Proceedings

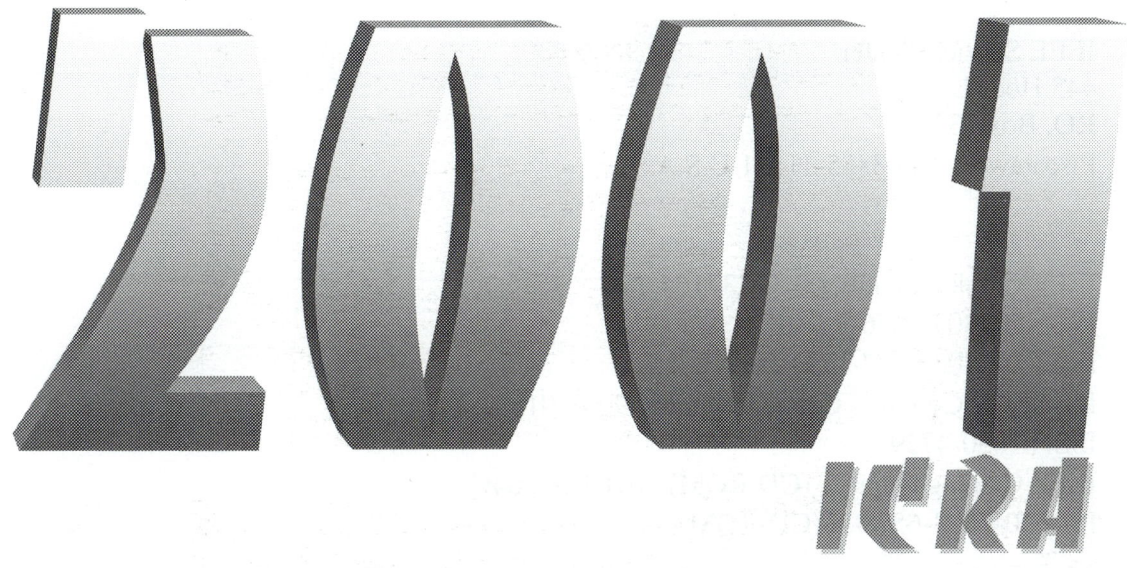

IEEE International Conference on Robotics and Automation

May 21~26, 2001
COEX
Seoul, Korea

Sponsored by

IEEE Robotics and Automation Society

Volume Ⅰ

Pages 1-1052

ICRA2001 PROCEEDINGS

Additional copies may be ordered from:

IEEE Service Center
445 Hoes Lane
P.O. Box 1331
Piscataway, NJ 08855-11331 U.S.A.

IEEE Catalog Number 01CH37164
ISBN 0-7803-6576-3
ISBN 0-7803-6577-1 (Microfiche)
Library of Congress Catalog Number 90-640158
ISSN 1050-4729
IEEE Catalog Number (CD-ROM) : 01CH37164C
ISBN 0-7803-6578-X (CD-ROM)

Copyright and Reprint Permission:

Abstracting is permitted with credit to the source. Libraries are permitted to photocopy beyond the limit of U.S. copyright law for private use of patrons those articles in this volume that carry a code at the bottom of the first page, provided the per-copy fee indicated in the code is paid through Copyright Clearance Center, 222 Rosewood Drive, Danvers, MA 01923. For other copying, reprint, or republication permission, write to IEEE Copyrights Manager, IEEE Operations Center, 445 Hoes Lane, P.O. Box 1331, Piscataway, NJ 08855-1331. All rights reserved. Copyright ©2001 by the Institute of Electrical and Electronics Engineers, Inc.

Printed in Korea by Kyung Hee Information Printing Co., Ltd.

The Institute of Electrical and Electronics Engineers, Inc.

Foreword

The 2001 IEEE International Conference on Robotics and Automation will be held in Seoul, the capital city of Korea. This is the fourth conference to be held outside the USA, and the second to be held in Asia. Seoul is the economic and cultural center of Korea, and offers a unique setting combining tradition and modernism for a conference on high technology.

We are witnessing the beginning of the new millennium, and the theme of this conference reflects the timely objective, "Frontiers of Robotics and Automation in the New Millennium". This conference will provide an opportunity to promote new emerging concepts in robotics and automation technology.

A record number of papers totaling over 1150 have been submitted to the program committee, by authors from more than 50 countries. Of these, 678 papers were selected for presentation at the conference. Geographically, the accepted papers come in roughly equal proportions from North America, Europe, and Asia.

The conference will be running in 13 parallel sessions with three plenary speeches and two panel discussions. The scope of the plenary speeches ranges from medical robotics and machine tool technologies to nano-robotics. We will also have panel discussions on manufacturing automation and rehabilitation robotic systems. We believe these topics will surely be critical issues for the future robotics community.

A number of special efforts were made in the preparation of this conference. For the first time in the history of this conference, an on-line electronic system was used to handle paper submissions, reviews, and registration. This contributed greatly in eliminating paperwork and communication delays. The conference will be held at the same time with KOFA2001, a full-scale annual exhibition on factory automation. A series of workshops and symposiums related to robotics and automation were also scheduled around the conference dates, to enable participants to attend more than one meeting while attending ICRA2001. Special sessions on Korean industrial automation have been scheduled during the conference in order to provide an opportunity for visitors to become acquainted with the local robotics and automation activities. These special sessions will focus on technology developments in automation of the automobile industry, semiconductor industry, and steel industry in Korea.

We are very grateful for all the hard work of many people and would like to express the deepest thanks to all the supporting organizations.

All the best,

Wook Hyun Kwon

General Chairman

Beom Hee Lee

Program Chairman

ICRA2001 Organization

A. Advisory Committee

Steve Hsia, University of California, USA
Myoung Sam Ko, Seoul National University, Korea
Fumio Harashima, Tokyo Metropolitan Institute of Technology, Japan
Tzyh-Jong Tarn, Washington University, USA
Norman Caplan, Washington University, USA
C.S. George Lee, Purdue University, USA

B. Organizing Committee

General Chair
Wook Hyun Kwon, Korea

General Co-Chairs
Arthur Sanderson, USA
Paolo Dario, Italy
Toshio Fukuda, Japan

General Vice-Chair
Hyung Suck Cho, Korea

Local Arrangements Chair
Sang-Rok Oh, Korea

Tutorials and Workshops Chair
Sukhan Lee, Korea

Video Proceedings Chair
Wankyun Chung, Korea

Publications Chair
Myung Jin Chung, Korea

Finance Chair
Jin Young Choi, Korea

Program Chair
Beom Hee Lee, Korea

Program Co-Chairs
Junku Yuh, USA
Li-Chen Fu, ROC
Hideki Hashimoto, Japan
Ruediger Dillmann, Germany
Il Hong Suh, Korea

Program Vice-Chairs
Myoung Hwan Choi, Korea
Dae Won Kim, Korea
Sooyong Lee, USA

Publicity Chair
Jaehyun Park, Korea

Publicity Co-Chair
Shigeki Sugano, Japan

Exhibition Chair
Chongwoo Park, Korea

C. Program Committee

Program Chair
Beom Hee Lee

Program Co-Chairs
Junku Yuh, Li-Chen Fu, Hideki Hashimoto, Ruediger Dillmann
Il Hong Suh

Program Vice Chairs
Myoung Hwan Choi, Dae Won Kim, Sooyong Lee

Asia/Oceania

Fumihito Arai	Jin Oh Kim	Sang-Rok Oh
Yoshikazu Arai	Jong-Hwan Kim	Toru Omata
Suguru Arimoto	Jongwon Kim	Jun Ota
Minoru Asada	Munsang Kim	Chongkug Park
Zeungnam Bien	Myung-Soo Kim	Gwi Tae Park
Shi-Chung Chang	Sungbok Kim	Jaehyun Park
Dongwoo Cho	Hiroshi Kimura	Jong Hyeon Park
Hye-Kyung Cho	Hyeong-Seok Ko	Jong-Oh Park
Young-Jo Cho	Nak-Yong Ko	Hong Qiao
Jin Young Choi	Kiyoshi Komoriya	Akihito Sano
Myung Hwan Choi	Kazuhiro Kosuge	Homayoun Seraji
Hong Tae Chun	Tae-Yong Kuc	Yoshiaki Shirai
Myung Jin Chung	Yoshinori Kuno	Kwee-Bo Sim
Wan Kyun Chung	In So Kweon	Shigeki Sugano
Teruo Fujii	Dong-Soo Kwon	Il Hong Suh
Masahiro Fujita	Jang-Myung Lee	Atsuo Takanishi
In-Joong Ha	Jin S. Lee	Toshio Tsuji
Woonchul Ham	Ju-Jang Lee	Masaru Uchiyama
Shigeo Hirose	Ki-Dong Lee	Michael Yu Wang
Katsushi Ikeuchi	Suk Gyu Lee	Seiji Yamada
Hidenori Ishihara	Ser Yong Li	Yoshio Yamamoto
Koji Ito	Joonhong Lim	Byung-Ju Yi
Ray Austin Jarvis	Ren C. Luo	Kazuhito Yokoi
MuDer Jeng	Takafumi Matsumaru	Yasuyoshi Yokokohji
Seul Jung	Hirofumi Miura	Kyung Hyun Yoon
Makoto Kaneko	Seungbin Moon	Kazuya Yoshida
Kazuo Kiguchi	Yoshihiko Nakamura	Tsuneo Yoshikawa
DaeWon Kim	Boo Hee Nam	Bum-Jae You
Hakil Kim	Y. Narahari	Shin'ichi Yuta

Europe/Mediterranean

Hassane Alla	Eckhard Freund	Jean-Paul Laumond
Antonio Bicchi	Volker Graefe	Alessandro De Luca
John Billingsley	Gerd Hirzinger	Jean-Pierre Merle
Martin Buss	Radu Horaud	Francois Pierrot
Maria Chiara Carrozza	Wisama Khalil	Juergen Rossmann
Jose Angel Castellanos	Heikki Koivo	Alberto Rovetta
Raja Chatila	Krzysztof Kozlowski	Imre J. Rudas
Ruediger Dillmann	Christian Laugier	Guenther Schmidt

Bruno Siciliano
Stefano Stramigioli

Karim A. Tahboub
Spyros Tzafestas

America

Sunil K. Agrawal
Jorge Angeles
Tucker Balch
Antal K. Bejczy
Beno Benhabib
Wayne J. Book
John Canny
C. L. Philip Chen
Yilong Chen
Gregory S. Chirikjian
Howie Choset
James J. Clark
Alan A. Desrochers
Rajiv Dubey
Khaled M. Elleithy
Ronald S. Fearing
Bijoy K. Ghosh
Maria Gini
Andrew Goldenberg
William A. Gruver
Kamal Gupta
Vincent Hayward
John M. Hollerbach
Roberto Horowitz
Seth Hutchinson

Youcef-Toumi Kamal
Lydia Kavraki
Pradeep K. Khosla
Antti J. Koivo
David J. Kriegman
Andrew Kusiak
David Lane
C.S. George Lee
Kok-Meng Lee
Sooyong Lee
Ming C. Lin
J.Y.S(John) Luh
Peter B. Luh
Vladimir Lumelsky
Anthony Maciejewski
Constantinos Mavroidis
Max Q.-H. Meng
Bradley J. Nelson
Paul Oh
David E. Orin
Nikos Papanikolopoulos
Lynne E. Parker
Gordon Robert Pennock
Nancy S. Pollard
Alfred A. Rizzi

Bernard Roth
Daniela Rus
Arthur C. Sanderson
Nilanjan Sarkar
Zvi Shiller
Reid G. Simmons
Wesley E. Snyder
Tarek M. Sobh
Harry E. Stephanou
Tzyh-Jong Tarn
Russell H. Taylor
Jeffrey C. Trinkle
N. Viswanadham
Richard M. Voyles
Kenneth J. Waldron
Ian D. Walker
Louis L. Whitcomb
Peter Will
Ning Xi
Woosoon Yim
Junku Yuh
Yuan F. Zheng
MengChu Zhou

D. Video Proceedings Committee

Video Proceedings Committee Chair
Wankyun Chung

Video Proceedings Committee Members
Rajiv Dubey, USA
Jin-oh Kim, Korea

Kazuhito Yokoi, Japan
Dragomir N. Nenchev, Japan

Giorgio Cannata, Italy

CONTENTS

SESSION : WA01 — Motion Planning

Planning Motion Patterns of Human Figures Using a Multi-layered Grid and the Dynamics Filter 1
Shiller, Zvi and Yamane, Katsu and Nakamura, Yoshihiko

Extending Procedural Reasoning toward Robot Actions Planning 9
Ingrand, Felix F. and Despouys, Olivier

Let's reduce the Gap between Task Planning and Motion Planning 15
Guere, Emmanuel and Alami, Rachid

An Adaptive Framework for 'Single Shot' Motion Planning: A Self-tuning System
for Rigid and Articulated Robots 21
Vallejo, Daniel and Remmler, Ian and Amato, Nancy M.

Optimal Line-sweep-based Decompositions for Coverage Algorithms 27
Huang, Wesley H.

Global Nearness Diagram Navigation (GND) 33
Minguez, Javier and Montano, Luis and Simeon, Thierry and Alami, Rachid

SESSION : WA02 — Petri Nets and Related Methods in Agile Automation

Combining Fault Detection and Process Optimization in Manufacturing Systems Using
First-order Hybrid Petri Nets 40
Balduzzi, Fabio and Di Febbraro, Angela

A Petri Net Graphic Method of Reduction Using Birth-death Processes 46
Zemouri, Ryad A and Racoceanu, Daniel I and Zerhouni, Noureddine

Modeling and Analysis of Semiconductor Manufacturing Systems with Degraded Behaviors
Using Petri Nets and Siphons 52
Jeng, MuDer and Xie, Xiaolan

Developing Software Controllers with Petri Nets and a Logic of Actions 58
Simon, Carlo

Resource-oriented Petri Nets in Deadlock Avoidance of AGV Systems 64
Wu, Naiqi and Zhou, MengChu

Algebraic Deadlock Avoidance Policies for Conjunctive / Disjunctive Resource Allocation Systems 70
Park, Jonghun and Reveliotis, Spyros A.

SESSION : WA03 — Control for Dextrous Manipulation

On a Two-level Hierarchical Structure for the Dynamic Control of Multifingered Manipulation 77
Casalino, Giuseppe and Cannata, Giorgio and Panin, Giorgio and Caffaz, Andrea

Robust Manipulation of Deformable Objects by a Simple PID Feedback 85
Wada, Takahiro and Hirai, Shinichi and Kawamura, Sadao and Kamiji, Norimasa

Robust Stabilization of the Plate-ball Manipulation System 91
Oriolo, Giuseppe and Vendittelli, Marilena

Robotic Pinching by Means of a Pair of Soft Fingers with Sensory Feedback 97
Han, Hyun-Yong and Arimoto, Suguru and Tahara, Kenji and Yamaguchi, Mitsuharu and Nguyen, Pham Thuc Anh

Fast Dextrous Regrasping with Optimal Contact Forces and Contact Sensor-based Impedance Control 103
Schlegl, Thomas and Buss, Martin and Omata, Toru and Schmidt, Guenther

DLR-hand II : Next Generation of a Dextrous Robot Hand 109
Butterfass, Joerg and Grebenstein, Markus and Liu, Hong and Hirzinger, Gerd

SESSION : WA04 — Micromanipulation and Microsystems

Micromanipulation Using a Friction Force Field 115
Sin, Jeongsik and Winther, Tobias and Stephanou, Harry

On the Coarse/Fine Dual-stage Manipulators with Robust Perturbation Compensator 121
Kwon, SangJoo and Chung, Wan Kyun and Youm, Youngil

Development of Global Vision System for Biological Automatic Micro-manipulation System 127
Li, Xudong and Zong, Guanghua and Bi, Shusheng

A Flexible Experimental Workcell for Efficient and Reliable Wafer-level 3D Microassembly 133
Yang, Ge and Gaines, James A. and Nelson, Bradley J.

Mechanical Micro-dissection by Microknife Using Ultrasonic Vibration and Ultra Fine Touch Probe Sensor 139
Arai, Fumihito and Amano, Takaharu and Fukuda, Toshio and Satoh, Hiroshi

Design and Analysis of an Electromagnetically Driven Valve for a Glaucoma Implant 145
Bae, Byunghoon and Kim, Nakhoon and Kee, Hongseok and Kim, Seonho and Lee, Yeon and Park, Kyihwan

SESSION : WA05 — Trajectory Generation & Control

A Path Generation Algorithm of Autonomous Robot Vehicle Through Scanning of a Sensor Platform 151
Park, Tong-Jin and Han, Chang-Soo

Hybrid Control of Formations of Robots 157
Fierro, Rafael O and Das, Aveek K and Kumar, Vijay and Ostrowski, James P

Tracking Control of a Mobile Robot Using a Neural Dynamics Based Approach 163
Yuan, Guangfeng and Yang, Simon X. and Mittal, Gauri S.

Backward Line Tracking Control of a Radio-controlled Truck and Trailer 169
Altafini, Claudio and Speranzon, Alberto

The Scalar Epsilon-controller: A Spatial Path Tracking Approach for ODV, Ackerman,
and Differentially-steered Autonomous Wheeled Mobile Robots 175
Davidson, Morgan E and Bahl, Vikas

Normalized Energy Stability Margin and its Contour of Walking Vehicles on Rough Terrain 181
Hirose, Shigeo and Tsukagoshi, Hideyuki and Yoneda, Kan

SESSION : WA06 — Programming Architecture

Creating the Architecture of a Translator Framework for Robot Programming Languages 187
    Freund, Eckhard and Luedemann-Ravit, Bernd and Stern, Oliver and Koch, Thorsten

A Multi-processing Software Infrastructure for Robotic Systems 193
    Jones, Andrew H. and DeSouza, Guilherme N. and Kak, Avinash C.

Open Real-time Interfaces for Monitoring Applications within NC-control Systems 199
    Weck, Manfred and Kahmen, Andreas

Robot Behavior Engineering Using DD-designer 205
    Bredenfeld, Ansgar and Indiveri, Giovanni

Object-oriented Design Pattern Approach for Modeling and Simulating Open Distributed Control System 211
    Tomura, Toyoaki and Kanai, Satoshi and Uehiro, Kiyoshi and Yamamoto, Susumu

Implementation of Internet-based Personal Robot with Internet Control Architecture 217
    Han, Kuk-Hyun and Kim, Shin and Kim, Yong-Jae and Lee, Seung-Eun and Kim, Jong-Hwan

SESSION : WA07 — Visual Servoing Experimetal Issues

Computer Animation: A New Application for Image-based Visual Servoing 223
    Courty, Nicolas and Marchand, Eric

Design and Implementation of Visual Servoing System for Realistic Air Target Tracking 229
    Yau, Wei Guan and Fu, Li-Chen and Liu, David

Robust Visual Servoing: Examination of Cameras Under Different Illumination Conditions 235
    Bachem, A. and Mueller, T. and Nagel, H.-H.

Asymptotic Motion Control of Robot Manipulators Using Uncalibrated Visual Feedback 241
    Shen, Yantao and Liu, Yun-Hui and Li, Kejie and Zhang, Jianwei and Knoll, Alois

Image-based Visual Servoing on Planar Objects of Unknown Shape 247
    Collewet, Christophe and Chaumette, Frangois and Loisel, Philippe

Localization of a Mobile Robot Using Images of a Moving Target 253
    Kim, Byung Hwa and Roh, Dong Kyu and Lee, Jang Myung and Lee, Man Hyung

SESSION : WA08 — Mobile Robot

Extracting Navigation States from a Hand-drawn Map 259
    Skubic, Marjorie A and Matsakis, Pascal and Forrester, Benjamin and Chronis, George

On Eye-sensor Based Path Planning for Robots with Non-trivial Geometry/Kinematics 265
    Gupta, Kamal and Yu, Yong

Optimizing Schedules for Prioritized Path Planning of Multi-robot Systems 271
    Bennewitz, Maren and Burgard, Wolfram and Thrun, Sebastian

Autonomous Characterization of Unknown Environments 277
    Pedersen, Liam

Towards a Meta Motion Planner A: Model and Framework 285
    Adam, Amit and Rivlin, Ehud and Shimshoni, Ilan

Towards a Meta Motion Planner B: Algorithm and Applications 291
    Adam, Amit and Rivlin, Ehud and Shimshoni, Ilan

SESSION : WA09 — Flexible Automation (I)

Efficient Scheduling of Behavior-processes on Different Time-scales 299
    Birk, Andreas and Kenn, Holger

An Efficient Depalletizing System Based on 2D Range Imagery 305
    Katsoulas, Dimitrios K and Kosmopoulos, Dimitrios I

Automatic Generation of Assembly Instructions Using STEP 313
    Mok, Swee M. and Ong, Kenlip and Wu, Chi-haur

Furniture Polishing Robot Using a Trajectory Generator Based on Cutter Location Data 319
    Nagata, Fusaomi and Watanabe, Keigo and Izumi, Kiyotaka

Force-guided Assemblies Using a Novel Parallel Manipulator 325
    Morris, Daniel M. and Hebbar, Ravi and Newman, Wyatt S.

Path Planning for Robot-assisted Grinding Processes 331
    Wang, Y.T. and Jan, Y.J.

SESSION : WA10 — Control and Mechanism for Production Systems

Data Filtering and Regression in Estimating Slew Motion Timing Model 337
    Hua, Wei and Zhou, Chen

An Exact Representation of Effective Cutting Shapes of 5-Axis CNC Machining
Using Rational Bezier and B-Spline Tool Motions 342
    Xia, Jun and Ge, Q. Jeffrey

Design of Programmable Passive Compliance Shoulder Mechanism 348
    Okada, Masafumi and Nakamura, Yoshihiko and Ban, Shigeki

Kinematic Control of Parallel Robots in the Presence of Unstable Singularities 354
    O'Brien, John F. and Wen, John T.

Complex Behaviors from Local Rules in Modular Self-reconfigurable Robots 360
    Kubica, Jeremy and Casal, Arancha and Hogg, Tad

Control of a Suspended Load Using Inertia Rotors with Traveling Disturbance 368
    Yoshida, Yasuo

SESSION : WA11 — Haptic Devices and Teleoperation

Variable Position Mapping Based Assistance in Teleoperation for Nuclear Cleanup 374
    Manocha, Karan A. and Pernalete, Norali and Dubey, Rajiv V.

Human-centered Scaling in Micro-teleoperation 380
    Sano, Akihito and Fujimoto, Hideo and Takai, Toshihito

Haptically Augmented Teleoperation 386
    Turro, Nicolas and Khatib, Oussama and Coste-Maniere, Eve

Elements of Telerobotics Necessary for Waste Clean Up Automation 393
   Hamel, William R. and Kress, Reid L.

Design and Implementation of Remotely Operation Interface for Humanoid Robot 401
   Kagami, Satoshi and Kuffner, James J. and Nishiwaki, Koichi and Sugihara,
   Tomomichi and Inaba, Masayuki and Inoue, Hirochika

Model-based Teleoperation of a Space Robot on ETS-VII Using a Haptic Interface 407
   Yoon, Woo-Keun and Goshozono, Toshihiko and Kawabe, Hiroshi and Kinami, Masahiro and
   Yuichi, Tsumaki and Masaru, Uchiyama and Mitsushige, Oda

SESSION : WA12   Space and Underwater Robotics

Analysis on Impact Propagation of Docking Platform for Spacecraft 413
   Lee, Sang Heon and Yi, Byung-Ju and Kim, Soo Hyun and Kwak, Yoon Keun

Flyaround Maneuvers on a Satellite Orbit by Impulsive Thrust Control 421
   Masutani, Yasuhiro and Matsushita, Motoshi and Miyazaki, Fumio

Nonlinear Control Methods for Planar Carangiform Robot Fish Locomotion 427
   Morgansen, Kristi A. and Duindam, Vincent and Mason, Richard J. and Burdick, Joel W. and Murray, Richard M.

Quaternion-based Kinematic Control of Redundant Spacecraft/Manipulator Systems 435
   Caccavale, Fabrizio and Siciliano, Bruno

Zero Reaction Maneuver: Flight Velification with ETS-VII Space Robot and
Extension to Kinematically Redundant Arm 441
   Yoshida, Kazuya and Hashizume, Kenichi and Abiko, Satoko

A Novel Adaptive Control Law for Autonomous Underwater Vehicles 447
   Antonelli, Gianluca and Caccavale, Fabrizio and Chiaverini, Stefano and Fusco, Giuseppe

SESSION : WA13   New Tools for Personal Robotics

Ethological Modeling and Architecture for an Entertainment Robot 453
   Arkin, Ronald C., and Fujita, Masahiro and Takagi, Tsuyoshi and Hasegawa, Rika

Human-like Robot Head that has Olfactory Sensation and Facial Color Expression 459
   Miwa, Hiroyasu and Umetsu, Tomohiko and Takanishi, Atsuo and Takanobu, Hideaki

Acquiring Hand-action Models by Attention Point Analysis 465
   Ogawara, Koichi and Tanuki, Tomikazu and Kimura, Hiroshi and Ikeuchi, Katsushi

New Architecture for Mobile Robots in Home Network Environment Using Jini 471
   Lee, Byoung-Ju and Lee, Hyun-Gu and Lee, Joo-Ho and Park, Gwi-Tae

Implementing Tolman's Schematic Sowbug: Behavior-based Robotics in the 1930's 477
   Endo, Yoichiro and Arkin, Ronald C.

Generating Linguistic Spatial Descriptions from Sonar Readings Using the Histogram of Forces 485
   Skubic, Marjorie A. and Chronis, George and Matsakis, Pascal and Keller, James

SESSION : WP01   Robot Learning

Real-time Robot Learning 491
   Bhanu, Bir and Leang, Pat and Cowden, Chris and Lin, Yingqiang and Patterson, Mark

Learning Task-relevant Features from Robot Data 499
Vlassis, Nikos and Bunschoten, Roland and Krose, Ben

Autonomous Learning Algorithm and Associative Memory for Intelligent Robots 505
Kojima, Kazuhiro and Ito, Koji

Learning Hierarchical Partially Observable Markov Decision Process Models for Robot Navigation 511
Theocharous, Georgios and Rohanimanesh, Khashayar and Mahadevan, Sridhar

Towards Automatic Shaping in Robot Navigation 517
Peterson, Todd S. and Owens, Nancy E. and Carroll, James L.

Obstacle Avoidance Learning for a Multi-agent Linked Robot in the Real World 523
Daisuke, Iijima and Wenwei, Yu and Hiroshi, Yokoi and Yukinori, Kakazu

SESSION : WP02 — Discrete Event Manufacturing Systems (I)

Performance Modeling of Supply Chains Using Queueing Networks 529
Viswanadham, N and Raghavan, N. R. Srinivasa

Dynamic Scheduling Rule Selection for Semiconductor Wafer Fabrication 535
Hsieh, Bo-Wei and Chang, Shi-Chung and Chen, Chun-Hung

A Deadlock Prevention Policy for Flexible Manufacturing Systems Using Siphons 541
Huang, YiSeng and Jeng, MuDer and Xie, Xiaolan and Chung, ShengLuen

An Effective Search Strategy for Wafer Fabrication Scheduling with Uncertain Process Requirements 547
Lin, Ming-Hung and Fu, Li-Chen

Model-based Control for Reconfigurable Manufacturing Systems 553
Ohashi, Kazushi and Shin, Kang G.

Design of Reconfigurable Semiconductor Manufacturing Systems with Maintenance and Failure 559
Tang, Ying and Zhou, MengChu

SESSION : WP03 — Sensing for Dextrous Manipulation

Dynamic Contact Sensing of Soft Planar Fingers with Tactile Sensors 565
Kinoshita, Genichiro and Kurimoto, Yujin and Osumi, Hisashi and Umeda, Kazunori

Interpretation of Force and Moment Signals for Compliant Peg-in-hole Assembly 571
Newman, Wyatt S. and Zhao, Yonghong and Pao, Yoh-Han

Dome Shaped Touch Sensor Using PZT Thin Film made by Hydrothermal Method 577
Arai, Fumihito and Fukuda, Toshio and Itoigawa, Kouichi and Thukahara, Yasunori

Heuristic Vision-based Computation of Planar Antipodal Grasps on Unknown Objects 583
Morales, Antonio and Recatala, Gabriel and Sanz, Pedro J. and Pobil, Angel P. del

Feature-guided Exploration with a Robotic Finger 589
Okamura, Allison M. and Cutkosky, Mark R.

Identification of Contact Conditions from Contaminated Data of Contact Force and Moment 597
Mouri, Tetsuya and Yamada, Takayoshi and Iwai, Ayako and Mimura, Nobuharu and Funahashi, Yasuyuki

SESSION : WP04 — Nano / Micro Robotics

Three-dimensional Bio-micromanipulation Under the Microscope 604
  Arai, Fumihito and Kawaji, Akiko and Luangjarmekorn, Poom and Fukuda, Toshio and Itoigawa, Kouichi

Force Control System for Autonomous Micro Manipulation 610
  Tanikawa, Tamio and Kawai, Masashi and Koyachi, Noriho and Arai, Tatsuo and Ide, Takayuki and Kaneko, Shinji and Ohta, Ryo

Micro Fluid Device Using Thick Layer Piezo Actuator Prepared on Si Micro-machined Structure 616
  Lee, Sukhan and Chung, Jaewoo and Lim, Seungmo and Lee, Changseung

Microrobotic Cell Injection 620
  Sun, Yu and Nelson, Brad

Force Feedback-based Microinstrument for Measuring Tissue Properties and Pulse in Microsurgery 626
  Menciassi, Arianna and Eisinberg, Anna and Scalari, Giacomo and Anticoli, Claud

3D Nanorobotic Manipulations of Multi-walled Carbon Nanotubes 632
  Dong, Lixin and Arai, Fumihito and Fukuda, Toshio

SESSION : WP05 — Real-Time Sensing and Supervisory Control of Robotic Operations

Observations Concerning Internet-based Teleoperations for Hazardous Environments 638
  Hamel, William R. Hamel and Murray, Pamela

Internet-based Teleoperation 644
  Tarn, Tzyh-Jong and Brady, Kevin

Language Model Approach to Nonblocking Supervisor Synthesis for Nondeterministic Discrete Event Systems 650
  Park, Seong-Jin and Lim, Jong-Tae

A Bone-reaming System Using Micro Sensors For Internet Force-feedback Control 656
  Ho, Antony W. T. and Li, Wen J. and Elhajj, Imad and Xi, Ning and Mei, Tao

Modeling and Control of Internet Based Cooperative Teleoperation 662
  Elhajj, Imad H. and Xi, Ning and Fung, Waikeung and Liu, Yunhui and Hasegawa, Yasuhisa and Fukuda, Toshio

Supervisory Control for Systems of Vehicles in Path Networks 668
  Roszkowska, Elzbieta K.

SESSION : WP06 — Collision Avoidance and Sensor Based Control

Sensor Based Online Path Planning for Serpentine Robots 674
  Gevher, Mustafa and Erkmen, Aydan M. and Erkmen, Ismet

Reactive Behaviours of Mobile Manipulators Based on the DVZ Approach 680
  Cacitti, Alessio and Zapata, Rene

The System Development of Unmanned Vehicle for the Tele-operated System Interfaced with Driving Simulator 686
  Shim, Jae-Heung and Kim, Min-Seok and Park, Young-Hoon and Kim, Jung-Ha

Motion Planning for Humanoid Robots Under Obstacle and Dynamic Balance Constraints 692
  Kuffner, James J and Nishiwaki, Koichi and Kagami, Satoshi and Inaba, Masayuki and Inoue, Hirochika

Exact Cellular Decomposition of Closed Orientable Surfaces Embedded in R³ 699
   Atkar, Prasad N. and Choset, Howie and Rizzi, Alfred A. and Acar, Ercan U.

Obstacle Detection Using Adaptive Color Segmentation and Color Stereo Homography 705
   Batavia, Parag H. and Singh, Sanjiv

SESSION : WP07 — Collision Avoidance and Sensor Based Control

Visual Servoing Based on Multirate Sampling Control-application of Perfect Disturbance Rejection Control 711
   Fujimoto, Hiroshi and Hori, Yoichi

Stacking Jacobians Properly in Stereo Visual Servoing 717
   Martinet, Philippe and Cervera, Enric

Visual Servoing: Path Interpolation by Homography Decomposition 723
   Borgstadt, Justin A. and Ferrier, Nicola J.

Design and Tracking of Desirable Trajectories in the Image Space by Integrating Mechanical and Visibility Constraints 731
   Mezouar, Youcef and Chaumette, Francois

A Visual Servoing Algorithm Based on Epipolar Geometry 737
   Chesi, Graziano and Prattichizzo, Domenico and Vicino, Antonio

A Solution to the Adaptive Visual Servoing Problem 743
   Astolfi, Alessandro and Hsu, Liu and Netto, Mariana S. and Ortega, Romeo

SESSION : WP08 — Mobile Robot Design & Application

An Industrial Application of Behavior-oriented Robotics 749
   Birk, Andreas and Kenn, Holger

Automated Container-handling System for Container Production Nurseries 755
   Schempf, Hagen and Graham, Todd and Fuchs, Robert and Gasior, Chris

Actively Steerable Inpipe Inspection Robots for Underground Urban Gas pipelines 761
   Roh, Se-gon and Ryew, SungMoo and Yang, Jonghwa and Choi, Hyoukryeol

Design of Continuous Alternate Wheels for Omnidirectional Mobile Robots 767
   Byun, Kyung-Seok and Kim, Sung-Jae and Song, Jae-Bok

Holonomic Omni-directional Vehicle with New Omni-wheel Mechanism 773
   Damoto, Riichiro and Cheng, Wendy and Hirose, Shigeo

Dynamic Simulation of Actively-coordinated Wheeled Vehicle Systems on Uneven Terrain 779
   Hung, Min-Hsiung and Orin, David E

SESSION : WP09 — Flexible Automation (II)

A Methodology for Software Development Cost Analysis in Information-based Manufacturing 787
   Sun, High-Way and Zhou, MengChu and Wolf, Carl

Optimal Fixture Layout Design in a Discrete Domain for 3D Workpieces 792
   Wang, Yu M. and Pelinescu, Diana M.

Holonic Robot System: A Flexible Assembly System with High Reconfigurability 799
   Sugi, Masao and Maeda, Yusuke and Aiyama, Yasumichi and Arai, Tamio

New Architecture for Corporate Integration of Simulation and Production Control in Industrial Applications 806
   Freund, Eckhard and Hypki, Alfred and Pensky, Dirk H.

Vision-based Automatic Forming of Rheological Objects Using Deformation Transition Graphs 812
   Tokumoto, Shinichi and Saito, Takuya and Hirai, Shinichi

Off-line Error Prediction, Diagnosis and Recovery Using Virtual Assembly Systems 818
   Baydar, Cem M. and Saitou, Kazuhiro

SESSION : WP10 — Design and Planning for Production Systems

Network Distributed Virtual Design Using Coevolutionary Agents 824
   Subbu, Raj and Sanderson, Arthur C.

Multi-component Genetic Algorithm for Generating
Best Bending Sequence and Tool Selection in Sheet Metal Parts 830
   Thanapandi, Chitra Malini and Walairacht, Aranya and Ohara, Shigeyuki

A Hybrid Software Agent Model for Decentralized Control 836
   Balduzzi, Fabio E. and Brugali, Davide

A Time Window Based Approach for Job Shop Scheduling 842
   Chen, Haoxun and Luh, Peter B. and Fang, Lei

FMS Scheduling Based on Timed Petri Net Model and RTA* Algorithm 848
   Kim, YoungWoo and Inaba, Akio and Suzuki, Tatsuya and Okuma, Shigeru

Optimal Configuration and Partner Selection in Dynamic Manufacturing Networks 854
   Viswanadham, Nukala and Gaonkar, Roshan S and Subramaniam, Velusamy

SESSION : WP11 — Haptic Systems

Development of a Scaled Teleoperation System for Nano Scale Interaction and Manipulation 860
   Sitti, Metin and Aruk, Baris and Shintani, Hiroaki and Hashimoto, Hideki

Haptic Display Device with Fingertip Presser for Motion/Force Teaching to Human 868
   Kikuuwe, Ryo and Yoshikawa, Tsuneo

A Framework for Haptic Rendering System Using 6-DOF Force-Feedback Device 874
   Kim, Dae-Hyun and Ko, Nak Yong and Oh, Geum Kun and Kim, Young Dong

A New Haptic Interface Device Capable of Continuous-time Impedance
Display within Sampling-period: Application to Hard Surface Display 880
   Kawai, Masayuki and Yoshikawa, Tsuneo

Design Of A New 6-DOF Parallel Haptic Device 886
   Lee, J.H. and Eom, K.S. and Yi, B-J. and Suh, I.H.

Wearable Master Device Using Optical Fiber Curvature Sensors for the Disabled 892
   Lee, Kyoobin and Kwon, Dong-Soo

SESSION : WP12 — Underwater Robotic Vehicle Control

Maneuverability of a Flat-streamlined Underwater Vehicle 897
  Guo, Jenhwa and Chiu, F.C.

Closed Loop Time Invariant Control of 3D Underactuated Underwater Vehicles 903
  Aicardi, Michele and Casalino, Giuseppe and Indiveri, Giovanni

Interoperability and Synchronisation of Distributed Hardware-in-the-loop
Simulation for Underwater Robot Development: Issues and Experiments 909
  Lane, David M. and Falconer, Gavin J. and Randall, Geoph and Edwards, Ian

Planar Motion Steering of Underwater Vehicles by Exploiting Drag Coefficient Modulation 915
  Aicardi, Michele and Casalino, Giuseppe and Indiveri, Giovanni

Underwater Cable Following by Twin-Burger 2 920
  Balasuriya, Arjuna P. and Ura, Tamaki

A Virtual Collaborative World Simulator for Underwater Robots Using
Multi-dimensional, Synthetic Environment 926
  Choi, Song K. and Yuh, Junku

SESSION : WP13 — Applications of Robotics Techniques to Computational Biology and Chemistry

Capturing Molecular Energy Landscapes with Probabilistic Conformational Roadmaps 932
  Apaydin, Mehmet Serkan and Singh, Amit P. and Brutlag, Douglas L. and Latombe, Jean-Claude

Physical Geometric Algorithms for Structural Molecular Biology 940
  Donald, Bruce R. and Bailey-Kellogg, Chris and Kelley, Jack and Lilien, Ryan

A Motion Planning Approach to Folding: From Paper Craft to Protein Folding 948
  Song, Guang and Amato, Nancy M.

Ligand Binding with OBPRM and User Input 954
  Bayazit, O. Burchan and Song, Guang and Amato, Nancy M.

Molecular Docking: A Problem With Thousands of Degrees of Freedom 960
  Teodoro, Miguel and Phillips, George and Kavraki, Lydia

Programmable Assembly at the Molecular Scale: Self-assembly of DNA Lattices (Invited Paper) 966
  Reif, John H. and LaBean, Thomas H. and Seeman, Nadrian C.

SESSION : WE01 — Motion Planning and Learning

A Simple Algorithm for Determining of Movement Duration in
Task Space without Violating Joint Angle Constraints 972
  Jin, Young G. and Choi, Jin Y.

Design of Jerk Bounded Trajectories for On-line Industrial Robot Applications 979
  Macfarlane, Sonja E. and Croft, Elizabeth A.

Control of Uncertain Flexible Joint Manipulator Using Adaptive Takagi-Sugeno Fuzzy Model Based Controller 985
  Park, Chang-Woo and Hyun, Chang-Ho and Park, Min-Sick and Lee, Chang-Hun

Sensor Based Planning for Rod Shaped Robots in Three Dimensions: Piece-wise Retracts of $R^3 \times S^2$ 991
Lee, Ji Yeong and Choset, Howie and Rizzi, Alfred A

Actor-Q Based Active Perception Learning System 1000
Shibata, Katsunari and Nishino, Tetsuo and Okabe, Yoichi

Inverse Kinematics Learning by Modular Architecture Neural Networks with Performance Prediction Networks 1006
Oyama, Eimei and Chong, Nak Young and Agah, Arvin and Maeda, Taro and Tachi, Susumu

SESSION : WE02 — Teleoperation

Virtual Repulsive Force Field Guided Coordination for Multi-telerobot Collaboration 1013
Chong, Nak-Young and Kotoku, Tetsuo and Ohba, Kohtaro and Tanie, Kazuo

The Visual Acts Model for Automated Camera Placement During Teleoperation 1019
Brooks, Bernard G and McKee, Gerard T and Schenker, Paul S

Sliding-mode-based Impedance Controller for Bilateral Teleoperation under Varying Time-delay 1025
Cho, Hyun C. and Park, Jong H. and Kim, Kyunghwan and Park, Jong O.

Ground-space Bilateral Teleoperation Experiment Using ETS-VII Robot Arm with Direct Kinesthetic Coupling 1031
Imaida, Takashi and Yokokohji, Yasuyoshi and Doi, Toshitsugu and Oda, Mitsushige and Yoshikawa, Tsuneo

A Numerical SC Approach for a Teleoperated 7-DOF Manipulator 1039
Tsumaki, Yuichi and Fiorini, Paolo and Chalfant, Gene and Seraji, Homayoun

Bilateral Controller Design for Telemanipulation in Soft Environments 1045
Cavusoglu, Murat Cenk and Sherman, Alana and Tendick, Frank

SESSION : WE03 — Part Handling

Orienting Parts by Inside-out Pulling 1053
Berretty, Robert-Paul M. and Goldberg, Ken and Overmars, Mark H. and van der Stappen, A. Frank

Compliant Grasping Force Modeling for Handling of Live Objects 1059
Lee, Kok-Meng and Joni, Jeffry and Yin, Xuecheng

Design of Robot Gripper Jaws Based on Trapezoidal Modules 1065
Zhang, Tao and Goldberg, Ken

Attractive Regions Formed by Constraints in Configuration Space-attractive Regions
in Motion Region of a Polygonal or a Polyhedral Part with a Flat Environment 1071
Qiao, Hong

A Geometric Approach to Designing a Programmable Force Field
with a Unique Stable Equilibrium for Parts in the Plane 1079
Sudsang, Attawith and Kavraki, Lydia

Effect of Conformability on Grasp Static Stability 1086
Hurtado, Jose F. and Melkote, Shreyes N.

SESSION : WE04 — Navigation

Distributed Sensor Fusion for Object Position Estimation by Multi-robot Systems 1092
Stroupe, Ashley W. and Martin, Martin C. and Balch, Tucker

Stereo Ego-motion Improvements for Robust Rover Navigation 1099
    Olson, Clark F. and Matthies, Larry H. and Schoppers, Marcel and Maimone, Mark W.

Autonomous Vehicle Positioning with GPS in Urban Canyon Environments 1105
    Cui, Youjing and Ge, Shuzhi S.

A Hybrid Approach for Robust and Precise Mobile Robot Navigation with Compact Environment Modeling 1111
    Tomatis, Nicola and Nourbakhsh, Illah and Arras, Kai and Siegwart, Roland

A Study of Autonomous Mobile System in Outdoor Environment 1117
    Takiguchi, Junichi and Hashizume, Takumi

A Visual Landmark Recognition System for Topological Navigation of Mobile Robots 1124
    Mata, Mario and Armingol, Jose M and de la Escalera, Arturo and Salichs, Miguel A

SESSION : WE05 — Nonlinear and Optimal Control

Quadratic Programming in Control of Brushless Motors 1130
    Aghili, Farhad and Buehler, Martin and Hollerbach, John M.

A New Optimization Algorithm for Singular and Non-singular Digital Time-optimal Control of Robots 1136
    Hol, Camile W.J. and van Willigenburg, Gerard and van Henten, Eldert J. and van Straten, Gerrit

On the Optimality and Performance of PID Controller for Robotic Manipulators 1142
    Choi, Youngjin and Chung, Wan Kyun

Robust Nonlinear Reduced-order Dynamic Controller Design and its Application to a Single-link Manipulator 1149
    Zhou, Jianying and Zhou, Rujing and Wang, Youyi

Decomposition-based Friction Compensation Using a Parameter Linearization Approach 1155
    Liu, Guangjun

Transition to Nonlinear H-inf Optimal Control From Inverse-optimal Solution for Euler-lagrange System 1161
    Park, Jonghoon and Chung, Wan Kyun

SESSION : WE06 — Visual Sensing and Control

VLSI Processor for Reliable Stereo Matching Based on Adaptive Window-size Selection 1168
    Masanori, Hariyama and Toshiki, Takeuchi and Michitaka, Kameyama

Range Estimation from a Pair of Omnidirectional Images 1174
    Bunschoten, Roland and Krose, Ben

Image-based Path-planning Algorithm on the Joint Space 1180
    Noborio, Hiroshi and Nishino, Yutaka

Automatic Extraction of Visual Landmarks for a Mobile Robot under Uncertainty of Vision and Motion 1188
    Moon, Inhyuk and Miura, Jun and Shirai, Yoshiaki

Hand Pose Recovery with a Single Video Camera 1194
    Kwon, Kihwan and Zhang, Hong and Dornaika, Fadi

A Performance Criterion for the Depth Estimation of a Robot Visual Control System 1201
    Chang-Jia, Fang and Shir-Kuan, Lin

SESSION : WE07 — Visual Tracking

Robust View-based Visual Tracking with Detection of Occlusions 1207
  Ito, Ken and Sakane, Shigeyuki

Real-time Color-based Tracking via a Marker Interface 1214
  PylkkoRiekki, Jukka P and R?ing, Juha J

Optimal Moving Windows for Real-time Road Image Processing 1220
  Choi, Sung Yug and Lee, Jang Myung

Model-based Object Tracking Using Stereo Vision 1226
  Ginhoux, Romuald and Gutmann, Jens-Steffen

Video-frame Rate Detection of Position and Orientation of Planar Motion Objects using One-sided Radon Transform 1233
  Tsuboi, Tatsuhiko and Masubuchi, Akihiro and Hirai, Shinichi and Yamamoto, Shinya and Ohnishi, Kazuhiko and Arimoto, Suguru

Proposal of an Adaptive Vision-based Attentive Tracker for Human Intended Actions 1239
  Ho, Minh Anh T. and Yamada, Yoji and Sonohara, Takayuki and Morizono, Tetsuya and Umetani, Yoji

SESSION : WE08 — Mobile Manipulation

Motion Generation for Formations of Robots a Geometric Approach 1245
  Belta, Calin A. and Kumar, Vijay

Manipulability Analysis for Mobile Manipulators 1251
  Bayle, Bernard and Fourquet, Jean-Yves and Renaud, Marc

Motion Planning for Active Acceleration Compensation 1257
  Decker, Michael W. and Dang, Anh X. and Ebert-Uphoff, Imme

Inverse Kinematics along a Geometric Spline for a Holonomic Mobile Manipulator 1265
  Altafini, Claudio

Sub-optimal Trajectory Planning of Mobile Manipulator 1271
  Mohri, Akira and Furuno, Seiji and Iwamura, Makoto and Yamamoto, Motoji

Utilization of Inertial Effect in Damping-based Posture Control of Mobile Manipulator 1277
  Kang, Sungchul and Komoriya, Kiyoshi and Yokoi, Kazuhito and Koutoku, Tetsuo and Tanie, Kazuo

SESSION : WE09 — Parallel Manipulators

Comparison Study of Architectures of Four 3 Degree-of-freedom Translational Parallel Manipulators 1283
  Tsai, Lung-Wen and Joshi, Sameer A

An Improved Design Algorithm Based on Interval Analysis for Spatial Parallel Manipulator with Specified Workspace 1289
  Merlet, Jean-Pierre

Singularity Analysis of CaPaMan: A Three-degree of Freedom Spatial Parallel Manipulator 1295
  Ottaviano, Erika and Gosselin, Clement M. and Ceccarelli, Marco

Algebraic Elimination-based Real-time Forward Kinematics
of the 6-6 Stewart Platform with Planar Base and Platform 1301
    Lee, Tae-Young and Shim, Jae-Kyung

New Methodology for the Forward Kinematics of 6-dof Parallel Manipulators Using Tetrahedron Configurations 1307
    Song, Se-Kyong and Kwon, Dong-Soo

Investigation of the Deficiencies of Parallel Manipulators
in Singular Configurations through the Jacobian Nullspace 1313
    Chan, Vincent K and Ebert-Uphoff, Imme

SESSION : WE10 — Rapid Prototyping and Design Automation

Computer-aided Synthesis of Higher Pairs via Configuration Space Manipulation 1321
    Kyung, Min-Ho and Sacks, Elisha P.

Combining Control Design Tools-from Modeling to Implementation 1327
    Ridderstrom, Christian and Ingvast, Johan and Wikander, Jan

The Development of a New Adaptive Slicing Algorithm for Layered Manufacturing System 1334
    Luo, Ren C. and Chang, Yi Cheng and Tzou, Jyh Hwa

An Efficient Scanning Pattern for Layered Manufacturing Processes 1340
    Yang, Yong and Fuh, Jerry Y. H. and Loh, Han Tong and Wang, Yun Gan

Model-based Design of an ECU with Data-and Event-driven Parts Using Auto Code Generation 1346
    Orehek, Martin and Robl, Christian

Development of a Sheet-based Material Handling System for Layered Manufacturing 1352
    Wei, Tao and Choi, Sangeun and Newman, Wyatt S.

SESSION : WE11 — Simulation and VR

Realistic Force Reflection in a Spine Biopsy Simulator 1358
    Kwon, Dong-Soo and Kyung, Ki-uk and Kwon, Sung Min and Ra, Jong Beom

Simulation of a Manual Gearshift with a 2 DOF Force-feedback Joystick 1364
    Frisoli, Antonio and Avizzano, Carlo A and Bergamasco, Massimo

Non-linear and Anisotropic Elastic Soft Tissue Models for Medical Simulation 1370
    Picinbono, Guillaume and Delingette, Hervi and Ayache, Nicholas

Hybrid Dynamic Simulation of Rigid-body Contact with Coulomb Friction 1376
    Son, Wookho and Trinkle, Jeffrey C. and Amato, Nancy M.

Force Saturation, System Bandwidth, Information Transfer, and Surface Quality in Haptic Interfaces 1382
    Kilchenman, Marcia and Goldfarb, Michael

Virtual Teaching Based on Hand Manipulability for Multi-fingered Robots 1388
    Kawasaki, Haruhisa and Nakayama, Kanji and Mouri, Tetsuya and Ito, Satoshi

SESSION : WE12 — Space Robotics

Command Data Compensation for Real-time Tele-driving System on Lunar Rover: Micro-5 1394
    Kunii, Yasuharu and Suhara, Masaya and Kuroda, Yoji and Kubota, Takashi

Attitude Control of a Space Robot with Initial Angular Momentum 1400
Matsuno, Fumitoshi and Saito, Kei

Proposal of A SkilMate Finger for EVA Gloves 1406
Yamada, Yoji and Morizono, Tetsuya and Sato, Shuji and Shimohira, Takahiro and
Umetani, Yoji and Yoshida, Tetsuji and Aoki, Shigeru

Fuzzy Rule-based Reasoning for Rover Safety and Survivability 1413
Tunstel, Edward and Howard, Ayanna and Seraji, Homayoun

Solar Navigational Planning for Robotic Explorers 1421
Shillcutt, Kimberly J and Whittaker, William L

Control of a Robotic Gripper for Grasping Objects in No-gravity Conditions 1427
Biagiotti, Luigi and Melchiorri, Claudio and Vassura, Gabriele

SESSION : WE13 — Frontiers of Robotics (I)

Swinging from the Hip: Use of Dynamic Motion Optimization in the Design of Robotic Gait Rehabilitation 1433
Wang, Chia-Yu E and Bobrow, James E and Reinkensmeyer, David J

The Switched Reluctance Motor Drive for the Direct-drive Joint of the Robot 1439
Chen, Hao and Zhang, Dong

Development of a Micro Transfer Arm for a Microfactory 1444
Maekawa, Hitoshi and Komoriya, Kiyoshi

Optimally Designed Electrostatic Microactuators for Hard Disk Drives 1452
Jung, Sunghwan and Choi, Jae-Joon and Lee, Changho and Park, Jihwang and Park, Changsu and Min, Dongki and Kim, Chulsoon

Dynamics of Gel Robots made of Electro-active Polymer Gel 1457
Otake, Mihoko and Kagami, Yoshiharu and Inaba, Masayuki and Inoue, Hirochika

Discrete-time Multirate Stabilization of Chained Form Systems: Convergence, Robustness, and Performance 1463
Conticelli, Fabio and Palopoli, Luigi

SESSION : TA01 — Recent Advances in Randomized Motion Planning

Decomposition-based Motion Planning:
A Framework for Real-time Motion Planning in High-dimensional Configuration Spaces 1469
Brock, Oliver and Kavraki, Lyida E.

Disassembly Sequencing Using a Motion Planning Approach 1475
Sundaram, Sujay and Remmler, Ian and Amato, Nancy M

Quasi-randomized Path Planning 1481
Branicky, Michael S. and LaValle, Steven M. and Olson, Kari and Yang, Libo

Motion Planning in Environments with Dangerzones 1488
Sent, Danielle and Overmars, Mark H.

Computer Aided Motion: Move3D within MOLOG 1494
Simeon, Thierry and Laumond, Jean-Paul and Van Geem, Carl and Cortes, Juan

Customizing PRM Roadmaps at Query Time 1500
Song, Guang and Miller, Shawna and Amato, Nancy M.

SESSION : TA02 — Assembly Planning

A Hierarchical Model for Coordinated Planning and Control of Automated Assembly Manufacturing 1506
Gyorfi, Julius S. and Wu, Chi-haur

Planning Motion Compliant to Complex Contact States 1512
Ji, Xuerong and Xiao, Jing

Direct Teaching and Error Recovery Method for Assembly
Task Based on a Transition Process of a Constraint Condition 1518
Toshio, Fukuda and Masaharu, Nakaoka and Tsuyoshi, Ueyama and Yasuhisa, Hasegawa

Deterministic Path Planning for Planar Assemblies 1524
Sacks, Elisha P.

Generating a Configuration Space Representation for Assembly Tasks from Demonstration 1530
Chen, Jason R and Zelinsky, Alex

C-space Cell Defragmentation for General Translational Assembly Planning 1537
Schwarzer, Fabian and Schweikard, Achim

SESSION : TA03 — Medical Sensor Technologies

Laser-pointing Endoscope System for Intra-operative 3D Geometric Registration 1543
Hayashibe, Mitsuhiro and Nakamura, Yoshihiko

A User Interface for Robot-assisted Diagnostic Ultrasound 1549
Abolmaesumi, P. and Salcudean, S.E. and Zhu, W.H. and DiMaio, S.P. and Sirouspour, M.R.

A Study of Natural Eye Movement Detection and Ocular Implant Movement
Control Using Processed EOG Signals 1555
Gu, Jason and Meng, Max and Cook, Albert and Faulkner, Gary

Recovery of Distal Hole Axis in Intra Medullary Nail Trajectory Planning 1561
Zhu, Yonggen and Phillips, Roger and Griffiths, John G and Viant, Warren and Mohsen, Amr and Bielby, Mike

Remote Ultrasound Diagnostic System 1567
Mitsuishi, Mamoru and Warisawa, Shin'ichi and Tsuda, Taishi and Higuchi, Takuya and Koizumi,
Norihiro and Hashizume, Hiroyuki and Fujiwara, Kazuo

Biorobotics: An Instrument for an Improved Quality of Life. An Application for the Analysis Neuromotor Diseases 1575
Rovetta, Alberto F.

SESSION : TA04 — Localization (I)

Edge-based Features from Omnidirectional Images for Robot Localization 1579
Vlassis, Nikos and Motomura, Yoichi and Hara, Isao and Asoh, Hideki

An Optimal Pose Estimator for Map-based Mobile Robot Dynamic Localization:
Experimental Comparison with the EKF 1585
Borges, Geovany A. and Aldon, Marie-Josi and Gil, Thierry

Robot Localization in Nonsmooth Environments: Experiments with a New Filtering Technique 1591
Antoniali, Fabio M. and Oriolo, Giuseppe

Data Fusion of Four ABS Sensors and GPS for an Enhanced Localization of Car-like Vehicles 1597
   Bonnifait, Philippe and Bouron, Pascal and Crubille, Paul and Meizel, Dominique

Self-localization for Mobile Robots by Matching of Two Consecutive Environmental Range Data 1603
   Jeong, In-Soo and Cho, Hyungsuck

Deriving and Mmatching Image Fingerprint Sequences for Mobile Robot Localization 1609
   Lamon, Pierre and Nourbakhsh, Illah and Jensen, Bjoern and Siegwart, Roland

SESSION : TA05 — Behavior and Flight Control

Autonomous Helicopter Control Using Reinforcement Learning Policy Search Methods 1615
   Bagnell, J. Andrew and Schneider, Jeff G.

Robust Attitude Stabilization of Spacecraft Using Minimal Kinematic Parameters 1621
   Park, Yonmook and Tahk, Min-Jea

Spatio-temporal Case-based Reasoning for Behavioral Selection 1627
   Likhachev, Maxim and Arkin, Ronald

Autonomous Helicopter Hover Using an Artificial Neural Network 1635
   Buskey, Gregg D and Wyeth, Gordon W and Roberts, Jonathan M

Flight Control System for a Micromechanical Flying Insect: Architecture and Implementation 1641
   Schenato, Luca and Deng, Xinyan and Sastry, Shankar

Behavior Control of Robot Using Orbits of Nonlinear Dynamics 1647
   Sekiguchi, Akinori and Nakamura, Yoshihiko

SESSION : TA06 — Object Tracking

Object Tracking with a Pan-tilt-zoom Camera Application to Car Driving Assistance 1653
   Clady, Xavier and Collange, Franois and Jurie, Frdric and Martinet, Philippe

A Real-time Color-based Object Tracking Robust to Irregular Illumination Variations 1659
   You, Bum-Jae and Lee, Yong-Beom and Lee, Seong-Whan

Tracking Multiple Moving Targets with a Mobile Robot Using Particle Filters and
Statistical Data Association 1665
   Schulz, Dirk and Burgard, Wolfram and Fox, Dieter and Cremers, Armin B.

Object Tracking Using the Gabor Wavelet Transforms and the Golden Section Algorithm 1671
   He, Chao and Dong, Jianyu and Zheng, Yuan F. and Ahalt, Stanley C.

Robust Object Tracking Using an Adaptive Color Model 1677
   Jang, Gi-jeong and Kweon, In-so

Fast and Robust Tracking of Multiple Moving Objects with a Laser Range Finder 1683
   Kluge, Boris and Koehler, Christian and Prassler, Erwin

SESSION : TA07 — Vision-Based Mobile Robot Control

Spatial Navigation Principles: Applications to Mobile Robotics 1689
   Suluh, Anthony and Sugar, Thomas G and McBeath, Michael

Navigation Strategies Referring to Insect Homing in Flying Robots 1695
    Kobayashi, Hiroshi and Kikuchi, Kohki and Ochi, Kazuhiro and Onogi, Yu

Visual Servoing for a Scale Model Autonomous Helicopter 1701
    Chriette, Abdelhamid and Hamel, Tarek and Mahony, Robert

Vision-based Control of Mobile Robots 1707
    Burschka, Darius and Hager, Gregory D

Real-time Vision-based Control of a Nonholonomic Mobile Robot 1714
    Das, Aveek K and Fierro, Rafael and Kumar, Vijay and Southall, John B and Spletzer, John R and Taylor, Camillo J

A Vision System for Landing an Unmanned Aerial Vehicle 1720
    Sharp, Courtney S and Shakernia, Omid and Sastry, S Shankar

SESSION : TA08 — Mobile Robot Navigation

Evaluation of Path Length Made in Sensor-based Path-planning with the Alternative Following 1728
    Horiuchi, Yohei and Noborio, Hiroshi

A Fast Path Planning-and-tracking Control for Wheeled Mobile Robots 1736
    Lee, T. H. and H.K., Lam and F.H. Frank, Leung and K.S. Peter, Tam

The Science Autonomy System of the Nomad Robot 1742
    Wagner, Michael D and Apostolopoulos, Dimitrios and Shillcutt, Kimberly and Shamah, Benjamin and Simmons, Reid and Whittaker, William

Parallelizing Planning and Action of a Mobile Robot Based on Planning-action Consistency 1750
    Miura, Jun and Shirai, Yoshiaki

Vision Based Navigation for an Unmanned Aerial Vehicle 1757
    Sinopoli, Bruno and MIcheli, Mario and Donato, Gianluca and Koo, John

Disconnection Proofs for Motion Planning 1765
    Basch, Julien and Guibas, Leonidas J and Hsu, David and Nguyen, An Thai

SESSION : TA09 — Spherical Motors and Accurate Positioning

Theory, Design, and Implementation of a Spherical Encoder 1773
    Stein, David and Chirikjian, Gregory S. and Scheinerman, Edward R.

Research of Spherical Direct Drive Actuators Control Systems 1780
    Sokolov, Sergey M. and Trifonov, Oleg V. and Yaroshevskiy, Victor S.

Optimization of Power Grasps for Multiple Objects 1786
    Yoshikawa, Tsuneo and Watanabe, Tetsuyoh and Mutsuo, Daito

An Approach to Basic Design of the PM-type Spherical Motor 1792
    Ebihara, Daiki and Katsuyama, Norikazu and Kajioka, Morimasa

Multi-degree-of-freedom Spherical Permanent Magnet Motors 1798
    Wang, Jiabin and Mitchell, Ken and Jewell, Geraint, W. and Howe, David

Task Priority Based Mode Shaping Method for In-phase Design
of Flexible Structures Aiming at High Speed and Accurate Positioning 1806
   Yoshikawa, Tsuneo and Ueda, Jun

SESSION : TA10 — Advances in Semiconductor Manufacturing Automation

A Colored Petri Net-based Approach to the Design of 300mm Wafer Fab Controllers 1813
   Park, Jonghun and Reveliotis, Spyros and Bodner, Douglas and Zhou, Chen and McGinnis, Leon F.

Identification of Potential Deadlock Set in Semiconductor Track Systems 1820
   Yoon, Hyun Joong and Lee, Doo Yong

The Modeling and Control of the Cluster Tool in Semiconductor Fabrication 1826
   Huang, Han-Pang and Wang, Che-Lung

The Development of Holonic Information Coordination Systems with Security Considerations and
Error-recovery Capabilities 1832
   Cheng, Fan-Tien and Yang, Haw-Ching and Lin, Jen-Yu and Hung, Min-Hsiung

Priority-based Tool Capacity Allocation in the Foundry Fab 1839
   Yu, Chih-Yuan and Huang, Han-Pang

Two-level Optimization Method for Optimal Control of a Class of Hybrid Systems 1845
   Ji, H. zhang and Kwon, Wook Hyun and Likuan, Zhao

SESSION : TA11 — Haptics and VR

Necessary Spatial Resolution for Realistic Tactile Feeling Display 1851
   Asamura, Naoya and Shinohara, Tomoyuki and Tojo, Yoshiharu and Shinoda, Hiroyuki

Finger Posture and Shear Force Measurement Using Fingernail Sensors: Initial Experimentation 1857
   Mascaro, Stephen A and Asada, Harry H

Time Domain Passivity Control of Haptic Interfaces 1863
   Hannaford, Blake and Ryu, Jee-Hwan

Human Interface for Maneuvering Nonholonomic Systems 1870
   Arai, Hirohiko

Virtual Reality Tools for Internet Robotics 1878
   Belousov, Igor R. and Chellali, Ryad and Clapworthy, Gordon J.

Robotic Mapping of Friction and Roughness for Reality-based Modeling 1884
   Lloyd, John E. and Pai, Dinesh K.

SESSION : TA12 — Teleoperation in space and Internet

Tele-manipulation of a Satellite Mounted Robot by an On-ground Astronaut 1891
   Oda, Mitsushige and Doi, Toshitsugu and Wakata, Koichi

The Control Oriented QoS: Analysis and Prediction 1897
   Wang, Qing Peng and Da Long, Tan and Ning, Xi and Yue Chao, Wang

Breaking the Lab's Walls: Tele-laboratories at the University of Pisa 1903
Bicchi, Antonio and Coppelli, Alessandro and Quarto, Francesco and Rizzo, Luigi and Balestrino, Aldo

UA Telehand: An Integrated Robotic Hand/Simulator System for Tele-manipulation via the Internet 1909
Guan, Yisheng and Ho, Teresa and Zhang, Hong

An Internet Robotic System Based Common Object Request Broker Architecture 1915
Jia, Songmin and Kunikatsu, Takase

Multimedia and Virtual Reality Techniques for the Control of ERA, the First Free Flying Robot in Space 1921
Freund, Eckhard and Rossmann, Juergen

SESSION : TA13 — Automobile Industry Automation in Korea

SESSION : TP01 — Modeling and Path Planning for Mobile Robots

Spanning-tree Based Coverage of Continuous Areas by a Mobile Robot 1927
Gabriely, Yoav and Rimon, Elon

Path Planning with Incremental Roadmap Update for Large Environments 1934
Li, Tsai-Yen and Chang, Chih-Ching

Probabilistic Roadmaps-Putting it all Together 1940
Dale, Lucia K and Amato, Nancy M

An Information Theoretical Approach to View Planning with Kinematic and Geometric Constraints 1948
Yu, Yong and Gupta, Kamal

A Pursuit-evasion BUG Algorithm 1954
Rajko, Stjepan and LaValle, Steven M.

Estimating Consistency of Geometric World Models Through Observation of a Localization Process 1961
Pavlin, Gregor and Braunstingl, Reinhard

SESSION : TP02 — Learning Systems

Protosymbol Emergence Based on Embodiment: Robot Experiments 1968
Karl, MacDorman F. and Koji, Tatani and Yoji, Miyazaki and Masanao, Koeda and Yoshihiko, Nakamura

Learning Momentum: Integration and Experimentation 1975
Lee, James B and Arkin, Ronald C

A Flexible, Hierarchical and Distributed Control Kernel Architecture
for Rapid Resource Integration of Intelligent Building System 1981
Chung, Wen-Ya and Fu, lichen and Huang, Shih-Shinh

Learning From Observation Using Primitives 1988
Bentivegna, Darrin C. and Atkeson, Christopher G.

A Framework for the Adaptive Transfer of Robot Skill Knowledge Using Reinforcement Learning Agents 1994
Malak, Richard J and Khosla, Pradeep K

Structurally and Procedurally Simplified Soft Computing for Real Time Control 2002
Tar, Jszsef K. and Rudas, Imre J. and Bito, Janos F. and Andersson, Paul, H. and Torvinen, Seppo, J.

SESSION : TP03 — Medical Surgery Robotics

Study on Robot-assisted Minimally Invasive Neurosurgery and its Clinical Application 2008
Liu, Da and Wang, Tianmiao and Wang, Zigang and Zesheng, Tang and Tian, Zengmin and Du, Jixiang and Zhao, Quanjun

Heartbeat Synchronization for Robotic Cardiac Surgery 2014
Nakamura, Yoshihiko and Kishi, Kousuke and Kawakami, Hiro

A Safe Robot System for Craniofacial Surgery 2020
Engel, Dirk and Raczkowsky, Joerg and Woern, Heinz

Control Algorithms for Interactive Shaping 2025
Hein, Andreas and Lueth, Tim C.

EndoBot: A Robotic Assistant in Minimally Invasive Surgeries 2031
Kang, Hyosig and Wen, John T.

Mobile Virtual Endoscope System with Haptic and Visual Information for Non-invasive Inspection Training 2037
Ikuta, Koji and Iritani, Koji and Fukuyama, Junya

SESSION : TP04 — Localization(II)

Precise 3-D Localization by Automatic Laser Theodolite and Odometer for Civil-engineering Machines 2045
Bouvet, Denis and Garcia, Gaetan and Gorham, Barry J and Bitaille, David

Vision-based Mobile Robot Localization And Mapping Using Scale-invariant Features 2051
Se, Stephen and Lowe, David and Little, Jim

Morphological Neural Networks for Vision Based Self-localization 2059
Raducanu, Bogdan and Grana, Manuel and Sussner, Peter

Robust Localization Algorithms for an Autonomous Campus Tour Guide 2065
Thrapp, Richard and Westbrook, Christian and Subramanian, Devika

Robust Localization Using Context in Omnidirectional Imaging 2072
Paletta, Lucas and Frintrop, Simone and Hertzberg, Joachim

Optimal Exploratory Paths for a Mobile Rover 2078
Lorussi, Federico and Marigo, Alessia and Bicchi, Antonio

SESSION : TP05 — Control of Underactuated Systems

Observer Based Kinematic Tracking Controllers for a Unicycle-type Mobile Robot 2084
Lefeber, Erjen and Jakubiak, Janusz and Tchon, Krzysztof and Nijmeijer, Henk

Stabilization of a PR Planar Underactuated Robot 2090
De Luca, Alessandro and Iannitti, Stefano and Oriolo, Giuseppe

Local Accessibility and Stabilization of an Underactuated Crawling Robot with Changing Constraints 2096
Matsuno, Fumitoshi and Ito, Kazuyuki and Takahashi, Rie

Hybrid Control for the Pendubot 2102
Zhang, Mingjun and Tarn, Tzyh-Jong

An Automatic Control System for Ship Harbour Manoeuvres Using Decoupling Control 2108
LE, Minh-Duc

Optimal Control of Underactuated Manipulators Via Actuation Redundancy 2114
Maciel, Benedito C O and Bergerman, Marcel and Terra, Marco H

SESSION : TP06 — Mapping & Reconstruction

Plane Segment Finder: Algorithm, Implementation and Applications 2120
Okada, Kei and Kagami, Satoshi and Inaba, Masayuki and Inoue, Hirochika

Correcting Polyhedral Projections for Scene Reconstruction 2126
Ros, Llums and Thomas, Federico

Efficient Car Recognition Policies 2134
Isukapalli, Ramana and Greiner, Russell

Automatic Training of a Neural Net for Active Stereo 3D Reconstruction 2140
Neubert, Jeremiah J. and Hammond, Anthony and Do, Yongtae and Hu, Yu Hen and Guse, Nils

Experiments in Robust Bistatic Sonar Object Classification for Local Environment Mapping 2147
Sillitoe, Ian P. and Lundin, Magnus and Caselli, Stefano and Ferraro, Domenico

Calibration and 3D Measurement from Martian Terrain Images 2153
Vergauwen, Maarten and Pollefeys, Marc and Van Gool, Luc

SESSION : TP07 — Camera Calibration

Machine-vision-based Estimation of Pose and Size Parameters from a Generic Workpiece Description 2159
Kececi, Ferit and Nagel, Hans-Hellmut

A New Technique for Camera Self-calibration 2165
Ji, Qiang and Dai, Songtao

An Iterative Approach to the Hand-eye and Base-world Calibration Problem 2171
Hirsh, Robert L. and DeSouza, Guilherme N. and Kak, Avinash C.

Camera Calibration With Genetic Algorithms 2177
Ji, Qiang and Zhang, Yongmian

SmartView: Hand-eye Robotic Calibration for Active Viewpoint Generation and Object Grasping 2183
Motai, Yuichi and Kosaka, Akio

Camera Self-calibration from Ellipse Correspondences 2191
Ji, Qiang and Hu, Rong

SESSION : TP08 — Grasp Analysis

Kinematic Feasibility Analysis of 3D Grasps 2197
Guan, Yisheng and Zhang, Hong

Rigid Body Analysis of Power Grasps: Bounds of the Indeterminate Grasp Force 2203
Toru, Omata

Enveloping Grasp Feasibility Inequality 2210
Park, Jonghoon and Harada, Kensuke and Makoto, Kaneko

Computation of Fingertip Positions for a Form-closure Grasp 2217
Ding, Dan and Liu, Yun-Hui and Zhang, Jianwei and Knoll, Alois

Whole-arm Impedance of a Serial-chain Manipulator 2223
Mochiyama, Hiromi

Dynamics and Control of Whole Arm Grasps 2229
Song, Peng and Yashima, Masahito and Kumar, Vijay

SESSION : TP09 — Flexible Manufacturing

C'mon Part, do the Local Motion! 2235
Reznik, Dan S. and Canny, John F.

Minimal Trap Design 2243
Agarwal, Pankaj K. and Collins, Anne D. and Harer, John L.

Error Compensation of Workpiece Localization 2249
Xiong, Z.H. and Li, Z.X.

Robotized Peg-in-hole Task Involving the Needle-like and Pill-like Objects with Tight Tolerances 2255
Borovac, Branislav A. and Nagy, Laszlo and Begovic, Edvard and Nikolic, Milan and Aleksandar, Popadic and Andric, Dejan

Robotic Manufacturing of Complete Dentures 2261
Zhang, Yongde and Zhao, Zhanfang and Lu, Jilian and Tso, Shiu Kit

Force Unloading of a Flexible Manipulator Link 2267
Cheboxarov, Victor V

SESSION : TP10 — Biped Robots

Sensory Redundant Parallel Mobile Mechanism 2273
Shoval, Shraga and Shoham, Moshe

Body Trajectory Generation for Legged Locomotion Systems Using a Terrain Evaluation Approach 2279
Bai, Shaoping and Low, Kin Huat

Gait Analysis of a Human Walker Wearing Robot Feet as Shoes 2285
Sardain, Philippe and Bessonnet, Guy

Feedforward and Deterministic Fuzzy Control of Balance and Posture During Human Gait 2293
Kubica, Eric and Wang, David and Winter, David

Real-time 3D Walking Pattern Generation for a Biped Robot with Telescopic Legs 2299
Kajita, Shuuji and Matsumoto, Osamu and Saigo, Muneharu

Walking Human Avoidance and Detection from a Mobile Robot Using 3D Depth Flow 2307
Okada, Kei and Kagami, Satoshi and Inaba, Masayuki and Inoue, Hirochika

SESSION : TP11 — Dynamic Simulation & Haptic Interaction

KAIST Interactive Bicycle Simulator 2313
Kwon, Dong-Soo and Yang, Gi-Hun and Lee, Chong-Won and Shin, Jae-Cheol and Park, Youngjin and Jung, Byungbo and Lee, Doo Yong

Randomized Parallel Simulation of Constrained Multibody Systems for VR/Haptic Applications 2319
Bicchi, Antonio and Pallottino, Lucia and Bray, Marco and Perdomi, Pierangelo

Development of the PNU Vehicle Driving Simulator and its Performance Evaluation 2325
Park, Min Kyu and Lee, Min Cheol and Yoo, Ki Sung and Son, Kwon and Yoo, Wan Suk and Han, Myung Chul

A Transparency-optimized Control for a 6-DOF Parallel-structured Haptic Device 2331
Kim, Hyung Wook and Eom, Kwang Sik and Suh, Il Hong and Yi, Byung-Ju

LEM-An Approach for Real Time Physically Based Soft Tissue Simulation 2337
Costa, Ivan F. and Balaniuk, Remis

Performance of Pinching Motions of Two Multi-DOF Robotic Fingers with Soft-tips 2344
Arimoto, Suguru

SESSION : TP12 — Underwater Robotic Vehicle Control and Behavior-Based Control

A Novel Neuro-fuzzy Controller for Autonomous Underwater Vehicles 2350
Kim, Tae Won and Yuh, Junku

An Experimental Undulating-fin Device Using the Parallel Bellows Actuator 2356
Sfakiotakis, Michael and Lane, David M. and Davies, Bruce J.

An Autonomous Underwater Vehicle Control with a Non-regressor Based Algorithm 2363
Yuh, Junku and West, Michael E. and Lee, Pan-Mook

Execution Control of the NGC Tasks for ROVs 2369
Coletta, Paolo and Bono, Riccardo and Bruzzone, Gabriele and Caccia, Massimo and Veruggio, Gianmarco

3D Force Control System Design for a Hydraulic Parallel Bellows Continuum Actuator 2375
O'Brien, Des J. and Lane, David M.

Behavior-based Control of a Non-holonomic Robot in Pushing Tasks 2381
Emery, Rosemary and Balch, Tucker

SESSION : TP13 — Semiconductor Industry Automation in Korea

SESSION : TE01 — Kinematics

Fully Autonomous Calibration of Parallel Manipulators by Imposing Position Constraint 2389
Rauf, Abdul and Ryu, Jeha

Closed-form Inverse Kinematics Solver for Reconfigurable Robots 2395
Chen, I-Ming and Gao, Yan

The Continuous Inverse Kinematic Problem for Mobile Manipulators: A Case Study in the Dynamic Extension 2401
Jakubiak, Janusz and Tchon, Krzysztof

Singularity Analysis of Three-legged Parallel Robots Based on Passive-joint Velocities 2407
  Yang, Guilin and I-Ming, Chen and Wei, Lin and Jorge, Angeles

Kinematic Modeling of Mobile Robots by Transfer Method of Augmented Generalized Coordinates 2413
  Kim, Whee Kuk and Kim, Do Hyung and Yi, Byung-Ju and You, Bum Jae

Improved Concept for Derivation of Velocity Profiles for Elevator Systems 2419
  Gagov, Zavarin and Cho, Young Cheol and Kwon, Wook Hyun

SESSION : TE02 — Robot Design

Mechanical Design of a Talking Robot for Natural Vowels and Consonant Sounds 2424
  Nishikawa, Kazufumi and Asama, Kouichirou and Hayashi, Kouki and Takanobu, Hideaki and Takanishi, Atsuo

Design and Implementation of Software Research Platform for Humanoid Robotics: H6 2431
  Kagami, Satoshi and Nishiwaki, Koichi and Sugihara, Tomomichi and
  Kuffner, James J. and Inaba, Masayuki and Inoue, Hirochika

A New Ultralight Anthropomorphic Hand 2437
  Schulz, Stefan and Pylatiuk, Christian and Bretthauer, Georg

Development of the Crown Motor 2442
  Hitoshi, Kimura and Shigeo, Hirose and Koji, Nakaya

Design Methodology for a Novel Planar Three Degrees of Freedom Parallel Machine Tool 2448
  Wang, Jinsong and Tang, Xiaoqiang and Duan, Guanhong and Li, Jianfeng

Active Hose: An Artificial Elephant's Nose with Maneuverability for Rescue Operation 2454
  Tsukagoshi, Hideyuki and Kitagawa, Ato and Segawa, Mitsuru

SESSION : TE03 — Modeling and Planning for Dextrous Manipulation

Real-time Tracking Meets Online Grasp Planning 2460
  Kragic, Danica and Miller, Andrew T. and Allen, Peter K.

Stability Analysis of 3D Grasps by a Multifingered Hand 2466
  Yamada, Takayoshi and Koishikura, Tarou and Mizuno, Yuto and Mimura, Nobuharu and Funahashi, Yasuyuki

Planning of Graspless Manipulation by Multiple Robot Fingers 2474
  Maeda, Yusuke and Kijimoto, Hirokazu and Aiyama, Yasumichi and Arai, Tamio

Joint Solutions of Many Degrees-of-freedom Systems Using Dextrous Workspaces 2480
  Agrawal, Sunil K. and Kissner, Lea and Yim, Mark

Tactile Display Which Presents Shear Deformation on Human Finger 2486
  Morita, Hideyuki and Kwon, Guiryong and Fukuda, Toshio and Matsuura, Hideo

Rolling Based Manipulation Under Neighborhood Equilibrium 2492
  Harada, Kensuke and Kaneko, Makoto

SESSION : TE04 — Robot Design and Control Architecture

Optimal Dynamics of Constrained Multibody Systems. Application to Bipedal Walking Synthesis 2499
  Chesse, Stephane and Bessonnet, Guy

Inverse Kinematics and Dynamics of the 3-RRS Parallel Platform 2506
    Li, Jianfeng and Wang, Jinsong and Chou, Wusheng and Zhang, Yuru

A Middleware for Open Control System in the Distributed Computing Environment 2512
    Lee, Wongoo and Park, Jaehyun

Design and Control of the BUAA Four-fingered Hand 2517
    Zhang, Yuru and Han, Z and Zhang, H. and Shang, X. and Wang, T. and Guo, W. and Gruver, W. A.

Open Robot Control Software: The OROCOS Project 2523
    Bruyninckx, Herman

Algebra and Dynamics of Manufacturing-like Processes 2529
    Canuto, Enrico and Tonani, Sergio

SESSION : TE05 — Control of Robotic System

Approximate Jacobian Feedback Control of Robots with Kinematic Uncertainty and its Application to Visual Servoing 2535
    Cheah, Chien C and Li, Kai and Arimoto, Suguru and Kawamura, Sadao

Well Structured Robot Positioning Control Strategy for Position Based Visual Servoing 2541
    Bachiller, Margarita and Adan, Antonio and Feliu, Vicente and Cerrada, Carlos

A Model-based Anti-swing Control of Overhead Crane with High Hoisting Speeds 2547
    Lee, Ho-Hoon and Choi, Seung-gap

Model-based Control of Hydraulically Actuated Manipulators 2553
    Honegger, Marcel and Corke, Peter

Disturbance Attenuation in Robot Control 2560
    Choi, Chong-Ho and Kwak, Nojun

Control of Robot Manipulators with Consideration of Actuator Performance Degradation and Failures 2566
    Liu, Guangjun

SESSION : TE06 — Robot Assistants

Physical and Affective Interaction between Human and Mental Commit Robot 2572
    Shibata, Takanori and Tanie, Kazuo

Integration of Tactile Sensors in a Programming by Demonstration System 2578
    Zvllner, Raoul D. and Rogalla, Oliver and Dillmann, R|diger

Acquisition of Fuzzy Control Based Exercises of a Rings Gymnastic Robot 2584
    Yamada, Takaaki and Watanabe, Keigo and Kiguchi, Kazuo and Izumi, Kiyotaka

Human-friendly Interaction for Learning and Cooperation 2590
    Kristensen, Steen and Horstmann, Sven and Klandt, Jesko and Lohnert, Frieder

Dynamic Gestures as an Input Device for Directing a Mobile Platform 2596
    Ehrenmann, Markus and L|tticke, Tobias and Dillmann, R|diger

Robot Personality Based on the Equations of Emotion defined in the 3D Mental Space 2602
Miwa, Hiroyasu and Umetsu, Tomohiko and Takanishi, Atsuo and Takanobu, Hideaki

SESSION : TE07 — Sensor Based Control

Visual Tracking Using Snake for Object's Discrete Motion 2608
Kim, Won and Lee, Ju-Jang

Task-based Compliance Planning for Multi-fingered Hands 2614
Kim, Byoung-Ho and Yi, Byung-Ju and Oh, Sang-Rok and Suh, IL Hong

A Study on Feature-based Visual Servoing Control of Eight Axes-dual Arm Robot
by Utilizing Redundant Feature 2622
Han, S. H and Choi, J. W and Son, K and Lee, M. C and Lee, J. M and Lee, M.H

A New Exoskeleton-type Masterarm* with Force Reflection Based on the Torque Sensor Beam 2628
Kim, Yoon Sang and Lee, Sooyong and Cho, Changhyun and Kim, Munsang and Lee, Chong-Won

Computer Vision Based Object Detection and Recognition for Vehicle Driving 2634
Liu, Cheng-Yi and Fu, Li-Chen

The LMS Hand: Force and Position Controls in the Aim of the Fine Manipulation of Objects 2642
Gazeau, Jean-Pierre and Zeghloul, Saod and Arsicault, Marc and Lallemand, Jean-Paul

SESSION : TE08 — Human Augmentation

Augmenting Human Performance in Motion Planning Tasks - The Configuration Space Approach 2649
Ivanisevic, Igor and Lumelsky, Vladimir

Natural and Manmade Shared-control Systems: An Overview 2655
Tahboub, Karim A.

Adaptive Annotation Using a Human-robot Interface System Partner 2661
Yamashita, Masaya and Sakane, Shigeyuki

Active Human-mobile Manipulator Cooperation Through Intention Recognition 2668
Fernandez, Vicente and Balaguer, Carlos and Blanco, Dolores and Salichs, M.A.

Realization of Skill Controllers for Manipulation of Deformable Objects Based on Hybrid Automata 2674
Kazuaki, Hirana and Tatsuya, Suzuki and Shigeru, Okuma and Kaiji, Itabashi and Fumiharu, Fujiwara

Human-robot Cooperative Handling Using Virtual Nonholonomic Constraint in 3-D Space 2680
Takubo, Tomohito and Arai, Hirohiko and Tanie, Kazuo

SESSION : TE09 — Mobile Robot Navigation

On the Stability and Design of Distributed Manipulation Control Systems 2686
Murphey, Todd D. and Burdick, Joel W.

A Destination Driven Navigator with Dynamic Obstacle Motion Prediction 2692
Yu, Huiming and Su, Tong

Computation of Time Optimal Movements for Autonomous Parking of Non-holonomic Mobile Platforms 2698
Kondak, Konstantin and Hommel, Guenter

Visual Avoidance of Moving Obstacles Based on Vector Field Disturbances 2704
Bianco, Giovanni M. and Fiorini, Paolo

A Systematic Approach to Parameter Design for a Coupled Oscillator Controller 2710
Qi, Baohua and Anderson, Gary T.

A Controllability Test and Motion Planning Primitives for Overconstrained Vehicles 2716
Murphey, Todd D. and Burdick, Joel W.

SESSION : TE10 — Robot Control

Design of High Speed Planar Parallel Manipulator and Multiple Simultaneous Specification Control 2723
Kang, Bongsoo and Chu, Jiaxin and Mills, James K.

Two-time Scale Force and Position Control of Flexible Manipulators 2729
Siciliano, Bruno and Villani, Luigi

Control of Underactuated Manipulators Using Variable Period Deadbeat Control 2735
Mita, Tsutomu and Nam, Taek-Kun

Analytical Time Optimal Control Solution for a 2 Link Free Flying Acrobots 2741
Mita, Tsutomu and Nam, Taek-Kun and Hyon, Sang-Ho

Flexible Motion Realized by Forcefree Control: Pull-out-work by Articulated Robot Arm 2747
Kushida, Daisuke and Nakamura, Masatoshi and Goto, Satoru and Kyura, Nobuhiro

Mobile Robot Navigation Using a Sensor-based Control Strategy 2753
Victorino, Alessandro C. and Rives, Patrick and Borrelly, Jean-Jacques

SESSION : TE11 — IEEE Education Forum

SESSION : TE12 — Underwater Vision and Autofocusing

Application of Extended Covariance Intersection Principle for Mosaic-based Optical Positioning and
Navigation of Underwater Vehicle 2759
Xu, Xun and Negahdaripour, Shahriar

A 2-D Visual Servoing for Underwater Vehicle Station Keeping 2767
Lots, Jean-Francois and Lane, David M and Trucco, Emanuele and Chaumette, Francois

Controlling the Manipulator of an Underwater ROV Using a Coarse Calibrated Pan/Tilt Camera 2773
Marchand, Eric and Chaumette, Francois and Spindler, Fabien and Perrier, Michel

Positioning an Underwater Vehicle through Image Mosaicking 2779
Garcia, Rafael and Batlle, Joan and Cufi, Xevi and Amat, Josep

Feature Extraction and Data Association for AUV Concurrent Mapping and Localisation 2785
Tena Ruiz, Ioseba and Petillot, Yvan R. and Lane, David M. and Salson, Cedric

Practical Issues in Pixel-based Autofocusing for Machine Vision 2791
Ng, Kuang-Chern and Poo, Aun-Neow and Ang, Marcelo H

SESSION : TE13 — Steel Industry Automation in Korea

SESSION : FA01 — Robot Modeling

Using a Scale: Self-calibration of a Robot System with Factor Method 2797
Zhuang, Hanqi and Meng, Yan

The Cable Array Robot: Theory and Experiment 2804
Gorman, Jason J. and Jablokow, Kathryn W. and Cannon, David J.

Theoretical and Experimental Investigation of Stratified Robotic Finger Gaiting 2811
Goodwine, Bill and Wei, Yejun

Efficient Dynamic Simulation of Robotic Systems with Hierarchy 2818
Esposito, Joel M and Kumar, Vijay

O(N) Forward Dynamics Computation of Open Kinematic Chains Based on the Principle of Virtual Work 2824
Yamane, Katsu and Nakamura, Yoshihiko

Six Methods to Model a Flexible Beam Rotating in the Vertical Plane 2832
Piedboeuf, Jean-Claude

SESSION : FA02 — Identification of Robotic Systems

State Estimation and Parameter Identification of Flexible Manipulators Based on Visual Sensor and Virtual Joint Model 2840
Tsuneo, Yoshikawa and Atsuharu, Ohta and Katsuya, Kanaoka

Combining Internal and External Robot Models to Improve Model Parameter Estimation 2846
Verdonck, Walter and Swevers, Jan and Chenut, Xavier and Samin, Jean-Claude

Parameter Identification and Passivity Based Joint Control for a 7DOF Torque Controlled Light Weight Robot 2852
Albu-Schaeffer, Alin and Hirzinger, Gerd

Identifiable Parameters for Parallel Robots Kinematic Calibration 2859
Besnard, Sebastien and Khalil, Wisama

Identification of Joint Stiffness with Bandpass Filtering 2867
Pham, Minh Tu and Gautier, Maxime and Poignet, Philippe

Pose-and-twist Estimation of a Rigid Body Using Accelerometers 2873
Parsa, Kourosh and Angeles, Jorge and Misra, Arun K.

SESSION : FA03 — Collision Avoidance

Using Coded Signals to Benefit from Ultrasonic Sensor Crosstalk in Mobile Robot Obstacle Avoidance 2879
Shoval, Shraga and Borenstein, Johann

Tracking Control of Multiple Mobile Robots : A Case Study of Inter-robot Collision-free Problem 2885
Jongusuk, Jurachart and Mita, Tsutomu

Fast Optical Flow Estimation and its Application to Real-time Obstacle Avoidance 2891
Song, Kai-Tai and Huang, Jui-Hsian

A Novel Collision Detection Method Based on Enclosed Ellipsoid 2897
Ju, Ming-Yi and Liu, Jing-Sin and Shiang, Shen-Po and Chien, Yuh-Ren

Kinetic Collision Detection: Algorithms and Experiments 2903
Guibas, Leonidas J and Xie, Feng and Zhang, Li

Application of a Blind Person Strategy for Obstacle Avoidance with the Use of Potential Fields 2911
Lopes, Ernesto P. and Aude, Eliana P. and Silveira, Julio T. and Serdeira, Henrique and Martins, Mario F.

SESSION : FA04   Architecture for Multi Agent Systems

System Architecture for Versatile Autonomous and Teleoperated Control of Multiple Miniature Robots 2917
Rybski, Paul E and Papanikolopoulos, Nikolaos

Towards a Team of Robots with Reconfiguration and Repair Capabilities 2923
Bererton, Curt A. and Khosla, Pradeep K.

An Extension of the Plan-merging Paradigm for Multi-robot Coordination 2929
Gravot, Fabien and Alami, Rachid

Optical Guidance System for Multiple Mobile Robots 2935
Paromtchik, Igor E. and Asama, Hajime

Evaluating Control Strategies for Wireless-networked Robots Using an Integrated Robot and Network Simulation 2941
Ye, Wei and Vaughan, Richard T and Sukhatme, Gaurav S and Heidemann, John

Pursuit-evasion Games with Unmanned Ground and Aerial Vehicles 2948
Vidal, Rene and Rashid, Shahid and Sharp, Cory and Shakernia, Omid and Kim, Jin and Sastry, Shankar

SESSION : FA05   Fuzzy Control

A New Fuzzy-logic Anti-swing Control for Industrial Three-dimensional Overhead Cranes 2956
Lee, Ho-Hoon and Cho, Sung-Kun

Fuzzy-sliding Mode Control of Automatic Polishing Robot System with
the Self Tuning Fuzzy Inference Based on Genetic Algorithm 2962
Go, Seok Jo and Lee, Min Cheol and Park, Min Kyu

A Fuzzy Compensator for Uncertainty of Industrial Robots 2968
Chen, Wuwei and Mills, James K. and Chu, Jiaxin and Sun, Dong

Fuzzy Logic Based Tuning of Sliding Mode Controller for Robot Trajectory Control 2974
Ryu, Se-Hee and Park, Jahng-Hyon

Autonomous Helicopter Control Using Fuzzy Gain Scheduling 2980
Kadmiry, Bourhane and Bergsten, Pontus and Driankov, Dimiter

Efficient Neuro-fuzzy Control Systems for Autonomous Underwater Vehicle Control 2986
Lee, C. S. George and Wang, Jeen-Shing

SESSION : FA06 — Robot Cooperation (I)

An Architecture for Tightly Coupled Multi-robot Cooperation 2992
  Chaimowicz, Luiz and Sugar, Thomas and Kumar, Vijay and Campos, Mario

Adaptive Relocation of Environment-attached Storage Device by Multiple Robots 2998
  Umetani, Tomohiro and Mae, Yasushi and Inoue, Kenji and Arai, Tatsuo

Decentralized Control of Multiple Mobile Robots Transporting a Single Object in
Coordination without Using Force/Torque Sensors 3004
  Kume, Youhei and Hirata, Yasuhisa and Kosuge, Kazuhiro and Asama, Hajime and Kaetsu, Hayato and Kawabata, Kuniaki

Map-based Control of Distributed Robot Helpers for Transporting an Object in Cooperation with a Human 3010
  Hirata, Yasuhisa and Takagi, Takeo and Kosuge, Kazuhiro and Asama, Hajime and Kaetsu, Hayato and Kawabata, Kuniaki

A Distributed Ladder Transportation Algorithm for Two Robots in a Corridor 3016
  Asahiro, Yuichi and Chang, Eric C.-H. and Mali, Amol and Suzuki, Ichiro and Yamashita, Masafumi

Coordination of Multiple Mobile Manipulators 3022
  Sugar, Thomas and Desai, Jaydev P. and Kumar, Vijay and Ostrowski, James P.

SESSION : FA07 — Dextrous Manipulation

Analysis on Detaching Assist Motion (DAM) 3028
  Kaneko, Makoto and Shirai, Tatsuya and Harada, Kensuke and Tsuji, Toshio

Fundamentals and Analysis of Compliance Characteristics for Multi-fingered Hands 3034
  Kim, Byoung-Ho and Yi, Byung-Ju and Oh, Sang-Rok and Suh, IL Hong

Stiffness Control for Geared Manipulators 3042
  Tischler, Neil A and Goldenberg, Andrew A

A New Theory in Stiffness Control for Dextrous Manipulation 3047
  Chen, Shih-Feng and Yanmei, Li and Imin, Kao

A Review of Modeling of Soft-contact Fingers and Stiffness Control for Dextrous Manipulation in Robotics 3055
  Li, Yanmei and Kao, Imin

Multi-metric Comparison of Optimal 2D Grasp Planning Algorithms 3061
  Bone, Gary M. and Du, Yonghui

SESSION : FA08 — Navigation & Exploration

A Rule-based Fuzzy Traversability Index for Mobile Robot Navigation 3067
  Howard, Ayanna and Seraji, Homayoun and Tunstel, Edward

Vision Based Navigation for Mobile Robots in Indoor Environment by Teaching and Playing-back Scheme 3072
  Tang, Lixin and Yuta, Shin'ichi

Inertially Assisted Stereo Tracking for an Outdoor Rover 3078
  Nickels, Kevin M. and Huber, Eric

Safe Navigation on Hazardous Terrain 3084
  Howard, Ayanna and Seraji, Homayoun and Tunstel, Edward

An Exploration and Navigation Approach for Indoor Mobile Robots Considering Sensor's Perceptual Limitations 3092
Romero, Leonardo and Morales, Eduardo and Sucar, Enrique

Autonomous Exploration Using Multiple Sources of Information 3098
Moorehead, Stewart J and Simmons, Reid and Whittaker, William

SESSION : FA09 — Flexible Automation (III)

Sensorless Machine Tool Condition Monitoring Based on Open NCs 3104
Plapper, Volker and Weck, Manfred

PLC Based Coordination Schemes for a Multi-robot System 3109
Moon, Chanwoo and Lee, Beomhee and Kim, M.S

Scheduling and Optimization for a Class of Single-stage Hybrid Manufacturing Systems 3115
Zhang, Jihui and Kwon, Wook Hyun and Likuan, Zhao

A High-performance Network Infrastructure and Protocols for Distributed Automation 3121
Kume, Shinji and Rizzi, Alfred A

Graph-based Surface Merging in CAD-guided Dimensional Inspection of Automotive Parts 3127
Sheng, Weihua and Xi, Ning and Song, Mumin and Chen, Yifan

An Operations Planner for Integrated Shop Floor Material Processing and Material Handling Operations 3133
Steele, Jay W and Cannon, David J. and Wysk, Richard A

SESSION : FA10 — Multi Legged Robots (I)

Extended Virtual Passive Dynamic Walking and Virtual Passivity-mimicking Control Laws 3139
Asano, Fumihiko and Hashimoto, Minoru and Kamamichi, Norihiro and Yamakita, Masaki

Unified Model Approach for Planning and Control of Mobile Manipulators 3145
Tan, Jindong and Xi, Ning

Dynamic Compliant Quadruped Walking 3153
De Lasa, Martin and Buehler, Martin

Combined Use of Ground Learning Model and Active Compliance
to the Motion Control of Walking Robotic Legs 3159
Zhou, Debao and Low, K.H.

Development of MEL HORSE 3165
Takeuchi, Hiroki

Feedforward and Feedback Dynamic Trot Gait Control for a Quadruped Walking Vehicle 3172
Kurazume, Ryo and Hirose, Shigeo and Yoneda, Kan

SESSION : FA11 — Rehabilitation Robotics (I)

General Danger-evaluation Method of Human-care Robot Control and Development of Special Simulator 3181
Ikuta, Koji and Nokata, Makoto and Ishii, Hideki

Back and Forward Moving Scheme of Front Wheel Raising for Inverse Pendulum Control Wheel Chair Robot ... 3189
Takahashi, Yoshihiko and Takagaki, Tsuyoshi and Kishi, Jun and Ishii, Yohei

A Relaxation System Adapting to User's Condition 3195
Hasegawa, Yasuhisa and Ootsuka, Takeo and Fukuda, Toshio and Arai, Fumihito and Kawaguchi, Mitsuo

Pressure Distribution Image Based Human Motion Tracking System Using Skeleton and
Surface Integration Model 3201
Harada, Tatsuya and Sato, Tomomasa and Mori, Taketoshi

Development of an Intelligent Guide-stick for the Blind 3208
Kang, Sung Jae and Kim, Young Ho and Moon, In Hyuk

Accurate Femoral Canal Shaping in Total Hip Arthroplasty Using a Mini-robot 3214
Yoon, Y.S. and Lee, J.J. and Kwon, D.S. and Won, J.H. and Hodgson, AJ and Oxland, T

SESSION : FA12 — Manipulation in Underwater Environment

Dexterous Underwater Object Manipulation Via Multirobot Cooperating Systems 3220
Casalino, Giuseppe and Angeletti, Damiano and Bozzo, Tommaso and Marani, Giacomo

Self-tuning Position and Force Control of an Underwater Hydraulic Manipulator 3226
Clegg, Andrew C. and Dunnigan, Matthew W. and Lane, David M.

Fault Diagnosis on Autonomous Robotic Vehicles with Recovery:
An Integrated Heterogeneous-knowledge Approach 3232
Hamilton, Kelvin and Lane, David M. and Taylor, Nick. K. and Brown, Keith

Control of Underwater Manipulators Mounted on an ROV Using Base Force Information 3238
Ryu, Jee-Hwan and Kwon, Dong-Soo and Lee, Pan-Mook

Application of Adaptive Disturbance Observer Control to an Underwater Manipulator 3244
Yuh, Junku and Zhao, Side and Lee, Pan-Mook

A Distributed Environment for Virtual and/or Real Experiments for Underwater Robots 3250
Ridao, Pere and Batlle, Joan and Amat, Josep and Carreras, Marc

SESSION : FA13 — Parallel Mechanisms (I)

H4 Parallel Robot: Modeling, Design and Preliminary Experiments 3256
Pierrot, Frangois and Marquet, Fridiric and Company, Olivier and Gil, Thierry

Robust Parallel Robot Calibration with Partial Information 3262
Daney, David and Emiris, Ioannis Z.

On the Stiffness and Stability of Gough-Stewart Platforms 3268
Svinin, Mikhail and Hosoe, Shigeyuki and Uchiyama, Masaru

Eclipse-II: A New Parallel Mechanism Enabling Continuous 360-degree Spinning
Plus Three-axis Translational Motions 3274
Kim, Jongwon and Hwang, Jae-Chul and Kim, Jin-Sung and Park, F.C.

Stiffness Estimation of a Tripod-based Parallel Kinematic Machine 3280
Huang, Tian and Mei, Jiangping and Zhao, Xingyu and Zhou, Lihua

Analysis and Design Criteria for a Redundantly Actuated 4-legged Six Degree-of-freedom Parallel Manipulator 3286
Yi, Byung-Ju and Cox, Daniel and Tesar, Delbert

SESSION : FP01 — Motion and Trajectory Planning

Motion Planning for a Bi-steerable Car 3294
Sekhavat, Sepanta and Hermosillo, Jorge and Rouchon, Pierre

Kinematic Controllability and Decoupled Trajectory Planning for Underactuated Mechanical Systems 3300
Bullo, Francesco and Lynch, Kevin M.

Generation of Optimal Trajectory for Real System of an Under-actuated Manipulator 3308
Luan, Nan and Ming, Aiguo and Kajitani, Makoto

Balancing the Selection Pressures and Migration Schemes in Parallel Genetic Algorithms for Planning Multiple Paths 3314
Oh, Sang-Keon and Kim, Cheol-Taek and Lee, Ju-Jang

Trajectory Planning for a Four-wheel-steering Vehicle 3320
Wang, Danwei and Qi, Feng

Optimal Scheduling Techniques for Cluster Tools with Process-module and Transport-module Residency Constraints 3326
Rostami, Shadi and Hamidzadeh, Babak

SESSION : FP02 — Kinematics and Design of Robot

On the Computation of the Direct Kinematics of Parallel Spherical Mechanisms Using Bernstein Polynomials 3332
Bombin, Carlos and Ros, Lluis and Thomas, Federico

Polyhedral Single Degree-of-freedom Expanding Structures 3338
Agrawal, Sunil K. and Kumar, Saravana and Yim, Mark and Suh, John

Geometry of Dynamic and Higher-order Kinematic Screws 3344
Stramigioli, Stefano and Bruyninckx, Herman

A Novel 4-DOF Parallel Manipulator and its Kinematic Modelling 3350
Chen, Wen-Jia and Zhao, Ming-Yang

On a New Generation of Torque Controlled Light-weight Robots 3356
Hirzinger, Gerd and Albu-Schaeffer, Alin and Haehnle, Matthias and Schaefer, Ingo and Sporer, Norbert

A Closed Form for Inverse Kinematics Approximation of General 6R Manipulators Using Genetic Programming 3364
Chapelle, Frederic and Bidaud, Philippe

SESSION : FP03 — Trajectory Planning & Control

Fault Tolerant Control Strategy for OmniKity-III 3370
Jung, Myung-Jin and Kim, Jong-Hwan

Balancing a Humanoid Robot Using Backdrive Concerned Torque Control and Direct Angular Momentum Feedback 3376
Kajita, Shuuji and Yokoi, Kazuhito and Saigo, Muneharu and Tanie, Kazuo

Motion Planning of Legged Vehicles in an Unstructured Environment 3383
Eldershaw, Craig and Yim, Mark

Control of Posture and Tool Location Using a Single Force/Torque Sensor 3390
Neumann, Mathias

Periodic Control for a Blimp-like Dynamical Robot 3396
Zhang, Hong and Ostrowski, James P.

Real-time Collision-free Path Planning and Tracking Control of a Nonholonomic
Mobile Robot Using a Biologically Inspired Approach 3402
Yang, Simon X. and Yuan, Guangfeng and Meng, Max and Mittal, Gauri S.

SESSION : FP04 — From Microrobotics to Nanorobotics

Layered Nanoassembly of Three-dimensional Structures 3408
Requicha, A. A. G. and Meltzer, S and Resch, R. and Lewis, D. and Koel, B. E. and Thompson, M. E.

A De-coupled Vibratory Gyroscope Using a Mixed Micro-machining Technology 3412
Lee, Byeung Leul and Lee, Sang Woo and Jung, Kyu Dong and Choi, Joon Hyock
and Chung, Taek Ryong and Cho, Yong Chul

Distributed Control System for an Active Surface Device 3417
Ku, Peng-Jui and Safaric, Riko and Winther, Tobias K. and Stephanou, Harry E.

Three-legged Wireless Miniature Robots for Mass-scale Operations at the Sub-atomic Scale 3423
Martel, Sylvain and Sherwood, Mark and Helm, Chad and Garcia de Quevedo, William
and Fofonoff, Timothy and Dyer, Robert and Bevilacqua, John

Flexible Micro-processing by Multiple Micro Robots in SEM 3429
Aoyama, Hisayuki and Fuchiwaki, Ohmi

Micro Force Sensing in a Micro Robotic System 3435
Fahlbusch, Stephan and Fatikow, Sergej

SESSION : FP05 — Neural and Adaptive Control

Adaptive Control of Flexible Joint Manipulator 3441
Yim, Woosoon

Adaptive Control Design Using Delayed Dynamical Neural Networks for a Class of Nonlinear Systems 3447
Yu, Wen-Shyong and Wang, Gwo-Chuan

Experimental Studies of Neural Network Impedance Force Control for Robot Manipulators 3453
Jung, Seul and Yim, Sun Bin and Hsia, T. C.

Nonlinear Model Based Coordinated Adaptive Robust Control of Electro-hydraulic
Robotic Arms via Overparametrizing Method 3459
Bu, Fanping and Yao, Bin

A Stable Neural Adaptive Force Controller for a Hydraulic Actuator 3465
Boubaker, Daachi and Abedelaziz, Benallegue and Nacer, K. M'Sirdi

Chattering-free Dynamical TBG Adaptive Sliding Mode Control of Robot Arms with Dynamic
Friction for Tracking in Finite-time 3471
Parra-Vega, Vicente

SESSION : FP06 — Robot Cooperation (II)

Human-robot Cooperation with Mechanical Interaction Based on Rhythm Entrainment
-Realization of Cooperative Rope Turning- 3477
  Maeda, Yusuke and Takahashi, Atsushi and Hara, Takayuki and Arai, Tamio

Position and Force Tracking of a Two-manipulator System Manipulating a Flexible Beam Payload 3483
  Sun, Dong and Liu, Yun-Hui

Inverse Dynamics Analysis and Trajectory Generation of
Incompletely Restrained Wire-suspended Mechanisms 3489
  Yanai, Noritaka and Yamamoto, Motoji and Mohri, Akira

Adaptive Generation of Desired Velocity Field for Leader-follow
Type Cooperative Mobile Robots with Decentralized PVFC 3495
  Yamakita, Masaki and Suh, Jin-Ho

Representing and Discovering the Configuration of Conro Robots 3503
  Castano, Andres and Will, Peter

Hand-over of Unstable Object between Multiple Manipulators 3510
  Hoshino, Tasuku and Furuta, Katsuhisa

SESSION : FP07 — Modeling and Vision Application

Velocity Kinematic Modeling for Wheeled Mobile Robots 3516
  Shin, Dong Hun and Park, Kyung Hoon

On Stability of the Resolved Acceleration Control 3523
  Campa, Ricardo and Kelly, Rafael and Garcia, Eloisa

Toward Automated Inspection of Textile Surfaces: Removing the Textural Information
by Using Wavelet Shrinkage 3529
  Fujiwara, Hisanaga and Zhang, Zhong and Hashimoto, Koichi

Measuring Data Based Non-linear Error Modeling for Parallel Machine Tool 3535
  Yu, Xiaoliu and Zhao, Mingyang and Fang, Lijin and Wang, Honggua and Wang, Qiyi

Toward Multiview Registration in Frame Space 3542
  Sharp, Gregory C. and Lee, Sang W. and Wehe, David K.

Uncalibrated Vision Based on Structured Light 3548
  Fofi, David and Salvi, Joaquim and Mouaddib, El Mustapha

SESSION : FP08 — Discrete Event Manufacturing System (II)

An On-line Production Scheduler Using Neural Network and
Simulator Based on Manufacturing System States 3554
  Kim, Ki-Tae and Jang, Seong-Yong and Yoo, Byung-Hoon and Park, Jin-Woo

Modeling, Scheduling, and Prediction in Wafer Fabrication Systems
Using Queueing Petri Net and Genetic Algorithm 3559
  Wen, Hung-We and Fu, Li-Chen and Huang, Shih-Shinh

An Optimal Deadlock Avoidance Policy for Manufacturing System, with Flexible
Operation Sequence and Flexible Routing 3565
    Xing, Keyi and Lin, Feng and Hu, Baosheng

An Unified State Avoidance Policy to Solve Forbidden State Problems in Generalized Controlled Petri Nets 3571
    Cho, Young C. and Kwon, Wook Hyun

An Analysis of Network-based Control System Using CAN(Controller Area Network) Protocol 3577
    Jeon, Jong Man and Kim, Dae Won and Kim, Hong Seok and Cho, Yong Jo

Petri Net Controller Synthesis for Discrete Event Systems Using Weighted Inhibitor Arc 3582
    Wu, Weimin and Su, Hongye and Hu, Jianbo and Chu, Jian

SESSION : FP09 — Mapping & Localization

Sensor Fusion for Mobile Robot Dead-reckoning with a Precision-calibrated Fiber Optic Gyroscope 3588
    Chung, Hakyoung and Ojeda, Lauro and Borenstein, Johann

Greedy Mapping of Terrain 3594
    Koenig, Sven and Tovey, Craig and Halliburton, William

Optimal Landmark Pattern for Precise Mobile Robots Dead-reckoning 3600
    Amat, Josep and Aranda, Joan and Casals, Alicia and Fernandez, Xavier

On Mobile Robot Localization from Landmark Bearings 3605
    Shimshoni, Ilan

Guaranteed 3-D Mobile Robot Localization Using an Odometer, an Automatic Theodolite and
Indistinguishable Landmarks 3612
    Bouvet, Denis and Garcia, Gaetan

Integration of Schema-based Behaviors and Variable-resolution Cognitive Maps for Stable Indoor Navigation 3618
    Cho, Hye-Kyung and Cho, Young-Jo and You, Bum-Jae

SESSION : FP10 — Multi-Legged Robots (II)

A Fourier Perspective in Multi-legged Systems 3624
    Machado, Jose A. T. and Rodrigues, Carlos M. B.

Optimal Motion Primitives for a 5 DOF Experimental Hopper 3630
    Albro, Juanita V. and Bobrow, Jim E.

Immobilization Based Control of Spider-like Robots in Tunnel Environments 3636
    Shapiro, Amir and Rimon, Elon and Shoval, Shraga

Biomimetic Design and Fabrication of a Hexapedal Running Robot 3643
    Clark, Jonathan E. and Cham, Jorge G. and Bailey, Sean A. and Froehlich, Edward M.

Proprioception Based Behavioral Advances in a Hexapod Robot 3650
    Komsuoglu, Haldun and McMordie, Dave and Saranli, Uluc
    and Moore, Ned and Buehler, Martin and Koditschek, Daniel E.

A Simulator to Analyze Creeping Locomotion of a Snake-like Robot 3656
    Ma, Shugen and Li, Wen J. and Wang, Yuechao

SESSION : FP11 — Rehabilitation Robotics (II)

Visual Servoing for a User's Mouth with Effective Intention Reading in a Wheelchair-based Robotic Arm 3662
Song, Won-Kyung and Kim, Dae-Jin and Kim, Jong-Sung and Bien, Zeungnam

Fuzzy-neuro Control of an Exoskeletal Robot for Human Elbow Motion Support 3668
Kiguchi, Kazuo and Kariya, Shingo and Watanabe, Keigo and Fukuda, Toshio

Human Interface Using PC Display with Head Pointing Device for Eating Assist Robot and
Emotional Evaluation by GSR Sensor 3674
Takahashi, Yoshihiko and Hasegawa, Naoya and Takahashi, Katsumi and Hatakeyama, Takuro

A Neuro-based Adaptive Training Method for Robotic Rehabilitation Aids 3680
Tsuji, Toshio and Harada, Kensuke and Kaneko, Makoto

Development of an Above Knee Prosthesis Using MR Damper and Leg Simulator 3686
Kim, Jung-Hoon and Oh, Jun-Ho

Estimation of Forearm Movement from EMG Signal and Application to Prosthetic Hand Control 3692
Morita, Satoshi and Kondo, Toshiyuki and Ito, Koji

SESSION : FP12 — Autonomous Vehicles

Self-calibration of a Stereo Vision System for Automotive Applications 3698
Broggi, Alberto and Bertozzi, Massimo and Fascioli, Alessandra

Robust Lane Keeping from Novel Sensor Fusion 3704
Lee, Sukhan and Kwon, Woong

Path Planning for Newly Developed Microrover 3710
Kubota, Takashi and Kuroda, Yoji and Kunii, Yasuharu and Yoshimitsu, Tetsuo

Motion Planning in Dynamic Environments: Obstacles Moving along Arbitrary Trajectories 3716
Shiller, Zvi and Large, Fred and Sekhavat, Sepanta

Smooth Path Planning for Cars 3722
Fraichard, Thierry and Ahuactzin, Juan-Manuel

Map Building for a Terrain Scanning Robot 3728
Najjaran, Homayoun and Kircanski, Nenad and Goldenberg, Andrew, A.

SESSION : FP13 — Parallel Mechanisms (II)

Control of Impact Disturbance by Redundantly Actuated Mechanism 3734
Lee, Sang Heon and Yi, Byung-Ju and Kim, Soo Hyun and Kwak, Yoon Keun

Dragline Automation 3742
Ridley, Peter R. and Corke, Peter I.

Analysis and Control of Redundant Parallel Manipulators 3748
Liu, G.F. and Wu, Y.L. and Wu, X.Z. and Yiu, Y.K. and Li, Z.X.

Characterization of the Analytical Boundary of the Workspace for 3-6 SPS Parallel Manipulator 3755
Wang, Qizhi and Wang, Dongsheng and Tan, min

Nonlinear Control of a Hydraulic Parallel Manipulator 3760
Sirouspour, M. R. and Salcudean, S. E.

On the Dynamics of Parallel Manipulators 3766
Yiu, Y.Ko. and Cheng, H. and Xiong, Z.H. and Li, Z.X. and Liu, G.F.

SESSION : FE01 — Environmental Navigation with Obstacle Detection and Avoidance

Real-time Action Acquisition for Autonomous Mobile Robots Based
on Information Criterion for Environment 3772
Fujisawa, Kae and Hayakawa, Soichiro and Aoki, Takeshi and Suzuki, Tatsuya and Shigeru, Okuma

A Method for Obstacle Avoidance and Shooting Action of the Robot Soccer 3778
Wong, Ching-Chang and Ming-Fong, Chou and Chin-Po, Hwang and Cheng-Hsin, Tsai

Flexible Exploitation of Space Coherence to Detect Collisions of Convex Polyhedra 3783
Mirolo, Claudio and Pagello, Enrico

An Integrated Mobile Robot Path (Re)Planner and Localizer for Personal Robots 3789
Kim, Jinsuck and Amato, Nancy M. and Lee, Sooyong

A Recursive Algorithm of Obstacles Clustering for Reducing Complexity of Collision Detection in 2D Environment 3795
Chen, Jin-Liang and Liu, Jing-Sin and Lee, Wan-Chi

Collision Prediction and Avoidance Amidst Moving Objects for Trajectory Planning Applications 3801
Bernabeu, Enrique J. and Tornero, Josep and Tomizuka, Masayoshi

SESSION : FE02 — Robotic Actuations

Prototyping Pneumatic Group Actuators Composed of Multiple Single-motion Elastic Tubes 3807
Hirai, Shinichi and Masui, Tomohiro and Kawamura, Sadao

New Control Strategy of Variable Reluctance Direct Drive Motor for Robotic Applications 3813
Yang, Bing and Dianguo, Xu and Zongpei, Wang

A Comparison of Certain Quasi-velocities Approaches in PD Joint Space Control 3819
Herman, Przemys3aw and Kozlowski, Krzysztof

Improvement of Response Properties of MR-fluid Actuator by Torque Feedback Control 3825
Takesue, Naoyuki and Furusho, Junji and Sakaguchi, Masamichi

Reduced-cable Smart Motors Using DC Power Line Communication 3831
Liu, Chun-Hung and Wade, Eric and Asada, H. Harry

Development of PZT and PZN-PT Based Unimorph Actuators for Micromechanical Flapping Mechanisms 3839
Sitti, M. and Campolo, D. and Yan, J. and Fearing, R.S. and Su, T. and Taylor, D. and Sands, T.S.

SESSION : FE03 — Control of Flexible Robots

Alleviation of Chattering in Variable Structure Control Signal for Flexible One-link Manipulator 3847
Thomas, Susy and Kim, Jong-Hwan

Gauge Based Collision Detection Mechanism for a New Three-degree-of-freedom Flexible Robot 3853
Garcia, Andres H. and Feliu, Vicente B. and Somolinos, Jose Andres S.

Input Preshaping Vibration Suppression of Beam-mass-cart Systems Using Robust Internal-loop Compensator 3859
Kim, Bong Keun and Park, Sangdeok and Chung, Wan Kyun and Youm, Youngil

Robust Control for a Flexible-link Manipulator Using Sliding Mode Techniques and
Nonlinear H_infinity Control Design Methods 3865
Hisseine, Dadi and Lohmann, Boris

Model-free Regulation of Multi-link Smart Materials Robots 3871
Ge, Shuzhi Sam and Lee, Tong Heng and Wang, Zhuping

Good Vibrations: A Vibration Damping Setpoint Controller for Continuum Robots 3877
Gravagne, Ian A. and Rahn, Christopher D. and Walker, Ian D.

SESSION : FE04 — Microrobots

Virtual Insect Flight Simulator (VIFS): A Software Testbed for Insect Flight 3885
schenato, luca and deng, xinyan and wu, wei chung and sastry, shankar

PZT Actuated Four-bar Mechanism with Two Flexible Links for Micromechanical Flying Insect Thorax 3893
Sitti, Metin

Towards Flapping Wing Control for a Micromechanical Flying Insect 3901
Yan, Joseph and Wood, Robert J and Avadhanula, Srinath and Sitti, Metin and Fearing, Ronald S

Remotely Controllable Mobile Microrobots Acting as Nano Positioners
and Intelligent Tweezers in Scanning Electron Microscopes (SEMs) 3909
Schmoeckel, Ferdinand and Woern, Heinz

Efficiency of Swimming Microrobots Using Ionic Polymer Metal Composite Actuators 3914
Laurent, Guillaume and Piat, Emmanuel

A Magnetic Parallel Motion Hand for Micro Grasping and Processes 3920
Nakamura, Tatsuya and Shimamura, Koichiro and Andou, Takayuki

SESSION : FE05 — Force and Impact Control

Position and Force Control by Reaction Compensation 3926
Tanner, Herbert G. and Kyriakopoulos, Kostas J.

The Position/ Force Control with Self-adjusting Select-matrix for Robot manipulators 3932
Zhang, Hui and Zhiqiang, Zheng and Qing, Wei and Wenseng, Chang

On the Stiffness Control and Congruence Transformation Using the Conservative Congruence Transformation (CCT) 3937
Li, Yanmei and Kao, Imin

An Experimental Study of Planar Impact of a Robot Manipulator 3943
Pagilla, Prabhakar R and Yu, Biao

Precise Control of Industrial Robot Arms Considering Trajectory Allowance under Torque and Speed Constraints 3949
Munasinghe, Sudath R and Nakamura, Masatoshi and Goto, Satoru and Kyura, Nobuhiro

Control of a Heavy-duty Robotic Excavator Using Time Delay Control
with Switching Action with Integral Sliding Surface 3955
Lee, Sung-Uk and Chang, Pyung Hun

SESSION : FE06 — Coordination of Multi Agent Systems

Formation Constrained Multi-agent Control 3961
Egerstedt, Magnus and Hu, Xiaoming

Planning and Obstacle Avoidance for Mobile Robots 3967
Papadopoulos, Evangelos and Poulakakis, Ioannis

Scalable Dynamical Systems for Multi-agent Steering and Simulation 3973
Goldenstein, Siome and Karavelas, Menelaos and Metaxas, Dimitris and Guibas, Leonidas and Goswami, Ambarish

An Adaptive Path Planning Algorithm for Cooperating Unmanned Air Vehicles 3981
Cunningham, Christopher T. and Roberts, Randy S.

Dynamic Task Assignment in a Multiagent/Multitask Environment Based on Module Conflict Resolution 3987
Uchibe, Eiji and Kato, Tatsunori and Hosoda, Koh and Asada, Minoru

Artificial Immune-based Swarm Behaviors of Distributed Autonomous Robotic Systems 3993
Sun, Sang-Joon and Lee, Dong-Wook and Sim, Kwee-Bo

SESSION : FE07 — Sensors in Robotics

Quick Primitives Extraction Using Inertia Matrix on Measures Issue from an Ultrasonic Network 3999
Canou, Joseph and Novales, Cyril and Poisson, Gerard and Marche, Pierre

Sensing Odour Sources in Indoor Environments without a Constant Airflow by a Mobile Robot 4005
Lilienthal, Achim and Zell, Andreas and Wandel, Michael and Weimar, Udo

A Force/Moment Sensor for Intuitive Robot Teaching Application 4011
Choi, Myoung H and Lee, Woo W

Learning to Locate an Odour Source with a Mobile Robot 4017
Duckett, Tom and Axelsson, Mikael and Saffiotti, Alessandro

Tactile Differentiator 4023
Kaneko, Makoto and Bessho, Yoshiharu and Tsuji, Toshio

Reconstruction of Curved Surfaces Using Active Tactile Sensing and Surface Normal Information 4029
Liu, Fuming and Hasegawa, Tsutomu

SESSION : FE08 — Redundant Manipulators

Kinematic/Static Dexterity Measure of Manipulators with Limit-driven Characteristics of Actuators 4035
MA, Shugen

Multiple Tasks Kinematics Using Weighted Pseudo-inverse for Kinematically Redundant Manipulators 4041
Park, Jonghoon and Choi, Youngjin and Chung, Wan Kyun and Youm, Youngil

Redundant Resolution with Minimum Joint Elastic Deflection for Elastic Joint Redundant Robots 4048
Zhao, jing and Zhang, yue-ming

Redundant Manipulator Infinity-norm Joint Torque Optimization
with Actuator Constraints Using a Recurrent Neural Network 4054
Tang, Wai Sum

A Novel Quantitative Measure of Redundancy for Kinematically Redundant Manipulators 4060
  Hung, Min-Hsiung and Cheng, Fan-Tien and Ting, Jen-Kuei

An Approach to Torque Optimizing Control for a Redundant Manipulator 4066
  Chung, Chi Youn and Kim, M. S. and Lee, Beom Hee

SESSION : FE09 — Programming

Modelling Manufacturing Control Software 4072
  Storoshchuk, Orest and Wang, Shige and Shin, Kang G.

Automated Programming of an Industrial Robot through Teach-by Showing 4078
  Myers, Donald R. and Pritchard, Michael J. and Brown, Mark D.J.

Developmental Software Environment that is applicable to Small-size Humanoids and Life-size Humanoids 4084
  Kanehiro, Fumio and Inaba, Masayuki and Inoue, Hirochika and Hirukawa, Hirohisa and Hirai, Shigeoki

Reconfigurable Software for Open Architecture Controllers 4090
  Wang, Shige and Shin, Kang G.

Programming by Demonstration: Removing Suboptimal Actions in a Partially Known Configuration Space 4096
  Chen, Jason R and Zelinsky, Alex

A Framework for Using Discrete Control Synthesis in Safe Robotic Programming and Teleoperation 4104
  Rutten, Eric P.

SESSION : FE10 — Walking Robots

Online Mixture and Connection of Basic Motions for Humanoid Walking Control by Footprint Specification 4110
  Nishiwaki, Koichi and Sugihara, Tomomichi and Kagami, Satoshi and Inaba, Masayuki and Inoue, Hirochika

Climbing Stairs with EP-WAR2 Biped Robot 4116
  Figliolini, Giorgio and Ceccarelli, Marco

Goal-oriented Biped Walking Based on Force Interaction Control 4122
  Silva, Filipe M. and Machado, Josi A.

Human-like Actuated Walking that is Asymptotically Stable Without Feedback 4128
  Mombaur, Katja D and Bock, Hans Georg and Longman, Richard W and Schloeder, Johannes P

Reflex Control of Biped Robot Locomotion on a Slippery Surface 4134
  Park, Jong Hyeon and Kwon, Ohung

To Wards the Design of a Biped Jogging Robot 4140
  Gienger, Michael and Loeffler, Klaus and Pfeiffer, Friedrich

SESSION : FE11 — Frontiers of Robotics (II)

Evolution of Generative Design Systems for Modular Physical Robots 4146
  Hornby, Gregory S. and Lipson, Hod and Pollack, Jordan B.

Autonomous Flight Experiment with a Robotic Unmanned Airship 4152
  Ramos, Josue J G and Paiva, Ely C and Azinheira, Jose R
  and Bueno, Samuel S and Maeta, Silvio M and Mirisola, Luiz G B and Bergerman, M

Autonomous Robotic Meteorite Identification in Antarctica 4158
    Pedersen, Liam and Wagner, Michael and Apostolopoulos, Dimitrios and Whittaker, William R

Psychological Analysis on Human-robot Interaction 4166
    Kanda, Takayuki and Ishiguro, Hiroshi and Ishida, Toru

Robotic Antarctic Meteorite Search: Outcomes 4174
    Apostolopoulos, Dimitrios and Pedersen, Liam and Shamah, Benjamin and Shillcutt, Kimberly and Wagner, Michael and Whittaker, William

Skyworker: A Robot for Assembly, Inspection and Maintenance of Large Scale Orbital Facilities 4180
    Skaff, Sarjoun and Urmson, Chris and Whittaker, William

SESSION : FE12 — Humanoid Robots

Learning a Coordinate Transformation for a Human Visual Feedback Controller Based on Disturbance Noise and the Feedback Error Signal 4186
    Oyama, Eimei and Chong, Nak Young and Agah, Arvin and MacDorman, Karl F.

Hormone-controlled Metamorphic Robots 4194
    Salemi, Behnam and Shen, Wei-Min and Will, Peter

Stability of Coupled Hybrid Oscillators 4200
    Klavins, Eric and Koditschek, Daniel E.

Imitation and Primitive Symbol Acquisition of Humanoids by the Integrated Mimesis Loop 4208
    Inamura, Tetsunari and Nakamura, Yoshihiko and Ezaki, Hideaki and Toshima, Iwaki

Towards the Development of a Humanoid Arm by Minimizing Interaction Forces Through Minimum Impedance Control 4214
    Desai, Jaydev P. and Howe, Robert D.

Humanoids Walk with Feedforward Dynamic Pattern and Feedback Sensory Reflection 4220
    Qiang, Huang and Nakamura, Yoshihiko and Inamura, Tetsunari

SESSION : FE13 — Robot Sensing and Data Fusion

Multilevel Multisensor Based Decision Fusion for Intelligent Animal Robot 4266
    Luo, Ren C. and Henry Phang, S. H. and Su, Kuo L.

Localization by Voronoi Diagrams Correlation 4232
    Blanco, Dolores and Boada, Beatriz L. and Moreno, Luis

A Counter Example to the Theory of Simultaneous Localization and Map Building 4238
    Julier, Simon and Uhlmann, Jeffrey

Drift-free Attitude Estimation for Accelerated Rigid Bodies 4244
    Rehbinder, Henrik and Hu, Xiaoming

A Probabilistic Approach to Hough Localization 4250
    Iocchi, Luca and Mastrantuono, Domenico and Nardi, Daniele

Building Topological Models for Navigation in Large Scale Environments 4256
    Van Zwynsvoorde, Dominique and Simeon, Thierry and Alami, Rachid

Planning Motion Patterns of Human Figures Using a Multi-layered Grid and the Dynamics Filter

Zvi Shiller[1], Katsu Yamane[2], Yoshihiko Nakamura[2]

[1] Department of Mechanical and Aerospace Engineering
University of California
Los Angeles, CA 90095

[2] Department of Mechano-Informatics
University of Tokyo
7-3-1 Hongo, Bunkyo-ku, Tokyo, 113-8656 Japan

Abstract

This paper presents a practical motion planner for humanoids and animated human figures. Modeling human motions as a sum of rigid body and cyclic motions, we identify body postures that represent the rigid-body part of typical motion patterns. This leads to a model of the configuration space that consists of a multi-layered grid, each layer corresponding to a single posture. A global search through this reduced configuration space yields a feasible path and the corresponding postures along the path. A velocity profile is calculated along the optimal path, subject to the speed and acceleration limits assumed for each posture. Cyclic motions, generated from "primitive" cyclic motion patterns for each posture, are then added to the trajectory produced by the path planner. This "kinematic" motion is then modified by the dynamics filter [16] to result in dynamically consistent behavior. Examples are presented which demonstrate the use of this planner in an office environment.

1 Introduction

Recent research on motion planning for humanoids and animated human figures has been driven by the computer-game and movie industries and by several visible humanoid projects in Japan [2] and in the US. Unlike earlier research, which focused on generating stable motions [14, 8], the focus of recent works has been on synthesizing "natural" looking, or human-like, motions for interactive animation [4, 1, 16, 6] and for real-time control of humanoids [16, 13, 11]. Because of the large number of degrees of freedom involved in synthesizing full motions of humanoids or human figures, current approaches to generating natural-looking motions rely heavily on motion captures [10, 6, 16]. Most motion planners thus first generate a path for the center body, and then fit motion capture data to this path. The focus in the latter procedure is usually on kinematic "correctness," i.e. producing motions that are smooth and continuous in all joints. Such motions, however, are usually dynamically "incorrect," in that they do not satisfy the character's own dynamics and physical constraints due to parametric differences between the original data and the edited motion.

Ensuring dynamic feasibility, or consistency with the equations of motion, is one step towards producing natural-looking animations and physically feasible motion patterns for humanoids. As an effort to this goal, the authors have proposed the dynamics filter [16]. It computes the closest joint motions to the original data that are dynamically feasible and satisfy all physical constraints. The dynamics filter can be used to produce motion patterns reflecting changes in the mass distributions or the geometry of body parts (a fat/thin or tall/short body, heavy boots, a backpack).

Despite the extensive use of character animation in games and movies, only few planners have been developed to account for obstacle avoidance. One such planner was presented in [6] for on-line animation of biped characters. It generates the shortest path, which is tracked by a simple feedback controller. Its efficiency stems from reducing the motion planning problem to a disk moving in a 2D space (the walking surface) by representing the character by a bounding cylinder. The velocity profile along the path is modulated by a PD controller, using the average velocity of the motion capture data as the reference speed. This model requires that the obstacles be spaced sufficiently apart to allow a forward-facing walk, which precludes situations where it is not possible. Applications where greater attention to obstacle avoidance is necessary, include humanoids moving in realistic environments, such as construction sites, power plants, and damaged structures. Other environments, more likely to be used for character animation, include crowded offices, hospitals, class rooms, and movie theaters, where the character is forced to squeeze between closely spaced furniture or people. Addressing environments, where a single motion pattern (or posture) such as forward-facing motion is insufficient, is the focus of this paper.

Sensor-based obstacle avoidance of a humanoid was addressed in [15]. This work focused on planning the foot steps to ensure dynamic stability while avoiding vision-sensed obstacles. In [5], randomized trees were used to plan collision-free grasping of a 7 DOF arm. Both planners address the detailed motions of the legs and hands, and are hence impractical for full-body gross-motion planning.

This paper presents a gross-motion planner that considers several, yet potentially unlimited, number of motion postures by modeling human walk as consisting of rigid body and of cyclic motions. The "rigid-body" motion represents the net motion of the body's main parts. Absent any cyclic components, rigid-body motion consists therefore of the body posture that is characteristic of each motion pattern, such as forward-facing walk or a crawl. The number of steady-state (sustainable over a period of time) postures is assumed small and typical of a given human figure. This allows an efficient representation of the "rigid-body" configuration space by modeling each posture by one layer of a multi-layered graph.

The planner first selects a path and motion postures by searching over the multi-layered graph. It then computes the velocity profile along the path by modeling the human figure as a point mass, propelled by a single force in the direction of motion. The "rigid-body" postures are then substituted with "kinematic" cyclic motions of all body parts. The kinematic cyclic motions are generated from "primitive" cyclic motion patterns, consisting of a single cycle produced from motion capture data. The dynamics filter [16] then filters the kinematic motion to ensure that the resulting motion reflects the actual parameters of the human figure and satisfies its equations of motion. We have imple-

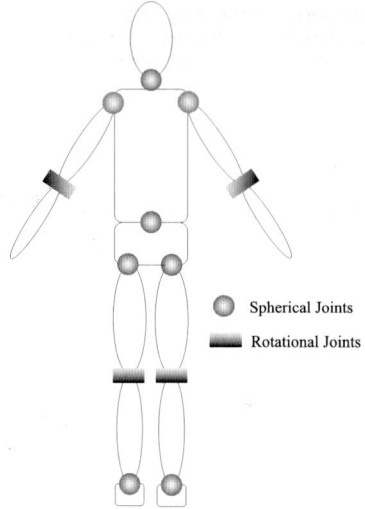

Figure 1: Human figure model

mented the planner for the three most common motion patterns: a forward-facing walk, a walk sideways, and crawling. While crawling is not commonly used in office environments or in movie theaters, it is often the pattern of choice in search and rescue operations. The planner is demonstrated for a human figure moving in a realistic office environment.

2 Modeling
2.1 Human figures

We consider a full-body human figure, with 13 links and 34 degrees of freedom (including the 6 DOF of the base body), as shown in Figure 1. For gross-motion planning, it is convenient to describe human motions as consisting of rigid body motion and cyclic motion [3]. This allows separating the motion planning problem into first determining the rigid body trajectory, and then attaching to it the cyclic motion. Such decomposition of the problem is necessary to overcome the inherent difficulty of planning in the large configuration space.

Planning for rigid body motions in $\mathcal{R}^3$ can be further simplified by recognizing that typically, the body attains a small number of sustainable postures, some shown in Figure 2. These postures are essentially approximations of the actual motion with all harmonics filtered out. We assume a small number of such postures, although more can be added with no difficulty. Reducing the number of possible configurations to a distinct set greatly simplifies the motion planning problem without significantly loosing motion fidelity.

To simplify collision checking, we represent each motion posture by a bounding polyhedron, as shown in Figure 3. Each polyhedron is then replaced with a cylinder, of the same height, and a radius equal to its

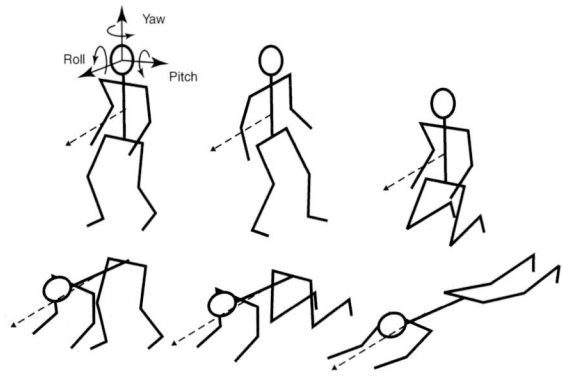

Figure 2: Various motion postures

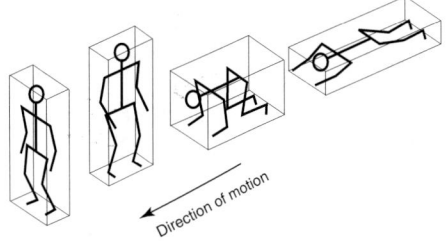

Figure 3: Bounding polyhedra of motion postures.

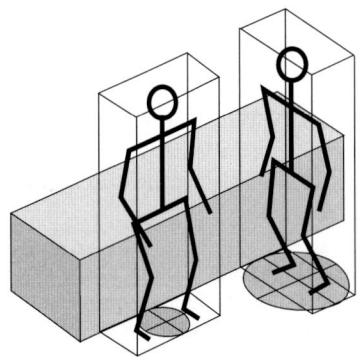

Figure 4: Replacing the polyhedron with a cylinder.

width (measured perpendicular to the direction of motion), as shown in Figure 4. This allows us to ignore rotation, thus reducing the configuration space of each pattern to 2D. For polyhedra that are wider than they are long, such as for the forward-facing walk, this approximation is conservative, and hence does not represent any difficulty. For polyhedra that are longer than they are wide, such as in sideways walk and crawl, this approximation does not pause a problem for straight-line motions, since the free space along the straight line provides room for the long, and unmodeled, part of the polyhedron. The problem arises at the turning points, where a free space for the cylinder does not ensure sufficient free space for the polyhedron. We circumvent this difficulty conservatively by permitting rotations only at points where sufficient free-space does exist, as discussed later.

Denoting the configuration space of the rigid body motion by G, and the configuration space of the full-body cyclic motion by H, the total configuration space is $Q = G \times H$. By planning for the rigid body motions, we actually plan in G.

Denoting the configuration space spanned by each motion pattern by $C_i \subset \mathcal{R}^2$, we effectively approximate G by $C \subset G$, $C = \bigcup C_i, (i = 1, \ldots, m)$, where m is the number of postures considered. C represents an efficient approximation of the *attainable* configuration space of the rigid body motion. Any inaccuracies introduced with this crude representation will be eliminated at the time the cyclic motion is added to the rigid body motion and filtered by the dynamics filter, as discussed later.

2.2 Obstacles

The obstacles are assumed known, or identified before motion planning begins. The obstacles can be of any shape, although polyhedral shapes are easier to handle.

For each reduced configuration space C_i, we generate the free subspace, $C_{ifree} \subset C_i$, by evaluating collisions at the grid points of a uniform grid, drawn onto the motion surface. At each grid point, the clearance to all obstacles within the height range of the cylinder, corresponding to the particular motion posture, is calculated. Points with clearance smaller than the cylinder's radius are marked "occupied." The union of all "occupied" points represents a conservative approximation of the occupancy map of all obstacles in C_i. The union of all free subspaces produces the total free-space for this problem: $C_{free} = \bigcup C_{ifree}, (i = 1, \ldots, m)$.

3 The Graph

Before constructing the search graph, we first clarify the problem we are actually solving. This problem is different from the standard shortest path problem because in addition to the path length we must account for the various motion postures. Since not all postures are created equal–some are more favorable than others, we assign to each posture a "priority factor" ν, with $\nu_1 = 1$ for the forward-facing walk, $\nu_2 = 1.1$ for sideways walk, and higher values for more taxing postures. We also wish to stay away from obstacles when possible, or equivalently, follow segments of the Voroni diagram. The problem addressed in this paper can therefore be stated as follows:

Denoting the start point $S = (s_x, s_y, i)$ (position x, y and posture i), the goal point $G = (g_x, g_y, j)$, and the free space C_{free}, compute the rigid-body motion, represented by the path $P \subset C_{free}$, from $S \in C_{free}$ to

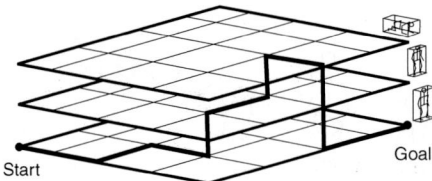

Figure 5: A multi-layered grid.

$G \in C_{free}$ that minimizes the cost function:

$$J = \int_0^{s_f} p(s)\delta(s,i)ds \quad (1)$$

where $p(s)$ is a potential function, reciprocal of the minimum distance to obstacles at each point, $\delta(s,i) = \nu_i$, i is the index of the selected posture, and s is the arc length along the path.

Minimizing (1) selects a short path that stays away from obstacles, using the highest-priority posture feasible. We now construct a weighted graph through which the shortest path solves the stated problem.

The reduced configuration space of each posture, $C_i \subset \mathcal{R}^2$ is represented by a uniform grid, projected onto the motion surface. Each grid point is a node $n_i(x, y, z)$ on the corresponding undirected graph $E_i, (i = 1, \ldots, m)$, with x, y, z being the three attributes of each node, (x, y) representing its location in a fixed coordinate frame attached to the motion surface, and z representing an assigned potential value. Each node in E_i is connected to eight of its nearest neighbors. Nodes in $C_{ifree} \subset C_i$ are assigned a value reciprocal to the distance to the nearest obstacle:

$$n_i(z) = \min_k \frac{d_0}{1 + d_{jmin}} \quad (2)$$

where k is the number of obstacles, d_0 is a constant, and d_{jmin} is the minimum distance of that node to the j-th obstacle. The value z is effectively a potential function, with lows coinciding with the Voroni diagram. This potential is used in the path search to verify sufficient room around turns.

The cost along the edge $e_i((x,y)_j, (x,y)_k) \in E_i$ between nodes $n_i((x,y,z)_j) \in E_i$ and $n_i((x,y,z)_k) \in E_i$ is the Euclidian distance, treating z as the "height" coordinate:

$$c_{j,k} = \sqrt{(x_j - x_k)^2 + (y_j - y_k)^2 + (z_j - z_k)^2}. \quad (3)$$

Minimizing this cost minimizes the path length and maximizes its distance to neighboring obstacles.

The graphs for the individual postures, $E_i, (i = 1, \ldots, m)$ are interconnected to form the total graph E by directed edges at nodes that are in the free space of both graphs. Thus, $e_{i,j,x,y}$ is an edge, connecting E_i to E_j by connecting the node $n_i(x,y), x,y \in C_{ifree}$ to node $n_j(x,y), x,y \in C_{jfree}$. The resulting graph, E, thus consists of interconnected 2D slices of the configuration space, as shown schematically in Figure 5.

Note that the graphs of the individual postures are connected to other postures according to the transition from one posture to the other. The cost for the edges between the graphs is assigned a positive value from a high priority posture to a lower priority posture (such as the sideways walk), and a zero value in the reverse direction. This encourages the search to select the preferred posture, and to return to that posture once such a transition is possible. The cost for each posture is scaled by ν to ensure that the path does not stay in a low priority posture unless absolutely necessary.

4 Path Search

A search for the shortest path through graph E accomplishes the stated objective by virtue of the cost assigned to the edges in each sub-graph E_i, and by the cost assigned to the edges connecting between the individual sub-graphs. The search itself is quite standard and deserves no special discussion, except to address the issue of orientation that was ignored when we generated the free space. This is resolved by penalizing for rotations, unless the minimum clearance to the nearest obstacle at a given node exceeds d_{turn} (see Figure 6):

$$d_{turn} = \frac{1}{2}\sqrt{(\frac{L}{2})^2 + W^2} \quad (4)$$

where W is the width and L is the length of the polyhedron. The minimum clearance is inferred from the potential value calculated in (2). Rotation is identified by a change in direction between the parent of a particular node and the parent of that parent. Obviously, this conservative treatment of rotations may miss possible solutions in tight spaces, but such approximations allow to reduce the configuration space to a small multiple of 2D spaces.

No other special attempt was made to speed up computation since the Dijkstra's algorithm [7] we used was sufficiently fast (a fraction of a second on a slow notebook PC) for a graph that accounts for three postures. The majority of the computation time was spent on processing the nodes and the collision checking. This can benefit from recent randomized tree approaches [5], however, ultimate efficiency was out of the scope of this paper.

The grid path is smoothed by a B-spline to establish curvature information for the computation of the velocity profile, as discussed next.

5 Velocity Profile

A properly selected velocity profile may add "physical reality" to the motion patterns. For this purpose,

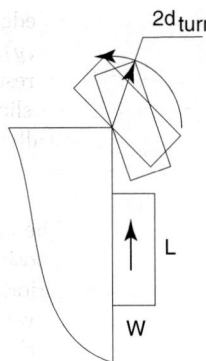

Figure 6: Minimum clearance during a turn.

we model the human figure as a point mass, pushed by a propelling force in the direction of motion. The computation of the velocity profile is formulated as an optimization problem that maximizes speed (or minimizes time), subject to constraints on the propelling force, the limit of friction force between the figure and the ground, and the speed limits for each posture.

This problem is solved efficiently using an existing algorithm for minimizing motion time of dynamic systems along specified paths [12]. The force and speed limits for the individual postures are assigned empirically to reflect typical human behavior. The velocity profile and path are used as inputs to the process of generating kinematic cyclic motions from primitive patterns produced from motion capture data.

6 Substituting Primitive Cyclic Motion Patterns

The reference trajectory generated by the motion planner consists of the projection of the body center on the walking surface, the motion posture, and the velocity vector tangent to the path. At each point, we determine the joint motions by substituting the rigid body postures with kinematic cyclic motion for each posture. The kinematic cyclic motions are motions of the limbs and the body that are scaled in time to reflect the specific motion pattern (posture) and speed along the rigid-body trajectory. They are "kinematic" since they do not necessarily satisfy physical and dynamic constraints. The kinematic motions are periodic cycles of a single cycle, we call "primitive," produced from motion capture data. The transitions between two different postures are generated by linearly interpolating the two motions around the transition time.

This procedure generates motions that may look smooth, but that are not necessarily physically feasible. Modifying the resulting motion patterns to be dynamically correct generally adds realism to the final motion. Ensuring dynamic correctness is done by the dynamics filter.

7 The Dynamics Filter

The reference trajectory, even when generated from a library of captured data, is likely to be physically infeasible due to differences between the dynamics of the human figure and the dynamics of the subject of the motion capture. The dynamics filter [16] is a computational procedure, intended to produce physically feasible motions from a given reference trajectory that may not be physically feasible. It considers the full dynamics of the human figure and is applicable to motions of any type.

The dynamics filter simulates the full-body dynamics of human figures that are driven by a stabilizing controller, using the kinematic motion patterns as the reference input. The stabilizing controller generates joint motions that satisfy system dynamics and are closest to the reference input. The performance of the dynamics filter depends largely on the performance of the stabilizing controller. We have successfully implemented a stabilizing control algorithm, however, the challenge remains to design a controller that is robust to variations in the motion patterns.

The current controller first computes the desired accelerations of the whole body, assuming all generalized coordinates are actuated. The implemented control law accounts locally for the generalized coordinates, and globally for the Cartesian position and orientation of the head. The computed accelerations of the generalized coordinates (joints) that are physically infeasible due to unactuated joints are then modified through an optimization procedure that ensures that the resulting motion is physically feasible. The controller finally produces the joint torques that correspond to the modified accelerations. The dynamic behavior of the whole body, driven by the joint torques, is then simulated.

The dynamics filter was applied to various motions, including a normal walk, jump, and karate kick [16]. We also used the dynamics filter to generate kinematically synthesized motion, such as a long walk and a turn using only one sequence of short-walk data from motion capture. Typical computation time for the controller and the dynamic simulator is approximately 50ms per frame using a common PC, which is the order of 20 times slower than real-time.

The equations of motion used in the optimization and the dynamic simulation are derived using the method developed in [9] for structure-varying kinematic chains. This method was designed to handle kinematic chains whose link connectivity changes dynamically among any possible open and closed configurations. This technique is essential for the problem addressed in this paper since the interaction between the human figure and the environment creates kinematic chains of various topologies.

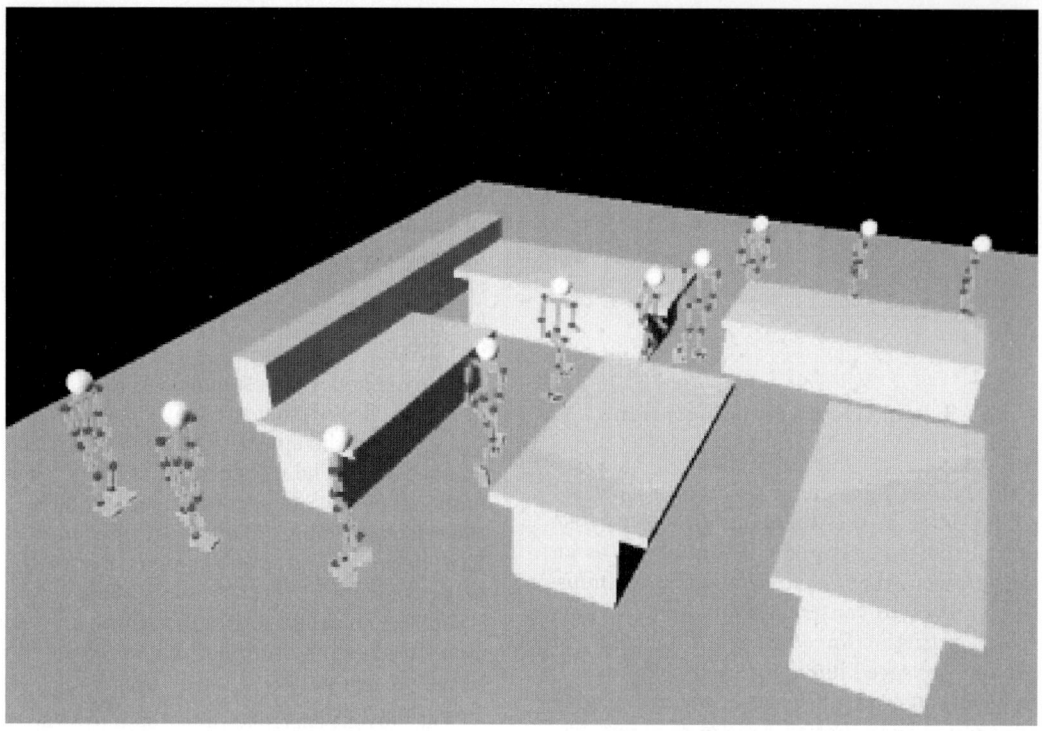

Figure 7: Primitive cyclic motion pattern substituted into the path

8 Examples

The following examples demonstrate the use of the motion planner and the dynamics filter for a humanoid moving in a typical office environment. Figure 7 shows the kinematic cyclic motion fitted to the path generated by the planner from the lower left to the upper right corners of the room. The kinematic motion is not dynamically feasible because of slippage of the feet along the turns. This is partly due to the slower speeds dictated by the velocity profile (not shown) than the speed of the original captured motion data, and partly due to the fact that the original data did not consider turns. This slippage was eliminated by filtering the kinematic motion through the dynamics filter, as shown in Figure 8. Note the slight bending of the back caused by the deceleration before each turn.

Bringing the desks closer together to force crawling and sideways motions resulted in the path and the kinematic cyclic motions shown in Figure 10. The velocity profile along this path is shown in Figure 9. The upper curve is the velocity limit curve, which reflects velocity constraints due to the path curvature and the simplified (point mass) dynamics. The lower curve is the actual velocity profile, which avoids crossing the limit curve and obeys the maximum speed limits assumed for each motion posture. This velocity profile adds "physical reality" to the resulting motion by suggesting typical human behavior.

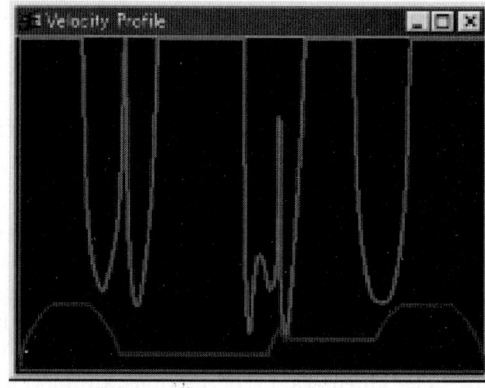

Figure 9: Velocity profile with crawl

9 Conclusions

A practical motion planner for humanoids and human figures has been presented. It considers the most common motion postures, each represented by a 2D grid. The interconnected configuration space is a small multiple of several 2D slices. This space can be efficiently searched for an optimal path that minimizes distance, maximizes clearance to obstacles, and maintains a hierarchical order of motion patterns. The speed along the path produced by the motion planner is maximized to reflect realistic behavior, subject to speed constraints on each motion pattern and simplified robot dynamics. The resulting trajectory is fitted with motion capture data, which is then filtered through the dynamics filter to ensure that the resulting motion satisfies robot dynamics and physical constraints. Numerical experiments were presented, which demonstrate the richness of the motions produced by the motion planner and by the dynamics filter.

Acknowledgments

This work is supported by the CREST project (PI: Y. Nakamura, University of Tokyo), the Japan Science and Technology Corporation.

References

[1] H. Ko and N.I. Badler: " Animating human locomotion with inverse dynamics," *IEEE Transactions on Computer Graphics*, vol.16, no.2, pp.50–59, 1996.

[2] H. Inoue, S. Tachi et al.: " HRP: Humanoid robotics project of MITI," *Proceedings of the First IEEE-RAS International Conference on Humanoid Robots*, Cambridge, MA, September 2000.

[3] J. Ostrowski and J.W. Budrick: " The Mechanics and Control of Undulatory Locomotion," *Int. Journal of Robotics Research*, vol.17, no.7, pp.683–701, July 1998.

[4] J.K. Hodgins, W.L. Wooten, D.C. Brogan, and J.F. O'Brian: "Animating human athletics," *SIGGRAPH*, vol.29, pp.71–78, 1995.

[5] J.J. Kuffner and S.M. LaValle: "RRT-Connect: An efficient apporoach to single-query path planning," *IEEE International Conference on Roboics and Automation*, pp.995–1001, San Francisco, April 2000.

[6] J.J. Kuffner Jr.: "Goal-directed navigation for animated characters using real-time path planning and control," *Proc. CAPTECH '98 : Workshop on Modelling and Motion Capture Techniques for Virtual Environments*, Springer-Verlag, 1998.

[7] E.L. Lawler: *Combinatorial Optimization*, Holt, Rinehart and Winston, New York, 1976.

[8] H. Miura and I. Shimoyama: "Dynamic walk of a biped," *Int'l Journal of Robotics Research*, pp.60–74, 1984.

[9] Y. Nakamura and K. Yamane: "Dynamics Computation of Structure-Varying Kinematic Chains and Its Application to Human Figures," *IEEE Transactions on Robotics and Automation*, vol.16, no.2, pp.124–134, 2000.

[10] Z. Popovic: "Editing Dynamic Properties of Captured Human Motion," *Proceedings of IEEE International Conference on Robotics and Automation*, pp.670–675, San Francisco, CA, April 2000.

[11] Q. Huang, K. Kaneko et al: " Balance control of a biped robot combining off-line pattern with real-time modification," *Proceedings of IEEE International Conference on Robotics and Automation*, pp.3346–3352, San Francisco, CA, April 2000.

[12] Z. Shiller and H.H. Lu: "Computation of path constrained time optimal motions with dynamic singularities," *ASME Journal of Dynamic Systems, Measurements and Control*, vol.114, no.1, pp.34–40, March 1992.

[13] S. Kagami, K. Nishiwaki et al.: " A fast generation method of a dynamically stable humanoid robot trajectory with enhanced ZMP constraint," *Proceedings of the First IEEE-RAS International Conference on Humanoid Robots*, Cambridge, MA, September 2000.

[14] M. Vukobratovic, A.A. Frank, and D. Juricic: "On the stability of biped locmotion," *IEEE Transactions of Biomedical Engineering*, pp.25–36, 1979.

[15] M. Yagi and V. Lumelsky: " Biped robot locomotion in scenes with unknown obstacles," *IEEE International Conference on Roboics and Automation*, pp.995–1001, April 1999.

[16] K. Yamane and Y. Nakamura: "Dynamics Filter — Concept and Implementation of On-Line Motion Generator for Human Figures," *Proceedings of IEEE International Conference on Robotics and Automation*, pp.688–695, San Francisco, CA, April 2000.

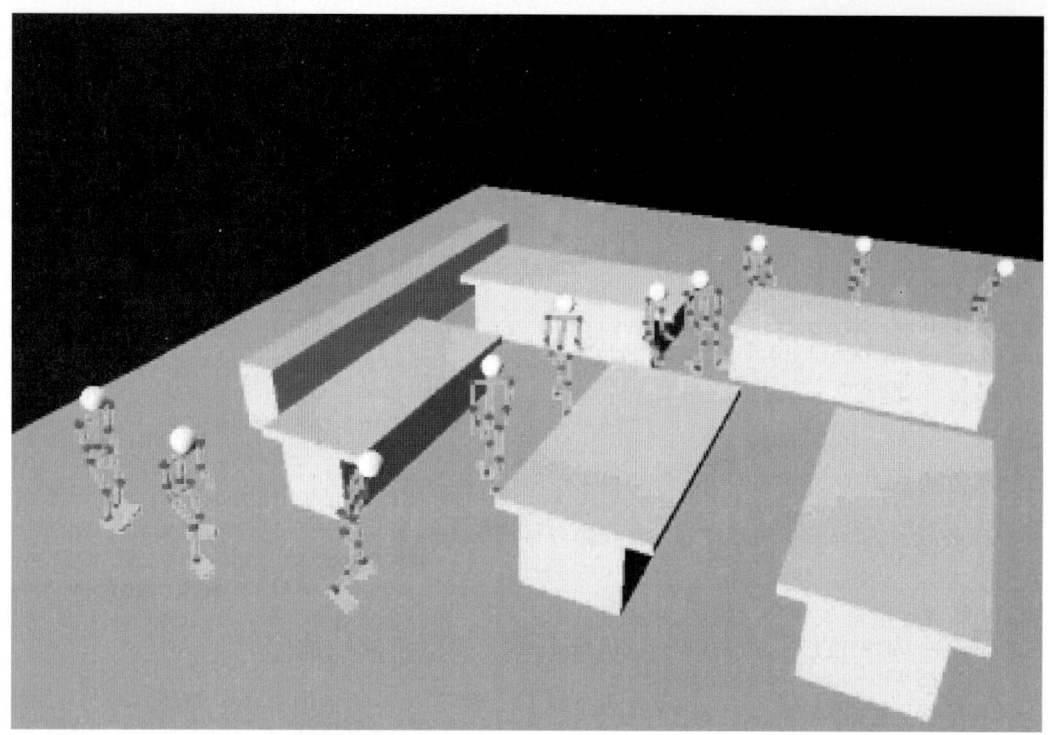

Figure 8: Filtered motion pattern obtained through the dynamics filter

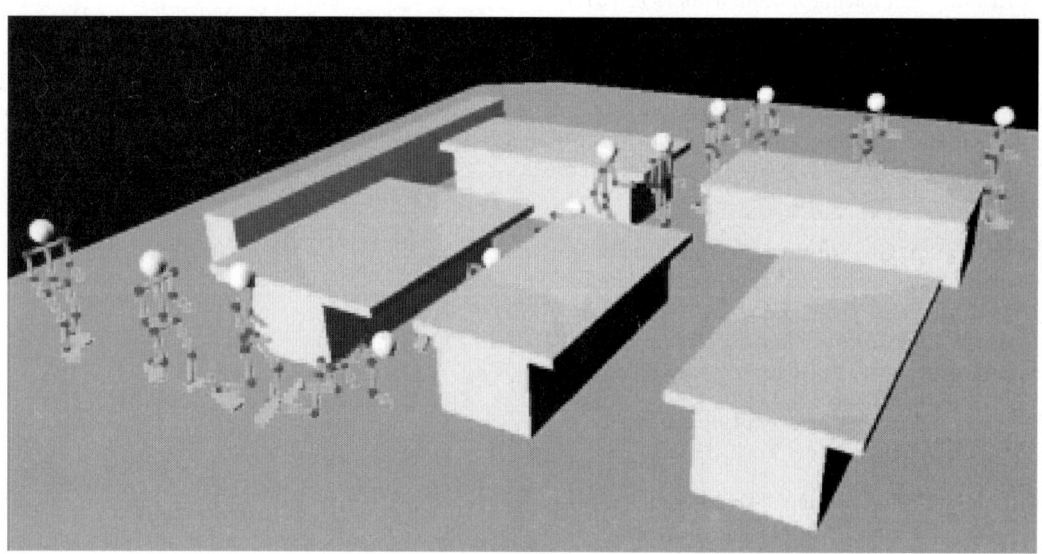

Figure 10: Postures attached to the trajectory with crawl

Extending Procedural Reasoning toward Robot Actions Planning

Félix Ingrand and Olivier Despouys *
LAAS/CNRS,
7 avenue du Colonel Roche,
F-31077 Toulouse Cedex 04, France
{felix,despouys}@laas.fr
Tel: +33 561 337 804
Fax: +33 561 336 455
http://www.laas.fr/~felix

Abstract

Procedural Reasoning Systems (PRS) have inspired some works in execution control and supervision, in particular in robotics applications. However, one of the main drawback of the PRS is the lack of planning capabilities, even though some of PRS execution mechanisms and knowledge representations are related to planning. In this paper, we propose some extensions to PRS, to integrate both execution and planning in robotics domain. In particular, we propose two planning mechanisms:

- plan synthesis to complement existing operational procedures by building new plans from operators;
- anticipation planning, which anticipates on the plans and procedures execution, to advise the execution for the best option to take when facing choices, and to forecast problems that may arise due to unforeseen situations.

These new extensions rely on a common language to represent plans, planning operators and operational procedures. In particular, the description we propose makes transitions between planning activities and execution seamless.
This work is used in two complex real-world problems: planning and control for our autonomous mobile robots (e.g. for delivery tasks), and for the management of the transition phases of a blast furnace.

1 Introduction

Procedural reasoning, originally developed at SRI International [9], has lead to a number of implementations and extensions and inspired a number of works in robot plan execution control [8, 12]. It appears to be a powerful and efficient tool to perform plans refinement and execution as well as the supervision of the robot [1]. Moreover, as described in [1] and [10], PRS have been used in conjunction with external planners, in particular the temporal planner IxTet [7]. Although this tandem supervisor/planner presents some clear advantages, there are situations where such planner is not available or not adapted. For this reason, we wanted to extend PRS toward planning.
One should note that some of the execution mechanisms and knowledge representations used in PRS are already related to planning:

- the procedures (sometimes called plans) given by the user can be triggered upon the posting of a goal or of an objective, i.e. an internal or external request to get the robot to achieve a particular state,

- moreover, the steps which compose the body of the procedure are subgoals the robot will try to achieve (by all available means if necessary),
- achieving a sequence of subgoals to fulfill a higher level goal can be seen as a reduction schema in hierarchical planning,
- using meta level procedures to decide which is the best method to fulfill a particular goal can also be seen as a form of heuristic planning.

On the other hand, PRS lacks some of the functionalities and representations which characterize most planning systems:

- PRS does not provide any form of implicit backtrack[1]. As a consequence, the user has to "program" explicit backtracking using conditional instructions and such in the body of procedures and plans.
- The procedure representation (originally named KA), with its invocation and context parts, is not really adapted to STRIPS (or Graphplan) like plan synthesis. It lacks explicit effects representation[2] and the context is only a filtering condition (it does not provide achievable or feasible condition, as defined in PDDL [14]).
- The body part of KA has some similarities with HTN [16, 18]. Indeed, the subgoals which compose the body are somehow similar to those in an HTN. However, these subgoals are interleaved with some control programming structures (such as if-then-else, while, etc), which prevent the complete KA from being easily used by HTN algorithms.

So if PRS are currently far from doing any substantial planning at all, some ingredients to do so are present and one should take advantage of it:

- Providing the procedure/plan representation is "augmented" with the appropriate information (mainly on the expected effects of its execution), we could search for a combination of these procedures to reach a given world state.
- Similarly, while executing procedures, we could anticipate on their execution (by simulating them) to examine in advance the outcomes of various possible execution paths and thus decide which one seems to be the most promising.
- At last, such anticipation could also foresee possible dead-ends or inefficient path and try to advise the system to opportunistically establish the feasible, yet unsatisfied, preconditions.

---

*Part of this work is funded under a contract with Usinor/Sollac, SACHEM Project.

[1]With good reason as the system runs online and cannot undo what has been done to reach the current situation.
[2]The effects field presented in some PRS papers is in fact just a shortcut for facts to be asserted to/retracted from the database upon successful KA execution.

These activities are definitely relevant to planning in robotic domain. Our goal is thus to integrate these approaches in a system called **Propice-Plan**.

Related work

This is not the first attempt to integrate some planning mechanism in PRS. In fact there has been some work [17, 15] done around SRI PRS and SIPE to design a common knowledge representation (ACT) and to enable the two systems to cooperate seamlessly. However, this approach uses an inter-langage to share common knowledge between heterogenous functions.

More generally, the goal of combining more tightly execution and planning has been addressed by other works. 3T [4], based on RAPS [6] and the Adversarial Planner, provides similar features in addition to a reactive skills manager. In [13], the author presents Propel, which provides a unified representation for anticipation planning and execution, but here also, the operators seem to be used by either planning or execution, but not both. In XFRM [2], adaptation is provided during execution by a set of pre-defined transformation rules.

The paper is organized as follows. Section 2 describes the changes made to the KA representation (now called OP for Operational Plans or Procedures) which is used to model plans, procedures, but also planning operators. Section 3 presents the overall organization of **Propice-Plan**, it describes the various modules which compose it and how they work together. Anticipation planning is further detailed in section 4: the heuristic search and the opportunistic precondition achievement. To illustrate how **Propice-Plan** works, an example is presented in section 5, followed some experimental results in section 6 and the conclusion.

2 Operational Procedures Representation

One of the goals of this work is to provide a unified representation for operational procedures (also named OPs), both used by supervision and planning. This representation can also be regarded as an extension of the procedure representation in PRS and its successors (UM-PRS, Propice,...), and it is inspired to some extend by the PDDL formalism [14]. The following example is related to the control of a mobile robot from its current room to a connex one (see Fig. 1).

▶ **Notation**
Symbols prefixed with '$' denote logical variables. Variable typing is allowed: (syntax `<variable>:<type>`).
Modal operators may be used for goals to achieve (!), tests (?), or waiting (∧).
=>, ~> (respectively +>, −>) assert or retract facts systematically (respectively under some condition indicated by the first expression).

▶ **Declarative Information**
The following fields represent the context of use of an OP and its known effects:

Invocation field. This corresponds to the main effect of an OP. An OP may be goal-triggered (its invocation part is prefixed with '!', like in the example above) or event-triggered (without '!').
Call field. This field is available to uniquely identify the OP and to bind all variables upon calling (indeed, different OPs may have the same invocation part). It allows the system to call the OP directly (thus avoiding the standard triggering/filtering mechanism).
Context field. This field gathers all preconditions required to apply the OP. As suggested in the PDDL formalism, we provide

```
(defop |Go To Room|
    :invocation (! (go-to-room $target $location $report))
    :call (<> (go-to-next-room $target:room $next:room
                               $location:room $door:door
                               $report:report
                               $dir:direction $dir0:direction))
    :context ((? (connex $location $next $door $dir0))
              (? (robot-dir $dir))
              (! (in-gripper nil)))
    :effects ((=> (robot-room $next))
              (~> (robot-room $location)))
    :resource ((battery 12) (duration 6))
    :priority 5
    :properties ()
    :body
        ((if (! (observed-door-status $door opened))
            (=> (robot-dir $dir0))
            (if (! (move-next $next $location))
                (! (= $report "OK"))
             else
                (! (= $report "FAILURE"))))
         else
            (=> (robot-dir (select-new-dir $dir0)))
            (! (= $report "DOOR")))))
```

Figure 1: Example of OP.

filtering conditions (prefixed with ?), *feasible conditions* (prefixed with !) which may be validated, if necessary, by triggering the execution of another OP prior to the new one. At last, conditions prefixed with # must remain true during the OP execution.
Effect field. This field contains the expected effects (apart from the one expressed in the invocation field) of a successful OP execution. Note that it is possible to use conditional effects.
Resource field. This field corresponds to an estimation of the resources required to perform the current procedure (including time).
Priority field. Some procedures are related to emergency situations, for which an autonomous robot is allowed to suspend its current activities in order to preserve its safety (low battery charge, etc.). This field is thus intended to qualify the relative criticity for OPs. Priorities are also inherited from calling procedures to callees.
Properties field. This field collects additional user-defined informations related to its use in the application.

2.1 Executive Information

An OP also contains informations related to what has to be done to execute it. It is either an *action* field (linked to an external execution code), or a body field describing the successive executions steps: it is a sequence of subgoals to satisfy (test, achieve, wait) if they are not already established in the current situation, combined with conditional constructs and loops (if-then-else, while, repeat), and possibly executed in parallel (expressed with //).
The body described in the example can therefore be interpreted as follows: "Determine if the door `$door` is opened, point toward the direction `$dir0`, and move. If the move is successful, set `$report` to OK, otherwise failure. If the door is closed, point toward another direction, and set `$report` to DOOR."
Applicability
Based on the field definitions above, we define some terms used in the following sections:
An OP is *relevant* for a goal (or a fact) if its invocation part unifies with it.
An OP is *applicable* if it is *relevant* and there exists a valid unifi-

cation of the `context`.

An OP is *potentially applicable* if it is *relevant* and there exists a unification satisfying all the *filtering conditions* of its `context`.

An OP is *non applicable* if it is *relevant* and there is an unsatisfied *filtering conditions* in its `context`.

2.2 Discussion

The OP representation extends the one defined for former PRS-based agents with fields that make it more suitable for robotics applications (resource, priority). In addition, it corresponds to a unique and global representation for:

Procedures This is the classical view of OPs. The new representation like the old one allows this possibility. Note that these procedures now have a real context and effect (possibly conditional) fields. As a consequence, action procedures, i.e. those which call an external C function, can be simulated in anticipation (without really calling the C function). The other procedures can be simulated going through their body "execution".

Planning operators Most OPs are given by the user to perform a particular goal, which in most situations also corresponds to an action (at some abstraction level). If those OPs are provided with the proper context and effect information (i.e. the declarative part), then they can be used as an operator to fulfill a higher level goal. This is what the system does when synthesizing new plans.

Plans Newly synthesized plans produced by **Propice-Plan** are added to the OP library. In this case, the executive part of the OP is the plan, with the successive steps representing the subgoals to reach to fulfill the overall goal.

3 Propice-Plan

Propice-Plan extends **Propice** (a C-PRS clone) by adding explicit planning capabilities. Figure 2 shows how the different components of **Propice-Plan** interact.

The original **Propice-Plan** algorithm (i.e. the classical procedural reasoning part) is handled by the Execution Module ($\mathcal{E}m$). Its algorithms are heavily inspired from the original PRS and we invite the reader to check [9, 8] for a more detailed account of the way it works. Roughly, it executes the OPs in a structure containing their execution state in response to events or to explicit goals given by the user. One of the main difference between **Propice** and **Propice-Plan** is its ability to request a new plan from the planning module ($\mathcal{P}m$) when a failure occurs. Another difference is that it explicitly takes into account the recommendations made by the anticipation module ($\mathcal{A}m$) (either OP choice or unification choice).

The $\mathcal{P}m$ implements the first planning method added in **Propice-Plan**. It is based on a Graphlan-like planner [3] (namely IPP [11]). When asked to find a plan to achieve a particular goal (presumably failed by the $\mathcal{E}m$), it uses the database as initial state and the set of OPs as planning operators (in particular their declarative information) to find and produce a new OP (the plan). Figure 3 shows an OP synthesized by IPP in response to an request from the supervisor to achieve the goal (! (on-shelf manip-shelf tube0)). Note that the facts which end up being those on which the plan is logically based are given in the context part.

The second planning method introduced in **Propice-Plan** is called anticipation planning and is implemented in the Anticipation Module ($\mathcal{A}m$) (see Figure 2).

The $\mathcal{A}m$ is similar to the $\mathcal{E}m$, except that it simulates assertions and actions in advance with the real time. Thus, it uses the same algorithm than the $\mathcal{E}m$, except for these differences:

```
(defop |ipp on-shelf_manip-shelf_tube0 (0)|
   :invocation (! (on-shelf manip-shelf tube0))
   :call (ipp_on-shelf_manip-shelf_tube0_0)
   :context ((? (shelf-room manip-shelf manip-room))
             (? (on-shelf store-shelf tube0))
             ... (? (robot-room manip-room)))
   :body ((<> (go-to-any-room store manip-room))
          (<> (pick-up-tube-from-shelf tube0 store-shelf store))
          (<> (carry-to-any-room manip-room store tube0))
          (<> (put-tube-on-shelf manip-shelf tube0
                                manip-room)))
   :properties ((op-type synthesised) (planning useless)))
```

Figure 3: Example of a plan returned by the $\mathcal{P}m$.

- It executes OP in its own database structure (initialized and resynchronized from time to time with the real world execution database).
- It simulates action OP execution (using their effects field) instead of really executing them. This is done by asserting/retracting the effects in/from the $\mathcal{A}m$ database.
- Unlike the $\mathcal{E}m$, it does not "stop" when the goal has been reached, it tries other alternative paths which may have been left "opened" (such as alternative variables bindings, or alternative execution paths in an OP). In fact, the $\mathcal{E}m$ builds a structure (the anticipation structure) which will hold trace of this anticipated execution.

If the $\mathcal{A}m$ detects an execution problem to come (unsatisfied, yet feasible condition, or non-explicit failures such as loop detection), it may modify the anticipation structure, to have these conditions established in time according to a *conservative* strategy

During a normal execution, both the $\mathcal{E}m$ and the $\mathcal{A}m$ run concurrently, with possible advices from the $\mathcal{A}m$ to the $\mathcal{E}m$. The $\mathcal{E}m$ has the highest "priority" to remain reactive, while the $\mathcal{A}m$ gets all the remaining cycles (which would be lost otherwise).

To perform anticipation planning we had to address the issues of handling non deterministic actions (such as perception action). This is done using some hypothetical reasoning (i.e. developing various hypothesis corresponding to the various possible outcomes).

4 Anticipation Planning

Anticipation Planning is the most original part of **Propice-Plan**. We describe here how it work and the focus on the two key features of this approach.

The initial state of the anticipation process is based on the current environment state. Then, when evaluating all possible outcomes for a choice point, the $\mathcal{A}m$ will develop different states accordingly; these will in turn lead to different options when the next choice point will be processed. Therefore, the projections examined by the $\mathcal{A}m$ is structured as a *projection tree*. Then, assigning instantiated OPs to choice points is equivalent to searching a path in this tree, and search control is performed wrt a specific criterion (either the adequacy to user-defined heuristics, or the minimisation of actions to perform).

In return, the $\mathcal{E}m$ synchronizes the anticipation process with the actual environment state by updating the database (and therefore the projection tree root). Obsolete paths are automatically cut off by the anticipation structure. Updates are processed every time the $\mathcal{E}m$ steps across a choice point even if it followed the $\mathcal{A}m$'s advise to take unexpected events into account.

A specific processing is required when loops and conditional branching are simulated. Indeed, the condition value may be un-

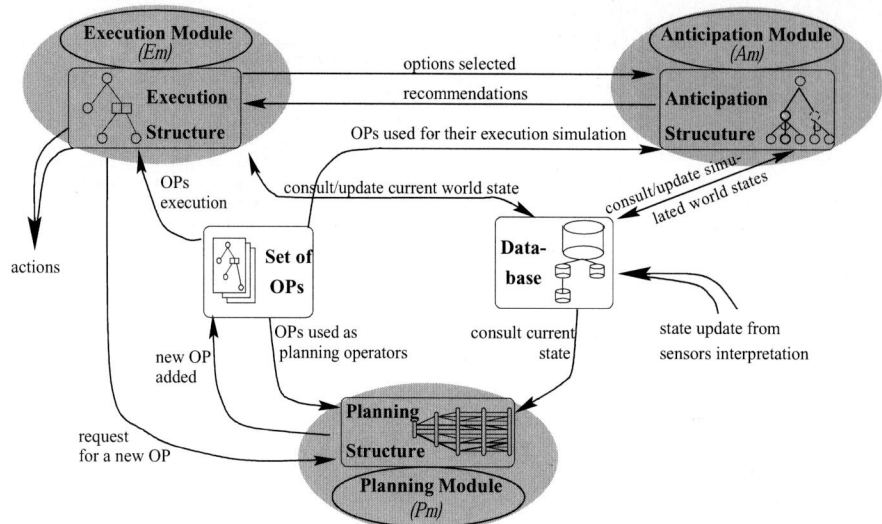

Figure 2: Organization of the various components.

predictable due to non determinism (if (? (& (nitrogen-flow $nf) (< $nf 12))) ...). Then, there is no mean to foresee which execution branch will be performed by the $\mathcal{E}m$. Both branches are then simulated and labeled with the corresponding constraints (here: one with a nitrogen flow value less than 12, and one for the opposite).

Heuristic Choice For a given goal, the $\mathcal{E}m$ may use execution heuristics to choose among various applicable and relevant OPs for the most trustworthy one. However, the selected operational plan may be more expensive than others (in terms of execution duration, resources,...). The first role of the $\mathcal{A}m$ is thus to evaluate such choice points before the $\mathcal{E}m$, with anticipation heuristics looking for the most cost-effective OPs. The corresponding results are gathered in the anticipation structure which contains all the projections examined so far, which will be consulted later by the $\mathcal{E}m$ when it faces this choice point. One key feature of the anticipation is that it is based on an interruptible interpreter; in addition, the more time is spent anticipating, the more execution paths are evaluated, and the better the advice. Thus, anticipation is used as an *anytime* algorithm by th $\mathcal{E}m$.

Opportunistic Precondition Achievement The operational plans in the set of OPs cover a wide range of expected situations including emergency ones. Yet, if the $\mathcal{E}m$ keeps executing and refining OPs without any anticipation, some preconditions required at a lower execution level may lead to inefficient plan, as they were unforeseen at the higher level.

The $\mathcal{A}m$ detects an OP inadequacy if its execution simulation failed. If it occurs that the relevant OPs for this goal are only potentially applicable, the $\mathcal{A}m$ will try to adapt the plan being anticipated. This is done by inserting, at the best place in the current plan, the proper OP which establishes the missing precondition(s). This adaptation is *complete*, and occurs according to an *opportunistic* and *conservative* strategy, to modify the original plan as little as possible and minimize harmful interactions due to these modifications (see [5] for further details).

5 Illustration

Although not fully detailed, the following example is intended to clarify the role played by the three modules in **Propice-Plan**. The domain is as follows.

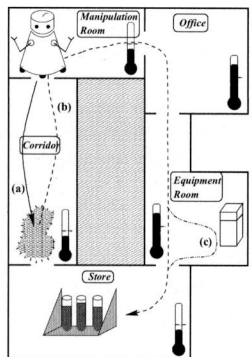

Figure 4: The laboratory robot domain.

In a chemistry laboratory, a robot is used to convey test tubes from a storage area to the manipulation room, via a corridor. The temperature in each room of this course is very low, to prevent the contents of the tubes from reacting. On contrary, the office and the equipment room, used by the staff of the laboratory, are warmer. Thus, is it necessary to use a refrigerated box (located in the equipment room) to convey tubes through those rooms, or their contents may be spoiled.

In this problem, the robot is given simple OPs to manipulate the refrigerated box, the tubes, and to go through doors to change rooms. An additional OP describes the main steps to accomplish its mission: from the manipulation room, move to the corridor, and then to the store, pick up a test tube, and bring it back to the manipulation room via the corridor[3]; at this point, though it knows the connections between the different rooms, the robot has no explicit route for his mission via the office nor the equipment room yet.

Initially, the robot is about to execute its mission, and starts from the manipulation room, but the way between the corridor and the store is blocked up. Of course, the robot has no information about the blockade yet (contingent event). Its route is represented by the solid arrow **(a)** in the figure. The $\mathcal{E}m$ therefore executes the

---
[3]Due to the lack of room, those OPs are not described.

OP corresponding to the mission, but will fail when attempting to reach the store; this could not have been detected by the $\mathcal{A}m$, since the robot had no information concerning the blockade before. As a result, the $\mathcal{P}m$ is in charge of finding a solution to achieve the current goal (ie: go to the store), from the corridor. The corresponding OP, synthesized by the $\mathcal{P}m$, corresponds to the dashed route **(b)** in the figure, that is: go back to the manipulation room, and move successively to the office, the equipment room, and finally to the store.

The $\mathcal{E}m$ then loads this OP and starts executing it. Meanwhile, the $\mathcal{A}m$ simulates this execution before the actual one.

However, this new sequence (see fig.5) is not feasible yet, because the tube cannot be conveyed through a warm room: it must be isolated first. This new flaw will lead to the insertion of a new sub-goal to get the refrigerated box before leaving the store. Now, suppose the robot is still on his way to the store, and it is currently moving through the office. The best moment to pick up the box is when it drives along the equipment room (because it minimizes the number of additional OPs): see arrow **(c)**; otherwise, it would reach the store, then step back to the equipment room, get the box, and finally come again to the store. **Propice-Plan** will thus insert additional sub-goals to make it feasible to bring the tube in the manipulation room; thus, the sub-goals sequence {*(! (location office)), (! (location equip)), (! (pickup box)), (! (location store))*} will replace the original {*(! (location corridor)), (! (location store))*} one. This is the most important gain provided by anticipation planning, which makes global plans lighter than the ones that would be obtained by a system simply combining execution and plan synthesis, and therefore adapting a procedure *only after an actual failure*.

6 Experimental results

In this section, we provide a few results to show, on a simple example, the benefits of to the integration of supervision and anticipation in **Propice-Plan**. The domain is a perfect labyrinth[4], where the robot has to find its way out as quickly as possible. In other words, the robot must interleave acting (to escape) and planning (to find the exit).

We realized experiments for various labyrinths sizes (from 5x5 up to 20x20 cells), and with different quantums allowed to anticipation planning: for one execution cycle, there was either 0, or 5, or more anticipation cycles (up to 250). For any experiment, the robot was given a heuristic function to follow the left wall to escape. In addition, execution and planning have access to the environment topology. No resource was define, nor particular execution priority for the OPs.

Due to space limitations, we only present the results for a series of experiments on a 15x15 labyrinth (see Figure 6).

All durations are expressed in milliseconds. The *execution time* corresponds to the cumulated duration for the execution of all the actions until the robot finds the exit; the *anticipation time* is related to the construction of the projection tree; last, the *advices and updates time* corresponds to the time spent to cut branches in the tree (after each choice selection by the supervisor), and to extract the most promising action to advise the supervisor.

This experiment can be analyzed in two ways :

▶ **Behavior:**

As one could expect, the more anticipation is performed, the better the overall behavior. In all experiments, the optimal strategy (wrt the visited cells) was obtained for 50 anticipation cycles per execu-

---
[4] *Perfect labyrinths* are such that there exists exactly one path from any cell to another.

tion cyle (including hard 20x20 labyrinths). For small labyrinths, the whole path was even found by the anticipation before executing the first move.

▶ **Time:**

As far as durations are concerned, it is obvious that the construction of the projections tree, its update and the search for the best advice has a cost that grows with the ratio of anticipation wrt execution. However, it seems there exists a trade-off for which the total time spent for solving the problem is minimal (the more time spent to anticipate, the better the overall behavior, and therefore less time is wasted executing sub-optimal moves); analog results appeared for all labyrinths sizes, where the optimal ratio varies with the domain size. Yet, too much anticipation is also a waste of time, since **Propice-Plan** develops too many possible projections before resynchronizing them wrt the actual action performed by the supervisor.

7 Conclusion

Propice-Plan, a system resulting of some planning extensions made to **Propice** proved rather effective in two real-world applications – autonomous mobile robot programming, and the management of the transition phases of a blast furnace – which called for such extensions to better handle situations:

- where all OPs have been attempted to perform a goal, and one tries to synthesize a new plan using available OPs as planning operator,
- where anticipating on the actual OPs execution, shows in advance some dead ends or preferred paths.
- where anticipation detects a more opportunistic spot to achieve a precondition which will be required in a near future.

The system has been implemented and the tests reported here show that a robot endowed with some anticipation perform better than without.

Nevertheless, some problems remains to be solved and some mechanisms could be improved:

- Logical soundness of OPs. We have redefined or added new fields in OPs to allow these new planning mechanisms to take place, but still nothing yet can prove that in most cases, an OP fulfills its declared effects. The $\mathcal{E}m$ has the last word as it always checks that requested goal get really achieved at run-time.
- The hypothetical reasoning, which takes place when encountering non-deterministic action (such as perception actions) needs to be improved.
- Time is always an important factor in execution control. So far we have not included any specific mechanism to handle time during the planning phases, but we are thinking about including some of the ideas and algorithms provided in [7].

Our goal in providing these new mechanisms is to take advantage of the large body of OPs available in most **Propice** execution control applications to better handle the problems execution may encounter. Planning is indeed a good candidate to make execution more robust and suitable to specific situations.

References

[1] R. Alami, R. Chatila, S. Fleury, M. Ghallab, and F. Ingrand. An Architecture for Autonomy. *International Journal of Robotics Research*, 17(4):315–337, April 1998.

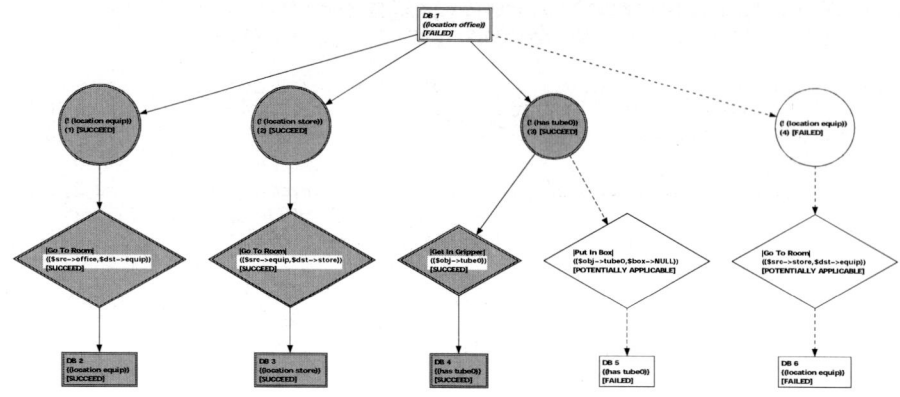

Figure 5: A projection tree example.

Quantum Anticipat. per Exec. cycles	0	5	15	25	50	250
Execution time	17903	16901	2410	2410	2410	2410
Advices & updates time	0	3582	4091	3825	4655	5596
Anticipation time	0	19130	38297	25799	24527	50077
Total time	17903	39613	44798	32034	**31593**	58083
# visited cells	221	198	**73**	**73**	**73**	**73**

Figure 6: Results for a 15x15 labyrinth.

[2] M. Beetz and D. McDermott. Improving Robot Plans During Their Execution. In *Proceedings of AIPS94*, June 1994.

[3] A. Blum and M. Furst. Fast Planning Through Planning Graph Analysis. *Artificial Intelligence*, 90:281–300, 1997.

[4] R.P. Bonasso, R.J. Firby, E. Gat, D. Kortenkamp, D.P. Miller, and M.G. Slack. Experiences with an Architecture for Intelligent, Reactive Agents. *Journal of Experimental and Theoretical AI*, January 1997.

[5] O. Despouys and F. Ingrand. Propice-plan: Toward a unified framework for planning and execution. In *Proceedings of European Conference on Planning*, 1999.

[6] R. J. Firby. Task Networks for Controlling Continuous Processes. In *Proceedings of the Second International Conference on AI Planning Systems, Chicago IL*, June 1994.

[7] M. Ghallab and A. Mounir-Alaoui. The Indexed Time Table Approach for Planning and Acting. In Rodriguez and Seraji, editors, *Procs. NASA Conference on Space Telerobotics*, volume 5, pages 321–332. JPL Publications 89-7, Feb. 1989.

[8] F. F. Ingrand, R. Chatila, R. Alami, and F. Robert. PRS: A High Level Supervision and Control Language for Autonomous Mobile Robots. In *IEEE International Conference on Robotics and Automation, St Paul, (USA)*, 1996.

[9] F. F. Ingrand, M. P. Georgeff, and A. S. Rao. An Architecture for Real-Time Reasoning and System Control. *IEEE Expert, Knowledge-Based Diagnosis in Process Engineering*, 7(6):34–44, December 1992.

[10] J. Gout and S. Fleury and H. Schindler. A New Design Approach of Software Architecture for an Autonomous Observation Satellite. In *5th International Symposium on Artificial Intelligence, Robotics and Automation in Space*, Noordwijk, the Netherlands, June 1999.

[11] J. Koehler, B. Nebel, J. Hoffmann, and Y. Dimopoulos. Extending Planning Graphs to an ADL Subset. In *Proceedings of European Conference on Planning*, 1997. See also: http://www.informatik.uni-freiburg.de/~koehler/ipp.html.

[12] J. Lee, M. J. Huber, E. H. Durfee, and P. G. Kenny. UM-PRS: An Implementation of the Procedural Reasoning System for Multirobot Applications. In *Proceedings of the Conference on Intelligent Robotics in Field, Factory, Service, and Space*, pages 842–849, Houston, Texas (USA), March 1994.

[13] R. Levinson. A General Programming Langage for Unified Planning and Control. *Artificial Intelligence*, 76:281–300, 1994.

[14] D. McDermott, M. Ghallab, A. Howe, C. Knoblock, A. Ram, M. Veloso, D. Weld, and D. Wilkins. PDDL – The Planning Domain Definition Langage. Technical Report CVC TR-98-003/DCS TR-1165, Yale Center for Computational Vision and Control, October 1998. Available on request to drew.mcdermott@yale.edu.

[15] K. L. Myers. Towards a Framework for Continuous Planning and Execution. In *Proceedings of the AAAI Fall Symposium on Distributed Continual Planning*, 1998.

[16] D. Nau, Y. Cao, A. Lotem, and H. Muñoz-Avila. SHOP: Simple Hierarchical Ordered Planner. In *Procs. IJCAI*, 1999.

[17] D. E. Wilkins, K. L. Myers, J. D. Lowrance, and L. P. Wesley. Planning and Reacting in Uncertain and Dynamic Environments. *Journal of Experimental and Theoretical AI*, 6:197–227, 1994.

[18] Q. Yang. *Intelligent Planning: A Decomposition and Abstraction Based Approach*, pages 163–171. Springer-Verlag Publisher, 1997.

Let's reduce the gap between task planning and motion planning

Emmanuel Guéré, Rachid Alami

LAAS-CNRS
7, avenue du Colonel-Roche 31077 Toulouse Cedex - France
{guere,rachid}@laas.fr

Abstract

In the stream of research that aims to speed up practical planners, we propose a new approach to task planning based on Probabilistic Roadmap Methods (PRM). Our contribution is twofold. The first issue concerns the development of ShaPer, a task planner that is able to deal efficiently with large problems. Shaper "captures" the structure of the task space. The second contribution involves promising results on robot task planning. This is obtained through an analysis of the task space structure that exhibits the relation between task and geometric reasoning for a given robot task. To illustrate such an approach, we solve a complex problem where motion and task planning are closely interleaved.

1 Motivations

Task planning has often used examples borrowed from robotics like, for instance, Pick&Place scenarii. However, the effective use of practical task planners in robotics has always been limited to domains where it was possible to establish a clear and impermeable hierarchy between a high-level task planner and a lower level where geometric problems are dealt with. This is clearly insufficient if one wants to tackle realistic robotics problems. For instance, a plan for building a stack of objects may be substantially modified if one adds an obstacle or changes the shape of the robot[10, 8].

We have proposed in the early nineties a geometrical formulation of the manipulation problem[2]. We formulated the problem as a series of motion planning problems in presence of movable obstacles. We showed that it was possible to compute regions in the global configuration of the system where a grasp or a release action may cause a qualitative change in the topology of the free space allowing to access new states in the task space of a given manipulation problem. Such regions correspond to links between various "slices" of the global configuration space[1].

While the formulation was satisfactory and gave a deep understanding of the manipulation problem, its effective application has been limited to environments with a small number of degrees of freedom[1].

Indeed, we faced three problems. The first one was due to the limitations of the motion planners of that (old) times. It was unrealistic to try to solve motion problems with more than 3 degrees of freedom. The second problem was the absence of an operational link between task planning and motion planning. The third problem which was also discouraging was the slowness of task planners.

We are convinced that the recent and independent advances in motion and task planning have reached a level where it becomes realistic and fruitful to investigate the links between them and to devise paradigms that effectively involve the two aspects in close relation and not simply through a gross and somewhat artificial hierarchical decomposition. This paper is a first step toward this goal.

Even though task planners have made very substantial progress [3] over the last years, they are still limited in their use. There are also domains, like in robotics which heavily influence the structure of the task space; learning such a structure will certainly help in building an efficient planner in a given domain. However, the structure of the environment (at least the "useful" one) heavily depends not only on the environment but also on the actions that can be performed. Our aim is to develop a generic planner that will exhibit and learn the "structure" of a given domain. This is the reason why we propose to investigate approaches based on Probabilistic Roadmap (PRM). PRM basically "captures" the space "topology" through random configuration generation and connectivity tests between states using a local planner. PRM obtains good results in robot path planning because it is relatively easy to test the validity of a randomly generated configuration and because there exist good metrics and numerous very efficient local planners. PRM can even obtain excellent results when careful techniques are devised in order to construct a compact graph and to "direct" the search toward non-explored regions[9].

Our contribution is twofold. The first issue concerns ShaPer[1], a task planner that is able to deal efficiently with large problems. We will show that, even if problems are modeled in first-order logic, there exists a topology for task planning domains. With a PRM like algorithm, we show how we can devise a task planner which builds a graph of the task space.

The second contribution involves promising preliminary results on joining task and geometrical reasoning. Indeed the environment definition and the robot configurations can be modeled in first-order logic. ShaPer's expressiveness coupling with a PRM like algorithm is also able to deal with geometrical constraints.

Both contributions are illustrated through a prototype implementation presented on a problem where motion and task planning are closely interleaved.

2 ShaPer: An efficient task planner

In this section, we briefly present ShaPer, a new version of the task planner proposed in [6]. Shaper is based on STRIPS formalism [4][2]. It performs in two steps: first an *accessibility* graph is generated *off-line*, and then the planner solves *online* task planning problems by extracting a plan from the learned graph. Even though usual task planners are limited by a combinatorial explosion, ShaPer is able to learn the *accessibility* graph for substantially large domains; this is possible because this graph only contains "relevant" states with respect to the state space "tolopology".

2.1 Relevant states

A new state is declared "relevant" if it allows to access to an unknown state (i.e. there is no other state in the current graph $\mathcal{G}$ with the same *shape*). We say that two states S and g have the same *shape* iff there exists a substitution σ such that $S = \sigma(g)$. Consider for instance the two block-world states g and S:

[1]ShaPer: Shape based Planner.
[2]An action o consists of three parts: P_o represents the preconditions, A_o the add-list (new facts - $A_o \cap P_o = \phi$) and D_o the Del-list (facts: not true when o is applied - $D_o \subseteq P_o$) represent the effects. o is applicable to a state S if and only if $P_o \subseteq S$. In this case $o(S) = (S - D_o) \cup A_o$.

```
Relevant(S', G, Max_Trials): Boolean
  ForEach g ∈ G
    If ∃σ_{g,S'} Then
      Return false /*S' shape already in G */
  trial ← 0
  found ← true
  While trial < Max_Trials Do
    Randomly choose Γ a local plan /*P_Γ ⊆ S'*/
    ForEach S'' ∈ G Do
      If L(S'', Γ(S')) Then
        found ← false /*Already in G*/
        trial ← trial + 1
        Break
    If found Then
      Return true /*S' adds information*/
  Return false
```

Table 1: Relevance of a state.

state g: A / B C   state S: A / B C

with $g = \{Clear(A), On(A,B), OnTable(B), Clear(C), OnTable(C)\}$ and $S = \{Clear(A), OnTable(B), Clear(B), OnTable(C), On(A,C)\}$. The state g, when A is substituted by A, B by C and C by B is equal to S ($\sigma = \{A/A, C/B, B/C\}$). Thus, if P is a sequence of actions applicable to g, then $\sigma(P)$ is applicable to S and $(\sigma(P))(S) = \sigma(P(g))$. Such a graph $\mathcal{G}$ also contains all accessible shapes (of a given connexity). Table 1 presents an algorithm to test the relevance of a state.

2.2 An accessibility graph learning

In motion planning, PRM generates a random configuration, checks its validity (if it is in $\mathcal{C}_{free}$) and tries to connect it to the current graph. In task planning, there is no general way to check the validity of a state. This is the reason why we do not generate states randomly. Instead of that, we generate random valid plans starting from a given initial valid state (see table 2 [3]).

[3]We develop a graph with only one connected component including the initial state S_{begin} by applying valid local plans.

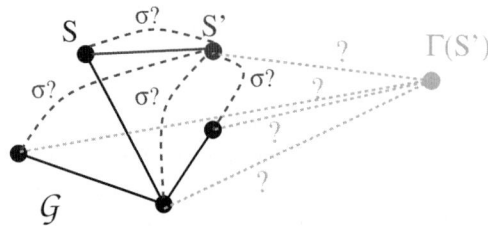

Figure 1: Is S' a relevant state for the graph $\mathcal{G}$?

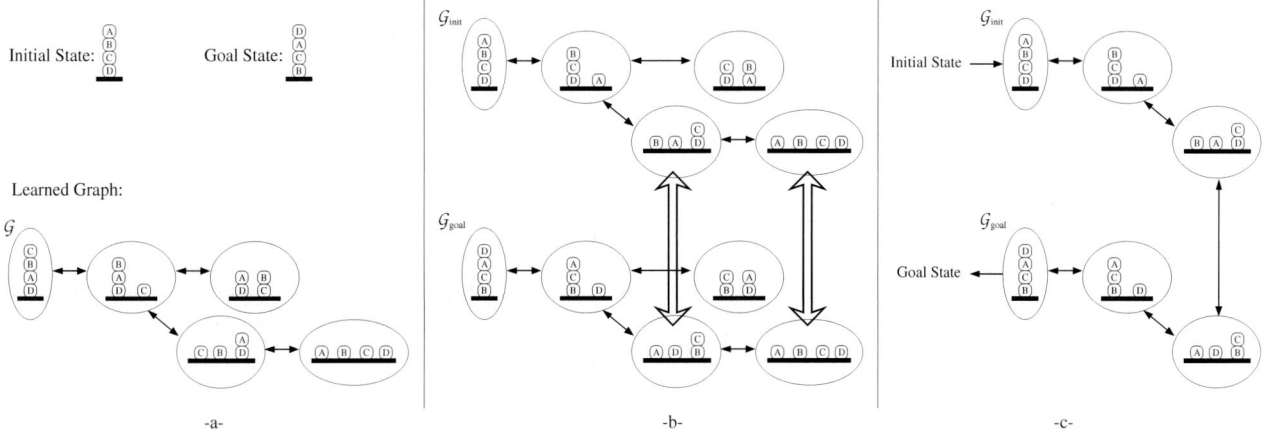

Figure 2: Solution extraction in block-world domain. a. The graph is generated *off-line* ; Initial and goal states are defined *online*. b. Create $\mathcal{G}_{init}$ and $\mathcal{G}_{goal}$ from the substitutions of S_{init} and S_{goal} in $\mathcal{G}$; connect $\mathcal{G}_{init}$ to $\mathcal{G}_{goal}$. c. Search a path in the graph.

After each random generation of a state S' by applying a random valid plan Γ, we must check the relevance of the new node $S' = \Gamma(S)$. To do so, we first look for a substitution σ between S' and $\mathcal{G}$. If it is not the case, S' is potentially relevant; we must then verify that S' allows to access to a new state S'' that is not directly accessible from the current $\mathcal{G}$ (i.e. $\forall g \in \mathcal{G}. \neg \mathcal{L}(S'', g)$ - see Table 1 and figure 1).

Owing to the accessibility property of shapes (i.e. if $S = \sigma(g)$ and P a plan such that $S' = P(S)$ then $\sigma(S') = (\sigma(P))(g))$, even if the graph is limited to one substitution, we are able to extract sound plans.

2.3 Solution extraction

Note that the planner can not directly use the graph $\mathcal{G}$ to extract a solution. Indeed, the initial state S_{init}, the goal state S_{goal} and the graph $\mathcal{G}$ may correspond to "different substitutions" (see, for instance, figure 2.a).

So, the solution extraction[4] consists in four steps: 1. Build the graphs $\mathcal{G}_{init}$ and $\mathcal{G}_{goal}$ using S_{init} and S_{goal} substitutions; 2. Connect[5] $\mathcal{G}_{init}$ and $\mathcal{G}_{goal}$ with a local planner; 3. Connect S_{init} to $\mathcal{G}_{init}$ (S_i) and S_{goal} to $\mathcal{G}_{goal}$ (S_g); 4. Search for a path between S_i and S_g.

This new algorithm answers the problem of randomly generating a valid state in the method presented in [6].

[4]Note that two states S_1 and S_2 with the same *shape* are not necessary in the same connexity (i.e. there exists no path between $\mathcal{G}_1$ and $\mathcal{G}_2$). In such case, we can say that there exists no solutions.

[5]Note that although connection is always possible (if a plan exists), in restricted domains, to connect $\mathcal{G}_{init}$ and $\mathcal{G}_{goal}$, the planner may need to find intermediate substitutions.

2.4 Results for classical planning domains

Table 3 presents some problem resolutions for the block-world and the gripper domain. We compare ShaPer to IPP-v4.0 [7] one of the fastest task planners. The problems that have been used in the block-world domain are defined by: the initial state corresponds to the highest possible tower and the goal state to the same tower except for the top block that is put at the bottom. In the gripper domain, the problem consists in moving all balls from a table to another with a robot which can pick two balls at a time.

3 Predicate independence

While the previous section presented the general framework of our planner, there are still efforts to be dedicated in order to reduce the graph size. For instance, one can take advantage of predicate independence. Indeed, given two independent predicates f and h, and F (resp. H) the accessibility graphs constructed from f (resp. h), we would have, with the algorithm described above, $Card(\mathcal{G}) = Card(F) \cdot Card(H)$, whereas the construction of $Card(\mathcal{G}) = Card(F) + Card(H)$ would have been sufficient.

3.1 When?

Two predicates f and h are said to be independent, in a given domain, iff $\forall o \in \mathcal{O}. f \in A_o \cup D_o \rightarrow h \notin A_o \cup D_o$. Similarly, we define classes of independent operators: two operators o_1 and o_2 are independent iff $(D_{o_1} \cup A_{o_1}) \cap (D_{o_2} \cup A_{o_2}) = \phi$ Note that even if o_1 and

o_2 are independent, we can have $P_{o_1} \cap (A_{o_2} \cup D_{o_2}) \neq \phi$. This means that o_1 may need preconditions from other predicate classes. For instance, in a domain where the robot moves blocks from a table to another, *block-position* is independent of *robot-position*, but to pick a block (operator of the *block-position* class) the robot needs to be *near* the table.

3.2 How?

The algorithm presented in table 4 builds a more compact *accessibility* graph by taking into account predicate independence.

To better understand this algorithm, we run[6] it on a very simple domain. The environment is composed of three locations (L_1, L_2 and L_3 with $Table(L_1)$ and $Table(L_3)$), two blocks (B_1 and B_2) and a robot which can *move*, *pick* or *place* a block (the *Pick & Place* actions are symmetric):

Move(X:Location, Y:Location)
    Pre: At(X), Connect(X,Y)
    Add: At(Y)
    Del: At(X)

Pick(X:Block, Y:Location)
    Pre: At(Y), On(X,Y), HandEmpty, Table(Y)
    Add: Hold(X)
    Del: On(X,Y), HandEmpty

After predicate decomposition, we obtain two classes: $E_1 = \{On, HandEmpty, Hold\}$ and $E_2 = \{At\}$[7].

Let us first decompose the initial state into E_1 and E_2: $e_1^1 = \{On(B_1, L_1), On(B_2, L_1), HandEmpty\}_{E_1}$, $e_2^1 = \{At(L_1)\}_{E_2}$ and $c = \{Table(L_1), Table(L_3), Connect(L_1, L_2), Connect(L_2, L_1), Connect(L_2, L_3), Connect(L_3, L_2)\}_{Const}$. Now we apply the algorithm step by step (see figure 3 for the state description):

---
[6]To simplify explanations, we present a deterministic version of table 4. The deterministic version develops all possible actions in all possible classes.

[7]Note that the predicates *Table* and *Connect* are not in any classes because no operator modifies them.

```
Learn_Graph(S_init, Coverage)
    G ← {S_init}
    cover ← 0
    While cover < Coverage Do
        Randomly choose S ∈ G
        Randomly choose Γ a valid local plan
        Given S' = Γ(S)  /*P_Γ ⊆ S */
        If Relevant(S',G,Max_Trials) Then
            G ← G ∪ {S'}
            cover ← 0
        Else cover ← cover + 1
```

Table 2: Learn the shape graph.

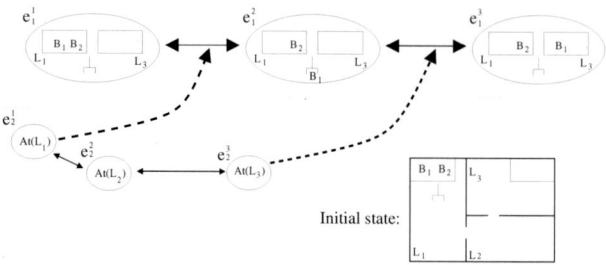

Figure 3: Algorithm with independent predicates explanation (dashed arrows represent preconditions).

1. No action is applicable to $e_1^1 \cup c$.
2. Apply $Pick(B_1, L_1)$ to $e_1^1 \cup e_2^1 \cup c$: create e_1^2 with precondition e_2^1.
3. Don't apply $Pick(B_2, L_1)$ to $e_1^1 \cup e_2^1 \cup c$ because of the substitution with e_1^2.
4. Apply $Move(L_1, L_2)$ to $e_2^1 \cup c$: create e_2^2.
5. No action is applicable to $e_1^2 \cup c$.
6. Don't apply $Place(B_1, L_1)$ to $e_1^2 \cup e_2^1 \cup c$ because of the substitution with e_1^1.
7. Apply $Move(L_2, L_3)$ to $e_2^2 \cup c$: create e_2^3.
8. No action is applicable to $e_1^2 \cup c$.
9. Apply $Place(B_1, L_3)$ to $e_1^2 \cup e_2^3 \cup c$: create e_1^3 with precondition e_2^3.
10. No *Move* applicable because of substitutions.
11. No action is applicable to $e_1^3 \cup c$.
12. Don't apply $Pick(B_2, L_1)$ to $e_1^3 \cup e_2^1 \cup c$ because of the substitution with e_1^2.
13. Don't apply $Pick(B_1, L_3)$ to $e_1^3 \cup e_2^3 \cup c$ because of the substitution with e_1^3.
14. No more action is applicable. The graph is complete.

The graph built with independent predicates (Table 4) is equivalent to the first one (Table 2). Consequently the solution extraction can be made as de-

Problem	IPP CPU	ShaPer		
		Graph learning		Extraction
		nb_{node}	CPU	CPU
12 blocks	171.9	77	1.58	0.01
15 blocks	-	176	16.73	0.06
20 blocks	-	627	173.79	0.51
10 balls	56.9	30	0.07	0.01
20 balls	-	60	0.55	0.01
50 balls	-	150	14.04	0.15

Table 3: Graph learning and solution extraction for some classical domains

```
Learn_Graph(S_init, Coverage)
    cover ← 0
    E_1 × ... × E_n ← Decompose_Predicates(S_init)
    While cover < Coverage Do
        ForEach E_i Do
            Randomly choose e ∈ E_i
            Randomly choose o ∈ O_i with P_o ⊆ e
            If ¬∃σ Then
                Add (e − D_o) ∪ A_o to E_i
                cover ← 0
            Else cover ← cover + 1
            ForEach e in completion E of E_i Do
                Randomly choose o ∈ O_i with P_o ⊆ e
                If ¬∃σ Then
                    Add (e − D_o) ∪ A_o to E_i with precondition on E
                    cover ← 0
                Else cover ← cover + 1
```

Table 4: Learning the shape graph with predicate independence.

scribed previously (with the independence transformation).

4 A step toward geometric reasoning

Now we show how we use the *accessibility* graph learning algorithm to integrate abstract reasoning (task planning) and geometrical constraints (motion planning). This is illustrated through an example.

4.1 Introduction of motion planning

In order to model numerical facts (e.g. Cartesian coordinates of the robot), we extend STRIPS formalism as following:

- numerical facts: $At(7.5, 1.3)$ models the robot position in Cartesian coordinates and $Size(Tank, 2.0)$ models the fact that the robot width is 2.0 when it holds the Tank;

- intersection and inequality in preconditions.

Move(X_1: Real,Y_1: Real,X_2: Real,Y_2: Real, S: Real)
    Pre: $At(X_1,Y_1)$, $Robot\_Size(S)$,
            $\forall d \in \mathcal{E}_{nv}.d \cap Disc([(X_1,Y_1),(X_2,Y_2)],S) = \phi$
    Add: $At(X_2,Y_2)$
    Del: $At(X_1,Y_1)$
Pick(X:Real, Y:Real, Z:Block, T:Table, S:Real, S':Real)
    Pre: $At(X,Y)$, $HandEmpty$, $On(Z,T)$, $Size(Z,S)$,
            $Robot\_Size(S')$, $Dist((X,Y),T) \leq Arm_{lenght}$
    Add: $Hold(Z)$, $Robot\_Size(S)$
    Del: $On(Z,T)$, $HandEmpty$, $Robot\_Size(S')$

Owing to the interaction between predicates of the different classes (e.g. *On* and *At*), Shaper is able to build a graph for the robot position (and accessibility

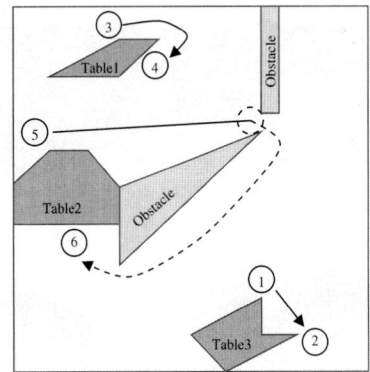

Figure 4: Environment description and "Grasp" classes explanations

via motion) that depends on the robot width, instead of generating one graph per robot width. In this case, edges are labeled by the *Robot_Size* predicate (see the previous section).

Figure 5 presents a learned graph for the tank of water example (a 2D representation of the environment is presented in figure 4). The goal is to have a glass of water on $Table_3$ (initial position of the robot). To do this, the robot must move a tank of water (initially on $Table_1$) and a glass (initially there is one on $Table_1$ and another one on $Table_2$) to $Table_3$ where a pump allows to fill the glass from the tank.

What is of interest here is the capability of this scheme to provide an effective way to deal with intricate links between the logics of the task and its geometric counterpart.

First, because there is a narrow passage which prevents the robot to go from one side to the other while holding a big object (the tank), ShaPer maintains two classes of robot positions for picking objects on $Table_2$. This fact has drastic consequences on robot plans that need to transfer the tank. This is the reason why a STRIPS plan fails (i.e. it deals with a too high level environment model).

Figure 5 compares a plan obtained by using a classical two-level task planner and motion planner with a plan produced by ShaPer. Even if one interleaves a task planner and a motion planner (to avoid failure during execution), ShaPer's expressiveness allows to find the shortest plan (e.g. pick the *Glass* from $Table_2$ instead of $Table_1$).

4.2 Object grasping

In the geometric context, we must define the fact that the robot is able to *pick* or *place* an object. Indeed, *Move* action does not allow to know the table

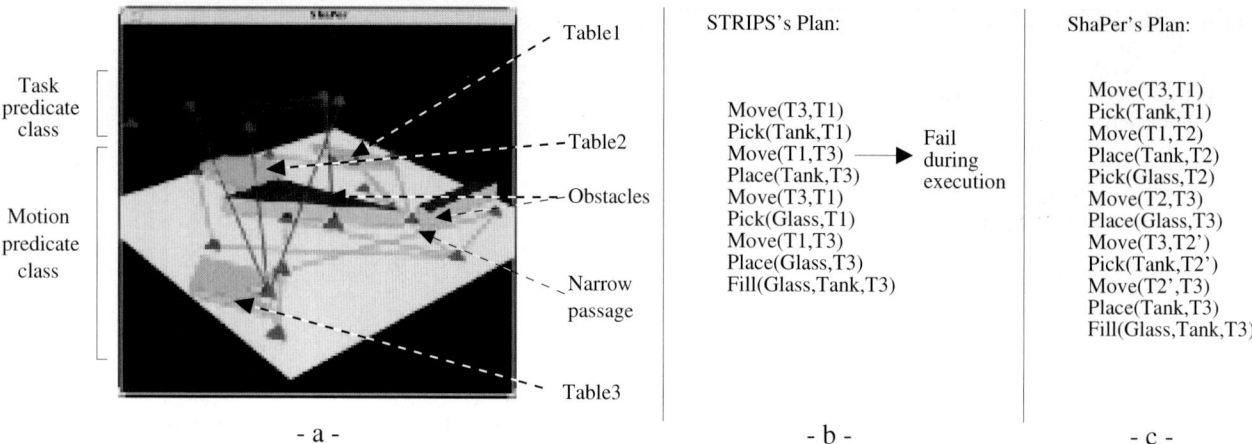

Figure 5: Tank of water problem. a. Learned graph with predicate independence. b. Plan in classical STRIPS model. c. Plan when the planner takes into account geometrical constraints.

proximity. A *near* precondition can, for instance, be defined based on robot-table distance.

Such a method allows infinite grasp positions; ShaPer allows to group them into several "grasp" classes (if necessary). Note that their number depends on the local planner ($\mathcal{L}$) capabilities. Figure 4 shows two distinct cases: i) $\mathcal{L}$ is a straight-line. The node n_2 is locally accessible from n_1, so n_2 is in the same class as n_1, it is not added. In the same way n_3 and n_4 define two classes; n_5 and n_6 define two classes. ii) $\mathcal{L}$ is a more powerful motion planner. n_4 is accessible from n_3, so there is only one class. However n_5 and n_6 define two classes because the robot width does not allow to pass through the narrow passage. Consequently the two nodes are necessary to capture the task topology.

5 Conclusion and future work

We have proposed a new planning algorithm based on an accessibility graph learning. ShaPer allows to demonstrate promising capabilities of such a method. Indeed, it is able to deal efficiently with complex task problems and geometric constraints. An example, where task and motion planning are closely interleaved, shows that ShaPer is more expressive than a hierarchical decomposition in a high level where task planning is performed and a lower level where geometric problems are dealt with.

Our future work will concern further investigations on the following aspects: i) improvement of the reasoning on robot manipulation [1] and ii) extension to deal with uncertainty especially by including perception actions [5].

References

[1] R. Alami, J.P. Laumond, and T. Siméon. Two manipulation planning algorithms. In *The First Workshop on the Algorithmic Foundations of Robotics, A.K. Peters Pub., Boston, MA.*, 1994.

[2] R. Alami, T. Siméon, and J.P. Laumond. A geometrical approach to planning manipulation tasks. the case of discrete placements and grasps. In *In Robotics Research: The Fifth International Symposium, Tokyo (Japan)*, pages 113–123, 1990.

[3] A.L. Blum and M.L. Furst. Fast planning through planning graph analysis. *Artificial Intelligence*, pages 281–300, 1997.

[4] R.E. Fikes and N.L. Nilsson. Strips: A new approach to the application of theorem proving to problem solving. *Artificial Intelligence*, 2:189–208, 1971.

[5] E. Guéré and R. Alami. A possibilistic planner that deals with non-determinism and contingency. *Proc. 16th Inter. Joint Conf. on Artificial Intelligence (IJCAI'99)*, 1999.

[6] E. Guéré and R. Alami. An accessibility graph learning approach for task planning in large domains. *ECAI 14th Workshop on Planning and Configuration*, 2000.

[7] J. Koehler, B. Nebel, J. Hoffmann, and Y. Dimopoulos. Extending planning graphs to an adl subset. In *4th European Conference on Planning (ECP'97)*, 1997.

[8] J.-C. Latombe. Robot motion planning. In *Kluwer Press*, 1991.

[9] T. Simeon, J.P. Laumond, and C. Nissoux. Visibility-based probabilistic roadmaps for motion planning. *(Submitted to Advanced Robotics Journal) A short version appeared in IEEE IROS*, 1999.

[10] P. Tournassoud, T. Lozano-Perez, and E. Mazer. Regrasping. In *IEEE, International Conference on Robotics and Automation, Raleigh*, 1987.

An Adaptive Framework for 'Single Shot' Motion Planning: A Self-Tuning System for Rigid and Articulated Robots*

Daniel Vallejo
Department of Computer Systems Engineering
Universidad de las Américas-Puebla
Cholula, Puebla 72820 Mexico
dvallejo@mail.udlap.mx

Ian Remmler   Nancy M. Amato
Department of Computer Science
Texas A&M University
College Station, TX 77843
{irr5509,amato}@cs.tamu.edu

Abstract

This paper describes an enhanced version of our previously proposed adaptive framework for single shot motion planning. This framework is versitile, and particularly suitable for crowded environments. Our iterative strategy analyzes the characteristics of the query and adaptively selects planners whose strengths match the current situation.

Contributions in this paper include an automatic method for setting and adaptively tuning planner characterizations, and reducing the reliance on programmer expertise present in the original framework. The adaptive refinement enables the system to evolve parameters specifically suited for particular classes of applications. The system now supports articulated robots, which were not supported previously. Our experimental results in complex 3D CAD environments show that our strategy solves queries that none of the planners could solve on their own.

1 Introduction

Automatic motion planning has application in many areas such as robotics, virtual reality systems, and computer-aided design. When many motion planning queries will be performed in an environment, it may be useful to preprocess the environment (as *roadmap* motion planning methods [18, 15] do, for example) in order to reduce the difficulty of subsequent queries.

However, if the `start` and `goal` configurations are known *a priori*, and only a few queries will be performed, it is generally not worthwhile to perform an expensive preprocessing stage. In this case, a more directed search of the free configuration space could lead to faster solution times. Motion planning methods that operate in this fashion are often called *single shot* methods. Single shot methods are also useful in dynamic environments in which other techniques, such as roadmap methods, are not suitable.

One of the first randomized planning methods is the Randomized Path Planner (RPP) of Barraquand and Latombe [5], which is a single shot planner. This potential field method uses random walks to attempt to escape local minima, which works well when C-space is relatively free, but is not as effective in cluttered scenarios [15]. Some success has been achieved in adapting the general probabilistic roadmap (PRM) strategy to solve single queries by restricting attention to 'useful' regions of C-space [14, 20]. A related idea is to use a sample of free points to specify promising subgoals [7, 11, 12]. Other approaches that have been used are to limit the part of the C-space that is explored. The planner in [16] finds paths for six-dof manipulators using heuristic search techniques. A skeleton of the C-space is built incrementally using a local opportunistic strategy in [8]. The Ariadne's clew algorithm [7] uses genetic algorithms to help search for a path in high-dimensional C-spaces. The Rapidly-exploring Random Tree (RRT) [17] method grows a tree from a specified start configuration. In [9] a hybrid approach is considered which utilizes a potential function (similar to RPP) on queries, but also saves information from past attempts in a graph to aid future queries in the same environment.

Although much progress has been made, good single shot solutions are needed for problems in cluttered environments. In [22], we present an iterative strategy for single shot motion planning. At any given time, we have a set of sub-queries, any one of which would solve the original motion planning problem, and a set of planning algorithms which can be applied to the sub-queries. We select a sub-query and a planner to be used on it, based on an evaluation process meant to select the best suited local planning method and the most promising sub-query. Anywhere from one to all of the local planners in the bank may be executed when solving a particular query – the goal is to use

---

[1]This research supported in part by NSF CAREER Award CCR-9624315 (with REU Supplement), NSF Grants IIS-9619850 (with REU Supplement), ACI-9872126, EIA-9975018, EIA-9805823, and EIA-9810937, and by the Texas Higher Education Coordinating Board under grant ARP-036327-017. This work was performed while Vallejo was studying at Texas A&M where he was supported in part by a Fulbright-CONACYT scholarship.

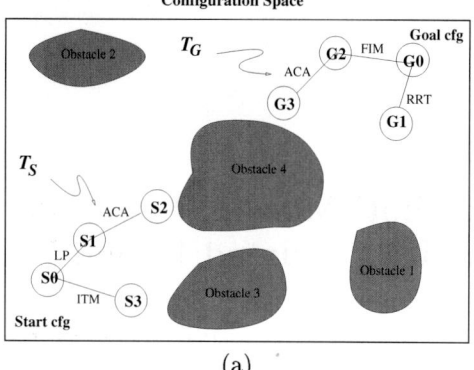

```
Repeat
 <qp,alg> := ExtractBest(PQ);
 PartialPath := Alg(qp);
 if ( qp.start in T_S)
   T_S := T_S + PartialPath; T := T_S;
 else T_G := T_G + PartialPath; T := T_G;
 if (T_S != T_G)
   last := LastCfg(PartialPath);
   for each (cfg in T)
     qp1 := <last,cfg>;
     qp2 := <cfg,last>;
     for each (alg in AlgoBank)
       insert( <qp1,alg>, score(qp1,alg), PQ);
       insert( <qp2,alg>, score(qp2,alg), PQ);
until ( T_S == T_G )
```

(a)   (b)

Figure 1: (a) A graph generated during Single Shot planning. Edges are labeled with the planning algorithm that made the connection, and (b) pseudo-code description of single shot loop.

them cooperatively.

Contributions in this paper include an automatic methodology for adaptively tuning the planner characterizations, thus reducing the reliance on programmer expertise present in the original framework. The adaptive refinement enables the system to evolve parameters specifically suited for particular classes of applications. In addition, the current system has been extended to articulated robots, which were not supported in the previous system. Our experimental results in complex 3D CAD environments show that our strategy solves queries that none of the planners could on their own. For completeness, the system in its entirety is described here, with more emphasis placed on the new additions.

2 Single Shot Framework

The framework is outlined in Figure 1. To find a path from the Start to the Goal configuration, we grow a tree T_S from the Start configuration and another tree T_G from the Goal configuration. In each iteration, we attempt to generate a path that connects T_S and T_G, and therefore also the Start and Goal configurations. In each iteration, we consider all potential query pairs with one configuration in T_S and one in T_G, and all the algorithms in the bank, and select the query pair and algorithm combination that is most likely to make a connection. Then, the selected algorithm is executed on the query pair and we add the (partial) path returned to T_S or T_G. If the sub-query fails to solve the problem, then we generate query pairs of the form (last, cfg) and (cfg, last), where last is the last configuration of the partial path returned by the sub-query, and cfg is a configuration in a different connected component from last. Each such pair is evaluated, and promising <QueryPair, Algorithm> tuples are added to the priority queue. The process continues in this way, growing T_S and T_G, until they can be connected or until some maximum number of iterations is reached.

2.1 Evaluation Criteria

The most important operation in each iteration is selecting the "best" <QueryPair, Algorithm> tuple. First, we calculate characteristics of the potential query pairs. Second, based on the query pair and algorithm characteristics, a score is assigned to each <QueryPair, Algorithm> tuple. These scored tuples are stored in a priority queue so that the "best" <QueryPair, Algorithm> tuple can be selected in each iteration.

2.1.1 Query properties and characteric values

For a given query pair, evaluation criteria are used to characterize the **local properties** of the space near the start and goal configurations and the **global properties** of the space between them (details can be found in [3]).

The evaluation begins by determining general characteristics of individual configurations. This is done by sampling additional configurations near the configuration of interest. For each characteristic property, we compute a numerical value (normalized to [0, 1]). Currently, for each configuration c, we consider the following **local properties**:

L_1: Clearance (distance to nearest obstacle),
L_2: Translation (free translational space near c),
L_3: Rotation (free rotational space near c),
L_4: Free (free space near c).

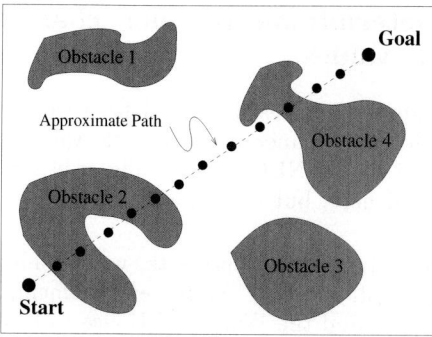

Figure 2: The approximate path P.

Next, the space between the configurations $c1$ and $c2$ is evaluated to analyze the type of planning required for the query. This is done in part by evaluating configurations sampled at a coarse resolution on a straight-line connecting $c1$ and $c2$ (see Figure 2). Currently, the **global properties** (normalized to $[0,1]$) considered are:

G_1: Distance between $c1$ and $c2$.
G_2: Proportion of free configurations in P,
G_3: Number of obstacles in collision with P.

2.1.2 Planner characteristics

We assign each algorithm values corresponding to the characteristics which describe the ideal situation in which the planner should be used. Each planning algorithm is assigned values s_{L_i} and g_{L_i}, $i = 1..4$, representing local characteristics of ideal start and goal configurations, respectively, and values q_{G_i}, $i = 1..3$, representing the global characteristics of a query on which that algorithm should be used.

In the original system described in [22], these values were pre-set using programmer expertise. As described in Section 4.3, our current system provides an automated mechanism for deriving, and automatically updating, the values for each planner. We consider two general scenarios, environments that are relatively "OPEN" and cluttered environments that are more "CONFINED," and derive default characteristic values for each. We have made this distinction since we believe that the selection of the planners for a particular sub-query depends mainly on the global characteristics of the environment considered. Thus, we ask the user to identify the queries as either "OPEN" or "CONFINED" when initiating the single shot query.

We also provide a mechanism for updating the characteristic values assigned to a planner dynamically. If a planner makes "good" progress when trying to solve a sub-query, the characteristic values of that query are averaged with the current values assigned to the planner. If the planner reduces the original distance between the start and goal configurations by at least half, then the characteristic values of the sub-query are averaged with those of the planner. In this way, the program can "learn" from good experiences.

2.1.3 Scoring

The score for a `<QueryPair, Algorithm>` tuple is a weighted average in which we give decreasing importance to the local properties of the start configuration (S), the global properties of the query (Q), and the local properties of the goal configuration (G).

First, for each component (S, Q, G), we sum the differences between the computed `QueryPair` value and the algorithm's assigned value. The score for the tuple $<(s,g), A>$ is:

$$\text{score}((s,g), A) = S(s, A) + \frac{1}{2}Q(s, g, A) + \frac{1}{4}G(g, A)$$

with the best possible score being zero, representing a perfect match between query pair (s,g) and planner A.

3 Single Shot Algorithm Bank

The algorithm bank should include a diverse set of algorithms so that at least one of them will perform well in each possible situation. See [21] for details.

Our algorithm bank includes several *local planning* methods that are commonly used in PRMs. They are relatively simple, deterministic methods that are fast but not very powerful. The methods currently in our bank are: straight-line in C-space, rotate-at-s (for rigid bodies), and simple A^*-like planners. All methods are described in detail in [1].

The *directed expansion methods* attempt to grow a connected component of configurations from the start to the goal. The iterative spread (ISM), translational (ITM) and rotational (IRM) methods try to 'spread' away from the start, and towards the goal when possible. ACA is a variant of the Ariadne's Clew algorithm [7] which uses random walks instead of the originally proposed Manhattan paths as its local planning method.

The *random expansion methods*, which grow a connected component of configurations from the start configuration, are useful in crowded situations. The *random walk* (RWM) method performs a random walk of a given number of steps from a start configuration. It is useful for escaping from "local minima." Our bank also includes a simple version RRT of the

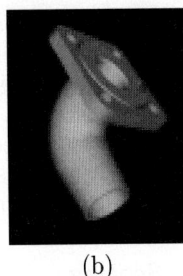

Figure 4: CAD models (a) alpha, (b) flange.

Rapidly-exploring Random Tree algorithm proposed in [17].

The *subgoal generation methods* generate subgoals for navigating around obstacles. They are suitable candidates when the `start` or `goal` is in open space and there are obstacles between them. The *first intersection method* (FIM) tries to go directly from the `start` to the `goal` using the straightline planner. If an obstacle interferes, it samples configurations around it in a manner similar to OBPRM [2]. The *recursive midpoint method* (RMM) splits the query at its midpoint into two sub-queries. This continues until each sub-query results in a collision-free connection.

4 Experimental Setup

Here we describe the implementation details of our system, the environments that we consider, and the experimental methodology used to obtain the characteristic values for the planners.

4.1 Implementation details

Our system consists of approximatelly 6,000 lines of C++ code, and is built on top of the OBPRM library ([1, 2, 4]), which consists of about 25,000 lines of C++ code. For collision detection, our system supports RAPID [13], V-Clip [19], and CSTK [23, 24]. The experiments were conducted on an HP V2200 system.

4.2 Environments studied

For our tests we used several artificial environments with rigid and articulated robots, and complex CAD-type environments with only rigid robots. The artificial environments (Figure 3) were designed so that it would be advantageous to use multiple planners. We tried to include representative environments, some relatively open, and some relatively confined. The CAD models (Figure 4) represent more realistic motion planning problems.

4.3 Determining planner characteristic values

The simulation studies described here were used to set the default planner characteristic values for the OPEN and CONFINED cases. A digest of the results is presented here, but complete results can be found in [21].

Of the artificial environments, we considered the Sandwich-Wall and the Stairs environments to be CONFINED, and the Walls and House to be OPEN. For each environment, we used a visualization tool to select about 20 representative configurations in open space, in narrow corridors, etc. We then considered all possible pairs of these configurations as start and goal. For each query pair, we recorded the characteristic values, whether each planner was successful, and the running time of each query. To find our default characteristic values, we averaged the characteristic values of the successful queries for each planner. The standard deviations were all relatively small, indicating that these averages are fairly good choices.

The values for the various planners determined by our experiments for the CONFINED case are shown in Table 1. We chose to distinguish between OPEN and CONFINED environments because, although similar trends were displayed in the two types of environments, some differences in the magnitudes of the characteristic values were noted. Moreover, programmer expertise can usually be relied upon to classify a query and environment as being OPEN or CONFINED. Thus, when initiating the single shot program, the programmer must specify whether the query involves an OPEN or CONFINED environment.

Our experiments also enable us to study the benefits of the characteristics and the individual planners. To see if any of the characteristics were redundant we calculated the correlations between all the planners. Only the translation proportion and free proportion showed a strong correlation. This indicates that it may be sufficient to consider only one of those characteristics. We consider both characteristics anyway because we believe there could be some environments in which they might differ.

We also calculated the correlation between the query successes for the planners. Little correlation was observed, indicating that none of the planners is completely redundant. In some environments, the RRT and the RWM planners seem to be highly correlated, and in these cases one should probably employ only RWM since it is faster than RRT. Something similar occurs with ITM, RRT and RWM. In this case we should perhaps employ the cheaper ones (ITM and RWM), and only the more expensive (RRT) if it is needed. Representative correlation results can be seen in Table 2.

 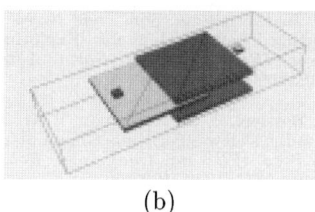 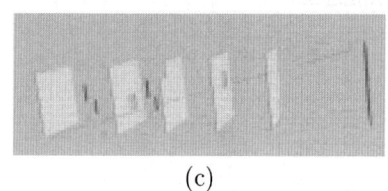 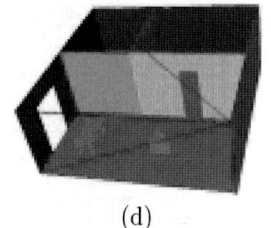

(a) (b) (c) (d)

Figure 3: Artificial environments: (a) sandwich-wall, (b) stairs, (c) walls, (d) house-piano.

Alg	Characteristic Values									
	Start				Query			Goal		
	L_1	L_2	L_3	L_4	G_1	G_2	G_3	L_1	L_2	L_3	L_4
LP	.06	.93	.60	.61	.17	.93	.08	.11	.93	.64	.64
ITM	.06	.94	.58	.58	.17	.82	.17	.10	.94	.65	.64
IRM	.06	.93	.58	.58	.18	.77	.19	.10	.94	.62	.62
ACA	.06	.93	.57	.57	.17	.79	.18	.10	.94	.64	.64
RRT	.06	.93	.57	.57	.17	.83	.16	.09	.94	.65	.65
RWM	.07	.93	.58	.58	.17	.81	.17	.10	.94	.64	.64
FIM	.06	.93	.60	.60	.17	.77	.19	.09	.93	.59	.59
RMM	.06	.94	.56	.55	.30	.39	.30	.08	.94	.56	.55

Table 1: Planner Characteristic Values (Confined)

Alg	LP	ITM	IRM	ACA	RRT	RWM	FIM	RMM
LP	1	.68	.67	.70	.70	.68	.65	.07
ITM	.68	1	.83	.87	.94	.93	.80	.06
IRM	.67	.83	1	.86	.85	.85	.77	.06
ACA	.70	.87	.86	1	.89	.90	.78	.08
RRT	.70	.94	.85	.89	1	.96	.79	.06
RWM	.68	.93	.85	.90	.96	1	.80	.07
FIM	.65	.80	.77	.78	.79	.80	1	.01
RMM	.07	.06	.06	.08	.06	.07	.01	1

Table 2: Planner Correlation Matrix (Confined)

5 Experimental Results

In this section we examine the performance of our prototype implementation of our single shot planner. We consider queries in both artificial environments and in complex mechanical CAD models. We compare the single shot planner with an obstacle-based probabilistic roadmap method (OBPRM), and with the versions of the rapidly-exploring random tree method (RRT) and the Ariadne's Clew Algorithm (ACA) that we implemented as subroutines in our algorithm bank (see Section 3).

Artificial Environments (Rigid Bodies)			
Env/Robot	Method	Time (s)	Solved
Sandwich Wall/block	OBPRM	10	yes
	RRT	864	NO
	ACA	946	NO
	SS:(ACA,RRT,RMM)	136	yes
Stairs/block	OBPRM	3,865	yes
	RRT	7,893	NO
	ACA	19,419	NO
	SS:(FIM,ACA,IRM,ACA, IRM,ITM,FIM,FIM)	3,061	yes
Walls/Stick	OBPRM	446	yes
	RRT	421	yes
	ACA	8	yes
	SS:(LP,RRT,ACA)	78	yes
House/Piano	OBPRM	70	yes
	RRT	2,014	NO
	ACA	16	yes
	SS:(IRM,RMM)	46	yes

Table 3: Artificial Environment Results

From Table 3, Table 4, and Table 5 we can see that the single shot system works well for both rigid and

Artificial Environments (Articulated)			
Env.	Method	Time (s)	Solved
Sandwich Wall	OBPRM	549	NO
	RRT	352	NO
	ACA	196	yes
	SS:(ACA)	196	yes
Stairs cline2-4	OBPRM	1,472	NO
	RRT	636	yes
	ACA	285	yes
	SS:(IRM,IRM,ITM, RWM,ACA)	302	yes
Walls	OBPRM	1,903	NO
	RRT	1,236	yes
	ACA	468	yes
	SS:(LP,ACA)	517	yes

Table 4: Artificial Environment Results (Articulated)

articulated robots in the artificial environments, and rigid robots in the CAD environments. The framework generally selected algorithms appropriate for the query, and the various planners do seem to work together to construct solutions that are competitive with the best of the individual methods. While there are cases when other methods outperform the single shot system, our performance is generally good. This indicates that the single shot system tends to avoid the pathological behavior of any of the individual methods.

6 Conclusion and Future Work

We have described a framework which enables multiple algorithms to work together to solve a query, and

CAD Environments			
Env.	Method	Time (s)	Solved
Alpha1.5	OBPRM	3,495	yes
	RRT	5,555	NO
	ACA	1,699	yes
	SS:(ACA)	1,699	yes
Alpha1.2	OBPRM	94,150	NO
	RRT	49,455	NO
	ACA	83,254	NO
	SS:(ACA,ACA,ITM,IRM)	68,934	yes
Flange.85	RRT	17,275	NO
	ACA	233	yes
	SS:(ACA)	233	yes

Table 5: CAD Environment Results

have shown that this yields better results on average than any algorithm alone. In this paper, we have extended our previous work [22] by showing that an automatic characterization of the planning algorithms can be used to effectively match algorithms with particular queries. We have also proposed a methodology for adaptively tuning these values so that the system can evolve over time for particular classes of applications. In addition, we have shown the generality of our approach by applying it to various types of environments and robots. There are still many interesting issues to be studied. For example, determining the best planning algorithms for the algorithm bank – perhaps some current algorithms should be removed, and others might be added. Similar questions exist for the characteristic values used for the queries and algorithms.

Acknowledgement

The alpha puzzle was designed by Boris Yamrom of the Computer Graphics & Systems Group at GE's Corporate Research & Development Center. GE also provided us with the Flange environment.

References

[1] N. M. Amato, O. B. Bayazit, L. K. Dale, C. V. Jones, and D. Vallejo. Choosing good distance metrics and local planners for probabilistic roadmap methods. In *Proc. IEEE Int. Conf. Robot. Autom. (ICRA)*, pages 630–637, 1998.

[2] N. M. Amato, O. B. Bayazit, L. K. Dale, C. V. Jones, and D. Vallejo. OBPRM: An obstacle-based PRM for 3D workspaces. In *Proc. Int. Workshop on Algorithmic Foundations of Robotics (WAFR)*, pages 155–168, 1998.

[3] N. M. Amato, C. V. Jones, and D. Vallejo. An adaptive framework for 'single shot' motion planning. Technical Report 98-025, Dept. of Computer Science, Texas A&M University, Nov 1998.

[4] N. M. Amato and Y. Wu. A randomized roadmap method for path and manipulation planning. In *Proc. IEEE Int. Conf. Robot. Autom. (ICRA)*, pages 113–120, 1996.

[5] J. Barraquand and J.-C. Latombe. Robot motion planning: A distributed representation approach. *Int. J. Robot. Res.*, 10(6):628–649, 1991.

[6] O. B. Bayazit, G. Song, and N. M. Amato. Enhancing randomized motion planners: Exploring with haptic hints. In *Proc. IEEE Int. Conf. Robot. Autom. (ICRA)*, pages 529–536, 2000.

[7] P. Bessiere, J. M. Ahuactzin, E.-G. Talbi, and E. Mazer. The ariadne's clew algorithm: Global planning with local methods. In *Proc. IEEE Int. Conf. Intel. Rob. Syst. (IROS)*, volume 2, pages 1373–1380, 1993.

[8] J. Canny and M. C. Lin. An opportunistic global path planner. In *Proc. IEEE Int. Conf. Robot. Autom. (ICRA)*, pages 39–48, 1990.

[9] S. Caselli and M. Reggiani. ERPP: An Experience-based Randomized Path Planner. In *Proc. IEEE Int. Conf. Robot. Autom. (ICRA)*, pages 1002–1008, 2000.

[10] H. Chang and T. Y. Li. Assembly maintainability study with motion planning. In *Proc. IEEE Int. Conf. Robot. Autom. (ICRA)*, pages 1012–1019, 1995.

[11] P. C. Chen and Y. K. Hwang. SANDROS: A dynamic graph search algorithm for motion planning. *IEEE Trans. Robot. Automat.*, 14(3):390–403, 1998.

[12] B. Glavina. Solving findpath by combination of directed and randomized search. In *Proc. IEEE Int. Conf. Robot. Autom. (ICRA)*, pages 1718–1723, 1990.

[13] S. Gottschalk, M.C. Lin, and D. Manocha. Obb-tree: A hierarchical structure for rapid interference detection. Technical Report TR96-013, University of N. Carolina, Chapel Hill, CA, 1996.

[14] D. Hsu, J-C. Latombe, and R. Motwani. Path planning in expansive configuration spaces. In *Proc. IEEE Int. Conf. Robot. Autom. (ICRA)*, pages 2719–2726, 1997.

[15] L. Kavraki, P. Svestka, J. C. Latombe, and M. Overmars. Probabilistic roadmaps for path planning in high-dimensional configuration spaces. *IEEE Trans. Robot. Automat.*, 12(4):566–580, August 1996.

[16] K. Kondo. Motion planning with six degrees of freedom by multistrategic bidirectional heuristic free space enumeration. *IEEE Trans. Robot. Automat.*, 7(3):267–277, 1991.

[17] J. J. Kuffner and S. M. LaValle. RRT-Connect: An Efficient Approach to Single-Query Path Planning. In *Proc. IEEE Int. Conf. Robot. Autom. (ICRA)*, pages 995–1001, 2000.

[18] J. C. Latombe. *Robot Motion Planning*. Kluwer Academic Publishers, Boston, MA, 1991.

[19] B. Mirtich. V-clip: Fast and robust polyhedral collision detection. Technical Report TR97-05, Mitsubishi Electric Research Lab, Cambridge, MA, 1997.

[20] M. Overmars and P. Svestka. A probabilistic learning approach to motion planning. In *Proc. Workshop on Algorithmic Foundations of Robotics*, pages 19–37, 1994.

[21] D. Vallejo. *An Adaptive Framework for 'Single Shot' Motion Planning*. PhD thesis, Dept. of Computer Science, Texas A&M University, December 2000.

[22] D. Vallejo, C. V. Jones, and N. M. Amato. An adaptive framework for 'single shot' motion planning. In *Proc. IEEE Int. Conf. Intel. Rob. Syst. (IROS)*, 2000. To appear.

[23] P. Xavier. Fast swept-volume distance for robust collision detection. In *Proc. IEEE Int. Conf. Robot. Autom. (ICRA)*, pages 1162–1169, 1997.

[24] P. G. Xavier and R. A. LaFarge. A configuration space toolkit for automated spatial reasoning: Technical results and ldrd project final report. Technical Report SAND97-0366, Sandia National Laboratories, 1997.

Optimal Line-sweep-based Decompositions for Coverage Algorithms

Wesley H. Huang
Department of Computer Science
Rensselaer Polytechnic Institute
Troy, New York 12180

Abstract

Robotic coverage is the problem of moving a sensor or actuator over all points in a given region. Ultimately, we want a coverage path that minimizes some cost such as time. We take the approach of decomposing the coverage region into subregions, selecting a sequence of those subregions, and then generating a path that covers each subregion in turn. In this paper, we focus on generating decompositions based upon the planar line sweep.

After a general overview of the coverage problem, we describe how our assumptions lead to the optimality criterion of minimizing the sum of subregion altitudes (which are measured relative to the sweep direction assigned to that subregion). For a line-sweep decomposition, the sweep direction is the same for all subregions. We describe how to find the optimal sweep direction for convex polygonal worlds.

We then introduce the minimal sum of altitudes (MSA) decomposition in which we may assign a different sweep direction to each subregion. This decomposition is better for generating an optimal coverage path. We describe a method based on multiple line sweeps and dynamic programming to generate the MSA decomposition.

1 Introduction

Coverage algorithms have received much attention in the past several years because of demining operations in several parts of the world. Landmines must be detected and removed in order to make these areas safe for public activity, and we would like to use robots to carry out the demining. Such a robot must pass a specialized sensor over all points in an area. These robots should be guaranteed to cover the entire area and should perform this task efficiently.

Coverage algorithms also have much broader applicability. Like demining, some applications require that a sensor be passed over all points in a given area. These applications include mapping (creating image mosaics), inspection, and search and rescue operations. Other applications require some sort of actuator to be passed over a given area: spraying coatings (such as paint), vacuum cleaning, lawnmowing, and various agricultural tasks. An application such as snow removal requires not only coverage but consideration of all the snow that is collected!

There have been several coverage algorithms published in the robotics literature, including online and offline algorithms, and single and multiple robot algorithms; however, none address the cost of the path generated to cover the given area.

Efficiency is important in many coverage applications. Time is critical in search and rescue operations, and in industrial applications, even a small improvement in efficiency can result in large cost savings through improved cycle times or reduced material use.

A coverage algorithm must generate what we will call a *coverage path*, i.e. a detailed sequence of motion commands for a robot that sweep the given sensor or actuator over a specified region. An optimal coverage algorithm would return a coverage path that minimized, for example, the time required to execute that path.

Several existing algorithms take the following basic approach to generating a coverage path: the region to be covered is decomposed into subregions, a traveling-salesman algorithm is applied to generate a sequence of subregions to visit, and a coverage path is generated from this sequence that covers each subregion in turn. These algorithms all use a single line sweep in order to decompose the coverage region into subregions, and these subregions are individually covered using a back and forth motion in rows perpendicular to the sweep direction. All subregions use the same sweep direction.

After finishing one row, the robot must turn around to start the next row, and we claim that minimizing the number of these turns is the most important factor in an efficient solution. The number of turns is directly related to the altitude of the subregion (measured along the sweep direction), so our optimality criterion is to minimize the sum of subregion altitudes.

In Section 3, we show that the optimal line sweep decomposition for convex polygonal worlds is generated by sweeping a line parallel to one of the boundary edges. We have also shown this result holds for nonconvex polygonal worlds.

In Section 4, we show that by allowing different sweep directions to be assigned to each subregion of a decomposition, we can produce a lower sum of subregion altitudes and thus a cheaper coverage path. We propose an algorithm to solve this minimal sum of altitudes (MSA) decomposition problem which uses multiple line sweeps and dynamic programming.

1.1 Previous work

Two similar and relatively recent algorithms for coverage are by Choset and Pignon [2] and Hert *et al.* [6].

Choset and Pignon describe an offline planning algorithm for polygonal worlds which explicitly performs a line sweep decomposition (the "Boustrophedon" decomposition) and creates a sequence of subregions (cells) using an heuristic traveling-salesman algorithm. This work includes experiments on a synchro-drive mobile robot. Hert *et al.* describe an online algorithm for nonpolygonal worlds which implicitly uses a line sweep decomposition and an heuristic Traveling Salesman algorithm. This work is described in the context of an autonomous underwater vehicle that creates an image mosaic of the ocean floor.

Schmidt and Hofner [9] describe a floor cleaning robot which has nonholonomic constraints. They use an offline planning algorithm to generate a coverage path based on a line sweep decomposition. A vocabulary of "basic motion macros" are used to maneuver the robot (i.e. for the portions of the coverage path that are not straight lines).

Kurabayashi *et al.* [8] describe an offline algorithm for planning coverage paths for multiple robots. It appears to generate a single coverage path, based on both "direction parallel" and "contour-parallel" motion. Zelinsky *et al.* [10] describes a grid-based coverage algorithm.

More recent results include: Gabriely and Rimon [5], who formulated a coverage algorithm based on traveling about the perimeter of a minimum spanning tree that fills the coverage region; Butler *et al.* [1], who created a distributed algorithm for multiple robots to cover an unknown rectilinear environment; and Choset *et al.* [3] who have extended the Boustrophedon decomposition to higher dimension Euclidean spaces.

2 The Coverage problem

We make the following assumptions about the three elements that describe a coverage problem:

- *coverage region* — the region to be covered is planar (or can be embedded in the plane), is connected, and is defined by an outer perimeter and holes its interior. In this paper we will assume that both the perimeter and holes are polygonal.

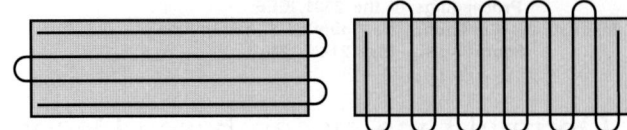

Figure 1: The number of turns is the main factor in the cost difference of covering a region along different sweep directions.

- *robot* — may have nonholonomic constraints, and the shape of the robot can be unrelated (in shape and size) to the sensor/actuator pattern. The starting or ending position of the robot may be specified, or we may insist that the robot end in the same place it starts.

- *sensor/actuator pattern* — the sensor or actuator has a one or two dimensional "coverage" pattern which sweeps out a two dimensional area as the robot moves. Common sensor/actuator patterns include a circle (for radially limited sensors), a rectangle (e.g. a video camera), a line (snowplow). We assume the sensor/actuator pattern does not move relative to the robot.

A coverage algorithm must return:

- *coverage path* — a detailed sequence of motion commands for the robot.

In the rest of this section, we discuss several issues regarding the general coverage problem and describe the assumptions upon which the remainder of the paper is based.

2.1 Optimal coverage

Our approach to the coverage problem is to decompose the coverage region, determine a sequence of subregions, and generate a coverage path that covers each region and then moves on to the next. We seek an optimal solution in this class of solutions.

The planar line sweep, upon which our decompositions are based, divides the coverage region into monotone subregions. These subregions can be easily and efficiently covered by back and forth motion along rows perpendicular to the sweep direction.

We reason that the time to cover a subregion in this manner consists of the time to travel along the rows plus the time to turn around at the end of the rows. Covering a subregion for a different sweep direction results in rows of approximately the same total length; however, there can be a large difference in the number of turns required as illustrated in Figure 1. Furthermore, turns take a significant amount of time: the robot must slow down, make the turn, and then accelerate.

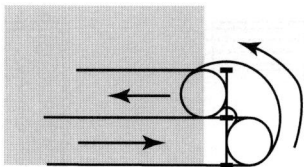

Figure 2: A lawnmower-like robot must drive outside the boundary in order to turn around efficiently.

We therefore wish to minimize the number of turns, and this is proportional to the altitude of the subregion measured along the sweep direction.

An additional cost we have not yet addressed is that of traveling from one subregion to another. In selecting a sequence of subregions, we can take into account the cost of traveling from one subregion to another, but in general, the decomposition of the coverage region cannot be independent of choosing a sequence of subregions to visit and generating a coverage path from that sequence.

We shall put these concerns aside for now by assuming that any gain from choosing a good decomposition is much larger than the variation in the total cost of traveling from subregion to subregion.

Under these assumptions, a decomposition that minimizes the sum of the subregion altitudes (as measured along the sweep direction) will produce an optimal coverage path. This is the problem we address in this paper, first for when the sweep direction must be the same for all subregions and then for when the sweep direction may be different in each subregion.

2.2 Boundaries & obstacles

We differentiate between the boundaries of the coverage region and obstacles within or outside the region (i.e. areas in which the robot cannot travel). Sometimes the two may be coincident, and sometimes they may be independent. For example, when spray painting, a "mask" may define the boundary of the coverage region, but it is permissible to spray beyond this boundary. Alternatively, when mowing a lawn, it is not acceptable to mow over any flowers surrounding the lawn.

A combination of the boundaries, obstacles, sensor/actuator pattern, and the robot may make it technically impossible for a robot to cover a region. For example, a circular robot cannot reach all the way into a convex corner when the boundary and "boundary obstacles" are coincident.

These observations lead us to the following assumption: the coverage region given as input to our algorithm has sufficient room between the boundary and any obstacle (beyond the perimeter or inside a hole) to turn around. This can be accomplished by assuming that the first time the robot encounters a hole or the perimeter, it makes one or more complete circuits around the boundary to create this "buffer zone."

This assumption also helps us deal with nonholonomic constraints of a robot platform. Figure 2 illustrates a simple 180 degree turn for a lawnmower-like robot (differential drive with a circular sensor/actuator pattern in front of the wheels).

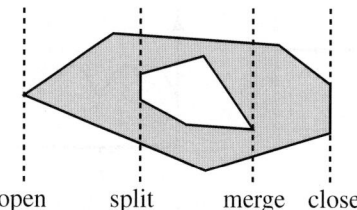

Figure 3: Illustration of the four main types of events in a line sweep from left to right.

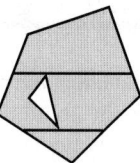 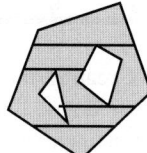

Figure 4: Introducing convex holes increases the subregion altitude sum by the altitude of the holes.

3 Optimal line sweep decompositions

Current coverage algorithms can be improved by determining the optimal sweep direction for a planar line sweep decomposition. We address this problem for convex worlds and then briefly discuss nonconvex worlds.

3.1 Convex worlds

A planar line sweep decomposes a region into monotone subregions by adding a diving line at certain "events" as the sweep line moves across the region. (A subregion is monotone with respect to a sweep direction if a line perpendicular to the sweep direction intersects the region to form a connected set of points.) Figure 3 shows the four types of events in a line sweep. For convex worlds (i.e. where the perimeter and the holes are convex polygons), where will be one OPEN event, one CLOSE event for the perimeter, and there will be one SPLIT and one MERGE event for every hole. It is easy to see, as illustrated in Figure 4, that the sum of subregion altitudes is simply the altitude of the perimeter plus the altitude of all holes.

To determine the optimal sweep direction, we can express the sum of subregion altitudes as a function of sweep

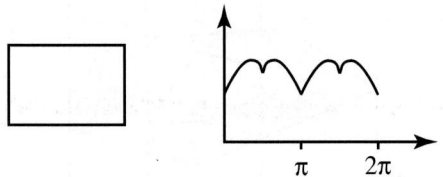

Figure 5: An example diameter function.

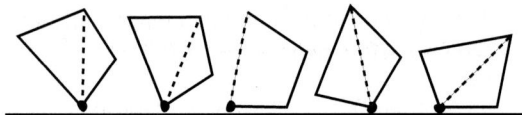

Figure 6: Creating the diameter function by rolling the polygon. The dotted line represents the chord used to determine the diameter or altitude of the polygon.

direction and then minimize this expression.

3.1.1 Altitude sum in terms of diameter functions

We can use the *diameter function* $d(\theta)$ to describe the altitude of the polygon along the sweep direction. For a given angle θ, the diameter of a polygon is determined by rotating the polygon by $-\theta$ and measuring the height difference between its highest and lowest point. An example diameter function is shown in Figure 5. The altitude of a polygon for a sweep direction at an orientation of α is $d(\alpha - \frac{\pi}{2})$.

We can then express the sum of subregion altitudes as:

$$S(\theta) = d_P(\theta) + \sum_i d_{H_i}(\theta) \qquad (1)$$

where $d_P(\theta)$ is the diameter function of the perimeter, and $d_{H_i}(\theta)$ is the diameter function of hole i. The optimal decomposition is then determined by the the sweep direction $\alpha = \theta + \frac{\pi}{2}$ that minimizes S.

3.1.2 Form of diameter functions

The form of a diameter function can be understood by considering its height as it rolls along a flat surface. Assume we start with one edge resting on the surface. As illustrated in Figure 6, we can draw a "chord from the pivot vertex to another vertex of the polygon, and the height of the polygon will be determined by this vertex. Whenever the polygon has rolled on to the next side or when an edge at the top of the polygon becomes parallel to the surface, we will change to a different chord (from a different pivot vertex or to a different top vertex). Therefore, a diameter function (for an n sided polygon) has the following form:

$$d(\theta) = \begin{cases} k_1 \sin(\theta + \phi_1) & \theta \in [\theta_0, \theta_1] \\ k_2 \sin(\theta + \phi_2) & \theta \in [\theta_1, \theta_2] \\ \vdots \\ k_{2n} \sin(\theta + \phi_{2n}) & \theta \in [\theta_{(2n-1)}, \theta_{2n}] \end{cases} \qquad (2)$$

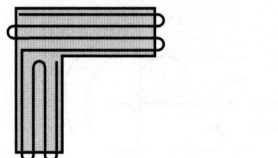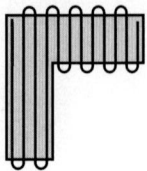

Figure 7: A simple example where assigning different sweep directions to subregions produces a better coverage path, i.e. one with fewer turns.

where $\theta_0 = 0$ and $\theta_{2n} = 2\pi$. The diameter function is piecewise sinusoidal; its "breakpoints" θ_i occur when an edge of the rotated polygon is parallel to the horizontal. This corresponds to when the sweep direction is perpendicular to an edge of the polygon. (The sweep line and the rows for covering the polygon would then be horizontal.)

Note that the diameter function only draws from the sine curve in the interval $\theta \in (0, \pi)$. Therefore, $d'' < 0$ everywhere except at the breakpoints.

3.1.3 Minimizing the altitude sum

The minimum of the function $S(\theta)$ must lie either at a critical point ($S'(\theta) = 0$) or at a breakpoint of one of its constituent diameter functions.

However, for any critical point in between breakpoints:

$$S''(\theta) = d_P''(\theta) + \sum_i d_{H_i}''(\theta) < 0 \qquad (3)$$

which means that it corresponds to a maximum! Therefore, the minimum must lie at a breakpoint of one of the component diameter functions. Since these breakpoints correspond to when the sweep direction is perpendicular to an edge of a hole or the perimeter, the minimum can be determined by testing each of these sweep directions.

3.2 Nonconvex worlds

With a nonconvex perimeter and nonconvex holes, a planar line sweep still places dividing lines at SPLIT and MERGE events, but now the perimeter and any obstacle can both produce any of the four types of events, and the sum of subregion altitudes is greater than the sum of diameters of the holes and the perimeter.

We have shown that the same result for convex worlds also holds for nonconvex worlds — the line sweep which minimizes the sum of subregion altitudes is along a sweep direction perpendicular to one of the sides of the perimeter or of a hole. For more details, see Huang [7].

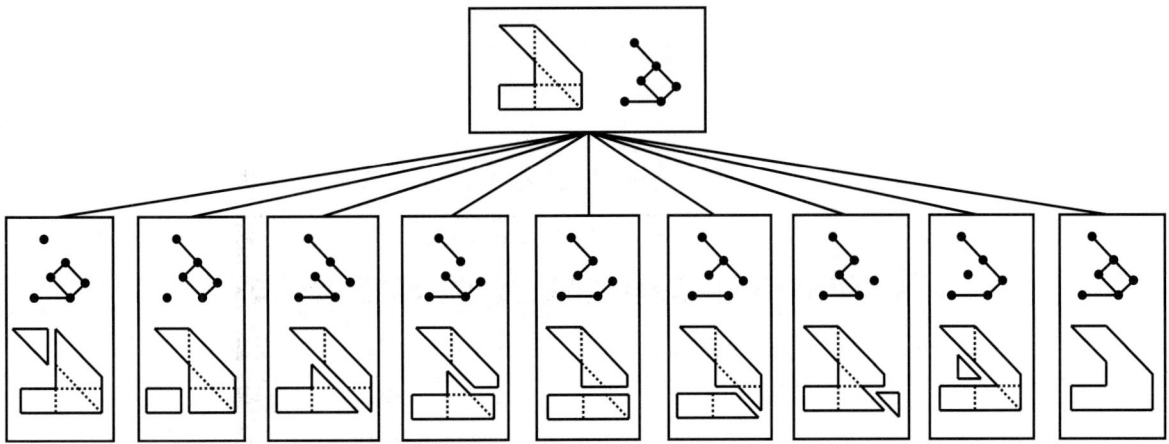

Figure 8: First stage of the dynamic programming problem decomposition. The top box shows a coverage region, its initial decomposition into cells, and the corresponding adjacency graph. There are 8 ways that this graph can be decomposed into two separate connected graphs; for each, the corresponding split of the coverage region is shown. The rightmost choice represents covering all cells as a single region.

4 MSA decomposition

Figure 7 shows a simple example where assigning a different sweep direction to each subregion results in a better decomposition. In this section, we address the general minimal sum of altitudes (MSA) decomposition: forming subregions and assigning sweep directions such that the sum of subregion altitudes (along their respective sweep directions) is minimized.

We propose an algorithm which is based on performing multiple line sweeps to decompose the coverage region into cells and then applying dynamic programming to combine cells into larger subregions and to assign a sweep direction for each subregion.

4.1 Decomposition of the coverage region

For each edge orientation (of the perimeter or a hole), we perform a line sweep using a sweep direction perpendicular to such edge. Each line sweep is done independently, but we overlay all decompositions, in effect taking the dividing lines introduced by all line sweeps. The resulting cells are monotone with respect to all sweep directions under consideration. We must additionally extend all nonconvex edges until they hit a boundary.

We hypothesize that the optimal MSA decomposition can be formed from combinations of the cells from this initial decomposition and now turn to combining these cells into subregions and assigning a sweep direction to each subregion.

4.2 Dynamic programming formulation

From the initial decomposition, we create an adjacency graph (each node represents a cell, and two nodes are connected if they share an edge). This adjacency graph may

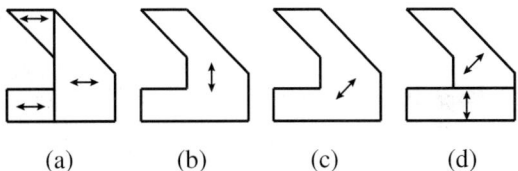

Figure 9: Figures (a) through (c) show the three line sweep decompositions of this region, of which (b) is the best with a subregion altitude sum of 7.0. Figure (d) shows an optimal MSA decomposition produced by our algorithm which has a sum of 5.5.

have cycles, even if there are no holes in the coverage region.

The basis of our dynamic programming formulation is to either split this graph in two, thus creating two smaller subproblems, or to try to unite all the cells corresponding to nodes in the graph and cover them as one large region.

At the start, we have a graph G which is the adjacency graph from the initial decomposition. If we split the graph, we create two (individually) connected subgraphs G_1 and G_2. These subgraphs together contain all the edges from G except those that connect a node from G_1 to a node in G_2.

We define the minimum sum of altitudes to be:

$$S(G) = \min \left\{ C(G), \min_i S(G_1^i) + S(G_2^i) \right\} \quad (4)$$

where i iterates over all possible ways to split the graph G into two connected subgraphs and $C(G)$ returns the cost of covering all cells corresponding to nodes in G as one subregion. When there is only one node in the graph, $S(G) = C(G)$.

The function $C(G)$ must consider all the directions under consideration to determine the cost for covering the cells in G as a single region. For some (or possibly all)

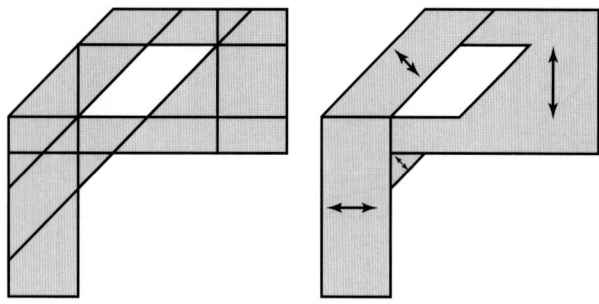

Figure 10: The left figure shows a coverage region and its initial decomposition. The right shows the MSA decomposition produced by our algorithm.

coverage directions, this region may not be monotone, in which case we assign the cost $+\infty$ for those directions. $C(G)$ returns the minimum over all sweep directions under consideration.

Figure 8 shows an example of the first level of decomposing a problem.

4.3 Results

We have implemented the MSA decomposition using the CGAL library [4]. Figure 9 shows the three line sweep decompositions and the optimal solution produced by our MSA decomposition algorithm. Figure 10 shows a more complicated example which took 15 minutes to generate on a Sun Ultra 10 workstation. The present algorithm is rather limited in the complexity of the environment input it can decompose in reasonable time.

There are two factors that indicate an exponential complexity for this algorithm. First, in creating the initial decomposition, each sweep direction contributes dividing lines that divide many other cells. This produces a large number of cells which results in an adjacency graph with many nodes.

Second, the dynamic programming phase must examine all connected subgraphs of 1 to n nodes. For a fully connected graph of n nodes, there are $2^n - 1$ such subgraphs, but it is not possible to have a fully connected adjacency graph (except in trivial cases). We may therefore expect many fewer than 2^n subproblems; however, the number of subproblems is most likely still exponential. The number depends not only on the geometry of the graph but also its maximum degree.

5 Conclusions

We have given a general overview of the coverage problem and described our assumptions and formulation for optimal coverage. We follow the basic approach of decomposing the coverage region into subregions, selecting a sequence of subregions, and generating a coverage path that covers each subregion in turn. This paper introduces a measure of optimality for coverage paths which we translated into a measure of optimality for the coverage region decomposition. The ultimate objective is to decompose a coverage region into subregions so that the sum of subregion altitudes is minimized.

We have shown that for polygonal worlds, the optimal line sweep decomposition uses a sweep direction perpendicular to an edge of the boundary. The more general minimum sum of altitudes (MSA) decomposition, however, can produce a better decomposition by allowing a different sweep direction to be assigned to each subregion. We have given an algorithm that performs multiple line sweeps to decompose the coverage region into cells, and then uses dynamic programming to combine these cells into larger subregions and to assign a sweep direction to each subregion.

References

[1] Z. Butler, A. Rizzi, and R. Hollis. Complete distributed coverage of rectilinear environments. In *Fourth International Workshop on the Algorithmic Foundations of Robotics*, 2000.

[2] H. Choset and P. Pignon. Coverage path planning: the boustrophedon cellular decomposition. In *Proceedings of the International Conference on Field and Service Robotics*, December 1997.

[3] H. Choset, E. Acar, A. Rizzi, and J. Luntz. Exact cellular decompositions in terms of critical points of Morse functions. In *IEEE International Conference on Robotics and Automation*, 2000.

[4] Computational Geometry Algorithms Library (CGAL). http://www.cgal.org/

[5] Y. Gabriely and E. Rimon. Spanning-tree based coverage of continuous areas by a mobile robot. Submitted to *Annals of Mathematics and Artificial Intelligence*.

[6] S. Hert, S. Tiwari, and V. Lumelsky. A terrain-covering algorithm for an auv. *Autonomous Robots*, 3(2–3):91–119, June–July 1996.

[7] W. Huang. The minimal sum of altitudes decomposition for coverage algorithms. Rensselaer Polytechnic Institute Computer Science Technical Report 00-3, June 2000.

[8] D. Kurabayashi, J. Ota, T. Arai, and E. Yoshida. Cooperative sweeping by mobile robots. In *IEEE International Conference on Robotics and Automation*, pages 1744–1749, 1996.

[9] G. Schmidt and C. Hofner. An advanced planning and navigation approach for autonomous cleaning robot operations. In *IEEE/RSJ International Conference on Intelligent Robots and Systems*, volume 2, pages 1230–1235, 1998.

[10] A. Zelinsky, R. A. Jarvis, J. C. Byrne, and S. Yuta. Planning paths of complete coverage of an unstructured environment by a mobile robot. *International Journal of Robotics Research*, 13(4):315–, 1994.

GLOBAL NEARNESS DIAGRAM NAVIGATION (GND)

J.Minguez L.Montano T.Simeon[†] R.Alami[†]
Computer Science and Systems Engineering [†]LAAS (CNRS)
CPS, University of Zaragoza, SPAIN Toulouse, FRANCE
{jminguez,montano}@posta.unizar.es {nic,rachid}@laas.fr

Abstract

This paper presents the Global Nearness Diagram (GND) navigation system for mobile robots. The GND generates motion commands to drive a robot safely between locations, whilst avoiding collisions. This system has all the advantages of using the reactive scheme Nearness Diagram (ND), while having the ability to reason and plan globally (reaching global convergence to the navigation problem). This framework has been extensively tested using an holonomic mobile base equipped with a laser range-finder. Experiments in unknown, unstructured, dynamic and complex environments are reported to validate the system.

1 Introduction

The development of a robust navigation system, which can work in different environments, and can adapt to everyday situations, is still an open research area in the field of robotics.

The construction of environmental models is highly coupled to navigation. This task is dependent on the natural and environmental conditions.

Focusing our attention on motion generation, we can divide navigation systems into three categories [22]: Model-based navigation systems, Hybrid systems and Reactive schemes.

- The *Model-based navigation systems* construct a model of the environment used directly to extract the motion commands. This model is based on the specific characteristics of the world (indoor/outdoor, static/dynamic, structured/unstructured...).
- The *Reactive navigation schemes* are restricted to the iteration between perception (usually the system inputs), and action (usually the system outputs). This constrains their solutions to a local section of the environment, and non optimal solutions are obtained. On the other hand, these reactive schemes have been demonstrated to be extremely well-adapted to very complex and dynamic environments, which model-based navigation systems cannot cope with.
- The *Hybrid systems* integrate both schemes in the sense that each one works independently, but they interact to perform the navigation task.

The difference between reactive systems and hybrid systems is that reactive schemes deal directly with perceptions, in order to generate motion commands, while hybrid systems build a model that interacts with the reactive scheme.

The main difference between model-based systems and hybrid systems, is the motion commands generation process. The former builds a model which is directly used to generate the motion commands. On the other hand, hybrid systems have two very well distinguished tasks: the model builder and the reactive navigation scheme, the latter generating the motion commands.

We focus our attention on reactive schemes, and their evolution over the last years. Early reactive navigation methods, firstly attempted to solve problems related to their internal behavior and drawbacks. Secondly, reactive methods evolved in order to deal with their lack of global reasoning and planning (towards hybrid methods), see Section 2 for an extended discussion on this topic.

In this paper, we present the evolution of the reactive method Nearness Diagram Navigation (ND) [1], towards the Global Nearness Diagram (GND). The GND is a navigation scheme that assures global convergence to the reactive navigation problem, inside the physical limits imposed by a model dynamically built. The GND is shown to be a very powerful navigation system, because it has all of the advantages of the reactive method ND, while incorporating global reasoning, which allows it to avoid trap situations.

The paper is organized as follows: Section 2 presents the related work, and the system requirements is introduced in Section 3. Sections 4 and 5 present two navigation systems (mND and mpND), and Section 6 shows how they cooperate to form the complete navigation system (GND). In Section 7 a comparison with other methods is presented, and in Section 8 we draw our conclusions.

2 Related Work

A reactive navigation scheme (also known as collision avoidance approach), is an algorithm that takes as input perceptions of the environment. The outputs are the motion commands that drive the robot towards the final location, while avoiding collisions. The reader is directed to [1] for an extended discussion and taxonomy of these methods.

The evolution in time followed by the common reactive navigation methods can be divided into two steps. In the first one, the methods evolved to cope with their internal drawbacks and limitations (eliminate oscillations, local trap situations, unstable motion, motion constraints, robot geometry...). In the second step, the methods evolved to deal with their lack of global reasoning, trying to increase its local nature (in the direction of hybrid methods).

The methods as they evolved cannot be considered to be purely reactive, because they go farther than only dealing with perceptions. On the other hand, they cannot be considered exclusively hybrid, because the devices introduced to increase the local nature, do not make complete sense outside of the navigation field, and are completely oriented

---

This work was partially supported by projects CICYT TAP97-0992-C02-01, DPI2000-1272 and Departamento de Educación y Cultura de la Diputación General de Aragón (Ref P29/98).

to improve the reactive method behavior.

We will discuss the evolution of five techniques (see Fig. 1): Potential Field Methods (PFM [2]), Vector Field Histogram (VFH [8]), Dynamic Window Approach (DWA [13]), Elastic Band (EB [15])) and Nearness Diagram (ND [1]).

Potential Field Methods (PFM)

The PFM [2] are obstacle avoidance methods that make a physical analogy to generate collision free motion. The obstacles and goal generate forces that are respectively repulsive and attractive. The motion commands are computed from these forces. The PFM technique has been widely used and studied by a large number of researchers [3], [4], [5], [17], [23], among others. These methods are in the first step of the evolution, because some inherent limitations settle in [6], are still a subject of research.

Vector Field Histogram (VFH)

The VFH [8] is an obstacle avoidance method, that selects the motion direction from a precalculated set of solutions (valleys), switching among three different laws. Later VFH+ [9] was presented, where internal problems and drawbacks of the original VFH were overcame. VFH+ took into account the width of the robot and the robot trajectory. Less oscillatory results were obtained, and it was possible to commit to a direction due to improved motion selection. Recently, the VFH* [10] was presented which basically deals with the local nature of the VFH+. VFH* uses a look-ahead verification to analyze the consequences of heading towards the candidate directions. The consequences are quantified by cost functions, allowing for the selection of the one which minimize some criteria. Trap situations are avoided by calculating a number of steps in advance of the algorithm's execution.

Elastic Band EB

The Elastic Band [15] and [16] is a framework that provides many of the benefits of reactive systems without sacrificing global planning. A path is provided by a global planner. Incremental adjustments to the path are based on the sensory data while maintaining the path in the free space. The concept of *bubble* is introduced to implement the elastic band efficiently. Later, [17] introduced a new formalism of this concept in a *Reed and Shepp* metric system, taking into account the kinematic constraints of the robot. Recently, the elastic strip (ES) framework has been presented [18] and [19]. Here, several local replanning operations are integrated in this framework to deal with moving obstacles, to improve the behavior in dynamic environments. While the elastic strips can be used to obstacle avoidance for mobile robots, it has been shown to work extremely well in high-dimensional configuration spaces.

Dynamic Window Approach (DWA)

In the mid-90's, some researches made an effort to incorporate vehicle dynamics into the collision avoidance problem, choosing motion commands rather than a travel direction (SAFA [11] and CVM [12]). But it was the DWA [13], the method that won more popularity in the scientific community. The DWA formulates the problem as a constrained optimization in the velocity space. Constraints are derived from physical limitations of the robot's velocities and from the sensor data (that indicates the presence of obstacles). The original DWA was formulated to synchro-drive robots. Recently, the GDWA [14] was presented, where the original DWA was reformulated to holonomic mobile robots,

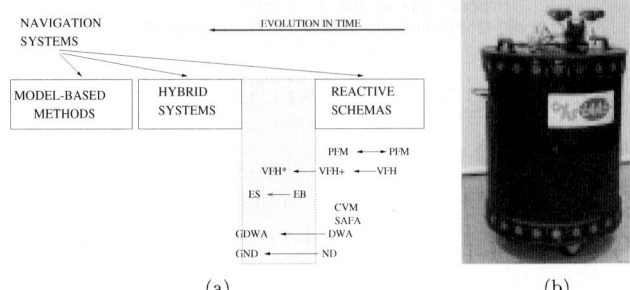

Fig. 1. a) Reactive schemes evolution. b) Nomad XR4000 platform.

and the cost function was slightly modified to improve the robot behavior [18]. Moreover, connectivity of the space was explored, allowing trap situations to be avoided.

Nearness Diagram (ND)

The ND [1] is a reactive scheme that performs a high level information extraction and interpretation of the environment. Subsequently, this information is used to generate the motion commands. In a first step, the ND extracts a description of regions which are free of obstacles, selects one of them, evaluates the robot security and chooses one of the five general situations defined. Secondly, it generates the motion commands with one of the five laws adapted to the general situations. The contribution of the ND scheme among other reactive methods can be seen in [1]. This paper describes the evolution from the ND towards the GND.

3 System Requirements

The aim of this work is to create a navigation system that drives the robot robustly among locations. We have identified three requirements, that have to be accomplished when designing a navigation system that executes motion tasks in an autonomous way:

1. **Information integration**: it is necessarily to integrate information from different perceptions into a model of the environment. Two reasons motivate this: firstly, it gives a framework to have an incremental global reasoning. Secondly, past perceptions may be used to avoid obstacles not perceived at the current moment (sensor constraints).

2. **Dynamic environments reaction**: when the environment changes dynamically, the description of the environment has to model instantaneously this change. If not, the robot will avoid parts of the space known to be free of obstacles, or will not avoid perceived obstacles.

3. **Trap situations solution**: There are a lot of obstacle configurations that produce trap situations. The most typical is the U-shape obstacles and they are common for all the reactive methods. Moreover, there are some symmetric environments where the reactive methods can produce alternate solutions. These environments create cyclic behaviors in the robot motion.

The evolution from the purely reactive navigation system ND, through to the final navigation system GND, has gone a way to accomplish these three requirements.

4 Mapping ND (mND)

In Section 3, three requirements were outlined in order to design an effective navigation system. We now go on to present the mND. It consists of a navigation system that uses a dynamically built model of the environment, and a reactive scheme to generate the motion commands. With this method we want to fulfill the first two requirements, related to information integration and the dynamic environments reaction.

No assumptions about the environment are made (static/dynamic, structured/unstructured...). The sensor used to perceive the environment is a laser range-finder.

The model of the environment is constructed by merging the information in an occupancy-grid that represents the working space. The laser sensory data is introduced directly into the grid model without any pre-processing, and is updated at each servo tic.

The grid has three types of cells: *occupied*, *free* and *unknown*. A point measured by the laser gives an *occupied* cell in the grid. Lines between the sensor and the measured points are projected to the map as *free* cells. Initially, all the cells of the map are set to *unknown* (never perceived). The Bresenham algorithm [20] is used to optimally update the map, in order to achieve real-time performance.

The occupancy-grid represents a finite subsection of the environment centered around the robot. A local region is defined in the center of the grid. When the robot escapes from this local region, the entire grid moves to encompass the robot within the local region. This allows the robot to move within the local region without having to move the complete grid. Grid displacements are always multiples to the cell's dimension, and rotation is not needed. Thus, error propagation associated to the measures in the cells is avoided.

Once the model has been built, it is used as input by the reactive navigation method, instead of directly using perception (see Fig. 4). For robust navigation, the approach relies on the fact that the robot's surroundings are constantly sensed, and that the map is updated at high rate.

It is important to remark that the last perception introduced in the map has no odometric errors with respect to the robot's location, and only sections not perceived accumulate them. Moreover, spurious measures are eliminated from map while introducing the new perceptions. Assuming that the robot performs instantaneous forward motions (that usually coincides with the main visibility sensor direction), and little slippage occurs during motion, this framework results in a very adequate method of **integrating the information** at different times (always in the obstacle avoidance context).

Moreover, the **the environment's dynamic** is reflected in the model as it is perceived, which is a consequence of updating the entire area covered by the last perception.

4.1 Experimental Results

We have extensively tested this navigation system on the XR4000 platform in LAAS (CNRS) shown in Fig. 1b. This base moves with an omnidirectional translational velocities of up to $1.2\frac{m}{sec}$, and accelerations of up to $1.5\frac{m}{sec^2}$. It is equipped with a SICK laser range-finder with a field of view of 180°, a range of 32 meters, and an accuracy of up to 3cm.

To perform the experiments, the dimensions of the map are 10 by 10 meters, and cell dimensions are 5 by 5 centimeters, which gives a grid of 200 by 200 cells. The process of updating the map with a laser measure (360 points), and moving the grid when necessary, takes approximately 100ms. The ND takes less than 50ms which gives a cycle-time of 150ms. These times are well suited to real-time collision avoidance. The maximum translational velocity set for the experiments was $v_{max} = 0.5\frac{m}{sec}$, and the maximum rotational one was $w_{max} = 1.57\frac{rad}{sec}$.

Fig. 2a presents a real experiment where a human walked between the robot and the selected passage. In this case, the environment's dynamic has to be automatically introduced because:

- If the human is not automatically integrated into the model, the reactive method will not have time to react.
- If the last of the robot's perceptions of the human are not eliminated from the model, the passage will remain closed, and the reactive method will avoid the free space.

From left to right, in Fig. 2b the robot moves towards the center of the passage. In Figs. 2c,d,e the human appears in the scene. In Fig. 2f the human enters in the security zone and the robot starts an avoidance manoeuvre, while moving towards the passage. In Fig. 2g, the human completely blocks the passage and the robot continues to avoid him. In Fig. 2h, the human has moved passed the passage, which appears now open for the robot to enter. It now turns towards the passage while continuing to avoid the human. In Fig. 2i the human has finally left the security zone, and so the robot recovers its motion towards the center of the passage.

The experiment shows that the human is automatically integrated into the model, so the reactive scheme avoids it instantaneously. Moreover, past human's perceptions are automatically eliminated, and the passage remains open after the person moved passed it. As consequence, the reactive scheme instantaneously directs the robot through the passage.

5 Mapping-Planning ND (mpND)

The mND is a framework which integrates the information in a model of the environment. The reactive scheme generates the motion commands to avoid collisions. Two advantageous properties are extracted from this coupling:

1. The model integrates past perceptions, thus the reactive method is able to avoid obstacles not perceived at the current moment.
2. The model reflects the environment's dynamics when it is perceived, so the reactive method reacts instantaneously to change.

However, due to the lack of global reasoning in the system, it still has limitations when dealing with trap situations.

The mpND is a navigation system that uses the mND scheme, but exploits the information of the connectivity of the space with a planning algorithm. With this method, we want to fulfill the third requirement, as stated in Section 3 (trap situations).

A minima-free navigation function NF1 [21] is built, using a wave propagation technique, over the configuration space calculated from the grid model. Finally, a path free of collisions, that connects the initial and final configurations, is obtained by a gradient-search technique. The main reasons for the use of this planning algorithm are: its grid-based navigation function (adapted to the grid-model); and its simplicity and efficiency, which allows for the computa-

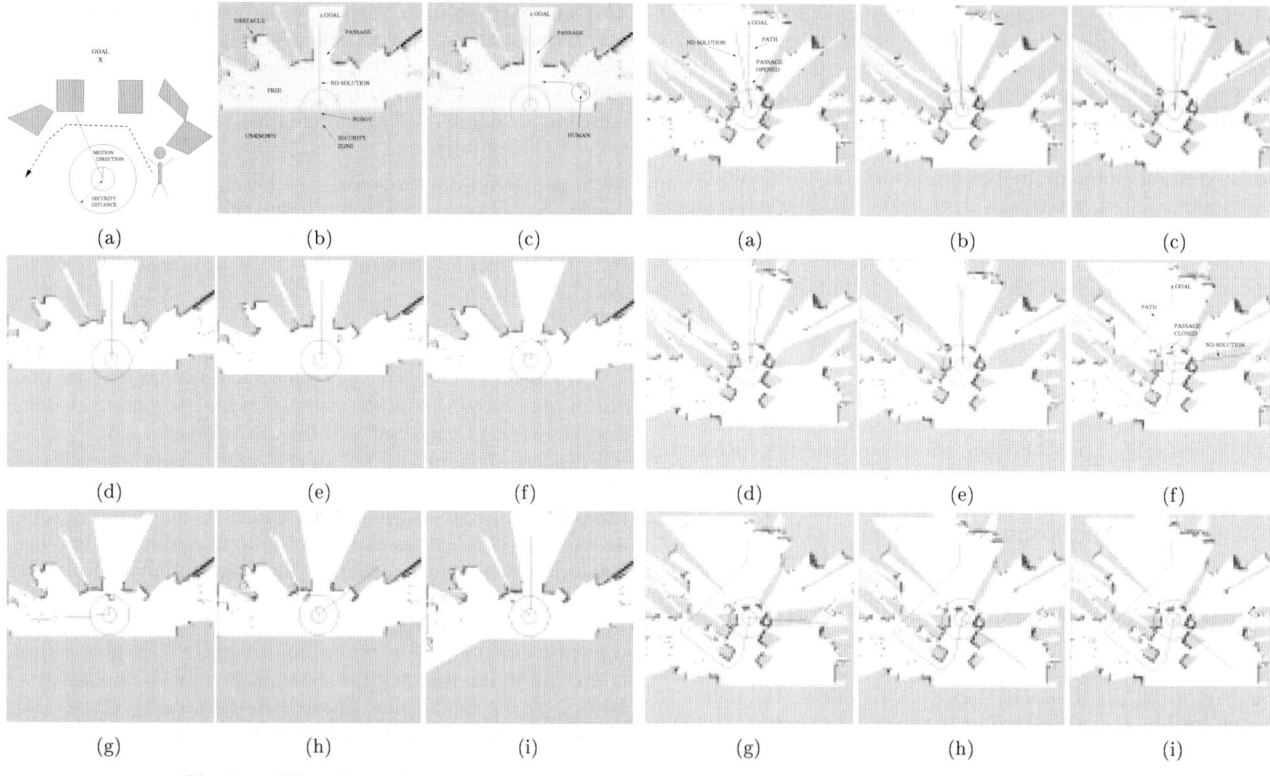

Fig. 2. mND real experiment

Fig. 3. mpND real experiment.

tion of this navigation function at each servo-tic during the robot control loop.

The path solution gives two important pieces of information:

1. *If it is or not possible to reach the final configuration from the actual robot configuration (global reasoning).*
2. *The* **instantaneous path direction** *in order to reach the final configuration (global reasoning).* The instantaneous path direction is the main direction of the first part of the path (in our current implementation the first meter is used to calculate it).

The case of unavailable trajectories will be discussed in next section. Once the path is calculated, it has to be linked to the reactive navigation method. The ND is modified to drive the robot towards the instantaneous path direction (when repeated at each servo tic assures convergence to the goal location), instead of directing the robot towards the goal location. See in Fig. 4 the complete mpND navigation scheme.

There are no obstacle configurations that produces **trap situations** (when a solution exists inside the grid-model), because the instantaneous path direction has the information needed to get the robot out of these situations. The reactive method only has to direct the robot towards this instantaneous direction to avoid trap situations. Moreover, the symmetries of the environment do not produce cyclic behaviors, because the possible alternate solutions are discriminated by the instantaneous path direction.

5.1 Experimental Results

The same platform and settings of the mND (Subsection 5.1) are used here. From mND to mpND, only the NF1 module is added, which introduces a time penalty of 100 ms. The complete servo-tic is then 250 ms, which is well-suited to achieve real-time performance.

Fig. 3 shows a real experiment where the robot is forced to fall into a trap situation. The navigation system has to react automatically to this situation and drive the robot out of it.

In Fig. 3a can be seen that the robot had to cross a passage to reach the final location. While the robot was traveling through the passage (Figs. 3b,c,d,e), a human blocked it. The robot was then inside of a big U-shape obstacle, what produced a trap situation, see Fig. 3f. Automatically, the new path calculated (and thus the instantaneous path direction) pointed out of the U-shape configuration. The ND generates the motion commands to follow this direction, see Fig. 3g,h,i. The result was that the robot was able to get out of the trap situation.

6 Global Nearness Diagram Navigation (GND)

Some researchers have signalled that it is possible to generate motion with a classical planner. But other researchers have used this result to generate reactive motion, using the planner in an iterative way. From our point of view this reasoning is not always valid. There exist two situations, where a planning algorithm does not find a solution, in order to connect the initial and final robot's configurations. If these situations appears, it is not possible to close the motion control loop, because it is not possible to generate the motion commands to follow the path.

Two situations produces this case:

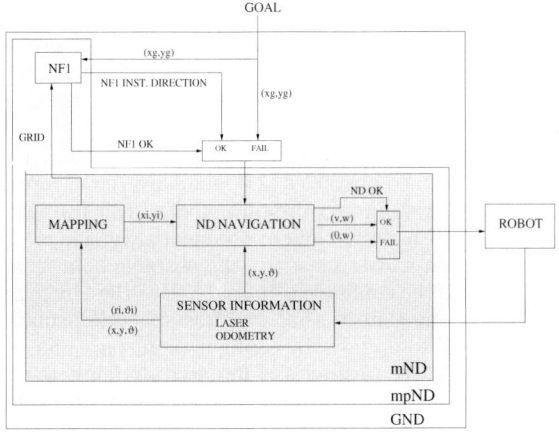

Fig. 4. GND navigation scheme.

1. *The final configuration is not in free space C_{free} [21]* (final location in collision with an obstacle). This is a very typical situation in unknown environments, where goals are iteratively placed for exploration. When the environment is incrementally discovered, the goal can be within an obstacle. In dynamic environments a mobile object can move, or even stop, at the goal location. Even in static and completely known environments, this situation appears when the goal moves to within an obstacle due to the robot drift.
2. *The robot or the goal are completely surrounded by an obstacle.*

One could think that these situations could be avoided by replacing the goal location. From our point of view, it is not the task of the navigation system to modify a final location imposed by an external agent, because the consequences can drastically determine the success of the global task.

Due to this shortcoming, we realize that when these situations occurs, the system should be able to continue its navigation task (close the motion control loop). To cope with this limitation, we developed the Global Nearness Diagram Navigation (GND). It combines the two schemes presented (mND and mpND) to achieve the complete navigation task, see Fig. 4 for the complete GND system.

The GND works as follows. First the mpND is used until a failure flag is produced in the NF1 module (a path connecting the initial and final configurations of the robot does not exist). Then the mND takes control and generates the motion commands. Now there are two possibilities: 1) the path is available (the control is passed to the mpND); 2) there is a ND failure. This last situation happens when there is no free walking area to move (the robot is completely surrounded by obstacles). The motion commands stop and rotates the robot about its center. This behavior updates the map in all directions. This continues until the environment changes, and the control can be transmitted to the mND or mpND.

The GND inherits the properties of the mND related to the **information integration** and the **dynamic environment reaction**. Moreover, the mpND is used when possible, thus avoiding the problem of **trap situations**. The motion commands in the GND are generated by the ND, which avoids many of the problems of other reactive schemes. These properties together make the GND a navigation system which is very well adapted to deal with unknown, dynamic, unstructured, dense and very complex environments.

6.1 Experimental Results

The GND system has been extensively tested with the XR4000 platform in LAAS (CNRS). In all the experiments the environment was unknown, and was incrementally explored. We have chosen two illustrative experiments to show the system behaving in dynamic, unstructured and complex environments (see Fig. 5). Due to the difficulty in reflecting the environment's dynamics, we decided to show the complete robot path and some snapshots of the experiment, rather than directly showing the sensory data and blurring thus the graphics.

Experiment 1: This experiment was designed to show the robot getting out consecutively of three trap situations, produced by changes in the environment's structure. The first snapshot shows the initial state of the robot and the environment, where the robot had to cross a passage to reach the goal location. When the robot arrived at the end of the passage, the right passage was opened (the robot could not see it), and the main passage was closed. Automatically the robot turned to get out of this trap situation. This part of the experiment can be seen step by step also in Fig. 3. When the robot was leaving the passage, it perceived the right passage and reacted to move the robot inside. Then the human closed this passage. The robot automatically reacted to get out of this new trap situation, and ended getting out of the global trap situation. Subsequently the robot resumed the motion towards the goal location.

Experiment 2: This experiment shows the robot in a typical populated environment. The first snapshot shows the initial state of the robot and the environment, where the robot had to cross a room to reach the final location. Humans were walking, building and modifying the environment randomly to disturb the robot's motion. During the experiment, the robot got trapped and had to move back to find the solution. The snapshots shows the highly dynamic nature of the environment.

7 Comparison with other methods

We next discuss the improvements of the GND over other navigation systems. We have chosen the more recent methods of the four techniques introduced in Section 2: Potential Field Methods (in a general fashion), the Vector Field Histogram (VFH* [10]), the Elastic Strip [19] and the Global Dynamic Window Approach (GDWA [14]).

The reader is directed to [1] for a comparison in purely reactive terms. The discussion here is oriented towards the three requirements introduced in Section 3: information integration, dynamic environments reaction and trap situations solution.

Potential Field Methods PFM

Many special solutions to the inherent limitations of PFM [6] still appear in the literature [23]. We have decided to orient the discussion with the PFM in a general fashion. In reactive terms, the ND (and thus the GND) solves all the inherent limitations of PFM except the trap situations (see [1] for a detailed discussion). Moreover, the GND avoids the trap situations when possible, solving also these last undesirable situations.

Vector Field Histogram VFH*

The VFH* uses a grid map of the environment [7] to integrate information. To discuss the drawbacks and advantages of each approach, in terms of information integration and dynamic environments reaction, is out of place, mainly due to the different nature of the sensors used (ultrasonic sensor in [7] and laser in GND). In terms of trap situations, the VFH* uses a look-ahead verification to analyze the consequences of heading towards a candidate direction. As far as we understand, when using a look-ahead verification to increment the local nature of the method, it is necessarily to fix a maximum number of steps (named goal depth in VFH*), which translates in a maximum distance inspected (named total projected distance in VFH*). To select this distance could be a trade off between the speed and the validity of the method, because it represents the maximum reach of the local nature of the method (measured in robot distance traveled). On the other hand, the GND assures global convergence inside of the map grid used.

The main advantage of a look-ahead verification is when dealing with platforms with low computational capabilities. The look-ahead then is well adapted because even by reducing the projected distance, good results can be obtained. Running the GND in real time requires high computational capacity, otherwise one can reduce the size of the grid, which drastically affects the reach of the solution.

Elastic Strip ES

The elastic strip framework [19] has been shown to work very well in high-dimensional configuration spaces. The discussion here is focused in low-dimensional configuration spaces, that is the case of this paper. Two strategies were introduced in [18] and [19] to deal with dynamic environments. The former is to impose constraints on the internal forces acting on two adjacent configurations, to allow a mobile obstacle to pop through the elastic strip. The second one is to maintain a set of alternative routes to chose when the elastic becomes invalid.

The elastic framework is based in the existence of a path that is not always available (see Section 6). As shown in [18], the elastic strip framework could fail in very tight or cluttered environments, where the GND is well-adapted due to the properties of the reactive method ND.

Global Dynamic Window Approach GDWA

The evolution of the GND has been inspired by the GDWA [14], [18]. We direct the comparison to the advantages/disadvantages of the model used, and to some implementation details.

- *Model*: The GDWA uses an occupancy-grid that represents the configuration space of the robot, and remains fixed in a global reference. We understand that the motivation to represent the configuration space is to not completely rebuild it at each step of the algorithm. While keeping the occupancy grid fixed in space, the same navigation function can be reused for every robot location, as long as the environment does not change. The GND uses an occupancy-grid to represent directly the working space, which moves centered around the robot's position.

To use a model that moves with the robot means that the dimension of the model does not depend on the distance traveled (the robot was at all times surrounded by the grid). Moreover, it ensures that the instantaneous surroundings of the robot are directly represented by the model. We would like to signal that the operation to displace a complete grid can be very efficiently implemented in terms of memory (in our current implementation takes about 10ms a 200 by 200 cells, i.e. 10m by 10m grid). On the other hand, the solution has to be found inside of the grid.

In GDWA, the data measured by the laser is translated into configuration space obstacles [21], that are represented in an occupancy grid. From our point of view, this framework does not represent fully the environment's dynamics. The reason is that in a laser measure, the obstacle point can be translated into the configuration space. But the line joining this point to the sensor, that has free space information, cannot be used to update the configuration space. So, only configuration space obstacles is updated, but not the free space. The consequence is that the robot avoids parts of the space that are free of obstacles (see experiment in Section 4, where the passage will remain blocked to the GDWA after the person crosses it). This information of the free space is lost in the GDWA while it is exploited by the GND.

- *Implementation details*: The GDWA computes the navigation function for a subsection of the configuration space, referred to as localized navigation function (localized N-F1). The subsection of the configuration space is increased while looking for a solution. This heuristic saves a lot of computational time currently lost by the GND (that computes the NF1 for the complete grid), and should be added to the GND. The multi-resolution GDWA is also presented [18], to deal with the impact of computational complexity of the size of the occupancy grid used.

The GDWA extracts information from the NF1 by examining the neighborhood of the grid cells that corresponds to the robot location. As shown in [18] and [14] some unnatural behaviors were found, because the border effects of the NF1, and because the solution can only be multiple of 45°. The GND calculates the complete path using a gradient-search technique. Subsequently, the path is tightened using a recursive algorithm and the instantaneous path direction is calculated. Using the instantaneous path direction, avoids the border effects of the NF1 and the constrained solutions, and thus no unnatural behaviors were found.

8 Conclusions

This paper presents a new navigation system that links global information with a local reactive scheme to generate motion. The GND uses the sensory information to build a grid representation of the environment. A planning algorithm is used to extract global information from the model. Finally, the reactive scheme ND uses the computed path and the model to generate collision free motion while directing the robot towards the final location.

We also present how some reactive techniques evolved in the last years, to discuss the advantages/disadvantages of this system among other existing methods. Experimental results in unknown, unstructured, dynamic and complex environments are also shown.

Acknowledgement

We thank R. Alami and T. Simeon of LAAS/CNRS (FRANCE) for accepting J.Minguez in their working group. Moreover, we thank S. Fleury of LAAS/CNRS (FRANCE), for her help with algorithm implementation on the XR4000 Nomad platform.

References

[1] J.Minguez L.Montano "Nearness Diagram Navigation. A New Real Time Collision Avoidance Approach" *IEEE/RSJ International Conference on Intelligent Robots and Systems (IROS 2000)*, Takamatsu, Japan, 2000.

[2] O.Khatib "Real-Time Obstacle Avoidance for Manipulators and Mobile Robots" *The International Journal of Robotics Research*, 5(1), 1986.

[3] B.H.Krough "A Generalized Potential Field Approach to Obstacle Avoidance Control" *Robotics Research: The Five Years and Beyond. SME Conference Proceedings* Bethlehem, 1984.

[4] R.B. Tilove "Local Obstacle Avoidance for Mobile Robots based on the method of Artificial Potentials" *Research Publication GMR 6650*, 1989.

[5] J.Borenstein Y.Koren "Real-time Obstacle Avoidance for Fast Mobile Robots" *IEEE Transactions on Systems, Man, and Cybernetics*, Vol. 19, No. 5, Sept./Oct., pp.1179-1187.

[6] Y.Koren J.Borenstein "Potential Field Methods and Their Inherent Limitations for Mobile Robot Navigation" *IEEE International Conference on Robotics and Automation*, pp. 1398-1404, California, USA, 1991.

[7] J.Borenstein and Y.Koren "Histogramic In-Motion Mapping for Mobile Robot Obstacle Avoidance" *IEEE Transactions on Robotics and Automation*, pp 535-539, 1991.

[8] J.Borenstein and Y.Koren "The Vector Field Histogram (VFH)-Fast Obstacle Avoidance for Mobile Robots" *IEEE Transactions on Robotics and Automation*, 7(3), 1991.

[9] I.Ulrich and J.Borenstein "VFH+: Reliable Obstacle Avoidance for Fast Mobile Robots" *IEEE International Conference on Robotics and Automation*, pp1572, Leuven, Belgium, 1998.

[10] I.Ulrich and J.Borenstein "VFH*: Local Obstacle Avoidance with Look-Ahead Verification" *IEEE International Conference on Robotics and Automation*, pp2505, San Francisco, USA, 2000.

[11] W.Feiten R.Bauer G.Lawitzky "Robust Obstacle Avoidance in Unknown and Cramped Environments" *IEEE International Conference on Robotics and Automation*, pp2412, 1996.

[12] R.Simmons "The Curvature-Velocity Method for Local Obstacle Avoidance" *IEEE International Conference on Robotics and Automation*, pp3375, Minneapolis, USA, 1996.

[13] D.Fox W.Burgard S.Thrun "The Dynamic Window Approach to Collision Avoidance" *IEEE Transactions on Robotics and Automation*, 4:1, 1997.

[14] O.Brock O.Khatib "High-Speed Navigation Using the Global Dynamic Window Approach" *IEEE International Conference on Robotics and Automation*, pp341, Michigan, USA, 1999.

[15] S.Quinlan O.Khatib "Elastic Bands: Connecting Path Planning and Control", *IEEE International Conference on Robotics and Automation*, Vol 2, pp. 802-807, Atlanta, USA, 1993.

[16] Sean Quinlan "Real-Time Modification of Collision Free Paths" *PHD thesis, Stanford University* December, 1994.

[17] Maher Khatib "Sensor-based motion control for mobile robots" *PHD thesis, LAAS-CNRS* December, 1996.

[18] Oliver Brock "Generating Robot Motion: The Integration of Planning and Execution' *PHD thesis, Stanford University* November, 1999.

[19] O.Brock and O.Khatib "Real-Time Replanning in High-Dimensional Configuration Spaces Using Sets of Homotopic Paths" *IEEE International Conference on Robotics and Automation*, pp 2328, San Francisco, USA, 2000.

[20] J.Foley, A.Van Dam, S.Feiner, J.Hughes, "Computer Graphics, principles and practice" *Addison Wesley*, edition 2nd, 1990. The system programming series.

[21] J.C.Latombe "Robot Motion Planning" *Kluwer Academic Publishers*, 1991.

[22] D.Maravall J.de Lope and F.Serradilla "Combination of Model-based and Reactive Methods in Autonomous navigation" *IEEE International Conference on Robotics and Automation*, pp 2328, San Francisco, USA, 2000.

[23] L.Chenqing M.Ang H.Krishman L.Yong "Virtual Obstacle Concept for Local-minimum-recovery in Potential-field Based Navigation" *IEEE International Conference on Robotics and Automation*, pp 983, San Francisco, USA, 2000.

Fig. 5. a) Experiment 1. b) Experiment 2.

Combining Fault Detection and Process Optimization in Manufacturing Systems Using First–Order Hybrid Petri Nets

Fabio Balduzzi, Angela Di Febbraro
Dipartimento di Automatica e Informatica
Politecnico di Torino, Torino, Italy
{balduzzi,difebbraro}@polito.it

Abstract— This paper describes a hierarchical two–level modeling and control framework for real–time manufacturing processes, based on hybrid Petri nets, and capable of integrating monitoring and fault detection techniques along with performance optimization procedures. At the higher level, First-Order Hybrid Petri Nets are used to generate the asynchronous concurrent events characterizing the nominal optimal system behavior, also providing the machine production rates. At the lower level, a real–time scheduler and a discrete Petri net communicate with the field–bus devices to force the machine to produce at the rates provided by the upper level, and monitor the occurring events by keeping track of part movements. Whenever a process failure is detected, the diagnosis procedure is triggered, and a recovery sequence is imposed.

I. INTRODUCTION

In automated manufacturing systems, unscheduled resource down–times, as well as process failures, are the factors that most affect the system productivity. Thus, the relevant control systems must provide integrated capabilities of fault detection and diagnosis, along with recovery procedures [8], in order to increase both the system reliability and productivity. The aim of monitoring the system functioning is to detect process failures, whereas diagnosis results in failure source identification and analysis.

Significant works about fault detection and diagnosis in discrete event systems, with special reference to manufacturing processes, can be found in the related literature (see, for instance, [6] and [9]). The idea of exploiting Petri net capabilities for fault monitoring and detection in manufacturing systems has already been proposed by some authors. Interesting examples are [4], [5], [10], and [11].

In this paper, a hybrid PN-based model to represent the dynamic behavior of a manufacturing system is proposed. The continuous time–driven dynamics arises from approximating the discrete movements of parts in the system resources by their average flows, whose values are first–order fluid quantities, i.e., the machine production rates. The event–driven dynamics models the discrete transitions of the system through a sequence of operational states, named *macro-states*, at the occurrence of such events as machine starvation, blockage, breakdown, and repair. Such a system can be efficiently described by using the special class of hybrid Petri nets named *First-Order Hybrid Petri Nets* (FOHPN) [2], [3], which allows a compact and convenient way of representing concurrency and parallelism aspects, and yet makes performance optimization easier.

With respect to previous works, a major novelty yielded by the proposed approach is the introduction of a hybrid Petri net model for fault detection and diagnosis. Moreover, this paper extends previous works by defining a unifying hierarchical modeling and control architecture that combines performance optimization with fault detection and monitoring capabilities.

A two–level control architecture that combines fault detection and process optimization is proposed. Fault detection is on–line accomplished by comparing the actual system response with the optimal control sequences provided by the FOHPN model. Process optimization is accomplished by a decision maker/planning system which provides the optimal control sequence according to a given cost criterion and the current configuration of the shop–floor.

At the higher level, a FOHPN model is used to generate the asynchronous concurrent events as they occur in the nominal optimal system behavior. In the proposed model, the firing speeds of the continuous transitions, representing the machine production rates, are time–varying, and provided by the decision maker which optimizes the system performance. At the lower level, a real–time scheduler performs according to the values of the machine production rates provided by the upper level, which, in turn, act as time–varying reference values within the time intervals named *macro–periods* between the occurrences of two subsequent *macro–events*. Typically, for each operation, the real–time scheduler decides which resources are going to perform it and the starting time of the corresponding activities.

Beside the real–time scheduler, a discrete Petri net model is used to communicate with the field-bus devices (PLC's, sensors, actuators, etc.) by tracking the discrete movements of parts, thus monitoring the occurring events. The information coming from such a PN is compared with the output of the FOHPN model of the system to conduct fault detection. As a result, a fault is revealed by a deviation of the actual system behavior from its nominal optimal behavior. In other words, a difference between the process behavior observed by the discrete PN controller through the sensors and the one modeled by the FOHPN is interpreted as a consequence of a process failure. In that case, the controller triggers the diagnosis procedure, and then imposes a recovery sequence.

Actually, in the proposed framework two different types of process failures are integrated. A higher–level failure model, embedded within the FOHPN model of the system, as detailed in subsection III-B, is used to account for machine failures occurring after a given production volume, either stochastic or deterministic (as in the case of preventive maintenance programs). At the lower–level, a process failure is modeled as a deviation of the observed behavior from the expected optimal behavior, which already includes maintenance programs. As a result, the lower–level reveals those kinds of faults which are really unforeseeable in presence of adequate maintenance operations of the system components prone to failures.

II. Basics of FOHPN

For the reader's convenience, the Petri net formalism used in this paper is briefly reported in this section, referring to [7] and [1] for a detailed description of place/transition Petri nets.

A *First–Order Hybrid Petri Net* (FOHPN) [2], [3] is a structure $N = (P, T, Pre, Post, \mathcal{D}, \mathcal{C})$. The set of *places* $P = P_d \cup P_c$ is split into the two sets of *discrete* places P_d (represented as circles) and *continuous* places P_c (represented as double circles). The set of *transitions* $T = T_d \cup T_c$ is partitioned into the two sets of discrete transitions T_d and continuous transitions T_c (represented as double boxes). The set T_d is further partitioned into a set of *immediate* transitions T_I (represented as bars), a set of *deterministic timed* transitions T_D (represented as black boxes), and a set of *exponentially distributed timed* transitions T_E (represented as white boxes), $T_d = T_I \cup T_D \cup T_E$.

The *pre-* and *post-incidence functions* that specify the arcs are $Pre, Post : \begin{cases} P_d \times T \to \mathbb{N} \\ P_c \times T \to \mathbb{R}_0^+ \end{cases}$, where $\mathbb{R}_0^+ = \mathbb{R}^+ \cup \{0\}$. To deal with *well-formed nets*, as it is preferable, it is required that for all $t \in T_c$ and for all $p \in P_d$, $Pre(p,t) = Post(p,t)$. The *incidence matrix* of the net is defined as $\mathbf{C}(p,t) = Post(p,t) - Pre(p,t)$. The function $\mathcal{D} : T_d \to \mathbb{R}^+$ specifies the timing associated with timed discrete transitions. A constant firing delay $\delta_j = \mathcal{D}(t_j)$ is associated with the deterministic timed transition $t_j \in T_D$, whereas an average firing delay $\frac{1}{\lambda_j}$, where λ_j is the parameter of the corresponding exponential distribution, is associated with an exponentially distributed timed transition $t_j \in T_E$.

The function $\mathcal{C} : T_c \to \mathbb{R}_0^+ \times \{\mathbb{R}^+ \cup \{\infty\}\}$ specifies the firing speeds associated with continuous transitions. For any continuous transition $t_j \in T_c$, let $\mathcal{C}(t_j) = (V_j', V_j)$, with $V_j' \leq V_j$, where V_j' represents the *minimum firing speed* (mfs) and V_j represents the *maximum firing speed* (MFS). In the following, unless explicitly specified, the mfs of a continuous transition is considered to be $V_j' = 0$.

Like in any hybrid model, also in a FOHPN system two different kinds of dynamics can be distinguished: a time–driven dynamics, associated with the firing of continuous transitions, and an event–driven one, associated with the firing of discrete transitions. The continuous evolution of the net strictly depends on the *instantaneous firing speeds* (IFS) of the continuous transitions, denoted as $v_j(\tau)$, for any $t_j \in T_c$. A *marking*

$$\mathbf{m} : \begin{cases} P_d \to \mathbb{N} \\ P_c \to \mathbb{R}_0^+ \end{cases}$$

is a function that assigns to each discrete place a non-negative number of tokens, and to each continuous place a fluid volume. The value of the marking at time τ is denoted as $\mathbf{m}(\tau)$, whereas m_i denotes the marking of place p_i.

The continuous evolution of the net strictly depends on the *instantaneous firing speeds* (IFS) of the continuous transitions, denoted as $v_j(\tau)$, for any $t_j \in T_c$. It follows that the evolution in time of the marking of a place $p_i \in P_c$ can be defined by:

$$\dot{m}_i(\tau) = \sum_{t_j \in T_c} C(p_i, t_j) v_j(\tau). \qquad (1)$$

Indeed, Equation 1 holds in the assumption that at time τ no discrete transition is fired and that all speeds $v_j(\tau)$ are continuous in τ. The enabling state of a continuous transition t_j defines its admissible IFS v_j as indicated in [2], [3], where also the enabling rules of transitions are described. The set of all the admissible IFS vectors can be characterized as follows.

Definition II.1. (admissible IFS vectors)
Let $\langle N, \mathbf{m} \rangle$ be a FOHPN system, i.e., an FOHPN N with a marking $\mathbf{m}$. Let $T_\mathcal{E}(\mathbf{m}) \subset T_c$ ($T_\mathcal{N}(\mathbf{m}) \subset T_c$) be the subset of continuous transitions enabled (not enabled) at $\mathbf{m}$, and $P_\mathcal{E} = \{p_i \in P_c \mid m_i = 0\}$ be the subset of empty continuous places. Any *admissible IFS vector* $\mathbf{v}$ at $\mathbf{m}$ is a feasible solution of the following set of linear

inequalities:

$$\begin{cases} (a) & V_j - v_j \geq 0 & \forall t_j \in T_\mathcal{E}(\mathbf{m}) \\ (b) & v_j - V'_j \geq 0 & \forall t_j \in T_\mathcal{E}(\mathbf{m}) \\ (c) & v_j = 0 & \forall t_j \in T_\mathcal{N}(\mathbf{m}) \\ (d) & \sum_{t_j \in T_\mathcal{E}} C(p, t_j) v_j \geq 0 & \forall p \in P_\mathcal{E}(\mathbf{m}) \end{cases} \quad (2)$$

■

Constraints (2.a), (2.b), and (2.c) result from the firing rules of continuous transitions. Constraints (2.d) follow from (1), because if a continuous place is empty its fluid content cannot decrease. Each admissible IFS vector $\mathbf{v}$ represents a particular operational mode of the system described by the net.

III. Monitoring and Control

A control-oriented model of a manufacturing system must take into account both constraints imposed by the physical characteristics of buffers and machines, to guarantee safe operations, and constraints resulting from part routing and scheduling. We propose to model such constraints via a single formalism to be used by the FOHPN system model.

Properly speaking, given a cost criterion, we characterize all the process behaviors, allowed by the system configuration during normal operation, that can achieve a desired goal. This can be done by solving a sequence of linear programming problems, one for each macro–period, thus obtaining the control sequence $u^o = \{\mathbf{v}^o(o), \mathbf{v}^o(1), \ldots\}$ of the optimal IFS of the continuous transitions, i.e., the machine production rates to be expected for each macro–period. Such a sequence defines the reference signal for the real–time scheduler that is in charge of forcing the machines to produce at those rates. To keep track of the discrete flows of parts through the machines, the real–time scheduler makes use of a discrete Petri net which acts as a process interface between the field–bus devices and the upper-level planning system.

During the normal functioning, the FOHPN used for performance optimization and the discrete PN used for monitoring the discrete flows of parts "almost" provide the same values $\mathbf{v}(k)$ for the machine production rates. More specifically, let $\hat{v}_i(k)$ be the production rate of machine i observed during the k–th macro–period, and let $v_i(k)$ be the IFS of the continuous transition t_i representing the production of machine i in the FOHPN model. If $|v_i(k) - \hat{v}_i(k)| \leq \Gamma(k)$, where $\Gamma(k)$ is a fixed nonnegative threshold, we say that the system is operating under nominal optimal conditions. When a process failure occurs, the resources at the shop–floor level cannot execute the service required by the upper–level

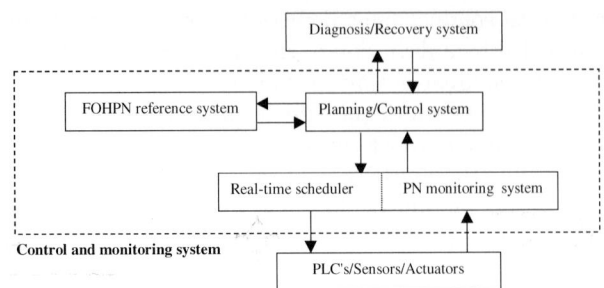

Fig. 1. The control and monitoring architecture.

controller, and therefore cannot satisfy the requirements for optimal behavior, so it is $|v_i(k) - \hat{v}_i(k)| > \Gamma(k)$.

The first function of a monitoring system concerns the mentioned detection issues of unforeseen situations. The proposed method is based on the comparison between the actual behavior of the process observed through the sensors and tracked through a discrete PN, and the expected optimal behavior described by the FOHPN model. The controller interprets a difference as a consequence of a process failure, and then triggers a diagnosis procedure, so as to find the causes of the observed deviations. A diagnosis system usually makes use of different information, such as the current configuration of the shop–floor, the activity during which the failure has been detected, and so on. It should be noted that some of these data can be deduced from the IFS vector $\mathbf{v}(k)$ and the markings of the two Petri nets.

Once found the cause of a failure, the monitoring system has to elaborate a recovery action in order to correct the failure. Diagnosis procedures can be performed, for instance, via *Fault Model and Effect Analysis* (FMEA) studies, or by means of fault tree analysis, which is a deductive form of failure analysis [12]. In this paper, the first function of the monitoring system, i.e., the fault detection step, is addressed within a Petri net framework that performs it with the goal of optimizing the system performance.

A. Architecture of the control and monitoring system

To implement the proposed control and monitoring system, the architecture shown in Figure 1 is defined, made up of the 5 sub–systems described as follows.

The *FOHPN Reference System* (FRS) represents the model of the manufacturing system which, as mentioned above, embeds both the higher–level machine failure model and the part routing constraints.

The *Planning/Control System* (PCS) provides the optimal values of the IFS of the FRS, i.e., the average machine production rates. Its goal is achieved by solving a sequence of linear programming problems, one for each macro–period, in the fulfilment of a linear set of constraints describing the current configuration of the shop-floor in terms of machine and buffer availabilities. It also collects data from the PN Monitoring System in order to detect process failure. In case of failures, it alerts the Diagnosis/Recovery System to activate the diagnosis procedure.

The *Real–Time Scheduler* and the *PN Monitoring System* are used to communicate with the field–bus devices, such as sensors, PLCs, and actuators. The former module receives the average machine rates to be expected for the next macro–period by the PCS and forces the machines to produce at those rates by routing and sequencing parts accordingly. The latter module tracks the discrete flows of parts through the machines providing the information set required by the PCS to perform process failure detection.

The *Diagnosis/Recovery System* performs procedures of diagnosis and recovery according to the information received from the PCS.

The *PLC's/Sensors/Actuators* block represents the field–bus devices which communicate with the shop–floor resources.

The main reason why the system is modeled independently of the controller is that the FOHPN model can serve as a simulator for testing new control policies. Thus, this model can be used to define the set of behaviors in which the system is expected to be optimally operating according to a given cost criterion. In order to show the functioning of the proposed control and monitoring architecture, a simple example is reported.

B. Example: a three stage transfer line

Consider a three stage transfer line producing a single product class and consisting of three unreliable machines M_1, M_2, and M_3. Parts coming from an external source are queued in the unlimited-capacity input buffer B_1, and finished products are collected at the exit of machine M_3 in the unlimited-capacity output buffer B_4. Machines M_1 and M_2 (resp., M_2 and M_3) are connected by a buffer B_2 (resp., B_3) with capacity C_{B_2} (resp., C_{B_3}). Since machines are unreliable, an operation–dependent failure model is taken into account, assuming that a machine fails after a given production volume has been processed since the previous repair operation.

Figure 2 shows the FOHPN model of the described production system, built in a modular way by composing the FOHPN models of elementary manufacturing components. The continuous transition t_{M_i} models the production of machine M_i, i=1,2,3, whose maximum production rate is represented by the corresponding maximum firing speed V_{M_i}, while the continuous places p_{B_i} and $\bar{p}_{B_i}$ (the corresponding co–buffer), i=1,2, model the finite buffers. Finally, transition t_{in} with a given minimum and maximum firing speeds (V'_{in}, V_{in}) is used to model the external arrival of parts.

The failure model of machine M_i is described as follows. The continuous place $p_{R,i}$ is initially marked with w_i, representing the production volume that will be processed by the machine before failing. After machine M_i has processed the fluid quantity w_i, the continuous place $p_{F,i}$ will be marked by w_i. Then, the immediate transition $t_{d,i}$ is enabled and fires, emptying place $p_{F,i}$ and removing a token from place $p_{u,i}$, thus disabling transition t_{Mi}. At the same time a token is added to the discrete place $p_{d,i}$, thus enabling the stochastic transition $t_{R,i}$. Place $p_{d,i}$ represents the condition of the machine under repairing, since, when it is marked, transition t_{Mi} is disabled, i.e., $v_{Mi} = 0$. The machine is down until the repair event occurs, i.e., transition $t_{R,i}$ fires, enabling t_{Mi} again.

Initially, all the buffers are assumed empty and all the machines are assumed operating at their maximum production rate. The proposed control architecture relies on a solver to implement linear programming, and on a real–time scheduler to dispatch parts according to the optimal sequence $u^o = \{\mathbf{v}^o(o), \mathbf{v}^o(1), \ldots\}$ of machine production rates.

Let us first define the control problem to be solved to optimize a given performance measure. The solver embedded in the PCS provides the optimal machines production rates according to the constraints defined by the current macro–state, by solving a linear programming problem, at the occurrence of each macro–event. At each step, the FOHPN model can be used to perform sensitivity analysis, and adjust the optimal myopic solution accordingly [2].

Let N be the final macro–event, and $T = \cup_{k=0}^{N-1}[\tau_k, \tau_{k+1})$ a finite time horizon, where τ_k, for $k = 0, 1, 2, \ldots$, are the occurrence times of the macro–events. The length of the k–th macro–period is $\Delta_k = (\tau_{k+1} - \tau_k)$. Let us define the instantaneous firing speed vector $\mathbf{v}(k) = [v_{in}(\tau_k), v_{M_1}(\tau_k), v_{M_2}(\tau_k), v_{M_3}(\tau_k)]^T$, and let $J = \mathbf{c}^T \mathbf{v}(k)$ be the performance function to be optimized. Finally, let $V_{in} = 10$, $V'_{in} = 5$, $V_{M_1} = 10$, $V_{M_2} = 5$, $V_{M_3} = 20$.

The chosen optimization objective is the maximization of the machine utilization; thus, the coefficient vector of the performance index is $\mathbf{c} = [0, 1, 1, 1]^T$. We show how to derive an optimal control policy that myopically maximizes the machine utilization over a finite time horizon, and how to perform fault detection. We describe the system evolution within the first two macro–periods: MP1, when all buffers are empty; MP2, buffer

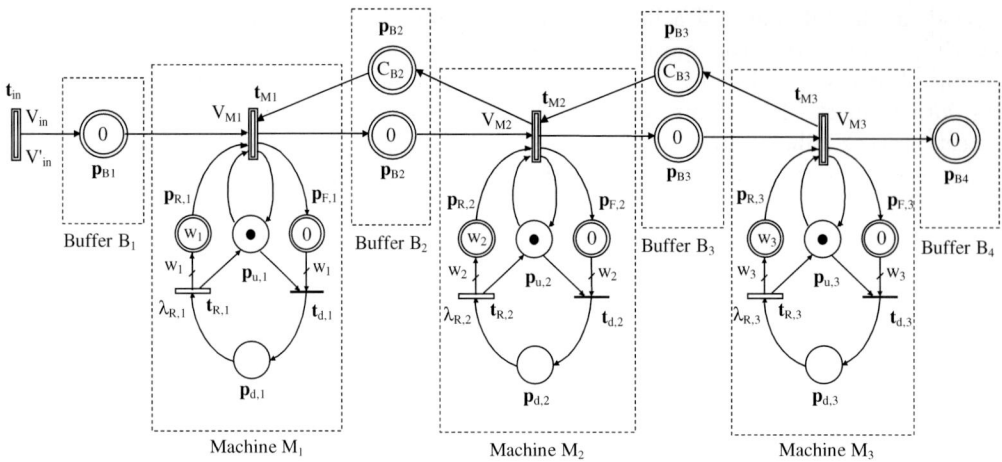

Fig. 2. The FOHPN model of a three stage transfer line.

B_2 becomes full.

(**MP1**). At the beginning of the first macro–period of length Δ_0 — at time τ_0 all buffers are empty and all machines are operational — we define and solve the following linear optimization problem:

$$\max_{\mathbf{v}(0)} \sum_{i=1}^{3} v_{M_i}(\tau_0)$$

subject to
$$\begin{cases} V'_{in1} \leq v_{in1}(\tau_0) \leq V_{in1} & \\ v_{M_i}(\tau_0) \leq V_{M_i}, & i=1,2,3 \\ v_{M_{i+1}}(\tau_0) - v_{M_i}(\tau_0) \leq 0, & i=1,2 \end{cases}$$

The solver provides the optimal solution as the vector $\mathbf{v}^o(0) = [10, 10, 5, 5]^T$, which represents an optimal control policy for the machine production rates that may be adopted during the first macro–period. In particular, throughout the interval Δ_0, B_2 keeps filling at a rate equal to 5, whereas B_3 is empty. Now, by executing the net with the IFS vector $\mathbf{v}^o(\tau_0)$ provided by the planning system, we generate the events of the discrete event production system under the optimal nominal behavior.

Let us now assume that the next macro–event occurs at time

$$\tau_1 = \frac{C_{B_2}}{v_1^o(\tau_0) - v_2^o(\tau_0)}$$

i.e., when buffer B_2 gets full. During time interval $\Delta_0 = [\tau_0, \tau_1)$, the task of the real–time scheduler is to force the machines to produce at rates $v_i^o(\tau_0)$. Then, the monitoring system, represented by the discrete Petri net (shown in Figure 3 as regards machine M_i), will track the discrete flows of parts, thus providing the information required by the PCS to compare the actual system behavior with the optimal one, in order to detect and diagnose process failures.

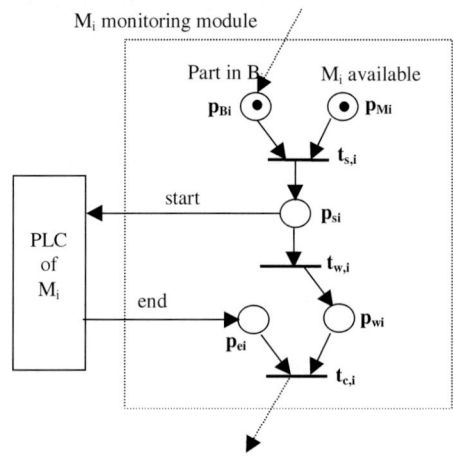

Fig. 3. The PN monitoring module of machine M_i.

In particular, this PN issues commands to the local controllers, and receives from the components of the production system information regarding task completion. As an example, if a part is queued at the input buffer B_i and machine M_i is not under repairing, i.e., places p_{Bi} and p_{Mi} are marked, then $t_{s,i}$ may fire, and send a signal to the PLC of M_i to start loading and processing the part. When the process is completed, that PLC will send back a signal that allows the firing of transition $t_{c,i}$, and then the process is repeated for all the downstream machines. At the same time, the control system makes use of this PN to collect such data as the starting time τ_s^j and completion time τ_c^j for each part on each machine, which correspond to the firing instants of the immediate transitions $t_{s,i}$ and $t_{c,i}$, respectively. Such data are filtered according to the length of the current macro–period, and the following quantity is elaborated:

$$\hat{v}_i(\tau_0) = \frac{1}{\eta(0)} \sum_{j=1}^{\eta(0)} \frac{\alpha^j}{\tau_c^j - \tau_s^j} \quad (3)$$

where α^j are suitable weighting coefficients and $\eta(0)$ is the number of parts completed during Δ_0. Under normal operating conditions, the real–time scheduler drives the production system, together with the Petri net monitoring module. The previously described FOHPN model generates the macro–events as they should occur in the plant operating under optimal conditions.

At the end of the current macro–period, the information set available through the monitoring system and that used by the FOHPN model of the system are compared by the PCS to conduct fault detection. The condition to be evaluated is $|v_i^o(\tau_0) - \hat{v}_i(\tau_0)| \leq \Gamma_i(\tau_0)$, $i = 1, 2, 3$, where $\Gamma_i(\tau_0)$ represents the relevant admissible deviation during MP1 for M_i. If it is not true, the PCS runs the diagnosis/recovery procedure. Otherwise, the production process is going on under nominal optimal conditions, thus guaranteeing maximization of machines utilization. The solver will be restarted according to the current system configuration in order to provide the optimal production rates to be expected for the next macro–period.

(**MP2**). At the beginning of the second macro–period of length Δ_1 — at time τ_2 buffer B_2 is full, B_3 is empty and all machines are operational — we define and solve the following constrained linear optimization problem:

$$\max_{\mathbf{v}(1)} \sum_{i=1}^{3} v_{M_i}(\tau_1)$$

subject to
$$\begin{cases} V'_{in1} \leq v_{in1}(\tau_1) \leq V_{in1} \\ v_{M_i}(\tau_1) \leq V_{M_i}, & i = 1, 2, 3 \\ v_{M_i+1}(\tau_1) - v_{M_i}(\tau_1) \leq 0, & i = 1, 2 \end{cases}$$

The solver provides the optimal solution $\mathbf{v}^o(1) = [10, 5, 5, 5]^T$, which represents an optimal control policy for the machine production rates that may be adopted during the second macro–period. In particular, throughout the interval Δ_1, B_2 remains full, while B_3 remains empty. Now, the same monitoring and control actions described for the macro–period **MP1** are applied within this macro–period.

Then, the process repeats until the end of the control horizon, i.e., the occurrence of the final macro–event. Upon each macro–event occurrence, the optimization problem to be solved may be different, obviously, but it always has the same structure as the problems faced in the first two macro–periods.

IV. Conclusions

In this paper, a unifying hierarchical modeling and control architecture is defined, presenting some innovative features. First novelty stands in the introduction of a hybrid Petri net model for fault detection and diagnosis. The chosen class of hybrid Petri nets, First-Order Hybrid Petri Nets, which has already proven valuable in modeling manufacturing systems, embeds the representation of failures due to maintenance schedules. Moreover, the proposed model integrates performance optimization with fault detection and monitoring capabilities, also allowing the sensitivity analysis of given performance measures with respect to variations of selected decision variables. Work is in progress to refine the proposed two–level model, and to apply it to more complex production systems.

References

[1] M. Ajmone Marsan, G. Balbo, G. Conte, S. Donatelli, G. Franceschinis, *Modelling with Generalized Stochastic Petri Nets*, Wiley Series in Parallel Computing, Wiley, 1995.

[2] F. Balduzzi, A. Giua, G. Menga, "First–Order Hybrid Petri Nets: a Model for Optimization and Control," *IEEE Trans. Robotics and Automation*, Vol. 16, N. 4, pp. 382–399, 2000.

[3] F. Balduzzi, A. Di Febbraro, A. Giua, C. Seatzu, "Decidability Results in Hybrid Petri Nets," *J. of Discrete Event Dynamic Systems*, Vol. 11, N. 1-2, 2001.

[4] C. Gao, X. He, H. Wang, P. Li, "Modeling, Safety Verification and Optimization of Operating Procedures in Process Systems Using Hybrid Petri Nets," *Proc. IEEE Int. Conf. Systems, Man, and Cybernetics*, October 1999, pp. 854–859.

[5] C.-H. Kuo, H.-P. Huang, "Failure Modeling and Process Monitoring for Flexible Manufacturing Systems Using Colored Timed Petri Nets," *IEEE Trans. Robotics and Automation*, Vol. 16, N. 3, pp. 301–312, 2000.

[6] F. Lin, "Diagnosability of Discrete Event Systems and Its Applications," *Journal of Discrete Event Dynamic Systems*, Vol. 4, N. 2, pp. 197–212, 1994.

[7] T. Murata, "Petri Nets: Properties, Analysis and Applications," *Proceedings IEEE*, Vol. 77, No. 4, pp. 541–580, 1989.

[8] R. Patton, P. Frank, R. Clark, *Fault Diagnosis in Dynamic Systems*, Prentice Hall, 1989.

[9] M. Sampath, R. Sengupta, S. Lafortune, K. Sinamohideen, D. Teneketzis, "Diagnosability of Discrete Event Systems," *IEEE Trans. Automatic Control*, Vol. 40, pp. 1555–1575, 1995.

[10] V.S. Srinivasan, M.A. Jafari, "Fault Detection/Monitoring Using Time Petri Nets," *IEEE Trans. Systems, Man, and Cybernetics*, Vol. 23, N. 4, pp. 1155–1162, 1993.

[11] L. Tromp, A. Benveniste, M. Basseville, "Fault Detection and Isolation in Hybrid Systems: a Petri–Net Approach," *Proc. 1999 IFAC World Congress*, Bejing, P.R. China, pp. 79–84.

[12] N. Viswanadham, Y. Narahari, *Performance Modeling of Automated Manufacturing Systems* Prentice Hall, NJ, 1992.

A Petri Nets Graphic Method of Reduction Using Birth-Death Processes

R. Zemouri, D. Racoceanu, N. Zerhouni

Laboratoire d'Automatique de Besançon (LAB), UMR CNRS 6596
25, rue Alain Savary, 25000 Besançon (France)
daniel.racoceanu@ens2m.fr

Abstract

Stochastic Petri nets are powerful tool for performance evaluation of concurrent systems like parallel computing, communication network and production systems. In many practical applications, performance evaluation using this model is very difficult because of the great dimension of the marking space. In this paper, we present a new graphical method for the reduction of stochastic Petri nets, applied for safe production system modeling. The approach is based on the principle of places interactivity in the model, function of transition firing rates. The reduction of the Petri net is applied directly on the graphical model after a simple analysis of places efficiency by using mathematical techniques of birth-death processes. Thus, the problem of the model dimension is solved since our method is independent of the marking graph.

Key Words

Stochastic Petri Nets, Markov Process, Birth-Death Process, Production System Safety, Maintenance Process, Singular Perturbations.

1 Introduction

A complete and effective performances analysis of any production system requires the use of random parameters. Stochastic models like stochastic Petri nets (SPN) find also a great applicability for modeling and simulation [1],[5],[11]. However, real industrial systems are often complex and the associate algorithms become complicate and can diverge. There are many simplification methods to reduce the dimension of such kind of model. Every place of the model represents a state of the studied system, so most existing reduction methods do not conserve physical meaning of the model and imply an important loss of information. Some researches try to keep the meaning of the reduced model, by eliminating insignificant states. Firstly, we present a brief survey of such existing methods of SPN simplification.

2 Overview of stochastic Petri nets simplification methods

2.1 Graphical methods of simplification

These methods of simplification allow a transformation of SPN into a more simple one, by keeping some important properties of the initial model, like liveness and boundedness, but it is not always possible to give a physical interpretation to this reduction [3], [4], [6]:

- Fusion of places;
- Transition fusion;
- Elimination of places and transitions in loop.

2.2 Analytic simplification methods

One of these methods is based on a development proposed by Racoceanu [9], concerning the application of the singular perturbation method for the reduction of Markov chains. An extension of this technique is used afterward for the simplification of the stochastic Petri net [2], [10]. This method consists in decoupling the slow and fast dynamics of the SPN. The meaning of slow and fast models of this category of methods is very interesting for the study of the transient behavior of the marking probabilities. The default of this method is not to take into account the steady state behavior of the system. Besides, the method uses the Markov process associated to the SPN marking. This graph has frequently a great number of states and the results of the study can diverge and become quickly illegible. Racoceanu [8], [10] initialized the study of such a random model simplification in steady state. It seems interesting to study this problem in the case of the SPN by introducing a graphic reduction method.

3 New graphic method of SPN simplification

3.1 Studied systems

Our study concerns vivacious, bounded stochastic Petri nets, $M(P_i) \leq N$ ($i \in \{1,..., n\}$, $N \in \mathbb{N}$) defined by $<P, \Gamma, \Lambda, Pre, Post, M_0>$ such as:

$P = \{P_1, ..., P_n\}$ is the finite set of places,
$\Gamma = \{T_1, ..., T_m\}$ the finite set of transitions,
$\Lambda = \{\mu_1, ..., \mu_m\}$ the set of transition firing rates,
Pre, Post : $P \times \Gamma \to \{0,1\}$ incidence applications
$M_0 : P \to \{0, 1, ..., N\}$ the initial marking.

3.2 Places interactivity

Having applied several series of rate using the method of Amodeo, we observe that token attraction or rejection depend on the value of the firing rates. To estimate their impact on the marking in steady-state regime, we study the interactivity of every place with the rest of the model, according to its input and output rates. We will be also able to know if the studied place tends to attract or to reject tokens. This is illustrated in Figure 1, where μ_s and μ_e are respectively the sum of output and input firing rates of the place P_1. The place P_r is a generic place that represents the rest of the model. So the study of the interactions between the place P_1 and the other places is reduced to the study of its behavior reporting to the generic place P_r.

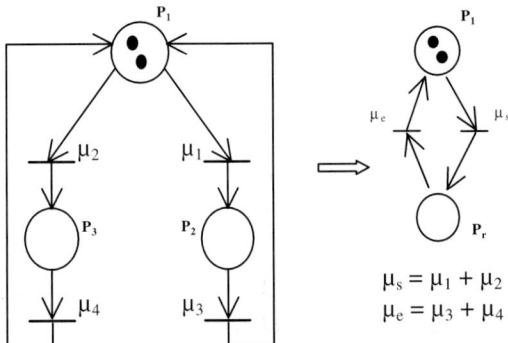

Figure 1. Interaction of the place P_1 with the rest of the model

3.3 Use of birth-death process theory

We can generalize the idea of the study of a place with regard to the rest of the model. We obtain thus a reduced two places stochastic Petri net, called generic Petri net (Figure 2), with:

- the studied place P,
- the place P_r representing the rest of the model,
- μ_s and μ_e the sums of all input and output firing rates of the place P

- N, the upper bound of number of tokens, which can contain the place P.

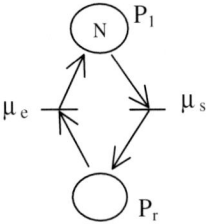

Figure 2. Generic SPN

After a transient period, during which each place is going to attract or to reject tokens according to the values of the firing rates, the system evolves towards one of markings M_i represented in Figure 3:

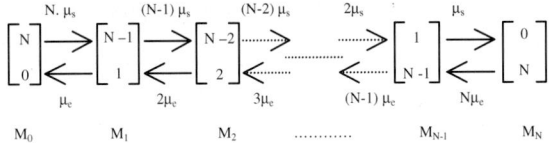

Figure 3. Generic SPN markings graph

Leaving the state M_i, the system can evolve only towards two M_{i-1} or M_{i+1} states with the respective firing rates $(i)\mu_e$ and $(N-i)\mu_s$. This particular characteristic is that of birth-death processes. These processes are a special case of a Markov chain in which the process makes transitions only to the neighboring states of its current position [7]. Thus, we can use all the analytical power of these processes to obtain a mathematical expression of places communication.

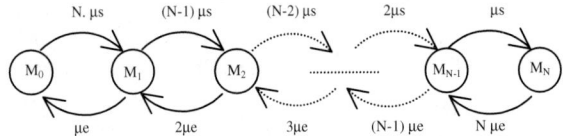

Figure 4. Birth-death process associated to the generic SPN

Birth and death rates of our N+1 states system are the following:

$$\begin{cases} \lambda_n = (N - n)\, \mu_s \\ \mu_n = n\, \mu_e \end{cases} \text{, with } n = 0,...,N \quad (1)$$

In such a system, the probability π_n being in a state n has the next form:

$$\pi_n = \frac{\rho^n.(1-\rho)}{1-\rho^{N+1}} \quad , \text{with} \quad \rho = \mu_S/\mu_e \quad (2)$$

This result allows us to deduct the steady-state average marking $M(P_i)$ of the place P_i:

$$M(P_i) = \sum_{n=0}^{N} M_n(P_i).\pi_n \quad (3)$$

with $M_n(P_i)$, the marking of the place P_i when the system is in the n state.

If we apply this formula for the place P of the model of figure 2, we obtain:

$$M(P) = N - \frac{(1-\rho)}{1-\rho^{N+1}} \cdot \sum_{n=0}^{N} n.\rho^n \quad (4)$$

Thus, we find an expression of the steady-state mean marking, according to the capacity N of the place P and to its input and output firing rates.

This result shows that the relation between the input and output firing rates is very important and has a strong influence on the steady-state behavior of the system. According to the rate ratio ρ, places can have two different possible evolutions:

- place P_i tends to attract tokens: $M(P_i) \gg 0$ and so P_i can not be eliminated because the corresponding state of the system is significant.
- place P_i tends to reject the tokens ($M(P_i) \approx 0$), so the state corresponding to this place is in no way significant and consequently, the place P_i can be neglected.

3.4 Use of the results for two-weighting scale decoupling of the SPN

From the expression of the average marking of the place P, we obtain the steady-state efficiency of this place:

$$R(P) = \frac{M(P)}{N} = 1 - \frac{(1-\rho)}{(1-\rho^{N+1})N} \cdot \sum_{n=0}^{N} n.\rho^n \quad (5)$$

Figure 5 illustrate the evolution of the efficiency R(P) function of the rate ratio ρ between output and input firing rate of the place P. The relation $R = F(\rho)$ depends on the upper marking bound N of the place P:

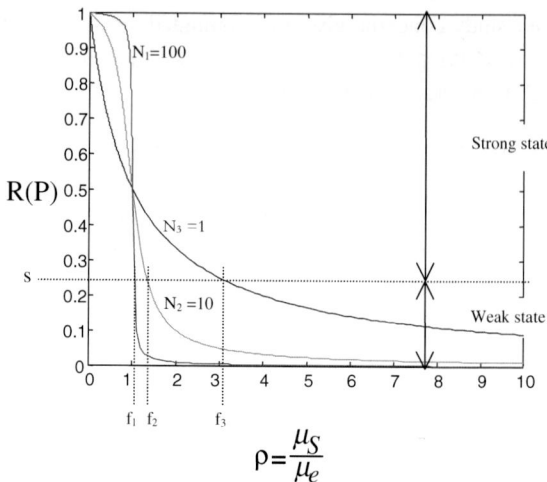

Figure 5. Evolution of the efficiency R of the place P, function of the rate ρ and the capacity N

All the curves pass by the point (1, 0.5), corresponding to an efficiency R=50% and an output/input rate ratio $\rho = 1$. According to $\rho=\mu_s/\mu_e$, R(P) can evolve in two manners:

1st case :

$$\mu_s \gg \mu_e \Rightarrow \rho = \frac{\mu_s}{\mu_e} \to \infty \quad (6)$$

and consequently,

$$\lim_{\rho \to \infty} M(P) = \lim_{\rho \to \infty} (N - \frac{1-\rho}{1-\rho^{N+1}} \sum_{n=0}^{N} n\rho^n) = 0 \quad (7)$$

$$\lim_{\rho \to \infty} R(P) = 0$$

In this case, the place P does not contain tokens in steady-state regime, so the system converges to the state M_N, corresponding to the marking $[0\ N]^t$. The place P tends to reject tokens rather than keeping them. The action associated to P appears rarely. We can qualify these states as *weak* or unstable *states*. Their elimination can only simplify our network without a consequent loss of information.

2nd case:

$$\mu_e \gg \mu_s \Rightarrow \rho = \frac{\mu_s}{\mu_e} \to 0 \quad (8)$$

and by next:

$$\lim_{\rho \to 0} M(P) = \lim_{\rho \to 0} (N - \frac{1-\rho}{1-\rho^{N+1}} \sum_{n=0}^{N} n\rho^n) = N \quad (9)$$

so:

$$\lim_{\rho \to 0} R(P) = 1$$

In this case, we remark that the place P has an efficiency of 100%, corresponding to a steady-state marking of N tokens. The state M_0 corresponding to the marking $[N\ 0]^t$ has a strong occurrence probability. The physical state or action corresponding to this place is very significant (*strong state*) and must be considered.

On the other hand, the efficiency R(P) tends quickly to zero for great capacities (N) when ρ is greater than 1, and evolves slowly to attempt 10% for a value $\rho = 10$ for the 1-bounded places (N=1).

Conclusion:

According to the interaction between the places, we conclude that the Petri net performances depend on the firing rates values. We calculate the average marking of places. This expression allows us to estimate the dynamics of every place in steady-state regime by estimating its efficiency, and so to see the effect of every place in the distribution of tokens in the model: some will tend to attract tokens towards them, and the others will reject them. Figure 5 give the place efficiency evolution function of output/input firing rates ratio ρ, for various values of N. In order to evaluate the weight of a place, we define an efficiency threshold s, $s \in [0,1]$ that shares places in two corresponding groups called strong and weak. Corresponding to this threshold, we have a bound f of rate ratio ρ ($f \in R^+$) such as:

$$s = 1 - \frac{(1-f)}{(1-f^{N+1})N} \sum_{n=0}^{N} n.f^n$$

This bound depends essentially on the capacity N of the place. It is close to 1, for places having an important capacity, and grows with the decrease of N.

We obtain thus for the 1st case:
$$\rho < f \Rightarrow \mu_s < f.\mu_e \Rightarrow R(P) > s \Rightarrow M(P) > s.N \quad (10)$$

The efficiency of the place P is important. The state or the action associated to this place is significant and so it cannot be eliminated.

2nd case:
$$\rho > f \Rightarrow \mu_s > f.\mu_e \Rightarrow R(P) < s \Rightarrow M(P) < s.N \quad (11)$$

The place P has a weak efficiency. The state or the action associated to this place is insignificant, so it can be eliminated without important loss of information in the model.

This method allows a simplification of the SPN by operating directly on the graphic model. The only necessary calculation is that of the efficiency on every place. After the reduction, the reduced model still corresponds to the real system. The places eliminated are negligible in the system evolution.

3.5 Firing rates calculus for the reduced SPN

Having listed and eliminated weak places of the system with our method, we obtain a reduced model with new firing rates. In Figure 6, for example, P_5 represents a weak place and P_1, P_2, P_3 and P_4 strong ones, the token flow between places P_i (i-th input place of P_5) and P_j (j-th output place of P_5) depends on the smallest rate between these places. After the simplification (Fig.6.b), we obtain the new firing rate μ_{ij}:

$$\mu_{ij} = \min(\mu_i, \mu_j)$$

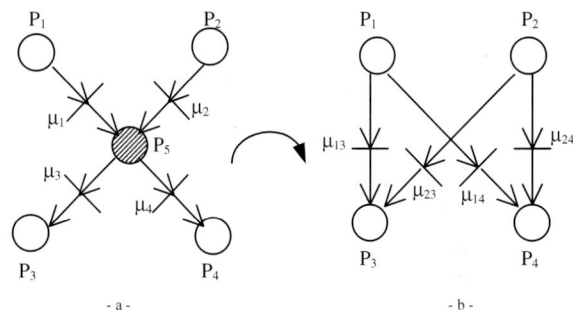

Figure 6. Calculus of firing rates of the reduced SPN in the case of multiple inputs/outputs

with:
μ_{ij}, the firing rate between the places P_i and P_j (Fig.6.b),
μ_i, the firing rate between P_i and P_5 (Fig.6.a),
μ_j, the firing rate between P_5 and P_j (Fig.6.a).

4 Industrial application

To estimate the performances of our method, we apply those results to a maintenance model elaborate in the frame of a global project with our automotive industrial partner. The studied model corresponds to the maintenance flow of spot-welding tongs between the robotized (flow-shop) production line and the repair shop. The various phases of this circuit are the following:
- Waiting zone after breakdown in a special place of the production line,
- Clips transit to the repair shop,
- Waiting zone for reparation,
- Test bed,
- Transit of clips to the different sectors of the flow-shop.

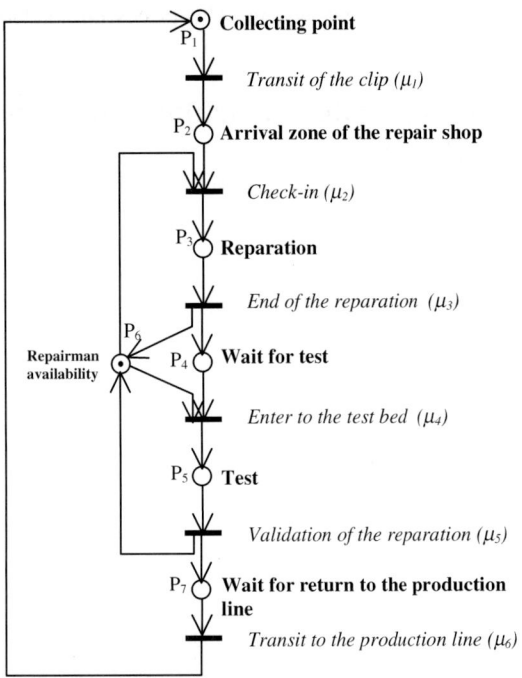

Figure 7. Maintenance flow modeling of a spot-welding tong.

A random transition period is necessary for every stage, so every transition can be characterized by a firing rate. Thus, we obtain the stochastic Petri net of Figure 7, with μ_i the different firing rates. This algorithm was programmed with the Matlab software and can be programmed with any other mathematical software. The table 1 gives the results of the SPN simplification using two methods: singular perturbations, and graphical method for different management cases (so different firing rates).

Table 1. Results using singular perturbations and graphic reduction methods

μ_1	1				1				10				5				
μ_2	20				20				15				90				
μ_3	2				80				90				5				
μ_4	5				5				5				100				
μ_5	4				90				100				5				
μ_6	40				4				6				40				
	μ_e	μ_s	ρ	R(P)	μ_e	μ_s	ρ	R(P)	μ_e	μ_s	ρ	R(P)	μ_e	μ_s	ρ	R(P)	
P1	40	1	0,03	0,98	4	1	0,25	0,8	6	10	1,67	0,38	40	5	0,13	0,89	
P2	1	20	20	0,05	1	20	20	0,05	10	15	1,5	0,4	5	90	18	0,05	
P3	20	2	0,1	0,91	20	80	4	0,2	15	90	6	0,14	90	5	0,06	0,95	
P4	2	5	2,5	0,29	80	5	0,06	0,94	90	5	0,06	0,95	5	100	20	0,05	
P5	5	4	0,8	0,56	5	90	18	0,05	5	100	20	0,05	100	5	0,05	0,95	
P6	6	5	4,17	0,19	170	25	0,15	0,87	190	20	0,11	0,9	10	190	19	0,05	
P7	4	40	10	0,09	90	4	0,04	0,96	100	6	0,06	0,94	5	40	8	0,11	

○ place eliminated by singular perturbations

▨ R(P) < 0,2 - elimination by graphical method

At first, we apply the singular perturbation method [2], [9] to the model and we compare the results with those given by our graphic method. The column R(P) gives the efficiency of the corresponding place according to N and ρ (the rate ratio between output and input sum of firing rates). For an efficiency threshold of 20%, we obtain a bound f =4 for the coefficient ρ. Hachured cells (tab 1) indicate the places with efficiency lower than 20%, and the surrounded cells represent the places, which are eliminated by the singular perturbation method. We remark that all places eliminated by the singular perturbation techniques have values of efficiency R(P) lower than 20%. The graphical method seems more complete than the singular perturbation method, thanks to the exhaustive elimination of all weak states. In this sense, we see (tab 1) that some weak states escaped from the elimination by singular perturbations.

To illustrate the meaning of the reduced model in our application, we take for example, the second column data of Table 1. The places eliminated by our method are:

P_2 : waiting in the flow–shop,
P_3 : reparation,
P_5 : the test bed.

Thus the reduced stochastic Petri net so obtained is the following:

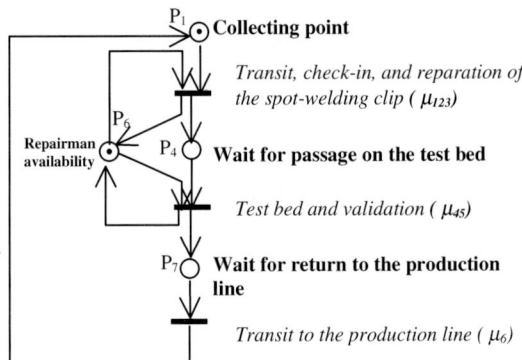

Figure 8. Reduced SPN

The reparation delay and the test duration of spot-welding clips are relatively small with a very short waiting time in the repair shop. The major part of the cycle is distributed among:

- Waiting in the collecting points of the production (assembly) line, (P_1)
- Waiting for passage on the test bed, (P_4)
- Waiting for return (P_7).

The firing rates have the following values:

$\mu_{123} = \min(\mu_1, \mu_2, \mu_3) = 1$
$\mu_{45} = \min(\mu_4, \mu_5) = 5$
$\mu_6 = 4$

Solutions for this type of maintenance strategy will be directed to
- an optimization of the transit times, especially since the mechanic is often available (P_6),
- a control of the passage of spot-welding clips in the test bed, because the time of the test is very short, the operator is often available, but clips waiting time is too long (P_4).

This technique allows us to watch the performances of the repair shop by establishing a dashboard having the efficiency R of each place as a performance indicator. The optimum to reach is the equilibrium between the flow of the spot-welding tongs in all the places of the model. This equilibrium corresponds to an equality between the input and output rates, which gives a communication coefficient of $\rho = 0.5$. Thus we can localize the part of the circuit to be optimized. We can also test this model to evaluate the impact of any amelioration operation on the performances and the flow of the system.

Let us note that the method of singular perturbation becomes impracticable with an important number of places or tokens, because of the enormous number of states of the marking graph, which gives an untreatable Markov generator. This problem does not occur in our method, because we don't need to establish the marking graph; a simple calculation of $R(P) = M(p)/N$ is enough. The application of this method can be extended to the simplification of other more complex systems (with divergent or convergent component) being able to be represented by a stochastic Petri net. We can study for example a rare but catastrophic situation (example of a fatal breakdown) as shown in Figure 9:

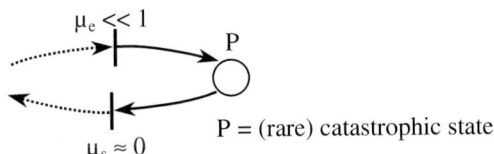

Figure 9. The study of a rare catastrophic state

The rate ratio ρ of this place tends to zero. So this place represents a strong state (Figure 5) and will not be eliminated with this method.

5 Conclusion

In this article, we developed a graphic method of stochastic Petri nets reduction, based on the principle of the interactivity of places. This technique puts in evidence the places that attract or reject the tokens of the model. Our method has the advantage to simplify the model by a simple calculation of the efficiency on every place in steady-state regime. Our future researches will deal with the development of a calculation methodology of a threshold between the weak and strong values of markings, according to the application chosen field.

6 References

[1] Alain J., Stochastic petri nets. Belgium French Netherland'Summer School on Discret Event Systems, Spa – Belgium, 1993.

[2] Amodeo L., Contribution à la simplification et à la commande des réseaux de Petri stochastiques. Application aux systèmes de production, PhD, Belfort, France, 1999.

[3] Brams G., Réseaux de Petri, théorie e pratique - Tome 1 :Théorie et analyse – Tome 2 : Modélisation et applications, Edition Masson, 1983.

[4] David R., et Alla H., Du Grafcet aux réseaux de Petri, Edition HERMES (2e édition), 1992.

[5] Florin G., Stochastic Petri nets : Properties, applications and tools, Microelectronics and Reliability, 31(4), 1991, pp.669-697.

[6] Murata T., Petri nets Properties, analysis and applications, IEEE, 77(4), 1989.

[7] Ng Chee Hock, Queueing Modelling Fundamentals, JOHN WILEY & SONS 1996.

[8] Racoceanu D., A. El Moudni, M. Ferney, N. Zerhouni., On a New Method of Markov Chain Reduction. Mathematical Modelling of Systems, vol. 1, no 3, pp. 83-101, 1995.

[9] Racoceanu D., Contribution à la modélisation et à l'analyse des chaînes de Markov à échelles de temps et échelles de pondération multiples. Application à la gestion d'un système hydro-énergétique, PhD, Belfort, France, 1997.

[10] Racoceanu D. and Zerhouni N., Use of Singular Perturbations for the Reduction of Manufacturing System Models, IFAC Congress MCPL-2000, Grenoble, France, 2000.

[11] Ruegg A., Processus stochastiques, Presses polytechniques romandes, 1989.

[12] Zhou, M.C. and J. Ma, Reduction and Approximation of Stochastic Petri Nets with Multiple Input Multiple-Output Modules, Preprints of 12th IFAC World Congress, vol. 4, 159-163, Sydney, Australia, 1993.

Modeling and Analysis of Semiconductor Manufacturing Systems with Degraded Behavior Using Petri Nets and Siphons

MuDer Jeng
Department of Electrical Engineering
National Taiwan Ocean University
Keelung 202, Taiwan, ROC

Xiaolan Xie
LGIPM and INRIA/MACSI Team
ENIM-Ile du Saulcy
F-57045 Metz Cedex 1, France

Abstract

Degraded behavior, such as reworks, failures, and maintenance, of a semiconductor manufacturing system (SMS) is not negligible in practice. When modeled by Petri nets, degraded behavior may be represented as initially-unmarked elementary circuits, interpreted as *local processing cycles*. Most existing "well-behaved" net classes for manufacturing have problems of describing such cycles and thus may have difficulties in modeling SMSs. In this paper, we extend the class of nets in [6] into the class of *RCN* merged nets* that model SMSs with such cycles. To model an SMS, we first describe the behavior of each resource type using a state-machine module, called RCN. Any RCN can be constructed as a connection of acyclic sub-nets called *blocks*, where one of them denotes the normal behavior of the resource type and the others denote its degraded behavior. Next, an RCN* merged net for the entire system is built by fusing all modules, conforming to three constraints, along their common *transition sub-nets*, which represent their synchronization. In the analysis of RCN* merged nets, we prove that their liveness and reversibility depend on the absence of unmarked siphons, which are structural objects that mixed integer programming can check rapidly. Examples are given to illustrate the proposed approach.

1. Introduction

Semiconductor manufacturing systems (SMSs) belong to a class of discrete manufacturing systems that contain complicated production procedures and a large number of shared resources. In particular, an SMS often reveals degraded behavior, which is not negligible in practice. For example, if in the photolithography stage a wafer is not properly coated with photo-resistors, it is necessary to remove the resistors and re-do the coating. On the other hand, when a wafer is waiting to be processed in some machine, the machine may not be available due to some unexpected fault or expected maintenance operation. These are three examples of degraded behavior -- reworks, failures, and maintenance. The purpose of this paper is to present a new class of Petri nets for the modeling and analysis of SMSs with such behavior.

Petri nets [16] have become a powerful tool for manufacturing after three decades of research and development. Recently, some researchers have applied them to SMSs. Kim and Desrochers [9] investigated the automatic generation and performance evaluation of a net model for an SMS based on the technology process flows and fabrication facilities. In [1, 13], Petri nets and heuristic search were adopted for modeling and scheduling an SMS. In [5, 11, 12], the modeling and analysis of several types of equipment in SMSs was performed using Petri nets. Cheng et al. [2] generated and anlyzed a net model for a semiconductor manufacturing execution system from IDEF0 and state diagrams. In the above-mentioned approaches, degraded behavior of SMSs was not explicitly considered at the net structure level. Yoon and Lee [14] exploited a very simple failure-repair net model for lot dispatching in a wafer fabrication facility. Degraded behavior is explicitly considered in [16], where a class of modularly composed Petri nets was proposed for modeling, qualitative analysis, and performance evaluation of a real-world IC fabrication system under the assumption that there exists exactly one global buffer. Combined with the uniformization technique, the class of nets was adopted for analytically evaluating SMSs in [8].

Since an SMS is very complex, analyzing its model for qualitative properties is a computation-intensive job. As a result, our focus in this paper is to explore a class of "well-behaved" nets for SMSs that can explicitly describe degraded behavior at the net structure level and whose properties can be checked within reasonable time. Previous "well-behaved" net classes may have difficulties in applications to SMSs because some are inflexible to represent shared resources (e.g. [10]), some take significant computation time in qualitative analysis (e.g. [15]), and others have problems to denote degraded behavior (e.g. [4, 6]). In terms of net structure, degraded behavior may be represented as initially-unmarked elementary circuits, interpreted as *local processing cycles*. Nevertheless, this does not mean that the last category of net classes mentioned above [4, 6] cannot describe such cycles. Through the transformation or reduction technique, they can denote local processing cycles that involve only one resource type as in [14]. In this paper, we present a new net class called *RCN* merged nets* that can describe local processing cycles that involve more than one resource type. Comparing to the nets in [8], RCN* merged nets do not require the assumption that there exists exactly one global buffer.

To model an SMS, we first describe the behavior of

each resource type (e.g., wafer, machine, buffer, AGV, etc.) using a state-machine module, called RCN. The concept of resource used here encompasses both the resources for fabricating the parts, e.g., wafers, and the parts themselves. Any RCN can be constructed as a connection of acyclic sub-nets called *blocks*, where one of them denotes the normal behavior of the resource type and the others denote its degraded behavior. The connection of any two blocks is accomplished by using degrading and restoration arcs. Next, an RCN* merged net for the entire system is built by fusing all modules, conforming to three constraints, along their common *transition sub-nets*, which represent their synchronization. In fact, we have generalized the nets in [6] into RCN* merged nets by relaxing one of the original constraints.

We show that liveness and reversibility of RCN* merged nets depend on the absence of unmarked siphons, which are structural objects much easier to check than the state-space based conditions. It is worth mentioning that a mathematical programming approach was proposed in [3] to avoid explicit enumeration of siphons. It allows one to check large Petri net models using powerful mathematical programming software packages. It is reported that this approach can check large nets of about 200 places and 200 transitions in a small amount of time. A net with such size can model significantly complex SMSs.

The remaining paper is organized as follows: Section 2 defines RCN* merged nets. Section 3 discusses their qualitative properties. Concluding remarks and future research are presented in Section 4. The appendix briefly summarizes the Petri net theory.

2. RCN* Merged Nets

As described above, the class of RCN* merged nets is constructed in a resource oriented and modular manner as the class of nets in [6]. The main construction steps include the modeling of the behavior of each resource type using Petri nets, and the integration of resource net modules by taking into account interactions among resource types. Each resource type is modeled as an RCN, constructed as a connection of blocks, and the integration is realized through merging of the RCNs along their common transition sub-nets, where the common elements of any two RCNs are denoted as having the same labels. Here the concept of resources includes both the resources for fabricating parts, e.g., wafers, and the parts themselves.

The following definition of an RCN was given in [6, 7].

Definition 1: An RCN $G_\alpha = (P_\alpha, T_\alpha, F_\alpha, M_{\alpha 0})$ is a strongly connected state machine where there exists exactly one place $p_{r\alpha} \in P$, called *resource place*, such that $M_{\alpha 0}(p_{r\alpha}) \neq 0$. The remaining places are called *operation places*.

Place $p_{r\alpha}$ denotes the availability of the resource type R_α that G_α models. In most manufacturing systems, there exists a state such that all resources are available and there is no work-in-process. We choose such a state as the initial state. It is then reasonable that only the resource places are initially marked. The portion of G_α excluding $p_{r\alpha}$ represents operations that require R_α.

Remark 1: An operation place with its input transition(s) and output transition(s) models some operation op_x being executed, the start of op_x, and the end of op_x, respectively, as shown in Fig. 1. Each such transition(s)-place-transition(s) structure represents an operation with a specific resource requirement. We can build an RCN by exploiting such net structures and by considering the precedence relationship among the operations, i.e., the input transitions for some operation are the output transitions for its previous operations (see Example 1 later).

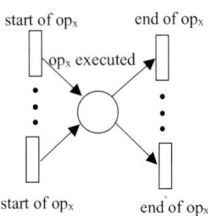

Fig. 1. Modeling of an operation.

The above definition of an RCN does not explicitly specify how to represent the degraded behavior of a resource type. Here we define degraded behavior as having local processing cycles:

Definition 2: A *local processing cycle* is a circuit that contains operation places only, while a *global processing cycle* is thus a circuit that contains at least one resource place.

The above definition applies to individual RCNs as well as integrated nets, as presented later. Note that the nets in [6] prohibit the modeling of local processing cycles at both the module and the integrated net levels due to one of their restrictions that the removal of all resource places from an integrated net results in an acyclic net. This prevents any RCN of the nets in [6] from containing local processing cycles.

To model the normal and degraded behavior of a resource type, any RCN can be constructed as a connection of blocks as shown in Fig. 2, where the blocks, which are defined below, are acyclic sub-nets and the double circle denotes the resource place p_r. Block 1 denotes the normal behavior of a semiconductor resource type while block x represents its degraded behavior due to, for example, (x-1) times of reworks, failures, or maintenance.

Definition 3: The blocks of an RCN G are defined as follows:

Step 1: Let $T_I^1 = p_r\bullet$ and $T_O^1 = \bullet p_r$ be respectively called the sets of *input* and *output transitions* of *block 1*, denoted as $B^1(G)$. Then $B^1(G) = (P^{B1}, T^{B1}, F^{B1}, M^{B1}_0)$ is a net defined as the maximal circuit-free union of elementary paths from T_I^1 to T_O^1 (of course, excluding p_r).

Step 2: Repeat the following step for $x = 2, \ldots$ until no new block is found.

Step 3: Let $T_I^x = \bigcup_{p \in P^{B(x-1)}} p\bullet \setminus \bigcup_{y=1}^{x-1} T^{By}$ and $T_O^x = \bigcup_{p \in P^{B(x-1)}} \bullet p \setminus \bigcup_{y=1}^{x-1} T^{By}$ be respectively called the sets of *input* and *output* transitions of *block x*, denoted as $B^x(G)$. Then $B^x(G) = (P^{Bx}, T^{Bx}, F^{Bx}, M^{Bx}_0)$ is a net defined as the maximal circuit-free union of elementary paths from T_I^x to T_O^x.

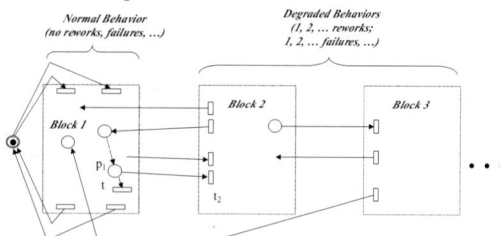

Fig. 2. The blocks within an RCN.

In the following, places in $\bullet T_I^x \cap P^{B(x-1)}$ will be called *interface places* (e.g., p_1 in Fig. 2) of block $(x-1)$ with $x > 1$. Arcs (p, t) from block $B^{x-1}(G)$ to $B^x(G)$ will be called *degrading arcs* and arcs (p, t) from block $B^x(G)$ to $B^{x-1}(G)$ will be called *restoration arcs*. Of course, any degrading arc connects an interface place to an input transition of the next block while a restoration arc connects an output transition to the previous block.

Example 1: Fig. 3, where the black dot denotes the initial state, shows a typical photolithography process for some wafer type W using online steppers. Wafers are initially coated with photo-resister circuit patterns. Then, the photo resistors are exposed and developed. Finally, they are inspected to see if their dimensions are correct or not. If not, the resistors on the wafers will be removed and the wafers are re-coated with new resistors. During the process, wafers are usually stored in buffers between two consecutive processing steps. Inline steppers are modern wafer-processing machines that can successively handle coating, exposure, and development in three chambers, respectively. In addition, they can perform resistor removal and re-coating in the coating chamber.

The process flow for W has degraded behavior for re-coating. Based on Fig. 3, we can model W as the RCN G_W in Fig. 4 (w denotes the initial marking of the resource place), where a dashed block models an operation, and blocks 1 and 2 (shaded) denote the normal process and the rework process respectively. In Fig. 4, we are aware of a local processing cycle p3t4p4t5p5t6p6t9p9t10t11p3 due to the degraded behavior for re-coating. [17] describes a system for such a proces. The system contains a set of inline steppers L, two types of buffers B1, B2, and a set of inspection machines E. Due to space limitation, the detailed modeling of the system using the proposed approach is left out.

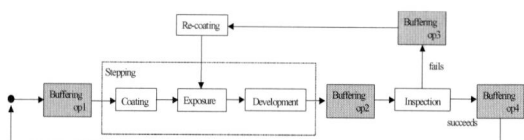

Fig. 3. A typical photolithography process for some wafer type W.

The following lemma shows an important property of blocks.

Lemma 1: Each block $B^x(G)$ in an RCN G is acyclic. Further, there exist a path in $B^x(G)$ from any given input transition to an output transition and a path from an input transition to any given output transition.

To build a system net, RCNs are merged along their common elements. Since each RCN is constructed according to Remark 1, it is clear that the common elements of any two RCNs form a transition sub-net, as given below [6, 7]:

Definition 4: A transition subnet $G_\beta = (P_\beta, T_\beta, F_\beta, M_{\beta 0})$ of a Petri net G is a subnet of G such that input transitions and output transitions of any place $p \in P_\beta$ (P_β can be empty) are transitions in T_β. In other words, the places of a transition sub-net have no arc connection outside the subnet.

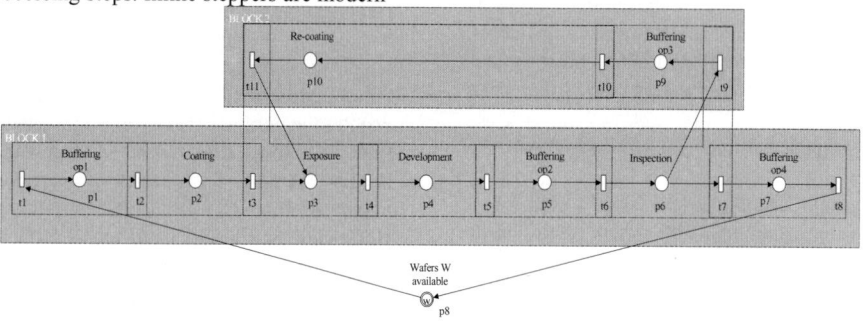

Fig. 4. RCN G_W for the wafer type W.

Definition 5: Given a set of n RCNs $\{G_s | G_s = (P_s, T_s, F_s, M_{s0}), s = 1,..., n\}$ built following Remark 1, an RCN* merged net $G = (P, T, F, M_0)$ is their union, i.e., $P = P_1 \cup P_2 ... \cup P_n$, $T = T_1 \cup T_2 ... \cup T_n$, $F = F_1 \cup F_2 ... \cup F_n$, and $M_0(p) = M_{s0}(p)$ if $p \in P_s$, satisfying the following three restrictions.

Restriction 1: At each common transition, there exists at most one input place that is an operation place.

Restriction 2: Common transition sub-nets should not include resource places.

Restriction 3: *(A)* In the integrated model G, for each place p with degrading outgoing arcs, there exist a non-degrading outgoing arc (p, t_1) such that $\bullet t_1 \subseteq \bullet t_2$ for any degrading outgoing arc (p, t_2). *(B)* The net G' derived from G by removing resource places has source and sink transitions, also called *global input and output transitions*, such that there exists a path in G' from an input transition to any node of G'. *(C)* The net G# derived from G by removing the resource places and degrading arcs is an acyclic graph.

Restrictions 1 and 2 also hold for the nets in [6]. We briefly repeat the explanations given in [6] for completeness. Restriction 1 excludes the modeling of the synchronization of parallel processes or the assembly of several components. Thus, it is restrictive and its relaxation is an issue of future research. Restriction 2 is natural for our resource-oriented approach since each resource type is modeled as exactly one RCN. That is, each resource place appears in exactly one RCN.

Restriction 3 here is more general than that for the nets in [6] because the integrated model can contain circuits that do not include the resource places as long as the "AC (asymmetric choice)" condition (i.e., $\bullet t_1 \subseteq \bullet t_2$ above) is satisfied. In other words, all such circuits, i.e., local processing cycles, include such transitions t_2.

Restriction 3 is reasonable as explained below. In most manufacturing systems, local processing cycles are due to operation constraints (e.g., reworks, failures, and maintenance) rather than resource constraints (e.g., the availability of resources). Thus, the decision of firing t_1 (generating a global processing cycle) or t_2 (generating a local processing cycle) at an interface place p is usually made in an "FC (free-choice)" manner, i.e., $\{p\} = \bullet t_1 = \bullet t_2$. In general, such decisions are based on information independent of the availability of resources such as product data, product quality or degrading state of equipments. In other words, no resources are involved in such a decision. For the same set of resources used at t_1 and t_2, this decision can be reasonably generalized to "extended FC ", i.e., $\bullet t_1 = \bullet t_2$. For example, in a semiconductor manufacturing system, a wafer may be placed in the same general buffer when either it is processed for the next operation, denoted by t_1, or reworked for a previous operation, denoted by t_2. In this condition, a local processing cycle is still, in effect, generated by operation constraints at the operation place p. We further extend the "extended FC" condition to the "AC" condition, leading to Restriction 3A. In the latter condition, a local processing cycle is more constrained to be created since a rework, failure, or maintenance operation may involve a larger set of resources than a normal operation does. For example, an extra robot is required for failure repair or maintenance.

As for Restrictions 3B and 3C, we consider that local processing cycles are usually the degrading behaviors of single resource (i.e., RCN) since they may be interpreted as reworks, failures or maintenance. Consequently, Restriction 3C excludes circuits that result from the normal operational behaviors without degrading arcs. For example, the net in Fig. 5 is not an RCN* merged net due to the circuit $t_1p_3t_2p_2t_1$ that contains all operation places, which are initially unmarked, where dotted blocks are RCNs, and t_1 and t_2 are the common transitions. Restriction 3B restricts processing cycles involving more than one RCN to be controllable via global input transitions. Local cycles such as the one of Fig. 6 create confusion between normal behaviors and degrading behaviors. In Fig. 6, t_1 belongs to the normal behavior and t_2 the degrading behavior in the left RCN while the reverse happens in the right RCN. Such confusion leads to a deadlock situation as proved in Lemma 2, given later.

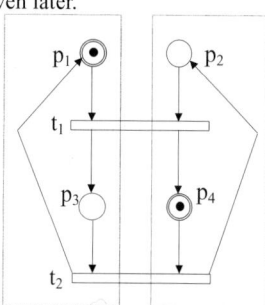

Fig. 5. A non-RCN* merged net.

By construction, an RCN* merged net G is state machine decomposable. Each RCN is a state machine component. The number of tokens in any state machine component remains constant whatever transition firings. Hence,

Property 1: G is conservative and structurally bounded. The following lemmata are due to Restriction 3:

Lemma 2: G contains an empty siphon at the initial marking and hence is not live if Restriction 3B does not hold.

Lemma 3: Under Restrictions 1 and 3B, any siphon in G contains at least one resource place.

3. Qualitative Properties

This section discusses two important qualitative properties, reversibility and liveness, of an RCN* merged net G, which satisfies Restrictions 1-3.

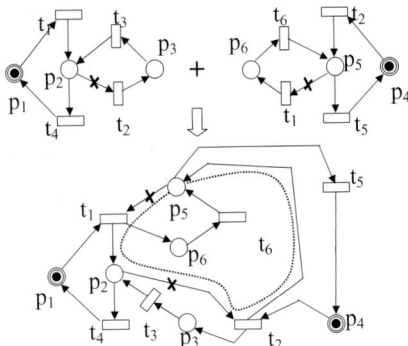

Fig. 6. A net with an inter-RCN cycle.

Theorem 1: Under Restrictions 1-3, G is reversible iff no siphons in G can become unmarked.

Notice that the condition of Theorem 1 can be checked using the following property obtained in [3]:

Property 2: A siphon S can never become empty if either it contains a marked trap or $F(S) > 0$ with $F(S) = \min\{\sum_{p \in S} M(p) \mid M = M_0 + CY, M \geq 0, Y \geq 0\}$.

Beside the above property, we can exploit the mathematical programming approach proposed in [3] to avoid explicit enumeration of siphons. Experimental results reported in [3] show that using powerful mathematical programming software packages, the approach is able to check large nets of about 200 places and 200 transitions in a small amount of time.

Under the reversibility, the liveness of the RCN* merged net G is reduced to its potential liveness, a property much easier to check.

Theorem 2: Under Restrictions 1-3, G is live and reversible iff no siphon in G can become unmarked and every transition can fire at least once, i.e., every transition is potentially firable.

The following results can be used to check the potential firability of the transitions.

Restriction 4: At any common transition, there is at most one output place that is an operation place.

Theorem 3: Under Restrictions 1-4, any transition in G is potentially firable.

The proof of this theorem is similar to that of Theorem 3 in [6], it is omitted and only a sketch of proof is given. The proof is based on the existence of an elementary path from a global input transition to any transition t composed of only operation places. Restrictions 1 and 4 ensure that transitions on this path do not have other operation input/output places. Firability of these transitions can then be easily proved.

Consider the case where Restriction 4 is not satisfied. It is more difficult to check the potential firability in this case since any transition not conforming to Restriction 4 creates parallel processes, i.e., more than one output operation place. The potential firability is not always true as shown in Fig. 7 where the RCN* merged net is obtained by composing two RCNs sharing t_1, t_2, and t_3.

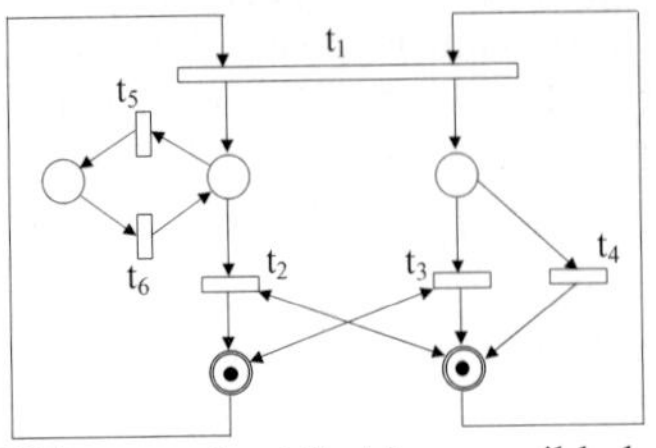

Fig. 7. RCN* merged net that is reversible but not live.

In the following, we use structural properties of the net to verify the potential firability. The basic idea is as follows: First, the RCN* merged net G is unfolded into an acyclic net that has the same potential firability for each transition. That is, if one net has a firing sequence to enable a transition t, the other has a firing sequence to enable t too, and vice versa. Second, we remove the non-determinism of the unfolded net by choosing for each operation place an output transition, and study the liveness of the resulting nets.

Definition 6: An unfolded net g^u of a net G' derived from G by removing resource places is constructed as follows:

Step 1: Let T_{in} and T_{out} be respectively the sets of *global input* and *output transitions*. Then $g^u = g^{u1} = (P^{u1}, T^{u1}, F^{u1}, M^{u1}_0)$ is a net defined as the union of all paths without degrading arcs from T_{in} to T_{out}.

Step 2: Repeat the following steps for $x = 2, \ldots$ until all degrading arcs of G' are considered.

Step 3: For all places p in $g^{u,x-1}$, add their degrading outgoing arcs (p, t) to the net g^u. Let T_{in}^x be the set of transitions related to new degrading arcs (p, t).

Step 4: Let g^{ux} be the net defined as the union of all paths without degrading arcs from T_{in}^x to T_{out}. Merge g^u and g^{ux} to construct the new g^u through the merging of transitions in T_{in}^x while the other common places and transitions between g^u and g^{ux} are not merged (i.e., they are duplicates of those nodes in g^u).

Lemma 4 The unfolded net g^u of the net G' derived from G by removing resource places is acyclic.

The unfolding eliminates the local processing cycles by separating normal behaviors and degrading behaviors. To complete the unfolding, we introduce the resource requirement for each transition t.

Definition 7: A unfolded net G^u of an RCN* merged net G is derived from the unfolded net g^u by adding resource places $\{p_{r1}, p_{r2}, \ldots\}$ and by adding an arc (p_{ri}, t) (resp. (t, p_{ri})) for each copy of transition t if the arc exists in G.

Lemma 5: G^u belongs to the class of nets in [11].

Lemma 6: Any transition t in G is potentially firable if t is potentially firable in G^u.

The following concepts are used for removing the non-determinism of the unfolded net for verifying the potential firability.

Definition 8: A Forward-Conflict-Free (FCF)

component G_1 of a net g^u is a sub-net generated by a sub-set T_1 of transitions having the following properties: (1) each place in G_1 has *one output transition* and at least one input transition; (2) a sub-net generated by T_1 is the net consisting of transitions in T_1, all of their input and output places, and their connecting arcs.

Definition 9: Let $G_1 = (P_1, T_1, F_1, M_{10})$ be an FCF component of the net g^u. $f(G_1)$ is defined as the sub-net of G^u generated by transitions in T_1. Clearly, $f(G_1)$ can be derived from G_1 by adding resource places that are input or output places of transitions in T_1.

As a result of Lemmata 5 and 6, we can exploit the properties of the class of nets in [6] to verify the potential firability of each transition in an RCN* merged net G.

Theorem 4: A transition t in G is potentially firable if there exists an FCF component G_1 in g^u that contains t and that no siphon in $f(G_1)$ can become unmarked.

4. Conclusions

In this paper, we have presented a new class of modularly composed Petri nets called RCN* merged nets that model semiconductor manufacturing systems with degraded behaviors such as reworks, failures, and maintenance. RCN* merged nets are generalized from the class of nets in [6]. Degraded behaviors are denoted as local processing cycles, which are elementary circuits that are initially unmarked. To model degraded behaviors, an RCN module can be built as a connection of blocks, where one block denotes the normal behavior and the others denote the degraded behaviors. The liveness and reversibility of RCN* merged net are proved to depend on the absence of unmarked siphons. Future research is directed to the extension to manufacturing systems with assembly/disassembly operations and degrading behaviors.

References

1. S. Cavalieri and O. Mirabella, "A PN-based scheduler for a flexible manufacturing semiconductor manufacturing system," *Proc. IEEE Int. Conf. Emerging Tech. & Factory Automat.*, pp. 724-729, 1996.
2. F.-T. Cheng, H.-C. Yang, T.-L. Kuo, C. Feng and M. D. Jeng, "Modeling and analysis of equipment managers in manufacturing execution systems for semiconductor packaging," *IEEE Trans. Syst. Man Cybern.*, vol. 30, part B, no. 5, pp. 772-782, Oct. 2000.
3. F. Chu and X. L. Xie, "Deadlock analysis of Petri nets using siphons and mathematical programming" *IEEE Trans. Robotics Automat.*, vol. 13, pp. 793-804, 1997.
4. J. Ezpeleta, J. M. Colom and J. Martinez, "A Petri net based deadlock prevention policy for flexible manufacturing systems," *IEEE Trans. Robotics Automat.*, vol. 11, pp. 173-184, 1995.
5. J. W. Janneck and M. Naedele, "Modeling a die bonder with Petri nets: a case study," *IEEE Trans. on Semicond. Manufact.*, vol. 11, no. 3, pp. 404-409, Aug. 1998.
6. M. D. Jeng and F. DiCesare, "Synthesis using resource control nets for modeling shared-resource systems," *IEEE Trans. Robotics Automat.*, vol. 11, pp. 317-327, 1995.
7. M. D. Jeng and X. L. Xie, "Analysis of modularly composed nets by siphons," *IEEE Trans. on Syst., Man, and Cybern.*, part A, vol. 29, no. 4, pp. 399-406, July 1999.
8. M. D. Jeng, X. L. Xie and W. Y. Hung, "Markovian timed Petri nets for performance analysis of semiconductor manufacturing systems," *IEEE Trans. Syst. Man Cybern.*, vol. 30, part B, no. 5, pp. 757-771, Oct. 2000.
9. J. Kim and A. A. Desrochers, "Modeling and analysis of semiconductor manufacturing plants using time Petri net models: COT business case study," *Proc. IEEE Int. Conf. Syst. Man Cybern.*, pp. 3227-3232, 1997.
10. I. Koh and F. DiCesare,"Modular transformation methods for generalized Petri nets and their applications in automated manufacturing systems," *IEEE Trans. Syst. Man Cybern.*, vol. 21, pp. 963-973, 1991.
11. S.-Y. Lin and H.-P. Huang, "Modeling and emulation of a furnace in IC fab based on colored-timed Petri net," *IEEE Trans. on Semicond. Manufact.*, vol. 11, no. 3, pp. 410-420, Aug. 1998.
12. R. S. Srinivasan, "Modeling and performance analysis of cluster tools using Petri nets," *IEEE Trans. on Semicond. Manufact.*, vol. 11, no. 3, pp. 394-403, Aug. 1998.
13. H. H. Xiong and M. Zhou, "Scheduling of semiconductor test facility via Petri nets and hybrid heuristic search," *IEEE Trans. on Semicond. Manufact.*, vol. 11, no. 3, pp. 384-393, Aug. 1998.
14. H. J. Yoon and D. Y. Lee, "A control method to reduce the standard deviation of flow time in wafer fabrication," *IEEE Trans. on Semicond. Manufact.*, vol. 13, no. 3, pp. 389-392, Aug. 2000.
15. M. C. Zhou and F. DiCesare, "Parallel and sequential mutual exclusions for Petri net modeling for manufacturing systems with shared resources," *IEEE Trans. Robotics Automat.*, vol.7, pp. 515-527, 1991.
16. M. C. Zhou and M. D. Jeng, "Modeling, analysis, simulation, scheduling, and control of semiconductor manufacturing systems: a Petri net approach," *IEEE Trans. on Semicond. Manufact.*, vol. 11, no. 3, pp. 333-357, Aug. 1998.
17. M. D. Jeng and X. L. Xie, "Modeling and Analysis of Semiconductor Manufacturing Systen with Degraded Behavior Using Petri Nets and Siphons" submitted to *IEEE Trans. Robotics Automat.*, 2001.

Developing Software Controllers with Petri Nets and a Logic of Actions

Carlo Simon
Institute of Software Technology
University of Koblenz-Landau, Germany
simon@uni-koblenz.de

Abstract

The presented software development process bases on extended timestamp nets (Petri net with an underlying time concept) and a logic of actions which allows to specify processes. By means of an example, the phases of the development process are discussed and the most important terms of the theory are explained. Since we can do direct and indirect proofing in the logic, we verify the developed controller with respect to process-like specifications. Moreover, also situation-like specifications can be tested. In this way, we put the machine virtually into operation.

Keywords: *Petri nets, Extended Timestamp nets, Logic of Actions, Controller Development, Software Engineering*

1. Introduction

In [6], a *software development process* for *controller programs* is presented which bases theoretically on *Petri nets* [5] and a *Logic of Actions* [3]. Since the main focus of [6] is on theoretical aspects and proofing techniques, only a small practical example is considered there.

This paper explains the software development process by means of a *pick-and-place operation* which is typical for mechanical engineering. Therefore, it is practically relevant and the specifications developed can be used for real production processes.

The controller developed for the pick-and-place operation has been tested in a virtual production environment. For a real machine, only the mathematical functions used would be different but not the models themselves.

This paper is organized as follows: after an overview over our software development process, we describe a machine which is able to realize a pick-and-place operation and demonstrate each step of our software development process by means of this example. Thereby, we point out the advantages of our approach.

2. The Software Development Process

Our software development process is a *phase model* where each phase defines *milestones*, *activities* to achieve them, and *models* to express the results. In this sense, it is leaned on *waterfall models* [1] like other software engineering approaches in computer science.

First, a *requirements analysis and specification* of the machine to be controlled and the processes to be realized is *verbalized*. In order to get a better comprehension of the machine and how it manipulates workpieces, in a *design and specification* phase the dynamic behavior of the *(uncontrolled) machine* and the *workpiece* is visualized. During this phase we benefit from *extended timestamp nets* (ETs nets) [4] which are expressive enough to specify even complex technical correlations. However, since models of the dynamic behavior of complex machines are complex, too, we use a *pictograph* representation to abbreviate our models. By using the Pascal-like programming language LAP ($\hat{=}$ Logic of Actions for Programming), we *formalize* the *controller specification*. In LAP we can specify sequences, conditions, iterations, and concurrent execution. Access-operations on system resources (actors and sensors) are conceptualized, too. LAP bases on a *Logic of Actions* (LA) where *actions* represent *elementary processes* which are used to build *complex processes*, and where *modules* - the formulas of LA - specify process sets. This theoretical base allows us to represent LAP programs by Petri nets. *Joining* them with the Petri net representation of the uncontrolled machine and the workpiece results in a Petri net model of the *controlled machine*. *Analyzing* and *simulating* this model allows us to *virtually put the machine into operation*. Here, we can profit from the formal definition of LAP again, because we can use *direct* and *indirect proofing techniques* of LA to verify the *soundness* and *completeness* of our implementations. If we are satisfied with our results, we *generate code* for programmable controllers from our models which is *put into operation* in reality.

In practical use, the phases will be iterated in order to *examine* and *evaluate* the results of each phase, and to *correct*, *optimize*, and *extend* them.

3. Requirements Analysis and Specification

In order to demonstrate the use of the above explained software development process, we consider a machine (figure 1 shows a scheme) which can realize pick-and-place operations in the following way: a box placed on the left lower base has to be picked up and being placed on the right higher base. The grab can be opened and closed (detected by sensors I0.1 and I0.2), the picker arm can be raised, lowered, and moved horizontally. Sensor I0.3 detects a fully raised arm, I0.4 that the picker arm is on the right base's level, and I0.5 that it is on the lower base's level. Sensors I0.6 and I0.7 detect that the picker arm is above one of both bases. Starting a production process can be initiated with the aid of a start button (input I0.0). By setting actors O0.0 (open) and O0.1 (close) the grab is controlled. The picker arm is raised by actor O0.2, lowered by O0.3, moved to the left by O0.4, and moved to the right by O0.5.

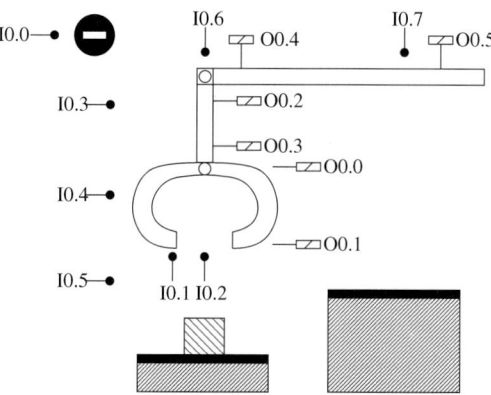

Figure 1: Scheme of a picker arm

In the following, we consider how to develop a controller for this machine in accordance with our software development process. We used the Petri net tool Poseidon developed at the University of Koblenz-Landau for assistance during each step.

4. Design and Specification of Machine and Workpiece

In order to be able to verify a controller program before putting a machine into operation, we need a precise model of the uncontrolled machine to verify the controller's specification against. All former approaches for modeling technical systems have shown that even reputed small machines imply a large number of (internal) states and transitions between them. One can deal with this amount of states by dividing the entire modeling task into sub-tasks. A good strategy is to model the components of a machine first and to relate these components' models to each other, afterwards. We chose ETs nets to model the dynamic behavior of components. We explain this net class during we develop a model of the component which horizontally moves the picker arm.

Like other Petri nets, ETs nets consist of *places*, *transitions*, and *arcs*. We represent the intrinsic states *idle (i)*, *moving left (ml)*, and *moving right (mr)* for horizontally moving by tokens on equally named places. Transitions allow us to change these intrinsic states from i to ml (transition *left (l)*) and back (transition *stop left (sl)*). Two transitions r and sr perform the respective task for the other direction. Figure 2 shows this first basic structure.

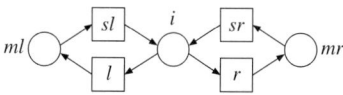

Figure 2: Intrinsic states for the horizontal motion

ETs nets are high-level nets where tokens are carrier for information which is organized in tuples. The first tuple value is the token's timestamp. It is generated by firing transitions and designates the moment a token was put on its place. The timestamps of the tokens on all preplaces of a transition are used to determine the time window during which a transition is enabled. We will discuss this later.

In the elementary net for horizontally moving a picker arm shown in figure 2, neither sensor nor actor representatives are considered. Since both, sensors and actors, have a state they should be modeled by a marked place where a marking-tuple of such a place consists of a timestamp and the current state of the sensor or actor. In the case of binary sensors or actors this state is encoded by 0 or 1.

Firing transitions l and sl does not only change the intrinsic state of the picker arm but comes along with changing the actor for moving to the left which is related to output O0.4 (cf. figure 1). Firing transitions r and sr comes along with changing actor O0.5. Figure 3 visualizes this. Tuples at arcs which consist of variable and constant elements are used to reference the places' markings. Variables are instantiated when transitions fire. In this case, the current moment is assigned to the special variable τ which always specifies the first tuple value at an outgoing arc of a transition. Consequently, after firing such a transition the timestamp of all produced tokens corresponds to the moment of firing. Token values irrelevant for firing a transition are signed by a don't-care symbol (_).

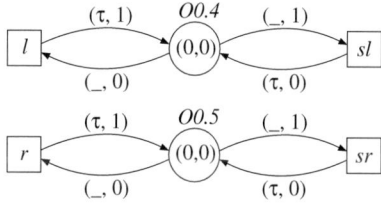

Figure 3: Setting and resetting of actors

The marking shown in figure 3 specifies that at moment zero actors O0.4 and O0.5 are disabled, and consequently only transitions l and r are enabled, because for them the second tuple value at their incoming arcs corresponds to the respective token value. Firing *l* replaces the marking of O0.4 by a new token with a new timestamp and an actor value 1. Firing *r* is comparable with respect to O0.5.

The principles we explained so far are comparable to firing a transition in a Predicate/Transition net [2] except the special variable τ and the interpretation of the first token value as the token's timestamp. Now, we consider at which moments a transition is enabled. For this purpose, we first improve the model of the intrinsic states for the horizontal motion (figure 2) and consider how and when sensor signals change, afterwards.

A token on one of the places *i*, *ml*, or *mr* means that the picker arm is idle, moving to the left, or moving to the right. However, this is an insufficient model if we are also interested in the current position of the picker arm. This can be modeled if tokens on *i*, *ml*, and *mr* contain additional information: a timestamp and the position of the picker arm at the moment designated by the timestamp.

In ETs nets, transitions are only enabled within time windows which are restricted due to the timestamps of the preplaces' tokens and intervals at the incoming arcs of transitions. After its adjacent place is marked, the permeability of an incoming arc of a transition which is labeled [*r*;*l*] is *retarded* for at least *r* and is *limited* at the most for *l* units of time. For example, if a token on a place has a timestamp 5 and an arc from this place to a transition is labeled [4; 6] this arc is permeable from 9 to 11. A transition is enabled when all its incoming arcs are permeable simultaneously. The chronology of events is taken for granted.

The moments at which the picker arm starts or stops to move is not determined by the picker arm itself, but is in the responsibility of an operator or an controller, and so these transitions could fire at any moment form the point of the picker arm. In our model, we represent this by intervals $[0; \omega]$ at the incoming arcs of all transitions. We interpret this as follows: at any moment - from zero up to infinitely many units of time - after the preplace of one of our transitions is marked, this transition is enabled.

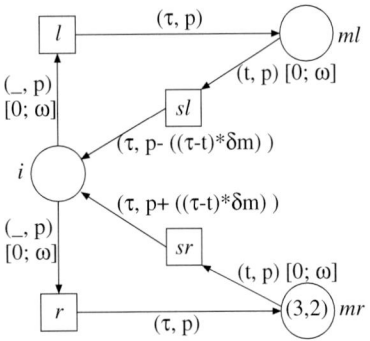

Figure 4: Determining the position of the picker arm

If transitions *l* and *r* fire, the picker arm starts to move to the left or to the right, but its position at this moment is unchanged. Therefore, place *ml* (or *mr*, respectively) is marked by a token with a new timestamp τ and the position *p* of the picker arm before this was moved.

If transitions *sl* or *sr* fire, the new picker arm position has to be computed from the position *p* at which moving started, the duration of motion $(\tau{-}t)$ where *t* is the moment moving started and τ the moment moving stopped, and the distance δm the picker arm moves per unit of time. For example, if transition *sr* fires at moment 6 under the marking shown in figure 4 and if $\delta m = 2$, a token $(6, 8)$ is placed on *i* which specifies that moving the picker arm is stopped at moment 6 at position 8.

While the firing of transitions *l*, *r*, *sl*, and *sr* is not temporary restricted in the picker arm model, this does not hold for the transitions which change sensor signals. In our machine model of figure 1, sensors I0.6 and I0.7 determine the left and the right border for the horizontal motion.

Now, let δv units of length be the distance over which the picker arm can horizontally move, δl the distance over which the left, and δr the distance over which the right sensor is enabled. Figure 5 shows these parameters, which allow us to model the enabling of sensors next, in a scheme.

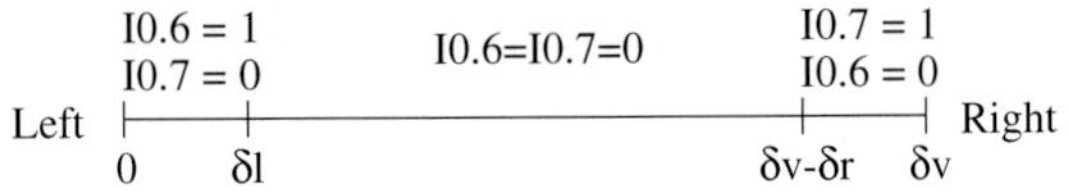

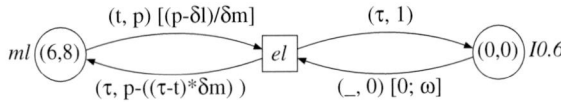

Figure 5: Parameters for the horizontal motion

We consider sensor I0.6, which we represent by a place *I0.6*, in more detail. A tuple on *I0.6* contains a timestamp and the current sensor state (0 or 1). I0.6 can get enabled while the picker arm moves to the left, and disabled while it moves to the right. If moving to the left starts at position *p* (where $p > \delta l$), it takes $((p-\delta l)/\delta m)$ units of time until I0.6 is enabled. In a net model of this behavior, a transition for changing the second tuple value of a token on place *I0.6* from zero to one has to fire $((p-\delta l)/\delta m)$ units of time after place *ml* gets marked by a token with position value *p*. Figure 6 shows a Petri net for this behavior, where transition *el* ($\widehat{=}$ enable left) models the enabling of sensor I0.6 and $[((p-\delta l)/\delta m)]$ is an abbreviation for $[((p-\delta l)/\delta m);((p-\delta l)/\delta m)]$

Figure 6: Enabling of a sensor

In this net, arc (*ml*,*el*) is permeable only $((p-\delta l)/\delta m)$ units of time after *ml* gets marked. If we demand that transitions in an ETs net have to fire before their incoming arcs loose their permeability, the net mirrors exactly the behavior of a real sensor. If the position from which moving to the left starts is less than δl, the interval value at arc (ml,el) is negative, and so this arc is never permeable.

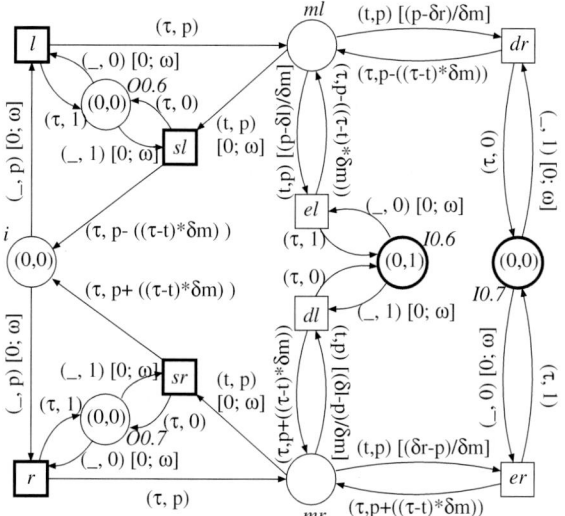

Figure 7: Model for horizontal motion

Figure 7 shows an overall model for horizontally moving the picker arm. Transition dl disables I0.6, er enables I0.7, and dr disables I0.7.

A Petri net model of the grab can be derived from figure 7 by changing the parameters of the net. If we extend it by an additional place for a third sensor and by transitions for enabling and disabling this sensor, we get a model for vertically moving the picker arm.

Modeling technical systems with ETs nets allows to find precise, formal descriptions of such systems. However, several information of component models are needless on a higher level when we combine them to a model of the overall machine. Therefore, we abstract such an ETs net by an interface which we use for the integration step only.

In accordance with [7], we distinguish two kinds of information interchanged between components: *event signals* indicate (the firing of) transitions in a component and *condition signals* are used to request the current (partial) state of components. So, an interface of an ETs net (modeling a technical component) comprises places and transitions of this net only. We subsume such an interface in a pictogram which we supplement by an image of the modeled component. Descriptive names are used for the vertices of the net, and such vertices are omitted which are not needed for combining this component with its environment. Figure 8 shows a pictogram for the horizontal motion.

For modeling the change of a workpiece during a pro-

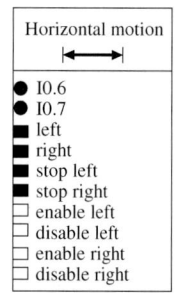

Figure 8: Pictogram for horizontal motion

duction process, we distinguish discrete states of the workpiece only. Although this implies that we could use condition/event nets (the most simple Petri nets), we use ETs nets, because we want to integrate our models later.

Like component nets, we can also represent workpiece nets by pictograms where a single pictogram represents a sequence of steps a workpiece passes through. Alternatives in the production process as well as assemble and disassemble processes are distributed over several pictograms.

Figure 9 shows a pictograph which contains three pictograms specifying the horizontal motion of the workpiece. The effect of our previously modeled component on the workpiece is specified with the aid of event arcs. Condition arcs, usually represented by labeled directed arcs, are not needed throughout this small model. They are used to model dependencies between components.

By combining all Petri nets specifying a pictograph of a machine and resolving all event and condition arcs in accordance with [6], we get an exact and formal Petri net *model of the uncontrolled machine and its influence on a workpiece*. Hereby, we replace condition signals by loops between the participated vertices. An event signal from an event source to an event target is resolved by extending the preset of the target transition by a time window for firing which is only open at moments the source transition fires.

Due to the limited space, we do not show the overall model here. However, it can be derived from our considerations we made above.

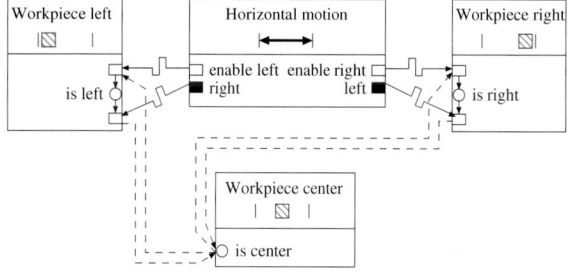

Figure 9: Pictograph for horizontal motion

5. Specification of the Controller

A controller for a machine has to perform two tasks: it has to prevent a machine from getting damaged and it has to realize the intended production process. The first task is usually performed by a so called security controller. In this paper, we focus on the second task.

While a machine is conceptualized, its production processes must be described in detail. Nowadays, such descriptions are usually verbalized and not formally specified. As a first consequence, they might be unprecise. However, a major disadvantage of this is that an informal description of a controller cannot be verified using formal methods.

In [3], a *logic of actions (LA)* is introduced which is extended in [6] by time. The central concept of LA are *processes* consisting of *actions* which *occur* or are *forbidden*. In non-elementary processes actions are organized *before*, *after*, or *coincidentally* to other actions. For example, if a, b and c are actions, then $((a \otimes b) \ominus c)$ is a process where a occurs before b and where both occur after c.

Modules specify process sets: process sets of elementary modules contain exactly one elementary process, and the processes of non-elementary modules are build up by concatenating or selecting the processes of their sub modules. In addition to process relations before, after, and coincident also *alternatives* are expressible. For example, the process set of $[a \otimes [b \boxplus c]]$ is $\{(a \otimes b), (a \otimes c)\}$ and $\boxplus$ is interpreted as an xor-operator.

In the timestamp logic of actions (TiLA) proposed in [6], the execution of actions in modules can be deferred.

While LA modules are implemented by (simple) Petri nets, TiLA modules are implemented by ETs nets. Each such implementation is an initially unmarked Petri net with start transition s (the only transition with empty preset) and goal transition g (the only transition with empty postset). Actions of an implemented module are identified with transitions of the net. *Processes of* such *a net* are firing sequences reproducing the empty initial marking and where s and g fire once. A process P of a module is implemented by a net process Q, if in Q all transitions identified with actions fire in the same sequence as specified in P. In TiLA, the implementation term is extended by the requirement that all transitions fire exactly at the same moments at which their corresponding actions occur. Finally, a net is an implementation of a module, if it implements exactly all processes of the module and no other processes.

Figure 10 shows implementations of simple LA modules. $\{(a \otimes b), (a \ominus b), (a \ominus b)\}$ is the process set of $[a \triangle b]$.

In imperative programming languages like Pascal processes are specified, too. *LAP* is a Pascal-like programming language which also allows to specify concurrent processes and contains commands to interact with a machine. All its *control flow concepts* are mapped on TiLA concepts. Therefore, we can also implement LAP programs by ETs

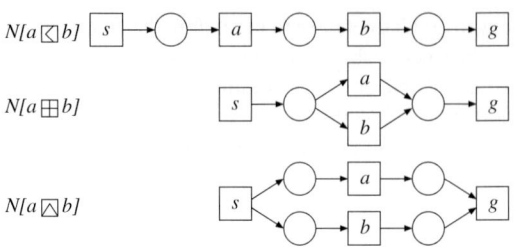

Figure 10: Petri net representation of modules

nets. This implementation together with an interpretation of the commands to interact with the machine allows us to integrate it with the ETs net model of the uncontrolled machine (and its influence on workpieces). This integration step results in an ETs net *model of the controlled machine* which we can use for verification purposes.

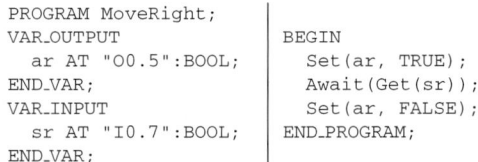

Figure 11: LAP-Program for moving to the right

Before we consider the verification of the controlled plant's model, we have a look at the short LAP program shown in figure 11. A LAP program starts with the declaration of the interface variables. Afterwards, the control flow of the program is specified where constructs like alternatives, iterations, and parallel splitting can also be used. The Set-command changes actors, the Get-command requests sensor information. The vertices of figure 7 and that part of the interface shown in figure 8 drawn with bold lines emphasize the actors and sensors accessible by a controller.

Figure 12 shows an implementation of the LAP program. To determine the ETs net of the controlled machine, we union this net and the net shown in figure 7 where we fuse transitions *S1* and *r* as well as *S2* and *sr*. Transitions *G* and *nG* are connected to place *I0.7* by a loop and are extended by a firing condition so that *G* is enabled if *I0.7* designates a sensor signal 1, and *nG* is enabled otherwise.

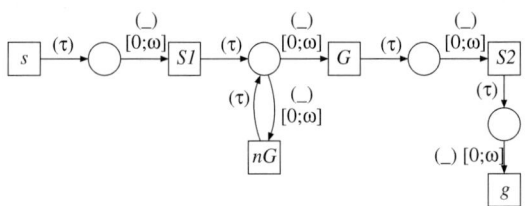

Figure 12: Petri net implementation of the LAP program

6. Virtually Putting a Machine into Operation

In the previous sections we developed a model of the uncontrolled machine and specified a controller to realize our production process. The used models can be developed independently from each other comparable to nowadays machine engineering where the machine and its controller are usually developed independently, too. By integrating both models we derive a model of the controlled machine. This step is comparable to *downloading* the finished software onto the machine controller. *Simulating* this model is like testing during the machine is put into operation. Since we work with models only, we call this step *virtually putting the machine into operation*. Due to the fact that our models are mathematically precise, we can even *verify* several properties and are not restricted to testing.

A module M_1 *fulfills* a module M_2, if each process of M_1 is also a process of M_2 after the process sets were made comparable. Making two modules comparable is called *completion* and means that we extend processes of M_1 by propositions concerning actions of M_2 which are not actions of M_1. M_1 *contradicts* M_2, if the union of their mutually completed process sets is empty. By *direct proofing* we show that a module fulfills another; *indirect proofing* shows contradictions. An LA *realization R is sound* with respect to an LA *specification S* iff R fulfills S, and R is *complete* with respect to S iff S fulfills R. Since LA modules can be implemented by Petri nets, we can now transfer *soundness* and *completeness* on a Petri net level and in this way apply these terms to the net model of the controlled machine.

In [6], net operations for the most important LA terms (completion, fulfill, contradict) are discussed. Finding processes in the participated nets is reduced to finding *T-invariants* and checking whether they are *realizable*, because Petri net processes reproduce the (empty) initial marking. Checking whether an implementation contradicts a certain specification can even be reduced on the question whether there exists a process at all, and this question can already be answered positively if there exists no T-invariant.

Since our method bases on LA, we have a powerful mathematical tool set to verify our models against *process-like specifications*. Moreover, we can also verify *situation-like specifications* with the aid of usual Petri net methods.

7. Implemented Software

Our previously developed specification of the controller contains all information relevant for implementing a controller for a any desired target environment. Since we do this specification with the aid of LAP which is based formally on LA, we have a precise comprehension on how single commands and complex control flows affect the machine. Therefore, we do not have to find a controller which satisfies our specification manually but *generate code* directly from this specification. Since such a generation algorithm can be implemented for several target systems, our software development process is applicable to any currently available programmable controller.

8. Implemented Production Process

Finally, the *machine* has to be *put into operation* and the controller software has to be *tested* in its target system. Since most investments into the engineering and the hardware of the machine have already been made at this moment, each day of this phase is exceptionally expensive which is why this phase should be as brief as possible.

In our software development process in this phase *correctness*, *adequacy*, and *consistency* of all models developed is validated. Still errors cannot be ruled out, but since the controller is checked against models of the machine's dynamics in earlier phases the risk of errors in the logical structure of the controller is reduced. Moreover, errors resulting from manually implemented software are ruled out, because the controller code is generated from the specification automatically. Consequently, our software development process should cause a remarkable reduction of the effort which has to be made to put the machine into operation.

Acknowledgement

This work was supported by the German Research Council (DFG) in the special program KONDISK - Analysis and Design of Technical Systems with Continuous Discrete Dynamics - under grants LA 1042/1-1.

References

[1] **B.W. Boehm**: *Software Engineering*. IEEE Transactions On Computers 25(12), 1976.

[2] **H.J. Genrich, K. Lautenbach**: *System Modelling with High-Level Petri Nets*. Theoretical Computer Science, 13, 1981.

[3] **M. Fidelak**: *Integritätsbedingungen in Petri-Netzen*. PhD-Thesis, University of Koblenz-Landau, 1993.

[4] **K. Lautenbach, C. Simon**: *Erweiterte Zeitstempelnetze zur Modellierung hybrider Systeme*. Technical Report Informatik 3-99, University Koblenz-Landau, Mai 1999.

[5] **J.L. Peterson**: *Petri-Net Theory and the Modelling of Systems*. Prentice-Hall, Englewood Cliffs, New Jersey, 1981.

[6] **C. Simon**: *A Logic of Actions and Its Application to the Development of Programmable Controllers*. PhD-Thesis, University of Koblenz-Landau, 2000.

[7] **R.S. Sreenivas, B.H. Krogh**: *On Condition/Event Systems with Discrete State Realizations*. Discrete Event Dynamic Systems: Theory and Application 1, 1991.

Resource-Oriented Petri Nets in Deadlock Avoidance of AGV Systems

Naiqi Wu
Department of Mechatronics Engineering
Guangdong University of Technology
Guangzhou 510090, P. R. CHINA
nqwu@yahoo.com

Mengchu Zhou
Dept. of Electrical and Computer Engineering
New Jersey Institute of Technology
Newark, NJ 07102-1982, USA
Zhou@njit.edu

ABSTRACT-This paper presents a colored resource-oriented Petri net (CROPN) modeling method to deal with conflict and deadlock arising in Automated Guided Vehicles (AGV) systems. The work can be viewed as the continuation of some of the authors' previous work. Some unique features in AGV systems require the further investigation into their deadlock avoidance using CROPN models. The proposed approach can easily handle both bidirectional and unidirectional paths. Bidirectional paths offer additional flexibility, efficiency and less cost than unidirectional paths. Yet they exhibit more challenging AGV management problems. By modeling nodes with places and lanes with transitions, one can easily construct a CROPN model for dynamic AGV systems with changing routes. A control policy suitable for real-time control implementation is then proposed.

I. Introduction

It is well known that deadlock may occur due to limited resources in automated manufacturing systems (AMSs), leading to a system wide standstill. This paper classifies deadlock into two types. One is caused by the competition by parts for such manufacturing resources as machines and buffers. The other type is due to the competition for nodes and lanes by AGVs when multiple AGVs are used in material handling systems (MHS). Several methods are developed to synthesize a live Petri net for AMS so that the resulting PN controllers are deadlock-free [1-2]. They belong to deadlock prevention techniques. Deadlock detection and recovery techniques allow deadlock to occur and then detect and recover from it [3-4]. Deadlock avoidance techniques dynamically assign the resources in the system such that the system is deadlock-free [5-12].

The above mentioned studies addressed the first type of deadlock in AMS but not the second one. When multiple AGVs are used, some problems may arise, e.g., blocking, conflict, deadlock, and collision [13-14]. To improve the productivity and resource utilization, it is desired to have as many parts as possible in a system. However, the more parts in AMS, the more likely deadlocked. The parts need to be delivered from a place to another by MHS. Often, only a few AGVs are available in an AGV system mainly due to their high-cost and easily satisfied transportation demand. The total number of AGVs likely remains constant. While the route of each part type is known in advance, the routes of an AGV frequently change based on real-time transportation requests. The theory for handling the first type of deadlock needs to be further studied for the second type.

The most widely used technique for vehicle management of AGV systems is zone control. Guided paths are divided into several disjoint zones. A zone can accommodate only one AGV at a time. The guided paths may be unidirectional or bidirectional. The deadlock problem for an AGV system with unidirectional guided paths is studied using the zone control [15-16]. Their algorithms predict deadlock based on the current routings of the AGVs, and then make decisions to avoid deadlocks. However, there is no effective method to deal with the conflict and deadlock in an AGV system with bidirectional paths. AGVs need to compete for not only zones as in unidirectional system but also lanes, increasing the control complexity.

One way to solve this problem is to carefully design the configuration of an AGV system so that its management is simplified. For example, tandem configuration [17] partitions all the stations into non-overlapping, single vehicle, closed loops with additional *pickup* and *deposit* locations provided as an interface between adjacent loops. Each station is assigned to only one loop, and each loop is served by exactly one AGV. A segmented flow approach [18] partitions the paths into non-overlapping segments. Each segment is comprised of one or more zones and is served by a single AGV. Transfer buffers are located at both ends of each segment and serve as interface devices between the segments. The buffers are designed to be able to serve both sides of the segments simultaneously. Therefore, the only possible conflicts are the use of the interface buffers. Their drawback is that a load might have to be handled by two or more AGVs before reaching its destination [14]. In addition, extra pickup and deposit stations are required to interface with each other, increasing the system scale and cost.

The conflict detection problem is studied for bidirectional systems using colored PN [14]. An approach is presented to detect the competition for a lane by AGVs in the adjacent zones. However, this is not enough to control the system. A conflict may occur when two AGVs in non-adjacent zones compete for some lanes. Furthermore, when such a conflict occurs, it in fact leads to deadlock. Thus, a more comprehensive approach is needed, motivating this present work.

A colored resource-oriented Petri net (CROPN) was originally developed to deal with the first type deadlock in AMS [10]. This paper further develops it to model AGV systems to derive a deadlock avoidance policy. The next section presents CROPN modeling of AGV systems. Sections 3-5 present conflict and deadlock-free conditions and control policy in different cases. Section 6 shows the application of the proposed approach. Section 7 concludes the paper.

II. SYSTEM MODELING WITH CROPN

A. Finite Capacity PN

A PN is a particular kind of directed graph containing places and transitions. It can be denoted by a quadruple PN = (P, T, I, O), where $P = \{p_1, p_2, ..., p_m\}$ is a finite set of places, $T = \{t_1, t_2, ..., t_n\}$ is a finite set of transitions, $P \cup T \neq \emptyset$, $P \cap T = \emptyset$, I: $P \times T \rightarrow \{0, 1\}$ is an input function, and O: $P \times T \rightarrow \{0, 1\}$ is an output function. We use $^\bullet t$ ($^\bullet p$) to denote the set of input places of transition t (the set of input transitions of place p) and $t^\bullet$ ($p^\bullet$) the set of output places of t (the set of output transitions of p).

A marking or state M: $P \rightarrow N = \{0, 1, 2, ...\}$ describes the distribution of tokens in PN. $M(P) = (M(p_1), M(p_2), ..., M(p_m))^T$

is an m×1 vector. Marking M_0 is referred to as the initial marking of a PN, representing the initial state of its modeled system.

In finite capacity PN, a place is limited to hold a finite number of tokens. They are adopted since a zone (also called node in this paper) and lane allow only a limited number of AGV at a time. Let $K(p)$, $p \in P$, denote the capacity of p, which is the maximum number of tokens that p can hold at any time.

Definition 2.1: A transition $t \in T$ in a finite capacity PN = (P, T, I, O) with marking M is said to be enabled if

$$M(p) \geq I(p, t), \forall p \in P \quad (1)$$

and

$$K(p) \geq M(p) - I(p, t) + O(p, t) \quad (2)$$

Definition 2.1 means that t is enabled and can fire if all the places in •t have enough tokens and all the places in t• have enough free spaces. When condition (1) is satisfied, t is process-enabled. When (2) is satisfied, t is resource-enabled. Thus, t is enabled only if it is both process and resource-enabled. Firing an enabled transition $t \in T$ at M changes M into M' according to

$$M'(p) = M(p) - I(p, t) + O(p, t) \quad (3)$$

A sequence of firings results in a sequence of markings. A marking M is said to be reachable from M_0 if there exists a sequence of firings that transforms M_0 to M. The set of all possible markings reachable from M_0 is denoted by $R(M_0)$.

B. Modeling Resources in a System

One important problem is how to model the resources in a system. Often, places in PN are used to model resources. In order to reduce the complexity of the resultant PN, there is a one-to-one relation between the resources in the system and places in CROPN [10]. There are two types of resources in an AGV system with bidirectional paths: nodes and lanes. An example semiconductor plant's AGV system is shown in Fig. 1 consisting of 11 nodes and 14 bi-directional lanes. This paper models nodes by places and lanes by transitions, respectively although both may be viewed as resources. A node in an AGV system is an intersection of several paths or a station where an AGV can load a part to a machine or pick up a part. Thus, we can regard a node as the resource just as a machine or buffer in CROPN presented in [10]. In Fig. 2, p represents a node, a token in it represents an AGV in the node. Suppose that a node can hold only one AGV at a time. Then $K(p) = 1$. Place p's multiple inputs and outputs represent that AGV can enter and leave the node along different paths. Since only one AGV can travel on a lane at a time, all the arcs are single if they exist.

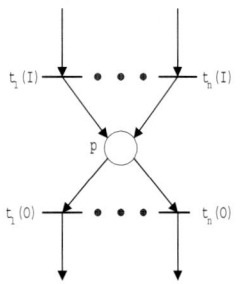

Fig. 2. PN model for a node.

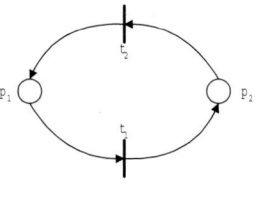

Fig. 3. PN model for a bidirectional lane between two nodes

A lane in an AGV system is a path between two adjacent nodes. An AGV can travel on a bidirectional lane in both directions. It is modeled by two transitions as shown in Fig. 3. A token representing an AGV can flow from $p_1 \rightarrow p_2$ by firing t_1 or $p_2 \rightarrow p_1$ by firing t_2, implying that the lane is assigned to an AGV from $p_1 \rightarrow p_2$ or $p_2 \rightarrow p_1$. Note that when p_1 and p_2 are both marked, neither transition is enabled according to enabling condition (2), thereby avoiding any collision on the lane. If a lane is one-way, e.g., from p_1 to p_2 only, then only t_1 is needed to model the lane.

C. Modeling the Traveling Processes of AGVs

An AGV can pick up a part in a node and travel on a path through one or more nodes. After reaching the destination node, it unloads the part. In general, it begins from a node and stops at another. According to different tasks assigned to AGVs, their routings change dynamically. The process of delivering a part from a node to another can be modeled by a sequence of places and transitions. Figure 4 shows the PN models for two processes of AGVs in Fig. 1 (V_1: $8 \rightarrow 7 \rightarrow 6 \rightarrow 11 \rightarrow 2 \rightarrow 1$ and V_2: $10 \rightarrow 9 \rightarrow 11 \rightarrow 4 \rightarrow 5 \rightarrow 6 \rightarrow 7$). We call such a model a subnet that models an AGV travelling from one node to another. Each subnet begins and ends with places (maybe the same place). Let $PN_i = (P_i, T_i, I_i, O_i)$ denote the subnet for V_i. Only its beginning place is initially marked with a token, i.e., its initial marking $M_{i0}=(1, 0, ..., 0)^T$. When its ending place is marked, V_i reaches its destination. The final marking, also called destination marking, $M_{id}=(0, ..., 0, 1)^T$.

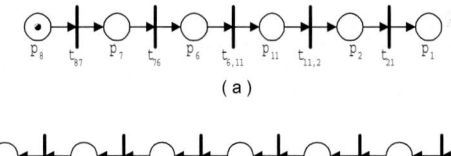

Fig. 4. Two example subnets for single AGV delivery processes.

Definition 2.2: PN = (P, T, I, O) for an AGV system is said to be the union of two subnets $PN_i = (P_i, T_i, I_i, O_i)$ and $PN_j = (P_j, T_j, I_j, O_j)$ for AGV V_i and V_j, if

$$P = P_i \cup P_j, \text{ and } T = T_i \cup T_j \quad (4)$$

$$M_0(p) = \begin{cases} M_{i0}(p), & \text{if } p \in P_i \\ M_{j0}(p), & \text{if } p \in P_j \\ M_{i0}(p) + M_{j0}(p), & \text{if } p \in P_i \cap P_j. \end{cases} \quad (5)$$

$$I(p, t) = \begin{cases} I_i(p, t), & \text{if } p \in P_i, t \in T_i \\ I_j(p, t), & \text{if } p \in P_j, t \in T_j \\ 0, & \text{otherwise} \end{cases}$$

$$O(p, t) = \begin{cases} O_i(p, t), & \text{if } p \in P_i, t \in T_i \\ O_j(p, t), & \text{if } p \in P_j, t \in T_j \\ 0, & \text{otherwise} \end{cases}$$

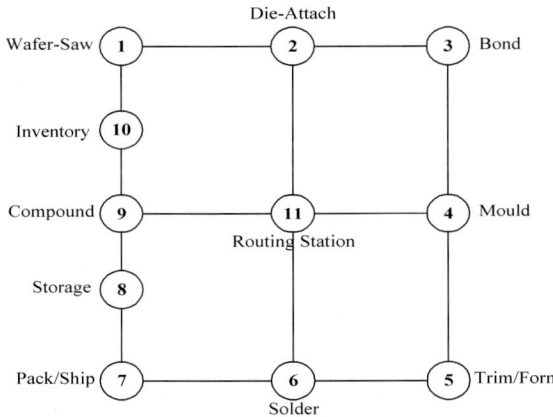

Fig. 1. An AGV system in a semiconductor plant

If $p_1, p_2 \in P_i \cap P_j$, the union of PN_i and PN_j merges two transitions $p_1 \to p_2$ ($p_2 \to p_1$) in PN_i and $p_1 \to p_2$ ($p_2 \to p_1$) in PN_j into one because they stand for the same lane. Thus, there is at most one transition between two places along the same direction in the union PN. By defining the union of multiple subnets in the same way, we obtain the PN model for the whole AGV system, which describes the traveling processes of all AGVs. The union PN of the two subnets in Fig. 4 is shown in Fig. 5.

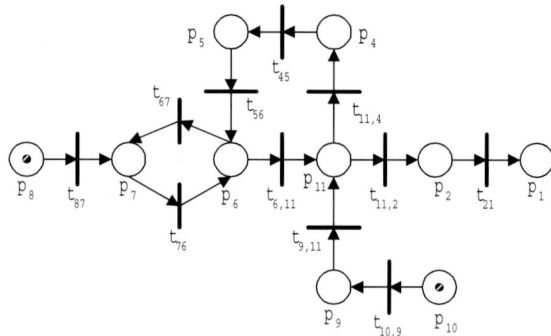

Fig. 5. The union of two subnets

Consider a token, i.e., AGV V_1 or V_2 in p_{11} (Node 11) in Fig. 5. It enables either $t_{11,2}$ (V_1) or $t_{11,4}$ (V_2) depending on which AGV the token represents. Only one of two transitions should be enabled. Unfortunately, ordinary PN cannot describe this. To model this feature, colors are introduced. Different from most existing work where colors are first defined for tokens in places [19], this work defines the colors for transitions first.

Definition 2.3: Define the color of transition $t_i \in T$ as b_i.

It states that each transition in the model is associated with a unique color, i.e., $b_i \neq b_j$ if $i \neq j$.

Definition 2.4: If $t_i \in p^\bullet$, define the color of a token in p that enables t_i as b_i.

For example, in Fig. 5, if the token in p_{11} stands for AGV V_1, it enables $t_{11,2}$. Thus its color is $b_{11,2}$. Note that we do not use colors to identify token types (or AGV), instead to describe the node for each AGV to go to. In this way, we model the dynamical processes of an AGV system. The resulting PN is named colored resource-oriented PN (CROPN).

Observing the CROPN shown in Fig. 5, we call $U = \{p_{11}, t_{11,4}, p_4, t_{45}, p_5, t_{56}, p_6, t_{6,11}, p_{11}\}$ a circuit and $Y = \{p_6, t_{67}, p_7, t_{76}, p_6\}$ a cycle denoted by $\{p_6, p_7\}$ for short. A "cycle" is a special "circuit" which models a bidirectional lane between two nodes as shown in Fig. 3. Let U denote a circuit and Y a cycle. Consider the PN model in Fig. 3. When two tokens with opposite directions are in p_1 and p_2, respectively, there is a conflict for the lane and at the same time a circular wait is formed. In other words, in a cycle, a conflict and deadlock occur simultaneously since two AGVs compete for both nodes and lanes in it. It is clear that if there is no deadlock in a cycle, there is no conflict in it either. Dislike a cycle, AGVs in a circuit compete for only nodes. In this sense, if we can avoid deadlocks by using the CROPN, we avoid conflicts as well.

Let M_{id} denote the destination marking of a subnet$_i$ and M_d the destination marking for the union CROPN, respectively. The below equation is similar to (5) for two subnets' union and can be easily extended to the union of multiple subnets.

$$M_d(p) = \begin{cases} M_{id}(p), & \text{if } p \in P_i \\ M_{jd}(p), & \text{if } p \in P_j \\ M_{id}(p) + M_{jd}(p), & \text{if } p \in P_i \cap P_j. \end{cases} \quad (6)$$

Definition 2.5: If a CROPN of an AGV system with initial marking M_0 can reach the destination marking M_d, the CROPN is deadlock-free.

Note that the CROPN of an AGV system is constructed dynamically according to the dispatching and routing of the AGVs. Each time when tasks are assigned to AGVs and their routings are determined, a CROPN is constructed. If every time we can control the CROPN such that it is deadlock-free, the system is conflict and deadlock-free. Note that it is easy to construct a CROPN dynamically since the configuration of an AGV system is known in advance.

III. DEADLOCK AVOIDANCE IN CIRUITS

A. Deadlock-Free Condition in Circuits

Let $P(U) \subset P$ denote the set of places on circuit U and $T(U) \subset T$ the set of transitions on U. Further, let $K(U) = \sum_{p \in P(U)} K(p)$ denote U's capacity.

Definition 3.1: If a token in place $p \in P(U)$ has color b_i whose corresponding transition $t_i \in T(U)$ is on circuit U, it is a cycling token, otherwise a leaving token of U.

After firing t_i, a cycling token remains in U while a leaving token leaves U, releasing a space in U. Let $M(U)$ denote the total number of cycling tokens and $M'(U)$ the total number of tokens in U at the current marking M. Clearly, $M(U) \leq M'(U)$.

Definition 3.2: $S(U) = K(U) - M'(U)$ is called the numbers of free spaces available in U at the current marking M.

Circuits are interactive if they share common places. For example, p_2 is shared by $U_1 = \{p_1, t_1, p_2, t_6, p_4, t_3, p_1\}$ and $U_2 = \{p_2, t_4, p_5, t_7, p_6, t_5, p_3, t_2, p_2\}$ in Fig. 6(a), and p_2 and p_5 shared by $U_3 = \{p_1, t_1, p_2, t_4, p_5, t_6, p_4, t_3, p_1\}$ and $U_4 = \{p_2, t_4, p_5, t_7, p_6, t_5, p_3, t_2, p_2\}$ in Fig. 6(b). If U_i and U_j are interactive and there is a free space in the common places, then both $S(U_i) > 0$ and $S(U_j) > 0$. Such common space is shared by U_i and U_j. In Fig. 6(a), if p_2 is empty, U_1 and U_2 share a space. Let $P_{ij} = P(U_i) \cap P(U_j)$ be the set of common places of U_i and U_j, $P_i = P(U_i) - P_{ij}$, $P_j = P(U_j) - P_{ij}$.

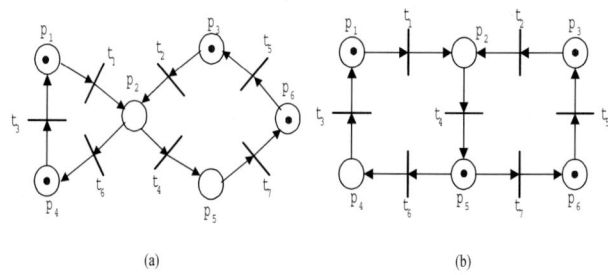

Fig. 6. PN models with two circuits.

Definition 3.3: Each of U_i and U_j is said to have a non-shared free space if 1) two spaces are in P_{ij}; 2) one space in P_i and another in P_{ij}; 3) one space in P_i and another in P_j; or 4) one space in P_{ij} and another in P_j.

By Definition 3.3 we mean that when the condition is met every circuit can be "allocated" a free space. For example, in Fig. 6(a), if places p_2 and p_6 are empty and all other places are occupied, then the spaces in p_2 and p_6 can be allocated to U_1 and U_2, respectively such that each of U_1 and U_2 has a non-shared space. The situation is similar if there are more than two circuits. If the free spaces in the PN in marking M are distributed such that every circuit can be allocated one or more free spaces, then we say that every circuit has at least one non-shared space.

Assume that the number of AGVs is less than the number of places in the CROPN of an AGV system since otherwise no vehicle can move. Also assume that no token (an AGV) stays in a place (node) forever since otherwise it should be moved out of the system. A sufficient condition for deadlock-free operation of an AGV system in the existence of circuits can be stated below.

Theorem 3.1: The PN model for the AGV system is live if every U_i in it has at least one non-shared space in any marking M.

Proof: See [20].

Note that in the PN in Fig. 6(a) if there is only one free space it may be deadlocked. For example, if only p_2 is empty (the two circuits share the only free space) and the token in p_3 needs to go to p_4. Firing t_2 leads to deadlock. It is the condition in Thoerem 3.1 that allows the token in p_2 in Fig. 6(a) to go to either circuit in the net. Thus the net is made to be live.

B. Control Law for Circuits

Before we present the control law we need some definitions.

Definition 3.4: A transition t in PN is said to be controlled if firing t or not is determined by a control law when t is both process and resource-enabled.

When a controlled transition t can fire according to the control law, t is said to be control-enabled.

Definition 3.5: A PN is said to be a controlled PN if at least one transition in it is controlled.

Therefore, a controlled transition in a PN can fire if it is process, resource and control-enabled. To effectively avoid deadlocks in an AGV system, the resources in the system should be assigned appropriately when more than one AGV complete for the same node.

Definition 3.6: t is said to be the input transition of circuit U if $t \notin U$ but $t^\bullet \in U$.

A circuit U in the CROPN may have two or more input transitions, we use $T_I(U) \subset T$ to denote the set of input transitions of U. Then from Theorem 3.1 we have:

Theorem 3.2: The PN model for the AGV system is deadlock-free if the condition given in Theorem 3.1. is satisfied by a) Initial marking, and b) a marking reached due to the firing of $t \in T_I(U_i)$ for any circuit U_i.

By this control law, we need to observe the state of the system on-line, calculate S(U) for every U, and control the firing of transitions in $T_I(U)$.

To implement the control law, we need to identify all the circuits in the CROPN of an AGV system and this must be done in real-time. If a PN has many circuits, it may be time consuming to do this. The below result from Theorem 3.1 can be used such that we need identify only a fewer number of circuits.

Corollary 3.1: In a CROPN of an AGV system, a necessary condition for a deadlock to occur in circuit U at marking M is

$$S(U) = 0 \qquad (7)$$

Corollary 3.1 means that deadlock can occur in U only when the number of tokens in U equals the number of places on U. This implies that the number of AGVs must be greater than or equal to the number of nodes on a circuit. Considering only fewer AGVs in the system than the number of nodes, deadlock can occur in only those circuits whose place count is less than or equal to the number of AGVs. Because the configuration and the number of AGVs in the system are known in advance, we can identify all these circuits in advance. For example, in the system shown in Fig. 1, a circuit can be deadlocked only if there are at least four AGVs. When there are four AGVs, the possible circuits in which deadlock can occur are $\{2 \to 3 \to 4 \to 11 \to 2\}$, $\{2 \to 11 \to 4 \to 3 \to 2\}$, $\{4 \to 5 \to 6 \to 11 \to 4\}$ and $\{4 \to 11 \to 6 \to 5 \to 4\}$. Thus, when a CROPN is constructed for a system, we need to ensure only those circuits in the CROPN to be deadlock-free. In fact for any two-AGV system, e.g., the one in Fig. 1, a circuit is guaranteed to have three or more places (otherwise becoming a cycle); thereby it will never be deadlocked.

IV. DEADLOCK AVOIDANCE IN THE CYCLES

A. Single Cycle

A single cycle is formed by two AGVs competing for a lane between two adjacent nodes, one AGV travels in one direction, the other in the opposite direction. The CROPN shown in Fig. 5 contains a cycle $Y = \{p_6, p_7\}$. If both AGVs take the lane, a collision (conflict) occurs. If both enter the nodes (both p_6 and p_7 have a token in the cycle Y in Fig. 5), a deadlock occurs. Thus, we have the below result.

Lemma 4.1: It is deadlock-free in a single cycle formed by routings of two AGVs in a CROPN iff there is at most one token in the cycle in any marking M.

Clearly, if there is no deadlock in a single cycle, two AGVs never occupy the two adjacent nodes in the cycle simultaneously. Thus, the conflict of using the lane never occurs.

B. Cycle Chain

Definition 4.1: In a CROPN of an AGV system, if p_1 and p_2 form a cycle, p_2 and p_3 form a cycle, …, and p_{n-1} and p_n form a cycle (n>2), these n places form a cycle chain.

A cycle chain example is shown in Fig. 8. It is formed by the AGV routings V_1: $p_8 \to p_2 \to p_3 \to p_4 \to p_5 \to p_{10}$ and V_2: $p_{11} \to p_5 \to p_4 \to p_3 \to p_2 \to p_1 \to p_7$. Firing t_{82} in Fig. 8 leads the token standing for V_1 in p_8 to p_2. Firing $t_{11,5}$ leads the token standing for V_2 to p_5. Now although either of cycles $Y_1 = \{p_2, p_3\}$ and $Y_2 = \{p_4, p_5\}$ has only one token, deadlock is inevitable. To progress, t_{23} and t_{34}, t_{54} and t_{43} or t_{32} and t_{54} may fire, leading to two tokens in cycle Y_2, Y_1 or $Y_3 = \{p_3, p_4\}$. All these situations are deadlocked.

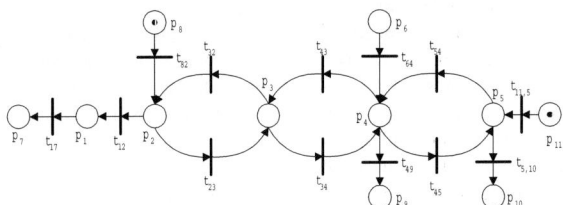

Fig. 8. The CROPNs containing a cycle chain.

It should be pointed out that a number of cycles form a cycle chain if there exist two AGVs that go through all the cycles in different direction. For example in Fig. 8, if three AGVs have routes $p_8 \to p_2 \to p_3 \to p_4 \to p_9$, $p_3 \to p_4 \to p_5 \to p_{10}$, and $p_{11} \to p_5 \to p_4 \to p_3 \to p_2 \to p_1 \to p_7$, then Y_1 and Y_3 form a cycle chain; and Y_2 and Y_3 form another.

Lemma 4.2: In a CROPN of an AGV system, if there is a cycle chain formed by the routes of two AGVs, the chain is deadlock-free iff there is at most one token in it in any marking.

A cycle chain may be formed by routes of more than two AGVs. Assume that there is another AGV V_3 with route $p_6 \to p_1 \to p_2 \to p_3 \to p_9$. Then two cycle chains $H_1 = \{Y_1, Y_2, Y_3\}$ and $H_2 = \{Y_4, Y_1\}$ as shown in Fig. 9 are formed by the routes of these three AGVs, where $Y_4 = \{p_1, p_2\}$.

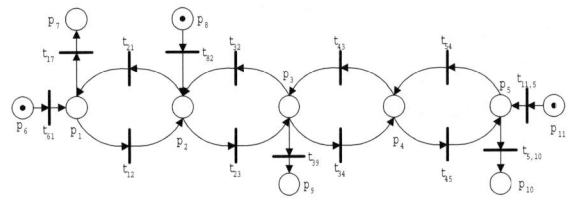

Fig. 9. A cycle chain formed by routes of three AGVs

It should be pointed out that the routes of more than two AGVs can generate more complicated structures than cycle chains. The cycle chains formed by more than two AGVs may have overlaps. For example, in Fig. 9, H_1 and H_2 overlap at Y_1.

Definition 4.2: A subnet of CROPN for AGV systems is called an interactive cycle chain if it is formed by two or more cycle chains and each of the chains overlaps with at least one of the other chains.

We call an interactive cycle chain subnet a cycle chain subnet for short and denote it w. Assume that a w is formed by n cycle chains H_1, H_2, $\cdots$, and H_n and there are m overlap segments D_1, D_2, $\cdots$, and D_m in w. Let $F_{Di}(w) = \{V_{fDi}, V_{bDi}, V_{cDi}; D_i\}$ and $F_{Hi} = \{V_{fHi}, V_{bHi}, V_{cHi}; H_i\}$, where V_f is the set of AGVs that move forward on D_i or H_i, V_b is the set of AGVs that move backward on D_i or H_i, and V_c is the set of AGVs that cross D_i or H_i. Further we let

$$\delta(F_\bullet(w)) = \begin{cases} 1, & \text{if Card}(V_f) > 0 \text{ and Card}(V_b) > 0 \\ 0, & \text{otherwise} \end{cases} \quad (8)$$

and $g(w) = \sum_{i=1}^{n} \delta(F_{Hi}(w)) + \sum_{i=1}^{m} \delta(F_{Di}(w))$. We assume, without loss of generality, that there are no same cycle chains in a w, or $H_i \neq H_j$, if $i \neq j$, then we have the following result.

Theorem 4.1: A subnet of cycle chains formed by n cycle chains in a CROPN of an AGV system is deadlock-free iff the following condition holds in any marking M.

$$g(w) = 0 \quad (9)$$

Notice that for cycle chains Lemmas 4.1 and 4.2 are the special cases of Theorem 4.1. Thus we can use condition (9) in Theorem 4.1 to avoid deadlock and conflict in all cycle chains.

A cycle chain can also be formed by the route of a single AGV. In fact, if an AGV goes to some nodes and comes back by the same path (with opposite direction), a cycle chain is formed. Clearly, no conflict and deadlock will occur in such a cycle chain since no other AGV enters the cycle chain.

Definition 4.2: t is said to be the input transition of cycle chain H if $t \notin H$ but $t^\bullet \in H$.

Let $T_I(H) \subset T$ denote the set of input transitions of H. To avoid deadlock and conflict in cycle chains is to control the firings of transitions in $T_I(H_i)$ such that condition (9) is always satisfied. It is easy to calculate $g(H_i)$ and thus simple to implement. An algorithm of complexity $o(|T|^2)$ can be derived to identify the set of all single cycles and all cycle chains in the CROPN to facilitate the real-time implementation [20].

V. DEADLOCK AVOIDANCE IN THE OVERALL SYSTEM

In a CROPN of an AGV system as the overall system, if there is no interaction between circuits and cycle chains, then we can control the overall system by using the control laws presented in the last two sections to control circuits and cycle chains, respectively. However, the circuits and cycle chains may interact with each other. This makes the problem more complicated. A subnet with interaction of a circuit and cycle chain is shown in Fig. 10. This subnet contains the cycle chain made of places p_{1-5} and a circuit $U = \{p_2, t_{2,12}, p_{12}, t_{12,4}, p_4, t_{43}, p_3, t_{32}, p_2\}$. Places p_{2-4} are on both the cycle chain and U.

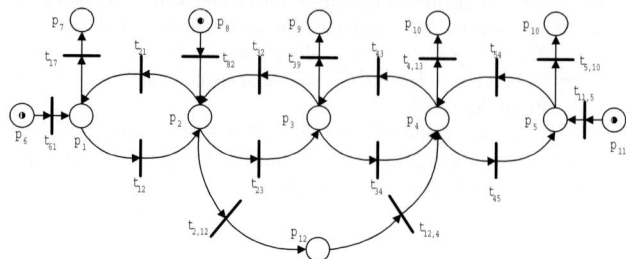

Fig. 10. A subnet with interaction of a circuit and cycle chain

Let us consider condition (9). By this condition, a shared path can be used only in one direction, as analogous to a time-shared system. It requires switching from one direction to another appropriately. In Fig. 10, if the direction from p_5 to p_1 is *active* (i.e., some AGV will occupy them at some time), then t_{43} and t_{32} are active, and so is U. If the direction from p_1 to p_5 is active, t_{23} and t_{34} are active but this time U is not. It is easy to show that when the circuit is active, circular wait may occur. Otherwise no circular wait will occur. Thus, we can prove the below result using Theorems 3.1 and 4.1.

Theorem 5.1: A subnet in a CROPN of an AGV system with interaction of circuits and cycle chains is deadlock-free if the following conditions hold in any marking.
1) The conditions given in Theorem 3.1 and (9) hold, if the circuits are active and
2) The condition (9) holds, if the circuits are not active.

Because in a two AGV system a circuit will never be deadlocked we don't need to consider the deadlocks in circuits. However, if there are multiple AGVs and we make the AGV assignments so that there are only cycle chains and circuits, then result given in Theorem 5.1 can be applied to solve the problem.

VI. ILLUSTRATIVE EXAMPLE

Consider the AGV system in Fig. 1. Two requests and assignments of AGVs are given one after another:
1) V_1: $8 \to 7 \to 6 \to 11 \to 2 \to 1 \to 10$
   V_2: $10 \to 9 \to 11 \to 4 \to 11 \to 9$
2) V_1: $10 \to 1 \to 2 \to 3 \to 4 \to 11 \to 9$
   V_2: $9 \to 11 \to 4 \to 5 \to 6 \to 7 \to 8$

The CROPN for Case 1 is shown in Fig. 11. There are two cycles $Y_1 = \{p_4, p_{11}\}$ and $Y_2 = \{p_{11}, p_9\}$, forming a cycle chain. It is easy to see that the cycle chain is due to the route of AGV V_1. Thus, there will be no conflict and deadlock in it. In fact, even transition $t_{10,9}$ is fired first and the token representing V_2 enters into p_9 in the cycle chain, t_{87}, t_{76} and $t_{6,11}$ can still fire and the token in p_8 (representing V_1) can enter p_{11}. Then we can fire $t_{11,2}$ and the token leaves p_{11} for p_2. If the token representing V_2 is in p_{11}, then $t_{6,11}$ cannot fire according to the transition enabling and firing rule. Therefore, both AGVs can reach the destination with no deadlock.

The CROPN for Case 2 is shown in Fig. 12. There are also two cycles $Y_1 = \{p_4, p_{11}\}$ and $Y_2 = \{p_{11}, p_9\}$, forming a cycle chain H. H is due to the routes of V_1 and V_2. Thus transition firings have to follow the control law specified in Theorem 4.1 to avoid deadlock. Theorem 4.1 requires that $g(H)=0$. Since V_2 is already in H, we have to limit another token to enter H. This leads to the below transition firing order. Transitions $t_{10,1}$, t_{12} and t_{23} can fire any time. t_{34} can fire only after $t_{9,11}$, $t_{11,4}$ and t_{45} fire. This way no conflict and deadlock will occur.

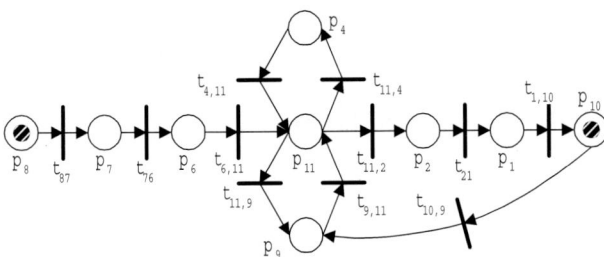

Fig. 11. The CROPN for Request 1

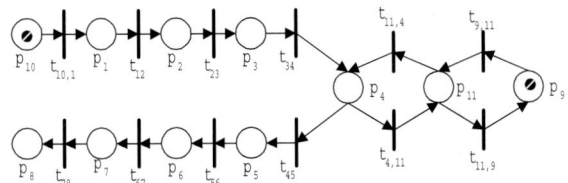

Fig. 12. The CROPN for Request 2.

VII. CONCLUSIONS

This paper has developed a PN model, called CROPN, to describe systematically the AGV travelling processes in an AGV system. The conflict and deadlock situations are well modeled by the circuits, cycles, cycle chains, and their interactions in CROPN. An effective control policy for conflict and deadlock avoidance is proposed. It can be used in the real-time control. It should be noted that the literature results, e.g., in [7-8], may apply to the discussed AGV systems but lead to much weaker results, e.g., only one AGV can be allowed in the system [20]. Besides conflict and deadlock, blocking is another important issue to be addressed in the operation in the AGV system. Blocking may be caused by inappropriate resource assignment in the traveling processes of the AGVs. It can also be caused by inappropriate routing. When a vehicle reaches its destination it may stay there until a new task is assigned to it. During its stay, it may block other vehicles. To solve this problem, it may be necessary to do the routing and control concurrently and develop sophisticated techniques for that. This is the future work to be done. The extension of the results should be performed for the cases where a node allows to park multiple AGVs. In CROPN, this implies that a place's capacity is more than one. The theory needs to be extended to the cases where multiple AGV's routes may form very complicated chains.

Acknowledgement

This work is partially supported by the Chinese NSF under grant 69974011 and New Jersey State Commission on Science and Technology. Anonymous reviewers helped us improve this paper's presentation.

References

1. M. -D. Jeng, and F. DiCesare (1995). "Synthesis Using Resource Control Nets for Modeling Shared Resource Systems," *IEEE Trans. Robotics and Automation,* 11(3), pp. 317-327.
2. M. C. Zhou, F. DiCesare, and A. Desrochers, Hybrid methodology for synthesis of Petri nets models for manufacturing systems, *IEEE Trans. on Robotics and Automation*, vol. 8, no. 3, 350-361, 1992.
3. R. A. Wysk, N. S. Yang, and S. Joshi, Detection of deadlocks in flexible manufacturing cells, *IEEE Trans. on Robotics and Automation*, vol. 7, 853-879, 1991.
4. Y. T. Leung and G. Shen, Resolving deadlocks in flexible manufacturing cells, *Journal of Manufacturing Systems*, vol. 12, no. 4, 291-304, 1993.
5. Z. A. Banaszak and B. H. Krogh, Deadlock avoidance in flexible manufacturing systems with concurrently competing process flows, *IEEE Trans. on Robotics and Automation*, vol. 6, 724-734, 1990.
6. E. Roszkowska and J. Jentink, "Minimal restrictive deadlock avoidance in FMS," In: *Proc. of The Second European Control Conf. ECC '93*, Groningen, Holland, volume 2, pages 530-534, 1993.
7. Fanti, M. P., B. Maione, S. Mascolo, and B. Turchiano, "Event-based feedback control for deadlock avoidance in flexible production systems," *IEEE Trans. Robotics Automat.*, Vol. 13, pp. 347-363, 1997.
8. A. Lawley, Deadlock avoidance for production systems with flexible routing, *IEEE Trans. on Robotics and Automation*, vol. 15, no. 3, 497-509, 1999.
9. Viswanadham, Y. Narahari, and T. L. Johnson, Deadlock prevention and deadlock avoidance in flexible manufacturing systems using Petri net models, *IEEE Trans. on Robotics and Automation*, vol. 6, 713-723, 1990.
10. N. Q. Wu, Necessary and sufficient conditions for deadlock-free operation in flexible manufacturing systems using a colored Petri net model, *IEEE Trans. on Systems, Man, and Cybernetics, Part C*, vol. 29, no. 2, 182-204, 1999.
11. K. Y. Xing, B. S. Hu, and H. X. Chen, Deadlock avoidance policy for Petri net modeling of flexible manufacturing systems with shared resources, *IEEE Trans. on Automatic Control*, vol. 41, no. 2, 289-295, 1996.
12. A. Reveliotis and P. M. Ferreira, Deadlock avoidance policies in automated manufacturing cells, *IEEE Trans. on Robotics and Automation*, vol. 12, no. 5, 1-13, 1996.
13. J. Malmbog, A model for the design of zone-control automated guided vehicle systems, *Int. J. of Production Research*, 28, 1741-1758, 1990.
14. L. Zeng, H.-P. Wang, and S. Jin, Conflict detection of automated guided vehicles: a Petri net approach, *Int. J. of Production Research*, 29, 865-879, 1991.
15. C.-C. Lee and J. T. Lin, Deadlock prediction and avoidance based on Petri nets for zone-control automated guided vehicle systems, *Int. J. of Production Research*, vol. 33, 3249-3265, 1995.
16. M.-S. Yeh and W.-C. Yeh, Deadlock prediction and avoidance for zone-control AGVS, *Int. J. of Production Research*, 36(10), 2879-2889, 1998.
17. Y. A. Bozer and M. M. Srinivasan, Tandem AGV systems: a partitioning algorithm and performance comparison with conventional AGV systems, *European Journal of Operational Research*, vol. 63, no. 2, 173-192, 1992.
18. Sinriech and J. M. A. Tanchoco, An introduction to the segmented flow approach for discrete material flow systems, *Int. J. of Production Research*, 33(12), 3381-3410, 1995.
19. N. Viswanadham and Y. Narahari, Colored Petri net models for automated manufacturing systems, *Proc. of IEEE Conference on Robotics and Automation*, Raleigh, NC, 1985-1990, 1987.
20. N. Wu and M. C. Zhou, "A Petri Net Method for Deadlock Avoidance of AGV systems," Working paper, ECE Dept., New Jersey Institute of Technology, September 2000.

Algebraic deadlock avoidance policies for conjunctive / disjunctive resource allocation systems

Jonghun Park and Spyros A. Reveliotis
School of Industrial & Systems Engineering
Georgia Institute of Technology
Atlanta, GA 30332

Abstract

Operational flexibility of deadlock avoidance policies (DAPs) is an essential requirement for achieving high resource utilization of the underlying deadlock-prone resource allocation system. This paper presents computational methods for synthesizing highly flexible DAPs for the class of conjunctive / disjunctive resource allocation systems (CD-RAS), which generalizes the resource allocation model that has been typically studied in the past, by allowing multiple resource acquisitions and flexible routings. Specifically, a linear programming method to compute DAPs for CD-RAS is developed, and subsequently a series of computational tools are provided to enhance the policy permissiveness / flexibility. These methods are based on (i) the pertinent exploitation of the policy parameterization, (ii) the observation that the considered class of policies is closed under policy disjunctions, and (iii) the systematic relaxation of the policy-imposed constraints through Petri net structural analysis. The presented results are demonstrated through an example modeling an agile automation system.

1 Introduction

Deadlock avoidance in sequential resource allocation systems (RAS) is a well-established problem in the Discrete Event System (DES) literature. Briefly, the issue is to develop real-time control policies / supervisors which will constrain the system behavior in a strongly connected component of its underlying state-space that contains the system initial empty state. Ideally, one would like to restrain the system to the *maximal* strongly connected component containing the RAS empty state, obtaining thus, the *optimal* deadlock avoidance policy (DAP), that provides maximum operational flexibility. Since, however, it has been established in the literature [16, 8] that the identification of this maximal strongly connected component is an NP-complete problem [7], a large body of the past results have focused on the development of suboptimal but polynomially computable DAPs, that seek to enhance their efficiency through pertinent exploitation of the available information on the system structure and dynamics. Furthermore, in an effort to manage the complexity underlying this policy development, [17] proposed a RAS taxonomy, in which the various sequential RAS are classified in four major classes, on the basis of the complexity of the admitted resource allocation structure. The two extreme cases of this taxonomy are the *Single-Unit (SU)-RAS*, in which every process stage requires only a single unit of a single resource for its successful execution, and the *Conjunctive / Disjunctive (CD)-RAS*, in which every process stage poses a finite number of alternative resource requests, each of which involves an arbitrary number of units from an arbitrarily structured resource set. Hence, the SU-RAS class represents the simplest RAS configurations, while the class of CD-RAS can incorporate more general resource allocation schemes, allowing for, both, multiple resource acquisitions and flexible routings.

The largest and more insightful body of results on deadlock avoidance in sequential RAS concerns the class of SU-RAS; e.g., [1, 16, 6, 5, 3, 9, 12]. Yet recently, there have been a series of results [2, 18, 15] that have addressed the liveness and deadlock avoidance problems in the CD-RAS context, effectively generalizing the insights developed for the SU-RAS class to this more complex RAS environment. The work presented in this paper seeks to complement and extend these theoretical developments, by developing a series of computational methods that can support automatic synthesis of highly flexible DAPs for CD-RAS. More specifically, the presented work first generalizes the CD-RUN DAP, originally presented in [15], to a broader DAP class, to be known as the *G-RUN* DAP, that can take better advantage of the routing flexibility inherent in the underlying RAS. It is also shown that given any CD-RAS configuration, an efficient G-RUN DAP realization can be easily obtained by solving a linear program [11] that is polynomially sized with respect to the given RAS configuration. Subsequently, the flexibility of the proposed class of DAPs is further enhanced through (i) the pertinent exploitation of the policy parameterization, (ii) the observation that the considered class of policies is closed under policy disjunctions, and (ii) the systematic relaxation of the policy-imposed constraints through Petri net structural analysis. The paper concludes by summarizing the major contributions of the presented work, and outlining possible directions for its extension.

2 Petri Net-Based Modeling and Analysis of CD-RAS

In the following discussion it is assumed that the reader is familiar with the basic Petri net structural and behavioral concepts; an excellent introduction on Petri net theory can be found in [10, 4].

CD-RAS and its PN-based Model The CD-RAS is formally is defined by a set of *resource types* $\mathcal{R} = \{R_i, i = 1, \ldots, m\}$, and a set of *job types* $\mathcal{J} = \{J_j, j = 1, \ldots, n\}$. Every resource type R_i is further characterized by its *capacity* $C_i \in Z^+$, where Z^+ is the set of positive integers. Processing requirements of job type J_j are defined by a set of *stages*, partially ordered through a set of precedence constraints. Each job stage p is associated with a *conjunctive* resource allocation requirement, expressed by an m-dimensional vector $(a_{ip})^{i=1,\ldots,m}$, where $a_{ip} \in \{0\} \cup Z^+, i = 1, \ldots, m$, indicates how many units of resource R_i are required to support the stage execution.

In order to model the resource allocation dynamics taking place in CD-RAS by a Petri net, first, we represent the process flow of each job type J_j by a particular net structure known as *Simple Sequential Process* (S^2P) [5]. This net structure is formally defined by an ordinary strongly connected state machine $\mathcal{N}_j = (P_{S_j} \cup \{p_{0_j}\}, T_j, W_j)$ such that (i) $P_{S_j} \neq \emptyset$, $p_{0_j} \notin P_{S_j}$, (ii) every circuit of $\mathcal{N}_j$ contains $\{p_{0_j}\}$, and (iii) $\forall p \in P_{S_j}$ s.t. $\{p_{0_j}\} \in p^{\bullet\bullet}$, $p^{\bullet\bullet} \cap P_{S_j} = \emptyset$. In the S^2P net, each place $p \in P_{S_j}$ is called a *process place*, and it corresponds to a job stage of J_j. Place p_{0_j} represents the *idle place*, since its marking represents the jobs of type J_j waiting to be loaded to the RAS. For notational convenience, we denote the set of resources supporting the execution of process place $p \in P_{S_j}$ by $Q_p = \{R_i \in \mathcal{R} \mid a_{ip} > 0\}$. The PN modeling of the CD-RAS is completed by interconnecting the S^2P nets through a set of *resource places*, P_R, which model the availability of the various resource types. The resulting PN class is referred to as *System of Simple Sequential Processes with General Resource Requirements*, and it will be denoted by S^3PGR^2. Formally, it is defined as follows:

Definition 1 *A well-marked S^3PGR^2 net is a marked PN $\mathcal{N} = (P, T, W, M_0)$ such that*

1. $P = P_S \cup P_0 \cup P_R$, where $P_S = \bigcup_{i=1}^n P_{S_i}$ s.t. $P_{S_i} \cap P_{S_j} = \emptyset, \forall i \neq j$, $P_0 = \bigcup_{i=1}^n \{p_{0_i}\}$ s.t. $P_0 \cap P_S = \emptyset$, and $P_R = \{r_1, \ldots, r_m\}$ s.t. $(P_S \cup P_0) \cap P_R = \emptyset$.

2. $T = \bigcup_{i=1}^n T_i$.

3. $W = W_S \cup W_R$, where $W_S : ((P_S \cup P_0) \times T) \cup (T \times (P_S \cup P_0)) \to \{0,1\}$ s.t. $\forall j \neq i, ((P_{S_j} \cup P_{0_j}) \times T_i) \cup (T_i \times (P_{S_j} \cup P_{0_j})) \to \{0\}$, and $W_R : (P_R \times T) \cup (T \times P_R) \to Z^+$.

4. $\forall i, i = 1, \ldots, n$, the subnet $\mathcal{N}_i$ generated by $P_{S_i} \cup \{p_{0_i}\} \cup T_i$ is an S^2P net.

5. $\forall r \in P_R$, $\exists$ *a unique minimal p-semiflow* y_r *s.t.* $\|y_r\| \cap P_R = \{r\}$, $\|y_r\| \cap P_0 = \emptyset$, $\|y_r\| \cap P_S \neq \emptyset$, *and* $y_r(r) = 1$. *Furthermore,* $P_S = \bigcup_{r \in P_R}(\|y_r\| - P_R)$.

6. $\mathcal{N}$ is pure and strongly connected.

7. $\forall p \in P_S$, $M_0(p) = 0$; $\forall r \in P_R$, $M_0(r) \geq max_{p \in \|y_r\|} y_r(p)$; and $\forall p_{0_i} \in P_0$, $M_0(p_{0_i}) \geq 1$.

We remark that the S^3PGR^2 net structure, originally proposed in [15], models the same of class of RAS behavior as the S^4PR net proposed in [18].

Liveness Analysis of S^3PGR^2 Net It turns out that the liveness of S^3PGR^2 nets is strongly related to the absence / non-development of a specially marked siphon in a modified reachability space characterizing the net dynamics [15]. Here we review this key result, providing first a series of definitions that are necessary for its formal statement. For a more extensive discussion, the reader is referred to [15].

Definition 2 *Consider an S^3PGR^2 net $\mathcal{N} = (P_0 \cup P_S \cup P_R, T, W, M_0)$. A set of places $S \subseteq P_0 \cup P_S \cup P_R$ is said to be a resource-induced deadly marked siphon at $M \in R(\mathcal{N}, M_0)$ iff (i) $^\bullet S \subseteq S^\bullet$; (ii) $\forall t \in {}^\bullet S$, t is disabled by some $p \in S$; (iii) $S \cap P_R \neq \emptyset$; and (iv) every place in $S \cap P_R$ is a disabling place.*

Definition 3 *Given a well-marked S^3PGR^2 net $\mathcal{N} = (P_0 \cup P_S \cup P_R, T, W, M_0)$ and $M \in R(\mathcal{N}, M_0)$, the modified marking $\overline{M}$ is defined by*

$$\overline{M}(p) = \begin{cases} M(p) & \text{if } p \notin P_0 \\ 0 & \text{otherwise} \end{cases} \quad (1)$$

Furthermore, the set of all modified markings induced by the reachable markings is defined by $\overline{R(\mathcal{N}, M_0)} = \{\overline{M} \mid M \in R(\mathcal{N}, M_0)\}$.

Theorem 1 *Let $\mathcal{N} = (P, T, W, M_0)$ be a well-marked S^3PGR^2 net. The net is live iff the space of modified reachable markings, $\overline{R(\mathcal{N}, M_0)}$, contains no resource-induced deadly marked siphon.*

3 G-RUN: An efficient algebraic DAP for CD-RAS

Algebraic DAPs Algebraic DAP's have been proposed in the past as a mathematically elegant and computationally powerful solution to the problem of deadlock avoidance in sequential RAS, able to provide a viable trade-off between computational tractability and operational efficiency [1, 16, 9, 12, 13]. In the context of S^3PGR^2 nets, an algebraic DAP is represented by the following system of linear inequalities:

$$\mathbf{A} \cdot M_S \leq \mathbf{f} \quad (2)$$

In Equation (2), $\mathbf{A} = [\bar{\alpha}_{ip}]_{p \in P_S}^{i=1,\ldots,N}$ is an $N \times |P_S|$ matrix such that $\bar{\alpha}_{ip} \geq 0, i = 1, \ldots, N, \forall p \in P_S$, and N is polynomially related to the number of the system resource types, m; M_S is a vector representation of the system resource allocation state, provided by the projection of the net marking M on the subspace defined by the process place subset P_S; $\mathbf{f} = (f_i)_{i=1,\ldots,N}$ is an N-dimensional vector of positive integers. As a control law, Equation 2 implies that the RAS state represented by vector M_S is *admissible* iff Equation 2 is satisfied.

Given an S^3PGR^2 net, the control logic imposed by the algebraic DAP, can be embedded to the behavior of the original uncontrolled net through of a *control subnet* constructed as follows: (i) Each constraint in Equation 2 is represented by a control place w_i, with initial marking $M_0(w_i) = f_i, i = 1,\ldots,N$; let $P_W = \{w_1, w_2, \ldots, w_N\}$. (ii) Subsequently, the control places $w_i \in P_W$ are connected to the uncontrolled net transitions as follows: By convention, we define $\bar{\alpha}_{ip} = 0, \forall p \in P_0$. Also, $\forall t \in T$, let $\{p\} \equiv {}^\bullet t \cap (P_S \cup P_0)$, and $\{q\} \equiv t^\bullet \cap (P_S \cup P_0)$. Then, $\forall w_i \in P_W, W(w_i, t) = \bar{\alpha}_{iq} - \bar{\alpha}_{ip}$, if $\bar{\alpha}_{iq} - \bar{\alpha}_{ip} > 0$; $W(t, w_i) = \bar{\alpha}_{ip} - \bar{\alpha}_{iq}$, if $\bar{\alpha}_{iq} - \bar{\alpha}_{ip} < 0$; $W(t, w_i) = W(t, w_i) = 0$, otherwise. The resulting net will be called as CS^3PGR^2 (*Controlled S^3PGR^2*). It is easy to see that, in the CS^3PGR^2 net, the control places $w_i \in P_W$ play the role of additional "logical" resources, establishing a new effective resource set $P_R \cup P_W$. Hence, we obtain the following lemma.

Lemma 1 *Given an S^3PGR^2 net, the CS^3PGR^2 net implementing an algebraic DAP on it belongs to the class of S^3PGR^2 nets.*

G-RUN DAP Consider a well-marked S^3PGR^2 net $\mathcal{N} = (P_0 \cup P_S \cup P_R, T, W, M_0)$, and let $o_i = o(R_i), i = 1, \ldots, m$, where $o() : \mathcal{R} \to \{1, \ldots, m\}$ is any partial order imposed on resource set $\mathcal{R}$. Given a place $p \in P_S$, an *i(mmediate)-neighborhood* N_p of p is defined by $N_p \subseteq p^{\bullet\bullet} \cap P_S$. Furthermore, a function $g() : \{p \in P_S \mid p^{\bullet\bullet} \cap P_0 = \emptyset\} \to \{N_p \mid p \in P_S \wedge p^{\bullet\bullet} \cap P_0 = \emptyset\}$ determines uniquely a *community* Ψ of $\mathcal{N}$, defined by $\Psi = \{(p,q) \mid p \in P_S \wedge p^{\bullet\bullet} \cap P_0 = \emptyset \wedge q \in N_p\}$. Let the set of all communities of $\mathcal{N}$ be denoted by $\mathcal{C}_\mathcal{N}$, and $\rho_p^{min} = \min\{o_i \mid a_{ip} > 0, i = 1, \ldots, m\}, \forall p \in P_S$. Then, an algebraic DAP, $(\mathbf{A} = [\bar{\alpha}_{ip}]_{p \in P_S}^{i=1,\ldots,N}, \mathbf{f} = (f_i)_{i=1,\ldots,N})$, is a G-RUN DAP iff $N = m$; $f_i = C_i, i = 1, \ldots, m$; and $\exists\, o() : \mathcal{R} \to \{1, \ldots, m\}$ and some community $\Psi \in \mathcal{C}_\mathcal{N}$, s.t.

$$\bar{\alpha}_{ip} \geq \bar{\alpha}_{iq} \quad \forall (p,q) \in \Psi, \forall R_i \in \mathcal{R} \quad (3)$$
$$\text{s.t. } o_i \geq \rho_p^{min}$$

$$\bar{\alpha}_{ip} = 0 \quad \forall p \in P_S, \forall R_i \in \mathcal{R} \quad (4)$$
$$\text{s.t. } o_i < \rho_p^{min}$$

$$C_i \geq \bar{\alpha}_{ip} \geq a_{ip} \quad \forall p \in P_S, \forall R_i \in \mathcal{R} \quad (5)$$

$$\bar{\alpha}_{ip} \in \{0\} \cup Z^+ \quad \forall p \in P_S, \forall R_i \in \mathcal{R} \quad (6)$$

The correctness of G-RUN DAP can be established by showing that the modified reachable state space of the corresponding CS^3PGR^2 net will present no resource-induced deadly marked siphons, since then, Lemma 1 and Theorem 1 immediately imply the liveness of the CS^3PGR^2 net. For the proof details, the reader is referred to [14].

Theorem 2 *A well-marked CS^3PGR^2 net $\mathcal{N} = (P_0 \cup P_S \cup P_R \cup P_W, T, W, M_0)$ implementing a G-RUN DAP, $(\mathbf{A} = [\bar{\alpha}_{ip}]_{p \in P_S}^{i=1,\ldots,m}, \mathbf{C})$ where $\mathbf{A}$ satisfies Constraints (3) - (6), is live.*

It is interesting to notice that, given an S^3PGR^3 net and any partial ordering imposed on the resource set, there always exists a feasible solution to Constraints (3) - (6): $\bar{\alpha}_{ip} = C_i, i = 1, \ldots, m, \forall p \in P_S$. This DAP enforces the liveness of the underlying CD-RAS by admitting only one job at a time, and therefore, it will be the most restrictive DAP. Based on the introductory discussion, we are naturally interested in a least restrictive – i.e., *optimal* – realization of G-RUN DAP. Since, however, the exact characterization of DAP permissiveness is an intractable problem [9, 13], in the subsequent discussion, the optimality of a G-RUN DAP realization $\mathbf{A}$ on a given S^3PGR^2 net $\mathcal{N}$ and for some given partial resource ordering $o_i, i = 1, \ldots, m$ and community Ψ, will be heuristically assessed through the following optimization problem:

$$\min G(\mathbf{A}; \mathcal{N}, o, \Psi) = \sum_{i=1,\ldots,m} \sum_{p \in P_S} \bar{\alpha}_{ip} \quad (7)$$

$$\text{s.t. Equations (3)} - (6)$$

The motivating idea for the objective function of Equation (7) is that the minimization of the total amount of additional resource allocations[1] imposed by the policy in order to enforce liveness will eliminate allocations that are unnecessary under the DAP structure defined by the triplet $<\mathcal{N}, o, \Psi>$, which, in turn, will lead to more permissive policy implementations.

Since the feasible region defined by Equations (3) – (6) is non-empty and bounded, the above optimization problem will always have a finite optimal solution. Furthermore, Constraint (6) may be omitted without violating the integrality of the optimal solution, since the network structure underlying Constraint (3) and the integrality of $C_i, i = 1, \ldots, m$ and $a_{ip}, i = 1, \ldots, m, p \in P_S$, imply that all the extreme points of the polyhedron defined by Constraints (3) - (5) will be integral [11]. Hence, given a triplet $<\mathcal{N}, o, \Psi>$, an optimal realization of the G-RUN DAP can be computed by solving the *linear programming (LP)* problem defined by Constraints (3) - (5) with the objective function (7).

Example 1 As an example of the G-RUN DAP implementation, consider the CD-RAS depicted in Figure 1, consisting of three resource types R_1, R_2, and

---
[1] Otherwise thought as reservations.

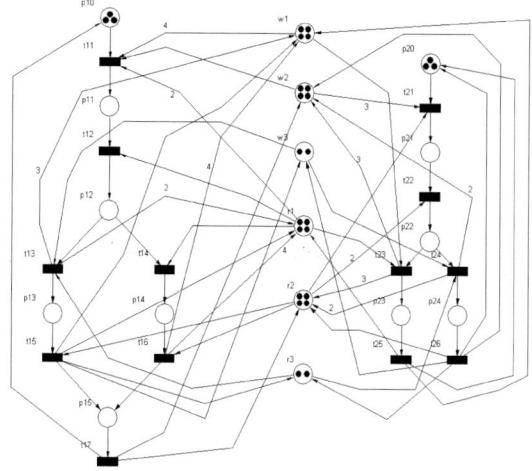

Figure 1: The CS^3PGR^2 net implementing the G-RUN DAP for the CD-RAS configuration of Example 1

R_3, with capacities $C_1 = C_2 = 4$, $C_3 = 2$, and supporting two job types J_1 and J_2. Job type J_1 (resp., J_2) is defined by the set of partially ordered job stages $\{p_{11}, p_{12}, p_{13}, p_{14}, p_{15}\}$ (resp., $\{p_{21}, p_{22}, p_{23}, p_{24}\}$). The conjunctive resource requirements associated with the various job stages are as follows: $a_{p_{11}} = (2, 0, 0)$, $a_{p_{12}} = (3, 0, 0)$, $a_{p_{13}} = (1, 0, 1)$, $a_{p_{14}} = (4, 0, 0)$, $a_{p_{15}} = (0, 1, 0)$, $a_{p_{21}} = (0, 1, 0)$, $a_{p_{22}} = (0, 3, 0)$, $a_{p_{23}} = (1, 0, 0)$, and $a_{p_{24}} = (0, 1, 1)$. We take an arbitrary partial resource ordering $o_1 = 2$, $o_2 = 3$, and $o_3 = 1$, and a community $\Psi = \{(p_{11}, p_{12}), (p_{12}, p_{14}), (p_{13}, p_{15}), (p_{14}, p_{15}), (p_{21}, p_{22}), (p_{22}, p_{23})\}$, and then solve the linear program defined by Equations (3) - (5), (7). The resulting G-RUN DAP is expressed by Equation (8), and its CS^3PGR^2 net implementation is also depicted in Figure 1.

$$\begin{bmatrix} 4 & 4 & 1 & 4 & 0 & 0 & 0 & 1 & 0 \\ 1 & 1 & 1 & 1 & 1 & 3 & 3 & 0 & 1 \\ 0 & 0 & 1 & 0 & 0 & 0 & 0 & 0 & 1 \end{bmatrix} \cdot M_S \leq \begin{bmatrix} 4 \\ 4 \\ 2 \end{bmatrix} \quad (8)$$

4 Enhancing the Policy Flexibility

Optimization of G-RUN Parameters According to the definition of G-RUN DAP, the policy-imposed constraints are computed based on an arbitrary partial ordering of the resource set and an arbitrary community. Obviously, the permissiveness of the resulting policy realization is affected by the partial ordering and the community selected for the policy synthesis. However, given an S^3PGR^2 net with m resource types and $|P_S|$ process places, there are $O(m^m \cdot |P_S|^{|P_S|})$ correct G-RUN DAP implementations in the worst case, and therefore, it is worth investigating a systematic computational method for obtaining efficient policy implementations by optimizing the order and community selection. Using the objective function defined in Equation (7), the resulting optimization problem can be formulated as a

$$\min \sum_{i=1,\ldots,m} \sum_{p \in P_S} X_{ip}$$

s.t.
Definition of Y_i:
$$\sum_{i=1}^m Y_i = \frac{m^2+m}{2}$$
$$Y_i \geq 1 \qquad \forall R_i \in \mathcal{R}$$

Definition of Y_p^{min}:
$$Y_i \geq Y_p^{min} \qquad \forall p \in P_S, \forall R_i \in Q_p$$
$$\sum_{R_i \in Q_p} Z_{ip} \geq 1 \qquad \forall p \in P_S$$
$$\nu(1 - Z_{ip}) \geq Y_i - Y_p^{min} \qquad \forall p \in P_S, \forall R_i \in Q_p$$

Definition of W_{ip}:
$$(\nu+1)W_{ip} \geq Y_i - Y_p^{min} + 1 \qquad \forall p \in P_{NT}, \forall R_i \in \mathcal{R}$$
$$\nu(1 - W_{ip}) \geq Y_p^{min} - Y_i \qquad \forall p \in P_{NT}, \forall R_i \in \mathcal{R}$$

Definition of I_{pq}:
$$\sum_{q \in F_p} I_{pq} \geq 1 \qquad \forall p \in P_{NT}$$

Definition of E^1_{ipq}:
$$I_{pq} \geq E^1_{ipq} \qquad \forall p \in P_{NT}, \forall q \in F_p, \forall R_i \in \mathcal{R}$$
$$W_{ip} \geq E^1_{ipq} \qquad \forall p \in P_{NT}, \forall q \in F_p, \forall R_i \in \mathcal{R}$$
$$E^1_{ipq} + 1 \geq I_{pq} + W_{ip} \qquad \forall p \in P_{NT}, \forall q \in F_p, \forall R_i \in \mathcal{R}$$

Definition of E^2_{ipq}:
$$I_{pq} \geq E^2_{ipq} \qquad \forall p \in P_{NT}, \forall q \in F_p, \forall R_i \in \mathcal{R}$$
$$1 - W_{ip} \geq E^2_{ipq} \qquad \forall p \in P_{NT}, \forall q \in F_p, \forall R_i \in \mathcal{R}$$
$$E^2_{ipq} \geq I_{pq} - W_{ip} \qquad \forall p \in P_{NT}, \forall q \in F_p, \forall R_i \in \mathcal{R}$$

Definition of G-RUN:
$$X_{iq} - X_{ip} \leq \mu(1 - E^1_{ipq}) \qquad \forall p \in P_{NT}, \forall q \in F_p, \forall R_i \in \mathcal{R}$$
$$X_{ip} \leq \mu(1 - E^2_{ipq}) \qquad \forall p \in P_{NT}, \forall q \in F_p, \forall R_i \in \mathcal{R}$$
$$C_i \geq X_{ip} \geq a_{ip} \qquad \forall R_i \in \mathcal{R}, \forall p \in P_S$$

Variable domains:
$$Y_i \in Z^+ \qquad \forall R_i \in \mathcal{R}$$
$$Z_{ip}, W_{ip}, E^1_{ipq}, E^2_{ipq}, I_{pq} \in \{0, 1\} \qquad \forall p \in P_S, q \in P_S, \forall R_i \in \mathcal{R}$$

Figure 2: The MIP formulation for the order and community optimization problem

mixed integer programming (MIP) problem. Figure 2 presents a MIP formulation for the problem of selecting simultaneously an optimal order and community pair. In Figure 2, $P_{NT} = \{p \in P_S \mid p^{\bullet\bullet} \cap P_0 = \emptyset\}$, $F_p = \{q \in P_S \mid q \in p^{\bullet\bullet}\}$, μ is a big-M parameter set to $\max\{C_i \mid i = 1, \ldots, m\}$, and ν is also a big-M parameter set to $\frac{m^2-m}{2}$. The variables used for the MIP formulation in Figure 2 are defined as follows: $X_{ip} \equiv \bar{\alpha}_{ip}$; $Y_i \equiv o_i, \forall R_i \in \mathcal{R}$; $Y_p^{min} \equiv \rho_p^{min}$; $Z_{ip} \equiv 1$ iff $Y_i = Y_p^{min}$, $\forall p \in P_S, \forall R_i \in Q_p$; $W_{ip} \equiv 1$ iff $Y_i \geq Y_p^{min}$; $I_{pq} \equiv 1$ iff $(p, q) \in \Psi$; $E^1_{ipq} \equiv 1$ iff $W_{ip} = 1 \land I_{pq} = 1$; and $E^2_{ipq} \equiv 1$ iff $W_{ip} = 0 \land I_{pq} = 1$. The efficacy of the proposed approach is demonstrated in the following example.

Example 2 Applying the MIP formulation of Figure 2 to the CD-RAS considered in Example 1, we obtain an optimal ordering o^* : $o_1^* = 3$, $o_2^* = 2$, $o_3^* = 1$ and community $\Psi^* = \{(p_{11}, p_{12}), (p_{12}, p_{13}), (p_{13}, p_{15}), (p_{14}, p_{15}), (p_{21}, p_{22}), (p_{22}, p_{24})\}$, with an optimal objective value

$G(\mathcal{N}, o^*, \Psi^*, \mathbf{A}) = 23$. The solution of the aforementioned MIP formulation provides also an optimized $\mathbf{A}$ matrix for the policy implementation under the new ordering and community pair, (o^*, Ψ^*), that is shown in Equation (9).

$$\begin{bmatrix} 3 & 3 & 1 & 4 & 0 & 0 & 0 & 1 & 0 \\ 0 & 0 & 1 & 0 & 1 & 3 & 3 & 0 & 1 \\ 0 & 0 & 1 & 0 & 0 & 0 & 0 & 0 & 1 \end{bmatrix} \cdot M_S \leq \begin{bmatrix} 4 \\ 4 \\ 2 \end{bmatrix} \quad (9)$$

The reader can verify that $R(\mathcal{N}; \mathcal{P}^0) \subset R(\mathcal{N}; \mathcal{P}^*)$, where $\mathcal{P}^*$ and $\mathcal{P}^0$ denote the G-RUN implementations represented by Equation (9) and Equation (8), respectively.

Efficient Disjunctive G-RUN Implementations
As it was observed in [16], given a sequential RAS with reachable state space $\mathcal{S}$, and k correct DAP implementation for it, $\mathcal{P}_i$, $i = 1, \ldots, k$, with policy-defining conditions $\mathcal{H}_i(\cdot)$, $\mathcal{S} \to \{0, 1\}$, and policy-admissible state spaces $\mathcal{S}(\mathcal{P}_i)$, the policy disjunction $\bigvee_{i=1}^k \mathcal{P}_i$, defined by the condition $\mathcal{L}(\cdot) \equiv \bigvee_{i=1}^k \mathcal{H}_i(\cdot)$, $\mathcal{S} \to \{0,1\}$, is another correct DAP that satisfies $\mathcal{S}(\bigvee_{i=1}^k \mathcal{P}_i) = \bigcup_{i=1}^k \mathcal{S}(\mathcal{P}_i)$, and therefore, it can be more permissive than any of its constituent policies. In the context of G-RUN implementation, this observation arises the problem of developing an algorithmic procedure for selecting an a priori sized set of ordering and community pairs of a given CD-RAS, in a way that tends to increase the permissiveness of the resulting disjunctive G-RUN implementation.

Next, we develop a heuristical approach to this problem, which is based on the following observation: Given a set of preselected ordering and community pairs, $< o_1, \Psi_1 >, \ldots, < o_l, \Psi_l >$, a new ordering and community pair, $< o_{l+1}, \Psi_{l+1} >$, is likely to contribute a larger number of states in the sub-space $\mathcal{S}(\bigvee_{i=1}^{l+1} \mathcal{P}(o_i, \Psi_i))$, if the pattern of reservations expressed by the $\mathbf{A}_{\mathcal{P}(o_{l+1}, \Psi_{l+1})}$ is differentiated as much as possible from the "cumulative" reservation pattern established by policies $\mathcal{P}^i \equiv \mathcal{P}(o_i, \Psi_i)$, $i = 1, \ldots, l$. By expressing this cumulative reservation pattern by the matrix $\bar{\bar{\mathbf{A}}}^l$ with elements $\bar{\bar{\alpha}}_{ip}^l = \max_{q=1}^l \bar{\alpha}_{ip}^q$, this differentiation is quantified by solving the MIP of Figure 2, with its objective modified as follows:

$$\min \sum_{i=1,\ldots,m} \sum_{p \in P_S} (\bar{\bar{\alpha}}_{ip}^l + \epsilon) X_{ip} \quad (10)$$

The term $\bar{\bar{\alpha}}_{ip}^l$ in the cost coefficients of the objective function of Equation 10 implements the idea discussed above, since the expression $\sum_{i=1,\ldots,m} \sum_{p \in P_S} \bar{\bar{\alpha}}_{ip}^l X_{ip}$ is the "inner product" (element-wise) of matrices $\bar{\bar{\mathbf{A}}}^l$ and $\mathbf{A}_{\mathcal{P}^{l+1}}$. The second cost term, ϵ, should be taken such that $0 < \epsilon < 1$, and it intends to penalize unnecessarily large reservations incurred by the new policy $\mathcal{P}^{l+1}$. Finally, we notice that the quantification of the (dis-)similarity of the reservation patterns through the "inner product" $\bar{\bar{\mathbf{A}}}^l \cdot \mathbf{A}_{\mathcal{P}^{l+1}}$ gives the proposed methodology the name of *(G-RUN implementation based on) "orthogonal" disjunction*. The complete algorithm for

1. $\bar{\bar{\mathbf{A}}} := \mathbf{0}$; $l = 1$;
2. **while** $l \leq k$ **do**
   (a) Solve the MIP of Figure 2 with the modified objective function in Equation 10
   (b) Compute $\mathbf{A}_{\mathcal{P}^l}$ based on the new order and community found.
   (c) $\bar{\bar{\mathbf{A}}}' := \bar{\bar{\mathbf{A}}} \vee \mathbf{A}_{\mathcal{P}^l}$, where $\vee$ denotes the (element-wise) max operator.
   (d) **if** $\bar{\bar{\mathbf{A}}}' \neq \bar{\bar{\mathbf{A}}}$ **then** $\bar{\bar{\mathbf{A}}} := \bar{\bar{\mathbf{A}}}'$; $l := l + 1$;
      else exit;

   endwhile

Figure 3: Order and community selections for k-disjunctive G-RUN implementation

selecting (up to) k orderings and communities for disjunctive G-RUN implementation according to the above ideas, is given in Figure 3.

Example 3 We demonstrate the idea of orthogonal disjunctive G-RUN implementation, by applying it on the CD-RAS of Example 1, with $k = 2$ and $\epsilon = 0.1$. In this case, $\mathbf{A}_{\mathcal{P}^1}$ is given by Equation (9). For $l = 2$, we solve the MIP of Figure 2 with the modified objective function of Equation (10), where $\bar{\bar{\mathbf{A}}}^1 = \mathbf{A}_{\mathcal{P}^1}$. The obtained order and community are $o^2 : o_1^2 = 3, o_2^2 = 1, o_3^2 = 2$, and $\Psi^* = \{(p_{11}, p_{12}), (p_{12}, p_{13}), (p_{13}, p_{15}), (p_{14}, p_{15}), (p_{21}, p_{22}), (p_{22}, p_{24})\}$. The corresponding policy defining matrix $\mathbf{A}_{\mathcal{P}^2}$ and the resulting policy constraints are:

$$\begin{bmatrix} 3 & 3 & 1 & 4 & 0 & 0 & 0 & 1 & 0 \\ 0 & 0 & 0 & 0 & 1 & 3 & 3 & 0 & 1 \\ 0 & 0 & 1 & 0 & 0 & 1 & 1 & 0 & 1 \end{bmatrix} \cdot M_S \leq \begin{bmatrix} 4 \\ 4 \\ 2 \end{bmatrix} \quad (11)$$

Furthermore, $\bar{\bar{\mathbf{A}}}^2$ is obtained by:

$$\bar{\bar{\mathbf{A}}}^2 \equiv \bar{\bar{\mathbf{A}}}^1 \vee \mathbf{A}_{\mathcal{P}^2} = \begin{bmatrix} 3 & 3 & 1 & 4 & 0 & 0 & 0 & 1 & 0 \\ 0 & 0 & 1 & 0 & 1 & 3 & 3 & 0 & 1 \\ 0 & 0 & 1 & 0 & 0 & 1 & 1 & 0 & 1 \end{bmatrix} \quad (12)$$

The disjunctive G-RUN implementation admits any state M_S of the considered S^3PGR^2 net that satisfies at least one of the constraint sets given in Equations (9) and (11).

G-RUN Constraint Relaxation by means of PN Structural Analysis The work presented in this subsection seeks to systematically relax the right-hand-side (rhs) vector in the constraints of Equation (2), in a way that ensures the liveness of the resulting controlled net, CS^3PGR^2. The mechanism for this relaxation is provided by the association of non-liveness in CS^3PGR^2 nets to the presence of resource-induced deadly marked siphons in the modified marking space. Specifically, first we develop a computational tool able to verify that the modified reachability space of a CS^3PGR^2 net does not contain resource-induced deadly marked siphons. This tool maintains the convenient form of a mixed integer programming (MIP) formulation, and in the light of Theorem 1, it provides a sufficient condition for CS^3PGR^2 net liveness. Subsequently, the availability of the MIP-based liveness test for CS^3PGR^2 nets allows the organization of a systematic search over the

$$G(M) = \min \sum_{p \in P} v_p$$

s.t.

$$f_{pt} \geq \frac{\overline{M}(p) - W(p,t) + 1}{SB(p)} \quad \forall W(p,t) > 0$$

$$f_{pt} \geq v_p \quad \forall W(p,t) > 0$$

$$z_t \geq \sum_{p \in {}^\bullet t} f_{pt} - |{}^\bullet t| + 1 \quad \forall t \in T$$

$$v_p \geq z_t \quad \forall W(t,p) > 0$$

$$\sum_{r \in P_R \cup P_W} v_r \leq |P_R \cup P_W| - 1$$

$$\sum_{t \in r^\bullet} f_{rt} - |r^\bullet| + 1 \leq v_r \quad \forall r \in P_R \cup P_W$$

$$v_p, z_t, f_{pt} \in \{0,1\} \quad \forall p \in P, \forall t \in T$$

Figure 4: The IP formulation for finding the maximal resource-induced deadly marked siphon

space of vectors, $\mathbf{f}$, that constitute *"meaningful"* rhs vectors in Equation (2), for maximal elements leading to correct policy implementations.

We proceed to the development of these results by first presenting an IP formulation in Figure 4 which, given a modified marking $\overline{M} \in \overline{R(\mathcal{N}, M_0)}$ of a CS^3PGR^2 net, computes the maximal resource-induced deadly marked siphon S, such that (i) $S \cap (P_R \cup P_W) \neq \emptyset$ and (ii) every place in $S \cap (P_R \cup P_W)$ is a disabling place. In Figure 4, $SB(p)$ denotes a structural bound for the markings of place $p \in P$, and $v_p = 1$ indicates that place p does not belong to the maximal resource-induced deadly marked siphon. [13] has also established that the infeasibility of the IP formulation of Figure 4 constitutes a necessary and sufficient condition for the non-existence of resource-induced deadly marked siphons in a given modified marking $\overline{M}$. This condition can be extended, in principle, to a test for the non-existence of resource-induced deadly marked siphons over the entire space $\overline{R(\mathcal{N}, M_0)}$ of a CS^3PGR^2 net $\mathcal{N} = (P, T, W, M_0)$, by: (i) turning marking vector $\overline{M}$ in the IP formulation of Figure 4 into a variable, (ii) introducing an additional set of variables, M, representing the net reachable markings, and (iii) adding two additional sets of constraints, the first one linking variables M and $\overline{M}$ according to the logic of Equation 1, and the second one ensuring that the set of feasible values for the variable vector M is equivalent to the PN reachability space $R(\mathcal{N}, M_0)$. Unfortunately, however, any system of linear inequalities exactly characterizing the set $R(\mathcal{N}, M_0)$ is of exponential complexity with respect to the net size. On the other hand, a superset of the reachability space $R(\mathcal{N}, M_0)$ is provided by the system *state equation*:

$$M = M_0 + \Theta \bar{x} \quad (13)$$

$$M \geq 0, \bar{x} \in Z^+ \quad (14)$$

The above remarks lead to a *sufficient* condition for the non-existence of resource-induced deadly marked siphons, in the entire space $\overline{R(\mathcal{N}, M_0)}$ of a given CS^3PGR^2 net $\mathcal{N}$. Furthermore, in the light of Theorem 1, this condition constitutes a *sufficient* condition for liveness of CS^3PGR^2 nets. Hence, we get the following theorem the proof of which can be found in [13].

Theorem 3 *Let $\mathcal{N} = (P, T, W, M_0)$ be a well marked CS^3PGR^2 net. If the MIP defined by the IP of Figure 4 with the additional constraints Equations (13)–(14) and Equation (1), is infeasible, then $\mathcal{N}$ is live.*

The above liveness criterion for CS^3PGR^2 nets can support the search for maximal elements in the space of *"meaningful"* rhs vectors, $\mathbf{f}$, for Equation 2, in a way similar to that presented in [12], for the flexibility enhancement of the original RUN DAP under the SU-RAS configuration. Specifically, given a CS^3PGR^2 net $\mathcal{N} = (P, T, W, M_0)$, controlled by a G-RUN implementation that is expressed by the system of linear inequalities $\mathbf{A} \cdot M_S \leq \mathbf{f}_0 (\equiv \mathbf{C})$, the search space is defined by the lattice $\{\mathbf{f} \in (Z^+)^m \mid \mathbf{f}_0 \leq \mathbf{f} \leq \bar{\mathbf{f}}\}$, where the (not necessarily tight) upper bound $\bar{\mathbf{f}}$ is computed by the following linear programs:

$$\forall i \in \{1, \ldots, m\}, \quad \bar{f}[i] =$$
$$\max_{\{x_p \,:\, p \in P_S \wedge \bar{\alpha}_{ip} > 0\}} \sum_{\{p \in P_S \mid \bar{\alpha}_{ip} > 0\}} \bar{\alpha}_{ip} x_p$$

s.t.

$$\sum_{\{p \in P_S \mid \bar{\alpha}_{ip} > 0\}} a_p[i] x_p \leq C_i, \forall i \in \{1, \ldots, m\}$$

$$x_p \geq 0, \quad \forall p \in P_S : \bar{\alpha}_{ip} > 0 \quad (15)$$

The search algorithm that identifies in the lattice defined above the *maximal* elements that lead to correct policy implementations, is based on the fact that any particular selection for the rhs vector $\mathbf{f}$ essentially defines the initial marking of the control places, $w_i \in P_W$, in the CS^3PGR^2 net modeling the controlled system behavior. A brief statement of this algorithm is as follows: Starting from the upper bound $\bar{\mathbf{f}}$, generate the arborescence of the elements defined by the '$\leq$' order; at every generated node solve the corresponding MIP formulation defined by the IP of Figure 4 with the additional constraints Equations (13)–(14) and Equation (1); terminate the search along each path when an element $\mathbf{f}^* \in \{\mathbf{f} \in (Z^+)^m \mid \mathbf{f}_0 \leq \mathbf{f} \leq \bar{\mathbf{f}}\}$ satisfying the condition of Theorem 3, is identified.

Example 4 Consider the G-RUN implementation of Example 1, which is represented by the system of linear inequalities given in Equation (8). Application of the LP formulation of Equation (15) to this policy implementation gives $\bar{\mathbf{f}} = (8, 15, 2)^T$. Subsequently, the application of the search algorithm outlined above results in the unique maximal element $\mathbf{f}^* = \{(7, 8, 2)^T\}$. Hence, a correct relaxed policy implementation is defined by the following set of constraints on the system state:

$$\begin{bmatrix} 4 & 4 & 1 & 4 & 0 & 0 & 0 & 1 & 0 \\ 1 & 1 & 1 & 1 & 1 & 3 & 3 & 0 & 1 \\ 0 & 0 & 1 & 0 & 0 & 0 & 0 & 0 & 1 \end{bmatrix} \cdot M_S \leq \begin{bmatrix} 7 \\ 8 \\ 2 \end{bmatrix} \quad (16)$$

It should be noted that the policy represented by Equation 16 admits all RAS states admitted by the policy of Equation 8, and furthermore, it admits state

$M_S = (0,1,0,0,0,1,1,0,0)^T$ (equivalently, marking $M = (2,0,1,0,0,0,1,1,1,0,0,1,0,2,3,1,2)^T$ of the CS^3PGR^2 net) which is not admitted by the policy of Equation 8. Therefore, the new relaxed policy is more permissive than the original one, and we can conclude that the proposed scheme provides an effective method to enhance the permissiveness of the original G-RUN definition.

5 Conclusions

This paper studied the DAP synthesis problem for the class of CD-RAS, that allows multiple resource acquisitions and flexible routings. The effective characterization of liveness for the class of S^3PGR^2 nets, modeling the CD-RAS class, through the notion of resource-induced deadly marked siphon, allowed the development of G-RUN DAP, which is represented by a system of linear inequalities that can be efficiently synthesized using linear programming. Furthermore, the second part of the paper investigated a series of techniques for enhancing the flexibility of the G-RUN implementation on any given CD-RAS. More specifically, the proposed methodologies included (i) a series of heuristical ideas that sought the pertinent selection of the policy parameters to be employed during the policy instantiation, and (ii) the systematic relaxation of the policy-imposed constraints while maintaining the liveness of the controlled system behavior, by exploiting the aforementioned siphon-based characterization of S^3PGR^2 liveness.

Future work will seek to experimentally assess the statistical significance of the heuristical ideas underlying the definition of the objective function in Equation (7) and some of the policy optimizing techniques presented in Section 4, and it will also try to address the optimal ordering and community selection problem for more traditional performance-related objectives, such as throughput maximization and cycle time reduction.

References

[1] Z. A. Banaszak and B. H. Krogh. Deadlock avoidance in flexible manufacturing systems with concurrently competing process flows. *IEEE Transactions on Robotics & Automation*, 6(6):724–734, 1990.

[2] K. Barkaoui, A. Chaoui, and B. Zouari. Supervisory control of discrete event systems based on structure theory of petri nets. In *IEEE International Conference on Systems, Man, & Cybernetics*, pages 3750–3755. IEEE, 1997.

[3] F. Chu and X-L. Xie. Deadlock analysis of petri nets using siphons and mathematical programming. *IEEE Transactions on Robotics & Automation*, 13(6):793–804, 1997.

[4] J. Desel and J. Esparza. *Free Choice Petri Nets*. Cambridge University Press, 1995.

[5] J. Ezpeleta, J. M. Colom, and J. Martinez. A petri net based deadlock prevention policy for flexible manufacturing systems. *IEEE Transactions on Robotics & Automation*, 11:173–184, 1995.

[6] M. P. Fanti, B. Maione, S. Mascolo, and B. Turchiano. Event-based feedback control for deadlock avoidance in flexible production systems. *IEEE Transactions on Robotics & Automation*, 13:347–363, 1997.

[7] M. R. Garey and D. S. Johnson. *Computers and Intractability : A Guide to the Theory of NP-Completeness*. W. H. Freeman, New York, 1979.

[8] M. Lawley and S. Reveliotis. Deadlock avoidance for sequential resource allocation systems : Hard and easy cases. to appear in *International Journal of Flexible Manufacturing Systems*, 2000.

[9] M. Lawley, S. Reveliotis, and P. Ferreira. A correct and scalable deadlock avoidance policy for flexible manufacturing systems. *IEEE Transactions on Robotics & Automation*, 14(5):796–809, 1998.

[10] T. Murata. Petri nets: Properties, analysis and applications. *Proceedings of the IEEE*, 77(4):541–580, 1989.

[11] M. Padberg. *Linear Optimization and Extensions*. Springer, 2nd edition, 1999.

[12] J. Park and S. A. Reveliotis. Algebraic synthesis of efficient deadlock avoidance policies for sequential resource allocation systems. *IEEE Transactions on Robotics & Automation*, 16(2):190–195, 2000.

[13] J. Park and S. A. Reveliotis. Deadlock avoidance in sequential resource allocation systems with multiple resource acquisitions and flexible routings. accepted for publication in *IEEE Transactions on Automatic Control (available at http://www.isye.gatech.edu/~spyros/)*, 2000.

[14] J. Park and S. A. Reveliotis. Liveness-enforcing supervisors for resource allocation systems with reworks, forbidden states, and uncontrollable events. submitted to *IEEE Transactions on Robotics & Automation (available at http://www.isye.gatech.edu/~spyros/)*, 2000.

[15] J. Park and S. A. Reveliotis. A polynomial-complexity deadlock avoidance policy for sequential resource allocation systems with multiple resource acquisitions and flexible routings. In *Proceedings of the IEEE International Conference on Decision & Control*, pages 2663–2669. IEEE, 2000.

[16] S. A. Reveliotis and P. M. Ferreira. Deadlock avoidance policies for automated manufacturing cells. *IEEE Transactions on Robotics & Automation*, 12(6):845–857, 1996.

[17] S. A. Reveliotis, M. A. Lawley, and P. M. Ferreira. Polynomial complexity deadlock avoidance policies for sequential resource allocation systems. *IEEE Transactions on Automatic Control*, 42(10):1344–1357, 1997.

[18] F. Tricas, F. García-Vallés, J. M. Colom, and J. Ezpeleta. An iterative method for deadlock prevention in fms. In *Proceedings of the 5th Workshop on Discrete Event Systems*, 2000.

On a Two-Level Hierarchical Structure for the Dynamic Control of Multifingered Manipulation

Giuseppe Casalino, Giorgio Cannata, Giorgio Panin, Andrea Caffaz

DIST, University of Genova - Via Opera Pia 13, 16145 Genova, Italy

E-Mail: {pino, cannata, gpanin, caffaz}@dist.unige.it

Abstract

The problem of grasping and manipulating rigid objects using a multifingered robotic system is dealt with in this paper.

A hierarchical two-level closed-loop trajectory tracking control strategy is presented here, in order to achieve the manipulation task, together with a complimentary force control, for the object grasping, which makes use of a recent formulation of grasping forces decomposition.

1 Introduction

Manipulating a rigid object using multiple robots or multifingered robotic hands can actually be considered as one of the most representative and complete tasks, among those generally considered within the robotic literature.

Within this framework, in fact, various interrelated problems commonly arise, ranging from: static and kinematic of grasping, including rolling contact situations [1], [2]; grasp planning and contact forces allocation [1], [3], [4] still including the cases of rolling contact situations, stability of grasping and related structural properties [5]; till arriving to the dynamics and control of multifingered manipulation, where all the previous problems are (or should be) comprehensively taken into account and consequently treated, to the control purposes, within a unifying approach.

To this respect, early attempts aiming to establish such a unifying approach date back to the well known work of Khatib [6] on operational space methods, then followed by various works, whose results have been successively reorganized within the book [7].

Within the here proposed approach, the motion tracking to be performed by the manipulated object is first considered in terms of the definition of a closed loop (kinematic only) upper level control law, accomplishing the task of generating, at each time instant, the reference velocity (linear and angular) that should be assigned to the fixed body frame, in case that the body itself could be considered as a pure kinematic entity (i.e. no mass, no inertia).

Then, via simple rigid body transformations, such reference velocity for the body frame is in turn real time translated into the corresponding set of reference linear velocities at the body contact points; thus coinciding with the reference linear velocities to be assigned to the corresponding fingertips (obviously assuming the persistency of the contact conditions). Such fingertips reference velocities are in turn real-time traslated into a related set of joint reference velocities for the robotic fingers, via the use of any right inverse of the linear part of the Jacobian matrix of the overall set of fingers.

Finally, such real time evaluated joint reference velocities are given as input to a suitable, lower level, closed loop joint velocity controller, accomplishing the task of guaranteeing their asymptotic tracking.

Naturally enough, to the aim of maintaining the contacts during motion, a complimentary force control law is also provided, not affecting the motion controlled by the above mentioned upper and lower levels control loops.

The present paper is organized as follows: in Section 2, the overall system dynamic equations are derived in the form of a set of DAE (Differential Algebraic Equations), thus mantaining the separation between the object and fingers dynamics. In Section 3, the proposed hierarchical control law for trajectory tracking is derived via the use of Lyapunov arguments, and the convergence to zero of the errors is proven, together with its stability properties. Finally, in Section 4 a simple complimentary force control law for maintaining the contacts without affecting the motion is also derived, on the basis of some of the recent force decomposition results presented in [4].

Simulations and experimental results, made on a 4-fingered and 16-degrees of freedom robotic hand (the "*DIST-Hand*" research activity) manipulating an object, can be found at our Robotics and Automation Lab internet site: *www.graal.dist.unige.it*.

2 Overall System Equations

The system considered here consists in a tree-structured robotic system composed by a number h of rigid end-effector tips, which all together handle a rigid body.

The contact model considered is the ideal *point contact with friction* (see [8]); in other words, it is assumed that the h contact points remain fixed on the object surface (and, off course, that the h end-effector tips are point-like and they do not slip during the manipulation task).

Within this system, first consider the subsystem represented by manipulated object and its assigned fixed frame $<b>$, whose orientation and position with respect to a given (inertial) base frame $<0>$ are, at each time instant, respectively given by the corresponding orthogonal matrix R and distance vector $\mathbf{x}$ (the latter projected on $<0>$), generally grouped within the homogenous transformation matrix

$$T_b \triangleq \left[\begin{array}{c|c} R & \mathbf{x} \\ \hline \mathbf{0} & 1 \end{array} \right] \quad (1)$$

By letting $\mathbf{v}$ and $\boldsymbol{\omega}$ be the projections (on frame $<0>$) of the linear and angular velocities of $<b>$ with respect to $<0>$, respectively, we can first of all recall the *kinematic* equations of such a rigid body; i.e., the well known equations

$$\left\{ \begin{array}{l} \dot{\mathbf{x}} = \mathbf{v} \\ \dot{R} = [\boldsymbol{\omega} \wedge] R \end{array} \right. \quad (2)$$

the second one representing the so-called "*Strapdown Differential Equation*", which relates the time derivative of R with vector $\boldsymbol{\omega}$ via the skew-symmetric matrix $[\boldsymbol{\omega} \wedge]$ (i.e. the cross-product operator matrix).

Note that throughout the following, and with a little abuse of notation, we shall always term the vector collection of the body velocities $\mathbf{v}, \boldsymbol{\omega}$ with the compact symbol

$$\dot{\mathbf{q}}_b \triangleq col(\mathbf{v}, \boldsymbol{\omega}) \quad (3)$$

where the notational abuse stands, obviously, from the well known fact that, generally speaking, the angular velocity $\boldsymbol{\omega}$ *is not* the time derivative of any pre-existing rotation vector.

As a second aspect of the object kinematics, it can be also reminded that, owing to its rigidity and the assumed invariance of the h contact points on it, a direct *linear* relationship actually exists between the velocities of such contact points and the body velocities.

More precisely, by letting $\mathbf{X} \triangleq col(\mathbf{x}_1, \mathbf{x}_2, ..., \mathbf{x}_h)$ be the vector collection of the h contact points on the body, each one projected on $<0>$, it is actually an easy task to verify that

$$\dot{\mathbf{X}} = J_b(R) \dot{\mathbf{q}}_b \quad (4)$$

being $J_b(R)$ the so-called Jacobian matrix of the body at the contact points, having the form

$$J_b(R) = \left[\begin{array}{c|c} I_3 & -[\mathbf{s}_1 \wedge] \\ \hline \cdots \\ \hline \cdots \\ \hline I_3 & -[\mathbf{s}_h \wedge] \end{array} \right] \quad (5)$$

where each vector $\mathbf{s}_i$ results projected on the inertial frame $<0>$, via the matrix R.

In particular, we can make such dependence on R explicit, by simply expressing each $\mathbf{s}_i$ in terms of its *constant* projection $\boldsymbol{\sigma}_i$ on the object frame $<b>$ (that is, $\mathbf{s}_i = R\boldsymbol{\sigma}_i$, and consequently $[\mathbf{s}_i \wedge] = R [\boldsymbol{\sigma}_i \wedge] R^T$), thus obtaining

$$J_b(R) = R^{(h)} G R^{(2)T} \quad (6)$$

being $R^{(k)}$ the $k-th$ order block diagonal organization of matrix R, and G the now *constant* Jacobian matrix of the body, projected on the body frame $<b>$, and built on the basis of constant vectors $\boldsymbol{\sigma}_i, i = 1, ..., h$.

Passing now to consider the *dynamic* aspects of the manipulated rigid object, let us first define $\mathbf{F} \triangleq col(\mathbf{f}_1, \mathbf{f}_2, ..., \mathbf{f}_h)$ as the vector collection of the h contact forces (each one projected on $<0>$), acting on the object at the corresponding contact points. Moreover, consider its relevant wrench reduction to the origin $\mathbf{x}$ of body frame $<b>$ (i.e., its equivalent representation in terms of the resultant force $\mathbf{f}$ applied to point $\mathbf{x}$, and resulting torque $\boldsymbol{\tau}$ evaluated with respect to $\mathbf{x}$, both projected on $<0>$), given by the dual relationship

$$\mathbf{p} \triangleq col[\mathbf{f}, \boldsymbol{\tau}] = J_b^T(R) \mathbf{F} \quad (7)$$

Then, by recalling the well known general form assumed by the *body dynamic* equations, whenever written with respect to any given point $\mathbf{x}$ on it (i.e. not necessarily coinciding with its mass center), projected on the inertial frame $<0>$, and by keeping into account relationship (7), we can consequently write such dynamic equations directly in the form

$$A_b(R) \ddot{\mathbf{q}}_b + B_b(R, \dot{\mathbf{q}}_b) \dot{\mathbf{q}}_b + C_b(R) = J_b^T(R) \mathbf{F} \quad (8)$$

where matrix $A_b(R)$ (positive definite) represents the so called *generalized inertia* matrix of the body, evaluated at point $\mathbf{x}$ on it, and where vectors $B_b(R, \dot{\mathbf{q}}_b) \dot{\mathbf{q}}_b$ and $C_b(R)$ keep into account the centrifugal-Coriolis and gravitational effects, respectively (being such effects also evaluated at point $\mathbf{x}$ on the body).

At this point, also consider the subsystem represented by the robotic manipulating structure. Then, by denoting with $\mathbf{q}_r$ the vector of its joint coordinate, and letting $\mathbf{Y} \triangleq col(\mathbf{y}_1, \mathbf{y}_2, ..., \mathbf{y}_h)$ be the vector collection of its h end effector tip points (each one projected on $<0>$) we can primarily recall their general *kinematic* behaviour, expressed by the relationship

$$\dot{\mathbf{Y}} = J_r(\mathbf{q}_r) \dot{\mathbf{q}}_r \quad (9)$$

being $J_r(\mathbf{q}_r) \triangleq col[J_1(\mathbf{q}_1), J_2(\mathbf{q}_2), ..., J_h(\mathbf{q}_h)]$ the column collection of the h linear Jacobian matrices (one for each end-effector tip) of the robotic fingers.

Moreover, by keeping into account the assumed contact constraint conditions between the body and the fingertips, i.e. the equality constraint

$$\mathbf{X} = \mathbf{Y} \qquad (10)$$

by differentiating it with respect to time (and keeping into account (4) and (9)) it follows that

$$\dot{X} = \dot{Y} \iff J_b(R)\,\dot{\mathbf{q}}_b = J_r(\mathbf{q}_r)\,\dot{\mathbf{q}}_r \qquad (11)$$

which makes explicit the linear constraint conditions existing between the body velocities and the joint velocities of the robotic structure.

Finally, by also recalling that the set of the h counteracting contact forces $-\mathbf{f}_1, ..., -\mathbf{f}_h$ (each one acting on the corresponding end-effector tip) globally reflect on the manipulating structure as an equivalent set of additional *joint* torques $\boldsymbol{\eta}$, expressed by the dual relationship

$$\boldsymbol{\eta} = -J_r^T(\mathbf{q}_r)\mathbf{F} \qquad (12)$$

we can consequently write down the *dynamic* equations of the manipulating structure itself, directly in the form

$$A_r(\mathbf{q}_r)\,\ddot{\mathbf{q}}_r + B_r(\mathbf{q}_r, \dot{\mathbf{q}}_r)\,\dot{\mathbf{q}}_r + C_r(\mathbf{q}_r) = \mathbf{m} - J_r^T(\mathbf{q}_r)\mathbf{F} \qquad (13)$$

where matrix $A_r(\mathbf{q}_r)$ (positive definite) represents the inertia matrix of the whole multifinger structure, and where vectors $B_r(\mathbf{q}_r, \dot{\mathbf{q}}_r)\,\dot{\mathbf{q}}_r, C_r(\mathbf{q}_r)$ keep into account for the corresponding centrifugal-Coriolis and gravitational effects, respectively. Moreover, vector $\mathbf{m}$ represents the set of *external input* joint torques, which can be applied to the robotic structure for control purposes.

The set of equations (8) (body dynamics), (13) (robot dynamics) and also (2) (for representing R, since not given by a pure integration process) completely defines the dynamics of the *entire*, contact constrained, system.

Due to the presence of the algebraic constraints (10), such equations take the form of a set of *DAE* (Differential-Algebraic Equations), which obviously maintain their validity within the fulfilment of the so-called "static friction conditions" for each one of the contacting forces $\mathbf{f}_i, i = 1, ...h$ (i.e., each $\mathbf{f}_i$ acting toward the body and located inside the corresponding *friction cone*).

Throughout the sequel, we shall very often refer to the above set of the DAE's (8), (13) and (10) by using the following, more compact representation (arguments avoided for ease of notations)

$$\begin{cases} A\,\ddot{\mathbf{q}} + B\,\dot{\mathbf{q}} + C = \boldsymbol{\mu} + J^T\mathbf{F} \\ X = Y \end{cases} \qquad (14)$$

where we have posed

$$\begin{aligned} \dot{\mathbf{q}} &\triangleq col(\dot{\mathbf{q}}_b, \dot{\mathbf{q}}_r) \\ \boldsymbol{\mu} &\triangleq col(\mathbf{0}, \mathbf{m}) \end{aligned} \qquad (15)$$

and consequently

$$\begin{aligned} A &\triangleq blockdiag(A_b, A_r) \\ B &\triangleq blockdiag(B_b, B_r) \\ C &\triangleq col(C_b, C_r) \\ J^T &\triangleq col(J_b^T, -J_r^T) \end{aligned} \qquad (16)$$

Finally observe that, as a consequence of the above definitions, also the previously noted existing velocity constraint (11) (second form) can be more concisely rewritten as

$$J\,\dot{\mathbf{q}} = \mathbf{0} \qquad (17)$$

thus showing that the set of admissible overall velocities $\dot{\mathbf{q}}$ always belong to the null space of the above defined overall Jacobian matrix J.

3 Definition of the Two-levels Hierarchical Trajectory Tracking Control Law

Within the considered overall system, first of all suppose the existence of at least *three distinct* contact points among the h assumed ones (thus $h \geq 3$) which are distributed on the body surface as a consequence of the grasping process.

Then, consider the fixed body frame $<b>$ and consequently note how, under the above conditions, its actual position $\mathbf{x}$ and attitude R can always be algebraically reconstructed via the knowledge (from the actual posture of the robotic structure) of both contact points absolute positions $\mathbf{x}_1, ..., \mathbf{x}_h$ and constant vectors $\boldsymbol{\sigma}_1, ..., \boldsymbol{\sigma}_h$ on the body, with simple geometric considerations.

While keeping in mind the above considerations, let us now introduce a "goal frame" $<a>$, characterized by a known absolute time evolution

$$T_a(t) \triangleq \left[\begin{array}{c|c} R_a(t) & \mathbf{x}_a(t) \\ \hline \mathbf{0} & 1 \end{array} \right] \qquad (18)$$

and, consequently, consider the problem of making the fixed body frame $<b>$ eventually tracking such given frame $<a>$.

Since $\mathbf{x}_a, R_a$ in (18) are, at each time instant, known quantities, as well as $\mathbf{x}$ and R in (1) (due to the above introduced assumption and related considerations), it then primarily follows that a couple of *directly measurable* error vectors representing the mismatch between frames $<b>$ and $<a>$ actually exist (to be possibly considered for control purposes), which are respectively defined as

$$\begin{cases} \mathbf{d} \triangleq \mathbf{x} - \mathbf{x}_a, \text{ distance error} \\ \boldsymbol{\rho} \triangleq \mathbf{r}\theta, \text{ rotation error} \end{cases} \qquad (19)$$

with the second one (projected on frame $<0>$) representing the so called "eigenaxis vector" of frame $<b>$

with respect to $<a>$. This is the vector whose unit versor $\mathbf{r}$ and relevant positive component $\theta \in [-\pi, \pi]$ respectively specify, at each time instant, the axis and the corresponding angle of rotation which are ideally needed for transporting $<b>$ to the actual attitude $R(R_a)^T$, with respect to $<a>$, starting from a condition where $<b>$ itself is thought to be parallel to frame $<a>$.

As it is well known, rotation error vector $\boldsymbol{\rho}$ can always be evaluated via the use of the following formulas ("versors lemma")

$$\begin{cases} [\mathbf{i}_a \wedge] \mathbf{i}_b + [\mathbf{j}_a \wedge] \mathbf{j}_b + [\mathbf{k}_a \wedge] \mathbf{k}_b = 2\mathbf{r} sen\theta; \\ (\mathbf{i}_a^T \mathbf{i}_b) + (\mathbf{j}_a^T \mathbf{j}_b) + (\mathbf{k}_a^T \mathbf{k}_b) = 1 + 2\cos\theta \end{cases} \quad (20)$$

being $(\mathbf{i}_a, \mathbf{j}_a, \mathbf{k}_a)$ and $(\mathbf{i}_b, \mathbf{j}_b, \mathbf{k}_b)$ the columns of rotation matrices R_a and R, respectively.

Since the considered tracking problem is equivalent to that of making both error vectors $\mathbf{d}$ and $\boldsymbol{\rho}$ eventually approaching zero for increasing time, then, in order to investigate about such possibility (if any), we may start by first considering the following positive definite scalar quantity

$$V = \frac{1}{2}(\mathbf{d}^T \mathbf{d} + \boldsymbol{\rho}^T \boldsymbol{\rho}) \triangleq \frac{1}{2} \mathbf{e}^T \mathbf{e} = \frac{1}{2} \|\mathbf{e}\|^2 \quad (21)$$

as a possible candidate Lyapunov function, measuring (one half of) the squared norm of the body global error vector

$$\mathbf{e} \triangleq col(\mathbf{d}, \boldsymbol{\rho}) \quad (22)$$

Then, by differentiating V with respect to time, while keeping into account that

$$\dot{\boldsymbol{\rho}} = \dot{\mathbf{r}}\,\theta + \mathbf{r}\,\dot{\theta} \quad (23)$$

with $\dot{\mathbf{r}}$ obviously orthogonal to $\boldsymbol{\rho}$, and with $\dot{\theta}$ given (after some algebra) by the expression

$$\dot{\theta} = \mathbf{r}^T(\boldsymbol{\omega} - \boldsymbol{\omega}_a) \quad (24)$$

being $\boldsymbol{\omega}_a$ the angular velocity of the target frame $<a>$, we directly get

$$\dot{V} = \mathbf{d}^T(\dot{\mathbf{x}} - \dot{\mathbf{x}}_a) + \boldsymbol{\rho}^T(\boldsymbol{\omega} - \boldsymbol{\omega}_a) = \mathbf{e}^T(\dot{\mathbf{q}}_b - \dot{\mathbf{q}}_a) \quad (25)$$

where we have obviously posed

$$\dot{\mathbf{q}}_a \triangleq col(\dot{\mathbf{x}}_a, \boldsymbol{\omega}_a) \quad (26)$$

At this point, by assuming *also* the knowledge of the velocity reference vector $\dot{\mathbf{q}}_a$, and provided that we could directly assign a body velocity vector $\dot{\mathbf{q}}_b$ of the form

$$\dot{\mathbf{q}}_b = \overline{\dot{\mathbf{q}}}_b \triangleq -\Pi \mathbf{e} + \dot{\mathbf{q}}_a; \Pi > 0 \quad (27)$$

which would, in turn, require a joint velocity vector $\dot{\mathbf{q}}_r$ of the form (consider the constraint (11), for $\dot{\mathbf{q}}_b = \overline{\dot{\mathbf{q}}}_b$, assume J_r full row rank, and then invert with respect to $\dot{\mathbf{q}}_r$)

$$\dot{\mathbf{q}}_r = \overline{\dot{\mathbf{q}}}_r \triangleq J_r^\# J_b \overline{\dot{\mathbf{q}}}_b = J_r^\# J_b(-\Pi \mathbf{e} + \dot{\mathbf{q}}_a) \quad (28)$$

being $J_r^\#$ any right inverse of Jacobian matrix J_r (e.g., in this case, the Moore-Penrose *pseudoinverse*); then, by substituting (27) into (25), it is readily seen that we could consequently make

$$\dot{V} = -\mathbf{e}^T \Pi \mathbf{e} \leq -\underline{\pi} \|\mathbf{e}\|^2 < 0 \quad (29)$$

being $\underline{\pi}$ the minimum eigenvalue of the positive definite *gain* matrix Π.

From (21), (29), the asymptotic convergence of the error vector $\mathbf{e}$ toward zero would then follow directly, with a convergence rate no slower than $\exp(-\underline{\pi} t)$.

In practice, however, since we cannot generally ensure the exact fulfilment of requirement (28) for all the time instants (this is due, at least, to the overall system dynamics), we are naturally led to consider $\overline{\dot{\mathbf{q}}}_r$, as given by (28) (and consequently $\overline{\dot{\mathbf{q}}}_b$, as given by (27)) as nothing more than appropriate reference signals that should be tracked "at best" by the actual robot joint velocities; being however aware of the fact that, generally, we shall have

$$\dot{\mathbf{q}}_r = \overline{\dot{\mathbf{q}}}_r + \delta \overline{\dot{\mathbf{q}}}_r = J_r^\# J_b(-\Pi \mathbf{e} + \dot{\mathbf{q}}_a) + \delta \overline{\dot{\mathbf{q}}}_r \quad (30)$$

which, in turn, will imply (substitute into constraint (11) while keeping (28) into account, and then invert with respect to $\dot{\mathbf{q}}_b$)

$$\dot{\mathbf{q}}_b = (-\Pi \mathbf{e} + \dot{\mathbf{q}}_a) + J_b^+ J_r \delta \overline{\dot{\mathbf{q}}}_r \triangleq \overline{\dot{\mathbf{q}}}_b + \delta \overline{\dot{\mathbf{q}}}_b \quad (31)$$

with J_b^+ the (now unique) *left* inverse matrix of the body Jacobian J_b, and the error term $\delta \overline{\dot{\mathbf{q}}}_b$ obviously defined.

This will, in turn, lead to a time derivative $\dot{V}$ taking on the perturbed form (no more guaranteed to be unconditionally negative definite)

$$\begin{aligned} \dot{V} &= -\mathbf{e}^T \Pi \mathbf{e} + \mathbf{e}^T J_b^+ J_r \delta \overline{\dot{\mathbf{q}}}_r \leq \\ &\leq -\underline{\pi} \|\mathbf{e}\|^2 + (\overline{\sigma}_r / \underline{\sigma}_b) \|\mathbf{e}\| \left\| \delta \overline{\dot{\mathbf{q}}}_r \right\| \end{aligned} \quad (32)$$

being $\underline{\sigma}_b$ the (constant) minimum singular value of matrix J_b, and $\overline{\sigma}_r$ the maximum singular value of matrix J_r, this one evaluated among all possible postures.

As an additional comment to the above considerations, we can also observe that, by substituting the above representations (30), (31) (it is sufficient to consider only the first and second right hand sides, respectively), within the velocity constraints (11), we get the conditions

$$J_b \overline{\dot{\mathbf{q}}}_b = J_r \overline{\dot{\mathbf{q}}}_r; J_b \delta \overline{\dot{\mathbf{q}}}_b = J_r \delta \overline{\dot{\mathbf{q}}}_r \quad (33)$$

thus bringing into evidence that the same constraints (11) also holds *separately* for both reference signals $\overline{\dot{\mathbf{q}}}_b, \overline{\dot{\mathbf{q}}}_r$ and the corresponding tracking errors $\delta \overline{\dot{\mathbf{q}}}_b, \delta \overline{\dot{\mathbf{q}}}_r$.

As a consequence, by simply letting

$$\overline{\dot{\mathbf{q}}} \triangleq col(\overline{\dot{\mathbf{q}}}_b, \overline{\dot{\mathbf{q}}}_r); \delta \overline{\dot{\mathbf{q}}} \triangleq col(\delta \overline{\dot{\mathbf{q}}}_b, \delta \overline{\dot{\mathbf{q}}}_r) \Rightarrow \dot{\mathbf{q}} = \overline{\dot{\mathbf{q}}} + \delta \overline{\dot{\mathbf{q}}} \quad (34)$$

and recalling the definition (16), we can trivially rewrite the above additional conditions (33) in the same concise form as in (17); that is

$$J \overline{\dot{\mathbf{q}}} = \mathbf{0}; J \delta \overline{\dot{\mathbf{q}}} = \mathbf{0} \quad (35)$$

thus showing that also $\overline{\dot{\mathbf{q}}}$ and $\delta \overline{\dot{\mathbf{q}}}$ separately belong to the null space of the overall Jacobian matrix J.

At this point, in order to face with the additional problem of tracking the joint velocity reference signal $\dot{\bar{\mathbf{q}}}_r$, let us now introduce the quadratic form

$$U = \frac{1}{2}(\dot{\mathbf{q}} - \dot{\bar{\mathbf{q}}})^T A (\dot{\mathbf{q}} - \dot{\bar{\mathbf{q}}}) = \frac{1}{2} \delta \dot{\bar{\mathbf{q}}}^T A \delta \dot{\bar{\mathbf{q}}} \quad (36)$$

which results in a positive definite candidate Lyapunov function of the sole vector $\delta \dot{\bar{\mathbf{q}}}_r$, once we keep into account that, due to (33) and definition (16), we have

$$U = \frac{1}{2} \delta \dot{\bar{\mathbf{q}}}_r^T [A_r + (J_b^+ J_r)^T A_b (J_b^+ J_r)] \delta \dot{\bar{\mathbf{q}}}_r \triangleq \\ \triangleq \frac{1}{2} \delta \dot{\bar{\mathbf{q}}}_r^T \Sigma \delta \dot{\bar{\mathbf{q}}}_r \quad (37)$$

where, being Σ positive definite, it is $U > 0$.

The existing analytical advantages in choosing a candidate Lyapunov function of the above form (written as in (36)) instead of a simpler one, will readily appear in the following.

In fact, by differentiating (36) with respect to time along (14), after some simple algebra we primarily get

$$\dot{U} = \delta \dot{\bar{\mathbf{q}}}^T A (\ddot{\mathbf{q}} - \ddot{\bar{\mathbf{q}}}) + \frac{1}{2} \delta \dot{\bar{\mathbf{q}}}^T \dot{A} \delta \dot{\bar{\mathbf{q}}} = \\ = \delta \dot{\bar{\mathbf{q}}}^T (\boldsymbol{\mu} - A\ddot{\bar{\mathbf{q}}} - B\dot{\bar{\mathbf{q}}} - C + J^T \mathbf{F}) + \frac{1}{2} \delta \dot{\bar{\mathbf{q}}}^T \dot{A} \delta \dot{\bar{\mathbf{q}}} = \\ = \delta \dot{\bar{\mathbf{q}}}^T (\boldsymbol{\mu} - A\ddot{\bar{\mathbf{q}}} - B\dot{\bar{\mathbf{q}}} - C) + \delta \dot{\bar{\mathbf{q}}}^T J^T \mathbf{F} + \\ - \delta \dot{\bar{\mathbf{q}}}^T B \delta \dot{\bar{\mathbf{q}}} + \frac{1}{2} \delta \dot{\bar{\mathbf{q}}}^T \dot{A} \delta \dot{\bar{\mathbf{q}}} \quad (38)$$

Then, by exploiting the well known property of matrices $B, \dot{A}$ (i.e. $\mathbf{z}^T B \mathbf{z} = 1/2(\mathbf{z}^T \dot{A} \mathbf{z})$) and also noting that the second term in the right hand side of (38) is actually zero (due to the second of (35)) we can consequently simplify (38) as follows

$$\dot{U} = \delta \dot{\bar{\mathbf{q}}}^T (\boldsymbol{\mu} - A\ddot{\bar{\mathbf{q}}} - B\dot{\bar{\mathbf{q}}} - C) \quad (39)$$

thus yelding (by making explicit the dependence on $\delta \dot{\bar{\mathbf{q}}}_r$ while recalling definitions (16))

$$\dot{U} = \delta \dot{\bar{\mathbf{q}}}_r^T [\mathbf{m} - (A_r \ddot{\bar{\mathbf{q}}}_r + B_r \dot{\bar{\mathbf{q}}}_r + C_r) + \\ - (J_b^+ J_r)^T (A_b \ddot{\bar{\mathbf{q}}}_b + B_b \dot{\bar{\mathbf{q}}}_b + C_b)] \quad (40)$$

At this point, by assuming also the knowledge of the body acceleration reference vector $\ddot{\bar{\mathbf{q}}}_a$ (which, as it can be shown, allows the on-line evaluation of vector signals $\ddot{\bar{\mathbf{q}}}_b, \ddot{\bar{\mathbf{q}}}_r$) from (40) we can readily see that, with the adoption of a joint torque control signal having the form

$$\mathbf{m} = -\Lambda \delta \dot{\bar{\mathbf{q}}}_r + (A_r \ddot{\bar{\mathbf{q}}}_r + B_r \dot{\bar{\mathbf{q}}}_r + C_r) + \\ + (J_b^+ J_r)^T (A_b \ddot{\bar{\mathbf{q}}}_b + B_b \dot{\bar{\mathbf{q}}}_b + C_b) \quad (41)$$

with $\Lambda > 0$, and where all the appearing quantities are actually measurable ones (in particular refer to (31) for $\dot{\mathbf{q}}_b$ appearing inside B_b); so, we can actually make

$$\dot{U} = -\delta \dot{\bar{\mathbf{q}}}_r^T \Lambda \delta \dot{\bar{\mathbf{q}}}_r < 0 \quad (42)$$

which guarantees the asymptotic convergence of $\delta \dot{\bar{\mathbf{q}}}_r$ (and, then, of $\delta \dot{\bar{\mathbf{q}}}_b$ due to (31) or, equivalently, to the second of (33) solved for $\delta \dot{\bar{\mathbf{q}}}_b$) toward zero.

As it concerns the convergence rate, it is instead sufficient to observe that, by factorizing the positive definite matrix Σ in terms of its square root matrix H; that is

$$\Sigma = H^T H \quad (43)$$

and introducing the auxiliary vector

$$\mathbf{z} \triangleq H \delta \dot{\bar{\mathbf{q}}}_r \Leftrightarrow \delta \dot{\bar{\mathbf{q}}}_r = H^{-1} \mathbf{z} \quad (44)$$

we can actually rewrite (37) as

$$U = \frac{1}{2} \|\mathbf{z}\|^2 \quad (45)$$

and consequently (42) as

$$\dot{U} = -\mathbf{z}^T (H^{-T} \Lambda H^{-1}) \mathbf{z} \leq -(\underline{\lambda}/\overline{\sigma}) \|\mathbf{z}\|^2 \quad (46)$$

being $\underline{\lambda}, \overline{\sigma}$ the minimum and maximum eigenvalue (the latter evaluated among all possible postures) of positive definite matrices Λ and Σ, respectively. Then, since (45), (46) together imply that auxiliary vector $\mathbf{z}$ converges to zero with a rate no slower than $\exp[-(\underline{\lambda}/\overline{\sigma})t]$, we can immediately conclude that the same also holds for $\delta \dot{\bar{\mathbf{q}}}_r$, due to the fact that (from the second of (44))

$$\|\delta \dot{\bar{\mathbf{q}}}_r\| \leq (1/\underline{\sigma})^{1/2} \|\mathbf{z}\| \quad (47)$$

being $\underline{\sigma}$ the minimum eigenvalue of the matrix Σ among all possible postures. Naturally enough, the same behaviour also extends to $\delta \dot{\bar{\mathbf{q}}}_b$, as a consequence of the existing condition (see (31), or equivalently the second of (33) solved for $\delta \dot{\bar{\mathbf{q}}}_b$, and keep into account the upper bound adopted in (32))

$$\|\delta \dot{\bar{\mathbf{q}}}_b\| \leq (\overline{\sigma}_r / \underline{\sigma}_b) \|\delta \dot{\bar{\mathbf{q}}}_r\| \quad (48)$$

Finally, the ensured exponential convergence of $\delta \dot{\bar{\mathbf{q}}}_r$ toward zero, also guarantees the asymptotic convergence to zero of the tracking error $\mathbf{e}$ (as just established by (32), whenever considered with an exponentially zeroing term $\delta \dot{\bar{\mathbf{q}}}_r$).

As an additional comment to the previous developments, it is last worth noting how the proposed control law (41) can actually be considered as a special case of the following more general form

$$\mathbf{m} = -\Lambda \delta \dot{\bar{\mathbf{q}}}_r + (A_r \ddot{\bar{\mathbf{q}}}_r + B_r \dot{\bar{\mathbf{q}}}_r + C_r) + \\ + (J_b^+ J_r)^T (A_b \ddot{\bar{\mathbf{q}}}_b + B_b \dot{\bar{\mathbf{q}}}_b + C_b) + J_r^T \boldsymbol{\Phi}; \quad (49)$$

$$\boldsymbol{\Phi} \in Ker(J_b^T)$$

where the presence of the additional term $J_r^T \boldsymbol{\Phi}$ (with $\boldsymbol{\Phi}$ as specified) *cannot* however affect the motion of $\delta \dot{\bar{\mathbf{q}}}_r$ (and, then, of $\delta \dot{\bar{\mathbf{q}}}_b$), otherwise imposed by the simpler control law (41).

In fact, by simply substituting (49) into (40), instead of (41), it is straightforward verifying that, with respect to (42), the corresponding expression for $\dot{U}$ simply modifies itself with the addition of the, however *null*, term

$$\delta \dot{\bar{\mathbf{q}}}_r^T J_r^T \boldsymbol{\Phi} = \delta \dot{\bar{\mathbf{q}}}_b^T J_b^T \boldsymbol{\Phi} = \mathbf{0} \quad (50)$$

where the equalities directly follow from kinematic constraints (33), and the assumed choice for $\boldsymbol{\Phi}$ in (49).

As it will be better clarified in the next section, the presence of the additional term $J_r^T \boldsymbol{\Phi}$, with $\boldsymbol{\Phi} \in Ker(J_b^T)$, will actually play an important role within the definition of an "ideal" complimentary force control law for contact constraints maintenance, in presence of dry friction.

4 Definition of the Contact Force Control Law

In the previous section, an "ideal" trajectory tracking control law for the overall system has been devised, under the assumption of having the contact constraints conditions always satisfied during motion. Naturally enough, since this requires the set of contact forces $\mathbf{F} \triangleq col(\mathbf{f}_1, \mathbf{f}_2, ..., \mathbf{f}_h)$ with elements each one directed toward the body, while remaining each one located inside the corresponding friction cone, the obvious need for a complimentary control law, capable of maintaining such conditions during motion, immediately arises.

In order to possibly devise such an additional control law, a preliminar discussion concerning the structure attained by the contact forces $\mathbf{F}$, under validity of contact constraints and joint torques $\mathbf{m}$ of the form (49), is however needed.

To this aim, let us start by first considering the general expression characterizing $\mathbf{F}$ under the action of unstructured input joint torques $\mathbf{m}$. Such an expression can be easily obtained from the general (DAE) dynamic model (14) by first twice differentiating the relevant constraints conditions (to do this, simply differentiate once the compact form (17)), thus obtaining the condition

$$J \ddot{\mathbf{q}} + \dot{J} \dot{\mathbf{q}} = \mathbf{0} \tag{51}$$

Then, by substituting in it the expression for $\ddot{\mathbf{q}}$ obtained from the first equation in (14), and in turn solving for $\mathbf{F}$, we get

$$\begin{aligned}\mathbf{F} &= -(JA^{-1}J^T)^{-1}[JA^{-1}(\boldsymbol{\mu} - B\dot{\mathbf{q}} - C) + \dot{J}\dot{\mathbf{q}}] = \\ &= -(JA^{-1}J^T)^{-1}[-J_r A_r^{-1} \mathbf{m} + J_r A_r^{-1}(B_r \dot{\mathbf{q}}_r + C_r) + \\ &\quad - J_b A_b^{-1}(B_b \dot{\mathbf{q}}_b + C_b) + \dot{J}\dot{\mathbf{q}}]\end{aligned} \tag{52}$$

At this point, by structuring $\mathbf{m}$ as given by (49), we get the equation below

$$\begin{aligned}\mathbf{F} = -(JA^{-1}J^T)^{-1}[&-J_r \ddot{\bar{\mathbf{q}}}_r - J_r A_r^{-1}(B_r \dot{\bar{\mathbf{q}}}_r + C_r) + \\ &-(J_r A_r^{-1} J_r^T)(J_b^+)^T(A_b \ddot{\bar{\mathbf{q}}}_b + B_b \dot{\bar{\mathbf{q}}}_b + C_b) + \\ &+ J_r A_r^{-1}(B_r \dot{\mathbf{q}}_r + C_r) - J_b A_b^{-1}(B_b \dot{\mathbf{q}}_b + C_b) + \\ &+ \dot{J}\dot{\mathbf{q}} + J_r A_r^{-1} \Lambda \delta \dot{\bar{\mathbf{q}}}_r - (J_r A_r^{-1} J_r^T)\boldsymbol{\Phi}]\end{aligned} \tag{53}$$

which can be rewritten as

$$\begin{aligned}\mathbf{F} = -(JA^{-1}J^T)^{-1}[&(-J_r \ddot{\bar{\mathbf{q}}}_r + \dot{J}\dot{\mathbf{q}}) + \\ &+ J_r A_r^{-1}(\Lambda + B_r)\delta \dot{\bar{\mathbf{q}}}_r - (J_r A_r^{-1} J_r^T)\boldsymbol{\Phi} + \\ &- (J_r A_r^{-1} J_r^T)(J_b^+)^T(A_b \ddot{\bar{\mathbf{q}}}_b + B_b \dot{\bar{\mathbf{q}}}_b + C_b) + \\ &- (J_b A_b^{-1} J_b^T)(J_b^+)^T(B_b \dot{\mathbf{q}}_b + C_b)]\end{aligned} \tag{54}$$

where we have simply added together the second and the fourth term in (53), and where the last term in (54) has been obtained from the fifth of (53) by simply exploiting the well known identity $J_b^+ J_b = I$.

We can actually furtherly process the first term of (54) by simply rewriting it as

$$\begin{aligned}-J_r \ddot{\bar{\mathbf{q}}}_r + \dot{J}\dot{\mathbf{q}} &= (-J_r \ddot{\bar{\mathbf{q}}}_r + \dot{J}\dot{\bar{\mathbf{q}}}) + \dot{J}\delta\dot{\bar{\mathbf{q}}} = \\ &= -J_b \ddot{\bar{\mathbf{q}}}_b + \dot{J}\delta\dot{\bar{\mathbf{q}}} = \\ &= -(J_b A_b^{-1} J_b^T)(J_b^+)^T A_b \ddot{\bar{\mathbf{q}}}_b + \dot{J}\delta\dot{\bar{\mathbf{q}}}\end{aligned} \tag{55}$$

where the second right hand side directly follows from differentiation of the first of conditions (35), while the third comes from the identity $J_b^+ J_b = I$.

As a consequence, by substituting (55) in (54) while keeping into account definition (31), and furtherly noting the following equalities

$$\begin{aligned}(JA^{-1}J^T) &= (J_b A_b^{-1} J_b^T) + (J_r A_r^{-1} J_r^T); \\ (J_r A_r^{-1} J_r^T)\boldsymbol{\Phi} &= [(J_b A_b^{-1} J_b^T) + (J_r A_r^{-1} J_r^T)]\boldsymbol{\Phi} = \\ &= (JA^{-1}J^T)\boldsymbol{\Phi}\end{aligned} \tag{56}$$

where the second follows directly from condition $\boldsymbol{\Phi} \in Ker(J_b^T)$ in (49), we finally get the expression

$$\begin{aligned}\mathbf{F} = (J_b^+)^T(A_b \ddot{\bar{\mathbf{q}}}_b + B_b \dot{\bar{\mathbf{q}}}_b + C_b) + \\ -(JA^{-1}J^T)[J_r A_r^{-1}(\Lambda + B_r)\delta\dot{\bar{\mathbf{q}}}_r + \\ + J_b A_b^{-1}(\Lambda + B_b)\delta\dot{\bar{\mathbf{q}}}_b + \dot{J}\delta\dot{\bar{\mathbf{q}}}] + \boldsymbol{\Phi};\end{aligned} \tag{57}$$

$$\mathbf{F} \triangleq \overline{\mathbf{F}} + \delta\overline{\mathbf{F}} + \boldsymbol{\Phi}$$

From the above final expression, we may note how the set of contacting forces $\mathbf{F}$ can be consequently interpreted as composed by three different terms:

- a first one $\overline{\mathbf{F}}$, clearly coinciding with the *minimum norm* forces required for assigning to the body the reference velocity $\dot{\bar{\mathbf{q}}}_b$ (consider the body dynamic equation (8) for $\dot{\mathbf{q}}_b = \dot{\bar{\mathbf{q}}}_b$ and solve it with respect to $\mathbf{F}$ minimum norm), in turn tending toward the desired one $\dot{\mathbf{q}}_a$ (see in fact (31) while recalling that both trajectory and velocity errors converge to zero);

- a second term $\delta\overline{\mathbf{F}}$ tending to zero with the velocity errors, which results from the initial transient mismatch between $\dot{\mathbf{q}}_b$ and $\dot{\bar{\mathbf{q}}}_b$;

- and a final third term $\boldsymbol{\Phi}$, only composed by arbitrary internal forces (i.e., lying inside $Ker(J_b^T)$), which result applied to the body as a consequence of the previously accounted presence of the arbitrary term $J_r^T \boldsymbol{\Phi}$ within control law (49).

Naturally enough, since *any* set of contacting forces $\mathbf{F}$ acting on the body can also be expressed (uniquely) as the sum of two *orthogonal* terms relevant to *pure motion forces* $\mathbf{F}_m \in Span(J_b)$ and internal forces $\mathbf{F}_o \in Ker(J_b^T)$, respectively; that is, in the form

$$\mathbf{F} = \mathbf{F}_m + \mathbf{F}_o = P_b\mathbf{F} + (I - P_b)\mathbf{F} \quad (58)$$

with P_b the matrix *orthogonal projector* onto $Span(J_b)$, given by

$$P_b \triangleq J_b(J_b^T J_b)^{-1} J_b^T \quad (59)$$

It immediately follows that, by applying the above decomposition to our case (57), we get the specification

$$\mathbf{F} = \mathbf{F}_m + \mathbf{F}_o = (\overline{\mathbf{F}} + P_b\delta\overline{\mathbf{F}}) + [(I - P_b)\delta\overline{\mathbf{F}} + \mathbf{\Phi}] \quad (60)$$

clearly showing that only the transient force term $\delta\overline{\mathbf{F}}$ generally splits into the two considered component parts, being the remaining terms $\overline{\mathbf{F}}, \mathbf{\Phi}$ already lying inside the pure motion forces and internal forces subspaces, respectively.

Within (60), furtherly note how the first term also represents the minimum norm forces actually needed for assigning to the body the actual velocity $\dot{\mathbf{q}}_b$ (consider, in fact, the body dynamic equation (8), solve it with respect to $\mathbf{F}$ minimum norm, recall that such solution necessarily belongs to $Span(J_b)$, and consequently conclude about its equality with the first term of (60)).

Still from (60) finally note how the problem of actually having each vector component $\mathbf{f}_i; i = 1, ..., h$ of $\mathbf{F}$ confined inside the corresponding friction cone, should consequently be solved by solely acting on $\mathbf{F}_o$, against $\mathbf{F}_m$, by suitably modulating the arbitrary vector $\mathbf{\Phi}$ (not influencing the motion), since all the remaining terms are already imposed by the motion status independently forced by the application of the previously structured control law (49).

In order to be able to eventually define a complimentary control law for the fulfilment of the contact constraint conditions, in the following we shall also make widely use of a suitable parametrization for the set of internal forces $\mathbf{F}_o$ acting on the body, which has been recently introduced in [4]. Such parametrization actually corresponds to what is hereafter explained.

Let $\mathbf{n}_i; i = 1, ..., h$ be the inward unitary normals to the body at the contact points, each one projected on the body frame $<b>$ (then *constant* in time). Moreover, define the so called *normal matrix* $N \triangleq blockdiag(\mathbf{n}_1, ..., \mathbf{n}_h) \in \Re^{3h \times h}$, also constant in time, and let $\boldsymbol{\alpha} \triangleq [\alpha_1, ..., \alpha_h]^T$ be an h-dimensional vector of scalar parameters. Furtherly denote with $\mathbf{\Psi}_o$ the projection on the body of the set $\mathbf{F}_o$ of (base frame projected) internal forces acting on the body.

Then, it can be shown that under very mild assumptions (namely, $h \geq 3$), such set $\mathbf{F}_o$ of forces always admits a parametrized representation of the form

$$\mathbf{F}_o = R^{(h)}\mathbf{\Psi}_o = R^{(h)}[(N + D)\boldsymbol{\alpha} + S\boldsymbol{\lambda}] \quad (61)$$

where $R^{(h)}$ is the h-th order block-diagonal organization of the body rotation matrix R (here used for simply transferring projections from body frame $<b>$ to base frame $<0>$), while $(N + D) \in \Re^{3h \times h}$, $S \in \Re^{3h \times (2h-6)}$ are full rank constant matrices, only depending on both constant normal matrix N and body Jacobian matrix G projected on the body frame. In particular, for a given configuration of contact points on the body, matrices D and obviously N are unique, while S can be changed only as a consequence of a change of basis in the space of the additional $(2h - 6)$-dimensional parameter vector λ.

For the structural constant matrices $(N+D)$ and S it can also be shown that, while entirely spanning the internal force space of the body, they also decompose it into two orthogonal subspaces: the first spanned by $(N + D)$, and the other one obviously spanned by S. Moreover, the vector force components separately due to $D\boldsymbol{\alpha}$ and $S\boldsymbol{\lambda}$ are each one *tangent* to the body surface, at the corresponding contact point. Consequently, only the vector forces due to $N\boldsymbol{\alpha}$ have non-zero components along the normals $\mathbf{n}_i; i = 1, ..., h$ at the contact points (which are obviously represented, globally, by the h-order vector $\boldsymbol{\alpha}$ itself).

With respect to parametrization (61) we shall also assume of having preliminarily determined (i.e. via *off-line* computations) at least one finite norm $[\underline{\boldsymbol{\alpha}}^T, \underline{\boldsymbol{\lambda}}^T]^T$ (supposed existing), satisfying the following conditions:

$$\mu\underline{\boldsymbol{\alpha}}_i \geq |D_i\underline{\boldsymbol{\alpha}} + S_i\underline{\boldsymbol{\lambda}}|, i = 1, 2, ..., h \quad (62)$$

where $0 < \mu < 1$ is the (assumed common) static friction coefficient relevant to the contact, while D_i, S_i are the matrices obtained from D and S, respectively, by "slicing" them into blocks, each one constituted by three sequential rows. As it can be easily realized, satisfying (62) simply corresponds to have preliminarily off-line established the structure of a *particular* set of (body frame projected) internal forces, taking on the form

$$\underline{\mathbf{\Psi}}_o = [(N + D)\underline{\boldsymbol{\alpha}} + S\underline{\boldsymbol{\lambda}}] \quad (63)$$

and such that the corresponding (non null) force components $\underline{\mathbf{\Psi}}_i; i = 1, 2, ..., h$ are all located inside the relevant friction cone. As a matter of fact, a finite norm parameter vector $[\underline{\boldsymbol{\alpha}}^T, \underline{\boldsymbol{\lambda}}^T]^T$ satisfying (62) can be computed using very simple linear programming techniques.

Together with the above defined particular set of internal forces $\underline{\mathbf{\Psi}}_o$ projected on body frame, let us now also consider the following related (strictly) positive parameter

$$\underline{\beta} \triangleq \min_i \{\underline{\beta}_i; i = 1, 2, ..., h\} > 0 \quad (64)$$

where $\underline{\beta}_i$ represents the distance exhibited by each force component $\underline{\Psi}_i; i = 1, 2, ..., h$ of $\underline{\Psi}_o$ with respect to the surface of the corresponding friction cone. Then, on the basis of (63), (64), let us finally define the following *reference* set of body frame projected internal force vectors:

$$\overline{\Psi}_o \triangleq \frac{1}{\underline{\beta}} \underline{\Psi}_o \quad (65)$$

which results in a set of body frame projected internal forces characterized by a *unitary* minimum distance from the friction cones surfaces.

Given this result, let us first confine ourselves to consider only the sets of base frame projected internal forces $\mathbf{F}_o$, which are strictly aligned with the above defined reference one $R^{(h)}\overline{\Psi}_o$, after projection on the base frame $<0>$.

Moreover, consider decomposition (60) (first right hand side) and, after definition

$$g(\mathbf{F}_m) \triangleq \max_i \{|\mathbf{f}_{m,i}|; i = 1, 2, ..., h\} \quad (66)$$

simply note that a sufficient condition for having each vector component $\mathbf{f}_i; i = 1, 2, ..., h$ of $\mathbf{F}$ confined within the corresponding friction cone during time, is actually that of having the internal force part $\mathbf{F}_o$ of $\mathbf{F}$ of the form (worst case design)

$$\mathbf{F}_o = \gamma g(\mathbf{F}_m) R^{(h)} \overline{\Psi}_o; \gamma > 1 \quad (67)$$

thus leading to the following expression for the additional internal force term $\mathbf{\Phi}$ in (49) (see the expressions for $\mathbf{F}_m$ and $\mathbf{F}_o$ in (60), second right hand side)

$$\mathbf{\Phi} = \gamma g(\overline{\mathbf{F}} + P_b \delta \overline{\mathbf{F}}) R^{(h)} \overline{\Psi}_o - (I - P_b) \delta \overline{\mathbf{F}}; \gamma > 1 \quad (68)$$

which, consequently, represents the previously searched complimentary control law for contact constraints maintenance during motion.

Apparently, such control law operates in real-time by first compensating for the presence of the independently imposed (transient) internal force term $(I - P_b)\delta \overline{\mathbf{F}}$, while simultaneously and suitably moving the resultant internal force set $\mathbf{F}_o$ along the base frame projection of $\overline{\Psi}_o$, for maintaining the validity of sufficient condition (67) during time. This is done on the basis of the knowledge of terms $\overline{\mathbf{F}}, \delta \overline{\mathbf{F}}$, which are, in turn, known functions of measurable quantities of posture and motion type (as established by their expression given by the first two terms of (57), respectively).

Finally, last but not all least, also note how complimentary control law (68) actually results to be of *open-loop type* with respect to its controlled output $\mathbf{F}_o$.

Note, however, that in case of a possible use of fingertips tactile sensors measuring $\mathbf{F}$, a closed loop force control law of the form

$$\dot{\mathbf{\Phi}} = K[\gamma g(\mathbf{F}_m) R^{(h)} \overline{\Psi}_o - \mathbf{F}_o]; K > 0 \quad (69)$$

could also be used, where $\mathbf{F}_o$ and $\mathbf{F}_m$ are on-line evaluated from measurements of $\mathbf{F}$, via the application of (58) and (59). As it can be easily realized, applying (69) simply corresponds to control the output $\mathbf{F}_o$ (seen as given by the second term of (60)), via an *integral* control action dealing with $(I - P_b)\delta \overline{\mathbf{F}}$ as with a transient disturbance.

It is however clear how further investigations are actually needed for getting insight into the possible effectiveness of (69).

5 Conclusions

In this paper a mathematical development for a two-level closed loop hierarchical control architecture, plus a complimentary force control law, has been done. Convergence to zero of the trajectory errors and stability porperties have also been proven via Lyapunov arguments.

References

[1] J. Kerr, *An Analysis of multifingered Hands*, PhD thesis, Dept. of Mechanical Engineering, Stanford University, 1984

[2] Y. Nakamura, K. Nagai, T. Yoshikawa, *Dynamics and stability in coordination of multiple robotic mechanisms*, Int. Journal of Robotics Research, 8(2), 1989

[3] B. Mishra, J. T. Schwartz and M. Sharir, *On the existence and synthesis of multifingered positive grips*, Algorithmica, 2:541-558, 1987

[4] M. Aicardi, G. Cannata, G. Casalino, *Contact Force Canonical Decomposition and the Role of Internal Forces in Robust Grasp Planning Problems*, Int. Journal of Robotics Research, vol. 15 n. 4, Aug. 1996

[5] D. J. Montana, *The kinematics of contact and grasp*, Int. Journal of Robotics Research, 7(3), 1988

[6] O. Khatib, *A Unified Approach for Motion and Force Control of Robot Manipulators: The Operational Space Formulation*, IEEE Journal on Robotics and Automation, Feb. 1987

[7] R. M. Murray, Z. Li, S. Sastry, *A Mathematical Introduction to Robotic Manipulation*, CRC Press 1993

[8] J. K. Salisbury, M. T. Mason, *Robot Hands and the Mechanics of Manipulation*, Cambridge: MIT Press, 1985

Robust Manipulation of Deformable Objects By A Simple PID Feedback

T.Wada[†], S.Hirai[‡], S.Kawamura[‡], and N.Kamiji[*]

[†] Dept. of Intelligent Mechanical Systems, Faculty of Engineering, Kagawa University
[‡] Department of Robotics, Faculty of Science and Engineering, Ritsumeikan Univ.
[*] Kamiji Ltd.
Email: wachan@robot.club.ne.jp   URL: http://www.eng.kagawa-u.ac.jp/~wada/

Abstract

Robust manipulation strategies of deformable objects will be presented. Manipulation of deformable objects can be found in many fields such as garment industry and food industry. Guidance of multiple points on a deformable object is a primitive operation in the manipulation of deformable objects. In this guidance, the points often cannot be manipulated directly. A model of the manipulated deformable object is needed in order to perform these operations. It is, however, difficult to build a precise model of a deformable object. Thus, we need a robust control scheme that allows us to realize the operations successfully despite of discrepancy between a manipulated deformable object and its model.

In this paper, we will firstly derive a mathematical model of deformable objects for their manipulation. Second, indirect simultaneous positioning operations of deformable objects are formulated. Then, we will propose a PID feedback control law with the rough object model to realize the manipulation. Furthermore, we will propose a simple PID feedback control law without deformation model. The validity and the robustness of the proposed manipulation method will be shown through simulation results.

1 Introduction

There exist many manipulative tasks that deal with deformable objects such as textile fabrics, rubber parts, paper sheets, and food products. Most these operations strongly depend on skilled human workers. We define manipulation of deformable objects as controlling of deformation of deformable objects as well as their positions and orientations in this paper. For example, a positioning operation called *linking* is involved in the manufacturing of seamless knitted products as shown in Fig.1 [1]. In linking of fabrics, knitted loops at the end of a fabric must be matched to those of another fabric so that the two fabrics can be sewed seamlessly. This operation is now done by skillful humans and automatic linking is required in manufacturing of knitted products. In this research, we describe the manipulations of deformable objects including linking by positioning of multiple points on the objects. Then, we regard the manipulations as the operations in which multiple points on a deformable object should be guided to the final locations simultaneously as shown in Fig.2. In many cases these points cannot be, however, manipulated directly. Thus, the guidance of positioned points must be performed by controlling some points except the positioned points. This operation is referred to as *indirect simultaneous positioning* [2]. In this paper, we will focus on indirect simultaneous positioning as a fundamental operation of manipulation of deformable objects.

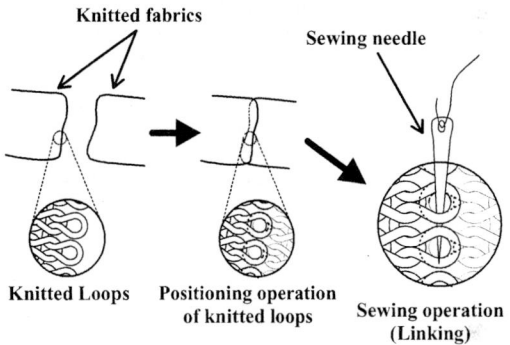

Figure 1: Linking of knitted fabrics

Some researches on manipulations of deformable objects have been conducted. For automated manufacturing of textile fabrics, many researches have been done [3]. Ono et al. [4] have derived a strategy for unfolding a fabric piece based on cooperative sensing of touch and vision. In these researches, since their approaches are for a specific task, thus it is difficult to apply the results to other different tasks in a systematic manner. Some researches have tried to deal with more general deformable object in systematic manners as follows. Hirai et al. [5] have proposed a method for modeling linear objects based on their potential energy and analyzed their static deformation. Wakamatsu et al. [6] have analyzed grasping of deformable objects and have introduced bounded force closure. In this approach, control of manipulative operations is out of consideration. Howard et al. [7] have proposed a method to model elastic objects by the connections of springs and dampers. A method to estimate the coefficients of the springs and dampers has been developed by

recursive learning method for grasping. This study has focused on model building. Thus, control problems for manipulative operations have not been investigated. Sun et al. [8] have studied on the positioning operation of deformable objects using two manipulators. They have focused on the control of the object position while deformation control is not discussed.

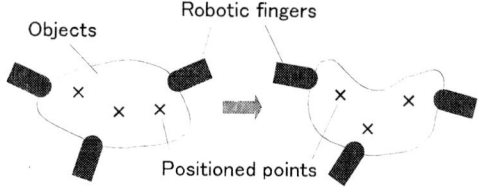

Figure 2: Indirect positioning of deformable object

In order to realize indirect simultaneous positioning, object model is important. However, it is difficult to build an exact model of the deformable objects in general due to nonlinear elasticity, friction, hysteresis, parameter variations, and other uncertainties. This is a main difficulty in manipulating deformable objects. To solve this dilemma, we have proposed to derive a robust manipulation strategy based on a coarse object model [2].

Based on this basic idea, we have proposed an iterative control scheme for the positioning [2]. It is, however, difficult to increase the speed of convergence in this approach. Also, it is difficult to maintain a stable grasp in grasping task as shown is Fig.2 because information of the deformation can only be utilized discretely [9]. Thus, it is desired to realize a new control law in which the deformation characteristics are utilized continuously.

Thus, we will realize the manipulation of deformable objects by a simple PID feedback control that can utilize information of the deformation. In this control law, we only use a velocity relationship between positioned points and manipulation points. Then, the locations of the positioned points are measured in real-time, and the location is fed back to the robotic finger using the velocity relationship.

In this article, we will firstly build a coarse model of deformable objects. Next, indirect positioning will be formulated based on the coarse model. In the formulation, we will derive a relationship between velocity of the positioned points and that of manipulation points based on the equilibrium equations at the positioned points. Then, we will propose a feedback control law with an approximate model. Furthermore, We explore feasibility of utilizing a feedback controller with only rough kinematic model, that is, without the deformation model. Simulation results will show the validity and robustness of the proposed control law.

2 Formulation of Manipulation of Deformable Objects

2.1 Modeling of Deformable Objects

First of all, a model of deformable objects is derived. On modeling of deformable objects, many researches have been conducted. For example in the area of computer graphics, cloth deformation is animated by Terzopoulos [10] or Louchet, Provot and Crochemore [11] and other many researchers. Our research goal is to realize robust manipulation of deformable objects. Therefore, we employ more simple deformation model. We model the object by connections of simple springs similar with Naster and Ayache [12]. For simplicity, we deal with two-dimensional deformable objects such as textile fabrics. We discretize the object by mesh points. Each mesh point is connected by vertical, horizontal, and diagonal springs as shown in Fig.3. In the model, we assume that the object deforms in a two-dimensional plane. In order to formulate the manipulation of deformable objects, object model must have the ability to describe translation, orientation, and deformation of the object simultaneously. Thus, position vector of the mesh points is utilized. Position vector of the (i,j)-th mesh point is defined as $\boldsymbol{p}_{i,j} = [x_{i,j}, y_{i,j}]^T$ $(i = 0, \cdots, M; j = 0, \cdots, N)$. Coefficients k_x, k_y, k_θ are spring constants of horizontal, vertical, and diagonal springs. Assume that no moment exert on each mesh point. Then, the resultant force exerted on mesh point $\boldsymbol{p}_{i,j}$ can be described as eq.(1).

$$\boldsymbol{F}_{i,j} = \sum_{k=1}^{8} \boldsymbol{F}_{i,j}^k = -\frac{\partial U}{\partial \boldsymbol{p}_{i,j}} \qquad (1)$$

U denotes whole potential energy of the object. Then, function U can be calculated by sum of all energies of springs [2]. Here, we assume that the shape of the object is dominated by eq.(1). Then, we can calculate the deformation of the object by solving eq.(1) under given constraints. Note that the following discussions are valid even if the object has an arbitrary three-dimensional shape by modeling the object similarly. Details have been reported in [2].

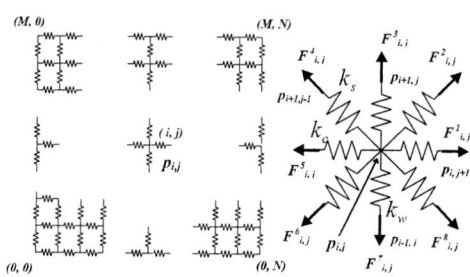

Figure 3: Spring model of deformable object

2.2 Problem Description

Here, we classify mesh points $\boldsymbol{p}_{i,j}$ into the following three categories(see Fig.4) in order to formulate indirect simultaneous positioning.

manipulation points: are defined as the points that can be manipulated directly by robotic fingers. ($\triangle$)

positioned points: are defined as the points that should be positioned indirectly by controlling manipulation points appropriately. ($\bigcirc$)

non-target points: are defined as the all points except the above two points. (others in Fig.4)

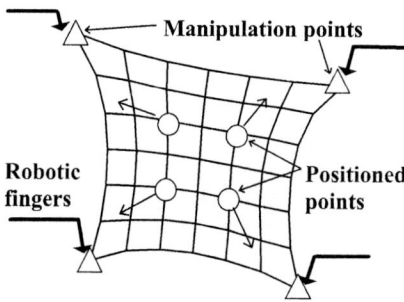

Figure 4: Classification of mesh point

Let the number of manipulation points and of positioned points be m and p, respectively. The number of non-target points is $n = (M+1) \times (N+1) - m - p$. Then, $\bm{r}_m$ is defined as a vector that consists of coordinate values of the manipulation points. Vectors $\bm{r}_p$ and $\bm{r}_n$ are also defined for positioned and non-target points in the similar way. Eq.(1) can be rewritten as eqs.(2),(3) using $\bm{r}_m$, $\bm{r}_p$, and $\bm{r}_n$.

$$\frac{\partial U(\bm{r}_m, \bm{r}_n, \bm{r}_p)}{\partial \bm{r}_m} - \bm{\lambda} = \bm{o}, \quad (2)$$

$$\begin{bmatrix} \frac{\partial U(\bm{r}_m, \bm{r}_n, \bm{r}_p)}{\partial \bm{r}_p} \\ \frac{\partial U(\bm{r}_m, \bm{r}_n, \bm{r}_p)}{\partial \bm{r}_n} \end{bmatrix} = \bm{o} \quad (3)$$

where a vector $\bm{\lambda}$ denotes a set of forces exerted on the object at the manipulation points $\bm{r}_m$ by robotic fingers.

Note that the external forces $\bm{\lambda}$ can appear only in eq.(2), not in eq.(3). This implies that no external forces are exerted on positioned points and non-target points. These equations represent characteristics of indirect simultaneous positioning of deformable objects.

Let us consider the following task:

[Indirect Simultaneous Positioning (ISP)]
Assume that the configuration of robotic fingers and the positioned points on an object are given in advance. In addition, the robotic fingers pinch the object firmly. Then, the positioned points $\bm{r}_p$ are guided to their desired location $\bm{r}_p^d$ by controlling manipulation points $\bm{r}_m$ appropriately.

2.3 Velocity Relationship

Let us derive velocity relation among positioned points and manipulated points. We can obtain the following equation by differentiating eq.(3) by time with $\bm{r}(t) = [\bm{r}_m^T(t), \bm{r}_n^T(t), \bm{r}_p^T(t)]^T$:

$$A(\bm{r})\dot{\bm{r}}_m + B(\bm{r})\dot{\bm{r}}_n + C(\bm{r})\dot{\bm{r}}_p = \bm{o} \quad (4)$$

where

$$A(\bm{r}) \triangleq \begin{bmatrix} \frac{\partial^2 U(\bm{r})}{\partial \bm{r}_m \partial \bm{r}_p} \\ \frac{\partial^2 U(\bm{r})}{\partial \bm{r}_m \partial \bm{r}_n} \end{bmatrix} \in R^{(2p+2n) \times 2m},$$

$$B(\bm{r}) \triangleq \begin{bmatrix} \frac{\partial^2 U(\bm{r})}{\partial \bm{r}_n \partial \bm{r}_p} \\ \frac{\partial^2 U(\bm{r})}{\partial \bm{r}_n \partial \bm{r}_n} \end{bmatrix} \in R^{(2p+2n) \times 2n},$$

$$C(\bm{r}) \triangleq \begin{bmatrix} \frac{\partial^2 U(\bm{r})}{\partial \bm{r}_p \partial \bm{r}_p} \\ \frac{\partial^2 U(\bm{r})}{\partial \bm{r}_p \partial \bm{r}_n} \end{bmatrix} \in R^{(2p+2n) \times 2p}.$$

Vector $\dot{\bm{r}}_m$ is defined as a velocity of the manipulation points. Vectors $\dot{\bm{r}}_n$ and $\dot{\bm{r}}_p$ are defined in the similar way. By transforming eq.(4), eq.(5) is obtained.

$$A(\bm{r})\dot{\bm{r}}_m + G(\bm{r})[\dot{\bm{r}}_n^T, \dot{\bm{r}}_p^T]^T = \bm{o}, \quad (5)$$

where $G(\bm{r}) \triangleq [B(\bm{r})\ C(\bm{r})] \in R^{(n+p) \times (n+p)}$. Note that $\det G(\bm{r}) \neq 0$ [14]. Then, velocity relation among manipulation points and positioned points can be derived as eq.(6) by transforming eq.(5).

$$\dot{\bm{r}}_p = J(\bm{r})\dot{\bm{r}}_m, \quad (6)$$

where $J(\bm{r}) \triangleq -S_L G^{-1}(\bm{r}) A(\bm{r}) \in R^{p \times m}$ and $S_L \triangleq [0, I] \in R^{p \times (n+p)}$. Note that the kinematic relationship among positioned points and manipulation points is included in matrix $J(\bm{r})$ as well as the deformation model.

We can obtain the following theorems. The proofs of these theorems have been reported in [13].

Theorem 1 *There exist infinitesimal displacements of manipulated points $\delta\bm{r}_m$ corresponding to arbitrary infinitesimal displacements $\delta\bm{r}_p$, if and only if,*
rank$[A\ B] = 2p + 2n$ **is satisfied.**

In addition, Theorem 1 needs the following result.

Result 1 *The number of the manipulated points must be greater than or equal to that of the positioned points in order to realize any arbitrary displacement $\delta\bm{r}_p$, that is, $m \geq p$.*

Therefore, we assume $m = p$, the number of robotic fingers equals to that of positioned points in this paper. Thus, Jacobian matrix $J(\bm{r})$ also become a square one.

3 Control Law

In this section, we propose simple PID feedback control laws in order to realize robust manipulation of deformable objects.

3.1 PID Feedback with Approximate Model

In the manipulation of a deformable object, the velocity of the positioned points and that of manipulation points have to satisfy the relationship described by eq.(6). Thus, the following feedback control law is derived as similar as the control of robot manipulators in task space:

$$u = -\hat{J}^T K_P(r_p - r_p^d) - K_V \dot{r}_m - \hat{J}^T K_I \int (r_p - r_p^d) d\tau, \quad (7)$$

where u denotes the input torque or force to actuators of a robot manipulator and $\hat{J}^T$ denotes the Jacobian matrix including some errors in deformation characteristics and the kinematic relation among points. In addition, we assume that a robot has prismatic joints when deriving eq.(7) for the sake of the simplicity. Note that we can derive the similar control law even if the manipulator has rotational joints. Eq.(7) is similar with the robot control in task space with uncertain transposed Jacobian by Cheah et al [15].

3.2 Simple PID Feedback without Deformation Model

In calculation of eq.(7), some errors are acceptable while we need a deformation model of the object with a certain accuracy. However, in the case that the deformation is not large, we may not need deformation model but require a rough kinematic relation among positioned points and manipulation points.

We will propose a method to describe kinematic relationship among manipulation points and positioned points. Before explaining the detail, we show a simple example in order to demonstrate the basic idea of the method. Fig.5 illustrates a manipulation of springs in a line. There exist two manipulation points and two positioned points as shown in the figure. The springs have spring constants k uniformly. Now we calculate J as follows:

Equations of equilibrium at x_2 and x_3 are given by

$$k(x_2 - x_1) + k(x_2 - x_3) = 0,$$
$$k(x_3 - x_2) + k(x_3 - x_4) = 0 \quad (8)$$

By transforming eqs.(8), we obtain eq.(9).

$$\begin{bmatrix} x_2 \\ x_3 \end{bmatrix} = \frac{1}{3} \begin{bmatrix} 2 & 1 \\ 1 & 2 \end{bmatrix} \begin{bmatrix} x_1 \\ x_4 \end{bmatrix} \quad (9)$$

Then, we can obtain

$$J = \frac{1}{3} \begin{bmatrix} 2 & 1 \\ 1 & 2 \end{bmatrix} \quad (10)$$

Note that matrix J of eq.(9) is not diagonal. This means that there exist couplings among the points since these points are connected one another. But, we human sometimes ignore such couplings. Thus, now the kinematic problems are simplified as follows.

1. We will ignore the coupling between x_2 and x_3.

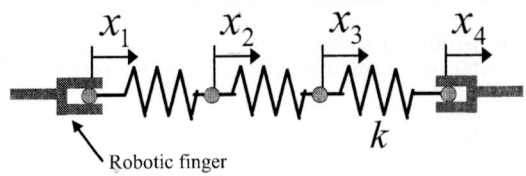

Figure 5: One-dimensional positioning

2. We will guide x_2 and x_3 by controlling x_1 and x_4, respectively. Namely, we will employ the following equation as a controller.

$$\begin{bmatrix} u_1 \\ u_4 \end{bmatrix} = -K_P I \begin{bmatrix} x_2 - x_2^d \\ x_3 - x_3^d \end{bmatrix} - K_V \begin{bmatrix} \dot{x}_1 \\ \dot{x}_4 \end{bmatrix}$$
$$- K_I I \int \begin{bmatrix} x_2 - x_2^d \\ x_3 - x_3^d \end{bmatrix} d\tau, \quad (11)$$

where u_1 and u_4 denote the forces exerted on x_1 and x_4, respectively, x_2^d and x_3^d are desired locations of x_2 and x_3, respectively.

We can prove that the control law, eq.(11) realize $x_2 \to x_2^d$ and $x_3 \to x_3^d$ as $t \to \infty$.

Here, we extend the above idea into two-dimensional positioning. Let us consider the situation shown in Fig.6. The error of a positioned point is fed back to the motion of a manipulation point nearest to the positioned one, as illustrated in Fig.6. Namely, we will ignore the coupling among positioned points and manipulated points. For instance, we have the following control input in the case of Fig.6.

$$\begin{bmatrix} u_{0,3} \\ u_{1,0} \\ u_{3,2} \end{bmatrix} = -S K_P \begin{bmatrix} p_{1,1} - p_{1,1}^d \\ p_{1,2} - p_{1,2}^d \\ p_{2,2} - p_{2,2}^d \end{bmatrix} - K_V \begin{bmatrix} \dot{p}_{0,0} \\ \dot{p}_{0,3} \\ \dot{p}_{3,2} \end{bmatrix}$$
$$- S K_I \int \begin{bmatrix} p_{1,1} - p_{1,1}^d \\ p_{1,2} - p_{1,2}^d \\ p_{2,2} - p_{2,2}^d \end{bmatrix} d\tau, \quad (12)$$

$$S \triangleq \begin{bmatrix} O_2 & I_2 & O_2 \\ I_2 & O_2 & O_2 \\ O_2 & O_2 & I_2 \end{bmatrix} \in R^{6 \times 6} \quad (13)$$

where $u_{0,3}$, $u_{1,0}$, and $u_{3,2}$ denote forces by robotic fingers exerted on $p_{0,3}$, $p_{1,0}$, and $p_{3,2}$, respectively. Matrices O_2 and I_2 denote a zero matrix of 2×2 and an unit matrix of two dimension, respectively. Matrix S defined by eq.(13) represents the correspondence of positioned points to manipulation points. Now, general form of the proposed control law is given as follows:

$$u = -S K_P(r_p - r_p^d) - K_V \dot{r}_m - S K_I \int (r_p - r_p^d) d\tau, \quad (14)$$

where S denotes the correspondence of the positioned points to the manipulation points. At present, it is difficult to determine the correspondence matrix S theoretically. Matrix S is related to the location of manipulation points on the object. Determination of matrix S is an important future work.

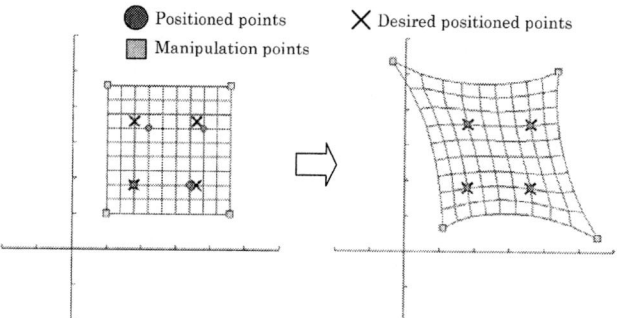

Figure 7: Simulation results

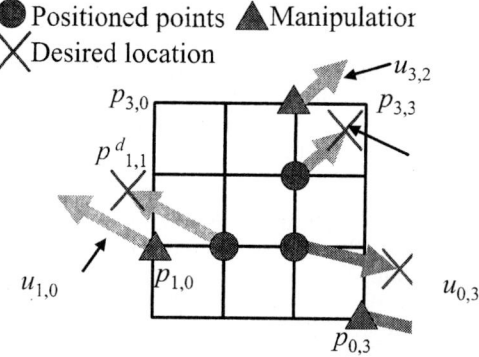

Figure 6: Simplified kinematic relation

4 Simulation Results

In this section, simulation results are illustrated in order to show the validity and the robustness of the proposed control law. Especially, the simple PID feedback control law without deformation model described in eq.(14) is investigated here.

We simulate the deformation of the 2-dimensional object from the left side of Fig.7 to the right side of the figure. Fig.7, the dimension of the object is $180[mm] \times 180[mm]$. The object is descritized into 100 lattice points. The positioned points and the manipulation points are located as shown in Fig.7. In this simulation, we define positioned points and manipulation points as follows:

$$r_p = [p_{2,2}^T, p_{2,6}^T, p_{6,3}^T, p_{6,7}^T]^T \quad (15)$$
$$r_m = [p_{0,0}^T, p_{0,9}^T, p_{9,0}^T, p_{9,9}^T]^T \quad (16)$$

The error of $p_{2,2}$ is fed back to $p_{0,0}$, $p_{2,6}$ to $p_{0,9}$, $p_{6,3}$ to $p_{9,0}$, and $p_{6,7}$ to $p_{9,9}$. Namely, matrix S in eq.(14)is a unit matrix of 6 dimension in this case. The desired location of positioned points is given as follows:

$$r_p^d = [90, 90, \ 180, 90, \ 90, 180, \ 180, 180]^T \quad (17)$$

Recall that we have ignored masses and dampers of an object when we derived eqs.(7) and (14). But in the simulations, we suppose that the deformable objects have masses and dampers as well as springs. Namely, we employ the control law eq.(14) that is derived by ignoring mass and dampers while an object in simulations has masses and dampers. Thus, we also investigate whether the control law eq.(14) works effectively with the objects that has masses,

dampers, and springs through the simulations. Assume that each lattice point has mass $m = 0.01[kg]$ and each mesh has damping coefficient $b = 5[Ns/m]$, and spring constant $k = 10[N/m]$. Each robotic finger has $10[kg]$ mass. All feedback gains are fixed through the simulations.

Fig.8 shows the convergence of the error norms. We can see that the error norm converges to zero. Thus, we can conclude that the manipulation of the deformable object can be realized without the deformation model. On the other hand, we can imagine that there exist limitations on the deformation if matrix J is replaced by S. Namely, the deformation model with a certain accuracy is required in the case of very large deformation. We, thus, would like to investigate how much error is permitted theoretically and experimentally. This is one of important future works.

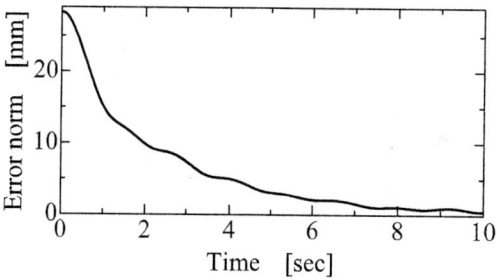

Figure 8: Error norm in simulation

5 Conclusions

In this paper, manipulation of deformable objects has been formulated. We have proposed a PID feedback control law with the approximate model and a simple PID feedback control law without deformation model. The validity and the robustness of the simple feedback without deformation model have been shown through the simulation results.

The simulation results show that the error can converge if the deformation is not large, that is, the differences between Jacobian J and constant matrix S is not large. Furthermore, a deformation model with a certain accuracy is required in order that the error converges to zero if the deformation is large. We will theoretically clarify how much deformations are acceptable when $\hat{J}$ and S are given. It is difficult to build matrix S in the case that the number of positioned points and that of manipulation points are large. This problem is essentially related to the location of the manipulation points on an object. This is also an important future work.

Acknowledgment

This research was funded in part by FANUC FA and Robot Foundation.

References

[1] Wada, T., Hirai, S., Hirano, T., Kawamura, S., "Modeling of Plain Knitted Fabrics for Their Deformation Control", Proc. of IEEE Int. Conf. on Robotics and Automation, pp.1960–1965, 1997

[2] Wada, T., Hirai, S., Kawamura, S., "Planning and Control of Indirect Simultaneous Positioning Operation for Deformable Objects", Proc. of IEEE Int. Conf. on Robotics and Automation, pp.2572–2577, 1999

[3] Taylor, P.M. et al.(Ed.), "Sensory Robotics for the Handling of Limp Materials", Springer-Verlag, 1990

[4] Ono, E., Kita, N., Sakane, S., "Strategy for Unfolding a Fabric Piece by Coorperative Sensing of Touch and Vision", Proc. of Int. Conf. on Intelligent Robots and Systems, pp.441–445, 1995

[5] Hirai, S., Wakamatsu, H., and Iwata, K., "Modeling of Deformable Thin Parts for Their Manipulation", Proc. IEEE Int. Conf. on Robotics and Automation, pp.2955–2960, 1994

[6] Wakamtatsu, H., Hirai, S., Iwata, K., "Static Analysis of Deformable Object Grasping Based on Bounded Force Closure", Proc. IEEE Int. Conf. on Robotics and Automation, pp.3324–3329, 1996

[7] Howard, A.M. and Bekey, G.A., "Recursive Learning for Deformable Object Manipulation", Proc. of Int. Conf. on Advanced Robotics, pp.939–944, 1997.

[8] Sun, D., Liu, Y., Mills, J.K., "Cooperative Control of a Two-Manipulator System Handling a General Flexible Object", Proc. of Int. Conf. on Intelligent Robots and Systems, pp.5–10, 1997

[9] Wada, T., Hirai, S., Mori, H., Kawamura, S., "Robust Manipulation of Deformable Objects Using Model Based Technique", Proc. of Int. Workshop on Articulated Motion and Deformable Objects, in Lecture Notes in Springer Computer Science vol. 1899, pp.1–14, 2000

[10] Terzopoulos, D., Platt, J., Barr, A., Fleischer, K., "Elastically Deformable Models", Proc. of Siggraph 87, Computer Graphics, 1987, Vol.21, No.4, pp.205–214

[11] Louchet, J., Provot, X., Crochemore, D., "Evoluntionary Identification of Cloth Animation Models", Computer Animation and Simulation'95: Proc. of the Eurograpics Workshopin Maastricht, pp.44–54, 1995

[12] Nastar, C. and Ayache, N., "Frequency-Based Nonrigid Motion Analysis: Application to Four Dimensional Medical Images", IEEE Trans. on Pattern Analysis and Machine Intelligence", Vol.18, No.11, pp.1067–1079, 1996

[13] Wada, T., Hirai, S., Kawamura, S., "Indirect Simultaneous Positioning Operations of Extensionally Deformable Objects", Proc. of Int. Conf. on Intelligent Robots and Systems, pp.1333–1338, 1998

[14] Wada, T., Hirai, S., Kawamura, S., "Analysis and Planning of Indirect Simultaneous Positioning Operation for Deformable Objects", Proc. of Int. Conf. on Advanced Robotics, pp.141–146 1999

[15] Cheah, C.C., Kawamura, S., Arimoto, S., Lee, K., "PID Control of Robotic Manipulator with Uncertain Jacobian Matrix",Proc. ICRA'99, pp.494–499,1999

Robust stabilization of the plate-ball manipulation system

Giuseppe Oriolo   Marilena Vendittelli

Dipartimento di Informatica e Sistemistica
Università di Roma "La Sapienza"
Via Eudossiana 18, 00184 Roma, Italy
{oriolo,venditt}@dis.uniroma1.it

Abstract

We consider the plate-ball system as a typical example of manipulation by rolling contacts. While there exist techniques for planning motions of this nonholonomic mechanism in nominal conditions, our objective in this paper is the robust execution of maneuvers in the presence of model perturbations. To this end, we adopt an iterative steering paradigm based on the use of a nilpotent approximation of the system. Simulation results are reported to confirm the robustness achieved with the proposed feedback controller.

1 Introduction

Rolling manipulation has recently attracted the interest of robotic researchers as a convenient way to achieve dexterity with a relatively simple mechanical design (see [1–3] and the references therein). In fact, the nonholonomic nature of rolling contacts between rigid bodies can guarantee the controllability of the manipulation system (hand+manipulated object) with a reduced number of actuators. More in general, this is another example of the minimalistic trend in the field of robotics, aimed at designing devices of reduced complexity for performing complex tasks.

The archetypal example of rolling manipulation is the plate-ball system [4–7]: the ball (the manipulated object) can be brought to any contact configuration by maneuvering the upper plate (the first finger), while the lower plate (the second finger) is fixed. Despite its mechanical simplicity, the planning and control problems for this device already raise challenging theoretical issues. In fact, in addition to the well-known limitations coming from its nonholonomic nature (e.g., the lack of smooth stabilizability), the plate-ball system is neither flat nor nilpotentizable; therefore the classical techniques (e.g., see [8]) for planning and stabilization of nonholonomic systems cannot be applied.

To this date, only the planning problem has been attacked with some success; e.g., see the symbolic algorithm of [5] (which contains an error but admits a suitable modification) and the numerical algorithm of [3]. Like for any planner based on open-loop control, however, the successful execution of maneuvers is not preserved in the presence of perturbations — some sort of feedback is necessary to induce a degree of robustness.

In this paper, we prove that robust stabilization of the plate-ball mechanism can be simply achieved through iterative application of an appropriate open-loop control law designed for the nilpotent approximation of the system. This paradigm, based on the theoretical results in [9], has already been effectively used for the stabilization of *general* (i.e., non-flat) nonholonomic systems, such as off-hooked trailer vehicles [10] or underactuated robots in the absence of gravity [11].

The paper is organized as follows. In Sect. 2, the model of the plate-ball system is given together with its nilpotent approximation. Section 3 describes our stabilization strategy, which makes use of a contracting open-loop control (Sect. 3.1) within an iterative scheme (Sect. 3.2). The robust performance of the method is confirmed by simulation in Sect. 4.

2 The plate-ball system

Consider the system shown in Fig. 1, consisting of a spheric ball of radius ρ rolling between two horizontal plates. The lower plate is fixed, while the upper is actuated and can translate horizontally.

2.1 Kinematic model

Denote by u and v the coordinates (latitude and longitude, respectively) of the contact point on the sphere, by x, y the cartesian coordinates of the contact point on the lower plane, and by ψ the angle between the x axis and the plane of the meridian through the contact point (see Fig. 1). We assume $-\pi/2 < u < \pi/2$ and $-\pi < v < \pi$, so that the contact point belongs always

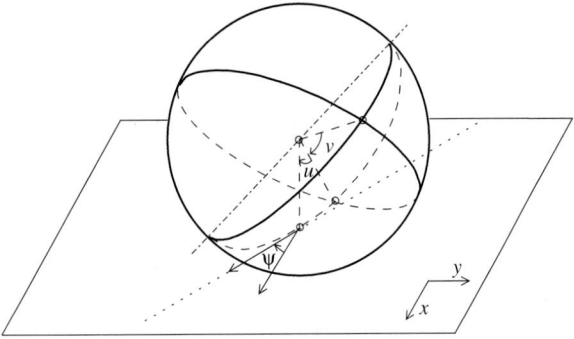

Figure 1: The plate-ball system. The upper plate is not shown in the figure for the sake of clarity.

to the same coordinate patch for the sphere.

The manipulation system is completely described by the kinematics of contact between the sphere and the lower plate [4]:

$$\begin{pmatrix} \dot{u} \\ \dot{v} \\ \dot{\psi} \\ \dot{x} \\ \dot{y} \end{pmatrix} = \begin{pmatrix} \cos\psi/\rho \\ -\sin\psi/\rho\cos u \\ \tan u \sin\psi/\rho \\ 1 \\ 0 \end{pmatrix} w_x + \begin{pmatrix} -\sin\psi/\rho \\ -\cos\psi/\rho\cos u \\ \tan u \cos\psi/\rho \\ 0 \\ 1 \end{pmatrix} w_y, \quad (1)$$

where w_x and w_y are the cartesian components of the translational velocity of the sphere, which we assume to be directly controlled[1].

In view of the nilpotent approximation procedure, it is convenient to perform the input transformation

$$\begin{pmatrix} w_x \\ w_y \end{pmatrix} = \begin{pmatrix} -\sin\psi\cos u & \cos\psi \\ -\cos\psi\cos u & \sin\psi \end{pmatrix} \begin{pmatrix} w_1 \\ w_2 \end{pmatrix}, \quad (2)$$

obtaining the triangular system

$$\begin{pmatrix} \dot{u} \\ \dot{v} \\ \dot{\psi} \\ \dot{x} \\ \dot{y} \end{pmatrix} = \begin{pmatrix} 0 \\ 1/\rho \\ -\sin u/\rho \\ -\sin\psi\cos u \\ -\cos\psi\cos u \end{pmatrix} w_1 + \begin{pmatrix} 1/\rho \\ 0 \\ 0 \\ \cos\psi \\ -\sin\psi \end{pmatrix} w_2. \quad (3)$$

Note that the input transformation (2) is always defined, except for $u = \pm\pi/2$ which is however outside our coordinate patch.

2.2 Nilpotent approximation

Nilpotent approximations [12, 13] of nonlinear systems are high-order local approximations that are useful

---

[1] Recall that the translational velocity of the sphere is half the translational velocity of the upper plane.

when tangent linearization does not retain controllability, as in nonholonomic systems. In particular, the computation of (approximate) steering controls can be performed symbolically, thanks to the closed-form integrability of the nilpotent system, which is polynomial and triangular by construction.

Thanks to the particular structure of our iterative steering strategy (see Sect. 3), it is sufficient to compute the nilpotent approximation at configurations of the form $\bar{q} = (0, 0, 0, \bar{x}, \bar{y})$. Applying the procedure in [13] to system (3), one obtains the so-called *privileged* coordinates by the following change of variables

$$\begin{aligned} z_1 &= \rho v \\ z_2 &= \rho u \\ z_3 &= \rho^2 \psi \\ z_4 &= -\rho^3 u + \rho^2(x - \bar{x}) \\ z_5 &= \rho^3 v + \rho^2(y - \bar{y}). \end{aligned} \quad (4)$$

This transformation is globally valid due to the fact that the degree of nonholonomy is 3 everywhere.

The approximate system is then computed by differentiating eqs. (4) and expanding the input vector fields in Taylor series up to a suitably defined order:

$$\begin{aligned} \dot{\hat{z}}_1 &= w_1 \\ \dot{\hat{z}}_2 &= w_2 \\ \dot{\hat{z}}_3 &= -\hat{z}_2 w_1 \\ \dot{\hat{z}}_4 &= -\hat{z}_3 w_1 \\ \dot{\hat{z}}_5 &= \frac{1}{2}\hat{z}_2^2 w_1 - \hat{z}_3 w_2. \end{aligned} \quad (5)$$

The approximation is polynomial and triangular; in particular, the dynamics of z_1 and z_2 is exact.

Another nilpotent approximation for the plate-ball system is given in [14].

3 The stabilization strategy

Assume that we wish to transfer the plate-ball system from q^0 to q^d, respectively the initial and desired contact configuration. Without loss of generality, we assume that $q^d = (0, 0, 0, 0, 0)$; this can always be achieved by properly defining the reference frames on the sphere and the lower plane.

Our objective is to devise a stabilization strategy which is robust w.r.t. the presence of model perturbations (e.g., on the sphere radius ρ). To this end, it is necessary to embed some form of feedback in the scheme. A natural way to realize this is represented by the iterative steering (IS) paradigm [9].

The essential tool of this method is a contracting open-loop control law, which can steer the system closer to the desired state q^d in a finite time. If such control is Hölder-continuous w.r.t. the desired reconfiguration, its iterated application (i.e., from the state reached at the end of the previous iteration), guarantees exponential convergence of the state to q^d. The resulting control is a time-varying law which depends on a sampled feedback action. A certain degree of robustness is also achieved: a class of non-persistent perturbations is rejected, and the error is ultimately bounded in the presence of persistent perturbations.

3.1 A contracting open-loop control

To comply with the IS paradigm outlined above, we must design an open-loop control which steers system (1) (or system (3)) from q^0 to a point closer in norm to $q^d = (0,0,0,0,0)$. Since the plate-ball manipulation system is controllable [5], such an open-loop control certainly exists. However, the necessary and sufficient condition for flatness [15] are not satisfied; equivalently, the system cannot be put in chained form, as already noticed in [3]. Therefore, we cannot use conventional techniques for generating the required open-loop control.

A possibility is to use the planning method of [3]; however, such numerical method is computationally intensive and therefore unsuitable for the real-time iteration of the open-loop control. Moreover, a symbolic expression of the control would be needed for guaranteeing the continuity properties required by the IS approach. We therefore settle for an approximate (but symbolic) solution; this is on the other hand consistent with the IS framework, which only requires the error to contract at each iteration.

Our open-loop controller requires two phases:

I. Drive the first three variables u, v and ψ to zero. This amounts to steering the ball to the desired contact configuration regardless of the variables x and y, i.e., of the cartesian position of the contact point. Denote by $q^I = (0,0,0,x^I,y^I)$ the contact configuration at the end of this phase.

II. Bring x and y closer to x^d and y^d (in norm), while guaranteeing that u, v and ψ return to their desired zero value.

Since the first three equations of (3) can be easily transformed in chained form (see Appendix), phase I can be performed in a finite time T_1 by choosing one of many available steering controls (see [8]). However, the latter should comply with the Hölder-continuity requirement w.r.t. the desired reconfiguration; relevant examples are given in [9].

For the second phase, a possible choice is to perform a cyclic motion of period T_2 on u, v and ψ, giving final values $x(T_1+T_2) = x^{II}$, $y(T_1+T_2) = y^{II}$ closer to zero than $x(T_1) = x^I$, $y(T_1) = y^I$. To design a control law that produces such a motion, we shall exploit the nilpotent approximation of the plate-ball system.

Consider the nilpotent approximation (5) at q^I. The synthesis of a control law that transfers in a finite time T_2 the state $\hat{z}$ from $z^I = 0$ to z^{II} (respectively, the images of q^I and $q^{II} = (0,0,0,x^{II},y^{II})$, computed through eqs. (4)) can be done as follows. Choose the open-loop control inputs as

$$w_1 = a_1 \cos \omega t + a_2 \cos 4\omega t \quad (6)$$
$$w_2 = a_3 \cos 2\omega t, \quad (7)$$

with $a_1, a_2, a_3 \in \mathbb{R}$ and $\omega = 2\pi/T_2$. The integration of eqs. (5) gives

$$\begin{aligned} z_1(T_2) = z_2(T_2) &= z_3(T_2) = 0 \\ z_4(T_2) &= k_1 a_1^2 a_3 \\ z_5(T_2) &= k_2 a_2 a_3^2, \end{aligned} \quad (8)$$

having set $k_1 = -T_2^3/32\pi^2$ and $k_2 = T_2^3/128\pi^2$.

In order to obtain $z_4(T_2) = z_4^{II}$ and $z_5(T_2) = z_5^{II}$, coefficients a_1 and a_2 in (6–7) must be chosen as

$$a_1 = \sqrt{\frac{z_4^{II}}{k_1 a_3}} \qquad a_2 = \frac{z_5^{II}}{k_2 a_3^2}. \quad (9)$$

Substitution of eq. (9) in eq. (8) proves that the value of a_3 is immaterial as long as (i) $a_3 \neq 0$ when $z_4^{II} \neq 0$ or $z_5^{II} \neq 0$, and (ii) $\text{sign}(a_3) = -\text{sign}(z_4^{II})$. Therefore, denoting by $||\cdot||$ denotes the euclidean norm, we let

$$a_3 = -\text{sign}(z_4^{II}) \cdot \left\| \begin{pmatrix} z_4^{II} \\ z_5^{II} \end{pmatrix} \right\|^{1/2r} \quad r > 1, \quad (10)$$

This choice, guarantees for a_1, a_2 and a_3 the Hölder-continuity property required by the IS paradigm.

The other condition to be met by our two-phase open-loop control is contraction from q^0 to q^{II}. It is easy to show that, with a suitable definition of norm, such condition is satisfied. This is true in spite of fact that the use of the nilpotent dynamics (5) for computing $z_4(T_2)$ and $z_5(T_2)$ induces an approximation error[2] on x and y, which increases with the required reconfiguration. In fact, the contraction property can be preserved by requiring a sufficiently small contraction.

---

[2]Note that u, v and ψ return to zero, as verified by integration of the first three equations of the original system (3). Thus, the open-loop controls (6–7) are exactly cyclic in u, v and ψ.

3.2 Iterative steering

We now clarify the use of the proposed open-loop controller within the iterative steering framework.

Starting from the initial contact configuration, apply the open-loop control of phase I for the required time T_1. Using the values x^I, y^I at the end of this phase, the desired z_4^{II} and z_5^{II} are generated as

$$z_4^{II} = \beta_1 z_4^d \qquad z_5^{II} = \beta_2 z_5^d, \qquad (11)$$

where $\beta_1 < 1$, $\beta_2 < 1$ are the chosen contraction rates and z_4^d, z_5^d are the images of $x^d = 0$, $y^d = 0$ computed inverting eqs. (4), in which $\bar{x} = x^I$, $\bar{y} = y^I$.

At this point, eqs. (9–10) are used to compute coefficients a_i, and the phase II open-loop controls (6–7) are applied to system (3). After $T_1 + T_2$ seconds from the initial time, the system state is sampled and the two-phase control procedure is repeated.

The values of z_4^{II} and z_5^{II} are updated at each iteration using eq. (11) (with constant β_1, β_2). In fact, as transformation (4) depends on the approximation point, the same is true for z_4^d, z_5^d. Note also that:

- Since the conditions of the IS paradigm [9] have been satisfied, it is guaranteed that the manipulation system state q exponentially converges to the desired contact configuration q^d.

- In the absence of perturbations, there is no need to repeat phase I after the first iteration.

- In perturbed conditions, it is necessary to analyze the structure of the perturbation itself. If certain requisites (see [9, Th. 2]) are met, the perturbation will be rejected on the simple basis of the stable behavior of the nominal system.

4 Simulation results

Two simulations are now presented to show the effectiveness of the proposed stabilization strategy: in the first, perfect knowledge of the system is assumed (*nominal* case), while in the second we have included a perturbation on the ball radius ρ (*perturbed* case).

In the first simulation, the radius $\rho = 1$ is exactly known and phase I has already been executed. The initial and desired configurations are $q^0 = (0, 0, 0, 0.5, 0.5)$ and $q^d = (0, 0, 0, 0, 0)$, respectively. In each iteration, the open-loop control (6–7) is applied with $T_2 = 1$ sec, $r = 1.5$ in eq. (10), and contraction rates $\beta_1 = \beta_2 = 0.4$ in eq. (11).

Figures 2 and 3 illustrate the exponential convergence of the state variables along the iterations. The complete cartesian path of the contact point is shown in Fig. 4: note how the path of the single iterations 'shrinks' with time. The contraction of the positioning error is visible in Fig. 5, which reports the path of the contact point during iterations 1, 4, 7 and 10.

In the second simulation, q^0, q^d as well as the control parameters are the same of the previous simulation, but a 10% perturbation on the value of the ball radius has been introduced; only its nominal value $\rho = 1$ is known and can be used for computing the control law. The theoretical framework of the IS paradigm (see [9, Th. 2]) guarantees that this kind of perturbation will be rejected by the iterative steering scheme.

Figures 6 and 7 confirm that exponential convergence is preserved despite the perturbation — only at a slightly smaller rate. The cartesian path of the contact point is very similar to the nominal case, as shown in Fig. 8, although Fig. 9 reveals that the paths in the single iterations are deformed.

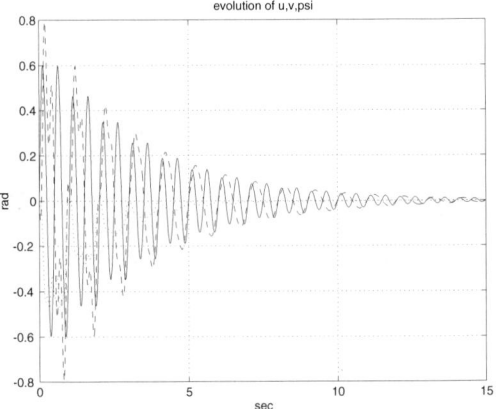

Figure 2: Nominal system: Evolution of u (solid), v (dashed) and ψ (dotted)

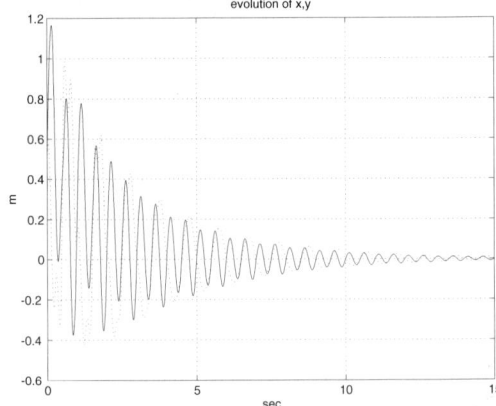

Figure 3: Nominal system: Evolution of x (solid) and y (dotted)

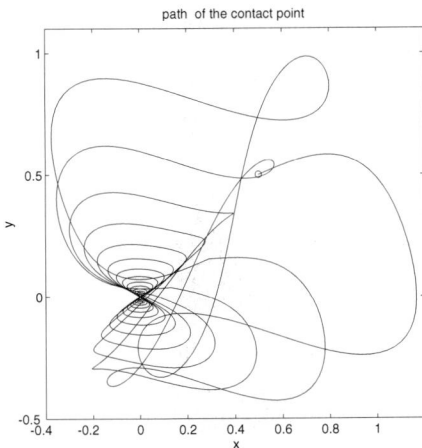

Figure 4: Nominal system: Cartesian path of the contact point (the small circle indicates q^0)

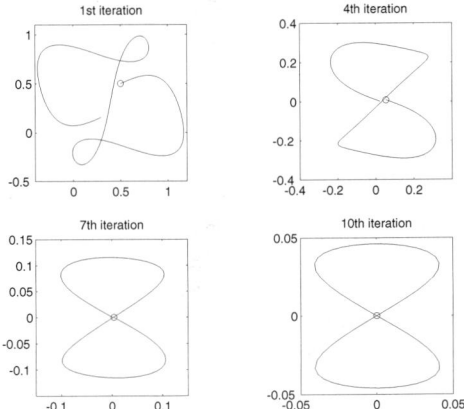

Figure 5: Nominal system: Cartesian paths of the contact point during the 1st, 4th, 7th and 10th iterations (the small circle indicates the starting configuration of each iteration). Notice the different scale in the plots.

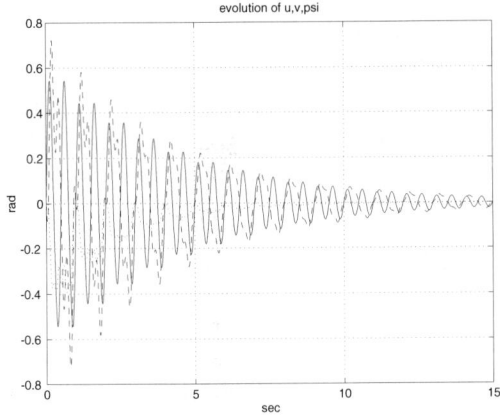

Figure 6: Perturbed system: Evolution of u (solid), v (dashed) and ψ (dotted)

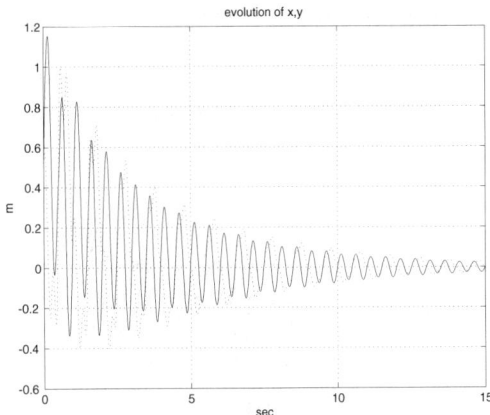

Figure 7: Perturbed system: Evolution of x (solid) and y (dotted)

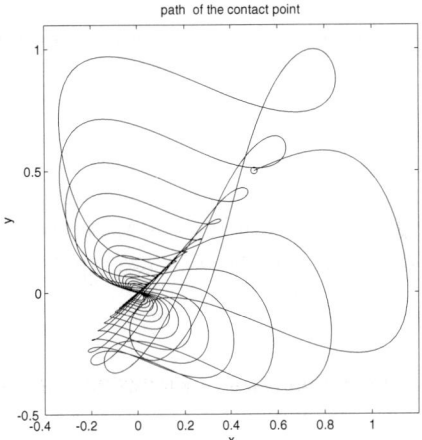

Figure 8: Perturbed system: Cartesian path of the contact point (the small circle indicates q^0)

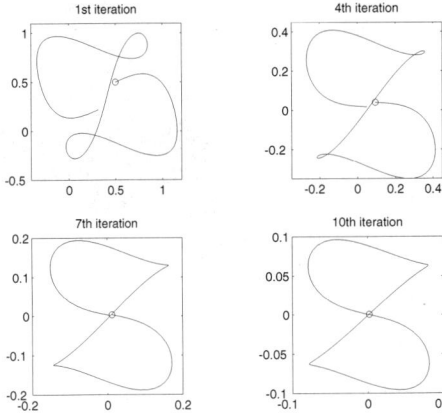

Figure 9: Perturbed system: Cartesian paths of the contact point during the 1st, 4th, 7th and 10th iterations (the small circle indicates the starting configuration of each iteration).

5 Conclusions

We have presented a feedback method for executing robust maneuvers with a plate-ball manipulation device in the presence of perturbations. Beside its practical interest, this problem is challenging from a theoretical viewpoint because the considered nonholonomic system is outside the class for which well-established planning and control techniques exist.

The proposed solution is based on an iterative steering scheme, which makes use of a nilpotent approximation of the system for designing the open-loop control law to be applied repeatedly. The performance of the algorithm, which can be established relying on the iterative steering theoretical framework, has been confirmed by simulations, both in the nominal case and in the presence of a perturbation on the ball radius.

Another advantage of the proposed technique, which could be useful for performing manipulation in the presence of obstacles, is the possibility of shaping the system trajectory during the generic iteration through the choice of the open-loop control. Finally, we point out that the same iterative approach may be successfully applied to other manipulation systems, such as the impulsive manipulator based on tapping described in [16].

Appendix

The first three equations of system (3) can be put in chained form by the following coordinate change

$$\begin{aligned} x_1 &= -v \\ x_2 &= \sin u \\ x_3 &= \psi \end{aligned}$$

and input transformation

$$\begin{aligned} v_1 &= -w_1/\rho \\ v_2 &= \cos u\, w_2/\rho. \end{aligned}$$

References

[1] R. M. Murray, Z. Li, and S. S. Sastry, *A Mathematical Introduction to Robotic Manipulation*, CRC Press, 1994.

[2] A. Bicchi and R. Sorrentino, "Dexterous manipulation through rolling", *IEEE Int. Conf. on Robotics and Automation*, pp. 452–457, 1995.

[3] A. Marigo and A. Bicchi, "Rolling bodies with regular surface: controllability theory and applications", *IEEE Trans. on Automatic Control*, 2000.

[4] D. J. Montana, "The kinematics of contact and grasp", *Int. J. of Robotics Research*, vol. 7, no. 3, pp. 17–32, 1988.

[5] Z. Li and J. Canny, "Motion of two rigid bodies with rolling constraint", *IEEE Trans. on Robotics and Automation*, vol. 6, pp. 62–72, 1990.

[6] V. Jurdjevic, "The geometry of the plate-ball problem", *Arch. for Rational Mechanics and Analysis*, vol. 124, pp. 305–328, 1993.

[7] R. W. Brockett and L. Dai, "Non-holonomic kinematics and the role of elliptic functions in constructive controllability", in *Nonholonomic motion planning*, Z. Li and J. F. Canny, Eds., pp. 1–21. Kluwer Academic Publishers, 1993.

[8] J.-P. Laumond (Ed.), *Robot Motion Planning and Control*, Springer-Verlag, 1998.

[9] P. Lucibello and G. Oriolo, "Robust stabilization via iterative state steering with an application to chained-form systems", *Automatica*, vol. 37, pp. 71–79, 2001.

[10] M. Vendittelli and G. Oriolo, "Stabilization of the general two-trailer system", *2000 IEEE Int. Conf. on Robotics and Automation*, pp. 1817–1822, 2000.

[11] A. De Luca, R. Mattone, and G. Oriolo, "Stabilization of an underactuated planar 2r manipulator", *Int. J. of Robust and Nonlinear Control*, pp. 181–198, 2000.

[12] H. Hermes, "Nilpotent and high-order approximations of vector field systems", *SIAM Review*, vol. 33, pp. 238–26, 1991.

[13] A. Bellaïche, "The tangent space in sub-riemannian geometry", in *Sub-Riemannian Geometry*, A. Bellaïche and J.-J. Risler, Eds., pp. 1–78. Birkhäuser, 1996.

[14] A. A. Agrachev and Y. L. Sachkov, "An intrinsic approach to the control of rolling bodies", *38th Conference on Decision and Control*, pp. 431–435, 1999.

[15] M. Fliess, J. Lévine, P. Martin, and P. Rouchon, "Flatness and defect of non-linear systems: Introductory theory and examples", *Int. J. of Control*, vol. 61, pp. 1327–1361, 1995.

[16] W. H. Huang and M. T. Mason, "Experiments in impulsive manipulation", *Proc. 1998 IEEE Int. Conf. on Robotics and Automation*, pp. 1077–1082, 1998.

Robotic Pinching by Means of a Pair of Soft Fingers with Sensory Feedback

H.-Y. Han, S. Arimoto, K. Tahara, M. Yamaguchi and P.T.A. Nguyen

Department of Robotics, Faculty of Science and Engineering,
Ritsumeikan University,
Nojihigashi 1-1-1, Kusatsu, Shiga, 525-8577, Japan
E-mail: han@se.ritsumei.ac.jp

Abstract

This paper proposes a pair of single or multi D.O.F. robot fingers with soft and deformable tips that can pinch an object stably in a dynamic sense with the aid of real-time sensory feedback. To realize dynamic stable pinching, a practical method of using optical devices is proposed for measuring both the maximum displacement of finger-tip deformation and the relative angle between the object surface and each of finger links. It is shown theoretically and by computer simulation that the overall closed-loop system of a pair of two single-D.O.F. fingers with soft tips with real-time sensory feedback of the difference between centers of two area-contacts at both sides of the object becomes asymptotically stable. This means that the pair achieves dynamic stable grasping (pinching). In the case of a pair of 1 D.O.F. and 2 D.O.F. fingers with soft tips, it is shown that the proposed method of closed-loop feedback of the difference between centers of two area-contacts and the rotational angle of the object can establish not only dynamic stable grasping but also regulation of the posture of the object.

1 Introduction

Human pinches stably and manipulates dexterously an object by using a thumb and other fingers. Motivated from this fact, a variety of multi-fingered robot hands have been proposed [1]. However, most of them have been used in open-loop control, where a certain amount of comprehensive plannings of grasp and manipulation is indispensable [1]. According to the vast literature referenced in the book [2] and the survey [1] there are many papers that have dealt with dynamics of multi-fingered hands contacting rigidly and pointwise with a rigid object, but there is a dearth of papers that analyze dynamics of a set of fingers with soft and deformable tips contacting an object with area-contacts. In fact there are little papers that attempt to derive a mathematical model expressing overall dynamics of motion for such a multi-fingered robot hand with soft tips grasping an object, though it is pointed out [3] that area-contacts with the aid of soft tips are important and useful in practice.

The purpose of this paper is to propose 1) a pair of dual 1 D.O.F. fingers with soft tips that can realize dynamic stable grasping (pinching an object) with the aid of sensory feedback of the difference between centers of area-contacts arising at both sides of the object and 2) another pair of 1 D.O.F. and 2 D.O.F. robot fingers with soft tips that can achieve not only stable grasping but also regulation of the object posture (set-point control of rotation angle of the object) with the aid of sensory feedback of both the difference of centers of area-contacts and the rotation angle of the object. To do this, a simple method of measuring both the maximum displacement of finger-tip deformation and the relative angle between the object surface and each of finger links is presented, which is based on the use of optical devices. The rotational angle of the object can be calculated in real time by means of the above measurement data and other measurements of joint angles that can be sensed by optical encoders or potentio-meters. In section 3, we present the full dynamics of horizontal motion of a pair of 1 D.O.F. and 2 D.O.F. fingers with soft tips pinching a rigid object. It is assumed as in previous papers [4]~[7] that motion of the overall system is confined to a horizontal plane and the effect of gravity is ignored (see Fig.1). At the same time it is assumed that the shape of the finger-tips is hemispherical and the distributed pressure arisen from deformation of each finger-tip can be lumped-parameterized into a single representative reproducing force that has a direction to the center of curvature of the finger-tip from the center of contact area between the finger-tip and one side surface of the object. In section 4, it is proved theoretically that the pair of such soft fingers can realize both dynamic stable grasping and regulation of the posture of the object. In section 5, computer simulation results are shown, which confirm the effectiveness of the proposed methods of sensory feedback. In section 6, an experimental setup is shown, which will be used in carrying out experiments to show practical usefulness of such pairs of soft fingers in pinching a variety of things.

2 Measurements of Maximum Displacement of Deformation and Rotation Angle of the Object

To control the object by soft fingers as being pointed out by the previous section, it is important to measure the maximum displacement of deformation of each soft finger-tip and the rotation angle of the object. In this paper a simple method for measuring of such physical values at the same time is proposed by using optical devices.

Firstly let us consider a pinched object by the finger-tip as shown in Fig. 1. Two optical sensors are located closely to both sides of the soft finger-tip. We assume that a finger-tip has hemispherical shape and is made of a soft and deformable material such as silicone gel. The object is rigid and its contact surface has a plane at the range between sensors on the finger-tip. Then the rotation angle θ of the grasped object can be calculated from the equation

$$\theta = \arctan((l_1 - l_2)/d) + q \qquad (1)$$

where d, l_1, l_2, and q denote the distance between sensors, distances from sensors to the object, and the angle of finger link respectively. Simultaneously, the maximum displacement of deformation of the soft finger-tip Δx can be given as

$$\Delta x = \\ r - \frac{1}{\sqrt{(y_{s2} - y_{s1})^2 + (x_{s2} - x_{s1})^2}} |(y_{s2} - y_{s1})x_{01} \\ - (x_{s2} - x_{s1})y_{01} - (y_{s2} - y_{s1})x_{s1} + (x_{s2} - x_{s1})y_{s1}| \quad (2)$$

where r and (x_{01}, y_{01}) denote the initial radius of finger-tip and the center point of soft finger-tip. (x_{s1}, y_{s1}) and (x_{s2}, y_{s2}) denote the coordinates of measured points by optical sensors on the surface of the object. These values can be calculated from the located points of optical sensors on the finger link and the values of l_1 and l_2 easily. The second term in eq.(2) means the length of a line segment perpendicular to a contact line(the surface of the object) from the center point of finger-tip.

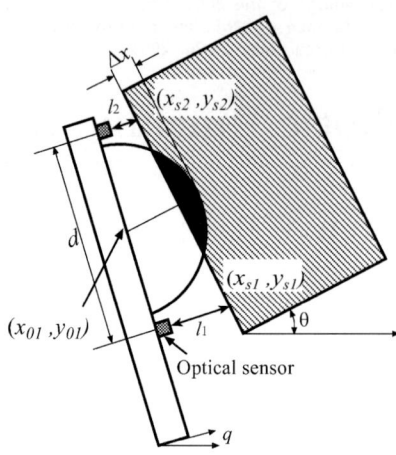

Figure 1: A setup of a finger with a hemispherical soft tip and two optical sensors

3 Dynamics of Pinch Motion

Firstly we derive the dynamics of pinch motion made by using two robot fingers with 1 D.O.F. and 2 D.O.F. whose tips are made of soft material. All symbols and coordinates in the overall system are defined in Fig. 2. There are four equations of geometric constraint as shown:

$$\begin{aligned}
Y_1 &= c_1 - r_1\varphi_1 \\
&= c_1 - r_1(\pi + \theta - q_{11} - q_{12}) \qquad (3) \\
Y_2 &= c_2 - r_2\varphi_2 \\
&= c_2 - r_2(\pi - \theta - q_{21}) \qquad (4) \\
x &= x_1 + \frac{l}{2}\cos\theta - Y_1\sin\theta \\
&= x_2 - \frac{l}{2}\cos\theta - Y_2\sin\theta \qquad (5) \\
y &= y_1 - \frac{l}{2}\sin\theta - Y_1\cos\theta \\
&= y_2 + \frac{l}{2}\sin\theta - Y_2\cos\theta \qquad (6)
\end{aligned}$$

The eqns.(3) and (4) are generated from tight area-contacts between each surface of finger-tips and the surface of the rigid object(see Fig. 3). Here, Y_1 and Y_2 in eqns.(3) and (4) should be expressed using position variables $z = (x, y, \theta)^T$, $q_1 = (q_{11}, q_{12})^T$, $q_2 = q_{21}$ and q_0 is a fixed angle of the 1 D.O.F. finger as follows:

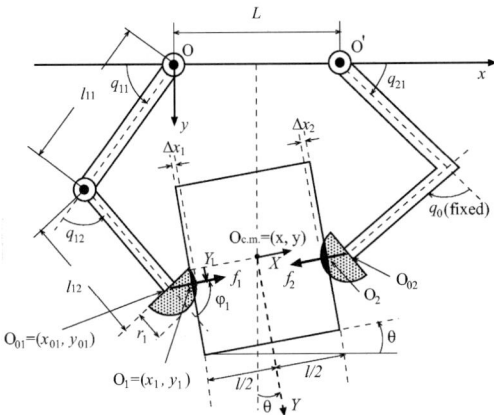

Figure 2: A setup of two fingers(1 D.O.F. and 2 D.O.F.) with soft tips pinching a rigid object

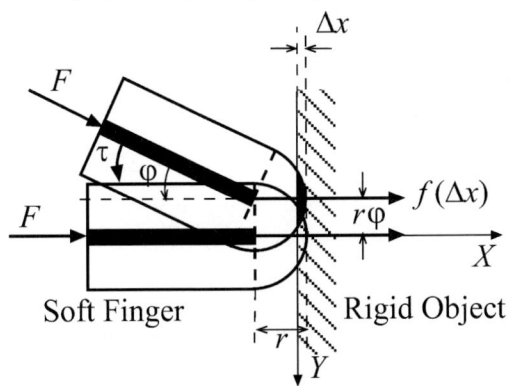

Figure 3: The center of contact area moves on the object surface by inclining the last link against the object.

$$Y_1 = (x_{01} - x)\sin\theta + (y_{01} - y)\cos\theta \quad (7)$$
$$Y_2 = (x_{02} - x)\sin\theta + (y_{02} - y)\cos\theta \quad (8)$$

where

$$x_{01} = -l_{11}\cos q_{11} - l_{12}\cos(q_{11} + q_{12}) \quad (9)$$
$$y_{01} = l_{11}\sin q_{11} + l_{12}\sin(q_{11} + q_{12}) \quad (10)$$
$$x_{02} = L + l_{21}\cos q_{21} + l_{22}\cos(q_{21} + q_0) \quad (11)$$
$$y_{02} = l_{21}\sin q_{21} + l_{22}\sin(q_{21} + q_0) \quad (12)$$

The eqns.(5) and (6) mean that the loop starting from the origin O of the first joint center on the left finger and return to it through the $O_{01}, O_1, O_{c.m.}, O_2, O_{02}, O'$ is closed, as shown in Fig.2. The Lagrangian of the overall system is given as

$$L = K - P + S \quad (13)$$

where K is the kinetic energy, P is the potential energy and S is related to constraints. These are defined as

$$K = \sum_{i=1,2} \frac{1}{2}\dot{q}_i^T H_i(q_i)\dot{q}_i + \frac{1}{2}\dot{z}^T H \dot{z} \quad (14)$$

$$P = \sum_{i=1,2} \int_0^{\Delta x_i} f_i(\xi) d\xi \quad (15)$$

$$S = \sum_{i=1,2} \lambda_i(Y_i - c_i + r_i\varphi_i) \quad (16)$$

where $H = \text{diag}(M, M, I)$, M and I denote the mass and inertia moment of the rigid object. Δx_i ($i = 1, 2$) denote the maximum displacement of deformation and λ_i ($i = 1, 2$) are Lagrange's multipliers. Here, by applying Hamilton's principle or by using the variational form

$$\int_{t_1}^{t_2} \left\{ \delta(K - P + S) + u_1^T \delta q_1 + u_2^T \delta q_2 \right\} dt = 0 \quad (17)$$

the Lagrange equation of the overall system is obtained, which is described as follows:

$$\left\{ H_i(q_i)\frac{d}{dt} + \frac{1}{2}\dot{H}_i(q_i) \right\} \dot{q}_i + S_i(q_i, \dot{q}_i)\dot{q}_i$$
$$+ \frac{\partial \Delta x_i}{\partial q_i} f_i - \lambda_i \frac{\partial \Phi_i}{\partial q_i} = u_i, \quad i = 1, 2 \quad (18)$$

$$H\ddot{z} + \sum_{i=1,2} \left(\frac{\partial \Delta x_i}{\partial z} f_i - \lambda_i \frac{\partial \Phi_i}{\partial z} \right) = 0 \quad (19)$$

$$\Phi_i = Y_i - c_i + r_i\varphi, \quad i = 1, 2$$

where eqn.(18) denotes dynamics of each of the two robot fingers. $\Phi_i = 0$ for $i = 1, 2$ express tight area-contact constraints. The eqn.(19) expresses dynamics of the rigid object. Here, it is possible to show that the overall dynamics of eqns.(18) and (19) satisfy passivity, i.e.,

$$\dot{q}_1^T u_1 + \dot{q}_2^T u_2 = \frac{d}{dt}\left\{ \sum_{i=1,2} \frac{1}{2}\dot{q}_i^T H_i(q_i)\dot{q}_i \right.$$
$$\left. + \frac{1}{2}\dot{z}^T H \dot{z} + \sum_{i=1,2} \int_0^{\Delta x_i} f_i(\zeta) d\zeta \right\}$$
$$= \frac{d}{dt}(K + P) = \frac{d}{dt}V(t) \quad (20)$$

whose integral over $[0, t]$ leads to

$$\int_0^t (\dot{q}_1^T u_1 + \dot{q}_2^T u_2) d\tau = V(t) - V(0) \geq -V(0) \quad (21)$$

Clearly the input-output pair (u_1, u_2) and $(\dot{q}_1, \dot{q}_2)$ satisfies passivity, where V signifies the total energy of the overall system, that is,

$$V = K + P \quad (22)$$

4 Dynamic Stable Grasping and Posture Control

By means of sensing described above, it is possible to design sensor feedback signals for closed-loop control of a pair of robot fingers, which are given as

$$u_{fi} = -k_{v_i}\dot{q}_i + \frac{\partial \Delta x_i}{\partial q_i}f_d$$

$$+(-1)^i f_d \left\{ \frac{r_i}{r_1 + r_2}(Y_1 - Y_2)e_i \right\}$$

$$+(-1)^i \gamma \left\{ \frac{\partial \Phi_i}{\partial q_i}\left(\frac{\dot{Y}_1 - \dot{Y}_2}{l - \Delta x_1 - \Delta x_2}\right) \right.$$

$$\left. + \frac{r_i}{r_1 + r_2}(\dot{Y}_1 - \dot{Y}_2)e_i \right\} \quad (23)$$

$$u_{\theta i} = (-1)^i \frac{\partial \Phi_i}{\partial q_i}\left(\frac{\beta \Delta \theta}{l - \Delta x_1 - \Delta x_2} + \alpha \dot{\theta}\right) \quad (24)$$

$$u_i = u_{fi} + u_{\theta i} + \Delta u_i, \quad i = 1, 2 \quad (25)$$

where $\Delta \theta = \theta - \theta_d$, $\alpha > 0$, $\beta > 0$, $\gamma > 0$, $e_1 = (1,1)^{\mathrm{T}}$, $e_2 = 1$, and θ_d signifies a prescribed desired angle of the object (see Fig. 2). Here, the coefficients $\frac{\partial \Delta x_i}{\partial q_i}$, $\frac{\partial \Phi_i}{\partial q_i}$, $\frac{\partial \Delta x_i}{\partial z}$ and $\frac{\partial \Phi_i}{\partial z}$ are given as (see [7])

$$\frac{\partial \Delta x_1}{\partial q_1} = \begin{pmatrix} l_{11}\sin(q_{11} - \theta) + l_{12}\sin(q_{11} + q_{12} - \theta) \\ l_{12}\sin(q_{11} + q_{12} - \theta) \end{pmatrix}$$

$$\frac{\partial \Delta x_2}{\partial q_2} = \begin{pmatrix} l_{21}\sin(q_{21} + \theta) + l_{22}\cos(q_{21} + q_0 + \theta) \end{pmatrix}$$

$$\frac{\partial \Phi_1}{\partial q_1} = \begin{pmatrix} l_{11}\cos(q_{11} - \theta) + l_{12}\cos(q_{11} + q_{12} - \theta) - r_1 \\ l_{12}\cos(q_{11} + q_{12} - \theta) - r_1 \end{pmatrix}$$

$$\frac{\partial \Phi_2}{\partial q_2} = \begin{pmatrix} l_{21}\cos(q_{21} + \theta) + l_{22}\cos(q_{21} + q_0 + \theta) - r_2 \end{pmatrix}$$

$$\frac{\partial \Delta x_1}{\partial z} = \begin{pmatrix} -\cos\theta \\ \sin\theta \\ -Y_1 \end{pmatrix}, \quad \frac{\partial \Delta x_2}{\partial z} = \begin{pmatrix} \cos\theta \\ -\sin\theta \\ Y_2 \end{pmatrix}$$

$$\frac{\partial \Phi_1}{\partial z} = \begin{pmatrix} -\sin\theta \\ -\cos\theta \\ \Delta x_1 - \frac{l}{2} \end{pmatrix}, \quad \frac{\partial \Phi_2}{\partial z} = \begin{pmatrix} -\sin\theta \\ -\cos\theta \\ -\Delta x_2 + \frac{l}{2} \end{pmatrix}$$

which can be computed in real-time on the basis of measurements of θ and q_i (which can be measured by optical encoders implemented in joint actuators). The third and fourth terms of u_{fi} are introduced for balancing the moments $f_d Y_1$ and $f_d Y_2$ by means of PD feedback with respect to $y = \dot{Y}_1 - \dot{Y}_2$. It should be remarked that $\dot{Y}_1 - \dot{Y}_2$ can be calculated in real-time from θ and q_i based on the form of eq.(7) minus eq.(8). It is implicitly assumed that $l - r_1 - r_2 > 0$ and therefore it holds that $l - \Delta x_1 - \Delta x_2 > 0$. It has been shown [6] that the closed-loop system when $u_i = u_{fi} + u_{\theta i} + \Delta u_i$ is substituted into eq.(18) satisfies passivity, i.e.,

$$\int_0^t (\dot{q}_1^{\mathrm{T}} \Delta u_1 + \dot{q}_2^{\mathrm{T}} \Delta u_2) \mathrm{d}\tau$$

$$= V(t) - V(0) + \int_0^t W(\tau) \mathrm{d}\tau \geq -V(0) \quad (26)$$

where

$$V(t) = K + \sum_{i=1,2} \int_0^{\delta x_i} \{f_i(\xi + \Delta x_{di}) - f_d\} \mathrm{d}\xi$$

$$+ \frac{f_d}{2(r_1 + r_2)}(Y_1 - Y_2)^2 + \frac{\beta}{2}\Delta\theta^2 \quad (27)$$

$$W(t) = -\sum_{i=1,2} k_{v_i}\|\dot{q}_i\|^2 - \alpha(l - \Delta x_1 - \Delta x_2)\dot{\theta}^2$$

$$- \frac{\gamma}{r_1 + r_2}(\dot{Y}_1 - \dot{Y}_2)^2 \quad (28)$$

Then, it is possible to prove that the closed-loop system with $\Delta u_1 = 0$ and $\Delta u_2 = 0$ satisfies asymptotic stability, that is,

$$\theta(t) \to \theta_d, \quad Y_1(t) - Y_2(t) \to 0, \quad f_i(t) \to f_d$$

as $t \to \infty$.

In this paper the details of the proof are omitted (see our future paper [8]). The novelty of the results is to show that the proposed sensory feedback scheme realizes stable grasping in a dynamic sense differently from the classical concept of stable grasping [9] standing from the static sense.

5 Computer Simulation of Motion under Geometric Constraints

Initial conditions of the robot-fingers and rigid object are shown in Fig. 4 and parameters chosen in the design of control inputs are shown in Table. 1. The results of computer simulation based on the use of a Constraint Stabilization Method are shown in Figs. 5~7, where it is assumed that the relation between the reproducing force f and the maximum displacement Δx_i is described as:

$$f = k(\Delta x_i)^2, \quad i = 1, 2 \quad (29)$$

where k is a constant value.

Table 1: Parameters of control inputs

f_d[N]	θ_d[deg]	k[N/mm^2]	k_{v_i}	α	β	γ
1.0	-4.6	0.25	0.1	0.0	0.5	5.9

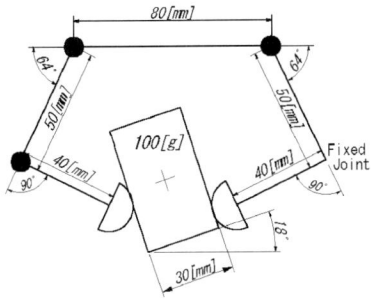

Figure 4: Initial condition of robot-fingers and a rigid object

In Figs. 5 and 6, both of reproducing forces $f_i (i = 1, 2)$ converge to the desired force and the settling time of convergence is about 0.5[s]. In Figs. 7 and 8, $Y_1 - Y_2$ also converges to 0 and the posture of the rigid object converges to the desired angle. These settling times are also about 0.5[s]. Figs. 9~12 show how the moment feedbacks works effectively even if the initial value of $Y_1 - Y_2$ is relatively big. Thus, stable grasping and posture control of a rigid object is achieved by using the proposed design of feedback control signals.

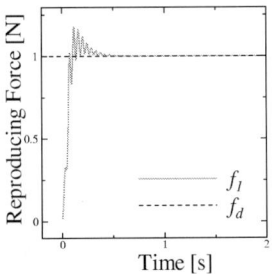

Figure 5: Reproducing force of 2 D.O.F. finger ($i = 1$)

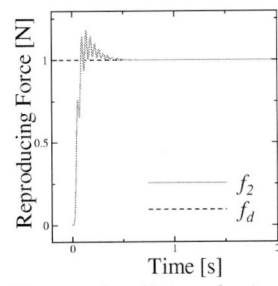

Figure 6: Reproducing force of 1 D.O.F. finger ($i = 2$)

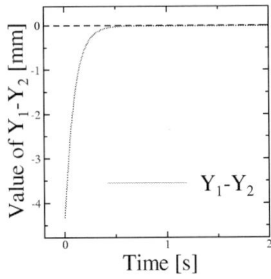

Figure 7: Vanishing phenomena of the couple of force

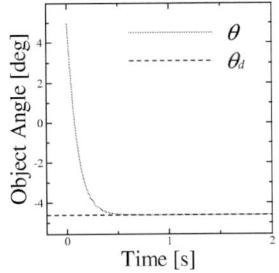

Figure 8: Posture of the rigid object

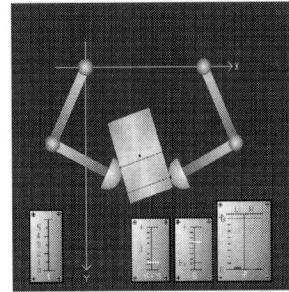

Figure 9: Start (0[sec])

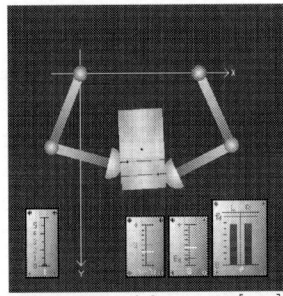

Figure 10: After 0.25[sec]

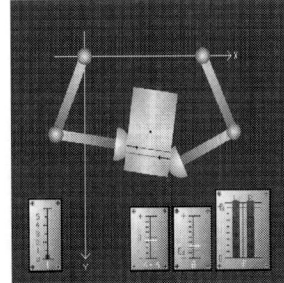

Figure 11: After 0.5[sec]

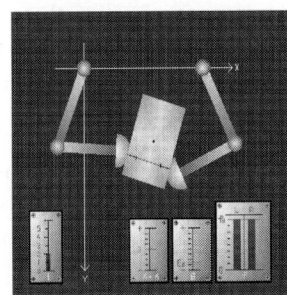

Figure 12: Finish (2[sec])

6 Experimental Results

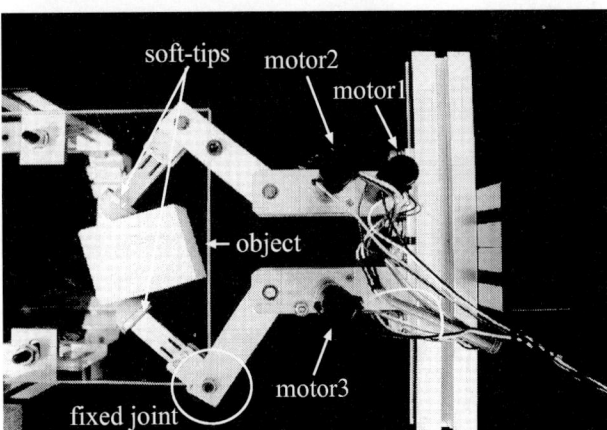

Figure 13: Experimental setup of a pair of robot fingers

An experimental setup in Fig. 13 and some results of experiments using the setup. The motion of this system is confined to a horizontal plane. In the system, timing belt are used to transmit the power and the ratios of reduction gears of actuators are set relatively small, in fact about 16:1 in the motor1, 3 and 4:1 in the motor2 in order to reduce the effect of any friction. These actuators are installed at the base frame in order to make up the total of link-weights as small as possible. The power of actuators is 3[W] and finger-tips are made of silicone gel. The tips have hemispherical shape of radius 10[mm]. Both fingers are constructed of 2 links but the second joint of 1 D.O.F. finger is fixed. The length of the first link is about 50[mm] and the second one is 40[mm]. The object is put on the acrylicresin board in order to decrease the effect of friction caused by the gravity. One of the results of grasping experiments is shown in Figs. 14 and 15. In this experiment, we used only the stable grasping control signal, which was computed in real-time from the equation of (7) minus (8) by using the measurement data on θ, In this experiment tactile sensing for measurement of f_1 and f_2 was not carried out. Instead, the desired contact force $f_d = 1.5$[N] was specified. It is actually observed that the sensory feedback of $Y_1 - Y_2$ as described in eq. (23) with $\gamma = 0$ leads to secure grasping of the object and the motion of the object stops often 0.2 second as seen in Fig. 15. However, as seen in Fig. 14, the value of $Y_1 - Y_2$ decreases quickly towards the value of zero, but eventually there remains a small offset. One of the principal reasons is that there is a small but ignorable static friction caused by various kinds of friction sources in gear heads, motors themselbes, area contacts between finger-tips and surfaces of the object and another contact between the bottom of the object and the acrylicresin board. Nevertheless, stable grasping was maintained even in the case that a small distarbance was artificially exerted to the object.

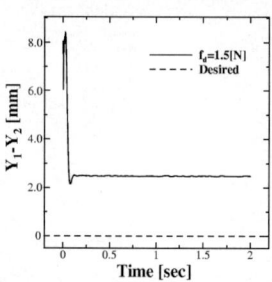

Figure 14: Vanishing phenomena of the couple of force

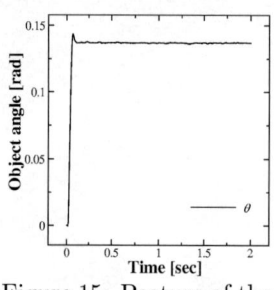

Figure 15: Posture of the rigid object

7 Conclusions

By solving numerically the equation of constrained dynamics of pinching motion by means of a pair of two 1 D.O.F. and 2 D.O.F. fingers with soft and deformable tips, it is shown that a designed closed-loop sensory feedback scheme can realize stable grasping in a dynamic sense and regulation of the rotation angle of the object to the desired one simultaneously. It should be pointed out that a pair of dual 1 D.O.F. robots with soft tips can also realize stable grasping in a dynamics sense though it is not possible to regulate the posture of the object. The detailed analysis of other cases including a set of 2 D.O.F. and 3 D.O.F. fingers with soft tips will be presented in a separate paper [8].

References

[1] K.B. Shimoga, "Robot grasp synthesis algorithms: A survey", Int. J. of Robotics Research, **15**(3), pp.230-266, 1996.

[2] R.M. Murray, Z. Li and S.S. Sastry, "A Mathematical Introduction to Robotic Manipulation", CRC Press, Boca Raton and Tokyo, 1994.

[3] M.R. Cutkosky, "Robotic Grasping and Fine Manipulation", Klumer Academic, Dordrecht, Netherlands, 1985.

[4] S. Arimoto, P. T. A. Nguyen, H. -Y. Han, and Z. Doulgeri, "Dynamics and control of a set of dual fingers with soft tips", Robotica, **18** (1), pp.71-80, 2000.

[5] P. T. A. Nguyen, S. Arimoto, and H.-Y. Han, "Computer simulation of dynamics of dual fingers with soft-tips grasping an object", Proc. of the 2000 Japan-USA Symp. on Flexible Automation, Ann Arbor, Michigan, July 2000.

[6] S. Arimoto, K. Tahara, M. Yamaguchi, P. T. A. Nguyen, and H. -Y. Han, "Principle of superposition for controlling pinch motions by means of robot fingers with soft-tips", Proc. of the IFAC Symp. on Robot Control 2000, pp.421-426, Wien, Austry, Sept. 21-23, 2000.

[7] K. Tahara, M. Yamaguchi, P. T. A. Nguyen, H. -Y. Han and S. Arimoto, "Stable grasping and posture control for a pair of robot fingers with soft tips", Proc. of the International Conference on Machine Automation (ICMA2000), pp.33-38, Osaka, Japan, Sept. 27-29, 2000.

[8] S. Arimoto, K. Tahara, M. Yamaguchi, P. T. A. Nguyen, and H. -Y. Han, "Principle of superposition for controlling pinch motions by means of robot fingers with soft tips", Robotica, **19** (1), pp.21-28, 2001.

[9] V. Nguyen, "Constructing stable grasps", Int. J. of Robotics Research, **8**(1), pp. 26-37, 1989.

Fast Dextrous Regrasping with Optimal Contact Forces and Contact Sensor-Based Impedance Control

Thomas Schlegl*   Martin Buss**   Toru Omata***   Günther Schmidt*

* Institute of Automatic Control Engineering, Technische Universität München, D-80290 Munich, Germany, Th.Schlegl@ieee.org
** Control Systems Section, Technische Universität Berlin, D-10587 Berlin, Germany, M.Buss@ieee.org
*** Department of Precision Machinery Systems, Tokyo Insitute of Technolgy (TIT), 4259 Nagatsuta, Midoriku Yokohama, Kanagawa 226, Japan, Omata@pms.titech.ac.jp

Abstract

This paper presents an approach to fast object manipulation by dextrous regrasping of multi-fingered hands. The approach is based on a real-time grasping force optimization algorithm (GFO) and a fingertip impedance control scheme. Both the controller and the GFO make use of 6D contact force sensor data at run-time. The latter keeps contact forces as small as possible while considering friction limits at the contact points during (re)grasping. The impedance controller is used to impose optimized contact forces onto the grasped object while simultaneously enabling active control of the fingertip positions. Experiments demonstrate the robustness of the approach and the increase in task speed during multi-fingered manipulation.

1 Introduction

Regrasping objects with a dextrous hand is in general characterized by slow task execution which is mainly caused by a high computational effort needed for grasp force planning. An object pose which changes during manipulation requires adaptation of fingertip contact forces to compensate for an external object force. If a grasp alters structurally due to regrasping, an object is to be fixed by a varying number of fingers. This fact has to be considered when planning contact forces. With high object manipulation speed and frequent structural grasp changes contact forces must be adjusted in real-time. Beside balancing external object forces unilateral friction constraints at the contact points have to be considered.

To satisfy all these requirements is not a trivial task. By now, no closed analytical solution for the proper choice of grasping forces is known. Several optimization problems were formulated, see [1,3]. One scheme is based on gradient flows on the smooth manifold of positive definite matrices, simplified versions of which were applied in experiments [4]. An approach to compensate disturbances based on force sensor information was suggested in [5].

In this paper we present a comprehensive real-time implementation of the semidefinite program for constant and time-varying grasps using 6D force sensor data [6]. Optimized contact forces are imposed on the grasped object by a fingertip impedance control scheme which additionally realizes desired fingertip positions. Experiments conducted with the TIT-hand (Figure 5) show an increased grasp robustness against disturbance forces as well as a higher achievable manipulation speed.

The paper is organized as follows: Section 2 reviews the semidefinite program for (re)grasping. The real-time algorithm using contact force sensor data is given in Section 3. Section 4 shows the combination of GFO and fingertip impedance control. Results from experiments are reported in Section 5.

2 Grasping Force Optimization
2.1 Basic Considerations

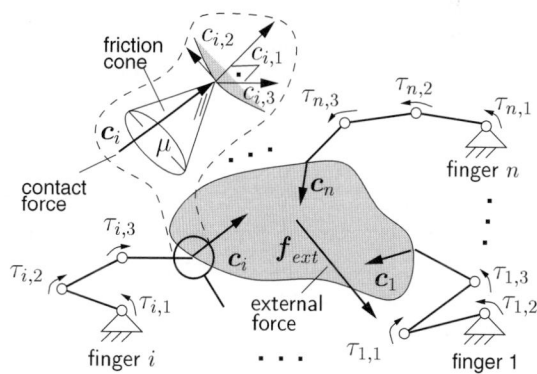

Figure 1: Multi-fingered grasp with friction

We assume a grasp with n point contacts as shown in Figure 1. The mapping of the contact force $\boldsymbol{c} = [\boldsymbol{c}_1^T \ldots \boldsymbol{c}_i^T \ldots \boldsymbol{c}_n^T]^T$ into a generalized force acting on the grasped object is given by the grasp matrix $\boldsymbol{W}(t)$ varying with time due to rolling effects and regrasping. The objective is to find a set of optimal contact forces balancing an external force

$$f_{ext} + W c = 0 . \quad (1)$$

Satisfying unilateral friction constraints $c_{i,1} > 0$ and $(c_{i,2}^2 + c_{i,3}^2)^{\frac{1}{2}} < \mu c_{i,1}$ with μ as friction coefficient is equivalent to the positive definiteness of a matrix $P = \mathrm{diag}(P_1, \ldots, P_i, \ldots, P_n)$ with

$$P_i = \begin{bmatrix} \mu c_{i,1} & 0 & c_{i,2} \\ 0 & \mu c_{i,1} & c_{i,3} \\ c_{i,2} & c_{i,3} & \mu c_{i,1} \end{bmatrix} . \quad (2)$$

Balancing f_{ext} and further constraints resulting from the special structure of P are summarized in

$$A \, \mathrm{vec}(P) = q . \quad (3)$$

The constraint matrix A can be constructed such that all elements are constant except a submatrix $\Delta(t)$ which comprises the elements of $W(t)$, see [2].

A cost index $\Phi : \mathcal{P}(n) \to \mathbb{R}$, $\Phi(P) = \mathrm{trace}(W_p P + W_b P^{-1})$ can be formulated whose first righthand term weights the linear costs of P and whose second term tends to infinity as P approaches the friction cone limits. It is solved by applying a recursive gradient flow method on

$$s = (I - A^{\#} A) \, \mathrm{vec}(W_p - P^{-1} W_b P^{-1}) \quad (4)$$

which is a gradient of the cost index $\Phi(P)$ projected in the tangential plane of linear constraints (3).

As the problem is convex a gradient flow converges at the unique equilibrium.

2.2 Regrasping with Optimal Contact Forces

One way to change a grasp is to lift off, move and recontact fingers. These discrete transitions of a finger result in structural changes both in the mechanical system and in the grasp controller. Switching from an n-fingered grasp to a reduced $(n-1)$-fingered grasp will cause non-smooth force trajectories, reducing the performance of the manipulation system. Even if a grasping force optimization scheme is used, grasp stability may be lost.

Therefore, it is necessary to *prepare* the system for the structural changes resulting from regrasping. Before lifting off a finger its contact force is to be decreased and distributed to other fingers remaining in contact in an unloading phase. At the instant when the regrasping finger starts moving the contact force should be almost zero in magnitude. On the other hand, as soon as the moving finger contacts the object again, its contact force is smoothly increased to a new final value in a loading phase. This type of regrasping behavior can be implemented through the GFO by adapting the weights of the regrasping finger in the cost index Φ and extending the linear constraints (3), see [7].

3 Real-Time Optimization
3.1 Use of Contact Force Sensor Data

The optimization algorithm needs to know the external force f_{ext} acting on the grasped object and the grasp matrix W. Based on the intendend manipulation task both variables can be calulated off-line and be provided to the control system at run-time out of a task database.

However, intended manipulaton may not succeed under presence of disturbances which affect grasp parameters ($\leadsto \bullet$), such as

- unknown forces acting on the object ($\leadsto f_{ext}$)
- imprecisely modeled object geometry ($\leadsto W$)
- regrasping and rolling effects ($\leadsto f_{ext}, W$)

On-line calculation of f_{ext} and W from known contact sensor data is in general superior to off-line planning and increases flexibilty of the control architecture. A

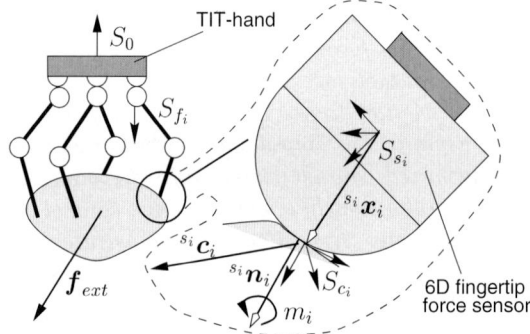

Figure 2: Contact force sensor information

6D contact force sensor at each fingertip of the TIT-hand, see Figure 2, provides contact point locations $^{s_i}x_i$, contact forces $^{s_i}c_i$, the contact normal vectors $^{s_i}n_i$ and the moment m_i around $^{s_i}n_i$, where $^{s_i}\bullet$ denotes the quantity $\bullet$ to be specified with respect to the sensor coordinate system S_{s_i}.

All measurements are related to the hand palm frame S_0. The external force acting on the object then follows to

$$^0 f_{ext} = \begin{bmatrix} \sum_{i=1}^{n} {}^0 c_i \\ \sum_{i=1}^{n} ({}^0 X_i \, {}^0 c_i + {}^0 n_i m_i) \end{bmatrix} \quad (5)$$

and the grasp matrix to

$$W = \begin{bmatrix} I_3 & \cdots & I_3 \\ {}^0 X_1 & \cdots & {}^0 X_n \end{bmatrix} \mathrm{diag}({}^0 R_{c_i}({}^0 n_i)) \quad (6)$$

with ^0X_i as special skew-symmetric matrices $\in \mathbb{R}^{3 \times 3}$ comprising the elements of ^0x_i. ^0X_i are used to express the cross products as vector matrix products, i.e. $^0X_i \, ^0c_i \equiv \, ^0x_i \times \, ^0c_i$.

As W and f_{ext} are calculated from contact sensor data at run-time rolling motion of the contact points is considered as well as a change of contact points due to regrasping.

3.2 Recursive Gradient Flow Algorithm

For real-time application the optimization problem min Φ is solved by a time-discrete gradient flow method as indicated in Figure 3.

Given the external force $\boldsymbol{f}_{ext}$, the grasp matrix $\boldsymbol{W}$, the weights for the cost index $\boldsymbol{W}_p$, $\boldsymbol{W}_b$, the duration $t_{r/i}$ of loading and unloading phases and the desired grasp state, i.e. information about contacting and moving fingers, the pseudo-inverse $\boldsymbol{W}^\#$ of $\boldsymbol{W}$ can be calculated. It is used to obtain a particular solution $\boldsymbol{c}_p$ of (1). To find a suitable initial value we increase the homogeneous contact forces until the initial $\boldsymbol{c}_0$ is inside friction limits using the increment vector $[1\ 0\ 0\ \ldots\ 1\ 0\ 0]^T$ and the mapping $\boldsymbol{I} - \boldsymbol{W}^\#\boldsymbol{W}$ to suppress any particular components which may be contained in the increment.

Next, recursive optimization is initiated with the calculation of a search direction $\boldsymbol{s}_{red}$ extracted from the projected gradient $\boldsymbol{s}$. For that purpose the projection operator $\boldsymbol{I} - \boldsymbol{A}^\#\boldsymbol{A}$ is calculated a priori *symbolically* depending on the elements of the inverse of $\boldsymbol{\Delta}$ which is $\in \mathbb{R}^{6\times 6}$ during both three and four fingered grasps or $\in \mathbb{R}^{7\times 7}$ during the (un)loading phases. An approximative line search is used to find a suitable stepsize α which for updating $\boldsymbol{c}_k$.

The algorithm terminates when the decrease in cost becomes zero near the optimal solution which is unique due to the convexity of the optimization problem. A high flexibility of the algorithm is guaranteed by explicit symbolic dependence of the gradients on the grasp matrix. The algorithm computes in 3-5ms on an $i686$-PC with 800MHz.

read $\boldsymbol{f}_{ext}$, $\boldsymbol{W}$, desired grasp state, $t_{r/i}$, $\boldsymbol{W}_p$, $\boldsymbol{W}_b$
determine mode: STABLE, REDUCE, MOVE, INCREASE
calculate $\boldsymbol{W}^\#$
$\boldsymbol{c}_p := \boldsymbol{W}^\# \boldsymbol{f}_{ext}$
update $w_{p,j}(t)$, $w_{b,j}(t)$ and $b(t)$
$\boldsymbol{c} := 0$
increase internal forces
$\boldsymbol{c} := \boldsymbol{c} + (\boldsymbol{I} - \boldsymbol{W}^\#\boldsymbol{W})[1\ 0\ 0\ 1\ 0\ 0\ \ldots]^T$
until $\boldsymbol{c}$ satisfies friction limits
$\boldsymbol{c}_k := \boldsymbol{c}_0 := \boldsymbol{c}$ is valid initial
calculate $\Delta(\boldsymbol{W})$, Δ^{-1}
calculate $\boldsymbol{s}_{red}$
$\alpha := \alpha_0$
while $\Phi(\boldsymbol{c}_k - \alpha\boldsymbol{s}_{red}) > \Phi(\boldsymbol{c}_k) \vee P(\boldsymbol{c}_k - \alpha\boldsymbol{s}_{red}) < 0$
$\alpha := \gamma\alpha$
$\boldsymbol{c}_{k+1} := \boldsymbol{c}_k - \alpha\boldsymbol{s}_{red}$, $k := k+1$
until $\|\Phi_{k+1} - \Phi_k\| < \epsilon$

Figure 3: Flowchart of on–line algorithm

4 Regrasping Control

For the experiments reported in the next section the GFO was included into a fingertip impedance control scheme.

4.1 Fingertip Impedance Control

Each finger is controlled by an impedance controller separately. Figure 4 shows the control loop for *one finger*. Its basic part is formed by a trajectory control law for the position of the ith fingertip, i.e.

$$\dot{\boldsymbol{\theta}}_i^d = {}^{f_i}\boldsymbol{J}^{-1}\, {}^0\boldsymbol{R}_{f_i}^T \left(\boldsymbol{K}_{I,i}\int \boldsymbol{e}_{x_i}dt + \boldsymbol{K}_{P,i}\boldsymbol{e}_{x_i} + \boldsymbol{K}_{V,i}\boldsymbol{e}_{\dot{x}_i}\right)$$

where ${}^{f_i}\boldsymbol{J}^{-1}$ is the inverse Jacobian of the fingertip position with respect to a finger base frame and ${}^0\boldsymbol{R}_{f_i}$ the rotation matrix from finger base to hand palm coordinates. $\dot{\boldsymbol{\theta}}_i^d$ is the desired joint velocity commanded to an analog actuator control loop with gain $\boldsymbol{K}_C$. A linear impedance model

$$\boldsymbol{M}_i \Delta^{f_i}\ddot{\boldsymbol{x}}_i(t) + \boldsymbol{D}_i \Delta^{f_i}\dot{\boldsymbol{x}}_i(t) + \boldsymbol{K}_i \Delta^{f_i}\boldsymbol{x}_i(t) = {}^{f_i}\boldsymbol{c}_i - {}^{f_i}\boldsymbol{c}_i^d$$

with $\boldsymbol{M}_i$, $\boldsymbol{D}_i$, and $\boldsymbol{K}_i$ as mass, damping, and spring parameters generates position and velocity offsets $\Delta^{f_i}\boldsymbol{x}$ and $\Delta^{f_i}\dot{\boldsymbol{x}}$. These are used to adapt the desired values ${}^0\boldsymbol{x}_i^d$ and ${}^0\dot{\boldsymbol{x}}_i^d$ of the fingertip position and velocity according to the difference between the desired and actual value of the contact force. This scheme results in a compliant behavior of the finger around the desired tip position ${}^0\boldsymbol{x}_i^d$ and velocity ${}^0\dot{\boldsymbol{x}}_i^d$.

For optimization of grasping forces measurements of all fingers are concatenated (denoted by $\oplus$) and mapped into palm coordinates S_0 using the forward kinematics maps ${}^0\mathsf{T}_{s_{1..4}}$ and rotation matrices ${}^0\boldsymbol{R}_{s_{1..4}}$ of the corresponding finger. The optimized forces ${}^{f_{1..4}}\boldsymbol{c}^*$ are then separated into subvectors comprising the contact forces of the respective fingers which is denoted by $\ominus$. These are used as desired contact forces for the impedance model.

In standard impedance controllers a priori planned forces are commanded to the impedance model by the reference generator which is depicted by a dashed line in Figure 4. This information is *not used* in the current approach.

4.2 Control Switching

During a manipulation task the GFO module is switched to different modes according to the desired discrete contact states $\boldsymbol{q}^d$ of the fingers, see [7].

When the desired contact state changes the GFO switches to an unloading phase. The contact force of the respective finger is reduced to zero and optimally distributed to still contacting fingers during this transition phase. The GFO then only considers contacting fingers for force optimization.

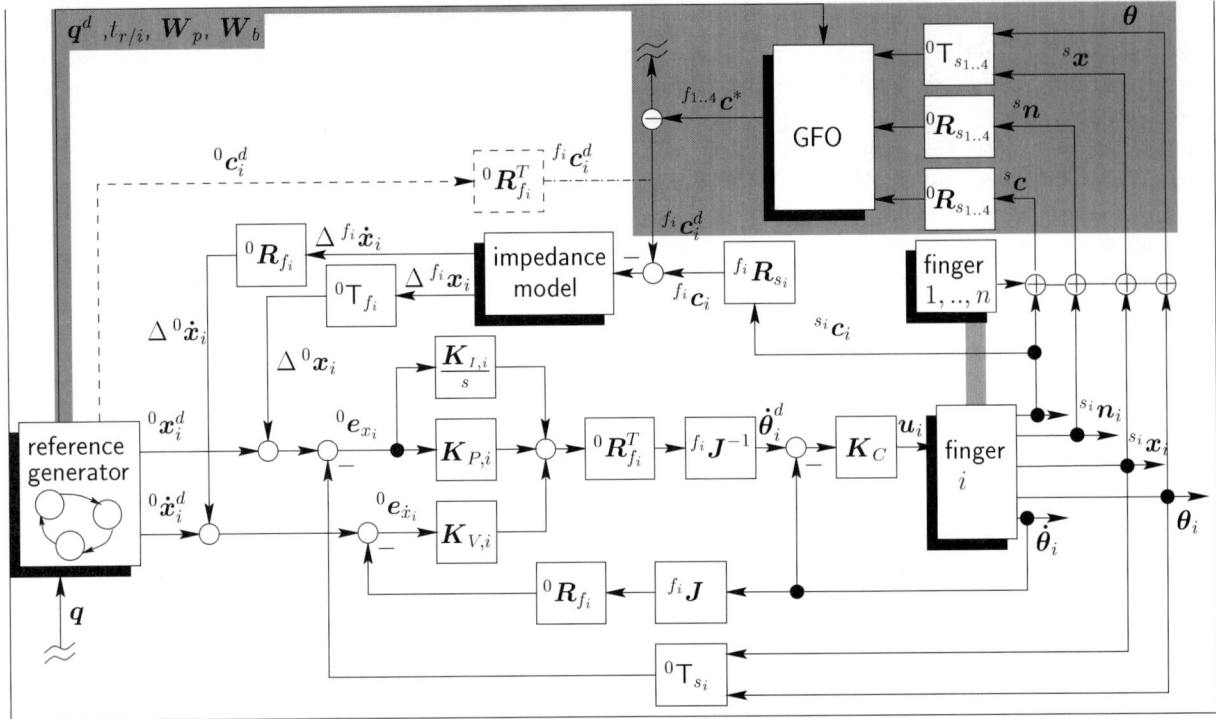

Figure 4: Fingertip impedance control law with grasping force optimization module

When the reference generator commands a moving finger to recontact the GFO is switched to a loading phase of duration $t_{r/i}$ during which the finger is smoothly reintegrated in the grasp again. When this operation is complete a full grasp mode is activated. The impedance control loops of the fingers are not directly affected by changes in the desired contact state.

5 Experimental Results

To demonstrate the efficiency of the proposed approach results of two different experiments are presented. In the first setup a cubic sample object with a mass of $81g$ is manipulated. It is initially grasped with four fingers in contact, see Figure 5. Results obtained by deliberately slow GFO performance are compared with the ones achieved with fast GFO.

In the second case a light bulb of mass $38g$ is screwed with high speed into a socket, i.e. fast object screw motion and quick regrasping. The setup is depicted in Figure 9.

A video clip showing the experiments described below is included in the Conference Video Proceedings.

Experiment 1a: Grasping a cube A wooden cube is constantly grasped by four fingers. The GFO is executed with a *purposely slow* sample time of $100ms$. During the task a time-varying disturbance force is acting on the object. Its maximum amplitude is about five times the object gravitational force. This is de-

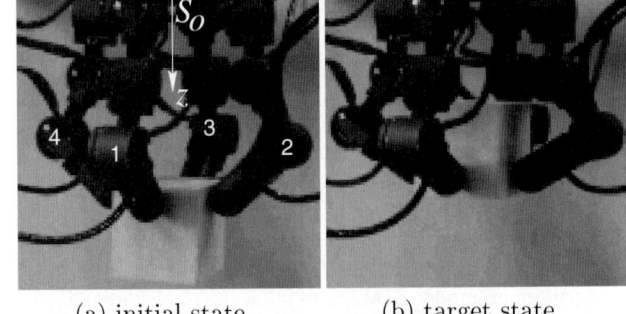

(a) initial state   (b) target state

Figure 5: TIT-hand grasping the sample object

picted in Figure 6(a). Figure 6(b) shows the friction reserve $\mu c_{1,1}$ (solid line) and the magnitude of tangential forces $(c_{1,2}^2 + c_{1,3}^2)^{\frac{1}{2}}$ (dashed line) as measured from the contact force sensor of finger 1. Obviously, the magnitude of the tangential force reaches the friction reserve indicating unintended slip of the fingertip. Results of a similar experiment with *fast GFO* ($4ms$) and an acting disturbance force of nearly eight times the object gravitational force is shown in Figure 7. Here, we achieve improved performance, since the friction reserve is not exceeded. Note that in both cases the GFO generates desired gasping forces $^{f_i}c_i^d$ within friction limits. However, slow GFO does not immediately compensate f_{ext} which results in a position error of the object and the fingertips, see Figure 6(c). The impedance controller compensates for this error

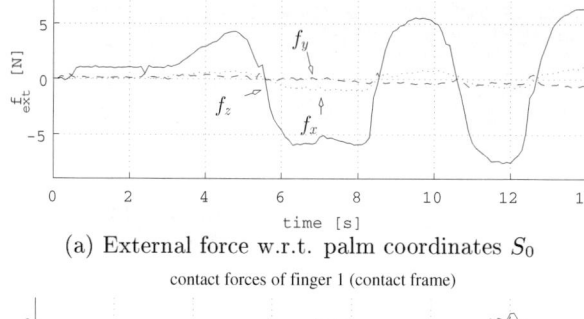

(a) External force w.r.t. palm coordinates S_0

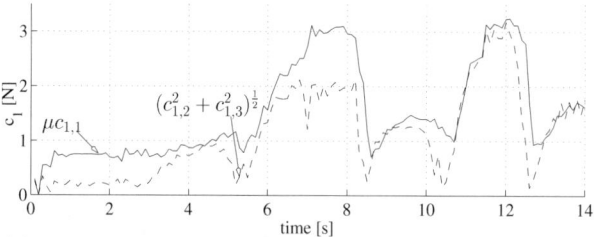

(b) Friction reserve (solid) and tangential force (dashed)

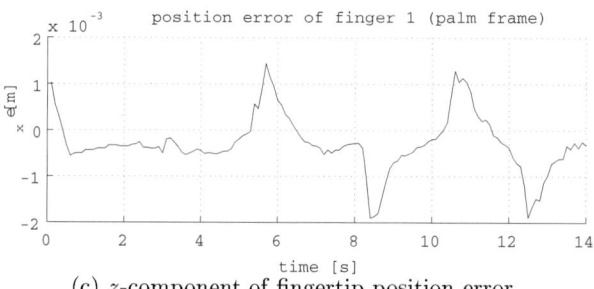

(c) z-component of fingertip position error

Figure 6: Grasping with slow optimization (100ms)

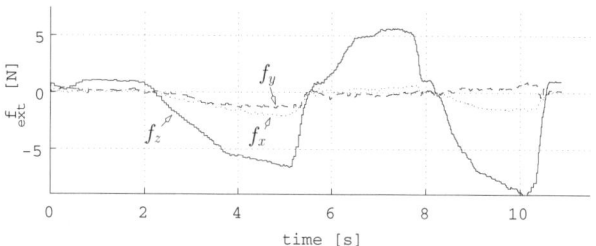

(a) External force w.r.t. palm coordinates S_0

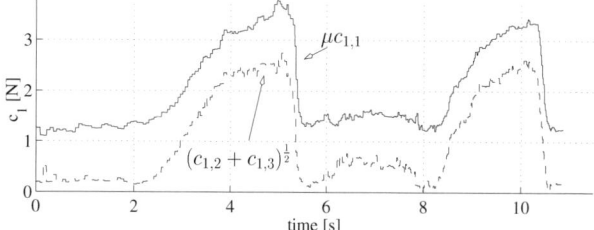

(b) Friction reserve (solid) and tangential force (dashed)

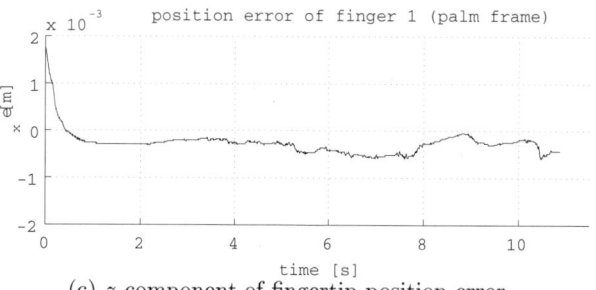

(c) z-component of fingertip position error

Figure 7: Grasping with fast optimization (4ms)

by generating contact forces *which are not necessarily located inside the friction cone*. The greater the position error the more optimized forces are overridden. Fast compensation for f_{ext} generates a smaller position error as demonstrated by Figure 7(c). This behavior results in an increased friction reserve at the contact points.

Experiment 1b: Regrasping a cube First, the contact force of finger 1, see Figure 5, is reduced to zero during $t_{r/i} = 1s$ beginning at $t = 55.7s$. The results are shown in Figure 8. Figure 8(c) depicts the change in the contact force of another finger indicating the changing force distribution as commanded by the GFO.

Next, the free finger is moved to a new contact point location about $\Delta z = 1.5 cm$ w.r.t. S_0 while a disturbance force is acting on the grasped object. After reaching the new location the contact force is increased.

During this task the GFO needs between 2 to 17 iterations for opimization (average: 11 iterations). A similar task with a heuristic choice of contact forces was not capable of completing the task in case of a disturbance force.

Experiment 2: Fast manipulation of a bulb The bulb is grasped by the TIT-hand, see Figure 9, and moved in z_0-direction until contact between bulb and socket is established. Because the grasp is disturbed by interaction forces between bulb and socket the weights in the cost index are adjusted to produce a rather tight grasp.

The bulb is rotated with respect to the z_0-axis about 0.63rad in 0.14s. This is followed by consecutive regrasping of all fingers where each finger regrasping task needs 0.75s. After that the bulb is again rotated. This procedure is repeated about 20 times until the bulb is tightly screwed into the socket after 80s.

The sample time of the GFO was varied between 4ms and 1s. The following table shows the results of a series of experiments to insert and screw the bulb into the socket.

4ms	20ms	50ms	75ms	125ms	0.25s	0.5s	1s
ok	ok	ok	fail	fail	fail	fail	fail

From 4ms to 50ms the task could be accomplished successfully. For a sample time $> 75ms$ the object was pushed out of the grasp during regrasping. The cause for this phenomenon is that the grasp forces were not

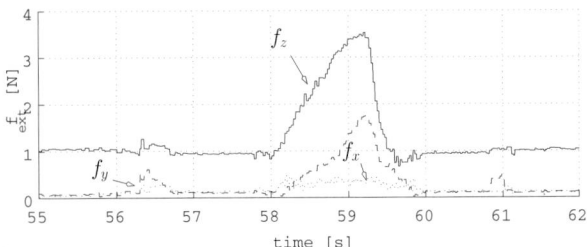
(a) External force w.r.t. palm coordinates S_0

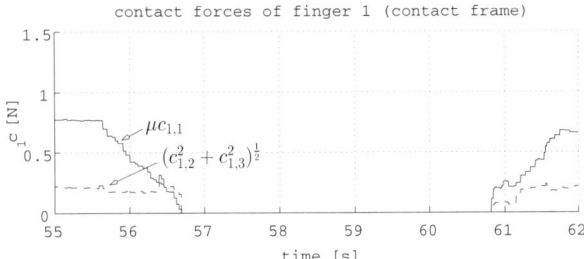
(b) Friction reserve and tangential force of finger 1

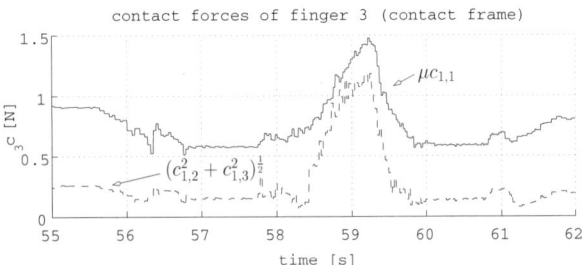
(c) Friction reserve and tangential force of finger 3

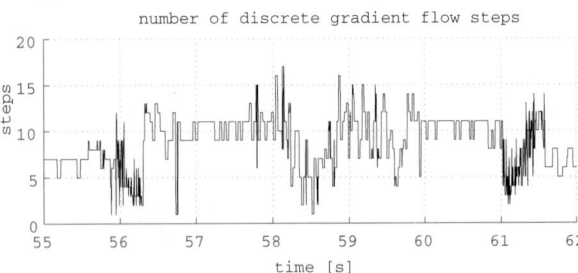
(d) Number of iterations during optimization

Figure 8: Regrasping of finger 1

adjusted fast enough to the changing orientation of the object after rotation and new grasp states with a different number of contacting fingers. This renders the grasp unstable with long sample times of the GFO.

6 Conclusions

This paper presented results of object regrasping and fast object manipulation using on-line grasping force optimization and an impedance controller. Fast GFO compensates external disturbance forces acting on the object without loosing grasp stability. Current information about the grasp is derived from fingertip force sensors which provide exact information on grasp matrix and object force. Rolling motion of fingertips is also considered.

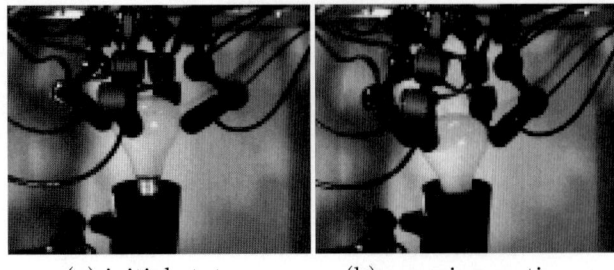

(a) initial state (b) screwing motion

Figure 9: TIT-hand screwing bulb in socket

A significant improvement of fast grasping force optimization is the increase of manipulation velocity shown in the bulb experiments. Only fast adjustment of grasping forces enables fast dextrous object manipulation. Many (re)grasping experiments could not be performed successfully without using the proposed fast GFO together with contact force sensor information.

Acknowledgments

The authors greatly appreciate the support by the German Academic Exchange Service (DAAD) making the first author's stay at TIT possible. This work was supported, in part, by the German Research Foundation (DFG) in the framework of the "KONDISK" priority research program.

References

[1] K. B. Shimoga, "Robot Grasp Synthesis Algorithms: A Survey," *Intl. Journal of Robotics Research*, vol. 15, June 1996.

[2] Z. Li, Z. Quin, S. Jiang, and L. Han, "Coordinated Motion Generation and Real–Time Grasping Force Control for Multifingered Manipulation," in *Proc. of the IEEE Intl. Conf. on Robotics & Automation*, 1998.

[3] T. Yoshikawa and K. Nagai, "Analysis of multi-fingered grasping and manipulation," in *Dextrous Robot Hands* (S. T. Venkataraman and T. Iberall, eds.), New York: Springer, 1990.

[4] M. Buss and K. Kleinmann, "Multi-Fingered Grasping Experiments Using Real-Time Grasping Force Optimization," in *Proc. of the IEEE Intl. Conf. on Robotics & Automation*, 1996.

[5] H. Maekawa, K. Tanie, and K. Komoriya, "Dynamic Grasping Force Control Using Tactile Feedback for Grasp of Multifingered Hand," in *Proc. of the IEEE Intl. Conf. on Robotics & Automation*, 1996.

[6] T. Omata and M. A. Farooqi, "Reorientation Planning for a Multifingered Hand based on Orientation States Network Using Regrasp Primitives," in *Proc. of the IEEE/RSJ Intl. Conf. on Intelligent Robots & Systems IROS*, 1997.

[7] M. Buss and T. Schlegl, "Multi-Fingered Regrasping Using On-Line Grasping Force Optimization," in *Proc. of the IEEE Intl. Conf. on Robotics & Automation*, 1997.

DLR-Hand II: Next Generation of a Dextrous Robot Hand

J. Butterfaß, M. Grebenstein, H. Liu and G. Hirzinger

German Aerospace Research Center (DLR)
Institute of Robotics and Mechatronics
P.O. Box 1116, D-82230 Wessling, Germany
Joerg.Butterfass@dlr.de

Abstract — *this paper outlines the 2nd generation of multisensory hand design at DLR. The results of the use of DLR's Hand I were analyzed and enabled - in addition to the big efforts made in grasping technology - to design the next generation of dextrous robot hands. An open skeleton structure for better maintenance with semi shell housings and the new automatically reconfigurable palm have been equipped with more powerful actuators to reach 30N on the fingertip. Newly designed sensors as the 6 DOF fingertip force torque sensor and integrated electronics together with the new communication architecture which enables a reduction of the cabling to the hand to only 12 lines outline the electronics concept. The Cartesian impedance control of all the fingers completes the new hand with its 13 DOF to what it is: the next step to autonomous and humanoid grasping*

1 Introduction

In 1997 DLR developed one of the first articulated hands with completely integrated actuators and electronics (fig. 1) [4]. This well known hand has been in

Figure 1. DLR's Hand I

use for several years and has been a very useful tool for research and development of grasping. The experiences with Hand I accumulated to a level that enabled us to design a new hand according to a fully integrated mechatronics concept which yields a reasonably better performance in grasping and manipulation and therefore accelerates further developments.

2 Design Philosophy

In order to achieve the goal of maximum flexibility and performance our philosophy is the miniaturization and complete integration of all components of the hand and also the massive reduction of cabling.
As on DLR's Hand I the main aspects in developing the new hand were maximum performance to improve autonomous grasping and fine manipulation possibilities and the use of fully integrated actuators and electronics without a forearm. This is the only possibility to use an articulated hand on different types of robots which are not specially prepared to be used with hands. Hands with forearms [5] or hands with just grasping abilities allow for a much smaller and thus more anthropomorphic design due to the possibility of using the additional space in the forearm for actuators and electronic components, but restrict the usability with e.g. industrial robots. Farther displacement of those components as known from the MIT-Utah Hand [3] nearly disables the use on mobile robots.
Furthermore the hand must be easy to maintain and use and even economically rebuild-able in case of any damage by daily research usage. For a good overview of more than fifty dextrous robot hands see [6].

3 The DLR Hand II

3.1 Open Skeleton Design

Due to maintenance problems with Hand I and in order to reduce weight and production costs the fingers and base joints of Hand II were realized as an open skeleton structure. The open structure is covered by 4 semi shells and one 2-component fingertip housing realized in stereolitography and vacuum mold (fig. 2). This enables us to test the influence of different shapes of the outer surfaces on grasping tasks without redesigning finger parts.

Figure 2. Finger with semi shell housings and rubber skin fingertip.

3.2 Kinematic Design of DLR's Hand II

The design process started on an anthropomophic base by evaluation of different workspace/manipulability measures like those of Salisbury [7] or Yoshikawa [9] to get optimal ratios of link lengths of one finger. The desired objects to be manipulated and technological restrictions resulted in absolute link lengths.

The second step was to get suitable hand kinematics. The main target developing Hand II beside the ability for fine manipulation has been the improvement of the grasping performance in case of precision- and power-grasp. Therefore the design of Hand II was

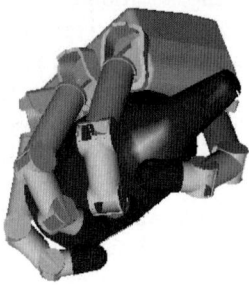

Figure 3. Optimization of kinematics with scalable hand-model.

based on performance tests with scalable virtual models as seen in fig. 3. Soon it turned out very important to be able to change the position of the 4th finger and the thumb as well. To perform power-grasps it is absolutely necessary to have a nearly parallel position of the second, third and forth finger as seen in fig. 4. On the other hand performing precision-grasps

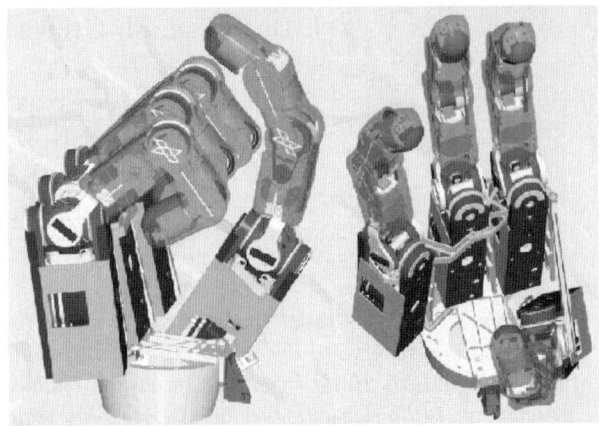

Figure 4. Simulation of Hand II in power-grasp- and fine manipulation configuration.

and fine-manipulation requires huge regions of intersection of the ranges of motion and the opposition of thumb and ring finger (fig. 4). Therefore Hand II was designed with an additional minor degree of freedom which enables to use the hand in 2 different configurations. This degree of freedom is a slow motion type to reduce weight and complexity of the system. The motion of the first and the fourth finger are both realized with just one brushed dc motor using a spindle gear.

The realized finger positions for both types of grasping were designed virtually and mapped to each other using the positions of 2nd and 3rd finger. Realizable kinematics were calculated and imported to the two virtually found configurations and optimized unless the actual configuration with an overall number of 13 DOF was found (fig. 4).

3.3 Actuator System

The three independent joints (there is one additional coupled joint) of each finger are equipped with appropriate actuators. The actuation systems essentially consist of brushless dc-motors, tooth belts, harmonic drive gears and bevel gears in the base joint. The configuration differs between the different joints. The base joint with its two degrees of freedom is of differential bevel gear type, the harmonic drive gears for geometric reasons being directly coupled to the motors. The differential type of joint (fig. 5) allows to use the full power of the two actuators for flexion or extension. Since this is the motion where most of the

Figure 5. Differential bevel gear of the new basejoint.

available torque has to be applied, it allows to use the torque of both actuators jointly for most of the time. This means that we can utilize smaller motors.

The actuation system in the medial joint is designed to meet the conditions in the base joint when the finger is in stretched position and can apply a force of up to 30 N on the finger tip. Here the motor is linked to the gear by the transmission belt (see [8] for advantages). The motor in the medial joint has less power than the motors in the base joint, however there is an additional reduction of 2:1 by the transmission belt. Thus we achieve the torque which corresponds to the torque created by the two motors in the base joint for an external force of 30 N on the finger tip. The harmonic drives used are of the same type for all joints, since the smallest appropriate type can stand the torque for both types of actuation.

3.4 Sensor Equipment

A dextrous robot hand for teleoperation and autonomous operation needs (as a minimum) a set of force and position sensors. Various other sensors add to this basic scheme (see table). Each joint is equipped with strain gauge based joint torque sensors and specially designed potentiometers based on conductive plastic. Besides the torque sensors in each joint we designed a tiny six dimensional force torque sensor for each finger tip which will be explained more precisely in 3.5. The potentiometers, each with an analogous filter of third order, would not be absolutely necessary, since one may calculate the joint position from the motor position, however they provide us with a more accurate information of joint position, and they can by the way eliminate the necessity of referencing the fingers after power up. In case of not using the potentiometers one would have to consider the elasticity of the transmission belt and the harmonic drive. With the potentiometer we achieve a resolution for the joint angles of $1/10°$, this means approximately 10 bits for the joint.

sensor type	count/finger	range (resolution)
joint position	3	110° (10 bit); 120° (10 bit)
joint torque	3	2.4; 4.8 Nm (11 bit)
force/torque	1	10-40N;150Nmm(11 bit)
motor speed	3	
temperature	6	0-125 °C (8 bit)

Since the base joint is of differential type, one has to calculate the joint position of the base joint from the potentiometer values. There was no way to measure the joint position directly due to space restrictions. For increasing the controllability of the actuators we appreciate speed sensors. Like in DLR's first generation hand we utilize so called Tracking Converters [1]. In contrast to the old version the complete calculation is done by software since there is enough computing power available now. The sensor itself is basically a position sensor with very high resolution, where the speed can be calculated by differentiation of the position signal. Each motor is equipped with two linear Hall effect sensors which are used for commutation of the motors as well. These sensors supply two sinusoidal signals with a phase shift of 120°. The position within the magnetic cyle of the motor is calculated from these signals. By additionally counting the cycles the position can be calculated. This type of sensor gives us just a relative position of the motor and has thus to be referenced after power up.

3.5 Force Torque Fingertip Sensor

A tiny six dimensional force torque sensor (20 mm in diameter and 16 mm in height) as shown in Fig. 7 with full digital output has been developed for the fingertip.

The force and torque measure ranges are 10 N for F_x and F_y, 40 N for F_z, 150 Nmm for M_x, M_y and M_z respectively. Also a 200 % mechanical overload protection is provided in the structure.

Measuring principle: The mechanical structure (as shown in Fig. 6) of the sensor is composed of two sensitive parts, one is a round plate (base element) with three symmetrical sensitive beams, another is a cantilever beam.

There are three elastic beams in the base element, which are sensitive to the M_x, M_y and F_z by measuring strain ϵ_1, ϵ_2 and ϵ_3. The cantilever beam is a rectangular pipe with a very thin wall. By using specialized torsion shear strain gauges it can measure the

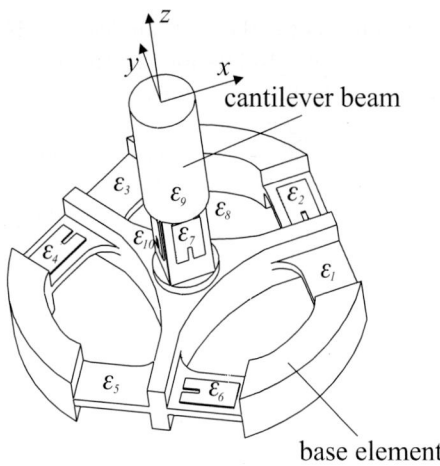

Figure 6. Mechanical structure of six dimensional force torque sensor.

F_x, F_y and M_z.

$$\begin{aligned}
F_x &= K_1(\epsilon_7 - \epsilon_9) \\
F_y &= K_2(\epsilon_8 - \epsilon_{10}) \\
F_z &= K_3\left[(\epsilon_1 - \epsilon_2) + (\epsilon_3 - \epsilon_4) + (\epsilon_5 - \epsilon_6)\right] \\
M_x &= K_4(\epsilon_3 - \epsilon_4 - \epsilon_5 + \epsilon_6) \\
M_y &= K_5\left[\epsilon_1 - \epsilon_2 - \frac{1}{2}(\epsilon_3 - \epsilon_4 + \epsilon_5 - \epsilon_6)\right] \\
M_z &= K_6(\epsilon_7 - \epsilon_8 + \epsilon_9 - \epsilon_{10})
\end{aligned}$$

Where K_i are the calculation parameters.

Figure 7. Six dimensional force torque sensor in fingertip.

Internal electronics: By using the latest technology the sensor with internal electronics can provide force and torque data at very high bandwidth and with very low noise. Signals from foil strain gage bridges are amplified and converted to digital representations of the force and torque applied to the sensor. All low level analog signals and the A/D converter (12bit) are within the sensor body, shielded from electromagnetic interference by the metal sensor body.

3.6 Integrated Electronics

One major goal of the design of the new DLR Hand was to fully integrate the electronics needed in the fingers and the palm in order to minimize weight, the amount of cables needed for a multisensory hand, and to increase the reliability by minimizing the amount of cables moved crossing the joints. In case of joints with a single degree of freedom we solved the problem of reliability by using flexible printed circuit boards (PCB) with appropriate bending space within the links (see top of Fig. 8). Tests showed after 100,000 cycles no

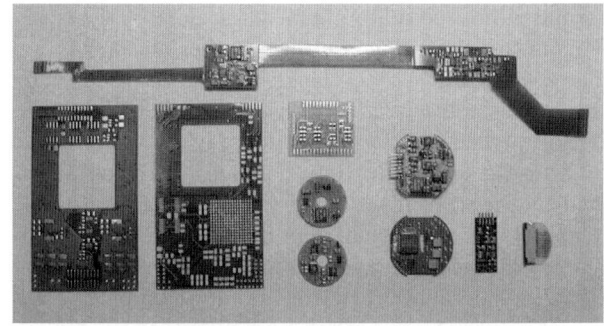

Figure 8. Set of PCBs in one finger.

visible or measurable effect on the flexible PCB.

In each link at least one serial ADC with 8 channels and 12 bit resolution converts the sensor signals as near as possible to the sensor circuitry into digital data. Thus only digital data is crossing any joint of the finger.

The power converters for driving the motors are located directly beside the motors and they are galvanically decoupled from the sensor electronics in order to minimize any noise induced by the running motors. Moreover the different fingers are galvanically decoupled from each other, too.

3.7 Communication Architecture

The control of the fingers and the hand is done by an external computer. In order to use the hand freely on different manipulators and to reduce cables and the possibility of noise in the sensor signals, we decided to design a fully integrated serial communication system. Each finger holds one communication controller in its base unit (see fig. 9). This controller is responsible for the collection and distribution of all information of interest. Furthermore it does some reasonable signal processing.

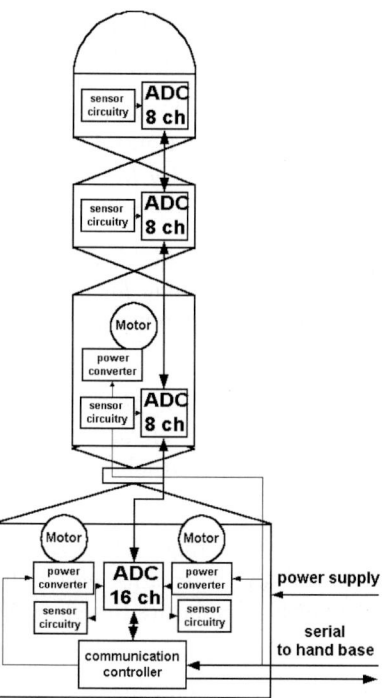

Figure 9. Electronics and communication in a finger.

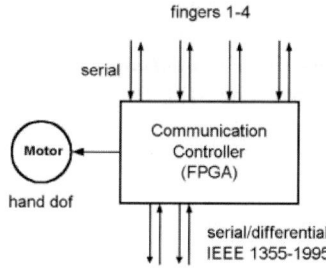

Figure 10. The communication controller in the hand base links the fingers to external computers.

It collects the data of all five ADCs per finger with together 40 channels of 12 bit resolution each and transmits these data to the communication controller in the hand base (see fig. 10). On the other hand it distributes the data from the control scheme to the actuators for finger control.

The communication controller in the hand base links the serial data stream of each finger to the data stream of the external control computer. By this hardware architecture we are able to limit the number of external cables of DLR's Hand II to a four line power supply and an eight line communication interface since the data is transmitted via differential lines. This interface even provides the possibility of using a quick-lock adaptor for autonomous tool exchange. Reducing external cabling from 400 (in Hand I) to 12 here, is one of the major steps forward in our new hand.

4 Cartesian Impedance Control

Salisbury [7] implemented cartesian stiffness control by using fingertip force sensors. The stiffness control scheme has the disadvantage of not being able to actively control the complete system dynamics, especially the system damping parameter. Hogan [2] introduced the impedance control scheme, which will improve system dynamic characteristics greatly. Using this control scheme we may control the complete impedance property of a finger:

$$F_{ext} = M_d \ddot{x}_k + B_d \delta \dot{x}_k + K_d \delta x_k$$

The equation above is central to the cartesian impedance control of the finger, where M_d, B_d, K_d are the 3×3 diagonal desired target impedance parameters of the finger. The selection of parameters will determine the directions in which large impedance is desirable and the directions in which small impedance is desirable. $\delta x_k = x - x_d$ is a 3×1 vector of position errors, while x, x_d, and F_{ext} are 3×1 vectors of actual positions, desired positions and external forces in the fingertip coordinate system, respectively. An essential variable in the compliant motion is the generalized force, F_{ext}, exerted by the end-effector. It is possible to select the coordinate system in view of the environment constrained in such a way that matrix M_d, B_d and K_d are diagonal. Each diagonal element is associated with a definite coordinate direction. For example, if a particular diagonal entry of the stiffness matrix K_d is small, then the contact force in this direction is small. On the other hand, when a specific diagonal entry is chosen as a large number, indicating a stiff spring, even a small displacement in this direction results in a considerable force. A large stiffness coefficient requires an accurate position control. The desired environmental generalized force may be generated indirectly by controlling the corresponding positions accurately.

When a robot hand performs any fine manipulation, there is always need that the fingertip should be soft in the direction normal to the contact surface and hard tangential to the contact surface. Therefore the impedance should be adaptable to the orientation of the fingertip. Therefore, a cartesian impedance controller has been built as shown in Fig 11. In steady state, all measured and desired velocity and acceleration values are zero. This induces that the value of the steady state torque is the stiffness multiplied by the steady state deformation $\delta\theta$, and the fingertip behaves like a programmable spring.

5 Conclusion and Future Work

DLR's work on one side aims at the development of robonaut systems for space applications and on the other side at the terrestrial use of ultralight weight arms and multifinger hands on mobile platforms. The DLR Hand II, for us, is a big step forward in approaching the goal of hands with human-like size and performance. Currently one new hand is in use in our lab meanwhile the second hand is in the stage of assembly. Future work will be a redesign to get the hand ready for manufacturing in a small series und on the other hand to get the hand simplified and qualified for space applications. Parallel to this work various efforts towards autonomous grasping and manipulation with the new hand are in progress.

References

[1] Butterfaß, J.; Hirzinger, G.; Knoch, S.; Liu, H.: *DLR's Multisensory Articulated Hand Part I: Hard- And Software Architecture*, Proceedings of the IEEE Int. Conference on Robotics and Automation, Leuven, Belgium, 1998, pp. 2081-2086.

[2] Hogan N.: *Impedance Control: An Approach to Manipulator: Part I-III*, Trans. ASME J. Dyn. Syst., Meas., Contr., Vol. 107, 1985, pp. 1-24.

[3] Jacobsen, S.C.; Wood, J.E.; Knutti, D.F.; Biggers, K.B.: *The UTAH/M.I.T. Dextrous Hand: Work in Progress*, The International Journal of Robotics Research, Vol. 3, No. 4, 1994.

[4] Liu, H.; Butterfaß, J.; Knoch, S.; Meusel, P.; Hirzinger, G.: *A New Control Strategy for DLR's Multisensory Articulated Hand*, Control Systems, Vol. 19, No. 2, April 1999, pp. 47-54.

[5] Lovchik, C.S.; Diftler, M.A.: *The Robonaut Hand: A Dexterous Robot Hand for Space*, Proceedings of the IEEE Int. Conference on Robotics and Automation, Detroit, 1999, S. 907-912.

[6] Reynaerts, D.: *Control Methods and Actuation Technology for Whole-Hand Dextrous Manipulation*, Dissertation, Katholieke Universiteit Leuven, 1995.

[7] Salisbury, J. K., Craig, J.J.: *Articulated hands: force control and kinematics issues*, The International Journal of Robotics Research, Vol. 1, 1982, pp. 4-17.

[8] Townsend, W. T., Salisbury, J. K.: *Mechanical Bandwidth as a Guideline to High-Performance Manipulator Design*, Proceedings of the IEEE Int. Conference on Robotics and Automation, Scottsdale, USA, 1989, pp. 1390-1395.

[9] Yoshikawa, T.: *Foundations of Robotics, Analysis and Control*, The MIT Press, 1990.

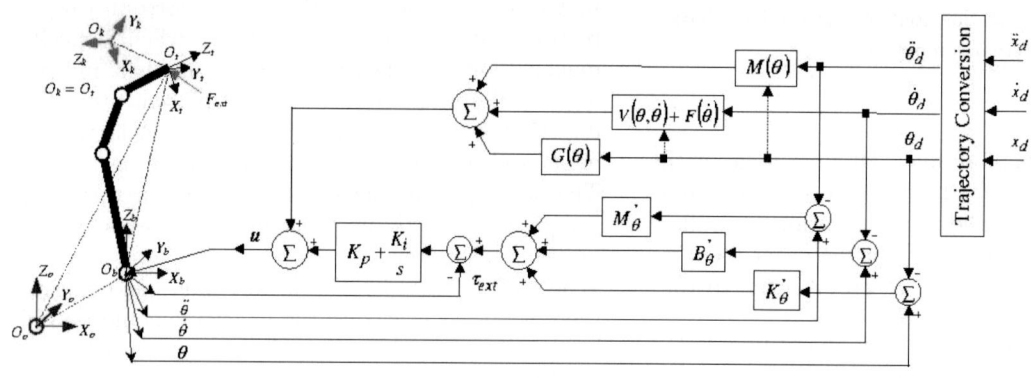

Figure 11. Block diagram of cartesian impedance control.

Micromanipulation Using a Friction Force Field

Jeongsik Sin, Tobias Winther and Harry Stephanou

Center for Automation Technologies
Rensselaer Polytechnic Institute
Troy, New York 12180, USA
sinj@cat.rpi.edu

Abstract

Most common requirements for micromanipulation are high precision, parallel manipulation, cost, etc. In this paper, we examine a manipulation method that may be useful for these requirements. The presented method uses an actively generated friction force field as a driving force of manipulation, and the method can have several actuation modes depending on the sliding condition of the friction force field. Planar dynamics regarding the proposed method were analyzed, and the design of the force field was investigated.

1 Introduction

The most common type of manipulation method is "pick & place" manipulation using a robot. Other manipulation methods such as pushing and tapping have been introduced too. As the object being manipulated becomes smaller and smaller, those methods show some limitations, and it even becomes difficult to pick or make contact at the right position.

We propose a manipulation concept and its implementation method, which may be useful for the manipulation of small planar objects. In this method, an object is placed in an actively generated friction force field and its positioning is performed by the design of the force field. To generate the force field, we used a 2-dimensional array of pneumatic suction nozzles. Some of them are moved by linear actuators so that normal force and motion create a friction force field based on the Coulomb friction law.

The suggested concept is expected to have several advantages over conventional methods. Using the force field to handle small objects has the advantage of not requiring precision gripping or a particular contact configuration which may be difficult for small objects. A specific friction force field can be generated in a bounded area under an object so that the generated friction force does not interfere with other friction fields. This may be useful when multiple objects are manipulated at the same time.

In this paper, we explain the manipulation mechanism and the design of the force field required for positioning. The planar dynamics of an object are analyzed when the friction force field is applied. Since the dynamic behavior is different for different shaped objects, we simulate trajectories for rotationally symmetric and non-symmetric cases. The design method to compute the force profile for the required friction field was considered in the case of a linear force field.

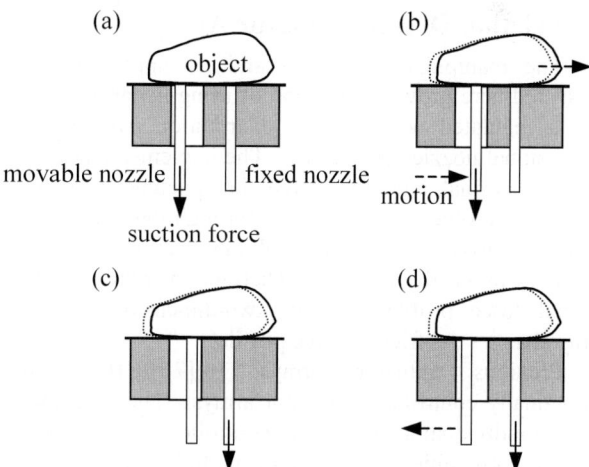

Figure 1. Concept of the proposed manipulation. (a) and (b): steps for moving forward, (c) and (d): steps for preparation of next motion.

2 Related Work

Pushing was first considered by [1] as an object manipulation method. It founded the framework for pushing mechanisms, where frictional forces play a dominant role. As a result, it was possible to predict the sense of rotation through the comparison of force rays exerted on the push contact point.

In [2], Lynch extended the pushing method to multiple contacts pushing. He considered kinematic and force constraints to determine possible instantaneous motions of a sliding object. The analysis resulted in a new manipulation primitive: stable rotational pushing, which may be useful for fine pushing operation.

Huang et al.[3] examined impulsive manipulation in which a manipulator strikes an object and lets it slide. He solved this manipulation as two problems; the inverse sliding problem which finds an initial kinematic condition to send the object to a desired position, and the impact problem to find a strike that achieves the condition. Huang et al.[4] also researched a tapping mechanism which is one particular form of impulsive manipulation. This work suggests a planning method to position a rotationally symmetric object with multiple tapping.

While the above works presented manipulation methods using conventional manipulators, the following works addressed new types of devices for manipulation. Luntz [5] analyzed the velocity field of a planar wheel array. The velocity field was used to position an object on the wheel array. A micro ciliary array [6][7] consists of ciliary legs which have a bimetal structure using differential thermal expansion. These arrays can be used as part feeder. In [8], Böringer analyzed the physical force field for positioning an object. Air jets [9] and magnetic forces [10] were also used as actuation forces in distributed manipulation.

3 Friction Driven Actuator Array

The manipulation method and its implementation presented here provide a method of using linear actuators and a regulated normal force distribution over a two-dimensional nozzle (grid) array. The implemented device supplies normal attractive force, i.e. pneumatic suction force, to the object on the device through the nozzles. It generates driving force by Coulomb friction when the actuators move the nozzles. Therefore variation of the normal force profile over the two-dimensional nozzle array changes the friction force profile.

Previous actuator arrays [6][7][9][10] used individually controllable actuator arrays. The movement of each individual actuator were controlled for continuous motion of one object or motion of multiple objects. They are generally microfabricated to allow for a large number of actuators in one device. The difference of the suggested device here is that this device only uses one actuator unit per axis (one for X and one for Y direction) which is shared by all nozzle elements. Instead of individual control of each actuator, individual regulation for the normal force profile makes planar motion possible.

The implemented device (Figure 2) consists of two piezoelectric actuators for X and Y motion, two linear motion guides, two sets of nozzle arrays, and solenoid valves to switch the pneumatic suction force individually. One set nozzle arrays is fixed relative to the device surface on which the objects are placed, the other set is actuated by the actuator relative to this surface.

For this device, there are two different actuation modes: uniform friction and non-uniform friction actuation. The first actuation mode uses uniform normal force profile and its magnitude is sufficient to hold the object so that there is no sliding between the nozzles and the object during the motion. The following steps are carried out for this mode as in Figure 1:

- Movable nozzle applies normal force to the object, while fixed nozzle does not.
- The actuator moves the movable nozzle to one extreme position (e.g. to the far right).
- The fixed nozzle applies attractive normal force to the object while the movable nozzle releases the force.
- The actuator moves the movable nozzle to the other extreme position (e.g. far left to original position).
- These steps are repeated until the object is close to the desired position. At that point the object will only be moved part of a step.

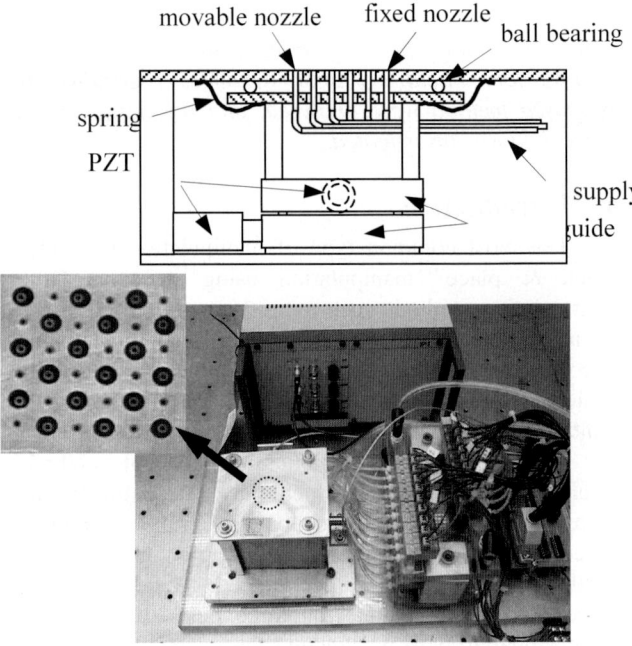

Figure 2. Implemented device

The time to apply and release the force is synchronized to the repeated actuation cycle of the actuators. The object motion is the same as the actuator's in terms of travel distance, resolution and speed. Rotational motion can be achieved by applying the force at a fixed nozzle which is located at the center of rotation while the force and motion are applied to the moving nozzle at the end of the object.

However, there can be practical situations where the nozzles can not supply the force required to hold the object, due to heavier weight or friction coefficient. In this case the object will slide relative to the nozzles, and the sliding friction is the driving force. The second actuation mode uses a non-uniform friction force field to control the sliding direction of the object.

The friction force during the sliding on the actuator array is a complex phenomenon in itself. There can be a combination of Coulomb friction, viscous friction, static friction, Stribeck effect and so on. In the following analysis of the object dynamics, we assumed that the object is sufficiently large so that any stiction force is negligible compared to the external normal force. In addition, instead of applying a complex friction model, the simple Coulomb friction model was applied because the detailed friction mechanism is still unknown and the purpose of this study is to drive the object manipulation method with the applied friction force field.

4 Dyanmics of Manipulation

4.1 Sliding Mode

When there are two interface surfaces on an object, sliding may occur on either one or both surfaces depending on the friction forces applied on the object. In this manipulation method, an object can slide against the device surface, nozzle or both surfaces.

Figure 3 shows a free body diagram of an object on the device surface with one nozzle actuation. During the actuation of the nozzle with the force F, a normal force f_n is applied on the object generating a friction force f_t on the object. If f_s is a friction force against the motion of the object, there can be four sliding cases depending on the magnitude of these forces.

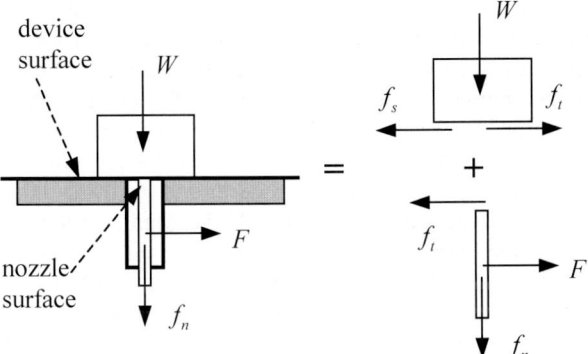

Figure 3. Free body diagram of an object on a device surface with one nozzle actuation

Case 1: $F \leq f_t$ and $f_t \leq f_s$
There is no sliding on either surface.
Case 2: $F \leq f_t$ and $f_t > f_s$
There is sliding between the object and the device surface.

Case 3: $F > f_t$ and $f_t \leq f_s$
There is sliding between the object and the nozzle.
Case 4: $F > f_t$ and $f_t > f_s$
There is sliding on both surfaces.

Cases 1, 2 and 3 are obvious. Case 4 can be verified using the following reasoning. The object cannot slide on the nozzle surface only, since $f_t > f_s$ should allow sliding on the device. The object cannot slide on the device surface only, because maximum force transferred to the object is f_t and the $F - f_t$ still exerts on the nozzle. Therefore there should be sliding on both surfaces. Then, the equation of motion for the object is $\ddot{x} = (f_t - f_s)/m$.

If we assume that (1) the nozzle surface is ideally flat and at the same level as the device surface, (2) nozzle's supporting force against the object is negligible due to the small area compared to the device surface, then $f_t = \mu_t f_n$ and $f_s = \mu_s (f_n + W)$. Where μ_t is friction coefficient between the object and the nozzle surface, μ_s is friction coefficient between the object and the device surface. The equation of motion and sliding distance of object are dependent on μ_t, μ_s, and f_n but not on the input force F if $F > f_t$.

The first actuation mode uses Case 2 sliding mode and a uniform normal force is applied. The second actuation mode uses Case 4 sliding mode and a non-uniform normal force field is applied. The dynamics of Case 4 sliding mode are analyzed in the next section.

4.2 Planar Dynamics on Non-Uniform Friction Field

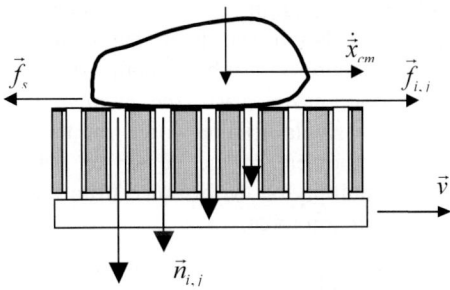

Figure 4. Free body diagram for object dynamics

There are two types of friction forces exerted on the object. An actively generated friction force is the driving force for object motion and is applied between the nozzle and the object. The i, j_{th} element of the nozzle supplies a normal force $n_{i,j}$. $f_{i,j}$ is a friction force on that nozzle. Its magnitude is proportional to the supplied normal force

$n_{i,j}$ and friction coefficient μ_t. The force direction is determined from the velocity difference between the actuator velocity v and the object velocity on the nozzle

$$\vec{f}_{i,j} = \mu_t n_{i,j} \frac{\vec{v} - \dot{\vec{x}}_{i,j}}{|\vec{v} - \dot{\vec{x}}_{i,j}|} \qquad (1)$$

The friction force df_s is the drag force during sliding between the object area dA and the device surface. The direction is opposite to the object moving direction. If the manipulated object is planar two-dimensional object, the force is

$$d\vec{f}_s = -\mu_s \left(\rho g t + p(x,y)\right) \frac{\dot{\vec{x}}}{|\dot{\vec{x}}|} dA \qquad (2)$$

Where μ_s, ρ, g and t denote the friction coefficient, object density, gravitational acceleration, and object thickness. $p(x,y)$ is a continuous pressure distribution function due to the discrete normal force $n_{i,j}$. It satisfies the following force and moment equilibrium condition with the supplied normal forces but it is statically indeterminate in most practical applications.

$$\begin{aligned} \int_A p(x,y) dA &= \sum\sum n_{i,j} \\ \int_A \vec{x} \times p(x,y) dA &= [Nx_c \quad Ny_c]^T \end{aligned} \qquad (3)$$

By combining equations (1) and (2) with summation over the nozzles and integration over the area, the total force applied on the object during the motion is computed.

$$\vec{F} = \sum\sum \vec{f}_{i,j} + \int_A d\vec{f} \qquad (4)$$

Friction forces usually can be reduced to a single force and its position is independent of the direction of motion. [1] found that there exists a center of friction during pure translation and it is the centroid of the pressure distribution. When both rotational and translation motion takes place, the moment is not just derived from a single friction force on the center of friction. The moment at the center of mass can be calculated by multiplying the distance from the center of mass with the friction force.

$$\vec{\tau}_{i,j} = (\vec{x}_{i,j} - \vec{x}_{cm}) \times \mu_t n_{i,j} \frac{\vec{v} - \dot{\vec{x}}_{i,j}}{|\vec{v} - \dot{\vec{x}}_{i,j}|} \qquad (5)$$

The moment due to the surface drag friction is computed at the center of mass also.

$$d\vec{\tau}_s = -(\vec{x} - \vec{x}_{cm}) \times \mu_s \left(\rho g t + p(x,y)\right) \frac{\dot{\vec{x}}}{|\dot{\vec{x}}|} dA \qquad (6)$$

Similarly, combining the two equations gives the total moment exerted on the center of mass.

$$T = \sum\sum \vec{\tau}_{i,j} + \int_R d\vec{\tau}_s \qquad (7)$$

Applying equations (4) and (7) with the equations of motion, $F = m\ddot{x}_{cm}$ and $T = I\ddot{\theta}$, the object motion can be simulated.

In planar manipulation, objects with rotationally symmetric shapes (circular, rectangular, etc.) and uniform pressure distribution are simpler than other shapes. Those cases can be treated as two separate problems: pure translation and pure rotation. Sliding friction cancels out except in the translation direction and the equation of motion can be solved for two dof (x, θ). This makes the object move in straight line.

For rotationally non-symmetric objects, friction force does not cancel out and the moment varies depending on the angular position. The integration of force and moment over the object region can be calculated in the local coordinate X' in which the origin is at the center of mass. If the rotational coordinate transformation matrix R transforms the object velocity to the local coordinate X' is the friction drag force exerted at the center of mass is

$$\vec{F}_s = -R^{-1} \iint \mu_s \left(\rho g t + p(x',y')\right) \frac{\dot{\vec{x}}'}{|\dot{\vec{x}}'|} dx'dy' \qquad (8)$$

Where, θ is the orientation of the object and $\dot{\vec{x}}' = R\dot{\vec{x}}$. In the same way, the moment is calculated in the local coordinate system.

$$T_s = -\iint \vec{x}' \times \mu_s \left(\rho g t + p(x',y')\right) \frac{\dot{\vec{x}}'}{|\dot{\vec{x}}'|} dx'dy' \qquad (9)$$

4.3 Inverse Sliding

The inverse sliding problem has been stated as the problem of finding the initial condition required for the object to come to rest at the desired final configuration as in the impulsive method [3]. Since finding the initial condition that satisfies the solution (i.e. the desired position) of the equation of motion is not algebraically derivable, a search method is used to find it. If the object shape is rotationally symmetrical, the search domain is monotonic, therefore relatively simple search methods can be applied. However if the object is of a non-symmetrical shape, the search domain may be non-monotonic and therefore advanced search methods are needed to find the global optimum [4].

In this manipulation method, the initial conditions consist of the force and torque that need to be applied on the object. The inverse sliding problem can be solved using the same method as in [3] [4]. However the inverse sliding solutions may not be applicable for precision positioning because the equation of motion uses simplified friction model and object shape.

5 Design of Friction Field

Based on the required force and moment (F_d, T_d) found from the inverse sliding problem a distributed normal force profile can be designed. An object initially stays at a certain position, the required force and moment equations become

$$\vec{F}_d = \sum\sum \mu_t n_{i,j} \frac{\vec{v}}{|\vec{v}|}, \quad T_d = \sum\sum \vec{x}_i \times \mu_t n_{i,j} \frac{\vec{v}}{|\vec{v}|} \quad (10)$$

The normal force is assumed to be a linear force profile over nozzles in a two-dimensional surface. Where, a, b, and c denote the parameters in an equation of plane.

$$n_k + a x_k + b y_k + c = 0 \quad (11)$$

By using a linear profile, the matrix form of the profile is an equation of the center of pressure $\vec{x}_{cn}$ and sum of normal force N as presented in [5]. Therefore, the problem of finding the force profile is reduced to finding the magnitude of N and its center $\vec{x}_{cn}$.

$$\vec{N}^T = NB^T\left(BB^T\right)^{-1}\left(\begin{bmatrix}1\\0\\0\end{bmatrix} + \begin{bmatrix}0&0\\1&0\\0&1\end{bmatrix}\vec{x}_{cn}\right) \quad (12)$$

where, $\vec{N}^T$ is a column vector of normal force n_k, k is serialized index from i and j, and matrix B is defined as

$$B = \begin{bmatrix}1 & \cdots & 1\\ x_1 & \cdots & x_n\\ y_1 & \cdots & y_n\end{bmatrix} \quad (13)$$

The matrix form for equation (10) including one-dimensional array of the force profile is,

$$\vec{F}_d = \mu_t \vec{V} \vec{N}^T, \quad T_d = \mu_t \left(\vec{X}\begin{bmatrix}0 & 1\\ -1 & 0\end{bmatrix} v\right)^T \vec{N}^T \quad (14)$$

where, $\vec{X} = [\vec{x}_1,...,\vec{x}_n]$, $\vec{x}_k = [x_i \quad y_j]^T$

Combining (14) with (12) gives magnitude N and center of the force profile $\vec{x}_{cn}$.

6 Experiment and Simulation

A simple experiment was carried out using uniform pressure to verify the manipulation concept of the first actuation mode. The object manipulated was a small piece of silicon wafer (1mm×4mm) that covers 2 nozzles (one fixed nozzle and one movable nozzle). The operating condition was −7 psi pressure on each nozzle (ID 508μm) with 10Hz of 10μm stroke of the piezoelectric actuator. Figure 5 shows the translation and rotational motion results after operation for 10sec. The translation distance d and rotational angle θ were measured as 0.96mm and 10°, which are less than the expected values (1mm and 21°) respectively. Rotational motion showed more error due to relatively weak suction force comparing to the friction force on the surface.

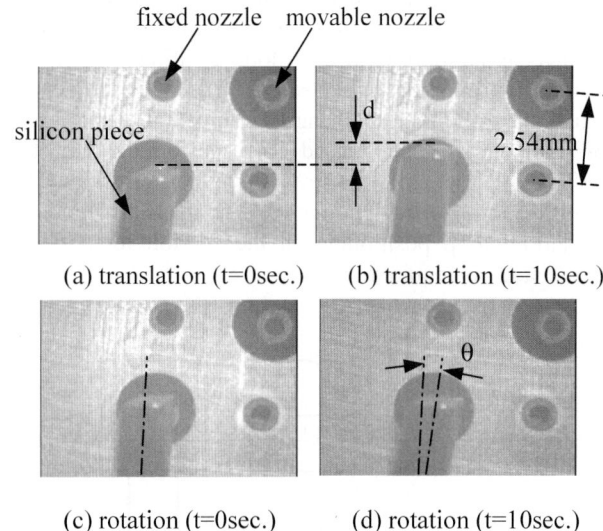

(a) translation (t=0sec.)   (b) translation (t=10sec.)

(c) rotation (t=0sec.)   (d) rotation (t=10sec.)

Figure 5. Experimental result of translation/rotational motion of the first actuation mode.

An object trajectory was simulated when a friction force field was applied as in the second actuation mode. The solution was computed through integration of the equations of motion. To avoid the singularities at $|\dot{\vec{x}}| = 0$, the continuous object area was discretized as a small rectangular area, and the force and moment were computed by summation except at the singularity points. The continuous pressure distribution function $p(x,y)$ in the equation (2) and (6) was assumed to be negligible.

As a simulation condition, the linear suction force of Figure 6 (b) was applied around the center of three objects (a) through nine nozzles. A step velocity input of 100mm/sec was applied to the piezoelectric actuator in x-direction for 0.1 sec. Figure 6 (c) shows simulation results of the trajectories for three different shapes.

7 Discussion

The first actuation mode with uniform force field provides positioning such as pick and place manipulation. The positioning accuracy that can be achieved is the same as the actuators accuracy. Experimental results show successful handling of objects beyond the stroke of the actuator. The second actuation mode may be useful when there exists sliding between the nozzle and object due to heavier object or reduced normal force. This actuation can be push style, exerting a normal force during relatively slow motion of actuator, or impulsive style exerting a normal force during fast motion of the actuator.

In the second actuation mode, the inverse sliding and the design of the force profile have been considered for trajectory planning purpose. The inverse solution may be critical for the impulsive method if it uses a low bandwidth conventional manipulator to apply force on the object. Its trajectory relies on the positioning accuracy of the inverse solution as in open loop positioning. In the manipulation method presented here, a higher frequency actuation can be used to change the direction of the force and moment. Then, closed loop positioning can be applied to compensate the inverse solution errors due to simplification of the friction model and mass distribution. For closed loop positioning on the friction force field, further study is required.

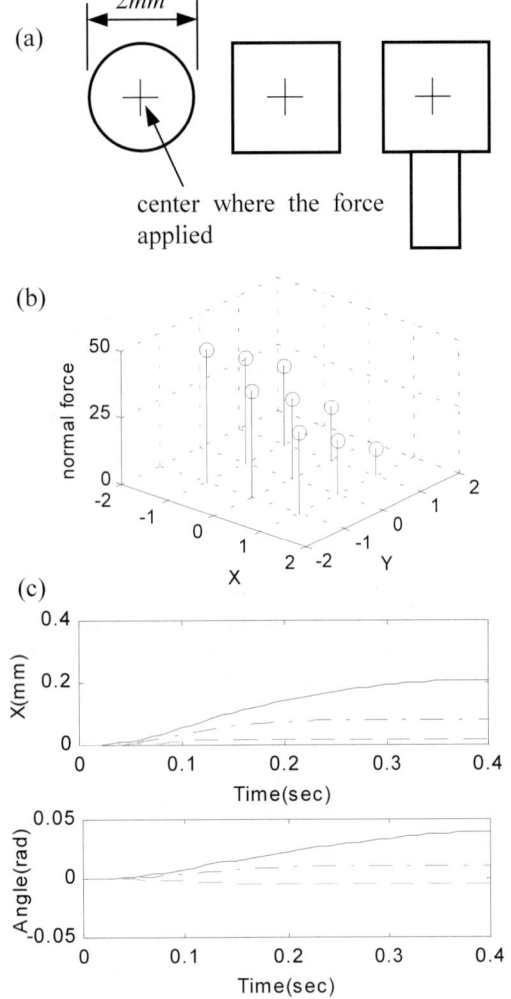

Figure 6. Simulation of an object trajectory when step input actuation was applied (a) three object shapes used in simulation: circular disk, square and asymmetric (b) applied suction force profile that centered at origin (c) trajectory after step input actuation: solid (disk), centered (square), and dashed (asymmetric)

Acknowledgement

This paper is based upon work supported in part by the New York State Office of Science, Technology and Academic Research

Reference

[1] M. Mason, "Mechanics and planning of manipulator pushing operations", International Journal of Robotics Research, Vol.5, No.3, 1986, pp.53-71.

[2] Kevin M. Lynch, "The mechanics of fine manipulation by pushing", Proceedings of IEEE International Conference on Robotics and Automation, 1992.

[3] W. Huang, E. Krotkov, M. Mason, "Impulsive manipulation", Proceedings of IEEE International Conference on Robotics and Automation, 1995, pp.120-125.

[4] W. H. Huang and M. T. Mason. "Mechanics, Planning, and Control for Tapping", Third International Workshop on the Algorithmic Foundations of Robotics, Houston TX, March 1998.

[5] J. Luntz, W., Messner, H. Choset, "Discrete actuator array vectorfield design for distributed manipulation", Proceedings of IEEE International Conference on Robotics and Automation, 1999, pp.2253-2241.

[6] K. Böhringer, B. Donald, N. MacDonald, "Single-crystal silicon actuators for micro manipulation tasks", Proceedings of IEEE Workshop on Micro Electro Mechanical Structures (MEMS), San Diego, CA, Feb. 1996, pp.7-12.

[7] J. Suh, R. Darling, K. Böhringer, B.Donald, H. Baltes, "CMOS Integrated ciliary actuator array as a general-purpose micromanipulation tool for small objects", Journal of microelectromechanical systems, Vol. 8, No. 4, 1999, pp.483-496.

[8] K. Böhringer, B. Donald, N. Macdonald, "Programmable force fields for distributed manipulation, with applications to MEMS actuator arrays and vibratory parts feeders", The International Journal of Robotics Research, Vol. 18, No.2, Feb. 1999, pp.168-200.

[9] S. Konishi, H. Fujita, "A conveyance system using air flow based on the concept of distributed micro motion systems", Journal of microelectromechanical systems, Vol. 3, No. 2, 1994, pp.54-58.

[10] C. Liu, T. Tsao, Y. Tai, W. Liu, P. Will, C. Ho, "A micromachined permalloy magnetic actuator array for micro robotics assembly systems", Transducers '95 Dig. 8th Int. Conf. On Solid-State Sensors and Actuators/Eurosensors IX, Vol. 1, Stockholm, Sweden, Jun. 1995, pp.328-331.

On the Coarse/Fine Dual-Stage Manipulators with Robust Perturbation Compensator

SangJoo Kwon
Ph.D. Student

Wan Kyun Chung
Professor

Youngil Youm
Professor

Department of Mechanical Engineering
Pohang University of Science & Technology(POSTECH), Pohang, KOREA
Tel: +82-54-279-2844; Fax: +82-54-279-5899; E-mail:{ksj,wkchung,youm}@postech.ac.kr

Abstract

A dual-stage, fast and fine robotic manipulator is presented. By adopting merits of both coarse and fine actuator, a desirable system having the capacity of large workspace with high resolution of motion is enabled. We constructed an ultra precision XY manipulator with dual-stage structure where the PZT driven fine stage is mounted on the motor driven XY positioner and applied it to fine tracking controls and micro-teleoperations as a slave manipulator. We describe essential merits of the compound actuation mechanism and the control strategy to successfully utilize it with proper servo system design. Through experimental results, the effectiveness of the coarse/fine manipulation by the dual-stage manipulator is shown.

1 Introduction

Two major sources of hindering control performance in systems with conventional actuators(hereafter "coarse actuators") such as electrical motors or hydraulic actuators may be the nonlinear friction in low speed motions and the resonance mode in high frequency motions. On the other hand, "fine actuators" such as piezoelectric actuators(PZT) are free of friction problem and much higher resolution of motion is possible but it is limited in motion range, at most, to several hundreds of microns. As a trial to overcome limitations of conventional actuators, several compound actuation systems with dual-stage construction were investigated for some applications [1–5]. The purpose of compound actuation is to increase the performance of the system by adopting merits of both coarse and fine actuators. In dual-stage systems, coarse and fine actuator are complementary to each other. That is, the coarse actuator offers a large workspace and actuation power while the fine actuator enables high resolution of motion and so defects of a actuator can be compensated by the merit of the other actuator. There were lots of studies where the concept of dual-stage compound actuation is applied. For example, we can find several works on macro/micro robot manipulators [6,7] and dual-stage XY positioning tables [3–5]. However, in these days, the research on the dual-stage servo seems to be the most active in hard disk drives(HDD) [1,2], where the 2nd actuator fabricated using the micro machining technology enables high speed track following.

In this paper, based on our former result [5], we propose a dual-stage control algorithm using a novel perturbation compensator and show experimental works using the dual-stage manipulator constructed for micro-teleoperations as well as fine tracking controls. Above all, we describe the system configuration with some mechanical considerations for compound actuation mechanism in section 2. A dual servo algorithm which are requisite for successful dual-stage manipulations are given in section 3, where a new perturbation compensator is proposed. Section 4 demonstrates experimental results including micro-tele operation. Finally, section 5 concludes this paper.

2 System Description

For fine tracking control applications or microscopic manipulation in micro-tele operating systems, we constructed a dual-stage manipulator where the fine stage is mounted on the coarse positioner as shown in Fig. 1, where the bottom coarse positioner offers a large work area of $200 \times 100\ mm$ in (x_1, y_1) coordinate plane and the top fine positioner enables nano level resolution of motion with travel range of $100 \times 100\ \mu m$ in (x_2, y_2) plane. The position sensors of coarse and fine stage are encoder(2500 $pulse/rev$) and capacitive gap sensor, respectively. The *absolute* accuracy of each positioner in closed-loop motion is about 20 μm in coarse positioner and at least submicron in fine positioner. As a result, by the cooperative manipulation of both postioning stages, a fast and fine manipulation with large workspace is possible. Figure 2 denotes the system configuration of the dual-stage manipulator in closed-loop, where the coordinates of the end-point of the dual-stage are $(x = x_1 + x_2, y = y_1 + y_2)$. So, there exists a *redundancy* of motion in each direction. In the dual-stage manipulator, the fine positioner is driven by piezoelectric actuator(PZT) with flexure hinge mechanism and the coarse positioner is driven by BLDC motors with ball-screw mechanism. So, the fine stage has much faster dynamics than the coarse stage and its high frequency redundant motion (x_2, y_2) makes it possible to compensate considerable tracking errors which are unavoidable in the motion of coarse stage.

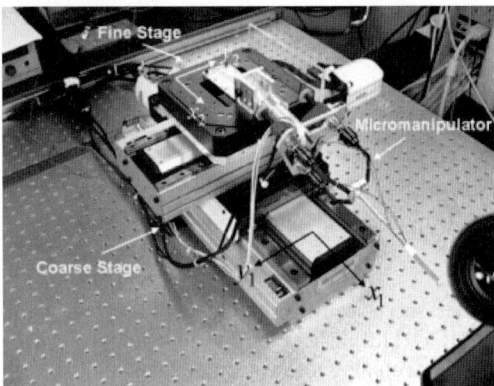

Fig. 1: Coarse/Fine dual-stage manipulator with micromanipulator.

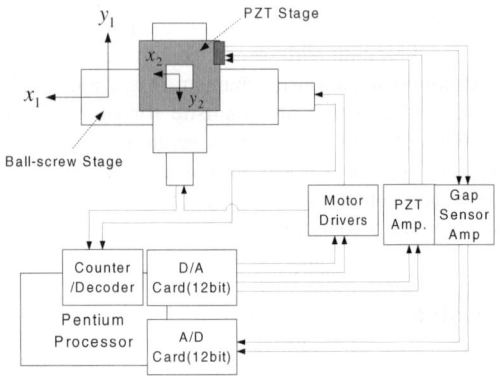

Fig. 2: Closed-loop structure of the dual-stage manipulator.

Modeling

Considering the mathematical model of the dual-stage manipulator in Fig. 1, the ball-screw driven positioner can be simply described as a mass-damper system: $H_n \ddot{z}_1 + B_n \dot{z}_1 = u_1(t)$, where the coordinate is $z_1 = [x_1 \ y_1]^T$, the control input, $u_1(t)$ corresponds to bipolar voltage input with $-5 \sim 5 \ Volt$. Nominal parameters, inertia and viscous damping coefficient, are identified as $H_n[Volt/(m/s^2)] = (0.2020, \ 0.1665)$, $B_n[Volt/(m/s)] = (2.25, \ 1.35)$ for (x_1, y_1) axis, respectively. On the other hand, the PZT driven fine positioner accepts unipolar input with $0 \sim 10 \ Volt$ and its motion is produced by elastic deformation of the flexure hinge and so there exists no friction due to the hard contact. So, the mass-spring model such as $m\ddot{z}_2 + kz_2 = u_2(t)$ is adequate for the fine stage, where the driving force(u_2) has the relationship to unipolar voltage input(V) as $u_2 = K\frac{dV}{dt}$, i.e., the generated force in the PZT is proportional to the rate of change of input voltage.

Mechanical considerations

Although the dual-stage manipulator offers an opportunity to enhance system performance, it also causes some problems which should be properly handled so as to effectively achieve its goal. Above all, when mounting a 2nd fine stage, the overall off-axis errors such as runout from the ideal straight line may become worse than the single-stage case. If only internally built-in sensors(*e.g.*, encoder of coarse positioner and gap sensor of the fine positioner) are available, this is a critical problem to achieve high positioning accuracy in absolute coordinate frame. So, in dual-stage case, more strict mechanical calibrations through external measurements are required to compensate such mounting errors. If some external feedback sensors detecting the position of the end-point are introduced, this problem can be greatly relaxed. After mechanical problems due to the dual-stage mechanism are satisfactorily settled down, we need a proper control strategy to effectively utilize its capacity.

3 Dual-Stage Servo Design

There are some approaches in controller design for dual-stage systems [1, 2, 6, 7]. These are usually classified into MIMO design and SISO design depending on the plant dynamics. When the cross coupling effect between two stages is not negligible, MIMO design is reasonable [1, 2]. However, if this is not the case as our dual-stage manipulator, SISO design is sufficient. Considering the characteristics of both actuators in the coarse/fine dual-stage manipulator, a way to obtain the maximum effect by the dual-stage actuation is let the coarse positioner track nominal trajectory and the fine positioner compensate coarse tracking errors or high frequency perturbations beyond the performance of the coarse actuator. Then, by applying the coarse stage tracking error as a reference input to the fine stage, control inputs to both actuators are naturally resolved. Figure 3 denotes the dual-servo structure, where $C_1(s)$ and $C_2(s)$ denote tracking controllers for the respective coarse and fine positioner to follow given reference trajectory. For robust performance of each stage under all sorts of perturbation, we propose, a robust controller with a novel perturbation compensator in Fig. 4, the details of which are described in the next section.

3.1 A novel robust tracking controller

Although considerable amount of tracking errors in the coarse positioner can be compensated by the fine stage motion, the overall performance of the dual-stage manipulator depends on the control performance of each positioner. In our dual-stage manipulator in Fig. 1, the positioning performance of coarse positioner is mainly disturbed by nonlinear friction and the fine stage performance is largely affected by the hysteresis effect. Considering the small travel range of the fine stage, the error level in coarse stage should be as small as possible. So, by applying a compensation

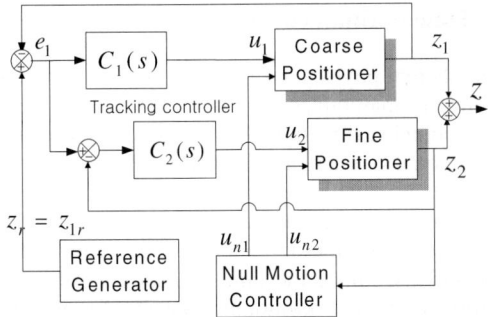

Fig. 3: Control loop of the dual-stage manipulator.

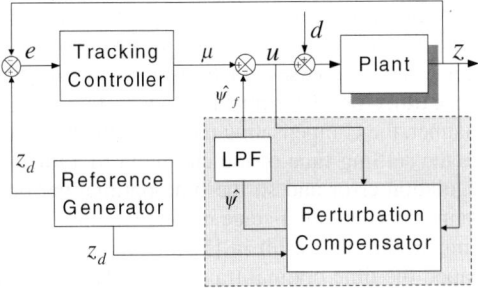

Fig. 4: A robust tracking controller with perturbation compensator.

scheme to reject such nonlinear hindering effects, a better performance of the dual-stage manipulator is possible.

For this purpose, we propose a new perturbation compensator which improves existing methods such as in [8–10]. First, let's consider the mass-damper system as a nominal model of robotic systems:

$$H_n \ddot{z} + B_n \dot{z} = u(t) + \psi(t), \quad (1)$$

where $\psi(t)$ denotes the perturbation to the nominal dynamics, which has components as

$$\psi(t) = (H_n - H)\ddot{z}(t) + f_c(t) + d(t) \quad (2)$$

where $f_c(t)$ includes all the coupling effects and model uncertainty and $d(t)$ is the external disturbances. Then, the perturbation of a plant can be *equivalently* expressed as

$$\psi(t) = \psi_{eq}(t) = H_n \ddot{z}(t) + B_n \dot{z}(t) - u(t). \quad (3)$$

While, investigating a class of perturbation observers in [8–10], all of them are based on the intuitive concept that the current perturbation can be monitored simply by

$$\hat{\psi}(t) = \psi_{eq}(t - L)$$
$$= H_n \ddot{z}(t - L) + B_n \dot{z}(t - L) - u(t - L) \quad (4)$$

under the assumption that $\psi(t) \approx \psi(t - L)$, where one-step delay in signals as much as the control interval(L) is inevitable for the causality between input and output($\hat{\psi}$).

By applying this input, the dynamic behavior of the inner loop in Fig. 4 will be changed to $H_n \ddot{z} + B_n \dot{z} = \mu(t) + \tilde{\psi}(t)$ where $\mu(t)$ is the tracking control input of outer loop and $\tilde{\psi}(t) = \psi(t) - \hat{\psi}(t)$ is the perturbation compensation error. In tracking controls, since the reference trajectory can be readily utilized, we can construct a different observer such as

$$\hat{\psi}_{ff}(t) = H_n \ddot{z}_d(t - L) + B_n \dot{z}_d(t - L) - u(t - L) \quad (5)$$

by replacing feedforward signals, $(z_d, \dot{z}_d, \ddot{z}_d)$ of a reference trajectory for feedback signals in (4). This feedforward type observer will estimate the deviation of the plant behavior from the desired dynamics according to the reference trajectory. The feedback type observer (4) requires full states of position outputs but the feedforward type observer is free of this problem.

Then, the perturbation compensator in Fig. 4 is constructed by hierarchically combining these two observers of (4) and (5), where, above all, a large amount of perturbation is compensated by the feedforward observer (5) and secondly, the *residue* of the perturbation still not attenuated, $\tilde{\psi}_1(t) = \psi(t) - \hat{\psi}_{ff}(t)$ is again compensated by the following feedback type observer.

$$\hat{\psi}_{fb}(t) = \tilde{\psi}_1(t - L) = \psi_{eq}(t - L) - \hat{\psi}_{ff}(t - L)$$
$$= H_n \ddot{z}(t - L) + B_n \dot{z}(t - L) - u(t - L)$$
$$- \hat{\psi}_{ff}(t - L) \quad (6)$$

As a result, we obtain a perturbation compensator in a discrete form as

$$\hat{\psi}(k) = \hat{\psi}_{ff}(k) + \hat{\psi}_{fb}(k)$$
$$= \psi_{eq}(k - 1) + \hat{\psi}_{ff}(k) - \hat{\psi}_{ff}(k - 1) \quad (7)$$

where the feedforward compensation by $\hat{\psi}_{ff}(k)$ and the hierarchical attenuation of the residual perturbation by $\hat{\psi}_{fb}(k)$ using feedback signals enables a reduced norm bound of the compensation error comparing with other perturbation observers in [8–10]. While, the sensor noise contained in the observer signals, which is unavoidable in real systems, should be attenuated through a low pass filter(LPF) as shown in Fig. 4 so that the filtered signal, $\hat{\psi}_f(s) = Q(s)\hat{\psi}(s)$ does not excite unmodeled resonance modes. As a feedback controller for nominal tracking performance of the outer loop, any PD type controller is acceptable.

Theorem 1: If the perturbation compensator (7) is applied to the plant (1) and the following conditions are satisfied: i) the plant is time-invariant during a control interval, ii) the full state is available(even by numerical differentiation), iii) the *change* of external disturbances during control intervals is bounded, and iv) the nominal inertia parameter, H_n satisfies the following condition for real parameter, $H(k)$ for all samples k:

$$0 < H_n < 2H(k), \quad (8)$$

then, the inner loop of the Fig. 4 is stable in such a manner that the perturbation compensation error, $\tilde{\psi}(t)$ is well bounded in a sufficiently small value.

Proof: By applying total input $u(t) = \mu(t) - \hat{\psi}(t)$ to the plant (1), we get an acceleration expression as

$$\ddot{z} = H_n^{-1}\left(\tilde{\psi}(t) + \mu(t) - B_n \dot{z}\right). \quad (9)$$

From this relationship, it is expected that the acceleration feedback term in the perturbation observer (6) will produce a dynamics of perturbation compensation error. If the full state is available and so the equivalent perturbation (3) faithfully represents the real perturbation, we can let $\psi_{eq}(t) = \psi(t)$. Then, the perturbation compensator (7) is equivalent to $\hat{\psi}(k) = \psi_{eq}(k-1) + \hat{\psi}_{ff}(k) - \hat{\psi}_{ff}(k-1) = \psi(k-1) + \Delta\hat{\psi}_{ff}(k)$. Using Eq. (2), the compensation error is described in discrete form as

$$\tilde{\psi}(k) = \psi(k) - \hat{\psi}(k) = \psi(k) - \psi(k-1) - \Delta\hat{\psi}_{ff}(k)$$
$$= [H_n - H(k)]\ddot{z}(k) - [H_n - H(k-1)]\ddot{z}(k-1)$$
$$+ \Delta f_c(k) - \Delta\hat{\psi}_{ff}(k) + \Delta d(k) \quad (10)$$

By letting $H(k) = H(k-1)$ under the assumption that plant is time-invariant during control intervals and substituting Eq. (9) into the above, a compensation error dynamics is derived as

$$\tilde{\psi}(k) = [1 - H_n H^{-1}(k)]\tilde{\psi}(k-1) + \eta(k) \quad (11)$$

where the driving function,

$$\eta(k) = [1 - H_n H^{-1}(k)][B_n \Delta \dot{z}(k) - \Delta\mu(k)]$$
$$+ H_n H^{-1}\left[\Delta f_c(k) - \Delta\hat{\psi}_{ff}(k) + \Delta d(k)\right](12)$$

where Δ denotes the change during a control interval(L) as $\Delta(\bullet)(k) = (\bullet)(k) - (\bullet)(k-1)$. First, we consider the boundedness of $\eta(k)$ in Eq. (12). Since mechanical system is a passive system, the velocity is always bounded, i.e., $\dot{z}(k) \in \mathcal{L}_\infty$ and since the control input is limited to the actuator power limit, $\mu(k) \in \mathcal{L}_\infty$. Then, also we have $\Delta\dot{z}(k), \Delta\mu(k) \in \mathcal{L}_\infty$. Considering characteristics of mechanical systems, the function $f_c(k)$ is composed of all bounded signals such as Coriolis force, centripetal force, and nonlinear friction, then $f_c(k) \in \mathcal{L}_\infty$, and so $\Delta f_c(k) \in \mathcal{L}_\infty$. Since the desired trajecoty, $\dot{z}_d, \ddot{z}_d \in \mathcal{L}_\infty$ and the input, $u(t) \in \mathcal{L}_\infty$ in Eq. (5), $\hat{\psi}_{ff}(k) \in \mathcal{L}_\infty$ and $\Delta\hat{\psi}_{ff}(k) \in \mathcal{L}_\infty$. As a result, *if the change of external disturbances is bounded*, i.e., $\Delta d(k) \in \mathcal{L}_\infty$, the driving function $\eta(k)$ of Eq. (12) will be well bounded. While, from the dynamic equation (11), the condition to guarantee $\tilde{\psi}(k) \in \mathcal{L}_\infty$ for $\eta(k) \in \mathcal{L}_\infty$ is readily determined using *Jury's test* for a discrete polynomial as $\|1 - H_n H^{-1}(k)\| < 1$ (i.e., $0 < H_n < 2H(k)$). So, the compensation error will be bounded in a sufficiently small value as far as the condition (8) is satisfied and the driving function $\eta(k)$ is well bounded. (Q.E.D.)

3.2 Time optimal control

An inherent physical property of dual-stage systems(or macro/micro manipulators) is that the effective inertia of the end-point in a direction of motion is less than that of the fine positioner(or micro manipulator) [5,7]. In other words, the dual-stage actuation enables larger acceleration of end-point than the single-stage actuation. Then, we can expect an enhanced time-optimal performance by dual-stage manipulations. Since the time-optimal control is a critical problem in most positioning systems, for example, in track seeking mode of hard disk drives or in high speed chip mounting device in semiconductor industry, the dual-stage manipulator has an obvious advantage in that point of view. We proposed a practical sub-optimal strategy where the fine actuator compensates tracking errors of the coarse actuator while the coarse actuator follows the time optimal reference trajectory determined for a target point in [5]. In this manner, the compensating motion of the fine stage can reduce the settling time of the end-point to a tolerance of the regulation error and so the time-optimal performance can be increased by dual-stage manipulations.

Summarizing the result in [5] for the completeness of this paper, the time optimal trajectory(TOT) for the mass-damper system, $\ddot{z} + a\dot{z} = b\nu(t)$(where ν is the normalized input, $|\nu| \le 1$), is determined as follows for the initial condition $\{z, \dot{z}\}(t(0)) = \{0, 0\}$ and the target condition $\{z, \dot{z}\}(t_f) = \{z_f, 0\}$.

i) When $t \le t_s$(switching time): the accelerated interval with the positive max. input($\nu = +1$)

$$t_s = \ln\left\{1 - \sqrt{1 - \exp\{-(a^2/b)z_f\}}\right\} \quad (13)$$
$$z(t) = (b/a)t + (b/a^2)\left(e^{-at} - 1\right) \quad (14)$$

ii) When $t_s < t \le t_f$(targeting time): the decelerated interval with the negative max. input($\nu = -1$)

$$t_f = t_s + (1/a)\ln\{1 + (a/b)\dot{z}(t_s)\} \quad (15)$$
$$z(t) = z(t_s) - (b/a)(t - t_s)$$
$$+ (1/a)[\dot{z}(t_s) + (b/a)]\left[1 - e^{-a(t-t_s)}\right] (16)$$

While, the velocity and acceleration profile can be readily determined by differentiating the position trajectory. For the dual-stage manipulator, if the reference trajectory (13)–(16) is applied to the coarse stage and the tracking error of the coarse stage is effectively compensated by the fine stage motion, we can expect an enhanced time-optimal performance.

3.3 Null motion control

As denoted in Fig. 2, adding a 2nd stage in the same direction produces actuation redundancy. So, there exist infinite number of control solutions for a fixed target point.

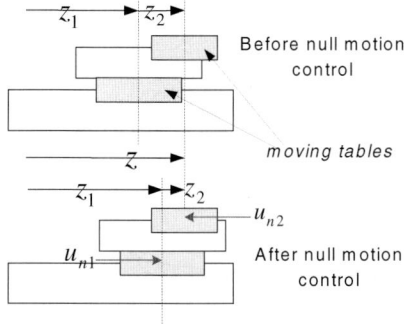

Fig. 5: Effect of the null motion control.

the end-point of both axes($x = x_1 + x_2, y = y_1 + y_2$) are considerably reduced and the tracked contour is so clean.

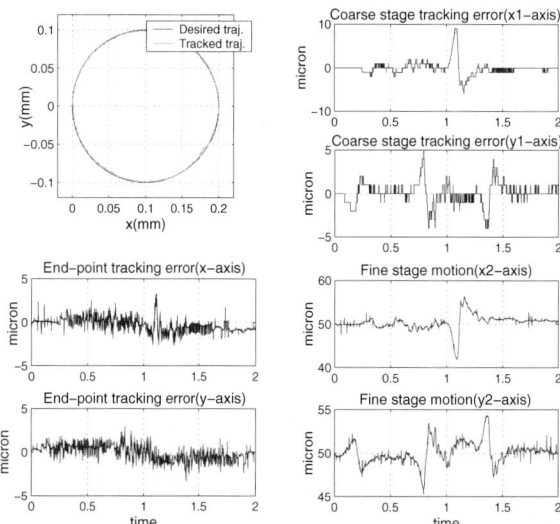

fig. 6: Circle tracking(100 μm radius) using dual-stage manipulator: Tracked contour, End-point tracking errors, Coarse stage tracking errors, and Fine stage motions, respectively.

Then, it is necessary to determine a *best* control among the candidates. Furthermore, since the travel range of a fine stage is usually very small, it may be easily saturated to the boundary. Therefore, the reference input to the fine stage should be properly limited so as not to be over its travel limit.

As well as the main control input, in the systems with actuator redundancy, a null motion control input which enables some desirable manipulation can be determined. In [5], we determined a null motion control input to accelerate the restoring action of the fine stage to its neutral position based on so-called the "dynamic consistency" [7]. This kind of input is necessary to reduce the possibility for the fine stage to be saturated and so to maximize the its capacity to compensate the coarse stage tracking errors. Figure 5 conceptually explains the effect of the null motion control, that is, the relative motion(z_2) of the fine stage to the coarse stage can be considerably reduced while the position of the end-point(z) is not changed.

4 Experimental Study

To confirm the effectiveness of the coarse/fine dual-stage manipulator, we have performed experiments on fine tracking control of planar motion, time optimal control of linear motion, and micro-teleoperations, as well. We applied the dual-servo algorithm in Fig. 3 to the dual-stage manipulator constructed. As a feedback controller, the robust tracking controller developed in Fig. 4 was applied to the coarse positioner but a simple PI controller to the fine stage. In all experiments, the control frequency was 1000 Hz under Windows 98 OS environment and the velocity/acceleration required in the algorithm were obtained through filtered derivatives of position outputs.

Fine trajectory tracking

Figure 6 shows the result in the fine tracking of circular trajectory with 100 μm radius. As shown, the coarse stage tracking errors are well compensated by the fine stage motions(the travel range is 0 $\sim$ 100 μm and the neutral position is 50 μm). As a result, the overall tracking errors in

Time-optimal performance

As remarked in the former section, a merit of the dual-stage manipulator is that the time-optimal performance can be enhanced using dual-stage manipulations. Applying the time-optimal trajectory in (13)–(16) to the coarse positioner as a reference trajectory and compensating the coarse tracking errors by fine stage motions, a faster regulation to the target point is possible than the single manipulation case by the coarse positioner. Figure 7 is the experimental result for the X-axis of the dual-stage manipulator when the target point is $x_f = 30\ mm$ and motor torques are limited to 1.2 $Volt$. Comparing the tracking errors in Fig. 7(a) and Fig. 7(c), the regulation time to the target point has been fairly reduced.

Micro-teleoperation

The constructed dual-stage manipulator with the proposed dual-stage servo is applied to micro-teleoperations. Figure 1 corresponds to the slave manipulator composed of the dual-stage manipulator and the micromanipulator, where the dual-stage manipulator follows XY motions by the master arm and the micromanipulator provides 3 DOF angular motions. In teleoperations, to achieve microscopic motions in the slave manipulator, the macroscopic motions of the master arm generated by the operator were scaled down as much as 1/200 in the slave manipulator. Figure 8 and 9 are the results of micro-teleoperation during 25 sec. and they denote the effectiveness of the dual-stage manipulator for fine tracking motions. As shown, the fine stage successfully compensates the tracking errors occurred in the coarse stage and so the tracking errors in the end-point are considerably reduced.

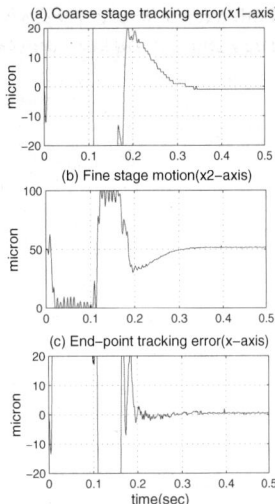

Fig. 7: Time-optimal control using coarse/fine manipulation: Coarse stage tracking errors, Fine stage motions, and End-point tracking errors, respectively.

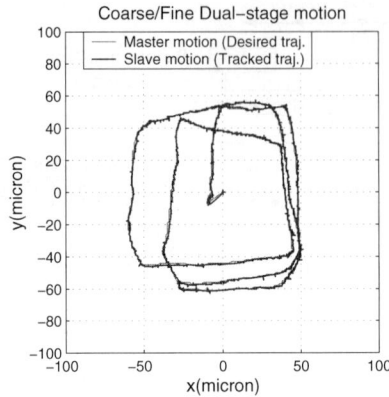

Fig. 9: (Micro-tele operation) Tracking master arm motion by the dual-stage manipulator.

REFERENCES

[1] Yen, J.-Y., Hallamasek, K., and Horowitz, R., 1990, "Track Following controller Design for a Compound Disk Drive Actuator," *ASME J. of Dyn. Sys. Meas. and Contr.*, Vol. 112, Sep., pp. 391-402.

[2] Li, Y. and Horowitz, R., 2000, "Track Following controller Design of MEMS Based Dual-Stage Servos in Magnetic Hard Disk Drives," Proc. of *2000 IEEE Int. Conf. on Robot. and Auto.*, pp. 953-958.

[3] Staroselsky, S. and Stelson, K. A., 1988, "Two-Stage Actuation for Improved Accuracy of Contouring," *'88 American Contr. Conf.*, pp. 127-132.

[4] Lee, C.W. and Kim, S.-W., 1997, "An Utraprecision stage for Alignment of wafers in Advanced microlithography," *Precision Engineering*, Vol. 21, Sep., pp. 113-122.

[5] Kwon, S. J., Chung, W. K., and Youm, Y., 2000, "Robust and Time-Optimal Control Strategy for Coarse/Fine Dual Stage Manipulators," Proc. of *2000 IEEE Int. Conf. on Robot. and Auto.*, pp. 4051-4056.

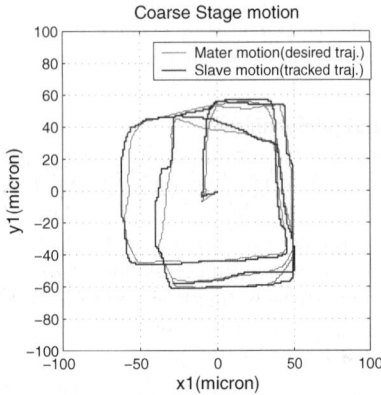

Fig. 8: (Micro-tele operation) Tracking master arm motion by the single coarse positioner.

5 Concluding Remarks

In this paper, we presented the effectiveness of the coarse/fine dual-stage manipulation though experimental studies. The dual-stage manipulator enables fast and fine manipulation with large workspace by adopting merits of both conventional manipulator and fine positioner with high speed and high resolution. We proposed a novel perturbation compensator for robust performance with dual-stage control strategy to effectively utilize and properly control the system. Through experimental studies on fine tracking control, time-optimal control, and also micro-teleoperation, we demonstrated that the coarse/fine dual-stage manipulator with the proposed control structure is effective in enhancing the robust performance of the system.

[6] Sharon, A., Hogan, N., and Hardt, D. E., 1993, "The macro/micro manipulator: An improved architecture for robot control," *Robotics and Computer Integrated Manufacturing*, Vol. 10, No. 3, pp. 209-222.

[7] Khatib, O., 1995, "Inertial properties in robotic manipulation: An object-level framework," *Int. J. of Robot. Res.*, Vol. 13, No. 1, Feb. 1995, pp. 19-36.

[8] Youcef-Toumi, K. and Ito, O., 1990, "A Time Delay Controller for Systems With Unknown Dynamics," *ASME J. Dyn. Sys., Meas., and Contr.*, Vol. 112, Mar., pp. 133-142.

[9] Ohnishi, K., Shibata, M., and Murakami, T., 1996, "Motion Control for Advanced Mechatronics," *IEEE/ASME Trans. on Mechatronics*, Vol. 1, No. 1, Mar. pp. 56-67.

[10] Yao, B., Al-Majed, M., Tomizuka, M., 1997, "High-Performance Robust Motion Control of Machine Tools: An Adaptive Robust Control Approach and Comparative Experiments," *IEEE/ASME Trans. on Mechatronics*, Vol. 2, No. 2, Jun. pp. 63-76.

Development of Global Vision System for Biological Automatic Micro-Manipulation System

Xudong Li, Guanghua Zong and Shusheng Bi

Robotics Research Institute, Beijing University of Aeronautics and Astronautics
37# Xueyuan Road, Haidian Dist., Beijing 100083, China
lixudong263@263.net

Abstract

An automatic micro manipulation system (AMMS), which aims at various kinds of operations in the field of bioengineering has been developed. Visual servo control is usually employed to improve the performance of the system. However, the visual information can not be obtained as fast as expected because there is a great amount of computation in image processing. This hinders the formation of closed-loop control system with visual servo. In this paper, efforts are made to improve the performance of the vision system in two aspects: algorithm design and hardware implementation. Methods of correlation based pattern matching and morphological operation based image processing technique are used to implement fast visual information abstraction, thus forming the real-time visual servo control system. The system performance and the result of an experiment on gene injection are given in the paper.

1. Introduction

In recent years, much attention has been given to micromanipulation system for manipulating various kinds of microscopic objects [1]~[4]. In the field of bioengineering and gene engineering, operators often need to manipulate microscopic objects such as oosperm, embryo, chromosome, etc. At present, such operations (e.g. gene injection) are accomplished manually with the help of a commercial system called MicroManipulator. During the operation, the operator must watch the objects through a microscope (usually optical microscope). Because of the eyestrain, the efficiency and the surviving rate are rather low. The main objective of our research is to accomplish these operations automatically. An automatic micromanipulation system (AMMS) that can finish such task has been developed. By using AMMS the authors have succeeded in accomplishing the process of gene injection, that is, injecting gene into a oosperm of the mouse.

Closed-loop control is usually employed to improve the performance of the system. However, because of the peculiarity in biological micromanipulation (described in the followed section), one should use the global vision system, instead of ordinary sensors to obtain the positions of the microscopic objects and manipulating tools (micro tube, micro injector, etc.)[5]. Moreover, the visual information can not be obtained as fast as expected because of the great amount of computation in image processing. This hinders the realization of closed-loop control system with visual servo. This paper will propose a global vision system for AMMS. The vision system has the ability of recognizing and tracking at least 3 targets and working out their position information simultaneously. This information is used as position feedback to realize real-time visual servo control.

In section 2 and section 3, a brief introduction to the peculiarity of biological micromanipulation and the prototype of AMMS will be presented as a background. The setup and implementation of the vision system will be described in detail in section 4. The system testing results and the experimental result of gene injection will be given in section 5. Finally, a conclusion will be offered in section 6.

2. Peculiarity of biological micro-manipulation

The biological micromanipulations have their particularities, which are as follows:

(a) The manipulated objects are living things, so they are sensitive to strong light. Exposing in strong light for a long time may affect the surviving rate.

(b) The manipulated objects may move randomly and continuously in a small range during the process of manipulation, mainly because of the disturbance from the movement of manipulating tools.

(c) The contact force is about $10^{-6} \sim 10^{-3}$ N.[6]

(d) The manipulated objects and the manipulating tools are all very small. The diameter of the oosperm is about 90 microns and the diameter of micro injector tip is about 1micron.

(e) System model is not constant. System models calibrated under different temperature or different time are not the same.

These mean that one can not use ordinary position sensors to position the manipulating tools as well as objects. So, during the manipulation, the accurate positions of objects can not be exactly known and their motions are out of control. Taking advantage of global vision system to obtain the position information to form visual servo control is a better way to overcome this obstacle.

3. Prototype of AMMS

Figure 1 and figure 2 show the setup of the AMMS. The part within dashed line box in figure 2 shows the setup of the vision system.

Fig. 1 The photo of the AMMS.

The left side is main controller and the right side is image processing computer. The microscope mounted with the two manipulators is in the middle.

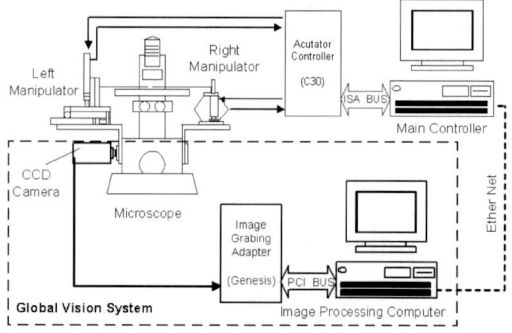

Fig. 2 The sketch of the AMMS.

The AMMS consists of three modules: executive module, control module and sensory module. The vision system is a part of sensory module.

The left and right manipulators form the executive module. The two manipulators use two different mechanisms to trade off between high accuracy and wide workspace, as shown in figure 3. The left manipulator is a 3-DOF XYZ stage with a workspace of about $10 \times 10 \times 10 mm^3$ and with an accuracy of 1micron. Each DOF is driven by a DC motor (DC-Mike of PI) with embedded encoder as position sensor. A micro tube is mounted on it to track and fix the oosperm.

And the right manipulator is a 3-DOF parallel mechanism (modified Delta mechanism) with a relatively narrow workspace of about $200 \times 400 \times 400 microns^3$ and with a relatively high accuracy of 0.1micron. Each link is driven by a piezotranslators (HVPZT of PI) with strain gauge as position sensor. A micro injector is mounted on it to perform the gene injection.

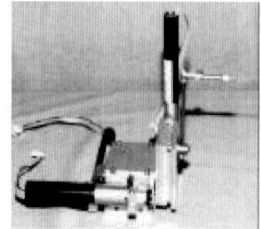 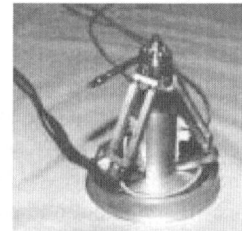

Fig. 3 Photos of left manipulator (left) and right manipulator (right) mounted with actuators, sensors and manipulating tools.

The control module consists of a main controller (PC, PII-333) and an actuator controller (TMS320C30 based DSP system). The sensory module consists of a global vision system and the actuator level sensors, that is, encoders and strain gauges.

4. Implementation of vision system

4.1 The setup of vision system

As shown in figure 2, the vision system consists of four parts: optical microscope, CCD camera, image grabbing and processing board, and image processing computer.

The optical microscope is employed because it may do less harm to the living objects, which are exposed in the light during the manipulation. In this study, the total magnification of the microscope is fixed to 100. At this magnification, the valid area of grabbed image is about $600 \times 450 microns^2$ with a resolution of 0.825micron/pixel.

During the manipulation, the scene of the "micro world" under the microscope will be transmitted to the image processing computer by CCD camera and the image board.

The image processing computer works out the positions of three targets by processing the CCD image data. The position information is then transferred to the main controller through the Ethernet to realize visual servo control.

4.2 Recognition algorithms

The main task of the vision system is to find out the accurate positions of three objects in grabbed image. Figure 4 is the typical image processed by the vision system. Objects that will be recognized are also shown. The recognition results will be represented in local image coordinates.

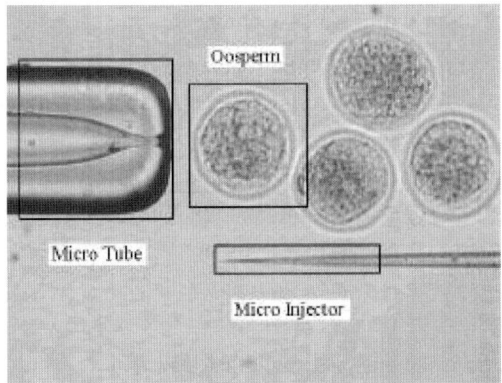

Fig. 4 A typical image grabbed and processed by vision system with size of 768×576 *pixels*. Three objects, micro tube, micro injector and oosperm, are marked by three rectangle respectively.

The flow chart of the vision system is shown in Figure 5. It is clear that the calculation speed depends on the image grabbing rate and the speed of recognition. The grabbing rate depends on the system of the camera. The camera we used is in PAL system and its grabbing rate is 25 frames per second (FPS). So, the speed of vision system eventually depends on the speed of recognition. The faster the recognition performs, the faster the system runs.

4.2.1 Recognition of oosperm and micro tube

Recognizing the oosperm and the micro tube is virtually a matching problem. Two kinds of matching methods are often used: Feature matching method and pattern matching method [8].

Feature matching method needs to abstract features (e.g. contour [6][7], or amplitude spectra [4]) first and then perform matching in the feature space. Because the feature abstracting is time-consuming, this kind of method is difficult to realize in real time. On the other hand, pattern matching method performs directly on original image. With the help of appropriate hardware, it is easy to fast realize. In this paper, we use correlation based pattern matching to realize the recognition of the oosperm and micro tube in real time.

Step. 1:
  Initialize the vision system, allocate the buffers and establish communications with the main controller.
Step. 2:
  Set searching parameters and define three patterns through the man-machine interface.
Step. 3:
  Grab the first frame of image into grabbing buffer.
Step. 4:
  If grabbing has not finished yet, wait until it finishes.
Step. 5:
  Transfer the grabbed image from grabbing buffer to processing buffer and begin to grab the next frame into grabbing buffer.
Step. 6:
  Set the searching region for each pattern, and recognize the three objects.
Step. 7:
  Transfer the results to main controller.
Step. 8:
  If the task has not finished, go to Step. 4.
Step. 9:
  Stop grabbing, end the current task and release the allocated buffers.

Fig. 5 The flow chart of the vision system.

First, define the rectangular image block of the oosperm and micro tube as patterns, as shown in Figure 4. Then calculate the correlation between the pattern and the corresponding image region at every pixel in grabbed images. The correlation represents the likelihood between the pattern and the corresponding region. The point where the correlation reaches local maximum value is considered the matching point.

The correlation R_{uv} at pixel (u, v) can be calculated by using the formula:[9]

$$R_{uv} = \frac{N \sum_{i=u-m+1}^{u+m} \sum_{j=v-n+1}^{v+n} M_{(i+m-u)(j+n-v)} S_{ij} - (\sum_{i=u-m}^{u+m} \sum_{j=v-n}^{v+n} S_{ij}) \sum_{i=1}^{m} \sum_{j=1}^{n} M_{ij}}{\sqrt{\left[N \sum_{i=u-m}^{u+m} \sum_{j=v-n}^{v+n} S_{ij}^2 - \left(\sum_{i=u-m}^{u+m} \sum_{j=v-n}^{v+n} S_{ij} \right)^2 \right] \left[N \sum_{i=1}^{m} \sum_{j=1}^{n} M_{ij}^2 - \left(\sum_{i=1}^{m} \sum_{j=1}^{n} M_{ij} \right)^2 \right]}}$$

Where, S is the grabbed image of the size of $[s] \times [t]$ *pixels*. M is pattern of the size of $[2m] \times [2n]$ *pixels*; N is the number of pixels in M. In our system $m=n=64$, $s=768$, $t=576$.

In order to avoid *"Edge Effect"*, calculations are only performed at the point (u, v) that lies in the rectangular region $[m, s-m] \times [n, t-n]$. At the same time, we use R_{uv}^2

instead of R_{uv}, to estimate the matching point. This avoids the slow square-root operation and is helpful to improve the performance.

4.2.2 Recognition of micro injector

Unlike the oosperm and the micro tube, micro injector appears as a long and narrow region in the image. It is not suitable to use pattern matching method to recognize the micro injector because of the low efficiency and poor accuracy. We use a method, which is based on binary morphological operation, to recognize the micro tube.

Fist, binarize the captured image S. There will be quite a few tiny blobs in the binarized image B. These blobs result from the noise in image and they are harmful to the successive processing. We remove them by using two successive morphological operations: dilating after eroding, which is known as "open" operation. The definitions of the two operations are as follows:

Eroding B with T:

$$er(B,T) = \{\vec{x} | (T+\vec{x}) \subset B\} \quad (2)$$

Dilating B with T:

$$dil(B,T) = \{\vec{x} | \vec{x} = \bigcup_{\vec{y} \in T}(B+\vec{y})\} \quad (3)$$

Where T is the predefined structure element, which is shown in Figure 6. Both two operations can be realized in form of convolution.

1	1	1
1	1	1
1	1	1

Fig. 6 Structure element

Then we pick the blob that corresponds to the micro tube out. Define a measure $E = W/H$, where W and H is the equivalent width and height of a blob. Here "equivalent" means that the blob has the same area and perimeter values as a rectangle with width of W and height of H.

Assume the micro injector is in shape of cylinder and we can get a threshold E_0 from the actual dimensions of the micro injector. For every blob in B, if its E value is greater than E_0, we consider that it corresponds to the micro injector, otherwise, it is erased. After the picking, there is only one blob in the binary image if the E_0 is carefully selected.

Finally, a scan-line method is used to obtain the coordinates of the most left point inside the blob. This point corresponds to the tip of the micro injector.

4.2.3 Improving the performance of vision system

We try to improve the performance of the system in two aspects.

First, we carefully design the algorithms so that they can be accomplished by convolution. In the algorithms mentioned above, the operations (the correlation, the dilation and the erosion) can all be realized in form of discrete convolution, which can be fast execute on DSP based hardware after the instruction level optimizing.

Second, we limit the matching into a relatively smaller searching region by considering the continuous motion constraint. Suppose the object moves with a constant velocity of v pixel/s, we can say that:

- 0 frame: initial position is (x_0, y_0), (which is given by operator through man-machine interface;) then,

- 1st frame: the searching region is $[x_0-v/25, x_0+v/25] \times [y_0-v/25, y_0+25/v]$; Suppose the matching position is (x_1, y_1), then,

- 2nd frame: the searching region is $[x_1, x_1 + x_1-x] \times [y_1, y_1 + y_1-y]$; Suppose the matching position is (x_2, y_2), then,

- $(i+1)^{th}$ frame: searching region is $[x_i, 2x_i-x_{i-1}] \times [y_i, 2y_i-y_{i-1}]$, (for i >1).

Where 25 is the grabbing rate of PAL system.

Compared with searching in the whole image (768×576 *pixels*), this can greatly reduce the computation and improve the performance of the vision system.

4.2.4 Implementation of the vision system

The recognition algorithms mentioned above are realized by using Microsoft Visual C++ and Matrox Image Library. Figure 7 is its man-machine interface. Through the interface, the operator can easily accomplish almost all kinds of operations in the field of bioengineering with few man-machine interaction.

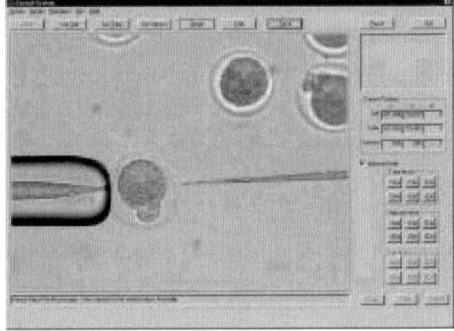

Fig. 7 The man-machine interface

4.3 calibration of the vision system

The vision system is integrated into the AMMS to realize visual servo control. This can improve the accuracy and degree of automatic operation of the system.

An image based look-and-move control strategy is employed in AMMS. The desired state is given in the image space. However, the input to the system is given in the actuator space. So, we need to calculate the Jacobian that can map the parameters from actuator space to image space.

The Jacobian matrix can be decomposed as the product of the other two Jacobians which map the parameters between the actuator space and right manipulator space, as well as image space and the right manipulator space, respectively. The first one can be gotten by kinematics analysis and the second one can be gained by calibrating the resolution of the vision system.

We use a micrometer to calibrate the resolution of the vision system. On the micrometer, there are reticles with equal spacing of 10 microns. The micrometer is placed on the stage of the microscope, and an image of the micrometer under a specified objective lens is taken through the vision system. Then, count the number of pixels between a pair of reticles. The distance between the two reticles can be read out from the micrometer directly. Suppose the readout distance is D and the corresponding pixel number is N, we can get the resolution K of the form

$$K = D/N \qquad (4)$$

Usually, the width of a reticle in the image is several pixels, so, a thinning operation must be performed fist to transform the width of the reticle to one pixel. Our vision system has a resolution of 0.825micron/pixel.

5. Experiment results

5.1 System testing

The global vision system was tested under the real environment and the results are shown in Table 1.

Table 1. The performance of the global vision system

Number of objects	Time (sec/200frame)	Average velocity (FPS)
an oosperm	12	16.7
a micro tube	12	16.7
a micro injector	17	11.8
All of above	22	9.1

The right manipulator tracks a circle with a diameter of 50 pixels (about 41microns) with and without visual servo, and the trajectories are shown in Figure 8 (a) and (b). The trajectories are recorded by saving the local image coordinates sequence of the tip of right manipulator into a file during tracking. From the figure, it is obvious that the tracking effect with visual servo becomes better.

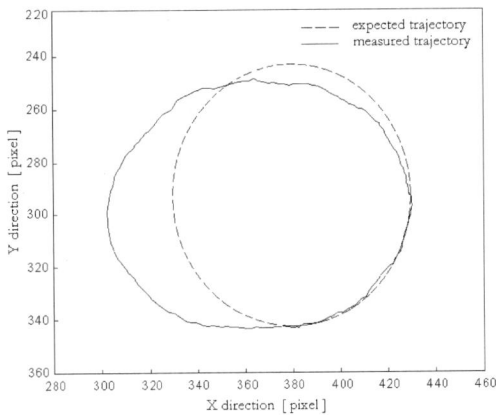

Fig. 8 (a) Experiment on tracking a circle without visual servo.

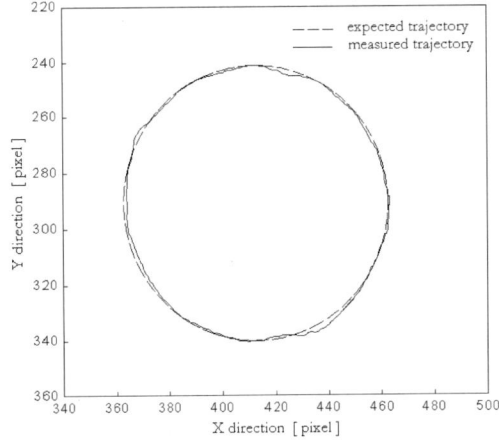

Fig. 8 (b) Experiment on tracking a circle with visual servo

5.2 Experiment on gene injection

Gene injection is the most often performed operation in the field of bioengineering. The manually operation can be accomplished though the following steps:

Fist, select the oosperm to be injected and fixed it in a suitable place by using a micro tube. Second, manipulate a micro injector to penetrate into the right place (nuclear) and inject genes. Third, release the oosperm and begin the next injection.

An experiment on gene injection is successfully carried out using AMMS and a series of key pictures taken from the process are shown in Figure 9. Where, the diameter of the oosperm is about 90 microns.

6. Conclusions

AMMS makes it possible to manipulate microscopic object with few human participation. The global vision system is one of the keys to the automatic manipulation.

The fast visual information abstracting results from the carefully designed algorithm and skills in implementation. Future work should be focused on stereo vision and to extend the system into other fields.

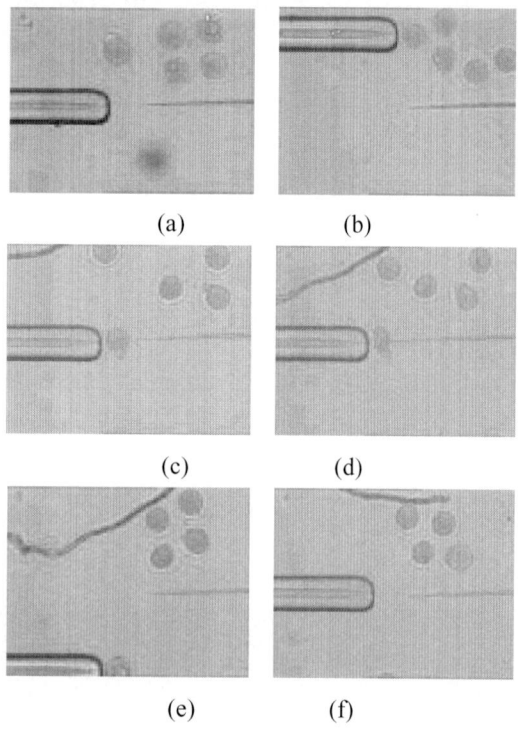

Fig. 9 Experiment on gene injection.

(a). Initial status, the operator uses mouse to click the oosperm to be injected. (b). The micro tube is tracking and attracts the oosperm automatically. (c). The micro tube carries the oosperm back into the workspace of the right manipulator (micro injector). (d). The operator selects the right nuclear by clicking it using mouse. Then the micro injector accomplishes injection automatically. (e). Releasing the oosperm. (f). The micro tube moves back to the initial position and ready to manipulate another oosperm

7. Acknowledgements

This paper is supported by the National Natural Science Foundation of China (NSFC) (Grant Number:59975002) and the Hi-Tech Research and Development Program of China (HTRDP).

Special thanks would be given to Mr. Zhao wei of Beijing University of Aeronautics and Astronautics for drawing the trajectories of tracking circle.

8. References

[1] P. Kallio, M. Lind, Q. Zhou and H. N. Koivo, "A 3 DOF Piezohydraulic Parallel Micromanipulator", Proceedings of the IEEE ICRA, pp.1823-1828, Leuven, May 1998.

[2] M.C. Carrozza, P. Dario, A. Menciassi and A. Fenu, "Manipulation Biological and Mechanical Micro-objects Using LIGA-Microfabircated End-Effectors", Proceedings of the IEEE ICRA, pp.1811-1816, Leuven, May 1998.

[3] T. Tanikawa, T. Arai, Y. Hashimoto. "Development of Vision System for Two-Fingered Micro- Manipulation", Proceedings of the Intelligent Robotics and System, pp.1051-1056, Grenoble, September 1997.

[4] S. Fatikow, J. Seyfried, S. Fahlbusch, A. Buerkle and F. Schmoeckel, "Development of a micro robot- based Microassembly Station", Proceedings of the Ninth International Conference on Advanced Robotics, pp.205-210, Tokyo, October 1999.

[5] F. Arai, T. Sugiyama, T. Fukuda, H. Iwata, K, Itoigawa, "Micro Tri-axial Force Sensor for 3D Bio-Micromanipulation". Proceedings of the IEEE ICRA, pp.2744-2749, Detroit, May 1999.

[6] T. Kasaya, H. Miyazaki, S. Saito and T. Sato, "Micro Object Handling Under SEM by Vision-Based Automatic Control", Proceedings of the IEEE ICRA, pp.2189-2196, Detroit, May 1999.

[7] M. Kass, A. Witkin and D. Terzopoulos. "Snakes: active contour models", International Journal of Computer Vision, Vol.1, No.4, pp.321-331, 1988.

[8] Y. Zhou, B.J. Nelson, B. Vikramaditya. "Fusing Force and Vision Feedback for Micromanipulation", Proceedings of the IEEE ICRA, pp.1220-1225, Leuven, May 1998.

[9] Matrox Image Processing Group, Matrox Imaging Library User Guide, Matrox Electronic Systems Ltd., 1997, Chap. 12, pp.154-157

A Flexible Experimental Workcell for Efficient and Reliable Wafer-Level 3D Microassembly

Ge Yang   James A. Gaines   Bradley J. Nelson

Department of Mechanical Engineering
University of Minnesota
Minneapolis, Minnesota 55455 USA
Email: gyang, jgaines, nelson@me.umn.edu

Abstract

This paper reports on an experimental microassembly workcell developed for efficient and reliable 3D assembly of large numbers of micromachined thin metal parts into micromachined holes in 4 inch silicon wafers. The major objective is to integrate techniques of microgripper design, microscopic imaging and high precision motion control to build a prototype system for industrial applications. The workcell consists of a multiple-view imaging system, a 4 DOF micromanipulator with high resolution rotation control, a large working space 4 DOF precision positioning system, a flexible microgripper, and control software system. A piezoelectric force sensing unit is developed to be integrated with the manipulator system to enhance pickup reliability. Operations are partially guided by a human operator through a graphical user interface (GUI). This system provides a highly flexible testbed for wafer-level 3D microassembly.

1. Introduction

In recent years, significant progress has been made in MEMS-based mesoscale and microscale[1] sensors, actuators, and highly integrated microsystems. Currently, many technologies used in MEMS fabrication come from silicon-based IC manufacturing. These technologies are restricted by the materials allowed and the requirements on process compatibility. Fabrication of complex 3D structures and hybrid systems is often difficult. These limitations can be partially overcome by using "nontraditional" techniques such as laser cutting, micro wire-EDM and micromilling[2][11]. Diverse materials, incompatible fabrication processes and the pursuit of hybrid integrated systems require assembly operations at meso and microscales, which is often referred to as microassembly.

Although a large portion of current microassembly research is devoted to the assembly of objects with dimensions of several hundred microns or less, it is also important to study the assembly of mesoscale parts, which can be machined using a variety of methods and are less susceptible to the limitations of silicon-based techniques. Due to the dimensions of these parts, mesoscale devices can generate forces and motions of more useful magnitudes. In some cases, development of MEMS also requires the assembly of microscale and/or mesoscale parts into macroscale systems. One example is wafer-level microassembly. Here the vastly different dimensions involved can be an obstacle.

Progress made in microassembly research has led to the development of highly integrated assembly systems [1][3][4][8][10]. In order to place microassembly technology into industrial applications, which have strict requirements on reliability and efficiency, several system-level aspects must be directly addressed.

(1) Part Transfer and Pick-and-Place

A good part transfer tool should help to reduce the complexity and increase the speed and robustness of subsequent microassembly operations. Besides existing transfer tools used in the IC industry, such as waffle packs and Gel-Pak vacuum release trays, it is often necessary to develop special transfer tools for microassembly [7].

Due to the different governing physics of microassembly [5], the development of robust and efficient pick-and-place operations is often challenging. The development of highly reliable microgrippers that use various gripping forces is critical[9].

(2) Manipulation

3D microassembly tasks typically require between four and six degrees-of-freedom. Since the manipulator end effector is normally monitored under a microscope, it must be small

---

1. We refer to 1μm to 100μm as "microscale" and 100 μm to 1 cm as "mesoscale".

in size and suitable in shape in order to remain in the camera's field-of-view, to minimize occlusion, and to avoid collisions. On the other hand, the assembly tolerance is typically on the order of microns. This places a lower limit on the motion resolution of the manipulator system. Building a highly flexible six DOF robot suitable for microassembly remains a challenging topic. However, we can decompose the requirement on DOF to reduce the complexity of the manipulator for certain applications.

In addition to its role in part *transfer*, a reliable microgripper is also critical for the efficiency and robustness of assembly *manipulations*. Due to the constraints posed by microscopes, the structure and actuator of microgripper often needs to be small. Sometimes, the end effector for part pickup and manipulation must be separated, as is the case for the application presented in this paper.

(3) Scale Factor

Microassembly tasks in industrial applications typically require large ranges of high precision motion with high repeatability. Many of these requirements are quite similar to those of automated IC fabrication equipment in which a coarse-to-fine approach is often used. A typical resolution of one micron can be implemented at reasonable cost using off-the-shelf high precision motion control systems. In many cases, this is sufficient for coarse positioning in microassembly.

(4) Assembly Tolerance

Different requirements on microassembly tolerance leads to vastly different assembly techniques and sensor configurations. For small microassembly tolerances, microscopic vision feedback combined with a high precision manipulator may be sufficient. For even smaller tolerances, the integration of force and vision feedback is necessary.

(5) System Integration

In general, the most important issue is system reliability and efficiency. Since it is still difficult to implement fully visually servoed microassembly, a human operator is often needed. In this case, it is often important to provide sufficient user assistance and sensing feedback, especially for the prevention of collision and for error recovery. An intuitive and convenient way for the operator to monitor and intervene in the assembly process is essential.

Other factors to be considered in system integration includes flexibility, expandability and connectivity. Due to the significant difference in the dominating physics, assembly planning and scheduling can be different from macroscopic applications [6].

This paper reports on the development of a microassembly workcell built for wafer scale microassembly of mesoscale parts with some microscale dimensions and tolerances. More specifically, the workcell is developed for the efficient and reliable assembly of hundreds of micromachined thin metal parts into DRIE etched holes in a 4 inch silicon wafer.

2. ASSEMBLY TASK ANALYSIS

The assembly task is to pick up thin metal parts from a vacuum release tray and insert them into vertically etched holes in a 4 inch silicon wafer (Figure 1a). The holes come in arrays and may not be regularly distributed on the wafer. In general, each assembly operation is a classic rectangular-peg-into-a-rectangular-hole problem.

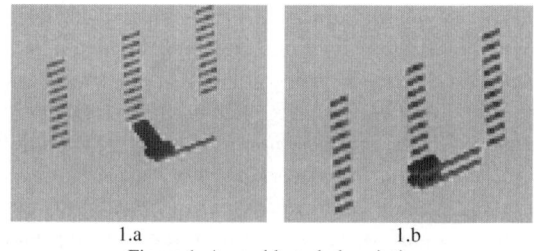

Figure 1. Assembly task description

(1) General Assembly Strategy

It is normally easier to transfer the metal parts horizontally to the workcell (Figure 3). Since the holes are vertically etched into the wafer, we choose to set the wafer on a vertical wafer mount such that the holes are also in a horizontal direction (Figure 1 & 3). This configuration does not require the flipping of the thin metal parts and is advantageous for both reliability and efficiency. Since each part has a thickness of less than $100\mu m$ and is shipped on a vacuum release tray, a vacuum pickup tool is chosen (Figure 2).

(2) Assembly Tolerance

Each metal part is approximately half a millimeter in width and less than $100~\mu m$ in thickness at its tip. The assembly clearance is approximately $20~\mu m$ in the horizontal direction and $10\mu m$ in the vertical direction.

(3) Distance Between Adjacent Holes

Another constraint comes from the distance between vertically adjacent holes, which is less than $200\mu m$, as illustrated in Figure 1b. The assembly must be performed in one specific direction to avoid collision.

(4) Working Space

The holes are distributed on an 4 inch wafer. This requires that the assembly station have a commensurate working space. A coarse-fine configuration is used (Figure 3). The large range coarse motion is provided by the 4 DOF coarse positioning unit. Small range fine adjustment is provided by a micromanipulator, which has a range of 2.5 cm and a maximum resolution of 0.04 μm for each of its X-Y-Z axes.

(5) Decomposition of Manipulation DOF

The assembly task requires 6-DOF manipulation. Instead of using a 6-DOF micromanipulator, we decompose the requirement on DOF and implement them on separate structures. Figure 2 shows the adapted Sutter manipulator used with an add-on rotation control unit actuated by a precision microstep motor. This configuration provides the yaw motion for assembly. The manipulator does not provide roll motion. Instead, it is implemented on the wafer mount, also actuated by a microstep motor (Figure 3). The pitch movement of the metal part after pickup is not automatic and is implemented by manual adjustment and calibration before assembly, as shown in Figure 2.

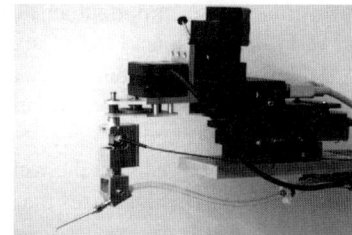

Figure 2. Adapted Sutter manipulator with vacuum pickup tool, rotation control unit and force sensing unit

3. WORKCELL DESCRIPTION

The entire workcell consists of a large working space 4-DOF coarse positioning unit, a micromanipulator unit, a multiple-view imaging unit, a microgripper unit and a control software unit. The microgripper unit and control software unit is introduced in Section 5 and Section 7, respectively.

3.1 Large Working Space Positioning Unit

Each wafer is oriented in the vertical direction on the wafer mount (Figure 3). This unit has 4-DOF. Planar motions in the horizontal direction are provided by an open frame high precision XY table with a travel of 32 cm

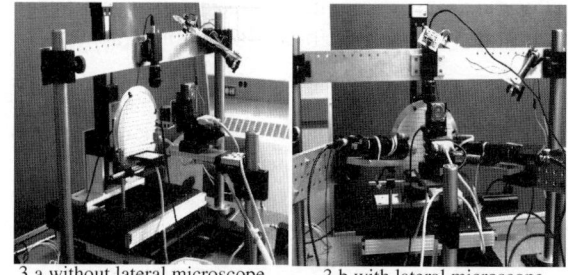

3.a without lateral microscope   3.b with lateral microscope
Figure 3. General view of the microassembly workcell

and a repeatability of 1 micron in both directions. Position feedback with a resolution of $0.1\mu m$ is provided by two linear encoders mounted on the table. A dual-loop PID plus feedforward control scheme is used for each axis. The internal speed loop is closed on the rotary encoder on the motor. The external loop is closed on the linear encoder.

Vertical movement is provided by a linear slide. It has a travel of 20 cm (8 inch) and a repeatability of $5\mu m$. It is controlled using a PID plus feedforward algorithm. The XY table is actuated by two API MBT-N232 AC servo motors, while the vertical linear slide is actuated by an API MBT-N231 AC servo motor. Each of these motors is driven by an API DS-3402i-E intelligent drive unit.

The wafer mount provides rotational control to implement roll movement. It is actuated by an Oriental 5-phase microstep motor with a maximum resolution of 0.0028 degree/step. In assembly operations, the human operator uses visual feedback through a microscope to perform fine adjustments. All motions are coordinated and commanded by a host computer.

3.2 Micromanipulator Unit

Fine movements needed for microassembly are implemented by an adapted high precision Sutter manipulator (Figure 2). For part orientation, a specially designed rotation mechanism is added onto the manipulator. It is also actuated by an Oriental 5-phase microstep motor with a maximum rotation resolution of 0.0028 degree/step. The structure is designed to minimize occlusion and to maximize working space. Again, in assembly operations, the operator uses visual feedback to perform fine adjustment.

3.3 Multiple-View Imaging Unit

Four different views are provided to the human operator: a global view of the entire assembly scene, a vertical view for part pickup and two lateral microscopic views for the fine adjustment of the position and orientation during micro-assembly operations (Figure 11). Each view uses a

CCD camera with a matching optical system. All images are captured by a Matrox Corona PCI frame grabber.

Microscopic stereo vision feedback is needed for precise 3D alignment. Two Navitar TenX zoom microscope systems with Mitutoyo ultra long working distance 5X objectives are used to provide two lateral views for fine adjustment. The resolving power of the objective is $2\mu m$.

The vertical view uses a long working distance zoom lens to guide pickup operation. A fourth global view is implemented using a miniature Marshall board camera to monitor the status of the entire assembly scene.

Several factors must be considered in setting up this multiple-view system. Firstly, due to the limited working distance of the microscope objective, the space between the wafer mount and those objectives is quite limited. It is therefore important to avoid collision among the objectives, the wafer mount and the manipulator. Secondly, since multiple views with vastly different magnifications are concentrated in this limited space, lighting must be controlled separately.

4. ASSEMBLY PROCESS FLOW

A simplified description of the assembly process flow consists of the following steps:

Step 1. Preprocessing
Read the wafer layout description file and the vacuum release tray layout description file. Generate the trajectory for pickup and assembly.

Step 2. Part Pickup
The XY positioning table automatically moves the part to be assembled under the vacuum pickup tool. The operator controls the manipulator to pick up the part.

Step 3. Move To Assembly Position
The XY positioning table and the vertical linear slide move automatically so that the hole for assembly on the wafer moves into the field of view of the two lateral microscopes. The manipulator moves the part automatically to a predetermined position close to the hole such that the part becomes observable by the lateral microscopes.

Step 4. Assembly Operation
The operator makes fine adjustments of the pose of the part and the wafer and guides the insertion of the part into the hole. Then the part is released.

Step 5. Manipulator Returns To Initial Position

Step 6. Start Next Assembly Operation
Get the position of the next part and the next hole. Update the assembly progress status. Go back to step 2.

5. MICROGRIPPER DESIGN AND GRIPPING SCHEME

5.1 Part Pickup Using Vacuum Gripper

Vacuum is a versatile method for part handling in microassembly [4][13]. Since the thickness of the metal part is less than $100\mu m$, vacuum pickup is a suitable choice for this task and has proven to be highly reliable in experiments.

5.2 The Need for an Additional Gripper

Although the vacuum tool is versatile in part *pickup*, it is limited in part *manipulation*. The major problem is that it can not provide sufficient constraint force, especially friction force, required in assembly operations. Another problem is that the orientation of the part after vacuum pickup is not consistent. Misalignment can only be accurately compensated after the part comes into the field-of-view of the lateral microscopes. This fine adjustment process can be rather time-consuming and error-prone.

Friction forces become more important in this assembly task for several additional reasons. First of all, the assembly can be more tolerant of alignment errors due to compliance if the microgripper can constraint the part with sufficient friction force. Secondly, since the assembly clearance is small, the part will contact the wafer substrate after it moves into the hole. This increases the required the friction force.

Figure 4. Microgripper working principle

Our solution is to use another gripper to clamp the part from the left and right sides. This additional gripper has two major functions. First, it forces the part into a repetitive orientation. Second, it provides additional friction force sufficient for the assembly. The working principle of the microgripper is illustrated in Figure 4. The actual microgripper with its control circuits is shown in Figure 5a. The complete two-gripper setup is shown in Figure 5b. Figure 6 shows the gripping scheme. The microgripper shown in Figure 5a is actuated by a 5mm Smoovy motor driving a precision lead screw. The final opening of the tweezer is adjustable.

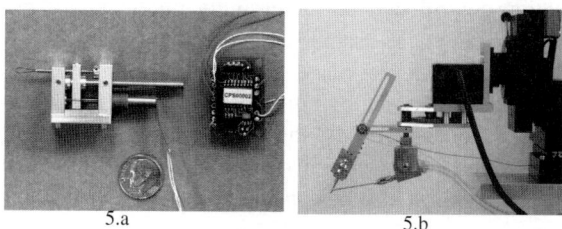

Figure 5. The lateral microgripper

This gripping scheme has been experimentally tested. A significant friction force between the part and the microtweezers was observed in experiments. The rough surface of the microtweezer, which is machined from spring steel by wire-EDM (Figure 8), and the relatively high holding force generated by the Smoovy motor and lead-screw combine to create this high horizontal constraint force.

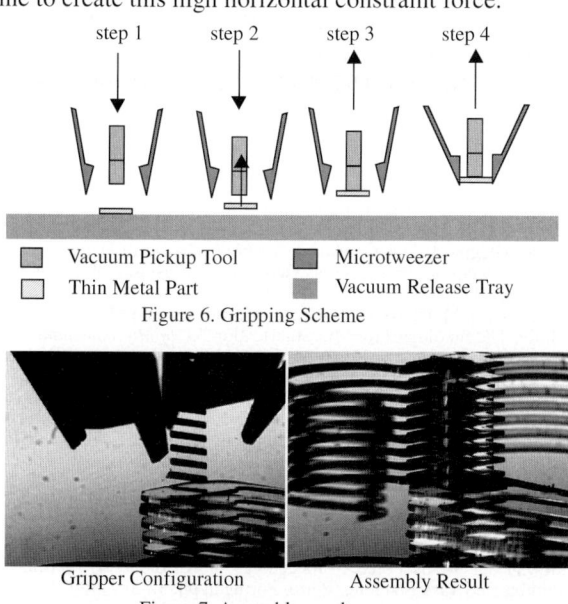

Figure 6. Gripping Scheme

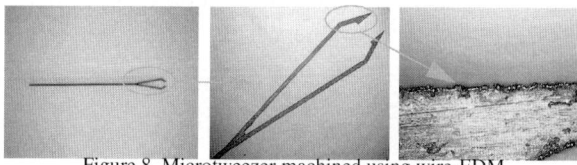

Figure 7. Assembly result

6. ENHANCING PICK-AND-PLACE RELIABILITY THROUGH FORCE SENSING

In pick-and-place operations, it is important to know when the microgripper comes into contact with parts or substrates. This feedback is important to minimize system cycle time and to avoid part damage. A piezoelectric force sensing unit was developed for this purpose. Its internal structure is shown in Figure 9. Its installation on the micromanipulator is shown in Figure 2.

Figure 8. Microtweezer machined using wire-EDM

Figure 9. Force sensing unit internal structure

One major design objective is to make this force sensing unit an independent module such that it can be integrated with different grippers. The force sensing unit is built around a high sensitivity PCB 209C1 piezoelectric force sensor. Its sensitivity is calibrated to 540.5mV/N. Its discharge time constant is approximately 1 second. A micro linear slide is used to guide the motion of the connection shaft, and to withstand torques and forces which are not vertical to the sensing surface.

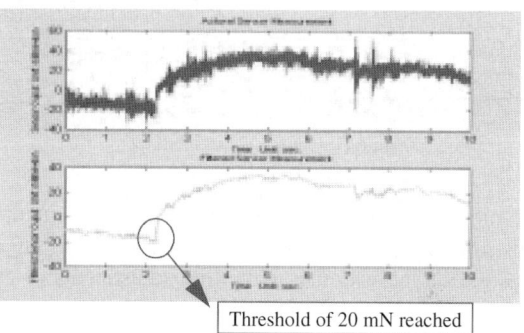

Figure 10. Comparison of measured force signals

It was found that under typical speeds for part pickup, the vibration of the manipulator is reflected in the sensor output. This significantly limits the attainable force resolution. A second order low pass filter with a roll over frequency of 30 Hz is used to eliminate these noises. Figure 11 compares the sensor output with and without this filtering. After contact force threshold is reached, a warning sound signal is generated. The large acceleration after reaching threshold shown in Figure 10 is due to the operator's reaction to reverse the manipulator motion.

One advantage of the piezoelectric force sensor is that it can withstand a relatively large static force so that a large structure can be connected to this module.

7. Microassembly Workcell Software Design

The software system consists of a multiple view control module, an assembly scheduling module and a motion control module (Figure 12). It provides dynamic vision feedback to the operator throughout assembly operations. It also provides motion control capability for manual position/orientation adjustment, emergency stop and error recovery. The status of the assembly process, e.g. position of next target, numbers and regions of parts assembled, is constantly monitored and recorded.

The image display part of GUI allows the user to quickly switch between different camera views and zoom in each view. One, two and four views can be displayed at the same time. The system supports simultaneous dynamic display in two views.

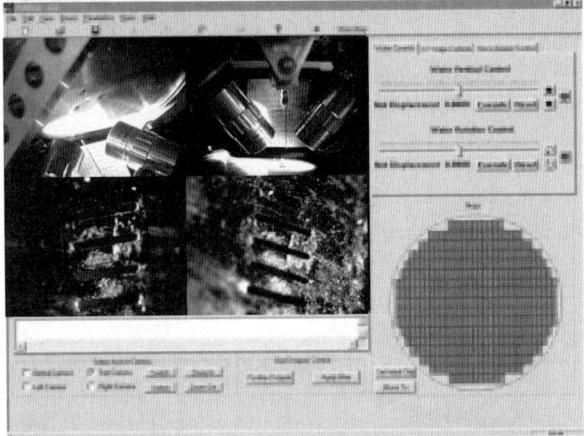

Figure 11. Graphical user interface

The motion control unit supports part alignment operations, microgripper control, and fine position/orientation adjustment. The user can make fine adjustments of every DOF of the coarse movement system with adjustable resolution, speed and range. This is essential for exception handling and error recovery.

The system keeps track of the progress of the assembly process. If the user click on the desired position on the wafer layout map, the system will move so that the area of interest comes into the field of view of the lateral microscopes.

8. CONCLUSION

An experimental microassembly workcell has been developed for wafer-level 3D microassembly. Research in several areas will be important for future development of wafer-level 3D microassembly systems. Continuing trends in the IC industry towards smaller line widths (currently 0.13μm) and larger wafer sizes (currently 12 inches) has driven underlying motion control technology to implement fast and highly repetitive motion with nanometer range repeatability within a large working space. These motion control technologies are able to meet the motion control requirements for wafer-level microassembly. Force feedback is also an important complement to vision feedback in high precision 3D microassembly. Currently, multiple DOF force measurement and control at mN to μN level remains difficult. In addition, better vibration isolation techniques will become important. Manipulators with five to six DOF, submicron repeatability and high precision rotation control will be increasingly important for more complex 3D microassembly. Finally, microgripper design for reliable pick-and-place and manipulation must continue to be addressed by investigating the use of different gripping forces and by integrating force sensing into microgrippers. The minimization of occlusion under the microscope must also be considered.

Acknowledgements

The authors would like to thank their colleagues in the Advanced Microsystems Lab for assistance in developing this workcell, and Ted Koury, Yu Zhou, and Bob Aldaz of Advantest America, Inc. for their advice on the microassembly system configuration. This research was supported in part by the National Science Foundation through Grant Numbers IRI-9612329, IRI-9702777, IIS-9996061, and IIS-9996062, the Office of Naval Research through Grant Numbers N00014-97-1-0668 and N00014-99-1-0144, Seagate Technology, Inc. and Advantest America, Inc.

References

[1] S. Allegro, *Automatic Microassembly by Means of Visually Guided Micromanipulation*, Ph.D. thesis, Swiss Federal Institute of Technology, Lausanne, 1998.
[2] P. Dario, M. Carrozza, N. Croce, M. Montesi and M. Cocco, "Non-traditional Technologies for Microfabrication," *J. of Micromechanics and Microengineering*, vol. 5, No. 2, pp. 64-71, 1995.
[3] G. Danuser, *Quantitative Stereo Vision for the Stereo Light Microscope*, Ph.D. thesis, Swiss Federal Institute of Technology, Zurich, 1997.
[4] R. Eberhardt, T. Scheller, G. Tittelbach, and V. Guyenot, "Automated Assembly of Microoptical Components", *Proceedings of SPIE, Microrobotics and Microsystem Fabrication*, vol. 3202, pp. 117-127, Pittsburgh, USA, 1997
[5] R. Fearing, "Survey of Sticking Effects for Micro-Parts," *IROS'95*, Pittsburgh, USA, pp.212-217.
[6] J. Feddema, P. Xavier, and R. Brown, "Micro-Assembly Planning with van der Waals Force," Proceedings of IEEE Int. Symposium on Assembly and Task Planning, Porto, Portugal, pp. 32-38, 1999.
[7] R. Grimme, W. Schmutz, D. Schlenker, M. Schuenemann, A. Stock, and W. Schaefer, "Modular Magazine for the Suitable Handling of Microparts in Industry," SPIE Microrobotics and Microsystem Fabrication, vol. 3202, pp. 157-167, Pittsburgh, USA, 1997.
[8] R. Hollis and J. Gowdy, "Miniature Factories for Precision Assembly" Int. Workshop on Microfactories, Tsukuba, Japan, pp. 9 - 14, 1998.
[9] A. Menciassi, *Microfabricated Grippers for Micromanipulation of Biological and Mechanical Objects*, Ph.D. thesis, Scuola Superiore Sant'Anna, Pisa, Italy, 1999.
[10] H. Morishita and Y. Hatamura, "Development of Ultra Micro Manipulator System Under Stereo SEM Observation", *IROS'93*, Yokohama, Japan, pp.1717-1721, 1993
[11] H. Suzuki, N. Ohya, N. Kawahara, M. Yokoi, S. Ohyanagi, T. Kurahashi and T. Hattori, "Shell-body Fabrication for Micromachines," *J. of Micromechanics and Microengineering*, vol. 5, No.1, pp.36-40, 1995.
[12] G. Thornell, M. Bexell, J. Schweitz, and S. Johansson, "Design and Fabrication of a Gripping Tool for Micromanipulation," *Sensors and Actuators A.*, vol. 53, pp. 428-433, 1996
[13] W. Zesch, M. Brunner and A. Weber, "Vacuum Tool for Handling Microobjects with a Nanorobot", *ICRA'97*, Albuquerque, USA, pp. 1761-1766.

Mechanical Micro-dissection by Microknife Using Ultrasonic Vibration and Ultra Fine Touch Probe Sensor

Fumihito Arai* Takaharu Amano** Toshio Fukuda*** Hiroshi Satoh*

*Department of Micro System Engineering, Nagoya University
**Department of Mechano-Infofmatics and Systems, Nagoya University
***Center for Cooperative Research in Advanced Science and Technology, Nogoya University
Furo-cho 1, Chikusa-ku, Nagoya 464-8603, JAPAN.

Abstract

Recently, research works on the bio-engineering actively carried out, and the useful tool for the micromanipulation is also developed. Purpose of this research is to develop the minute size knife for the cutting of the minute object such as the cell. In this paper, we report on the microknife that cuts off the object by using ultrasonic vibration of the sharp needle. We employed the multilayer piezoelectric actuator for generating the ultrasonic vibration for cutting. And we made the touch probe sensor for microknife by the PZT thin film which is made by the hydrothermal method. We evaluated the performance of cutting and the touch probe sensor by the experiments.

1 Introduction

Recently research works on the biotechnology advanced greatly, and the bio micromanipulation, which is available for various operations of the biological cell under the microscope, is actively used. In one of the micro operations, dissection technology that cuts off the submillimeter order sample is required. The glass needle, the metal edge, the silicon edge[1] and laser beam are conventionally used for these micro cutting operations.

As the tool for the bio-micromanipulation, the glass tool is conventionally used for various micro operations like perforation of the biological cell, cutting off and so on. The glass needle is effective for the micro operation because it is easy to make the tip diameter to be submicron order, but there are defects; such as low strength and fragility. The strength of the tool is important to cut off the hard object sharply. The strength of the knife is higher than that of the needle, so it is effective to cut off in one direction. However, it is necessary to change the direction of the edge, when we need to change the cutting direction. For the micromanipulator, it is difficult to change the orientation while keeping the position accuracy. In the case of the needle, it is not necessary to change the posture of the tool and the object. So it is possible to cut off in the various directions on the stage in narrow work spaces.

In this paper, we propose a new method for cutting off the object in the various directions without changing the posture of the tool and object. This method is applied the ultrasonic vibration to the needle, which is set perpendicular for the object on the surface. The vibration cutting is effective in the reduction of cutting resistance. It is used in the ultrasonic cutter and the ultrasonic vibration cutting[2] and so on.

There is the other system using laser which can cut off the object in various directions on the stage, however the laser system needs exclusive equipments. The microknife is set at the end effector of conventional micro manipulation system. We report the result of cutting of the onion epidermal cell, to see the effectiveness of the proposed microknife.

The micromanipulation under the microscope is difficult to carry out, since recognition of the height direction is difficult and the image of the microscope is two-dimensional in spite of the operation in the three-dimensional space. So it often occurs that the tool pierce the target and cell, or the tool collides with the stage and the very thin tip is broken.

The similar problem occurs in microknife operation. It is quite difficult to adjust the depth position to insert the needle in the cell. In such case, the cutting condition is not constant. To solve these problems and to improve the operability, we made the touch probe sensor for the microknife by the PZT thin film which is made by the hydrothermal method, and report the experiment result of this sensor.

2 Microknife

2.1 Principle of Microknife

The structure of the microknife is shown in Fig.1. The multilayer piezoelectric actuator is assembled as vibrating actuator for vertical direction. The microknife is constructed in this vibration unit, fixing the needle unit, and the junction unit for the manipulator. The junction unit has the gradient of 30 degrees, and the needle is bent for 90 degrees as the tip becomes perpendicular for the surface of the object.

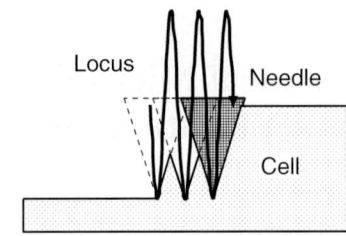

Figure 2: Locus of Needle(Side View)

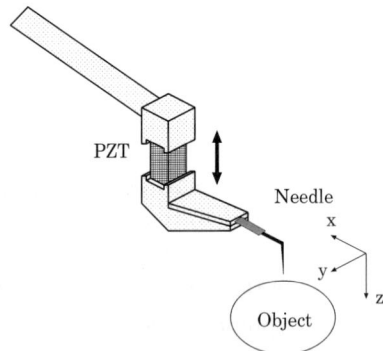

Figure 1: Microknife

The amplitude of the tool (the tip of the pipe fixing the needle) is about 100 nm, when the driving voltage of the multilayer piezoelectric actuator is 20 V_{p-p} at the frequency 35kHz. We determined the vibration frequency referring to the one of the ultrasonic cutter.

We used the probe needle, which is used for inspecting the IC wafer, as the needle for cutting. This is made by the electropolishing. Tip radius is about 2.5μm and material is the hardened tungsten. The tip is so sharp and hard that it is effective to cut off the cell. It is thought that the locus of the tip of the needle which vibrates in the vertical direction when the microknife cut off the cell is shown in Fig.2. In Fig.2, we think that the needle cut off the cell with ultrasonic vibration for top and bottom by the side edge.

The proposed microknife can cut off the object without changing the posture of the tool and the object. Normally, it is quite difficult for the conventional mechanical cutting system to change the cutting direction without changing the tip position, since precise orientation control is difficult in micromanipulation works. However, the proposed method can change the cutting direction easily by translational control of the stage.

2.2 Cutting Experiment

We cut off the onion epidermal cell as the object using the microknife equipment. Configuration of the micromanipulation system for maicroknife is shown in Fig.3. We come down the needle to the surface of the cell, and move it in the horizontal direction by the micromanipulation system whose driving resolution is 40 nm, and drive range is 100 μm[3]. The driving voltage for the multilayer piezoelectric actuator is 20 V_{p-p}, and the frequency is 35kHz. The mobile speed of the tool in the horizontal direction is 50μm/s.

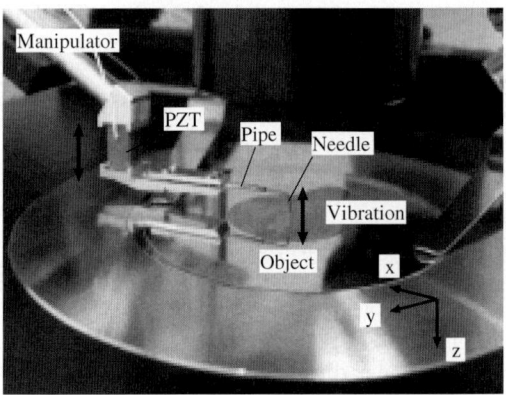

Figure 3: Photo of Microknife System

The result of cutting off the cell without vibration is shown in Fig.4. The results of cutting off the cell with vibration are shown in Fig.5 and Fig.6.

The cut off position is corrugated in Fig.4, so the damage of the cell is big. When the needle does not vibrate, it is difficult to stick the needle into the cell, so we forcefully cut and the result in Fig.4 was obtained. The needle is frequently broken. In Fig.5 and Fig.6 , the cut off line is narrow and it is thought the damage

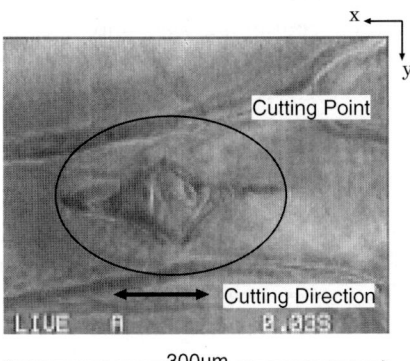

Figure 4: No vibration

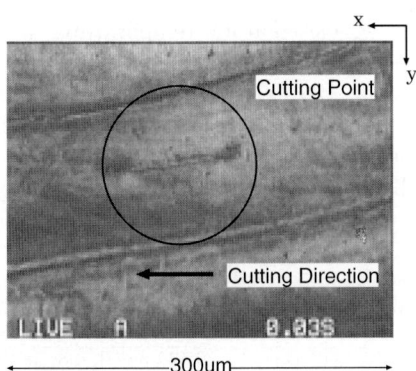

Figure 5: Cutting in x-direction with vibration

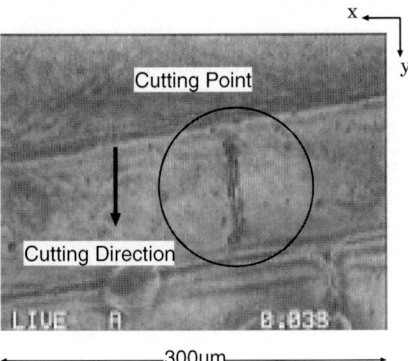

Figure 6: Cutting in y-direction with vibration

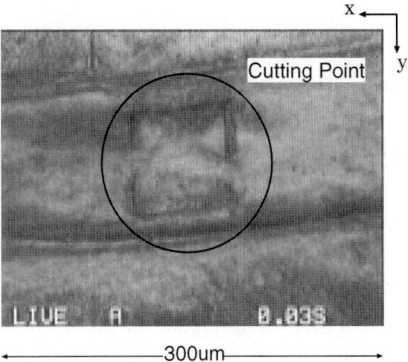

Figure 7: Cutting along square trajectory with vibration

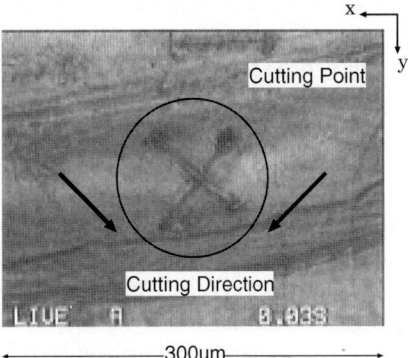

Figure 8: Cutting in crisscross with vibration

of the cell is small. However, if the depth of the needle height position is different, the condition of the cut off position is correspondingly different. The cell was not be cut or the cut off position is corrugated, when the height of the needle was not set properly.

In Fig.7, the cell was continuously cut off in the square trajectory. In Fig.8 the cell was cut off in the crisscross. Sharp cut was realized similarly.

3 Touch Probe Sensor

3.1 Principle of Sensor

The tool of the microknife is composed of the needle for the cutting which is tightened in the pipe and the tool which fixes the pipe.

We deposited the PZT thin film on the surface of this pipe by the hydrothermal method and use this

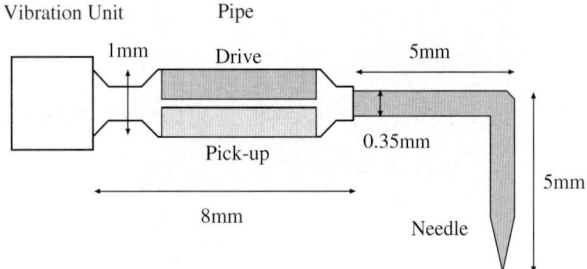

Figure 9: Outline of Touch Probe Sensor

pipe as the touch probe sensor in Fig.9. The principle of this sensor is like the tapping mode AFM[4] [5].

This sensor has two electrodes as shown in Fig.9. One electrode exists on the opposite side of the pipe. We use the upper electrode for the driving actuator, and the lower electrode for the pick-up sensor.

3.2 Hydrothermal Method

PZT thin film has been produced by Sol-Gel method, Screen-Press, Sputtering, MOCVD, and so on. Those methods are difficult to make PZT thin film on a three dimensional structure. And we have to keep high temperature more than 500°C. Using hydrothermal method, we can avoid those problems. PZT thin film deposited by the hydrothermal method was researched before[6] [7] [8]. The princple of hydrothermal method is that the titanium sample which is deposited PZT thin film, lead nitrate(Pb^{2+}), zirconium oxychloride(Zr^{4+}), titanium tetrachloride(Ti^{4+}), and potassium hydroxide are reacted in autoclave under the condition of high pressure and temparature for a long time. This method consists of two processes, that is, nucleation process and crystal growth process.

The nucleation process is the first process. The titanium sample are reacted about 24hour, 140 ℃. Then PZT nuclei are formed on a titanium sample.

The next process is the crystal growth process. Pb^{2+}, Zr^{4+}, and Ti^{4+} react for 24hour, at 120 ℃. The PZT crystal is subsequently grown from the nuclei that is made by the first process.

The advantage of the hydrothermal method compared with the others are following.

- PZT thin film can be fabricated on the three-dimensional structure, because this method grow PZT thin film under the solution.

- PZT thin film can be fabricated only on the titanium substrate.

In nucleation process, the titanium substrate reacts with the solution containing lead and zirconium to form PZT nuclei on the surface. So we can fabricate PZT thin film selectively.

- Thickness of PZT film can be controlled by repeating crystal the growth process

This method consists of two processes, nucleation process and crystal growth process. By repeating the crystal growth process, we can control thickness of PZT film from about 10 μm to 100 μm.

- This method doesn't need polarization process, because PZT thin film is polarized during this method.

- PZT thin film is deposited at relatively low temperature below 200°C.

PZT films are deposited at low temperature, the nucleation process is about 140 degrees, the crystal growth process is about 120 degrees, in autoclave. So, this method can decrease thermal stress compared with the conventional method.

The condition of hydrothermal method is shown in table.1

Table.1
The reaction condition of hydrothermal method

Process	Material	Concentoration
nucreation	$Pb(NO_3)_2$	0.520 mol/l, 4.8g in 28ml H_2O
	$ZrOCl_2 8H_2O$	0.535mol/l, 2.1g in 8ml H_2O
	$TiCl_4$	1.603mol/l, 1.4ml
at 140 ℃	KOH	4N, 44ml
24hours		
crystal	$Pb(NO_3)_2$	0.780mol/l, 7.2g in 28ml H_2O
growth	$ZrOCl_2 8H_2O$	0.803mol/l, 3.0g in 8ml H_2O
	$TiCl_4$	1.603mol/l, 2.12ml
at 120 ℃	KOH	4N, 44ml
24hours × 2		

Fig.10 shows the photograph of the PZT crystal grown on the surface. The PZT crystal grew on the titanium substrate. The grain size of PZT crystal is about 5 to 10 μm.

We measured the piezoelectric constant of PZT thin film made by the hydro thermal method in this time. The value of d_{31} was $6.87 \times 10^{-12}[V/m]$. This is generally lower than that of general piezoelectric ceramics.

3.3 Analysis of Sensor

We performed modal analysis on this touch probe sensor using ANSYS that is the finite element analysis software. We show the analysis model in Fig.11

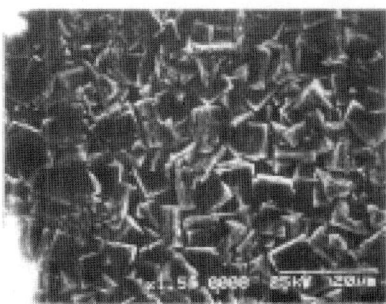

Figure 10: PZT crystal

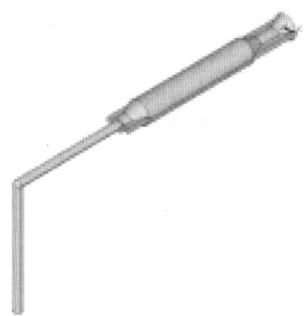

Figure 11: Analysis Model

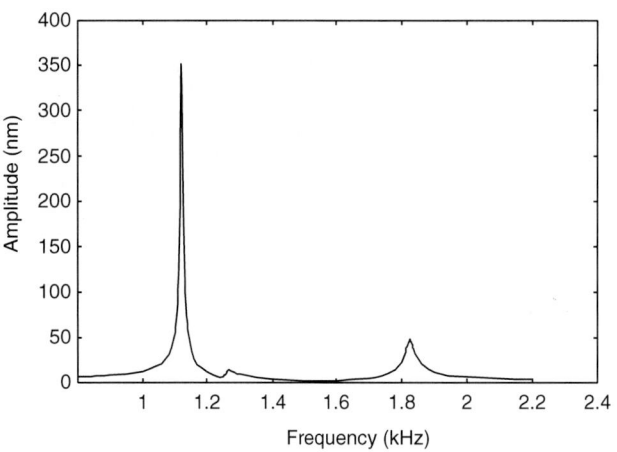

Figure 12: Resonant Frequency

At the 1st and 3rd resonance frequency, the needle and the pipe vibrates in vertical direction. At the 2nd resonance frequency, the needle mainly vibrates in horizontal direction. The value of the resonance frequency is that 1st : 0.558 kHz, 2nd : 0.612 kHz, 3rd : 1.588 kHz. The object is set under the needle, so we think that the vertical vibration is desirable. The PZT thin film is deposited on the pipe. However, in this analysis, PZT thin film was not included in the model of the modal analysis.

3.4 Experiment

We measured the amplitude of the tip of the pipe and the resonant frequency. We use a one electrode for driving. The driving voltage was kept at $2V_{p-p}$. The vibration velocity of the pipe was measured by using the laser Doppler vibrometer. Fig.12 shows this result.

The value of the resonance frequency is that 1st : 1.12 kHz, 2nd : 1.27 kHz, 3rd : 1.82 kHz. The measured value of resonance frequency was higher than the analyzed one, because PZT thin film was deposited on the pipe. The amplitude of the 1st resonance frequency is bigger than that of the 2nd and the 3rd resonance frequency, and Q value is quite high. So we think that the 1st resonance mode is suitable for this sensor.

We measured the shift of the phase of pick-up signal when the needle contacts the object by using the scheme in Fig.13. We used one electrode for driving and another one for picking up the signal. The driving voltage was kept at 2 V_{p-p}, and the frequency was 1.12 kHz. The sensor comes down to the cell in speed at 1 $\mu m/s$ by micromanipulation system. The driving resolution of micromanipulation system is 40 nm (P1-Polxtec, P-762 Multi-Axis Alignment System), and the driving range is 100 μm.

We use the cantilever beam with the strain gauge for evaluation of the contact force, and detection of the contact.

Fig.14 shows the result of measurement. We could observe the phase shift of pick-up signal when the needle contacts the object. When the tip of the needle contacted the cantilever beam, the output signal of the strain gauge increased. At the same time, it was observed that the output of touch sensor greatly changed. It is thought that each areas corresponds to (A) for the free vibration, (B) for the tapping vibration and the contact.

4 Conclusion

We proposed the method to cut off the cell using the needle with the ultrasonic vibration, and succeeded

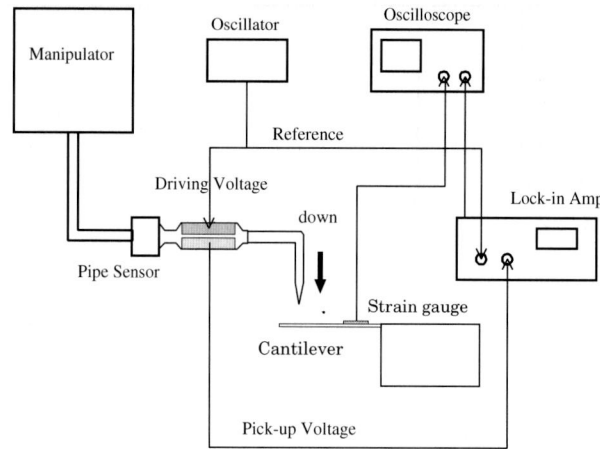

Figure 13: Set up for Measurment

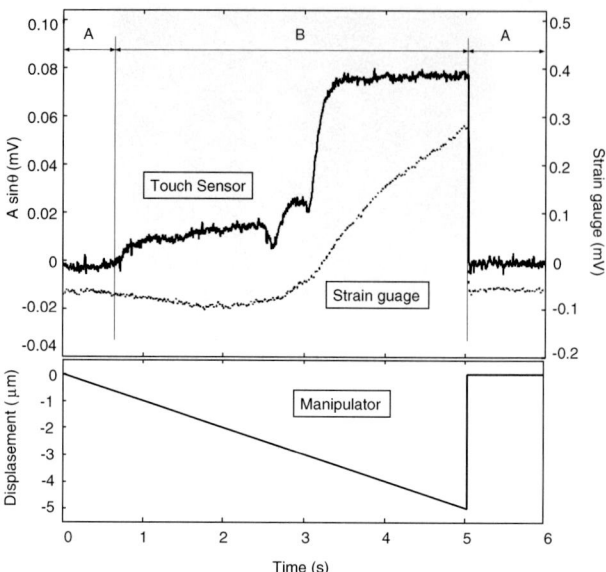

Figure 14: Relationship between the pick-up signal phase of the sensor and the displacement of the tool

in the experiment to cut off the onion epidermal cell with the developed equipment. The cut off condition is obviously different in the case that ultrasonic vibration was used. The cut off edge is sharpe and the damage of the cell is light. We confirmed that it is possible that the microknife can cut off in various directions on the stage without changing the posture of the tool or the object. It is a great advantage of the needle shaped microknife.

In addition, we made the touch probe sensor for the microknife by the hydrothermal method, and evaluated the characteristics of this sensor. It is possible to detect the contact state with the object.

In future, we will develop the micromanipulation system for cutting with the touch probe sensor, and analyze the cutting phenomenon in micro region. We also plan to make the smaller microknife that has sensor and actuator by using the hydro thermal method, and develop the micromanipulation system with the function of SPM and the ultrasonic cutter.

References

[1] T.Irita,S.Hara,Y.Suzuki and Y.Hiramoto:'Thin-film Microknife for Biological Samle',National Convention of the Institute of Electrical Engineers of Japan,pp277-278,1997(in Japanese)

[2] E.Shamoto and T.Moriwaki,'Study on Elliptical Vibration Cutting',Annals of the CIRP,43/1,pp35-38,1994

[3] F.Arai,T.Sugiyama,P.Luangjarmekorn,A.Kawaji, T.Fukuda,K.Itoigawa and A.Maeda:'3D Viewpoint Selection and Bilateral Control for Bio-Micromanipulation', Int.Conf.on Robotics and Automation,pp947-952,2000

[4] G.Binnig,C.F.Quate,and Ch.Gerber: 'Atomic force microscope' ,Phys.Rev.Lett.,56,pp930-933,1986

[5] T.Kanda,T.Morita,M.K.Kurosawa,T.Higuchi: 'A Flat Type Touch Sensor Using PZT Thin Film Vibrator' ,Transducers,pp1508-1511,1999

[6] H.Satoh,F.Arai,H.Ishihara,T.Fukuda,H.Iwata, K.Itoigawa:'New PZT Actuator Using Piezoelectric Thin Film on Parallel Plate Structure', Proc.of.Int.Sym.on.Micromechatronics and Human Science(MHS), pp79-84,1997

[7] Ohba,Arita,Sakai and Daimon: 'Lead Zirconate Titanate Film Synthesized from Sulfate Slurry' ,MRS in Japan,1996

[8] M.K.Kurosawa,H.Yasui,T.Kanda and T.Higuchi: 'Performance of Hydrothermal PZT Film on High Intensity Operation',Proc.of.the IEEE International Workshop on Micro Elctro Mechanical Systems (MEMS),pp56-61,2000

Design and Analysis of an Electromagnetically Driven Valve for a Glaucoma Implant

Byunghoon Bae*, Nakhoon Kim*, Hongseok Kee**, Seonho Kim**, Yeon Lee**, and Kyihwan Park*

*Department of Mechatronics, Kwangju Institute of Science and Technology,
Gwangju, Korea, Email: khpark@kjist.ac.kr

**Lee yeon ophthalmic hospital, Gwangju, Korea
Email: oculand@hananet.net

Abstract— Glaucoma is an eye disease which is caused by abnormal high IOP (Intra Ocular Pressure) in the eye. The use of implants is increasing in these days in order to treat the glaucoma. However, though conventional implants have a capability of pressure regulation, they cannot maintain IOPs desired for different patients. To solve these problems, it is needed to develop a new implant which is capable of controlling the IOP actively and copes with the personal difference of patients. An active glaucoma implant consists of the valve actuator, pressure sensor, controller, and power supply. In this paper, firstly, we make an analysis of the operation of the conventional implant using a bond graph, and show its defects and limitations of the conventional valve analytically. Secondly, we design and analyze an electromagnetically driven valve actuator composed of membrane, air-core solenoid, permanent magnet, and covering frame. To save the electrical power for operating the valve actuator, appropriate actuating methods are suggested and an optimal design of the valve actuator is carried out. Finally, using simulations, the possibility of the proposed valve actuator is investigated.

Keywords— glaucoma, IOP, glaucoma implant, valve actuator

I. Introduction

In the human's eye, the aqueous humor which is fluid like water is produced consistently and pressure is maintained by this flow of fluid, called Intra Ocular Pressure(IOP). The the aqueous humor is produced at the processus ciliaries, located in the posterior chamber and is flowing from the posterior chamber through the pupil to the anterior chamber, and is drained away via the trabecular system into the channel of Schlemn which is in the chamber corner. Finally, the aqueous humor passes the channel of Schlemn through the sclera into the venous system. Glaucoma is developed if too much aqueous humor is produced or if the the aqueous humor encounters with an abnormal fluid resistance somewhere in the venous system[1].

In USA, about 1.5 million patients have surgical operation and glaucoma patients are increasing year by year. In South Korea, many surgical operations have been taken in hospital. However, there has been no innovative advance in surgical operation or implant device during past 30-40 years. Especially, due to the environmental problem and frequent use of the steroid it is expected that the number of patients increases.

There are three methods to treat glaucoma - using medicines, surgical operation, and implant device. If the condition of the patient gets serious, it is recommended to use the implant devices which decreases the IOP by draining out the the aqueous humor compulsory. Otherwise, he/she will lose the sight because the high IOP destructs the optic nerve. Normal IOP is about 2250Pa(17mmHg), but the IOP of a serious glaucoma patient is 8000-9000Pa. There are conventional implant devices such as Ahmed valve and Molteno valve [7]. These implants have fluidic resistances composed of a silicon tube and membranes to regulate the IOP. These implants have a capability of pressure regulation, however, they cannot maintain the IOPs desired for different patients, due to the fixed resistance. In many cases, too much the aqueous humor is drained out, which causes hypotony. This is also the cause of blindness[6], [7]. Moreover, after 1-2 year, fibrosis is formed at the wall of the silicon tube, so the the aqueous humor cannot flow well. Therefore, patients must have surgical operations every two years. Additionally, the big size of the implants gives fear to patients.

To solve these problems, it is needed to develop a new active implant which is small and capable of controlling the IOP actively so as to cope with the personal difference among patients, and help the convienient surgical operation. The active implant proposed in this work consists of the valve actuator, pressure sensor, controller, and power supply. For clarity, the implant is defined to indicate the whole system and valve actuator indicates just the valve part of the whole system.

This paper is composed of the following contents. In chapter II, we make an analysis of the operation of a conventional implant using a bond graph[2] and show its defects and limitation. In chapter III, we investigate design considerations which include actuation principles, resistance elements, control methods, and energy supplies. Among these, we focus on a design of the proper type of a valve actuator and control method for power saving. In chapter IV, the model of the electromagnetically driven valve actuator is achieved by using bond graph modeling. In chapter V, we simulate the IOP assuming the actuator is implanted in the eye of patient and investigate the possibility of the use of the proposed valve. In chapter VI, the conclusion of this paper is made

II. MODELING OF THE PASSIVE VALVE

The conventional implant is analyzed to find the important design factors required for developing a new active implant. Figure 1 Shows the appearance and implant method of the Ahmed valve. Assumptions are made for a simple modeling,

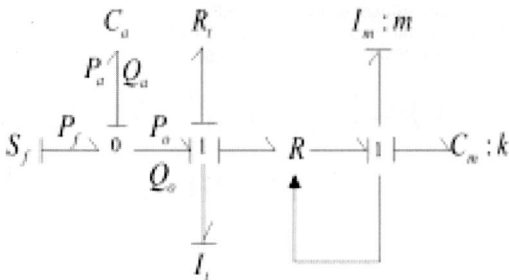

Fig. 3. A bond graph model of the Ahmed valve

and R_t can be neglected. Because C_a is related to the anterior chamber of a patient, it cannot be controlled. R depends on the design specification of the valve actuator. Figiure 4 is the simplified bond graph of Fig. 3.

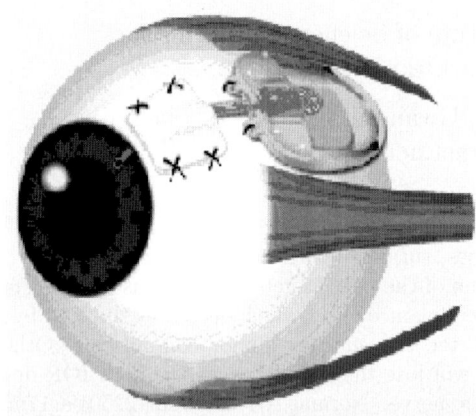

Fig. 1. The appearance of the Ahmed valve and implant method

1. The pressure drop occurs only inside the implant device.
2. The the aqueous humor is drained out to the veins where the pressure is same as the atmosphere pressure.
3. The aqueous humor flows with the constant flow rate.

The flow rate of the aqueous humor is almost constant as 2.5ml/min for the wide variation of the IOP. Therefore, the flow rate of the aqueous humor can be assumed to be an ideal flow source. Considering the structure of the Ahmed valve composed of two membranes and a silicon tube, an equivalent mechanical model can be drawn like Fig. 2 , and a bond graph model[2] is shown in Fig. 3

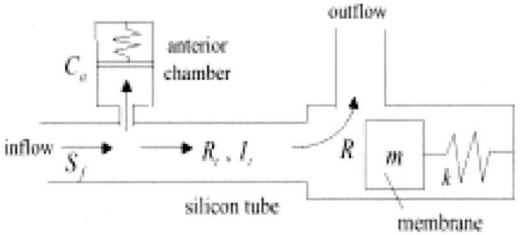

Fig. 2. An equivalent mechanical model of the Ahmed valve

S_f represents the flow source of the the aqueous humor which is uniformly produced, C_a is the compliance of the anterior chamber, R_t and I_t are the resistance and inertia of the fluid in the silicon tube, respectively. I_m and C_m are the inertia and compliance of the membrane, respectively. R_v is the resistance of the orifice varied by the gap length. Because the flow rate in the silicon tube is very small, I_t

Fig. 4. A simplified bond graph of the Ahmed valve

When the membrane is open, the small value of R makes Q_o increase, then Q_a decreases because $Q_a = Q_f - Q_o$. Hence P_a becomes low. Since P_o is the same as P_a, the low P_a indicates the low P_o. Hence, Q_o decreases, then Q_a increases. Because at a steady state, Q_o and P_a is regulated to have certain values, the Ahmed valve is an open loop stable system. However, the P_a cannot be guaranteed to be desired normal pressure.

The differential equation of the model is,

$$\dot{P_a} = \frac{1}{C_a}(Q_f - Q_o) = \frac{1}{C_a}(Q_f - \frac{P_a}{R}) \quad (1)$$

The solution is,

$$P_a(t) = e^{-\frac{t}{RC_a}}P_a(0) + RQ_f - RQ_f e^{-\frac{t}{RC_a}} \quad (2)$$
$$= (P_a(0) - RQ_f)e^{-\frac{t}{RC_a}} + RQ_f \quad (3)$$

where $P_a(0)$ is an initial IOP of a patient. The first term of Eqn. 2 is a transient-state term related to the decaying rate of the IOP, and the second term is related to a steady state. At a steady state ,

$$P_a(\infty) = (R_t + R_v)Q_f \quad (4)$$

The value of R is uncontrollable in the passive valve such as the Ahmed valve because it is determined when manufactured. The fixed resistance R makes the IOP converge to a certain value. However, since it is controllable in the active valve, the active implant can cope with the desired IOP which is different to peoples.

III. DESIGN CONSIDERATION OF THE VALVE ACTUATOR

Design of the active valve actuator includes the selection of actuation principle, variable resistance and control method to save power consumption, and energy supply, as shown in Fig. 5. In this work, actuation principle, resistance element, and control method are mainly discussed.

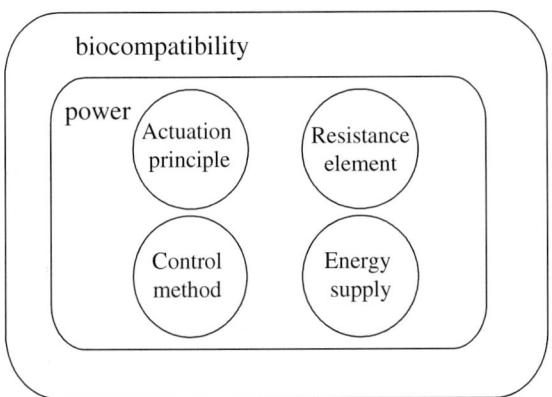

Fig. 5. The relation of design factors

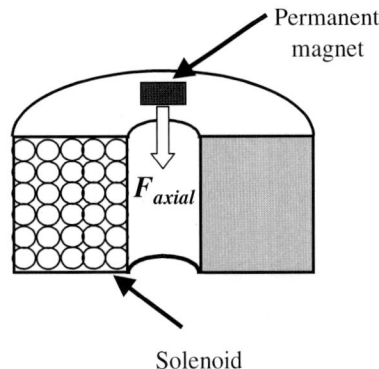

Fig. 6. The actuation principle used in the proposed valve actuator

A. Design specification of the valve actuator

Some design guidelines of the active implant are provided from [3]

1. Maximum size; $5 \times 5 \times 3 [mm^2]$
2. The tube diameter must be over $50 \mu m$ to avoid clogging of the tube due to the protein component of the aqueous humor.
3. Long durability : 5~10 years
4. Fail-safe design
5. Reliability
6. Simplicity

B. Actuation principle

The operation of the active implant can be performed by using piezoelectric and SMA(Shape Memory Alloy) materials, and thermal, thermopneumatic, electrostatic, and electromagnetic forces. Piezoelectric material and SMA need high voltage and high temperature for operation, respectively. Therefore, they are not suitable to implant into human's eye. Thermal and thermopneumatic method also use high temperature, so, they are not suitable for implant. Valve operation using electrostatic method has on-off operation, i.e. discrete operation. Electromagnetic method needs permanent magnet and solenoid to generate force. Due to the size of magnet and solenoid, The actuator size may be a little large, but actuation stroke is long and power consumption is not so high. In this work, we select firstly electromagnetic principle and propose a structure which consists of an air-core solenoid and a magnet which use the axial force as shown in Fig. 6

C. Resistance element

At a steady state, the outflow Q_o is equal to the inflow Q_f, from the Fig. 4, i.e.

$$Q_f = Q_o = \frac{P_a}{R} \quad (5)$$

Eqn. 5 is an IOP sensitivity over resistance. If this sensitivity is high for small variation of R, power consumption can be reduced. Therefore, we select the resistance element which gives a high sensitivity. There are variable resistance elements shown in Fig. 7. The (b)type was known to have a high sensitivity in [3], [4] by comparing the sensitivity of all types of Fig. 7.

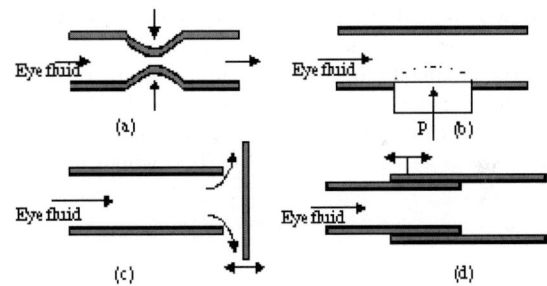

Fig. 7. Various resistance elements

D. Control method

To keep the desired IOP, the orifice of the implant of Fig. 7(b)must be regulated to have a constant gap length. To do this, the power should be continuously supplied. However, this control method should be avoided because it requires a big power supplying capacity. Hence, it is important to consider control methods which reduce the power supply.

D.1 PWM method

A PWM method is a method by which power is supplied during the time only when the current pulse is on. We can change the IOP area by adjusting the current on or off time. Figirue 8 shows the controlled IOP using the PWM

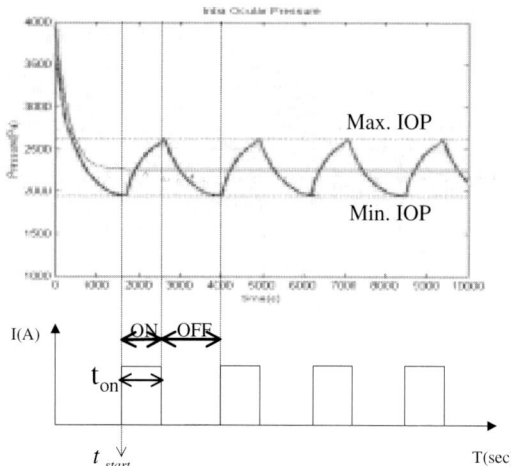

Fig. 8. IOP control using PWM method for power supply

method in a normally open valve.
When the IOP is lower than the minimum IOP, the control input is supplied and the IOP increase. When the IOP is higher than the maximum IOP, the control input is off and the IOP decreases because the valve is open. The average IOP is maintained between the maximum and minimum IOPs. The electric consumed energy is $W = v \cdot i \cdot t_{on}$, where v is the supplied voltage, i is current, t_{on} is the current on-time.

D.2 Impulse method

An impulse method is a method that the valve can be controlled so that its state of open or close position can be maintained until another pulse current is re-supplied. Since the supplied current has a similar shape of impulse, this method is called as an impulse method. Fig. 9 shows the controlled IOP and supplied voltage. The electric consumed energy is $W = \sum_{k=1}^{n} v(k) \cdot i(k)$, where n'th is on-time.

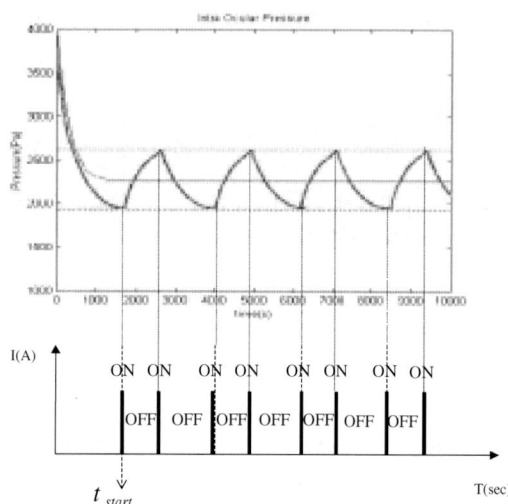

Fig. 9. IOP control using impulse method for power saving

Fig. 10 shows the typical example of a structure manufactured by using MEMS technology to which the impulse method can be applied[8].
The structure has a buckling membrane. At each on and off state of the valve, the membrane keeps its mechanical shape.

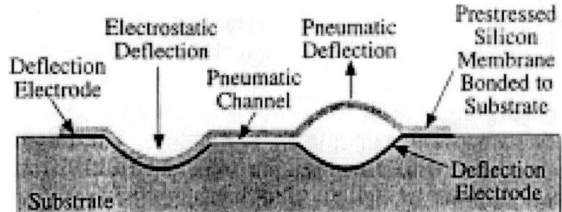

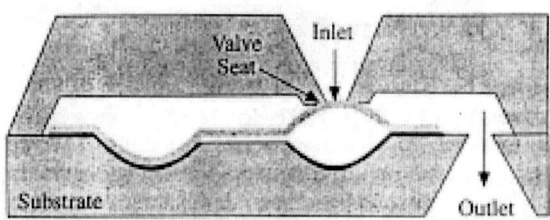

Fig. 10. A buckled membrane possible for the impulse method

We can save more power by considering the operating point of the valve with these two methods. The operating point is set to the position where the stable IOP is obtained without applying the current. One thing to note in determining the operating point is the gap length should be large so that some protein element of the aqueous humor doesn't clog it.

IV. MODELING OF THE VALVE ACTUATOR

Figure 11 shows the proposed valve actuator, where all design factors are graphically defined. Focusing on the power saving problems, we optimize the parameters d_m, i_m, c_r, and c_z which makes the axial force maximize under the design specifications. The result are d_m=3.44mm, i_m=0.8mm, c_r=0.78mm (13turns×0.06mm diameter wire), and c_z=0.78mm(13turns×0.06mm diameter wire).
To simplify the model, we make following assumptions.

1. The movement of the membrane is not influenced by the pressure of the aqueous humor flow.
2. The flow of aqueous humor is a laminar flow.

The first assumption allow us to model the valve axtuator using the one port R instead of the two-port R as shown in Fig. 12

F is the actuator force, z, p, I_m, C_m, and R_m are the membrane displacement, momentum, inertia, compliance, and damping, respectively. V_a is the volume of the anterior chamber.

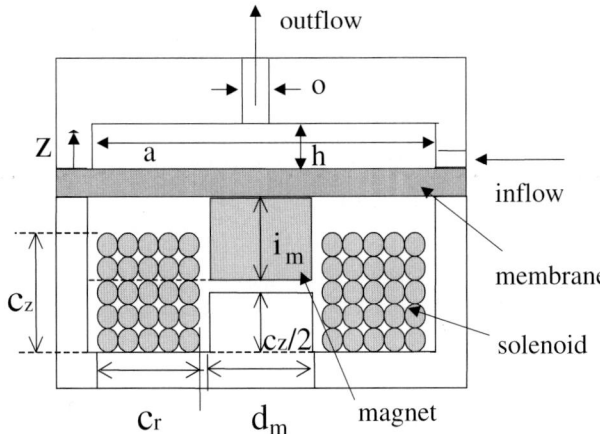

Fig. 11. The proposed valve actuator

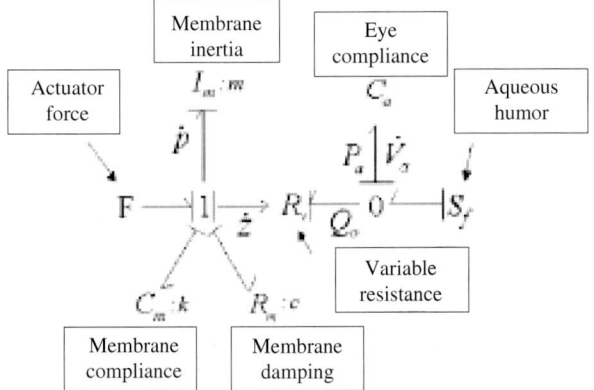

Fig. 12. A bond graph modeling of the proposed valve

The compliance of an eye C_a is known to be [1]

$$C_a = \frac{1}{(aP_a + b)(K_{eq} + K_f e^{-m_f t} + K_s e^{-m_s t})} \quad (6)$$

where,

$$a = 0.022\mu l^{-1}, \ b = 0.208 mmHg/\mu l, \ K_{eq} = 0.15,$$
$$K_f = 0.38, \ K_s = 0.47, \ m_f = 1.803 min^{-1}, \quad (7)$$
$$m_s = 0.072 min^{-1}$$

The compliance is highly nonlinear. V_a is about 0.25ml in a normal eye.

The material of the membrane is a silicon rubber. The size of the membrane is $4mm \times 4mm \times 100\mu m$. The young's modulus and poisson's ratio of the membrane are known to be $\sim 1[MPa]$ and 0.5, respectively. We use the maximum young's modulus($1[MPa]$) in here.

The variable resistance R_v is obtained analytically[5].

$$R_v = 2\frac{a\mu(1 - 4\frac{ah(\pi o + h - z)}{\pi o(4a+h)(h-z)})(\pi o + h - z)^2}{\pi^3 o^3 (h-z)^3 (\frac{\pi^3 o^3 (h-z)^3 (4a+h)^3}{64 a^3 h^3 (\pi o + h - z)^3} - 1)} \ [Ns/m^5] \quad (8)$$

where $\mu [Pa \cdot s]$ is the viscosity of the fluid and other parameters are defined at Fig. 11.

The state equations are

$$\dot{p} = F - kz - c\dot{z} \quad (9)$$
$$\dot{z} = \frac{p}{m} \quad (10)$$
$$\dot{V_a} = S_f - \frac{P_a}{R_v} \quad (11)$$

Let's take the Fourier transform of first equation of Eqn. 9

$$F = [m(j\omega)^2 + c(j\omega) + k] \cdot z \quad (12)$$

Because the flow rate is very small, operating frequency of the valve, w is very small, and the Eqn. 12 can be approximated to $F = k \cdot z$.

If we set the output variables as the membrane displacement and the IOP,

$$y_1 = z \quad (13)$$
$$y_2 = P_a = \frac{V_a}{C_a} = \frac{1}{C_a}\int (S_f - \frac{P_a}{R}) dt \quad (14)$$

V. Simulation of the Valve Actuator

For the simulation of in-vitro environment, firstly, we investigate the variance of the IOP using the PWM method. The initial IOP of patient is assumed to be 4000Pa. The outlet diameter(o) and orifice gap length(h) are $50\mu m$ and $100\mu m$, respectively. The rising ratio of the IOP is varied for the different membrane displacements. The falling ratio of the IOP is same in all cases, because the IOP depends on only passive resistances element.

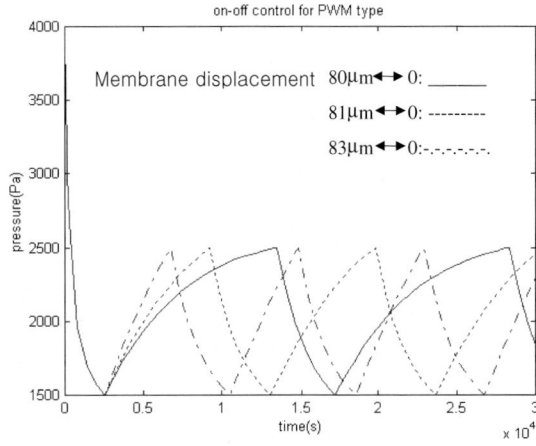

Fig. 13. Simulation result using PWM method

In the next simulation, we consider the case that the operating point is set to the position where the stable IOP is obtained without applying current. The orifice gap length is set to be $50\mu m$ when no current is supplied with the other size taken same as before.

From Fig. 14, we discover the valve actuator is operated 5 times in one day. The supplied voltage and corresponding current are v=0.0273[V] and i=10[mA]. The operated time

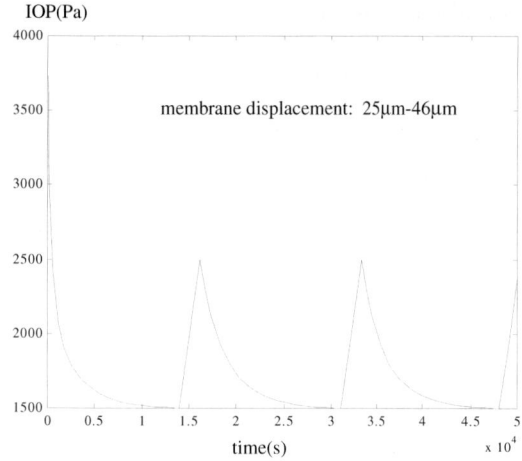

Fig. 14. Simulation result considering the operating point

in one day is about 12500[secs](3hr 30minutes). Therefore, the consumed power in one day is supposed to be $0.956[mW \cdot h]$. This result shows that power is much reduced.

VI. Conclusions and further researches

Glaucoma is an eye disease due to a high IOP, to make one blind, finally. Among current treatment of glaucoma, The method using implant is used in these days for their long acting advantage. However, conventional implants cannot make the IOPs the normal IOPs in all patients, and a big problem is to drain out too much the aqueous humor, to cause hypotony. In this paper, we did following contributions

1. Analyzed the operation of the conventional implant and shows its defects and limitations.
2. Established the design factors and design the electromagnetically driven valve actuator.
3. Modeled the valve actuator.
4. Simulate the variance of the IOP for in-vitro environment and investigate the possibility of the valve actuator.

Further researches are manufacturing of the proposed valve(on-going), investigation of the actuator performance, certification of the pressure-down effect of the manufactured valve in in-vitro experiment, and finally, certify the possibility of the proposed valve actuator by in-vivo experiment.

References

[1] R. Collins, T. J. van der Werf, Mathematical models of the dynamics of the human eye, Springer Verlag, 1980
[2] Ronald C. Rosenberg, Dean C. Karnopp, Introduction to Physical System Dynamics, 1983
[3] Cristina R. Neagu, A Medical Microactuator based on an Electrochemical Principle, Ph.D. Dissertation, University of Twente, Netherland, 1996
[4] D. J. IJntema, Feasibility Study for a Micro Machined Eye Pressure Regulator for Glaucoma Patients, Report, University of Twente, Netherland, 1992
[5] Peter Gravesen, Jens Branebjerg and Ole S. Jensen, "Microfluidics-a review", J. Micromech. Microeng. 3. 168-182. 1993
[6] D. L. Eisenberg, Edward. Y. K, G. Hafner, and J. S. Schauman, "In vitro flow properties of glaucoma implant devices", Ophthalmic surgery and lasers, Vol. 30, No. 8, pp. 662-667, 1999
[7] J. Antonio, A. Mermoud, L. LaBree, and D. S. Minckler, "In vitro and in vivo flow characteristics of glaucoma drainage implants", Ophthalmology, Vol. 102, No. 6, June 1995
[8] Wagner, B., Quenzer, H. J., Hoerschelmann, S., Lisec, T., and Juerss, M., " Bistable Microvalve with Pneumatically Coupled Membranes,", Proc. of the 9th Annual Workshop on MEMS, San Diego, CA, Feb. 11-15, 1996, pp. 384-388

A Path Generation Algorithm of Autonomous Robot Vehicle Through Scanning of a Sensor Platform

Tong-Jin Park* and Chang-Soo Han**
Department of Precision Mechanical Engineering Hanyang University, Korea*
Department of Mechanical Engineering Hanyang University, Korea**

Abstract-- In this paper, a path generation through use of a sensor platform is proposed. The sensor platform is composed of two electric motors which make panning and tilting motions. An algorithm for computing a real path and an obstacle length is developed by using a scanning method that controls rotation of the sensors on the platform. An Autonomous Robot Vehicle (ARV) can recognize the given path by adapting this algorithm. A scanning algorithm for recognizing environments around the ARV is applied to the sensor platform. The path generation algorithm is composed of two parts. One is recognizing a path pattern; the other is used to avoid an obstacle. An optimal controller is designed for tracking the reference path which is generated by recognizing the path pattern. Using the optimal controller and the path generation algorithm, the ARV is constructed. Based on the results of actual experiments, this algorithm for an ARV proved sufficient for path generation by small number of sensors and for a low cost controller by using the sensor platform with a scanning method.

Keywords -- ARV (Autonomous Robot Vehicle), Sensor platform, Optimal Control, Scanning algorithm, Path tracking, Path-generation,

I. INTRODUCTION

Conditions under which an Autonomous Robot Vehicle (ARV) recognizes a path and generates drive path instructions occur whenever there is a demand for driving flexibility. The ARV should generate a proper path for obstacle avoidance and path recognition which then creates a path for reaching a goal point. Thus a study of a path generation algorithm contains path planning, obstacle avoidance, self-localization, and a self-organizing map. Recent research has been conducted which investigated recognition of objects while reducing the number of sensors. Matthies and Shafer(1987) studied recognition using a stereo camera. Miller and Wagner(1987) investigated the path generation algorithm along with the input of an infra-red sensor using a rotating sensor platform. Appropriate control of the ARV actuators based on kinematic modeling has also been the subject of research. Sarkar(1996) constructed a study of dynamic feedback control based on an ARV kinematic model with two steerable wheels. This kinematic model has the advantage that the ARV can be improved as the ARV actuator control is achieved by controlling a kinematic model. The algorithm for path generation can, therefore, be developed more easily.

In this paper, a path generation algorithm is proposed. For a more effective method of sensor input, application of the actuator to the sensor platform is investigated, and for the path generation algorithm it is the recognition of a path pattern or an obstacle by rotating the sensor and about an optimal controller for path tracking. To apply the path generation algorithm, first of all, the sensor platform and the ARV are designed and experimented. The type of ARV's driving is two-wheel drive (2WD), which is steered by difference angular velocity of both wheels.

II. DESIGN OF THE ARV WITH SENSOR PLATFORM

A. Design of sensor platform

In the present study, a sensor platform is designed to enhance sensor inputs for more effective path recognition.

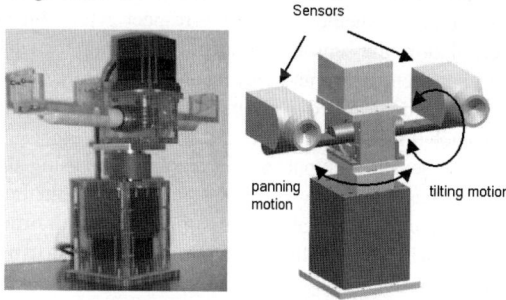

Fig. 1 Sensor Platform

Fig. 1 shows the sensor platform that is operated by two electric motors. The number of sensors for path generation may be reduced, so that sensor inputs are more effective with the ARV controller. As the sensor inputs become larger, they require more accompanying filters or a larger quantity of managing data. Consequently, a higher level of control is needed or there would be increased difficulty in driving at high speeds.

The sensor platform can be scanned by two electric motors with a transmission mechanism. To recognize the direction of the right and left paths, the platform utilizes a panning motion. To direct the up and down paths, a tilting motion is simultaneously available. The sensor platform is composed of the two sensor-decks able to apply various sensors. In the present study, only the panning motion of the sensor platform itself is allowed to generate data for the algorithm governing the horizontal recognition of objects and their respective paths.

B. Sensor

Sensor data obtained for measurement when the analog voltage ranged between 0[V]~10[V] contains many noise components. A sensor which is rotating in the platform has more noise while operating than in a motionless platform. A filtering method should be required for reducing the noises. In this study, the low and high pass filter is applied for clearing and reducing the noise. Fig. 2 shows the results of filtered sensor data.

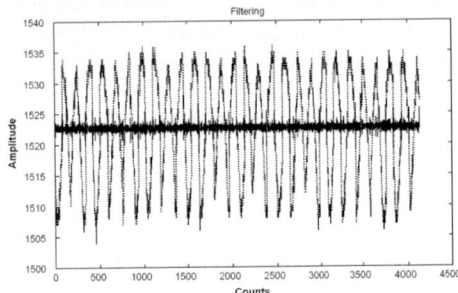

Fig. 2 The filtered sensor data

When comparing filtered data with pure sensor data, the noise levels are remarkably lower. In order to obtain get information regarding the distance from an object to its monitoring sensor, the relationship between the voltage and the distance is derived using a linear interpolation method. With respect to linear interpolation, the sensor's outputs are obtained for each 5cm between 0.2m and 3m apart and the sensor's outputs are determined to the means about sensor output from each distance.

The sensor's outputs are interpolated by using the mean values. Fig. 3 shows a linear interpolation method and the results interpolated using a first and fourth order equation with the resulting mean values obtained using measured sensor data.

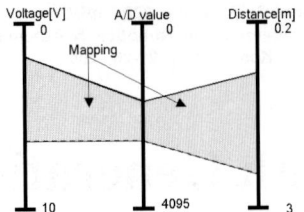

(a) Relationships of sensor input and recognition input

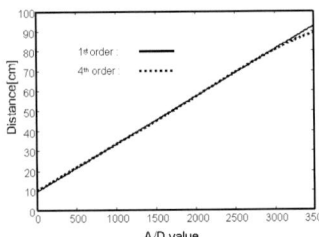

(b) Interpolation of the sensor data

Fig. 3 Linear interpolation method of sensor data

In the present study, the first and fourth order equation results are almost identical. After comparative evaluation, the first order interpolation equation was selected for the reduction of calculated amounts.

C. The ARV

In producing the ARV, the most important part is the performance of the actuators. The AC servomotor is used for easy control and for its capacity. This 24-volt motor is located in the wheel. It is often used in autonomous vehicles in the industrial field.

Fig. 4 shows the structure and design of the ARV.

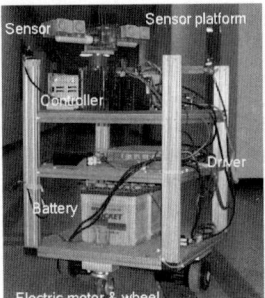

Fig. 4 The Autonomous Robot Vehicle

As seen in Fig. 4, the driving motor is composed of one body with wheel. The driving motor and wheel sets are located on both sides, and two caster wheels are located at the front and rear of the ARV. The maximum speed of the motor is 45cm/min and it is powered by a 24-volt DC battery. The battery has a 2-hour continuous drive capacity at maximum speed. Fig. 5 shows the ARV hardware structure and the internal signal flowchart for controlling the ARV.

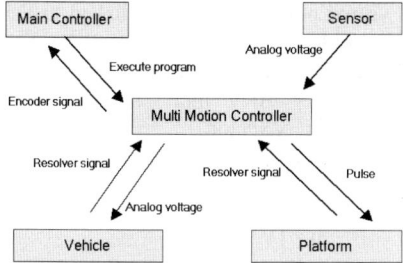

Fig. 5 Hardware block diagram of the ARV controller

The Multi-Motion Controller (MMC) transfers the rotating angle signal of each wheel to the main controller, and transfers the control algorithm output to the motor's driver through voltage signals.

III. PATH GENERATION ALGORITHM

A. Path algorithm 1 – Recognition of navigating path pattern

Before initiating navigation, the ARV creates an algorithm for recognizing path patterns. First, the ARV is assumed to be stationary. Sensor data is obtained once for each sampling period. Each data set collected is a relative variable obtained from the sensor frame on the ARV. Fig. 6 represents the coordinated relationship between the sensor and the object frame.

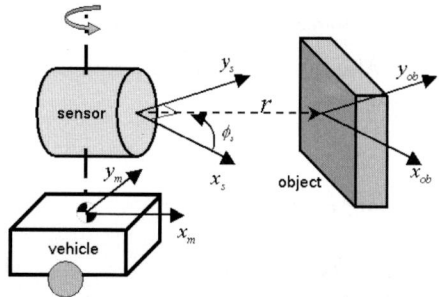

Fig. 6 Coordinates of sensor and object

The mathematical formulae regarding the recognition objects to be measured using a polar coordinates is as follows:

$$x_{ob/m}(k) = r(k)\cos\phi_s(k) \quad (1)$$

$$y_{ob/m}(k) = r(k)\sin\phi_s(k) \quad (2)$$

Where, k means k^{th} sampling time. These formulae are converted coordinates for representing any given obstacle with respect to the absolute rectangular coordinates. The approximate shape of an obstacle is modeled by linking each point indicated to its respective obstacle coordinate. If this relational theory is applied to an actual driving situation, the center position of the ARV is of critical importance. Related formulae are as follows:

$$x_{ob}(k) = x(k) + r(k)\cos\phi_s(k) \quad (3)$$

$$y_{ob}(k) = y(k) + r(k)\sin\phi_s(k) \quad (4)$$

Through the scanning motion of the sensor platform, the measured size of the object is found using from Eq. (1) to Eq. (4). Initial results of the path recognition obtained by the ARV are shown in Fig. 7. All the shapes of the paths that can be driven by the ARV are modeled using a combination of lines.

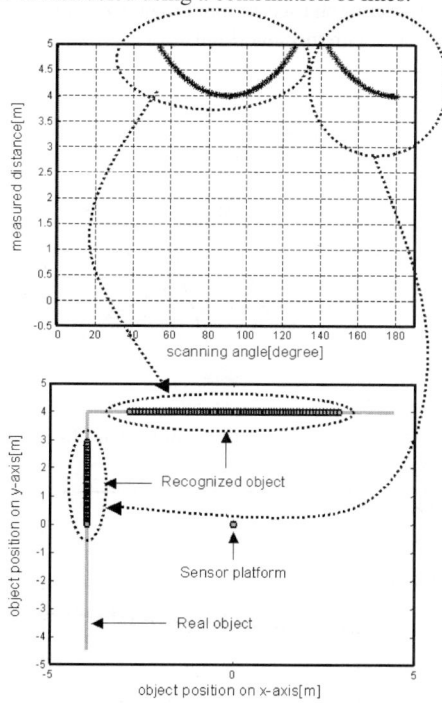

Fig. 7. In front of and on the right side walls recognition

B. Path algorithm 2 – Obstacle avoidance

Path algorithm 2 is used to recognize the shape of an obstacle in the navigation path of the ARV. This is the algorithm which measures the approximate shape of the object employing the scanning method.

The equation measures the obstacle size according to the input obtained from the following:

$$s(k) = \sqrt{(x_{ob}(k+1) - x_{ob}(k))^2 + (y_{ob}(k+1) - y_{ob}(k))^2} \quad (5)$$

A decision algorithm regarding whether the object is moving or not is derived from Eq. (6) to Eq. (7). Thus, an object which is not moving is represented by the following:

$$\dot{x}_{ob/m}[k] = \dot{x}_m[k] \quad (6)$$

$$\dot{y}_{ob/m}[k] = \dot{y}_m[k] \quad (7)$$

A moving object is represented by the following:

$$\dot{x}_{ob/m}[k] \neq \dot{x}_m[k] \quad (8)$$

$$\dot{y}_{ob/m}[k] \neq \dot{y}_m[k] \quad (9)$$

Information regarding where a measured object is moving is obtained by Eq. (5)~Eq. (9). With an obstacle which has a complicated shape, it is necessary for the sensor's inputs to ascertain information quickly. In this case, sensor data should be sampled with in a suitable period. This algorithm is constructed by recognizing the size of the complicated object.

Fig. 8 and Fig. 9 suggest an algorithm which can roughly recognize the approximate shape of an obstacle by interpolating the sensor input data, and can recognize the obstacle with this algorithm.

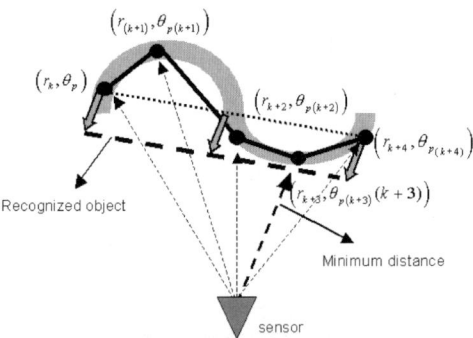

Fig. 8 Recognition of a complicated obstacle

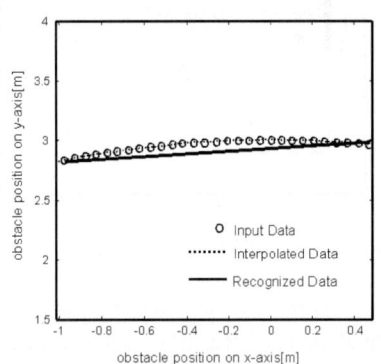

Fig. 9 Recognition result plot of a complicated obstacle

IV. DESIGN OF THE CONTROLLER

Fig. 10 shows a modeling of the ARV for an optimal controller.

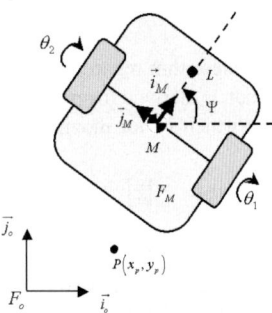

Fig. 10 Modeling of the ARV

A virtual point, L, which is linked to the ARV's frame, is located at a distance l from the direction of wheel's axis of the ARV navigation.

Modeling of the ARV represents the configuration vector and auxiliary vector:

$$\mathbf{x} = \begin{bmatrix} x & y & \Psi \end{bmatrix}^T \tag{10}$$

$$\mathbf{u} = \begin{bmatrix} v & \dot{\Psi} \end{bmatrix}^T \tag{11}$$

In Eq. (11), v is the tangential velocity of the ARV along the $\vec{i}_M$ axis. The control vector $\mathbf{u}$ is constrained by the driving wheel's velocity and represented by the following equations:

$$\mathbf{u} = \begin{bmatrix} \dfrac{r}{2} & \dfrac{r}{2} \\ \dfrac{r}{2R} & -\dfrac{r}{2R} \end{bmatrix} \begin{bmatrix} \dot{\theta}_1 \\ \dot{\theta}_2 \end{bmatrix} \tag{12}$$

Therefore, the ARV's state equation is expressed by control input $\mathbf{u}$. Eq. (13) yields:

$$\dot{\mathbf{x}} = \mathbf{A}(\mathbf{x}) \cdot \mathbf{u} \tag{13}$$

To obtain the matrix $\mathbf{A}(\mathbf{x})$, the relative velocity of any point $P(x_p, y_p)$ with respect to fixed frame F_o is described by Eq. (14):

$$\vec{V}_{P/F_M} = \vec{V}_{P/F_O} - \vec{V}_{M/F_O} - \vec{\omega}_{F_M/F_O} \times \overrightarrow{MP} \tag{14}$$

Here, $\vec{\omega}_{F_M/F_O}$ is the instantaneous rotational velocity of the ARV's frame F_M with respect to the fixed frame F_O. Also, $\vec{\omega}_{F_M/F_O}$ and $\overrightarrow{MP}$ are represented in the following equations:

$$\vec{\omega}_{F_M/F_O} = \dot{\Psi} \cdot \vec{k}_o \tag{15}$$

$$\vec{\omega}_{F_M/F_O} \times \overrightarrow{MP} = y\dot{\Psi} \cdot \vec{i}_M - (l+x)\dot{\Psi} \cdot \vec{j}_M \tag{16}$$

$$\vec{V}_{M/F_O} = v \cdot \vec{i}_M \tag{17}$$

Therefore, Eq. (14) is transformed to the next equation:

$$\begin{bmatrix} \dot{x} \\ \dot{y} \end{bmatrix} = \begin{bmatrix} -1 & y \\ 0 & -(l+x) \end{bmatrix} \begin{bmatrix} v \\ \dot{\Psi} \end{bmatrix} + \begin{bmatrix} \cos\Psi & \sin\Psi \\ -\sin\Psi & \cos\Psi \end{bmatrix} \begin{bmatrix} \dot{x}_p \\ \dot{y}_p \end{bmatrix} \tag{18}$$

Where the point P is the same as the origin of the fixed frame F_O, it turns to be $\dot{x}_p = \dot{y}_p = 0$. In Eq. (18), that is, $\mathbf{A}$ matrix is expressed by next equations:

$$\mathbf{A}(\mathbf{x}) = \begin{bmatrix} -1 & y \\ 0 & -(l+x) \\ 0 & 1 \end{bmatrix} \tag{19}$$

Eq. (19) shows that the values of the open loop control depend on the initial position $\mathbf{x}$. However, the matrix $\mathbf{A}$ is not square, this system is not controllable. This model should rather be considered as a reference model following control which is a path tracking algorithm.

Ψ_r is the ARV's direction when it is navigating along the recognized path navigation algorithm and Ψ_t is the direction of ARV when it is on a deviated path. The direction error between Ψ_r and Ψ_t is as defined in Eq. (20):

$$\tilde{\Psi} = \Psi_t - \Psi_r \quad (20)$$

The state vector is shown in Eq. (21):

$$\mathbf{x} = \begin{bmatrix} x & y & \tilde{\Psi} \end{bmatrix} \quad (21)$$

$\tilde{v}(t)$ is the velocity error of the ARV between its velocity along the reference path and its velocity deviation from the reference path. In this case, the control input, u is will be the same as in Eq. (22):

$$\mathbf{u} = \begin{bmatrix} \tilde{v} \\ \dot{\tilde{\Psi}} \end{bmatrix} = \begin{bmatrix} v_t - v_r \\ \dot{\Psi}_t - \dot{\Psi}_r \end{bmatrix} \quad (22)$$

The system equation is as follows:

$$\dot{\mathbf{x}} = \mathbf{F}\mathbf{x} + \mathbf{G}\mathbf{u} \quad (23)$$

where,

$$\mathbf{F} = \begin{bmatrix} 0 & \dot{\Psi}_r & \dfrac{v_r \cdot \cos\tilde{\Psi} + l \cdot \dot{\Psi}_r \cdot \sin\tilde{\Psi} - 1}{\tilde{\Psi}} \\ \dot{\Psi}_r & 0 & \dfrac{-v_r \cdot \sin\tilde{\Psi} + l \cdot \dot{\Psi}_r \cdot \cos\tilde{\Psi} - 1}{\tilde{\Psi}} \\ 0 & 0 & 0 \end{bmatrix}$$

$$\mathbf{G} = \begin{bmatrix} -1 & y \\ 0 & -(l+x) \\ 0 & 1 \end{bmatrix}$$

The optimal control method to be used when choosing a control input with which the ARV can track a reference path involves using a quadratic optimal control method. The performance index is selected as in Eq. (24):

$$J = \int_0^{+\infty} \left(\mathbf{X}^T \mathbf{Q} \mathbf{X} + \mathbf{U}^T \mathbf{R} \mathbf{U} \right) dt \quad (24)$$

Control input, which minimizes a performance index, is as follows:

$$\mathbf{U} = \mathbf{K}\mathbf{X} \quad (25)$$

where, $\quad \mathbf{K} = -\mathbf{R}^{-1}\mathbf{G}^T \mathbf{P}$

In Eq. (25), matrix **P** can be found by solving the following Riccati equation:

$$(\mathbf{F} - \mathbf{K}\mathbf{G})^T \mathbf{P} + \mathbf{P}(\mathbf{F} - \mathbf{K}\mathbf{G}) + \mathbf{Q} + \mathbf{K}^T \mathbf{R} \mathbf{K} = \mathbf{0} \quad (26)$$

where,

$$\mathbf{F} = \begin{bmatrix} 0 & 0 & 0 \\ 0 & 0 & -v_r \\ 0 & 0 & 0 \end{bmatrix}, \quad \mathbf{G} = \begin{bmatrix} -1 & 0 \\ 0 & 0 \\ 0 & 1 \end{bmatrix}$$

$$\mathbf{P} = \begin{bmatrix} 1 & 0 & 0 \\ 0 & \dfrac{\sqrt{1+2|v_r|}}{|v_r|} & -sign(v_r) \\ 0 & -sign(v_r) & \sqrt{1+2|v_r|} \end{bmatrix}$$

By substituting a matrix, P, from Eq. (26) into Eq. (25), control gain matrix, K, can be obtained:

$$\mathbf{K} = \begin{bmatrix} 1 & 0 & 0 \\ 0 & sign(v_{A1}) & -\sqrt{1+2|v_r|} \end{bmatrix} \quad (27)$$

The system is not controllable when the AGV is motionless $\left(v_r = \dot{\Psi}_r = 0 \right)$ on a reference path. However the system becomes controllable as soon as the reference ARV starts moving. Using the control gain matrix **K**, the path tracking algorithm is constructed.

V. PATH GENERATION SIMULATION

To verify the performance of the controller of the ARV, it is necessary to determine if it tracks in compliance with the recognized path navigation algorithm. Initially various operational environments are detected by using a scanning algorithm. First, a reference path is generated using sensor data, and then the ARV tracks using a generated algorithm.

Fig. 11 shows the ARV navigation to the path generation after recognition of the path pattern and the ARV tracking the reference path after avoiding an obstacle which the ARV has recognized for size.

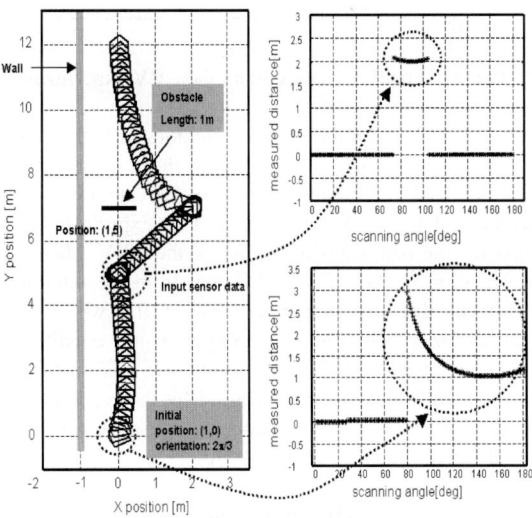

Fig. 11 Simulation of the navigation path algorithm

VI. EXPERIMENT

The control input is the angular velocity in both wheels. Fig. 12 shows that the ARV tracks to the reference path after recognizing the path in accordance with user input.

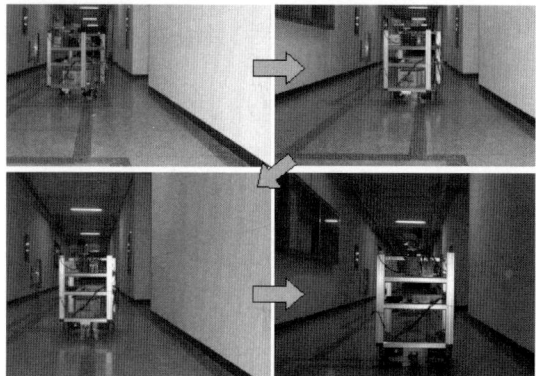

Fig. 12 Photos of the ARV's trajectory tracking the reference path

By using the scanning method of the sensor, the ARV recognizes the path pattern and navigates accordingly.

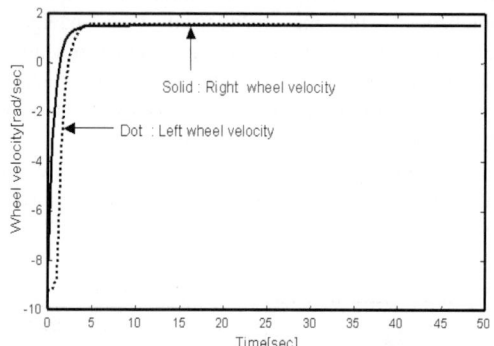

Fig. 13 Wheel velocities about the ARV experiment

Fig. 13 shows the angular velocities of both wheels during convergence of the ARV to the reference path. As the results show, the wheel angular velocity is well controlled for tracking the path. This experiment is a test for the path tracking ability of the ARV when the orientation of the ARV is different from the reference path. In this experiment, control of the wheel was satisfactorily achieved under experimental conditions, but there appeared to be some error. As a road surface is not perfectly flat, the error caused by the road surface can be regarded as the unmodeled errors between the road surface and the ARV's wheels.

VII. CONCLUSIONS

In this paper, a navigation algorithm and controller for the path generation of the two-wheel drive ARV was proposed. Simulations and experiments showed the validity of the control algorithm when applied to real vehicle. The results of the experiments show that the recognitions of a path pattern are constructed by the scanning method using the sensor platform. The path tracking is performed by the optimal controller based on a kinematic modeling. A scanning method using only one sensor showed a recognizing a path pattern adequately. The path generation algorithm was able to confirm through the experiments.

The conclusions of this study are as follows:
1. The path pattern and obstacle magnitude were verified by the sensor platform without the use of many sensors.
2. The ARV tracks to the generated path with the optimal controller.
3. The scanning algorithm and the optimal controller show the variability of the path generation algorithm applying these to constructed the ARV with the sensor platform.

VIII. REFERENCES

[1] C. samson and K.Ait-Abderrahim, "Feedback Control of Nonholonomic Wheeled Cart in Catesian Space", IEEE International Conference on Robotics and Automation, pp. 1136-1141, April 1991.

[2] X. Yun and Nilanjan Sarkar, "Dynamic Feedback Control of Vehicle with Two Steerable Wheels", IEEE International. Conference on Robotics and Automation, pp. 3105-3110, April 1996.

[3] Yongji Wang, J.A. Linnett and J. W. Roberts. "A unified approach to inverse and direct kinematics for four kinds of wheelded mobile robots and its applications", Proceedings of the 1996 IEEE Interenational Conference on Robotics and Automation, Minneapolis, Minnesota-April, pp. 3458-3465, April 1996.

[4] Kanayama, Y. Kimura, Y. Miyazaki and Noguchi T. "A Stable Tracking Control Method for an Autonomous Mobile Robot", Proceedings of the IEEE Conference On Robotics and Automation, pp. 384-389, 1990.

[5] G.L.Miller and E.R. Wagner, " An optical Rangefinder for Autonomous Robot Care Navigation", Proceedings of the SPIE, Vol.851, Mobile Robots II, 132-144, 1987.

[6] Larry Matthies and Steven A. Shafer, " Error Modeling in Stereo Navigation", IEEE Journal of Robotics And Automation. Vol. RA-3, No. 3, pp. 239-248, 1987.

[7] Barry Steer, "Trajectory Planning for a Mobile Robot", International Journal of Robotics Research, Vol. 8, No. 5, pp. 3-14, 1989.

[8] J.H. Jang and C.S. Han, "The State Sensitivity Analysis of the Front Wheel Steering Vehicle: In the Time Domain," KSME International Journal, Vol. 11, No. 6, 1997.

[9] M. Krstic, l. kanellakopoulos and P. Kokotovic, "Nonlinear and Adaptive Control Design," Jon Wiley & Sons, Inc, 1995.

[10] Luis E. Aguilar, T. Hamel and P. Soueres, September 7-11, "Robust Path Following Control for Wheeled Robots via Sliding Mode," Proceedings of the IROS 97, pp. 1389-1395, 1997.

[11] Y. Zhao and M. Reyhanoglu, "Nonlinear Control of Wheeled Mobile Robots", Proceedings of the 1992 IEEE/RSJ International Conference on Intelligent Robots and Systems, pp. 1967-1973, July 7-10 1992.

[12] Koh, K. C. and Cho, H. S, "A Steering Control Method for Wheel-driven Mobile Robot", Korea Automatic Control Conference 1991, pp. 787-792, 1991.

Hybrid Control of Formations of Robots

R. Fierro, A. K. Das, V. Kumar, and J. P. Ostrowski

GRASP Laboratory – University of Pennsylvania, Philadelphia, PA, 19104, USA
{rfierro, aveek, kumar, jpo}@grasp.cis.upenn.edu

Abstract

We describe a framework for controlling a group of nonholonomic mobile robots equipped with range sensors. The vehicles are required to follow a prescribed trajectory while maintaining a desired formation. By using the leader-following approach, we formulate the formation control problem as a hybrid (mode switching) control system. We then develop a decision module that allows the robots to automatically switch between continuous-state control laws to achieve a desired formation shape. The stability properties of the closed-loop hybrid system are studied using Lyapunov theory. We do not use explicit communication between robots; instead we integrate optimal estimation techniques with nonlinear controllers. Simulation and experimental results verify the validity of our approach.

1 Introduction

Research activity in multi-robotic systems has increased substantially in the last few years. Topics include cooperative manipulation [9], multi-robot motion planning, collaborative mapping and exploration [2], software architectures for multi-robotic systems [12], and formation control [6]. Areas of application include, undersea and space exploration, surveillance, target acquisition, and service robotics for mention just a few. Researchers in multi-robotic systems are facing new challenges and open issues that require deeper investigation. For instance, we need to address stability and robustness of multi-agent hybrid systems and develop the methodology and the software that will enable robots to exhibit deliberative and reactive behaviors, and to learn and adapt to unstructured, dynamic environments and new tasks, while providing performance guarantees.

This work considers the problem of formation control. Formation control of multiple autonomous vehicles arises in many scenarios of current interest. For example, in military applications and intelligent vehicle highway systems (IVHS) vehicles need to maneuver while keeping a prescribed formation. To be more specific, we consider a team of n nonholonomic mobile robots that are required to follow a prescribed trajectory while maintaining a desired formation. A robot designated as the *reference* robot follows a trajectory generated by a high-level planner. By using the *leader-following* approach, we split the formation control problem into:

Continuous-state robot control: Control algorithms are designed based on I/O feedback linearization. Each robot can maintain a prescribed separation and bearing from its adjacent neighbors. Explicit inter-robot communication is avoided by using optimal estimation techniques.

Discrete-state formation control: A desired formation is achieved by sequential composition of basic maneuvers (control algorithms). Switching rules are formulated based on sensor constraints.

The paper is organized as follows. In section 2, we provide some mathematical preliminaries and present a brief description of the set of controllers we use in our work. The sequential composition of behaviors and the formation switching strategy are addressed in section 3. Section 4 presents simulation and experimental results. Finally, some concluding remarks and future work ideas are given in section 5.

2 Formation Control

In this section, we describe a formation of n robots as a tuple $\mathcal{F} = (r, \mathcal{H})$ where r is a set of variables describing the relative positions of the robots with respect to the reference robot, and $\mathcal{H}$ is a formation graph describing the control strategy used by each robot. Thus, $\mathcal{F}$ is a dynamical system evolving in continuous-time on the interval $T = [t_0, t_N] \subset \mathbb{R}^+$. The configuration space for $\mathcal{F}$ is $\mathcal{C} = SE(2)^n$.

A formation change can be accomplished by using the compositional control approach introduced in [3]. The main idea is to define a set of controllers $U = \{\xi_1, \ldots, \xi_p\}$ for each robot. Let Φ_j and Ω_j be the domain and goal of controller ξ_j, respectively. It is said that controller ξ_i *prepares* controller ξ_k (denoted $\xi_i \succ \xi_k$) if $\Omega_i \subseteq \Phi_k$. For a given suitably designed set of controllers U, a switching strategy can be found such that the team of robots achieves a desired formation $\mathcal{F}^d$ from any initial formation $\mathcal{F}_0$. Thus, the control problem of formations of robots can be formulated as a hybrid system whose continuous dynamics change in a *controlled* fashion [7]. Let $g \in SE(2)$ denote the reference robot's trajectory. The kinematics of the nonholonomic i-robot are given by

$$\dot{x}_i = v_i \cos\theta_i, \qquad \dot{y}_i = v_i \sin\theta_i, \qquad \dot{\theta}_i = \omega_i \quad (1)$$

In the next subsections we describe briefly three controllers used for formation control purposes. The first two are adopted from [6]. We derived here a third controller that takes into account obstacles.

2.1 Separation Bearing Control

In the *Separation Bearing Controller* (denoted $SB_{ij}C$), robot R_j follows R_i with a desired separation l_{ij}^d and a desired relative bearing ψ_{ij}^d, see Figure 1. The control velocities for the follower are given

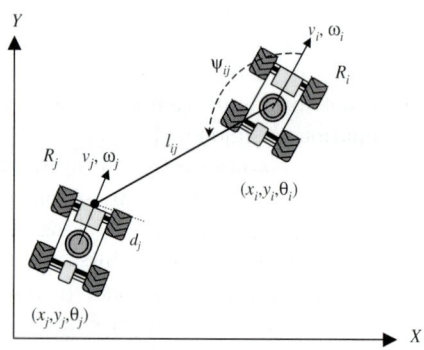

Figure 1: The *Separation Bearing Controller*.

by

$$v_j = s_{ij}\cos\gamma_{ij} - l_{ij}\sin\gamma_{ij}(b_{ij}+\omega_i) + v_i\cos(\theta_i - \theta_j) \quad (2)$$

$$\omega_j = \frac{s_{ij}\sin\gamma_{ij} + l_{ij}\cos\gamma_{ij}(b_{ij}+\omega_i) + v_i\sin(\theta_i-\theta_j)}{d} \quad (3)$$

where

$$\begin{aligned}\gamma_{ij} &= \theta_i + \psi_{ij} - \theta_j, \quad s_{ij} = k_1(l_{ij}^d - l_{ij}),\\ b_{ij} &= k_2(\psi_{ij}^d - \psi_{ij}), \quad k_1, k_2 > 0\end{aligned} \quad (4)$$

The closed-loop linearized system becomes

$$\dot{l}_{ij} = k_1(l_{ij}^d - l_{ij}), \quad \dot{\psi}_{ij} = k_2(\psi_{ij}^d - \psi_{ij}), \quad \dot{\theta}_j = \omega_j \quad (5)$$

2.2 Separation Separation Control

In the *Separation Separation Controller* (denoted $S_{ik}S_{jk}C$), robot R_k follows R_i and R_j with desired separations l_{ik}^d and l_{jk}^d, respectively. See Figure 2. In this case, the control velocities for the follower become

$$v_k = \frac{s_{ik}\sin\gamma_{jk} - s_{jk}\sin\gamma_{ik} + v_i\cos\psi_{ik}\sin\gamma_{jk}}{\sin(\gamma_{jk}-\gamma_{ik})}$$
$$\quad - \frac{v_j\cos\psi_{jk}\sin\gamma_{ik}}{\sin(\gamma_{jk}-\gamma_{ik})} \quad (6)$$

$$\omega_k = \frac{-s_{ik}\cos\gamma_{jk} + s_{jk}\cos\gamma_{ik} - v_i\cos\psi_{ik}\cos\gamma_{jk}}{d\sin(\gamma_{jk}-\gamma_{ik})}$$
$$\quad + \frac{v_j\cos\psi_{jk}\cos\gamma_{ik}}{d\sin(\gamma_{jk}-\gamma_{ik})} \quad (7)$$

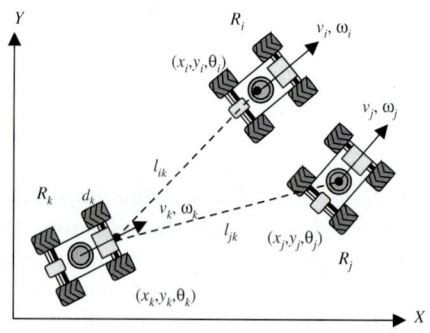

Figure 2: The *Separation Separation Controller*.

The closed-loop linearized system is

$$\dot{l}_{ik} = k_1(l_{ik}^d - l_{ik}), \quad \dot{l}_{jk} = k_1(l_{jk}^d - l_{jk}), \quad \dot{\theta}_k = \omega_k \quad (8)$$

2.3 Separation Distance-To-Obstacle Control

In the *Separation Distance-To-Obstacle Controller* (denoted SD_OC), the outputs of interest are the separation l_{ij} between the follower robot and leader, and the distance δ from an obstacle to the follower. We define a virtual robot R_0, as shown in Figure 3, which moves on the obstacle's boundary with linear velocity v_0 and orientation θ_0. For this case the kinematics of R_j become

$$\begin{aligned}\gamma_{0j} &= \theta_0 - \theta_j \quad (9)\\ \dot{l}_{ij} &= v_j\cos\gamma_{ij} - v_i\cos\psi_{ij} + d\omega_j\sin\gamma_{ij}\\ \dot{\delta} &= v_j\sin\gamma_{0j} - d\omega_j\cos\gamma_{0j}\\ \dot{\theta}_j &= \omega_j\end{aligned}$$

where γ_{ij} is given in (4) and $\delta = \inf \|x_j - x_{Obs}\|$. Feedback I/O linearization is possible as long as

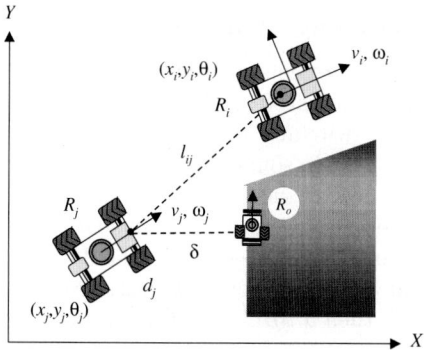

Figure 3: *Separation Distance–To–Obstacle Control*.

$d\cos(\gamma_{0j} - \gamma_{ij}) \neq 0$, *i.e.*, the controller is not defined whether $d = 0$ or $\gamma_{0j} - \gamma_{ij} = \pm k\frac{\pi}{2}$. The latter occurs when vectors $\vec{\delta}$ and $\vec{l}_{ij}$ are collinear. The velocity inputs for R_j are given by

$$v_j = \frac{s_{ij}\cos\gamma_{oj} + s_{oj}\sin\gamma_{ij} + v_i\cos\psi_{ij}\cos\gamma_{oj}}{\cos(\gamma_{oj} - \gamma_{ij})} \tag{10}$$

$$\omega_j = \frac{s_{ij}\sin\gamma_{oj} - s_{oj}\cos\gamma_{ij} + v_i\cos\psi_{ij}\sin\gamma_{oj}}{d\cos(\gamma_{oj} - \gamma_{ij})} \tag{11}$$

Thus, the linearized kinematics become

$$\begin{aligned}\dot{l}_{ij} &= k_1(l_{ij}^d - l_{ij}) \equiv s_{ij} \\ \dot{\delta} &= k_0(\delta_0 - \delta) \equiv s_{0j} \\ \dot{\theta}_j &= \omega_j\end{aligned} \tag{12}$$

where k_1, k_0 are positive controller gains, and δ_0, l_{ij}^d are desired distances to the obstacle and reference robot, respectively.

2.4 Stability Analysis

In this section, we provide stability results for the *SBC* and *SSC*, respectively. Proofs are omitted here due to space constraints. Details are discovered in [8].

Theorem 2.1 *Assume that the reference linear velocity along the trajectory $g(t) \in SE(2)$ is lower bounded i.e., $v_i > V_{\min} > 0$, the reference angular velocity is also bounded i.e., $\|\omega_i\| < W_{\max}$, and the initial relative orientation $\|\theta_i(t_0) - \theta_j(t_0)\| < c_1\pi$ for some positive constant $c_1 < 1$. If the control velocities (2)–(3) are applied to R_j, then system (5) is stable and the output system error of the linearized system converges to zero exponentially.*

□

Theorem 2.2 *Assume that the reference linear velocity along the trajectory $g(t) \in SE(2)$ is lower bounded i.e., $v_i > V_{\min} > 0$, the reference angular velocity is also bounded i.e., $\|\omega_i\| < W_{\max}$, the relative velocity $\delta_v \equiv v_i - v_j$ and orientation $\delta_\theta \equiv \theta_i - \theta_j$ are bounded by small positive numbers ε_1, ε_2, and the initial relative orientation $\|\theta_i(t_0) - \theta_k(t_0)\| < c_2\pi$ for some positive constant $c_2 < 1$. If the control velocities (6)–(7) are applied to R_k, then system (8) is stable and the output system error of the linearized system converges to zero exponentially.*

□

Remarks The two output variables in (5) and (8) converge to the desired values arbitrarily fast (depending on k_1 and k_2). The main difficulty arises in considering the internal dynamics, for instance θ_k in (8), which depends on the controlled velocity ω_k. The orientation error can be expressed as

$$\dot{e}_\theta = \omega_i - \omega_k \tag{13}$$

After some work, we have

$$\dot{e}_\theta = -\frac{v_i}{d}\sin e_\theta + \eta(\boldsymbol{u}, e_\theta) \tag{14}$$

where $\boldsymbol{u}$ is a vector that depends on the output system error and reference angular velocity ω_i. $\eta(\cdot)$ is a nonvanishing perturbation for the *nominal* system in (14) which is (locally) exponentially stable. By using stability of perturbed systems [10], it can be shown that system (14) is stable, thus the stability results in Theorems 2.1 and 2.2 follow.

3 A 3-Robot Formation Control Case

We illustrate our approach using three nonholonomic mobile robots $R_{1,2,3}$ moving in an obstacle-free environment. First, R_1, the *reference* robot, follows a given trajectory $g(t) \in SE(2)$. Second, R_2, the *leader* robot, follows R_1 with $SB_{12}C$. Finally, R_3, the *follower*, has to maintain a specified distance from R_1 and R_2, *i.e.*, $S_{13}S_{23}C$. However, R_3 may change its control behavior depending on its position with respect to R_1 and R_2. Thus, for any arbitrary initial configuration, R_3 may follow R_1 or R_2 with $SB_{13}C$ or $SB_{23}C$. Eventually, R_3 will switch between different control behaviors in order to reach the desired formation. The palette of controllers becomes $U = \{U_2 \cup U_3\}$, and $U_2 = \{SB_{12}C\}$, $U_3 = \{SB_{13}C, SB_{23}C, S_{13}S_{23}C\}$. The finite set of discrete formation modes $Q = \{q_1, q_2, q_3\}$ is illustrated in Figure 4.

Assume that $q_3 \in Q$ is the desired formation $\mathcal{F}^d$, and $\mathcal{F}_0$ is an initial formation. The hybrid system is designed using the compositional control approach outlined in section 2. Let $\{\Phi_1, \Omega_1\}$, $\{\Phi_2, \Omega_2\}$, and $\{\Phi_3, \Omega_3\}$ be the {domain, goal} of $SB_{13}C$, $SB_{23}C$, $S_{13}S_{23}C$, respectively. We design the controllers such that $\Omega_1 \subseteq \Phi_3$ and $\Omega_2 \subseteq \Phi_3$, then $SB_{13}C \succ S_{13}S_{23}C$, similarly $SB_{23}C \succ S_{13}S_{23}C$. In the next section, we formalize this approach by using Lyapunov stability theory to show that under reasonable assumptions $\mathcal{F}^d$ is achieved in a stable manner from any $\mathcal{F}_0$.

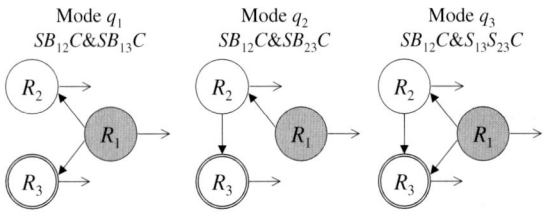

Figure 4: Formation modes for the 3-Robot case.

The closed-loop formation modes are given by
Mode q_1: $SB_{12}C\&SB_{13}C$

$$\begin{aligned} \dot{l}_{12} &= k_1(l_{12}^d - l_{12}) & \dot{l}_{13} &= k_1(l_{13}^d - l_{13}) \\ \dot{\psi}_{12} &= k_2(\psi_{12}^d - \psi_{12}) & \dot{\psi}_{13} &= k_2(\psi_{13}^d - \psi_{13}) \\ \dot{\theta}_2 &= \omega_2 & \dot{\theta}_3 &= \omega_3 \end{aligned} \quad (15)$$

Mode q_2: $SB_{12}C\&SB_{23}C$

$$\begin{aligned} \dot{l}_{12} &= k_1(l_{12}^d - l_{12}) & \dot{l}_{23} &= k_1(l_{23}^d - l_{23}) \\ \dot{\psi}_{12} &= k_2(\psi_{12}^d - \psi_{12}) & \dot{\psi}_{23} &= k_2(\psi_{23}^d - \psi_{23}) \\ \dot{\theta}_2 &= \omega_2 & \dot{\theta}_3 &= \omega_3 \end{aligned} \quad (16)$$

Mode q_3: $SB_{12}C\&S_{13}S_{23}C$

$$\begin{aligned} \dot{l}_{12} &= k_1(l_{12}^d - l_{12}) & \dot{l}_{13} &= k_1(l_{13}^d - l_{13}) \\ \dot{\psi}_{12} &= k_2(\psi_{12}^d - \psi_{12}) & \dot{l}_{23} &= k_1(l_{23}^d - l_{23}) \\ \dot{\theta}_2 &= \omega_2 & \dot{\theta}_3 &= \omega_3 \end{aligned} \quad (17)$$

Actually, the gains k_1 and k_2 can be different in each mode. For simplicity, we use the same values in our simulations and experiments. Let the system error be defined as

$$\begin{aligned} e_1 &= l_{13}^d - l_{13}, & e_2 &= \psi_{13}^d - \psi_{13}, & e_3 &= \theta_1 - \theta_3 \\ e_4 &= l_{23}^d - l_{23}, & e_5 &= \psi_{23}^d - \psi_{23}, & e_6 &= \theta_2 - \theta_3 \\ e_7 &= l_{12}^d - l_{12}, & e_8 &= \psi_{12}^d - \psi_{12}, & e_9 &= \theta_1 - \theta_2 \end{aligned}$$

For every mode, we have $e_{ijk} \equiv [e_i\, e_j\, e_k]^T$ where e_{ij} and e_k correspond to the outputs of interest and the internal dynamics, respectively. Moreover, if the assumptions in theorems 2.1 and 2.2 hold, then each formation mode (15)–(17) is stable. Now, we need to prove that for a given switching strategy S_w, the hybrid system is stable, i.e., given any initial formation $\mathcal{F}_0$, a desired $\mathcal{F}^d$ is achieved in finite time.

3.1 Switching Strategy

Our robots are equipped with an on-board omni-directional vision system. The sensor constraints determine the switching sequence S_w. R_3 may detect R_1, R_2 or both. In some cases, neither R_1 nor R_2 are within the field of view of R_3. Figure 5 depicts the switching boundaries in Cartesian space. Notice the triangle inequality $l_{ik} + l_{jk} > l_{ij}$ should be satisfied. If R_i with $i = 1, 2, 3$ were collinear, SSC would not be defined, then a SBC should be utilized.

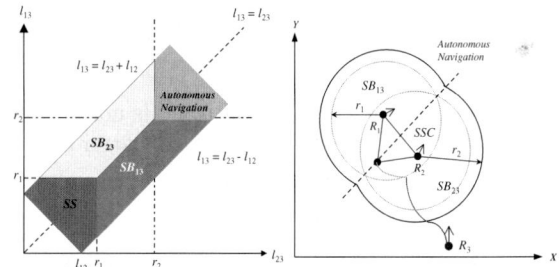

Figure 5: Switching boundaries based on sensor constraints.

The formation control objective is to drive R_3 to a region where it can detect both R_1 and R_2. Thus, the switching control strategy for R_3 can be summarized as follows

If $(l_{13} < l_{23})\&(l_{23} > r_1)\&(l_{13} < r_2)$ Then $SB_{13}C$
If $(l_{13} > l_{23})\&(l_{13} > r_1)\&(l_{23} < r_2)$ Then $SB_{23}C$
If $(l_{13} < r_1)\&(l_{23} < r_1)$ Then $S_{13}S_{23}C$
If $(l_{13} > r_2)\&(l_{23} > r_2)$ Then $AutonNavig$

3.2 Stability Analysis

Since a palette of controllers and a switching strategy are given, we need to *verify* that the hybrid system is stable provided that each mode shares a common equilibrium point $\boldsymbol{x}_0 \in \Omega_3$. One way to solve this verification problem is to find a common Lyapunov function, thus the switched system is stable for any arbitrary fast switching sequence. This is in general a difficult task. A number of approaches have been proposed in the literature to confront this problem (see [11] and the references therein). In our 3-robot formation example, it turns out that under some reasonable assumptions, there may exist a common Lyapunov function. Therefore, the equilibrium point is stable, and the system error of the desired formation mode converges to zero. However, the property of exponential convergence is lost in the switching process. Let $\bar{V}_3(\boldsymbol{e}) = V_3 + V_{12}$ be a Lyapunov function candidate for the desired formation

$\mathcal{F}_3$ in (17), and

$$V_3 = \frac{1}{2}\left[e_1^2 + e_4^2 + e_3^2\right], \quad V_{12} = \frac{1}{2}\left[e_7^2 + e_8^2 + e_9^2\right] \quad (18)$$

V_{12} is a Lyapunov function candidate for subsystem $SB_{12}C$ i.e., R_2 follows R_1 using a separation-bearing controller. If the assumptions in theorem 2.1 are satisfied, then $\dot{V}_{12} \leq 0$. Moreover, if the assumptions in theorem 2.2 are satisfied for subsystem $S_{13}S_{23}C$, then $\dot{V}_3 \leq 0$. We impose an additional constraint on our hybrid system which is R_2 has already reached its equilibrium point. Thus, we only need to consider V_3 in (18) for studying the stability of the switched system $\mathcal{F}_q$. By definition V_3 is a Lyapunov function for $\mathcal{F}_3$. We would like to show that, V_3 is also a Lyapunov function for $\mathcal{F}_1$ and $\mathcal{F}_2$.

Let us consider formation mode $\mathcal{F}_1$. $SB_{13}C$ makes $e_1 \to 0$ and $e_2 \to 0$ exponentially as $t \to \infty$. But we need to show that $\dot{V}_4 = e_4\dot{e}_4 \leq 0$ or $(l_{23}^d - l_{23})\dot{l}_{23} \geq 0$. The main idea here is to pick ψ_{13}^d such that $l_{23} \to l_{23}^d$ as $e_2 \to 0$. Then, we have

$$\psi_{13}^d = \cos^{-1}\left(\frac{l_{12}^{d\,2} + l_{13}^{d\,2} - l_{23}^{d\,2}}{2l_{12}^d l_{13}^d}\right) + \psi_{12}^d \quad (19)$$

Using the inequality constraint imposed by the geometry of the problem i.e., $l_{23}^d < l_{12}^d + l_{13}^d$, it is easy to show that $\dot{V}_4 = e_4\dot{e}_4 \leq 0$. Then V_3 is a Lyapunov function for $\mathcal{F}_1$ (similarly for $\mathcal{F}_2$).

It is well known that Lyapunov methods provide conservative stability regions, since we always consider the worst case. Simulation results reveal that the desired formation is achieved even when some of the assumptions discussed here are not satisfied e.g., positions and orientation of R_2 and R_3 are randomly initialized.

4 Simulation and Experimental Results

We simulate the switching strategy outlined in section 3.1. As it can be seen in Figure 6, after some mode switching and obstacle avoidance the 3-Robot system reaches the desired formation. The parameters for simulation are: $v_1 = 0.5$ m/s, R_1: $(0, 0, 30°)$, R_2: $(1.5, 0, 0°)$, R_3: $(0.2, 2, 30°)$, $\omega_1 = 0.1\sin(0.2t)$, $l_{12}^d = l_{13}^d = l_{23}^d = 1$m, and $\psi_{12}^d = 90°$.

4.1 The Experimental Setup

The mobile robots we use for the experiments are shown in Figure 7. Each robot has an onboard omnidirectional vision system, a wireless video transmitter, and a battery pack. The receiver (located at the host NT computer) feeds the signal to a frame grabber that is able to capture video at full frame rate (30 Hz.) for image processing.

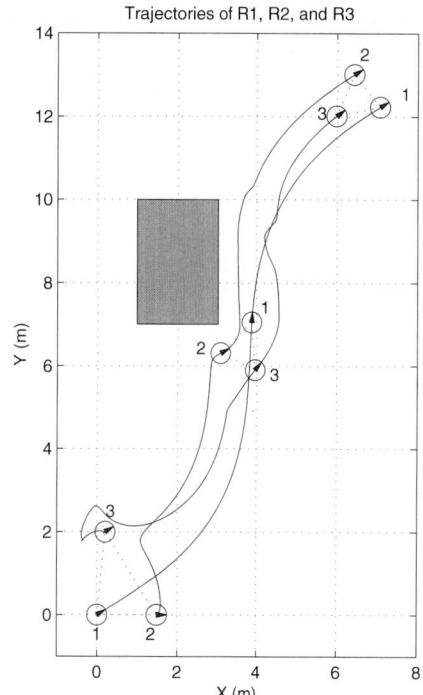

Figure 6: 3-Robot case formation control.

Figure 7: The experimental setup.

The formation controllers described here require reliable estimation of the linear velocity $v_i(t)$ and angular velocity $\omega_i(t)$ of the leader robot R_i, and relative orientation $(\theta_i - \theta_j)$. The omni-directional vision system provides the range ρ_{ij} and the angle β_{ij} of the observed leader. This information is used by the *velocity estimator* that is based on an extended Kalman filter [4]. Control velocities for the follower robot are computed and sent to the driving and steering servomotors. Figure 8 presents experimental results for the separation bearing control (*SBC*). The desired separation and bearing are $l_{ij}^d = 0.6$ m and $\psi_{ij}^d = 180°$, respectively. The reference robot follows a circular path. The robustness of the system is verified when we manually hold the follower for a few seconds at $t \approx 65$ s.

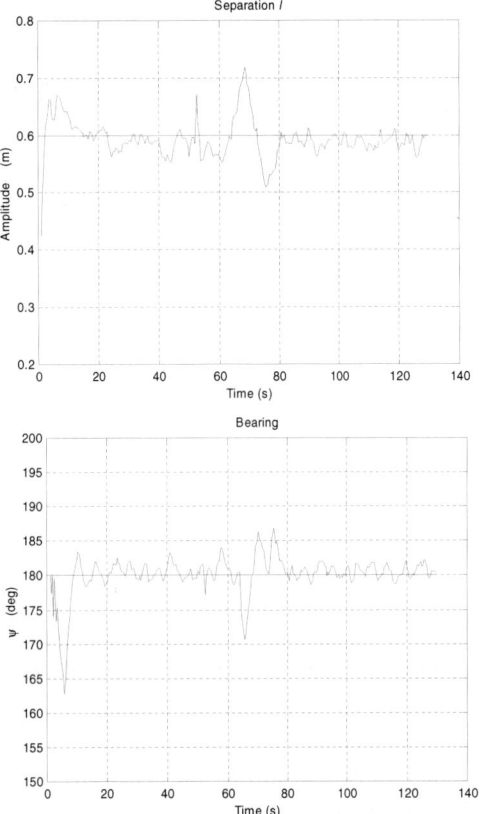

Figure 8: Measured separation and bearing.

Acknowledgements

Research Supported by the DARPA ITO MARS Program, Grant No. 130-1303-4-534328-xxxx-2000-0000

5 Conclusions

In this paper, we have presented a hybrid system approach for formation control. We have designed a suite of controllers for leader following and obstacle avoidance. These individual controllers are sequentially composed in order to achieve a desired formation. Simulation and experimental results verify the validity of our approach. Velocity estimation techniques based on an EKF have been integrated in the closed loop system. Estimation of leader's velocities is required, since there is no inter-agent communication. Experiments are being extended to more complex scenarios where robots need to exhibit a variety of behaviors such as localization, target acquisition, collaborative mapping and formation keeping. The controllers presented in this work are valid for $SE(2)$. Currently, we are investigating similar controllers for $SE(3)$.

References

[1] T. Balch, "Social potentials for scalable multi-robot formations", *Proc. IEEE Int. Conf. Robot & Automat.*, San Francisco, CA, April 2000, pp. 73–80.

[2] W. Burgard, M. Moors, D. Fox, R. Simmons, and S. Thrun, "Collaborative multi-robot exploration," *Proc. IEEE Int. Conf. Robot. and Automat.*, San Francisco, CA, April 2000, pp. 476–481.

[3] R. Burridge, A. Rizzi, and D. Koditschek, "Sequential composition of dynamically dexterous robot behaviors," *Int. J. Robot. Research*, vol. 18, no. 6, pp. 534–555, June 1999.

[4] A. K. Das, R. Fierro, V. Kumar, B. Southall, J. Spletzer, and C. Taylor, "Real-time vision-based control of a nonholonomic mobile robot," *Proc. IEEE Int. Conf. Robot. and Automat.*, Seoul, Korea, May 2001.

[5] A. De Luca, G. Oriolo, and C. Samson, "Feedback control of a nonholonomic car-like robot," in *Robot Motion Planning and Control*, J.-P. Laumond (ed.), London: Springer-Verlag, 1998, pp. 171–253.

[6] J. Desai, J. P. Ostrowski, and V. Kumar, "Controlling formations of multiple mobile robots," *Proc. IEEE Int. Conf. Robot Autom.*, Leuven, Belgium, May 1998, pp. 2864–2869.

[7] R. Fierro and F. L. Lewis, "A framework for hybrid control design," *IEEE Trans. Syst., Man, Cyber.*, vol. 27-A, no. 6, pp. 765–773, Nov. 1997.

[8] R. Fierro, P. Song, A. K. Das, and V. Kumar, "Cooperative control of robot formations," Submitted to Cooperative Control and Optimization Series, Kluwer, 2001.

[9] W. Kang, N. Xi, and A. Sparks, "Formation control of autonomous agents in *3D* workspace," *Proc. IEEE Int. Conf. Robot. Automat.*, San Francisco, CA, April 2000, pp. 1755-1760.

[10] H. K. Khalil, *Nonlinear Systems*, Upper Saddle River, NJ: Prentice Hall, 2nd ed., 1996.

[11] D. Liberzon and A. S. Morse, "Basic problems in stability and design of switched systems," *IEEE Control Systems*, vol. 19, no. 5, pp. 59–70. Oct. 1999.

[12] L. E. Parker, "Current state of the art in distributed robot systems," *Distributed Autonomous Robotic Systems 4*, L. E. Parker, G. Bekey, and J. Barhen (eds.), Springer, pp. 3–12, 2000.

[13] D. Stilwell and B. Bishop, "A framework for decentralized control of autonomous vehicles," *Proc. IEEE Int. Conf. Robot. Automat.*, San Francisco, CA, April 2000, pp. 2358–2363.

[14] H. Yamaguchi and J. W. Burdick, "Asymptotic stabilization of multiple nonholonomic mobile robots forming groups formations," *Proc. IEEE Int. Conf. Robot. Automat.*, Leuven, Belgium, May 1998, pp. 3573–3580.

Tracking Control of a Mobile Robot using a Neural Dynamics based Approach*

Guangfeng Yuan, Simon X. Yang† and Gauri S. Mittal
School of Engineering, University of Guelph
Guelph, Ontario, N1G 2W1, Canada

Abstract

In this paper, a novel tacking control approach is proposed for real-time navigation of a nonholonomic mobile robot. The proposed tracking controller is based on the error dynamics analysis of the mobile robot and a neural dynamics model derived from Hodgkin and Huxley's membrane model of a biological system. The stability of the control system and the convergence of tracking errors to zeros are guaranteed by a Lyapunov stability theory. Unlike many tracking control methods for mobile robot where the generated control velocities start with large initial velocities, the proposed neural dynamics based approach is capable of generating smooth, continuous robot control signals with zero initial velocities. In addition, it can deal with the situation with a very large tracking error. The effectiveness and efficiency are demonstrated by comparison and simulation studies.

1 Introduction

Real-time tracking control of a mobile robot is a very important issue in mobile robotics. Due to slippage, disturbance, noise, vehicle-terrain interaction and sensor errors, it is very difficulty to avoid the errors between the desired and actual robot paths. How to effectively control a mobile robot to precisely track a desired trajectory is still an open question in robotics. In a 2-dimensional (2D) Cartesian workspace, the location of a mobile robot can be uniquely determined by three variables, the spatial position (x_c, y_c) of the robot center C and its orientation θ_c with respect to C, which is referred as a *posture* of the mobile robot (see Fig. 1).

There has been many studies on tracking control of a mobile robot in recent years (e. g., [1]-[8]). The existing tracking control methods for a mobile robot can be classified into five categories: (1) sliding mode [1]; (2) linearization [2]; (3) backstepping [4, 5], (4) neural networks [6]; (5) fuzzy systems [7]; and (6) neuro-fuzzy systems [8]. The control algorithm using sliding mode is complicated and computationally expensive. The generated velocity command with respect to time is not a smooth curve [1], which may lead to discontinuousness in the robot velocities. The linearization based methods (e.g., [2]) requires a small initial error between the target and actual robot positions. The backstepping based tracking controllers (e.g., [4, 5]) are the most commonly used approach. They are very simple and the system stability is guaranteed by a Lyapunov stability theory. In addition, some of the backstepping based controllers can deal with arbitrarily large initial error. However, the generated robot velocity commands using those conventional control approaches start with a very large value, and suffers from velocity jumps when sudden tracking errors occur, i.e., the required accelerations and forces/torques are infinitely large at the velocity jump points, which is not practically possible. Fierro and Lewis [3] proposed a novel controller based on backstepping technique for a mobile robot by generating torque signals using computed torque control or a three-layer neural network based control, which can solve the impractical problem of large initial velocities. But both control methods are computationally complicated. In addition, the computed torque control required the exact robot model that mostly is not available, while the neural network require on-line learning in order to make the robot perform properly. Recently Zhang et al. [4] proposed a controller based on backstepping and neural network, where the backstepping is used for tracking control, while the neural network is for compensating the robot dynamics. However, the mobile robot also starts with a very large initial velocity, and the algorithm is computationally expensive. The fuzzy rules based tracking control approaches (e.g., [7, 8]) can solve the problem of large initial robot velocities,

---

*This work was supported by Natural Sciences and Engineering Research Council (NSERC) of Canada.
†All correspondence should be addressed to S. X. Yang. Email: syang@uoguelph.ca

but it is very difficulty to formulate the fuzzy rules, which are usually obtained by trial and error. The existing neural networks based tracking control algorithms (e.g., [3, 6, 8]) for a mobile robot require either on-line and/or off-line training procedures before the controllers are capable of controlling the robot properly, which also add computational cost.

In this paper, inspired by the unique features of the neural dynamics in Hodgkin and Huxley's membrane model [9] for a biological neural system, and based on the error dynamics analysis of a mobile robot that is similar to the integrator backstepping, a novel tracking controller is proposed for real-time navigation of a mobile robot. The control signal consists of a component from a neural dynamics model, and a component from the error dynamics that is similar to a proportional (P) control part. The stability of the control system and the asymptotical convergence of tracking errors to zero are rigorously proved using a Lyapunov stability analysis. Distinct from the previous neural networks based approaches, no learning procedure is needed in the proposed control algorithm. The proposed controller is capable of generating smooth, continuous control signals with *zero* initial robot velocities. In addition, it can deal with arbitrarily large tracking errors between the target and current robot postures. To the best of our knowledge, it is the first time that a neural dynamics based model is proposed for tracking control of a mobile robot, where no learning is needed.

2 Model Algorithm

2.1 Nonholonomic Mobile Robot and its Error Dynamics

In a 2D workspace, a nonholonomic mobile robot can be described by two coordinate systems: the world coordinate system $\{X, 0, Y\}$ and the local coordinate system $\{D, C, L\}$, where D is the driving direction (longitudinal direction), L is the lateral direction (latitudinal direction), and C is the robot center point. The world coordinate system is fixed to the Cartesian workspace and the local coordinate system is attached to the mobile platform. A robot posture in the world coordinate system can be *uniquely* determined by a vector $P_c = [x_c\ y_c\ \theta_c]^T$, where (x_c, y_c) denotes the spatial position of the robot center C, and θ_c is the robot orientation angle with respect to C (see Fig. 1).

A freely movable mobile robot that is referred as holonomic mobile robot has three degrees of freedom (d.o.f.), x_c, y_c, and θ_c. However, because of the kine-

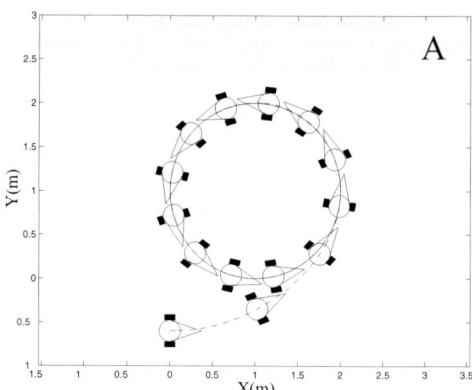

Figure 1: *Model of a nonholonomic mobile robot*

matical constraint, the degrees of freedom for a nonholonomic mobile robot reduces to two. On the conditions of non-slipping, the kinematic constraint of a nonholonomic mobile robot is given as

$$\dot{y}_c \cos\theta_c - \dot{x}_c \sin\theta_c = 0. \qquad (1)$$

From the motion control perspective, a mobile robot has 2-d.o.f., v_c and w_c, where v_c is the linear velocity and w_c is the angular velocity of the mobile robot. For a nonholonomic mobile robot, the relationship between velocity in the world coordinate system $\dot{P}_c = [\dot{x}_c\ \dot{y}_c\ \dot{\theta}_c]^T$ and velocity $v = [v_c\ w_c]^T$ in the local coordinate system can be described by a Jacobian matrix as

$$\begin{bmatrix} \dot{x}_c \\ \dot{y}_c \\ \dot{\theta}_c \end{bmatrix} = \dot{P}_c = \begin{bmatrix} \cos\theta_c & 0 \\ \sin\theta_c & 0 \\ 0 & 1 \end{bmatrix} \begin{bmatrix} v_c \\ w_c \end{bmatrix}. \qquad (2)$$

The reference path of a nonholonomic mobile robot provides the target robot posture $P_d = [x_d\ y_d\ \theta_d]^T$. in the world coordinate system. The tracking error in the local coordinate system is defined as $E_p = [e_D\ e_L\ e_\theta]^T$, where e_D, e_L and e_θ are the errors in the driving direction, lateral direction and the orientation, respectively. The relationship between the tracking errors in the world and local coordinate systems can be obtained by geometrical projection transformation as

$$E_p = \begin{bmatrix} e_D \\ e_L \\ e_\theta \end{bmatrix} = \begin{bmatrix} \cos\theta_c & \sin\theta_c & 0 \\ -\sin\theta_c & \cos\theta_c & 0 \\ 0 & 0 & 1 \end{bmatrix} \begin{bmatrix} x_d - x_c \\ y_d - y_c \\ \theta_d - \theta_c \end{bmatrix}. \qquad (3)$$

The error dynamics of the mobile robot can be derived from the time derivative of the above posture error equation in (3) as

$$\begin{bmatrix} \dot{e}_D \\ \dot{e}_L \\ \dot{e}_\theta \end{bmatrix} = \begin{bmatrix} w_c e_L - v_c + v_d \cos e_\theta \\ -w_c e_D + v_d \sin e_\theta \\ w_d - w_c \end{bmatrix}. \qquad (4)$$

2.2 Tracking Control Problem

The function of a tracking controller in this paper is to implement a mapping between the known information (e.g., the desired information and the sensory information) and the velocity commands designed to achieve the robot's task. The controller design problem can be described as: given the desired robot posture $P_d(t) = [x_d(t)\ y_d(t)\ \theta_d(t)]^T$ and the desired velocities $v_d(t)$ and $w_d(t)$, i.e., the desired state of the mobile robot defined as

$$X_d(t) = [x_d(t)\ y_d(t)\ \theta_d(t)\ v_d(t)\ w_d(t)]^T, \quad (5)$$

design a control law for the linear velocity v_c and angular velocity w_c, which drive the robot to move, such that the actual robot state

$$X_c(t) = [x_c(t)\ y_c(t)\ \theta_c(t)\ v_c(t)\ w_c(t)]^T, \quad (6)$$

will precisely tracking the desired robot state $X_d(t)$,

$$\lim_{t \to \infty} X_c(t) = X_d(t), \quad (7)$$

i.e., the tracking error converges to zero while time approaches infinitely.

2.3 The Proposed Controller

The biological neural model has a certain property that can be used to solve the problem of sudden speed jumps. A typical shunting model derived from the Hodgkin and Hulexy's [9] membrane model is given as

$$\frac{dy_i}{dt} = -A_i y_i + (B_i - y_i)S_i^e(t) - (D_i + y_i)S_i^i(t), \quad (8)$$

where i is the neuron index, y_i is the membrane potential, A_i represents the passive decay rate, B_i and D_i are the upper and lower bounds of the membrane potential, and S_i^e and S_i^i are excitatory and inhibitory inputs to the ith neuron.

By analyzing the backstepping technique based tracking controller proposed in [3, 4, 5] the sharp speed jumps are caused by the suddenly changes in tracking errors. So a biological inspired tracking controller is proposed to solve the problem. Substituting $A_i = A$, $B_i = B$, $D_i = D$, $y_i = v_s$ $S_i^+ = f(e_D)$, $S_i^- = g(e_D)$ in Eq. (8), a velocity dynamics equation with respect to the error in the driving direction is obtained as

$$\frac{dv_s}{dt} = -A v_s + (B - v_s)f(e_D) - (D + v_s)g(e_D), \quad (9)$$

where v_s is the velocity component that will be used in tracking controller design, A is the passive rate of the velocity, B and D are the upper boundary and lower boundary of the velocity, respectively, Functions $f(x)$ is a linear-above-threshold function defined as $f(x) = \max\{x, 0\}$ and the non-linear function $g(x)$ is defined as $g(x) = \max\{-x, 0\}$. The proposed tracking control law for the linear and angular velocities are given as

$$v_c = v_s + v_d \cos e_\theta, \quad (10)$$
$$w_c = w_d + k_2 v_d e_L + k_3 v_d \sin e_\theta, \quad (11)$$

where the k_2 and k_3 are positive parameters. At the start period, we need that the speed of the mobile robot increases exponentially and reaches the desired speed v_0. According to the time response property of the first order system, the reference velocity can be defined as $v_d(t) = v_0(1 - e^{-t/\tau})$, where v_0 is the desired speed, τ is the time constant.

When the proposed tracking controller is used together with a path planner, the output of the path planner is the desired robot posture P_d. The current posture can be obtained from measurement. The system architecture of the proposed tracking controller is shown in Fig. 2. The error vector E_p in the local coordinate systems is obtained through a a transformation matrix T_e from the posture error between the current and desired postures in the world coordinate system. The input of the path tracker is the error vector and the desired velocities. The output of the path tracker is the steering commands in linear velocity v_c and angular velocity w_c.

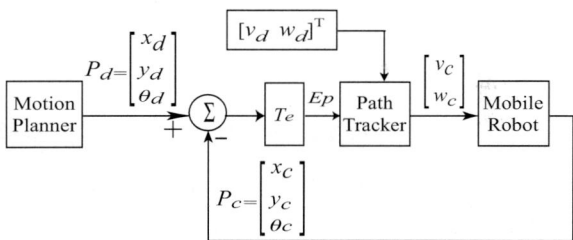

Figure 2: *System architecture of the proposed tracking controller*

2.4 Stability Analysis

The tracking control system proposed in this paper is asymptotically stable, and the tracking errors converge to zeros. In the shunting equation (9), the excitatory input item $(B - v_s)f(e_D)$ forces the output of the shunting model to stay below the upper bound B while the inhibitory input item $(D + v_s)g(e_D)$ guarantees that the output of the shunting model stay above lower bound $-D$. Therefore, the output of the shunting model is bounded in the finite interval $[-D, B]$. To rigorously prove the asymptotical stability and the convergence of the proposed tracking control system, a Lyapunov function candidate is chosen as,

$$V(t) = \frac{1}{2}(e_D^2 + e_L^2) + \frac{1}{k_2}(1 - \cos e_\theta)$$
$$+ \frac{1}{2k}(e_v^2 + e_w^2) + \frac{1}{2B}v_s^2, \quad (12)$$

where v_s is the velocity component defined in Eq. (9), and e_v and e_w are the auxiliary velocity errors that will be defined later in this section. It is obvious that $V(t) = 0$ if and only if $e_D = 0, e_L = 0, e_\theta = 0, e_v = 0, e_w = 0$, and $v_s = 0$.

From Eq.s (4), (9), (10), and (11), the time derivative of the Lyapunov function L becomes

$$\dot{V}(t) = \dot{e}_D e_D + \dot{e}_V e_V + \frac{1}{k_2}\dot{e}_\theta \sin e_\theta$$
$$+ \frac{1}{k}(\dot{e}_v e_v + \dot{e}_w e_w) + \frac{1}{B}\dot{v}_s v_s$$
$$= -v_s e_D - \frac{k_3}{k_2}v_d \sin^2 e_\theta + \frac{1}{k}(\dot{e}_v e_v + \dot{e}_w e_w)$$
$$+ \frac{1}{B}[-A - f(e_D) - g(e_D)]v_s^2$$
$$+ \frac{1}{B}[Bf(e_D) - Dg(e_D)]v_s, \quad (13)$$

If we choose the constants $B = D$ in the shunting equation, Eq. (13) can be rewritten as

$$\dot{V}(t) = -\frac{k_3}{k_2}v_d \sin^2 e_\theta + \frac{1}{k}(\dot{e}_v e_v + \dot{e}_w e_w)$$
$$+ \frac{1}{B}[-A - f(e_D) - g(e_D)]v_s^2$$
$$+ [f(e_D) - g(e_D) - e_D]v_s. \quad (14)$$

From to the definition of $f(e_D)$ and $g(e_D)$, if $e_D \geq 0$, then $f(e_D) = e_D$ and $g(e_D) = 0$. Thus we have,

$$[f(e_D) - g(e_D) - e_D]v_s = [e_D - 0 - e_D] = 0. \quad (15)$$

Similarly, if $e_D < 0$, then $f(e_D) = 0$ and $g(e_D) = -e_D$. Thus we have

$$[f(e_D) - g(e_D) - e_D]v_s = [0 - (-e_D) - e_D] = 0. \quad (16)$$

Therefore, Eq. (14) can be rewritten as

$$\dot{V}(t) = -\frac{k_3}{k_2}v_d \sin^2 e_\theta + \frac{1}{k}(\dot{e}_v e_v + \dot{e}_w e_w)$$
$$+ \frac{1}{B}[-A - f(e_D) - g(e_D)]v_s^2 \quad (17)$$

To prove the convergence of the robot velocities, the auxiliary velocity error variables, e_v and e_w in Eq. (12), are define as

$$e_v = v_c - v_d, \quad (18)$$
$$e_w = w_c - w_d. \quad (19)$$

Considering the nonlinear feedback acceleration control input as

$$a_v = \dot{v}_d + k(v_d - v_c), \quad (20)$$
$$a_w = \dot{w}_d + k(w_d - w_c), \quad (21)$$

where k is a positive parameter, and a_v and a_w are the actual accelerations of the robot, i.e., namely, $a_v = \dot{v}_c$, $a_w = \dot{w}_c$. Note that if the second term on the right hand of Eq.s (20) and (21) is crossed out, the control law is usually called perfect velocity tracking, which is not possible for a real robot. By doing time-derivative on both sides of Eq.s (18) and (19), and using Eq.s (20) and (21), we have

$$\dot{e}_v = \dot{v}_c - \dot{v}_d = a_v - \dot{v}_d = -ke_v, \quad (22)$$
$$\dot{e}_w = \dot{w}_c - \dot{w}_d = a_w - \dot{w}_d = -ke_w. \quad (23)$$

By substituting Eq. (22) and (23) into Eq. (17), we have

$$\dot{V} = -\frac{k_3}{k_2}v_d \sin^2 e_\theta - (e_v^2 + e_w^2)$$
$$+ \frac{1}{B}[-A - f(e_D) - g(e_D)]v_s^2. \quad (24)$$

Obviously we have $-(e_v^2 + e_w^2) \leq 0$. Since parameters k_2 and k_3 are positive numbers, and the desired linear velocity v_d is a positive constant, we have $-(k_3/k_2)v_d \sin^2 e_\theta \leq 0$. From the definition of functions $f(e_D)$ and $g(e_D)$, we have $f(e_D) \geq 0$ and $g(e_D) \geq 0$. In addition, the parameters A and B are nonnegative constants. Thus we can $[-A - f(e_D) - g(e_D)]v_s^2/B \leq 0$. Therefore, the time derivative of the Lyapunov function candidate $\dot{V}$ in Eq. (24) along all the system trajectories is not larger than zero, i.e. $\dot{V} \leq 0$. The proposed tracking control system for a mobile robot is stable.

The velocity error dynamic equations for mobile robot satisfy the following relationship: $v_c \to v_d$, $w_c \to w_d$ as $t \to \infty$. Obviously, by assuming that $v_d > 0$, $\dot{V}(t) \leq 0$ and the entire error E_p, e_v, e_w, v_s are bounded. From Eq.s (4), (18), (19) and $\dot{V}(t) \leq 0$, we can infer that $\|E_p\|$, $\|\dot{E}_p\|$, $\|e_v\|$, $\|\dot{e}_v\|$, $\|e_w\|$, $\|\dot{e}_w\|$, $\|v_s\|$ and $\|\dot{v}_s\|$ are bounded. Thus we have $\|\ddot{V}(t)\| < \infty$. Since $V(t)$ does not increase and converges to certain constant value. By Barbalat's lemma, $\dot{V}(t) \to 0$ as $t \to \infty$, from which we can deduce that $e_v \to 0, e_w \to 0, v_s \to 0$ as $t \to \infty$. By using Eq. (9) and the input-output property of the shunting model, we can infer that if output converges to some constant value (zero), the input is supposed to go to a constant value (zero), namely, $e_D \to 0$ as $v_s \to 0$. From the first term in Eq. (24), $\frac{k_3}{k_2}v_d \sin^2 e_\theta = 0$, we have $e_\theta \to 0$ as $t \to \infty$. From Eq. (4), as $e_\theta \to 0$, we have $w_d - w_c = 0$. From Eq. (11), $e_w \to 0$ as $t \to \infty$, thus $k_2 v_d e_v = 0$. From the assumption $v_d > 0$, we have $e_V \to 0$ as $t \to \infty$. Thus the equilibrium point is $E_p = 0, e_v = 0, e_w = 0$. Therefore, the tracking control system is asymptotically stable. From the Jacobean transformation in Eq. (2), it is obvious that $x_d - x_c = 0$ and $y_d - y_c = 0$ from the results: $e_D = 0$ and $e_L = 0$. Therefore, we have

$$\lim_{t \to \infty}(X_d(t) - X_c(t)) = 0 \qquad (25)$$

Therefore, the proposed control algorithm is asymptotically stable and the tracking errors are guaranteed to converge to zeros.

3 Simulation Studies

3.1 Tracking a Straight Path

The desired robot path is a straight line described as $y = 5$ and $x = 0$; the robot starts at $(0, 3.2, 0)$. The desired cart is supposed to forward speed set out at point $(0, 5, 0)$. So the initial error is $(0, 1.8, 0)$. Time varies from 0 to 10s; the parameters are chosen as: $k_2 = 5$, $k_3 = 2$, $A = 5$, $B = 3$, $D = 3$, $v_d = 1$, $w_d = 0$. During the beginning period, the parameters used in the first order exponent equation for linear speed are as: $k = 1$, $\tau = 0.5$. The sampling time is 0.01 second. The actual path is denoted with dash dot line and the desired path is the solid line as shown in Fig. 3A. The smooth and reasonable variation of speeds is demonstrated in Fig. 3B. The errors of longitude, lateral and orientation converge to zero as time approaches to infinity shown in Fig. 3C. Fig. 3D shows the change of the orientation of the mobile robot.

3.2 Tracking a Circular Path

Consider a circular path, $x^2 + y^2 = 1$, as shown in Fig. 4, and assume that the desired linear and angular velocities are 1 and 1 respectively. The desired virtual car proceeds at such velocities and sets out from point $(1, 0, 0)$. The actual robot starts at point $(0, -.6, 0)$. this means that the initial error is $(-1, -0.6, 0)$. During the beginning period, the parameters used in

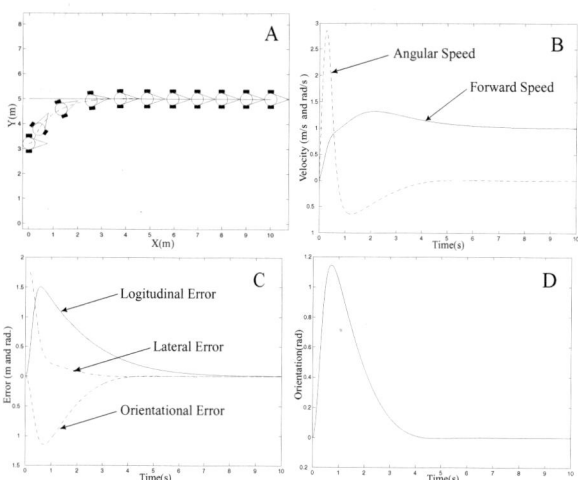

Figure 3: *Tracking a straight path*. A: *dynamic tracking performance;* B: *The generated linear and angular velocities;* C: *The tracking errors in the longitudinal and lateral directions, and in orientation;* D: *The varying orientation of the mobile robot.*

the first order exponent equation for linear speed are: $k = 1$, $\tau = 0.05$. The parameters used in the controller are: $A = 5$, $B = 3$, $D = 3$, $K_2 = 2$, $k_3 = 4$. the sampling time is equal to 0.01 second. As illustrated in Fig. 4A, the robot chooses a perfect angularity to track the circular path. Fig. 4B shows the smooth and reasonable velocity curves versus time. The speeds first increase and then track the desired speeds. The tracking errors approach to zero as time goes to infinity shown in Fig. 4C, Fig. 4D shows that the orientation increases until it equals 2π and then increases from 0. The robot advances about more than one and half circles shown in Fig. 4D during 10 seconds.

3.3 Comparison to a Conventional Backstepping Model

In this section, the backstepping technique based control law is compared to the biological neural network based tracking controller proposed in this paper. As we can see in the following equation (26), the suddenly changing errors result in the speed jump as shown in the Fig. 5B. The backstepping technique based tracking control law used by [3] is given as

$$\begin{bmatrix} v_c \\ w_c \end{bmatrix} = \begin{bmatrix} c_1 e_D + v_d \cos e_\theta \\ w_d + c_2 v_d e_L + c_3 v_d \sin e_\theta \end{bmatrix}, \qquad (26)$$

where c_1, c_2, and c_3 are the parameters. The tracking control law defined in Eq. (26) is simulated here using the same parameters as the controller proposed in this

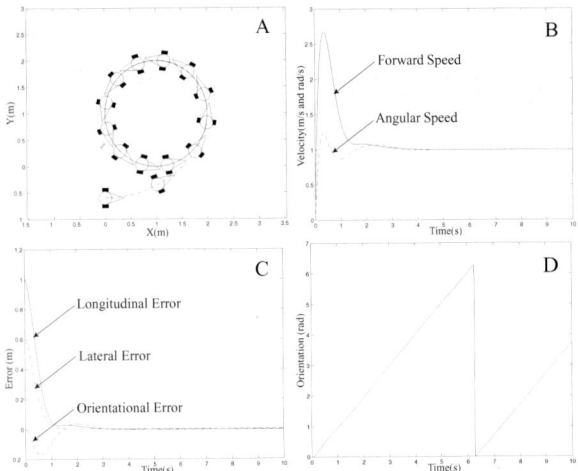
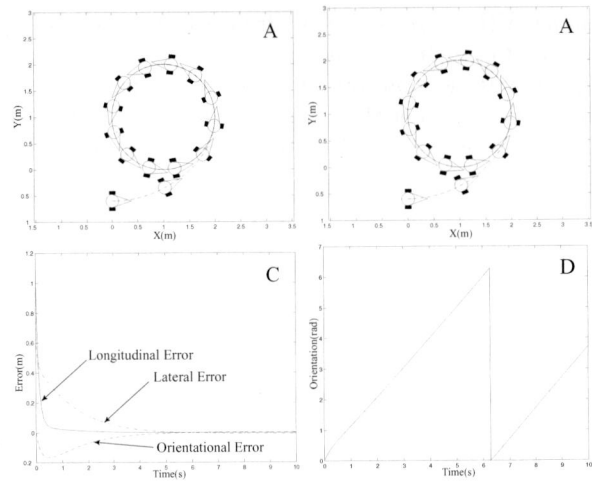

Figure 4: *Tracking a circular path. A: The dynamic tracking performance; B: The generated linear and angular velocities; C: The tracking errors in the longitudinal and lateral directions, and in orientation; D: The varying orientation of the mobile robot.*

Figure 5: *Tracking a straight path with a conventional backstepping controller. A: The dynamic tracking performance; B: The generated linear and angular velocities; C: The tracking errors in the longitudinal and lateral directions, and in orientation; D: The varying orientation of the mobile robot.*

paper ($c_1 = 10, c_2 = 2, c_3 = 4$). The Fig. 5 shows that the results of the tracking controller defined by Eq. (26). Fig. (5B) demonstrates that the speed changes suddenly rather than gradually from zero when time equals zero. So this does not hold in practice. In comparison to Fig. (4B), the proposed path tracking controller can obtain a reasonable results. What' more, the proposed controller in this paper tracks the circle path quicker than the controller proposed in [3] by comparison of Fig. (4A) and Fig. (5A).

4 Conclusions

In this paper, a novel neural dynamics based tracking controller is proposed, which is capable of generating smooth and continuous velocity commands with zero initial value. The tracking control system is asymptotically stable, and the tracking errors are guaranteed to converge to zeros. The proposed controller resolves the problem of sharp speed jumps at beginning and when sudden tracking errors occur.

References

[1] J. M. Yang and J. H. Kim: Sliding mode control for trajectory tracking of nonholonomic wheeled mobile robots. *IEEE Trans. on Robotics and Automation*, 15 (3): 578-587, 1999.

[2] D. H. Kim and J. H. Oh: Tracking control of a two-wheeled mobile robot using input-output linearization. *Control Engineering Practice*, 7 (3): 369-373, 1999.

[3] R. Fierro and F. V. Lewis: Control of a nonholonomic mobile robot: backstepping kinematics into dynamics. *J. of Robotic Systems*, 14 (3): 149-163, 1997.

[4] Q. Zhang, J. Shippen and B. Jones: Robust backstepping and neural network control of a low-quality nonholonomic mobile robot. *Intl. J. of Machine Tools and Manufacture*, 39 (7): 1117-1134, 1999.

[5] M. Ahmadi, V. Polotski and R. Hurteau: Path Tracking Control of Tracked Vehicles. *Proc. of IEEE Intl. Conf. on Robotics and Automation*, San Francisco, USA, pp. 2938-2943, 2000.

[6] V. Boquete, R. Garcia, R. Barea and M. Mazo: Neural control of the movements of a wheelchair. *J. of Intelligent and Robotic Systems*, 25 (3): 213-26, 1999.

[7] A. Ollero, A. G. Cerezo and J. V. Martinez: Fuzzy supervisory path tracking of mobile robots. *Control Engineering Practice*, 2 (2): 313-19, 1994.

[8] K. C. Ng and M. M. Trivedi: A neural-fuzzy controller for real-time mobile robot navigation. *Proc. of the SPIE - The Intl. Soc. for Optical Engineering*, 2761: 172-183, 1996.

[9] A. V. Hodgkin and A. F. Huxley: A quantitative description of membrane current and its application to conduction and excitation in nerve. *J. of Physiology London*, 117: 500-554, 1952.

Backward line tracking control of a radio-controlled truck and trailer

Claudio Altafini*

Optimization and Systems Theory
Royal Institute of Technology
Stockholm, Sweden
altafini@math.kth.se

Alberto Speranzon

Signals, Sensors and Systems
Royal Institute of Technology
Stockholm, Sweden
albspe@s3.kth.se

Abstract

A control scheme is proposed for the backward line tracking problem of a truck with trailer. It combines two different regulators, one for backward motion and the other for forward, in a switching scheme that assures convergence to the desired line. The scheme has been implemented and successfully used to reverse a radio-controlled vehicle.

1 Introduction

This paper describes a feedback control scheme used to stabilize the backward motion of the radio-controlled truck and trailer shown in Figure 1. The vehicle re-

Figure 1: The radio-controlled truck and trailer

produces in detail the geometry of a full-scale lorry; it has four axles, actuated front steering and actuated second axle to govern the longitudinal motion. Like the real one it presents saturations on the steering angle and on the two relative angles between the bodies. It is equipped with potentiometers and differential encoders so that full state feedback is possible. Our control task is to drive the system backward along a preassigned straight line, avoiding jack-knife effects on the angles. There is a moderate literature on backward steering control of multiple wheeled vehicles reporting on experimental results achieved with different control techniques and with different kinds of vehicles, mainly especially built laboratory mobile robots, see for example [4, 8, 12]. Numerous papers treat the backing problem with tools spanning from neural network [9], fuzzy control [5, 12], learning, genetic algorithms and expert systems [3, 10]. Only a few works make use of more theoretical tools steaming from the literature on control of kinematic vehicles (overviewed for example in [7]), see [6, 11]. According to such formalism, our system is a general 3-trailer, general because of the kingpin hitching between the second axle and the dolly (see [1]). From a system theory point of view, the control problem is quite challenging: it is an unstable nonlinear system with state and input constraints. The "reduced" control goal of stabilization along a line (instead of a point) allows to consider a system with controllable linearization so that local asymptotic stability can be achieved via Jacobian linearization. Still, the combination of instability and saturations makes the task impossible with a single controller. The scheme we use consists of a switching controller with a logic variable (the sign of the longitudinal velocity input v) that allows switching between the two different modes, backward (open-loop unstable) and forward (open-loop stable), each of them governed by a linear state feedback designed via linear quadratic techniques on the Jacobian linearizations. Since v is a control input for the system, it becomes the natural choice for a logic variable that switches between two regimes governed by two different feedback controllers. In order to automatically select the logical value of v, a suitable partition of the state space has to be given. The crossing of

---

*This work was supported by the Swedish Foundation for Strategic Research through the Center for Autonomous Systems at KTH

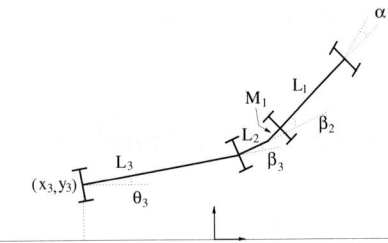

Figure 2: The kinematic model of the truck and trailer of Fig. 1

the switching surfaces of the partition and the direction of crossing provide the feedback information to v. In synthesis, the switching can be seen as an extra feedback loop around the two different closed loop modes. The switching surfaces and the switching logic are designed in such a way that the desired equilibrium inside the backward motion regime is given the character of global attractor from all the initial conditions in a prespecified domain.

2 Kinematic equations and linearization

The differential equations describing the kinematic of the vehicle under exam (see Figure 2 for notation) can be found for example in [1]:

$$\dot{x}_3 = v \cos \beta_3 \cos \beta_2 \left(1 + \frac{M_1}{L_1} \tan \beta_2 \tan \alpha \right) \cos \theta_3$$

$$\dot{y}_3 = v \cos \beta_3 \cos \beta_2 \left(1 + \frac{M_1}{L_1} \tan \beta_2 \tan \alpha \right) \sin \theta_3$$

$$\dot{\theta}_3 = v \frac{\sin \beta_3 \cos \beta_2}{L_3} \left(1 + \frac{M_1}{L_1} \tan \beta_2 \tan \alpha \right) \quad (1)$$

$$\dot{\beta}_3 = v \cos \beta_2 \left(\frac{1}{L_2} \left(\tan \beta_2 - \frac{M_1}{L_1} \tan \alpha \right) - \quad (2)$$
$$- \frac{\sin \beta_3}{L_3} \left(1 + \frac{M_1}{L_1} \tan \beta_2 \tan \alpha \right) \right)$$

$$\dot{\beta}_2 = v \left(\frac{\tan \alpha}{L_1} - \frac{\sin \beta_2}{L_2} + \frac{M_1}{L_1 L_2} \cos \beta_2 \tan \alpha \right) \quad (3)$$

The two inputs of the system are the steering angle α and the longitudinal velocity at the second axle v. Call $\mathbf{p} = [y_3\ \theta_3\ \beta_3\ \beta_2]^T$ the configuration state obtained neglecting the longitudinal component x_3.

In a compact way, the state equations are written as:

$$\dot{\mathbf{p}} = v(\mathcal{A}(\mathbf{p}) + \mathcal{B}(\mathbf{p}, \alpha)) \quad (4)$$

The sign of v decides the direction of motion. $v < 0$ corresponds to backward motion. The entire state is measured via two potentiometers on the relative angles β_2 and β_3 and a pair of encoders on the two wheels of the rearmost axle. No inertial measure is available.

State and input saturations Both the relative angles β_2 and β_3 present hard constrains:

$$|\beta_2| \leq \beta_{2_s} = 0.6 \text{ rad} \quad (5)$$
$$|\beta_3| \leq \beta_{3_s} = 1.3 \text{ rad} \quad (6)$$

These limitations are due to the front and rear body touching each other and to the dolly touching the wheels. They are particularly critical since for back-up maneuvers the equilibrium point is unstable and jack-knife effects appear on both angles. Also the input has a saturation:

$$|\alpha| \leq \alpha_s = 0.43 \text{ rad} \quad (7)$$

The steering servosystem tolerates very quick variations, so we do not assume any slew rate limitation in the steering signal.

Jacobian linearization along straight lines The system (4) is homogeneous in the longitudinal input v. Fixing v as a given nonnull function means having a drift component which gives a nonvanishing term to the differential equations of the system. The steering angle α can be used to give asymptotic stability to the system along a trajectory. The equilibrium point of $\mathbf{p}$ is the origin $\mathbf{p}_e = 0$ and it corresponds to a nominal value of the steering input $\alpha_e = 0$. The linearized system is

$$\dot{\mathbf{p}} = v(A\mathbf{p} + B\alpha) \quad (8)$$

where

$$A = \begin{bmatrix} 0 & 1 & 0 & 0 \\ 0 & 0 & \frac{1}{L_3} & 0 \\ 0 & 0 & -\frac{1}{L_3} & \frac{1}{L_2} \\ 0 & 0 & 0 & -\frac{1}{L_2} \end{bmatrix} \quad B = \begin{bmatrix} 0 \\ 0 \\ -\frac{M_1}{L_1 L_2} \\ \frac{L_2 + M_1}{L_1 L_2} \end{bmatrix} \quad (9)$$

3 Controllers for backward and forward motion

3.1 Backward controller

Consider the straight line backing case. The linearization (8) is open-loop unstable: the characteristic polynomial of the uncontrolled system is

$$\det(sI - vA) = s^2 \left(s + \frac{v}{L_2}\right)\left(s + \frac{v}{L_3}\right) \quad (10)$$

Since (8) is controllable, the origin of the nonlinear system (4) locally can be made an asymptotically stable equilibrium by linear state feedback. We treat it as a linear quadratic optimization problem and in the weight assignment we use the rule of thumb of trying to have decreasing closed-loop bandwidths when moving from the inner loop to the outer one in a nested loopshaping design. In fact, the relative displacement y_3 comes after a cascade of two integrators from the relative angles as can be seen on the linearization (9). It turns out that such a heuristic reasoning is very important in the practical implementation in order to deal with the saturations.

It is in general difficult to draw conclusion on the invariance properties of the flow of a nonlinear system. If in addition one takes into account the state and input constraints (5)-(7), then an analytic description becomes almost impossible. Therefore, in order to obtain estimates of the region of attraction of the linear controller

$$\alpha = -K_B \mathbf{p} \qquad K_B = [k_{B_1} \ldots k_{B_4}] \quad (11)$$

and of the contractivity of the resulting integral curves, we rely on the numerical simulation of the closed-loop behavior of the original nonlinear system (4) paired with the linear controller (11)

$$\dot{\mathbf{p}} = \mathcal{F}_B(\mathbf{p}) = v\big(\mathcal{A}(\mathbf{p}) + \mathcal{B}(\mathbf{p}, -K_B\mathbf{p})\big) \quad (12)$$

In order to obtain a graphical representation of the results, in the following we neglect the y_3 component of the state space which is by far the less critical one with the LQ controller under use.

The cloud of initial conditions that represents the region of attraction closely resembles an ellipsoid in $\hat{\mathbf{p}} = [\theta_3 \; \beta_3 \; \beta_2]^T$ space. The fitting of an ellipsoid $\hat{\mathcal{E}}$ strictly contained in the set of succesfull initial conditions can be done by direct investigation, see Figure 3. The principal axes $\hat{\mathbf{q}} = [q_1 \; q_2 \; q_3]^T$ of the ellipsoid are related to $\hat{\mathbf{p}}$ by an orthogonal transformation:

$$\hat{\mathbf{p}} = \hat{R}_{\mathcal{E}} \hat{\mathbf{q}} \qquad \hat{R}_{\mathcal{E}} \in SO(3)$$

Calling ε_1, ε_2 and ε_3 the semiaxes of $\hat{\mathcal{E}}$, the ellipsoid is given by the algebraic equation

$$\hat{\mathcal{E}} = \left\{ \frac{q_1^2}{\varepsilon_1^2} + \frac{q_2^2}{\varepsilon_2^2} + \frac{q_3^2}{\varepsilon_3^2} = 1 \right\} \quad (13)$$

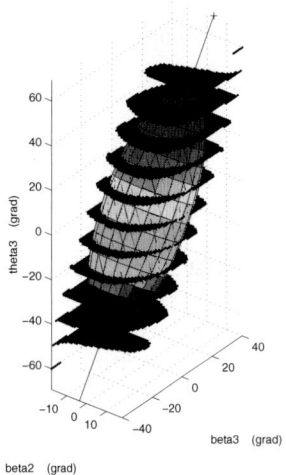

Figure 3: The succesful initial conditions and the fitted ellipsoid $\hat{\mathcal{E}}$.

Taking into account also the y_3 component of the initial conditions, the ellipsoid $\mathcal{E} \in \mathbb{R}^4$ is given by

$$\mathcal{E} = \left\{ \frac{q_1^2}{\varepsilon_1^2} + \frac{q_2^2}{\varepsilon_2^2} + \frac{q_3^2}{\varepsilon_3^2} + \frac{q_4^2}{\varepsilon_4^2} = 1 \right\} \quad (14)$$

with $\varepsilon_4 \gg \varepsilon_i$, $i = 1, 2, 3$. In D the difference with respect to Figure 3 can hardly be appreciated.

From Figure 3, we draw the qualitative conclusion that for the closed-loop nonlinear system $\mathcal{E}$ is a positively invariant set.

3.2 Stabilization for forward motion

When $v > 0$, in (10) the two unstable poles move on the open left half of the complex plane. Considering the subsystem $\hat{\mathbf{p}}$ means neglecting one of the two poles in the origin. The origin of $\hat{\mathbf{p}}$ is asymptotically stabilizable by linear feedback and this time convergence for the nonlinear system is a less critical problem. The reason for neglecting y_3 when moving forward is again the same: the closed loop mode relative to y_3 has a natural time constant higher of several orders of magnitude when compared to the other states.

Assume for example $v = 1$. Extracting from (9) the three dimensional system $(\hat{A}, \hat{B})$, linearization around the origin of (1)-(3), it is possible to choose a linear feedback

$$\alpha = \hat{K}_F \hat{\mathbf{p}} \quad (15)$$

such that the closed loop system $\dot{\hat{\mathbf{p}}} = v(\hat{A} - \hat{B}\hat{K}_F)\hat{\mathbf{p}}$ is asymptotically stable and has three distinct real modes. The practical rule here for the selection of the eigenvalues is to try to have all 3 closed-loop poles of the same order of magnitude. The unavoidable input saturation will not destroy stability anyway.

A forward feedback on y_3 is of no practical interest because of the long time constant of the y_3 mode.

4 Switching controller

The region of attraction of the backward controller is only a subset $\mathcal{E}$ of the entire domain D. Starting from outside $\mathcal{E}$, it is necessary to first drive forward for example with a controller like (15) until the system enters inside $\mathcal{E}$ and only then switch to backward motion. When reversing, the main manifestation of a destabilizing perturbation is a jack-knife effect on the relative angles. Just like on a full-scale truck and trailer vehicle, the only way to recover from such a situation is to move forward and try again. So, in order to guarantee stability of the backward motion in D and not only inside the ellipsoid $\mathcal{E}$ for the nominal model and in order to cope with the perturbations, one single controller is not enough. The switching variable between the two controllers is the longitudinal velocity v. For example we assume that $v \in \{-1, +1\} \triangleq \mathcal{I}$. The backward regime is selected by $v = -1$ and the forward one by $v = +1$. Since the longitudinal input v is a control input, if we assume that $v \in \mathcal{I}$ then v becomes a controlled logic variable. Moreover, if the selection of the logic value of v is made according to a partition of the state space, the overall system with multiple controllers becomes a feedback controlled system. This is feasible in our case since we have on-line full state information available.

4.1 Selection of the two switching surfaces

Two are the switching surfaces that delimit the partition of the state space, and their crossing in a prescribed direction by the flow of the system induces a sign change in v. This in its turn causes the inversion of the direction of motion and induces the activation of the corresponding linear state feedback controller. These switching surfaces, call them $\mathcal{S}_{-+}$ and $\mathcal{S}_{+-}$ have to be chosen such that they give to the point $\mathbf{p} = 0$ of the backward motion the character of global attractor (in D).

Since in both regimes the origin is the closed-loop local asymptotically stable equilibrium point, we choose both $\mathcal{S}_{-+}$ and $\mathcal{S}_{+-}$ as closed hypersurfaces in $\mathbb{R}^4$ containing the origin in their interior.

The switching surface from forward to backward motion: $\mathcal{S}_{+-}$ From Section 3.1, $\mathcal{S}_{+-}$ has to be contained inside $\mathcal{E}$. The simplest choice is to consider $\mathcal{S}_{+-} = \mathcal{E}_\rho$ for some ρ such that $\frac{1}{2} < \rho < 1$. The trade-off is the following:

- if $\mathcal{S}_{+-}$ is large ($\rho \to 1$) the system will be sensitive to disturbances and more easily destabilized by perturbations (meaning more switches can occur);

- if $\mathcal{S}_{+-}$ is small ($\rho \to \frac{1}{2}$) the forward regime will be very long, which is often unacceptable for practical implementations.

Ellipsoids smaller that $\mathcal{E}_{\frac{1}{2}}$ are also not recommendable for other reasons, like the possibility of being completely "jumped over" in case of relevant sensor error.

The switching surface from backward to forward motion: $\mathcal{S}_{-+}$ Such a switching surface has to "tell" the system that backing is not going well and the trailers need to be realigned. The choice is quite flexible, the only constraint is that $\mathcal{S}_{+-}$, $\mathcal{S}_{-+}$ and the sides of D must not intersect. In particular the set distance between $\mathcal{S}_{+-}$ and $\mathcal{S}_{-+}$ gives the hysteresis between the two regimes. If this distance is positive, problems like chattering will be avoided. One simple choice for $\mathcal{S}_{-+}$ is for example to use a cube in $\mathbb{R}^4$ which is a rescaling of D by a factor less than 1.

Control logic for v D is divided into three nonintersecting regions:

- $\mathcal{C}_- =$ region inside $\mathcal{S}_{+-}$ where $v = -1$;

- $\mathcal{C} =$ region between $\mathcal{S}_{-+}$ and $\mathcal{S}_{+-}$ where v can be either $+1$ or -1;

- $\mathcal{C}_+ =$ region outside $\mathcal{S}_{-+}$ ($\mathcal{C}_+ = D \cap (\mathcal{C} \cup \mathcal{C}_-)^\perp$) where $v = +1$.

Changes on v occur only at crossing with the rules of Table 1.

4.2 Convergence for the nominal and perturbed system

For the nominal system we can assert the following:

change in v	switching surface	crossing direction
$+1 \to -1$	$\mathcal{S}_{+-}$	$\mathcal{C} \to \mathcal{C}_-$
$-1 \to +1$	$\mathcal{S}_{-+}$	$\mathcal{C} \to \mathcal{C}_+$

Table 1: Switching rules for v.

Theorem 1 *Under the assumptions of invariance of $\mathcal{E}$, the system (4) with the two controllers (11) and (15) respectively for the cases $v = -1$ and $v = +1$ and with the feedback rule of Table 1 for $v \in \mathcal{I}$, asymptotically converges to the origin in backward motion from any initial condition in D.*

Proof From the analysis of Section 3 and looking at the switching rules of Table 1, the following order relation is the only possible one for the system:

$$\begin{array}{ccccc} \mathcal{C}_+ & \to & \mathcal{C} & \to & \mathcal{C}_- \\ v=+1 & & v=+1 & & v=-1 \end{array}$$

In fact, from any $\mathbf{p}_0 \in D$, the controller (15) steers the system inside $\mathcal{S}_{+-}$ and $\mathcal{S}_{+-}$ is a positively invariant set for the controller (11) ∎

So for the nominal system the switching surface $\mathcal{S}_{-+}$ is never in use. Due to the unstable equilibrium point, the effect of perturbations is critical in $\mathcal{C}_-$. Since the whole stabilization developed here occurs along a trajectory, we cannot expect the disturbances affecting the system to be vanishing at the equilibrium point of (12). When a perturbation is large enough to pull the state out of $\mathcal{E}$ the system diverges. Trying to quantify the amplitude of the destabilizing perturbations and consequently trying to infer total stability for a class of bounded perturbations is very hard in our situation because of the input saturation involved. The destabilized system keeps driving backwards until it hits the $\mathcal{S}_{-+}$ surface. After that, it inverts the direction of motion and try again to converge inside $\mathcal{S}_{+-}$ with the forward controller.

As said above, if the $\mathcal{S}_{-+}$ and $\mathcal{S}_{+-}$ do not touch each other, then degenerate switching phenomena (normally referred to as Zeno chattering) do not occur. Furthermore, also the different pole placement philosophy adopted in the two controllers (11) and (15) (in one the critical mode, the θ_3 mode, is slow, in the other it is instead faster) is meant to avoid a chattering type of behavior (like keep moving the system back and forth between the same points on $\mathcal{S}_{-+}$ and $\mathcal{S}_{+-}$) which can happen if the two closed-loops resemble each other.

5 Experimental results

The controller for the truck and trailer shown in Figure 1 was implemented using a commercial version of PC/104 with an AMD586 processor and with an acquisition board for the sensor readings. For the relative angles β_2 and β_3 we used AD Converter provided with the acquisition board while the distance from the target line (y_3) and the angle of the trailer with respect this line (θ_3) were measured using a Digital Input/Output Port. These inputs were read by a Interrupt Service Routine called at a frequency of 2kHz. This frequency and the 500 pulse shaft ensured a maximum speed of about 0.2 m/s which was sufficient for this application. The error on the relative angles is about 3-4 degrees due to a very high resistance of the potentiometers with respect the voltage applied (5V). To reduce it the values of the angles were averaged over 5 measures. The error on the angle of the trailer, due to the encoders, is given by the resolution of the encoder itself and it is less than 0.7 degrees while the error on the linear distance (along x_3 and y_3) is about 0.4 mm. The controller was written in C and used at a frequency of about 10Hz since the velocity of the system was very low.

Fig. 4-6 present the result of a simple real manouvre. The switching scheme used is the two-state automaton described in Section 4. The states and input are plotted versus the distance travelled. The manouvre is shown in Fig. 7 using the experimental data. The vehicle starts with saturated relative angles and first drives forward in order to realigne itself, then reverse along the reference line. The transient in y_3 is very long and only a part of it is shown. Notice that since the θ_3 mode is slower than those of the relative angles most of the forward motion is needed to get θ_3 inside the ellipsoid $\mathcal{S}_{+-}$. Finally, in Figure 6 it is instructive to compare the activity of the feedback input when the open loop system is stable (upper plot) and when it is unstable (lower plot).

References

[1] C. Altafini. Controllability and singularities in the n-trailer system with kingpin hitching. Proc. 14th IFAC World Congress, vol. Q, p.139-145, Beijing, China.

[2] P. Bolzern, R.M. De Santis, A Locatelli and D. Masiocchi. Path-tracking for articulated vehicles with off-axle hitching. *IEEE Transaction on Control Systems Technology*, 6:515-523, 1998.

[3] D.F. Hougen, M. Gini and J. Slagle. Rapid unsupervised connectionist learning for backing a robot with

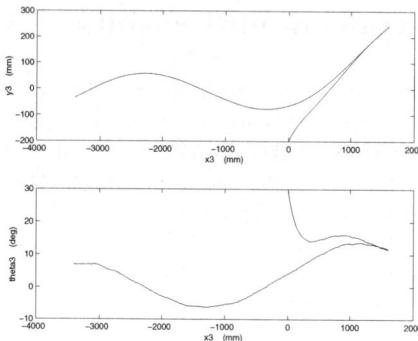

Figure 4: y_3 displacement and θ_3 angle.

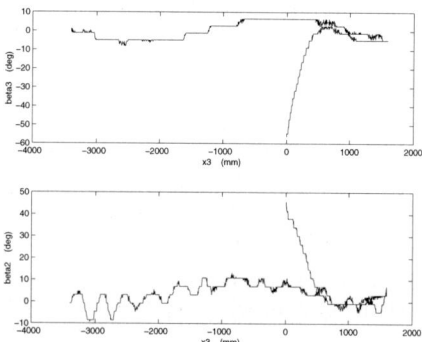

Figure 5: Relative angles β_3 and β_2.

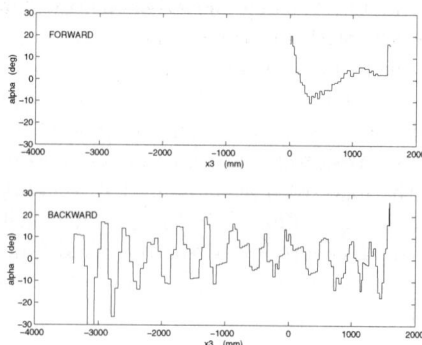

Figure 6: Steering input α (forward and backward are plotted separately to avoid confusion as they consistently overlay).

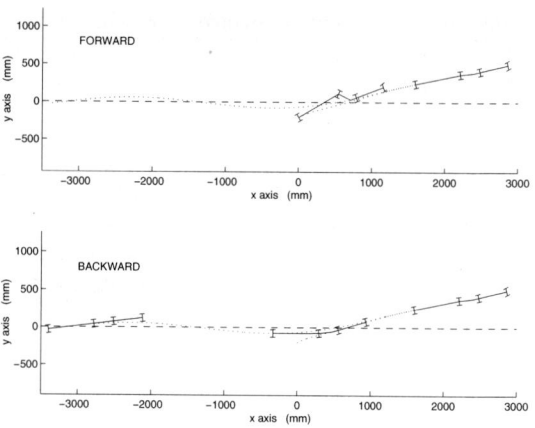

Figure 7: Schetch of the motion of the vehicle for the data of Fig. 4-6. The dotted line represents the path followed by the (x_3, y_3) point.

two trailers. Proc. 1997 IEEE Int. Conf. on Robotics and Automation, p.2950-2955, Albuquerque, NM.

[4] D.H Kim and J.H Oh. Experiments of backward tracking control for trailer system. Proc. 1999 Int. Conf. on Robotics and Automation, p.19-22, Detroit, MI.

[5] G.S. Kong and B. Kosko. Adaptive fuzzy systems for backing up a truck and trailer. *IEEE Trans. on Neural Networks*, **3**:211-223, 1992.

[6] F. Lamiraux and J.P. Laumond. A practical approach to feedback control for a mobile robot with trailer. Proc. 1998 IEEE Int. Conf. on Robotics and Automation, p.3291-3296, Leuven Belgium.

[7] J.P. Laumond (ed). Robot Motion Planning and Control, Lecture notes in control and information sciences, Springer-Verlag, 1998.

[8] Y Nakamura, H. Ezaki, Y. Tan; W. Chung. Design of steering mechanism and control of nonholonomic trailer systems. Proc. 2000 IEEE Int. Conf. on Robotics and Automation, p.247 - 254, San Francisco, CA.

[9] D. Nguyen and B-Widrow. The truck backer-upper: an example of self-learning in neural networks. Proc. of SPIE vol.1293, pt.1, p.596-602, 1990.

[10] R. Parra-Loera and D.J. Corelis. Expert system controller for backing-up a truck-trailer system in a constrained space. Proc. of the 37th Midwest Symposium on Circuits and Systems p.1357-1361, 1995.

[11] M. Sampei, T. Tamura, T. Kobayashi and N. Shibui. Arbitrary path tracking control of articulated vehicles using nonlinear control theory. *IEEE Transaction on Control Systems Technology*, **3**, 125-131, 1995.

[12] K. Tanaka, T. Taniguchi and H.O. Wang. Trajectory control of an articulated vehicle with triple trailers. Proc. 1999 IEEE International Conference on Control Applications, p.1673-8.

The Scalar ϵ-Controller: A Spatial Path Tracking Approach for ODV, Ackerman, and Differentially-Steered Autonomous Wheeled Mobile Robots

Morgan Davidson and Vikas Bahl
Center for Self-Organizing and Intelligent Systems (CSOIS)
Utah State University, UMC 4160
Logan, UT 84322-4160
email: morg@ece.usu.edu

Abstract

Path tracking algorithms for wheeled mobile robots (WMRs) are frequently parametric in the sense that they are time-based. This has the potential of introducing lag-related errors, and is not a direct approach. A spatial path tracking control algorithm, the ϵ-controller (C_ϵ), is developed in this paper. It is based solely on static path geometry with position feedback. The C_ϵ is applied in simulation to three different WMR steering configurations to illustrate the performance and generality of this new approach. Actual results are found to parallel the simulated results.

1 Introduction

The path tracking, or path following problem is prevalent in the WMR community [1, 2, 3, 4, 5]. Frequently, a WMR is required to follow a path in space with a certain speed in order to complete a certain mission. The desired path is often parameterized in time to provide time-varying position set-points [3, 4, 5]. Since the locus of desired set-points follow (in time) the trajectory (in space), it is inferred that a controller that can track the parametric set-points in time will effectively track the desired trajectory in space. Because this is an indirect approach, the robot dynamics can generate unexpected results, especially in the presence of external disturbances such as operation of a WMR on a slope or in cases of actuator saturation etc. Positional errors may cause the WMR to "cut corners". Additionally, this parameterization approach is not well-suited to real time changes in desired speed along the path due to the requirement of re-calculating the parametric equations for the remaining portion of the path each time a new desired speed is given. If in fact the requirement of "following the path" is more important than the position of the WMR along the path at a given time, then a different approach should be used which is time independent and is completely spatial.

Motivated by previous work on spatial path tracking [1, 2], a novel control law is developed in this paper which, while maintaining the WMR on the path in space, impels it toward the desired endpoint of the path with the desired speed. This path tracking controller (dubbed the ϵ-controller) is based completely on static inputs - the geometry of the path - and the desired speed along with feedback of the current position of the WMR. An important follow-on topic is the transformation of the output of the C_ϵ to actuator commands. The MakeSetPoints (MSP) algorithm presented in this paper carries out this transformation, and is specific to the steering kinematics configuration of the WMR. In this paper, three MSP algorithms are developed. These correspond to Omni-Directional Vehicle (ODV) steering [4, 6], Ackerman steering, and Differential steering. Computer simulations as well as actual results demonstrate the effectiveness of the C_ϵ in the aforementioned steering configurations. Results are compared and contrasted, and in conclusion future work is outlined.

2 Controller Notation

The development of C_ϵ is carried out in a right-handed inertial Cartesian coordinate system (ICS). The vector quantity generated by C_ϵ is referenced to this coordinate system (CS), which is shown with axes subscripted with I in Figure 1. A right-handed body-fixed CS (BFCS) is also employed which translates and rotates with the robot [4]. Yaw (ψ) is the rotational relation between ICS and BFCS and is encapsulated in the Euler rotation matrix $\mathbf{R}_\psi$ [7].

Qualitatively, the desired path-tracking behavior of the robot is such that it begins near the initial point of the desired path and travels along the path to the final point. This performance expectation is entirely spatial. The desired path is shown in Figure 1 as the dash-dot line. Velocity along the path (and therefore the time taken to complete the path) is secondary, i.e., although the desired velocity is specified it is not explicitly regulated. Hence, the desired speed (V_d) along the path can be allowed to vary continuously throughout the execution of the path without affecting the robot trajectory in space. In this development, all desired paths

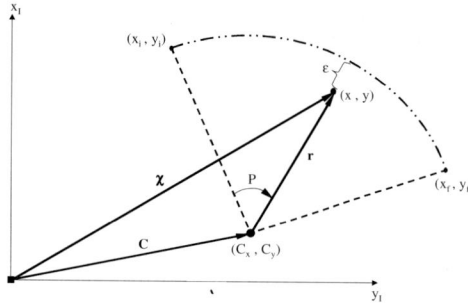

Figure 1: Path and Path Related Quantities

are composed of arc segments. Straight lines are a special case and can be closely approximated as an arc segments with a very large radius. Thus any complex trajectory can be built up from basic arc segments, so there is no loss of generality in possible combinations of paths. Any path geometry in this type of architecture can be described by its initial point $\chi_i = [\ x_i\ y_i\ ]^T$, final point $\chi_f = [\ x_f\ y_f\ ]^T$, signed radius R (following the right-hand rule a positive sign on R indicates clockwise direction of travel). These quantities define the location of the center point of the arc **C** (in this paper vectors and matrices are represented by "bold faced" symbols). In Figure 1 the vector $\chi = [\ x\ y\ ]^T$ is the actual position of the robot in ICS. The vector from the center of the arc to the actual position is the vector **r** in ICS defined as

$$\mathbf{r} = \chi - \mathbf{C} = \begin{bmatrix} x - C_x \\ y - C_y \end{bmatrix} \quad (1)$$

The angle P in Figure 1 represents the progress made toward the final point on the path, and is always positive in the direction of desired travel. It provides for coupling control of the angular motion of an ODV robot to the spatial path tracking. This topic is beyond the scope of this paper and will be discussed elsewhere.

The central quantity in this work is ϵ, the deviation from the path. It is the distance from χ to the nearest point on the arc as seen in Figure 1. It is a scalar, and is always measured in the radial direction (except in the trivial case where the robot is at the center of the arc). When ϵ is positive, the robot lies "inside" the desired arc. When ϵ is negative, the robot is "outside" the arc, and as ϵ goes to zero the robot approaches the desired path. From Figure 1, it is clear that

$$\epsilon = |R| - \|\mathbf{r}\|. \quad (2)$$

2.1 ϵ-Controller Development

The nonlinear path tracking controller C_ϵ developed in this paper is quite general and can control the linear motion of ODV, Ackerman, and Differential steering configurations of a mobile robot without changes in form. It is the outermost controller in the cascade architecture shown in Figure 2. The input vector $\mathbf{U_p}$ contains the parameters of the path command as discussed earlier:

$$\mathbf{U_p} = [\chi_i\ \chi_f\ R\ V_d]^T \quad (3)$$

Based upon the deviation from the path, C_ϵ generates the inertial velocity setpoints ($\mathbf{V_I^*}$) specifying corrective linear motion (in this paper superscript " * " denotes setpoint). Based on this $\mathbf{V_I^*}$ and kinematics, MSP generates the desired drive velocities and steering angle set-points for each wheel subsystem. The low-level wheel controllers C_w provide actuator control signals **E** to the robot. We define $\mathbf{dV_I^*}$ as the sum of normal and tangential velocity components $\mathbf{V_n}$ and $\mathbf{V_t}$ respectively. It is used for computing the *direction* of $\mathbf{V_I^*}$, as shown in Figure 3. To handle the case of negative path radii, we let

$$\mathbf{dV_I^*} = \mathbf{V_n} + \mathbf{V_t}\ \text{sign}(R) \quad (4)$$

where the sign term effects a reversal of direction on the vector quantity $\mathbf{V_t}$ based on the direction of the desired path. The normal velocity component $\mathbf{V_n}$ is

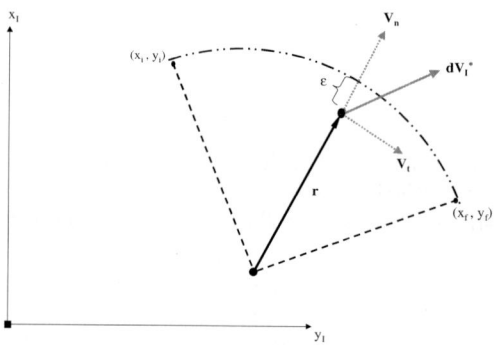

Figure 3: ϵ-Controller Quantities

the corrective action taken by the C_ϵ to minimize ϵ. Thus C_ϵ conceptually acts as a regulator operating on ϵ. This regulator could consist of any of several control methods, but for this paper we use P and PID algorithms. For the P control case, for example, we have

$$\mathbf{V_n} = K_p\ \epsilon\ \left(\frac{\mathbf{r}}{\|\mathbf{r}\|}\right) \quad (5)$$

where K_p is the proportional controller gain. Multiplication by the normalized vector **r** gives the desired *direction* of $\mathbf{V_n}$ in the ICS.

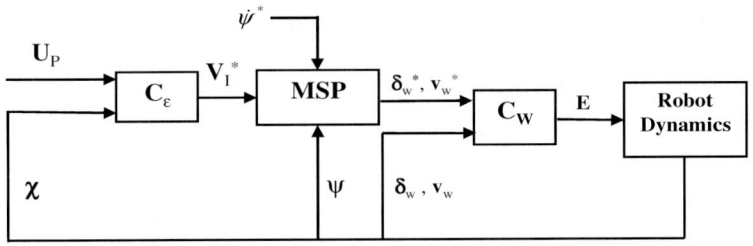

Figure 2: Cascade Control Architecture

In the tangential direction, we define $\mathbf{V}_t$ to be

$$\mathbf{V}_t = (V_d - \|\mathbf{V}_n\|)(\frac{\mathbf{r}_t}{\|\mathbf{r}_t\|}) \quad (6)$$

where

$$\mathbf{r}_t = \begin{bmatrix} -(y - C_y) \\ x - C_x \end{bmatrix} \quad (7)$$

and the quantity $(V_d - \|\mathbf{V}_n\|)$ is never allowed to be negative. This choice of constraints makes $\|\mathbf{V}_t\|$ a function of ϵ. The result is a tangential velocity which approaches zero as ϵ increases and approaches V_d as ϵ goes to zero. Multiplication by the normalized vector $\mathbf{r}_t$ gives the desired *direction* of $\mathbf{V}_t$ in the ICS.

For the case of P control (see Equation 5), Equations 4 through 7 can be expressed more compactly in the matrix form

$$\mathbf{dV}_I^* = \mathbf{\Lambda} \ \mathbf{V}_\epsilon \quad (8)$$

Where $\mathbf{\Lambda} \in \Re^{2 \times 2}$ is given by:

$$\mathbf{\Lambda} = \frac{1}{\|\mathbf{r}\|} [\ \mathbf{r} \ | \ \mathbf{r}_t \ ] \quad (9)$$

as $\|\mathbf{r}\| = \|\mathbf{r}_t\|$. The vector $\mathbf{V}_\epsilon$ is given by:

$$\mathbf{V}_\epsilon = [\ K_p\epsilon \ \ (\ V_d \ - \ K_p|\epsilon| \ )\text{Sign}(R) \ ]^T \quad (10)$$

with the quantity $(V_d - K_p|\epsilon|)$ only allowed a positive or zero value (saturation at zero). Finally, $\mathbf{V}_I^*$ is computed as

$$\mathbf{V}_I^* = V_d \frac{\mathbf{dV}_I^*}{\|\mathbf{dV}_I^*\|} \quad (11)$$

The effect of Equation 11 is a path tracking controller which, when the robot is far from the path, directs it to the closest point on the path in the radial direction. As the robot approaches the path, the controller generates a desired velocity vector tangent to the path. Note that the ϵ-controller presented here is a nonlinear controller operating on the scalar ϵ. This controller effectively turns the two-dimensional path tracking problem into a scalar regulation problem with respect to ϵ.

3 The MakeSetPoint (MSP) Algorithms

The MSP algorithm converts inertial velocity commands into body-fixed wheel velocity and steering angle set-points. It uses the assumption that the wheels do not slip, satisfying the relative motion equation of rigid body kinematics. In this paper we will develop the MSP algorithms for three different steering types of WMR, namely, Omni-Directional (MSP_{ODV}), Ackeraman (MSP_{ACK}), and Differentially Steered robots (MSP_{DS}).

3.1 MSP_{ODV}

WMRs with an omni-directional steering system as shown in Figure 4(a) (such as the T-Series of Robots developed at Utah State University (USU)), have distinct advantages over other types of steered robots in terms of their maneuverability and stability [6]. ODV maneuverability is derived from a fully independent omni-directional steering system allowing independent motion control in the longitudinal, lateral, and rotational directions. The MSP_{ODV} algorithm ensures that kinematically correct wheel velocities and steering angle set-points are passed to the actuator controllers so that the wheels do not "fight each other" while carrying out any desired maneuver.

The input to MSP_{ODV} is $\mathbf{V}_I^*$ (obtained from C_ϵ), ψ (obtained from an onboard fiber optic gyroscope) and the angular velocity setpoint $\dot{\psi}^*$ or ω^* (obtained from a separate yaw controller). The vector $[\mathbf{V}_I^* \ \omega^*]^T$ is rotated into the BFCS using Euler's rotation matrix resulting in the body fixed velocity vector $\mathbf{v}_B^* = \begin{bmatrix} v_{Bx}^* \\ v_{By}^* \\ \omega^* \end{bmatrix}$. Components of individual drive motor velocity ($\mathbf{v}_w^*$) and steering motor angle set-points (δ_w^*) are obtained using relative motion equation of rigid body kinematics, which can be embodied in the robot geometry matrix $\mathbf{G}_{\text{ODV}}$, given by

$$\mathbf{G}_{\text{ODV}} = \begin{bmatrix} 1 & 0 & 1 & 0 & 1 & 0 & 1 & 0 \\ 0 & 1 & 0 & 1 & 0 & 1 & 0 & 1 \\ r_y & r_x & -r_y & r_x & r_y & -r_x & -r_y & -r_x \end{bmatrix} \quad (12)$$

where r_x is one half the track length of the robot and

r_y is one half the wheel base of the robot (see Figure 4(a) and (b)). More details about controlling omnidirectional WMR using this type of architecture can be found in [4].

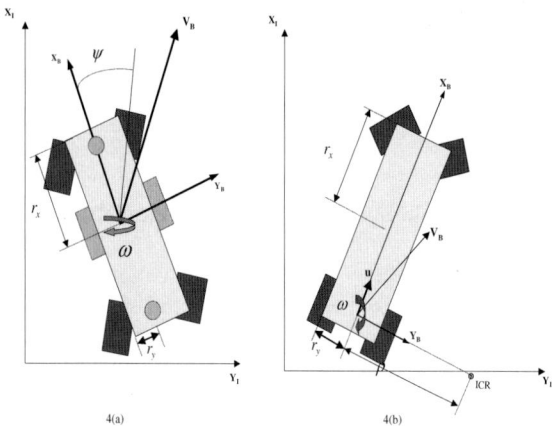

Figure 4: ODV, Differential and Ackerman Steering Modes

3.2 MSP_{ACK}

Ackerman steering (front-wheel-steering, rear-wheel-fixed) is probably the most common steering system that one comes across in daily life. Essentially all passenger vehicles on roads utilize some type of Ackerman mechanism. In this type of system all the wheels are required to travel different distances around their respective arcs when cornering. They are positioned such that they are perpendicular to the line connecting them with the instantaneous center of rotation (ICR). The robot has a linear velocity u (assumed always positive for this paper) at the origin of the BFCS and is constrained to lie along the X_B axis (see Figure 4(b)). The kinematically correct steering angles and drive velocities are obtained from the MSP_{ACK} algorithm, which also ensures that the wheels follow the desired radius to make the turn. The input to MSP_{ACK} is $\mathbf{V}_I^*$ (obtained from C_ϵ).

Unlike omni-directional WMRs, which can allow rotation in the horizontal plane to be independent of its translational motion, orientation of the robot in Ackerman steering mode is slaved to the linear motion of the vehicle. The MSP_{ACK} algorithm first rotates $\mathbf{V}_I^*$ into the BFCS using Euler's rotation matrix to get a pseudo body-fixed velocity setpoint vector($\mathbf{Pv}_B^*$). Note that as the angular velocity ω in this steering mode is path dependent, the Euler's rotation matrix $\mathbf{R}_\psi \in \Re^{2\times 2}$ is given by

$$\mathbf{R}_\psi = \begin{bmatrix} \cos\psi & \sin\psi \\ -\sin\psi & \cos\psi \end{bmatrix}. \quad (13)$$

ψ in the above equation is the actual orientation of the robot in ICS and is path dependent. The expression for $\mathbf{Pv}_B^*$ can thus be written as

$$\mathbf{Pv}_B^* = \mathbf{R}_\psi \mathbf{V}_I^* = \begin{bmatrix} Pv_{Bx}^* \\ Pv_{By}^* \end{bmatrix} \quad (14)$$

and the linear velocity (u) of the robot is chosen to be $|Pv_{Bx}^*|$. This is because the y-component of the body fixed velocity vector in an Ackerman steered WMR is always zero as the ICR for the robot lies along the Y_B axis.

Next the algorithm computes δ_{avg} (steering angle of an imaginary wheel lying on the midpoint of the line connecting the two front wheels) as

$$\delta_{avg} = \tan^{-1}\left(\frac{Pv_{By}^*}{Pv_{Bx}^*}\right) \quad (15)$$

Finally the angular velocity (ω^*) of the robot is computed as

$$\omega^* = \frac{u}{\rho} \quad (16)$$

where ρ is the distance between the ICR and the origin of the BFCS as shown in Figure 4(b). ρ is computed as

$$\rho = \frac{2r_x}{\tan^{-1}(\delta_{avg})} \quad (17)$$

Hence the body-fixed velocity vector $\mathbf{v}_B^*$ is $\begin{bmatrix} u \\ 0 \\ \omega^* \end{bmatrix}$, which, when multiplied by the vehicle geometry matrix $\mathbf{G}_{ACK}$ [4], gives the components of the desired steering motor angle and drive motor velocity set-points. The robot geometry matrix for this steering mode with four wheels as in Figure 4(b) is

$$\mathbf{G}_{ACK} = \begin{bmatrix} 1 & 0 & 1 & 0 & 1 & 0 & 1 & 0 \\ 0 & 1 & 0 & 1 & 0 & 1 & 0 & 1 \\ r_y & 2r_x & -r_y & 2r_x & r_y & 0 & -r_y & 0 \end{bmatrix}. \quad (18)$$

3.3 MSP_{DS}

Differentially steered robots are another class of WMR, an example of which is the popular Pioneer^{TM} WMR. In this type of system two wheels mounted on a single axis are independently powered and controlled, thus providing both drive and steering. Additional passive wheels (casters) are usually provided for support. If both drive wheels turn at equal speed in the same direction, the robot moves in a straight line. If one wheel turns faster than the other, the robot follows a curved path. If the wheels turn at equal speed, but in opposite directions, the robot spins in place. Hence steering the robot is just a matter of varying the speeds of the drive wheels. The MSP_{DS} algorithm derives the kinematically correct drive speeds so that the WMR can move in the desired direction.

MSP$_{DS}$ computes $\mathbf{Pv}_B^*$ from $\mathbf{V}_I^*$ (obtained from C$_\epsilon$) in the same fashion as in MSP$_{ACK}$ algorithm. The linear velocity (u) of the robot is again $|\mathrm{Pv}_{Bx}^*|$ with the y-component of the velocity being zero. Angular velocity ω^* in this steering mode is

$$\omega^* = \tan^{-1}(\frac{\mathrm{Pv}_{By}^*}{\mathrm{Pv}_{Bx}^*}) \qquad (19)$$

Hence $\mathbf{v}_B^*$ is $\begin{bmatrix} u \\ 0 \\ \omega^* \end{bmatrix}$, which, when multiplied by $\mathbf{G}_{DS}$, gives the components of the desired steering motor angle and drive motor velocity set-points. The $\mathbf{G}_{DS}$ matrix for this steering mode with two wheels (lightly shaded) as shown in Figure 4(a) is

$$\mathbf{G}_{DS} = \begin{bmatrix} 1 & 0 & 1 & 0 \\ 0 & 1 & 0 & 1 \\ r_y & 0 & -r_y & 0 \end{bmatrix}. \qquad (20)$$

Note that in this steering mode we take into account only two wheels and the position of the castors are not considered because they are passive wheels provided for support and don't contribute in determining the steering angles or the drive velocities.

Finally, the exact form of $\mathbf{G}$ in all three steering modes will depend on the ordering that is used to number the wheels.

4 Simulated And Experimental Results

A Matlab/Simulink simulation for all three steering modes was built to test the spatial path tracking algorithm presented above. The model parameters for these simulations were obtained from a validated model [4]. Random transmission delay noise was added to represent real world effects resulting from the distributed multi-processor controller implementation used in the test vehicle. Figure 5 shows five different spatial trajectories that all three WMR's were required to traverse in simulation. Trajectory 1 has the largest radius (representing a straight line), trajectory 5 has the smallest radius and the radii of trajectories 2 through 4 are in decreasing order of their magnitude. Also note that the radii of trajectories 2 and 3 are chosen negative, thus requiring the WMRs to traverse them in the counterclockwise direction. Figures 6 through 8 demonstrate the effectiveness of C$_\epsilon$ on ODV, Ackerman and Differentially-steered WMRs respectively in making them traverse the desired paths. These figures were generated from the data obtained from an actual WMR. Note that only a limited number of results are shown in the interest of space. It is also interesting to note that in the case of the ODV robot, C$_\epsilon$ is merely a proportional regulator, while for the other two cases we used complete PID regulation with conditional integration. This is not to say that a proportional regulator would not work in the Ackerman and Differentially-steered WMR. On the contrary, a proportional controller has been found in simulation to stabilize the robot in both these latter cases. The steady-state value of ϵ in these cases was, however, much larger with mere proportional control. The aforementioned coupling of heading to yaw in the Ackerman and Differential steering cases slows the response of the vehicle to changes in heading. Figure 9 shows the effectiveness of C$_\epsilon$ in bringing the WMR's to the desired path when the initial condition of ϵ is large.

5 Conclusions

In this paper a completely spatial path-tracking controller is developed and applied to the motion control of three very different types of WMR. The simulation results obtained are very encouraging and comparable to the actual results. Future work in this direction will include investigations into stability for all three cases based on Lyapunov criteria or dissipative dynamical system theory. In addition, the desired trajectory of the WMR in space represents a sort of reduced order manifold in the state space. With this interpretation, the ϵ-controller could be considered in some sense a sliding mode controller. Exploration of this avenue of thought along with the comparision of C$_\epsilon$ to contemporary path tracking methodologies will be additional points of discussion. Finally, different methods of regulating ϵ will be used in place of PID.

Acknowledgments

The authors would like to thank Dr. Kevin L. Moore, the Director of CSOIS, for his support and efforts to implement the ideas presented in this paper.

References

[1] D. Cripps, "Proprietary report submitted to project sponsor on obstacle detection and avoidance," in *Center for Self-Organizing and Intelligent Systems, Utah State University, Logan, Utah*, September 1999.

[2] D. Cripps, "Spatially-robust vehicle path tracking using normal error feedback," in *Submitted to Unmanned Ground Vehicle Technology III, SPIE* 15th *Annual Aerosense Symposium*, April 2001.

[3] W. E. Dixon, D. M. Dawson, F. Zhang, and E. Zergeroglu, "Global exponential tracking control of a mobile robot system via a PE condition," in *IEEE Transactions on Systems, Man, and Cybernetics - Part B: Cybernetics*, vol. 30, pp. 129–142, February 2000.

[4] K. L. Moore, M. Davidson, V. Bahl, S. Rich, and S. Jirgal, "Modelling and control of a six-wheeled autonomous robot," in *Proceedings of the 2000 American Control Conference*, (Chicago,Illinois), June 2000.

[5] A. G. O. .Mutambara and H. F. .Durrant-Whyte, "Estimation and control of a modular wheeled mobile robot," in *IEEE Transactions on Control Systems Technology*, vol. 8, pp. 129–142, January 2000.

[6] M. Davidson, V. Bahl, and C. Wood, "Utah State University's T2 ODV mobility analysis," in *Proceedings of SPIE Conference on Ground Robotic's*, (Orlando, Florida), April 2000.

[7] J. J. Craig, *Introduction to Robotics:Mechanics and Control 2nd Edition*. Reading, Massachusetts: Addison-Wesely, 1978.

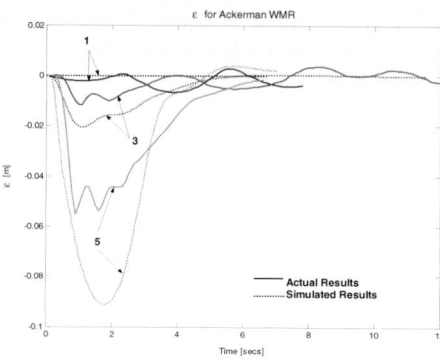

Figure 7: ϵ-Controller Response for Ackerman Steered WMR

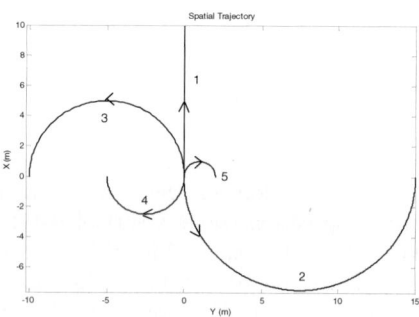

Figure 5: Desired Spatial Trajectory to be Traversed by the WMR

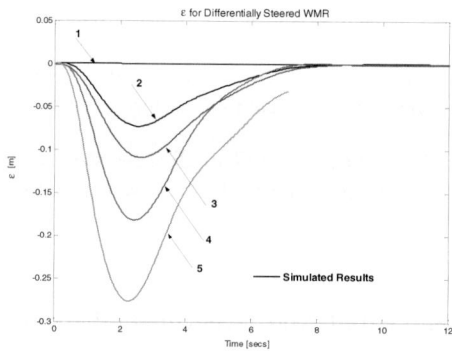

Figure 8: ϵ-Controller Response for Differentially Steered WMR

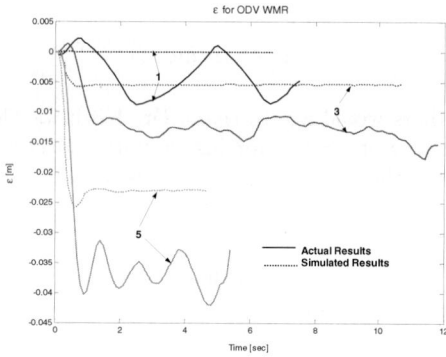

Figure 6: ϵ-Controller Response for ODV WMR

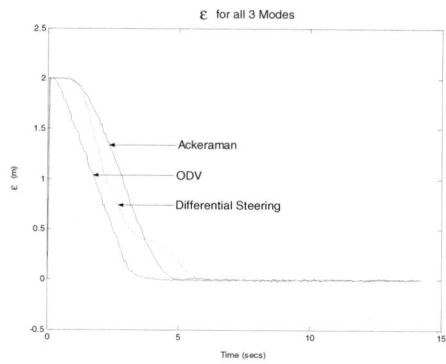

Figure 9: ϵ-Controller Response with Large Initial Position Error

Normalized Energy Stability Margin and its Contour of Walking Vehicles on Rough Terrain

Shigeo Hirose   Hideyuki Tsukagoshi   Kan Yoneda

Mechanical and Control Eng., Tokyo Institute of Technology
2-12-1 Ohokayama Meguro-ku Tokyo 152-8552, Japan

Abstract

Some stability criteria for walking vehicles on rough terrain are discussed in this paper. Several criteria for stability margin were proposed up to now, but those can be roughly divided into three categories. This paper compares them each other and concludes that a stability criterion based on energy consideration is the most reasonable for practical use through simple experiments. But the existing stability margin is apparently inadequate because it varies with weight of the vehicle with the same posture, while weight doesn't affect its resistance to the tumble. Therefore, the improved stability margin normalized by weight is proposed here. At the same time, a helpful tool to derive the desirable posture for walking vehicles is described which can maximize the proposed stability margin on rough terrain.

1. Introduction

Although the mobile efficiency of walking vehicles is lower than that of wheel type vehicles or crawler type ones, walking vehicles have some desirable characteristics as follows: (i)They can move on rough terrain by selecting the stable footholds discretely. (ii)They can have a stable posture even on a slope by shifting their center of gravity to a suitable position. (iii)Their legs can be used as powerful manipulators. (iv)The freedom of the body can contribute to perform heavy tasks, when some tools are attached to the body. Because of the above potentiality, the application to manipulative operations by walking vehicles is being studied actively at present[1], to say nothing of the application to mobile operations on rough terrain. We have also paid attention to these advantages, and have developed a quadruped walking vehicle for operations on steep slopes, called TITAN VII shown in Fig.1[1], which is expected to walk around at the construction site and to perform several operations to prevent the landslide instead of human workers.

In order to put walking vehicles into practical use for the above application, they must manage to keep walking slowly but as stably as possible, even if they traverse on rough terrain including slopes. From this point of view, it is indispensable to define the most reasonable stability criterion for walking vehicles and to design a gait maximizing its margin. Needless to say, several stability criteria have been proposed up to now. However, the postures maximizing each stability margin have never been compared each other before and the difference among them has never been studied from the physical point of view. Then, first, this paper compares the proposed stability criteria on the same basis, and proposes the improved one, called "the normalized energy stability margin," which can be the most practical criterion for walking vehicles on rough terrain. Second, it introduces a new tool of "the S_{NE} contour" to prepare for the design of the most stable posture and the most stable gait.

Fig.1 TITAN VII climbing a slope with its posture maximizing the stability margin to step over obstacles by using a vision sensor.

2. General Study of the Stability Criteria
2.1 Classification of the Existing Stability Criteria

It can be considered that roughly six tumble stability criteria for walking vehicles were proposed as follows.
1) "Stability Margin"[2]: It evaluates the distance between the projection of the center of gravity on the ground and the border of the polygon formed by the supporting feet of the walking vehicle on the plane.
2) "Tumble Stability Margin[3]": When the walking vehicle tumbles around the line connecting two support feet, it evaluates the

absolute value of the moment divided by its weight which generates around the line to avoid tumbling. It corresponds to the "Stability Margin" ignoring the dynamic effect when the walking vehicle is on the level ground.

3) "Gradient Stability Margin[4]":It evaluates the inclination of the walking vehicle at which it starts tumbling owing to gravity, when it gets inclined little by little from the level ground.

4) "Tipover Stability Margin[5]": It is similar to the criterion of the "Gradient Stability Margin," but all the external forces including gravity are considered to work on the center of gravity of the walking vehicle.

5) "Energy Stability Margin[6]":In the process of tumbling, the center of gravity passes over the point at which it possesses the maximum potential energy under the field of gravity. This criterion evaluates the stability by the magnitude of the difference between its maximum potential energy and its initial one.

6) "Dynamic Energy Stability Margin[7]": It is similar to the criterion of the "Energy Stability Margin," but all the external forces including gravity are considered to work on the center of gravity of the walking vehicle.

Furthermore, the stability margin of the arbitrary posture at a moment is given as the minimum value evaluated by each criterion around all the yaw angles of its body.

The criteria of 2), 4), and 6) add the dynamic effect to the criteria of 1), 3), and 5) respectively. This paper focuses on the static stability on rough terrain for the practical use, therefore, the criteria of 1), 3), and 5) are only compared each other after this. The criterion 1) was proposed on the assumption that walking vehicles were on the level ground, but it is contradictory on the rugged slope shown in Fig.2 (1), where the walking vehicle touches to the ground by its two feet in 2D. Then let's consider the most stable point for the center of gravity on the line at the constant height from the ground. According to the "Stability Margin," the most stable posture is the one shown in Fig.2(2), where the center of gravity is situated just in the middle. It is true that this posture can maximize the minimum moment(M_g) by gravity around each foot which works to stabilize the body. However, once the body is affected by a disturbance force from the horizontal direction, the moment around the downhill foot(M_{d2}) becomes larger than that around the uphill one(M_{d1}) and it is easier for the walking vehicle to tumble down the slope, even if the sizes of the disturbance force are kept equal. Consequently, the "Stability Margin" doesn't give us a right result, when all the support feet are not on the same level plane.

Considering the fact that the moments(M'_{d1}, M'_{d2}) caused by the horizontal disturbance forces differ in each support foot, the posture should be changed so that M'_{d1} and M'_{d2} generates in proportion to M'_{g1} and M'_{g2} respectively. Such a posture is expressed by Fig.2(3) at which the resultant vector of both gravity and the horizontal force vectors passes over each support foot. As a result, it corresponds to the posture maximizing the "Gradient Stability Margin," which evaluates the inclination of the waling vehicle on rough terrain when it starts tumbling by the instant disturbance force.

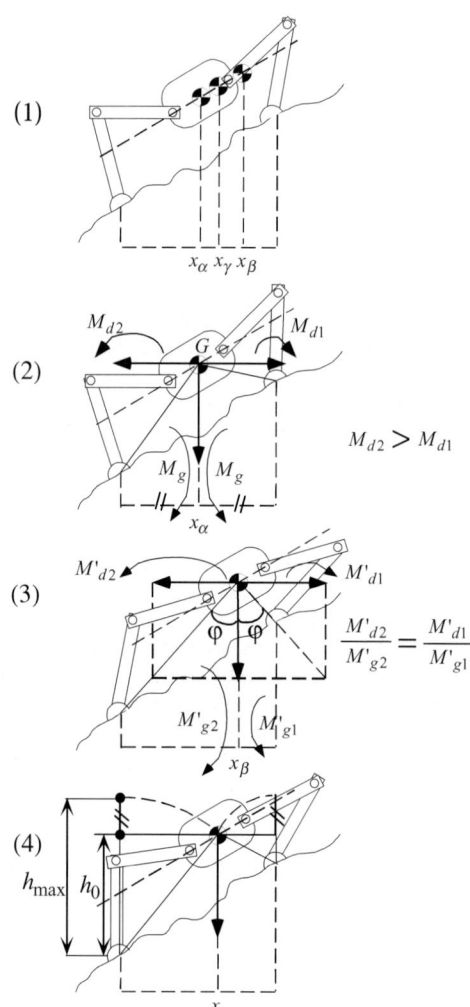

Fig.2 The relationship between the posture and stability criterion. (1) The walking vehicle under consideration. (2) Posture maximizing the Stability Margin. (3) Posture maximizing the Gradient Stability Margin.(4) Posture maximizing the Energy Stability Margin.

From these points of view, another consideration comes up to our mind, which regards the cause of the tumble as not the instant force but as the energy working on the body. More specifically, the center of gravity won't reach its highest position in the process of its rotating around the support foot, if its kinematic energy by the disturbance is completely consumed by increasing its potential energy. In other words, a large difference between the potential energy at the initial position of the center of gravity and one at its highest position can evaluate the stability of walking vehicles from the energy point of view. This difference was proposed as the "Energy Stability Margin."

This criterion shows us that the posture in Fig.2(2) is easier to tumble down the slope because the lifted distance of the center of gravity for the downhill side is less than that for the uphill side, while the posture in Fig.2(3) is easier to tumble to the uphill side. As a result, the most stable posture maximizing the Energy Stability Margin is one in Fig.2(4) which divides it equally into both sides. Eventually, these criteria lead to a different optimal position of the center of gravity respectively, as shown in Fig.2(1), but which one is the most reasonable for the practical use?

2.2 Comparative Experiment

According to the consideration in 2.1, it seems that the "Energy Stability Margin" is apparently more valid than the "Stability Margin" and the "Gradient Stability Margin." However, when the walking vehicle is affected by the disturbance force and one of the legs is lifted up, it can be guessed that the lifted foot hits the ground with the large impulse and another leg is lifted up again and again, and finally the walking vehicle leads to tumble. Judging from it, it might be necessary to come up with the intermediate criterion combining the "Energy Stability Margin" and the "Gradient Stability Margin."

In order to verify the above presumption, the validity of each criterion was considered through the next simple experiment. The experimental devices are composed of a small model of walking vehicle and a shock generating device as shown in Fig.3. The small model is 1.88[kg] in weight and 225[mm] in length between backward and forward. It is equipped with the adjustable weight at the height of 145[mm] to change the position of the center of gravity. The shock generating device consists of the sliding slope and the spring. The sliding slope carrying the model can slide on the level ground with the maximum inclination of $\theta = 13.6°$, and it is pulled by a hand from one side against the spring force. Once it is released suddenly, it crashes into the elastic material on the other side, which generates the horizontal disturbance force. In this experiment, the sliding slope with the model was crushed repeatedly with the increase of the spring extension little by little, and the spring extension at which the model started tumbling was regarded as the size of the disturbance force. Furthermore, the model was regarded as stable, so long as it didn't tumble after having its foot lifted up. The result of this experiment is shown in Fig.4, whose horizontal axis expresses the position of the center of gravity and its vertical expresses the initial length of the extended spring. According to this result, the position of the center of gravity, which is the least likely to tumble to both the uphill side and the downhill one, is $x = 151[mm]$. Mathematically, the positions of the center of gravity corresponding to the "Stability Margin," the "Gradient Stability Margin," and the "Energy Stability Margin" on the small model are given as $x_\alpha = 112.5[mm]$, $x_\beta = 174.0[mm]$, and $x_\gamma = 153.[mm]$ respectively. In the presumption before the experiment, it was expected that there would be the optimal position for the center of gravity between the posture maximizing the "Energy Stability Margin" and one maximizing the "Gradient Stability Margin."

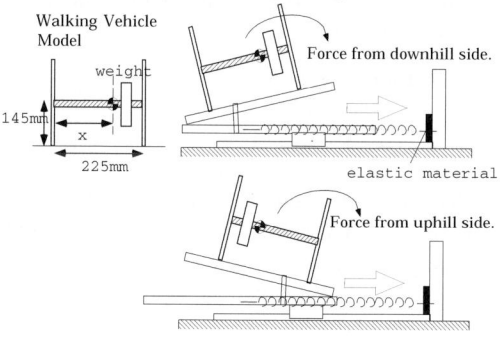

Fig.3 The experimental device.

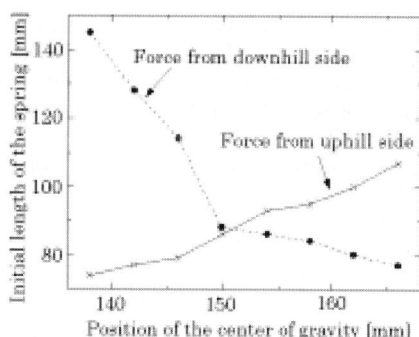

Fig.4 The experimental results showing the conditions to start tumbling.

Judging from the above results, however, it can be concluded that it is better to evaluate the stability by means of the "Energy Stability Margin" by itself.

Apart from the experiment in Fig.3, another one shown in Fig.5 was carried out, at which two small models taking the different posture were put on the vibrating slope and were swung from left to right. In this experiment, the same conclusion as Fig.3 was obtained.

Fig.5 Another experiment. The model maximizing "Stability Margin" in front tumbles earlier than one maximizing "Energy Stability Margin."

2.3 Proposal of "NE Stability Margin"

The above consideration made it clear that "Energy Stability Margin" proposed by Klein was the most desirable stability criterion for walking vehicles on rough terrain, and its basic concept is physically reasonable. However, some inconvenient aspects are remained in this criterion when it is used without any modification.

According to the definition of "Energy Stability Margin", walking vehicles would become more stable in proportion to its weight, even though their posture doesn't change at all. This result is reasonable in the sense that the potential energy to keep the original posture increases proportionally. But at the same time, the disturbance acting on the center of gravity also becomes large with the increase of weight, therefore, the increase of weight does not necessarily leads to the increase of stability. For example, the stability of walking vehicles affected by their miss stepping and their sudden stops has no relation to their weight. On account of this reason, the static stability criterion should be expressed by the dimension of length without including weight, that is, it should be defined as just the vertical distance between the initial position of the center of gravity and its highest position in the process of tumbling. Then, "Energy Stability Margin" normalized by weight, "Normalized Energy Stability Margin" or "NE Stability Margin" for short, is newly introduced in this paper, as expressed in the following equation.

$$S_{NE} = h_{mzx} - h_0 \quad (1)$$

Although "NE Stability Margin" doesn't change essentially from "Energy Stability Margin", it has a few advantages as follows. i)Stability can be evaluated in proper way when such a disturbance as mentioned before occurs. ii)As it is expressed by not the unit of [J] but the unit of [mm], it is convenient to derive a gain by means of the geometric way. iii)When walking vehicles are on the ground, "NE Stability Margin" corresponds to "Stability Margin" in the case where the center of gravity touches to the ground, which has the continuous relationship and is easier to understand intuitively.

Furthermore, "NE Stability Margin" can be made dimensionless if it is normalized by the total size. But the height of the center of gravity is expected to affect stability greatly, therefore, the dimension of length shall be remained in the criterion.

3. S_{NE} Contour and its characteristics
3.1 Derivation of the S_{NE} contour

When walking vehicle tumble, their center of gravity rotates around the support line l_s as shown in Fig.6(b). Judging from this phenomenon, "NE Stability Margin" S_{NE} can be derived from the vertical difference between the initial position of the center of gravity G and its highest position G_{max}, when G rotates around the point P on l_s. At this time, the S_{NE} contour is defined as the set of all the points P which possess the same S_{NE} on the contact plane.

Then, it will be explained here how to derive the S_{NE} contour for walking vehicles which stands on the contact plane as shown in Fig.6(a). The height of the initial center of gravity from P in the absolute coordinate frame, $PG|_{z0}$ and its maximum height $PG_{max}|_{z0}$ are expressed in the following equation.

$$\overrightarrow{PG}|_{z0} = h\,\cos\theta + d\,\sin\theta_e \quad (2)$$

$$\overrightarrow{PG_{max}}|_{z0} = \sqrt{h^2+d^2}\,\cos\theta_s \quad (3)$$

Notice that θ_s and θ_e represent the angle between l_s and the level ground and the angle between l_e and the level ground respectively, and they are given as the following equations.

$$\sin\theta_e = -\cos(\eta-\xi)\cdot\sin\theta \quad (4)$$

$$\cos\theta_s = \sqrt{1-\sin^2(\eta-\xi)\cdot\sin^2\theta} \quad (5)$$

In this case, S_{NE} can be defined as follows.

$$S_{NE} = \overrightarrow{PG_{max}}\big|_{z0} - \overrightarrow{PG}\big|_{z0} \qquad (6)$$

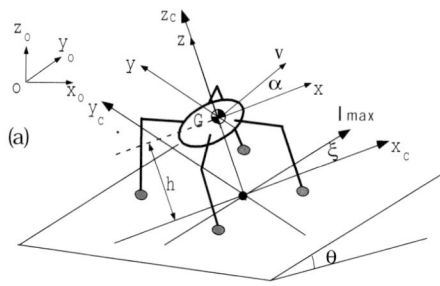

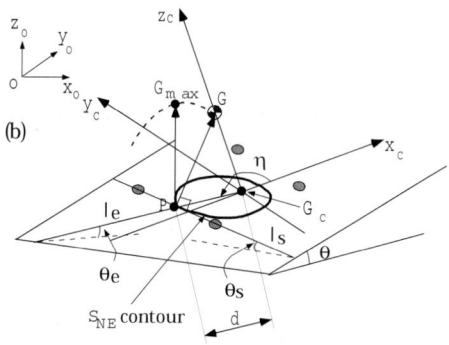

Fig.6 3D posture of the walking vehicle on a slope.
(a) : The body and contact surface coordinates.
(b) : The introduction of the S_{NE} contour.

The S_{NE} contour can be given as the set of points $(d\cos\eta, d\sin\eta, 0)$ satisfying equation (6), which are calculated in the coordinate frame of $\Sigma_c(G_c x_c y_c z_c)$. Therefore, it can be drawn on the contact plane by plotting all the points P keeping the distance of d from G_c for the direction of η in the range of $0 \le \eta < 2\pi$. Furthermore, the equation (6) gives the solution of d as follows.

$$d = \frac{-B \pm \sqrt{B^2 - AC}}{A} \qquad (7)$$

In the above equation, A, B, and C are given as $A = \sin^2\theta_e - \cos^2\theta_s$, $B = (S_{NE} + h\cos\theta)\sin\theta_e$, and $C = (S_{NE} + h\cos\theta)^2 - h^2\cos^2\theta_s$ respectively. This equation has two solutions for a d, but the solution in the range of $d < 0$ is neglected here. If both two solutions are in the rage of $d > 0$, it means that there are two points of P along the same emitted axis.

Fig.7 shows the contour for $\theta = 15°$. Here, the distance h between the center of gravity and the contact plane is kept 380[mm], and the maximum gradient line l_{max} corresponds to the axis x_c.

The number at the side of each contour expresses S_{NE} and its unit is [mm]. The S_{NE} contours for $\theta = 0°$ form the concentric circles, which are similar to ones derived by the criterion of "Stability Margin." On the other hand, the S_{NE} contours for $\theta = 15°$ results in forming the unique curves on the contact plane which are expanded to the downhill side.

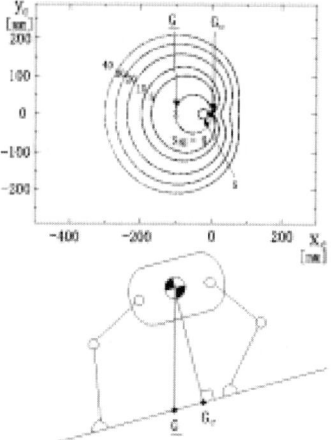

Fig.7 S_{NE} contour in the case of 15 degrees.

3.2 Characteristics of the S_{NE} contour

The S_{NE} contours on slopes in Fig.7 include two singular points. One of them is the intersection G_c between the contact plane and the line perpendicular to it through G, and the other is the intersection $\underline{G}$ between the contact plane and the line vertical to it through G. These singular points possess the following characteristics individually.

(1) G_c is the centeral point of every S_{NE} contour, therefore, the support line through G_c can be drawn for any direction. Among all the support lines through G_c, the support line perpendicular to l_{max} can make S_{NE} maximize and such S_{NE} shall be expressed as $S_{NE}(G_c \max)$. As the suppourt line is inclined with its passing through G_c, S_{NE} decreeses and it becomes $S_{NE} = 0$ when it corresponds to the maximum gradient line.

(2) $\underline{G}$ is the point above which the center of gravity exists, therefore, there is no chance when the center of gravity is situated at the higher point than its beginning, as long as it is rotated around the point $\underline{G}$. In other words, every support line through $\underline{G}$ makes S_{NE} equal to 0 regardless of its direction.

Two singular points with these characteristics results in drawing next two kinds of unique S_{NE} contour on slopes as follows.

i) Interestingly, the S_{NE} contour of $S_{NE}=0$ on slopes forms not a point but a circle whose diameter is $\underline{GG_c}$, while it forms a point on the level ground. This phenomenon can be explained in the following way by using Fig.8. The emitted axis G_cP perpendicularly intersects the support line l_s. In addition, the S_{NE} contour with respect to l_s connecting P and $\underline{G}$ is kept 0 because of (2). As a result, all the points P satisfying this condition become the set of points forming $\angle \underline{G}PG_c = 90°$. On the other hand, there exists a geometric constraint that the angle of circumference becomes $90°$ if its string corresponds to the radius. Therefore, the set of such points P forms a circles whose diameter is $\underline{GG_c}$.

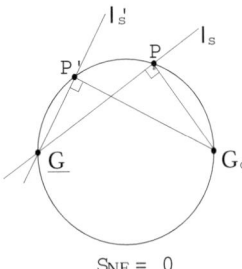

Fig.8 Extended figure of $S_{NE}=0$ in Fig.7.

ii) The S_{NE} contours satisfying $0 < S_{NE} < S_{NE}(G_c \max)$ are drawn inside the circle whose diameter is $\underline{GG_c}$, and the S_{NE} contours satisfying $0 < S_{NE} < \infty$ are drawn outside it. The 3D characteristic of G's rotation leads to the generation of two contours possessing the same S_{NE}, which is equivalent to the fact that the quadratic equation (7) has two solutions. In Fig.7, the S_{NE} contour of $S_{NE}(G_c \max) \approx 10$ and the S_{NE} contour of $S_{NE}=5$ exist inside and outside the circle of $S_{NE}=0$.

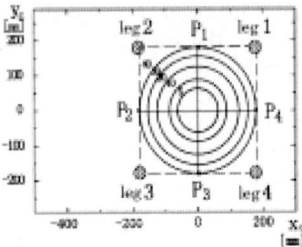

Fig.9 (a) Posture with $S_{NE}=40$ for all directions on 0 degree slope.

Finally, the application of the S_{NE} contour to derive a stable posture is described. For example, when four feet are on the level ground as shown in Fig.9(a), $S_{NE}=40$ is maintained. In order to maintain the same stability $\theta = 15$, it is clear that four feet should be situated at dotted circles in Fig.9(b).

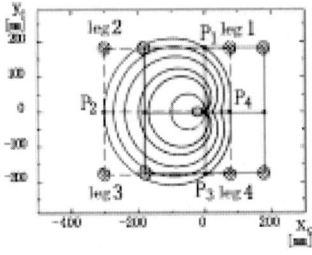

Fig.9(b) Improved posture on 15 degree slope with the dotted line which has the same S_{NE} contour as (a).

4. Conclusions

In this paper, the existing static stability criteria for walking vehicles on rough terrain were compared systematically, and "NE Stability Margin" was newly proposed as the most proper one. In addition, the contour for the stability was proposed as the tool to derive the desirable posture and its derivation was described. The proposed tool here is assured to be helpful to design the most stable gait for quadruped walking vehicles, called "intermittent crawl gait"[8].

References

[1] S. Hirose, K. Yoneda, H. Tsukagoshi : ``TITAN VII: Quadruped Walking and Manipulating Robot on a Steep Slope,'' IEEE International Conference on Robotics and Automation, 494/500(1997).

[2] McGhee,R.B. and Frank,A.A,:``On the Stability Properties of Quadruped Creeping Gaits,'' Math.Biosciences,3 331/351(1968)

[3] K.Yoneda, S.Hirose:``Tumble Stability Criterion of Integrated Locomotion and Manipulation,'' IROS96 Proceedings, 870/876(1996)

[4] Hirose, Iwasaki, Umetani:"Static Stability Criterion for Walking Vehicles," 21st SICE Symposium (In Japanese),253/254(1978)

[5] E.G.Papadopoulos and D.A.Rey:``A New Measure of Tipover Stability Margin for Mobile Manipulators,"IEEE International Conference on Robotics and Automation, 3111/3116 (1996)

[6] DOMINIC A. MESSURI and CHARLES A. KLEIN: ``Automatic Body Regulation for Maintaining Stability of a Legged Vehicle During Rough-Terrain Locomotion," IEEE JOURNAL OF ROBOTICS AND AUTOMATION, vol.RA-1, No.3, SEPTEMBER (1985)

[7] A.Ghasempoor and N.Sepehri:``A Measure of Machine Stability for Moving Base Manipulators," IEEE International Conference on Robotics and Automation, 2249/2254(1995)

[8] H.Tsukagoshi, S.Hirose:``Intermittent Crawl Gait for Quadruped Walking Vehicles on Rough Terrain," Clawar'98, 323/328(1998)

Creating the Architecture of a Translator Framework for Robot Programming Languages

Eckhard Freund, Bernd Lüdemann-Ravit, Oliver Stern, Thorsten Koch

Institute of Robotics Research (IRF), University of Dortmund
Otto-Hahn-Str. 8
D-44227 Dortmund, Germany
email: {freund,luedeman,stern,koch}@irf.de

Abstract

This paper presents a novel approach to facilitate the development and maintenance of translators for industrial robot programming languages. Such translators are widely used in robot simulation and offline programming systems to support programming in the respective native robot language. Our method is based upon a software architecture, that is provided as a complete translator framework. For the developer of a new translator, it offers convenient strategies to concentrate on robot specific language elements during the design and implementation process: fill-in templates, libraries for common functionality, design patterns etc., all tied up with a general translation scheme. In contrast to other compiler construction tools, the developers need not care about the complex details of a whole translator. As a matter of principle, the architecture offers a complete default translator (except for the grammar). Robot specific elements can be held in separate units - outside of the actual translator - to facilitate maintenance and feature extension. The most probable changes in the translator product life cycle are restricted to the adaptation of these units. Several translators built upon this framework are in actual use in the commercial robot simulation system COSIMIR® to support native language robot programming, as well as in the widely used robot programming system COSIROP to verify the syntax of robot programs.

1 Introduction

There exist two different basic approaches to develop industrial robot programs within a robot simulation system. The first one uses a general engineering language which is translated into robot specific code for execution on robot controllers (post-processors). The second technique is characterized by programming the robot in its native robot programming language inside the simulation system [1].

Modern robot simulation systems should support native robot programming of industrial robots for two main reasons. First, robot programs executed in industrial environments are characterized by frequent changes, e.g. due to product or production cell changes. These changes are most often carried out directly on the shop floor. Simulation systems which support native robot programming can process these modified programs for further analysis, simulation or reprogramming. Secondly, the acceptance of a robot simulation system increases with the support for native robot programming. The users need not learn a new programming language, but can program in the robot language he is familiar with.

The main difficulty for robot simulation systems to support native robot programming is the fact that there exist several hundred different proprietary robot programming languages [2]. Each robot manufacturer has developed his own language. The robot programming language does even vary between different version of a robot controller. Some manufacturers support even more than one programming language with the same controller, e.g. Mitsubishi Electric [3].

Another problem is the history of the language design. The manufactures have changed, extended and modified their languages in the course of time. As a result, these languages are difficult to handle, because there is no strict systematics (e.g. lack of grammars) within the languages. Besides, modern production cells are very heterogeneous. Industrial robots from different manufactures and robot controllers with different software versions are found within the same workcell.

Consequently, the effort to support more than one language within a robot simulation or programming system is extremely high, and the development process is very expensive. In this paper, we present a framework to minimize the time-to-market for the support of new industrial robot programming languages within a robot simulation system, allowing the implementation of new translators on customer demand.

2 Requirements

To support different robot programming languages in a simulation system, all these languages are translated into the same "assembler-like" intermediate code, e.g. IRDATA [4] or ICR [5]. This code is interpreted by a virtual robot controller inside the simulation system. This method is a modified version of the proposal in [1] (see figure 1).

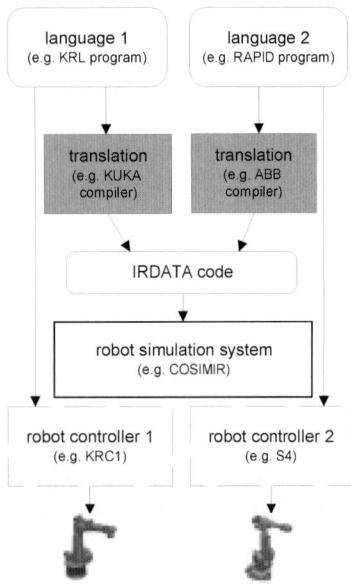

Figure 1: Native language support

Our experience has shown that the development time for such translators varies by about 30% (see table 1), depending on the complexity of the language. The IRL (Industrial Robot Language) translator was used as the reference for the values in the table.

Language (Manufacturer)	Development time
BAPS (Bosch)	80%
IRL (Standard)	100%
MBA3 (Mitsubishi)	90%
MRL (Mitsubishi)	70%
VAL II (Unimation)	100%

Table 1: Previous development times (in % of IRL)

Therefore, a software architecture is needed to support the programmer in the development process. The aim of our translation framework is to reduce the development and maintenance time.

We have identified the following main requirements for such an architecture:

(A) Common functionality for different languages should be unified and standardized in libraries.

(B) Language specific issues should be completed by filling in templates rather than programmed anew.

(C) Common functionality must be adaptable to the needs of language specific issues.

(D) It should be possible to use libraries of native and intermediate code to facilitate maintenance and to reduce the necessary amount of code for the translator.

So, the developer of a translator for a specific robot programming language should be put into the position to concentrate on the language specific issues and to neglect the implementation of recurring mechanisms.

3 System Structure

Our framework implements a complete translator for a general (higher) programming language, except for the grammar. The overall structure of the system consists of five main units, which correspond to the classic phases of compilation (see figure 2).

This concept is based on an advanced paradigm of translation: the translation steps are executed sequentially on an internal representation (abstract syntax tree, AST) of the complete robot program. In older compilers, each translation step was executed only on a single command. This means that only the necessary information to translate one command was held in memory [6], resulting in reduced memory consumption. The advantage of the newer concept is in the design of the translation system: the translation steps can be separated much more easily, thus operating independently on the AST [7]. This concept is therefore better suited to serve as a basis for the requirements shown above.

All robot programming languages have built-in elements, such as predefined variables (e.g. actual position), data types (e.g. positions), functions (e.g. frame manipulation) and statements (e.g. movement commands). These built-in elements distinguish a robot programming language from a "general" programming language. Some built-in elements are explicitly defined in the so called "system files" of the corresponding robot controller. These files are usually coded in the robot programming language.

The declarations of built-in elements, which cannot be found inside the system files, should be put into the built-in language elements files.

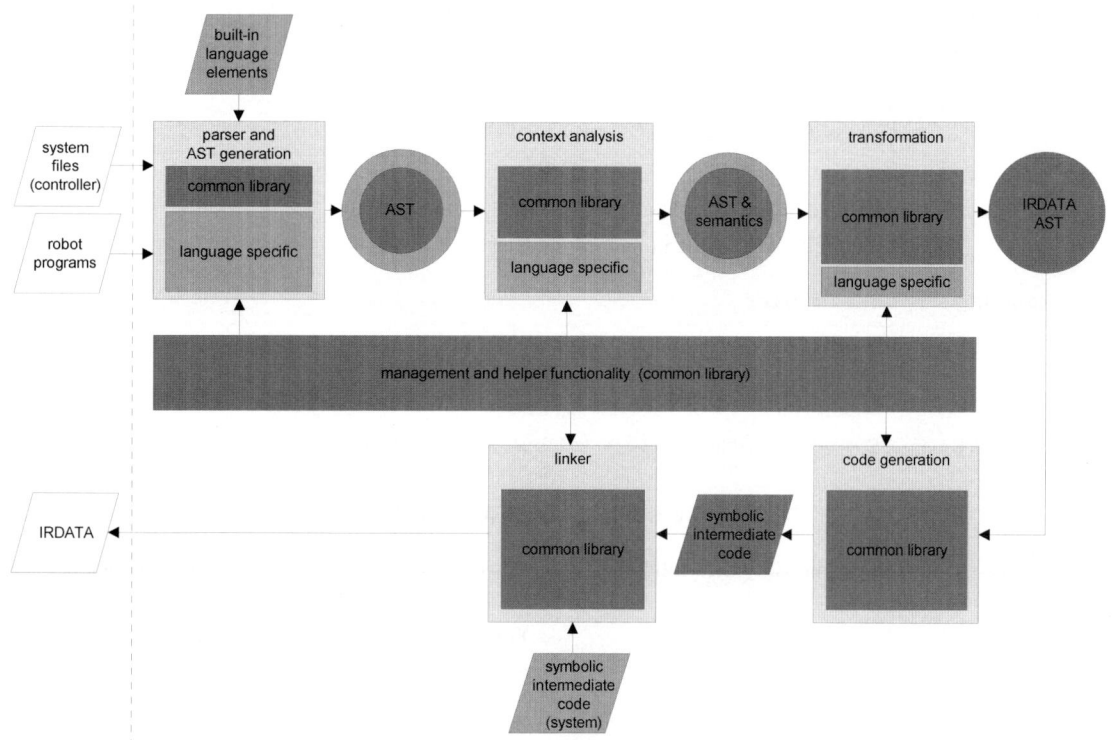

Figure 2: Structure of the translator framework

The translator can then use this information to perform a complete syntax and context analysis, and to generate a standardized call into the intermediate system code library, where the functionality has to be implemented in symbolic intermediate code.

The advantage of this approach is the fact, that for changing the semantics and/or implementation of one of these built-in elements, the translator itself need not be touched. Thus, considerable time in the test, correction and feature extension phases of the translator life cycle is saved. This facilitates maintenance enormously (see requirement (D)).

Due to the development history of some industrial robot programming languages, it is not always possible to formalize the declaration of built-in elements in one of these files. Such built-in elements must be added to the language specific AST, because they need a specific treatment inside the translator. The following example, a palletizing function, is taken from the Mitsubishi MELFA BASIC IV (MBA4) language [7]. The parameters are not allowed to be in brackets, which would be necessary for a declaration in ordinary syntax.

```
10   P1 = PLT 3,5
```

3.1 Design of the System Units

The main characteristic of this architecture is the division of each system unit and each data structure of the AST into a common part and a language specific part. In the common part, everything is united that can be found in "ordinary" programming languages (e.g. program flow instructions). A default processing for this part of the AST is offered to the developer of a new translator. In the language specific part, everything is concentrated that belongs to a specific robot language, and that cannot be declared in the built-in elements or system files.

Every portion of the AST is divided into a common library part and a language specific part in this manner. The division was made analogously for declarations, statements, expressions, actual and formal parameter, routines, etc. The following example illustrates the division of the statements with an indirect call (common library) and a linear movement (language specific part). We use GENTLE [7] notation, the PROLOG like programming language that was used to implement the framework.

```
-- common part
   'action' AnalyzeStmts(STMT -> STMT)
(I)      'rule' AnalyzeStmts(Stmt -> A_Stmt):
             CS_Pre_AnalyzeStmt(Stmt -> A_Stmt)
(II)     'rule' AnalyzeStmts(cc_callindir(Pos, …) -> cc_callindir(Pos, …):
(III)    'rule' AnalyzeStmts(cs_stmt(Pos, CsStmt) -> cs_stmt(Pos, A_CsStmt)):
             CS_AnalyzeStmt(CsStmt -> A_CsStmt)

-- language specific part
(i)      'action' CS_Pre_AnalyzeStmt(STMT -> STMT)
(ii)     'action' CS_AnalyzeStmt(CS_STMT -> CS_STMT)
             'rule' CS_AnalyzeStmt(mba4_mvs(Pos, …) -> mba4_mvs(Pos, …)):
             …
```

Figure 3: Context analysis example

The common AST part comprises the following elements:

```
'type' STMT
  -- common part
  cc_callindir(Pos:    POS,
               Expr:   EXPR,
               AParam: APARAMLIST)
  -- language specific statements
  cs_stmt(Pos:  POS,
          Stmt: CS_STMT)
```

The language specific AST part includes the following:

```
'type' CS_STMT
  -- linear movement
  mba4_mvs(Pos:    POS,
           Target: EXPR,
           …)
```

The common part includes one predicate (here, "cs_stmt") where all language specific statements can be included. So, common units can work on the common AST, while language specific units can process the language specific part. Thus, no common unit can see any language specific parts of the AST.

Every processing step ("action") on the AST in a common unit is structured in the same manner, see figure 3 for an example from the context analysis:

(I) One predicate to overwrite the standard processing of common AST parts, if desired (requirement (C)).

(II) Several common predicates that define a default behavior for the common AST processing (requirement (A)).

(III) One predicate to handle all language specific AST parts (requirement (B)).

For (I) and (III), there are corresponding processing actions (i) and (ii) in the language specific unit.

Action (i) is filled-in in the language specific unit, if the processing of the common AST is insufficiently covered by the common rules. Usually, this action is empty, so the default processing is used. Thus, it is possible to add some processing steps to the common rules from (II), that are carried out before the default behavior. But also, the complete default processing can be overwritten.

Action (ii) must be implemented for all language specific statements, that exist inside the language specific AST, and that are not declared in the built-in elements or system files (e.g. "mba4_mvs()").

3.2 Recipe for new Translators

To develop a new translator for a – maybe even exotic - robot programming language, the following main steps have to be performed:

1. Define a grammar. This has to be done in any case. Except for expressions, there is no common grammar part available. For expressions, design patterns from the common library can be used. Due to the compiler construction kit used here [7], this step is well supported.

2. Determine the built-in elements of the language (check the manuals).

    2.1. Analyze the controller's system files for declarations of built-in elements.

2.2. If other built-in elements can be declared in the native robot language, declare them in the built-in elements files.

2.3. Add the remaining built-in elements to the language specific AST.

3. Fill in the templates in the language specific units (context analysis, transformation) of the framework to handle the parts of the language specific AST. (ii)

4. Add language specific handling of built-in data types inside expressions to the context analysis unit. (ii)

5. Add type compatibility and conversion issues to the context analysis unit, especially for assignment, actual and formal parameters, and function return values. (ii)

6. Analyze the language for elements, that are treated in another way than in the common part, and override the corresponding behavior inside the language specific system units. (i)

7. Implement built-in statements and functions inside the symbolic intermediate system code file.

4 Results

In table 2, 3 and 4, we show the robot programming languages that have been developed based on the framework presented above. The linker was added to the tables although it is not a robot programming language. The reason is, that the linker itself uses common parts of the system units parser and AST generation, context analysis and code generation.

As can be seen, the framework handles robot programming languages with different characteristics equally well: PASCAL-like languages (KRL, RAPID), BASIC-like languages (MBA4), NC-like languages (ROSIV) and proprietary style languages (V+).

Table 2 and 3 show the gain of time in development and maintenance for these robot programming languages. The "estimated time" reflects the estimated time to develop, respectively to maintain a translator for this robot programming language without the framework, based on our experiences from table 1. The "development time", respectively the "actual time" reflects the real time, that was needed to develop/maintain the translator. The development times of the KRL and RAPID translators contain the time to develop the framework, otherwise they would be much shorter. The overall reduction in development time is about 70% to 80% on average. The gain in maintenance time is also about 80%. In total, the gain is higher than the proportion of use of common functionality (see table 4). The reason is, that the framework offers templates to fill in, rather than forcing the developer to program things anew.

Language (manufacturer)	Estimated time	Develop. time	Relative gain
KRL (KUKA)	100%	75%[1]	- 25 %
RAPID (ABB)	100%	83%[1]	- 17 %
V+ (Adept)	100%	29%	- 71 %
ROSIV (Reis)	90%	25%	- 72 %
MBA4 (Mitsub.)	100%	17%	- 83 %
Linker	25%	3%	- 88 %

Table 2: Compiler development time (in % of IRL)

Table 4 shows the share of common and specific parts in each system unit for the translators. The management and helper functionality (see figure 2) is not listed, because all translators use almost 100% of it. Due to this omission, the total share of common and specific parts cannot be calculated from the unit entries. Evidently, the share of common functionality increases with every translation step. On average, the translators use 2/3 of common functionality and 1/3 of language specific code.

Language	Estimated time	Actual time	Rel. gain
KRL	100%	30%	- 70 %
RAPID	100%	15%	- 85 %
V+	80%	25%	- 69 %
ROSIV	75%	15%	- 80 %
MBA4	130%	30%	- 77 %
Linker	100%	35%	- 65 %

Table 3: Maintenance time (in % of IRL)

Another conclusion can also be drawn: developers, who have a longer framework experience, can achieve a further gain (e.g. >80% for MBA4). But even framework novices can achieve a high gain in productivity (e.g. about 70% for V+).

---

[1] The framework presented in this article has been developed as a side product of the KRL and RAPID translators. So, both translators could already profit from its advantages.

	Syntax		Context		Transformation		Code Generation		Total	
	common	specific	common	specific	common	specific	common	specific	common	specific
KRL	59.4%	40.6%	56.8%	43.2%	80.6%	19.4%	93.8%	6.2%	70%	30%
RAPID	51.5%	48.5%	56.3%	43.7%	71.4%	28.6%	89.6%	10.4%	66%	34%
V+	48.1%	51.9%	49.0%	51.0%	76.1%	23.9%	92.2%	7.8%	64%	36%
ROSIV	59.8%	40.2%	65.5%	34.5%	80.7%	19.3%	94.4%	5.6%	65%	35%
MBA4	46.7%	43.3%	66.7%	33.3%	73.2%	26.8%	89.2%	10.8%	67%	33%
Linker	49.6%	50.4%	29.8%	60.2%	-	-	30.1%	69.9%	65%	35%

Table 4: Programming effort with framework (in percent of total lines of code per step)

5 Summary

This article presents a software architecture to facilitate the development of translators for robot programming languages. This framework does not share the aims of other general compiler construction tools, like e.g. [7], [8]. Such systems can be used to implement our framework. Residing on top of these construction kits, our system abstracts the development of robot language translators. This abstraction from application specific languages offers the developer e.g. fill-in templates, common functionality, design patterns and recipes to build translators. He can therefore concentrate on the robot specific programming language elements, instead of a complete programming language in general, as it has been the case up to now [9]. We have shown that in this manner, our framework achieves an enormous reduction of development time.

The framework encapsulates almost all built-in elements of a specific language in separate files. These built-in elements usually cause most of the code changes within the product life cycle. Our approach allows to perform these changes without changing binary executables. Therefore, a considerable reduction of maintenance time has been achieved.

6 Commercial Applications

The translators that were built on basis of this software architecture, have been integrated with the commercial robot simulation system COSIMIR® (Cell Oriented Simulation of Industrial Robots) [10]. This 3D-simulation system has been developed at the IRF and is in global, practical use.

COSIROP is the combined online and offline programming software for Mitsubishi robots. It includes our translators to check the syntax of robot programs before they are downloaded to the robot controller.

References

[1] R. Dillmann, M. Huck: "Informationsverarbeitung in der Robotik," p. 124, Springer-Verlag, 1991.

[2] S. Hesse: "Industrieroboterpraxis: automatisierte Handhabung in der Fertigung," p. 201 ff, Vieweg, 1998

[3] "CRx Controller: Instruction Manual," *BFP-A5992*, Mitsubishi Electric Corporation, 1999

[4] "IRDATA - Industrial Robot Data," *DIN 66314*, Beuth-Verlag, 1994

[5] "ICR - Intermediate Codes for Robots," *ISO/CD 10562-2*

[6] A. V. Aho, R. Sethi, and J. D. Ullman: "Compilers: Principles, Techniques and Tools," Addison-Wesley, 1987

[7] F. W. Schröer: "The GENTLE Compiler Construction System," Oldenbourg-Verlag, 1997

[8] E. M. Gagnon, L. J. Hendren: "SableCC, an Object-Oriented Compiler Framework," *Proc. on Technology of O-O Lang.*, p. 140-154, Aug. 1998

[9] M. Rackovic: "Construction of a Translator for Robot-Programming Languages," *Jnl. on Intell. & Robotic Systems*, p. 209-232, Feb. 1996

[10] E. Freund, J. Rossmann: "Systems Approach to Robotics and Automation," *IEEE Int. Conf. on Robotics and Automation*, p. 3-14, May 1995

A Multi-processing Software Infrastructure for Robotic Systems

Andrew H. Jones   Guilherme Nelson DeSouza   Avinash C. Kak

Purdue University

{ajones, gdesouza, kak}@ecn.purdue.edu

Abstract

Robotic systems and their software design are based on the same principles found in other computer applications. However, while many computer systems to date are deeply rooted in concepts such as multi-processing, multi-threads, modularity, etc., robotic systems are still limited by mono-processing, centralized designs. This paper addresses these limitations by proposing a new philosophy for robotic system software design – software infrastructure – that allows for the design of an efficient, modular, fault-tolerant, and distributed software architecture. Two examples of this infrastructure applied to mobile robot navigation are also discussed.

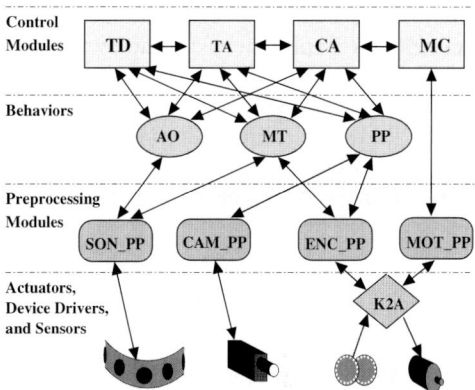

Figure 1: Behavior Based Navigation System

1 Introduction

In the past fifteen years, specially since the works of [4, 1], researchers in robotic systems began to realize the importance of decomposing a centralized control architecture into smaller distributed units. These new distributed architectures became widely accepted [10, 5, 3], and much work has been dedicated to refine the high-level architecture. Despite some effort dedicated to message passing [11], there has been little attention to solutions addressing the practical issues in the low level infrastructure necessary for a robust, generic, and modular software architecture in the domain of robotic systems.

The resources necessary to implement this type of infrastructure have become available in computer science and have been employed in many areas [12]. While robotic control architectures are executing on the same computer systems, many of the software systems developed for robots are still designed and implemented using early monolithic computer models. Many robotic systems today use concepts such as behaviors [10, 1], subsumption [4], tiered-level [3], etc., to deal with the multi-facet requirements of the specific applications for which they were designed. However, most of these systems neglected the more profound changes that would add characteristics such as modularity, portability, encapsulation, and parallelism (distributed processing).

In this paper, we present a software infrastructure which is the foundation of some of our lab's systems, most notably in our mobile robot and visual servoing systems. This infrastructure combines concepts such as threads, processes, pipes, and software wrappers in order to provide a portable and modular environment. This allows the robot's resources (motors, cameras, sonars, range scanners, etc.) to be accessed concurrently by the different application modules operating in a distributed network of processors. Also at the end of the paper, we briefly introduce two mobile robot applications – a behavior-based navigation system (Fig. 1), and an object-oriented navigation system (Fig. 4)– that are being developed on top of this multi-processing software foundation.

2 Terminology and Background

In this section, we briefly present the definitions for some of the terms used throughout the paper. We understand that some of these terms are found in the computer science literature with a much more intricate definitions [12] than the one presented here. However, our goal in this section is to recapitulate some of these definitions for the target reader of this paper, which may not necessarily be familiar with these terms.

Resource - A physical device that needs to be controlled or accessed. For example: motors of a mobile robot; image grabber for cameras; etc.;

Process - An executing instance of a software function (or set of software functions) that run concurrently in the CPU. For example: a word processor and a web browser operating in different windows on the same computer where each program is a different process for the system;

Thread - A thread is frequently referred to as a *light-weight* process. In fact, a process can be made of many different threads, each one concurrently running inside the same process. Two threads inside a process can share the same variables. Spell-checkers, printing functions, and help assistants are possible examples of threads in a word processor;

Parent/Child Process - The relationship between processes and its implications to the system are not important in this paper. However, it is important to say that a parent process can create as many child processes as necessary, and those, in turn, can create their own child processes;

Server/Client - A *server* is a process or a set of processes that execute a specific task for the system. This task is only performed when a request is sent to the server by the *client* of this service. In our case, most of the servers' tasks is to provide access to a specific resource;

Software Wrapper - A set of functions or routines that give access to a server. The wrapper is the interface between processes and it is the wrapper that creates the *abstraction layer* that frees one process from knowing the details about how (or where) the other processes run;

Pipes - A *pipe* is a mechanism provided by the operating system with which processes can communicate and exchange information. In this paper, we use the term *pipe* indistinctively for intra-process, inter-process, or network (socket) communications.

Modularity/Encapsulation - A system is said to use encapsulation when all functions, routines, data structures, etc. relating to one and only one specific task of the system is enclosed in a single independent *module*. A module is a self-contained, executable program file.

3 System Infrastructure

As mentioned before, our motivation in proposing this software infrastructure is to bring some well known concepts found in distributed systems, real-time systems, etc., to the robotics domain.[1] In those systems, characteristics such as modularity, encapsulation, and portability, among others, are deemed necessary. For example, it's not uncommon for a particular task in a distributed system to migrate from one computer to another, depending on issues such as *fault-tolerance*, *CPU load*, etc. For this migration to happen, this same task must be well encapsulated in a single module so that it can be easily ported to another computer in the network.

In the robotics domain we found no different needs. The computer hardware for a mobile robot, for example, can be constructed using many independent CPUs performing tasks such as: a) controlling actuators; b) executing navigation software; c) supervising sensors; etc. Many of the above tasks need to communicate with one another regardless of the task allocation among the CPUs. In order to distribute the tasks, each task must run as processes, which are encapsulated in modules. Each module can be installed in its own CPU (or share CPU with other modules) completely independent of the other modules. However, since all these tasks need to cooperate in order to perform the overall system function, we need to provide a mechanism to exchange information between tasks. This mechanism, which is based on pipes and software wrappers, is the main focus of this paper. As it will be clear in the following example, this mechanism allows the design of a system with all the desired characteristics.

3.1 Example

In order to clarify these concepts, we present a simple example (Fig. 2) using our mobile robot, PETER, and its motion command service, K2A. Our mobile robot consists of a PC-based system mounted on top of a K2A Cybermotion base. The PC-based system running the Linux operating system contains two image grabbers, a network card, and a wireless interface to the internet. PETER also has an onboard controller for the active-vision stereo head, a controller for a laser range scanner, and an interface to five ultrasonic transceivers. Since the PC-based system is connected to the internet using a wireless bridge, any workstation in the network is a potential candidate to run any software task of our navigation system.[2]

To perform a simple navigation task, a basic system could be designed using the following four modules:

K2A Server: this module receives requests from various clients. These requests include *set-speed*, *turn*, *move*, *read-position*, *read-status* (e.g. battery voltage), *reset-deadman*, etc. Once a request is received, the server executes the command, monitors its completion, and returns a status along with the appropriate data regarding the current command.

Navigation Client: this module performs basic navigation decisions. It communicates with the K2A server to issue motion commands such as turn, set drive speed, etc.

---

[1] It is important to emphasize that despite the fact that these concepts have been widely applied in other domains, they are fairly new in the robotics domain.

[2] We usually run simulations and other matlab programs to display range maps in any of our Sun or PC workstations.

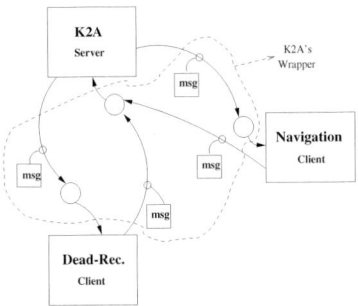

Figure 2: The clients call the K2A wrapper to exchange messages through their pipes

Dead-reckoning Client: this module reads the current position and orientation of the robot from the K2A server in order to keep track of its location within the environment.

System Status Client: this module monitors the status of the on-board batteries in the K2A and informs the K2A by resetting the deadman timer that the main CPU is in good standing.

As the reader can infer, the above three clients can concurrently send requests to the K2A server. Those requests are sent by invoking the K2A wrapper which communicates through the input pipe of the K2A server. However, the K2A server does not know which client sent the request. Therefore, the client also provides, through the wrapper, a return address to its own reply pipe.

3.2 Wrappers

As mentioned earlier, the motivation behind the wrapper is the need for the system to be encapsulated into independent, self-contained modules. With that in mind, the wrapper behaves as an abstraction layer that interfaces with the server. In simple words, the wrapper is a set of specific functions and routines that constitutes the protocol with which the client passes information to the server. Examples of the functions in the wrapper for the K2A include ROBOT_OPEN and ROBOT_MOVE. The client uses ROBOT_OPEN to establish a connection with the server while it uses ROBOT_MOVE to specify a drive speed and a turn angle. For all functions in the wrapper, the client needs to provide a reply pipe to which the server can respond. Within the wrapper, each function constructs a message which is sent forward to the server. This message is atomic, contains a specified number of fields, and can vary in length based on the command.

Upon reception of a message from its wrapper, the server identifies the command to be executed, and extracts all other necessary information that the message may contain to carry out the request. The server then executes the request and returns the appropriate data to the client by constructing a return message which contains the requested data and the status regarding the command executed.

3.3 Pipes, Processes, and Messages

Atomic message passing is a vital mechanism for exchanging data between processes. This is especially true in our robotic system where many processes are servers while other processes are clients overseeing the use of resources or services. Pipes are handled by the operating system in an efficient manner and offer a fast mechanism of communication between independent processes that may be running on different interconnected computers. The access to this mechanism resembles the normal open, read, and write commands for files. But unlike files, pipes are internal data structures stored in local memory or distributed in the network, not stored on slow disk drives. Another advantage with pipes is that when a process is waiting for data from its input pipe, it goes into idle mode. That is, the waiting process does not consume CPU time.

As mentioned before, pipes are only part of the picture. The message itself is a very important aspect of useful interprocess communication and the message structure deserves further discussion.

The message structure is intimately related to the relationship between processes. It is not unusual for processes to become client and server to each other at the same time. In fact, the relationship between client and server is only momentary. In a large system, processes can grow so interconnected that messages may need to pass from any one process to any other process. One could choose to define a unique message structure for each process. But this choice would force each process to know the specific message structure of the processes with which it communicates. As the system escalates in the number of processes, storing and using such specific message structures becomes more and more complex. Therefore, it is desirable to clearly define a single message structure that is versatile enough to handle all possible data types and data lengths needed for each process in the system. An example of a single message structure will be presented in subsection 4.1.

Atomicity is another aspect of message passing that a flexible, distributed topology requires. An atomic message is a message that is fully contained in a single data structure and is sent in a single write operation. If these conditions were not satisfied, two or more messages could be entangled with each other in the input pipe of the receiving process. For example, process A sends a message, '123456', where each digit is written separately into the pipe of process C, violating message atomicity. Concurrently, process B sends a message, 'FOO', to process C. In this situation, process C could receive an entangled and indecipherable messages, such as '123FOO456', which is unacceptable. However, if processes A and B sent atomic messages, process C would be guaranteed to receive the messages '123456' and 'FOO'. Note that the way in which the read operation from the pipe is preformed does not violate message atomicity.

Finally, another consideration in creating a flexible topology is forcing the message to contain information about the process that originated the message. In particular, the message must contain either the reply pipe, or the *process-id* of the sending process so that the receiving process can reply appropriately.

All these considerations – use of pipes, unique message structure, and atomic message passing – form the foundation of our system infrastructure. Together they provide flexibility, efficiency, portability, modularity, and encapsulation to any robotic system.

4 Infrastructure Applications

Up to this point, we have described a software infrastructure at the operating system level. In section 3, we demonstrated how one could construct a simple software architecture for a simple application of control and communication with the motion command service, K2A. But to demonstrate the true expandability and modularity of the proposed infrastructure, we will briefly discuss two system architectures being developed for a mobile robot: a Behavior Based navigation system and an Object Oriented navigation system.

4.1 Behavior-Based System

A behavior based architecture consists of many specific, task-oriented modules called behaviors that execute independently. For mobile robots, each behavior obtains input data that originates from some sensor or group of sensors and produces an output. This output combines in some fashion with outputs from other behaviors to determine vehicle motion within an environment. It is not the purpose of this paper to address behavior based navigation systems such as in [8, 10, 3, 1]. Instead, the main focus of this work is explaining the mechanisms embedded in the infrastructure and how these mechanisms allow the behaviors to communicate with each other. But, before discussing this communication, a brief description of the architecture is needed.

Architecture Overview Our architecture, Task Learning Architecture using Behaviors, TLAB (pronounced T-lab), naturally breaks into three units or levels of modules: *preprocessing modules*, *behaviors*, and *control modules* as illustrated in Fig.1. The *preprocessing modules* are strict[3] servers that encapsulate all functions and data necessary to serve a particular

---

[3]A strict server is a process that is never a client of other servers.

sensor or group of sensors. Many times these preprocessing modules are running on dedicated CPUs or microcontrollers. The *behaviors* map sensory data from appropriate preprocessing modules to a set of outputs for various control modules. As a module, each behavior is totally encapsulated and can operate at its own rate. This is important because reactive behaviors such as Avoid Object (AO) must be able to quickly affect the motion of the vehicle, while a passive behavior such as Plan Path (PP) affects the motion less frequently. The *control modules* decompose a high level task such as "navigate to room-D" into a sequence of safe, basic motion commands for the mobile robot. As Fig.1 indicates, the control modules combine the behavior outputs to carry out this task.

Architecture Development The proposed infrastructure of messages and independent modules developed as a result of the limitations in attempting to implement TLAB under other possible infrastructures. The first possible infrastructure considered was a mono-process, mono-thread system. But given the large scale of this system and the need for each behavior to operate at a different rate, a single module that encapsulates the entire system would be very complex if not impossible. The second system design attempt was using multiple processes within a single module. In this single-module system, the Task Descriptor (TD) was the only parent process, while the rest of the system was created as a set of separate child processes. This design proved to be a poor choice for a number of reasons. First, because of operating system constraints; when one of the child processes failed (halted or exited prematurely), the entire system would fail. This alone creates an unacceptable solution for any robotic system which requires fault-tolerance. Second, due to the close relation of parent/child processes, debugging of a child process was nearly impossible. Third, child processes could not communicate with each other except through the parent process, so the parent became the bottleneck for message passing. These reasons lead us to an improved infrastructure, the one proposed, which consists of multi-processes distributed among independent modules.

Once the decision was made to create the system of behaviors by using independent modules, a message passing structure had to be clearly defined. Due to the interconnectedness of this particular system where any one module might need to communicate with any other module, all the messages within the system needed to be formulated using an identical and flexible structure. Given the above considerations, we arrived at a structure that is composed of two parts: a header and a body. See Fig.3 for a pictorial view of the message structure. The *header* is the first part of all messages while the *body* contains specific data that can vary in length. The header is based on the Unix message pro-

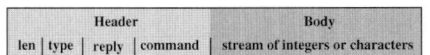

Figure 3: Message Structure for Behavior System

tocol [12] and contains the information such as: *len*, *type*, *reply*, and *command*. The field *len* contains the total length of the message body. The field *type* specifies both the type of data – character or integer – in the body and the message priority. The field *reply* passes the unique identification of the reply pipe. Finally, the action to be taken by the process receiving the message is provided in the *command* field. Note that no module actually fills the message. This task is performed by the wrapper of the recipient module.

Since the message varies in length, the receiving module reads from its input pipe in two operations. It first reads the header of the message to obtain the length of the body, then it can read the entire body without checking for an end-of-message marker.

Advantages of this Architecture Unfortunately, all further details about this architecture have to be omitted from this paper. However, we must list the reasons why this infrastructure offers an efficient and flexible solution for a behavior based architecture. These reasons are as follows:

Fault-tolerance - a global supervisor module can be instated to monitor all modules in the system. Because of the modularity of the design, upon the detection of a fault, the supervisor can take actions that range from re-prioritizing the scheduling of slow-responding modules, to the re-instantiation of unexpectedly terminated modules.

CPU Efficiency - due to the intrinsic characteristics of pipes, processes waiting for messages do not consume CPU time. Even for large scale systems with expanding number of processes, there is only minor impact on the CPU load.

Dissemination of Error Messages - by using a single, identical message structure, even the lowest-level modules can directly inform the highest-level modules of errors in the resources.

Elasticity - because of the encapsulation properties of this infrastructure, modules can be dynamically created and destroyed as necessary with little or no impact on other modules. For example, if a given task does not require ultrasonic transducer information, the corresponding preprocessor module can be destroyed.

4.2 Object-Oriented System

Most of the successful robot indoor navigation systems reported to date are constrained either by some kind of map of the environment [7, 2], by some artificial landmarks [6], or by the appearance of the environment where the robot is to navigate [9, 13]. In either

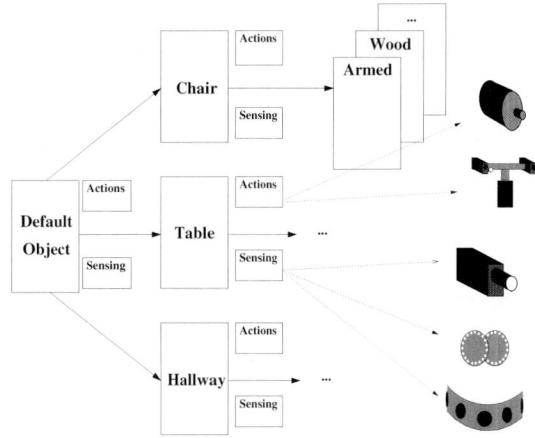

Figure 4: Object-Oriented Navigation System

case, information about the environment is stored in the robot memory[4] in a way that the system presents no generic structure that can be used in different and unexpected situations.

We believe that some of the limitations experienced with the navigation systems today derive from this lack of generic information about the environment. For that reason, we are developing a navigation system that is based on three central ideas: a) a Semantic Occupancy Map (SOM) that stores a description of each observed object; b) an object driven navigation scheme that embeds a generic representation of the environment inside directions such as *go down the* **hallway** *and turn at the third* **door**; and c) a task subsumption architecture that allows the robot to take specific actions associated with each observed object.

In our system, the meaning of the word *object* is two-fold. An object is any element of the environment, such as: a table, a chair, a hallway, etc. But an object is also this element's representation under the Object-Oriented paradigm (OO). Objects can belong to classes of objects and inherit properties from more generic objects. For example, Fig. 4 depicts three basic objects – chair, table, and hallway – and their relationship with the generic object class, *Default Object*.

Some of the properties that an object may inherit are the class methods. Every object has a set of methods that we grouped as *Action* and *Sensing*. The Action group includes all the methods that in one way or another will cause parts of the robot to move – active vision stereo head, motors, etc. The Sensing group consists of the methods used by the Action group to monitor the execution of its actions by means of the robot sensors – odometers, image grabbers, sonar, etc. The so called *Task Subsumption Stacker* selects which objects will be active at a given time. This decision

---

[4]This information is either pre-loaded in the robot memory or it is acquired during a learning phase.

is based on the current configuration of the environment, which is stored in the SOM, and since all objects have access to the SOM, they can check the position and description of other objects in the environment. The objects take actions that help the robot to recognize new objects and that lead the robot to a different position toward the robot's goal.

It is not the scope of this paper to drudge through the details of each of the ideas above. Instead, we wish only to emphasize the importance of the proposed infrastructure for this navigation scheme. This importance derives from the fact that each object instance, by nature of OO, is encapsulated in a module. The intrinsic property of objects as being modules lends to the ability of dynamically instantiating objects (create/destroy) based on the observed environment. Finally, since a method is defined within the scope of each object and the servers can be implemented as objects, the wrappers of a service becomes the method itself. Therefore, all the above discussions on distribution, modularity, multi-processing, and encapsulation present in our proposed infrastructure also offers an efficient and flexible solution for the object-oriented architecture.

5 Conclusions

We have presented a new philosophy for robotic system software design – software infrastructure – that allows the development of efficient, distributed, fault-tolerant control architectures. This infrastructure makes use of widely accepted concepts such as multi-processing, software wrappers, pipes, message passing, etc. The application of these concepts leads to the design of systems that are intrinsically characterized by encapsulation, modularity, and fault-tolerance.

Two examples, in which we applied this infrastructure, were also introduced. These examples constitute two completely different paradigms of robotic control, but the proposed infrastructure proved to be indispensable in both cases.

The same philosophy presented here has also been applied and tested by integrating a previously developed system [7] on top of this new infrastructure. Other systems developed in the lab, such as visual servoing for tracking moving objects, are also benefiting from this infrastructure.

Acknowledgement

The authors would like to thank Ford Motor Company for supporting this work.

References

[1] R. C. Arkin, "Motor schema-based mobile robot navigation," International Journal of Robotics Research Vol. 8, No. 4, pp. 92-112, 1989.

[2] S. Atiya and G. D. Hager, "Real-time vision-based robot localization," IEEE Trans. on R&A, Vol. 9, No. 6, pp. 785-800, Dec. 1993.

[3] R. P. Bonasso, R. J. Firby, E. Gat, D. Kortenkamp, D. Miller, M. Slack, "Experiences with an Architecture for Intelligent, Reactive Agents", JETAI, Vol. 9, pp. 237-256, 1997.

[4] R. A. Brooks, "A Robust Layered Control System for a Mobile Robot," IEEE Journal of R&A, Vol. RA-2, No. 1, pp. 14-23, 1986.

[5] A. A. D. de Medeiros, R. Chatila, and S. Fleury, "Specification and Validation of a Control Architecture for Autonomous Mobile Robots", in Proc. IEEE IROS, pp. 162-169, 1996.

[6] M. R. Kabuka and A. E. Arenas, "Position verification of a Mobile Robot Using Standard Pattern," IEEE Journal of R&A, Vol. RA-3, No. 6, pp. 505-516, Dec. 1987.

[7] A. Kosaka and A. C. Kak, "Fast vision-guided mobile robot navigation using model-based reasoning and prediction of uncertainties," Computer Vision, Graphics, and Image Processing – Image Understanding, Vol. 56, No. 3, pp. 271-329, 1992.

[8] J. Kosecka, H. Christensen, and R. Bajcsy, "Experiments in behavior composition", Robotics and Autonomous Systems, Vol. 19; pp. 287-298; 1997.

[9] T. Ohno, A. Ohya and S. Yuta, "Autonomous navigation for mobile robots referring pre-recorded image sequence," in Proc. of 1996 IEEE IROS, Vol. 2, pp. 672-679, Nov. 1996.

[10] J. Rosenblatt and D. Payton, "A Fine-Grained Alternative to the Subsumption Architecture for Mobile Robot Control", in Proc. of IEEE/INNS International Joint Conference on Neural Networks, Vol. 2, pp. 317-323, 1989.

[11] R. Simmons, "Structured Control for Autonomous Robots", IEEE Trans. on R&A, Vol. 10, No. 1, pp. 34-43, 1994.

[12] W. R. Stevens, UNIX Network Programming: Interprocess Communications, Vol. 1, Ed. 2, Prentice Hall PTR, Upper Saddle River, NJ, 1999.

[13] J. Weng, and S. Chen, "Vision-guided navigation using SHOSLIF," Neural Networks, Vol. 11, No. 7-8, pp. 1511-1529, Oct-Nov. 1998.

Open Real-Time Interfaces for Monitoring Applications within NC-Control Systems

Manfred Weck, Andreas Kahmen

Laboratory for Machine Tools and Production Engineering (WZL)
RWTH Aachen University of Technology
A.Kahmen@wzl.rwth-aachen.de

Abstract

Process monitoring is the key to an increasing degree of automation and consequently to an increasing productivity in manufacturing. The realization of process monitoring functionality demands an extension of the control system. The prerequisite for these extensions are open interfaces in the NC-kernel. Nowadays controls with open NC-kernel interfaces are available on the market. However these interfaces are vendor specific solution that do not allow the reuse of monitoring software in different controls. To overcome these limitations a vendor neutral open real-time interface for the integration of process monitoring functionality into the NC-kernel is presented in this paper. Additionally two implementations of the real-time interface for different target systems are described.

1 Introduction

Machine tools with rising functionality are demanded by the market these days. Innovative concepts and rationalization measures are supposed to lower the costs for all participants in the production process. Thus new control functionality is necessary to gain more flexibility and quality in production. Furthermore users wish to be independent of single suppliers.

These demands cannot be fulfilled solely by control vendors. It should rather be the task of technology suppliers to offer products that increase the functionality of controls. For that reason, open control systems are getting more and more popular lately [1,2].

In terms of open controls, different levels of openness have to be distinguished. They can be subdivided into "outer openness" and "inner openness" categories. Controls of the category outer openness aim at a unification and standardization of external control interfaces, like a harmonization of the NC-programming, an extendible Man Machine Interface (MMI) or a standardized interface for drives. Controls of inner openness allow access to control internal data and methods as well as the integration of additional software into the control.

Nowadays many controls offer open interfaces in the MMI. This openness is mostly achieved with a PC-based system architecture of the MMI. A few control vendors even took a further step. They provide libraries in the NC-kernel of their control that enable third parties to integrate software at certain points of the control internal data flow. Data access or even the manipulation of data is thereby possible.

Up to now all open control interfaces are vendor specific. Hence porting third party software modules is impossible.

This paper will present a solution for the mentioned limitations. For this purpose the "Application Interface", which is a new interface for integrating third party software modules into various controls, is introduced. The Application Interface focuses on integrating applications in the field of process monitoring, but it can also be applied to less time critical applications.

2 Architectures of process monitoring systems

Nowadays process monitoring applications are state of the art in industrial production. They are used to ensure safe processes at any time. Mass production is the main field of process monitoring applications. The need for monitoring functionality depends on the degree of automation. Monitoring of tool breakage and tool wear for drilling, milling, turning and threading are the most prevalent applications. Nowadays external sensors are the main source for the acquisition of input data for monitoring applications. However, control internal signals, such as the torque values of digital drives are used more and more.

Today the PLC is the most important link between the control of a machine tool and external components of process monitoring applications. For the data exchange between time-critical applications and the control, fieldbusses, like CAN, Profibus or Interbus, are the preferred communication media. However, for less time-critical applications, that mostly communicate with the control via the MMI, usually ethernet is chosen. Digital and analogue I/Os are still widely used for hardware interfaces. Furthermore, vendor specific solutions in terms of communication media as well as protocols that are developed for specific demands are still applied in certain monitoring applications.

In figure 1 the architecture of a sensor based application for tool monitoring is illustrated. The current cutting forces, which are measured with an external sensor, are preprocessed in an external analyzing unit. Here the cutting forces are also compared to their upper limits. If they exceed these limits, an error message is sent to the PLC where a suitable reaction is triggered.

Due to the specific software interfaces that are used for the communication between the PLC and the external components, porting the monitoring application to different controls or even machines with different PLC programs is impossible without changing the interfaces. However, the adaptation of interfaces always implies additional costs as well as the risk of failures.

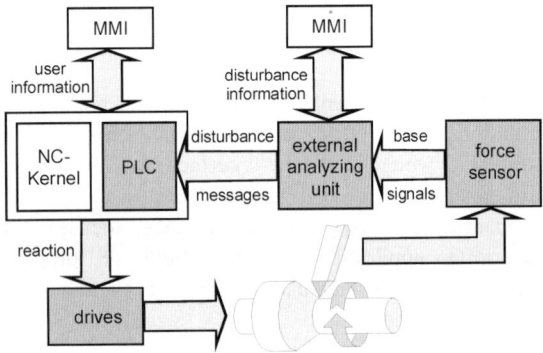

Figure 1: Architecture of an application for tool monitoring based on external sensors

The advantage of systems using external sensors is that they can be optimized for specific requirements of monitoring tasks. On the other hand, extensive additional components as well as costly adaptations and tuning are necessary. For this reason a trend towards the integration of process monitoring applications into the control can be noticed [3, 4]. Through the integration of monitoring applications, control internal information, such as the current of drives or position values, can be used as data origin for monitoring algorithms.

Because of the integration of process monitoring applications, additional hardware for external sensors, preprocessing and analyzing units as well as their electrical connections can be reduced. Figure 2 illustrates the architecture of a control integrated process monitoring system.

A data acquisition module filters and amplifies the base signals, which are provided by external sensors or are acquired from the drives. Disturbances in the signals are then eliminated and process data is prepared to be evaluated in an analyzing module. The analyzing module hands over information about disturbances to a reaction module which triggers an appropriate action in the NC-kernel.

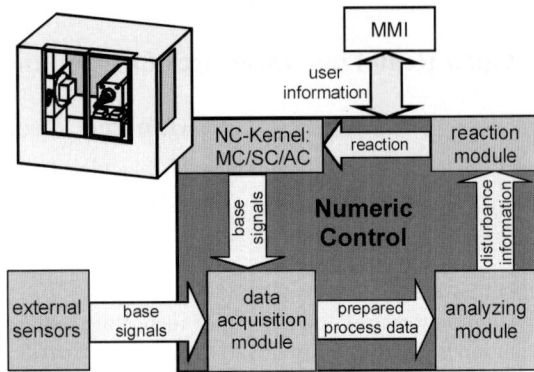

Figure 2: Architecture of a control integrated process monitoring system

As pointed out in the introduction, controls with an open NC-kernel are already available on the market. This is the prerequisite for the integration of monitoring functionality into the control. However, the interfaces for the integration are still vendor specific. For that reason the requirement for an adaptation of interfaces remains.

Both examples have demonstrated that porting process monitoring functionality without adaptation of interfaces is so far impossible. This can only be achieved with the development of vendor neutral open interfaces.

3 Requirements on a vendor neutral open real-time interface for process monitoring

Process monitoring functionality requires different levels of real-time performance. For example, monitoring tool wear is less time critical than monitoring tool breakage. The demanded performance of a real-time interface for process monitoring has to meet the requirements of the most critical application. Furthermore it has to be ensured that the interface can be implemented on different hardware platforms (e.g. vendor specific control platforms, DSPs, micro-controllers, etc.). The interface has to be scalable in functionality because not every control platform provides the same features, and finally, the integration of the interface must not affect the stability of the control system.

4 Concept of the Application Interface

For a simple integration of process monitoring systems into controls, the monitoring functionality should be split into small modular units such as preprocessing, analyzing or reaction determination units. As illustrated in figure 2, each unit should be encapsulated in a separate software module. These software modules are called Process Modules. The interfaces of the Process Modules have to follow the specification of the Application Interface. Through this, a seamless integration into the control should be feasible.

During the design of the Application Interface, existing standards or de facto standards were used wherever they were applicable. This approach should also apply to the implementation of the interface in different controls.

Figure 3 illustrates the concept of the Application Interface and the integration of Process Modules. The Application Interface is a uniform interface for data access in the NC-kernel or on external hardware platforms. It comprises a thin communication layer that hides the vendor specific communication mechanisms inside the control. The communication connection between the Application Interface and the control internal data structure is realized individually, depending on communication protocols (e.g. OSACA, CORBA, vendor specific protocols, etc.) and communication media (e.g. ethernet, fieldbus, etc.) that are either already available in the control or that can be adapted easily. Through this, the real-time performance of the interface can also be scaled by choosing an appropriate communication protocol and communication media for a specific control platform.

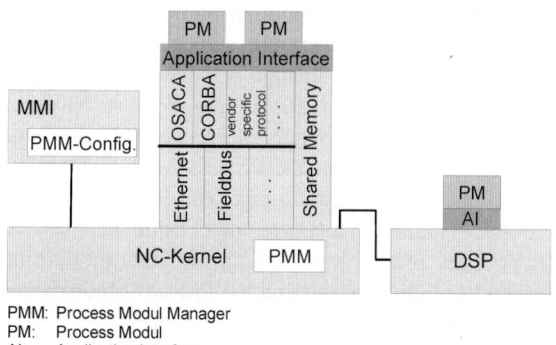

PMM: Process Modul Manager
PM: Process Modul
AI: Application Interface

Figure 3: Concept of the Application Interface and the integration of process monitoring applications

To meet the mentioned requirements concerning the real-time performance and the ability to be integrated on various hardware platforms, the Application Interface has been designed as a low level interface using C as programming language. C compilers are provided for almost any hardware platform. Furthermore a C-interface offers efficient methods for data exchange, which fulfill the requirements concerning the real-time performance.

In the following, the concept of the Application Interface, which is composed of the three parts "Communication System", "Reference Architecture" and "Process Module Manager", will be detailed.

Communication System:
It is the task of the Communication System to offer mechanisms for the data exchange between different applications and the control. In terms of data transfer the Communication System is the core of the Application Interface. The communication mechanisms have a significant influence on the performance of the interface. That is why the realization of efficient mechanisms for data exchange is essential.

A high flexibility of the Application Interface is achieved by defining a small set of methods that can access and handle different communication objects. This small set of methods covers the basic functionality of reading and writing data values as well as invoking and receiving function calls. Handling different communication objects via one method implies an identification of the objects, which is needed for their distribution in the interface. For this reason all communication objects used by the Process Modules for exchanging data as well as for invoking and receiving function calls have to be registered in the Application Interface during the start up phase of the monitoring system. The descriptions of the objects, which are used for the registration, have to follow the specification in the reference architecture which will be introduced later.

After the registration has been finished, the data exchange between Process Modules and the control system can be handled with simple get and set services for reading and writing data values. For invoking function calls in the control system, a method for requesting a service execution can be used.

For receiving function calls from the control system throughout normal operation of the monitoring system, the function pointers of the corresponding methods in the Process Modules have to be registered in the Application Interface during the start up phase.

The communication via the Application Interface follows the client-server-principle. This means that data or executable functions are offered by server objects to clients anywhere in the control system.

Reference Architecture:
The interoperability of Process Modules requires a clear specification of the server communication objects offered by the Process Modules and the Application Interface in addition to the access functions described in the previous paragraphs. This specification has to include non-ambiguous object names and the description of the objects´ attributes. The Reference

Architecture has to be understood as an open specification that can be extended whenever new Process Modules are developed or existing Process Modules are upgraded with new functionality. It is important, to make sure that the specification is always consistent and compatible to former versions.

In the Reference Architecture, the communication objects are subdivided into objects for data exchange and objects for handling of function calls. The specification of objects for data exchange has to provide the description of the objects' attributes. These attributes are the object name, a verbal description of the object's purpose, the identification of the object's internal use, the description of its data type, the object's access rights, the time layer that contains information about the period in which the data can change, the scaling which includes the object's measurement unit and the range specifying the limits of the object's values.

For the communication objects that handle function calls, the logical correlation between the states of the respective Process Module and the functions that are allowed to be called in certain states has to be part of the specification, too. To establish such a correlation, the functions that can be called in a Process Module as well as the states of a module (e.g. READY, ACTIVE, etc.) are organized in so called function groups. Each function group represents an image of a certain operation mode of the Process Module (e.g. the start up mode or the normal operation mode). Therefore in the specification of the communication objects for handling function calls, each function group has to be described with all states and corresponding methods. The methods can also be called with passing arguments as well as return arguments. These arguments have to be specified following the description of the communication objects for data exchange.

The structure of the presented Reference Architecture is closely related to the reference architecture developed in the European Project OSACA (Open System Architecture for Controls within Automation Systems) [5]. Therefore the basic elements of the OSACA reference architecture, like the specification of data types and the formal description templates, have been adopted.

Process Module Manager:
The Process Module Manager is a part of the Application Interface that offers functionality for the coordination of the process monitoring system. Its functionality comprises the execution of the system start up, including configuration and parameterization, as well as the system coordination during normal operation.

By being part of the Application Interface, the Process Module Manager is also part of the control platform. Not every control platform offers all features that are needed for an implementation of the full functionality that a Process Module Manager could provide. That is why conformance classes have been introduced that define the scale of functionality offered by a Process Module Manager for a certain control platform. For the definition of conformance classes, the functionality is subdivided into different levels. At the moment three different levels are defined:
- an elementary level which comprises the basic mechanisms to establish communication connections,
- a second level that offers mechanisms for a dynamic configuration of the monitoring system in addition to the functionality of the elementary level and
- a third level that additionally offers functionality for a full parameterization of the system and for monitoring the output of the Process Modules.

Further conformance classes can be defined at a later time.

For a successful start up of the monitoring system, a certain order has to be kept for the execution of methods that build up communication connections. In the first start up phase, the server communication objects for the data exchange and the objects for the reception of function calls in the Process Modules have to be registered. When all Process Modules have finished the first phase, the Process Module Manager can initiate the registration of client objects. After all Process Modules have returned from the second registration phase, the start up of the monitoring system has been finished and the Process Modules can be switched to normal operation mode.

The mechanisms for building up communication connections are part of the functionality defined in the elementary conformance class. A Process Module Manager based on conformance class two, additionally offers mechanisms for a dynamic configuration of the monitoring system and its communication connections. For a dynamic configuration, the set up of the monitoring system has to be specified in a configuration file that is parsed during start up. This specification of the process monitoring system has to include information about the Process Modules that have to be started and the client communication objects that have to be registered.

A Process Module Manager implemented based on conformance class three provides mechanisms for a parameterization of the monitoring system in addition to the functionality defined in the conformance classes one and two. The parameter values can contain limits for data generated in Process Modules, priority values for the Process Modules as well as indications for strategies that are used to shut down the process in case of a disturbance.

These parameters are needed as a basis for mechanisms that ensure safety operations of the process monitoring system. One of these mechanisms analyzes the data generated in the Process Modules. If a trend towards the limit of a certain data value can be observed, the machine tool operator is notified. If this trend moves too fast for a notification of the operator, the Process Module Manager initiates a controlled shut down of the process. The strategy of the shutdown has to be included in one of the parameterization values. It is important to choose an adequate strategy in order to avoid endangering the operator or damaging the machine tool or the workpiece (e.g. in case of a tool break while drilling, a different strategy has to be chosen than if a break occurs while finishing a free form surface).

The priority values for Process Modules are needed to prevent deadlocks if more than one Process Module attempt to set the same data value for the NC-kernel (e.g. if the feedrate has to be reduced because of a spindle overload detected by one module, but at the same time another module attempts to rise the feedrate to avoid chattering).

Similarly to the system configuration the parameter values are specified in a parameterization file that is parsed during the start up of the monitoring system.

5 Implementations of the Application Interface

This section presents the implementations of the Application Interface in a Siemens control with open NC-kernel (Sinumerik 840D) and in an OSACA conform control, the WZL-NC developed at WZL.

In figure 4 the Application Interface for the Sinumerik 840D is illustrated. As stated before, the Process Module Manager is part of the Sinumerik 840D control platform. It organizes the start up of the process monitoring system according to the start up of the Sinumerik control system. The Application Interface builds up the communication connections between the Process Modules and the control internal data. In this example the implementation of the Process Module Manager follows the elementary conformance class. Therefore the communication connections are coded in the Process Modules which restricts the flexibility of a dynamic connection management.

The implementation of the Process Modules follows the specification of the Application Interface. Consequently the Process Modules can be ported on any control platform that offers an Application Interface.

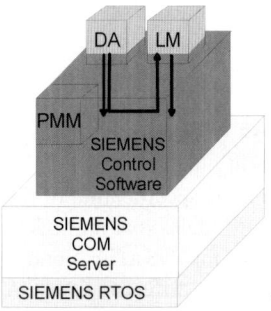

DA: Data Acquisition Module
LM: Logging Module
PMM: Process Module Manager

Figure 4: Implementation of an Application Interface for a Siemens Sinumerik 840D control

For an OSACA control, the Process Module Manager can be provided as Architecture Object (AO). It is again part of the control software, because for OSACA controls, the control functionality is encapsulated in discrete AOs. The encapsulation of the Process Module Manager's functionality offers the option of porting it to different OSACA conform controls.

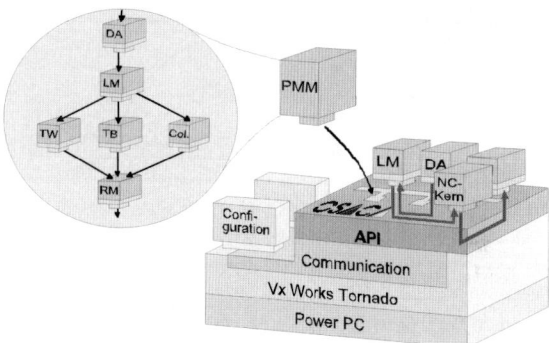

Col.: Collision Detection   RM: Response Module
DA: Data Acquisition Module   TB: Tool Breakage
LM: Logging Module   TW: Tool Wear
PMM: Process Module Manager

Figure 5: Implementation of an Application Interface for an OSACA control

The WZL-NC offers all features for an implementation of the full functionality that a Process Module Manager can provide (conformance class three). That is why a dynamic connection management and an individual parameterization has been realized for this control. Furthermore security measures for the monitoring of data generated in the Process Modules, the prevention of deadlocks and the execution of controlled shutdowns have been implemented to ensure a safe operation of the process monitoring system.

By linking the Process Modules with the Application Interface, the Process Modules are upgraded to be complete OSACA AOs. Through this they can be operated on any OSACA conform control platform. Figure 5 illustrates the implementation and the use of an Application Interface in an OSACA control.

6 Portability of Process Modules

Figure 6 illustrates the portability of Process Modules that are implemented following the specification of the Application Interface. Process Modules that are linked with an Application Interface for OSACA control platforms can be ported between different OSACA platforms without any change. In figure 6 the Data Acquisition Module can be integrated either on the OSACA control or on the OSACA platform in the MMC of the Siemens control. For other system modifications, such as porting a Process Module from the NC-kernel of a Sinumerik 840D to an OSACA control, an Application Interface for the control platform that the Process Module is supposed to be integrated on has to be linked to the Process Module. Nevertheless in both cases the implementation of the Process Modules does not have to be changed.

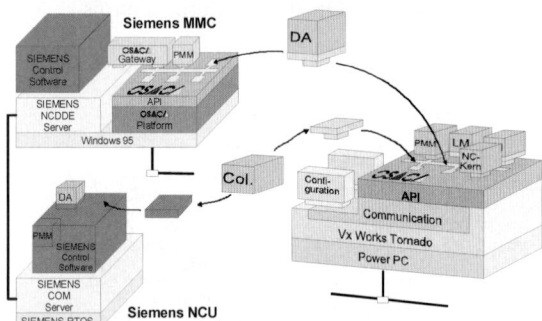

Figure 6: Portability of Process Modules following the specification of the Application Interface

7 Conclusion

These days, machine tools with rising productivity are demanded by the market. To reach this aim, the degree of automation has to be increased. For this reason, process monitoring functionality is essential.

The use of external sensors is the state of the art in acquiring information for process monitoring. Because of the availability of controls with open NC-kernel, which is the prerequisite for the integration of process monitoring functionality into the control, a trend towards control integrated solutions can be noticed. However, due to the vendor specific open interfaces, the software modules cannot be ported to different controls. Therefore vendor neutral open interfaces for the integration of third party software into the NC-kernel have to be developed.

In this paper the Application Interface, a real-time interface for the integration of third party software modules, is introduced. It focuses on application in the field of process monitoring, but it can also be applied to less time critical applications. The Application Interface is composed of three parts: a Communication System that handles the data exchange, a Reference Architecture for the specification of communication objects and a Process Module Manager for the configuration and parameterization of the monitoring system.

Implementations of the Application Interface for a control with vendor specific open NC-kernel interfaces and for an OSACA conform control have been described. Finally, the portability of monitoring modules has been explained by means of exchanging Process Modules between different control platforms.

8 References

1 *N. N.:* User´s manual, SINUMERIK 840D, OEM-package NCK. Publication of the Siemens AG, Erlangen 1997

2 *N.N.:* Publication of the Robert Bosch GmbH, Erbach 1999

3 *Kaever, M.; Weck, M.:* Intellignet Process Monitoring for Rough Milling Operations Based on Digital Drive Currents and Machine Integrated Sensors. In: Manufacturing and Technology, Vol. 1, 1997

4 *Hartmann, M.; Kilsmark, P.; Tönshoff, H.; et.al.:* The Future of Monitoring. In: VDI Berichte 1179, Überwachung von Zerspan- und Umformprozessen, VDI Verlag, Düsseldorf 1995

5 *N.N.:* OSACA Handbook Version 1.0.1, OSACA Association, Stuttgart 1997

Robot Behavior Engineering using DD-Designer

Ansgar Bredenfeld, Giovanni Indiveri
GMD - Institute AiS
D-53754 Sankt Augustin, Germany
+49-2241-14-2841
{bredenfeld,indiveri}@gmd.de

Abstract

We present a novel robotic software development framework for rapid simulation and programming of mobile robot teams. The framework uses a specification-centred generative approach. A robot control program is specified in an intuitive graphical representation of a hyper-graph of typed data processing elements. This specification is automatically refined to all design artifacts required in a robotic software development environment: simulation models, robot control programs, team communication infrastructure, documentation, and real-time monitoring. Although our framework was originally intended to design behavior-based robot control programs using the Dual Dynamics architecture, we demonstrate the flexibility of the environment by an example which integrates a "classical" control schema into a Dual Dynamics behavior system. Experimental validation was done on GMD-Robots, our team of mid-size league RoboCup robots.

Keywords - *Programming and architecture, rapid prototyping and design automation, behavior-based robotics, robotic software development framework*

1. Introduction

Autonomous mobile robots present fascinating research perspectives for today, and likewise fascinating prospects for applications in the future. The challenge is to achieve a behavior of autonomous robots that is both complex and reliable. We have identified three research areas which are crucial to achieve these objectives:

- Formal models of complex behavior systems.

- Methods, tools and development environments for very fast design-redesign cycles.

- Fast real-time processing for control and inter-robot communication.

With these requirements in mind, designing autonomous mobile robots is a demanding, interdisciplinary challenge. Developing software for complex robots is done in many iterated design-redesign cycles, often with substantial modifications in all required design artifacts: specification, simulation models, robot control programs and documentation.

Frameworks [1], component software [2] and generative programming are the most promising engineering approaches to speed-up the construction of novel design environments and the construction of robotic software using these environments. In this paper, we present our DD-Designer framework which has been designed and implemented exploiting these leading-edge technologies.

Our RoboCup [3] mid-size league robots are the test bed for our ideas, techniques and environment [4]. We use our development environment to program and operate this team of soccer playing robots. We present the control program of the goal keeper as demonstrative example in this work.

This paper is structured as follows. In section 2 we introduce the main idea behind our reactive, behavior-based approach called Dual Dynamics. In sections 3 to 5 we present the architecture, realization and the environment of our robotic software development framework. In section 6 we take a user perspective and demonstrate the flexibility of our framework by giving a robot controller example. Section 7 discusses related work and the last sections conclude the paper with an outlook to future work.

2. Dual Dynamics

Our approach to robot programming is based on a mathematical model for robot behaviors which was developed at GMD [5]. It integrates central aspects of a reactive, behavior-based approach to robot control [6][7], methods from continuous-time nonlinear control (like in [8]), and a dynamical systems representation of actions and goals.

In our approach robot behaviors are specified through differential equations, forming a global dynamical system made of behavior subsystems which interact through specific coupling and bifurcation-induction mechanisms. Behaviors are organized in levels where higher levels have slower dynamics than lower levels. Since the activation of behaviors (activation dynamics) is separated from their actuator control laws (target dynamics), the approach was named "Dual Dynamics" [5]. An important feature of DD is that it allows for robust and smooth changes between different behavior modes, which results in very reactive, fast and natural motions of the robots.

3. DD-Designer

The successful design of robot software requires means to specify, to produce code and to simulate as well as to run and debug the robot software in real-time on physical robots. In this section we give an overview of DD-Designer, the central specification tool of our robotic software development framework.

3.1 Data processor network

As abstraction of a robot control architecture we use a hyper-graph of data processing elements. Figure 1 shows a screen shot of a robot control program as appearing to the user. The interpretation of data processing elements depends on the robot control architecture. In our Dual Dynamics architecture, data processing elements are either sensor filters, activation dynamics of higher-level behaviors, activation dynamics of elementary behaviors or target dynamics of elementary behaviors.

Data processing elements are connected by states which are classified. In the Dual Dynamics architecture we have sensor variables as outputs of sensor filters, activations as output of elementary or higher-level behaviors and target variables calculated by the target dynamics of elementary behaviors. The final actuator control signal is a weighted sum of all its target variables.

Seen from an abstract point of view, DD-Designer is a graphical specification editor for data processor hyper-graphs where processors are of different type and have to fulfill constraints regarding their connectivity with other types of data processing elements.

3.2 Data processor scheduling

All data processing elements have to be scheduled in order to be executed in a simulator or on the real robot. At present, we use a static schedule for the execution sequence of data processing elements which is executed on our robots in a main loop at a rate of 50 Hz. The schedule of the sensor filters is ruled by data dependencies in the sensor filter network. Unavoidable cycles in the data flow are reported to the user and are

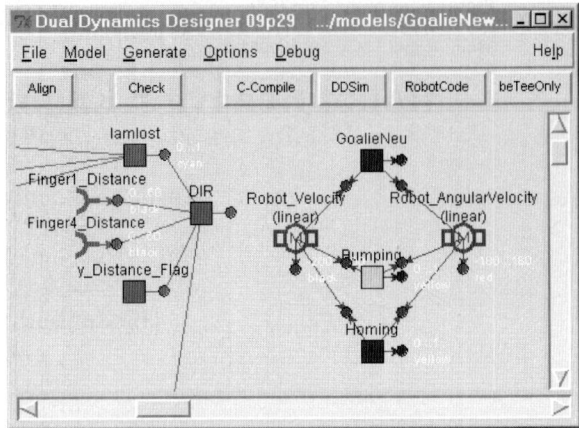

Figure 1. DD-Designer: sensors, sensor filters, behaviors, actuators entered as network of typed data processing elements

subject to be re-scheduled by a suitable user modification of the network. The static schedule of behaviors is ruled by the behavior hierarchy; we first schedule all higher-level behaviors and afterwards the elementary behaviors. Schedule construction takes care of avoiding unallowed relationships, for example, that activation variables from lower-level behaviors affect higher level ones, which is forbidden in the Dual Dynamics architecture [5].

3.3 Processor equations

The functionality of data processing elements is further detailed using an equation editor. The specification language uses common control flow statements known from C and Java. Data types comprise scalars, vectors and static multi-dimensional arrays. We restrict arrays to be static in order to prepare migration of robot control programs to field-programmable gate arrays (FPGA) in the future.

Besides common control and data flow, our language supports the formulation of differential equations. They are simply written by specifying an assignment to a derivative term like $\dot{\alpha} = -\alpha$. Derivative terms are converted to difference equations without any further interaction required by the user.

Since equations are parsed and stored as a fine-grained syntax tree, we can offer convenient editing, searching and browsing functionality like known from state-of-the-art integrated software development environments.

3.4 Interface abstraction

The data processing elements of a robot control program have to be connected to the sensoric input and the actuator output. Sensor and actuator abstractions are supported as prominent modelling constructs. This explicit interface abstraction has three major benefits.

The first is to have a clear cut edge between the robot control program and robot platform resp. simulator specific realizations of sensors and actuators. In this framework, the control program can be re-used in the simulation and on the real robot regardless of the implementation language used in the simulator or on the real robot. By example, the simulator is written in Java whereas the real robot is programmed in C++. Interface abstraction in combination with data processor equations that are independent of a dedicated target language maximize the re-usability of our robot behavior systems and avoids all problems related to manual code migration.

The second benefit of explicit interface abstraction is a simplification of the back-annotation across different implementation targets, namely the simulation code, the real robot code and our real-time monitoring tool. This enhances the system integration process and allows to keep track of all the relations and links among the design artifacts. As an example, we are able to configure the real-time monitoring tool via the specification and we can re-use the monitoring tool for both simulation and real-time observation of our robots. The enabling factor is explicit modelling of sensors, actuators and states (activations, sensor variables, target variables, actuator variables).

The third benefit is simplification of extensions. Adding a new sensor to the DD-Designer framework is simply done by adding an entry in a textual sensor configuration file and providing sensor wrapper functions for the simulator and the real robot. The same extension process is used for adding new functions to the framework. This is done by adding the signature of the new function and providing a simple wrapper to the native implementation in the target language. This "plug-and-play" concept makes each extension directly visible in the DD-Designer editor. Model checking, automatic documentation, and code generation for the target languages is then directly available for such extensions.

3.5 Team communication support

Communication between robots is a prerequisite for the realization of message-based cooperation in a team of robots. Therefore, a team communication feature is already supported at the specification level. Specifying team communication is as simple as tagging a state - which connects data processing elements - to be publically visible in a team.

Team communication is realized with a black-board like mechanism, which allows to share public states in a given team of robots. In the simulator, team communication is emulated, on the real robot the communication layer uses TCP/IP. The communication protocol between the robots is automatically synthesized for a given team. If a robot joins or leaves the team, its states are simply not distributed any longer to its team members. A given robot accesses public states of other robots using a dynamic state vector. If no team member provides values for a state, this vector remains empty. This mechanism turns out to be very flexible for varying size teams.

3.6 Data processor refinement

The central specification of the robot behavior system is the hyper-graph of state connected data processing elements. This representation allows to design Dual Dynamics models on a high level of abstraction and to synthesize all design artifacts required to make the models operative in practice: a documentation, a simulation model, executable control programs for our RoboCup robots and a parameter set for our real-time monitoring and analysis tools.

4. Realization of the DD-Designer

DD-Designer was constructed using the model-based rapid software prototyping environment APICES [9]. In previous work [10][11] we concentrated on the construction and evolution of the DD-Designer from the very first prototype to the full-fledged design tool it is now. Here, we summarize very shortly this aspect of the DD-Designer. We generate a design tool framework from an UML class diagram with APICES. Figure 2 shows the simplified UML class diagram of the DD-Designer without attributes or generated methods. The generated tool framework consists of an efficient C++ implementation of the class diagram embedded in the scripting language Tcl. In addition, it comprises a re-configurable interactive graphical instance editor.

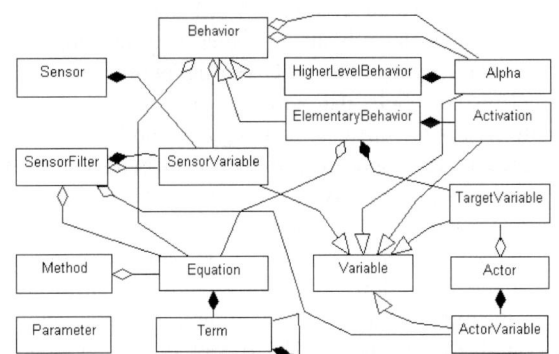

Figure 2. UML class diagram of the DD-Designer

5. DD-Designer Environment

The DD-Designer is the central tool of an integrated design and simulation environment for mobile robot teams. Besides the DD-Designer, it consists of the simulator *DDSim* and the real-time monitoring tool *beTee*.

DD-Designer allows to specify a Dual Dynamics model in terms of sensors, actuators, sensor pre-processing elements and a hierarchy of coupled behaviors. Each of the processing elements is formulated using a combination of control data flow and differential equations. This specification is the basis for an automatic refinement of all code artifacts required by the tools in our design environment.

DDSim allows to simulate a communicating team of robots on a field. It provides simulations of the ball and the sensors of the robots. This includes laser scanner simulation and an emulation of the vision system used by our robots. Since each robot may operate different behavior systems, it is possible to benchmark behavior systems against each other in a team of robots.

beTee is a real-time monitoring tool for tracing arbitrary internal states of the simulated or (via wireless LAN and TCP/IP) the physical robot. In addition, it serves as a control cockpit for the robots.

6. Controller design using the DD-Designer

The DD-Designer war originally conceived as a design tool for robot control programs relying on the Dual Dynamics scheme [5]. In order to illustrate the basic underlying ideas of the Dual Dynamics (DD) control architecture and the use of the DD-Designer, we describe the design of a control schema of the goal keeper of the GMD RoboCup robot team.

In synthesis the name Dual Dynamics refers to a two layer control architecture: the lower one consists of task-oriented controllers (or behaviors) as by example "avoid obstacle", "move to a point", "track a target" that are designed independently of each other with standard tools of automatic control theory. Given the purely kinematic model of the robots, the output of the generic behavior controller *i* is the linear u_i and angular velocity w_i to be assigned to the robot. The higher-level behavior layer consists of a dynamic system regulating the interactions between behaviors. This kind of architecture was inspired by biological considerations [5] in consideration of which the activation dynamics should generally have more than one locally stable equilibria and bifurcations among them may occur. The output of the activation dynamics relative to behavior *i* is a scalar number α_i between 0 and 1, such that the total input to the actuators of the robot at each time are given by

$$u_{tot} = \sum_{i=1}^{n} \alpha_i u_i \text{ and } w_{tot} = \sum_{i=1}^{n} \alpha_i w_i.$$

For the present purpose it might suffice to notice that the activation dynamics may be thought of (and used) simply as a smoothing mechanism to switch among tasks. In the goal keeper case, by example, the behaviors or tasks were three: "avoid obstacle", "home" and "keep goal". Their activation dynamics were very simply given by

$$\gamma_i \dot{\alpha}_i = -\alpha_i + f_i,$$

being γ_i a positive constant, f_i in $\{0,1\}$ for every *i* and $\sum_{i=1}^{n} f_i = 1$. This condition insures that the rise of an activation variable implies the decrease of the others, i.e., that except for a transient period only one behavior is active at a time.

The transitions among the f_i variables are governed by a finite state machine that guarantees the satisfaction of the constraint $\sum_{i=1}^{n} f_i = 1$.

The "keep goal" behavior should allow the robot to move along the line of the goal and to stop in front of the ball. With reference to figure 3, the steering law is designed such that the candidate Lyapunov function

$$V = \frac{1}{2}(\phi^2 + (y - y_r)^2)$$

has semi-negative definite time derivative. Noticing that $\dot{y} = u\sin\phi$, and $\dot{\phi} = w$ being *u* and *w* the linear and angular velocities,

$$\dot{V} = \phi w + (y - y_r)u\sin\phi$$

such that imposing $\dot{V} = -h\phi^2$, $h>0$ constant gain, implies

$$w = -h\phi - (y - y_r)u\frac{\sin\phi}{\phi}$$

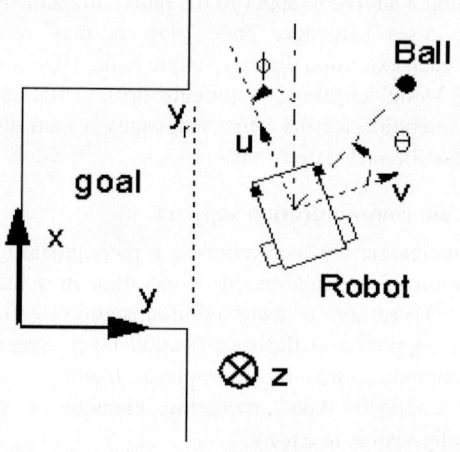

Figure 3. Control law for RoboCup goal keeper

which is well defined on the whole state space.

La Salle's Invariance Principle may be applied to V showing that as long as u is not identically null both ϕ and $(y-y_r)$ will be globally asymptotically driven to zero.

ϕ and $(y-y_r)$ are estimated by onboard infrared sensors and u is taken to be proportional to θ (see figure 3) which is measured by the vision system of our robots. This control law for u and w is the target dynamics of the "keep goal" behavior.

The corners of the goal are detected by other infrared sensors: the variable f_i relative to the activation dynamics of the "keep goal" behavior is put to zero when a corner is reached while the f_i variable within the "avoid obstacle" activation dynamics is put to 1. The angular and linear velocities relative to the "avoid obstacle" behavior are both null.

If the robot should be hit by opponent robots thus losing its position in front of the goal, a homing behavior is triggered in order to recover the correct position and the keep goal behavior is thus restarted.

By using the integrated DD-Designer environment the design process of the described behavior system was performed in very short time. Going from the basic idea through the simulation to the running robot took less than a week. The robustness and effectiveness of this approach has been experimentally verified during the 4th RoboCup World Championships in Melbourne where the GMD Robots team reached the quarter finals. Figure 4 shows our self-constructed robot platform.

7. Related Work

We identified related work in two domains: languages for mobile robot programming and robotic software development frameworks.

Several specific languages for mobile robot programming with different emphasis and focus have been proposed in the past. PDL [12] (Process Description Language) is a language developed by Steels and Vertommen for defining and experimenting with networks of dynamical processes for behavior-based control. Similar to the DD-Designer, PDL is targeted to specify process networks for behavior-based robot control. In contrast to PDL, the DD-Designer offers a visual language to construct constrained networks of typed data processing elements. COLBERT [13] is a language for writing reactive robot controllers in C++. Unfortunately, it is dedicated to robot controllers having the Saphira [14] architecture. CES [15] is a programming language extension of C++

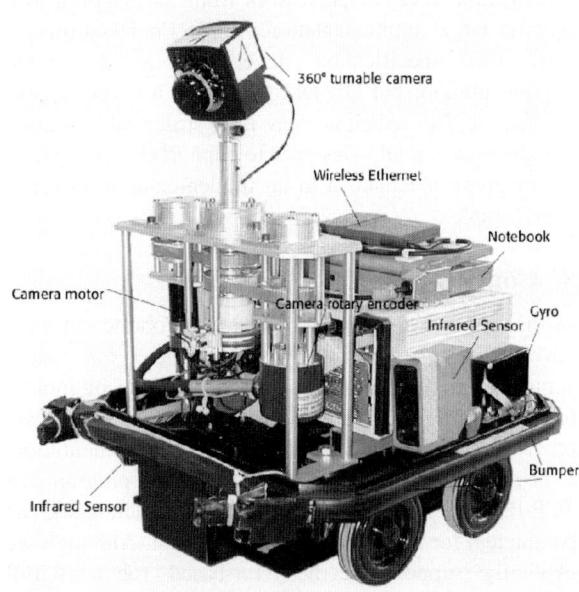

Figure 4. GMD RoboCup mid-size league robot

specifically targeted towards mobile robot control. It adds two ideas to C++: computing with probability distributions and built-in mechanisms for learning from examples. The DD-Designer follows a different extension approach. The user specifies a robot behavior system explicitly in a graphical editor. Since the specification is mapped to an implementation language as C++ or Java, the suggested approach has two advantages: first it allows to generated robot control programs in C++ and simulation models in Java for DDSim without any manual code migration. This important benefit stems from the rigorous separation of abstract control specification from concrete target implementations. Second, with the advent of powerful embedded real-time Java platforms, we will simply re-target our robot behavior systems to Java without any cumbersome and error-prone code migration from C++ to Java.

In the domain of robotic software development frameworks integrated design and simulation environments, by example COSIMIR [16], exists. This environment is not targeted to behavior-based mobile robots but to manufacturing robotics. Interesting work on a robotic software development framework for humanoids is PredN [17]. As our environment it starts from a hyper-graph representation of the robot control specifications. This representation is mapped to a multi-threaded real-time communication platform. In comparison, the architecture of PredN is very similar to ours. Both systems clearly separate the hyper-graph specification level of processors from its mapping to a specific target implementation. While PredN allows to map the specification level to a distributed implementation, the DD-Designer does not yet support this feature. Nevertheless, there is no structural limitation in extending the DD-Designer to support this feature, as the mapping mechanism to an implementation target is configurable.

8. Conclusions

In this paper, we presented the novel robotic software development framework DD-Designer for rapid simulation and behavior-based programming of mobile robot teams. From an abstract hyper-graph of typed data processing elements we generate HTML documentation, Java simulation code, C++ robot control programs, a TCP/IP based team communication infrastructure and the parameters for a real-time monitoring tool. Although we explicitly support the behavior-based robot control architecture Dual Dynamics, the system allows to easily integrate other control architectures. The environment is in productive use within the GMD team of mid-size league RoboCup robots.

9. Future work

It is worthwhile to investigate an adaption of the DD-Designer to other robot control architectures. Since the DD-Designer is constructed using the UML-class diagram based rapid-prototyping tool APICES [9], it can be seen as an instance of a design tool family. One aspect of our future work is to look at abstractions of robot control architectures using "architecture templates". If we factor out these templates, we will be able to tailor the DD-Designer to different robot control architectures by changing the meta-model (Fig. 2) in APICES. This would allow to compare alternative architectural approaches within a single, integrated environment for simulation and real robot programming.

10. References

[1] M. E. Fayad, D. C. Schmidt. Object-Oriented Application Frameworks - Introduction, Communications of the ACM 40(10), 1997, pp. 32-38

[2] C. Szyperski. Component Software – Beyond Object-Oriented Programming, Addison-Wesley, 4th printing, 1999

[3] H. Kitano, M. Asada, Y. Kuniyoshi, I. Noda, and E. Osawa. RoboCup: The Robot World Cup Initiative. In Proceedings of the IJCAI-95 Workshop on Entertainment and AI/ALife, 1995

[4] A. Bredenfeld, T. Christaller, W. Goehring, H. Guenther; H. Jaeger, H.-U. Kobialka, P. Ploeger, P. Schoell, A. Siegberg, A. Streit, C. Verbeek, J. Wilberg. 'Behavior engineering with "dual dynamics" models and design tools', in RoboCup-99: Robot Soccer World Cup III (M. Veloso, E. Pagello, H. Kitano, eds.), LNAI 1856, pp. 231-242, 2000

[5] H. Jaeger and T. Christaller. 'Dual dynamics: Designing behavior systems for autonomous robots', Artificial Life and Robotics, 2:108-112, 1998

[6] R. C. Arkin. Behavior-Based Robotics (Intelligent Robots and Autonomous Agents), MIT Press, 1998

[7] R. A. Brooks. Intelligence without reason. A.I. Memo 1293, MIT AI Lab, 1991 (http://www.ai.mit.edu/)

[8] M. Aicardi, G. Casalino, A. Bicchi, and A. Balestrino. Closed-loop steering of unicyle-like vehicles via Lyapunov techniques. IEEE Robotics and Automation Magazine, March:27-35, 1995

[9] A. Bredenfeld. APICES - rapid application development with graph pattern. in Proc. of the 9th IEEE Int. Workshop on Rapid System Prototyping (RSP 98), Leuven, Belgium, pages 25-30, 1998

[10] A. Bredenfeld. Co-Design Tool Construction Using APICES, in Proc. of the 7th ACM/IEEE Int. Workshop on Hardware/Software Codesign, pp. 126-130, May 1999

[11] A. Bredenfeld. 'Integration and Evolution of Model-Based Prototypes', in Proc. of the 11th IEEE Int. Workshop on Rapid System Prototyping (RSP 2000), Paris, France, June 21-23, 2000

[12] L. Steels. The PDL Reference Manual. VUB AI-Lab AI Memo 92-5, 1992

[13] K. Konolige. "COLBERT: A Language for Reactive Control in Saphira", in German Conference on Artificial Intellgence, Freiburg, 1997

[14] K. Konolige, K. Myers. The Saphira Architecture for autonomous mobile robots, In AI-based Mobile Robots: Case studies of successful robot systems (D. Kortenkamp, R. P. Bonasso, and R. Murphy, eds.), MIT Press, 1996

[15] S. Thrun. Towards Programming Tools for Robots That Integrate Probabilistic Computation and Learning, in Proc. of the IEEE Int. Conf. on Robotics and Automation (ICRA-2000) San Francisco, CA, April 2000

[16] E. Freund, J. Rossmann. Projective virtual reality: Bridging the gap between virtual reality and robotics. IEEE Transaction on Robotics and Automation; 15:3: 411-422, June 1999

[17] O. Stasse and Y. Kuniyoshi. PredN : Achieving efficiency and code re-usability in a programming system for complex robotic applications, in Proc. of the IEEE Int. Conf. on Robotics and Automation (ICRA-2000) San Francisco, CA, April 2000

Object-Oriented Design Pattern Approach for Modeling and Simulating Open Distributed Control System

Toyoaki Tomura
Asahikawa National College of Technology, Japan
e-mail : tomura@asahikawa-nct.ac.jp

Satoshi KANAI
Hokkaido Univ. School of Engg.

Kiyoshi Uehiro
Motorola Japan Ltd., Japan
e-mail : rnm175@email.sps.mot.com

Susumu Yamamoto
Motorola Japan Ltd., Japan

Abstract

An open distributed control system (DCS), which consists of a large number of devices and a single open network interconnecting those devices, is now used in many automation areas. One critical problem of the DCS, however, is that because there is massive traffic on the network, the system integrator must carefully tune the network traffic after the construction of the system to assure its control performance. Hence a DCS simulator is strongly needed. In our study, we propose an object-oriented design pattern approach as a uniform, rapid and accurate method of modeling and simulating a DCS. Two special design patterns are proposed: Statechart Pattern which defines classes and the state-transition execution mechanism for realizing the dynamic behavior of devices, and Device-Constructor Pattern which defines classes and their instantiation mechanism for realizing the structure of devices composed of many kinds of sensors and actuators. The systematic procedures from those patterns to the executable Java code of the simulation model are also discussed. Furthermore, the effectiveness of our approach was investigated through actually developing an in-house Java-based DCS simulator.

1. Introduction

In recent years, the DCS based on open networks has been rapidly replacing traditional centralized control systems in factory, process and building automation. As shown in Figure 1, the DCS generally consists of a large number of devices composed of sensors, actuators and local controllers, and of a high-speed open network interconnecting those devices. Now we have ControlNet, Foundation Fieldbus, LonWorks, Device-Net, and etc. [1] as examples of such a high-speed open network. The DCS is capable of making the control system more scalable, and its construction and wiring cost much less. For these reasons, the DCS is expected to be used more widely in automation areas in the future.

One critical problem of the DCS is that the number of devices tends to be large and massive communication traffics among those devices concentrates on one single network. This causes unacceptable time-delay or fatal loss of data which assure reliable control performance. To avoid this problem, the system integrator must optimize the network traffic by carefully tuning the several communicative timing parameters of each device after the

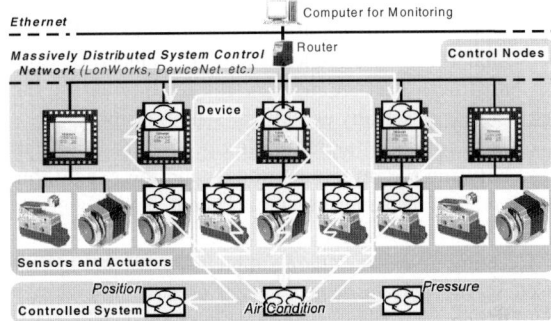

Figure 1 Typical Structure of Open DCS

construction of the system. However, this task, requiring extra time and cost, becomes a bottleneck for rapid development of the DCS. Therefore, the system integrator strongly needs simulation techniques which enable the integrator to make uniform models of the structure and dynamic behavior of each device, to quickly implement the models as executable codes, and accurately estimate the dynamic behavior of the network traffic among the devices through combining those models.

In order to satisfy those requirements, object-oriented methodology has been used in modeling several distributed systems. SEMATECH has proposed *CIM Application Framework* [2] for semiconductor production. It specifies the reference model of the structure of manufacturing resource classes and association among them, and of the dynamic behavior of objects. OSE has proposed *OSEC (Open System Environment for Controllers)* [3] for open NC controller. It specifies the reference model of structure of function block classes of NC controller and association among them. SEMI's *Common Device Model* [4] specifies a reference model of static structure and dynamic behavior of each device composed of various sensors, actuators and local controllers in *UML* [5]. Those approaches provide the system integrators with object-oriented reference models that can be used in several technical domains, and are very useful for uniform analysis and description of specification of each model to be used for DCS simulation. However, the approaches still have some problems which prevent us from efficiently building executable models necessary for DCS simulation in our study. The problems are as follows:

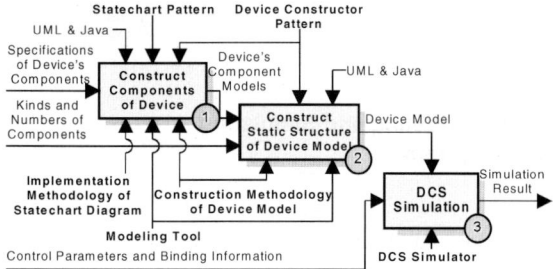

Figure 2 Overview of Modeling and Simulation of DCS

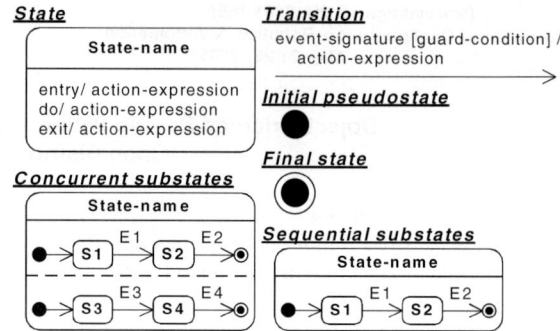

Figure 3 Notation of Statechart Diagram

1) The reference models can only define the minimum required structure of classes and association among them, or state-transitions, and they do not have a detailed mechanism for implementing them. In additon, they fail to specify mechanisms for creating objects dynamically, for uniform management of the objects of various concrete subclasses inherited from abstract reference models, and for interacting the objects to actually execute the state-transition.
2) The concrete implementing procedure to bridge between the reference models and the executable codes of the simulation model is not explicitly given.

Recently, "*Design Pattern*" approaches [6] and [7] are used in the object-oriented design process. The design pattern approach specifies reusable mechanisms for collaboration and communication among classes or among objects to solve common object-oriented problems in a general domain. Generally, the approach classifies design patterns into three categories: patterns for creating objects, patterns for defining the structure of objects, and patterns for defining the behavior of the objects. Although the design pattern approach can solve the first problem mentioned above by combining several basic patterns, it still fails to solve the second one. In order to solve the two problems at the same time, we propose a new object-oriented design approach for modeling and simulating a DCS. The purposes of our study are:

1) We propose two design patterns for modeling a DCS, that is, *Statechart* which defines classes and state-transition execution mechanism for realizing the dynamic behavior of devices, and *Device-Constructor* which defines classes and their instantiation mechanism for realizing the static structure of devices.
2) We also discuss a systematic implementation procedure to bridge from those patterns to the executable Java code of the simulation model.
3) We confirm the effectiveness of our proposed approach based on those patterns through actually developing an in-house Java-based DCS simulator for building automation.

2. Modeling and Simulation of DCS

Figure 2 shows a process overview of modeling and simulation of the DCS based on the proposed two patterns. In each process, the system integrator's tasks are as follows:

1) To construct the static structures and dynamic behaviors of the devices components by referring to the specifications of the components. The Device Constructor and Statechart patterns are used.
2) To construct the static structure of a device model by using the Device-Constructor pattern.
3) To connect the device model with the open DCS simulation model (which can simulate its communication protocol), and then evaluate network properties such as traffic on the network and packet collision frequencies.

In order to execute those processes, Motorola Japan, Inc. has developed a modeling tool for the device model and the DCS simulator using the proposed design patterns. This simulator is actually used in designing and evaluating the actual DCS governed by LonWorks for a heating-cooling system in the building automation area.

3. Statechart Pattern

In order to model a device, the system designer must develop all the device-specific component models with dynamic behavior. A statechart diagram [5] can describe the detailed dynamic behavior of a device-specific component model. In order to implement the statechart diagram systematically and easily, the system designer needs a design pattern and an imple-mentation methodology based on it. In this section, we describe the Statechart pattern and the implementation methodology with the pattern, which are proposed in our study.

3.1 Statechart Diagram

The statechart diagram describes the dynamic behavior in common with all the objects of a class. In particular, it describes possible sequences of states and actions through which an object can proceed during its life as a result of reacting to discrete events. The notation of the statechart diagram are substantially those of Harel's statechart [8] with extensions to make them object-oriented. Figure 3 shows the notation of the statechart diagram.

Table 1 Patterns for Implementing Statechart Diagram

Patterns \ Elements	Conditional Statements Pattern	State Pattern	StateTable Pattern	Statechart Pattern
State	Case Statement in a Switch Statement	Object of Subclass of Class *State*	*State* Object	*State* or *StateMachine* Objcet
Transition	Changing Value of State Variable	Creating and Deleting *State* Object	*Transition* Object	*Transition* Object
Event	Event Variable	Event Variable of Class *State*	Event Variable	String (Argument of Method)
Action and Activity	Method of Class *Context*	Overrided Method of Class *State*	Method of Class *Context*	Interface
Current State	State Variable	Attribute of Class *State*	State Variable	Attribute of Class *StateMachine*
easyness of implementation	Good	Bad	Good	Excellent!
Support of Substates	Bad	Not Bad	Bad	Excellent!
Corresponsibility to Java Code	Bad	Bad	Bad	Excellent!
Performance of Execution	Bad	Not Bad	Good	Not Bad

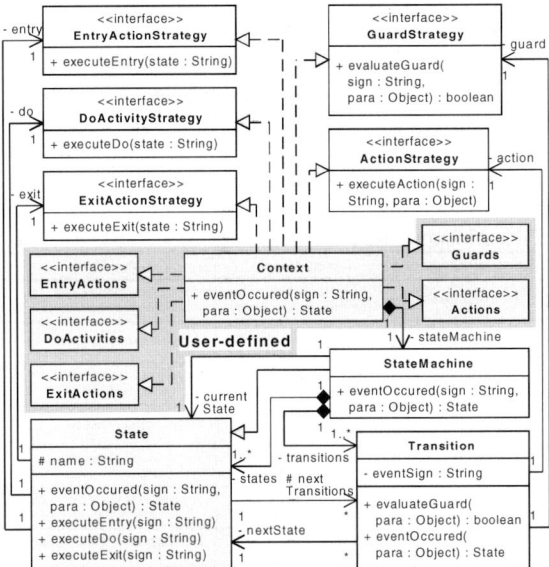

Figure 4 Structure of Statechart Pattern

3.2 Previous Design Patterns

Table 1 shows design patterns for implementing the dynamic behavior described by statechart diagram [6], [7], [9]. These patterns except the Statechart pattern have critical problems in easiness of implementation, support of substates, and corresponsibility to the Java code. In order to solve these problems, the system designer needs a new design pattern which meets the following needs:

- The pattern has a mechanism for state-transition execution including transition in substates.
- The pattern can be easily implemented to the executable Java code by referring to a mapping rule.
- The implemented executable Java code can be easily modified by the system designer later.

3.3 Structure of Statechart Pattern

Figure 4 shows the structure of the Statechart pattern which is proposed in our study. The classes and interfaces in Figure 4 are defined as follows:

- *Context* is the class which has the dynamic behavior described by the statechart diagram. The object has one *StateMachine* object.
- *StateMachine* has two sets of states and transitions. The object is a either of statechart diagram, sequential substates, or one substates in concurrent substates.
- *EntryActions*, *DoActivities*, and *ExitActions* represent three sets of entry-actions, do-activities, and exit-actions respectively. Each action in a state is defined in these interfaces as an empty method.
- *Guards* and *Actions* represent two sets of guard-conditions and actions. Each condition and each action in transition are defined in these interfaces as an empty method.
- *EntryActionStrategy*, *DoActivityStrategy*, and *ExitActionStrategy* represent three logics in order to execute actions according to current states.

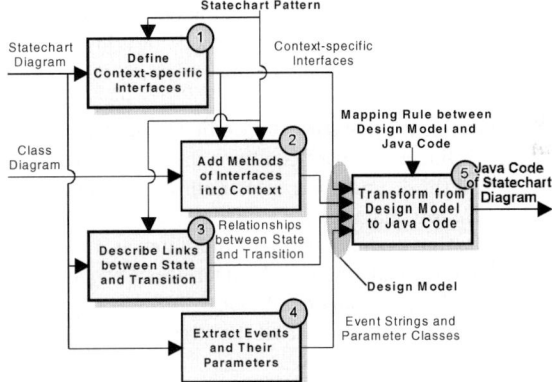

Figure 5 Implementation Methodology of Statechart Diagram with Statechart Pattern

- *GuardStrategy* and *ActionStrategy* represent two logics in order to evaluate conditions and execute actions according to occurred events.

3.4 Implementation Methodology of Statechart Diagram with Statechart Pattern

Figure 5 shows the implementation methodology of statechart diagrams with the Statechart pattern. It consists of the following five processes:

1) Defining the five context-specific interfaces,
2) Adding all the methods of the defined interfaces to the class *Context*,
3) Describing all the relationships between state and transition,
4) Extracting all the events and their parameters, and
5) Transforming from the design model to the Java code by referring to the mapping rule.

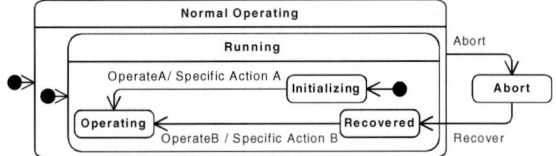

Figure 6 An Example of Statechart Diagram

Table 2 Mapping Rule between Design Model and Java Code

Statechart Diagram	Design Model	Java Code
State and Substates	Node in Tree Structure of States	*State* Object and *StateMachine* Object
Transition	Transition Relationship between States	*Transition* Object
Initial Pseudostate	Node in Tree Structure of States	Variable of Class *StateMachine* (i. e. *State* Object)
Final State	Node in Tree Structure of States	*State* Object
Event and Its Parameter	Event String and Class of Parameters	Event String and Class of Parameters
Guard Condition	Empty Method of Interface *Guards*	Implemented Method of class *Context*
Action (Included inside State)	Empty Method of Interfaces *Actions*, etc.	Implemented Method of class *Context*

As an example of implementation of a statechart diagram, we implement the statechart diagram shown in Figure 6 by applying the implementation methodology. In Figure 5, each process can be divided into the following steps:

1.1) Define the five context-specific interfaces.
1.2) Define each guard condition and each action as a method, and then add them to the interface *Guards* and *Actions*.
1.3) Define each action in all the states as a method, and then add them to the interfaces *EntryActions*, *DoActivities*, and *ExitActions*.
2.1) Add the methods of all the strategy interfaces to the class *Context*.
2.2) Add the methods of all the interfaces defined in the first process to the class *Context*.
3.1) Identify the tree structure of states, all the transition relationships between states, and all the initial pseudostates.
4.1) Extract event names from all the transitions.
4.2) If an event has one or more parameters, define a set of parameters as a class.
5.1) Transform from the design model to the Java code by referring to the mapping rule shown in Table 2.

Figure 7 shows the executable Java code of the statechart diagram shown in Figure 6.

4. Device-Constructor Pattern

In order to construct the static structure of a device model, we describe the Device-Constructor pattern and the construction methodology of the device model with it.

4.1 Previous Design Patterns

Figure 8 (a) shows the common device model. When the component class inherits any one of the three abstract classes, the configuration of the device components may

```
public class SensorA extends Sensor implements ActionStrategy,
  ActionsOfSensorA {
  StateMachine stateMachine;
  public SensorA(String name) {
    stateMachine = new StateMachine(name);
    stateMachine.createSequentialStateMachine(
      "Normal Operating");
    stateMachine.createState("Abort");
    stateMachine.createSequentialStateMachine(
      "Normal Operating", "Running");
    stateMachine.createState("Running", "Initializing");
    // ........
    stateMachine.connectState("Normal Operating", "Abort",
      "Abort");
    stateMachine.connectState("Abort", "Recovered", "Recover");
    // ........
    stateMachine.setInitialState("Normal Operating");
    stateMachine.setInitialState("Running");
    stateMachine.setInitialState("Initializing");
  }
  public void executeAction(String event, Object parameters) {
    if(event.equals("OperateA"))
      specificActionA(event, parameters);
    else if(event.equals("OperateB"))
      specificActionB(event, parameters);
    else
      return;
  }
  public void specificActionA(String event, Object parameters) {
    // Describe Specific Action A
  }
  public void specificActionB(String event, Object parameters) {
    // Describe Specific Action B
  }
}
```

Figure 7 An Executable Java Code of Statechart Diagram

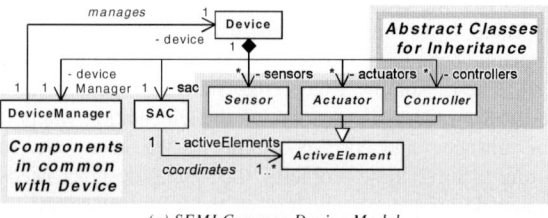

(a) *SEMI Common Device Model*

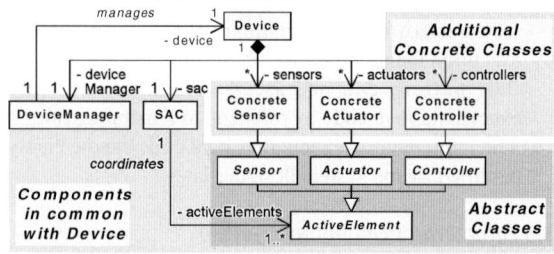

(b) *Direct Composition of Concrete Classes by Class Device*

Figure 8 Common Device Model and Extended Model

become ambiguous. In order to solve this problem, an extended model can be considered as shown in Figure 8 (b). Because the class *Device* manages the component classes directly, the configuration of the components will be clear. However, the system designer must define the association between the class *Device* and the component class. The system designer must implement the procedure for creating the component objects into the class *Device*'s constructor. In order to solve these problems, the system designer needs a design pattern which meets the following requirements:

1) The pattern can define new component classes without associating between the class *Device* and them.
2) The pattern provides interfaces for creating the component objects without depending on the kinds of components and the numbers of their objects.
3) Interfaces provided by the pattern are very simple.

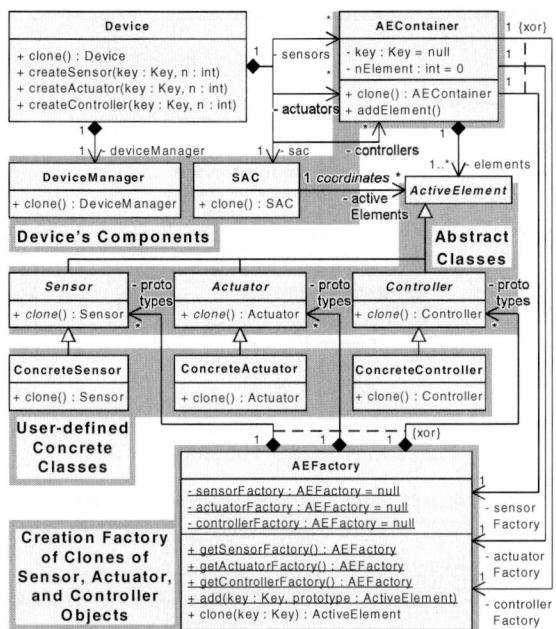

Figure 9 Structure of Device-Constructor Pattern

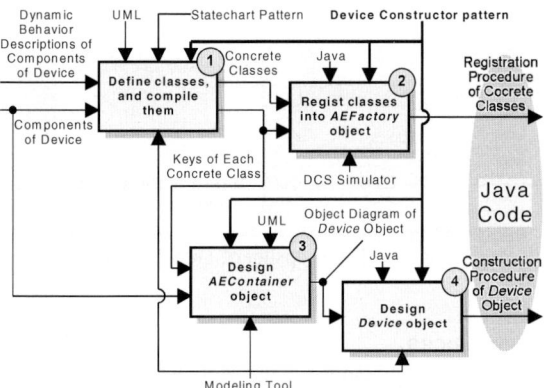

Figure 10 Construction Methodology of Device Model with Device-Constructor Pattern

4.2 Structure of Device-Constructor Pattern

Figure 9 shows the structure of the Device-Constructor pattern, which is proposed in our study. In the Device-Constructor pattern, the classes are defined as follows:

- **Device** represents the device model defined by the SEMI Common Device Model.
- **AEContainer** represents the container for all the objects of a particular concrete class.
- **AEFactory** represents the factory for creating the cloned objects of the concrete classes.
- **ConcreteSensor**, **ConcreteActuator**, and **ConcreteController** represent the concrete classes of three abstract classes *Sensor*, *Actuator* and *Controller*.

The Device-Constructor pattern partially extends the following recent design patterns; *Abstract Factory*, *Factory Method*, *Prototype*, *Singleton*, and *Composite*.

```
AEFactory f = AEFactory.getSensorFactory();
Sensor s = new SensorA();
// Regist class SensorA
  f.add(new Key("SensorA"), s);
                // Key String for SensorA
Actuator a = new ActuatorA();
// Regist class ActuatorA
  f.add(new Key("ActuatorA"), a);
                // Key String for ActuatorA
```
(a) Registration Procedure of Concrete Classes

```
Device device = new Device();
// create two SensorA objects
device.createSensor(
  new Key("SensorA"), 2);
       // Key String   number of objects
// create three ActuatorA objects
device.createActuator(
  new Key("ActuatorA"), 3);
       // Key String   number of objects
```
(b) Construction Procedure of Device Object

Figure 11 Registration Procedure of Concrete Classes and Construction Proccedure of *Device* Object

The Device-Constructor pattern meets the requirements described in section 4.1 by the following mechanisms:

- The object of the class *AEFactory* contains the objects of each concrete class with a key string.
- The object of the class *AEContainer* contains all the objects of a particular concrete class.
- In order to create device components, the system designer must describe the sequence of class *Device* methods with the key string of the concrete class and the number of objects as two arguments.

4.3 Construction Methodology of Device Model with Device Construtor Pattern

Figure 10 shows the construction methodology of the device model with the Device-Constructor pattern. As an example of the device model, we consider a device model which consists of two sensors called *Sensor A*, and three actuators called *Actuator A*. The methodology consists of the following four processes:

1) By referring to the device components, define each component as a concrete class of the abstract classes *Sensor*, *Actuator*, and *Controller*, and then compile them. The dynamic behavior of the concrete class is implemented by using the Statechart pattern. Then allocate unique strings to each concrete class.
2) By referring to the defined concrete classes and their key strings, define the registration procedure of the concrete classes into the *AEFactory* object, as shown in Figure 11 (a).
3) In order to define the static structure of the device model, design the *AEContainer* objects which contain all the objects of a particular concrete class by drawing an object diagram as shown in Figure 12.
4) By referring to Figure 12, define the construction procedure of *Device* object shown in Figure 11 (b).

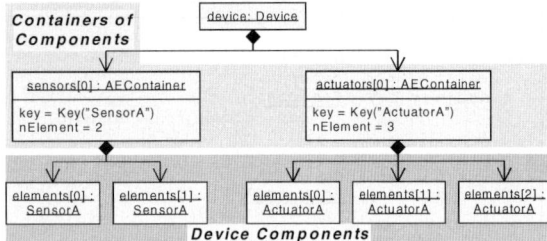

Figure 12 Object Diagram of Example of Device Model

5. Modeling Tool and DCS Simulator

In order to construct device models in the DCS and simulate its dynamic behavior, Motorola Japan Inc. has developed the modeling tool of the device model and the DCS simulator. In the modeling tool, the Device-Constructor pattern is used for defining the static structure of the device model, and the Statechart pattern is used for implementing the dynamic behavior of the device model as the Java code. Figure 13 shows the snapshot of the modeling tool. In Figure 13, five kinds of device models are modeled, and are registered into the list of devices. By modeling the device models actually, we confirmed that two design patterns were very useful for rapidly developing the device models. Figure 14 shows the snapshot of the DCS simulator. The system designer can observe the performance of the network and the log of packets. Where the protocol of LonWorks is implemented as the network protocol of the DCS. By comparing the log of packets with the message sequence among the device models, we confirmed that two design patterns were very useful for simulating the dynamic behavior of the control network with the device models. By embedding two design patterns into the modeling tool and the DCS simulator, the period of time required for developing these applications was enormously shortened from four months to two months. Furthermore, because the newly defined component classes are imported into the DCS simulator alone, we consider that the efficiency of device modeling in the modeling tool is very high.

6. Conclusions

The conclusions of our study are summarized as follows:

1) We have proposed two design patterns for modeling a DCS. The *Statechart* pattern defines classes and the state-transition execution mechanism for realizing the dynamic behavior of devices. The *Device-Constructor* pattern defines classes and their instantiation mechanism for realizing the static structure of devices.

2) We have discussed the systematic implementation procedures to bridge from those patterns to the executable Java code of the simulation model.

3) We have confirmed the effectiveness of our approach based on these patterns through actually developing of an in-house Java-based DCS simulator for the building automation.

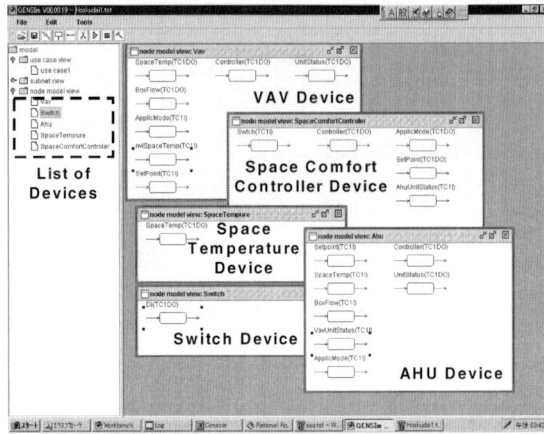

Figure 13 A Snapshot of Modeling Tool

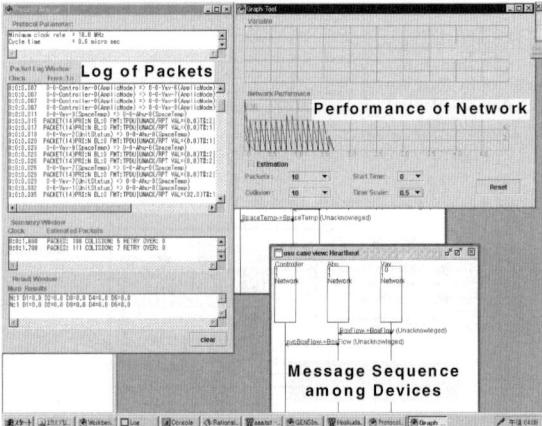

Figure 14 A Snapshot of DCS Simulator

References

[1] Piggin, R., Young K., and McLaughlin, R., "Current Fieldbus Standards Situation - A Europian View", *Assembly Automation*, Vol. 19, No. 4, 286-289, 1999.

[2] SEMATECH, "CIM Application Framework, Specification 1.2", 1995.

[3] OSE, "OSEC Architecture Version 2.0", 1996.

[4] SEMI, "SEMI E54.1-0298 Standard for Sensor/ Actuator Network Common Device Model", 1996.

[5] Object Management Group, Inc., "OMG Unified Modeling Language Specification Version 1.3", *Object Management Group, Inc.*, 1999.

[6] Gamma, E.et al., "Design Patterns Elements of Reusable Object-oriented Software", *Addison Wesley Longman, Inc.*,1995.

[7] Grand, M., "Patterns in Java - A Catalog of Reusable Design Patterns, Illustrated with UML, Volume 1", *John Wiley & Sons*, 1998.

[8] Harel, D., and Politi, M., "Modeling Reactive Systemswith Statecharts", *McGraw-Hill*, 2000.

[9] Douglass, B. P., "Real-Time UML - Developing Efficient Objects for Embedded Systems", *Addison Wesley Longman, Inc.*, 1998.

Implementation of Internet-Based Personal Robot with Internet Control Architecture

Kuk-Hyun Han, Shin Kim, Yong-Jae Kim, Seung-Eun Lee and Jong-Hwan Kim

Department of Electrical Engineering and Computer Science,
Korea Advanced Institute of Science and Technology (KAIST),
Kusong-dong, Yusong-gu, Taejon, 305-701, Republic of Korea
{khhan, skim, yjkim, selee and johkim}@vivaldi.kaist.ac.kr
http://vivaldi.kaist.ac.kr

Abstract

This paper describes the implementation of an internet-based personal robot with novel direct internet control architecture which is insensitive to the inherent internet time delay. The personal robot can be controlled by using a simulator provided at a local site. However, a large internet time delay may make some control inputs distorted. Moreover, since it is affected by the number of the internet nodes and loads, this delay is variable and unpredictable. The proposed control architecture guarantees that the personal robot can reduce the path error and the time difference between a virtual robot at the local site and a real robot at the remote site. Simulations and experimental results in the real internet environment demonstrate the effectiveness and applicability of the internet-based personal robot with the proposed internet control architecture.

1 Introduction

Recently it has been observed that many researchers take interest in internet robotics because of the merits of internet which enables users to access any systems on the network cheaply. The robot arm control system [1] through a Web browser was designed, and TeleGarden system [2] and Mars Pathfinder [3] were developed. The sensor-based mobile robot system [4] which can be controlled by using a Web browser and the internet-based supervisory architecture [5] were reported. The concept of a personal tele-embodiment [6] and an intelligent telerobot [7] were introduced recently. Most of them have the supervisory control scheme which enables users to issue high level commands. The internet time delay is variable and unpredictable so that the design of a direct control scheme which enables users to control the motion of the robot continuously may not be easy. The direct control scheme [8] on the internet was proposed, but the modeling of the internet time delay was not adequate.

This paper describes the implementation of an internet-based personal robot with novel internet control architecture which guarantees that the personal robot can reduce the path error and the time difference between the actions of a virtual robot at the local site and a real robot at the remote site. An internet user can control the real robot using a simulator provided at the local site, and can have information on the real environment at the remote site since the simulator has a virtual environment. The path error and the time difference in the internet-based personal robot system are caused by the unpredictable internet time delay and the difference between the real environment of the remote site and the virtual environment of the local site. It is not easy to model internet time delay, hence a control architecture that is insensitive to time delay, is needed. Main components of the proposed internet control architecture are a command filter to recover the information loss of control commands, a path generator and a path-following controller to reduce the time difference between the real robot and the virtual robot. The difference between the real robot and the virtual robot model of the simulator can be overcome by a posture estimator. The problem caused by the difference between the two environments can be solved by applying a virtual environment supervisor to the control architecture. We have already proposed the control architecture with the virtual environment supervisor [9], and it will not be mentioned in this paper. The graphic user interface(GUI) implemented with Java and the practical applications of the internet-based personal robot were described in [10].

This paper is organized as follows. In Section 2, the developed internet-based personal robot system, the modeling of a mobile robot, and the characteristics of internet time delay are described. In Section 3, the novel internet control architecture is designed step by step, which is insensitive to internet time delay. Section 4 presents the simulation results of the proposed internet control architecture. Real experiments with the developed internet-based personal robot are provided to show the effectiveness and applicability of the proposed internet control architecture. Concluding remarks follow in Section 5.

2 Internet-based Personal Robot

2.1 System description

The internet-based personal robot (IPR), a kind of service robot, can be used for a person's convenient life in a house/office or any indoor environment. It has a personal computer (PC) as a main part, and it can obtain information about environmental changes by using vision cameras, sonar sensors, etc. Actuators enable the robot to move and to carry out some physical work. It has a wireless LAN system for the internet remote control. A user can control the IPR using a simulator provided at a local site. It has the intelligence to gather the data from the sensors and to process them to decide its action.

The overall system consists of computers at the local sites, internet, wireless LAN system and the IPR. Users can access the IPR located at the remote site via internet using a computer at the local site. The wireless LAN system connects the IPR to the internet.

At the local site to control the IPR, a remote control architecture of the IPR system should be designed considering the inherent internet time delay. A basic remote control architecture of the IPR system considered in this paper is described in the following. A user controls a virtual robot in a simulator provided at the local site, and the virtual robot uses a virtual environment for obstacle avoidance. The command signals given by the user are sent to the IPR at the remote site via the internet. The IPR moves like the virtual robot, and avoids obstacles using sensor information. The posture of the virtual robot in the simulator can be updated by feedback of the IPR posture information through the internet.

The developed IPR has a square body of size $45cm \times 52cm \times 75cm$ as shown in Figure 1. The weight is about $70Kg$. It is a 4-wheeled drive with two fixed wheels and two auxiliary off-centered orientable wheels. It consists of a personal computer (Pentium II 333Mhz), a wireless LAN (Samsung MagicWave, 2Mbps), a head with two vision color cameras, a 12.1 inch TFT monitor, a speaker, a microphone, sonar sensors (10 pairs), a 12V 100Ah battery (5hr 80Ah), and two AC servo motors (LG Industrial Systems, 200W). Two cameras of the head part can rotate around a vertical and a horizontal axis under the command of the three DC motors. The IPR is connected to the internet through the wireless LAN, and it works as a server. The user can connect to the IPR using a Web browser or a TCP/IP application anywhere and also the user can give motion commands to the IPR.

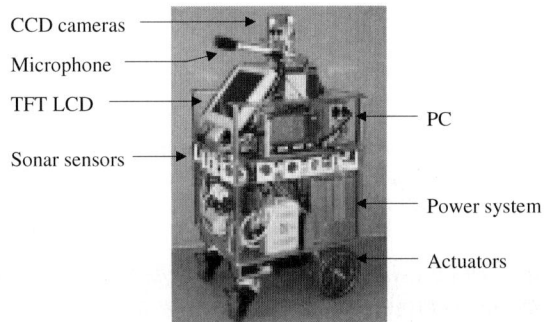

Figure 1: The IPR hardware

2.2 Modeling of the IPR

The modeling of IPR is needed for the implementation of the simulator. Two fixed and two auxiliary off-centered orientable wheeled mobile robots with non-slipping and pure rolling are considered. The velocity vector $\mathbf{u} = [v \;\; \omega]^T$ consists of the translational velocity of the center of driving wheel axis and the rotational velocity with respect to the center of driving wheel axis. The velocity vector $\mathbf{u}$ and a posture vector $\mathbf{P_c} = [x_c \;\; y_c \;\; \theta_c]^T$ are associated with the robot kinematics as follows:

$$\dot{\mathbf{P}}_c = \begin{bmatrix} \dot{x}_c \\ \dot{y}_c \\ \dot{\theta}_c \end{bmatrix} = \begin{bmatrix} \cos\theta_c & -h\sin\theta_c \\ \sin\theta_c & h\cos\theta_c \\ 0 & 1 \end{bmatrix} \begin{bmatrix} v \\ \omega \end{bmatrix} = \mathbf{J}(\theta_c)\,\mathbf{u} \quad (1)$$

$$\mathbf{u} = \begin{bmatrix} v \\ \omega \end{bmatrix} = \begin{bmatrix} \frac{1}{2} & \frac{1}{2} \\ \frac{1}{L} & -\frac{1}{L} \end{bmatrix} \begin{bmatrix} v_R \\ v_L \end{bmatrix} \quad (2)$$

where v_R is the right wheel velocity, v_L is the left wheel velocity, h is the displacement from the center of robot to the wheel axis, and L is the distance between the two wheels.

2.3 Internet time delay

A large internet time delay may make some control inputs distorted. Moreover, since it is affected by the number of the internet nodes and loads, this delay is variable and unpredictable. Figure 2 shows the influence of the internet time delay on the control information. The received data at the remote site was distorted severely, and the information of the sine function was almost lost, when the test function used was $y(t) = 5\sin(0.2\pi t) + 5$.

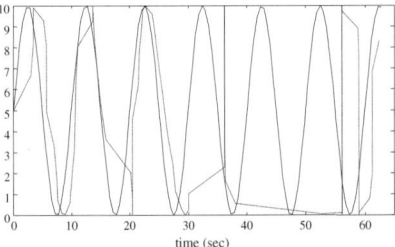

Figure 2: Influence of the internet time delay

3 Internet Control Architecture

A user can control the IPR at the remote site through internet using a simulator provided at the local site. The user regards the status of the virtual IPR at the local site as that of the real IPR at the remote site. Since the user cannot recognize the environment of the remote site, it is expected that the real IPR moves as the virtual IPR does. However, because of time delay we have to compensate for the path error and the time difference between the real IPR and the virtual IPR, which increase as time goes on.

In this section, a novel internet control architecture is designed step by step to minimize the effect of internet time delay. The proposed architecture is completed in three design steps, and its effectiveness is verified through simulations and experiments.

The internet control architecture-I consists of a user interface, simulator, virtual environment and posture estimator, which can be devised from the basic concept as shown in Figure 3. In the figure, $\mathbf{u}_r(i)$ is the ith control command $[v_r(i) \ \omega_r(i)]^T$ from a user, $\mathbf{u}_r^d(i)$ the ith control command passed through the internet, $\mathbf{P}_c(i)$ the ith robot posture, $\mathbf{P}_c^d(i)$ the ith robot posture passed through the internet, $\hat{\mathbf{P}}_c(i)$ the ith estimated posture, and $\mathbf{P}_c^s(i)$ the ith posture of the virtual robot. In order to correct the posture error between the virtual robot and the real robot, the real robot generates feedback signals such as posture information of the real robot, to the simulator. The architecture-I can be considered as a basic structure.

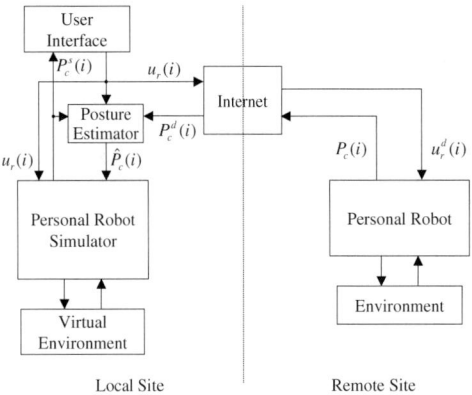

Figure 3: Internet control architecture-I (Step 1)

User Interface which can be implemented by Java, C++, etc., enables a user to control a remote IPR. *Posture estimator* estimates the current posture of the virtual IPR based on the feedback information of the real IPR. *Personal robot simulator* is the same as the virtual mobile robot at the local site. *Virtual environment* has the information of the real environment so that it enables the virtual robot to avoid obstacles. *Personal robot* is the same as the real mobile robot at the remote site. *Environment* is a circumstance where the real IPR is working.

The internet control architecture-I has a weak point that the information loss of control commands increases when internet time delay occurs. The *posture estimator* can recover the information loss eventually, but the time required for the recovery becomes too long. The architecture which can get rid of the cause of the information loss is needed. Figure 4 shows the internet control architecture-II, where a *command filter* is introduced. The *command filter* can recover the information loss of control commands caused by the internet time delay. It means that the filter reduces the path error between the real robot and the virtual robot. The function of the *command filter* is shown in Figure 5. Command signals received at the same time after the internet time delay T_d are regenerated with the sampling time T in the *command filter*. The *command filter* consists of two modules such as a *command queue* and a *command generator*. The *command filter* and the two modules can be defined by DEVS (Discrete Event Systems Specifications) formalism [9]. The *command filter* receives a control command, and stores it in the *command queue*. The *command gener-*

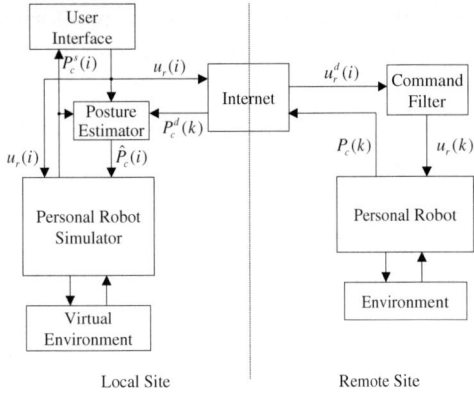

Figure 4: Internet control architecture-II (Step 2)

ator pulls out the command from the *command queue* and outputs it at each sampling time T.

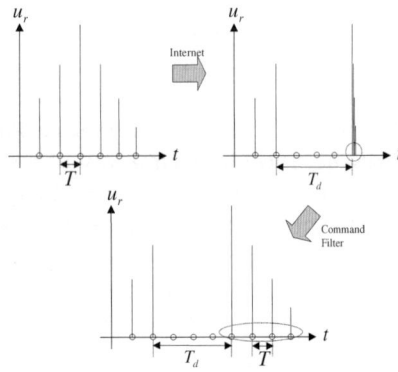

Figure 5: Function of *command filter*

The internet control architecture-II can recover the information loss of control commands, though internet time delay exists, but it still has the serious problem that the time difference between the real robot and the virtual robot increases, as internet time delay T_d is accumulated in the *command filter*. In order to solve this problem, the internet control architecture-III is finally designed.

The internet control architecture-III guarantees that the path error and the time difference between the real IPR at the remote site and the virtual IPR at the local site can be reduced. The proposed architecture includes a *path generator* and a *path-following controller*. The *path generator* restores the moving path of the virtual robot. The *path-following controller* guarantees that the real robot follows the generated path. The time difference between the real robot and the virtual robot can be reduced by the *path generator* and the *path-following controller*. As the control input of the real robot is separated from the control command passed through the internet by the two modules, the *command generator* in the *command filter* can be modified by replacing sampling time T with the processing time T_p which is shorter than T. The processing time is the computing interval for generating a path segment for one control command.

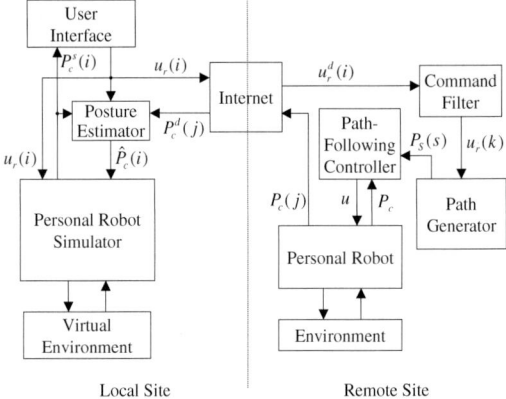

Figure 6: Internet control architecture-III (Step 3)

Figure 6 shows the internet control architecture-III, where $\mathbf{P}_S(s)$ is the moving path of the virtual robot, $\mathbf{u}$ is the control input of the *path-following controller*, $\mathbf{P}_c$ is the current posture of the real robot, and $\mathbf{P}_c(j)$ is the jth robot posture to be fed back to the simulator.

In this paper, the *path-following controller* is implemented with the uni-vector field navigation method [11]. The uni-vector field makes the mobile robot converge to a desired path.

4 Simulations and Experiments

4.1 Simulation results

Simulations were performed in the real internet environment. The physical distance between the local site and the remote site was about $300Km$ in Korea. Figure 7 shows the path error and the time difference between the two robots.

It should be noted that the architecture-I had cumulative errors. Because of this, it might be inconvenient for the user to control the robot in the simulation environment by the architecture-I. By the architecture-II the path error could be reduced, but the time difference increased continuously. However, by the architecture-III, the time difference as well as the path error were very small, although the internet time delay was quite variable.

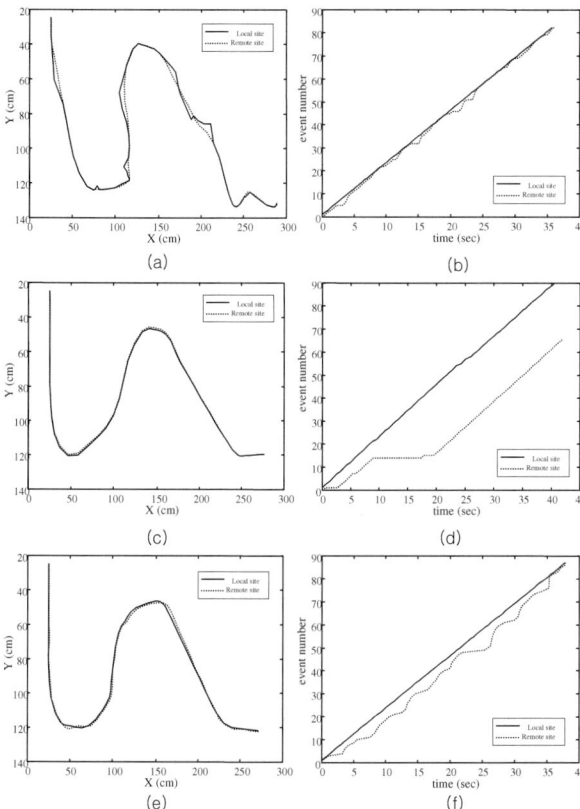

Figure 7: Simulation results. The path error and the time difference. (a),(b) architecture-I. (c),(d) architecture-II. (e),(f) architecture-III.

4.2 Experimental results

Experiment was performed with the developed IPR system equipped with an overhead CCD camera for global positioning.

In this experiment, the proposed internet control architecture was implemented as a TCP/IP application version, and the physical distance between the local site and the remote site was about $300Km$, which was the same condition as that of computer simulations. The *Posture Estimator* was removed from the architecture so as to find out the exact characteristics of each control structure. Figure 8 shows the path error and the time difference between the two robots. In the experimental results of the internet control architecture-I, the path error was caused by the information loss of control commands. The information loss of control commands made the real robot path different from the virtual robot path. In the results by the internet control architecture-II, the time difference between the actions of the two robots increased continuously, since internet time delay accumulated as time goes on. In the results by the internet control architecture-III, the path error and the time difference were quite small. The experimental results demonstrated the effectiveness and the applicability of the proposed internet control architecture-III as the simulation results did.

5 Conclusions

An internet-based personal robot with novel internet control architecture was developed. The proposed architecture was insensitive to internet time delay and guaranteed that the path error and the time difference between a real IPR and a virtual IPR could be reduced. Simulations and experimental results in a real internet environment demonstrated the effectiveness and the applicability of the proposed internet control architecture.

Acknowledgements

The authors would like to thank the support given by MIC(Ministry of Information and Communication), Korea, to the development of the internet-based personal robot.

References

[1] K. Taylor and B. Dalton, "Issues in Internet Telerobotics," in *Int. Conf. on Field and Service Robotics*, Dec. 1997.

[2] C. Sutter and J. Wiegley, "Desktop Teleoperation via the World Wide Web," in *Proc. IEEE Int. Conf. Robot. Automat.*, pp. 654-659, May 1995.

[3] R. Volpe, J. Balaram, T. Ohm and R. Ivlev, "The Rocky 7 Mars Rover Prototype," in *Proc. IEEE/RSJ Int. Conf. on Intelligent Robots and Systems*, pp. 1558-1564, Nov. 1996.

[4] T. M. Chen and R. C. Luo, "Remote Supervisory Control of An Autonomous Mobile Robot Via World Wide Web," in *Proc. IEEE Int. Symposium on Industrial Electronics*, vol. 1, pp. ss60-ss64, July 1997.

[5] K. Brady and T. J. Tarn, "Internet-Based Remote Teleoperation," in *Proc. IEEE Int. Conf. Robot. Automat.*, pp. 65-70, May 1998.

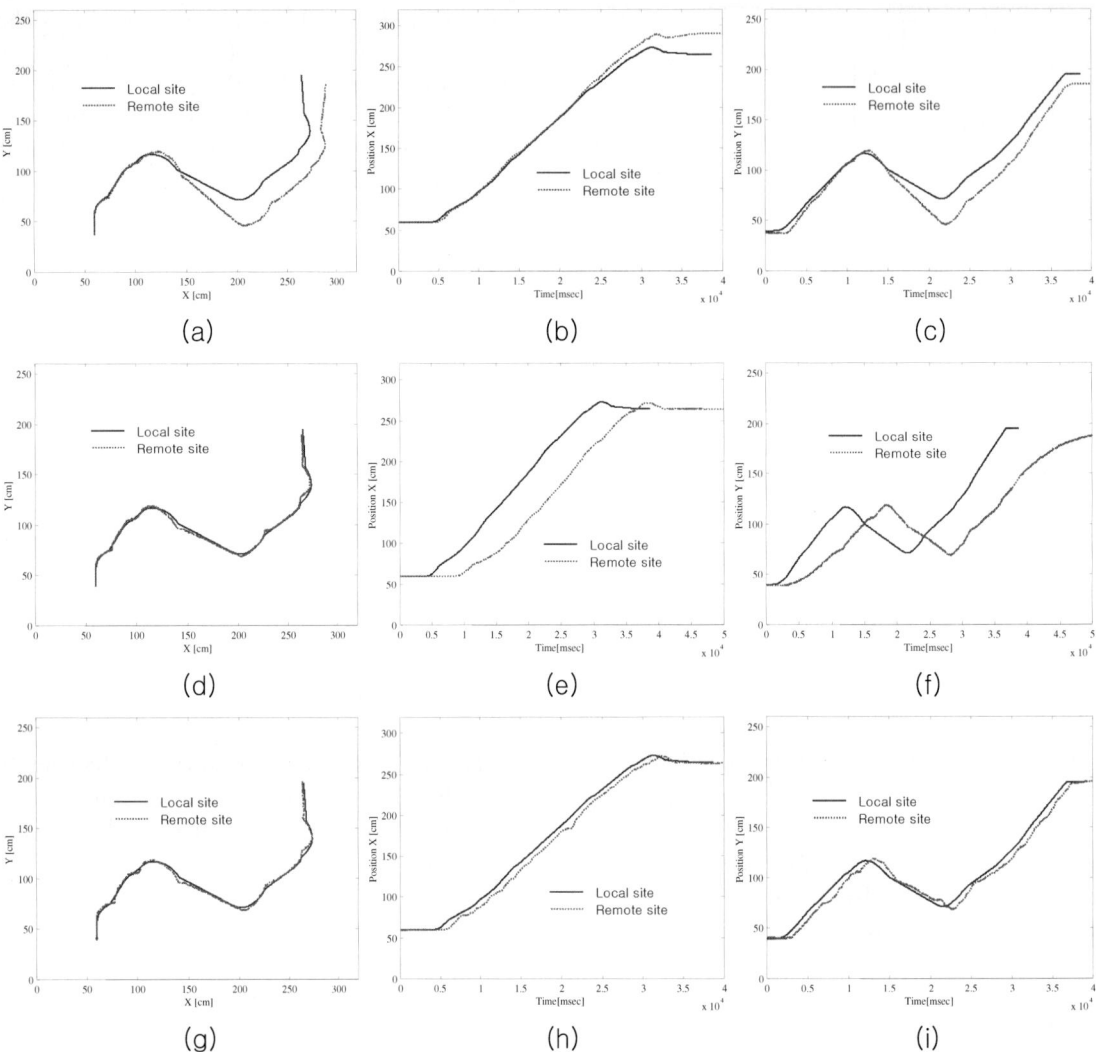

Figure 8: Experimental results. The path error and the time difference. (a),(b),(c) architecture-I. (d),(e),(f) architecture-II. (g),(h),(i) architecture-III.

[6] E. Paulos and J. Canny, "Designing Personal Tele-embodyment," in *Proc. IEEE Int. Conf. Robot. Automat.*, pp. 3173-3178, May 1998.

[7] E.P.L. Aude, G.H.M.B. Caneiro, H. Serdeira, J.T.C. Silveira, M.F. Martins and E.P. Lopes, "CONTROLAB MUFA: A Multi-Level Fusion Architecture for Intelligent Navigation of a Telerobot," in *Proc. IEEE Int. Conf. Robot. Automat.*, pp. 465-472, May 1999.

[8] R. Oboe and P. Fiorini, "A Design and Control Environment for Internet-Based Telerobotics," *Int. Journal of Robotics Research*, vol. 17, no. 4, pp. 433-449, Apr. 1998.

[9] K.-H. Han, S. Kim, Y.-J. Kim and J.-H. Kim, "Internet Control Architecture for Internet-Based Personal Robot," *Autonomous Robots Journal*, Kluwer Academic Publishers, vol. 10, no. 2, pp. 135-147, Mar. 2001.

[10] K.-H. Han, S. Kim, Y.-J. Kim and J.-H. Kim, "Internet-Based Personal Robot System using Map-Based Localization," in *Proc. the 32nd Int. Sym. on Robotics*, Apr. 2001.

[11] J.-H. Kim, K.-C. Kim, D.-H. Kim, Y.-J. Kim and P. Vadakkepat, "Path Planning and Role Selection Mechanism for Soccer Robots," in *Proc. IEEE Int. Conf. Robot. Automat.*, pp. 3216-3221, May 1998.

Computer animation: A new application for image-based visual servoing

Nicolas Courty, Éric Marchand
IRISA - INRIA Rennes
Campus de Beaulieu, 35042 Rennes Cedex, France
Email {Eric.Marchand}@irisa.fr

Abstract

This paper presents a new application for image-based visual servoing: computer graphics animation. Indeed, the control of a virtual camera in virtual environment is not a trivial problem and usually required skilled operators. Visual servoing, a now well known technique in robotics and computer vision, consists in positioning a camera according to the informations perceived in the images. Using this method within computer graphics context leads to a very intuitive approach of animation. Furthermore, in that case a full knowledge about the scene is available. It allows to easily introduce constraints within the control law in order to react automatically to modifications of the environment. In this paper, we apply this approach in two different contexts: highly reactive applications (virtual reality, video games) and the control of humanoid avatars.

1 Overview

Issues and related work. For now more than 10 years, visual servoing has been successfully and widely used to achieve various robotic tasks from assembly/disassembly tasks to docking or navigation tasks [5, 7]. A good review and introduction to visual servoing can be found in [8]. In this paper we proposed the use of this powerful framework for a new class of applications: computer graphics animation. Indeed, camera control in a virtual environment raised many difficult issues. Mainly the camera has to position itself wrt. its environment (first important issue) but it also has to react in an appropriate and efficient way to modifications of this environment (second important issue). Dealing with the first issue, even with a full knowledge of the scene, as in computer graphics, this positioning task is not a trivial problem [1]. Indeed it requires to precisely control the six degrees of freedom (d.o.f) of the camera in the 3D space. The second issue, that can be seen as the introduction of constraints in the camera trajectory, is even difficult. In order to be able to consider unknown or dynamic environments and to control in real time the motion of the camera, these constraints must be properly modeled and "added" to the positioning task. Image-based visual servoing has proved to be an efficient solution to these two problems.

Image-based control also received attention in computer graphics. The main difference wrt. computer vision or robotics is that the problem is no longer ill-posed. Indeed, in that case a full knowledge about the scene is available. Furthermore, even in an interactive context, the past and current behavior of all the objects are fully known. Ware and Osborn [21] consider various metaphors to describe a six d.o.f. camera control including "*eye in hand*". Within this context, the goal was usually to determine the position of the "eye" wrt. its six d.o.f in order to see an object or a set of objects at given locations on the screen. To control such a virtual device, people may consider user interfaces such as 3D mouse or six d.o.f joystick. Obtaining smooth camera motions required a skilled operator and has proved to be a difficult task. The classical lookat/lookfrom/vup parameterization is a simple way to achieve a focusing task on a world-space point. However specifying a complex visual task within the lookat/lookfrom framework is quite hopeless. Attempts to consider this kind of problem have been made by Blinn [1], however it appears that the proposed solutions are dedicated to specific problems and hardly scaled to more complex tasks. The image-based control have been described within the computer graphics context by Gleicher and Witkin in [6], they called it "*Through-the-lens camera control*". They proposed to achieve very simple tasks such as positioning a camera with respect to objects defined by static "virtual" points. This technique, very similar to the visual servoing framework, consider a local inversion of the nonlinear perspective viewing transformation. A constrain optimization is used to compute the camera velocity from the desired motion of the virtual points in the image. The image Jacobian is considered only for point features.

Interesting attempts to solve the introduction of constraints received great attention in both computer vision (e.g., [19]) and computer graphics [4] community. The resulting solutions are often similar. Each constraints is defined mathematically as a function of the camera parameters (location and orientation) to be minimized using deterministic (gradient approaches) or stochastic (simulated annealing) optimization processes. These approaches feature numerous drawbacks. First they are usually time consuming (the search space is of dimension six) and the optimization has to be considered for each iteration of the animation process (i.e., for each new frame). It is then difficult to consider these technics for reactive applications such as video-games. Visual servoing allows the introduction of constraints in the camera trajectory. Control laws taking into account "bad" configurations can thus to be considered [15, 14]. It combines the regulation of the vision-based task with the minimization of cost functions which reflect the constraints imposed on the trajectory. As the camera trajectory that ensure the task and the constraints is computed locally, it can be handled in real-time as required by the considered application.

Proposed system and contributions. We aimed at the definition of basic camera trajectories for virtual movie directors as well as the automatic control of a camera for reactive applications such as video games. We assume that we fully know the model of the scene at the current instant. Within this context, we present a complete framework, based on visual servoing, that allows the definition of positioning tasks wrt. a set of "virtual visual features" located within the environment (these features can be points, lines, spheres, cylinders, etc.). When the specified task does not constrain all the camera degrees of freedom, the method allows the introduction of secondary tasks that can be achieved under the constraint that the visual task is itself achieved. Furthermore the considered features are not necessarily motionless. Using this approach we present solutions to various non-trivial problems in computer animation. Some of these tasks are more concerned with reactive applications (target tracking and following, obstacles and occlusions avoidance) while others deal with the control of digital actors.

2 Image-based camera control

Image-based visual servoing consists in specifying a task as the regulation in the image of a set of visual features[5][7]. Embedding visual servoing in the task function approach allows the use of general results helpful for the analysis and the synthesis of efficient closed loop control schemes.

Control issues Let us denote $\mathbf{P}$ the current value of the set of selected visual features used in the visual servoing task and measured from the image at each iteration of the control law. To ensure the convergence of $\mathbf{P}$ to its desired value $\mathbf{P_d}$, we need to know the interaction matrix (also called image Jacobian) $\mathbf{L_P}$ defined by the classic equation [5]:

$$\dot{\mathbf{P}} = \mathbf{L_P}(\mathbf{P},\mathbf{p})\mathbf{T_c} \qquad (1)$$

where $\dot{\mathbf{P}}$ is the time variation of $\mathbf{P}$ due to the camera motion $\mathbf{T_c}$. The parameters $\mathbf{p}$ involved in $\mathbf{L_P}(\mathbf{P},\mathbf{p})$ represent the depth information between the considered objects and the camera frame.

A vision-based task $\mathbf{e}$ is defined by:

$$\mathbf{e} = \mathbf{W}^+\mathbf{C}(\mathbf{P} - \mathbf{P_d}) + (\mathbf{I} - \mathbf{W}^+\mathbf{W})\mathbf{e_2} \qquad (2)$$

where $\mathbf{C}$, called combination matrix, has to be chosen such that $\mathbf{CL_P}(\mathbf{P},\mathbf{p})$ is full rank about the desired trajectory $q_r(t)$. It can be defined as $\mathbf{C} = \mathbf{WL_P^+}(\mathbf{P},\mathbf{p})$ ($\mathbf{L}^+$ denotes the pseudo inverse of $\mathbf{L}$). In that case, we set $\mathbf{W}$ as a full rank matrix such that Ker $\mathbf{W}$ = Ker $\mathbf{L_P}$. If the vision-based task does not constrain all the n robot degrees of freedom, a secondary task $\mathbf{g_s}$ can also be performed. $\mathbf{e_2}$ is the gradient of a cost function h_s to be minimized ($\mathbf{e_2} = \frac{\partial h_s}{\partial \mathbf{r}}$). This cost function is minimized under the constraint that $\mathbf{P} = \mathbf{P_d}$. The two projection operators $\mathbf{W}^+$ and $\mathbf{I}-\mathbf{W}^+\mathbf{W}$ guarantee that the camera motion due to the secondary task is compatible with the regulation of $\mathbf{P}$ to $\mathbf{P_d}$.

To make $\mathbf{e_1}$ decrease exponentially and behave like a first order decoupled system, we get:

$$\mathbf{T_c} = -\lambda\mathbf{e} - \alpha\mathbf{T_0} - (\mathbf{I} - \mathbf{W}^+\mathbf{W})\frac{\partial\mathbf{e_2}}{\partial t} \qquad (3)$$

where:
- $\mathbf{T_c}$ is the camera velocity;
- λ is the proportional coefficient involved in the exponential convergence of $\mathbf{e}$;
- the last term $\mathbf{T_0}$ (the pure target motion) allows to fully suppress the tracking errors if $\alpha = 1$.

3 Reactive viewpoint planning

The positioning tasks that can be considered within the framework presented in the previous section are quite simple. As we did not consider the environment, the target was assumed to be "alone". We now present a method that makes it possible to achieve far more complex tasks in dynamic *"cluttered environments"*. We will propose a purely reactive framework in order to avoid undesirable configurations in an animation context.

3.1 Avoiding obstacles

Obstacle avoidance is a good example of what can be easily given within the proposed framework. Let us assume that the camera is moving in a cluttered environment while gazing on a visual target. The goal is to ensure this task while avoiding all the obstacles in the scene.

There are in fact multiple solutions to this problem: one solution is to planify a trajectory that avoids the obstacles using a trajectory planning process. Another solution is to consider a secondary task that uses the redundant d.o.f of the camera to move away from obstacles. This function will tend to maximize the distance between the camera and the obstacle. A good cost function to achieve the goal should be maximum (infinite) when the distance between the camera and the obstacle is null. The simplest cost function is then given by:

$$h_s = \alpha \frac{1}{2\|C - O_c\|^2} \qquad (4)$$

where $C(0,0,0)$ is the camera location and $O_c(x_c, y_c, z_c)$ are the coordinates of the closest obstacle to the camera, both expressed in the camera frame (note that any other cost function that reflects a similar behavior suits the problem). If $O_s(x_s, y_s, z_s)$ are the coordinates of the obstacle within the scene frame (or reference frame) and $M_c(RT)$ the homogenous matrix that describes the camera position within this reference frame, the obstacle coordinates within the camera frame are given by $X_c = R^T X_s - R^T T$.

The components of the secondary task are given by:

$$\mathbf{e_2} = -(x_c, y_c, x_c, 0, 0, 0)^T \frac{h_s^2}{\alpha} \quad \text{and} \quad \frac{\partial\mathbf{e_2}}{\partial t} = 0 \qquad (5)$$

Multiple obstacles can be handled considering the cost function $h_s = \sum_i \alpha \frac{1}{\|C - O_{c_i}\|^2}$.

3.2 Avoiding occlusions

The goal here is to avoid the occlusion of the target due to static or moving objects (with unknown motion). The virtual camera has to perform adequate motion in order to avoid the risk of occlusion while taking into account the desired constraints between the camera and the target. Related work are proposed by [10]. There are actually many situations that may evolve in an occlusion. The first and most simple case is a moving object that crosses the camera/target line (see Figure 1.a). Two other similar cases may be encountered: in the first one (see Figure 1.b) the target moves behind another object in the scene while in the second one (see Figure 1.c) the camera follows an undesirable trajectory and is hidden behind an object.

We will now present a general image-based approach that make it possible to generate adequate camera motion automatically to avoid occlusions [14]. In a second time we will see a simple method to determine the risk of occlusion in order to weight adequately the camera response (i.e. its velocity).

Figure 1: Occlusion issues (a) occlusion due to a moving object (b) occlusion due to the target motion (c) occlusion due to the camera motion

Automatic generation of adequate motions Let us consider $\mathcal{O}$ the projection in the image of the set of objects in the scene which may occlude the target T: $\mathcal{O} = \{O_1, \ldots O_n\}$. According to the methodology presented in paragraph ?? we have to define a function h_s which reaches its maximum value when the target is occluded by another object of the scene. In fact this occlusion problem can be fully defined in the image. If the occluding object is closer than the target, when the distance between the projection of the target and the projection of the occluding object decreases, the risk of occlusion increases.

We thus define h_s as a function of this distance in the image:

$$h_s = \frac{1}{2}\alpha \sum_{i=1}^{n} e^{-\beta(\|T - O_i\|^2)} \qquad (6)$$

where α and β are two scalar constants. α sets the amplitude of the control law due to the secondary task. The components of $\mathbf{e_2}$ and $\frac{\partial \mathbf{e_2}}{\partial t}$ involved in (3) are then:

$$\mathbf{e_2} = \frac{\partial h_s}{\partial \mathbf{r}} = \frac{\partial h_s}{\partial \mathbf{P}} \frac{\partial \mathbf{P}}{\partial \mathbf{r}}, \qquad \frac{\partial \mathbf{e_2}}{\partial t} = 0$$

Computing $\frac{\partial h_s}{\partial \mathbf{P}}$ is seldom difficult. $\frac{\partial \mathbf{P}}{\partial \mathbf{r}}$ is nothing but the image Jacobian $L_\mathbf{P}$.

Let us consider the case of a single occluding object here considered as a point. The generalization to other and/or to multiple objects is straightforward. We want to see the target T at a given location in the image. Thus we will consider the coordinates $\mathbf{P} = (X, Y)$ as its center of gravity. If we also consider the occluding object $\mathcal{O}$ by a point $\mathbf{P}_\mathcal{O} = (X_\mathcal{O}, Y_\mathcal{O})$, defined as the closest point of $\mathcal{O}$ to T, we have:

$$h_s = \frac{1}{2}\alpha e^{-\beta \|\mathbf{P} - \mathbf{P}_\mathcal{O}\|^2}$$

and $\mathbf{e_2}$ is given by:

$$\mathbf{e_2} = \frac{\partial h_s}{\partial \mathbf{r}} = \frac{\partial h_s}{\partial X}\mathbf{L}_X + \frac{\partial h_s}{\partial Y}\mathbf{L}_Y \qquad (7)$$

with

$$\frac{\partial h_s}{\partial X} = -\alpha\beta(X - X_\mathcal{O})e^{-\beta\|\mathbf{P} - \mathbf{P}_\mathcal{O}\|^2}$$

and

$$\frac{\partial h_s}{\partial Y} = -\alpha\beta(Y - Y_\mathcal{O})e^{-\beta\|\mathbf{P} - \mathbf{P}_\mathcal{O}\|^2}$$

In fact $\mathbf{e_2}$ as defined in (7) is an approximation of $\frac{\partial h_s}{\partial \mathbf{r}}$. Indeed $\mathbf{L_P} = [\mathbf{L}_X, \; \mathbf{L}_Y]^T$ is the image Jacobian related to a physical point. In our case, since the point is defined as the closest point of $\mathcal{O}$ to T, the corresponding physical point will change over time. However considering $\mathbf{L}_X$ and $\mathbf{L}_Y$ in (7) is locally a good approximation.

Risk of occlusion Using the presented approach to compute the camera reaction is fine if the occluding object moves between the camera and the target [14] as depicted in Figure 1. Indeed, in that case occlusion will occur if no action is taken. However, it is neither necessary nor desirable to move the camera in all the cases (if the occluding object is farther than the target). A key point is therefore to detect if an occlusion may actually occur. In that case we first compute a bounding volume $\mathcal{V}$ that includes both the camera and the target at time t and at time $t + ndt$ assuming a constant target velocity (see Figure 2 and Figure 3). An occlusion will occur if an object is located within this bounding box. The time-to-occlusion may be computed as the smallest n for which the bounding box is empty. If an object $\mathcal{O}$ of the scene is in motion, in the same way, we consider the intersection of the volume $\mathcal{V}$ with a bounding volume that includes $\mathcal{O}$ at time t and at time $t + ndt$.

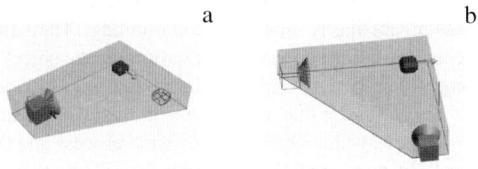

Figure 2: Computing the risk of occlusion

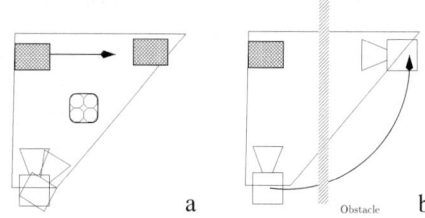

Figure 3: Detection of a future (a) occlusion (b) collision with an obstacle

Let us point out two other interesting issues:

- Obstacle avoidance may be considered in this context. Indeed, if an obstacle is on the camera trajectory, it will be located in the created bounding box (see Figure 3.b). The system will therefore forbid the camera to move in that direction.

- Some cases are more difficult to handle. A good example is a target moving in a corridor. In that case, the only solution to avoid the occlusion of the target by one of the walls and to avoid the contact with the other wall is to reduce the camera/target distance. This can only be done if the z axis is not controlled by the primary task [13].

In conclusion, let us note that in this paragraph, we have just proposed a method to detect and quantify the risk of occlusion. The method proposed in paragraph 3.2 must be, in all cases, used to generate the adequate motion that will actually avoid occlusion. The time-to-occlusion computed here will in fact be used to set the parameter α (see equation (6)) that tunes the amplitude of the response to the risk.

4 Digital actors control

Another possible application of visual servoing in computer animation is the control of digital actors (also called virtual humanoids or avatars). Achieving humanoids control through a virtual vision process is interesting for multiple reasons:

- the sensing process is more selective since the humanoid do not have access to the whole environment data-base.
- a consequence of this selectivity process is that each avatar is more independent since it has not access to the same information that the other avatars.
- the avatar is more autonomous and its behavior will be more realistic.

Mainly two approaches have been proposed to simulate this vision process. In the former the scene observed by the digital actors is rendered and information is extracted from the resulting image using image processing algorithm [20, 17, 9]. In these approach we have only access to a limited amount of information and image processing is usually time consuming. The latter approach consider a direct access (*direct sensing*) to the object of the environment data-base in which all interesting informations are encoded, the actors are then omniscient. In this two approaches, there are few interactions between the vision process and the actor control. In many case the actor sees then it acts in a strong sequential process that is very different from the real perception process.

We think that visual servoing is a good way to achieve the low level control of such digital actors. It will allow, as in the previous experiments, a fast reactivity to external stimuli. Another interesting point in this approach is that the specification of the humanoid motions are, here again, done in the 2D image space, allowing simple automatic generation by behavioral engines ruling the humanoid. With respect to the previous experiments, the motion of virtual camera (the eyes of the humanoid) is no longer free. The eyes can be considered as mounted on the end-effector of a highly redundant robot.

In this paragraph we will show how to modify the previous control law to consider such particular robot. We show how to deal with joint limits in order to achieve realistic motions.

Control in the articular frame. In Section 2, the control laws have been expressed in the operational space (i.e., in the camera frame). However, in order to combine a visual servoing with the avoidance of joint limits, we have to directly express the control law in the articular space.

This leads to the definition of a new interaction matrix such that:
$$\dot{\mathbf{P}} = \mathbf{H_P}\, \dot{\mathbf{q}} \qquad (8)$$
Since we have $\mathbf{T_c} = \mathbf{J}(\mathbf{q})\,\dot{\mathbf{q}}$, where $\mathbf{J}(\mathbf{q})$ is nothing but the robot Jacobian, we simply obtain:
$$\mathbf{H_P} = \mathbf{L_P}\,\mathbf{J}(\mathbf{q}) \qquad (9)$$
The vision-based task $\mathbf{e_1}$ is then defined by:
$$\mathbf{e_1} = \mathbf{C}(\mathbf{P} - \mathbf{P_d}) \qquad (10)$$
where $\mathbf{C}$ can be defined as $\mathbf{C} = \mathbf{W}\mathbf{H_P^+}$.

Joint limits avoidance. Joint limits avoidance is a fundamental process that has to be implemented to achieved realistic motion. The most classical way to solve the joint limits avoidance problem is to define the secondary task as the gradient of a cost function h_s ($\mathbf{e_2} = \frac{\partial h_s}{\partial \mathbf{q}}$). This cost function must reach its maximal value near a joint limits and its gradient must be equal to zero when the cost function reaches its minimal value. Several cost functions h_s which reflect this desired behavior have been presented in [18, 2, 12]. An example of such a cost function is given by:
$$h_s = \beta \sum_{i=1}^{n}(q_i - (\frac{q_{i_{min}} + q_{i_{max}}}{2}))^2 \qquad (11)$$
where $q_{i_{min}}$ and $q_{i_{max}}$ are the minimum and maximum allowable joint values for the ith joint. The parameter β that sets the amplitude of the control law due to the secondary task is very important. If β is too small, the change in the configuration will occur when $(\mathbf{I} - \mathbf{W^+}\mathbf{W})\mathbf{e_2}$ will become large wrt. the primary task. It may be too late and may produce some overshoot in the effector velocity. If β is too large, it will result in some oscillations. Therefore β is usually set based on trial and errors.

To cope with this problem we have used a fully automatic approach proposed in [3]. A good solution to achieve the avoidance task is to cut any motion on axes that are in critical area or that moves the robot toward it. Considering that $\mathbf{q}_k$ is one of these axes, we have to compute a velocity $\dot{\mathbf{q}}_k = 0$. This can be done by iteratively solving a system of linear equation. Our goal in this paper is not to describe this approach here. Let us just say this method provides a complete solution to ensure that, if a solution exists, the joints in critical situation will not encounter their limits.

5 Results

In this section some results are presented to illustrate our approach. Most of the images are generated in "real-time" (i.e. less than 0.1 s/frame without texture-mapping) on a simple SUN Ultra Sparc (170Mhz) using Mesa GL.

5.1 Avoiding occlusions: museum walkthrough.

In this example, we applied the proposed methodology to a navigation task in a complex environment. The target to be followed is moving in a museum-like environment. This "museum" has two rooms linked by stairs. The experiment goal is to keep the target in view (i.e. to avoid occlusions) while considering *on-line* the modifications of the environment (i.e. other moving objects).

We do not address in this paper the definition of the target trajectory. Finding a path for the target is a planning problem on its own. Solutions are proposed in, e.g. [4][11]. Most of these approaches are based on a global path planning strategy (usually based on potential field approach).

In this example, we consider a focusing task wrt. an image centered virtual sphere that has to be centered in the image. This task constrains 3 d.o.f of the virtual camera (i.e. to achieve the focusing task and to maintain the radius constant in the image). The reader can refer to [5] for the complete derivation of the image Jacobian related to a sphere. Figure 4 shows the camera trajectories for various applied strategies while target and camera are moving in the first room of the environment. Obstacles appear in yellow. The target trajectory is represented as a red dotted line, while the trajectory of another moving object is represented as a blue dotted line. The red trajectory represents the simplest strategy: just focus on the object. As nothing is done to consider the environment, occlusions and then collisions with the environment occur. The blue trajectory only considers the avoidance of occlusions by static objects; as a consequence, the occlusion by the moving object occurs. The green trajectory considers the avoidance of occlusions by both static and moving objects.

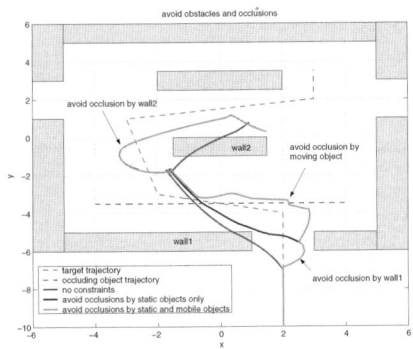

Figure 4: Museum walkthrough: camera trajectories for various strategies

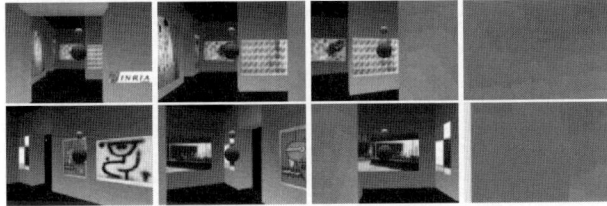

Figure 5: Museum Walkthrough. The occlusions/obstacles avoidance process is not considered. This leads to multiple occlusions of the target and multiple collisions with the environment.

Figure 5 shows the views acquired by the camera if no specific strategy is considered to avoid occlusion of the target and obstacle avoidance. This leads to multiple occlusions of the target and multiple collisions with the environment. In Figures 6 the control strategy considers the presence of obstacles. This time, the target always remains in the field of view, and at its desired position in the image. The collisions with the wall and the occlusions of the target are correctly avoided. Let us note that the environment is not flat, and neither the target nor the camera move within a plane (the target "gets down" stairs on last row of Figure 6 last row). Tracking and avoidance process perform well despite the fact that the target moves in 3D. On the bird's eye view the yellow volume (associated to the camera-target couple) corresponds to the bounding volumes used to predict the occlusions.

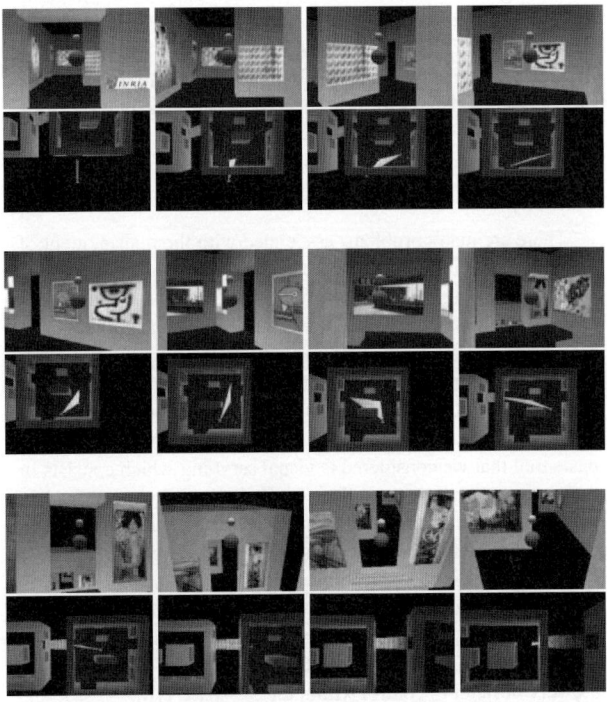

Figure 6: Museum Walkthrough: camera views and corresponding bird's eye views

5.2 Humanoid control

For the digital actor we currently consider the animation of the torso of an humanoid with 9 degrees of freedom: pelvis (3 d.o.f), spine (1 dof), neck (3 dof), eyes (2 dof). The modeling of this humanoid robot has been done using the Denavit-Hertenberg parameterization. For the moment, we have only consider simple positionning task with respect to the environment considering the joint limits avoidance issue.

Positionning wrt. a sphere. The first experiment (see Figure 7) deals with a positioning task wrt. a sphere that has to be seen centered in the image. Three dof are used to achieve this task, six dof are then free to deal with the joint limits.

Figure 7: Humanoid control: positioning wrt. a sphere

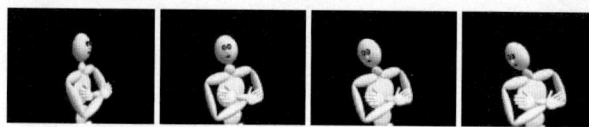

Figure 8: Humanoid control: tracking a point

Tracking task. The second experiment deals with a tracking task. It allows to demonstrate the capabilities of the joint limits avoidance algorithm. A point-object is crossing the scene. On Figure 8, we can see that, at the beginning, mainly the eyes are moving, when they move near their joint limits the motion is automatically transferred to the neck, then to the pelvis.

6 Conclusion

There are many problems associated with the management of a camera in a virtual environment. It is not only necessary to be able to carry out a visual task (often a focusing task or more generally a positioning task) efficiently, but it is also necessary to be able to react in an appropriate and efficient way to modifications of this environment. Furthermore, if we consider digital actors it is necessary to act in a realistic way. We chose to use techniques widely considered in the robotic vision community. The basic tool that we considered is visual servoing which consists in positioning a camera according to the information perceived in the image. This image-based control constitutes the first novelty of our approach. The task is indeed specified *in a 2D space*, while the resulting camera trajectories are *in a 3D space*. It is thus a very intuitive approach of animation since it is carried out according to what one wishes to observe in the resulting images sequence. This is specially true for the control of humanoid that are very difficult to control within the 3D space [16].

However, this is not the only advantage of this method. Indeed, contrary to previous work [6], we did not limit ourselves to positioning tasks wrt. virtual points in static environments. In many applications (such as video games) it is indeed necessary to be able to react to modifications of the environment, of trajectories of mobile objects, etc. We thus considered the introduction of constraints into camera control. Thanks to the redundancy formalism, the secondary tasks (which reflect the constraints on the system) do not have any effect on the visual task. To show the validity of our approach, we have proposed and implemented various classic problems from simple tracking tasks to more complex tasks like occlusion or obstacle avoidance or joint limits avoidance. The approach that we proposed has real qualities, and the very encouraging results obtained suggest that the use of visual control for computer animation is a promising technique. The main drawback is a direct counterpart of its principal quality: the control is carried out in the image, thus implying loss of control of the 3D camera trajectory or of the upper part of the humanoid body trajectory. This 3D trajectory is computed *automatically* to ensure the visual tasks but is not controlled by the animator. For this reason, one can undoubtedly see a wide interest in the use of these techniques within real-time reactive applications.

Acknowledgment. The authors wish to thank François Chaumette for his valuable comments.

Animations on-line. Most of the animations presented in this paper can be found as mpeg film on the VISTA group WWW page (http://www.irisa.fr/vista then follow the "demo" link).

References

[1] J. Blinn. Where am I ? what am I looking at ? *IEEE Computer Graphics and Application*, pages 76–81, July 1998.

[2] T.-F. Chang and R.-V. Dubey. A weighted least-norm solution based scheme for avoiding joints limits for redundant manipulators. *IEEE Trans. on Robotics and Automation*, 11(2):286–292, April 1995.

[3] F. Chaumette and E. Marchand. A new redundancy-based iterative scheme for avoiding joint limits: Application to visual servoing. In *IEEE Int. Conf. on Robotics and Automation*, volume 2, pages 1720–1725, San Francisco, CA, Avril 2000.

[4] S.M. Drucker and D. Zeltzer. Intelligent camera control in a virtual environment. In *Graphics Interface'94*, pages 190–199, Banff, Canada, 1994.

[5] B. Espiau, F. Chaumette, and P. Rives. A new approach to visual servoing in robotics. *IEEE Trans. on Robotics and Automation*, 8(3):313–326, June 1992.

[6] M. Gleicher and A. Witkin. Through-the-lens camera control. In *ACM Computer Graphics, SIGGRAPH'92*, pages 331–340, Chicago, July 1992.

[7] K. Hashimoto. *Visual Servoing : Real Time Control of Robot Manipulators Based on Visual Sensory Feedback*. World Scientific Series in Robotics and Automated Systems, Vol 7, World Scientific Press, Singapor, 1993.

[8] S. Hutchinson, G. Hager, and P. Corke. A tutorial on visual servo control. *IEEE Trans. on Robotics and Automation*, 12(5):651–670, October 1996.

[9] J.J. Kuffner and J.C Latombe. Fast synthetic vision, memory, and learning models for virtual humans. In *Computer Animation'99*, pages 118–127, Genève, Suisse, mai 1999.

[10] M. LaValle, H.-H. González-Baños, C. Becker, and J.-C. Latombe. Motion strategies for maintaining visibility of a moving target. In *Proc. IEEE International Conference on Robotics and Automation, ICRA'97*, volume 1, pages 731–736, Albuquerque, NM, April 1997.

[11] T.Y. Li, J.-M. Lien, S.-Y. Chiu, and T.-H. Yu. Automatically generating virtual guided tours. In IEEE Comput. Soc, editor, *Proc. of Computer Animation 1999*, pages 99–106, Geneva, Switzerland, May 1999.

[12] E. Marchand, F. Chaumette, and A. Rizzo. Using the task function approach to avoid robot joint limits and kinematic singularities in visual servoing. In *IEEE/RSJ Int. Conf. on Intelligent Robots and Systems, IROS'96*, volume 3, pages 1083–1090, Osaka, Japan, November 1996.

[13] E. Marchand and N. Courty. Image-based virtual camera motion strategies. In *Graphics Interface Conference, GI2000*, pages 69–76, 2000.

[14] E. Marchand and G.-D. Hager. Dynamic sensor planning in visual servoing. In *IEEE Int. Conf. on Robotics and Automation*, volume 3, pages 1988–1993, Lueven, Belgium, May 1998.

[15] B. Nelson and P.K. Khosla. Integrating sensor placement and visual tracking strategies. In *IEEE Int. Conf. Robotics and Automation*, volume 2, pages 1351–1356, San Diego, May 1994.

[16] Baerlocher P. and Boulic R. Task-priority formulations for the kinematic control of highly redundant articulated structures. In *Proc. of IROS'98*, Victoria, Canada, October 1998.

[17] O. Renault, N. Magnenat-Thalmann, and D. Thalmann. A vision-based approach to behavioural animation. *Journal of Visualization and Computer Animation*, 1(1):18–21, 1990.

[18] C. Samson, M. Le Borgne, and B. Espiau. *Robot Control: the Task Function Approach*. Clarendon Press, Oxford, United Kingdom, 1991.

[19] K. Tarabanis, P.K. Allen, and R. Tsai. A survey of sensor planning in computer vision. *IEEE trans. on Robotics and Automation*, 11(1):86–104, February 1995.

[20] D. Terzopoulos and T.M. Rabie. Animat vision: Active vision with artificial animals. In *Fifth International Conf. on Computer Vision (ICCV'95)*, pages 801–808, Cambridge, MA, June 1995.

[21] C. Ware and S. Osborn. Exploration and virtual camera control in virtual three dimensional environments. In *Proc. 90 Symposium on Interactive 3D Graphics*, pages 175–183, March 1990.

Design and Implementation of Visual Servoing System for Realistic Air Target Tracking

Wei Guan Yau[1], Li-Chen Fu[1,2], and David Liu[1]

Department of Electrical Engineering[1]
Department of Computer Science and Information Engineering[2]
National Taiwan University, Taipei, Taiwan, R.O.C.
r7921018@ms.cc.ntu.edu.tw

Abstract

In this paper, a real-time visual tracking system based on our proposed motion estimation algorithms is developed. The proposed motion estimation algorithm is used to predict the location of target and then generate a control input so as to keep the target stationary in the center of image. The work differs from previous ones in that it is capable to decouple the estimation of motion from the estimation of structure. The major contribution of this work is that simple, none computation intensive, correspondence-free, and numerically stable 3D motion estimation algorithms are developed. The robust target detection method in simple environment and a time reduction of SSD method in complex environment are minor contributions. The visual tracking system can achieve at a rate of 30 Hz. The robustness of the visual tracking system is validated by a number of experiments.

1. INTRODUCTION

Recently, research on motion estimation has been widely conducted [16, 26, 20, 21, 18, 1, 2, 13, 12]. However, one cannot estimate the target motion without knowledge of its structure or of the camera motion. In light of this observation, an adaptive method [23, 25, 11, 14, 15] or fuzzy controller is used to overcome this problem. On the other hand, the Interacting Multiple Model (IMM) algorithm is another famous target tracking framework [9, 28, 27, 4, 8].

The methods described above however are either too complicated in form or have their own drawbacks, which in turn increase the computational burden or deteriorates the tracking performance. Therefore, this paper proposes another close form solution for estimating the object motion during the tracking phase. This yields better performance than the previous methods.

Target detection algorithms such as Sum of Squared Difference (SSD) [23, 24, 26], and Normalized Cross-Correlation (NCC) [3, 5] are primitive methods. A small area of the image is used as template, which is later searched throughout the interested region in the next frame for finding the target location. However, these methods are computationally expensive and the SSD method is sensitive to the changes of the illumination. Sawasaki [5] used a specific hardware to speed up the computation. Hager and Belhumeur [19] also used the SSD method, but they provided an illumination-insensitive tracking algorithm. However, they need the images under various lighting conditions as the illumination basis. Another template matching method called contour matching [17, 6, 22] uses a non-rigid contour as model and is more robust to cluttered background.

Another well-known algorithm is motion-based recognition [10]. Motion-based tracking systems have the advantage of tracking a moving object regardless of its shape [29] [30]. This method is suitable for real-time implementation. However, they are easily corrupted in outdoor environment when the background contains other moving objects such as waving tree leaves.

Several existing target detection algorithms have already demonstrated appealing performance [17, 6, 22, 7, 29]. However, capability of real-time implementation is our main consideration. Therefore, we will design a simple, efficient, real-time monocular visual tracking system in this paper.

The remainder of this paper is organized as follows. Section 2 describes the target detection methods which are operated in simple and complex environment. The trajectory tracking algorithm is illustrated in Section 3. The implementation of these methods and the experimental results are presented in Section 4. Finally, conclusions are drawn in Section 5.

2. TARGET DETECTION METHODS

2.1 Target Detection in Simple Environment

Detecting the target by finding the centroid of edges is prone to fail as shown in Fig. 1. To solve this problem, we calculate the centroid of edges in a small attention window instead of the whole image as shown in Fig. 2. But at first we have to determine the location and the size of the attention window.

To determine the location of the attention window is critical for containing the target. Therefore, we use the direction of optical flow to provide the location of the attention window, which is the so-called optical flow based attention window as shown in Fig. 3. In particular, for fast sampling rate systems, the current centroid of the target can be used as the center location of the attention window in the next frame. Then we determine the size of the attention window. Assume the size of the attention window is $m \times n$, and the target has a maximum velocity $\mathbf{v}_m$. Then the size of attention window can be determined as follow:

$$m \geq |u_m| \quad \text{and} \quad n \geq |v_m| \quad (1)$$

where $[u_m \ v_m]^T$ is the optical flow induced by $\mathbf{v}_m$. The concept of appropriately determining the size of the attention window not only reduces the computation time but also makes this method more robust.

Fig. 1. Finding centroid of edges in whole image.

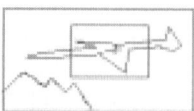

Fig. 2. Finding centroid of edges in attention window.

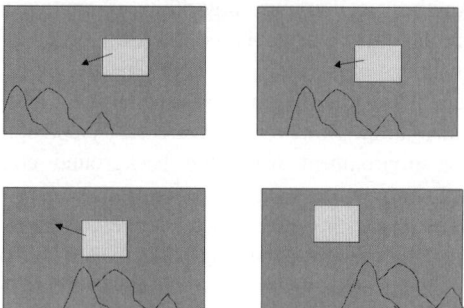

Fig. 3. Optical flow based attention window.

2.2 Target Detection in Complex Environment

The SSD measure having the moving edges as candidate is used to detect the target in complex environment. The method of finding the moving edges is based on motion energy detection [29] except for the background compensation algorithm. The background compensation algorithm used in [29] is based on Kanatani's relationship which is valid only when the pan/tilt angles between successive frames are not larger than $3°$.

We propose another background compensation algorithm that can compensate for arbitrary pan/tilt angles if the object is still visible in the current frame after pan/tilt motion. Denote Δx and Δy be the displacement in x and y direction respectively. From [25], Δx and Δy are

$$\Delta x = f \tan(\phi) \quad (2)$$

$$\Delta y = [-\Delta x \sin(\phi) - f \cos(\phi)] \tan(\theta) \quad (3)$$

where ϕ and θ are the pan and tilt angles. By using the actual coordinate system $\{F_a\}$ in Fig. 4, the relationship between pixel position in the previous and the current frame is:

$$x_{t-1} = x_t - \Delta x \quad (4)$$

$$y_{t-1} = y_t - \Delta y \quad (5)$$

where $[x_{t-1} \ y_{t-1}]^T$ and $[x_t \ y_t]^T$ are the pixel position in the previous and the current frame. Figure 5 and Fig. 6 show image subtraction with and without the proposed background compensation algorithms. Executing a logical AND operation between subtracted image and edge image can detect the moving edges as shown in Fig. 7.

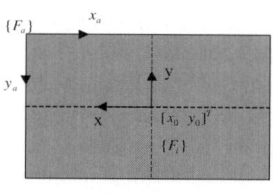 

Fig. 4. Image coordinate system. Fig. 5. Image subtraction with compensation.

Fig. 6. Image subtraction without compensation. Fig. 7. Moving edges in $3.5°$ pan and tilt angles.

In the SSD algorithm, assume the attention window size is $m \times n$ and the template size is $T_a \times T_b$, then the template is shifted $(m - T_a) \times (n - T_b)$ times. Obviously the computation is redundant if we have knowledge about the location of the target in the attention window. In light of this observation, moving edges are chosen as the candidates for the SSD measure. Let set $\Im$ of the candidates be the set of points:

$$\Im = \{\mathbf{x} = [x \ y]^T \mid I_m(x, y) > 0\} \quad (6)$$

where $I_m(x, y)$ is the brightness value of the result of motion energy detection. Thus, any $\mathbf{x} \in \Im$ is a candidate for the SSD measure. Obviously, the number of elements which belong to the set $\Im$ is less than $(m - T_a) \times (n - T_b)$. Accordingly, not only the computation time is reduced significantly, but also the noise is reduced and thus the confidence of SSD measure is increased.

If the object is static, then this concept can be extended by using the edges as the candidates for the SSD measure. Similarly, define the set

$$\mathfrak{I}_e = \{\mathbf{x} = [x\ y]^T \mid G(x,y) > 0\} \quad (7)$$

where $G(x,y)$ is the gradient magnitude at the location $\mathbf{x}$ in the attention window. Thus, any $\mathbf{x} \in \mathfrak{I}_e$ is a candidate for the SSD measure.

Table 1 shows the comparison of the processing time among these methods.

Table 1. The processing time among the different target detection methods.

Different methods / Processing time	SSD	SSD having moving edges as candidate	SSD having edges as candidate
In simple environment	540 ms	30 ms	58 ms
In complex environment	550 ms	30 ms	228 ms

3. DESIGN OF TRAJECTORY TRACKING ALGORITHM

We will make clear our main concept and notion for designing the trajectory tracking algorithm in this section. Our goal here is to design a simple, none computation intensive, and numerically stable algorithm. Our main idea is that if the target motion is known, then the camera motion which will eliminate the relative motion of the target may be generated such that the tracking is achieved. In particular, if the motion of target is known well, then the "perfect tracking" may be achieved, i.e., image position of the target can always be kept at the center of image.

It is apparent that the estimation of the target motion is the first step in our proposed algorithm. In general, the motion of target is not known a *priori*. However, it can be estimated through the change of the image position which is called "optical flow" or image displacement. Using the observation of image displacement or optical flow which induced by target motion, one can estimate the motion of target if the camera is stationary. However it is not the case for the tracking phase since both the camera and target are moving during the tracking phase and therefore the optical flow is subject to both camera and target motion. In light of this observation, we must modify the optical flow equation such that the optical flow is induced only by the target motion and irrelevant to the camera motion.

We can then estimate the target motion by using the modified optical flow. Motion estimation task becomes peculiarly difficult in trajectory tracking since neither the structure (depth) nor the target motion is known. Accordingly, we proposed a motion estimation algorithm which is independent of the estimation of structure. In other words, our algorithm is capable to decouple the estimation of motion from the estimation of structure. Thus, the structure (depth) need not be known or estimated before solving the motion estimation task. To this end, the weak perspective projection which is used to alleviate this problem, gives a good approximation when the size and the depth variation of the object are small compared with the distance between the object and the camera.

The second step in our proposed algorithm is prediction. The prediction task is to predict where the target image location in the current frame will 'move to' in the next frame. The last step is tracking. Using the predicted target position, we can then calculate a feasible camera motion in order to track the moving target. The whole process of these algorithms makes up of three steps and the flow chart is shown in Fig. 10. At any time instant k, the target image position at this time instant k and at the previous time instant $k-1$ are as the input of the motion estimation module. Then, it output the motion parameters to the prediction and filtering module. After that, the prediction and filtering module will calculate the predicted target image position in the next frame and eventually the tracking module generate the desired camera motion which will achieve tracking.

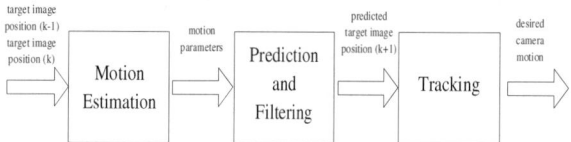

Fig. 8. Flow chart of our proposed algorithm.

For the lack of the space, the equation of this algorithm is summarized in Fig. 11 and the details can be found in [31].

Trajectory Tracking Algorithm

1. The target induced optical flow is calculated as follow:

$$u_o(k-1) = x(k) - x(k-1) + \left(x_{predict}(k-1) - x_o\right)$$
$$v_o(k-1) = y(k) - y(k-1) + \left(y_{predict}(k-1) - y_o\right)$$

where $[x_o\ y_o]^T$ are the image center and $[x_{predict}(k-1)\ y_{predict}(k-1)]^T$ are the predicted target image position.

2. Select three different edge points $[x_1\ y_1]^T$, $[x_2\ y_2]^T$, $[x_3\ y_3]^T$ which lie around the centroid of target and use the following equation to estimate the target motion parameters.

$$\mathbf{v} = \mathbf{G}^{-1}\mathbf{c}$$

where

$$G = \begin{bmatrix} f & 0 & -x_1(k-1) & -\frac{x_1(k-1)y_1(k-1)}{f} & \frac{x_1^2(k-1)+f^2}{f} & -y_1(k-1) \\ 0 & f & -y_1(k-1) & -\frac{y_1^2(k-1)+f^2}{f} & \frac{x_1(k-1)y_1(k-1)}{f} & x_1(k-1) \\ f & 0 & -x_2(k-1) & -\frac{x_2(k-1)y_2(k-1)}{f} & \frac{x_2^2(k-1)+f^2}{f} & -y_2(k-1) \\ 0 & f & -y_2(k-1) & -\frac{y_2^2(k-1)+f^2}{f} & \frac{x_2(k-1)y_2(k-1)}{f} & x_2(k-1) \\ f & 0 & -x_3(k-1) & -\frac{x_3(k-1)y_3(k-1)}{f} & \frac{x_3^2(k-1)+f^2}{f} & -y_3(k-1) \\ 0 & f & -y_3(k-1) & -\frac{y_3^2(k-1)+f^2}{f} & \frac{x_3(k-1)y_3(k-1)}{f} & x_3(k-1) \end{bmatrix}$$

$$\mathbf{v} = \begin{bmatrix} \frac{T_x(k-1)}{Z_c(k-1)} \\ \frac{T_y(k-1)}{Z_c(k-1)} \\ \frac{T_z(k-1)}{Z_c(k-1)} \\ \omega_x(k-1) \\ \omega_y(k-1) \\ \omega_z(k-1) \end{bmatrix}, \quad \mathbf{c} = \begin{bmatrix} u_0^1(k-1) \\ v_0^1(k-1) \\ u_0^2(k-1) \\ v_0^2(k-1) \\ u_0^3(k-1) \\ v_0^3(k-1) \end{bmatrix}.$$

3. Evaluate the predicted target centroid coordinate by assuming $\mathbf{T}(k) \approx \mathbf{T}(k-1)$, $\boldsymbol{\omega}(k) \approx \boldsymbol{\omega}(k-1)$, and $Z(k) \approx Z(k-1)$ for short sampling time as follow:

$$\begin{bmatrix} u_o(k) \\ v_o(k) \end{bmatrix} = \begin{bmatrix} f & 0 & -x_c(k) \\ 0 & f & -y_c(k) \end{bmatrix} \begin{bmatrix} \frac{T_x(k)}{Z(k)} \\ \frac{T_y(k)}{Z(k)} \\ \frac{T_z(k)}{Z(k)} \end{bmatrix} + \begin{bmatrix} -\frac{x_c(k)y_c(k)}{f} & f+\frac{x_c^2(k)}{f} & -y_c(k) \\ -\left(f+\frac{y_c^2(k)}{f}\right) & \frac{x_c(k)y_c(k)}{f} & x_c(k) \end{bmatrix} \begin{bmatrix} \omega_x(k) \\ \omega_y(k) \\ \omega_z(k) \end{bmatrix}$$

$$x_c(k+1) = x_c(k) + u_o(k)$$
$$y_c(k+1) = y_c(k) + v_o(k)$$

where $[x_c \; y_c]^T$ is the target centroid coordinate in image plane.

4. Use the following equations to compute the accommodated tracking command:

$$\phi = \tan^{-1}\left(\frac{x_o - x_c(k+1)}{f}\right),$$

$$\theta = \tan^{-1}\left(\frac{y_o - y_c(k+1)}{-\left(x_{org} - x_c(k+1)\right)\sin\phi - f\cos\phi}\right),$$

where ϕ is the pan angle and θ is the tilt angle.

Fig. 9. Trajectory Tracking Algorithm

4. IMPLEMENTATION AND EXPERIMENTS

This section illustrates the performance of the trajectory tracking algorithms under a variety of circumstances. All experiments were performed on live image sequences which were grabbed by a JAI MCL-1500 DSP color camera and then processed by the MATROX CORONA image processing card in the Pentium III-450 PC. The control command is then sent to the ADVANTECH stepping motor control card so as to generate an appropriate pan/tilt motion that achieves the goal of tracking.

4.1 Validation of the Trajectory Tracking Algorithm

To validate the performance of the proposed trajectory tracking algorithm, we will use the target detection method in simple environment as described in Section 2. Timing of the control cycle indicates that the proposed visual tracking system can perform frame rate (30 Hz) tracking of image regions with size 640×480 pixels.

- **Indoor Experiment**

An aircraft model is used as our target. At any time instant, the location of the target in the image plane is fed to the trajectory tracking algorithm to predict the future position as shown in Fig. 10(a) and 10(b). The tracking error is shown in Fig. 10(c). Note that the center of the image plane is $[320 \; 240]^T$. It is clear that the predicted target position is close to the true target position. This implies that our proposed algorithm provides accurate prediction of the maneuvering target trajectory.

- **Outdoor Experiment**

We use a remote controlled helicopter as our target. The helicopter is first near the camera. Then, it flies away, turns around and flies back toward the camera as shown in Fig. 11. The data of these image sequences are shown in Fig. 12. It demonstrates that the algorithm is robust against fast varying target size. Finally, it is interesting to notice Fig. 12(a) and 12(b), which show that the predicted target position well matches the motion of target.

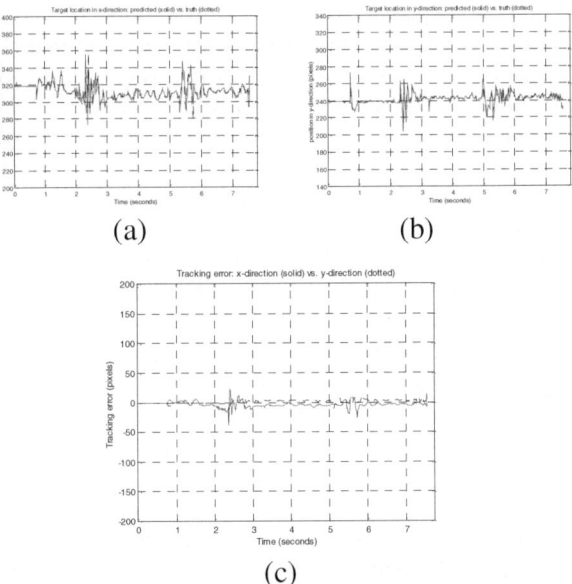

Fig. 10. Experimental results of indoor experiment.

Fig. 11. Image sequence.

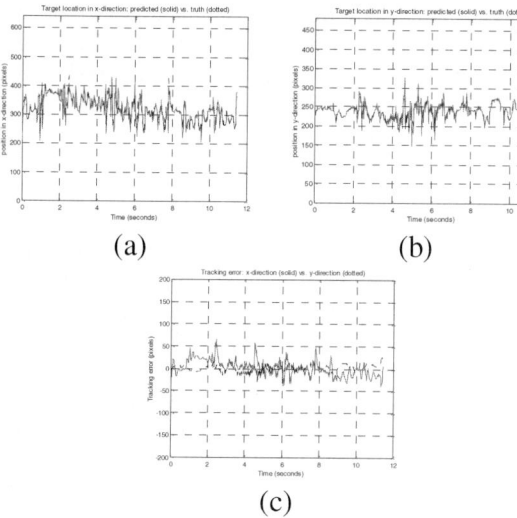

(a)　　　　　　　　(b)

(c)

Fig. 12. Experimental results of outdoor experiment.

4.2 Target Tracking in Complicated Scene

Image sequences in Fig. 13 demonstrate that we can successfully track an aircraft model which passes through highly cluttered scenes.

Fig. 13. Complicated environment scene image sequence.

5. CONCLUSION

In this paper, we have proposed a real-time visual tracking system. The whole system consists of a target detection method and a trajectory tracking algorithm.

We have proposed two different target detection methods which operate in simple and complex environment, respectively. In simple environment, the centroid of edges in the attention window is recognized as the location of target. The proposed method has been proved robust in general conditions if the target is not occluded by other objects. On the other hand, the SSD measure which has the moving edges as candidate is used to detect the target in the complex environment. The capability of tracking in highly cluttered scenes has been validated by the experiment. The computational burden of both algorithms is modest such that our system can be implemented real-time.

The centroid of target which is detected by the proposed target detection method is used as the tracking information which is needed by the trajectory tracking algorithm. The trajectory tracking algorithm based on proposed motion estimation algorithm provides accuracy of the prediction of the trajectory of the maneuvering target. The robustness of proposed visual tracking system is validated by a number of experiments.

REFERENCES

[1] G. Adiv, "Determining Three-Dimensional Motion and Structure from Optical Flow Generated by Several Moving Objects," *IEEE Trans. Pattern Analysis and Machine Intelligence*, vol. PAMI-7, no. 4, pp. 384-401, Jul 1985.

[2] S. Soatto, R. Frezza, and P. Perona, "Motion Estimation via Dynamic Vision," *IEEE Trans. Automatic Control*, vol. 41, no. 3, pp. 393-413, Mar 1996.

[3] Robert J. Schilling, *Fundamentals of Robotics Analysis and Control*, Englewood Cliffs, N.J. : Prentice-Hall, c1990.

[4] Y. Bar-Shalom, K. C. Chang and H. A. P. Blom, "Tracking a Maneuvering Target Using Input Estimation Versus the Interacting Multiple Model Algorithm", *IEEE Trans. on Aero. and Electro. Sys.*, vol. AES-25, no. 2, pp. , Mar 1989.

[5] N. Sawasaki, T. Morita, and T. Uchiyama, "Design and Implementation of High-Speed Visual Tracking Systems for Real-Time Motion Analysis," *IEEE Int. Conf. Pattern Recognition*, vol. 3, 478-483, 1996.

[6] A. Blake, R. Curwen, and A. Zisserman, "A Framework for Spatiotemporal Control in the Tracking of Visual Contours," *Int. J. Computer Vision*, vol. 11, no. 2, pp. 127-145, 1993.

[7] Andrew Blake and Michael Isard, *Active Contours,*

London : Springer-Verlag, c1998.

[8] H. A. P. Blom and Y. Bar-Shalom, "The Interacting Multiple Model Algorithm for Systems with Markovian Switching Coefficients", *IEEE Trans. Automatic Control*, vol. 33, no. 8, pp. 780-783, Aug 1988.

[9] K. J. Bradshaw, I. D. Reid, and D. W. Murray, "The Active Recovery of 3D Motion Trajectories and Their Use in Prediction," *IEEE Trans. Pattern Analysis and Machine Intelligence*, vol. 19, no. 3, pp. 219-233, Mar 1997.

[10] C. Cedra and M. Shah, "Motion-based recognition: a survey," *Image and Vision Computing*, vol. 13, no. 2, pp. 129-155, Mar 1995.

[11] L. Chen and S. Chang, "A Video tracking system with adaptive predictors," *Pattern Recognition*, vol. 25, no. 10, pp. 1171-1180, Oct 1992.

[12] B. Sabata and J. K. Aggarwal, "Estimation of Motion from a Pair of Range Images: A Review," *CVGIP: Image Understanding*, vol. 54, no. 3, pp. 309-324, Nov 1991.

[13] Z. Duric, J. A. Fayman, and E. Rivlin, "Function From Motion," *IEEE Trans. Pattern Analysis and Machine Intelligence*, vol. 18, no. 6, pp. 579-591, June 1996.

[14] J. T. Feddema, C. S. G. Lee, "Adaptive image feature prediction and control for visual tracking," *IEEE Trans. Systems, Man and Cybernetics*, vol. 20, no. 5, pp. 1172-1183, Sept-Oct 1990.

[15] J. T. Feddema, C. S. G. Lee, "Adaptive image feature prediction and control for visual tracking with a moving camera," *Proc. IEEE Int. Conf. Systems, Man and Cybernetics*, pp. 20-24, Nov 1990.

[16] D. B. Gennery, "Visual Tracking of Known Three-Dimensional Objects," *Int. J. Computer Vision*, vol. 7, no. 3, pp. 243-270, 1992.

[17] G. D. Hager and P. N. Belhumeur, "Efficient Region Tracking With Parametric Models of Geometry and Illumination," *IEEE Trans. Pattern Analysis and Machine Intelligence*, vol. 20, no. 10, pp. 1025-1039, Oct 1998.

[18] D. J. Heeger and A. D. Jepson, "Subspace Methods for Recovering Rigid Motion I: Algorithm and Implementation," *Int. J. Computer Vision*, vol. 7, no. 2, pp. 95-117, 1992.

[19] Horace H. S. Ip and D. G. Shen, "An affine-invariant active contour model (AI-snake) for model-based segmentation," *Image and Vision Computing*, vol. 16, pp. 135-146, 1998.

[20] T. S. Jebara and A. Pentland, "Parametrized Structure from Motion for 3D Adaptive Feedback Tracking of Faces," *Proc. IEEE Computer Society Conf. on Computer Vision and Pattern Recognition*, pp. 144-150, 1997.

[21] T. Joshi, "Structure and Motion Estimation from Dynamic Silhouettes under Perspective Projection," *Int. J. Computer Vision*, vol. 31, no. 1, pp. 31-50, 1999.

[22] J. S. Park and J. H. Han, "Contour matching: a curvature-based approach," *Image and Vision Computing*, vol. 16, pp. 181-189, 1998.

[23] N. P. Papanikolopoulos and P. K. Khosla, "Adaptive Robotic Visual Tracking: Theory and Experiments," *IEEE Trans. Automatic Control*, vol. 38, no. 3, pp. 429-445, Mar 1993.

[24] S. Nassif and D. Capson, "Real-Time Template Matching Using Cooperative Windows," *IEEE Canadian Conf. Electrical and Computer Engineering*, vol. 2, pp. 391-394, 1997.

[25] D. Maravall and L. Baumela, "Adaptive control of a video camera for the automatic detection and Tracking of Mobiles," *Proc. IEEE Int. Conf. Systems, Man and Cybernetics*, pp. 53-58, Oct 1993.

[26] L. Matthies, T. Kanade, and R. Szeliski, "Kalman Filter-based Algorithms for Estimating Depth from Image Sequences," *Int. J. Computer Vision*, vol. 3, pp. 209-236, 1989.

[27] E. Mazor, A. Averbuch, Y. Bar-Shalom and J. Dayan, "Interacting Multiple Models methods in Target Tracking: A Survey", *IEEE Trans. Aero. and Electro. Sys.*, vol. 34, no. 1, Jan 1998.

[28] A. Munir & D. P. Atherton, "Maneuvring target tracking using different turn rate models in the interacting multiple model algorithm," *Proc. 34th IEEE Conf. Decision & Control New Orleans*, pp. 2747-2751, Dec 1995.

[29] D. Murray and A. Basu, "Motion Tracking with an Active Camera," *IEEE Trans. Pattern Analysis and Intelligence*, vol. 16, no. 5, pp. 449-459, May 1994.

[30] D. Murray and A. Basu, "Active Tracking," *Proc. 1993 IEEE/RSJ Int. Conf. on Intelligent Robots and Systems Yokohama*, Japan, pp. 1021-1028, Jul 1993.

[31] W. G. Yau, "Design and Implementation of Visual Servoing System for Realistic Air Target Tracking," Master's thesis, Dept. Elec. Eng., National Taiwan University, 2000.

Robust Visual Servoing: Examination of Cameras Under Different Illumination Conditions

A. Bachem, T. Müller, and H.-H. Nagel

Institut für Algorithmen und Kognitive Systeme,
Universität Karlsruhe (TH)
Postfach 6980, 76128 Karlsruhe, Germany
Phone: +49 (0)721 608-4323; Fax: +49 (0)721 608-6116
E-mail: {bachem | thmu | nagel}@ira.uka.de

Abstract— Experience has shown that changes in illumination conditions may influence the robustness of robot control based on machine vision even if such illumination changes appear practically negligible to a human. Recently, CMOS cameras have become available with a larger dynamic range than conventional (high quality) CCD-cameras. The use of such higher dynamic range cameras for robot control by machine vision is compared with the use of conventional cameras under various illumination conditions. The task to be performed by the robot consists in the removal of a battery from the engine compartment of a used car. In all cases, the servo task could be accomplished successfully. The stop of CCD cameras had to be adjusted manually, however, whereas an automatic adaptation to the current lighting conditions could be implemented for the other camera type, thus allowing to deal automatically with significant changes of illumination.

Keywords— Automatic Disassembly, High Dynamic Range Cameras, Illumination Dependence.

I. Introduction

Feed-back based on machine vision can significantly extend application areas and reduce the dependency of robot manipulation on a-priori knowledge. Experience has shown, however, that machine vision must be robust against short-term changes of ambient illumination conditions due to, e. g., clouds covering the sun: a malfunction of machine vision may cause critical situations for the control of a manipulator.

In order to become less dependent on ambient illumination conditions, usually great care is taken to insure sufficient artificial lighting for machine vision in a robot work cell. Parts of the work cell may be shadowed during critical phases by a moving manipulator or workpiece. A large number of illumination sources can thus become necessary. Since light sources may malfunction occasionally, failsafe illumination conditions are by no means easy to guarantee. A different approach consists in the utilization of higher dynamic range CMOS (HDRC) cameras.

Various attempts have been reported in general in order to increase independence from illumination conditions, for example hardware solutions ([Skribanowitz & König 99], [Schneider et al. 99], [Schanz et al. 00]), modifications of image illuminance ([Sakaue et al. 95], [Alter-Gartenberg 96]) or by multiple exposures [Robertson et al. 99]. Apart from such general investigations, higher dynamic range cameras have been studied in connection with machine vision applied to lane detection, tracking, and obstacle recognition in the field of autonomous driving or driver assistance systems – see, e. g., [Stiller et al. 97], [Goldbeck & Hürtgen 99]. So far, the authors did not find any references regarding the exploitation of CMOS cameras for machine vision control of manipulators in the context of disassembly tasks.

In the remainder of this contribution, we report about a systematic comparison between CMOS and CCD cameras for a complex task. The successful removal of a car battery from the engine compartment of a used car provides a criterion by which we assess different illumination conditions and their effect on machine-vision-based robot control.

II. Experimental setup

The working cell at our laboratory comprises an 'observer robot' and a 'manipulation robot' – see Fig. 2(b). The observer robot carries a stereo camera setup with two conventional CCD cameras. A separation of observation and manipulation subtasks as well as the assignment of each subtask to a different robot allow to position the cameras such that workpieces and tools involved in a manipulation can be recorded without major difficulties. In particular, significant occlusions of workpieces by tools or vice versa can be avoided by a suitable choice of the observer robot's pose.

A. Setup with conventional CCD cameras

Images recorded by a CCD camera are grabbed by a dedicated special purpose hardware (MiniVista, see Fig. 1) which also calculates Edge Elements (EEs) for each image half-frame ('field', 584 columns × 264 rows)

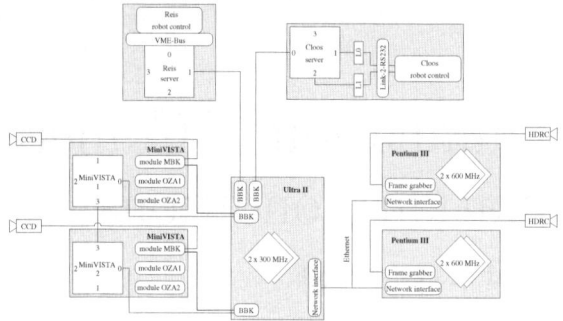

Fig. 1. Hardware configuration: Either the CCD cameras (connected via MiniVista modules) or the HDRC cameras (connected to frame grabbers in standard PCs) are used. Both robots are linked via transputer channels directly to BBK SBUS cards in the Sun Ultra-2.

on the fly. The greyvalues and EEs associated with the first half-frame of each CCD image frame are sent via a transputer link to a Sun Ultra-2 workstation for further evaluation and machine-vision-based control of the manipulation robot.

B. Setup of HDRC cameras

Due to a limitation on the admissible cable length, the stereo setup of HDRC cameras could not (yet) be mounted rigidly on the hand of the observer robot. For comparison purposes, this setup is placed on a tripod at the same position otherwise taken by the CCD camera setup on the robot's hand (see Fig. 2). In addition, the relative position of the two cameras is nearly the same for both stereo setups. This was checked by projection of object models into the image plane of each camera. Image clipping is rather similar, too, for both setups.

HDRC cameras provide a logarithmic compression of the input irradiance w. r. t. the output signal ([Seger et al. 93], [Seger et al. 99]). For dark areas, already small differences in brightness can thus be detected whereas in brighter image areas a large change of input irradiance just causes a small change of output greyvalues.

The full-frame images (resolution 640 × 480) of a HDRC camera are transmitted directly to a framegrabber on a 600 MHz Pentium-III dualprocessor system (one for each camera). The CMOS-cameras require special framegrabber cards which prevents us from connecting them to the fast Minivista EE-extraction and communication with the Sun Ultra-2 double processor.

Since we utilize HDRC *color* cameras which happened to be at our disposal, the intensity component has to be extracted, followed by an EE-computation roughly equivalent to the corresponding computations on the MiniVista. The EE-images provided by each Pentium-III double-processor are sent to the Sun

(a)

(b)

Fig. 2. Part of the robot's working space before and after a manipulation under closed-loop control by machine vision: (a) Using the HDRC cameras (on a tripod) for the dismantling operation. The gripper is shown in its initial position. (b) Looking towards the engine compartment with the CCD camera setup rigidly mounted on the robot's hand (image center), the car battery has just been placed onto the desk after removal from the engine compartment by the manipulation robot.

Ultra-2 workstation via TCP/IP network sockets (see Fig. 1).

III. COMPARISON OF CCD AND HDRC CAMERAS

The two camera setups are compared under a variety of illumination conditions. The performance criterion for this comparison consists in the successful removal of a car battery from the engine compartment of a used car by the manipulation robot under machine-vision-based control.

The *internal* camera calibration (intrinsic camera parameters) was determined off-line using the same calibration configuration for both camera setups.

A. Servoing by machine vision

The servoing task to be performed comprises the following steps:

1. automatic localization of the gripping tool and of the car battery,
2. tracking of the moving gripper,
3. closed-loop control of the robot's hand carrying the gripper to its goal pose relative to the estimated pose of the car battery,
4. closing the gripper, lifting the car battery, taking it from the engine compartment.

The final deposition of the car battery, once outside the engine compartment, onto a desk next to the manipulation robot is performed under open loop control. Two images representing the start and end of the closed-loop control removal phase as well as one image during the final open-loop deposition phase are shown in Fig. 3.

The poses of both objects are estimated initially by a model-based localization approach described in [Müller & Nagel 2000]. We thus do not need to know the exact relative pose of these objects with respect to the stereo setup, i. e. we do *not* require the determination of an extrinsic calibration of the cameras with respect to the scene.

During closed-loop control, the relative pose of *both* objects is estimated continuously with respect to the *same* camera coordinate system (see [Tonko & Nagel 2000]). We thus do *not* need an *exact* calibration of the manipulation robot with respect to the cameras on the observer robot (or on the tripod) in order to move the gripper by machine-vision-control to its relative *goal* pose.

Polyhedral CAD models have been created for the car battery and the gripper. Straight-line 'model segments' are derived by projection of such a model from the currently estimated scene position into the image plane and by subsequent hiddenline removal.

EEs ('contour points') are defined by image locations at which the gradient assumes a maximum in gradient direction, and exceeds a specific threshold. In Figures 3 (a, b), 5 and 6, EEs are plotted in black.

For object tracking, a Kalman Filter is used to estimate object poses in the camera coordinate system at each frame time (see [Tonko & Nagel 2000]). No prediction is performed inside the Kalman Filter (i. e. the 'motion model' relative to the camera coordinate system implies relative stationarity of the scene) nor will robot commands be used for predicting the relative gripper pose. Motion of the gripper is thus modelled just by system noise. Since the limited bandwidth between the Pentium-IIIs for EE-extraction and the Sun Ultra-2 forces us at the moment to move the gripper only slowly, this simple approach turned out to be sufficient.

(a)

(b)

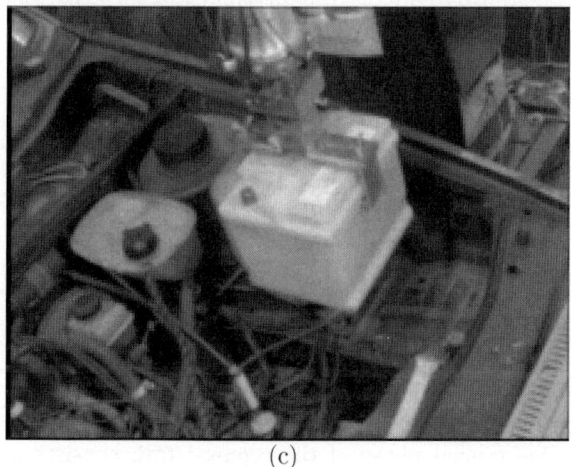

(c)

Fig. 3. HDRC images (right camera) for servoing: (a) shortly after the localization, during gripper approach for grasping; visible model-segments of the CAD model are shown in white, EEs assigned to the model segments are shown in black, (b) at the end of the machine-vision-based closed-loop control removal phase, (c) during final removal of the car battery from the engine compartment (without further closed-loop control).

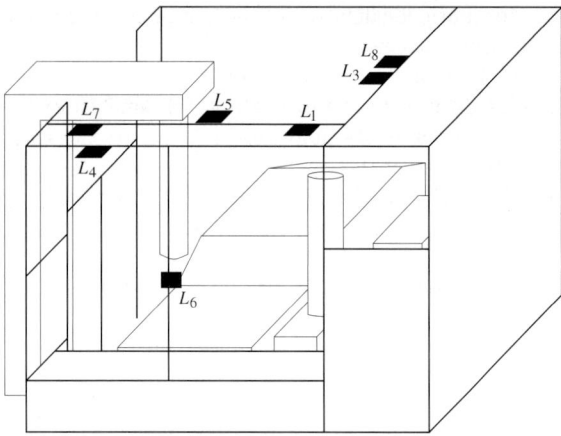

Fig. 4. Alignment of the light sources in our robot working cell. Only the spotlights L_4, L_5, L_6, and L_7 have been used for the experiments reported here.

B. Varying the illumination

Our robot working cell provides various illumination sources as sketched in Fig. 4. In addition, we could choose between the general ceiling illumination being switched on or off: since the ceiling is high ($> 10\ m$) above the ground floor, its switching state did not influence the experiments significantly. Natural daylight illumination through a large window front right next to the working cell – see Fig. 2(b) – turned out to be much more significant. In fact, the entire investigation reported here had been caused by initially unexplainable failures of visual servoing. Upon scrutiny, short-term variations of daylight intensity – due to fast-moving clouds which temporarily occluded the sun – increasingly appeared to be the root cause (unless, of course, a programming error, or an occasionally malfunctioning contact, or an inappropriately chosen parameter, or ... intervened).

During our investigations, visual servoing experiments were run using one of the following illumination conditions:
1. no additional artificial light,
2. one (L_4),
3. two (L_4 and L_7), or
4. three spotlights (L_5, L_6, L_7)

mounted at the robot working cell, see Fig. 4.

C. CMOS camera adaptation to illumination

Significant changes of illumination conditions during visual servoing can be caused, e. g., by varying daylight, by shadows of moving objects inside or outside the working cell, or by turning artificial light sources on or off.

Gain and offset of the image signal from HDRC-cameras can be set electronically and can even be changed during image grabbing. The sensor output function is given by

$$U_a = U_{offset} + u_{gain} \cdot \log(1 + \frac{E}{E_{dark}})$$

E: input irradiance [W/m^2]
E_{dark}: minimum detectable irradiance [W/m^2]
U_a: sensor output voltage [V]
U_{offset}: sensor output voltage [V] for $E = 0$
u_{gain}: slope of output signal in logarithmic range [V/decade]

So far, an adaptation algorithm is implemented which attempts to exploit the 10-bit range for image output greyvalues, but prevents that this range becomes saturated. This adaptation scheme allows to react quickly to changing illumination conditions even while closed-loop control continues.

IV. Experimental results

If there is sufficient ambient light (daylight from the window front behind the dismantling robot, see Fig. 2(b)), it is possible to find a *single* setting for the lens stop of the CCD cameras such that the visual servoing task succeeds even if two or three additional spotlights are turned on in the working cell. If, however, the illumination is varied more – no spotlights at all, only the daylight; or all spotlights switched on simultaneously – visual servoing can only be performed successfully by an *interactive change* of the lens stop of the *CCD cameras*. In addition, we lowered the minimum threshold for the gradient magnitude during EE-extraction in order to obtain a sufficient number of usable EEs. In an attempt to be fair, this threshold has been adjusted interactively in *both* cases such that between 2000 and 3500 EEs could be associated although adjustment of this EE-extraction threshold appeared necessary only for the CCD cameras. Fig. 5 illustrates representative examples recorded by the right CCD camera for different choices of additional illumination sources.

While the admissible brightness range is quite narrow for CCD cameras given a fixed stop, the HDRC cameras can cope with the full illumination range *without requiring manual adjustment of the lens stop*; near the limits of the illumination range indicated in the preceding paragraph, the automatic adaptation of gain and offset for the CMOS cameras will be activated – see Fig. 6.

One crucial phase of the removal task consists in the *initial localization*. This localization phase requires enough EEs which can be successfully associated with model-segments related to the gripper.

Another critical phase occurs when the gripper passes the upper boundary of the car battery on its way towards the goal gripping pose: during this maneuver, the total tolerance between fingers of the gripper and the upper boundary of the car battery

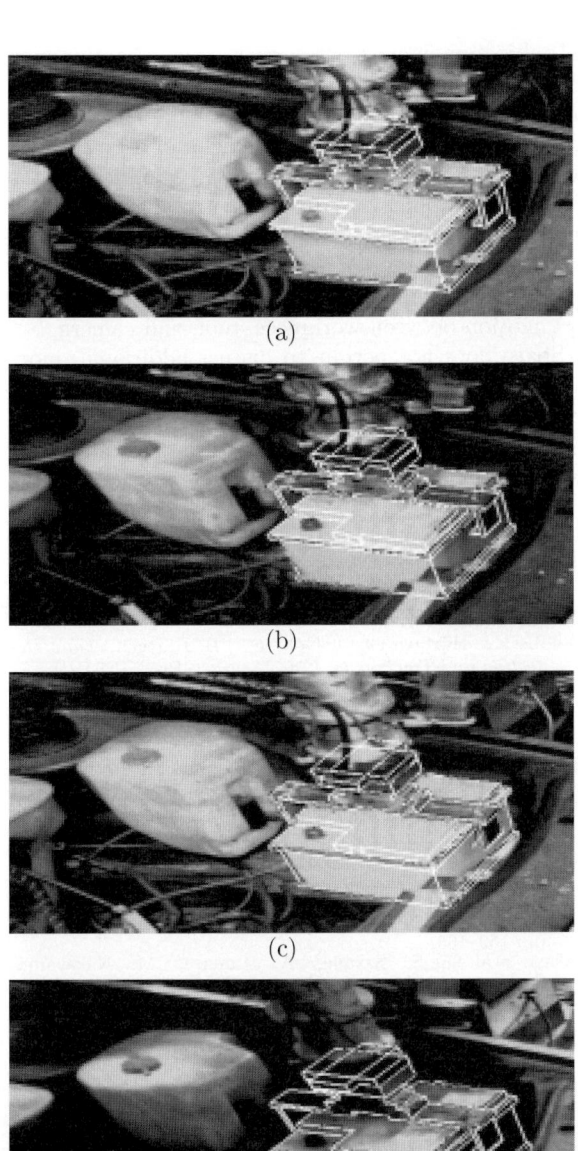

Fig. 5. Images of the right CCD camera for various illumination conditions, showing the gripper just before the fingers will be closed and the car battery will be lifted. Lighting at the working cell, setting of the camera's lens stop, and threshold parameter for EE-extraction:
(a) three halogen spotlights (L_5, L_6, L_7), two mounted at the top and one in the front of the robot's working space (lens stop: 5.6, threshold for EE extraction: 15);
(b) two halogen spotlights (L_4, L_7) mounted at the top of the robot's working space (lens stop: 4, threshold for EE extraction: 15),
(c) one halogen spotlight (L_4) mounted at the top of the robot's working space (lens stop: 2.8, threshold for EE extraction: 5),
(d) no additional illumination, just the lights at the ceiling and the late afternoon daylight (lens stop: 1.4, threshold for EE extraction: 5).

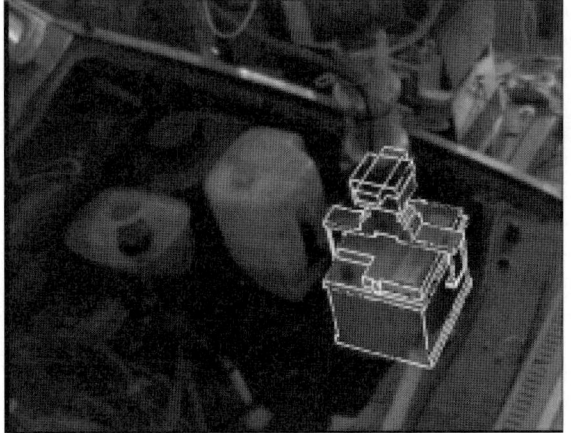

Fig. 6. Analogous to Fig. 5(a), (top panel), but for the right CMOS camera without adaptation of the lens stop. Similarly, the bottom panel corresponds approximately to Fig. 5(d), but without lights at the ceiling, just the afternoon daylight.

amounts to less than 2 cm. During this part of the approach maneuver, the fingers of the gripper move into a region shaded by the car body and the robot arm, which significantly reduces the contrast between gripper fingers and fore-/background, thereby complicating a reliable extraction of relevant EEs. In addition, one finger of the gripper becomes partially occluded by the car battery when the gripper approaches its goal position (see Fig. 3b). This is particularly critical since the finger contours have to be extracted reliably in order to determine the required translation perpendicular to the fingers and the rotation around an axis joining the center points of the two fingers.

As Fig. 7 illustrates, the aforementioned (manual or automatic) adaptations served to cover visible model-segments of the gripper by associated EEs at a rate between 60 and 90 %.

V. CONCLUDING DISCUSSIONS

Based on the experiments performed so far, we were forced to the conclusion that (very good) CCD cam-

Camera	no add. illumination	L_4	L_4, L_7	L_5, L_6, L_7
right CCD	76.3	88.9	61.7	68.9
left CCD	74.5	90.3	72.1	77.0
right CMOS	70.1	65.2	69.3	69.0
left CMOS	75.6	76.2	75.7	77.5

Fig. 7. Percentage of visible model segments covered by EEs associated during the Kalman Filter update step (and thereby contributing to the update of the pose estimate for the moving gripper). The numbers given represent an average over the first 300 images recorded during each removal experiment for the illumination conditions indicated in the last four columns of this table. Data obtained for the linear CCD cameras are directly compared with data obtained for the log-compressed CMOS camera.

eras enable a successful removal of the battery under machine vision control despite considerable variation of illumination conditions – *provided* one can assure that the system parameters (in our case, the lens stop and the threshold for EE-extraction) are adjusted appropriately. As expected, the CMOS cameras required less *interactive* adjustments. One could argue, however, that automatic control of the lens stop for CCD-cameras would allow an acceptable *system* performance over a much larger illumination range than without adjustment of the lens stop.

This may sound trivial, but nevertheless provides important stimuli for further experiments. First of all, our investigations demonstrate the importance of comparing a *system performance* and not only some isolated attributes of an experimental arrangement: an entire system may offer options which can be exploited in order to compensate for weaknesses in certain components or configurations. Second, such compensatory adjustments will show their effect predominantly if one compares two systems for a non-trivial task such as in our investigations. The successful performance of such a task depends on many factors – and thereby offers several options to compensate for local weaknesses.

Third, we do not claim that one does not need CMOS cameras (provided one can put one's hand on them at all). Clearly, CMOS cameras offer a wider potential. The surprise to us came in the extent by which we could compensate the *expected* weakness of CCD cameras – during the experiments performed *so far*. Based on the experience reported here, clearly additional experiments recommend themselves in order to explore the potential of CMOS cameras more fully. We expect to overcome some of the deficiencies in our experimental arrangement, for example the limited admissible cable length for the connection of CMOS cameras to the required framegrabber cards. This would allow us to place the CMOS cameras directly onto the observer robot and thus remove one (possibly, but – due to our efforts – not very power-ful) objection against our current setup. Another step clearly requires to increase the transmission bandwidth between the Pentium double processors and the Sun Ultra-2 which executes the tracking and robot control programs. Such a step should allow us to increase the temporal sampling rate during control and thereby to increase the speed with which the manipulation robot can move. It may be that the advantage of CMOS cameras will become more pronounced for faster relative motion between workpiece, tool, and camera.

Space does not permit to discuss additional, more complicated, aspects. After all, technology has become available to *systematically* explore the performance of an entire system capable of non-trivial task under real-world boundary conditions.

References

[Alter-Gartenberg 96] R. Alter-Gartenberg: *Nonlinear Dynamic Range Transformation in Visual Communication Channels*. IEEE Transactions on Image Processing 5:3 (1996) 538–546.

[Goldbeck & Hürtgen 99] J. Goldbeck, B. Hürtgen: *Lane Detection and Tracking by Video Sensors*. Proc. IEEE/IEEJ/JSAI International Conference on Intelligent Transportation Systems ITSC'99, 5 – 8 October 1999, Tokyo, Japan, pp. 74–79.

[Müller & Nagel 2000] Th. Müller, H.-H. Nagel: *Parallelizing a CAD-Model-Based Object Localization for Initialization of Tracking in Visual Servoing*. Proc. IEEE 1st International Conference on Cluster Computing CLUSTER 2000, 28 November – 1 December 2000, Chemnitz, Germany, IEEE Computer Society, Los Alamitos, CA, pp. 399-400.

[Robertson et al. 99] M.A. Robertson, S. Borman, R.L. Stevenson: *Dynamic Range Improvement Through Multiple Exposures*. Proc. International Conference on Image Processing ICIP'99, 24 – 28 October 1999, Kobe, Japan, Vol. 3, pp. 159–163.

[Sakaue et al. 95] S. Sakaue, A. Tamura, M. Nakayama, S. Maruno: *Adaptive Gamma Processing of the Video Cameras for the Expansion of the Dynamic Range*. IEEE Transactions on Consumer Electronics 41:3 (1995) 555-562.

[Schanz et al. 00] M. Schanz, C. Nitta, A. Bußmann, B.J. Hosticka, R.K. Wertheimer: *A High-Dynamic-Range CMOS Image Sensor for Automotive Applications*. IEEE Journal of Solid-State Circuits 35:7 (July 2000) 932–938.

[Schneider et al. 99] B. Schneider, P. Rieve, M. Böhm: *Image Sensors in TFA (Thin Film on ASIC) Technology*. In B. Jähne, H. Haußecker, P. Geißler (Eds.): 'Handbook of Computer Vision and Applications', Volume 1 – Sensors and Imaging, Academic Press, San Diego, CA, 1999, pp. 237–270.

[Seger et al. 93] U. Seger, H.-G. Graf, M.E. Landgraf: *Vision Assistance in Scenes wit Extreme Contrast*. IEEE Micro 13:1 (February 1993) 50–56.

[Seger et al. 99] U. Seger, W. Apel, B. Höfflinger: *HDRC-Imagers for Natural Visual Perception*. In B. Jähne, H. Haußecker, P. Geißler (Eds.): 'Handbook of Computer Vision and Applications', Volume 1 – Sensors and Imaging, Academic Press, San Diego, CA, 1999, pp. 223–235.

[Skribanowitz & König 99] J. Skribanowitz, A. König: *First Application Results of a Dedicated Vision Chip for Image Sequence Processing and Visual Inspection*. Third International Conference on Knowledge-Based Intelligent Information Engineering Systems, 31 August – 1 September 1999, Adelaide, Australia, pp. 385–388.

[Stiller et al. 97] C. Stiller, W. Pöchmüller, B. Hürtgen: *Stereo Vision in Driver Assistance Systems*. IEEE Conference on Intelligent Transportation Systems ITSC'97, 9–12 November 1997, Boston, Mass., pp. 888–896.

[Tonko & Nagel 2000] M. Tonko, H.-H. Nagel: *Model-Based Stereo-Tracking of Non-Polyhedral Objects for Automatic Disassembly Experiments*. International Journal of Computer Vision 37:1 (2000) 99-118.

Asymptotic Motion Control of Robot Manipulators Using Uncalibrated Visual Feedback*

Yantao Shen and Yun-Hui Liu
Dept. of Automation and Computer-Aided Engr.
The Chinese University of Hong Kong, Hong Kong

Kejie Li
Dept. of Mechatronic Engr.
Beijing Institute of Technology, China

Jianwei Zhang and Alois Knoll
Faculty of Technology
University of Bielefeld, Germany

Abstract

To implement a visual feedback controller, it is necessary to calibrate the homogeneous transformation matrix between the robot base frame and the vision frame besides the intrinsic parameters of the vision system. The calibration accuracy greatly affects the control performance. In this paper, we address the problem of controlling a robot manipulator using visual feedback without calibrating the transformation matrix. We propose an adaptive algorithm to estimate the unknown matrix on-line. It is proved by Lyapunov approach that the robot motion approaches asymptotically to the desired one and the estimated matrix is bounded under the control of the proposed visual feedback controller. The performance has been confirmed by simulations and experiments.

1 Introduction

Visual feedback is an important approach to improve the control performance of robot manipulators [1]. In order to conduct high precision visual feedback manipulation, some key parameters such as the homogeneous transformation matrix between the robot base frame and the vision frame in position-based methods [2]-[5], or those concerning image Jacobian in image-based methods [6]-[11] should be calibrated accurately besides the intrinsic parameters of the vision system. However, an accurate calibration requires substantial efforts and time, and is also impossible in some cases such as when the vision system is mounted on a mobile platform (robot). For this reason, tremendous efforts have been recently made to visual feedback control with uncalibrated vision system. Papanikolopoulos et

---

*This work is supported in part by Hong Kong Research Grant Council under the grants CUHK4173/00E and the Germany/HongKong joint research scheme. Email: yhliu @acae.cuhk.edu.hk

al. [7] proposed an algorithm based on on-line estimation for the relative distance of the target with respect to the camera. This algorithm obviates the need for off-line calibration of the eye-in-hand robotic system. Yoshimi et al. [10] utilized a simple geometric property, that is rotational invariance under a special setup of system for a peg-in-hole alignment task, to estimate image Jacobian. Hosada et al. [8] and Jägersand et al. [9] employed the Broyden updating formula to estimate the image Jacobian. The methods in [11] by Kim et al. do not use depth in the feedback formulation. However, the methods above considered kinematics only and neglected dynamic effect of the robot manipulator. To achieve high-lever performance for a manipulator-vision system, the controller must incorporate the dynamics of the manipulator.

In this paper, we address the design of a position-based visual feedback controller for motion control of a robot manipulator when the homogeneous transformation matrix between the robot base frame and the vision frame is not calibrated. It is assumed that the intrinsic parameters of the vision system have been calibrated accurately and the vision system can measure the 3D position and orientation of the end-effector of manipulator. Based on an important observation that the visual Jacobian matrix can be represented as a product of a known matrix, which depends on the kinematics of the manipulator, and the unknown rotation matrix R between the robot base frame and the vision frame, we propose a simple adaptive algorithm to estimate the unknown matrix on-line. The controller can be considered as a combination of an on-line calibration and the real-time control. It is proved with a full consideration of dynamics of the system by Lyapunov approach that this controller yields asymptotic convergence of the motion error to zero and the estimated matrix is bounded. The performance of controller has been verified by simulations and experiments.

2 Kinematics and Dynamics

2.1 The Coordinate Frames

Fig. 1 shows a typical set-up of a robot workcell using a visual feedback. Three coordinate frames, namely the robot base frame $\sum_B$, the end-effector coordinate frame $\sum_E$, and the vision frame $\sum_V$ are defined, respectively. Here, the ^VT_B is generally represented as: $^VT_B = \begin{bmatrix} R & p \\ 0 & 1 \end{bmatrix}$, where $R \in \Re^{3\times 3}$ denotes the rotation matrix. $p \in \Re^3$ is position of the origin of $\sum_B$ with respect to $\sum_V$. Assume that the intrinsic parameters of the vision system have been calibrated, and that the vision system can measure the 3D position and orientation of robot in real-time. However the homogeneous transformation matrix ^VT_B of the robot base frame with respect to the vision frame is unknown.

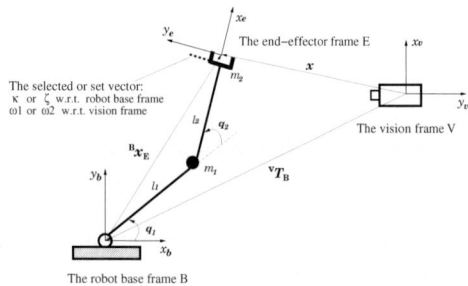

Figure 1: The coordinate frames.

2.2 Kinematics of System

Denote by $^Bx_E = [^bx_{e1}, ^bx_{e2}, ..., ^bx_{e6}]^T$ the position and orientation of the end-effector with respect to the robot base frame. The first three components of ^Bx_E denote the position, and the last three are the roll, pitch and yaw angles representing the orientation. Let $x \in \Re^6$ denotes the position and orientation of the end-effector with respect to the vision frame. Denote by q the joint angles of the robot. From the forward kinematics, we have

$$^B\dot{x}_E = J(q)\dot{q} \qquad (1)$$

where $^B\dot{x}_E \in \Re^6$ is the velocity of the end-effector. $J(q)$ is the Jacobian matrix of the robot. $\dot{q}$ denotes the joint velocity. According to the relation of kinematics, we also have

$$\dot{x} = \underbrace{\begin{bmatrix} R & 0 \\ 0 & R \end{bmatrix}}_{A} J(q)\dot{q} \qquad (2)$$

where $\dot{x} \in \Re^6$ denotes the velocity of the end-effector with respect to the vision frame. The matrix $AJ(q)$ is called visual Jacobian matrix. Assuming that $J(q)$ is square and nonsingular, we then have

$$\dot{q} = J^{-1}(q) \underbrace{\begin{bmatrix} R^T & 0 \\ 0 & R^T \end{bmatrix}}_{A^T} \dot{x} \qquad (3)$$

Differentiating the equation (3) results in

$$\ddot{q} = J^{-1}(q)A^T\ddot{x} + \frac{d}{dt}(J^{-1}(q)A^T)\dot{x} \qquad (4)$$

Note that R in equations (2)$\sim$(4) is unknown if no calibration is performed.

2.3 The System Dynamics

The dynamics of robot in the joint space can be represented as

$$H(q)\ddot{q} + \underbrace{\left(\frac{1}{2}\dot{H}(q) + S(q,\dot{q})\right)}_{C(q,\dot{q})}\dot{q} + G(q) = \tau \qquad (5)$$

where $H(q)$ is the symmetric and positive definite inertia matrix. $S(q,\dot{q})$ denotes a skew symmetric matrix. $G(q)$ is the gravity force. The τ represents the joint input of the manipulator.

3 Position Control

In this section, we consider the problem of moving the end-effector of robot from a position x to a desired one x_d. Firstly, we adopt the popular PD plus gravity compensation scheme for position control:

$$\tau = G(q) - K_v\dot{q} - K_pJ^T(q)\underbrace{\begin{bmatrix} \hat{R}^T & 0 \\ 0 & \hat{R}^T \end{bmatrix}}_{\hat{A}^T}\Delta x \qquad (6)$$

where $\hat{R}^T$ denotes an estimated value of transposed matrix of R. $\hat{A}^T$ represents the estimated value of transposed matrix of A. $\Delta x = x - x_d$ denotes the position error with respect to the vision frame. K_p and K_v are the positive scalar constant gains. Substituting this control law into equation (5) results in the following closed-loop dynamics equation:

$$H(q)\ddot{q} + C(q,\dot{q})\dot{q} = -K_v\dot{q} - K_pJ^T(q)\hat{A}^T\Delta x \qquad (7)$$

Note that the right side of equation (7) can be re-written as follows:

$$H(q)\ddot{q} + C(q,\dot{q})\dot{q}$$
$$= -K_v\dot{q} - K_pJ^T(q)\hat{A}^T\Delta x - K_pJ^T(q)(\hat{A}^T - A^T)\Delta x$$
$$= -K_v\dot{q} - K_pJ^T(q)\hat{A}^T\Delta x - K_pJ^T(q)Y(\Delta x)\Delta\Theta \quad (8)$$

where we arrange the elements of R into a 9×1 vector Θ. $\hat{\Theta}$ is an estimated value of Θ and $\Delta\Theta = \hat{\Theta} - \Theta$. The $Y(\Delta x)$ is a regressor matrix without depending on any element of $\hat{R}$ and R. Obviously, based on the fact that R is an orthonormal matrix, we draw into the following two approaches for leading the estimated rotation matrix to tend to the actual rotation matrix. In the first approach, we select a vector κ, and then calculate

$$(\hat{R}^T\hat{R} - I)\kappa = (\hat{R}^T\hat{R} - R^TR)\kappa$$
$$= \hat{R}^T(\hat{R} - R)\kappa + (\hat{R}^T - R^T)\underbrace{R\kappa}_{w_1}$$
$$= Y_2(\hat{\Theta}, \kappa, w_1)\Delta\Theta \quad (9)$$

where I denotes a 3×3 identity matrix. $Y_2(\hat{\Theta}, \kappa, w_1)$ is a 3×9 regressor matrix which does not depend on the elements of $\Delta\Theta$. If the vector κ is defined with respect to the robot base frame, w_1 is a vector with respect to the vision frame. When the vector κ is so properly selected that its coordinates with respect to the robot base frame and the vector w_1 with respect to the vision frame can be measured by the encoders and the vision system respectively, the matrix $Y_2(\hat{\Theta}, \kappa, w_1)$ can be calculated without using unknown R.

In the second approach, we define

$$\hat{R}\zeta - \underbrace{R\zeta}_{w_2} = Y_2(\zeta)\Delta\Theta \quad (10)$$

where ζ is a vector on the robot. w_2 is a vector with respect to the vision frame, so w_2 can be measured by the vision system.

Based on the two approaches in eqs (9) or (10), we propose the following updated laws to calculate the estimated $\hat{\Theta}$, respectively:

$$\Delta\dot{\hat{\Theta}}^T = -\underbrace{[(\hat{R}^T\hat{R} - I)\kappa]^T}_{\Delta\Theta^T Y_2^T(\hat{\Theta},\kappa,w_1)} B_2 Y_2(\hat{\Theta},\kappa,w_1)$$
$$+ \frac{1}{B_1}\dot{q}^T K_p J^T(q) Y(\Delta x)$$

or

$$\Delta\dot{\hat{\Theta}}^T = -\underbrace{[\hat{R}\zeta - w_2]^T}_{\Delta\Theta^T Y_2^T(\zeta)} B_2 Y_2(\zeta)$$
$$+ \frac{1}{B_1}\dot{q}^T K_p J^T(q) Y(\Delta x) \quad (11)$$

where B_1 and B_2 are the positive constant gains. From the adaptive laws, we have

$$\dot{q}^T K_p J^T(q) Y(\Delta x)\Delta\Theta =$$
$$\Delta\dot{\hat{\Theta}}^T B_1 \Delta\Theta + \Delta\Theta^T Y_2(\bullet)^T B_1 B_2 Y_2(\bullet)\Delta\Theta \quad (12)$$

where $Y_2(\bullet)$ represents either $Y_2(\hat{\Theta},\kappa,w_1)$ or $Y_2(\zeta)$ respectively in two approaches.

Theorem 1: *Under the control of the proposed controller in equation (6), the robot manipulator system yields*
- *the estimated matrix $\hat{R}$ is bounded, and*
- *asymptotic convergence of position error Δx with respect to the vision frame to zero as the time approaches to the infinity.*

A Proof can be referred to [15].

4 Trajectory Tracking Control

In this section, we consider the problem of controlling the end-effector of the robot to trace a given time-varying trajectory $(x_d(t), \dot{x}_d(t), \ddot{x}_d(t))$ with respect to the vision frame. Firstly we define the following nominal reference with respect to the vision frame

$$\dot{x}_r = \dot{x}_d - \lambda\Delta x \quad (13)$$

$$\ddot{x}_r = \ddot{x}_d - \lambda\Delta\dot{x} \quad (14)$$

where λ is a postive constant. $\Delta x = x - x_d$ and $\Delta\dot{x} = \dot{x} - \dot{x}_d$ denote the position and velocity errors with respect to the vision frame, respectively. The error vector is given by

$$s = \dot{x} - \dot{x}_r = \Delta\dot{x} + \lambda\Delta x \quad (15)$$

Referring to the nominal reference and the error vector, the joint space nominal reference and the error vector are given as follows:

$$\dot{q}_r = J^{-1}(q)\hat{A}^T \dot{x}_r \quad (16)$$

$$s_q = \dot{q} - \dot{q}_r = \dot{q} - J^{-1}(q)\hat{A}^T \dot{x}_r \quad (17)$$

From equation (3), we can also re-write s_q as:

$$s_q = J^{-1}(q)(A^T - \hat{A}^T)\dot{x} + J^{-1}(q)\hat{A}^T s \quad (18)$$

Using the computed-torque method, we propose the following control law for trajectory tracking:

$$\tau = -K_p J^T(q)\hat{A}^T s - K_v s_q + G(q)$$
$$+ H(q)\underbrace{(J^{-1}(q)\hat{A}^T \ddot{x}_r + (J^{-1}(q)\hat{A}^T)\dot{\dot{x}}_r)}_{\ddot{q}_r} + C(q,\dot{q})\dot{q}_r \quad (19)$$

where K_p and K_v are the positive scalar constant gains. Substituting the control law into equation (5) results in the following closed-loop dynamics equation:

$$H(q)\dot{s}_q + C(q,\dot{q})s_q = -K_p J^T(q)\hat{A}^T s - K_v s_q \quad (20)$$

An adaptive law is necessary to update the estimated matrix $\hat{A}$. Consider the following equation

$$\hat{A}^T \dot{x} - \underbrace{A^T \dot{x}}_{v_f} = Y_3(\dot{x})\Delta\Theta \quad (21)$$

where $Y_3(\dot{x})$ is a regressor matrix without depending on elements of R. The vector v_f is the velocity of the end-effector with respect to the robot base frame and can be measured by the encoders.
Furthermore,

$$\dot{x}^T(\hat{A}-A)K_p\hat{A}^T s = Y_4(\dot{x}, K_p, \hat{A}, s)\Delta\Theta \quad (22)$$

Note that the regressor matrix $Y_4(\dot{x}, K_p, \hat{A}, s)$ does not depend on the elements of $\Delta\Theta$.
Then, the following adaptive law is proposed:

$$\Delta\dot{\Theta}^T = -\underbrace{(\hat{A}^T\dot{x} - v_f)^T}_{\Delta\Theta^T Y_3^T(\dot{x})} B_3 Y_3(\dot{x}) - \frac{1}{B_4} Y_4(\dot{x}, K_p, \hat{A}, s) \quad (23)$$

where B_3 and B_4 are the positive constant gains. From this equation, we have

$$\Delta\dot{\Theta}^T B_4 \Delta\Theta = -\Delta\Theta^T Y_3^T B_3 B_4 Y_3 \Delta\Theta - Y_4 \Delta\Theta \quad (24)$$

Theorem 2: *Under the control of the proposed controller in equation (19), the robot manipulator system yields*
- *the error $(\hat{R}-R)$ is bounded, and*
- *asymptotic convergence of trajectory tracking errors $\Delta x(t)$ and $\Delta \dot{x}(t)$ with respect to the vision frame to zero as the time approaches to the infinity.*

Proof: Define the following nonnegative scalar function

$$V = \frac{1}{2}\{s_q^T H(q) s_q + \Delta\Theta^T B_4 \Delta\Theta\} \quad (25)$$

By multiplying the closed-loop dynamics equation (20) from the left-hand side by s_q^T derives

$$s_q^T H(q)\dot{s}_q + s_q^T \frac{1}{2}\dot{H}(q)s_q = -s_q^T K_p J^T \hat{A}^T s - s_q^T K_v s_q \quad (26)$$

From equation (18), note that the term in the equation (26)

$$-s_q^T K_p J^T \hat{A}^T s = \dot{x}^T(\hat{A}-A)K_p\hat{A}^T s - (\hat{A}^T s)^T K_p(\hat{A}^T s) \quad (27)$$

Differentiating the nonnegative scalar function in equation (25) and then substituting eqs (24), (26) and (27) into it, we obtain

$$\dot{V} = -\Delta\Theta^T Y_3^T B_3 B_4 Y_3 \Delta\Theta \\ -s_q^T K_v s_q - (\hat{A}^T s)^T K_p(\hat{A}^T s) \quad (28)$$

For the positive constants K_v, K_p, B_3 and B_4, $\dot{V}$ is nonpositive and hence V never increases. It states that V is a Lyapunov function. From Barbalat's lemma, we can say the error $\hat{R}-R$ is bounded. Also, we have asymptotic convergence of the trajectory tracking errors Δx and its successive derivative $\Delta \dot{x}$ with respect to the vision frame to zero. □

5 Simulations

In this section, we show the performance of the proposed trajectory tracking controller by simulations. We conducted the simulations on a two-link planar arm with the physical parameters $m1 = m2 = 1, l1 = l2 = 2$, as shown Fig. 1. The arm base frame is located at $(-10, -10)$ with respect to the vision frame. In the simulations, the rotation matrix $R(\theta) = \begin{bmatrix} cos\theta & -sin\theta \\ sin\theta & cos\theta \end{bmatrix}$ is a function of θ. The end-effector of the arm is required to follow the following desired trajectory

$$x_d(t) = \begin{bmatrix} -0.8sin(\omega t) - 9.0 \\ 0.8cos(\omega t) - 8.5 \end{bmatrix}$$

with respect to the vision frame. All units in the simulations used are in the SI system.

The simulation results are plotted in Fig.2. In the simulations, the real rotation matrix $R(\theta = \pi/2)$, the initial estimation of the rotation matrix $\hat{R}(\theta = \pi/5)$; the positive gains are $K_p = 40, K_v = 30, B_3 = 50, B_4 = 200$; $\lambda = 3, \omega = 1$; the initial position of the end-effector is $x_0 = (-8.5, -9.0)$. As shown in Fig.2, the results confirmed asymptotic convergence of the tracking errors and the bounded $\hat{R}$.

6 Experiments

We have implemented the controller in the five-fingered robot hand system developed at the Chinese University of Hong Kong using DSP's and workstations. Each finger of the robot hand has three revolute joints driven by AC motors through a harmonic drive of 80:1 reduction ratio. The joint angles of the finger are measured by high-precision encoders with

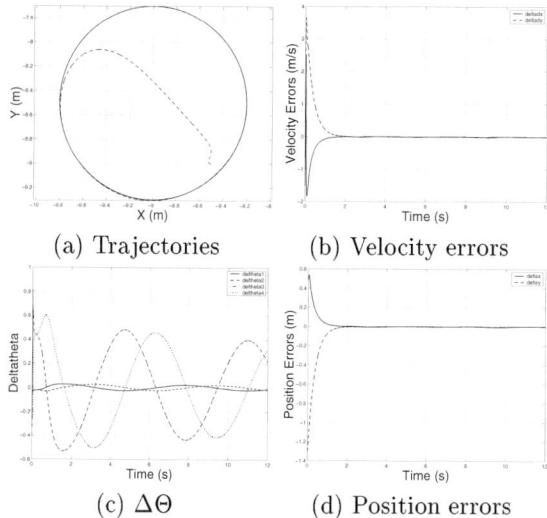

(a) Trajectories (b) Velocity errors

(c) $\Delta\Theta$ (d) Position errors

Figure 2: The example of simulation.

resolutions of 30720 pulse/turn. The joint velocities are obtained by differentiating the joint angles. In the experiments, one finger of the robot was employed as a 3DOF arm. About 2m away from this 3 DOF finger, we set an OPTOTRAK/3020 position sensor system to measure in real-time the 3D positions of markers mounted at the fingertip with resolution of 1:200000 (power axis). Like the joint velocities, we can also obtain the velocities of the fingertip by differentiating the positions detected by OPTOTRAK sensors. In order to communicate the data, an interface board is installed in a PC with Intel80486 CPU between the robot hand and OPTOTRAK system. To compensate for the frictions at the joints, we adopt the following friction model:

$$F = K_{vf}\dot{q} + [K_{df} + (K_{sf} - K_{df})dexp]sgn(\dot{q}) \quad (29)$$

where $dexp = diag\{exp(-|\dot{q}_i|/\alpha)\}$. $K_{vf} = diag\{0.1, 0.1, 0.1\}$ is the coefficient matrix of viscous friction, $K_{sf} = diag\{0.6, 0.4, 0.5\}$ is the coefficient matrix of static friction and $K_{df} = diag\{0.4, 0.2, 0.35\}$ is the coefficient matrix of dynamic friction. $\alpha = 0.001$ is a small positive parameter. The sampling time of the experimental system is 2.36 ms. Note that all units in the experiments are in the SI system. Fig. 3 shows the robot manipulator and the vision system.

6.1 Position Control

Two experiments have been conducted to validate the proposed position control scheme using the different approaches in the equation (9) and the equation (10),

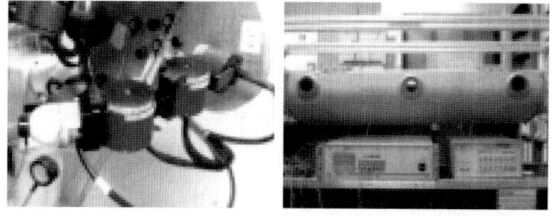

(a) Robot manipulator (b) OPTOTRAK/3020

Figure 3: The robot manipulator and the vision system.

respectively. We set two sensor markers of OPTOTRAK as a vector whose distance is 0.039m at the robot fingertip. The initial estimation of the rotation matrix $\hat{R}$ between the finger base frame and the vision frame is given as a 3×3 identity matrix. The gains are chosen as: $K_p = 250, K_v = 10, B_1 = 600, B_2 = 100$. Fig. 4(a) and (b) show the experimental results of two approaches, respectively. As shown in Fig. 4, the results ascertain the effectiveness of the two proposed control algorithms.

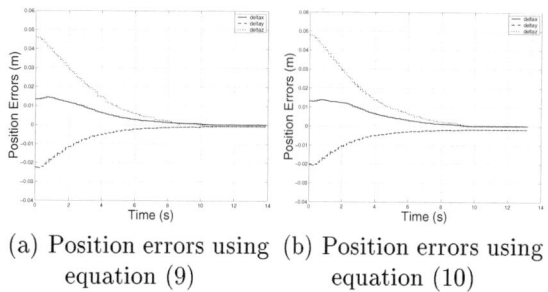

(a) Position errors using equation (9) (b) Position errors using equation (10)

Figure 4: The experimental results of position control.

6.2 Trajectory tracking Control

In this experiment, the end-effector of the robot is required to trace a given trajectory

$$x_d(t) = \begin{bmatrix} 0.006cos(\omega t) + 0.028sin(\omega t) - 0.163 \\ -0.027cos(\omega t) + 0.002sin(\omega t) - 0.257 \\ 0.012cos(\omega t) - 0.01sin(\omega t) - 1.802 \end{bmatrix}$$

with respect to the vision frame. We set a marker to detect the change of position of the end-effector. The initial estimation of the rotation matrix is also set as the identity matrix. The initial position is $x_0 = (-0.140071, -0.237624, -1.792197)$; the parameters are set to: $K_p = 50, K_v = 30, B_3 = 100, B_4 = 500, \lambda = 3, \omega = 1$. The results in Fig. 5 confirmed good convergence of the trajectory tracking errors.

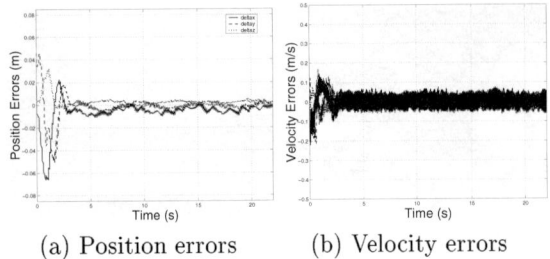

(a) Position errors (b) Velocity errors

Figure 5: The experimental results of the tracking controller.

7 Conclusions

In this paper, we proposed a motion controller using visual feedback without calibrating the homogeneous transformation matrix between the robot base frame and the vision frame before executing task. Differing from other approaches, the controller is developed by considering a full dynamics of the system and using an adaptive algorithm to estimate the unknown matrix on-line. The proposed adaptive algorithm is based on an important observation, that is the visual Jacobian matrix can be represented as a product of a known matrix, which depends on the kinematics of the manipulator, and the unknown rotation matrix R between the robot base frame and the vision frame. This controller greatly simplifies the implementation process of a robot-vision workcell and is especially useful when a pre-calibration is impossible. Simulation and experimental results verified the performance of asymptotic convergence of the new controller. The future work is to extend this method to the image-based visual feedback control.

References

[1] S. Hutchinson, G. D. Hager, and P. I. Corke, "A Tutorial on Visual Servo Control," *IEEE Transaction on Robotics and Automation*, Vol. 12, No. 5, pp. 651-670, 1996.

[2] P. K. Allen, A. Timcenko, B. Yoshimi, and P. Michelman, "Automated Tracking and Grasping of a Moving Object with a Robotic Hand-Eye System," *IEEE Transaction on Robotics and Automation*, Vol. 9, No. 2, pp. 152-165, 1993.

[3] Y. Yokokohj, M. Sakamoto, and T. Yoshikawa, "Vision-Aided Object Manipulation by a Multifingered Hand with Soft Fingertips," *Proceedings of IEEE International Conference on Robotics and Automation*, pp. 3201-3208, 1999.

[4] H. H. Fakhry and W. J. Wilson, "A Modified Reslovedd Acceleration Controller for Position-Based Visual Servoing," *Mathematical and Computer Modeling*, Vol. 24, No. 5/6, pp. 1-9, 1996.

[5] W. J. Wilson, C. C. Williams Hulls, and G. S. Bell, "Relative End-Effector Control Using Cartesian Position Based Visual Servoing," *IEEE Transaction on Robotics and Automation*, Vol. 12, No. 5, pp. 684-696, 1996.

[6] K. Hashimoto, T. Kimoto, T. Ebine, and H. Kimura, "Manipulator Control with Image-Based Visual Servo," *Proceedings of IEEE International Conference on Robotic and Automation*, pp. 2267-2272, 1991.

[7] N. P. Papanikolopoulos and P. K. Khosla, "Adaptive Robotic Visual Tracking: Theory and Experiments," *IEEE Transaction on Automatic Control*, Vol. 38, No. 3, pp. 429-445, 1993.

[8] K. Hosada and M. Asada, "Versatile Visual Servoing without Knowledge of True Jacobain," *Proceedings of IEEE/RSJ International Conference on Intelligent Robots and Systems*, pp. 186-191, 1994.

[9] M. Jägersand, O. Fuentes, and R. Nelson, "Experimental Evalution of Uncalibrated Visual Servoing for Precision Manipulation," *Proceedings of International Conference on Robotics and Automation*, pp. 2874-2880, 1997.

[10] B. H. Yoshimi and P. K. Allen, "Active, Uncalibrated Visual Servoing," *Proceedings of IEEE International Conference on Robotics and Automation*, pp. 156-161, 1994.

[11] D. Kim, A. A. Rizzi, G. D. Hager, and D. E. Koditschek, "A "Robust" Convergent Visual Servoing System," *Proceedings of IEEE/RSJ International Conference on Intelligent Robots and Systems*, Vol. 1, pp. 348-353, 1995.

[12] E. Malis, F. Chaumette, and S. Boudet, "Positioning a Coarse-Calibrated Camera with respect to an Unknown Object by 2D 1/2 Visual Servoing," *Proceedings of IEEE International Conference on Robotics and Automation*, pp. 1352-1359, 1998.

[13] T. J. Tarn, A. K. Bejczy, A. Isidori, and Y. Chen, "Nonlinear Feedback in Robot Arm Control," *Proceedings of 23rd IEEE Conference on Decision and Control*, pp. 736-751, 1984.

[14] Y. H. Liu, K. Kitagaki, T. Ogasawara, and S. Arimoto, "Model-Based Adaptive Hybrid Control for Manipulators Under Multiple Geometric Constraints," *IEEE Transaction on Control System Technology*, Vol. 7, No. 1, pp. 97-109, 1999.

[15] Y. T. Shen, Y. H. Liu, and K. J. Li, "Asymptotic Position Control of Robot Manipulators Using Uncalibrated Visual Feedback," *Proceedings of IEEE/RSJ International Conference on Intelligent Robots and Systems*, pp. 435-440, 2000.

Image-based visual servoing on planar objects of unknown shape

Christophe Collewet

Cemagref
17 Avenue de Cucillé
35044 Rennes Cedex, France
christophe.collewet@cemagref.fr

François Chaumette

IRISA / INRIA Rennes
Campus Universitaire de Beaulieu
35042 Rennes Cedex, France
francois.chaumette@irisa.fr

Philippe Loisel

Cemagref
17 Avenue de Cucillé
35044 Rennes Cedex, France
philippe.loisel@cemagref.fr

Abstract

This paper proposes a way to achieve positioning tasks by visual servoing, for any orientation of the camera, when the desired image of the observed object cannot be precisely described. The object is assumed to be planar and motionless but no knowledge about its shape or pose is required. To simplify the problem, first, we treat the case of a threadlike object and then we show how our approach can be generalized to an object with three particular points. The control law is based on the use of 2d visual servoing and on an estimation of a 3d parameter. Experimental results relative to objects of unknown shape are given to validate the approach. In addition, an algorithm to estimate the depth between the object and the camera is provided which leads to the dimensions of the object.

1 Introduction

Visual servoing is now a classical technique in robot control (see [1] for a description of the different approaches). Nevertheless, in the most often case of an "eye-in-hand" system [2, 3, 4], we still cannot achieve positioning tasks with regard to deformable or not well known objects. Such a case appears when we have to treat applications for example in surgical domain, agri-food industry, agriculture or in unknown environments (underwater, space). Indeed, except manufactured goods for which a model often exists, we rarely have a precise description of the object or of the desired visual features.

In the case of a 3d visual servoing two approaches exist. The first one and the most often used, is based on the computation of the pose object/camera and then requires a model of the object. Therefore, this approach cannot be used in our case. The second one is based on 3d reconstruction by dynamic vision [5, 6] but these techniques are currently not accurate enough with regard to the errors of reconstruction. On the other hand, active vision [7, 8, 9, 10, 11] can limit such errors. However, this approach has only been used on very simple objects. Moreover, let us point out that those 3d reconstruction techniques are sensitive to the calibration of the system.

On the other hand, lots of work in 2d visual servoing have shown that the closed-loop system is little sensitive to calibration problems [4, 12, 13]. Similar results have been obtained concerning the recent 2 1/2d approach [14]. However, the 2d approach as well as the 2 1/2d cannot cope with the objects being studied. Indeed, let us consider an accurate positioning task related to such objects, a raw ham for example. Even though these objects are part of a same class, they are different enough not to use unique desired visual features for all the objects in the class. In such applications the desired visual features have to be considered as unknown. Few authors relate such cases. In [15], thanks to dynamic visual features a positioning task consisting in moving the camera in front of a planar object of unknown shape can be achieved. However, such an approach needs particular motion parameters estimation leading currently to high computation duration and, consequently, to a low control scheme rate. Moreover, this approach does not well suit for positioning tasks since a motion is necessary.

The method described in our paper is based on the use of points as visual features, it can achieve positioning tasks when any orientation is required and for the same objects as those used in [15]: planar objects of unknown shape. Our approach needs no assumption about the shape of the observed object but we have to assume the object motionless. The approach is based on the use of 2d visual servoing and on an estimation of a 3d parameter that will be precised below. We will see that this structure maintains the object of interest in the field of view of the camera, does not need a calibrated camera although it provides dimensions of the object and an estimation of the depth. Unfortunately, it is sensitive to the calibration of the robot. Finally, our method combines in part the advantages of both 2d visual servoing and 3d reconstruction.

This paper is organized as follows: in Section 2, we deal with a threadlike object to achieve the particular positioning task consisting in moving the robot in front of this object. Next, we show in Section 3 how this approach can be readily generalized to an object with three particular points and for any orientation of the camera. Finally, experimental results are given to validate the algorithm.

2 Case of a simple object: the segment

2.1 Task specification

In this section, we propose to move a robot in front of a threadlike object by visual servoing. In order to achieve applications as those described in the previous section, we suppose the length of the object unknown as well as its pose. The object is described by a segment $[m_1, m_2]$ with $\underline{Om_1} = (x_1, y_1, z_1)^T$ and $\underline{Om_2} = (x_2, y_2, z_2)^T$, expressed in the camera frame which is centered in the optical center O. The optical axis of the camera is the z axis of the camera frame (see Figure 1). This object projects on the image plane by a perspective projection as a segment $[M_1, M_2]$ with $\underline{OM_1} = (X_1, Y_1, f)^T$ and $\underline{OM_2} = (X_2, Y_2, f)^T$ according to:

$$\underline{OM} = \frac{f}{z} \underline{Om} \quad (1)$$

where f is the focal length assumed to be equal to 1.

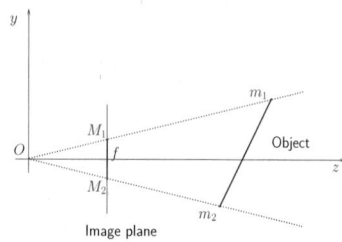

Figure 1: Projection of the object on the image plane

The task is achieved, *i.e.* the camera is in front of the object, when the optical axis is orthogonal to $[m_1, m_2]$ and the image centered with respect to the y axis. In this case, we have $z_1 = z_2 = z^*$ and $Y_1 = -Y_2$.

Even though the object is very simple, the task is not obvious to achieve since an infinity of object poses leads to a centered image without achieving the task. Nevertheless, a way to achieve it is to center the image and then, to scan other positions. This is a typical case where the *task function approach* described in [2] has to be used. So, we will consider as the *main task* the one which maintains centered the image, and the *secondary task* the one which scans other positions while ensuring the image centered.

2.2 Control law

2.2.1 Main task

A way to center the image is to choose as visual features the following vector:

$$\underline{s} = (X_1, X_2, Y_1 + Y_2)^T \quad (2)$$

and as the desired visual features the vector:

$$\underline{s}^* = (0, 0, 0)^T \quad (3)$$

We immediately obtain the interaction matrix at the desired position [1, 2]:

$$L_{\underline{s}^*}^T = \begin{pmatrix} -\frac{1}{\widehat{z^*}} & 0 & 0 & 0 & -1 & Y_1 \\ -\frac{1}{\widehat{z^*}} & 0 & 0 & 0 & -1 & -Y_1 \\ 0 & -\frac{2}{\widehat{z^*}} & 0 & 2(1+Y_1^2) & 0 & 0 \end{pmatrix} \quad (4)$$

where $\widehat{z^*}$ is an approximation of z^*.

2.2.2 Secondary task

At this step, we want to equate z_1 to z_2 while ensuring the image centered. Therefore, such a motion has to belong to Ker $L_{\underline{s}^*}^T$ to ensure $\underline{s} = \underline{s}^*$. Thereafter, we can apply a control law of the following form to achieve the task:

$$T_c = -\lambda L_{\underline{s}^*}^{T+}(\underline{s} - \underline{s}^*) + T_s \quad (5)$$

with:

$$T_s = \begin{pmatrix} 0 & \widehat{z^*}\omega_x(1+Y_1^2) & 0 & \omega_x & 0 & 0 \end{pmatrix}^T \quad (6)$$

which belongs to Ker $L_{\underline{s}^*}^T$ and in which ω_x has to be determined.

If $\underline{s}$ is in the neighborhood of $\underline{s}^*$, only T_s generates a motion. Intuitively, it is easy to see that the length of the segment $l = Y_1 - Y_2$ may supply useful information. In particular, we think it takes a maximum when the camera is in front of the object. In order to verify this assumption and to obtain all the parameters needed for (5), we want to know how l changes during the motion of the camera. Besides, we think that the trajectory of the camera can supply interesting information.

2.3 Modeling

2.3.1 Modeling of the length of the segment

Let us consider the points $m_1 = (y^*, z^*)^T$, $m_2 = (-y^*, z^*)^T$ and $m_0 = (0, z_0)^T$ expressed in the camera frame when the camera is at the desired position. We consider that the motion during the maximization of l results from a x axis rotation of angle θ_x, centered in m_0 (see Figure 2). Therefore, m_1 and m_2 can be expressed as: $m_1{}' = \mathcal{R}(m_1 - m_0) + \mathcal{T}$ and $m_2{}' = \mathcal{R}(m_2 - m_0) + \mathcal{T}$ which yields the projection $M_1 = (X_1, Y_1)^T$ and $M_2 = (X_2, Y_2)^T$ of $m_1{}'$ and $m_2{}'$ respectively. Then, we search the solution of $Y_1 = -Y_2$ with respect to z_0 to perform a motion which belongs to $\text{Ker } L^T\big|_{\underline{s}=\underline{s}^*}$, as seen in 2.2.2. This expression is complicated. Nevertheless, when $v = y^*/z^*$ is low, it can be expressed at 2nd order by:

$$z_0 = z^* \qquad (7)$$

while l becomes simply at 3rd order:

$$l_m = 2v \cos \theta_x \qquad (8)$$

2.3.2 Modeling of the trajectory of the camera

Thanks to (7), we can derive the trajectory of the camera with respect to the desired camera position (see figure 2):

$$\begin{cases} z &= z^*(1 - \cos \theta_x) \\ y &= z^* \sin \theta_x \end{cases} \qquad (9)$$

Thus the trajectory is simply a circle centered in $(0, z^*)$ with a radius z^*:

$$y^2 - (z - z^*)^2 = z^{*2} \qquad (10)$$

2.3.3 Application to control

During a motion, the form of which is given by the control law (5), we proceed to an on-line learning of l, supposed to be modelized by (8). In practice, the function we need is $l_m(\Theta_x - \Theta_{xc})$ with $\Theta_x - \Theta_{xc} = \theta_x$ (remember that (8) is obtained with respect to the desired position) and thereafter, Θ_{xc} is the unknown value where l_m takes its maximum. $l_m(\Theta_x - \Theta_{xc})$ yields a linear expression in $\cos \Theta_x$ and $\sin \Theta_x$ that provides Θ_{xc} thanks to a least squares algorithm. Moreover, to ensure an exponential decay of the angular error $\Theta_x - \Theta_{xc}$, we impose:

$$\omega_x = K_x(\Theta_x - \Theta_{xc}) \qquad (11)$$

with K_x a positive value.

At the begining of the motion, when the estimation supplies a not accurate enough value of Θ_{xc}, we fix ω_x to a constant value ω_0.

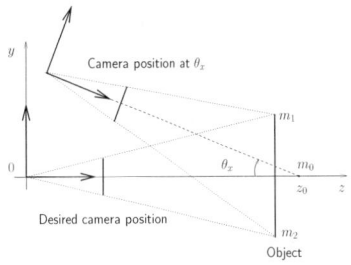

Figure 2: Modeling of the length of the segment with respect to θ_x

remark: Note that Θ_x is obtained thanks to the odometry of the robot and therefore does not depend of the calibration of the camera. Of course, it depends on the calibration of the robot.

On the other hand, it is thus possible to extract $\widehat{z^*}$ from the trajectory of the robot thanks again to an on-line nonlinear algorithm. It must lead to low errors $\|\underline{s} - \underline{s}^*\|$ during the motion defined by (5). Moreover, thanks to the perspective transformation, 3d features can be obtained as will be shown in Section 3.2.

After having treated the case of segments, we will show, in the next section, how our method can be generalized to a planar object of unknown and complex shape when at least three feature points can be extracted from the image.

3 Case of a planar object with at least three feature points

3.1 Specifying the main task

In section 2, the task has been achieved by maximization of the length of the object image. In the case of a 2d object, a natural transposition consists in maximizing in the image its area S. However, obtaining its analytical expression in the general case seems to be out of reach. Thus, we simplify the approach by assuming that three feature points exist and then the image becomes simply a triangle with vertices $M_1(X_1, Y_1)$, $M_2(X_2, Y_2)$ and $M_3(X_3, Y_3)$. In the same way we have modelized l in 2.3.1, we have tried to express a modeling of S. Unfortunately, this study failed. Nevertheless, thanks to the results of section 2, we can maximize sequentially l then $h = X_3 - X_1$. Thus, we impose $Y_1 = -Y_2$ with regard to the maximization of l and $X_1 = -X_3$ with regard to the maximization of h. If we want to ensure that vectors of the kernel of the interaction matrix related to $\underline{s} = (X_1 + X_3, Y_1 + Y_2)^T$ are of the form of (6), we have to impose $Y_1 = Y_3$ and $X_1 = X_2$. Unfortunately these two constraints cannot be satisfied simultaneously in the case of any triangle. Therefore, we have arbitrarily chosen to impose $X_1 = X_2$ yielding:

$$\underline{s} = (X_1 - X_2, X_1 + X_3, Y_1 + Y_2)^T \quad (12)$$

$$\underline{s}^* = (0, 0, 0)^T \quad (13)$$

$$L_{\underline{s}^*}^T = \begin{pmatrix} 0 & 0 & 0 & 2X_1Y_1 & 0 & 2Y_1 \\ -2/\widehat{z^*} & 0 & 0 & X_1(Y_1 - Y_3) & -2(1 + X_1^2) & Y_1 + Y_3 \\ 0 & -2/\widehat{z^*} & 0 & 2(1 + Y_1^2) & 0 & -2X_1 \end{pmatrix} \quad (14)$$

and the two vectors which belong to Ker $L_{\underline{s}^*}^T$:

$$\underline{T}_{sl} = (-\widehat{z^*}\omega_x X_1 Y_3, \widehat{z^*}\omega_x(1 + X_1^2 + Y_1^2), 0,$$
$$\omega_x, 0, -\omega_x X_1)^T \quad (15)$$

$$\underline{T}_{sh} = \left(-\widehat{z^*}\omega_y \left(1 + X_1^2\right) \ 0 \ 0 \ 0 \ \omega_y \ 0 \right)^T \quad (16)$$

We will see that even if (15) is different from (6), satisfactory results are obtained. The control law is thus:

$$T_c = -\lambda L_{\underline{s}^*}^{T^+} (\underline{s} - \underline{s}^*) + T_{sl} + T_{sh} \quad (17)$$

In the next section we present the behavior of this control law first on a simple object and next on a more complicated object.

3.2 Experimental results

3.2.1 Experiments on a simple object

The first experiment consists in moving the camera in front of an object with 3 feature points. To simplify the image processing, a binary object has been used. Figures 3a (4a), 3b (4b), 3c (4c) and 3d (4d) depict respectively the behavior of the components of T_c, the error in the sensor $\|\underline{s} - \underline{s}^*\|$, the desired and current angular values and the measured and modelized image length during the maximization of l (h). Figure 5 summarizes the different phases to achieve the task. Figure 3d (4d) confirms the theorical results about the modeling of $l = f(\theta_x)$ ($h = f(\theta_y)$). Moreover, on Figure 3c (4c) we remark that we quickly obtain the desired angular value which refines during the estimation and finally stabilizes. In addition, thanks to the least squares algorithm, lets us point out that the noise has not a lot of effect on this value. In other respects, we have applied DeMenthon's method [16] to obtain the poses of the camera before and after servoing with respect to the object (4 points are then required). The initial pose was around $(25^\circ, 19^\circ, 11^\circ)$ according to the x, y and z axes of the camera. The positioning error was around 0.5°. Therefore, these results are satisfactory.

Besides, as seen in 2.3.2 we have access to $\widehat{z^*}$. This algorithm yielded a value of 706.1 mm while DeMenthon's method gave in desired position $z_1 = 707.9$ mm, $z_2 = 708.6$ mm and $z_3 = 709.9$ mm. The consequence of a good value for $\widehat{z^*}$ can be seen on Figure 4b since the error decreases suddenly. Thereafter, once the task is achieved, the camera is in front of the object and $\widehat{z^*}$ is known. Then, according to (1) 3d features can be reconstructed. For example for l and h the following values have been obtained: $\widehat{l} = 13.06$ cm and $\widehat{h} = 13.05$ cm instead of $l = h = 13$ cm.

To validate those first results, 20 other experiments has been led for different initial positions (into [-4°, 22°] x [8°, 33°] x [-25°, 36°] with respect to the rotations of axes x, y and z). Very good results have been obtained concerning as well the positioning task as the reconstruction of 3d features. They are summarized in the table 1 in which the function $Q(x)$ means the absolute value of the relative error on the measure of x in percent.

Table 1: Results on 20 experiments.

x	$\overline{m} = 1/N \sum_i x_i$	$\max |x_i|$	σ_x	$\min |x_i|$
γ	-0.009	0.439	0.007	0.017
β	-0.489	0.751	0.156	0.133
$Q(l)$	0.576	1.103	0.136	0.246
$Q(h)$	0.562	1.104	0.140	0.242
$Q(z)$	0.562	1.109	0.145	0.240

In addition, errors have been introduced in the intrinsic parameters of the camera: we have added an error of 10 % on these parameters and we have neglected the radial distorsion of the lens. Experimental results, consisting in achieving positioning tasks, showed that non significant difference exists concerning the positioning error between a calibrated camera and a coarse calibrated one. This result agrees with the remark made in section 2.3.3.

The second experiment consists in positioning the camera not in front of the object but for a particular desired pose. Since we know the angular value Θ_{xc} (Θ_{yc}) (see 2.3.3) which leads to move the robot in front of the object, by adding specified constant values any orientations can be reached (see Figures 6 and 7). In this case we obtained (-15.2°, -19.0°) instead of (-15°, -20°). Again, these results are satisfactory.

3.2.2 Experiment on a complicated object

In this third experiment, we consider the case of a raw ham. This object is moving along a conveyor and the goal is to place the camera in front of it to perform an inspection of its cut by a machine vision. To do this, we used the method described in [17] to select and track the 3 feature points we need. It is based on SSD matching and assumes translational frame-to-frame displacements. The experimental results are depicted on Figures 8 and 9. They lead to similar conclusions as the previous experimental results. However, the algorithm of extraction of the visual features is more noisy and a higher positioning error is obtained but, this time, difficult to measure. Again, 3d features can be obtained to characterize the cut of the ham.

4 Conclusion

We have presented in this paper a way to achieve any 2d visual servoing positioning task in the case of a planar and motionless object of complex shape thanks to simple geometric visual features. Experimental results have shown that accurate positionings can be obtained ($\approx 0.5^o$) even if the camera were not calibrated. In addition, thanks to the trajectory of the camera, the depth between the camera and the object can be obtained yielding precise 3d reconstruction (≈ 0.6 %). Thus, our method combines in part the advantages of both 2d and 3d visual servoing in the sense that 3d information can be obtained with a coarse calibrated camera.

Finally, we think that this work contributes to an expansion of the application area of visual servoings in the sense that now, complex objects or objects of unknown shape can be treated even if the desired image is not precisely known.

References

[1] S. Hutchinson, G. D. Hager, and P. I. Corke, "A tutorial on visual servo control," *IEEE Trans. on Robotics and Automation*, vol. 12, no. 5, pp. 651–670, October 1996.

[2] B. Espiau, F. Chaumette, and P. Rives, "A new approach to visual servoing in robotics," *IEEE Trans. on Robotics and Automation*, vol. 8, no. 3, pp. 313–326, June 1992.

[3] N. Papanikolopoulos, P. K. Khosla, and T. Kanade, "Vision and control techniques for robotic visual tracking," in *IEEE Int. Conf. on Robotics and Automation, ICRA'91*, Sacramento, USA, April 1991, pp. 857–864.

[4] H. Hashimoto, T. T. Kimoto, T. Ebine, and H. Kimura, "Manipulator control with image-based visual servo," in *IEEE Int. Conf. on Robotics and Automation, ICRA'91*, Sacramento, California, USA, April 1991, pp. 2267–2272.

[5] J. K. Aggarwal and N. Nandhakumar, "On the computation of motion from sequences of images, a review," *Proc. of IEEE*, vol. 76, pp. 917–935, August 1988.

[6] G. Adiv, "Inherent ambiguities in recovering 3d motion and structure from a noisy flow field," *IEEE Trans. on Pattern Analysis and Machine Intelligence*, vol. 11, no. 5, pp. 477–489, May 1989.

[7] Y. Aloimonos, I. Weiss, and A. Bandopadhay, "Active vision," *Int. Journal of Computer Vision*, vol. 1, no. 4, pp. 333–356, January 1987.

[8] R. Bajcsy, "Active perception," *Proc. of the IEEE*, vol. 76, no. 8, pp. 996–1005, August 1988.

[9] G. Sandini and M. Tistarelli, "Active tracking strategy for monocular depth inference over multiple frames," *IEEE Trans. on Pattern Analysis and Machine Intelligence*, vol. 12, no. 1, pp. 13–27, January 1990.

[10] K. Kutulakos and C. Dyer, "Recovering shape by purposive viewpoint adjustment," *Int. Journal of Computer Vision*, vol. 12, no. 2, pp. 113–136, February 1994.

[11] F. Chaumette, S. Boukir, P. Bouthemy, and D. Juvin, "Structure from controlled motion," *IEEE Trans. on Pattern Analysis and Machine Intelligence*, vol. 18, no. 5, pp. 492–504, May 1996.

[12] B. Espiau, "Effect of camera calibration errors on visual servoing in robotics," in *Third Int. Symp. on Experimental Robotics, ISER'93*, Kyoto, Japan, October 1993.

[13] G. D. Hager, W. C. Chang, and A. S. Morse, "Robot hand-eye coordination based on stereo vision," *IEEE Control Systems Magazine*, vol. 15, no. 1, pp. 30–39, February 1995.

[14] E. Malis, F. Chaumette, and S. Boudet, "2 1/2d visual servoing," *IEEE Trans. on Robotics and Automation*, vol. 15, no. 2, pp. 238–250, April 1999.

[15] A. Crétual and F. Chaumette, "Positioning a camera parallel to a plane using dynamic visual servoing," in *IEEE/RSJ Int. Conf. on Intelligent Robots and Systems, IROS'97*, Grenoble, Septembre 1997, pp. 43–48.

[16] D.F. DeMenthon and L.S. Davis, "Model-based object pose in 25 lines of code," *International Journal of Computer Vision*, vol. 15, pp. 123–141, 1995.

[17] J. Shi and C. Tomasi, "Good features to track," in *IEEE Int. Conf. on Computer Vision and Pattern Recognition, CVPR'94*, Seattle, USA, June 1994, pp. 593–600.

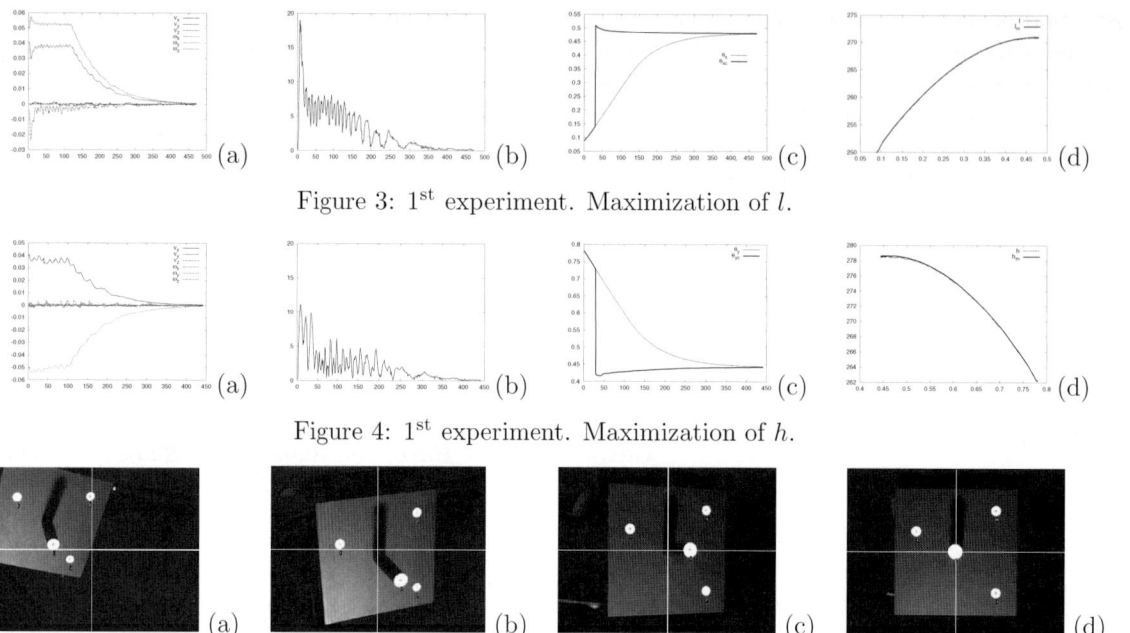

Figure 3: 1st experiment. Maximization of l.

Figure 4: 1st experiment. Maximization of h.

Figure 5: 1st experiment. (a) Initial Position. (b) Position after "centering". (c) Position after maximization of l. (d) Final position.

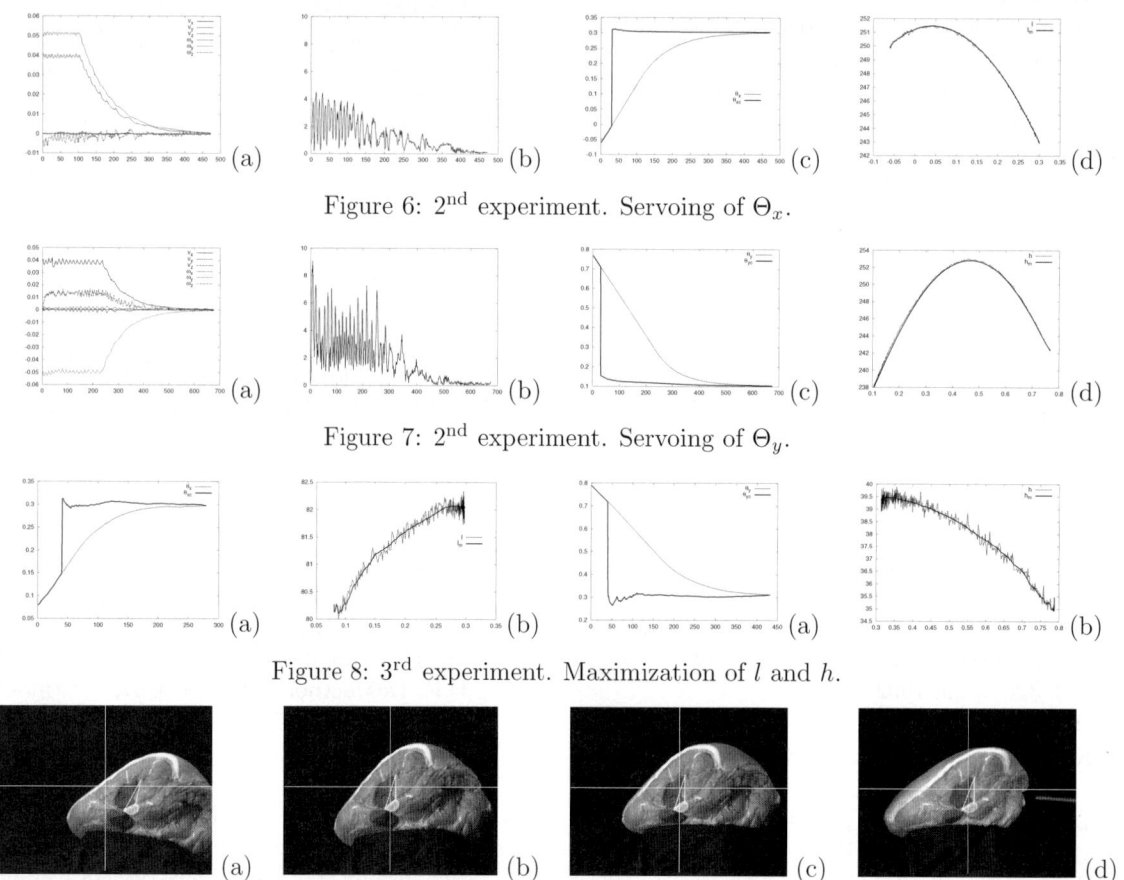

Figure 6: 2nd experiment. Servoing of Θ_x.

Figure 7: 2nd experiment. Servoing of Θ_y.

Figure 8: 3rd experiment. Maximization of l and h.

Figure 9: 3rd experiment. (a) Initial Position. (b) Position after "centering". (c) Position after maximization of l. (d) Final position.

252

Localization of a Mobile Robot using Images of a Moving Target

B. H. Kim[1], D. K. Roh[1], Jang M. Lee[1], M. H. Lee[2], K. Son[2], M. C. Lee[2], J. W. Choi[2], and S. H. Han[3]

1. Dept. of Electronics Engineering
Pusan National University
Pusan, 609-735, Korea
Tel. : 82-51-510-2378
E-mail : jmlee@hyowon.cc.pusan.ac.kr

2. School of Mechanical Engineering
Net Shape & Die Manufacturing ERC
Pusan National University, Pusan, Korea

3. Dept. of Mechanical Engineering
Kyungnam University, Masan, Korea

Abstract

In this paper, the localization of a mobile robot using images of a moving target is introduced. An example of this kind of task can be given from the military purpose: a tank is targeting the enemy tank while it is moving. Or we may need a mobile robot to catch a moving cockroach. For both of these cases, we have neither the position nor velocity of the target. If there is a stereo camera system on the mobile robot, the localization of the mobile robot is relatively easy. However, for the sake of convenience, most of cases, a single camera will be utilized for the purpose and we are going to show that the goal can be achieved. Typical objects are stored in the database for the localization of the mobile robot. With a fixed camera, a perspective camera model, and given object database, an image frame can provide the pose (distance and orientation) of the object with respect to the camera. Utilizing the consecutive image frames and motion estimation technology, the relative pose of the object with respect to the camera can be obtained accurately and during the process, calibration of camera with respect to the world frame, i.e., localization of a mobile robot, is gradually performed. This localization scheme is demonstrated by the experiments.

I. Introduction

A mobile robot is used in various field such as the autonomous gantry vehicle(AGV) for conveying materials in production line, a document-delivering robot in a office room, a nursing robot in a hospital, a human-assistant robot for helping a physically handicapped person, a service robot in a waiting room of airport or station, and a fixing robot in a nuclear reactor or spacecraft of severe environments.

Most mobile robot in production line moves along the guided line of the ground in the working area by the guiding system.

If we develop and use the autonomous mobile robot without guided line in a random path, we can obtain the effectiveness of the increase in production because we take appropriate measure regardless of the change in equipment setting or variation in production plan. It is necessary for the mobile robot to move autonomously recognizing the current position of the robot itself in the case of moving a random path for the given work[1].

In this paper, the localization of a mobile robot is proposed utilizing position and velocity information of a moving target. Due to the uneven ground, it is not accurate enough to provide sufficient information of a mobile robot from only position sensor data attached at the driving wheel. In this study, we attached a CCD camera at the top of a mobile robot and we developed a algorithm of a pose (position/orientation) recognition using position data of moving target saved in a database[3,4].

For this, in section II we present an algorithm of estimating position and orientation to the object which lies in the image plane of a mobile robot using camera perspective model and object's database[7]. In section III we suggests a method of localization of a camera with respect to the world frame using the pose(position/orientation) between camera and object from current position, the pose between camera and object from previous position, and estimated motion vector minimizing errors among these data[5,6].

In existing study of others, the initial position of the object is known in a position estimation of moving object and motion vector estimation, and the position of a mobile robot is commonly recognized matching camera's image and various environment data saved in map in the study of the algorithm of position recognition of a mobile robot. But, in this paper we don't know the initial position of target position, and we estimate the position utilizing a database and we focus on the work of object capturing or tracking independent of a surrounding environment. In such a course, localization of a mobile robot is gradually performed.

II. Pose Estimation of an Object

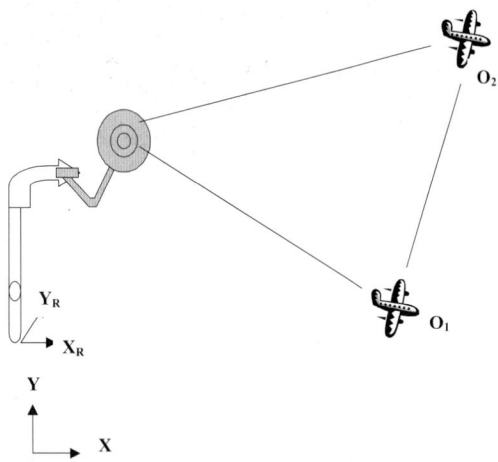

Fig. 1. Localization of a mobile robot.

From Fig. 1, the position of the camera, X, is initially given by the dead-reckoning sensors of mobile robot. By the captured image and database for the airplane, the position/orientation of the airplane, O_1, will be estimated.

We derive some equations on the image processing for the mobile robot. We begin with considering kinematics for actuator of active camera.

A. Kinematics for actuator of active camera system

In our active camera system, it has the ability of panning and tilting as shown in Fig. 2. We define the position and posture of camera about the base frame. According to the Denabit-Hatenberg convention, we can obtain the homogeneous matrix after establishing the following coordinate system and representing the parameter as shown in Eq. (1).

$$^0H_4 = {^0H_1} \cdot {^1H_2} \cdot {^2H_3} \cdot {^3H_4} = \qquad (1)$$

$$\begin{bmatrix} \cos(\alpha)\cos(\beta) & -\cos(\alpha)\sin(\beta) & \sin(\alpha) & l_2\sin(\alpha) + l_3\cos(\alpha)\cos(\beta) \\ \sin(\beta) & \cos(\beta) & 0 & l_s\sin(\beta) \\ -\sin(\alpha)\cos(\beta) & \sin(\alpha)\cos(\beta) & \cos(\alpha) & l_1 + l_2\cos(\alpha) - l_3\sin(\alpha)\cos(\beta) \\ 0 & 0 & 0 & 1 \end{bmatrix}$$

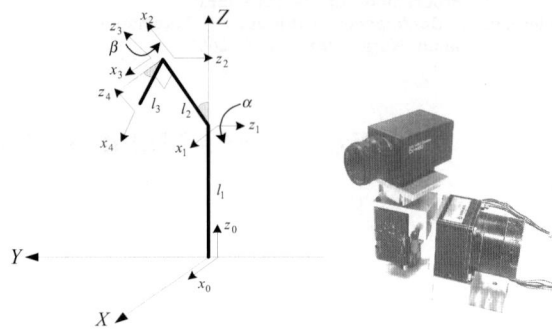

Fig. 2. The two degrees of freedom of the camera platform (Left) and what it looks like (Right)

The position of camera by homogeneous matrix is presented as follows

$$\begin{aligned} x_{ccd} &= l_2\sin(\alpha) + l_3\cos(\alpha)\cos(\beta) \\ y_{ccd} &= l_3\sin(\beta) \\ z_{ccd} &= l_1 + l_2\cos(\alpha) - l_3\sin(\alpha)\cos(\beta) \end{aligned} \qquad (2)$$

It can be obtained a tilted angle and rotated angle by panning and tilting of camera as shown in Eq. (3). θ_T is angle between vector x_4 transformed to value of absolute coordinate system and base frame X-Y coordinate after. θ_R is angle between vector y_4 transformed to value of absolute coordinate system and base frame X-Y coordinate.

$$\theta_T = \tan^{-1}\left(\frac{\sin(\alpha)\cos(\beta)}{\sqrt{\cos^2(\alpha)\cos^2(\beta) + \sin^2(\beta)}}\right)$$

$$\theta_R = \tan^{-1}\left(\frac{\sin(\alpha)\sin(\beta)}{\sqrt{\cos^2(\alpha)\sin^2(\beta) + \cos^2(\beta)}}\right) \qquad (3)$$

B. Relation between camera and real coordinate

Now that the position where the obstacle meets the bottom can be detected, we can estimate the obstacle position kinematically from the tilt angle and rotation angle of the camera mounted on the mobile robot. Those results are in Eq. (4)-(6).

$$\begin{bmatrix} j' \\ k' \end{bmatrix} = \begin{bmatrix} \cos(\theta_R) & -\sin(\theta_R) \\ \sin(\theta_R) & \cos(\theta_R) \end{bmatrix} \begin{bmatrix} j - \frac{P_x}{2} \\ k - \frac{P_y}{2} \end{bmatrix} \qquad (4)$$

$$\hat{r}_0 = z_{ccd} \cdot \cot(\theta_T - \frac{k'}{p_y}\theta_{ry}) \quad \hat{\theta}_0 = \frac{j'}{p_x}\theta_{rx} \qquad (5)$$

$$\hat{x}_0 = r_{ccd} \cdot \cos(\beta) + \hat{r}_0 \cdot \cos(\beta + \hat{\theta}_0)$$
$$\hat{y}_0 = r_{ccd} \cdot \sin(\beta) + \hat{r}_0 \cdot \sin(\beta + \hat{\theta}_0) \quad (6)$$
$$(r_{ccd} = \sqrt{x_{ccd}^2 + y_{ccd}^2})$$

where j', k' are the rotated image by moving of camera, $\hat{r}_0$, $\hat{\theta}_0$ is the estimated values of distance, direction respectively and $\hat{x}_0, \hat{y}_0$ is x, y position of obstacle based on the coordinate of a mobile robot.

C. Inverse kinematics for placing the center of image to the desired position

In case of using active camera, we can obtain visual information of area that we want to know but it is required inverse kinematics to obtain information for desired area. Eq. (7) is the inverse kinematics equations that describe the relation to place the center of image to the desired position.

$$\beta_d = \tan^{-1}\left(\frac{y_d}{x_d}\right)$$

$$\alpha_d = \cos^{-1}\left(\frac{-l_1 l_2 + \sqrt{l_1^2 l_2^2 - (l_1^2 + r_d^2)(l_1^2 - r_d^2)}}{(l_1^2 + r_d^2)}\right)\left(\frac{1}{\sin(\beta_d)}\right) \quad (7)$$

where α_d, and β_d are the attitude of actuator respectively, x_d and y_d are the desired position that we want to place. r_d is $\sqrt{x_d^2 + y_d^2}$ and link parameters l_1, l_2 of camera actuator respectively.

III. Localization of the Mobile Robot

In Fig. 1, now we have two values O_2: One is obtained directly from the image at O_2 and the other is obtained by the vector summation of $O_1 + (O_2 - O_1)$ which is the summation of the initial position vector and the motion vector. Here let us hypothesize that if there is any difference in between these two O_2's, it came from the mal-calculation of position, X from the dead-reckoning sensors. So in this section, we searched a true value of X in the neighborhood of the initial X and found a right X that minimizes the error between the two O_2's.

In this section, we introduce the Extended Kalman Filter(EKF) formalism which is applied to localization of the mobile robot[9].

A. Unifying the Problems

In all of these previously listed problems, we are confronted with the estimation of an unknown parameter $a \in R^n$ given a set of k possibly nonlinear equations of the form

$$f_i(x_i, a) = 0 \quad (8)$$

where $x_i \in R^m$ and f_i is a function from $R^m \times R^n$ into R^p. The vector x_i represents some random parameters of the function f_i in the sense that we only measure an estimate $\hat{x}_i$ of them, such that

$$\hat{x}_i = x_i + v_i \quad (9)$$

where v_i is a random error. The only assumption we make on v_i is that its mean is zero, its covariance is known, and that it is a white noise

$$E[v_i] = 0$$
$$E[v_i v_i^t] = \Lambda_i \geq 0$$
$$E[v_i v_j^t] = 0 \quad \forall i \neq j.$$

These assumptions are reasonable. If the estimator is biased, it is often possible to subtract its mean to get an unbiased one. If we do not know the covariance of the error (or at least an upper bound of it), the estimator is meaningless. If two measurements $\hat{x}_i$ and $\hat{x}_j$ are correlated, we take the concatenation of them $\hat{x}_k = (\hat{x}_i, \hat{x}_j)$ and the concatenated vector function $f_k = [f_i^t, f_j^t]^t$. The problem is to find the optimal estimate $\hat{a}$ of a given the function f_i and the measurements $\hat{x}_i$.

B. Linearizing the Equations

The most powerful tools developed in parameter estimation are for linear systems. We decided to apply these tools to a linearized version of our equations. This is the EKF approach that we now develop.

For each nonlinear equation $f_i(x_i, a) = 0$ we need to know an estimate $\hat{a}_{i-1}$ of the sough parameter a, and again a measure S_i of the confidence we have in this estimate.

Actually, we model probabilistically the current estimate $\hat{a}_{i-1}$ of a by assuming that

$$\hat{a}_{i-1} = a + w_i \quad (10)$$

where w_i is a random error. The only assumptions we make on w_i are the same as for v_i, i.e.,

$$E[w_i] = 0$$
$$E[w_i w_i^t] = S_i \geq 0$$

where S_i is a given non-negative matrix. Here again, no assumption of gaussianness is required.

Having an estimate $\hat{a}_{i-1}$ of the solution, the equations are linearized by a first-order Taylor expansion around $(\hat{x}_i, \hat{a}_{i-1})$

$$f_i(x_i, a) = 0 \approx f_i(\hat{x}_i, \hat{a}_{i-1}) + \frac{\partial \hat{f}_i}{\partial x}(x_i - \hat{x}_i) + \frac{\partial \hat{f}_i}{\partial a}(a - \hat{a}_{i-1}) \quad (11)$$

where the derivatives $\partial \hat{f}_i / \partial x$ and $\partial \hat{f}_i / \partial a$ are estimated at $(\hat{x}_i, \hat{a}_{i-1})$.

Equation (11) can be rewritten as

$$y_i = M_i a + u_i \quad (12)$$

where

$$y_i = -f_i(\hat{x}_i, \hat{a}_{i-1}) + \frac{\partial \hat{f}_i}{\partial a} \hat{a}_{i-1}$$

$$M_i = \frac{\partial \hat{f}_i}{\partial a}$$

$$u_i = \frac{\partial \hat{f}_i}{\partial x}(x_i - \hat{x}_{i-1})$$

Equation (12) is now a linear measurement equation, where y_i is the new measurement, M_i is the linear transformation, u_i is the random measurement error. Both y_i and M_i are readily computed from the actual measurement $\hat{x}_i$, the estimate $\hat{a}_{i-1}$ of a, the function f_i, and its first derivative. The second-order statistics of u_i are derived easily from those of v_i

$$E[u_i] = 0$$

$$W_i \cong E[u_i u_i^t] = \frac{\partial \hat{f}_i}{\partial x} \Lambda_i \frac{\partial \hat{f}_i^t}{\partial x}.$$

C. Recursive Kalman Filter

When no gaussianness is assumed on the previous random errors u_i, v_i, and w_i, the Kalman filter equations provide the best (minimum variance) linear unbiased estimate of a. This means that among the estimators which seek a_k as a linear combination of the measurements $\{y_i\}$, it is the one which minimizes the expected error norm squared

$$E[(\hat{a}_k - a)^t (\hat{a}_k - a)]$$

while verifying

$$E[\hat{a}_k] = a.$$

The recursive equations of the Kalman filter which provide a new estimate $(\hat{a}_i, S_i)$ of a from $(\hat{a}_{i-1}, S_{i-1})$ are as follows:

$$\hat{a}_i = \hat{a}_{i-1} + K_i(y_i - M_i \hat{a}_{i-1}) \quad (13)$$

$$K_i = S_{i-1} M_i^t (W_i + M_i S_{i-1} M_i^t)^{-1} \quad (14)$$

$$S_i = (I - K_i M_i) S_{i-1} \quad (15)$$

or equivalently

$$S_i^{-1} = S_{i-1}^{-1} + M_i^t W_i^{-1} M_i. \quad (16)$$

When moving object is located at (P_x, P_y), it is represented with respect to the robot coordinate frame as equation (17)

$$x_i = [x_{bi} \quad y_{bi}]^t \quad (17)$$

where the robot position is as follows

$$a = [x_r \quad y_r \quad \theta_r]^t. \quad (18)$$

And the constraint equation is

$$f_i(x_i, a) = \begin{bmatrix} x_{bi} - (P_x - x_r)\cos\theta_r - (P_y - y_r)\sin\theta_r \\ y_{bi} + (P_x - x_r)\sin\theta_r - (P_y - y_r)\cos\theta_r \end{bmatrix}. (19)$$

Derivatives of upper equation are as follows

$$\frac{\partial \hat{f}_i}{\partial a} = \begin{bmatrix} \cos\theta_r & \sin\theta_r & (P_x - x_r)\sin\theta_r - (P_y - y_r)\cos\theta_r \\ -\sin\theta_r & \cos\theta_r & (P_x - x_r)\cos\theta_r + (P_y - y_r)\sin\theta_r \end{bmatrix}$$

$$\frac{\partial \hat{f}_i}{\partial x} = \begin{bmatrix} 1 & 0 \\ 0 & 1 \end{bmatrix}. \quad (20)$$

IV. Experiments

A. System Configuration

Fig. 3 shows the mobile robot, ZIRO, which is developed by my laboratory on our own. A mobile part of the robot has the structure of a wheeled vehicle and the two differential wheels control linear velocity and angular velocity of the robot. We also use a DC motor for each wheel and a ball-caster for an assistant wheel. Since the robot can get a great torque output of the reduction gear, it uses independent joint control, which treats dynamic characteristics of the robot as disturbance. Two encoders, a gyro sensor and a vision sensor are used for the navigation control. The odometry sensor is attached to each motor, and the gyro sensor(ENV-05D) is used for recognizing the orientation of the robot by measuring the rotational velocity and it is applied to velocity control and position calculation of the robot. A CCD camera is fixed on the top of the mobile robot for the recognition of a moving object and environments, and it has the structure of 2 DOF active camera for free vision control and active object recognition.

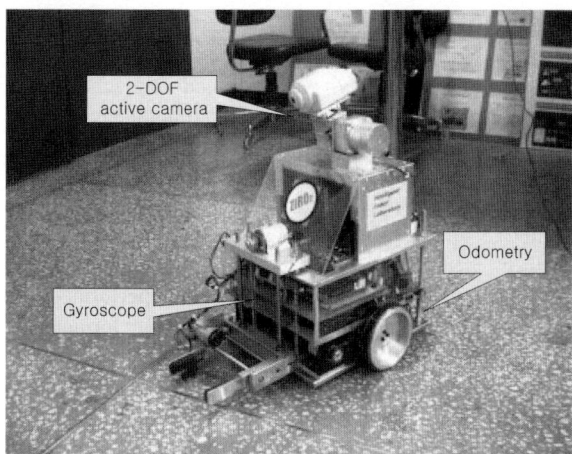

Fig. 3. A mobile robot, ZIRO.

A Camera is designed for a pan/tilt movement and uses step motor for driving. Each motor is controlled by an 8-bit micro controller, and we implemented variable speed control and holding function and therefore the camera can move smoothly and precisely. If a camera's pose is determined in the main controller, the value of its pose is transferred to each axis controller via ISA bus and distributed control is achieved. We used color frame grabber for capturing a camera's image that contains object's position information.

The developed mobile robot has many sensors such as a vision sensor, a gyro sensor and a encoder to accomplish autonomous navigation. A real-time execution is needed in this system, but if we process these many works in one computer, much load is required to execute these tasks and it is difficult to get real-time processing.

In our mobile robot, we realized a real-time distributed control. So, each independent controller can performs sensor interface, position and velocity control of each axes and motion planning, and integration of these controllers are implemented using CAN(Controller Area Network) [2,8]. In recent distributed control system, it is general to adopt a standard real-time communication network such as CAN to interface between main controller and lower controller aiming at open architecture, small size and easy maintenance.

Fig. 4 shows the overall organization of developed mobile robot system. There are totally six controllers and they are divided into one main controller and five lower controller. A main controller uses Pentium MMX-233 PC board where frame grabber card is interfaced via PCI bus for capturing images, and CAN controller card for controller networking and pan/tilt control interface card is interfaced via ISA bus for the movement of an active camera. The main controller periodically acquires image information and robot state information, executes image processing and robot movement planning, and transfers command to lower controller.

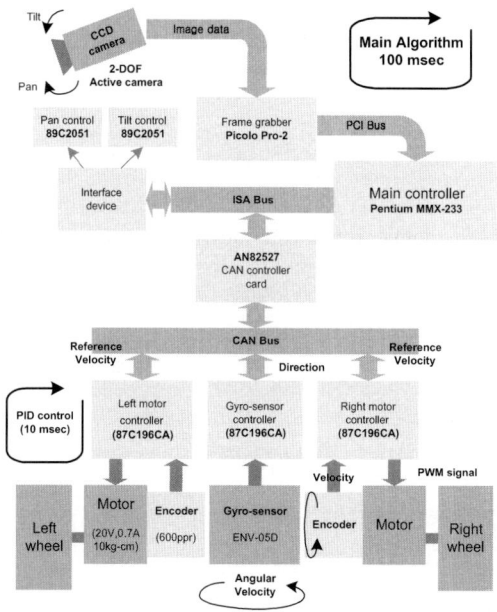

Fig. 4. The overall organization of robot system.

B. Experiment

In our experiment, we added a noise to the robot position and estimated the position of the moving object. We detects the position of the moving target using the images of 50 sequences and estimates the position of mobile robot using EKF.

Noise covariance in this experiment is as follows

$$S_i = \begin{bmatrix} 0.005^2 & 0 \\ 0 & 0.005^2 \end{bmatrix} \quad (21)$$

The robot's moving path is

$$Y = -0.1(X-500)^2 + 1200, X = 400-600, Z = 100 \quad (22)$$

Simulation results are as follows and these figures are shows the x, y, θ errors.

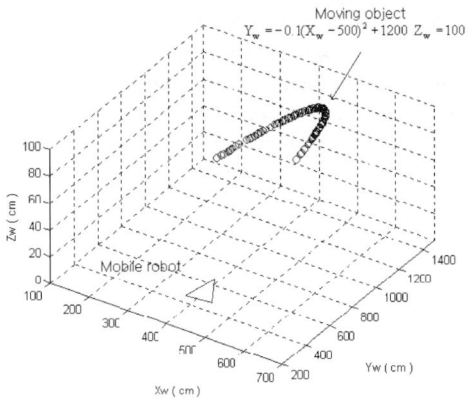

Fig. 5. mobile robot and moving object : parabolic motion.

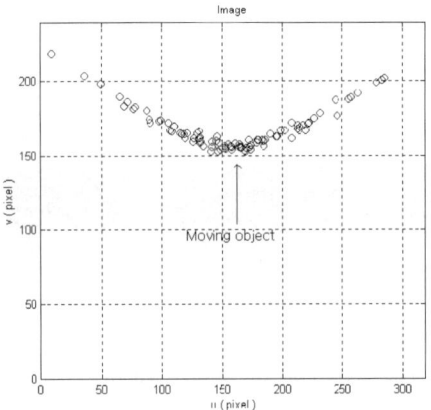

Fig. 6. image coordinates of a moving object.

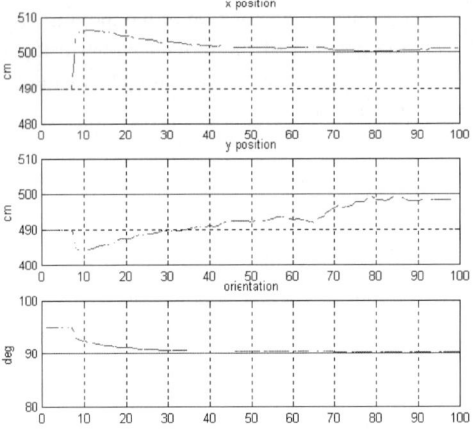

Fig. 7. position estimation.

V. Conclusions

Through this study, we presented the scheme of autonomous localization of a mobile robot using images of a moving target and this method is generally applied to the field of precise robot control.

Utilizing the consecutive image frames and Kalman Filter technique, the relative pose of the object with respect to the camera obtained accurately and localization of a mobile robot is gradually performed.

The control technique of a mobile robot formulated in our laboratory can applied to the mobile robot working in a nuclear reactor, spacecraft or underwater exploration.

References

[1] M. Y. Han and Jang M. Lee, "Precision Control of a Mobile/Task Robot Using Visual Information," *Journal of The Korea Institute of Telematics and Electronics*, in Korean, vol. 34, no. S-10, pp. 71-79, 1997.

[2] J. W. Park, D. K. Roh, J. H. Park, H. R. Hur, J. M. Lee, "Implementation of a Mobile Robot with Distributed Control Structure using CAN," *Conference of The Korea Institute of Telematics and Electronics*, in Korean, vol. 8, no. 2, pp. 251~255, 1999.

[3] Roger Pissard-Gibollet and Patrick Rives, "Applying Visual Servoing Techniques to Control a Mobile Hand-Eye System," *Proc. IEEE Int. Conf. on Robotics and Automation*, pp. 166-171, 1995.

[4] Hanqi Zhuang and Zvi S. Roth, "Camera-Aided Robot Calibration," CRC Press, 1996.

[5] F. Moscheni, F. Duifaux, and M. Kunt, "Object tracking based on temporal and spatial information," *IEEE Int'l. Conf. Acoustic, Speech, Signal Processing*, pp. 1914-1917, 1996.

[6] N.C. Mohanty, "Computer tracking of moving point targets in space," *IEEE Trans. Pattern and Machine Intelligence*, vol. PAMI-3, pp. 606-611, Sept. 1981.

[7] Stephane Betge-Brezetz, Raja Chatila, and Michel Devy, "Object-based Modeling and Localization in Natural Environment ," *IEEE Int'l. Conf. Robotics and Automation*, pp. 2920~2927, 1995.

[8] M. Dani Baba and E. T. Power, "Scheduling Performance in Distributed Real-Time Control System," *Proc. of 2nd International CAN Conf.*, pp. 7/2~7/11, 1995.

[9] R. E. Kalman, "A New Approach to Linear Filtering and Prediction Problems," *Trans, ASME, J. Basic Eng*, Series 82D, pp. 35-45, Mar. 1960.

Extracting Navigation States from a Hand-Drawn Map

Marjorie Skubic, Pascal Matsakis, Benjamin Forrester and George Chronis
Dept. of Computer Engineering and Computer Science
University of Missouri-Columbia
email: skubicm@missouri.edu

Abstract

Being able to interact and communicate with robots in the same way we interact with people has long been a goal of AI and robotics researchers. In this paper, we propose a novel approach to communicating a navigation task to a robot, which allows the user to sketch an approximate map on a PDA and then sketch the desired robot trajectory relative to the map. State information is extracted from the drawing in the form of relative, robot-centered spatial descriptions, which are used for task representation and as a navigation language between the human user and the robot. Examples are included of two hand-drawn maps and the linguistic spatial descriptions generated from the maps.

1 Introduction

Being able to interact and communicate with robots in the same way we interact with people has long been a goal of AI and robotics researchers. Much of the robotics research has emphasized the goal of achieving autonomous robots. However, this ambitious goal presumes that robots can accomplish human-like perception, reasoning, and planning as well as achieving human-like interaction capabilities.

In our research, we are less concerned with creating autonomous robots that can plan and reason about tasks, and instead we view them as semi-autonomous tools that can assist a human user. The user supplies the high-level and difficult reasoning and strategic planning capabilities. We assume the robot has some perception capabilities, reactive behaviors, and perhaps some limited reasoning abilities that allow it to handle an unstructured and possibly dynamic environment.

In this scenario, the interaction and communication mechanism between the robot and the human user becomes very important. The user must be able to easily communicate what needs to be done, perhaps at different levels of task abstraction. In particular, we would like to provide an intuitive method of communicating with robots that is easy for users that are not expert robotics engineers. We want domain experts to define their own task use of robots, which may involve controlling them, guiding them, or even programming them.

In ongoing research on human-robot interaction, we have been investigating the use of spatial relations in communicating purposeful navigation tasks. Linguistic, human-like expressions that describe the spatial relations between a robot and its environment provide a symbolic link between the robot and the user, thus comprising a type of navigation language. The linguistic spatial expressions can be used to establish effective two-way communications between the robot and the user, and we have approached the issue from both perspectives.

From the robot perspective, we have studied how to recognize the current (qualitative) state in terms of egocentric spatial relations between the robot and objects in the environment, using sensor readings only (i.e., with no map or model of the environment). Linguistic spatial descriptions of the state are then generated for communication to the user. See our companion paper [1] for details on the approach used.

In this paper, we focus on the user perspective, and offer one approach for communicating a navigation task to a robot, which is based on robot-centered spatial relations. Our approach is to let the user draw a sketch of an environment map (i.e., an approximate representation) and then sketch the desired robot trajectory relative to the map. State information is extracted from the drawing on a point by point basis along the sketched robot trajectory. We generate a linguistic description for each point and show how the robot transitions from one qualitative state to another throughout the desired path. A complete navigation task is represented as a sequence of these qualitative states based on the egocentric spatial relations, each with a corresponding navigation behavior. We assume the robot has pre-programmed or pre-learned, low-level navigation behaviors that allow it to move safely around its unstructured and dynamic environment without hitting objects. In this approach, the robot does not have a known model or map of the environment, and the user may have only an approximate map. Thus, the navigation task is built upon relative spatial states, which become qualitative states in the task model.

The idea of using linguistic spatial expressions to communicate with a semi-autonomous mobile robot has been proposed previously. Gribble *et al* use the framework of the Spatial Semantic Hierarchy for an intelligent wheelchair [2]. Perzanowski *et al* use a combination of gestures and linguistic directives such as "go over there" [3]. Shibata *et al* use positional relations to overcome ambiguities in recognition of landmarks [4]. However, the idea of communicating with a mobile robot via a hand-drawn map appears to be novel.

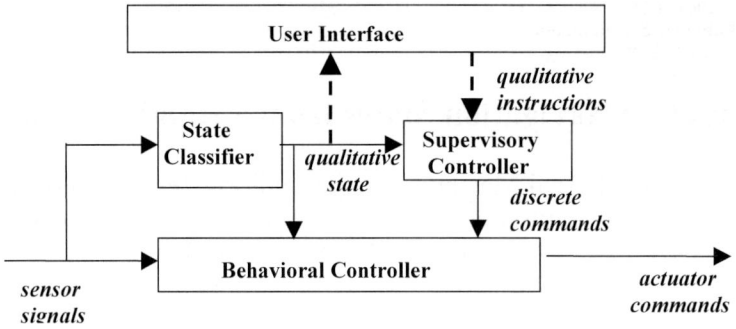

Figure 1. The User Interface and Robot Control Architecture.

The strategy of using a sketch with spatial relations has been proposed by Egenhofer as a means of querying a geographic database [5]. The hand-drawn sketch is translated into a symbolic representation that can be used to access the geographic database.

In this paper, we show how egocentric spatial relations can be extracted from a hand-drawn map sketched on a PDA. In Section 2, we discuss background material on the human-robot interaction framework. In Section 3, we show the method for extracting the environment representation and the corresponding states from the PDA sketch. Experiments are shown in Section 4 with two examples of hand-drawn maps and the spatial descriptions generated. We conclude in Section 5.

2 Framework for Human-Robot Interaction

Much of our research efforts in human-robot interaction have been directed towards extracting robot task information from a human demonstrator. Figure 1 shows the framework for the robot control architecture and the user interface.

2.1 Robot Control Components

We consider procedural tasks (i.e., a sequence of steps) and represent task structure as a Finite State Automaton (FSA) in the *Supervisory Controller*, following the formalism of the Discrete Event System (DES) [6]. The FSA models behavior sequences that comprise a task; the sensor-based qualitative state (QS) is used for task segmentation. The change in QS is an event that corresponds to a change in the behavior. Thus, the user demonstrates a desired task as a sequence of behaviors using the existing behavior primitives and identifiable QS's, and the task structure is extracted in the form of the FSA. During the demonstration, the QS and the FSA is provided to the user to ensure that the robot is learning the desired task structure. With an appropriate set of QS's and primitive behaviors, the FSA and supervisory controller is straightforward. Also, this task structure is consistent with structure inherently used by humans for procedural tasks, making the connection easier for the human. We have used this approach in learning force-based assembly skills from demonstration, where a qualitative contact state provided context [7]. For navigation tasks, spatial relations provide the QS context.

With the *State Classifier* component, the robot is provided with the ability to recognize a set of qualitative states, which can be extracted from sensory information, thus reflecting the current environmental condition. For navigation skills, robot-centered spatial relations provide context (e.g., *there is an object to the left front*). Adding the ability to recognize classes of objects provides additional perception (e.g., *there is a person to the left front*).

The robot is also equipped with a set of primitive (reactive) behaviors and behavior combinations, which is managed by the *Behavioral Controller*. Some behaviors may be preprogrammed and some may be learned off-line using a form of unsupervised learning. The user can add to the set of behaviors by demonstrating new behaviors which the robot learns through supervised learning, thus allowing desired biases of the domain expert to be added to the skill set.

Note that this combination of discrete event control in the Supervisory Controller and the "signal processing" in the Behavioral Controller is similar to Brockett's framework of hybrid control systems [8].

2.2 User Interface

As shown in Figure 1, the interface between the robot and the human user relies on the qualitative state for two-way communications. In robot-to-human communications, the QS allows the user to monitor the current state of the robot, ideally in terms easily understood (e.g., *there is an object on the right*). In human-to-robot communications, commands are segmented by the QS, termed qualitative instructions in the figure (e.g., *while there is an object on the right, move forward*).

The key to making the interactive robot training work is the QS, especially in the following ways: (1) the ability to perceive an often ambiguous context based on sensory conditions, especially in terms that are understandable for the human trainer, (2) choosing the right set of QS's so as to communicate effectively with the trainer, and (3) the ability to perform self-assessment, as in knowing how well the QS is identified which helps in knowing when to get further instruction. Spatial relationships provide powerful cues for humans to make decisions; thus, it is plausible to investigate their use as a qualitative state for robot tasks, as well as a linguistic link between the human and the robot.

3 Extracting Spatial Relations States

The interface used for drawing the robot trajectory maps is a PDA (e.g., a PalmPilot). The stylus allows the user to sketch a map much as she would on paper for a human colleague. The PDA captures the string of (x,y) coordinates sketched on the screen and sends the string to a computer for processing (the PDA connects to a PC through a serial port).

The user first draws a representation of the environment by sketching the approximate boundary of each object. During the sketching process, a delimiter is included to separate the string of coordinates for each object in the environment. After all of the environment objects have been drawn, another delimiter is included to indicate the start of the robot trajectory, and the user sketches the desired path of the robot, relative to the sketched environment. An example of a sketch is shown in Figure 2, where each point represents a captured (x,y) screen pixel.

For each point along the trajectory, a view of the environment is built, corresponding to the radius of the sensor range. The left part of Figure 3 shows a sensor radius superimposed over a piece of the sketch. The sketched points that fall within the scope of the sensor radius represent the portion of the environment that the robot can sense at that point in the path.

The points within the radius are used as boundary vertices of the environment object that has been detected. They define a polygonal region (Figure 3, step (a)) whose relative position with respect to the robot (assimilated to a square) is represented by two histograms (Figure 3, step (b)): the histogram of constant forces and the histogram of gravitational forces [9][1]. These two representations have very different and interesting characteristics. The former provides a global view of the situation and considers the closest parts and the farthest parts of the objects equally. The latter provides a more local view and focuses on the closest parts.

The notion of the histogram of forces, introduced by Matsakis and Wendling, ensures processing of raster data as well as vector data, offers solid theoretical guarantees, allows explicit and variable accounting of metric information, and lends itself, with great flexibility

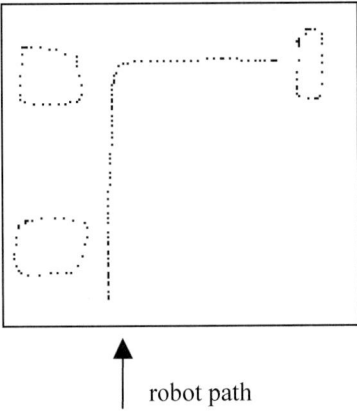

Figure 2. A sketched map on the PDA. Environment objects are drawn as a boundary representation. The robot path starts from the bottom.

to the definition of fuzzy directional spatial relations (such as *to the right of, in front of*, etc.).

For our purposes, it also allows for a low-computational handling of heading changes in the robot's orientation and makes it easy to switch between a world view and an egocentric robot view. The heading is computed as the direction formed by the current point and the second previous point along the sketched path. A pixel gap in the heading calculation serves to smooth out the trajectory somewhat, thereby compensating for the discrete pixels.

The histogram of constant forces and the histogram of gravitational forces associated with the robot and the polygonal region are used to generate a linguistic description of the relative position between the two objects. The method followed is the method described in [10][11] (and applied to LADAR image analysis).

First, eight numeric features are extracted from the analysis of each histogram (Figure 3, step (c)). They constitute the "opinion" given by the considered histogram. The two opinions (*i.e.*, the sixteen values) are then combined (Figure 3, step (d)). Four numeric and two symbolic features result from this combination. They feed a system of fuzzy rules that outputs the expected linguistic description.

The system handles a set of adverbs (like *mostly, perfectly*, etc.) which are stored in a dictionary, with other terms, and can be tailored to individual users. Each description generated relies on the sole primitive directional relationships: *to the right of, in front of, to the left of,* and *behind*.

The spatial description is generally composed of three parts. The first part involves the primary direction (e.g., *the object is mostly to the right of the robot*). The second part supplements the description and involves a secondary direction (e.g., *but somewhat to the rear*). The third part indicates to what extent the four directional relationships are suited to describing the relative position between the robot and the object (e.g., *the description is satisfactory*). In other words, it indicates to what extent it is necessary to utilize other spatial relations (e.g., *surrounds*).

Figure 4 shows the linguistic description generated for some point on the robot path. In this example, a secondary direction is not generated because the primary direction clause is deemed to be adequate. Figure 5 shows a second example along the robot path, with the three-part linguistic spatial description generated for that point.

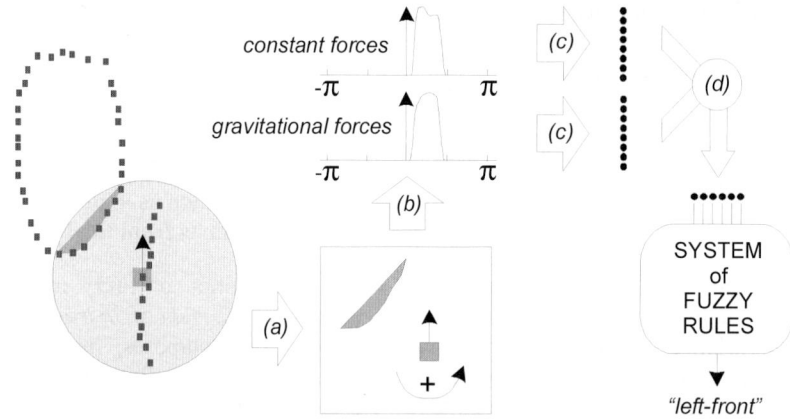

Figure 3. Synoptic diagram. (a) Construction of the polygonal objects. (b) Computation of the histograms of forces. (c) Extraction of numeric features. (d) Fusion of information.

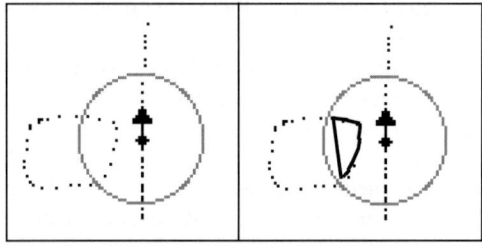

"*Object is to the left of the Robot (the description is satisfactory)*"

"*Object is mostly to the left of the Robot but somewhat to the rear (the description is satisfactory)*"

Figure 4. Building the environment representation for one point along the trajectory, shown with the generated linguistic expression.

Figure 5. Another example with a three-part linguistic spatial description generated.

4 Experiments

Experiments were performed on two hand-drawn maps to study the linguistic spatial descriptions generated. The first map is shown in its raw (pixel) state in Figure 2. The user first draws the three objects in the bottom left, top left and top right locations. Then, she draws a desired robot trajectory starting from the bottom of the PDA screen.

Representative spatial descriptions are shown in Figure 6 for several points, labeled 1 through 11, along the sketched robot trajectory. The assessment was always satisfactory so it is not specified on the figure.

Note that the heading is also calculated and used in determining the robot-centered spatial relations. The sensor radius was set to 22 pixels.

At position 1, part of the object **A** is detected to the left-front of the robot, according to the generated linguistic description. As the robot proceeds through positions 2, 3, and 4, the parts of **A** that are within the 22 pixel radius are processed and the corresponding linguistic descriptions are shown in the figure. At position 5, there is nothing within the sensor radius of the robot, so no linguistic descriptions are generated. At points 6, 7, and 8 we observe a sharp right turn. The corresponding parts of the second object **B** that fall within the sensor radius at each point are expressed in linguistic terms. At point 9, the robot is again between objects and nothing is within the sensor radius. Finally, part of the last object **C** is detected to the front of the robot at position 10, and at position 11 an extension of the part of **C** also falls within the radius to the right of the robot.

Figure 7 shows the second map sketched on the PDA. To experiment with a different scaling factor, the sensor radius was set to 30 pixels. Several spatial descriptions are shown in Figure 8. All linguistic descriptions were accepted as satisfactory. An interesting variation in this second experiment is the simultaneous detection of two different objects, namely **A** and **B**. For positions 3, 4, and 5, we show the linguistic descriptions while the robot passes between **A** and **B**.

These experiments indicate the feasibility of using spatial relations to analyze a sketched robot map and trajectory, but much work remains to be done. The limited resolution of the PDA screen results in abrupt changes of the robot heading, which can affect the accuracy of the description generated. The current algorithm for building the object representation for the map cannot handle all cases (e.g., concave objects). Also, we need to study the granularity of the spatial descriptions generated. While they are descriptive for human users, they may be too detailed for use in navigation task representation. The next step is to perform further experiments and extract the corresponding navigation behavior to study the granularity issue. Future work will also address the discrepancy between the sketched trajectory and the actual robot path taken, as indicated by the sensory data.

5 Concluding Remarks

In this paper we have proposed a novel approach for human-robot interaction, namely showing a robot a navigation task by sketching an approximate map on a PDA. The interface utilizes spatial descriptions that are generated from the map using the histogram of forces. The approach represents a first step in studying the use of spatial relations as a symbolic language between a human user and a robot for navigation tasks.

Acknowledgements

The authors wish to acknowledge support from ONR, grant N00014-96-0439 and the IEEE Neural Network Council for a graduate student summer fellowship for Mr. Chronis. We also wish to acknowledge Dr. Jim Keller for his helpful discussions and suggestions.

References

[1] M. Skubic, G. Chronis, P. Matsakis and J. Keller, "Generating Linguistic Spatial Descriptions from Sonar Readings Using the Histogram of Forces", in *Proceedings of the 2001 IEEE International Conference on Robotics and Automation.*

[2] W. Gribble, R. Browning, M. Hewett, E. Remolina and B. Kuipers, "Integrating vision and spatial reasoning for assistive navigation", In *Assistive Technology and Artificial Intelligence,* V. Mittal, H. Yanco, J. Aronis and R. Simpson, ed., Springer Verlag, , 1998, pp. 179-193, Berlin, Germany.

[3] D. Perzanowski, A. Schultz, W. Adams and E. Marsh, "Goal Tracking in a Natural Language Interface: Towards Achieving Adjustable Autonomy", In *Proceedings of the 1999 IEEE International Symposium on Computational Intelligence in Robotics and Automation*, Monterey, CA, Nov., 1999, pp. 208-213.

[4] F. Shibata, M. Ashida, K. Kakusho, N. Babaguchi, and T. Kitahashi, "Mobile Robot Navigation by User-Friendly Goal Specification", In *Proceedings of the 5th IEEE International Workshop on Robot and Human Communication*, Tsukuba, Japan, Nov., 1996, pp. 439-444.

[5] M.J. Egenhofer, "Query Processing in Spatial-Query-by-Sketch", *Journal of Visual Languages and Computing*, vol. 8, no. 4, pp. 403-424, 1997.

[6] P.J. Ramadge and W.M. Wonham, "The control of discrete event systems", Proceedings of the IEEE, vol. 77, no. 1, pp. 81-97, Jan., 1989.

[7] M. Skubic and R.A. Volz, "Acquiring Robust, Force-Based Assembly Skills from Human Demonstration", *IEEE Transactions on Robotics and Automation*, vol. 16, no.6, Dec., 2000, pp. 772-781.

[8] R.W. Brockett, "Hybrid models for motion control systems," in *Essays on Control: Perspectives in the Theory and Its Applications*, H.L. Trentelman and J.C. Willems, Eds., chapter 2, pp. 29--53. Birkhauser, Boston, MA, 1993.

[9] P. Matsakis and L. Wendling, "A New Way to Represent the Relative Position between Areal Objects", *IEEE Trans. on Pattern Analysis and Machine Intelligence*, vol. 21, no. 7, pp. 634-643, 1999.

[10] P. Matsakis, J. M. Keller, L. Wendling, J. Marjamaa and O. Sjahputera, "Linguistic Description of Relative Positions in Images", *IEEE Trans. on Systems, Man and Cybernetics*, to appear.

[11] J. M. Keller and P. Matsakis, "Aspects of High Level Computer Vision Using Fuzzy Sets", *Proceedings, 8th IEEE Int. Conf. on Fuzzy Systems*, Seoul, Korea, pp. 847-852, 1999.

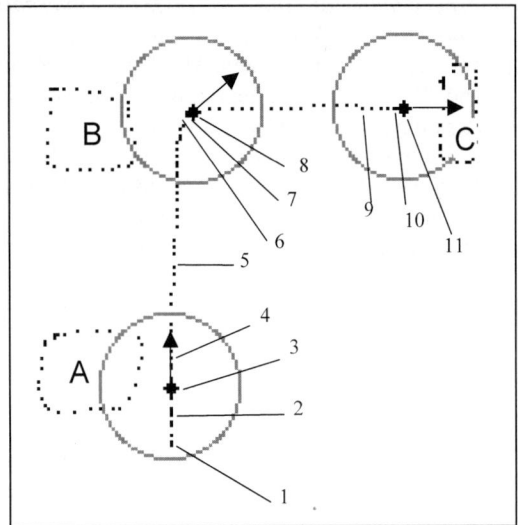 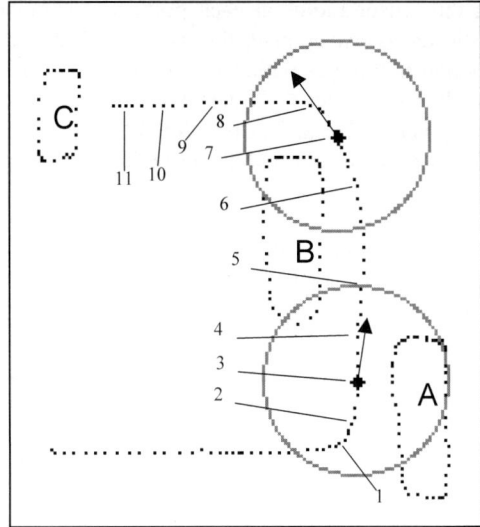

1. **Object A** is to the left-front of the **Robot**.
2. **Object A** is mostly to the left of the **Robot** but somewhat forward.
3. **Object A** is to the left of the **Robot** but extends forward relative to the **Robot**.
4. **Object A** is to the left of the **Robot**.
5. None
6. **Object B** is mostly to the left of the **Robot** but somewhat to the rear.
7. **Object B** is behind-left of the **Robot**.
8. **Object B** is mostly behind the **Robot** but somewhat to the left.
9. None
10. **Object C** is in front of the **Robot**.
11. **Object C** is in front of the **Robot** but extends to the right relative to the **Robot**.

Figure 6. Representative spatial descriptions along the sketched robot trajectory for the PDA-generated map 1.

1. **Object A** is in front of the **Robot** but extends to the right relative to the **Robot**.
2. **Object A** is to the right of the **Robot**.
3. **Object A** is to the right of the **Robot** but extends to the rear relative to the **Robot**.
   Object B is to the left-front of the **Robot**.
4. **Object A** is mostly to the right of the **Robot** but somewhat to the rear.
   Object B is mostly to the left of the **Robot** but somewhat forward.
5. **Object A** is mostly behind the **Robot** but somewhat to the right.
   Object B is to the left of the **Robot**.
6. **Object B** is to the left of the **Robot** but extends to the rear relative to the **Robot**.
7. **Object B** is mostly to the left of the **Robot** but somewhat to the rear.
8. **Object B** is to the left of the **Robot** but extends to the rear relative to the **Robot**.
9. **Object B** is behind-left of the **Robot**.
10. **Object C** is in front of the **Robot**.
11. **Object C** is in front of the **Robot** but extends to the left relative to the **Robot**.

Figure 8. Representative spatial descriptions along the sketched robot trajectory for the PDA-generated map 2, showing the simultaneous detection of two different objects.

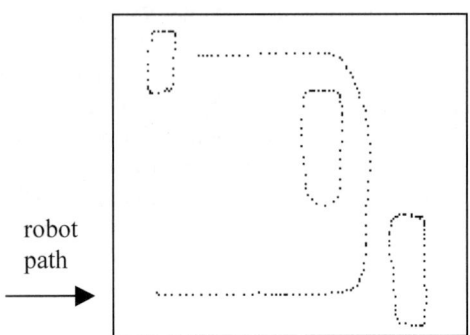

robot path →

Figure 7. The sketched map used for the second experiment. The robot path starts from the bottom left.

On Eye-Sensor based Path Planning for Robots with non-trivial Geometry/Kinematics

Kamal K. Gupta
kamal@cs.sfu.ca

Yong Yu
yongyu@cs.sfu.ca

School of Engineering Science
Simon Fraser University
Burnaby, B.C. V5A 1S6 Canada

Abstract

We formally pose and explore some novel issues that arise for eye-sensor based motion planning for robots with non-trivial geometry/kinematics. The key issue is that while the sensor senses in physical space, the planning takes place in configuration space, and the two spaces are distinctly different for robots with non-trivial geometry/kinematics. This leads to some very interesting, fundamental yet novel issues. In particular, we introduce several novel notions: notion of s-reachability, notion of s-completeness that characterizes completeness for sensor-based planning algorithms, notion of explorability of configuration space, and notion of observability of physical space. We give sufficient conditions for a (discrete) eye-sensor based planner to be s-complete.

1 Introduction

Motion planning (MP) in robotics can be divided into two categories [7]: (i) model-based MP where the environment is assumed to be completely known, and the task for the robot is to reach a desired goal configuration [13], and (ii) sensor-based MP where the environment is unknown and the task for the robot, equipped with a sensor, is to explore the environment and reach a given goal configuration. A variety of robot-sensor systems have been used in the latter category, ranging from mobile robots with vision/range ("eye" type) sensing to manipulator arms with "skin" type sensors [14]. In this paper, we formalize the problem of eye-sensor based motion planning for robots with non-trivial[1] geometry/kinematics. This class of robots is broad and includes robots ranging from a simple polygonal mobile robot to complex articulated manipulators. The key issue is that while the sensor senses in physical space, the planning takes place in configuration space, and the two spaces are distinctly different for robots with non-trivial geometry/kinematics. Some recent work including our own [2, 3, 6, 4, 5] has presented implemented sensor-based motion planners and view planning algorithms for exploration with eye-in-hand systems (a manipulator arm with a wrist mounted range camera). However, some of the underlying fundamental issues have not yet been explored. This paper addresses them. For instance, it may even be impossible to decide if a given goal configuration is reachable. An example that illustrates this is shown in Figure 1 — a planar eye-in-hand system consisting of a 2-link robot arm, equipped with an "eye" sensor at its end-effector that gives distance or range of the objects (from the sensor). The arm is required to plan and execute collision-free motions in an environment initially unknown to the robot. Sensor field of view is indicated by the triangular region; in addition, the sensor has an additional degree of freedom and can rotate. White region around the initial robot configuration (in left image, shown in dark gray) is known to be free at the start, light gray region is free space unknown to the robot, and dark gray areas are obstacles unknown to the robot. The robot in goal configuration is shown in black. It is not possible for a sensor-based planner to determine if the given goal configuration is reachable or not, since the robot is not able to sense the unknown region (which is free but the robot does not know it). Right image shows the physical space that the robot is able to see. At best, the planner could terminate with "goal not reachable with the given sensor and initial free region". The key intuitive idea here is that in the sensor-based case, the robot must be able to sense the physical space *before* it can occupy it. We formalize such notions in this paper. Although our focus is on eye type sensors, most of the notions are applicable to sensor-based planning in general. In particular, we introduce several novel notions: notion of s-reachability, notion of s-completeness that characterizes completeness for sensor-based planning algorithms, notion of explorability of configuration space, and notion of observability of physical space. We give sufficient conditions for a discrete eye-sensor based planner to be s-complete. Finally we discuss some related open problems.

To the best of our knowledge, planning research with eye sensors has used mobile robots, often modelled as a point. Most of these algorithms are applicable to planar case and deal with 2-dimensional polygonal world [8], or more recently 3-dimensional case [11, 12]. [9] presents an algorithm for a convex polygon translating in unknown polygonal world. However, the assumed sensor model is that of entire visibility polygon from a given point.

2 Physical Sensors from C-space Perspective

From a broad motion planning perspective sensors provide — directly or indirectly — distances to objects (or ob-

---
[1] Non-trivial geometry/kinematics implies that the physical space and C-space are different. Idealized cases such a point or circle robots are considered to have trivial geometry/kinematics from this point of view.

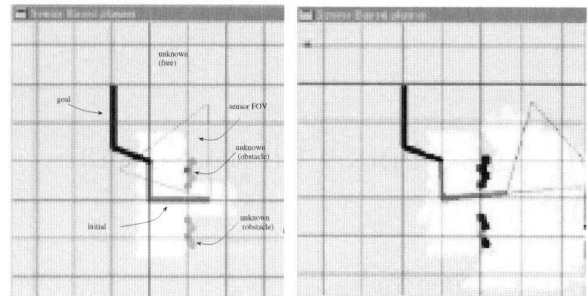

Figure 1: A 2-link robot arm with an eye sensor (a range camera) mounted at the wrist. From the initial configuration (shown in gray in left image) it is not possible for a sensor-based planner to determine if the given goal configuration (shown in black) is reachable.

ject features) in the environment and can be roughly categorized into two broad types[2]: (i) "eye" type sensor, and (ii) "skin" type sensor. For our purposes, we assume that an eye sensor senses distance of objects from a single vantage point (or a reference frame) in physical space (implicitly also classifying which points are in free-space and which are obstacles). The actual implementation may consist of sonar, or laser range finder, or stereo vision. The skin sensor senses if *any* part of the robot is in contact with obstacles (skin-contact sensor), or a small distance away from the obstacles (skin-range sensor), i.e., it senses along the entire surface geometry of the robot. A skin sensor often consists of proximity or range sensors distributed along the entire manipulator geometry [10, 14].

Interestingly, there is a fundamental computational distinction between these two types of sensors. The action of a skin (skin-contact, for example) sensor has a "natural" interpretation in C-space since it can determine if the current configuration of the robot (and possibly a small neighborhood[3]) belongs to a C-obstacle or C-free. For eye type sensors, however, this is not true, i.e., a sensing action may not render an adjacent C-space point known or unknown (i.e., generally only some part of the region that the robot would have occupied in the adjacent configuration would be sensed by the robot). Note that for a point holonomic robot, this fundamental distinction between an eye sensor and a skin sensor disappears, since the eye sensor can also sense the current robot configuration (a point) and a neighborhood around it.

Another distinction between an eye sensor and a skin sensor is that of discrete time versus continuous time sensing. Eye sensors, generally speaking involve processing images, which is computationally intensive and slow. Therefore, often, the sensing action is triggered at discrete instants of time.[4] Skin sensors, on the other hand, involve proximity sensors, and can be carried out continuously, as the robot moves.

Real sensors that sense in physical space (either eye or skin) normally satisfy two properties in physical space[11]: (i) they sense in a "local" neighborhood and (ii) the sensing action is memoryless, i.e., the region sensed by the sensor in physical space does not depend on what it has sensed before. Note, however that these two properties do not carry over to C-space. Islands or "disconnected" components may suddenly appear in the C-space as a result of a new scan; Clearly, which points in C-space become free as a result of a scan critically depends on what has been sensed before.

Several other "abstract" sensors such as those that assume a form of data that is either too abstract, too complex, or too far from current practice (unrealistic), have been considered, but are not focus of our work. For instance, some works assume that the sensor returns (in a polygonal environment), the entire polygonal scene within the sensor range [8]; others may ssume that the sensor returns "features" in C-space [16].

3 Formal Problem Statement
3.1 General Notation

Let $\mathcal{A}$ denote the robot (including all the physical parts), $\mathcal{S}$ denote the sensor, and $\mathcal{AS}$ denote the combined robot and sensor system. Let $\mathcal{P}$ denote the physical space (normally R^p, $p = 2$ or 3). Let $\mathcal{B}$, a closed set, denote the static obstacles. $\mathcal{A}(q) \subset \mathcal{P}$ denotes the physical space occupied by the entire robot (including the body of the sensor) at configuration q. We assume that the sensor body is "absorbed" into the robot $\mathcal{A}$. Hence, we can treat the physical sensor as an abstract reference frame without any physical body. We will attach a frame $\mathcal{F}_\mathcal{S}$ to the sensor. For simplicity, we also assume that the sensor has no "independent" degrees of freedom, i.e., $\mathcal{F}_\mathcal{S}$ depends only on q. The sensor senses in the physical space $\mathcal{P}$ and sensing is defined by a function[5] $s : \mathcal{C} \to \mathcal{V}_\mathcal{S} \subset R^p - int(\mathcal{B})$, where int denotes the interior of a set. Semantically, it implies that with the sensor frame, positioned at $\mathcal{F}_\mathcal{S}(q)$, the sensor can determine the p-dimensional co-ordinates of all points in $\mathcal{V}_\mathcal{S}$. With a slight abuse of notation, we will use $\mathcal{V}_\mathcal{S}(q) \subset \mathcal{P}$ to denote the sensed free region (an open set) when robot scans at configuration q. We will assume "line of sight" type sensors, hence, the actual sensed entity is a $p - 1$ dimensional set composed of obstacle boundaries (surfaces) or the boundary of the sensing region. Let $\mathcal{V}_{\mathcal{S}obs}(q)$ denote the obstacle boundaries sensed by the sensor. Then, $\mathcal{V}_\mathcal{S}(q)$ and $\mathcal{V}_{\mathcal{S}obs}(q)$ is the information obtained at configuration q. Figure 2 illustrates the sensing function. $\mathcal{P}_{free}^{(i)}(\mathcal{P}_{obs}^{(i)}) \subset \mathcal{P}$ is the cumulative free (obstacle) physical space acquired (or seen) by the sensor, and $\mathcal{P}_{unk}^{(i)} \subset \mathcal{P}$ is the physical space remaining unknown, at a certain stage i of the planning process (for instance, after the ith scan has been taken and processed). $\mathcal{P}_{free}^{(i)}$ is an open set. $\mathcal{C}_{free}^{(i)}, \mathcal{C}_{obs}^{(i)}, \mathcal{C}_{unk}^{(i)}$ denote the corresponding free C-space (an open set), obstacles in C-space,

---
[2]Nature itself seems to have favoured these two modalities of sensing for navigation/obstacle avoidance in animal kingdom.

[3]The nature of this neighborhood would depend on the kinematics of the robot. For a car-like non-holonomic robot, this local neighborhood would lie in the tangent space of C-space; for a holonomic robot, this local neighborhood would be a "ball" of appropriate size in the C-space.

[4]It is noteworthy that bulk of the attentive visual processing in humans also occurs at "discrete" times in between saccade movements.

[5]Ideally s should be defined as a function of $\mathcal{F}_\mathcal{S}$ and not q, since $q \to \mathcal{F}_\mathcal{S}$ is a many to one mapping. However for simplicity, we use q here. It is indeed possible to use $\mathcal{F}_\mathcal{S}$ in our formulation using an approach similar to [1].

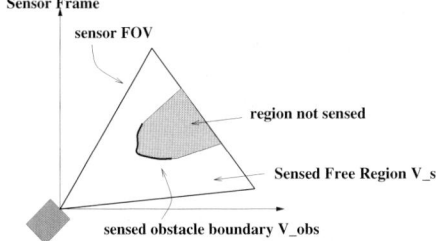

Figure 2: An illustrative sensor model showing the sensed region. White region is $\mathcal{V}_\mathcal{S}(q)$, dark curve is the sensed obstacle boundary $\mathcal{V}_{\mathcal{S}obs}(q)$. The triangular region (open set) is the sensor field of view (FOV).

and still unknown C-space. $\mathcal{P}_{free}$ is the entire free physical space and $\mathcal{C}_{free}$ is the entire free C-space. Similarly, $\mathcal{P}_{obs}$ and $\mathcal{C}_{obs}$ are obstacles in physical space and C-space. A path π^i is a continuous curve in $\mathcal{C}_{free}^{(i)}$.

3.2 Sensor-based Reachability: s-feasible Path

We now introduce some key concepts for sensor-based motion planning problem. Our presentation first assumes that the sensor is scanning continuously as the robot moves. The corresponding concepts are then naturally extended to the discrete time case.

Recall that for the model-based case (known environment, *a priori*), a configuration q is reachable if there exists a collision-free path from the start configuration q_0. We use the term m-reachable to denote the reachability in model-based case. In the sensor-based case, for robots with non-trivial geometry/kinematics, what the sensor can sense and where the robot can go are intricately tied. The robot can only move into those regions already discovered to be free by the sensor. The path that the robot takes to a given configuration is, in general, different from model-based ones. Such paths are fundamental in analysis of sensor-based MP and we call them s-feasible paths.[6]

s-feasible Path: Let the initial configuration $\mathcal{AS}$ be q_0, and $\mathcal{A}(q_0) \subset \mathcal{P}_{free}^{(0)}$. Let $\pi_{[q_0,q_1]} \subset \mathcal{C}$ denote a path from q_0 to q_1. q_a is a point on $\pi_{[q_0,q_1]}$ and let a one dimensional half-open set $\pi_{[q_0,q_a)}$ represent the sub-segment of path $\pi_{[q_0,q_1]}$ from q_0 to q_a (include q_0, but exclude q_a), and let $\pi_{[q_0,q_a]}$ represent the closed sub-segment of path $\pi_{[q_0,q_1]}$ from q_0 to q_a (include both q_0 and q_a). Then, $\bigcup_{q\in\pi_{[q_0,q_a]}} \mathcal{A}(q)$ is the volume swept by the robot as it moves along $\pi_{[q_0,q_a]}$, and $\bigcup_{q\in\pi_{[q_0,q_a)}} \mathcal{V}_\mathcal{S}(q)$ is the total sensing volume sensed all along the path segment before the robot reaches q_a. If $\forall q$ along the path, the volume swept by the robot is a subset of the total sensing volume previously sensed, then this path is a feasible path. Formally, if $\forall q_a \in \pi_{[q_0,q_1]}$,

$$\bigcup_{q\in\pi_{[q_0,q_a]}} \mathcal{A}(q) \subset (\bigcup_{q\in\pi_{[q_0,q_a)}} \mathcal{V}_\mathcal{S}(q)) \bigcup \mathcal{P}_{free}^{(0)} \quad (1)$$

then path $\pi_{[q_0,q_1]}$ is an s-feasible path. If the above inclusion does not hold, the path is non-feasible. We will refer to this inclusion condition as swept volume inclusion (SVI) condition.

Figure 3(a) shows an example of an s-feasible path for a rectangular robot. This robot has a sensor which can sense within a cone with a fixed direction toward the right of the robot (shown as dotted lines). In order to sense region A, so that the robot can move to point e, the robot has to move to point c to sense. Path $a \to b \to c \to d \to e$ is an s-feasible path. Without the back-trace segment $b \to c \to d$, path $a \to b \to e$ is not an s-feasible path, as shown in (b).

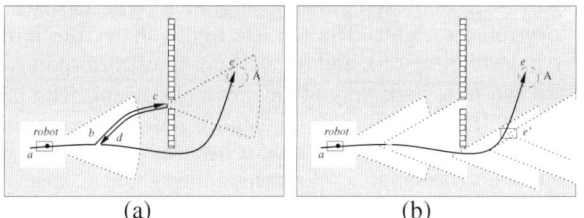

Figure 3: The path in (a) is s-feasible. In (b), it is not s-feasible.

s-reachable Configuration: A configuration q_1 is s-reachable from q_0 iff there exists an s-feasible path for $\mathcal{AS}$ to move from q_0 to q_1. Otherwise, we say q_1 is not s-reachable from q_0.

The goal configuration in Figure 1 is not s-reachable. Having defined s-reachability, we can now formally state the sensor-based planning problem.

3.3 The Problem

Our key assumptions are: (i) the robot environment is static, with $\mathcal{AS}$ being the only moving entity, and (ii) there is an implicit assumption that the robot knows its current configuration (using joint encoders, for example), and (iii) a small region of physical space, $\mathcal{P}_{free}^{(0)}$, surrounding the robot in its initial configuration is free. This latter requirement is important because otherwise even the very first move may not be possible after the initial scan. $\mathcal{AS}$ starts off at configuration q_0. As the robot moves, the sensor continuously senses the free region $\mathcal{V}_\mathcal{S}$ and obstacle boundaries $\mathcal{V}_{\mathcal{S}obs}$. Different motion planning problems can be formulated. We state the salient start-goal problem:

Problem: Given a robot-sensor, an initial known free region $\mathcal{P}_{free}^{(0)}$, an initial configuration q_0 ($\mathcal{A}(q_0) \in \mathcal{P}_{free}^{(0)}$), and a goal configuration $q_{(goal)} \in \mathcal{C}$, determine an s-feasible path from q_0 to $q_{(goal)}$. Report failure if no such path exists.

4 Observability, Explorability, s-completeness

Let $\mathcal{C}_{\mathcal{S}reach} \subset \mathcal{C}_{free}$ denote the set of all s-reachable configurations from the given initial configuration q_0 of the robot. In Figure 4 (bottom right), the white region containing the initial robot configuration, is $\mathcal{C}_{\mathcal{S}reach}$. This denotes

---

[6]Our definition of s-feasibility (and other notions) is based on the sensed free region and can only be checked after execution. A different (weaker) notion of s-feasibility may be defined based on sensed region (the entire field of view) without knowing the free space. Of course, in this case, s-feasible path would not imply feasible but rather that it can be attempted by the robot with its on-board sensors. Thanks to one of the anonymous referees for pointing this out.

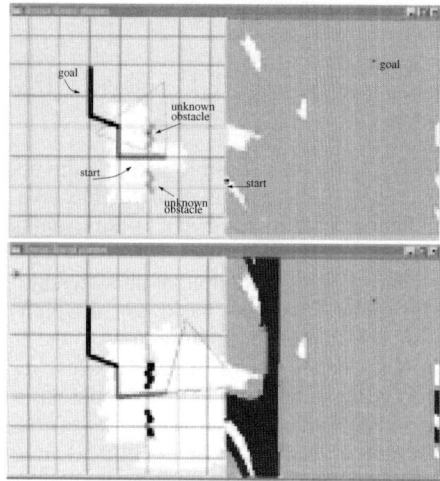

Figure 4: Same example as in Figure 1. The C-space is shown on the right. The top left figure shows the initial $\mathcal{P}^0_{free}$ (white region), and $\mathcal{P}^0_{unk}$ (light gray corresponds to unknown free, dark gray to unknown obstacles). The bottom row (left figure) shows the entire *observable* physical space – white (free) and black (obstacles). The gray region is not observable. Correspondingly, the C-space is shown on the (bottom) right. The entire gray region is unexplorable, the white (free) and black (obstacle) regions being *explorable* C-space.

the entire C-space where the robot can go given the initial $\mathcal{P}^{(0)}_{free}$ and the sensor constraints.

Observable Physical Space: Given an environment $\mathcal{P}$,
$$\mathcal{P}_{observ} = \left\{ \bigcup_{q \in \mathcal{C}_{Sreach}} (\mathcal{V}_S(q) \bigcup \mathcal{V}_{Sobs}(q)) \right\} \subset \mathcal{P} \text{ is called}$$
the observable subset of this environment. As the name indicates, this is the subset of the environment that the robot-sensor system is able to sense (or observe). In Figure 4 (bottom left), the white and black regions combined form $\mathcal{P}_{observ}$. It is composed of two parts: the observable free space, denoted by $\mathcal{P}_{free-observ}$ (first union term in the expression above) and observable obstacles, denoted by $\mathcal{P}_{obs-observ}$ (second union term). In Figure 4 (bottom left), the white region is $\mathcal{P}_{free-observ}$, the black region is $\mathcal{P}_{obs-observ}$, and the gray region is unobservable physical space.

Explorable Configuration:[7] A configuration q_e is explorable if $\mathcal{A}(q_e) \subset \mathcal{P}_{free-observ}$ OR $\mathcal{A}(q) \bigcap \mathcal{P}_{obs-observ} \neq \emptyset$. The explorable C-space, denoted by $\mathcal{C}_{expl}$, is the set of all explorable configurations. In Figure 4 (bottom right), any configuration q belonging to the white (free) or black (obstacle) regions is explorable. The gray region is unexplorable C-space. Note that the desired goal configuration lies in the

---

[7]Generally speaking, one could define two types of *explorability*: *weak explorability* and *strong explorability*. If the entire $\mathcal{A}(q)$ can be "seen" by the sensor, i.e., $\mathcal{A}(q) \subset \mathcal{P}_{observ}$, q is explorable in a strong sense. The weak explorable configuration and strong explorable configuration are different when $q \in \mathcal{C}_{obs}$. The former requires to determine the status of q only while the latter requires that the entire $\mathcal{A}(q)$ is observable. We use explorable in the weak sense.

unexplorable C-space, and is therefore, not s-reachable as mentioned earlier in the introduction.

In model-based approaches, there is the notion of completeness (let us call it m-completeness). An algorithm is m-complete if it can find a collision-free path to an arbitrary goal if there exists one (i.e., the goal is reachable), otherwise it reports failure if no such path exists. Note however, that in sensor-based planning, as we discussed earlier, even if there is a collision-free path, the arm may not be able to find it — part of the physical space may not be observable given the particular robot-sensor system. Therefore, for sensor-based motion planning, the planner can, in general, decide only if a goal is s-reachable. The only meaningful completeness for sensor-based motion planning, therefore is s-completeness, as defined below.

s-completeness: An s-complete algorithm for the start–goal problem should return an s-feasible path to the goal if such a path exists, for any given environment and any initial free region $\mathcal{P}^{(0)}_{free}$. Otherwise it should report that the goal is not s-reachable.

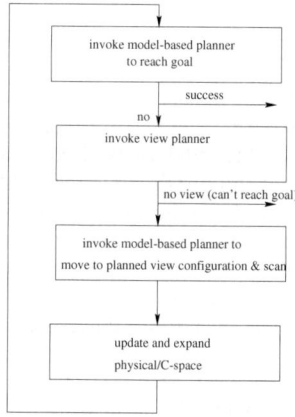

Figure 5: A general framework for sensor-based MP.

5 A General Framework for discrete Eye-sensor based MP

Figure 5 shows a "high-level" framework for eye-sensor based MP (See [3, 6, 5] for a specific sensor-based planning algorithm SBIC-PRM[8] based on this framework) assuming that the sensor is a discrete time sensor. This framework assumes that a "model-based" planner is available. This underlying model-based MP algorithm is used to plan paths within the known part of the environment to further sense the unknown part of the environment. We call this problem of determining the next sensing action, i.e., which region to scan and from where to scan, the view planning problem (See [4] for an information theoretical approach to solve this problem). Having taken a new scan, the robot "updates" its internal representation (of physical and configuration space) and re-invokes the model-based planner. This process is repeated until[9] the given goal is declared reachable by the

---

[8]Acronym for Sensor Based Incremental Construction of Probabilistic Road Map

[9]The termination condition would depend on whether it is a start-goal problem or exploration problem. This one is for start-goal problem.

model-based planner, or the goal is declared unreachable. Now we discuss the capabilities of a sensor-based planner developed within this framework.

Before more discussion of the completeness issue, we need to introduce the concept of *touchable boundary* (denoted by $\mathcal{E}$) in physical space. Touchable boundary is essentially that part of the boundary between unknown physical space $\mathcal{P}_{unk}^{(i)}$ and the cumulative region ($\subset \mathcal{P}_{free}^{(i)}$) occupiable by the robot at all configurations belonging to the closure of entire $\mathcal{C}_{free}^{(i)}(q_0)$ (the connected component of $\mathcal{C}_{free}^{(i)}$ containing q_0), i.e. $\mathcal{E}^{(i)} = (\bigcup_{q \in cl(\mathcal{C}_{free}^{(i)}(q_0))} \mathcal{A}(q)) \bigcap \mathcal{P}_{unk}^{(i)}$, where $cl()$ stands for the closure of an open set.[10]

Our concern is only the part of $\mathcal{E}$ which is visible from the sensor from some configuration in $\mathcal{C}_{free}^{(i)}(q_0)$. Let $\mathcal{E}_\mathcal{V}^{(i)} = \mathcal{E}^{(i)} \bigcap (\bigcup_{q \in \mathcal{C}_{free}^{(i)}(q_0)} \mathcal{V}_\mathcal{S}(q) \bigcup \mathcal{V}_{\mathcal{S}obs}(q))$. Note $\mathcal{V}_\mathcal{S}(q) \bigcup \mathcal{V}_{\mathcal{S}obs}(q)$ is the visible set of $\mathcal{P}$ from configuration q. $\mathcal{E}_\mathcal{V}$ is essentially the part of boundary between $\mathcal{P}_{unk}^{(i)}$ and $\mathcal{P}_{free}^{(i)}(q_0)$ that is touchable by a part of the robot if the robot were to move within the entire $cl(\mathcal{C}_{free}^{(i)}(q_0))$, and is visible from $\mathcal{C}_{free}^{(i)}(q_0)$.

We now define a complete view planner as follows: if there exists a point $x \in \mathcal{E}_\mathcal{V}$, then a complete view planner can find a configuration q_v, such that point x can be sensed from configuration q_v, i.e., $x \in \mathcal{V}_\mathcal{S}(q_v) \bigcup \mathcal{V}_{\mathcal{S}obs}(q_v)$. Otherwise, it reports $\mathcal{E}_\mathcal{V} = \emptyset$.

Assuming that the underlying model-based planner and the view planner are complete, would the overall sensor-based planner (as in Figure 5) be s-complete? Unfortunately, not quite as we discuss below.

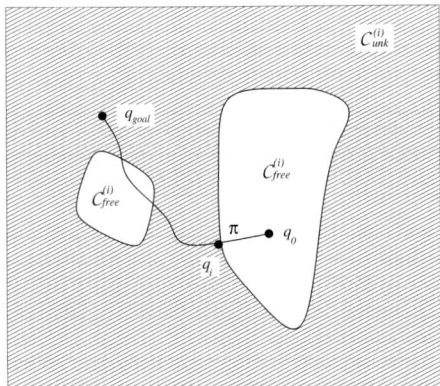

Figure 6: The status of the C-space when the algorithm stops without finding any s-feasible path, and there exists is an s-feasible path π. π intersects the boundary between $\mathcal{C}_{unk}^{(i)}$ and $\mathcal{C}_{free}^{(i)}(q_0)$ at q_i.

Lemma 1 *A sensor-based planner with a complete model-based planner and a complete view planner may end in one of the following three states: (1) declares goal is reached, or (2) declares goal is not s-reachable, and (3) loops infinitely sensing new portions of $\mathcal{E}_\mathcal{V}$.*

Proof:
Case (1) will occur if the robot has sensed enough physical space so that the goal configuration belongs to $\mathcal{C}_{free}^{(i)}(q_0)$. Once this happens, the underlying model-based planner, since it is complete will return a path to the goal configuration. Case (3) may occur since it may take the robot infinite number of scans to cover the entire $\mathcal{E}_\mathcal{V}$.

We now show case (2), i.e., the planner will never falsely declare that the goal is not s-reachable. The proof is by contradiction. Assume that there exists an s-feasible path π starting from q_0 to q_{goal} and the algorithm stops and the second stop condition is satisfied, i.e., no s-feasible path is found (see Figure 6). When the algorithm stops, known free C-space is $\mathcal{C}_{free}^{(i)}$, and known physical space is $\mathcal{P}_{free}^{(i)}$, and etc. Both $\mathcal{C}_{free}^{(i)}$ and $\mathcal{P}_{free}^{(i)}$ are open sets. Since π is an s-feasible path, π can not intersect $\mathcal{C}_{obs}$. π can not lie in $\mathcal{C}_{free}^{(i)}$ completely either (otherwise, underlying model-based planner would find this path). Therefore, π must intersect $\mathcal{C}_{unk}^{(i)}$. Let q_i be the first intersection of π and the boundary between $\mathcal{C}_{unk}^{(i)}$ and $\mathcal{C}_{free}^{(i)}(q_0)$. Since $q_i \in \mathcal{C}_{unk}^{(i)}$, part of $\mathcal{A}(q_i)$ should intersect $\mathcal{P}_{unk}^{(i)}$, i.e., $\mathcal{A}_{unk}(q_i) \neq \emptyset$ ($\mathcal{A}_{unk}(q_i)$ is the part of $\mathcal{A}(q_i)$ in unknown physical space). Since q_i is on the boundary of $\mathcal{C}_{unk}^{(i)}/\mathcal{C}_{free}^{(i)}(q_0)$, $q_i \in cl(\mathcal{C}_{free}^{(i)}(q_0))$, and then $\mathcal{A}_{unk}(q_i) \subset \mathcal{E}$. On the other hand, on this s-feasible path π, by definition, $\bigcup_{q \in \pi_{[q_0,q_i]}} \mathcal{A}(q) \subset \{\bigcup_{q \in \pi_{[q_0,q_i]}} \mathcal{V}_\mathcal{S}(q) \bigcup \mathcal{P}_{free}^{(0)}\}$. This implies that the sensor can sense entire $\mathcal{A}_{unk}(q_i)$ on the path segment of π from q_0 to q_i (before reaching q_i). Specifically, $\mathcal{A}_{unk}(q_i) \subset \{\bigcup_{q \in \pi_{[q_0,q_i]}} \mathcal{V}_\mathcal{S}(q) \bigcup \mathcal{P}_{free}^{(0)}\}$. Because the way q_i is chosen, (it is the first intersection point on π with $\mathcal{C}_{unk}$,) $\pi_{[q_0,q_i]} \subset \mathcal{C}_{free}^{(i)}(q_0)$. Hence, $\mathcal{A}_{unk}(q_i)$ is visible from $\mathcal{C}_{free}^{(i)}(q_0)$. Therefore, $\mathcal{A}_{unk}(q_i) \subset \mathcal{E}_\mathcal{V}$. Therefore, there is still a subset of $\mathcal{E}_\mathcal{V}$ which is scannable when the algorithm stops. This contradicts the stop condition. Therefore, the algorithm can not stop without finding path π.

But for case (3), the planner would have been s-complete. Indeed, the inability of the planner to terminate is due to the fact that *infinite* number of scans may be needed by the discrete view planner. We now define a stronger notion of s-reachability (and other corresponding notions), intrinsically appropriate for a discrete time sensor.

6 Discrete Sensing: Discrete s-feasible

Assuming that the sensor is a discrete time sensor, the notions presented so far are easily extended. A path $\pi_{[q_0,q_1]}$ is discrete s-feasible if there exists a finite number m of configurations q_{v_i} along the path (ordered such that $q_{v_{i-1}}$ occurs before q_{v_i}) such that the volume swept by the robot in reaching q_{v_i} is a subset of the cumulative volume sensed by the robot at all the previous sensing configurations q_{v_j},

---
[10]This is for holonomic robots. For more general case including nonholonomic robots, $\mathcal{C}_{free}^{(i)}(q_0)$ could be replaced by $\mathcal{C}_{reach}^{(i)}$, the reachable portion of $\mathcal{C}_{free}^{(i)}$.

$j = 1, \ldots, i-1$. Formally, if $\forall q_{v_i}$, $i = 1, \ldots, m$,

$$\bigcup_{\pi_{[q_0, q_{v_i}]}} \mathcal{A}(q) \subset \left(\bigcup_{j=1, i-1} \mathcal{V}_\mathcal{S}(q_{v_j}) \right) \bigcup \mathcal{P}_{free}^{(0)} \qquad (2)$$

then path $\pi_{[q_0, q_1]}$ is a discrete s-feasible path. If there exists no such finite set of view configurations, the path is not discrete s-feasible.

A configuration q is discrete s-reachable from a given initial configuration q_0 if there exists a discrete s-feasible path to q from q_0. We can similarly extend notions of observability, explorability and s-completeness to the discrete sensing case.

Suppose now that there are a finite number n of discrete (and ordered) s-reachable configurations[11], denoted by $\mathcal{Q}_\mathcal{V} = \{q_{v_i}, i = 1, \ldots, n\}$ that can see the entire observable workspace. We now define a complete view planner (in the discrete case) as follows: if there exists a point $x \in \mathcal{E}_\mathcal{V}$, then a complete view planner will find the next configuration q_v from a set $\mathcal{Q}_\mathcal{V}$ such that point x can be sensed from configuration q_v, i.e., $x \in \mathcal{V}_\mathcal{S}(q_v) \bigcup \mathcal{V}_{\mathcal{S}obs}(q_v)$. Otherwise, it reports $\mathcal{E}_\mathcal{V} = \emptyset$. Indeed with this stronger definition of complete view planner, the sensor-based planner in Lemma 1 will always terminate with case (1) or (2), and hence would be s-complete. The condition on view planner, however, is quite strong and requires that the view planner be able to find a finite number of s-reachable configurations from which the entire observable physical space can be seen.

7 Open Issues/Problems and Discussion

The existence of configurations which are not s-reachable but may be m-reachable (reachable if the environment were known *a priori*) is due to a combination of (i) the SVI condition, i.e., the robot can only move to and scan from those regions already discovered to be free by the sensor; (ii) sensor constraints such as occlusion, finite field of view, etc; and (iii) non-trivial geometry/kinematics of the robot as discussed in the introduction.

Our conjecture is that if the sensor is able to sense an open neighborhood (in C-space) around q, it would be necessary and sufficient to guarantee that the set of s-reachable configurations is the same as the set of m-reachable configurations. For point robot, [11] shows that sensor-based MP is solvable as long as the robot possesses the ability to "explore" the entire environment. We are currently investigating if similar results could be obtained for robots with non-trivial geometry/kinematics. A related issue is to determine conditions (on robot-sensor system) under which a restricted class of environments would be completely observable, or the corresponding C-space would be completely explorable.

There are interesting design issues here as well. For instance, given a sensor $\mathcal{S}$, how should it be "mounted" on the robot $\mathcal{A}$, so that navigability in a given class of environments is "maximized". The weaker notion of s-feasibility suggested in footnote 6 may be more useful for such purposes.

---
[11] This set may not be unique.

References

[1] Juan-Manuel Ahuactzin and Kamal Gupta. A roadmap based approach to inverse kinematics of redundant robots: Theory and experiments. *IEEE Transactions on Robotics and Automation*, 15(4):653–669, 1999.

[2] E. Kruse, R. Gutsche, and F. Wahl. Effective, iterative, sensor based 3-d map building using rating functions in configuration space. In *Proceedings of IEEE International Conference on Robotics and Automation*, pages 1067 – 1072, 1996.

[3] Y. Yu and K. Gupta. On sensor-based roadmap: A framework for motion planning for a manipulator arm in unknown environments. In *Proceedings of IEEE/RSJ International Conference on Intelligent Robot and System*, pages 1919 – 1924, 1998.

[4] Y. Yu and K. Gupta. An information theoretical approach to view planning with kinematic and geometric constarints. In the same Proceedings. A related version also appeared in *Proceedings of 31st International Symposium on Robotics*, pages 306– 311, 2000.

[5] J. Ahuactzin and A. Portilla. A basic algorithm and data structure for sensor-based path planning in unknown environment. In *Proceedings of IROS*, 2000.

[6] Y. Yu and K. Gupta. Sensor-based roadmaps probabilistic roadmaps: Experiments with an eye-in-hand system. *Advanced Robotics*, 14(6), pages 515-537, 2000. A version also appeared In *Proceedings of IEEE/RSJ International Conference on Intelligent Robot and System*, pages 1707–1714, 1999.

[7] Kamal K. Gupta and Angel del Pobil, editors. *Practical Motion Planning in Robotics: Current Approaches and Future Directions*. John Wiley, 1998.

[8] N.S.V. Rao, S. Kareti, W. Shi and S. S. Iyengar. Robot navigation in unknown terrains: introductory survey of non-heuristic algorithms. *Technical Report*, Oak Ridge National Lab. ORNL/TM-12410:1-58, July 1993.

[9] Subir Ghosh and Joel Burdick. Exploring an unknown polygonal environment with a sensor based strategy. *Technical Report*, Tata Institute of Fundamental Research, Bombay, India.

[10] H. Choset and J.W Burdick. Sensor based planning for a planar rod robot. In *Proceedings of 1996 IEEE International Conference on Robotics and Automation*, pages 3584–3591, 1996.

[11] K. Kutulakos, V. Lumelsky, and C. Dyer. Vision-guided exploration: A step toward general motion planning in three dimensions. In *Proceedings of IEEE International Conference on Robotics and Automation*, pages 289–296, 1993.

[12] I. Kamon and E. Rimon. Range-sensor based navigation in three dimensions. In *Proceedings of IEEE International Conference on Robotics and Automation*, pages 163–169, 1999.

[13] J. C. Latombe. *Robot Motion Planning*. Kluwer Academic Publications, 1991.

[14] V.J. Lumelsky and E Cheung. Real-time collision avoidance in teleoperated whole-sensitive robot arm manipulators. *IEEE Transactions on Systems, Man and Cybernetics*, 23(1):194–203, Jan.-Feb 1993.

[15] P. Renton, M. Greenspan, H. Elmaraghy, and H. Zghal. Plan-n-scan: A robotic system for collision free autonomous exploration and workspace mapping. *Journal of Intelligent and Robotic System*, 24:207 – 234, 1999.

[16] E. Rimon and J.F Canny. Construction of c-space roadmaps using local sensory data – what should the sensors look for? In *Proceedings of IROS*, pages 117–124, 1994.

Optimizing Schedules for Prioritized Path Planning of Multi-Robot Systems

Maren Bennewitz[†]   Wolfram Burgard[†]   Sebastian Thrun[‡]

[†]Department of Computer Science, University of Freiburg, 79110 Freiburg, Germany
[‡]School of Computer Science, Carnegie Mellon University, Pittsburgh PA, USA

Abstract

The coordination of the motions of the robots is one of the fundamental problems for multi-robot systems. A popular approach to avoid planning in the high-dimensional composite configuration space are prioritized and decoupled techniques. While these methods are very efficient, they have two major drawbacks. First, they are incomplete, i.e. they sometimes fail to find a solution even if one exists, and second, the resulting solutions are often not optimal. They furthermore leave open how to assign the priorities to the individual robots. In this paper we present a method for optimizing priority schemes for such prioritized and decoupled planning techniques. Our approach performs a randomized search with hill-climbing to find solutions and to minimize the overall path lengths. The technique has been implemented and tested on real robots and in extensive simulation runs. The experimental results demonstrate that our method is able to seriously reduce the number of failures and to significantly reduce the overall path length for different prioritized and decoupled path planning techniques and even for large teams of robots.

1 Introduction

Path planning is one of the fundamental problems in mobile robotics. As mentioned by Latombe [9], the capability of effectively planning its motions is "eminently necessary since, by definition, a robot accomplishes tasks by moving in the real world."

In this paper we consider the problem of motion planning for multiple mobile robots. This problem is significantly harder than the path planning problem for single robot systems, since the size of the joint state space of the robots grows exponentially in the number of robots. Therefore, the solutions known for single robot systems cannot directly be transferred to multi-robot systems.

The existing methods for solving the problem of motion planning for multiple robots can be divided into two categories [9]. In the *centralized* approach the configuration spaces of the individual robots are combined into one composite configuration space which is then searched for a path for the whole composite system In contrast, the *decoupled* approach first computes separate paths for the individual robots and then resolves possible conflicts of the generated paths.

While centralized approaches (at least theoretically) are able to find the optimal solution to any planning problem for which a solution exists, their time complexity is exponential in the dimension of the composite configuration space. In practice one is therefore forced to use heuristics for the exploration of the huge joint state space.

Many methods use potential field techniques [2, 3, 17] to guide the search. These techniques apply different approaches to deal with the problem of local minima in the potential function. Other methods restrict the motions of the robots to reduce the size of the search space. For example, [16, 8, 10] restrict the trajectories of the robots to lie on independent roadmaps. The coordination is achieved by searching the Cartesian product of the separate roadmaps.

Decoupled planners determine the paths of the individual robots independently and then employ different strategies to resolve possible conflicts. According to that, decoupled techniques are incomplete, i.e. they may fail to find a solution even if there is one. A popular decoupled approach is planning in the configuration time-space [6] which can be constructed for each robot given the positions and orientations of all other robots at every point in time. Techniques of this type assign priorities to the individual robots and compute the paths of the robots based on the order implied by these priorities. The method presented in [18] uses a fixed order and applies potential field techniques in the configuration time-space to avoid collisions. The approach described in [7] also uses a single priority scheme and chooses random detours for the robots with lower priority.

Another approach to decoupled planning is the path coordination method which was first introduced in [14]. The key idea of this technique is to keep the robots on their individual paths and let the robots stop, move forward, or even move backward on their trajectories in order to avoid collisions (see also [4]). To reduce the complexity in the case of huge teams of robots [12] recently presented a technique to separate the overall coordination problem into sub-

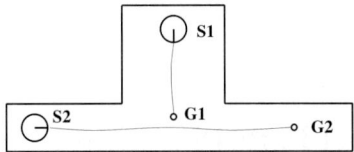

Figure 1: Situation in which no solution can be found if robot 1 has higher priority than robot 2

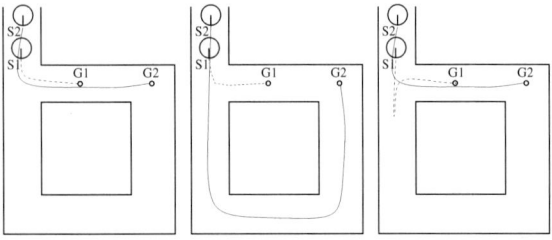

Figure 2: Independently planned optimal paths for two robots (left), suboptimal solution if robot 1 has higher priority (center), and solution resulting if the path for robot 2 is planned first (right).

problems. This approach, however, assumes that the overall problem can be divided into very small sub-problems, a serious assumption which, as our experiments described below demonstrate, is often not justified. In general, therefore, a prioritized variant has to be applied.

The methods described above leave open how to assign the priorities to the individual robots. In the past, different techniques for selecting priorities have been used. [5] applied a heuristic which assigns higher priority to robots which can move on a straight line from the starting point to their target location. In [1] all possible assignments are considered. Due to its complexity this approach has only been applied to groups of up to three robots.

For decoupled and prioritized methods the order in which the paths are planned has a serious influence on whether at all a solution can be found and if so, how long the resulting paths are. Figure 1 shows a situation in which no solution can be found if robot 1 has a higher priority than robot 2. Since the path of robot 1 is planned without considering robot 2, it arrives at its target location marked G1 before robot 2 has passed the t-junction. Thus, it blocks the way of robot 2 which can no longer reach its designated target point G2. However, if we change the priorities and plan the trajectory of robot 2 before that of robot 1, then robot 1 considers the trajectory of robot 2 during path planning and this way waits until robot 2 has passed by.

Another example is shown in Figure 2 (left). If we start with robot 1 then every planner has to choose a large detour for robot 2 (see Figure 2 (center)), because robot 1 blocks the corridor. However, if the path of robot 2 is planned first, then we can obtain a much more efficient solution (see Figure 2 (right)).

These two examples illustrate that the priority scheme has a serious influence on whether a solution can be found and on how long the resulting paths are. Unfortunately the problem of finding the optimal schedule is NP-hard for most of the decoupled approaches. For example, the Job-Shop Scheduling Problem with the goal to minimize maximum completion time with unit processing time for each job [11] can be regarded as a special instance of the path coordination method.

In this paper we present a randomized and hill-climbing technique which starts with an initial priority scheme and optimizes this by swapping two randomly chosen robots. This way it can significantly increase the number of problems for which a solution can be found. Additionally it is able to reduce the overall path length. Furthermore, our approach has any-time characteristic which means that it can return the best solution found so far at any point in time and whenever it is interrupted. Our technique has been implemented and tested on real robots. In extensive experiments it has been proven to be very effective even for large teams of robots and using two different decoupled path planning techniques.

The paper is organized as follows. The following section describes the prioritized and decoupled path planning techniques we apply our algorithm presented in Section 3 to. Section 4 contains experimental results illustrating the capabilities of our approach.

2 Prioritized A^*-based Path Planning and Path Coordination

The basic algorithm to compute optimal paths for single robots applied throughout this paper is the well-known A^* search procedure. The next section briefly describes the variant we are using. To represent the environment of the robots we apply occupancy grids [13] which separate the environment into a grid of equally spaced cells and store in each cell $\langle x,y \rangle$ the probability $P(occ_{x,y})$ that it is occupied. In the remainder of this section we then present the key ideas of decoupled prioritized path planning and discuss how the A^* procedure can be utilized to plan the motions of teams of robots by this approach.

2.1 A^*-based Path Planning

The A^* procedure simultaneously takes into account the accumulated cost of reaching a certain location $\langle x,y \rangle$ from the starting position as well as the estimated cost of reaching the target location $\langle x^*,y^* \rangle$ from $\langle x,y \rangle$. In our case, the cost for traversing a cell $\langle x,y \rangle$ is proportional to its occupancy probability $P(occ_{x,y})$. Furthermore, the estimated cost for reaching the target location is approximated by $c \cdot ||\langle x,y \rangle - \langle x^*, y^* \rangle||$ where c is chosen as the minimum occupancy probability $P(occ_{x,y})$ in the map and $||\langle x,y \rangle - \langle x^*, y^* \rangle||$ is the straight-line distance between $\langle x,y \rangle$ and $\langle x^*, y^* \rangle$. Since this heuristic is admissible, A^* determines the cost-optimal path from the starting position to the target location.

2.2 Decoupled Path Planning for Teams of Robots

In this paper we consider decoupled and prioritized path planning approaches which plan the paths in the configuration time-space. Such approaches proceed as follows. First, one computes for each robot the path without considering the paths of the other robots. Then one checks for possible conflicts in the trajectories of the robots (we regard it as a conflict between two robots if their distance is less than δ where $\delta = 1.2m$ in our current system). Conflicts between robots are resolved by introducing a priority scheme. A priority scheme determines the order in which the paths for the robots are planned. The path of a robot is planned in its configuration time-space computed based on the map of the environment and the paths of the robots with higher priority.

Our system applies the A^* procedure to compute the cost-optimal paths for the individual robots, in the remainder denoted as the independently planned optimal paths for the individual robots. We also apply A^* search to plan the motions of the robots in the configuration time-space. In this case the cost of traversing a location $\langle x,y \rangle$ at time t is determined by the occupancy probability $P(occ_{x,y})$ plus the probability that one of the other robots with higher priority covers $\langle x,y \rangle$ at that time.

In this paper we consider two different strategies: A^*-based planning in the configuration time-space as well as a restricted version of this approach denoted as the path coordination technique [12]. It differs from the general A^*-search in that it only explores a subset of the configuration time-space given by those states which lie on the initially optimal paths for the individual robots. The path coordination technique thus forces the robots to stay on their initial trajectories. The overall complexity of both approaches is $O(n \cdot m \cdot \log(m))$ where n is the number of robots and m is the maximum number of states expanded by A^* during planning in the configuration time-space (i.e. the maximum length of the OPEN-list).

Due to the restriction during the search, the path coordination method is more efficient than the general A^* search. Its major disadvantage, however, lies in the fact that it fails more often. A typical example is shown in Figure 2. Whereas the path coordination method fails independently of the planning order, the general A^* procedure is able to find a solution in both cases.

3 Optimizing Priority Schemes

As already mentioned above, prioritized and decoupled approaches to multi-robot path planning are incomplete and sub-optimal. However, as the examples given in Figures 1 and 2 illustrate, the order in which the paths are planned has a significant influence on whether a solution can be found and on how long the resulting paths are. This raises the question of how to find a priority scheme for which the decoupled approach does not fail and how to find the order of the robots leading to the shortest paths.

Recently, randomized search techniques have been used with great success to solve constraint satisfaction problems or to solve satisfiability problems [15]. Our algorithm presented here is a variant which performs a randomized and hill-climbing search in order to optimize the planning order for decoupled and prioritized path planning of teams of mobile robots. It starts with an arbitrary initial priority scheme Π and randomly exchanges the priorities of two robots in this scheme. If the new order Π' results in a solution with shorter paths than the best one found so far, we continue with this new order. Since hill-climbing approaches like this frequently get stuck in local minima, we perform random restarts with different initial orders of the robots. The complete algorithm is listed in Table 1.

Table 1: The algorithm to optimize priority schemes.

```
FOR tries := 1 TO maxTries BEGIN
    select random order Π
    if (tries = 1)
        Π* := Π
    FOR flips := 1 TO maxFlips BEGIN
        choose random i, j with i<j
        Π' := swap(i, j, Π)
        if moveCosts(Π') < moveCosts(Π)
            Π := Π'
    END FOR
    if moveCosts(Π) < moveCosts(Π*)
        Π* := Π
END FOR
return Π*
```

Please note that an additional advantage of our randomized optimization approach lies in its any-time character. The procedure can be terminated at any point in time and return the currently best priority order whenever it is interrupted.

Figure 4 shows a typical application example carried out with our robots Albert and Ludwig shown in Figure 3 in our office environment. In this example we used the general $A*$ procedure in the configuration time-space. While Ludwig starts at the left end of the corridor of our lab and has to move to right end, Albert has to traverse the corridor in the opposite direction. If the path of Ludwig is planned before that of Albert, the system fails because Albert cannot reach

Figure 3: The mobile robots Albert (left) and Ludwig (right).

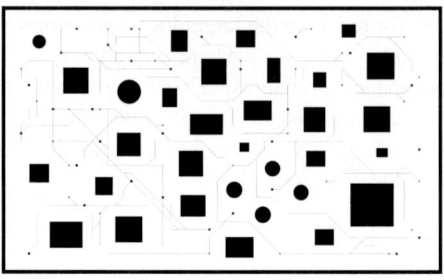

Figure 6: Paths resulting after priority optimization.

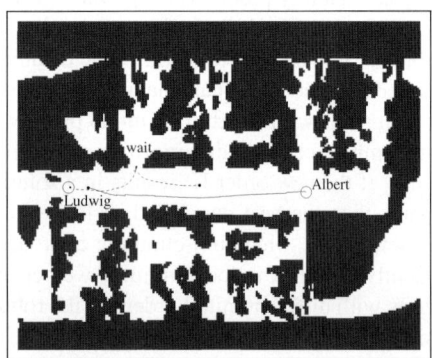

Figure 4: Real world application of A^*-based planning in the configuration time-space.

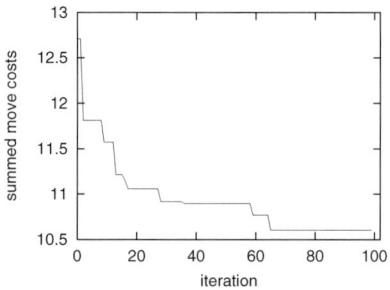

Figure 7: Summed move costs plotted over time.

its target point if Ludwig stays on its optimal trajectory. If we alter the planning order, our system is able to find a solution. In this case, Ludwig is moved into a doorway in order to let Albert pass by. Please note, that no solution can be found in this situation if the path coordination technique would be used. The resulting trajectories are shown in Figure 4 including the position where Ludwig waited to let Albert pass by.

Figure 5 shows a simulated situation with 30 robots. By applying our algorithm using the general A^* procedure we obtain the paths depicted in Figure 6. In this and all

experiments described below we used a value of 10 for `maxFlips` and `maxTries`. Figure 7 plots the evolution of the summed move costs of the best solution found so far during these 100 trials. Obviously, compared to the initial solution shown in Figure 5 with summed move costs of 12.7, the final solution illustrated in Figure 6 has move costs of 10.9 which corresponds to a reduction of 15%. Please note, that there is a huge number of conflicts between the robots in this example. As a result, the whole trajectory graph is a single connected component. Accordingly, the decomposition technique presented in [12] cannot be applied.

4 Experimental Results

The algorithm described above has been tested thoroughly in extensive simulation runs. To evaluate the general applicability, we applied our method to the two decoupled and

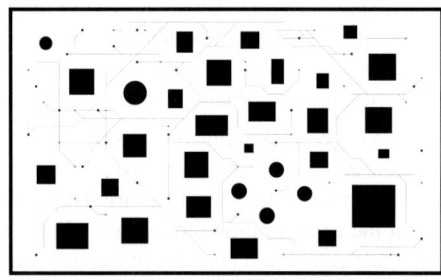

Figure 5: Independently planned paths for 30 robots.

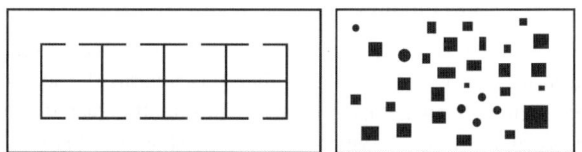

Figure 8: Environments used for the simulation runs.

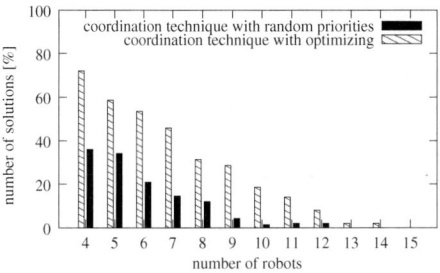
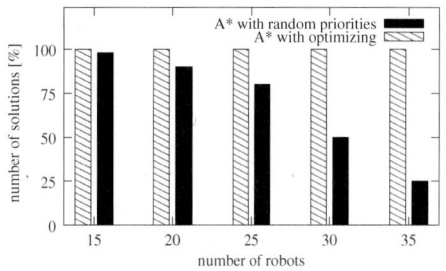

Figure 9: Reducing the number of failures for the path coordination technique by optimizing priority schemes.

Figure 10: Reducing the number of failures for A^*-based planning in the configuration time-space by optimizing priority schemes.

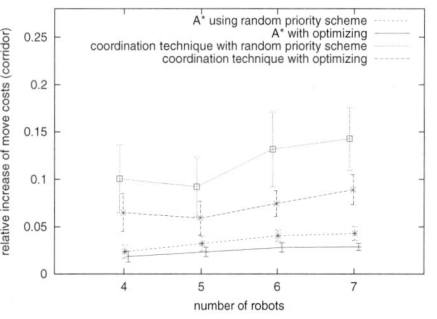

prioritized path planning techniques described above. The current implementation is highly efficient. It requires less than 0.2 seconds to plan a collision-free path for one robot in all environments described below. Throughout the experiments we used the two different environments depicted in Figure 8. Whereas the map on the left of Figure 8 is a typical corridor environment, the map on the right is corresponds to an unstructured environment.

4.1 Reducing the Number of Failures

The first set of experiments is designed to illustrate that the overall number of failures can be reduced significantly using our optimization technique. Figure 9 summarizes the results we obtained using the path coordination technique for different numbers of robots in over 600 runs in the unstructured environment. In each experiment we randomly chose the starting and target locations of the robots and determined whether the coordination technique is able to find a solution given a randomly chosen initial priority scheme. Then we optimized this scheme using our algorithm described above. For example, if 4 robots are used then the path coordination technique fails in more than 60% of the cases. Our approach, in contrast, yields a solution in more than 70% of all situations. For more than 15 robots, however, it is nearly impossible to find configurations for which the path coordination method can find a solution. One reason is that starting or goal locations often lie too close to the trajectories of other robots so that they cannot pass by any more.

Additionally, we applied our approach to 100 randomly chosen situations to which we used the A^*-based planning in the configuration time-space. The result of this experiment is shown in Figure 10. As this figure shows, A^* has a significantly higher success rate even for larger numbers of robots. However, even using this approach and a single priority scheme, the number of solutions decreases monotonously. Our optimization technique, in contrast, was always able to find a solution.

Figure 11: Relative increase of the move costs compared to the sum of the optimal move costs for the independently planned paths in the corridor environment.

4.2 Minimizing the Overall Path Lengths

The second set of experiments is designed to demonstrate that our optimization technique is able to significantly reduce the overall path length. For different numbers of robots we performed a series of experiments in both environments shown in Figure 8. Again we randomly chose starting and target locations and then computed the paths for the robots using the two decoupled and prioritized path planning techniques with and without our optimization technique. We then measured the average path length and compared it to the average length of the optimal paths for the individual robots[1].

Figure 11 shows the relative increase of the move costs for four to seven robots in the corridor environment (in contrast to the other experiments, the starting and target locations were chosen from a given set of hand-selected positions in the map). As can be seen, our approach reduces the overall path length for the path coordination technique and for A^*-based planning in the configuration time-space. Additionally, it illustrates that the latter approach results in more

---
[1] Please note that throughout this paper we take the optimal paths for single robot planning problems as reference, since computing the optimal solution for the whole problem is not tractable in practice at least for larger numbers of robots.

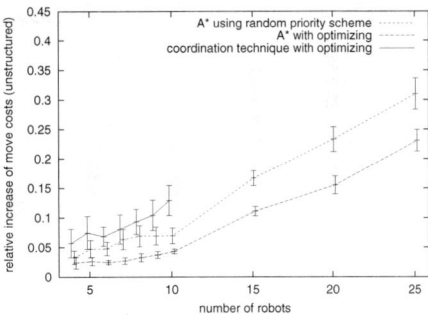

Figure 12: Increased move costs compared to the sum of the optimal move costs for the independently planned paths in the unstructured environment.

efficient paths than the coordination technique.

We performed similar experiments for the unstructured environment. Figure 12 summarizes the results we obtained for over 300 runs. It shows the relative increase of the move costs for different numbers of robots. Since the path coordination technique using the initially chosen priority scheme failed in most of the cases, we omit the corresponding results here. As the figure shows, our optimization technique applied to A^*-based planning in the configuration time-space yields the best results.

5 Conclusions

In this paper we presented an approach to optimize the priorities for decoupled and prioritized path planning methods for groups of mobile robots. Our approach is a randomized method which repeatedly reorders the robots to find a sequence for which a plan can be computed and to minimize the overall path lengths. It is an any-time algorithm since it can be stopped at any point in time and can always return its currently best estimate. The approach has been implemented and tested on real robots and in extensive simulation runs for two different decoupled path planning techniques and for large numbers of robots. The experiments demonstrate that our technique significantly decreases the number of failures in which no solution is found for a given planning problem. Additionally, its application leads to a significant reduction of the overall path length.

References

[1] K. Azarm and G. Schmidt. A decentralized approach for the conflict-free motion of multiple mobile robots. In *Proc. of the IEEE/RSJ International Conference on Intelligent Robots and Systems (IROS)*, pages 1667–1674, 1996.

[2] J. Barraquand, B. Langois, and J. C. Latombe. Numerical potential field techniques for robot path planning. *IEEE Transactions on Robotics and Automation, Man and Cybernetics*, 22(2):224–241, 1992.

[3] J. Barraquand and J. C. Latombe. A monte-carlo algorithm for path planning with many degrees of freedom. In *Proc. of the IEEE International Conference on Robotics & Automation (ICRA)*, 1990.

[4] Z. Bien and J. Lee. A minimum-time trajectory planning method for two robots. *IEEE Transactions on Robotics and Automation*, 8(3):414–418, 1992.

[5] S. J. Buckley. Fast motion planning for multiple moving robots. In *Proc. of the IEEE International Conference on Robotics & Automation (ICRA)*, 1989.

[6] M. Erdmann and T. Lozano-Perez. On multiple moving objects. *Algorithmica*, 2:477–521, 1987.

[7] C. Ferrari, E. Pagello, J. Ota, and T. Arai. Multirobot motion coordination in space and time. *Robotics and Autonomous Systems*, 25:219–229, 1998.

[8] L. Kavraki, P. Svestka, J. C. Latombe, and M. Overmars. Probabilistic road maps for path planning in high-dimensional configuration spaces. *IEEE Transactions on Robotics and Automation*, pages 566–580, 1996.

[9] J.C. Latombe. *Robot Motion Planning*. Kluwer Academic Publishers, Boston, MA, 1991. ISBN 0-7923-9206-X.

[10] S. M. LaValle and S. A. Hutchinson. Optimal motion planning for multiple robots having independent goals. In *Proc. of the IEEE International Conference on Robotics & Automation (ICRA)*, 1996.

[11] E. L. Lawler, J. K. Lenstra, A. H. G. Rinnooy Kan, and D. B. Shmoys. Sequencing and scheduling: Algorithms and complexity. Technical report, Centre for Mathematics and Computer Science, 1989.

[12] S. Leroy, J. P. Laumond, and T. Simeon. Multiple path coordination for mobile robots: A geometric algorithm. In *Proc. of the International Joint Conference on Artificial Intelligence (IJCAI)*, 1999.

[13] H.P. Moravec and A.E. Elfes. High resolution maps from wide angle sonar. In *Proc. IEEE Int. Conf. Robotics and Automation*, pages 116–121, 1985.

[14] P. A. O'Donnell and T. Lozano-Perez. Deadlock-free and collision-free coordination of two robot manipulators. In *Proc. of the IEEE International Conference on Robotics & Automation (ICRA)*, 1989.

[15] B. Selman, H. Levesque, and D. Mitchell. A new method for solving hard instances of satisfiability. In *Proc. of the National Conference on Artificial Intelligence (AAAI)*, 1992.

[16] P. Sveska and M. Overmars. Coordinated motion planning for multiple car-like robots using probabilistic roadmaps. In *Proc. of the IEEE International Conference on Robotics & Automation (ICRA)*, 1995.

[17] P. Tournassoud. A strategy for obstacle avoidance and its application to multi-robot systems. In *Proc. of the IEEE International Conference on Robotics & Automation (ICRA)*, pages 1224–1229, 1986.

[18] C. Warren. Multiple robot path coordination using artificial potential fields. In *Proc. of the IEEE International Conference on Robotics & Automation (ICRA)*, pages 500–505, 1990.

Autonomous Characterization of Unknown Environments

Liam Pedersen*
The Robotics Institute
Carnegie Mellon University
Pittsburgh PA 15213, USA

Abstract

Key to the autonomous exploration of an unknown area, by a scientific robotic rover is the ability of the vehicle to autonomously recognize objects of interest and generalize about the region. This paper presents a Bayesian framework under which a mobile robot can learn how different classes of objects are distributed over a geographical region, using imperfect observations and non-random sampling. This yields dramatic improvements in classification accuracy by exploiting the interdependencies between objects in an area and allows the robot to autonomously characterize the region. This is demonstrated with data from Carnegie Mellon University's Nomad robot in Antarctica, where it traversed the ice sheet, classifying rocks in its path.

Introduction

For many tasks involving the robotic searching and exploration of an area to find and identify objects, it is necessary to characterize the operating environment. To classify objects using onboard sensors, the likely candidates and potential false targets must be known in advance. In addition, to optimally classify objects their prior probabilities of being encountered must be known.

Unfortunately, it is difficult to know *a priori* the relative chances of finding different objects in an unexplored area. Consider the problem of classifying rocks from a robot for the purpose of geological exploration. There are many possible rock types, some hard to distinguish from each other. A geological map, if available, only indicates the most common rock type over a very large area. It does not indicate all rock types or their relative probabilities, and ignores small-scale variations. The latter is very important, as the kinds of rocks present can change significantly over a short distance, such as when crossing geological strata.

The standard approach to classifying objects is to consider each one independently, and classify it based on observations. However, objects in an area may be correlated, exploiting this can significantly increase classification accuracy. It is common for objects of the same type to be clustered together.

This paper will show a Bayesian approach to using the dependencies between objects distributed over an area by learning the statistical prior probabilities for different objects as a function of position, and specifically from the perspective of exploring an area with a mobile robot. Classification is thus improved not only by exploiting spatial dependencies, but also through improved knowledge of the priors. Furthermore, a map of prior probabilities over a geographic area is itself a useful summary with which to characterize the region and recognize gross properties.

This work is applicable to a variety of tasks involving the classification of objects distributed across an area, such as geological exploration, landmine removal, or soil profiling for agriculture. However, this analysis focuses on the problem of identifying the abundances of different rock types in an area of the Antarctic ice sheets using a scientific robotic vehicle, Nomad (Figure 1), built at Carnegie Mellon University to look for meteorites in Antarctica [1].

Figure 1 *Nomad robot in Antarctica, investigating rocks with a spectroscopic sensor to classify them.*

Autonomously identifying rocks with a robot is a challenging task with a high error rate [2]. It is highly relevant to the next generation of planetary rovers for exploring Mars, intelligently selecting samples for return to Earth, and incorporates the issues that arise when classifying objects from a mobile robot.

Robotic learning of the environment

Using a robotic vehicle to explore an area introduces unique issues.

- Sampling is not random. Rocks are examined by the robot as it traverses a path through the terrain of interest, leaving areas unexplored (Figure 2). The uncertainty about what is found in these areas must be noted, and

---
* Email: pedersen+@ri.cmu.edu

used to constrain subsequent changes in beliefs about the area as new data is added.

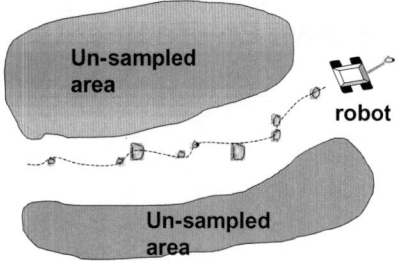

Figure 2 *Modus operandi of a robotic explorer. Only selected samples are visited in a large area, and they are not randomly distributed, but lie along the robots path.*

- Relatively few rock samples are examined. The Nomad robot (Figure 1) obtained measurements of no more than 50 rocks in 2 days. To make matters worse, there are many possible rock classes [3]. The rock probabilities must be therefore be initially coarsely defined and subsequently improved if and when data becomes available.

- Rock samples cannot usually be identified with complete certainty. Rather, when sensor data is obtained from a rock sample, only the likelihoods of different rock classes generating that data is known.

- The probabilities of different rock types are conditioned on geographical position, a continuous 2D (or 3D) quantity.

While machine learning and statistical estimation are mature fields, little prior work directly addresses the problem of characterizing a geographical area for the purposes of classification. The evidence grids of [4] are used to model the likelihoods of obstacles in an area. However, they fail to account for statistical dependencies between objects and require that space be discretized into a grid.

[5] and [6] survey strategies for either autonomously searching an area, given a prior description of the geographical distribution of targets, or exploiting the knowledge that targets tend to cluster together. They do not address how this information is obtained in the first place.

Representing the rock priors over an area

Consider the parameter $\bar{\theta}_x = [\theta_{x1},...,\theta_{xM}]$, representing the relative proportions of each rock type present at a geographic location x, and the random variable $\mathbf{R_x}$ the rock type (class labels 1, 2... M) of a rock sample found at there. Therefore

$$\theta_{x1},...,\theta_{xM} \in [0,1], \sum_{i=1}^{M} \theta_{xi} = 1 \quad (1)$$

$$P(\mathbf{R_x} = \mathbf{k} \mid \bar{\theta}_x) = P(\mathbf{R_x} = \mathbf{k} \mid \theta_{xk}) = \theta_{xk} \quad (2)$$

$$P(\mathbf{R_x} = \mathbf{k}) = E\{\theta_{xk}\} \quad (3)$$

$\bar{\theta}_x$ is itself a random variable and depends on position. Furthermore, knowing its distribution allows the determination of the rock type priors at x (3). Therefore, the problem of learning these priors is solved by learning the distribution of $\bar{\theta}_y$ at all positions y, given a sequence of robot sensor observations $\{\mathbf{O}_{x_i} | i=1,2...N\}$ made at locations $\{x_i \mid i = 1,2...N\}$. To do this it is necessary to model the statistical relationships between $\bar{\theta}_y$, the rock types of each rock sample, and the associated observations. Consequently we will show how to compute $p(\bar{\theta}_y | \mathbf{O}_{x_i})$, the posterior representation of the rock priors over the area, and how to use it to improve classification.

Pseudo Bayes network generative model

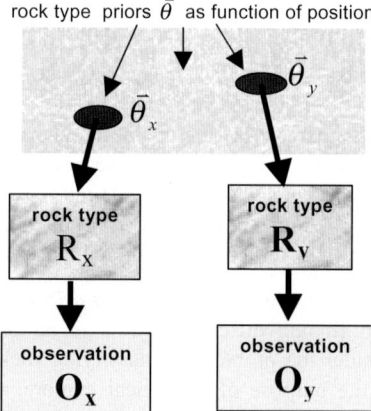

Figure 3 *Pseudo Bayes network generative model of the statistical relationships between position (x,y), relative proportions of different rock types at those positions $(\bar{\theta}_x, \bar{\theta}_y)$, type of rock samples found (R_x, R_y), and the observations on those samples (O_x, O_y).*

It is reasonable to assume that the rock type of a sample at any given location x is conditionally dependent only on the local rock ratios $\bar{\theta}_x$, and sensor observations of a rock depend only upon its type (Figure 3). Therefore, for any positions x,y, and rock type $k \in \{1,2,...,M\}$

$$P(R_x = k \mid \bar{\theta}_x, \bar{\theta}_y, R_y) = P(R_x = k \mid \bar{\theta}_x) = \theta_{xk} \quad (4)$$

No assumptions are yet made on the relationship between $\bar{\theta}_x$ and $\bar{\theta}_y$, which is not usefully expressible in terms of a standard Bayes network diagram.

Furthermore, (4) implies that

$$P(R_x = k \mid \bar{\theta}_y) \quad (5)$$

$$= \iint P(R_x = k \mid \bar{\theta}_x, \bar{\theta}_y) p(\bar{\theta}_x \mid \bar{\theta}_y) d\bar{\theta}_x$$

$$= \int \theta_{xk} \left\{ \int p(\bar{\theta}_x \mid \bar{\theta}_y) d\theta_{x1}..d\theta_{xk-1}, d\theta_{xk+1}..d\theta_{xM} \right\} d\theta_{xk}$$

$$= \int \theta_{xk} p(\theta_k \mid \bar{\theta}_y) d\theta_{xk}$$

$$= E\{\theta_{xk} \mid \bar{\theta}_y\}$$

a result that will be used subsequently. See [7] for more on the Bayes network representation of statistical relationships.

Geographical models

Consider the case when rocks can be identified with complete certainty. Suppose a rock at position x is determined to be of type k. Then, using Bayes rule, the posterior density of $\bar{\theta}_y$ is given by

$$p(\bar{\theta}_y \mid R_x = k) = \frac{P(R_x = k \mid \bar{\theta}_y) p(\bar{\theta}_y)}{P(R_x = k)} \quad (6)$$

$$= \frac{E\{\theta_{xk} \mid \bar{\theta}_y\} p(\bar{\theta}_y)}{E\{\theta_{xk}\}}$$

To clarify the functional relationships it is convenient to define

$$M_k(\bar{\theta}_y; x, y) := \begin{cases} \theta_{xk} & , x = y \\ E\{\theta_{xk} \mid \bar{\theta}_y\} & , x \neq y \end{cases} \quad (7)$$

Then, the posterior density $\bar{\theta}_y$ given the definitive observation is simply expressed in terms of $M_k(\bar{\theta}_y; x, y)$

$$p(\bar{\theta}_y \mid R_x = k) = \frac{M_k(\bar{\theta}_y; x, y) p(\bar{\theta}_y)}{E\{\theta_{xk}\}} \quad (8)$$

The function $M_k(\bar{\theta}_y; x, y)$ is fundamental to determine how a rock find at the location x affects the rock ratio's at all other locations y (including x). In recognition of the functions importance it will be henceforth referred to as a *geographical model*, as it describes the statistical relationships between samples at different geographic positions.

Properties of $M_k(\bar{\theta}_y; x, y)$

The geographical model $M_k(\bar{\theta}_y; x, y)$ is a complicated multi-variate function, determined by the underlying geological (or other) mechanism by which different rock types are distributed across an area. Nonetheless, applying reasonable assumptions it is possible to constrain it and make it tractable without going into the details of the (unknown) underlying mechanism.

- *Small influence at a distance*

Observations at a distant location should have little or no relation to the rock ratios at the current location. This assumption can be formalized as

$$M_k(\bar{\theta}_y; x, y) \xrightarrow[|x-y| \to \infty]{} E\{\theta_{xk}\} \quad (9)$$

Following this it also reasonable to assume some finite cut-off distance, beyond which measurements have no effect:

For some distance D > 0 (10)

$$|x - y| > D \Rightarrow M_k(\bar{\theta}_y; x, y) = E\{\theta_{xk}\}.$$

- *Decreasing influence with distance*

Not only should observations at locations distant to each other be largely independent, but the dependence between observations should never increase when the distance between them is increased. While motivated by the previous statement, this is a stronger assumption. Formally:

For any locations x, y, z s.t. $|x\text{-}y| > |x\text{-}z|$ (11)

$$\left| M_k(\bar{\theta}_y; x, y) - E\{\theta_{xk}\} \right| \leq \left| M_k(\bar{\theta}_z; x, z) - E\{\theta_{xk}\} \right|.$$

- *Smoothness assumption*

$M_k(\bar{\theta}_y; x, y)$ should vary smoothly with x and y. This is consistent with natural laws and not very restrictive. However, relaxing this slightly to allow a discontinuity at the finite cut-off distance may be convenient.

- *Spatial invariance and isotropy*

This assumption requires that $M_k(\bar{\theta}_y; x, y)$ be solely a function of the distance $|x\text{-}y|$ between samples, and not depend on their individual positions or the direction from one to the other. However, this is inconsistent with the previous assumptions, which require explicit dependence on x. Nonetheless, if this dependence can be explicitly accounted for and functionaly separated from the rest of $M_k(\bar{\theta}_y; x, y)$, then spatial invariance and isotropy are desirable and reasonable properties, provided that the geographic area of interest is not too large.

- *Conjugacy requirement*

This is the most restrictive and useful of the assumptions, requiring that for all positions x,y, and rock types k, the prior distribution of rock ratios $p(\bar{\theta}_y)$ is the same class of distribution as the posterior $p(\bar{\theta}_y | R_x = k)$. Conjugate prior distributions are computationally convenient and the prior is easily interpretable as earlier measurements.

It is natural to represent $p(\bar{\theta}_y)$ by a Dirichlet distribution [7] with parameters $\alpha_{y1}\ldots\alpha_{yM}$:

$$p(\bar{\theta}_y) = \text{Dirichlet}(\bar{\theta}_y; \alpha_{y1} \ldots \alpha_{yM}) \quad (12)$$

$$= \frac{\Gamma(\alpha_{y1}+\ldots+\alpha_{yM})}{\Gamma(\alpha_{y1})\ldots\Gamma(\alpha_{yM})} \theta_{y1}^{\alpha_{y1}-1}\ldots\theta_{yM}^{\alpha_{yM}-1}$$

Ensuring that $p(\bar{\theta}_y | R_x = k)$ is also a Dirichlet distribution requires that

$$M_k(\bar{\theta}_y; x, y) = \quad (13)$$
$$Z_k(x,y) \theta_{y1}^{\beta_1(x,y,k)} \ldots \theta_{yM}^{\beta_M(x,y,k)}$$

which guarantees from (8) that

$$p(\bar{\theta}_y | R_x = k) = \quad (14)$$

$\text{Dirichlet}(\bar{\theta}_y; \alpha_{y1}+\beta_1(x,y,k), .., \alpha_{yM}+\beta_M(x,y,k))$

The small influence at a distance assumption (9) implies that

$$Z_k(x,y) \xrightarrow[|x-y|\to\infty]{} E\{\theta_{xk}\} \quad (15)$$
$$\beta_i(x,y,k) \xrightarrow[|x-y|\to\infty]{} 0$$

Introducing the finite cutoff of (10) implies the limits are attained when x and y are separated by a finite distance. The assumption of decreasing influence with distance (11) means that Z_k and $\beta_1\ldots\beta_M$ never get further from their limits as $|x-y|$ increases. They are smooth functions as $M_k(.;x,y)$ is assumed smooth.

Note that computation of the posterior distribution requires only that $\beta_1(x,y,k)\ldots\beta_M(x,y,k)$ be specified. $Z_k(x,y)$ is implicitly defined by normalization of the posterior in (8). Furthermore, $Z_k(x,y)$ accounts for the spatial dependence of $M_k(.;x,y)$, making it possible to assume spatial invariance and isotropy for $\beta_1(x,y,k)\ldots\beta_M(x,y,k)$. That is, $\beta_j(x, y, k) = \beta_j(|x-y|, k)$.

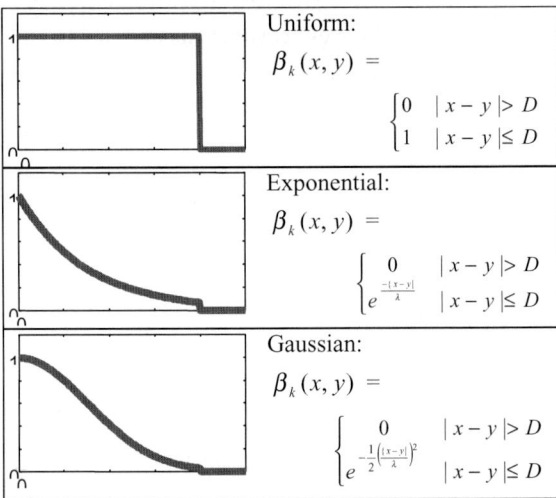

Figure 4 *Possible formulae for the functions $\beta_k(x,y)$. While ad hoc, they satisfy all the restrictions on $\beta_k(x,y)$, including spatial invariance and isotropy. The uniform formula weights all samples within an area equally, whilst the others have steadily decreasing influence with distance. Choice of the constants D and λ involves a trade-off between spatial resolution and power to generalize. Large values allow the rapid learning of the rock ratios over a large area from a few samples but are less effective in learning regional variations, which will be smoothed over. Ideally, they should be comparable to the average separation between samples.*

For co-located samples $x = y$, hence $M_k(\bar{\theta}_y; x, y) = \theta_{yk}$, implying that

$$\beta_k(x,x,k) = 1 \text{ and } \beta_j(x,x,k) = 0 \text{ (for every } j \neq k) \quad (16)$$

Because $\beta_1(x,y,k)\ldots\beta_M(x,y,k)$ approach zero and may not increase as distance $|x - y|$ increases, it follows that $\beta_j(x,y,k) = 0$ for all x, y and $j \neq k$. Therefore the notation for the β's is redundant; it is sufficient to denote the remaining nonzero term $\beta_k(x,y,k)$ as $\beta_k(x,y)$.

This implies that a *definitive* rock find (identified with 100% certainty) will not increase the assumed probabilities of finding other related but distinct rock types in the area.

To summarize

$$M_k(\bar{\theta}_y; x, y) = Z_k(x,y)\theta_{yk}^{\beta_1(x,y)} \quad (17)$$

and

$$p(\bar{\theta}_y | R_x = k) = \quad (18)$$

Dirichlet($\bar{\theta}_y$; α_{y1},..., $\alpha_{yk} + \beta_k(x,y)$, ..., α_{yM})

To learn the distribution of rock ratios at a particular point given definitive observations in the area, sum the contributions of each observation using (18) and an appropriate formula for the β's (Figure 4).

Learning from uncertain observations

Regrettably, it is rarely the case that rock samples can be autonomously identified with certainty. Otherwise, there would be little reason to learn the priors.

Consider an observation O_x made on a rock sample R_x at position x, with likelihoods $w_{xk}=P(O_x|R_x=k)$. It can be shown that

$$p(\bar{\theta}_y | O_x) \propto \sum_k [P_{xk} Z_k(x,y) \theta_{xk}^{\beta_k(x,y)}] p(\bar{\theta}_y) \quad (19)$$

where

$$P_{xk} = P(R_x = k | O_x) \quad (20)$$

$$= \frac{P(O_x | R_x = k) E\{\theta_{xk}\}}{\sum_j P(O_x | R_x = j) E\{\theta_{xj}\}} \quad \text{(Bayes rule)}$$

$$= \frac{w_{xk} \alpha_{xk}}{\sum_j w_{xj} \alpha_{xj}}$$

As $p(\bar{\theta}_y)$ is a Dirichlet distribution, it follows that

$$p(\bar{\theta}_y | O_x) = \quad (21)$$

$$\sum_{k=1}^{M} P_{xk} \text{ Dirichlet}(\bar{\theta}_y ; \alpha_{y1},..,\alpha_{yk}+\beta_k(x,y),.., \alpha_{yM})$$

This is not a Dirichlet distribution, but a mixture model of Dirichlet distributions. It violates the conjugacy requirement and is intractable as more observations are made. Subsequent observations would produce mixture models with M^2, M^3, and so on terms. To maintain a closed form model with a bounded number of parameters it is necessary to approximate

$$P(\bar{\theta}_y | O_x) \quad (22)$$

Dirichlet($\bar{\theta}_y$; $\alpha_{y1}+P_{x1}\beta_1(x,y),..,\alpha_{yM}+P_{x1}\beta_M(x,y)$)

This is equivalent to computing the posterior distribution of $\bar{\theta}_y$ given the definite observations $\{\mathbf{R}_x=k | k=1...M\}$ at x, each weighted according to their probability $P(\mathbf{R}_x=k|\mathbf{O}_x)$ given the current observation and priors on $\bar{\theta}_y$. This exactly equals the true posterior whenever the observations identify the rock with complete certainty. The approximation is worst when the observations do not favor any rock type. The latter occurs when an observation is made that is equally likely for all possible rock types, in which case the probability of each rock type is equal to the prior. The approximation (21) of the posterior distribution on $\bar{\theta}_y$ then correctly has the same mean as the prior but a slightly decreased variance, which is incorrect as nothing has been learned.

Multiple rock finds

Assuming the prior rock ratio density at any point x is Dirichlet, with parameters α_{xk}^{prior}, and an observation is made on a particular sample. Then, the rock type probabilities for that sample are given by (20), and the rock ratio densities at all other points given by (22). If observations are made at all sample locations then (20) and (22) are combined to create the following coupled simultaneous equations, relating the rock type probabilities at each sample and the posterior rock ratio distribution at each sample given the rock type probabilities at the other samples:

$$P_{xk} = \frac{w_{xk}\alpha_{xk}}{\sum_j w_{xj}\alpha_{xj}}, \quad \alpha_{xk} = \alpha_{xk}^{prior} + \sum_{y \neq x} P_{yk}\beta_k(y,x) \quad (23)$$

The α_{xk}^{prior}'s define the assumed (Dirichlet) distributions on $\bar{\theta}_x$ at all sample locations x prior to any measurements.

Note that the order in which samples are examined does not affect the computed probabilities and rock ratio distributions.

Experimental results

Results obtained by applying (23), using Gaussian geographical models (Figure 4) to simulated data have shown statistically robust improvements in classification accuracy and generated consistent probability maps. However, there are too many arbitrary parameters to set when simulating data for it necessarily to be a good indicator of performance. There is no substitute for *real* data, gathered in the field by a robot.

In December 1998 the Nomad robot was dispatched to Patriot Hills, Antarctica where it traversed a moraine and acquired spectral data, using a fiber optic reflectance spectrometer operating in the visible range, from 51 rocks in its path, along with their spatial coordinates (Figure 5). These rock samples were recovered and subsequently classified by a geologist to ascertain ground truth.

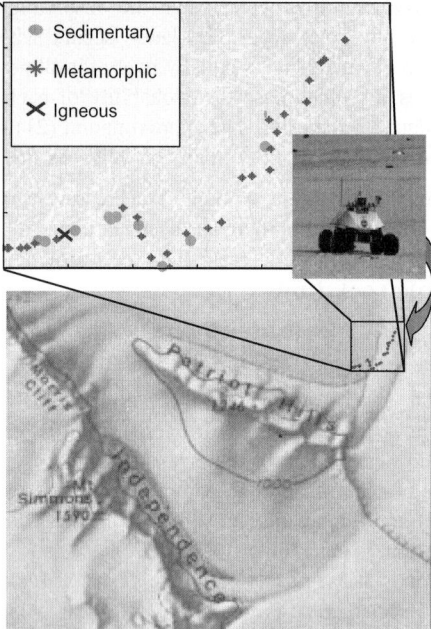

Figure 5 *Patriot Hills, Antarctica, data collection site. The Nomad robot traversed a 600m x 300m area, collecting spectroscopic data from each rock along it's path. The samples have been grouped into according to their formation process: sedimentary, metamorphic, igneous and extraterrestrial (meteorites, although none are present in the sample), each of which encompasses many non-overlapping rock types.*

Each sample spectrum was processed and run through a Bayes network classifier developed for rock and meteorite identification [3] to determine the conditional likelihood of that spectrum for each of 25 different rock types. This data was sequentially entered into a statistical model (22) of the possible rock ratio's at each sample position, using a Gaussian geographical model (Figure 4) and assuming initial distributions with α_{xk}^{prior}'s = 0.1 at all locations. At each iteration the rock type probabilities for every sample entered so far were recomputed and the most likely formation (sedimentary, metamorphic, igneous and extraterrestrial) deduced. Comparing with the known formation processes, the cumulative number of classification errors Figure 6(ii) as each sample is entered are determined.

Note the occasional *reduction* in the total number of misclassifications as more samples are examined and the model becomes more precise. This would not be possible if each rock sample was examined independently, as in Figure 6(i) and Figure 6(v) where the rocks are respectively classified with the assumption of uniform fixed rock type priors everywhere, and the priors fixed to the known data set rock type ratios. These indicate the worst and the best that the rock classifier in [3] can do under different assumptions on the rock type priors over the entire region if each rock is classified independently and the priors are assumed the same at each sample location.

Note that classification performance improves for various values of λ, indicating robustness to the exact form of the geographic models. Further computations using a uniform geographic model empirically confirm this. Furthermore, for large values of λ, Bayes optimal classification performance is *exceeded*. While from this data set it is hard to tell if the improvement is statistically significant, it is not inconsistent as the system is exploiting dependencies between samples as well as learning their geographical distributions.

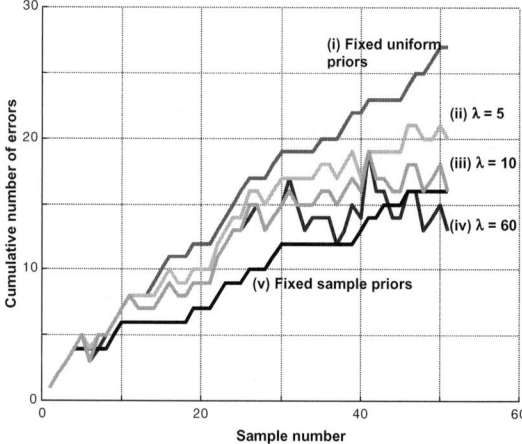

Figure 6 *Cumulative number of misclassifications as samples from the Antarctic field data are examined in the order they were encountered by the robot. In (i) samples are each independently classified assuming uniform rock type probabilities. In (ii)-(iv), the rock type probabilities are learned as samples are acquired, using Gaussian geographic models with λ values of 5, 10 and 60 respectively (c.f. Figure 4). Curve (v) indicates performance when rocks are again classified independently, using the known fraction of each rock type in the data set as the priors. This is the best performance possible for the independent classification of the rocks.*

With all the rock data entered into the model it is also possible to compute the learned rock type probabilities everywhere in an area (Figure 7, Figure 8, and Figure 9), not just at the samples. Compare these to the actual distribution of the rock samples along the robots path in Figure 5 to verify that they are indeed consistent with what was seen by the robot. The regions dominated by sedimentary and metamorphic rocks are clearly identified. Conversely, nothing is learned about the areas distant from the robot path, since a geographic model with a finite cut-off is assumed. Furthermore, the low density of igneous rocks (and meteorites, which have a probability map almost identical to Figure 9) is learned, even in the area dominated by metamorphic rocks, which are often confused with igneous rocks.

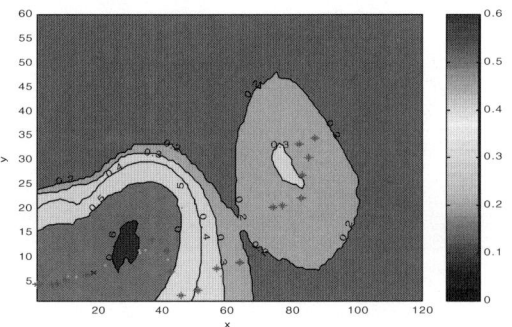

Figure 7 *Learned probability of sedimentary rocks across the explored region of the data collection site in Antarctica. The original sample positions and types are indicated by the colored dots (c.f. Figure 5).*

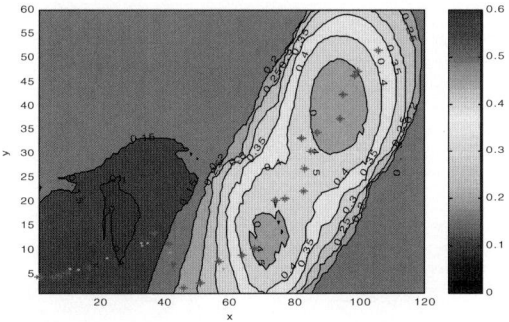

Figure 8 *Learned probability of metamorphic rocks.*

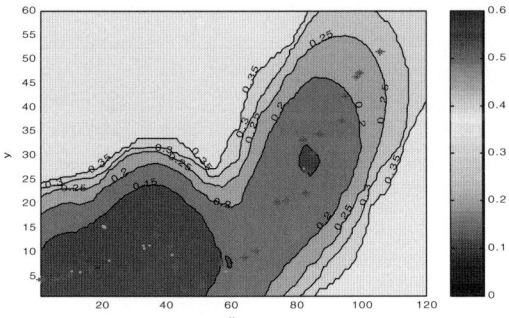

Figure 9 *Learned probabilities of igneous rocks.*

Conclusions

The results of Figure 6 are very significant. They show that learning the probabilities gives a clear improvement in classification over assuming uniform priors everywhere and classifying each sample independently. In fact, performance approaches and possibly exceeds the optimum, achieved when the average priors over the region are known beforehand. The latter might occur with more data and pronounced regional variations that can be exploited by this approach.

This is equivalent to a human geologist who looks at many rocks, constantly re-evaluating all the previous rocks seen every time another rock is looked at. Unfortunately, because of the lack of prior work in this exact area it is not possible to compare this with other methods.

A weakness of the approach proposed here is the ad hoc nature of the geographical models. While based upon reasonable assumptions they are still under-constrained. Nonetheless, empirical evidence suggests that the improvements in classification are robust to changes in the assumed geographical models. Choosing models with a wide footprint (large cut-off distance and low decay rate) results in faster convergence and generalizes over a larger area, but also less ability to capture and exploit small scale variations. Further work is needed to determine the optimal geographic models from the data.

All rock type probabilities are re-computed every time another rock sample is examined, and do not depend on the order in which they are found. Therefore, this method is robust to unlucky sequences of samples not representative of the area. Except for the approximation (22), the learning algorithm is Bayesian, and should converge (in the probabilistic sense) to the correct probabilities as more data is added.

Computationally requirements are minimal. They increase with the number of samples squared. In practice, the matrices in (23) are sparse and, depending on the robot path, complexity is order N^2 each time a new sample is added, where N is the average number of samples within a circle whose radius is the geographical model cut-off distance.

Acknowledgements

Invaluable assistance was provided by William Cassidy, Ted Roush, Andrew Moore, Martial Hebert, Dimitrious Apostolopoulos, Peter Cheeseman and all the participants in CMU's RAMS program. This work was supported by a grant from NASA.

References

[1] D. Apostolopoulos, M. Wagner, B. Shamah, L. Pedersen, K. Shillcutt, W. Whittaker, "Technology and Field Demonstration of Robotic Search for Antarctic Meteorites", *International Journal of Robotic Research*, special Field & Service Robotics issue, in press December 2000.

[2] L. Pedersen, D. Apostolopoulos, W. Whittaker, T. Roush and G. Benedix "Sensing and data classification for robotic meteorite search" Proceedings of *SPIE Photonics East Conference*. Boston, USA, 1998.

[3] L. Pedersen, D. Apostolopoulos, W. Whittaker "Bayes Networks on Ice: Robotic Search for Antarctic Meteorites", *Neural Information Processing Symposium 2000*, Denver, Colorado, 2000

[4] A. Elfes, "Occupancy rids: a stochastic spatial representation for active robot perception", *Autonomous Mobile Robots: Perception, Mapping and Navigation,* S. Iyengar, A. Elfes, Eds. IEEE Comp. Soc. Press, 1991.

[5] E. Gelenbe, N. Schmajuk, J. Staddon, J. Reif, "Autonomous search by robots and animals: a survey", *Robotics and Autonomous Systems* 22, 1997, pp 23-34.

[6] E. Gelenbe, Y. Cao, "Autonomous search for mines", *European Journal of Operational Research* 108, 1998, pp319-333.

[7] A. Gelman, J. Carlin, H. Stern, D. Rubin, *Bayesian Data Analysis,* Chapman & Hall, 1995 edition.

[8] Pearl, J., "Probabilistic Reasoning in Intelligent Systems: Networks of Plausible Inference", Morgan Kaufman, 1988.

Towards a Meta Motion Planner A: Model and Famework

Amit Adam
Dept. of Mathematics
amita@tx.technion.ac.il

Ehud Rivlin
Dept. of Computer Science
ehudr@cs.technion.ac.il

Ilan Shimshoni
Dept. of Industrial Engineering
ilans@ie.technion.ac.il

Technion – Israel Institute of Technology
Haifa 32000 – Israel

Abstract

We address the problem of rating or comparing navigation algorithms, or more generally navigation packages. For a given environment a navigation package consists of a motion planner and a sensor to be used during navigation. The ability to rate or measure a navigation package is important in order to address issues like sensor customization for an environment and choice of a motion planner in an environment.

We develop a framework under which we can rate a given navigation package. Based on the navigation package, a partially observable Markov decision process (POMDP) is defined. Next an optimal policy to be used in this POMDP is searched for. The performance achieved under the resulting policy serves to measure the navigation package.

This paper presents the motivations for solving the problem, the model we use and the framework which we have developed. An accompanying paper [1] presents the algorithm which we use and some results.

1 Introduction

When given a navigation task in an environment, a robot will usually employ certain sensors and a certain motion planner. In this and a companion paper [1] we discuss the problem of choosing the best combination of motion planner and sensor. This paper focuses on the motivations for solving this problem, the model which we use for the problem, and the framework under which this problem may be approached. In the companion paper [1] we present an algorithm to be used for solving the problem and demonstrate results of applying this framework and algorithm to some sample problems.

We will start with an example which will clarify our motivation and the terms we will use.

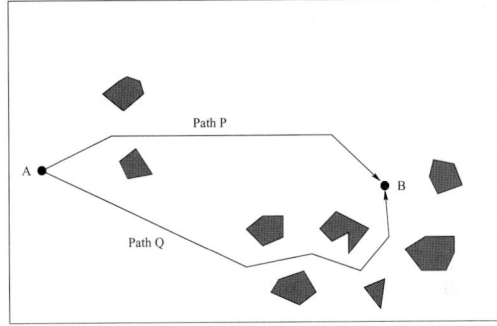

Figure 1: An example of an environment

1.1 A Motivating Example

Consider the environment shown in Fig. 1. Suppose that our robot has to move from point A to point B, without colliding with the obstacles. In order to compute its course, the robot employs a motion planning algorithm. A large number of motion planning algorithms may be employed [11], and in general each will compute a different *nominal path* from A to B. Paths P and Q in the figure are examples of two nominal paths which were generated by two different motion planners.

Should our robot use path P or should it choose path Q ? In principal, both paths are good and should be suitable for navigation. We might want to consider the shorter path between the two. However, other factors also come into play. The first is uncertainty in positioning. For various reasons (see for example [3]), we cannot expect the robot to be able to follow accurately a nominal path which was planned by the motion planner. As the robot attempts to execute a path, uncertainty in its position will develop. This uncertainty will result for example in a certain chance of hitting an obstacle. In our example, it is likely that for path P this chance is lower than for path Q. Thus, in addition to path length, other spatial characteristics such as distance from obstacles, number of turns, where in the environment the path actually traverses (e.g in the bottom region, in the center ...) etc. may also affect our choice of nominal path.

In order to reduce the uncertainty in position, the robot invokes from time to time a sensor. Many types of sensors are available [3]. Let us assume specifically that our robot is equipped with a camera and that it updates its position by comparing the current image it obtains from the camera, with a database of images of the environment that is stored in its memory. Let us assume that this database of images is given. A natural question that we now ask is the following: along which path, P or Q, will the robot be better able to compare the images it will acquire with the specific images stored in its database ? For example, it may be possible that the images in the database were all photographed from points in the bottom half of the environment. In this case, we expect the vision sensor to be more useful along path Q rather than along path P.

Once we have chosen the nominal path to follow, we are faced with the question of actually executing this path. The robot has to decide for example when it wants to invoke its sensor. It would not be efficient to invoke the sensor too often. On the other hand, we cannot let uncertainty in position grow too much (because of collisions with obstacles for example). The amount of uncertainty we are willing to allow may also vary along each different nominal path (because of variable distance to obstacles for example). Thus for each nominal path we expect to find a different sensing strategy which should be used along the path.

2 Navigation Packages

The example above shows us that a number of factors or elements interact with each other in the navigation problem. The factors we have identified are the environment itself, nominal paths in the environment, uncertainty in the position of the robot and a sensor which is used to reduce this uncertainty.

It is convenient to represent the sensor abstractly by a performance map defined on the environment. For every point p in the environment, this map states the expected performance or accuracy of the sensor if invoked at the point p. For example, in our image database sensor, this accuracy will be related to the ability to compare the image seen from the point p to one of the images in the database. By representing the sensor as a performance map we free ourselves from referring to the actual implementation of the sensor.

We use the term "navigation package" to denote a combination of all the factors mentioned above. Because the environment and robot are usually given and fixed, we will sometimes think of the navigation package as consisting of a motion planning algorithm producing nominal paths, and a sensor, which is represented abstractly by its performance map.

3 Rating Navigation Packages

In this work we discuss the problem of rating or comparing different navigation packages. The interaction between the nominal path, the performance of the sensor across the environment, the spatial features of the environment, and the uncertainty which develops in the position of the robot, result in certain measurable consequences. These may be the cost of sensing, the chances of bumping into obstacles, the chances of reaching the goal, the lengths of the paths taken etc. It is therefore natural to try to *rate* or *measure* the quality of different navigation packages for a given environment.

By being able to compare navigation packages, many interesting problems may be tackled:

- What is the best motion planner to be used with a given environment and sensing capability ? In our example, this was the question of choosing between path P and path Q. Note that not only the environment but also the sensing capability are important in making the decision.

- Suppose the sensor can be customized for a given environment. What is the preferred customization ? In our example, "customizing" may mean the choice of images to be stored in the database. How should we photograph the environment in order for the image database to constitute a useful sensor ? The answer to this question may depend on the nominal paths we plan to use.

- Given a specific sensing capability in an environment, should we use different motion planners in different regions of the environment ? For each region of the environment, the motion planner which is best combined with that region's sensing capability and spatial characteristics, should be chosen.

Notice that in each of these examples, the choice we actually have to make is between different navigation packages. In the first example the sensor is given and we change the nominal paths, thus yielding different navigation packages. In the second example different "customizations" yield different sensor performance maps, thus defining different navigation packages. Therefore it is important to be able to compare different navigation packages.

We note that the problems we have discussed are especially relevant when working with vision sensors. One reason is that the performance of vision-based sensors is highly variable and dependent on location. For example, occlusions that interfere may degrade the performance in a very well defined region. The second reason is that the issue of customization is also very relevant. For example, we may choose what images to store in the database, or we may choose where to place visual landmarks.

4 Difficulty of Rating a Navigation Package

Assume that we are given a navigation package for a certain environment consisting of a motion planner which plans nominal paths, and a sensor which would be used to update the position of the robot along the path. The first thing we should understand is how to augment the nominal path with a sensing strategy. This strategy will be employed upon actual traversal of the path. A sensing strategy is required since sensing may be a costly operation which we would like to avoid as much as possible. Some considerations that should be addressed by this strategy are:

- The performance of the sensor at the place where sensing is invoked - there is no point in invoking the sensor where it is useless.

- The required accuracy in position. In some places an accurate update may be required (for example near obstacles) while in other places accuracy may be compromised.

- The cost of sensing - how often should the sensor be invoked.

Note that the robot is uncertain of its position. As the current position is unknown, the robot cannot use the map which describes sensor performance as function of position in a straight forward way.

All of these considerations make it clear that finding a reasonable sensing strategy is not a trivial task. However, it would only be fair to rate a navigation package based on its performance under the *best* sensing strategy available for this package. Therefore, we should first find the strategy under which the robot should act when using the navigation package. The navigation package will then be rated based on the performance achieved by using it in conjunction with the strategy found.

5 Our Framework and Related Work

Before we describe our framework, let us first describe related work by other authors. Many authors have noted that the performance of a sensor varies across the environment. In [12] regions of perfect sensing have been called landmarks or "islands of truth". A map describing the sensor performance has been termed Sensory Uncertainty Field (SUF) in [18]. Quite a few works have shown the benefits of using such a map. Use of the map in motion planning has been demonstrated in [18, 7, 10]. In [15] a related notion is the information content of the environment at each configuration. Another similar idea motivated by visual servoing is described in [16]. In [4, 5] sensing management issues are discussed and the sensor performance map is a factor taken into account. Finally, in a previous work [2] we have addressed the issue of computation of such a map for vision-based sensing.

The motion planning approach taken in previous works (for example [18, 15]) is to search for a path in free space by minimizing a function which takes into account both the length of the path and the sensory uncertainty along the path. This approach involves some arbitrary decision on how to trade off between sensory uncertainty and path length. These two different factors are usually combined into one objective function by introducing an arbitrary scale factor between the two.

Both our problem and our method of solution are different. As explained above, we would like to find a method for comparing two different navigation packages. We have chosen the partially-observable Markov decision process (POMDP) framework in order to address this problem. In our framework, every navigation package defines a POMDP which expresses the sensing capability, the underlying nominal path, the developing uncertainty in position and the environment. An optimal policy under which to act in this POMDP is searched for. The expected performance under this policy is the rating of the navigation package. We are then able to compare different navigation packages by comparing the performance in the POMDPs they have defined, under the policies that were found.

As a by-product of the process of rating a navigation package, our method yields a policy under which the robot should act while navigating in the environment. This policy tells the robot how to move and when to invoke the sensor. The policy takes into account all the aspects of the problem: the performance of the sensor, the costs of sensing, probabilities of undesired events like losing the way or bumping into an obstacle etc. This is in contrast with previous approaches (for example [18, 15]) in which the highest localization accuracy possible was desired, but eventually compromised in order to account for the length of the path.

6 The POMDP Framework - A Brief Review

We will now describe the Partially Observable Markov Decision Process (POMDP) framework. We begin with the notion of Markov Decision Process (MDP).

Assume that an agent operates in a certain world. We assume that time is discrete and at each point in time the world is in a certain state s belonging to the set of possible states S. The agent chooses at each step an action a from the set of actions A. A reward function $R : S \times A \to \mathcal{R}$ determines the reward $R(s, a)$ the agent accepts when performing action a while the state is s. Apart from affecting the reward, the action a also affects the next state of the system as follows.

The state of the world in the next epoch is a random variable with a distribution which depends on both the current state s and the chosen action a. Let x_t denote the state at time t, and let a_t denote the action chosen at time t. We let $T(s, a, s')$ denote the state transition probabilities:

$$T(s, a, s') = \Pr(x_t = s' | x_{t-1} = s \text{ and } a_{t-1} = a)$$

Let $r_t = R(s_t, a_t)$ denote the reward collected at time t. The performance measure that is to be maximized is the sum of these rewards. The sum is either over a finite number of steps (so called finite horizon)

$$R = \sum_{t=1}^{N} r_t$$

or we may also look at an infinite horizon. In this case future rewards are discounted by a discount factor $0 < \lambda < 1$:

$$R = \sum_{t=1}^{\infty} \lambda^t r_t$$

In the Markov decision process (MDP) the agent always knows the state of the world x_t. The goal of the agent is to choose a policy or mapping between states and actions, which will maximize its expected total reward.

The POMDP framework is similar to the MDP framework except that the current state is unknown. Instead, a set O of possible observations exists, from which an observation o is given to the agent. The observation which the agent "measures" is a random variable which is distributed according to

$$\nu(o|s', a) = \Pr \left(\begin{array}{c} \text{after executing } a \text{ and reaching state} \\ s' \text{ the observation will be } o \end{array} \right)$$

Based on the dynamics of the system which are governed by $T(s, a, s')$, and on the measurements governed by $\nu(o|s', a)$, the agent maintains a "belief function" $b : S \to \mathcal{R}$. The belief function b is the probability distribution of the current state. The beliefs are propagated according to the dynamics and updated after each observation by using Bayes' rule.

A policy is a mapping from the belief state to actions: given the current belief b, what is the best action to choose ? As in the MDP, the goal is to find a policy which will maximize the expected total reward.

A more comprehensive introduction to POMDPs may be found in [9, 8].

In principle a POMDP is a particular case of MDP where the state space is the belief space. However the belief space is much too large (in fact infinite) for standard techniques for solving MDPs to be used. Therefore special algorithms have been proposed for POMDPs. Examples of such algorithms may be found in [9] and the references therein. However, even for problems with tens of states, finding the optimal policy exactly is computationally prohibitive. Therefore various researchers have worked on algorithms for finding approximations of the optimal policy. A review of approximation techniques and some new algorithms may be found in [8]. Other algorithms for finding approximate policies are presented in [13, 14, 19]. In the robotics context, a successful experience in which the search for the optimal policy was bypassed is reported in [17]. There actions were chosen according to the most probable action rule: at each state the correct action was defined, and the most probable action according to the current belief was chosen.

7 Description of the POMDP

We now turn to phrase our problem of rating navigation packages in terms of the POMDP framework. Let us assume that our environment is quantized to a grid. On the grid initial and goal configurations are marked. At each point on the grid the robot may choose one of six actions. It may either move in one of the four directions (up, right, down or left), or it may stop, or it may invoke its sensor to get an update of its position. Thus the state space S is the set of possible (discrete) locations, and the set of actions is

$$A = \{ \text{up, right, down, left, update, stop} \}$$

After either a movement command or a "stop" command, no observation is given. We mark this as being given the "nil" observation. If the action taken was the "update" action, then the observation returned is a state $\hat{s}$, which is a random variable describing the result of invoking the sensor. The distribution of this observation varies with the true position the robot is in. This way we model the variation in the performance of the sensor across the environment. Thus the set of observations is

$$O = \{ \text{nil} \} \cup S$$

The uncertainty which develops in the position of the robot is governed by the dynamics of its movements. Let $T(s, a, s')$ denote the probability that performing the action a while being in s will lead the robot to s'. For $a \in \{ \text{update, stop} \}$ we have

$$T(s, a, s') = \begin{cases} 1 & s = s' \\ 0 & s \neq s' \end{cases}$$

while for movement actions, $T(s, a, s')$ should model factors such as control inaccuracies.

We allow some states to be defined as occupied by obstacles. Bumping into an obstacle is modeled by allowing $T(s, a, s')$ to be possibly non-zero for an obstacle state s'. After hitting an obstacle s', the robot bounces back to free space.

We must now specify how different actions in different states are rewarded. The first step is to use the motion planner to generate the nominal paths from every (free) state to the goal state. This defines a *preferred* action at every free state - namely the first movement command according to the motion planner. We reward movement according to the motion plan with (say) 1 unit. Movement which is not according to the nominal motion plan is rewarded with a smaller $w < 1$ reward. Thus, the motion planner is incorporated into the POMDP by encouraging movements according to the motion plan. We will later see another way in which the motion planner affects the POMDP.

All other factors that are part of the navigation process should also affect the rewards in the POMDP. We will now consider them.

The cost of sensing is brought into account by the reward u given for the sensing action. This u may in general depend on the state s in which the sensing action is performed.

If the robot chooses to stop while it is in the goal position, a large reward g is collected. Stopping in non-goal positions results in a lower reward d.

To summarize, for a non-obstacle state s we have:

$$R_1(s, a) = \begin{cases} 1 & a \text{ is movement according to motion plan} \\ w & a \text{ is another movement} \\ u & a = \text{update} \\ g & a = \text{stop and } s \text{ is the goal} \\ d & a = \text{stop and } s \text{ is not the goal} \end{cases}$$

In order to discourage bumping into obstacles, we give a (possibly negative) reward ob for any action performed while in an obstacle state:

$$R(s, a) = \begin{cases} R_1(s, a) & s \text{ is not an obstacle state} \\ ob & s \text{ is an obstacle state} \end{cases}$$

Choice of parameters: The different parameters above reflect the preferences one might have. For instance, if it is critical not to bump into obstacles then one has to lower the parameter ob. As another example, the cost of sensing is another parameter that is clearly dependent on the specific scenario one encounters. Although having a large number of parameters is undesirable, the POMDP has to express a large number of such different aspects of the problem. The POMDP framework is general enough to allow expression of these different factors, and this is done by choosing different rewards.

8 Discussion

In this part of the paper we have introduced navigation packages and have demonstrated that many interesting problems may be phrased as problems of rating navigation packages one with respect to the other. We have described a framework under which this problem may by treated. We conclude this part of the work with some additional remarks.

Introducing the Motion Planner As discussed earlier, many other researchers have worked on integrating the sensor performance map into the motion planning stage. The approach in most of these works was to strive for the highest possible sensing capability along the path, but compromise on this requirement in favor of path length. An underlying motion planner was not part of the approach. Instead, a motion planning algorithm was devised to find paths that optimize a criterion which includes both the path length and the sensing accuracy along the path.

We argue that a broader viewpoint must be used. Why should we strive for the highest possible sensing accuracy ? Instead, we strive for a *sufficient* sensing accuracy required for completing the nominal path. As a basis for the path we take a nominal path generated by a motion planner. This path is augmented with sensing operations. Unnecessary sensing will yield sub-optimal rewards in the POMDP and therefore will be discouraged.

We remark that Erdmann has addressed the issue of minimal sensing capability required by a motion plan in [6]. In that work, the robot does not need to know exactly at what state it is in, but only which action will advance it towards its goal. The distance from the goal is measured in terms of a nominal motion plan. It is argued that the sensor should be designed to answer this question ("What action will now move me towards the goal ?") and in some sense this is the minimal sensing needed.

The Roles of the Motion Planner The nominal motion planner which is part of the navigation package serves as a "guiding force" in the POMDP. This is due to the preferred rewards being given to movement actions that are in accordance with the motion planner.

In addition, as is shown in the companion paper [1], the nominal motion planner has an important role in the search for an optimal policy. It gives rise to the initial value function from which we try to search for the optimal value function.

As is shown in the companion paper, the nominal motion planners may greatly affect the policy which is found for the POMDP. Thus, although the nominal motion plan is not used "as is", it still has an important role and it is certainly possible that one nominal motion plan would be much better to use than another.

In the next part of the paper [1] we present an algorithm which searches for the optimal policy in the POMDPs and some example applications.

Acknowledgment A. Adam and I. Shimshoni were supported in part by Israeli Ministry of Science Grants no. 9766 and no. 2104.

References

[1] A. Adam, E. Rivlin, and I. Shimshoni. Towards a Meta Motion Planner B: Algorithm and Applications. Proceedings of ICRA 2001.

[2] A. Adam, E. Rivlin, and I. Shimshoni. Computing the sensory uncertainty field of a vision based localization sensor. In *Proc. IEEE Int. Conf. on Robotics and Automation*, pages 2993–2999, 2000.

[3] J. Borenstein, H.R. Everett, L. Feng, and D. Wehe. Mobile robot positioning: Sensors and techniques. *Journal of Robotic Systems*, 14(4):231–249, 1997.

[4] T. Celinski and B. McCarragher. Achieving efficient data fusion through integration of sensory perception control and sensor fusion. In *Proc. IEEE Int. Conf. on Robotics and Automation*, pages 1960–1965, 1999.

[5] T. Celinski and B. McCarragher. Improving sensory perception through predictive correction of monitoring errors. In *Proc. IEEE Int. Conf. on Robotics and Automation*, pages 2608–2613, 1999.

[6] Michael Erdmann. Understanding action and sensing by designing action-based sensors. *Int. Journal of Robotics Research*, 14(5):483–509, October 1995.

[7] T. Fraichard and R. Mermond. Path planning with uncertainty for car-like robots. In *Proc. IEEE Int. Conf. on Robotics and Automation*, pages 27–32, 1998.

[8] M. Hauskrecht. *Planning and control in stochastic domains with imperfect information*. PhD thesis, MIT, 1997.

[9] L. P. Kaelbling, M. L. Littman, and A. R. Cassandra. Planning and acting in partially observable stochastic domains. *Artificial Intelligence*, 101:99–134, 1998.

[10] A. Lambert and N. L. Fort-Piat. Safe actions and observation planning for mobile robots. In *Proc. IEEE Int. Conf. on Robotics and Automation*, pages 1341–1346, 1999.

[11] Jean-Claude Latombe. *Robot Motion Planning*. Kluwer Academic Publishers, 1991.

[12] Anthony Lazanas and Jean-Claude Latombe. Motion planning with uncertainty: a landmark approach. *Artificial Intelligence*, 76:287–317, 1995.

[13] M. L. Littman, A. R. Cassandra, and L. P. Kaelbling. Learning policies for partially observable environments: Scaling up. In *Proc. of 12'th Int. Conf. on Machine Learning*, pages 362–370, 1995.

[14] R. Parr and S. Russell. Approximating optimal policies for partially observable stochastic domains. In *Proc. of 14'th IJCAI*, 1995.

[15] N. Roy, W. Bugard, D. Fox, and S. Thrun. Coastal navigation - mobile robot navigation with uncertainty in dynamic environments. In *Proc. IEEE Int. Conf. on Robotics and Automation*, pages 35–40, 1999.

[16] R. Sharma and H. Sutanto. A framework for robot motion planning with sensor constraints. *IEEE Transactions on Robotics and Automation*, 13(1):61–73, February 1997.

[17] R. Simmons and S. Koenig. Probabilistic navigation in partially observable environments. In *Proc. of 14'th IJCAI*, 1995.

[18] H. Takeda, C. Facchinetti, and J.C. Latombe. Planning the motions of a mobile robot in a sensory uncertainty field. *IEEE Transactions on Pattern Analysis and Machine Intelligence*, 16(10):1002–1017, October 1994.

[19] S. Thrun. Monte Carlo POMDPs. In *Proc. of NIPS*, 1999.

Towards a Meta Motion Planner B: Algorithm and Applications

Amit Adam
Dept. of Mathematics
amita@tx.technion.ac.il

Ehud Rivlin
Dept. of Computer Science
ehudr@cs.technion.ac.il

Ilan Shimshoni
Dept. of Industrial Engineering
ilans@ie.technion.ac.il

Technion – Israel Institute of Technology
Haifa 32000 – Israel

Abstract

In a companion paper [1] we have developed a framework for rating or comparing navigation packages. For a given environment a navigation package consists of a motion planner and a sensor to be used during navigation. The ability to rate or measure a navigation package is important in order to address issues like sensor customization for an environment and choice of a motion planner in an environment.

In this paper we present the algorithm which we use in order to rate a given navigation package. Under the framework which was presented in [1], a partially observable Markov decision process (POMDP) is defined. The algorithm searches for an optimal policy to be employed in this decision process.

We briefly review the problem and the framework, develop the algorithm and present experimental results.

1 Introduction

The navigation problem consists of planning and executing a path between two different points in an environment. Many different factors are involved in this problem, of which two important ones are the motion planner and the sensors which will be used by the robot. In a companion paper [1] we have raised and discussed the problem of choosing between different possible combinations of motion planners and sensors. Solving this problem would enable for example to choose among different sensor customizations - for example different placement of visual landmarks in the environment. We have presented a framework in which the basic idea was that each combination of motion planner and sensor (which we have termed a "navigation package") defines a partially observable Markov decision process or POMDP. The navigation package is then rated by the expected payoff which can be obtained in the decision process, while acting under the best possible policy (i.e. the policy which maximizes the expected payoff).

This paper complements [1] by introducing a simple reinforcement learning algorithm which finds a suitable policy for a given POMDP. We present the algorithm, some results which were obtained by using it and a concluding discussion of the work.

2 Approximating an Optimal Policy

Following [1] each navigation package defines a POMDP. The state space consists of a quantization of the environment. The possible actions are movement in different directions, update of the position (by invoking a sensor) or stopping. After every action a reward is collected, the amount of which depends both on the state the robot was in and on the action taken. At each state there is a preferred action (leading to a maximal reward) which is movement in the direction specified by the nominal motion plan. However, since the robot is unsure of its position, it cannot always choose this preferred action. At all times a belief function $b(\cdot)$ is maintained by the robot. This function is the probability distribution of the current state. It is updated after each step based on the dynamics of the robot (where it moved) and on the observations which are generated. In our model an observation is generated only when the robot invokes its sensor.

Our goal now is to develop an algorithm for finding a policy which will maximize the expected payoff. A policy is a mapping which associates an action with each belief function. Let us begin with the way we represent the policy. A standard way of representing a policy is through its value function or Q-values (details may be found in [3] for example). The function $Q(b(\cdot), a)$ represents the total reward the agent may expect if it performs the action a and then continues optimally, when the current state is distributed by $b(\cdot)$. This is a very convenient representation for a policy: using the value function Q, an agent with a current belief function $b(\cdot)$, chooses the optimal action simply by

$$a^* = \arg\max_{a \in A} Q(b(\cdot), a)$$

(A is the set of possible actions).

When using the value function approach to represent a policy, we have to address two issues. The first is the issue of storing and representing the value function Q. The second issue is how to actually compute the value function.

Representing the Q-function is not straight forward since one of the arguments Q accepts is a belief function which comes from an infinite space. Therefore we cannot store the values Q obtains on every possible pair $(b(\cdot), a)$ of belief function and action. Instead, we discretize the belief space into a finite set of beliefs

$$B_d = \{b_1(\cdot), \ldots, b_N(\cdot)\}$$

and store the values Q obtains on the finite set $B_d \times A$. In other words, Q is represented by a finite lookup table.

The second issue which we now address is how to actually compute the function Q. When the state space is not very small, exact computation of the optimal value function is computationally infeasible (see [3, 2] for example). Therefore we strive to approximate the value function. By finding an approximate value function and using it, we obtain not the optimal policy but an approximation for the optimal policy. We have chosen to use reinforcement learning in order to compute an approximation for the value function.

The reinforcement learning algorithm starts with an initial approximation $\hat{Q}$ for the value function Q. Each iteration of learning involves simulation of actions which were chosen on the basis of the current approximation $\hat{Q}$. The "empirical" rewards obtained in the simulation are used to update the expected value of taking the action - in other words to update the current approximation $\hat{Q}$. More specifically, each learning iteration has the following structure:

A Learning Iteration Starting from Belief $b_0(\cdot)$

- Let $\hat{Q}$ be the current approximation of the value function

- Draw a random state s distributed according to $b_0(\cdot)$

- Let $\tilde{b}(\cdot) = b_0(\cdot)$

- Repeat l times:

  1. Let $b_d(\cdot)$ be the discretized value of $\tilde{b}(\cdot)$.
  2. Based on the current belief $\tilde{b}(\cdot)$ and the current value function $\hat{Q}$, choose the best action a^*.
  3. Based on simulation, update the current value of $\hat{Q}(b_d(\cdot), a^*)$.
  4. From the current state s jump to a new state with probabilities governed by the current state s and the action a^*. Let s now denote the new state.
  5. Obtain an observation o.
  6. Based on the action taken and the observation made, update the belief function. Let $\tilde{b}(\cdot)$ now denote the updated belief function.
  7. Return to step 1 (loop l times).

Let us elaborate on some of the steps.

Step 2: By default we choose the best action based on our current belief and current value function:

$$a^* = \arg\max_{a \in A} \hat{Q}(b_d(\cdot), a)$$

However, recall that $\hat{Q}$ is not the true value function. Therefore we sometimes (say with a 0.1 probability) choose a random action instead. By doing this we maintain a constant "exploring" behaviour.

Step 3: In this step we *learn* the value of $\hat{Q}(b_d(\cdot), a^*)$. This is done by simulating the action a^* at states s which are drawn from the distribution $b(\cdot)$. By simulating the action from a state, we get an immediate reward $R(s, a^*)$. In addition we obtain an observation o. Based on o we update the belief function. Using the updated belief we can look up the expected optimal future reward, based on the current $\hat{Q}$. We add this value to the immediate reward we got, and this is the empirical value obtained from acting a^* while the belief was $b(\cdot)$. We repeat this simulation of acting a^* a number of times, and take the average of empirical values we have obtained. This average value is the updated value of $\hat{Q}(b_d(\cdot), a^*)$.

The above learning iteration is repeated, each time beginning with the same $b_0(\cdot)$. By doing this we obtain updates of the values of $\hat{Q}$, mostly in those areas of the belief space that will actually be traversed when we use the resulting value function. This is a result of the fact that we choose at each step the best action based on the current approximation $\hat{Q}$.

The reinforcement learning algorithm which we have used is rather standard [5]. Reinforcement learning has been applied in various other works on POMDPs including [4, 6, 7]. Although we have used a very basic version of this algorithm, we have obtained very reasonable policies as will be shown in section 3.

Quantizing the Belief Space and Initializing the Value Function Recall that we have to quantize the belief space to a finite set B_d of beliefs. We have found the following quantization useful. For a given belief function we first compute the expected state and check what is the probability of being in that state (recall that our world is represented by a grid of possible positions). We then compute the probability of being in a neighboring state to the expected state. The two probabilities, namely the probability of being in the expected state and the probability of being in a neighboring state, are quantized and serve as a representation of the belief function. In the experiments which will be described below, we used 10 quantization levels for each probability.

Finally, what function serves as the initial $\hat{Q}$ with which we start the reinforcement learning iterations? Let us first define $Q_{opt}(s)$ to be the value which may be obtained by a robot with perfect control which starts at state s and moves to the goal according to the motion plan. In other words, we assume that the robot always ends up in the position it intended to reach after a movement command. In this case sensing actions are not needed. Now, for a belief function b which after quantization has probability 1 at a given state s, the initial value of $\hat{Q}(b,a)$ is $Q_{opt}(s)$ assuming that a is indeed movement according to the motion plan. If a is another action, then we take $\hat{Q}(b,a)$ to be 0. If the belief b has probability p at the expected state after quantization, then we take the initial value to be $\hat{Q}(b,a) = pQ_{opt}$ for the motion-plan action a and 0 for other actions. This value function is even more optimistic than MDP-based approximations for the POMDP value function, which are sometimes used [2]. Let us call this initial approximation for the value function the "perfect robot value function". Note that this initialization is clearly dependent on the underlying motion planner: the function describes the values obtainable by a perfect robot which acts according to the nominal motion plan.

3 Results

We now present results that were obtained by using the algorithm which we have now described. The results we present were obtained for different environments, sensors and motion planners. Each environment is represented by a grid. Some of the squares in the grid are obstacles. The robot moves between the free cells in the grid. At each time step it may move one square to the left, to the right, up or down. With each movement, position uncertainty grows, since with a probability of 20% the robot ends up in a square which is nearby the square it intended to reach (see Fig. 1).

The robot may invoke its sensor at any time. The sensor returns a grid position which is the estimated current position. The accuracy of this estimate may vary and depends on the actual position of the robot. We have used 3 accuracy levels, depicted in Fig. 2.

For every scenario tested, a value function was computed using the reinforcement learning algorithm de-

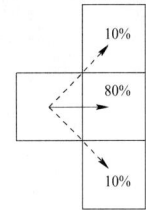

Figure 1: Position uncertainty resulting from a movement to the right. Similar uncertainty develops from movements in other directions

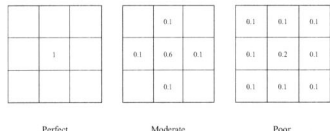

Figure 2: Sensor performance. The robot is in the middle square. The sensor estimates the current position with the probabilities depicted. We use three sensor accuracy levels.

scribed previously. The policy resulting from this value function was used in 1000 runs of the simulated robot. In all runs the robot started in the initial configuration, with the belief function being the initial configuration with probability 1 (i.e. no uncertainty in position). Each run consisted of 30 steps of action/observation. This number of steps is sufficient to permit reaching the goal. The results we show for each scenario are based on the results collected in those 1000 runs.

We start with the environment shown in Fig. 3. The asterisk marks the initial configuration and the + marks the goal configuration. A roadmap-based motion planner has been used to plan paths from every free configuration to the goal position. The directions of motion in each configuration as determined by this motion planner are shown in part (a) of the figure.

Let us first present the consequences of moving without any updates. We let the robot use the "perfect robot" value function. The policy associated with this value function is to perform a sequence of motions and then stop, without ever invoking the sensor. Fig. 4 presents the results from 1000 runs using this policy. In part (a) of the figure we see a histogram of the actions performed at each time step. We see that in all 1000 runs we had 14 motion actions and then the robot stopped for the next 16 time steps. Due to inaccurate control, this policy has led to the goal configuration in only about 20% of the runs. This is shown in part (b)

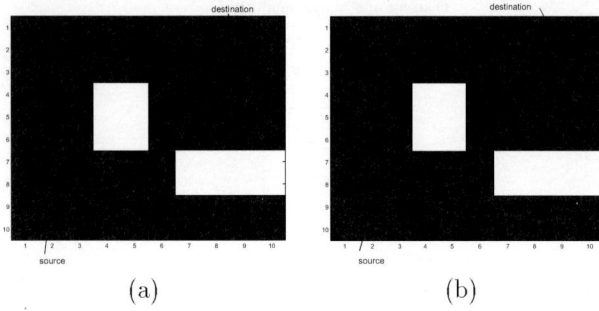

(a)   (b)

Figure 3: First environment. (a) Roadmap-based motion planner. (b) Visibility graph based motion planner

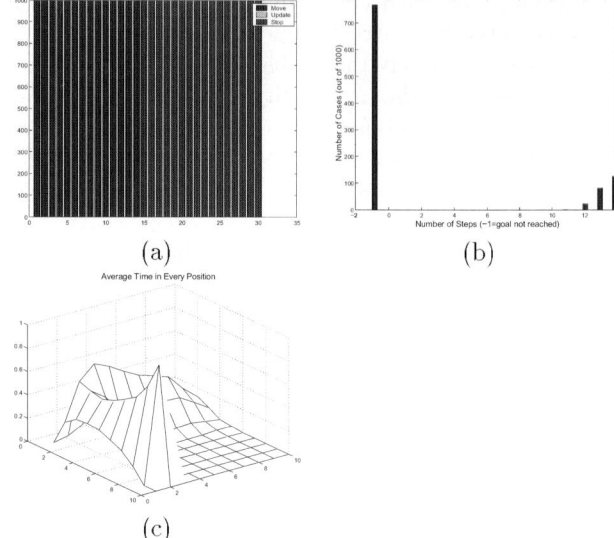

Figure 4: Results of using the "perfect robot" value function on the first environment. (a) Histogram of the actions performed in each time step. (b) The time it took to reach the goal. Note that in nearly 80% of the runs the goal was not reached. (c) The average "presence" of the robot along the path.

of the figure. Part (c) of the figure shows the average number of time steps (per run) the robot has spent in every configuration. Note that in the initial configuration this number is 1. Then due to inaccurate control the presence of the robot is "spread out" on a wide strip around the nominal path. Notice that around the goal configuration the spread is rather wide.

Next we used the value function which was computed using the reinforcement learning algorithm for 100000 learning iterations. We assumed perfect sensing capability across the environment. Fig. 5 shows the results. Part (a) of the figure shows that the policy executed consists of movement actions for 8 steps and then the robot invokes its sensor. Part (b) of the figure shows that now the goal has been reached in almost all of the runs. In part (c) we see the "presence" of the robot along the path. Compare this figure with with Fig. 4(c). Part (d) shows the positions in which the sensor was invoked. Note the effect of position update on the "presence" of the robot as seen in part (c) of the figure. Near the final goal the policy calls for an update since stopping in a non-goal position is much less rewarding than stopping in the goal. Part (e) of the figure shows a histogram of the number of times an obstacle was hit during a run. This, together with the histogram in part (b), are "operational" criteria which might be of interest.

We now change the sensor in this scenario. A sensor which does not operate properly in part of the environment is introduced in Fig. 6. We now have a new navigation package, consisting of the first environment with the roadmap-based motion planner (see Fig. 3), and the new non-perfect sensor which we have now defined.

The results for this environment are shown in Fig. 7. Notice in parts (a) and (d) of the figure how the policy has changed to invoke the sensor earlier in the path. This is due to the fact that it makes no sense to invoke the sensor where it performs poorly.

For the same environment and sensor, we now consider changing the motion planner. Fig. 3(b) shows a second motion planner for this environment. A new navigation package is now defined and a value function was computed for it. Results of using this value function are presented in Fig. 8.

We see that the time in goal and the chances of reaching the goal are quite similar (compare Figs. 8(b) and 5(b)). However, the chances for colliding with an obstacle are higher when using the second motion planner (Figs. 8(e) and 5(e)). Therefore, we might conclude the first navigation package is better than the second.

Fig. 9 shows a different environment with two motion plans. The first is based on a potential field planner and the second on the visibility graph. The sensor we have used on this environment (with both motion plans) is abstractly represented by the performance map shown in Fig. 10.

The first navigation package on this environment used the potential-field based motion plan (depicted in Fig. 9(a)). The results obtained by using this navigation package are shown in the top row of Fig. 11. The bottom row shows the results for the second motion plan. We can see that when using the potential-field based planner, the robot had a chance of approximately 8% of not reaching the goal. When it did reach the goal,

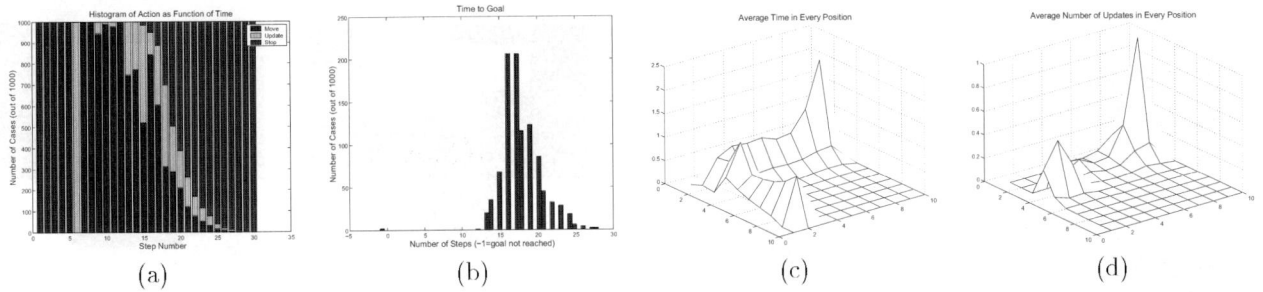

Figure 7: First environment, sensor as in Fig. 6. (a) Histogram of the actions performed in each time step. Notice earlier sensing as compared with Fig. 5(a). (b) The time it took to reach the goal. (c) The average "presence" of the robot along the path. (d) Places where the sensor was invoked. Compare with Fig. 5(d).

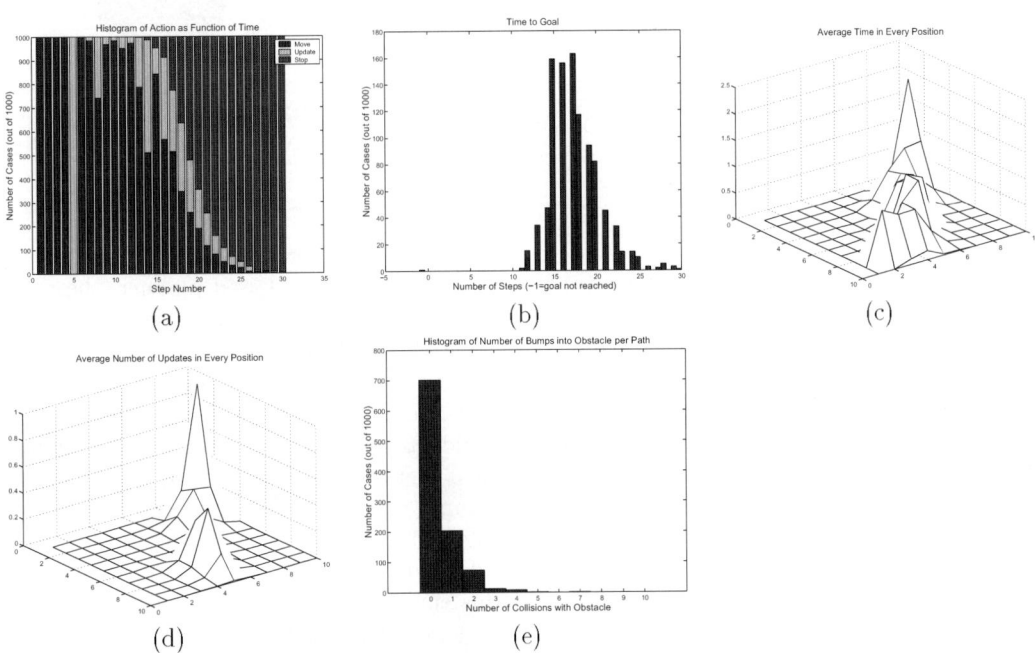

Figure 8: First environment with a second motion planner (as in Fig. 3(b)). (a) Histogram of the actions performed in each time step. (b) The time it took to reach the goal. (c) The average "presence" of the robot along the path. (d) Places where the sensor was invoked. (e) Histogram of the number of times an obstacle was hit along the path.

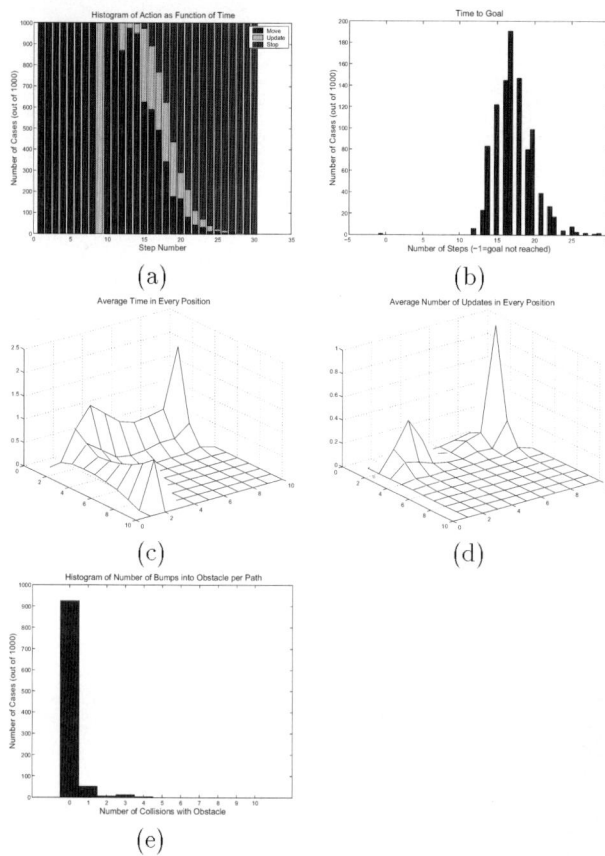

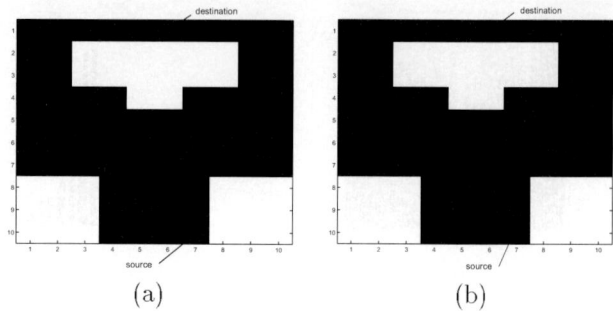

Figure 9: Second environment with two motion plans on it. (a) Potential field based motion plan. (b) Visibility graph based motion plan.

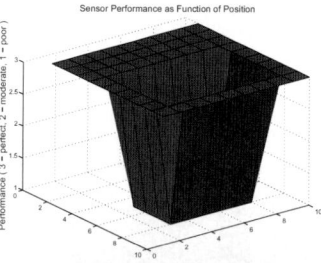

Figure 10: Sensor used on the second environment

Figure 5: First environment with perfect sensing. (a) Histogram of the actions performed in each time step. (b) The time it took to reach the goal. Note that now in almost all of the runs the goal was reached. (c) The average "presence" of the robot along the path. (d) Places where the sensor was invoked. (e) Histogram of the number of times an obstacle was hit along the path.

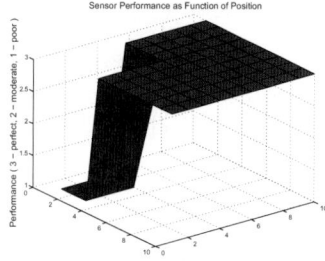

Figure 6: Sensor performance in the first environment.

it usually took over 20 time steps. However, when using the visibility graph based planner, the robot had only a 2% chance of not reaching the goal, and when it did reach the goal, it usually took less time steps. Looking at the chances of colliding with an obstacle, we see that using the visibility graph planner our chances of colliding are above 60% (in less than 400 runs of the 1000 there were no collisions), while the potential field planner gives us a chance of collision of approximately 45%. Depending on our preferences, we may now decide which of the motion planners should be used in this environment with this specific sensor.

The last example relates to the environment shown in Fig. 12. We have used this environment and motion planner in conjunction with a perfect sensor. In this example we want to illustrate the effects of choosing different rewards. We have used two sets of rewards, where the second set penalizes more severely collisions with obstacles and uses of the sensor. The results from the first set of rewards are shown in parts (a)-(c) of Fig. 13, while the results that were obtained with the second set of rewards are shown in parts (d)-(f) of the figure. As may be seen, the original motion plan has been used in part (a) of the figure, but it has been abandoned completely in the second case (part (d)) because it called

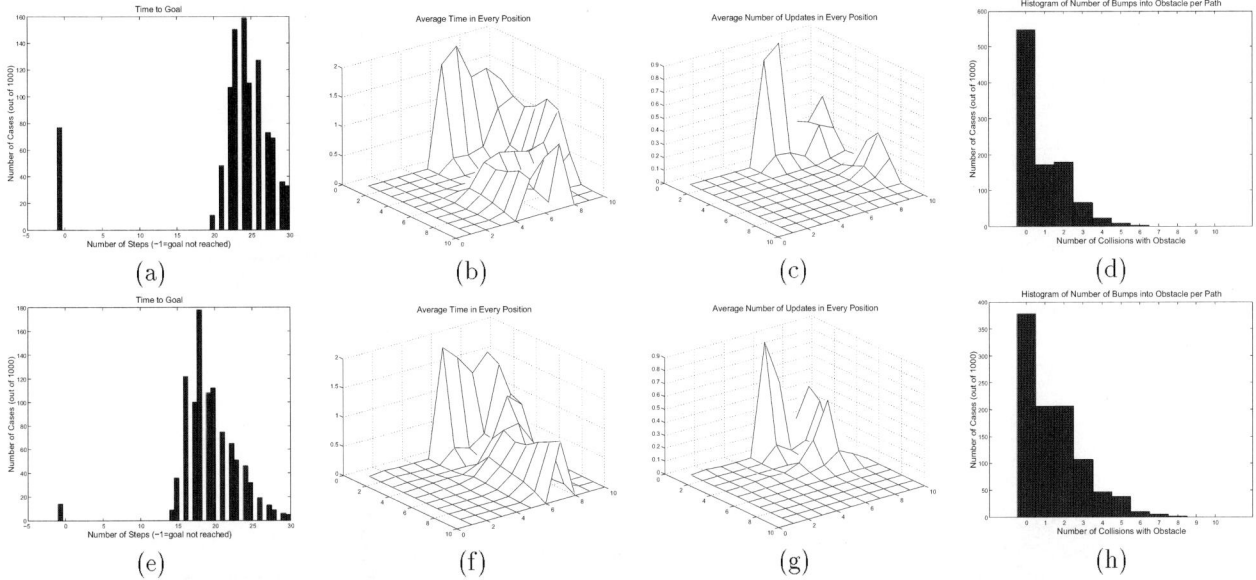

Figure 11: Second environment. Top row refers to a potential field planner and bottom row to a visibility graph motion planner. (a),(e): The time it took to reach the goal. (b),(f): The average "presence" of the robot along the path. (c),(g): Places where the sensor was invoked. (d),(h): Histogram of the number of times an obstacle was hit along the path.

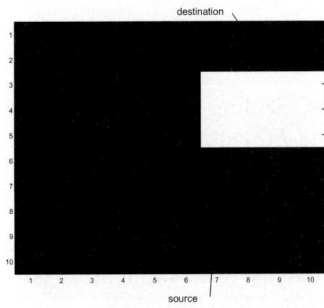

Figure 12: Third environment

for travel near an obstacle, which is very undesirable under this choice of rewards.

4 Discussion and Conclusions

The navigation problem involves a number of different aspects and factors such as the environment, the sensors and the algorithms. We have chosen the term navigation package to denote a specific combination of these factors. In this paper and a companion paper [1] we have discussed the problem of comparing navigation packages.

The algorithm we have presented in this paper uses a simple version of reinforcement learning. We have applied this simple algorithm to a number of test environments and have shown that it produces rather "reasonable" policies.

Besides enabling the comparison of different navigation packages, the POMDP framework allows us to augment a nominal path with sufficient sensing. In contrast to other approaches, we recognize the fact that in some cases the robot may not need to know its exact position. An optimal policy will choose sensing actions only in cases where reduction of the position uncertainty is actually required.

We believe that this work is a first step towards a *meta-algorithm* for choosing between different navigation algorithms and/or sensors. In order to completely achieve this goal, more research is required. Firstly, the POMDP framework we have used can compare two different navigation packages for navigating from a given initial position to a final position. If we want our meta-algorithm to customize a sensor for an environment for example, we must consider all possible combinations of initial and goal positions. Solving a POMDP for every possible combination is not feasible currently. Therefore our framework may not be used in a straight-forward manner for such tasks.

Additionally, solving large POMDPs (thousands of states) even approximately is still beyond our capability. For larger environments techniques such as multi-resolution or division of the environment into sub-

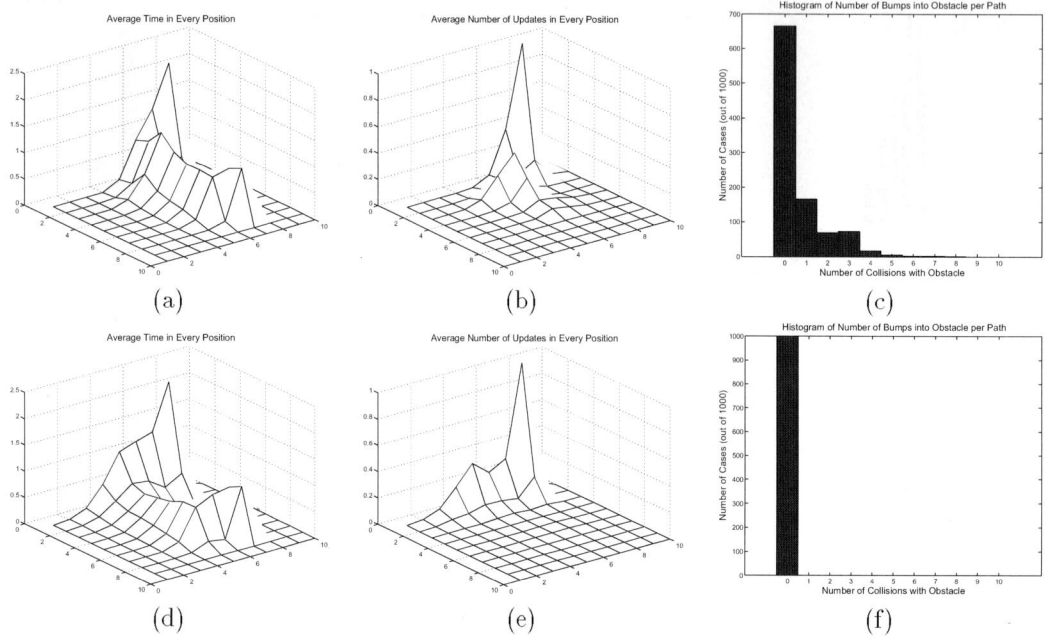

Figure 13: Using different rewards, environment as in Fig. 12. The bottom row shows the results of using a policy which was found for rewards which penalize collisions and position updates more severely. (a),(d): Paths taken. Notice in part (d) how the robot first moves to the left in order to keep a safe distance from the obstacle region. In part (a) the robot "sticks" to the nominal motion plan. (b),(e): Where the sensor was invoked. (c),(f): Collisions with obstacles.

environments may be required. Our future research will focus on these issues in order to enable the use of a meta algorithm for navigation in larger and more "real life" environments.

Acknowledgment A. Adam and I. Shimshoni were supported in part by Israeli Ministry of Science Grants no. 9766 and no. 2104.

References

[1] A. Adam, E. Rivlin, and I. Shimshoni. Towards a Meta Motion Planner A: Model and Framework. Proceedings of ICRA 2001.

[2] M. Hauskrecht. *Planning and control in stochastic domains with imperfect information.* PhD thesis, MIT, 1997.

[3] L. P. Kaelbling, M. L. Littman, and A. R. Cassandra. Planning and acting in partially observable stochastic domains. *Artificial Intelligence*, 101:99–134, 1998.

[4] M. L. Littman, A. R. Cassandra, and L. P. Kaelbling. Learning policies for partially observable environments: Scaling up. In *Proc. of 12'th Int. Conf. on Machine Learning*, pages 362–370, 1995.

[5] T. Mitchell. *Machine Learning*. McGraw-Hill, 1997.

[6] R. Parr and S. Russell. Approximating optimal policies for partially observable stochastic domains. In *Proc. of 14'th IJCAI*, 1995.

[7] S. Thrun. Monte Carlo POMDPs. In *Proc. of NIPS*, 1999.

Efficient Scheduling of Behavior-Processes on Different Time-Scales

Andreas Birk and Holger Kenn
Vrije Universiteit Brussel, Artificial Intelligence Laboratory,
Pleinlaan 2, 1050 Brussels, Belgium

Abstract

In behavior-oriented robotics, the control of a system is distributed over various processes or behaviors running in virtual parallel. But the assignment of processing power to different processes is a non-trivial task. Existing approaches to this problem like rate-monotonic scheduling focus on the fulfillment of deadlines, i.e., upper time-bounds. For behavioral control, the periodicy of processes is also of interest. Here, a novel scheduling algorithm for behavioral processes at different time-scales is introduced, which ensures time-optimal and periodically balanced execution.

1 Introduction

Behavior-oriented robotics has matured in the last 15 years from a scientific critic of "classical" AI [Bro86, Bro91, Ste91] into a wide robotics field [Ark98] including a range of applications [Bir98]. As pointed out in [Ste94], the notion of behavior is used within a wide range of interpretations. Nevertheless, the common property of all behavior-oriented systems is that the overall control of the system is distributed over various processes running in virtual parallel. A major technical problem with this approach is that it needs *scheduling*, i.e., a scheme to assign processing power to different behavior-processes such that they seem to run in virtual parallel.

Existing behavior-oriented programming languages like the subsumption architecture [Bro86, Bro90] or motor schemas [Ark87, Ark92] came out of early scientific work in the field of behavior-oriented robotics. Accordingly, they did not incorporate any considerations on efficiency or software-engineering, forcing the user to do a lot of hand-tailoring for each particular application, especially in respect to scheduling. As a consequence, these languages are not widely distributed. Instead, the complete software environment for every behavior-oriented project around the globe is usually developed from scratch.

Scheduling is a major research topic in a completely different scientific area, namely the field of real-time systems [BW97, Mel83, You82], which provides a wide range of solutions to the problem. One approach for behavior-oriented robotics accordingly is to use a real-time operating-system and programming-language and to build the control-behaviors on top. The problem is that standard real-time approaches, especially the widely used rate-monotonic scheduling [LL73, LSD89], focus on the fulfillment of deadlines, i.e., upper time bounds for the execution of the processes. For control, it is in addition of interest that behavior-processes are executed as regular as possible, i.e., the periodicity plays a significant role.

Here, a novel scheduling algorithm, the so-called B-scheduling, is presented which handles behaviors running on different time-scales represented through so-called exponential effect priorities. These special priorities allow our special algorithm to achieve both desired features for behavior-scheduling, namely time-optimal and periodically balanced execution.

2 Behaviors and Time-Scales

Behavioral processes in general can span very different time-periods. The pulse-width-modulation (PWM) as speed-control has for examples to operate for some DC-motors in the 20 kHz range, i.e., on a time-basis of $5 \cdot 10^{-5}$ seconds. A behavior monitoring batteries in contrast operates on a scale of minutes. Some adaptive or learning behaviors could operate on much higher scales like hours or even days.

We hypothesize that in general it is desirable to span several orders of magnitude of time-periods. A linear priority scheme is not suited for this. Therefore, so-called *exponential effect priorities* are introduced here. The idea is that for each increase in a priority-value by one, the periodicity is halved.

In the remainder of this paper the following naming conventions are used: the set of processes: $\mathcal{P}=$

$\{p_0, ..., p_{N-1}\}$, the priority-value of process p_i: $pv[p_i]$, the set of processes with priority k or the k-th priority class: PC_k, and the highest used priority-value: $maxpv$.

The exact semantic of a priority-value $pv[p_i]$ of process p_i within *exponential effect priorities* is:

- $pv[p_i] = 0 \iff p_i$ is executed with the maximum frequency f_0

- $pv[p_i] = n \iff p_i$ is executed with the frequency f_n which is half the frequency of the previous priority-class, i.e., $f_n = f_{n-1}/2$

3 The Chores of Scheduling

For solving the task of finding a suitable order of execution of the processes, we use a *cyclic executive scheduling* approach [BW97]. This means there is a so-called *major cycle*, which is constantly repeated. The major cycle consists of several so-called *minor cycles*. Each minor cycle is a set of processes, which are executed when the minor cycle is activated.

The general problem of finding a suitable schedule within this approach is NP-hard as it can be reduced to the Bin-Packing-problem in a straightforward manner. We present an extremely efficient, namely linear-time algorithm, which is based on the restriction to exponential-effect-priorities. As motivated above, we do not see this as a limitation, but even as a feature.

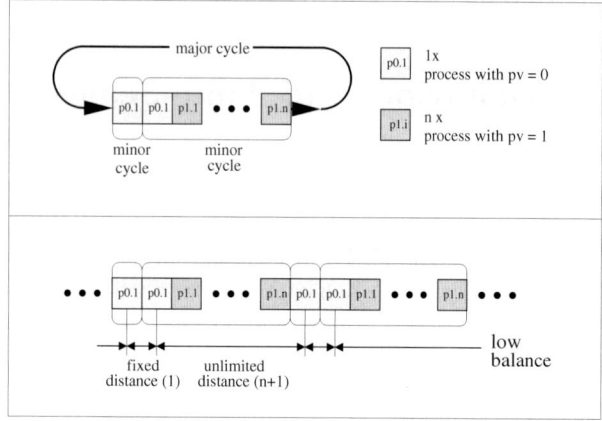

Figure 2: The simple scheduler S_1 leads to a so-called unbalanced execution. One minor cycle can consist of a single process $p0.1$ while a second minor cycle contains unlimited many other processes. Hence, the execution of $p0.1$ is not evenly spread.

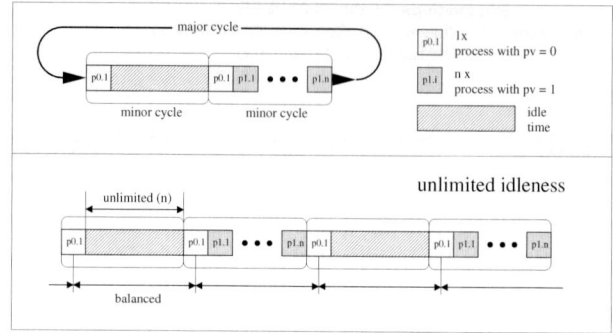

Figure 3: Adding idle-time to balance the schedule made by S_1 can lead to an unlimited waste of time.

Before the algorithm for behavior scheduling or *B-scheduling* is presented, an example is considered to illustrate the problems involved in scheduling. Figure 1 shows a simple algorithm, which schedules behaviors based on their priorities. The major cycle is simply a loop proceedings in rounds. The minor cycle simply executes all processes of priority-class PC_k in every round which is a multiple of 2^k.

The major problem with this algorithm is illustrated in figure 2. Assume there is a single process $p0.1$ with priority 0 and n processes $p1.i$ with priority 1. So, $\#PC_0 = 1$ and $\#PC_1 = n$. The first minor cycle consists of $p0.1$. As S_1 executes all processes of a priority-class together, the second minor cycle includes all processes with priority 0 and priority 1, i.e., this minor cycle has $n+1$ processes. From a naive viewpoint, we can simply say that the processes are badly distributed.

In a more formal approach, the so-called *balance* of a schedule S is defined as

$$balance(S) = \min \frac{\min\ dist(p_i, x)}{\max\ dist(p_i, y)}$$

with $dist(p_i), z)$ is the number of processes which are executed between start of the execution of p_i in cycle z and its next execution in cycle $z + 2^{pv[p_i]}$. If the balance is one then the schedule manages an equidistant spreading of every process over the cycles. If the balances is close to zero than there is at least one process which is very unevenly executed.

A small balance is very bad. As illustrated in the above example, a process with low priority-value, i.e., a process which should be executed very often, has to wait for an unbounded time-period. This is also

```
1   /* Execute the Major Cycle */
2   for(round = 0; round < n_mic; round = round + 1) {
3       /* Execute the Minor Cycle */
4       ∀p_id ∈ P: {
5           if(round modulo 2^{pv[p_i]} == 0) {
6               execute p_id
7           }
8       }
9   }
```

Figure 1: A simple scheduler S_1 which is very inefficient.

expressed by the balance of S_1 which is in this case:

$$balance(S_1) = \frac{1}{n} = 0 \text{ for } n \to \infty$$

The balance of S_1 can be improved by adding idle-time as illustrated in figure 3. This way, the balance can even be tuned to reach the optimum of one. But this is bought at the cost of an unlimited waste of time. The *idleness* as the sum of idle-times in a major cycle is now unbounded.

In general, a schedule S is *time-optimal* if and only if the idleness is zero.

4 Properties of B-Scheduling

The workload WL within a major cycle can be computed as the sum of the occurrences of each process, i.e.,:

$$WL = \sum_{0 \le i \le maxpv} \#PC_i \cdot 2^{maxpv-i}$$

The number n_{mic} of minor cycles per major cycle is determined by the highest priority-value $maxpv$ as the process or the processes with this priority has/have to be executed once per major cycle. It follows that the average number av of processes per minor cycle has to be

$$av = WL \,/\, n_{mic} \text{ with } n_{mic} = 2^{maxpv}$$

For an even distribution of the workload, the actual number of processes in a minor cycle has to be equal to the average number av. Unfortunately, av is not necessarily an integer. Therefore, we define

$$perfect = \lceil av \rceil \text{ and } dirty = \lfloor av \rfloor$$

A so-called *perfect minor cycle* has perfectly many processes, whereas the number of processes in a *dirty minor cycle* accordingly is dirty. A *bad minor cycle* includes more than *perfect* or less than *dirty* many processes. B-scheduling computes a schedule S_B such that

1. S_B is time-optimal

2. the major cycle consists only of perfect and dirty minor cycles

3. the processes are distributed over the cycles in an optimal manner, i.e., p_i is executed in cycle $c + 2^{pv[p_i]}$ if and only if p_i is executed in cycle c

It follows from properties 2 and 3 that S is well balanced as

$$balance(S_B) = \frac{dirty + 1}{perfect + 1}$$
$$= 1 \text{ for } av \to \infty$$

The worst-case balance of S_B is 1/2 when only two processes are used and one is more frequent than the other. In general, the balance becomes better the more work-load is handled in each minor cycle. It can be shown that any time-optimal schedule S has to consist of a set of dirty and perfect cycles. Hence, S_B provides the best balance that is possible.

```
1   /* Initialization */
2   /* computing the initial wait-values for each process p_id */
3   quicksort(P)
4   pc = 1
5   start = 0
6   n_slots = 1
7   ∀i ∈ {0, ..., maxpv − 1} : {
8         start = 2 · start
9         n_slots = 2 · n_slots
10        ∀id with pv[p_id] = pc : {
11              wait[p_id] = reverse((start + id) modulo n_slots)
12        }
13        start = (start + #{p_id | pv[p_id] = pc}) modulo n_slots
14        pc = pc + 1
15  }
```

Figure 4: The initialization of B-scheduling.

```
1   /* Execute the Major Cycle */
2   for(round = 0; round < n_mic; round = round + 1) {
3         /* Execute the Minor Cycle */
4         id = 0
5         done = 0
6         while( (done < perfect) ∧ (id < #P) ) {
7               if(wait[p_id] == 0) {
8                     execute p_id
9                     wait[p_id] = 2^{pv[p_id]}
10                    done = done + 1
11              }
12              id = id + 1
13        }
14        ∀p_id ∈ P : if(wait[p_id] > 0) : wait[p_id] = wait[p_i] − 1
15  }
```

Figure 5: The execution of a B-schedule.

name	p1.1	p1.2	p1.3	p2.1	p2.2	p3.1	p4.1	p4.2	p4.3
$pv[]$	1	1	1	2	2	3	4	4	4
$2^{pv[]}$	2	2	2	4	4	8	16	16	16
wait	0	1	0	1	3	0	4	12	2

Table 1: A set of processes with their priority-values $pv[]$, their according waiting-time between executions, and their initial wait values calculated with the algorithm shown in figure 4. The wait values lead to the schedule shown in table 2 when the B-scheduler 5 is invoked.

5 The Heart of B-Scheduling

Figure 4 and figure 5 show the critical parts of B-scheduling in a pseudo-code. An important variable in both parts is $wait[p_{id}]$. It specifies for each process p_{id} how long it has to wait in number of cycles until it is executed again. During the execution of a B-schedule (figure 5), wait is constantly decremented in each cycle. When a process p_{id} is executed, its wait $wait[p_{id}]$ is set to $2^{pv[p_{id}]}$. Therefore, the execution of p_{id} is spread evenly other the minor cycles in the major cycle.

minor cycle number	processes within the cycle
0	p1.1 p1.3 p3.1
1	p1.2 p2.1
2	p1.1 p1.3 p4.3
3	p1.2 p2.2
4	p1.1 p1.3 p4.1
5	p1.2 p2.1
6	p1.1 p1.3
7	p1.2 p2.2
8	p1.1 p1.3 p3.1
9	p1.2 p2.1
10	p1.1 p1.3
11	p1.2 p2.2
12	p1.1 p1.3 p4.2
13	p1.2 p2.1
14	p1.1 p1.3
15	p1.2 p2.2

Table 2: A simple example of a major cycle computed with B-scheduling. The notation $pX.Y$ denotes process number Y within priority-class PC_X. Note that there is no straight-forward distribution of dirty and perfect cycles, i.e., minor cycles which consist in this example of either two or three processes.

The dynamic execution part of a B-schedule (figure 5) is more or less straightforward. The "real magic" is done in the static initialization of the $wait$-values (figure 4). Note that the initial value of $wait[p_{id}]$ determines in which minor cycle p_{id} will be executed for the first time. So, computing suited initial waits produces a B-schedule. Note, that the number of $wait$-values is equal to the number of processes $\#\mathcal{P}$. So, the complete schedule which is of size $O(2^{\#\mathcal{P}})$ is represented in a single variable in each processes, i.e., in the overall size $O(\#\mathcal{P})$.

Before discussing the initialization of the $wait$-values in more detail, a special command from figure 4 has to be explained. The `reverse()` is used to reverse the bit-order of a binary number. More concretely, let $B_n = [b_0, ..., b_{n-1}]$ and $R_n = [r_0, ..., r_{n-1}]$ denote two binary numbers, each represented as array of bits b_i, respectively r_i. The function `reverse()` is then defined as:

$$\texttt{reverse}(B_n) = R_n \text{ with } r_i = b_{n-i}$$

The main idea when computing suited initial $wait$-values is as follows. Imagine a set $\mathcal{S}$ of natural numbers with a cardinality equal to a power of 2. Let $S(start, d)$ denote a sequence which begins at the number $start$ and "jumps" further to numbers x which are distance d away, i.e., $x = (k \cdot d) \texttt{modulo} \#\mathcal{S}$ with $k \in \mathbb{N}$. When $start$ and d are powers of 2, S is called harmonic. It holds that for each harmonic list S, we can create two harmonic lists S_1 and S_2 such that $S = S_1 \cup S_2$, namely:

- $S_1 = S(start, 2 \cdot d)$
- $S_2 = S(start + d/2, 2 \cdot d)$

The overall set $\mathcal{S}$ can be expressed as $S(0, 1)$. It can recursively be divided in smaller lists and sublist.

When computing the initial $wait$-values, the goal is to distribute processes such, that the minor cycles are equally filled up. Each execution process of class PC_k

can be seen as a list $S(start, 2^{maxpv-k})$ of minor cycles. The first value for *start* is zero, i.e., the first slot in the first minor cycle is used. The distance d is $2^{maxpv-pv[p_0]}$. From then on, further lists can be computed. The difficulty is to keep track of the *start* position. Especially, so-to-say left-overs, i.e., empty lists not used up by class PC_{k-1}, have to be used when the class PC_{k-1} is handled.

Table 1 shows as an example a set of processes with their priority-values $pv[]$, their according waiting-time $2^{pv[]}$ between executions, and their initial wait values calculated with the algorithm shown in figure 4. The interested reader can try to find a time-optimal, well balanced schedule of the processes (of course without using the pre-computed wait-values). The time-optimal, well balanced schedule computed by B-scheduling is shown in table 2.

6 Conclusion

Behavior-oriented robotics is based on the distribution of control over various processes running in virtual parallel. Scheduling these processes, i.e., assigning processing power to them, is a non-trivial task. Existing approaches from the field of real-time systems mainly focus on the fulfillment of deadlines. For control, it is also important that behavior-processes are executed as regular as possible in addition. Here, the novel algorithm of B-scheduling is presented which handles behaviors running on different time-scales, represented through so-called exponential effect priorities. B-scheduling ensures both desired properties for behavior-control, namely time-optimal and periodically balanced execution of processes.

References

[Ark87] R. C. Arkin. Motor schema based navigation for a mobile robot. In *Proc. of the IEEE Int. Conf. on Robotics and Automation*, pages 264–271, 1987.

[Ark92] Ronald C. Arkin. Cooperation without communication: Multiagent schema-based robot navigation. *Journal of Robotic Systems*, 9(3):351–364, April 1992.

[Ark98] Ronald C. Arkin. *Behavior-Based Robotics*. The MIT Press, 1998.

[Bir98] Andreas Birk. Behavior-based robotics, its scope and its prospects. In *Proc. of The 24th Annual Conference of the IEEE Industrial Electronics*. IEEE Press, 1998.

[Bro86] Rodney A. Brooks. A robust layered control system for a mobile robot. In *IEEE Journal of Robotics and Automation*, volume RA-2 (1), pages 14–23, April 1986.

[Bro90] Rodney A. Brooks. The behavior language; user's guide. Technical Report A. I. MEMO 1227, Massachusetts Institute of Technology, A.I. Lab., Cambridge, Massachusetts, April 1990.

[Bro91] Rodney Brooks. Intelligence without reason. In *Proc. of IJCAI-91*. Morgan Kaufmann, San Mateo, 1991.

[BW97] Alan Burns and Andy Wellings. *Real-Time Systems and Programming Languages*. Addison-Wesley, 1997.

[LL73] C. L. Liu and James W. Layland. Scheduling algorithms for multiprogramming in a hard-real-time environment. *Journal of the ACM*, 20(1):46–61, January 1973.

[LSD89] I. Lehoczky, L. Sha, and Y. Ding. The rate monotonic scheduling algorithm: Exact characterization and average case behavior. In IEEE Computer Society Press, editor, *Proceedings of the Real-Time Systems Symposium - 1989*, pages 166–171, Santa Monica, California, USA, December 1989. IEEE Computer Society Press.

[Mel83] Mellichamp. *Real-Time Computing*. Van Nostrand Reinhold, New York, 1983.

[Ste91] Luc Steels. Towards a theory of emergent functionality. In Jean-Arcady Meyer and Steward W. Wilson, editors, *From Animals to Animats. Proc. of the First International Conference on Simulation of Adaptive Behavior*. The MIT Press/Bradford Books, Cambridge, 1991.

[Ste94] Luc Steels. The artificial life roots of artificial intelligence. *Artificial Life Journal*, 1(1), 1994.

[You82] SJ Young. *Real Time Languages*. Ellis Horwood, 1982.

An efficient Depalletizing System based on 2D Range Imagery

D.K. Katsoulas[1], D.I. Kosmopoulos[2]

[1] University of Freiburg, Institute for Pattern Recognition and Image Processing, Freiburg i.Br., D-79085

[2] National Technical University of Athens, Department of Electrical Engineering, Division of Computer Science, Athens, GR-157 73

Email: [1]dkats@informatik.uni-freiburg.de, [2]dkosmo@telecom.ntua.gr

Abstract

In this paper we propose an efficient approach towards the solution of the depalletizing problem, based on active vision. We describe a system comprising an industrial robot and a time of flight laser sensor, which performs the depalletizing task in real time, and independently of lighting conditions. In our case, the target objects are solid boxes of known identical dimensions, neatly layered but with arbitrary orientation within a layer, which are all placed on a platform. The layered structure of the target platform allows for two-dimensional imagery. The system locates the position of the boxes by tracking one of the corners they expose to the laser source. The system locates the desired corners by applying the scan line approximation technique [11], adapted to fit the needs of our application, to the two-dimensional input data. The advantages of our system over existing applications are its simplicity, robustness, speed and ease of installation.

1. Introduction

This paper addresses the depalletizing problem (or bin picking problem) in the context of which, a number of objects, of arbitrary dimensions, texture and type must be automatically located, grasped and transferred from a pallet, to a specific point defined by the user. The objects can be either placed on the pallet in a structured manner, for example, in the case of boxes, when they are placed on layers, or they can be jumbled. The requirement of a robust and generic automated depalletizing system stems primarily from the car and the food industries, where boxes and sacks of various dimensions, texture, and weight, either neatly placed, or jumbled, are laid on a pallet. By the term pallet we mean a rectangular platform like a table. A grouping of the configurations met at target pallets which are usually encountered at distribution centers or automobile industry factories has as follows: neatly placed identical sacks, jumbled identical sacks, neatly placed identical cardboard boxes, neatly placed identical boxlike objects, neatly placed boxlike objects or cardboard boxes or both, with varying dimensions, jumbled cardboard boxes, with varying dimensions. In this paper, we are addressing the third case, according to which neatly placed cardboard boxes are grasped and unloaded, but what is notable is that the technique used, could form a basis for the realization of a more general solution.

An automated system for depalletizing is of great importance because it undertakes a task that is very boring, strenuous and sometimes quite dangerous for humans. Furthermore, the robust and fast transfer of goods from an originating pallet to a target position (pallet, conveyor belt etc.) can accelerate significantly the logistics-processes in the industry and the warehouses. Thus much time and labor is saved and the costs are minimized.

1.1 Related Work

In general, the existing systems can be sorted in the following categories: systems incorporating no vision at all, and systems incorporating vision. The majority of the systems employed in industrial depalletizing applications so far, do not contain any vision modules. They usually employ preprogrammed gantry robots for bulk depalletizing tasks (e.g. [22], [23], [25], [26], [28]). Such systems may be much more effective in strictly controlled environments, (since they are fast and accurate), but they fail in adverse environments e.g. in a distribution center, where the pallet's position is not well-defined and the objects to be picked may be arbitrarily jumbled due to human intervention. In the initial attempts to use visual data, 2D techniques were employed, combined with additional sensors in the gripper to obtain the third dimension [14], [15]. These systems were limited to simple objects with special surface properties, e.g. identical cylindrical pieces of metal with a ground surface. Similar techniques are still used in the industry [24], where a feature-based approach identifies the objects to be picked by locating on them patterns such as letters or logos. These logo-tracking techniques are as well used in three-dimensional imagery [27]. The disadvantage of these methods is that they have all the problems of camera-based identification. The matching can fail in the case that the pattern does not appear, e.g. due to poor lighting conditions, reflections or dust. Furthermore, this method needs training for all patterns and is inappropriate in case that there is no obvious feature. Attempts incorporating range imagery seem much more promising. In [2] a structured light range sensor with additional sensors (force and proximity sensors) on the gripper were successfully employed, to deal with unloading piles of postal parcels with very good results. Even if the vision methods used did not allow for high accuracy, the idea of usage of range imagery and sensor fusion, resulted in a relatively efficient system. Nevertheless, the vision algorithm employed can only deal with planar objects. Its extension towards more complicated configurations is not straightforward. The recognition algorithms endorsed on the other hand, can take from 15 to 45 seconds to detect the position of only one graspable object at a time. Additionally, the disturbing of the pallet by the robot when the recognition system fails, may damage the target objects. The system of Vayda and Kak [20] deals with depalletizing of jumbled cylinders and parallelepipeds of unknown dimensions with the help of a range sensor. The authors attempt a complete scene understanding, via processes originating from artificial intelligence. According to these methods, size and pose estimation of target objects are facilitated by virtually extending their dimensions in the direction away from the sensor until they physically contact other objects in the scene The adoption of such methods results in an time-inefficient system, since the interpretation of complex scenes can take up to 20 minutes [2] in non-specialized hardware. The system closer to ours, is the one developed by Chen and Kak [5]. In both systems, a hypothesis generation for the existence of a target is based on Feature Sets (vertices in 3D in [5], corners in 2D here) detection. The hypothesis verification is performed by hardware in our case.

A direct comparison in grasping accuracy between the existing systems and ours is not possible, since no detailed accuracy measurements are provided by the systems' constructors. In the context of layered depalletizing of boxes, our framework surpasses all the existing systems as far as the speed, the ease of installation and the robustness are concerned because 2D - and thus simpler- image processing is applied.

A detailed description of our approach, as well as concrete experimental results follow in the subsequent paragraphs.

2. Depalletizing using a laser sensor

The system comprises a vision system that is responsible for the detection of the object's position and an industrial robot, which grasps the boxes. This system is based on one time-of-flight laser sensor, which is mounted on the robot's arm. The sensor is integrated on the tool, which performs the boxes' grasping (gripper) and is seamlessly attached to the robot's flange (figure 1). The laser sensor has specific advantages over cameras, since the measurements are not affected by environmental conditions like target objects' surface reflectance or lighting. The input data of the vision subsystem, is a set of two-dimensional points, which are defined as the intersection of the objects in front of the laser sensor, with the sensor's scanning plane, the range of which is adjustable. The integration of the laser sensor on the robot-hand allows for accurate data acquisition. Upon the input data, efficient algorithms that will be subsequently described are performed, which detect the boxes on the pallet. The system is characterized by speed, robustness, accuracy, low cost and ease of installation.

2.1 System configuration

A side view of the whole system and the pallet is presented in figure 1. The origin of the base coordinate system (B; x_b, y_b, z_b) is located on the base of the robot. The relative position and orientation of (B; x_b, y_b, z_b) to the world coordinate system (W; x_w, y_w, z_w) is known a-priori [16]. We attach the coordinate systems (T; x_t, y_t, z_t), (S; x_s, y_s, z_s), (F; x_f, y_f, z_f) to the robot tool (gripper) the sensor and the robot flange correspondingly as depicted in figure 1. We also attach the (C; x_c, y_c, z_c) to the identified corner of the box so that x_c, y_c are parallel to the box sides. The sensor provides the corner coordinates in the (S; x_s, y_s, z_s), and the tool moves using the coordinates of the box centre transformed in (T; x_t, y_t, z_t). The scanning plane is defined by the axes x_s, y_s.

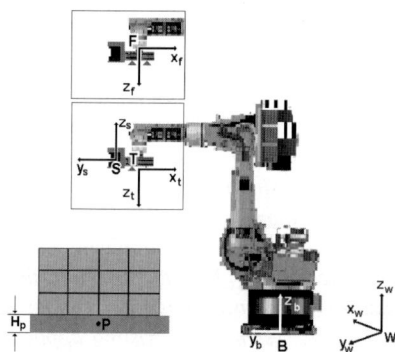

Figure 1. The world (W; x_w, y_w, z_w), robot-base (B; x_b, y_b, z_b), robot-tool (T; x_t, y_t, z_t), robot-flange (F; x_f, y_f, z_f) and sensor (S; x_s, y_s, z_s) coordinate systems.

If ^IT_J denotes the homogeneous transform matrix from (I; x_i, y_i, z_i), to (J; x_j, y_j, z_j) then ^TT_S is obtained by calibration (see experimental results section), ^ST_C is calculated from the measurement data, and ^TT_F is given by the tool manufacturer.

The approximate position of the pallet P in (B; x_b, y_b, z_b), the height of the pallet basis H_p, the layer height L_h and the maximum pallet height Hh_{max} are also required. These parameters are defined during the system's installation phase.

2.2 Operation

The flow diagram of the system is illustrated in figure 2. Initially the height of the upper layer of the heap is estimated. The sensor is positioned at distance T_h (figure 3b) from the ground where:

$$T_h = Hh_{max} + H_p - L_h/2 \qquad (1)$$

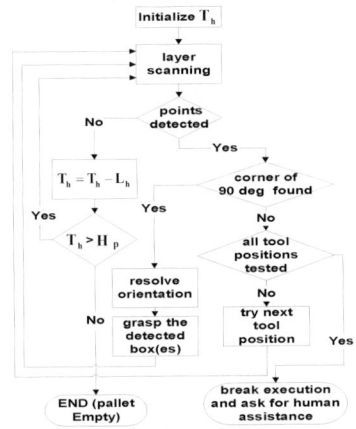

Figure 2. The flow diagram of the overall process

The scanning plane is parallel to the ground and z_s has the direction of z_b. The vector y_s forms an angle of approximately 45 degrees with the pallet's side (figure 3a) in order to achieve higher accuracy (section 3) in the frequent case that the boxes are neatly placed on the layers of the pallet.

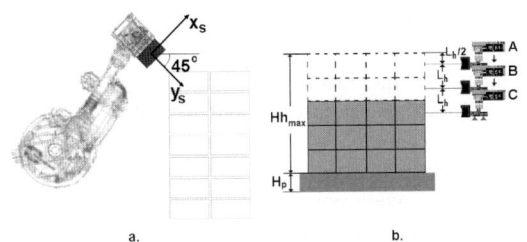

Figure 3 (a) The initial robot position – top view (b) The heap height estimation procedure – side view

The sensor performs a layer scanning and if no points are detected its height is decreased by L_h. The procedure is repeated until points are detected in the scanning area, e.g. at position C in figure 3b. From the acquired range data a scan line is derived and the box corners are extracted. In the event that no box corners could be extracted (figure 4a) the sensor is moved and rotated in order to acquire a better view of the boxes (figure 4b). Two such predefined movements are performed and if even now no corners can be detected, the system asks for human intervention. Otherwise the detected boxes are grasped and the layer unloading operation continues until the current layer is empty (no points appearing in the viewing range of the sensor). Then the tool goes down to the next layer and the process is repeated until the height of the tool is less than H_p (that is, when the pallet is empty).

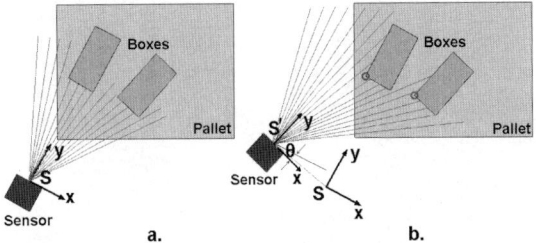

Figure 4 (a) The corner identification problem. (b) The sensor is moved and rotated and the corner(s) is (are) identified

Every time the sensor identifies a corner the grasping procedure is initiated. From the detected corner position, the centre of gravity of the box's upper surface G is calculated in $(S; x_s, y_s, z_s)$, and then it is transformed to $(T; x_t, y_t, z_t)$. The resulting frame is used to position the gripper onto G. If we attach the coordinate system $(C; x_c, y_c, z_c)$ to the identified corner of the box then $^CX_{G1} = (W/2, L/2, 0, 1)$ (case A) or $^CX_{G2} = (L/2, W/2, 0, 1)$ (case B) according to the orientation of the box (figure 5). In $(T; x_t, y_t, z_t)$ the point G will be given by:

$$^TX_G = {^TT_S} \cdot {^ST_C} \cdot {^CX_G} \qquad (2)$$

The orientation of the box is not known in advance and unfortunately it is very difficult to use the laser sensor for finding it, because in many cases the gap between the neighbouring boxes is 2-3mm (smaller than the sensor's resolution). Therefore, we send the gripper to the point where the gripper touches the topside of the box assuming that the orientation case is B (figure 5). If the actual case is B the gripper will sense low pressure and it is guided to $^CX_{G2}$. In the case A the gripper senses high pressure (no edge) and the gripper is sent to $^CX_{G1}$. Of course this solution is not optimal because much time is wasted in the attempt to find the orientation of the box. Later we will try to eliminate this procedure by employing a camera for this purpose.

The system in operation is displayed in figure 6.

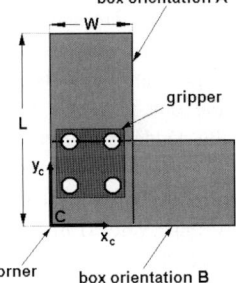

Figure 5. The gripper test position and the orientation cases in box-grasping

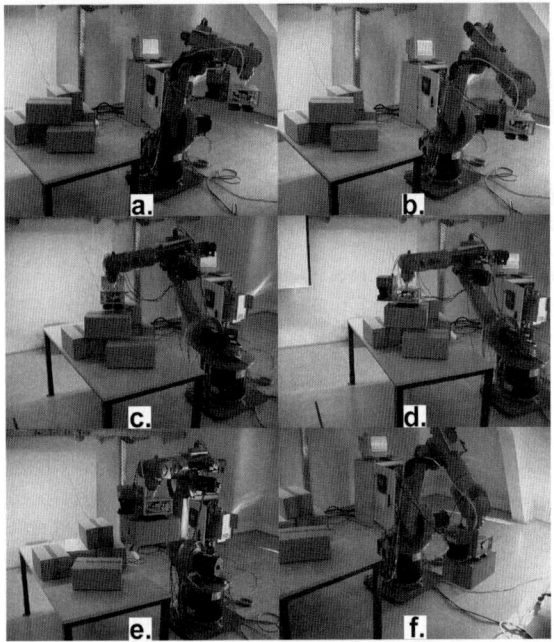

Figure 6 The system in operation: (a)-(b) heap height estimation (b) box localization (c) determination of orientation (d) box grasping (e) box picking (f) box placement.

2.3 Identification of box corners from 2D data

This is the core of the system. The input of the process is a set of planar points which comprise a two dimensional representation of the current layer of the pallet. The target of the process is the localization of the 90 degrees corner of the boxes to be grasped by the robot. In general terms we have to deal with the range image segmentation problem in two dimensions. From the vast literature having to do with range image segmentation based on 2D information (e.g. [1], [4], [7], [11], [21]), an algorithm should be selected, which should be efficient, accurate and able to deal with three-dimensional range data, when, in a future system, the objects on the pallet are not layered. The algorithm should have some sort of qualitative capabilities as well in order to detect the corners' size. An algorithm that satisfies the requirements stated above, is the one proposed by Jiang and Bunke [11], based on the Duda and Hart scan line splitting method [6], which detects and evaluates crease edges (discontinuities in range normal vectors.) and step edges (discontinuities in range values) in range data by using a scan line approximation technique. The particular algorithm was selected because of its simplicity and straight-forwardness, its potential to accurately deal, when combined with an edge grouping technique [18], with 3D configurations [9], [10], [11], [12], and finally its time-efficiency [17], which allows for rapid target detection when no image processing hardware is employed.

The scan line approximation algorithm splits the scan line to segments that can be accurately represented by functions. In figure 7a, a subset of a scan line is approximated with two curved segments namely $f_1(x), f_2(x)$.

From the edge points x_1 and x_2 the midpoint $\bar{x} = (x_1 + x_2)/2$ is extracted, with the help of which the jump and crease edge strengths are calculated as follows:

$$\text{Jump Edge Strength} = \left| f_1(\bar{x}) - f_2(\bar{x}) \right| \quad (3)$$

$$\text{Crease Edge Strength} = \cos^{-1}\left(\frac{(-f_1'(\bar{x}),1)\cdot(-f_2'(\bar{x}),1)}{\left\|(-f_1'(\bar{x}),1)\right\|\cdot\left\|(-f_2'(\bar{x}),1)\right\|}\right) \quad (4)$$

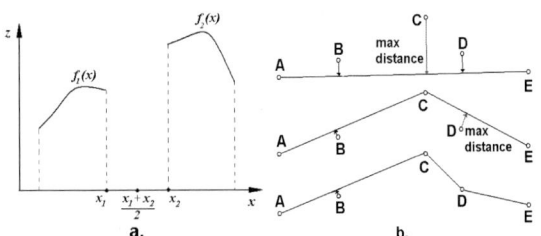

Figure 7 (a) Edge Strength Definition (b) The scan line splitting method

The accuracy in the edge strength value depends on the selected approximation functions. Our application induces the selection of linear functions. Thus, initially we represent each set of 2D points with a linear segment by means of the Duda and Hart [6] splitting method (figure 7b). If we suppose that our scan line comprises the points with labels A-E, a linear segment is initially estimated from the end points and the maximum distance of the line to every point of the scan line is calculated. If no point has a distance greater than T_{split} from the approximation curve, then the process stops for the particular scan line segment, since it is satisfactorily approximated. If the maximum distance is bigger than a predetermined threshold T_{split}, then the whole process is repeated recursively e.g. for the scan lines AC and CE.

Problems of the splitting method are the frequent over-segmentation of the scan line and the not optimal recovery of the real edge position [11]. The first problem can be solved with an appropriate selection of the threshold T_{split} in combination with a merge process. Over-segmentation problems are solved when the threshold T_{split} is increased. However, arbitrary increase in the value of T_{split} produces the under-segmentation phenomenon. In order to adjust the value of T_{split} we must have an a priori estimation of the noise level in the scan line. A solution to the problem is proposed in [3], according to which for every point of the scan line the previous and the next point are considered. Upon these points a straight-line segment is fitted. If the end points of the group modeled with the line, have an absolute difference in their y values bigger than two times the sensor's resolution R plus its random error E (both defined by the vendor) the point initially considered is regarded as belonging to a range discontinuity region and thus discarded. Otherwise, the approximation error is calculated. Given N_p the number of points of the smooth regions and $\epsilon_{1,3}(p)$ the RMSE of the linear segment corresponding to the point p, the image's quality measure ρ is calculated as

$$\rho = \frac{1}{N_p} \sum_p \epsilon_{1,3}(p) \quad (5)$$

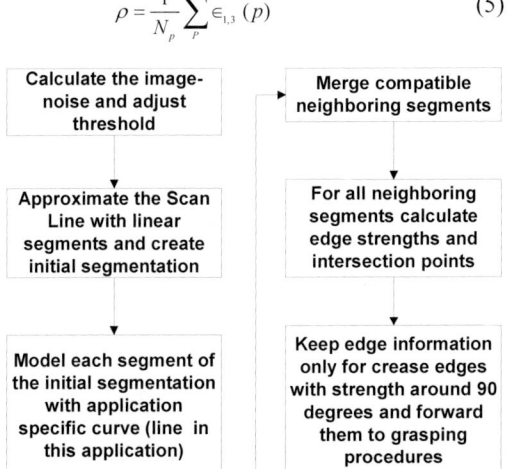

Figure 8. The layer scanning flow diagram

In our approach the threshold T_{split} was set to the value: $T_{split} = \rho + R$ where R is the resolution of the sensor. As mentioned, the second problem of this initial segmentation is that the edges generated do not correspond with accuracy to the real edge points. In order to realize higher accuracy, the initial segments produced are being again approximated. A least square fit is performed to the points comprising the segments originating from the splitting method. Due to the fact that, in some cases the over-segmentation problem could not be alleviated without a merging step, such a step is introduced by checking whether the angle of the normal vectors (AON), of least square modeled neighboring segments is lower than an input parameter.

Figure 8 describes in detail the adopted segmentation method. The corner position and orientation and the AON, are then forwarded to the grasping procedures, which move the gripper and grip the box(es).

3. Experimental results

Experimental setup As already mentioned, the system incorporates an industrial robot, namely the model KR 15/2 manufactured by KUKA GmbH, a square vacuum-gripper, which grips the boxes from their top side, and a time of flight laser sensor namely the model LMS200–30106, manufactured by SICK GmbH, with resolution 10mm, random error of 5mm and acquisition time 13ms. The sensor emits a beam, every f degrees, where -α≤f≤α [19]. In our experiments, f=0.25 degrees and α=50 degrees. The system uses a Pentium PC (400 MHz, 256MB RAM), in which the vision software resides, as well as the necessary hardware needed for the grasping procedure operation. The system from the software point of view, comprises the vision module (implemented in C++), which accepts data from the laser sensor and calculates the position and orientation of the box(es), and the Robot controlling module (implemented in KUKA KRL), which requests correction frames from the vision module and moves the robot.

The inputs of the system during the initialization phase are the position of the center of the pallet P (figure 1) relative to the robot Base B coordinate system, the maximum heap height Hh_{max} (figure 3b), the dimensions H_p and L_h, the coordinate transformation matrices $^T T_S$, $^T T_F$, the sensor's resolution and its random error and finally the segments merging threshold. Only the latter should be extracted by experimentation, (in our experiments set to 2 degrees). A pallet of 1100mm×600mm was used with H_p =1000mm. The target objects were card boxes of 250mm×350mm with L_h =150mm. The sensor was placed 300 mm away from a corner of the pallet.

Calibration In order to calculate the transform matrix $^T T_S$ we executed a calibration procedure (offline). In this phase, we had to deal with the 3D – 3D absolute orientation problem, which is elegantly solved in [8]. According to this solution, when the coordinates of N 3D points relative both to the sensor and the tool reference frame are known, the transformation from the tool to the sensor can be determined by adopting a singular value decomposition approach. We used 5 3D points whose coordinates were known in both (S; $\mathbf{x}_s$, $\mathbf{y}_s$, $\mathbf{z}_s$) and (T; $\mathbf{x}_t$, $\mathbf{y}_t$, $\mathbf{z}_t$). In order to obtain these points, we used a solid box. The positions of the suction pads, which allow for gripping the box from the centre of its topside, were marked. We placed the box in various positions on the pallet and more specifically on the four corners and the center of it. For each position of the box, the position of one of the square box's corners (consequently the box's center) was calculated in (S; $\mathbf{x}_s$, $\mathbf{y}_s$, $\mathbf{z}_s$). For each of the corners, we manually moved the robot's arm, in such a way that the suction pads fitted the marks on the topside of the target box. We noted the tool coordinates provided by the robot's console in (T; $\mathbf{x}_t$, $\mathbf{y}_t$, $\mathbf{z}_t$) and in this respect we corresponded the coordinates of two coordinate systems. We have developed a software tool, which eases the above operation by helping the user to perform all the necessary actions. From the above, it is evident that the installation procedure is simple and can be executed within a reasonable amount of time. In all other systems known to the authors, the system's setup is much more time consuming and strenuous, since it requires sophisticated calibration procedures, training, or elaborate hardware installations on the ceiling of the installation site, or above the pallet.

Algorithm evaluation We verified the algorithm's performance regarding the accuracy and the speed. In order to estimate accuracy, a layer scanning was performed 50 times on a layer comprising up to 5 boxes. Many configurations of boxes were tried (some machine segmentation (MS) outputs are depicted in figure 9). Afterwards, two human operators examined the data. The operators identified the boxes, fitted lines to selected points on the sides of the boxes (they excluded noisy points) and calculated the corner's position, orientation and magnitude (ground truth–GT). The difference in measurements between MS and GT is displayed in (Table 1).

	Corner magnitude error (degrees)	Box orientation error (degrees)	Corner Position error (mm)		MS-GT corner distance (mm)
			X	Y	
Average	1.60	0.072	1.787	1.568	2.521
Standard deviation	1.80	0.137	1.191	1.257	1.531

Table 1. Segmentation algorithm's accuracy

The speed of the segmentation algorithm was estimated by executing the layer scanning operation 10000 times. 50 different boxes' configurations were tried, each one comprising up to five boxes. For each such configuration the system ran 200 times. The results are depicted in table 2. The time needed for a complete segmentation of the scan line, which comprises about 300 2D points is 1.56 ms. If we add the 13 ms of the scan line acquisition time to this quantity we come to the conclusion that the next graspable box can be detected in less than 15 ms, due to the fact that more than one boxes (1.8 on the average according to these experiments) can be detected in the scan line. This is significantly faster than any other solution to the problem proposed, up to our knowledge.

	Time (ms)/hit
Image Quality Calculation	0.41
Scan Line Splitting	0.79
Segment Approximation	0.10
Segments Merging	0.02
Edge Strength Calculation	0.24
Edge Detector (overall process)	1.56

Table 2. Processing time

Overall system's accuracy and robustness We placed the model box used in the calibration phase to various positions on the pallet (figure 10) with the depicted orientation. The position, and orientation of the box corner were measured in the tool coordinate system. Then we directed manually the tool so that the suction pads coincided with the marks on the box. The difference between the values observed on the robot's console and the calculated ones gave the system's overall accuracy measurements. For each position 20 measurements were executed.

The system proved its robustness by unloading the boxes every time for positions 1 to 17. In the cases of 16 and 17 the system was able to locate the box only after the automatic rotation and translation of the tool (due to the problem described in figure 4). Table 3 displays the results for positions 1 to 15. It is evident that as the exposed sides of the box face the laser beam with the same angle (45 degrees) the detection becomes more accurate. In the experiments conducted, in which the robot arm was executing linear movements with the maximum speed and acceleration [16], each box was grasped in less than 3 seconds, on the average.

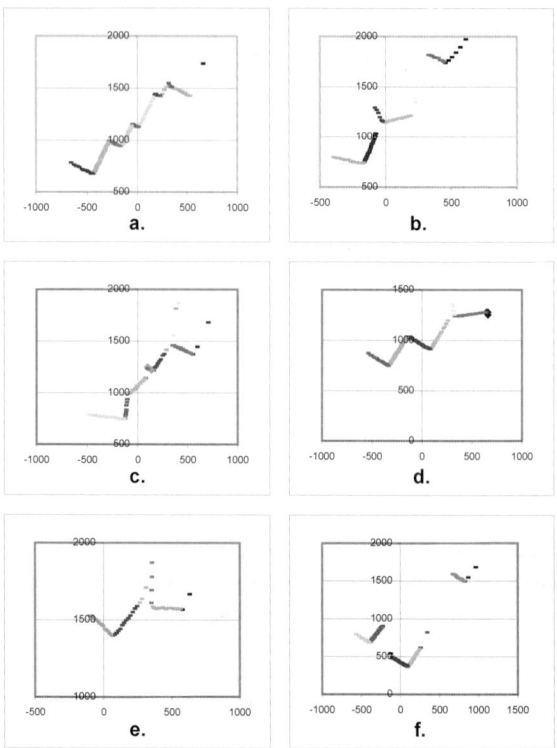

Figure 9 (a)-(f) The output of the segmentation procedure in some typical cases (all dimensions in millimeters).

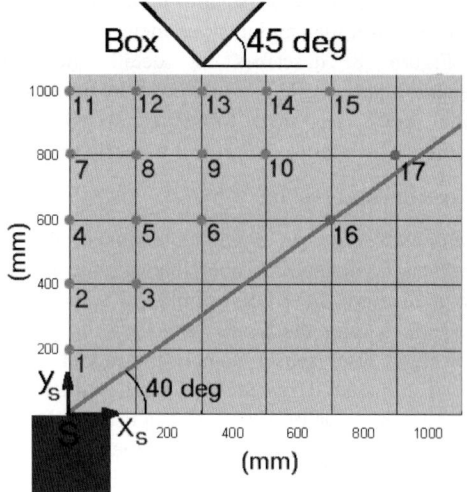

Figure 10 The measurement positions

Position	Position error (mm)		Orientation error (deg)	
	mean	st. dev	Mean	st. dev
1	14.64	1.66	0.45	0.90
2	14.50	1.85	1.59	1.08
3	15.55	1.03	2.20	1.11
4	14.72	2.48	1.70	1.33
5	15.53	1.92	1.88	0.94
6	15.84	1.55	1.06	0.58
7	15.09	2.95	0.42	0.57
8	15.30	3.06	2.80	1.48
9	16.22	4.57	2.55	1.50
10	18.32	5.23	2.95	1.77
11	14.86	1.62	2.05	1.71
12	16.35	2.44	1.39	1.14
13	16.57	3.42	2.40	1.34
14	17.55	4.41	2.81	1.84
15	18.25	5.35	2.88	1.80

Table 3 The system accuracy for the positions 1-15.

4. Conclusions and future work

In this paper a novel robotic system for depalletizing boxes was introduced and demonstrated. The task was executed using an industrial robot and a laser sensor for hand-eye coordination. The proposed solution to this problem is remarkably simple, fast and efficient. The boxes were always identified, independently of their appearance and of the illumination conditions. This is a major advantage against the camera-based solutions.

A variation of Jiang - Bunke range edge detection algorithm was employed and the corners were identified with mean accuracy better than 20mm in position and 5 degrees in orientation (mainly due to the systematic error of the sensor). This performance is sufficient for tasks like unloading a pallet to a conveyor belt but better accuracy would be probably desirable if the target position is another pallet. Therefore currently we try to implement a fine-localization procedure using a camera mounted on the robot's tool. The camera integration aims also to resolve the box orientation as mentioned in section 2.4.

In our future research we also plan to solve the problem of grasping piled boxes and piled sacks. Due to the fact that no layering exists, three-dimensional information should be extracted. Three-dimensional information will be obtained by collecting the scan lines acquired during a movement of the hand of the robot. Edge detection will be applied to the two dimensional scan lines and an edge grouping technique will be employed to extract desired feature sets. The algorithm presented, is able to deal not only with planar objects, but also with curved ones. According to this framework, the scan line approximation technique, should utilize curved functions [13] and not linear as in the application presented in this paper.

5. Acknowledgment

We gratefully thank Dr. Lambis Tassakos, INOS GmbH, for giving a hint to employ 2D techniques to deal with this problem and for supporting the first steps of this research. Many thanks are also due to Dr. Jens Schick and Dr Leonidas Bardis for their useful ideas and their overall support. We also wish to thank Professor Hans Burkhardt, University of Freiburg, Institute for Pattern Recognition and Image Processing and Dr. Costas Tzafestas for their fruitful comments on the manuscript.

References

[1] E. Al-Hujazi, A. Sood, "Range image segmentation with applications to robot bin-picking using vacuum gripper", IEEE transactions on Systems Man and Cybernetics 20(6), 1313-1325, 1990.

[2] A.J. Baerveldt, "Robust Singulation of Parcels with a Robot System using multiple sensors", Ph.D. Thesis, ETH No 10348, Swiss federal institute of technology, Zurich, 1993.

[3] P.J. Besl and R.C. Jain, "Segmentation Through Variable-Order Surface Fitting", IEEE Transactions on Pattern Analysis and Machine Intelligence, 10(2), 167–192, Feb. 1988.

[4] P. Boulanger et al., "Detection of depth and orientation discontinuities in range images using morphology", 10th International Conference in Pattern Recognition, B, 729-732, 1990.

[5] C.H.Chen and A.C.Kak, "A Robot Vision System for recognizing 3-D Objects in Low Order Polynomial Time", IEEE Transactions on SMC, Vol.19, No.6, 1989

[6] R.O. Duda, P.E. Hart, "Pattern Classification and Scene Analysis", Wiley, New York, 1972.

[7] S. Ghosal and R. Mehrota, "Detection of composite edges", IEEE Transactions on Image Processing 3(1), 14-25, 1994.

[8] Haralick et al., "Pose Estimation From Corresponding Point Data", IEEE Computer Society Workshop on Computer Vision, Miami, FL, pp. 258–263, 1987.

[9] A. Hoover et al, "An experimental comparison of Range Image Segmentation Algorithms", IEEE Transactions on Pattern Analysis and Machine Intelligence, 18(7), 673-687, June 1996.

[10] X. Jiang, "Optimality analysis of edge detection algorithms for range images", Image Analysis and Processing, Lecture Notes in Computer Science, Springer, 1310, 182–189, 1997.

[11] X. Jiang, and H. Bunke, "Edge detection in range images based on scan line approximation", Computer Vision and Image Understanding, 73(2), 183–199, 1999.

[12] X.Y. Jiang et al., "A methodology for evaluating edge detection techniques for range images", 2nd Asian Conference on Computer Vision, Singapore, 2, 415-419, 1995.

[13] X.Y. Jiang et al., "High-level feature based range image segmentation", Image and Vision Computing 18(10) 817-822, 2000.

[14] R.B. Kelley, et al., "A robot system which acquires Cylindrical work-pieces from bins", IEEE Transactions. On Systems Man and Cybernetics, 12(2), March/April 1982.

[15] R.B. Kelly, "Heuristic Vision Algorithms for Bin-Picking", 14th Symposium on Industrial Robots, Gothenburg, Sweden, October, 1984.

[16] KUKA Roboter GmbH, KR 15/2 manual.

[17] E. Natonek, "Fast, Range Image Segmentation for Servicing Robots" IEEE International Conference on Robotics and Automation, 406-411, IEEE, May 1998.

[18] L. Thurfjell et al, "A new three-dimensional connected components labelling algorithm with simultaneous object feature extraction capability", Computer Vision Graphics and Image Processing, 51(4), July, 357-364, 1992.

[19] SICK GmbH, Measurement Software Tool (MST) technical description.

[20] A.J. Vayda and A.C. Kak, "A robot vision system for recognition of generic shaped objects", Computer Vision Graphics and Image Processing 54(1), 1-46, July 1991.

[21] M.A. Wani, B.G. Batchelor, "Edge region based segmentation of range images", IEEE Transactions on Pattern Analysis and Machine Intelligence 16(3), 314-319, 1994.

[22] www.abcpackaging.com/model108.html

[23] www.alvey.com/Products_Main_Sub_1.html

[24] www.automatedconcepts.com/home.html

[25] www.cat.it/english/index.htm

[26] www.iandh.com/pallet.html

[27] www.isra.de/frame/home_engl/produkt/index.html

[28] www.ocme.it/Maine.html

Automatic Generation of Assembly Instructions using STEP

Swee M. Mok*
Motorola Inc.
1301 East Algonquin Road
Schaumburg, IL 60196
swee.mok@motorola.com

Kenlip Ong* and Dr. Chi-haur Wu
Dept. of Electrical and Computer Engineering
Northwestern University
Evanston, IL 60208
kenlip,chwu@ece.nwu.edu

Abstract

An automatic method for generating assembly instructions using CAD files is presented in this paper. Algorithms for extracting geometrical information of objects stored in a non-proprietary format, ISO-10303, STEP-CAD data file are explained. The developed algorithms form an important link between design and manufacturing. In our previous work, we designed a hierarchical assembly model that allows designers to evaluate products for manufacturing cost using a structured assembly coding system (SACS). Each SACS code represents a sequence of assembling/disassembling operations for mating two parts. The developed algorithms in this paper will generate the required SACS codes based on the geometrical and topological information extracted from the STEP files of a product in a CAD environment. Based on the generated SACS codes, the proper assembly operations can then be derived for assembling the designed product. An example of assembly with simple parts is presented to verify the method.

1. Introduction

In recent years, the typical product life cycles have shrunk for consumer products. The time for putting a new production line together has also shrunk as a result. Therefore, to implement a cost-effective automation project successfully, a product must be designed for automation before actual manufacturing. As a result, the traditional paradigm of using automation to streamline production as an afterthought is no longer acceptable. A cost-effective automation solution must be implemented on day one of production for it to be profitable. In other words, a CAD/CAM approach that treats design-to-manufacturing as an integrated process is required. To tackle this problem, a product designer must be made aware of issues faced by manufacturing. Therefore, a tool that can assist the designer to determine a product's manufacturability will be very useful. In addition, automatic generation of assembly and disassembly instructions that can be executed by production equipment will also help in solving the CAD/CAM integration problem.

In our previous work [1-4], we had proposed a system that can assist product designers to determine manufacturing cost during the product design phase. The system is a virtual environment that uses a generic assembly and disassembly (GENAD) model to capture information related to assembly and disassembly operations [1-2]. In turn, the required assembly operations are generated based on a developed structured assembly coding system (SACS) [3,4]. In the mean time, this model can also estimate manufacturing and de-manufacturing cost based on the list of SACS codes. Cost estimation is possible because each SACS code is a well-defined operation that can be carried out by robotic equipment with a cost value associated to it. A brief description of various SACS codes is given in Appendix A.

The assignment of SACS codes to an assembly or disassembly operation is critical in making correct cost prediction in the described virtual environment. In this paper, we will present a method that can automatically generate such codes by examining the product's CAD data file. To allow the developed method to be applied to a large number of CAD systems, an International Standards Organization (ISO) standard CAD format, ISO-10303: Standard for the Exchange of Product Model Data (STEP), is used [5-10].

Figure 1 shows the hierarchical structure used by STEP to represent objects [10]. At the top level, each object is defined as a closed-shell. A closed-shell is created by combining a set of advanced-faces. Each advanced-face is created by using a set of faces (or extensions for faces such as face_bounds). A face is created by edge-loops that are the equivalent of oriented-edges but tagged with a direction flag. The direction flag is used to simplify the creation of

---

* S. Mok and K. Ong are currently Ph.D. and M.S. students of ECE at Northwestern University, Evanston, IL.

higher-level objects that have the correct geometrical properties but with a different vector (direction) than the lower-level object. Each oriented-edge is created by combining basic objects such as circles, planes, edge-curves, lines, vertices, and direction vectors. The creation of each object is achieved in STEP by calling pre-defined functions with scalars or by referencing previously defined objects. In other words, higher complexity objects are defined only after the basic objects.

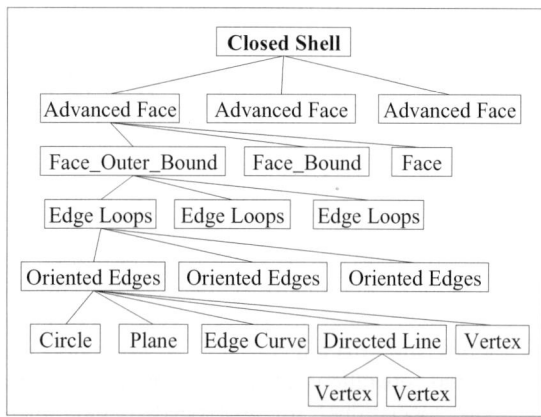

Figure 1. A diagram illustrating the hierarchical structure of STEP for defining objects.

Based on a standard STEP file generated from a CAD station for a designed product, we have developed algorithms to extract geometrical and topological information from the STEP data [6-10] and generate the SACS codes for assembling the designed product. To verify our algorithms, a STEP file of a simple assembly with five parts was generated for our analysis using commercially available CAD software called Professional Engineer (ProE).

2. STEP to SACS – Automated Assembly

In this section, we will present a method, illustrated in Fig. 2, to generate SACS codes from STEP CAD data files. In our method, a STEP file is the input to the system. The file is parsed and searched to retrieve all the relevant geometrical data of parts encoded in it. A conversion process is then applied to all parts that make up the product so that they reference a single coordinate system. This is a necessary step to recreate the topological relationships among all the product's parts when they are assembled. In the subsequent steps, a boundary table (BT) and a feature table (FT) are created. The boundary table contains geometrical boundary information of all objects. By using the BT information, a near optimal assembly tree based on a bottom-up design (see section 2.1) can be generated.

Then our algorithm will use the FT to generate SACS codes for the assembly tree. The FT is made up of defined features, such as holes, protrusions, grooves, planes, and their relationships to each other.

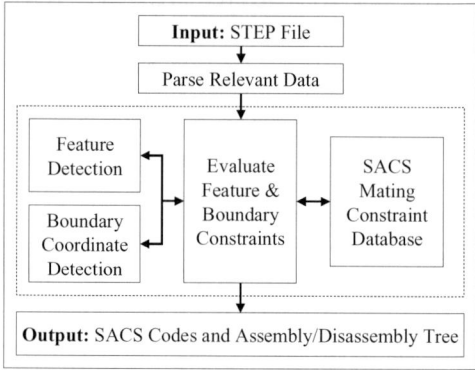

Figure 2. A diagram illustrating the process for generating SACS codes from a STEP file.

2.1 Boundary Table

The maximum and minimum coordinate points on the x-, y-, and z-axis are the boundary points of an object. This gives us a set of six points for each object. They are obtained by analyzing each object in its assembled location with respect to its reference coordinate frame. In other words, we would analyze a product in its assembled form in order to obtain the boundary points of all its parts. This is an important step for simplifying the problem of trying to relate many different frames of reference; in this way, each part will be referenced to the same frame. The STEP file has all the necessary coordinate transformations [5-9] to reference every part's vertices to one single reference frame. After this coordinate transformation, all parts would be in the assembled form for us to extract the boundary points.

After the boundary points for all parts are obtained, our algorithm will compare every part with each other geometrically to find an efficient assembly sequence using a bottom up approach. This means that a part with the lowest boundary point (e.g. we can use the z-axis for comparison) should be assembled first with its overlapping parts. The overlapping geometrical information among parts will tell us how to assemble the product in an efficient way. Therefore, the BT will help us generate an efficient assembly tree for assembling the product.

2.2 Feature Table

There are four types of features that the FT algorithm can look for in a STEP file: plane, hole (circular or otherwise), protrusion (circular or otherwise), and groove. These features will be used to determine the proper SACS codes for mating two parts. We will present our algorithms for feature extraction from a STEP file below.

To detect a plane on a part, we search for the command, ADVANCED_FACE, from the STEP file. The syntax of STEP files can be found in the listed references [5-9]. If it is found and it has only one feature defined by the STEP file, it is implied that there is a smooth plane. For example, a part with a plane feature has the following lines in a STEP file:

```
....
#141=AXIS2_PLACEMENT_3D(",#138,#139,#140);
#142=PLANE(",#141);
....
#149=FACE_OUTER_BOUND(",#148,.F.);
#150=ADVANCED_FACE(",(#149),#142,.F.);
......
```

It is found that line#150 has only one feature at line#149. The Boolean flag at the end of line#150 is used to indicate if the reference normal axis is in the same direction as that of the normal in the plane of line#142.

To search for a hole on a part, it is necessary to detect two or more features on two separate faces of an object, and that the reference axes match with each other. For example, another part with a hole feature has the following lines in a STEP file:

```
....
#101=CARTESIAN_POINT(",(0.E0,0.E0,0.E0));
#102=DIRECTION(",(0.E0,0.E0,1.E0));
#103=DIRECTION(",(1.E0,0.E0,0.E0));
#104=AXIS2_PLACEMENT_3D(",#101,#102,#103);
#105=PLANE(",#104);
....
#115=FACE_OUTER_BOUND(",#114,.F.);
....
#121=FACE_BOUND(",#120,.F.);
....
#122=ADVANCED_FACE(",(#115,#121),#105,.F.);
....
#151=CARTESIAN_POINT(",(0.E0,0.E0,3.E0));
#152=DIRECTION(",(0.E0,0.E0,1.E0));
#153=DIRECTION(",(1.E0,0.E0,0.E0));
#154=AXIS2_PLACEMENT_3D(",#151,#152,#153);
#155=PLANE(",#154);
....
#165=FACE_OUTER_BOUND(",#164,.F.);
....
#169=FACE_BOUND(",#168,.F.);
....
#170=ADVANCED_FACE(",(#165,#169),#155,.T.);
....
```

It is found that both line#122 and line#170 have two features (#115, #121) and (#165, #169), respectively. Our algorithm searches the reference axes of the two features to find matches as in the case of line #102 matching with #152 and line #103 matching with #153. This means that this object has a hole feature.

Next, we need to determine whether the hole feature is circular or otherwise. The following lines generated in our STEP files shows how line #121 was formed:

```
....
#5=CIRCLE(",#4,1.E0);
....
#10=CIRCLE(",#9,1.E0);
....
```

```
#79=VERTEX_POINT(",#77);
#80=VERTEX_POINT(",#78);
....
#116=EDGE_CURVE(",#79,#80,#5,.T.);
#117=ORIENTED_EDGE(",*,*,#116,.T.);
#118=EDGE_CURVE(",#80,#79,#10,.T.);
#119=ORIENTED_EDGE(",*,*,#118,.T.);
#120=EDGE_LOOP(",(#117,#119));
#121=FACE_BOUND(",#120,.F.);
```

It can be seen that line #121 ultimately codes for two circle statements, line#5 and line#10. In STEP, one instance of the circle statement will only encode a semi-circle [5-9]. Therefore, when our algorithm found a face formed by two circle statements, we detected a circular shape.

To detect a protrusion, it is necessary for our algorithm to detect two or more features on one ADVANCED_FACE of the object, and just one feature on another ADVANCED_FACE with the same principle axis. As such, protrusion detection is identical to hole detection with the exception that one of the advanced faces would only possess one feature. For example, if line #170 was altered such that it appeared as follows:

```
#170=ADVANCED_FACE(",(#169),#155,.T.);
```

the object would be a protrusion. Circular protrusions can similarly be found by looking for the presence of two consecutive circle statements in the STEP file.

For detecting a groove in an object as shown in Figure 3, we check for a concave feature. First, we identify the individual normal reference axis for every side of an object. The algorithm then proceeds in a clockwise direction to cross-multiply the normal axes successively, using the right-hand rule, to obtain the angles between the planes. When there is a cross-product angle larger than 180 degrees, it means there is a possibility of a groove present in the object. If there are even (2, 4, etc.) consecutive angles larger than 180 degrees, we found a groove feature.

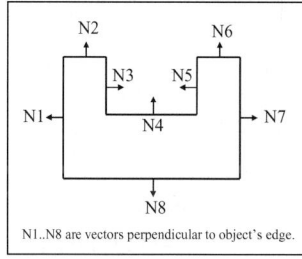

Figure 3. An object with a groove can be detected by analyzing the cross-product angle of its edges.

The four types of basic features just described cover a wide range of object shapes. Based on the detected features in mating parts and the assembly-tree generated from the boundary table for the product, our algorithm will output the appropriate SACS codes. For example, if we have a peg that needs to be assembled into a hole in a square block, the following logic will apply. First, the peg and its corresponding hole's principle axes are extracted from the STEP file. Second, the boundary points of both the peg and square block are obtained. Third, an analysis of the peg in its assembled position (in the block) is performed. This will then allow us to determine that the block has a lower overall minimal boundary at its lower face compared to the peg (unless the hole is drilled through the block). We can then conclude that the block has to be placed first onto a fixture (such as a flat surface) before the peg is inserted. An insertion operation in this case can be deduced by the fact that the peg is a protrusion, and the block's hole has a similar dimension to match the peg's radius. Moreover, the principle axes of both the peg and hole match in opposite directions after assembly as defined by SACS [3].

Our algorithms presented here are obtained by studying the geometrical and topological properties of assembled parts. We will now go through a simulation in the next section to verify our method.

3. Simulation and Analysis

To verify our algorithms, an assembly with five simple parts, as shown in Fig. 4, is simulated using ProE for our analysis. The product consists of five parts labeled: A, B, C_1, C_2, and D. Both part A and part B have two drilled holes for part C_1 and C_2 to fit through. In addition, part A has a large hole in its center that is occupied by part D after assembly. The assembled product is a module that will be mated to a larger assembly.

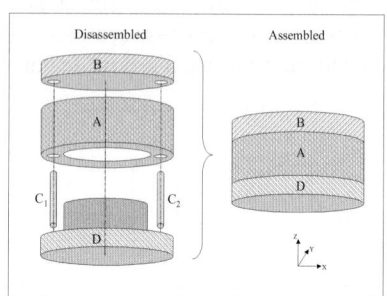

Figure 4. Diagram of the un-assembled and assembled product used in simulation.

As described in section 2, the product's STEP file is first parsed to retrieve all relevant geometrical information. A transformation is made on all parts to ensure that they are referenced to a single coordinate system. The assembled product as illustrated on the right-hand side of Fig. 4 is thus obtained.

A boundary table (BT) is first created to record the maximum and minimum height of each part, as shown in Table 1. The name of each part is listed in column 1. The minimum and maximum values of each part are then recorded for the x-, y-, and z-axis in the remaining six columns. These results will be used to determine which two parts should be assembled first, and so on.

Parts Name	X (min)	X (max)	Y (min)	Y (max)	Z (min)	Z (max)
Part_A	-8	8	-8	8	1	5
Part_C_1	-6.5	-7.5	-0.5	0.5	1	6
Part_C_2	6.5	7.5	-0.5	0.5	1	6
Part_D	-8	8	-8	8	0	5
Part_B	-8	8	-8	8	5	6

Table 1. Boundary information of the five parts.

A feature table (FT), as shown in Table 2, is also generated. Column one lists the five parts of the product. When a feature is detected, such as a circular hole, the table is updated in column two. Notice that there could be multiple detected features that are of the same geometrical type as is in the case of part A. Each detected feature has a reference point and it is recorded in column three; this is the location where the feature is located in the global reference coordinate system. In column four, the reference axis for the feature found is recorded. Column five contains additional information for the detected feature. For example, part A has a hole with a radius of six units which is the large inner cavity. In addition, it also has two smaller holes located at (7,0,0), and (-7,0,0), each with a radius of 0.5 units.

Parts Name	Feature detected	Ref. Point of feature	Ref. Axis of feature	Additional data
Part_A	Circular hole	(0,0,0)	(0,0,1)	Radius 6
	Circular hole	(7,0,0)	(0,0,1)	Radius 0.5
	Circular hole	(-7,0,0)	(0,0,1)	Radius 0.5
Part_C_1	Circular protrusion	(7,0,0)	(0,0,1)	Radius 0.5
Part_C_2	Circular protrusion	(-7,0,0)	(0,0,1)	Radius 0.5
Part_D	Circular protrusion	(0,0,0)	(0,0,1)	Radius 6
Part_B	Circular hole	(7,0,0)	(0,0,1)	Radius 0.5
	Circular hole	(-7,0,0)	(0,0,1)	Radius 0.5

Table 2. Feature table for the five parts.

At this point, we have enough information from the BT and FT to deduce the assembly tree and its SACS codes for the product. First, we will establish the assembly tree configuration. The minimum and maximum boundary values of each part in the z-axis are compared. Part D has the smallest minimum boundary value and is thus selected first for assembly. To choose the second part (that will mate with part D), parts A, C_1 and C_2 are considered since they overlap or make contact with part D after assembly. Part A is selected since both C_1 and C_2 have a higher maximum boundary value - a selection rule defined by us. Parts C_1, C_2, and B are then considered for assembly but both part C_1 and part C_2 have minimal boundary values. As a result, we arbitrarily choose C_1 followed by C_2. Finally, part B is chosen as the last part to be assembled onto part DAC_1C_2 to form the final product DAC_1C_2B. The comparisons are shown in Table 3. The assembly tree is illustrated on the right-hand side of Fig. 5.

Assembly Parts	Axis	Minimum compare	Maximum compare	Feature detected	Possible Operation	SACS Code
D, A	x	A = D	A = D	3 Circular holes on A 1 Protrusions on D Identical reference axis & reference point Insertion detected	Rz	0123
	y	A = D	A = D			
	z	A > D	A = D			
DA, C1	x	C1 > DA	C1 < DA	1 Cir. protrusion on C1 2 matching holes in DA Different reference axis Insertion detected	Rz	0123
	y	C1 > DA	C1 < DA			
	z	C1 > DA	C1 > DA			
DAC1, C2	x	C2 > DAC1	C2 < DAC1	1 Cir. protrusion on C1 2 matching holes in DAC1 Different reference axis Insertion detected	Rz	0123
	y	C2 > DAC1	C2 < DAC1			
	z	C2 > DAC1	C2 < DAC1			
DAC1C2, B	x	B = DAC1C2	B = DAC1C2	2 Circular holes in B 2 Cir. protrusions on DAC1C2 Identical reference axis 2 Insertions detected	NULL	0023
	y	B = DAC1C2	B = DAC1C2			
	z	B > DAC1C2	B = DAC1C2			

Table 3. A set of generated SACS codes for assembling the product.

To generate SACS codes for mating part D to A, C_1 to DA, and so on, the topological constraints between the mating parts are analyzed. Obviously, the parts to consider come from the previously generated assembly tree. For example, the constraints between part D and part A will result in an assembled part that will only have one degree of rotation freedom about the z-axis (Rz). This constraint is also known as a "fit" mating constraint [1]. The appropriate SACS code for this type of mating operation is 0123, as described in [1] and also as shown in row three of Fig. A in Appendix A. The BT shows that part A has an outer radius of eight units and a hole in the middle with a radius of six units. It also has a minimum and a maximum height of one and five units, respectively. Part D has a smaller minimum boundary value of zero and the same maximum height as part A. This tells us that part A can indeed be inserted into part D resulting in a combined part called DA with minimum and maximum boundary values of zero and five, respectively. Part C_1 and part C_2 have a minimum and maximum boundary value (height) of one and six units, respectively. Their SACS codes for assembly into part DA and DAC_1 are both 0123. Finally, part B is evaluated to have an "against-fit" mating operation with DAC_1C_2 since it has two simultaneous insert operations with parts C_1 and C_2; thus, preventing it from any further movement after assembly. From Fig. A in Appendix A, a SACS code of 0023 is obtained. The results are summarized in Table 3.

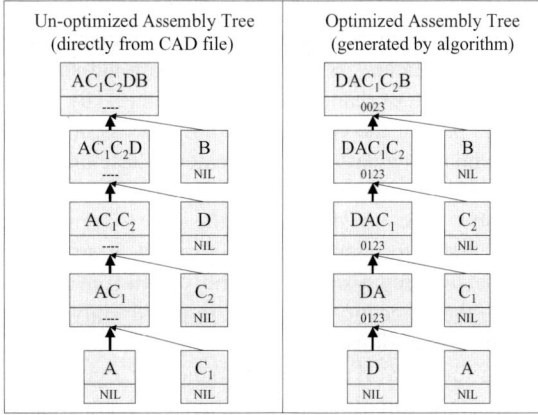

Figure 5. Two versions of product assembly trees.

From the FT and BT analysis, a near optimal product assembly tree for assembling the product is created, as shown on the right-hand side in Fig. 5. A SACS code is needed for each assembly operation. Part A (moving) is the first part to be assembled onto part D (stationary – thick arrow in Fig. 5). Parts C_1, C_2, and B are then subsequently assembled into the previously assembled modules.

In comparison, another assembly tree obtained directly from the CAD-STEP file is shown on the left-hand side of Fig 5. This sequence was our parts' design sequence in the ProE CAD environment. As most CAD designers, we didn't consider the actual assembly sequence during the design stage. As a result, this assembly tree is not in the proper assembly sequence. Therefore, it requires an excessive amount of tooling; making it more expensive or even impossible to assemble in manufacturing. For example, part AC_1 at the bottom of the tree means that part C_1 can fall out of part A if part AC_1 is not oriented correctly (90 degrees from vertical) when part C_2 or part D is being inserted in the subsequent assembly steps. This experience taught us that, to obtain a better assembly tree, the CAD designer must design each product in a bottom-up fashion so that the CAD system can record a much better assembly procedure. To solve this potential problem of linking CAD to CAM, we developed the described algorithms to sort out a CAD STEP file and to automatically create a near optimal assembly tree for manufacturing plus generating the required assembly operations represented by SACS codes.

4. Conclusion

In our previous work, we proposed an integrated CAD/CAM system [1,2] for handling the process of product design to automatic assembly and disassembly. Within the system, a set of Structured Assembly Coding System (SACS) codes is used for encoding assembly operations. In this paper, we presented a method for solving the problem of generating such SACS codes from standard STEP files automatically. STEP is an international standard for CAD data exchange, as described in [5-10].

As part of the code generation process, a feature table and a boundary table is created. These tables are then used by our algorithms to generate a product's assembly tree to determine the most appropriate SACS codes for assembling the product. A study was conducted using a simple assembly with five parts to verify the concept. All parts required only z-axis insertion operations.

In future work, we will expand our algorithm to cover more complex products that have part-mating operations in the x-, y-, and z-axis directions. We will also take parts' tolerance and clearance into account to reflect real assembly operations more accurately.

Acknowledgment

The authors wish to thank Dr. Iwona Turlik and Dr. Thomas Babin from Motorola Inc. for their support. This work was partly supported by the Motorola Center for Telecommunications at Northwestern University.

References

1. Mok, Swee, Chi-haur Wu, D.T. Lee, "A System for Analyzing Automatic Assembly and Disassembly Operations," IEEE International Conference on Robotics and Automation, San Francisco, California, Apr. 27, 2000.
2. Mok, Swee, Chi-haur Wu, D.T. Lee, "A Hierarchical Workcell Model for Intelligent Assembly and Disassembly," IEEE International Symposium on Computational Intelligence in Robotics and Automation (CIRA99), Monterey, California, Nov. 8-9, 1999.
3. Wu, Chi-Haur, Myong Gi Kim, "Modeling of Part-Mating Strategies for Automating Assembly Operations for Robots," IEEE Trans. On Systems, Man, and Cybernetics, Vol. 24, No. 7, July 1994, pp. 1065-1074.
4. Kim, M. G., C. H. Wu, "Formal part mating model for automating assembly operations," in Proc. The 1990 IEEE Int. conf. Syst., Man, Cybern., Los Angeles, CA, Nov 1990.
5. U.S. Product Data Association, "An American National Standard: Product Data Exchange using STEP (PDES) Part 1 – Overview and Fundamental Principles (ANS US PRO/IPO-200-001-1994," U.S. Product Data Association, 2722 Merrilee Drive, Suite 200, Fairfax, VA 22031.
6. U.S. Product Data Association, "An American National Standard: Product Data Exchange using STEP (PDES) Part 41 – Integrated Generic Resources: Fundamental of Product Description and Support (ANS US PRO/IPO-200-041-1994," Trident Research Center, Suite 204, 5300 International Blvd., N. Charleston, SC.
7. U.S. Product Data Association, "An American National Standard: Product Data Exchange using STEP (PDES) Part 42 – Integrated Generic Resources: Geometric and Topological Representation (ANS US PRO/IPO-200-042-1994," Trident Research Center, N. Charleston, SC.
8. U.S. Product Data Association, "An American National Standard: Product Data Exchange using STEP (PDES) Part 43 – Integrated Generic Resources: Representation Structures (ANS US PRO/IPO-200-043-1994," Trident Research Center, N. Charleston, SC.
9. U.S. Product Data Association, "An American National Standard: Product Data Exchange using STEP (PDES) Part 203 – Application Protocol: Configuration Controlled Design (ANS US PRO/IPO-200-203-1994," Trident Research Center, N. Charleston, SC.
10. Al-Timimi Kais, MacKrell John, "STEP: Toward Open Systems," CIMdata Inc., 1996, ISBN: 1-889760-00-5.

Appendix A

The Structured Assembly Coding System (SACS) [3,4] can encode all mating-operations for assembling two primitive parts in a three-dimensional task space. SACS uses force compliance to perform assembly operations between a *stationary* part and a *moving* part. The topological constraints between two mating parts, a moving part and a stationary part, are used to determine the assembly or disassembly process. The basic SACS coding system uses four levels (the first 4 columns in Figure A) to code 24 defined types of part-mating (the fifth column in Figure A) operations for primitive parts [1]. In column six of Figure A, seven types of operations that have level-0 or "hole" features are shown. Levels 1, 2, and 3 are also defined for plane, groove, and handlers, respectively, but they are not shown in Figure A. Handlers are objects such as fixtures, robots, end-effectors, pallets and feeders. Finally, each SACS code is associated with assembly and disassembly cost, as shown in columns 7 and 8, and they are each a function of variables including cycle time, tooling cost, and assembly yields.

						Cost	
Level 0	Level 1	Level 2	Level 3	Part Mating Type	Code	Assembly	DisAssem.
0 (Hole)	0R	P	Z	against-fit (null)	0023	c0023	d0023
		P	W	screw-fit (null)	0024	c0024	d0024
	1R	P	Z	fit (Rz)	0123	c0123	d0123
		V	Y	side-fit (Ry)	0112	c0112	d0112
	2R	V	Y	e-fit (RxRz)	0212	c0212	d0212
		P	Z	e-side-fit (RxRy)	0223	c0223	d0223
	3R	V	Z	sph-fit (RxRyRz)	0313	c0313	d0313

Key:
V=Vertical(1) P=Parallel(2) Z=Z-Axis(3) W=Screw(4)

Figure A. Some of the mating operations defined by the Structured Assembly Coding System.

Furniture Polishing Robot Using a Trajectory Generator Based on Cutter Location Data

Fusaomi Nagata[*1] Keigo Watanabe[*2] and Kiyotaka Izumi[*3]

[*1]Interior Design Research Institute, Fukuoka Industrial Technology Center,
Agemaki-405-3, Ohkawa, Fukuoka 831-0031, JAPAN
nagata@fitc.pref.fukuoka.jp

[*2]Department of Advanced Systems Control Engineering, Graduate School of Science and Engineering,
[*3]Department of Mechanical Engineering, Faculty of Science and Engineering,
Saga University, Honjomachi-1, Saga 840-8502, JAPAN
{watanabe, izumi}@me.saga-u.ac.jp

Abstract

In this article, a furniture polishing robot, which uses a trajectory generator based on cutter location data, is proposed. The position and orientation of the polishing tool is suitably controlled by referring the output from the trajectory generator. The trajectory generator yields a zigzag path, whirl path or their combinational path according to the shape of each workpiece. Using the trajectory generator, the present polishing robot doesn't need the conventional complicated teaching process. The contact force acting between the polishing tool and the workpiece is controlled through a desired impedance model to follow in Cartesian space. The effectiveness and promise of the proposed polishing robot are demonstrated through some experiments concerning polishing tasks using an industrial robot JS-10 with a PC based controller.

1. Introduction

In manufacturing industry of wooden furniture, CAD/CAM system and NC machine tool are currently being used generally and widely, and these advanced systems have drastically rationalized the design and manufacturing process of furniture. However, the polishing process after NC machining has hardly automated yet, and can not help depending on skilled workers who can carry out both compliant force control and skillful trajectory control for the polishing tool. The skilled workers usually use air driven rotational polishing tools. In using these types of tools, keeping contact with the object from normal direction with desired contact force is the most important to obtain good finishing of the surface. Furthermore, when conducting the polishing task, it should be noted that skilled workers move the polishing tool along the object's surface using zigzag path, whirl path or their combinations.

Generally, since the repetitive position accuracy at the top of industrial robots is 0.1 [mm] or its neighborhood, it is so difficult to finish the surface of the wooden furniture using only position control strategy. In the polishing process of the wooden parts, the accuracy of 10 [μm] or less is finally required. Especially, when the robot contacts to an object, several factors that decrease the total stiffness of the system are included. They are called as clearance, strain and deflection, all of which exist in not only the robot itself but also force sensor, polishing tool, and so on. Therefore, it is meaningless to discuss the position accuracy of the top of the polishing tool attached to the top of the robot arm. If only the position control is used to a polishing task, where the polishing tool and the object contact to each other, then both the stiffness of the robot itself and the total stiffness including the polishing tool must be high. Further, the robots must deal with wooden materials which tend to be warped on the condition of temperature and humidity.

Up to now, although several papers have described that force control strategy is indispensable to realize any polishing robots [1-3], there still remain a significant problem that should be developed to practical applications in the future. Such problem is how to construct an easy and safe teaching system to provide the continuous trajectory of the attached polishing tool according to the object's shape. Generally, the desired trajectory which is indispensably required for the position/orientation control, is obtained through a well-known teaching process using a teaching pendant. However, it is complicated and time-consuming to implement. Especially, in teaching for the polishing robot, the object with curved surface demands a large number of teaching points.

In this paper, we try to develop a furniture polishing robot which can suitably generate the desired trajectory as performed by skilled workers. To achieve the goal, a trajectory generator based on cutter location data (CL data) is first proposed. If the object is designed by CAD/CAM system, the CL data can be used as the desired trajectory. The CL data are the basic tool path for the multi-axis NC machine tool, generated from the main processor of CAM system. The trajectory generator allows the polishing robot to accomplish the polishing task without any teaching process. On the other hand, the contact force acting between the polishing tool and the object is controlled by an impedance model following force control which can be easily applied to open architectural industrial robots. The effectiveness of the proposed method is proved through some experiments using an industrial robot JS-10 with a PC based controller.

2. Trajectory Generator Based on Cutter Location Data

Profiling control is the basic control strategy for polishing robots and it is performed by both force control and position/orientation control. The desired trajectory is indispensable so that the position/orientation of the tool can be controlled suitably and rapidly. The basic polishing task needs a trajectory that the polishing tool can profile the object keeping contact from the normal direction. When the robot executes a motion, the trajectory is generally computed or obtained in advance, e.g. by teaching. However the teaching is complicated and time consuming.

In manufacturing industry of furniture, almost all the workpieces are fortunately designed with CAD/CAM system, so that the CL data can be referred as the information of each shape. In this section, a trajectory generator based on the CL data is discussed to exclude the conventional teaching process. The trajectory generator creates not only the desired trajectory in the direction of position/orientation control but also feedforward quantity in the direction of force control. The polishing tool is moved along the desired trajectory to keep contact with the workpiece from normal direction. In the following, it is described how to put the profiling control into practice using the trajectory generator.

The CL data are composed of position vector and its normal vector. The n-th step $CL(n) \in \Re^6$ is written by

$$CL(n) = \begin{bmatrix} CL_p^T(n) & CL_o^T(n) \end{bmatrix}^T \quad (1)$$

where $CL_p^T(n)=[x(n)\ y(n)\ z(n)]$ and $CL_o^T(n)=[\alpha(n)\ \beta(n)\ \gamma(n)]$ are the position vector and normalized orientation vector, respectively. The trajectory generator yields a desired trajectory $r(k)$ at the discrete time k given by

$$r(k) = \begin{bmatrix} r_p^T(k) & r_o^T(k) \end{bmatrix}^T \quad (2)$$

where $r_p(k)=[r_x(k)\ r_y(k)\ r_z(k)]^T$ and $r_o(k)=[r_\alpha(k)\ r_\beta(k)\ r_\gamma(k)]^T$ are the position and orientation components, respectively. $r(k)$ is calculated using both the CL data and profiling velocity $v(k)$ given by

$$v(k) = \begin{bmatrix} v_x(k) & v_y(k) & v_z(k) & 0 & 0 & 0 \end{bmatrix}^T \quad (3)$$

where it is assumed that the magnitude of $v(k)$, $|v|$, is constant.

An example of profiling control is shown in **Fig. 1**. In this case, assuming $r(k) \in [CL(n), CL(n+1)]$ we obtain $r(k)$ through the following procedure. First of all, a direction

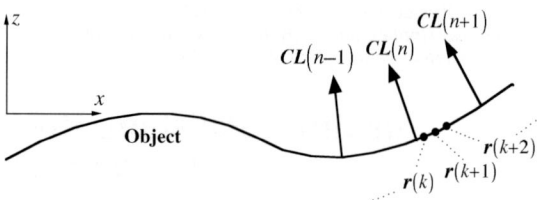

Fig. 1 Relation between CL data and an object

vector $D = \begin{bmatrix} D_p^T & D_o^T \end{bmatrix}^T$ is defined as

$$D = CL(n+1) - CL(n) \quad (4)$$

where $D_p=[D_x\ D_y\ D_z]^T$ and $D_o=[D_\alpha\ D_\beta\ D_\gamma]^T$ are the position and orientation components, respectively. The each directional profiling velocity is obtained by

$$v_i(k) = |v(k)| \frac{D_i}{|D_p|} \quad (i=x,y,z) \quad (5)$$

Using a sampling width Δt, the desired position $r_p(k)$ is given by

$$r_i(k) = r_i(k-1) + v_i(k)\Delta t \quad (i=x,y,z) \quad (6)$$

Next, the rotational component $r_o(k)$ is considered. We define two angles $\theta_1(n)$, $\theta_2(n)$ as shown in **Fig. 2**. $\theta_1(n)$ and $\theta_2(n)$ are the tool angles of inclination and rotation, respectively. Each component of $CL_o(n)$ is represented by

$$\alpha(n) = \sin\theta_1(n)\cos\theta_2(n) \quad (7)$$
$$\beta(n) = \sin\theta_1(n)\sin\theta_2(n) \quad (8)$$
$$\gamma(n) = \cos\theta_1(n) \quad (9)$$

The desired tool angles $\theta_{r1}(k)$, $\theta_{r2}(k)$ of inclination and rotation at the discrete time k can be calculated as

$$\theta_{ri}(k) = \theta_i(n) + \{\theta_i(n+1) - \theta_i(n)\}\frac{|r_p(k) - CL_p(n)|}{|D_p|} \quad (10)$$

where $i = 1, 2$. If Eq. (10) is substituted into Eqs. (7), (8),

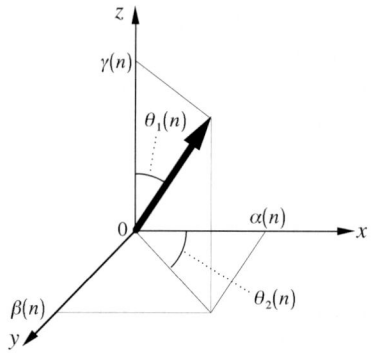

Fig. 2 Tool vector $CL_o(n)$ in robot base coordinate system

(9), we finally obtain

$$r_\alpha(k) = \sin\theta_{r1}(k) \cos\theta_{r2}(k) \quad (11)$$
$$r_\beta(k) = \sin\theta_{r1}(k) \sin\theta_{r2}(k) \quad (12)$$
$$r_\gamma(k) = \cos\theta_{r1}(k) \quad (13)$$

In the direction of force control, $r_p(k)$ is used for the feedforward control. On the other hand, in the direction of position/orientation control, $r_p(k)$ and $r_o(k)$ are used for desired trajectory of the polishing tool. The orientation vector $r_o(k)$ is transformed into Z-Y-Z Euler angles which can be given to the PC based servo controller used.

3. Design of Profiling Control System

3.1 Velocity Based Impedance Model Following Force Control with Integral Action

So far, a trajectory generator based on cutter location data has been discussed. Regarding the force control, we use the impedance model following force control that can be easily applied to industrial robots with an open architecture controller. The desired impedance equation for Cartesian-based control of a robot manipulator is designed by

$$M_d(\ddot{x} - \ddot{x}_d) + B_d(\dot{x} - \dot{x}_d) + SK_d(x - x_d) = SF + (E - S)K_f(F - F_d) \quad (14)$$

where $x \in \Re^6$, $\dot{x} \in \Re^6$ and $\ddot{x} \in \Re^6$ are the position, velocity, and acceleration vectors, respectively. $M_d \in \Re^{6\times6}$, $B_d \in \Re^{6\times6}$ and $K_d \in \Re^{6\times6}$ are the coefficient matrices of the desired mass, damping and stiffness, respectively. $F \in \Re^6$ is the raw force-moment vector acting between the end-effector and its environment defined by $F^T = [f^T \ n^T]$, where $f \in \Re^3$ and $n \in \Re^3$ are the force and moment vectors, respectively. $K_f \in \Re^{6\times6}$ is the force feedback gain matrix. x_d, $\dot{x}_d$, $\ddot{x}_d$ and $F_d^T = [f_d^T \ n_d^T]$ are the desired position, velocity, acceleration and force-moment vectors; S and E are the switch matrix diag$(S_1, \ldots, S_6)$ and identity matrix. It is assumed that M_d, B_d, K_d and K_f are positive-definite diagonal matrices. Note that if $S = E$, Eq. (14) becomes an impedance control system in all directions; whereas if $S = 0$, it becomes a force control system in all directions.

If the force control is used in all directions, $X = \dot{x} - \dot{x}_d$ gives

$$\dot{X} = -M_d^{-1}B_d X + M_d^{-1}K_f(F - F_d) \quad (15)$$

In general, (15) is solved as

$$X = \exp(-M_d^{-1}B_d t) X(0)$$
$$+ \int_0^t \exp\{-M_d^{-1}B_d(t-\tau)\} M_d^{-1}K_f(F - F_d) d\tau \quad (16)$$

In the following, we consider the form in the discrete time k using a sampling time Δt. It is assumed that M_d, B_d, K_f, F and F_d are constant at $\Delta t(k-1) \leq t < \Delta tk$. Defining $X(k) = X(t)|_{t=\Delta tk}$, it can be obtained by

$$X(k) = \exp(-M_d^{-1}B_d\Delta t) X(k-1)$$
$$- \{\exp(-M_d^{-1}B_d\Delta t) - E\} B_d^{-1}K_f \{F(k) - F_d\} \quad (17)$$

Remembering $X = \dot{x} - \dot{x}_d$ and setting $\dot{x}_d = 0$ in the direction of force control, the recursive equation of velocity command in terms of Cartesian space is derived by

$$\dot{x}(k) = \exp(-M_d^{-1}B_d\Delta t) \dot{x}(k-1)$$
$$- \{\exp(-M_d^{-1}B_d\Delta t) - E\} B_d^{-1}K_f \{F(k) - F_d\} \quad (18)$$

where $x(k)$ is composed of position vector $[x(k) \ y(k) \ z(k)]^T$ and orientation vector $[\phi(k) \ \theta(k) \ \psi(k)]^T$ expressed by Z-Y-Z Euler angles. Profiling control is the basic strategy for polishing or sanding and it is performed by both force control and position/orientation control. However, it is so difficult to realize a stable profiling control under such environments that have unknown dynamics or shape [4, 5]. Undesirable oscillations and non-contact state tend to occur. To reduce the influences, an integral action is added to Eq. (18), which yields

$$\dot{x}(k) = \exp(-M_d^{-1}B_d\Delta t) \dot{x}(k-1)$$
$$- \{\exp(-M_d^{-1}B_d\Delta t) - E\} B_d^{-1}K_f \{F(k) - F_d\}$$
$$+ K_i \sum_{n=1}^{k} \{F(n) - F_d\} \quad (19)$$

where $K_i = \text{diag}(K_{i1}, \ldots, K_{i6})$ is the integral gain. Further, we use Eq. (2) as a feedforward control in the direction of force control, which gives

$$\dot{x}(k) = \exp(-M_d^{-1}B_d\Delta t) \dot{x}(k-1)$$
$$- \{\exp(-M_d^{-1}B_d\Delta t) - E\} B_d^{-1}K_f \{F(k) - F_d\}$$
$$+ K_i \sum_{n=1}^{k} \{F(n) - F_d\} + \frac{r(k) - r(k-1)}{\Delta t} \quad (20)$$

Since polishing tools used in the manufacturing industry of furniture have a tendency to cause high frequency vibration, the measured force data have to be smoothed. To obtain the smoothed force sensor data $\hat{F}(k)$, we use the following moving average technique:

$$\hat{F}(k) = \frac{1}{m} \sum_{i=k-m}^{k} F(i) \quad (21)$$

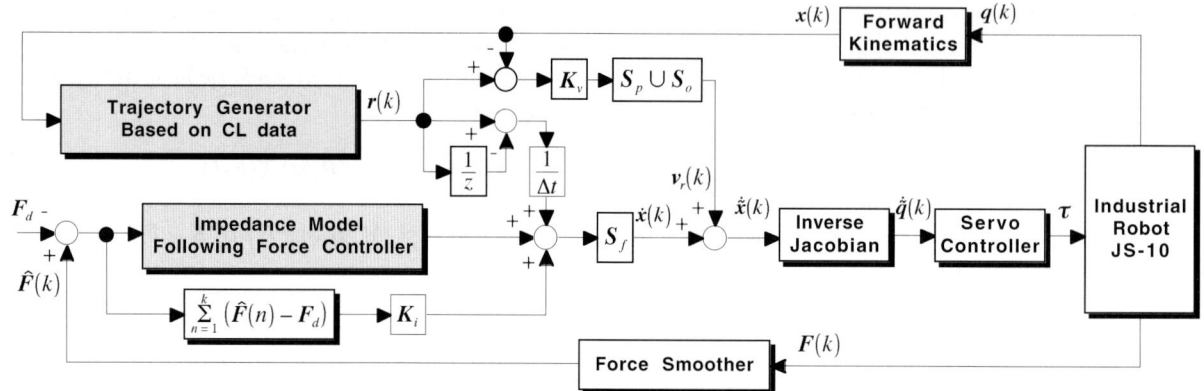

Fig. 3 Block diagram of impedance controlled polishing robot system using a trajectory generator based on CL data

where m is the sampled number for the data window. Substituting Eq. (21) into Eq. (20), the following force control strategy is finally obtained.

$$\dot{x}(k) = \exp(-M_d^{-1}B_d\Delta t)\,\dot{x}(k-1)$$
$$- \left\{\exp(-M_d^{-1}B_d\Delta t)-E\right\}B_d^{-1}K_f\left\{\hat{F}(k)-F_d\right\}$$
$$+ K_i \sum_{n=1}^{k}\left\{\hat{F}(n)-F_d\right\} + \frac{r(k)-r(k-1)}{\Delta t} \quad (22)$$

The control input $\dot{x}(k)$ given by Eq. (22) is substituted into the reference of servo system every sampling time so that the contact force is controlled to be F_d.

3.2 Position and Orientation Control

The block diagram of the impedance model following force controller using the trajectory generator is shown in **Fig. 3**. Here, matrices S_p=diag($S_{p1}, S_{p2}, S_{p3}, 0, 0, 0$), S_o=diag($0, 0, 0, S_{o1}, S_{o2}, S_{o3}$) and S_f=diag($S_{f1}, S_{f2}, S_{f3}, S_{f4}, S_{f5}, S_{f6}$) make each switch of position control, orientation control and force control to be active, respectively. When one value is presented, each controller is in effect, whereas when zero value is presented, each controller is without effect. Note that each matrix has a following relation:

$$S_p \cup S_o \cup S_f = E \quad (23)$$

The interpretation of the block diagram shown in **Fig. 3** is as follows: In the direction of force control, Eq. (22) generates a velocity command $\dot{x}(k)$, which includes the feedforward quantity from the trajectory generator. On the other hand, in the direction of position/orientation control, the position error between $r(k)$ and $x(k)$ generates another velocity command $v_r(k)$ with the velocity transformation gain K_v=diag($K_{v1},...,K_{v6}$). The each directional velocity command is added each other, which yields $\dot{\tilde{x}}(k)$. $\dot{\tilde{x}}(k)$ is transformed

Photo 1 Experimental setup using an industrial robot

into the joint angle velocity $\dot{q}(k)$ with the inverse Jacobian, then $\dot{q}(k)$ is given to the reference of servo system installed in the open architecture controller. Note that the velocity command given by Eq. (22) is updated every sampling rate Δt, e.g., 10 [msec]. Thus, if the object is manufactured using the NC machine tool, we can use the CL data as prior information of each environmental shape.

4. Experiments on Polishing Task

In this section, the ability of the proposed method shown in **Fig. 3** is examined through some experiments on polishing task using an industrial robot JS-10 with a PC based controller. **Photo 1** shows the experimental setup. On this system, several Windows API (Application Programming Interface) functions, such as kinematics, coordinate transformation for force sensor, and servo control with position command, can be used. In order to avoid the problem of singularities, the position of each workpiece in robot base coordinate system should be considered according to its size and shape. In manufacturing industry of furniture, skilled workers usually use some kinds of handy-type vibrational tools which are driven by air power. In this experiment, a similar type tool with a polish

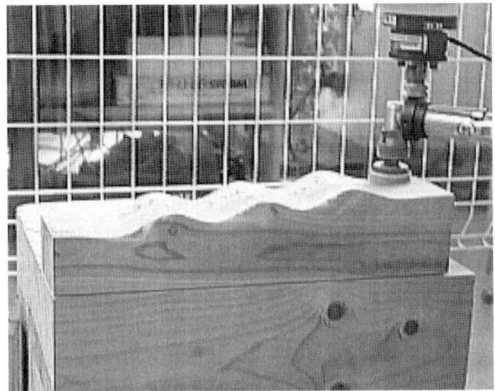

Photo 2 Polishing scene using a zigzag path

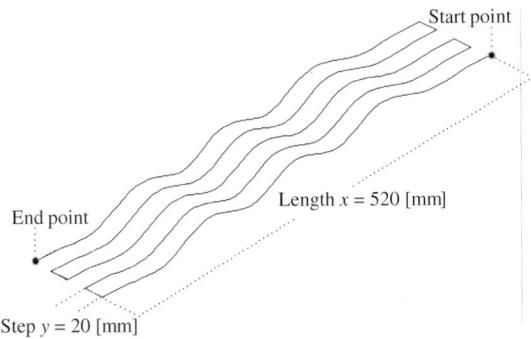

Fig. 4 CL data composed of zigzag path

Table 1 Parameters used in experiments

Desired contact force $\|f_d\|$	0.7 [kgf]
Profiling velocity $\|v\|$	10 [mm/s]
Desired inertia coefficient M_{d1}, M_{d2}, M_{d3}	0.01 [kgf·s²/mm]
Desired damping coefficient B_{d1}, B_{d2}, B_{d3}	20 [kgf·s/mm]
Force feedback gain K_{f1}, K_{f2}, K_{f3}	1
Integral control gain K_{i1}, K_{i2}, K_{i3}	0.002
Velocity transformation gain K_{v1}, K_{v2}, K_{v3}	0.08
Velocity transformation gain K_{v4}, K_{v5}, K_{v6}	0.014
Sampling width Δt	10 [msec]

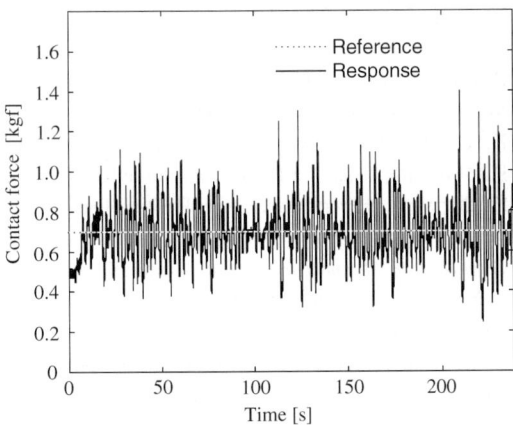

Fig. 5 Force control result

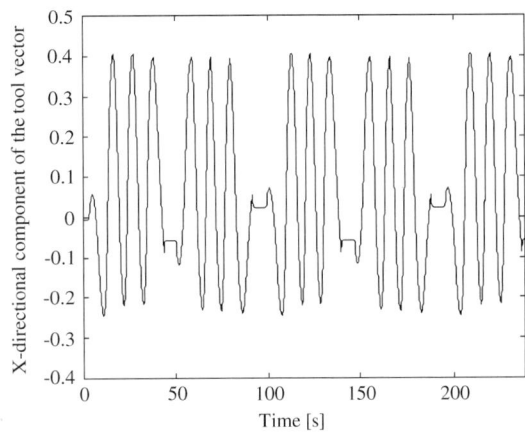

Fig. 6 Orientation control result

paper is selected and attached to the top of the robot arm via a 6-DOF force/torque sensor 67M25A provided by Nitta corporation. The diameter of the tool is 50 [mm] and paper roughness is #220. In order to obtain a good finishing in using these types of tools, it is the most important to keep contact with the workpiece from normal direction of its curved surface and to stably control the contact force to be reference value.

4.1 In the case of using a zigzag path

As a first case study, a workpiece as shown in **Photo 2** was polished. The workpiece is designed and manufactured using CAD/CAM system and 5-axis NC machine tool respectively, so that we can refer the CL data created from the CAM modules. The CAD/CAM system used in the experiments is UNIGRAPHICS provided by Electronic Data Systems Corporation. The CL data used are composed of zigzag paths as shown in **Fig. 4**. The x-directional length x and y-directional step y are 520 [mm] and 20 [mm], respectively. In this experiment, the polishing tool comes into contact with the workpiece and profiles its curved surface with a desired constant force 0.7 [kgf] using the zigzag path. The parameters used are shown in **Table 1**.

Photo 3 Another workpiece polished with a whirl path

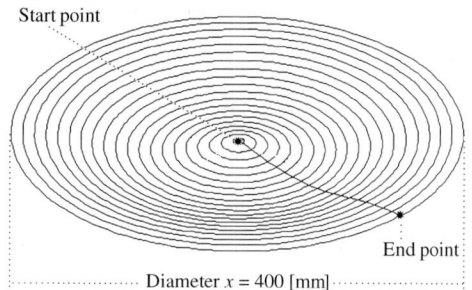

Fig. 7 CL data composed of whirl path

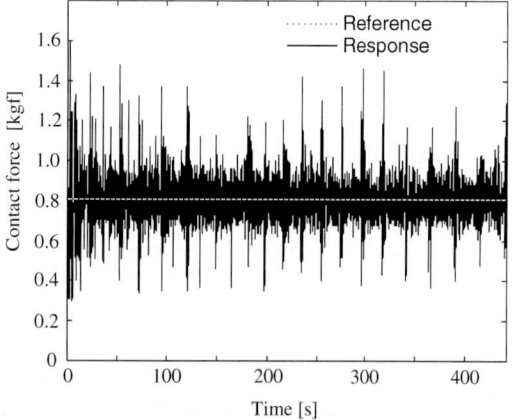

Fig. 8 Force control result

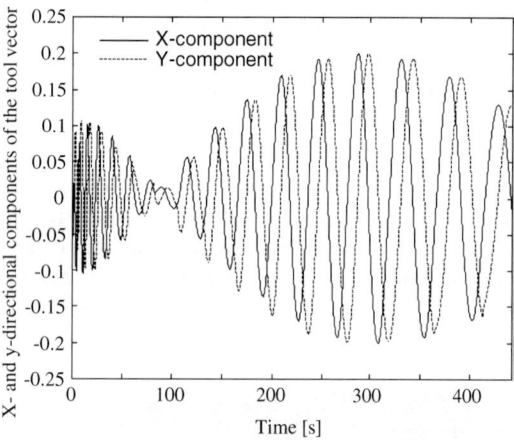

Fig. 9 Orientation control result

Figure 5 shows the z-directional force control result in tool coordinate system in the case of using Eq. (22). As can be seen, a desirable response is obtained around the reference value. Of course, the orientation of the polishing tool was simultaneously controlled to keep contact with the workpiece from normal direction of the curved surface as shown in **Fig. 6**. The vertical line of the **Fig. 6** represents the x-directional component of the normalized tool vector in robot coordinate system.

4.2 In the case of using a whirl path

Another case study was carried out using a workpiece as shown in **Photo 3**. The workpiece was manufactured using a whirl path as shown in **Fig. 7**. The polishing tool was whirlingly moved from the start point to the end point (from the center of the path to outside) with profiling velocity 20 [mm/s]. The desired contact force was set as 0.8 [kgf]. **Figure 8** shows the force control result. From this figure, it is observed that the contact force was controlled around the reference value, although some spikes appeared caused by the vibration of the polishing tool. **Figure 9** shows the orientation control result. The solid line and dotted line are the x- and y-directional components of the normalized tool vector in robot coordinate system.

5. Conclusion

This paper presented a furniture polishing robot system which uses an impedance model following force controller and a trajectory generator based on cutter location data. The experiments demonstrated that the proposed polishing robot can successfully accomplish the polishing task of wooden materials without complicated teaching. Furthermore, the surface accuracy of polished workpieces was so good condition as well as polished by skilled workers. The measurements evaluated by arithmetical mean roughness method were less than 2μm.

References

[1] F. Ozaki, M. Jinno, T. Yoshimi, K. Tatsuno, M. Takahashi, M. Kanda, Y. Tamada and S. Nagataki, "A Force Controlled Finishing Robot System with a Task-Directed Robot Language," *Journal of Robotics and Mechatronics*, Vol. 7, No. 5, pp. 383-388, 1995.

[2] F. Pfeiffer, H. Bremer and J. Figueiredo, "Surface Polishing with Flexible Link Manipulators," *European Journal of Mechanics, A/Solids*, Vol. 15, No. 1, pp. 137-153, 1996.

[3] Y. Takeuchi, D. Ge and N. Asakawa, "Automated Polishing Process with a Human-like Dexterous Robot," *Procs. of IEEE International Conference Robotics and Automation*, pp. 950-956, 1993.

[4] K. Takahashi, S. Aoyagi and M. Takano, "Study on a Fast Profiling Task of a Robot with Force Control Using Feedforward of Predicted Contact Position Data," *Procs. of the 4th Japan-France Congress & 2nd Asia-Europe Congress on Mechatronics*, Vol. 1, pp. 398-401, 1998.

[5] F. Nagata, K. Watanabe and K. Izumi, "Profiling Control for Industrial Robots Using a Position Compensator Based on Cutter Location Data," *Journal of the Japan Society for Precision Engineering*, Vol. 66, No. 3, pp. 473-477, 2000 (in Japanese).

Force Guided Assemblies Using a Novel Parallel Manipulator

Daniel M. Morris, Ravi Hebbar, Wyatt S. Newman (wsn@po.cwru.edu)
Department of Electrical Engineering and Computer Science
Case Western Reserve University
Cleveland, Ohio 44106

Abstract: In this paper we present recent experimental results using a novel parallel manipulator operating under Natural Admittance Control performing difficult mechanical assemblies. The example assemblies are components from an automotive automatic transmission. Success depends on appropriate force responsiveness, rather than precise position control. The robotic assemblies are shown to be gentler than human assemblies while competitive with human assembly rates. The results show the potential for robotic automation in application areas that are currently limited to manual assembly.

1. Introduction: In automotive manufacturing, use of robots is widespread in spray painting and spot welding tasks. However, mechanical assembly is still dominantly manual. The barrier to robotic mechanical assembly is that this task domain requires appropriate responsiveness to contact forces, rather than reproducible position control.

The need for force responsiveness in mechanical assembly has long been recognized (see, e.g., [1] for a review). While this has been an active area of research for over 25 years, industrial applications of force controlled robotic assembly remain rare. In part, this is due to the fact that industrial robots are designed to achieve high repeatability, and both industrial controllers and industrial robot programming languages are restricted to precision position control.

Under a National Institute of Standards and Technology (NIST) Advanced Technology Program (ATP) effort by an industrial consortium for Flexible Robotic Assembly for Powertrain Applications (FRAPA), a new robot has been designed and built specifically for enhanced force responsiveness. This robot, the "ParaDex", is shown in Fig 1. Designed by MicroDexterity Systems, Inc., it was first publicly described at ICRA 2000 [2,3]. It is a unique mechanism that incorporates aspects of the Stewart Platform, the Modified Stewart Platform [4], and the Merlet-INRIA mechanism [5] as well as innovations not included in previous designs.

An open-architecture controller for this robot, designed at Sandia National Laboratories, allows for implementation of custom control algorithms. Force sensors integrated within the connecting links of the closed-chain mechanism are accessible to the controller at high speed with low latency. Similarly, direct control of the motor efforts is also available at high rates and low latency. These capabilities make the system well suited for force control.

Force-responsive control algorithms were implemented on this robot at Case Western Reserve University. Specifically, the technique used was Natural Admittance Control [6-9]—a flavor of Impedance Control [10] that suppresses Coulomb friction while emulating a passive physical equivalent system, thus guaranteeing contact stability.

The combination of these elements—a robot designed for force control, an open-architecture controller receptive to high-speed, low-latency implementation of custom control algoritms, and use of appropriate low-level compliant-motion control algorithms—has resulted in a remarkably effective mechanical assembly system. This paper presents results obtained using this system to perform relatively difficult mechanical assemblies within an automotive transmission. It will be shown that the robotic assembly system can meet or exceed the current performance of manual assembly.

2. ParaDex Manipulator: Like the Stewart platform, the ParaDex manipulator has six actuators that allow control of a distal platform in 6 degrees of freedom. Like the INRIA prototype [5], the ParaDex has an active link to passive link connection for each kinematic chain between the base and platform. The motors are not mounted on the active links and therefore do not have to accelerate most of their own mass, yielding a lower inertia-to-payload ratio than a typical Stewart platform. In addition, the linear actuators are designed to have low Coulomb friction. The passive links cross over one another in space in three pairs, like the Modified Stewart Platform, to increase the available horizontal force. In addition, the links are curved to allow better spacing of the universal joints that connect the passive links and the distal platform. This yields a more uniformly strong force-torque capacity over the workspace [3].

In addition, the ParaDex has two important features that make it more useful for robotic assembly. First, it has a central, redundant tool-roll actuator. Because of this actuator, the tool-flange rotation about its own normal, which is vertical in the reference position, may be decoupled from the movement of the linear actuators. Second, force sensors are integrated within the passive links. Such placement of these sensors achieves feedback which is inherently more stable than use of endpoint force feedback [9].

3. Control Algorithm: Our foundation for achieving high-performance assembly that responds to environmental contact forces is an *impedance controller* [11] operating at a low level in the control hierarchy [12].

For purposes of mechanical assembly, simply using a "force control" algorithm is not sufficient; also needed is the ability for the robot to remain stable when its dynamics are coupled with those of a passive environment [11,9]. This is the property of an *interaction controller* [13]. Natural Admittance Control offers this capability, while rejecting Coulomb friction, an important consideration for dexterous manipulation.

Our controller is comprised of a 700 Mhz PC running the QNX real-time OS and Delta Tau's Turbo PMAC-2 PC, a 32-axis motion-control card.

A description of the Natural Admittance Control (NAC) algorithm follows, based on the explanation in [12]. The concept of virtual springs and dampers is illustrated in Figure 2. An attractor is defined in the Cartesian space of the robot. The attractor is defined as a position and orientation, as well as a velocity and angular velocity. Virtual springs and dampers are defined in software and used to generate virtual forces on the robot. These virtual forces depend on the differences between the position, orientation, velocity, and angular velocity of a reference point relative to the robot, and the corresponding states of the attractor. The reference point may be on the robot itself (e.g. tool tip), or have a fixed location with respect to the robot in tool-space coordinates. Remote center of compliance (RCC) behavior is implemented in our algorithm by defining this reference position to be at fixed offset along the gripper z axis.

In our implementation, the orientational springs and dampers are defined as linear springs and dampers that attach to the unit span vectors and exert virtual forces on the vectors, which results in virtual moments exerted on the tool. The virtual springs and dampers between one set of those axes are shown in Figure 2.

The robot has six linear actuators, which completely specify platform position and orientation, plus a seventh redundant rotational axis controlling tool-flange rotation. The direction in which the tool-roll actuator's normal axis points is constrained by the platform. The relation between rotation of the central-axis actuator and rotation of the tool flange is a function of platform position and orientation [14], but the actuator rotational angle itself is completely decoupled from the distal platform. Therefore, the Natural Admittance Control problem with respect to tool z-rotation may be considered as a separate, 1-dof system.

To achieve desirable interaction behavior, we define an idealized model of the robot dynamics, including realistic responsiveness to external forces, then control the robot to mimic this model. The model dynamics are computed in real time, in response to sensed contact forces. Ultimately, the contact stability of the controlled system depends on the choice of model. (See [9] for details). While achieving stability with this control method is still something of an art, it was found empirically that the ParaDex robot was relatively easy to stabilize under this control approach. The resulting performance was remarkably responsive to end-effector forces, yet the robot was stable in contact with rigid environments (e.g., steel fixtures).

4. Search Algorithm: Achieving gentle and stable interactions with stiff environments is a necessary condition for effective force-responsive robotic assembly, but this property alone is insufficient. It is also necessary to guide the robot towards a parts-mating solution. In our approach, this is done by specifying a trajectory for the virtual attractor.

For our example assemblies, insertion is to be performed in the vertical direction. This requires a hunt for alignment of the parts in the horizontal plane. Our "blind" search for parts mating consists of three strategies: 1) the attractor is moved downwards, resulting in a force preload in the vertical direction due to stretching of the vertical virtual spring; 2) the attractor is moved in the horizontal plane; and 3) the part to be inserted is rotated back and forth about its z-axis by defining corresponding movement of the rotational attractor.

While an assembly is being attempted and the parts are in contact, the attractor invokes a search in the horizontal plane in a spiral pattern (see [12,15,16] for various implementations).

While the attractor is being moved in the horizontal plane, the angular attractor of tool rotation follows a trapezoidal angular-velocity trajectory, causing corresponding rotation of the tool-roll axis of the ParaDex.

To evaluate the effectiveness of the above combination of novel robot, open controller, NAC force feedback and virtual attractor trajectories, we compared human performance in several assemblies to performance of the robotic system.

5. Experimental Setup: For our evaluations, parts were assembled by the ParaDex and by humans. Some additional assembly experiments were performed with human/robot cooperation in which the human provided manual guidance and gravity assist was provided by the ParaDex. For assemblies involving the ParaDex, motions of the grasped parts were sensed via the robot's encoders. A JR[3] force/torque sensor was used to measure contact forces and moments during assembly—both for robotic and human assembly trials.

One of the performance metrics was assembly time. Assembly time was defined as the time span from initial contact between parts until the grasped part had been fully inserted. For the robotic assemblies, the initial and final times were detected precisely by the force and position sensors. For the human assemblies, the assembly times were measured with a stopwatch. Four different assembly examples were evaluated. A benchmark was set for each assembly task by having four volunteers manually complete 30 time trials of each task.

In human contact-force experiments, the human used the JR[3] sensor mounted in series between a pneumatic gripper and an acrylic handle. The goal during each human experiment was to complete the assembly as fast as possible. The contact forces obtained are therefore representative of those during the timed trials.

Our experiments included four assembly examples from automotive automatic transmissions: a forward clutch assembly, a rear planet gear assembly, a front planet gear assembly, and a reverse clutch assembly. The two clutch assemblies were similar. Each contained multiple toothed spline rings that could move in the horizontal plane and rotate about the vertical. (See [17] for further details of the forward clutch). During assembly, gear teeth of the mating part had to align with each of the five spline rings sequentially during insertion. The two planetary assemblies were also similar to each other. In these assemblies, a subassembly containing multiple planet gears had to be inserted into an external ring gear, meshing all planets with the ring gear.

Three of these assemblies are shown in Figure 3. In sequence the rear planet gear subassembly (3.b) is inserted into the rear ring gear (3.a); the front planet gear subassembly (3.c) is inserted into the the rear planet gear subassembly; and finally the reverse clutch (3.d) is assembled onto this stack..

6. Experimental Results:

6.1. AX4N Reverse Clutch Insertion: This assembly and the rear planet gear assembly were easier for the robot than for humans. A light contact force (12N) was specified for robotic assembly of the reverse clutch assembly. The reverse clutch weighed about 30 N, and thus the robot exerted a gravity-compensating upwards force on the grasped part during insertion. This behavior, which seemed to be difficult for people to emulate, was favorable for fast and gentle assembly of this component.

The reverse clutch contained four toothed spline rings that could move about in the horizontal plane and rotate about the vertical. Each spline ring had forty-nine teeth, which had to mesh with teeth on the fixtured subassembly (see Figure 3.d.). A soft virtual vertical spring (0.30 N/mm) was chosen, resulting in a very gentle assembly. In addition, RCC behavior was seen to have a helpful effect on the reverse clutch assembly.

Thirty 30 trials of the reverse clutch insertion using the ParaDex were compared against 30 manual assembly trials each by four volunteers. In preparation for each assembly, the spline rings in the cylinder were deliberately misaligned by hand. All trials were successful. The JR3 was not used for any of these time trials. On the average, the robot was faster then the volunteers (2.1 sec vs 3.0 sec). The standard deviation of robotic assembly times was also lower than that of manual assembly (0.46 sec vs 1.76 sec)..

Insertions of the AX4N reverse clutch were also performed by the ParaDex and by humans with a force/torque sensor in series for the purpose of comparing contact forces.

Figure 4 shows vertical position and forces in one reverse clutch insertion as performed by the ParaDex. The ParaDex handled this assembly task very gently. The impact forces barely exceeded 50 N. The steady contact forces (just following the downward-pointing peaks in Figure 4) barely exceeded 10 N.

The corresponding assembly-force signature for a manual assembly is shown in Fig 5. The sustained forces were about 60 to 75 N. A large impact of approximately 500 N was observed at completion of the assembly. The time shown is roughly 2.5 seconds (first impact at about 1.4 seconds and final impact at about 3.8 seconds).

6.2. Forward Clutch Assembly:
This assembly was similar to the reverse clutch assembly, in that multiple, movable spline rings had to engage with a gear component. In this case, though, the spline rings were contained within the fixtured subassembly, and the grasped component to be inserted was a monolithic toothed hub.

A substantial downward force and strong vertical spring were found to be useful in this assembly task. The force threshold for this assembly was set at 50 N and a vertical spring of 5 N/mm was selected. In addition, it was observed that the assembly performance was better without the use of remote center of compliance (RCC) behavior. The clutch hub weighed 5 N and the gripper used in the assembly weighed about 10 N.

Based on 30 time trials with the ParaDex and a total of 120 time trials with 4 volunteers, the ParaDex was observed to be somewhat slower than humans on average for this assembly (1.8 sec. vs 1.5 sec). However, humans would not be able to maintain the pace of these short-duration, maximum-speed time trials, and thus the ParaDex may be expected to have a higher sustained throughput than humans in this assembly task.

In addition, the assembly-force tests showed a maximum impact force for the robotic assembly of approximately 300N, whereas the peak impact force for manual assembly was nearly 500N. If the robot were permitted to be similarly aggressive in contact forces, it might also achieve faster assembly times, though the lower impact forces might be more desirable than the modest throughput gains.

6.2. AX4N Rear Planet Gear Assembly Insertion: The AX4N rear planet gear assembly insertion task was easier than the forward clutch assembly task for the ParaDex. It only required rotational alignment of the four planet gears with the rear ring gear. In addition, the inside of the rear ring gear cylinder was slightly chamfered just above the helical teeth. The robot's compliance allowed it to exploit use of this chamfer. After insertion, the rear planet gear assembly is free to spin about its z-axis as the planet gears follow the track of the rear ring gear.

The spiral search for this assembly was trivial. It was still carried out, but its main purpose was to provide a small force disturbance to the system in the unlikely event of a jam, and it had the effect of "wiggling" the part to be inserted. Maintaining a light downward force and rotating the tool were the most important parts of this assembly task. The threshold contact force used was 10 N, which is less than the weight of the mechanism itself, 20 N. Like the reverse clutch example, the robot provided partial gravity compensation during assembly. In addition, a soft vertical spring of 0.30 N/mm was chosen.

Again, 30 trials of robotic assembly were recorded. The same assembly was also performed 30 times each by four volunteers. The mean robotic assembly time was significantly faster than the humans (1.15 sec vs. 2.3 sec).

Additionally, the ParaDex was significantly more gentle than the volunteers. The peak impact was around 180 N for the robotic assembly, compared to approximately 600N for manual assembly.

6.3. AX4N Front Planet Gear Assembly Insertion: The front planet gear assembly insertion was more difficult than the rear planet gear assembly insertion because, unlike the latter, the shape of the parts did not act as a guide to proper assembly. In addition, the grasped component included tabs that had to be aligned with slots in the fixtured component after the planets had meshed with their corresponding ring gear. Aligning the tabs was easy during manual assembly, as this operation benefited from use of vision. The ParaDex system used here did not have vision and had to search blindly for the tab/slot alignment.

Based on 30 robotic assembly trials and 120 human trials, the robot was slower, on average, than the humans (2.75 sec vs. 1.93 sec). With the addition of machine vision for the final alignment of the tabs and slots, the ParaDex would be more competitive with humans.

Although the ParaDex was slower than the volunteers in performing this assembly, it was significantly gentler. The peak impact force during robotic assembly was less than 100N, whereas the impact force during manual assembly was nearly 400N.

7. **Conclusions**: Results of our assembly experiments are summarized in Figures 6 and 7. Figure 6 shows the mean assembly times and standard deviations of assembly times for the ParaDex vs human volunteers. The ParaDex was faster than humans in two of the four examples, and slower in the other two cases. All 120 assembly trials with the ParaDex were successful. Overall, the ParaDex was competitive with humans in terms of assembly times. It should be noted that this comparison is based on volunteers' times while performing limited-duration, motivated speed trials. Exhaustion, boredom and sustainable pace were not considered. The sustainable throughput of the ParaDex is expected to be superior to manual assembly.

More dramatically, as shown in Fig 7, it is clear that the robot was much gentler than humans in all four cases. It should be noted, though, that the force data is still anecdotal; more force-signature experiments should be conducted before a statistically valid conclusion can be stated. Nonetheless, the limited data suggests that the robot is gentler than humans. Such a conclusion could be a significant issue in manufacturing quality, since deformed or chipped components could lead to early failure in service.

To achieve the results reported here, our system included several important features. First, the assembly robot was intentionally designed for force-responsive behavior. Virtues included relatively low Coulomb friction, relatively low inertia, and force sensors located near the actuators rather than near the end effector. The robot controller was also supportive of high-performance impedance control. It offered high sampling rates and direct, low-latency access to the force sensors and the motor forces, and it supported development of custom real-time control algorithms. The specific force-feedback algorithm implemented, Natural Admittance Control, achieved responsive and stable interaction dynamics with arbitrarily stiff environments. The use of virtual springs, virtual dampers and virtual attractors offered sufficient flexibility to perform fast yet gentle assembly of all four cases examined.

We conclude that exploitation of force feedback for robotic assembly may now have matured to where industrial applications are viable. In our investigation, we have considered components that have proven difficult to assemble automatically by conventional means, and we have shown that force-responsive robotic assembly is arguably superior to manual assembly.

Acknowledgments: This work was performed under the support of the U.S. Department of Commerce, National Institute of Standards and Technology, Advanced Technology Program, Cooperative Agreement Number 70NANB7H3024. The ParaDex robot was designed by Steve Charles and Robert Stoughton of MicroDexterity Systems, Inc., in collaboration with David Kozlowski of Sandia National Laboratories. The authors are grateful for the extensive support from Ford Motor Co. in this project, including resources and technical input, with special thanks to David Gravel, Frank Masler and Valerie Bolhouse of Ford's Advanced Manufacturing Technology Development Center.

References:

1. Whitney, D. E., "Historical Perspective and State of the Art in Robot Force Control," Int. J. of Robotics Research, Vol 6, No 1, Spring 1987, pp 3-14.
2. Wang, Y., Newman, W. S., and Stoughton, R. S., "Workspace Analysis of the ParaDex Robot," *Proc. of IEEE Int. Conf. on Robotics and Automation*, San Francisco, CA., Vol. 3., pp. 2392-2397, Apr. 2000.
3. Charles, S., Stoughton, R. S., Kozlowski, D., Morris, D., Hebbar, R., and Newman, W. S., "The ParaDex Manipulator: a Novel, Parallel-Actuated Robot Design for Mechanical Assembly Applications", *Video Proc. of IEEE Int. Conf. on Robotics and Automation*, San Francisco, CA., Apr. 2000.
4. Stoughton, R. S., and Arai, T., "A Modified Stewart Platform Manipulator with Improved Dexterity," *IEEE Transactions on Robotics and Automation*, Vol. 9 [2], pp. 166-173, Apr. 1993.
5. Mouly, N., Merlet, J.-P., "Singular Configurations and Direct Kinematics of a New Parallel Manipulator," *Proc. of IEEE Int. Conf. on Robotics and Automation*, Vol. 1, pp. 338 –343, 1992.
6. Mathewson, B. B., and Newman, W. S., "Integration of Force Strategies and Natural Admittance Control," *Proc. of the Int. Mechanical Engineering Congress and Exposition,* Vol. 1, Nov. 1994.
7. Newman, W. S., and Zhang, Y., "Stable Interaction Control and Coulomb Friction Compensation Using Natural Admittance Control," *Journal of Robotic Systems,* Vol. 11, No. 1, pp. 3 - 11, 1994.
8. Glosser, G. D., and Newman, W. S., "Implementation of a Natural Admittance Controller on an Industrial Manipulator," *Proceedings of IEEE Int. Conf. on Robotics and Automation,* May, 1994.
9. Newman, W. S., "Stability and Performance Limits of Interaction Controllers," *Transactions of the ASME Journal of Dynamic Systems, Measurement, and Control,* Vol. 114, No. 4, pp. 563-570, 1992.
10. Hogan, N., "On the Stability of Manipulators Performing Contact Tasks," *IEEE Journal on Robotics and Automation, Vol. 4* [6], pp. 677-686, Dec. 1998.
11. Hogan, N. "Impedance Control: An Approach to Manipulation: Parts I, II, III," *ASME Journal of Dynamic systems, Measurement, and Control,* Vol. 107, pp. 1-24, 1985.
12. Chhatpar, S. R., *Experiments in Force-Guided Robotic Assembly*, M.S. thesis, Department of Electrical Engineering and Applied Physics, Case Western Reserve University, Jan. 1999.
13. Colgate, J. E., *The Control of Dynamically Interacting Systems*, Ph.D. thesis, Department of Mechanical Engineering, MIT, 1988.
14. Chuckpaiwong, I., *Reflexive Collision Avoidance for a Novel Parallel Manipulator*, M.S. thesis, Department of Electrical Engineering and Applied Physics, Case Western Reserve University, Jan. 2001.
15. Huang, L. E., *Impedance Control for Accommodation of Position Uncertainty in Robotic Assembly*, M.S. thesis, Department of Electrical Engineering and Applied Physics, Case Western Reserve University, May 1998.
16. Morris, D. M., *Experiments in Mechanical Assembly Using A Novel Parallel Manipulator*, M.S. thesis, Department of Electrical Engineering and Applied Physics, Case Western Reserve University, Jan. 2001.
17. Newman, W. S., Branicky, M. S., Chhatpar, S., Huang, L., and Zhang, H., "Impedance Based Assembly," *Video Proc. of IEEE Int. Conf. On Robotics and Automation*, Detroit, MI, 1999.

Figure 1. The ParaDex Manipulator

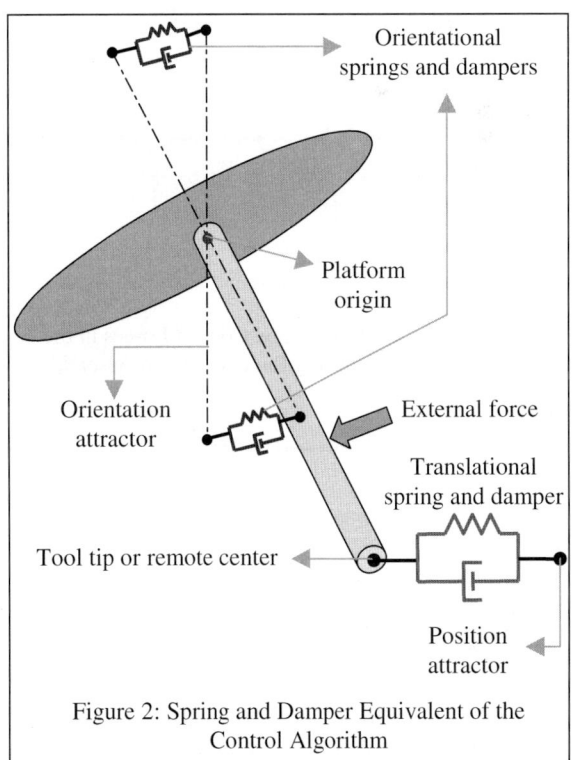

Figure 2: Spring and Damper Equivalent of the Control Algorithm

Figure 3. (a) Rear ring gear cylinder. (b) Rear planet gear assembly. (c) Front planet gear assembly. (d) Reverse clutch (shown inverted).

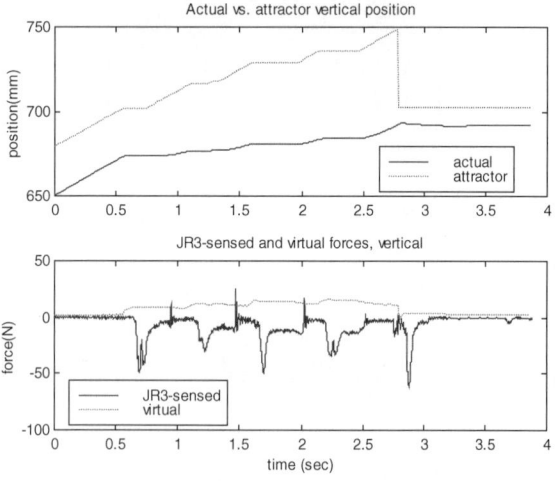

Figure 4. Robot Position and Contact Forces in the Vertical Direction during Reverse Clutch Assembly

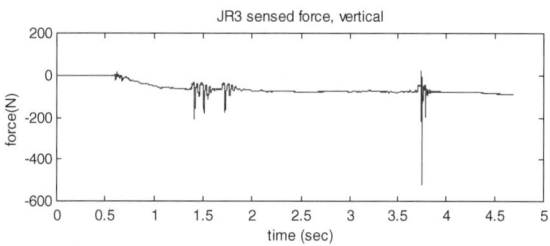

Figure 5. Contact Forces in the Vertical Direction during Reverse Clutch Assembly by a Human Volunteer

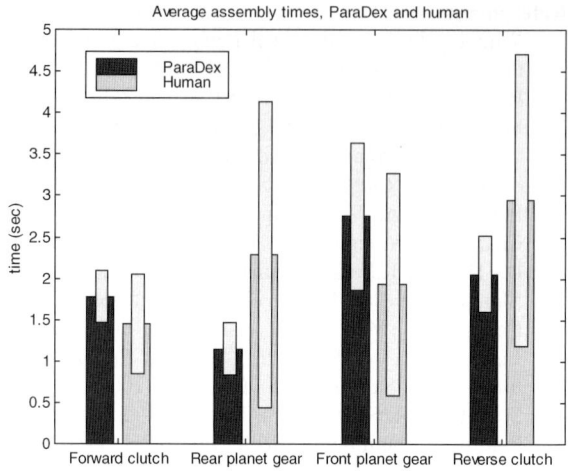

Figure 6. Average assembly times for ParaDex compared with times by human volunteers

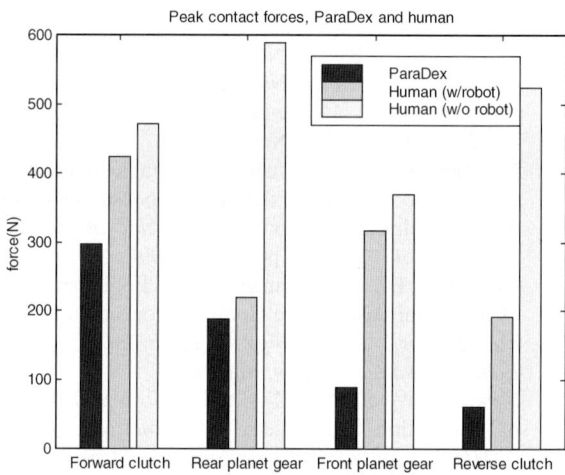

Figure 7. Largest forces recorded during assemblies by ParaDex compared with human volunteers.

RESULTS	Mean Assembly Time	Standard Deviation
ParaDex	2.06 sec	0.46 sec
All Volunteers	2.95 sec	1.76 sec

Table 1. Reverse clutch insertion task –timing results for the ParaDex and human volunteers

Path Planning for Robot-Assisted Grinding Processes

Y.T. Wang and Y.J. Jan
Department of Mechanical Engineering, Tamkang University
151 Ying-Chuan Rd., Tamsui, Taipei, Taiwan 25137
ytwang@mail.tku.edu.tw

Abstract

Path planning for a robot-assisted surface finishing system with an active torque controller is presented in this paper. We utilize a dexterous manipulator to attain the desired position and orientation in three-dimensional space during finishing processes. A single-axis active controller consists of a dc motor and a software observer is attached to the robot wrist and used to actuate a pneumatic hand-grinder. The torque observer is designed to sense the grinding contact force based on the driving current and output position of the motor. Zigzag and fractal paths on curved surfaces are designed for the grinding processes. In order to determine an ideal grinding condition, Taguchi's method for experimental design is utilized. We choose four grinding conditions, namely, path pattern, grinding contact pressure, tool diameter, and feed rate. Tendencies of these factors can be found from the experiments. In this research, the prototype of a robot-assisted finishing system is constructed and tested on a Tatung A530 robot. The experimental results show that the robot-assisted finishing system functions well under a variety of grinding conditions.

1. Introduction

Finishing processes of die-cast manufacturing include grinding, honing, lapping, polishing lapping, and polishing. These processes are time-consuming and monotonous operations that strongly rely on skilled human-workers. To automate these processes and achieve desired surface roughness, it is important to control the grinding path and contact force, as well as to choose suitable feed-rate and tool diameter. Among them, to generate a suitable tool-path and to control the contact force are two major challenge issues. For example, to polish free-form surfaces of an object requires a delicate machine to follow complicated polishing paths. In this case, a polishing system based on a robot manipulator is more effective than that on a NC machining center in order to follow the curved free-form surfaces. Different robot grinding path on the specimen will affect the surface roughness. On the other hand, during finishing operations, the tool comes into physical contact with the workpiece and causes contact forces between them. It is difficult to control these contact forces which depend on the cutting depth, feed rate, grinding-wheel speed and material properties.

Many researchers have proposed automated systems for grinding of dies, deburring of castings, and removing of weld beans etc [5-8]. Usually, a grinding tool is mounted on a NC machining center or a robot manipulator and a multi-dimensional force sensor is included in the system to improve finishing accuracy. It is troublesome to handle the multi-dimensional motion and force control system in run-time processes, besides the passive-type force sensors are expensive in price and sensitive to a noise.

We propose an automated finishing system for polishing free-form surfaces of dies specimens. Two types of tool path, zigzag and fractal, are used in this research for comparison. To simplify the force-control action, only the contact force normal to the polishing surface is concerned and a software-type torque observer is used to replace the role of a hardware sensor. We decide four important grinding conditions, namely, path pattern, grinding contact pressure, tool diameter, and feed rate. In order to determine the best combination of the four grinding conditions, Taguchi's method for experimental design is utilized.

2. Finishing robot system

The developed robot-assisted surface finishing system consists of a 5-axis articulated industrial robot, an end-effector, a robot controller, a xy-table for setting the metal mold, and a personal computer for sensory processing essential to a contact force control. The system configuration is shown as Figure 1. The system utilizes a dexterous manipulator, Tatung A530, to attain the desired position and orientation of the end-effector in three-dimensional space. A dc motor is attached to the robot wrist and used to actuate the polishing tool. The torque observer is designed to sense the applied torque based on the driving current and output position of the dc motor. A pneumatic hand-grinder is serially mounted on the observer-motor. We control the motion and contact force of the hand-grinder to perform the desired finishing action. The robot follows a desired tool path and drives the hand-grinder to come in contact with the workpiece. The single-axis torque observer can sense the contact force and direct the hand-grinder to apply a desired contact pressure on the workpiece. The kinematic analysis, path planning, and torque control algorithm for the robot and torque observer are derived in the following sections.

In order to polish a workpiece with free-form surfaces, a 7-d.o.f. mobility is provided by the finishing system formed by a Tatung A530 robot and a xy-table as shown in Figure 1. The workpiece is placed on the xy-table. During the process, the manipulator drives the polishing tool to follow a programmed path and attain a desired orientation. The robot performs rotation in three rotational angles and translation in z-axis, while the xy-table translates in x- and x-axis. There is one more d.o.f. mobility provided by the dc observer-motor. We control the current command and the angular position of the motor in order to ensure that the tool is kept at a desired contact angle and contact pressure with the workpiece. Usually, the contact angle, θ_f, is measured from the surface at contact point in the drive-feed direction as shown in Figure 2.

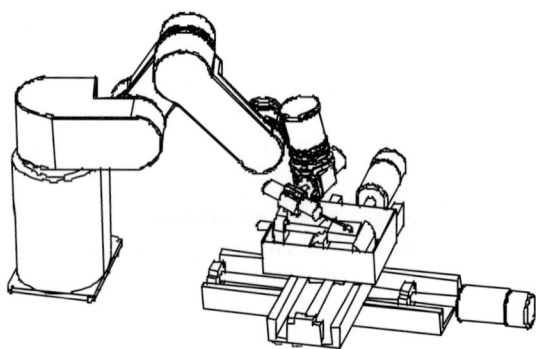

Figure 1 Finishing Robot System

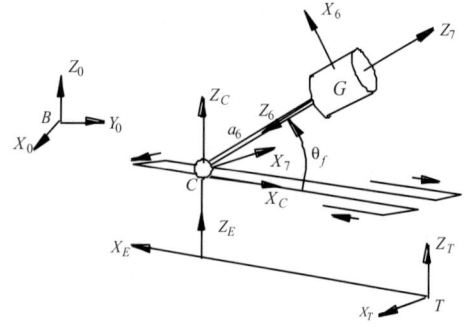

Figure 2 Coordinate systems for the contact area

Assume that the contact angle is retained at a magnitude of θ_f, and the contact at point C is a point contact, then the homogenous transformation matrix from the robot-base coordinate, B, to the coordinate of the contact point, C, can be expressed as

$$A_C^B = A_T^B A_C^T = A_P^B A_C^P$$

where T and P are located at the coordinate origins of the base of the xy-table and the robot end-effector, respectively. During the finishing process, the desired position coordinate of the contact point C and the orientation of the tool relative to surfaces of the specimen are generated. The transformation from the robot-base frame to the end-effector is determined by

$$A_P^B = A_T^B A_C^T (A_C^P)^{-1} = A_C^B (A_C^P)^{-1} \qquad (1)$$

where A_C^P could be found by the knowledge of θ_f and tool length. From Equation (1), we can find the desired joint angles of A530 robot by an inverse kinematic method. These joint angles are the inputs of the robot motion controller.

3. Grinding Path Planning

We use B-spline function to describe the surface of a specimen,

$$Q(u,v) = \sum_{i=1}^{n+1} \sum_{j=1}^{m+1} B_{i,j} N_{i,k}(u) M_{j,l}(v)$$

Where $B_{i,j}$ is a position matrix; $N_{i,k}(u)$ and $M_{j,l}(v)$ are two basis functions in u and v axes respectively. Typical curved surfaces of a specimen are shown in Figures 3 and 4. We can formulate different types of tool path for the robot to perform a task on this workpiece. Two types of tool path, zigzag and fractal, are used in this research for comparison. A zigzag is a path with repeated switching of directions as shown in Figure 3. Figure 5 depicts the top-view of a specimen ground by zigzag tool path. On the other hand, a fractal tool path is generated based on the Hilbert "£v" pattern. A L-system method [4] using logical symbols is adopted here to generate the Hilbert "£v" pattern. Definitions of the logical symbols are stated as following:

F: *Drawing a fixed-length line from current position to new position*
-: *Turning an angle of 90° in CCW direction*
+: *Turning an angle of 90° in clockwise direction*
L: +F-F-F+
R: -F+F+F-

Generation rules of the Hilbert curve are: (a) substituting "+RF-LFL-FR+" into L, "-LF+RFR+FL-" into R when the order increases by one, (b) repeating the processes in every increment of order, and (c) the zero-order starting from the L-operation.

Figure 4 depicts a 3rd-order fractal tool path based on the generation rules, while the top-view of the ground specimens by fractal tool path are shown in Figure 5-7. We found that the fractal path has an advantage of consistency in direction [4].

We develop a PID position controller for the motion of each axis of the xy-table. The control block diagram is shown in Figure 8 and the transfer function is given by

$$\frac{P(s)}{P^*(s)} = \frac{K_t K_d s^2 + K_t K_p s + K_t K_i}{Js^3 + (B + K_t K_d)s^2 + K_t K_p s + K_t K_i}$$

Where $P^*(s)$ and $P(s)$ represent the position command and output, respectively; J is the inertia and B is the coefficient of viscous friction; K_t is the drive torque constant; K_p, K_i, K_d are gains of the PID controller. For the xy-table, point-to-point motion is programmed by using a trapezoid velocity profile as shown in Figure 9.

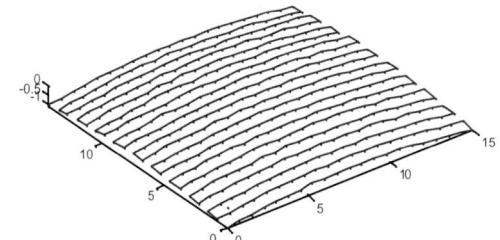

Figure 3 Zigzag tool path

Figure 4 Third-order fractal tool path

Figure 5 Ground zigzag tool path

Figure 6 Ground 3rd-order fractal path

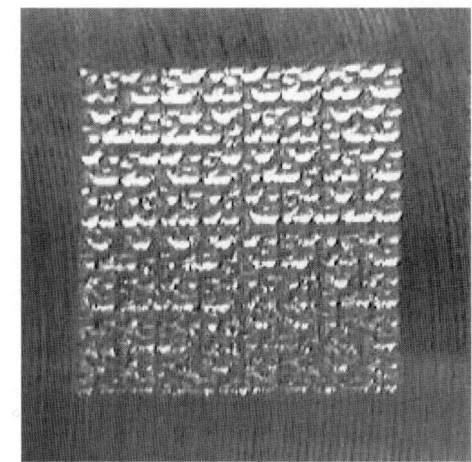

Figure 7 Ground 4th-order fractal path

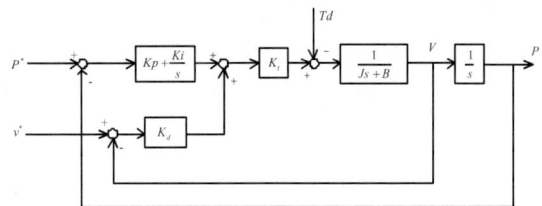

Figure 8 PID position control

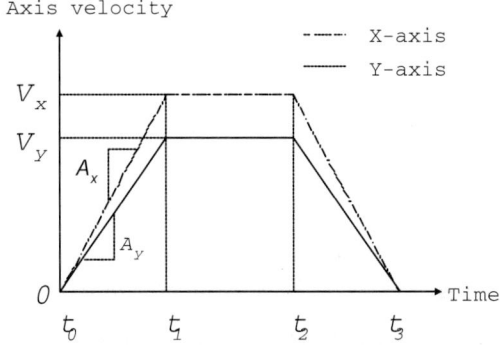

Figure 9 Trapezoid velocity profile

4. Contact torque control

In the finishing process, an incorrect CAD data or tool wear will cause a contour error. In this case, a pure position controller would cause the tool to be no contact with or over-cut the workpiece. A suitable torque controller is always needed in a finishing process. In this research, we propose to control only the contact pressure normal to the contact surface. Then the torque controller could be simplified and decoupled from the robot motion controller.

We model the contact behavior between the tool and the specimen as a linear rotational spring in Figure 10,

$$T_f = k_e \theta \quad (2)$$

where T_f is the torque exerted by the observer motor on the surface of the workpiece; k_e is the rotational spring constant; θ is the angular displacement of the motor. The dynamic equation of the end-effector system with an observer motor and a hand-grinder can be expressed as

$$T_{em} = J\ddot{\theta} + k_e\theta + T_L \qquad (3)$$

where T_{em} and T_L are the applied torque and external load, respectively. J is the inertia of the system. From Equation (2), we have

$$\theta = k_e^{-1}T_f$$
$$\ddot{\theta} = k_e^{-1}\ddot{T}_f$$

Equation (3) can be rewritten as the following equation with the new variable T_f,

$$T_{em} = Jk_e^{-1}\ddot{T}_f + T_f + T_L \qquad (4)$$

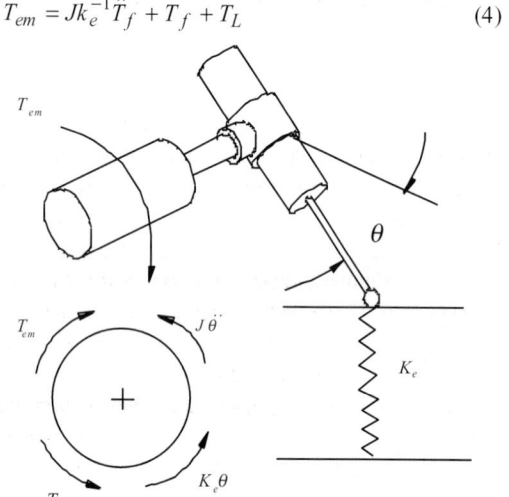

Figure 10 Linear spring model

In order to control the applied torque, a proportional-integral-derivative (PID) controller is considered

$$T_{em} = Jk_e^{-1}\left[\ddot{T}^* + k_{fp}e_f + k_{fi}\int e_f dt + k_{fd}\dot{e}_f\right] + T_f + T_L \qquad (5)$$

where T^* denotes the torque command; e_f is the torque error, $e_f = T^* - T_f$; k_{fp}, k_{fi} and k_{fd} are the gain values of the PID controller. In this case, T_L is not suitable to be put in the controller because it is not easy to be measured on-line. The term $(T_f + T_L)$ in Equation (5) is replaced by a feed-forward command T^*. The resultant controller is a PID-plus-feedforward (PIDFF) controller,

$$T_{em} = Jk_e^{-1}\left[\ddot{T}^* + k_{fp}e_f + k_{fi}\int e_f dt + k_{fd}\dot{e}_f\right] + T^* \qquad (6)$$

The block diagram of the PIDFF controller is shown in Figure 11. Where k_t is the torque constant of the motor; $\hat{J}$, $\hat{k}_e$ and $\hat{k}_t$ are the estimated values of the system parameters J, k_e and k_t, respectively. In the feedback loop, we replace sT_f by $k_e\omega$ in order to reduce the signal noise to the controller. Where ω is the angular speed of the motor. In the finishing processes, the contact torque will be kept at a constant value, i.e. $\dot{T}^* = \ddot{T}^* = 0$. From Equations (4) and (6), we find the error equation,

$$\ddot{e}_f + k_{fd}\dot{e}_f + \frac{e_f}{Jk_e^{-1}} + k_{fi}\int e_f dt = T_L$$

$$\dddot{e}_f + k_{fd}\ddot{e}_f + k_{fp}\dot{e}_f + k_{fp}\dot{e}_f + k_{fi}e_f = \dot{T}_L \qquad (7)$$

We can see that the value of the steady-state error vanishes as time approaches infinity. If the estimated values of the system parameters are correct, we can derive the transfer function which represents the relationship between the output torque, T_f, and the command torque, T^*,

$$\frac{T_f}{T^*} = \frac{(Jk_{fp} + k_e)s + Jk_{fi}}{Js^3 + Jk_{fd}s^2 + (Jk_{fp} + k_e)s + Jk_{fi}} \qquad (8)$$

In the case of no disturbance, the output will track the command exactly at steady state, as we can see from Equations (7) and (8). If there exists a disturbance, T_L, the steady-state error will disappear as time approaches infinity. The PIDFF controller provides a solution to the torque control problem during the finishing process. The analytical and experimental works in this paper are based on the concept of the PIDFF controller.

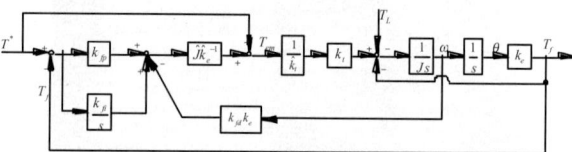

Figure 11 Block diagram of the PIDFF controller

5. Torque observer

We utilize a torque observer to estimate the contact torque during the finishing processes. The torque observer is a linear Luenberger observer [9] as shown in Figure 12. The block diagram is composed of two parts: the upper loop is the motor system with a torque controller and the lower one is the torque observer. The torque observer estimates the contact torque based on the information of the torque command and the output position of the system.

According to Figure 12, the observed torque can be determined as

$$\hat{T}_f = \frac{\frac{\hat{k}_t}{k_t}(k_{od}s^2 + k_{op}s + k_{oi})}{\hat{J}s^3 + k_{od}s^2 + k_{op}s + k_{oi}}T_f$$

$$+ \frac{\hat{J}(k_{od}s^3 + k_{op}s^2 + k_{oi}s)}{\hat{J}s^3 + k_{od}s^2 + k_{op}s + k_{oi}}\left[(\frac{J}{\hat{J}}\frac{\hat{k}_t}{k_t} - 1)\omega\right] \qquad (9)$$

Where k_{op}, k_{oi} and k_{od} are gains of the observer. If the parameters, $\hat{k}_t$ and $\hat{J}$, are correctly estimated, Equation (9) can be simplified as

$$\frac{\hat{T}_f}{T_f} = \frac{(k_{od}s^2 + k_{op}s + k_{oi})}{\hat{J}s^3 + k_{od}s^2 + k_{op}s + k_{oi}} \quad (10)$$

In steady state, the torque observer can sense the contact torque exactly.

The parameters of the dc observer-motor system are listed in Table 1. The controller is equipped with a PCL-726 D/A interface card and a PCL-833 encoder card by Advantech [1,2]. The D/A card has a 12-bits resolution to represent the output current in the range of ±5A. The basic unit of the current command of the motor drive is calculated as

$$10A \times \frac{1}{2^{12}} = \frac{10}{4095} \cong 2.44\,mA$$

The resolution of the observer is

$$T = i \times k_t = 2.44\,mA \times 0.185\frac{N \cdot m}{Amp}$$
$$= 0.4514\,mN \cdot m \times \frac{1\,kgf}{9.81\,N} \cong 4.601 \times 10^{-2}\,kgf \cdot mm$$

The radius between the motor axis and the contact point is 95mm. Then, the resolution of the applied pressure is

$$4.601 \times 10^{-2}/95\,mm = 0.48\,gf$$

This is the smallest force can be applied by the active torque controller. On the other hand, the maximum output-torque is limited by the rate-current of the motor which value is 1 ampere in this case. The maximum output torque can be generated by the controller is determined as

$$T = 0.185\frac{N \cdot m}{Amp} \times 1\,Amp \times \frac{1\,kgf}{9.81\,N} \times \frac{1000mm}{1\,m} \cong 18.9\,kgf \cdot mm$$

which is about 199 gf for this system.

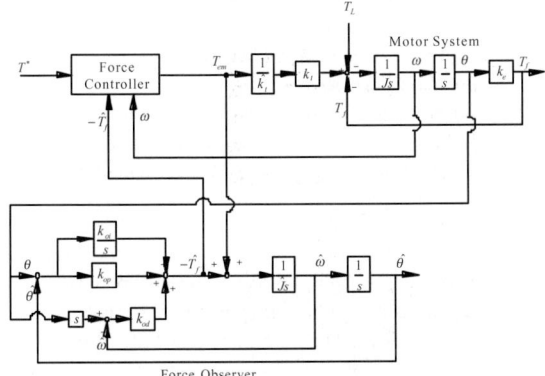

Figure 12   Torque controller with a Luenberger observer

Table 1   Parameters of the observer-motor

Rated Power	60Watt
Rated Voltage	75V
Rated Current	1.2A
Torque Constant	0.185N·m/A
Rotor Inertia	$1.72656 \times 10^{-5} kg \cdot m^2$
Weight	0.8kgf

6. Taguchi's Parametric Design

The integrated robot-assisted finishing system is shown in Figure 13. The system includes a Tatung A530 robot for implementing the position and orientation of finishing processes. The torque observer is a 60*watt* dc motor which is serial-connected to the wrist of the robot. The pneumatic hand-grinder is equipped at the front end of the torque observer. The motion of the manipulator is controlled by a single-board controller. The estimated inertia and contact stiffness of the hand-grinder system are

$$\hat{J} \cong 0.0011\,kgm^2$$
$$\hat{k}_e = 11.95\,Nm/rad$$

The gains of the controller and observer are given as

$$k_{fp} = 299.3636;\quad k_{fi} = 226981;\quad k_{fd} = 183$$
$$k_{op} = 5.28;\quad k_{oi} = 70.4;\quad k_{od} = 0.132$$

The purpose of Taguchi's parametric design is to determine the best programming of the four grinding factors, namely, tool diameter, path pattern, federate, and grinding contact pressure. In this paper, Taguchi's L_{18} orthogonal table is used as shown in Table 2. Taguchi's L_{18} table has the property that interactions are distributed uniformly among all factors and the major trend of each factor can still be seen.

Experimental results are expressed in terms of the signal to noise ratio (or S/N ratio) η, which is defined by the decibel value of surface roughness R_a,

$$\eta = -10\log(R_a)^2 = -20\log(R_a)$$

The surface roughness R_a is defined by the arithmetic mean values of surface heights

$$R_a = \frac{a + b + c + d + \dots}{n}$$

where $a, b, c, d, \dots$ are absolute values of surface heights measured in μm.

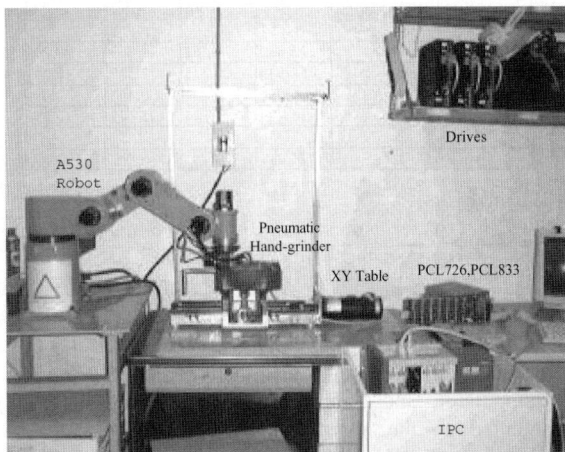

Figure 13   Automated Surface Finishing System

In this experiment, we have four grinding factors with several levels shown as follows,

Tool Diameter   3mm, 4mm
Path Pattern   zigzag(200μm), 5th-order fractal, 4th-order Fractal
Feedrate   12.5mm/s, 25mm/s, 37.5mm/s
Grinding Pressure 0.556N, 1.111N, 1.667N

Note that, only the scale of tool diameter is divided into two levels, scale of all the other factors are divided into three levels. Results for the various factors are also shown in the last two columns of Table 2. According to the method in reference [3], contribution of each factor is calculated, and the results are shown in Table 3. From this table, we know that among the four factor which may affect surface roughness, their contribution in descending order are path pattern, grinding contact pressure, tool diameter, and feed rate. The signal-to-noise ratio for each of the grinding factors can be determined from Table 2. The results show that the best combination of the grinding factors is $A_1B_1C_1D_3$, which includes 4-mm tool diameter, 5th-order fractal path, 12.5mm/s federate, and 1.667N grinding pressure.

Table 2 Experiment design and results

	tool diameter (mm)	path pattern	feed-rate (mm/sec)	grinding pressure (N)	Surface roughness Ra(μm)	SN Ratio η_i
1	4	5th-order fractal	12.5	0.556	5.000	6.021
2	4	5th-order fractal	25	1.111	3.500	9.119
3	4	5th-order fractal	37.5	1.667	3.750	8.519
4	4	4th-order fractal	12.5	0.556	8.250	1.671
5	4	4th-order fractal	25	1.111	6.000	4.437
6	4	4th-order fractal	37.5	1.667	6.000	4.437
7	4	zigzag	12.5	1.111	4.250	7.432
8	4	zigzag	25	1.667	5.250	5.597
9	4	zigzag	37.5	0.556	6.750	3.414
10	3	5th-order fractal	12.5	1.667	4.500	6.936
11	3	5th-order fractal	25	0.556	6.250	4.082
12	3	5th-order fractal	37.5	1.111	5.000	6.021
13	3	4th-order fractal	12.5	1.111	7.500	2.499
14	3	4th-order fractal	25	1.667	6.750	3.414
15	3	4th-order fractal	37.5	0.556	10.500	-0.424
16	3	zigzag	12.5	1.667	6.750	3.414
17	3	zigzag	25	0.556	9.250	0.677
18	3	zigzag	37.5	1.111	6.500	3.742

Table 3 Contribution of various factors

Factors	Degree of freedom (DF)	Sum of squares (SS)	Variance (V)	Pure sum of squares (TS)	Percent contribution (%)
Tool diameter	1	22.862	22.862	22.862	20.2%
Path pattern	2	52.549	26.275	52.549	46.5%
Federate	2	0.453	0.227	0.453	0.4%
Grinding pressure	2	33.487	16.744	33.487	29.6%
Error		3.656			3.3%
sum	7	113.008			100.0%

7. Conclusions

This paper presents the development of a robot-assisted surface finishing system with an active torque controller. This system utilizes a dexterous manipulator to attain the desired position and orientation of finishing processes in three-dimensional space. A torque observer is attached to the tool frame of the robot manipulator, and a pneumatic hand-grinder is serially mounted on the observer. The function of the active torque controller in the system includes observing the contact torque, applying a desired contact pressure in the normal direction of the workpiece surface, and adjusting the contact angle between the hand-grinder and the surface of the workpiece. In this research, we construct the prototype of a robot-assisted finishing system. The experimental results show that the developed torque observer and controller system functions well under a variety of grinding conditions.

Taguchi's method is used to determine the effects of the following factors: tool diameter, path pattern, federate, and grinding contact pressure. Tendencies of these factors are found. From the experimental results, we know that among the four factor which may affect surface roughness, their contribution in descending order are path pattern, grinding contact pressure, tool diameter, and feed rate.

Acknowledgement

This work was supported by the National Science Council in Taiwan under grant no. NSC89-2212-E-032-008.

References

[1] Advantech, 1994a, PCL-726 six-channel D/A output card user's manual.

[2] Advantech, 1994b, PCL-833 3-axis quadrature encoder and counter card user's manual.

[3] Belavendram, N., Quality by Design, Prentice Hall, 1995.

[4] Chen, A.C.-C., K.C. Tai, C.H. Chen, Y.C. Wang, P.H. Lo, and Y.T. Wang, 1998, Generation of Fractal Tool Paths for Automated Surface Finishing Processes, Pacific Conference on Manufacturing, Brisbane, Australia.

[5] Furukawa, T., D.C. Rye, M.W.M.G. Dissanayake, and A.J. Barratt, 1996, Automated polishing of an unknown three-dimensional surface, Robotics & Computer-Integrated Manufacturing, Vol.12, No.3, pp.261-270.

[6] Jenkins, H.E. and T.R. Kurfess, 1996, Design of A Robust Controller for A Grinding System, IEEE Transactions on Control Systems Technology, Vol. 4, No.1, pp.40-48.

[7] Kunieda, M., T. Nakagawa, and T. Higuchi, 1984, Development of a Polishing Robot for Free Form Surface, Proceedings of the 5th International Conference on Production Engineering, pp.265-270, Tokyo.

[8] Kurfess, T.R., D.E. Whitney, and M.L. Brown, 1988, Verification of a dynamic grinding model, Journal of Dynamic systems, Measurement and Control, Vol.110, n4, pp.403-409.

[9] Luenberger, D.G., 1971, An Introduction to Observers, IEEE Transactions on automatic control, Vol. AC-16, No. 6, pp.596-602.

[10] Perdikaris, G.A. and K.V. VanPatten, 1982, Computer Schemes for Modeling, Tuning, and Control of DC Motor Drive Systems, Proceedings of the Power Electronics International Conference, pp.83-96.

Data Filtering and Regression in Estimating Slew Motion Timing Model

Wei Hua and Chen Zhou
School of Industrial and Systems Engineering
Georgia Institute of Technology
Atlanta, GA 30332
chen.zhou@isye.gatech.edu

Abstract

In many high-speed automated equipment, such as placement machines in surface mount technology lines, the most important thing is to move certain device to specific positions quickly. The trajectory is not critical. These machines utilize slew motions. The process of such machines can be divided into a set of sequential tasks moving between programmed points. In order to operate the machines efficiently, it is important to know the time it would take to move from any point to any other point, i.e. its timing model. This paper presents a method to estimate the timing model. The method is based on a data filtering and curve fitting. The work was done on a placement machine but the method is applicable to other high-speed machines with slow motion.

1 Introduction

The most time demanding process in a surface mount technology (SMT) line is to place hundreds or even thousands of components on each of the printed circuit board. A fast placement machine can achieve a rated speed between 10 to 20 chips per second.

The actual placement rate depends on the machine design as well as the sequencing of the components and the assignment of feeder locations. If the feeder's position and the placement sequence are not well arranged, the placement heads have to travel unnecessarily long distance to complete each pick and placement task and result in low placement rate. Minimizing placement time can significantly improve the line productivity in high volume production environment such as in personal computer and cellular phone industry.

The feeder allocation and placement sequence are two important decision problems in placement process optimization. The decision process is called placement programming in SMT industry. The problem can be modeled as Traveling Salesman Problems (TSP). In such problems, the picking head of the placement machine is the traveler. The physical address of each component on the board is a point the traveler needs to visit. The solution of TSP requires the travel time from a point to another, i.e. the timing model of the traveler. In other words, we need to predict accurate travel time of the revolver head between any pair of point given the physical address in order to find the best feeder arrangement and placement sequence.

Users can choose from two types of optimization software to do the placement programming. The first type is vendor exclusive packages, which normally support the vendor's own equipment. They tend to provide better optimization result for the intended machines but does not support equipment from other vendors. Moreover, the vendor software tends to have limited functionality in manufacturing process management since equipment vendors may not specialize in process management. Another type of optimization software is general assembly process management software. It normally supports machines from multiple vendors and provides additional management functionality and interface to other system modules. However, the placement rate it can achieve is usually lower than those controlled by vendor exclusive software. The main of this weakness is the lack of accurate timing models of the placing head.

In SMT industry, many users picked the best-in-class equipment and resulted in multi-vender assembly environment. It is more convenient to use general assembly process management software to manage the whole line even the whole workshop. However, the inferior result due to the lack of accurate timing model can hinder the throughput and line efficiency. In high volume productions such as cellular phone, automotive electronics and computer industry, the loss can be significant.

In recent years, most placement machine starts to be equipped with a standard interface - Generic Equipment Model (GEM). The GEM interface provides event-reporting protocol in which the equipment reports time stamps to a computer upon completion of tasks. The task completion times can be used to identify timing model of the placement machine as well as other slew motion equipment.

Zuhlke *et al* introduced the possibility and challenge of robots application in Micro-Assembly industry [1]. They intensified the importance of assembly accuracy and the requirement of high accuracy assembly equipment. Swevers *et al* presented an experimental method to identify robot dynamics [2]. They first built and combined internal and external dynamic models of the robot. The researchers then designed the optimal excitation trajectories for model parameter identification. Khalil *et al* compared different parameter calibration methods [4]. All methods he mentioned are carried out by solving a system of nonlinear equations. The difference between them is the type of endpoint sensing, the endpoint constraints, and number of equations. In addition to the nonlinear equation system, Omodei *et al* propose method to solve the parameter calibration problem, linear equations or Kalman Filter [4]. Kusiak and Tseng discussed the parameter design problem as a data-mining process [5]. They choose rough set theory instead of regression analysis and neural network.

In this paper, we present data filtering and a linear-regression methods to model and estimate timing models for placement machines, or other machines with complex combination of slew motions. The method is relatively simple.

2 The Operation of Placement Machines

Placement machines are specialized high-speed picking and placing manipulators. We will discuss the method with a type of revolver head, five-axis gantry placement machine. The method applies to other type of placement machine and general slew motion mechanisms. The schematic system is shown in Figure 1.

A turret revolver has several nozzles on the placing head. During assembly process, it first performs a picking cycle. In a picking cycle, each nozzle on the revolver picks one component from corresponding feeder. We call each pick a picking segment. If there are 12 nozzles, a picking cycle can have maximum of 12 picking segments.

The revolver then performs a placement cycle. In a placement cycle, the revolver moves and rotates each nozzle to the right position, rotate the nozzle (θ) and place each component by moving downward. We call each placement a placing segment. Both the sequences of picking segments and placing segment are decided by the optimization software.

To summarize, this type of machine has five independent axes: x, y, θ, ϕ, z. x, y, z are Cartesian axis of the working space. Where θ is for turret rotation and ϕ is for nozzle rotation.

2.1 Timing Characteristics

Each picking segment and placing segment consists of several independent slew motions of different parts of the machine, gantry in x and y, revolver rotation θ, revolver vertical motion (z) and nozzle rotation ϕ. The x, y, θ, and ϕ motions are concurrent while z is sequential after the completion other four movements, shown in Figure 2. What GEM interface can provide is a time stamp at the end of a pick or placement segment. The motion of a segment in time can be depicted in Figure 2.

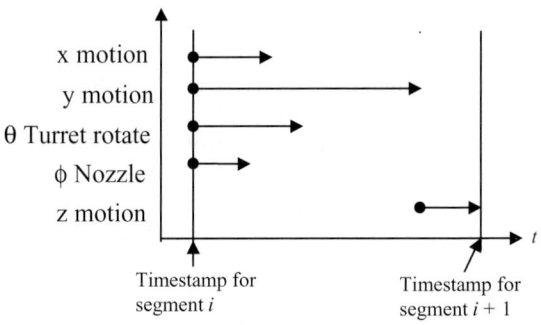

Figure 2. Timing of one motion task.

Let f_i and f_j, be any two slot positions in a feeder and p_u and p_v be any two locations on the circuit board. The objective of timing model is to predict the travel times between any position pairs in $\{f_i, f_j\}$ or $\{p_u, p_v\}$.

2.2 Timing Model of a Single Axis

Consider the kinematics of a sigle axis x. Let v be the maximum motor speed, and a be acceleration rate, the speed curve as a function of time t is shown in Figure 3.

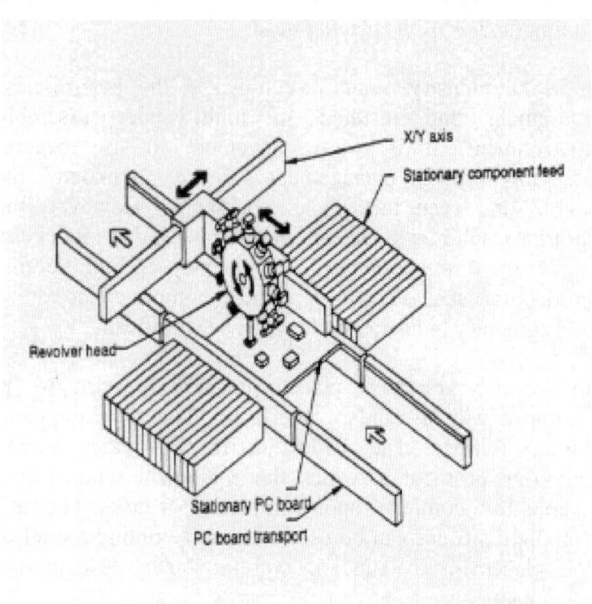

Figure 1. Schematic of a placement machine.

338

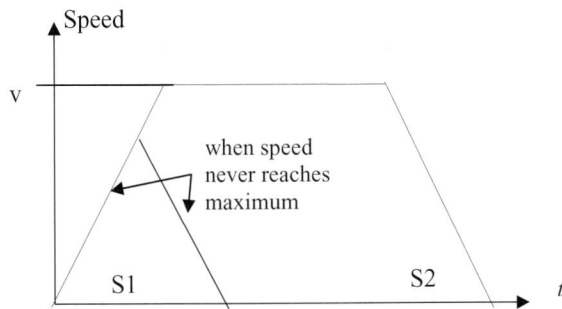

Figure 3. The speed change as a function of time.

Here, we are interested in the travel time as a function of travel distance $T(x)$, which is the integration of the speed function, i.e.,

$$T(x) = \begin{cases} \sqrt{4x/a} & \text{if } x \leq v^2/a \\ v/a + x/v & \text{otherwise} \end{cases}$$

The schematic of this function is shown in Figure 4.

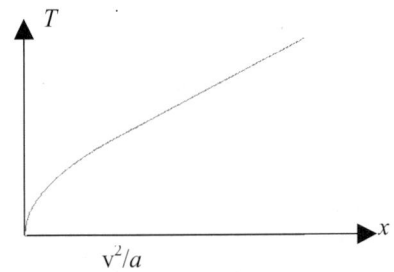

Figure 4. Travel time T as a function of distance x.

$T(x)$ is nonlinear and has discontinuity with respect to second derivative. It has one input and one output. There are only two parameters in the model. If we have a group of travel time of the revolver head on this axis and the address of starting and ending point on the corresponding direction, this is a rather simple regression problem.

2.3 Timing Model for Multiple Axes

However, the time and distance pair for individual axis is not directly available through GEM. What GEM can get is time stamps at the end of a picking or placing segment. Each segment is a combination of several axes' movement. Some movements are concurrent while others are sequential, as shown in Figure 3.

The time required of the segment equal to the travel time in z direction plus the maximum of the other four movements. The travel time of z direction movement is unknown. However, in the special case of placement machine, it can be considered as a constant since the travel distance on Z direction is always the same during each component picking and placing segment. Each of the other four movements has its own maximum speed and acceleration rate, i.e. the 'v' and 'a'. The system has four inputs, i.e. the corresponding distance in direction x, y, θ, and ϕ. There are nine parameters need to be estimated, i.e.,

$$T = T(x, y, \theta, \phi) + T(z)$$
$$= \max\{T_{v,a}(x), T_{v,a}(y), T_{v,a}(\theta), T_{v,a}(\phi)\} + T_{const.}(z)$$

Because of the maximum function in this model, it will be very difficult to explore the model as a whole using regression method.

Through GEM, we get a tuple $\{t_i, x_i, y_i, \theta_i, \phi_i\}$ from each picking or placing segment. Next step, we will try to decouple these tuples. One good feature of the maximum function is that once an entry dominates the result, other entry will not affect it anymore. The challenge is to find which motion dominates in the tuple.

Without loosing generality, consider a single axis x, there are two cases in terms of dominance:

$$\begin{cases} T(x) \geq \max\{T(y), T(\theta), T(\phi)\} & \text{(a)} \\ T(x) < \max\{T(y), T(\theta), T(\phi)\} & \text{(b)} \end{cases}$$

We can project partial tuples $\{t_i, y_i\}$ on t-y plot. For each tuple point when condition (a) holds, x dominates y, θ, ϕ. We have $T = T(x) + T(z)$ or a point in the relationship $T(x_i) = t_i - T(z_i)$. Since $T(z_i)$ can be approximated with a constant, we get a relationship between travel time and x coordinate. When (b) holds, $T > T(x) + T(z)$. One (or more) of the other axle dominate the segment. This observation does not provide relationship between travel time and x coordinate. However, it can provide relationship between travel time and one (or more) of the dominate coordinate.

The above analysis can be depicted in a T - x plot shown in Figure 5. Where T(z) is the expected travel time in the z-direction movement. When condition (a) holds, the projection of tuple $\{t_i, x_i, y_i, \theta_i, \phi_i\}$ on the T - x plot is on the lower boundary. Those points on the boundary contains the information of function $T(x)$. When condition (b) holds, the travel time is dominated by another axis and the point projection on T - x plot will be above the lower boundary. The point is not helpful to determine the relationship between travel time and x coordinate. Therefore, the function of $T(x)$ can be extracted from the lower boundary of T - x plot. The functions of T(y), T(θ), and T(ϕ) are also determined accordingly.

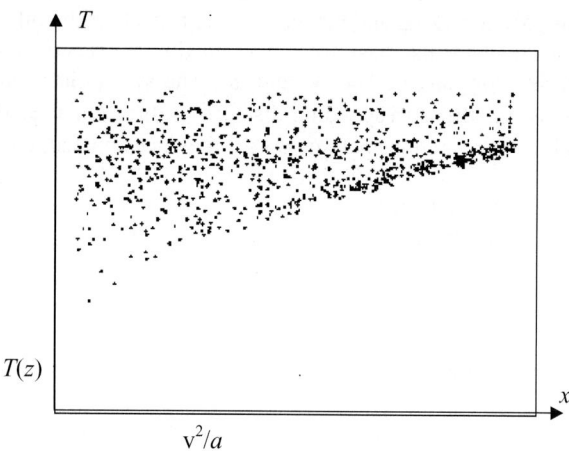

Figure 6: travel time plotted on T - y chart

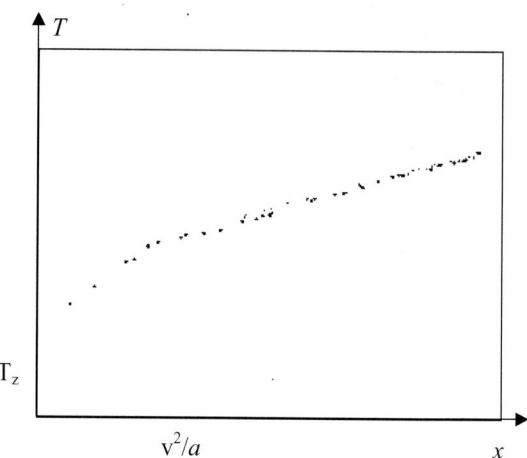

Figure 7: The dominant Points

2.4 Data Filtering for Each Axis

The problem is then to identify the points on boundary and fit a curve for the boundary points. Without loosing generality, we will use x coordinate as an example. Assume we observed N gantry movements, and collect the allied coordination in the following vector:

$$\{(x_1, t_1), (x_2, t_2), ..., (x_N, t_N)\},$$

where each data pair indicate one x-axis distance and the corresponding travel time, x motion dominants or not.

First, we need to remove those points not on the lower boundary using the following algorithm.

Sort (x, t) tuples with increasing value of x, assign index j to each tuple to get (x_j, t_j). Set $j = 1$

loop   If $x_{j+1} = x_j$ and
        if $t_{j+1} > t_j$, remove tuple (x_j, t_j).
        Else remove tuple (x_{j+1}, t_{j+1}).
    Else (i.e. $x_{j+1} > x_j$)
        if $t_{j+1} \leq t_j$, remove tuple (x_j, t_j).
    If not the end of list $j = j + 1$.
    Else, end.
Go to loop.

After this procedure we will get a series of non-decreasing points in the x-t plane as shown in Figure 7:

These points (or data pairs) will be recorded as (x_1, t_1), (x_2, t_2), ..., (x_n, t_n), where $x_1 < x_2 < ... < x_n$ and $t_1 < t_2 < ... < t_n$.

We know that remaining points all follow the model

$$T = \begin{cases} \sqrt{4x/a} + t_z + e & \text{if } x \leq v^2/a \\ v/a + x/v + t_z + e & \text{otherwise} \end{cases}$$

2.5 Piecewise Linear Regression

The next step is to identify the parameters of this model. This model has two parts:

a) a non-linear function of 'x' with two unknown parameters 'a' and 't_z' and
b) a linear function of 'x' with unknown parameter 'a', 't_z' and 'v'.

The data points should be divided into two sets S_1 and S_2. We can then apply regression analysis on each data set to estimate the model parameters. However, the dividing point of $x = v^2/a$ is unknown. We developed a searching procedure:

(1) Choose an initial dividing point x_L. There are several ways to choose x_L. We can choose it arbitrarily between x_1 and x_n. Then we will need to decide the searching direction later. Or we can choose x_L equal to x_{n-1}, then S_2 will include at least two points for the requirement of regression analysis. Then the x_L will decrease in the following searching procedure. The first data set S_1 will include all points (x_1, t_1) to (x_m, t_m), where $x_m < x_L$. And the second data set S_2 will include the rest points (x_{m+1}, t_{m+1}) to (x_n, t_n), and $x_L = x_{m+1}$.

(2) Perform regression on data set S_2. In this data set, $T_i = x_i/v + v/a + t_z + e$.

T_i is a linear function of x_i, i.e. $T_i = \alpha x + \beta_1 + e$. where

$$\beta_1 = v/a + t_z, \quad (i)$$
$$\alpha = 1/v, \quad (ii)$$

Standard regression procedure can generate the estimation of β_1 and α.

(3) Consider data set S_1. In this data set, $T_i = \sqrt{4x_i/\alpha} + t_x + e$.

$$T_i - T_j = \frac{2}{\sqrt{a}}(\sqrt{x_i} - \sqrt{x_j}) + \varepsilon$$

Let $T' = T_i - T_j$ And
$x' = \sqrt{x_i} - \sqrt{x_j}$

With a simple transformation, Then T' becomes a linear function of x', i.e. $T' = \beta_2 x'$. where

$$\beta_2 = 2/\sqrt{a} \quad (iii)$$

The regression procedure will generate the estimation for β_2.

(4) From (i), (ii) and (iii), we can calculate the estimation value of the unknown parameters as:
$a = 4/\beta_2^2$
$v = 1/\alpha$
$t_z = \alpha - v/a = \alpha - \beta_2^2/(4\alpha)$

(5) Adjust the jointing point to v^2/a with the given a and v. Stop when the difference of v^2/a between two consecutive iterations is small enough or go to (2) and begin the next searching iteration.

2.6 Numerical Example:

We construct a numerical example to test the algorithm. The system has three axes with dynamic feature following model. And the parameter is given as follows:

V_1	v_2	v_3	a_1	a_2	a_3	T_z
10000	10000	10800	1500	1500	216000	2

We apply the algorithm on the first axis under different noise levels ($\delta(\varepsilon)/E(T_z)$) and randomly chosen initial jointing point. The algorithm converges quickly in all cases. The result is listed as follows:

$\sigma/E(T)$	# of runs	α	v	t_z	Average # of Iteration
1%	10	1480 (-1.3%)	10003 (+0.3%)	2.07 (+3.5%)	4
3%	10	1478 (-1.5%)	9770 (-2.3%)	1.95 (-2.5%)	3.2
10%	10	1392 (-7.2%)	9345 (-6.5%)	1.82 (-9%)	3.5

3 Conclusions

The operation of placement machine can be considered as a combination of slew motions, i.e. a series of parallel or consequential movements. The parameter of these movements can be estimated by time stamps on certain points in the process. Parallel movements can be decoupled by selecting the dominant observation points. The simple algorithm we presented can effectively estimate the parameters of slew motions in the given kinematics models.

4 References

[1] Detlef Z., Ralf F., and Johannes H., "Stepwise into Micro-Assembly", Proceedings of ICAR '97, pp 259 – 263, Monterey CA, July 1997.

[2] Swevers, J., Ganseman, C., Chenut, X., and Samin, J. C., "Experimental Identification of Robot Dynamics for Control", Proceedings of the 2000 IEEE ICRC, pp 241 – 246, San Francisco, April 2000.

[3] Khalil, W. Besnard, S. and Lemoine, P. "Comparison Study of the Geometric Parameter Calibration Methods," Int J. of Robotics and Automation, vol 15, no. 2, pp 56 – 66, 2000.

[4] Omodei, A. Legnami, G. and Adamini, R. "Three Methodologies for the Calibration of Industrial Manipulators: Experimental Results on a SCARA Robot", Jounal of Robotic Systems, vol 17, no. 6, pp 291 – 307, 2000.

[5] Kusiak, A., and Tseng, T.L., "Data Mining in Engineering Design: A Case Study", Proceedings of the 2000 IEEE ICRC, pp206 – 211, San Francisco, April 2000.

AN EXACT REPRESENTATION OF EFFECTIVE CUTTING SHAPES OF 5-AXIS CNC MACHINING USING RATIONAL BÉZIER AND B-SPLINE TOOL MOTIONS

J. Xia
Department of Mechanical Engineering
State University of New York
Stony Brook, New York 11794-2300

Q. J. Ge[1]
Department of Mechanical Engineering
State University of New York
Stony Brook, New York 11794-2300

ABSTRACT

Presented in this paper is a new approach to 5-axis CNC tool path generation for sculptured surface machining with a flat-end cutter. Rational Bézier and B-spline motions are used to plan cutter motions so that an exact representation of the effective cutting shape can be obtained. The exact representation leads to an accurate computation of the scallop curve generated by two adjacent tool paths. Two examples are given to show how this result can be used to accurately plan and verify tool paths for 5-axis CNC milling of sculptured surfaces.

1 INTRODUCTION

It is well known that CNC machining of sculptured surfaces using flat-end cutting tools and 5-axis machines offers the benefits of higher material removal rates, better accessibility, and reduced number of set-ups. The quality of the machined surface is determined by geometric factors as well as kinematic, dynamic, and thermal properties of the machine tool. A vast amount of research has been done in the area of 5-axis tool path planning in order to achieve the competing goals of higher accuracy for manufactured surface and reduced machining time. Much of the existing work on CNC tool path generation focuses on geometric issues such as scallop heights, local and rear gouging (Vichers and Quan, 1989; Marciniak, 1991; Jensen and Anderson, 1992; Choi et al., 1993; Chen et al., 1993; Menon and Voelcker, 1993; Li and Jerard, 1994; Kim and Chu, 1994; Suresh and Yang, 1994; Lee and Chang, 1995; Lin and Koren, 1996; Sarma and Dutta, 1997; Lee, 1997 and 1998; Lee and Ji, 1997; Rao et al., 1997; Lo, 1999; Rao and Sarma, 2000).

The present paper deals also with the geometry of CNC tool path. Instead of focusing on a particular instant of the tool motion and studying local geometric issues at the instant, this paper uses the recently developed rational Bézier and B-spline motions (Ge and Ravani, 1991; Jüttler and Wagner, 1996; Srinivasan and Ge, 1998) for CNC tool path generation. The main advantages of using such freeform motions for tool path representation include (a) the entire tool path can be represented using a much more compact set of control positions of the freeform motion as opposed to a huge data set of discrete cutter positions; (b) since the tool motion representation is analytic, it may provide a framework for including kinematic and dynamic factors of the machine tool in tool path generation (Ge, 1996). Furthermore, Xia and Ge (1999) has shown that the boundary surfaces of the swept volume of a flat-end cylindrical cutter undergoing a rational Bezier or B-spline motion can be represented exactly. The present paper extends this recent result to obtain an exact representation of the effective cutting shape. This leads to an accurate computation of the scallop curve generated by two adjacent rational Bezier or B-spline tool paths. These results, when combined with existing approaches to 5-axis CNC tool path planning, would make these methods much more reliable.

The organization of the paper is as follows. The Section 2 reviews the kinematics fundamentals required for the development of the paper. Section 3 presents an analytic representation of the exact swept section of a flat-end cutting tool under a rational Bézier motion. Section 4 presents an example to show how this result can be used to plan iso-parametric rational Bézier and B-spline tool paths such that the resulting scallop heights of the entire manufactured surface do not exceed the specified scallop height. Section 5 presents another example in which near-constant scallop rational B-spline tool paths are obtained.

2 RATIONAL BÉZIER MOTIONS

This paper follows Ge and Ravani (1991) and Xia and Ge (1999) and uses dual quaternions to represent spatial motions. A dual quaternion $\hat{\mathbf{Q}}$ consists of a pair of quaternions $\mathbf{Q}$ and $\mathbf{R}$, where $\mathbf{Q}$, is a quaternion of rotation and $\mathbf{R}$ is another quaternion associated with the translation component. Details on quaternions and dual quaternions can

---

[1]Q. J. Ge is currently on sabbatical leave in the Department of Automation and Computer Aided Engineering, The Chinese University of Hong Kong. Email: qjge@yahoo.com

be found in McCarthy (1990).

In dual-quaternion representation, rigid transformations of homogeneous point coordinates, $\tilde{\mathbf{P}} = (\tilde{P}_1, \tilde{P}_2, \tilde{P}_3, \tilde{P}_4)$ and $\mathbf{P} = (P_1, P_2, P_3, P_4)$, are given by the following quaternion equations:

$$\tilde{\mathbf{P}} = \mathbf{QPQ}^* + P_4(\mathbf{RQ}^* - \mathbf{QR}^*) \quad (1)$$

where "$*$" denotes the conjugate of a quaternion.

Given a set of dual quaternions $\hat{\mathbf{Q}}_i$, the following rational Bézier representation

$$\hat{\mathbf{Q}}(t) = \sum_{i=0}^{n} B_i^n(t) \hat{\mathbf{Q}}_i \quad (2)$$

defines a Bézier curve in the space of dual quaternions. The dual-quaternion curve corresponds to a rational Bézier motion whose point trajectories are rational Bézier curves.

The trajectory of a point $\mathbf{P}$ under a rational Bézier motion (2) can be rewritten in matrix form as

$$\tilde{\mathbf{P}}^{2n}(t) = [H^{2n}(t)]\mathbf{P}$$

where

$$[H^{2n}(t)] = \sum_{k=0}^{2n} B_k^{2n}(t)[H_k], \quad (3)$$

$$[H_k] = \sum_{i+j=k} \frac{C_i^n C_j^n}{C_k^{2n}} ([H_i^+][H_j^-] + [H_j^-][H_i^{0+}] - [H_i^+][H_j^{0-}]). \quad (4)$$

In the above, the matrices $[H_i^+], [H_j^-], [H_i^{0+}], [H_j^{0-}]$ are given by

$$[H_i^+] = \begin{bmatrix} Q_{i,4} & -Q_{i,3} & Q_{i,2} & Q_{i,1} \\ Q_{i,3} & Q_{i,4} & -Q_{i,1} & Q_{i,2} \\ -Q_{i,2} & Q_{i,1} & Q_{i,4} & Q_{i,3} \\ -Q_{i,1} & -Q_{i,2} & -Q_{i,3} & Q_{i,4} \end{bmatrix},$$

$$[H_j^-] = \begin{bmatrix} Q_{j,4} & -Q_{j,3} & Q_{j,2} & -Q_{j,1} \\ Q_{j,3} & Q_{j,4} & -Q_{j,1} & -Q_{j,2} \\ -Q_{j,2} & Q_{j,1} & Q_{j,4} & -Q_{j,3} \\ Q_{j,1} & Q_{j,2} & Q_{j,3} & Q_{j,4} \end{bmatrix},$$

$$[H_i^{0+}] = \begin{bmatrix} 0 & 0 & 0 & R_{i,1} \\ 0 & 0 & 0 & R_{i,2} \\ 0 & 0 & 0 & R_{i,3} \\ 0 & 0 & 0 & R_{i,4} \end{bmatrix},$$

$$[H_i^{0-}] = \begin{bmatrix} 0 & 0 & 0 & -R_{j,1} \\ 0 & 0 & 0 & -R_{j,2} \\ 0 & 0 & 0 & -R_{j,3} \\ 0 & 0 & 0 & R_{j,4} \end{bmatrix}.$$

It is clear from (3) that the point trajectory is a rational Bézier curve of degree $2n$.

3 THE EFFECTIVE CUTTING SHAPE

The circular boundary of the base of a cylindrical cutting tool is called the *cutting circle* of the tool. When the tool follows a trajectory, it traces out a swept volume. Xia and Ge (1999) presented an exact representation of all boundary surfaces of a cylindrical cutter under rational Bézier and B-spline motions. Another related work is by Jüettler and Wagner (1999).

This section focuses on the swept surface generated by a cutting circle under a rational Bézier motion. In particular, we investigate the profile of the swept surface in the cutting plane which is a plane normal to the direction of motion at a given instant. The profile is referred to as *effective cutting shape* or *swept section* in CNC machining literature. A scallop is an uncut volume left between two adjacent tool paths. The scallop height δ is defined at the maximum of the height of the volume measured from the designed surface $\mathbf{S}(u, v)$. The highest point, called the scallop point, traces out a curve along the tool path, called the scallop curve. The distance between two neighboring scallop curves is called step-over distance or path interval (Sarma and Dutta, 1997). When a cutter moves along a tool path without changing its orientation, the effective cutting shape is an ellipse obtained as the intersection of the cutting plane with the cylindrical cutter. Traditionally, the ellipse has been used to represent the effective cutting shape whether or not there is change in tool orientation. However, as pointed out by Sarma (2000a), when there is change in the tool orientation, the effective cutting shape can deviate significantly from the ellipse for points that are away from the contact point (CC). Since a scallop point is away from CC and is traditionally computed as the intersection of two ellipses, the error in estimating the scallop point and the step-over distance are even greater.

We now consider the problem of obtaining an exact representation of the effective cutting shape of a cutting circle under a rational Bézier motion. First, we represent a circular arc of sweep angle 180° with a quadratic Bézier curve with one control point at infinity (see, for example, Piegl and Tiller 1995). Without loss of generality, we assume that the coordinate system is chosen such that the circle is on XZ plane with radius R and its center is the origin of the coordinate system. Then the half circle below the

X-axis is represented as:

$$\mathbf{P}(s) = \sum_{i=0}^{2} B_i^2(s)\mathbf{P}_i \quad (5)$$

where $B_i^2(s)$ are quadratic Bernstein polynomials. and $\mathbf{P}_0 = (R, 0, 0, 1)$, $\mathbf{P}_1 = (0, 0, -R, 0)$ and $\mathbf{P}_2 = (-R, 0, 0, 1)$ are the homogeneous coordinates of the three Bézier control points. Similarly, the circular arc above x-axis can be represented by the same formula as (5), but with $\mathbf{P}_1 = (0, 0, R, 0)$. When the arc as shown by (5) is under a rational Bézier motion as defined by (2), the swept surface of the arc can be represented in the following tensor product form:

$$\begin{aligned}\mathbf{P}(s,t) &= [H^{2n}(t)]\mathbf{P}(s) \\ &= \sum_{k=0}^{2n}\sum_{i=0}^{2} B_k^{2n}(t)B_i^2(s)[H_k]\mathbf{P}_i \\ &= \sum_{k=0}^{2n}\sum_{i=0}^{2} B_k^{2n}(t)B_i^2(s)\mathbf{P}_{ik} \quad (6)\end{aligned}$$

where $\mathbf{P}_{ik} = [H_k]\mathbf{P}_i$ and the matrix $[H_k]$ is given by (4).

Now turn our attention to a CNC tool path on a designed surface. Let $\mathbf{n}$, $\mathbf{t}$, $\mathbf{b}$ denote the normal, tangent, and binormal vectors associated with the Frenet frame at the contact point CC of the tool path. The binormal plane defined by $\mathbf{n}$ and $\mathbf{b}$ captures the instantaneous position of the tool motion. Let $\mathbf{M}$ denote a four-dimensional vector whose coordinates are the homogeneous coordinates of the binormal plane. Then the intersection of the plane with the swept surface (6) defines the swept section at the instant t. The swept section is a planar curve $s(t)$ on the swept surface such that $\mathbf{P}(s,t) \cdot \mathbf{M} = 0$, i.e.,

$$\sum_{k=0}^{2n}\sum_{i=0}^{2} B_k^{2n}(t)B_i^2(s)(\mathbf{P}_{ik} \cdot \mathbf{M}) = 0.$$

This leads to

$$\sum_{i=0}^{2} B_i^2(s)f_i = 0, \quad (7)$$

where $f_i = \sum_{k=0}^{2n} B_k^{2n}(t)(\mathbf{P}_{ik} \cdot \mathbf{M})$. Solving the quadratic equation (7), we obtain

$$s(t) = \frac{f_0 - f_1 \pm \sqrt{f_1^2 - f_0 f_2}}{f_0 + f_2 - 2f_1}. \quad (8)$$

When $f_1^2 - f_0 f_2 \geq 0$, the swept surface (6) intersects with the plane $\mathbf{M}$. Otherwise there is no intersection between them. The curve of intersection, $\mathbf{C}(t) = \mathbf{P}(s(t), t)$, obtained by substituting (8) into (6) is the exact representation of the effective cutting shape of a flat-end cylindrical cutter under a rationl Bézier motion of degree $2n$.

4 PLANNING ISO-PARAMETRIC TOOL MOTIONS FOR BOUNDED SCALLOP HEIGHT

In this section we describe a procedure for planning an iso-parametric tool motions for a given sculptured surface $\mathbf{S}(u, v)$ such that the resulting scallop curves do not exceed the specified scallop height δ. This procedure is used as an example to show how the exact representation of the cutting shape can be used in conjunction of rational Bézier and B-spline motion for tool path planning. In the following, without loss of generality, we consider only the iso-parametric paths defined by fixing $u = u_i$ to obtain $\mathbf{S}(u_i, v)$.

4.1 Cutter position generation and interpolation

For a given iso-parametric curve $\mathbf{S}(u_i, v)$, we first discretize it into a set of n points $\mathbf{S}(u_i, v_j)$ with $v_j (j = 0, \cdots, n-1, v_0 = 0, v_{n-1} = 1)$. We use an existing method such as the one proposed by Lee (1998) to generate a set of cutter positions on the $\mathbf{S}(u_i, v_j)$ and the associated tool-frames T_{ij}. We then convert these tool positions from matrix representation to quaternion representation $\hat{\mathbf{Q}}_{ij}$. After that, we use a piecewise Bézier dual quaternion curve such as a cubic B-spline dual quaternion curve $\hat{\mathbf{Q}}(u_i, v)$ to interpolate the dual quaternions $\hat{\mathbf{Q}}_{ij}$ in a manner similar to curve interpolation in CAGD (see, for example, Farin, 1997). See also Srinivasan and Ge (1998).

Once we obtained a rational Bézier or B-spline motion, we can proceed to compute the swept volume of a cylindrical cutter using the method presented in Xia and Ge (1999). For each tool motion segment $[v_i, v_{i+1}]$, we check the the deviation of the swept surface from the designed surface to determine if further subdivision within the segment $[v_i, v_{i+1}]$ is needed. The following factors need to be examined:

- global gouging between the swept volume segment and surface segment $[v_i, v_{i+1}]$;
- over-cut situation when the wept surface generated by the base of the cutter contributes to the boundary surfaces of the swept volume;
- If none of the above two situation arises, then check the distance from $\mathbf{S}u_i, v_j$ and the swept surface to see if it is within the specified error δ.

If any of above three cases exists, the segment $[v_i, v_{i+1}]$

should be divided into smaller ones to obtain a more refined interpolating rational B-spline motion.

4.2 Estimation of step-over distance

Once we have obtained a rational B-spline motion $\hat{\mathbf{Q}}(u_i, v)$ such that its swept surface approximates the designed surface near the iso-parametric curve $\mathbf{S}(u_i, v)$. The next step is to obtain the step size Δu_i such that the resulting scallop height generated by $\hat{\mathbf{Q}}(u_i, v)$ and $\hat{\mathbf{Q}}(u_{i+1}, v)$ (where $u_{i+1} = u_i + \Delta u_i$) is no larger than the specified scallop height δ.

The step size Δu_i is related to the step-over distance l_i of two adjacent tool motions by

$$l_i = \Delta u_i \|\mathbf{S}_u\| \tag{9}$$

where $\mathbf{S}_u = \partial \mathbf{S}/\partial u$ and $\|\mathbf{S}_u\|$ denote the maximum length of the derivative vector for $u \in [0, 1]$. For non-isoparametric motions, the step-over distance is related to step sizes Δu_i and Δv_j by (Lo, 1999):

$$l_{ij}\mathbf{b} = \mathbf{S}_u \Delta u_i + \mathbf{S}_v \Delta v_j, \tag{10}$$

where $\mathbf{S}_v = \partial \mathbf{S}/\partial v$ and Δv_j is the step size in v direction.

Given a specified maximum scallop height δ, the step-over distance can be estimated using a formula proposed by Lin and Koren (1996):

$$l_i = \sqrt{\frac{8\delta}{\kappa_e + \sigma \kappa_b}} \tag{11}$$

where κ_e is the effective cutting curvature of the effective cutting shape at CC, κ_b is the effective surface curvature of the designed surface $\mathbf{S}(u, v)$ at CC, and $\sigma = 1$ for convex surface and $\sigma = -1$ for concave surface. Since the curvatures vary when CC changes along the tool path $\mathbf{S}(u_i, v)$, the step-over distance l_i is a function of v. For the example presented in this paper, we choose the value of l_i when $v = 1/2$ for the purpose of simplicity.

The curvature κ_b is a property of the designed surface and can be obtained using formulas presented in Faux and Pratt (1981). The effective cutting curvature has been traditionally computed using the elliptic approximation of the cutting shape, which can be significantly different from true cutting curvature. In this paper, since we have developed an exact representation of the cutting shape, we can compute the cutting curvature exactly. We note that recently Sarma (2000b) presented formulas for computing the exact cutting curvature using the incline and tilt angles of the cutting tool. The use of exact cutting curvature gives better estimation for the step-over distance.

Once the step-over distance is obtained, one can use Eq. (9) to obtain an estimate for Δu_i. After that one can generate a neighboring rational B-spline tool motion $\hat{\mathbf{Q}}(u_i + \Delta u_i, v)$. For a given instant v, one can compute the locations of the cutting planes associated with $\hat{\mathbf{Q}}(u_i, v)$ and $\hat{\mathbf{Q}}(u_i + \Delta u_i, v)$. The intersection of these two planes yield a line which intersect the cutting shape in one of the planes to obtain the scallop point $\mathbf{m}$. As v varies in $[0, 1]$, the scallop point traces out the scallop curve $\mathbf{m}(u_i, v)$. The distance between the two curves $\mathbf{m}(u_i, v)$ and $\mathbf{S}(u_i, v)$ defines the scallop function $h(v)$. If $h(v) \leq \delta$ for all $v \in [0, 1]$, then the estimate l_i is good. Otherwise, a new estimate needs to be generated using a search routine. In the example presented in this paper, a simple binary search is used.

5 PLANNING NEAR-CONSTANT SCALLOP RATIONAL B-SPLINE TOOL MOTIONS

In the case of iso-parametric tool path planning, we know the cutting direction because CC points move along an iso-parametric curve of $\mathbf{S}(u, v)$. In near constant scallop case, we do not know the next CC point, so we do not know the cutting direction. If we assume the cutting direction is along a parametric curve at each CC point, the maximum effective cutting radius can not be achieved. In the following, we propose a method which compromises between the goals of maximum cutting radius and required scallop height. Four steps are included:

1. Generate a set of cutter locations $\mathbf{T}_{ij}$ based on the local geometry of surface $\mathbf{S}(u, v)$ by using Eq. (11) and (10).
2. Obtain quaternions $\hat{\mathbf{Q}}_{ij}$ transforming the cutter from its original location to $\mathbf{T}_{ij}$ and associated knot sequence u_{ij}, v_{ij}. Then a two-parameter rational B-spline motion $\hat{\mathbf{Q}}(u, v)$ is generated to interpolate all $\hat{\mathbf{Q}}_{ij}$.
3. Fine tune the control quaternions of $\hat{\mathbf{Q}}(u, v)$ so that the surface swept by the circular edge under $\hat{\mathbf{Q}}(u, v)$ matches the design surface. Swept volume of cylindrical cutter under two-parameter motion $\hat{\mathbf{Q}}(u, v)$ (see Xia and Ge, 2000) is needed to avoid global interference and make sure the swept surface generated by the base plane of the cutter does not contribute to the boundary surfaces of the swept volume, which would cause over-cut. If necessary, we raise the degree of $\hat{\mathbf{Q}}(u, v)$ such that there are more control quaternions can be used provide additional flexibility for adjusting the two-parameter motion.
4. Assuming that $\hat{\mathbf{Q}}_I(t)$ is known, obtain a one-parameter motion $\hat{\mathbf{Q}}_{i+1}(t)$ from $\hat{\mathbf{Q}}(u, v)$ such that the scallop

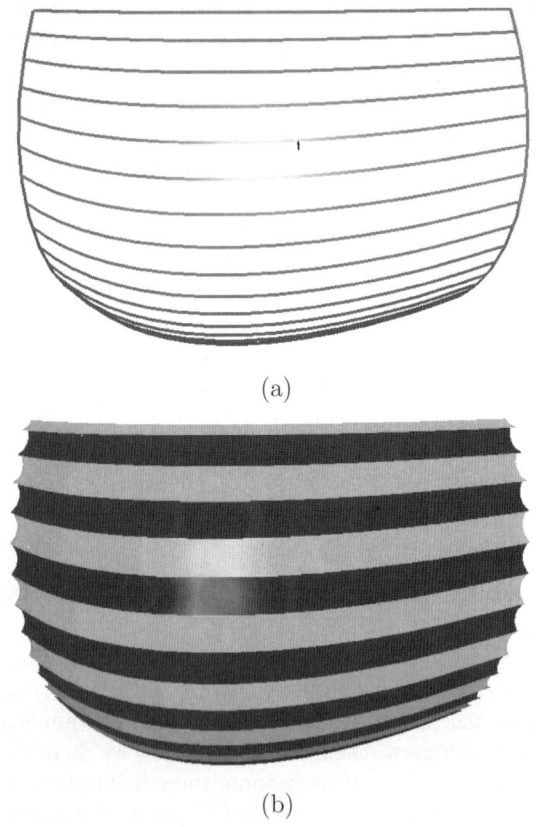

Figure 1. Iso-parametric (a)tool paths (b)manufactured surface.

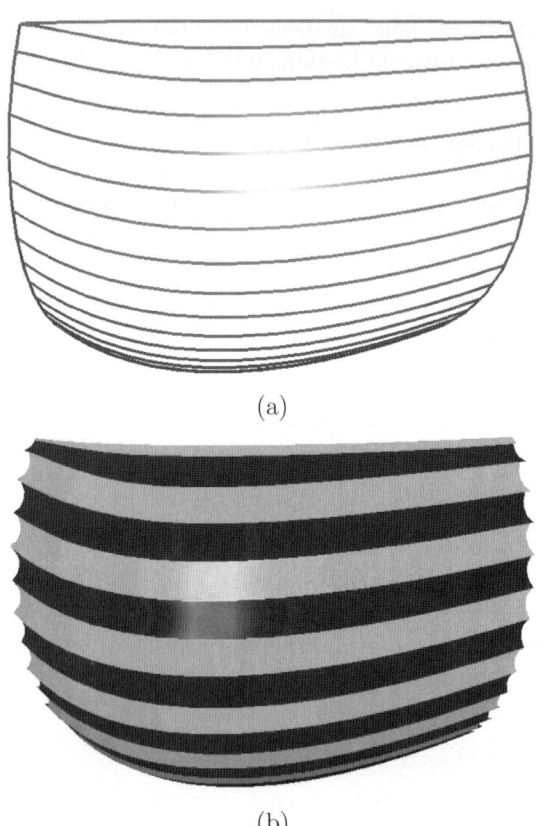

Figure 2. Iso-scallop (a)tool paths (b)manufactured surface.

height generated by cutter under $\hat{\mathbf{Q}}_i(t)$ and $\hat{\mathbf{Q}}_{i+1}(t)$ is within the range.

This 4th step above requires more explanation. The basic idea is as follows. First find a set of discrete points $\mathbf{P}$ on the swept surface of the cutting circle under the existing one-parameter motion $\mathbf{Q}_i(t)$ such that these points maintain constant distance δ from the designed surface $\mathbf{S}(u,v)$. Then find a set of discrete cutter locations from the two-parameter $\mathbf{Q}(u,v)$ such that cutter bottom edges contain points $\mathbf{P}$. Finally interpolate all cutter positions to obtain $\mathbf{Q}_{i+1}(t)$.

As an example, we implemented the algorithms for planning C^2 rational B-spline tool paths for CNC machining of the following design surface:

$$\mathbf{S}^x(u,v) = -2 + 6v + 2v^3,$$
$$\mathbf{S}^y(u,v) = 6u,$$
$$\mathbf{S}^z(u,v) = 6u + 6v - 6u^2 - 6v^2,$$

where $u,v = [0,1]$. Figure 1(a) shows the tool path and 1(b) is the resulting manufactured surface for iso-parametric case, Figure 2(a) shows the tool path and 2(b) is the manufactured surface for near constant scallop case.

6 CONCLUSIONS

In this paper we have developed a method for the exact representation of the effective cutting shapes for a flat-end cutter under rational Bézier or B-spline motion. We presented two examples to demonstrate how the result can be used to plan iso-parametric and near constant scallop rational tool motions.

ACKNOWLEDGMENT

This work was supported by NSF grant DMI-9800690 to the State University of New York at Stony Brook.

REFERENCES

Chen, Y.D., Ni, J. and Wu, S.M., 1993, Real-time CNC tool path generation for machining IGES surfaces, *ASME Journal of Engineering for Industry*, 115, pp. 480-486.

Choi, B.K., Park, J.W. and Jun, C.S., 1993, Cutter location data optimization in 5- axis machining, *Computer Aided Design*, vol. 25, No. 6, pp. 377-386. Farin, G., 1997, *Curves and Surfaces for Computer Aided Geometric Design*, Academic Press.

Faux, I.D., and Pratt, M.J., 1981, *Computational geometry for design and manufacturing*, John Wiley and Sons, New York.

Ge, Q.J., and Ravani, B., 1991, Computer aided geometric design of motion interpolants, *ASME Journal of Mechanical Design*, 116(3):756-762.

Ge, Q. J., 1996, Kinematics-driven geometric modeling: A framework for simultaneous tool-path generation and sculptured surface design, *Proc. 1996 IEEE Robotics and Automation Conference*, Vol. 2, pp 1819-1824, Minneapolis, MN.

Ge, Q.J., and Sirchia, M., 1999, Computer aided geometric design of two parameter freeform motions, *ASME J. of Mechanical Design*, 121:502-506. Jensen, C.G. and Anderson, D.C., 1992, Accurate tool placement and orientation for finish surface machining, *ASME Concurrent Engineering*, PED-Vol. 59, pp. 127-145.

Jerard, R.B., and Drysdale, R.L., 1988, Geometric simulation of numerical control machinery, *ASME Computers in Engineering*, Vol.2, p 129-136.

Jüttler, B., and Wagner, M.G., 1996, Computer Aided Design With Spatial Rational B-Spline Motions, *ASME J. of Mechanical Design*, 119(2):193-201. Jüttler, B., and Wagner, M.G., 1999, Rational motion-based surface generation,. *Computer Aided Design*. Vol. 31(3): 203-13.

Kim, B.H. and Chu, C.N., 1994, Effect of cutter mark on surface roughness and scallop height in sculptured surface machining, *Computer Aided Design*, vol. 26, no. 3, pp. 179-188.

Lee, Y.S. and Chang T.C., 1995, 2-phase approach to global tool interference avoidance in 5-axis machining, *Computer Aided Design*, vol. 27, no. 10, pp. 715-729.

Lee, Y.S., 1997, Filleted endmill placement problems and error analysis for multi- axis CNC machining, *Transactions of the NAMRI/SME*, pp. 129-134. Lee, Y.S., 1998, Non-isoparametric tool path planning by machining strip evaluation for 5-axis sculptured surface machining, *Computer Aided Design*, vol. 30, no. 7, pp. 559-570.

Lee, Y.S. and Ji., H., 1997, Surface interrogation and machining strip evaluation for 5-axis CNC die and mold machining, *international Journal of Production Research*, vol. 35, no. 1, pp. 225-252.

Li, S.X. and Jerard, R.B., 1994, 5-axis machining of sculptured surfaces with a flat-end cutter, *Computer Aided Design*, vol. 26, no. 6, pp. 422-436.

Lin, R.S. and Koren, Y., 1996, Efficient tool-path planning for machining freeform surfaces, *ASME Journal of Engineering for Industry*, vol. 118, pp. 20-28.

Lo, C-C., 1999, Efficient Cutter-path planning for five-axis surface machining with a flat-end cutter, *Computer Aided Design*, vol. 31, pp. 557-566.

Marciniak, K., 1991, *Geometric Modeling for Numerically Controlled Machining*, Oxford University Press.

McCarthy, J.M., 1990, *Introduction to Theoretical Kinematics*, MIT.

Menon, J.P. and Voelcker, H.B., 1993, Toward a comprehensive formulation of NC verification as a mathematical and computational problem, *Journal of Design and Manufacturing*, vol. 3, no. 4, p 263-278.

Rao, N., Ismail, F. and Bedi, S., 1997, Tool path planning for five-axis machining using the principle axis method, *International Journal of Machine Tools and Manufacture*, vol. 37, no. 7, pp. 1025-1040.

Rao, A., and Sarma, R., 2000, On local gouging in five-axis sculptured surface machining using flat-end tools. *Computer Aided Design*, 32:409-420.

Sarma, R., and Dutta, D., 1997, The geometry and generation of NC Tool Paths, *ASME Journal of Mechanical Design*, 119(2):253-258.

Sarma, R., 2000a, Flat-Ended Tool Swept Sections for Five-Axis Machining of Sculptured Surfaces, *ASME Journal of Manufacturing Science and Engineering*, Vol. 122, pp. 158-165.

Sarma R., 2000b, On local gouging in five-axis sculptured surface machining using flat-end tools, *Computer Aided Design*, 32:409-420.

Srinivasan, L., and Ge, Q. J., 1998, Fine tuning of rational B-spline motions, *ASME Journal of Mechanical Design*, 120(1):46-51.

Suresh, K. and Yang D.C.H., 1994, Constant scallop-height machining of freeform surfaces, *ASME Journal of Engineering for Industry*, vol. 116, pp. 253-259.

Vickers, G.W. and Quan, K.W., 1989, Ball-mills versus end-mills for curved surface machining, *ASME Journal of Engineering for Industry*, vol. 111, pp. 11-26.

Xia, J. and Ge, Q.J. 1999, On the exact representation of the boundary surfaces of the swept volume of a cylinder undergoing rational Bézier and B-spline motions, *Proc. 1999 ASME Design Automation Conference*, Las Vegas, Paper No. DETC99/DAC-8607. Also accepted for publication by *ASME Journal of Mechanical Design*.

Xia, J. and Ge. Q.J., 2000, On the exact computation of the swept surface of a cylindrical surface under two-parameter rational Bézier motions, *Proceedings of ASME/DETC 2000*, Paper No. DFM-14039.

Design of Programmable Passive Compliance Shoulder Mechanism

Masafumi OKADA[*1], Yoshihiko NAKAMURA[*1*2] and Shigeki BAN[*1]

[*1]Dept. of Mechano-Informatics, University of Tokyo
7-3-1 Hongo Bunkyo-ku Tokyo, 113-8656 Japan
[*2]CREST Program, Japan Science and Technology Corporation
e-mail : okada@ynl.t.u-tokyo.ac.jp

Abstract

Design of mechanical compliance would be one of the most important technical foci in making humanoid robots really interactive with the humans. For safety insurance the mechanical compliance should be developed to humanoid robots. The introduction of the passive compliance to humanoid robots has large possibility to achieve the human skill by using the dynamical energy stored in the compliant members. The programmable passive compliance plays an important role to cope with the changing environments and task execution. In this paper, we evaluate the effectiveness of the passive compliance for the realization of the human skill and design a programmable passive compliance mechanism 'PPC cybernetic shoulder' which is the four degree of freedom shoulder mechanism for humanoid robots using a closed kinematic chain. The programmability of the PPC cybernetic shoulder is evaluated by experiments.

Key Words : *Skill of compliance, Prgrammable passive compliance, The cybernetic shoulder, Humanoid robot*

1 Introduction

Humanoid robots that share the space and environments with human should have compliance for human friendliness, safety issue and relief of impacts. There are two strategies to develop the robot compliance. One is active compliance on which many researches have been reported [1]~[6], the other is passive compliance. The active compliance is realized by actuators. The compliance of robot joints is developed using control theories such as impedance matching method. It has high programmability of compliance, however cannot cope with fast responses because of the low resolution of sensors, a long sampling time of control and noises of sensors. The passive compliance means mechanical compliance of members of robot arm or some special joint mechanisms. This compliance works effectively in all frequency (both fast and slow responses) but its programmability is low. For the safety issue, the passive compliance is important because there are many humans in the environments of the humanoid robots.

Our research focuses on the 'Skill of Compliance', which means (1) tuning of passive compliance, (2) planning of swing pattern and (3) design of the control law. In the casting of fishing, for example, the potential energy is accumulated in the rod by taking the swing and the large kinetic energy is obtained by discharging the potential energy in the instant to throw the prickle farer. In this motion, the passive compliance of the rod is tuned, the swing pattern of the rod and the force control of our arm are well designed.

So far, we have developed the cybernetic shoulder[7] that is the three degree-of-freedom mechanism for humanoid robots. It has human-like motion and passive compliance using the closed kinematic chain. In this paper, we design the programmable passive compliance mechanism for the cybernetic shoulder (Programmable Passive Compliance Cybernetic Shoulder) that is useful to filling up the low programmability of the passive compliance, and obtain the compliance ellipsoid[4] of this mechanism that is helpful to design the swing up pattern. The programmability of the designed mechanism is evaluated by experiments.

2 Passive Compliance

2.1 Compliance, control law and swing pattern

In this section, we show the skill of the passive compliance. Consider the two links manipulator in the horizontal plane shown in Fig.1. One joint is actuated and another is free joint that has passive compliance. ℓ_i are the length of links (we set $\ell_1 = 0.3$ [m], $\ell_2 = 0.5$ [m]), s_i are the positions of the center of gravity of

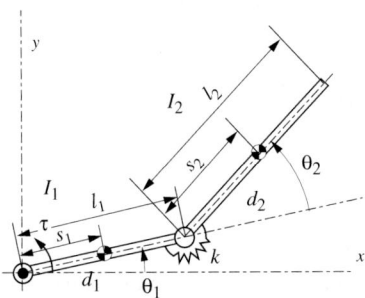

Figure 1: Two links manipulator in the horizontal plane

links ($= \ell_i/2$), I_i are the inertias of links, d_i are the coefficients of the viscosity of joints ($d_1 = 0.3$ [Nms/rad], $d_2 = 1.0$ [Nms/rad]), θ_i are the rotation angles of the links, k is the spring constant of the passive joint and τ is the torque of the motor. θ_1 is controlled by PD controller K as shown in Fig.2. P is the two links ma-

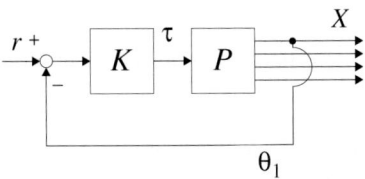

Figure 2: Control system of the two links manipulator

nipulator, r is the reference signal for θ_1 and X is as follows.

$$X = \begin{bmatrix} \theta_1 & \dot{\theta}_1 & \theta_2 & \dot{\theta}_2 \end{bmatrix}^T \quad (1)$$

The dynamics of the two links manipulator is as fallows.

$$M(\theta_2)\ddot{\Theta} + C(\Theta, \dot{\Theta}) = U \quad (2)$$

$$\Theta = \begin{bmatrix} \theta_1 & \theta_2 \end{bmatrix}^T \quad (3)$$

$$M = \begin{bmatrix} a + 2b\cos\theta_2 + c & b\cos\theta_2 + c \\ b\cos\theta_2 + c & c \end{bmatrix} \quad (4)$$

$$C = \begin{bmatrix} -b\cos\theta_2(2\dot{\theta}_1^2 + \dot{\theta}_2^2)\dot{\theta}_2 & b\sin\theta_2 \cdot \dot{\theta}_1^2 \end{bmatrix}^T \quad (5)$$

$$U = \begin{bmatrix} \tau - d_1\dot{\theta}_1 & -k\theta_2 - d_2\dot{\theta}_2 \end{bmatrix}^T \quad (6)$$

$$a = m_1 s_1^2 + m_2 \ell_1^2 + I_1 \quad (7)$$

$$b = m_2 s_2 \ell_1 \quad (8)$$

$$c = m_2 s_2^2 + I_2 \quad (9)$$

Setting the reference signal as

$$r(t) = -\sin(2\pi t), \quad 0 \leq t \leq 1 \quad (10)$$

we get the optimal spring constant k_{opt} which minimizes the following cost function J.

$$J = \sum_{i=1}^{2} w_i J_i \quad (11)$$

$$J_1 = \max_t (\dot{\theta}_1(t)\tau(t)) \quad (12)$$

$$J_2 = \frac{1}{\max_t (v_y(t))} \quad (13)$$

$$v_y(t) = \dot{\theta}_1 \ell_1 \cos\theta_1 + (\dot{\theta}_1 + \dot{\theta}_2)\ell_2 \cos(\theta_1 + \theta_2) \quad (14)$$

$$w_1 = 1, \quad w_2 = 500 \quad (15)$$

J_1 aims at reduction of the actuator power. J_2 aims

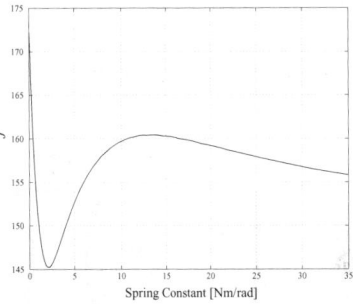

Figure 3: Value of J versus spring constant k

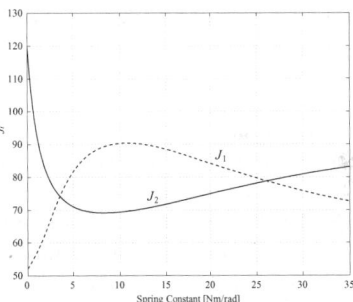

Figure 4: Value of J_i versus spring constant k

at maximizing the velocity of the end of the arm along with y axis. Maximization of the velocity means that the two links manipulator can throw fastball. Though the optimized spring constant depends on the motor controller (control law) and reference signal in equation (10) (swing pattern), we optimize only the spring constant (compliance) in one situation that fixes control law and swing pattern. The values of J and J_1, J_2 due to the spring constant k are shown in Fig.3, 4 respectively, which are given from the numerical simulations. These figures show that the optimal spring constant k_{opt} is given as

$$k_{opt} = 2.15 \quad (16)$$

and the maximum velocity is 6.19 [m/s]. These results show that by using the passive compliance, the two links manipulator can throw the faster ball by small consumption of the motor energy.

2.2 Programmable passive compliance mechanism

Because the optimal spring constant given in the previous section depends on the weight of links and trajectory of the reference signal, the spring constant should be changed adaptively, which is achieved by the programmable passive compliance. Figure 5 shows the example of the programmable passive compliance mechanism using a closed kinematic chain. There

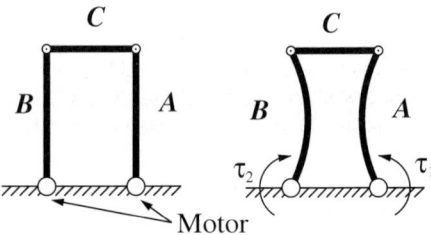

Figure 5: Programmable passive compliance mechanism

are two redundant actuators. When the members A and B have a nonlinear relationship between the strain and stress, the compliance of the position C can be changed by giving tension to members A and B. These types of PPC mechanisms have been developed [6, 8, 9]. The drawbacks of these mechanisms are as follows.

Development of the multi-DOF mechanism
If we develop the multi-degree of freedom mechanism assembling the single degree of freedom mechanism, it gets heavy weight and large volume.

Control of redundant actuators The programmable passive compliance is realized by two redundant actuators whose outputs should be exactly same. Otherwise the joint may rotate or has an oscillation.

To overcome these problems, we develop the programmable passive compliance mechanism using a closed kinematic chain.

3 Programmable Passive Compliance (PPC) Cybernetic Shoulder

3.1 Design and mechanism

We have designed the cybernetic shoulder[7] that is the three DOF shoulder mechanism for humanoid robots. The passive compliance mechanisms using closed kinematic chain have been developed. The model of the cybernetic shoulder is shown in Fig.6. β

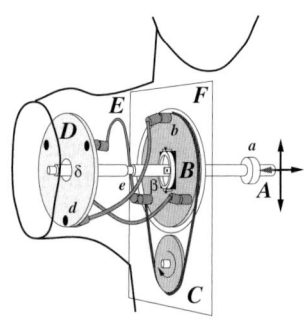

Figure 6: The cybernetic shoulder

and δ are two degree of freedom gimbal mechanisms, d is a three degree of freedom ball joint, b is a two degree of freedom universal joint, a is a four degree of freedom joint of spherical and prismatic motion, and e is a prismatic joint. Moving point A within vertical plane alters the pointing direction of the main shaft G, which determines, along with the constraints due to the free curved links E between points b and d, the direction of the normal vector of D. The rotation about the normal of D is mainly determined by the rotation of C through B and G. Note that the rotation of C is coupled with the pointing direction of D when B and D are not parallel. Based on this mechanism, we design the PPC cybernetic shoulder shown in Fig.7. The advantages of this mechanism are as follows.

PPC mechanism We replace the prismatic joint e in Fig.6 with a linear actuator (4.5[W] DC motor and ball screw) as shown in Fig.8. By changing the length of L in ΔL, the internal force is applied to members E, which causes the programmable passive compliance when E have nonlinear relationship between strain and stress.

Compactness and small backlash The universal joints on the point b and d are replaced with elastic universal joints as shown in Fig.9. It has the

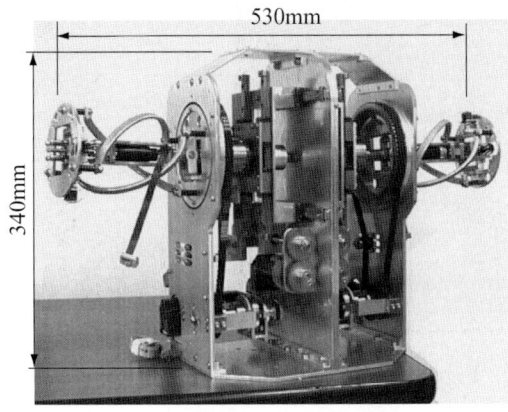

Figure 7: The PPC cybernetic shoulder

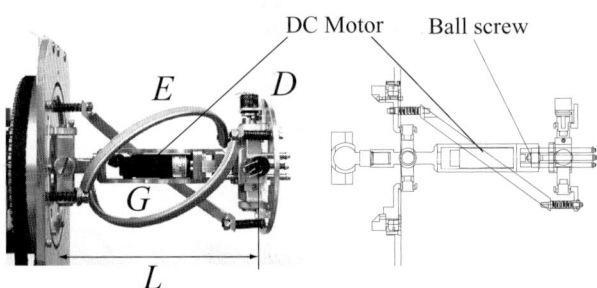

Figure 8: PPC mechanism

Figure 9: Elastic universal joint

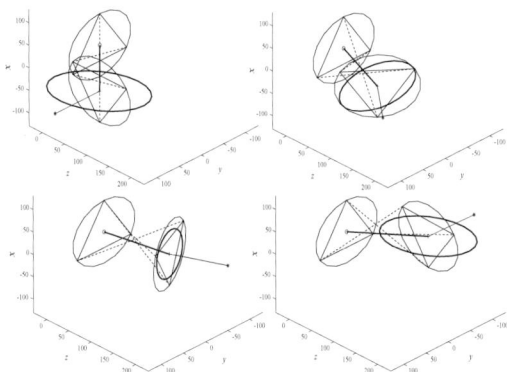

Figure 10: Compliance ellipsoid of the cybernetic shoulder

same structure as a flexible coupling. This is for the compactness and the small backlash.

Multi-DOF compliance Because the end disk D has a gimble mechanism on its center, the PPC cybernetic shoulder has two degree of freedom compliance around the rotation axis of the gimble mechanism. Because the center rod G is rigid, the PPC cybernetic shoulder has high stiffness for any other degree of freedom of compliance.

3.2 Compliance ellipsoid of the cybernetic shoulder

The compliance ellipsoid [4] is helpful for the foundation of the swing pattern and motor control law.

Consider the compliance matrix C defined as

$$C = JK^{-1}J^T \quad (17)$$

Here, J is the jacobian matrix and K is the spring constant matrix. Using the singular value decomposition of C,

$$C = USV^T \quad (18)$$
$$= [\ U_1\ \cdots\ U_n\ ]\,\mathrm{diag}\{\ s_1\ \cdots\ s_n\ \}V^T \quad (19)$$

the compliance ellipsoid is defined in the n dimensional space whose axes are $s_i U_i (i = 1, 2, \cdots n)$. In this paper, we consider the two-dimensional compliance ellipsoid of the cybernetic shoulder. Figure 10 shows the compliance ellipsoid in accordance with the motion of the cybernetic shoulder. These ellipsoids are calculated in each orientation using equation (19).

3.3 Evaluation of the programmability

Each occasion is defined as Table 1. In this section, we evaluate the programmability of the passive compliance on PPC cybernetic shoulder. We set two configurations of the PPC cybernetic shoulder as shown in Fig.11. By cutting the 500[g] weight hung from the end of the arm, the external force is applied. The torque of the external force becomes 0.539 [Nm]. Two cases are adopted on each configuration, in one case $\Delta L = 0$ [mm], in another case $\Delta L = -3$ [mm]. The responses of each case are shown in Fig.12 and 13.

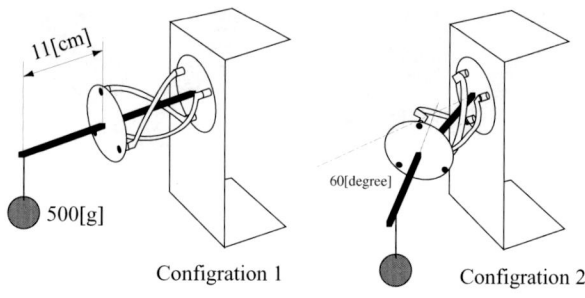

Figure 11: Configurations of the PPC cybernetic shoulder

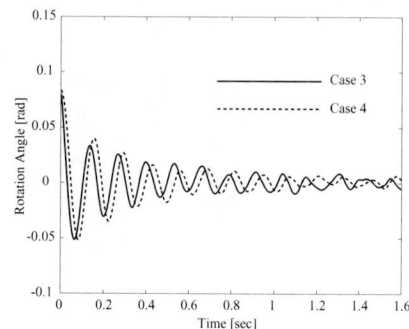

Figure 13: Responses on configuration 2

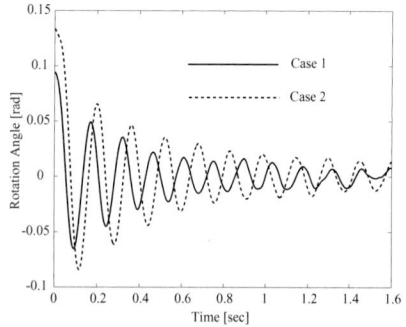

Figure 12: Responses on configuration 1

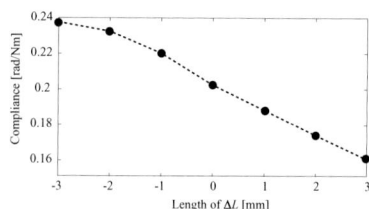

Figure 14: PPC due to ΔL

In this prototype, the members E are rigid but joints (elastic joints) have compliance. The passive compliance of this mechanism is caused by the joint compliance. The compliance on each case is as follows which is calculated from the rotation angle in time zero.

Case 1 : 0.202 [rad/Nm]
Case 2 : 0.237 [rad/Nm]
Case 3 : 0.156 [rad/Nm]
Case 4 : 0.170 [rad/Nm]

In configuration 2, the compliance cannot be changed so much. In configuration 1, we measure the passive compliance by small resolution of changing ΔL. Figure 14 shows the compliance due to ΔL in the configuration 1. The shorter L yields the higher compliance.

Table 1: Definition of the experimental set

	$\Delta L = 0$ [mm]	$\Delta L = -3$ [mm]
Configuration 1	Case 1	Case 2
Configuration 2	Case 3	Case 4

The elastic universal joints have high compliance for yaw and pitch direction but have low compliance on thrust direction, that yield the passive compliance of the PPC cybernetic shoulder. The more dominant the thrust compliance becomes, the lower the passive compliance of the PPC cybernetic shoulder becomes.

3.4 Design of elastic members

The programmable passive compliance in the previous section was realized adopting the elastic universal joint in Fig.9. In this section, we design the nonlinear elasticity as a property of link E in Fig.8.

Figures 15 and 16 illustrate the idea and design of nonlinear elastic link. Link E has a series of holes with different diameters and small cuts. As seen in Fig.16, the largest hole has the minimum thickness and, therefore, bends first when bending moment is applied until the cut C-shaped hole deforms and becomes closed. If the bending moment exceeds, it further deforms the second largest C-shaped hole and so on. Since the thickness of halls are different, the elastic coefficient of link E changes in a discrete manner. Figure 17 shows the result of measurements of the fabricated link, which clearly shows the nonlinear discrete elasticity. The spring constant in each area is as Table 2. The diameters, thickness, and width of cut would need

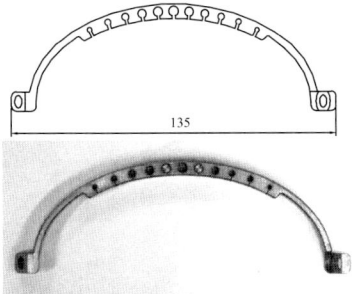

Figure 15: Elastic link

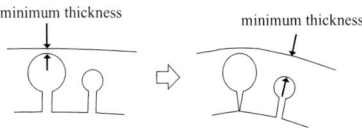

Figure 16: Change of the spring constant

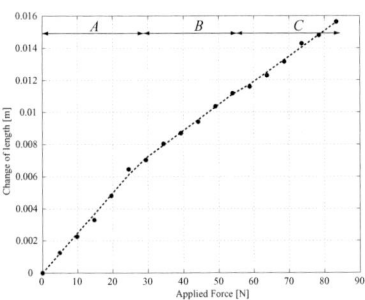

Figure 17: Applied force and change of length

Table 2: Tunable spring constant

Area	Spring constant [N/m]
A	4.009×10^3
B	6.203×10^3
C	6.303×10^3

more careful design using FEM numerical analysis if it requires a specific shape of elastic curve.

The redundant actuator of "DC Motor" in Fig.8 determines length L. When it shortens, it is accompanied by the bends of three of link E, which determines the mechanical compliance of the PPC cybernetic shoulder mechanism.

4 Conclusions

In this paper, we discuss on the skill of compliance, which is tuning of the passive compliance, planning of the swing pattern and design of the control law, and design the programmable passove compliance cybernetic shoulder. The results are as follows.

1. By using the passive compliance mechanism, robots can throw a ball faster by small actuator power.
2. We design the programmable passive compliance cybernetic shoulder which is the shoulder mechanism for humanoid robots.
3. The programmable passive compliance cybernetic shoulder has high programmability of the passive compliance by using the elastic joint and elastic link.

This research is supported by the Research for the Future Program, the Japan Society for the Promotion of Science (Project No. JSPS-RFTF96P00801).

References

[1] R.P.C.Paul and B.Shimano "Compliance and Control, Proc. of the 1976 Joint Automatic Control Conference, pp.694–699, 1976.

[2] H.Hanafusa and H.Asada "Stable Pretension by a Robot Hand with Elastic Fingers, Proc. of the 7th International Symposium on Industrial Robots, pp.361–368, 1977.

[3] N.Hogan "Mechanical Impedance Control in Assistive Devices and Manipulators, Proc. of the 1980 Joint Automatic Control Conference, pp.TA10-B, 1980.

[4] J.K.Salisbury "Active Stiffness Control of a Manipulator in Cartesian Coordinates, Proc. of the IEEE Conference on Decision and Control, 1980.

[5] N.Hogan "Impedance Control: An Approach to Manipulation: Part 1~3, ASME Journal of Dynamic Systems, Measurement and Control, Vol.107, pp.1–24, 1985.

[6] K.F.L-Kovitz, J.E.Colgate and S.D.R.Carnes: Design of Components for Programmable Passive Impedance, Proc. of IEEE International Conference on Robotics and Automation, pp.1476–1481, 1991.

[7] M.Okada and Y.Nakamura: Development of the Cybernetic Shoulder – A Three DOF Mechanism that Imitates Biological Shoulder-Motion –, Proc. of IEEE/RSJ International Conference on Intelligent Robots and Systems , Vol.2, pp.543-548 (1999)

[8] T.Morita and S.Sugano: Design and Development of a new Robot Joint using a Mechanical Impedance Adjuster, Proc. of IEEE International Conference on Robotics and Automation, pp.2469-2475 (1995)

[9] H.Kobayashi, K.Hyodo and D.Ogane: On Tendon-Driven Robotic Mechanism with Redundant Tendons, Int. J. of Robotics Research, Vol.17, No.5, pp.561-571 (1998)

Kinematic Control of Parallel Robots in the Presence of Unstable Singularities

John F. O'Brien and John T. Wen

Center for Automation Technology
Department of Electrical, Computer, & Systems Engineering
Rensselaer Polytechnic Institute, Troy, NY 12180

Abstract

Parallel mechanisms frequently contain an unstable type of singularity that has no counterpart in serial mechanisms. When the mechanism is at or near this type of singularity, it loses the ability to counteract external forces in certain directions. This paper considers an interesting alternative to redundant actuation through the application of additional kinematic constraint via passive joint braking in the close neighborhood of unstable poses. The brake has the advantage of not requiring additional actuator/sensor pairs or mechanism architecture redesign, and of being implemented only when necessary to stabilize the system. The kinematic analysis of braked mechanisms is supported with results using a six degree-of-freedom (DOF) parallel machining mechanism.

1 Introduction

Parallel robots provide a stiff connection between the payload and the base structure, with pose accuracies that are superior to serial chain manipulators. The principal drawbacks concerning parallel robots are their limited workspace, and the complexity of singularity analysis [1]. In contrast to serial chain manipulators, singularities in parallel mechanisms have different manifestations. This issue has been studied in the multi-finger grasping context in [2,3] and more recently for general parallel mechanisms in [4,5,6]. In [4], the singularities are separated into two broad classifications: end-effector and actuator singularities. The former is comparable to the serial arm case, where the end-effector losses a degree-of-freedom in the task space. The latter is defined when a certain task wrench cannot be resisted by active joint torques. Or equivalently, the task frame can move even when all the active joints are locked. These are called the unstable configurations in [6] which correspond to unstable grasps in the multi-finger grasp literature. The unstable type of singularity is obviously unattractive, as unpredictable task motion could result.

Problems associated with unstable singularity in practical robotic applications have been reported. In [7], a new parallel machining center is proposed. After the mechanism was built, it was discovered that the workspace contains unstable singularities – severely compromising the utility of the mechanism. A solution proposed in [7] introduces a new active degree of freedom to the mechanism. In [8,9], the activation of passive joints and implementation of additional parallel kinematic chains are proposed as solutions to the unstable singularity problem.

This paper proposes an alternative to redundant actuation to eliminate unstable singularity in parallel mechanisms, namely the application of additional kinematic constraint in the close neighborhood of unstable singularity by locking a passive joint.

2 Differential Kinematics

This section considers the differential kinematics of general rigid multibody systems. Consider a general mechanism subject to kinematic constraints. The generalized coordinate (with the constraints removed) is denoted by θ. The active joints' angles are denoted by θ_a and passive ones by θ_p. We order the angles so that $\theta^T = [\theta_a^T, \theta_p^T]$. Consider a general constraint (written in terms of the joint velocity vector)

$$J_C(\theta)\dot{\theta} = 0. \qquad (1)$$

Let the velocity of the task frame be

$$v_T = J_T(\theta)\dot{\theta}. \qquad (2)$$

Partition J_C and J_T according to the dimension of θ_a and θ_p:

$$J_C = \begin{bmatrix} J_{C_a} & J_{C_p} \end{bmatrix} \quad J_T = \begin{bmatrix} J_{T_a} & J_{T_p} \end{bmatrix}.$$

Then (1) can be used to solve for $\dot{\theta}_p$:

$$\dot{\theta}_p = -J_{C_p}^\dagger J_{C_a}\dot{\theta}_a + \tilde{J}_{C_p}\xi \qquad (3)$$

where $\text{col}(\tilde{J}_{C_p})$ spans the null space of J_{C_p}, and ξ is arbitrary. Substituting into (2), we have

$$v_T = (J_{T_a} - J_{T_p}J_{C_p}^\dagger J_{C_a})\dot{\theta}_a + J_{T_p}\tilde{J}_{C_p}\xi. \qquad (4)$$

Define the manipulability Jacobian as

$$\overline{J}_T \doteq J_{T_a} - J_{T_p}J_{C_p}^\dagger J_{C_a}. \qquad (5)$$

In this paper, we will not address mechanisms that are under- or redundantly actuated, thus $\overline{J}_T$ is square. There are two cases of singularities:

1. Unmanipulable Singularity: This corresponds to configurations at which $\overline{J}_T$ loses rank.

2. Unstable Singularity: This corresponds to configurations at which $J_{T_p}\tilde{J}_{C_p} \neq 0$.

It may happen that $J_{T_p}\tilde{J}_{C_p} = 0$ but $\tilde{J}_{C_p} \neq 0$. This corresponds to the existence of self motion involving only passive joints in the mechanism. At an unmanipulable singularity, the composite Jacobian $\overline{J}_T$ loses rank. At an unstable singularity, the maximum singular value of $\overline{J}_T$ approaches infinity. Thus, it is evident that the condition number of $\overline{J}_T$ can be used in conjunction with its maximum singular value to determine the closeness to and type of singularity.

3 Kinematic Control Algorithm

A common set-point kinematic control problem is presented. Choose

$$\dot{\theta}_a(t) \ni (R(t), p(t)) \rightarrow (R_d, p_d). \qquad (6)$$

The pair $(R(t), p(t))$ denotes the pose of the mechanism end-effector defined by the 3×3 orthonormal rotation matrix and the 3×1 position vector expressed in a common reference frame. Dropping the independent variable from the notation, the position error function is defined as

$$e_p = \frac{1}{2}\| p - p_d \|^2. \qquad (7)$$

Obtaining an error function for the rotational pose component is somewhat less straightforward. The following is one of several candidate functions:

$$e_R = \frac{1}{2}\| q \|^2 \qquad (8)$$

where q is the vector quaternion of the orientation error. A positive definite Lyapunov function is constructed via the sum of (7) and (8).

$$V = e_p + e_R \qquad (9)$$

The time derivative of (9) is

$$\dot{V} = \Delta p^T \dot{p} + \frac{1}{2}\sin\frac{\phi}{2}\cos\frac{\phi}{2}k^T\omega = \Delta p^T\dot{p} + s^T\omega \qquad (10)$$

where ϕ is the relative angle about the unit length axis of rotation k, and the 3×1 vector of reals, ω, is the task space angular rate. Applying the assumption that the mechanism is not at an unstable singularity (but could be in the close neighborhood), equation (10) is rearranged and combined with (4).

$$\dot{V} = \begin{pmatrix} s \\ \Delta p \end{pmatrix}^T \begin{pmatrix} \omega \\ \dot{p} \end{pmatrix} = \begin{pmatrix} s \\ \Delta p \end{pmatrix}^T \overline{J}_T \dot{\theta}_a \qquad (11)$$

The manipulable variable, $\dot{\theta}_a$, is now driven by an appropriate control function. Candidate control laws are:

$$\dot{\theta}_a = -\overline{J}_T^\dagger K_p \begin{pmatrix} s \\ \Delta p \end{pmatrix} \qquad (12)$$

and,

$$\dot{\theta}_a = -\overline{J}_T^T K_p \begin{pmatrix} s \\ \Delta p \end{pmatrix} \qquad (13)$$

where K_p is a diagonal matrix of proportional gains. Both (12) and (13) are asymptotically stable controllers with the stipulation that the mechanism is fully manipulable over the entire trajectory.

4 Passive Joint Braking Near Unstable Singularity

It is proposed in this section to provide additional kinematic constraint to a parallel mechanism by seizing a passive joint when it enters a "brake" space, defined as the set of all mechanism configurations in the workspace that satisfy

$$(R, p) \in \{SE(3) : cond(\overline{J}_T) \geq m_1, \overline{\sigma}(\overline{J}_T) \geq m_2\} \qquad (14)$$

where m_1 and m_2 are real scalars corresponding to selected threshold values. Mechanism poses that satisfy (14) are in the close neighborhood of unstable singularity.

When the pair (R, p) satisfies (14), a passive joint is seized. In this condition, the constraint matrix J_C

of (1) has a row of zeros appended to it with a unit element corresponding to the locked passive joint.

$$J_{Cl} = \begin{pmatrix} & J_C & \\ 0 & \cdots & 1 & \cdots & 0 \end{pmatrix} \quad (15)$$

If J_{C_p} is fat, then the passive partition of J_{Cl} is square or fat, and there is no additional constraint on $\dot{\theta}_a$. A square J_{C_p} matrix is a more likely situation, as passive joint self motion would be possible only at singularity. If J_{C_p} is square, the passive partition of J_{Cl} is tall, and the set of feasible $\dot{\theta}_a$ that satisfy (1) is reduced.

To find the differential kinematic description of the locked mechanism, we premultiply (1) by the annihilator of J_{Cp}

$$0 = \widetilde{J_{Cp}} J_{Ca} \dot{\theta}_a \quad (16)$$

where J_{Cp} and J_{Ca} now represent the partition of the constraint matrix of the locked mechanism, J_{Cl}. This indicates a constraint on the active joint space, specifically $\dot{\theta}_a \in \mathcal{N}(\widetilde{J_{Cp}} J_{Ca})$. If (16) is satisfied, the passive joint rate and task velocity can be found.

$$\dot{\theta}_p = -J_{Cp}^\dagger J_{Ca} \dot{\theta}_a \quad (17)$$

$$v_T = (J_{Ta} - J_{Tp} J_{Cp}^\dagger J_{Ca}) \dot{\theta}_a \quad (18)$$

subject to $0 = \widetilde{J_{Cp}} J_{Ca} \dot{\theta}_a$.

Select a new matrix P such that $sp\{col(P)\} = \mathcal{N}(\widetilde{J_{Cp}} J_{Ca})$. The active joint variable may be expressed as

$$\dot{\theta}_a = P\xi \quad (19)$$

where ξ is an arbitrary variable that maps to the admissible space of $\dot{\theta}_a$. The matrix P is tall and orthonormal, thus (5) can be expressed with the constraint on $\dot{\theta}_a$.

$$\overline{J}_{Tl} = (J_{Ta} - J_{Tp} J_{Cp}^\dagger J_{Ca}) P P^T \quad (20)$$

which is degenerate. Applying the transpose control law (13) with an appropriately selected gain matrix to the braked mechanism results in a negative semi-definite derivative Lyapunov function.

$$\dot{V} = -\begin{pmatrix} s \\ \Delta p \end{pmatrix}^T \overline{J}_{Tl} \overline{J}_{Tl}^T K_p \begin{pmatrix} s \\ \Delta p \end{pmatrix} \quad (21)$$

The braked mechanism is stable in the Lyapunov sense.

4.1 Passive Brake Selection

Instability is a sufficient condition for passive joint self motion, mathematically described by $\mathcal{N}(J_{Cp}) \neq \{0\}$. An abuse of Matlab notation is used to develop a selection algorithm.

$$[a, b] = \max \underline{v}(J_{Cp}) \implies \dot{\theta}_p(b) = 0 \quad (22)$$

where $\underline{v}$ represents the minimum right singular vector. The input direction given by (22) represents the largest projection onto the minimum input direction of J_{Cp}. Certainly (22), which considers only passive joint rate amplitude, represents only one of many possible algorithms.

It seems feasible to blend other performance characteristics into the selection algorithm. While (22) should provide the best performance in terms of normed passive joint rate, the added constraint may result in large *active* joint rates depending on the required instantaneous task velocity. An alternative to (22) may be posed as a discrete optimization problem that minimizes the following cost function evaluated over all n such that (14) is satisfied

$$J(n) = w_1 \overline{\sigma}(J_T(n))^2 + w_2 \|\dot{\theta}_a(n)\|^2. \quad (23)$$

where $\overline{\sigma}(J_T(n))$ is the maximum singular value of the composite Jacobian. This system represents a trade off between braking performance and active joint effort. Additional terms associated with task axis coupling and other quantities can be applied to (23).

4.2 Stuck Poses in the Brake Space

A zero value of (21) implies $s \to \{0\}$, $\Delta p \to \{0\}$, or $\begin{pmatrix} s \\ \Delta p \end{pmatrix} \in \mathcal{N}(J_{Tl}^T)$. The first condition indicates that the set-point has been reached, while the second implies the error function requires instantaneous task velocity that is inadmissible, and the controller is "stuck". An interim control algorithm or kinematic constraint must be implemented in this case.

One possible solution to the "stuck" pose problem is to introduce an interim kinematic constraint in the form of an alternate passive joint brake. A subset of the brake space defined as

$$(R, p) \in \{SE(3) : |\frac{\Delta x}{\|\Delta x\|} \cdot \underline{U}| \geq m_3\} \quad (24)$$

where $\Delta x = \begin{pmatrix} s \\ \Delta p \end{pmatrix}$, m_3 is a real scalar of positive sign (usually slightly less than unity), and $\underline{U}$

is the minimum left singular vector of $\overline{J}_{Tl}$ which is inadmissible task velocity direction. When the pose satisfies (24), the mechanism is approaching a stuck configuration. The "nominal" brake is released, and another passive joint is seized, serving as an "interim" brake. A new manipulability Jacobian corresponding to the new constraint is found using (19) and (20).

A possible shortcoming of the multiple brake algorithm is if the interim brake also results in a "stuck" condition. This corresponds to either

$$\mathcal{N}(\overline{J}_{Tn}^T) = \mathcal{N}(\overline{J}_{Ti}^T) \qquad (25)$$

or

$$\mathcal{N}(\overline{J}_{Tl_n}^T) = \mathcal{N}(\overline{J}_{Tl_i}^T) \qquad (26)$$

where the subscript n corresponds to the manipulability Jacobian associated with the nominal brake, and i the interim brake.

5 Illustrative Analytical Example

The Seoul National University has designed and fabricated a novel 6-DOF parallel mechanism called *Eclipse* for five-face machining [7]. The *Eclipse* replaces the nominal serial structure working in concert with a rotating table with a single mechanism, able to provide continuous spindle rotations from the horizontal to the vertical faces.

The mechanism consists of three parallel PPR kinematic chains that are fixed to the end effector by spherical joints. The circular and linear prismatic joints are active, while the three revolute joints mounted on the prismatic joints are passive. In contrast to the majority of parallel mechanisms, the Eclipse workspace is quite spacious, especially in rotation.

The Eclipse exhibits both unmanipulable and unstable singularity. The former occurs when an end effector spherical joint is directly over the center of the circular guide, while the later occurs when the end-effector pitches through approximately $\frac{\pi}{3}$. The unstable singularity is particularly troublesome in that it represents a barrier that cannot be avoided when pitching the spindle to the horizontal, as shown in the simulation results of figures 1 and 2. Note the barrier where the manipulability Jacobian is ill-conditioned with large maximum singular value at approximately $\frac{\pi}{3}$. This is indicative of unstable singularity. The rise in condition number at $\frac{\pi}{2}$ indicates unmanipulable singularity.

The singularity barrier problem is addressed through the use of a passive joint brake that activates in the close neighborhood of the barrier in concert with a kinematic control algorithm described in

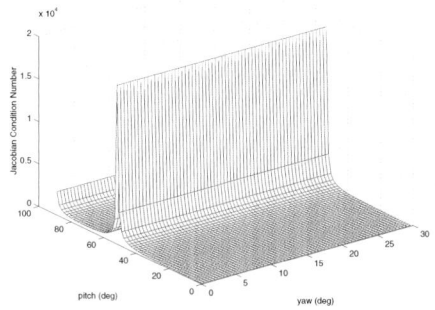

Figure 1: Eclipse Unstable Singularity Barrier: Condition Number

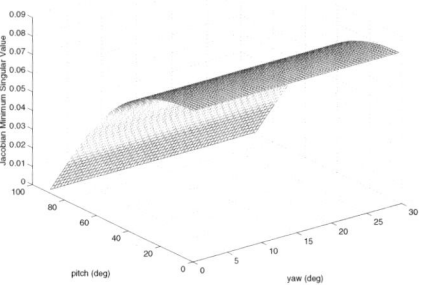

Figure 2: Eclipse Unstable Singularity Barrier: Minimum Singular Value

section 3. The controller is tasked to pitch the end-effector from 0 to $\frac{\pi}{2}$. The passive joint to be seized was selected via (22). The minimum left singular vector of J_{C_p} is $\underline{v} = \begin{bmatrix} -0.9998 & 0.014 & 0.014 \end{bmatrix}$, thus the revolute joint on the first chain is selected. Indeed, this is the only effective brake to apply, as the unstable direction is rotation about the line segment that connects the centers of the spherical joints of the second and third fingers. The position and orientation error components are shown in figures 3 and 4, and the composite Jacobian maximum singular values are shown in figure 5.

The transient in x-position shown in figure 3 is indicative of the coupling between that axis and the pitch axis due to the mechanism's unmanipulability in the braked condition. While the normed pose error is still asymptotically decreasing, this small error in the position may be unacceptable for precision machining operations. As is the case for many parallel robot applications, the *Eclipse* mechanism is redundant. The tool tip axis is fixed on the yaw-axis of the mechanism, and the pose control problem is five DOF. The 90 degree tilting opera-

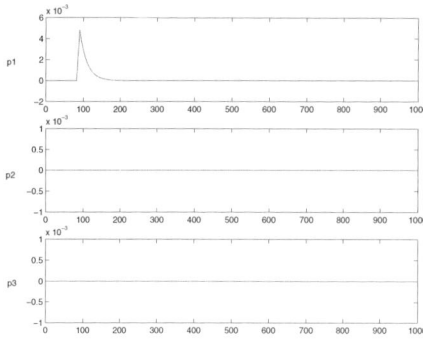

Figure 3: Eclipse Kinematic Controller with Brake: Position Error Components

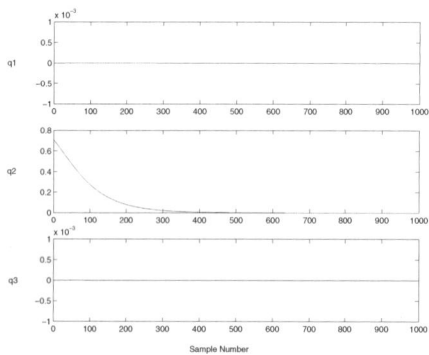

Figure 4: Eclipse Kinematic Controller with Brake: Error Vector Quaternion

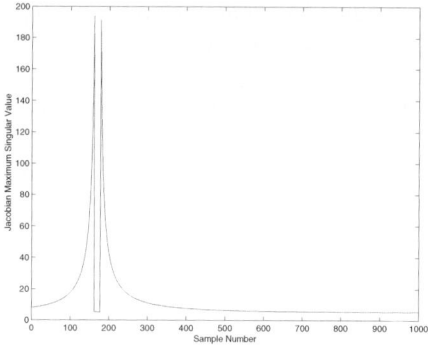

Figure 5: Eclipse Kinematic Controller with Brake: Jacobian Maximum Singular Value

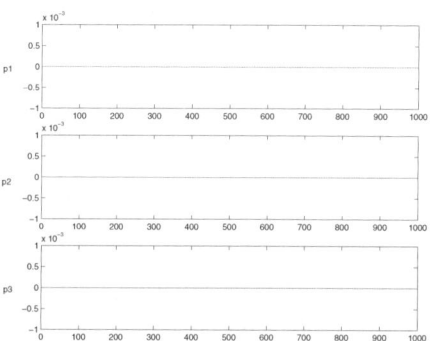

Figure 6: Eclipse Kinematic Controller with Brake: Position Error Componenets (90 deg roll)

tion is revisited with an initial 90 degree slew on the redundant yaw axis (equivalent to a 90 degree roll operation). When the mechanism enters the brake space, the minimum left singular vector of J_{C_p} is $\underline{v} = \begin{bmatrix} 0.3765 & -0.2084 & -0.9027 \end{bmatrix}$, indicating greater distribution of passive joint self motion amongst the three revolute joints at this pose. By weighting transient position error, the joint on the first chain is selected for braking, and the resulting error tracking is as shown in figures 6 and 7. Note that for this 90 degree tilting operation through the unstable singularity barrier, the error transient due to task axis cross-coupling in the braked condition is not present. This is a good example of effective exploitation of mechanism redundancy to effect a desired performance despite transient unmanipulability.

6 Conclusions and Future Work

A method of eliminating kinematic instability in parallel mechanisms has been presented. By locking a passive joint when the mechanism is in a defined "brake" space, kinematic instability is traded for local unmanipulability. Kinematic set point control algorithms maintain asymptotic stability with the brake algorithm in place, except when the error vector moves into the null direction of the degenerate transpose manipulability Jacobian. A method of brake switching has been suggested to avoid stuck configurations. Analytical results of the passive braking algorithm have been presented as evidence of its effectiveness.

The passive joint brake is suggested as an inexpensive alternative to redundant actuation. In addition to requiring less hardware, it has the added advantage of only being implemented when necessary. It is, of course, not a panacea for the unstable singularity problem. The analytical results in this paper suggest proper scrutiny must be applied to task axis cross-coupling issue in the braked condition. Mechanism redundancy can be used to mitigate this, and further investigation into this will be conducted in future work, along with dynamic analysis and brake design issues.

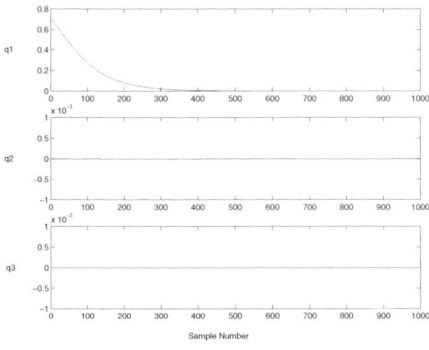

Figure 7: Eclipse Kinematic Controller with Brake: Error Vector Quaternion (90 deg roll)

Acknowledgment

This work is supported in part by the Center for Advanced Technology in Automation, Robotics & Manufacturing under a block grant from the New York State Science and Technology Foundation, the National Science Foundation (Grant IIS-9820709), and a U.S. Department of Energy Integrated Manufacturing Predoctoral Fellowship.

References

1. J.-P. Merlet,"Parallel manipulators: state of the art and perspective," in *Robotics, Mechatronics, and Manufacturing Systems* (T. Takamori and K. Tsuchiya, eds.), Elsevier, 1993.

2. A. Bicchi, C. Melchiorri, and D. Balluchi, "On the mobility and manipulability of general multiple limb robots," *IEEE Transactions on Robotics and Automation*, vol. 11, pp. 215-228, April 1995.

3. P. Chiacchio, S. Chiaverini, L. Sciavicco, and B. Siciliano, "Global task space manipulability ellipsoids for multiple-arm systems," *IEEE Transactions on Robotics and Automation*, vol 7, pp. 678-685, October 1991.

4. F. Park, and J. Kim, "Manipulability and singularity analysis of multiple robotic systems: A geometric approach," in *Proc. 1998 IEEE International Conference on Robotics and Automation*,Leuven, Belgium, pp. 1032-1037, May 1998.

5. A. Bicchi and D. Prattichizzo, "Manipulability of co-operating robots with passive joints," in *Proc 1998 IEEE International Conference on Robotics and Automation*, Leuven, Belgium, pp. 1038-1044, May 1998.

6. J. Wen and L. Wilfinger, "Kinematic manipulability of general constrained rigid multibody systems," in *Proc. 1998 IEEE International Conference on Robotics and Automation*, Leuven, Belgium, pp. 1020-1025, May 1998.

7. S. Ryu, C. Park, J. Kim, J. Hwang, J. Kim, and F. Park," Design and performance analysis of a parallel mechanism-based universal machining center," technical report, Seoul National University, 1998.

8. J. O'Brien and J. Wen, "Redundant Actuation for Improving Kinematic Manipulability," in *Proc. 1999 IEEE International Conference on Robotics and Automation*, Detroit, MI, May 1999.

9. S. Ryu, J. Kim, J. Hwang, C. Park, H. Cho, K. Lee, U. Cornel, F.C. Park, J. Kim, "Eclipse: An over-actuated parallel mechanism for rapid machining," *Proceedings of the ASME IMECE*, 1998.

Complex Behaviors from Local Rules in Modular Self-reconfigurable Robots

Jeremy Kubica, Arancha Casal,[*] Tad Hogg

Xerox Palo Alto Research Center, Palo Alto, CA 94304

Abstract

We demonstrate how simple local rules, inspired by social insects, produce complex dynamic behaviors required for locomotion and navigation in modular self-reconfigurable robots. We show how systems made up of many modules respond dynamically to their environment, such as obstacles during navigation. We present control algorithms tested on simulation experiments of TeleCube, a new modular robot developed at Xerox PARC.

1 Introduction

Modular self-reconfigurable (MSR) robots [2, 8, 3, 7, 6, 4, 5] consist of many simple identical modules, that can attach and detach from one another to change their overall connectivity. This means these systems can dynamically adapt their shape to suit the needs of the task at hand, e.g., for manipulation, locomotion, and the creation of static structures.

From a planning and control viewpoint, modular self-reconfigurable robots pose interesting challenges. Because typical systems consist of very large numbers of modules (hundreds or thousands), centralized control and planning schemes must be abandoned in favor of distributed ones. This observation and the fact that each module is a self-contained unit with its own processing, sensing and actuation means that the large body of work on distributed multi-agent control is particularly relevant.

Social insects, such ants and termites, can be viewed as powerful problem solving systems with sophisticated collective intelligence[14, 15]. Despite the fact that a single insect has very limited intelligence, the group of insects, as a whole, can achieve remarkably complex behaviors needed to coordinate, build and maintain the colony and its nest[16, 17]. Social insects can provide a fitting computational metaphor for modular self-reconfigurable robots, as both systems share basic characteristics: each module has limited capabilities but the ensemble must be able to achieve complex tasks in a distributed fashion, hence the intelligence must lie primarily in the interactions between modules.

Much of the work published to date on modular robots has concentrated on planning the reconfiguration between two static pre-defined shapes, and has often involved small numbers of modules[8, 3, 5, 6, 7, 9, 10]. Lately attention has been turned to the use of Genetic Algorithms to control MSR robots[11, 12] However, dynamically-adaptable robot structures involving hundreds of modules can also be achieved as an "emergent" result of biologically-inspired local rules[1]. This paper further extends that work, in the context of a new type of modular robot, TeleCube, developed and built at Xerox PARC. Specifically, we explore the use of *simple local* rules to produce control algorithms for a common mobile robot scenario: reconfigure, locomote and navigate through obstacles in the environment to reach a target object.

The assumptions we make are as follows:

- Modules have limited computational capabilities, and run a simple finite-state machine (FSM) to switch among different "modes". State transitions are driven by the local state, the states of neighboring modules, their locations, and some external sensor information.

- Communication is limited to immediate neighbors, and a limited number of bits are exchanged at each step.

The following section describes the robot platform used in our work. The remainder of this paper presents control algorithms for MSR robots that coordinate their actions locally to achieve emergent, global behaviors.

---

[*]Contact information: acasal@parc.xerox.com

Figure 1: Two TeleCube modules shown with arms retracted and expanded.

2 TeleCubes

2.1 Hardware

TeleCubes is a new MSR robot in which each module is a 3D cube that can prismatically extend each of its six faces independently up to a factor of two times its fully retracted configuration. A similar 2D version was previously developed at Dartmouth [7]. Figure 1 shows a TeleCube module and its telescoping faces. This mechanical design, adopted for its simplicity, has the characteristic of internal motion: modules bound inside a group of modules can move to change their positions. This capability was not possible for a previous MSR, called Proteo[3], in which modules are restricted to motions on the surface of the group. Thus the TeleCube modules allow investigating a different range of behaviors that was possible for Proteo[1].

Other existing modular robot designs include [7, 6, 8, 4, 5, 2]. Even though MSR robots like TeleCube and Proteo require specific attention to their particular motion primitives, the methods presented here are not tied to any specific design, being applicable to MSR robots in general.

2.2 Module Movement

Motions for a TeleCube module are only possible along the directions of the cube's faces. In order to change its position within the robot, a module must "slide" to it, which in general requires breaking and making connections with neighboring modules along any number of its six faces. This is illustrated in Figure 2.

All of the modules use the same motion primitive. The modules first disconnect from all neighbors in directions perpendicular to the direction of motion. Then the actual move is preformed by expanding the "back" arm and contracting the "front" arm, where front and back are determined by the direction of motion. Performing the dual expansion and contraction has the advantage of allowing the module to effectively slide forward on its arms, instead of pushing or pulling

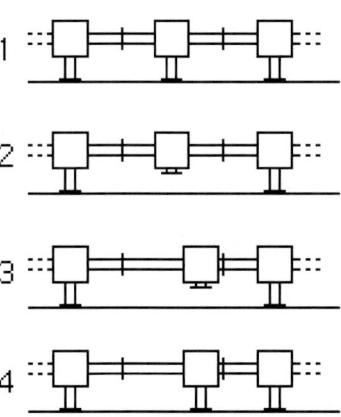

Figure 2: Detail of individual module motion, showing a side view of 3 modules. The middle module moves by contracting its bottom arm (2), then simultaneously expanding and contracting two horizontal arms (3), and finally reextending the bottom arm(4).

large chains of modules. Finally, the module will attempt to connect to any modules to which it is now aligned.

There are two other basic motion primivites, which involve more than one module: dragging modules and cooperative moves. These are illustrated in Figure 3. Dragging can occur when it is possible to pull along a neighbor module that is not connected to anyone else. Cooperative moves are used to help maintain connectedness and are accomplished by two neighbors actively moving at the same time in the same direction without breaking connections.

2.3 Simulator

The experiments were carried out on a simulated version of the TeleCube system. The simulator, written in Java, accounted for the motions of the modules in three dimensions and typically ran with hundreds of modules. Each module moves once per time step,

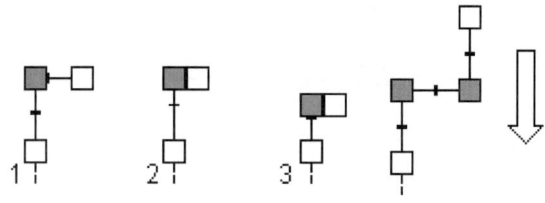

Figure 3: The three steps of a drag move and a cooperative move.

but in order to simulate an asynchronous system the order in which the modules move is random and different each turn. Each of the modules is given a finite strength, allowing the module to move itself and drag up to four other modules, with at most one module per face except for the two faces along the direction of motion.

The simulator makes several simplifying assumptions. First, the module arms can only occupy two states, fully expanded and fully contracted. Second, the modules arms are infinitely rigid, so no drooping occurs when an arm is extended. Both of these assumptions simplify alignment of modules after a move. Lastly, the simulator assumed infinite friction between the module and the floor, effectively eliminating sliding effects.

In addition, a global connectivity check was used to insure disconnecting from a neighbor would not result in separation from the rest of the ensemble. This check may be easily preformed in hardware, and was implemented as a depth-first search of the nodes in the software.

3 Control Approach

Our approach takes advantage of the distributed, homogenous nature of MSR robots. It also scales well as the number of modules increases, whereas conventional centralized schemes become computationally intractable. It has the following characteristics:

There is no central control processor, or designated leader module, or offboard planner. Global motion results, or "emerges", from purely local control rules executed at every module and through communication between neighboring modules alone.

Each module runs a local FSM control "script", consisting of simple if-then rules which may act probabilistically. This use of randomization is helpful in preventing modules from becoming indefinitely stuck in unproductive configurations.

We make use of the following concepts[1]:

The **Mode** of a module is its present FSM state, which determines the rules of behavior of a module.

Seeds The **seed** mode is the catalyst for starting new structures and focusing the movement of modules to specific spots.

Scents are the means of global communication among modules. Scents are propagated through the system in a distributed breadth-first fashion [1]. As scents are propagated they act as global gradients to guide motion.

Each module is also capable of communicating with any neighbors in all six directions. This communication falls into three basic categories: messages, queries, and commands. Messages, which include scents and impulses, are information actively sent from one module to another. An impulse is a direct message to a neighbor. A scent, explained above, (which can be either positive or negative) is a message that gets propagated to other modules and whose strength or value changes with each propagation. Queries are requests sent to neighboring modules for basic types of information, including arm length and module state. Finally, commands are forcible messages instructing a neighbor to either contract an arm, extend an arm, or to move in a given direction.

4 Results

We present results for a typical mobile robot task: reach a target object by navigating through an environment with obstacles. MSR robots present the additional requirement of reconfiguring into the appropriate shape as the task progresses.

Figure 4 shows a group of modules reconfiguring into a snake and moving towards a goal location. As it encounters obstacles, it reconfigures to be able to turn and navigate them, finding a narrow passage to squeeze through, again as a snake, and finally reaches the target. From the point of view of control, the task requires three distinct behaviors: Reconfiguration, Locomotion (including turning) and Path Planning (with obstacle avoidance).

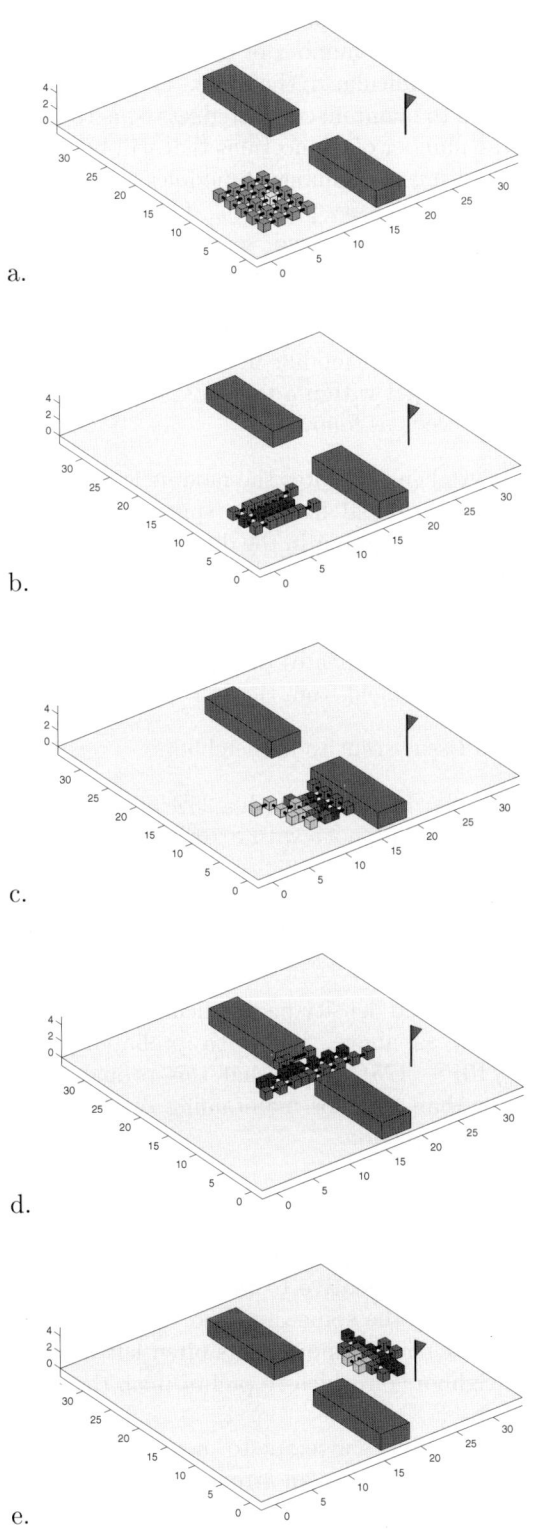

Figure 4: Reconfiguration and movement toward a goal. Module colors represent different modes as described in the text.

The remainder of this paper describes in detail the local control rules required by these behaviors. Although not presented in this paper due to space constraints, this approach has also been used to achieve 3D manipulation of an object fully surrounded by hundreds of modules. This kind of "internal" robot manipulation is a unique capability of the TeleCube module design.

4.1 Modes and Scents

The behaviors are handled via four FSM states or modes (**undefined**, **tissue**, **seed**, and **spine**) and seven scents (assignment, reset, spine-direction, tissue, obstacle, suggested-move and distance-from-spine). In Figure 4, the modules are colored according to their mode: dark gray for **undefined**, light gray for **tissue** modules beyond their distance threshold from the **spine**, red for **tissue** within the distance threshold, blue for **spine** and yellow for **seed**.

A module performs the actions below before any of its current mode-specific actions.

- If an assignment scent is received with a greater priority than the present mode, switch to that mode (mode priorities are, from low to high: **undefined**, **tissue**, **seed**, **spine**).

- If a reset message is received, transition to the **undefined** mode and end turn.

- Propagate a decayed version of any tissue and obstacle scents received to neighbors.

- Propagate spine-direction scents to neighbors.

4.2 Reconfiguration

The Snake configuration is not a set assignment of module positions, but rather a configuration with a given width and height.

The reconfiguration begins when one node is chosen at random to be a **seed** (the yellow module in Figure 4a). This choice does not need to be internal, but could also be triggered by an outside stimulus.

The **seed** module acts as a catalyst to form a "spine". The spine is a straight line of modules oriented along one of the three global axis, serving as the center of the snake structure. The **seed** module, which immediately turns into a **spine** module, serves as the first module in the spine. The spine then spreads out from the original module in the two directions parallel the spine direction. In addition, the

spine modules emit scents in the perpendicular directions, where receiving modules switch to the **tissue** mode and keep propagating the spine scent gradient. The **tissue** modules try to move in towards the spine as if attracted by a magnetic field. The process of modules following the gradient towards the spine forms a thinner and thinner snake structure and terminates when all of the tissue nodes are within a given threshold from the spine (less than 3 modules wide). Figures 4a and b show the snake being formed from an original square configuration of the modules, but any initial shape could have been used.

The rules for each mode are as follows:

Seed Modules: The module in the **seed** mode catalyses the formation of the spine. Its rules are, in order, as follows:

- If there are no neighbors perpendicular to the direction of the spine but there is a neighbor in the direction opposite the spine direction, send a seed assignment scent to that neighbor, transition to undefined state, and end turn.

- Emit spine assignment scents in directions parallel to the spine direction.

- Emit tissue assignment scents in directions perpendicular to the spine direction.

- Transition to the **spine** mode.

Spine Modules: A module in the **spine** state makes up part of the line of modules forming the spine. A module in this mode proceeds as follows:

- If a suggested-move impulse is received, move in that direction.

- Verify that the spine is not misaligned by checking that no neighbor in a perpendicular spine direction is in the **spine** mode. If there is such a neighbor transition both modules to the **undefined** mode and end the turn.

- Emit spine assignment scents in directions parallel to the spine direction.

- Emit tissue assignment scents in directions perpendicular to the spine direction.

- Maintain uniform density as follows: Check the distances to neighbors along the spine directions by taking the sum of the two arm lengths. If the module is farther from or closer to a neighbor than a distance of one fully extended arm, then move towards or away from the neighbor respectively with a probability of $1 - ((N_c/8) + .75)$ where N_c is the number of the module's connections perpendicular to the spine. This probability is biased to maintain connectedness by accounting for the number of connections that will be broken and limiting the number of modules moving each turn due to density problems. In the event that a movement is triggered, send a suggested-move message to all neighbors that will be disconnected from before moving, and then move.

Tissue Modules: A module in the **tissue** state simply attempts to get within a threshold distance of the spine. It proceeds as follows:

- If the gradient indicates the module is within the distance threshold from the spine and the distance to the neighbor in the direction towards the **spine** is zero, then with a 10% probability, try to join the spine.

- If the gradient indicates the module is within the distance threshold from the spine, end turn.

- Emit a tissue scent to all neighbors.

- Move toward the spine with a 25% probability, or make a random move with a 10% probability.

- Maintain good density along the spine direction as described in the **spine** mode to maintain uniform density.

- Maintain good density perpendicular to the spine direction as above, but with probability $1 - ((N_c/10) + .475)$. Note that this probability is greater than that for maintaining density along the spine direction.

Experimentally it was found that reconfiguration moved faster and more robustly when there was a large number of density movements perpendicular to the spine direction relative to the number of density movements along the spine direction. The main reason being that parallel movements often left modules without neighbors on which to pull or push their way to the spine.

A **tissue** module's attempt to move closer to the spine can be visualized as an attempt for a module to move down a level, where the levels consist of parallel lines of modules. This move turns out to be both complex and very common in the snake reconfiguration. Therefore the motion was given as a set sequence of actions that includes movements, communications

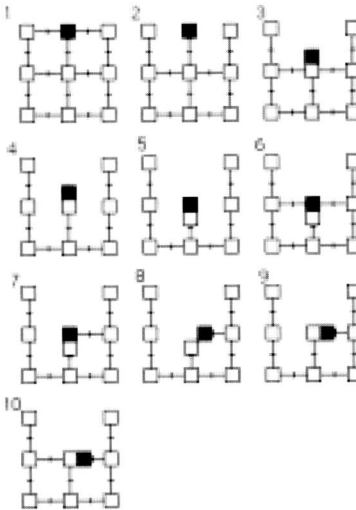

Figure 5: Sequence of actions needed to move down by a row.

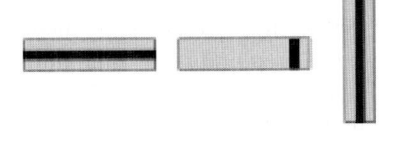

Figure 6: Snake with old spine direction (left), before reconfiguration (center) and after reconfiguration (right).

with neighbors, and forced moves. The sequence is shown in Figure 5 for the darkened module trying to move into the row below it.

Reconfiguration continues until the tissue scent is no longer detected.

4.3 Locomotion

Once the modules have reached a snake configuration, locomotion of the snake structure begins (Figures 4b-e). Specifically, the locomotion is triggered by the lack of the tissue scent. Thus, when all of the modules have moved within the distance threshold, the scent decays and movement begins. Locomotion along the direction of the spine is accomplished through the use of a movement scent. The foremost module in the direction of motion that is in the spine state acts as the leader for the structure. It initiates the movement by sending all of its neighbors movement scents. When a module receives it, it will move if possible, and in turn propagate the scent to its neighbors. After responding to movement scent, a module will wait a set number of time steps, until its "energy level" has returned to normal, before responding to the next one. The energy cost of a move is set at three, and the module recovers one energy unit per turn. This delay is used to insure the modules move in waves, resulting in a centipede-like gait of expansions and contractions.

Turning is accomplished by changing the location and direction of the spine and reconfiguring into a new snake structure around the new spine (Figure 6). A turn begins when any module sends out a reset scent. The reset scent carries with it a direction of the new spine and also information on where the new **seed** module should be relative to the module that released the reset. Once the new **seed** receives the reset scent it transitions to the **spine** mode and begins forming the new spine. All of the other modules transition to the **undefined** state and wait for assignment scents from the spine. Upon completion of the reconfiguration, a snake is formed pointing in the new direction. Because the new spine can only form at angles perpendicular to the old spine, the snake is limited to making right angle turns only.

4.4 Path Planning

In the example of Figure 4, the robot has to navigate its way around two obstacles and the sourrounding walls. In order to make it to the goal flag, it has to detect an opening between the obstacles and squeeze past it. We assume the modules have contact sensors to detect when an obstacle is hit. Also, we assume perfect dead-reckoning for localization purposes. To plan a path towards a goal location, the modules track the moves they have made from their original coordinates. In this way, they can determine their global position and relative position to the given goal location.

The modules use local, reactive path planning. At the beginning of the simulation, modules are only given information about their coordinates and the location of the goal module. At the onset of locomotion, the modules are arrayed in a snake configuration along a spine that is oriented in a given direction. The foremost **spine** module in the spine direction becomes the effective "head" of the snake, handling the path planning decisions. This module's control is limited to sending movement impulses and reset impulses. Movement impulses force the modules in the snake to take a step in the spine direction. Reset impulses cause the modules to reset and reconfigure along a new spine direction specified in the reset. Thus these two signals can be viewed as step and turn commands.

At its simplest level, path planning consists of repeated attempts to minimize distance to the goal object. If taking a step in the spine direction will serve to decrease this distance then the path planning will take a step in that direction. Otherwise, the snake will make a right-angle turn in the direction that will get it closer to the goal object. When to make a turn is determined by the alignment of the module that is going to be the new **seed** module (the second **spine** module from the front) with the object. Thus moving towards a goal at its simplest involves two directions and a single right-angle turn.

Obstacles in the environment pose additional constraints. When the snake hits an obstacle in the direction of travel it will take a right-angle turn in the direction that will move it closer to the goal. It will then continue in this direction until it hits another obstacle (such as a wall), or aligns itself with the obstacle and waits for an opening it can turn into (see Figure 4c,d). In cases where the goal is directly behind an obstacle, the snake should move away from the obstacle until it can turn around it. Using an obstacle scent the snake tracks where the **tissue** modules on the outside of the spine are touching the obstacle. The snake will only try to turn towards the goal when there is no longer an obstacle in the direction that it wants to turns in. Thus the robot will follow a wall if it hits one.

This approach to dealing with obstacles brings with it the complexity of determining when the snake is truly able to turn. Factors such as the decay rate of the obstacle scent and the contour of the obstacle can lead to undesirable behaviors such as repeatedly trying to turn into the obstacle as the snake moves along its surface and overshooting an opening.

Local path planning has the drawbacks of local minima. A prime example is the case of an L shaped obstacle. If the goal is near the vertex of the L and the snake is inside the L, it will repeatedly turn between two directions to try to minimize the distance to the goal. This can be prevented by adding a timeout, after which the snake performs a random move. Because the same module may not be doing the path planning each time, this knowledge must be shared among all modules in the snake.

5 Conclusions

This paper presents an approach to the control of modular self-reconfigurable robots inspired by social insects. We show how the use of simple local rules to switch between "modes" allows completion of a complex task, involving reconfiguration, locomotion and navigation in an environment with obstacles.

Complex interactions are produced through the propagation of information and local action guided by simple rules. This approach replaces an emphasis on global planning and centralized control with distributed behavior, local control and communications. We show results in a simulation of TeleCube, a robot developed at Xerox PARC. The control primitives can easily be scaled up in the number of modules, and down in the size of individual modules, and are general enough to fit most modular robot designs.

As the results in this paper and [1] show, the social insect programming metaphor is proving a viable approach to the tremendous growth of complexity in software and control, of which modular robotics is a case in point.

References

[1] H. Bojinov, A. Casal and T. Hogg, "Emergent Structures in Modular Self-Reconfigurable Robots", Proc. ICRA2000.

[2] Yim, M., "Locomotion With A Unit-Modular Reconfigurable Robot", Stanford University PhD Thesis, 1994.

[3] Yim, M., Lamping, J, Mao, E., Chase J.G., "Rhombic Dodecahedron Shape for Self-Assembling Robots", Xerox PARC, SPL TechReport P9710777, 1997.

[4] Murata S., Kurakawa, H., Kokaji, S., "Self-Assembling Machine", Proc. of IEEE ICRA'94.

[5] Murata S. , Kurakawa, H., Yoshida, E., Tomita, K., Kokaji, S., "A 3-D Self-Reconfigurable Structure", Proc. IEEE ICRA'98.

[6] Kotay, K., Rus, D., Vona, M., McGray, C., "The Self-reconfiguring Robotic Molecule: Design and Control Algorithms", Algorithmic Foundations of Robotics, 1998.

[7] Rus,D., Vona,M., "Self-reconfiguration Planning with Compressible Unit Modules", Proc. IEEE ICRA'99.

[8] Pamecha, A., Ebert-Uphoff, I., Chirikjian G.S.,"Useful Metrics for Modular Robot Motion Planning", IEEE

Transactions on Robotics and Automation, Vol.13, No.4, 1997.

[9] Casal, A., Yim, M. "Self-Reconfiguration Planning for a Class of Modular Robots", Proc. SPIE Symposium on Intelligent Systems and Advanced Manufacturing, Sept. 1999, Vol.3839.

[10] K. Hosokawa, T. Tsujimori, T. Fujii, H. Kaetsu, H. Asama, Y. Kuroda, I. Endo: "Self-Organizing Collective Robots with Morphogenesis in a Vertical Plane", Proc. IEEE ICRA'98.

[11] Bennett,F.H., Rieffel,E.G., "Design of Decentralized Controllers for Self-Reconfigurable Modular Robots Using Genetic Programming".

[12] Lipson, H. and J. B. Pollack, "Towards Continuously Reconfigurable Self-Designing Robotics" in *Proceedings of the 2000 IEEE International Conference on Robotics and Automation*, 2000.

[13] Hackwood, S., Beni G., "Self-organization of Sensors for Swarm Intelligence", Proc. IEEE ICRA'92.

[14] M. Resnick, *Turtles, Termites and Traffic Jams, Explorations in Massively Parallel Microworlds*, The MIT Press, 1994.

[15] E. Bonabeau, M. Dorigo, G. Theraulaz, *Swarm Intelligence, From Natural to Artificial Systems*, Oxford University Press, 1999.

[16] Johnson, George, "Mindless Creatures Acting 'Mindfully' ", *New York Times* on the Web, March 23, 1999.

[17] Gordon, Deborah, "Ants At Work: How an Insect Society is Organized ", W.W. Norton and Company, 2000.

Control of a Suspended Load Using Inertia Rotors with Traveling Disturbance

Yasuo Yoshida

Department of Mechanical Engineering
Chubu University
1200 Matsumoto-cho, Kasugai-shi, Aichi 487-8501, Japan
yyoshida@isc.chubu.ac.jp

Abstract

Rotational control and swing suppression of a crane suspended load model with traveling disturbance are studied. A rotational free rigid body suspended by a single rope is controlled using three inertia rotors. The end-supporting-point of the single rope is forced to be traveled as disturbance. Control angles and angular velocities are derived from measured data of fiber optic gyros installed on the suspended load. Rotor angular velocities controlling the suspended load are obtained by integrating the computed digital sliding-mode control feedback accelerations based on the coupling system's dynamics. Experiments and simulations investigate simultaneous control of the load's rotational orientation and swing suppression for traveling disturbance.

1. Introduction

A crane suspended load in a construction work is frequently rotated and swung by wind pressure or inertia force accompanied with movement of the crane. Most studies to control the suspended load have focused on suppressing swing of the load as a lumped mass and manipulating the crane. For example, Sakawa and Nakazumi [1] treated a rotary crane and Lee [2] discussed the simultaneous traveling and traversing trolley. The suspended load is actually a rigid body different from above-mentioned a lumped mass model and possible to be rotated. Kanki et al. [3] developed an active control device using gyroscopic moment to control the rotation of a crane suspended load.

There are few studies to control both rotation and swing of the suspended load as a rotational free rigid body. Yoshida and Mori [4] studied simultaneous control of rotation and swing of a pendulum having a rotational free rigid body using inertia rotors and recently Yoshida and Yajima [5] studied for a rotational free rigid body model suspended by a single rope with initial swing disturbance.

This paper presents rotational control and swing suppression of a suspended load model with traveling disturbance. The rotational free rigid body is suspended by a single rope and controlled using three inertia rotors, where the end-supporting-point of the single rope is forced to be traveled as disturbance. The suspended load as a rigid body has non-actuating (passive) three-degree-of-freedom and three inertia rotors have individually actuating (active) one-degree-of-freedom. The model system has six-degree-of-freedom and passive degrees of freedom are controlled by active degrees of freedom based on the coupling system's dynamics.

2. Dynamic Model and Controller

Figure 1 shows static state of the suspended load model, where the coordinate $\hat{x}_0\hat{y}_0\hat{z}_0$ is nonmoving base frame and $\hat{x}\hat{y}\hat{z}$ is moving frame fixed with the load. The rope-end-point travels to $\hat{x}_0$ direction. Three inertia rotors installed on the load. The purpose of this study is to control the motion of the load using inertia rotors. Fig.2 shows moving state, where the suspended load is represented with the rope-end-point displacement x, zxy Euler angles $\theta_1, \theta_2, \theta_3$ and zxz Euler angles $\bar{\theta}_1, \bar{\theta}_2, \bar{\theta}_3$. Angular velocities of the load are $\omega_x, \omega_y, \omega_z$ with respect to $\hat{x}, \hat{y}$ and $\hat{z}$ axes.

Angles and angular velocities of inertia rotors are

$\theta_4, \theta_5, \theta_6$ and $\omega_z + \dot\theta_4, \omega_x + \dot\theta_5, \omega_y + \dot\theta_6$, and torques are τ_1, τ_2, τ_3 with respect to $\hat{z}, \hat{x}$ and $\hat{y}$ axes.

Based on zxy Euler angles $\theta_1, \theta_2, \theta_3$ of the suspended load and $\theta_4, \theta_5, \theta_6$ of inertia rotors, the equation of motion can be written as

$$\begin{bmatrix} M_{cc} & M_{cf} \\ M_{fc} & M_{ff} \end{bmatrix} \begin{bmatrix} \ddot{q}_c \\ \ddot{q}_f \end{bmatrix} + \begin{bmatrix} h_c \\ h_f \end{bmatrix} + \begin{bmatrix} X \\ 0 \end{bmatrix} \ddot{x} = \begin{bmatrix} 0 \\ \bar{\tau} \end{bmatrix}, \quad (1)$$

where

$$M_{cc} = \begin{bmatrix} m_{11} & m_{12} & m_{13} \\ m_{12} & m_{22} & 0 \\ m_{13} & 0 & m_{33} \end{bmatrix}, \quad M_{cf} = M_{fc}^T = \begin{bmatrix} m_{14} & m_{15} & m_{16} \\ m_{24} & m_{25} & 0 \\ 0 & 0 & m_{36} \end{bmatrix},$$

$$M_{ff} = \begin{bmatrix} I_{R3} & 0 & 0 \\ 0 & I_{R1} & 0 \\ 0 & 0 & I_{R2} \end{bmatrix},$$

$q_c = [\theta_1\ \theta_2\ \theta_3]^T$, $q_f = [\theta_4\ \theta_5\ \theta_6]^T$,

$h_c = [h_1\ h_2\ h_3]^T$, $h_f = [h_4\ h_5\ h_6]^T$, $X = [x_1\ x_2\ x_3]^T$,

$\bar{0} = [0\ 0\ 0]^T$, $\bar{\tau} = [\tau_1\ \tau_2\ \tau_3]^T$.

q_c shows controlled (non-actuating) angle state vector and q_f free (actuating) angle state vector. Mass matrix $M_{cc}, M_{cf}(=M_{fc}^T)$ are function of θ_2, θ_3. h_c, h_f are vector of Coriolis, centrifugal and gravity terms and function of $\theta_2, \theta_3, \dot\theta_i (i=1\sim 6)$. X is vector of traveling disturbance and function of $\theta_1, \theta_2, \theta_3$.

The system equation of controlled zxy Euler angle's state vector q_c is linearized by using the control input vector u_c as

$$\ddot{q}_c = u_c = [u_1\ u_2\ u_3]^T. \quad (2)$$

A digital sliding-mode controller by which the controlled angles $\theta_1, \theta_2, \theta_3$ of the suspended load follow the desired values $\theta_{1d}, \theta_{2d}, \theta_{3d}$ is designed from equation (2). Digital state equation of control angle error is obtained using sampling time T and time steps $k(=1,2,\cdots)$ as

$$E_{i,k+1} = pE_{ik} - q\bar{u}_{ik} \quad (i=1\sim 3) \quad (3)$$

$$E_{ik} = \begin{bmatrix} e_{ik} \\ \dot{e}_{ik} \end{bmatrix}, \quad p = \begin{bmatrix} 1 & T \\ 0 & 1 \end{bmatrix}, \quad q = \begin{bmatrix} 0.5T^2 \\ T \end{bmatrix}$$

$e_{ik} = \theta_{idk} - \theta_{ik}$, $\dot{e}_{ik} = \dot\theta_{idk} - \dot\theta_{ik}$, $\bar{u}_{ik} = u_{ik} - \ddot\theta_{idk}$

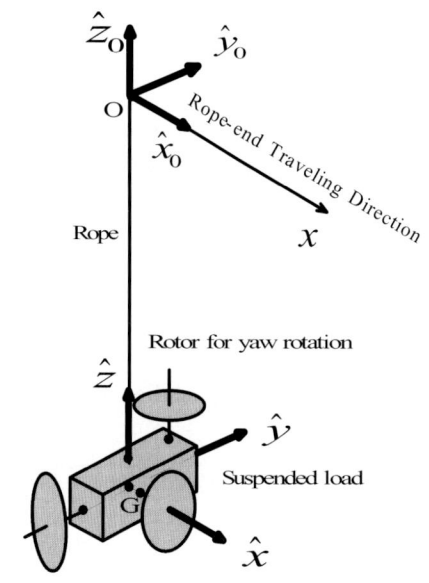

Fig.1 Suspended load model using inertia rotors with traveling disturbance.

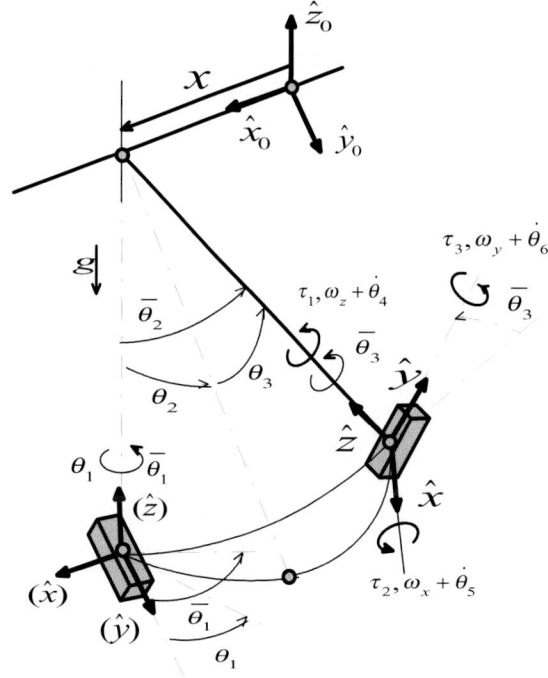

Fig.2 Coordinates and angles of the suspended load model.

Switching function σ_{ik} is defined with scalar vector $S = [s_1 \ s_2]$ as follows

$$\sigma_{ik} = SE_{ik} \quad (i = 1 \sim 3), \tag{4}$$

where, $\sigma_{ik} = 0$ gives switching line. Sliding-mode control input $\overline{u}_{ik}$ can be represented as follows, where $\overline{u}_{eqik}$ is equivalent input on switching line and $\overline{u}_{nlik}$ is nonlinear input acting to reach switching line,

$$\overline{u}_{ik} = \overline{u}_{eqik} + \overline{u}_{nlik}, \tag{5}$$

$$u_{eqik} = (Sq)^{-1} S(p - I) E_{ik},$$

$$u_{nlik} = \eta (Sq)^{-1} \sigma_{ik} \quad for \quad 0 < \eta < 2.$$

From equation (2) to (5), sliding-mode controller is obtained as

$$u_{ik} = K_1 e_{ik} + K_2 \dot{e}_{ik} + \ddot{\theta}_{idk} \ (for \ i = 1 \sim 3), \tag{6}$$

$$K_1 = \frac{(\eta/T)}{0.5T + (s_2/s_1)}, K_2 = \frac{1 + (\eta/T)(s_2/s_1)}{0.5T + (s_2/s_1)}.$$

Estimated accelerations of inertia rotor's angles q_f are taken as

$$\ddot{q}_f = \hat{u}_f = [\hat{u}_{4k} \ \hat{u}_{5k} \ \hat{u}_{6k}]^T. \tag{7}$$

After putting equation (2), (7) into equation (1), upper part of the equation of motion (1) becomes as

$$M_{cc} u_c + M_{cf} \hat{u}_f + h_c + X\ddot{x} = \overline{0}. \tag{8}$$

Above equation (8) shows the dynamic coupling between controlled angles q_c and free angles q_f. Then, estimated accelerations $\hat{u}_f$ are given as

$$\hat{u}_f = -M_{cf}^{-1}(M_{cc} u_c + h_c + X\ddot{x}). \tag{9}$$

By integrating equation (9) with the sampling time T, manipulating velocities of inertia rotors $\dot{q}_f = \overline{v}_k = [v_{4k} \ v_{5k} \ v_{6k}]^T$ are obtained as

$$\overline{v}_k = \overline{v}_{k-1} + \hat{u}_f T. \tag{10}$$

Fiber optic gyros installed on the suspended load can measure the angles $\overline{\theta}_{13} (\equiv \overline{\theta}_1 + \overline{\theta}_3), \theta_2, \theta_3$ and the angular velocities $\omega_x, \omega_y, \omega_z$.

zxy Euler angles $\theta_1, \theta_2, \theta_3$ and zxz Euler angles $\overline{\theta}_1, \overline{\theta}_2, \overline{\theta}_3$ are mutually convertible. zxz Euler angles are obtained by measured angles as

$$\overline{\theta}_1 = \overline{\theta}_{13} - \overline{\theta}_3,$$
$$\overline{\theta}_2 = \cos^{-1}(C_2 C_3), \tag{11}$$
$$\overline{\theta}_3 = A\tan 2(-C_2 S_3, S_2).$$

Under the condition that θ_2, θ_3 are directly measured by fiber optic gyros, θ_1 must be calculated based on equation (11) as

$$\theta_1 = A\tan 2(\overline{C}_1 \overline{S}_3 + \overline{S}_1 \overline{C}_2 \overline{C}_3, -\overline{S}_1 \overline{S}_3 + \overline{C}_1 \overline{C}_2 \overline{C}_3), \tag{12}$$

where $\overline{C}_i$ and $\overline{S}_i$ are defined as $\cos\overline{\theta}_i$ and $\sin\overline{\theta}_i$ for $i = 1, 2, 3$, respectively.

Angular velocities $\dot{\theta}_1, \dot{\theta}_2, \dot{\theta}_3$ are obtained using $\omega_x, \omega_y, \omega_z$ and θ_2, θ_3 as

$$\begin{bmatrix} \dot{\theta}_1 \\ \dot{\theta}_2 \\ \dot{\theta}_3 \end{bmatrix} = \begin{bmatrix} -S_3/C_2 & 0 & C_3/C_2 \\ C_3 & 0 & S_3 \\ T_2 S_3 & 1 & -T_2 C_3 \end{bmatrix} \begin{bmatrix} \omega_x \\ \omega_y \\ \omega_z \end{bmatrix}, \tag{13}$$

where T_2 is $\tan\theta_2$. Fig.3 shows the control system.

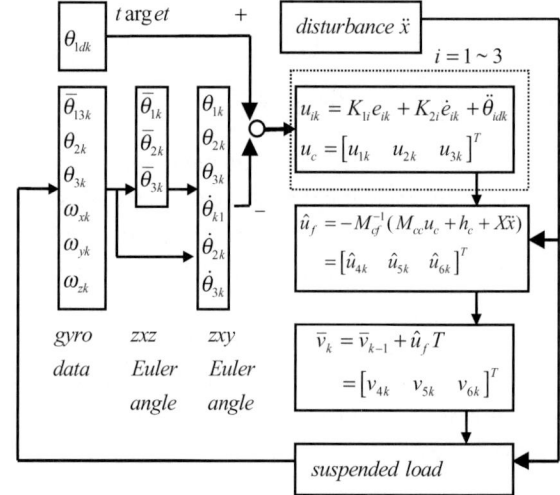

Fig.3 Control system.

Simulation is performed for the sampling time between t (=step k) and t+T (=step k+1). Following simultaneous differential equations are solved using Runge-Kutta numerical integration.

$$\begin{bmatrix} \ddot{q}_c \\ \ddot{q}_f \end{bmatrix} = \begin{bmatrix} -M_{cc}(M_{cf} \overline{u}_f + h_c + X\ddot{x}) \\ \overline{0} \end{bmatrix}, \tag{14}$$

where $\overline{u}_f = (\overline{v}_k - \overline{v}_{k-1})/T$.

Since the manipulating velocities are held and constant at the step k, angular accelerations of free angle $\ddot{q}_f$ are zeros in the time between sampling time T.

However, the manipulating velocities change at each time step and therefore accelerations of free angle are approximately taken as $\bar{u}_f$ that is the change of the manipulating velocity between k-1 and k steps. $\bar{u}_f$ assures the dynamic coupling in the simulation.

3. Experimental Results

Fig.4 shows the experimental suspended model photograph. Length between support point and the load's center of gravity is $l = 1.22m$, mass of the load including inertia rotors $m = 20.9kg$, moments of inertia of the load $I_x = 1.36 kgm^2$, $I_y = 0.676 kgm^2$, $I_z = 1.17 kgm^2$ and inertia rotors $I_{R1} = I_{R2} = I_{R3} = 0.0377 kgm^2$. Maximum angular velocities of servomotors driving inertia rotors are $\dot{\theta}_4 = 173 rpm$, $\dot{\theta}_5 = 184 rpm$, $\dot{\theta}_6 = 168 rpm$. Sliding-mode digital gains are $K_1 = 80, K_2 = 16$ using sampling time $T = 0.05 \sec$ and $\eta = 0.5, s_2 / s_1 = 0.1$, selected from simulation results using $\eta = 0 \sim 2$, $s_2 / s_1 = 0 \sim 10$. And, digital poles of closed system obtained from equation (3) and (6) using these digital gains are $z = 0.5$ and 0.6, therefore the control system is stable. Simulation is used by Runge-Kutta numerical integration in time by steps of size $0.005 \sec$. Fig.5 shows the linear traveling of the end-supporting-point of the single rope as disturbance for the suspended load. Fig.5(a) is traveling displacement moving $0.25m$ within $1.5 \sec$. Fig.5(b) shows velocity and acceleration, that is, accelerating to 0.5 sec, uniform velocity to 1.3 sec and decelerating to 1.5 sec. Initial angles of the suspended load are manually positioned and it is difficult to give the accurate same initial angles to the experiments of both without and with controls. But the differences of initial zxy Euler angles are small between $\theta_1 = 25°$, $\theta_2 = 1°$, $\theta_3 = 0°$ in the case of experimental values without control and $\theta_1 = 27°$, $\theta_2 = 0°$, $\theta_3 = 0°$ with control. Zero desired values $\theta_{1d}, \theta_{2d}, \theta_{3d}$ are given for controlled angle $\theta_1, \theta_2, \theta_3$.

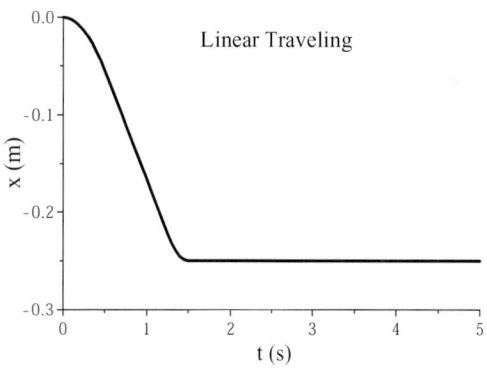

(a) Traveling displacement.

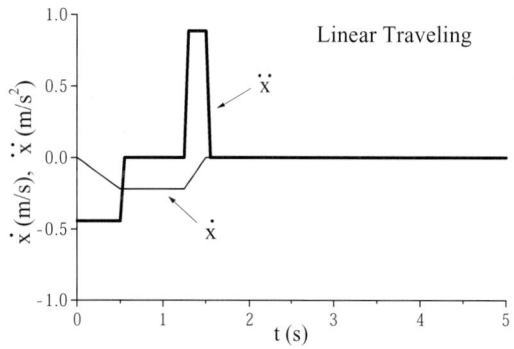

(b) Traveling velocity and acceleration.

Fig.5 Linear traveling disturbance.

Fig.4 Experimental suspended model's photograph.

Fig.6 shows the experimental time response of controlled rotational angle θ_1 within 30 sec. Thin line indicates the case of without control and thick line with control, and the following figures show in similar ways. Thin line of without control moves from initial angle $25°$ to maximum $-66°$ at 12 sec and thereafter largely fluctuates. This phenomenon is explained by Yoshida and Mori [5] that both centrifugal force caused by swing and the difference between moment of inertia I_x and I_y generate yaw rotation torque. Thick line of with control moves rapidly from initial angle to zero desired value within 2 sec and hereafter keep desired value.

Fig.7 shows the experimental swing angle's components. Fig.7(a) shows controlled angle θ_2. Thin line of without control enlarges vibration amplitude to maximum $6°$ between 5 sec to 15 sec and and thick line with control keeps within $1.5°$ value from initial time to 30 sec. Fig.7(b) shows controlled angle θ_3. Both cases of without and with control are same till the time 15 sec, but after that time thin line of without control remains $4°$ constant vibration amplitude and thick line with control damps to zero desired value.

Fig.8 shows the response of swing angle (zxz Euler angle) $\bar{\theta}_2$. The amplitudes in the case of with control are suppressed under half of those of without control.

Fig.9 shows swing displacement x, y for the fixed coordinate when the load is at rest. The vibration displacement x is induced by travel disturbance.

Thin line of x without control leaves vibration amplitude at 30 sec, but thick line of x with control damps to zero desired value.

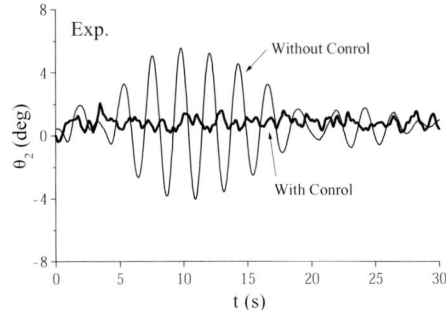

(a) Time response of θ_2.

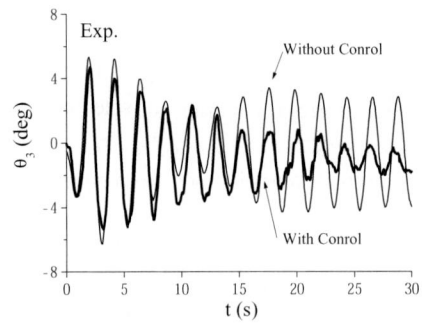

(b) Time response of θ_3.

Fig.7 Experimental swing angle's components.

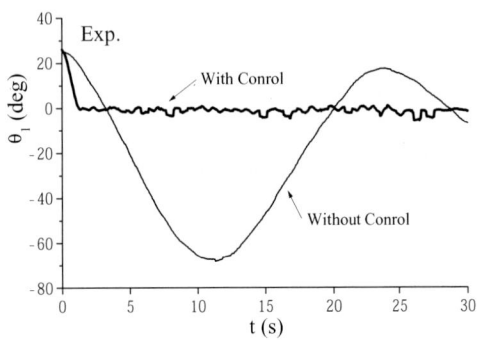

Fig.6 Experimental time response of rotational angle θ_1.

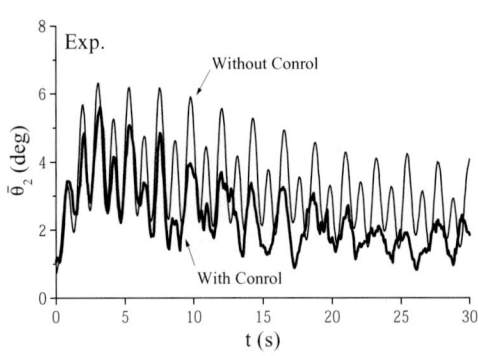

Fig.8 Experimental swing angle $\bar{\theta}_2$.

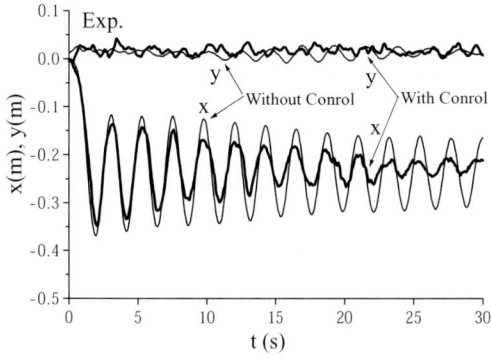

Fig.9 Experimental swing displacements x, y.

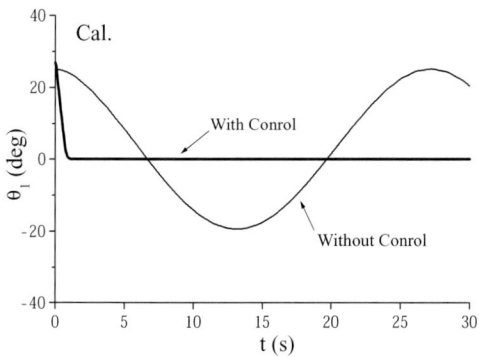

Fig.10 Simulation time response of rotational angle θ_1.

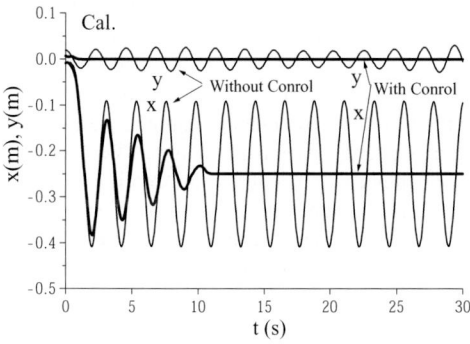

Fig.11 Simulation swing displacements x, y.

Fig.10 shows simulation time response of rotational angle θ_1 corresponding to experimental result of Fig.6. Torsional spring coefficient of the rope is considered as $0.1 Nm/rad$ in simulation. Fluctuation tendency of thin line without control shows the same pattern of experiment and thick line with control moves rapidly from initial angle to zero value same as experiment.

Fig.11 shows simulation swing displacements x, y corresponding to experimental result of Fig.9 and these are similar as experimental ones.

4. Conclusions

A rotational free rigid body suspended by a single rope model with traveling disturbance is controlled using three inertia rotors. Velocity-command-type control system is developed for inertia rotors by integrating the computed feedback accelerations of digital sliding-mode control, based on the coupling system's dynamics. The experimental and simulation results show that inertia rotors can control the rotation and swing of the suspended load for traveling disturbance.

References

[1] T. Sakawa and A.Nakazumi, "Modeling and Control of a Rotary Crane", ASME Journal of Dynamic Systems, Measurement, and Control, Vol.107, pp.200-205, 1985.

[2] H. H. Lee, "Modeling and Control of a Three-Dimensional Overhead Crane", ASME Journal of Dynamic Systems, Measurement, and Control, Vol. 120, pp.471-476, 1998.

[3] H. Kanki and Y. Nekomoto et al., "Development of Suspender Controlled by CMG (in Japanese with English abstracts)", Proceedings of JSME Dynamics and Design Conference '95, No.95-8 (1), B, pp.34-37, 1995.

[4] Y. Yoshida and K. Mori, "Simultaneous Control of Attitude and Swing of a Pendulum Having a Rotational Free Body (in Japanese with English abstracts)", Trans. of JSME, Series C, Vol.64, No.628, pp.4660-4665, 1998.

[5] Y. Yoshida and M. Yajima, "Control of a Suspended Load Using Inertia Rotors", Proceedings of the 1999 ASME Design Engineering Technical Conferences, Symposium on Motion and Vibration, DETC99/MOVIC-8418, pp.1-6, Las Vegas, 1999.

Variable Position Mapping Based Assistance in Teleoperation for Nuclear Cleanup

K.A. Manocha[1], N. Pernalete[2], and R. V. Dubey[2]

Abstract

Radioactive tank waste remediation and decontamination and decommissioning (D&D) of contaminated Department of Energy (DOE) facilities, and other nuclear cleanup tasks require extensive remote handling technologies. The unstructured nature of these tasks and limitations of the current sensor and computer decision-making technologies prohibit the use of completely autonomous systems for remote manipulation.

This paper presents a new methodology in which model-based computer assistance is incorporated into human controlled teleoperator systems.

This approach implies a form of assistance function in which the human input is enhanced rather than superseded by the computer. A specific task of cutting a pipe with a saw is chosen as an example to demonstrate the implementation of the assistance functions in D&D size reduction tasks and the results are presented.

1. Introduction

Environmental restoration and waste management challenges in the United States Department of Energy (DOE), and around the world, involve radiation or other hazards. Over the past 40 years, more than 600 waste transfer pits have been constructed for use in the transfer of liquid high-level waste. During the many years of operation, the pits have become contaminated with radioactive material from leaks and spills. The pits have also been used to accumulate worn out components and parts.

The DOE's office of Environmental Management created in 1994 the Tank Focus Area to develop tank waste remediation technologies. One of the needs this area has identified is the use of remote technology to enhance cleaning, decontamination, and reconfiguration operations in radioactive jumper pits used to place pumps and jumper lines for transferring waste. Current methods for modifying, operating, cleaning and decontaminating these pits are labor intensive and costly.

One of the high-priority teleoperation tasks identified as necessary in the tank pits is the *size reduction*, which consists of cutting up jumpers and angles using gripper held tools, such as band saws, shears or rotary wheel cutters.

Traditional teleoperated systems are difficult to operate and make simple manipulation operations tedious and time consuming, and thus, increase the costs and operator fatigue. In addition, these systems are highly dependent on the human operator for safety. The approach presented in this work involves a form of assistance that is provided by adjusting system parameters that are not under direct control by the operator, such as impedance parameters and workspace mappings between the master and slave manipulators. This strategy will enhance clean-up operations in this environment by reducing personal exposure and increasing the efficiency of task execution.

2. Background

A number of approaches to merge human decisions with computer assistance have developed in recent years including virtual constraint strategies that have been used to assist an operator in maneuvering a slave manipulator, such as those by Joly and Andriot [1] and Kosuge et al. [2]. In their work, virtual mechanisms have been used to constrain the manually controlled manipulator's motion on a desired surface or to be pulled into alignment with a task. Aigner [3] integrated potential field effects and remote control of a manipulator providing teleoperation assistance. Potential fields were used to produce velocity commands, which, when added to those generated by the input device, maneuvered the manipulator away from walls or around obstacles automatically.

The current work is an extension of work conducted by Chan [4], wherein experiments were conducted using sensor data to adjust stiffness and damping of the slave to suit various task requirements [5]. These ideas lead the investigation into the alteration of other human independent parameters in a telerobotic system, specifically the position and velocity mapping parameters between the master and slave manipulators. The operator uses an input device to control the motion of the manipulator. Information from sensors, such as force/torque, ultrasonic, range, and image processing, as well as available environmental models, is collected. The testing system for this took form of a telerobotic controller. Everett [6] implemented this controller on a testbed consisting of a seven-degree-of-freedom Robotics Research Corporation manipulator and a six-degree-of-freedom force-reflecting Kraft hand controller.

This work prompted the need for assistance algorithms, which use this available information to alter parameters, such as position and velocity mappings and dynamic parameters in impedance control implementations,

---

[1]K.A. Manocha is with Corning Inc.,Telecommunications Products, MS-5, 310 N. College Rd., Wilmington, NC 28405, USA Email: manochaka@corning.com [2] N. Pernalete and R.V. Dubey are with the Robotics and Intelligent Machines Laboratory, University of South Florida, ENB 118, 4202 E. Fowler Avenue, Tampa, FL 33620, USA. Email: dubey@eng.usf.edu and pernalet@eng.usf.edu.

on-line. The result is a passive form of assistance, which leaves the operator in control of the motion of the manipulator, but assisted to the extent that the sensor and model information may be relied upon.

3. Assistance Function Development

A. Assistance Concept Description

The underlying idea behind the assistance function concept is the generalization of position and velocity mappings between master and slave manipulators of a telerobotic system. This concept was conceived as a general method for introducing computer assistance in task execution without overriding an operator's command to the manipulator.

The assistance functions can be classified as regulation of position, velocity, contact force, and joint configuration of the manipulator. All of these assistance strategies are accomplished by modification of system parameters.

This research was focused on the implementation of the position assist strategy based on model information, although it can be used with the same type of information provided by sensors.

In the position assistance strategy, the mapping between master and slave velocities is linear, and thus there is a corresponding position mapping as well. The position assistance is achieved by scaling in which the slave workspace is enlarged or reduced as compared to the master workspace. The mapping from master to slave manipulators is done by virtual linear fixtures and virtual planar fixtures based position functions. In the linear and planar virtual fixtures, the motion of the manipulator is constrained to lie along a given line or plane, respectively. This type of assistance can be useful in drilling and sawing tasks, where the manipulator must maneuver a tool along a particular direction.

Before the assistance can be applied, a model of the environment around the desired object is made.

B. Creating the Model

Cutting is a common task performed in a pit, in which equipment like pipes, tees, and elbows of given lengths and diameters need to be cut. Before starting to cut, a model of the environment around the desired object to be cut is made. In the real system implementation [8], a robot manipulator (Schilling manipulator) with two stereo heads will be used to capture live images of the environment. These live images are converted to a static model using the Envision software. The cameras can be tilted and zoomed so that the required frame is captured. Once the frame is selected, the live image is made into a static one and the model is built on this static image.

These standard equipment dimensions are present in the Envision software [8] in a database called the Master Part list. If the part is not present, the part is added to the master part list.

The software overlaps the objects on the static image with parts from the Master Part List for all those cases where a close dimensional match is found. In this way, accuracy of model is maintained in spite of imperfect sensory mapping.

C. Manual Part Placement Calculations

One of the primary functions before the calculations begin is the manual modeling process, in which part models from a library of templates are placed in a virtual environment. Positions and orientations of base frames of the parts are calculated using the pan and tilt angles chosen through adjustment of the sensor head position by an operator and the distance measured by a laser range finder. The calculations carried out to achieve this result are described in this section.

In all three part-modeling processes, it is assumed that the operator places the laser point along the geometric centerline of the image of the pipe of interest. Accuracy of this position is determined by the operator's estimation ability and the resolution of the pan-tilt motion. The point that is placed is at the centerline of the image from the camera's viewpoint, while calculations assume that the point is at the centerline of the image from the viewpoint of the laser range finder, introducing some small error in calculations.

The following section describes the mathematical procedure necessary to calculate the base frame locations of each of the primary types of parts.

D. Pipe Placement

Coordinate frames associated with the placement of pipe sections are shown in Figure 1. In the manual modeling process for pipes, the operator is required to orient the pan-tilt head such that the laser spot is placed at the ends of the pipe section to be modeled. Each of the two spot locations chosen by the operator must be along the axis of the image of the pipe, regardless of the pipe's position or orientation.

In the first step the coordinates of each of the points chosen is calculated so that the vector connecting the two points, $^{PT}L_1$, may be specified. Ideally, this vector is parallel to the axis of the pipe, and the unit vector in this direction is used as the z-vector in the part surface frame, Σ_s. The cross product between this vector and the direction to the first point chosen, $^{PT}p_1$, defines a tangent to the pipe cross-section, which is used to determine the y unit vector of the pipe surface frame, Σ_s. The final vector necessary to define the base frame is the x unit vector, which is determined by taking a cross product of the z and y vectors and points toward the centerline of the pipe. The base frame is then determined by translating the surface frame along the x-direction a distance of half the pipe diameter so that the new base frame, Σ_p, lies on the centerline of the pipe.

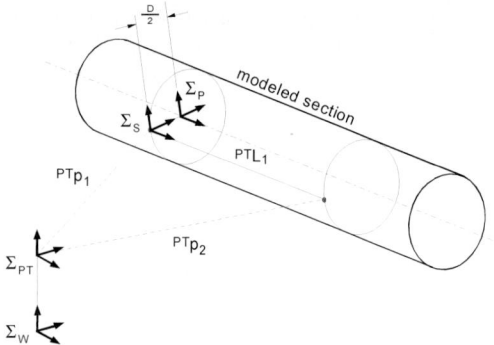

Figure 1: Pipe placement coordinate frames

E. Tee Placement

Coordinate frames associated with the placement of tees are shown in

Figure 2. In the manual modeling process for tees, the operator is required to orient the pan-tilt head such that the laser spot is placed at the end of the horizontal section, and on points somewhere along the pipes connected to the vertical and horizontal sections. As usual, each of the three spot locations chosen must be along the axis of the image of the cylinders.

In the first step the coordinates of each of the three points chosen is calculated so that the vector $^{PT}L_1$ along the pipe connected to the horizontal section of the tee may be specified. Ideally, this vector is parallel to the axis of the horizontal section of the tee, and the unit vector in this direction is used as the z-vector in the part surface frame, Σ_s. Another vector $^{PT}L_2$ is calculated using the direction between the point at the end of the tee horizontal section and the point along the vertical section of pipe. The cross product of vectors $^{PT}L_1$ and $^{PT}L_2$ is used to determine the y unit vector of the tee surface frame, Σ_s. The final vector necessary to define the base frame is the x unit vector, which is determined by taking a cross product of the z and y vectors and points toward the centerline of the horizontal section of the tee. The base frame is then determined by translating the surface frame along the x-direction a distance of half the tee diameter so that the new base frame, Σ_p, lies on the centerline of the horizontal section of the tee.

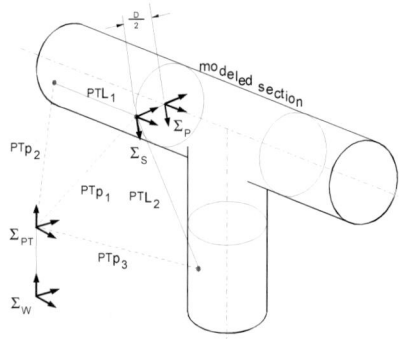

Figure 2: Tee placement coordinate frames

F. Virtual Planar Fixture Based Position Function

This is a position assist function, which ensures the movement of the end-effector in a specified plane. One of the applications of this assistance function is cutting a pipe with a saw.

This assistance function when used in the cutting operation ensures that the cutting tool like the band saw moves only in the plane in which cutting is desired. This function works by scaling by a nominal factor the velocity vector in the desired plane, and by a small factor the velocity perpendicular to the desired cutting plane.

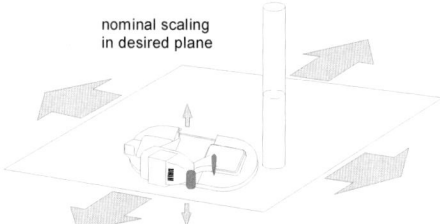

Figure 3: Virtual Planar Fixture Based Function

G. Definition of the Task

The task of cutting a pipe can be divided into subtasks such as: moving the saw towards the pipe, cutting the pipe, and finally returning the saw back to base.

The specific subtask of cutting a pipe with a saw will be used as the nominal task to demonstrate the use of assistance functions. An operator working from a remote location achieves this task, where the data about the position and orientation of the pipe and possible obstacles are available through the previously built model.

H. Function Elaboration

The object to be cut (pipe) is selected by the operator. As all the dimensions are known, direction ratio of the pipe axis (a_3, b_3, and c_3) is available to the model. Also known are the co-ordinates of a point on the pipe axis (x_R, y_R, z_R). The equation of the pipe axis can thus be defined.

Using the model, the operator selects the cutting point on the axis surface, represented by x_2, y_2, and z_2. The point of intersection of the perpendicular between the cutting point and the pipe axis, x_1, y_1, and z_1 is the origin of the cutting plane frame. The pipe axis is the Z-axis and the line joining the two points is the X-axis. Y-axis is defined based on X and Z-axes.

The General equation of the pipe axis is:

$$\frac{(x-x_R)}{a_3} + \frac{(y-y_R)}{b_3} + \frac{(z-z_R)}{c_3} = K$$

Since x_1, y_1, z_1 lie on this axis:

$$x_1 = x_R - K a_3 \quad (1)$$
$$y_1 = y_R - K b_3 \quad (2)$$
$$z_1 = z_R - K c_3 \quad (3)$$

The plane perpendicular to pipe axis and passing through x_1, y_1, z_1 and x_2, y_2, z_2 is:
$$(x_2 - x_1)a_3 + (y_2 - y_1)b_3 + (z_2 - z_1)c_3 = 0 \quad (4)$$
Solving K from equations (5)-(8)

$$K = \frac{a_3(x_R - x_2) + b_3(y_R - y_2) + c_3(z_R - z_2)}{(a_3^2 + b_3^2 + c_3^2)} \quad (5)$$

The direction ratio of X-axis is defined as a_1, b_1, c_1 where
$$\begin{aligned} a_1 &= x_2 - x_1 \\ b_1 &= y_2 - y_1 \\ c_1 &= z_2 - z_1 \end{aligned} \quad (6)$$

Direction ratios of Y-axis can thus be calculated:
$$a_1a_2 + b_1b_2 + c_1c_2 = 0; \quad (7)$$
$$a_3a_2 + b_3b_2 + c_3c_2 = 0; \quad (8)$$

Let $r = a_2/c_2$ and $s = b_2/c_2$. From equations (7) and (8),
$$a_1r + b_1s + c_1 = 0; \quad (9)$$
$$a_3r + b_3s + c_3 = 0; \quad (10)$$

From equations (9) and (10):
$$s = \frac{(c_3a_1 - c_1a_3)}{(b_1a_3 - b_3a_1)} \quad (11)$$

$$r = \frac{(-c_1 - b_1s)}{a_1} \quad (12)$$

$$a_2 = r; \; b_2 = s; \; c_2 = 1; \quad (13)$$
$$suma_i = a_i^2 + b_i^2 + c_i^2 \quad (14)$$

The direction cosines are defined as:
$$a_{ij} = a_i / suma_i; \; b_{ij} = b_i / suma_i; \; c_{ij} = c_i / suma_i \quad (15)$$

The transformation matrix between the cutting and the base frame is:
$$^0R_c = \begin{bmatrix} a_{11} & b_{11} & c_{11} \\ a_{21} & b_{21} & c_{21} \\ a_{31} & b_{31} & c_{31} \end{bmatrix} \quad (16)$$

Let the velocity vector in the base frame be:
$$V_{BASE} = \begin{bmatrix} v_x \\ v_y \\ v_z \end{bmatrix} \quad (17)$$

Velocity in the cutting plane is then:
$$V_{CUTPLANE} = [^0R_c]^T * V_{BASE} \quad (18)$$

Since velocity components in the Z-axis of the cutting frame need to be scaled down, a scaling factor m between zero and 1 is selected. The scaling matrix is then:
$$SCALE = \begin{bmatrix} 1 & 0 & 0 \\ 0 & 1 & 0 \\ 0 & 0 & m \end{bmatrix} \quad (19)$$

The modified velocity in the cutting plane is then:

$$V_{SCALEDCUTPLANE} = SCALE * V_{CUTPLANE} \quad (20)$$

The modified velocity in the base frame is then:
$$V_{SCALEDBASE} = {}^0R_C * V_{SCALEDCUTPLANE} \quad (21)$$

While the previously described procedure addresses the basic algorithm for the cutting operation, the need for movement along the Z-axis and adjustable scaling becomes inevitable in the case of binding of the saw. Fault recovery for binding was addressed by providing adjustable scaling, which could be initiated when desired, by recalling the scale function matrix as in (19).

4. Simulation Experiments

The cutting operation requires the implementation of the Virtual Planar Position assistance function. This is the function tested using the software development environment GHOST by Sensable Technologies. The master input device used here is the six-degree-of-freedom PHANToM hand controller, which drives the slave end effector. GHOST provides a means for monitoring the PHANToM's position and rotation information over time, as well as for providing force feedback, graphics, and other functions. The developed application consisted of two objects: a saw and a pipe. Each of these is defined as a solid object within GHOST, and each can have its transformation matrix modified as desired.

The operator inputs the information needed to initialize the saw coordinate axes, in addition to the scaling parameters.

The simulation is run for a variety of scaling values, ranging from the ideal complete scaling of the velocity to non-scaled velocity, to ascertain the effects of scaling on the performance of the system. Though the simulation for the virtual linear fixture is not developed here, the scaling results would help predict the multiple constraining effects. The graphical display for this is shown in Figure 4

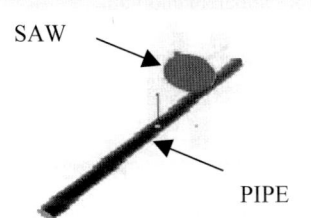

Figure 4: Window for the GHOST graphical simulation

In the figure, the disc represents the saw or the end effector (slave) position, and the cone represents the master. This display with both master and slave positions helps visualize the position and velocity of the slave with respect to the master at all times and thus allows a more comprehensive analysis. For all simulations, a diagonal orientation of the pipe was assumed to avoid any singularities due to the orientation.

The scaling is tested at three different levels, classified on degree of constraints (scale factor) applied, namely: No motion constraint (1), total motion constraint (0), and partial motion constraint (between 0 and 1).

5. Results

Depending on the scale factor chosen, several simulation experiments in the GHOST environment were run. Note that the plotted curves are a set of stacked values obtained from the inputs by three different operators, thus avoiding any operator errors.

No Motion Constraint: Here the scale factor is kept as 1, i.e. no scaling is applied along the pipe axis to scale any velocities in undesired directions (normal to the cutting plane, Z in this case). The effects are as seen in Figure 5 below.

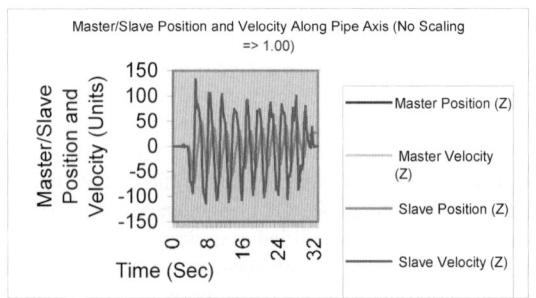

Figure 5: Position/velocity of master/slave for no scaling

These indicate the typical unwanted inputs along the pipe axis, which are encountered during a normal cutting operation. Though only the slave position and velocity is visible due to no scaling and thus, superimposing of identical graphs, it can be seen that the slave parameters can typically rise to high values, which is extremely undesirable in any task. Thus, the need for scaling of the velocity becomes apparent and vital.

Total Motion Constraint: The scale factor in this case is set to zero. The effects can be seen in Figure 6.

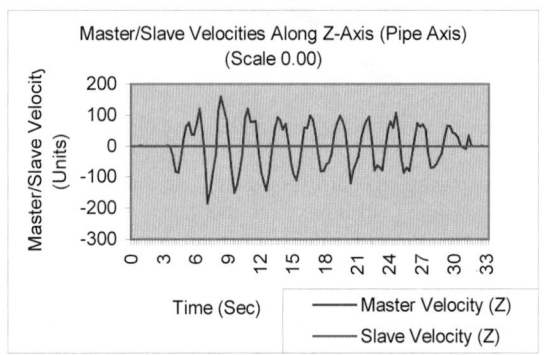

Figure 6: Velocities along the pipe axis for the master and slave for complete scaling

The velocity along the undesired directions is completely scaled. Though, this can be considered as an ideal case, this approach has certain limitations in practical implementation. The most prominent limitation is the case of binding of the saw. A minimal amount of movement is always desirable to allow the smooth cutting operation, which certainly cannot proceed along a complete straight line due to pipe material structure. The fault recovery mode is introduced by providing an adjustable scaling function, which can be initiated when desired. Following experiments support this need.

Partial Motion Constraint: As the need for optimal scaling is already apparent, the experiments conducted henceforth included minimal scaling of the velocities along the pipe axis. The operator before task execution optimally selects the scaling parameter, which is completely adjustable. The selection can be based on the task requirements and task environment constraints. This leaves the operator with complete control of the operation, which is clearly aided by the assist functions provided.

The Figures 7 and 8 demonstrate the master and slave position and velocity along the pipe axis respectively. It can well be observed that an adjustable scaling provides with the best results. Apart from the above-mentioned factors, this also assists the operator in maintaining precision along the cutting pipe direction.

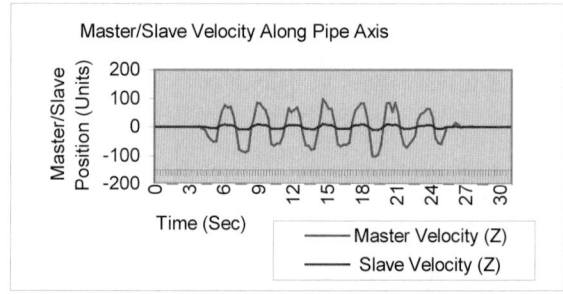

Figure 7: Master and Slave velocity along pipe axis for adjustable scaling.

Other intangible advantages include the reduction of the operator's fatigue and increased efficiency, due to shorter execution times and precise assisted alignment of the tool.

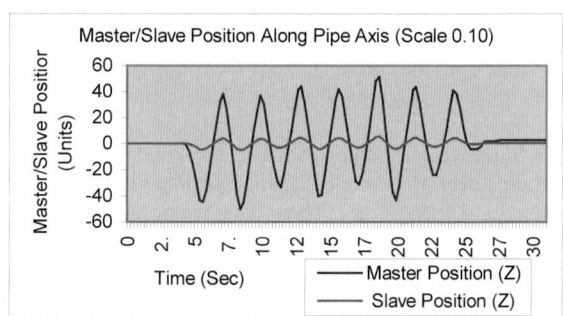

Figure 8: Master/Slave position along pipe axis for adjustable scaling.

Finally, sets of simulations were run for scaling values varying from 0 to 1 in steps of 0.05. In this case, two modes of operation are present. The first mode allows the operator to go "ballistic" in the workspace at a desired velocity whenever the saw is far from the pipe. The second mode is a "fine-tuning", which is implemented once the center of the saw is sensed to be close enough to the center of the pipe, as the cutting operation begins.

In this case, the times in which the mode is being switched to fine-tuning and the cutting operation is completed are recorded as "switching mode time" and "Final time" respectively.

The results can be observed in Figure 9, in which the need of scaling becomes even more apparent given the fact that total or no motion constraints resulted in longer execution times for the cutting task.

It is important to note that even though the research was focused on the position based assist function; a form of velocity based assist function was also introduced to guide the operator towards the pipe. This form was introduced during the "ballistic" mode of operation, and it was noticed that the time to reach the pipe at the cutting point was considerably shorter when the velocity based function assistance was provided.

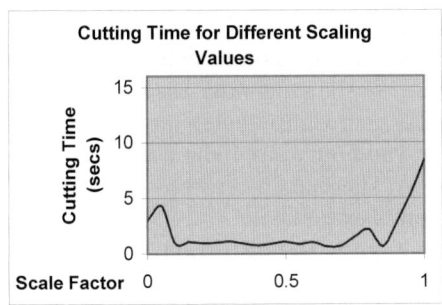

Figure 9: Cutting Time for several Scaling Values

6. Conclusions

General forms of assistance, which can be used and made available for specific Department of Energy (DOE) tasks in dismantlement and decommissioning (D&D) nuclear clean up operations, were analyzed and described.

Pipe cutting was the task that was identified for the demonstration of the assistance function strategy. The graphical simulation for the demonstration of the saw cutting operation was developed using the GHOST software development toolkit. The six-degree-of-freedom PHANToM hand controller was used as the input device for the simulation. The results from the simulation clearly indicate the efficacy of the assistance strategy.

It was evident that the assistance provided both tangible benefits such as precision in task execution, reduced times as well as intangible benefits such as reduced operator fatigue due to aided motion and reduced times.

In addition, a very commonly encountered issue of fault recovery of binding of the saw was also addressed by providing adjustable non-uniform scaling along the cutting axis. Moreover, it was observed that the adjustable scaling resulted in better performance as compared to complete motion constraining.

7. Future Work

Usually, the geometry of the task environment is highly unstructured and uncertain. Likewise, the precision and accuracy of the requisite geometric knowledge varies from task to task, as does the extent of the task space itself. A significant fraction of the tasks to be performed are complex by any standard. To address these issues, rotational motion assistance is being addressed. Encouraging results from assistance in translational motion further support this.

References

[1] Luc D. Joly and Claude Andriot, "Imposing motion constraints to a force reflecting telerobot through real-time simulation of a virtual mechanism," in Proceedings of the 1994 IEEE International Conference on Robotics and Automation, San Diego, CA, May 1994, pp. 357-362.

[2] Kazuhiro Kosuge, Koji Takeo, and Toshio Pukuda, "Unified approach for teleoperation of virtual and real environment manipulation based on reference dynamics " in Proceedings of the 1995 IEEE International Conference on Robotics and Automation, Nagoya, Japan, May 1995, pp. 938-943.

[3] Peter Aigner and Brenan McCaxragher, "Human integration into robot control utilizing potential fields," in Proceedings of the 1997 IEEE International Conference on Robotics and Automation, Albuquerque, NM, Apr. 1997, pp. 291-296.

[4] T. F. Chan. *"Design and experimental studies of a generalized bilateral controller for a teleoperator system with a six degree-of-freedom master and a seven degree-of-freedom slave"*. Doctor of Philosophy in Mechanical Engineering, University of Tennessee, May 1996.

[5] Rajiv V. Dubey, Steve E. Everett, and Tan Fang Chan, "Variable damping and stiffness control of a bilateral telerobotic system," IEEE Control Systems Magazine, vol. 17, no. 1, pp. 37-45, Feb. 1997

[6] S. E. Everett. *"Human-Machine Cooperative Telerobotics Using Uncertain Sensor and Model Data"*. Doctor of Philosophy in Mechanical Engineering, University of Tennessee, August 1998.

[7] Tank Focus Area, Pacific National Laboratory. http://www.pnl.gov/tfa.

[8] Human-Machine Cooperative Telerobotics. Topical Report. Federal Energy Technology Center, U.S. Department of Energy, Morgantown, WV. November 1998.

Human-Centered Scaling in Micro-Teleoperation

Akihito Sano, Hideo Fujimoto and Toshihito Takai

Department of Mechanical Engineering
Nagoya Institute of Technology
Gokiso-cho, Showa-ku, Nagoya 466-8555, JAPAN
{sano, fujimoto}@vier.mech.nitech.ac.jp

Abstract

In micro-teleoperation, such as micro tele-surgery, a scaling problem is one of the most important problems. The human operator must be able to identify the micro-environment and execute well the micro-teleoperation. In this study, a human-centered scaling is suggested. In the developed micro-teleoperaion system, the human operator can choose the optimum scaling parameters in consideration of his ability of perception and motor control.

1 Introduction

Human being can master a dextrous micro manipulation by the experience and the training for years. Since this ability is unfathomable, it is impossible to substitute completely the ability for the robot by the present technique. However, it is very important to develop effective and suitable methods that can easily execute the micro manipulation and reduce the exhaustion. In the medical field, many new potential uses of telerobotics and master-slave technology have been explored.

Recently, a surgical teleoperator, known as daVinci system, has been designed to provide enhanced dexterity to doctors performing minimally invasive surgical procedures [1]. The doctor performs virtually the teleoperation from the surgeon's console. We know that haptic senses are fundamentally significant for the complex operations. Therefore, we believe that a force-reflecting teleoperation based on the bilateral control is well suited to the micro-operation. Further, the computer network may allow the teleoperation to expand and be popular [2]. For the design of bilateral controller which is required for the performance specifications in the frequency region, the H_∞ control theory and μ-synthesis/analysis are very effective.

In micro-teleoperation, such as micro tele-surgery, scaling problem is one of the most important problems. Colgate [3] has proposed an impedance shaping bilateral manipulation and derived the general condition based upon the approach using a structured singular value. Many researches tend to discuss a general design methodology or scheme. However, we know that there is an optimum diameter of grasping cylinder, such as the grip of tennis racket, for the arch formed by our fingers and palm. This is geometrical matching. In the same way, it is supposed that the dynamical or impedance matching is unconsciously done between the human and the scaled environment. In this study, the human-centered scaling in micro teleoperation is examined.

In Section 2, the distortion of intensive property, the developed micro-teleoperator and the task are introduced. In Section 3, the design of the time varying H_∞ controller by which the human operators can choose on-line the scaling parameters is discussed. In Section 4, the human characteristics concerning the perception and the recognition are investigated. In Section 5, the significance of human-centered optimum scaling is confirmed, and the adjustment and the matching of human's impedance are discussed.

2 Micro-Teleoperation

2.1 Distortion of Intensive Property

Goldfarb [4] has developed a design methodology, which incorporates dimensional analysis to define a necessary and sufficient set of similarity conditions. The physical relationship requires three fundamental dimensions, which are length [L], mass [M] and time [T]. Now, the arguments upon which the force depends for a given geometry are density $\rho[L^{-3}M]$, viscosity $\mu[L^{-1}MT^{-1}]$, shear modulus $G[L^{-1}MT^{-2}]$, geometrical scale $l[L]$, displacement $x[L]$, and time $t[T]$. If ρ, G, and l are selected as the dependent variables, the resulting dimensionless representation of the physical relationship can be written as:

$$\frac{f}{Gl^2} = \Phi\left[\left(\frac{\mu}{l\sqrt{\rho G}}\right), \left(\frac{t}{l}\sqrt{\frac{G}{\rho}}\right), \left(\frac{x}{l}\right)\right] \quad (1)$$

The dimensionless groups must remain invariant.

The bilateral control architecture imposes additional constraints on geometric and kinematic scaling ξ_x, force scaling ξ_f, and time invariance. The constraints can be expressed as:

$$\frac{l_m}{l_s} = \frac{x_m}{x_s} = \xi_x, \quad \frac{f_m}{f_s} = \xi_f, \quad \frac{t_m}{t_s} = 1 \quad (2)$$

where the subscript s indicates a variable of the micro environment at the slave site and the subscript m indicates a variable of the scaled environment at the master si

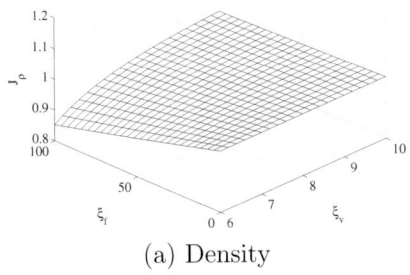

(a) Density

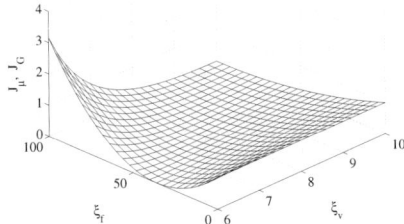

(b) Viscosity and shear modulus

Fig.1: Distortions of intensive properties by scaling

Combining the dimensionless and bilateral control constraints yields the analytic constraints of the scaling problem in the generalized form [4].

$$\frac{\sqrt{\rho_m G_m}}{\mu_m} = \frac{\sqrt{\rho_s G_s}}{\xi_x \mu_s}, \quad \frac{\rho_m}{G_m} = \frac{\rho_s}{\xi_x^2 G_s}, \quad G_m = \frac{G_s \xi_f}{\xi_x^2} \quad (3)$$

Finally, the distortions of intensive (mechanical) properties that are independent of size are given as:

$$J_\rho := \frac{\rho_m - \rho_s}{\rho_s}, \quad J_\mu := \frac{\mu_m - \mu_s}{\mu_s}, \quad J_G := \frac{G_m - G_s}{G_s} \quad (4)$$

Figure 1 shows the distortions J_ρ, J_μ and J_G of intensive properties by scaling. In this study, the ranges of scaling parameters are chosen as:

$$\xi_x \in [6, 10], \quad \xi_f \in [1, 100]$$

A primary objective is to ideally preserve the dynamic similarity and concomitantly preserve the intensive properties. The geometric/kinematic scaling factor is often determined by workspace criteria. On the other hand, for example, the force scaling factor is selected to minimize the distortions of intensive properties [4].

2.2 Micro-Teleoperator and Task

Figures 2(a) and 2(b) show the macro-micro teleoperator with one-degree-of-freedom. The weight of the master and the slave are 4.5 and 2.7×10^{-2} kg respectively. The viscosity coefficient of them are 19.95 and 1.979 Ns/m. The master and the slave are driven through the linear direct-drive motor and the linear voice-coil motor independently. The drive units are used in the force control mode.

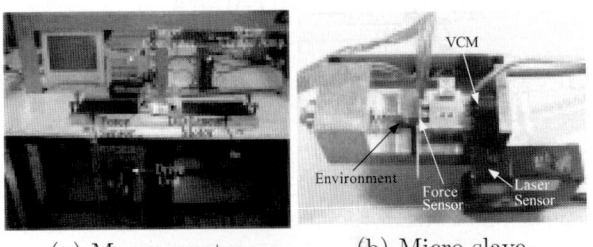

(a) Macro master (b) Micro slave

(c) Operation style with scaling adjuster

Fig.2: Macro-micro teleoperator (experimental setup)

Two force sensors are able to measure the operation force imposed on the master by the operator and the reaction force from the slave environment. The resolution of the force sensors and the position sensor is 0.5 mN and 1 μm respectively. In this study, the real time OS (VxWorks) is adopted.

Figure 2(c) shows the operation style using the adjuster (left-hand side) of scaling parameters. The operator can choose on-line the optimum scaling parameters for the given micro-operation. We focus on the force-reflecting bilateral control, since the haptic senses is very important in the micro-operation. Accordingly, as the operating task, the human operator executes the positioning at the point where he perceives a just-noticeable-difference of the nonlinear spring K_e by the haptic senses without referring to the visual information (see Fig.3).

$$K_e = \begin{cases} 647\,\text{N/m} & (x_s < 1.3 \times 10^{-3}\,\text{m}) \\ 2490\,\text{N/m} & (x_s \geq 1.3 \times 10^{-3}\,\text{m}) \end{cases}$$

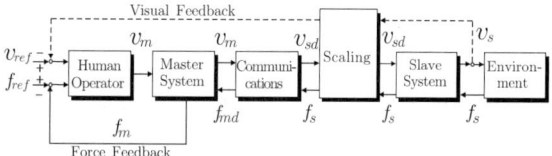

Fig.3: Force-reflecting micro-teleoperation

3 Design of H_∞-Controllers

3.1 Bilateral System

Figure 4(a) illustrates a block diagram of the bilateral control system with the scaling parameters. The transfer functions of the master and the slave are represented by P_m and P_s. Thus

$$P_m = \frac{1}{M_m s + Q_m}, \quad P_s = \frac{1}{M_s s + Q_s} \quad (5)$$

where M_m and M_s denote the mass of master and slave, Q_m and Q_s denote the viscosity coefficients. The slave is in contact with the (assumed known) external environment S_e ($S_e = \frac{\bar{K}_e}{s}$, $\bar{K}_e$ is a nominal stiffness of nonlinear spring.)

The blocks e^{-sT_1} and e^{-sT_2} represent the time delay from the slave to the master and, vice-versa respectively. The scales $\xi_v (= \xi_x)$ and ξ_f represent the velocity scaling factor and the force scaling factor respectively. u_z and u_v are the applied forces on the master and the slave. v_m and v_s are the resultant velocities. In addition, f_m and f_s represent the operation force and the reaction force respectively. A design of the H_∞ controllers K_z, K_v and K_f will be mentioned in Section 3.3.

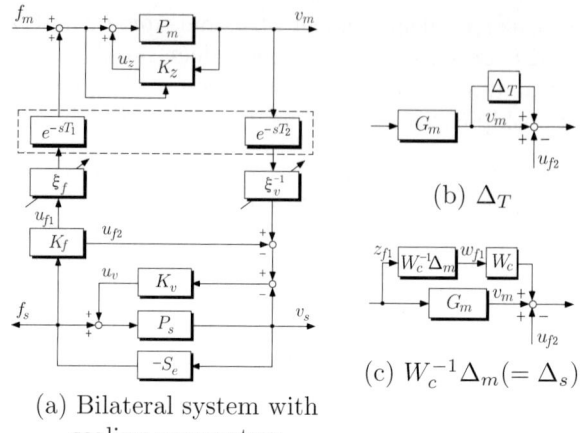

(a) Bilateral system with scaling parameters

(b) Δ_T

(c) $W_c^{-1}\Delta_m (= \Delta_s)$

Fig.4: Bilateral system and perturbation model

3.2 Time Delay and Perturbation

In Fig.4(a), the left delay block e^{-sT_1} can be moved around the loop to the forward path of the loop in order to lump the delays into one block e^{-sT}. T represents the round trip time delay. Consequently, the bilateral system can be reconfigured so that the time delay is reflected as a perturbation Δ_T to the system as shown in Fig.4(b) [5]. Let

$$\Delta_T(s) := e^{-sT} - 1 \quad (6)$$

To reduce the conservatism[1], the perturbation Δ_T can be moved to surround the master system G_m. G_m represents the master of which impedance is regulated by the controller K_z. Then, the new perturbation Δ_m is defined as:

$$\Delta_m = \Delta_T G_m \quad (7)$$

Since G_m is strictly proper by adequate design, there is a bandpass filter W_c such as

$$||W_c^{-1}\Delta_m||_\infty < 1 \quad (8)$$

Finally, the system of Fig.4(b) can be redrawn as that in Fig.4(c), since Δ_m can be expressed as $W_c W_c^{-1}\Delta_m$. Now, the perturbation is $W_c^{-1}\Delta_m$, and it has norm < 1.

Since it can be considered that the time delay reflects as the perturbation to the system, the compensation of time delay results in the robust stability problem of the system with the perturbation. Thus, the following condition is required

$$||\Delta_s Q||_\infty < 1 \quad (9)$$

---
[1] The H_∞ norm $||\Delta_T||_\infty$ equals 2 for every $T > 0$, so it would be compensating for all perturbations of norm ≤ 2, not just the time delay.

for the stability. $W_c^{-1}\Delta_m$ is replaced by Δ_s. Q is denoted as the transfer function from w_{f1} to z_{f1} as shown in Fig.4(c).

3.3 Force-Reflecting Bilateral Control

The design of the time varying H_∞ controller by which the operator can choose on-line the appropriate scaling parameters for the given micro operation is discussed based on the framework of the gain scheduling [2]. Figure 5 illustrates a design algorithm of controllers.

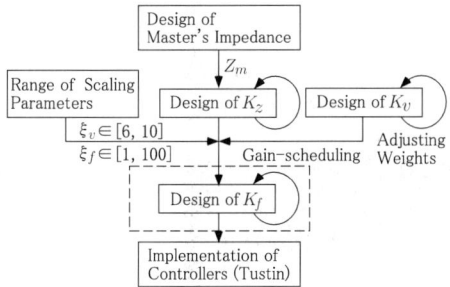

Fig.5: Design algorithm of controllers

In this study, it is assumed that the maximum time delay incurred in the communication channel is 0.2 seconds total. The range of scaling parameters was referred in Section 2.1. First of all, by introducing the adjustable parameters α and β, a desired impedance Z_m of the master is defined as $\alpha M_m s + \beta Q_m$. As seen from Eq.(7), Δ_m can be suppressed by adequately regulating the master's dynamics G_m. Thus, α and β are set to 1.0 and 5.0 respectively. The design specifications are the following: (1) minimizing the impedance error and (2) satisfying prespecified saturation limits for applied force u_z.

The controller K_v is designed to control the velocity of the slave. The controller K_f is designed to compensate the instability caused by the time delay. Figure 6 illustrates a generalized plant. G_s represents the slave which is in contact with the environment S_e and the velocity of the slave is controlled by K_v. The design specifications for K_f are taken as the following: (1) stabilizing for up to the prespecified amount of time delay, (2) tracking the slave velocity to the scaled master velocity, (3) minimizing the model-matching error $u_{f1} - S_M \xi_v^{-1} v_m$ ($S_M = \frac{\bar{K}_e}{s+10^{-3}}$), and (4) satisfying prespecified limits for the control inputs u_{f1} and u_{f2} (omission in Fig.6).

For the derived Δ_m, the bandpass filter W_c was chosen. By appropriate design of weighting functions (W_{f1}, W_{f2},...), it is possible to design the controller. The controller $K_f(\xi_v, \xi_f)$ has been designed based on gain-scheduling procedure by utilizing the LMI Control Toolbox (*hinfgs*) in MATLAB.

4 Human Characteristics

In this section, the human characteristics concerning the perception and the recognition are investigated. The scaling parameters ξ_v, ξ_f that are related to the pushing distance and the resistance force respectively influence greatly the perception and the attainment of the mentioned task in Section 2.2. The compliance discrimination relies on the total work (force times displacement) performed during operation, as well as the final force value [6].

First, a compliance resolution is examined within the range of scaling. In Fig.7, the horizontal axis denotes ξ_v, and the vertical axis denotes ξ_f. The solid line represents an absolute threshold and the broken line represents a just-noticeable-difference of nonlinear spring. Unfortunately, the compliance resolution in the operation through the master arm tends to increase, since the dynamics of master cannot be completely canceled. However, the prespecified range of scaling is sufficiently wide.

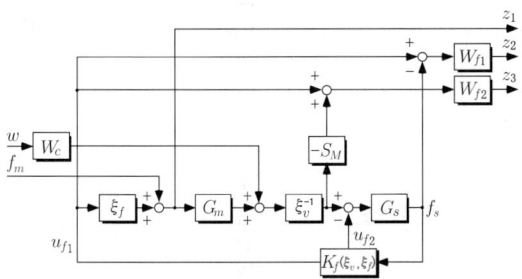

Fig.6: Generalized plant for $K_f(\xi_v, \xi_f)$

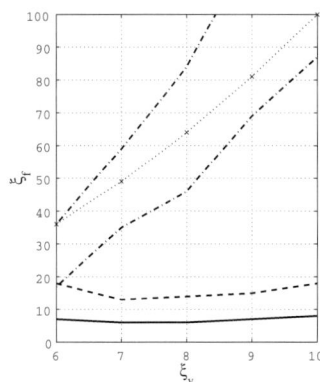

Fig.7: Compliance resolution and mental image

The human operator makes the mental image of dynamics from the visual information of scaled environment. Next, the mental image dynamics is examined. In the experiment, both the scaled information and the resistance force calculated from the scaled environment are displayed. The subject answers a just-incongruous-compliance, such as it is too hard or too soft, compared with the displayed image. In Fig.7, the dashed-dotted lines indicate the derived upper and lower bounds concerning stiffness. As seen from Eqs.(3) and (4), J_G becomes zero on the dotted line in Fig.7 when ξ_f is equal to $\xi_v^2(=\xi_x^2)$. A sense of incongruity caused by the difference between the haptic display and the dynamics imaged by operator yields the workload of operation.

5 Human-Centered Scaling

5.1 Optimum Scaling for Human Operator

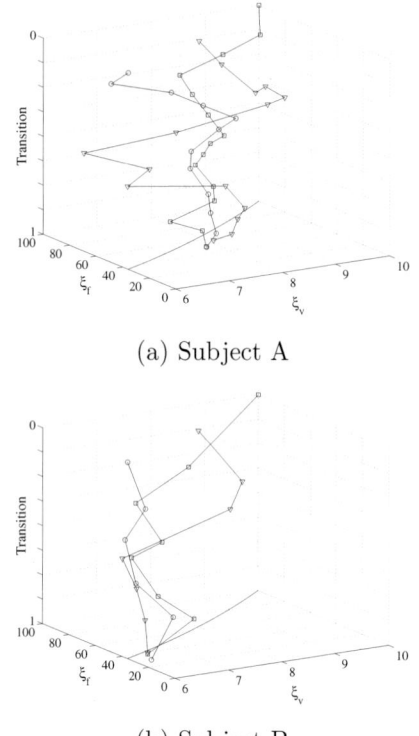

(a) Subject A

(b) Subject B

Fig.8: Transition of on-line scaling by operators

By using the proposed micro-teleoperation system, the operator can on-line choose the optimum scaling parameters so as to get on well with the given micro-operation. The time delay is virtually generated in the computer system. Figure 8 illustrates the transition of on-line scaling by two subjects. The vertical axis denotes the normalized trial number. In the experiments, three sets $(\xi_v, \xi_f)_0 = \{(8, 64), (6, 36), (10, 100)\}$ of initial values are selected at which ξ_f is equal to ξ_v^2.

As seen from Fig.8, each subject has chosen the almost same pair of scaling parameters in spite of the different initial values. And, the parameters that were selected after the satisfactory search are close to the preserving line of shear modulus. Furthermore, subject B prefers the compliant environment to the hard one compared with subject A.

Figure 9 illustrates the variation of force. The operation force f_m (master) is denoted as solid line, and the force $-\xi_f \times f_s$ (slave) is plotted as broken line. As seen from Fig.9, although the time delay exists, the system maintains stability.

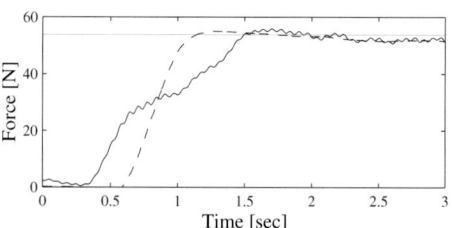

Fig.9: Variation of force

5.2 Adjusting and Matching of Impedance

It is supposed that the operator is adequately adjusting the impedance of his arm. In this section, the human's impedance in the direction of operation at the hand coordinate is estimated by superposing the continuous perturbed vibration [7]. If the perturbed vibration u_p is sufficiently faster than the variation of velocity, acceleration and operation force which vary voluntarily, the signals ($\bar{x}_m, \dot{\bar{x}}_m, \ddot{\bar{x}}_m$ and $\bar{f}_m$) caused by perturbation are sampled from the band-pass filter F (10-20 Hz: order of 4). The amplitude and the frequency of u_p were set to 18 N and 15 Hz respectively so that the given task was not disturbed.

Now, the hand-impedance can be approximately expressed as:

$$M_{op}\ddot{\bar{x}}_m + Q_{op}\dot{\bar{x}}_m + K_{op}\bar{x}_m = \bar{f}_m \qquad (10)$$

where M_{op}, Q_{op} and K_{op} are inertia, viscosity and stiffness at the hand coordinate respectively. In order to estimate the hand-impedance, a least square method with time-varying forgetting factor is applied.

Table 1 shows the average of selected scaling parameters of subject A~C and the estimated hand-impedance in the steady state. As seen from Table 1, it seems that the impedance matching is unconsciously done between the human ($\hat{Q}_{op}, \hat{K}_{op}$) and the selected environment (k_s), as well as the geometrical matching.

Table 1: Experimental results

		Subject A	Subject B	Subject C
ξ_v		7.80	6.42	7.46
ξ_f		50.3	37.8	50.0
k_s	[N/m]	4172	3809	4336
J_G		3.00×10^{-2}	0.69×10^{-2}	1.03×10^{-2}
$\hat{M}_{op}$	[Kg]	0.20	0.20	0.40
$\hat{Q}_{op}$	[Ns/m]	150	130	180
$\hat{K}_{op}$	[N/m]	500	450	500

6 Conclusions

In this study, human-centered scaling in micro tele-operation was discussed. The results of this study are summarized as follows:

1. The compensation of time delay resulted in the robust stability problem of the system with perturbation. The design of the variable controller by which the operator is able to choose on-line the appropriate scaling parameters for the given micro operation could be realized in the framework of the gain scheduling.

2. The task in which haptic sense becomes a key factor was produced in order to demonstrate the significance of direct force-reflection from the environment. The human operator could regulate the optimum scaling parameters so as to get on well. Of course, although the time delay existed, the system has maintained the stability.

3. The human characteristics within the range of scaling were investigated. The pair of selected scaling parameters preserves the intensive property. The scaled dynamics without the sense of incongruity was selected. Furthermore, it seems that the impedance matching was unconsciously done between the human and the scaled environment.

4. We emphasize that the individual variation should be thought highly from the *human-centered* point of view. In the future, the development of man-machine system such as the micro-teleoperator, such as micro tele-surgery, will change from the standard to the intimate based on the information such as the scaling parameters and the intention [8] etc..

References

[1] G.S. Guthart and J.K. Salisbury: "The Intuitive^{TM} Telesurgery System: Overview and Application," *Proc. of the 2000 IEEE Int. Conf. on Robotics & Automation*, pp.618–621, 2000.

[2] A. Sano, H. Fujimoto, and T. Takai: "Network-Based Force-Reflecting Teleoperation," *Proc. of the 2000 IEEE Int. Conf. on Robotics & Automation*, pp.3126–3131, 2000.

[3] J.E. Colgate: "Robust Impedance Shaping Telemanipulation," *IEEE Transactions on Robotics and Automation*, Vol.9, No.4 pp.374–384, 1993.

[4] M. Goldfarb: "Dimensional Analysis and Selective Distortion in Scaled Bilateral Telemanipulation," *Proc. of the 1998 IEEE Int. Conf. on Robotics & Automation*, pp.1609–1614, 1998.

[5] G.M.H. Leung, B.A. Francis, and J. Apkarian: "Bilateral Controller for Teleoperators with Time Delay via μ−Synthesis," *IEEE Transactions on Robotics and Automation*, Vol.11, No.1, pp.105–116, 1995.

[6] G.C. Burdea: "Force and Touch Feedback for Virtual Reality," *John Wiley & Sons, Inc*, pp.13–39, 1996.

[7] H. Gomi and T. Konno: "On-Line Estimation of Human-Arm Stiffness by Continuous Perturbation," *Proc. of the 10th Bioengineering Conference*, pp.411–412, 1998.

[8] H. Fujimoto, A. Sano and A. Sakakibara: "Operational Assistance Based on Adjustment of Impedance That Reflects Human's Intention Concerning Difficulty," *Trans. of The Institute of Systems, Control and Information Engineers*, Vol.13, No.12, pp.552–559, 2000.

Haptically Augmented Teleoperation

Nicolas Turro
Inria/Stanford
Nicolas.Turro@sophia.inria.fr

Oussama Khatib
Robotics group
Stanford University
ok@robotics.stanford.edu

Eve Coste-Maniere
CHIR Medical Robotics group
INRIA Sophia Antipolis
Eve.Coste-Maniere@sophia.inria.fr

Abstract

1 Introduction

Recent progresses in robotic-aided telesurgery help the surgeons to achieve comparable performances during laparoscopic operations and classical open sky procedures. In state of the art systems ([3]), the master console provides a lightweight mechanism whose ergonomy approaches real surgical tools; slave robots are becoming sufficiently dextrous to reproduce the surgeon's motion with great fidelity; and optical systems and video restitution provide a 3D feedback convincing enough to really immerse the surgeon inside the body of its patient. Thus, the main drawbacks of laparoscopy (limited field of view, awkward tools..) can be overcome.

Existing systems, such as the JPL's RAMS system ([1]), already propose tremor elimination and scaling of motions, which are considerably useful for applications like micro-surgery in cardiac operations or eye surgery. The purpose of this article is to go beyond the strict master/slave scheme and actually *enhance* the operator's capabilities during teleoperation. We focused our work on the implementation of three types of constraints for the operators movements:

- constrained movement (along a curve or on a predefined surface)
- virtual obstacle avoidance
- geometric constraints to limit the robots workspace

Constraints for the operator's movements can be implemented mechanically, as described in [7]. Our approach uses a haptic master robot, and consists in adding constraint forces to its control scheme. Constraint forces are computed according to attractive or repulsive potential fields placed around constraints, as defined in [4].

This article presents the principles of these control enhancements which, we believe, will raise dramatically the level of safety and precision that surgeon can achieve, but those principles can also be applied to a wide variety of teleoperation applications (eg. human friendly robotics, cooperative robots...). We also describe one implementation on an experimental platform composed by a Phantom haptic device for the master device and a PUMA arm as the slave manipulator.

2 Principles

In this section, we will first explain the core control scheme of the teleoperation, which reproduces movements of the master device on the slave robot, and also provides the operator with some feedback of the interaction between the robot and its environment. Then we will explain how constraints can be superposed to this control scheme.

2.1 Teleoperation - Slave robot control

In robotic-aided laparoscopy surgery, the master and slave system have vastly different geometry and dynamic properties. The master device is designed to be lightweight and ergonomic, whereas the slave robot must match the geometric constraints of the laparoscopy (operate thru fixed 'ports' into the human body), and usually are much more massive and voluminous than the master device. Thus, the master device controls the position of the end effector of the slave using *operational space control*: the master device absolute position is used to control the position of the end effector of the slave robot. We compute the force F_s^* to apply to the end effector using a PD controller whose goal position is the master position x_m with a scaling factor S and an offset $x_0 = Sx_{s_0} - x_{m_0}$:

$$F_s^* = -K_p(x_s - Sx_m + x_0) - K_v(\dot{x}_s - \dot{x}_m). \quad (1)$$

Then, the torque τ_s to send to the motors is computed using the dynamic (Mass Matrix Λ and gravity g) of the slave robot. Computing the dynamics of the robot is critical in our approach : A good position and velocity servoing could be achieved without it but would involve high gains K_p and K_v, making the robot stiff. Using the dynamics allows us to achieve the same performances, but with lower gains, making the robot more compliant and sensitive to external forces.

$$\tau_s = J^t \Lambda F_s^* + g. \quad (2)$$

2.2 Teleoperation feedback - Master control

On the master device, we provide a force feedback F_m computed using the offset between the master arm position and the position of the slave robot end effector. This way, when external forces (from environment) are applied to the slave robot, those forces will be felt on the master device because the error on the slave tracking will increase. This method doesn't require any force sensor. However, there is constantly an error between the master and the slave position, mostly due to friction on the slave and also to its inertia. Since we do not want to feel a constant drag on the master in free-space movements, we use a cubic function for the computation of F_m:

$$F_m = -K(x_m - \frac{1}{S}(x_s + x_0))^3$$

the gain K is chosen small to exploit the cubic function near zero, on its flat part. So, small tracking errors will be minimized whereas real contact forces will be amplified. We assume that the system is naturally damped (because the manipulator is handling the master device). Furthermore, since our master device is very light, and well balanced, its mass can generally be neglected and its dynamic does not have to be computed. Thus, the torques to send to the master device are computed directly from F_m using the master's Jacobian matrix.

2.3 Teleoperation feedback - Constraints

The main added value of our work is to augment this basic teleoperation scheme with some constraints in order to increase the system safety and

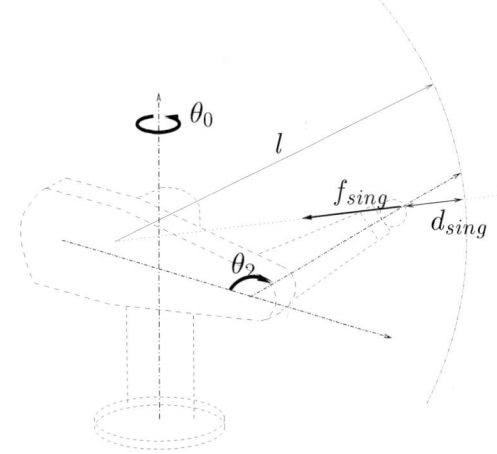

Figure 1: Singularity repulsive force

the users dexterity. All those constraints are based on *potential field*, and we define several possibilities to introduce them in the basic control scheme described above, depending on the nature of the constraint.

2.3.1 Slave-side constraints

Some constraints are critical for the safety of the slave robot, so it is preferable to plug them inside the *slave* controller, in case of a network or master device failure. For example, the master and the slave workspace are different most of the time: the joints limits and singularities are not the same. We propose to enforce those limitations using *repulsion fields*. Those fields will prevent the slave robot in given directions. Then, the basic control scheme described above will haptically render those slave constraints on the master device.

The repulsion fields can be either in the *joint space* or in the *operational space*. For example, our PUMA slave robot has a singularity on its second joint (elbow singularity) when the angle θ_2 (figure 1) reach the value π, or the end effector reaches a sphere whose radius is the length l of the arm fully extended.

For stability and safety reasons, we want to avoid this configuration, so we add to the slave control force F^* in equation (1) an *operational space repulsion force* f_{sing} whose direction is indicated on figure 1. The amplitude of $f_{(sing)}$ is proportional to the distance $\frac{1}{d_{sing}^3}$ between the end effector and a sphere whose radius is the lenght l of the maximal extension of the robot (corresponding to $\theta_2 = \pi$).

Furthermore, due to mechanical reasons, the angle θ_0 must stay between -180 and 180 degrees. A natural way to enforce this property is to add a repulsive torque on the corresponding joint's control. Since this torque τ_{sing} is in the *joint space*, we add it to τ_s (eq. (2)).

Since the amplitude of those repulsing forces grows faster than the PD control force and torque, the slave robot will not reach the undesired configurations. Meanwhile, the operator will 'feel' fast-growing force feedback if he tries to move the slave in those configurations. Thus, imposing constraints on the slave robot, in its control scheme, both in operational and joint space, actually constraints the operator's motions and makes them safer.

2.3.2 Virtual constraints

In this section, we will explain how potential fields can be used on the master's control, in order to help him to carry out some complex tasks, like moving on a perfect line, or help him stay out of some predefined zones.

Master servo loop constraints We propose to help the operator move the slave end effector on a predefined surface or curve by the following scheme. We project the operators cartesian position on the desired trajectory. We call this point a *Proxy*. The curve, or the control surface to follow is surrounded by an *attractive potential field* whose amplitude increases with the distance between the master end effector and the proxy. Then we apply the corresponding attractive force F_m^c on the haptic device's end effector. By choosing the appropriate gains, the operator will easily move on the unconstrained directions, but will have to fight high torques on its master device to go away from it. Subsequently, the slave robot, following the master device will move according to the predefined constraints. Moreover, to further enforce the constraint, we can use the potision of proxy to control the slave, instead of the real position of the master. Figure 2 displays the proxy and the corresponding force, when the constrained motion is along a line Δ.

This location of the constraints gives the better haptic feedback, since it runs at the speed of the master controller. However, despite ongoing work ([5] complex virtual interactions are very hard to computed in haptical real-time (in the thousand Hz range).

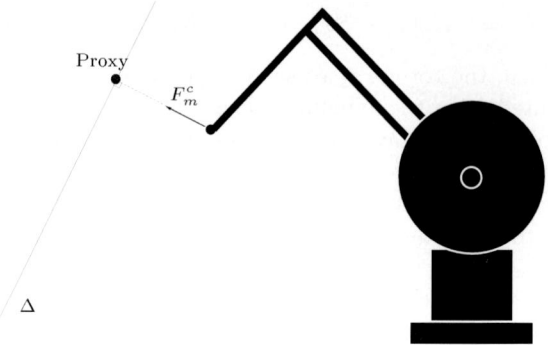

Figure 2: Attractive potential along a line

Master side virtual environment Constraining the movements of the end effector on a predefined curve or surface is simple enough to be incorporated to the servo loop, and running at the same frequency as the control F_m. However, we would like to put constraints on the whole robot movements (not only its end effector) and interact with complex virtual scenes (represented by thousands of triangles, for example) leading to much more complex computations.

To fulfill this requirement, we propose to integrate a model of the real slave robot inside a virtual environment (figure 4). In this 3D environment, the robot will follow the moves of the real one, but will also interact with models of real objects as well as purely virtual obstacles. We compute this interaction asynchronously with respect to the master's control, usually at a much slower rate.

This virtual world can also be displayed on the master screen, providing a convenient visual feedback to the operator. Should the slave and the master robots be in separate rooms, this virtual visual feedback is a good complement to a classical video display of the slave, since it can provide any point of view and any zoom of the real scene.

To model the interaction of the robot in its virtual environment, we define this environment with a set of n convex objects O_i (figure 3). Each object O_i is surrounded with a predefined repulsive potential field whose amplitude and range of action can be parameterized. Then, we compute the resulting forces of this potential field on each rigid moving part B_j of our robot. For each body B_j, we compute the shortest distance d_{ij} between the body and any obstacle. This distance computation will also provide the point of application p_{ij} and the direction of the partial virtual interaction force f_{ij}^{vp} between the part B_j of the robot and the ob-

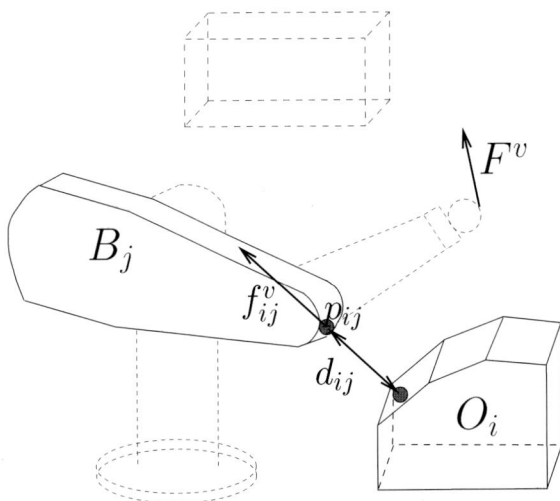

Figure 3: Interaction with the virtual environment: each Obstacle O_i produces a repulsion force f_{ij}^v on each part B_j of the robot, resulting in a force F_m^v at the end effector

ject O_i. Once this is done for all couples of objects and robot bodies, the final virtual interaction force applied by the environment will be :

$$F_m^v = J^{t^{-1}} \sum_{i=1}^{n} \sum_{j=1}^{m} J_{p_{ij}}^t f_{ij}^{vp} \qquad (3)$$

where $J_{p_{ij}}$ is the intermediate Jacobian of the slave robot at the point p_{ij}. Once computed, this force F_m^v will be sent to the master servo loop, and be added to F_m. Due to the slow update frequency of this forces, the amplitude of f_{ij}^{vp} should not vary too fast in order to preserve the haptic feedback stability and is likely to need experimental tuning.

3 Experimental results

In this section we describe our implementation of those principles on the robotic platform of the Stanford Robotic Manipulation Group.

3.1 Hardware architecture

The master device is a Phantom with 6 degrees of freedom. On the used model, only the first three joints of this phantom can be read and controlled. Thus our experiments will deal only with positions of the end effector, not its orientation. The slave robot is a PUMA 560.

We use a linux PC quadri-processor Pentium Pro 200 MHz to control the Phantom, run the display and compute the virtual constraints on the master device. The slave controller runs on a PC Pentium II 333 MHz, using the QNX realtime OS. Both PCs are linked together thru a switch, using a dedicated 100Mb/s ethernet line. This experimentals setup prefigures experiments on real medical robots.

3.2 Software implementation

On the Master PC, we use three separate processes to perform the different tasks, thus exploiting the physical processors. The three process communicate using UNIX UDP sockets. As a consequence, the distance computations described in section 2.3.2 is running asynchronously with the control, typically, much slower.

We implemented this distance computation using the Proximity Query Package (PQP) [2], freely provided by the department of Computer Science of the university of North Carolina. For our tests, we modelled the shoulder and the elbow of the puma using respectively 46 and 26 triangles. A simple environment was modelled using two sets of 22 and 12 triangles.

The Master PC sends its position to the slave PC at the same rate as its servo loop, always providing the freshest value to the slave. In order to achieve optimal performance, this communication uses INET UDP sockets with fixed length 200 bytes packets. This size being smaller than the ethernet MTU (which is 1500 bytes), each position update is completely sent on the network in one shot, avoiding data fragmentation.

The slave controller runs at a slower rate than the master controller, and sends its positions at the same rate as its control loop, which is slower than the masters one. In order to avoid accumulation of data coming from the master, we set the buffer size in the IP stack to the length of one of the packet we receive. Using this technique, we do not need to flush the socket each time we read it: each new update from the master erases any older data present in the socket buffer.

The graphical display of the scene is implemented using the client/server architecture described in [8], using *Mesa* implementation of openGL on a 3DFX graphical adapter (figure 4).

3.3 Performance

In the most complex case, moving on a line, with all virtual obstacles enabled, the master controller runs at 5000 Hz, which is appropriate to achieve a good feedback on the master device.

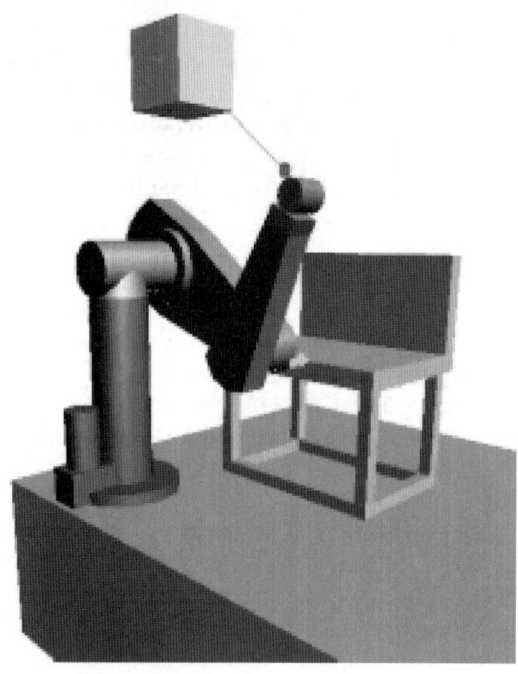

Figure 4: 3D graphical display of the scene, including the robot, real objects (the bench) and virtual ones (the floating cube).

A servo rate of 600 Hz was proven sufficient to run smoothly the slave PUMA 560.

The interaction with the virtual environment was running at 200 Hz, beeing limited by the CPU power. A faster computation would greatly improve the haptic feedback.

3.4 Feedback

The following plots display the feedback (amplitude in the vertical direction) felt by the operator on the master device, during different interactions :

3.4.1 Real Contact

Figure 5 displays the force feedback on the master robot when the slave end effector makes two consecutive contacts with an horizontal plane (z= -0.06 m). When the robot moves in free space, the feedback force is almost zero. When the robot hits the plane, the feedback on the master device rapidly increases as the error between the slave and the master augment. Contact with any part of the robot (not necessarily its end effector) would give a similar feedback to the operator, in the direction of the applied force.

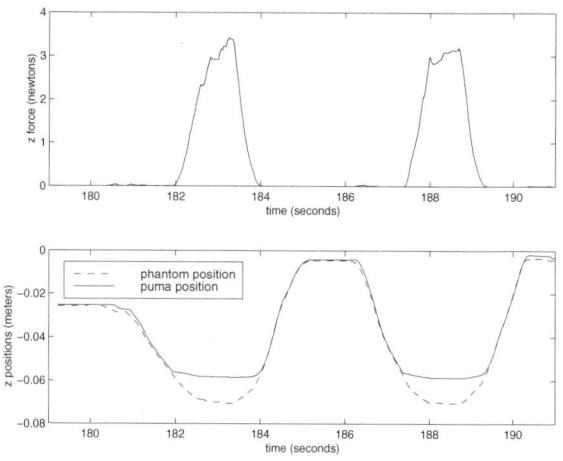

Figure 5: Force Feedback during a real contact on a plane $z = -0.06m$

3.4.2 Virtual Obstacles: Repultion fields

Contact with a virtual obstacle results in a feedback profile displayed in figure 6. The best results for computing f_{ij}^{vp} in equation (3) were achieved using a cubic function of the distance to the obstacle. As in the previous section, the obstacle is a z=-0.06m plane. The effects of the repulsion field are effectively felt at 1cm from the obstacle (about 2 Newtons), but the feedback profile is much less steep than the one associated with a real contact. Because of the relatively slow computation of the interaction with the virtual world (200 Hz), it was necessary to use low gains for the computations of f_{ij}^{vp} (eq. (3)) to ensure the master control stability. Small steps that can be observed on the plots are also due to this slow rate of computation.

3.4.3 Movement along a line: Attractive fields

Figure 7 display the force feedback during a constrained motion along an horizontal line (z = 0.018 m). Here, the feedback is much more stronger than in the two other exemples: when the user is moving along the constraint line (between start and $t = 48$ and $t = 50$ to the end), the feedback force is almost zero, but when the operator tries to go away from the line(between $t = 48$ and $t = 50$), the maximum saturated force of five newtons is quickly reached (for an error of 5 milimeters). This clearly shows the advantages of including the constraint forces computation inside the master servo loop whenever the computation time is compatible with 'haptic'

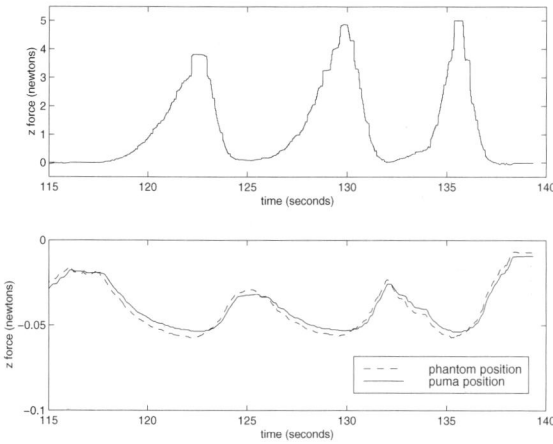

Figure 6: Force feedback during interaction with a virtual obstacle

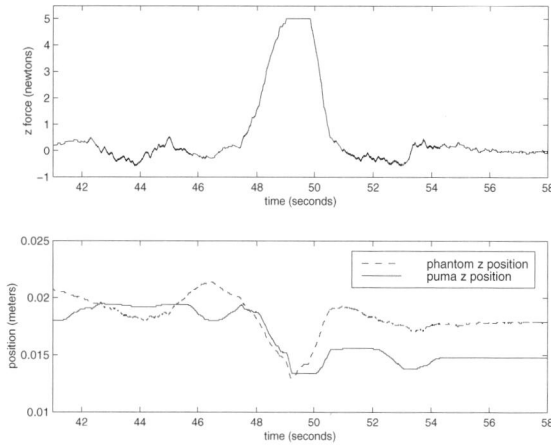

Figure 7: Force feedback during a constrained line movement: trying to go away from the constraint triggers a strong feedback

real-time.

4 Summary and future trends

This work provides a control framework summarized in figure 8 to enrich robotic teleoperation with various types of constraints. Those constraints will increase the safety and the dexterity of the operator in tele-surgery as well as various tele-exploration applications. This framework was extensively experimented, providing a demonstration stable enough to be run tens of times for our visitors.

However, this first haptic teleoperation experiment raises several issues that will be investigated in the future.

Preliminary experiments show that in the case of network delays greater than 200 milliseconds, this control framework is not sufficient, and must be complemented, for example, with a wave transmission scheme ([6]). The ability to cope with delays longer than 100 milliseconds opens the gates of teleoperation over the internet, for example.

Our current virtual interaction is implemented without using the proxy principles as defined in [8], because those principles model only a moving point in a virtual word, and we need to compute the interaction of whole 3D bodies (the robots parts). But as a consequence, the notion of inside or outside obstacle is not present in our implementation. In the near future, we plan to integrate previous work that better model interaction with virtual obstacles.

We also plan to use this framework to control a real medical teleoperation platform, using a patient 3d model for the virtual environment. On such platform, the slave robot will be more precise than our PUMA and will have less friction, leading to better performance for our control framework.

References

[1] Steve Charles, Hari Das, Tim Ohm, Curtis Boswell, Guillermo Rodriguez, Robert Steele, and Dan Istrate. Dexterity-enhanced telerobotic microsurgery. In *Proc. of the 8th Int. Conf. on Advanced Robotics (ICAR '97)*, 1997.

[2] Stefan Gottschalk, Ming Lin, and Dinesh Manocha. Obb-tree: A hierarchical structure for rapid interference detection. In Holly Rushmeier, editor, *Computer Graphics Proceedings, Proceedings of SIGGRAPH, Annual Conference Series, New Orleans, Louisiana*, pages 171–180, August 1996. http://www.cs.unc.edu/ geom/SSV/.

[3] Gary S. Guthart and J. Kenneth Jr. Salisbury. The intuitive telesurgery system: Overview and application. In *Proc. of the 2000 IEEE Int. Conf. on Robotics and Automation*, volume I, pages 618–621, 2000.

[4] Oussama Khatib. Real-time obstacle avoidance for manipulators and mobile robots. *The International Journal of Robotics Research*, 5(1):90–98, Spring 1986.

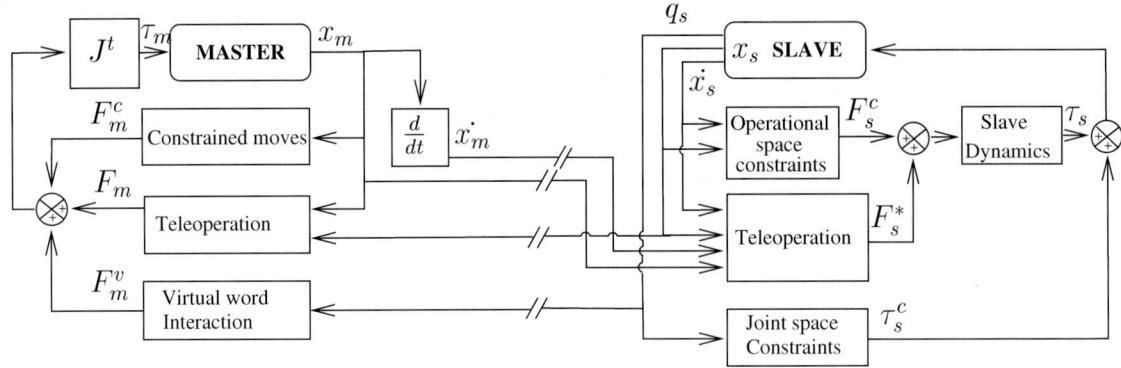

Figure 8: Complete control scheme

[5] Jean-Christophe Lombardo, Marie-Paul Cani, and Fabrice Neyret. Real-time collision detection for virtual surgery. In *Computer Animation, Geneva*, May 1999.

[6] Gnter Niemeyer and Jean-Jacques E. Slotine. Towards force-reflecting teleoperation over the internet. In *Proc. of the 1998 IEEE Int. Conf. on Robotics and Automation*, pages 1909–1915, 1998.

[7] Schneider O., Troccaz J., Chavanon O., and Blin D. Padyc : a synergistic robot for cardiac puncturing. In *Proc. of the 2000 IEEE Int. Conf. on Robotics and Automation*, volume III, pages 2883–2888, 2000.

[8] Diego C. Ruspini, Krasimir Kolarov, and Oussama Khatib. The haptic display of complex graphical environments. In *SIGGRAPH*, pages 345–352, 1997.

Elements of Telerobotics Necessary for Waste Clean Up Automation

Dr. William R. Hamel and Dr. Reid L. Kress
Mechanical & Aerospace Engineering and Engineering Science Department
207 Dougherty Engineering Building
University of Tennessee
Knoxville, Tennessee 37996-2210
USA
Phone: (423)-974-6588, Fax: (865) 974-5274, whamel@utk.edu

ABSTRACT

A promising way to achieve increased remote worksystem efficiency is to layer telerobotic technologies onto teleoperated remote systems. The research being reported here will enable the teleoperation baseline to be supplemented with operator-selective telerobotic modes of operation that allow automatic performance of subtasks that are either repetitive, require high precision, or involve extreme patience. Before subtask automation can be exploited, however, it is necessary to explicitly represent the 3-D geometry of the task space scene surrounding the remote worksystem. The Robot Task Space Analyzer (RSTA) is a tool for remote equipment operators that combines infrared laser and visible stereo imaging, human-interactive modeling and computer-based object recognition to build 3-D models of the immediate work zone in which a robot system is operating. This paper presents the hardware and software design of the RTSA system. Human factors aspects the system operation and design are discussed.

1. INTRODUCTION

Environmental restoration and waste management (ER&WM) challenges in the United States, and around the world, involve radiation and other hazards which will necessitate the use of remote operations to protect human workers from dangerous exposures. Remote operations carry the implication of greater costs since remote work systems are inherently less productive than contact human work due to the inefficiencies/complexities of teleoperation. To reduce costs and improve quality, much attention has been focused on methods to improve the productivity of combined human operator/remote equipment systems; the achievements to date are modest at best. The most promising avenue in the near term is to supplement conventional remote work systems with robotic planning and control techniques borrowed from manufacturing and other domains where robotic automation has been used. Practical combinations of teleoperation and robotic control will yield telerobotic work systems that should outperform currently available remote equipment. Moving in this direction may not be as daunting as it appears to many since the basic hardware and software features of most modern remote manipulation systems can readily accommodate the functionality required for telerobotics. Further, several of the additional system ingredients necessary to implement telerobotic control such as computer vision, 3D object and workspace modeling, automatic tool path generation and collision-free trajectory planning are at a mature level of development.

Practical and reliable implementation of telerobotic systems in Environmental Restoration and Waste Management (ER&WM) contexts is an unrealized objective, despite the potential payoff of telerobotics. This can be attributed to several formidable technical challenges unique to field automation. Almost always the geometry of the task environment is highly unstructured and uncertain. Likewise, the precision and accuracy of the requisite geometric knowledge varies from task to task, as does the extent of the task space itself. A significant fraction of the tasks to be performed are complex by any standards. These factors put full automation of ER&WM tasks beyond the reach of current technology. However,

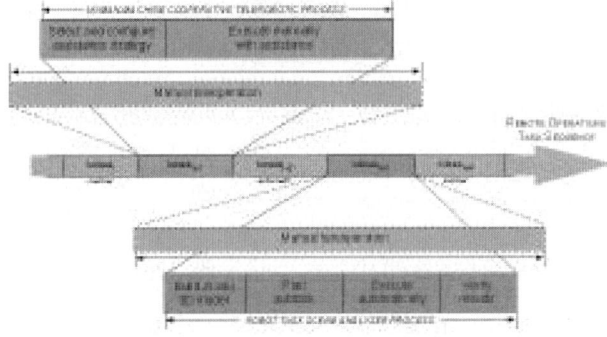

Figure 1. Telerobotics Operations Cycle

there are certain subtasks that are amenable to automatic planning and execution by interjecting telerobotic subtasks into the overall sequence. Implementation of telerobotic capability in a typical ER&WM application will involve operational sequences such as that as depicted in Figure 1.

The type of operation implied by Figure 1 puts emphasis on the human-machine interaction and cooperation. In the case of RTSA, it is believed that human-interactivity is foundational for ultimate task space modeling efficiency as well as seamless maneuvering between manual and automated operations.

2. HUMAN COOPERATIVE TELEROBOTIC SYSTEM CONCEPT

A telerobotic remote work system must provide a very flexible operator environment for discretionary use of subtask automation. The features of the human-machine interface must facilitate synergy that falls into the realm of human cooperation. Human cooperation is used here in the sense of allowing the human operator to control and intercede as dictated by task conditions and execution results. As depicted in Figure 2, the human cooperative telerobot must provide a seamless relationship between manual and autonomous subtask execution. This means that the operator must be able to obtain a rich understanding of the task environment within the current workspace of the mobile system. This is called the "task space." Sensory capabilities to characterize the task space are necessary to provide in situ representation of the tasks to the operator and the robot system (within robot coordinates). Some type of task planner is needed so that

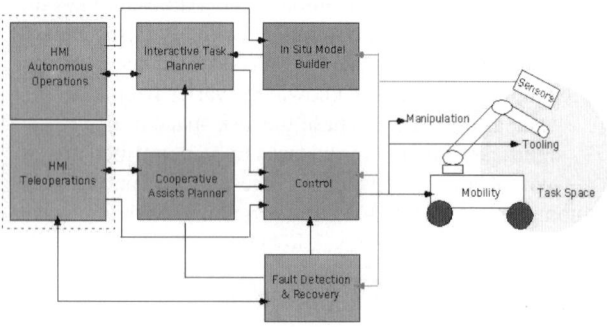

Figure 2. Elements of a Human Cooperative Telerobot

selected subtask executions can be codified for autonomous execution. This planner must make it possible for manual and autonomous subtasks to be interleaved within an overall work task. Finally, a robust form of fault detection and isolation must be provided to notify the operator of the occurrence of a fault, and to facilitate the analysis of the fault scenario. The emphasis in the fault detection and recovery capability is in the operational space of the robot rather than its internal hardware and software space (which must also be monitored). Operational space faults are those associated with task execution and the performance of tools, manipulators, sensors, etc. One distinction between operational and system faults is that an operational fault can occur without hardware or software failures simply because a tooling process did not work (e.g., a saw blade became jammed).

The remote work efficiency can also be improved by enhancing the nature of manual operations by providing computer-based control functions that allow the operator to accomplish manual tasks faster and more effectively. The cooperative assists planner shown in Figure 2 is associated with this element of the telerobot and is the subject of another research activity.

Several aspects of the human cooperative telerobot are being pursued in research at the Mechanical Engineering Robotics and Electromechanical Systems Laboratory at the University of Tennessee in support of the D&D robotics work at the Oak Ridge National Laboratory and the National Energy Technology Laboratory. This paper will focus on the subsystem that is needed to acquire three-dimensional data about the task space. It is called the robot task space analyzer.

3. ROBOT TASK SPACE ANALYZER CONCEPT

Automation of a task requires complete quantitative data about the task/subtasks to be performed, the manipulation systems, and the tooling devices to be used. Task space scene analysis (TSSA) refers to the process by which the remote work system gathers geometrical and other types of information that are necessary to characterize, analyze, and plan the automated task execution [1,2]. For example, in a dismantlement scenario the task may be to remove a segment of process piping using remote manipulators and cutting tools. If such a task is to be automated, it is necessary to describe the location and orientation of each piping element with respect to the remote work system. This data representation, or model, must be complete and accurate to an extent dictated by the specific tool being used: positioning of a shear demands less accuracy than maintaining the proper standoff for a plasma arc torch. Once a sufficient model is available, planning the manipulator and tooling motions can be defined, and the cutting can be automatically executed. The RTSA is a system that performs TSSA, and is in essence a model builder of the near field of view of the mobile work system. Unlike the notion of world model building, RTSA functions in the region of "space" in the near field that is within the sphere of influence of the remote work system where the current task operations are to be performed. RTSA performs an integral step in the telerobotics operations cycle and it must exhibit a level of efficiency that allows telerobotic execution to provide performance benefits over conventional teleoperational execution.

As depicted in Figure 1, telerobotic execution requires a "programming" phase and an "execution" phase for each task to be performed. The programming phase is the RTSA function plus task planning; it is the most important part of the operation since subsequent execution is fully automatic and can progress at the full operating speed of the remote hardware. Therefore, RTSA is an enabling technology that determines the ultimate overall performance of any telerobotics concept.

3.1 Functional Architecture

The RTSA has three major components: 1) a panoramic viewing system, 2) a manual model building component, and 3) an automated model building component. These components are based on the work previously done with Human-Interactive Stereo [1,2], Artisan [3], and a manual model building component where the operator input is used exclusively. From a panoramic view (PV) of the task scene, the operator selects a region of interest (ROI) and assigns the building pipes and fittings models in the ROI to be done manually (manual background) or to be done by an AutoScan method (automated background). When the user chooses to have a ROI analyzed automatically, the stereo AutoScan or the range AutoScan function would be chosen along with specified classes of objects to be found in the ROI. In its current implementation, RTSA contains object classes for standardized process piping components and a custom object tool. The class of objects describes the schedule and size of the piping and whether it is welded, flanged, or screwed piping; this also includes the fitting or fittings to be found including tees, elbows, and pipes.

Human-interactive stereo was used as the foundation for the stereo AutoScan function included in RTSA. Human-interactive stereo [1,2] uses a pair of black and white cameras to capture images of the task scene. Once the images were displayed, the operator indicated corresponding points of a pipe segment in each image. From points at each end of the pipe segment, stereo calculations could be made to construct a 3D model of the pipe including its size, position, and orientation.

Artisan provides the foundation for the range AutoScan capability included in RTSA. CMU's Artisan [3,4] is a perception system that automatically creates three-dimensional models of the area in which a robot works. An operator begins a session with Artisan by instructing the system to acquire range data of the scene using a scanning laser range finder or structured light sensor. Special filtering algorithms are applied to the range image to further reduce noise (while preserving the range discontinuities) and the images are displayed on the operator's workstation. Since the sensor field of view is usually larger than the area the operator wishes to work on, he restricts the system's attention to a particular region of interest by drawing a box around it. Next, he indicates what objects Artisan should expect to find in the region of interest by selecting from a menu of pre-defined object types and sizes. Artisan then creates a Cartesian mesh from the range data in the region of interest thus defining a 3-D surface representation of the data.

Two different object recognition algorithms have been developed for Artisan. The first method (Quadric/Planar Segmentation and Matching, or "QPSM") segments the 3-D surfaces into planar and quadric patches and matches the resulting scene description to analogous descriptions of object models in a database (developed off-line from CAD descriptions of objects). The other method (Free-form Object Recognition Method or "FORM") is based on a technique known as geometric indexing. In this case a collection of 3-D surface points is transformed into a set of 2-D representations, called spin images, that describe the spatial relationship of each point to all the others. The stack of spin images representing the scene data are then compared to stacks of spin images of models in the database to arrive at a few number of plausible correspondences. Each of these is further refined using a modified iterative closest point (ICP) algorithm that outputs the optimal estimate of the recognized object's dimensions, location and orientation in the task space. For each object recognized, the operator can either accept or reject what Artisan found in the data. Each accepted object appears in the World Model window in the location that Artisan has calculated. This process of range data collection, processing and user interaction continues until the operator is satisfied with the 3-D model of the robot's work space.

By selecting ROIs, the operator limits the volume of information required to be analyzed by either of the background AutoScan algorithms and increases their collective efficiency. While the AutoScan algorithms are being executed in the background, the operator can build models of the pipes/fittings manually in other ROIs in the foreground. The operator's list of ROI's assigned to be analyzed manually is known as the manual queue.

The structure of the RTSA flows naturally from the desire to automatically develop models with the AutoScan methods and the need to have operator input. With three paths available for the creation of the task space model, the operator is both an administrator and an active participant.

Administratively, the operator separates the scene into ROIs and assigns the ROIs to be sent either to the manual queue or to an AutoScan method. By allowing the operator to assign parts of the scene to an AutoScan method, the operator's knowledge of an AutoScan method's past successes and failures will aid in the his decision to use AutoScan. Under certain scene conditions, such as occlusions and poor lighting, the operator can decide which method to use on specific regions. During manual modeling the operator designates the placement of the object with the laser range pointer and then approves the object placement by making small adjustments in translation and rotation of the on screen model as is done when the operator approves the results of an AutoScan algorithm.

The operator's input, in the form of manually placing objects in a ROI, is essential. The operator's skill for recognizing objects in a ROI as well as the intuitive ability to place and orient those objects makes him the most robust avenue to creating a model of the scene. The operator also acts as backup to the AutoScan methods as each object can be tweaked into the correct position and orientation if the AutoScan method does not produce modeling results of sufficient accuracy. In the event that an AutoScan method fails by missing an object or by placing an erroneous object, the operator can complete a partially modeled ROI or delete those objects that don't belong. By displaying a visual representation of the model in front of the stereo images, the operator can approve or disapprove of the model built by an AutoScan method.

The stereo AutoScan algorithm uses a pair of images taken from a set of black and white, charge coupled device (CCD) cameras with servo lenses mounted on a pan-tilt head. The stereo head points to the appropriate ROI and acquires a set of stereo images. The stereo images are supplied to the stereo AutoScan algorithm with the desired class of objects to be found. Unlike the previous work done in Human-Interactive Stereo, a model of the class of objects already exists so certain parameters such as pipe diameter and elbow radius are already known. Standard piping and fittings for various pressure ratings and line sizes have been included in an object library within RTSA. Automated object recognition and positioning is greatly simplified with the limitation to the class of standard piping. The algorithm finds the location and orientation of the objects of interest (OOI) in the task scene such as pipes, elbows, and tees.

3.2 System Implementation

The RTSA implementation philosophy is intended to reduce the ultimate recurring costs of systems by maximizing the use of low-cost PC-based software and hardware. The overall hardware architecture is shown in Figure 3. Initially, the background range AutoScan image acquisition is implemented on a separate Silicon Graphics Workstation in order to

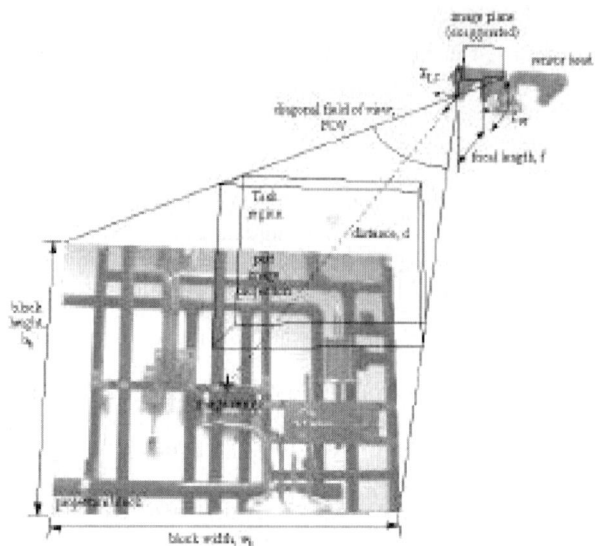

Figure 3. RTSA Hardware Configuration

satisfy budget and schedule constraints. All of the foreground and background modeling functions are implemented within the two dual-PC workstations. In future designs the laser range camera can be directly interfaced to one of the dual PC's.

The computer controlling the stereo head is a Dell 400 workstation with dual 300MHz Pentium II™ processors running Windows NT™. The intensive video processing associated with RTSA is performed by an Elsa™ Gloria-XL video card. The stereo head is controlled through a set of four serial ports; one serial port was required for each of the servo lenses, one for the pan and tilt drives, and one for the laser range pointer. The images are acquired through a Matrox™ Meteor RGB/PPB frame grabber. An RGB frame grabber was chosen so that the black and white images could be captured on different channels – red, green, or blue – of the frame grabber with inherent synchronization. The other computer used is a Dell 400 workstation as well and uses dual 333MHz processors. This computer is used for software development and stereo AutoScan algorithm execution. Given the operating system and experience with the C language, the choice of development tools used to write the program is Microsoft Visual C++™ 5.0. The Microsoft developer's environment was found to be an effective program development environment for this application.

One of the most critically important aspects of RTSA is to provide the operator (and the computer model) with an effective 3D model representation and visualization medium. A commercially available 3D package was chosen for use with RTSA called Envision™ VP. Envision is a 3D kinematic modeling package most often used in simulating the motion of manipulators and virtual path planning.

3.3 Sensor Configurations

The stereo sensor head consists of two Newport™ drives and stepping motors in a pan/tilt arrangement, two Panasonic™ CCD cameras, two Electonique-Informatique Applications™ (EIA) servo lenses, and a SICK™ laser range pointer. The Newport drives and stepper motors allow sensor head pointing with a step size of one one-thousandth of a degree. The CCD cameras are Panasonic GP-MF552 units that produce black and white images with 640 X 480 resolution. The EIA servo lenses, Model X6, allow the digital control of focus, zoom, and aperture. The laser range pointer is a SICK model DME 2000 that measures the phase of a returning laser beam to determine distance to a reflecting surface and is mounted as close to the tilt drive as possible to minimize any deflection in the bracket that the added weight of the laser range pointer might induce.

The laser range camera used in initial RTSA is the Minolta Vivid 7000 laser range camera. The range camera has 8X zoom and a maximum field of view of thirty degrees in its zoomed out configuration. The range of the camera is 600 to 3000 millimeters. The Minolta uses a structured light approach to calculate the distance to an object in the range of 600 to 3000 millimeters. Communication is via a Small Computer Systems Interface (SCSI).

4. RTSA GRAPHICAL USER INTERFACE

The graphical user interface (GUI) is the "connection" between the computer and the operator and is one of the most important aspects of the system. All of the operator input required by RTSA goes through the GUI, and all the information required by the operator is displayed by the GUI. If RTSA is less easy to use than to control the manipulators under teleoperation control, the operator will most likely choose to complete all the tasks in teleoperation mode. In the following discussion, the flow of information at the GUI to and from RTSA are discussed. The GUI windows and their hierarchy are shown in Figure 4.

4.1 Defining Regions of Interest (ROI)

After calibrating the sensors and obtaining the desired panoramic view of the task space of interest, the operators task is to subdivide the task analysis into region of interest (ROI) that contain objects that must be modeled. Refer to step 3 in Figure 4. The object of splitting the task scene PV into ROIs is to allow different processes to work on different parts of the scene at the same time. The ROIs are a way for the operator to keep track of which modeling method is being executed in which part of the scene. Also, ROIs speed up the AutoScan methods, i.e., each ROI can be analyzed more quickly than the entire scene. Duplication of efforts that would occur, if both AutoScan methods were used to analyze the same object, is not a concern.

4.2 Placing Objects Manually

The manual placement of object models must be the most intuitive and practical part of the program as it will determine if a task scene is modeled in a timely matter. Refer to step 4a in Figure 4. The position and orientation of each object model must be designated by the operator using the laser pointer along with the dimensions of the model being placed. For the pipe, the orientation around the pipe axis does not matter and the length can vary; so two points in space (as defined with the laser pointer) can accurately define the end points of a pipe and with the diameter information from the model fully define the pipe. In the case of an elbow, the dimensions of an elbow are known, but all orientation axes must be defined requiring three points. As for the tee, like the elbow the tee is of fixed dimensions and requires three points (from the laser pointer) to describe its orientation.

4.3 Choosing Object Class Information

The definition of the object class information is necessary for the AutoScan methods. Refer to step 4b in Figure 4. This information is necessary because the AutoScan methods are model-based; so a correct mode representation of the OOI must be supplied to the algorithms. For example, since the RTSA operator would know the difference between a three inch pipe and a two inch pipe, the AutoScan methods need not waste time attempting to determine the size pipe that is not in the scene and can eliminate from consideration OOIs that appear to be of diameters different from that specified by the operator.

4.4 Information Required from the RTSA

As mentioned earlier, the result of the program is the model of a task space scene. For a model of the task space scene to be built, the operator needs to be presented with pertinent information from the RTSA. For example, in the placement of points (i.e., the laser pointer spots) when defining the location of objects manually, the operator needs an interactive screen to zoom in on the OOI and position the laser range finder dot on that object. Also, the validation of the correct placement of object models from either the manual identification of points or the AutoScan algorithms requires a view of the task space scene. This task space scene view needs to incorporate the object models' placement information so the operator can visually inspect the object models' placement in comparison to the actual location of the objects. The last example is that the ROI information needs to be presented to the operator so that effort is not wasted on modeling an

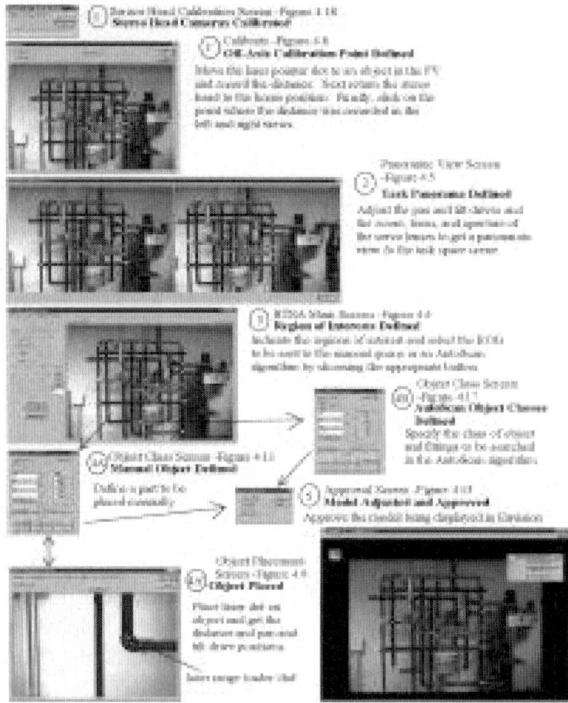

Fig. 4. Operational Flow

area that has already been modeled or on assigning an area to be modeled by two methods. By supplying the operator with pertinent information, the operator's efficiency in modeling the task space scene can be maximized. In its current form, the RTSA GUI involves seven separate windows and requires no keyboard input if the mouse is used to control the stereo sensor head pan and tilt motions. Work continues on the evaluation and streamlining of the GUI. The goal is eliminate all keyboard operations and to minimize the number of windows while keeping their structures simple.

5. EXPERIMENTAL EVALUATIONS

As discussed at the outset, the quantitative performance of the RTSA process is critically important with regard to the practicality of remote telerobotics. In the interest of quantitative evaluation, a structured modeling environment and experimental scheme has been developed.

5.1 Test Mock-Up

The task space scenes shown in Figure 4 show the task space mock-up that was constructed for RTSA testing. The mock-up provides the density and size of process piping objects that one would expect in a typical task space scene. The mock-up was constructed from conventional piping components including some stainless steel items. Image properties such as occlusions, surface colors, and surface spectral characteristics are realistic. A precise Envision 3D graphical model ($\pm$ 0.25 inches) of the mock-up was constructed and is used to compare RTSA modeling results with "ground truth." The graphical model is "calibrated" relative to the true mock-up position using a theodolite with range measurement capability (i.e., Hewlett Packard Total Station™). The Total Station is also used to establish the coordinates of the sensor head relative to the task mock-up. Coordinate transformations were developed to allow the RTSA modeling results and the graphical model to be expressed in terms of a coordinate frame located at the base position of the actual mock-up. This allows RTSA results to be superimposed with the graphical model to provide an excellent visualization of the model correspondence with the real world. The standoff distance between the sensor head tripod and the task mock-up was approximately 16 feet.

5.2 Testing

The goal of testing was to determine the speed and accuracy with which a typical task could be manually modeled. Timing data between novice and expert users was compared to determine the ease of using the interface, and also comparisons between joystick and mouse control of the pan/tilt.

The novice group consisted of 3 graduate students from the psychology department and the expert group consisted of 2 mechanical engineering students and 1 mechanical engineering faculty member who had experience with the RTSA program.

The modeling task consisted of placing a seven-item section from the mockup, which included one (1) tee, one (1) elbow, and five (5) pipes. Subjects were given the opportunity to practice modeling each type of object with RTSA until they felt comfortable with their ability to understand and use the system. They were also given a choice of using the joystick or mouse controls, and allowed to try each beforehand.

Each subject modeled the test section of the mockup twice. The start-point in the mockup (where modeling began) was randomly selected by the experimenter. Subjects were instructed to model the given section as quickly and accurately as possible, and to use the translation and rotation features of RTSA to adjust any parts they thought needed it.

The data collected were based on the location error and time to complete the task. The errors in locating each component were recorded from Envision for 1) where the RTSA manual modeling initially placed the component, and 2) the location of the part after being adjusted by the subject. The time was recorded to complete the initial placement and adjustment phases for each component.

5.3 Results

When given a choice of using the joystick or mouse, all subjects preferred using the joystick, with some using the mouse for fine-tuning and smaller adjustments.

Accuracy: Average error in the initial placement of a part was 65.6 mm (s=32.52mm), representing an error rate of 5-7% of the entire distance for any given dimension (x, y, or z) of a part. Error for each part and dimension, along with relevant interactions, are presented in the figures below.

Error was significantly greater in the Y dimension for novices. (F=11.877, p=.00). Error in the Z dimension was significantly greater for tee in both novice and expert groups.(F=22.457, p=.00). Adjustment did not always improve accuracy, and in the case of our subjects increased error by an average of 9.87 mm.

The average time to completely model the test section with RTSA, including adjustment, was 6.6 minutes (396 seconds). Without adjustment, the average time for modeling the section was 5.06 minutes (303.5 seconds). Both novices and experts showed improvement with practice, with novices making the largest improvement and reaching performance levels of the experts by the second time through (novices from 500 to 343 seconds, and experts 405 to 345 seconds).

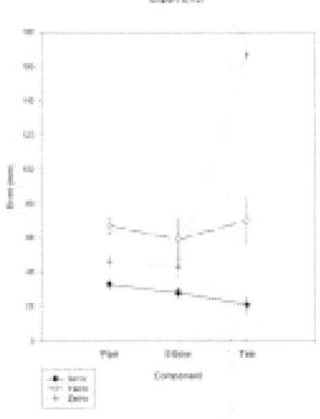

Figure 5. RTSA Manual Modeling Test Results

5.4 RTSA Design Implications

The joystick was the preferred device for controlling larger, sweeping movements of the pan/tilt, with the mouse controls preferred for finer adjustments. Alternate input devices should be consistently available or some hybrid device should be employed that allows for the two types of movements.

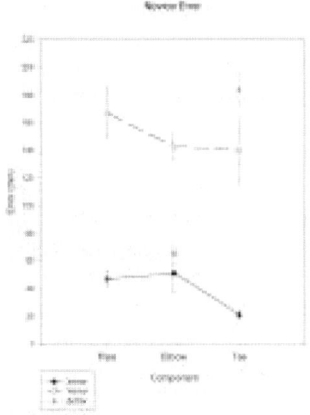

The ability to fine tune, or trim, object location within the Envision interface proved to be beneficial only in certain situations. Its utility lies in the ability to judge when a part has been placed in an entirely incorrect position. This factor accounts for the outcome that improvement was not reflected in the data because subjects would delete objects that were very obviously incorrect from the adjustment/wallpaper view and would place it again. For finer judgments, the RTSA system tends to place parts in a more accurate location than when subjects adjusted the parts using the wallpaper image. However, Envision adjustment tools need to be simplified and more intuitive, since novice subjects made consistently greater and more frequent errors with them.

6. CONCLUSIONS AND FUTURE WORK

A key objective in the development of the RTSA was to use earlier work in task space scene analysis as a foundation for the development of an in situ geometrical modeling system which is a practical tool that typical remote equipment operators could use comfortably. It is believed that the current RTSA design achieves this important goal. Care has been taken in the design of the human-machine interface to assure its simplicity and ease of use. A minimum number of simple windows have been used. Background autonomous modeling functions have been incorporated with the hope that their parallel execution with operator-based manual modeling will accelerate the modeling process. The modeling "mainstay" however is the human interactive stereo image manual mode of operation in which the human operator's powerful skills for scene analysis and object recognition will assure that any scene complexity can be handled.

Experimental results indicate that RTSA has the ability to construct models of a task space scene layer on the order of minutes, and with accuracy's in the range of a few inches. This subtask time fits within the subtask execution operation cycle shown in Figure 1. These time results encourage the further development of RTSA and the automation of subtasks in telerobotic operations.

Future work will involve the implementation and detailed evaluation of a complete RTSA system. Full-scale experiments will be performed at the Remote Technology Assessment Facility at the Oak Ridge National Laboratory. These tests will be performed on several task mock-ups with multiple subjects and trials. In addition, the RTSA system will be integrated with a to-be-determined version of the Dual Arm Work Platform to achieve a comprehensive and working telerobotic system.

ACKNOWLEDGEMENTS

The guidance and direction of Mr. Vijay Kothari, the contracting officer's representative at the U.S. DOE National Energy Technology Laboratory has been especially appreciated.

The RTSA Project is a joint research activity between the University of Tennessee - Knoxville and the Carnegie Mellon University. At UTK, the RTSA team included Professors Mongi Abidi, Steven Everett, Stephen Handel and Phillip Smith and graduate students T. C. Widner, S. Hale, X. Fu, S. Singh, S. Mitchell, Ambrose Ononye, Geoffrey Yerem, Susan Yoder, Ge Zhang, Sewoong Kim. At CMU, Jim Osborn, Dr. Scott Thayer, Dr. Martial Hebert and Owen Carmichael were involved in the development of the range AutoScan functions.

REFERENCES

[1] S. Thayer, C. Gourley, P. Butler, H. Costello, M. Trivedi, C. Chen and S. Marapane. Three-dimensional sensing, graphics, and interactive control in a human-machine system for decontamination and decommissioning applications, *In Proceedings of the Sensor Fusion V Conference*, pages 74-85. SPIE, Nov. 1992.

[2] S. Thayer, "Design and Analysis of Human-Machine and Automatic Vision Systems for Cylinder Detection and Localization", Masters Thesis, University of Tennessee, August, 1993.

[3] A. Johnson, P. Leger, R. Hoffman, M. Hebert, and J. Osborn, 3-D Object modeling and recognition for telerobotic manipulation. In *Proceedings of the 1995 IEEE International Conference on Intelligent Robots and Systems*, volume 1, Aug. 1995.

[4] A. Johnson and M. Hebert, Recognizing objects by matching oriented points. Technical report CMU-RI-TR-96-04. The Robotics Institute, Carnegie Mellon University, May 1996.

Design and Implementation of
Remotely Operation Interface for Humanoid Robot

Satoshi Kagami James J. Kuffner Jr. Koichi Nishiwaki Tomomichi Sugihara
Takashi Michikata Takuma Aoyama Masayuki Inaba Hirochika Inoue
Dept. of Mechano-Informatics, Univ. of Tokyo.
7-3-1, Hongo, Bunkyo-ku, Tokyo, 113-8656, Japan.
Email: kagami@jsk.t.u-tokyo.ac.jp

Abstract

This paper describes a design of remote operation interface for humanoid robot with following three functions: 1) Body DOFs control interface through 3D robot model in a virtual environment, and it has two types of online stablilizing software for maintaining the body balance, 2) Environmental recognition display interface for sensors, sound and 3D vision, 3) Voice and face recognition interface for interacting with operator and a human in front of the robot body. Then, implementation and experiments on our humanoid robot H6 are denoted.

1 Introduction

Recently research on humanoid type robot is active field in robotics society, and many elemental functions are proposed. Especially bipedal dynamic walking, soft skin, 3D vision, motion planning and other topics are very much progressing. However in order to achieve a humanoid robot which work in a human world together with human being, not only elemental functions but also integration of these functions will be a important problem. At present, many humanoid robots are developed, but almost all robots are designed for bipedal locomotion experiments. To satisfy both locomotion and high-level behavior by integrating tactile sensing/3D vision based perception and motion, robot should have a good functionality for mechanism, hardware and software. Especially full body behavior with many contact points to the environment will be a important problem, however such a motion requires more sophisticated body design.

So far, we developed child size full body humanoid "H6" (tall:1370mm, weight:51kg) for a vision/tactile and motion coupling behavior research [1]. Dynamically stable walking pattern generation, motion planning, 3D vision functions are studied by using this H6 [2, 3]. However, these functions are still remaining primitive for a autonomous behavior of the humanoid type robot. Therefore network operation task is adopted, because it requires low-level autonomy for stability and task execution. Currently Humanoid Robot Project(HRP:MITI Japan) has sub-project for network through operation of a humanoid robot [4], however instead of developing low-level autonomy, it is strongly depend on a virtual reality technology and availability of human being who is controlling the robot.

This low-level autonomy will be required for a higher level autonomous behavior research on a humanoid type robot. Therefore, not only tele-operated application, but also this autonomy will be useful for developing from low-level function to high-level autonomous behavior in various environment.

In this paper, humanoid robot H6 is controlled remotely in order to achieve a humamanoid robot which work in a human world together with human being.

2 Network Operation Interface for Humanoid Robot

There exists three major requirements for a network operated humanoid type robot interface.

The first requirement is interface for controlling the robot body joints. In order to control humanoid robot remotely, it is impossible to control a robot by adjusting each joint manually. Even having a master-slave controller, it is very hard to control the robot by satisfying dynamically stable condition. Therefore, low-level autonomy and sophisticated robot DOF control interface are required for controlling a humanoid type robot.

The second requirement is interface for recognizing the

Figure 1: H6 Control Interface

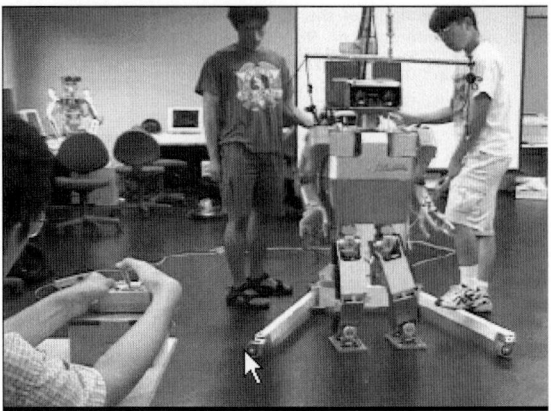

Figure 2: H6 Joystick Control Interface

environment. When robot is working by controlling through network in a real environment, camera has relatively narrow view angle so that it is hard to have a 3D recognition of the environment. Several method are proposed for a tele-operated vision based robot, i) reconstructing a 3D environment [5], ii) using virtual reality technology [4], and so on.

The third requirement is interface for human being who is a) in front of the remotely controlled humanoid robot, and b) in front of the remote operation interface. So far, several interface methods are proposed, however no methods is examined on a tele-operated humanoid type robot.

Therefore, requirements for a network operation interface of humanoid robot are as follows.

- Robot control interface
- Environment Interface
- Human Interaction Interface

3 Robot Control Interface

Robot control interface is divided into two major functions. The former is a interface for working mostly using its hand. For this purpose, we propose a layered system as follows: 1) virtual puppet interface for intuitively control the DOF, and 2) "Autobalancer" which compensate a given motion to dynamically stable one online. The latter is a interface for walking and mostly using its legs. We propose a online walking pattern generation method by combining pre-calculated discrete patterns.

3.1 Virtual Puppet Interface

Since humanoid type robot has many degrees of freedom, it is hard to control dynamic stability by directly controlling its joints. Many research issues have been proposed especially for a robot arm, however there are relatively few researches have been proposed with a humanoid shape robot. So far we have proposed master-slave type humanoid robot interface "Puppet" which has the same DOF arrangement with the real robot. This master-slave based methods have an advantage of the intuitive control, however, either birateral or uni-rateral methods are difficult to apply to the human shape body, since body configuration may changed dramatically in case of birateral method, and since body configuration may be completely different with the real-body in case of uni-rateral method.

Instead of using real device for control the robot body, we propose a virtual puppet interface concept. Robot configuration and its behavior can be simulated with environmental model and it can be displayed on a current 3D CG hardware. Fig.1 shows a virtual puppet interface in a virtual environment, and human operator can control end-effector position and orientation by mouse.

3.2 Autobalancer

"AutoBalancer" reactively generates the stable motion of a standing humanoid robot on-line and from the given motion pattern [6]. The system consists of two parts, one is a planner for state transition from the relationship between legs and ground, and the other is a dynamic balance compensator which solves the balance problem as a second order nonlinear programming optimization by introducing several conditions. The latter can compensate for the centroid position and the tri-axial moments of any standing motion, using all joints of body in real-time. The complexity of AutoBalancer is $O((p+c)^3)$, where p is number of DOFs and c is number of condition equations. Therefore, any motion input from virtual puppet interface will be dynamically compensated using this "Autobalancer".

3.3 Walk Interface

3.3.1 Walk Trajectory Generation

For walking, it is also hard to control every DOF interactively. We have been proposed a offline dynamically stable trajectory generation method for a humanoid robot [2]. From a given input motion and the desired ZMP trajectory, the algorithm generates a dynamically stable trajectory using the relationship between the robot's center of gravity and the ZMP. A simplified robot model is introduced that represents the relationship between its center of gravity and ZMP. Then it is shown that horizontal shift of the torso satisfies given desired ZMP trajectory.

Let z axis be the vertical axis, and x and y axis be the other component of sagittal and lateral plane respectively. First, we introduce a model of humanoid type robot by representing motion and rotation of the center of the gravity (COG). Let total mass of the robot be m_{total}, total center of the gravity be $\boldsymbol{r}_{cog} = (r_{cog_x}, r_{cog_y}, r_{cog_z})$, and total force that robot obtains be $\boldsymbol{f} = (f_x, f_y, f_z)$. ZMP $\boldsymbol{p}_{cog} = (p_{cog_x}, p_{cog_y})$ around point $\boldsymbol{p} = (p_x, p_y, h)$ on the horizontal place $z = h$ is defined as a point where moment around point $\boldsymbol{p}$ be $\boldsymbol{T} = (0, 0, Tz)$. Then following differential equation is obtained.

$$\boldsymbol{p}_{cog}^{err}(t) = \boldsymbol{r}_{cog}^{err}(t) - \frac{m_{total} r_{cog_z}(t) \ddot{\boldsymbol{r}}_{cog}^{err}(t)}{f_z^o(t)} \quad (1)$$

Here let $\boldsymbol{p}_{cog}^{err}$ be an error between ideal ZMP $\boldsymbol{p}_{cog}^*$ and current ZMP $\boldsymbol{p}_{cog}$, and $\boldsymbol{r}_{cog}^{err}$ be the an error between

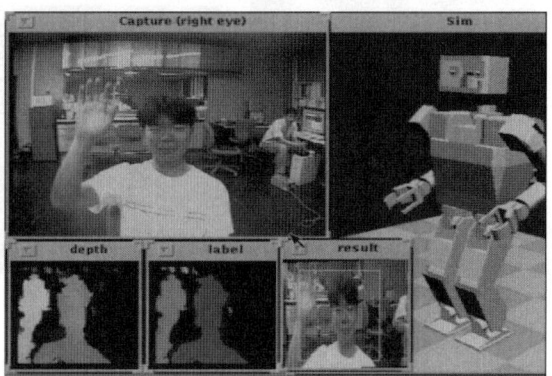

Figure 3: Environment Interface of H6

ideal center of gravity trajectory $\boldsymbol{r}_{cog}^*$ and current trajectory $\boldsymbol{r}_{cog}$.

Finally a convergence method is adopted to eliminate approximation errors arising from the simplified model.

3.3.2 Online Mixture and Connection of Pre-designed Motions

However, in order to control robot online, offline method is impossible to adopt. Therefore online generation of desired walking pattern is proposed. Utilizing the characteristics of ZMP dynamically stable mixture of pre-designed motion is carried out to generate desired walking motion. Eleven pre-designed motions are generated by previous offline trajectory generation method for translation motions. The mixture requires low calculation cost, so that desired walking pattern can be generated in real time. User only have to designate the direction and speed of the motion from the pointing device. Fig.2 shows a joystick control experiment of humanoid H6.

4 Environment Interface

In order to control remotely the humanoid robot, not only the body degrees of freedoms interface, but also environment present interface is important. Humanoid robot must work by its hand and walk in the complex environment, and may be disturbed from the environment (or human being).

There are two kind of environment information. One is the robot internal sensor informations that indicate robot current physical state. The other is 3D vision informations and it is important to the operator to

 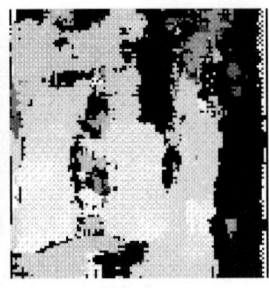  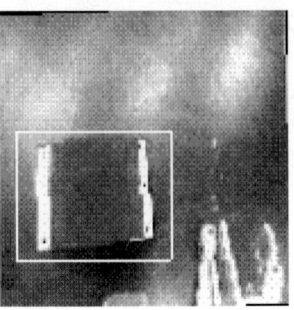

Figure 4: H6 Find out a Step from the Plane Segment Finder in Walking Time

control the robot. With vision recognition software, operator can only say/designate for example "grasp this object" or "climb this step".

4.1 Sensor Interface

Internal state of the robot, tactile information and sound data should be present to the operator. By using virtual puppet interface that is mentioned in previous section, these informations are overlayed on the virtual robot(Fig.3).

4.2 Depthmap Generation System

Real-time 3D Vision functions are fundamentally important for a robot that behaves in real-world. Recently there are several real-time 3D depth map generation systems have been proposed in computer vision field (ex. [7,8]) and some commercial products are also available (ex. [9]). However, these solutions requires special hardware. Since onbody real-time system is required for mobile robotics (or other camera moving) applications, it is hard to make a onbody system using such extra hardwares.

In order to solve this problem, we proposed a real-time depth map generation system using only standard PC hardware and simple image capture card [10]. Four key issues are adopted to achieve real-time and to obtain accurate range data, as follows: 1) recursive (normalized) correlation technique, 2) cache optimization, 3) online consistency checking method, 4) applying MMX/SSE(R) multimedia instruction set. results are denoted.

4.3 Plane Segment Finder

3D plane information is very useful in a artificial environment. In order to find out a 3D plane, we proposed

Figure 5: H6 Voice Control

a Plane Segment Finder by combining depth map generation and 3D hough transformation method [11]. The process includes: 1) precise depth map generation, 2) 3D hough transformation in order to find the plane segment candidates, 3) fit the candidates into the depth map so that plane regions and non plane regions can be distinguished, and 4) track segmented plane in order to achieve real-time plane segmentation. 3D plane is described by parametric notation as follows:

$$\rho = (x_0 \cos(\phi) + y_0 \sin(\phi)) \cos(\theta) + z_0 \sin(\theta) \quad (2)$$

The algorithm requires $O(M^3)$ calculation cost, however by limiting the search space (for example orientation of the plane), its cost reduces. Fig.4 shows the step finding experiment.

5 Human Interaction Interface

In order to make humanoid robot work in a human world by human operator, human interaction interface is important. Two interface is required, a) in front of the remotely controlled humanoid robot, and b) in front of the remote operation interface. Two functions are adopted, A) voice function and B) face recognition function.

5.1 Voice Interface

Since humanoid robot has many degrees of freedom, noises happen while it is working. Therefore, voice recognition software should have a function to resist its noises. Adopted voice recognition software is developed by Dr.Hayamizu at ETL, and this software has advantages that it can run on onbody processor (it runs on Linux) and programmer can very easily to manage its dictionary. Using this advantage, task based dictionaries which contain only several words are prepared, and it is robust in terms of noises. Speech software is a commercial software (Fujitsu) and it also runs on Linux. Fig.5 shows a voice command based walking experiment.

5.2 Human Finding and Face Recognition

In order to find out and recognizing a human being, human finding software and face recognition software are developed. It find out a human existence by segmenting the depth image from the depthmap generation software. Fig.6 shows a human segmentation result.

Then, the human region image is sent to commercial neural-net based face recognition software (MIRO) to recognize. Therefore, the robot automatically recognize a human. Fig.7 shows a face recognition experiment.

6 Conclusion

In this paper, remote operation interface for humanoid robot is discussed and based on a motivation to develop a low-level autonomy of the humanoid robot, three key issues are denoted, 1) control interface, 2) environmental interface, and 3) interaction interface. Then as for control interface, virtual-puppet interface combining with Autobalancer and walk interface are denoted. Operator can control humanoid robot to manipulate the object and to walk to the desired direction. Every designated motion is filtered by online dynamically stabilize functions and its efficiency is shown throughout the experiment of our humanoid robot H6. As for environmental interface, internal robot state display and 3D vision functions are denoted. With this interface, operator can simply designated the target object to grasp or the steps to climb.

Finally as for interaction interface, voice recognitionspeech and human findingface recognition functions are denoted. Operator can react to the human beings around the robot without too much spend a effort to communicate with them, and can concentrate to the robot task.

We examined this remote operation interface by using humanoid robot H6, and confirmed its efficiency for object handling task and walk in a complex environment task. Using this interface, only PC console with microphone and joystick is enough to control the humanoid robot, even humanoid robot has many degrees of freedom and must satisfy dynamic balance. Remote operation interface of the humanoid type robot has an application in a hazardous environment, however, we also believe that remote operation task is good for developing a low-level autonomy of the humanoid robotics.

Acknowledgments

This research has been supported by Grant-in-Aid for Research for the Future Program of the Japan Society for the Promotion of Science, "Research on Micro and Soft-Mechanics Integration for Bio-mimetic Machines (JSPS-RFTF96P00801)" project and several grants of Grant-in-Aid for Scientific Research.

References

[1] K. NISHIWAKI, T. SUGIHARA, S. KAGAMI, F. KANEHIRO, M. INABA, and H. INOUE. Design and development of research platform for perception-action integration in humanoid robot : H6. In *Proc. of IEEE/RSJ International Conference on Intelligent Robots and Systems (IROS'00) (to be appear)*, 2000.

[2] S. KAGAMI, K. NISHIWAKI, T. KITAGAWA, T. SUGIHARA, M. INABA, and H. INOUE. A fast generation method of a dynamically stable humanoid robot trajectory with enhanced zmp constraint. In *Proc. of IEEE International Conference on Humanoid Robotics (Humanoid2000)*, 2000.

[3] J. J. Kuffner, S. KAGAMI, M. INABA, and H. INOUE. Dynamically-stable motion planning for humanoid robots. In *Proc. of IEEE International Conference on Humanoid Robotics (Humanoid2000)*, 2000.

[4] H. Inoue and S. Tachi and K. Tanie and K. Yokoi and S. Hirai and H. Hirukawa and K. Hirai and S. Nakayama and K. Sawada and T. Nishiyama and O. Miki and T. Itoko and H. Inaba and M. Sudo. HRP: Humanoid Robotics Project of MITI. In

  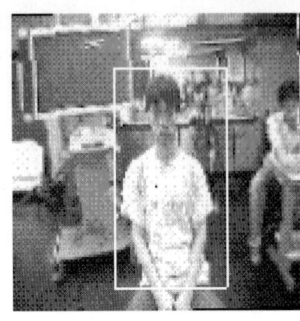

Figure 6: H6 Find out a Human from the Depth in Walking Time, Left:Depth map, Center:Distance Labeling, and Right:Bounding Box

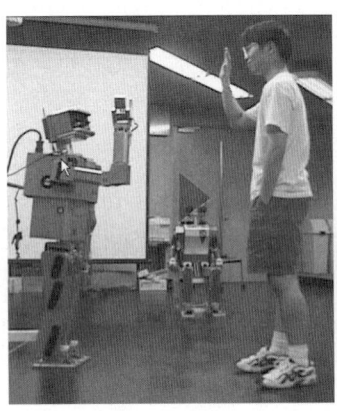 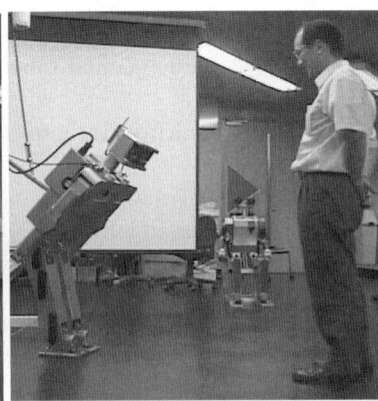 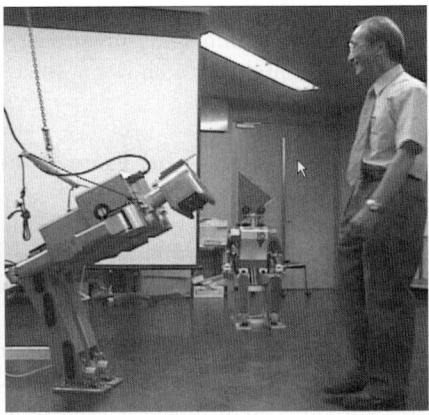

Figure 7: H6 Interacts with Human Being

Proc. of IEEE International Conference on Humanoid Robotics (Humanoid2000), 2000.

[5] M. Maimone, L. Matthies, J. Osborn, E. Rollins, J. Teza, and S. Thayer. A Photo-Realistic 3D Mapping System for Extreme Nuclear Environments: Chornobyl. In *Proc. of International Conference on Intelligent Robots and Systems (IROS'98)*, Vol. 3, pp. 1521–1527, 1998.

[6] S. KAGAMI, F. KANEHIRO, Y. TAMIYA, M. INABA, and H. INOUE. Autobalancer: An online dynamic balance compensation scheme for humanoid robots. In *Proc. of Fourth Intl. Workshop on Algorithmic Foundations on Robotics (WAFR'00)*, pp. SA–79–SA–89, 2000.

[7] K. Konolige. Small Vision Systems: Hardware and Implementation. In Y. Shirai and S. Hirose, editors, *Robotics Research: The Eighth International Symposium*, pp. 203–212. Springer, 1997.

[8] T. Kanade, A. Yoshida, K. Oda, H. Kano, and M. Tanaka. A Stereo Machine for Video-rate Dense Depth Mapping and Its New Applications. In *Proc. of the 1996 International Conference on Computer Vision and Pattern Recognition*, pp. 196–202, Jun 1996.

[9] Point Grey Research Inc. *Triclops Stereo Vision System*. http://www.ptgrey.com.

[10] S. KAGAMI, K. OKADA, M. INABA, and H. INOUE. Design and implementation of onbody realtime depthmap generation system. In *Proc. of International Conference on Robotics and Automation (ICRA'00)*, pp. 1441–1446, 2000.

[11] S. Kagami, K. Okada, M. Inaba, and H. Inoue. Plane segment finder. In *5th Robotics Symposia*, pp. 381–386, 2000.

Model-Based Teleoperation of a Space Robot on ETS-VII Using a Haptic Interface

Woo-Keun Yoon*, Toshihiko Goshozono***, Hiroshi Kawabe*, Masahiro Kinami*,
Yuichi Tsumaki*, Masaru Uchiyama*, Mitsushige Oda** and Toshitsugu Doi***

*Graduate School of Engineering, Tohoku University, Japan
**National Space Development Agency of Japan
***Toshiba Corporation, Japan

Abstract

In our previous works, we have proposed a mixed force and motion commands-based space robot teleoperation system and a compact 6-DOF haptic interface to achieve an effective manual teleoperation. Until now, the effectiveness of this system and the haptic interface has been confirmed by the experiments in our laboratory. The most important features of this teleoperation system are robustness against modeling errors and ability to realize the operator exerted force at a remote site. In this paper, elements of this system and the haptic interface have been employed to teleoperate the 6-DOF manipulator mounted on the Engineering Test Satellite VII (ETS-VII) which is a real space robotic system. Surface-tracking and peg-in-hole tasks with modeling errors have been executed in this experiment. The results show that our space robot teleoperation system including the haptic interface has been able to execute these tasks in real space environment without any big problems.

1 Introduction

The communication time delay is one of the biggest problems for a space robot to be teleoperated from the ground. In a contact task, a force feedback to the operator is an effective control method. Under the time delay, however, the force feedback system is unstable and can not readily display the force to the operator [1]. In order to solve this problem, the model-based teleoperation system has been proposed [2], [3]. The virtual world model is used in the above system, however, does not match exactly with the real world. As a result, it is difficult to execute the remote-site tasks effectively. Some precise model matching ways have also been proposed [4], [5]. However, it is impossible to completely remove the modeling errors between the virtual world and the real one.

Until now, various real space robotic projects like the space robot technology experiment (ROTEX) [6], the Manipulator Flight Demonstration (MFD) and ETS-VII [7] have been conducted and many state-of-art technologies were developed for the space robot teleoperation.

In our previous works, we have proposed a mixed force and motion commands-based space robot teleoperation system with robustness against the modeling errors [8]. Furthermore, we have also developed a new and compact 6-DOF haptic interface to be used as a master arm [9]. The haptic interface can display the force to the operator. This force is calculated on the model. In this paper, we introduce a new control method which improves our model-based teleoperation system. This new method is applied to the ETS-VII manipulator teleoperation. Using these elements, surface-tracking and peg-in-hole tasks with modeling errors are performed. The effectiveness of our space teleoperation system is verified by the above tasks carried out in a real space robotic system.

2 Experimental system
2.1 System elements

This experimental system consists of the space and the ground systems. An overview of the system is shown in Fig. 1. The space system is composed of ETS-VII owned by National Space Development Agency of Japan (NASDA) and Tracking and Data Relay Satellite (TDRS) owned by National Aeronautics and Space Administration (NASA). The ground system is composed of NASDA's satellite operation system and our operator support system. The operator support system consists of a mixed force and motion commands-based space robot teleoperation system and a master controller which is a 6-DOF compact haptic interface. The communication time delay in this communication loop is about 6 seconds. The main components of this experimental system are described as follows.

- Slave system

  The robotic arm mounted on ETS-VII, shown in Fig. 2, is used as the slave manipulator [7]. The

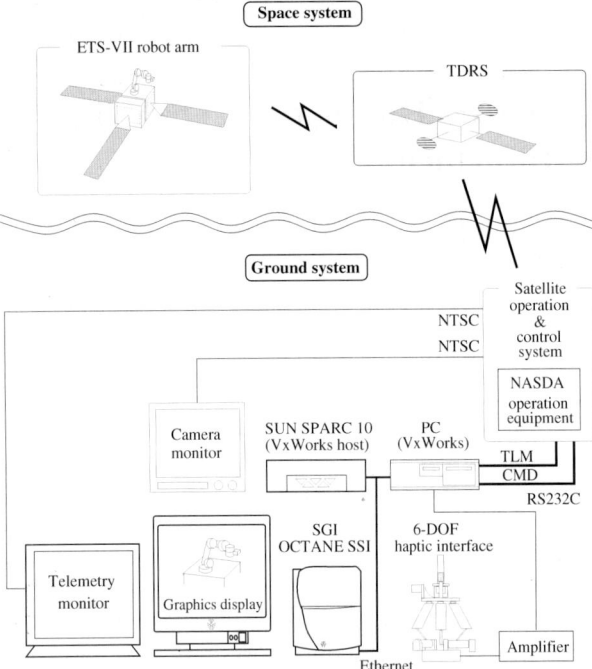

Figure 1: Overview of the system.

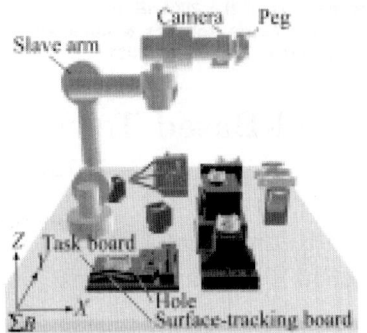

Figure 2: Satellite based slave system.

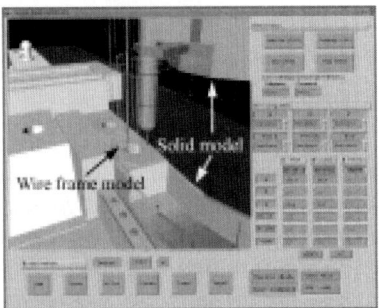

Figure 3: 3D graphics and GUI.

peg is installed at the tip of this manipulator. A camera is also attached at the end of this manipulator. Task bard (TB) is mounted on the ETS-VII deck. Σ_B in Fig. 2 is the arm's coordinate system.

- Graphics computer

  This graphics computer displays the following:

  - the virtual models of the slave arm (virtual arm) and the TB in the solid graphics,
  - the reference tip position of the slave arm, which sends to the ETS-VII, in the wire-frame model,
  - the numerical data of the tip positions and the tip forces of the virtual arm,
  - the GUI for input of the control commands and parameters.

  A real image of this display is shown in Fig. 3. The reason to use the solid and the wire-frame models is given in detail in section 2.2. The concepts of the virtual beam and the virtual grip are utilized in this graphics for the operator support [10], [11].

- Master arm

  A compact 6-DOF haptic interface is used as a master device [9]. A small six-axis force/torque sensor is installed in it to compensate the non-backdrivability feature of the high-ratio reduction gears. The master device, the virtual arm and the slave arm are commanded by this force/torque sensor data.

- Telemetry terminal display

  The numerical data of the positions and the forces of the slave arm are displayed at this terminal.

2.2 Control methods

The main feature of our mixed force and motion commands-based space robot teleoperation system is that the tip velocity commands are generated by the operator exerted forces on the master device and the motion information of both the virtual and the slave arms. The system also incorporates an automatic function to change between the contact and the non-contact modes. However, we cannot manipulate the velocity control and change the control modes between the contact and the non-contact for the manipulator on ETS-VII. Therefore, the master and the virtual arms are controlled by the end-tip velocity and the slave arm is controlled by the position under compliance control. Here, the stiffness gains of the slave arm are 20 kg and 40 kgm^2/rad, the inertia gains are 2795 Ns/m and 2262 Nms/rad and the viscous gains are 200 N/m and 20 Nm/rad. We control the translation only while the orientation is fixed as it is in the initial state. The maximum velocities at the tip of the master, the virtual arms and the reference position of the slave arm are set 2.0 m/s on the ground. However, the maximum velocity of the manipulator on ETS-VII is set 50.0 m/s at the space system computer. The control methods of all these arms are as follows:

- Master arm

408

- Contact
$$\dot{x}_m = k_f(f_{ref} - f_{va}) \quad (1)$$
- Non-contact
$$\dot{x}_m = k_v f_{ref} \quad (2)$$
$$f_{ref} = f_m + k_d(\dot{x}_{va} - \dot{x}_m) + k_p(x_{va} - x_m) \quad (3)$$
- Virtual arm
  - Contact
$$\dot{x}_{va} = k_f(f_m - f_{va}) \quad (4)$$
  - Non-contact
$$\dot{x}_{va} = k_v f_m \quad (5)$$
- Slave arm
$$f_{sa} = m_{sa}\ddot{e} + c_{sa}\dot{e} + k_{sa}e \quad (6)$$
$$e = x_s - x_{sa} \quad (7)$$
$$x_s = x_{va} + \frac{f_m}{k_{sa}} \quad (8)$$

where

x_m, x_{va}: the tip position of the master and the virtual arms respectively,

x_s, x_{sa}: the reference tip position and tip position of the slave arm respectively,

$\dot{x}_m$, $\dot{x}_{va}$: the reference tip velocities of the master and the virtual arms respectively,

f_m: the force/torque sensor data of the master arm,

f_{va}: the virtual refraction/reaction force,

f_{ref}: the position restraint force for the certification of the master arm backdrivability,

f_{sa}: the force/torque sensor data of the slave arm,

k_v, k_f: the velocity and force gains respectively,

k_p, k_d: the position and damping gains for f_{ref} respectively,

k_{sa}, m_{sa}, c_{sa}: the stiffness, inertia and viscous gains of the slave arm respectively.

The change of modes between the contact and the non-contact is carried out automatically realizing a contact in the virtual world.

Second term on the right side of eqn. (8) is added in the tip position of the virtual arm to realize of the operator's force. Using eqn. (8), the slave arm can realize a force equal to the reference force if there are no modeling errors. With modeling errors, the slave arm cannot realize the exact forces. However, the slave arm can compensate for the modeling errors without generating large forces.

Using above equations, the position of the virtual arm is different form the reference position of the slave arm in a contact task. In order to distinguish between these arm positions, the solid and the wire-frame models in the graphics, shown in Fig. 3, are needed.

Generally, the force which the operator inputs is unstable. It is thought that the movement of the slave arm becomes unstable using the original force of the operator. However, the motion of the slave arm is not affected, since a tip speed limitation of the reference position is 2.0 mm/s.

3 Details of the experiments

The safety, effectively and robustness against modeling errors for our teleoperation system are evaluated in the experiments. In the experiments, the virtual model with and without the artificially modeling errors is introduced. The virtual model without the artificially introduced modeling errors dose include a small modeling errors as it is only developed by the draft of the ETS-VII and NASDA's experimental data. Moreover, all virtual models include the dynamic modeling errors. The contents of the experiments are given below.

1. Surface-tracking task

   Surface-tracking task is carried out using the peg and the tracking-surface board, shown in Fig. 2. At first, an operator presses the peg up to 20 N along the z-direction keeps on the tracking-surface board. The operator checks this force at the telemetry monitor and tracks the surface maintaining this force. The details of the surface-tracking task are described as follows:

   - Experiment 1-1
     The operator performs this task with the virtual model without the artificially introduced modeling errors. This model is named "virtual model-1."
   - Experiment 1-2
     The operator performs this task with the virtual model with the artificially introduced modeling errors of +10 mm in both the x and the z-directions. This model is named "virtual model-2."

   In the experiment 1-2, the operator already knows about the scale of the modeling errors before executing the task.

2. Peg-in-hole task

   Peg-in-hole task is also carried out using this set-up, shown in Fig. 2. The peg and the hole diameters are 18.0 mm and 18.4 mm respectively. The details of the peg-in-hole task are described as follows:

   - Experiment 2-1
     The operator performs this task with the virtual model-1, same as in the experiment 1-1.
   - Experiment 2-2
     The operator performs this task with the virtual model having the artificially introduced modeling errors of +5 mm to the x-direction and +10 mm to the z-direction respectively. This model is named "virtual model-3."

   In the experiment 2-2, the operator knows about the introduction of the modeling errors, but dose not know about the scale of these errors before executing the experiment.

4 Results of the experiments

In the experiment 1-1 and 2-1, these tasks is performed smoothly. Therefore, the details of these tasks are left out.

4.1 Surface tracking

The results of the experiment 1-2 are shown in Fig. 4 and Fig. 5. We can confirm the effects of the introduced modeling errors from the peg trajectory in Fig. 4. When the operator presses the peg up to 20 N along the z-direction, the difference between the reference position and the virtual arm is 100 mm in the z-direction. The force deviance due to the modeling errors, which is +10 mm in the z-direction, are 2 N.

In Fig. 5, the x and the z positions of the virtual and the slave arms deviate a little in the first and greatly later. The reasons for this situation are the shape of the tracking-surface board, the relative positions of the corresponding arms, the velocity limits of the two arms and an ignored frictional force in the virtual model. The peg positions of the virtual and the slave arms, and the reference position of the slave arm are shown in Fig. 6. In moving up the surface, all these arms move at 2.0 mm/s. While changing the viewpoint, the virtual arm and the reference position of slave arm are locked. The slave arm, however, keeps moving towards the reference position. Therefore, the slave arm passes by the current position of the virtual arm. In moving down the surface, the frictional force against the slave arm becomes very small because the reaction force of the slave arm becomes small, and hence the actual force working for motions of the slave arm becomes large. Therefore, the slave arm can move even at over 2.0 mm/s because its limit is at 50.0 mm/s. However the virtual arm's motion is limited to 2.0 mm/s. Therefore, the virtual arm is keeping at 2.0 mm/s. As a result, the position error between the slave and the virtual arms becomes larger during this motion.

In Fig. 5, the x-direction force of the slave arm is widely different from that of the master arm. This situation is due to the psychological effects on the operator, the shape of the tracking-surface board and the velocity limit of both the master arm and the reference position for the slave arm. At first, we would describe about the master arm in detail. Before starting the surface-tracking task, both the slave and the virtual arms stop while the master arm is exerting a force of 20 N against the surface. When the tracking-surface task starts, the operator exerts a large force in the x-direction to move the both arms. Therefore the master arm generates a large reaction force. The same situation happens after changing the viewpoint. The operator maintains a large force during this task because the velocity limit of the master arm is very small. Next we would describe about the slave arm using Fig. 6. In moving up the surface, the right side of the slave arm's peg is in contact with the surface. In moving down the surface, the left side is in contact. Therefore, the sign of the slave arm's force in the x-direction is changed while moving from the upward to downward motion along the surface.

In Fig. 5, the master arm always exerts to −20 N in the z-direction. The slave arms achieves to −20 N in the z-direction while moving up the surface. However, this force decreases in moving down the surface. This is why the slave arm goes on moving down the surface.

From these results, it can be said that this task could be performed successfully without generation of big forces that could disturb the execution in both the experiment 1-1 and 1-2.

4.2 Peg-in-hole

The results of the experiment 2-2 are shown in Fig. 7 and Fig. 8. We can confirm the presence of the modeling errors in Fig. 7.

In this experiment 2-2, the operator knows about the presence of the modeling errors in the virtual model but dose not know how about the scale and character. Therefore, the operator has to teleoperate carefully.

1. 0 ∼ 15 s: The peg is moved right above the hole in the virtual world.

2. 15 ∼ 25 s: The peg of the virtual arm is inserted a bit into the virtual TB. But as the force of the slave arm is generated in the z-direction, the operator can notice that the slave arm could not insert the peg in the hole.

3. 25 ∼ 45 s: The operator starts to search for the hole in the real world. f_m/k_s which is the second term on the right side of eqn. (8) is utilized for this purpose. When the virtual peg inserted a bit into the hole of the virtual TB, the operator inputs the x and the y forces and controls the wire-frame model, shown in Fig. 3, for the hole search. In this while, the virtual and the master arms hardly move. During the search, the z-direction force of the slave arm is quite high. If the slave arm can reach the hole and insert peg in it, this force will go to zero. Therefore, the operator keeps on searching noticing only the z-direction force. This situation is shown in Fig. 8.

4. 45 s: The operator discovers the spot where the z-direction force becomes zero and decides that this might be the position of the hole.

5. 45 ∼ 65 s: The operator inserts the virtual peg into the virtual hole. At the same time, the wire-

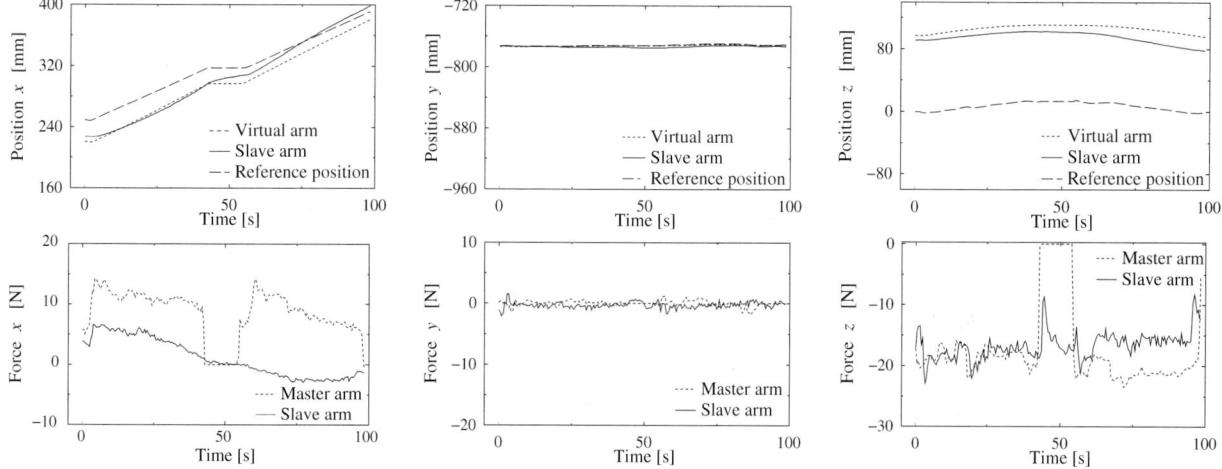

Figure 5: surface-tracking task with the artificially introduced modeling errors.

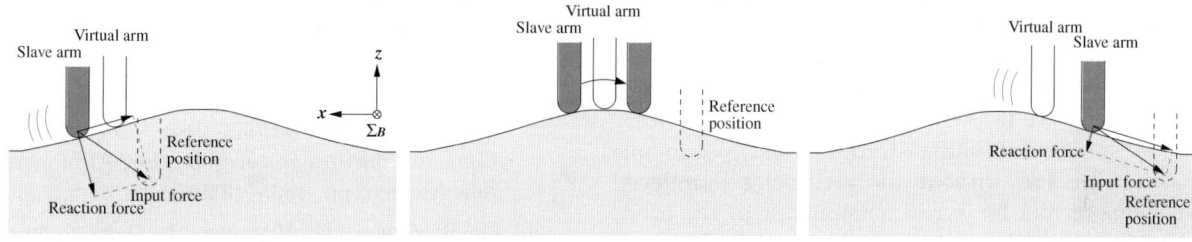

(a) Moving up the surface.   (b) Changing the viewpoint.   (c) Moving down the surface.

Figure 6: Relation to the virtual arm, the slave arm and the reference position of the slave arm.

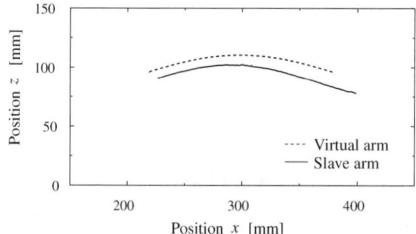

Figure 4: surface-tracking task with the artificially introduced modeling errors.

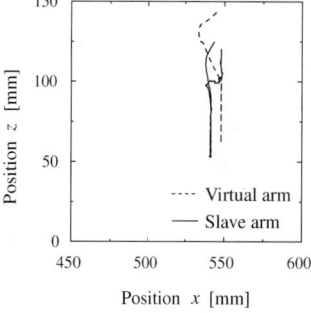

Figure 7: Peg-in-hole task with the artificially introduced modeling errors.

frame model also moves to get into the hole maintaining its position where the hole was discovered, and the operator inserts the slave peg into the real hole.

6. 65 ~ 85 s: In order to confirm the complete peg insertion, the operator presses the peg to over 5 N.

7. 85 ~ 125 s: The peg is pulled out.

8. 125 ~ s: The peg is moved to the finish point in the free space.

The peg-in-hole task could be performed very easily in the experiment 2-1. Although the operation became quite complicated in the experiment 2-2, the operator could carry out the task successfully without generating large disturbance forces.

5 Conclusions

In our previous work, we have proposed a mixed force and motion commands based space robot teleoperation system. The features of this system are robustness against modeling errors and ability to realize the operator exerted force at the remote site. All arms in our system are controlled by the velocity command and incorporated an automatic function to change between the contact and the non-contact modes. However, we cannot manipulate the velocity control and change the control modes between the contact and the non-contact for the ETS-VII manipulator. Therefore,

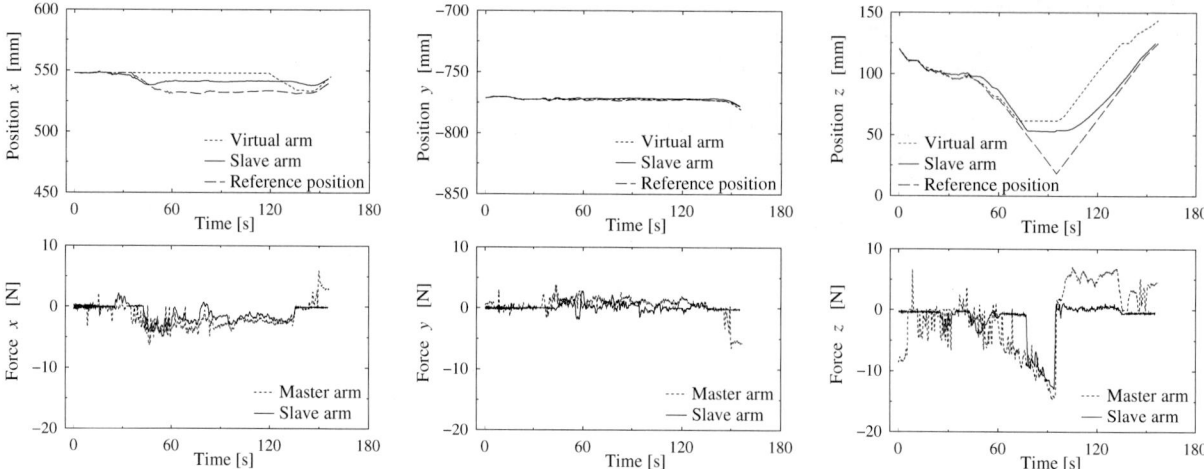

Figure 8: Peg-in-hole task with the artificially introduced modeling errors.

we introduced a new control method which was improved our model-based teleoperation system. And this method was applied to the ETS-VII manipulator teleoperation. A compact haptic interface was used as a master device in these experiments.

We performed surface-tracking and peg-in-hole tasks with and without the artificially introduced modeling errors for a real manipulator in the orbit. The surface-tracking task was carried out safely with and without the modeling errors. The peg-in-hole task could also be carried out successfully though the operation with the artificially introduced modeling errors became very complicated.

From these results, we could confirm that our teleoperation system can be applied easily to the real space robotics, and our system dose have some level of the robustness against the modeling errors.

References

[1] W. R. Ferrel, "Delayed Force Feedback," *IEEE Trans. on Human Factors Electronics*, Vol. HFE-8, pp. 449–455, 1966.

[2] S. Tachi and T. Sakaki, "Impedance Controlled Master-slave Manipulation System. Part I. Basic Concept and Application to the System with a Time Delay," *Advanced Robotics*, Vol. 6, No. 4, pp.483–503, 1992.

[3] T. Kotoku, "A Predictive Display with Force Feedback and its Application to Remote Manipulation System with Transmission Time Delay," *Proc. of the 1992 IEEE/RSJ Int. Conf. on Intelligent Robotics and Systems*, Raleigh, North Carolina, pp. 239–246, 1992.

[4] E. Oyama, N. Tsunemoto, S. Tachi and Y. Inoue, "Remote Manipulation Using Virtual Environment," *Proc. of the Second Int. Symp. on Measurement and Control in Robotics*, Tsukuba, Japan, pp. 311–318, 1992.

[5] W. S. Kim, D. B. Genney and E. C. Chalfant, "Computer Vision Assisted Semi-Automatic Virtual Reality Calibration," *Proc. 1997 IEEE Int. Conf. on Robotics and Automation*, Albuquerque, New Mexico, pp. 1335–1340, 1997.

[6] G. Hirzinger, B. Brunner, J. Dietrich and J. Heindl, "Sensor-Based Space Robotics-ROTEX and Its Telerobotic Features," *IEEE Trans. on Robotics and Automation*, Vol. 9, No. 5, pp. 649–663, 1993.

[7] M. Oda, "Space Robot Experiments on NASDA's ETS-VII Satellite," *Proc. 1999 IEEE Int. Conf. on Robotics and Automation*, Detroit, Michigan, pp. 1390–1395, 1999.

[8] Y. Tsumaki and M. Uchiyama, "A Model Based Space Teleoperation System with Robustness against Modeling Errors," *Proc. 1997 IEEE Int. Conf. on Robotics and Automation*, Albuquerque, New Mexico, pp. 1594–1599. 1997.

[9] Y. Tsumaki, H. Naruse, D. N. Nenchev and M. Uchiyama, "Design of a Compact 6-DOF Haptic Interface," *Proc. 1998 IEEE Int. Conf. on Robotics and Automation*, Leuven, Belgium, pp. 2580–2585, 1998.

[10] M. Uchiyama, S. Kaneda amd K. Kitagaki, "A Teleoperated Force Control System for Space Robots," *J. of the Robotics Society of Japan*, Vol. 9 ＜No. 7 ＞pp. 849–856 ＜1991, (in Japanese).

[11] W.K. Yoon, Y. Tsumaki and M. Uchiyama, "An Experimental System for Dual-Arm Robot Teleoperation in Space with Concepts of Virtual Grip and Ball," *Proc. of the Ninth Int. Conf. on Advanced Robotics*, Tokyo, Japan, pp. 225–230, 1999.

Analysis on Impact Propagation of Docking Platform for Spacecraft

Sang Heon Lee[*], Byung-Ju Yi[**], Soo Hyun Kim[*], and Yoon Keun Kwak[*]

*Department of Mechanical Engineering, KAIST,
373-1, Kusong-dong, Yusong-gu, Taejon 305-701, Korea*

**School of Electrical Engineering and Computer Science, Hanyang University
1271 Sa-1 dong, Ansan, Kyunggi-do 425-791, Korea*

Abstract

When spacecrafts dock with each other, impact will rise at the docking point. It is assumed in this paper that a docking platform having the shape of Stewart platform is attached at the one of the docking spacecrafts to reduce the effect of impact. Although the impact between two spacecrafts at the contact point of the upper plate of the platform is very large, the impact is absorbed by the docking platform and the negligible impact is just transmitted to the spacecraft. To show this, the modeling of the internal impulses at the joints of multi-body system is introduced and the internal impulses of the docking platform are analyzed by the modeling method.

1 Introduction

Impact due to collision between the systems occurs at the contact point during the transition from free motion to constrained motion. The interacting force at the contact point is difficult to control, and the modeling and control of impact has been considered as an important issue [1][2].

Methods to evaluate the impulse have been proposed by several researchers. For a robot system with kinematic redundancy, it is feasible that changing the manipulator configuration can reduce the undesirable effects of the impact. Walker [3, 4] introduced the external impulse model for serial-type robotic manipulator, and proposed a method to reduce the effect of impact by utilizing the self-motion of a kinematically redundant manipulator. Liao and Leu [5] presented the Lagrangian external impact model to derive an impact equation for an industrial manipulator.

When the multi-body system collides with environment, the joints of the system also experience impulsive forces or moments. Zheng and Henami [6] derived the internal impulse model at the joints by using Newton-Euler equations, but their model was confined to the serial-type manipulators. General robot system may contain not only serial-chains, but also closed-chains as well as hybrid-chains. Wittenburg [7] provides a general methodology for modeling external and internal impulses. However, his approach is not directly applicable to modeling of the impulse of robot manipulator systems or general multi-body systems since it is derived in an implicit form. Therefore, in this work, we employ an explicit internal impulse model derived for general classes of multi-body mechanisms [11].

Berthing is understood as a way for typical birds to land on the ground or any particular surface. The foot of birds is composed of several toes. When birds lay land on the ground, the external impulse experienced at the contacting points is distributed to the toes. Majority of the external impulse is converted to momentum of the toes and finally the internal impulse experienced at the roots of the toes becomes relatively small. A biomimetic approach resembling the behavior of birds can be also employed in the docking problem between spacecrafts. A parallel type manipulator can be effectively employed as a way to minimize the internal impulse.

2 External Impulse on Robotic System

Most generally, the impact is partially elastic in the range of $0 < e < 1$, where e is the coefficient of restitution. The component of the increment of relative velocity along a vector n that is normal to the contact surface is given by [7]

$$(\Delta \boldsymbol{v}_1 - \Delta \boldsymbol{v}_2)^T \boldsymbol{n} = -(1+e)(\boldsymbol{v}_1 - \boldsymbol{v}_2)^T \boldsymbol{n}. \quad (1)$$

where $\boldsymbol{v}_1$ and $\boldsymbol{v}_2$ are the absolute velocities of the colliding bodies immediately before impact, and $\Delta\boldsymbol{v}_1$ and $\Delta\boldsymbol{v}_2$ are the velocity increments immediately after impact.

When a robot system interacts with environment, the dynamic model of general robot systems given in Eq. (1) is given as [8-10]

$$\boldsymbol{T}_a = [\boldsymbol{I}_{aa}^*]\ddot{\boldsymbol{\phi}}_a + \dot{\boldsymbol{\phi}}_a^T [\boldsymbol{P}_{aaa}^*]\dot{\boldsymbol{\phi}}_a - [\boldsymbol{G}_a^{v_I}]^T \boldsymbol{F}_I, \quad (2)$$

where $\boldsymbol{F}_I$ is the impulsive external force at the contact point, and $[\boldsymbol{I}_{aa}^*]$ and $[\boldsymbol{P}_{aaa}^*]$ denote the inertia matrix and the inertia power array referenced to the independent joint set, respectively. The independent joint set $\boldsymbol{\phi}_a$ denotes the minimum coordinate set, which is required to describe the system's motion or kinematics. $\boldsymbol{T}_a$ is the torque vector at the independent joint set.

$[\boldsymbol{G}_a^{v_I}]$ defined in

$$\boldsymbol{v}_I = [\boldsymbol{G}_a^{v_I}]\dot{\boldsymbol{\phi}}_a \quad (3)$$

denotes the 1st-order KIC (Kinematic Influence Coefficient) relating the contact point's velocity $\boldsymbol{v}_I$ to the independent joint velocities [10].

Integration of the dynamic model given in Eq. (2) over contacting time interval gives

$$\int_{t_0}^{t_0+\Delta t} \boldsymbol{T}_a \, dt = \int_{t_0}^{t_0+\Delta t}[\boldsymbol{I}_{aa}^*]\ddot{\boldsymbol{\phi}}_a \, dt + \int_{t_0}^{t_0+\Delta t}\dot{\boldsymbol{\phi}}_a^T[\boldsymbol{P}_{aaa}^*]\dot{\boldsymbol{\phi}}_a \, dt \\ - \int_{t_0}^{t_0+\Delta t}[\boldsymbol{G}_a^{v_I}]^T \boldsymbol{F}_I dt. \quad (4)$$

Since the positions and velocities are assumed finite all the time during impact, the integral term involving $\dot{\boldsymbol{\phi}}_a^T[\boldsymbol{P}_{aaa}^*]\dot{\boldsymbol{\phi}}_a$ becomes zero as Δt goes to zero, as does the term involving actuation input $\boldsymbol{T}_a$. Thus, we obtain the following simple expression

$$\Delta \dot{\boldsymbol{\phi}}_a = [\boldsymbol{I}_{aa}^*]^{-1}[\boldsymbol{G}_a^{v_I}]^T \hat{\boldsymbol{F}}_I \quad (5)$$

where $\hat{\boldsymbol{F}}_I = \int_{t_0}^{t_0+\Delta t} \boldsymbol{F}_I dt$ is defined as the external impulse at the contact point. Assuming that the robot impacts on a fixed solid surface, the absolute velocity and the velocity increment of the fixed surface are always zero ($\boldsymbol{v}_2 = \Delta\boldsymbol{v}_2 = 0$). Impulse always acts on the contact point along the normal vector $\boldsymbol{n}$ if we assume no friction exists on the contacting surface. Thus, we have the magnitude of the external impulse

$$\hat{F}_I = \frac{-(1+e)(\boldsymbol{v}_I)^T \boldsymbol{n}}{\boldsymbol{n}^T ([\boldsymbol{G}_a^{v_I}][\boldsymbol{I}_{aa}^*]^{-1}[\boldsymbol{G}_a^{v_I}]^T)\boldsymbol{n}}. \quad (6)$$

where

$$\hat{\boldsymbol{F}}_I = \hat{F}_I \boldsymbol{n}. \quad (7)$$

3 Internal Impulse of Serial-chain Module

Consider a subsystem of the serial-chain module shown in Fig. 1. $^j\boldsymbol{r}_j$ is the position vector directing from mass center C_j to the contact point. The superscript in front of variable means the coordinate frame in which the variable is expressed. Assume that the contact point is located at the center of the joint. $^j\boldsymbol{F}_j$ and $^j\boldsymbol{\tau}_j$ are the impulsive force and the impulsive moment at the jth contact point, respectively. These impulsive forces and moments between interacting links are equal in magnitude, opposite in direction. Based on Newton-Euler's equation, the dynamics of the jth link with respect to the jth coordinate frame can be expressed as

$$m_j{}^j\dot{\boldsymbol{v}}_{C_j} = -{}^j\boldsymbol{F}_j + [{}_{j+1}^{j}\boldsymbol{R}]^{j+1}\boldsymbol{F}_{j+1} + {}^j\boldsymbol{f} \quad (8)$$

$$[^{C_j}\boldsymbol{I}]^j\dot{\boldsymbol{\omega}}_j + {}^j\boldsymbol{\omega}_j \times [^{C_j}\boldsymbol{I}]^j\boldsymbol{\omega}_j = -{}^j\boldsymbol{\tau}_j + [{}_{j+1}^{j}\boldsymbol{R}]^{j+1}\boldsymbol{\tau}_{j+1} \\ - {}^j\boldsymbol{r}_j \times {}^j\boldsymbol{F}_j + {}^j\boldsymbol{r}_{j+1} \times [{}_{j+1}^{j}\boldsymbol{R}]^{j+1}\boldsymbol{F}_{j+1} + {}^j\boldsymbol{\tau} \quad (9)$$

where m_j and $[^{C_j}\boldsymbol{I}]$ are the mass and the moment of inertia of the jth link, respectively. The inertia matrix $[^{C_j}\boldsymbol{I}]$ is expressed in the jth coordinate frame. The velocity of the jth mass center and the angular velocity of jth link are denoted as $^j\boldsymbol{v}_{C_j}$ and $^j\boldsymbol{\omega}_j$, respectively. $^j\boldsymbol{f}$ and $^j\boldsymbol{\tau}$ represent the non-impulsive forces and moments acting on the link j. $[{}_{j+1}^{j}\boldsymbol{R}]$ is the rotation matrix transforming $(j+1)$th coordinate frame to jth coordinate frame.

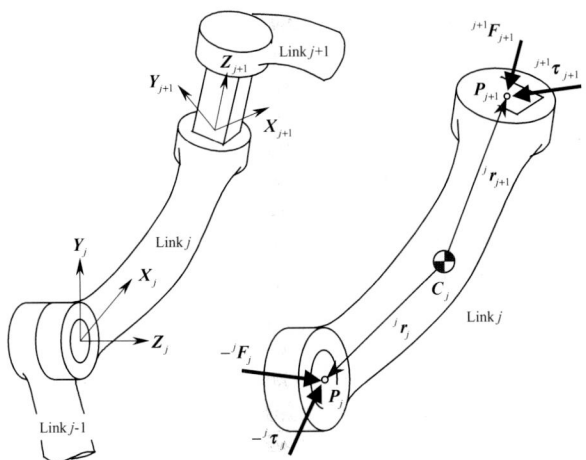

Fig. 1. Link and joint of a serial-chain

By integrating the dynamic equations with respect to time, the relationship between the velocity increment ($\Delta^j v_{C_j}$ and $\Delta^j \omega_j$) and the impulse ($\hat{F}$ and $\hat{\tau}$) in the jth coordinate frame are obtained as

$$m_j \Delta^j v_{C_j} = -{}^j\hat{F}_j + [{}^j_{j+1}R]\,{}^{j+1}\hat{F}_{j+1}, \qquad (10)$$

$$\begin{aligned}[{}^{C_j}I]\Delta^j \omega_j =& -{}^j\hat{\tau}_j + [{}^j_{j+1}R]\,{}^{j+1}\hat{\tau}_{j+1} \\ & -{}^j r_j \times {}^j\hat{F}_j + {}^j r_{j+1} \times [{}^j_{j+1}R]\,{}^{j+1}\hat{F}_{j+1}.\end{aligned} \qquad (11)$$

If the joint j connecting the jth link to the (j-1)th link is a revolute joint and the Z_j-axis is coincident with the rotating axis, the constraint moment along the Z_j-axis does not exist. Therefore, the impulse moment vector at the joint j should be modified to

$$\begin{bmatrix}1 & 0 \\ 0 & 1 \\ 0 & 0\end{bmatrix}{}^j\hat{\tau}_j = \begin{bmatrix}1 & 0 \\ 0 & 1 \\ 0 & 0\end{bmatrix}\begin{pmatrix}{}^j\hat{\tau}_{jx} \\ {}^j\hat{\tau}_{jy}\end{pmatrix},$$

while the impulse force

$${}^j\hat{F}_j = \begin{pmatrix}{}^j\hat{F}_{jx} & {}^j\hat{F}_{jy} & {}^j\hat{F}_{jz}\end{pmatrix}^T$$

is still unchanged. When the joint (j+1) is prismatic, the Z_{j+1} axis is collinear with the moving axis and the constraint force along this axis does not exist. So the impulse at the joint (j+1) must be modified to

$$\begin{bmatrix}1 & 0 \\ 0 & 1 \\ 0 & 0\end{bmatrix}{}^{j+1}\hat{F}_{j+1} = \begin{bmatrix}1 & 0 \\ 0 & 1 \\ 0 & 0\end{bmatrix}\begin{pmatrix}{}^{j+1}\hat{F}_{(j+1)x} \\ {}^{j+1}\hat{F}_{(j+1)y}\end{pmatrix},$$

but the impulse moment

$${}^{j+1}\hat{\tau}_{j+1} = \begin{pmatrix}{}^{j+1}\hat{\tau}_{(j+1)x} & {}^{j+1}\hat{\tau}_{(j+1)y} & {}^{j+1}\hat{\tau}_{(j+1)z}\end{pmatrix}^T$$

need not be modified. To generalize, two types of constraint matrices $[C_{Fj}]$ and $[C_{Tj}]$ are defined. When the joint j is revolute, $[C_{Fj}]$ is a 3×3 identity matrix and

$$[C_{Tj}] = \begin{bmatrix}1 & 0 & 0 \\ 0 & 1 & 0\end{bmatrix}^T.$$

If the joint is prismatic, $[C_{Fj}]$ and $[C_{Tj}]$ are exchanged with each other.

Thus, the Eqs. (10) and (11) can be represented as

$$m_j [C_{Fj}]^T \Delta^j v_{C_j} = [C_{Fj}]^T \Big(-[C_{Fj}]\,{}^j\hat{F}_j \\ + [{}^j_{j+1}R][C_{F(j+1)}]\,{}^{j+1}\hat{F}_{j+1} \Big), \qquad (12)$$

$$[C_{Tj}]^T [{}^{C_j}I]\Delta^j \omega_j = [C_{Tj}]^T \Big(-[C_{Tj}]\,{}^j\hat{\tau}_j \\ + [{}^j_{j+1}R][C_{T(j+1)}]\,{}^{j+1}\hat{\tau}_{j+1} - [{}^j r_j^j][C_{Fj}]\,{}^j\hat{F}_j \quad (13) \\ + [{}^j r_j^{j+1}][{}^j_{j+1}R][C_{F(j+1)}]\,{}^{j+1}\hat{F}_{j+1} \Big).$$

where $[{}^j r_j^{j+1}]$ is the skew-symmetric matrix corresponding to '${}^j r_{j+1} \times$'. The transposes of constraint matrices are premultiplied to the both sides of Eqs. (12) and (13) to adjust the number of rows to the rank of the equation. Let N be the number of links. Then, the impulses ${}^{N+1}\hat{F}_{N+1}$ and ${}^{N+1}\hat{\tau}_{N+1}$ experienced at the end of Nth link are zero vectors.

The increments of the velocity of the jth mass center and the angular velocity of the jth link are obtained by pre-multiplying the 1st-order KICs to the velocity increments of the joints $\Delta\dot{\phi}$:

$$\Delta^j v_{C_j} = [{}^0_j R]^T [G_\phi^{C_j}]\Delta\dot{\phi} \qquad (14)$$

$$\Delta^j \omega_j = [{}^0_j R]^T [G_\phi^{\omega_j}]\Delta\dot{\phi} \qquad (15)$$

where $[G_\phi^{C_j}]$ and $[G_\phi^{\omega_j}]$ denote the 1st-order KICs relating the velocity of mass center v_{C_j} and the angular velocity ω_j of the jth link to the joint velocity vector $\dot{\phi}$, respectively. Recalling the impulse model given in Eq. (5), the velocity increments of joints of serial-chain module are expressed as

$$\Delta \dot{\phi} = [I^*_{\phi\phi}]^{-1} [G^I_\phi]^T \begin{pmatrix} \hat{F}_I \\ \hat{\tau}_I \end{pmatrix}. \quad (16)$$

where $[G^I_\phi]$ given by $[G^I_\phi] = \begin{bmatrix} [G^{v_I}_\phi] \\ [{}_I G^{\omega_I}_\phi] \end{bmatrix}$ denotes the 1st-order KIC relating the 6×1 velocity vector at the contact point to the joint velocity vector, and $[I^*_{\phi\phi}]$ denotes the inertia matrix of the serial chain module. The external impulse moment $\hat{\tau}_I$ does not exist if there is no external moment applied to the contact point or no friction on the contact surface. However, when the serial-chain is regarded as a subsystem of a hybrid mechanism, $\hat{\tau}_I$ always exists. So, we included $\hat{\tau}_I$ in the impulse model of a serial-chain module for generalization.

Substituting Eqs. (14) through (16) into the left-hand sides of Eqs. (12) and (13) yields ($j = 1, 2, ...N$)

$$m_j [C_{Fj}]^T \Delta^j v_{C_j} = \left(m_j [C_{Fj}]^T [{}^0_j R]^T [G^{C_j}_\phi] \right)$$
$$\times [I^*_{\phi\phi}]^{-1} [G^I_\phi]^T \begin{pmatrix} \hat{F}_I \\ \hat{\tau}_I \end{pmatrix} = [V_j] \begin{pmatrix} \hat{F}_I \\ \hat{\tau}_I \end{pmatrix} \quad (17)$$

$$[C_{Tj}]^T [{}^{C_j}I] \Delta^j \omega_j = \left([C_{Tj}]^T [{}^{C_j}I][{}^0_j R]^T [G^{\omega_j}_\phi] \right)$$
$$\times [I^*_{\phi\phi}]^{-1} [G^I_\phi]^T \begin{pmatrix} \hat{F}_I \\ \hat{\tau}_I \end{pmatrix} = [W_j] \begin{pmatrix} \hat{F}_I \\ \hat{\tau}_I \end{pmatrix}. \quad (18)$$

If the kth link experiences the external impulses $\hat{F}_I$ and $\hat{\tau}_I$, then Eqs. (12) and (13) are modified as

$$m_k [C_{Fk}]^T \Delta^k v_{C_k} = [C_{Fk}]^T \left(-[C_{Fk}]^k \hat{F}_k \right.$$
$$\left. + [{}^k_{k+1}R][C_{F(k+1)}]^{k+1} \hat{F}_{k+1} + [{}^k_0 R]\hat{F}_I \right). \quad (19)$$

$$[C_{Tk}]^T [{}^{C_k}I] \Delta^k \omega_k = [C_{Tk}]^T \left(-[C_{Tk}]^k \hat{\tau}_k \right.$$
$$+ [{}^k_{k+1}R][C_{T(k+1)}]^{k+1} \hat{\tau}_{k+1} - [{}^k r^k_k][C_{Fk}]^k \hat{F}_k$$
$$+ [{}^k r^{k+1}_k][{}^k_{k+1}R][C_{F(k+1)}]^{k+1} \hat{F}_{k+1} \quad (20)$$
$$\left. + [{}^k_0 R]\hat{\tau}_I + [{}^k r^u_k][{}^k_0 R]\hat{F}_I \right).$$

Eqs. (12), (13), (19) and (20) can be expressed in an augmented form as follows

$$[D] \begin{pmatrix} \hat{F} \\ \hat{\tau} \end{pmatrix} = [A] \begin{pmatrix} \hat{F} \\ \hat{\tau} \end{pmatrix} + [B] \begin{pmatrix} \hat{F}_I \\ \hat{\tau}_I \end{pmatrix} \quad (21)$$

where $\hat{F} = ({}^1\hat{F}_1^T, \cdots, {}^N\hat{F}_N^T)^T$ and $\hat{\tau} = ({}^1\hat{\tau}_1^T, \cdots, {}^N\hat{\tau}_N^T)^T$. The block matrix $[D]$ consists of $[V_j]$ and $[W_j]$, and the block matrix $[B]$ has all zero matrices except the kth and the $(N+k)$th sub-matrices

$$[D] = \begin{bmatrix} [V_1] \\ \vdots \\ [V_N] \\ [W_1] \\ \vdots \\ [W_N] \end{bmatrix}, \quad [B] = \begin{bmatrix} 0 & 0 \\ \vdots & \vdots \\ [C_{Fk}]^T[{}^k_0 R] & 0 \\ 0 & 0 \\ \vdots & \vdots \\ [C_{Tk}]^T[{}^k r^u_k][{}^k_0 R] & [C_{Tk}]^T[{}^k_0 R] \\ 0 & 0 \\ \vdots & \vdots \end{bmatrix} \begin{matrix} k \\ \\ (N+k) \end{matrix}$$

and $[A]$ given by $[A] = \begin{bmatrix} [A_{FF}] & 0 \\ [A_{TF}] & [A_{TT}] \end{bmatrix}$, where the sub-matrices, $[A_{FF}]$, $[A_{TT}]$, and $[A_{TF}]$ are defined as follows:

$[A_{FF}]=$
$$\begin{bmatrix} -[C_{F1}]^T[C_{F1}] & [C_{F1}]^T[{}^1_2 R][C_{F2}] & 0 & 0 & 0 & 0 \\ 0 & -[C_{F2}]^T[C_{F2}] & \ddots & 0 & 0 & 0 \\ 0 & 0 & \ddots & [C_{F(j-1)}]^T[{}^{j-1}_j R][C_{Fj}] & 0 & 0 \\ 0 & 0 & 0 & -[C_{Fj}]^T[C_{Fj}] & \ddots & 0 \\ 0 & 0 & 0 & 0 & \ddots & [C_{F(N-1)}]^T[{}^{N-1}_N R][C_{FN}] \\ 0 & 0 & 0 & 0 & 0 & -[C_{FN}]^T[C_{FN}] \end{bmatrix}$$

$[A_{TT}]=$
$$\begin{bmatrix} -[C_{T1}]^T[C_{T1}] & [C_{T1}]^T[{}^1_2 R][C_{T2}] & 0 & 0 & 0 & 0 \\ 0 & -[C_{T2}]^T[C_{T2}] & \ddots & 0 & 0 & 0 \\ 0 & 0 & \ddots & [C_{T(j-1)}]^T[{}^{j-1}_j R][C_{Tj}] & 0 & 0 \\ 0 & 0 & 0 & -[C_{Tj}]^T[C_{Tj}] & \ddots & 0 \\ 0 & 0 & 0 & 0 & \ddots & [C_{T(N-1)}]^T[{}^{N-1}_N R][C_{TN}] \\ 0 & 0 & 0 & 0 & 0 & -[C_{TN}]^T[C_{TN}] \end{bmatrix}$$

$$[A_h] =$$

$$\begin{bmatrix} -[C_{j1}]^T[r_1^h][C_{j1}] & [C_{j1}]^T[r_1^{j-1}]_2^2[R][C_{j2}] & 0 & 0 & 0 & 0 \\ 0 & -[C_{j2}]^T[r_2^2][C_{j2}] & \ddots & 0 & 0 & 0 \\ 0 & 0 & \ddots & [C_{j(j-1)}]^T[r_{j-1}^{j-1}]_j^j[R][C_{jj}] & 0 & 0 \\ 0 & 0 & 0 & -[C_{jj}]^T[r_j^j][C_{jj}] & \ddots & 0 \\ 0 & 0 & 0 & 0 & \ddots & [C_{j(N-1)}]^T[r_{N-1}^{N-1}]_N^N[R][C_{jN}] \\ 0 & 0 & 0 & 0 & 0 & -[C_{jN}]^T[r_N^N][C_{jN}] \end{bmatrix}$$

Therefore, the internal impulses at the joints of serial-chain systems are evaluated as

$$\begin{pmatrix} \hat{F} \\ \hat{\tau} \end{pmatrix} = [A]^{-1}([D] - [B]) \begin{pmatrix} \hat{F}_I \\ \hat{\tau}_I \end{pmatrix}. \tag{22}$$

4 Internal Impulse of One-Module Closed-chain Systems

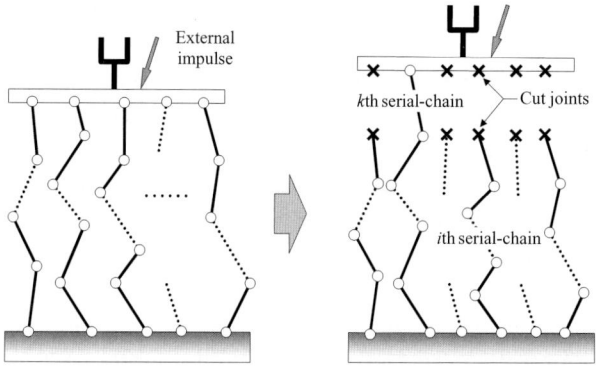

Fig. 2. A closed-chain module and its open-tree structure

In this section, the internal impulse model for closed-chain systems is introduced. It is assumed that the external impulse and the velocity increments at the joints are already evaluated according to Eqs. (5) and (6). To obtain the internal impulse model at the joints for a closed-chain system, we consider the open-tree structure of such system by cutting joints. If the system has only one closed-chain like a four-bar or five-bar mechanism, the open-tree structure can be made by cutting one joint. If the system consists of several closed-chains, at least one joint of each closed-chain should be cut. In the example given in Fig. 2, the cut joints are located at the top platform, because it is convenient to model the kinematics of the serial-chain. However, other cut joints may be chosen if there were other advantages.

The internal impulses at the cut joints are derived based on the dynamic models of the serial-chains. In Fig. 2, the joint set of the ith serial-chain is noted as $_i\phi$ and the set of the cut joints is noted as ϕ_h. The impulsive forces and the impulsive moments at the cut joints are represented as F_h and τ_h, respectively. These impulsive forces and moments acting on the both sides of the cut joint are equal in magnitude, opposite in direction. So the force and moment acting on one serial-chain are set as positive and those on the other are set as negative. The cut joint belonging to the ith serial-chain may be more than one, and the sets of the impulsive forces and the impulsive moments acting on the cut joints of the ith serial-chain are noted as $_iF_h$ and $_i\tau_h$, respectively.

The dynamic model of the ith serial-chain of the open-tree structure is given by

$$_iT = [_iI^*_{\phi\phi}]_i\ddot{\phi} + _i\dot{\phi}^T[_iP^*_{\phi\phi\phi}]_i\dot{\phi} \\ - [_iG^h_\phi]^T \begin{bmatrix} [_iC_F] & 0 \\ 0 & [_iC_T] \end{bmatrix} \begin{pmatrix} _iF_h \\ _i\tau_h \end{pmatrix}. \tag{23}$$

where $[_iC_F]$ and $[_iC_T]$ represent the constraint matrices for the cut joints. $[_iG^h_\phi]$ is the block matrix defined from

$$\begin{pmatrix} _iv_h \\ _i\omega_h \end{pmatrix} = [_iG^h_\phi]_i\dot{\phi}, \tag{24}$$

where $_iv_h$ is the velocity set for the cut joints of the ith serial-chain and $_i\omega_h$ is the angular velocity set for the link connected to the cut joints of the ith serial-chain. By integrating the dynamic model of Eq. (23) over the infinitesimal time interval, we have

$$\Delta_i\dot{\phi} = [_iI^*_{\phi\phi}]^{-1}[_iG^h_\phi]^T[_iC_h] \begin{pmatrix} \hat{F}_h \\ \hat{\tau}_h \end{pmatrix}, \tag{25}$$

where $[_iC_h]$ is the block matrix consisting of $[_iC_F]$ and $[_iC_T]$.

Premultiplying $[_iG^h_\phi]$ and $[_iC_h]^T$ consecutively to the both sides of Eq. (25) yields

$$[_iD_h]\begin{pmatrix}\hat{F}_I\\ \hat{\tau}_I\end{pmatrix}=[_iA_h]\begin{pmatrix}_i\hat{F}_h\\ _i\hat{\tau}_h\end{pmatrix}, \quad (26)$$

where

$$[_iA_h]=[_iC_h]^T[_iG_\phi^h][_iI_{\phi\phi}^*]^{-1}[_iG_\phi^h]^T[_iC_h]. \quad (27)$$

The left-hand side of Eq. (25) is obtained as follows

$$[_iC_h]^T[_iG_\phi^h]\Delta_i\dot{\phi}=[_iC_h]^T[_iG_\phi^h][^iG_a^\phi]$$
$$\times[I_{aa}^*]^{-1}[G_a^I]^T\begin{pmatrix}\hat{F}_I\\ \hat{\tau}_I\end{pmatrix}=[_iD_h]\begin{pmatrix}\hat{F}_I\\ \hat{\tau}_I\end{pmatrix} \quad (28)$$

according to the following relationships.

$$\Delta_i\dot{\phi}=[^iG_a^\phi]\Delta\dot{\phi}_a \quad (29)$$

$$\Delta\dot{\phi}_a=[I_{aa}^*]^{-1}[G_a^I]^T\begin{pmatrix}\hat{F}_I\\ \hat{\tau}_I\end{pmatrix} \quad (30)$$

where $[G_a^I]$ denotes the 1st-order KIC relating the 6×1 velocity vector at the contact point to the velocity vector at the independent joint set of a closed-chain system.

When a link of the *k*th serial-chain is influenced by the external impulses, the dynamic model of the *k*th serial-chain becomes

$$\Delta_k\dot{\phi}=[_kI_{\phi\phi}^*]^{-1}[_kG_\phi^h]^T[_kC_h]\begin{pmatrix}_k\hat{F}_h\\ _k\hat{\tau}_h\end{pmatrix}$$
$$+[_kI_{\phi\phi}^*]^{-1}[_kG_\phi^I]^T\begin{pmatrix}\hat{F}_I\\ \hat{\tau}_I\end{pmatrix}. \quad (31)$$

Applying the same procedure (i.e., consecutive premultiplying $[_iG_\phi^h]$ and $[_iC_h]^T$ to Eq. (31)) gives

$$[_kD_h]\begin{pmatrix}\hat{F}_I\\ \hat{\tau}_I\end{pmatrix}=[_kA_h]\begin{pmatrix}_k\hat{F}_h\\ _k\hat{\tau}_h\end{pmatrix}+[_kB_h]\begin{pmatrix}\hat{F}_I\\ \hat{\tau}_I\end{pmatrix}, \quad (32)$$

where

$$[_kB_h]=[_kC_h]^T[_kG_\phi^h][_kI_{\phi\phi}^*]^{-1}[_kG_\phi^I]^T$$

By aggregating the relation given in Eqs. (26) and (32) for all serial-chains, we have

$$[D_h]\begin{pmatrix}\hat{F}_I\\ \hat{\tau}_I\end{pmatrix}=[A_h]\begin{pmatrix}\hat{F}_h\\ \hat{\tau}_h\end{pmatrix}+[B_h]\begin{pmatrix}\hat{F}_I\\ \hat{\tau}_I\end{pmatrix}, \quad (33)$$

where $[A_h]$ and $[D_h]$ are the block diagonal matrices composed of $[_iA_h]$ and $[_iD_h]$, respectively. $[B_h]$ has all zero matrices except the block matrix corresponding to the *k*th serial-chain. The number of equations is larger than the number of variables, so the rows of Eqs. (26) and (32) should be carefully extracted in accordance with the mobility of each serial-chain so that the matrix $[A_h]$ of Eq. (33) should be a nonsingular square matrix. From Eq. (33), the internal impulses of all the cut joints of the system is obtained from Eq. (35). Next, the impulses at the remaining uncut joints should be obtained. All the serial-chains for the open-tree structure are treated as independent serial-chains, and the internal impulses and the internal impulse moments are evaluated by Eq. (22) for each serial-chain.

5 Docking at Space

Freely floating multi-body system is considered here, which is a simplified spacecraft docking with an artificial satellite as shown in Fig. 3. Assuming that the artificial satellite is approaching the stationary spacecraft and a docking platform is installed at the top of the spacecraft to absorb the shock when the two systems collide. The docking platform takes the shape of Stewart platform as shown in Fig. 4. The platform has 6 legs and each leg has three joints, which are universal, prismatic, and spherical joints. The legs are joined to the base plate by universal joints, and the length of the leg can be varied by virtue of a prismatic joint. The connection to the upper plate is a spherical joint

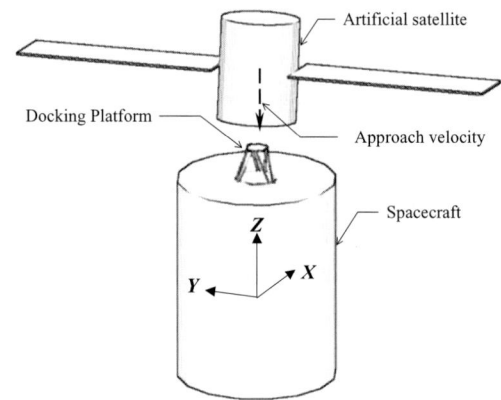

Fig. 3.   Spacecraft docking with an artificial satellite

The condition of simulation is that the spacecraft is stationary and artificial satellite is approaching at the velocity of $5 cm/s$ along opposite direction of Z-axis as shown in Fig. 3. The kinematic and dynamic parameters for the artificial satellite, the spacecraft, and the docking platform are summarized in Table 1. The satellite is tilted $10°$ about Y-axis, so the contact point is located on the edge of the upper plate. When the coefficient of restitution is 0.8, the magnitude of the external impulse evaluated from the external impulse model is
$$\hat{F}_I = \begin{pmatrix} -0.045 & 0 & -0.257 \end{pmatrix}^T = 0.261\, Ns.$$
The external impulse is illustrated as the dark gray arrow of Fig. 4. Once the external impulse is obtained, the internal impulses at the joints of the legs can be calculated by the previously introduced modeling method. The internal impulses at the joint of the platform are described in Table 2 and illustrated in Fig. 4. The relative magnitude of the internal impulses is shown as the dark arrows. It is noted that the internal impulses are too small compared to the external impulse. The internal impulse moments are not presented because the amounts of them are negligible.

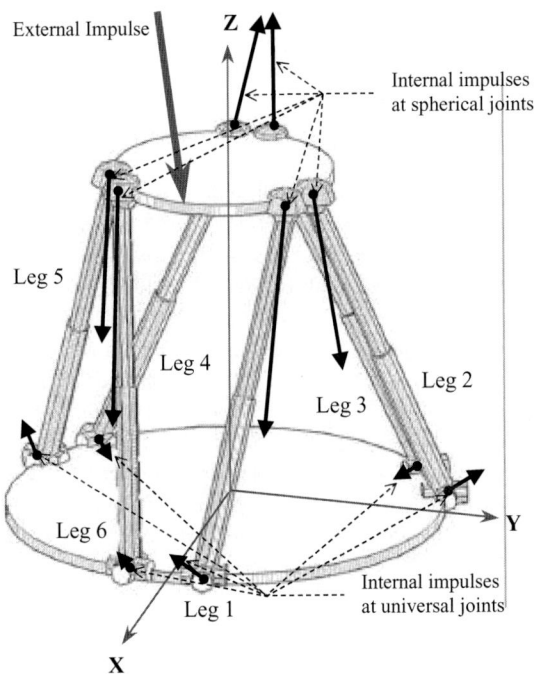

Fig. 4. External and internal impulses of the docking platform

It is shown that the external impulse is transmitted dispersively to the internal impulses at the spherical joints connected to the legs, and that the internal impulses are reduced as compared with the external impulse. It can be observed that the internal impulses at the spherical joints of leg 3 and 4 are acting upwardly and the directions are reverse to that of the external impulse. This is due to the contribution of the internal impulsive force to the increase of internal impulsive moment.

Table 1. Parameters of the docking system

Body	Kinematic parameter		Dynamic parameter	
	Radius (m)	Length (m)	Mass (kg)	Inertia [X,Y,Z] ($kg·m^2$)
Spacecraft	4.0	10.0	400	[-4.93, 4.93, 3.20]×10^3
Satellite	2.0	5.0	180	[3.08, 3.08, 2.00]×10^3
Upper plate	0.5	0.02	6	[0.31, 0.31, 0.63]
Platform base	1.0	–	–	–
Link of the leg	0.02	0.8	1	[0.05, 0.05, 0.00]

Table 2. Internal impulses of the docking platform ($×10^{-3} Ns$)

Leg No.	1	2	3	4	5	6
Spherical	60.4	44.6	36.2	36.2	44.6	60.4
Prismatic	2.8	1.3	1.6	1.6	1.3	2.8
Universal	10.5	5.2	5.9	5.9	5.2	10.5

The trend of the internal impulses shows that the magnitude of the internal impulse at the proximal joint is generally less than those of the distal joints. It is deduced that the external impulse experienced at the contact point is transmitted from the distal joint to the proximal joint in a decreasing fashion, since some part of the impulse is transformed into the momentum changes of the links and the remainder is transmitted to the next joint.

The impulse forces at the prismatic joints are very small because the directions of the impulses transmitted from the spherical joints are nearly the same with the longitudinal directions of the legs. The upper link of a leg can move easily along the direction of the impulse and the most part of the impulse is absorbed into the momentum increase of the link, so the very small part of the impulse is transmitted.

On the contrary, the impulses at the universal joints of the base plate become larger than those of the prismatic joints although the impulses transmitted from the prismatic joints are very small. This is also due to the internal impulse moment caused by the internal impulse.

Finally, the impulse acting on the spacecraft is 0.0173 $N \cdot s$, and it is 6.62 % of the external impulse. About 94 % of the impulse is absorbed by the momentum change of the mechanism. Therefore, the satellite can berth smoothly onto the spacecraft.

From the above result, it is observed that the impact propagation is deeply associated with the structure and the configuration of the shock absorber and the careful design and pose control is needed to absorb the impact effectively according to circumstances.

6 Conclusion

Analysis on the impulses acting at the joints when the robot system collides with environment has not attracted much attention. In this work, we employed the internal impulse model was derived as compact closed-form solutions that facilitate the analysis of the internal impulses. It was shown through simulation that the proposed modeling method can successfully evaluate the internal impulses of general multi-body systems.

Specifically, an artificial satellite docking with a spacecraft was treated. By the introduced impact analysis method, the internal impulses of the docking platform are evaluated and analyzed. From the analysis results, it was observed that the internal impulse has strong inter-relationship with momentum change of the linkages of the system. Thus, in order to minimize the internal impulse, the design and the motion planning of the given linkage system becomes crucial.

Besides this example, the proposed algorithm is applicable to analysis of internal impact propagation of general closed-chain mechanisms. Furthermore, the analysis results can give the fundamental data for the design of a shock absorbing mechanism.

Acknowledgement

This work was supported in part by the Brain Korea 21 project.

References

[1] R. M. Brach, "Classical planar impact theory and the tip impact of a slender rod", *International Journal of Impact Engineering*, Vol. 13, No. 1, pp. 21-33, 1993.

[2] G. Ferretti, G. Magnani, and A. Zavala Rio, "Impact modeling and control for industrial manipulators", *IEEE Control System Magazine*, Vol. 18, No. 4, pp. 65-71, 1998.

[3] I. D. Walker, "Impact configurations and measures for kinematically redundant and multiple armed robot systems", *IEEE Transactions on Robotics and Automation*, Vol. 10, No. 5, pp. 670-683, 1994.

[4] I. D. Walker, "The use of kinematic redundancy in reducing impact and contact effects in manipulation", *Proceeding of 1990 IEEE Conference on Robotics and Automation*, pp. 434-439, 1990.

[5] H-T. Liao and M. C. Leu, "Analysis of impact in robotic peg-in-hole assembly", *Robotica*, Vol. 16, No. 3, pp. 347-356, 1998.

[6] Y-F. Zheng and H. Henami, "Mathematical modeling of a robot collision with its environment", *Journal of Robotic Systems*, Vol. 2, No. 3, pp. 289-307, 1985.

[7] J. Wittenburg, *Dynamics of systems of rigid bodies*, B.G. Teubner, Stuttgart, 1977.

[8] R. A. Freeman and D. Tesar, "Dynamic modeling of serial and parallel mechanisms/robotic systems, Part I-Methodology, Part II-Applications", *Proceedings on 20th ASME Biennial Mechanisms Conference, Trends and Development in Mechanisms, Machines, and Robotics*, Orlando, FL, DE-Vol. 15-3, pp. 7-27, 1988.

[9] Y. Nakamura and M. Ghodoussi, "Dynamic computation of closed-link robot mechanisms with nonredundant and redundant actuators", *IEEE Transaction on Robotics and Automation*, Vol. 5, No. 3, pp. 294-302, 1989.

[10] H. J. Kang, B-J. Yi, W. Cho, and R. A. Freeman, "Constraint-embedding approaches for general closed-chain system dynamic in terms of a minimum coordinate set", *The 1990 ASME Biennial Mechanism Conference*, Chicago, IL, DE-Vol. 24, pp. 125-132, 1990.

[11] S. H. Lee, B. J. Yi, S. H. Kim, and , Y. K. Kwak, "Modeling and Analysis of Internal Impact for General Classes of Robotic Mechanisms," *2000 IEEE/RSJ Int. Conference on Intelligent Robots and Systems*, pp. 1955-1962, 2000

Flyaround Maneuvers on a Satellite Orbit by Impulsive Thrust Control

Yasuhiro Masutani   Motoshi Matsushita   Fumio Miyazaki

Graduate School of Engineering Science, Osaka University

Toyonaka, Osaka 560-8531, Japan

`masutani@me.es.osaka-u.ac.jp`

Abstract

Close circumnavigation is an important function indispensable for servicing satellites. We discuss the bi-elliptic flyaround maneuver by impulsive thrust control for a small and low cost servicing satellite flying around a target satellite. An optimal feedback control scheme for the thrust is proposed to maintain this trajectory in the presence of disturbances. The extended Kalman filter is employed to estimate state variables which are not available for measurement. Simulation results that verify the trajectory keeping capability of the proposed thrust control are presented.

Keywords: satellite, flyaround maneuvers, impulsive thrust control, optimal regulator, extended Kalman filter

1 Introduction

In order to provide services for an Earth-orbiting satellite, it is desirable to be able to use a fully autonomous servicing satellite that flies around it, observes it carefully, and performs distant operations. In practice, NASA developed a prototype remotely controlled free-flying television camera, AERCam Sprint (Autonomous Extravehicular Activity Robotic Camera Sprint), so as to inspect large scale structures such as the International Space Station and experimentally examined its performance in a Space Shuttle mission STS-87 (Space Transportation System-87) in 1997 [1][2]. NASDA is planning to develop "Eyeball Satellite" flying around its mother satellite and sending inspected data to the Earth [3].

Close circumnavigation is an important function indispensable for servicing satellites. Conventional flyaround maneuvers including these two examples use circular, elliptical, or rectilinear trajectories relative to the target satellite. Future space service missions are driving the need for small, low cost satellites whose motion relative to the target satisfy the following conditions;

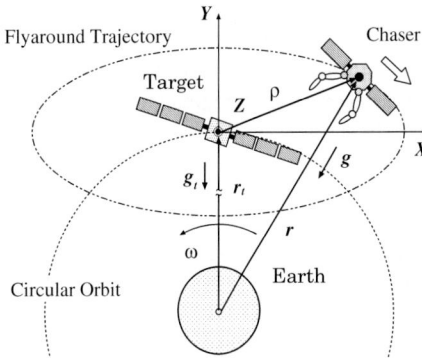

Fig. 1 Modeling of Flyaround Maneuver

- The flyaround period is variable (especially shorter than the Earth-orbiting period of the target).
- The distance from a target satellite does not vary much.
- The amount of fuel consumption is small.
- The trajectory can be maintained under disturbances by simple thruster control.

In this paper, we discuss the bi-elliptic flyaround maneuver that meets all of the requirements mentioned above and propose an impulsive thrust control scheme to hold the trajectory under disturbances.

This paper is organized as follows. In **Sec.2**, we investigate the bi-elliptic flyaround orbit for a servicing satellite (chaser) flying around an objective satellite (target) using the linearized relative orbital dynamics known as Hill's equation. The bi-elliptic flyaround maneuver can be accomplished using impulsive thrust two times within a flyaround period to partly combined two elliptic trajectory. An optimal feedback control scheme for the thrust is given in **Sec.3** to maintain this trajectory in the presence of external and internal disturbances. **Sec.4** describes how to estimate state variables necessary for the feedback control using the extended Kalman filter, assuming that the direction of the target is the only variable the chaser can observe. Simulations in **Sec.5** illustrate the performance of the proposed flyaround maneuver. Concluding remarks are made in **Sec.6**.

2 Modeling of Flyaround Maneuvers

In this section, we will examine the motion of the chaser with respect to the target and propose a bi-elliptic flyaround maneuver for the chaser.

2.1 Hill's Equation

Fig. 1 presents the vector positions of the chaser and target satellites at some time with respect to the center of the Earth, $\boldsymbol{r}$ and $\boldsymbol{r}_t$. The position of the chaser with respect to the target is $\boldsymbol{\rho}$. An orthogonal coordinate frame is attached to the target and moves with it. The vector position of the chaser yields

$$\boldsymbol{r} = \boldsymbol{r}_t + \boldsymbol{\rho} \qquad (1)$$

Differentiating this equation with respect to an inertial coordinate frame results in

$$\ddot{\boldsymbol{r}} = \ddot{\boldsymbol{r}}_t + \ddot{\boldsymbol{\rho}} + 2(\boldsymbol{\omega} \times \dot{\boldsymbol{\rho}}) + \dot{\boldsymbol{\omega}} \times \boldsymbol{\rho} + \boldsymbol{\omega} \times (\boldsymbol{\omega} \times \boldsymbol{\rho}) \qquad (2)$$

where

- $\ddot{\boldsymbol{r}}$ the inertial acceleration of the chaser
- $\ddot{\boldsymbol{r}}_t$ the inertial acceleration of the target
- $\ddot{\boldsymbol{\rho}}$ the acceleration of the chaser relative to the target

Now, let $\ddot{\boldsymbol{r}} = \boldsymbol{g} + \boldsymbol{A}$, where $\boldsymbol{g}$ is the gravitational acceleration and $\boldsymbol{A}$ is the acceleration applied by thrust whose components are given by (A_x, A_y, A_z).

Assuming that the target-to-chaser distance is much smaller than the orbit radius of the target ($\boldsymbol{r}_t$), the linearized equation of Eq.(2) called "Hill's equation" [4][5] becomes

$$\ddot{x} = -g_t x/r_t + 2\omega \dot{y} + \dot{\omega} y + \omega^2 x + A_x \qquad (3)$$
$$\ddot{y} = 2g_t y/r_t - 2\omega \dot{x} - \dot{\omega} x + \omega^2 y + A_y \qquad (4)$$
$$\ddot{z} = -g_t z/r_t + A_z \qquad (5)$$

When the target is in a circular orbit, $\dot{\omega} = 0$ and $\omega = \sqrt{g_t/r_t}$, these equations result in

$$\ddot{x} = 2\omega \dot{y} + A_x \qquad (6)$$
$$\ddot{y} = -2\omega \dot{x} + 3\omega^2 y + A_y \qquad (7)$$
$$\ddot{z} = -\omega^2 z + A_z \qquad (8)$$

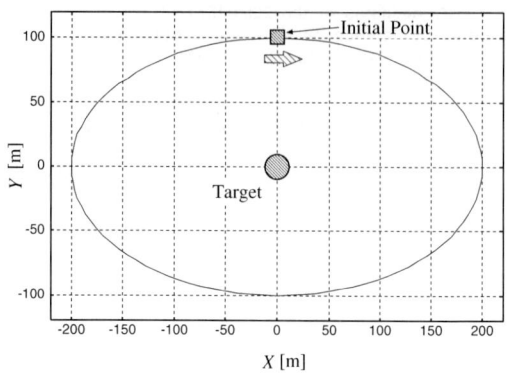

Fig. 2 Elliptic Flyaround Maneuver

2.2 Single Elliptic Flyaround Maneuver

It is well-known that the elliptic flyaround maneuver can be achieved over an orbit period by an initial velocity impulse, which is a sort of solution of Eqs.(6)-(8) in case that no external forces act ($A_x = A_y = A_z = 0$). The solution in the orbital plane is of the form

$$\begin{bmatrix} x(t) \\ y(t) \end{bmatrix} = y_0 \begin{bmatrix} 2\sin \omega t \\ \cos \omega t \end{bmatrix} \qquad (9)$$

where the initial condition is given by

$$[x(0), y(0), \dot{x}(0), \dot{y}(0)] = [0, y_0, 2y_0\omega, 0]$$

This solution means that the major axis of the ellipse is equal in magnitude to twice its minor axis as shown in Fig. 2 and that its period is equal to the Earth-orbiting period $2\pi/\omega$. As a result, although this elliptic trajectory needs no fuel consumption except for the initial thrust theoretically, it is inadequate to servicing satellites because the distance between the chaser and target changes a lot as time passes by and its flyaround period is fixed.

A circular flyaround at a constant radius can also be performed. However, it requires a continuous application of thrust, which makes it difficult to realize low-cost servicing satellites.

2.3 Bi-elliptic Flyaround Maneuver

In this section, we consider a bi-elliptic trajectory shown in Fig. 3. This trajectory is obtained by solving Eqs.(6)-(8) taking account of compatible connection of

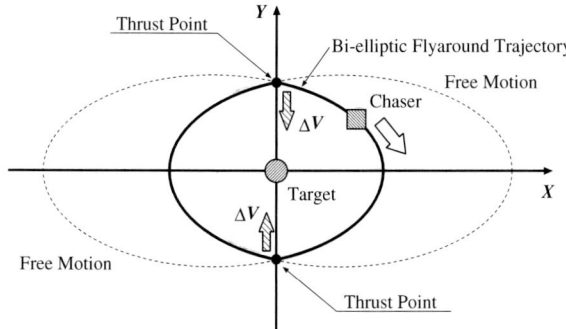

Fig. 3 Bi-elliptic Flyaround Maneuver

two elliptic trajectories, which can be written as:

$$\begin{bmatrix} x(t) \\ y(t) \end{bmatrix} = \begin{cases} \frac{y_0}{\sin \frac{T}{4}} \begin{bmatrix} -2\cos\frac{T}{4} + 2\cos(-\omega t' + \frac{T}{4}) \\ +\sin(-\omega t' + \frac{T}{4}) \end{bmatrix} \\ \qquad\qquad\qquad\qquad (0 < t' < \frac{T}{2}) \\ \frac{y_0}{\sin \frac{T}{4}} \begin{bmatrix} +2\cos\frac{T}{4} - 2\cos(-\omega t' + \frac{3T}{4}) \\ -\sin(-\omega t' + \frac{3T}{4}) \end{bmatrix} \\ \qquad\qquad\qquad\qquad (\frac{T}{2} < t' < T) \end{cases} \quad (10)$$

where T is the flyaround period of the chaser and $t' = \mathrm{mod}(t, T)$.

The flyaround maneuver in this case needs the discontinuous change in velocity two times within a flyaround period given by

$$\Delta \boldsymbol{V} = \begin{bmatrix} \Delta v_x \\ \Delta v_y \end{bmatrix} = \begin{bmatrix} 0 \\ \pm y_0 \frac{2\omega}{\tan \frac{T}{4}} \end{bmatrix} \quad (11)$$

This can be achieved by applying two thrust impulses in the direction of the y-axis at time $t' = 0$ and $t' = T/2$. It is possible to change the flyaround period T by adjusting the magnitude of velocity impulse $\Delta \boldsymbol{V}$.

3 Trajectory Keeping by Optimal Feedback Control

The bi-elliptic maneuver proposed in the previous section is an approximation based on ignoring certain effects that need to be included in practical situations. A satellite in flyaround trajectory experiences small but significant perturbations due to atmospheric drag and impulsive thrust approximation, that is, the losses due to the finite duration of thrust. In order to maintain the flyaround orbit, it is necessary to modify the magnitude of velocity impulse based on the observed perturbation. In this section, we consider the standard linear quadratic (LQ) state feedback strategy for the thrust control.

3.1 Discrete-time State-space Representation

Since the impulsive thrust is applied every half of the flyaround period $T/2$, we convert the equation of the bi-elliptic trajectory given by Eq.(10) to a discrete-time state equation with the sampling period $T/2$. We put even and odd numbers at the beginning of the right and left half of the bi-elliptic trajectory to represent sampling instants respectively.

Let us consider state transition on the half of the bi-elliptic trajectory. Given some conditions of state variables $\boldsymbol{\zeta}_i \in \mathbf{R}^4$ and the velocity impulse $\boldsymbol{\nu}_i \in \mathbf{R}^2$ at time $t = iT/2$, the state variables $\boldsymbol{\zeta}_{i+1}$ after one sampling period yield

$$\boldsymbol{\zeta}_{i+1} = \boldsymbol{A}\boldsymbol{\zeta}_i + \boldsymbol{B}\boldsymbol{\nu}_i \quad (i = 0, 1, 2, \cdots) \quad (12)$$

where

$$\boldsymbol{\zeta}_i = \left[x(i\tfrac{T}{2}), y(i\tfrac{T}{2}), \dot{x}(i\tfrac{T}{2}), \dot{y}(i\tfrac{T}{2}) \right]^T \quad (13)$$

$$\boldsymbol{\nu}_i = \left[\Delta v_x(i\tfrac{T}{2}), \Delta v_y(i\tfrac{T}{2}) \right]^T \quad (14)$$

$$\boldsymbol{A} = \begin{bmatrix} 1 & -6\sin\frac{\omega T}{2} + 3\omega T & \frac{4}{\omega}\sin\frac{\omega T}{2} - \frac{3T}{2} & \frac{2}{\omega}(1 - \cos\frac{\omega T}{2}) \\ 0 & 4 - 3\cos\frac{\omega T}{2} & \frac{2}{\omega}(\cos\frac{\omega T}{2} - 1) & \frac{1}{\omega}\sin\frac{\omega T}{2} \\ 0 & 6\omega(1 - \cos\frac{\omega T}{2}) & 4\cos\frac{\omega T}{2} - 3 & 2\sin\frac{\omega T}{2} \\ 0 & 3\omega\sin\frac{\omega T}{2} & -2\sin\frac{\omega T}{2} & \cos\frac{\omega T}{2} \end{bmatrix} \quad (15)$$

$$\boldsymbol{B} = \begin{bmatrix} \frac{4}{\omega}\sin\frac{\omega T}{2} - \frac{3T}{2} & \frac{2}{\omega}(1 - \cos\frac{\omega T}{2}) \\ \frac{2}{\omega}(\cos\frac{\omega T}{2} - 1) & \frac{1}{\omega}\sin\frac{\omega T}{2} \\ 4\cos\frac{\omega T}{2} - 3 & 2\sin\frac{\omega T}{2} \\ -2\sin\frac{\omega T}{2} & \cos\frac{\omega T}{2} \end{bmatrix} \quad (16)$$

This equation is obtained by solving Hill's equation of Eqs.(6)-(8).

On the other hand, state transition is symmetric with respect to the interchange of coordinates between the right and left half of the trajectory. Thus, in order to use a common nominal state for the right and left half, the state transition Eq.(12) is rewritten as follows:

$$\tilde{\boldsymbol{\zeta}}_{i+1} = \tilde{\boldsymbol{A}}\tilde{\boldsymbol{\zeta}}_i + \tilde{\boldsymbol{B}}\tilde{\boldsymbol{\nu}}_i \quad (i = 0, 1, 2, \cdots) \quad (17)$$

where

$$\tilde{\boldsymbol{A}} = -\boldsymbol{A} \quad (18)$$

$$\tilde{\boldsymbol{B}} = -\boldsymbol{B} \quad (19)$$

$$\tilde{\zeta}_i = \begin{cases} +\zeta_i & (i=0,2,4,\cdots) \\ -\zeta_i & (i=1,3,5,\cdots) \end{cases} \quad (20)$$

$$\tilde{\nu}_i = \begin{cases} +\nu_i & (i=0,2,4,\cdots) \\ -\nu_i & (i=1,3,5,\cdots) \end{cases} \quad (21)$$

The nominal operating condition $\tilde{\zeta}^*$ and $\tilde{\nu}^*$ are derived from the results in **Sec.2.3** as follows:

$$\tilde{\zeta}^* = \left[0,\ y_0,\ 2\omega y_0,\ \frac{\omega y_0}{\tan\frac{T}{4}}\right]^T \quad (22)$$

$$\tilde{\nu}^* = \left[0,\ -\frac{2\omega y_0}{\tan\frac{T}{4}}\right]^T \quad (23)$$

Let us introduce small deviations $\Delta\tilde{\zeta}_i$ and $\Delta\tilde{\nu}_i$ such that

$$\Delta\tilde{\zeta}_i = \tilde{\zeta}_i - \tilde{\zeta}^* \quad (i=0,1,2,\cdots) \quad (24)$$
$$\Delta\tilde{\nu}_i = \tilde{\nu}_i - \tilde{\nu}^* \quad (i=0,1,2,\cdots) \quad (25)$$

Substituting Eqs.(24) and (25) in Eq.(17), we obtain the following discrete-time state-representation:

$$\Delta\tilde{\zeta}_{i+1} = \tilde{A}\Delta\tilde{\zeta}_i + \tilde{B}\Delta\tilde{\nu}_i \quad (26)$$

3.2 Optimal Control Design

The problem of maintaining the nominal trajectory can be solved by the standard LQ state feedback approach. We consider the problem of determining the optimal control vector $\Delta\tilde{\nu}_i$ for the system described by Eq.(26) and the performance index given by

$$J = \sum_{k=1}^{\infty}(\Delta\tilde{\zeta}_k^T Q \Delta\tilde{\zeta}_k + \Delta\tilde{\nu}_k^T R \Delta\tilde{\nu}_k) \quad (27)$$

where Q is a positive-semidefinite real symmetric matrix and R is a positive-definite real symmetric matrix. It is well-known that the optimal control that minimizes the performance index is

$$\Delta\tilde{\nu}_i = -F\Delta\tilde{\zeta}_i \quad (28)$$

where the optimal feedback gain matrix F is given by

$$F = (\tilde{B}^T S \tilde{B} + R)^{-1}\tilde{B}^T S \tilde{A} \quad (29)$$

The matrix S in Eq.(29) is the solution to the algebraic Riccati equation

$$\tilde{A}^T S \tilde{A} - S - \tilde{A}^T S \tilde{B}(\tilde{B}^T S \tilde{B} + R)^{-1}\tilde{B}^T S \tilde{A} + Q = O \quad (30)$$

Note that the discrete-time system of Eq.(26) is controllable on the overall bi-elliptic trajectory.

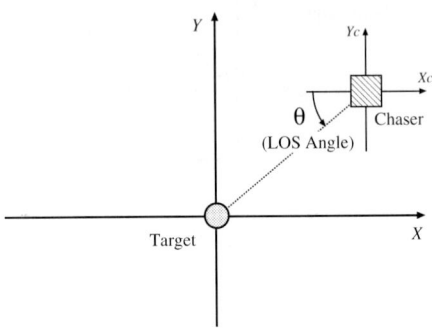

Fig. 4 Observation of the Target from the Chaser

4 State Estimation Using Extended Kalman Filter

The optimal control strategy given in the previous section is based on the assumption that the full state vector is available for measurement. In flyaround maneuvers, however, the entire state vector is not available for measurement. For example, Global Positioning System (GPS) which is used to measure the relative position among satellites is not available to a target satellite in trouble. Radar observations suitable for position measurement in long range is not applicable to close circumnavigation. In this paper, we assume that the line of sight (LOS) shown in Fig. 4, that is, the direction of the target relative to the chaser, is the only state variable available for measurement. This may be a practicable assumption, referring to currently available optical devices.

4.1 Extended Kalman Filter Design

From Hill's equation of Eqs.(6) and (7), the equation of motion for a chaser satellite can be expressed in the general state space form

$$\dot{\xi}(t) = A_c\xi(t) + B_c u(t) + Gv(t) + F_D \quad (31)$$
$$\eta(t) = c(\xi(t)) + w(t) \quad (32)$$

where

$$\xi(t) = [x(t), y(t), \dot{x}(t), \dot{y}(t)]^T \quad (33)$$
$$u(t) = [A_x(t), A_y(t)]^T \quad (34)$$

$$A_c = \begin{bmatrix} 0 & 0 & 1 & 0 \\ 0 & 0 & 0 & 1 \\ 0 & 0 & 0 & 2\omega \\ 0 & 3\omega^2 & -2\omega & 0 \end{bmatrix} \quad B_c = \begin{bmatrix} 0 & 0 \\ 0 & 0 \\ 1 & 0 \\ 0 & 1 \end{bmatrix} \quad (35)$$

$$G = \begin{bmatrix} 0 & 0 \\ 0 & 0 \\ 1 & 0 \\ 0 & 1 \end{bmatrix} \quad F_D = \begin{bmatrix} 0 \\ 0 \\ f_D \\ 0 \end{bmatrix} \quad (36)$$

$$c(\boldsymbol{\xi}(t)) = \theta = \tan^{-1} \frac{y(t)}{x(t)} \quad (37)$$

and F_E includes disturbances due to atmospheric drag and other deterministic noises, and $v(t), w(t)$ represent the process and measurement noise processes which are assumed white with the following characteristics:

$$E[\boldsymbol{v}(t)] = \boldsymbol{0} \quad (38)$$
$$E[w(t)] = 0 \quad (39)$$
$$E[\boldsymbol{v}(t)\boldsymbol{v}(\tau)^T] = \boldsymbol{Q}\delta(t-\tau) \quad (40)$$
$$E[w(t)w(\tau)^T] = R\delta(t-\tau) \quad (41)$$
$$E[\boldsymbol{v}(t)w(\tau)^T] = \boldsymbol{0} \quad (42)$$

Considering the available discrete-time measurement for the output η, we convert the system equation of Eqs.(31) and (32) to a discrete-time system equation with sampling period T_s.

$$\boldsymbol{\xi}[n+1] = \boldsymbol{A}_d \boldsymbol{\xi}[n] + \boldsymbol{B}_d \boldsymbol{u}[n] + \boldsymbol{G}\boldsymbol{v}[n] \quad (43)$$
$$\eta[n] = c(\boldsymbol{\xi}[n]) + w[n] \quad (44)$$

where

$$\boldsymbol{A}_d = \begin{bmatrix} 1 & 0 & T_s & 0 \\ 0 & 1 & 0 & T_s \\ 0 & 0 & 1 & 2\omega T_s \\ 0 & 3\omega^2 T_s & -2\omega T_s & 1 \end{bmatrix} \quad \boldsymbol{B}_d = \begin{bmatrix} 0 & 0 \\ 0 & 0 \\ T_s & 0 \\ 0 & T_s \end{bmatrix}$$
$$(45)$$

Since the output model of Eq.(44) is nonlinear, we have to use the Extended Kalman Filter (EKF) to produce estimates of the states [6]. Linearizing the output model near the estimated point $\widehat{\boldsymbol{\xi}}[n|n-1]$, we obtain the linearized output equation

$$\eta[n] = \boldsymbol{C}[n]\boldsymbol{\xi}[n] + w[n] \quad (46)$$

where

$$\boldsymbol{C}[n] = \left.\frac{\partial c}{\partial \boldsymbol{\xi}}\right|_{\boldsymbol{\xi}=\widehat{\boldsymbol{\xi}}[n|n-1]} \quad (47)$$
$$= \left[-\frac{y}{(x^2+y^2)}, \frac{x}{(x^2+y^2)}, 0, 0\right]_{\boldsymbol{\xi}=\widehat{\boldsymbol{\xi}}[n|n-1]} \quad (48)$$

The EKF algorithm has the form

$$\widehat{\boldsymbol{\xi}}[n|n] = \widehat{\boldsymbol{\xi}}[n|n-1] + \boldsymbol{K}[n](\eta[n] - c(\widehat{\boldsymbol{\xi}}[n|n-1])) \quad (49)$$

$$\widehat{\boldsymbol{\xi}}[n+1|n] = \boldsymbol{A}_d \widehat{\boldsymbol{\xi}}[n|n] + \boldsymbol{B}_d \boldsymbol{u}[n] \quad (50)$$

where the Kalman gain $\boldsymbol{K}$ is the solution of the equations

$$\boldsymbol{K}[n] = \boldsymbol{P}[n|n-1]\boldsymbol{C}[n]^T(\boldsymbol{C}[n]\boldsymbol{P}[n|n-1]\boldsymbol{C}[n]^T + R')^{-1}$$
$$(51)$$
$$\boldsymbol{P}[n|n] = \boldsymbol{P}[n|n-1] - \boldsymbol{K}[n]\boldsymbol{C}[n]\boldsymbol{P}[n|n-1] \quad (52)$$
$$\boldsymbol{P}[n+1|n] = \boldsymbol{A}_d \boldsymbol{C}[n|n]\boldsymbol{A}_d^T + \boldsymbol{G}_d \boldsymbol{Q}' \boldsymbol{G}_d^T \quad (53)$$

In these equations, the values for R and $\boldsymbol{Q}$ are assumed to be identified as R' and $\boldsymbol{Q}'$.

5 Simulation Results

The proposed bi-elliptic flyaround maneuver has been tested in simulation. Here, we present a result demonstrating that the proposed control scheme with the EKF makes it possible to maintain the bi-elliptic flyaround orbit under disturbances.

5.1 Thrusting Method — VIC

Several thrusting methods can be employed to generate the velocity impulse. Considering the short distance between the chaser and target (less than a few hundred meters), we use the Velocity Increment Cutoff (VIC) method in the following simulation.

Procedures in VIC to generate the velocity impulse are

1. Calculating the desired velocity $\boldsymbol{v}_d = \boldsymbol{\zeta}_{3,4} + \boldsymbol{\nu}$, where the actual velocity impulse $\boldsymbol{\nu} = [\Delta v_x, \Delta v_y]^T$ is calculated from Eqs.(25) and (21) and the optimal deviation of the velocity impulse $\Delta \widetilde{\boldsymbol{\nu}}$ is given by the feedback law Eq.(28).

2. Comparing $\boldsymbol{v}_d$ and the current velocity estimated with the EKF for each component, then giving constant thrusts $\pm U_x, \pm U_y$ in the direction of the x-axis and y-axis.

3. Repeating 2 until the sign of the difference does not change every time the estimated state is updated at intervals of T_s.

which are illustrated in Fig. 5.

5.2 Flyaround Example

Fig. 6 gives a block diagram of the chaser's bi-elliptic trajectory keeping control system together with VIC. All parameters required to perform simulations are shown in TABLE I. Fig. 7 shows a simulation result for the flyaround bi-elliptic maneuver with the flyaround period 59% of the Earth-orbiting period of the target.

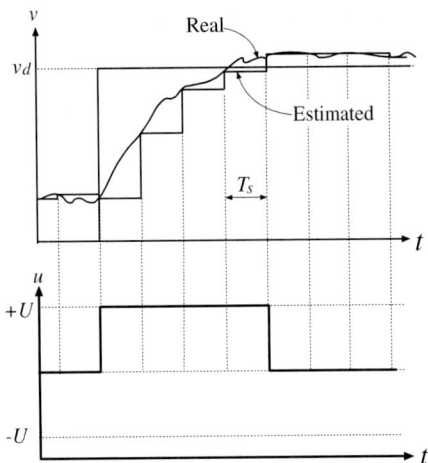

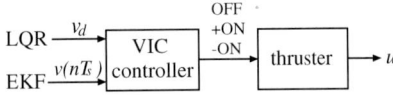

Fig. 5 VIC Controller

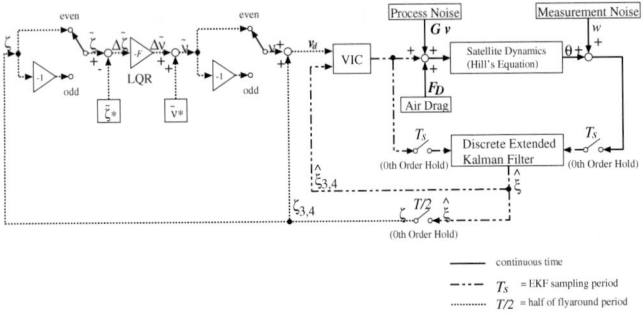

Fig. 6 Block Diagram of Bi-Elliptic Flyaround System

TABLE I Parameters of Bi-elliptic Flyaround Simulation	
Radius of Target Orbit r_t [km]	6900
Angular Velocity ω [rad/s]	1.1015×10^{-3}
Flyaround Period T [s]	3365
Normalized Flyaround Period k	0.59
EKF Sampling Period T_s [s]	0.1
S.D. of Process Noise $\sqrt{Q}/\alpha_{in}$	0.01
S.D. of Measurement Noise $\sqrt{R_\theta}$ [deg]	0.1
LQR Weight q : $Q = qI$	1.0
LQR Weight r : $R = rI$	1.0
Mass(Chaser) m [kg]	500
Air Drag f_D/m [m/s^2]	2.5×10^{-8}
Thruster Ability U_x [N]	5
Thruster Ability U_y [N]	5

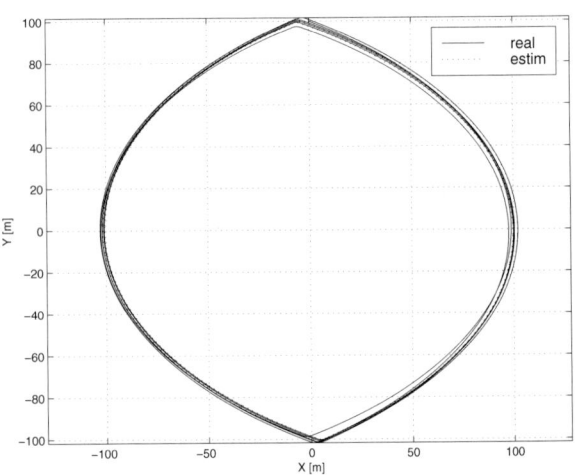

Fig. 7 Result of Bi-elliptic Flyaround Simulation

6 Conclusion

The bi-elliptic flyaround maneuver for servicing satellites was studied, and an optimal control scheme with the EKF was proposed to maintain this trajectory in the presence of external and internal disturbances. In the simulation performed, the proposed control scheme was found to be successful for the trajectory keeping problem in a practical situation. Analyzing the observability of the system in more detail will be pursued in future work.

References

[1] Howie Choset, David Kortenkamp, "Path Planning And Control for Free-Flying Inspection Robot In Space", Journal of Aerospace Engineering, April, 1999.

[2] WWW page on AERCam/Sprint, http://tommy.jsc.nasa.gov/projects/Sprint/

[3] WWW page on Eyeball Satellite http://sentan.tksc.nasda.go.jp/hypersat/eyeball.html

[4] Vladimir A. Chobotov (ed.), "Orbital Mechanics", AIAA, 1991.

[5] John E. Prussing, Bruce A. Conway, "Orbital Mechanics", Oxford University Press, 1993.

[6] A. Gleb (ed.), "Applied Optimal Estimation", MIT Press, 1974.

Nonlinear Control Methods for Planar Carangiform Robot Fish Locomotion

Kristi A. Morgansen[†‡], Vincent Duindam[‡], Richard J. Mason[‡],
Joel W. Burdick[‡], Richard M. Murray[†‡] *

[†]Control and Dynamical Systems [‡]Mechanical Engineering
California Institute of Technology, Mail Code 107-81, Pasadena, CA, 91125
{kristi,vincent,murray}@cds.caltech.edu, {jwb,mason}@robotics.caltech.edu

Abstract: *This paper considers the design of motion control algorithms for robot fish. We present modeling, control design, and experimental trajectory tracking results for an experimental planar robotic fish system that is propelled using carangiform-like locomotion. Our model for the fish's propulsion is based on quasi-steady fluid flow. Using this model, we propose gaits for forward and turning trajectories and analyze system response under such control strategies. Our models and predictions are verified by experiment.*

1 Introduction

This paper investigates the control of fish-like robots that propel themselves by changes in their shape rather than by the use of propellers and maneuvering surfaces. The study of underwater locomotion has long been a subject of interest to the biological community [3, 6, 13]. In the past several years, the robotics and engineering communities have been inspired by this research to construct mechanisms that mimic the behavior of swimming lifeforms. The motivation for this work comes from the high maneuverability that fish demonstrate over conventional propeller-driven underwater vehicles.

Some of the most impressive swimmers in nature propel themselves by the *carangiform* style of swimming. In carangiform swimming, the front two-thirds of the fish's body moves in a largely rigid way, with the propulsive body movements being confined to the rear third of the fish's body—primarily the tail. Carangiform movement is one of the easiest to replicate from a mechanical design perspective. Previous work in this area has come from the robopike and robotuna projects at MIT and Draper Laboratories [1, 2, 15].

Prior robotic fish research has focused on the issue of propulsion efficiency and fluid flow effects and has been primarily empirical. For example, the approach taken in the MIT and Draper Laboratories projects focused on a parameterized kinematic model of the mechanical fish rather than detailed models of the robot fluid-body interaction. The parameters in the model were determined by extensive experimental trials so as to minimize overall system drag. The end result of this effort produced a reproduction of a tuna that swims untethered in open water. However, the accuracy and robustness with which these vehicles can track a trajectory is unknown. Others have recently studied the vortices shed by a pitching and heaving plate [5].

Our work differs from these studies in one main respect: we focus on the issue of motion planning and control. A suitable model for the fluid-body interaction is clearly a prerequisite for control system analysis and design. In previous years, Caltech work in this area has focused on studying the fluid-body interaction using two approaches. The first approach involves a reduced Lagrangian formulation where a fish is taken to be a rigid body with the action of the tail represented by a point vortex of independently controlled position and strength [4]. The second approach is based on a highly simplified quasi-static lift and drag model of the forces on the fish body and tail [8, 9]. Previous papers presented the development of a robotic testbed for planar carangiform locomotion [8]. An updated version of this testbed is used in this paper to verify our approach. Prior papers also presented comparisons between the behavior of the experimental system and the restriction of the model to the case of forward propulsion [9]. Some simple turning maneuvers were also discussed.

In this paper we use methods from nonlinear control theory to generate system inputs which allow our fish-like robot to track simple trajectories. The model for this work is based on a modest extension of the simple quasi-static lift and drag approach previously

---

*This work was supported in part by the National Science Foundation through an Engineering Research Center grant and through NSF grant CMS-9502224.

proposed. While this model does not capture every detail of the fluid-body interaction, it does serve several useful purposes. First, it allows us to write a description of our system in a control affine form where the control inputs enter linearly. We can therefore apply known methods to analyze control performance. Importantly, we note that the linearized equations are not controllable, precluding the use of standard linear methods for trajectory tracking. However, as we will show, application of recent results in nonlinear control theory demonstrate that we can generate a variety of gaits and achieve trajectory tracking for this system. In particular, our approach predicts "wriggling" sequences that cause turning. These maneuvers would be difficult to find by intuition or by direct adoption of the maneuvers of real fish whose mechanical structure is sufficiently different than our robot fish. Second, it is not absolutely necessary to capture all fluid flow details for the purposes of control design, since control feedback can compensate for reasonable modeling errors. In fact, an important conclusion of our work is that the detailed computational or experimental models of fluid flow that have been developed in prior work [16] are not completely necessary for the purposes of robot fish control system design.

2 Experimental Apparatus

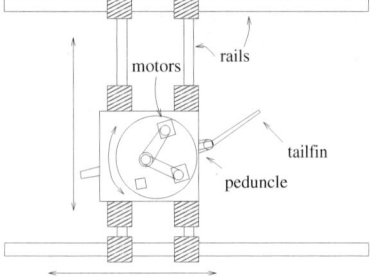

Figure 1: Photograph from rear and schematic of top view of fish design.

To motivate the fluid-body model that is developed in Section 3, we first review our prototype carangiform robot fish testbed. Our robot is a simple ap-

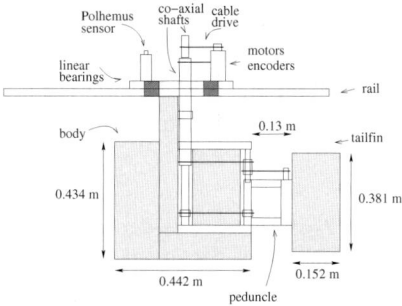

Figure 2: Photograph and schematic of side view of fish design.

proximation to a carangiform-type fish consisting of three links: a flat rectangular "body", a flat rectangular "tail", and an open brace "peduncle" connecting the two (see Figs. 1 and 2). The three-link mechanism is suspended from a passive trolley in a water tank 4 ft wide by 4 ft deep by 36 ft long. The trolley consists of two orthogonal sets of rails and a rotating platform, all supported on low friction bearings. The robot is attached to a shaft that runs through the center of the rotational platform. Thus the robot is free to move in a plane, but the point of body rotation will be determined by the location of the shaft connection to the robot rather than by the robot's center of mass. During tail flapping, the trolley mechanism allows the fish to propel itself and its supporting carriage around the tank. Our assumption of planar motion is not unrealistic and is a typical assumption for the study of fish locomotion [13, 16].

The tail and peduncle joints are independently controlled by transmitting torques from two DC motors through a steel cable-drive system. Joint angles are continuously measured via optical shaft encoders mounted on each motor. Position and orientation of the body are measured with a Polhemus sensor attached to the carriage. This sensor is based on magnetic field measurements and has been calibrated to compensate for field distortion effects due to the presence of steel in the lab.

The fish "body" and "tail" consist of a combination of flat plexiglas plates with dimensions as shown in Fig. 2. The peduncle is an open rigid brace which connects the body to the tail, and we assume it has little hydrodynamic effect. The mass of the entire robot and trolley system is 30 kg, and the inertia of the body with the tail fully extended is 0.5038 kg·m². Because of the tail's small mass relative to the body, we assume that variations in the moment of inertia due to tail motion are negligible. By rotating the peduncle and tail joints, the tail moves back and forth, and its velocity relative to the fluid induces lift, drag, and virtual fluid mass reaction forces which are transmitted to the body via the peduncle.

The nonlinear control methods in which we are interested require that we either begin with a passively stable system or apply closed-loop control to stabilize the system. For our application, the term "passively stable" implies that when the fish is placed in a constant velocity flow, the body and tail will tend to align with the flow–i.e., a "weather-vane" effect. We state without proof that a sufficient condition for passive stability of our system is that the trolley shaft be attached to the body no more than half the body length from the body's front end. Essentially this condition ensures that the drag experienced by the portion of the body behind the shaft is larger than the drag on the front portion of the body. This force imbalance produces moments which cause the body to align with, and pointing into, the flow. Our current design was modified from the original [9] to be stable in this way.

3 A Control Affine Model

This section develops a simplified model for the robot described in Section 2. As discussed in [9], our three-link mechanism is a reasonably general planar approximation to carangiform locomotion, and therefore small modifications of this model should have general utility in the analysis of carangiform swimming.

As we will show, the joint velocities will enter our system equations quadratically. In order to produce a control-affine model where controls enter linearly, we must then take our control inputs to be the angular accelerations of the joints. The result will be a system with ten-dimensional state space and two-dimensional control space. We neglect three-dimensional fluid effects and assume that we can restrict our attention to a plane parallel to the floor. We assume that the forces applied to the system come from quasi-static lift effects, drag effects, and added fluid mass effects on the tail and body.

Fig. 3 shows a simple diagram of the robot with the principal acting forces. Without loss of generality, we assume that the point at which the body is attached

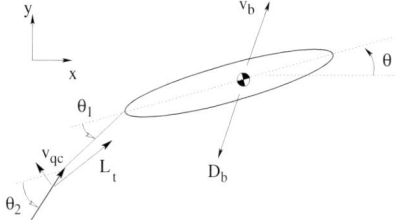

Figure 3: Diagram for control-affine model.

to the trolley shaft coincides with the body center of mass. The body position is given relative to an inertial frame by $[x, y]$ where the positive direction is taken to the right. The body orientation, measured relative to an inertial x-axis, is denoted by θ. When measured relative to the body longitudinal axis, the peduncle and tail angles are denoted by $[\theta_1, \theta_2]$, whereas these angles are denoted by ψ_1 and ψ_2 when they are measured relative to the inertial x-axis. The body has length l_b, the distance between the center of rotation and the peduncle joint is l_{cort}, the peduncle has length l_p, the tail has length l_t, the depth of the body in the water is h_b and the depth of the tail is h_t. The unit vector along the tail plate, $l_{e,t}$ is given by

$$l_{e,t} = [-\cos(\psi_2), -\sin(\psi_2), 0].$$

The angle between the body translational velocity at the center of rotation, $[\dot{x}, \dot{y}, 0]$, and the inertial x-axis is $\alpha_0 = \tan^{-1}(\dot{y}/\dot{x})$.

From standard quasi-steady airfoil theory we know that the lift on a flat plate is given by

$$L = \pi \rho A \left(v_{qc} \times l_e \right) \times v_{qc},$$

and the moment about the quarter chord point by

$$\tau = -\pi \rho \frac{l_f^2}{4} \left(\dot{x}_m \dot{y}_m \cos(2\psi_2) + \frac{1}{2} \left(\dot{y}_m^2 - \dot{x}_m^2 \right) \sin(2\psi_2) \right)$$

where A is the surface area of the plate, ρ is the density of the fluid, v_{qc} is the velocity of the plate relative to the flow at a point that is one-quarter of the chord length from the plate's leading edge, and $\dot{x}_m$ and $\dot{y}_m$ are the velocity of the tail at its midpoint. We assume here that the tail does not stall and that the angle of attack never exceeds ninety degrees (which would result in the leading edge becoming the trailing edge). The velocity at the quarter chord point of the tail is

$$v_{qc,t} = \begin{bmatrix} \dot{x} + l_{cort} s_\theta \dot{\theta} - l_p s_{\psi_1} \dot{\psi}_1 - (l_t/4) s_{\psi_2} \dot{\psi}_2 \\ \dot{y} - l_{cort} c_\theta \dot{\theta} + l_p c_{\psi_1} \dot{\psi}_1 + (l_t/4) c_{\psi_2} \dot{\psi}_2 \\ 0 \end{bmatrix}$$

where $s_{(\cdot)} = \sin(\cdot)$ and $c_{(\cdot)} = \cos(\cdot)$.

The drag on a pitching and heaving flat plate is approximated as the projection of the blunt body drag onto the velocity direction. Because the incremental velocity element corresponding to angular rotation depends on position, we must integrate the incremental drag over the plate. For a plate of length l rotating about a point at a distance a from one end of the plate, the drag integration can be evaluated as

$$D = \frac{1}{2}\rho C_D h \int_{a-l}^{a} \|V_a \times r - s\dot{\theta}\| V ds$$

where C_D is the drag coefficient, V_a is the translational velocity at a, r is a unit vector aligned with the plate and $V = V_a + \dot{\theta} r$. Similarly, the moment of the drag force about the point of rotation is given by

$$M_D = \frac{1}{2}\rho C_D h \int_{a-l}^{a} \|V_a \times r - s\dot{\theta}\| (V_a \times r - s\dot{\theta}) s ds.$$

Additionally, we must account for the added mass effect that results from accelerating a rigid body through a fluid. For a flat plate, the effect of moving through water adds a mass of quantity $m_w = 1/4\pi l^2 h$ along the lateral direction of the body-fixed mass matrix where l is the length of the plate and h is its depth.

Collecting these terms results in the equations

$$\begin{bmatrix} \ddot{\psi}_1 \\ \ddot{\psi}_2 \\ m_x \ddot{x} \\ m_y \ddot{y} \\ I_\theta \ddot{\theta} \end{bmatrix} = \begin{bmatrix} u_1 \\ u_2 \\ L_{t,x} + D_{b,x} \\ L_{t,y} + D_{b,y} \\ [x_t, y_t] \times [L_{t,x}, L_{t,y}] + M_{D,b} + \tau \end{bmatrix} \quad (1)$$

where $L_{t,x}$ describes the x-component of the lift force on the body, $D_{b,y}$ describes the tail drag in the y directions, etc. Drag on the tail and lift on the body are small compared to body drag and tail lift and have not been included in this model. To simplify notation, we will sometimes utilize generalized coordinates $[q, \dot{q}]$ where $q = [\psi_1, \psi_2, x, y, \theta]$.

4 Nonlinear Control Background

4.1 Controllability and Accessibility

Given a system of the form $\dot{x} = f(x, u)$, the first question we should ask is whether the system has a controllable linearization either at a point or about a trajectory. While our system does not possess linear controllability at a point, we believe that linearization about a trajectory can be achieved but have not yet been able to verify this result. We must then resort to the use of nonlinear methods. The general form of a two input nonlinear system in control affine form, where controls enter linearly, is

$$\frac{d}{dt}\begin{bmatrix} q \\ \dot{q} \end{bmatrix} = f(q, \dot{q}) + g_1(q, \dot{q}) u_1 + g_2(q, \dot{q}) u_2. \quad (2)$$

The vector $f(\cdot)$ is referred to as the system drift, and $g_1(\cdot)$ and $g_2(\cdot)$ are termed the control vector fields. The Lie bracket of two vector fields h_i and h_j is denoted $[h_i, h_j] = ad_{h_i} h_j$ and is defined to be $[h_i, h_j] = \frac{\partial h_j}{\partial x} h_i - \frac{\partial h_i}{\partial x} h_j$. Given a set $\mathcal{H}$ of C^∞ vector fields on a manifold M, the Lie algebra of $\mathcal{H}$, $L(\mathcal{H})$ is the set of all Lie brackets of elements of $\mathcal{H}$ and of all the Lie brackets of vector fields generated by Lie bracketing. The family $\mathcal{H}$ satisfies the *Lie algebra rank condition* (LARC) at a point $p \in M$ if the Lie algebra of $\mathcal{H}$ evaluated at p, $L(\mathcal{H})(p)$, is the whole tangent space of M at p. The set $\mathcal{H}$ is said to satisfy the *accessibility property* from a point p if, for every $T > 0$, the set of points reachable from p in time $\leq T$ is nonempty. The following is a standard result [14]:

Proposition 1 *Let $\mathcal{H}$ be a family of C^∞ vector fields on a C^∞ manifold M. Then the LARC at p implies accessibility from p.*

A control system $\dot{x} = f(x) + \sum_{i=1}^{m} g_i(x) u_i$ is *small time locally controllable* (STLC) at p if the vector fields $\{f, g_1, \ldots, g_m\}$ satisfy the accessibility property at p and p is contained in the set of points reachable from p in time T for every $T > 0$. This definition requires that a nonlinear system with nonzero drift term satisfies the STLC condition for points p such that $f(p) = 0$. For mechanical systems described with generalized coordinates $[q, \dot{q}]$ these points are simply those for which $\dot{q} = 0$.

4.2 Second-Order Linearly Uncontrollable Systems

Based on the structure of the Lie algebra of our system, we will use some recent results from the control of second-order linearly uncontrollable systems [10, 11] to motivate a choice of control functions. As a starting point, we consider the class of driftless nonholonomic systems which can be written as

$$\dot{x} = \sum_{i=1}^{m} g_i(x) u_i, \quad x \in \mathbb{R}^n \quad (3)$$

and which satisfy m constraints of the form

$$\omega_i(x) \dot{x} = 0, \quad 1 \leq i \leq m.$$

These driftless nonholonomic systems can be extended to two more or less general classes of second-order systems by cascading either the inputs or outputs through a set of integrators. In particular, when the inputs are passed through a set of integrators, the resulting systems will take the form

$$\dot{\xi} = u, \quad \dot{x} = \sum_{i=1}^{m} g_i(x)\xi_i, \quad \xi \in \mathbb{R}^m. \qquad (4)$$

If the underlying nonholonomic system (3) is controllable, then as shown in [10], the cascaded system satisfies the accessibility property with the vector fields shown in the following proposition:

Proposition 2 *Consider a system of the form (4) where the underlying nonholonomic system (3) is controllable with the set of vector fields*

$$\mathcal{G} = \left\{ g_i, ad_{g_{i_1}} g_{i_2}, \ldots, ad_{g_{i_1}} \cdots ad_{g_{i_r}} g_{i_{r+1}} \right\}.$$

Then the system (4) is accessible with the set of vector fields $\{g_i, \tilde{\mathcal{G}}\}$*, all brackets in* $\mathcal{G}$ *are zero and*

$$\tilde{\mathcal{G}} = \{ad_f g_i, ad_{ad_f g_{i_1}} ad_f g_{i_2},$$
$$\ldots, ad_{ad_f g_{i_1}} \cdots ad_{ad_f g_{i_r}} ad_f g_{i_{r+1}} \}.$$

The structure of the Lie brackets of a system determines which control function combinations will generate motions along the different basis directions of the state space. The control vector fields specify which states are linearly influenced by given control functions, and $ad_f g_i$ corresponds to the effect of the system's inertial response after control u_i is applied. The bracket of two control vector fields corresponds to infinitesimal periodic switching between the corresponding controls. Recursive application of these rules determines the combinations of controls that generate motion in different directions of the state space. Practically speaking, with this analysis we can determine which "wrigglings" of the fish joints will generate motion in a given direction.

The notion of generating motion along the direction of Lie bracket vector fields using periodic control functions has led to a variety of nonlinear control methods based on the use of amplitude-modulated time varying sinusoidal control functions with integrally related frequencies. We list the general relations here and refer the reader to [7, 10, 12] for details and examples. As mentioned above, the state space directions corresponding to the control vector fields are directly controlled with the system inputs. Motions corresponding to brackets of two vector fields are generated by a switching between the appropriate controls which can be accomplished using sinusoids ninety degrees out of phase. Second level bracket directions can be produced by cosines of one frequency along two of the control directions and a cosine of twice that frequency along the third direction. A control vector formed from p brackets can be produced using p controls with cosines at a single frequency ω and one cosine at a frequency of $p\omega$. To summarize, for vector fields generated from Lie brackets of control vector fields, motion can be generated with the following relations:

$$\begin{aligned}
g_i &\rightarrow \alpha_i(t) \\
ad_{g_{i_1}} g_{i_2} &\rightarrow u_{i_1} = \alpha_{i_1}(t)\sin(\omega t) \\
& u_{i_2} = \alpha_{i_2}(t)\cos(\omega t) \\
ad_{g_{i_1}} ad_{g_{i_2}} g_{i_3} &\rightarrow u_{i_1} = \alpha_{i_1}(t)\cos(\omega t) \\
& u_{i_2} = \alpha_{i_2}(t)\cos(\omega t) \\
& u_{i_3} = \alpha_{i_3}(t)\cos(2\omega t) \\
&\vdots
\end{aligned}$$

As shown in [10] for systems with drift, relations which will generate motion along the appropriate directions come by replacing each vector field in the above table with the appropriately corresponding term from the set of vector fields $\tilde{\mathcal{G}}$ constructed from Prop. 2.

Using these results and some knowledge of the Lie bracket structure of a nonlinear system, we can generate motion along desired directions and ultimately track given trajectories.

5 System Analysis and Experimental Results

For our model, the drift vector is zero when $[\dot{\psi}_1, \dot{\psi}_2, \dot{x}, \dot{y}, \dot{\theta}] = [0,0,0,0,0]$. However, at any point with zero velocity, all of the Lie brackets become zero. Thus, our system is not STLC. But if we consider the problem of trajectory tracking rather than stabilization or moving between two points, then we need only be concerned with satisfying the conditions for accessibility. I.e., we simply need to show that from a given point with nonzero velocity, we can move to another point with nonzero velocity.

Assuming nonzero velocity, we have the following correspondence for the joint positions and velocities:

$$\begin{aligned}
\dot{\psi}_1 &\rightarrow g_1, & \psi_1 &\rightarrow ad_f g_1 \\
\dot{\psi}_2 &\rightarrow g_2, & \psi_2 &\rightarrow ad_f g_2
\end{aligned}$$

From (1), we can see that the elements of the vectors g_1 and g_2 are either zero or one. Thus the term $ad_{g_1} g_2$ is identically zero, and our system possesses the characteristics of (4). Due to the functional form of any higher level brackets, we are not able to determine linear independence of terms by symbolic calculation. We can, however, evaluate vector fields at particular points in the state space and argue by continuity that characteristics of the vector fields at those points must hold in a neighborhood of the points. In particular, if the robot is moving straight ahead from the origin with no sideways or rotational velocity, we have the following correspondence:

$$\dot{x} \rightarrow ad_{ad_fg_1} ad_f g_2$$
$$\dot{\theta} \rightarrow ad_{ad_fg_1} ad_{ad_fg_1} ad_f g_2$$
$$x \rightarrow ad_f ad_{ad_fg_1} ad_f g_2$$
$$\theta \rightarrow ad_f ad_{ad_fg_1} ad_{ad_fg_1} ad_f g_2$$

where the pairing is determined by the largest nonzero entry in the displayed vector field. Generally more than one entry will be nonzero, but in these cases, the given value dominates. Unfortunately, higher order brackets that have been tested have not produced independent motion in the y direction as well. One can argue that this coupling of y with the x and θ directions occurs because one cannot simply move the robot sideways but would need to achieve a parallel parking behavior. We are currently more interested with forward and turning gaits, so we will leave this issue to be addressed in the future and will restrict our attention to forward and rotational motions.

Given the state-bracket relations above and the results from the preceding section, we expect that we can achieve forward propulsion with system controls of the form

$$u_1 = \alpha_1 \sin(\omega t), \quad u_2 = -\alpha_2 \cos(\omega t)$$

and rotation with controls of the form

$$u_1 = \alpha_1 \cos(\omega t), \quad u_2 = \alpha_2 \cos(2\omega t).$$

The simulated response of our system model to each of these sets of controls with $\alpha_1 = \alpha_2 = 0.4$ in both cases, $\omega = 8$ for forward propulsion and $\omega = 3.5$ for turning gives the results shown respectively in Figs. 4 and 5. By simulation, we mean a numerical integration of the equations of motion. In each simulation, system parameters such as inertia and mass are taken from measurements on the experimental robot. In Fig. 4, the simulation shows that when the controls are turned on with the fish starting from rest, the body turns slightly. As the body velocity increases to a constant value, the fish orientation oscillates about this perturbed value, and the fish travels in a straight line. The turning gait, shown in Fig. 5, is primarily produced from the effects of added mass on the tail. This simulation predicts that when the fish is started from rest with this gait, the response of the body to the forces on the tail will be to initially pull backwards and to the side (due to scooping motions of the tail) before

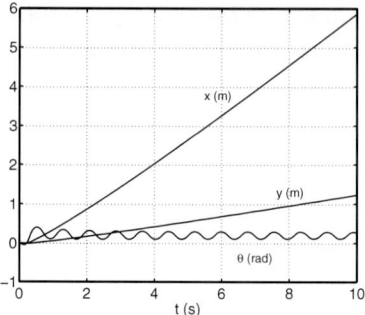

Figure 4: Simulated model response for forward propulsion.

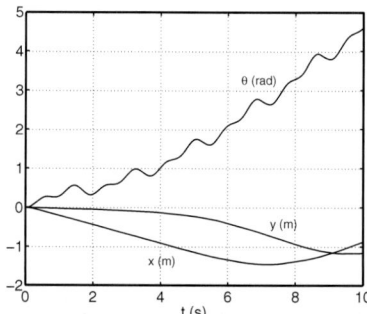

Figure 5: Simulated model response for turning.

settling into a circular motion. In both of these cases, we would like to draw attention to the fact that our model is based on quasi-static approximations of lift and drag, and we expect discrepancies between simulation and experiment during periods of large body acceleration such as when we start the body from zero velocity.

Considering the number of assumptions and simplifications made in this model, we do not expect that our robot will exactly produce these motions. Indeed, comparison of the above simulations and the following experiments demonstrates scaling discrepancies up to an order of magnitude. One likely source for these errors is the bearings in the trolley which are assumed to produce no stiction. Regardless, we do expect that the underlying nonlinear structure of the simulation and experiment will be the same. If this assumption is correct, then our robot should demonstrate the following behavior. First, for a forward propulsive gait, the robot should move forward without significant turning. Second, for a turning gait, the robot should turn significantly without large translational motions. Third, because g_1 and g_2 each appear once in the Lie bracket assumed to produce forward motion, the forward motion should scale linearly with the amplitude of u_1 and linearly with the amplitude of u_2. Fourth, because g_1 appears twice in the Lie bracket associated with turning and g_2 appears once, turning should scale linearly with the amplitude of u_2 and quadratically with the amplitude of u_1.

In Figs. 6-8 we show recorded experimental results for application of controls of the form $u_1(t) = \alpha_1 \sin(8t)$, $u_2(t) = -\alpha_2 \cos(8t)$ with $\alpha_i = \{0.1, 0.2, 0.3, 0.4\}$. The typical response shown in Fig. 6 demonstrates the same behavior as the simulation where the body moves in a straight line and oscillates slightly about that direction. The average orientation is not a fixed value due to system disturbances from motion of the water and tension in the vehicle power and communication cables. Our prediction was that

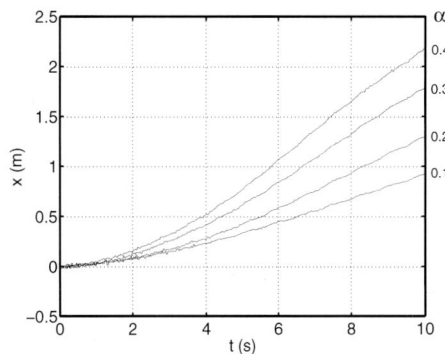

Figure 8: Experimental data for forward propulsion with $\alpha_1 = 0.4$ and $\alpha_2 = \{0.1, 0.2, 0.3, 0.4\}$.

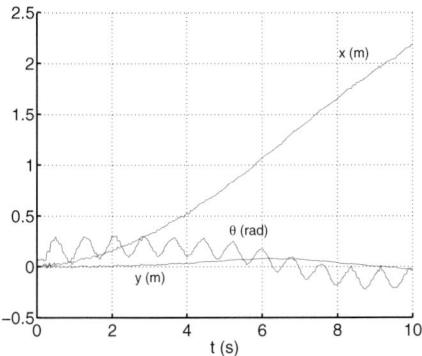

Figure 6: Experimental data for forward propulsion with $\alpha_1 = \alpha_2 = 0.4$.

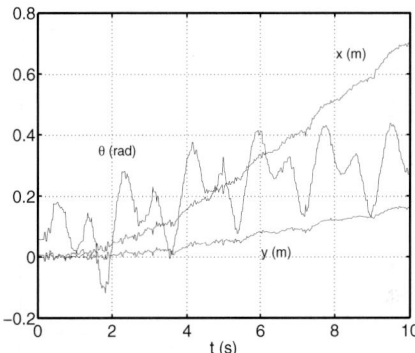

Figure 9: Experimental data for turning motion with $\alpha_1 = \alpha_2 = 0.4$.

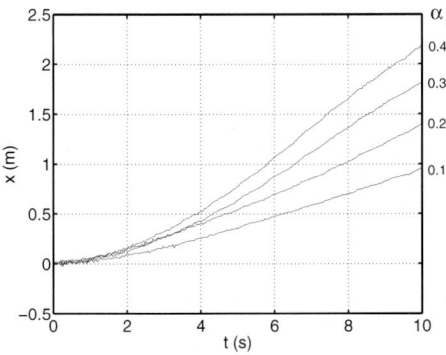

Figure 7: Experimental data for forward propulsion with $\alpha_1 = \{0.1, 0.2, 0.3, 0.4\}$ and $\alpha_2 = 0.4$.

forward motion in this direction corresponds to a single bracket and that the net motion should scale linearly and symmetrically with the amplitudes of each of the controls. Clearly, our experimental results do validate this hypothesis.

We now consider turning maneuvers. Interestingly, our analysis predicts turning maneuvers for inputs of the form $u_1(t) = \alpha_1 \cos(3.5t)$, $u_2(t) = \alpha_2 \cos(7t)$. Figs. 9-11 show experimental results for application of these controls with $\alpha_i = \{0.1, 0.2, 0.3, 0.4\}$. This data illustrates two things. First, the fish actually turns as predicted by the theory. Unlike the simulation, however, the body is not pulled backward by the initial tail kick. Stiction in the bearings and inaccurate model drag coefficients could account for this behavior. Second, for a turning motion, our model predicts that a second level bracket where g_1 appears twice and g_2 appears once should produce the desired behavior. As discussed above, because g_1 appears twice, we expect net motion to scale quadratically with α_1. Because g_2 appears once we expect net motion to to scale linearly in the amplitude of u_2. The amplitudes that we are using for our inputs are all less than one, so doubling α_1 with α_2 unchanged should produce less net increase in θ than doubling α_2 with α_1 unchanged. The data in the figures does indeed support this expectation with the net turn much more strongly affected by changes in α_2 than in α_1.

6 Conclusions and Future Work

We have developed a control affine model to which we are able to apply nonlinear control methods to pro-

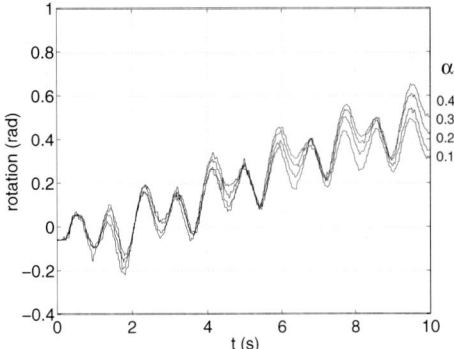

Figure 10: Experimental data for turning motion with $\alpha_1 = \{0.1, 0.2, 0.3, 0.4\}$ and $\alpha_2 = 0.4$.

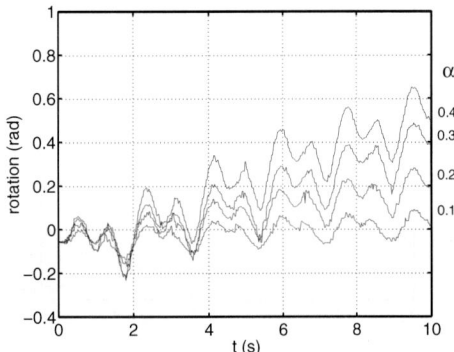

Figure 11: Experimental data for turning motion with $\alpha_1 = 0.4$ and $\alpha_2 = \{0.1, 0.2, 0.3, 0.4\}$.

duce forward propulsion and turning gaits. The qualitative trends of our experiments do correspond well to the theoretically predicted behavior. A key aspect of this work is the fact that we are able to produce these results with an extremely simplified model that captures only the most basic of system effects. Also, we must emphasize that these methods are open-loop: no state feedback is used to produce the motions.

Several avenues of investigation are now available for exploration. Due to the complicated nature of the equations of motion, we have had to resort to numerical methods for part of our analysis. This situation is not unique to our system, and appropriate numerical tools for evaluating system controllability characteristics must be developed. Given that we can generate motions in two independent state directions, we wish to track trajectories in those directions. Open-loop methods exist for this task if the system model is known exactly and if system initial conditions can be dictated exactly. In general neither situation exists. Our current research effort is directed toward the construction of feedback control functions which enable tracking in the presence of inexact system models.

References

[1] J. M. Anderson and P. A. Kerrebrack. The vorticity control unmanned undersea vehicle–an autonomous vehicle employing fish swimming propulsion and maneuvering. In *Proc. 10th Int. Symp. on unmanned untethered submersible technology*, pages 189–195, Durham, NH, September 1997.

[2] D. Barrett, M. Grosenbaugh, and M. S. Triantafyllou. The optimal control of a flexible hull robotic undersea vehicle propelled by an oscillating foil. In *Proc. 1996 Symp. Aut. Underwater Vehicle Tech.*, pages 1–9, 1996.

[3] S. Childress. *Mechanics of Swimming and Flying*. Cambridge University Press, Cambridge, 1981.

[4] S. D. Kelly, R. J. Mason, C. T. Anhalt, R. M. Murray, and J. W. Burdick. Modelling and experimental investigation of carangiform locomotion for control. In *Proc. of the 1998 Amer. Cont. Conf.*, pages 1271–1276, 1998.

[5] P. Y. Li and S. Saimek. Modeling and estimation of hydrodynamic potentials. In *Proceedings of the 1999 CDC*, pages 3253–3258, 1999.

[6] J. L. Lighthill. *Mathematical Biofluiddynamics*. SIAM, Philadelphia, 1975.

[7] W. Liu. An approximation algorithm for nonholonomic systems. *SIAM J. Control. Opt.*, 35(4):1328–1365, July 1997.

[8] R. J. Mason and J. W. Burdick. Construction and modelling of a carangiform robotic fish. In *1999 International Symposium on Experimental Robotics*, 1999.

[9] R. J. Mason and J. W. Burdick. Experiments in carangiform robotic fish locomotion. In *Proceedings of the 2000 ICRA*, pages 428–435, 2000.

[10] K. A. Morgansen. *Temporal patterns in learning and control*. PhD thesis, Harvard University, 1999.

[11] K. A. Morgansen and R. W. Brockett. Nonholonomic control based on approximate inversion. In *Proc. of the 1999 American Control Conference*, pages 3515–3519, 1999.

[12] R. M. Murray and S. Sastry. Nonholonomic motion planning: Steering using sinusoids. *IEEE Trans. Aut. Cont.*, 38(5):700–716, 1993.

[13] J. N. Newman and T. Y. Wu. Hydrodynamical aspects of fish swimming. In T. Wu, C. Brokaw, and C. Brennen, editors, *Swimming and Flying in Nature, Vol 2.*, pages 615–634. Plenum Press, New York, 1975.

[14] H. J. Sussmann and J. Jurdjevic. Controllability of nonlinear systems. *J. Diff. Eqns.*, 12:95–116, 1972.

[15] M. S. Triantafyllou and G. S. Triantafyllou. An efficient swimming machine. *Scientific American*, pages 64–70, March 1995.

[16] M. J. Wolfgang, J. M. Anderson, M. A. Grosenbaugh, D. K. P Yue, and M. S. Triantafyllou. Near-body flow dynamics in swimming fish. *J. Experimental Biology*, 202(17):2303–2307, 1999.

Quaternion-Based Kinematic Control of Redundant Spacecraft/Manipulator Systems

Fabrizio Caccavale and Bruno Siciliano

°PRISMA Lab, Dipartimento di Informatica e Sistemistica
Università degli Studi di Napoli Federico II
Via Claudio 21, 80125 Napoli, Italy
{caccaval,siciliano}@unina.it

Abstract

This paper addresses kinematic control of a redundant space manipulator mounted on a free-floating spacecraft. Redundancy of the system with respect to the number of task variables for spacecraft attitude and manipulator end-effector pose is considered. Also, the problem of both spacecraft attitude and end-effector orientation representation is tackled. A nonminimal singularity-free representation of rigid body orientation is adopted: the unit quaternion; this allows avoiding representation singularities and formulating the inverse kinematics algorithms in terms of geometrically meaningful variables. Depending on the nature of the task for the spacecraft/manipulator system, a number of closed-loop inverse kinematics algorithms are proposed. Case studies are developed for a system of a spacecraft with a six-joint manipulator attached.

1 Introduction

When considering a robotic system composed by a space manipulator mounted on a free-floating spacecraft, the reaction effects of the manipulator motion on the spacecraft can be taken into account by proper kinematic and dynamic modeling of the whole system. To the purpose, a general description of the reaction effects is the generalized Jacobian concept established in [1]; the ordinary manipulator Jacobian is modified by the addition of a term accounting for the spacecraft Jacobian and the relative weight between the spacecraft and manipulator inertia.

The generalized Jacobian approach discloses the problem of coordinated motion between manipulator and spacecraft to a potential of inverse kinematics solutions that resemble those developed for ground-fixed manipulator. Among them, resolved motion rate control can be carried out as well as exploitation of redundant degrees of freedom with respect to tasks defined for either the spacecraft only or the overall spacecraft/manipulator system [2],[3].

The goal of this paper is to present solution algorithms that are logically derived from closed-loop inverse kinematics schemes for constrained redundant manipulators in the framework of task space augmentation with task priority [4]. Then, the inverse kinematics algorithms are used to compute the joint trajectories corresponding to a given end-effector trajectory, which then constitute the reference inputs to some joint feedback control scheme (i.e, the so-called kinematic control problem).

When six-degree-of-freedom tasks are considered, the problem of properly describing the attitude of the free-floating base and the orientation of the manipulator end effector is of concern. In order to overcome the typical drawback concerned with a minimal representation of orientation (Euler angles), the unit quaternion is adopted which provides a geometric description of the rotation in terms of an equivalent angle and axis. The resulting schemes extend those in [5] to solve the problem of kinematic control of spacecraft/manipulator systems.

A number of case studies are developed by simulating the motion of a system composed by a free-floating spacecraft with a six-joint manipulator attached.

2 Spacecraft/Manipulator System Modeling

Consider a system composed by an n-degree-of-freedom manipulator with rigid links mounted on a rigid body spacecraft free-floating in a zero-gravity environment. In the following, q will denote the $(n \times 1)$ vector of joint variables. The (3×1) vector p_s represents the position of a spacecraft-fixed coordinate frame Σ_s with respect to an inertial reference frame;

R_s is the (3×3) rotation matrix expressing the spacecraft attitude, i.e., the orientation of Σ_s with respect to the inertial frame. Let also the manipulator end-effector pose be described by the (3×1) vector p_e expressing the end-effector position and the (3×3) rotation matrix R_e expressing the orientation of a frame Σ_e attached to the end-effector with respect to the same inertial reference frame. Hereafter, a superscript will denote the frame to which a quantity (vector or matrix) is referred; the superscript is dropped whenever a quantity is referred to the inertial frame.

The kinematic equation relating the joint and spacecraft variables to the end-effector position can be written as

$$p_e = p_s + R_s \Delta^s p_{es}(q), \qquad (1)$$

where $\Delta^s p_{es} = R_s^T(p_e - p_s)$ is the position of the end-effector frame relative to the spacecraft frame; the end-effector orientation can be described by the rotation matrix

$$R_e = R_s {}^s R_e(q), \qquad (2)$$

where ${}^s R_e$ is the rotation matrix expressing the relative orientation between the end-effector frame and the spacecraft frame. Notice that $\Delta^s p_{es}(q)$ and ${}^s R_e(q)$ represent the usual direct kinematics equations of a ground-fixed manipulator with respect to its base frame (represented by Σ_s).

In view of solving the inverse kinematics for such a system, it is convenient to consider differential kinematics in lieu of eqs. (1),(2). Let $v_e = [\dot{p}_e^T \; \omega_e^T]^T$ be the (6×1) vector of generalized end-effector velocity, where $\dot{p}_e$ and ω_e denote the linear and angular velocity, respectively; ${}^s J_{es}$ is the usual $(6 \times n)$ manipulator geometric Jacobian relating the joint velocities $\dot{q}$ to the end-effector velocity relative to the spacecraft frame ${}^s v_{es} = [(\Delta^s \dot{p}_{es})^T \; (\Delta^s \omega_{es})^T]^T$, where $\Delta \omega_{es}^s = R_s^T(\omega_e - \omega_s)$. Let also $v_s = [\dot{p}_s^T \; \omega_s^T]^T$ be the (6×1) vector of generalized spacecraft velocity. Then, by differentiating (1) and (2), it follows

$$v_e = J_s(q, R_s) v_s + R_s {}^s J_{es}(q) \dot{q}, \qquad (3)$$

with

$$J_s(q, R_s) = \begin{bmatrix} I_3 & -S(R_s \Delta^s p_{es}) \\ O_3 & I_3 \end{bmatrix}, \qquad (4)$$

where I_l and O_l denote $(l \times l)$ identity and null matrices, respectively, and $S(\cdot)$ is the (3×3) skew-symmetric matrix operator performing the cross product. Since the spacecraft position is of no concern, $\dot{p}_s$ can be eliminated from (3); by following the guidelines in [1] and assuming null initial velocity for the system's center of mass, the following expression can be derived from (3)

$$v_e = \bar{J}_s(q, R_s) \omega_s + \bar{J}_e(q, R_s) \dot{q}, \qquad (5)$$

where the matrices $\bar{J}_s$ and $\bar{J}_e$ depend on the Jacobian matrices J_s and ${}^s J_{es}$ in (3).

It is assumed that no external forces or torques act on the center of mass of the system, that is no devices (reaction wheels or thrusters) are employed to change spacecraft attitude. In view of momentum conservation of the system, it is

$$M_s \omega_s + M_e \dot{q} = 0, \qquad (6)$$

where M_s is a (3×3) matrix related to the spacecraft inertia and M_e is a $(3 \times n)$ matrix related to the manipulator inertia [1].

Equations (5) and (6) are fundamental for analyzing the motion of the system composed by the robotic manipulator mounted on the free-floating spacecraft. Since $M_s \omega_s$ represents the spacecraft rotational momentum, M_s is a non-singular matrix; then, solving (6) for ω_s and substituting in (5) allows eliminating the dependence on the spacecraft attitude changes, i.e.

$$v_e = J_G \dot{q} \qquad (7)$$

where the matrix

$$J_G = \bar{J}_e - \bar{J}_s M_s^{-1} M_e \qquad (8)$$

is termed the *generalized Jacobian* for the spacecraft/manipulator system [1].

The attractive feature of eq. (8) is its formal analogy with the well-known differential kinematics equation for ground-fixed manipulators. The manipulator Jacobian $\bar{J}_e$ is modified by the presence of a term accounting for the relative inertial weight between the spacecraft and the manipulator. The larger the spacecraft inertia, the smaller the reaction caused by the manipulator motion; in the limit of a very massive spacecraft, the generalized Jacobian will tend to the manipulator Jacobian.

3 Kinematic Control with Redundancy Resolution

Robot *kinematic control* consists of solving the motion control problem into two stages, i.e., the desired end-effector trajectory is transformed via inverse kinematics into the corresponding joint trajectories, which then constitute the reference inputs to some joint space control scheme [6]. This approach differs from operational space control in the sense that manipulator kinematics is handled outside the control loop thus allowing the problem of kinematic singularities and/or redundancy to be solved separately from the motion control problem. The key point of kinematic control is the solution to the inverse kinematics problem.

Let $\boldsymbol{p}_d(t)$ and $\boldsymbol{R}_d(t)$ denote the desired end-effector trajectory. The inverse kinematics problem consists of computing the corresponding joint trajectories $\boldsymbol{q}(t)$ which are solution of the direct kinematics equation. In order to overcome the typical numerical drift concerned with discrete-time implementation of integration of (7), a *closed-loop inverse kinematics* (CLIK) algorithm can be devised which acts upon an error characterizing the displacement between the desired and the end-effector trajectory computed via the direct kinematics (1) and (2).

Let

$$\Delta \boldsymbol{p}_{de} = \boldsymbol{p}_d - \boldsymbol{p}_e(\boldsymbol{q}) \tag{9}$$

denote the end-effector position error. The end-effector orientation error depends on the choice of the orientation description. In the case of Euler angles, the error is simply

$$\Delta \boldsymbol{\varphi}_{de} = \boldsymbol{\varphi}_d - \boldsymbol{\varphi}_e(\boldsymbol{q}) \tag{10}$$

where $\boldsymbol{\varphi}_d$ and $\boldsymbol{\varphi}_e$ denote the desired and current quantities, respectively. Notice that the computation of $\boldsymbol{\varphi}_e(\boldsymbol{q})$ requires the extraction of the Euler angles from the end-effector rotation matrix $\boldsymbol{R}_e(\boldsymbol{q})$ via inversion formulæ, which suffer from representation singularities though.

The joint velocity algorithmic solution is given by

$$\dot{\boldsymbol{q}} = \boldsymbol{J}_A^{-1}(\boldsymbol{q}) \begin{bmatrix} \dot{\boldsymbol{p}}_d + k_{Pp}\Delta \boldsymbol{p}_{de} \\ \dot{\boldsymbol{\varphi}}_d + k_{Po}\Delta \boldsymbol{\varphi}_{de} \end{bmatrix} = \boldsymbol{J}_A^{-1}(\boldsymbol{q})(\dot{\boldsymbol{x}}_d + \boldsymbol{K}_P \Delta \boldsymbol{x}_{de}) \tag{11}$$

where $\boldsymbol{K}_P = \text{diag}\{k_{Pp}\boldsymbol{I}_3, k_{Po}\boldsymbol{I}_3\}$ is a positive definite matrix gain, $\Delta \boldsymbol{x}_{de} = [\Delta \boldsymbol{p}_{de}^T \; \Delta \boldsymbol{\varphi}_{de}^T]^T$, and $\boldsymbol{J}_A$ is the analytical Jacobian [6] that is related to the geometric Jacobian by the relationship

$$\boldsymbol{J}_A = \begin{bmatrix} \boldsymbol{I}_3 & \boldsymbol{O}_3 \\ \boldsymbol{O}_3 & \boldsymbol{T}^{-1}(\boldsymbol{\varphi}) \end{bmatrix} \boldsymbol{J}_G \tag{12}$$

where $\boldsymbol{T}$ is so that $\boldsymbol{\omega} = \boldsymbol{T}(\boldsymbol{\varphi})\dot{\boldsymbol{\varphi}}$. Note that $\boldsymbol{J}_A$ is singular at a representation singularity, where it is not possible to describe an arbitrary set of Euler angles time derivatives with a set of joint velocities.

In view of (9),(10), substituting (11) into (7) gives

$$\Delta \dot{\boldsymbol{p}}_{de} + k_{Pp}\Delta \boldsymbol{p}_{de} = \boldsymbol{0} \tag{13}$$

$$\Delta \dot{\boldsymbol{\varphi}}_{de} + k_{Po}\Delta \boldsymbol{\varphi}_{de} = \boldsymbol{0}. \tag{14}$$

It is easy to see that system (13) is exponentially stable, implying that $\Delta \boldsymbol{x}_{de}$ converges to zero. Likewise, system (14) is exponentially stable as long as no representation singularity occurs.

In order to devise an inverse kinematics scheme based on the unit quaternion, a suitable end-effector orientation error shall be defined [5]. Let $\{\eta_d, \boldsymbol{\epsilon}_d\}$ and $\{\eta_e(\boldsymbol{q}), \boldsymbol{\epsilon}_e(\boldsymbol{q})\}$ represent the unit quaternions associated with $\boldsymbol{R}_d$ and $\boldsymbol{R}_e(\boldsymbol{q})$, respectively ($\eta$ is the scalar part and $\boldsymbol{\epsilon}$ is the vector part of the quaternion). The mutual orientation can be expressed in terms of the unit quaternion $\{\eta_{de}, {}^e\boldsymbol{\epsilon}_{de}\}$ that can be extracted from the rotation matrix ${}^e\boldsymbol{R}_d = \boldsymbol{R}_e^T \boldsymbol{R}_d$ expressing the mutual orientation between the desired and the actual end-effector frame. The end-effector orientation error can be defined as

$$\boldsymbol{\epsilon}_{de} = \boldsymbol{R}_e \, {}^e\boldsymbol{\epsilon}_{de} \tag{15}$$

which has been conveniently referred to the base frame.

The joint velocity algorithmic solution is given by

$$\dot{\boldsymbol{q}} = \boldsymbol{J}_G^{-1}(\boldsymbol{q}) \begin{bmatrix} \dot{\boldsymbol{p}}_d + k_{Pp}\Delta \boldsymbol{p}_{de} \\ \boldsymbol{\omega}_d + k_{Po}\boldsymbol{\epsilon}_{de} \end{bmatrix} = \boldsymbol{J}_G^{-1}(\boldsymbol{q})(\boldsymbol{v}_d + \boldsymbol{K}_P \boldsymbol{e}_{de}) \tag{16}$$

where $\boldsymbol{e}_{de} = [\Delta \boldsymbol{p}_{de}^T \; \boldsymbol{\epsilon}_{de}^T]^T$. It is worth pointing out that in (16), differently from the previous Euler angles-based algorithm in (11), the geometric Jacobian appears in lieu of the analytical Jacobian, and thus representation singularities cannot occur.

It can be recognized that eq. (13) still holds for the position error. As for the orientation error, in view of (15), substituting (16) into (7) gives

$$\Delta \boldsymbol{\omega}_{de} + k_{Po}\boldsymbol{\epsilon}_{de} = \boldsymbol{0} \tag{17}$$

with $\Delta \boldsymbol{\omega}_{de} = \boldsymbol{\omega}_d - \boldsymbol{\omega}_e$. It should be observed that now the orientation error equation is not homogeneous in $\boldsymbol{\epsilon}_{de}$ since it contains the end-effector angular velocity error instead of the time derivative of the orientation error. This, in turn, allows expressing the inverse kinematics algorithm via geometrically meaningful quantities, i.e., angle/axis representations and angular velocities. The stability of the system (17) can be analyzed by resorting to a Lyapunov argument [5].

In the framework of inverse kinematics algorithms, it is important to recognize the presence of redundant degrees of freedom in the system with respect to the required task. As pointed out in [2], three cases of redundancy can be distinguished in connection with the number of degrees of mobility (joint variables) n versus the number of degrees of freedom characterizing the assigned task (task space variables) $m_s + m_e$, where

m_s and m_e refer to the spacecraft and manipulator task, respectively. If $n < m_s + m_e$, the manipulator can be redundant with respect either to the spacecraft task ($n > m_s$) or to the end-effector task ($n > m_e$), but redundancy will not allow specifying a coordinated task for the spacecraft and the end-effector. If $n = m_s + m_e$, the available redundancy can be exploited to coordinate the motion of the spacecraft with that of the end-effector. If $n > m_s + m_e$, it is possible to introduce additional constraints to be satisfied along with spacecraft/manipulator motion coordination. In the following it is assumed that $n \geq m_s + m_e$.

Hence, the Jacobian matrix to be considered in (7) is in general obtained by eliminating some rows of J_G corresponding to the relaxed task variables, i.e. its dimensions become $(m_e \times n)$ with $m_e \leq 6$ and $n \geq 6$. This implies that suitable strategies have to be pursued to manage both the presence of a non-square Jacobian matrix and the redundant degrees of freedom in the inverse kinematics algorithms previously defined.

By applying the well-known Jacobian transpose algorithm to the robotic system described by eq. (7), the joint velocity solution can be computed as $\dot{q} = J_G^T K_P e_{de}$. The tracking error is upper-bounded – the larger the elements of K_P the smaller the error norm– and asymptotic convergence at steady-state is ensured. The essential advantage of this algorithm resides in the avoidance of pseudoinversion of J_G, which is more time-consuming and may be computationally ill-posed.

With the above solution, however, the resulting spacecraft attitude varies as the manipulator end-effector moves along the trajectory. It is then advisable to exploit the redundant degrees of freedom $n - m_e \geq m_s$ to impose a desired time evolution of spacecraft attitude R_{s_d}. This is a typical case of task space augmentation for redundant manipulators. Since a conflict may arise between the end-effector and the constraint (spacecraft) task, an order of priority should be assigned. By revisiting the solution algorithm proposed in [4], the joint velocity solution (16) can be computed as

$$\dot{q} = J_G^\dagger (v_d + K_P e_{de}) + (I_n - J_G^\dagger J_G) J_C^T K_C e_C \quad (18)$$

where the matrix $(I_n - J_G^\dagger J_G)$ projects the additional joint velocity contribution on the null space of the generalized Jacobian so as to avoid interference of the constraint task with the end-effector task, which has then been given higher priority. Further, e_C is the error for the constraint task and J_C is the associated Jacobian, while K_C is a positive definite (diagonal) matrix playing the same role as K_P. It is worth noticing that the first term of solution (18) is based on the pseudoinverse and not on the transpose of J_G, since $J_G^\dagger$ is to be computed anyhow for the second term; to the purpose, it is assumed that J_G is non-singular. Note also that a feedforward velocity term is added like in conventional resolved motion rate algorithms and the feedback term serves as a correction term to avoid numerical drift or deviations along the trajectory due to discrete-time implementation of the algorithm.

Regarding the constraint task, if $n - m_e = m_s$, then $e_C = \epsilon_{s_ds}$ is extracted from $R_s^T R_{sd}$ and

$$J_C = -M_s^{-1} M_e \quad (19)$$

from (6); in this case, manipulator redundancy is exploited to reach the desired spacecraft attitude R_{s_d} while tracking the desired end-effector trajectory. On the other hand, if $n - m_e > m_s$, then the constraint task can encompass also an additional constraint, such as mechanical joint range, obstacle avoidance etc.

With solution (18), the constraint task is not guaranteed to be satisfied along the whole motion execution. Therefore, for those applications where the constraint task is judged to be more important than the end-effector task, the order of priority can be switched with obvious transposition of subscripts in (18).

An interesting case is that when it is desired to keep the spacecraft attitude constant during manipulator motion ($\omega_{s_d} = 0$), e.g. not disturbing the orientation of some antenna for communication between spacecraft and earth. According to the above technique, the joint velocity solution can be computed as

$$\dot{q} = J_C^\dagger K_C e_C + (I_n - J_C^\dagger J_C) J_G^T K_P e_{de}. \quad (20)$$

If no other constraint is imposed and $n - m_e = m_s$, then J_C is given as in (19); it is assumed that J_C is non-singular, otherwise a transpose should be employed. By using the expression for J_G given in (8), one can write

$$J_G = \bar{J}_e + \bar{J}_s J_C; \quad (21)$$

then computing the matrix $J_G(I_n - J_C^\dagger J_C)$ leads – after ordinary algebraic manipulation– to

$$J_G(I_n - J_C^\dagger J_C) = \bar{J}_e(I_n - M_e^\dagger M_e), \quad (22)$$

which coincides with the so-called *fixed-attitude-restricted* (FAR) Jacobian introduced in [7].

As a consequence, solution (20) provides end-effector trajectories not changing spacecraft attitude which are computed via the transpose of the fixed-attitude-restricted Jacobian –the projector ($I_n -$

$J_C^\dagger J_C$) is symmetric– and then can be simplified into

$$\dot{q} = J_C^\dagger K_C e_C + (I_n - M_e^\dagger M_e)\bar{J}_e^T K_P e_{de}. \quad (23)$$

This result is argued to be quite important in view of occurence of algorithmic singularities of the above Jacobian [2], whereas an algorithm based on a pseudoinverse of such Jacobian might suffer from numerical problems.

4 Case Studies

In order to study the performance of the above inverse kinematics algorithms, a number of case studies were developed for a system composed by a free-floating spacecraft with a six-joint manipulator attached.

The first case study is aimed at showing the reaction effects induced by the motion of the manipulator on the spacecraft. The spacecraft orientation is initially aligned with the reference frame, while a smooth trajectory is commanded to the manipulator end-effector position and orientation which involves crossing of a representation singularity for the Euler angles description; the duration of the motion is 2 s.

The joint velocity algorithmic solution based on (16) has been implemented in discrete time using Euler integration rule with 0.002 s time interval. The matrix gain has been set to $K_P = \mathrm{diag}\{1000I_3, 850I_3\}$.

The results are displayed in Fig. 1 in terms of the time history of the spacecraft orientation (Euler angles) and angular velocity, manipulator joint angles and velocities, and in Fig. 2 in terms of the time history of the end-effector position and (quaternion) orientation errors. It can be recognized how the spacecraft attitude varies while the end-effector successfully tracks the imposed trajectory. Also, the joint motion is smooth and the errors vanish at steady state.

The same case study has been executed by adopting the joint algorithmic solution based on (11) with the same data as above.

The results are displayed in Fig. 3 in terms of the time history of the spacecraft orientation (Euler angles) and angular velocity, manipulator joint angles and velocities, and in Fig. 4 in terms of the time history of the end-effector position and (Euler angles) orientation errors. It can be recognized how the representation singularity affects both the spacecraft and the manipulator joint motion and provokes a considerable error on the end-effector orientation.

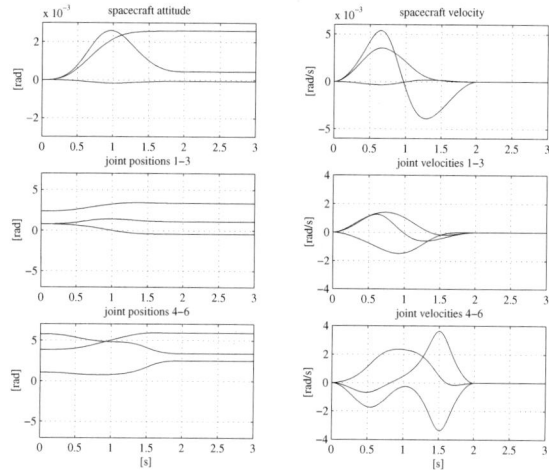

Figure 1: Time history of the spacecraft orientation and angular velocity, the manipulator joint angles and velocities using the CLIK algorithm based on the unit quaternion.

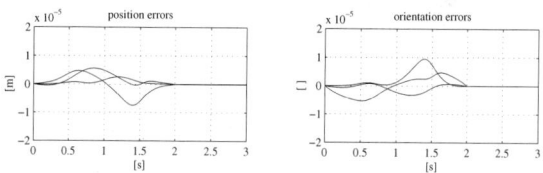

Figure 2: Time history of the end-effector position and orientation errors using the CLIK algorithm based on the unit quaternion.

The second case study is aimed at showing the potential of utilizing kinematic redundancy. This is obtained by relaxing the end-effector position, while the constraint task consists of keeping the spacecraft attitude constant during the motion. The spacecraft orientation is initially aligned with the reference frame, while a smooth trajectory is commanded to the manipulator end-effector position and orientation with the same interpolation as above.

The joint velocity algorithmic solution based on (20) has been implemented in discrete time using Euler integration rule with 0.002 s time interval. The matrix gains have been set to $K_P = 1500I_3$ and $K_C = 850I_3$.

The results are displayed in Fig. 5 in terms of the time history of the (end-effector orientation) constraint errors and the spacecraft orientation errors.

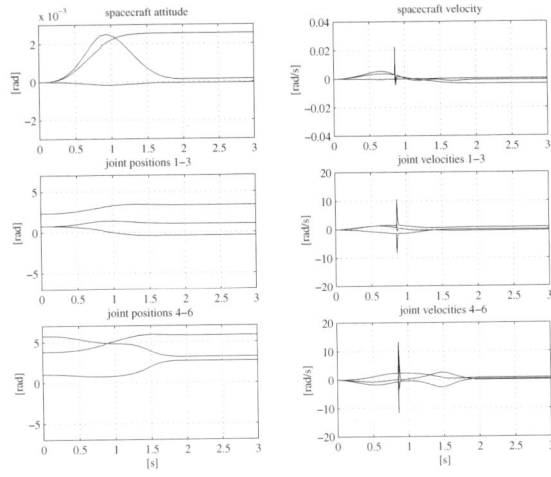

Figure 3: Time history of the spacecraft orientation and angular velocity, the manipulator joint angles and velocities using the CLIK algorithm based on the Euler angles.

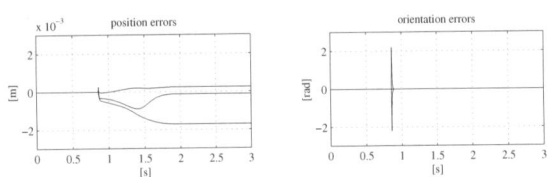

Figure 4: Time history of the end-effector position and orientation errors using the CLIK algorithm based on the Euler angles.

It can be recognized how the spacecraft orientation tracking performance is further improved, at the expenses of a limited end-effector orientation error both during the transient and at steady state; such error is due to the occurrence of an algorithmic singularity caused by a conflict between the (end-effector orientation) constraint task and the spacecraft orientation task.

5 Conclusion

Kinematic control schemes for redundant free-floating robotic systems have been presented in this work. A key point has been the adoption of the unit quaternion to represent orientation errors. A number of case studies on a system composed by a spacecraft and a six-joint manipulator have revealed the effectiveness of the various solutions, including the occurrence of a representation singularity and the requirement of maintaining constant spacecraft attitude.

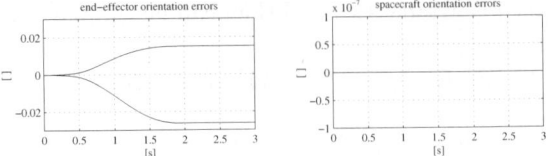

Figure 5: Time history of the (end-effector orientation) constraint errors and the spacecraft orientation errors using the CLIK algorithm with redundancy resolution.

Acknowledgments—This work was supported by *MURST* and *ASI*.

References

[1] Y. Umetani and K. Yoshida, "Resolved motion rate control of space manipulators with generalized Jacobian matrix," *IEEE Transactions on Robotics and Automation*, vol. 5, pp. 303–314, 1989.

[2] D. Nenchev, Y. Umetani, and K. Yoshida, "Analysis of a redundant free-flying spacecraft/manipulator system," *IEEE Transactions on Robotics and Automation*, vol. 8, pp. 1–6, 1992.

[3] B. Siciliano, "Closed-loop inverse kinematics algorithms for redundant spacecraft/manipulator systems," *Proceedings of the 1993 IEEE International Conference on Robotics and Automation*, Atlanta, GA, pp. 95–100, 1993.

[4] P. Chiacchio, S. Chiaverini, L. Sciavicco, and B. Siciliano, "Closed-loop inverse kinematics schemes for constrained redundant manipulators with task space augmentation and task priority strategy," *International Journal of Robotics Research*, vol. 10, pp. 410–425, 1991.

[5] S. Chiaverini and B. Siciliano, "The unit quaternion: A useful tool for inverse kinematics of robot manipulators," *Systems Analysis Modelling Simulation*, vol 35, pp. 45–60, 1999.

[6] L. Sciavicco and B. Siciliano, *Modeling and Control of Robot Manipulators*, 2nd Edition, Springer-Verlag, London, UK, 2000.

[7] D. Nenchev, K. Yoshida, and Y. Umetani, "Analysis, design and control of free-flying space robots using fixed-attitude-restricted Jacobian matrix," in *Robotics Research – 5th International Symposium*, H. Miura and S. Arimoto (Eds.), MIT Press, Cambridge, MA, pp. 251–258, 1990.

Zero Reaction Maneuver: Flight Validation with ETS-VII Space Robot and Extension to Kinematically Redundant Arm

Kazuya Yoshida, Kenichi Hashizume and Satoko Abiko

Department of Aeronautics and Space Engineering, Tohoku University,
Aoba 01, Sendai 980-8579, JAPAN,
yoshida@astro.mech.tohoku.ac.jp

Abstract

This paper presents the experimental results and post-flight analysis of Reaction Null-Space based reactionless manipulation, or Zero Reaction Maneuver (ZRM). The concept has been developed with an insight into the motion dynamics of free-flying multibody systems and its practical availability is clearly demonstrated with ETS-VII, a Japanese space robot. The ZRM is proven particularly useful to remove the velocity limit of manipulation due to the reaction constraint and the time loss due to the waiting for the attitude recovery. The existence of the ZRM is very limited for a 6 DOF manipulator arm mounted on a free-flying base, but it is discussed that more operational freedom is obtained with a kinematically redundant arm.

1. Introduction

The Engineering Test Satellite VII (ETS-VII), Figure 1, developed and launched by National Space Development Agency of Japan (NASDA) has been successfully flown and carried out a lot of interesting orbital robotics experiments with a 2 meter-long, 6 DOF manipulator arm mounted on this un-manned spacecraft.

The ideas for the rescue or service to a malfunctioning satellite by a free-flying space robot has been discussed since early 80s (for example [1]), but very few attempts have ever done in orbit. The maintenance missions of the Hubble Space Telescope and the retrieval of the Space Flyer Unit are such important examples carried out with the Space Shuttle Remote Manipulator System. However, in these missions the manipulator was manually operated by a well-trained flight crew. Autonomous target capture by an un-manned space robot is a big challenge for space robotics community for many years, and very recently, essential parts of this technology have been successfully verified and demonstrated in orbit by ETS-VII.

The mission objective of ETS-VII is to test robotics technology and demonstrate its utility for un-manned orbital operation and servicing tasks. The mission consists of two subtasks, autonomous

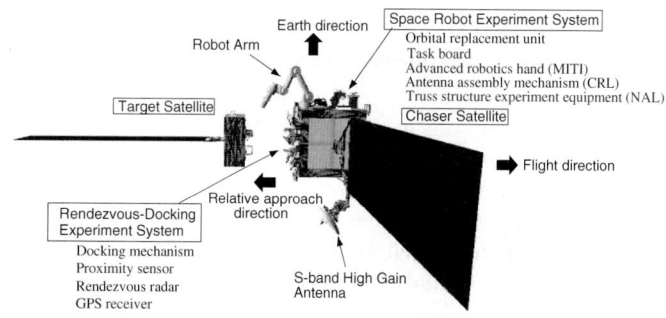

Figure 1: The Engineering Test Satellite VII

rendezvous/docking (RVD) and robot experiments (RBT). The robot experiments include a variety of topics such as: (1) teleoperation from the ground with large time-delay, (2) robotic servicing task demonstrations such as ORU exchange and deployment of a space structure, (3) dynamically coordinated control between the manipulator reaction and the satellite attitude, and (4) capture and berthing of a target satellite. Early reports on some of these experiments were made in [2][3][7][9], for example.

The initially planned flight experiments were successfully completed by the end of May 1999. But since the ETS-VII was still operational in a good condition, an extensive mission period was set till the end of December 1999. In this period the opportunity was opened for academic proposals and four research groups from Japanese universities were given the time to do their own flight experiments. The groups and their topics are (A) Tohoku University 1: several different dynamic control methods, which is elaborated in this paper, (B) Tohoku University 2: teleoperation using a 6DOF haptic interface device, (C) Tokyo Institute of Technology: identification of the vibratory dynamics, and (D) Kyoto University: teleoperation with a bilateral force feedback control.

The present authors carefully prepared for the above flight experiments (A) and have successfully obtained invaluable flight data, where the focus was made on the dynamic characteristics of the base/arm coupling and coordination. Specific research subjects and corresponding flight data have been reported in [8]:

(a) manipulation in the inertial space using the Generalized Jacobian Matrix,

(b) reactionless manipulations based on the Reaction Null-Space,

(c) non-holonomic path planning and operation for terminal endpoint control,

(d) coordinated control between the manipulator arm and the base satellite by feedback control with offset attitude commands.

Above all, the results of the reactionless manipulation, or *Zero Reaction Maneuver*, are so clear and obvious that the attitude of the base satellite has kept very close to zero while the manipulator arm makes motion from a given point to another with following the reactionless path obtained from the Reaction Null-Space theory. The Zero Reaction Maneuver should be very useful for future space operations, because one of the reasons why the current manipulator motion in space is so slowly is due to the restriction on the base reaction, and this restriction would be removed.

This paper focuses on the background theory, flight data analysis, and further discussion for more practical usage of the Zero Reaction Maneuver by a manipulator arm with more DOF.

The paper is organized as follows. In Section 2, the formulation of dynamics, particularly about the Reaction Null-Space is briefly reviewed. In Section 3, the flight data of the extended ETS-VII flight experiments are presented, and Section 4 extends the discussion to the case with a redundant arm.

2. Dynamics and Control of a Free-Flying Space Robot

A unique characteristics of a free-flying space robot is found in its motion dynamics. According to the motion of the manipulator arm, the base spacecraft moves due to the action-to-reaction principle or the momentum conservation. The reaction of the arm disturbs its footing base, then the coupling and coordination between the arm and the base becomes an important issue for successful operation. This is a main difference from a terrestrially based robot manipulator and a drawback to make the control of a space manipulator difficult. Earlier studies for the modeling and control of such a free-flying robot are collected in the book [4].

2.1. Basic equations

The equation of motion of a free-flying space robot as a multibody system is, in general, expressed in the following form:

$$\begin{bmatrix} \boldsymbol{H}_b & \boldsymbol{H}_{bm} \\ \boldsymbol{H}_{bm}^T & \boldsymbol{H}_m \end{bmatrix} \begin{bmatrix} \ddot{\boldsymbol{x}}_b \\ \ddot{\boldsymbol{\phi}} \end{bmatrix} + \begin{bmatrix} \boldsymbol{c}_b \\ \boldsymbol{c}_m \end{bmatrix}$$
$$= \begin{bmatrix} \mathcal{F}_b \\ \boldsymbol{\tau} \end{bmatrix} + \begin{bmatrix} \boldsymbol{J}_b^T \\ \boldsymbol{J}_m^T \end{bmatrix} \mathcal{F}_h \quad (1)$$

where we choose the linear and angular velocity of the base satellite (reference body) $\dot{\boldsymbol{x}}_b = (\boldsymbol{v}_b^T, \boldsymbol{\omega}_b^T)^T$ and the motion rate of the manipulator joints $\dot{\boldsymbol{\phi}}$ as generalized coordinates. The formulation is not limited to a single, serial-link manipulator arm, but in this paper, we suppose one serial manipulator system with n Degrees-Of-Freedom (DOF) is mounted on a base body. The symbols used here are defined as follows:

$\boldsymbol{H}_b \in R^{6 \times 6}$: inertia matrix of the base.

$\boldsymbol{H}_m \in R^{n \times n}$: inertia matrix for the manipulator arms (the links except the base.)

$\boldsymbol{H}_{bm} \in R^{6 \times n}$: coupling inertia matrix.

$\boldsymbol{c}_b \in R^6$: velocity dependent non-linear term for the base.

$\boldsymbol{c}_m \in R^6$: that for the manipulator arms.

$\mathcal{F}_b \in R^6$: force and moment exert on the centroid of the base.

$\mathcal{F}_h \in R^6$: those exert on the manipulator hand.

$\boldsymbol{\tau} \in R^n$: torque on the manipulator joints.

Especially in the free-*floating* situation, the external force/moment on the base, which can be generated by gas-jet thrusters, and those on the manipulator hand are assumed zero; i.e. $\mathcal{F}_b = \boldsymbol{0}$, $\mathcal{F}_h = \boldsymbol{0}$. The motion of the robot is governed by only internal torque on the manipulator joints $\boldsymbol{\tau}$, and hence the linear and angular momenta of the system $(\mathcal{P}^T, \mathcal{L}^T)^T$ remain constant.

$$\begin{bmatrix} \mathcal{P} \\ \mathcal{L} \end{bmatrix} = \boldsymbol{H}_b \dot{\boldsymbol{x}}_b + \boldsymbol{H}_{bm} \dot{\boldsymbol{\phi}} \quad (2)$$

2.2. Angular momentum

The integral of the upper set of the equation (1) gives the momentum conservation, as shown in Equation (2), which is composed of the linear and angular momenta. The linear momentum has further integral to yield the principle that the mass centroid stays stationary or linearly moves with a constant velocity.

The angular momentum equation, however, does not have the second-order integral hence provides the first-order non-holonomic constraint. The equation is expressed in the form with the angular velocity of the base $\boldsymbol{\omega}_b$ and the motion rate of the manipulator arm $\dot{\boldsymbol{\phi}}$ as:

$$\tilde{\boldsymbol{H}}_b \boldsymbol{\omega}_b + \tilde{\boldsymbol{H}}_{bm} \dot{\boldsymbol{\phi}} = \mathcal{L} \quad (3)$$

where $\mathcal{L}$ is the initial constant of the angular momentum, and the inertia matrices with a tilde are those modified from Equation (2). $\tilde{\boldsymbol{H}}_{bm} \dot{\boldsymbol{\phi}}$ represents the angular momentum generated by the manipulator motion.

2.3. Manipulation with zero disturbance to the base

From a practical point of view, the attitude change is not desirable, then the manipulator motion planning methods to have minimum attitude disturbance on the base are also well studied. An ultimate goal of those approaches is completely zero disturbance, and such operation is found with an insight into the angular momentum equation.

The angular momentum equation with zero initial constant $\mathcal{L} = \boldsymbol{0}$ and zero attitude disturbance $\boldsymbol{\omega}_b = \boldsymbol{0}$:

$$\tilde{\boldsymbol{H}}_{bm} \dot{\boldsymbol{\phi}} = \boldsymbol{0} \quad (4)$$

yields the following null-space solution:

$$\dot{\boldsymbol{\phi}} = (\boldsymbol{I} - \tilde{\boldsymbol{H}}_{bm}^{+} \tilde{\boldsymbol{H}}_{bm}) \dot{\boldsymbol{\zeta}} \quad (5)$$

The joint motion given by this equation is guaranteed to make zero disturbance on the base attitude. Here the vector $\dot{\boldsymbol{\zeta}}$ is arbitrary and the null-space of the inertia matrix $\tilde{\boldsymbol{H}}_{bm}$ is termed *Reaction Null-Space* (RNS) [5].

The rank of this null-space projector is $n - 3$, while for the ETS-VII the manipulator arm has 6 DOF, i.e. $n = 6$, then there remains 3 DOF for additional criterion to specify $\dot{\boldsymbol{\zeta}}$.

By the way, the manipulator hand motion observed in the satellite base frame $\dot{\boldsymbol{x}}_h = (\boldsymbol{v}_h^T, \boldsymbol{\omega}_h^T)^T$ are expressed as:

$$\begin{bmatrix} \boldsymbol{v}_h \\ \boldsymbol{\omega}_h \end{bmatrix} = \begin{bmatrix} \boldsymbol{J}_v \\ \boldsymbol{J}_\omega \end{bmatrix} \dot{\boldsymbol{\phi}} \quad (6)$$

using conventional Jacobian matrices.

If paying attention to either upper set or lower set of Equation (6), each of them has 3 DOF, then can be a good candidate for this motion constraint.

Let us consider to combine Equations (4) and the lower set of (6):

$$\begin{bmatrix} \tilde{\boldsymbol{H}}_{bm} \\ \boldsymbol{J}_\omega \end{bmatrix} \dot{\boldsymbol{\phi}} = \begin{bmatrix} \boldsymbol{0} \\ \boldsymbol{\omega}_h \end{bmatrix} \quad (7)$$

The solution for $\dot{\boldsymbol{\phi}}$ gives the manipulator motion to generate zero reaction on the base, while the orientation change of the hand, $\boldsymbol{\omega}_h$, is specified or constraint.

In case of the ETS-VII, where the combined inertia and Jacobian matrix

$$\boldsymbol{G} = \begin{bmatrix} \tilde{\boldsymbol{H}}_{bm} \\ \boldsymbol{J}_\omega \end{bmatrix} \quad (8)$$

is a 6 by 6 square matrix. Its solution is then obtained with a matrix inversion as follows, as far as the matrix is not singular.

$$\dot{\boldsymbol{\phi}} = \begin{bmatrix} \tilde{\boldsymbol{H}}_{bm} \\ \boldsymbol{J}_\omega \end{bmatrix}^{-1} \begin{bmatrix} \boldsymbol{0} \\ \boldsymbol{\omega}_h \end{bmatrix} \quad (9)$$

This solution belongs to the Reaction Null-Space given by Equation (5), but unique since any freedom is not left over due to the kinematic constraint introduced in (7).

Note again that the manipulator motion given by this unique solution yields both zero reaction on the base and the specified orientation change of the hand. However, any specification or constraint has not been made on the translational motion of the hand: eventually it moves and the motion trace forms a line. This *resultant* motion of the hand is calculated by plugging Equation (9) into the upper set of (6).

$$\boldsymbol{v}_h = \boldsymbol{J}_v \boldsymbol{G}^{-1} \begin{bmatrix} \boldsymbol{0} \\ \boldsymbol{\omega}_h \end{bmatrix} \quad (10)$$

Again, $\boldsymbol{v}_h$ is a unique solution.

For the Zero Reaction Maneuver tested on the ETS-VII, we prepared several motion paths from or to a given point in the operational space. The motion trace is obtained by a numerical integration of Equation (10) with a constant, non-zero $\boldsymbol{\omega}_h$. As both $\boldsymbol{v}_h$ and $\boldsymbol{\omega}_h$ are dependent in a complex way, several different $\boldsymbol{\omega}_h$ are tried and the resultant $\boldsymbol{v}_h$ are checked in advance simulation, then a most suitable operation is picked up for the flight experiment.

2.4. Singularity Consistent Inversion

In the above process, the operation to obtain $\boldsymbol{v}_h$ includes the inversion of $\boldsymbol{G}$. Unlike conventional Jacobians or inertia matrices, the combined matrix $\boldsymbol{G}$ may involve many singular points in a non-intuitive manner. And at or around a singular point, the inversion is not defined or yields unstable solution.

In order to obtain stable solutions near the singularity, a good computational method using the following equation is developed [6]:

$$\dot{\boldsymbol{\phi}} = k \cdot \mathrm{adj}(\boldsymbol{G}) \dot{\boldsymbol{x}} \quad (11)$$

where k is an arbitrary scaler and $\dot{\boldsymbol{x}}$ stands for $(\boldsymbol{0}, \boldsymbol{\omega}_h^T)^T$. If k is chosen as $k = 1/\det(\boldsymbol{G})$, then the computation becomes same as with the conventional inverse, and a finite k works to bound the magnitude of $\dot{\boldsymbol{\phi}}$ in the vicinity of the singularity.

We employed this method in the practical computation.

3. Flight Experiments

The extended flight experiment proposed by Tohoku University was carried out on September 30, 1999, using three successive *flight paths*. The flight path is a control of the base satellite using reaction wheels. Even under the control, the attitude disturbance is observed when the base receives the manipulator reaction, since the control torque of reaction wheels is relatively small. The attitude control here mainly works for the recovery after the attitude disturbed.

Figure 2 depicts a typical flight data to compare the conventional and reactionless manipulations. The top graph shows the velocity norm of the manipulator hand. The middle shows the reaction momentum induced by the manipulation. And the bottom shows the attitude motion. The graphs include three sets of manipulator motion, the first one is the conventional PTP manipulation generating a relatively large momentum and attitude disturbance, while the other two are the RNS based reactionless manipulation yielding very small, almost zero reaction and disturbance.

It should be noted that, not only the maximum attitude change is remarkably different, but the time for the recovery is also very different. This waiting time for the attitude recovery in the conventional manipulation is not negligible and degrade the efficiency of the operation in practice. However, the reactionless manipulation, or Zero Raction Maneuver, provides almost zero attitude disturbance and almost zero recovery time, thus assures a very high operational efficiency.

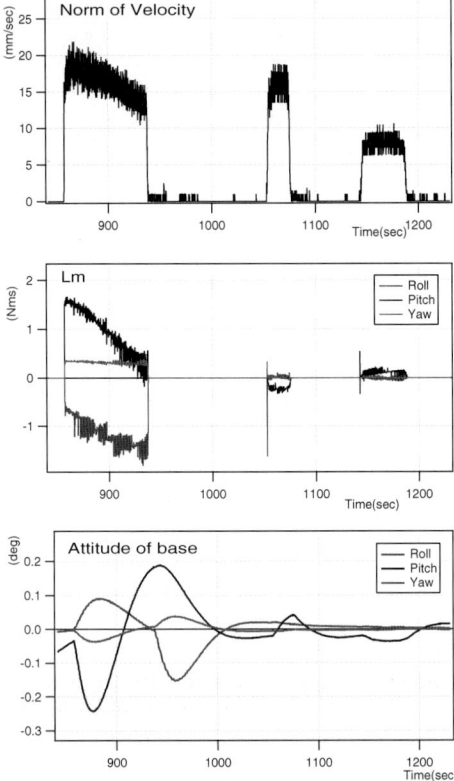

Figure 2: ETS-VII flight data for the RNS based reactionless manipulation

communication window between Tsukuba Space Center, NASDA, and ETS-VII via TDRS, a US data relay satellite located in GEO above the pacific ocean. In each flight path almost net 20 minutes operation (command uplink) and dense telemetry (including video downlink) are allowed.

For our experiments, the manipulator motion trajectories were carefully prepared in a motion file and the safety was preliminary checked on an offline simulator. During the experiment, the motion file is uploaded to ETS-VII at 4 Hz frequency as an isochronous command and the manipulator arm in space is controlled to follow these given trajectories.

In this paper, the focus is made on the experimental result of the RNS based reactionless manipulation only.

3.1. RNS based reactionless manipulation

In the RNS experiment, several sets of reactionless trajectories were prepared using Equation (10) with (11). We prepare the trajectories to go to or from a useful control point such as a standard approach point (150 [mm] right above the corresponding optical marker) to an onboard ORU or a target satellite, and compared with the motion by conventional PTP trajectories.

The experiment was carried out under the attitude

4. Extension to the Case with a Redundant Manipulator Arm

The flight experiments have been carried out under the practical constraints of an existing flight system, ETS-VII. Of particular since ETS-VII has a 6 DOF manipulator arm, non-redundant in trivial sense, the trajectories for Zero Reaction Maneuver are too much constraint to perform Point-To-Point operation from a given initial point A to an arbitrary goal point B. In this section, we discuss how this characteristics could change if we would have a 7 or more DOF manipulator arm.

Let us recall Equations (9) and (10). When $n = 6$ these equations represents a fully determined system, and the integration of (10) yields a one-dimensional line in the operational space.

Now let us consider the cases with $n > 6$. In these cases, the matrix G becomes non-square, $6 \times n$, then a general solution for $\dot{\phi}$ is given with a pseudo inverse component and a null-space component:

$$\dot{\phi} = G^{\#} \begin{bmatrix} 0 \\ \omega_h \end{bmatrix} + (I - G^{\#}G)\dot{\xi} \quad (12)$$

$$v_h = J_v G^{\#} \begin{bmatrix} 0 \\ \omega_h \end{bmatrix} + J_v(I - G^{\#}G)\dot{\xi} \quad (13)$$

In case of $n = 7$, the solution for $\dot{\phi}$ stays on a plane in the configuration space given by mutually perpendicular two vectors, the first term and the second term

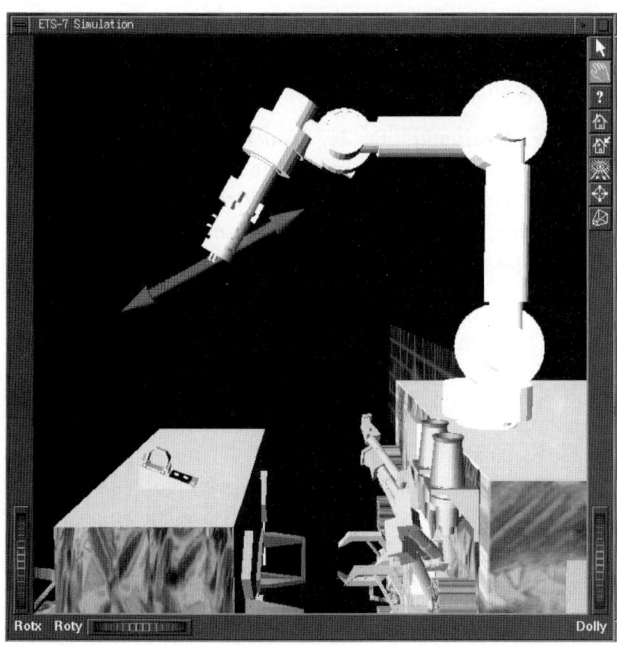

Figure 3: A direction for a reactionless path when $n = 6$

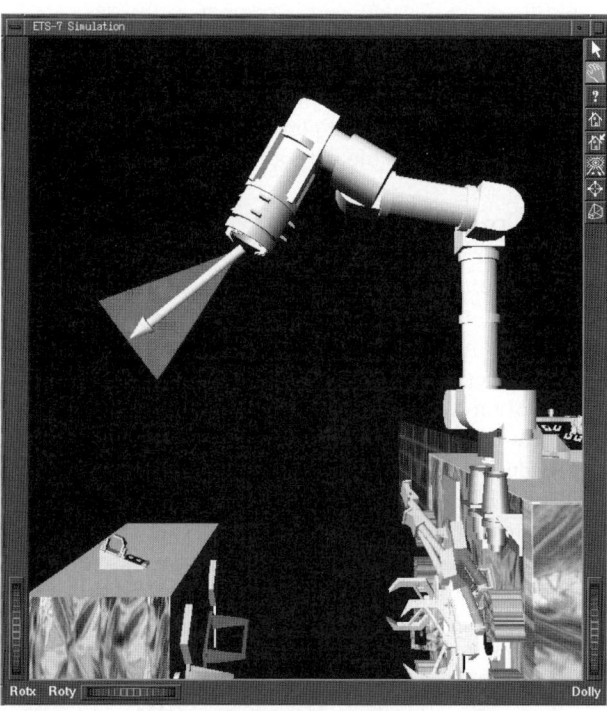

Figure 4: A plane for reactionless paths when $n = 7$

of (12). And the solution for v_h stays on a plane in the operational space given by two vectors, the first term and the second term of (13).

Note that in both spaces, the second term vector has variable magnitude according to the variety of $\dot{\xi}$ but its direction is all the same, invariant for a given configuration. Also in the operational space, the two vectors are not necessary perpendicular but the summation of two stays on a plane which includes both vectors.

Practical situations are compared with computer graphic images. Figure 3 depicts a case with a 6 DOF manipulator arm. For an arbitrary input of ω_h there exists a single instantaneous direction for the hand to move while keeping zero attitude change for the base and the given motion for the hand orientation. Figure 4 depicts a case with a 7 DOF manipulator arm. In this case, there are choices of the instantaneous motion direction of the hand from a plane illustrated here, for an arbitrary input of ω_h. The size of the lateral component depends on $\dot{\xi}$ of Equation (13).

Figure 5 illustrates an application to a motion planning toward a specified terminal point, such as a fixture mounted on a free-floating target satellite. In case with a 6 DOF manipulator arm we cannot expect that a terminal point always stays on a reactionless path [10], but as illustrated here, Zero Reaction Maneuver with some operational flexibility would be possible with a 7 DOF manipulator arm.

However, it should be note that there still remains constraint and some degrees of difficulty in the operation. We need ω_h as a driving input, and we may need non-holonomic planning to locate the hand with

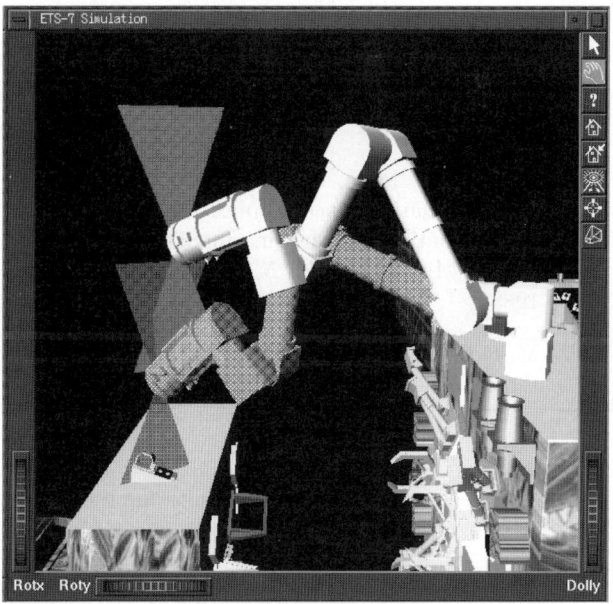

Figure 5: Example of the motion planning to a given goal with Zero Reaction Maneuver ($n = 7$)

a proper orientation, which is left for further investigation.

It is inferred that the possible directions of the hand motion would form a three-dimensional space when the arm has 8 DOF. And the Zero Reaction Maneuver without any operational constraint, except singularities, would be possible with a 9 DOF arm.

5. Conclusions

This paper summarizes the experimental results and post-flight analysis of Reaction Null-Space based reactionless manipulation, or Zero Reaction Maneuver. This concept has been developed with an insight into the motion dynamics of free-flying multibody systems and its practical availability is demonstrated by the extended flight experiments carried out in September 1999 on ETS-VII, a Japanese space robot.

It is clearly verified with the Zero Reaction Maneuver that the attitude disturbance of the base is kept almost zero during manipulator tasks, and it is particularly useful to remove the velocity limit of manipulation due to the reaction constraint and the time loss due to the waiting for the attitude recovery.

Such Zero Reaction Maneuvers (ZRM) are very specified for a 6 DOF manipulator arm mounted on a free-flying base, but it is clarified that more operational freedom is obtained with a kinematically redundant arm. An instantaneous ZRM direction forms a specific vector with a 6 DOF arm, but is can be chosen from a plane with a 7 DOF arm and from a 3-D space with a 8 DOF arm. Fully arbitrary ZRM, in the sense that zero attitude change of the base, can be achieved with a 9 DOF arm.

References

[1] D. L. Akin, M. L. Minsky, E. D. Thiel and C. R. Curtzman, "Space Applications of Automation, Robotics and Machine Intelligence Systems (ARAMIS) phase II," *NASA-CR-3734 – 3736*, 1983.

[2] M. Oda et al, "ETS-VII, Space Robot In-Orbit Experiment Satellite," *Proc. 1996 IEEE Int. Conf. on Robotics and Automation,* pp.739–744, 1996.

[3] There are a number of papers reporting the ETS-VII flight experiments in *Proc. 5th Int. Symp. on AI, Robotics and Automation in Space, i-SAIRAS'99,* June 1999, ESTEC, Netherlands.

[4] *Space Robotics: Dynamics and Control,* edited by Xu and Kanade, Kluwer Academic Publishers, 1993.

[5] K. Yoshida, D. N. Nenchev and M. Uchiyama, "Moving base robotics and reaction management control," *Robotics Research: The Seventh International Symposium,* Ed. by G. Giralt and G. Hirzinger, Springer Verlag, 1996, pp. 101–109.

[6] Y. Tsumaki, D. N. Nenchev and M. Uchiyama, "Jacobian Adjoint Matrix Based Approach to Teleoperation," *Proc. Int. Symp. of Microsystems, Intelligent Materials and Robots,* Sendai, Japan, pp.532-535, 1995.

[7] N. Inaba and M. Oda, "Autonomous Satellite Capture by a Space Robot," *Proc. 2000 IEEE Int. Conf. on Robotics and Automation,* pp.1169–1174, 2000.

[8] K. Yoshida, D. N. Nenchev, N. Inaba and M. Oda, "Extended ETS-VII Experiments for Space Robot Dynamics and Attitude Disturbance Control," *22nd Int. Symp. on Space Technology and Science,* ISTS2000-d-29, Morioka, Japan, May, 2000.

[9] K. Yoshida, K. Hashizume, D. N. Nenchev, N. Inaba and M. Oda, "Control of a Space Manipulator for Autonomous Target Capture –ETS-VII Flight Experiments and Analysis–," *AIAA Guidance, Navigation, and Control Conference & Exhibit,* AIAA2000-4376, Denver, CO, August, 2000.

[10] D. N. Nenchev and K. Yoshida, "Point-To-Point and Reactionless Motions of a Free-Flying Space Robot," *Proc. 2001 IEEE Int. Conf. on Robotics and Automation,* submitted.

A Novel Adaptive Control Law for Autonomous Underwater Vehicles

Gianluca Antonelli Fabrizio Caccavale[†] Stefano Chiaverini Giuseppe Fusco

Dipartimento di Automazione, Elettromagnetismo,
Ingegneria dell'Informazione e Matematica Industriale
Università degli Studi di Cassino
Via G. Di Biasio 43, 03043 Cassino (FR), Italy
{antonelli,chiaverini,fusco}@unicas.it

[†] Dipartimento di Informatica e Sistemistica
Università degli Studi di Napoli Federico II
Via Claudio 21, 80125 Napoli, Italy
caccaval@unina.it

Abstract

In this paper a new adaptive control law for Autonomous Underwater Vehicles (AUVs) is presented. Usually, the control laws are designed with respect to a mathematical model expressed either in the earth-fixed frame or in the vehicle-fixed frame. These two approaches, however, do not take into account the different origin of the effects that can affect the steady state errors: namely, the restoring generalized forces and the ocean current. With the use of a suitable adaptive action those effects can be properly taken into account as will be shown in this paper.

1 Introduction

The position and attitude tracking control problem for Autonomous Underwater Vehicles (AUVs) and Remotely Operated Vehicles (ROVs) is addressed in this paper. A new adaptive control law is proposed and the use of adaptive/integral actions is discussed.

AUVs performing station-keeping tasks, e.g., AUVs carrying a robot to perform a manipulation task, need to be controlled in 6 degrees of freedom (dofs). The control problem can be formulated in a similar way that a rigid robot in 6 dofs. However, the presence of the hydrodynamic effects significantly increases the complexity of the control problem for AUVs. In detail, two are the main causes of steady state errors: the presence of external disturbances such as the current, and the imperfect compensation of the gravity and buoyancy. To overcome these problems most control laws contain an adaptive or integral action to guarantee null steady state error. Experimental results on ODIN (Omni-Directional Intelligent Navigator) are provided in [3]. References [10] and [11] proposed several quaternion-based control laws. Experiments on ODIN in 6 dofs have been realized in [2], where a control law is proposed based on a quaternion attitude representation, and in [13], where a fault tolerant control law is implemented. Several other control laws use adaptive actions to compensate for the disturbances [4, 5, 8, 15]; however, most of them only verify the approach by reduced order simulations and 6-dofs experiments are not common [14]. Among the few papers that explicitly take into account the current, [7] compensates it in vehicle-fixed coordinates while in [12] the current is supposed to be measurable and used in the definition of the sliding manifold. An overview of control techniques for AUVs is reported in [6].

All the above papers do not consider the different nature of the disturbances; hence, the adaptive or integral action is defined either in the earth-fixed frame or in the vehicle-fixed frame. In this paper the use of a suitable adaptive action is proposed that can significantly reduce the tracking error is investigated. The resulting control scheme is validated in 6-dofs simulations.

2 Modeling

Let define as $\Sigma_i, \{O-\boldsymbol{xyz}\}$ a reference frame that we will suppose earth-fixed and inertial and as $\Sigma_v, \{O_v-\boldsymbol{x}_v\boldsymbol{y}_v\boldsymbol{z}_v\}$ a vehicle-fixed frame. The versor $\boldsymbol{z}$ is considered parallel to the gravity, $\boldsymbol{x}_v$ is parallel to the vehicle fore aft direction and $\boldsymbol{z}_v$ is aligned with $\boldsymbol{z}$ when the vehicle is at the surface. Moreover, $\boldsymbol{\eta} = [\boldsymbol{\eta}_1^{\mathrm{T}} \; \boldsymbol{\eta}_2^{\mathrm{T}}]^{\mathrm{T}}$, where $\boldsymbol{\eta}_1 = [x \; y \; z]^{\mathrm{T}} \in \mathbb{R}^3$ is the vector of vehicle position coordinates in a earth-fixed reference frame, $\boldsymbol{\eta}_2 = [\phi \; \theta \; \psi]^{\mathrm{T}} \in \mathbb{R}^3$ is the vector of Euler-angles coordinates expressing the vehicle orientation in the earth-fixed reference frame. $\boldsymbol{\nu} = [\boldsymbol{\nu}_1^{\mathrm{T}} \; \boldsymbol{\nu}_2^{\mathrm{T}}]^{\mathrm{T}}$, $\boldsymbol{\nu}_1 \in \mathbb{R}^3$ is the vector of vehicle linear velocity expressed in the vehicle-fixed reference frame, $\boldsymbol{\nu}_2 \in \mathbb{R}^3$ is the vector of vehicle angular velocity expressed in the vehicle-fixed reference frame.

The vehicle-fixed velocity $\boldsymbol{\nu}$ and the time derivative of the earth-fixed vehicle coordinates are related by the following:

$$\boldsymbol{\nu}_1 = \boldsymbol{R}_I^B \dot{\boldsymbol{\eta}}_1, \qquad (1)$$

$$\boldsymbol{\nu}_2 = \boldsymbol{T}(\boldsymbol{\eta}_2) \dot{\boldsymbol{\eta}}_2. \qquad (2)$$

where $\boldsymbol{R}_I^B$ is the rotation matrix expressing the transformation from the earth-fixed frame to the vehicle-fixed frame, the matrix $\boldsymbol{T}(\boldsymbol{\eta}_2) \in \mathbb{R}^{3\times 3}$, expressed in terms of Euler angles, is given, e.g., in [6].

The vehicle's orientation can also be expressed by mean of the unit quaternion to avoid the occurrence of representation singularities. Defining by $\mathcal{Q} = [\eta \quad \boldsymbol{\varepsilon}]^\text{T} \in \mathbb{R}^4$ the quaternion, the differential kinematics can be expressed in terms of

$$\dot{\mathcal{Q}} = \boldsymbol{E}(\mathcal{Q})\boldsymbol{\nu}_2, \qquad (3)$$

where the expression of $\boldsymbol{E} \in \mathbb{R}^{4\times 3}$ can be found, e.g., in [6]. The inverse mapping is simply given by

$$\boldsymbol{\nu}_2 = 4\boldsymbol{E}^\text{T}(\mathcal{Q})\dot{\mathcal{Q}}. \qquad (4)$$

The equations of motion of an AUV can be written in vehicle-fixed reference frame in the form [6]:

$$\boldsymbol{M}\dot{\boldsymbol{\nu}} + \boldsymbol{C}(\boldsymbol{\nu})\boldsymbol{\nu} + \boldsymbol{D}(\boldsymbol{\nu})\boldsymbol{\nu} + \boldsymbol{g}(\boldsymbol{R}_I^B) = \boldsymbol{\tau} \qquad (5)$$

where $\boldsymbol{M} \in \mathbb{R}^{6\times 6}$ is the mass matrix including the added mass, $\boldsymbol{C}(\boldsymbol{\nu})\boldsymbol{\nu} \in \mathbb{R}^6$ is the vector of Coriolis and Centripetal terms including the effects of the added mass, $\boldsymbol{D}(\boldsymbol{\nu})\boldsymbol{\nu} \in \mathbb{R}^6$ is the vector of friction and hydrodynamic damping terms, $\boldsymbol{g}(\boldsymbol{R}_I^B) \in \mathbb{R}^6$ is the vector of gravitational and buoyant generalized forces, $\boldsymbol{\tau} \in \mathbb{R}^6$ is the vector of forces and moments acting on the vehicle.

It can be proven that the inertia matrix is symmetric and positive definite, the matrix $\boldsymbol{C}(\boldsymbol{\nu})$ is skew-symmetric and $\boldsymbol{D}(\boldsymbol{\nu})$ is positive definite.

For a ground-fixed serial chain of rigid bodies the property of linearity in the dynamic parameters holds. In the case of underwater vehicles, by adopting a suitable mathematical model for the hydrodynamic forces, eq. (5) can be rewritten the form:

$$\boldsymbol{\Phi}(\boldsymbol{R}_B^I, \boldsymbol{\nu}, \dot{\boldsymbol{\nu}})\boldsymbol{\theta} = \boldsymbol{\tau} \qquad (6)$$

with $\boldsymbol{\Phi} \in \mathbb{R}^{6 \times n_\theta}$, being n_θ the size of the vector of parameters $\boldsymbol{\theta}$. Notice that n_θ depends on the adopted model for the hydrodynamic generalized forces. An estimate is given in [12] where, due to the hydrodynamic terms, $n_\theta > 100$.

Focusing our attention on the terms that can affect the steady state of the vehicle, only the gravity and buoyancy have to be considered. Let define as $\boldsymbol{g}^I = [0 \quad 0 \quad 9.81]^\text{T}$ m/s$^2 \in \mathbb{R}^3$ the gravity vector expressed in the earth-fixed frame, m the vehicle mass, $\rho = 1000$ kg/m^3 the water density, ∇ the vehicle volume, $W = m\|\boldsymbol{g}^I\|$ and $B = \rho\nabla\|\boldsymbol{g}^I\|$. Moreover, the center of gravity and buoyancy are defined, in the vehicle-fixed coordinates, as $\boldsymbol{r}_G = [x_G \quad y_G \quad z_G]^\text{T}$ and $\boldsymbol{r}_B = [x_B \quad y_B \quad z_B]^\text{T}$ respectively. Notice that all the above quantities are constant. It can be observed that the vector $\boldsymbol{g}(\boldsymbol{R}_B^I)$ can be represented as $\boldsymbol{g}(\boldsymbol{R}_B^I) = \boldsymbol{\Phi}_R(\boldsymbol{R}_B^I)\boldsymbol{\theta}_R$ where the matrix $\boldsymbol{\Phi}_R(\boldsymbol{R}_B^I) \in \mathbb{R}^{6\times 4}$ is simply given by:

$$\boldsymbol{\Phi}_R(\boldsymbol{R}_B^I) = \begin{bmatrix} \boldsymbol{R}_I^B \boldsymbol{z} & \boldsymbol{O}_{3\times 3} \\ \boldsymbol{0}_{3\times 1} & \boldsymbol{S}(\boldsymbol{R}_I^B \boldsymbol{z}) \end{bmatrix}, \qquad (7)$$

where $\boldsymbol{S}(\cdot)$ is the 3×3 matrix operator performing the cross product. The (4×1) vector of dynamic parameters is then composed by $\boldsymbol{\theta}_R = [W - B \quad y_G W - y_B B \quad z_G W - z_B B \quad x_G W - x_B B]^\text{T}$ and it is constant.

Control of marine vehicles cannot neglect the effects of the ocean current. Let us assume that the ocean current, expressed in the earth-fixed frame, $\boldsymbol{\nu}_c^I$ is constant and irrotational, i.e., $\boldsymbol{\nu}_c^I = [\nu_{c,x} \quad \nu_{c,y} \quad \nu_{c,z} \quad 0 \quad 0 \quad 0]^\text{T}$ and $\dot{\boldsymbol{\nu}}_c^I = \boldsymbol{0}$. Its effects can be added to the dynamics of a rigid body moving in a fluid simply considering the *relative* velocity in vehicle-fixed frame $\boldsymbol{\nu}_r = \boldsymbol{\nu} - \boldsymbol{R}_I^B \boldsymbol{\nu}_c^I$ in the derivation of the Coriolis, centripetal and the damping terms in the equation (5).

In [2, 7], the current is assumed to be not measurable, its effect is modelled as a time-varying disturbance acting on the vehicle-fixed generalized forces. The current is generally considered irrotational and constant in the earth-fixed frame. Its effect on the vehicle can be modeled as a constant disturbance in the earth-fixed frame that is further projected on to the vehicle-fixed frame. Let define $\boldsymbol{\theta}_C \in \mathbb{R}^7$ be the vector of constant parameters contributing to the generalized forces due to the current. Then the current disturbance is modelled as $\boldsymbol{\Phi}_C(\boldsymbol{R}_B^I)\boldsymbol{\theta}_C$ where $\boldsymbol{\Phi}_C(\boldsymbol{R}_B^I) \in \mathbb{R}^{6\times 7}$, is given by

$$\boldsymbol{\Phi}_C(\boldsymbol{R}_B^I) = \begin{bmatrix} \boldsymbol{R}_I^B & \boldsymbol{O}_{3\times 4} \\ \boldsymbol{O}_{3\times 3} & \boldsymbol{E}^\text{T} \end{bmatrix}. \qquad (8)$$

Notice that a differential quaternion-based representation has been assumed for the moment components of the disturbance.

The two regressors $\boldsymbol{\Phi}_R$ and $\boldsymbol{\Phi}_C$ can be grouped. Notice, however, that the vertical component of the current and the parameter $W - B$ have the same dynamical effect. This means that they can be merged in linear combination in order to achieve the following formulation:

$$\boldsymbol{M}\dot{\boldsymbol{\nu}} + \boldsymbol{C}(\boldsymbol{\nu})\boldsymbol{\nu} + \boldsymbol{D}(\boldsymbol{\nu})\boldsymbol{\nu} + \boldsymbol{\Phi}_P(\boldsymbol{R}_B^I)\boldsymbol{\lambda} = \boldsymbol{\tau} \qquad (9)$$

where $\boldsymbol{\Phi}_P \in \mathbb{R}^{6\times 10}$, is given by

$$\boldsymbol{\Phi}_P = \begin{bmatrix} \boldsymbol{O}_{3\times 3} & \boldsymbol{R}_I^B & \boldsymbol{O}_{3\times 4} \\ \boldsymbol{S}(\boldsymbol{R}_I^B \boldsymbol{z}) & \boldsymbol{O}_{3\times 3} & \boldsymbol{E}^\text{T} \end{bmatrix}, \qquad (10)$$

and $\boldsymbol{\lambda} = [\boldsymbol{\lambda}_1^\text{T} \quad \boldsymbol{\lambda}_2^\text{T}]^\text{T}$. The vector $\boldsymbol{\lambda}_1 \in \mathbb{R}^3$ collects the parameters contributing to the restoring moment (*vehicle-fixed* disturbances), i.e., the elements 2, 3 and 4 of the vector $\boldsymbol{\theta}_R$ defined by eq. (7), and $\boldsymbol{\lambda}_2 \in \mathbb{R}^7$ collects the parameters contributing to the generalized forces caused by the current as well as for the restoring linear force (*earth-fixed* disturbances).

It is possible to rewrite the dynamic model in terms of an earth-fixed frame instead of the vehicle-fixed frame as shown above. In this case the state variables are $\boldsymbol{\eta}$, $\dot{\boldsymbol{\eta}}$ and $\ddot{\boldsymbol{\eta}}$:

$$\boldsymbol{M}^\star(\boldsymbol{R}_B^I)\ddot{\boldsymbol{\eta}} + \boldsymbol{C}^\star(\boldsymbol{R}_B^I, \dot{\boldsymbol{\eta}})\dot{\boldsymbol{\eta}} + \boldsymbol{D}^\star(\boldsymbol{R}_B^I, \dot{\boldsymbol{\eta}})\dot{\boldsymbol{\eta}} + \boldsymbol{g}^\star(\boldsymbol{R}_B^I) = \boldsymbol{\tau}^\star, \qquad (11)$$

the expression of $M^\star$, $C^\star$, $D^\star$, $g^\star$ and $\tau^\star$ can be found, e.g., in [6].

Properties of the model at rest
When the rigid body is at rest, i.e., $\eta = cost.$, $\nu = \dot{\nu} = 0$, or, equivalently, $\eta = cost.$, $\dot{\eta} = \ddot{\eta} = 0$, two forces have to be compensated: namely, the restoring generalized forces and the current.

The restoring generalized forces are given by the gravity and the buoyancy. As pointed out above, this term is function of the orientation of the body (R_B^I), the difference between gravity and buoyancy ($W - B$) and a linear combination of the first moment of gravity/buoyancy of the vehicle (mr_G, $\rho \nabla r_B$). Notice that the difference between gravity and buoyancy only affects the linear force acting on the vehicle (see eq. (7)) which is constant in the earth-fixed frame. On the other hand, the two vectors of the first moment of inertia affect the moment acting on the vehicle and are constant in the vehicle-fixed frame. Usually, the vehicle is designed to be slightly positive buoyant and with the center of buoyancy along z_v for stability purposes [2, 12] (see Figure 1). The larger is the metacentric height of the vehicle, the larger are the corresponding restoring moments.

The current is supposed to be irrotational and constant in the earth-fixed frame. Although this is not strictly true, it is a common simplification. Its effect on the vehicle is not a linear force parallel to the current itself due to the coupling effects of the hydrodynamic terms. Moreover, the presence of a coupling moment can also be observed. However, we made the reasonable hypotheses that the *main* contribution is experienced along the current direction.

As it will be shown in the following, by taking into account the different nature of the restoring and current forces the performance of the control law can be significantly improved.

3 Integral actions

Several control laws have been presented in literature to solve the tracking problem of AUVs. All the proposed control laws are based on error variables defined either in the vehicle-fixed frame or in the earth-fixed frame. In this Section, we will discuss the drawbacks experienced in considering an adaptive, or integral, action totally conceived in one frame.

3.1 Earth-fixed frame based errors

Let consider as error variables the following:

$$\tilde{\eta} = \eta_d - \eta \quad (12)$$
$$\dot{\tilde{\eta}} = \dot{\eta}_d - \dot{\eta} \quad (13)$$

where the subscript d denotes the desired value of the corresponding variable. A possible integral action is:

$$\tau_{int}^\star = K_{int}^\star \int_0^t \tilde{\eta}(\sigma) d\sigma \quad (14)$$

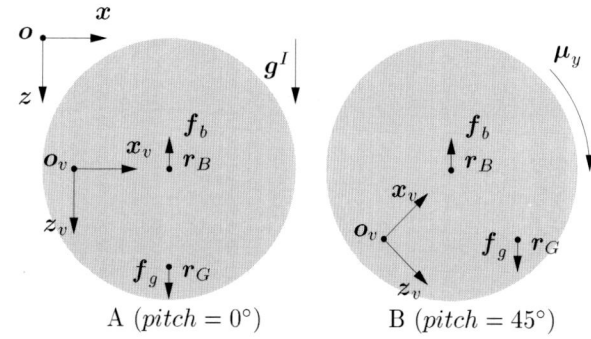

Figure 1: Different configurations and corresponding restoring forces and moments. Planar view parallel to the xz earth-fixed frame.

where $K_{int}^\star \in \mathbb{R}^{6 \times 6}$ is a positive definite matrix of gains. At steady state, with null tracking error, the integral action is in charge of compensating the restoring and current effects on the vehicle dynamics. While this action suitably compensates for the current effect, it is not appropriate to compensate for the restoring effect. In Figure 1 a vehicle is shown in different postures. Let suppose that the vehicle, starting from configuration A, is driven to configuration B and, after a while, back to configuration A. In configuration A the integral action (14) will not give any contribution to the moment, since the centers of gravity/buoyancy are aligned. In configuration B, however, at steady state, the integral action will compensate exactly for the moment generated by the misalignment with the gravity. When the vehicle is driven back to configuration A, the closed loop system has to wait that the integral action is discharged to obtain a null steady state error, thus affecting the bandwidth of the control.

3.2 Vehicle-fixed frame based errors

A different definition of the error is obtained considering the vehicle-fixed variables:

$$\tilde{y} = \begin{bmatrix} R_I^B(\eta_{1,d} - \eta_1) \\ \tilde{\varepsilon} \end{bmatrix} \quad (15)$$
$$\tilde{\nu} = \nu_d - \nu \quad (16)$$

where $\tilde{\varepsilon}$ is the quaternion based attitude error [2]. A possible integral action is then given by:

$$\tau_{int} = K_{int} \int_0^t \tilde{y}(\sigma) d\sigma \quad (17)$$

where $K_{int} \in \mathbb{R}^{6 \times 6}$ is a positive definite matrix of gains.

The drawback of this integral action can be experienced in the presence of a current. Let suppose that the vehicle is at steady state in configuration A (see Figure 2). The control action, thus, is compensating for the current that, in this particular configuration, is parallel to the y_v axis.

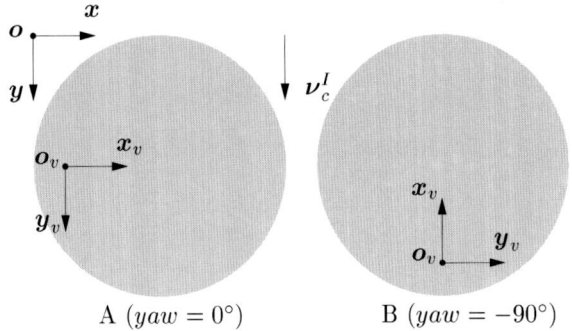

Figure 2: Different configurations with respect to a constant current. Planar view parallel to the xy earth-fixed frame.

When the vehicle is rotated to configuration B, the compensation action is still acting in the vehicle-fixed coordinates, i.e., along $\boldsymbol{y}_v$ and $\boldsymbol{x}$. The steady state in position B will be reached only when the integral action will *rotate* the compensation vector in order to be parallel to $\boldsymbol{x}_v$ and $\boldsymbol{y}$. During this transient, thus, the integral term is acting as a disturbance.

Moreover, for this specific control action, the same drawback as for the restoring forces compensation can be observed. In fact the integral action along the moment does not make use of the mathematical model, in other words it does not make use of the knowledge of the equations representing the restoring force. However, this can be avoided resorting to a simple model-based adaptive action as the one proposed in, e.g., [2, 8, 10]:

$$\boldsymbol{\tau}_{int} = \boldsymbol{\Phi} \cdot \hat{\boldsymbol{\theta}} \quad (18)$$
$$\dot{\hat{\boldsymbol{\theta}}} = \boldsymbol{K}_\theta^{-1} \boldsymbol{\Phi}^{\mathrm{T}} (\tilde{\boldsymbol{\nu}} + \boldsymbol{\Lambda} \tilde{\boldsymbol{y}}) \quad (19)$$

where $\boldsymbol{K}_\theta > \boldsymbol{O}$ (its dimension depends on the model used to retrieve the regressor) and $\boldsymbol{\Lambda} > \boldsymbol{O}$.

In this case, the adaptation is computed with respect to the constant vector $\boldsymbol{\theta}$, i.e., the value of $\boldsymbol{\theta}$ is filtered by the regressor matrix $\boldsymbol{\Phi}$. In configuration A (Figure 1), thus, the estimation of $\boldsymbol{\theta}$ is different from the null value while the compensation is null due to the structure of the regressor.

Mathematically, both drawbacks could be avoided by resorting to an earth-fixed, model-based, adaptive control law. The velocity error should then make use of the current measurements. From the practical point of view, this approach cannot guarantee a fine positioning for the following reasons: the model is an approximation of the hydrodynamic effects, the measure of the current is not exact and local vortex can make this measure too noisy.

It must be noted that, if the control law makes use of a model-based adaptive action, null position error at steady state, in the presence of a current, cannot be guaranteed. The dynamic model, in fact, does not depend on the absolute vehicle position; hence, null vehicle linear velocity with a non-null position error would not *excite* a corresponding adaptive control action. This can be easily understood from eq. (5), in which all the terms would be null in this case no matter the values of the dynamic parameters are. The reason is in the mathematical model used to derive the control law, which that does not consider external disturbances such as, e.g., the current. A solution is proposed in next Section.

4 Proposed control law

Our aim is to take into account the different nature of the steady state disturbances by an appropriate use of the integral action and the kinematic transformation between the frames. Moreover, an approach appealing for practical implementation is required, i.e., no complete knowledge of the dynamic model is available.

Let define as $\hat{\boldsymbol{\lambda}} \in \mathbb{R}^{10}$ the vector of parameters to be adapted whose meaning has been clarified in the modeling Section.

Let define as $\boldsymbol{s} = \tilde{\boldsymbol{\nu}} + \boldsymbol{\Lambda}\tilde{\boldsymbol{y}}$, with $\boldsymbol{\Lambda} = \mathrm{blockdiag}\{\lambda_p \boldsymbol{I}_{3\times 3}, \lambda_o \boldsymbol{I}_{3\times 3}\}$, $\boldsymbol{\Lambda} > \boldsymbol{O}$. The control law is given by:

$$\boldsymbol{\tau} = \boldsymbol{K}_D \tilde{\boldsymbol{\nu}} + \boldsymbol{K}_P \tilde{\boldsymbol{y}} + \boldsymbol{\Phi}_P \hat{\boldsymbol{\lambda}} \quad (20)$$
$$\dot{\hat{\boldsymbol{\lambda}}} = \boldsymbol{K}_\lambda^{-1} \boldsymbol{\Phi}_P^{\mathrm{T}} \boldsymbol{s}, \quad (21)$$

where $\boldsymbol{K}_D \in \mathbb{R}^{6\times 6}$, $\boldsymbol{K}_P \in \mathbb{R}^{6\times 6}$ and $\boldsymbol{K}_\lambda \in \mathbb{R}^{10\times 10}$ are positive definite design matrices of gains.

Notice that the control law is a PD action plus a suitable integral term that take into proper consideration the origin of the disturbances. Without current measurements and without exact knowledge of the vehicle dynamics is possible to compensate efficiently the steady state disturbances. The stability analysis is investigated in [1].

5 Simulations

The simulations have been run on a mathematical model of ODIN (Omni-Directional Intelligent Navigator), an AUV built at the Autonomous Systems Laboratory (ASL) of the University of Hawaii. It has a near-spherical shape with diameter of ≈ 0.63 m. Its dry weight is ≈ 125 Kg; it is, thus, slightly positive buoyant. Despite the small metacentric height, and thus small restoring moment, the proposed control law exhibits improvement of tracking performances with respect to other controllers.

The simulations have been run considering the equations given by (5) taking into proper consideration the current, i.e., computing the relative velocity $\boldsymbol{\nu}_r$.

The desired trajectory is designed in order to show the benefits of the proposed approach. Under a current of 0.3 m/s the vehicle is commanded to move in pitch of 20 deg and yaw of 90 deg in 10 s (see Figure 3) while keeping constant the position.

Three simulations have been run. First, an earth-fixed based control law is implemented [11]. Second, a vehicle-

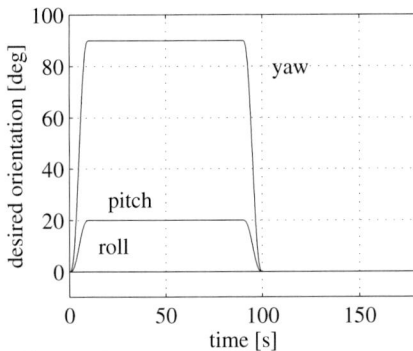

Figure 3: Desired orientation vs time.

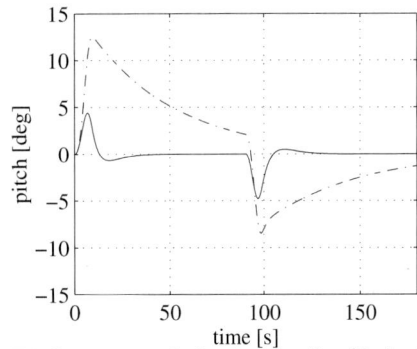

Figure 5: Pitch error, earth-fixed controller (dashed) and proposed controller (solid).

fixed based control law is implemented [2]. Finally, the proposed *mixed* control law (20)-(21) is compared with the previous two. The control laws being different it is not possible to use the same gains. A fine tuning has been implemented for each of the control laws before the comparison.

Comparison between earth-fixed and proposed control law

The position tracking error is similar. The current mainly affects the vehicle position and both the control laws are estimating the disturbance given by the current in the earth-fixed frame. This gives a prompt reaction to the vehicle rotation (see Figure 4).

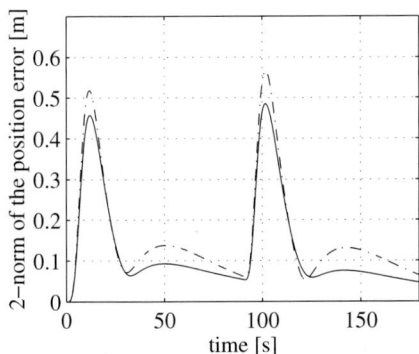

Figure 4: 2-norm of the position error, earth-fixed controller (dashed) and proposed controller (solid). In this case, the two tracking errors are similar.

Observing the pitch error in Figure 5, however, it is possible to notice the different behavior of the control laws. The earth-fixed based controller is estimating the effect of the gravity on the vehicle moment, this means that, from the rest position to the inclined position, the adaptation to the restoring moment has to be waited for. In the proposed approach, however, once that a *good* estimation of the vector λ_1 is obtained, the compensation of the restoring moment is effectively performed.

Due to the structure of the vehicle, two parameters (one for each implemented controller) are significantly different from zero when the vehicle is not at rest, i.e., $\phi \neq 0$ and $\theta \neq 0$. In detail, for the earth-fixed controller this is the fifth element of the integral action, for the proposed controller this is the third element of the first moment of buoyancy: $\rho \nabla r_{B,3}$. In Figure 6 those two variables are reported (note that they do not have the same measurement unit). It can be noted that the parameter of the proposed controller is almost constant with respect to the variation of the parameter relative to the earth-fixed action.

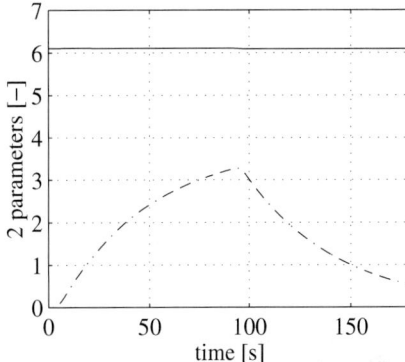

Figure 6: Time vs adaptive actions, earth-fixed controller (dashed) and proposed controller (solid).

Comparison between vehicle-fixed and proposed control law

A symmetric reasoning can be done for the adaptive, vehicle-fixed controller. The adaptation with respect to the restoring moment is the same as the proposed controller. The performances, thus, are similar as it can be observed from the pitch error plot in Figure 7.

A difference arises in comparing the tracking behavior in the position variables as shown in Figure 8. The adaptive action, in the vehicle-fixed frame, after the rotation of 90° acts as a disturbance until the adaptation does not compensate properly the effect of the current. This can be seen in Figure 9 where the polar plot of the current compensation is shown. The proposed controller is working in the earth-fixed frame, thus compensating the current along its direction. The adaptive, vehicle-fixed controller however, is *rotating* its action with the vehicle, thus causing a larger tracking error.

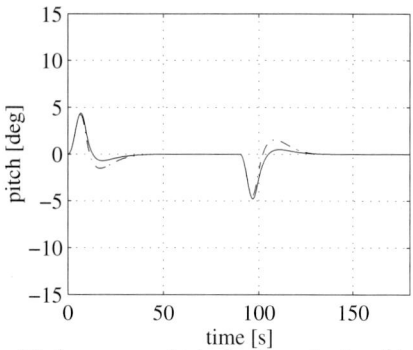

Figure 7: Pitch error, vehicle-fixed controller (dashed) and proposed controller (solid).

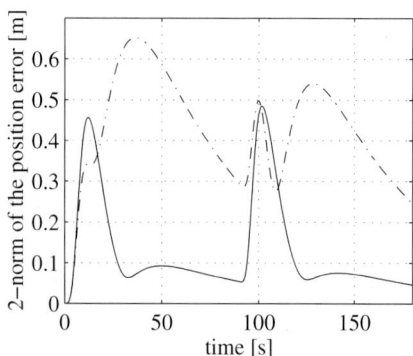

Figure 8: 2-norm of the position error, vehicle-fixed controller (dashed) and proposed controller (solid).

In Table 1 the integral of the position/orientation errors of the different control laws along the simulated trajectory is shown. The values have been normalized with respect to that obtained with the proposed controller. It can be noted that the proposed controller is more efficient also for the errors handled with the same approach. This can be justified by observing that the mathematical model is coupled; a disturbance not compensated in one direction, thus, affects also the other directions.

6 Conclusions

A new adaptive control law for Underwater Vehicles has been presented in this paper. With the use of a suitable adaptive action the effects of the the restoring generalized forces and the ocean current can be properly taken into account. The controller performance has been verified by comparison with known control schemes in simulation.

control law	position error	orientation error
earth-fixed	1.23	**2.20**
adaptive, vehicle-fixed	**3.29**	1.28
proposed controller	1.00	1.00

Table 1: Normalized integral of the position/orientation errors of the different control laws along the simulated trajectory.

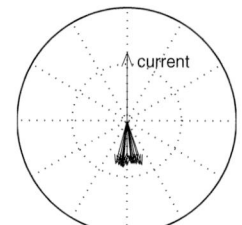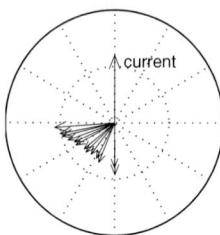

Figure 9: earth-fixed polar representation of the current compensation of the proposed controller (left) and of the adaptive, vehicle-fixed controller (right). Notice that the compensation is *rotating* with the vehicle-fixed frame.

References

[1] G. Antonelli, F. Caccavale, S. Chiaverini, and G. Fusco, "Adaptive Control Laws for Autonomous Underwater Vehicles," *PRISMA Tech. Report 00-01*, Napoli, Sept. 2000.

[2] G. Antonelli, S. Chiaverini, N. Sarkar, M. West, "Adaptive Control of an Autonomous Underwater Vehicle: Experimental Results on ODIN," *IEEE Trans. on Control System Tech.*, in press.

[3] S.K. Choi and J. Yuh, "Experimental Study on a Learning Control System with Bound Estimation for Underwater Robots," *Proc. IEEE Int. Conf. Rob. and Aut.*, Minneapolis, MN, pp. 2160–2165, 1996.

[4] M.L. Corradini and G. Orlando, "A Discrete Adaptive Variable-Structure Controller for MIMO Systems, and Its Application to an Underwater ROV," *IEEE Trans. on Control Systems Tech.*, vol. 5, pp. 349–359, 1997.

[5] R. Cristi, F.A. Pappulias, and A. Healey, "Adaptive Sliding Mode Control of Autonomous Underwater Vehicles in the Dive Plane," *IEEE J. of Oc. Eng.*, vol. 15, pp. 152–160, 1990.

[6] T. Fossen, *Guidance and Control of Ocean Vehicles*, John Wiley & Sons, Chichester, UK, 1994.

[7] T. Fossen and J. Balchen, "The NEROV Autonomous Underwater Vehicle," *OCEANS '91 Conf.*, Honolulu, HI, Oct. 1991.

[8] T. Fossen and S.I. Sagatun, "Adaptive Control of Nonlinear Systems: A case Study of Underwater Robotic Systems," *J. of Robotic Systems*, vol. 8, pp. 393–412, 1991.

[9] T. Fossen and S.I. Sagatun, "Adaptive Control of Nonlinear Underwater Robotic Systems," *Proc. IEEE Int. Conf. Rob. and Aut.*, Sacramento, CA, pp. 1687–1694, Apr. 1991.

[10] O.-E. Fjellstad and T.I. Fossen, "Quaternion Feedback Regulation of Underwater Vehicles," *Proc. IEEE Conf. on Control Appl.*, Glasgow, UK, pp. 857–862, 1994.

[11] O.-E. Fjellstad and T.I. Fossen, "Position and Attitude Tracking of AUVs: A Quaternion Feedback Approach," *IEEE J. of Oc. Eng.*, vol. 19, pp. 512–518, 1994.

[12] A.J. Healey and D. Lienard, "Multivariable Sliding Mode Control for Autonomous Diving and Steering of Unmanned Underwater Vehicles," *IEEE J. of Oc. Eng.*, vol. 18, pp. 327–339, 1993.

[13] T.K. Podder, G. Antonelli, and N. Sarkar, "An Experimental Investigation into the Fault-Tolerant Control of an Autonomous Underwater Vehicle" *J. of Advanced Robotics*, in press.

[14] L.L. Whitcomb, "Underwater Robotics: Out of the research laboratory and Into the Field," *Proc. IEEE Int. Conf. Rob. and Aut.*, San Francisco, CA, pp. 709–716, Apr. 2000.

[15] J. Yuh and K.V Gonugunta, "Learning Control of Underwater Robotic Vehicles," *Proc. IEEE Int. Conf. Rob. and Aut.*, Atlanta, GE, pp. 106–111, May 1993.

Ethological Modeling and Architecture for an Entertainment Robot

Ronald C. Arkin*, Masahiro Fujita**, Tsuyoshi Takagi**, Rika Hasegawa**

This paper presents a novel method for creating high-fidelity models of animal behavior for use in robotic systems based on a behavioral systems approach, and describes in particular how an ethological model of a domestic dog can be implemented with AIBO, the Sony entertainment robot.

I. INTRODUCTION

Ethology is the science of studying the behavior of animals in their natural environment. While much attention has been paid in robotics to neuroscientific models of behavior (e.g., [1,17]), less attention has been paid to realistic ethological models other than in simulated studies. It is our contention that ethology provides great insights into the design of practical robotic systems.

In this paper, a behavior systems methodology is presented drawing on work from both psychology and ethology. A specific ethological model is created for *Canis Familiaris*, the domestic dog. The modeling process itself is extensible to other animal species.

This model is then transformed into an ethological controller suitable for implementation on AIBO, Sony's entertainment robot (Figure 1) [7]. The underlying architecture in support of this model is discussed.

Figure 1. AIBO - Sony's Entertainment Robot

*College of Computing, Georgia Tech, Atlanta, GA, U.S.A.
**Sony Digital Creatures Laboratory, Kitashinagawa, Tokyo, Japan

The notion of robotic pets is not new. In the late 1980s, a robot pet called Petster was marketed for around $80 U.S.D. It purred, snarled, and responded to clapping by running around. It was marketed as a mail order item apparently without much success. A more sophisticated robotic pet's construction was advocated in a book by DeCosta almost a decade earlier. Its motivation was for use in apartments that banned real dogs. It had 3 pairs of sonar sensors mounted on a rigid head. There was no control box; rather an artificial tonal language was used. It understood both the speech of the owner and used its own synthesizer to communicate. It barked for a recharge, which the owner had to provide. Neither of these systems looked more than superficially at actual canine behavior as the basis for their design.

II. CANINE BEHAVIOR

The domestic dog is one of the most studied mammals in existence. An extensive range of literature resources is available describing its behavior in gory detail from almost every aspect. Perhaps the two richest veins relative to the needs of a roboticist are found in the work of Scott [4] and Fox [5,12]. Scott's work in particular has led to the development of an ethogram: a categorization of all of the exhibited behaviors of the dog. This ethogram provides the basis for the model used in this work. Other ethograms [5,18] for the dog exist, but Scott's is the most comprehensive.

A behavior pattern is a unique and independent piece of behavior having a complete adaptive function. The main behavioral classes for the dog can be characterized as shown in Table 1. The last two entries are behavioral subsystems that have been added above and beyond Scott's original ethogram but are included for implementation purposes.

Investigative (searching/seeking)
Sexual
Epimeletic (care and attention giving)
Eliminative (excretion and urination)
Et-epimeletic (attention getting or care soliciting)
Ingestive (food and liquids)
Allelomimetic (doing what others in group do)
Comfort-seeking (shelter-seeking)
Agonistic (associated with conflict)
Miscellaneous Motor
Play
Maladaptive

Table 1: Main Behavioral Subsystems of Dog

Extensive research was conducted on the pertinent canine behavior literature (the literature is voluminous but [e.g., 5,8,9] are good starting points), then summarized in a series of proprietary reports for Sony Corporation.

III. BEHAVIOR SYSTEMS APPROACH

Clearly the organization for such a behavioral control system for the dog is a complex task. To manage this complexity, a concurrent top-down and bottom-up approach to its development has been undertaken. In particular the behavior systems approach developed by the psychologist William Timberlake [6] has provided the structure and terminology for the methods used here. Critical features of a behavior system are:

1. Motivational processes that prime other structures and help organize and maintain their sequence of expression.
2. Perceptual-motor structures (modules) that relate specific stimuli to specific responses. The response components are often sequentially and temporally related and are elicited, initiated, controlled and terminated by stimuli.

This is a *functional,* not physiological, model. (which is good for the purposes of robot design). Four hierarchical levels are represented using this approach:

- *Systems* - These are assumed to be at least partially independent from other systems. They represent a collection of motivational states that prime underlying subsystems and modules.
- *Subsystems* - Coherent strategies that serve the general function of the system. Activation of a subsystem sensitizes an animal to particular stimuli and potentiates responses (Figure 2).
- *Modes* - Motivational substates that are related to the sequential and temporal organization of action patterns with respect to terminal stimuli (e.g., Figure 3).
- *Perceptual-motor modules* - Respond to particular stimuli with particular response components (e.g., Figure 4).

The basic unit of output is an action-pattern: a coherent, recognizable, relatively stereotypical movement (although some variability may be present). The environment is involved in the definition of an action pattern, as well as limb and posture position and temporal patterning. Action patterns vary in strength with particular modules and may be controlled by several modules. Learning can occur through the refinement, combination and reassembling of action patterns. Of note is the fact that there is a relationship between the notion of action pattern and that of motor schema [11] as found in schema theory and utilized in previous designs in our research [10].

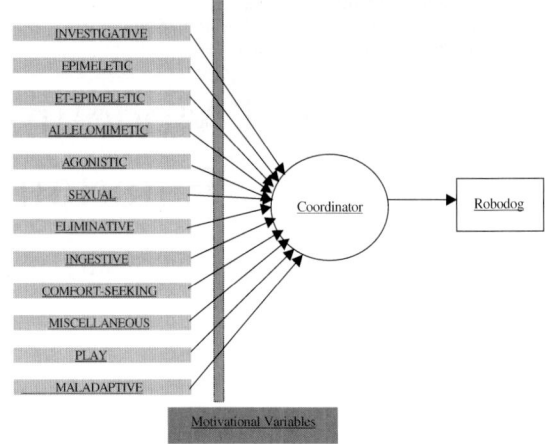

Figure 2: Complete Set of Subsystems

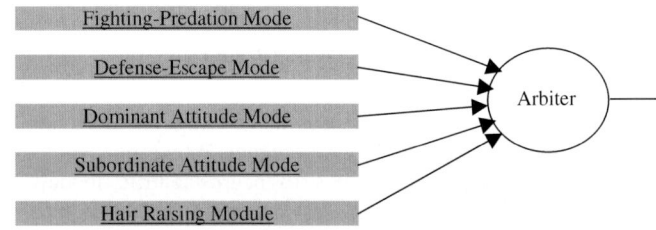

Figure 3. Modules comprising Agonistic Subsystem

Stimulus = threat or dominant animal present + attack
 + escape route/area present + high fear
(the escape areas may include corners of rooms)

Response = run(fast, towards escape route/area)
 + ear-position(both, back)

Figure 4. Example: run-away module

Scott and Fuller's ethogram represents a format suitable for incremental implementation in a robotic dog. The higher-level behavioral systems are decomposed into collections of subsystems that ultimately ground into a set of robotic primitives. From the bottom-up, a collection of parameterizable robot dog primitives have been specified (taxes, reflexes, and fixed action patterns) that map onto the higher-level behavioral systems.

The advantage of this method lies in the reuse and composability of the constituent subsystems and primitives, enabling easy personalization of the robotic pet to different owners. Specifications for motivational variables that affect these components are included to allow emotional state to dynamically alter the embodied control system.

An idealized potential input device inventory should include the following considerations in regards to fidelity for biological dogs:
- **Auditory**
  1. Verbal communication from owner
  2. Non-verbal communication from other robots/animals
  3. Environmental sounds
- **Visual**
  1. Displays of owner
  2. Displays of other animals/robots
  3. Specific behavioral cues (colors, shapes, etc.)
- **Tactile**
  1. Petting
  2. Collision
  3. Holding
- **Force** - to detect restricted movement
- **Olfactory**
- **Dermal (temperature) Taste**
- **Vestibular**

Not all of these are currently available on AIBO.

A preliminary motor output set was developed to enable a mapping of the behaviors onto the actuators. This list of output primitives serves as the set of response functions for the developed behaviors [11]. The list is too extensive to reproduce here, but includes sets of actions organized under major classes of actuation such as tail, head, legs, body, vocalization, locomotion, and elimination.

Two of the smaller behavioral subsystems are reproduced with their behavioral modules listed in Figure 5. For all 12 subsystems there are over 140 total behavioral modules.

Investigative
- Investigative locomotion module
- Head in air sniffing module
- Sniffing conspecific module
- Alert looking module
- Nosing/sniffing scented objects module
- Scent tracking module
- Encounter module
- Pointing module

Et-epimeletic
- Whining for attention module
- Yelping for attention module
- Tail wagging friendly module
- Licking face/hands module
- Pawing module
- Jumping up module
- Pup rooting for nipple module
- Close following of caregiver module

Figure 5. Behavior modules within two of the 12 major subsystems

The following motivational variables were recommended for eventual implementation based on literature studies of canine emotionality and motivation (e.g., [19]):

Guilt	Frustration
Boredom	Anger/aggression
Fear	Patience
Submission / dominance	Spite
Happiness/love	Sadness/depression
Jealousy (pack competition)	Tense/relaxed
Alertness	Sexual
Hunger/thirst	Fatigue
Pain	Gregariousness
Elimination needs	Loyalty
Temperature	Somnolence

Table 2. Motivational Variables

IV. COORDINATION FUNCTIONS

Numerous approaches regarding choice of the coordination functions for the behaviors were investigated. While it was determined that a motivational space approach [20] would likely be the best in the long-term, it was decided that using a model developed initially by Ludlow [15,16] to model biological action-selection and then later imported to computer graphics applications by Blumberg [13,14] would serve well in the short-term.

Ludlow [15,16] introduced the notions of lateral inhibition and the fatiguing of behaviors later utilized by others in the action-selection community. Blumberg's Hamsterdam project [15] speaks to *persistence*, avoiding behavioral dithering (the rapid oscillation between different behaviors) through the incorporation of lateral inhibition and fatigue. It also incorporates *time-sharing* for low priority behaviors to execute in the presence of a high priority behavior. The coordination mechanism is a winner-take-all approach that incorporates both releasing mechanisms (external stimuli) and endogeneous variables (internal data). It can model ethologically observed motivational isoclines. Strict homeostatic modeling is not required however. Blumberg's later work extended this model into the ALIVE system and was based on a dog-like graphical character [14]. Much of his work is drawn from the animation community, but very little from or to robotic systems functioning in the real world.

The overall action-selection mechanism is summarized as follows:
1. At the top-level behavior group competition occurs by updating releasing mechanism values and combining the result with the motivation and interest levels. Inhibition is then applied from other competing behaviors. This process is iteratively repeated until only one behavior has a non-zero value. This is the active behavior within that group.

2. Behaviors within the selected group that are not active can issue secondary commands (i.e., they can execute if it doesn't cause a problem with the selected primary behavior).
3. The process is recursive through the hierarchy until a single behavior is selected to issue the motor commands.

A variant of this approach has been used for initial implementation on AIBO, where a recursive descent through the behavioral systems hierarchy, as described in the previous Section, occurs until one single behavior is selected for execution.

V. IMPLEMENTATION

In order to verify the advantages of the ethological approach, we implemented the model described in the previous sections. We focus on checking if the following advantages hold in the actual robotic implementation.

(1) The fusion of internal motivations and external stimuli
(2) The coordination of behaviors via lateral inhibition
(3) Computational efficiency with a layered architecture

In order to simplify and shorten development time, we implemented a subset of the whole model with limited perception (recognition targets) as follows:

(1) Only 3 subsystems shown in Fig. 8 are realized in part
(2) Only 3 objects, WATER, FOOD, and MASTER, can be recognized by color classification

Fig.6 shows the implemented software architecture on the AIBO. As described in the previous sections, roughly speaking, there are 3 major parts, Release Mechanism, Motivation Creator, and Action Selection Module.

The Release Mechanism component computes its output $RM[I]$ (see Fig. 7) using environmental perceptual results such as the distance to a recognized object. As itemized above, we only use the color camera signal for this purpose and only 3 objects can currently be recognized.

Motivation Creator computes its output $Mo[I]$ (see Fig. 7) using an Instinct and Emotional Model, which has 6 internal variables: nourishment, moisture, bladder, tiredness, curiosity, and affection. Furthermore, another 6 variables act to keep the 6 internal variables within some bounded values. These are called instinct variables, which include hunger, thirst, eliminate, tiredness, curiosity, and affection. The output of the Motivation Creator $Mot[I]$ is computed using these instinct variables.

In the Action Selection Module, a behavior variable $V[I]$ is computed using a function of $RM[I]$ and $Mot[I]$ as shown in the graph of Fig.7. The computation is carried out from behaviors in the higher layer. The lateral inhibition to avoid the behavioral dithering described in the previous section is also carried out here so that the system can select one behavior. From the highest layer (subsystems) to the lowest layer (primitive modules), the computations are performed to select a proper action command that is sent to a Finite-State-Machine where the specific sequences on how to complete the command are described.

Thus, the action to be executed is selected based on the value $V[I]$, which is affected by both $Mot[I]$ related to the internal variables and $RM[I]$ related to the external stimuli. For example, even if the robot has high motivation for ingestive behavior, without the relevant external stimuli, then the robot doesn't select the ingestive behavior, and vice versa.

Figure 8 shows a layered and tree structured architecture for subsystems, modes, and primitive modules. Figure 9 shows the implemented behavior tree, where 3 subsystems, investigative, ingestive, and play, are implemented. Investigative means investigative behaviors such as walk around (locomotion), ingestive means ingestive behaviors such as eating or drinking, and play means interactive behaviors with a human such as giving a paw.

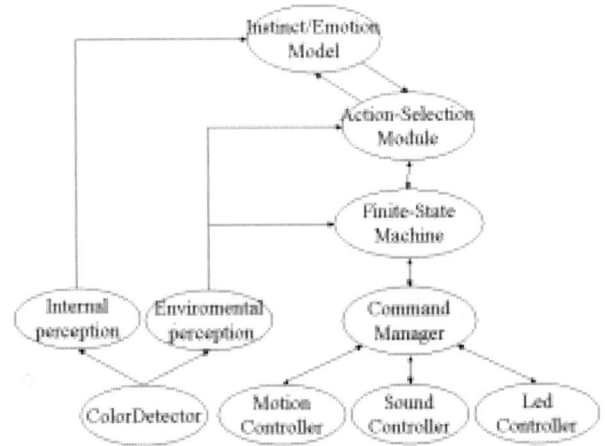

Figure 6. Software architecture

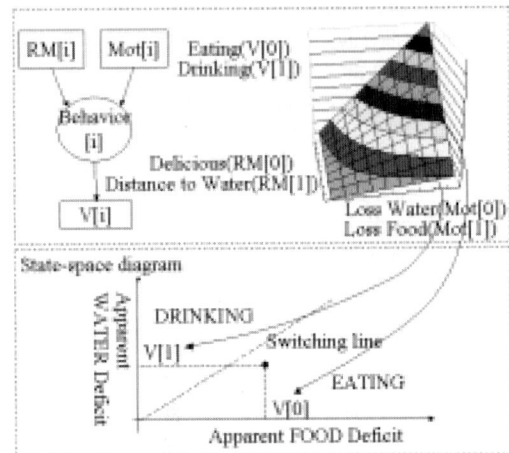

Figure 7. State-space Diagram

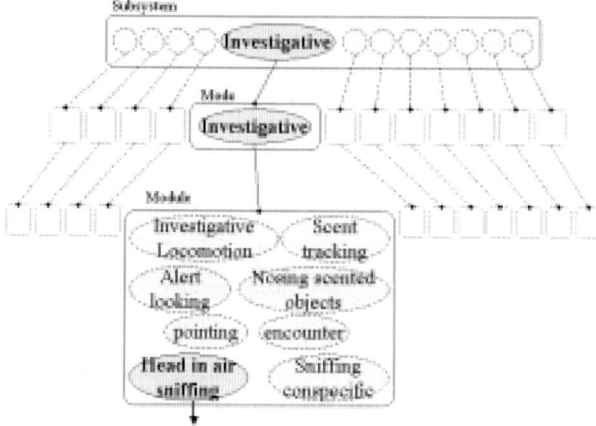

Figure 8. Behavioral Tree (Whole)

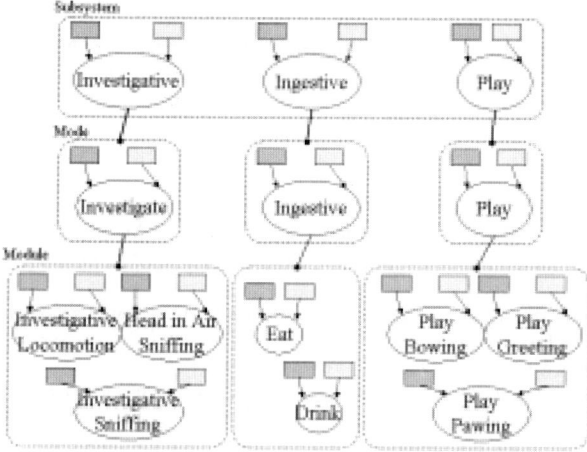

Figure 9. Behavioral Tree (Implemented)

VI. EXPERIMENTS AND RESULTS

In order to verify if the advantages of this approach are achieved, we build a test field as shown in Fig. 10. For easy recognition, we make red, blue, and green circles with 12-cm diameter, which correspond to FOOD, WATER, and MASTER respectively. The field is 120cm square and is surrounded by walls. We placed the robot described in the previous section in the field and measure the RM[I], Mot[I], V[I], selected behavior, and so on.

Figures 11-15 show various time sequences of some relevant measurements. Figure 11 shows the Time-Instinct variable graph. Figure 12 and 13 show Time-Motivation variable graphs corresponding to Mot[I] of subsystems and modules. Figure 14 shows a Time-Release Mechanism (RM[I]) variable graph, and Figure 15 shows the time sequence of selected behaviors. Here, the 6 internal variables are decreased as time passes but increased while the corresponding behavior is executed.

Comparing Figure 11 with Figure 15, we can observe an increase of the instinct variables as well as its decrease when the corresponding action is selected. Moreover, comparing Figures 12 and 13 with Figure 15, we can observe that the corresponding action is not selected (as expected) even when higher Motivation variable Mot[I] is found in some time periods. Then, comparing between Figures 14 and 15, in this period, the Release Mechanism value RM[I] is small, so not enough external stimuli is presented within that period.

During such a period, the system selected "investigative" behavior. Thus, the motivation variables or the internal variables and the external stimuli affect the action selection mechanism in this system, as anticipated.

We encountered a problem, when an action cannot be selected properly. For example, when "hunger" motivation is large, and WATER exists, then the highest layer selects "ingestive" behavior. Because WATER doesn't produce a big Release Mechanism value for the eating behavior, there is no action that has both of larger RM[I] and Mot[I] in the lowest layer of the selected ingestive subsystem. This can be avoided by designing a proper tree structure.

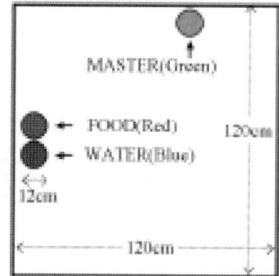

Figure 10. Field

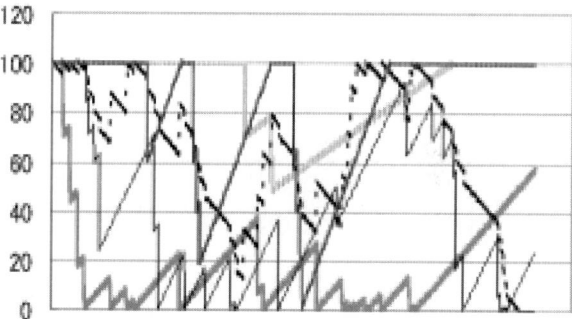

Figure 11. Instinct-Time graph

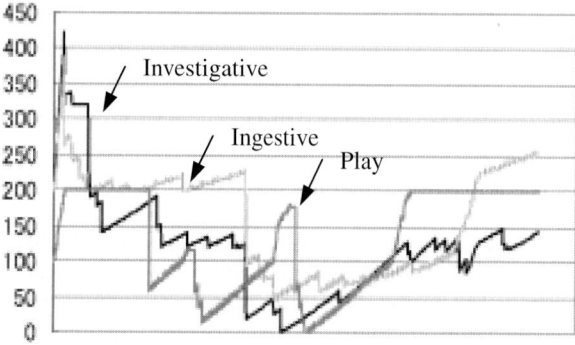

Figure 12. Motivation-Time graph for subsystem

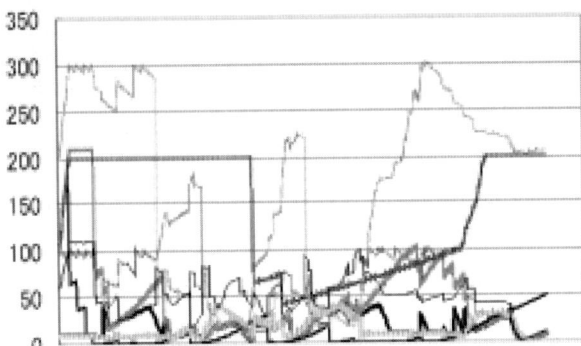

Figure 13. Motivation-Time graph for Module

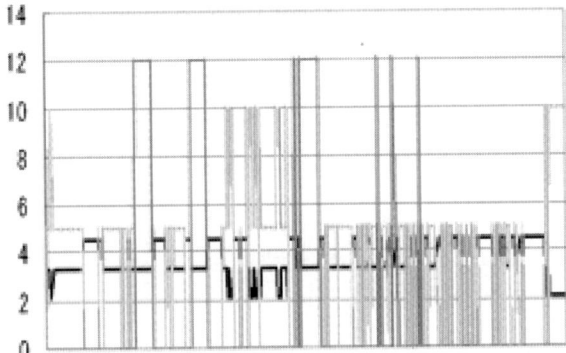

Figure 14. Release Mechanism-Time graph

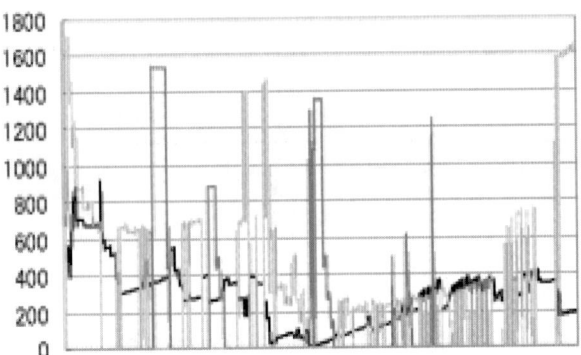

Figure 15. Behavior-Time graph

VII. SUMMARY AND CONCLUSIONS

An ethological controller derived directly from canine behavior has been developed for use in Sony's AIBO. It employs the behavioral systems approach as developed by Timberlake [6] as applied to Scott's ethogram of canine behavior [4]. In the short term, behavioral coordination is accomplished through a variant of the Ludlow [15]-Blumberg [14] action-selection mechanism.

This model has been successfully implemented in part on AIBO. We implemented 3 subsystems with 8 primitive actions with this model. The complex action selection mechanism is observed, which is caused by both internal variables and external stimuli. Moreover, efficient computation is performed by the layered, tree-structured architecture.

References

[1] Beer, R., *Intelligence as Adaptive Behavior: An Experiment in Computational Neuroethology,* Academic Press, N.Y., 1990.
[3] DeCosta, F., *How to Build Your Own Working Robot Pet*, Tab Books, Blue Ridge Summit, PA, 1979.
[2] Engelberger, J.F., *Robotics in Service*, MIT Press, 1989.
[4] Scott, J.P. and Fuller, J.L., *Genetics and the Social Behavior of the Dog*, University of Chicago Press, Chicago, IL, 1965.
[5] Fox, M., *The Dog: Its Domestication and Behavior*, Garland, New York, 1978.
[6] Timberlake, W. and Lucas, G., "Behavior Systems and Learning: From Misbehavior to General Principles", in *Contemporary Learning Theories: Instrumental Conditioning Theory and the Impacts of Biological Constraints on Learning*, S. Klein and R. Mowrer (eds.), LEA Associates, Hillsdale, NJ, 1989.
[7] Sony Robot Reference User's Manual.
[8] Coren, S., *The Intelligence of Dogs*, Bantam, New York, 1994.
[9] Buytendijk, F., *The Mind of the Dog*, Arno Press, New York, 1973.
[10] Arkin, R.C., "Motor Schema-based Mobile Robot Navigation", *International Journal of Robot Research*, Vol. 8, No. 4, pp. 92-112.
[11] Arkin, R.C., *Behavior-based Robotics*, MIT Press, 1998.
[12] Fox, M., *Integrative Behavior of Brain and Behavior in the Dog*, University of Chicago Press, Chicago, 1971.
[13] Blumberg, B., "Action-Selection in Hamsterdam: Lessons from Ethology", *From Animals to Animats 3*, ed. Cliff et al, MIT Press, 1994, pp. 108-117.
[14] Blumberg, B., Old Tricks, New Dogs: Ethology and Interactive Creatures, Ph.D. Thesis, MIT Media Lab, Massachusetts Institute of Technology, Cambridge, MA 1996.
[15] Ludlow, A., "The Behaviour of a Model Animal", *Behaviour*, LVIII, 1-2, pp. 131-172, 1976.
[16] Ludlow, A., "The Evolution and Simulation of a Decision-maker", in *Analysis of Motivational Processes*, F. Toates, and T. Halliday, Academic Press, New York, 1980.
[17] Touretzky, D. and Saksida, L., "Operant Conditioning in Skinnerbots", *Adaptive Behavior*, Vol. 5, No. 3-4, 219-247, 1997.
[18] Serpell, J. (Ed.), *The Domestic Dog: Its Evolution, Behaviour, and Interactions with People*, Cambridge University Press, 1995.
[19] Milani, M., The Body Language and Emotion of Dogs, Quill, New York, 1986.
[20] McFarland, D. (ed.), *Motivational Control Systems Analysis*, Academic Press, London, 1974.
[21] Timberlake, W. and Lucas, G., 1989, Behavior Systems and Learning: From Misbehavior to General Principles, in Contemporary Learning Theories: Instrumental Conditioning Theory and the Impact of Biological Constraints, ed. Klein S. & Mowrer, R. Lawrence Erlbaum Associates, Inc. Hillsdale NJ.
[22] McFarland, D. & Sibley, R., 1975, The behavioral final common path, Philosophical Transactions of the Royal Society, B. 270.

Human-like Robot Head that has Olfactory Sensation and Facial Color Expression

Hiroyasu Miwa[*], Tomohiko Umetsu[*], Atsuo Takanishi[**, ***], Hideaki Takanobu[**]

*Graduate School of Science and Engineering, Waseda University
**Department of Mechanical Engineering, Waseda University
***Humanoid Robotics Institute, Waseda University
#59-308, 3-4-1 Ookubo, Shinjuku-ku, Tokyo, 169-8555 Japan
Tel: +81-3-5286-3257, Fax: +81-3-5273-2209
takanisi@mn.waseda.ac.jp
http://www.takanishi.mech.waseda.ac.jp/

Abstract

The authors have been developing human-like head robots in order to develop new head mechanisms and functions for a humanoid robot that has the ability to communicate naturally with a human by expressing human-like emotion. We were able to four sensations (the visual, auditory, cutaneous and olfactory sensations) of a human's five senses by developing WE-3RIV (Waseda Eye No.3 Refined IV). In this paper, we targeted olfactory sensation and facial color that humans use to communicate with each other. A human's olfactory sensation recognizes over a thousand smells. WE-3RIV can recognize the smell of alcohol, ammonia and cigarette smoke. Our olfactory sensor consists of four semiconductor gas sensors. WE-3RIV was able to recognize the recognition of certain smells within a few seconds of being presented with the smell stimuli by using sensor output and changing rate of the sensor output. Further, we developed the facial color expression using red EL sheets. By adding the facial color expression, the expressiveness of the robot and the recognition rate of a robot's facial expression were improved.

1. Introduction

Many animals have almost all their sensory abilities in their heads. A human has all five senses, the visual, auditory, olfactory, taste and tactile sensations, in the head (Fig. 1.1, Fig. 1.2 [12]). Humans or animals express their mental states by facial expressions or eyes motions. Humans also obtain a lot of information from not only the meaning of the word but also the facial expressions or gaze of one's conversation partner over and above the actual meanings of the words used. The head has an input function and an output function with regard to the environment. Thus, the head is the most important part of the human body in communication. Therefore, the authors aimed at both the input and output functions of the head for studying a humanoid

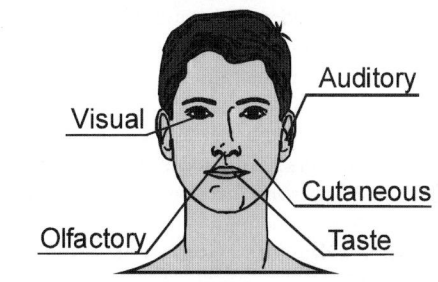

Fig. 1.1 Sense Organs on the Head

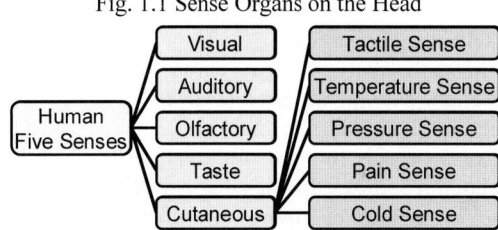

Fig. 1.2 Human Five Senses

robot.

The authors have been developing a human-like head robot in order to elucidate a human vision system from an engineering point of view, and to develop new head mechanisms and functions for a humanoid robot having the ability to communicate naturally with a human.

The head robot is an area of active research in the field of robotics. Brooks developed a head robot which expresses facial expression using eyes, eyelids, eyebrows and a mouth. It communicates with humans using visual information from CCD cameras [1]. Hara developed a head robot that uses fourteen Action Units of Ekman [2][3]. It can express the Six Basic Facial Expressions as quickly as a human and can recognize a human's facial expression and express the same facial expression [4][5]. Regarding other research on sensors for the five human senses, visual and auditory sensation have been studied heavily, but olfactory sensation have not received much attention.

The authors developed the human-like head robot

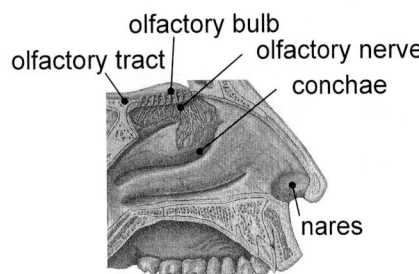

Fig. 2.1 Section of Nasal Cavity [13]

WE-3. In addition, we were able to create coordinated head-eye motion with V.O.R (Vestibular-Ocular Reflex) [6][7], depth direction that uses the angle of convergence between the two eyes [8]. Then, we were able to adjust to the brightness of an object by adding eyelids and human-like expression by adding eyebrows, lips and a jaw [9][10]. We added auditory sensation and tactile sensation to the robot [11]. We also introduced the Equations of Emotion and a mental model that has three independent parameters [10][11].

We added olfactory sensation and a facial color expression function to our new model WE-3RIV. We also improved and rebuilt our mental model. In this paper, we describe the newly added olfactory sensation and facial color expression function.

2. Human's Olfactory Sensation and Facial Color

2.1 Olfactory Sensation

A human's olfactory organ exists on the olfactory epithelium deep in the nose (Fig. 2.1). Molecules of smell are breathed from the nare, and changed to electrical signals by the olfactory cells. Human's cerebrum recognizes that signal as a specific smell. The olfactory epithelium covers the olfactory cleft, and consists of olfactory cells, supporting cells, and basal cells. There are six cilia at the end of an olfactory cell that extend to the surface of the olfactory epithelium. A dog, which has a keen sense of smell, has about two hundred million olfactory cells and about a hundred cilia for each cell. A human has about ten million olfactory cells [12].

The molecules of smell naturally valy widely. For example, a rose gives out over ten kinds of smell (alcohol, aldehyde, ester, hydrocarbon and etc.) For recognizing this enormous variety of smell, a human has various receptor proteins. Humans recognize the visible spectrum by four photoreceptive proteins, but humans recognize smells with over a thousand olfactory receptor proteins [13].

2.2 Facial Color

The color of human's skin depends on the pigmentation of the cuticle, the cutis, and the blood color that has penetrated the cuticle [14]. The former doesn't depend on emotion, but the latter does. The change of blood color that penetrated the cuticle is caused by a change of blood value and oxygen density.

The relationship between the body state and facial color has been researched in the field of medicine. For example, a liver affection can cause jaundice and an iron deficiency anemia causes pallor. We often diagnose our physical condition by our facial color. On the other hand, the relation between emotion and facial color isn't active because it is difficult to rouse a subject's emotion. Watanabe and Kuroda have researched the change of facial color when a human is happy, angry and drunk. The hue and saturation of the bridge is higher than the cheeks and glabella [15]. Therefore, we considered that the change of facial expression is different in each area.

3. Robot's Olfactory Sensation and Facial Color

3.1 Olfactory Sensation

We examined how many smells a robot can recognize keeping in mind though a human can recognize over a thousand smells. The following are the conditions which are necessary for a robot.

(1) A human feels pleasantness or unpleasantness
(2) Influencing a robot's behavior
(3) Safety of handling
(4) Facility to purchase and familiar to human

We chose the following smell. (1) Alcohol; it is a good smell and the robot was able to reveal a new state, "drunken". (2) Ammonia; it is an irritating smell. (3) Cigarette Smoke; it is an unpleasant smell. WE-3RIV was able to recognize these three smells.

The olfactory sensation of WE-3RIV consists of a

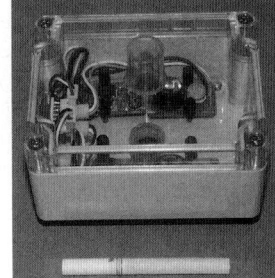

Fig. 3.1 Olfactory Sensor Box

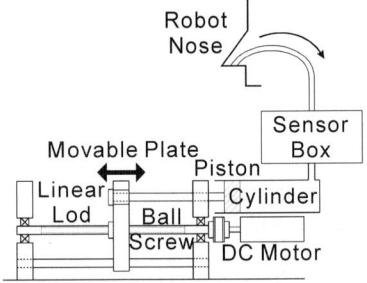

Fig. 3.2 Lung Mechanism of Olfactory Sensor

sensor component and a lung component as in the following explanation.

Table 3.1 Semiconductor Gas Sensor

Sensor Name	Target Gas
SB-19	Inflammable Gas
SB-31	Solvent
SP-32	Alcohol
SP-AQ2	the Whole

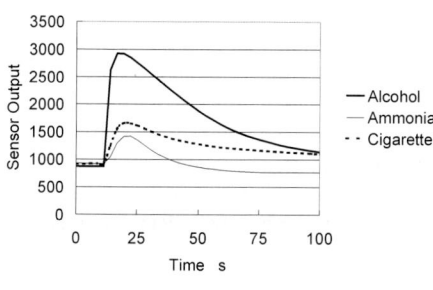

(a) SB-30

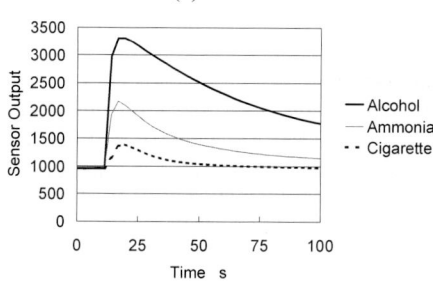

(b) SP-32

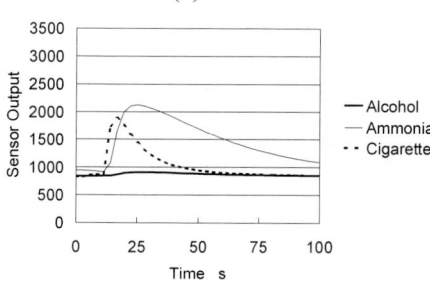

(c) SB-19

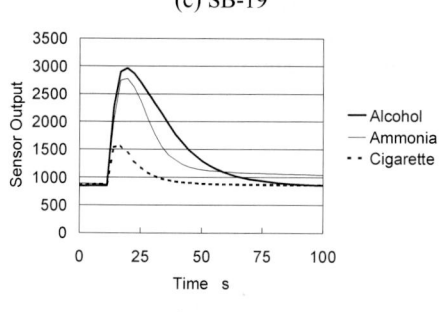

(d) SP-AQ2

Fig. 3.3 Semiconductor Gas Sensor

3.1.1 Sensor Component

We used the semiconductor gas sensors, SB-30, SP-32, SB-19 and SP-AQ2, manufactured by FIS Inc. They can quickly detect an ingredient and their response is stable. We kept these four sensors in a sensor box as shown in Fig. 3.1, and set it at the back of the robot connected by flexible air hoses between the nose and the lung. WE-3RIV can recognize the smell of alcohol, ammonia and cigarette smoke. Table 3.1 shows the target gas of each sensor.

3.1.2 Lung Component

The lung component consists of four parallel sets of cylinders, pistons, and a DC motor which drives the pistons. WE-3RIV breathes by moving the movable plate using a DC motor and a ball screw. The air moves into the nose, passes through the sensor box, and arrives at the lung. Then, the air is directly breathed out. The lung volume is 3700 $[cm^3]$. The maximum airflow is 6100 $[cm^3/s]$. The maximum airflow acceleration is 12800 $[cm^3/s^2]$. Fig. 3.2 shows the mechanism of the lung component.

3.1.3 Recognition Algorithm

WE-3RIV was able to recognize certain smells within a few seconds of being presented with the smell stimuli using its sensor output and the changing rate of the sensor output. We used a 12[bit] A/D board to sample the sensor output. Fig. 3.3 shows the output of the four sensors for each smell stimuli. SB-30 and SP-32 strongly reacted to alcohol. SB-19 strongly reacted to ammonia and cigarette smoke. And, SP-AQ2 strongly reacted to alcohol and ammonia.

The following is the algorithm for recognizing the kind of smell. Fig. 3.4 shows the flow chart of the recognition algorithm.

(1) WE-3RIV sampled the output of gas sensor 30 times in the sampling period 400[μs]. We call this average "sensor output."

(2) WE-3RIV distinguishes the smell's level from the average of the four sensors' output S. In the case of (a) and (b), WE-3RIV doesn't distinguish the kind of smell.

(a) $0 < S < 1200$: Level 1: The smell level is odorless level because S is under the threshold.

(b) $1200 < S < 1400$: Level 2: The smell level is unspecified level because S is over the threshold and under a recognition value.

(c) $1400 < S < 2400$: Level 3: The smell level is distinguishable level because S is over the recognition value.

(d) $2400 < S < 4095$: Level 4: The smell level is too strong level because S is too high.

(3) WE-3RIV calculates the changing rates of the sensor output. We defined d1 as the changing rate of SB-30, d2 as changing rate of SB-19 and d3 as changing rate of SP-AQ2.

(4) WE-3RIV distinguishes the kind of smell by comparing the changing rates.
  (a) d1 > 800, d2 < 90, d3 < 100: Alcohol
  (b) d1 < 800, d2 < 90, d3 > 100: Ammonia
  (c) d1 < 800, d2 > 90, d3 > 100: Cigarette Smoke
  (d) The others: Unknown Smells
(5) If the smell's level is Level 4, WE-3RIV expresses disgust or turns its face away from smells.

3.2 Facial Color Expression

We recently added a facial color expression function to the robot skin. We used a red EL (Electro Luminescent) sheet that is a thin and light device and doesn't influence the other devices on the skin, such as the FSR (Force Sensitive Resistor), which is used to detect the external forces for tactile sensation. We assigned them to the forehead, around the eyes, the cheeks and the nose of the robot as shown in Fig. 3.5. Now the robot skin has both sensing and expression functions, which have a four-layer structure as shown in Fig. 3.6. We assigned the six basic facial expressions and a neutral facial expression. Moreover, we have added the "drunken" and the "shame" facial expressions to WE-3RIV, as shown in Fig. 3.7, in addition to the six basic facial expressions. These new facial expressions are applied to the EL.

4. Robot System

Fig. 4.1 and Fig. 4.2 present the hardware overview of WE-3RIV. WE-3RIV has 26-DOF (Degrees of Freedom) as shown in Table 4.1 and has sensors shown in Table 4.2 as its sensory organs for extrinsic stimuli. The followings are descriptions of each component.

4.1 Neck, Eyeballs, Eyelids, Eyebrows and Lips

The maximum angular velocity of each axis is similar to a human's with 160[deg/s] for the neck, 600[deg/s] for the eyeballs and 900[deg/s] for the eyelids. Furthermore, this robot can blink within 0.3[s], which is as fast as a human blinks. The eyebrows and lips consist of springs and wires [11].

4.2 Visual, Auditory and Cutaneous Sensation

WE-3RIV recognizes the target position, pursues the target, adapts to the brightness in regard to the visual sensation, and localizes the sound from the loudness and the phase difference in 3D space with regard to the auditory sensation. Further, WE-3RIV has a tactile sense and a temperature sense as human cutaneous sensation, and recognizes not only the magnitude of a force but also the difference in touching behavior such as "push," "hit" and "stroke" [11].

4.3 Total System

Fig. 4.3 shows the total system configuration of WE-3RIV. We used four computers (PC/AT compatible.) An Ethernet system connects PC1 and the other three computers. Table 4.3 shows the functions of each PC.

PC1 obtains the outputs of the semiconductor gas sensors using a 12 [bit] A/D board to recognize smell. PC1 determines WE-3RIV's mental state according to the information of the visual, auditory, cutaneous, and olfactory sensations.

PC2 controls the DC motors of the eyeballs, the neck, the eyelids, the jaw and the lung according to the visual

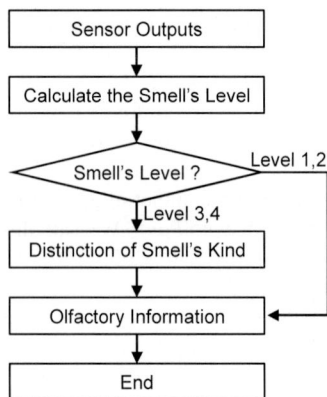

Fig. 3.4 Smell Recognition Algorithm

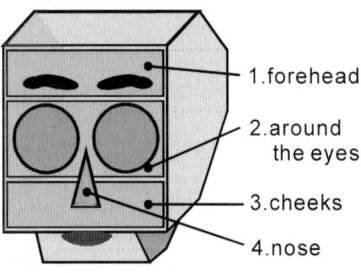

Fig. 3.5 EL Sheet on the Robot Face

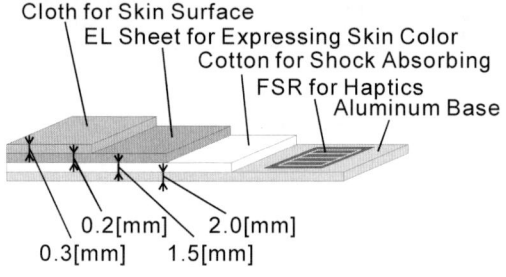

Fig. 3.6 Robot Skin

(a) Drunken   (b) Shame
Fig. 3.7 New Facial Expressions

and mental information sent from PC1. In addition, PC2 obtains information from the temperature sensors through a 12 [bit] A/D board, and transmits this to PC1.

PC3 controls the eyebrows, lips and facial color. To control the facial color, PC3 sends the reference input to EL sheets through the 12 [bit] D/A board and the inverter circuit.

PC4 calculates the direction of the sound from loudness and the sound pressure difference between the right and left, and transmits this information to PC1.

5. Experimental Evaluation

5.1 Olfactory Sensation

The authors evaluated the effect of a robot's olfactory sensation in recognizing certain smells. We presented smells of alcohol, ammonia and cigarette smoke for 1.5[s]. We used vodka with 40[%] degrees of alcohol as a sample of alcohol. We repeated this 50 times. We measured the number of times the robot correctly recognized the smells. The experimental results are shown in Table 5.1. Because the average of the recognition rates is 93[%], we confirmed that the robot can recognize the smells of alcohol, ammonia and cigarette and the olfactory sensation and algorithm of olfactory sensation are effective.

5.2 Facial Color

The authors evaluated the effect of a facial color expression function. We showed 31 testers (men: 31, average age: 23) the photographs of the six basic facial expressions exhibited by WE-3RIV. We examined the recognition rate of those facial expressions, and compared the recognition rate of WE-3RIV (with facial color) to the previous robot WE-3RIII (without facial color). The experimental results are presented in Table 5.2.

The recognition rates of the "happiness", "disgust" and "surprise" facial expressions are higher than those of WE-3RIII (Avg. 89.3[%]). For the "anger" state, the

Table 4.1 DOF Configuration

Part	DOF
Neck	4
Eyes	4
Eyelids	4
Eyebrows	8
Lips	4
Jaw	1
Lung	1
Total	26

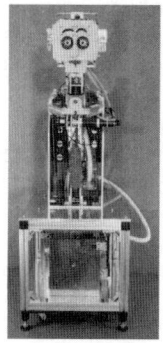

Fig. 4.1 WE-3RIV (whole view)

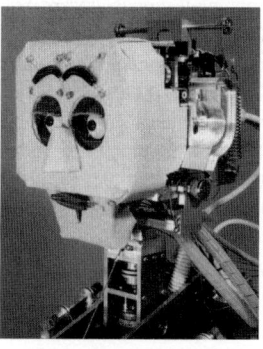

Fig. 4.2 WE-3RIV (head part)

Table 4.2 Sensors on WE-3RIV

Sensation		Device	Quantity
Vision		CCD Camera	2
Auditory		Microphone	2
Cutaneous	Tacticle	FSR	28
	Temperature	Sensor IC	4
Olfactory		Semiconductor Gas Sensor	4

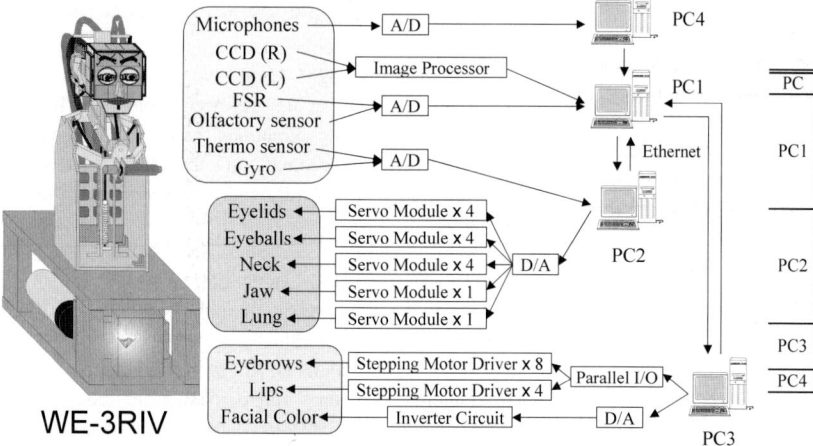

Table 4.3 Functions of PC

PC	OS	Function
PC1	Windows 95	Visual Sensation
		Tacticle Sensation
		Olfactory Sensation
		Eyelashes
		Mental State
PC2	MS-DOS	Neck Motion
		Eyeballs Motion
		Eyebrows Motion
		Lung Motion
		Temprature Sensation
PC3	MS-DOS	Facial Expression
		Face Color
PC4	Windows 98	Auditory Sensation

Fig. 4.3 System Configuration of WE-3RIV

Table 5.1 Experimental Result of Smell Recognition

Subject	Correct Recognition Rate [%]
Vodka	92
Ammonia	96
Cigarette	92

Table 5.2 Experimental Result of Facial Expression System

Facail Expressions	Recognition Rate[%]	
Facial Color	OFF	ON
Happiness	78.9	80.7
Disgust	78.9	87.1
Surprise	89.5	100.0
Sadness	78.9	64.5
Anger	68.4	90.3
Fear	21.1	51.6

recognition rate is very high (90.3[%]). The recognition rate for the "fear" state greatly rose, too, but it is still lower than the other states (51.6[%]). We determined that the robot needs not only red facial color but also a pale facial color for expressing a more effective "fear" state. In the "sadness" state, the recognition rate is low (64.5[%]). We have considered that the robot needs some neck motion to accompany the facial expression in expressing a more effective "sadness" state. As a whole, the average of all the recognition rates of WE-3RIV's facial expression became 79.0[%] while WE-3RIII's was 69.3[%].

Therefore, we have confidently confirmed the effect of facial color expression by these specific increases in the five different emotional states excepting the "sadness" state.

6. Conclusions and Future Work

(1) We have developed the human-like head robot "WE-3RIV" (Waseda Eye No.3 Refined IV), which has olfactory sensation and facial color expression function.
(2) The sub-system that used the semiconductor gas sensors was developed as olfactory sensation. And, WE-3RIV can recognize the smells of alcohol, ammonia and cigarette smoke.
(3) The sub-system that used the EL sheets was developed as facial color expression function. The expressiveness of the robot and the recognition rate of the robot's facial expression were improved.

We determined that the robot would recognize more kinds of smell and express a pale facial color as more effective facial expressions.

Acknowledgement

A part of this research was done at the Humanoid Robotics Institute (HRI), Waseda University. The authors would like to express thanks to ATR (Advanced Telecommunications Research Institute International), HITACHI, Ltd., MINOLTA Co., Ltd., OKINO Industries, Ltd., SANYO ELECTRIC Co., Ltd., SHARP Corp., SMC Corp., SONY Corp., for their financial support for HRI.

References

[1] Cynthia Breazeal, Brian Scassellati: Infant-like Social Interactions between a Robot and a Human Caretaker
[2] Paul Ekman, Wallace V.Friesen: Facial Action Coding System, Consulting Psychologists Press Inc., 1978
[3] Tsutomu Kudo, P.Ekman, W.V. Friesen; Hyojo Bunseki Nyumon -Hyojo ni Kakusareta Imi wo Saguru - (Japanese), Seishin Shobo, 1987
[4] Hiroshi Kobayashi, Fumio Hara, et al: Study on Face Robot for Active Human Interface - Mechanisms on Face Robot and Facial Expressions of 6 Basic Emotions -, the Journal of the Robotics Society of Japan Vol.12 No.1, pp.155-163, 1994
[5] Hiroshi Kobayashi, Fumio Hara: Real Time Dynamic Control of 6 Basic Facial Expressions on Face Robot, the Journal of the Robotics Society of Japan Vol.14 No.5, pp.677-685, 1996
[6] Kazutaka Mitobe, et al.: Consideration of Associated Movements of head and Eyes to optic and Acoustic Stimulation, The institute of electronics, information and communication engineers, Vol.91, pp.81-87, 1992
[7] Laurutis V.P. and Robinson D.A.: The vestibulo-ocular reflex during human saccadic eye movements, J. Physiol., 373, pp.209-233, 1986
[8] Atsuo Takanishi, et al.: Development of an Anthropomorphic Head-Eye Robot with Two Eyes, Proceedings of the IEEE/RSJ International Conference on Intelligent Robots and Systems, pp.799-804, 1997
[9] Atsuo Takanishi, et al.: Development of an Anthropomorphic Head-Eye System for a Humanoid Robot-Realization of Human-like Head-Eye Motion Using Eyelids Adjusting to Brightness-, Proceedings of the IEEE International Conference on Robotics and Automation, pp.1308-1314, 1998
[10] Atsuo Takanishi, et al.: Development of an Anthropomorphic Head-Eye Robot WE-3RII with Autonomous Facial Expression Mechanism, Proceedings of the IEEE International Conference on Robotics and Automation, pp.3255-3260, 1999
[11] Atsuo Takanishi, et al.: "An Anthropomorphic Head-Eye Robot expressing Emotions based on Equations of Emotion", Proceedings of the IEEE International Conference on Robotics and Automation, pp.2243-2249, 2000
[12] Kyoji Tazaki, et al: Shin Seirikagaku Taikei Vol.9 (Japanese), IGAKU-SHOIN Ltd, 1989
[13] Yutaka Kurioka, Mitsuo Tonoike: Nioi no Oyou Kogaku (Japanese), Asakura Shoten, p.15, 1994
[14] Shinji Oshima: Jintai no Kouzo to Kinou, Sinsicho-sya
[15] Tsutomu Kuroda, Tomio Watanabe; Analysis and Synthesis of Facial Color Using Color Image Processing, Journal of the Japan Society of Mechanical Engineers Series C Vol.63 No.608, pp.217-222, 1997

Acquiring hand-action models by attention point analysis

Koichi Ogawara   Soshi Iba [†]   Tomikazu Tanuki [††]   Hiroshi Kimura [†††]   Katsushi Ikeuchi

Institute of Industrial Science, Univ. of Tokyo, Tokyo, 106-8558, JAPAN
[†] The Robotics Institute, Carnegie Mellon University, Pittsburgh PA, USA
[††] Research Division, Komatsu Ltd. Kanagawa, 254-8567, JAPAN
[†††] Univ. of Electro-Communications, Tokyo, 182-8585, JAPAN

{ogawara, ki}@iis.u-tokyo.ac.jp, iba+@cmu.edu
tomikazu_tanuki@komatsu.co.jp, hiroshi@kimura.is.uec.ac.jp

Abstract

This paper describes our current research on learning task level representations by a robot through observation of human demonstrations. We focus on human hand actions and represent such hand actions in symbolic task models. We propose a framework of such models by efficiently integrating multiple observations based on attention points; we then evaluate the produced model by using a human-form robot.

We propose a two-step observation mechanism. At the first step, the system roughly observes the entire sequence of the human demonstration, builds a rough task model and also extracts attention points (APs). The attention points indicate the time and the position in the observation sequence that requires further detailed analysis. At the second step, the system closely examines the sequence around the APs, and obtains attribute values for the task model, such as what to grasp, which hand to be used, or what is the precise trajectory of the manipulated object.

We have implemented this system on a human form robot and demonstrated its effectiveness.

1 Introduction

One of the most important issues in robotics is how to program robot behaviors. Several methodologies for programming robots have been proposed. We can classify them into the following three categories: static textual programming, manipulation by a human through a control device, and automatic programming. The former two methods require human intervention throughout the entire task. In contrast, automatic programming is intended to reduce human aid and to generate an entire robot program automatically. Given the necessary initial knowledge, robots try to acquire their behavior automatically from observation, simulation or learning.

Our research goal is automatic acquisition of robot behavior, in particular, hand-actions, from observation based on the automatic programming approach. We divide the acquisition process of human tasks into two levels: task level, e.g, what-to-do and behavior level, e.g., how-to-do it. This paper covers the former one, task level acquisition, while the latter one is presented in [4]. In Chapter 2, we discuss the necessity of integration of multiple observations. In Chapter 3, we introduce the concept of attention points and present a method for constructing a task model by two kinds of attention point (AP) analyses. In Chapters 4 and 5, we describe implementation details for each attention point analysis. In Chapter 6 we present experimental results. Chapter 7 contains our conclusions and remarks on future work.

2 Acquisition of human task

Ikeuchi, Suehiro and Kuniyoshi et al. studied vision based task acquisition [1, 2]. In their research, the acquisition system observed a human performing an assembly task and constructed high-level task models. Then, using those constructed models, a robot performed the same task. Kimura et al. proposed task models which could be used to realize cooperation between a human and a robot [3]. In this scheme, the robot first observes sequential human operations, referred to as events, by vision and analyzes mutual event dependencies (pair of pre-conditions and results) in the tasks. The robot is able to change its assistant behavior according to the current event observed and the knowledge of what is to be done next, derived from the task model, and to generate a large number of cooperative patterns from a single a task model. However, these models depend on one (typically a single camera) or a few sensors and are constructed through one-time observation, therefore they are not suitable for close analysis.

Our approach utilizes multiple observations which vary in sensor variety and granularity for efficient analysis. By analyzing each observation sequentially or repeatedly, we can determine the necessary part in the human demonstration where the level of detail in the subsequent analysis should be changed and can then accumulate each result to build the task model efficiently. Integration of observations enables us to build heterogeneous task models in which accuracy is enhanced locally. We introduce the concept of attention point (AP) as a key of integration and propose a two-step analysis based on APs as a method of constructing a human task model.

3 Attention point
3.1 Two-step analysis

Integration of multiple observations is accomplished by two-step analysis. At the first step, the system roughly analyzes the input modalities and recognizes the outline of the entire human demonstration (rough task model). At the same time, the system also extracts APs. APs, which require close observation to learn a particular behavior, are defined around specific time and position along a sequence of a human demonstration.

At the second step, the system closely examines the demonstration around each AP to enhance the task model. This sequence can be the same observation data or another one. In the latter case, the system synchronizes two observation data which are derived from different demonstrations of the same task.

We employ a type of task models similar to Ikeuchi's. We decomposed a hand-action task as a sequence of discrete hand-actions, during which a human performs some action by manipulating objects, and we symbolized possible hand-actions as "Action Symbols", which indicate what-to-do information. The task model also includes several attributes for "Action Symbol", detailed information to achieve that "action," such as which hand to use or which object to grasp (Table 1). In the proposed two-step approach, this "Action Symbol" is obtained from the rough analysis at the first step. Then, from the detailed analysis around the APs previously determined, those attributes are obtained at the second step.

Table 1: Task Model

Attributes	Priority	Value
Action Symbol	3(high)	Power Grasp, Precision Grasp Release, Pour, Hand Over
Object Model	3	Shape and Color histogram
Hand	2	Right, Left, Both
Position	1	Absolute Position in 3D space
Time Stamp	1(low)	Absolute Time (start and stop time)

We propose two different kinds of AP analyses in the following sections.

3.2 Integration of sensors separated in space

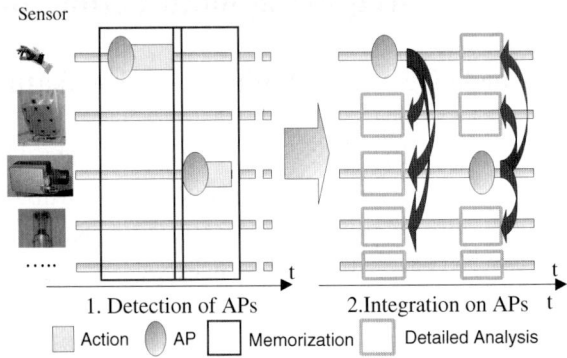

Fig. 1: Two Steps Analysis using Attention Point

When several input sensors are available simultaneously, it is generally ineffective to precisely analyze all the data along the entire human demonstration. So the system temporally records all the raw data available and employs a two-step analysis of a human task (Fig.1).

To realize the two-step analysis, we utilized the short-term memorization method. At one observation sequence, the system first analyzes the input data given by the set of modalities that require the cheapest computation. It extracts "action symbol," and APs as the boundaries of each segmented action while recording all the data around each AP on storage devices.

After an observation sequence completed, i.e., after one demonstration was finished, the system acquires the recorded data corresponding to each AP from the storage devices and applies a detailed analysis on them off-line. This process obtains the remaining attributes in the task model.

3.3 Integration of sensors separated in time

The method described above requires temporal sets of recorded input data; as the number of sensors and work time increases, the amount of unused data expands. And also, for some sensors, it is not advisable to adopt a specific sensor configuration at all times because of range, speed, precision trade-off.

So we propose another two-step analysis in which the system requires quantitative evaluations of a number of demonstrations for the same task. The system roughly analyzes the demonstration and extracts APs at the first observation. Then the system changes the sensor configuration if necessary and examines the second demonstration around the APs to enhance the task model. For the synchronization issues, the system can predict the hand motion from the first observation and, by watching for the appearance of the predicted motion at each AP in the second observation, the multiple observations can be synchronized.

4 AP analysis for sensors separated in space

Our system employs a pair of data gloves and a 9-eye real-time stereo vision system. We can acquire depth and color images from the stereo vision system and can acquire hand motion (finger shape, absolute position and orientation) from the data gloves. The image processing is much more time-consuming as opposed to the processing of the data gloves; thus we adopted the AP analysis described in Section 3.2. We utilized the data gloves to extract APs and "action symbols;." then, to determine attributes of the task model, the system analyzes depth and color images around those Aps.

Fig.2 and Fig.3 show the flow of the AP based two-step analysis. The subsequent sections describe the outline of the analysis. Please refer [5] for details.

4.1 Rough analysis by gesture spotting

We set up a task domain for a specific hand-action task and built a finite set of "action symbols," which represents all the possible hand motions that appeared in that task domain. "Action symbols" are combinations of finger actions and local hand motions. For now, we classify possible finger actions into three actions: "Power Grasp", "Precision Grasp"[6] and "Release," and described human hand actions as a finite set of "Action Symbols" which are combinations of above finger actions and local hand motion. By excluding hand actions composed of independent finger motion, we can segment the entire hand-action task into meaningful "Action Symbols".

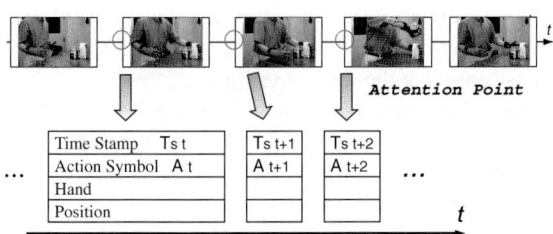

Fig. 2: Rough Analysis by Gesture Spotting

To obtain "Action Symbols" in the task model, we aim to spot human gestures from hand-actions performed by a human demonstrator as shown in Fig.2. In this experiment, we chose "transferring content of container" as a task domain and selected five gestures as possible hand actions (Table2). APs are defined as the starting point of each gesture. To extract these "Action Symbols," we employ datagloves and a gesture spotting technique based on Hidden Markov Models (HMMs).

Table 2: Gesture definitions

Gesture	Primitives	Action
Grip	cls+sp	Power-grasp from open position
Pick	prc+sp	Precision-grasp from open position
Pour	cls+roll+sp	Power-grasp, and roll the wrist
Hand-over	prc+forw+sp	Precision-grasp, move forward, and back
Release	opn+sp	Open a grasp hand
Garbage	gb	A filler model for spotting
Start,End	sil	Silence at the start and end

We utilize a pair of data gloves (CyberGlove 18-DOF each), and 6-DOF position sensors (Polhemus) as input devices for the HMM-based gesture spotting module. So, 24 dimensions and their differentials are the input to the HMM module for each hand. The second column of Table2 indicates the defined HMM primitives for each gesture. Each primitive is defined as 5-state left-right HMMs. *sil* is a silent state used at the time of training, *sp* is a short pause which tends to occur at the end of the gesture corresponding to an action symbol, and *gb* is a garbage collector trained on arbitrary non-gesture movement. By sharing primitives, each action symbol requires a small number of training data with better efficiency.

Left and right single-hand gestures are spotted separately in a parallel manner, while two-handed gestures are spotted by combining results from the analysis of both hands. Our system can sample the data from a pair of data gloves in 30Hz and can spot gestures corresponding to action symbols in parallel without delay.

4.2 Attention point analysis by vision

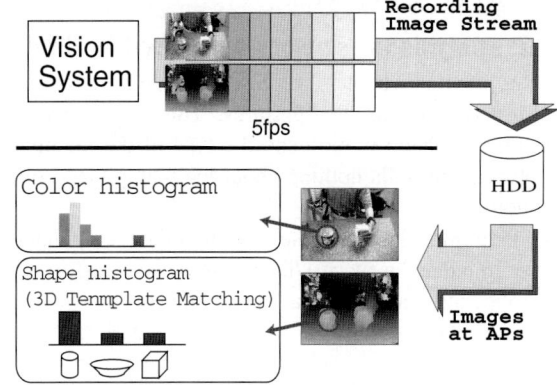

Fig. 3: AP analysis by vision

Computation time for image processing is rather time-consuming. So we first record all the raw data from the vision system around APs. These recorded data are synchronized with the data-glove analysis and the correspondence between them is easily made. After the first analysis is finished and extracts APs, the system fetches the corresponding images and extracts the information about the manipulated objects (Fig. 3).

By analyzing just before each AP, we can obtain the images in which the target object is not occluded by the hand. The object is modeled by calculating shape and color histograms.

We assume that a human task is demonstrated on a table whose geometric information is known. By extracting depth regions corresponding to each object on the table, we calculate shape histogram as a list of goodness of matching between an object extracted in the depth image and each object model in the database. This goodness of matching is obtained by using the 3D Template Matching(3DTM)[7] technique.

3D Template Matching, a technique for localization, finds the precise position and orientation of the target object in depth data. This process is calculated by projecting the corresponding 3D model into the 3D space generated from a depth image and calculates goodness of matching between the 3D model and the 3D data by summing up weighted distance between each center point of the meshes in the template model and the closest 3D point. 3DTM adopts M robust estimator to eliminate the effect of outliers.

Color histogram is calculated as a normalized hue histogram which counts pixels with large saturation value among the area of the object on the color image. These depth and color images are produced at 5 fps (up to 30 fps) synchronously by the 9-eye multi-baseline stereo vision system.

The system registers this histogram information in the attribute slot of the task model.

5 AP analysis for sensors separated in time

In the previous chapter, we described the hand-action model in terms of classified gestures. This model gives a good notion of hand motion and the type of the manipulated objects, but tells nothing about the manipulated object's motion.

In oder to model the delicate motion of the manipulated object or to judge the success/failure of the task performed by the robot automatically, a task model must contain some information about precise position and orientation of the manipulated object at particular parts in the entire task.

We developed an efficient method to acquire the precise trajectory based on repeated observations and APs. In this chapter, we present the method, which uses the zoom stereo system.

5.1 Repeated observation

To acquire the precise trajectory of the object, we combined two kinds of two-step analyses as shown in Table3.

Table 3: Process of repeated observation

	Zoom	Model	Time	description
1	x1	Coarse	0.6s	extraction of APs
2	x2	Coarse	1.4s	tracking the object in real-time
3	x3	Fine	2.0s	tracking the object in off-line

 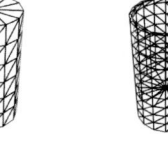

Real object   Coarse model (168 polygons)   Fine model (640 polygons)

Fig. 4: 3D Model

Process $1 \rightarrow 2$ adopts the method described in Section 3.2 (repeated observation), while Process $2 \rightarrow 3$ adopts the method described in Section 3.1.

Fig.4 shows the object and its CAD model used in this experiment.

5.1.1 Extraction of APs

At first, the zoom configuration is set to x1(default) and the system roughly tracks the object for each hand action using the 3DTM method. To estimate the initial position and time of the object to be tracked, we utilized data-gloves used in the previous chapter; the gloves were enhanced by tactile sensors to classify the grasping. The system can detect the grasping motion directory from tactile sensors, so it roughly estimates the initial object position (from the polhemus sensor) at the time of grasping.

We used the coarse object model during tracking, because precise position and orientation is not important. The system gets the rough trajectory and also gets the APs as the initial position and time of the tracking.

Fig.5 shows the tracking result (intensity images overlaid with the wire-frame model).

5.1.2 Tracking in repeated observation

At this stage, the system demands the repeated demonstration of the same task. In this stage, the doubly zoomed cameras cannot put the entire action in sight in a fixed orientation; therefore, driving of the pan/tilt moving mechanism synchronized with image processing in real-time is necessary to track the target object to be kept in the center of the view.

All the depth and intensity images are recorded during tracking. These images are used in the third process below.

Fig. 5: Tracking at the first stage

Fig. 6: Tracking at the second stage

This tracking is also processed by 3DTM with the coarse object model, because precise localization is not important.

Fig.6 shows the tracking result (intensity images overlaid with the wire-frame model).

5.1.3 Estimation of the precise trajectory

At the third stage, the system fetches the recorded images and localizes the object in each scene to estimate the precise trajectory with the fine model. This image fetching is the same technique as that described in chapter 4.

This process is executed off-line. To localize the object precisely, we developed a method to combine the 3DTM and 2DTM. 3DTM is a method for localizing the 3D model in the 3D points obtained from the depth data[7]. 2DTM is the edge-based localization method between the 3D model and the estimated 3D edges of the contour of the object, which are derived from the intensity image[7].

2DTM is sensitive to the edges in the image background and does not offer a good guess about z position (parallel to the viewing direction) of the model because of the approximation of z position of the 3D edge. But, at the final stage of the localization, 2DTM offers a good guess about the position and orientation perpendicular to the viewing direction.

So, we first adopt the 3DTM only to localize the object to the approximate position and then we adopt 2DTM & 3DTM combined method to localize the object to the exact position as shown in Table4.

2DTM and 3DTM are calculated in the same 3D space by M-estimator (Lorentzian) with different weight. Sigma is the parameter to reduce the effect of the outliers.

Fig.7 shows the tracking result. The upper row shows the intensity images and the lower row shows the disparity images. The contour of the object's model is overlaid in each image.

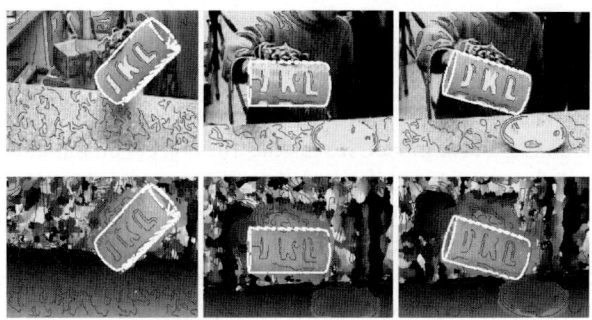

Fig. 7: Tracking at the third stage

Table 4: 2DTM & 3DTM combined localization

Method	Sigma[mm]
3DTM	10.0
3DTM	4.0
3DTM	2.0
3DTM & 2DTM	2.0
2DTM	1.0

5.2 Experimental result

Our stereo vision consists of zoom lens cameras. This is actually digital zooming but, when capturing images, the stereo system re-samples each pixel at the ratio of one-quad, so we can expect that to doubling the power of zooming value will not reduce the quality of the image captured by our stereo system.

Stereo processing is done on a hardware chip and the system can acquire a depth image and the corresponding intensity image (280×200) in 15fps at most.

The average tracking rate of the first stage is 0.6 [sec/frame] on our Pentium3 500MHz PC. Similarly, the tracking rates of the second stage and third stage are 1.4 [sec/frame] and 2.0 [sec/frame], respectively.

The difference in the rate between the first and the second stages is mainly due to the construction time of the Kd-tree used in localization. We restricted the search area of 3DTM to be very close to the object, the inside of the rectangle shown in Fig.5 and Fig.6, so the search area of the first stage is relatively smaller and the construction time is short.

The difference in frame-rate between the second and the third stages is due to the difference in the number of iterations in the localization process and the level of detail of the model.

6 Performance by robot

We have developed a human-form robot as an experimental platform for learning and performing human hand-action tasks[9]. The robot has similar capabilities and body parts to those of humans, including vision, dual arms and upper torso.

When the robot is to perform the same task after constructing a task model, it searches for objects on the table and, for each object, it calculates mean square distance between the shape and color histogram of the object on the table and those in the model database. The smallest value determines the best matching objects. In this way, the robot recognizes the object. Once the recognition of the current environment is done, the robot sequentially executes the action corresponding to each "Action Symbol" in the task model adapting to the current environment condition.

Fig.8 shows the experimental result in which the robot performed the same task successfully.

Fig. 8: Experiment

7 Conclusion

We proposed a novel method of constructing a human task model by attention point (AP) analysis. Attention points relate and integrate multiple observations and construct a locally enhanced task model of human demonstration. AP analysis consists of two steps. In the first step, action segment and APs are extracted. Then, at the second step, by closely examining human demonstration only around APs, the system extracts the attribute values and improves the model.

By reducing unnecessary analysis, the system can construct the task model efficiently. Efficiency is important when we consider human-robot cooperation tasks in which the robot must respond to the action taken by both a human and the robot itself in relatively short time.

We presented two kinds of AP analyses, one for integration of sensors available simultaneously and the other for integration of sensors derived from different observations of the same task by repeated demonstration.

To realize the first AP analysis, we proposed a short-term memorization method, which records all the raw input data around each AP to be processed at the second step. And also, we proposed a localization method which combines 2DTM and 3DTM to track and localize a moving object robustly.

The future work is to solve the problem of integrating the trajectory information into the current task model for training the robot itself automatically. We are also planning to combine this task level acquisition with the behavior level acquisition method [4]. First, the task level acquisition constructs task models to perform the entire task to be adapted to the environment. It also extracts special APs that require behavior level acquisition. Second, the behavior level acquisition analyzes those APs closely and obtains a suitable motion sequence (sub-skill). This two-layer approach should extend the capabilities of the learning robot that can acquire a human task through observation.

Acknowledgment

This work is supported, in part, by Japan Society for the Promotion of Science (JSPS) under the grant RFTF 96P00501, and, in part, by Japan Science and Technology Corporation (JST) under Ikeuchi CREST project.

References

[1] K. Ikeuchi and T. Suehiro: "Toward an Assembly Plan from Observation Part I: Task Recognition With Polyhedral Objects," *IEEE Trans. Robotics and Automation*, 10(3):368–384, 1994.

[2] Y. Kuniyoshi, M. Inaba, and H. Inoue: "Learning by watching," *IEEE Trans. Robotics and Automation*, 10(6):799–822, 1994.

[3] H. Kimura, T. Horiuchi and K. Ikeuchi: "Task-Model Based Human Robot Cooperation Using Vision," *IROS '99*, 2:701–706, 1999.

[4] J. Takamatsu, H. Tominaga, K. Ogawara, H. Kimura and K. Ikeuchi: "Symbolic Representation of Trajectories for Skill Generation," *IEEE ICRA*, 4:4077-4082, 2000.

[5] K. Ogawara, S. Iba, T. Tanuki, H. Kimura, K. Ikeuchi: "Recognition of Human Task by Attention Point Analysis," *IEEE/RSJ IROS*, 3:2121-2126, 2000.

[6] M. R. Cutkosky, "On Grasp Choice, Grasp Models, and the Design of Hands for Manufacturing Tasks," *IEEE Trans. on Robotics and Automation*, 5(3):269–279, 1989.

[7] M. D. Wheeler: "Automatic Modeling and Localization for Object Recognition", *Ph.D Thesis*, CMU, 1996.

[8] K. M. Knill and S. J. Young: "Speaker Dependent Keyword Spotting for Accessing Stored Speech," *Cambridge University Engineering Dept., Tech. Report*, No. CUED/F-INFENT/TR 193, 1994.

[9] K. Ogawara and J. Takamatsu and S. Iba and T. Tanuki and H. Kimura and K. Ikeuchi: "Acquiring hand-action models in task and behavior levels by a learning robot through observing human demonstrations," *IEEE Conf. on Humaniod Robots*, 2000.

New architecture for mobile robots in home network environment using Jini

Byoung-Ju Lee*, Hyun-Gu Lee*, Joo-Ho Lee** and Gwi-Tae Park*

*Department of Electrical Engineering, Korea University
5-1, Anam-Dong, Sungbuk-Ku, SEOUL, KOREA
{sseng, hglee99, gtpark}@elec.korea.ac.kr

**IIS #3, University of Tokyo
7-22-1, Roppongi, Minato-ku, Tokyo, Japan
leejooho@ieee.org

Abstract

Recently, many robotic researchers are much interested in designing mobile robots that can work in the networked and intelligent space. In this case, these robots perform their tasks by sharing many external resources around them. However, these researches only show the possibility to realize the robot. The robot has some problems. To solve the problems, we implemented a network protocol that links all resources around the robot.

In this paper, by using network protocol, Jini, we propose an advanced resource sharing architecture(RSA) for a mobile robot. We made several services with all resources around a robot by using Jini. There is a task managing service for a certain task, and the service uses other service to perform a task as well as make a new service of other services. To verify our architecture we built a simple experimental space. In this space we performed the experiments such as a detection of desired objects, path generation and motion control for a robot, object avoidance and so on. Experimental results show the mobile robot with our advanced RSA performs various tasks successfully.

1 Introduction

In the past decade, many robotic researchers have tried to make mobile robots fully intelligent, but there are still so many problems to be solved. For example, to recognize environment around it and to build a map for navigation, a robot needs so expensive sensors such as a gyroscope and laser range finder. Hence until now it is difficult to put service robots for practical use. Recently, several robotic researchers are much interested in mobile robots in the networked and thinking space[1]. Such kind

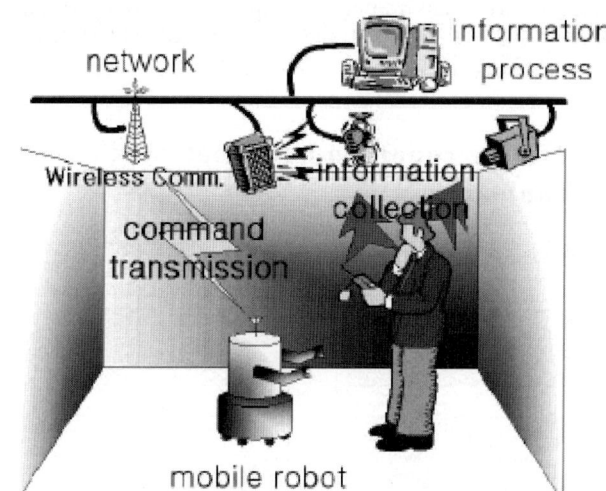

Fig.1 An example of mobile robots in a home network environment

of space can perceive what is happening in it, build a model of itself, communicate with its habitants and act based on things happening in it. Coen et al. suggested intelligent agents known as the Scatterbrain[2]. The agent's primary tasks are to link various components of the room and to connect them to internal and external stores of information. Fig.1 shows this scheme. Lee et al. tried to applied this paradigm to a mobile robot[3]. In this way a robot receives several data from external resources to build map for navigation. As a result, the robot does not need sensing devices. Since 1998 our laboratory has been interested in such an embedded space and mobile robots for it. By combining mobile robots and the embedded space, we could get many good features that could not be obtained before. For example, even if a robot does not have gyroscopes nor laser range finders the robot can get enough precise positions of itself and easily extend its functions because the robot shares the resources in the room [4][5]. However, these researches only show the possibility to realize such a robot. The robot has some problems such as its extension, modification and maintenance[10]. To solve the problems, we implemented a network protocol that links all resources around the robot. We used Jini as a network protocol. It was designed by the Sun microsystems for a protocol of networked consumer electronics[6]. Many vendors are developing networked consumer electronics with a network transceiver using Jini. As a result, if we use Jini, we can use so many devices such as TV, air conditioner and light as our resources.

In this paper, by using Jini, we propose an advanced resource sharing architecture(RSA) for a mobile robot in the home network environment. We made several services with all resources around a robot by using Jini. There is a task managing service for a certain task, and the service uses other service to perform a task as well as makes a new service of other services.

In chapter 2, we will explain mobile robots in the home network environment. Chapter 3 describes an advanced RSA. Chapter 4 shows results of experiments. Finally chapter 5 concludes this paper.

2 Mobile robots in home network environment

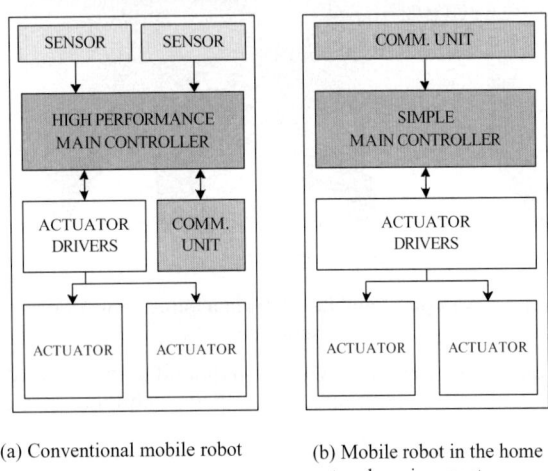

(a) Conventional mobile robot

(b) Mobile robot in the home network environment

Fig.2 Architecture of a mobile robot

As shown in Fig.2(a) a conventional architecture of mobile robots fundamentally consists of four components; main controller with high performance, sensor, actuators and communication unit. Robot with this architecture is so expensive due to sensors such as gyroscopes and laser range finders, and moreover its sensors get only limited data about its environment. On the other hand, Fig.2(b) shows an architecture of the mobile robot in the home network environment. The robot has only simple three components: main controller, actuators, and communication unit. The robot does not have to build a global map of its environment because a sensing devices for doing it is not on the robot. A map is built by external resources(e.g. CCD camera hanging on the wall and an external host computer connected with the camera). Because the sensing resources are able to watch a global area where a robot is located, they can calculate a position of the robot more precisely. Moreover, tasks such as a localization and decision making are done by an external resource so that we can make a robot very simply.

3 A new architecture of mobile robots using Jini

Networks must link all resources in RSA together. So a proper protocol is needed to construct RSA from several resources over the network. There are some technologies to construct systems from distributed objects over the network such as Jini [TM]and UPnP(universal plug and play). We selected Jini as a tool to construct RSA because it offers a simple infrastructures for delivering services over the networks and creating spontaneous interaction between programs that use these services, regardless of their hardware or software implementations. Moreover it is recently more popular than UPnP. Hence we can use so many resources in which Jini is implemented.

3.1 Protocol interface, Jini

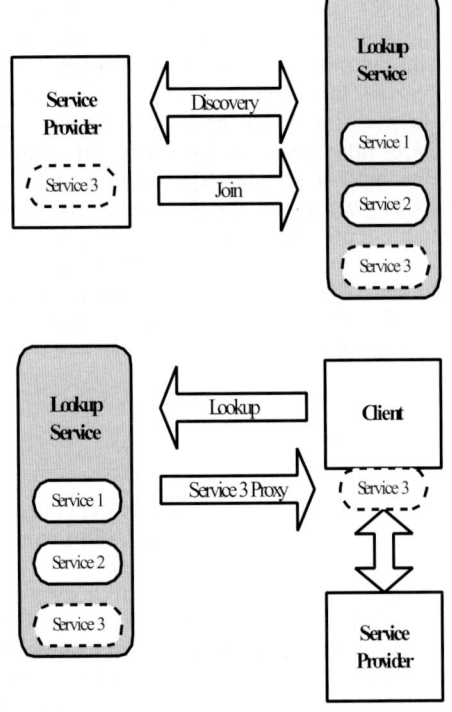

Fig.3 A run time infrastructure of a protocol, Jini

Jini is a set of APIs and network protocols that can help developer to build and deploy distributed systems that are organized as federations of services. Hardware devices, software, communications channels that reside on the network can be services. Jini defines a runtime infrastructure that resides on the network and provides mechanisms that enable you to add, remove, locate, and access services. The runtime infrastructure resides on the network in three places: in lookup services that sit on the network; in the service providers (such as Jini-enabled devices); and in clients. Lookup services are the central organizing mechanism for Jini-based systems. When new services become available on the network, they register themselves with a lookup service. When clients wish to locate a service to assist with some task, they consult a lookup service.

The runtime infrastructure uses one network-level protocol, called discovery, and two object-level protocols, called join and lookup. Fig.3 describes it. Discovery enables clients and services to locate lookup services. Join enables a service to register itself in a lookup service. Lookup enables a client to query a lookup service for services that can help the client to accomplish its goals.

3.2 An advanced resource sharing architecture

The most important characteristic of RSA is that external resources are regarded as local components on the robot themselves. As a result, a developer designs a robot without considering a network protocol between a robot and external resources. Additionally because Jini supplies PnP(plug and play), a developer doesn't have to consider a configuration of each resource, and also partial faults of a system does not effect their whole system[6].

The suggested architecture basically consists of physical components and logical components, as shown in Fig.4. The Physical components are devices such as mobile robots, cameras, display monitor, mic, speaker and external host PC, whereas the logical components are Jini services such as an image processing service, robot operation and robot control.

- Physical components

Jini service should run on each device to realize an ideal physical component. However, we cannot make Jini run on all devices, since the OS such as Window98 or Linux is required in order to run Jini. Hence some Jini services are run on a PC. For example, we use the PC as a Jini service for a image sensing resource and robot operation.

- Logical components

Logical components are classified into two logical components. One is a fundamental service, and the other is an inherited service that is made of fundamental services. Gray-colored rectangles in Fig.4 describe the ser-

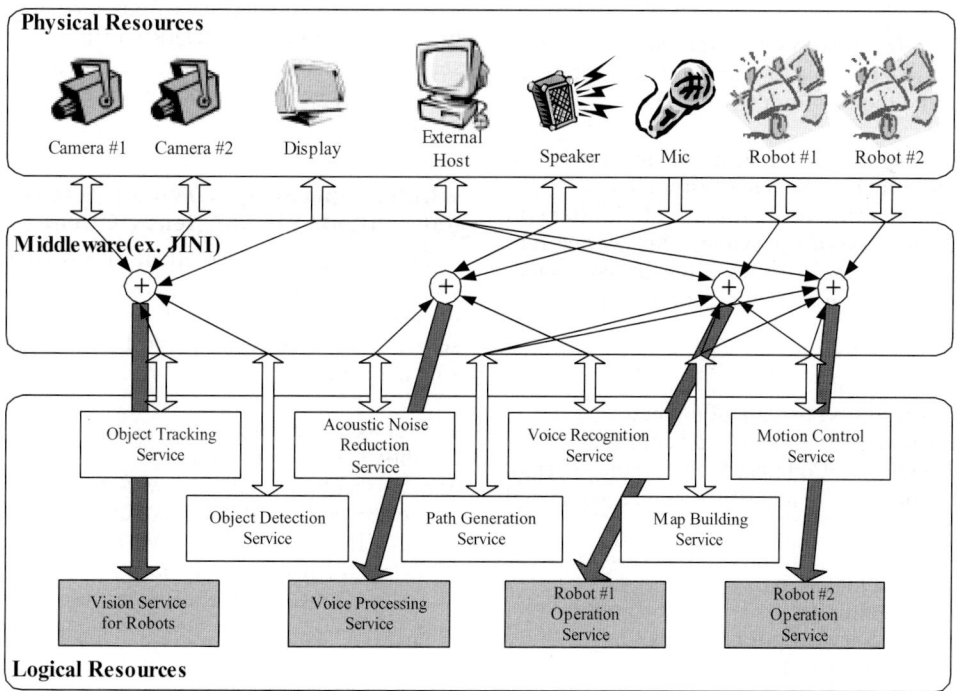

Fig.4 An advanced resource sharing architecture

vices.

We designed four inherited services that are basically required to operate two mobile robots in a home network environment. They consist of a vision service, robot operating service, voice processing service and supervising service.

- **Vision service : Jini service for image processing, VisionService (Service Provider)**

Jini service for a vision system acts as a service provider. It means that this service provides a service for other clients. In this system, its functions are to find a robot in environment, to localize the robot and to send a position data to other clients to request it. The following program written in Java is to do it.

```
1: class VisionServiceImpl {
2:    int robotX; // a x-axis coordination of the robot
3:    int robotY; // a y-axis coordination of the robot
4:    int robotDir; // a direction of the robot
5:    public void GetRobotInfo( int x, int y, int dir); //
      retrieve a position information of the robot

6:    public static void main() {
7:    new VisionServiceImpl();
      // Create a VisionServiceImpl instance;
8:    new JoinManager(.,this,..);
      // Join to the lookup service;
      //Wait for client's request;
9:    }
10:}
```

In line 8, a service provider registers an interface of this function to a lookup table of a proxy in their network and then waits for a request from any client.

- **Voice processing service : Jini service for voice processing, VoiceService (Service Provider)**

Jini service for a voice recognition and TTS(text to speech) also acts as a service provider. This service recognizes a voice and then identifies a human command. Additionally, It tells a man its states or processing results by TTS.

```
1: class VoiceServiceImpl {
2:    char *voiceStream; // raw data of a voice
3:    char *voiceCommand; // identified command
4:    public void GetVoiceCommand( char *voiceStream,
      char *voiceCommand); // identify a human command
      from raw data

5:    public static void main() {
6:    new VoiceServiceImpl();
      // Create a VoiceServiceImpl instance;
7:    new JoinManager(.,this,..);
      // Join to the lookup service;
      //Wait for client's request;
8:    }
9:}
```

- **Supervising service : Jini service for robot operation, VisionClient (Client, Server)**

Jini service for robot operation acts as a service client for image processing service and also service server for robot control service. This service receives information about robots and decides its task. This performed by following sequences. Firstly, a robot find an desired interface at a lookup table of a proxy in its network and if the interface is registered in the lookup table, it processes the interface. Otherwise, the robot should wait until the interface is registered by any service provider. The following program describes this process.

```
1:class VisionClient {
2:    int targetX; // a x-axis coordination of a target
3:    int targetY; // a y-axis coordination of a target

4:    public void GetRobotOperation( int x, int y); // retrieve a kind of an operating method of robots
5:    public static void main() {
6:    ServiceFinder sf = new ServiceFinder( VisionService.class ); // find the lookup service;
7:    VisionService vs = (VisionService) sf.getObject(); //get the matching service
8:    vs.GetRobotInfo();// use the service interface
9:    }
10:}
```

In line 6, a robot(client) finds a desired interface and in line 8, an interface, 'GetRobotInfo()' is declared in a service provider. Because it can find it at a lookup table, it uses the function supplied by a service provider, vision server.

- **Robot operating service : Jini service for robot control, VisionClient (Client)**

This acts as a service client of a service for robot operation. On the other hand previous services is performed on a host PC connected to a vision system, while this service is on robots and control its actuators, simple sensors and so on.

```
1:class RobotClient {
2:public static void main() {
3:    ServiceFinder sf = new ServiceFinder( VisionClient.class ); // find the lookup service;
4:    VisionClient vs = (VisionClient) sf.getObject();
      //get the matching service
5:    vs.GetRobotOperation();// use the service interface
6:    }
7:}
```

So far we explain a new architecture of mobile robots using Jini. In the next chapter, we will describe our robot system and show some experimental results.

4 Experiment

We performed two experiments. One is to evaluate an accuracy of a localization of mobile robot by a proposed vision system. The other is to verify a flexibility of a proposed architecture when a task is changed during its processing and a robot has a trouble.

4.1 Experimental system

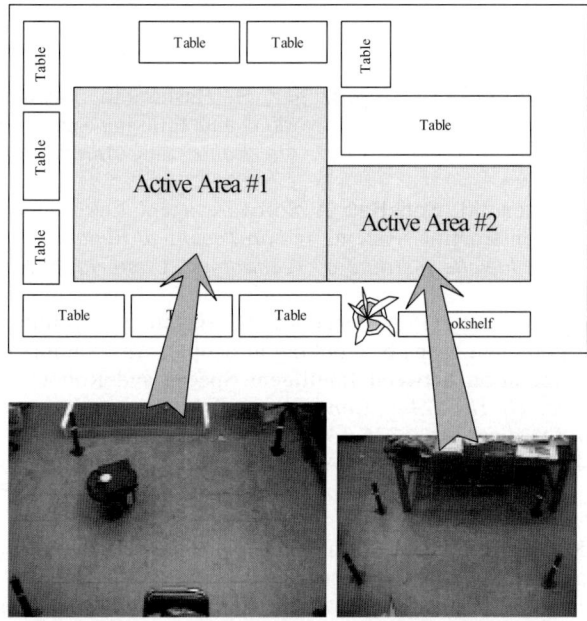

Fig.5 A diagram of a test field for a localization (above) and real pictures of the field (bottom)

Our system consists of two mobile robots, two vision system, PC and a voice processing system. The robot is 'Pioneer2 CE', and its main controller is a PC(AMD K6 450MHz). We use a PCI-type image capture card and CCD camera as a vision system. The robot and vision system are linked by a wireless LAN(IEEE 802.11). A running speed of the robot is about 40[cm/sec]. We divide a test field by 2 as shown in Fig.5. The size of each area is respectively about 4x3[m] and 3x2[m]. There are six waypoints in the areas.

Firstly, we make the robot go to the 'GOAL' point and pass through all subgoals in intended order as shown in Fig.6. Finally we make it come back to the 'START' point. We repeat this process seven times. We use PI-control for a path tracking and a simple calibration algorithm for compensating a distortion of a captured image[10].

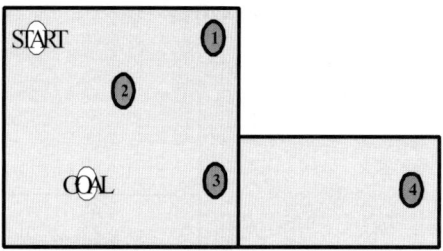

Fig.6 Running path of robot for a localization experiment

Second experiment is to explore a test field as shown in Fig.7, passing through all subgoals in the field by two robots. We design a inherited service for this task. Fig.8 describes this service.

The most important function of a task managing service of a field exploring service is to find the shortest paths of two robots. Additionally, it reorganizes itself in an error of any service.

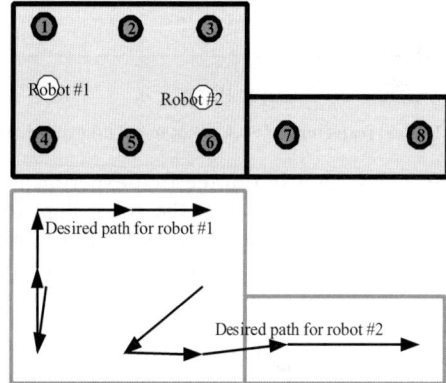

Fig.7 Test field for exploring experiment and desired paths of two robots

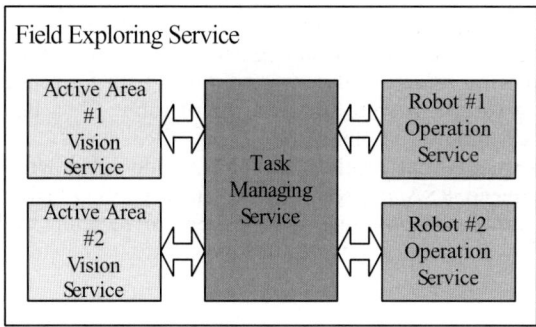

Fig.8 Field exploring service constructed by other inherited services

4.2 Experimental results

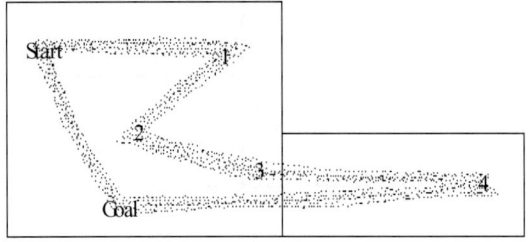

Fig.9 Real trajectory of a mobile robot in a localization experiment

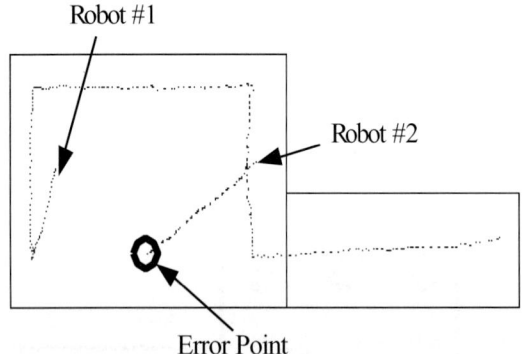

Fig.10 Real trajectory of each robot in a field exploring test

Fig.9 shows the result of the first experiment. The dotted lines are trajectories that the robot kept track of in processing the task. There are some position errors at each subgoal. We get a direction of the robot by a vector of its displacement. Hence, if a robot has a distorted direction during its stopping, it takes some time to compensate the distortion, but the error is decreased in running. The position error in the points is caused mechanical faults of the robot, slip, image calibration and so forth. But because this error is small enough not to effect an operation, we neglect it.

Fig.10 describes the result of the second experiment. In this experiment, when robot #2 was turned off at the subgoal 2, the task managing service reorganized a field exploring service and generated a new path for robot #1. As a result, in spite of the error of robot #2, a desired task could be completed.

Experimental results show the mobile robot with our advanced RSA performs successfully various tasks and the proposed architecture is proper to develop mobile robots for a home network environment.

5 Conclusion

The main idea of the proposed architecture is that RSA make application areas of Jini more various. Even though Jini is not necessary for a robot with RSA, we can get some advantages by using Jini. For example, firstly, developers can easily design the robots without considering a network programming such as a socket programming because external resources can be used as if they are local components of robots. Secondly, developers can save a development time because Jini is based on Java so that a number of classes with various functions are supported. Finally, nowadays, standard interfaces of Jini for consumer electronics are being discussed. If they are established, Jini will be used in various fields. It means that resources which robots can use will increase. Thus, robots can be utilized at various fields because all devices in the space (e.g. TV, air conditioner, lights, windows, gas valves and etc...) can be used as resources.

6 References

[1] J.H. Lee, G. Appenzeller, H. Hashimoto, Physical agent for a sensored, networked and thinking space, *In Proc. of IEEE Int. Conf. On Robot and Automation*, 1998.

[2] Coen, M. SodaBot: A Software agent Environment and Construction System. *AI Lab Technical Report 1493, Massachusetts Institute of Technology, Cambridge, MA*, 1994.

[3] J.H. Lee, G. Appenzeller, H. Hashimoto, Building Topological Maps by Looking at people: An Example of Cooperation between Intelligent Spaces and Robots, *In Proc. Of IEEE Int. Conference of Intelligent Robotics and Systems*, pp. 1326-1333, 1997.

[4] B.J. Lee, G.T. Park, A Design of a Physical Agent for a Cooperative Space, *Proc. Of KACC99, vol. B, pp. 168-171*, 1999

[5] B.J. Lee and G. T. Park, A Robot in Intelligent Environment : Soccer Robot, *Proceedings of the 1999 IEEE/ASME International Conference on Advanced Intelligent Mechatronics, pp. 73-78*, 1999

[6] W.Keith Edwards, Core Jini, *Prentice Hall PTR*, 1999

[7] Kwun Han, Manuela Veloso, Reactive Visual Control of Multiple Non-Holonomic Robotic Agents, *Proc. of IEEE Int. Conference on Robotics and Automation*, vol.4, pp.3510-3515, 1998

[8] T.Yoshikawa, J.Ueda, Analysis and Control of Master-salve systems with Time Delay, *Proc. Of IEEE Int. Conference of Intelligent Robotics and Systems*, pp. 1366-1373, 1996.

[9] T.Matsumaru, S.Kawabata, T.Kotoku, N.Matuhira, Task-based Data Exchange for Remote Operation System through a Communication Network, *Proc. of IEEE Int. Conference on Robotics and Automation*, pp. 557-564, 1999.

[10] H.G. Lee, Development of Object Oriented Architecture for an Autonomous Mobile Robot under Home Network Environment, 2000

Implementing Tolman's Schematic Sowbug: Behavior-Based Robotics in the 1930's

Yoichiro Endo   Ronald C. Arkin

Mobile Robot Laboratory
College of Computing
Georgia Institute of Technology
Atlanta, Georgia 30332-0280 U.S.A.

Abstract

This paper reintroduces and evaluates the schematic sowbug proposed by Edward C. Tolman, psychologist, in 1939. The schematic sowbug is based on Tolman's purposive behaviorism, and it is believed to be the first prototype in history that actually implemented a behavior-based architecture suitable for robotics. The schematic sowbug navigates the environment based on two types of vectors, orientation and progression, that are computed from the values of sensors perceiving stimuli. Our experiments on both simulation and real robot proved the legitimacy of Tolman's assumptions, and the potential of applying the schematic sowbug model and principles within modern robotics is recognized.

1 Introduction

In the field of cognitive science, the psychologist Edward C. Tolman is best known for introducing the concept of "cognitive map" [12] in the late 1940's. He studied how both rats and people store information regarding their physical locations with respect to the environment, in past, current, and future perspectives [1]. This paper, however, focuses on another research project of Tolman, the concept of a schematic sowbug, which was a product of his earlier work on purposive behaviorism. Tolman, born in 1886 in Massachusetts, proposed purposive behaviorism in the early 1920's. According to Innis [8] who studied Tolman's approach:

> *Initially, in Tolman's purposive behaviorism, behavior implied a performance, the achievement of an altered relationship between the organism and its environment; behavior was functional and pragmatic; behavior involved motivation and cognition; behavior revealed purpose.*

In other words, the goal was to investigate how high-level factors, such as motivation, cognition, and purpose, were incorporated into the tight connection between stimulus and response that the prevailing behaviorist view largely ignored. Tolman derived a formula to compute the value of a behavior (B) from environmental stimuli (S), physiological drive (P), heredity (H), previous training (T), and mutuality or age (A) [12]:

$$B = f(S, P, H, T, A)$$

From the inputs S, P, H, T, and A, the formula generates the output behavior B, by applying various laws, such as the laws of perception, law of motivation, and laws of learning (Figure 1).

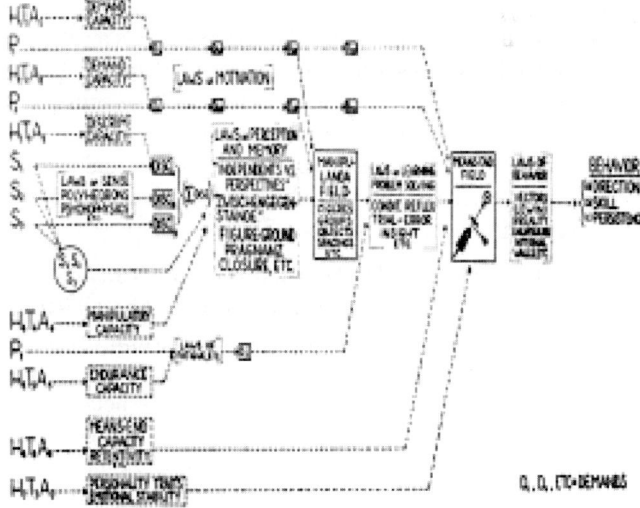

Figure 1: Tolman's Purposive Behaviorism: $B = f(S, P, H, T, A)$. (Reproduced from [12].)

Purposive behaviorism spoke to many of the same issues that modern behavior-based robotics architectures address [2]: how to produce intelligent behavior from multiple concurrent and parallel sensorimotor (behavioral) pathways, how to coordinate their outputs meaningfully, how to introduce the notion of goal-oriented behavior, how to include motivation and emotion, and how to permit stages of developmental growth to influence behavior.

1.1 Tolman's Schematics Sowbug Model

Based on his purposive behaviorism, Tolman proposed the concept of the schematic sowbug (Figure 2) in 1939. The schematic sowbug consists of various unique features that are explained in detail in [11]. The following are brief descriptions of these features, summarized from Tolman's writings:

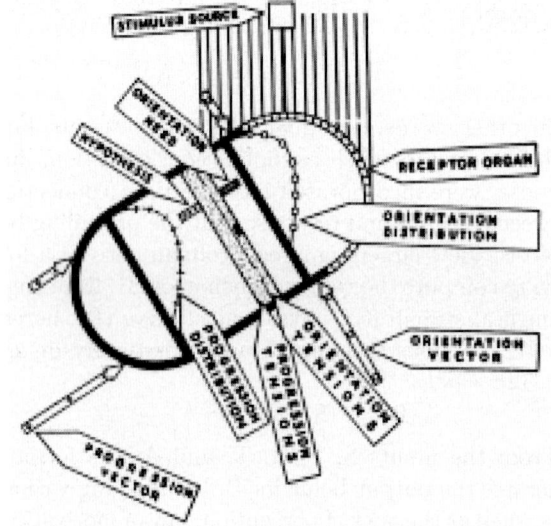

Figure 2: Tolman's Schematic Sowbug. (Reproduced from [12].)

- **Receptor Organ**: The *Receptor Organ* is a set of multiple photo-sensors that perceive light (or any given stimuli) in the environment. These sensors are physically mounted on the front end surface of the sowbug, forming an arc. An individual sensor outputs a value based on the intensity of the stimuli.

- **Orientation Distribution, Orientation Need, and Orientation Tensions**: The *Orientation Distribution*, shown as a line graph drawn inside the front-half of the sowbug, indicates the output values of the photo-sensors. The height of each node in the graph is the value of the corresponding photo-sensor in the *Receptor Organ*. For example, if there is a light source (or any given stimulus) on the left-hand side of the sowbug (as shown in Figure 2), the nodes on the left-hand side of the graph become higher than the ones on the right-hand side. The height of the nodes is also determined by a specific *Orientation Need* (a little column rising up from the stippled area). This stippled area is referred to as the *Orientation Tensions*. The *Orientation Need* is a product of the *Orientation Tension*, where the level of *Orientation Tension* corresponds to the degree of the motivational demand. For example, if the stimulus is a food object, *Orientation Tension* is determined by the degree of the sowbug's hunger. Moreover, it is assumed that if the sowbug is facing directly toward the stimulus, the *Orientation Need* decreases.

- **Orientation Vector**: The vectors pointing at the sides of the sowbug are the *Orientation Vectors*. The length of the right-hand side vector is the total sum of the left-hand side *Orientation Distribution*, and the length of the left-hand side vector is the total sum of the right-hand side *Orientation Distribution*. When an *Orientation Vector* is generated, the sowbug will rotate toward the direction it is pointing. For example, if there is only a right-hand side vector pointing towards the left, the sowbug will try to rotate in a counter-clockwise direction. If there are two vectors pointing toward each other, the net value (after summation) will be the direction the sowbug will try to rotate. The *Orientation Vector* will not cause translational movement of the sowbug, only rotational.

- **Progression Distribution, Hypothesis, and Progression Tensions**: The *Progression Distribution* is also shown as a line graph drawn inside the rear-half of the sowbug. The shape of the *Progression Distribution* is proportional to the shape of the *Orientation Distribution*. However, the height of the *Progression Distribution* is determined by the strength (or certainty) of a specific *Hypothesis*. For example, if the sowbug is reacting to a food stimulus, the level of *Hypothesis* is how much the sowbug believes "this stimulus source is really food." The level of a *Hypothesis* becomes higher the more the sowbug assumes that the stimulus is indeed food. A *Hypothesis* is a product of the *Progression Tensions* and the past experience relative to this specific stimulus.

- **Progression Vector**: *Progression Vectors* are located at the rear-end corners, left and right, of the sowbug, pointing toward the front. These vectors represent the velocities of the left-hand side and right-hand side motors of the sowbug, respectively. As for *Orientation Vector*, the length of the left-hand side *Progression Vector* is determined by the right-hand side of *Progression Distribution*, and the length of the right-hand side *Progression Vector* is determined by the left-hand side of *Progression Distribution*. In other words, if there is a stimulus on the left-hand side of the sowbug, it will generate a larger right-hand side vector, and try to move forward while turning to the left, similar to the notion described decades later by Braitenberg [4]. However, if the sowbug sustains negative experiences regarding the stimulus, the hypothesis then becomes weaker, and it will not move towards the stimulus.

The main behavioral characteristic of the schematic sowbug is its positive phototactic behavior. With the combination of the *Orientation Vector* and *Progression Vector*, the sowbug is expected to respond to the stimulus in the environment by orienting and moving towards it based on its *Orientation Need* and *Hypothesis*. Since both the *Orientation Need* and *Hypothesis* are determined by the internal state of the sowbug, which changes as the sowbug increases its experiences with the stimulus, the trajectory of the sowbug is not consistent for different trials even if the external conditions are setup same.

1.2 Remarks

Tolman acknowledges [11] that his schematic sowbug was inspired by Lewin's "psychological life space" [9] and Loeb's "tropism theory" [10], which was also studied by Blum [3]. Lewin's "psychological life space" indicates "the totality of facts which determine the behavior of an individual at a certain moment." Before Tolman invented the formula which can compute the behavior of animals (i.e., B = f (S, P, H, T, A)), Lewin, a psychologist, attempted to form an equation which outputs the behavior (B) of a person (P) for a given event (E):

$$B = f(P, E)$$

The word "tropism" in Loeb's "tropism theory" describes how plants and low-level organisms try to turn towards a light source. According to Fleming who translated Loeb's work [10], the origin of the word "tropism" comes from a Greek word "trope" for turning. Loeb, a biologist, studied animals' phototactic behavior by trying to figure out how photosensitive substances in animals' bodies undergo chemical alternations by light, and how they would effect the animals' motor behavior [10]. This study let Blum create a model of a phototactic animal by connecting its left-hand side photo-sensors to the right-hand side motor, and right-hand side photo-sensors to the left-hand side motor, and compared it to the statistical results taken from the experiments with cucumber beetles [3]. This again is very similar in spirit to Braitenberg's later descriptions of vehicles exhibiting similar phototactic behaviors [4]. Even though Tolman proposed his system a half-century before Braitenberg did, they were both inspired by Loeb's "tropism theory", and their systems should exhibit similar behaviors. However, Braitenberg's model was implemented with *Progression Vectors* only, while Tolman's model has both *Orientation* and *Progression Vectors* as Blum's model does.

From a roboticist's point of view, Tolman's schematic sowbug is remarkable because it was the first prototype that actually described a behavior-based robotics architecture in history, to the best of our knowledge. It was a half-century before Brooks developed the subsumption architecture [5] in the mid-1980's. However, it should be noted that Tolman's schematic sowbug is not a purely reactive architecture. Past training and internal motivational state will also affect the behavior.

2 Schematic Sowbug Implementation

In order to determine the feasibility of actually implementing the schematic sowbug model proposed by Tolman on a physical robot, a C++ program *eBug*[1] (Emulated Sowbug), which runs on RedHat Linux 6.2, was created. *eBug* emulates the basic features of the schematic sowbug, such as the *Orientation Vector*, *Progression Vector*, *Orientation Distribution*, and *Progression Distribution*. The program environment can control both a simulated sowbug that reacts to simulated stimuli (Figure 3), and a real robot that reacts to real color objects (Figure 4). The key features of *eBug* are explained in Section 2.1 and the configuration of the real robot experiment is explained in Section 2.2.

2.1 Key Features of eBug

The following key features are implemented in *eBug*:

- **Stimulus Source**: For the simulation mode, the user can place any number of stimuli on the screen, and change their intensities or types by clicking mouse buttons. For the real robot mode, the perceived color objects as seen by the robot are displayed on the screen.

- **Receptor Organ**: The nine circles located at the front end surface of the sowbug emulate the

---
[1] The source code is available from http://www.cc.gatech.edu/ai/robot-lab/research/ebug/ .

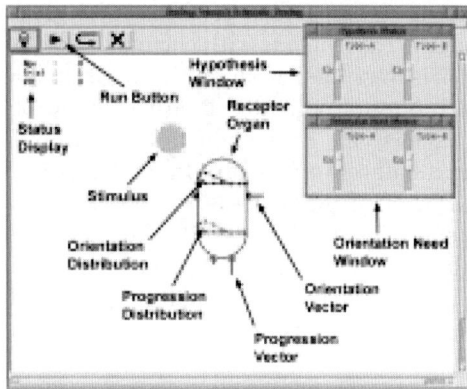

Figure 3: User interface of *eBug*.

properties of the photo-sensors in the schematic sowbug's *Receptor Organ*. Suppose that the number of the stimuli perceived by the photo-sensor is N_S. The intensity reading of the photo-sensor (I_P) is then computed from Equation 1, which takes the intensity of each stimulus (I_S), the angle between the normal of the sensor surface and the rays from the stimulus (θ_{PS}), the distance between the sensor and the stimulus (D_{PS}), and a constant k_P as its inputs. I_P is assumed to be proportional to the inverse of D_{PS}, and the values for I_S and k_P are arbitrary assigned.

$$I_P = k_P \sum_{i=1}^{N_S} I_{Si} \cos(\theta_{PSi}) \frac{1}{D_{PSi}} \quad (1)$$

- **Orientation Distribution**: The line graph drawn inside the sowbug's body is the *Orientation Distribution*. The height of each node o_P in the graph is a function of the corresponding sensor reading I_P and the *Orientation Need* value O (Equation 2). Moreover, as Tolman assumed, O will be set to be zero if a stimulus is in the direction of the sowbug's heading.

$$o_P = I_P O \quad (2)$$

- **Orientation Vector**: At each side of the sowbug, a set of vectors is pointing towards the sowbug, representing the *Orientation Vectors*. As shown in Equation 3, the right *Orientation Vector* value V_{OR} is the total sum of the left-hand-side *Orientation Distribution* (N_L nodes), and the left *Orientation Vector* value V_{OL} is the total sum of the right-hand-side *Orientation Distribution* (N_R nodes).

$$V_{OR} = \sum_{i=1}^{N_L} o_{PLi}, \quad V_{OL} = \sum_{i=1}^{N_R} o_{PRj} \quad (3)$$

- **Progression Distribution**: Another line graph drawn inside the sowbug's body represents the *Progression Distribution* of the schematic sowbug. The height of each node p_P in the graph is a function of the corresponding sensor reading I_P and the *Hypothesis* value H (Equation 4).

$$p_P = I_P H \quad (4)$$

- **Progression Vector**: The vectors pointing to the sowbug from the rear-end represent the *Progression Vectors*. As it was for the *Orientation Vectors*, the right *Progression Vector* value V_{PR}, which is the speed of the right motor, is the total sum of the left-hand-side *Progression Distribution*, and the left *Progression Vector* value V_{PL}, which is the speed of the left motor, is the total sum of the right-hand side *Progression Distribution* (Equation 5).

$$V_{PR} = \sum_{i=1}^{N_L} p_{PLi}, \quad V_{PL} = \sum_{i=1}^{N_R} p_{PRi} \quad (5)$$

- **Running Sowbug**: The sowbug will start reacting to the stimuli when the user clicks on the *Run Button* in the menu bar. The angular speed ω (counter-clockwise positive) is computed from Equation 6, which takes *Orientation Vectors*, *Progression Vectors*, and constants κ_ω and κ'_ω. The forward speed v is also computed from averaging the right and left *Progression Vectors*, and multiplying it with a constant k_v (Equation 7). The values for κ_ω, κ'_ω, and k_v are arbitrary assigned.

$$\omega = \kappa_\omega (V_{OR} - V_{OL}) + \kappa'_\omega (V_{PR} - V_{PL}) \quad (6)$$

$$v = k_v \frac{(V_{PR} + V_{PL})}{2} \quad (7)$$

The schematic sowbug in *eBug* can distinguish two different types of stimuli, red and green. When these two types of the stimuli are present at the same time, ω and v are calculated independently for each stimulus type, and are summed up at the end to obtain their total values (Equations 8 and 9).

$$\omega_{total} = \omega_{green} + \omega_{red} \quad (8)$$

$$v_{total} = v_{green} + v_{red} \quad (9)$$

- **Hypothesis Window**: This window allows the user to directly specify the sowbug's *Hypothesis* values as well as monitor their changes through the duration of an experiment. Two slider bars indicate the value of the *Hypothesis* for each of the two colored stimuli, green and red, (labeled "Type-A" and "Type-B", respectively).

- **Orientation Need Window**: This window allows the user to monitor the changes in the sowbug's *Orientation Need*. As in the *Hypothesis* window, two slider bars indicate the value of the *Orientation Need* for the two stimuli.

2.2 Real Robot Configuration

eBug runs not only in simulation but also on a real robot. When it is in the real robot mode, *eBug*, running on a Dell Precision 410 (Pentium III, 500MHz) desktop computer, remotely communicates with *HServer*, a component of the *MissionLab* system [6], running on a Toshiba Libretto 110CT (Mobile Pentium MMX, 233MHz) laptop that sits atop an ActivMedia Pioneer AT robot (Figure 4). *HServer* is a hardware server that contains drivers for various robots and sensors. After an on-board Sony EVI camera captures the images of the environment, a Newton Cognachrome Vision System [13] processes the images to identify the stimuli (color objects). *HServer* then sends the data that contain locations of the stimuli to *eBug* (Figure 5). *eBug* in return sends the movement commands to the robot via *HServer* based on the computed values of *Orientation Vectors* and *Progression Vectors*. *eBug* communicates to *HServer* through IPT [7], and *HServer* communicates to the robot hardware through serial links.

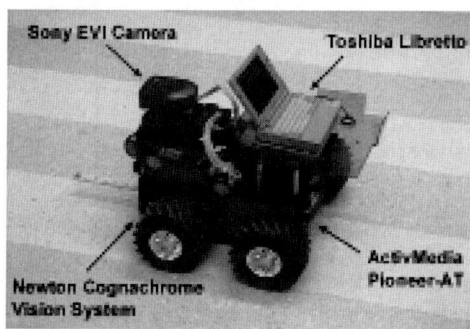

Figure 4: Robot Hardware.

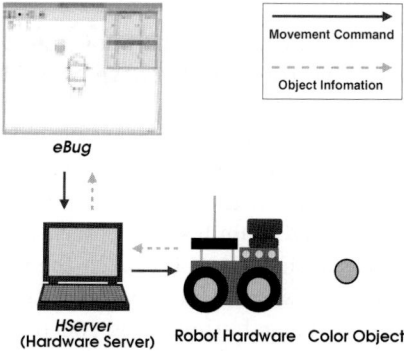

Figure 5: Information flow for the real robot configuration.

3 Evaluation

3.1 Simulation Experiment

Simulation experiments were conducted to verify the positive phototactic behavior of the sowbug and to investigate how the sowbug's trail changes from both positive and negative experiences with a stimuli.

3.1.1 Setup

As shown in Figure 6, two stimuli, green and red, were placed in front of and equally away from the sowbug. Initially, the sowbug's *Hypothesis* values of both stimuli were set to 50 percent. In other words, with 50 percent uncertainty, the sowbug initially "believes" that the green and red objects are both food objects. Note that the green object is labeled "+" and the red object is labeled "-". If the sowbug eats a positive stimuli, the *Hypothesis* value increases 10 percent. The *Hypothesis* value decreases 10 percent if it eats a negative stimuli. This experiment was, therefore, set up to observe how the sowbug learns that the green stimulus is indeed a food.

The movement trail of the sowbug was recorded during the eight trials of the experiment. Each trial ends when the sowbug either eats both stimuli or no longer makes any movement. Through the eight trials, the *Hypothesis* values were not reset by the user, and only altered by the sowbug itself. If the sowbug alters the *Hypothesis* value after eating a positive or negative stimulus, the value was kept for use in the subsequent trial. At the beginning of each trial, however, the sowbug was relocated to the same initial position, and the two stimuli were placed again at the same exact previous positions with the same positive/negative types.

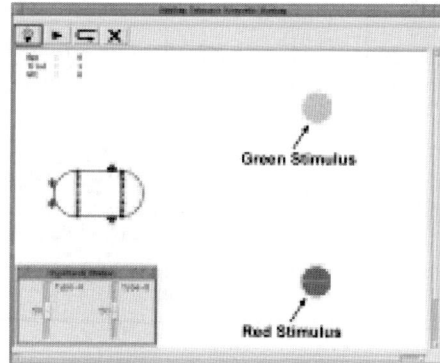

Figure 6: Schematic sowbug at the initial position in the simulation experiment.

3.1.2 Results

As can be observed from Figure 7, the simulated schematic sowbug exhibited positive phototactic behavior. Figure 8 shows that the sowbug began to "believe" that the green stimulus is indeed food, and started to "disbelieve" the red stimulus as a food source during the experiment. The captured screen images in Figure 10 show how the *Hypothesis* values affected the shape of sowbug's trail. In the beginning of the experiment, when the *Hypothesis* values for the two stimuli were very close, the sowbug approached the stimuli from the center. Later, when the sowbug started discerning that the green stimulus is indeed food, it began to take a path closer to the green stimulus, and when the *Hypothesis* value of the red stimulus became zero (at the sixth trial), the sowbug no longer approached the red stimulus, even though it oriented itself to the stimuli at the end.

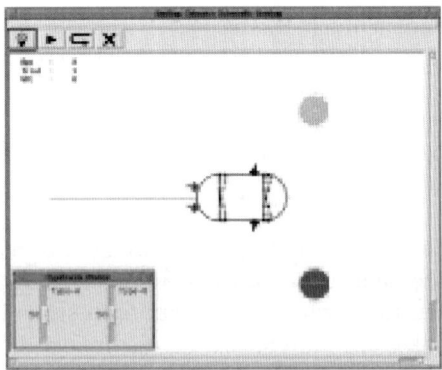

Figure 7: Screen capture from the simulation experiment. The schematic sowbug is approaching the stimuli during the first trial.

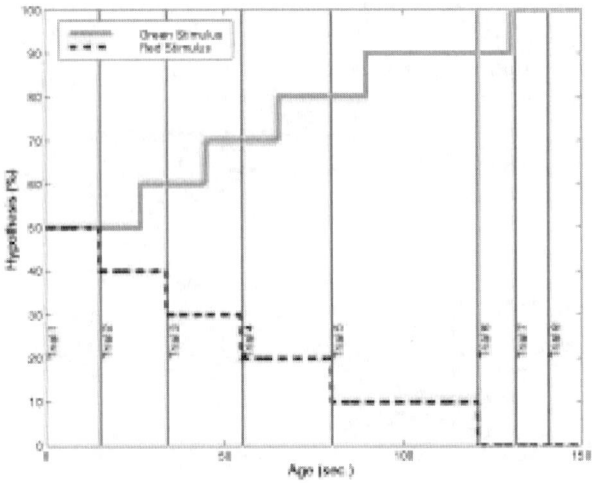

Figure 8: *Hypothesis* values during the simulation experiment.

3.2 Real Robot Experiment

In order to determine the potential of the schematic sowbug model for use in the robotics domain, an experiment similar to the one conducted in simulation was also performed on the real robot.

3.2.1 Setup

Two stimuli, green and red objects, that were equally distant from the robot's initial position were placed in front of the robot at the beginning of each trial (Figure 9). The Cognachrome Vision System was trained to identify those two colors. In this experiment, however, the *Hypothesis* values were set manually at the beginning of each trial. Three trials were recorded during the experiment. In the first trial, the *Hypothesis* values for both green and red stimuli were 50 percent. In the second trial, the *Hypothesis* value for the green stimulus was 100 percent and for the red stimulus it was 0 percent. In the third trial. the *Hypothesis* value for the green stimulus was 0 percent and for the red stimulus it was 100 percent. During this entire experiment, the locations of the color objects were not changed.

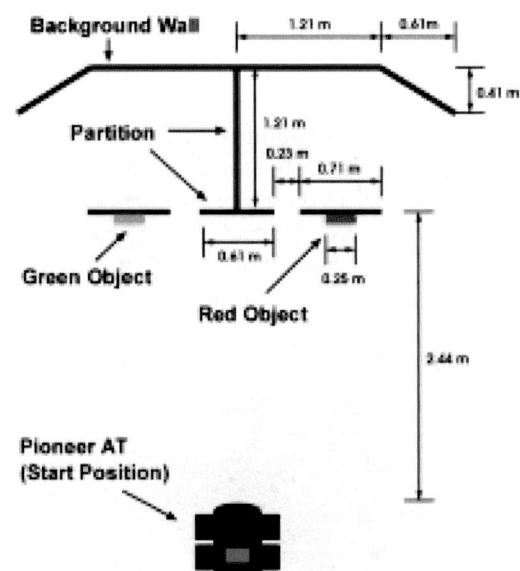

Figure 9: Robot at the initial position in the real robot experiment (top view).

3.2.2 Results

Captured images in Figure 11 show how the robot performed during the three trials of the experiment. In the first trial, when the *Hypothesis* values for both green and red stimuli are 50 percent, the robot chose to approach the red object. In the second trial, when the *Hypothesis* value for the green stimulus is 100 percent and for the red stimulus is 0 percent, the robot chose to approach the green object. In the third trial, when the *Hypothesis* value for the green stimulus is 0 percent and for the red stimulus is 100 percent, the robot chose to approach the red object.

4 Conclusion

Tolman's schematic model is the first instance in history, to our knowledge, of a behavior-based model suitable for implementation on a robot. It predates both Brooks' subsumption and Braitenberg's vehicles by approximately a half century. While useful conceptually as a model, it was the goal of this research to test indeed whether or not the model could be implemented on real robots.

The primary features of Tolman's schematic sowbug were successfully implemented in both simulation and on a real robot. The results of the simulation experiment were consistent with Tolman's assumption for the sowbug in which he expected to observe its phototactic behavior; when the stimulus was in the field, the sowbug rotated itself to face the stimulus; and if there was enough belief in the hypothesis, the sowbug moved towards the stimulus. It was also observed that, when the internal state (*Hypothesis*) of the sowbug is different, the sowbug produces different trajectories even though the external conditions are set up identically. The results from the real robot experiment proved that it is indeed possible to apply Tolman's schematic sowbug in robotics. Future research could expand the model as currently implemented more completely to draw together more closely psychological models of animal behavior and robotic systems.

5 Acknowledgments

The authors would like to thank William C. Halliburton for his assistance on interfacing *eBug* to the real robot. The Georgia Tech Mobile Robot Laboratory is supported from a variety of sources including DARPA, Honda R&D, C.S. Draper Laboratory, and SAIC.

References

[1] Anderson, R.E. "Imagery and Spatial Representation." *A Companion to Cognitive Science.* ed. Bechtel, E. and Graham, G., Blackwell Publishers Inc., 1998, pp. 204-211.

[2] Arkin, R.C. *Behavior-Based Robotics.* Cambridge, MIT Press, 1998.

[3] Blum, H.F. "An Analysis of Oriented Movements of Animals in Light Fields." *Cold Springs Harbor Symposia on Quantitative Biology.* Vol. 3, 1935, pp. 210-223.

[4] Braitenberg, V. *Vehicles: Experiments in Synthetic Psychology.* Cambridge, the MIT Press, 1984.

[5] Brooks, R. "A Robust Layered Control System for a Mobile Robot." *IEEE Journal of Robotics and Automation.* Vol. RA-2, No. 1, 1986, pp. 14-23.

[6] Endo, Y., MacKenzie, D.C., Stoytchev, A., Halliburton, W.C., Ali., K.A., Balch, T., Cameron, J.M., and Chen, Z. *MissionLab: User Manual for MissionLab version 4.0.* ed. Arkin, R.C., Georgia Institute of Technology, 2000.

[7] Gowdy, J. *IPT: An Object Oriented Toolkit for Interprocess Communications - Version 6.4.* Carnegie Mellon University, 1996.

[8] Innis, N.K. "Edward C. Tolman's Purposive Behaviorism." *Handbook of Behaviorism.* Academic Press, 1999, pp. 97-117.

[9] Lewin, K. *Principles of Topological Psychology.* trans. Heider, F. and Heider, G.M. McGraw-Hill, New York, 1936.

[10] Loeb, J. *The Mechanistic Conception of Life.* ed. Fleming, D. Cambridge, The Belknap Press of Harvard University Press, 1964.

[11] Tolman, E.C. "Prediction of Vicarious Trial and Error by Means of the Schematic Sowbug." *Psychological Review.* ed. Langfeld, H.S. Vol. 46, 1939, pp. 318-336.

[12] Tolman, E.C. *Behavior and Psychological Man.* University of California Press, 1951.

[13] Wright, A., Sargent, R., Witty, C., and Brown, J. *Cognachrome Vision System User's Guide.* Newton Research Labs, 1996.

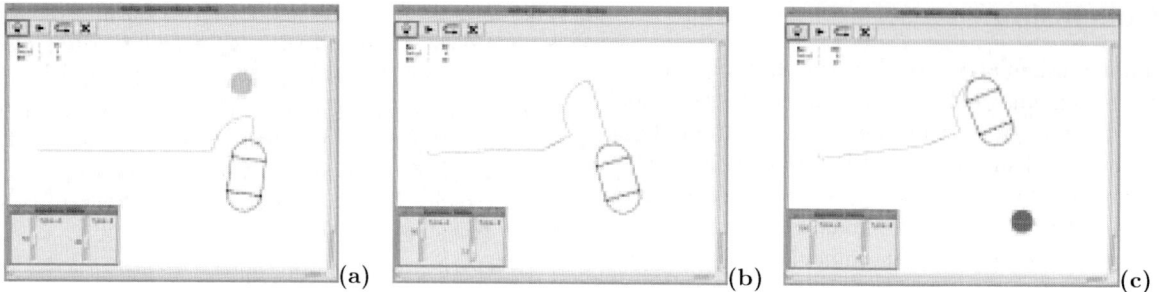

Figure 10: Screen captures from the simulation experiment: (a) Final position of Trial 1; (b) Final position of Trial 4; (c) Final position of the last trial (Trial 8).

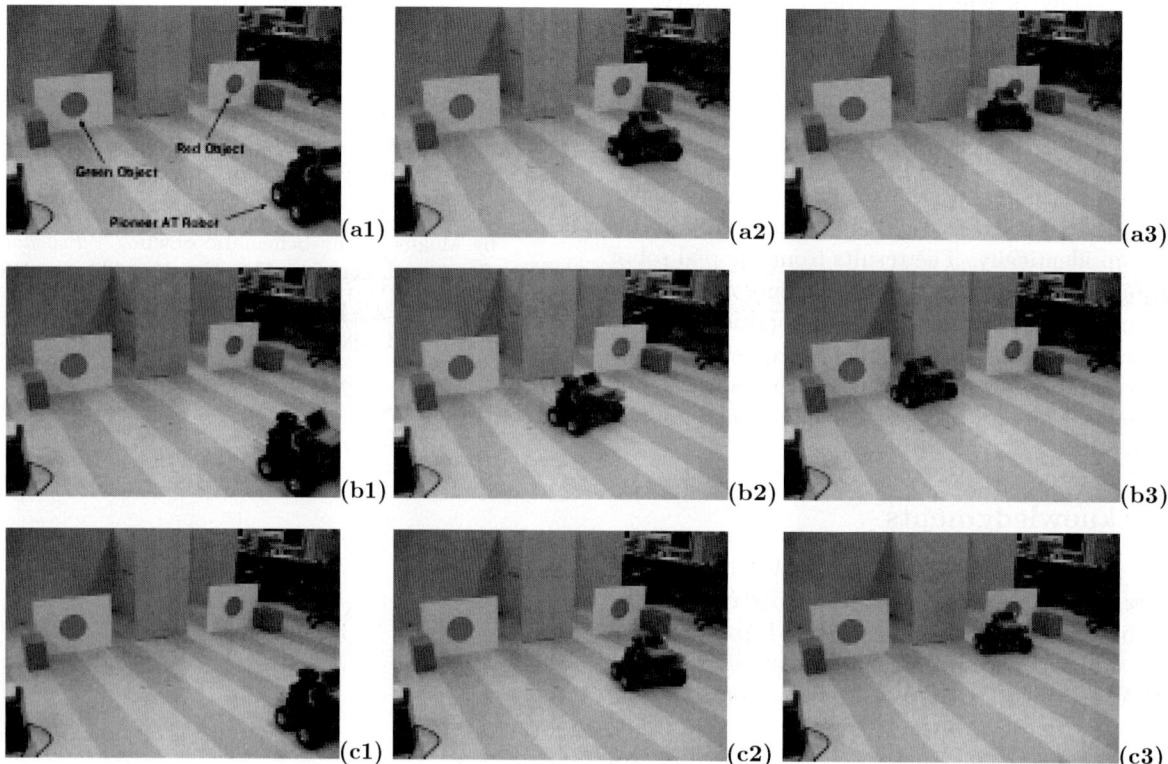

Figure 11: Sequence of images from the real robot experiment: (a1) - (a3) Trial 1. The *Hypothesis* values for both green and red stimuli are 50%; (b2) - (b3) Trial 2. The *Hypothesis* values for the green stimulus is 100% and for the red stimulus is 0%; (c2) - (c3) Trial 3. The *Hypothesis* values for the green stimulus is 0% and for the red stimulus is 100%.

Generating Linguistic Spatial Descriptions from Sonar Readings Using the Histogram of Forces

Marjorie Skubic, George Chronis, Pascal Matsakis and James Keller
Dept. of Computer Engineering and Computer Science
University of Missouri-Columbia
email: skubicm@missouri.edu

Abstract

In this paper, we show how linguistic expressions can be generated to describe the spatial relations between a mobile robot and its environment, using readings from a ring of sonar sensors. Our work is motivated by the study of human-robot communication for non-expert users. The eventual goal is to use these linguistic expressions for navigation of the mobile robot in an unknown environment, where the expressions represent the qualitative state of the robot with respect to its environment, in terms that are easily understood by human users. In the paper, we describe the histogram of forces and its application to sonar sensors on a mobile robot. Several environment examples are also included with the generated linguistic descriptions.

1 Introduction

Our work is motivated by the study of human-robot interaction and, in particular, the investigation of human-robot communication. The ultimate goal is to provide easy and intuitive interaction by naïve users, so that they can guide, control, and/or program a robot to perform some purposeful task. We consider the communication between the human user and the robot to be crucial to intuitive interaction by users that are not robotics experts. We further argue that good communications is essential both from the human to the robot (to command the robot to perform purposeful tasks) and also from the robot to the human (so that the user can monitor the robot's current state or condition). See also [1] and [2] for examples and further motivation on task-oriented dialogues between a robot and a human user.

In this paper, we show how linguistic expressions can be generated to describe the spatial relations between a mobile robot and its environment, using readings from a ring of sonar sensors. The eventual goal is to use these linguistic descriptions for navigation of the mobile robot in an unstructured, unknown, and possibly dynamic environment. We are not attempting to build an exact model of the environment, nor to generate a quantitative map. However, we do want to generate linguistic descriptions that represent the qualitative state of the robot with respect to its environment, in terms that are easily understood by human users.

The linguistic spatial descriptions provide a symbolic link between the robot and a human user, thus comprising a navigation language for human-robot interaction. The linguistic expressions can be used for two-way communications with the robot. First, in robot-to-human communication, they provide a qualitative description of the robot's current state (e.g., *there is an object to the left*, or *there is an object to the right front*).

Second, in human-to-robot communication, the human can command the robot to perform navigation behaviors based on the spatial relations (e.g., *while there is an object on the left, move forward*, or *if there is an object on the right front, turn left*, or even a high-level and very human-like directive such as *turn left at the second intersection*). A task can be represented and described as a sequence of qualitative "states" based on spatial relations, each state with a corresponding navigation behavior. We assume the robot has pre-programmed or pre-learned, low-level navigation behaviors that allow it to move safely around its unstructured and dynamic environment without hitting objects.

To accomplish both cases of communication, the robot must be able to recognize its state in terms of egocentric spatial relations between itself and objects in its environment, and it must be able to generate a linguistic description of the spatial relations. The main focus of this paper is the creation of these linguistic spatial descriptions from a ring of sonar sensors.

The idea of using linguistic spatial expressions to communicate with a semi-autonomous robot has been proposed previously. Gribble *et al* use the framework of the Spatial Semantic Hierarchy for an intelligent wheelchair [3]. Perzanowski *et al* use a combination of gestures and linguistic directives such as "go over there" [4]. Shibata *et al* use positional relations to overcome ambiguities in recognition of landmarks [5]. In [6], Stopp *et al* use spatial expressions to communicate with a 2-arm mobile robot performing assembly tasks. Spatial relations are used as a means of identifying an object in a geometric model. That is, the robot has a model of its environment, and the user selects an object from the model using relational spatial expressions.

The work presented here is an extension of spatial analysis previously applied to image analysis. Background material on the spatial analysis algorithms is included in Section 2. In Section 3, we show how the robot's sonar readings can be used to generate inputs for the spatial analysis algorithms. Specific test cases are shown in Section 4 along with a discussion of future work. Concluding remarks are found in Section 5. The interested reader is also referred to a companion paper on using spatial analysis to extract navigation states from a hand-drawn map [7].

2 Background on Spatial Relations

Freeman [8] proposed that the relative position of two objects be described in terms of spatial relationships (such as "above", "surrounds", "includes", etc.). He also proposed that fuzzy relations be used, because "all-or-nothing" standard mathematical relations are clearly not suited to models of spatial relationships. Moreover, "although the human way of reasoning can deal with qualitative information, computational approaches of spatial reasoning and object recognition can benefit from more quantitative measures" [9]. By introducing the notion of the histogram of angles, Miyajima and Ralescu [10] developed the idea that the relative position between two objects can have a representation of its own and can thus be described in terms other than spatial relationships. However, the representation proposed shows several weaknesses (*e.g.*, requirement for raster data, long processing times, anisotropy).

In [11][12], Matsakis and Wendling introduced the histogram of forces. Contrary to the angle histogram, it ensures processing of raster data as well as of vector data. Moreover, it offers solid theoretical guarantees, allows explicit and variable accounting of metric information, and lends itself, with great flexibility, to the definition of fuzzy directional spatial relations (such as "to the right of", "in front of", etc.). For our purposes, the histogram of forces also allows for a low-computational handling of heading changes in the robot's orientation and also makes it easy to switch between a world view and an egocentric robot view.

2.1 The Histogram of Forces

The relative position of a 2D object A with regard to another object B is represented by a function F^{AB} from $I\!R$ into $I\!R_+$. For any direction θ, the value $F^{AB}(\theta)$ is the total weight of the arguments that can be found in order to support the proposition "A is in direction θ of B". More precisely, it is the scalar resultant of elementary forces. These forces are exerted by the points of A on those of B, and each tends to move B in direction θ (Fig. 1). F^{AB} is called the *histogram of forces associated with* (A,B) *via* F, or the *F–histogram associated with* (A,B). The object A is the *argument*, and the object B the *referent*.

Note that throughout this paper, the referent is always the robot. Actually, the letter F denotes a numerical function. Let r be a real. If the elementary forces are in inverse ratio to d^r, where d represents the distance between the points considered, then F is denoted by F_r. The F_0–histogram (histogram of constant forces) and F_2–histogram (histogram of gravitational forces) have very different and very interesting characteristics. The former coincides with the angle histogram—without its weaknesses—and provides a global view of the situation. It considers the closest parts and the farthest parts of the objects equally, whereas the F_2–histogram focuses on the closest parts. Details can be found in [11][12].

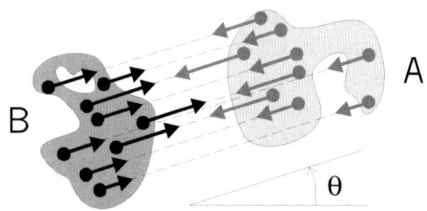

Figure 1. Computation of $F^{AB}(\theta)$. It is the scalar resultant of forces (black arrows). Each one tends to move B in direction θ.

2.2 Handling of Vector Data

In previous work, we generated the F_0 and F_2 histograms using raster image data. In this paper, we present the first application of histograms that uses vector data, *i.e.*, a boundary representation based on the objects' vertices.

In practice, the F-histogram associated with a pair (A,B) of objects is represented by a limited number of values (*i.e.*, the set of directions θ is made discrete). For any θ considered, the objects are partitioned by sorting both A and B vertices, following direction $\theta+\pi/2$. The computation of F^{AB} is of complexity $O(n \log(n))$, where n denotes the total number of vertices. It is translated into a set of assessments of predetermined algebraic expressions. Each assessment corresponds to the process of a pair of trapezoids. In the case illustrated by Figure 2, the scalar resultant of the forces represented by black arrows is Γ_0 for constant forces and is Γ_2 for gravitational forces:

$$\Gamma_0 = \varepsilon[(x_1+x_2)(z_1+z_2)+x_1z_1+x_2z_2] / [6\cos^2(\theta)]$$
$$\Gamma_2 = \varepsilon[f(x_1+y_1,x_2+y_2)-f(y_1,y_2)+f(y_1+z_1,y_2+z_2)$$
$$-f(x_1+y_1+z_1,x_2+y_2+z_2)]$$

where f denotes the function defined by:

$$\forall (r,s)\in I\!R_+^* \times I\!R_+^*, \ r\neq s \Rightarrow f(r,s) = [s\ln(s)-r\ln(r)]/(s-r)$$
$$\text{and} \ \forall r\in I\!R_+^*, \ f(r,r) = \lim_{s\to r} f(r,s) = 1+\ln(r)$$

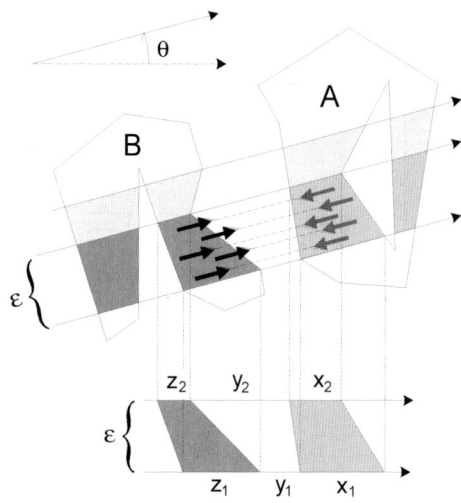

Figure 2. The evaluation of $F^{AB}(\theta)$ is based on the partitioning of the objects.

2.3 Linguistic Description of Relative Positions

In [13][14], Matsakis *et al.* present a system that produces linguistic spatial descriptions. The description of the relative position between any 2D objects A and B relies on the sole primitive directional relationships: "to the right of", "above", "to the left of" and "below" (imagine that the objects are drawn on a vertical surface). It is generated from F_0^{AB} (the histogram of constant forces associated with (A,B)) and F_2^{AB} (the histogram of gravitational forces). First, eight values are extracted from the analysis of each histogram: a_r(RIGHT), b_r(RIGHT), a_r(ABOVE), b_r(ABOVE), a_r(LEFT), b_r(LEFT), a_r(BELOW) and b_r(BELOW). They represent the "opinion" given by the considered histogram (*i.e.*, F_0^{AB} if r is 0, and F_2^{AB} if it is 2). For instance, according to F_0^{AB} the degree of truth of the proposition "A is to the right of B" is a_0(RIGHT). This value is a real number greater than or equal to 0 (proposition completely false) and less than or equal to 1 (proposition completely true). Moreover, according to F_0^{AB} the maximum degree of truth that can reasonably be attached to the proposition (say, by another source of information) is b_0(RIGHT) (which belongs to the interval $[a_0(RIGHT),1]$). F_0^{AB} and F_2^{AB}'s opinions (*i.e.*, the sixteen values) are then combined. Four numeric and two symbolic features result from this combination. They feed a system of 27 fuzzy rules and meta-rules that outputs the expected linguistic description. The system handles a set of 16 adverbs (like "mostly", "perfectly", etc.) which are stored in a dictionary, with other terms, and can be tailored to individual users. A description is generally composed of three parts. The first part involves the primary direction (*e.g.*, "A is mostly to the right of B"). The second part supplements the description and involves a secondary direction (*e.g.*, "but somewhat above"). The third part indicates to what extent the four primitive directional relationships are suited to describing the relative position of the objects (*e.g.*, "the description is satisfactory"). In other words, it indicates to what extent it is necessary to utilize other spatial relations (*e.g.*, "surrounds").

3 Egocentric Spatial Relations from Sonar Readings

In this section, we describe the application of the F_0 and F_2 histograms for extracting spatial relations from the sonar ring of a mobile robot. In our work, we have used a Nomad 200 robot with 16 sonar sensors evenly distributed along its circumference. The sensors' readings are used to build an approximate representation of the objects surrounding the robot. The vertices of each object are extracted and used to build the F_0 and F_2 histograms, as described in Section 2.2, which are then used to generate linguistic descriptions of relative positions between the robot and the environment objects (see Figure 3).

The first step in recognizing spatial relations from sonar readings is to build objects around the robot from the sonar readings. Let us consider a simple case of the robot and a single obstacle, shown in Figure 4. The sonar sensor S returns a range value (which is less than the maximum), indicating that an obstacle has been detected. In the case of Figure 4, all sonar sensors except S return the maximum value, which means that no other obstacle was detected. In this case, a single object is plotted as a trapezoid in the center of cone S. The depth of the obstacle cannot be determined from the sonar reading; thus, we use a constant arbitrary depth when building objects. We also represent the cylindrical robot as a rectangular object, because it is easier to process using vector data, since there are only 4 vertices in a rectangle. The bounding rectangle we build around the robot is also shown in Figure 4.

In the case of multiple sonar returns, we examine the sonar readings that are adjacent to each other. There is a question on whether adjacent sonar readings are from a single obstacle or multiple obstacles. Our solution to this issue is to determine if the robot can fit between the points of two adjacent sonar returns. If the robot cannot fit between two returns, then we consider these returns to be from the same object. Even if there are actually two objects, they may be considered as one for robot navigation purposes. In the case that the distance between the two points of the sonar returns is big enough to allow the robot to travel through, we consider separate objects. To form objects from multiple sonar returns we join the centers of the corresponding sonar cones.

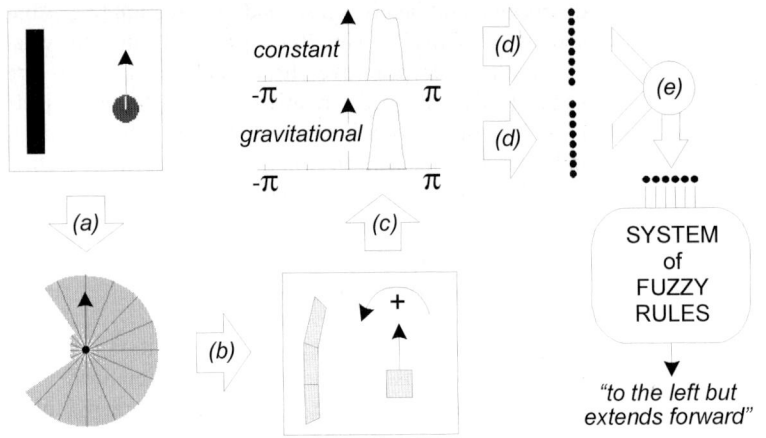
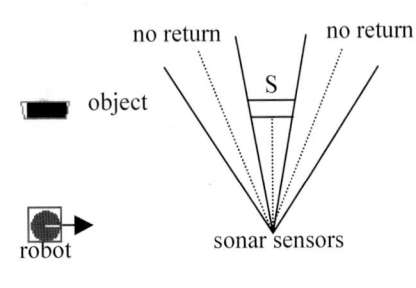

Figure 3. Synoptic diagram. (a) Sonar readings. (b) Construction of the polygonal objects. (c) Computation of the histograms of forces. (d) Extraction of numeric features. (e) Fusion of information.

Figure 4. A single object is formed from a single sonar reading.

The distance we compute to determine if two adjacent sonar returns are "close" or not can be expressed by the following formula (distance between two points in polar coordinates):

$$\sqrt{s_1^2 + s_2^2 - 2s_1 s_2 \cos(2\pi/c)}$$

where: s_1 is the return of sonar sensor S_1,
s_2 is the return of sonar S_2, adjacent to S_1,
c is a constant that determines the angle between the two sonar sensors S_1 and S_2.

For $c = 16$, the angle between the two sonar sensors is set to the real angle between them ($2\pi/16$), and the formula returns the exact distance between the points of the two sonar returns. However, for our application we used $c = 24$, for which the distance computed between the points of the adjacent sonar readings is shorter than the actual one.

This way, when the robot diameter is compared to the distance between two obstacles, the distance will be big enough for the robot to easily travel between the obstacles. Thus, we allow extra clearance to make sure that the robot can easily fit between two obstacles.

For example, consider the obstacle in Figure 5. Since the obstacle is relatively far from the robot, the distance between the sonar returns is rather big, and we cannot determine whether the obstacle continues between the three sonar readings, or we have three different obstacles. In this case, we plot three different objects until the robot gets closer to the obstacle and we have a better resolution of the obstacle, since more sensors would detect its presence. In the same figure we show the distance computed for $c = 16$, which is the distance between A and B, and for $c = 24$, which is the distance between C and D.

In Figure 6, we show the same obstacle at a closer distance to the robot. There are five adjacent sonar sensors that have returns from the obstacle in this case. The distance measure determines that all sonar returns are close together, for the object to be considered as one.

After building the objects around the robot based on the sonar sensor readings, we represent the relative position between each object and the robot by the histograms of constant and gravitational forces associated with the robot/object pair, as described in Section 2. We then generate an egocentric linguistic description, *i.e.*, from the robot's point of view. Thus, the descriptions also depend on the robot's orientation or heading. A change in robot heading is easily accomplished by shifting the histogram along its horizontal axis. In the next section we show some test cases that illustrate the function of the approach.

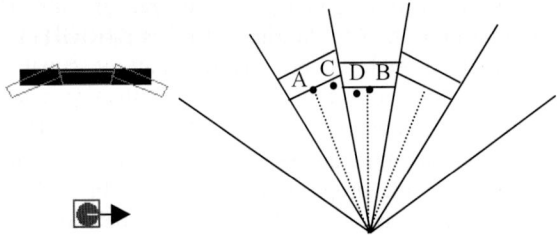

Figure 5. Three different objects are formed from 3 different sonar readings, if the readings are not "close" enough, according to the distance measure.

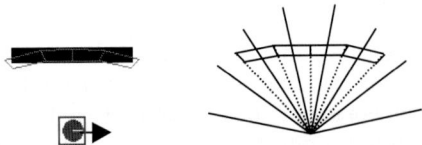

Figure 6. A single object is formed from 5 different sonar readings, if the readings are "close" enough.

4 Experiments and Discussion

The experiments included in this section were generated using the Nomad simulator. The program ran on the simulator at real-time speed. Processing of all obstacles, plotting of objects, processing of histograms and linguistic description generation is done faster than the robot can move, so there are no "delayed" results.

A simple case that demonstrates the functionality is shown in Figure 7. The sonar sensor readings are displayed on the right, the robot is shown as a circular model and an obstacle is drawn as a solid rectangle. For illustration, the software plots a hollow trapezoid based on the sonar readings, which should roughly coincide with the real obstacle, and it also plots the bounding rectangle that represents the robot. The software outputs the linguistic description, after executing the spatial analysis algorithm for all generated objects with respect to the robot. As described in Section 2, the linguistic expressions are generated in a three-part form: (1) "Object 1 is mostly to the left of the robot" (the primary direction), (2) "but somewhat forward" (the secondary direction), and (3) "the description is satisfactory" (the assessment indicating an adequate description).

"*Object 1* is mostly to the left of the robot but somewhat forward (the description is satisfactory)"

Figure 7. The robot detects one obstacle. The sonar sensor readings are shown on the right. The generated linguistic expression is shown in italics.

In Figure 8, we show a more complex case. Object 1 from Figure 7 remains at the same position. A new obstacle is introduced behind the robot, which is recognized as a single object (Object 2). The obstacle to the right of the robot however, is plotted as three different objects. Since there are only three sonar readings from the right obstacle, and they are far apart according to the distance measure, the readings may not be from a single obstacle. If more detail is needed, the robot may approach these three plotted objects to the right, to get a better resolution from more sonar sensors. This action may indeed reveal a passage through two of the three plotted objects or, if all sensors get returns that are close according to the distance measure, the three objects will prove to be the same one. Figure 8 shows the linguistic description generated for each object detected; in all cases, the assessment shows an adequate description.

Figure 9 shows the detection of two objects. The two obstacles to the left of the robot are so close together, that the robot cannot travel through them. Therefore, for navigation purposes these two obstacles are considered to be one object. Figure 9 shows the description generated, including a satisfactory assessment.

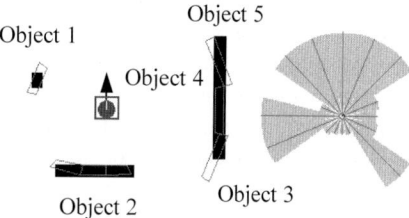

"*Object 1* is mostly to the left of the Robot but somewhat forward (the description is satisfactory)"

"*Object 2* is behind the Robot but extends to the left relative to the Robot (the description is satisfactory)"

"*Object 3* is mostly to the right of the Robot but somewhat to the rear (the description is satisfactory)"

"*Object 4* is to the right of the Robot (the description is satisfactory)"

"*Object 5* is mostly to the right of the Robot but somewhat forward (the description is satisfactory)"

Figure 8. The robot detects 5 obstacles.

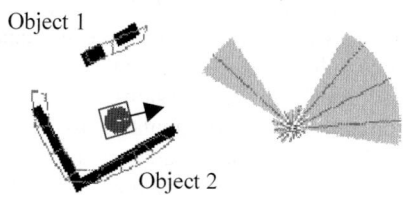

"*Object 1* is to the left of the Robot (the description is satisfactory)"

"*Object 2* is loosely to the right of the Robot and extends to the rear relative to the Robot (the description is rather satisfactory)"

Figure 9. The robot detects 2 obstacles.

The L-shaped object behind and to the right of the robot is an example of a rather satisfactory (*i.e.*, less satisfactory) linguistic description. The algorithm determines that for such a relative position there is not a really good description in terms of the four primitive directions only. It introduces the term "loosely" together with the classification of the whole description as "rather satisfactory" as opposed to "satisfactory" in all previous examples. This assessment indicates that we may need additional spatial relations (like "surrounds").

In the future, we plan to use more spatial relations for descriptions to include situations such as the one of Figure 9. A higher level of processing may generate such descriptions after considering the outputs of our current algorithm. For example, if there is an object to the right and an object to the left of the robot, then the robot is between the two objects.

We are also planning to introduce descriptions that indicate distance, in addition to relative position, such as *close* or *far*. These descriptions may be generated after processing the distance information that the sonar sensors return. Information from the robot's camera may also be combined with the sonar data to achieve more complete linguistic descriptions of the robot's environment (*e.g.*, recognize and label objects).

Temporal data may also be used for realization of corridors, rooms, etc. For example, if we have many consecutive linguistic descriptions of being between objects, then the robot could be traveling in a corridor. If we have consecutive descriptions of being surrounded, this could mean that the robot is in a room of a certain size.

5 Concluding Remarks

In this paper, we have shown how the histogram of forces can be used to generate linguistic spatial descriptions representing the qualitative state of a mobile robot in an unknown environment. Using the robot's sonar readings, a boundary approximation of the obstacles is made, and their vertices are used as input to the histogram of forces. The usage described in this paper represents the first application of F_0 and F_2 histograms that uses vector data instead of raster data.

Several examples have been presented which illustrate the linguistic expressions automatically generated. The approach is computationally efficient, and the spatial descriptions can be generated in real time. Note that although we have assumed an unknown environment and therefore must build an approximation of the environment from the sonar readings, the approach could also be used to generate linguistic descriptions for a robot in a known environment using a map. In either case, the linguistic expressions can be used to facilitate natural communication between a robot and a human.

Acknowledgements

The authors wish to acknowledge support from ONR, grant N00014-96-0439 and the IEEE Neural Network Council for a graduate student summer fellowship for Mr. Chronis.

References

[1] K. Morik, M. Kaiser and V. Klingspor, ed., *Making Robots Smarter*, Kluwer Academic Publishers, Boston, 1999.

[2] M. Skubic and R.A. Volz, "Acquiring Robust, Force-Based Assembly Skills from Human Demonstration", *IEEE Trans. on Robotics and Automation*, vol.16, no.6, Dec.,2000, pp.772-781.

[3] W. Gribble, R. Browning, M. Hewett, E. Remolina and B. Kuipers, "Integrating vision and spatial reasoning for assistive navigation", In *Assistive Technology and Artificial Intelligence*, V. Mittal, H. Yanco, J. Aronis and R. Simpson, ed., Springer Verlag, , 1998, pp. 179-193, Berlin, Germany.

[4] D. Perzanowski, A. Schultz, W. Adams and E. Marsh, "Goal Tracking in a Natural Language Interface: Towards Achieving Adjustable Autonomy", In *Proceedings of the 1999 IEEE Intl. Symp. on Computational Intelligence in Robotics and Automation*, Monterey, CA, Nov., 1999, pp.208-213.

[5] F. Shibata, M. Ashida, K. Kakusho, N. Babaguchi, and T. Kitahashi, "Mobile Robot Navigation by User-Friendly Goal Specification", In *Proceedings of the 5th IEEE Intl. Workshop on Robot and Human Communication*, Tsukuba, Japan, Nov., 1996, pp. 439-444.

[6] E. Stopp, K.-P. Gapp, G. Herzog, T. Laengle and T. Lueth, "Utilizing Spatial Relations for Natural Language Access to an Autonomous Mobile Robot", In *Proceedings of the 18th German Annual Conference on Artificial Intelligence*, Berlin, Germany, 1994, pp. 39-50.

[7] M. Skubic, P. Matsakis, B. Forrester and G. Chronis, "Extracting Navigation States from a Hand-Drawn Map", in *Proceedings of the 2001 IEEE Intl. Conf. on Robotics and Automation*.

[8] J. Freeman, "The Modelling of Spatial Relations", *Computer Graphics and Image Processing (4)*, pp. 156-171, 1975.

[9] I. Bloch, "Fuzzy Relative Position between Objects in Image Processing: New Definition and Properties Based on a Morphological Approach", *Int. J. of Uncertainty Fuzziness and Knowledge-Based Systems*, vol. 7, no. 2, pp. 99-133, 1999.

[10] K. Miyajima, A. Ralescu, "Spatial Organization in 2D Segmented Images: Representation and Recognition of Primitive Spatial Relations", *Fuzzy Sets and Systems*, vol. 65, no.2/3, pp. 225-236, 1994.

[11] P. Matsakis, *Relations spatiales structurelles et interprétation d'images*, Ph. D. Thesis, Institut de Recherche en Informatique de Toulouse, France, 1998.

[12] P. Matsakis and L. Wendling, "A New Way to Represent the Relative Position between Areal Objects", *IEEE Trans. on Pattern Analysis and Machine Intelligence*, vol. 21, no. 7, pp. 634-643, 1999.

[13] P. Matsakis, J. M. Keller, L. Wendling, J. Marjamaa and O. Sjahputera, "Linguistic Description of Relative Positions in Images", *IEEE Trans. on Systems, Man and Cybernetics*, to appear.

[14] J. M. Keller and P. Matsakis, "Aspects of High Level Computer Vision Using Fuzzy Sets", *Proceedings, 8th IEEE Int. Conf. on Fuzzy Systems*, Seoul, Korea, pp. 847-852, 1999.

Real-Time Robot Learning

Bir Bhanu, Pat Leang, Chris Cowden, Yingqiang Lin and Mark Patterson

Center for Research in Intelligent Systems
University of California, Riverside, CA 92521
{bhanu, pleang, ccowden, yqlin, mark}@vislab.ucr.edu

Abstract

This paper presents the design, implementation and testing of a real-time system using computer vision and machine learning techniques to demonstrate learning behavior in a miniature mobile robot. The miniature robot, through environmental sensing, learns to navigate a maze choosing the optimum route. Several reinforcement learning based algorithms, such as Q-learning, Q(λ)-learning, fast online Q(λ)-learning and DYNA structure, are considered. Experimental results based on simulation and an integrated real-time system are presented for varying density of obstacles in a 15×15 maze.

1. Introduction

Real-time robot learning has been a very challenging and active research area for several years [3]. Reinforcement learning has been applied to robot learning and machine intelligence tasks, and several different techniques have been developed [8, 9, 11]. In this paper, we consider a systematic comparison of some key reinforcement learning techniques, such as Q-learning, Q(λ)-learning, fast Q(λ)-learning, DYNA-based learning and an algorithm that combines heuristics with Q-learning. First, we compare these algorithms in simulation for navigation in a maze that may contain a number of obstacles. Second, we build a real maze and a miniature robot equipped with basic sensors and a vision camera that monitors the activity of the robot. We integrate a number of reinforcement learning techniques for real-time navigation and control of the miniature robot.

In Section 2, we present the related research using reinforcement learning. This is followed by a discussion of several reinforcement learning algorithms that we compare in Section 3. Here we also present the details of the robot, interface and the integrated learning system. Section 4 presents experimental results, both in simulation and in real time. Finally, Section 5 presents the conclusions of the paper.

2. Related Research

Asada et al. [1] present a method for vision based reinforcement learning where a robot learns to shoot a ball into a goal. The robots use a technique called "learning from easy missions," which reduces the learning time to about the linear order in the size of the state space. Yamagnchi et al. [13, 14] propagate learned behaviors of a virtual agent to a physical robot in order to accelerate learning in a physical environment. Reinforcement learning is used as the basis of the ball-pushing task. Suh et al. [10] extend Q-learning that incorporates a region-based reward to solve a structural credit assignment problem and a triangular type Q-value model. This may enable a robot to move smoothly in a real maze. Hailiu and Sommer [5] embed syntax rules and environment knowledge into the learner to achieve satisfactory performance with a reinforcement-learning algorithm. Huber [6] presents a hybrid architecture that applies reinforcement on top of an automatically derived, abstract, discrete event dynamic system supervisor. This reduces the complexity of the learning task and allows the incorporation of *a priori* knowledge. Bhanu and Peng [2] have also developed techniques to incorporate *a priori* knowledge into a reinforcement-learning framework for integrated image segmentation and object recognition.

As compared to the previous works, we present a systematic comparison of old and some new reinforcement-learning algorithms (for example, fast Q(λ), where complexity is O(| actions |)), and implement and evaluate them on a real miniature robot which has a bi-directional wireless link with the computer.

3. Technical Approach

Figure 1 shows various components of the integrated system for robot learning that we developed. It consists of a maze, an overhead camera, a robot, a vision module, a learning module and an interface module. In the following, first, we describe various reinforcement learning algorithms that we have considered. This is followed by the details of the real-time integrated system that we have developed.

3.1 Reinforcement-Learning Algorithms

The reinforcement algorithms we consider are Q-learning, Q(λ)-learning, fast Q(λ)-learning and some of their variants, such as incorporating DYNA structure and some environmental information into the learning algorithm.

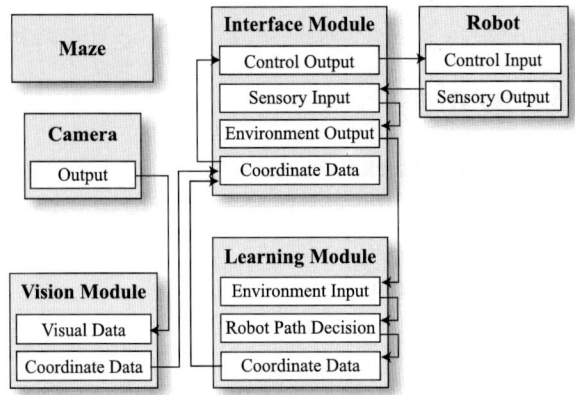

Figure 1. Various modules of the robot learning system.

- **Discrete-Time Finite Markov Random Process**

The discrete time finite Markov random process is a 4-tuple (S, A, P, R), where:

S is a finite set of states: $S = \{S_1, S_2, ..., S_n\}$
A is a finite set of actions taken at each state.
P is a probabilistic state-transition law. It is a function
$P_{ij}^a: S \times S \times A \to \Re$, where $\Re$ is the set of real nos.
$P_{ij}^a = P\{S_{t+1} = S_j | S_t = S_i, a_t = a\}$ for $S_i, S_j \in S$ and $a \in A$
P_{ij}^a is the probability to reach state S_j from state S_i when action a is taken at state i. S_t means at time t, the agent is at state S_t.
R: $S \times A \to \Re$ is the reward function mapping
$(S_i, a) \in S \times A$ to a scalar reinforcement $R(S_i, a)$.

The reward at time t is r_t, and a_t is the action taken at time t. A discount factor $\gamma \in [0, 1]$ discounts later against immediate rewards.

As an example, in the maze problem, the state is the cell in the maze. The actions that can be taken at each state are: traveling north, traveling west, traveling south and traveling east. The state-transition function is very simple and deterministic. For example, if the agent takes the action of going to the north, the next state is the cell above; if the cell is in the top row, or if the cell above is occupied by an obstacle, the agent stays in the original cell. If the agent achieves the goal, the reward is 100. Otherwise, the reward is 0.

- **ε-Greedy Policy**

There is a trade-off between exploration and focusing on the path that has already been found. In order to find a short path, the agent must explore the search space (e.g., maze) actively at the beginning. But as the exploration proceeds, we assume the agent can find a good (short) path, so the agent should gradually focus on the path that has been found.

We use ε-greedy policy to guide the exploration. With probability ε, the agent explores the search space, and with probability 1 - ε, it goes along the best path already found.

So the larger the value of ε, the more actively the agent explores the search space.

Initially, the value of ε is 0.9. Each time the agent finds a path, ε is decreased by 0.05. The minimum value of ε is greater than zero or a small positive number.

- **Q-learning**

Given (S_t, a_t, r_t, S_{t+1}), standard one-step Q-learning updates just a single Q-value $Q(S_t, a_t)$ as follows:

$Q(S_t, a_t) \leftarrow Q(S_t, a_t) + \alpha\, e_t'$ e_t' is given by:

$e_t' = (r + \gamma V(S_{t+1}) - Q(S_t, a_t)),\quad \gamma \in [0,1]$

where $V(S) = \max_a Q(S, a)$ and α is the learning rate.

The agent can take one of the actions at each cell. We associate each action, which can be taken at each cell, with a value. This associated value represents the "goodness" of the corresponding action. We refer this value as the Q-value.

Algorithm 1 (Q-learning):
1. Put the agent at the starting point.
2. While (not end) do
3.   Use ε-greedy policy to select an action a_t.
4.   Take the action to get (S_t, a_t, r_t, S_{t+1}).
5.   Apply Q-learning to update Q-value.
6.   if goal is reached then
7.   change the value of ε and put the agent back at the starting point.
    end
end

- **Q(λ)-learning**

Q-learning only considers the immediate reward. It propagates the reward backward only one step. Unlike Q-learning, Q(λ)-learning not only considers the immediate reward, it also takes the discounted future rewards into consideration. Q(λ)-learning updates the Q-values in the following way:

$Q(S_t, a_t) \leftarrow Q(S_t, a_t) + \alpha\, e_t^\lambda$

$e_t^\lambda = e_t' + \sum_{i=1}^{\infty} (\gamma \lambda)^i e_{t+i}$ where $\lambda \in [0,1]$

$e_{t+i} = (r_{t+i} + \gamma V(S_{t+i+1}) - V(S_{t+i}))$

- **Online Q(λ)**

The above updates in Q(λ)-learning cannot be made as long as future rewards are not known. However, we can compute them incrementally [7], by using eligibility traces. We define $\eta^t(S,a)$ as 1 if (S,a) occurred at time t, and 0 otherwise. Also,

define $l_t(S,a) = \sum_{i=1}^{t-1} (\gamma \lambda)^{t-i} \eta^i(S,a)$, we have

$Q(S,a) \leftarrow Q(S,a) + \alpha\, [e_t' \eta^t(S,a) + e_t\, l_t(S,a)]$

Based on the last expression, an online Q(λ)-learning algorithm is proposed in [9]. We can use Q(λ)-learning to replace the Q-learning in *Algorithm 1*.

- **Fast Online Q(λ)-learning**

Fast online Q(λ)-learning [12] is an improvement of Q(λ)-learning. At each step, it only updates a state-action pair *(S,a)* when it is needed, thus increasing the speed of learning. In Q(λ)-learning, all the *(S,a)* pairs in history list are updated at each step.

Note that the complexity of Q(λ) is O(|S||A|), but the complexity of fast Q(λ) is only O(|A|).

The algorithm is based on the observation that only Q-values needed at any given time are those for the possible actions given the current state. Hence, using "lazy learning", we can postpone updating Q-values until they are needed. Suppose some state-action pairs (SAP) *(S, a)* occur at steps $t_1, t_2, t_3, \ldots$. Let us abbreviate $\eta^t = \eta^t(S,a)$, where η^t is 1 for $t = t_1, t_2, t_3, \ldots$ and 0 otherwise, $\phi = \gamma\lambda$ and define:

$$\Delta_t = \sum_{i=1}^{t} e_i \varphi^i, \quad l'_t(S,a) = \sum_{i=1}^{t} \frac{\eta^t(S,a)}{\varphi^t}, \text{ we can have:}$$

$$\Delta Q(S,a) = \lim_{k \to \infty} \sum_{t=1}^{k} [e'_t \eta^t(S,a) + l'_t(S,a)(\Delta_{t+1} - \Delta_t)]$$

Based on the above expressions, we can build a fast online Q(λ)-learning algorithm. This algorithm relies on two procedures: the Local update procedure calculates exact Q-values once they are required; the Global update procedure updates the global variables and the current Q-value.

- **DYNA Structure**

With DYNA, we incorporate planning into learning process. It is proposed as a simple but principled way to achieve more efficient reinforcement learning in an autonomous agent.

DYNA algorithm:
1. *Initialize All Q(S,a) to 0 and the priority queue PQueue to empty.*
2. *While (not end)*
   (a) *x ← the current state.*
   (b) *Use ε-greedy to select an action a.*
   (c) *Carry out action a to get the experience (x, a, y, r), where x is the current state, a is the action taken by the agent, and y is the next state and r is the immediate reward.*
   (d) *Apply Q-learning to update Q-value.*
   (e) *Compute e = | r + γ V(y) – Q(x, a) |. If e ≥δ, insert (x, a, y, r) into PQueue with key e.*
   (f) *If PQueue is not empty, do planning:*
      i. *(x', a', y', r') ← **first_experience**(PQueue).*
      ii. *Update:*
           $Q(x',a') = Q(x',a') + \alpha (r' + \gamma V(y') - Q(x',a'))$
      iii. *For each predecessor x'' of x' do:*
           Compute e = | $r_{x''x'}$ + γ V(x') - Q(x'', $a_{x''x'}$) |.
           If e ≥δ, insert (x'', $a_{x''x'}$, x', $r_{x''x'}$) into PQueue with key e.

δ is a parameter of the algorithm. It is a threshold value.

After we insert a *(x, a, y, r)* into PQueue, we can extract a fixed number of *(x, a, y, r)* from PQueue (if it has) and use them to do planning.

- **Incorporating Environmental Information**

We try to allow the agent to remember and use some environmental information. In particular, during exploration in the maze, we make the following assumptions:

1) Whenever the agent runs into an obstacle, it receives a negative reward (-100) and uses it to update the Q-value associated with the action taken at the cell. So, the next time the agent is at this cell, it won't take this action, avoiding the obstacle.

2) After the agent goes from cell S_1 to S_2, it won't go back from S_2 to S_1 immediately.

3) The agent remembers the cells it has visited in a trial. It won't visit a cell twice in a particular trial. Here, a trial is for the agent to go from the starting point to the goal. During exploration, after the agent goes from S_1 to S_2 and finds itself stuck, i.e., the other 3 neighbors of S_2 are either boundaries or obstacles, it has two choices: to either give up this trial and return to the starting point or to return to S_1 and continue the current trial.

We can also incorporate the environmental information and DYNA structure together into the learning process.

3.2 Integrated Learning System
- **Maze**

The maze used in real experiments is an 8-foot square block made from 2 sheets of 4×8 inch square plywood and bordered by 2×4 boards. The surface is entirely black with ¼-inch white lines. The lines are made with white typographer's tape. These lines, placed horizontally and vertically, produce a grid whereby each cell is 6×6 square inches. There are a total of 225 cells. The only two colors that exist on the maze are black and white.

The obstacles are made from 6-inch cubed blocks of wood. The obstacles are all black in color. The sides have been lined with paper to produce a smooth surface. This smooth finish allows for better reflection of IR light. Experiments have shown that the rough texture of the wood creates uncertainties in object detection because of the many different angles by which IR light can be reflected from the obstacle's rough wooden surface. Uncertainties in the amount of reflected IR signal causes errors in its detection, and thus, errors in recognizing the presence of the obstacle.

- **Robot Design**

The robot is a miniature robot that stands 6 inches tall with a diameter of 4.5 inches. The robot is controlled by the Handy Board micro-controller. It is based on the Motorola 68HC11 microprocessor. Features of the controller include a 32K static RAM, four DC motor output ports, a variety of sensory inputs, and a 16×2 LCD display. The software used is Interactive C.

Figure 2. The miniature robot navigating the real maze in the presence of obstacles.

The robot is equipped with various sensors [4] to aid in communication with its environment and obstacle detection. It has four modulated IR proximity sensors mounted on its front, back, and on each of its sides for detecting objects within a 3-inch distance. A line sensor is mounted in the bottom front portion of the robot. It consists of three pairs of IR emitters and collectors. Detection of the ¼-inch white line depends on the amount of IR light that is reflected from the emitters to the collectors.

For added insurance, a touch sensor has been added in the front of the robot. Errors are inevitable, but the detection of these errors is invaluable. The touch sensor acts as a last resort by physically sensing the presence of an obstacle upon contact. In the event that the IR proximity sensor fails to detect the object, the touch sensor will eventually detect it, and thus, prevent the robot from constantly trying to run into the obstacles.

Two Tower Hobbies TS-53 servomotors propel the robot. These motors have been modified for continuous rotation, meaning that the position-sensing device has been removed as well the physical gear stopper, allowing the motor to turn continuously.

The communication between the robot and the host computer is done via an RF module. The RF module is made by Parallax, Inc. They are transmitter and receiver pairs, model TXAM315A and RXAM315A, respectively. Powered by 5V standard TTL output, these modules operate at 315 MHz. Conversion of the TTL signal to RS232 is done for communication with the computer.

The use of battery power to the robot has been completely eliminated due to problems with consistency. The robot was subjected to running for long hours at a fairly predictable level of performance. Batteries could not provide for this prolonged usage or consistency due to the power drainage over time. The solution was to connect it to a power cord so that any standard wall socket can serve as the power source. This would ensure continuous usage and power level consistency.

- **Vision Algorithms**

The vision system was implemented to ensure that overall activity would be error free. It ensures that the learning algorithm has correct information about the position of the robot. Before any command is sent to the robot, the vision system is called upon to determine the current position of the robot. If this position does not match the position that the learning algorithm thinks the robot is in, then a correction algorithm is invoked, in which case, the robot is automatically commanded to the correct position intended by the learning algorithm. Not until both vision and learning algorithm positions match will the program continue.

The vision hardware is the ITEX imaging system. It is a Modular Vision Computer (MVC) 150/40, and a CCD camera combined with a Sun Ultra 1 computer. The camera is mounted 10 feet above the center of the maze, and captures images at a rate of 30 frames per second. The image is captured in 256 gray scales. The image of the maze consists of 512×480 pixels with a resolution of 72 pixels per inch.

Software used to control the captured images was written in the C programming language. To detect the robot, a circular white disc is placed atop the robot. A white surface indicates the presence of a robot and a black surface indicates the maze floor. The software scans the image and looks at every sixth pixel. If the pixel value is within a certain prescribed threshold value for white, indicating the presence of a robot, then 4 surrounding pixels are also sampled. If all these 5 pixels are within the threshold value, then the robot has been located, and its corresponding position is relayed back to the learning algorithm for processing. The average runtime to sample an image is 0.10 seconds.

- **Interface**

The interface program was written in the C programming language. The program is run using the Sun Microsystems computer under a Solaris operating system. This is the main module that controls all other peripherals, which includes the robot, the RF communication between the robot and computer, the vision system, the learning algorithm, and the user interface. All information coming from these peripherals is routed and controlled by the interface program.

The program begins with the initialization of the peripherals. For the robot, it makes sure that the robot is ready to receive commands and that it is in the correct cell position for starting. In the event that the robot is not initialized, the interface scheme will allow for a user interface mode in which the user can command the robot, test its sensors, or test its response before the main learning program should begin. The vision system is also initialized, meaning that it sets up the hardware, initializes the display and calibrates the maze size. The RF communication is also set up to receive signals through its RS232 serial port at a rate of 300 baud.

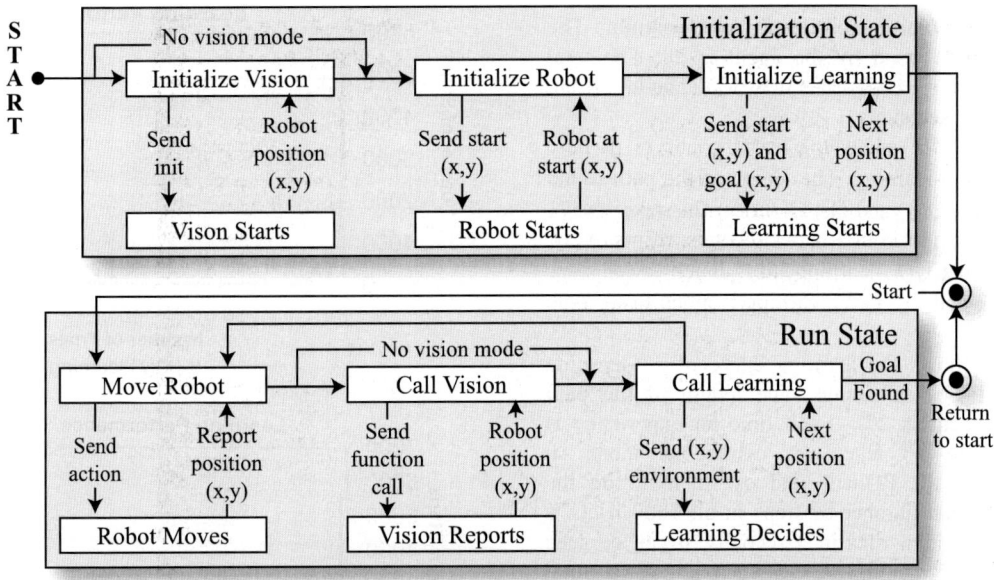

Figure 3. Details of initialization procedure and the run loop.

The sequence of events to control the robot is as follows. First of all, it queries the robot for environmental information. This means that it gathers information from the robot about surrounding obstacles in its 4 cardinal directions relative to the robot. Vision information is next queried for the robot's position. The intended position, the position assumed by the interface program, of the robot is compared against the actual position gathered from the vision. Discrepancies between the intended position and the actual position are handled by a position-correction algorithm, which relocates the robot to the correct position. Once both positions agree, the learning algorithm is called to determine the next position. The interface next commands the robot to this new position and this entire sequence of events is repeated until the goal is found. From goal position, the robot is directed to its starting position using a gradient-descent algorithm. From there, the next trial will begin. When the maximum number of trials has been reached, the shortest path is calculated based upon the type of learning algorithm used in the interface program. The entire event is done in real-time. Figure 3 shows the details of initialization procedure and the run loop.

4. Experimental Results
4.1 Simulation of Reinforcement-Learning Algorithms

A maze is a two-dimensional array with a cell as the starting point and some cells as a goal or goals. There may also be some cells occupied by obstacles. The agent can move from a cell to any one of its 4 direct neighbors with the restriction that the agent cannot move out of the maze, or into a cell occupied by an obstacle. The task of the agent is to find a good path from the starting point to the goals efficiently. A good path should be a short path, if it is not the shortest one.

Figure 4 is an example maze. The size of the maze is 18×12. The light gray cell on the left side is the starting point; the four cells on the upper right side are the goal; the dark gray cells are obstacles and the rest are empty cells. One of the paths from the starting point to the goals is also shown and it consists of a sequence of mid gray cells. The path shown here is one of the shortest paths with length 27 from the starting point to the goals.

The initial value of ε is 0.9. Each time the agent finds a path, ε is decreased by 0.05. The minimum value of ε is 0.1. The number of trials is 1000, i.e., the algorithm stops after the agent goes from the starting point to the goals 1000 times. The values of the parameters are: α=0.5, γ=0.95, λ=0.9. Threshold value for eligibility trace is 0.0001 and δ=10. Figure 5 shows the learning curves. The horizontal axis represents the number of successful trials and the vertical axis represents the number of steps. Here, we only show the first 200 trials.

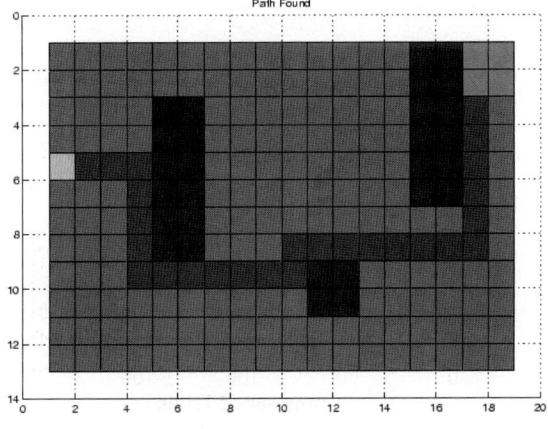

Figure 4. Simulated maze

Figure 5(a) is the learning curve of pure Q-learning. The length of the path found by the agent is 28. The total number of steps in all 1000 trials is 80156. The algorithm runs in less than 1 second.

Figure 5(b) and 5(c) show the learning curves of pure Q(λ) and fast Q(λ)-learning. The length of the path found by the agent is 30. Q(λ) and fast Q(λ)-learning take 48571 and 50684 steps, and run in 3 and 1 seconds, respectively. It can be seen that Q(λ)-learning takes much fewer steps than Q-learning and fast Q(λ)-learning is much faster than Q(λ)-learning.

Figure 5(d) shows the learning curve of Q-learning incorporating DYNA planning. The length of the path found by the agent is 27. Each time an experience is inserted into the PQueue, we extract at most 10 experiences from the PQueue and use them to do the planning. The total number of steps in all 1000 trials is 34145. The algorithm runs in 3 seconds. It can be seen after using DYNA planning, the agent takes fewer total number of steps than it takes in Q and Q(λ)-learning

Figure 5(e) shows the learning curves of incorporating some environmental information into Q-learning. In this experiment, the agent doesn't visit a cell twice in a trial, and if the agent gets stuck during exploration, it doesn't give up the current trial. The length of the path found by the agent is 28. The total number of steps is 45368. The algorithm runs in less than 1 second. It can be seen that after using some environmental information, the agent takes fewer total number of steps than it takes in pure Q-learning.

From the above curves, we can see that initially it takes the agent a large number of steps, because at that time, the agent has no knowledge about the maze; it can only explore the maze blindly. As learning proceeds and the agent gains more and more knowledge, the number of steps in a trial drops dramatically.

Although reinforcement learning can solve the search problem, it is expensive. It takes the agent many steps to find a good path, especially in the initial trials. It can be seen that Q-learning incorporated with DYNA or environmental information reduce about 50 percent of the number of steps taken by the agent. The combination of Q-learning and DYNA gave the best results.

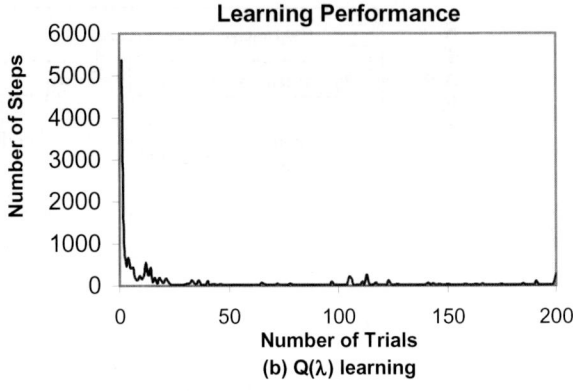
(b) Q(λ) learning

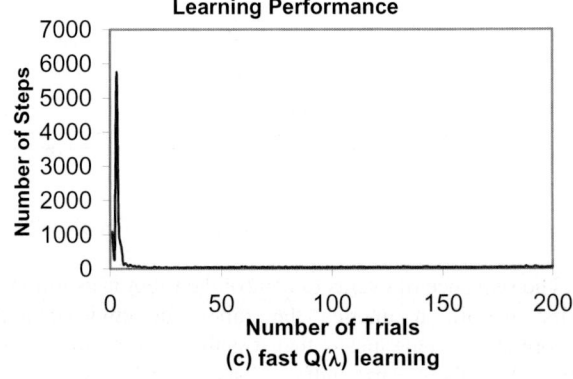
(c) fast Q(λ) learning

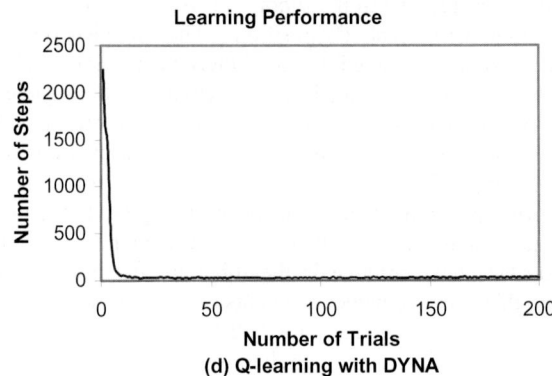
(d) Q-learning with DYNA

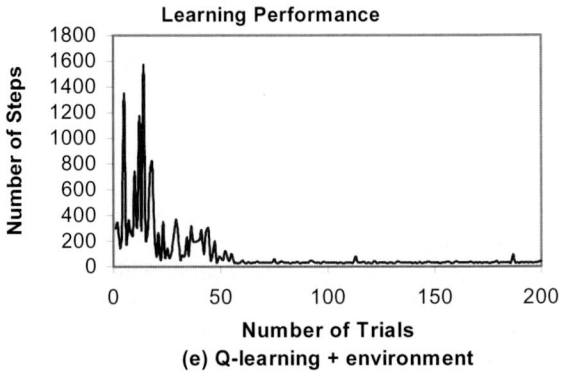
(e) Q-learning + environment

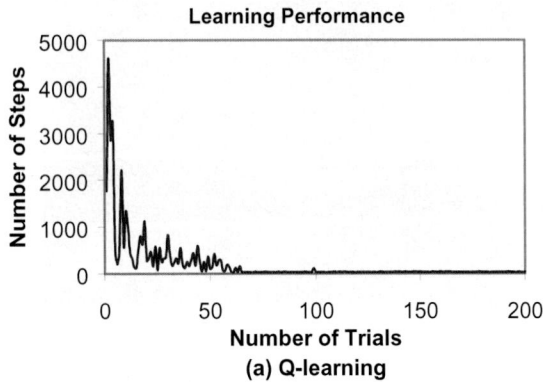
(a) Q-learning

Figure 5. Learning performance for various algorithms.

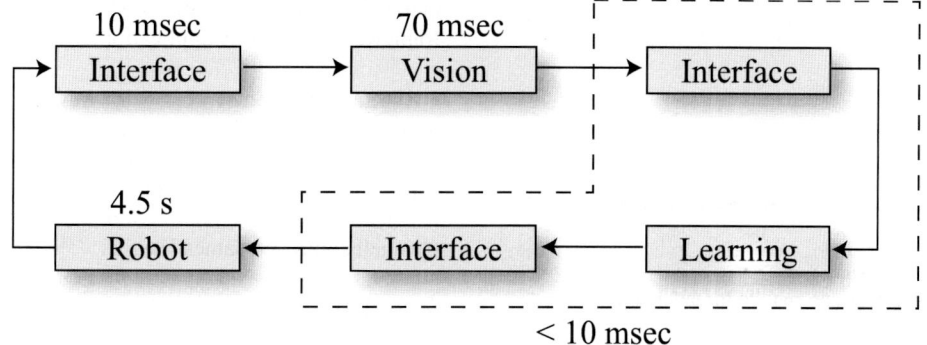

Figure 6. System-timing diagram.

4.2 Integrated Real-Time System

Figure 6 shows the time needed to complete one iteration of the learning process.

The results of the 15×15 maze with 28 obstacles are shown in Figure 7. The figure shows the decrement of actions taken as the number of trials increases. Before the first 8 trials, however, the number of actions taken oscillates tremendously, but as the number of trials increases, the graph begins to converge upon a set value. This value is the number of actions taken to produce the shortest path from start to goal.

In Figure 8, this same 15×15 maze result is graphed along with the results from a 6×6 maze with 12 obstacles. From the 6×6 maze, the results indicate the same type of oscillation found in the beginning trials but converges in later trials. 25 total trials were done for the 6×6 maze. After about 12 trials, the robot shows little oscillation in the number of actions taken and the values are at their minimum, meaning that the robot travels the shortest path nearly every time from start to goal.

In Figure 9, results from two 6×6 maze events were graphed against each other. The first event contained 6 obstacles while the second event contained 12 obstacles. Because the size of the maze was the same, similar results can be seen. In both events, the number of actions taken tends to fluctuate within roughly the same range. However, both experiments show that the 12-obstacle event tends to converge more quickly than the 6-obstacle event. With more obstacles on the maze, there are fewer cells for the robot to travel, and hence, it takes the robot a shorter time to go from the starting point to the goal. With fewer obstacles, there are more cells to explore, and so it requires the robot to take more actions.

Results for the Q and Q(λ) algorithms for a 6×6 maze are plotted against each other in Figure 10. From this graph, there appears to be few significant differences between the two learning algorithms. Both begin to converge upon the shortest path at approximately the same number of trials. Overall, the Q(λ) algorithm takes fewer actions during the entire experiment, which suggests that it is faster in finding the shortest path, but for a better comparison, more trials on a larger size maze should be done.

Q(λ)-learning was tested on a larger size maze of 15×15 with 28 obstacles. This experiment was conducted under the same parameters as the Q-Learning experiment shown in Figure 7, but with the Q(λ) algorithm. The results of this test are shown in Figure 11. The results show that in the Q(λ) algorithm, the number of steps in the later trials is lower than the Q-Learning algorithm. After 4 trials, the number of steps decreases below that of the Q-Learning algorithm. The number of steps shows a significant decrease after 10 trials.

Figure 12 shows two results of the Q(λ) experiments, one with 6 obstacles and the other with 12 obstacles. Similar to the results in Figure 9, these graphs show that the 12-obstacle experiment converges more quickly upon the shortest path than the 6-obstacle experiment, ascertaining the fact that with more obstacles, there are fewer cells to explore and hence, faster convergence upon the shortest path.

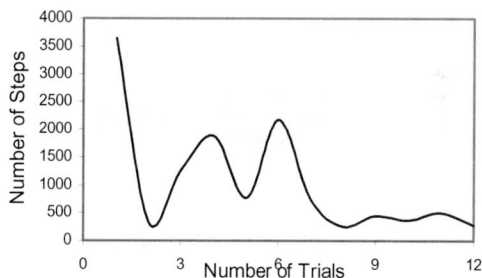

Figure 7. Q-learning results for 15x15 maze, 28 obstacles.

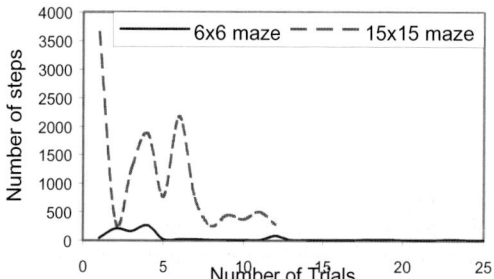

Figure 8. Q-learning results for different maze sizes.

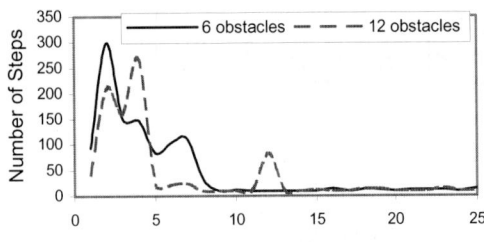

Figure 9. Q-learning results for 6x6 maze.

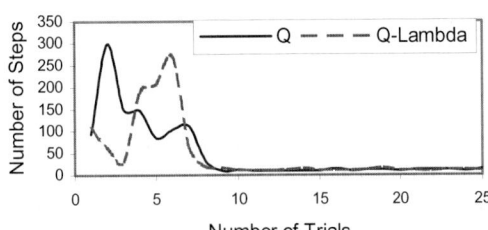

Figure 10. Results for 6x6 maze, 6 obstacles.

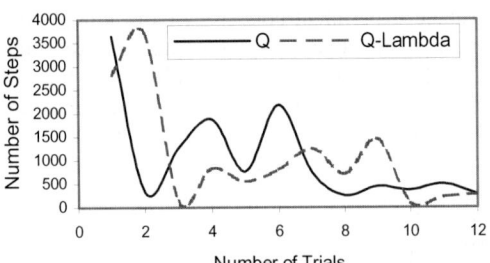

Figure 11. Results for 15x15 maze, 28 obstacles.

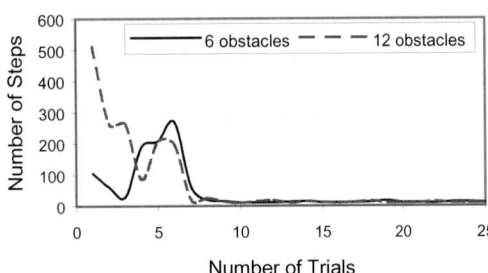

Figure 12. Q-lambda results for 6x6 maze.

5. Conclusions

In this paper, we have presented a systematic comparison of several reinforcement learning algorithms. We find that the fast Q(λ) and Q-learning with DYNA structure are the best. In the future, we plan to incorporate *a priori* task knowledge into these techniques and have hierarchical abstraction of the state space in a multi-agent framework for robotic applications.

• **Acknowledgments:** This work was supported by a NSF grant EIA-9610082-003. The authors would like to acknowledge the support they received from Roberto Carrillo, Vincent Hernandez, Tom Huyuh, Jing Peng, Michael Boshra, Sohail Nadimi, Stephanie Fonder, Grinnell Jones and James Harris.

6. References

[1] M. Asada et al., "Vision-based reinforcement learning for purposive behavior acquisition," *Proc IEEE Int. Conf. Robotics & Automation*, pp. 146-153, May 1995.

[2] B. Bhanu and Jing Peng, "Adaptive integrated image segmentation and object recognition," *IEEE Trans. on Systems, Man and Cybernetics*, Vol. 30, No. 4, pp. 427-441, November 2000.

[3] J. H. Connell and S. Mahadevan, *Robot Learning*, Kluwer Academic Publishers, 1993.

[4] H. R. Everett, *Sensors for Mobile Robots – Theory and Application*. A. K. Peters, Ltd, MA, 1995.

[5] G. Hailu, G. Sommer, "Embedding knowledge in reinforcement learning," *Proc. 8th Int. Conf. On Artificial Neural Networks*, pp. 1133-1138, September 1998.

[6] M. Huber, "A hybrid architecture for hierarchical reinforcement learning," *Proc. IEEE Int. Conf. On Robotics & Automation*, pp. 3290-3295, April 2000.

[7] J. Peng and R. J. Williams, "Incremental multi-step Q-learning," *Machine Learning*, vol. 22, No. 1-3, pp. 283 – 290, Kluwer Academic Publishers, January – March 1996.

[8] J. Peng and B. Bhanu, "Closed loop object recognition using reinforcement learning," *IEEE Trans. on Pattern Analysis and Machine Intelligence*, Vol. 20, No. 2, pp 139-154, February 1998.

[9] J. Peng and B. Bhanu, "Delayed reinforcement learning for adaptive image segmentation and feature extraction," *IEEE Trans. on System, Man and Cybernetics*, pp 482 – 488, August 1998.

[10] H. Suh, J. H. Kim and S. R. Oh, "Region based Q-learning for intelligent robot systems," *IEEE Int. Symp. on Computational Intelligence in Robotics & Automation*, pp. 172-178, July 1997.

[11] R. S. Sutton and A. G. Barto, *Reinforcement Learning*, MIT Press, 1998.

[12] M. Wiering and J. Schmidhuber, "Fast online Q(λ)," *Machine Learning*, Vol. 33, No. 1, pp. 105-115, October 1998.

[13] T. Yamagnchi et al., "Propagating learned behaviors from a virtual agent to a physical robot in reinforcement learning," *Proc. IEEE Int. Conf. On Evolutionary Computation*, pp. 855-859, May 1996.

[14] T. Yamagnchi et al., "Reinforcement learning for a real robot in a real environment," *European Conf. On Artificial Intelligence*, pp. 694-698, August 1996.

Learning task-relevant features from robot data

Nikos Vlassis   Roland Bunschoten   Ben Kröse

RWCP, Autonomous Learning Functions SNN
Computer Science Institute
Faculty of Science
University of Amsterdam
The Netherlands
{vlassis,bunschot,krose}@science.uva.nl
http://www.science.uva.nl/research/ias

Abstract

Feature extraction from robot sensor data is a standard way to deal with the high dimensionality and redundancy of such data. An automatic, commonly used way to learn such features from a set of robot observations is Principal Component Analysis (PCA). However, as we argued in previous work, PCA can yield features with little discriminatory power between robot positions, leading to suboptimal localization performance of the robot. In order to get optimal task-relevant features, PCA must be replaced by a supervised projection method.

In this paper we extend our previously proposed supervised linear feature extraction method in two ways: (i) the projection matrix is optimized simultaneously over all columns under the constraint of orthonormality, (ii) a Jacobi parametrization of the matrix allows the use of unconstrained nonlinear optimization algorithms. The new algorithm is more efficient and many times faster than the old version. We show experimental results in extracting features from panoramic images of a mobile robot. The results compare favorably to the PCA solutions.

1 Introduction

In several mobile robot applications where a model of the environment must be built and used for navigation, appropriate *landmarks* or *features* must be extracted from the raw robot sensor measurements prior to modeling. The rationale is that normally the dimensionality of these data is very high, making any statistical inference in the original space unrealistic.

The features that are extracted from robot sensor data can be classified as *local* or *global*. The former usually refer to location-dependent distinctive characteristics of the environment like doors, hallways, etc., (natural landmarks), or landmarks realized through specialized devices like beacons (artificial landmarks) [1]. On the other hand, a global feature is normally location-independent and aims at providing good robot localization on the average.

Recently there has been a growing interest in automatic procedures that *learn* such features from a set of data (see, e.g., [13]). Automatic learning of features is a natural objective because on the one hand it obviates the need for man interference in the feature extraction process, while on the other hand makes the process (potentially) environment independent.

Learning features from a set of robot observations is most often carried out with statistical methods, and the easiest and most commonly used is Principal Component Analysis (PCA) [10]. This is a global feature extraction method which projects a set of robot observations linearly to a low-dimensional subspace, computed by solving a matrix eigenvalue problem. The nice thing about PCA is that it combines many optimality properties and is very simple to implement [10]. Recent reports on the use of PCA on mobile robots are [8, 2, 6, 11, 15, 5].

However, when the robot observations are collected in a 'supervised' manner, i.e., when they are annotated in the sample with the position of the robot where each observation was taken, then, as argued in [16], PCA can be suboptimal. The reason is that PCA is an unsupervised feature extraction method that uses only the observed sensor vectors to compute the projection directions, and thus the extracted features can have little discriminatory power between robot positions. If feature extraction is to be used for tasks like robot localization and navigation, then PCA should be substituted by a *supervised* projection method [16].

In the current paper we extend the results in [16] in two main ways. First, in the above work the projection directions were learned in a greedy fashion, namely, a projection to an optimal direction was computed, then a second optimal direction was sought which was orthogonal to the first, etc. This strategy can be suboptimal and it is not difficult to devise artificial data sets that show this suboptimal behavior. In this paper we optimize the projection matrix (see below) simultaneously for all dimensions while keeping its columns pairwise orthonormal.

Second, we adopt an optimization strategy which obviates the need for constrained nonlinear optimization by parametrizing the projection matrix as a product of Jacobi matrices satisfying the orthogonality constraint during optimization. These two improvements make the method more efficient and much faster than the original version.

In the following we first describe the proposed method and then show experimental results from its application in panoramic image data collected by a mobile robot in a typical indoor environment. The average localization performance—evaluated through an appropriate risk function—when using the proposed method vs. PCA, and the visualization of the projected data manifold in the reduced subspace permit a quantitative and qualitative verification of our theoretical claims.

2 Feature extraction and the localization risk

For clarity of exposition and visualization we will limit our analysis to a robot that follows a predefined one-dimensional trajectory in its workspace. The results extend directly to the general case. For each position (offset) s of the robot on the trajectory we assume that the sensors provide an observation vector $\mathbf{x} \in \mathbb{R}^d$. For our analysis we assume a supervised training set $\{s_i, \mathbf{x}_i\}$, $1 \leq i \leq n$, of observations $\mathbf{x}_i$ collected at respective trajectory positions s_i.

Linear feature extraction amounts to reducing the dimensionality of the data $\mathbf{x}_i$ by linearly projecting them to a subspace $\mathbb{R}^q$, $1 < q < d$, multiplying them

with a $d \times q$ matrix $\mathbf{W}$ with orthonormal columns

$$\mathbf{y}_i = \mathbf{W}^T \mathbf{x}_i, \quad 1 \leq i \leq n, \quad \mathbf{W}^T \mathbf{W} = \mathbf{I}_q \quad (1)$$

where $\mathbf{I}_q$ stands for the q-dimensional identity matrix. Moreover, we assume a probabilistic model that associates robot locations with sensor observations. For an observation $\mathbf{x}$ that is projected through (1) to a feature vector $\mathbf{y}$ we assume a model for $p(s|\mathbf{y})$, the conditional density of the robot position s given $\mathbf{y}$.

To assess the quality of an individual projection we must define an appropriate *risk* function that measures the average localization performance of the robot using the extracted features $\mathbf{y}_i$. For this purpose it was proposed in [13] the risk function

$$R_L = \frac{1}{n} \sum_{i=1}^{n} \int |s - s_i| p(s|\mathbf{y}_i) ds, \quad (2)$$

i.e., the average over the training set mean absolute distance to the true—conditioned on the feature vector $\mathbf{y}_i$—location s_i. This risk penalizes position estimates that appear on the average far from the true position of the robot. The above formula was approximated in [13] from the training set with complexity $O(n^3)$.

In [16] we proposed an alternative risk which is $O(n^2)$. This risk is based on the simple observation that, for a given observation $\mathbf{x}_i$ which is projected through (1) to $\mathbf{y}_i$, the density $p(s|\mathbf{y}_i)$ will always exhibit a *mode* on $s = s_i$. Thus, an approximate measure of divergence from this mode is the Kullback-Leibler distance between $p(s|\mathbf{y}_i)$ and a unimodal density sharply peaked at $s = s_i$, giving the approximate estimate $-\log p(s_i|\mathbf{y}_i)$ plus a constant. Averaging over all points $\mathbf{y}_i$ we have to minimize the risk

$$R_K = -\frac{1}{n} \sum_{i=1}^{n} \log p(s_i|\mathbf{y}_i) \quad (3)$$

which can be regarded as the average negative log-likelihood of the data given the model of $p(s_i|\mathbf{y}_i)$ and the projection matrix $\mathbf{W}$.

From (3) we see that a nonparametric estimate of $p(s|\mathbf{y})$ is needed. For an appropriate sequence of weights $\lambda_j(\mathbf{y}), 1 \leq j \leq n$, such an estimate is [12]

$$p(s|\mathbf{y}) = \sum_{j=1}^{n} \lambda_j(\mathbf{y}) \phi_{h_s}(s - s_j) \quad (4)$$

where

$$\phi_{h_s}(s) = \frac{1}{\sqrt{2\pi} h_s} \exp\left(-\frac{s^2}{2h_s^2}\right) \quad (5)$$

is the univariate Gaussian kernel with bandwidth h_s, defining a local *smoothing* region around s. A weight function $\lambda_j(\mathbf{y})$ which satisfies the conditions in [12] and makes the above estimate a smooth function of the projection matrix $\mathbf{W}$ is

$$\lambda_j(\mathbf{y}) = \frac{\phi_{h_y}(\mathbf{y} - \mathbf{y}_j)}{\sum_{k=1}^n \phi_{h_y}(\mathbf{y} - \mathbf{y}_k)} \quad (6)$$

where

$$\phi_{h_y}(\mathbf{y}) = \frac{1}{(2\pi)^{q/2} h_y^q} \exp\left(-\frac{\|\mathbf{y}\|^2}{2h_y^2}\right) \quad (7)$$

is the q-dimensional spherical Gaussian kernel with bandwidth h_y. The two kernel bandwidths h_y and h_s are the only free parameters of the model $p(s|\mathbf{y})$ and their values affect the resulting projections. Substituting $p(s|\mathbf{y})$ from above into (3) we get a risk with complexity $O(n^2)$.

3 Model selection and optimization

3.1 Kernel smoothing

Using a nonparametric estimate of a density using (4) and (5)–(7) requires a choice for the smoothing parameters y_s and h_y. Our approach was to assign constant values to these two bandwidths during optimization. For projections to 2-d we set $h_y = n^{-2/7}$ which can be kept fixed during optimization after sphering the data (see next). This value is within the optimal bounds $O(n^{-1/3})$ and $O(n^{-1/4})$ given in [4, Sec. 4] for the related problem of projection pursuit regression, while it was found to give good results in practice. For the s-bandwidth we chose the Gaussian MISE optimal value $h_s = (3n/4)^{-1/5}$ [17, Ch. 3.2].

3.2 Sphering

A sphering of the data $\mathbf{x}_i$, namely, a normalization to zero mean and identity covariance matrix, makes the kernel bandwidth h_y independent of the projection. Then h_y can be kept constant during optimization leading to considerable computational savings. Sphering means a rotation of the data to their PCA directions and then standardization of the individual variances to one. To avoid modeling noise in the data, it is typical to ignore directions with small eigenvalues, and a heuristic method to do this is by putting a threshold to the ratio of the cumulative variance (added eigenvalues) to the total variance.

The numerically most accurate way to sphere the data is by singular value decomposition [9]. Let $\mathbf{X}$ be the $n \times d$ matrix whose rows are the data $\mathbf{x}_i$ after they have been normalized to zero mean. For $n > d$, we compute the singular value decomposition $\mathbf{X} = \mathbf{U}\mathbf{L}\mathbf{V}^T$ of the matrix $\mathbf{X}$ and form the matrix $\mathbf{A} = \sqrt{n}\mathbf{V}\mathbf{L}^{-1}$. The points $\mathbf{X}\mathbf{A}$ are then sphered [10].

For $n \leq d$ the data $\mathbf{x}_i$ lie in general in a $(n-1)$-dimensional Euclidean subspace of $\mathbb{R}^d$. In this case it is more convenient to compute the principal directions through eigenanalysis of $\mathbf{K} = \mathbf{X}\mathbf{X}^T$, the inner products matrix of the zero mean data. We compute its singular value decomposition $\mathbf{K} = \mathbf{U}\mathbf{L}\mathbf{V}^T$ and remove the last column of $\mathbf{V}$ and last column and row of $\mathbf{L}$ (the last eigenvalue of $\mathbf{K}$ will always be zero). Then we form the matrix $\mathbf{A} = \sqrt{n}\mathbf{V}\mathbf{L}^{-1}$. The points $\mathbf{K}\mathbf{A}$ are $(n-1)$-dimensional and sphered [7].

Moreover, all projections of sphered data $\mathbf{x}_i$ in the form of (1) give also sphered data $\mathbf{y}_i$ because

$$E[\mathbf{y}\mathbf{y}^T] = \mathbf{W}^T E[\mathbf{x}\mathbf{x}^T]\mathbf{W} = \mathbf{I}_q \quad (8)$$

due to the constraint of orthonormal columns of $\mathbf{W}$. This frees us from having to reestimate (co)variances of the projected data in each step of the optimization algorithm. In the following we assume that the data $\mathbf{x}_i$ have already been sphered and the position data s_i have been normalized to zero mean and unit variance.

3.3 Optimization

The smooth form of the risk R_K as a function of $\mathbf{W}$ allows the minimization of the former with nonlinear optimization. For constrained optimization we must compute the gradient of R_K and the gradient of the constraint function $\mathbf{W}^T\mathbf{W} - \mathbf{I}_q$ with respect to $\mathbf{W}$, and then plug these estimates in a constrained nonlinear optimization routine to optimize with respect to R_K [3].

An alternative approach which avoids the use of constrained nonlinear optimization, in a similar problem using kernel smoothing for discriminant analysis, has been recently proposed in [14]. The idea is to parametrize the projection matrix $\mathbf{W}$ by a product of *Jacobi* rotation matrices [9] and then optimize with respect to the angle parameters involved in each matrix. For projections from $\mathbb{R}^d$ to $\mathbb{R}^q$ this parametrization takes the form

$$\mathbf{W} = \prod_{o=1}^{q} \prod_{u=q+1}^{d} \mathbf{G}_{ou} \quad (9)$$

where $\mathbf{G}_{ou}$ is a Jacobi rotation matrix which equals $\mathbf{I}_d$ except for the elements $g_{oo} = \cos\theta_{ou}$, $g_{ou} = \sin\theta_{ou}$,

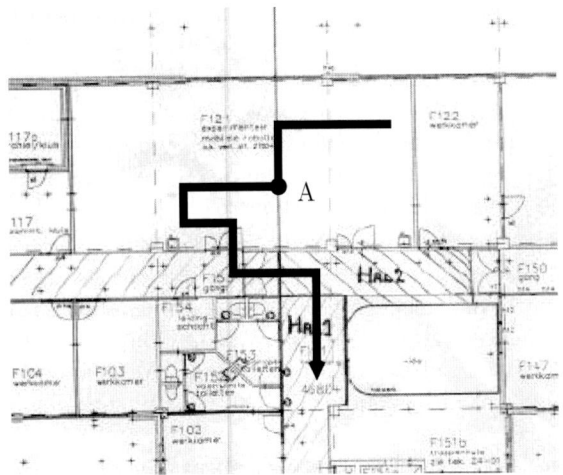

Figure 1: The robot trajectory.

Figure 2: Panoramic snapshot from position A.

$g_{uo} = -\sin\theta_{ou}$, and $g_{uu} = \cos\theta_{ou}$ for an angle θ_{ou} which depends on o and u. For simplicity we let in the above notation g_{oo}, g_{ou}, etc., denote the (o,o)-th, (o,u)-th, etc., elements of the matrix $\mathbf{G}_{ou}$, respectively. To ensure that $\mathbf{W}$ is $d \times q$, only the first q columns of the last matrix $\mathbf{G}_{qd}$ in (9) are retained, while multiplications must be carried from right to left to reduce the evaluation cost.

Multiplication with a matrix $\mathbf{G}_{ou}$ causes a rotation by θ_{ou} along the plane defined by the dimensions o and u, while the range of indices in (9) ensures that all rotations take place along planes defined by at least one non-projective direction, i.e., one among the $d - p$ remaining dimensions. This fact also reduces the total number of parameters from qd in the constrained optimization case (elements of matrix $\mathbf{W}$) to $q(d-q)$ here (angles θ_{ou}).

The derivative of the risk R_K with respect to an angle θ_{kl} is (we skip an analytical derivation here)

$$\frac{\partial}{\partial \theta_{kl}} R_K = \text{trace}\left\{ (\nabla_{\mathbf{W}} R_K)^T \left(\frac{\partial}{\partial \theta_{kl}} \mathbf{W} \right) \right\} \quad (10)$$

where the first term in the trace is

$$\nabla_{\mathbf{W}} R_K = \frac{1}{nh_y^2} \mathbf{X}^T [\mathbf{B} + \mathbf{B}^T - \text{diag}(\mathbf{1}^T \mathbf{B})] \mathbf{X} \mathbf{W} \quad (11)$$

where $\mathbf{X}$ is the $n \times d$ matrix of the sphered data, $\mathbf{1}$ is a column vector of all ones, $\text{diag}(\cdot)$ transforms a vector to a diagonal matrix, and $\mathbf{B}$ is the $n \times n$ matrix with elements

$$b_{ij} = \lambda_j(\mathbf{y}_i) - \frac{\phi_{h_y}(\mathbf{y}_i - \mathbf{y}_j)\phi_{h_s}(s_i - s_j)}{\sum_{k=1}^n \phi_{h_y}(\mathbf{y}_i - \mathbf{y}_k)\phi_{h_s}(s_i - s_k)}. \quad (12)$$

The second term of the trace is

$$\frac{\partial}{\partial \theta_{kl}} \mathbf{W} = \prod_{o=1}^{q} \prod_{u=q+1}^{d} \frac{\partial}{\partial \theta_{kl}} \mathbf{G}_{ou} \quad (13)$$

where

$$\frac{\partial}{\partial \theta_{kl}} \mathbf{G}_{ou} = \begin{cases} \mathbf{G}'_{ou} & \text{if } k = o \text{ and } l = u \\ \mathbf{G}_{ou} & \text{otherwise} \end{cases} \quad (14)$$

and $\mathbf{G}'_{ou}$ is the matrix $\mathbf{G}_{ou}$ with the ones substituted by zeros and the trigonometric functions substituted by their derivatives.

A point we should note is that the mixture density form of (4) and the additional trigonometric functions in (9) can make the landscape of the risk R_K have numerous local minima. For this reason, combining a gradient-free optimization method like, e.g., Nelder-Mead [9], with nonlinear optimization is requisite. Also an appropriate dimension reduction through sphering prior to optimization can significantly facilitate the search. In any case, the optimization algorithm must be applied many times and the solution with the minimum risk must be retained.

4 Experiments

We applied the above algorithm to data collected by a Nomad Scout robot following a predefined trajectory in our mobile robot lab and the adjoining hall as shown in Fig. 1. The omnidirectional imaging device which is mounted on top of the robot consists of a vertically mounted standard camera aimed upward looking into a spherical mirror. The data set contains 104 omnidirectional images (320×240 pixels) captured every 25 centimeters along the robot path. Each image is transformed to a panoramic image (64×256) and this set of 104 panoramic images constitutes the training set of our algorithm. A typical panoramic image shot at the position A of the trajectory is shown in Fig. 2.

In order to apply our supervised projection method, we first sphered the panoramic image data using the

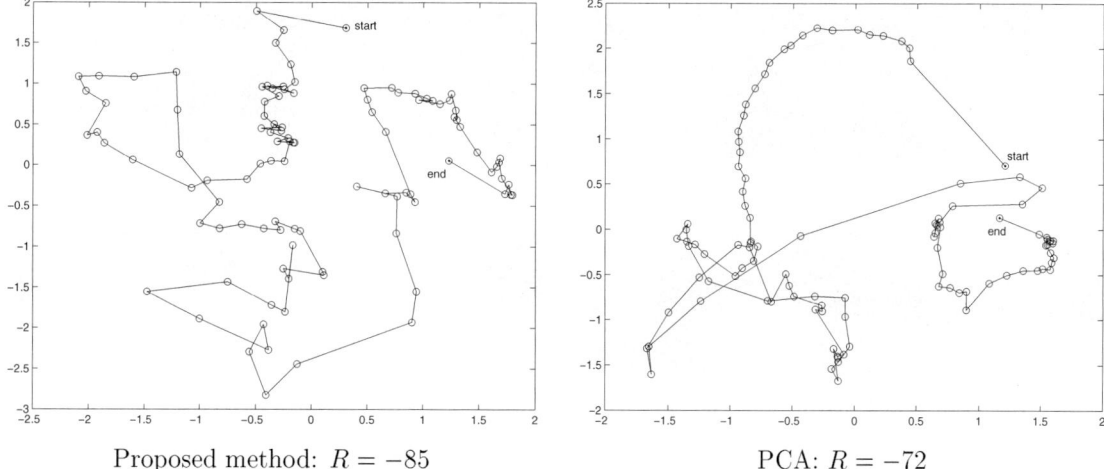

Figure 3: Projection of the sphered panoramic image data from 10-d to 2-d: using the proposed method (left), projection on the first two principal components (right). The 'start' and 'end' points are the projections of the panoramic images captured by the robot at the beginning and end, respectively, of its trajectory.

inner products matrix as explained above and kept the first 10 dimensions explaining about 60% of the total variance. Then we applied our method projecting the sphered data points from 10-d to 2-d. The resulting two-dimensional points are shown on the left part of Fig. 3. For optimization we ran several times a combined search using the Nelder-Mead algorithm with random initial values for the Jacobi angles in $[-\pi/2, \pi/2]$, together with nonlinear optimization with the BFGS algorithm [3, 9]. Running only BFGS required many more runs with random initial guesses to reach the global minimum, leading to comparable total expenses. Each execution of the optimization algorithm took a couple of seconds in a Sparc Ultra 5 machine.

On the right part of Fig. 3 we show the result of projecting the sphered 10-d points on the first two principal components of the data. We clearly see the advantage of the proposed method over PCA. The risk is smaller, while from the shape of the projected manifold we see that taking into account the pose information during projection can significantly improve the resulting features: there are fewer self-intersections of the projected manifold in our method than in PCA which, in turn, means better robot position estimation on the average.

Finally, in Fig. 4 we show the first two feature vectors (points in the original space of panoramic images) learned by our method and by PCA. In the PCA case these are the familiar first two eigenimages of the panoramic data which, as is normally observed in typical data sets, exhibit low spatial frequencies. We see that the proposed supervised projection method yields very different feature vectors than PCA, namely, images with higher spatial frequencies and distinct characteristics.

5 Conclusions

We proposed a method for learning task-relevant linear features from high-dimensional robot observations. Our method is supervised in the sense that the position of the robot in the sample is also taken into account during optimization. This makes the method superior to PCA which is unsupervised. We showed results of linear feature extraction from panoramic robot data when the robot was moving in a typical office environment. The results show clearly the superiority of the proposed method over PCA.

Our method can be useful in various robotic settings and is not limited to mobile robots. In particular, it can used in any case where global feature extraction from *supervised* robot observations is in order. The extension of the method to handle nonlinear features is possible (e.g., by using a neural network) but then additional issues have to be addressed (complexity of the network, overfitting, etc.). Besides, the wide use of PCA in robotic problems shows that linear feature extraction is still a viable approach in robotics.

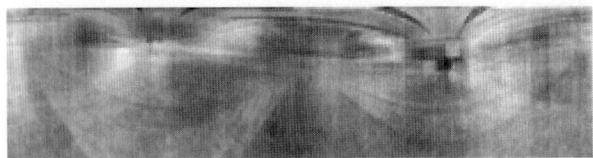
1st optimal feature vector

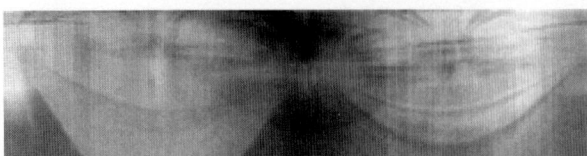
1st eigenvector

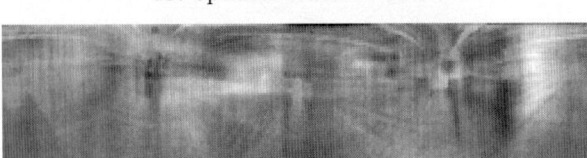
2nd optimal feature vector

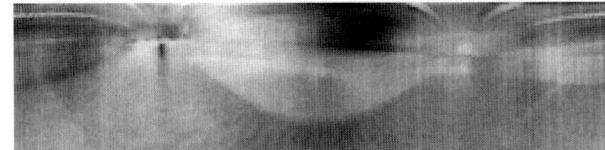
2nd eigenvector

Figure 4: The first two feature vectors using our method (left), and PCA (right).

References

[1] J. Borenstein, B. Everett, and L. Feng. *Navigating Mobile Robots: Systems and Techniques.* A. K. Peters, Ltd, Wellesley, MA, 1996.

[2] J. L. Crowley, F. Wallner, and B. Schiele. Position estimation using principal components of range data. In *Proc. IEEE Int. Conf. on Robotics and Automation*, Leuven, Belgium, May 1998.

[3] P. E. Gill, W. Murray, and M. Wright. *Practical Optimization.* Academic Press, London, 1981.

[4] P. Hall. On projection pursuit regression. *Ann. Statist.*, 17(2):573–588, 1989.

[5] M. Jogan and A. Leonardis. Robust localization using the eigenspace of spinning-images. In *Proc. IEEE Workshop on Omnidirectional Vision*, South Carolina, June 2000.

[6] B. Kröse and R. Bunschoten. Probabilistic localization by appearance models and active vision. In *Proc. ICRA'99, IEEE Int. Conf. on Robotics and Automation*, pages 2255–2260, Detroit, Michigan, May 1999.

[7] V. Kumar and H. Murakami. Efficient calculation of primary images from a set of images. *IEEE Trans. Pattern Analysis and Machine Intelligence*, 4(5):511–515, 1982.

[8] S. K. Nayar, H. Murase, and S. A. Nene. Learning, positioning, and tracking visual appearance. In *Proc. IEEE Int. Conf. on Robotics and Automation*, pages 3237–3244, San Diego, CA, 1994.

[9] W. H. Press, S. A. Teukolsky, B. P. Flannery, and W. T. Vetterling. *Numerical Recipes in C.* Cambridge University Press, 2nd edition, 1992.

[10] B. D. Ripley. *Pattern Recognition and Neural Networks.* Cambridge University Press, Cambridge, U.K., 1996.

[11] R. Sim and G. Dudek. Learning visual landmarks for pose estimation. In *Proc. IEEE Int. Conf. on Robotics and Automation*, Detroit, Michigan, May 1999.

[12] C. J. Stone. Consistent nonparametric regression (with discussion). *Ann. Statist.*, 5:595–645, 1977.

[13] S. Thrun. Bayesian landmark learning for mobile robot localization. *Machine Learning*, 33(1), 1998.

[14] K. Torkkola and W. Campbell. Mutual information in learning feature transformations. In *Proc. Int. Conf. on Machine Learning*, Stanford, CA, June 2000.

[15] N. Vlassis and B. Kröse. Robot environment modeling via principal component regression. In *Proc. IEEE/RSJ Int. Conf. on Intelligent Robots and Systems*, pages 677–682, Kyŏngju, Korea, Oct. 1999.

[16] N. Vlassis, Y. Motomura, and B. Kröse. Supervised linear feature extraction for mobile robot localization. In *Proc. IEEE Int. Conf. on Robotics and Automation*, pages 2979–2984, San Fransisco, CA, Apr. 2000.

[17] M. P. Wand and M. C. Jones. *Kernel Smoothing.* Chapman & Hall, London, 1995.

Autonomous Learning Algorithm and Associative Memory for Intelligent Robots

Kazuhiro Kojima

Koji Ito

Department of Computational Intelligence and Systems Science
Tokyo Institute of Technology
4259 Nagatsuta Midori-ku, Yokohama 226-8502
Tel:+81-45-924-5654, Fax:+81-45-924-5654
kojima@ito.dis.titech.ac.jp
ito@dis.titech.ac.jp

Abstract

In this paper, we propose autonomous learning algorithm based on the internal state of the associative memory for intelligent robots. The proposed associative memory model consists of structural unstable oscillators and a common field such as chemical concentration.

In computer simulations, we use the binary pattern as the stimuli. When the pattern memorized in the network is given to the network from the outer world, the internal state of the network becomes a periodic state. On the other hand, when the pattern has not been memorized is given to the network, the state becomes an intermittently chaotic and the output of the network travels around the input and some memorized patterns. This chaotic state is regarded as "I don't know" state. Further, when the proposed autonomous learning algorithm is applied to the proposed network, the network can learn only the novel patterns automatically without destroying the previously memorized patterns.

Keywords: autonomous learning algorithm, associative memory, time-dependent internal state.

1 Introduction

In recent years, the autonomous robots[1][2], e.g., robocup, robocup rescue and pet robot, has been studied. The autonomous robot interacts with the outer world for itself, that is, the environments and other robots. In other words, robot gets the stimuli from the outer world, e.g. visual, auditory and somatic senses. Further, these stimuli are transformed to the internal information to the robot. Furthermore, the robot takes action to the environments based on the internal information. This process is called as the perception-action circulation and shown in Fig.1. In this circulation, the transformation from the stimulus to the internal information is based on the history, memory of the robots. Hence, the memory is very important factor for the realization of the autonomous intelligent robots.

In this paper, we focused on the formation of memory system. We employ a kind of the Hopfield model[3], specially the determinate and continuous time type, as the memory system. Because, the dynamical and statistical properties of this model have been studied in a decade. Further, other memory models, e.g. Boltzmann machine, are come down to the Hopfield model.

In this model, the memories are embedded in the network such as each synaptic weight between the neurons i and j in advance. For retrieving the memory, an initial state that is a point inside of the basins is given to the network as the stimulus. Then, the state of the network converges to the relevant point attractor, that is, the network dynamics is based on the Liapunov stability theory. This stability is robust to the noise. However, due to this robustness, if novel stimuli are given to the network as the continuous external input, the network must converges to 1) one of the embedded patterns, 2) the input pattern, or 3) a confused pattern of them (the mathematical framework is explained in section 2.).

The main drawback of the memory system which works according to the Hopfield model is that, in the above three cases, after a short transient time the network goes to a time-independent state and it is impossible that the autonomous robot discriminates among

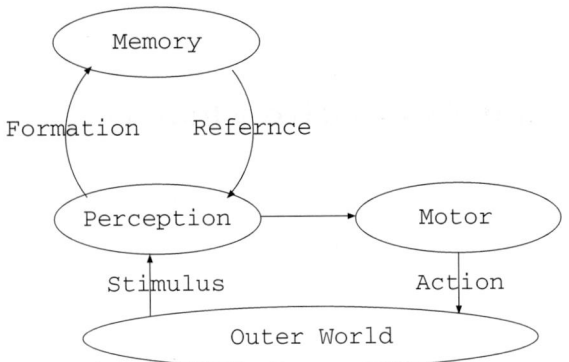

Figure 1: Perception-action circulation.

the three cases by oneself. But, it is very important to distinguish between the known stimuli and the unknown one obtained from the outer world. Because, if the stimulus is regarded as the unknown one, the autonomous robot can behave actively in the unknown outer world, e.g. active learning, searching, communicating, avoiding, etc.

For this problem, there are very interesting physiological studies. Freeman and coworkers[4] showed that when novel odor stimuli are given to rabbit, the olfactory bulb responses chaotically, which is known as "I don't know" state. Further, if the rabbit has learned the novel odor, the olfactory bulb changes to response periodically. These physiological facts suggest that brain and neural systems utilize chaotic dynamics and phase transition or bifurcation for the recognition process including the learning process.

From this point of view, we propose a dynamical memory network constructed from chaotic dynamics and an autonomous learning algorithm based on the internal state of the network. Although, there are many network based on chaotic dynamics[5][6][7][8], these studies focused on the dynamical associative process of each network. On the other hand, in this paper, we focus on the learning process of the network.

The paper is organized as follows. In Section 2, the mathematical framework of the conventional Hopfield model is explained and the new dynamical memory model is proposed. In Section 3, the condition of the simulations are presented. Section 4 presents the dynamical associative process. In Section 5, we propose the autonomous learning algorithm. The last two sections give discussion and conclusion.

2 Memory Dynamics

In this section, we show the mathematical framework of the conventional Hopfield model and disadvantage of it. Then, we propose a new dynamical associative memory model.

2.1 Conventional Memory Model

The Dynamics of the conventional Hopfield model is represented as following differential equation

$$\dot{u}_i = -u_i + h_i, \qquad (1)$$

where h_i is local field and defined as follows.

$$h_i = \sum_{j \neq i}^{N} J_{ij} s_j + I_i \qquad (2)$$

$$s_i = g(u_i) = \tanh(\beta u_i) \qquad (3)$$

In Eq.(3), $g(x)$ is an output function. In Eq.(2), I_i is the input signal for the neuron i and J_{ij} is the coupling weight in which the memories are embedded. For auto-associative memory, these weight matrix is obtained as symmetric matrix. Hence, the energy function is defined as follows.

$$E(s) = -\frac{1}{2} \sum_{i=1}^{N} \left(\sum_{j \neq i}^{N} s_i J_{ij} s_j - \frac{1}{\beta} \int_0^{s_i} g^{-1}(x) dx + I_i s_i \right) \qquad (4)$$

Its time derivative is

$$\frac{dE}{dt} = -\frac{1}{2} \sum_i \frac{ds_i}{dt} \left(\sum_{j \neq i}^{N} J_{ij} s_j - u_i + I_i \right) \qquad (5a)$$

$$= -\frac{1}{2} \sum_i \frac{ds_i}{dt} \frac{du_i}{dt} \qquad (5b)$$

$$= -\frac{1}{2} \sum_i^N \left(\frac{ds_i}{dt} \right)^2 \frac{du_i}{ds_i}. \qquad (5c)$$

Since $g^{-1}(s_i)$ is a monotone increasing function.

$$\frac{dE}{dt} \leq 0, \frac{dE}{dt} = 0 \rightarrow \frac{ds_i}{dt} = 0. \qquad (6)$$

Hence, the energy function E is the Liapunov function of Eq.(1) and the conventional Hopfield model is Liapunov stable.

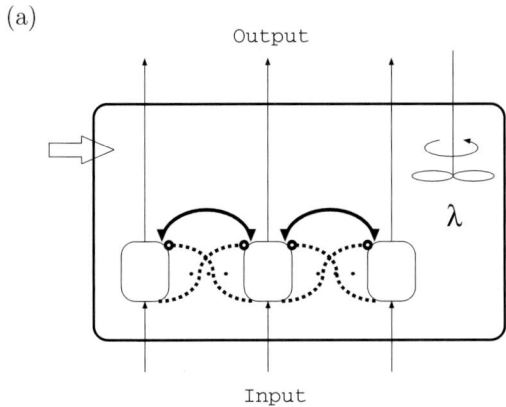 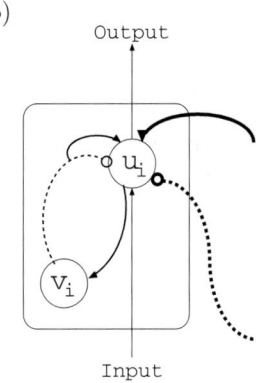

Figure 2: Schematic illustration of the network architecture. This network consists of the neural oscillators and the common field which stands for the concentration of a certain kind of chemical substances. Further, these oscillators are soaked in this common field.

2.2 Proposed Dynamical System

We need a dynamical memory system not based on the Liapunov stability in order to realize "I don't know" state. We proposed the new network based on an analogy between associative memory and Lorenz system[9].

The proposed network is shown in Fig. 2 and the evolution equations are represented by a set of ordinary differential equations as follows.

$$\dot{u}_i = -u_i + \alpha \left(\lambda - \frac{1}{N} \sum_{j=1}^{N} v_j^2 \right) v_i + h_i \quad (7a)$$

$$\dot{v}_i = -\sigma(v_i - u_i) \quad (7b)$$

$$\dot{\lambda} = b \left(1 - \lambda - \frac{\gamma}{N} \sum_{j=1}^{N} v_j^2 \right) \quad (7c)$$

where $\gamma = 2\sigma/b - 1$.

This network consists of the neural oscillators and the common field which stands for the concentration of a certain kind of chemical substances. Further, these oscillators are soaked in this common field. The schematic illustrations of the proposed network are shown in Fig.2.

Each neural oscillator consists of the excitatory u-neuron and the inhibitory v-neuron shown in Fig.2 (b). There are two type intra-connections from the inhibitory v-neuron to the excitatory u-neuron. These are the first order excitatory connection and the third order inhibitory connection. In Fig.2 (b), the former and the latter corresponds to the solid and the broken lines, respectively. Further, the intension of the first order excitatory connection depends on the common filed λ.

On the other hand, there are two types of interconnections among oscillators, one is the mutual inhibitory connections from the $v_j (j \neq i)$ neurons to the u_i neuron, the other is the mutual connections among u neurons which corresponds to the thick broken and the thick solid line in Fig.2 (a) and (b), respectively.

The mutual connections among u neurons induces the local filed h_i defined as follows.

$$h_i = \varepsilon_1 \sum_{\substack{j \neq i}}^{N} J_{ij} s_j + \varepsilon_2 I_i \quad (8)$$

$$s_i = \tanh(\beta u_i). \quad (9)$$

where J_{ij}, I_i and β is coupling weight, input and stepness parameter, respectively.

In Eq.(7a), when the parameter α is zero, the evolution equations are equivalent to Eq.(1), that is, the conventional Hopfield model. Then, the proposed model is an expanded model of the conventional Hopfield model.

The common filed is uniform for the space. Further, it is consumed by the oscillators and supplied from the outside. Hence, this system is an open system.

3 Simulations

We use binary patterns that each pixel consists of $\{-1, 1\}$ and its size is 49. These binary patterns are no-orthogonal set.

In order to investigate the network dynamics when

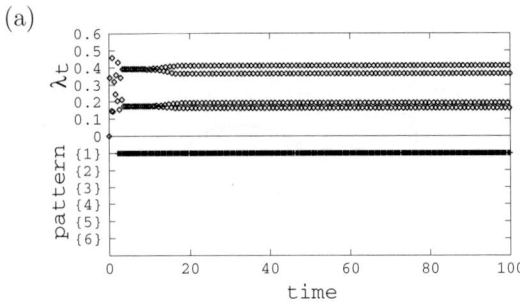

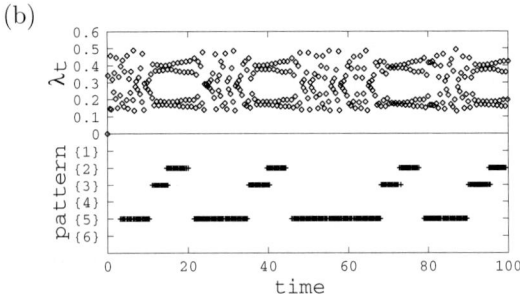

Figure 3: Temporal Association Process. In this Figure, when the Hamming distance is less than 0.2 or more than 0.8, dot is plotted.

known and unknown patterns are given to the network, the patterns $\xi^{(1)}, \xi^{(2)}, \xi^{(3)}$ and $\xi^{(4)}$ are embedded in the network in advance. Namely, the synaptic weights J_{ij} between u_i and u_j neurons are given by the superposition of auto-correlation matrix of the each pattern as follows.

$$J_{ij} = \frac{1}{N} \left(\sum_{\mu=1}^{4} (1-\delta_{ij}) \xi_i^{(\mu)} \xi_j^{(\mu)} \right) \quad (10)$$

where δ_{ij} is Kronecker's δ. It is simple auto-correlation matrix. we assume no special structure on the synaptic coupling weights J_{ij}.

Each parameter of the network was selected as follows: $k = 10.0$, $b = \frac{3}{8}$, $\beta = 50.0$, $\varepsilon_1 = 0.2$ and $\varepsilon_2 = 0.048$.

4 Dynamical Associative Process

Next, we show the dynamical behavior of λ_t and the output pattern of the network around this critical region. The parameter α was selected as $\alpha = 200$.

For measuring the output pattern of the network, we calculated Hamming distances defined as follow.

$$H_t^{(\mu)} = \frac{1}{2} \left(1 - \frac{1}{N} \sum_{i=1}^{N} \xi^{(\mu)} s_{i,t} \right) \quad (11)$$

When Hamming distance $H_t^{(\mu)}$ is 0 or 1, the output pattern is pattern $\xi^{(\mu)}$ or the inverse pattern of it, respectively.

These result are shown in Fig. 3. In this Figure, when the Hamming distance is less than 0.2 or more than 0.8, dot is plotted.

When the known pattern $\xi^{(1)}$ is given to the network, λ_t oscillated periodically with cycle 4, and the output pattern of the network oscillated between the relevant embedded pattern and it's inverse as shown Fig. 3 (a). On the other hand, when the unknown patterns $\xi^{(5)}$ is given to the network, λ_t became a turbulent state from oscillatory state with cycle 4 intermittently, and the output pattern of the network changed with the intermittent turbulence as shown Fig. 3 (b). Therefore, this turbulent state is considered as "I don't know" state.

5 Learning Process

In the above section, we showed the dynamical behaviors of the proposed network, that is, when the known pattern is given to the network, the state of the network is oscillatory state. On the other hand, when the unknown pattern is given to the network, the state of it is chaotic state. Now, under the phase of "I don't know", a learning rule was applied to the network. Then, what will be occurred in the network? In this section, we investigate the effect of a learning process for the proposed network. Further, we propose a new learning rule that depends on the internal state of the network.

5.1 Constant Learning

First, we applied the conventional Hebbian learning rule as following equation.

$$\Delta J_{ij} = \eta s_i s_j \quad (12)$$

Where η is the learning rate parameter.

The result of the constant Hebbian learning is shown in Fig.4. In this case, η is 1.0×10^{-4} and the initial synaptic coupling weights are given by Eq.(10).

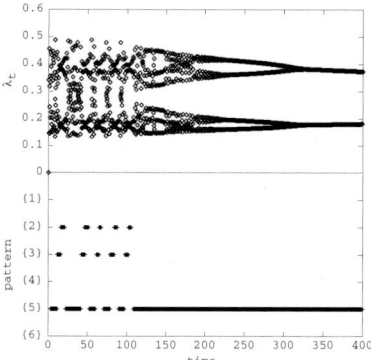

Figure 4: Inverse period-doubling bifurcation under the constant learning.

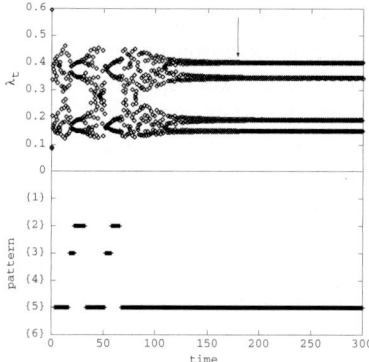

Figure 6: Inverse period-doubling bifurcation under the proposed learning rule.

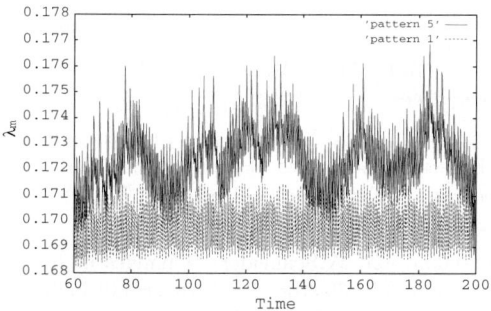

Figure 5: Temporal behavior of the time average $\lambda_m(t)$

The unknown pattern $\xi^{(5)}$ is given to the network. Then, we could observe the inverse period-doubling bifurcation of λ_t responses with the progress of the learning as shown Fig.4. It is very natural result, because the inverse period-doubling bifurcation smoothly connects from the disorder to the order.

But when the learning process progresses, the internal state of the network changes from the limit cycle to the fix point, so that previously memorized pattern are destroyed. Therefore, there is only the input pattern in the network as the memorized pattern. This state is the over learning. Next subsection, we propose a solution for avoiding the over learning.

5.2 Proposed New Learning Rule

As shown in Fig.4, the inverse bifurcation of λ_t corresponds to the index of the progress of the learning. Hence, we notice the periodicity of λ_t. In general, the calculation algorithm of the periodicity needs the memory or long sampling time, e.g. FFT and Lyapunov spectrum method. But the learning process is an on-line process.

We consider the history of the variable λ, that is the time average of it. It can be obtained to solve the following differential equation.

$$\tau \dot{\lambda}_m = -\lambda_m + \lambda \qquad (13)$$

where τ is a time constant. When the known and the unknown pattern is given to the network, the temporal behavior of the time average $\lambda_m(t)$ is represented in Fig.5, where the dash line and the solid line corresponds to the know and the unknown pattern, respectively. From this result, we can find an appropriate threshold that divides the state between the order and the disorder.

According to the threshold property, we propose a new learning rule that depends on the state of the network.

$$\Delta J_{ij} = \eta(\lambda_m) s_i s_j \qquad (14)$$

$$\eta(\lambda_m) = \begin{cases} 0 & : \lambda_m < V_L \\ \eta_0(\tanh(\lambda_m - V_L) + 1) & : \lambda_m \geq V_L \end{cases} \qquad (15)$$

where V_L is the threshold.

The result of the proposed learning rule is shown in Fig.6. In this simulation, we set the threshold parameter V_L to 0.172.

The unknown pattern $\xi^{(5)}$ is given to the network. Then, λ_t inversely bifurcates like as the above constant learning case. But the learning process is reduced gradually. Further, it completely stop around the arrow that is shown in Fig.6, so that we can avoid the over learning.

6 Discussion

In section 1, we explained the perception-action circulation. We pointed out the importance on the memory for the intelligent robot. However we couldn't consider the behavior of the autonomous robot in this paper. Hence one of future work is that action generator is integrated into the proposed model. Consequently, the perception-action circulation is closed between the robot and the outer world. Recently, reinforcement learning is studied for learning of the behavior under the unknown environments. In this algorithm, robot doesn't have the internal model of the environments. Hence, the new action is searched according to the random numbers, in a word, *try and error*.

We showed the dynamical association process in section 4. This process is regarded as the searching process. However, this searching process doesn't depend on randomness completely. Namely, when the current stimulus contradicts the memories, the randomness is generated. It seems that this randomness is used as a new random number generator based on the internal state of the robot. Further, it is suggested that there is the entrainment among perception, action and outer world through the body of robot[10][11].

In this paper, we used the binary pattern as the stimuli that obtained from the outer world. In the real world, however these could be represented by the real number. If the real value is transformed to the binaries, the topology such as a distance defined in the original real value space isn't equated with one defined in the binary-coded space. The same problem arises in the binary genetic algorithm. The improvement for dealing with the real value is one of the most important future works.

7 Conclusion

In this paper, we proposed an autonomous learning algorithm based on the internal state of the associative memory for intelligent robots. The proposed memory model have a common field such as chemical concentration, that has mutual interaction with each oscillator. In computer simulations, when the memorized pattern in the network is given to the network from the environment, the state of the common field becomes a periodic state. On the other hand, in the case where the non-memorized pattern is given to the network, the state of the common field becomes an intermittently chaotic state and the output of the network travels around the input and some memorized patterns.

Moreover, when the proposed autonomous learning algorithm is applied to the proposed network, the network can learn only the novel patterns automatically without destroying the previously memorized patterns.

Acknowledgments

A part of this research was supported by the Scientific Research Foundation of Ministry of Education (10450165) and "Research for the Future" Project (JSPS-RFTF96I00105) from the Japan Society for the Promotion of Science.

References

[1] R.A.Brooks, "Intelligence without representation", Artificial Intelligence, Vol. 47, 1991, pp.139-160.

[2] H.Kitano, "RoboCup-97:Robot Soccer World Cup I", Springer, 1998.

[3] J.J.Hopfield, "Neural Networks and Physical Systems with Emergent Collective Computational Abilities", Proceedings of Natl.Acad.Sci.USA, Vol. 79, 1982, pp. 2554-2558.

[4] W.J.Freeman and W.Schneider, "Changes in spatial patterns of rabbit olfactory EEG with conditioning to odors", Psychophysiology, Vol. 19, 1982, pp. 44-56.

[5] S.Nara, P.Davis and H.Totsuji, "Memory Search Using Complex Dynamics in a Recurrent Neural Network Model", Neural Networks, Vol. 6, 1993, pp. 963-973.

[6] K.Aihara, Y.Takebe and M.Toyoda, "Chaotic neural networks", Phys. Lett 144A, 1990, pp. 333-340.

[7] I.Tsuda, "Dynamic Link of Memory - Chaotic Memory Map in Nonequilibrium Neural Networks" Neural Networks, Vol. 5, 1992, pp. 313-326.

[8] M.Dauce, M.Quoy, B.Cessac, B.Doyon and M.Samu-elides, "Self-organization and dynamics reduction in recurrent networks stimulus presentation and learning", Neural Networks, Vol. 11, 1998, pp. 521-533.

[9] E.N.Lorenz, "Deterministic nonperiodic flow", J.Atmos.Sci., Vol. 20, 1963, pp.130-141.

[10] G.Taga, Y.Yamaguchi and H.Shimizu, "Self-organized control of bipedal locomotion in unpredictable environment", Biol. Cybern., Vol. 65, 1991, pp.147-159.

[11] Y.Kuniyoshi and L.Berthouze, "Neural learning of embodied interaction dynamics", Neural Networks, Vol. 11, 1998, pp.1259-1276.

Learning Hierarchical Partially Observable Markov Decision Process Models for Robot Navigation

Georgios Theocharous
theochar@cse.msu.edu

Khashayar Rohanimanesh
khash@cse.msu.edu

Sridhar Mahadevan
mahadeva@cse.msu.edu

Department of Computer Science and Engineering
Michigan State University
East Lansing, MI 48823

Abstract— We propose and investigate a general framework for hierarchical modeling of partially observable environments, such as office buildings, using Hierarchical Hidden Markov Models (HHMMs). Our main goal is to explore hierarchical modeling as a basis for designing more efficient methods for model construction and useage. As a case study we focus on indoor robot navigation and show how this framework can be used to learn a hierarchy of models of the environment at different levels of spatial abstraction. We introduce the idea of model *reuse* that can be used to combine already learned models into a larger model. We describe an extension of the HHMM model to includes actions, which we call hierarchical POMDPs, and describe a modified hierarchical Baum-Welch algorithm to learn these models. We train different families of hierarchical models for a simulated and a real world corridor environment and compare them with the standard "flat" representation of the same environment. We show that the hierarchical POMDP approach, combined with model reuse, allows learning hierarchical models that fit the data better and train faster than flat models.

I. INTRODUCTION

Recent work in artificial intelligence (AI) on sequential decision-making under uncertainty has adopted the framework of Partially Observable Markov Decision Processes (POMDP) [1], [2]. This framework allows modeling environments where the underlying states are "hidden", and only partially observable through noisy sensory observations and actions. POMDPs extend the well-known Hidden Markov Model (HMM) [3] to include actions and rewards. There are well-developed algorithms for both building and using POMDP models, such as the Baum-Welch procedure and dynamic programming. However, past work on POMDPs has been restricted to "flat" uniform scale models, and the model-learning and planning algorithms scale poorly with the size of the model. Several studies have shown that the HMM/POMDP framework can be used to program autonomous mobile robots to navigate in real office environments [1], [4], [2], but these systems use flat models. As the size of the environment grows, it becomes increasingly difficult to learn and use flat models, and it would be desirable to have a natural way of reusing previously learned sub-models.

In this paper, we address these limitations of earlier standard HMM/POMDP systems using the Hierarchical Hidden Markov Model (HHMM) framework [5]. Using the HHMM approach, a robot learns and uses a hierarchy of homogeneous representations of the environment in which each layer of the hierarchy maintains a probabilistic model of the environment defined at some resolution. The tree structure of HHMM models also provides a natural approach for rapid model learning by reusing previously learned sub-models. We extend HHMM models to include primitive actions, and show a procedure for converting hierarchical models into flat POMDPs.

Intuitively, hierarchical modeling should allow us to learn the environment in a modular fashion and therefore be able to learn faster, and even be able to discover relationships at abstract levels that would not be trivial in flat representations. Fortunately, many natural environments can be viewed at different levels of spatial abstraction. For example, an office environment modeled at a coarse level would consist of nodes for corridors, intersections and dead-ends; at a finer level, each model element would represent a fixed-length region of a corridor.

In this paper we demonstrate through experiments that the HHMM framework extended to the hierarchical POMDP case appears to have some compelling advantages as a basis for designing scalable spatial learning algorithms. We will provide empirical results from learning in both real-world and simulated robot navigation to illustrate the potential of the hierarchical POMDP framework.

II. HIERARCHICAL HIDDEN MARKOV MODELS

The Hierarchical Hidden Markov Model (HHMM) [5] generalizes the standard hidden Markov model (HMM) [3], by allowing hidden states to represent stochastic processes themselves. An HHMM is visualized as a tree structure in which there are three types of states, production states (leaves of the tree) which emit observations, and internal states which are (unobservable) hidden states that represent entire stochastic processes. Each production state is associated with an observation vector which maintains distribution functions for each observation defined for the model. Each internal state is associated with a horizontal transition matrix, and a vertical transition vector. The horizontal transition matrix of an internal state defines the transition probabilities among its children. The vertical transition vectors define the probability of an internal state to activate any of its children. Each internal state is also associated with a child called an *end-state* which returns control to its parent. The end-states do not produce observations and cannot be activated through a vertical transition from their parent. The HHMM is formally defined as a 5 tuple $\langle S, T, \Pi, Z, O \rangle$:

- S denotes the set of states. The functions $p(s)$ denotes the parent of state s. The function $c(s, j)$ returns the jth child of state s. The end-state child of an abstract state s is denoted by e^s. The set of children of a state s is denoted by C^s and the number of children by $|C^s|$. There are three types of states.
  - Production states
  - Abstract states
  - End-states
- $T^s : \{C^s - e^s\} \times C^s \to (0, 1)$ denotes the horizontal transition functions, defined separately for each abstract state. A horizontal transition function maps each child state of s into a probability distribution over the children states of s. We write $T^s(c(s,i), c(s,j))$ to denote the horizontal transition probability from the ith to the jth child of state s. As an example, in Figure 1, $T^{s4}(s7, s8) = 0.8$.
- $\Pi^s : \{C^s - e^s\} \to (0, 1)$ denotes the vertical transition function for each abstract state s. This function defines the initial distribution over the children states of state s, except from the end-state child e^s. For example, in Figure 1, $\Pi^{s4}(s6) = 0.5$.
- Z denotes the set of observations.
- $O : S^{product} \to (0, 1)$ denotes a function that maps every product state to a distribution over the observation set. We write $O(s, z)$ for the probability of observing z in state s.

Figure 1 shows a graphical representation of an example HHMM. The HHMM produces observations as follows:

1. If the current node is the root, then it chooses to activate one of its children according to the vertical transition vector from the root to its children.
2. If the child activated is a product state, it produces an observation according to an observation probability output vector. It then transitions to another state within the same level. If the state reached after the transition is the end-state, then control is returned to the parent of the end-state.
3. If the child is an abstract state then it chooses to activate one of its children. The abstract state waits until control is returned to it from its child end-state. Then it transitions to another state within the same level. If the resulting transition is to the end-state then control is returned to the parent of the abstract state.

Fine et al. [5] describe a hierarchical Baum-Welch algorithm that is able to re-estimate the model parameters λ (including transitions matrices, vertical vectors, and observation vectors) of an HHMM given observation sequences $M = z_1, z_2, ..., z_T$. We have extended this algorithm to be able to learn the parameters of an HHMM that includes actions, as we describe next.

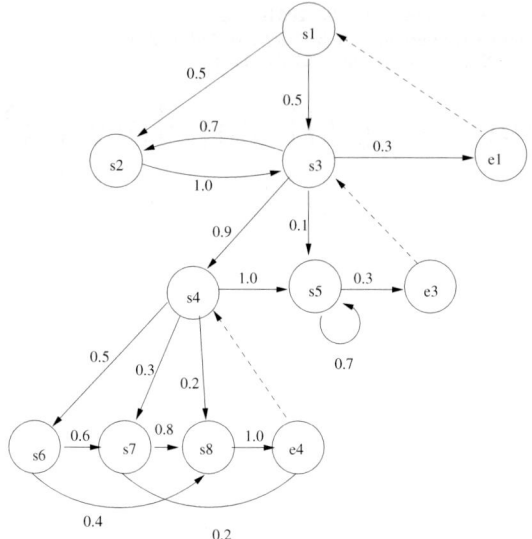

Fig. 1. An example hierarchical HMM. Only leaf (production) states (s2, s5, s6, s7, and s8) have associated observations.

III. HIERARCHICAL PARTIALLY OBSERVABLE MARKOV DECISION PROCESS MODELS

A. Model definition

Robot navigation is an example of a planning problem that requires extending the HHMM model to include both primitive and abstract actions, as well as reward functions for specifying goals. We call this extended model *hierarchical POMDP* and we formally define it as follows:

- S denotes the set of states which are exactly the same as an HHMM.
- A denotes the set of primitive actions. The primitive actions initiate and terminate in product states. The product states represent the lowest resolution of a physical environment and therefore every action has to start in some product state and end in some other product state. For example, in an indoor robot navigation environment, if abstract states are corridors and product states are 2 meter locations in the corridors, then every primitive action such as "go-forward" will always take the robot from one product state to another product state. However, the transition probability from one product state $s1$ to some other product state $s2$ is not simply a lookup operation into a global transition matrix as in flat POMDPs. In the hierarchical POMDP model we may have to look at more than a single horizontal transition matrix and even multiple vertical vectors. An example of this calculation is shown in Figure 2.
- $T^{s,a} : \{C^s - e^s\} \times C^s \to (0, 1)$ denotes the horizontal transition functions, which are defined separately for each abstract state s and action a. A horizontal transition function maps each child state of s and action pair into a probability distribution over the children states of s.
- $\Pi^s : \{C^s - e^s\} \to (0, 1)$ denotes the vertical transition function for each abstract state s. This function defines the initial distribution over the children states of state s.

- Z denote the set of observations as before.
- $O : S^{product} \to (0,1)$ is a function that maps every product state to a distribution over the observation set.
- $R : S^{product} \times A \to \Re$ denotes an immediate reward function defined on the product states.

Figure 2 shows an example hierarchical POMDP. A hierarchical Baum-Welch algorithm for the hierarchical POMDP can be defined by extending the hierarchical Baum-Welch algorithm for HHMMs. First we define the α variable as shown in Equation 1

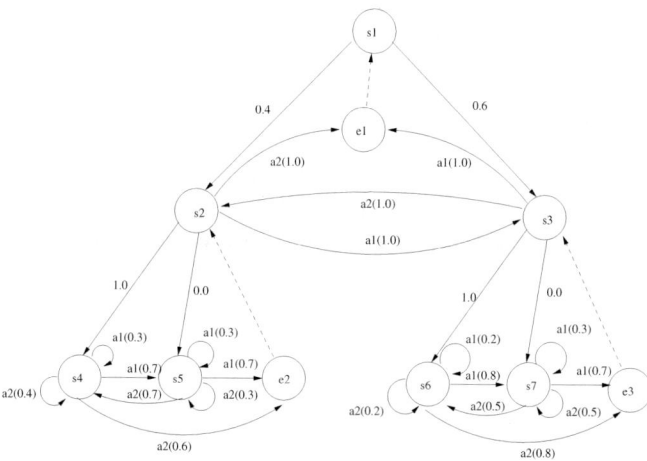

Fig. 2. An example hierarchical POMDP with two primitive actions, $a1$ and $a2$. To calculate the transition from state $s4$ to state $s4$ under action $a2$ we have to consider all non-zero probability paths. One path is $(s4, s4)$ and the other path is $(s4, e2, s2, e1, s1, s2, s4)$. Thus while $T^{s2,a2}(s4, s4) = 0.4$, the probability of transitioning from $s4$ to $s4$ under action $a_2 = 0.4 + 0.6 \times 1.0 \times 0.4 \times 1.0 = 0.64$.

$$\alpha(t, t+k, s) = \\ P(z_t, ...z_{t+k}, s \ finished \ at \ t+k \ | a_t, ...a_{t+k-1}, p(s) \ started \ at \ t, \lambda) \quad (1)$$

The α variables gives us the probability of the observation sequence $z_t...z_{t+k}$ and that state s finished at time $t+k$ given that actions $a_t, ...a_{t+k-1}$ were taken and the parent of s, $p(s)$ was started at time t. We say that a product state finishes at time t after the production of observation z_t and an abstract state finishes when control is returned to it from its child end-state after one of its children produces its last observation z_t. We also say that a product state s, started at time t, if at time t it produced observation z_t. An abstract state s starts at time t if at time t one of its children produced observation z_t but observation z_{t-1} was generated before s was activated by its parent or any other horizontal transition.

Note that using the α variable we can calculate the probability of an observation sequence given the model as shown in Equation 2.

$$P(M|\lambda) = \sum_{s=1|c(root,s) \neq e^{root}}^{|C^{root}|} \alpha(1, T, c(root, s)) \quad (2)$$

The backward variable β, defined in Equation 3, denotes the probability that a state s was entered at time t and that the observations $z_t...z_{t+k}$ were produced by the parent of s, which is $p(s)$ and actions $a_t, ...a_{t+k}$ were taken and p(s) terminated at time $t + k$.

$$\beta(t, t+k, s) = \\ P(z_t, ...z_{t+k}|a_t, ...a_{t+k}, s \ started \ at \ t, p(s) \ generated \\ z_t...z_{t+k} \ and \ finished \ at \ t+k, \lambda) \quad (3)$$

The next important variable is ξ, defined in Equation 4, which denotes the probability of making a horizontal transition from s to s' at time t.

$$\xi(t, s, s') = P(s \ finished \ at \ time \ t, \\ s' \ started \ at \ time \ t+1 \ | a_1, ...a_T, z_1, ...z_T, \lambda) \quad (4)$$

Another important variable is $\chi(t, s)$, shown in Equation 5, which defines the probability that state s was activated by its parent $p(s)$ at time t given the action and observation sequence

$$\chi(t, s) = P(p(s) \ started \ at \ time \ t, s \ started \ at \ time \ t| \\ a_1, ...a_T, z_1, ...z_T, \lambda) \quad (5)$$

Based on these variables we can re-estimate the model parameters. The vertical vectors are re-estimated in Equations 6 and 7.

$$\Pi^s(n) = \chi(0, n), \ if \ s = root, \ n \in C^s \quad (6)$$

$$\Pi^s(n) = \frac{\sum_{t=1}^T \chi(t, n)}{\sum_{i=1|c(s,i)\neq e^s}^{|C^s|} \sum_{t=1}^T \chi(t, c(s,i))}, \ if \ s \neq root, \ n \in C^s \quad (7)$$

The horizontal transition matrices are re-estimated in Equation 8. The re-estimation calculates the average number of times the process went from state s to state s' over the number of times the process exited state s under action a.

$$T^{p(s),a}(s, s') = \frac{\sum_{t=1|a_t=a}^T \xi(t, s, s')}{\sum_{t=1|a_t=a}^T \sum_{i=1}^{|C^{p(s)}|} \xi(t, s, c(p(s), i))} \quad (8)$$

The observation vectors are re-estimated in Equation 9. The re-estimation of the observation model calculates the average number of times the process was in state s and perceived observation z over the number of times the process was in state s.

$$O(s,z) = \frac{\sum_{t=1,z_t=z}^{T} \chi(t,s) + \sum_{t=2,z_t=z}^{T} \gamma_{in}(t,s)}{\sum_{t=1}^{T} \chi(t,s) + \sum_{t=2}^{T} \gamma_{in}(t,s)}$$

$$\text{where } \gamma_{in}(t,s) = \sum_{i=1 | c(p(s),i) \neq e^{p(s)}}^{|C^{p(s)}|} \xi(t-1, c(p(s),i), s) \quad (9)$$

B. Planning using Hierarchical POMDPs

Solving a POMDP means that we have to find a mapping from each "belief state" (probability distribution over states) to actions that will achieve the best long term sum of rewards [6]. The belief state is a sufficient statistic that summarizes the past history of observations and actions. There is an efficient Bayesian update procedure that can be used to calculate the belief state probability distribution over all the product states given a sequence of observations and actions. Unfortunately the number of belief states is infinite and exact solutions to large POMDPs are computationally infeasible. Fortunately, many heuristics solutions such as the most likely state heuristic (MLS) and the Q-MDP method are known to provide satisfactory approximate solutions for robot navigation [7]. We have extended these approximate methods to hierarchical POMDPs, by implementing hierarchical versions of these methods (e.g. for the MLS procedure, we compute the most likely abstract state, and then recursively the most likely product state).

Any hierarchical POMDP model with actions can be converted into an equivalent flat POMDP. The states of the equivalent flat POMDP are the product states of the hierarchical POMDP and are associated with a global transition matrix that is calculated from the vertical and horizontal transition matrices of the hierarchical POMDP. To construct this global transition matrix of the equivalent flat POMDP, for every pair of states (s_1, s_2), we need to sum up the probabilities of all the paths that will transition the system from $s1$ to $s2$ under some action a. One such example is shown in Figure 2. However, "flattening" a hierarchical model in this manner will destroy the ability to learn faster by reusing sub-models, and negate one of the primary advantages of the hierarchical approach.

IV. LEARNING HIERARCHICAL POMDPs FOR ROBOT NAVIGATION

We now describe a detailed set of experiments comparing hierarchical POMDP models with flat models in terms of learning speed and fit to the data. We have conducted experiments using both a real robot platform (a Nomad 200 robot) and a simulated environment. Using the Nomad 200 simulator we constructed a model of the second floor of the MSU Engineering building, shown in Figure 3. We also used a real indoor environment, shown in Figure 10.

Such topological maps can be automatically compiled into a Markov representation, either as a flat POMDP or as a hierarchical POMDP model. In our experiments, we used both "good" initial models as well as uninformed weak "ergodic" initial models. Figures 4, 5, and 6 show hierarchical and flat POMDP models. In a good initial model, we provide a priori the appropriate connectivity of the actions and we also initialize the observation models according to the location of a state. An observation consists of 5 components: the action taken and the probabilities that the robot has seen a wall or an opening on its four sides, front, left, back and right. The probabilities of the observations are computed by neural nets which take as input local occupancy grid maps around the robot (constructed from 16 sonars) and output the probability of a wall and opening for every direction [8]. For example, if a product state is facing in the corridor direction, then the initial observation model would be front (wall:0.4, opening:0.6), left (wall:0.6, opening:0.4), back (wall:0.4, opening:0.6), and right (wall:0.6, opening:0.4). In the case of uninformed ergodic initial models, we also provide good observation models, but every state is connected with every other state under the same parent for every action at all levels.

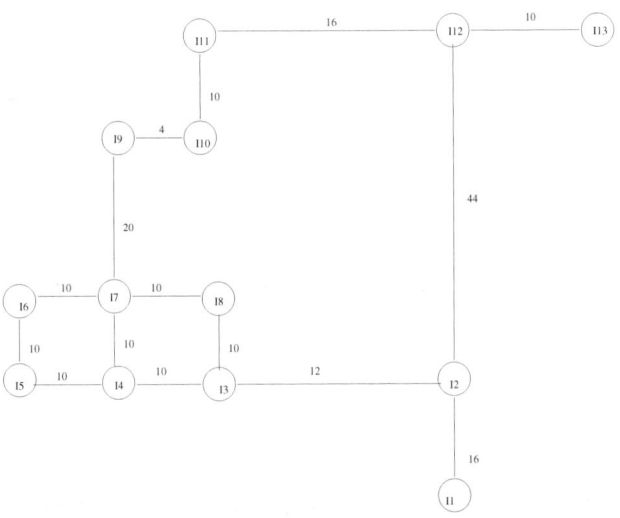

Fig. 3. A topological map of the 2nd floor of the Engineering building. Numbers indicate distances in meters. For each edge and each vertex there are four states at the second level of the hierarchical POMDP. States representing the (North-South or East-West) direction of corridors are expanded to a number of third level product states, each one representing two meters.

In the first experiment, we compared hierarchical and flat POMDPs, where we biased the training with a good initial model. We collected 26 "short" observation sequences, where a short sequence is one where the robot goes from one topological node to the next. We trained a hierarchical model, a "hierarchical with reuse" model (in which all the low level submodels were trained separately) and a flat model as shown in Figure 7. For the "hierarchical with reuse" model we also collected a separate sequence of observations for every abstract state. The result in Figure 7 shows that the hierarchical with reuse model converges a lot faster, and the fit to the the data is very good right from the beginning due to the fact that the submodels were already trained. We say a model converges when the mean

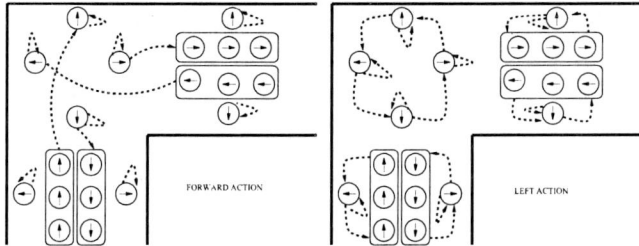

Fig. 4. These figures show how we represent a corridor environment as a hierarchical POMDP model. The circles represent product states and the solid arrows inside the circle indicate their orientation. The rectangles indicate abstract states (level 2) and the circles inside the rectangles are product states at level 3. The dashed arrows on the left show the transition matrix for the forward action at level 2 for a "good initial model". In the ergodic initial model (not shown), transitions are possible between all states under a given parent (for both abstract and product states). Before training we set all transitions to 0.5. The figure on the right shows the initial transition matrix for the "turn left" action. The "turn right" action is similar, except that the transitions are reversed.

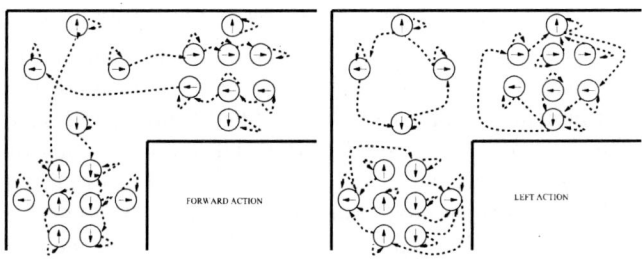

Fig. 5. These figures show the equivalent flat POMDP corridor model that can be automatically derived from the hierarchical POMDP model. The figure on the left show the transitions for the forward action and the one on the right shows the left action (both for good initial models).

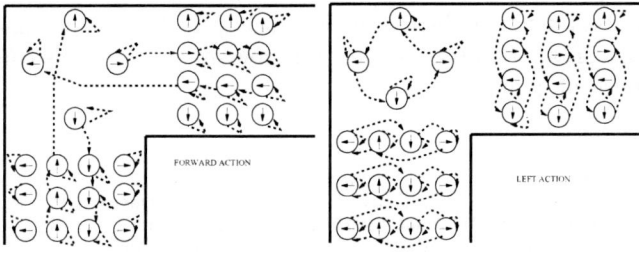

Fig. 6. These figures show how we represent a corridor environment as a flat POMDP model. The figure on the left shows the forward action and the figure on the right the turn left action (for good initial models).

square error across successive log likelihood values for all training traces remains stable within some threshold.

In the second experiment, we compared hierarchical and flat POMDPs, beginning with weak uninformed ergodic models. The training convergence is shown in Figure 8. In this experiment, we see that the hierarchical with reuse model fits the data significantly better than the hierarchical and the flat models.

In a third experiment we trained the same models as those used in experiment 2 using one long sequence of 150 observations as shown in Figure 9. We observe that the

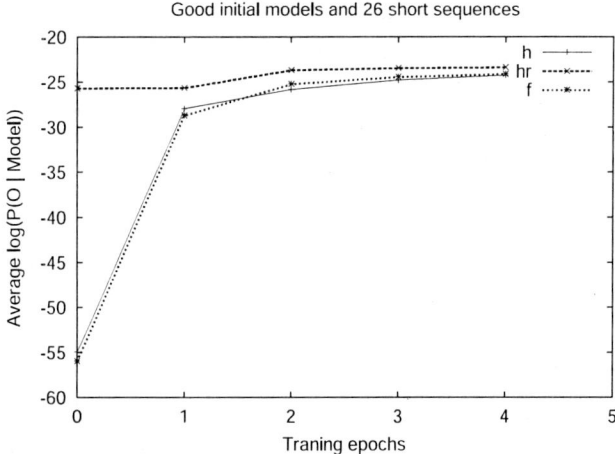

Fig. 7. The graph shows the training convergence between 3 different models. Model convergence is measured by mean square error across successive log likelihood values. The horizontal axis represents the number of training epochs and the vertical axis the average log likelihood fit of 26 different sequences of observations. h stands for a hierarchical model, hr stands for a hierarchical model whose initial model was created by first training separately the different ergodic models at level 3, and f stands for a flat model. Good initial models were provided for the training cases.

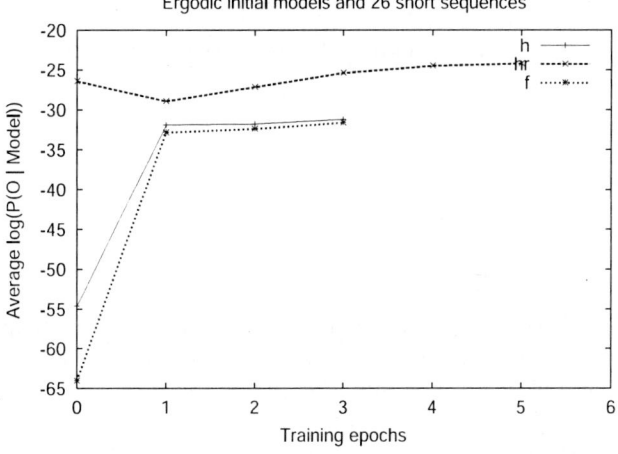

Fig. 8. The graph shows the training convergence between 3 different models. The X axis represents the number of training steps and the Y axis the average log likelihood fit of 26 different sequences of observations. h stands for a hierarchical model, where the starting model was ergodic, hr stands for a hierarchical model whose initial model was created by first training separately the different ergodic models at level 3, and f stands for a flat model. Here the starting model was also ergodic.

hierarchical with reuse model and its equivalent flat model fit the data better than all the other models. The flat POMDP provides a poorer fit to the data than the other models. Table I shows both the size (number of states) of the various models and the time (number of seconds) it takes for a training epoch in each experiment. The hierarchical models and their equivalent flat models train faster than the flat POMDP model, partly due to the smaller number of states in these models as compared to the flat model.

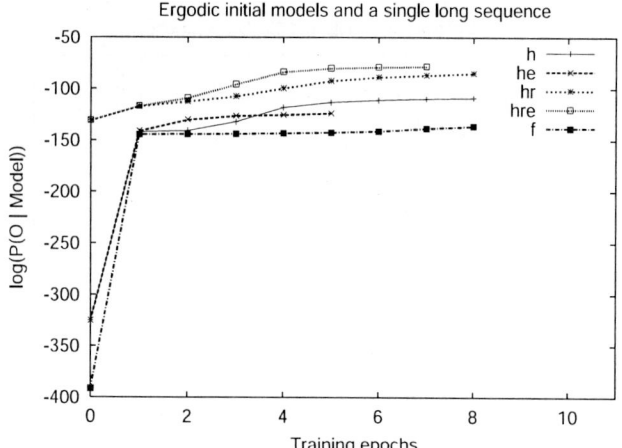

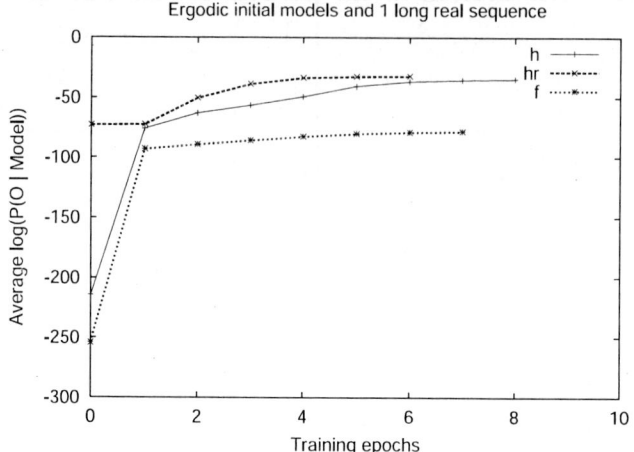

Fig. 9. The graph shows the training convergence between 5 different models for one long observation sequence. *h* stands for a hierarchical model, *he* stands for a flat model equivalent to the hierarchical, *hr* stands for a hierarchical with reuse model whose initial model was created by first training separately the different ergodic modules at level 3, *hre* is the equivalent flat model of the hierarchical with reuse model, and *f* stands for a flat model. For all cases, the starting models were ergodic.

	Figure 7, 8	Figure 9	Figure 11
Model	Time (Size)	Time (Size)	Time (Size)
h, hr	328 (346)	2200 (346)	300 (256)
he, hre	443 (286)	677 (286)	490 (232)
f	1416 (458)	21550 (458)	560 (410)

TABLE I

THIS TABLE SHOWS SIZE OF EACH MODEL (NUMBER OF STATES), AND THE TIME IT TAKES (NUMBER OF SECONDS) TO TRAIN PER EPOCH.

Fig. 11. The graph compares two hierarchical (*h* and *hr*) and one flat (*f*) model in terms of the goodness of fit and convergence (training epochs) for a real corridor environment.

VI. CONCLUSIONS AND FUTURE WORK

In this paper, we describe a framework for learning hierarchical models of partially observable indoor corridor environments. We presented a modified Baum-Welch algorithm for learning hierarchical POMDP models. Using the new algorithm we compared flat and hierarchical POMDP models for both simulated and real world robot navigation environments. Our experimental results show that the hierarchical POMDP model is smaller in size, and correspondingly faster to train, even starting from a weak ergodic initial model. Furthermore, hierarchical models provide a natural way of reusing previously learned submodels. The fit to the data of the hierarchical models was also better than the flat model. We are now investigating alternate ways of defining abstract actions using the hierarchical nature of our POMDP model, and how abstract actions can speed up planning.

V. REAL ROBOT EXPERIMENTS

We also did experiments in an actual indoor environment as shown in Figure 10, where we collected 1 long sequence of a 100 observations using the mobile robot shown in the figure. The training convergence is shown in Figure 11, where the hierarchical with reuse model fits the data better than the rest of the models. The flat POMDP model once again provides the worst fit to the data and took the longest time to train, due to its size.

REFERENCES

[1] Sven Koenig and Reid Simmons, "A Robot Navigation Architecture Based on Partially Observable Markov Decision Process Models," in *Artificial Intelligence Based Mobile Robotics:Case Studies of Successful Robot Systems*. MIT press, 1998.
[2] I. Nourbakhsh, R. Powers, and S. Birchfield, "Dervish: An office-navigation robot," in *AI Magazine 16(2):53-60*. 1995.
[3] Lawrence Rabiner, "Tutorial on Hidden Markov Models and Selected Applications in Speech Recognition," in *Proceedings of the IEEE*, February 1989, vol. 77.
[4] Hagit Shatkay and Leslie Kaebling, "Learning Topological Maps with Weak Local Odeometric Information," in *IJCAI97*, 1997.
[5] Shai Fine, Yoram Singer, and Naftali Tishby, "The Hierarchical Hidden Markov Model: Analysis and Applications," *Machine Learning*, vol. 32, no. 1, July 1998.
[6] Leslie Pack Kaebling, Michael L. Litman, and Anthony R. Cassandra, "Planning and acting in partially observable stochastic domains," *Artificial Intelligence*, vol. 101, 1998.
[7] Anthony R. Cassandra, Leslie Pack Kaelbling, and James A. Kurien, "Acting under uncertainty: Discrete bayesian models for mobile robot navigation," in *Proceedings of IEEE/RSJ International Conference on Intelligent Robots and Systems (IROS)*, 1996.
[8] Sridhar Mahadevan, Georgios Theocharous, and Nikfar Khaleeli, "Fast concept learning for mobile robots," in *Autonomous Robots Journal*, vol. 5, pp. 239-251. 1998.

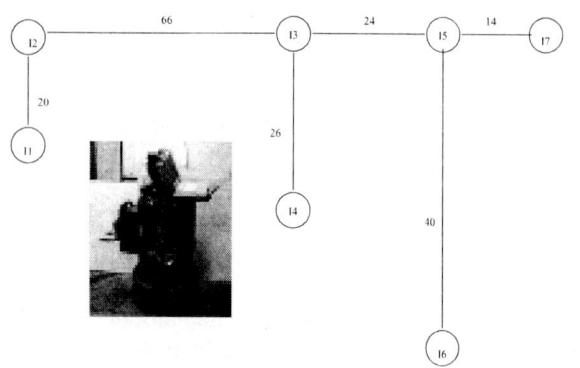

Fig. 10. The figure shows the topological map of a real indoor environment that is learned by PAVLOV (a Nomad 200 platform).

Towards Automatic Shaping in Robot Navigation

Todd S. Peterson, Nancy E. Owens & James L. Carroll
Computer Science Department
Brigham Young University
todd@cs.byu.edu, owens@cs.byu.edu, james@jlcarroll.net

March 1, 2001

Abstract

Shaping is a potentially powerful tool in reinforcement learning applications. Shaping often fails to function effectively because of a lack of understanding about its effects when applied in reinforcement learning settings and the use of inadequate algorithms in its implementation. Because of these difficulties current shaping techniques require some form of manual intervention. We examine some of the principles involved in shaping and present a new algorithm for automatic transferral of knowledge which uses Q-values established in a previous task to guide exploration in the learning of a new task. This algorithm is applied to two different but related robot navigation tasks.

1 Introduction

Reinforcement learning is an attractive method for developing robot behaviors because it does not require a model of the environment, nor does it require the designer to specify the exact behaviour of the robot in every situation that may occur. There are a few key concerns when applying reinforcement learning in robot settings: first, reinforcement learning can be very slow in large domains; and second, policies learned in reinforcement learning apply only to single tasks. One method for overcoming these concerns is to transfer knowledge gained from one task to the performance of a new task through a process called shaping. Unfortunately, current shaping methods require manual intervention.

1.1 Reinforcement Learning

Reinforcement learning [11] is a class of algorithms that uses temporal differences to learn optimal control policies (or near optimal policies in the case of learning with function approximation) for agents that are situated in a real or simulated environment.

A popular on-line algorithm for learning the optimal policy function is *Q-learning* [14]. In Q-learning the policy is formed by determining a Q-value for each state-action pair. A Q-value is the discounted expected on-line return for performing an action at the current state.

$$Q(s_t, a_t) = R(s_t, a_t) + \gamma max_a Q(s_{t+1}, a),$$

where $Q(s,a)$ is the Q-value for state s and action a, $R(s,a)$ is the reinforcement received from taking action a from state s and γ, where $0 < \gamma < 1$, is the discount factor. The update equation is

$$\Delta Q(s_t, a_t) = \alpha(R(s_t, a_t) + \gamma max_a Q(s_{t+1})),$$

where α is the learning rate.

In order to avoid learning sub-optimal policies the agent often explores his environment semi-randomly. One example of a semi-random exploration policy is to choose actions through a Boltzmann distribution:

$$prob(a_t) = \frac{e^{Q(s_t, a_t)/\tau}}{\sum_a e^{Q(s_t, a_t)/\tau}},$$

where τ effects the amount of randomness in the agents actions, and decays over time.

Reinforcement learning is particularly slow when applied directly to robot settings because of the extra time needed for physical interactions. One way of decreasing the learning time in robot settings is to use a simulator to speed up the interaction of the robot with its environment. Unfortunately, inherent noise in the sensors and the actuators of the robot prevents a simulator from exactly representing the dynamics of the environment or the robot's actuators. Function approximation is often used to increase the simulation's accuracy, but is not sufficient to completely solve this problem. Because of these and other modeling difficulties, tasks learned on simulators often transfer poorly to the real robot.

The second concern with applied reinforcement learning is that policies learned to achieve one task are not easily modified for new situations. Combining function approximation with reinforcement learning allows an agent to apply a previously learned policy to new environments provided that the new environments share some underlying environmental characteristics, but it does not facilitate the modification of a learned policy to match new goal conditions. If the underlying environmental characteristics do not match, or if goal conditions are sufficiently different, further training is required in order to achieve adequate performance.

1.2 Robot Shaping

Because of these and other difficulties, several researchers have used various shaping techniques to transfer knowledge from one problem domain to another. Shaping is a term used in animal psychology [8, 9] to describe a process in which an animal is trained to perform a complex behavior in stages. The animal is first trained to perform a very simple task, and is then retrained to perform similar, although slightly more difficult, tasks in gradual degrees until the desired behavior is attained. Shaping covers a variety of different approaches (ex. [2, 5, 7, 13]), each of which is somewhat successful. However, each of the techniques requires some form of manual intervention.

The first approach to shaping focuses on manually changing the reward structure of the problem so that the problem is easier to learn [13, 3, 5, 4]. For example a gradient reward was added to the terminating reward in our previous work on a hierarchical reinforcement learning strategy applied to a mine field navigation task [10]. This approach to shaping has shown to be effective in speeding up the learning process, but it requires intimate knowledge of the environment in order to appropriately place intermediate rewards.

Another approach to shaping is to change the physics of the problem in order to make the problem easier to learn. This approach was applied to a mountain car task [7]. The height of the mountain was gradually increased for each run, enabling the agent to learn to climb the mountain faster than if the task is learned on just the final height. This approach has also shown to be effective, but it is impractical to change the physics of a real-world environment, and also requires knowledge of how to appropriately change the physics. It can be practical to apply this approach to shaping in a simulator, then transfer the knowledge learned in a simulator into a robot or other physical system.

In the last approach to shaping, the policy learned for one task is modified through reinforcement learning in order to perform a new task [2]. The idea here is to only change the portion of the previously learned policy which is different than the original. In order for the task to be learned in less time than just learning the task from scratch, Bowling et. al. needed to fix the overlapping portion the task so that it could not be modified. This required manual intervention in order to determine which portions should be learned and which should remain fixed.

Although each of the shaping techniques described above allowed the agent to learn complex tasks faster than traditional reinforcement learning techniques, they all required some form of *design intervention*.[1] Part of the reason intervention is required is in the subtleties involved in appropriately transferring previously learned policies.

In this work we demonstrate the failure of a naive approach to shaping on a simple wall-following task. We then present an analysis of shaping in situations where the problem dynamics stay the same, but where the reward structure of the problem is different, and present an algorithm for applying shaping in this situation that doesn't require manual intervention. We then demonstrate the success of this technique on simplified simulations, and on a more realistic simulation of obstacle-avoidance and wall-following.

2 The wall-following task

Using a reinforcement learning agent in a simulated environment, we explored the application of shaping in learning a wall-following behavior. Specifically, the agent has as 8 sonars inputs, evenly spaced throughout 360 degrees. The agent is given a reward of 1.0 whenever the agent detects a wall with its left or right sonar and is within 35% of its maximum sonar range. The agent is given a reward of −1.0 for colliding with a wall.

The agent has a choice of five different actions, each of which represents the desired turn angle

---

[1] manual intervention by the designer of the system

for the robot. The actions include a left or right turn of 0, 20, or 60 degrees. The robot must continue moving forward as it turns, and cannot move backwards. The agent uses a CMAC [1] to approximate the Q-values for each action. The CMAC has 10 layers, each layer containing 3 bins per input.

A natural precursor to the wall-following task is the task of wall-avoidance. In this task the agent was trained to avoid collisions with walls. Specifically, the agent was given a reward of -1.0 whenever the agent collided with a wall, and a reward of 0.0 otherwise. Final Q-values from the learned wall-avoidance task were then used as the initial Q-values in a wall-following task.

We trained five agents directly on the wall-following task, and five additional agents first on wall-avoidance and then on wall-following. Figure 1 shows the average learning curve of the agents trained from scratch on the wall-following task. Figure 2 shows the average learning curve of the agents trained on the wall-following task through shaping. As can be seen in Figure 2 this approach to shaping was dismally inadequate.

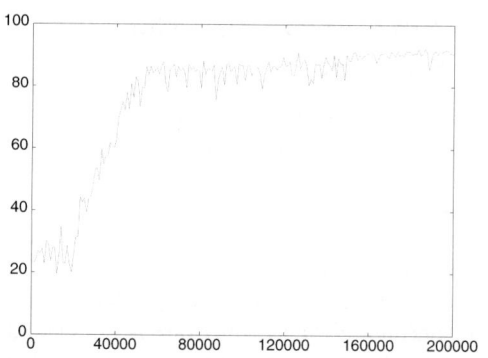

Figure 1: Average of 5 runs of learning wall-following from scratch. Shows the percentage of time the agent is performing wall-following.

Not only did the shaping attempt not learn the desired task more quickly than a "control" agent learning the task from scratch, but the agent failed to learn the task *at all* within 200,000 time steps (a more than generous allotment of time). Other researchers have also noticed this phenomenon [2].

This unexpected result lead us to a more careful examination of the principles involved in shaping, particularly in shaping situations which involve modified reward structures in a static environment. The hope was that a better understanding of the mechanics of shaping would lead to an effective algorithm which could be successfully applied to various shaping situations.

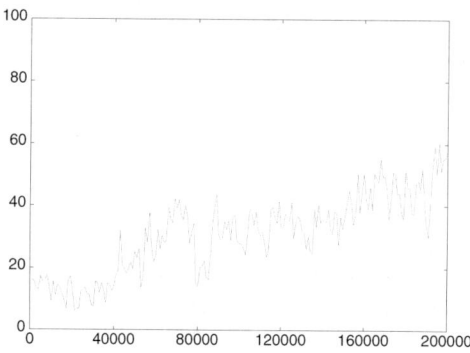

Figure 2: Average of 5 runs of learning wall-following through shaping. Shows the percentage of time the agent is performing wall-following.

3 Analysis

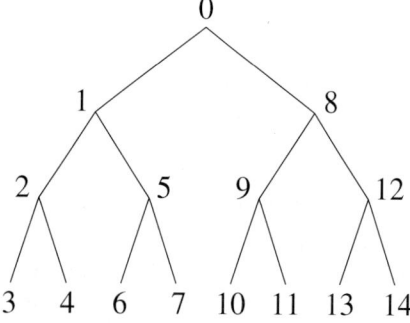

Figure 3: Simple analysis task. Agent starts task at node 0 and terminates at one of the nodes at the bottom of the tree.

We begin our discussion of shaping by considering a very simple decision task shown in Figure 3. In the task the agent starts at the top node of the tree. Each of the leaf nodes of the tree are termination nodes for the decision problem. One of the paths (unknown to the agent) terminates in a reward. The agent must learn to consistently choose the path which leads to the reward. This task is very easy and is learned in about 10 iterations using a best-first exploration strategy.

In the second phase of training, the task is changed by altering the location of the reward. If the reward remains close to its previous location (is moved perhaps one node to the right) then a large portion of the optimal policies of the first and second tasks are identical. Intuition suggests that information acquired during the first task should be easily applicable to the second, making this an ideal problem for shaping.

Several shaping approaches were tried, but each failed to significantly decrease learning time.

3.1 Direct Transfer of Q-values

If the agent trains on the initial task until all of the Q-values reach their optimal values, all of the Q-values in the whole tree will be zero, except for those on the branch leading towards the goal. When the agent is retrained on the task, using Q-values learned from the previous task as the initial Q-values for the new task (a technique which we call direct transfer of Q-values), the agent takes slightly *longer* to converge to an optimal policy.

The reason for this extra time is explained by considering that the agent is losing information about its environment faster than it is gaining information. If we assume a discount factor of .95 and a learning rate of 1.0 (which is sufficient for a deterministic world), and a best-first (deterministic) exploration strategy, the results of training are as follows. During the first episode of training the agent (incorrectly) assumes that the goal will be exactly where it was. The agent receives a reward of 0.0, and updates its Q-values to be zero. At this point, all of the Q-values actions leading towards leaf nodes are zero. This Q-value of 0.0 is propagated backwards each successive episode until the value of the root node is also updated to zero. The agent is then required to do a blind search for the reward.

We might assume that the learning rate or the discount factor is the problem here. However, similar problems arise with other learning rates. When a reward is moved from its original position the original policy must be unlearned. The difficulty is that if the learning rate is too low, the agent spends even *more* time unlearning the task than it would take to learn the initial task with a high learning rate.

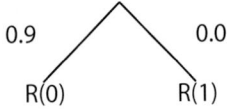

Figure 4: Simple decision problem. If Q-values are initialized as shown, the problem is difficult to learn.

For example consider the above simple decision problem. When the reward is moved just one position to the right the agent has a very difficult re-learning task. It must drop the left Q-value of .9 all the way down to say .49, while moving the 0 up to .5.

Thus the learning rate must balance the agent's need to unlearn incorrect old information, while preserving old information which was correct. Best performance was achieved with a learning rate near 0.5. Changing the discount factor had little effect.

3.2 Modified Prioritized Sweeping

To help unlearn old incorrect information we tried an algorithm which propagates the change in Q-values more quickly, (similar to prioritized sweeping [6].) The modified prioritized sweeping doesn't use a model of the environment, it simply keeps a history of the past n updated nodes, which will be the state's predecessors, and propagates updates back to these nodes.

Modified prioritized sweeping decreased the overall search time for both shaped and traditional reinforcement learning, but did not decrease the shaping time in comparison to the original learning time.

3.3 Alternate Exploration Strategies

We also tried several alternate exploration strategies [12] including recency-based, counter-based, and error-based exploration. None of these methods work in conjunction with direct transfer of Q-values for the same two reasons: First, if the learning rate is too high, correct information is overwritten as new Q-values are updated. Second, if the learning rate is low enough to prevent the overwriting of good information, it takes too long to unlearn the incorrect portion of the previously learned policy.

4 Memory-Guided Exploration

One solution to these problems is to keep the old Q-values in a separate place, and initialize the new Q-values to the same starting position as when learning from scratch. This is advantageous because the agent doesn't need to spend additional time unlearning incorrect information.

As learning commences, all of the Q-values are set to some initial value such as 0.5. Even when a small learning rate is used, the policy becomes correct with one update, even though Q-values may be far from their optimal values.

For example, in the decision problem shown in Figure 5 only one iteration is required to find the optimal policy, (even if the agent makes the wrong choice), as compared to many iterations in the decision problem shown in Figure 4.

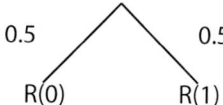

Figure 5: Simple decision problem. If Q-values are initialized to 0.5, the problem is learned in one iteration.

If the old Q-values are not used to initialize the new Q-values, how then are they used to facilitate shaping? If the old Q-values can be used to guide the agent in its initial exploration, the agent should be able to find the optimal policy much faster. The update of the new Q-values is done from scratch, and is unaffected by the old Q-values. In this way the agent is prevented from having to unlearn an incorrect policy.

One exploration policy that worked well was a version of choose best exploration with an evaluation function of:

$$Eval(s,a) = W_1 * Q(s,a) + W_2 * oldQ(s,a),$$

where W_1 and W_2 are weights that represent the amount of consideration that should be given to each value. $Q(s,a)$ represents the current Q-value of state s and action a, and $oldQ(s,a)$ represents the Q-values learned in the primary task. W_2 should eventually decay to 0.0 so that the final policy is based only on the new Q-values. This ensures that the agent will eventually learn the new task.

The effect of this exploration policy is to initially bias exploration near the previous policy. W_2 is set to be small enough that if the previous policy is incorrect, as Q-values are updated the exploration shifts toward the new Q-values. If the optimal policy is similar to the old policy then this will significantly reduce the time required to learn the new task. In cases where the primary and secondary tasks are dissimilar, shaping would normally be inappropriate. However, this shaping algorithm requires only a few more time steps to learn a dissimilar task than it would to learn it from scratch. Therefore, a robot encountering a completely unknown problem can safely choose to apply shaping if a primary policy has been learned previously.

5 Results

We applied memory-guided exploration to the simulated wall-following task. The initial exploration weights of $W_1 = 1.0$, $W_2 = 0.1$. W_1 remained constant and W_2 decayed to 0 by 50,000 steps.

Figure 6 shows the agent learning the wall-following task from scratch, and Figure 7[2] shows the agent using the memory-guided exploration algorithm.

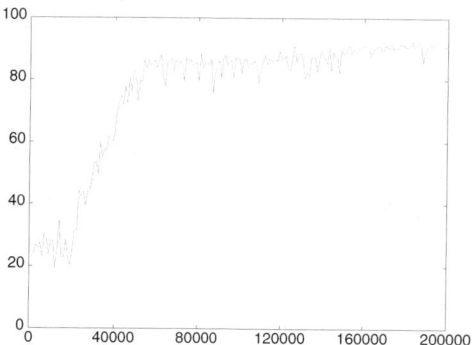

Figure 6: Average of 5 runs of learning wall-following from scratch. Shows the percentage of time the agent is performing wall-following.

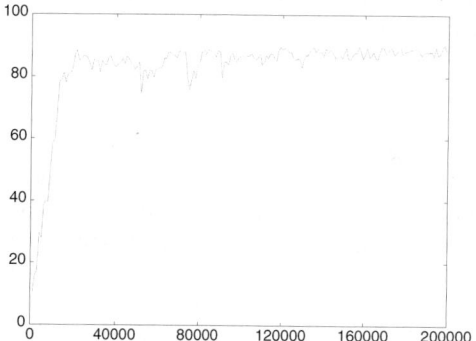

Figure 7: Average of 5 runs of agent shaping wall-following using memory-guided exploration. Shows the percentage of time the agent is performing wall-following.

Using the new exploration policy, the agent learned the same shaping task in 20,000 steps as compared to 50,000 steps to learn the task from scratch, and more than 200,000 steps with direct transfer of Q-values.

---

[2]The CMAC offsets were incorrectly restored during information transfer, further complicating the task. Direct transfer failed to compensate, while memory-based exploration succeeded. Preliminary data suggests that while direct transfer can perform better than shown here, memory-based exploration is still more effective.

6 Conclusion

Effective shaping in a reinforcement learning setting can be difficult. This happens for two main reasons, first the agent must unlearn a biased incorrect policy, and second as new Q-values are updated useful information about the old policy is lost. These difficulties can be overcome through the use of a memory-guided exploration policy. This policy allows the agent to initialize and update Q-values normally, while retaining knowledge of the previous policy which can be used to guide the agent's exploration.

This algorithm was shown to be effective in shaping an obstacle-avoidance behavior into a wall-following behavior. Memory-guided exploration attained the desired behavior much more quickly than either learning the task from scratch or the use of a naive shaping algorithm based on direct transfer of Q-values. This algorithm also differs from previous shaping approaches in that it is more easily automated.

7 Future Research

The successful development and application of this algorithm suggest several potential avenues of future research. Examination of our preliminary graphs indicates that the proficiency of an agent in learning a shaped task is dependent upon how fully the original task was learned. Further research must be done to determine whether this is the case for all shaping situations or only for certain shaping problems.

This exploration strategy could conceivably use memory of more than one past skill to explore in an unfamiliar situation. Thus, an agent with a repertoire of several previously learned skills could apply knowledge acquired in all of these tasks to a new problem. A filtering mechanism could be used to determine which of the previously-learned tasks are most similar to the current one.

One immediate application of this algorithm is the transfer of simulator-learned policies to real-world robots. Because of the difficulties discussed earlier, simulators rarely model a problem accurately, and policies which execute perfectly on simulators often fail to produce the desired results in the real world. However, simulated and real-world tasks have an extremely high degree of similarity. Agents trained on a simulator could use memory-guided exploration to apply simulated information to the real world. This approach would combine the quick learning rates achieved with simulators with the optimal behavior obtained through real world training.

References

[1] J. S. Albus. A new approach to manipulator control: The cerebellar model articular controller (cmac). *Trans. ASME, J. Dynamic Sys., Meas., Contr.*, 97:220–227, 1975.

[2] M. Bowling and M. Veloso. Reusing learned policies between similar problems. In *Proceedings of the AIIA-98 Workshop on new trends in robotics*. Padua, Italy, 1998.

[3] M. Dorigo and M. Colombetti. The role of the trainer in reinforcement learning. In *Proceedings of MLC-COLT '94 Workshop on Robot Learning, S.Mahadevan et al. (Eds)*, pages 37–45. New Brunswick, NJ, 1994.

[4] M. Dorigo and M. Colombetti. Precis of robot shaping: An experiment in behavior engineering. *Adaptive Behavior*, 5:3–4, 1997.

[5] M.J. Mataric. Reward functions for accelerated learning. In *Machine Learning: Proceedings of the Eleventh International Conference*. Morgan Kaufmann, CA, 1994.

[6] A. W. Moore and Christopher G. Atkeson. Prioritized sweeping: Reinforcement learning with less data and less real time. *Machine Learning*, 13:103–130, 1993.

[7] J. Randlov and P. Alstrom. Learning to drive a bicycle using reinforcement learning and shaping. In *Machine Learning: Proceedings of the Eleventh International Conference*. Morgan Kaufmann, CA, 1999.

[8] B. F. Skinner. *The Behavior of Organisms: An Experimental Analysis*. Prentice Hall, Englewood Cliffs, New Jersey, 1938.

[9] B. F. Skinner. *Science and Human Behavior*. Colliler-Macmillian, New York, 1953.

[10] R. Sun and T. Peterson. A hybrid model for learning sequential navigation. In *Proc. of IEEE International Symposium on Computational Intelligence in Robotics and Automation*, pages 234–239, San Diego, CA., 1997. IEEE.

[11] R. Sutton. Learning to predict by the methods of temporal differences. *Machine Learning*, 3(1):9–44, August 1988.

[12] Sebastian Thrun. Exploration and model building in mobile robot domains. In *Proceedings of the 1993 International Conference on Neural Networks*, 1993.

[13] D.S. Touretzky and L.M. Saksida. Operant conditioning in skinnerbots. *Adaptive Behavior*, 5(3/4):219–247, 1997.

[14] C. J. C. H. Watkins. *Learning from Delayed Rewards*. PhD thesis, University of Cambridge, 1989.

Obstacle Avoidance Learning for a Multi-Agent Linked Robot in the Real World

Daisuke Iijima, Wenwei Yu, Hiroshi Yokoi, and Yukinori Kakazu

Autonomous Systems Eng. Lab., Hokkaido Univ.
N-13, W-8, Kita-ku, Sapporo 060-8628, Japan
{iijima, yu, yokoi, kakazu}@complex.eng.hokudai.ac.jp

ABSTRACT

In order to achieve an autonomous system which can adaptively behave through learning in the real world, we constructed a distributed autonomous swimming robot that consisted of mechanically linked multi-agent and adopted adaptive oscillator method that was developed as a general decision making for distributed autonomous systems (DASs). One of the our aims by using this system is to verify whether the robot could complete a target approaching including obstacle avoidance. For this purpose, we introduced a modified Q-learning in which plural Q-tables are used alternately according to dead-lock situations. By using this system, as a result, the robot acquired stable target approaching and obstacle avoiding behavior.

1. INTRODUCTION

This work proposes a design method for an autonomous robot system consisting of mechanically linked plural identical agents that enable the robot to behave adaptively in the real world.

Much attention has recently been focused on distributed autonomous systems (DASs) due to limitation of canonical centralized systems when they are assigned more and more complicated tasks. A DAS does not have a central controller that supervises the whole system. It is composed of structural elements that interact cooperatively or competitively with each other by behaving autonomously but act as one system to enable the objective task to be performed. A DAS therefore has the following advantages : 1). adaptability to a complex environment based on the self-organizability of the control rules by dynamical interaction of structural elements, and 2). competence to complete the objective tasks by re-configuration of the control rules in the cases where there is a partial fault in the system.

Considering these advantages, the usefulness of DASs for robots has been studied by using distributively structured real module robots (Here, a module is called an "agent", defined as an independent decision-making-unit that can input and output to the outer world) [1], [2], [3]. In these works, self-organizability in morpho-generation or re-configuration of the system and movability on various morphologies were discussed; however, there was little discussion of the behavioral design in an actual dynamical environment.

One of the most important features of a multi-agent-linked robot is self-organizability of locomotion patterns, which enable the robot to adapt to complex environment by interactions among plural decision-making mechanisms and by various motions generated by multi-part actuators. Therefore, in this study, we proposed adaptive oscillator method (AOM) as a decision-making method based on interaction with other agents and with the outer world for the multi-agent linked robot systems. In this paper, we employed a learning method to realize the AOM. In order to verify this learnability of locomotion patterns, in this study, we constructed a distributed autonomous swimming robot according to the proposed design method and make it learn target-approaching behavior in a complex water environment.

In previous studies, we have confirmed obtainability of target approaching behavior in an environment that has no obstacles [4] and competence to complete an objective task in the case where there is a partial fault in the system [5] by the swimming robot. In this paper, as the next essential problem for autonomous robots, the problem of obstacle avoidance is dealt with. For learning efficiency, we propose the Switching-Q learning method in which the control shifts between the action rules obtained in the previous environment and action rules acquired later in an environment containing obstacles.

2. AGENT DESIGN

2.1 Design Conditions for Adaptive Behavior Learning
For a robot constructed by multi-agents, designing the system means designing agents. One difficult point in designing DASs is how to embed dynamics, that is, a mechanism by which action rules are generated by interaction among structural elements. In the following, a method for designing agents for autonomous acquisition of environment-oriented locomotion patterns is discussed by focusing on the features of a complex environment and the structure of a DAS.

2.1.1 Realization of Locomotion in a Complex Environment: The real world is a complex environment. Various physical force in the surrounding environment must be taken into consideration in order for a robot to be able to move in the real world. For instance, the movement of a fish in water is determined not only by the fin's motion at that time but also by many other factors, such as the force of inertia caused by past movement of the fish, water flow, and the flow generated by other moving objects near the fish. Therefore, a process by which locomotion patterns are obtained by dynamical interaction with the outer world appears to be necessary.

2.1.2 Effective Use of Multi-Agent Linked Structure: In a situation where multiple agents connected mechanically, an action of one agent is considerably affected by the other agents' actions. Then, it is necessary that objective functions of all agents are set identically because if so, it is expected that all agents cooperatively behave and effectively learn locomotion patterns.

Furthermore, one of the advantages of such a structure

is the easy expansion or reduction of scale. In order to ensure that this merit can be maintained, each agent must be designed as a "gradeless" and homogeneous unit in both terms of structure and function so that the system can adapt to any change in scale or arrangement.

2.2 Decision-Making : "Adaptive Oscillator Method"

The adaptive oscillator method (AOM) [4] was proposed as a decision-making method based on interaction with the outer world for multi-agent structured robot systems. A general form of the method is shown in equation (1):

$$g(\sum_n k_n \frac{d^n}{dt^n} y) + f(I) = e, \quad (1)$$

where
- y: solution of equation (1), i.e., vibrating action mode that the agent performs
- k_n: coefficient toward differentiated y, the weight of an n-dimensional value
- $g()$: function to transform outputs of the differential calculating term
- I: input vector including all sensory inputs and decision-making factors (the energy acquired by the outputted action of the AOM also being a component of this input vector)
- $f()$: function to transform all external inputs
- e: internal energy for each agent

The first term on the left side of equation (1) is a differential calculating term that enables the system to reflect the effects of long-term, high-dimensional derivatives because it is constructed as an AR (autoregressive)-related model. Hence, the system's action in the next step is determined according to time sequential data involving effects on inertia, current, waves, etc. The second term on the left side is an external-effect-input-term, and the right side of the equation is the internal energy of the agent. Thus, it is assumed that all parameters are smoothly learned by the effect on energy management (but, actually, in this paper, the energy value is not dealt with in order to simplify the problem).

3. EXPERIMENTAL SETTINGS

3.1 Distributed Autonomous Swimming Robot

In determining hardware structure, the type of arrangement among agents affects the movability of the robot. In this study, the robot is realized by linking all agents as a two-dimensional chained structure.

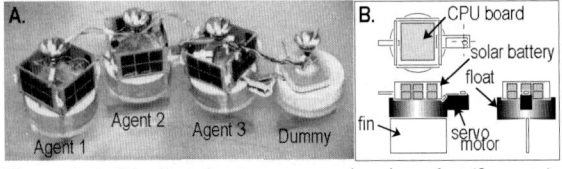

Figure 1: A, Distributed autonomous swimming robot (3 agents); B, Structure of an agent

As shown in Figure 1-A and -B, each round float (diameter : 15 cm) is taken as one agent, and a three-agents version is dealt with here. The total length of the robot system is 80 cm, and the total weight is 5 kg. Each agent has four solar panels that function as light sensors, a CPU, and an angular controlling servo motors (torque : 9.5 kg-cm) that located at the pivot of each joint between neighboring agents.

Two-channel serial ports are used for communication between agents. Moving force is generated by pushing water aside with fins. Additionally, for structural necessity, the robot has a dummy agent at one end of the chain. This dummy agent carries battery cells for driving the robot.

3.2 Other Experimental Settings

- Work Space

As the working space, water environment where physical force such as currents or waves remarkably affects motion of the robot is chosen.

- Communication Range

The communication range for each agent is limited to neighboring agents. That is, agents 1 and 3 (see Figure 1-A) do not know each other's situation.

- Negotiation in Decision Making

In decision making, the agents do not negotiate with each other. Cooperative behavior is expected to be obtained through the learning process.

- Synchronicity in Action

For simplification of the problem, all agents are assumed to be synchronized when they execute their actions.

- Introduction of Elementary Actions

For learning efficiency, oscillation patterns based on a sine function are preset as elementary actions. These patterns can be acquired by transforming the differential calculating term of equation (1) as follows.

In the differential calculating term, y is replaced with x, instead, the whole this term is replaced with y. Then, equation (2) is input to the differential calculating term (m is natural number), and we obtain the sine function as equation (3) (Infinity is introduced to n.).

$$k_n = \begin{cases} 0 & , n = 2m \\ \frac{-(-1)^{(n+1)/2}}{n!} & , n = 2m-1 \end{cases} \quad (2)$$

$$y = x - \frac{x^3}{3!} + \frac{x^5}{5!} - \frac{x^7}{7!} \cdots = \sin x \quad (3)$$

- Introduction of Learning Method: "Q-Learning"

Although there are several possible learning methods that could be used in this system, we employed the Q-learning method [6]. This method is an approximate dynamic programming method in which only value updating is performed based on local information. This fast step-by-step value updating and decision making can reduce the sequential effects of waves and inertia in a water environment. If the learning environment is based on the Markov Decision Process [7], it can be guaranteed that learning will proceed continuously, thereby guaranteeing the successful acquisition of target-approaching behavior.

The Q-table is constructed as follows. The sensing states S are divided into 4 states. Each state denotes one of the 4 directions by which the largest value is sensed. As shown in Figure 2-A, if the sensor on the right side receives the largest input, the representation of the agent's sensing state is defined as $[S:3]$. On the other hand, the action patterns A are divided into 12 patterns from the basic sine function that has an amplitude of 30 degrees. Twelve variations are achieved by using 3 variations of center angle and 4 variations of phase (Figure 2-B). If the actions chosen by all agents are proper, the robot can swim forward or backward in a straight line, forward and to the left or right, or backward and to the left or right. As shown in the figure, if the oscillation center angle of the chosen action is +30 degrees and its phase is 0, the representation of the agent's action pattern is defined as [A:3-2]. The execution term of each oscillation motion per one action is two periods.

current selection probability is maximum, α is the learning rate, γ is the discounting rate, r_t is the reward value, and T is the thermal parameter that decides the randomness of the action. In the experiments, we used $\alpha=0.1$, $\gamma=0.95$, $r_t=100$, and $T=4.0$ based on the experiences in previous tests [5].

$$\Delta Q(S_t, A_t) = \alpha(r_t + \gamma \max_b Q(S_{t+1}, b) - Q(S_t, A_t)) \quad (4)$$

$$P(A|S) = \frac{\exp(Q(S,A)/T)}{\sum \exp(Q(S,A)/T)} \quad (5)$$

4. EXPERIMENTS

4.1 Exp. 1: Target Approaching

We first performed a behavior acquisition test for target approaching, since the probability for achieving a task would be reduced and the learning efficiency would be much worse if an obstacle avoidance test were first performed. In the experiment, the robot was place in the water at one corner of a 2.6 m^2 square pool. The target, which was a light source, was set at the opposite corner. A reward was given to all agents only when the robot reached the target. The same trials were repeated.

A total of 405 trials were carried out. Figure 3 shows the results of the first to the 150th trial. This is a transition of the number of steps needed to reach the target.

After 100 trials, the number of steps needed to complete the task became relatively stable at about 15 steps.

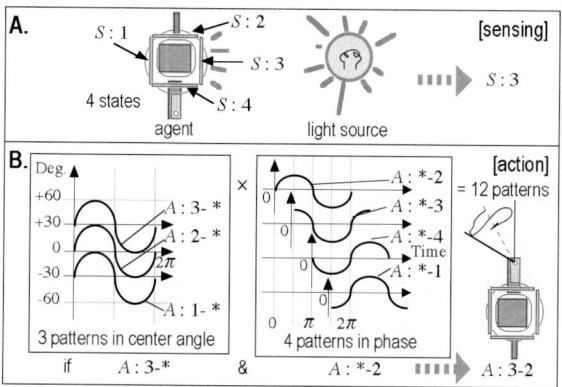

Figure 2: A, Definition of sensing states;
B, Definition of element action patterns

To set the Q-value, information collected by neighboring agents is introduced. However, in order to prevent the size of the state space exploding, we determined the Q-value using only the sensing states of the neighboring agents (number of the state spaces, 768).

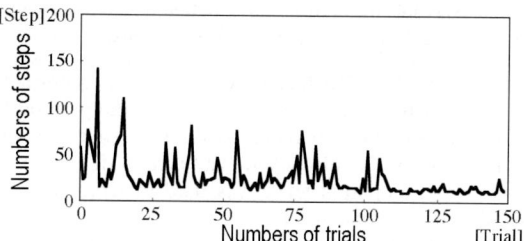

To update the Q-value, equation (4) is used. For action selection, the Boltzmann exploration depending on equation (5) is implemented. Here, S_t and A_t are the state and the action of the agent at time t, respectively, $Q(S_t, A_t)$ is the Q-value at that time, b is the action for which the

Figure 3: Numbers of steps needed to reach the target

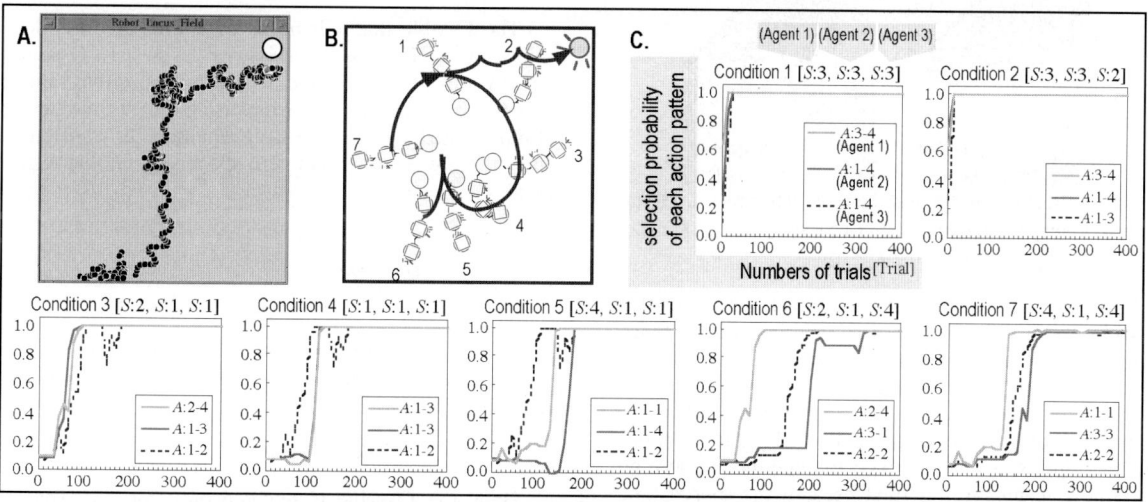

Figure 4: A, Replay behavior based on acquired action rules; B, Flows of stable motion patterns in replay behavior;
C, Transitions of selection probabilities of the seven motion patterns shown in Figure 4-B

Figure 4-A shows an example of the robot's behavioral locus (coordinates of gravity) observed by reusing the action rules acquired after the above trials. In the figure, the white circle at the corner is the position of the light source. It can be seen that the robot approached the light source from the left side of the figure.

Furthermore, Figure 4-B shows the flows of stable motion patterns in replay behavior, and Figure 4-C shows the transition of the selection probabilities of the motion patterns. In these figures, motion patterns in the most observable seven conditions are depicted. First, the robot acquired two motion patterns in certain sensing states, which are referred to as conditions 1 and 2. (In condition 1, the sensing states of the robot's three constituent agents (1, 2, and 3) can be expressed as [S:3, S:3, S:3] and the action patterns can be expressed as [A:3-4, A:1-4, A:1-4]. In a similar way, in condition 2, the sensing states can be expressed as [S:3, S:3, S:2] and the action patterns can be expressed as [A:3-4, A:1-4, A:1-3].) The robot gradually became able to reach the target from other situations by using the conditions as sub-goals. Then, finally, the robot learned sequences of motion patterns for reaching the target from any situation or position. Accordingly, it was verified that the robot acquired almost all of the basic motion patterns.

4.2 Exp. 2: Obstacle Avoidance

In the next obstacle avoidance test, the objective of this task was to solve obstacle avoidance problems as simply as possible by reusing previously obtained basic action rules without special measures, e.g., detailed environmental maps and correctly recognizable obstacle sensors, etc.

4.2.1 Setting of the Obstacles:
At first, we set up obstacles and observed the robot's behavior by using the previously acquired action rules in order to discuss what measure should be taken. The obstacle was a fence made of acrylic resin, and a transparent fence was used to avoid the perceptual aliasing problem [8].

The width of the entrance was 32 cm, about twice the diameter of each agent.

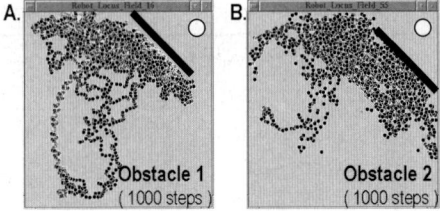

Figure 5: A, Replay behavior based on past action rules in Obstacle 1; B, Replay behavior based on past action rules in Obstacle 2

Figure 5-A shows one case of obstacle setting (called Obstacle 1). In this case, the robot approached the fence using the locomotion patterns for reaching the target from the left side, but after that, the robot was not able to approach to the target because of the fence. Figure 5-B shows another case of obstacle setting (called Obstacle 2). In this case, the robot was not able to go through the entrance no matter how many times if tried because the approaching angle was not appropriate for entering. In both of these cases, the robot could not reach the target even though 1000 steps were performed.

4.2.2 Proposed Method: "Switching-Q":
For cases involving complex problems, such as a robot's navigation learning, some hierarchical learning methods have been proposed ([9], [10], [11], etc.). It has been verified that such a hierarchical learning method works effectively for a centralized controlled systems, but the effectiveness of such a distributed controlled system is not guaranteed. Thus, a new method is needed for a distributed controlled system. In addition, the mechanism of the method should be constructed as simply as possible in order for the method to work effectively in the real world.

Based on the observations described in section 4.2.1, it is thought that one general solution for effective learning of motion in more complex environments by reusing previously obtained rules is a method in which the robot first moves by reusing past rules, and if the robot falls into a deadlock situation, it starts to learn by means of acquiring new action rules. For this purpose, a new learning method, which is described by equation (6), is proposed. The method has a deadlock checking mechanism and past and new rules. The control shifts between the past rules and new rules, depending on the deadlock situation.

$$Switching\_Q = \{Q_1, ..., Q_x, Deadlock\_Detector\} \quad (6)$$

The new Q-table for learning is assigned to Q_x, which has the maximum value of x. Here, only the case of $x=2$ is dealt with, so the actual Q-tables can be represented as follows.

$$Q_1(S_t, A_t): Past\_Qtable$$
$$Q_2(S_t, A_t): New\_Qtable$$

Switching-Q works as follows. First, for i steps from the start, the past Q-table is used unconditionally. During the trial, the log of the sensed values in recent j steps, $Sv_{t-j}, ..., Sv_t$ is memorized, and in each time step, slope l is calculated by the function $v()$, which calculates the linear approximation of their log data (equation (7)).

$$l = v(Sv_{t-j}, ..., Sv_t) \quad (7)$$

The Q-table that will be used in time step t is determined by l. If l is a positive value, the past Q-table is chosen at that time step, and if l is not a positive value, a new Q-table is chosen. Q_t which is chosen at time step t is represented by the following equation. In addition, i and j were set to 9 and 5, respectively, in this experiment.

$$Q_t = \begin{cases} Q_1(S_t, A_t), & 1 \leq t \leq i \\ Q_1(S_t, A_t), & t > i \cap l > 0 \\ Q_2(S_t, A_t), & t > i \cap l \leq 0 \end{cases} \quad (8)$$

If Q_2 is used, the Q-value at that time step is memorized as $Rule_{recent}$, which is the rule that has been used most recently (equation (9)).

$$Rule_{recent} = Q_2(S_t, A_t) \quad (9)$$

Here, the Q-value is updated only for Q_2. That is, $Rule_{recent}$ is updated every time when Q_2 is used. Finally, when the task is completed, the reward is given to the rule.

Moreover the reward is propagated to Q_2 gradually by equation (4).

The feature of Switching-Q is to active reuse of past experiences. One shortcoming, however, is that an optimal solution is difficult to obtain because the new Q-table tends to learn the parts that are weak points in the past Q-table. Furthermore, this method can be applied up to the case of $x=2$, but if x is 3 or greater, a new mechanism for switching Q_2 to Q_x is needed. Hence, further study is needed in cases where three or more action rules are necessary in the same sensing states.

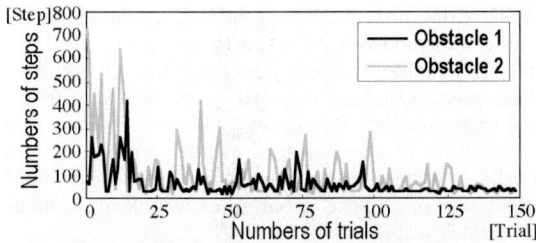

Figure 6: Numbers of steps needed to reach the target in Obstacle 1 and Obstacle 2

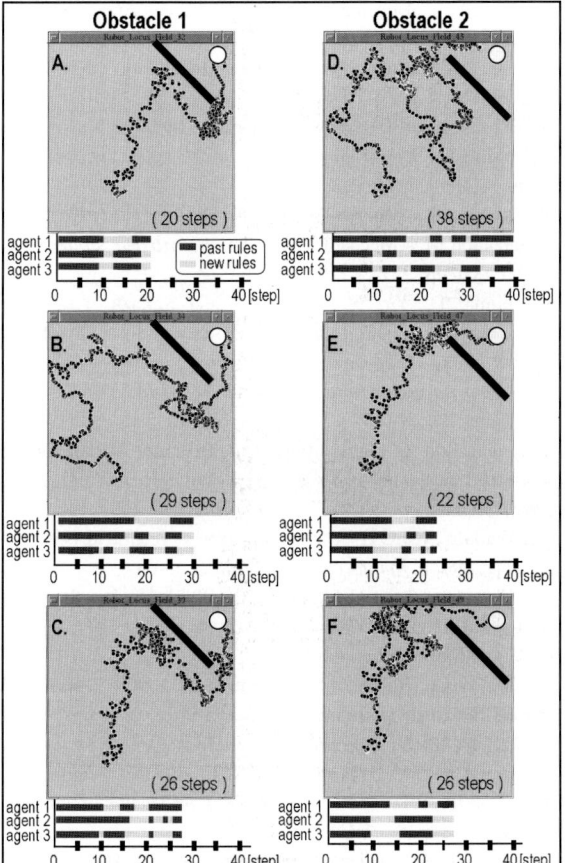

Figure 7: Replay behavior based on Switching Q in Obstacle 1 and 2, and Transitions of chosen action rules in the replay tests

4.2.3 Experimental Results: The procedure for experiment 2 was similar to that for experiment 1.

Figure 6 shows the numbers of steps needed to reach the target in Obstacle 1 and Obstacle 2. Just after the start of the tests, the robot sometimes needed over 400 steps in Obstacle 1 and over 700 steps in Obstacle 2 to reach the target. However, the number of steps gradually decreased, and after the 130th trial, the robot could complete the task in about 30 steps in both cases.

Figure 7 shows some results of replay tests in Obstacle 1 and Obstacle 2 performed by reusing the action rules acquired by Switching-Q. The results show that all achieve the tasks although there are various behavioral patterns.

Under each of the pictures showing the results of behavior, the transition of action rules chosen by each agent is shown. The dark-colored lines represent the past rules, and the light-colored lines represent the new rules. After the first nine steps that was assumed to use the past rules in section 4.2.2, it was observed that each agent individually switched from new to past rules, or from past to new rules, based on the deadlock checking.

5. DISCUSSION AND ANALYSIS

5.1 Behavior Learning in a Distributed System

First, we consider the mechanism of behavioral learning of simple target approaching. The main point of the discussion is how the motion patterns can be obtained as a whole system when each agent learns its action rules only from local information. Because each agent can not know the situation of the whole system directly, certain mechanism that can somehow extrapolate the unknown information ought to exist in the system.

Under the setting of AOM, each agent can know the local situation within the range of neighboring agents. Then, it is expected that "weak" network for sharing information is constructed among agents according as all agents know the neighboring situation. On the other hand, by the function of Q-learning, it is expected that action sequence will be learned in each agent. By this way, it is thought that the system has a possibility of making implicit connection of each state among agents and among time series.

In addition, according to the setting of same object function for cooperative learning and the condition of mechanical constraint, a schema in which an effect caused by one agent that precedently acquired motion patterns propagates to the other agents and the behavioral learning proceeds continuously is considered. Actually, we can see such behavior in Figure 4-C. Where, it is observed that the agents that are left behind raise their selection probabilities one after another, as if the agent that first learned led the others.

Table 1: Rate of Reused Rule Transition in Replay Experiment

$Q(S_t, A_t)$ $\downarrow$ $Q(S_{t+1}, A_{t+1})$	single Q-learning		Switching-Q	
	initial phase	non obstacle	Obstacle 1	Obstacle 2
Rate of Inter-agent (%)	7.3	90.5	64.6	63.6
Rate of Inner-agent (%)	17.1	91.0	68.1	65.9

To verify this propagation effect and also to investigate the reproducibility of the rules obtained, we calculate the rate of reused rule transition in replay experiment. Table 1 shows the statistical data, in which, "Rate of inter-agent" means the percentage of re-appeared one-time-step rule transition between two neighboring agents, with respect to all the rule transition appeared in replay experiment. On the other hand, "Rate of inner-agent" means that of rule transition inside the certain single agent. From this table, we can see that in the single Q-learning case, the corresponding rates of both cases were about 10% at initial phase of learning, while, after learning, the rates rose up to over 90%. That is, as a result of distributed learning, selection probabilities of actions so rise that some strong connections of rules among the agents or inside one individual agent were implicitly formed, consequently, the sequential motion patterns were acquired.

5.2 Mechanism of Behavioral Acquisition

Next, we discuss the mechanism of behavior acquisition for obstacle avoidance. The main point of the discussion is the role of Switching-Q.

For verifying this point, replay behavior performed by using only newly obtained action rules was observed (Figure 8). In case of Obstacle 1, it was observed that the robot went directly through the entrance. On the other hand, in case of Obstacle 2, the robot remained in the lower right area of the picture. Thus, it was thought that the new rules were obtained so as to complete the objective task by combining them with past rules or by using only new rules.

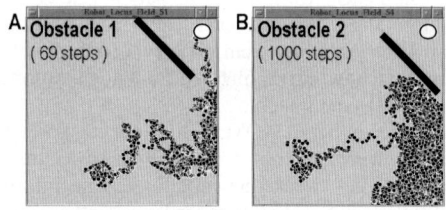

Figure 8: A, Replay behavior by new action rules in Obstacle 1; B, Replay behavior by new action rules in Obstacle 2

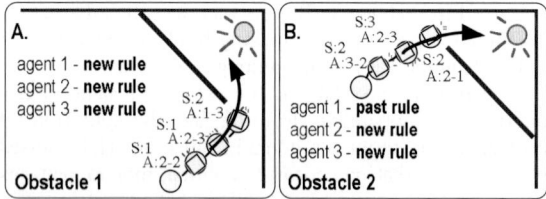

Figure 9: Obstacle avoiding behavior; A, 20th motion pattern in Figure 7-C; B, 34th motion pattern in Figure 7-D

In order to confirm this hypothesis, we focused on the behaviors of the robot in passing through the entrance in Obstacle 1 and Obstacle 2. Figure 9-A shows the entering behavior of the robot in the 20th step shown in Figure 7-C. Where, all agents used new action patterns because the previous locomotion patterns were not appropriate for the situation. On the other hand, Figure 9-B shows the entering behavior of the robot in the 34th step shown in Figure 7-D. Where, the robot was able to avoid the obstacle because the entering angle generated by combining the new rules with past rules was appropriate for passing through the entrance. Thus, each agent acquired its action rules in order to appropriately use those rules in various situations.

Furthermore, Table 1 shows that, in the Switching-Q case, the rates fall in all situations, comparing with the 90% of after-learning situation in Single-Q case. It is because that, for distributed agents, the transitions between new rule table and past rule table were not simultaneous. Nevertheless, the rates were over 60%, which is high, comparing with the initial phase in Single-Q case. Accordingly, it is confirmed that strong connections among rules were also constructed under Switching-Q.

7. CONCLUSION

In order to develop an autonomous system capable of adaptive behaving in the real world, we constructed a distributed autonomous swimming robot which consisted of mechanically linked multi-agent and adopted adaptive oscillator method as a general decision-making for DASs. And for learning, we proposed Switching Q-learning in which plural Q-tables are used alternately according to dead-lock situations. By this way, the robot acquired stable target approaching and obstacle avoidance behavior. Additionally, among agents, it was confirmed that implicit connections of acquired action rules were formed, though the learning was distributed.

REFERENCES

[1] Fukuda, T., Ueyama, T., Kawauchi, Y., and Arai, F. (1992). Concept of Cellular Robotic System (CEBOT) and Basic Strategies for its Realization, Computers Elect. Eng 18(1):11-39, Pergamon Press.

[2] Yim, M. (1993). A Reconfigurable Modular Robot with Many Modes of Locomotion, *Proc. of the JSME Int. Conf. on Advanced Mechatronics*, 283-288.

[3] Kokaji, S., Murata, S., and Kurokawa, H. (1994). Self Organization of a Mechanical System, *Distributed Autonomous Robotic Systems*, 237-242.

[4] Iijima, D., Yu, W., Yokoi, H., and Kakazu, Y. (1998). Autonomous Acquisition of Adaptive Behavior for a Distributed Floating Robot Based on the AHC Method, *Intelligent Engineering Systems through Artificial Neural Networks* 8:537-542.

[5] Iijima, D., Yu, W., Yokoi, H., and Kakazu, Y. (1999b). Distributed Robotic Learning: Adaptive Behavior Acquisition for Distributed Autonomous Swimming Robot in Real World, *Proc. of the 16th Int. Conf. on Machine Learning*, 191-199.

[6] Watkins, C.J.C.H., Dayan, P. (1993). Technical Note: Q-Learning, Sutton, *Reinforcement Learning*, 55-68.

[7] Christman, L., and Littman, M. (1995). Hidden State and Short-Term Memory, *Presentation at Reinforcement Learning Workshop*, Machine Learning Conf.

[8] McCallum, R.A. (1995). Instance-Based Utile Distinctions for Reinforcement Learning with Hidden State, *Proc. of the 12th Int. Conf. on Machine Learning*, 387-395.

[9] Lin, L.J. (1993). Scaling Up Reinforcement Learning for Robot Control, *Proc. of the 10th Int. Conf. on Machine Learning*, 182-196.

[10] Wiering, M., and Schmidhuber, J. (1997). HQ-Learning, *Adaptive Behavior* 6(2):219-246.

[11] Tani, J., and Nolfi, S. (1998). Learning to Perceive the World as Articulated: An Approach for Hierarchical Learning in Sensory-Motor Systems, *From Animals To Animats 5*, 270-279.

Performance Modeling of Supply Chains using Queueing Networks

N. Viswanadham
Mechanical and Production Engg.
National Univ. of Singapore, Singapore 119260
e-mail: mpenv@nus.edu.sg

N. R. Srinivasa Raghavan
Management Studies
Indian Institute of Science, Bangalore, India 560 012
e-mail: raghavan@mgmt.iisc.ernet.in

Abstract

Supply chain networks are formed out of complex interactions amongst several companies whose aim is to produce and deliver goods to the customers at the time and place specified by them. Computing the total lead time for customer orders entering such a complex network of companies is an important exercise. In this paper, we present analytical models for evaluating the average lead times of make-to-order supply chains. In particular, we illustrate the use of fork-join queueing networks to compute the mean and variance of the lead time. The existing literature on approximate methods of analysis of fork-join queueing systems assume heavy traffic and require tedious computations. We present two applications of a tractable approximate analytical method for lead time computations in a class of fork-join queueing systems. For the case where the arrivals are deterministic and service times are normally distributed, we present an easy to use approximate method. Specifically, we illustrate the use of the above method in setting service levels in assemble-to-order type supply chains.

1 Supply Chain Networks

Supply chain networks (SCNs) are formed out of complex interconnections amongst various manufacturing companies and service providers such as raw material vendors, original equipment manufacturers (OEMs), logistics operators, warehouse operators, distributors, retailers and customers (see Figure 1). One can succinctly define supply chain management(SCM) as the coordination or integration of the activities of all the companies involved in procuring, producing, delivering and maintaining products and services to customers located in geographically different places. Traditionally, each company performed marketing, distribution, planning, manufacturing and purchasing activities independently, optimizing their own functional objectives. SCM is a process-oriented approach to coordinating all organizations and all functions involved in the delivery process. The product moving through the SCN transits several organizations and each time a transition is made, logistics is involved. Also since each of the organizations is under independent control, there are interfaces between organizations and material and information flows depend on how these interfaces are managed. We define interfaces

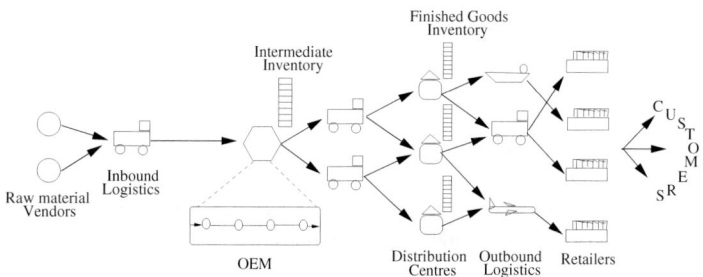

Figure 1: The supply chain network

as the procedures and vehicles for transporting information and materials across functions or organizations such as negotiations, approvals (so called paper work), decision making, and finally inspection of components/assemblies, etc. For example, the interface between a supplier and manufacturer involves procurement decisions such as price, delivery frequencies and nature of information sharing at the strategic level and the actual order processing and delivery at the operational level. The coordination of the SCN plays a big role in the over all functioning of the SCP. In most cases, there is an integrator for the network, who could be an original equipment manufacturer, coordinating the flow of orders and materials through out the network. Modeling and analysis of such a complex system is crucial for performance evaluation and for comparing competing supply chains. In this paper, we view a supply chain as a probabilistic network and present a modeling approach to compute performance measures such as lead time and work in process inventory. In particular, we investigate the use of queueing network models for computing the lead time and other performance measures. In the rest of this section we review the types of supply chain networks and the different order-fulfilment policies.

1.1 Operational Models

An important aspect of the supply chain operation is the supply chain planning and control methodology (SPC). A customer order for a product triggers a series of activities in the supply chain facilities, and these have to be synchronized so that the end customer order is satisfied. The SPC specifies the business model and hence determines the paths for the information and material flow in the supply chain. There are three broad models followed in practice: Make-to-stock (MTS), Make-to-order (MTO), Assemble-to-order

(ATO).

The crucial issues of when to order and how much to order define these policies. For instance in base stock policies, one unit (alternatively, a stock keeping unit, SKU) of inventory is replenished as soon as a unit of goods held at the facility is depleted. On the other hand, if the facility is following a reorder point based policy, it replenishes items as soon as a preset reorder level is reached, ordering each time such that a targeted level of inventory is reached. For a detailed discussion, refer [12].

1.2 Modeling of Supply Chains

Supply chain networks are discrete event dynamical systems (DEDS) in which the evolution of the system depends on the complex interaction of the timing of various discrete events such as the arrival of components at the supplier, the departure of the truck from the supplier, the start of an assembly at the manufacturer, the arrival of the finished goods at the customer, payment approval by the seller, etc. The state of the system changes only at discrete events in time. Over the last two decades, there has been a tremendous amount of research interest in this area. There are several classes of models that are useful in this context. These models can be used for either qualitative or quantitative analysis. Qualitative analysis yields results on stability and deadlock analysis. Quantitative methods, on the other hand, highlight the determination of system performance measures such as throughput and lead time. Series-Parallel graphs, Petri nets and queuing networks are fundamental models for DEDS. Discrete event simulation is a very general method and is widely followed. System dynamic models are also widely used for supply chain prformance evaluation [10].

1.3 Models for Lead Time Computation

The lead time of an order entering the supply chain is the total time it spends in the supply chain and is a crucial performance measure. All the models discussed above can be used to arrive at the total expected supply chain lead time and its variance. An estimate of the mean and variance of the supply chain lead time can help one in quoting reliably the delivery time for a given customer order. In this paper, we treat the lead time computation problem for a class of make-to-order supply chains as multi-class open generalized queueing networks. We utilize the existing efficient approximation algorithms for computing the expected supply chain lead time and variance. This is the subject matter of Section 2. We remark that existing literature on approximate methods of analysis of fork-join queueing (FJQ) systems assume heavy traffic [9] and require tedious computations. For a class of fork-join queueing systems, we present in Section 3, a new approximate method of analysis which we use in the analysis of supply chains. For the case of deterministic arrivals and normally distributed processing times, we present an easy to use approximate method, based on results of Clarke [3]. We report encouraging results for this class of fork-join systems, with some possible applications in the supply chain context. We conclude this paper in Section 4.

2 Approximate Analysis of Fork-Join Queueing Networks

In this section, we present an approximate method for the performance analysis of certain make-to-order supply chains. Consider the following supply chain:

- There is one end product or a product group that is made to order. Thus we can handle a single product or all products belonging to a particular family.

- This product (family) is manufactured from two major sub-assemblies supplied by two different suppliers. The inbound logistics is managed by the suppliers themselves.

- The sub-assemblies are then joined at a manufacturing plant. There is a synchronization delay at this manufacturing plant, for both the sub-assemblies to arrive.

- Since the end item is made-to-order, there is forking at the supplier end i.e. orders for the sub-assemblies are placed simultaneously with the suppliers. This is a structural feature that simplifies the analysis of the underlying FJQ system.

- The assembled end product (family) is then delivered to distributors.

We model such a SCN by a queueing network as shown in Figure 3. Observe that this queueing network has fork-join structure preceding a generalized queueing network. Once we analyze the fork-join structure, we can easily further the analyze using well known approximations [2]. We are interested in computing the end-to-end delay, or the total mean supply chain lead time. It is well known (see [5]) that FJQ systems are difficult to analyse exactly. Hence many approximations have been proposed in the literature (see the references in [6, 1]). Exact results are available only for the case where the arrivals are Poisson, with exponential service times and just two joining nodes. See [4, 8] for details. Most of the approximate methods assume the following:

- The processing times at the servers belong to the exponential family of distributions (exponential or Erlangian or Hyper-exponential).

- The buffer sizes at the various queues are all bounded.

- Studying the *mean* cycle time alone is sufficient.

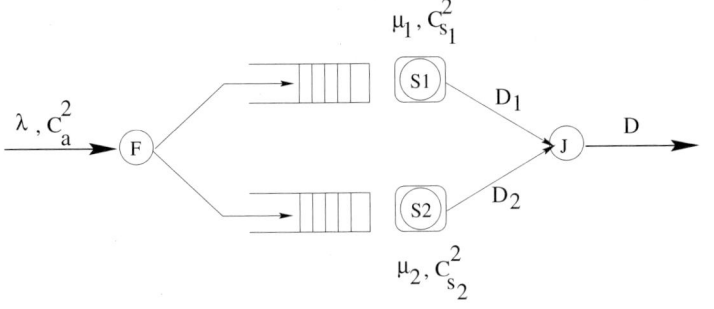

Figure 2: The fork-join structure

λ and C_a^2	The mean rate and SCV of the arrival process
μ_i	The mean service rate at server $S_i, i = 1, K$
$C_{s_i}^2$	The SCV of service time at server $S_i, i = 1, K$
ρ_i	Utilization of node $i, i = 1, K$
T_i	The sojourn time at server $S_i, i = 1, K$
T_i^*	The maximal value (order statistic) of $T_i, i = 1, K$
D_i	The departure process from server $S_i, i = 1, K$
$\sigma_{D_i}^2$	Variance of $D_i, i = 1, K$
$C_{D_i}^2$	SCV of $D_i, i = 1, K$
D	The departure process after the join node
T	The sojourn time in the fork-join structure
μ and σ^2	Mean and variance of D

Table 1: Notation for the approximation method

Since in the case of supply chains, it is common to encounter more general distributions, it is necessary to include general service times in the analysis. Also, the buffer sizes need not necessarily be bounded. Though mean cycle times capture the steady state behaviour, it is necessary to compute the variance of the cycle times too. Before we go into the complete analysis of the supply chain, we describe our approach for the fork-join structure with normally distributed service times and deterministic inter-arrival times. In the approximation that we will be developing in this section, we assume that the joining nodes have Gaussian service time distributions with their means at least thrice the standard deviation and in real life supply chains, this assumption is found to be adequate and reasonable [7].

2.1 Computing the SCV of Inter-departure Time

Consider two servers with general service times and infinite buffer sizes as shown in Figure 2. Let the arrival process to these be generally distributed, with a fork on every arrival. Let there be a join immediately after the two servers complete service. We are interested in computing the approximate departure process *after* the join, by its first two moments. Observe that the departure process from the join station is the arrival process to any downstream server. Once the moments of the above departure process are computed, the analysis of the remaining part of the queueing network is at hand. In Table 1 we detail the notation used. The average values will be denoted by $\mathbb{E}[.]$. Under conditions of stability, it is worth noting that when the fork-join structure is under steady-state, the departure rate out of the join node will be equal to the arrival rate at the fork node. Hence it is enough if we get an approximation for the variance of the departure process from the join node. In order to compute the second moment of $D_1 \ldots D_K$, we analyse servers $S_1 \ldots S_K$ as independent GI/G/1 queues, with the *same* arrival rates as the given external arrival rate, prior to forking. This analysis would give us the mean and SCV of inter departure times from each server. Towards this end, we use the first approximation for the mean cycle time and SCV, given in [2], pp-75. We ignore the effect of blocking of servers and make the (first) assumption that the mean inter departure rate is same as the mean inter arrival rate for each server. Thus, when there is a fork to the two servers, D_1 and D_2 will have the same expected values. We use:

$$C_{D_i}^2 = (1-\rho_i^2)\frac{C_a^2 - \rho_i^2 C_{s_i}^2}{1+\rho_i^2 C_{s_i}^2} + \rho_i^2 C_{s_i}^2, \quad i = 1..K \quad (1)$$

$$\rho_i = \frac{\lambda}{\mu_i}, \quad i = 1..K. \quad (2)$$

By our first assumption, $\mathbb{E}[D_1] = \ldots = \mathbb{E}[D_K] = \frac{1}{\lambda}$. On an average, it is the server (say, $\tilde{k}$) with the greatest mean flow time ($= \max_i T_i$) which is expected to delay the other jobs. Thus the server with the greatest average processing time contributes most to the variance of the inter departure process after the join node. Hence, we compute $C_{D_i}^2, i = \tilde{k}$ and choose this value as an approximation for the SCV of the inter departure process *after* the join node, i.e., $C_D^2 = C_{D_{\tilde{k}}}^2$. Thus the departure process *after* the join, viz., D is computed by its first two moments. We are now ready for the next stage of our aggregated approximate analysis of the entire supply chain.

2.2 Mean Waiting Time for the Fork-Join Structure

We know that $T_1 \ldots T_K$ can be computed by their first two moments using standard GI/G/1 analysis (see [2]). Thus the mean waiting time for the fork-join construct is given by $\max(T_1, \ldots T_K)$. Recently, some interpolation and diffusion approximations are available for symmetric fork-join systems with generally distributed arrival and service processes (see [11, 9]). We note that such methods again involve tedious numerical computations and are valid only under heavy traffic conditions. For the case where the joining servers have service times which are normally distributed, and when the arrival pattern is deterministic, we proceed as follows. We assume that the waiting times are independent of each other, and that they are normally distributed with the mean and standard deviation as computed. The first two moments of the maximum of n independent normally distributed random variables can be obtained using the approximation detailed in [3]. We use the same here and obtain the mean flow time at the fork-join stage. We reproduce Clarke's result for the two random variables case:

$$\mathbb{E}[D] = \mathbb{E}[D_1]\Phi(\alpha) + \mathbb{E}[D_2]\Phi(-\alpha) + a\phi(\alpha) \quad (3)$$

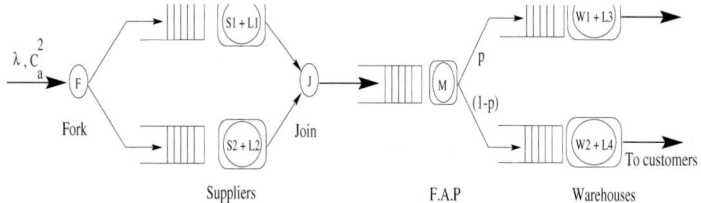

Figure 3: The supply chain network (FJQN) considered for study

P_{S_i} i=1..K, Mean processing times (secs): 200 to 380 in steps of 20
Standard deviation of processing times: $\frac{1}{4}$ of mean
Squared coefficient of variation of all processing times: 0.0625
Mean inter-arrival time seconds: 760, 475, and 422.2

Table 2: Input parameters (Case A) for the supply chain of Figure 3

$$\mathbb{E}^2[D] = (\mathbb{E}[D_1]^2 + \sigma_{D_1}^2)\Phi(\alpha)$$
$$+ (\mathbb{E}[D_2]^2 + \sigma_{D_2}^2)\Phi(-\alpha)$$
$$+ (\mathbb{E}[D_1] + \mathbb{E}[D_2])a\phi(\alpha) \quad (4)$$
$$\alpha = \frac{1}{a}(\mathbb{E}[D_1] - \mathbb{E}[D_2]) \quad (5)$$
$$a^2 = \sigma_{D_1}^2 + \sigma_{D_2}^2 - 2\sigma_{D_1}\sigma_{D_2}\varrho \quad (6)$$

In the above equations, ϕ is the standard normal density function and Φ is the corresponding cumulative distribution function. Also, ϱ is the coefficient of correlation between the variables D_1 and D_2, which is assumed to be 0 in our case, owing to the independence assumption. The above formulae can be easily extended for the n independent normal random variables case. Observe that max(a, b, c) = max(a, max(b, c)).

2.3 Numerical Results

For purposes of validation, we very briefly present the results of using our approximation on two single stage systems, one containing two servers and the other containing ten servers.

In order to test our approximation for the SCV of the departure process after the join node, and the mean flow time at the fork join structure, the input cases are illustrated in Table 2. In the table, the mean inter-arrival time is specified for varying utilization values of the server with the maximum service time of 380 seconds. The utilizations considered were 50, 80, and 90% respectively at the server with the maximum mean service time.

In this table, Case B refers to the case where the servers are identical with service times equal to 380 seconds, for both two and ten server systems. Case C refers to non-identical servers with same mean service times as Case A, but their standard deviations decrease from 50 seconds (for the server with mean service time 200s) to 5 seconds

N	ρ (%)	Mean flow time at FJ stage in seconds			
		Case A		Case B	
		Exact	Approximation	Exact	Approximation
2	50	382.96	382.07	433.04	433.58
	80	394.30	385.26	449.70	437.15
	90	444.88	414.78	514.31	476.16
10	50	450.15	460.00	526.05	524.77
	80	461.50	460.80	550.73	528.80
	90	498.21	477.68	670.85	587.52
		Case C		Case D	
10	50	380.00	381.6	459.36	447.48
	80	380.14	381.6	461.62	448.00
	90	380.30	381.6	474.49	450.53

Table 3: Validation results for single stage fork-join queueing systems; N: Number of servers, ρ: Utilization

(for the server with mean service time 380s) in steps of 5s. Case D is similar to Case C, but the service time standard deviations now decrease from 150s to 60s in steps of 10s as in Case C. The results are tabulated in Table 3. The maximum absolute error percentage from the table shown is found to be 12. We note that this occurs for Case B (with ten identical servers) and when the utilization value is 90%. Also, Clarke [3] showed that the approximation for the maximum value of n normally distributed random variables is error prone especially when the random variables considered have the same mean and variance. This is precisely the case when we consider identical servers in the fork-join stage.

On an average, the approximation was found to give absolute error percentages of less than 4%. The absolute error percentage was computed as difference of the approximated flow time from that computed from simulation, divided by the latter.

2.4 Setting Service Levels in a Two Echelon Assemble-to-order System

The approximation that we have developed, can also aid in computing the total costs in certain assemble-to-order supply chains, in the process 'determine' the service levels, which is defined as 1-{probability of stockout}. The two main cost components in such supply chains are the inventory and the delay costs. We present a simple illustration for this case. Consider the supply chain in Figure 3 which shows a two echelon supply chain with suppliers at the first echelon and the OEM as the second. Now, let us assume that the system is operated in a assemble-to-order fashion, rather than the make-to-order type which was considered in the earlier discussion. This would mean that we will have inventories of components bought from suppliers S1 and S2, that are assembled at M and sold to end customers on order. In the ensuing discussion, we omit the warehouses from the analysis. Name the components bought from S1 and S2 as A and B respectively. Let C be the finished goods. Let us assume that the components are ordered when M is out of stock. We associate probabilities of stock outs for A and

B. Let us assume that the suppliers manage the inbound logistics, and that the processing times of individual components at these suppliers are known, along with the logistics times. We can thus aggregate these two along with any interfaces present, by the servers S1+L1 and S2+L2 of Figure 3. For ease of exposition, let us assume that the probabilities of stock outs of A and B are the same, and equal to p. Also p^2 be the probability that M is out of stock of A and B simultaneously. This is usually a management set parameter, and hence a decision variable. (Our analysis is easily extendable to the case when A and B have different probabilities of stock outs.) The target inventories of A and B at M (respectively, I_A and I_B) are set based on the stock out probabilities. For instance, see [12] on how this can be done. This requires an assumption that the lead times from the suppliers (including processing at their factories, logistics and interface times) are normally distributed, to make matters tractable. Let D_1 and D_2 denote the waiting times from S1 and S2, respectively, whose mean and variance can be computed using GI/G/1 analysis of S1 and S2, with arrival rate equal to the external demand rate. Let D_3 be the waiting time at M. The following cases can occur:

- M is out of stock of A,
- M is out of stock of B, or,
- M is out of stock of both A and B.

Thus the total average lead time for arriving orders is obtained as:

$$D = p\{D_1 + D_2 + p * \mathbb{E}[max(D_1, D_2)]\} + D_3 \quad (7)$$

Similarly, the total average inventory in the system is given by:

$$I = (1-p)(I_A + I_B) + p\{L_1 + L_2 + p(L_1 + L_2)\} + L_3, \quad (8)$$

where L_1, L_2, and L_3 are the steady state average WIP at S1, S2, and M respectively, which are computed using GI/G/1 analysis of the respective servers. Observe that $max(D_1, D_2)$ is at hand, thanks to Clarke's method. All other values are calculable using simple approximate solutions of GI/G/1 queues. Hence one can, in principle, perform a total cost analysis as follows. Let H_1 be the holding costs of inventory and H_2 the delay costs. Thus the total cost of operating the supply chain in the assemble-to-order fashion, is given as

$$TC = H_1 * I + H_2 * D \quad (9)$$

For varying values of $\frac{H_2}{H_1}$, one can then compute the total cost, thus enabling the setting of the stock out probability p, in turn, the target inventories of A and B at M. We note here that, although H_1 is assumed to be the same for inventories

λ and C_a^2	10/day, 0.5
μ_{S_1} and μ_{S_2}	15/day, 20/day
$C_{s_1}^2$ and $C_{s_2}^2$	(0.8, 0.4)
μ_M and C_M^2	30/day, 0.2

Table 4: Input parameters for the assemble-to-order supply chain

of the components and the finished goods, in practice, it is possible to have the holding costs of finished goods to be, say, 20% higher than those of the components. Similarly for the delay costs H_2. This can be easily incorporated into our analysis by altering the total cost function suitably, although we dont do that here. The input case considered is shown in Table 4. As discussed above, the various variables are computed for the given input parameters, and we get the following expression for the total cost:

$$TC = H_1[0.370 + (1-p)(I_A + I_B) + 2.169p] + 2.169p^2 + H_2[0.037 + 0.217p + 0.174p^2] \quad (10)$$

The above equation can be used to determine the total cost given p. Alternatively, if p is a decision variable, we enumerate for various values of p and get the least cost solution. We know that

$$p = \mathbb{P}(DDLT_i \geq I_i), i = A, B, \quad (11)$$

where DDLT is the demand during lead time (i.e., the orders for finished goods that arrived even when the required components are on order) which is a random variable. This is obtained as the product of the arrival rate and the lead time for replenishment. Assuming now, that each arriving order for finished goods C requires one component each of A and B, we compute the following:

$$DDLT_i = \lambda * D_i, i = A, B \quad (12)$$
$$C_{DDLT_i}^2 = C_{D_i}^2, i = A, B \quad (13)$$

We now make the assumption that the DDLT computed above, is a Gaussian random variable. Using the definition of p above, we can easily determine I_A and I_B. For various values of the stock out probability p, and the ratio of $\frac{H_2}{H_1}$, we computed the total costs, the same being presented in Figures 4–5. The trend shown in the graphs is expected, because, as the probability of stock outs is allowed to decrease, the inventories go up and vice versa. The minimal total cost can thus be traced to an appropriate value for p, although it requires exhaustive enumeration.

3 Conclusions

Performance modeling intended for decision making in supply chains is a critical issue. In this paper, we have presented queueing network based models for analysing supply chain networks in a dynamic and stochastic setting.

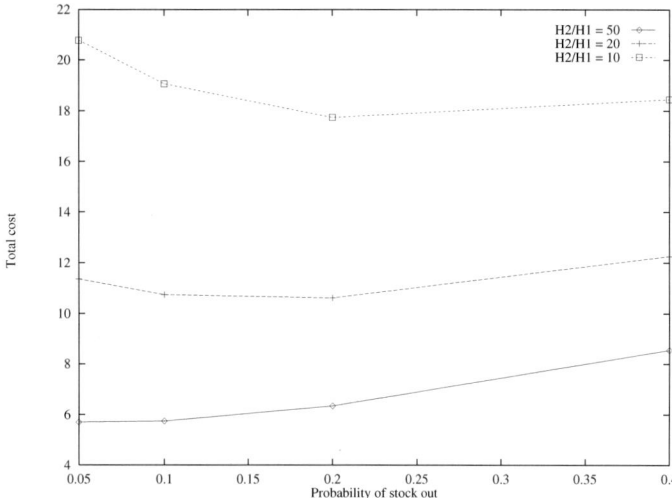

Figure 4: Total cost analysis of assemble-to-order system at high ratio of $\frac{H_2}{H_1}$

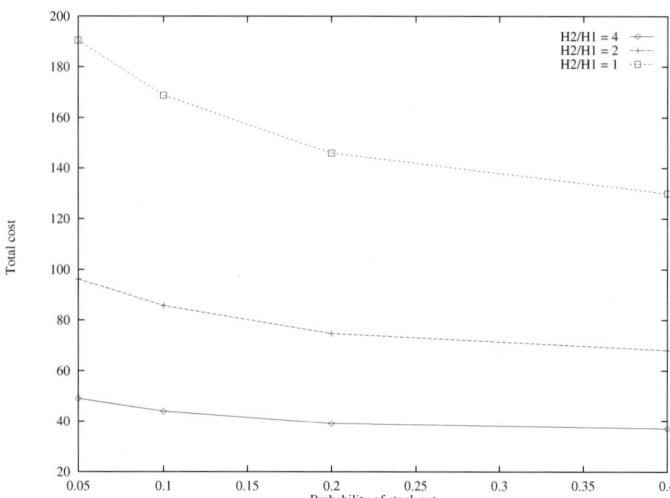

Figure 5: Total cost analysis of assemble-to-order system at low ratio of $\frac{H_2}{H_1}$

Specifically, we considered fork-join queueing systems and presented an approximate method for performance analysis. We presented a potential application of this method in setting service levels in certain assemble-to-order supply chains.

References

[1] F. Baccelli, W. A. Makowski, and D. Towsley. Acyclic fork-join queueing systems. *Journal of the ACM*, 36:615–642, 1989.

[2] J. A. Buzacott and G. Shantikumar. *Queueing Models of Manufacturing Systems*. Prentice Hall, 1993.

[3] C. E. Clarke. The greatest of a finite set of random variables. *Operations Research*, pages 145–161, March-April 1961.

[4] L. Flatto and S. Hahn. Two parallel queues created by arrivals with two demands: I. *SIAM Journal of Applied Mathematics*, 44:1041–1053, October 1984.

[5] C. Kim and A. K. Agrawala. Analysis of the fork-join queue. *IEEE Transactions on Computers*, 38(2):250–255, 1989.

[6] A. Kumar and R. Shorey. Performance analysis and scheduling of stochastic fork-join jobs in a multicomputer system. *IEEE Transactions on Parallel and Distributed Systems*, 4(10):1147–1154, 1993.

[7] H. L. Lee and C. Billington. The evolution of supply-chain-management models and practice at Hewlett Packard. *Interfaces*, 25(5):42–63, Sep-Oct 1995.

[8] R. Nelson and A. N. Tantawi. Approximate analysis of fork/join synchronization in parallel queues. *IEEE Transactions on Computers*, 37(6):739–743, 1988.

[9] V. Nguyen. Processing networks with parallel and sequential tasks: Heavy traffic analysis and Brownian limits. *The Annals of Applied Probability*, 3(1):28–55, 1993.

[10] N. R. Srinivasa Raghavan. *Performance Analysis and Scheduling of Manufacturing Supply Chain Networks*. Ph.D Thesis, Indian Institute of Science, Bangalore, 1998.

[11] S. Varma and A. Makowski. Interpolation approximations for symmetric fork-join queues. *Performance Evaluation*, 20(3):245–265, 1994.

[12] T. E. Vollman, W. L. Berry, and D. C. Whybark. *Manufacturing Planning and Control Systems*. The Dow Jones-Irwin/APICS Series in Production Management, Fourth Edition 1998.

Dynamic Scheduling Rule Selection for Semiconductor Wafer Fabrication

Bo-Wei Hsieh
Dept. of Electrical Engineering
National Taiwan University
Taipei, Taiwan, R.O.C.
bwhsieh@ac.ee.ntu.edu.tw

Shi-Chung Chang
Dept. of Electrical Engineering &
Grad. Inst. of Industrial Engineering
National Taiwan University
Taipei, Taiwan, R.O.C.
scchang@cc.ee.ntu.edu.tw

Chun-Hung Chen
Dept. of Systems Engineering &
Operations Research
George Mason University
Fairfax, VA 22030
cchen9@gmu.edu

Abstract

In this paper, we exploit the speed of an ordinal optimization (OO)-based simulation tool designed by Hsieh et al. to investigate dynamic selection of scheduling rules for semiconductor wafer fabrication (fab). Although a scheduling rule is a combination of loading wafer release and dispatching rules, this paper specifically focuses on dispatching when significant amount of wafers-in-process (WIPs) are held due to engineering causes and when major machine failures occur. Four prominent dispatching rules combined with the wafer release policy of workload regulation constitute a basic set of rule options. The dispatching rule may be weekly selected based on fab states over a four-week horizon. A total of 256 rule options are then evaluated and ranked by the OO-based simulation tool under the performance index of mean cycle time and throughput rate. Results demonstrate the value of dynamic rule selection for uncertainty handling, the insightful selection of good rules and the needs for further research.

1. Introduction

Major fab scheduling problems include how wafers should be released into a fab and how they should be dispatched among machines for processing. A popular practitioners' approach for scheduling the production in a fab is to select from the many empirical scheduling rules available for IC fabs [6]. To quickly select a good enough scheduling rule from a rule library, Hsieh et al. developed a fast simulation tool (Figure 1) based on the ordinal optimization (OO) and optimal computing budget allocation (OCBA) methods, which will be referred to as the OO-based method hereafter [2].

Operation objectives of a fab change dynamically as well as the machine and inventory states. To achieve competitive fab operations, such a dynamic nature intuitively may lead to the need for dynamic selection of a scheduling rule based on the changes of objectives and states. In addition to the finding of [5], experimental studies of static rule selection by Hsieh et al. have indicated that rule selections vary with factors of initial state, performance index (objective) and time horizon [2]. This motivates our further investigation about how the efficiency of the OO-based simulation may be exploited to facilitate dynamic rule selection.

Dynamic dispatching rule selection is essentially a stochastic optimal control problem. As a closed-loop solution is generally impossible to be obtained for stochastic optimal control of a complex system [1], we consider the open-loop feedback selection (OLFS) instead. At each decision point, OLFS uses whatever available system information to select a good rule for a coming period of time as if no further information will be received in the period. Although not truly dynamic, OLFS exploits feedback information and fast evaluation of rule options to select scheduling rules. In specific, OLFS is applied to selection of dispatching rules upon the occurrence of two significant uncertain events in fab operation: holding of a significant amount of WIPs due to engineering causes and failure of a major machine. The study exploits the speed of the OO-based simulation tool of [2] and adopts a 10-product, 60-step and 12-tool-group fab model, which is extended from the single-product model of Lu et al. [3]. The potential of the OO-based simulation for application to dynamic selection of dispatching rules is also assessed.

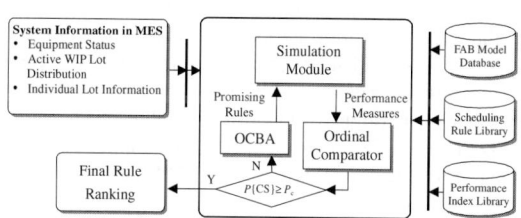

Figure 1 An OO-based Simulation Tool

The remainder of this paper is organized as follows. Section 2 describes the needs for dynamic selection of scheduling rules. The simulation model is described in Section 3. Dynamic selections of dispatching rules under significant WIP holding and major machine failures are given in Sections 4 and 5 respectively. Section 6 concludes this paper.

2. Dynamic Selection of Scheduling Rules

Reentrant feature of the fabrication line and uncertainties of machines are two critical characteristics that make scheduling problems challenging. As the circuitry is fabricated layer by layer onto a wafer and basic processing steps among layers may be similar, the production flow of each type of product may re-visit the same type of machines, i.e., the same machine group, a few times. Wafers of different product types as well as those of the same type but processed at different layers may compete for the finite capacity of a machine group.

Among the uncertainty factors in fab production management, machine failures and temporary holding of WIPs from processing due to engineering causes are known as the two most prominent perturbations that lead to significant state and/or objective changes. When an engineering hold event occurs, a certain amount of WIPs is held from production until the engineering problem is cleared. Such holding results in a sudden reduction of available WIPs at each stage and may lead to a shortage of WIP for processing at the stage. If a stage in short of WIP requires the processing by a bottleneck machine, WIP holding may then cause bottleneck capacity loss, i.e., the output volume may decrease. When the held WIPs are released back to the production line, they are very often expedited to meet the due dates or the output volume target. When an unscheduled machine failure occurs, the machine group of the failed machine may become a short-term bottleneck. Its loss of capacity may also result in a lower fab output than the original target.

In practice, minimizing the mean cycle time while keeping the output volume per week or month above a target level is usually a fab operation objective. The latter, however, is really the bottom line performance requirement of a fab. Since the occurrence of either of the two aforementioned events may reduce the output volume, fab operation objective might be shifted from cycle time reduction to output volume maximization. In the event of a long period of engineering hold, one may want to adjust the wafer release policy. First, increase wafer release during the period of holding so that there is enough workload in the fab to keep a good utilization of fab capacity and the output volume. Then reduce wafer release after the held WIPs are back to production. And finally put wafer release back to a normal level. On the contrary, wafer release may first need to be reduced under a long time of machine failure to avoid unnecessary increase of WIP levels and cycle times, and then return to the normal level after the machine is repaired.

As for the selection of dispatching rule under a given wafer release policy, there are two intuitive and common strategies in response to the aforementioned two events. One is to feed proper amounts of WIPs to the bottleneck machines so that their available capacity is prevented from starvation. The other is to use machines in processing available WIPs that can be effectively moved to reduce total waiting times incurred by the holding or failure events. The selection is challenging because of the re-entrant nature, where a machine failure may affect the processing of several stages and the productions at one time become the future re-entrant flows to individual machine groups. What the proper amounts of WIPs are to prevent capacity from loss and how to effectively move WIPs and feed the bottleneck machines obviously depend on the state(s) of a re-entrant line.

3. Simulation Model Description

In this paper, a 10-product model (named FAB) extended from the single-product model of Lu et al. [3] is adopted. There are three types of processing technologies: $T1$, $T2$ and $T3$, each having a specific sequence of processing stages. Among the ten product types, four product types use technology $T1$, three product types use $T2$, and the other three use $T3$. The model involves 12 failure-prone processing stations, each having one or more identical but independent machines. Wafers are moved among machines in the unit of lot, which consists of 24 wafers. Among the processing stations, Station 8 is modeled as a batch-processing machine group where each batch consists exactly of 6 lots. Processing times, times between failures and times to repair are exponentially distributed. The numbers of operation steps of $T1$, $T2$ and $T3$ are 60, 41 and 30 respectively. With release rates of 0.3 lots/hour for $T1$, 0.2 lots/hour for $T2$ and 0.12 lots/hour for $T3$, the capacity bottleneck machine is Station 6 whose percentage utilization is 95.3%. Detailed model parameters of FAB are given in Table 1.

In our experimental study, two performance indices are considered: per circuit layer mean cycle time (LMCT) and total throughput rate, which are among the most frequently used fab performance indices. There are four prominent dispatching rules and a representative wafer release policy considered in this study as listed in Table 2. Workload regulation release policy proposed by Wein [7], FSVCT dispatching rules proposed by Lu et al. [3], and the OSA rule proposed by Li et al. [4] are known to be good for reducing mean and variance of cycle time. We designed the LDF rule for controlling production smoothness and for tracking production targets [8].

Table 1 Plant Data of FAB

Station	# of Machines	# of Visits (T1)	# of Visits (T2)	# of Visits (T3)	MPT[1]	MTBF[2]	MTTR[3]	% Util
1	4	14	10	8	0.500	150	5	92.7%
2	3	12	9	7	0.375	200	9	82.3%
3	10	7	5	4	2.500	200	5	91.9%
4	1	1	0	1	1.800	200	1	76.1%
5	1	2	1	1	0.900	200	1	83.3%
6	2	3	2	2	1.200	200	6	95.3%
7	1	1	1	0	1.800	200	1	90.5%
8	4	8	6	4	0.800	150	5	84.8%
9	1	3	0	0	1.000	200	5	92.4%
10	9	5	4	1	3.000	130	5	84.4%
11	2	3	2	1	1.200	200	5	87.6%
12	2	1	1	1	2.500	200	5	79.9%

[1] MPT: Mean Processing Time (by hours)
[2] MTBF: Mean Time between Failures (by hours)
[3] MTTR: Mean Time to Repair (by hours)

Table 2 Scheduling Rules

Rule	Symbol	Description
Release policy	WR(C_p)	In a one-bottleneck system, whenever the expected work of type-p products in fab drops below C_p hours for the bottleneck machine, then release a new type-p lot into the fab.
Dispatching rules	FSVCT	Choose the lot with smallest $(a_n + C_p - \zeta_i)$, where p represents the index of product type, a_n is the release time of lot n, C_p is the mean cycle time, and ζ_i is the estimate of the remaining cycle time from buffer i.
	LDF	Let the completion of one wafer processing at a stage be a move. Choose a stage with the largest deviation of completed moves from the desired moves, where the desired number of moves of each product type at each stage is pre-specified. Then choose from the stage a lot which is released into the fab the earliest.
	OSA	Choose a step according to the following priorities: Priority I: step i such that $N_i(t) > \overline{N}_i$ and $N_{i+1}(t) < \overline{N}_{i+1}$; Priority II: step i such that $N_i(t) < \overline{N}_i$ and $N_{i+1}(t) < \overline{N}_{i+1}$; Priority III: step i such that $N_i(t) > \overline{N}_i$ and $N_{i+1}(t) > \overline{N}_{i+1}$; Priority IV: step i such that $N_i(t) < \overline{N}_i$ and $N_{i+1}(t) > \overline{N}_{i+1}$, where $N_i(t)$ is the WIP at time t at step i, $\overline{N}_i$ is the average WIP at step i. Choose a lot with the same priority using FSVCT.
	FIFO	Select the lot which arrived at the station the earliest.

4. Rule Selection under Engineering Holds

Consider the operations of FAB for the coming four weeks. The fab has been operated under the scheduling rule of the workload regulation release policy combined with FSVCT dispatching rule (WR-FSVCT) for one year. Now suppose that over the whole line, half of the technology-*T1* WIP belongs to one customer order and that customer orders an engineering hold for one week. The WR wafer release policy remains unchanged due to its capability of regulating the workload of the production line. Under the WR policy, the workload for the bottleneck machine group, Station 6, is set at a level of 106 hours, which is the long-term average workload to achieve a throughput rate of 0.62 lots/hr. At this throughput rate, the utilization of Station 6 is 95.31%. Recall that the machine of Station 9, whose utilization is 92%, is only used in processing products of technology T1. A little calculation reveals that the engineering hold may lead to 12% capacity loss of Station 9 over the week of holding.

As has been discussed at the beginning of this section, proper amounts of available WIPs should be supplied in priority to both the bottleneck station and Station 9 to prevent their capacity from loss. When the held wafers are released back, the utilization of Station 9 needs to be raised to about 95% and Station 6 to 97% for the later three weeks in order to catch up with the delayed work of the first week. Since FSVCT determines lot priority based on individual lot information rather than station information, continual application of FSVCT as the dispatching rule does not match the needs when a significant holding/release event occurs. The question is then how dispatching rules should be dynamically selected so that the performance requirements for LMCT and/or throughput rate can be well achieved.

The dispatching rule library now consists of only four rules: FSVCT, FIFO, LDF, and OSA, which will be referred to as rules A, B, C, and D respectively. By means of OLFS, weekly change of dispatching rules is investigated. Over a four-week horizon, there are therefore 256 (4x4x4x4) options. The OO-based simulations are then conducted to find good combinations of the four dispatching rules from the 256 options.

Simulation results listed in Tables 3 and 4 clearly indicate that throughput close to 0.62 lots/hr can be achieved by many options, and that to minimize LMCT, the dispatching rule should be changed from WR-FSVCT to WR-LDF. The former observation is due to the fact that the capacity loss of Station 9 is only 12% in the first week and it can be made up by a higher utilization of Station 9 for the rest three weeks. The best rule listed in Table 4, A-C-D-D, achieves the maximum throughput rate, which clearly shows the transient of dispatching rule selection over a four-week horizon from the originally used rule A. In the latter observation, although not the best in throughput rate performance, the option C-C-C-C obtains a throughput rate of 0.618 lots/hr, which is only 0.418% lower than that of rule option A-C-D-D. But The LMCT performance of C-C-C-C, 12.388 hour, is 10% shorter than the 13.690 hour of A-A-A-A. Computation time required for this rule selection experiment by using the OO-based simulation tool is about three hours. Our study shows that evaluation of the 256 rule combinations by using a regular simulation may take 150 to 300 hours of computation time, which is infeasible for such an application.

Table 3 Dynamic Rule Selection under Engineering Holds (ranked by LMCT)

Rank	Rule[4]	LMCT	%	Throughput
1	C-C-C-C	12.388	-	0.618
2	C-C-C-B	12.395	0.06%	0.616
3	C-C-C-D	12.492	0.84%	0.617
4	B-C-C-B	12.644	2.07%	0.602
5	B-C-C-D	12.653	2.14%	0.607
6	B-C-C-C	12.705	2.56%	0.600
7	A-C-C-C	12.733	2.78%	0.608
8	A-C-C-B	12.740	2.84%	0.607
9	D-C-C-B	12.759	2.99%	0.601
10	C-D-C-C	12.781	3.17%	0.608
131	D-D-D-D	13.628	10.01%	0.610
145	A-A-A-A	13.690	10.51%	0.612

[4] Rule-A: WR-FSVCT; Rule-B: WR-FIFO; Rule-C: WR-LDF; Rule-D: WR-OSA

Table 4 Dynamic Rule Selection under Engineering Holds (ranked by throughout)

Rank	Rule	Throughput	%	LMCT
1	A-C-D-D	0.621	-	13.514
2	A-A-B-A	0.621	0.00%	13.970
3	A-A-D-A	0.620	0.16%	13.995
4	A-C-C-A	0.619	0.32%	13.279
5	A-B-C-B	0.619	0.32%	14.043
6	A-B-B-D	0.619	0.32%	14.170
7	C-C-C-C	0.618	0.48%	12.388
8	D-D-C-C	0.618	0.48%	13.106
9	D-D-C-A	0.618	0.48%	13.605
10	D-B-A-D	0.618	0.48%	13.642
46	A-A-A-A	0.612	1.45%	13.690

It can be concluded from Table 3 that when engineering hold occurs, the switching from FSVCT to LDF rule leads to a superior LMCT while maintaining reasonable throughput rates. Conceptually, this is no surprise because under holding, the actual production of wafers of *T1* technology largely deviates from the desired targets over the whole line. LDF then gives a higher priority to available WIPs of *T1* technology than WIPs of other types. In so doing, available WIPs of *T1* move faster than ordinary to Station 9, which can reduce the capacity loss of Station 9 and the cycle times of available WIPs of *T1* technology. Such a gain compensates the LMCT increase for the held WIPs of *T1* technology. Similarly, at the bottleneck station, available WIPs of *T1* technology are given a higher priority of processing and re-enter the station faster. When the held WIPs are released, they are still given a higher priority and are expedited until the actual production of wafers of *T1* technology catches up its desired targets of individual steps. In contrast, the FSVCT rule prioritize WIPs of all types by their slack times; the held WIPs of *T1* technology get a high priority after being released because of resultant short slack times. But the available WIPs of *T1* technology does not get a higher priority during the holding period. So, when WIPs of *T1* technology are rushed to Station 9 in the later three weeks, congestion occurs and the FSVCT rule leads to a longer LMCT.

5. Rule Selection after Unusual Machine Failure

Again, consider the four-week operations of FAB model, where one machine of Station 9 goes into an unusual down situation at the beginning of a week and will need five days to be repaired. Although not frequently happened in the fab, this unusual event may lead to significant impact on fab performance. To complete the target 4-week workload of Station 9 in the remaining 23 days, the average utilization of Station 9 has to be more than 100%, which means that the capacity of Station 9 is not enough to complete the 4-week target after a 5-day failure. Station 9 then becomes a short-term capacity bottleneck instead of Station 6. Even 100% utilized for the remaining 23 days, Station 9 can only complete 89% of the target throughput which leads to at least 5% decrease in total throughput rate, i.e. the throughput rate of FAB over the four-week horizon is not greater than 95% of the target throughput. Under such a failure, WIPs belonging to technology *T1* cumulate at the steps processed by Station 9 and machines for processing downstream steps are starved. Due to the reentrant feature of IC fabrications, the shortage of *T1* products at downstream steps of Station 9 will propagate to the original bottleneck station (Station 6). Station 6 then allocates more resource to products of *T2* and *T3* technologies, which occupy 57% of capacity of Station 6 under normal states, to prevent capacity from loss and results in shorter LMCT of products of *T2* and *T3*, which are not processed by Station 9.

Under the failure of Station 9, throughput rate of the fab will decrease by a certain amount due to capacity loss of Station 9. For such an unusual event, the most important of all might be maintaining the throughput, which is again set as 0.62 lots/hour. Dispatching rule of LDF is supposed to be good in the aspects of maintaining throughput rate and reducing LMCT. LDF gives a higher priority to WIPs of *T1* technology after Station 9 is repaired to prevent available capacity of Station 9 from further loss and to reduce the LMCT of *T1* WIPs by expedition.

Table 5 Dynamic Rule Selection under Unusual Machine Failure (ranked by LMCT)

Rank	Rule[5]	LMCT	%	Throughput
1	C-C-C-C	14.527	-	0.553
2	C-C-C-D	14.702	1.20%	0.549
3	C-C-C-B	14.795	1.84%	0.546
4	B-C-C-C	14.833	2.11%	0.538
5	D-C-C-B	14.845	2.19%	0.543
6	D-C-C-C	14.850	2.22%	0.537
7	C-C-A-D	14.912	2.65%	0.546
8	B-C-C-D	14.919	2.70%	0.543
9	C-A-C-C	14.940	2.84%	0.543
10	A-C-C-C	14.966	3.02%	0.545
69	A-A-A-A	15.547	7.02%	0.545

[5] Rule-A: WR-FSVCT; Rule-B: WR-FIFO; Rule-C: WR-LDF; Rule-D: WR-OSA

Table 6 Dynamic Rule Selection under Unusual Machine Failure (ranked by throughout)

Rank	Rule	Throughput	%	LMCT
1	A-B-A-A	0.555	-	15.889
2	D-B-D-D	0.555	0.00%	16.419
3	C-C-C-C	0.553	0.36%	14.527
4	C-B-D-D	0.553	0.36%	15.946
5	D-A-A-A	0.552	0.54%	15.163
6	D-A-D-A	0.552	0.54%	15.865
7	D-A-A-B	0.551	0.72%	15.243
8	C-D-A-B	0.551	0.72%	15.843
9	B-D-A-A	0.551	0.72%	15.936
10	B-D-D-A	0.551	0.72%	16.092
59	A-A-A-A	0.545	1.80%	15.547

Simulation experiments similar to those of Section 4 are conducted to select a good combination of dispatching rules. Only the initial states are different. The initial state of this experiment is obtained by running the FAB simulation with a machine of Station 9 set to be down and to be repaired after five days. Simulation results listed in Tables 5 and 6 indicate that throughput rate decreases by more than 10% under the unusual machine failure for all rules, and that to minimize LMCT, dispatching rule should be changed from WR-FSVCT to WR-LDF. It is our expectation that the throughput rate decreases due to capacity loss of Station 9. If the operation objective is to maximize throughput rate, rule option A-B-A-A results in the best throughput rate of 0.555 lots/hr, shown in Table 6. However, the throughput rates of the top-ranking rules are not significantly different in this machine failure case as well as in the previous engineering hold case. Such observations imply that throughput rate is insensitive to dispatching rules under the WR release policy. Since we only calculate $P\{CS\}$ for the top-ranking option, the relative ranking among other options does not really have a significant statistical support under the insensitivity of throughput. Namely, the differences between the top-10 rule options for maximizing throughput and the top-10 rule options for minimizing LMCT (Tables 3-6) are not so big as they appear to be.

In Table 5, the best selection for LMCT performance is rule option C-C-C-C, whose LMCT is 14.527 hours, 7% shorter than the 15.547 hours of A-A-A-A, and throughput rate is 0.553 lots/hr. Therefore, under a capacity loss situation, rule option C-C-C-C not only performs the best in LMCT performance but also obtains a good throughput rate. Dispatching rule should be changed from WR-FSVCT to WR-LDF for the coming four weeks when the failure event occurs. Under both engineering holds and unusual machine failure events, CCCC (WR-LDF) is the only option that commonly appears in the top-10 rule combinations across Tables 3-6. This reveals a strong appeal of CCCC for handling these two unusual events in the fab. Note that such a conclusion may not be applicable to other problems. The main objective of the study here is to demonstrate that our OO-based simulation tool can determine a good dispatching rule very efficiently when any unexpected event occurs.

6. Conclusions

In this paper, selections of dispatching rules at the occurrence of significant WIP holding and major machine failure were investigated. We exploited the speed of the OO-based simulation tool to select a good scheduling rule for the coming four weeks, where rule changes weekly. Simulations yielded insightful results that dispatching rule should be switched from the slack time-based FSVCT to the deviation-from-target-based LDF to handle these unusual events. These observations justified that dispatching rule should be changed dynamically to handle these unusual events. Simulation time saving up to 50 times can be achieved by the OO-based simulation. However, the number of rule combinations grows combinatorially over time and the number of tool groups. Further research on option search method exploiting the OO-based simulation is thus needed for large problems with combinatorial complexity.

Acknowledgement

This work was supported in part by the National Science Council of the Republic of China under Grants NSC88-2212-E-002-065 and NSC 89-2212-E-002-040, and by NSF Grant DMI-9732173 and DMI-0002900, Sandia National Laboratory Grant BD-0618, and the George Mason University Research Foundation.

References

[1] D. P. Bertsekas, *Dynamic Programming and Stochastic Control*. New York: Academic Press, 1976.

[2] B. W. Hsieh, C. H. Chen, and S. C. Chang, "Fast Fab Scheduling Rule Selection by Ordinal Comparison-based Simulation," in *Proc. 1999 Int. Symposium*

Semicond. Manuf., pp.53-56, Oct. 1999.

[3] S. H. Lu, D. Ramaswamy, and P. R. Kumar, "Efficient Scheduling Policies to Reduce Mean and Variance of Cycle-Time in Semiconductor Manufacturing Plants," *IEEE Trans. Semicond. Manuf.*, vol. 7, no. 3, pp. 374-388, Aug. 1994.

[4] S. Li, T. Tang, and D. W. Collins, "Minimum Inventory Variability Schedule with Applications in Semiconductor Fabrication," *IEEE Trans. Semicond. Manuf.*, vol. 9, no. 1, pp. 145-149, Feb. 1996.

[5] S. C. Park, N. Raman, and M. J. Shaw, "Adaptive Scheduling in Dynamic Flexible Manufacturing System: A Dynamic Rule Selection Approach," *IEEE Trans. Robot. Automat.*, vol. 13, no. 4, pp. 486-502, Aug. 1997.

[6] M. Thompson, "Using Simulation-Based Finite Capacity Planning and Scheduling Software to Improve Cycle Time in Front End Operations," in *Proc. 1995 IEEE/SEMI Advanced Semicond. Manuf. Conf. and Workshop*, pp. 131-135, 1995.

[7] L. M. Wein, "Scheduling Semiconductor Wafer Fabrication," *IEEE Trans. Semicond. Manuf.*, vol. 1, no. 3, pp. 115-130, Aug. 1988.

[8] G. L. Wu, K. Wei, C. Y. Tsai, S. C. Chang, N. J. Wang, R. L. Tsai, and H. P. Liu, "TSS: a Daily Production Target Setting System for Foundry Fabs," in *Proc. 1998 Int. Symposium Semicond. Manuf.*, pp. 75-78, Oct. 1998.

A Deadlock Prevention Policy for Flexible Manufacturing Systems Using Siphons

YiSheng Huang[1]   MuDer Jeng[2]   Xiaolan Xie[3]   ShengLuen Chung[4]

[1]*Departmet of Electronic Engineering, FUSHIN Institute of Technology,*
Tou-Cheng 261, Taiwan Taiwan, ROC
yshuang@mail.fit.edu.tw

[2]*Department of Electrical Engineering, National Taiwan Ocean University,*
Keelung 202, Taiwan, ROC
b0162@ind.ntou.edu.tw

[3]*INRIA / MACSI Team ENIM-ILE DU SAULCY, 57045-Metz Cedex, France*
xie@loria.fr

[4]*Department of Electrical Engineering, National Taiwan University of Science and*
Technology, Taipei 106, Taiwan, ROC
slchung@event.ee.ntust.edu.tw

Abstract

In this paper, we present a new deadlock prevention algorithm for the class Petri nets. A new class of net that is extended from S^3PR, called ES^3PR where deadlocks are related to unmarked siphons. This method is an iterative approach by adding two kinds of control places called *ordinary control place* and *weighted control place* to the original model to prevent siphons from being unmarked. We have obtained the relation of the algorithm and the liveness and reversibility of the controlled net. Finally, a flexible manufacturing example is presented for illustrating the method.

1. Introduction

Petri nets (PNs) have been recognized as one of the most powerful tools for modeling FMS [12]. The increasing interest in PN's is stimulated by their analysis of the modeled systems. However, several fundamental problems remain open. One of them is the prevention of deadlocks. In essence, the deadlock prevention problem is an important issue in manufacturing systems because of two reasons. First, the existence of deadlocks causes a certain degree of disruption of production, and thus may significantly increase the production cost. Second, the computation of deadlocks is in general is an exponential-time problem.

Some deadlock prevention avoidance schemes for controlling an FMS have been proposed in prior work [1] [3-4][6-8]. Many of them adopted PN models as a formalism to describe FMS's and to develop deadlock avoidance policies. In [1][5-6], the deadlock avoidance problem is solved using the concept of siphons. In particular, Ezpeleta *et al.* [6] proposed a deadlock prevention control policy that is implemented by means of the addition to the initial PN model of some new elements such that the final model is live. That is, new places are added to the net imposing restrictions that prevent the presence of unmarked siphons (direct cause of deadlocks). Unfortunately, all output arcs of the new control places are added to source transitions of the resultant net if all places representing resources (we shall call them resource places) in the original net are removed. Since the source transitions denote the entry points of the processing of raw parts, this policy is rather conservative. In [3], the deadlock detection and avoidance is proposed using a graph-theoretic approach. The approach is simple because it only looks ahead one step. Unfortunately, the approach considers only single resources in a state. In [8], the authors use an algebraic polynomial kernel for analysis and deadlock avoidance for augmented marked graphs. A main drawback of the prevention method is that the algorithm directly controls source transitions. Other developments involving deadlock prevention avoidance were reported in [4][6][11].

In [2], the authors have developed a fast deadlock detection approach based on mixed integer programming (MIP) for structurally bounded nets whose deadlocks tied to unmarked siphons. Since no explicit enumeration of siphons is required, this formulation opens a new avenue for checking deadlock-freeness of large systems. Its computational efficiency is relatively insensitive to the initial marking and expected to be more efficient than classical state enumeration methods [9]. The MIP method is able to find a maximal siphon unmarked at a reachable marking. Based on this, we can formalize an algorithm that can efficiently obtain a minimal siphon from the result of the MIP method.

In [5], we have proposed a deadlock prevention algorithm for S^3PR modeled manufacturing systems. The algorithm is an iterative approach based on Chu and Xie's MIP method. At each iteration, the MIP technique is used to find an unmarked maximal siphon. Next, an unmarked minimal siphon is obtained from the maximal siphon. The algorithm consists of two main stages: one stage is called *siphons control*, the other stage is called *augmented siphons control*. This latter stage assures that there are no new siphons generated. We have shown the relation of the algorithm and the liveness and reversibility of the controlled net. Comparing with prior work, this approach provides better performance since control is not solely added to the entry point of jobs.

In this paper, we present a new deadlock prevention algorithm for a new class of nets that are extended from S^3PR called ES^3PR where deadlocks are related to unmarked siphons. This method is an iterative approach by adding two kinds of control places to the original model to prevent siphons from being unmarked. One is added a control place, called *ordinary control place*, to an original net G with its related arcs such that the controlled net is still an ordinary Petri net; the other is added a control place, called *weighted control place*, to an original net G with its related arcs such the controlled net become to a generalized Petri net. We have obtained the relation of the algorithm and the liveness and reversibility of the controlled net. Finally, a flexible manufacturing example is presented for illustrating the method.

The rest of the paper is organized as follows: Section 2 presents basic definitions and properties of Petri nets that are related to this paper. Section 3 presents our deadlock avoidance algorithm. Section 4 gives an example. Conclusions are presented in section 5.

2. Basic Definitions and Properties

Consider a generalized PN $G = (P, T, F, W, M_0)$ where P is the set of *places*, T is the set of *transitions*, $F \subseteq (P \times T) \cup (T \times P)$ is the set of directed arc, $W: F \rightarrow IN\setminus\{0\}$ gives the weights of arcs, and $M_0: P \rightarrow IN$ is the initial marking, where IN is the set of nonnegative integers. A net such that $W: F \rightarrow \{1\}$ is called an ordinary net, and can be denoted as $G = (P, T, F, M_0)$. The set of input (resp. output) transitions of a place p is denoted by $\bullet p$ (resp. $p\bullet$). Similarly the set of input (resp. output) places of a transition t is denoted by $\bullet t$ (resp. $t\bullet$). For any subset of places S, $\bullet S$ (resp. $S\bullet$) denotes the set of transitions with at least one output (resp. input) place belonging to S. A transition t is *enabled* and can *fire* under a marking M iff $M(p) \geq 1$, $\forall p \in \bullet t$. A transition t is said to be *live* if for any $M \in R(M_0)$, there exists a sequence of transitions firable from M that contains t. A PN is said to be live if all the transitions are live. A PN is said to be *deadlock-free* if at least one transition is enable at every reachable marking. A place p is said to be *bounded* if there exists a constant K suck that $M(p) \leq K$ for all $M \in R(M_0)$. A PN is said to be bounded if all the places are bounded. It is said to be *structurally bounded* if it is bounded whatever the initial marking is. A PN is said to be *reversible* if, for any $M \in R(M_0)$, M_0 is reachable from M. A subset of places S is called a *siphon* if $\bullet S \subseteq S\bullet$, i.e., any input transition of S is also an output transition of S.

It is called a *trap* if $\bullet S \supseteq S\bullet$. A siphon is *minimal* if it does not contain any other siphons [10].

Our deadlock prevention method uses the MIP deadlock detection approach proposed by Chu and Xie [2] since the MIP approach can deal with large systems within reasonable computation time according to their experimental study. The MIP approach is briefly summarized as follows: Given a siphon S, two binary indicators are introduced:

$$v_p = 1\{p \notin S\} \quad and \quad z_t = 1\{t \notin S\bullet\} \quad (1)$$

It is obvious that any p with $v_p = 1$ and any t with $z_t = 1$ will not be included in a siphon. Since S is a siphon, $v_p = 0 \Rightarrow z_t = 0$, $\forall t \in p\bullet$ and $z_t = 1 \Rightarrow v_p = 1$, $\forall p \in t\bullet$ which lead to

$$z_t \geq \sum_{p \in \bullet t} v_p - |\bullet t| + 1, \forall t \in T \quad (2)$$

$$v_p \geq z_t, \forall (t, p) \in F \quad (3)$$

$$v_p, z_t \in \{0, 1\}. \quad (4)$$

For a structurally bounded net, we have
$$v_p \geq M(p)/SB(p), \forall p \in P \quad (5)$$

where the structural bound SB is defined as: $SB(p) = max\{M(p) \mid M = M_0 + CY, M \geq 0, Y \geq 0\}$. Therefore, the *maximal* siphon unmarked at a given marking M can be determined by the following MIP problem and there exist siphons unmarked at M iff $G^{MIP}(M) < |P|$:

$$G(M) = Minimize \sum_{p \in P} v_p$$

such that constraints (1)-(4) are satisfied and
$$M = M_0 + CY, \quad M \geq 0, Y \geq 0 \quad (6)$$

Several important properties are presented below.

Property 1: The system (2)-(4) is properly speaking an integer linear system.

Property 2: Let S be a minimal siphon and let p and q be two places such that $p\bullet \subseteq q\bullet$ then $q \in S \Rightarrow p \notin S$.

Property 3: Let S be a minimal siphon. Then for any $p \in S$, there exists $t \in p\bullet$ such that $\bullet t \cap S = \{p\}$.

Property 4: Let S be a minimal siphon. Then the subnet induced by S and $\bullet S$ is strongly connected.

The following properties (see [2]) show the importance of siphons and traps in the detection of deadlocks.

Property 5: A siphon free of tokens at a marking remains token-free whatever the transition firings. A trap marked by a marking remains marked. For any marking such that no transition is enabled, the set of unmarked places forms a siphon.

Property 6: A Petri net is deadlock-free if no minimal siphon eventually becomes unmarked.

3. The Deadlock Prevention Algorithm

The deadlock prevention algorithm proposed in this paper is targeted at systems modeled by a class of ordinary PNs called ES^3PR. An ES^3PR is defined as follows:

Definition 1: An ES^3PR net G is a net that results from adding a set R of initially marked places (resource places) to a set of process nets such that:

(i) Each resource place r is associated with a set of operation places, $OP(r)$. This implies that these operation places require resource r.

(ii) For each input transition t of some $p \in OP(r)$, there exists an arc from r to t if $\bullet t \cap OP(r) = \phi$.

(iii) For each output transition t of some $p \in OP(r)$, there exists an arc from t to r if $t\bullet \cap OP(r) = \phi$

ES^3PRs properly include S^3PRs and partially overlap with RCN merged nets. Fig. 1 helps us to understand an ES^3PR net.

As described above, our deadlock avoidance algorithm is an iterative approach. At each iteration, the proposed method adds a control place to the original net G with its output arcs to the sink transitions of a minimal siphon, or to the source transitions of the resultant net G^* if all resource places are removed.

Based on the discussions of the above section, this section presents and illustrates our deadlock avoidance algorithm. Since the algorithm uses the MIP approach described in Section 2 to obtain maximal siphons unmarked at some markings. We also use an algorithm [5] to obtain unmarked minimal siphons from the maximal siphons.

For each minimal siphon S_i that can become unmarked, two cases are possible.

When an *ordinary control place* (*OC*) is added on an ordinary net, the resultant net is still an ordinary net (i.e. depicted in Fig. 2(a)); but when a *weighted control place* (*WC*) is added to an ordinary net, the resultant net becomes a generalized net (i.e. depicted in Fig. 2(b)).

As mentioned previously, our method consists of two kinds of control places: We add an *OC* to a net G with its output arcs to the *sink transitions* of a minimal siphon. However, we add a *WC* control place to a net G with its output arcs to the *source transitions* of the net. The principles and theorems are depicted as follows.

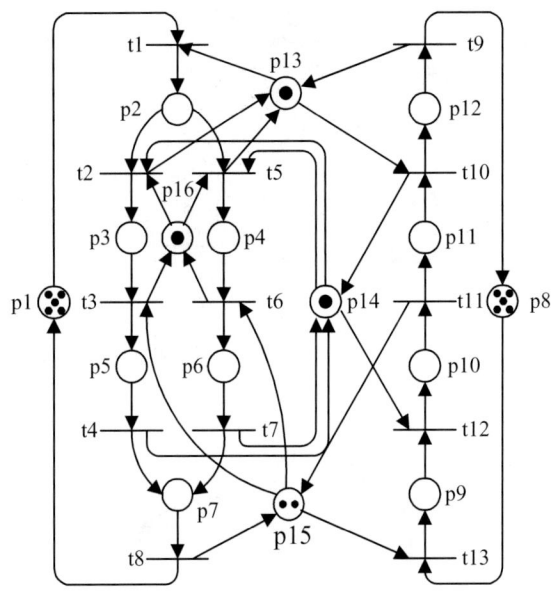

Fig. 1: An ES^3PR net.

Definition 2:

(i) An operation place $p \notin S$ is called stealing place if $p \in OP(r)$ for some resource place $r \in S$.

(ii) Let $O(S)$ be the set of stealing place of S.

(iii) Let $N_p = \sum_{r \in S \cap R} 1(p \in OP(r)) =$ Number of resource tokens of S stolen by a token in place p.

Case I: $N_p = 1, \forall\, p \in O(S)$

To prevent S from becoming unmarked, an OC is added as follows:

Definition 3: For each siphon S such that $N_p = 1, \forall\, p \in O(S)$, a new place OC is added such that:

(i) $M_0(OC) = \sum_{r \in S \cap R} M_0(r) - 1,$

(ii) For any input transition t of a place in $O(S)$, there exists an arc from OC to t if $\bullet t \cap O(S) = \phi$,

(iii) For any output transition t of a place in $O(S)$, there exists an arc from t to OC if $t \bullet \cap O(S) = \phi$.

Theorem 1: The net with the place OC is also an ES^3PR net where OC can be considered as a resource place with $OP(OC) = O(S)$.

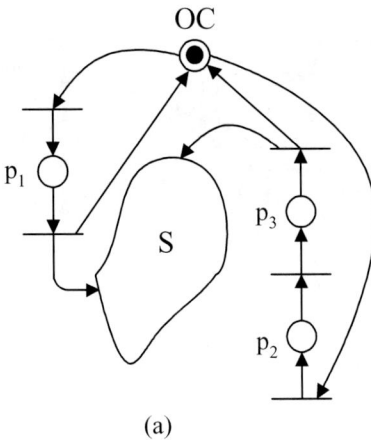

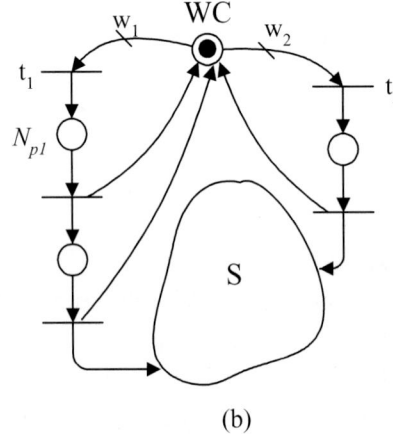

Fig. 2: Two different control places.

Case II: $N_p > 1, \forall p \in O(S)$

To prevent S from becoming unmarked, a WC is added as follows:

Definition 4: $W_p = \max\limits_{p' \in L_p} N_{p'}$, where L_p is the set of operation place in all paths from p to the sink transitions of G^* with obvious extension $N_p = 0$ for all $p \notin O(S)$.

Definition 5: $W_t = \max\limits_{p' \in L_t} N_{p'}$, where L_t is the set of operation place in all paths from p to the sink transitions of G^* with obvious extension $N_p = 0$ for all $p \notin O(S)$. Note that $W_t = W_p$, for the unique operation place $p \in t\bullet$.

Definition 6: For each siphon S such that $N_p > 1, \forall p \in O(S)$, a new place WC is added such that:

(i) $M_0(WC) = \sum\limits_{r \in S \cap R} M_0(r) - 1$,

(ii) For all source transition t of G^* such that $W_t > 0$, add an arc (WC, t) of weight W_t,

(iii) For any output transition t that is not a source transition G^*, let p be its input operation place. Add an arc (t, WC) of weight $W_p - W_t$ if $W_p > W_t$.

Theorem 2: The net obtained by adding WC is potentially live.

Theorem 3: The net obtained by adding WC is reversible iff no siphon S without WC cannot become unmarked.

As previously mentioned, the net has several features as follows:

(i) There may be more than one siphon that can become unmarked.
(ii) Adding OC may create a new unmarked siphon.
(iii) Adding WC will not create new unmarked siphons.

From the above features, they imply that an iterative process is needed to eliminate all existing/new unmarked siphons. As a result, our goal is to eliminate any unmarked siphons at each iteration. More detail steps are shown as follows:

Algorithm Deadlock_Prevention
Step 1:
Starting from an ES^3PR net G^0 an unmarked siphon S_1, is determined and a control place C_1 is added.
Let $(G^0 + C_1)$ denote the new net.

Step 2:
  Case I: C_1 is an OC.
  (i) Then, the new net $(G^0 + C_1)$ is an ES^3PR net.
  (ii) The next step determines an unmarked siphon S_2 of the net $(G^0 + C_1)$.
  (iii) The related control place C_2 is added to obtain net $(G^0 + C_1 + C_2)$.
  Case II: C_1 is an OW.
  (i) Then, the new net $(G^0 + C_1)$ is a generalized net.
  (ii) According to the *Theorem 15*, the next step determines a siphon S_2 in net G^0 that can become unmarked in $(G^0 + C_1)$. Note that the WC with its arcs added on the net cannot generate any new siphons.

(iii) The related control place C_2 is added to obtain net $(G^0 + C_1 + C_2)$.

For convenience to explain our policy, in the next section, we will show an example.

4. Example

Fig. 1 shows the net model of a flexible production system where two types of processes execute concurrently and share a set of common resources. The system net is an ES^3PR.

In Fig. 3, we obtain an *unmarked maximal siphon* $\{p3, p5, p6, p7, p10, p11, p12, p13, p14, p15, p16\}$ of the net. From the maximal siphon, an *unmarked minimal siphon* $\{p5, p6, p7, p10, p11, p14, p15\}$ by which the sub-net generated is shown in Fig.3. Hence, we can point out that $t2$, $t5$ and $t13$ are *sink transitions*, $\{p3, p4, p9\} \in O(S)$, and $M_0(p17) = M_0(p14) + M_0(p15) - 1 = 2$.

Note that p17 is considered a resource place in the algorithm. Similarity, we obtain an *unmarked maximal siphon* $\{p3, p4, p5, p6, p7, p10, p12, p13, p14, p15, p17\}$ of the net. From the maximal siphon, an *unmarked minimal siphon* $\{p3, p4, p5, p6, p12, p13, p14\}$ is obtained. Hence, we can point out that $t1$ and $t12$ are *sink transitions*, $\{p2, p10, p11\} \in O(S)$, and $M_0(p18) = M_0(p13) + M_0(p14) - 1 = 1$. In the next iteration, by the same way, we obtain an *unmarked maximal siphon* $\{p3, p5, p6, p7, p10, p11, p12, p13, p15, p17, p18\}$. From the maximal siphon, an *unmarked minimal siphon* $\{p3, p4, p13, p17, p18\}$ by which the sub-net generated is shown in Fig.4. Hence, we can point out that $t2$ and $t13$ are *sink transitions* and source transition as well. This situation is different from the former iteration. Since $N_{p2} = 2$, in this case, we add a weighted control place to the net. Specifically, we obtain $W_{t1} = 2$ and $W_{t13} = 1$, and the weight of arc $(p19, t1)$ is equal to 2 and the weight of $(p19, t13)$ is equal to 1. It is worth to note that $p19$ (WC) cannot generate a new siphon (*Theorem 3*).

Due to that $p19$ cannot be considered a resource place and generates a new siphon, in fact, add a restriction to the MIP method. The objective is to eliminate some unreachable states. The restriction is as follows:

$M(3) + M(4) + M(12) + M(13) + M(17) + M(18) \leq 3$

As a result, in the new iteration, we can obtain another minimal siphon $\{p5, p6, p7, p10, p11, p15, p17, p18\}$ and the controlled sub-net is depicted in Fig. 5.

Finally, we can obtain another minimal siphon $\{p3, p4, p10, p11, p17, p18\}$ and the controlled sub-net is depicted in Fig. 6.

In summary, we add three *OCs* control place and two *WCs* control place to the original net. After adding those control places, the resultant net is live and reversibe.

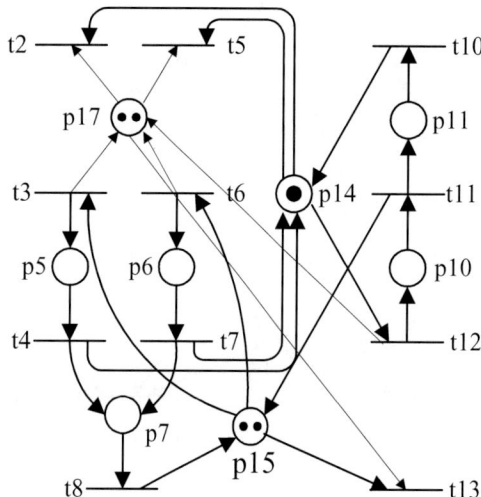

Fig. 3: A minimal siphon with p17 (*OC*) being the added control place.

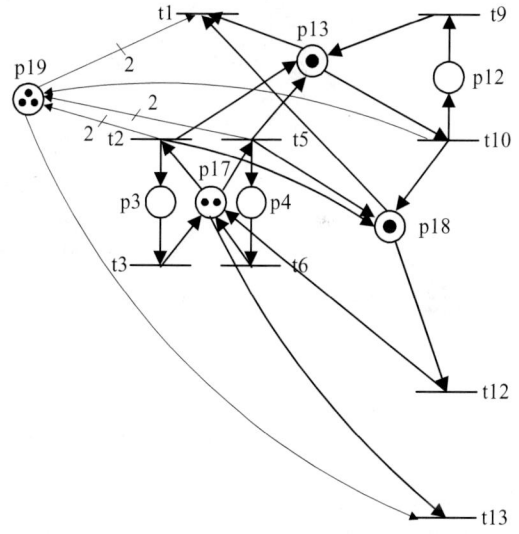

Fig. 4: A minimal siphon with p19 (*WC*) being the added control place.

5. Conclusions

This paper presents a deadlock prevention method for a class of FMS where deadlocks are caused by unmarked siphons in their PN models. The FMSs are modeled using ES^3PRs, which are extended from S^3PR nets.

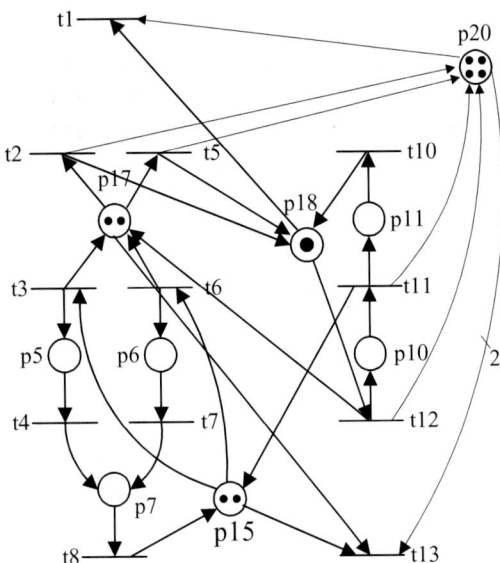

Fig. 5: A minimal siphon with *p20* (WC) control place.

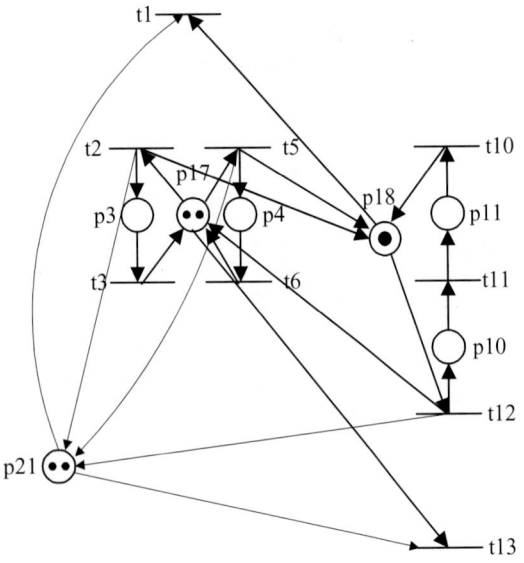

Fig. 6: A minimal siphon with *p21* (OC) control place.

The proposed deadlock control policy can be implemented by adding two kinds of control places to the original model. The algorithm main object is to decide which one kind of control place is suitable. More specific, the method used two kinds of control places to add on the original net G. We have obtained the relation of the algorithm and the liveness and reversibility of the controlled net. Finally, a flexible manufacturing example is presented for illustrating the method.

6. Acknowledgement

This work was supported in part by the National Science Council of Taiwan, ROC under Grant NSC89-2618-E-011-009.

References

1. K. Barkaoui, and I. B. Abdallah, "Deadlock avoidance in FMS based on structural therory of Petri nets," *Proc. INRIA/IEEE Symposium on ETFA*, vol. 2, no .4, pp. 499-510, 1995.
2. F. Chu and X. L. Xie, "Deadlock analysis of Petri nets using siphons and mathematical programming," *IEEE Trans. Robotics Automat.*, vol. 13, no. 6, pp. 793-804, 1997.
3. M. P. Fanti, G. Maione, Saverio Mascolo and B. Turchiano, "Event-Based Feedback Control for Deadlock Avoidance in Flexible Production Systems," *IEEE Trans. Robotics Automat.*, vol. 13, no. 3, pp. 347-363, 1997.
4. F. S. Hsieh and S. C. Chang, "Dispatching-driven deadlock avoidance controller synthesis for flexible manufacturing systems," *IEEE Trans. Robotics Automat.*, vol. 10, no. 2, pp. 196-209, 1994.
5. Y. S. Huang, M. D. Jeng, X. L. Xie, and ShengLuen Chung, " A deadlock Prevention Policy Based on Petri nets and Siphons," *to appear in International Journal of Production Research,* 2000.
6. J. Ezpeleta, J. M. Colom, and J. Martinez, "A Petri net based deadlock prevention policy for flexible manufacturing systems," *IEEE Trans. Robotics Automat.*, vol. 11, no. 2, pp. 173-184, 1995.
7. M. D. Jeng, " Petri nets for modeling automated manufacturing systems with error recovery," *IEEE Trans. Robotics Automat.*, vol. 13, no. 5, pp. 752-760, 1997.
8. M. D. Jeng, M. Y. Peng, and Y. S. Huang, " An algorithm for calculating minimal siphons and traps of Petri nets," *International Journal of Intelligent Control and Systems,* Vol. 3, No. 3, pp. 263-275.1999.
9. J. Park and S. A. Reveliotis, "Algebraic Synthesis of Efficient Deadlock Avoidance Policies for Sequential Resource Allocation Systems," *IEEE Trans. Robotics Automat.*, vol. 16, no. 2, pp. 190-195, 2000.
10. Murata T., "Petri nets: properties, analysis and application", *Proc. IEEE,* vol. 44, no .4, pp. 541-579, Apr. 1989.
11. N. Viswanadham, Y. Narahari, and T. L. Johnson, " Deadlock prevention and deadlock avoidance in flexible manufacturing systems using Petri net models," *IEEE Trans. Robotics Automat.*, vol. 6, no. 6, pp. 713-723, 1990.
12. M. C. Zhou and F. DiCesare, *Petri Net Synthesis for Discrete Event Control of Manufacturing Systems*, Boston, MA: Kluwer, 1993.

An Effective Search Strategy for Wafer Fabrication Scheduling with Uncertain Process Requirements

Ming-Hung Lin and Li-Chen Fu
Dept. of Computer Science and Information Engineering
National Taiwan University
Taipei, Taiwan, R.O.C
Email: lichen@ccms.ntu.edu.tw

Abstract

In this paper, we can decide the operation order with unknown potential order requirements. The scheduling architecture is proposed in this paper, First, the branch-and-bound search based on Markov chain method is proposed. The Markov chain gets the service rate records and arrival rate records from the MES(manufacturing execution system). We can get the possible beginning times of operations for each job via Markov chain. The information of the possible beginning time can help us to approximate the solution space. Thus, by the information of the possible beginning times of operations, a branch-and-bound search scheduler can be used to find a sub-optimal scheduling.

1 Introduction

Wafer fabrication is the most costly phase of semiconductor manufacturing [1, 2, 3]. Recent papers by Uzsoy et al. [2, 4], Johri [5], and Duenyas et al. [6] highlight the difficulties in planning and scheduling of wafer fabrication facilities. These papers also survey the literature on related topics. Effective shop-floor scheduling can be a major component of reduction in cycle time [7, 8, 9, 10, 11, 12, 16]. Yet in many wafer fabrications the product spends much more waiting time than actually being processed, so there is a large potential for reducing waiting time and a great benefit for doing so. People also considered other issues of wafer fabrication systems, such as batch processing system [14] and maintenance scheduling as well as staff policy [13]. It is well known in the scheduling literature that the general job shop problem is NP-hard, which lead to no efficient algorithm exists for solving the scheduling problems optimally in polynomial time for wafer fabrication.

In this paper, we assign the involved job to a machine at each wafer processing step so that the total completion time

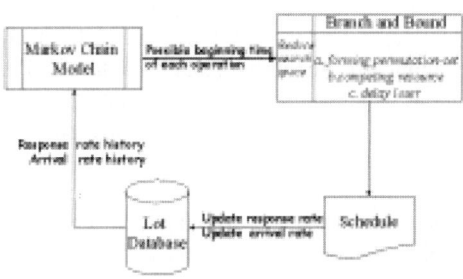

Figure 1: The Markov scheduling

of the processing is minimum subject to the constraints of some promised finishing time or due date. As we have mentioned earlier, we can get the possible beginning times of operations for each job via Markov chain. The information of the possible beginning time can help us to approximate the solution space. As shown in Figure 1, given such information, a branch-and-bound search scheduler can be used to find a sub-optimal scheduling. The Markov chain acquires the service rate records and arrival rate records from the MES (manufacturing execution system).

The organization of this paper is described as follows. In Section 2, some scheduling problems of wafer fabrication are illustrated. In Section 3, the detailed analytical Markov chain method is discussed. In Section 4, an embedded search method over the wafer fabrication system is employed. In Section 5, we demonstrate an example of using the proposed mechanism and analyze the performance. Finally, conclusions are provided in Section 6.

2 Problem Formulation

Suppose there is a set of n types of lots ($i = 1 \ldots n$). Each type has S_i wafer processing steps, and each wafer

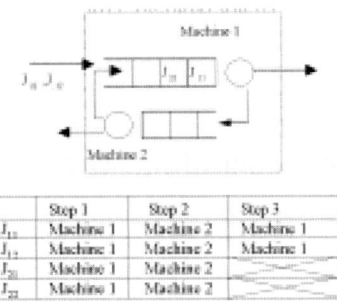

Figure 2: The processing steps of 4 jobs

Figure 3: A possible permutation schedule in Example 1

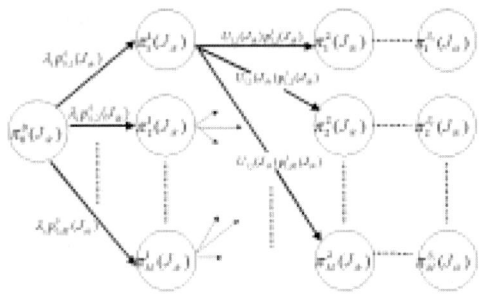

Figure 4: The Markov chain of state probability for the job J_{ik} which appears in the queue of the machine m for executing the l-th wafer processing step at time t.

processing step belongs to one kind of wafer processing operation. Each processing operation is to be performed on one of the machine group, each of which has the same corresponding wafer processing capability. Let job J_{ik} be the k-th arrival lot of the i-th type, and each job J_{ik} has its arrival time a_{ik}, due date d_{ik}, and weight α_{ik}. Presume that there are totally M machines in a wafer factory. The wafer processing operation $op(l)$ of the l-th processing step of job J_{ik} requires a mean processing rate $u_{op(l),m}(i)$ on machine m. Besides, the i-th type lot is released into the wafer factory with the mean arrival rate λ_i. Let $F^l(J_{ik})$ be the time at which job J_{ik} finishes the l-th wafer processing step. Since the following discussion will all in statistical paradigm, the arrival time, processing time, and weight are modeled as independent random variables. However, we still prefer to assume that machine availability is deterministic.

Let K_i be the total number of lots of type i, which include the lots that have been released into factory and the lots that will be released. The total completion time is defined to be the weighted sum of times that all the jobs take to finish their final processing steps as shown below:

$$f = \sum_{i=1}^{n} \sum_{k=1}^{K_i} \alpha_{ik} F^{S_i}(J_{ik}) \qquad (1)$$

Example 2.1

Consider the following numerical example with 4 jobs. Job J_{11} and J_{21} are the jobs corresponding to the first arrival lots of Type 1 and Type 2, respectively, whereas Job J_{12} and J_{22} are the jobs corresponding to the second arrival lots of Type 1 and Type 2, respectively. Presume that J_{11} and J_{21} have already been released into the wafer factory, but J_{12} and J_{22} have not been released into the wafer factory yet. As shown in Figure 2, from the manufacturing database Type 1 job needs three wafer operation steps and Type 2 job needs two. In this example, we suppose that the mean service time on the Machine 1 is 4 time units for each processing step of any job, and the mean service time on the Machine 2 is 5 time units for each processing step of any job. Moreover, we suppose that the mean arrival time for lots of Type 1 is 2 time units and the mean arrival time for lots of Type 2 is 3 time units. The scheduling problem is to determine the permutation schedule so that the total completion time f is the minimum. Figure 3 shows a possible permutation schedule where the arrows indicate the possible finishing times of the four jobs. □

3 Markov Chain Model

The job J_{ik} may be already in the factory or may wait outside the factory. Let $\pi_m^l(J_{ik})(t)$ be denoted as the state probability that the job J_{ik} appears in queue of the machine m for executing the l-th wafer processing step at time t. As shown in Figure 4, the state probability $\pi_m^l(J_{ik})(t)$ can be formulated in terms of the Markov chain. On the other hand, let $p_{s,t}^l(J_{ik})$ be the routing probability from the machine s to the machine t after the job J_{ik} finishes the l-th wafer processing step on the machine s. Let $U_{l,m}(J_{ik})$ be the mean service rate of the l-th processing step of the i-th type on the machine m.

In a practical factory, the transient solution is more meaningful than steady-state solution. Besides, there is a corresponding Markov chain as shown in Figure 4 for every job. Therefore, we need to compute the transient state probability $\pi_m^l(J_{ik})(t)$ for every job J_{ik}. Due to the special structure of this kind of Markov chain, it is possible to obtain a closed-form transient solution. We can derive a system of linear differential equations for this kind of Markov chain from differentiation and integration theory.

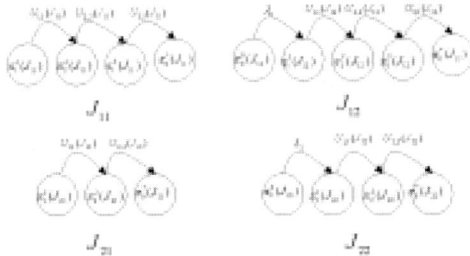

Figure 5: The Markov chains of the four jobs in Example 2.1

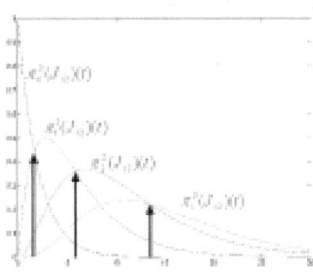

Figure 8: The transient state probability of job J_{12}

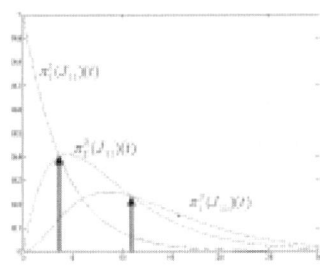

Figure 6: The transient state probability of job J_{11}

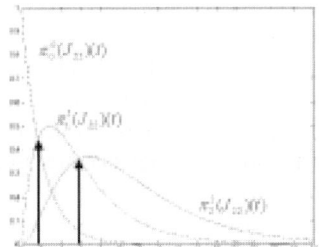

Figure 9: The transient state probability of job J_{22}

Example 3.1 *(Continued)*

Figure 5 shows the Markov chains of the jobs $J_{11}, J_{12}, J_{21}, J_{22}$ from Example 2.1. In this example, $\pi_0^*(J_{11}), \pi_0^*(J_{12}), \pi_0^*(J_{21}), \pi_0^*(J_{22})$ are the final states. We can apply elementary differentiation and integration rules and get closed-form solution for each transient state probability. The result of these transient state probabilities are shown in Figures 8 - 7 □

Precedence Constraints Considering processing-step precedence constraints, the operation $op(l)$ of the l-th processing step of job J_{ik} must start before the operation $op(l+1)$ of the $(l+1)$-th processing step on the same machine m, i.e.,

$$\pi_m^l(J_{ik})(t) > \pi_m^{l+1}(J_{ik})(t), \forall i = 1, \ldots, n, \forall k \in N$$

Thus, as shown in Figures 8- 7, we need to modify the value of $\pi_m^l(J_{ik})(t)$ where the value of $\pi_m^l(J_{ik})(t)$ is either 0 or 1 as follows, i.e.,

$$\pi_m^l(J_{ik})(t) = 1, \forall t,$$

if

$$\pi_m^{l-1}(J_{ik})(t) < \pi_m^l(J_{ik})(t)$$

or

$$\pi_m^l(J_{ik})(t) > \pi_m^{l+1}(J_{ik})(t);$$

whereas

$$\pi_m^l(J_{ik})(t) = 0, \forall t,$$

if

$$\pi_m^{l-1}(J_{ik})(t) > \pi_m^l(J_{ik})(t)$$

or

$$\pi_m^l(J_{ik})(t) < \pi_m^{l+1}(J_{ik})(t) \quad (2)$$

Resource Constraints In addition, each operation is assigned to only one machine, i.e.,

$$\sum_{m=1}^{M} \pi_m^l(J_{ik})(t) = 1, l = 1, \ldots, S_i, i = 1, \ldots, n, k = 1, \ldots, K_i \quad (3)$$

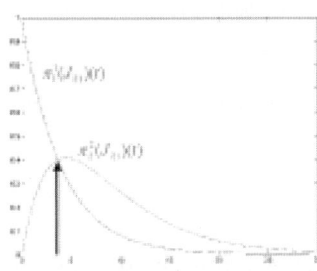

Figure 7: The transient state probability of job J_{21}

4 Branch-and-Bound Search Strategy

The scheduling problem can be viewed as finding a sub-optimal permutation schedule. We know that finding permutation schedules is NP-complete. In this paper, we use a branch-and-bound search strategy as a solution to solve such a problem in the environment of a wafer fab. It is worthwhile to note that the branch-and-bound method is based on the idea of *"intelligently"* enumerating all the feasible points in a combinatorial problem.

The qualification *"intelligent"* is important here because, as should be self-clear, it is impractical simply to examine all possible solutions. Perhaps a more sophisticated way of describing such approach is to say that we try to construct a proof to show that a solution is sub-optimal, based on successive partitioning of the solution space. The term *"branch"* in branch-and-bound method refers to the above-mentioned partitioning process, whereas the term *"bound"* refers to the lower bounds that are used to construct a proof for optimality without going through exhaustive search. We can visualize this process as a tree, where the root represents the original feasible region and each node represents a sub-problem. The natural way to branch is to choose the first job to be scheduled to each machine at the first level of the branching tree, the second job to each machine at the next level, and so on. For simplicity, let the processing operation $op(l)$ of the l-th processing step of job J_{ik} be denoted as $op_{ik,l}$. The following is an algorithm which tries to solve the scheduling problem through job assignment.

Algorithm 4.1 *The branch-and-bound search algorithm for sub-optimal scheduling:*

Begin

1. schedule-time $t := 0$;

2. for each machine m, let PS_m be the permutation-set of machine m at schedule-time t, $PS_m := 0$;

3. active-set $:= \{ \underbrace{0, \cdots, 0}_{M} \}$;

4. $Q := \infty$;

5. current-best $:=$ anything;

    For *each operation $op_{ik,l}$* **Do**

    Get mean of arrival rate and mean of processing time from manufacturing database and compute $\pi_m^l(J_{ik})(t)$;

    End-for

While *active-set is not empty* **Do**

Begin

(a) *choose a branching node x, node $x \in$ active-set;*

(b) *let node x be denoted as that $x = (\underbrace{ps_1, \cdots, ps_m, \cdots, ps_M}_{M})$, where $ps_m \in PS_m$;*

(c) *remove node x from active-set;*

    For *each operation $op_{ik,l}$* **Do**

    If $\pi_m^l(J_{ik})(t)=1$ **and** *operation $op_{ik,l} \neq ps_m$* **and** $op_{ik,l} \notin PS_g$ *(subject to Equation 3)*, $\forall g \neq m$ **then** *add operation $op_{ik,l}$ into the permutation-set PS_m of machine m.*

comment : *That the constraint $op_{ik,l} \notin PS_g$, $\forall g \neq m$, means that each operation is assigned to only one machine.*

    End-for

(d) *generate the all possible children of the node x*

(e) *child $i = (ch_1, \cdots, ch_m, \cdots, ch_M)$, where the winner operation $ch_m \in PS_m$, $i = 1, \ldots, n_k$, where $n_k = |PS_1| \times |PS_m| \times |PS_M|$;*

(f) *let the loser-set LS_i of the child i be the set of loser-lists $(ls_1, \cdots, ls_m, \cdots, ls_M)$, where the operation $ls_m \in PS_m - \{ch_m\}$, ls_m is also denoted as a loser-operation competing with the operation ch_m on the machine m;*

(g) *calculate the corresponding **lower-bound**, z_i;*

    For $i = 1, \ldots, n_k$ **Do**

    Begin

    If $z_i > Q$

    then *kill child i*

    else $Q := z_i$, *current-best := child i, add child i to active-set;*

    End

    For *each loser operation ls_m in a loser-list$(ls_1, \cdots, ls_m, \cdots, ls_M)$ of loser-set LS_i* **Do**

    A. *assume the winner operation ch_m is the operation $op_{ik,l}$ and $\frac{1}{u_{op(l),m}(i)}$ is the mean processing time of operation $op_{ik,l}$.*

    B. *assume the loser operation ls_m is the operation $op_{ab,c}$ of the job J_{ab}, shift the value of the state probabilities $\pi_m^c(J_{ab})(t)$, $\pi_m^{c+1}(J_{ab})(t)$, $\cdots$, $\pi_m^{S_a}(J_{ab})(t)$ to satisfy the processing-step constraints of the job J_{ab};*

    For *processing step y of the job J_{ab}, y from c to S_a*, **Do loop**

    Begin

C. suppose t from t1 to t2, $\pi_m^y(J_{ab})(t) = 1$;

D. let t0 be from t1 to t2 do
Begin

E. $\pi_m^y(J_{ab})(t0 + \frac{1}{u_{op(l),m}(i)}) = \pi_m^y(J_{ab})(t0);$

F. $\pi_m^y(J_{ab})(t0) = 0;$
End

G. suppose t from t3 to t4, $\pi_m^{y+1}(J_{ab})(t) = 1$;

H. **If** $t2 + \frac{1}{u_{op(l),m}(i)} \geq t3$ **then** continue the loop **else** stop the do loop;

End do loop

Comment: it means that the job J_{ab} is a loser when competing with the job J_{ik} at its c-th processing step. Thus, all successor operations of the job J_{ab} may be delay caused by the job J_{ik}. It also indicate that all successor operations of the job J_{ik} may be delay by a time-period $\frac{1}{u_{op(l),m}(i)}$ which caused by the winner operation $op_{ik,l}$ of the job J_{ik}.

End-for
End-for

(h) update the current schedule-time $t = t + 1$;
End-while
End

5 Simulation Results

There are 24 workstations which are divided into six types, and each of the workstations comprises several identical pieces of equipment. In the simulation model presented here, each lot entering the fab is associated with a specific process flow. The model contains two different process flows. Since the process flows are deterministic, the Markov Chain can be easy constructed.

The experiment duration is 20000 hours. Eight kinds of lot-release interval are examined in this experiment, namely, 42,52,62,72,82,92,102 and 112 hours. The experiment is conducted on a computer with Intel 500 MHz CPU and 128 MB RAM. The result of this experiment is compared with the simple heuristic rule FCFS. Figure 10 shows the mean production cycle time of Type 1 lot under the cases with different lot-release intervals. We can see that the proposed branch-and-bound search based on Markov chain method produces lower mean production cycle time in comparison with the simple scheduling

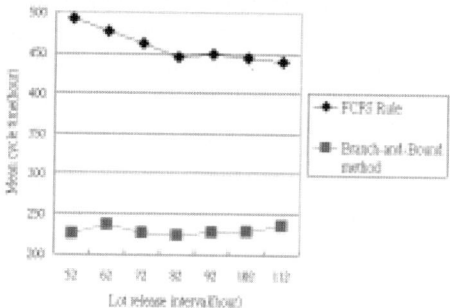

Figure 10: The mean production cycle time

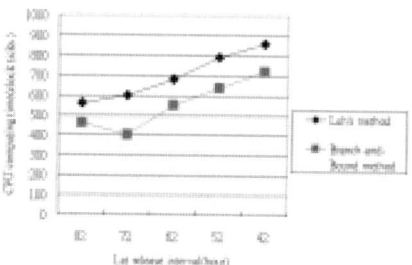

Figure 11: The computing time

based on FCFS rule. On the other hand, we provide another test case, the result of this test case is compared with Lagrangian relaxation and dynamic programming method [15]. Five kinds of lot-release interval are examined in this experiment, namely, 42,52,62,72 and 82 hours. Figure 11 shows the CPU computing time for two different scheduling strategies. We can see that the proposed branch-and-bound search based on Markov chain method needs less CPU computing time than the Lagrangian relaxation and dynamic programming method.

6 Conclusion

We can get the possible beginning times of operations for each job via Markov chain. Thus, by the information of the possible beginning times of operations, a branch-and-bound search scheduler can be used to find a sub-optimal scheduling. The results of the experiment show that the proposed branch-and-bound search based on Markov chain method produces lower mean production cycle time in comparison with the simple scheduling based on FCFS rule. In addition, the results of the experiment also show that the proposed branch-and-bound search based on Markov chain method needs less CPU computing time than the Lagrangian relaxation and dynamic programming method. Since the wafer production system is complex and the WIP level is very large, it is indeed difficult to obtain an optimal scheduling under the dynamic environment. Thus, the hereby proposed scheduling method is performed actually off-line. However, the scheduling solution by our method can be a good candidate of solution for dynamic scheduling.

References

[1] Hong Chen, J. Michael Harrison, Avi Mandelbaum, Ann Van Ackere, and Lawrence M. Wein, "Empirical Evaluation of a Queueing Network Model for Semiconductor Wafer Fabrication, " *Operation Research*, vol. 36, no. 2, pp. 202-215, 1988.

[2] Reha Uzsoy, Chung-Yee Lee, Louis A. Martin-Vega, "A Review of Production Planning and Scheduling Models in the Semiconductor Industry Part: System Characteristics, Performance Evaluation and Production Planning, " *IIE Transactions*, vol. 24, no. 4, pp. 47-60, 1992.

[3] S.M. Sze, "VLSI Technology", *McGraw-Hill* New York, 1983.

[4] Reha Uzsoy, Chung-Yee Lee, Louis A. Martin-Vega, "A Review of Production Planning and Scheduling Models in the Semiconductor Industry Part: Shop-Floor Control, " *IIE Transactions*, vol. 26, no. 5, pp. 44-55, 1994.

[5] Pravin K. Johri, "Practical Issues in Scheduling and Dispatching in Semiconductor Wafer Fabrication, " *Journal of Manufacturing Systems*, vol. 12, no. 6, pp. 474-485, 1993

[6] Lzak Duenyas, John W. Fowler, and Lee W. Schruben, "Planning and Scheduling in Japanese Semiconductor Manufacturing, " *Journal of Manufacturing Systems*, vol. 13, no. 5, pp. 323-332, 1994.

[7] Lawrence M. Wein, "Scheduling Semiconductor Wafer Fabrication, " *IEEE Transactions on Semiconductor Manufacturing* , vol. 1, no. 3, pp. 115-130, 1988.

[8] C. Roger Glassey and Mauricio G. C. Resende, "Closed-Loop Job Release Control for VLSI Circuit Manufacturing, " *IEEE Transactions on Semiconductor Manufacturing*, vol. 1, no. 1, pp. 36-46, 1988.

[9] Shu Li, Tom Tang, and Donald W. Collins, "Minimum Inventory Variability Schedule with Applications in Semiconductor Fabrication, " *IEEE Transactions on Semiconductor Manufacturing*, vol. 9, no. 1, pp. 145-149, 1996.

[10] Y. Narahari and L. M. Khan, "Modeling the Effect of Hot Lots in Semiconductor Manufacturing Systems, " *IEEE Transactions on Semiconductor Manufacturing*, vol. 10, no. 1, pp. 185-188, 1997.

[11] Yeong-Dae Kim, Jung-Ug Kim, Seung-Kil Lim, and Hong-Bae Jun, "Due-Date Based Scheduling and Contril Policies in a Multi-product Semiconductor Wafer Fabrication Facility, " *IEEE Transactions on Semiconductor Manufacturing*, vol. 11, no. 1, pp. 155-164, 1998.

[12] Yeong-Dae Kim, Dong-Ho Lee, and Jung-Ug Kim, "A Simulation Study on Lot Release Control, Mask Scheduling, and Batch Scheduling in Semiconductor Wafer Fabrication Facilities, " *Journal of Manufacturing Systems*, vol. 17, no. 2, pp. 107-117, 1998.

[13] Sarah A. Mosley, Tim Teyner, and Reha M. Uzsoy, "Maintenance Scheduling and Staffing Policies in a Wafer Fabrication Facility, " *IEEE Transactions on Semiconductor Manufacturing*, vol. 11, no. 2, pp. 316-323, 1998.

[14] C. Roger Glassey and W. Willie Weng, "Dynamic Batching Heuristic for Simultaneous Processing, " *IEEE Transactions on Semiconductor Manufacturing*, vol. 4, no. 2, pp. 77-82, 1991.

[15] Peter B. Luh, Dong Chen and Lakshman S. Thakur, " An Effective Approach for Job-Shop Scheduling with Uncertain Processing Requirements," *IEEE Transactions on Robotics and Automation* , vol. 15, no. 2, pp. 328-339, 1999.

[16] D. Y.Liao, S. C. Chang, K. W.Pei, C. M.Chang,"Daily Scheduling for R&D Semiconductor Fabrication," *IEEE Transactions on Semiconductor Manufacturing*, pp. 550-561, Nov. 1996.

Model-based Control for Reconfigurable Manufacturing Systems

Kazushi Ohashi[*1] and Kang G. Shin[*2]

*1: Industrial Electronics & Systems Laboratory
Mitsubishi Electric Corp.
8-1-1 Tsukaguchi-Honmachi
Amagasaki, Hyogo 661-8661, Japan
ohashi@fas.sdl.melco.co.jp

*2: Real-Time Computing Laboratory
Eelectrical Engineering and Computer Science
The University of Michigan
Ann Arbor, MI 48109-2122, U.S.A.
kgshin@eecs.umich.edu

Abstract: The manufacturing industry cannot stay competitive and survive in today's market without agile adaptation to rapidly changing customers' demands. This in turn requires manufacturing systems to be reconfigurable for timely introduction of new products in the market. Unfortunately, at present, the system designers cannot systematically and completely manage their design data, because manufacturing systems have gradually become too large and too complicated to manage. In order to reconfigure and reuse H/W and S/W components in manufacturing systems, and improve the engineering environment of system control design, we propose a model-based control design using state transition diagrams and a general graph description, while taking reconfiguration and reuse of design data into account. We demonstrate the utility of the proposed approach using a real application.

1. Introduction

User demands for products tend to change rapidly. To respond to such demands, manufacturers must modify the design and functions of products, and change their production schedule to be competitive. This in turn requires to quickly reconfigure their manufacturing systems like a cell or line. It is, however, difficult and time-consuming to change system control programs due mainly to their ad hoc, and sometimes proprietary, nature. Especially, many parts of the System Control Design (SCD) data are inter-related to one another, thus requiring significant time and effort to identify and determine which parts to be modified. As a result, it is very costly to modify the control programs, or sometimes costlier than re-developing them from scratch.

Since graphs are very useful to represent data relations, many researchers attempted to model manufacturing systems using various graph descriptions, such as state transition diagrams (STDs) [2,3,4,6,7] and IDEF [5]. For accurate system descriptions, however, these models usually consist of an excessive number of states when they are used to model realistic (large) systems. To cope with this difficulty, some of these models adopt a hierarchical structure, but they have not taken into account system reconfiguration and reuse of H/W, S/W and SCD data, while its unit of reconfiguration is coarse in others. The authors of [3] and [4] presented a model based on CIM-OSA, which is composed of event flow and resource behaviors at a control specification level, and described with an object diagram and a Petri-Net. This model is decomposed according to manufacturing system functionalities. Researchers have taken into account reconfiguration of a manufacturing system and reuse of components of SCD data. But the unit of replaceable parts is a resource object. [5] provided a model consisting of message flows based on IDEF3 and a resource configuration graph. In this model sequence control flows and synchronization between resources in the message flow can be understood easily, but it didn't consider data reconfiguration. [6] presented a model consisting of a manufacturing system's structure graph and production routes based on Colored Petri-Nets. This model adopts templates and allows use of variables. It includes the concept of time to estimate runtime, but it is difficult to identify synchronization parts between resources when system size is large. [7] proposed a cooperation model between resource behaviors and a system control model using hierarchical Petri-Nets. The authors demonstrated the use of the analysis tools to verify the properties of their models, and analyzed them both quantitatively and qualitatively. Their concept is similar to ours in terms of a cyclic description of resource behaviors, synchronization descriptions and their compositions. However, in their model it is difficult to detect the boundaries between resource behaviors in the final composed model.

We previously attempted to model manufacturing systems with a cyclic STD model [2]. Each equipment behavior is described with a cyclic STD, and the model of a target system is configured to connect synchronized transitions between STDs of resources. This modeling method is very simple, but the definitions of system sequence control are included implicitly, and hence, it is difficult to modify part of the control. Moreover, the number of redundant descriptions explodes when the

---

The work reported in this paper was done during K. Ohashi's visit to RTCL and supported in part by DARPA under Grant F33615-00-1706 administered by the US Airforce Wright Laboratory.

number of combinations between resources increases in workpiece path. Neither did it give any information about the redundancy in behavior descriptions.

In this paper we propose a model-based control design method in which one can build a control system easily and efficiently using a reconfigurable SCD data model. We (1) make the SCD data model reconfigurable from the designers' standpoints and (2) reduce redundant state descriptions. This model is described by combining three sub-models corresponding to process flow, workpiece path, and resource behavior. Each of them is a cyclic STD or a graph description, and is hierarchically-structured.

The paper is organized as follows. Section 2 describes the architecture of reconfigurable manufacturing systems. Section 3 introduces the notation in graph descriptions, and Section 4 presents a real application example. Section 5 describes data structures, and the paper concludes with Section 6.

2. Architecture

We will focus on how to change the type, number and location of resources, and process flow in SCD as easily as possible for designers who do not completely manage the SCD data. Such a robust model requires the following basic functionalities:

(1) Data definition from designer's viewpoints,
(2) Hierarchical and componential model structures.

The main objects in system control design are:

(1.1) Manufacturing (M-) process flow,
(1.2) Selection of resources,
(1.3) Workpiece path/flow between resources,
(1.4) Resource behavior (e.g., sequence of parts of a motion or process control program),
(1.5) Collaboration and synchronization (C&S) between resources,
(1.6) Layout (resource and workpiece base locations).

The model is composed of the following functionalities:

(2.1) Hierarchal skeleton structure of SCD data,
(2.2) Parameterized descriptions of motion or process behavior,
(2.3) Variable and parameter list.

Fig. 1 shows the proposed SCD model which consists of three sub-models and combinations thereof:

- M-process flow (for (1.1))
- Workpiece path (for (1.2), (1.3) and (1.5))
- Resource behavior (for (1.4) and (1.5))

M-process flow and resource behavior are described with STD and workpiece path using a graph description. The behaviors of each resource are described with a cyclic STD and can be understood easily. Parts of data in these models can be linked to each other at a low level (e.g., a place, transition or node, which is an element of each graph description). These parts together form a SCD model, but these graph descriptions only show the data structure of SCD. Actual process or motion data are linked to places or nodes. Moreover, process and motion data are decomposed into parameterized descriptions, and variable and parameter list. This way, redundant descriptions are reduced in decomposing the SCD data into components at lower levels than a resource, and sharing the same parts. As a result, the local data can be reconfigured easily.

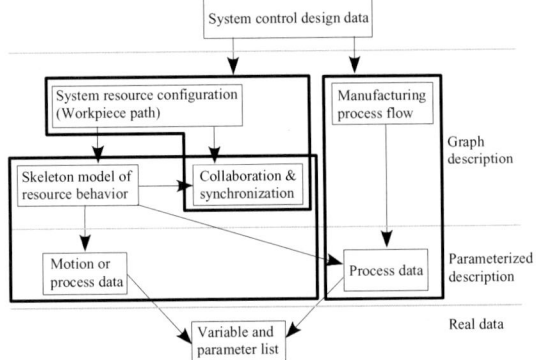

Fig. 1 The proposed model configuration

3. Notation and Modeling
3.1 M-process flow and resource behavior

M-process flow and resource behavior are described with STDs. A STD is constructed with a set of places, transitions and arrows. Graph models are constructed by connecting places and transitions with arrows. A place represents a part of manufacturing process and can have a hierarchical structure, and designers define and expand SCD data from an upper level to a lower level. Also, a place itself has data objects as shown in Fig. 2 where the line ended with a circle means a reference to another data object.

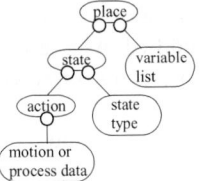

Fig. 2 Data structure of a place

A place only represents a node in graph descriptions. A combination of state, variable and parameter list represents the actual context of place, and finally identifies the place. "Variable list" means a list of variables defined by the user and can be defined locally, and "parameter list" means a list of resource variables obtained from the profiled data of resources. Also, a state is composed of action and state type data. Action data is motion or process execution data like a program and is parameterized. State type includes the data related to execution type of resources (type of M-process, execution code type of action, and so on). Likewise, SCD configuration and its actual detailed context are separated. State is reused at other places in the same SCD or other SCDs. M-process flow and resource behavior are represented below as a flow of such places.

M-process flow

The M-process flow shows a sequential flow of M-processes for each workpiece as shown in Fig. 3.

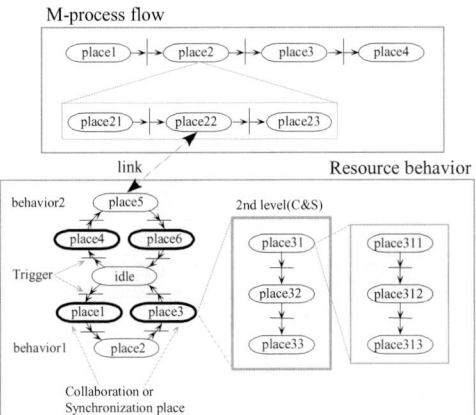

Fig 3 M-process flow and resource behavior

Each M-process is described by a place and can be defined hierarchically. All terminal places, which are elementary M-processes, can be linked to any place of any resource behavior. But on this phase of M-process flow definition, actual resources are not decided exactly until it is linked to resource behaviors, and state and variable lists are defined. Then the M-process flow itself is independent of resources. Users can reuse the M-process flow for a system that has the same M-process flow by modifying the context of places.

Resource behavior

The resource behavior represents a sequence of resource actions and is shown in Fig. 3. The resource behavior is defined based on the cyclic STD of the uppermost level. The cyclic STD is a conceptual behavior described with basic actions like "Loaded", "Unloaded" and "Processing." Some resources can use the same cyclic STD at this level, and we can provide a certain basic cyclic STD as a template. A resource can own multiple STDs each of which corresponds to a motion or process control program. Cyclic STDs, which are not independent within a resource, share a start place, which is called the "Idle" place. Upon completion, every resource behavior always returns to the "Idle" place. A place can be described hierarchically in the same way as an M-process flow. The difference between M-process flow and resource behavior is that all cyclic STDs have more than one collaboration and synchronization (C&S) place,

which is represented by a thick-lined oval, at the top level of a resource behavior. A C&S place represents collaboration or synchronization with any C&S places of other resources (e.g., "Loaded" and "Unloaded" place) as shown in Fig. 6. There always exists a C&S place next to the "Idle" place, because one behavior is selected by one of transitions between the "Idle" place and the activated place. If multiple actions need to be executed in parallel within a resource, it needs to own multiple behaviors, which do not share the "Idle" place.

Lower-level places of a C&S place are defined here or in the connection node of workpiece path. Details of collaboration and synchronization are defined at the second level of resource behavior as shown in Fig. 6.

3.2 Workpiece path

A workpiece path is specified with a graph description, and uses 4 types of nodes and 2 types of arrows as shown in Fig. 4.

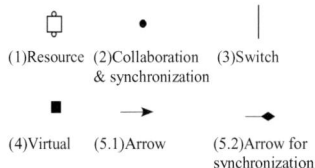

Fig. 4 Nodes in workpiece path

A resource node represents a real or conceptual resource and each small circle corresponds to the C&S place of a resource behavior. When the node represents a conceptual resource, it can be expanded hierarchically. Then, small circles mean C&S places of a complex behavior to the outside world. In case of an actual resource, it corresponds to a resource behavior.

A C&S node is linked to the C&S places of collaborating or synchronizing resource behaviors. The normal arrow of (5.1) in Fig. 4 is used for collaboration to link nodes, and the diamond-shape arrow of (5.2) is used for synchronization. Designers can define C&S conditions between resource behaviors at a lower level through this node as shown in Fig. 6.

A switch node represents the branch or join of resource nodes, and is linked to the designer-defined conditions.

A virtual node is used in the "detailed" connection node and represents real input and output places of approximate resource nodes locally. Usually, a connection between resources is described conceptually at the uppermost level, and all actual connection nodes don't appear at the upper level of workpiece path. But when a conceptual resource node is expanded, the number of connection nodes is not restricted to one (e.g., when multiple resources are used in one process as shown in Fig. 5). As an example, a workpiece path among resource nodes and the lower level of connection node CO2 are given in Fig. 5. Two resource nodes are sensors, and three virtual nodes are C&S places of the equipment in the expanded node CO2. Also "PM11.1. Loaded" means <Resource name>.<Behavior number>. <Place name> in this figure.

The relation between workpiece path and resource behavior is shown in Fig. 6. Details of synchronization can be defined through a connection node.

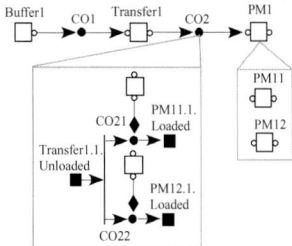

Fig. 5 An example of workpiece flow

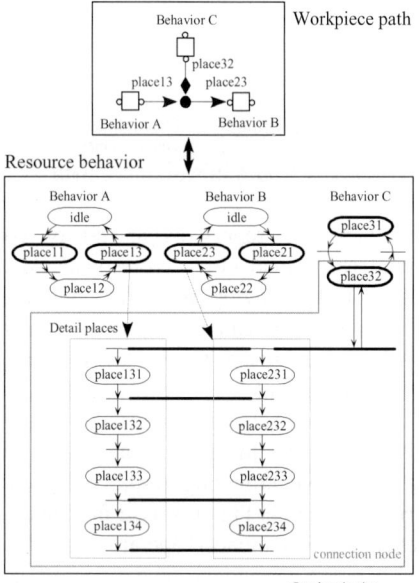

Fig. 6 Relation between workpiece path and resource behavior

4. An Example Application

The above model is applied to the SCD of a Plasma Display Panel (PDP) transfer cell as an example. This cell consists of elements of panels, three cassettes storing panels, two types of robots (GTR and RH-L3), two positioning tables (PT) (one of which is PT1 for input and the other is PT2 for output), and process machine (PM) evaporating MgO on panels. The graphical scene of this cell is given in Fig. 7 and its layout in Fig. 8. One cassette stores 15 panels. The types of work are to transfer panels from a cassette to PM, perform a chemical process on PM, and restore the panels to a cassette. PM draws in a carrier and 2 panels can be placed on the carrier each time. The workpiece path is cyclic as shown in Fig. 8. On PTs, the workpiece position is corrected. But position control is omitted to simplify the presentation. Each robot, PTs and PM have sensors, which check the existence of panels. The actual definition of each sub-model is explained below using the cell, and details are given using GTR and PT1.

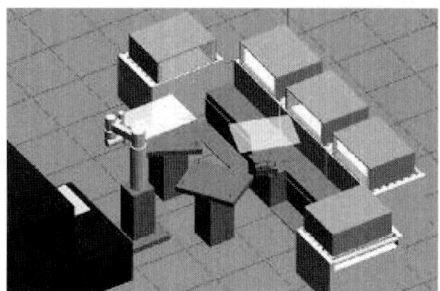

Fig. 7 PDP transfer cell (CG screen)

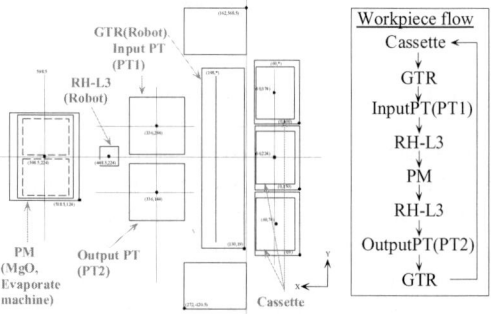

Fig. 8 Cell layout

4.1 M-process flow and resource behavior

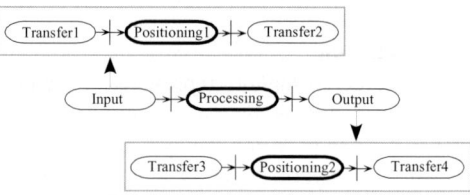

Fig. 9 M-process flow

This cell's process consists of transfer processes and a chemical process, and a process flow is described as shown in Fig. 9. The "Input" place shows how to transfer from buffer to PM, and the "Output" place represents a reverse-process for it. Each basic resource behavior can be described as shown in Fig. 10 where thick-lined ovals represent C&S places. The resource behavior is described with <Resource name>.<Behavior number> and * represents the corresponding resource or behavior number. "Wait*" or "Passive*" places denote the action of previous state being finished. An example of a place list of resource behaviors is shown in List 1. States can be shared by these multiple places.

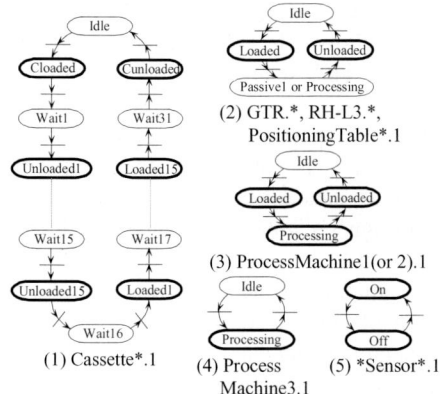

Fig. 10 Resource behaviors

List 1 Place list of GTR and PT1

	Behavior	Place		
		Level1	Level2	Level3
GTR	To PM (GTR.1)	Idle		
		Loaded (ToBuffer)	Move1	Move11(1-3)
				Move12(1-15)
			Approach1	
			Absorb1	
			Leave1	
		Passive1		
		Unloaded (To PT1)	Move2	
			Approach2	
			Release1	
			Leave2	
			Wait1	
Positioning Table1	ToPM (Positioning Table1.1)	Idle		
		Loaded	Wait1	
			Rotate0	
		Positioning		
		Unloaded	Wait2	
			Rotate90	

4.2 Workpiece path

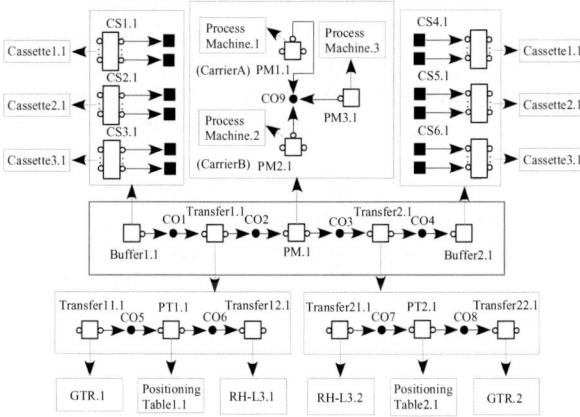

Fig. 11 Workpiece path

Workpiece path can be described hierarchically as shown in Fig. 11 where CO* is a C&S node's name. C&S node details are described as shown in Fig. 12 (CO7, 8 and 9 are omitted.). In an equipment resource node, the left-hand side circle represents the "Loaded" place and the right-hand side circle does the "Unloaded" place. In a sensor resource node, the upper-side circle represents the "Off" place and the bottom-side circle does the "On" place. Conditions stored in switch nodes are described as follows for the case of CO1 (Place1 and Place2 are variables):

Type, Priority
CS1.1.Unloaded1   // Priority 1
CS1.1.Unloaded2   // Priority 2
　　　　:

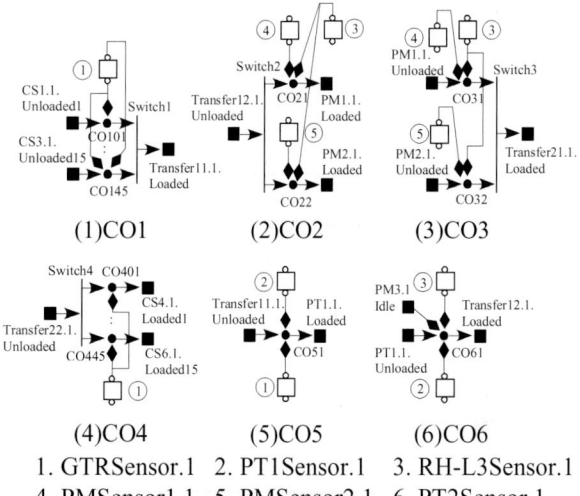

1. GTRSensor.1 2. PT1Sensor.1 3. RH-L3Sensor.1
4. PMSensor1.1 5. PMSensor2.1 6. PT2Sensor.1
Fig. 12 Detailed C&S nodes

List 2 An example of synchronization data on CO51

Transfer11	PT1	GTRSensor	PT1Sensor
Move2	Wait1		
Approach2			
Release1			
Leave2		<-N:Off	<-N:On
Wait1	Rotate0		

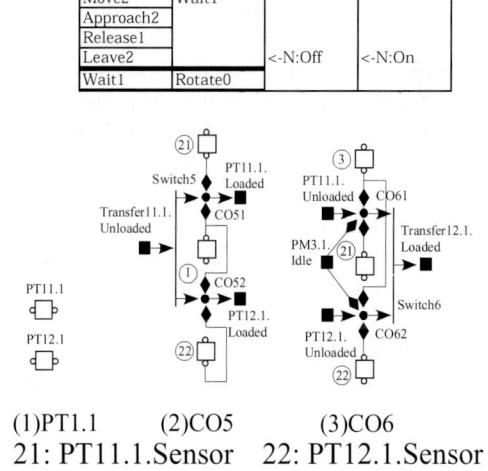

(1)PT1.1   (2)CO5   (3)CO6
21: PT11.1.Sensor   22: PT12.1.Sensor
Fig. 13 In a case of changing number of input PT

An example of synchronization condition data included in C&S nodes (CO5) is shown in List 2. Here the lines above or below a place name represent transitions, and thick lines denote collaboration. In sensor places, the

557

arrow is synchronization, and <-N means synchronization at the next transition of a pointed place.

If designers want to change the number of Input PTs to two (PT11 and PT12), they need to modify SCD data by changing resource node PT1 and C&S nodes, CO5 and CO6, as shown in Fig. 13. Then, the states used in PT1 can be reused in PT11 and PT12. State data is omitted here, but one can easily infer that the parameterized data on actions can be reused without increasing the amount of data too much.

5. Data Structure

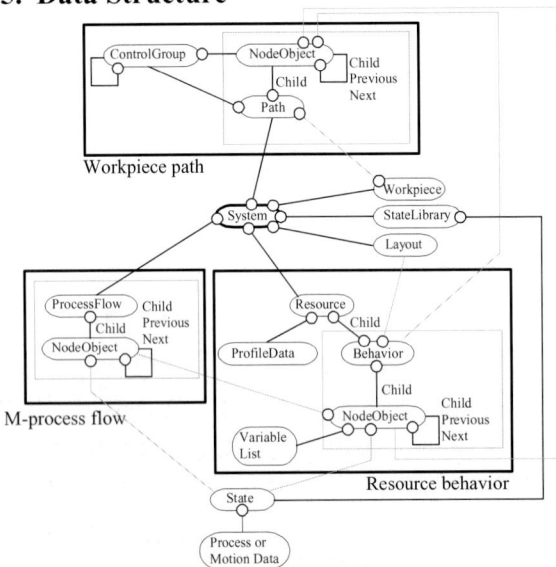

Fig. 14 Data structure of SCD model

The data structure to realize the proposed model is shown in Fig. 14. Ovals represent data objects based on object-oriented methods. The line ended with a circle denotes the reference of object data. Especially, gray lines represent the relations between sub-models and main data objects. Previous, next and child in the figure mean to own data lists (e.g., hierarchy). Places and nodes of sub-models are linked and an SCD model is constructed by their combination. We only give the meaning of some special data classes as follows.

- System: Top object in SCD data
- ProfileData: Resource parameter list
- ControlGroup: Administration of resource groups
- Behavior: Cyclic STD of resource
- NodeObject: Super class of every node (place, transition, resource node, etc.)

6. Conclusion

We proposed a modeling method for supporting the reconfiguration of a manufacturing system. Specifically, we presented a system control design model and described how to define the data as the first step. The model is then composed of 3 sub-models based on workpiece path, manufacturing process flow, and resource behavior. These sub-models are described with general graph descriptions and state transition diagrams, and decomposed into multiple data levels, such as skeleton data configuration, parameterized description, and variable and parameter lists. By separating control structure from actual data and components of the structure based on the designer's viewpoints, this model enables the data to be modified only locally and facilitates efficient reconfiguration of manufacturing systems. Also part of skeleton of control configurations, and process, motion and synchronization data can be reused. A Plasma Display Panel transfer cell is modeled as an example.

As future work, we are planning to
(1) verify the state-level system behavior by simulation using the model, and refine the model,
(2) investigate how to reconfigure the data and make execution data for the model., and
(3) expand the model to be able to deal with timing constraints, complex distributed control systems and description of exceptions.

Reference

[1] J. L. Peterson., "PETRI NET THEORY AND THE MODELING OF SYSTEMS", Prentice-Hall, 1981.
[2] K. Furusawa, T. Yoshikawa, and K. Ohashi, "Development of an Integrated Workcell Design and System Engineering (II) -Applying Line-Model to Real Systems-", *Proceedings of the 41th Annual Conference of ISCIE*, pp. 337-338, 1997 (Japanese)
[3] R. P. Monfared, and R. H. Weston., "The re-engineering and reconfiguration of manufacturing cell control systems and reuse of their components", *Proc. Instn. Mech. Engrs.*, vol. 211, pp. 495-508, 1997
[4] J. M. Edwards and M. Wilson, "A top down and bottom up approach to manufacturing enterprise engineering using the function view", *Int. J. Computer Integrated Manufacturing*, vol. 11, no. 4, pp. 364-376, 1998
[5] H. Cho and I. Lee, "Integrated framework of IDEF modeling methods for structured design of shop floor control systems", *Int. J. Computer integrated manufacturing*, vol. 12, no. 2, pp.113-128, 1999
[6] A. Zimmermann, S. Bode, and G. Hommel., "Performance and Dependability Evaluation of Manufacturing Systems Using Petri Nets", "Manufacturing and Petri Nets" of *17th Int. Conf. on Application and Theory of Petri Nets*, pp.235-250, 1996
[7] M. Heiner, P. Deussen., and J. Spranger., "A Case Study in Developing Control Software of Manufacturing Systems with Hierarchical Petri Net", "Manufacturing and Petri Nets" of *17th Int. Conf. on Application and Theory of Petri Nets*, pp.177-196, 1996

Design of Reconfigurable Semiconductor Manufacturing Systems with Maintenance and Failure

Ying Tang and MengChu Zhou
Dept. of Electrical & Computer Engineering
New Jersey Institute of Technology
University Heights, Newark, NJ07102
Email: zhou@njit.edu and yxt8958@njit.edu

Abstract

Due to expensive, highly complex and time-consuming processes, semiconductor-manufacturing systems have been given a special attention. In our previous work [5], a heuristic algorithm to design the reconfigurably automated production system was proposed. However, machine breakdowns and planned and unplanned maintenance were not considered. This paper extends that work and addresses the related design issues in reconfigurable back-end semiconductor manufacturing systems with failures and maintenance. Considering different conditions of machines, queuing network approaches are used to derive the throughput of machines to enhance the proposed virtual production line (VPL) design methodology in which the throughput was originally computed using the deterministic machine processing time. A priority is assigned to each idle machine according to its past performance and adaptive algorithms for reconfiguration are proposed.

1. Introduction

A Reconfigurable Manufacturing System (RMS) is one designed at the outset for rapid change in its structure, as well as its hardware and software components, in order to accommodate rapid adjustment of production capacity and functionality needed in response to new market demands [2]. The National Research Council in a recently released study identified RMS as the number one priority technology in manufacturing for the Year 2020 [9]. Due to the mammoth needs of the semiconductor market, semiconductor manufacturing companies are trying to necessitate the use of RMS, which can be reconfigured and reprogrammed to provide manufacturers with a rapid response capability. Qiu and Wysk introduced the concept of virtual production lines (VPL) to improve the flexibility of the back-end semiconductor manufacturing systems [4]. The fundamental issues for the discrete-event driven VPL design was presented in [5]. However, these methodologies were limited when applied to a system subject to machine failure and periodic maintenance. In the recent academic literature, researchers use Petri nets (PN) to model manufacturing system and catch such dynamics [3, 4]. Queuing theories are also adopted for these analysis and design [1, 7]. Despite these relatively modest activities, it is strongly emphasized that the variability of VPL is a decisive topic to investigate because of the complexity of the semiconductor manufacturing procedures. The primary goal of this paper is to present a method for the adaptive VPL design, which considers the performance of the failure prone machines and periodic maintenance. The rest of the paper is organized as follows. Section 2 considers a VPL with periodical maintenance; Section 3 analyses failure prone machines in a VPL; Section 4 focuses on adaptive algorithms for the VPL design; Section 5 gives an example, and Section 6 is our conclusion.

2. VPL with periodical maintenance

A VPL is organized as a sequence of workcells, each with one or more machines to handle processes in a stage [5]. Machines may have different capacity and efficiency in a workcell. Moreover, idleness of machines due to failure is harmful to the throughput and thus increases the production cost. Therefore, most of machines need to be maintained regularly for a period of time to achieve high availability. In this section, machines' periodic maintenance is considered first. Some notation for a VPL is defined as follows:

i: machine index
j: workcell index
k: machine class index

w:	VPL index
N_w:	total number of magazines in the wth work-order
$n_w(t)$:	number of finished magazines in the wth work-order at time t
M_k:	the kth class of machine pools
R_k:	the kth class of repair machine pools
I_k:	the kth class of idle machine pools
s_w:	number of workcells in the wth VPL
ϕ_k:	number of idle machines of the kth class
c_{ijw}:	the ith machine's capacity in the jth workcell of the wth VPL
τ_{ijw}:	the ith machine's processing time of a magazine in the jth workcell of the wth VPL
β_{ijw}:	the percentage time of the ith machine dedicated to the jth workcell of the wth VPL
υ_{ijw}:	the speed of the ith machine in j^{th} workcell of the wth VPL
ε_{jw}:	number of machines in the jth workcell of the wth VPL
a_{jw}:	average speed of the jth workcell in the wth VPL
$\underline{a}_w$:	the minimal real-time speed of workcells in the wth VPL
α_{ijw}:	the failure rate of the ith machine in the jth workcell of the wth VPL
p_{ik}:	the mean failure rate of the ith machine in the kth class
P_{jw}:	the allowed maximum failure rate of machines in the jth workcell of the wth VPL without getting immediate repair
χ_{ijw}:	the maintenance (repair) time for the ith machine in the jth workcell of the wth VPL
r_{ijw}:	the repair rate of the ith machine in the jth workcell of the wth VPL
ϖ_{jw}:	maintenance periodicity of machines in the jth workcell of the wth VPL
ι_{ijw}:	the maintenance rate of the ith machine in the jth workcell of the wth VPL
δ_{ik}:	the priority of the ith machine in the M_k
$\zeta_w(t)$:	slack time of the wth VPL at time t
$\varphi_w(t)$:	tardiness of the wth VPL at time t
$\psi_w(t)$:	earliness of the wth VPL at time t
E:	set of VPLs with earliness ($\psi_w(t)>0$)
D:	set of VPLs with tardiness ($\varphi_w(t)>0$)
T_w:	production time for the wth work-order

In this paper, it is assumed that machines may have different maintenance time and the maintenance periodicity of workcells is independent. The speed and the mean maintenance rate of the ith machine in the jth workcell of the wth VPL are calculated as:

$$\upsilon_{ijw} = \frac{c_{ijw}}{\tau_{ijw}} \quad (2.1)$$

$$\iota_{ijw} = \frac{\chi_{ijw}}{\varpi_{jw}} \quad (2.2)$$

Then, the average speed of the jth workcell is obtained.

$$a_{jw} = \sum_{i=1}^{\varepsilon_{jw}} \upsilon_{ijw}(1-\iota_{ijw})\beta_{ijw} \quad (2.3)$$

3. VPL with failure and maintenance

Even though periodical maintenance is a useful mean to prevent machines from failure, some malfunctions and exceptions are still inevitable, which in turn cause line imbalance and decrease the speed of workcells. In this section, each machine with failure is modeled as a state machine as shown in Fig. 1. Note that the qth workcell precedes the jth one. Meanwhile, this paper assumes that machines may break down only when they are busy.

Let π^I_{ijw}, π^B_{ijw} and π^F_{ijw} denote the probabilities of the idle, busy, and failure conditions of the ith machine in the jth workcell of the wth VPL, respectively.

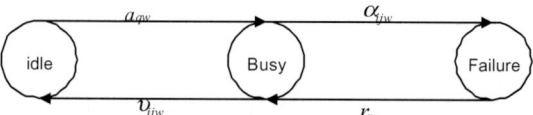

Fig. 1 State transition model of the ith machine in the jth workcell of the wth VPL

Then, the steady-state probabilities are obtained by solving the following linear equations [8]:

$$\Pi D_{ijw} = 0 \quad (3.1)$$

$$\sum_{t\in\{I,B,F\}} \pi^t_{ijw} = 1 \quad (3.2)$$

where $D_{ijw} = [(d_{yz})_{3\times 3}]_{ijw}$ is the transition rate matrix.

$$D_{ijw} = \begin{pmatrix} -a_{qw} & a_{qw} & 0 \\ \upsilon_{ijw} & -(\upsilon_{ijw}+\alpha_{ijw}) & \alpha_{ijw} \\ 0 & r_{ijw} & -r_{ijw} \end{pmatrix}$$

Then solving Equations 3.1 and 3.2 leads to π_{ijw}^I, π_{ijw}^B and π_{ijw}^F are obtained.

$$\pi_{ijw}^I = \frac{r_{ijw}\upsilon_{ijw}}{\Delta} \quad \pi_{ijw}^B = \frac{a_{qw}r_{ijw}}{\Delta} \quad \pi_{ijw}^F = \frac{a_{qw}\alpha_{ijw}}{\Delta}$$

where $\Delta = r_{ijw}\upsilon_{ijw} + a_{qw}\alpha_{ijw} + a_{qw}r_{ijw}$ and $\alpha_{ijw} = p_{hk}$ if the hth machine in M_k serves as the ith machine in the jth workcell of the wth VPL.

Due to periodic maintenance, the service speed of a machine is decreased from υ_{jw} to $\upsilon_{ijw}(1-\iota_{ijw})$ where ι_{ijw} is the maintenance rate. Thus, considering the effects of both failure and periodic maintenance of machines, equation 2.3 becomes:

$$a_{jw} = \sum_{i=1}^{\varepsilon_{jw}} \upsilon_{ijw}(1-\iota_{ijw})\beta_{ijw}\pi_{ijw}^B \quad (3.3)$$

4. Adaptive reconfiguration for the VPL

From the above analysis, it is found that failure of machines is a critical factor that affects their performance. Thus, to select a machine with higher speed and lower failure rate is very important. For example, assume that machines *1* and *2* both can serve jobs in a certain stage *j*, υ_{1jw} is larger than υ_{2jw} but the failure rate of machine 1 is higher than that of machine 2. Then, the system may frequently spend resources to perform the operation associated with machine *1* without success, and then perform with machine *2* with success. This will certainly decrease the system throughput. In the other words, if the system can select machine *2* prior to machine *1*, the system throughput may be improved. Introducing priority for machines and integrating them into VPL design should enable system to adapt for the best reconfiguration of VPLs.

Based on the past performance of machines, the values of δ_{hk} assigned to the hth machine in the M_k is decided as follows:

$\delta_{hk} = e$ if $\upsilon_{jw}(1-p_{hk})$ is the eth greatest value in { $\upsilon_{jw}(1-p_{hk}), h \in I_k$} (assuming the hth machine in M_k is idle and may serve as the ith machine in the jth workcell of the wth VPL). It is clear that the priority of a machine may change depending on the job it serves.

When a new job comes or other events (e.g., breakdown) happen, the system will update the idle machine pools. If the mean failure rate of a machine is higher than a pre-set number P_{jw} (assuming this machine will serve jobs in the jth workcell of the wth VPL), remove this machine from its corresponding idle machine pool and add it into R_k, where machines get immediate repairs.

At initialization, machines all work in good condition. Due to the statistical data from a real time semiconductor manufacturing system, the mean failure rate of a machine is assumed known. Thus, to configure a VPL for a new work-order, the system orders these machines in I_k in a decreasing order of their priority values according to work-order information. Then, machines are selected from these sorted idle machine pools according to their values from the highest to the lowest. The extended algorithm is as follows:

Algorithm 1(Adaptive reconfiguration):
(1) Set $X = E$, the set of VPLs with earliness, and assume the jth workcell in a VPL requires the kth class of machines.
(2) Evaluate the information of the new work-order and calculate the desire speed of the $(u+1)^{th}$ VPL:

$$a_{u+1} = \frac{N_{u+1} + s_{u+1} - 1}{Due(u+1)} \quad (4.1)$$

(3) Update idle machine pools: if the failure rate of an idle machine in the I_k is larger than $P_{j(u+1)}$, move it into its corresponding repair machine pool R_k; assign priorities to the remained machines in I_k and order them in decreasing priority values
(4) Evaluate the system status and calculate the maximum speed of each workcell in the $(u+1)^{th}$ VPL (assuming the arrival rate of jobs is larger than speed of any machine in a VPL):
i). First, calculate the maximum speed of workcells with maintenance:

$$a_{j(u+1)} = \sum_{i=1}^{\phi_k} \upsilon_{ijw}(1-\iota_{ijw})\beta_{ijw} \quad (4.2)$$

ii). Second, get the minimal speed of the $(u+1)^{th}$ VPL only considering workcells with maintenance, that is:

$\underline{a}_w = Min \{a_{jw}, j \in [1, s_w]$ and the jth workcell without failure$\}$

iii). Then, calculate speeds of workcells with failures and maintenance (Note that $\underline{a}_w$ is arrival rates of machines in such workcells):

$$a_{jw} = \sum_{i=1}^{\phi_k} \upsilon_{ijw}(1-\iota_{ijw})\beta_{ijw}\pi_{ijw}^B \quad (4.3)$$

(5) Re-get the minimal speed of the $(u+1)^{th}$ VPL according to $\underline{a}_w = Min \{a_{jw}, j=1, 2, ..., s_w\}$

(6) If $\underline{a}_{(u+1)} \geq a_{u+1}$, keep this new VPL running with speed $\underline{a}_{(u+1)}$. Then, choose the minimum number of machines ($\varepsilon_{j(u+1)}$) using Eq. 3.3 such that $a_{j(u+1)} \geq \underline{a}_{(u+1)}$ and they are the first $\varepsilon_{j(u+1)}$ machines in I_k, then update $\phi_k = \phi_k - \varepsilon_{j(u+1)}$, and calculate $\psi_w(t)$. If $\psi_w(t) > 0$, add this new VPL into set E and go to Step (10)

(7) If $X \neq \Phi$, do:
  a). Select the wth VPL in the set X, remove it from set X, and adjust its $\underline{a}_w$ as follows:

  $$\text{Min } \varepsilon_{jw} \quad j = 1 \text{ to } s_w \quad (4.4)$$

  Subject to:

  $$\psi_w \geq 0 \quad (4.5)$$

  $$a_w = \frac{N_w - n_w(t)}{Due(w) - Date(w)} \quad (4.6)$$

  $$a_{jw} \geq a_w \quad j = 1 \text{ to } s_w \quad (4.7)$$

  b). If Step a) succeeds, update ε_{jw}. Exclude the redundant machines from this VPL and add them into idle machine pool I_k;
  c). If $\psi_w = 0$, remove it from the set E;
  d). Return to Step (2)

(8) If there is at least one machine available at each stage of the $(u+1)^{th}$ work-order and $\sum_{w=1}^{u}(\psi_w - \varphi_w) - \varphi_{u+1} > 0$, configure this new VPL with a slow rate $\underline{a}_{(u+1)}$, put this new line into the set D and go to Step (10)

(9) Reject this new order, consider it later and exit.

(10) Calculate the production time for the $(u+1)^{th}$ work-order T_{u+1} and exit:

$$T_{u+1} = (N_{u+1} + s - 1) / \underline{a}_{(u+1)} \quad (4.8)$$

Because of tardiness, malfunctions and exceptions during operation, how to reallocate resources to minimize these negative impacts is extremely important. Considering these, an adjustment algorithm is presented based on our previous work [5]:

Algorithm 2 (Dynamical adjustment)
System adjusts VPLs when tardiness, malfunctions or exceptions happen:
(1). Repeat the following steps for all VPLs:
  a) Calculate $\zeta_w(t)$, $\varphi_w(t)$ and $\psi_w(t)$;
  b) If $\psi_w(t) > 0$, update $E = E \cup \{w\}$; otherwise if $\varphi_w(t) > 0$, update $D = D \cup \{w\}$;
(2). If $D \neq \Phi$, do:
  a) If $E \neq \Phi$, for each VPL in set E, Reallocate resources through Equations 4.4-4.7; and update ϕ_k and $\psi_w(t)$. If $\psi_w(t) = 0$, remove it from the set E;
  b) Arrange VPLs in D with an ascending order of earliest due dates. For each VPL in D, do:
    I) For $j = 1$ to s_w (assuming the jth workcell in the wth VPL requires the kth class of machines)
      i) If $\phi_k \neq 0$, update I_k using the same way as Step 3 in Algorithm 4.1, then, assign the first ϑ_j ($\vartheta_j \leq \phi_k$) machines in I_k to the jth workcell to increase its speed.
        for ($i = \varepsilon_{jw}$ to $\varepsilon_{jw} + \vartheta_j$) do:
          if ($\alpha_{ijw} = 0$) $a_{jw} = a_{jw} + \upsilon_{ijw}(1 - \iota_{ijw})\beta_{ijw}$
          else $a_{jw} = a_{jw} + \upsilon_{ijw}(1 - \iota_{ijw})\beta_{ijw}\pi^B_{ijw}$
      ii) Otherwise, check other VPLs in the system (i.e., x^{th} VPL and the vth workcell in the xth VPL requires the kth class of machines). If there is machine i such that its utilization is less than 0.5, this machine can be shared by wth VPL.
        for ($x = 1$ to u ($x \neq w$))
          for ($i = 1$ to ε_{vx})
            if ($\alpha_{ivx} = 0$)
              $a_{jw} = a_{jw} + \upsilon_{ivx}(1 - \iota_{ivx})(1 - \beta_{ivx})$
            else
              $a_{jw} = a_{jw} + \upsilon_{ivx}(1 - \iota_{ivx})(1 - \beta_{ivx})\pi^B_{ivx}$
    II) Calculate $\underline{a}_w = \text{Min } \{a_{jw}, j = 1, 2, ..., s_w\}$
    III) If $|\underline{a}_w - a_w| \leq 10^{-2}$, remove this VPL from D
    IV) Otherwise, update its tardiness φ_w. If $\varphi_w > 0$, this order will be delayed φ_w time to finish based on the present forecast.

5. An example

To better understand the above concepts and methods, the system throughput and machine utilization are introduced first [5].

Definition 5.1: The system throughput g is:

$$g = \frac{\sum_{w=1}^{u} N_w}{\text{Max } \{Com(w)\} - \text{Min } \{ST(w)\}} \quad (5.1)$$

where $Com(w)$ is the completion time of the wth work-order and $ST(w)$ is the actual start time of the wth work-order.

Definition 5.2: The machine utilization f is:

$$f = \frac{\text{Actual processing time of a machine}}{\text{Total production time}} \quad (5.2)$$

Then, to demonstrate our methodology, a simplified back-end semiconductor line is used to run three cases. In the conventional case, the system successively processes work-orders according to their order date and at each time, the line is predefined to manufacture certain type of products. In the second case, the system configures a VPL using the methodology presented in [5]. In the third case, the above algorithms are applied to configure and control a VPL. Through these cases, the system is already running and two new work-orders need to be made right now by the following sequential processes: Entry, saw, D/A, curing, plasma, wire bond, inspection and exit. For the simplicity, it is assumed that except in "Saw" and "Wire bond" workcells, machines are reliable. Workcells are equipped with identical machines with different failure rates and machines are fully dedicated to its corresponding workcell.

For these three cases, this paper considers the following instance: before configuring a VPL for the first new work-order, there are ten machines with a failure rate in I_5, which can process jobs in "Wire bond" workcell. Their failure rates are {0.004, 0.008, 0.01, 0.01, 0.01, 0.008, 0.008, 0.008, 0.04, 0.04}. Based on field engineers' experience, the P_{jw} depends on the ϖ_{jw}. In this example, the baseline for the highest failure rate is assumed 0.004 (1/min). On the fifth day of the first new work-order being processed, three machines in the "Saw" workcell break down. The next day after that, eighteen idle machines are added into I_5 since an old work-order was finished. These input data are shown in Tables 1 and 2.

Table 1: The input data for two work-orders

Work-order	EST	Due	$N_{w\,(mag.)}$
1st order	1/1/00	1/16/00	10^5
2nd order	1/3/00	1/13/00	10^4

Table 3 lists our computation results, which compare the changes of speeds of production lines, system throughput and utilization of "Wire bond" workcell, when two work-orders going through the three cases. It is clear to see that in the last two cases, the system can concurrently handle multiple work-orders and adjust the speed of these VPLs according to their conditions and system capacity. Thus, the systems' throughputs increase by 26.7%. In the third case, due to the priority introduced, system can monitor machines' performance and immediately repair machines with a high failure rate, that is $p_{ik} > 0.004$, instead of keeping such machines running until periodic maintenance. Thus, the second work-order was finished in advance by one day, and workcell utilization increases by 10.2% and 4.2% than those in the first and second cases, respectively.

6. Conclusion

This paper addresses the design issues in reconfigurable back-end semiconductor manufacturing systems. A queuing network model is used to analyze a workcell's throughput due to its failure, unplanned and planned maintenance. Because unexpected breakdowns of machines can degrade significantly the system performance, a priority value is introduced for each idle machine. Adaptive algorithms are developed based on this and allow the system to dynamically select machines and adjust the VPL. From the example, the approach is found to be effective in increasing system throughput and machine utilization. To test the proposed algorithms through more complicated cases is our future work.

Reference:

[1] Ciprut, P., Hongler, M. O. and Salama, Y., "On the variance of the production output of transfer lines," *IEEE Trans. On Rob. & Aut.*, Vol. 15, No. 1, 1999, pp. 33-43.

[2] Koren, Y., Hu, J.S., and Weber, T. W., "Impact of Manufacturing System Configuration on Performance," *Annals of the CIRP*, Vol. 47, pp. 369-372, 1998.

[3] Proth, J. M. and Xie, X. L., "Cycle time of stochastic event graphs: evaluation and marking optimization," *IEEE Trans. On Automatic Control*, Vol. 39, No. 7, 1994, pp. 1482-1486.

[4] Qiu, R. and Wysk, R., "Design and implementation of virtual production lines for discrete automated manufacturing systems," *The 14th Int. Federation of Automatic Control World Congress*, Beijing, P. R. China, July 5-9, 1999, pp. 455-460.

[5] Tang, Y., Zhou, M. C. and Qiu, R., "Design of virtual production lines in back-end semiconductor manufacturing systems," to appear in *Proc. of IEEE Int. Conf. On System, Man & Cybernetics*, Nashaville, Tennessee, Oct. 8-11, 2000.

[6] Xie, X. L., "Superposition properties and performance bonds of stochastic timed-event graphs," *IEEE Trans. On Automat. Contr.*, Vol. 39, No. 7, 1994, pp. 1376-1386.

[7] Vinod, B. and Altiok, T., "Approximating unreliable queuing networks under the

assumption of exponentiality," *Journal of the Operational Research Society*, 37(3), 1986, pp: 309-316.

[8] Zhou, M. C. and Venkatesh, K., *Modeling, Simulation, and Control of Flexible Manufacturing Systems-A Petri net Approach*, World Scientific, Singapore, 1998.

[9] *Visionary Manufacturing Challenges for 2020*, National academy press, Washington, D. C., 1998.

Table 2: The input data for workcells

	Saw	D/A	Cure	Plasma	Wire bond	Inspection
c_{ijw} (mag.)	1	1	24	2	1	1
τ_{ijw} (min.)	7	6	120	4	30	2.5
χ_{ijw} (hour)	1	1	1	1	1	1
ϖ_w (day)	7	7	7	7	7	7
ϕ_k	50	50	50	40	150	30

Table 3: The computation results

# of days	conventional case ($\underline{a}_w$ 1/min)		2nd case (previous algorithms [5])		3rd case (extended algorithms)	
	1st work-order	2nd work-order	1st work-order	2nd work-order	1st work-order	2nd work-order
1	4.742		4.742		4.659	
2	4.742		4.742		4.659	
3	4.742		4.639	0.123	4.639	0.318
4	4.742		4.639	0.123	4.639	0.318
5	4.429		4.429	0.123	4.429	0.318
6	4.429		4.639	0.719	4.639	0.915
7	4.429		4.639	0.719	4.639	0.915
8	4.429		4.639	0.719	4.639	0.915
9	4.429		4.639	0.719	4.639	0.915
10	4.429		4.639	0.719	4.639	0.915
11	4.429		4.639	0.719	4.639	0.915
12	4.429		4.639	0.719	4.639	0.915
13	4.429		4.639	0.719	4.639	
14	4.429		4.639	0.719	4.639	
15	4.429		4.639		4.639	
16	4.429					
17	4.429					
18		5.358				
19		5.358				
g	5789		7333		7333	
f	90.2% (the 5th workcell)		95.4% (the 5th workcell)		99.4% (the 5th workcell)	

Dynamic Contact Sensing of Soft Planar Fingers with Tactile Sensors

Genichiro Kinoshita, Yujin Kurimoto, Hisashi Osumi and Kazunori Umeda

Faculty of Science and Engineering
Chuo University
kino@robotics.elect.chuo-u.ac.jp

Abstract

This paper focuses on a method of estimating contact force dynamically between environment and an object, which is grasped by the soft planar fingers with tactile sensors. The external contact force acting on the grasped object causes the tactile sensors to generate the displacement distributions and the force distributions. We present a method of dynamic tactual image acquisitions for both distributions at the pseudo video rate. Experimental results show the measurement of the tactual image flows due to the deflection of an object and the magnitude and orientation of force vectors operating on the finger according to the external contact force. The contact position of the external force acting on the grasped object is estimated within 15 % from the tactual image distributions.

1 Introduction

The tasks of grasp and manipulation of an object are done by operations of fingers. The tactile sensors equipped on each finger have functions of acquisition for the 3-D tactual images caused by coming in contact with an object's surface.

As it is able to estimate the displacement distributions and the force distributions from these images, these distributions give the contact geometrical shape of an object and some parameters such as the location of contact, the magnitude of force at static contact[1],[2].

Dexterous manipulations require dynamically some of parameters such as the contact location between the object's suraface and the finger, the changes of contact force, the contact states of the object.

Two methods of real time sensing for the contact location are presented as tactile array extrinsic sensing and force-torque based intrinsic contact sensing[3].

The finger-shaped tactile sensor have been proposed for measuring the contact location and the normal force vector in contact with the surface of an object[4],[5],[6].

The implementation of an object reorientation strategy called a "twirling" has been carried out with multi-fingers[4],[5] manipulated by using the sensing data of the tactile sensor equipped on the finger-tip.

The curvature of contact object's surface is also estimated at the cylindrical tactile sensor[6],[7].

As the sensing surface of a finger is covered with compliant material[8], a contact force is given as blurred distribution which generates a small contact area around the contact point[9]. A model of compliant interaction between the hand and the object has been proposed[10].

For realization of both extrinsic and intrinsic sensing, it is able to determine some parameters like the shape of an object surface, its location and magnitude of contact force or external contact force, as the optical tactile sensor[11] takes the tactual images as the visual images.

In this paper, we present a model of tactile sensor and a tactual sensing behavior of fingers. First, the sensing principle of the optical tactile sensor and its sensing method are given. Second, we present a method of locating contact as the centroid of the force distribution[9] and the dynamic change of tactual image flows during the external force applied on the tip of grasped object.

2 Sensing principle of optical tactile sensor

2.1 A model of tactile sensor on the finger

On interaction between a soft planar finger and an object's surface, the finger makes contact with the surface at a contact point acting on a force.

A model of compliance interaction based upon the elastic rubber between the sensing surface of a finger equipped with a tactile sensor and an object's surface is proposed as shown in Fig.1.

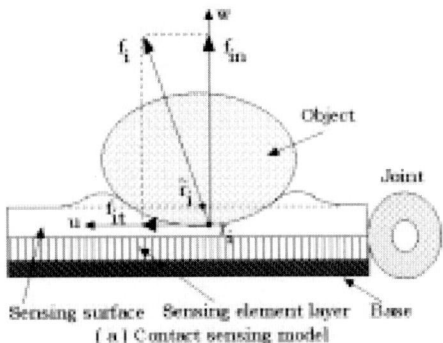

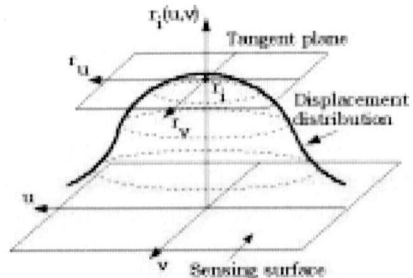

Figure 1: A model of tactual sensing

The displacement distribution and the force distribution are generated in the sensing surface, as the sensing surface of tactile sensor contacts with the object's surface. Let us suppose the displacement distribution is indicated with the parameter representation u and v at a contact point $\boldsymbol{r}_i$ by Eq.(1)

$$\boldsymbol{r}_i = (x(u,v),\ y(u,v),\ z(u,v)) \qquad (1)$$

It is possible to define a tangent plane with the tangent vector $\boldsymbol{r}_u$ and $\boldsymbol{r}_v$ given by Eq.(2) at a contact point $\boldsymbol{r}_i$:

$$\boldsymbol{r}_u = \frac{\partial \boldsymbol{r}_i}{\partial u}, \qquad \boldsymbol{r}_v = \frac{\partial \boldsymbol{r}_i}{\partial v} \qquad (2)$$

The normal vector of the tangent plane is defined as follows,

$$\boldsymbol{n} = \frac{\boldsymbol{r}_u \times \boldsymbol{r}_v}{|\boldsymbol{r}_u \times \boldsymbol{r}_v|} \qquad (3)$$

Let us define a force vector $\boldsymbol{f}_i$ acting on the tactual sensing surface of a soft planar finger equipped with a tactile sensor. Force vectors $\boldsymbol{f}_{in}$ and $\boldsymbol{f}_{it}$ are represented with the normal force component and the tangential force component at the contact point on the sensing surface, respectively. The displacement $\delta \boldsymbol{r}_i$ is defined by the displacement difference between the initial displacement $\tilde{\boldsymbol{r}}_i$ and the displacement $\boldsymbol{r}_i$ generated by applying a force as shown in Eq.(7). The displacement $\delta \boldsymbol{r}_{it}$ is also represented for its variation in the tangential direction. The following force conditions must be satisfied at a contact point between the sensing surface of the tactile sensor and the object's surface,

$$\boldsymbol{f}_i = \boldsymbol{f}_{in} + \boldsymbol{f}_{it} \qquad (4)$$
$$\boldsymbol{f}_{in} = K_{in}\delta\boldsymbol{r}_{in} \qquad (5)$$
$$\boldsymbol{f}_{it} = K_{it}\delta\boldsymbol{r}_{it} \qquad (6)$$
$$\delta\boldsymbol{r}_i = \boldsymbol{r}_i - \tilde{\boldsymbol{r}}_i \qquad (7)$$

where the coefficients K_{in} and K_{it} show the stiffness in the normal and tangential directions, respectively.

Let us suppose the tactile sensor is able to detect the normal force component and the normal displacement component. On tactual sensing, the normal displacement and force components are represented by $\delta r_{in}(u,v)$, $f_{in}(u,v)$ respectively and detected by the sensing elements with the coordinates (u,v) arranged within the array. The relationship between the force and displacement is given by the coefficient $K_{in}(u,v)$ as follows,

$$f_{in}(u,v) = K_{in}(u,v)\delta r_{in}(u,v) \qquad (8)$$

In this case, the normal force vector $\boldsymbol{f}_{in}$ operating at the contact point is represented as follows,

$$\boldsymbol{f}_{in} = \iint_S f_{in}(u,v)\,du\,dv \qquad (9)$$
$$= \iint_S K_{in}(u,v)\delta r_{in}(u,v)\,du\,dv \qquad (10)$$

where S describes the contact area on the sensing surface of the tactile sensor.

2.2 Configuration of the tactile sensors

The tactile sensor utilizes the principle[11] that the total internal reflection of a wave guide is lost when an elastic material contacts on the surface of the waveguide. A white silicon rubber sheet is used as the elastic material. One side of the sheet is composed of a matrix of circular cones, which is attached to the wave-guide without any pressure as shown in Fig.2. A black colored cloth sheet is covered over the other side of rubber sheet as the sensing surface.

When a pressure pattern is given on the sensing surface due to an object's shape, the circular cones at the contact area are squeezed and the area of contact between the circular cone and wave-guide becomes large. Light which is injected into the side of wave-guide from the plastic fibers is scattered at the area of contact on the wave-guide, because the condition of the total reflection doesn't stand any more. The patterns of the scattered light are taken visual images as the tactual images by the CCD camera which is set up behind the wave-guide as shown in Fig.2.

As the sensing surface of the tactile sensor is composed of the elastic rubber and the covered cloth, the CCD camera takes the contact visual images onto the wave-guide as the 3-D tactual images at psuedo video rate according to coming in contact with object's surface. Therefore, the acquisition rate of the tactual images may be based upon the rate of image processing.

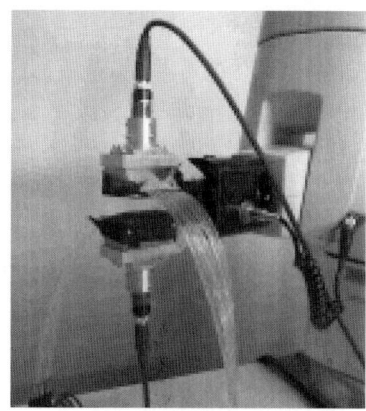

Figure 3: Photograph of the soft planar parallel two fingeres

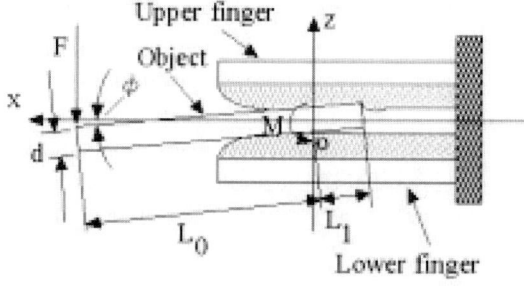

Figure 4: The soft planar parallel two fingers

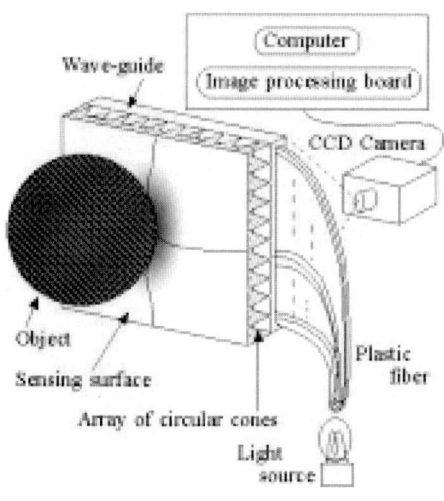

Figure 2: Configuration of tactile sensing

3 Prototype of soft planar parallel two fingers

A parallel two-fingered hand with a compact and high-density optical tactile sensor are developed as shown in Fig.3. The tactile sensor is able to take 3-D tactual images at pseudo video rate. Tactual images are taken during grasping an object by fingers, as the tactile sensor sets on the grasping side.

The grasping force can be easily controlled by motion of each finger. The size of each finger is $88mm$ in length and $30mm$ in width, and the tactual sensing surface is $30mm \times 55mm$ as shown in Fig.3, but tactual images capable to detect in the range $20mm \times 48mm$. The cross section of the finger tip is made round with an arc of a circle in a diameter $100mm$.

The tactile sensors on the finger are manufactured as shown in Fig.3. An acrylic plate of $5mm$ in thickness is used as the wave-guide as shown in Fig.2. The circular cones are arranged at equal spaces of $2mm$, and the matrix of the circular cones is 10×24, each circular cone corresponds to 200×480 pixels of the visual image at the CCD camera, and thus 20×20 pixels are assigned to one circular cone. The pressure applied to the circular cone is obtained as the average of the gray-level($8bit, 0-255$) of the 20×20 pixels. The maximum pressure applicable to one circular cone is about $120gf/cone$ (the contact area of a cone $= 3.1mm^2$). Sensitivity of the tactile sensor is $0.4gf/pixel$.

4 Experimental results

Let us consider the acquisitions of dynamic tactual images and the estimations of the internal force position and its magnitude experimentally, which are generated by an object grasped by soft planar parallel two fingeres. The experimental setup and the architecture of the tactual image processing show in Fig.3 and Fig.5, respectively.

4.1 Tactile sensing method of soft planar parallel two fingeres

As the soft planar parallel two fingeres with the tactile sensor have functions for discriminating the object's shape and the force component in the z axis direction, moments about the x and y axis, these functions of the tactile sensor are used for estimationg the shape of an object and finding the contact position of the applied external contact force F.

If the grasped object is a rod typed one and its orientation is parallel to the x axis as shown in Fig.4, the force balance between the grasped object and its environment is considered approximately on the x axis.

Let us suppose the parameters (i, j) with the i-th position in x axis ($i = -n \sim n$) and the j-th position in y axis ($j = -m \sim m$) at finger coordinate system. Moreover, we assume the parameters a_{ij} and b_{ij} indicate the displacement of the squeezed (i, j)-th cone at the upper and the lower fingers respectively.

The force F_u and F_ℓ acting on the upper and lower fingers with the grasped object are obtained from sum of the deflection of the squeezed cone at the contact area for each finger as shown in Fig.4. The coefficient

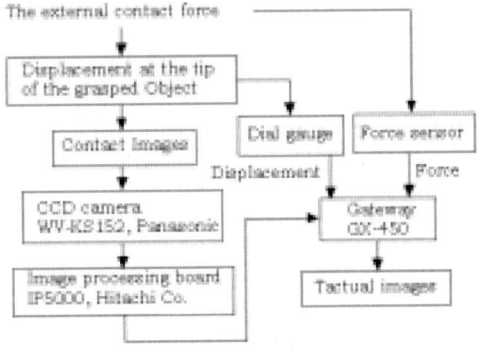

Figure 5: Architecture of the tactual image processing

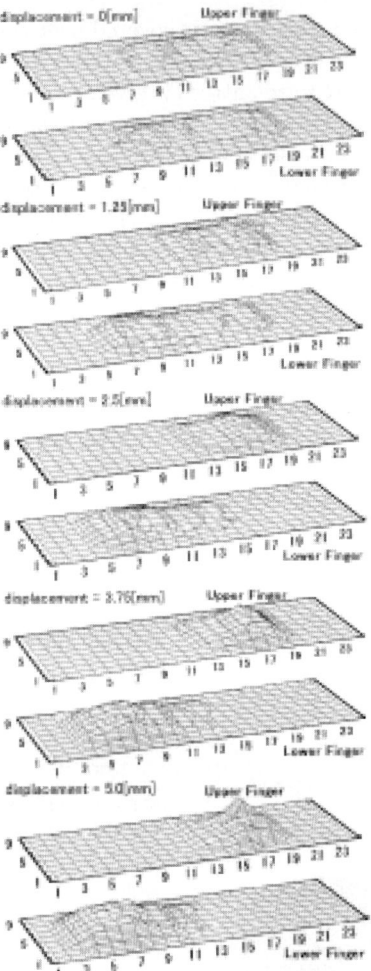

Figure 6: The changes of the tactual images induced according to the displacement at the tip of grasped object

k_{ij} is corresponding to $K_{in}(u, v)$ in Eq.(8).

$$F_u = \sum_{j=-m}^{m} \sum_{i=-n}^{n} k_{ij} a_{ij} \cos \phi \quad (11)$$

$$F_\ell = \sum_{j=-m}^{m} \sum_{i=-n}^{n} k_{ij} b_{ij} \quad (12)$$

where the angle ϕ indicates the tilt angle of an object applied with the external contact force as shown in Fig.4. The coefficient k_{ij} is the stiffness of sensing element at (i, j). The centroid of the displacement distribution is also estimated as the contact location.

The tactual sensors take the tactual images as the displacement distribution for a grasping object and estimate the forces with Eq.(11) and (12) from its distribution.

Let the parameter s be the distance between the centroid of displacement distribution caused with the edge of object at the upper finger and the origin of finger coordinate system, and also the parameter t be the distance between the centroid of the displacement distribution induced on the round surface of the lower finger and the origin of finger coordinate system.

The applied position L_0 of the external contact force is derived from the force balance between the grasped object and the external contact force as shown in Fig.4. We represent the applied position L_0 of the external contact force from the moment about the origin of finger coordinate system as follows:

$$L_0 = \frac{F_\ell \cdot t + F_u \cdot s}{F \cos \phi} \quad (13)$$

As the edge position of an object is recognized from the tactual image data, we indicate as the distance L_1 between edge position and origin in the finger coordinate system. Therefore, the rod length L between the edge position and external force applied position is given as follows:

$$L = L_0 + L_1 \quad (14)$$

If the edge of an object make in contact with the environment, then the external applied force is estimated by Eq.(15) because the length of an object is given in general.

$$F = \frac{F_\ell \cdot t + F_u \cdot s}{(L - L_1) \cos \phi} \quad (15)$$

4.2 Dynamic tactual image flows generated by the deflection of an object

After the calibration of the tactile sensor, we obtained the tactile images due to external contact force. In this experiment, the object is a square pillar which is $10mm \times 10mm$ in a cross section and $90mm$ in length, and a column $15mm$ in a diameter and same length. The mass of the object is neglected. The grasped object is applied the external contact force at the tip of a square pillar and its displacement is caused on $0mm$ to $5mm$.

The experimental results are represented as the tactual image flows as shown in Fig.6. As the external contact force is not applied and the amplitude of grasping inner force for holding the object is a very small, the tactual images are obtained as the displacement contours at the center region of the surface of both fingers in Fig.6. Although the left side in figure corresponds to the tip of finger, the displacement

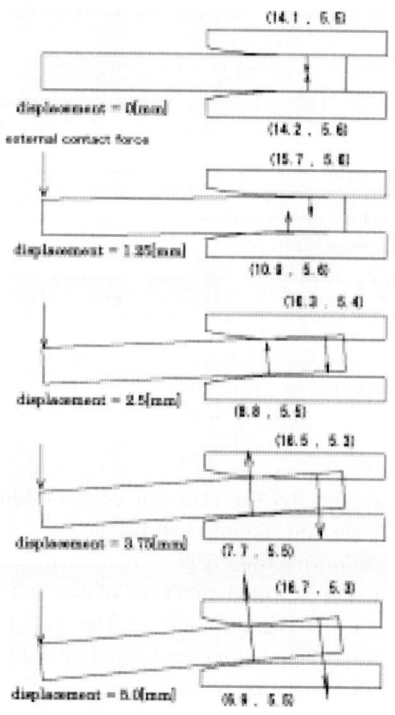

Figure 7: Estimation of the force vector on the finger and the external contact force

contours put in position of a center because of the roundness of a finger tip.

When some external contact force is applied at the tip of the grasped object, the flows of tactual images come about from the center to the right side at the upper finger and from the center to the left side at the lower finger according to the amplitude of external contact force.

On the other hand, the tactual image of the left side at the lower finger is obtained for being in contact with the behind of an object along the round surface. The close displacement contours result in large deflection of the distribution as shown in Fig.6. The normal vector of force acting on the sensor and its amplitude are calculated from the displacement distribution. Forces acting on the finger are illustrated with the change of it due to the external contact forces as shown in Fig.7.

4.3 Estimation of the applied position of external contact force from the tactual images

In the previous section, it is shown that the deflection of the grasped object caused by the external contact force is represented as the change of the tac-

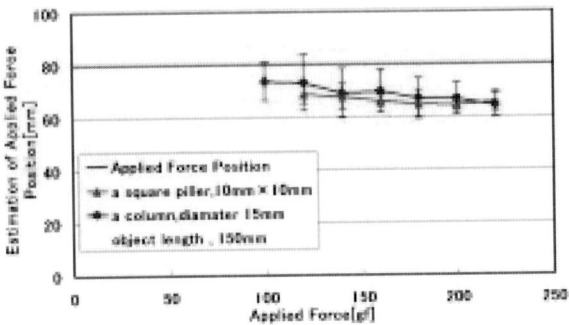

Figure 8: Estimation of the applied force position

tual images due to the state of contact between the object and the surface of each finger.

At the handling like a peg in hole task, it is necessary that the applied position of the external contact force is discriminated from the contact sensing between the grasped force and its environment on the operation of the robot handling.

The applied position of the external contact force is determined experimentally from Eq.(13) by using the tactual images. It is obtained the force contact positions were determined within 15 % accuracy as shown in Fig.8. The tactile sensor will be quite useful for the finger dexterous applications.

5 Conclusion

The soft planar parallel two-fingeres with the tactile sensors have developed for some tasks with the tactile sensings. It is clarified that the usage of tactile sensors is adequate for estimating the normal force and the contact position between the object and its environment.

The dynamic tactual images with the psuedo video rate are obtained for estimating the grasping states of the object during the operation of handling. It will be quite useful for the tactual feedback to attain the dexterous handling.

References

[1] R.D.Howe,"Tactile Sensing and Control of Robotic Manipulation," *Advanced Robotics*, Vol.8, No.3, pp.245-261,1994

[2] G.Kinoshita, E.Moutoh and K.Tanie,"Haptic Aspect Graph Representation of 3-D Solid Object Shapes by Tactile Sensing," *Proceedings of the 1992 IEEE Int. Conf. on Robotics and Automation*, pp.1912-1917, 1992

[3] J.S.Son, M.R.Cutkosky, and R.D.Howe, "Comparison of Contact Sensor Localization Abilities During Manipulation,"IROS'95, pp.96-103, 1995

[4] S.Begej,"Planar and Finger-shaped Optical Tactile Sensors for Robotic Applications,"*IEEE J. Robotics and Automation*, Vol.4, No.5 pp.472-484,1988

[5] H.Maekawa, K.Tanie, K.Komoria and M.Kaneko, "Development of a Finger-shaped Tactile Sensor and its Evaluation by Active Touch,"*Proceedings of the 1992 IEEE Int. Conf. on Robotics and Automation*, pp.1327-1334,1992

[6] R.S.Fearing and T.O.Binford,"Using a Cylindrical Tactile Sensor for Determining Curvature,"*IEEE Transactions on Robotics and Automation*, Vol.7, No.6, pp.806-817, 1991

[7] E.J.Nicolson and R.S.Fearing,"The Reliability of Curvature Estimates from Linear Elastic Tactile Sensors,"*Proceedings of the 1995 IEEE Int. Conf. on Robotics and Automation*,pp.1126-1133,1995

[8] M.Shimojo,"Spatial Filtering Characteristic of Elastic Cover for Tactile Sensor,"*Proceedings of the 1994 IEEE Int. Conf. on Robotics and Automation*,pp.287-292,1994

[9] A.Bicchi,"Intrinsic Contact Sensing for soft Fingers," *Proceedings of the 1990 IEEE Int. Conf. on Robotics and Automation*, pp.968-973,1990

[10] M.M.Svinin and C.V.Albrichsfeld, "Analysis of Constrained Elastic Manipulation,"*1995 IEEE/RSJ Int. Conf. on Intelligent Robots and Systems*, pp.414-421, 1995

[11] G.Kinoshita,Y.Sugeno,H.Osumi,K.Umeda,"High Compliant Sensing Behaviour of a Tactile Sensor,"*1999 IEEE/RSJ Int. Conf. on Intelligent Robots and Systems*,pp.826-831, Oct. 1999

[12] H.Osumi, T.Miyashita, K.Umeda and G. Kinoshita, "Sensing of Contact Point Between a Grasped Object and Environment by Tactual Behavior of a Parallel Two-Fingered Hand,"*The 1996 IEEE/SICE/RSJ Int. Conf. on Multisensor Fusion and Integration for Intelligent Systems*, pp.227-232,1996.

Interpretation of Force and Moment Signals for Compliant Peg-in-Hole Assembly

Wyatt S. Newman (wsn@po.cwru.edu), Yonghong Zhao, Yoh-Han Pao
Department of Electrical Engineering and Computer Science
Case Western Reserve University
Cleveland, Ohio 44106

Abstract: Interpretation of force and moment data is used to guide the searching phase of a circular peg-in-hole assembly. It is shown that the available measurements carry varying quality of information, ambiguous in much of the search space and relatively precise in a small region. By interpreting the history of sensory data acquired during an assembly attempt, reliable interpretation of the data is possible. A feature-based technique is described for interpreting sensory data. Experiments with robotic peg-in-hole assembly demonstrated a speed-up of nearly an order of magnitude relative to a blind search.

1. Introduction: In this paper, we present research on applying intelligent methods to deal with a class of robotic assembly tasks. Specifically, we focus on the simple task of inserting a circular part into a circular hole, guided by interpretation of sensory data acquired during searching.

At present, successful implementations of robotic assembly generally depend on generous clearances that are larger than the accumulated errors from part variations, fixturing, sensing, grasping and robot motion precision. The subset of assemblies that satisfy this error tolerance is small, and thus application of robotics to assembly is still far from meeting its potential.

Our motivation for this investigation is a current effort at CWRU in performing robotic assembly of automotive powertrain components. In these assemblies, which include meshing gears and splines in transmissions, internal components within subassemblies have position and orientation uncertainties that are much larger than the clearance between successfully mated parts. Locations of the internal components cannot be known *a priori*, and machine vision is not an option, since the internal parts are occluded during assembly. (See e.g. [1]).

Invoking force feedback for compliant motion control, we have successfully assembled a variety of these components using three different robot/controller combinations: an AdeptOne with a Cimetrix controller [2], a Kawasaki JS-10 with a Trellis controller [3], and a novel closed-chain manipulator—the "ParaDex"—with a custom PC-based controller [4]. In each case, the robot exerted a steady, gentle downwards force via the grasped component onto a fixtured sub-assembly. While exerting this downwards force, the robot hunted in x, y and z-rotation until alignment within tolerance was achieved, at which point the interacting parts "fell" into assembly. In these example assemblies, jamming during insertion was generally not a problem. Assembly rates were limited by the relatively simple planar searching phase.

While the above simple approach has been successful on a variety of assemblies (i.e. competitive with manual assembly rates and contact forces), the searching phase we have used is still far from optimal. To understand current limitations and opportunities for improvement, we consider the deliberately simplified case of inserting a round peg into a round hole. Our goal is to exploit sensory interpretation to achieve intelligent rather than blind searching. Specifically, we examine measurements of reaction forces and moments during manipulation in contact to guide parts into assembly.

2 Background

Sensory interpretation for the peg-in-hole problem has been researched extensively over the last 20 years. Kinematic analyses have resulted in successful passive techniques for assembly [5]. Active compliance, using sensory feedback, has also been pursued extensively. (See, e.g., [6]). More recently, intelligent techniques have been investigated for guiding parts into assembly based on higher-level interpretations of force and moment signals. Many of these works are strictly theoretical and remain to be validated empirically. Also, much of the research attention has been focussed on the jamming problem during insertion of pegs into holes with small clearances. This is not an issue for our components; instead, we focus on the need for improving the speed of convergence of the searching phase, from initial contact to alignment of the peg and hole. Speed of the searching phase is seldom reported in the literature. Inferring from results that are typically reported in terms of iterations of quasi-static position increments, the searching phase of most experimental efforts is impractically slow.

Several of the works most relevant to our presentation include the following. In 1989, Gottschlich and Kak presented a method invoking straight-line motion goals chosen based on interpretation of assembly states from sensor values [7]. In our approach, we adopt this philosophy of invoking straight-line searches and interpretation of sensory data based on such motion commitments. Gottschlich and Kak also presented a quasi-static force/moment balance analysis for a circular peg partially overlapping a circular hole where the surface of the peg is parallel to the assembly surface. While such force/moment analysis is directly relevant to our needs, we have found experimentally that the ideal condition of perfect parallelism does not occur. Rather, the contact conditions are better modeled by a tilted peg (whether intentionally tilted or not). Finally, these authors did not report convergence rates, which were presumably slow.

In 1990, Asada described a method for intelligent interpretation of force/moment data for guiding peg-in-hole assemblies using a neural network [8]. This relatively early venture in this area was followed by intensive research in intelligent control for assembly. As with many of the methods to follow, this work used interpretation of instantaneous measurements of forces and moments to compute incremental motion commands to drive the robot closer to the desired assembly state. In our experiments, however, we have found that dependence on instantaneous measurements results in erratic performance, due to sensor noise, dynamic effects and regions in which the sensor signals provide no useful information.

More recently, Cervera and del Pobil have presented novel approaches to intelligent control of robotic assembly, including use of self-organizing neural networks and recurrent neural networks [9,10]. Notable in these works is the observation that the force/moment information is inherently ambiguous. They point out that such ambiguity can be addressed by depending on a *history* of sensor readings rather than making incremental decisions based on incremental readings. In [9], response to a history of readings is encoded implicitly via recurrence in the neural net, creating a finite state automaton. They report a dramatic improvement over decisions based exclusively on current sensor readings. (However, only simulation results are presented).

McCarragher has pursued intelligent control methods in experimental settings, including consideration of dynamics, friction and noise [11,12]. These efforts have included use of Hidden Markov Models, clustering and discriminant functions. In [11], the Hidden Markov Model implicitly operates on a history of sensor readings.

Similar to the philosophical approaches of Cervera, delPobil and McCarragher, our own research presumes use of a history of sensor readings to guide robotic assembly. Our methods depend on sensor readings that include friction, noise, surface roughness and intermittent contact dynamics, and our emphasis is on speeding up the searching phase for initial insertion.

3 Analysis of Mappings from Moments onto Positions

Mathematically, one can consider the force/moment balance for static equilibrium between a component and a subassembly. In the case considered here, a right cylinder (i.e. a circular peg) is to be inserted into a circular hole in a flat plate. Neither the peg nor the hole is chamfered.

For a tilted peg, Figure 1 illustrates the region over which we can expect useful information from measurements of reaction moments. For conditions 1 and 3, the moments will provide no clues regarding the relative location of the hole. The contact conditions for these cases are identical near the hole vs far from the hole. In case 2, however, the contact conditions are significantly different, and this change will show up in reaction-moment measurements. The computed moments as a function of x and y peg displacement for a tilted peg are shown in Figure 2.

From mathematical analysis, one can deduce that, over a relatively large "capture" region (i.e., conditions of case 2), one can compute the relative displacement of the center of a peg from the center of a hole from a single measurement of the reaction moments M_x and M_y.

To test this analysis, we conducted an experiment with an AdeptOne robot instrumented with a JRRR force/torque sensor. The AdeptOne was controlled to be compliant in the vertical direction and precisely position-controlled in the horizontal plane. An 80mm diameter peg was grasped by the robot and used to contact a smooth plate containing a 100mm diameter hole. To consider non-parallel contact, the plate was tilted slightly (since the Adept was not capable of tilting the peg). The robot was controlled to contact approximately 600 x-y locations over a 100mm by 100mm region. To obtain data as close to ideal as possible, frictional and dynamic effects were suppressed by controlling the robot to approach each contact point slowly from above, and to establish steady-state contact at a prescribed vertical force before sampling the force and moment data. At each contact location, approximately 500 force/torque samples were acquired over a 1-second duration and averaged to suppress noise. While this painstaking data acquisition process is not representative of high-speed assembly conditions, the data is useful for evaluating best-case conditions for deducing assembly status from force/moment information. The measurement results are shown in Fig 3. While Fig 3 has some resemblance to the model data of Fig 2, it is obvious that even our unrealistically careful data acquisition process yields quite noisy data.

Ideally (as per the mathematical model), for conditions corresponding to peg and hole overlap like that of case 2 in Fig 1, there would be a unique mapping from (M_x, M_y) onto desired peg motion, (dx,dy), to achieve peg-in-hole insertion. For the synthetic data of Fig 2, this inversion is well behaved. For the physical data in Fig 3, however, this inversion is problematic.

Our data constituted a unique mapping from relative positions, (dx,dy), onto observed moments, (M_x, M_y). To obtain the desired inverse mapping from moment measurements onto hole coordinates, we needed to invert this function. To visualize the prospects for using moment measurements to guide robotic assembly, we performed analysis of our moment data using neural-network functional approximation and the k-means algorithm for unsupervised learning [13]. The mathematical inversion, (dx,dy) as a function of (M_x, M_y), was performed numerically by training a feedforward neural net. The measured moments (M_x, M_y) were defined as inputs to the net, and the corresponding relative hole coordinates (dx,dy) were defined as the target output values. The neural net weights were adjusted iteratively using backpropagation.

Performing the desired functional inversion by training a neural network was only partially successful. In some regions, the network returned reasonably precise estimates of the hole location as a function of moment values. But in other regions, the hole location estimate was poor. This was due to the fact that quite similar values (M_x, M_y) were observed for quite different values of peg coordinates. That is, mathematical analysis notwithstanding, the experimentally-observed function was not invertible.

Rather than dismissing attempts to perform functional inversion altogether, one can analyze over what regions inversion is possible. To achieve a good functional inversion, it should be true that a small neighborhood of inputs maps onto a small neighborhood of outputs, and further that a small neighborhood in output space corresponds to a small neighborhood in input space. To examine this condition, we analyzed the data via the k-means algorithm. In this analysis, robot coordinates (relative to the known hole location) were treated as input patterns, and the corresponding moment measurements were treated as output patterns. Input patterns were organized into sets (clusters) based on the similarity of their outputs. The resulting sets of clustered inputs were then analyzed to see if they were also similar to each other (i.e., within a small Euclidean distance).

In some regions of moment space, the bi-directional mapping was accurate, but in other regions, widely scattered position coordinates corresponded to similar moment values. From this analysis, we were able to map an approximate functional inversion from moments onto positions, augmented with a "quality" measure characterizing the credibility of the mapping as a function of (M_x, M_y).

This map is shown in Fig 4 (with the x and y sample values scaled to the range 0 to 1). In this figure, the (x,y) sample points are labeled as high-quality, medium-quality and low-quality data. The circular region in which there are no samples corresponds to the range of coordinates in which the peg would fall into the hole. Data points marked with circles and diamonds correspond to samples for which the moment measurements provide little or no information regarding the hole-center coordinates. These points include contact conditions like those of case 1 and case 3 in Fig 1.

Conversely, points marked with an "x" correspond to high-quality samples. The moments corresponding to these sample locations map unambiguously unto accurate predictions of the hole coordinates.

Points marked with triangles or squares carry limited information. They do not yield accurate estimates of the hole coordinates, but they can indicate a general direction in which to move in order to get closer to the hole.

4 Pattern Matching and Moment Maps for Guided Assembly

Our analysis of experimental data showed that there is potential for using moment information to guide peg-in-hole robotic assembly. However, the effectiveness of such an algorithm depends on the locations of the robot's samples. One approach we investigated was to trust only the hole-coordinate estimates produced by the most credible measurements. When only low-quality data was available, the robot was controlled to perform a blind search until higher-quality data was available. A limitation of this approach, though, is that the region of high-quality data is restricted to a fairly narrow band. Further, this region of high-quality information is even smaller for more challenging assemblies (i.e., as the clearance between the peg and hole gets smaller). An additional complication is that noise from dynamic effects during data acquisition makes the response to such a controller erratic.

However, the moment measurements do contain clues, and this information should be mined for intelligent control. To extract the information more effectively, instead of mapping individual measurements onto corresponding control commands, we acquire relatively long records of data from continuous motion in contact. Such scans are subsequently analyzed to recognize patterns that indicate the desired destination.

In principal, one could interpret the moment data in terms of an optimal pattern-matching problem. In searching, the robot could datalog samples of its x-y coordinates and corresponding Mx, My reaction-moment measurements. Subsequently, such a trajectory could be interpreted in terms of the pre-recorded moment maps. Given a postulated hole-center location, an interpolation of the measured (x,y) vs (Mx,My) could be used to compute a predicted stream of (Mx,My) values for the actual (x,y) trajectory. By iterating over the two unknown parameters—the x and y offset coordinates of the hole-center location—a best fit can be sought for matching a computed trajectory in moment space to the measured trajectory in moment space. Such a process would be relatively immune to noise, would avoid the problems of functional inversion, and would avoid excessive dependence on individual samples. We note, though, that the sample trajectory should pass through the region of high-quality moment data to produce good estimates of the hole coordinates. Further, the goodness-of-fit measure for test trajectories in sensor space should be weighted most heavily in the regions near the high-quality data.

The described general procedure for pattern matching could utilize the entire history of data acquired during an assembly attempt. In addition to weighting the importance of matching data in the high-information regions, it would also be appropriate to weight the most current data more strongly.

The above method could employ a variety of pattern-matching and optimization techniques for sensory interpretation. A simpler, faster subset of this approach is to perform pattern matching based on features. The searching trajectory can be designed intentionally to ease detection of such features. We next describe such an implementation.

5 Feature Detection and Interpretation of Moment Space

In our simplified version of pattern matching, the search trajectory was designed as follows. First, the peg was tilted a small, negative rotation about its x-axis. (Equivalently, the assembly surface was tilted positively about the robot's x-axis). The robot was then commanded to contact the assembly surface at a location estimated to be $(-r_{peg}, -r_{peg})$ = (x,y) relative to the hole center. This estimate could be quite coarse; an error as much as $+/-r_{peg}$ would still result in success. The robot was then commanded to drag the peg along the surface in the x direction, while maintaining a steady downwards contact force on the surface, for a distance of $2*r_{hole}$. During this scan, both the hole and the peg's contact point should be to the "left" relative to the direction of motion. If the y coordinate of the peg's trajectory is within a distance (r_{hole} + r_{peg}) of the y-coordinate of the hole center, the peg's contact point will shift as the peg partially dips into the hole during the scan. Consequently, the measured moment M_x will change as the peg passes near the hole.

Both Fig 2 (analytic) and Fig 3 (empirical) show that the M_x reaction moment peaks as a function of x along lines of constant y. This peak occurs when the peg y-coordinate is nearly equal to the hole y-coordinate. The *location* of this peak offers an estimate of the x coordinate of the hole center. The *magnitude* of this peak also offers an estimate of the y coordinate, but with lower confidence.

Figures 2, 3 and 4 are based on a peg of radius 40mm and a hole of radius 50mm, corresponding to an unrealistically large clearance. As the clearance gets smaller, the region of high information signals in Fig 4 also gets smaller. Thus, algorithms that depend on immediate response to instantaneous measurements will typically be forced to react to medium or low-information data. However, as the radial clearance decreases, the features (peaks and discontinuities) in the moment data become more pronounced. Figure 5 shows the computed moments for a tilted peg of radius=40mm in the vicinity of a hole of radius 41mm. Our assembly experiments were performed with this size of peg and hole (1mm radial clearance).

In addition to scans parallel to the *y* axis, scans parallel to the *x* axis also yield characteristic features in sensor space indicating the relative hole *y*-position. In our implementation, though, we moved the peg to the identified *x*-coordinate of the hole before executing the *y*-scan. As a result, the peg typically fell into the hole during the *y*-scan, terminating the assembly search without the need for *y*-scan data interpretation.

If the peg does not happen to fall into the hole during the *y*-scan, a strong, peaked *y*-force is produced as the peg passes near the hole, from which we can recognize the hole's *y* coordinate. Then, based on the resulting estimated *x* and *y* hole coordinates, a spiral search is performed near the estimated hole center. Ordinarily, such searches rapidly concluded in success. However, an upper-limit search time was specified, at which the spiral search was abandoned and the entire process was restarted from the initial x-scan.

6. Results

Figure 6 shows the peg's trajectory for an example assembly using our featured-based moment interpretation algorithm on an AdeptOne robot. In this trial, the assembly concluded during the *y* scan. This is the fastest type of conclusion we can expect. (The *x* scan is deliberately offset in *y*, so we do not expect insertion during the *x* scan). In the figure, the circle is the start point and the triangle marks the end point of the assembly. The arrowheads indicate the peg's direction of motion.

From 225 assembly trials, we found that 78% of the assemblies concluded during the *y* scan. The average assembly time for this type of termination was approximately 4.4 seconds. Most of the remaining assemblies concluded successfully while performing the spiral search. The average assembly time for this kind of termination was approximately 6.2 seconds. A small number of assemblies (2%) failed to conclude within the time limit of the spiral search. The composite average assembly time (including approximately 28 seconds total penalty for the 4 failed assemblies) was 4.77 seconds.

We can compare this result to a blind search with equivalent uncertainty and clearances. Our search space was 100mm by 50mm (5,000 mm^2). For a 41mm radius hole and a 40mm radius peg, the target area for success is 254 mm^2 out of 5,000, or 5% of the search area. One must impose a maximum search speed, based on the impact forces that can occur when the peg drops into the hole during fast scanning. Experimentally, we found a scan speed of roughly 30mm/sec to be acceptable for our test example with the peg, hole and AdeptOne robot. At this speed, a blind search in a raster pattern with 2mm spacing between scan lines could cover the search space exhaustively in 83 seconds. On the average, the hole would be found in half this time (roughly 40 seconds). Thus, the feature-based, sensory-guided search algorithm was nearly an order of magnitude faster than a blind search. This search-speed advantage would be further exaggerated for smaller clearances and/or larger search spaces.

7 Conclusions and Future Work

Intelligent approaches for using sensory information as clues to guide hole-searching in a robotic peg-in-hole assembly were investigated in this paper. We constructed and analyzed mathematical contact models for the peg-in-hole problem. It was shown that reaction-moment measurements can provide information regarding desired assembly coordinates. However, this information has variable quality, providing credible, accurate coordinates in some regions of moment space and little or no predictive value in other regions. It was shown that a neural-net functional inversion from sensor space to assembly coordinates can be augmented with a credibility measure to more effectively utilize the sensory data for guidance.

It was further advocated to collect entire streams of sensor data while moving parts in contact. Such measurements can then be analyzed in terms of pattern matching against a pre-recorded map of positions vs moments. A special-case version of such pattern matching was described and implemented in terms of characteristic features (in this case, strong peaks). Experimentally, this algorithm resulted in nearly an order of magnitude improvement in speed over a blind search for our simple example with a 50mm by 100mm search space and a 1mm radial clearance. The advantage of a sensory-guided search over a blind search should be more dramatic for more demanding clearances.

More importantly still, the advantage of sensory-guided searches over blind searches should be exaggerated in higher dimensions. Extending these results to 3-D searching (by including the dimension of z-rotation for non-circular pegs and holes) would have applicability to a wide class of industrially relevant assembly problems. Because sensory interpretation is all the more difficult in higher-dimensional searches, it will be important to identify key features in sensor space, as suggested by the 2-D experiments described here.

Acknowledgments: This work was performed under the support of the U.S. Department of Commerce, National Institute of Standards and Technology, Advanced Technology Program, Cooperative Agreement Number 70NANB7H3024.

References
1. Newman, W. S., et. al., *Force-Responsive Robotic Assembly of Transmission Components*, Proceedings, IEEE Int. Conf. on Robotics and Automation, May, 1999, pp 2096-2102.
2. Chhatpar, S.R., *Experiments in Force-Guided Robotic Assembly*. M. S. thesis, Department of Electrical Engineering and Applied Physics, CWRU, 1999.
3. Zhang, Cheng, *Towards a Practical Robotic Assembly System for Industrial Mechanical Assembly Applications*, M. S. thesis, Department of Electrical Engineering and Applied Physics, CWRU, 2000.
4. Morris, Daniel, *Experiments in Mechanical Assembly Using a Novel Parallel Manipulato*, M. S. thesis, Department of Electrical Engineering and Applied Physics, CWRU, 2000.
5. Whitney, D.E., *Quasi-Static Assembly of Compliantly Supported Rigid Parts, in Robot Motion*, Cambridge, MA; MIT Press, 1981, pp 439-471.
6. Whitney, D. E., *Historical Perspective and State of the Art in Robot Force Control*, Int. J. of Robotics Research, Vol 6, No 1, Spring 1987, pp 3-14.
7. Gottschlich, S.N. and Kak, A.C., *A Dynamic Approach to High-Precision Parts Mating*, IEEE Trans. Systems, Man and Cybernetics, Vol 19, no. 4, 1989, pp 797-810.
8. Asada, H. *Teaching and Learning of Compliance using Neural Nets: Representation and Generation of Nonlinear Compliance*, Proceedings of IEEE International Conference on Robotics and Automation, pp1237-1244, 1990.
9. Cervera, E.; del Pobil, A.P. *Sensor-Based Learning for Manipulation. Part I: Eliminating Sensor Ambiguities via Recurrent Neural Networks*, in Proceedings of IEEE International Conference on Robotics and Automation, Leuven, Bélgica, vol.3, pp. 2174-2179, 1998.
10. Cervera, E.; del Pobil, A.P.; Marta, E.; Serna, M.A. *A Sensor-Based Approach for Motion in Contact in Task Planning*, in Proceedings of IEEE/RSJ International Conference on Intelligent Robots and Systems, Vol2, pp468 –473, 1995.
11. Hovland, G.E.; McCarragher, B.J. *Frequency-domain Force Measurements for Discrete Event Contact Recognition*. In Proceedings of the IEEE international Conference on Robotics and Automation, pages 1166-1171, 1996.
12. Sikka, P.; McCarragher, B.J. *Monitoring Contact using Clustering and Discriminant Functions*. In proceedings of the IEEE International Conference on Robotics and Automation, pages1351-1356, 1996.
13. MacQueen, J. *Some Methods for Classification and Analysis of Multivariate Observations*, Proceedings of Fifth Berkeley Symposium on Math. State. And Prob., I, pp281-297, 1967.

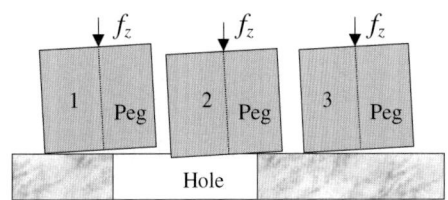

Figure 1. Different contact states for a tilted peg

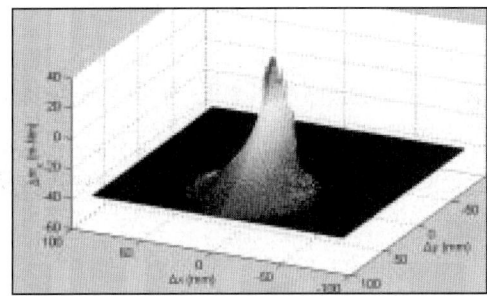

Figure 2.a Moments m_x (N-mm) vs. Δx, Δy (mm)

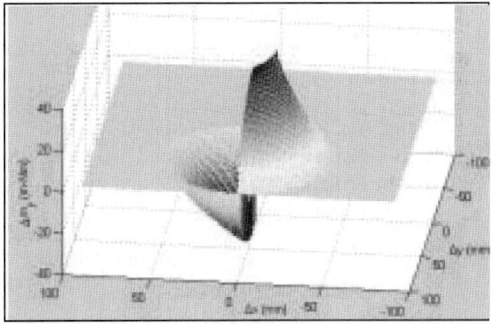

Figure 2.b Moments m_y (N-mm) vs. Δx, Δy (mm)

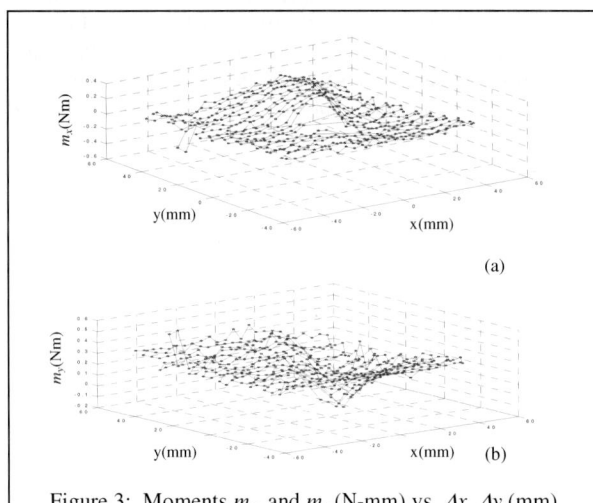

Figure 3: Moments m_x and m_y (N-mm) vs. Δx, Δy (mm)

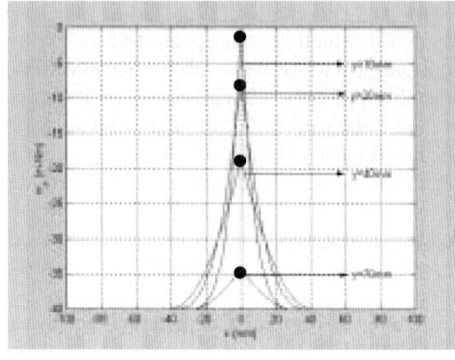

Figure 5: Computed moment m_x for scan along x at constant y offset. Peg is 40mm radius and hole is 41mm radius. Moment peak is aligned with hole's x coordinate

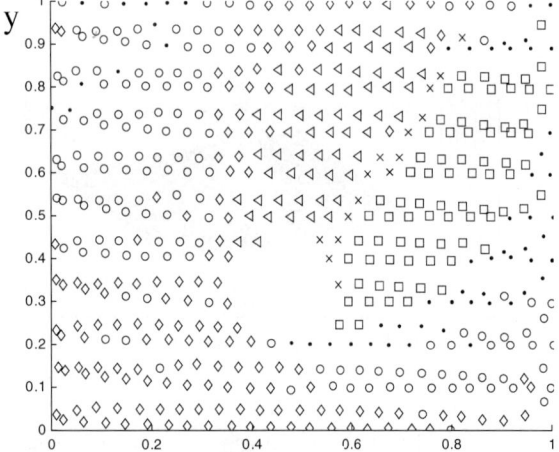

Figure 4: Clustering result in position space: "x" is high-information data; diamond and circle indicate low-information data; triangle and square indicate medium-information data

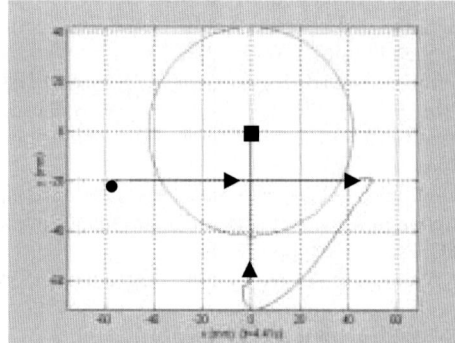

Figure 6: Trajectory of an example assembly

Dome Shaped Touch Sensor Using PZT Thin Film Made by Hydrothermal Method

Guiryong KWON*, Fumihito ARAI*, Toshio FUKUDA**, Kouichi ITOIGAWA***, and Yasunori THUKAHARA***

* Department of Micro System Engineering, Nagoya University,
 Furo-cho, Chikusa-ku, Nagoya 464-8603, Japan, Phone +81-(52)-789-3116
** Center for Cooperative Research in Advanced Science and Technology, Nagoya University,
 Furo-cho, Chikusa-ku, Nagoya 464-8603, Japan, Phone +81-(52)-789-4478
*** Tokai Rica Co.,LTD., Oguchi-cho, Niwa-gun, Aichi-Ken 480-0195, Japan, Phone +81-(578)-95-7042
 E-mail : kwon@robo.mein.nagoya-u.ac.jp, arai@mein.nagoya-u.ac.jp

Abstract

We propose a dome shaped touch sensor unit, which is small and is applicable in high temperature environment. The PZT thin film was made on the half-round Ti substrate, and the electrodes were deposited on its surface to form the sensor and driving actuator. The PZT thin film was made on the curved surface of the substrate by the hydrothermal method. The actuator part is driven by the high frequency voltage at the resonant frequency [6.2kHz] and the sensing part can detect the impedance change before and after the contact. The sensor works with the low voltage [5V]. It works under the high temperature [over 50°C]. The structure is simple and easy to miniaturize. This sensor can be used as the control touch-pad of the car instead of the conventional capacitive type computer touch pad, because it can resist high temperature and disturbance from the electrostatic inductance change. The basic property of the sensor is shown.

Keyword: Touch sensor, PZT thin film, Micro-sensor, Vibration, Hydrothermal method

1. Introduction

Today, a role of the automobile is not only a vehicle but also a moving office combined with the communication tools, such as a cellular phone and car navigation system with online information. Since the driver must concentrate on driving as much as possible, the role of the human interface or command input device is very important. There are a lot of research works on the intelligent human machine interface, such as a verbal communication system to reduce the burden. However, there is a permanent demand on the switch or key input device for secure operation. In future, the number of those input device will increase and more compact input device will be required. The conventional switch is based on the mechanical contact and has difficulty in miniaturization. So, we need a new switching device which is suitable for miniaturization and robust against the disturbance. The inside temperature of the car will range from 85 °C to –20 °C depends on the season and location. So, the sensor must resist such a wide temperature variation.

Recently various kinds of touch sensors were developed [1-5]. Touch sensors that have been developed until now can be classified as follows.
(1) Switch conductor type
(2) Resistance change type
(3) Polymer type
  (Ex: pressure sensitive conductor material)
(4) Capacitance change type
(5) Optical type (measurement of intensity of light)
(6) Thermal type
(7) Vibration type (Ex: piezo-electric material)

The switch conductor type sensor was fabricated by MEMS (Micro Electro Mechanical System) technology. But it has the problem of the structural rigidity. Nilsson [2] proposed the skin-type touch sensor, which adopted the resistance change. The main characteristic of this sensor is its structural flexibility. Niihara et al. developed a new polymer type touch sensor using rubber composite [3]. This sensor has good quality of sensing, but works under 50°C, so it is not suitable for usage in the hazard environment. A touch pad based on the capacitance change is very popular, but is not suitable for the car because of high inside temperature in summer. Yamada et al. developed a touch sensor, which can sense slip motion [4]. However, it has difficulty in sensing the contact state of the object. Optical type has the problem in miniaturization. Higuchi et al. fabricated a tactile prove sensor [5]. This sensor can also be used as the proximity sensor. However, it is not suitable for application of sensing soft object like a finger, since the shape of the contact point is so sharp.

In order to use the sensor in harshness environment such as a car for all seasons, we need to develop a new touch sensor, which is gentle for the human finger and not so fragile, can endure in high or low temperature, and strong for electrostatic disturbance.

In this paper, we propose a dome shaped touch sensor unit, which is small and is applicable in high temperature environment. The PZT thin film was made on the half-round Ti substrate, and the electrodes were deposited on its surface to form the sensor and driving actuator. Figure 1 shows the outline of the sensor. The PZT thin film was made on the curved surface of the substrate by the hydrothermal method. The actuator part is driven by the high frequency voltage around the resonant frequency [6.2kHz], and the sensing part can detect the impedance change before and after the contact. The sensor works with the low voltage [5V]. In general, the PZT and its composite shows a stable property under the temperature around 80°C. So, the PZT material is considered to be suitable for the touch sensor in harshness environment. It works under the high temperature [over 50°C]. The structure is simple and easy to miniaturize. This sensor can be used as the control touch-pad of the car instead of the conventional capacitive type computer touch pad, because it can resist high temperature and disturbance from the electrostatic inductance change. The basic concept of the sensor is shown in fig. 1.

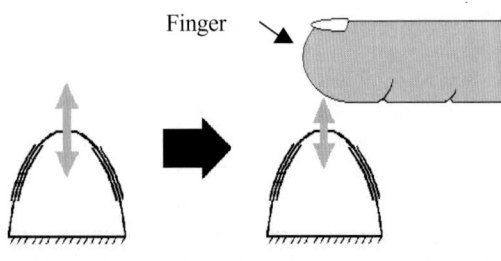

Fig. 1 Outline of the dome shaped touch sensor

2. Hydrothermal Method

In these days, many researchers have interest in piezo-eletric material, because it can transfer the mechanical deformation to the electric charge and transfer the electric charge to the mechanical deformation. Piezo-electric material has been applied to the various kinds of sensor, for example, a touch probe sensor [5] and vibration gyroscope [6]. Those sensors use the PZT thin film as a sensor and actuator. To form the PZT thin film on the dome shaped substrate, we used the hydrothermal method.

In the past, PZT thin film was made by the various methods such as the Sol-Gel method, Screen-Press, MOCVD, Sputtering and so on [6]. By these methods, it is difficult to make the PZT thin film on a three-dimensional structure. Moreover, reaction temperature is higher than 500°C. By using the hydrothermal method, we can avoid such manufacturing difficulties. The advantages of the hydrothermal method are summarized as follows [6].

- PZT thin film can be fabricated on the three-dimensional structure. This method grows the PZT thin film in the solution.
- Thickness of PZT film can be controlled by repeating the crystal growth process. This method consists of two processes, namely the nucleation process and crystal growth process, we can control the thickness of the film by repeating the crystal growth process.
- PZT thin film can be fabricated only on the titanium substrate.
- PZT thin film is deposited at relatively low temperature (below 200°C).
- This method doesn't need the polarization process.

The process to fabricate the PZT thin film by the hydrothermal method had been researched by Shimomura and Ohba et al. [7]. The improved process was proposed by Higuchi et al. [5].

In the autoclave, potassium hydroxide, lead nitrate, zirconium oxychloride, titanium substrate, and titanium tetrachloride are ionized in solution. Chemical reaction formulas in each solution are written as follows.

Potassium hydroxide

$$KOH \rightarrow K^+ + OH^- \qquad (1)$$

Lead nitrate

$$Pb(NO_3)_2 \rightarrow Pb^{2+} + 2NO_3^- \qquad (2)$$

$$Pb^{2+} + 2OH^- \rightarrow H^+ + HPbO_2^- \qquad (3)$$

Zirconium oxychloride

$$ZrOCl_2 \rightarrow ZrO^{2+} + 2Cl^- \qquad (4)$$

$$ZrO^{2+} + 4OH^- \rightarrow H^+ + Zr(OH)_5^- \qquad (5)$$

Titanium substrate

$$Ti + 4OH^- \rightarrow Ti(OH)_4 \qquad (6)$$

Titanium tetrachloride

$$TiCl_4 + 4OH^- \rightarrow Ti(OH)_4 + 4Cl^- \qquad (7)$$

Finally, the formulas of mixed solution in autoclave is written as follows.

$$HPO_2^- + 0.52Zr(OH)_5^- + 0.48Ti(OH)_4 \rightarrow Pb(Zr_{0.52}Ti_{0.48})O_3 + 2H_2O + 1.52OH^- \quad (8)$$

The concentration of KOH was the same (4 mol - 44ml) for both processes. But the other solutions are changed for each process. Table 1 shows the condition of the hydrothermal method for each process. Figure 2 shows the photograph of crystal shape after the growth process. The temperature of the nucleation process is 150 °C and that of the crystal growth process is 120 °C. The reaction temperature was controlled by the electric heating of the oil bath, which contains the silicon oil and the autoclave. The autoclave was rotated in the bath by the AC motor in order to blend the solvent inside it. The rotation velocity of the autoclave is 7 [rpm]. It is considered that the suitable rotation velocity ranges from 7 [rpm] to 10 [rpm][5],[7].

Table 1. Process condition of the hydrothermal method

	Nucleation Process	Crystal Growth Process
$Pb(NO_3)_2$	0.52mol 28ml	0.776mol 28ml
$ZrOCl_2 8H_2O$	0.815mol 8ml	1.179mol 8ml
$TiCl_4$	1.6mol 1.4ml	1.6mol 2.1ml

Fig. 2 Photograph of crystal shape by the hydrothermal method

3. Vibration Analysis of Curvature Beam

The shape of the sensor unit is half-round. There are 4 electrodes on inside and outside surface of it. The length of each electrode is one third of the length of the quarter of the circle length. The inside electrode is used as a sensor part and the outside electrode is used as a driving part. When the voltage is applied in the driving part, the top of the touch sensor moves up and down.

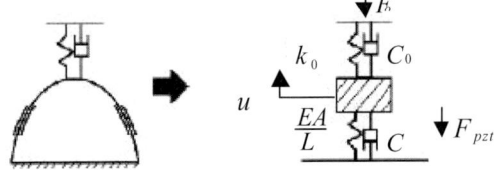

Fig.3 Model of the touch sensor

3.1 Simplified model

To derive the dynamical equation of this sensor, we assume the following conditions. (1) The touch sensor is symmetric and its top moves in vertical direction. (2) The touch sensor is modeled as the one-dimensional spring-mass model. (3) The finger is assumed as the visco-elastic material. The finger is fixed firmly and the mass of the finger does not move. Based on these assumptions, the model of this sensor is shown in Fig. 3. Then, the dynamical equation of motion of this sensor is written as follows.

$$m\ddot{u} + (c+c_0)\dot{u} + (\frac{EA}{L}+k_0)u = F_0 + F_{PZT} \quad (9)$$

Here, m is total mass. c and c_0 are damping coefficients. E, A, and L are the Young's modulus, the cross section of the plate, and the plate length of the sensor, respectively. F_{pzt} is force caused by the PZT thin film actuator. F_0 is the external force.

It is obvious that the impedance of the total system is changed before and after the contact of the finger. So, if the sensor is sensitive enough to detect the impedance change, it can detect the contact.

In order to analyze the property of the sensor, we need equivalent coefficients of the model including the finger. However, it is very difficult to obtain those parameters precisely. So, next we analyze the vibration property of the dome shaped structure by the finite element method (FEM) to design a sensitive sensor.

3.2 Vibration modal analysis by FEM

We need to know the vibration mode and resonant frequency to determine the driving frequency of the actuator. We did modal analysis by the FEM.

We considered the effect of the piezo-electric effect in vibration analysis by FEM. We used the ANSYS program. For input data, elastic modulus of substrate is 110Gpa, poisson's ratio is 0.33, and applied voltage of the driving part is 5[v]. Specific inductive capacity is set $e_{33}/e_0 = 1700$ and $e_{11}/e_0 = 1730$. Thickness of the titanium substrate is 25 μm. Since we assume the sensor is symmetric, we used the half-length model for the finite element analysis as shown in Fig. 4.

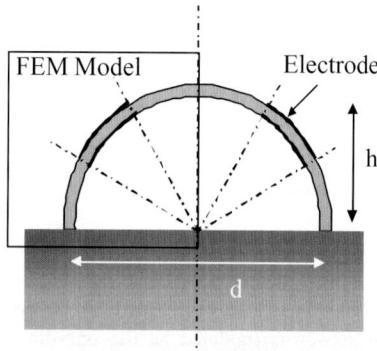

(a) Configuration of the model (side view)

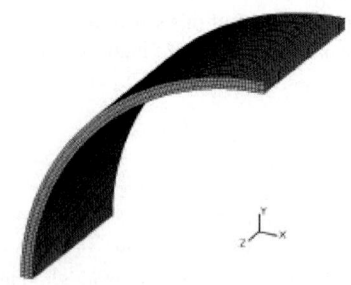

(b) Half-length model

Fig. 4 Model for finite element analysis

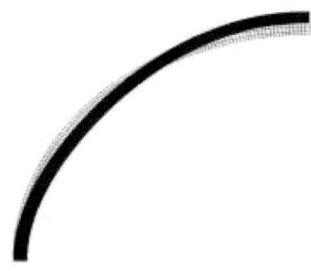

(a) First vibration mode (2.1 KHz)

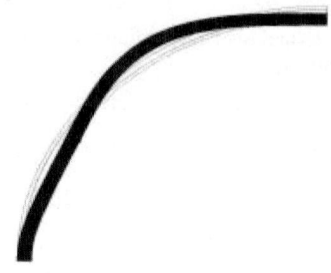

(b) Second vibration mode (6.2 KHz)

Fig. 5 Vibration mode of the sensor by FEM analysis

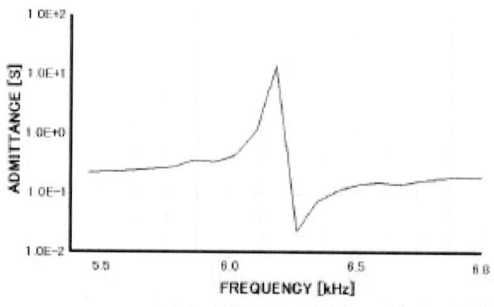

Fig. 6 Resonance point by FEM

Figure 5 shows the vibration mode by the FEM analysis. The first resonant frequency is about 2.1[kHz]. The second resonant frequency is about 6.2[kHz]. Figure 6 shows the admittance change according to the driving frequency. From this analysis, we can evaluate the second resonant frequency around 6.2 [kHz]. We used the 2-nd resonant frequency for the touch sensing.

5. Experiment Result

5.1 Touch sensing experiment

We fabricated the sensor unit by the hydrothermal method. The process condition of the hydrothermal method is shown in Table 1. The size of titanium substrate is 3×10×0.025 [mm]. The height of the curved sensor unit is 3.5 [mm] (h in Fig. 4 (a)), and its width is 6.1 [mm] (d in Fig. 4 (a)). After finishing the hydrothermal process, we made electrodes at both sides of sensor unit by the vacuum evaluator. Electrodes inside the sensor unit are used for the driving part, and electrodes outside the sensor unit are used for the sensing part. Sensing part is connected to the lock-in Amplifier. Figure 7 is the electric circuit diagram of sensor pad. Figure 8 is photograph of the sensor unit.

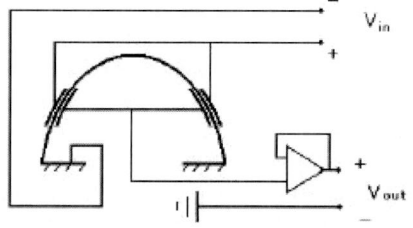

Fig. 7 Electric circuit of the sensor

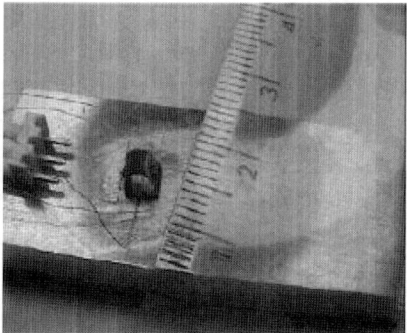

Fig. 8 Photograph of the sensor unit

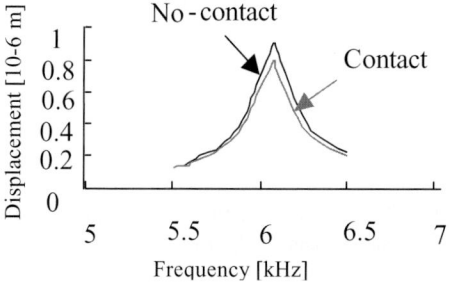

Fig. 9 Difference of displacement of the sensor before and after the contact

We tested the sensor with driving voltage 5 [V] and vibration frequency 6.2 [kHz]. The sensor at this condition was very sensitive. Table 2 is the experiment result of this sensor before and after the contact.

When the finger contacts the top of the sensor, the displacement and output voltage decreased together. At the same time, phase difference between the driving frequency and output frequency occurred. Figure 9 shows the difference of displacement according to the driving frequency when the finger touched the sensor.

We obtained the result in Fig. 9 by using the Laser Doppler vibration measurement device. The amplitude of vibration from the analysis and experiment is almost the same. The displacement amplitude is about 1 μm.

When we changed the thickness of the titanium substrate from 0.1[mm] to 0.01[mm], we found that output voltage of thin substrate is bigger and clearer than that of the thick one.

Table 2 Comparison of output signal before and after the contact

	Output Voltage [mV]	Phase Difference [deg]
Before contact	22.3	66.5
After contact	20.4	57.3

We also did an experiment of touching by the nonconductive material, like wood. The result of this experiment is almost the same as the previous experiment touching by the finger. This means that output voltage was influenced mainly by the variation amplitude. So, we conclude that the main factor to determine the sensitivity of the sensor is the variation amplitude.

Next, we linked three sensors with the parallel circuit. The sensor array is shown in Fig. 10. The sensor unit B located at the center was touched by the finger. The result is shown in Fig. 11. The output from the sensor B decreased after the contact. On the other hand, the output from the sensor A did not decrease after the contact. This means the sensor unit is independent. This result suggests us that we will be able to develop sensor array with fine special resolution in future.

Fig. 10 Photograph of parallel linked sensor array.

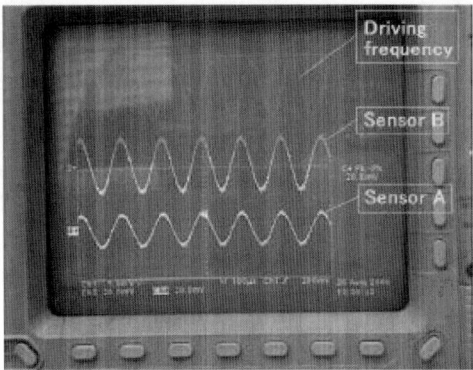

(a) Before contact

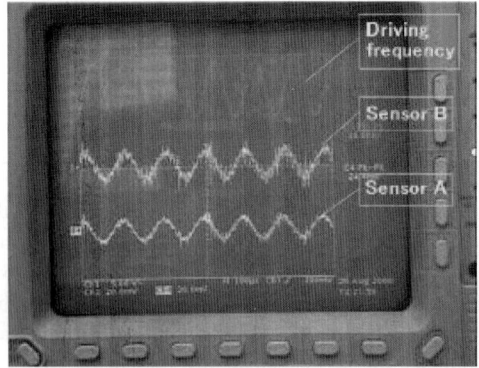

(b) After contact

Fig. 11 Difference of output signal before and after the contact (Oscilloscope monitor)

5.2 Force sensing experiment

We tested the relationship between the applied force and output voltage of the sensor. Figure 12 shows the experiment result. From this result, the proposed touch sensor is very sensitive for tactile sensing. On the other hand, this sensor is not suitable for force sensing, since the output decays suddenly when the force exceeds 0.3 mN. To use this sensor for force measurement, we need to modification.

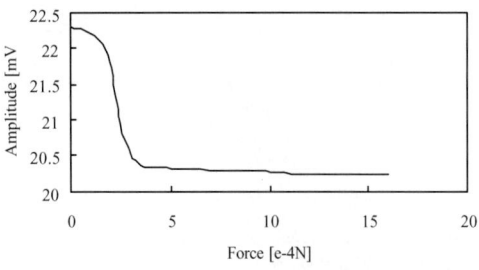

Fig. 12 Relationship between applied force and output voltage of sensor

5. Conclusion

We fabricated a new touch sensor using the PZT thin film, which is made by the hydrothermal method, in order to develop a new type touch sensor, which can be miniaturized easily and is applicable in hazard environment, such as a car. The prototype sensor detected the difference before and after the contact. The signal level was several mV. In future, we plan to evaluate this sensor in high and low temperature environment.

We also evaluated the response to the force input. The change of the signal was very sharp and it is not practical for force sensing of the finger at present. However, if we modify the configuration, the sensor has possibility to measure the force with good precision. We will expand the application of this sensor in future.

References

[1] H.Meixner and R. Jones, "Sensors, A Comprehensive Survey, Micro and Nanosensor Technology/Trends and Markets", VCH, Vol.8, pp.3-13, 1995

[2] M. Nilsson, "Tactile Sensing with Minimal Wiring Complexity", Proceeding of the IEEE ICRA, Vol.1, pp.293-298, 1999

[3] K. Niihara, Y.H. Choa, M. Hussain, Y. Hamahashi, H. Kawahara, Y.Okamoto and H.Nishida, "New Contact Sensor in Organic/Inorganic Composite System (in Japanese)", Materials Integration, Vol.12, No.5, pp.47-53, 1999.

[4] Y. Yamada, M. Furukawa, S. Nishi, K. Imai and Y. Umetani, "Primary Development of Viscoelastic Robot Skin with Vibrotactile Sensation of Pacinian / Non-Pacinian Channels ", Proceedings of the 3rd International Conference on Advanced Mechatronics, pp.879-885, 1998.

[5] T.Kanda, T.Morita, M.K.Kurosawa, T.Higuchi, "A Rod-Shaped vibro touch Sensor using PZT thin film", IEEE Transaction on Ultrasonics, Ferroelectrics, and Frequency Control, Vol. 46, No.4, pp.875-882, 1999

[6] T. Fukuda, H. Sato, F. Arai, H. Iwata, and K.Itoigawa, "Parallel Beam Micro Sensor/Actuator Unit Using PZT Thin Films and Application Examples", Proceeding of the IEEE ICRA, Vol.2, pp.1498-1503, 1998

[7] Y. Ohba, M. Miyauchi, T. Tsurumi, M. Daimon, "Analysis of Bending Displacement of Lead Zirconate Titanate Thin Film Synthesized by Hydrothermal Method", Journal of J.Appl. Phys., Vol.32, pp.4095-4098,1993

Heuristic Vision-Based Computation of Planar Antipodal Grasps on Unknown Objects

Antonio Morales, Gabriel Recatalá, Pedro J. Sanz, Ángel P. del Pobil

Department of Computer Science, Universitat Jaume I
Campus del Riu Sec, E-12071 Castellón (Spain)
{morales,grecata,sanzp,pobil}@inf.uji.es

Abstract

A key issue in robotics is the development of the ability to grasp unknown objects. This ability requires a grasp determination mechanism that, based on the analysis of the description of the object, determines how it can be stably grasped. In this paper, a grasp determination method is presented that computes a set of grasps that comply with the force-closure condition. Its input is a set of contours, extracted from vision data, describing the shape of the object. The internal holes of the object are taken into account, so the algorithm can find grasps on them. The algorithm also finds expansion and squeezing grasps, which are executed by opening and closing the gripper fingers.

Keywords: 2D grasping, visually-guided grasping, service robotics

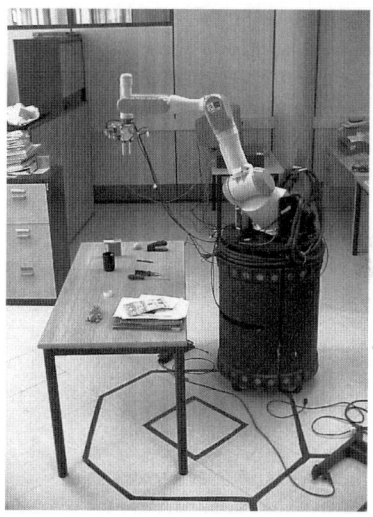

Figure 1: Experimental setup for the grasp determination method proposed in this paper.

1 Introduction

One of the basic problems in robotics is the development of manipulation skills, which include the ability to grasp both known and unknown objects. A key component of this ability is the mechanism used to determine how each object should be grasped.

Due to its versatility and inexpensiveness, the combination of a simple two-fingered gripper and a single camera, either in a camera-in-hand or in a fixed-camera configuration (see figure 1), has become popular on many works on grasping [5, 14, 13]. In this case, a planar (2D) description of the object is used in the grasp determination process. Therefore, the development of robust and efficient grasp determination algorithms based on 2D object descriptions and two-fingered grippers is fundamental for the application of these popular configurations to fields such as service or industrial robotics.

An important concept related to the analysis of grasps on planar objects is the *force closure* condition, introduced by Nguyen [10], which ensures the stability of the grasp assuming point contacts with friction. Grasps that comply with this condition are also referred to as *antipodal grasps*.

The force-closure condition has been used in many important papers on planar grasp determination, such as [3, 4, 9]. Nevertheless, there are several problems for the applicability of the above procedures to the on-line processing of unknown objects. First, they rely on predefined geometric models, so they cannot be used to handle unknown objects. In addition, they make use of analytical optimization methods that require a lot of computation time.

Other works on planar grasping have relied on vision for extracting the object description. Jarvis [7] and Hauck [5] based their methods on the analysis of

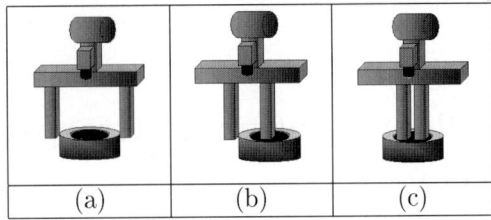

Figure 2: Types of grasps considered by the proposed method: (a,b) squeezing grasps; (c) expansion grasp.

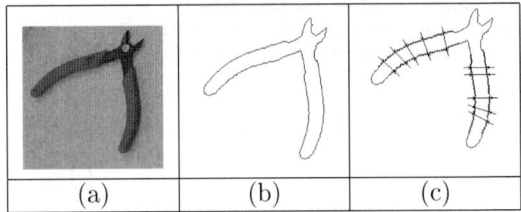

Figure 3: (a) Image of an object as observed by the vision system; (b) Input to the grasp determination method: contours of the object silhouette; (c) Output: set of grasps.

the skeleton, while Stanley [14] performed a quadtree resolution expansion of the image in order to extract the contour. In general, they do not require a previous identification of the object [15], since they are not based on models. Kamon [8] used visually-computed thresholds to select the grasping points on the contour of the object. Sanz [11, 12] associates thresholds to a set of grasp quality conditions, and relies on a heuristic to find the best grasp. These methods are heuristic in nature, since they estimate generic values of some parameters related to the force closure (friction, force to apply to the object,...) based on either vision or previous experiments.

Many of the above procedures perform a limited grasp determination, since they only consider grasps along the boundaries of the object; the holes in the object are ignored during the analysis, so some possible grasps are missed. Moreover, only *squeezing grasps* –those that are executed by closing the gripper fingers on the object– are considered, while, in some cases, *expansion grasps* –which are performed with an opening of the fingers– are also possible [3].

In this paper, an heuristic method for planar grasp determination is presented that considers both squeezing and expansion grasps, shown in figure 2. Furthermore, it takes into account the whole morphology of the object, including not only the external contour but also the internal ones. Unlike in other works, instead of selecting a single grasp, a set of grasps is produced. All these grasps comply with the force-closure condition. This set can then be used by other methods for selecting the most suitable grasp, taking into account application-specific criteria. A set of force-closure grasps is also produced by Sanz [13], but the scope is more limited, since neither expansion grasps nor grasps on internal contours are considered.

This method has been designed for a robotic system that consists, at least, of an arm with a parallel-jaw gripper and a camera, either fixed or in a camera-in-hand configuration, that observes the area reachable by the arm from a zenithal point of view (see figure 1). The objects to be grasped by the arm are assumed to be lying on a relatively-flat surface, perpendicularly to the optical axis of the camera. With this system, the objects will be grasped so that the *grasping line* –the line joining the points where the gripper fingers are placed– is parallel to this surface.

This paper is organized as follows. First, the proposed method is described. The results obtained are shown in section 3. Finally, section 4 provides some concluding remarks about the proposed method and the immediate future lines to extend this research.

2 Methodology

The proposed method takes as input a 2D description of the object, consisting of a list of its external and internal contours. Currently, a list of points is being used for representing these contours. The output is a list of grasps that, according to a set of criteria, comply with the force-closure condition. Each grasp is described by a pair of points, each one belonging to one of the contours and being the center of the region where the gripper fingers are placed. For each grasp, it is specified if it is an expansion or a squeezing grasp.

This method assumes that the objects are relatively flat, or that can be treated as an extrusion of their projected silhouettes. Figure 3 shows a sample image of an object, the corresponding description provided as input to the grasp determination method, and the set of grasps it produces.

The method relies on an external image-processing module for the extraction of the object shape description, such as the one described in [6]. As an observed scene will usually contain several objects, an additional module, which will be application-specific, is also required that selects the object to grasp.

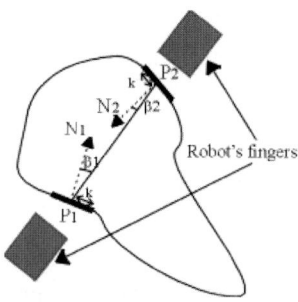

Figure 4: Geometric interpretation of the quality criteria used for the selection of grasps.

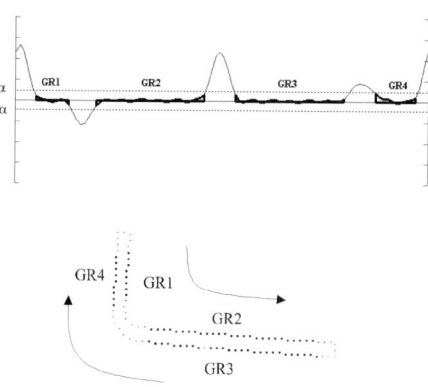

Figure 5: Grasping regions on an Allen key's contour (below), selected based on the analysis of the curvature function (above).

The grasps in the list produced by this method can then be processed by another module that will select the most suitable grasp according to criteria that will be specific to the task to be performed with the object. The execution of the selected grasp is left to the control system of the robot (e.g., the one used in [11]).

2.1 Grasp characterization

The quality assessment performed by this module is based on the following criteria:

- *Gripper adaptation criterion.* It evaluates the adaptation between the object and the gripper's fingers in terms of the estimated area of contact when the fingers are placed on the object in order to execute the grasp. This estimation is based on the curvature of the object's observed contour at the points where the fingers should be placed. This curvature should not exceed a threshold α (*curvature threshold*).

- *Force-closure criterion.* Its purpose is to ensure that the gripper does not cause the object to slide due to a torque when it closes its fingers to grasp it. Its evaluation is based on the concept of *friction cone* [10] and makes use of the angles (β_1 and β_2 in figure 4) between the normals (N_1 and N_2) to the object's contour at the grasping points and the *grasping line* ($\overline{P_1 P_2}$). Force closure [10] is achieved when the grasping line lies inside both friction cones. These angles should not exceed a threshold β (*angular threshold*).

The values of α and β have been determined empirically. Figure 4 illustrates how the above criteria are used in the evaluation of the quality of a grasp.

2.2 Grasp determination

The only *a priori* information the proposed method considers is the geometry of the gripper –in particular, the width of its fingers–. For the grasp determination, we assume contacts with friction between the object and the fingers, and consider the Coulomb friction model [10]. However, the static coefficient of friction between the object and the fingers is not known beforehand. This method involves three steps.

Step 1: Extraction of grasping regions

The purpose of this step is to find regions on the object contour that comply with the curvature threshold, which will be referred to hereinafter as *grasping regions*. In this step, a *curvature function* is computed at each point on the contour (see figure 5). For the calculus of this function, a neighborhood centered at each point is considered that is of the same size as the projection of the width of the robot's fingers on the image ($2k$ in figure 4).

The task of finding the grasping regions lies in the analysis of the curvature function and grouping consecutive points with curvature below the curvature threshold in a single region. Ideally, it should be possible to approximate these regions as straight lines. Nevertheless, in some objects, such as the one shown in figure 6, this procedure could produce large regions, covering portions of the contour that could not be reduced to straight lines. For this reason, the accumulated curvature along a grasping region is used to limit its size. The size of the grasping regions is also lower-bounded, since, being too small, they would not allow

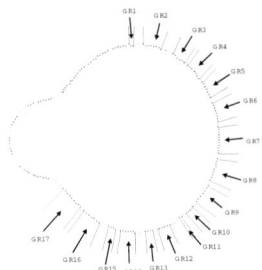

Figure 6: Grasping regions on a circular contour. The use of the accumulated curvature produces a set of small regions, instead of a large one covering most of the contour.

for tolerances in the positioning of the robot's fingers. Figure 5 illustrates the use of the curvature function to find grasping regions on the object's contour.

Unlike in other works [11, 12, 13], the contours corresponding to the internal holes of the object have also been considered for the grasp determination. Therefore, the grasping regions are extracted from both the external and the internal contours.

Step 2: Selection of compatible regions

Once the grasping regions have been found, the next step is to build a list of pairs of grasping regions where the robot's fingers could be placed to grasp the object. These regions are characterized by containing two points –one per region– such that, when the robot's fingers are placed at them, the grasp complies with the force-closure criterion. The regions in each pair will be termed hereinafter *compatible regions*.

The compatibility between two grasping regions is verified through the following conditions: (1) the angle between the normal vectors to each region is π; and (2) the projection of each region, in the direction of its normal, intersects with the other region, that is, the regions are confronted. With regard to the first condition, the angular threshold (β) is considered for tolerance in the angle between both vectors; therefore, the allowed range between these vectors is $\pi \pm 2\beta$. Figure 7 shows how these two conditions are checked.

This test is not performed only between regions within the same contour, but also between regions of different contours, so all combinations of regions are checked. This produces pairs of compatible regions such that the normals to each region point to the space between both regions, and pairs in which the normals point towards outside this space. The first ones correspond to squeezing grasps, which are executed by

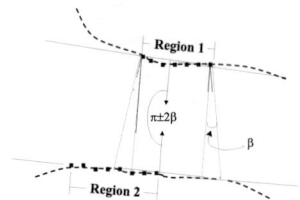

Figure 7: Compatibility test between grasping regions.

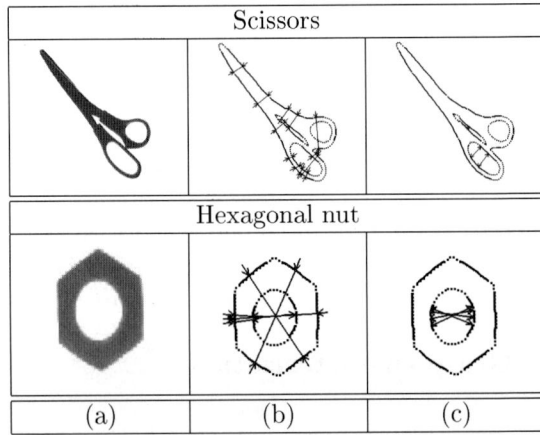

Figure 8: (a) Segmented images; (b) Squeezing grasps obtained; (c) Expansion grasps obtained.

closing the gripper fingers on the object and are the only ones considered by many works on grasp determination. The others correspond to expansion grasps.

Step 3: Grasp refinement

This step involves the selection of a pair of grasping points within each pair of compatible regions, and can be seen as a refinement of the compatible regions. This selection can be performed by exhaustively checking each point from each region and select the pair that best complies with the quality criteria. However, a faster procedure is being used to limit the check to the neighborhood of the pair of points in the center of the regions.

3 Results

Figures 8 and 9 show the performance of our method on a set of objects. All of them are common objects that can be found in a usual human environment and have not been specially designed for a robot manipulation task. The results show that our method is able

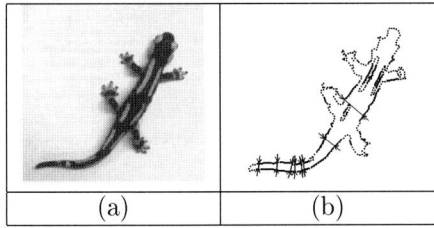

Figure 9: Toy salamander: (a) Original image; (b) Obtained grasps and grasping regions.

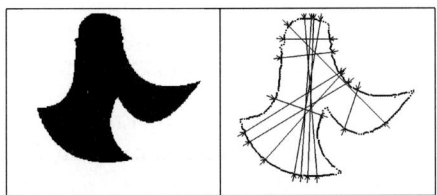

Figure 10: Synthetic image from [4].

to find a large variety of grasps over them, not only the most intuitive, but also some unexpected ones.

Grasps found on the internal contours are specially interesting. Some previous works ignore the holes in the object, what can result in losing some of the obvious grasps that can be observed in figure 8. Moreover, our approach can also obtain expansion grasps using the same principles as for squeezing ones.

Figure 10 shows the results obtained for an object proposed in [4]. This work is relevant since it proposes an analytical method for finding all the grasps that satisfy the force-closure constraint. Our methodology demonstrates that it can find most of these results with smaller computational cost (see table 1).

Note that it is possible obtain all the solutions proposed by Sanz [11, 12] from the group of grasps that our method generates. This could be done by applying an appropriate selection rule that chose the best grasp according to global stability conditions (i.e: the distance to the centroid of the grasping line).

Another interesting result can be observed in figure 11. There, a group of objects lying in a heap is shown. In this case, the grasping skills can be used for identifying and isolating the different objects by manipulating them. This follows the approach that in [2] has been called *active perception*. Our method offers a wide range of grasping possibilities to modify and manipulate the placement of the objects in the scene.

Finally, table 1 shows the computation times of our

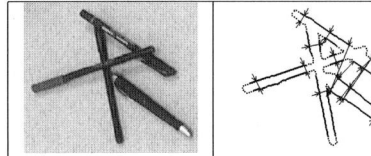

Figure 11: Scene with overlapped objects (pens).

	Image processing	Grasp Determination
Scissors	0.041 sec.	0.034 sec.
Hexagonal Nut	0.034 sec.	0.007 sec.
Toy Salamander	0.047 sec.	0.035 sec.
Figure from [4]	0.038 sec.	0.027 sec.
Heap	0.045 sec.	0.048 sec.

Table 1: Time for grasp determination, obtained with a 166 MHz Pentium PC.

method. For the image processing, the module described in [6] has been applied. As it can be seen, the time spent in grasp determination increases with the complexity of the contours. More precisely, it increases as more grasping regions are found, since more combinations must be checked. Furthermore, all the computations require only a few milliseconds, allowing real-time performance.

4 Discussion

A novel strategy to compute planar grasps on unknown objects has been presented. They can be squeezing and expansion grasps, and take into account external and internal regions, including holes. The short computing time, including image processing and grasp determination, in a state-of-the-art PC, makes fast response applications feasible. Furthermore, a satisfactory comparison with other works has been showed.

The main benefit of our strategy are the inexpensive resources required. It can be easily implemented in all kinds of industrial and service robots with a conventional parallel-jaw gripper and a camera-in-hand. Some potential applications could be food handling tasks or biomedical assistance systems. Another advantage of our method is its flexibility, since it offers a wide range of possible stable grasps that can be filtered depending on the different requirements of the applications. Furthermore, this method is able to find solutions –those that include internal contours or expansion grasps– that none of the other referred algorithms could find. In conclusion, the simplicity, flexibility and fast response of this proposed method make it suitable as a component for reactive and behavior-based architectures [1].

However, our method has several limitations that must be overcome. First, a filtering is needed of the grasps that do not fit some security conditions related with the dimensions and the maximum and minimum opening of the gripper fingers. Second, a mechanism for selecting the best grasp from the list depending on the application requirements must be defined.

In addition, our method is only a small component in a larger system, and the overall performance depends on many modules not directly related to the algorithm proposed here. For example, the image processing modules must generate a proper input, consisting of the contour description of the scene; and a control scheme must be able to execute the selected grasp. The performance of such modules can constrain the behavior of the grasp determination module.

Finally, we are currently introducing the use of other object representations such as active contours, that will allow the tracking and grasping of mobile objects. We are also extending this research toward 3D object descriptions by using stereo systems capabilities.

Acknowledgments

This work has been funded in part by the CICYT under projects TAP98-0450 and HA1999-0038, by the Generalitat Valenciana under project GV97-TI-05-8, by the Fundació Caixa-Castelló under project P1B97-06. The first author is supported by the Ministerio de Ciencia y Tecnología under a FPI program graduate fellowship.

References

[1] R. C. Arkin. *Behavior-Based Robotics*. The MIT Press, 1998.

[2] R. Bajcsy. Active perception and exploratory robotics. In Dario, Sandini, and Aebischer, editors, *Robots and Biological Systems: Towards a New Bionics?*, NATO ASI Series, pages 3–20. Springer Verlag, 1993.

[3] I.M. Chen and J.W. Burdick. Finding antipodal point grasps on irregularly shaped objects. In *Proc. of the IEEE Intl. Conf. on Robotics and Automation*, pages 2278–2283, May 1992.

[4] B. Faverjon and J. Ponce. On computing two-finger force-closure grasps of curved 2D objects. In *Proc. of the IEEE Intl. Conf. on Robotics and Automation*, pages 424–429, 1991.

[5] A. Hauck, J. Rüttinger, M. Sorg, and G. Färber. Visual determination of 3D grasping points on unknown objects with a binocular camera system. In *Proc. of the IEEE/RSJ Intl. Conf. on Intelligent Robots and Systems*, pages 272–278, Kyongju, Korea, 1999.

[6] J.M. Iñesta, P.J. Sanz, and A.P. del Pobil. An automatic transformation from bimodal to pseudobinary images. In A. del Bimbo, editor, *Image Analysis and Processing*, volume 1310 of *LNCS*, pages 231–238. Springer-Verlag, 1997.

[7] R. A. Jarvis. Automatic grip site detection for robotics manipulators. *Australian Computer Science Communications*, 10(1):346–356, 1988.

[8] I. Kamon, T. Flash, and S. Edelman. Learning to grasp using visual information. In *Proc. of the IEEE Intl. Conf. on Robotics and Automation*, pages 2470–2476, Minnesota, April 1996.

[9] X. Markenscoff, L. Ni, and C.H. Papadimitriou. The geometry of grasping. *The Intl. J. of Robotics Research*, 9(1):61–74, 1990.

[10] V.-D. Nguyen. Constructing force-closure grasps. *The Intl. J. of Robotics Research*, 7(3), 1988.

[11] P.J. Sanz, A.P. del Pobil, J.M. Iñesta, and G. Recatalá. Vision-guided grasping of unknown objects for service robots. In *Proc. of the IEEE Intl. Conf. on Robotics and Automation*, pages 3018–3025, Leuven, Belgium, 1998.

[12] P.J. Sanz, J.M. Iñesta, and A.P. del Pobil. Planar grasping characterization based on curvature-symmetry fusion. *Applied Intelligence*, 10:25–36, 1999.

[13] P.J. Sanz, V.J. Traver, G. Recatalá, and A.P. del Pobil. Towards a reactive grasping system for an industrial robot arm. In *IEEE Intl. Symp. on Computational Intelligence in Robotics and Automation*, pages 1–6, Monterey, California, 1999.

[14] K. Stanley, J. Wu, A. Jerbi, and W.A. Gruver. A fast two dimensional image based grasp planner. In *Proc. of the IEEE/RSJ Intl. Conf. on Intelligent Robots and Systems*, pages 266–271, Kyongju, Korea, 1999.

[15] S.A. Stansfield. Robotic grasping of unknown objects: A knowledge-based approach. *The Intl. J. of Robotics Research*, 10(4):314–326, August 1991.

Feature-Guided Exploration with a Robotic Finger

Allison M. Okamura
Haptic Exploration Laboratory
Department of Mechanical Engineering
The Johns Hopkins University
aokamura@jhu.edu

Mark R. Cutkosky
Dexterous Manipulation Laboratory
Department of Mechanical Engineering
Stanford University
cutkosky@cdr.stanford.edu

Abstract

Haptic exploration with robotic fingers is accomplished by feature-guided exploration, where information about surface features such as cracks and ridges is used to guide the finger in an exploratory procedure. A local exploration strategy uses contact trajectory information from a tactile sensor to identify features while moving over and around them. Using an algorithm based on the Voronoi diagram, an approximation of the medial axis of the feature is found, then pruned using an edge length threshold. Multiple feature skeletons are then used to create a global skeleton that partitions the surface into regions. The models resulting from these local and global explorations can be used to characterize objects for information storage and manipulation planning.

Figure 1: A three degree-of-freedom robotic finger, equipped with tactile sensor, exploring a flat surface with a single ridge feature.

1 Introduction

As humans, we are able both to manipulate and explore objects in our environment using the sense of touch. Klatzky and Lederman[12] identified haptic exploratory procedures that allow us identify object properties, such as local and global shape, surface texture, and stiffness. These properties are useful not only in building an object model, but also for planning future manipulation.

An object can be explored by a dexterous robot hand by alternating manipulation and exploration, whereby all or a subset of the fingers are used to manipulate, then a subset are used to hold the object while other fingers explore[19]. When some fingers are grasping while others are free to roll and slide over the surface of the object, there are regions of the surface that can be explored by each free finger. In this paper, we will confine ourselves to the exploration of a single finger within such a region, as shown in Figure 1. Multiple regions that have been explored and modeled can later be connected to form a complete object model. The main problems of interest in this work are how to explore the features in a local region, and create a map of features on the object surface.

In this paper, we begin with a short review of related research. The next section describes the high-level procedure for haptic exploration, as well as goals for exploration and assumptions about the environment. Then, the control for exploration is developed, including tactile sensor data interpretation. A method for local exploration of features is presented. Next, shape skeletons are examined as a way to store feature information and provide a global mapping of features on a surface. Finally, global exploration strategies are compared for different feature types and spacings.

1.1 Previous Work

Although there have been several investigations of haptic exploration, active or feature-guided exploration in three dimensions has received relatively little attention in the literature. The relevant areas of research are frameworks for haptic exploration, and object modeling from data.

In the area of haptic exploration, a system for integrating vision and touch for object recognition tasks was developed by Allen[1, 2, 3]. Stansfield[23] also used a combination of vision and touch to create a robotic perceptual system. Hemami and collaborators[10] developed a conceptual framework for tactually guided exploration and shape perception. They

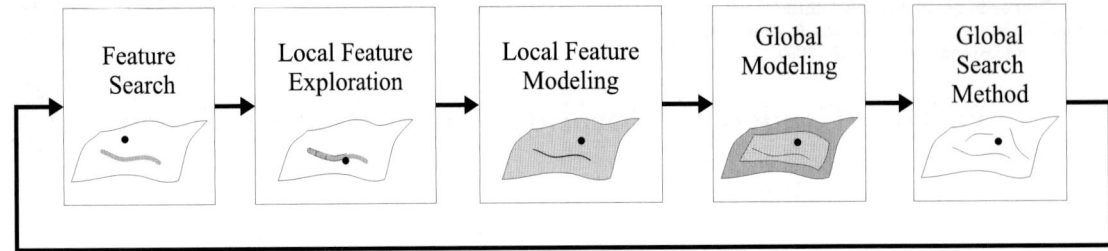

Figure 2: An overview of the phases for local and global haptic exploration. The black dot represents the contact point between the finger and the object.

have also investigated tactile servoing, hybrid control, and learning controllers for moving a robotic finger over globally unknown objects[8, 21]. The low-level hybrid control law used for moving a robotic finger over a surface in this paper is similar to that in Hemami's work.

In the area of object modeling from data, several researchers have investigated the development of modeling based on cloud sets of data[13, 22]. Most of the work in this area has focused on solid or surface models rather than local surface deformations. Shape skeletons have also been used to model objects from data points or pixels. Skeletons, known in 2D as the *medial axis* and in 3D as the *medial surface*, are geometric abstractions of curves, surfaces, or solids that are useful as lower-dimensional representations of objects. Many researchers have developed algorithms for creating the medial axis[4, 6, 15, 24, 25, 26].

2 A Procedure for Feature-Guided Exploration

This section describes a high-level procedure for feature-guided haptic exploration. This is termed "feature-guided" exploration because the discovery of a feature leads to changes in the fingertip path. Figure 2 shows the progression of exploration using several phases. A feature may be discovered on the surface of an object during a manipulation task, a random walk of the fingertip over the surface, or a specific multi-fingered exploratory procedure. For the purposes of this work, let us assume that one of these three finger motions results in the finger encountering a feature while rolling and sliding over the surface. This fingertip motion corresponds to the "feature search" phase of exploration as shown in Figure 2.

The first step in a feature encounter is identification of the feature type. A basic algorithm for feature detection was provided by the authors[18]. In this paper, that work is expanded to include active exploration of three-dimensional features. In order to identify feature type, the finger must go through a local exploratory procedure that will extract feature properties for type identification. Then the feature can be further explored by tracing to find its boundary. We assume in this work that the features are sparsely distributed so that, once a feature has been detected, the best way to obtain further information about the object is to continue exploring the current feature until it can be completely modeled. The procedures for determining feature type and boundary tracing correspond to the "local feature exploration" phase in Figure 2.

The next two blocks in Figure 2 refer to modeling steps. For each feature, the modeling process involves the development of a shape skeleton, or 2D medial axis, created in a coordinate system fixed on the surface of the object. This is known as a *feature skeleton*. Using these feature skeletons, a *global skeleton* can be used to model the layout of multiple features on a surface. Finally, as shown in the last block of Figure 2, a global exploration strategy may be developed by observing the spatial distributions of features that have already been explored. Then the finger resumes the search for a new feature.

The goals of active exploration in this work are to identify each feature type and obtain data that can be used to create a skeleton of its shape. In this high-level exploration strategy, we are considering a situation where a robotic finger has sufficient degrees of freedom to explore a surface actively in three dimensions. Due to the hardware used in the experiments for local feature exploration, we require two simplifying assumptions. The first is that the closest points on any two features are at least one fingertip diameter apart. This is so that the finger can trace the boundary of one feature without being distracted by other feature contacts. The second is that each surface feature is one of the basic types shown in Table 1. In general, it is possible to identify and trace any macro feature type, however, we simplify our approach by considering only a few examples. Compound features, such as a ridge that suddenly dips down and becomes a ravine, would be especially difficult to trace with only contact point sensing.

Feature Name	3D picture	2D picture
Convex "Cusp"		$k_1 = +$, $k_2 = +$
Step		$k_1 = +$, $k_2 = \varepsilon$; $k_1 = -$, $k_2 = \varepsilon$
Bump		$k_1 = -$, $k_2 = \varepsilon$; $k_1 = +$, $k_2 = +$

Table 1: A partial list of possible macro features. In the 2D picture (a plan view of the surface), the gray regions indicate curvature features, with the necessary principal curvatures labeled. In this table, + indicates a positive curvature feature, - indicates a negative curvature feature, and ε is no curvature feature ($|k_i| < \frac{1}{r_f}$), where k_1 and k_2 are the principal curvatures of the surface and r_f is the radius of the fingertip. It is assumed that $|k_1| \geq |k_2|$. The white regions between the curvature feature regions have a maximum width of r_f.

3 Local Exploration

Features may be detected when a finger's path moves the contact point over the feature, although complete feature type and shape identification may require further exploration. Thus, the robotic finger's trajectory is modified in order to collect additional data about the feature. This involves issues in control and tactile sensor data interpretation, and uses feature definitions developed by the authors[18]. In these definitions, features are identified as surface regions with curvature higher than that of the fingertip.

3.1 Hardware

Surface exploration requires a three-degree-of-freedom robotic finger with a tactile sensor. The tactile sensor can be any shape, but a spherical one is most useful for feature tracing[18]. The robotic finger used was the 3GM haptic interface from Immersion Corporation; the software for controlling it was easily modified to make it into a robotic finger. Because the 3GM is backdrivable and lightweight, open-loop force control and gravity compensation can be used in the device control. Details can be found in Goldenberg, et al.[9].

The tactile sensor used was the Optical Waveguide Tactile Sensor from Maekawa, et al.[14]. This tactile sensor uses an analog position-sensitive device (PSD) to measure the reflection of light from the point(s) of contact. It can report contact centroid and contact intensity, which is related to the amount of light reflected. Because the sensor is analog and fast, data can be obtained at the 1kHz rate used for controlling the finger. Figure 1 illustrates this robotic finger and tactile sensor exploring a flat surface with a ridge feature.

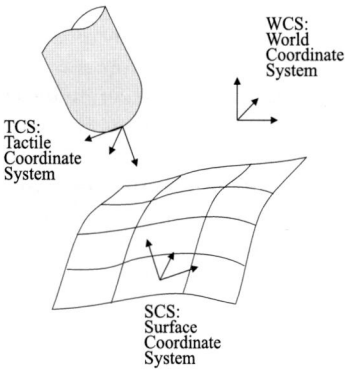

Figure 3: Three coordinate systems are used in local haptic exploration.

3.2 Control

If we are to use the feature definitions presented in [17, 18] for identification, a useful control law is one that causes the finger to travel parallel to the surface while maintaining contact. Thus, proportional-derivative (PD) control is used parallel to the surface and the normal force is controlled perpendicular to the surface. Tactile sensor feedback is used to determine the current surface normal and provide information for feature detection algorithms. Several coordinate systems must be considered in the control for surface tracking, as shown in Figure 3.

The combination of PD and normal force control allows the finger to slide over the surface and travel over features while following the contours of the object. The normal force is light enough so that friction does not impede the motion of the fingertip, but strong enough to allow the tactile sensor to measure the contact point. (This requires a minimal amount of

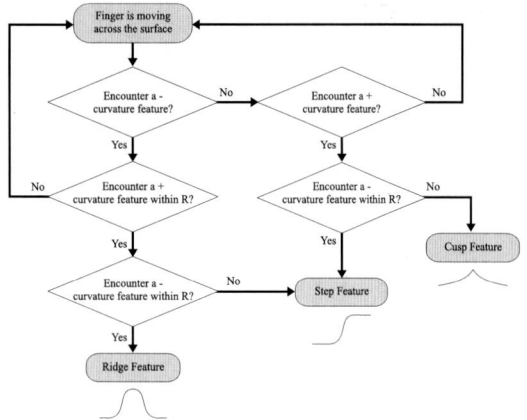

Figure 4: Flow chart for feature type identification. The possible features in this example are a cusp, a step, and a ridge.

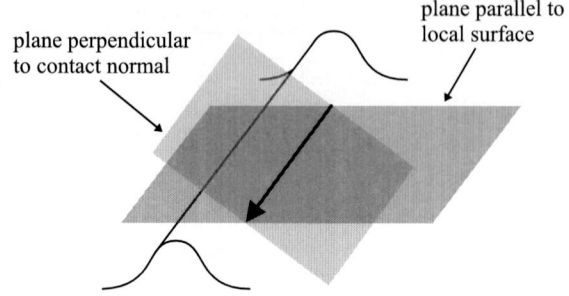

Figure 5: The direction of contact point travel during feature tracing is determined by the intersection of a plane parallel to the local surface and a plane orthogonal to the contact normal.

pressure.) Because friction during sliding caused erroneous readings from the tactile sensor[17], graphite was spread on the surface to lower the coefficient of friction so that higher normal forces could be applied. Tactile sensor data was used to determine the direction of the surface normal, so that as the surface curved, the normal force applied by the fingertip also changed direction. PD control caused the finger to move parallel to the surface with a particular velocity. (Desired positions are selected using the current position and desired velocity.) Gravity compensation was also applied.

3.3 Feature Type Detection

Let us assume that a finger has encountered a feature and detection has been performed as described in [18], so that curvature features are recorded. It is important to note that the direction of finger travel during a search may not correspond to the first principal direction of a feature, so the finger can actually travel over a feature without detecting it. Thus, it may be desired to perform type detection even if it is uncertain whether the curvature is sharp enough to warrant designation as a feature.

A flow chart can be used to describe a type detection algorithm, where the branching direction depends on previous discoveries. Because it is possible to define an infinite number of macro features, we will restrict the possible feature types for simplicity. In this example, let us consider three feature types: a convex cusp, a step, and and a ridge. Figure 4 shows an example of a feature type identification flow chart using these feature types. (Similar algorithms have been developed for other feature types[17].)

3.4 Feature Tracing

After the feature type has been identified, the geometry of the feature is further examined by tracing the feature boundary. Tracing is accomplished by moving in the plane orthogonal to the contact normal when the finger is on a negative curvature feature. With only contact location sensing, it is difficult to decide which direction to travel on this plane. However, when the nominal surface is gently curving, one can also include the constraint of staying in the plane of travel before the feature was encountered. The intersection of these two planes creates a path through space that also traces along a negative curvature feature, as shown in Figure 5. When tracing a feature, these planes should not be close to parallel, thus, the path of travel will always be defined. The direction along this path can be chosen arbitrarily.

Figure 6 shows the contact point data from a finger that has partially explored a feature and is beginning the tracing step. While performing this tracing, it is possible for the finger to lose contact with the feature. At intervals, the finger can re-approach the feature and find the curvature feature regions again. During the tracing, it is important to keep track of locations that represent the boundary of the feature. When the finger moves away from the feature in order to relocate a curvature feature, the extra contact data should be excluded from the set used for feature modeling. In this work, non-boundary data taken during a tracing phase was removed manually. However, this removal can be performed automatically by noting the increase in impedance force as the finger travels over a feature and allowing time for overshoot before the finger returns to tracing the feature.

4 Local and Global Feature Modeling

Using feature data points obtained by tracing, the skeleton, or medial axis, of the feature can be identified

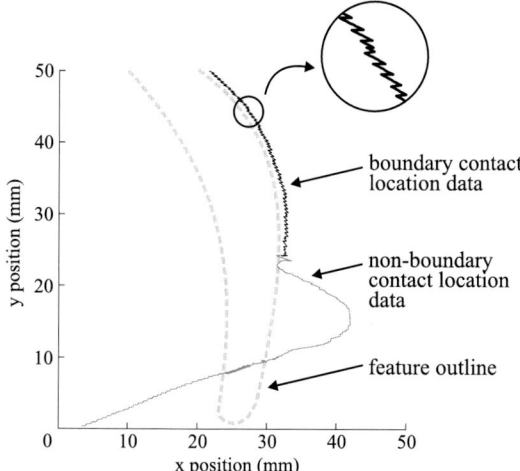

Figure 6: Contact location data was obtained using the system in Figure 1 as the finger approaches the feature, travels over it (overshooting), and begins tracing around the boundary, with a velocity command of 10 mm/s. Data that are not part of the feature boundary should be excluded from the feature modeling process.

for macro features. This skeleton is useful for modeling and object identification purposes, as well as for creating global maps of features and feature regions.

4.1 Feature Modeling with Shape Skeletons

The medial axis is defined as the loci of centers of locally maximal balls (in 3D) or disks (in 2D) inside an object. Each point on the medial axis, also known as the shape skeleton, is associated with the radius of a locally maximal ball or disk. These medial axis points, together with their associated radii, define the Medial Axis Transform (MAT) of an object. In this work, the medial axis is used to represent the locations and shapes of features and is the most important part of the transform. If the features are to be reconstructed, the entire MAT is needed. Only the 2D MAT will be considered here, as it is sufficient for describing the shapes and locations of features on a surface.

There are two major skeletonization methods that can be used to extract the medial axis transform of feature data, taken from medial axis algorithms for 2D image data[5]. The first is region-based, where the input is usually an array of filled image data. In a "thinning" approach, pixels of images are iteratively thinned, or equivalently, redundant pixels are successively deleted until a final skeleton is derived. The second method is based on the boundaries of images. The boundaries are extracted from the image by edge detection and skeletons are generated directly from the boundary data.

Because the feature data is obtained by tracing the boundary of the feature, the data that must be used in the MAT is the feature boundary. This is advantageous because boundary-based approaches are known to be more stable and less problematic than thinning methods[7]. Much of the overhead of the boundary approach is in the detection of a boundary of the image and removal of excessive small branches due to noisy data[11]. Because the portion of local exploration which is identified as a "tracing" phase is well-defined, the data taken during that phase can be separated from the rest of the tactile or position data. Thus, the boundary is already calculated, although we must still deal with the issue of noisy data, which is certainly the case as shown in Figure 6. Downsampling is also be used to speed up the skeletonization process, because a finger moving slowly with respect to the data collection rate obtains more contact points than needed.

Brandt and Algazi[5] and Ogniewicz[16] have used approaches in which boundary points are used to find Voronoi diagrams in order to generate a discrete medial axis. The algorithm begins with the creation of a Voronoi diagram from the boundary data points. The Voronoi diagram is the partitioning of a plane with n points into n convex polygons, where each polygon contains exactly one data point, and every point in a polygon is closer to this point than to any other data point. Ogniewicz calls the Voronoi diagram of a discrete set of boundary points the *discrete Voronoi medial axis (DVMA)*. Brandt showed formally that the DVMA approaches the continuous Voronoi diagram as the number of boundary samples increases (Figure 7).

Once the DVMA has been found, the skeleton must still be extracted by pruning away undesirable Voronoi polygon edges. There are many pruning algorithms that can be used for this purpose. Brandt[5] presents a method where polygon segments are deleted based on the absolute regeneration error and an empirically-determined threshold. For feature detection based on tactile and position data, however, there is a simpler pruning algorithm based on the length of the Voronoi polygon edges. Given the velocity of the robotic fingertip performing the exploration and the rate at which data is sampled, there will be a minimum distance between samples at the boundary of the feature. Thus, a length threshold can be used for removing undesirable polygon edges. The results of pruning using this method are shown in Figure 7. One can see that where there are data points that are particularly noisy and far away from the main feature shape, unwanted skeleton arcs appear. These could also be pruned away, provided that the features under consideration do not branch. Pruning is accomplished by removing the skeleton arcs which are smaller than a threshold length, defined using the noise present in the data. Some researchers[20] use the medial axis itself to determine if an object "branches," so the choice of threshold will depend on the application.

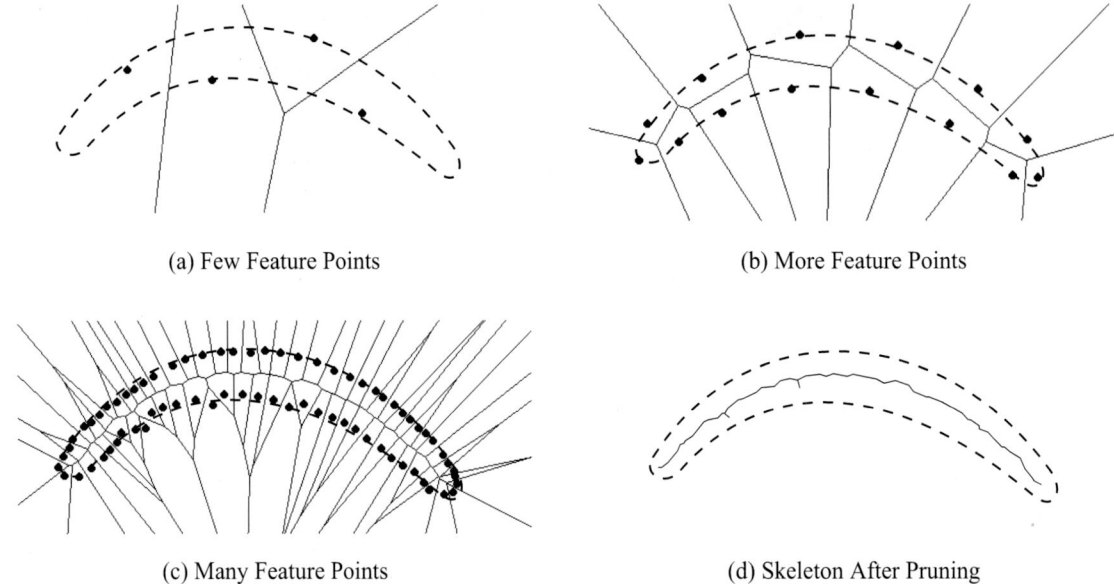

(a) Few Feature Points

(b) More Feature Points

(c) Many Feature Points

(d) Skeleton After Pruning

Figure 7: As the number of boundary samples increases, the discrete Voronoi medial axis (DVMA) approaches the continuous Voronoi diagram. The DVMA is then pruned to get get a more accurate object skeleton. In this example, the boundary samples are noisy and cause unwanted arcs in the final skeleton. These can be removed using a length threshold.

This basic algorithm results in a single medial axis for features that are not significantly curved on the surface. The threshold can be chosen to make the skeleton inside of the boundary samples completely connected for a given sample spacing. However, if the feature curves sharply on the surface, the algorithm will also find part of the skeleton outside the region of the feature. In this case, the connected skeleton with the most edges is taken as the medial axis[17]. With these feature skeletons created, we can now consider methods for creating a global skeleton.

4.2 Global Modeling with Shape Skeletons

After a local feature exploration has been completed, tracing data may be used to start building a global map of features. As additional features are discovered and explored, a *global skeleton* may be developed to partition the surface of the object into regions as shown in Figure 8. This global skeleton is similar to the concept of a *negative medial axis*, which is the medial axis formed from data points consisting of the previously defined feature medial axes.

Given the medial axis for each feature, the negative MAT can be calculated by taking the locally maximal balls between the medial axes. This amounts to taking the MAT of the feature medial axes. As shown in the previous section, algorithms based on the Voronoi diagram of sample data are particularly straightforward for boundary data. Thus, we will consider another simple algorithm based on the Voronoi diagram to de-

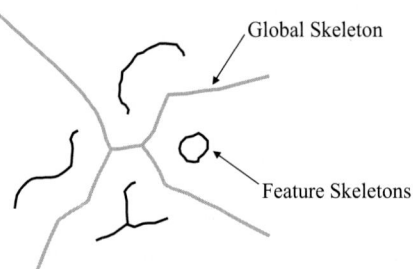

Figure 8: The global skeleton creates regions around features.

termine a simple approximation to the negative MAT of the features. The basic steps in this algorithm are shown in Figure 9.

The algorithm for obtaining the global medial axis begins with the points on the DVMAs of the features. The Voronoi diagram of these medial axes is constructed. Similar to the method for determining the feature medial axes, a number of pruning algorithms can be used to extract the skeleton from the Voronoi diagram. However, the nature of the contact and position data obtained during exploration allows for a new, simple extraction method.

One advantageous property of the local feature exploration algorithm is that features are naturally segmented during the exploration process. When the finger encounters a feature, that feature is explored fully and all the data points observed during this exploration can be attached to a structure for that feature.

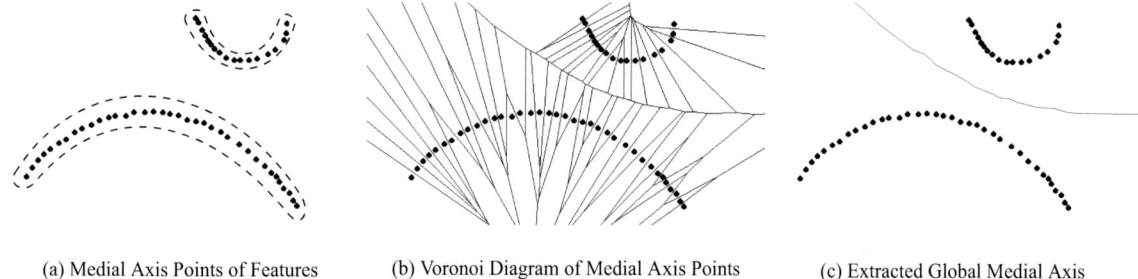

(a) Medial Axis Points of Features (b) Voronoi Diagram of Medial Axis Points (c) Extracted Global Medial Axis

Figure 9: The global skeleton is approximated using the Voronoi diagram of the medial axes of the features. It is extracted from the Voronoi diagram by considering only the polygon edges which correspond to medial axis points from two different features.

When a different feature is explored, the new data points are associated with the new feature. Therefore, each boundary sample is associated with a single feature. This property is used in the extraction of the global medial axis from the Voronoi diagram by considering only the Voronoi polygon edges which are shared by two sample points associated with different features. By using only these polygon edges, only segments which are equidistant from two different features are considered.

This technique also eliminates some medial axis arcs that might be created using the traditional negative MAT calculation. A true negative MAT calculation would include points that are equidistant from multiple points on a single feature. While this information may be useful for some applications, such as navigation, it would distort the description of feature spacing or the division of a surface into discrete regions surrounding different features. By using Voronoi polygon edges that are only shared between points on different features, we obtain a simpler extraction algorithm and a useful global medial axis.

5 Global Exploration

Planning using the global medial axis is left as future work. However, there are several key planning applications that will be described here. One important use of this partitioning is to evaluate the spacing of the features. The minimum radius of the disks along the medial axis represents the smallest clearance, and therefore the minimum spacing, among objects. Given different feature spacings, various global exploration or manipulation strategies may be invoked. In addition, the global medial axis may be such that it bounds regions on the surface. Larger regions that do not closely bound a feature (to within a fingertip diameter) may need to be explored further to ensure that no features were missed.

If the goal is to avoid features, the trajectory of the global skeleton depicts the safest navigation path that is the farthest from all features. Thus, the relationship between the size of the finger and the radius for each locally maximal disk tells us whether the finger can navigate without hitting any features.

6 Conclusion

This work addressed issues in the planning and execution of local and global exploratory procedures for finding features on a surface in 3D. Using the feature definitions presented in earlier work and local exploratory procedures, features can be identified by type and geometry. Feature type is identified by reactive local exploratory procedures, and feature geometry is recorded by tracing around or along the length of the feature.

The control for moving the finger over the surface at a constant velocity and normal force is presented. This control can be used both during the search for a feature and during feature tracing. Using tactile sensor data, the surface normal is continuously updated, and the motion and normal force trajectories are modified. During feature tracing, the contact location and nominal surface orientation are used to determine the direction of motion when only contact location is available for feedback.

Data obtained from this exploration may be used to model individual features and the layout of multiple features on a surface. Various methods for global exploration have also been discussed, their use depending on the goal of the exploration and spatial feature distribution.

The are many intriguing avenues for future work related to haptic exploration. Possibilities include planning methods for global exploration, improved robot finger and tactile sensor design specifically for the purpose of exploration, and reality-based modeling systems that use data obtained during exploration to create realistic virtual environments.

Acknowledgments

The authors acknowledge Dr. Hitoshi Maekawa for the MEL optical waveguide tactile sensor, and Dr. Bernard Roth for his review of this research. This work was supported by NSF Dissertation Enhancement Award 9724763 and Army STRICOM Grant M67004-97-C-0026 through Immersion Corporation.

References

[1] P. K. Allen. Mapping haptic exploratory procedures to multiple shape representations. *Proc. of the IEEE Int'l. Conf. on Robotics and Automation*, pages 1679–1684, 1990.

[2] P. K. Allen and P. Michelman. Acquisition and interpretation of 3-d sensor data from touch. *IEEE Trans. on Robotics and Automation*, 6(4):397–404, 1990.

[3] P. K. Allen, A. T. Miller, P. Y. Oh, and B. S. Leibowitz. Using tactile and visual sensing with a robotic hand. *Proc. of the IEEE Int'l. Conf. on Robotics and Automation*, pages 676–681, 1997.

[4] H. Blum. A transformation for extracting new descriptors of shape. *Models for Perception of Speech and Vsiual Form*, pages 362–380, 1967.

[5] J. W. Brandt and V. R. Algazi. Continuous skeleton computation by voronoi diagram. *CVGIP: Image Understanding*, 55(3):329–338, 1992.

[6] C.-S. Chiang. Algorithms for extracting the medial axis transform of 2d images. *Proc. of SPIE - The Int'l. Society for Optical Engineering, Visual Communications and Image Processing*, 2501(2):1173–1182, 1995.

[7] E. R. Davies and A. P. N. Plummer. Thinning algorithms: a critique and a new methodology. *Pattern Recognition*, 14(1-6):53–63, 1981.

[8] R. E. Goddard, Y. F. Zheng, and H. Hemami. Dynamic hybrid velocity/force control of robot compliant motion over globally unknown objects. *IEEE Trans. on Robotics and Automation*, 8(1):132–138, 1992.

[9] A. S. Goldenberg, E. F. Wies, K. Martin, and C. J. Hasser. Next-generation 3d haptic feedback system. Poster, American Society of Mechanical Engineers Haptics Symp., November 1998.

[10] H. Hemami, J. S. Bay, and R. E. Goddard. A conceptual framework of tactually guided exploration and shape perception. *IEEE Trans. on Biomedical Engineering*, 35(2):99–109, 1988.

[11] J.-H. Kao. *Process Planning For Additive/Subtractive Solid Freeform Fabrication Using Medial Axis Transform*. PhD thesis, Stanford University, Department of Mechanical Engineering, 1999.

[12] R. L. Klatzky and S. Lederman. Intelligent exploration by the human hand. In S. T. Venkataraman and T. Iberall, editors, *Dextrous Robot Hands*, pages 66–81. Springer-Verlag, 1990.

[13] C. T. Lim, G. M. Turkiyyah, M. A. Ganter, and D. W. Storti. Implicit reconstruction of solids from cloud point sets. *ACM SIGGRAPH Symp. on Solid Modeling and Applications*, pages 393–402, 1995.

[14] H. Maekawa, K. Tanie, and K. Komoriya. A finger-shaped tactile sensor using an optical waveguide. *Proc. of the IEEE Int'l. Conf. on Systems, Man and Cybernetics*, pages 403–408, 1993.

[15] C. W. Niblack, D. W. Capson, and P. B. Gibbons. Generating skeletons and centerlines from the medial axis transform. *Proc. of the Int'l. Conf. on Pattern Recognition*, 1:881–885, 1990.

[16] R. L. Ogniewicz. Skeleton-space: a multiscale shape description combining region and boundary information. *Proc. of IEEE Comp. Society Conf. on Computer Vision and Pattern Recognition*, pages 746–751, 1994.

[17] A. M. Okamura. *Haptic Exploration of Unknown Objects*. PhD thesis, Stanford University, Department of Mechanical Engineering, 2000.

[18] A. M. Okamura and M. R. Cutkosky. Haptic exploration of fine surface features. *Proc. of the IEEE Int'l. Conf. on Robotics and Automation*, 3:2930–2936, 1999.

[19] A. M. Okamura, M. L. Turner, and M. R. Cutkosky. Haptic exploration of objects with rolling and sliding. *Proc. of the IEEE Int'l. Conf. on Robotics and Automation*, 3:2485–2490, 1997.

[20] N. M. Patrikalakis and H. N. Gursoy. Shape interrogation by medial axis transform. *Proc. of the ASME Design Technical Conf., Design Engineering Division*, 3(1):77–88, 1990.

[21] K. Pribadi, J. S. Bay, and H. Hemami. Exploration and dynamic shape estimation by a robotic probe. *IEEE Trans. on Systems, Man, and Cybernetics*, 19(4):840–846, 1989.

[22] F. Solina and R. Bajcsy. Recovery of parametric models from range images: the case for superquadrics with global deformations. *IEEE Trans. on Pattern Analysis and Machine Intelligence*, 12(2):131–147, 1990.

[23] S. Stansfield. A robotic perceptual system utilizing passive vision and active touch. *Int'l. Journal of Robotics Research*, 7(6):138–161, 1988.

[24] D. W. Storti, G. M. Turkiyyah, M. A. Ganter, C. T. Lim, and D. M. Stal. Skeleton-based modeling operations on solids. *Proc. of the Symp. on Solid Modeling and Applications*, pages 141–154, 1997.

[25] A. Sudhalkar, L. Gursoz, and F. Prinz. Continuous skeletons of discrete objects. *Proc. of the Symp. on Solid Modeling and Applications*, pages 85–94, 1993.

[26] G. M. Turkiyyah, D. W. Storti, M. Ganter, H. Chem, and M. Vimawala. An accelerated triangulation method for computing the skeeltons of free-form solid models. *Computer Aided Design*, pages 5–19, 1997.

Identification of Contact Conditions from Contaminated Data of Contact Force and Moment

Tetsuya MOURI [*], Takayoshi YAMADA [**], Ayako IWAI [**],
Nobuharu MIMURA [***], and Yasuyuki FUNAHASHI [**]

[*] Virtual System Laboratory, Gifu University, Yanagido 1-1, Gifu 501-1193, Japan
[**] Department of Mechanical Engineering, Nagoya Institute of Technology,
Gokiso-cho, Showa-ku, Nagoya 466-8555, Japan
[***] Department of Biocybernetics, Niigata University, Ninomachi, Ikarashi, Niigata 950-2181, Japan
tmouri@vsl.gifu-u.ac.jp, yamat@eine.mech.nitech.ac.jp

Abstract

This paper discusses a method for identification of contact conditions from the information of 6-axes force sensor equipped with a robot hand. The previous paper has the following problems. (1) The noise of sensing force was not considered. (2) Hence, the estimates obtained by the previous method are biased from true value, and the identification of contact types depends on an unknown contact position. This paper thinks over the noise of force. We propose a method of removing the bias from the estimates. Hence, it is guaranteed that asymptotically unbiased estimates are obtained. The contact types can be judged by eigenvalues of a covariance matrix, which is derived from estimates of contact moment. The effectiveness of the algorithm is demonstrated by simulations.

1. Introduction

When a robot manipulates an object which is in contact with external environment and performs assembly tasks, it is required to recognize contact conditions. The contact conditions mean contact type, contact position, and contact force. For example, let us imagine the simple task such that a box is set on the table by a robot manipulator. If the manipulator is equipped only with position control, it cannot judge which contact type happens, and cannot accomplish the task. On the other hand, human can recognize the current contact type by only using a sense of forces and accomplish the task easily. Therefore, it is necessary to identify and control the contact conditions in order to endue a robot with the skill of human.

So far, there are various researches about contact conditions using force information. To our knowledge, Akella et al. [1] and Hyde et al. [5] proposed the method of controlling contact transition from free motion to constrained motion. Dütre et al. [3] identified geometrical uncertainties and detected topological transitions in contact situations. Tsujimura et al. [16], Kaneko et al. [6] and Ueno et al. [17] detected contact position by a probe, Active Antenna and Multiple Active Antenna, respectively. Bicchi et al. [2] identified contact position and contact normal direction in case of point contact type and soft finger one. Zhou et al. [19] inferred accuracy of the proposed approach within the limit that the contact type was point one. Salisbury et al. [14] proposed a method for identification of contact position on the surface of sensor using multiple active sensing. Nagata et al. [13] and Kitagaki et al. [7] proposed methods of identifying the contact position between unknown object and environment using active sensing. In these methods, however, it was assumed that the contact condition is point contact type with friction. Oussalah [12] proposed a method of identifying the contact normal by measured force and velocity without friction.

The authors [9][10][11][18] discussed identification of contact conditions including soft finger, line and plane contact type in addition to the point one. Ref. [9] showed the possibility of identifying the contact conditions between unknown grasped object and external environment. Ref. [10] expressed constraints on contact moment in an explicit form and reduced the identification of contact conditions into a linear problem. Ref. [18] treated the case where moment data were contaminated with noise. Ref. [11] introduced standard deviation of contact moment in order to express the restriction of contact moment. A method of identifying contact conditions was statistically provided by using the eigenvalues of covariance matrix of estimated contact moment. However, Ref. [11] did not consider the noise of force.

This paper treats the case when both force data and moment one are contaminated with noise. If we try to use the method of Ref. [11], the bias of estimates occurs and the identification of contact types depends on contact position. We provide a method of removing the bias and identifying contact conditions, where the covariance of noise is known from the sensor calibration beforehand. An algorithm for identification of contact conditions is proposed. The effectiveness of the algorithm is

2. Problem Formulation

In this paper, the system shown in Fig. 1 is considered. Unknown object is grasped by a robot hand and is made in contact with external environment.

2.1. Symbols

We define the following symbols, as shown in Fig. 1.

- o : origin of a force sensor equipped with the hand,
- c : contact point between the grasped object and external environment,
- Σ_o : sensor coordinate frame fixed at o,
- Σ_c : contact coordinate frame fixed at c,
- f_o, n_o : measured force and moment in Σ_o,
- f_c, n_c : contact force and moment in Σ_c,
- r_c : position vector of c in Σ_o,

where Σ_c is the same orientation as Σ_o.

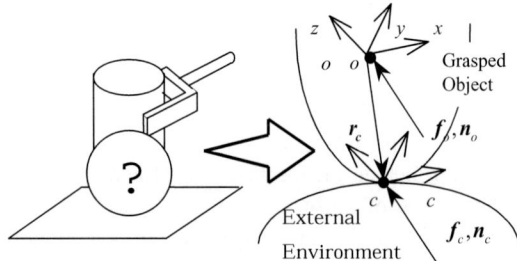

Fig. 1: Interaction between grasped object and external environment

2.2. Assumptions

We make the following assumptions for clarification of discussions.

- (A1) The object is firmly grasped by the robot hand and can be manipulated arbitrarily by the robot manipulator.
- (A2) The grasped object is in contact with external environment through point, soft finger, line, or plane contact type with friction. However the contact position and types are unknown.
- (A3) The contact position and type remain unchanged during identification process. And no slip occurs at the contact position c.
- (A4) The force f_o, the moment n_o, and the position o are measurable and controllable. The force f_c and the moment n_c are immeasurable.
- (A5) The measured force f_o and the moment n_o are contaminated with noise ε_f and ε_n, respectively. The noise ε_f and ε_n are independent of each other and occur randomly.
- (A6) The contact force f_c and the moment n_c occur randomly by active sensing.
- (A7) The number of times of active sensing motion is sufficient large, and our analysis is based on ergodic hypothesis.

Assumption (A2) is illustrated in Fig. 2. Contact types are classified by DOF of contact [8]. It is denoted by m.

- $m = 0$: Plane contact type with friction
- $m = 1$: Line contact type with friction
- $m = 2$: Soft finger contact type with friction
- $m = 3$: Point contact type with friction

The soft finger contact type does not exist between rigid objects.

From Assumption (A5), the covariance matrices of noise are given by

$$\mathrm{Cov}[\varepsilon_f] = \mathrm{diag}[\gamma_{fx}^2, \gamma_{fy}^2, \gamma_{fz}^2] =: \Gamma_f,$$
$$\mathrm{Cov}[\varepsilon_n] = \mathrm{diag}[\gamma_{nx}^2, \gamma_{ny}^2, \gamma_{nz}^2] =: \Gamma_n, \quad (1)$$

where $\mathrm{diag}[\bullet]$ implies a diagonal matrix, and γ means standard deviation of noise. Γ_f and Γ_n are known by the calibration of force sensor beforehand. Under these assumptions, the problem of identifying contact conditions (type, position, and force) is investigated.

The essential difference of this paper from Ref. [11] is consideration of the noise ε_f. In Section 3.3 and 3.4, we will make it clear that the estimates of contact conditions are biased, if the method of Ref. [11] is simply used. And we propose a revised method.

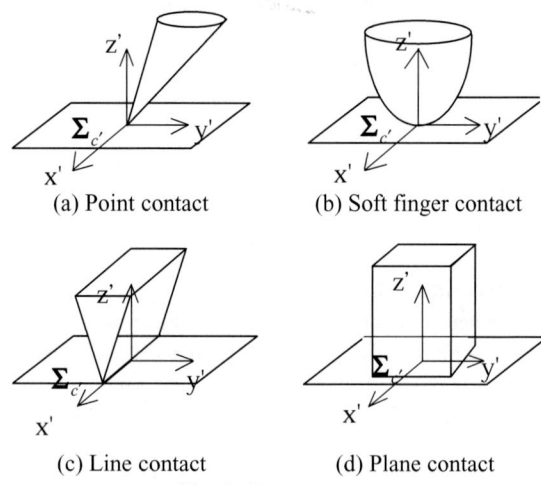

(a) Point contact (b) Soft finger contact

(c) Line contact (d) Plane contact

Fig. 2: Contact types

2.3. Constraint of Contact Moment

According to Ref. [11], we consider some appropriate frame $\Sigma_{c'}$ as shown in Fig. 2. The origin of $\Sigma_{c'}$ is fixed at c, and the direction of z' axis is normal to the contact plane. In case of line contact type, moreover, the contact line should be taken as x' axis.

Contact moment with respect to $\Sigma_{c'}$ is denoted by $n'_c = [n'_{cx}, n'_{cy}, n'_{cz}]^T$. We define the covariance matrix of n'_c.

$$\mathrm{Cov}[n'_c] = \mathrm{diag}[\sigma_{nx}^2, \sigma_{ny}^2, \sigma_{nz}^2] =: V, \quad (2)$$

where σ_{nj} ($j=x,y,z$) denotes a standard deviation of components of contact moment n'_c. From Assumption (A7), each contact type is characterized by the standard deviation σ_{nj}.

$$\left.\begin{array}{ll} \text{Point contact} & : \sigma_{nx}=\sigma_{ny}=\sigma_{nz}=0 \\ \text{Soft finger contact} & : \sigma_{nx}=\sigma_{ny}=0 \\ \text{Line contact} & : \sigma_{nx}=0 \\ \text{Plane contact} & : \text{no constraint} \end{array}\right\} \quad (3)$$

n_c is expressed by
$$n_c = R n'_c, \quad (4)$$
where
$$R := [r_x, r_y, r_z] \in \Re^{3\times 3}$$
is orientation of $\Sigma_{c'}$ with respect to Σ_c.

3. Identification of Contact conditions

We will identify contact conditions such as contact type, position, and force from the measured data contaminated with noise.

3.1. Force and Moment Equilibrium

The equilibrium equation of force and moment is represented by
$$r_c \times f_o + n_c = n_o. \quad (5)$$

In Eq. (5), the measurable parameters are f_o and n_o, while immeasurable ones are n_c and r_c from Assumption (A4). In order to clarify expression of the equation, we use the following notations.
$$a := f_o, \quad A := -[a\times], \quad b := n_o, \quad x := r_c. \quad (6)$$

From k times of active sensing motion, we have
$$b_i^* = A_i^* x^* + n_{ci}^*, \quad i=1,2,\cdots,k, \quad (7)$$

where the subscript "i" implies the i-th value, $\bullet^*$ means the true value of $\bullet$, and x remains unchanged by active sensing. From Assumption (A5), the measured force a_i and moment b_i are expressed by the following equations.
$$A_i = A_i^* + \mathrm{E}_{fi}, \quad b_i = b_i^* + \varepsilon_{ni}, \quad (8)$$

where $\mathrm{E}_{fi} := -[\varepsilon_{fi}\times]$. Substituting Eq. (8) into (7) yields
$$b_i = A_i x^* + n_{ci}^* + \varepsilon_{ni} - \mathrm{E}_{fi} x^*. \quad (9)$$

Note that the force data are contaminated with noise, and Eq. (9) includes the unknown term "$\mathrm{E}_{fi} x^*$". In the Section 3.3 and 3.4, we will find an effect of the term on estimates.

3.2 Generation of Deviated Data

We generate deviated force and moment, whose averages are equal to zero, because four contact types shown in Fig.2 are treated even when the averages of measured force and moment are not zero.

Deviated force and moment from the average are given by
$$\overline{a}_i := a_i - \frac{1}{k}\sum_{j=1}^{k} a_j, \quad \overline{b}_i := b_i - \frac{1}{k}\sum_{j=1}^{k} b_j,$$

$$\overline{n}_{ci} := n_{ci} - \frac{1}{k}\sum_{j=1}^{k} n_{cj}, \quad \overline{\varepsilon}_{fi} := \varepsilon_{fi} - \frac{1}{k}\sum_{j=1}^{k}\varepsilon_{fj},$$

$$\overline{\mathrm{E}}_{fi} := \mathrm{E}_{fi} - \frac{1}{k}\sum_{j=1}^{k}\mathrm{E}_{fj}, \quad \overline{\varepsilon}_{ni} := \varepsilon_{ni} - \frac{1}{k}\sum_{j=1}^{k}\varepsilon_{nj}. \quad (10)$$

From Eq. (10), Eq. (9) is rewritten by
$$\overline{b}_i = \overline{A}_i x^* + \overline{n}_{ci}^* + \overline{\varepsilon}_{ni} - \overline{\mathrm{E}}_{fi} x^*. \quad (11)$$

According to Ref. [11], contact moment with noise is represented by
$$\overline{v}_i := \overline{n}_{ci}^* + \overline{\varepsilon}_{ni} - \overline{\mathrm{E}}_{fi} x^*. \quad (12)$$

Eq. (11) becomes
$$\overline{b}_i = \overline{A}_i x^* + \overline{v}_i, \quad i=1,2,\cdots,k. \quad (13)$$

3.3. Identification of Contact Position

In this subsection we use the method of Ref. [11] for identification of contact conditions. Since an average of component of $\overline{v}_i$ is equal to zero, we identify x by minimizing a performance index
$$J = \frac{1}{k}\sum_{i=1}^{k}\left\|\overline{A}_i x - \overline{b}_i\right\|^2. \quad (14)$$

The estimate is given as
$$\left(\frac{1}{k}\sum_{i=1}^{k}\overline{A}_i^T \overline{A}_i\right)\hat{x}(k) = \left(\frac{1}{k}\sum_{i=1}^{k}\overline{A}_i^T \overline{b}_i\right). \quad (15)$$

Substituting Eqs. (8), (9) and (11) into (15) yields
$$\left(\frac{1}{k}\sum_{i=1}^{k}(\overline{A}_i^* + \overline{\mathrm{E}}_{fi})^T(\overline{A}_i^* + \overline{\mathrm{E}}_{fi})\right)\hat{x}(k)$$
$$= \left(\frac{1}{k}\sum_{i=1}^{k}(\overline{A}_i^* + \overline{\mathrm{E}}_{fi})^T(\overline{A}_i^* x^* + \overline{n}_{ci}^* + \overline{\varepsilon}_{ni})\right) \quad (16)$$

Here, from Assumptions (A5)~(A7), the estimate of contact position, $\hat{x}(\infty)$, is
$$\hat{x}(\infty) = \lim_{k\to\infty}\hat{x}(k)$$
$$= x^* + \begin{bmatrix} \dfrac{\gamma_{fy}^2 + \gamma_{fz}^2}{\sigma_{fy}^2 + \sigma_{fz}^2 + \gamma_{fy}^2 + \gamma_{fz}^2} & 0 & 0 \\ 0 & \dfrac{\gamma_{fz}^2 + \gamma_{fx}^2}{\sigma_{fz}^2 + \sigma_{fx}^2 + \gamma_{fz}^2 + \gamma_{fx}^2} & 0 \\ 0 & 0 & \dfrac{\gamma_{fx}^2 + \gamma_{fy}^2}{\sigma_{fx}^2 + \sigma_{fy}^2 + \gamma_{fx}^2 + \gamma_{fy}^2} \end{bmatrix} x^*,$$
$$\quad (17)$$

where the covariance of true value of measured force a_i^* is represented by
$$\mathrm{Cov}[a^*] = \mathrm{diag}[\sigma_{fx}^2, \sigma_{fy}^2, \sigma_{fz}^2]. \quad (18)$$

The bias of the estimate $\hat{x}$ from true value is given by the second term of Eq. (17).

In order to remove the bias from $\hat{x}$, we use Assumption (A5) in which the covariance of noise of force is known. We replace Eq. (15) with the following equation, and identify the contact position.

$$\left(\frac{1}{k}\sum_{i=1}^{k}\overline{A}_i^T\overline{A}_i - \Gamma\right)\hat{x}'(k) = \left(\frac{1}{k}\sum_{i=1}^{k}\overline{A}_i^T\overline{b}_i\right), \quad (19)$$

where

$$\Gamma := \operatorname{diag}[\gamma_{fy}^2 + \gamma_{fz}^2, \gamma_{fz}^2 + \gamma_{fx}^2, \gamma_{fx}^2 + \gamma_{fy}^2]. \quad (20)$$

Here, Eq. (19) coincides with the estimate by CLS [15]. From Assumptions (A5)~(A7), it is guaranteed that the estimate $\hat{x}'(k)$ is asymptotically unbiased.

$$\hat{x}'(\infty) = x^* \quad (21)$$

The estimated value of contact moment with noise is represented by

$$\hat{v}_i(k) := \overline{b}_i - \overline{A}_i \hat{x}'(k). \quad (22)$$

From Eqs. (8), (10) and (11), Eq. (22) is expressed by

$$\hat{v}_i(k) = \overline{A}_i^* x^* + \overline{n}_{ci}^* + \overline{\varepsilon}_{ni} - (\overline{A}_i^* + \overline{E}_{fi})\hat{x}'(k). \quad (23)$$

Using Assumptions (A5)~(A7) and putting k close to ∞,

$$\hat{v}_i(\infty) = \lim_{k\to\infty}\hat{v}_i(k) = \overline{n}_{ci}^* + \overline{\varepsilon}_{ni} - \overline{E}_{fi}x^* = \overline{v}_i. \quad (24)$$

So it is guaranteed that $\hat{v}_i(\infty)$ in Eq. (24) coincides with $\overline{v}_i$ in Eq. (12).

3.4. Identification of Contact Types

In this subsection we analyze the relation between contact types and the eigenvalues of covariance matrix obtained by estimated contact moment $\hat{v}_i$.

From Eq. (12), we define a covariance matrix of $\overline{v}_i$.

$$M := \operatorname{Cov}[\overline{v}] \quad (25)$$

From Eqs. (1) and (4), the matrix M becomes

$$M = M' + \Gamma_n + [x^* \times]\Gamma_f [x^* \times]^T, \quad (26)$$

where

$$M' = RVR^T. \quad (27)$$

M depends on contact position x, noise of force ε_f and moment ε_n. M' is independent of them.

Here, we define $N(k)$:

$$N(k) := \frac{1}{k}\sum_{i=1}^{k}\hat{v}_i(k)\hat{v}_i(k)^T - \Gamma_n - [\hat{x}(k)\times]\Gamma_f[\hat{x}(k)\times]^T. \quad (28)$$

When we put k close to ∞ in Eq. (28) from Assumption (A7),

$$N(\infty) = M'. \quad (29)$$

Performing the singular value decomposition, the matrix $N(k)$ is rewritten by

$$N(k) = [r_1, r_2, r_3]\begin{bmatrix}\lambda_1(k) & 0 & 0 \\ 0 & \lambda_2(k) & 0 \\ 0 & 0 & \lambda_3(k)\end{bmatrix}[r_1, r_2, r_3]^T, \quad (30)$$

where

$$\lambda_1(k) \geq \lambda_2(k) \geq \lambda_3(k).$$

Using the matrix $N(\infty)$ and Eq. (3), we can identify the contact conditions by the method of Ref. [11]. The distribution of eigenvalues in case of each contact type is illustrated in Fig. 3. If a threshold θ is set at a value larger than 0, we can judge which contact type happens.

$$\begin{cases}\theta > \lambda_1 & \text{for a point contact type} \\ \lambda_1 > \theta > \lambda_2 & \text{for a soft finger contact type} \\ \lambda_2 > \theta > \lambda_3 & \text{for a line contact type} \\ \lambda_3 > \theta & \text{for a plane contact type}\end{cases} \quad (31)$$

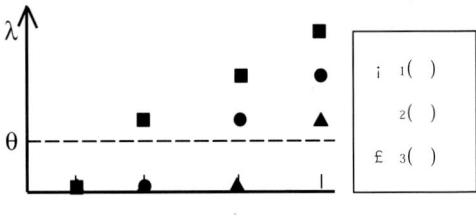

Fig. 3 Relation between contact types and eigenvalues

4. Identification of Unknown Parameters

In the previous section, the estimated contact position $\hat{x}'(k)$ and moment with noise $\hat{v}_i(k)$ were obtained, and it was shown that the contact types could be identified in case of the infinite times of active sensing motion. In that case, however, the estimate error cannot be ignored. Also, we did not consider the constraint of contact moment such as Eq. (3), when the contact position was estimated in Section 3.3.

Therefore, in this section, we investigate the finite time case and propose a method for more accurate identification of unknown parameters.

A performance index is represented by

$$J_{type} = \frac{1}{k}\sum_{i=1}^{k}\left\|\overline{A}_i x + \overline{n}_{ci} - \overline{b}_i\right\|^2. \quad (32)$$

4.1 Case of Point Contact Type

In case of point contact type, contact moment n_{ci} is restricted to zero. The index of Eq. (32) is represented by the same form as Eq. (14). Considering the noise of force, the estimate $\hat{x}$ is given by Eq. (19).

4.2. Case of Soft Finger Contact Type

In case of soft finger contact type, contact moment is restricted to the form:

$$n_{ci} = r_z n'_{czi} = [r_{z1}\ r_{z2}\ r_{z3}]^T n'_{czi}, \quad (33)$$

where r_z denotes the unit normal vector at contact point, and remains unchanged. If the element of maximum absolute value of r_z is defined as $r_{z\max}$, the moment can be rewritten by

$$\overline{n}_{ci} = \begin{cases}[1\ y_1\ y_2]^T\overline{z}_i & \text{for } r_{z\max} = r_{z1} \\ [y_1\ 1\ y_2]^T\overline{z}_i & \text{for } r_{z\max} = r_{z2} \\ [y_1\ y_2\ 1]^T\overline{z}_i & \text{for } r_{z\max} = r_{z3}\end{cases}. \quad (34)$$

For example, in case of $r_{z\max} = r_{z1}$, the index is represented by

$$J_s = \frac{1}{k}\sum_{i=1}^{k}\left\|\overline{A}_i x + y\overline{z}_i - \overline{b}_i\right\|^2, \quad y = [1\ y_1\ y_2]^T. \quad (35)$$

When we identify the unknown parameters by minimizing the index J_s, it is necessary to solve the following nonlinear equations.

$$\frac{\partial J_s}{\partial \boldsymbol{x}}=0, \frac{\partial J_s}{\partial y_1}=0, \frac{\partial J_s}{\partial y_2}=0, \frac{\partial J_s}{\partial z_i}=0. \quad (36)$$

Here, z_i is estimated as

$$\bar{z}_i = \hat{\boldsymbol{y}}^+ (\bar{\boldsymbol{b}}_i - \bar{A}_i \hat{\boldsymbol{x}}), \quad (37)$$

where

$$\boldsymbol{y}^+ = [y_1^+ \ y_2^+ \ y_3^+] := (\boldsymbol{y}^T \boldsymbol{y})^{-1} \boldsymbol{y}^T = \frac{[1 \ y_1 \ y_2]}{1+y_1^2+y_2^2}.$$

Considering the noise of force, Eq. (37) can be rewritten by

$$\bar{z}_i = \bar{z}_i^* - \hat{\boldsymbol{y}}^+ (\bar{\boldsymbol{\varepsilon}}_{ni} - \bar{E}_{fi} \hat{\boldsymbol{x}}). \quad (38)$$

Eq. (38) implies that $\bar{z}_i$ has the bias from true value.

From Eq. (38), $\boldsymbol{x}, y_1$, and y_2 are estimated by

$$\begin{bmatrix} \frac{1}{k}\sum \bar{A}_i^T \bar{A}_i - \Gamma & \frac{1}{k}\sum \bar{A}_{2i}^T \bar{z}_i + \hat{D}_2^T & \frac{1}{k}\sum \bar{A}_{3i}^T \bar{z}_i + \hat{D}_3^T \\ \frac{1}{k}\sum \bar{A}_{2i}\bar{z}_i + \hat{D}_2 & \frac{1}{k}\sum (\bar{z}_i)^2 - \hat{\zeta} & 0 \\ \frac{1}{k}\sum \bar{A}_{3i}\bar{z}_i + \hat{D}_3 & 0 & \frac{1}{k}\sum (\bar{z}_i)^2 - \hat{\zeta} \end{bmatrix} \begin{bmatrix} \boldsymbol{x} \\ y_1 \\ y_2 \end{bmatrix}$$

$$= \begin{bmatrix} \frac{1}{k}\sum(\bar{A}_i^T \bar{\boldsymbol{b}}_i - \bar{A}_{1i}^T \bar{z}_i) - \hat{D}_1^T \\ \frac{1}{k}\sum \bar{b}_{2i}\bar{z}_i - \gamma_{ny}^2 \hat{y}_2^+ \\ \frac{1}{k}\sum \bar{b}_{3i}\bar{z}_i - \gamma_{nz}^2 \hat{y}_3^+ \end{bmatrix}, \quad (39)$$

where

$$\bar{A}_i = \begin{bmatrix} \bar{A}_{i1} \\ \bar{A}_{i2} \\ \bar{A}_{i3} \end{bmatrix}, \bar{\boldsymbol{b}}_i = \begin{bmatrix} \bar{b}_{i1} \\ \bar{b}_{i2} \\ \bar{b}_{i3} \end{bmatrix},$$

$$\begin{bmatrix} \hat{D}_1 \\ \hat{D}_2 \\ \hat{D}_3 \end{bmatrix} := (\hat{\boldsymbol{y}}^T \hat{\boldsymbol{y}})^{-1} \begin{bmatrix} \gamma_{fx}^2(\hat{x}_2\hat{y}_2 - \hat{x}_3\hat{y}_1) \\ \gamma_{fy}^2(\hat{x}_3 - \hat{x}_1\hat{y}_2) \\ \gamma_{fz}^2(\hat{x}_1\hat{y}_1 - \hat{x}_2) \end{bmatrix} \times,$$

$$\hat{\zeta} := \hat{\boldsymbol{y}}^+ (\Gamma_n + [\hat{\boldsymbol{x}}\times]\Gamma_f [\hat{\boldsymbol{x}}\times])(\hat{\boldsymbol{y}}^+)^T.$$

Therefore, we summarize the result as follows [11].

Step 1: Initial values are set as the estimates in section 3.
Step 2: $\boldsymbol{x}, y_1, y_2$ are fixed and z_i is identified by minimizing J_s by Eq. (37).
Step 3: z_i is fixed and $\boldsymbol{x}, y_1, y_2$ are identified by minimizing J_s by Eq. (39).

Steps 2 and 3 are repeated until some convergence condition is satisfied. In fact, though the normal vector $\boldsymbol{r}_z$ is unknown, the eigenvector $\boldsymbol{r}_1$, which is close to $\boldsymbol{r}_z$, is obtained by Eq. (30). Therefore, we can set as $\boldsymbol{r}_z = \boldsymbol{r}_1$ and select the index J_s.

4.3. Case of Line Contact Type

In case of line contact type, contact moment is restricted to the form:

$$\boldsymbol{n}_{ci} = \begin{bmatrix} r_{y1} & r_{y2} & r_{y3} \\ r_{z1} & r_{z2} & r_{z3} \end{bmatrix}^T \begin{bmatrix} n'_{cyi} \\ n'_{czi} \end{bmatrix}. \quad (40)$$

The index is represented by Eqs. (32) and (40). In a similar way of soft finger contact type, we can identify unknown parameters. This formulation is omitted for lack of space.

4.4. Case of Plane Contact Type

In case of plane contact type, we cannot identify unknown parameters by the active sensing method.

4.5. Algorithm for Identification of Contact Conditions

Summarizing the result above, the algorithm for identification of contact conditions is proposed as follows.

Step 1: Command active sensing motion.
Step 2: Calculate deviated force and moment from Eq. (10).
Step 3: Estimate contact position $\hat{\boldsymbol{x}}$ from Eq. (19).
Step 4: Estimate contact moment $\hat{\boldsymbol{v}}$ from Eq. (22).
Step 5: Calculate matrix N from Eq. (28) and analyze the eigenvalues $\lambda_1, \lambda_2, \lambda_3$.
Step 6: Judge which contact type happens by Eq. (31).
Step 7: If it is judged as soft finger contact type, then unknown parameters are re-estimated by Eqs. (38) and (39).
Step 8: If it is judged as line contact type, then unknown parameters are re-estimated.
Step 9: Go to Step 1.

By using this algorithm, the contact conditions can be identified continuously by active sensing motion. Also, in case of plane contact type, unknown parameters cannot be identified, however contact type can be judged.

5. Simulation

In this section, we demonstrate the effectiveness of our proposed method by using some simulations. The measured force $\boldsymbol{f}_o$ and moment $\boldsymbol{n}_o$ are generated under the following conditions, where $\boldsymbol{n}_c$ follows Eq. (3).

$$\boldsymbol{x}^* = [2.00, 2.00, 2.00]^T, R = \text{diag}[1,1,1],$$
$$\text{Cov}[\boldsymbol{f}_c] = \text{diag}[2.00^2, 1.50^2, 1.00^2],$$
$$V = \text{Cov}[\boldsymbol{n}_c] = \text{diag}[2.00^2, 1.50^2, 1.00^2],$$
$$\Gamma_f = \text{Cov}[\boldsymbol{\varepsilon}_f] = \text{diag}[0.10^2, 0.20^2, 0.30^2],$$
$$\Gamma_n = \text{Cov}[\boldsymbol{\varepsilon}_n] = \text{diag}[0.10^2, 0.20^2, 0.30^2].$$

Figs. 4 and 5 show the distribution of eigenvalues calculated by the method of Ref. [11] and that of our proposed method, respectively. These figures are obtained by 100 trials and 100 times of active sensing motion per 1 trial.

In Fig. 4, the eigenvalues are not separated into two types of the variance, which means the variance of noise and that of moment. These eigenvalues depend on unknown contact position, as explained in Eq. (26). Hence, the threshold cannot be determined, and the contact type cannot be judged.

Compared with Fig. 4, the eigenvalues shown in Fig. 5 are clearly separated into the two types. From Eq. (31), the contact types can be judged if the threshold θ is selected between 0 and 1.0^2 (=min{ σ_{nx}^2, σ_{ny}^2, σ_{nz}^2 }). From Fig. 5, it can be set as about 0.4.

Figs. 6 and 7 show the estimates of contact position and normal direction in case of soft finger contact type, respectively. Figs. 6(a) and 7(a) show linear estimates, which are estimated by the method of Section 3. Figs. 6(b) and 7(b) show typed estimates which are re-identified as described in Section 4. Both linear estimates and typed ones are converged to a true value. However, Fig. 6 shows that the typed estimates are more accurate than the linear estimates.

6. Conclusion

The identification of contact conditions was investigated when the shape of grasped object is unknown and both force and moment data are contaminated with noise. First, it was shown that the bias of estimates occured if the previous method was used. Secondly, a method of obtaining an asymptotically unbiased estimate of contact position was established, where the variance of noise of force and moment were evident by the sensor calibration beforehand. The relation between contact types and the eigenvalues of covariance matrix of estimated contact moment was analyzed. Thirdly, an algorithm for more accurate identification of unknown parameters was proposed. Finally, the effectiveness of our proposed method was demonstrated by numerical examples. This method can easily estimate the normal direction in case of soft finger contact type and the direction of contact line in case of line contact type.

This method can be applied to the case where a walking robot recognizes contact conditions between floor and its own legs.

References

[1] Akella, P. N., and Cutkosky, M. R., "Contact Transition Control with Semiactive Soft Fingertips," *IEEE Trans. Robotics and Automation*, Vol. 11, No. 6, pp. 859-867, 1995.

[2] Bicchi, A., et al,. "Contact Sensing from Force Measurements," *Int. J. Robotics Research*, Vol. 12, No. 3, pp. 249-262, 1993.

[3] Dütre, S., et al., "Contact Identification and Monitoring Based on Energy," *Proc. IEEE Conf. Robotics and Automation*, pp. 1333-1338, 1996.

[4] Horn, R., A. and John, C. R., *Topics in Matrix Analysis,* Cambridge Univ. Press, 1991.

[5] Hyde, J. M., and Cutkosky, M. R., "Contact Transition Control: An Experimental Study," *Proc. IEEE Conf. Robotics and Automation*, pp. 363-368, 1993.

[6] Kaneko, M., et al., "Active Antenna," *Proc. IEEE Conf. Robotics and Automation*, pp. 2665-2671, 1994.

[7] Kitagaki, K., et al., "Methods to Detect State by Force Sensing in an Edge Mating Task," *Proc. IEEE Conf. Robotics and Automation*, pp.701-706, 1993.

[8] Mason, M. T. and Salisbury, J. K., *Robot Hands and the Mechanics of Manipulation*, Cambridge, MA, MIT Press, 1985.

[9] Mimura, N, and Funahashi, Y., "Parameter Identification of Contact Conditions by Active Force Sensing," *Proc. IEEE Conf. Robotics and Automation*, pp. 2645-2650, 1994

[10] Mimura, N., et al., "An Algorithm for Identification of Contact Conditions," *Trans. Jpn. Soc. Mech. Eng.*, Vol. 63, No. 610, Series C, pp. 2061-2068, 1997 (in Japanese).

[11] Mouri, T., et al., "Identification of Contact Conditions from Contaminated Data of Contact Moment," *Proc. IEEE Conf. Robotics and Automation*, pp. 585-591, 1999.

[12] Nagata, K., et al., "Pose Estimation of Grasped Object from Contact Force or Joint Data of Manipulator," *SICE*, Vol. 28, No. 7, pp. 783-789, 1992 (in Japanese)

[13] Oussalah, M., "Fuzzy linear regression for contact identification," *Proc. IEEE Conf. Robotics and Automation*, pp. 3616-3621, 2000

[14] Salisbury, J. K., et al., "Interpretation of Contact Geometries from Force Measurements," *Proc. 1st Int. Symp. on Robotics Research*, pp. 565-577, 1983.

[15] Stoica, P., and Soderstorm, T., "Bias Correction in Least Square Identification", *Int. J. Control*, 35, pp. 449-457

[16] Tsujimura, T., and Yabuta, T., "Object Detection by Tactile Sensing Method Employing Force/Torque Information," *IEEE Trans. Robotics and Automation*, Vol. 5-4, pp. 444-450, 1989.

[17] Ueno, N., and Kaneko, M., "Contact Localization by Multiple Active Antenna," *Proc. IEEE Conf. Robotics and Automation*, pp. 1942-1947, 1999.

[18] Yamada, T., et al., "Identification of Contact Conditions from Contaminated Data," *Trans. Jpn. Soc. Mech. Eng.*, Vol. 64, No. 618, Series C, pp. 584-489, 1998 (in Japanese).

[19] Zhou, X., et al, "Contact Localization Using Force/Torque Measurements," *Proc. IEEE Conf. Robotics and Automation*, pp. 1339-1344, 1996.

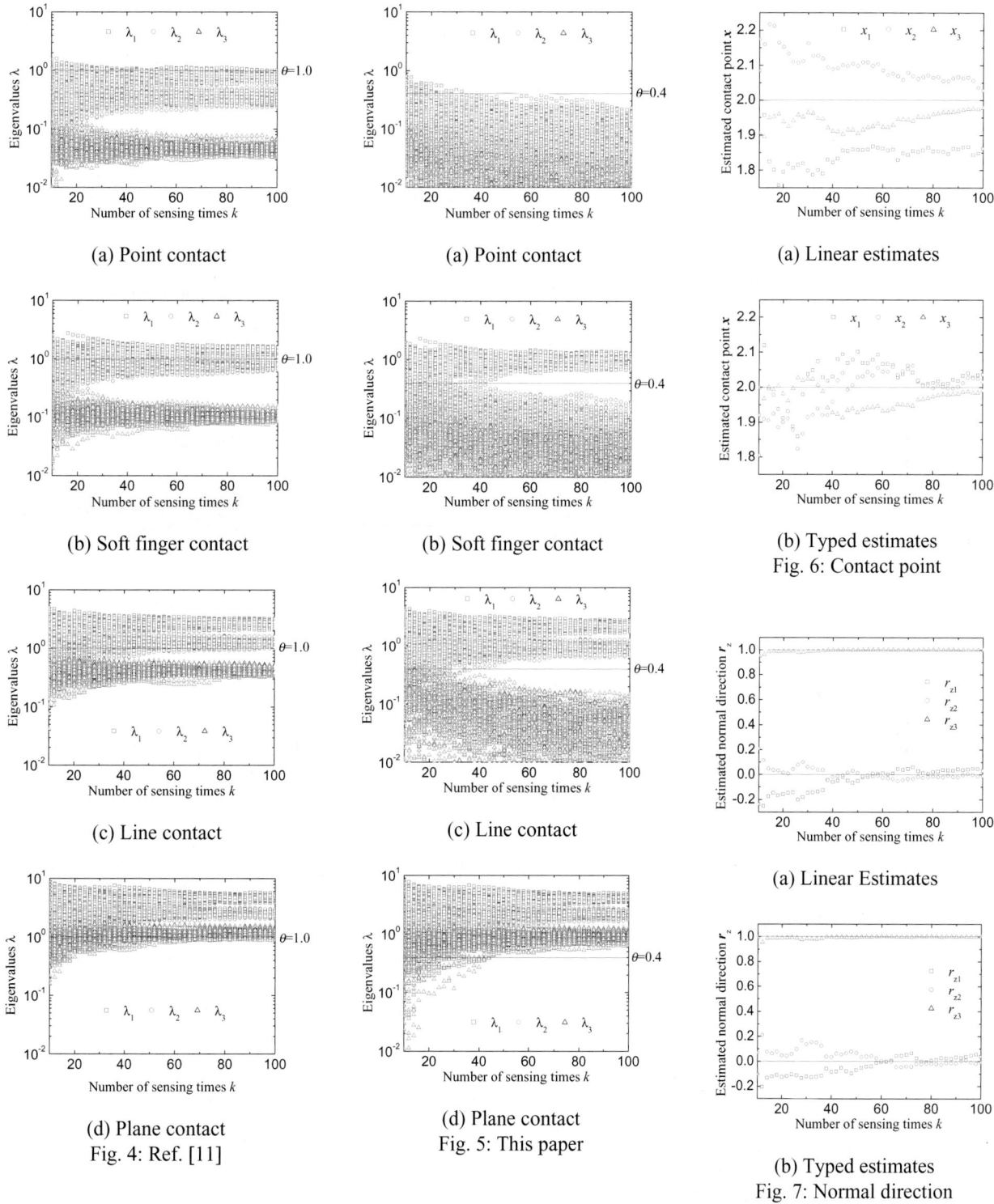

(a) Point contact

(b) Soft finger contact

(c) Line contact

(d) Plane contact
Fig. 4: Ref. [11]

(a) Point contact

(b) Soft finger contact

(c) Line contact

(d) Plane contact
Fig. 5: This paper

(a) Linear estimates

(b) Typed estimates
Fig. 6: Contact point

(a) Linear Estimates

(b) Typed estimates
Fig. 7: Normal direction

Three-Dimensional Bio-Micromanipulation under the Microscope

Fumihito Arai*, Akiko Kawaji*, Poom Luangjarmekorn*, Toshio Fukuda**, and Kouichi Itoigawa***

* Department of Micro System Engineering, Nagoya University
Furo-cho, Chikusa-ku, Nagoya 464-8603, JAPAN, http://www.mein.nagoya-u.ac.jp
** Center for Cooperative Research in Advanced Sci. & Tech., Nagoya University
*** Tokai Rika Co., LTD., Oguchi, Niwa-gun, Aichi 480-0195, JAPAN

Abstract

Bio-micromanipulation is important for Biology and Bio-engineering field. However, operation is very difficult, since the object is very small, kept in the liquid, and observed by the optical microscope. We have been developing a new micromanipulation system for anatomical operation of the micro object such as an embryo, cell, and microbe. The image of the microscope is two-dimensional, so it is hard to manipulate the target in the 3-D space. To improve the manipulation work, we proposed the 3-D bio-micromanipulation system combined with the Virtual Reality (VR) space. Here we proposed the 3-D modeling method of the object to present the 3-D visual information to the operator and improved the operation environment. In this system, we still have difficulty to change the orientation of the microscopic object by manipulating a mechanical manipulator. The Bio-aligner, we proposed, is a micro device for the posture control of an object. We developed a two-dimensional Bio-aligner by microfabrication. Here we show its fabrication process and basic rotating experiment with yeast cells.

Key words: Micromanipulation, biological application, VR, 3-D, MEMS, electrorotation

1. Introduction

Recently, dexterous micromanipulation of small biological objects, such as an embryo, cell, and microbe in high speed is demanded. Bio-micromanipulation is important for Biology and Bioengineering field. However, it is very difficult, since the object is very small, and is kept in the liquid. We have developed the contact type micromanipulation system that has multiple degrees of freedom for anatomical operation, such as nuclear transplantation and electrophysiological inspection, of the micro object such as an embryo, cell, and microbe [1-6].

Most of the cases, operators manage to manipulate the micro objects by the micromanipulators with the two-dimensional image from the optical microscope [7,8]. However, the total work is carried out in the three-dimensional space as shown in Fig. 1. So, operation of the micromanipulator is quite difficult and hard work. To improve the manipulation of the micro object, we proposed the three-dimensional (3-D) bio-micromanipulation system [3-5]. We have studied on the key technologies to realize this concept.
(1) Precise positioning of the micromanipulator in 3-D space [6]
(2) Bilateral control system using the multi-axial micro force sensor [3,4]
(3) VR modeling of the manipulation environment
(4) Intelligent user interface for 3-D manipulation [5]
   EX: 3-D free viewpoint selection in the VR space

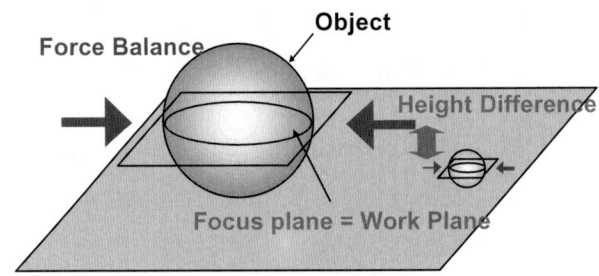

Fig. 1 2-D Bio-micromanipulation under the Microscope

The image of the microscope is two-dimensional, so it is hard to manipulate the target in the 3-D space. Depth of focus is very narrow. To expand the image in depth direction, we linked the VR space and the real space based on the precise 3-D calibration technique. There has been reported on the VR navigation for the micromanipulation [9], however, previous works didn't consider the matching between the VR space and the real space with consideration of the difference of refractive index. We realized absolute positioning accuracy of the micromanipulator around 1μm - 3μm within 40μm^3 working range [6] by calibrating the parameters related to the refractive index. The accuracy is enough for the cell manipulation works. Based on this achievement, we can match the VR space and the real operation space. Here we propose the high-speed 3-D modeling method of the target object to present the 3-D visual information to the operator to improve the operation environment.

In this system, we still have difficulty to change the orientation of the microscopic object by manipulating a mechanical manipulator. The Bio-aligner, we proposed, is a micro device for the posture control of an object [1,2]. We developed a two-dimensional Bio-aligner by microfabrication. Here we show its fabrication process and basic rotating experiment result with yeast cells.

2. Three-Dimensional Micromanipulation System

Figures 2 and 3 show a three-dimensional (3-D) bio-micromanipulation system integrated with the VR space. It has a micromanipulator, which has a 3-DOF narrow-range positioning device on a 3-DOF wide-range positioning device, and a micro tri-axial force sensor. Detail of this system is shown in the references [3-5].

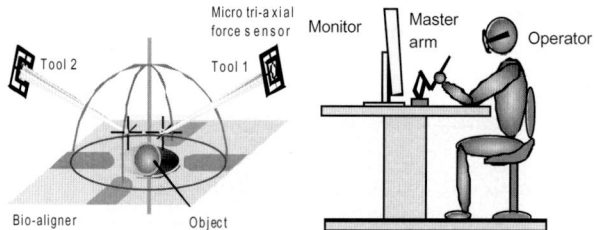

Fig. 2 Concept of the 3-D Bio-micromanipulation

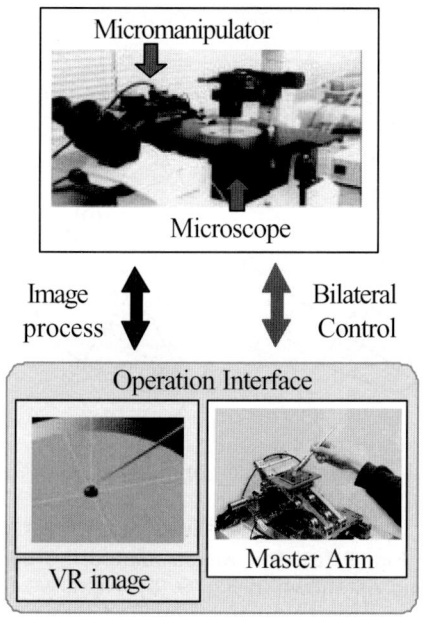

Fig. 3 3-D Bio-micromanipulation System

We have reported basic strategies to improve the operability. To improve the operation environment, we proposed to use the VR space, which is well matched with the real system. In this paper, we propose a high-speed modeling method of the target in the VR space.

3. How to Build a 3-D Model of the Target

3.1 Assumption of Modeling

Our target is an embryo, cell, and microbe. The shape of most of those objects is approximated as an ellipsoid. Here, we assume the target is symmetric to the equatorial plane. In most case, the target sinks in the base. We also assume the object is consistent, and the gravity and buoyancy are balanced at the center of the object. Under such assumptions, we need 2 kinds of parameters to build the approximated ellipsoid model. Those are the height from the base and the shape on the equatorial plane as shown in Fig. 4.

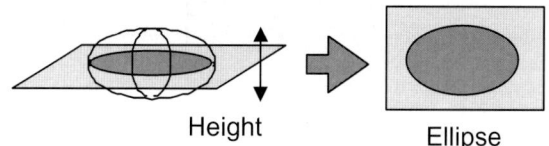

Fig. 4 Ellipsoid model

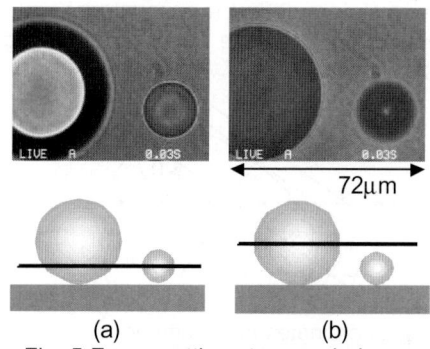

Fig. 5 Focus setting at several planes

3.2 Height Estimation

When we adjust the focus of high N.A. lens for the spherical object, we can get the image in the best contrast condition at the equatorial plane. For example, Fig. 5 shows two images of the different size spheres. The larger one is φ50μm glass bead and the smaller one is φ20μm glass bead. In case Fig. 5 (a), the smaller one is focused at its equatorial plane. In case Fig. 5 (b), the larger one is focused at its equatorial plane. It is clear to see that the contrast of the image is the best at the equatorial plane. By adjusting the focus, we can estimate the equatorial plane, and thus we can estimate the height of the approximated object as shown in Fig. 6.

In our microscope system, the position of the objective lens is controlled by the closed loop position feedback control. The position accuracy of the lens is sub-micron order. We calibrated the displacement of the lens and the corresponding displacement of the focused plane in the previous research works [6]. We considered the refractive index of the glass cover and the liquid. Based on the calibration, we can estimate the height of the target with high precision.

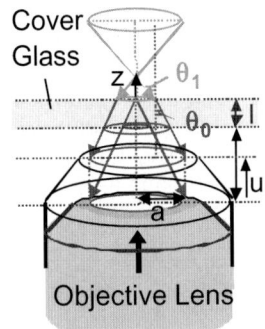

Fig. 6 How to calculate the height of the target

3.3 Modeling of the Target in VR Space

The shape of the equatorial plane is approximated as an ellipse. To model the target, we use the parameters in Fig. 7.

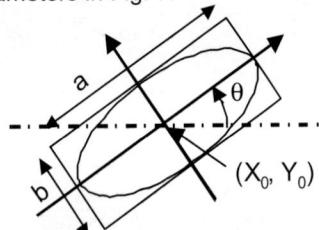

(X_0, Y_0): Coordinates of the center
a : Length of major axis
b : Length of minor axis
θ : Angle of rotation

Fig. 7 Parameters of equatorial plane

To get these parameters, we used the image processing and developed an object recognition algorithm. We used some of the image processing functions of MIL (Matrox Image Library) version 5.0.

The image taken from the microscope is shown in Fig. 8 (a). The target is chlorella and the length of major axis of one cell is about 6 μm. We used the 40x lens (oil immersion type) with the inverted microscope IX70 (OLYMPUS). The image is converted to the 640 pixels x 480 pixels of 256 gray-scale data. The distance between each pixel is 0.12μm.

At first, we used an edge detection function to get the edge points of 256 gray scale level as shown in Fig. 8 (b). Next, we used a binarization function to get the binary image as shown in Fig. 8 (c).

To model multiple objects in the same image independently, we grouped the points of the same target as follows. At first, we used a dilation function to join the close points, and then used an erode function. Next, we filled the closed area as shown in Fig. 8 (d). It is used for the object recognition phase.

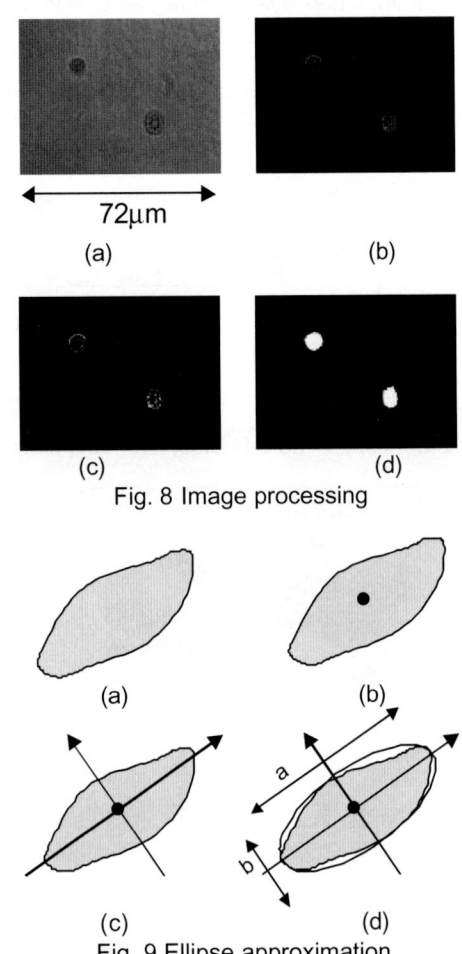

Fig. 8 Image processing

Fig. 9 Ellipse approximation

To approximate the ellipse from the processed image in Fig. 9 (a), first we calculated the centroid of each group as shown in Fig. 9 (b). Next, we used the principal component analysis to find direction of the major axis. In the case of two-dimensional data that has ellipse-like distribution, it is the direction of the major axis.

Let x_i is the coordinate of each points of a group and $\bar{x}$ is the coordinate of the centroid, a covariance matrix E is written as follows,

$$E = \frac{1}{n-1}\sum_{i=1}^{n}(x_i - \bar{x})(x_i - \bar{x}) \quad (1)$$

Let V is a vector of the major axis as follows,

$$V = \begin{pmatrix} V_x \\ V_y \end{pmatrix} \quad (2)$$

V will be an eigenvector of the maximum eigenvalue of E. Then the rotation angle is calculated as follows.

$$\theta = \tan^{-1}(V_y / V_x) \quad (3)$$

Base on this method, two axes were determined as shown in Fig. 9 (c). From these axes, we can calculate the orientation of the ellipse. Then, we determined the axis length "a" and "b" and an appropriate ellipse was fitted as shown in Fig. 9 (d).

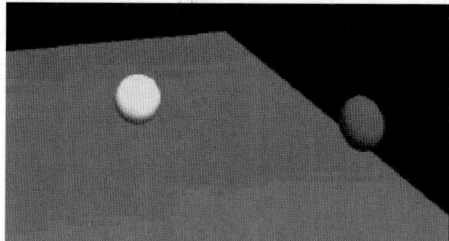

Fig. 10 VR Space

Table 1 Calculation Time

Operation	Time (ms)
Edge detection	80
Binarization	10
Grouping	60
Labeling	50
Scene recognition	310
Total	510

Finally, we made a VR space as shown in Fig. 10. The VR space was written by the OPENGL on WINDOWS NT. Calculation time of each process is shown in table 1. The updating time is fast. The proposed system is practical and useful in approaching the target point.

4. Bio-aligner

4.1 Concept and Principle of the Bio-aligner

In the biotechnology field, the positional precision of the tip of a micromanipulator is influenced to the success rate of phenotypic expression. We need to adjust the tip on the target precisely in a cell, microorganism and so on. Bio-aligner can change the position and orientation of the object in a liquid by using electric field [1,2].

The following are the techniques for micro-manipulation.
(1) Non-contact type
   (i) Laser tweezers
   (ii) Electromanipulation
      Coulomb force, Dielectrophoresis,
      Electrorotation, Traveling wave drive
   (iii) Ultrasonic Wave
(2) Contact type
   (i) Mechanical Manipulator
   (ii) Vibrating Stage

Among those methods, electrorotation are suitable for changing orientation of microscopic material. Electrorotation can easily extend to three-dimensional control. In our research work, we use electrorotation for the Bio-aligner.

The two-dimensional Bio-aligner has four electrodes in the center. These electrodes cross at right angles each other. We give alternating current voltages shifted each 90 degrees phase to the electrodes. These voltages will cause a rotating field in the center of the micro device. The polarization by the electric field is too slow to keep up with the rapidly rotating field, and the interaction between field and a dielectric body immersed in a non-conducting liquid gives rise to a torque to spinning. The size of the torque varies with frequency [10].

4.2 Fabrication Process of the Bio-Aligner

We used an inverted microscope for the observation of micro materials because of its wide working space for a mechanical micromanipulator as shown in Fig. 11. For that reason a substrate of the Bio-aligner had to meet observation conditions on an oil immersion objective of the inverted microscope. We used a borosilicate glass plate (Φ4inch, Thickness: 200μm) for the substrate. The process flowchart for making of Bio-aligner and its sketches are shown in Fig. 12. The pattern of electrodes was shaped by lift-off processing of photoresist. We used gold and chromium for electrical wiring. (Thickness: Au = 200nm, Cr = 100nm) Chromium is suitable for a middle layer between gold electrodes and a glass substrate because of its adhesive strength and abrasion resistance. The surface of electrodes was made by gold. The volume resistivity of gold is low compared with chromium. Metals are evaporated on the glass substrate. In our trial experiment, the line and space of electrodes could be formed finely up to ten micrometers width. The orientation control of a microorganism has to be done in the liquid. Electrodes were coated with an insulating film. We adopted Spin-on-Glass (SOG) for the electrical insulation. SOG can spread uniformly over the thin glass. The thickness of the insulator is about 1200nm.

Figure 13 gives a finished product of the Bio-aligner. Each one of nine squares in the center of the substrate is a single pattern of electrodes. The small rectangles around the patterns are bonding pads. Electric conductors were attached onto the pads with silver paste. Figure 14 shows a center of a pattern in a microscopic image.

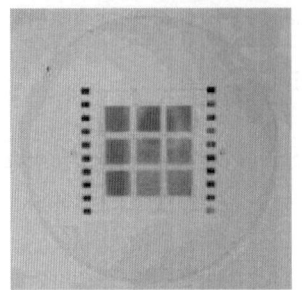

Fig. 13 Exterior of Bio-aligner

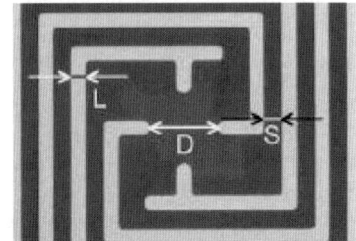

L: Width of Electrode Lines (20μm)
D: Distance between Center Electrodes (50μm)
S: Width of Space between Electrodes (20μm)

Fig. 14 Center of Pattern

Fig. 11 Bio-aligner on Inverted Microscope

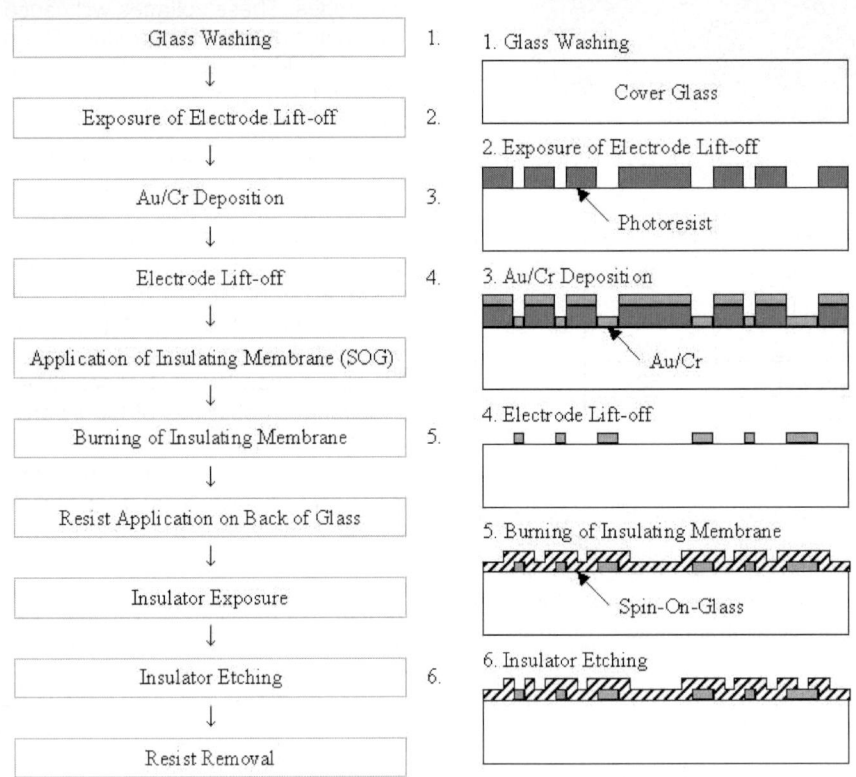

Fig. 12 Process for Making of Bio-aligner

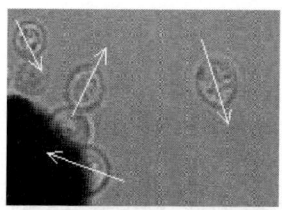

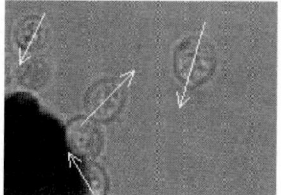

Fig. 15 Rotation of Yeast Cells

Fig. 16 Pearl Chains of Cells between Electrodes

4.3 Experiment

A sinusoidal oscillator impressed the alternating voltage of 7.5 V on an electrode. We changed its frequency from 1KHz to 15MHz in this experiment. Four electrodes generated sine wave voltage delayed 90 degrees each counterclockwise. The rotating objects we used were yeast cells. These were about 5μm across in diameter. Figure 15 gives rotating image of yeast cells nearby a center electrode of the Bio-aligner. The fastest rotation was observed when the frequency was from 70KHz to 80kHz. We estimated that the cell spinning were reversed in the band of higher frequency, because changing of induced surface charges on the cells was late furthermore from rapidly rotating field. However the rotating direction did not change. The phenomenon may be responsible for dielectrophoretic force that pulls down the cells to the surface of plane electrodes on the substrate. In Fig. 16, we can see pearl chains of cells formed by potential differences between the electrodes.

5. Conclusions

To improve the manipulation work, first we developed the 3-D bio-micromanipulation system with the VR space. We proposed the 3-D modeling method of the object to present the 3D visual information to the operator to improve the operation environment. The time to construct the VR space is fast and the proposed system is practical to reduce the approaching time to the target. The modeling of the target is not precise compared with the 3-D modeling based on the several sectional images taken by the confocal microscope. However, the proposed method is real time and easy to implement with low cost. Current dexterous manipulation work is performed with the fixed focus plane. However, if the operator tries to move the manipulator in wide area, the proposed system is useful to reduce the time to approach the target point.

Next we explained the necessity of the Bio-aligner and its operating principle. We then showed how to make the two-dimensional Bio-aligner. A rotating experiment of yeast cells was performed to confirm validity of this micro device. In our future, its function will be extended to three-dimensional rotation.

Acknowledgement

This research work was supported in part by a grant from NEDO, Japan.

References

[1] F. Arai, et al.: "Bio-Micro-Manipulation (New Direction for Operation Improvement)", Proc. of IEEE/RSJ Int. Conf. Intelligent Robots and Systems (IROS), VOL. 3, 1997, pp.1300-1305.
[2] F. Arai, et al.:"3D Position and Orientation Control Method of Micro Object by Dielectrophoresis", IEEE Proc. MHS'98, 1998, p.149-154
[3] F. Arai, et al.: " 3D Micromanipulation System under Microscope ", IEEE Proc. MHS'98, 1998, p.127-134
[4] F. Arai et al.:" Micro Tri-axial Force Sensor for 3D Bio-Micromanipulation", Proc. IEEE ICRA'99, Detroit, 1999, pp.2744-2749
[5] F. Arai, et al.: " 3D Viewpoint Selection and Bilateral Control for Bio-Micromanipulation", Proc. IEEE ICRA'00, SF, 2000, pp.947-952
[6] A. Kawaji, F. Arai, et al.:" Calibration for Contact Type Micromanipulation", Proc. of Int. Conf. Intelligent Robots and Systems (IROS), Vol.2, 1999, pp.715-720
[7] B. Vikramaditya and B.J. Nelson: " Visually Guided Microassembly Using Optical Microscopes and Active Vision Techniques ", Proc. IEEE Int'l Conf. on Robotics and Automation, Vol. 4, p.3172-3177, 1997.
[8] T. Tanikawa, T. Arai, and Y. Hashimoto: " Development of Vision System for Two-Fingered Micro manipulation", Proc. IEEE/RSJ Int. Conf. on Intelligent Robotics and Systems, VOL. 2, pp. 1051-1056, 1997.
[9] A. Sulzmann, et al.: " Virtual Reality and High Accurate Vision Feedback as Key Information for Micro Robot Telemanipulation", SPIE, Proc. Microrobotics: Components and Applications, Vol. 2906, Boston, p.38-57, 1996
[10] U. Zimmermann et al., "Electromanipulation of cells," pp.283-299, CRC Press (1996)

Force Control System for Autonomous Micro Manipulation

Tamio TANIKAWA†, Masashi KAWAI††, Noriho KOYACHI†, Tatsuo ARAI*,
Takayuki IDE**, Shinji KANEKO**, Ryo OHTA**, Takeshi HIROSE††

†Mechanical Engineering Laboratory, 1-2 Namiki, Tsukuba, Ibaraki
††Nihon University
*Osaka University
**Olympus Optical Cooperation
E-mail tamio@mel.go.jp

Abstract

A dexterous micro manipulation system was developed for applications such as assembling micro machines, manipulating cells, and micro surgery. We have proposed a concept of a two-fingered micro hand, and designed and built a prototype. We succeeded in performing basic micro manipulations, including the grasp, release, and rotation of a microscopic object. The micro hand is controlled with a position control only. An operator has to guess a micro force on grasping from a behavior of object in a microscope image. The accurate micro manipulation depends on a skill of the operator yet. For an easy manipulation and an automatic manipulation, it is necessary to measure the micro forces between the finger and the object. A micro force sensor has developed for a force control in micro manipulation on a corroboration research of Mechanical Engineering Laboratory and Olympus Optical Cooperation. Its resolution is 0.5 [nN] in theoretically. In this paper, we will mention the micro force sensor and to perform an automatic micro manipulation with installing the sensor and a force control. Basic experiment shows excellent micro capability.
Keywords: force sensor, force control, micro manipulation, automatic manipulation

1 Introduction

A dexterous micro manipulation system was developed for applications such as assembling micro machines, manipulating cells, and micro surgery. We have proposed a concept of a two-fingered micro hand, and designed and built a prototype. We succeeded in performing basic micro manipulations, including the grasp, release, and rotation of a microscopic object [1]. The micro hand is controlled with a position control only. An operator has to guess a micro force on grasping by a behavior of object from a microscope image. The accurate micro manipulation depends on a skill of the operator yet. For an easy manipulation and an automatic manipulation, it is necessary to measure the micro forces between the finger and the object.

Various approaches have been presented for using a master-slave system in micro manipulation [2-5], but there have been no published reports of actual experiments using a master-slave system for manipulation on the micron scale. The main difficulty is in developing a force sensor capable of sensing a micro force. For sensing such a small force, it is necessary that the sensor have a resolution order of [nN], as well as multiple sensing degrees. Therefore, it is currently difficult to create a genuine master-slave system.

A micro force sensor with high resolution for a force control in micro manipulation has developed on a corroboration research of Mechanical Engineering Laboratory and Olympus Optical Cooperation. Its resolution is 0.5 [nN] in theoretically. In this paper, we will mention the force sensor and to perform an automatic micro manipulation with installing the sensor and a force control. Basic experiment in micro manipulation shows effectiveness of the micro force sensor and force control in micro manipulation.

In section 2, the detail of a micro force sensor with high resolution for micro manipulation. Its resolution is 0.5 [nN] in theoretically. For an end-effector in a micro hand, a short glass needle is attached on the end of force sensor. It will be discuss about the end-effector in section 3. In section 4, overview of the two-fingered micro hand for micro manipulation will be discussed. With the system mentioned above, the automatic micro manipulation is performed for effectiveness of usage of the force sensor in section 5. Finally, some conclusions will be presented in section 6.

2 Micro force sensor

In a micro manipulation, a force between the end-effector and the object is very small. In order to control the micro force, a micro force sensor with high resolution is necessary. The estimated resolution order is [nN] in theoretically. A conventional strain gage has the resolution of more than [μN]. It is difficult to make the micro force sensor with high resolution, therefore the micro manipulation with force control is difficult.

A micro force sensor has been developed on a corroboration research of MEL and Olympus Opt. Co. The sensor is formed on a silicon wafer by using MEMS technique. Fig. 1 shows the fabrication process in the micro force sensor.

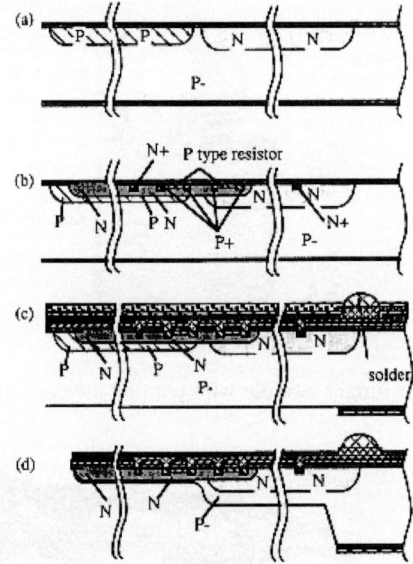

Fig. 1 Process sequence of micro force sensor fabrication

The sensor has two strain gages of semiconductor resistor. For high resolution of the force sensor, it is necessary that a plate on the two strain gages is very thinly fabricated. Here, it is well known in the Electro Chemical Etching that the etching is stopped at a depletion layer between P and N diffused layer [6]. In Fig. 1(a), N diffused layer is doped on the P-base, in order to obtain the depletion layer. At the point of sensor, the more thin plate is necessary. In order to obtain narrow depletion layer, P diffused layer with high concentration is doped, and then N diffused layer is doped. In Fig. 1(b), P diffused layer, P+ and N+ for the electric parts are doped. Then the SiN layer is patterned for an etching mask in Fig. 1(c). In Fig. 1(d), the Electro Chemical Etching is performed. At the part of sensor, the thin plate is obtained, since there is the narrow depletion layer between P and N diffused layer. At the interface between P- and N diffused layer, the depletion layer is wider than sensor part. The part is thicker than sensor part after the etching process. By control of doping depth and its concentration, the thickness can be partly controlled at the any point.

Also, for the high resolution of sensor, each strain sensor has an opposite gage factor. Even the force sensor has very small bending, the high output signal can be obtained.

Fig. 2 shows an overview of the micro force sensor with one axis type. Also, Fig. 3 shows the characteristic of the force sensor with 5[V] in source voltage to the bridge. In this graph, the applied force is bigger than the desired force magnitude [nN]. Currently, there is no idea to apply micro force with [nN] magnitude for a calibration of the force sensor. In this current calibration way, the end part of micro force is pushed to an electrical balance. By using this way, the relation of the sensor output and the applied force is obtained. As future work, a new idea for accurate calibration should be considered. From the result of Fig. 3, it is guessed the force sensor has the sensitivity of 272[μV/μN]. The resolution of 0.5[nN] is theoretically guessed with a strain amplifier having 2,000-power magnitude and a 14-bit AD converter.

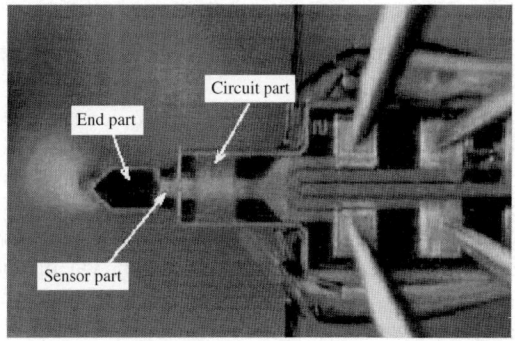

Fig. 2 Micro force sensor

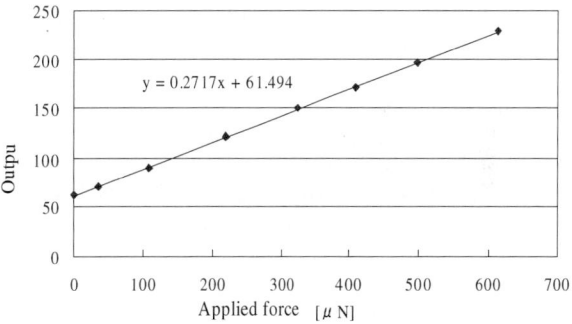

Fig. 3 Characteristic of the micro force sensor

3 End-effector with force sensor

The size of a target object is less than 20 microns. The force sensor cannot be directly used as an end-effector in a micro hand, since the end part in the force sensor is much bigger compared with the object size. Therefore, the short glass needle is attached on the end part by glue. Fig 4 shows whole view of the end-effector.

can be obtained by arranging two finger modules in series as illustrated in Fig. 6. The lower module provides global motion while the upper module provides local, relative motion of the two fingertips [1].

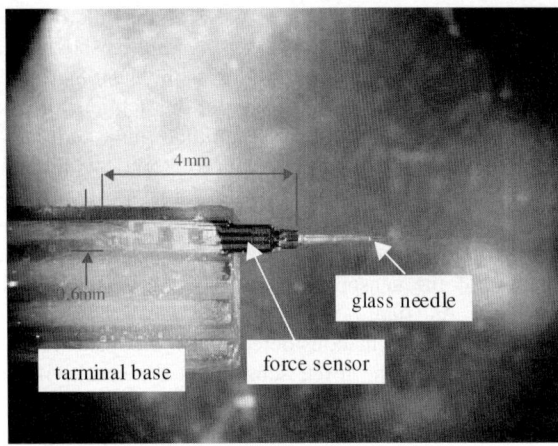

Fig. 4 Picture of the end-effector with the force sensor

4 Two-fingered micro hand with the end-effector

In order to manipulate microscopic object, multi degree motion and high positioning are necessary for a manipulator. In micro field, the adhesion forces, which are surface tension, electric force and friction force, are more dominant than inertia force and gravity. It is easy to grasp object because of sticky condition. On the other hand, to release is very difficult. One finger is enough for just grasp, however, considered with releasing, two fingers are suitable for micro manipulation. If each finger has multi degree motion as chopsticks manipulation, rotation of object can be easily performed also. So, two-fingered manipulator with capable of chopsticks motion is suitable for micro manipulation. A parallel mechanism is used to the manipulator to achieve multi degree motion and high positioning [7].

Fig. 5 shows a finger module with the parallel mechanism. This is one finger mechanism. Flexure hinges are used as passive joint. "R" and "P" mean a revolving joint and a prismatic joint respectively in Fig. 5. Three piezo electric devices are installed to the module. So, the end plate can be controlled with 3 degrees of motion [8][9].

A two-fingered micro hand is designed base on chopsticks motion strategy. In usage of chopsticks, one chopstick is fixed at the side of forefinger and the other at the side of thumb. Each chopstick has a different task. The chopstick at the side of forefinger moves in order to grasp, rotate, and release objects. The other chopstick works only for supporting. The motion of the thumb-side chopstick is linked with the motion of the hand. When global motion for positioning an object is required, the thumb-side chopstick will move in unison with the other chopstick. This is caused by an overall motion of the hand. Thus, each chopstick is controlled by a different strategy. A structure, which mimics the use of chopsticks,

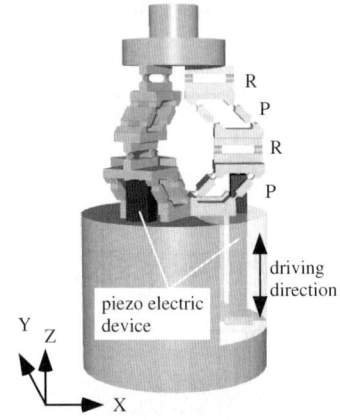

Fig. 5 Finger module with parallel mechanism

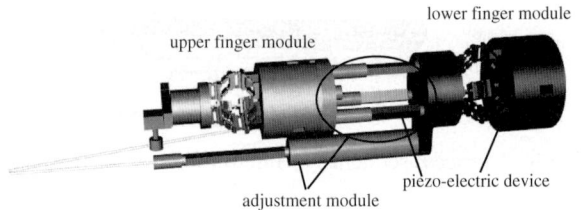

Fig. 6 Two-fingered micro hand

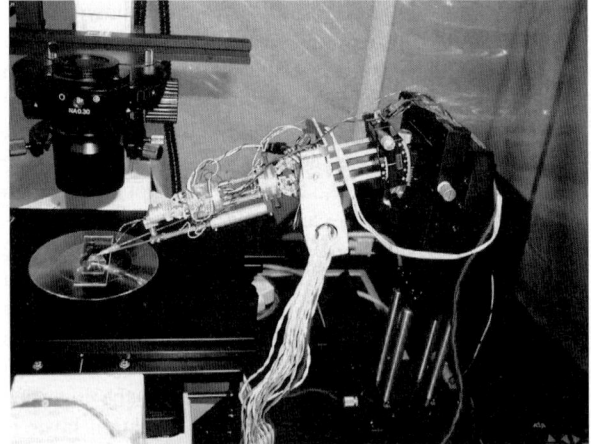

Fig. 7 Configuration of micro manipulation system

The micro hand is set on the side of a microscope shown in Fig. 7. The microscope (Olympus Co. IX70) has a maximum magnifying power of 450. Also, using a 5 power CCD camera in electric magnification, the 2250 power real-time image is obtained on a monitor. It is limit magnification in an optical microscope.

The end-effector with a force sensor is attached on the top of finger in the micro hand. Fig. 8 shows a photograph of the top of finger with the end-effector.

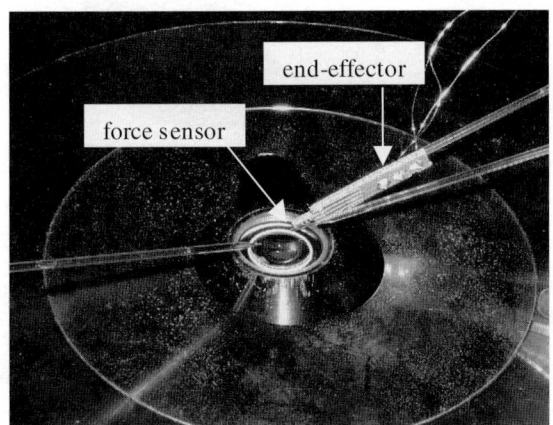

Fig. 8 The top of finger with the end-effector

5. Automatic control for grasping task in micro manipulation

5.1 Flow of automatic grasping task with one force sensor

A master-slave tele-operation system can be performed with the force. The effectiveness of the force sensor can be shown with estimating the feeling of the tele-operation. However, it is difficult to estimate the effectiveness quantitatively with the tele-operation system, since the subjective sense of an operator has to be estimated. On the other hand, an automatic manipulation can be performed with the force sensor also. Therefore, the quantitative effectiveness of the force sensor is shown with performing automatic manipulation. Here, a grasping task to the microscopic object is chosen as the automatic manipulation for the estimation.

Fig. 9 shows a configuration in the grasping task. The force sensor is attached on top of the one finger. The sensing direction of the sensor is closing direction to another finger. A glass ball with 5 [μm] in size is target object. The flow of the automatic grasping task is in blow.

Step 1: The fingers are positioned at over the target by an operator with remote control.
(After that, the automatic manipulation is started.)
Step 2: The fingers go down until contact with ground.
(The contact can be checked with output from the force sensor.)
Step 3: The fingers go up to the center of the target. Then the fingers are closed.
(It is assumed that the target size is known.)
Step 4: The target is grasped with keeping a constant grasping force by using output from the sensor. And then, the target is picked up to the start position in the step 1.

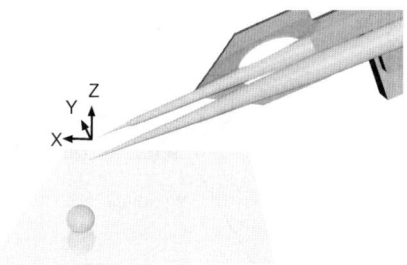

Fig. 9 Configuration in the grasping task

5.2 Result of automatic grasping task with one force sensor

Fig. 10 shows a result of experiment in the automatic grasping task. The "Z Position" is a height of finger position and the "Y Position" is a space between both fingers. The origin of "Z Position" and "Y Position" is start position in the automatic manipulation.

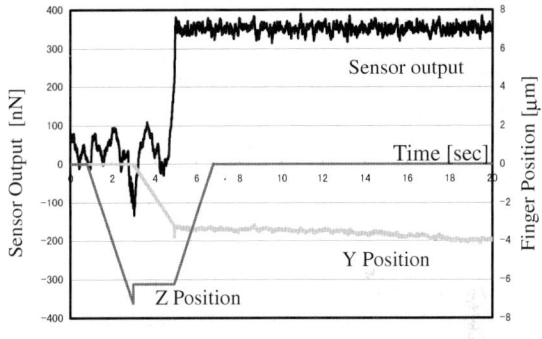

Fig. 10 Sensor output in grasping task with keeping a constant grasping force

It can be found to grasp the target with keeping a constant force in Fig. 10. This is the result when the automatic task is success. In this automatic task, the success rate is not so high. For example, although the "Sensor output" keeps a constant value, the space of both fingers "Y Position" gets narrow little by little. It means that the fingers slip on the target surface. If the grasping direction is accurately set to the gravity center of the target, fingers dose not slip on the target surface. In micro field, it is difficult to obtain the shape information of the target, particularly in depth direction, from microscope image. Because of this reason, the high success rate of the task cannot be obtained with the automatic control with keeping a grasp force.

In micro field, it is easy to keep the grasp condition because of adhesion forces. A special control is not so necessary to keep the grasp condition.

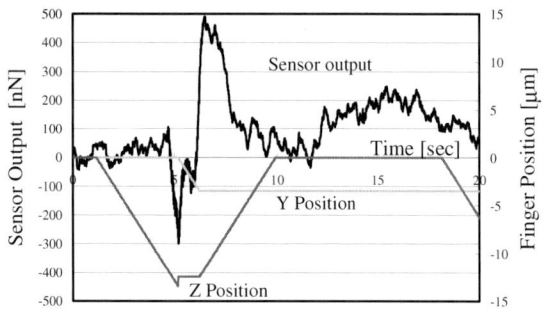

Fig. 11 Sensor output in grasping task with keeping a finger space

Fig. 11 shows the result of automatic grasping task with keeping a finger space during the grasping condition. In case of this automatic task, the force control is only applied when the fingers just grasp. The finger space "Y Position" is kept after going over a threshold of grasp force from touching the target. It can be found to reduce the grasp force when the picking motion is started, since there is no applied force from the ground. It means the closing direction is not faced to the center of gravity of the target. In this condition, if force control is applied during picking process as the previous automatic task, fingers are closed further more and the stable grasp cannot be performed. In this task, finger space is fixed during picking task. So, stable grasp can be performed with adhesion forces from the both side fingers. In this control way, the success rate can be more increased than a previous control way. It can be found that the force control is not always necessary during whole task. The force control is important when the condition just shifting to grasping in micro field.

5.3 Automatic grasping task with two force sensors

In previous sub-session, the automatic task is not performed so easily, since the fine target shape cannot be obtained from microscope image. In order to check where the target is grasped, the force control with two axes is necessary. However, it is difficult to fabricate a high-resolution force sensor having two axes sensing. Here, two force sensors with one axis can be attached on the each finger, since the micro hand has two fingers. Two force sensors with one axis are attached with different sensing direction for other. For example, one sensor is attached with face to the grasping direction in sensing direction, and another sensor is attached with rectangle direction for other. Fig 12 shows the configuration using two force sensors.

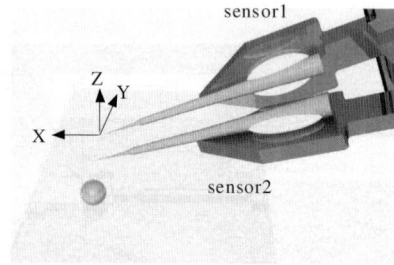

Fig. 12 Configuration in the grasping task with two force sensors

In order to estimate this way, the outputs from force sensors are checked when upper part and lower part of a target are grasped. A glass ball with 20[μm] in size is used as a target object.

5.4 Result of automatic grasping task with two force sensors

Fig. 13 shows the sensor outputs when the lower part of the target is grasped.

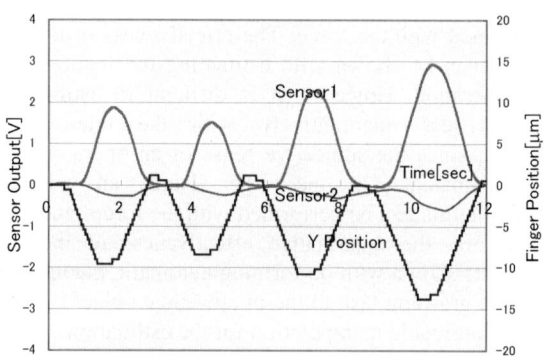

Fig. 13 Sensor outputs with grasping lower part of the glass ball

The output of "Sensor 1" shows grasping force on Y-axis and "Sensor 2" shows vertical force on Z-axis. It can be found from "Sensor 2" that the force is exerted to the finger as under direction force on Z-axis corresponding to the finger space "Y Position" and the grasping force "Sensor 1". It is considered that this force consists of the gravity force of target and adhesion forces between the target and the ground.

Fig. 14 shows the sensor output when the upper part of the target is grasped. Corresponding to the finger space and the grasp force, the vertical force on Z-axis changes. Its direction is opposite compared with the previous result Fig. 13. In this case, the applied force to the "Sensor 2" dose not return when the finger is opened. Because of grasping the upper part, the target is inserted and fixed into the space between the finger with "Sensor 2" and the ground. Therefore "Sensor 2" output dose not return, even the finger is opened. The increase of "Sensor 2" output means that the target more inserts into the space between the finger with "Sensor 2" and the ground with repetition of grasping. Under this condition, the picking is hardly success.

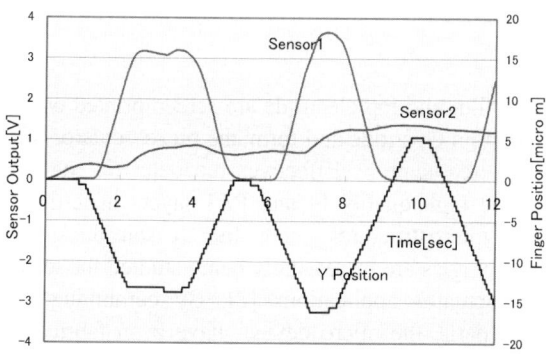

Fig. 14 Sensor outputs with grasping upper part of the glass ball

From the results of two experiments above, it can be clear that the difference of position between the grasping point and center of gravity of the target, as two force sensors are attached with different sensing direction.

6 Conclusions

The micro force sensor for micro manipulation with force control was developed in the corroboration research of MEL and Olympus Opt. Co. The force sensor has the resolution of 2[nN] in theoretically and one axis sensing. A short grass needle is attached on the end of the force sensor in order to obtain the end-effector for micro manipulation. For effectiveness of the force sensor, the automatic grasping task is performed with force control using the end-effector. It is found that the force control is not always necessary in the grasping task, since the grasping is basically easy with adhesion forces. The force control is important under the condition change when micro object is just grasped. Therefore, the success ratio of the task can be increased with non-force control for keeping grasp condition. Also, for more increasing the success rate, two force sensors are attached on each finger of micro hand with different sensing direction. By using this configuration, two axes force sensing can be perform and then it can be clear if fingers grasp to the center gravity of the target. From these experiments, the usage of the force sensor to micro manipulation was estimated.

References

[1] T. Tanikawa, T. Arai, "Development of a Micro-Manipulation System Having a Two-Fingered Micro-Hand", IEEE Transaction on Robotics and Automation, vol.15, No. 1, pp. 152 -162, Feb., 1999

[2] Y. Yokokohji, N. Hosotani and T. Yoshikawa, "Analysis of Maneuverability of Micro-Teleoperation Systems," Proceedings of IEEE Intnational Conference on Robotics and Automation, Vol. 1, pp. 237-242, 1994.

[3] T. Sakaki, Y. Inoue and S. Tachi, "Tele-Existence Virtual Dynamic Display Using Impedance Scaling with Physical Similarity," Proceedings of the 1993 JSME International Conference on Advanced Mechatronics, Tokyo, Japan, pp. 127-132, 1993.

[4] J. E. Colgate, "Power and Impedance Scaling in Bilateral Manipulation," Proceedings of IEEE International Conference on Robotics and Automation, Vol. 3, pp. 2292-2297, 1991.

[5] K. Kaneko, H. Tokashiki, K. Tanie, K. Komoriya, "Impedance Shaping based on Force Feedback Bilateral Control in Macro-Micro Teleoperation System," Proceedings of IEEE International Conference on Robotics and Automation, New Mexico, U.S.A, pp. 710-715, 1997.

[6] A. Stoffel, A. Kovacs, C. Reckleben, M. Schatle, "Electro Chemical Etching of thin membranes on silicon surfaces containing integrated circuit", The Electrochemical Society Proceeding Vol. 94-32, pp. 323-332, 1994

[7] Stewart, D., "A Platform with Six Degrees of Freedom," Prc. Institution of Mechanical Engineers, Vol. 180, Pt. 1, No.15, pp. 371-386,1965-66.

[8] HERVE, J.M., "Group Mathematics and Parallel Link Mechanisms", Proceedings of IMACS/SICE International Symposium on Robotics, Mechatronics and Manufacturing Systems '92 (Kobe, Japan), pp.459-464, 1992

[9] Arai, T., HERVE, J.M., Tanikawa, T., "Development of 3 DOF Micro Finger", Proceedings of the 1996 IEEE/RSJ International Conference on Intelligent Robots and Systems, November 4-8, Osaka, Japan, pp. 981-987, 1996.

Micro fluid device using thick layer piezo actuator prepared on Si micro-machined structure

Sukhan Lee, Jaewoo Chung*, Seungmo Lim, and Changseung Lee

System & Control Sector
Samsung Advanced Institute of Technology
P.O. Box 111, Suwon 440-600, Korea
*E-mail : jwchung@sait.samsung.co.kr

Abstract

The design and fabrication of a piezoelectric micro actuator on a silicon diaphragm is presented for application of micro ink jet. We developed the micro devices for the fluid transference that are fabricated using thick film technologies for ceramic materials and Si micro-machining. The performance of drop ejection was shown from the optimal design parameters. It is closely related to the precise machining of flow channel structure and piezoelectric actuator capacity. The features of micro fabrication and ink drop are described in this work.

1. Introduction

The future piezoelectric applications in micro devices will require PZT based films with several tens um thickness in micro-electromechanical system. The micro actuator based on piezoelectric mechanism can be utilized in many applications due to its high driving energy. The demands for the higher quality micro fluid devices have been intensified along with the explosion in recent years in MEMS as well as color digital images. Various types of ink jet heads have been produced and supplied to a variety of market [1]. For the generation of an inkjet, there is a variety of actuation principles, such as shear, push and bending modes, each of which has gained its market share. We have been developing piezo type micro devices by bending mode technology.

Using computer modeling, many variations in micro pumping devices can be designed and tested. Our devices comprised two major parts that are piezo thick film actuator and flow channel structure prepared by micro-machined Si materials. Si crystals are machined by the wet and dry etching processes forming the three-dimensional complex fluid flow channel and reservoir structures. A piezoelectric thick film and top electrode are screen printed on the Pt bottom electrode and form the piezo actuator after hear treatment. Effective piezoelectric devices require high quality Pt and PZT layers in terms of their microstructures. For this reason thick film technologies are increasingly being studied for sensor and actuator applications [2]. By combining two main parts, the micro devices have been fabricated. Several types of devices prepared with different kind of orifice density provided us successive in drop on demand ejection.

2. Device Fabrication

Piezoelectric thick film devices are multi-layer structures with Si substrates, a screen-printed top electrode and PZT materials, and sputtered bottom electrodes onto Ti seed later above Si oxides. Pt is chosen for the electrode and barrier layer avoiding diffusion between PZT and Si. After annealing in the furnace with special treatment, including low temperature sintering, slow heat experience, and the control of binder evaporation in raw materials, the green body changed to considerably high dense material. The resulting thickness of the PZT layer was 20 um and its composition showed near a morphotropic phase boundary. Between Si and PZT materials, there was not almost inter-diffusion that may degrade the dielectric constant, although the electrical properties are inferior to those of the bulk materials. The adhesion of the Pt bottom electrode is one of the important considerations in actuator applications because the PZT actuator deforms continuously during actuation. The Si etching process was carries out using combined TMAH wet and ICP dry etching. A schematic fabrication process of the fluid channel structure is shown in Fig. 1.
Thermal oxide on the front is etched by RIE and

coated with LPCVD low stress silicon nitride. After etching out the silicon nitride and oxide, Si and oxides is successively etched in TMAH and by RIE, respectively. Repeating the removing silicone nitride and oxide and Si step by step using the same way, the fluid channel structure is completed having with the micro fluid flow channel and reservoir. In an end step process, the orifice can be formed using the ICP etching process after photo-resist (PR) film patterned. The actuator with diaphragm can be bonded with this structure to form the micro fluid devices finally.

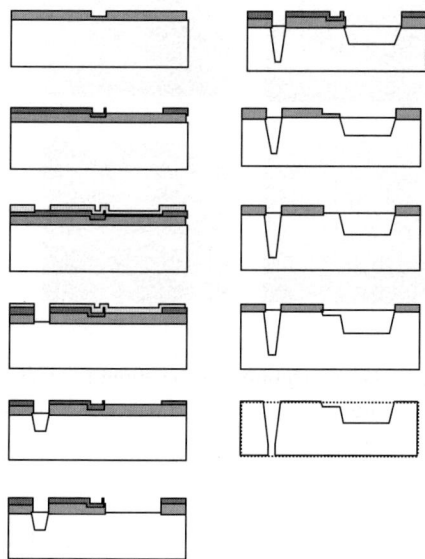

Fig. 1: Fabrication process of the micro fluid structure made of Si

3. Micro fluid pumping devices

The devices design is a complex and iterative process that attempts to satisfy many constants [3]. The abbreviated schematic of a front view cross-section for devices is shown in Fig. 2.

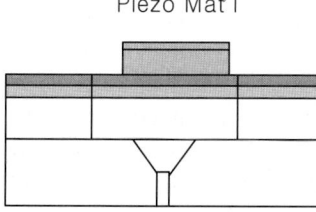

Fig. 2 ; Schematic of micro devices actuated by piezoelectric actuator fabricated with Si structure

The flow channel composes five parts such as, pressure chamber, manifold, outlet damper, restrictor and nozzle. Each pressure chamber is divided by upper plate. It provides the separation walls necessary to prevent cross talk effects. The manifold supplies ink into the pressure chamber. The outlet damper plays a role to avoid inclined fluidic flow as well as keeping straight drop ejection out of nozzle. The ink can be prevented from reverse flow into manifold when the chamber is pressurized with piezoelectric actuation. A piezo electric element is changing the volume of an ink chamber inducing drop ejection through the nozzles. The nozzle is intimately related to the speed of ink drop changing the size of its diameter. These actuators are fabricated as thick film of about 20 um on Si diaphragm forming the bi-morph structure. The expansion and contraction of the piezo materials makes the diaphragm to deflect, resulting in fires of micro droplets. To be able to determine the ratio of piezo material to diaphragm, we adopted the modeling using the equation [4] as follow ;

$$\delta = \frac{3l^2 d_{31} E \cdot g(m,n)}{T_p} = \frac{3l^2 d_{31} E}{T_s} \cdot \frac{g(m,n)}{n} \quad (1)$$

$$g(m,n) = \frac{mn^2(n+1)}{m^2 n^4 + 4mn^3 + 4mn + 1} \quad (2)$$

where,
l ; length of cantilever, E ; electric field, Tp ; Thickness of piezo material, Ts ; thickness of elastic diaphragm, m ; Y's Modulus ratio between piezo material and diaphragm, n ; thickness ratio between piezo material and diaphragm

The function of g(m,n) give us the plot as Fig. 3, From this relation, it is certain that the deflection is dependent on the size and physical properties.
In a case of diaphragm thickness fixed, it's thickness can be determined considering the stress of elastic diaphragm, the stability of controlled thickness, and

force necessary for actuator.

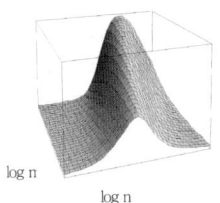

Fig. 3 ; g (m, n)

The deflection should be changed according to the thickness of piezo materials. Fig. 4. The maximum deflection was obtained at the thickness ratio about

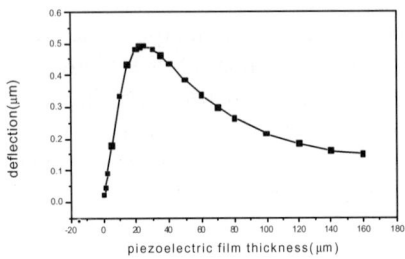

Fig. 4 : Deflection change in relation to thickness of piezo material at fixed thickness of diaphragm

1.25 keeping the diaphragm 20 um, width of the piezo device relative to the ink chamber. However, upon considering the displacement and displacement x pressure, which determine the volume and speed of ink droplet ejection [5], with the maximized value, we adopted 10 um diaphragm for the 20 um piezo actuator devices reflecting the competion in a process.

4. Drop ejection results

When the devices were fabricated with electrical connection, the piezoelectric actuator creates pressure wave in the chamber. The drop ejection is observed demanding on specific input pulse shape. The result is shown in Fig. 5. Each of drop velocity and volume was measured using stroboscope 3~5 m/sec, about 15 pl, respectively. The circle indicates the droplet fired form the orifice. Two drops from the next channel are ejected simultaneously to define the cross talk effect. There was not interaction during drop flight between two adjacent orifices. Two circles are shown for each channel because the real drop is reflected as same for each one. The fire frequency ranges from 1 kHz to 12 kHz.

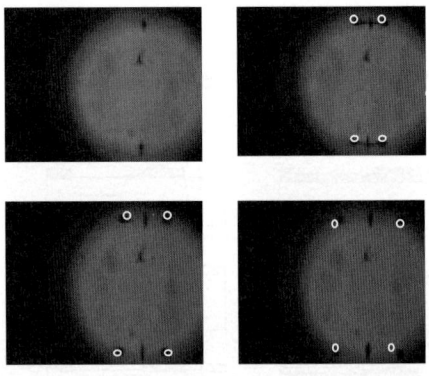

Fig. 5 ; Drop ejection ; (a) be about to eject from the orifice, (b) after 10 us, (c) 30 us, (d) 50 us

Higher resolution necessitates scaling jet designs to decrease drop volumes, while maintaining the high drop velocities needed for accurate drop displacement. The size of droplets may be reduced with lowering driving voltage, but lowering the driving voltage cause the droplets speed. To obtain higher performances, we are under development of the inkjet device with continuity to change many constants and control the process parameters to meet the high speed and large drop volume simultaneously.

5. Conclusion

The actuator and flow channel structure prepared by thick film process and Si micro-machined technologies, respectively, are fabricated as micro fluid pumping devices, so that micro size drops are ejected from the orifice. They showed straight flying from the orifice with spherical shape. Some satellites are also observed during firing resulting from the residual vibration generated in the meniscus. This will be improved to control driving pulse for cutoff the flight of drops. This piezo type actuator works on the inkjet head as well as on the necessary micro pumping parts in micro-system. The fabrication using Si materials for the flow channel structure has the advantages to make process as batch scale and control precise dimension with cleanness essential to the fluid flow behavior. The maintenance to have compatibility of thick film PZT actuator on the Si diaphragm through the complex process is also very important to show high performances for the micro fluid pumping.

References

[1] Hue P. Le, "Progress and Trends in ink jet printing Technology", Journal of Imaging Science and Technology, Vol. 42, 1, Jan./Feb., p49~62 (1998)

[2] Morten, B., G. D. Cisso, M. Prudenziati, Sensor and Actuator, A, 1994. 33~42,

[3] Ronald F. Burr, David A. Tence, and Sharon S. Berger, " Multiple Dot size Fluidics for phase change Piezoelectric Ink jets", Recent Progress in Ink jet Technologies II, edited by Eric Hanson, Society for Imaging Science and Technology, chap. 3, p192~198 (1999)

[4] M. R. Steel, F. Harrison, and P. G. Harper, " The piezoelectric bi-morph ; An experimental and theoretical study of its quasi-static response", J. Phys. D., vol. 11, p979~989 (1978)

[5] Minoru Usui, " Development of the new MACH, "Recent Progress in Ink jet Technologies II, edited by Eric Hanson, Society for Imaging Science and Technology, chap. 3, p199~202 (1999)

Microrobotic Cell Injection

Sun Yu   Bradley J. Nelson

Department of Mechanical Engineering
University of Minnesota
Minneapolis, Minnesota 55455
E-mail: yus, nelson@me.umn.edu

Abstract

Advances in microbiology demonstrate the need for manipulating individual biological cells, such as for cell injection which includes pronuclei injection and intracytoplasmic injection. Conventionally, cell injection has been conducted manually, however, long training, disappointingly low success rates from poor reproducibility in manual operations, and contamination all call for the elimination of direct human involvement. In this paper, we present a microrobotic system capable of performing automatic embryo pronuclei DNA injection autonomously and semi-autonomously through a hybrid visual servoing control scheme. After injection, the DNA injected embryos were transferred into a pseudopregnant foster female mouse to reproduce transgenic mice for cancer studies. Experimental results show that the injection success rate is 100%. The system setup, hybrid control scheme and other important issues in this application such as automatic focusing are discussed.

1. Introduction

Recent advances in microbiology, such as cloning, demonstrate that increasingly complex micromanipulation strategies for manipulating individual biological cells are required. Microrobotics and microsystems technology can play important roles in manipulating cells, a field referred to as biomanipulation.

Biomanipulation entails such operations as positioning, grasping, and injecting material into various locations in cells. Existing biomanipulation techniques can be classified as non-contact manipulation including laser trapping [3][4][5][22] and electro-rotation [1][17][21], and contact manipulation referred to as mechanical micromanipulation [11]. When laser trapping [3][4][5][22] is used for non-contact biomanipulation, a laser beam is focused through a large numerical aperture objective lens, converging to form an optical trap in which the lateral trapping force moves a cell in suspension toward the center of the beam. The longitudinal trapping force moves the cell in the direction of the focal point. The optical trap levitates the cell and holds it in position. Laser traps can work in a well controlled manner. However, two reasons make laser trapping techniques undesirable for automated cell injection. The high dissipation of visible light in aqueous solutions requires the use of high energy light close to the UV spectrum, raising the possibility of damage to the cell. Even though some researchers claim that such concerns could be overcome using wavelengths in the near infrared (IR) spectrum [5], the question as to whether the incident laser beam might induce abnormalities in the cells' genetic material still exists. One alternative to using laser beams is the electro-rotation technique. Electric-field-induced rotation of cells was demonstrated by Mischel [13], Arnold [2] and Washizu [21]. This non-contact cell manipulation technique is based on controlling the phase shift and magnitude of electric fields. These fields, appropriately applied, produce a torque on the cell. Different system configurations have been established for cell manipulation based on this principle [1][17], which can achieve high accuracy in cell positioning. However, it lacks a means to hold the cell in place for further manipulation, such as injection, since the magnitude of the electric fields has to be kept low to ensure the viability of cells. The limits of non-contact biomanipulation in the laser trapping and electro-rotation techniques make mechanical micro-manipulation desirable. The damage caused by laser beams in the laser trapping technique and the lack of a holding mechanism in the electro-rotation technique can be overcome by mechanical micro-manipulation.

Cell injection such as pronuclei DNA injection and intracytoplasmic sperm injection is one typical biomanipulation operation. A successful injection is determined greatly by applied force, injection speed and trajectory [11]. To date, there have been few quantitative measurements of membrane forces for cell injection, though MEMS devices are being developed for this purpose [6]. With the integration of a force sensor, real time sensor feedback on the forces exerted onto cells can be provided to the injection system, which can further improve the performance of the injection system compared to the control scheme using only visual feedback. This is currently under investigation.

Quantitative force measurements on cell membranes are not only significant for injection operations, but important for studies on properties of cell membranes, which holds potential for understanding key issues in cell injury and repair.

In this paper we present a visually servoed microrobotic system capable of performing automatic pronuclei DNA injection, which is a method for introducing DNA into embryos in order to create transgenic organisms. In Figure 1, a holding pipette holds a mouse embryo and an injection pipette performs the injection task. The objective of pronuclei injection is, in this case, to produce transgenic mice for use in cancer studies.

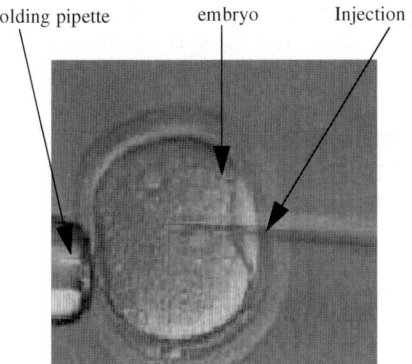

Figure 1. Cell injection of a mouse embryo. The embryo is approximately 50μm in diameter.

In Section 2, the procedure of embryo preparation is described. Section 3 presents the system setup. The visual servo control scheme is introduced in Section 4. Auto focusing is a critical issue in automated cell injection, and the implementation of auto focussing in this application is described in Section 5. We conclude the paper in Section 6.

2. Embryo Preparation

The embryos are collected in the Cancer Center at the University of Minnesota in accordance with standard embryo preparation procedure [8]. Three week old FVB/N female mice are injected with pregnant mare serum (PSM) to promote oval maturation. After approximately 45 hours the mice are injected with human chorionic gonadotropin (hCG) to promote synchronized ovulation. Then the superovulated female mice are mated to fertile male mice. Finally embryos are collected from the ampulla of female mice. A typical embryo is shown in Figure 2. The average diameter of the embryos is 50 μm.

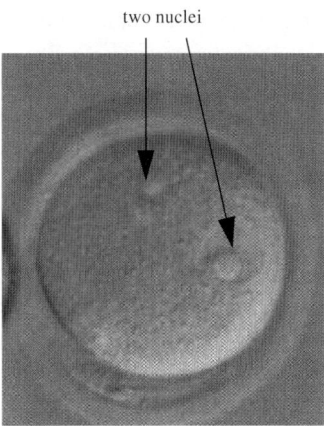

Figure 2. Embryo with two nuclei

For embryo pronuclei DNA injection, only embryos with two visible pronuclei can be selected for injection. DNA is deposited in one of the two nuclei. From the perspective of control, intracytoplasmic injection into embryos is less demanding than pronuclei DNA injection because intracytoplasmic injection only requires that a foreign genetic material be deposited within the embryo membrane, not necessarily in the nucleus.

3. System Setup

The autonomous embryo injection system is composed of an injection unit, an imaging unit, a vacuum unit, a microfabricated device, and a software unit. Figure 3 shows the system setup.

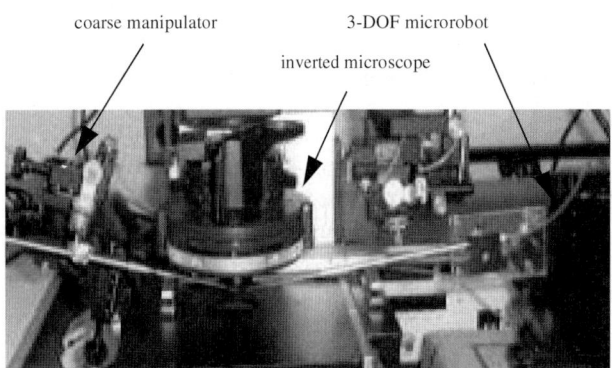

Figure 3. Autonomous embryo injection system

For embryo injection, vibration must be well controlled. Vibration not only causes difficulty in visually tracking features but also produces permanent and fatal harm in the injected location and the surrounding area. To avoid vibration, all units except the host computer and the vacuum units of our embryo pronuclei DNA injection system are placed on a floating table.

3.1. Injection Unit

The injection unit of the system includes a holding pipette, an injection pipette, two standard pipette holders, a high precision 3 DOF microrobot, and a coarse manipulator.

The injection and holding pipettes are both processed using a micropipette puller. The dimensions of the pipette tips are 1μm in inner diameter for the injection pipettes and 20μm in outside diameter for the holding pipettes. Both the holding pipettes and injection pipettes are held by pipette holders.

Extremely high precision motion control is required for successful embryo injection. A 3 DOF microrobot is used in which the XYZ axes each has a travel of 2.54 cm with a step resolution of 40nm. An injection pipette with a pipette holder is installed on the microrobot as shown in Figure 4

Figure 4. 3-DOF high precision microrobot

The holding pipette is installed on a micromanipulator that is a manually operated three dimensional coarse manipulator. The holding pipette holder and the injection pipette holder are both connected with Teflon tubing such that negative and positive pressure is provided to the tips of the pipettes for holding embryos and depositing DNA.

3.2. Imaging Unit

The imaging unit of the embryo injection system includes an inverted microscope, a CCD camera, a PCI framegrabber, and a host computer. An inverted microscope is used with a 400x objective. The CCD camera is mounted on port of the microscope. The framegrabber captures thirty frames per second. The tracking of image features, which is required for semi-autonomous teleoperation and autonomous injection, is performed on the host computer (a 450MHz Celeron) at 30Hz.

4. Embryo Injection

4.1. Automatic Embryo Injection

For embryo pronuclei DNA injection, focusing needs to be done precisely on the central plane of one of the two nuclei, the tip of the holding pipette, and the tip of the injection pipette. Failure to do so will cause the injection pipette to slide over the top of the embryo failing to puncture the nucleus membrane and possibly causing serious injure to the cell membrane.

4.1.1. Hybrid Control Scheme for Embryo Injection

Our hybrid control scheme consists of image-based visual servo control and precise position control. In image-based visual servo control, the error signal is defined directly in terms of image feature parameters. The motion of the microrobot causes changes to the image observed by the vision system. Although the error signal is defined on the image parameter space, the microrobot control input is typically defined either in joint coordinates or in task space coordinates. In formulating our visual servo system, task space coordinates are mapped into sensor space coordinates through a Jacobian mapping. Let x_T represent coordinates of the end-effector of the microrobot on the task space, and $\dot{x}_T$ represent the corresponding end-effector velocity. Let x_I represent a vector of image feature parameters and $\dot{x}_I$ the corresponding vector of image feature parameter rates of change. The image Jacobian, $J_v(x_T)$, is a linear transformation from the tangent space of task space T at x_T to the tangent space of image space I at x_I.

$$\dot{x}_I = J_v(x_T)\dot{x}_T \qquad (1)$$

where $J_v(x_T) \in \Re^{k \times m}$, and

$$J_v(x_T) = \left[\frac{\partial x_I}{\partial x_T}\right] = \begin{bmatrix} \frac{\partial x_{I1}(x_T)}{\partial x_{T1}} & \cdots & \frac{\partial x_{I1}(x_T)}{\partial x_{Tm}} \\ \cdots & \cdots & \cdots \\ \frac{\partial x_{Ik}(x_T)}{\partial x_{T1}} & \cdots & \frac{\partial x_{Ik}(x_T)}{\partial x_{Tm}} \end{bmatrix} \qquad (2)$$

m is the dimension of the task space T. The derivation of the model can be found in [20].

The state equation for the visual servo control system is as follows.

$$x(k+1) = x(k) + TJ_v(k)u(k) \qquad (3)$$

where $x(k) \in \Re^{2M}$ (M is the number of features being tracked); T is the sampling period of the vision system; and $u(k) = \begin{bmatrix} \dot{X}_T & \dot{Y}_T \end{bmatrix}$ is the microrobot's end-effector velocity.

The control objective of the system is to control the motion of the end-effector, i.e. the injection pipette, in order to place the image plane coordinates of the feature on the target in the switch area shown in Figure 5(a). The control strategy used to achieve the control objective is based on the minimization of an objective function that places a cost on errors in feature positions, $[x(k+1) - x_{switch}]$, and a cost on providing a visual control input $u(k)$.

$$E(k+1) = [x(k+1) - x_{switch}]^T Q [x(k+1) - x_{switch}] + u^T(k) L u(k) \qquad (4)$$

This expression is minimized with respect to the current control input $u(k)$. The result is the expression for the visual control input.

$$u(k) = -[TJ^T_v(k)QTJ_v(k)+L]^{-1}TJ^T_v(k)Q[x(k)-x_{switch}] \quad (5)$$

The weighting matrices Q and L allow the user to place more or less emphasis on the feature error and the control input. Methods for selecting these matrices can be found in [15].

When the visual servo controller guides the end-effector of the microrobot into the switching area, the control scheme switches to precise position control. The visual servo control and the precision position control jointly form the hybrid control scheme for automatic embryo pronuclei DNA injection. The complete hybrid control scheme is

$$U(k) = F_\sigma \begin{bmatrix} x(k)-x_{switch} \\ x_T(k)-x_{TD} \end{bmatrix} \quad (6)$$

where

$$F_\sigma = \left[-\sigma_1[TJ^T_v(k)QTJ^T_v(k)+L]TJ^T_v(k)Q \;\; \sigma_2 I \right] \quad (7)$$

I is a 2×2 unit matrix; and x_{TD} is the desired position on the task space T.

The switching condition is $\sigma_1\sigma_2 = 0$, and
$\sigma_1 = 1$ when $x(k) \notin (c,r)$;
$\sigma_2 = 1$ when $x(k) \in (c,r)$.
where (c,r) is the switching area shown in Figure 5(a).

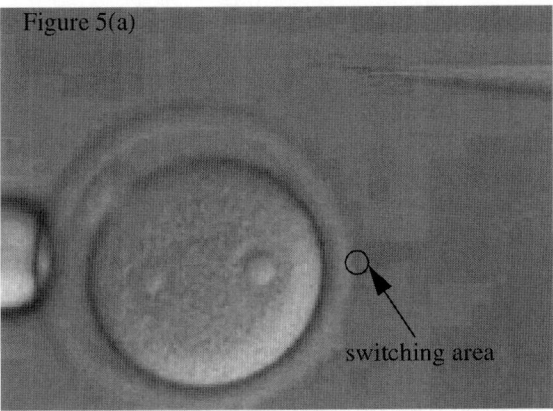
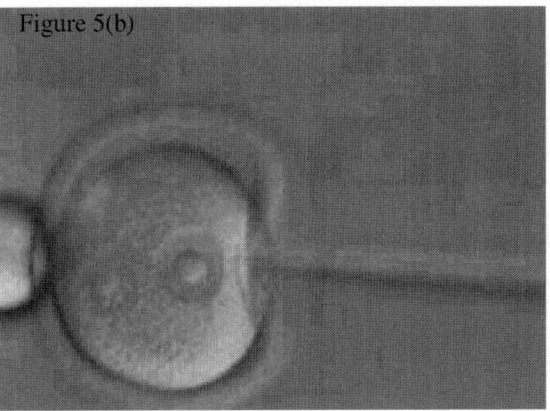

Figure 5. Embryo pronuclei DNA injection using hybrid control

The injection pipette is originally positioned away from the embryo shown in Figure 5(a). The hybrid controller guides the microrobot with the injection pipette into the nucleus of the embryo where DNA is deposited. Figure 5(b) shows the injection process.

4.1.2. SSD Tracking Algorithm

For visual servo control, we adopt the sum-of-squared-differences optical flow (SSD) tracking algorithm [16]. SSD is an effective method for tracking in a structured environment where image patterns do not change considerably between successive frames of images. In visual servoing of microrobot under a microscope, the predictable environment and controlled illumination make SSD a robust tracking method. It is desirable to select features with high gradients, such as edges and corners that are distinct from their neighboring regions. In embryo injections, we select the tip of the injection pipette as a feature.

The basic assumption of SSD tracking is that intensity patterns $I(x,y,t)$ in a sequence of images do not change rapidly between successive images $I(x,y,t+1)$. In implementing the algorithm, we acquire a template of 20x20 pixels $T_{20 \times 20}$ around the feature, i.e. the tip of the injection pipette. An SSD correlation measure is calculated for each possible displacement (dx,dy) within a 20x20 pixel search window in the new image $I(x,y,t+1)$

$$SSD(dx,dy) = \sum_{i,j \in N} [I(x_1+dx+i, y_1+dy+j) - T(x_1+dx+i, y_1+dy+j)]^2 \quad (8)$$

The distance (dx,dy) having the minimum SSD measure shown in (8) is assumed to be the displacement of the feature. The amount of processing depends greatly on the template size and the size of the search window. A large template will increase robustness, while a larger search window will handle larger displacements, provided frames of images can be processed in real time.

4.2. Teleoperated Embryo Injection

In teleoperation mode, a supervisor guides a cursor on a monitor (using a computer mouse) which the visual servoing system accepts as a control input to a visual servoing control law. Figure 6 shows the program interface that performs teleoperation injection.

The main teleoperation injection process flow is described as follows.

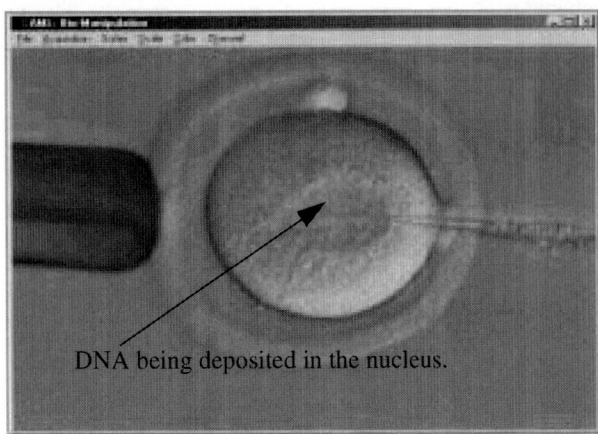

Figure 6. Teleoperated embryo injection

Step1: Focus on the nucleus of the embryo, the tips of the holding pipette and the injection pipette.
Step2: Control the vacuum unit to hold the embryo.
Step3: Guide the injection pipette to the edge of the embryo, well aligned with the larger nucleus of the embryo.
Step4: Control the injection pipette to move into the nucleus of the embryo.
Step5: Deposit DNA inside the nucleus of the embryo.
Step6: Move the injection pipette out of the embryo. This completes the teleoperation process.

5. Automatic Focusing

In our experiments, each batch of oocytes consisting of approximately ten egg cells is loaded from the incubator onto slides. Unloading and reloading cannot be completed unless the injection pipettes are moved out of focus and back into focus. In addition, when switching among the ten cells in one batch, the injection pipette must be moved out of the focus plane and brought back into focus before injection is conducted. Refocusing takes significant time and effort during manual operations, which makes automatic focusing attractive.

Various focus measures and metrics have been proposed and developed in the past [7][9][10][12][14][18][19]. These measures include Fourier transform, Tenengrad, high pass filtering, histogram entropy, gray level variance and sum-modulus-difference. In this application, since the injection pipettes only need to be moved in a plane perpendicular to the platform of the microscope when switching among cells or among batches, automatic focusing can be implemented by using a simple template matching technique.

$$error = \sum_{i,j \in N} [I(x_1+dx+i, y_1+dy+j) - T(x_1+dx+i, y_1+dy+j)]^2 \quad (9)$$

The error is minimized when the injection pipette is in

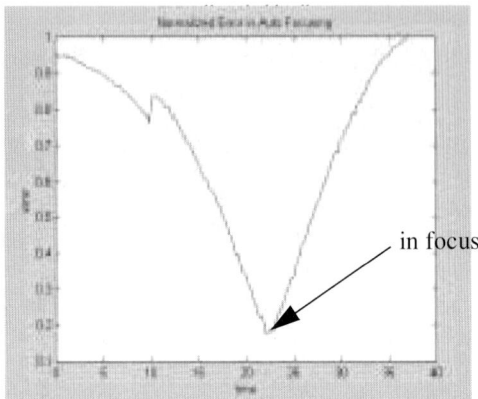

Figure 7. Error change in automatic focusing

the focus plane. Figure 7 shows the change of the error with the distance between each position of the injection pipette and the focus plane. Templates can be obtained while the injection pipettes are initially in focus. After switching among cells or batches, the microrobot controls the injection pipettes to move down, performing template matching to guide the pipettes into focus. Experimental results demonstrate that this technique performs well.

6. Experiment Results

Eight mouse embryos were collected for embryo pronuclei DNA injection. Three of the eight embryos were discarded due to abnormalities in the nuclei. The other five embryos were injected by the autonomous system and were then placed in an incubator for 45 minutes. The injected embryos were visually inspected for viability. All five injected embryos proved to be viable and were transferred into a foster female mouse to reproduce transgenic mice. In nineteen days, transgenic mice were reproduced. Experimental results demonstrate that the success rate for automatic injection is 100%. Figure 8 shows the mouse (with a clip) with the injected embryos implanted.

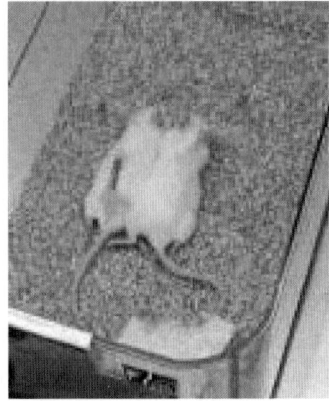

Figure 8. Mouse implanted with the injected cells

7. Conclusions

An autonomous embryo pronuclei DNA injection system was developed. Experimental results show that our success rate for embryo pronuclei DNA injection is 100%. Our system is not only capable of conducting pronuclei DNA injection, but is also suitable for performing intracytoplasmic injection. It is important to note that the task of embryo pronuclei DNA injection differs in two aspects from that of intracytoplasmic injection into embryos. In embryo pronuclei DNA injection, DNA must be deposited in one of the two nuclei, while for intracytoplasmic injection foreign genetic material such as sperm only needs to be deposited within the embryo, which makes the motion control task more difficult for pronuclei injection. Membrane material properties are also different. Oocyte membranes are much more elastic than embryo membranes, making membrane penetration more complex for intracytoplasmic sperm injection than for pronuclei injection. Therefore, force sensing is particularly important for obtaining quantitative measurements during intracytoplasmic injection. When the force sensor is integrated into the injection system, real time force feedback can be provided, which is expected to further improve the performance of the injection system. Integrating the force sensor [6] that is being developed into the injection system is under investigation.

8. Acknowledgements

The authors would like to thank Ms. Sandra Horn and Dr. David Largaespada of the Cancer Center at University of Minnesota for assistance in embryo preparation and in the experiments. We would also like to thank Prof. John Bischof and Prof. Ken Roberts for their invaluable discussions concerning this research.

9. References

[1] F. Arai, K. Morishima, T. Kasugai, T. Fukuda, "Bio-Micromanipulation (new direction for operation improvement)," Proceedings of the 1997 IEEE/RSJ International Conference on Intelligent Robot and Systems, IROS 1997, New York, 1300-1305.
[2] W.M. Arnold, U. Zimmermann, "Electro-Rotation: Development of a Technique for Dielectric Measurements on Individual Cells and Particles," Journal of Electrostatics, Vol. 21, No. 2-3, pp. 151-191, 1988.
[3] A. Ashkin, "Acceleration and Trapping of Particles by Radiation Pressure," Physical Review Letters, Vol. 24, No. 4, pp. 156-159, 1970.
[4] T.N. Bruican, M.J. Smyth, H.A. Crissman, G.C. Salzman, C.C. Stewart, J.C. Martin, "Automated Single-Cell Manipulation and Sorting by Light Trapping," Applied Optics, Vol. 26, No. 24, pp. 5311-5316, 1987.
[5] J. Conia, B.S. Edwards, S. Voelkel, "The Micro-robotic Laboratory: Optical Trapping and Scissing for the Biologist," Journal of Clinical Laboratory Analysis, Vol. 11, No. 1, pp. 28-38, 1997.
[6] E.T. Eniko, B.J. Nelson, "Three-Dimensional Microfabrication for a Multi-Degree-of-Freedom Capacitive Force Sensor Using Fibre-Chip Coupling," Journal of Micromech. Microeng., Vol. 10, pp.492 - 497, 2000.
[7] P. Grossman, "Depth from Focus, Pattern Recognition Letters," Vol. 5, pp.63-69, 1987.
[8] B. Hogan, R. Beddington, F. Costantini, E. Lacey, *Manipulating the Mouse Embryo: A Laboratory Manual*, Second Edition, Cold Spring Harbor Laboratory Press, 1994.
[9] B.K.P. Horn, *Robot Vision*, MIT Press, 1986.
[10] R.A. Jarvis, "A Perspective on Range Finding Techniques for Computer Vision," IEEE Trans. PAMI, Vol. 5, No. 2, pp.122-139, 1983.
[11] Y. Kimura, R. Yanagimachi, "Intracytoplasmic Sperm Injection in the Mouse," Biology of Reproduction, Vol. 52, No. 4, pp. 709-720, 1995.
[12] E.P. Krotkov, "Focusing," Int. J. of Computer Vision, pp.223-237, 1987.
[13] M. Mischel, A. Voss, H.A. Pohl, "Cellular Spin Resonance in Rotating Electric Fields," Journal of Biological Physics, Vol. 10, No. 4, pp. 223-226, 1982.
[14] H.K. Nayar, Y. Nakagawa, "Shape from Focus: An Effective Approach for Rough Surfaces," Proc. IEEE Int. Conf. on Robotics and Automation, pp.218-225, 1990.
[15] N.P. Papanikolopoulos, B.J. Nelson, P.K. Khosla, "Full 3-D Tracking Using the Controlled Active Vision Paradigm," IEEE Int. Symp. Intell. Contr., ISIC-92, 1992, pp. 267-274.
[16] N.P. Papanikolopoulos, "Selection of Features and Evaluation of Visual Measurements During Robotic Visual Servoing Tasks," Journal of Intelligent & Robotic Systems: Theory and Applications, Vol. 13, pp 279-304, 1995.
[17] M. Nishioka, S. Katsura, K. Hirano, A. Mizuno, "Evaluation of Cell Characteristics by Step-Wise Orientational Rotation Using Optoelectrostatic Micromanipulation," IEEE Transactions on Industry Applications, Vol. 33, No. 5, pp. 1381-1388, 1997.
[18] A.P. Pentland, "A New Sense for Depth of Field," IEEE Trans. PAMI, Vol. 9, No. 4, pp.523-531, 1987.
[19] J.F. Schlag, A.C. Sanderson, C.P. Neumann, F.C. Wimberly, "Implementation of Automatic Focusing Algorithms for a Computer Vision System with Camera Control," Tech. Report, CMU-RI-TR-83-14, 1983.
[20] B. Vikramaditya, B.J. Nelson, "Visually Guided Microassembly Using Optical Microscopes and Active Vision Techniques," IEEE Int. Conference on Robot and Automation, Albuquerque, NM, Apr. 21-27, 1997, pp. 3172-3177.
[21] M. Washizu, Y. Kurahashi, H. Iochi, O. Kurosawa, S. Aizawa, S. Kudo, Y. Magariyama, H. Hotani, "Dielectrophoretic Measurement of Bacterial Motor Characteristics," IEEE Transactions on Industry Applications, Vol. 29, No. 2, pp.286-294, 1993.
[22] W.H. Wright, G.J. Sonek, Y. Tadir, M.W. Berns, "Laser Trapping in Cell Biology," IEEE Journal of Quantum Electronics, Vol. 26, No. 12, pp. 2148-2157, 1990.

FORCE FEEDBACK-BASED MICROINSTRUMENT FOR MEASURING TISSUE PROPERTIES AND PULSE IN MICROSURGERY

A. Menciassi, A. Eisinberg, G. Scalari, C. Anticoli, M.C. Carrozza, P. Dario

Scuola Superiore Sant'Anna - MiTech Lab

via Carducci, 40 - 56127 Pisa (Italy)

e-mail: arianna@sssup.it

Abstract

Miniaturized and "smart" instruments capable of characterizing the mechanical properties of tiny biological tissues are needed for research in biology, physiology and biomechanics, and can find very important clinical applications for diagnostics and minimally invasive surgery (MIS). We are developing a set of robotic microinstruments designed to augment the performance of the surgeon during MIS. These microtools are intended to restore (or even enhance) the finger palpation capabilities that the surgeon exploits to characterize tissue hardness and to measure pulsating vessels in traditional surgery, but that are substantially reduced in MIS.

This paper describes the main features and the performance of a prototype miniature robotic instrument consisting of a microfabricated microgripper, instrumented with semiconductor strain-gauges as force sensors. For the (*in vitro*) experiments reported in this paper, the migrogripper is mounted on a workstation and teleoperated. An haptic interface provides force-feedback to the operator. We have demonstrated that the system can discriminate tiny skin samples based on their different elastic properties, and feel microvessels based on pulsating fluid flowing through them.

Keywords: tissue characterization, palpation, micromanipulation, microgrippers, force-feedback.

1. Introduction

Measuring the mechanical properties of cells and tiny biological tissues is important for research in biology, physiology and biomechanics. Cells of the body are exposed to mechanical stresses and strains throughout life, and this is critical to the health and functions of various tissues and organs of the body [1-2]. It is clear that microfabrication and microrobotics will provide very powerful means for the development of a novel generation of research instruments. In fact, a variety of laboratory apparatuses and microfabricated instruments have been developed recently for investigating cells and tissues properties by mechanical stimulation [3-5]. Miyazaki et al. [6] designed a micro tensile test system for cells and fine biological fibers and determined tensile properties of collagen fibers isolated from the patellar tendon. Sato et al. [7] measured the viscoelastic properties of cultured porcine aortic endothelial cells and analyzed them using a standard linear viscoelastic model. Wang and Coceani [8] set up an *in vitro* preparation of fetal lambs pulmonary arteries and veins to examine local factors responsible for hemodynamic control

In addition to basic research in biomechanics, miniaturized robotic instruments are fundamental tools for minimally invasive diagnostics and surgery. Perhaps the most critical factor in MIS is the severe reduction of sensory and dexterous manipulation capabilities of the surgeon. Restoring (or even augmenting) these capabilities by developing new "smart" surgical tools would have a major impact on the future of the whole field of MIS.

Palpation is a procedure that surgeons exploit regularly and "naturally" in traditional surgery in order to estimate tissue hardness and to locate blood vessels hidden beneath opaque tissues. This is very important because the accidental puncturing of blood vessels during MIS is a rather frequent life threatening complication.

Research on robotics palpation has received some attention recently. A tactile array system aimed at finding hidden arteries has been proposed by Howe et al. [9], and a commercial tool for MIS has been instrumented in order to enhance the surgeon's haptic perception of the manipulated tissue by Bicchi et al. [10].

The authors are investigating a new class of robotic microinstruments whose ultimate goal is to augment the performance of the surgeon during MIS. In previous papers we have presented some preliminary work on the fabrication, characterization and control of microgrippers for micromanipulation tasks [11-13]. We have also described some initial results on the use of microgrippers for characterizing soft tissues [14].

This paper discusses the main features and the performance of a prototype miniature robotic instrument consisting of a microfabricated microgripper, instrumented with semiconductor strain-gauges as force sensors. *In vitro* experiments have been carried out on the migrogripper mounted on a workstation and teleoperated. An haptic interface provides force-feedback to the operator. The system can discriminate tiny skin samples based on their different elastic properties, and feel microvessels based on pulsating fluid flowing through them.

2. The Modular Workstation

The apparatus we developed for testing the microrobotic system is modular and teleoperated. It comprises: a microfabricated instrumented probe which can exert controllable force-displacement cycles on soft tissues and measure force generated by pulsating flow in microvessels; a 3 d.o.f. motorized manipulator which moves the microgripper; a fiber optic microscope (50-200X) with monitor which allows the operator to visualize the sample and the microgripper position; a PC-based control unit; and a haptic interface (Phantom 1.0, SensAble Technologies Inc.) which provides force-feedback to the operator. A scheme of the apparatus is illustrated in Figure 1.

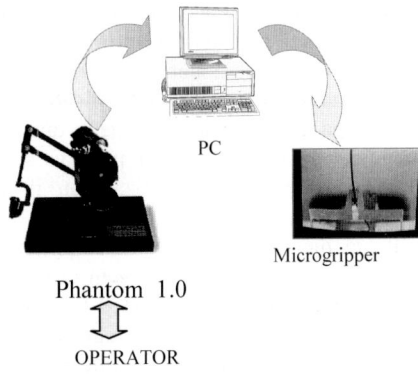

Figure 1 *Scheme of the apparatus*

2.1 The instrumented microprobe

The microprobe selected as the end-effector of the testing apparatus is a LIGA-fabricated microgripper made out of electroplated nickel coated with a thin gold layer [11]. The geometry of the microgripper (overall length 17 mm, overall width 7.5 mm, thickness 0.4 mm) is showed in Figure 2.

The microgripper exploits flexure joints in order to generate large displacement at the fingertips in a compact structure, relatively easy to fabricate and assemble by microfabrication technologies. The microgripper is actuated by a low voltage multi-layer PZT stack (TOKIN AE0203D16), whose maximum driving voltage is 150 V, and maximum displacement is about 17 μm). The voltage is supplied by a power amplifier which receives an input signal directly from a PC through an A/D interface (DAQ AT-A06-10, National Instruments Inc.). When the actuator pushes the rear part of the microgripper, the flexure joint-based deformable structure amplifies the displacement, thus producing much larger displacement at the tip.

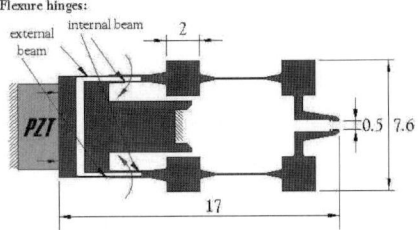

Figure 2. *Microgripper design*

The microprobe is instrumented with semiconductor strain gauges (ESU-025-1500 Entran Devices Inc.) [12]. In order to obtain accurate and repeatable measurements of force, the symmetry of the microgripper structure should be matched by a symmetrical configuration of the strain gauge sensors. A symmetrical configuration is also useful for thermal compensation and better signal to noise ratio [15]. The optimal symmetrical configuration consists of a full Wheatstone bridge sensor comprising four active strain gauges. The sensor can be used to implement a PI closed loop force control, as described in [12-13]. The four strain gauges have been mounted in two pairs on the microgripper, each pair located at a flexure joint: one strain gauge of the pair measures compression and the other measures tension. Figure 3 illustrates the approximate location of the strain gauges in the microgripper structure.

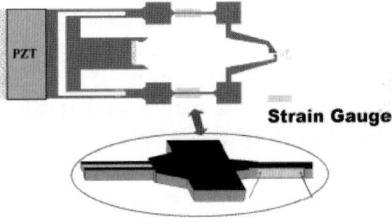

Figure 3. *Scheme of the microgripper showing the location of the strain gauge sensors*

As the fingertips grasp a tissue sample, the strain gauges measure the deformation of the microprobe structure. After proper calibration, the output signal of the strain gauge bridge can be read as a force signal. To this purpose, the strain gauge sensor system was calibrated by opening the microgripper fingertip against a load cell (Model GM2 3M, PTC Electronics Inc.; full scale 300 mN, accuracy 0.01 mN). Calibration tests showed the good linearity of the strain gauge sensors and indicated that the microstructural deformation, and hence the force exerted by the micro gripper, can be monitored rather accurately using the Wheatstone bridge configuration.

The output signal of the instrumented microgripper in the force range 0-6 mN is illustrated in Figure 4. The force signal measured by the instrumented microprobe is delivered to the Phantom haptic system, thus providing force feedback to the human operator.

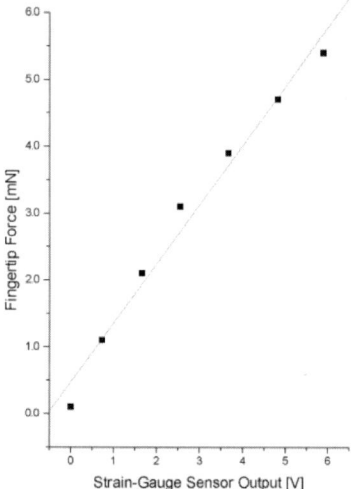

Figure 4. *Calibration of the strain gauge sensor mounted on the microgripper*

We have also developed a new version of microgripper, fabricated in superelastic alloy (Ni50.8Ti49.2) by wire micro Electro Discharge Machining (µEDM). µEDM allows to fabricate high aspect-ratio structures made out of different conducting and semiconducting materials with good surface finishing and without any thermal alterations even in the smallest features. Moreover, even when the above characteristics are not strictly required (in our design the aspect ratio is ~5), µEDM is an elective choice to machine hard materials which could not be machined by other technologies.
Superelastic alloy is the material of choice because of its favorable mechanical properties. This alloy allows to obtain a more robust and stiffer microprobe and to fabricate flexure joints which reach a large displacement amplification factor still within the elastic range of the material. A photograph of the new microgripper is shown in Figure 5.

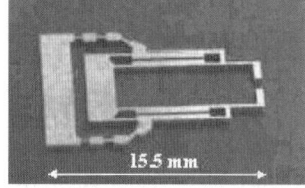

Figure 5. *The new superelastic alloy microgripper fabricated by micro Electro Discharge Machining*

2.2 The haptic interface

The system control is implemented by means of a PC, which interfaces the microgripper actuating and sensing circuits with the Phantom haptic interface. A graphic interface continuously displays the state of the system to the operator. The system is activated through the graphic interface and various modes of functionality can be selected.

For calibration purposes, the Phantom can be by-passed by retrieving the closure commands via a mouse or via an automated sine wave generator. In normal operation, the operator drives the actuators by the haptic interface and, at the same time, "feels" the grasping forces measured by the instrumented microprobe. By using this apparatus, different micro-samples of soft tissue have been tested and pulse in microvessels has been "felt". However, in order to extract quantitative parameters from the tested samples, signals must be processed. To this aim a method has been devised for the identification of the probe-sample system, as illustrated in the following paragraph.

3. Testing Methods and Experimental Results

While for research in biology and physiology accurate measurements of tissue mechanical properties are required, but on-line monitoring is not strictly necessary, most microsurgery tasks (e.g. the identification of microvessels embedded in the organ or tissue to operate) require an instrument which is able to provide some information to the operator in real time, even if the information is less accurate.
In order to evaluate the performance of the proposed microrobotic system in different measurement tasks, we have selected two experiments: one directed to test the elastic properties of micro samples of biological tissue, and the second to detect pulsating flow in microvessels. These experiments are described in the next two sections.

3.1 Testing the elastic properties of microsamples of biological tissue

Although the main intended application of the apparatus is for *in vivo* experiments in physiology and for microsurgery, in this phase we elected not to make tests with animals for ethical reasons. Instead, we used tiny samples (width about 100 µm) of human skin freshly excised from areas around the fingernails of three volunteers. In order to obtain information on the elastic properties of the microsamples, the following procedure was used: the microgripper grasped the sample and, whilst grasping, a small signal step voltage was applied to the piezo-actuator resulting in a sudden closure of the gripper. This results in small oscillations of the gripper fingertips at frequencies dependent on the resonant frequencies of the mechanical system ("gripper plus sample"). The step response of the system in idling conditions and for the three different skin samples is shown in Figure 6.

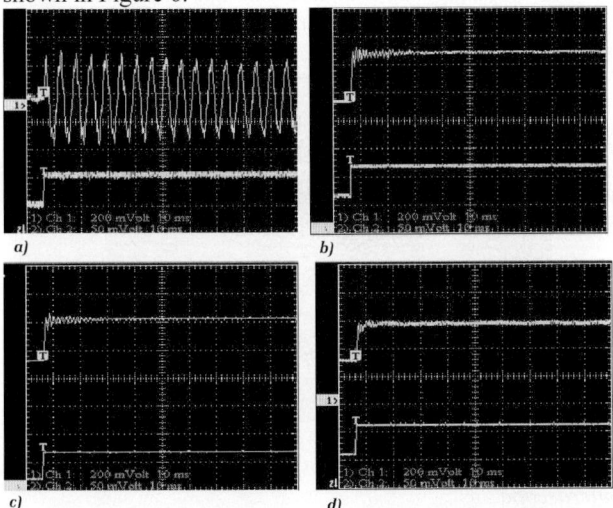

Figure 6. *Step response of the system in idling conditions (a) and when grasping three different skin samples (b, c, d)*

As clearly visible in Figure 6, the step response signals are not sufficient per se to discriminate the different skin samples. Additional processing is needed to provide the apparatus with the required discrimination capabilities. The method we elected to use for analyzing the experimental data presented in Figure 6 is based on system identification [16]. The microgripper system was identified by means of a small step impulse excitation, filtering and Fast Fourier Transform (FFT) of the output signals. The experimental data were frequency analyzed using MATLAB 5.3 (The MathWorks, Inc.). Results are illustrated in Figure 7.

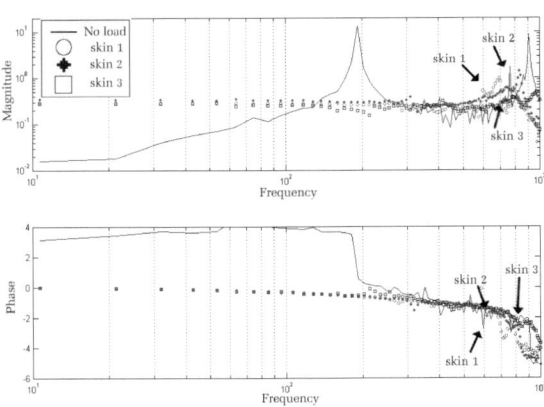

Figure 7. *Bode plots of the frequency response of the gripper microsystem in idling conditions (blue line) and when grasping three different skin samples*

In idling conditions the gripper exhibits two main resonant frequencies. When the gripper grasps a sample, the fundamental mode of oscillation is blocked, the second resonant frequency is shifted depending on the mechanical properties of the sample, and the damping increases due to internal friction. This behavior can be observed in the Bode diagrams of Figure 7 which refer to the idling condition and to the cases when three different skin samples are grasped. In the system identification procedure, the resonant frequency and the fading constant of the poles relative to the secondary mode of oscillation are calculated for each system "gripper plus sample" starting from the acquired oscillating signals. Figure 8 illustrates the Bode diagrams related to the identified systems for the cases of idling condition and when the three different skin samples are grasped.

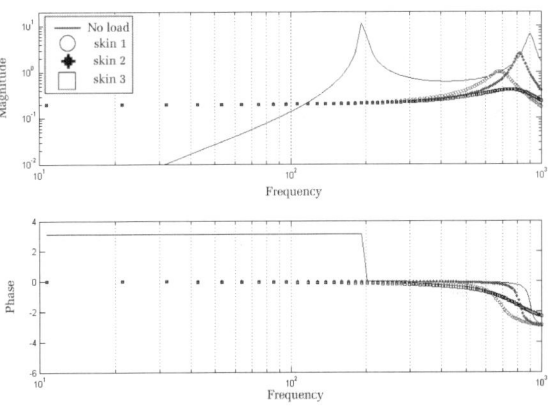

Figure 8. *Bode plots of the identified systems. The signals corresponding to the gripper grasping three different tissue samples are much easier to discriminate*

In idling conditions, the following experimental transfer function was obtained:

$$G(s) = K \cdot \frac{s^2}{\left(\frac{s^2}{w^2_{p0}} + 2\frac{\zeta_{p0}}{w_{po}} + 1\right) \cdot \left(\frac{s^2}{w^2_{p1}} + 2\frac{\zeta_{p1}}{w_{p1}} + 1\right)}$$

with :

$$\begin{cases} K = 10^{-13} \\ f_{p0} = 195 Hz \quad \zeta_{p0} = 0.00005 \\ f_{p1} = 900 Hz \quad \zeta_{p1} = 0.025 \end{cases}$$

When grasping skin samples, the following resonant frequencies were derived from the diagrams in Figure 8:

Skin sample 1: f = 690 Hz, ς = 0.1, K = 0.2
Skin sample 2: f = 800 Hz, ς = 0.25, K = 0.2
Skin sample 3: f = 820 Hz, ς = 0.04, K = 0.2

In order to extract information on the mechanical properties of each sample, the resonant frequency and the fading constant of the poles relative to the secondary mode of oscillation should be interpreted using models for the cases of the gripper alone (idling condition) and of the complete system "gripper plus sample". The system "gripper and grasped sample" has been modeled as a 2 d.o.f. system and this simplified model has been used to calculate the first two resonating frequencies of the system [14]. The characteristic equations of this system have been obtained and the relative frequencies of the modes of oscillation have been calculated. Finally, the poles of the system have been calculated by resolving the following second order polynomial equation:

$$P(\omega) = m_1 m_2 \omega^4 + [m_2(k_1 + k_2) + m_1(k_2 + k_c)]\omega^2 + \\ + [(k_1 + k_2)(k_2 + k_c) - k_2^2]$$

where ω is the resonant frequency of the system; k_1 and k_2 are the elastic constants of the microgripper (according to our 2 d.o.f. model); m_1 and m_2 are the characteristic masses of the system; k_c is the elastic constant of the sample.

The resonant frequencies ω_i of the system "microgripper and grasped sample" obtained experimentally (see Figure 6) corresponds to the roots of the equation above. By knowing the value of k_1, k_2, m_1 and m_2 from mechanical simulations of the microgripper, the experimental resonant frequencies can be related to the elastic constant (k_c) of each sample.

3.2 Testing pulse in microvessels

The same system (connected to the haptic interface) was used to to measure and "feel" pulse in microvessels. For ethical reasons, microvessels were substituted by polymeric microtubes which simulate microvessels (for example those in the coronary tree) with good approximation. Pulsating blood flow was simulated by a simple microfluidic circuit comprising a Micro Annular Gear Pump (mzr-2903, Mikrosysteme GmbH) and polymeric microtubes (Zeus Scientific Inc), with an external diameter of 800 µm and an internal diameter of 500 µm. The micro pump is PC-controlled by a software interface that allows dosing and continuos flow of fluid with an accuracy of more than 99.5% on delivered volume. Experimental parameters such as acceleration time, maximum speed and time interval among dosing pulses are variable in a wide range. Physiological blood flow in different sites and conditions can be simulated quite accurately in the microfluidic circuit, and therefore the performance of the proposed microrobotic system can be evaluated in realistic conditions. Water with a flow rate of 40 µl and a pulsating frequency of 8.3 Hz (reproducing physiological parameters in coronary system of mice) was supplied through the microtube grasped by the instrumented microgripper, as illustrated in Figure 9.

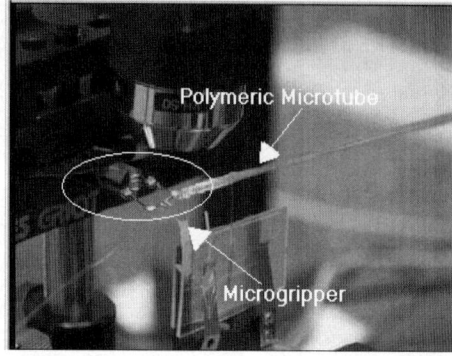

Figure 9. *Microgripper grasping a microtube under microscope*

The pulse signal was measured by the strain gauge sensor and also "felt" distinctly by the operator through the Phantom interface. Figure 10 shows the recorded strain gauge signal and the operator hand who "feels" the pulse signal through the haptic interface.

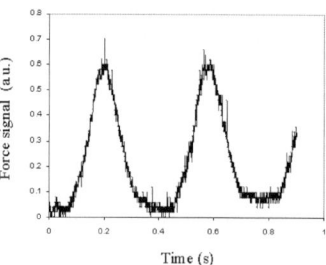 

Figure 10. *Pulsating microtube signal and haptic interface sensing*

4. Conclusion and Future Work

In this paper we have presented a technique suitable for measuring *in vivo* the mechanical properties of different tissues, although in a rather qualitative way. The proposed technique can be adopted for research and clinical purposes, as it does not require that the biological sample is excised and prepared by drying and fixation, like in traditional (and more accurate) techniques.

The main features and components of the apparatus developed in order to implement the proposed technique have been described. The key component of the apparatus is a LIGA microfabricated and instrumented microgripper which is used to grasp and measure the mechanical properties of the tissue sample. A method has been presented for the identification of the microgripper-sample system. Different samples of human skin have been tested *ex vivo* and successfully distinguished based on their mechanical characteristics. Furthermore the system has been used successfully to sense pulse in a microvessel.

Further work is in progress along the two different directions described in this paper.

The first direction aims to improve the performance of the microfabricated probe. In fact, a limitation of the LIGA microgripper used for the experiments presented in this paper is its low stiffness, comparable to that of some tissue micro-samples. When the probe is much stiffer than the tested samples, the discrimination accuracy of the apparatus improves considerably [14]. The new probe fabricated in superelastic alloy by using micro Electro Discharge Machining has these characteristics and thus it will be extensively used in the near future.

A second direction aims to further exploit the force feedback capabilities of the system. In fact, the haptic interface provides the most intuitive approach to tissue and microvessel palpation and provides real time, high bandwidth feedback signal. This feature may make the Phantom interface very useful to provide a "feeling" of tissue "hardness" to the operator, and even an "alarm signal" during delicate microsurgical operations. We intend to exploit these characteristics in order to develop a new a class of robotic microinstruments designed to augment the performance of the surgeon during MIS.

ACKNOWLEDGEMENTS

The work described in this paper has been supported in part by some Projects funded by the Commission of European Communities: "MEDEA" (Microscanning Endoscope with Diagnostic and Enhanced Resolution Attributes), BIOMED2 Contract No. BMH4-CT97-2399); "MINIMAN" (A Miniaturised Robot for Micro Manipulation"), ESPRIT Contract No. 33915. The Scuola Superiore Sant'Anna has provided the financial support for basic research on micromanipulation. The authors wish to thank Mr. C. Filippeschi and Mr. P. Francabandiera for their technical contributions to the paper.

REFERENCES

[1] F. Guilak, M. Sato, C.M. Stanford, R.A. Brand: "Editorial – Cell Mechanics", *Journal of Biomechanics* 33 (2000), pp. 1-2.

[2] T. Brown: "Techniques for Mechanical Stimulation of Cells in vitro: a Review", *Journal of Biomechanics* 33 (2000), pp. 3-14.

[3] M.E. Fauver, D.L. Dunaway, D.H. Lilienfeld, H. G. Craighead, G.H. Pollack: "Microfabricated Cantilevers for Measurement of Subcellular and Molecular Forces", *IEEE Transaction on Biomedical Engineering*, Vol. 45, No. 7, July 1998, pp. 891-898.

[4] G. Lin, K.S.J. Pister, K.P. Roos: "Surface Micromachined Polysilicon Heart Cell Force Transducer", *Journal of Microelectromechanical Systems*, Vol. 9, No. 1, March 2000, pp. 9-17.

[5] H. Miyazaki, Y. Hasegawa: "A Newly Designed Tensile Tester for Cells and Its Application to Fibroblasts", *Journal of Biomechanics* 33 (2000), pp. 97-104.

[6] H. Miyazaki, K. Hayashi: "Tensile Tests of Collagen Fibers Obtained from the Rabbit Patellar Tendon", *Biomedical Microdevices*, No. 2, 1999, pp. 151-157.

[7] M. Sato, N. Ohshima, R.M. Nerem: "Viscoelastic Properties of Cultured Porcine Aortic Endothelial Cells Exposed to Shear Stress", *Journal of Biomechanics* Vol. 29, No. 4 (1996), pp. 461-467.

[8] Y. Wang, F. Coceani: "Isolated Pulmonary Resistance Vessels From Fetal Lambs Contractile Behavior and Responses to Indomethacin and Endothelin-1", *Circulation Research* Vol 71, No 2 August 1992, pp. 320-330.

[9] R. Howe, W.J. Peine, D.A. Kontarinis, J.S. Son: "Remote Palpation Technology for Surgical Applications", *IEEE Engineering in Medicine and Biology Magazine*, 14 (3) (1995), pp 318-323.

[10] A. Bicchi, G. Canepa, D. De Rossi, P.Iacconi, E.P. Scilingo: "A sensorized Minimally Invasive Surgery Tool for Detecting Tissutal Elastic Properties", *Proc. of 1996 IEEE ICRA*, Minneapolis, Minnesota, April 1996.

[11] M.C. Carrozza, P. Dario, A. Menciassi, A. Fenu: "Manipulating Biological and Mechanical Micro-Objects with a LIGA-Microfabricated End Effector", *Proc. of 1998 IEEE ICRA*, Leuven, Belgium, May 16-20, 1998, pp. 1811-1816.

[12] M.C. Carrozza, A. Eisinberg, A. Menciassi, D. Campolo, S. Micera, P. Dario: "Towards a Force-controlled Microgripper for Assembling Biomedical Microdevices", *Journal of Micromechanics and Microengineering*, Vol. 10 (2000) pp. 271-276.

[13] A. Eisinberg, A. Menciassi, S. Micera, D. Campolo, M.C. Carrozza, P. Dario: "PI Force Control of a Microgripper for Assembling Biomedical Microdevices", *IEE Proceedings on Circuits, Devices and Systems.* (In press).

[14] A. Menciassi, G. Scalari, A. Eisinberg, C. Anticoli, P. Francabandiera, M.C. Carrozza, P. Dario: "An Instrumented Probe for *in vivo* Characterization of Soft Tissues". (Accepted for Publication on *Biomedical Microdevices*)

[15] A.L. Window and G.S. Holister, eds.: "Strain Gauge Technology", *Elsevier Applied Science,* London, UK (1989).

[16] A. Cohen: "Biomedical Signal Processing", CRC Press Inc., Boca Raton, FL, U.S.A. (1986).

3D Nanorobotic Manipulations of Multi-Walled Carbon Nanotubes

Lixin DONG*, Fumihito ARAI*, and Toshio FUKUDA**

*Department of Micro System Engineering, Nagoya University
**Center for Cooperative Research in Advanced Sci. and Tech., Nagoya University
Furo-cho, Chikusa-ku, Nagoya 464-8603, JAPAN
dong@robo.mein.nagoya-u.ac.jp, arai@mein.nagoya-u.ac.jp, fukuda@mein.nagoya-u.ac.jp

Abstract:
Multi-walled carbon nanotubes (MWNTs) are manipulated in 3D space with a 10-DOF nanorobotic manipulator, which is actuated with PZTs and Picomotors™ (New Focus Inc.) and operated inside a scanning electronic microscope (SEM). The coarse linear resolutions of the manipulator are better than 30nm (X, Y, Z stages actuated by Picomotors) and the rotary one 2mrad, while the resolutions of fine motions (actuated by PZTs) are within nano-order. AFM cantilevers are used as the end-effector. Several kinds of manipulations of MWNTs are performed with the developed manipulators in the assistance of dielectrophoresis and van der Waals forces. A single MWNT with estimated dimensions of $\phi 40nm \times 7\mu m$ has been picked up on an AFM cantilever. Another $\phi 50nm \times 6\mu m$ MWNT is placed between two cantilevers, and still another $\phi 40nm \times 8\mu m$ MWNT is bent between a cantilever and sample substrate. Carbon nanotube (CNT) junctions are basic building blocks for more complex devices based on CNTs. A cross-junction was constructed with two MWNTs of dimensions of $\sim\phi 40nm \times 6\mu m$ and $\sim\phi 50nm \times 7\mu m$, and a T-junction was made of two MWNTs with the dimensions of $\sim\phi 40nm \times 3\mu m$ and $\sim\phi 50nm \times 2\mu m$. Force measurements are performed and the flexural rigidity and Young's Modulus of an $\sim\phi 30nm \times 7\mu m$ MWNT are estimated to be $8.641 \times 10^{-20} Nm^2$ and 2.17TPa respectively. Such manipulations are essential for both the property research of CNTs and the fabrication of CNT-based NEMS.

Key words: 3D nanorobotic manipulations, CNTs, MWNTs, AFM cantilever, dielectrophoresis, SEM

1. INTRODUCTION

After Iijima[1] observed and identified the first carbon nanotube (CNT) in fullerene soot, many research works have been done on CNTs. Theoretical and experimental research works on carbon nanotubes showed that they have exceptional mechanical and electrical properties. Mechanically, CNTs are regarded as ultimate fibers, electronically as quantum wires, and chemically and biologically as probes and nano containers. Saito[2] and Hamada[3] firstly predicted the metallic/ semiconducting properties of CNTs. Single-walled nanotubes (SWNTs) were synthesized by Iijima and Ichihashi[4]. Individual multi-walled tubes (MWNTs) were measured respectively by Ebbesen[5], Dai[6] and L. Langer[7]. Dai[8] also tried to use an individual nanotube as an AFM tip. Thess[9] produced bulk and monodisperse 1.4-nm SWNTs. Tans[10] and Bockrath[11] observed single-electron effects in individual SWNTs and bundles. Devices such as a CNT quantum resistor and a room temperature, SWNT transistor have recently been reported[12-13]. The mechanical properties have been researched[14-17], among these researches, the strength and breaking mechanism of MWNTs under tensile load has been tackled by Fu with the help of a 3D nanomanipulator[16]. Some manipulations of CNTs in 2D plane by using an AFM with an excellent virtual reality interface are reported[18].

Nanomanipulation, or positional control at the nanometer scale, is the first step towards molecular nanotechnology. With the advancement of nanotechnology, the objects required to be manipulated become entering nano-scale. Although Atomic Force Microscope (AFM) is capable of suitably small motions (tenths of an angstrom), but cannot repeatedly position to, and hold, sub-angstrom accuracy due to hysteresis and drift. It is also currently limited to three degrees of freedom, with no rotational control. As a scanning probe, it works quite well; as a construction device for nanotechnology, it is practically limited to 2D plane.

It is much significant to manipulate nano scale objects in 3D space for constructing nano structures and devices. In order to realize such manipulations, robotic manipulators with nanometer resolutions will be useful tools. The basic requirements for a nanorobotic manipulator for 3D manipulations include a nano-scale positional resolution, a relative large working space, enough DOFs for 3D positioning of the end effectors, and usually with multi end-effectors for complex operations. One of the key techniques will be how to design the end effector because of the specialty of the physical phenomena in micro/nano order world[19]. It has been shown that it is extremely difficult to apply a micro clipper to realize the pick-and-place operations for nano-order objects because the electromagnetic interactions between the clipper and the objects become larger than gravity that cause the departure of the object from the clipper to be difficult[19]. Hence, it is a much more promising strategy to realize nano manipulation by controlling the interactions between tools and the objects rather than by using a gripper. Several strategies for controlling the interactions have been presented[19-21], and a teleoperated touching system has also been proposed[22]. Basic experiments have been reported[23-24].

In the following, a nanorobotic manipulator is firstly presented in section 2, the strategies for nanorobotic manipulations are then introduced in section 3, and some experiments on manipulations are reported in section 4. In section 5, force measurement method is introduced.

2. NANOROBOTIC MANIPULATORS

A set of nanorobotic manipulators is developed. As shown in Fig.1, there are 3 units with totally 10 DOFs including a 3-DOF unit 1 (X-Y-α stage, α is along the direction rotating with X axis) for placing sample substrates, a 1-DOF unit 2 (Z stage) for positioning an AFM cantilever and a 6-DOF PZT-actuated unit 3 for positioning another cantilever. Sample substrates can also be placed on unit 2 or 3 to facilitate manipulations. Unit 1 and unit 2 have linear strokes 6mm respectively and a 360 degree rotation. The linear resolutions for the coarse movements are 30nm (X, Y and Z stages) and the rotary one is 2mrad. Unit 3 is actuated with PZTs for compensating the stepping motions of units 1 and 2, which has 6-DOF with nano-order resolutions.

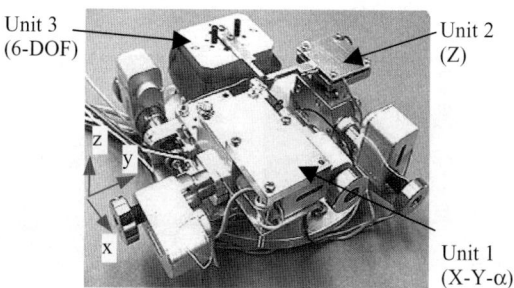

Fig. 1 A nanorobotic manipulator

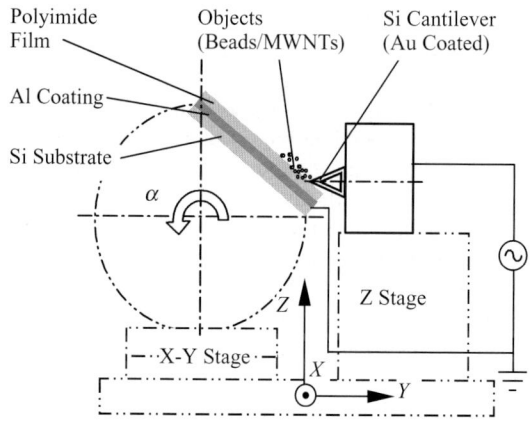

Fig.2 Dielectrophoresis between samples and cantilever

In unit 1, a rotary Picomotor (New Focus Inc.) α is amounted above an X-Y stage actuated by two translatory Picomotors, and is used for mounting a silicon substrate, which is used for placing objects to be manipulated. Over the Si substrate, a thin film of aluminum is coated as an electrode for applying electrical field to generate dielectrophoretic force. Another electrode can be wired to either the cantilever of units 2 or that of 3. To make the tip of cantilevers and the Si plate insulated to each other, a polyimide thin film is bonded over the aluminum film. The applying method of dielectrophoresis between the AFM cantilevers and samples are shown in Fig.2 (wiring of unit 3 is the similar as Z-stage and hence not shown). Note that the position of the sample substrate on unit 1 and the cantilever of unit 2 are exchangeable.

The 6-DOF 8-PZT-actuated Unit 3 is designed to compensate the stepping motion of Units 1 and 2. For getting a larger working space ($26\times22\times35\mu m^3$) and a higher resolution, bi-directive actuation and close-loop control are applied in this micro-moment manipulator. The superposition of actuators and sensors keep the compact volume of Unit 3 so as to place it into the SEM readily[25].

For getting a real-time observations as operating, the whole set of manipulators are mounted inside a SEM (JEOL JSM-5300) with a secondary electron detector, which has a relative large vacuum chamber. The resolution of the microscope is specified as 4nm at 30kV. However, the real time video resolution is a factor of two or three lower. All of the wires are connected through the isolated vacuum feed-throughs passing through the SEM chamber wall. All of the components and cables of the manipulators are properly shielded from the SEM observation region to minimize image distortion from charging effect.

It can be found that the manipulators meet all of the basic requirements for nano manipulations. It has totally 10 DOFs and three units with bi-cantilevers, the working space is $6\times6\times12mm^3$ with a 360° rotation, the coarse resolutions are 30nm and 2mrad while the fine ones are nano-scaled. It is amounted inside the vacuum chamber of a SEM, so the real time observations for the manipulations are realized and the forces between the cantilever tips and samples can also be measured with multiple exposure techniques.

3. PRINCIPLES OF NANOROBOTIC MANIPULATION

As dealing with micro objects, the interactions caused by quantum and electromagnetic effects can no longer be ignored, which is different from that in the macro world. For instance, the interactions between a bead with diameter of d and an infinite plate in the VHV inside SEM is mainly van der Waals force[22]

$$F_{vdw} = \frac{H}{6}\left\{\frac{d}{2z^2} + \frac{d}{2(z+d)^2} - \frac{1}{z} + \frac{1}{z+d}\right\} \quad (1)$$

On the other hand, the dielectrophoretic force F_{dep} is a function of the radius of objective bead r, the dielectric coefficient of circumstance ε_0, the dielectric coefficient of the objective bead ε and the electric field intensity E_0.

$$F_{dep} = 2\pi r^3 \frac{\varepsilon_0(\varepsilon-\varepsilon_0)}{\varepsilon+2\varepsilon_0}\nabla(E_0^2) \quad (2)$$

Because the dielectrophoretic force is a function of the gradient of the electric field intensity, it is easy to control it by changing the applied voltage. It is also easier to be realized than to control other kinds of adhesive forces. For generating a nonuniform electric field gradient, it is effective to apply a sharp pin such as an AFM cantilever vs. to a plate as two electrodes. Hence, it is possible to pick up an object placed in a plate such as that shown in Fig.2 if the dielectrophoretic force is greater than van der Waals forces acted on a particle. In order to realize such manipulations, two ways can be effectively used. One of them is to decrease the van der Waals forces as described in [19] and [20]. Another way is to control the intensity and gradient of the electric field.

As shown in Fig.3, a $\phi 1\mu m$ bead placing on an ideal plate (roughness b=0, hence b/z=0) inside SEM cannot be picked up when the voltage for generating dielectrophoretic force is 100V (region A in Fig.3) because $F_{dep}<F_{vdw}$ (in this case, the other kinds of adhesive forces can be ignored because they are largely less than van der Waals). The manipulation can be realized either by increasing the voltage (e.g. up to 500V) so as to enhance dielectrophoretic force, or by increasing the roughness (e.g. b/z=10 or even 100) of the plate so as to decrease van der Waals force subjected to the bead.

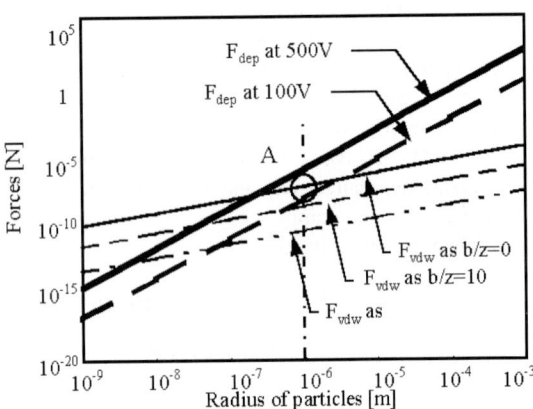

Parameters used in computations:

$H = 14.8\times 10^{-20}[J]$, $z = 0.4[nm]$, $\rho = 2300[kg/m^3]$
$\varepsilon_0 = 8.85\times 10^{-12}[F/m]$, $\varepsilon = 5\varepsilon_0$, $\sigma = 26.5[\mu C/m^2]$

Fig. 3 Controlling dielectrophoretic force and/or van der Waals force

4. MANIPULATION EXPERIMENTS

4.1 Picking-up, Placing and Bending Single MWNTs

With the developed nanorobotic manipulators, some kinds of manipulations of individual MWNTs have been tried. We aim at constructing 3D structures of CNTs and researching their mechanical and electronic properties at the same time, and here we show some preliminary results.

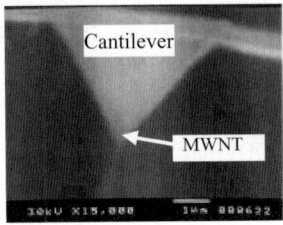

Fig.4 Pick up a single MWNT with an AFM cantilever

Fig.4 shows that a single MWNT is picked up on the AFM cantilever of unit 2, which has an estimated dimension of $\phi 40nm\times 7\mu m$. Fig.5 shows the MWNT is placed between two cantilevers, and Fig.6 shows a MWNT is bent. Such manipulations are essential for both the property research of CNTs and the fabrication of CNT-based NEMS.

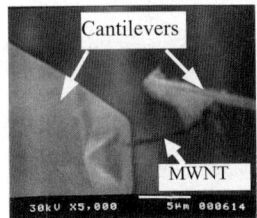 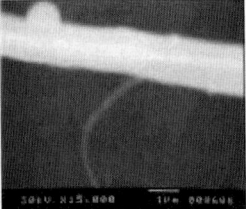

Fig.5 Place an MWNT   Fig.6 Bend of an MWNT

4.2 Construction of MWNT-Junctions

4.2.1 Types of Junctions

The possibility of connecting nanotubes of different diameter and chirality has generated considerable interest recently[26-30]. This is because of the possibility of the junctions being building blocks of nanoscale electronic devices. Although junctions are randomly found in CNT samples, it is significant to find the techniques to fabricate such basic structures. The difficulties for the construction of CNT junctions depend on the kinds of junctions. The types of CNT-junctions are determined by the kinds of CNTs, the configurations of CNTs and the conjunction methods:

(1) Kinds of CNTs
  1) Metallic SWNTs
  2) Semiconducting SWNTs
  3) Semimetallic SWNTs
  4) (Metallic) MWNTs
(2) Configurations
  1) V or I-junctions
  2) T-junctions
  3) Y-junctions
  4) X-junctions
  5) More complex (e.g. 3D) junctions

(3) Conjunction methods
 1) Van der Waals
 2) Electron beam soldering
 3) Chemical bonds
 4) Other methods

Here we show an X-junction and a T-junction of MWNTs jointed with van der Waals forces by the nanorobotic manipulator.

4.2.2 MWNT Junctions

(1) X-junction

An X-junction (cross-junction) was constructed with two MWNTs of dimensions of ~φ40nm×6μm and ~φ50nm×7μm. As shown in Fig.7, the two MWNTs are supported between the raw material of carbon nanotubes on the sample substrate and an AFM cantilever. Although it cannot be determined clearly that how the two MWNTs connected since the limitation of the SEM, it is reasonable to say that they are linked by van der Waals forces.

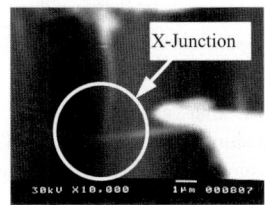
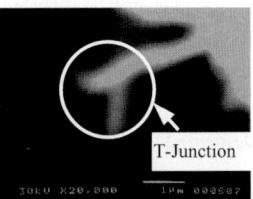

Fig.7 X-Junction of MWNTs   Fig.8 T-Junction of MWNTs

(2) T-junction

A T-junction was made of two MWNTs with the dimensions of ~φ40nm×3μm and ~φ50nm×2μm as shown in Fig.8. The T-junction is holding on the AFM cantilever. Similarly, it seems that the two MWNTs are jointed with van der Waals forces.

5. FORCE MEASUREMENTS

It is significant to understand the force information between the cantilever and the objects because it is necessary for the better control of the manipulator and for the investigation of the properties of the CNTs and CNT junctions as well as more complex CNT structures. We applied SEM Photos and calibrated AFM cantilevers to measure the forces with video or multiple exposure techniques.

5.1 Flexural Rigidity of MWNTs

By buckling a single MWNT, we tried to evaluate the stiffness of nanotubes by measuring the forces subjected to the nanotube and the deformations of the nanotube and cantilever. Fig.9 (a) and (b) show two continuous SEM image frames recording the bending process, while (c) and (d) depict the analytic models for (a) and (b) respectively, and (e) shows the forces subjected to the MWNT. According to Euler's Formula and force balance relations, the following equations can be obtained

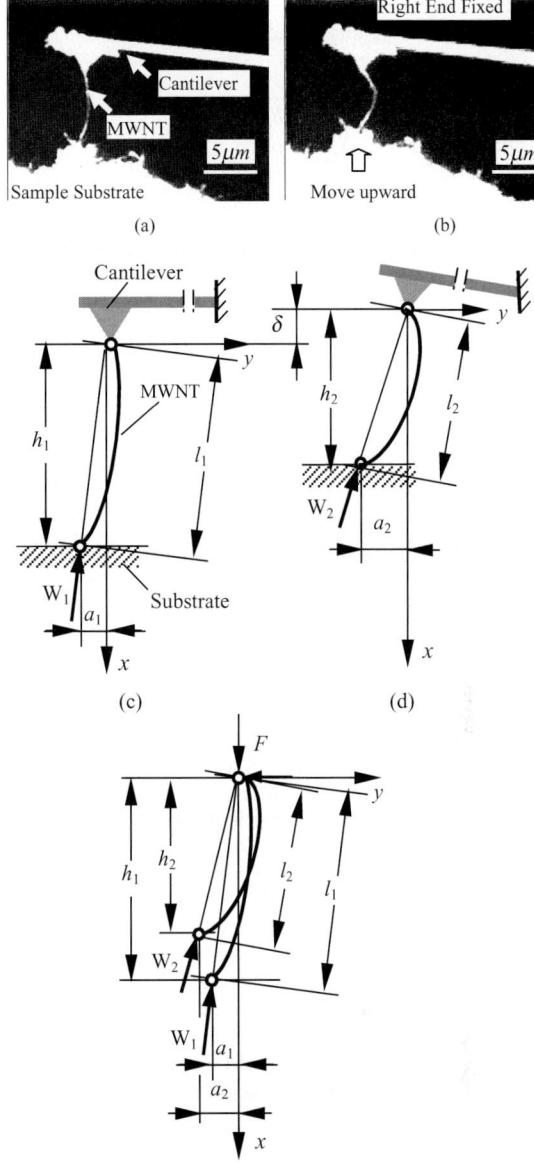

Fig.9 Bending of an MWNT and Force Measurement

$$\begin{cases} W_1 = \dfrac{\pi^2 E I_z}{l_1^2} \\ W_2 = \dfrac{\pi^2 E I_z}{l_2^2} \\ F = k\delta \\ F = W_2 \dfrac{h_2}{l_2} - W_1 \dfrac{h_1}{l_1} \end{cases} \quad (3)$$

where, W_1 and W_2 are buckling forces subjected to the nanotube by the substrate, F is the difference of reaction force of the cantilever between Fig.9 (a) and (b), E is Young's modulus, I_z is second moment of area, and other parameters and their values are listed in Table 1, where the values of a_1, a_2, h_1, h_2, δ and d are measured from Fig.9 (a) and (b), k is a given calibrated value, and l_1 and

l_2 are calculated values.

The flexural rigidity of the MWNT is hence

$$EI_z = \frac{k\delta}{\pi^2 \left(\frac{h_2}{l_2^3} - \frac{h_1}{l_1^3} \right)} \quad (4)$$

and finally we got the result

$$EI_z = 8.641 \times 10^{-20} \, \text{Nm}^2. \quad (5)$$

Table 1 Parameters and their values

Parameter	Value
a_1(refer to Fig.9(c))	0.690μm
a_2(refer to Fig.9(d))	0.690 μm
h_1(refer to Fig.9(c))	6.852 μm
h_2(refer to Fig.9(d))	5.755 μm
$l_1 = \sqrt{a_1^2 + h_1^2}$ —buckling length (Fig.9(c))	6.886 μm
$l_2 = \sqrt{a_2^2 + h_2^2}$ —buckling length (Fig.9(d))	5.796 μm
k—Stiffness of Cantilever Tip	0.03N/m
δ—Deformation of Cantilever Tip	0.244 μm
d –diameter of CNT	~30nm

The SEM limited us to get an accurate value of the diameter of the nanotube and detailed geometric structure of the nanotube, so it is difficult to obtain a relative accurate value of Young's modulus. But for getting a conservative estimation of Young's modulus, it is reasonable to hypotheses the MWNT to be solid cylinder and let d=30nm, then we got E=2.17TPa. It is a little bit larger than the mean value obtained in [14], where they applied thermal vibration method and got an average value E=1.8TPa but their data for individual nanotubes ranged from 0.4 to 4.15TPa. Hence, the result in Eq.(5) is reasonable.

5.2 Force Measurement by Multi Exposure Technique

Fig.10 shows the principle for this force measurement method. Two calibrated cantilevers are "fighting" with each other. As the left one moves upward 20.20μm, the tip of the right one deformed with the same distance. According to the stiffness of the cantilever, it can be known that the force between the two cantilevers is about 607.2nN.

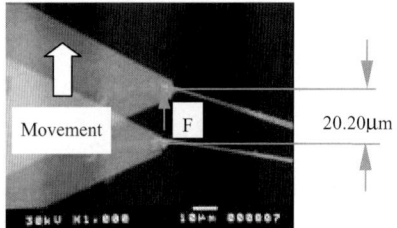

F=607.2nN ($k_{cantilever}$=0.03N/m)
Fig.10 Principle for Force Measurement
with Multiple Exposure Technique

Fig.11 shows that an MWNT is picked up onto an AFM cantilevers. In the process, the forces (mainly van der Waals forces) between the AFM cantilever and the sample substrate is estimated to be 314.9nN.

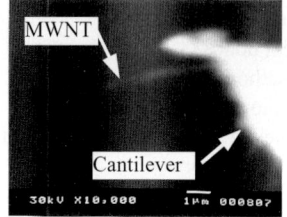

(a) MWNT Stretched onto AFM Cantilever

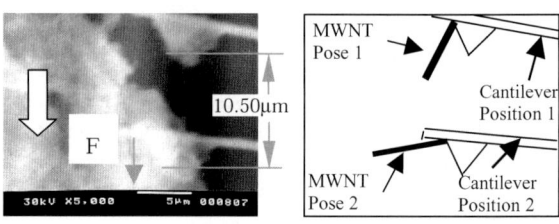

(b) Multiple Exposed Photo   (c) Outline of (b)
F=314.9nN ($k_{cantilever}$=0.03N/m)
Fig.11 Force Measurement as Stretching an MWNT

Fig.12 (a) and (b) show that an X-junction is deformed by pulling and pushing the upper MWNT. Fig.12 (c) is a multiple exposure photo that depicts the same process as shown in Fig.12 (a) and (b). From Fig.12 (c), the force occurred in this process is measured to be 54.6nN.

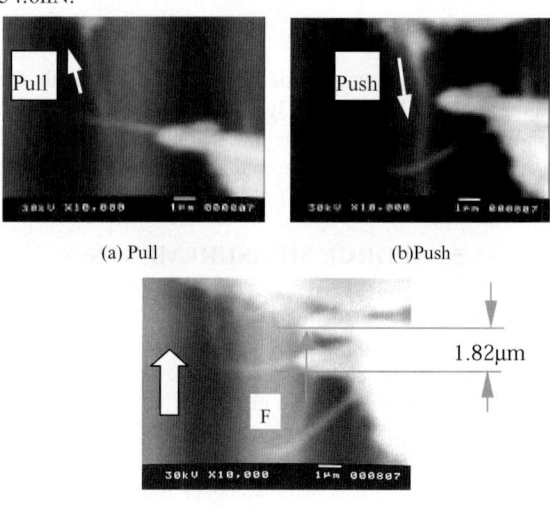

(a) Pull   (b) Push

(c) Multiple Exposed Photo of (a) and (b)
F=54.6nN ($k_{cantilever}$=0.03N/m)
Fig.12 Deformation and Force Measurement as Pulling and Pushing an MWNT X-Junction

6. CONCLUSIONS

A 10-DOF nanorobotic manipulator with two cantilevers has been constructed inside a SEM. By adjusting the voltage applied between the AFM cantilevers and the sample substrate, the dielectrophoretic force subjected to the objects was controlled effectively. 3D manipulations of MWNTs were realized in the assistance of dielectrophoretic forces being controlled, and force measurement performed. The manipulators being developed will be fundamental tools both for the research of the properties of nano scale particles and for the construction of nano-order device with nano building blocks such as carbon nanotubes. Force measurements are performed and the flexural rigidity and Young's Modulus of an $\sim\phi 30nm \times 7\mu m$ MWNT are estimated to be $8.641 \times 10^{-20} Nm^2$ and 2.17TPa respectively.

ACKNOWLEDGMENTS

We are grateful to Prof. Y. Saito at Mie University for providing us MWNT samples, Prof. Shinohara at Nagoya University for a helpful discussion, and the first author would like to thank Prof. R. Saito at University of Electro-Communications for presenting an instructive book on CNTs.

REFERENCES

[1] S. Iijima, Helical Microtubules of Graphitic Carbon, Nature, Vol.354, pp.56-58 (1991).

[2] R. Saito, G. Dresselhaus and M. S. Dresselhaus, Physical Properties of Carbon Nanotubes, Imperial College Press (1998).

[3] N. Hamada, S. I. Sawada and A. Oshiyama, New One-Dimensional Conductors: Graphitic Microtubules, Phys. Rev. Lett., Vol.68, pp.1579-1581(1991).

[4] S. Iijima and T. Ichihashi, Single-Shell Carbon Nanotubes of 1-nm Diameter, Nature, Vol.363, pp.603-605 (1993).

[5] T. W. Ebbesen, H. J. Lezec, H. Hiura, J. W. Bennett, H. F. Ghaemi and T. Thio, Electrical Conductivity of Individual Carbon Nanotubes, Nature, Vol.382, pp.54–56 (1996).

[6] H.J. Dai, E.W. Wong and C.M. Lieber, Probing Electrical Transport in Nanomaterials: Conductivity of individual Carbon Nanotubes, Science, Vol.272, pp.523-526 (1996).

[7] L. Langer, V. Bayot, E. Grivei, J.P. Issi, J.P. Heremans, C.H. Olk, L. Stockman, et al, Quantum Transport in A Multiwalled Carbon Nanotube, Phys. Rev. Lett., Vol.76, pp.479-482 (1996).

[8] H.J. Dai, J.H. Hafner, A.G. Rinzler, D.T. Colbert and R.E. Smalley, Nanotubes as Nanoprobes in Scanning Probe Microscopy, Nature, Vol.384, pp.147-150 (1996).

[9] Thess, R. Lee, P. Nikolaev, H.J. Dai, P. Petit, J. Robert, C.H. Xu, Y.H. Lee, S.G. Kim, A.G. Rinzler, D.T. Colbert, G.E. Scuseria, D. Tománek, J.E. Fischer and R.E. Smalley, Crystalline Ropes of Metallic Carbon Nanotubes, Science, Vol.273, pp.483-487 (1996).

[10] S. J. Tans, M. H. Devoret, H. J. Dai, A. Thess, R. E. Smalley, L. J. Geerligs and C. Dekker, Individual Single-Wall Carbon Nanotubes as Quantum Wires, Nature, Vol.386, pp.474–477 (1997).

[11] M. Bockrath, D. H. Cobden, P. L. McEuen, N. G. Chopra, A. Zettl, A. Thess and R. E. Smalley, Single-Electron Transport in Ropes of Carbon Nanotubes, Science, Vol.275, pp.1922–1925 (1997).

[12] S. Frank, P. Poncharal, Z. L. Wang and W. A. d. Heer, Carbon Nanotube Quantum Resistors, Science, Vol.280, pp.1744–1746 (1998).

[13] S. J. Tans, A. R. M. Verchueren and C. Dekker, Room-Temperature Transistor Based on a Single Carbon Nanotube, Nature, Vol.393, pp.49-52 (1998).

[14] M. J. Treacy, T. W. Ebbesen and J. M. Gibson, Exceptionally High Young's Modulus Observed for Individual Carbon Nanotubes, Nature, Vol.381, pp.678-680 (1996).

[15] E. W. Wong, P. E. Sheehan and C. M. Lieber, Nanobeam Mechanics: Elasticity, Strength, and Toughness of Nanorods and Nanotubes, Science, Vol.277, pp.1971-1975 (1997).

[16] M.F. Yu, O. Lourie, M.J. Dyer, K. Moloni, T.F. Kelley and R.S. Ruoff, Strength and Breaking Mechanism of Multiwalled Carbon Nanotubes Under Tensile Load, Science, Vol.287, pp.637-640 (2000).

[17] S. Ruoff, J. Tersoff, D. C. Lorents, S. Subramoney and B. Chan, Radial Deformation of Carbon Nanotubes by van der Waals Forces, Nature, Vol.364, pp.514-516 (1993).

[18] M. Guthold, M.R. Falvo, W.G. Matthews, S. Paulson, S. Washburn, D. A. Erie, R. Superfine, F. P. Brooks, Jr. and R. M. Taylor II, "Controlled Manipulation of Molecular Samples with the nanoManipulator", IEEE/ASME Trans. On Mechatronics, Vol.5, No.2, pp.189-198 (2000).

[19] F. Arai, D. Andou, T. Fukuda, et al., Micro Manipulation Based on Micro Physics -Strategy Based on Attractive Force Reduction and Stress Measurement, Proc. of IEEE/RSJ Int. Conf. on Intelligent Robotics and Systems, Vol.2, pp.236-241 (1995).

[20] F. Arai, D. Andou, Y. Nonoda, T. Fukuda, H. Iwata, and K. Itoigawa, Integrated Microendeffector for Micro-manipulation, IEEE/ASME Trans. Mechatronics, Vol.3, No.1, p.17-23 (1998).

[21] S. Saito, H. Miyazaki and T. Sato, Pick and Place Operation of Micro Object with High Reliability and Precision based on Micro Physics under SEM, Proc. ICRA'99, pp.2736-2743 (1999).

[22] M. Sitti, S. Horiguchi and H. Hashimoto, Tele-touch Feedback of Surfaces at the Micro/Nano Scale: Modeling and Experiments, Proc. of the IEEE/RSJ Int. Conf. on Intelligent Robots and Systems, pp.882-888 (1999).

[23] F. Arai, T. Noda, T. Fukuda and L.X. Dong, Basic Experiment on the Three Dimensional Nano-manipulation, Proc. of RoboMec'00 (JSME), No.2A1-62-080(2000) (in Japanese).

[24] L.X. Dong, F. Arai and T. Fukuda, 3D Nanorobotic Manipulators inside SEM, Proc. of RSJ'2000, pp.81-82(2000).

[25] L.X. Dong, F. Arai and T. Fukuda, 3D Nanorobotic Manipulation of Nano-order Objects inside SEM, Proc. of the 2000 Int'l Symp. on Micromechatronics and Human Science, pp.151-156 (2000).

[26] L. Chico, V.H. Crespi, L.X. Benedict, et al, Pure Carbon Nanoscale Devices: Nanotube Heterojunctions, Phys. Rev. Lett., Vol.76, pp.971-974 (1996).

[27] J.-C. Charlier, T. W. Ebbesen and Ph. Lambin, Structural and Electronic Properties of Pentagon-Heptagon Pair Defects in Carbon Nanotubes, Phys. Rev. B., Vol.53, pp.11108-11112 (1996).

[28] R. Saito, G. Dresselhaus and M. S. Dresselhaus, Tunneling Conductance of Connected Carbon Nanotubes, Phys. Rev. B., Vol.53, pp.2044-2050 (1996).

[29] S. Iijima, T. Ichihashi and Y. ando, Pentagons, Heptagons and Negative Curvature in Graphite Microtubule Growth, Nature, Vol.356, pp.776-778 (1992).

[30] M. Menon and D. Srivastava, Carbon Nanotube "T Junctions": Nanoscale Metal-Semiconductor-Metal Contact Devices, Phys. Rev. Lett., Vol.79, pp.4453-4456 (1997).

Observations Concerning Internet-based Teleoperations For Hazardous Environments

Professor William R. Hamel

Pamela Murray

Mechanical & Aerospace Engineering & Engineering Science Department
The University of Tennessee
Knoxville, Tennessee, USA
whamel@utk.edu

Abstract

Teleoperated systems are used to perform remote maintenance functions in hazardous environments usually with the fundamental objective of reducing, or eliminating, human worker exposure to dangers Researchers are now considering how might Internet capabilities be used in such operations. This paper discusses this general idea in terms of the practical needs, constraints, and concepts that are associated with the notion of Internet-based teleoperations.

1. Introduction

Teleoperated manipulators/systems are used to perform maintenance functions in hazardous environments usually with the fundamental objective of reducing, or eliminating, human worker exposure to dangers. Hazardous environments include undersea, nuclear, space, and toxic exposure situations. The hazards can be thought of as either intrinsic, or extrinsic. The undersea and space environments involve intrinsic hazards, namely extreme hydrostatic pressures in a liquid environment and the absolute vacuum of deep space respectively. Nuclear and toxic hazards are extrinsic in that the environments are non-hazardous for humans except for the presence of dangerous substances such as materials emitting ionizing radiation.

Regardless of the specific hazardous environment, the remote systems used to accomplish maintenance and support functions under such conditions are complex and expensive to construct and operate. The normal mode of operation of these systems is manual control, or teleoperation, and they are notoriously slow in comparison to the work rates that humans accomplish in benign environments. Over the past 25 years, major research efforts have been focused on improving the work efficiencies of these systems. Research is addressing techniques to increase the transparency of teleoperations and schemes to use subtask robotic automation. Major investments are going into the design and development of comprehensive integrated work systems that can move around in the remote environments to accomplish needed work [1]. Also, major resources are required to train human operators and to maintain such systems. Remote operations in any hazardous environments are extremely expensive and difficult to manage.

The Internet has become a common and pervasive worldwide service that touches almost every level of human endeavor. The performance properties of the Internet have increased to the point that it makes sense to consider to what degree high-fidelity teleoperators can be implemented across the Internet. Because of our background, the nuclear remote maintenance domain is used as a basis, however it is believed that the discussion is applicable to most of the hazardous environments where remote operations are used.

2. Basics

Teleoperated systems used in remote operations like any engineered system go through a cycle of development, testing and deployment. Internet-based concepts may be useful throughout this cycle or for a particular phase only. The basic steps of the cycle include:
- System development
- Testing and debugging
- Operations
    - Nominal operations
    - Off-nominal operations
- Personnel training
- Maintenance

During system development, testing and debugging, the full features of the remote maintenance system go through a process of design, implementation, and evaluation that involve intimate interaction between engineers and the system hardware/software. In the case of mature products, this phase is essentially systems integration.

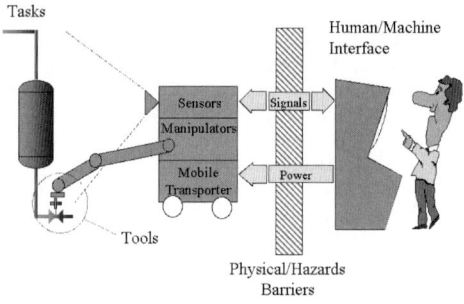

Figure 1, Basic Subsystems of a Teleoperated Remote System

A teleoperated remote system will generally involve subsystems for mobility, manipulation, tooling, sensing, and human/machine interfacing. Nominal operations involve the collective workings of these subsystems to accomplish remote operational goals. Any remote maintenance system will eventually experience some aspect of off-nominal operation that may be the result of unexpected environmental events or system malfunctions.

Because of their inherent complexity and the very nature of remote operations, operator training is a major challenge that requires the use of simulations and cold testing facilities that provide operators with comprehensive and realistic training. Such training typically will encompass all aspects of nominal and anticipated off-nominal operations.

Remote maintenance systems themselves will eventually experience equipment failures. Remote maintenance/operation of failed remote maintenance systems must be an integral part of their basic design and operational features. Hardware/software features must be provided for the analysis, recovery from, and correction of problems.

This paper discusses various issues pertaining to how Internet-based system concepts may be incorporated into these basic system functions.

3. Constraints

Remote operations in hazardous environments involve numerous technical and regulatory constraints that any system concept must satisfy. As depicted in Figure 1, information and power transmission paths must be provided between the remote maintenance system and the external safe area where human are located. With respect to Internet-based concepts, the following attributes of the information communications links are paramount:

- Data transmission time delay
- Noise
- Signal dropout likelihood
- Signal dropout recovery

The magnitude of the data transmission delay effects the utility of sensory feedback to the operator (i.e., refresh rate of remote television views) and the stability of any closed loop controls that cross the remote environment including teleoperations. Force-reflecting teleoperated manipulators provide kinesthetic feedback to operators, and necessary to accomplish more complex tasks. The servo controls between the master and slave subsystems are interconnected and require control sampling rates of 200 to 1000 Hz for stability and performance. Such systems are obviously very sensitive to even small amounts of time delay.

Noise corruption of transmitted information may erode system performance in a host of ways. Loss of signal transmission during system operations is critically important in terms of detecting, responding, and recovering gracefully without the loss of system control and stability. Obviously, an Internet-based system structure will involve superimposing additional complex communications processes directly into the data transmission path.

In addition to technical constraints, most remote operations involve regulatory restrictions. In the case of nuclear remote operations, these regulatory restrictions are particularly rigid with respect to nuclear, environmental and worker safety issues. Adherence to regulatory restrictions normally becomes matters of reliability and fail safe operational assurance. One might expect that there would be concern about operational system concepts in which specific subsystems are not under the direct control of the operators or if such subsystems include suitable back-up modes of operation. We will discuss concepts that may be useful in addressing this constraint later.

4. Internet Performance

The actual performance of the Internet is a key factor pertaining to its practical use in teleoperations in actual applications. Very little quantitative performance data is available, but a number of projects have been created recently to analyze performance.

The performance of a computer connection is based on the speed and reliability with which the data is transmitted over that connection. Currently, the Internet Traffic Report [2] and Internet Weather Report [3] provide an approximate measure of Internet performance by measuring the performance of connections among a small group of monitored sites distributed throughout the world. Delay and packet losses are monitored as measures of connection speed and reliability respectively. The minimum delay values generated represent delay of non-congested paths giving an indication of the baseline propagation and transmission delay. Higher values reflect congestion. The round-trip delay can also be calculated from this data if desired [4]. This round-trip delay is called response time or latency. This response time is the time it takes for data to travel from A to B and back to A. Packet loss is measured by sending a packet of data to a router that is supposed to send the packet back and then measuring the success of this operation [2].

The speed and reliability of the entire Internet cannot be measured at present. IPMA, the Internet Performance Measurement and Analysis Project [4] and Internet Protocol Performance Metrics at IETF [5] are at the forefront of developing metrics to analyze overall Internet delay and packet loss. IPPM proposes the use of one-way delay measurements because the path from the destination to source is not usually the same as that from source back to the destination, and one direction may be more important than the other for an application. A proposed derivative metric of packet loss is the loss pattern. This pattern measures the frequency and length of loss. The loss period measures the frequency and length of loss, and the loss distance metric measure the space between loss periods [5]. It will be critical to quantify the performance of a connection as many applications are connection dependent. The Internet Traffic Report figures from September 2000 show that the average response time is 272 ms and the average packet loss is 2%. These figures are based on tests between 74 routers around the world. The Internet Weather Report figures for the same period based on a large number of sites show that the average latency is 130.2 ms and that the average packet loss is 4.3%. The magnitudes of these results are significant with respect to the concept of high fidelity Internet teleoperations.

5. Concepts

An objective of this paper is to discuss how the Internet might be used to accomplish practical aspects of teleoperations in hazardous environments more effectively. As shown in Figure 2, an Internet-based teleoperated system structure is identical to the classical architecture except the signal transmission path between the operator and the system passes through the Internet in some sense.

Before we discuss the technical ramifications of passing signals through the Internet, let us consider the functional impacts. Probably, the greatest advantage for Internet architectures is the ability to connect field systems with expertise centers located at distant geographical sites. These expertise centers could include systems/equipment suppliers, centralized training facilities, and R&D facilities involved in technologies used in the robotic remote maintenance system. Via the e-connection, in principle, remote operators could tap into specialized expertise rapidly without travel and scheduling delays.

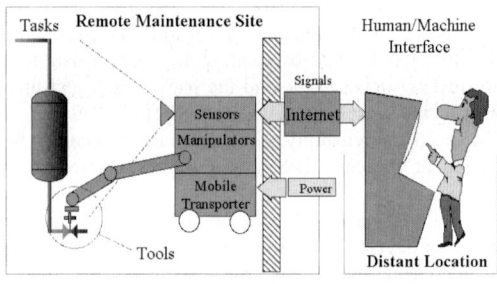

Figure 2, Internet-based Teleoperator System

Training costs could be reduced while effectiveness enhanced by using the best teachers available. Downtimes could conceivably be reduced through direct interaction (trouble shooting) with source designers and specialists.

This approach could ultimately reduce capital resource costs dramatically by allowing centralized training and maintenance facilities. The rudiments of the concept are depicted in Figure 3.

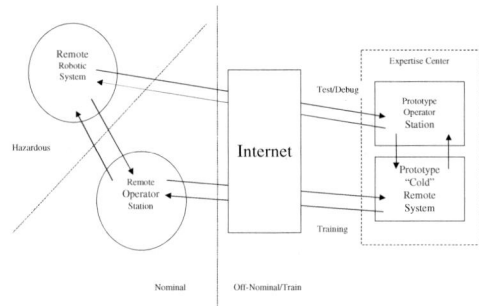

Figure 3, Functional Architecture of Training/Maintenance Support

Another conceptual approach is direct control of distributed remote maintenance systems from a centralized facility. In this scenario, for example, an electrical utility with multiple nuclear power plants would perform remote maintenance functions from a central technical center perhaps hundreds of kilometers from the operating plants. Remote operations require specialized skills and extensive training to achieve high remote work efficiency. Centralized remote control would reduce the number of operators and equipment necessary for large operations like utilities. Today, virtually all remote maintenance is performed using pure teleoperations in which remote manipulation position/force controls are closed across the signal transmission system (usually to allow electronics to remain outside of the radiation zone). With the human/machine interface at a distant location, it is necessary that either the signal communications link be robust with acceptable time delay, or dedicated data links with sufficient throughput be provided.

Delay is of concern because many applications cannot operate with variable delays. It is difficult to sustain high bandwidths with large delays. [5] In current Internet teleoperation systems, an invalid constant bounded delay is often assumed. [6] The time delay is especially critical as its length directly effects sensory feedback and closed loop control capabilities. The data transmission time delay seen over the Internet must be acceptable or a dedicated line must be used. Operator performance decreases significant as the time delay increases beyond 0.2 seconds [15]. For example, when the image update rate decreases from 30 frames/second to 3 frames/second, the task completion time increases by 100% [7]. The system becomes unstable when the frequency of the input is such that half a cycle is equal to the time delay. [8] Different data types have different levels of importance to the operator. It is more important to minimize the image delays even if the commands then experience a larger delay [6]. These image delays can be minimized by sending images that contain only the needed information at the optimal resolution. Then, more frequently, data can be sent containing the current status, position, and events [9].

Recently, considerable R&D is addressing the integration of the Internet with various engineering enterprises even including sharing research facilities among distributed research groups [10]. Brady and Tarn [11] discuss internet-based remote teleoperation including a theoretical foundation for modeling communications delays and a new supervisory control concept. Taylor and Dalton [12] present recent work with a web-operated telerobot that is also based on supervisory control. Most of the applications do not require the comparatively fast and deterministic communications coupling necessary for *practical* remote operations in hazardous environments.

In remote operations, it is recognized that teleoperations place the greatest demands on communications bandwidth and these demands are dominated by remote viewing requirements. With force reflecting systems incorporating multiple remote views with color or stereo, one can readily require continuous throughputs on the order of 200 to 500 MBaud. One scheme to alleviate high communications data rates is to incorporate supervised control and/or autonomy into the remote system [11]. In this way, the operator interacts with the remote system at a higher level requiring comparatively low command interactions. This means that the sensor and control capabilities of the remote system must be adequate to perform the required tasks. While the concept is straightforward, its implementation is another matter in the complex and unstructured task environments that are typical of hazardous environments. All researchers are proposing the use of supervisory control schemes to essentially compensate for the destabilizing effects of the communications time delay. The fundamental practical problem with this approach for hazardous environments is the complexity and uncertainty of the remote task environment. Very little a priori task information is available for programmed control at the remote site. Major research efforts are directed toward comprehensive telerobotics that can achieve autonomous subtask execution with high task uncertainty and stringent requirements for robust and deterministic operations [13]. To date, practical working systems have not been realized.

Probably the greatest concern about Internet teleoperations in hazardous environments is the non-deterministic system behavior that would result during packet loss or total dropout. One approach to this concern would be to incorporate a "smart" communications interface where the Internet connects with the remote system. If such an interface could reliably distinguish between significant amount packet loss or dropout from normal operations communications (admittedly a non-trivial capability), then controls actions could be taken to assure stable slave operation (e.g., servo about current state) until communications is restored. This could very well be an important research area.

6. Practical Experience

In 1996 a relevant experience occurred in which a remote excavator located at the Oak Ridge National Laboratory was teleoperated using commercial communications lines from a hotel in Washington, DC at a distance of over 1000 kilometers. The excavator was operated both day and night. The Telerobotic Small Emplacement Excavator (TSEE) was a joint development between the U.S. Army and Department of Energy to evaluate low cost remote digging equipment [14].

Figure 4, Teleoperated Small Emplacement Excavato (Courtesy of the Oak Ridge National Laboratory)

The TSEE is a conventional backhoe that can be operated remotely using the compact console shown in Figure 5. The TSEE was being used to simulate excavation of buried nuclear waste drums as shown in Figure 4 and this was the task that was performed in the experiment. High-speed T1 telephone lines were rented to connect the two sites. Sun workstations were used at each end to buffer and control data transfers. For a system like the TSEE, about 20 KBaud is necessary for control signals and about 60 MBaud (for 30 frames/s) for the single black and white remote television view being sent to the operator. With all of the routing and overhead associated with the communications hook-up the remote television update rate obtained was several seconds per screen refresh. The experiment was primarily

*Figure 5, TSEE Remote Control Console
(Courtesy of the Oak Ridge National Laboratory)*

a functional demonstration using non-radioactive materials, and did not involve any quantitative analysis. It was found that the remote view refresh rate was barely acceptable and severely hampered the operator in performing the remote handling. Most experts felt that a refresh rate in the range of 5 to 10 frames/sec would allow nearly normal teleoperations. These results are consistent with established data describing the limits of teleoperations under time delay. High fidelity teleoperations begin to breakdown when the time delays approaches 0.25 s [15]. Time delays greater than this value are difficult, if not impossible, for operators to compensate for, even with methodical "command and wait" strategies. During this experiment there were significant problems with noise and losses of the communications link. The demonstration results do however serve as a useful benchmark of remote teleoperations over large distances. The communications delays in this case were of course small compared to space applications such as terrestrial control of low earth orbit space remote operations.

7. Technical Issues

The broad concept of using the Internet as a data communications link to facilitate teleoperations operations in hazardous environment is subject to two key technical factors: 1) communications reliability, and 2) safe systems operations.

As the data communications bandwidth decreases, the range of admissible remote teleoperated tasks decreases. This effect can be offset by incorporating supervised autonomy into the remote system. But supervisory control (in its various forms) requires that a significant number of the remote tasks can be automated reliably. Because of the inherent complexity of typical remote maintenance tasks, very little subtask automation has actually been achieved. Data links must be capable of meeting the needs of teleoperations. The most demanding form of remote teleoperation is force reflection where the remote operator actively senses force interactions. Efficient force reflecting teleoperations require that the time delay be on the order of tenths of seconds maximum.

The assurance of safe operations is a very stringent and comprehensive regulatory requirements for this class of systems. As discussed earlier, safety analyses include hardware and software. Extensive fault and reliability analyses are performed to identify system vulnerabilities. Because the data link to the remote environment is essential to the operation of the remote equipment, the non-deterministic, other party ownership factors of the Internet will restrict the Internet teleoperations. The determinism and high reliability of the communications links for this class of systems are essential. It is very difficult to quantify reliability parameters at this stage, but the objective would be to achieve the level of reliability achievable with a hardwired remote system. One would imagine the dropout periods would have to be limited to a maximum of a couple control sample periods, or on the order of a few milliseconds. Until Internet specifications provide this level of performance, Internet teleoperations will probably be limited to the functional architecture shown in Figure 3. Reliable dropout detection and recovery schemes could alleviate many of these issues.

8. Conclusions

Internet teleoperations in hazardous environments such as those associated with nuclear operations depends on adequate and reliable data communications whether accomplished via the Internet or other means. Because of the complexity and non-determinism of the Internet, it is not likely that such concepts will be used in actual hazardous operations in the near future. Internet-based distant operations appear to be reasonable for terrestrial training and system test/debug type operations that do not involve physical operations in the hazardous environment.

As the Internet performance capabilities are expanded in the future, potential applications associated with e-maintenance in hazardous environments should be re-evaluated.

Acknowledgments

The content of this paper derives from many years at

the Oak Ridge National Laboratory where many aspects of nuclear remote maintenance were studied. Thanks to all DOE and ORNL colleagues.

References

[1] Noakes, M., Haley, D., and Willis, D., "The Selective Equipment Removal System Dual Arm Work Module," 7th ANS International Topical Meeting on Robotics and Remote Systems, Augusta, GA, April 27-May 1, 1997.

[2] http://www.internettrafficreport.com/

[3] http://www.mids.org/weather

[4] http://nic.merit.edu/ipma/

[5] http://io.advanced.org/IPPM/

[6] Y. Kunii, H. Hashimoto. Computer Networked Robotics. In IEEE/RSJ International Conference on Intelligent Robots and Systems, Oct. 1998.

[7] A. Rastogi. Design of an Interface for Teleoperation in Unstructured Environments using Augmented Reality Displays,1996,http://gypsy.rose.utoronto.ca/people/anu_dir/thesis/fchp1.fm.html.

[8] H. Hirukawa, I. Hara. The Web Top Robotics. IEEE/RSJ International Conference on Intelligent Robots and Systems, Oct. 1998.

[9] R. Siegwart, C. Wannaz, P. Garcia, R. Blank. Guiding Mobile Robots through the Web. Presented at IROS, Oct. 1998.

[10] Collins, J., et. al, "Data Express, a Gigabit Junction with the Next Generation Internet," IEEE Spectrum, February, 1999.

[11] Brady, K., and Tarn, T.J., "Internet-Based Remote Teleoperation," Proceedings of the 1998 IEEE International Conference on Robotics and Automation, Leuven, Belgium, May 1998.

[12] Taylor, K. and Dalton, B., "Internet Robots: A New Robotics Niche," IEEE Robotics and Automation Magazine, vol 7, no. 1, March 2000.

[13] Hamel, W. and Osborn, J., "Human-Interactive and Multi-Spectral Robot Task Space Modeling," 8th ANS International Topical Meeting on Robotics and Remote Systems, Pittsburgh, PA, April 25-29, 199

[14] Thompson, D, Burks, B. and Killough, S., "Remote Excavation Using the Telerobotic Small Emplacement Excavator" 5th ANS International Topical Meeting on Robotics and Remote Systems, pp. 465-470, Knoxville, TN, April 25-30, 1993.

[15] Johnsen, E. and Corliss, W., Human Factors Applications in Teleoperator Design and Applications, pp. 90-92, Wiley Interscience, 1971.

Internet-Based Teleoperation

Kevin Brady, System Engineering and Analysis Group, MIT Lincoln Laboratory,
244 Wood Street, Lexington, MA 02420-9108, USA, kbrady@ll.mit.edu

Tzyh-Jong Tarn, Systems Science and Mathematics Department, Washington University,
One Brookings Drive, St. Louis, MO 63130, USA, tarn@wuauto.wustl.edu

Abstract

A new method for controlling robots over the internet, where communication propagation delays exist, is presented. These delays are potentially destabilizing, and certainly degrade the human operator's intuition and performance. A state space formulation is presented, taking into account the time-varying non-deterministic nature of the control and observation delays. A model of the delay characteristics for the communication medium is also derived. Using the state space framework a general-purpose supervisory architecture is developed, allowing the projection of human "intelligence" to the remote environment via the telerobot. Dynamics of the robotic system, as well as the delay characteristics of the communication medium, become part of the design process. The design criteria of transparency, generality, and safety have been met and successfully tested in an experimental setup between Albuquerque, New Mexico and Washington University's Center for Robotics and Automation.

1. Introduction

There have been myriad designs for constructing telerobotic architectures where delays exist in the observation and control of telemanipulators. In general, though, these attempts can be grouped into three general approaches: predictive control, bilateral control, and teleprogramming. Many designs tend to blur these distinctions.

Predictive control is used in this paper to refer to a broad range of predictive approaches including traditional predictive control [1], Kalman filtering [2], and Internal Model Control [3,4]. Its role in remote teleoperation is to generate a model for the remote system to compensate for the delay. This predicted model may be used graphically to provide a pseudo real-time response to teleoperator commands. The Jet Propulsion Laboratory (JPL) [5,6,7] has done much work in using predictive displays to help control space-based robotic systems. One manifestation of their predictive paradigm is coined the "Phantom Robot" [5].

Much research has taken place in controlling telerobots using bilateral control. The intention is to control both the contact force and position (velocity) of the telerobot. As a practical matter, doing both simultaneously is not possible. Instead, bilateral control is modeled as a two-port where force is the input and velocity is the output for the teleoperator. Likewise, velocity is the input and force is the output for the telerobot. For more information see [8,9,10,11,12,13,14,15,16].

"Teleprogramming" is used in this paper to encompass a broad range of approaches that perform remote teleoperation in a similar way. This approach enables a cooperation between the remote planner and the robot on an abstract level in order to overcome the limitations of delay and bandwidth. The local controller has broad latitude in interpreting the general plan and implementing it. This supervisory control approach typically results in greater modularity and independence for the remote telemanipulator. For more information see [17,18].

Much research has been done in controlling systems with delays. They have a broad range of applications including process control, ship stabilization, and many biological problems such as population models. The difficulty is that traditional control methods typically cannot guarantee stability for such systems except for possibly small gains. One way out of this is to take the process model to be controlled and model it as part of the control. The first example of this approach is the well-known Smith predictor [19]. Utilizing a process model the time-delay is eliminated from the characteristic equation converting the closed-loop system into a delay-free one.

Olbrot [20,21] established conditions for the controllability, detectability, and stabilization of time-varying delayed systems. Manitius and Olbrot [22] came up with a feedback law to yield a finite spectrum for the closed loop time-delay system. Their research stems from the idea that time-delay systems are infinite dimensional systems. Assuming perfect system knowledge they develop feedback that includes an integral term on the feedback itself as well as the expected state feedback. Using standard Laplacian methods they show that the resulting system can be made finitely dimensional, while mapping the finite poles to the left-hand complex plane using standard pole-placement techniques. They further investigate the case of disturbances or imperfect system knowledge and ask two questions. First, does the system become infinitely dimensional? Second, what happens to the poles of the system? Their answer to the first question is, yes, the system does become infinitely dimensional. Their answer to the second question is more interesting. They find that they still have finite poles in the left-hand plane that are in the neighborhood of the poles assigned using standard techniques. Furthermore, the infinite poles come in far out in the left-hand plane. Their conclusion is that the system remains stable for reasonable disturbances/modeling errors.

Watanabe and Ito [23] came up with an observer for the delayed system with finite spectrum, comparing their results with those of the Smith predictor. Klamka [24] extended these results to the case of distributed delays. Later, Pandolfi [25] showed that the Watanabe-Ito observer was a simplified form of the compensator presented in his paper based on the theory of distributed control processes.

Much work has been also done over the years in the area of stability of systems with state delays, particularly the stability independent of the delay magnitude. Li and de Souza [26] use a Razumikhin [27] approach for achieving robust stability of such systems. Kojima et al [28] reformulate the problem as a H-infinity problem to achieve robust stabilization results. Clarke et al [29] use an adaptive predictive controller to achieve such ends. Numerous other results exist that come up with stability criterion for time-delay systems independent and dependent on the size of the delay. Other research concerning systems with state delays include [30,31,32].

The contribution of this work is a general-purpose architecture for projecting "intelligently" into a remote site via a telerobot. It is based on an analytic state-space formulation of the system. This formulation includes the appropriate delays in observation and control. An investigation of these delays is a prerequisite to finalizing the state space model. Upon investigation, these delays often tend to be time varying and non-deterministic. This mathematical analysis is general enough to encompass remote teleoperation using various communication mediums such as sonar, electromagnetic radiation, and the internet.

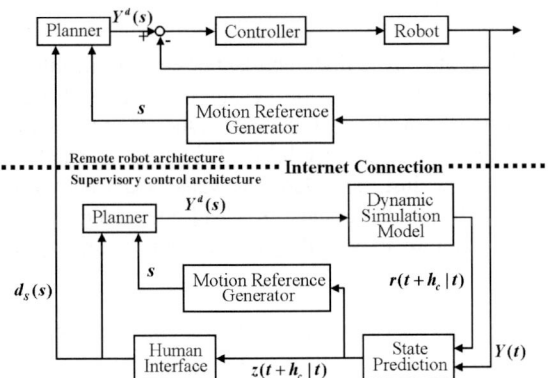

Figure 1: Robotic Architecture for Internet Based Operations with Time Delays.

2. Design
2.1. Software Architecture and Design

In order to meet the objectives laid out in the first section there are a number of intermediary goals. These include deriving a flexible supervisory control scheme and developing an interface that gives the teleoperator a sense of telepresence. Figure 1 is the general idea for how to meet the first intermediary goal. It is based on Event Based Planning and Control [33,34], whose basic design is on the top half of the Figure 1. The variable **s** is the reference variable that is used instead of time, while **Y^d(s)** and **Y^{ds}(s)** are the nominal and supervisory commands. The major problem is that the supervisory control is not located where the rest of the system is, and there is a delay in the exchange of information between the supervisory control and the rest of the architecture. The role of this section is to apply the state space model of the previous section to formulating this problem.

The section of the figure below the dotted line represents the supervisory controller, while the other half comprises of the robot, the controller, and the corresponding planner. The role of the supervisory controller is to deliver a sense of telepresence to the teleoperator. Telepresence is presented through the Human Interface. The state **$z(t+h_c|t)$** that is fed to the Human Interface represents the fusion of the time-forward observer and a dynamic model of the robot's remote environment. This allows the teleoperator to interact with the virtual environment contained within the Human Interface as if he/she were doing so in real-time without any delays.

Note that there is an Event Based Planner and Motion Reference Generator local to each side of the architecture. This distributed architecture gives greater flexibility and autonomy to the remote manipulator. The planned trajectory **Y^d(s)** is the same on either side. Through use of the time-forward observer presented in the previous section the reference variable **s** is coordinated on either side of the system.

2.1.1. TeleSupervisory Commands

Due to communication limitations it is often necessary to communicate with the remote workcell on an abstract level. TeleSupervisory Control, as presented here, aspires to present a canonical methodology for performing such supervisory control. It is unique in a number of ways, including:
- The first robotic teleprogramming technique based on a state space model.
- Considers the time-varying nature of the communication medium as part of the design.
- It is event-based, rather than time-based.

This set of supervisory commands is based on the joystick information provided through the human interface. The joystick commanded velocity inputs are discussed in the next section. The supervisory commands are based on Function Based Sharing Control [35] and are extended here for time-delayed systems.

2.1.2. Time-Synchronization

A predictive controller is included in the architecture in Figure 1. This prediction gives the teleoperator the ability to interact with a virtual model of the remote environment as if there were no time-delay. In order to facilitate this the observer model for the robot starts its virtual operation h_c seconds before the actual robot starts. This time offset gives the teleoperator enough time to react and instruct the remote telerobot to take evasive actions in the event that the teleoperator determines that a collision is about to take place.

2.2. Interface Design

Figure 2 illustrates the human interface that has been successfully tested. It consists of two computer interfaces. The first has a virtual model of the remote system. The human teleoperator may interact with the virtual model in Deneb's TGRIP with the expectations that h_c seconds later the same actions are taking place on the actual robot. A control panel allows the teleoperator to enable the appropriate functionality of the robot. Live video using SGI's Inperson that is delayed similiarly to the observed data is shown on a second computer. Finally, a joystick is used to enter supervisory commands.

An internet connection utilizing a UDP/IP link is used for conveying information between the remote sites.

2.3. Control

Event Based Planning [36,37] is used to provide the desired trajectory for this model. It is based on a reparameterization of the control so that it is indexed by an event reference variable **s** instead of by time. Details are not provided here, but are available in the references for those interested in further investigation. These references also provide details of the nonlinear feedback necessary to decouple the robotic system to enable the linear model presented in this subsection.

2.3.1. Time Delays and Bandwidth Limitations

$$h(t) = h_n + \overline{h}_d(t) + h_b(t)$$

The goal of this section is to describe the nature of the propagation delay that is incurred, describe its relationship to the bandwidth, and come up with a delayed model for the state space system. The uni-directional delay is designated $h(t)$. It is only for illustrative purposes and may denote the delay in either the forward or backward direction. The delay $h(t)$ may be broken down into three components as follows:

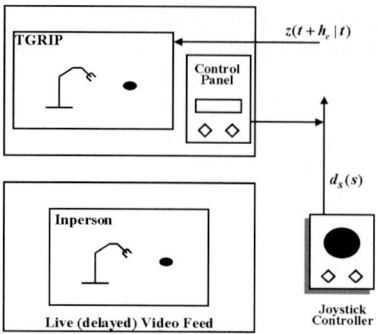

Figure 2: Human interface.

h_n is the nominal propagation delay. It represents the time that it takes for the signal to physically propagate without disturbance from its source to its destination across the communication medium. Its value may be experimentally determined and is non time-varying.

h_d is the disturbance delay. It represents the deviation from the expected delay that results from unknown disturbances or even loss of information. h_d is non-deterministic and time-varying.

$\overline{h}_d$ is a step function that is based on the function h_d. Since the data exchange is discrete the only time that the disturbance delay is relevant is when the information actually propagates

h_b is the bandwidth delay. Since information is exchanged at finite rate **b** across the communication medium a corresponding delay will be incurred that is known, but time varying. It is not known a priori absolutely, but the form of the function is known, as seen in Figure 3. The delay h_b is a sawtooth function and is bounded as follows:

$$\overline{h}_b = \frac{1}{b} \geq h_b(t) \geq 0$$

The choice of **b** is the first important design consideration of our system. A large value of **b** enables a larger exchange of information, as well as a reduction in the effective delay of the system. On the other hand, there may be bandwidth limitations to just how large a value may be chosen. Choosing **b** too large may overload the communication system, resulting in lost data and eventually an untimely shutdown.

Note that the delayed information is treated differently depending on if it is sent in the observer or controller direction. Observed information is processed as soon as it comes in. Mathematically, this means that the delay in the observer direction h_o is defined as $h_o(t) = h(t)$. The delay in the control direction is evaluated slightly differently in order to synchronize the control of the remote robot with a local simulated model. First, a control delay h_c is determined as a design consideration. As mentioned earlier, h_c value represents the difference in time between when a control is applied at the teleoperator and at the telerobot side of the system. It is chooses so that it is at least as large as the maximum value of the deterministic part of the delay model:

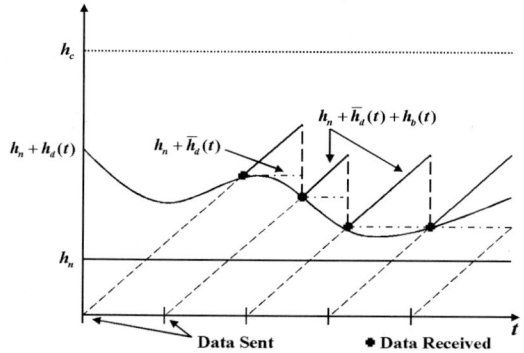

Figure 3: Nature of the time delays.

$$h_c \geq h_n + \overline{h}_b$$

Once the choice of h_c is made a function $\Delta h_c(t)$ can be used to model when $h(t)$ exceeds h_c:

$$\Delta h_c(t) = \begin{cases} h(t) - h_c & h(t) > h_c \\ 0 & h(t) \leq h_c \end{cases}$$

$h_c + \Delta h_c(t)$ now represents the time difference between when a control is generated by the teleoperator and enacted by the telerobot. The value of $\Delta h_c(t)$ is bounded by Δh_{max}. If this were not the case the system would effectively be open loop.

2.3.2. State Space Model

The state space model for a telerobotic system with delays in the communication channels is modeled as:

$$\dot{x}(t) = Ax(t) + B_1 u_1(t) + B_2 u_2(t - h_c - \Delta h_c(t))$$
$$y(t) = \sum_{j=1}^{N} C_j x(t - h_{o_j}(t))$$

where $x(t)$ is the task space vector $y(t)$ is the observation of $x(t)$. It is assumed that $u_1(t)$, $u_2(t)$, and $y(t)$ are measureable, though $x(t)$ is not. Matrices A, B_1, B_2, and C are time invariant and of appropriate dimension. The delay components $\Delta h_c(t)$ and $h_{oi}(t)$ are non-deterministic, time-varying, non-negative, and bounded. The prediction in Figure 1 is based on this state space model.

Note that the various components $h_{oi}(t)$ of the observation delay represent observed information that comes over different channels with possibly different delays. A good example of why such a model is needed is the ROTEX [39] project. NASA allocated two different channels for feeding back the position of the robot and sending back the video feed for the robot. Each of these channels had a very different delay. A distributed observer delay model would also be important if there were calculation delays. Rough estimates might be readily available, while more accurate filtered information is available but requires heavy computation and a resulting time lag.

2.3.3. Data Fusion

The architecture's goal is to provide an interface to the robot as if the teleoperator was present in the robot's workspace. A time-forward observer, as presented in the previous section, is an excellent first step to achieving this functionality. The robot though, may physically interact or make contact with the remote environment rather than simply move in free space. A hybrid force / position control is necessary for this reason. In order to achieve the robust prediction for the hybrid control of the remote manipulator a dynamic simulation of the robot's activities is necessary. This is achieved through fusing the results of the time-forward observer with the results of a dynamic simulation of the remote environment.

The distributed nature of the system's architecture allows the remote telemanipulator to react to unexpected circumstances without needing to wait for a delayed command from the teleoperator. In particular, the Event Based Planner/Controller will suspend forward motion upon contact with an object. This contact cannot be directly modeled in the time-forward observer, though it may be dynamically simulated.

3. Experimental Results

The result shown in Figure 4 here is based on a move along a straight line. The human operator determines that there is an object that may obstruct the move and issues a **STOP** command. After further investigation the manuever is allowed to proceed. The results are shown based on three approaches:

- Apply supervisory commands h_c seconds after they are issues (approach identified in this research).
- Apply supervisory commands as they show up.
- No prediction is used, supervisory commands are used as they show up.

It is readily apparent that the approach shown here shows a greater correlation between the state of the predictive model and the actual state of the robot. The advantage is that the closer to reality that the virtual model is the more likely that the human operator will attain the intuitiveness (telepresence) for the remote environment to act as an appropriate supervisory controller. Further experimental results are illustrated in reference [36].

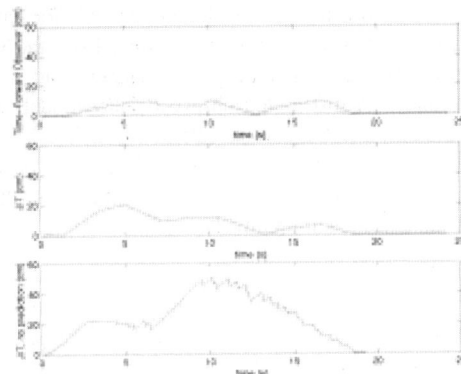

Figure 4: Experimental results: Move, stop, move.

4. Conclusions

This paper has presented a promising architecture for controlling remote telerobots. The goals of safety, efficiency, telepresence, and transparency have been met. Its effectiveness has been tested, including a live demonstration during a Plenary address at the 1997 IEEE Conference on Robotics and Automation in Albuquerque, New Mexico. Such architecture has ready application to remote controlled space-based and underwater robots. Additionally, it is highly relevant to the nascent field of internet-based control.

A delay model was developed and presented. This model is flexible enough to embrace the wide variety of possible communication mediums for remote teleoperation. Limitations in bandwidth and time-varying delays were discussed, and their affect on system design parameters. Based on the delay model a State Space Model for controlling, observing, and predicting the telemanipulator was developed.

Using the State Space Model the corresponding architecture has been developed. A human in the loop architecture was used to ensure the safe and intelligent operation at the other end. This required the development of a supervisory paradigm based on the state space model. The teleoperator was also immersed in a graphical predictive model to develop a degree of "telepresence" in the remote environment.

A dynamic model of the robot was fused with the state space time-forward observer. Using data fusion, the interaction with a virtual object in the remote environment was performed.

Experimental results show the efficacy of this approach. Very importantly, it is robust to varying time-delays; an area that has not been explored before. The predictive nature of the architecture, as well as the virtual model, allow for a degree of transparency to the user of the remoteness of the working environment and the delays in the communication channels. This is crucial for preventing the degradation of the teleoperator's intuition and performance.

5. References

[1] Ronald Soeterboek, Predictive Control: A Unified Approach, Prentice Hall, 1992.

[2] Ian b. Rhodes, A Tutorial Introduction to Estimation and Filteing, IEEE Transactions on Automatic Control, AC-16(6):688-706, December 1971.

[3] Naoto Abe, Practically Stability and Disturbance Rejection of Internal Model Control for Time-Delay Systems, CDC.

[4] Ning-Shou Xu and Zhi-Hong Yang, A Novel Predictive Structural Control Based on Dominant Internal Model Approach, IFAC, San Fransisco, USA, pages 451-456, 1996.

[5] Antal K. Bejczy, Paolo Fiorini, Won Soo Kim, and Paul S. Schenker, Toward Integrated Operator Interface for Advanced Teleoperation Under Time-Delay, Intelligent Robots and Systems, pages 327-348, 1995.

[6] Antal K. Bejczy, Won. S. Kim, and Steven C. Venema, The Phantom Robot: Predictive Displays for Teleoperation with Time Delays, IEEE International Conference on Robotics and Automation, Cincinatti OH, pages 546-551, May 1990.

[7] Won S. Kim and Antal K. Bejczy, Demonstration of a High-Fidelity Predictive / Preview Display Technique for Telerobotic Servicing in Space, IEEE Transactions on Robotics and Automation, 9(5):698-701, October 1993.

[8] Robert J. Anderson, Teleoperation with Virtual Force Feedback, Prodeecings of the 1993 SPIE International Symposium on Optical Tools for Manufacturing and Advanced Automation, Boston MA, September 1993.

[9] Robert J. Anderson, How to build a Modular Robot Control System using Passivity and Scattering Theory, IEEE Conference on Decision and Control, New Orleans, December 1995.

[10] Gunter Niemeyer and Jean-Jacques E. Slotine, Shaping the Dynamics of Force Reflecting Teleoperation with Time-Delays, IEEE International Conference on Robotics and Automation, Workshop on Teleoperation and Orbital Robotics.

[11] Dale A. Lawrence, Stability and Transparency in Bilateral Teleoperation, IEEE Transactions on Robotics and Automation, 9(5): 624-637, October 1993.

[12] H. Kazerooni, Tsing-Iuan Tsay, and Karin Hollerbach, A Controller Design Framework for Telerobotic Systems, IEEE Transactions on Control

Systems Technology, 1(1):50-62, March 1993.
[13] Joseph Yan and S. E. Salcudean, Teleoperation Controller Design Using H-infinity – Optimization with Applications to Motion Scaling, IEEE Transactions on Control Systems Technology, 4(3):244-258, May 1996.
[14] Gary M. H. Leung, Bruce A. Francis, and Jacob Apkarian, Bilateral Controller for Teleoperators with Time Delay via μ-synthesis, IEEE Transactions on Robotics and Automation, 11(1):105-116, February 1995.
[15] Won Kim, Force-Reflection and Shared Compliant Control in Operating Telemanipulators with Time-Delay, IEEE Transactions on Robotics and Automation, 8(2):176-184, April 1992.
[16] Sukhan Lee and Hahk Sung Lee, Modeling, Design, and Evaluation of Advanced Teleoperator Control Systems with Short Time Delay, IEEE Transactions on Robotics and Automation, 9(5):607-623, October 1993.
[17] Lynn Conway, Richard A. Bolz, and Michael W. Walker, Teleautonomous Systems: Projecting and Coordinating Intelligent Action at a Distance, IEEE Transactions on Robotics and Automation, 6(2):146-158, April 1990.
[18] Y. Wakita, S. Hirai, K. Machida, K. Ogimoto, T. Itoko, P. Backes, and S. Peters, Applications of Intelligent Monitoring for Super Long Distance Teleoperation, IROS, Osaka, Japan, November 1996.
[19] O. J. M. Smith, A Controller to Overcome Dead Time, ISA, 6:28-33, 1959.
[20] Andrzej W. Olbrot, On Controllability of Linear Systems with Time Delays in Contol, IEEE Transactions on Automatic Control, pages 664-666, October 1972.
[21] Andrzej W. Olbrot, Stabilizability, Detectability, and Spectrum Assignment for Linear Autonomous Systems with general Time Delays, IEEE Transactions on Automatic Control, AC-23(5):887-890, October 1978.
[22] Andrzej Z. Manitius and Andrzej W. Olbrot, Finite Spectrum Assignment Problemfor Systems with Delays, IEEE Transactions on Automatic Control, AC-24(4):541-553, August 1979.
[23] Keiji Watanabe and Masami Ito, An Observer for Linear Feedback Control Laws of Multivariable Systems with Multiple Delays in Controls and Outputs, Systems and Control Letters, 1(1):54-59, July 1981.
[24] Jerzy Klamka, Observer for Lineaqr Feedback Control of Systems with Distributed Delays in Controls and Outputs, Systems and Control Letters, 1(5):326-331, 1982.
[25] Luciano Pandolfi, Dynamic Stabilization of Systems with Input Delays, Automatica, 27(6):1047-1050.
[26] Xi Li and Carlos E. de Souza, Criteria for Robust Stability of Uncertain Linear Systems with Time-Varying State Delay, In 13th IFAC World Congress, San Fransisco, pages 1-6, June 1996.
[27] Jack K. Hale and Sjoerd M. Verduyn Lunel, Introduction to Functional Differential Equations, Springer-Verlag, 1993.
[28] Akira Kojima, Kenko Uchida, Etsujiro Shimemura, and Shintaro Ishijima, Robust Stabilizatin of a System with Delays in Control, IEEE Transactions on Automatic Control, 39(8):1694-1698, August 1994
[29] D. W. Clarke, E. Mosca, and R. Scattolini, Robustness of an Adaptive Predictive Controller, IEEE Transactions on Automatic Control, 39(5):1052-1056, May 1994.
[30] S. D. Brierley, J. N. Chiasson, E. B. Lee, and S. H. Zak, On Stability Independent of Delay for Linear Systems, IEEE Transactions on Automatic Control, AC-27(1):252-254, February 1982.
[31] Takehiro Mori, Criteria for Asymptotic Stability of Linear Time-Delay Systems, IEEE Transactions on Automatic Control, AC-30(2):158-161, February 1985.
[32] Takehiro Mori and H. Kokame, Stability of $x(t)=ax(t) + bx(t-\tau)$, IEEE Transactions on Automatic Control, 34(4)460-462, April 1989
[33] Ning Xi, Event-Based Motion Planning and Control for Robotic Systems, PhD Thesis, Washington University, September 1993.
[34] Ning Xi, Tzyh-Jong Tarn, and Antal K. Bejczy, Intelligent Planning and Control for Multi-Robot Coordination --- An Event-Based Approach, IEEE Transactions on Robotics and Automation, 12(3): 439-452, June 1996.
[35] Chuanfan Guo, Tzyh-Jong Tarn, Ning Xi, and Antal K. Bejczy, Fusion of Human and Machine Intelligence for Telerobotic Systems, Proceedings of the IEEE/RSJ International Conference on Robotics and Automation, pages 3110-3115, August 5-9, 1995.
[36] Kevin Brady, Time-Delayed Telerobotic Control, Doctor of Science Dissertation, 1997.
[37] Ning Xi, Event-Based Motion Planning and Control for Robotic Systems, PhD Thesis, Washington University, September 1993.
[38] Ning Xi, Tzyh-Jong Tarn, and Antal K. Bejczy, Intelligent Planning and Control for Multi-Robot Coordination --- An Event-Based Approach, IEEE Transactions on Robotics and Automation, 12(3): 439-452, June 1996.
[39] Gerd Hirzinger, K. Landzettel, and Ch. Fagerer, Telerobotics with Large Time-Delays – the ROTEX Experience, IROS, Munich, Germany, pages 571-578, September 12-16, 1994.

Language Model Approach to Nonblocking Supervisor Synthesis for Nondeterministic Discrete Event Systems

S.-J. Park and J.-T. Lim

Dept. of Electrical Engineering, KAIST,
Taejon, 305-701, Korea, E-mail: jtlim@stcon.kaist.ac.kr

Abstract

This paper considers the nonblocking supervisory control problems of nondeterministic discrete event systems (DESs) which are modeled as nondeterministic automata with ϵ-moves. Based on language models, this paper presents the necessary and sufficient conditions for the existence of a nonblocking supervisor to achieve a given language specification for a nondeterministic DES. The developed nonblocking supervisor always guarantees the absence of blocked states in a controlled nondeterministic system. Moreover, when the language specification does not satisfy the existence conditions, this paper provides the computational algorithm for finding the supremal language of the specification which satisfies the conditions. Furthermore, the results developed are demonstrated through the example of an assembly work station.

I. Introduction

Most of the research on the supervisory control focuses on deterministic DESs that are modeled as deterministic automata [1]. In a deterministic DES, given a state and an event that occurs in that state, the state reached after the occurrence of the event is uniquely known. However, such an assumption is not satisfied for nondeterministic DESs, in which unmodeled dynamics, partial observation, or inherent nondeterminism are present. In recent years, a great deal of attention is given to the supervisory control of nondeterministic DESs that are modeled as nondeterministic automata. The work of [3] introduces the trajectory model formalism for modeling of nondeterministic behavior, and the prioritized synchronous composition (PSC) to model the interaction of a system and a supervisor. In [4]-[7], it is shown that the supervisory control of nondeterministic DES can be achieved using trajectory models and PSC. In [5], in particular, it is shown that the usual notion of nonblocking, called language model nonblocking, may not be adequate in the setting of nondeterministic systems, and a stronger notion, called trajectory model nonblocking, is introduced. However, it is possible to have such a system that it is trajectory model nonblocking yet it gets blocked [5]. In [8] and [9], it is shown that, through a lifting procedure, a nondeterministic system is translated to a partially observed deterministic system, and the algorithms for the supervisory control of deterministic systems under partial observation can be adapted for synthesis of supervisors for nondeterministic systems. In particular, the work of [8] deals with the nonblocking supervisory control problem subject to trajectory model specifications, where a supervisor is nonblocking in the sense that every trajectory enabled by the supervisor is a prefix of a trajectory that ends at a marked state. However, for some nondeterministic systems represented as nondeterministic automata, even though there exist nonblocking supervisors in the sense of trajectory models, supervised systems may get blocked. In [10], the failure semantics formalism is presented for the supervisory control of nondeterministic DESs.

In [11], based on language models and PSC, robust and nonblocking supervisory control problems are developed for uncertain nondeterministic DESs which are modeled as sets of some possible nondeterministic automata without ϵ-moves. However, in the work, the nonblocking supervisor synthesis problem is missing. In this paper we address the nonblocking supervisory control problems of nondeterministic DESs which are modeled as nondeterministic automata with ϵ-moves. Specifically, we present the existence conditions of a nonblocking supervisor to achieve a given language specification and guarantee the absence of blocked states in a controlled system. Moreover, we address the nonblocking supervisor synthesis problem to find a supremal sublanguage of the specification which satisfies the conditions. The method presented in this paper deals with the nonblockingness problem on the basis of language models and PSC. It is shown that the method based on language models always guarantees that a nondeterministic DES controlled by a nonblocking supervisor developed does not get blocked. Furthermore, to show feasibility of the obtained results, we illustrate an example of an assembly work station performing peg-in-hole linking operations.

II. Preliminaries

Let the 5-tuple $G:=(\Sigma, Q_G, \delta_G, q_G^0, Q_G^m)$ represent a nondeterministic DES modeled as a nondeterministic automaton with ϵ-moves, where Σ is the set of events, Q_G is the set of states, $\delta_G : Q_G \times (\Sigma \cup \{\epsilon\}) \mapsto 2^{Q_G}$ is the transition function, where ϵ-moves represent internal events or silent transitions, i.e., ϵ-transitions, q_G^0 is the initial state, and $Q_G^m \subseteq Q_G$ is the set of marked states. We assume that G does not contain any cycle of ϵ-transitions. Let Σ^* denote the set of all finite strings over Σ, including the empty string ϵ. Note that a deterministic automaton is a special case of nondeterministic automata in which the cardinality of $\delta_G(q_G, \sigma)$ is less than or equal to 1. It is customary to call a subset of Σ^* a language over Σ. The ϵ-closure of $q \in Q$, denoted by $\epsilon_G^*(q) \subseteq Q_G$, is defined as follows: $q \in \epsilon_G^*(q)$; $q' \in \epsilon_G^*(q) \Rightarrow \delta_G(q', \epsilon) \subseteq \epsilon_G^*(q)$. Also, the set of refusal events at $q \in Q_G$, denoted by $R_G(q) \subseteq \Sigma$, is defined as $R_G(q) := \{\sigma \in \Sigma \mid \delta_G(q', \sigma) = \emptyset, \forall q' \in \epsilon_G^*(q)\}$. The extended transition function $\delta_G^* : Q_G \times \Sigma^* \mapsto 2^{Q_G}$ is defined as follows: $\delta_G^*(q, \epsilon) := \epsilon_G^*(q)$ and $\delta_G^*(q, s\sigma) := \epsilon_G^*(\delta_G(\delta_G^*(q, s), \sigma))$. The prefix closure of $L \subseteq \Sigma^*$ is defined as $pr(L) := \{u \in \Sigma^* \mid uv \in L \text{ for some } v \in \Sigma^*\}$. The closed behavior of G, denoted by $L(G)$, is defined as $L(G) := \{s \in \Sigma^* \mid \delta_G^*(q_G^0, s) \neq \emptyset\}$, and the marked behavior of G, denoted by $L_m(G)$, is defined as $L_m(G) := \{s \in \Sigma^* \mid \delta_G^*(q_G^0, s) \cap Q_G^m \neq \emptyset\}$.

Let $S = (\Sigma, Q_S, \delta_S, q_S^0, Q_S^m)$ be a nondeterministic automaton of a supervisor. In this paper, we assume that $Q_S^m = Q_S$. Also, let $A, B \subseteq \Sigma$ be the priority sets of G and S, respectively. In general, $A \cap B$ is the set of controllable events, $A - B$ is the set of uncontrollable events, $B - A$ is the set of driven events, and $\Sigma - (A \cup B)$ is assumed to be empty. Then, the prioritized synchronous composition (PSC) of a system G and a supervisor S means a closed-loop system represented as a nondeterministic automaton denoted by $G_A \|_B S := R := (\Sigma, Q_R, \delta_R, q_R^0, Q_R^m)$, where $Q_R = Q_G \times Q_S$, $q_R^0 = (q_G^0, q_S^0)$, $Q_R^m = Q_G^m \times Q_S^m$, and the transition function $\delta_R : Q_R \times (\Sigma \cup \{\epsilon\}) \mapsto 2^{Q_R}$ is defined as follows: for all $q_R = (q_G, q_S) \in Q_R$ and $\sigma \in \Sigma$, $\delta_R(q_R, \sigma) :=$

$$\begin{cases} \delta_G(q_G, \sigma) \times \delta_S(q_S, \sigma), \\ \quad \text{if } \delta_G(q_G, \sigma) \neq \emptyset, \delta_S(q_S, \sigma) \neq \emptyset \\ \delta_G(q_G, \sigma) \times \{q_S\}, \\ \quad \text{if } \delta_G(q_G, \sigma) \neq \emptyset, \sigma \in R_S(q_S), \sigma \notin B \\ \{q_G\} \times \delta_S(q_S, \sigma), \\ \quad \text{if } \sigma \in R_G(q_G), \sigma \notin A, \delta_S(q_S, \sigma) \neq \emptyset \\ \emptyset, \quad \text{otherwise,} \end{cases}$$

$$\delta_R(q_R, \epsilon) := [\delta_G(q_G, \epsilon) \cup \{q_G\}] \times [\delta_S(q_S, \epsilon) \cup \{q_S\}] - \{(q_G, q_S)\}.$$

Thus, an event occurs synchronously whenever both systems can participate. However, it can occur asynchronously whenever one of the systems can participate and the second system refuses it but has no priority over it. The closed behavior of $G_A \|_B S$ is defined as $L(G_A \|_B S) := \{s \in \Sigma^* \mid \delta_R^*(q_R^0, s) \neq \emptyset\}$, and the marked behavior of $G_A \|_B S$ is defined as $L_m(G_A \|_B S) := \{s \in \Sigma^* \mid \delta_R^*(q_R^0, s) \cap Q_R^m \neq \emptyset\}$.

For $\Sigma_d \subseteq \Sigma$, let $D(\Sigma_d)$ be a deterministic automaton with one state and self-loops labeled by every event in Σ_d. Then, the augmentation of G by Σ_d, denoted by G^{Σ_d}, is defined as a nondeterministic automaton $G^{\Sigma_d} := G_\emptyset \|_\emptyset D(\Sigma_d)$. The state set of G^{Σ_d} can be identified with the state set of G, and G^{Σ_d} is then obtained from G by adding self-loop at each $q_G \in Q_G$

labeled by every event in $\Sigma_d - \Sigma_G(q_G)$ where $\Sigma_G(q_G) := \{\sigma \in \Sigma \mid \delta_G(q_G, \sigma) \neq \emptyset\}$.

III. MAIN RESULTS

Let us define nonblocking languages and nonblocking supervisors respectively as follows:

Definition 1: $M \subseteq \Sigma^*$ is a nonblocking language w.r.t.(with respect to) a nondeterministic automaton G if, for any $s \in pr(M)$ and $q \in \delta_G^*(q_G^0, s)$, there exists $t \in \Sigma^*$ such that $st \in pr(M)$ and $\delta_G^*(q, t) \cap Q_G^m \neq \emptyset$.

Definition 2: S is a nonblocking supervisor for G if $L(G_A \|_B S)$ is a nonblocking language w.r.t. $G_A \|_B S$.

A nonblocking supervisor S for G always guarantees the absence of blocked states in a controlled system $G_A \|_B S$. Also, it satisfies the language property of the nonblockingness in [1], i.e., $pr(L_m(G_A \|_B S)) = L(G_A \|_B S)$.

Note that the following relationships between the behaviors of $G_A \|_B S$ and G^{B-A} hold: $L(G_A \|_B S) \subseteq L(G^{B-A})$ and $L_m(G_A \|_B S) \subseteq L_m(G^{B-A})$. They can be easily proved using the definition of PSC.

Let us consider the following problem: Given $K(\neq \emptyset) \subseteq L_m(G^{B-A})$, find conditions for the existence of a nonblocking supervisor S such that $L_m(G_A \|_B S) = K$. In the following lemma, we show a basic result for solving the problem.

Lemma 1: For $K(\neq \emptyset) \subseteq L_m(G^{B-A})$, there exists a supervisor S such that $L(G_A \|_B S) = pr(K)$ iff $pr(K)(A - B) \cap L(G^{B-A}) \subseteq pr(K)$.

Proof: (If) Let $S := det(pr(K))$, i.e., the deterministic automaton such that $L(S) = pr(K)$. Under the assumption that for $s \in \Sigma^*$, $s \in L(G_A \|_B S) \Leftrightarrow s \in pr(K)$, we need to show that $L(G_A \|_B S) = pr(K)$. Firstly, for $\sigma \in \Sigma$, let $s\sigma \in L(G_A \|_B S)$. Then, for some $(q_G, q_S) \in \delta_R^*(q_R^0, s)$, the following three cases can be considered: (Case a) $\delta(q_G, \sigma) \neq \emptyset$ and $\delta_S(q_S, \sigma) \neq \emptyset$, (Case b) $\delta_G(q_G, \sigma) \neq \emptyset$, $\sigma \in R_S(q_S)$, and $\sigma \notin B$, (Case c) $\sigma \in R_G(q_G)$, $\sigma \notin A$, and $\delta_S(q_S, \sigma) \neq \emptyset$. Then, by $S := det(pr(K))$ and the sufficient condition, it can be shown that $s\sigma \in pr(K)$ for any cases. Next, for $\sigma \in \Sigma$ let $s\sigma \in pr(K)$. Then, for some $(q_G, q_S) \in \delta_R^*(q_R^0, s)$, the following two cases can be considered: (Case a) $\delta_G(q_G, \sigma) \neq \emptyset$, (Case b) $\sigma \in R_G(q_G)$. Then, using the definition of PSC, it can be shown that $s\sigma \in L(G_A \|_B S)$ for any cases. Therefore, we conclude that $L(G_A \|_B S) = pr(K)$.

(Only if) Let $s \in pr(K)$, $\sigma \in A - B$, and $s\sigma \in L(G^{B-A})$. Then, $s \in L(G_A \|_B S)$ by $L(G_A \|_B S) = pr(K)$, and there exists $q = (q_G, q_S) \in \delta_R^*(q_R^0, s)$ such that $\delta_G(q_G, \sigma) \neq \emptyset$. Then, the following two cases can be considered: (Case a) $\delta_S(q_S, \sigma) \neq \emptyset$, (Case b) $\delta_S(q_S, \sigma) = \emptyset$. By $L(G_A \|_B S) = pr(K)$ and the definition of PSC, it can be shown that $s\sigma \in pr(K)$. Therefore, we conclude that $pr(K)(A - B) \cap L(G^{B-A}) \subseteq pr(K)$. ∎

For solving the above problem, the following theorem is necessary.

Theorem 1: For $K(\neq \emptyset) \subseteq L_m(G^{B-A})$, there exists a nonblocking supervisor S such that $L(G_A \|_B S) = pr(K)$ iff

(A1) $pr(K)(A - B) \cap L(G^{B-A}) \subseteq pr(K)$,

(A2) K is nonblocking w.r.t. G^{B-A}.

Proof: (If) From Lemma 1, $S := det(pr(K))$ satisfies $L(G_A \|_B S) = pr(K)$ by (A1). Hence, by (A2), $L(G_A \|_B S)$ is nonblocking w.r.t. G^{B-A}, and it can be shown that it is nonblocking w.r.t. $G_A \|_B S$.
(Only if) (A1) is satisfied by Lemma 1, and $L(G_A \|_B S)$ is nonblocking w.r.t. $G_A \|_B S$. Hence, $pr(K)$ is also nonblocking w.r.t. $G_A \|_B S$, and we can show that it is nonblocking w.r.t. G^{B-A}. ∎

We outline the complexisty analysis as follows: Let $\|K\|$ be the number of states in the minimal automaton realization of K, and m be the number of states of the automaton G. Then, the computational complexity for testing the condition (A2) is $O(\|K\|m^2)$ where n is the cardinality of G.

As a solution for the above problem, the following theorem provides the necessary and sufficient conditions for the existence of a nonblocking supervisor.

Theorem 2: For $K(\neq \emptyset) \subseteq L_m(G^{B-A})$, there exists a nonblocking supervisor S for G such that $L_m(G_A\|_B S)=K$ iff the conditions (A1), (A2) in Theorem 1 and (A3) $pr(K) \cap L_m(G^{B-A}) = K$ are satisfied.

Proof: (If) From Theorem 1, $S := det(pr(K))$ is nonblocking for G by (A1), (A2). Also, the forward inclusion of $L_m(G_A\|_B S) = K$ can be shown using (A3), and the reverse inclusion can be shown using $K \subseteq L_m(G^{B-A})$ and $Q_S^m = Q_S$. (Only if) Since S is nonblocking for G such that $L_m(G_A\|_B S) = K$, it holds that $pr(L_m(G_A\|_B S)) = L(G_A\|_B S) = pr(K)$. Hence, by Theorem 1, (A1) and (A2) are satisfied. Also, using $L_m(G_A\|_B S) = K$, it can be shown that (A3) is satisfied. ∎

Now let us consider the following problem: When $K \subseteq L_m(G^{B-A})$ does not satisfy the conditions (A1) and (A2), find a supremal sublanguage of K to satisfy the conditions. In other words, we consider the the nonblocking supervisor synthesis problem. Let $K_0:=K$ and $K_{i+1}:=SupC(\Phi(K_i))$, $i = 0, 1, 2, \cdots$, where $\Phi(K_i) := \{s \in K_i \mid \text{for any } t \in pr(s) \text{ and } q \in \delta_{G^{B-A}}^*(q_{G^{B-A}}^0, t), \exists u \in \Sigma^* \text{ s.t. } tu \in pr(K_i) \text{ and } \delta_{G^{B-A}}^*(q, u) \cap Q_{G^{B-A}}^m \neq \emptyset\}$ and $SupC(\Phi(K_i))$ is a supremal controllable sublanguage of $\Phi(K_i)$ w.r.t. $A-B$ and $L(G^{B-A})$, i.e., $SupC(\Phi(K_i)):=\bigcup\{M \mid M \subset \Phi(K_i) \text{ and } pr(M)(A-B) \cap L(G^{B-A}) \subseteq pr(M)\}$ [2]. Then the following lemmas hold:

Lemma 2: The limit $K' := \lim_{i \to \infty} SupC(\Phi(K_i))$ exists and $K' = SupC(\Phi(K'))$, i.e., K' is a fixpoint of the operator $SupC \circ \Phi$.

Proof: According to the definition of $SupC$ and Φ, the following relationship always holds for any K_i: $K_{i+1} \subseteq K_i$. Hence, $K_0 \supseteq K_1 \supseteq K_2 \supseteq \cdots$, so that the limit $K' := lim_{i \to \infty} SupC(\Phi(K_i)) = \bigcap_{i=0}^{\infty} K_i$ exists. Next, assume that $K' \supseteq SupC(\Phi(K'))$ but $K' \neq SupC(\Phi(K'))$. Then $K' \neq \bigcap_{i=0}^{\infty} K_i$, hence it is a contradiction. Thus, we conclude that $K' = SupC(\Phi(K'))$. ∎

Lemma 3: K' satisfies (A1) and (A2).

Proof: Since K' is a controllable sublanguage w.r.t. $A - B$ and $L(G^{B-A})$, it satisfies (A1). Also, since $K' = SupC(\Phi(K')) \subseteq \Phi(K') \subseteq K'$, it is true that $\Phi(K') = K'$. Thus, it satisfies (A2). ∎

The following theorem shows a solution of the synthesis problem:

Theorem 3: For any $K'' \subseteq K$ satisfying (A1) and (A2), $K'' \subseteq K'$, i.e., K' is the supremal sublanguage of K satisfying (A1) and (A2).

Proof: Let $K'' \subseteq M \subseteq K$. Then, we need to show that $K'' \subseteq SupC(\Phi(M))$. For any $s \in K''$, since K'' satisfies (A2), it is true that for any $t \in pr(s)$ and $q \in \delta_{G^{B-A}}^*(q_{G^{B-A}}^0, t)$, there exists $u \in \Sigma^*$ such that $tu \in pr(K'')$ and $\delta_{G^{B-A}}^*(q, u) \cap Q_{G^{B-A}}^m \neq \emptyset$. Also, since $K'' \subseteq M$, it holds that $tu \in pr(M)$. Hence, $s \in \Phi(M)$. Thus, $K'' \subseteq \Phi(M)$. Next, since K'' satisfies (A1), it is true that $K'' \subseteq SupC(\Phi(M))$. Therefore, since $K' = lim_{i \to \infty} SupC(\Phi(K_i))$ and $K_0 = K$, we conclude that $K'' \subseteq K'$, i.e., K' is the supremal sublanguage of K satisfying (A1) and (A2). ∎

The algorithm to find $\Phi(K_i)$ is as follows:

1. Initialize $M \subseteq \Sigma^*$ as $M := \emptyset$, and for all $s \in K_i$, repeat the following procedure:
2. Compute $\bigcup_{t \in pr(s)} \delta_{G^{B-A}}^*(q_{G^{B-A}}^0, t)$.
3. For all $q \in \bigcup_{t \in pr(s)} \delta_{G^{B-A}}^*(q_{G^{B-A}}^0, t)$, repeat the following procedure:
   (a) Let $q \in \delta_{G^{B-A}}^*(q_{G^{B-A}}^0, t')$ where $t' \in pr(s)$.
   (b) If $q \notin Q_{G^{B-A}}^m$ and there does not exist $\alpha \in \Sigma$ s.t. $\delta_{G^{B-A}}(q, \alpha) \neq \emptyset$ and $t'\alpha \in pr(K_i)$, then $M \leftarrow M \cup \{s\}$.

After all the computation, $\Phi(K_i)$ is found as $\Phi(K_i) = K_i - M$. Also, $SupC(\Phi(K_i))$ can be computed using the algorithm in [2]. Until a fixpoint K' is found, the algorithms are iteratively applied.

IV. Example

Let us consider an assembly work station performing peg-in-hole linking operations and welding operations for the pegs and the holes shown in Fig. 1. After a peg and a hole are linked, they are fixed through a welding operation. We assume that, due to the similarity of pegs and imperfection of a sensor, the sensed results for peg 1 and peg 2 may be incorrect.

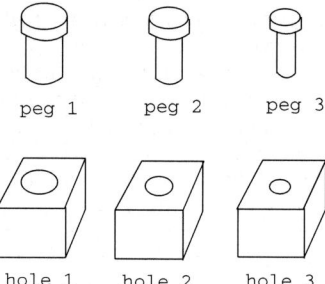

Figure 1. The pegs and holes.

For example, when peg 1 arrives in the station, it may be sensed as peg 2. The same assumption is also applied to the case of peg 2 and peg 3. Under the assumption, the assembly process is modeled as nondeterministic automaton G as shown in Fig. 2. The arrival of peg 1 or peg 2 in

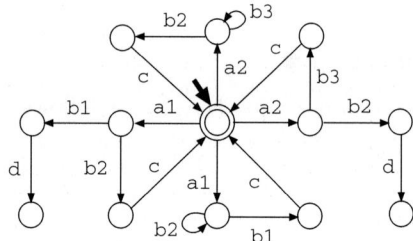

Figure 2. The nondeterministic system G.

the work station is labeled as the identical event $a1$, and the arrival of peg 2 or peg 3 is labeled as $a2$. When a peg arrives, hole 1 in a hole storage can be selected for a linking operation as a default (event $b1$). Moreover, a supervisor can request that hole 2 or hole 3 should be selected for linking with an incoming part (event $b2$ and $b3$, respectively). In the assembly process, if peg 1, peg 2, and peg 3 are linked with hole 1, hole 2, and hole 3, respectively, then the welding operations for the linked parts are assumed to be successfully completed (event c). Also, it is assumed that peg 1 is not linked with hole 2 and hole 3, and peg 2 is not linked with hole 3. If peg 2 is linked with hole 1, then loose linking of parts causes a failed welding operation (event d) due to abnormal welding situations such as the spattering fault that a tip of a torch nozzle is choked with the molten material called weld spatter. Then, the process must be stopped to clear the weld spatter in the tip. If peg 3 is linked with hole 1 or hole 2, then the system's behaviors are assumed to be analogous to the above case.

The set of events $\Sigma = \{a1, a2, b1, b2, b3, c, d\}$ is assumed to be categorized as follows: $A = \{a1, a2, b1, c, d\}$ and $B = \{b1, b2, b3\}$. Also, for $B - A$, the augmented system G^{B-A} is shown in Fig. 3. Let $K_1 = (a1\, b1\, c)^* \subseteq L_m(G^{B-A})$.

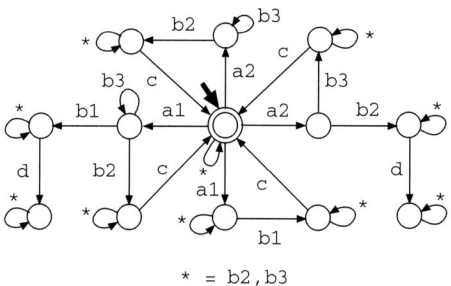

Figure 3. The augmented system G^{B-A}.

Then, the condition (A1) in Theorem 1 is not satisfied for the following reason: $a1\, b1 \in pr(K_1)$, $a1\, b1\, d \in pr(K_1)(A - B) \cap L(G^{B-A})$, but $a1\, b1\, d \notin pr(K_1)$. Also, for the states $q_1 \in \delta^*_{G^{B-A}}(q_G^0, a1\, b1)$, there does not exist $s \in \Sigma^*$ such that $\delta^*_{G^{B-A}}(q_1, s) \cap Q^m_{G^{B-A}} \neq \emptyset$. Hence, the condition (A2) in Theorem 1 is not also satisfied. Thus, there does not exist a nonblocking supervisor S for G such that $L(G_A\|_B S) = pr(K_1)$.

Let $K_2 = (a2\, b2\, (\epsilon + b3)\, c)^* \subseteq L_m(G^{B-A})$. Then, for the state $q_2 \in \delta^*_{G^{B-A}}(q_G^0, a2\, b2)$, there does not exist $s \in \Sigma^*$ such that $\delta^*_{G^{B-A}}(q_2, s) \cap Q^m_{G^{B-A}} \neq \emptyset$. Thus, there does not exist a nonblocking supervisor S for G such that $L(G_A\|_B S) = pr(K_2)$.

Let $K_3 = (a1\, b2\, (\epsilon+b1)\, c + a2\, b3\, (\epsilon+b2)\, c)^* \subseteq L_m(G^{B-A})$. Then, it is easily verified that K_3 satisfies all conditions in the developed theorems. Thus, there exists a nonblocking supervisor S for G such that $L_m(G_A\|_B S) = K_3$. Then, S is designed as the deterministic automaton $det(pr(K_3))$ shown in Fig. 4, and the closed-loop system $G_A\|_B S$ is also shown in Fig. 5. The supervisor S implements the following

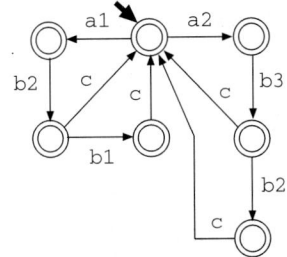

Figure 4. The supervisor S.

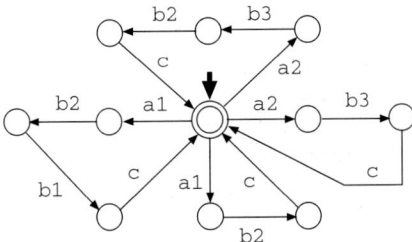

Figure 5. The closed-loop system $G_A \|_B S$.

control strategy: When peg 1 or peg 2 arrives, the supervisor requests that hole 2 should be selected for linking with the peg. If the peg is peg 2, the assembly operation is successfully completed. However, if the peg is peg 1, hole 2 is not linked with the peg. Next, the supervisor requests that hole 1 should be selected for linking with the peg. Since the peg is peg 1, the assembly operation is successfully completed. When peg 2 or peg 3 arrives, the supervisor requests that hole 3 should be selected for linking with the peg. If the peg is peg 3, the assembly operation is successfully completed. However, if the peg is peg 2, hole 3 is not linked with the peg. Next, the supervisor requests that hole 2 should be selected for linking with the peg. Since the peg is peg 2, the assembly operation is successfully completed.

V. Conclusions

In this paper, based on language models and prioritized synchronous compositions, we provide the necessary and sufficient conditions for the existence of a nonblocking supervisor to achieve a given language specification for a nondeterministic automaton with ϵ-moves. Moreover, we present the algorithm to find a supremal sublanguage of a given specification which satisfies the conditions. The nonblocking supervisor developed always guarantees the absence of blocked states in a controlled nondeterministic automaton.

References

[1] P. J. Ramadge and W. M. Wonham, "Supervisory control of a class of discrete event processes," *SIAM J. Contr. Optim.*, vol. 25, no. 1, pp. 206-230, 1987.

[2] W. M. Wonham and P. J. Ramadge, "On the supremal controllable sublanguage of a given language," *SIAM J. Contr. Optim.*, vol. 25, no. 25, pp. 637-659, 1987.

[3] M. Heymann, "Concurrency and discrete event control," *IEEE Contr. Syst. Mag.*, vol. 10, no. 4, pp. 103-112, 1990.

[4] M. A. Shayman and R. Kumar, "Supervisory control of nondeterministic systems with driven events via prioritized synchronization and trajectory models," *SIAM J. Contr. Optim.*, vol. 33, no. 2, pp. 469-497, 1995.

[5] R. Kumar and M. A. Shayman, "Nonblocking supervisory control of nondeterministic systems via prioritized synchronization," *IEEE Trans. Automat. Contr.*, vol. 41, no. 8, pp. 1160-1175, 1996.

[6] R. Kumar and M. Heymann, "Masked prioritized synchronization for interaction and control of discrete event systems," in *Proc. 36th IEEE Conf. Decision Contr.*, vol. 3, pp. 2952-2957, 1997.

[7] S. Jiang and R. Kumar, "Supervisory control of nondeterministic discrete event systems with driven events via masked prioritized synchronization," in *Proc. 38th IEEE Conf. Decision Contr.*, vol. 3, pp. 2212-2217, 1999.

[8] M. Heymann and F. Lin, "Nonblocking supervisory control of nondeterministic systems," Technion-Israel Institute of Technology, *Tech. Rep. CIS-9620*, 1996.

[9] M. Heymann and F. Lin, "Discrete-event control of nondeterministic systems," *IEEE Trans. Automat. Contr.*, vol. 43, no. 1, pp. 3-17, 1998.

[10] A. Overkamp, "Supervisory control using failure semantics and partial specification," *IEEE Trans. Automat. Contr.*, vol. 42, no. 4, pp. 498-510, 1997.

[11] S.-J. Park and J.-T. Lim, "Robust supervisory control of nondeterministic discrete event systems via language models and prioritized synchronization," *Submitted to IEEE Trans. Automat. Contr.*, 2000.

A Bone Reaming System Using Micro Sensors For Internet Force-feedback Control

Antony W. T. Ho[1], Imad Elhajj[2], Wen J. Li[1], Ning Xi[2], and Tao Mei[3]

[1]Center for Micro and Nano Systems, The Chinese University of Hong Kong
{wtho,wen}@acae.cuhk.edu.hk
[2]Dept. of Electrical and Computer Eng., Michigan State University
{elhajjim, xin}@cgr.msu.edu
[3]Institute of Intelligent Machines, Chinese Academy of Sciences
tmei@mail.iim.ac.cn

Abstract

The development of a medical surgical tool packaged with micro sensors for transmission of *Supermedia* information over the Internet is described in this paper. We define *Supermedia* as a set of communication media, which encompasses acoustic, force, visual, audio, temperature, tactile, and chemical (e.g., taste an smell) information, and which can be physically experienced by a communicator. In this project, we specifically develop *Supermedia* capability for a bone-reaming system that is used for intramedullary fixation procedure of fractured bone treatments. Thus far, transmission of temperature, force, and pressure information from MEMS sensors over the Internet has been demonstrated. Force-reflective control over the Internet using force information from a micro tip has also been shown. We have also packaged a MEMS pressure sensor inside a bone reaming guide-rod and proved that pressure variations inside a long cavity that simulated the environment inside a bone can be monitored, even with the guide-rod rotating up to 600rpm. This paper describes our experimental methods and gives the experimental results for these accomplishments.

1 Introduction

Internal fixation is a widely used treatment method to repair bone fracture, which is one of the most common orthopedic trauma. However, in reaming of the intramedullary canal to facilitate insertion of a guiding rod by creating a surgical passage, severe disturbances such as increased local cortical temperature and elevation of intramedullary pressure can occur (a pressure of 300mmHg was recorded in rabbit, and around 800mmHg in goat from experimental procedures [1].) The significantly high pressure can push fat content in bone marrow into the blood stream, which causes serious problem if the fat is transported along the blood vessel to a patient's heart and lung; in which case, the patient will suffer from fat embolism. When fat embolism syndrome occurs in post-traumatized patients, the mortality rate is very high.

Currently, pressure inside the intramedullary cavity is monitored by inserting pressure sensors (typically about 1cm diameter and 5cm length) into additionally drilled holes at the circumferential wall of the fractured bones (see Figure 1). These additional holes not only weaken the bone structure mechanically, but cause additional complexity for the internal fixation surgical procedures. One of the main aim of our project is to eliminate the additional drilling of holes for sensor insertion by packaging MEMS pressure sensors into the guide-rod used for guiding drill bits during the intramedullary canal boring. In addition, integrated temperature sensors will be packaged to monitor the temperature of the guide-rod, and thermal couples will be used to monitor the bone marrow temperature. This new system is very important in two aspects: 1) it will allow physicians to drill just one instead of the currently required minimal of two holes in the fractured bone; 2) it will allow physicians to have a supermedia-based control during a surgical procedure, which prevents over-pressurization or over-heating of the bone cavity. We believe that, by merging MEMS and tele-robotic technologies for this medical application, we will dramatically improve the safety and time consumed for the internal fixation procures.

2 Design of the New Bone Reaming System

In improving the existing system, our fundamental philosophy is that the modifications should have minimal effect or alteration on the existing medical procedures. Giving the dimensions of the components needed for the internal fixation procedures, we proposed to package MEMS sensors into the head section of the guiding rod. This modification will allow the new sensing system to be integrated with the drilling system, and hence, eliminate the need to drill additional holes on the bone structure. Moreover, this implementation will not require any additional medical procedures. Three basic engineering challenges need to be overcome before the new system can be realized: 1) package multiple sensors into a small volume (MEMS technology is essential here); 2) design a relatively long and hollow mechanical structure with small radius, which is able to handle the same load as the existing guide-rod; 3) send power to and receive signal from the sensors embedded in a rotating mechanical structure.

Our solutions to these challenges are presented below.

3 Conventional System

The existing bone reaming operation system is illustrated in Figure 1. A long medical drill with a hole in the center goes into the bone cavity by using a guide-rod as a guide. Physicians can control the depth and drilling direction of the drill by manipulating the guide-rod. The push-forward motion of the medical drill raises the bone marrow pressure similar to the piston-pushing effect. Currently, separate pressure-measuring devices are implemented by drilling other holes into the bone to monitor any change on the bone marrow pressure. The surgeon would temporarily stop the operation until the pressure drops back to safety level. The extra holes necessary for pressure sensors would cause additional and undesired damage to the patient's bone structure.

3.1 New Reaming System with MEMS Sensors

The proposed new system is shown in Figure 2. This new system, which has almost the same shape and physical dimensions as the existing guide-rod, will allow doctors to use it with existing operation tools, and thus, will permit them to learn how to operate the new system quickly.

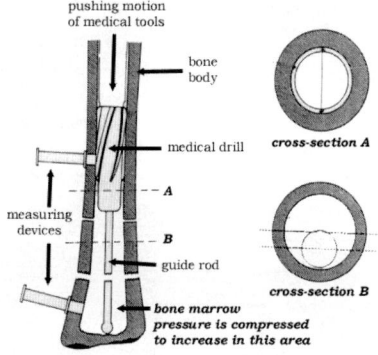

Figure 1. Reamer drill is pushed into bone cavity to enlarge volume for intramedullary nailing. A guide-rod is used to guide the reamer-drill's direction.

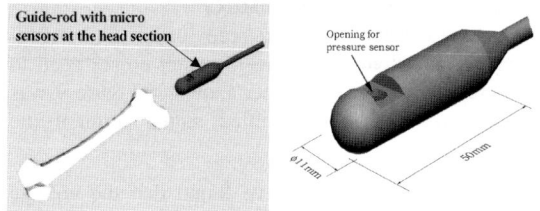

Figure 2. New system with same physical features as the existing guide-rod is inserted into bone cavity.

3.1.1 Packaging Design

For feasibility test, a miniature pressure sensor from Entran Devices Inc. was used. It was calibrated for a pressure range up to 50psi, which is the range reported in the actual medical operation conditions [1][2]. The Entran Devices MEMS piezoresistive pressure sensor is cased in a stainless steel body and was chosen for this project due to its ability to operate in a fluidic environment.

The longitudinal and cross-sectional views of the hollow head-part of the new guide-rod are shown in Figure 3 and Figure 4, respectively. In Figure 3, the left-most T-shaped channel is the opening passage for bone marrow pressure and other supermedia measurements. The MEMS pressure sensor is positioned at the end of the T-channel and has its thin stainless steel diaphragm pointing towards the channel. The sensor is adhered by a waterproof epoxy in a suitable position. Behind the MEMS sensor, a chamber is designed for incorporating the sensor's factory-made conditioning module and our own designed circuit module with functions of voltage stabilization and signal conditioning.

3.1.2 Electrical Signal Output

A special bearing-signal-transmission system was designed to obtain electrical signal from the sensors in the rod. The bearing is designed to be without an inner ring. It fits the rotating rod (3mm outer diameter) and has a flexible roller cage that is suitable for applications where the bearing is installed and removed frequently. The bearing will ease the rotational motion of the guide-rod during operation and simultaneously supply electrical power and transmit sensor signals. The design details and the mechanical analyses of the rod are given in [3]. A comparison of the stresses on the redesigned (hollow) rod and the original guide-rod is given in Table 1. It was shown in [3] that all stresses are within the yield limit of the stainless hollow rod.

Table 1. Theoretical comparison of the mechanical characteristics of the solid and hollow guide-rod.

Stress/Deflection	Ratio (hollow/solid)
Bending stress	1.52
Shear Stress	1.46
Torsional Shear Stress	1.011
Tip deflection	1.013

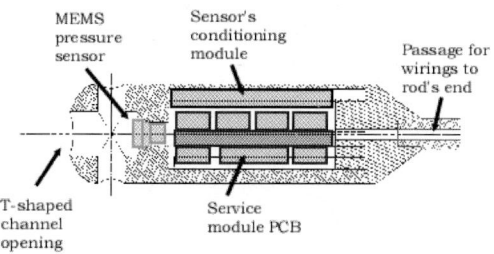

Figure 3. Longitudinal cross section diagram of the new guide-rod's head part. It shows the implanted components.

657

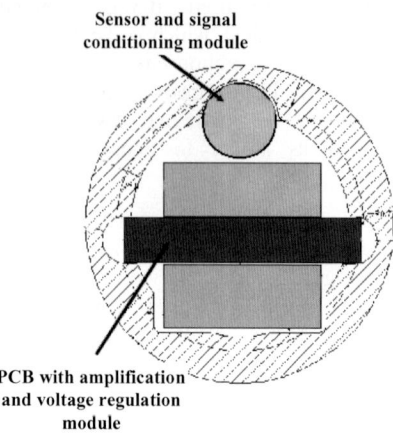

Figure 4. Cross sectional view of service module chamber of guide-rod's head part.

4 Experimental Results of the New Guide-Rod System

Experimental results of transmitting pressure signals from the new guide-rod are presented in this section. Pictures of the actual redesigned guide-rod components are shown in Figure 5 and Figure 6 below. For the prototype test, we packaged only a pressure sensor and its signal conditioning circuit into the guide-rod head (Figure 6). In reality, the current guide-rod head has enough volume to encase other supermedia sensors.

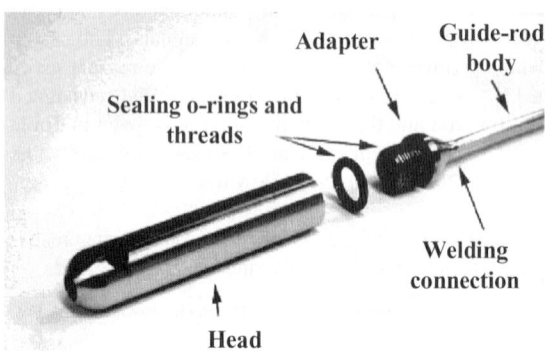

Figure 5. The redesigned guide-rod system.

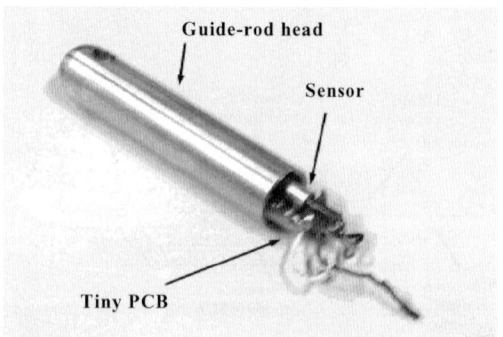

Figure 6. The micro sensor and the signal conditioning circuitry are encased in the guide-rod head.

4.1 Experimental Setup

The guide-rod system was put into a cylindrical pressure vessel as illustrated in Figure 7. An airflow valve was used to control the outflow of air from the vessel. An inlet of the vessel was connected to an air pump such that the pressure inside the vessel can be increased. Controlling the airflow valve can decrease the pressure in the vessel chamber.

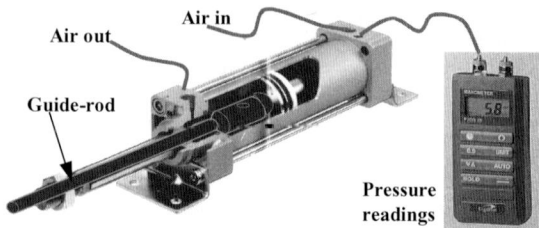

Figure 7. Exposed view of the cylindrical vessel used to simulate environment inside bone cavity.

To simulate the rotating motion of the guide-rod during a bone-reaming operation, a belt-drive system was connected to a rotation motor to rotate the guide-rod. The rotation speed of the guide-rod was measured by an optical encoder packaged with a connection system that read the sensor data during the rotation experiments. The entire experimental setup is shown in Figure 8.

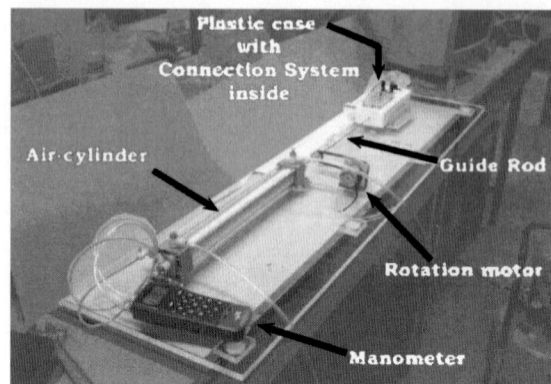

Figure 8. Experimental setup to test pressure variation inside a cylinder vessel using a rotating guide-rod.

4.2 Experimental Results

Since the sensor signal is read using a bearing-transmission system, mechanical contact noises need to be filtered. The low-pass filter shown in Figure 9 was used to process the transmitted signal through the rotating bearings.

In the current system, we have found that the voltage output as a function of pressure is very linear; however, the pressure signal output through the connection system decreases with increasing rotation speed (see Figure 10, Figure 11, and Figure 12).

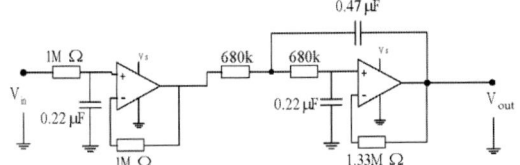

Figure 9. The low-pass filter circuit used to reject mechanical noise due to the rotation bearing.

We have speculated that rotating the guide-rod in a cylindrical cavity may induce Couette flow motion in the vessel [4], and causes drop of pressure as a function of rotation speed. However using Bernoulli's Equation to calculate the change of pressure $\Delta p = \rho r_i^2 \omega_i^2 / 2$ due to rotation at the guide-rod head surface r_i (5.5mm), where ρ is the fluid density (assuming air density of 1.2929kg/m^3), ω_i is the angular velocity of the rotating guide-rod (83.78 rad/s or 800rpm). The calculated value for this pressure gradient is 0.137Pa, which is much less than the 1% tolerance of the sensor's reading span, and hence cannot be detected by the sensor. Therefore, the radial pressure gradient due to the rotating guide-rod cannot be the cause for the observed experimental result. Other possible causes of the output degradation include 1) deflection of the sensor membrane due to centrifugal force and 2) signal rejection by the low-pass filter (Figure 9) at higher rotation speeds. These hypotheses are currently under investigation.

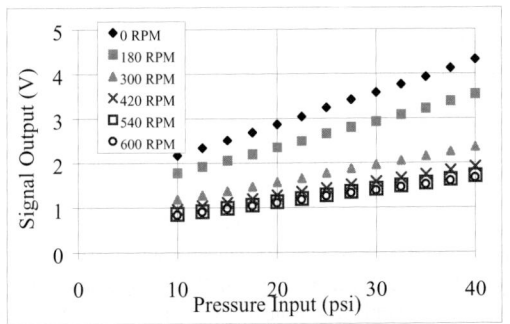

Figure 10. Voltage output versus pressure in the cylindrical vessel; the output also depends on rotation speed of the guide-rod.

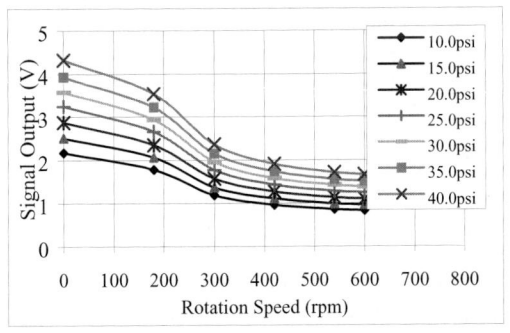

Figure 11. Same data as in Figure 10 except that pressure is used as a parameter in this case.

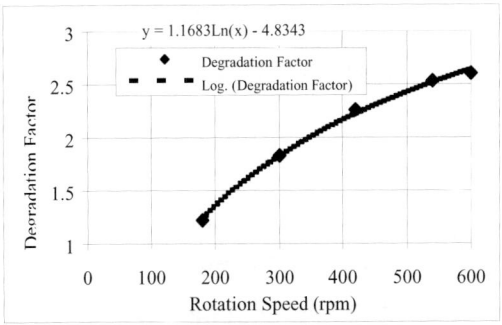

Figure 12. Calibrated degradation factor of the voltage output as a function of the guide-rod rotation speed.

5 Internet Force-Reflection Experiment

Some experiments were conducted in using the Internet to transmit micro sensor signal, which was then used to perform force-reflective control of an x-y table. Details of these experiments are given in [5], but the results are summarized in the following subsections. We are currently integrating the bone reaming guide-rod system with a force-reflective control system to demonstrate a teleoperated bone surgical system. Eventually, the teleoperated system will also include other supermedia information such as temperature and vision.

5.1 Event-based Internet Force-Reflective Control

Delay in communication links has several effects on the stability and synchronization of teleoperation systems, which is even more prevalent when force feedback is included. These effects are caused by the use of time as the reference variable; therefore, if a non-time based reference is used the system would become immune to delay. This suitable action or motion non-time reference variable is called event. The event-based controller design was first introduced in [6], then, several studies and applications followed [7]. The planning and control of the traditional time-based and the event-based schemes are shown in Figure 13.

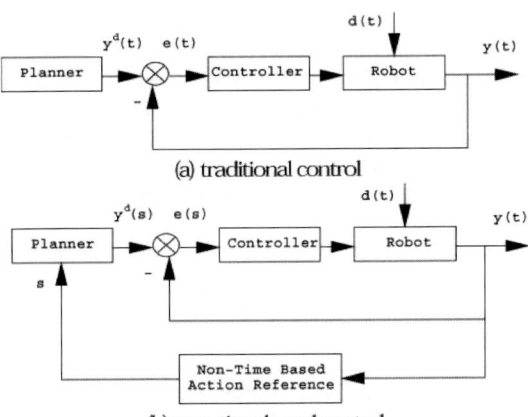

Figure 13. Comparison between traditional time-based and event-based planning and control.

Event-based control results in not only stability but also event synchronization. Because of delay, visual feedback does not reflect the current state of the system. By using event-based control, the force is event synchronized, which implies that the force always reflects the most up-to-date state of the system.

5.2 Internet Force-Reflection Experimental Results

5.2.1 Experimental Setup

We have developed Polyvinylidence fluoride (PVDF) piezoelectric micro tips as rate-of-force sensors for force-reflective control, and which can be eventually integrate into many medical tools [5]. The micro tips were laser-micromachined to geometries of 2.5mm long, 0.8mm at the triangular base, and with tip-radii of ~100µm. The output from a PVDF micro-tip sensor was amplified using an inverted amplifier with feedback gain of 50. Its signal was then feed to the 8255 analog-to-digital conversion (ADC) card connected to a PC for signal transmission to the Internet. The sensor tip was attached to an x-y computer-control positioning table which was housed in the Advanced Microsystems Laboratory (AML) of The Chinese University of Hong Kong. The x-y table could be controlled via the Internet by a force-reflection joystick in the Robotics and Automation Laboratory (RAL) at Michigan State University. This x-y positioning table will be replaced with a computer-controlled drilling system eventually. The AML sensor tip position was manipulated by the RAL joystick to contact a vibrating cantilever which vibrated at frequencies from 1 to 120Hz with amplitudes ranging from 100µm to 1mm. The RAL operator observed the AML tip position using a video-conferencing software. The force of the vibrating cantilever sensed by the tip was sent to RAL via the Internet. The force was then received and played by the force-reflective joystick. The operator then generated a new movement command that was sent to move the sensoing tip via the Internet.

5.2.2 Internet-Based Control Results

To emphasize the delay problem over the Internet, a sample of round trip delay between the RAL and AML is shown in Figure 14. It is clear that the delay is random with no specific pattern or model. If not dealt with, this delay might cause instabilities and de-synchronizations. Figure 15 shows a plot of the desired position increments in both directions and a plot of the played force with respect to the event. It is clear that the commands are random, which is typical of a teleoperation scenario. This makes approaches based on prediction of forces or virtual forces non-realistic. Therefore, actual force had to be sensed and fed back. Figure 16 presents plots of the force felt by the operator, the force sampled for the sensor and the error between them. As seen the force felt is closely following the one

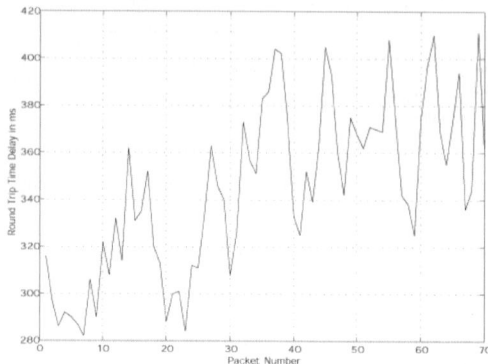

Figure 14. A sample of round trip delay between Hong Kong and Michigan State.

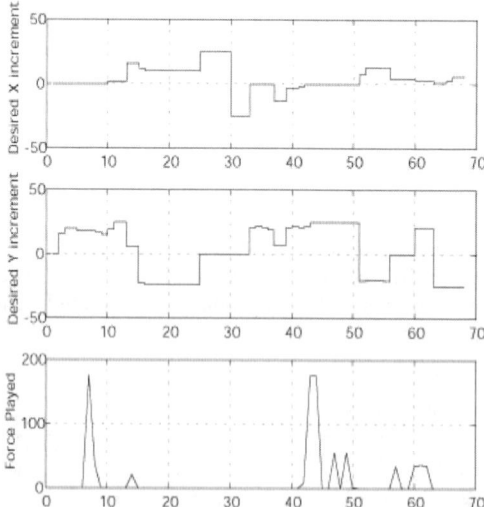

Figure 15. Plots of the desired position increments and the force felt by the operator.

sampled from the sensor. Although this is not occurring at the same time instant, since both plots are with respect to local and not global time, the system is still stable and event synchronized [7]. Despite the random time delay experienced between Hong Kong and Michigan Sate, the system performance is stable as seen from the error, which is constantly converging to zero and has a small value at all times. This implies that, for the given sampling frequency, the system is transparent – in the case that the operator was controlling the sensor from a local machine a similar force profile would have been experienced.

6 Conclusion

The development of a novel bone reaming system using micro sensors for Internet force-reflective teleoperations was described in this paper. A new guide-rod was developed for rotation signal transmission and pressure sensing with a micro sensor. An event-based control scheme over the Internet has been found experimentally stable over a telemanipulation distance from Michigan to Hong Kong. The integration of all components of the new reaming system is underway. The final product will improve the existing surgical system dramatically by greatly reduce the over all medical operation time and risk to patients while providing an opportunity for Internet tele-surgery.

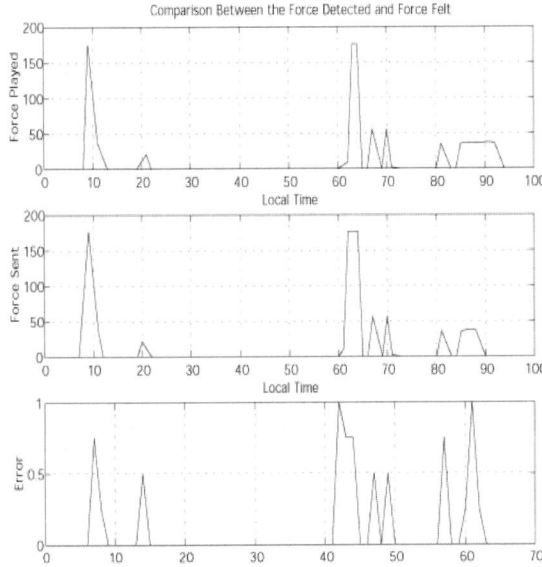

Figure 16. Comparison between the forces felt and the ones sent.

7 Acknowledgement

This work was funded by the Research Direct Grant (2050173) of The Chinese University of Hong Kong, and the NSF Grants IIS-9796300 and IIS-9796287 of Michigan State University. We would like to thank Carmen K. M. Fung and King W. C. Lai for their significant contributions in developing the Internet control software and electronic interface circuits for this project.

8 References

[1] Cheung E., "The Effect of Reaming on Intramedullary Pressure and Marrow Fat Embolisation", Thesis of M. Phil., The Chinese University of Hong Kong, 1997.K. M. Stürmer, "Measurement of intramedullary pressure in an animal experiment and propositions to reduce the pressure increase", Injury, Supplement 3, pp. S7-S21, 1993.

[2] Heim D., Schlegel U., and Perren S.M., "Intramedullary Pressure in Reamed and Unreamed Nailing of the Femur and Tibia", Injury, Supplement 3, pp S56-S63, 1993.

[3] Ho W.T., Lai W.C., Li W.J., Elhajj I., and Xi N., "Development of A Bone Reaming System Using Micro Sensors For Internet Force-feedback Control", Proc. of the Workshop of Service Robotics and Automation, City University, Hong Kong, June 19-21, 2000.

[4] White F.M., Fluid Mechanics, 3rd edition, McGraw-Hill, Inc. 1994.

[5] Lai W.C., Fung K.M., Li W.J., Elhajj I., and Xi N., "Transmission of Multimedia Information on Micro Environment via Internet", Proc. of IEEE IECON 2000, October 2000, Nagoya, Japan.

[6] N. Xi, ``Event-Based Planning and Control for Robotic Systems'', Doctoral Dissertation, Washington University, December 1993.

[7] Elhajj I., Xi N., and Liu Y.H., "Real-time control of Internet based teleoperation with force reflection", Proc. of IEEE International Conference on Robotics and Automation, pp.3284-3289, 2000.

Modeling and Control of Internet Based Cooperative Teleoperation *

I. Elhajj, N. Xi	W. K. Fung, Y. H. Liu	Y. Hasegawa, T. Fukuda
Dept. of Electrical and Computer Engineering Michigan State University East Lansing, MI 48824, U.S.A. {elhajjim, xin}@egr.msu.edu	Dept. of Mechanical and Automation Engineering Chinese University of Hong Kong Shatin; N.T., Hong Kong {wkfung, yhliu}@mae.cuhk.edu.hk	Center for Cooperative Research in Advanced Science and Technology Nagoya University Nagoya 464-8603, Japan {yasuhisa, fukuda}@mein.nagoya-u.ac.jp

Abstract

Robotic operations carried out via the Internet face several challenges and difficulties. These range from human-computer interfacing and human-robot interaction to overcoming random time delay and task synchronization. These limitations are intensified when multi-operators at multi-sites are collaboratively teleoperating multi-robots to achieve a certain task. In this paper, a new modeling and control method for Internet-based cooperative teleoperation is developed. Combining Petri Net model and event-based planning and control theory, the new method provides an efficient way to model the concurrence and complexity of the Internet-based cooperative teleoperation. It also provides an efficient analysis tool to study the stability, transparency and synchronization of the system. Furthermore, the new modeling and control method enables us to design an Internet-based cooperative telerobotic system that is reliable, safe and intelligent. This new method has been experimentally implemented in a three site test bed consisting of robotic laboratories in the USA, Hong Kong and Japan. The experimental results have verified the theoretical development and further demonstrated the advantages of the new modeling and control method.

1 Introduction

The Internet is maturing from a simple data link between computers to a world where operators in Hong Kong and Japan can bilaterally cooperate a robot at Michigan State University in the USA in real-time. This blooming of the Internet is a result of the Internet's availability, low cost and world wide coverage. These characteristics make the Internet the media of choice for remote robotic teleoperation. Specifically the Internet makes multi-operator at multi-site collaborative teleoperation feasible.

Multi-operator at multi-site collaborative teleoperation is a combination of teleoperation and collaboration, where several operators at different remote sites are trying to collaboratively control a robot or several robots to achieve a certain task [1]. The motivation behind tele-cooperation is the existence of tasks that require several different expertise to collaborate concurrently. In cases where these expertise are at different locations, real-time collaborative teleoperation via the Internet is the answer. An example of such a system is illustrated in Fig.1, where two operators are collaborating to bilaterally control a mobile manipulator.

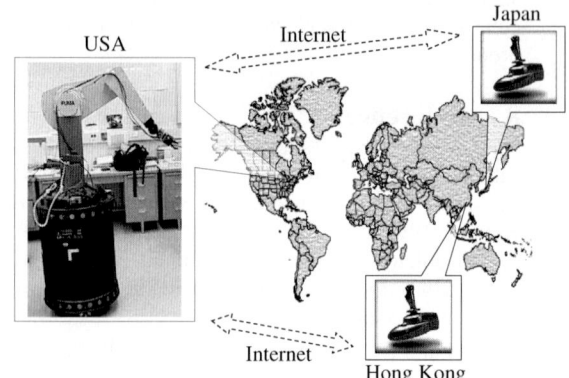

Figure 1: The structure of a multi-operator at multi-site collaborative teleoperation system.

However, real-time bilateral collaborative teleoperation via the Internet faces many difficulties. A major one is time delay, which can render a system unstable and asynchronous. Another difficulty is achieving efficient and safe collaboration where it is unclear what the other operator's intentions are. Extensive

---
*Research Partially supported under NSF Grant IIS-9796300, IIS-9796287 and EIA-9911077, and DARPA Grant DABT63-99-1-0014.

research has been done relating to time delay and stability in bilateral teleoperation [2]-[6]. Each of these approaches has its limitations: delay is taken to be fixed and not random, or the same in both directions, or having an upper bound. These limitations become critical when the Internet is used as a communication media. When it comes to the Internet no assumptions should be made regarding the time delay, because it can not be modeled with a specific simple statistical model and there is no upper bound on it [7] [8]. Because of the lack of simple models the inter-arrival times of packets can not be estimated. Therefore any assumption made about them would be limiting.

Another difficulty is the lack of an efficient and simple model for such systems. There is a need for a model that can capture the concurrence, non-determinism and logical behavior of such systems. This model should be easily analyzed to study the underlying system performance. Properties such as stability and synchronization can not be easily studied using the current models. Another difficulty is designing a control method based on such models that can ensure certain system qualities. So the challenge is to develop a model that can help solve these difficulties.

This paper presents a new model that can capture various properties of complex distributed robotic systems. This model is Petri Net based and thus can be easily analyzed to study different system properties. The model can also be used to design control methods that satisfy performance constrains. The new model can also be integrated with the event-based planning and control method [9]. The model discussed is that of a real-time bilateral collaborative teleoperation system, which was implemented and tested via the Internet.

2 Petri Net Model

The dynamic model of the system will be first presented to illustrate the operation and behavior of the multi-operator at multi-site collaborative teleoperation system illustrated in Fig.2. Each block will be discussed in detail and all terms are explained in Table1.

Human Operator: This is the most difficult to model, but a spring-like behavior may be assumed, as shown in [11] and as used in several instances in the literature [10]. Once the operators feel a force, they will generate a new joystick position according to the following:

$$X_m(s+1) = \frac{F_p(s)}{K_m} \qquad X_p(s+1) = \frac{F_m(s)}{K_p} \qquad (1)$$

where K_m and K_p are scaling constants, s is the *event* and $s \in \aleph$. $\aleph$ is the set of all positive integers, this implies that s represents the number of cycles the system has gone through. So s represents how many commands have been sent and executed. $F_m(s)$ and $F_p(s)$ are the applied forces, i.e. the forces that the operators feel. As seen in eq.1, $X_m(s+1)$ and $X_p(s+1)$ are related to $F_p(s)$ and $F_m(s)$, so $X_m(s+1)$ and $X_p(s+1)$ at event $s+1$ are generated by the previous force at event s. This results in an event-based system where each event is triggered by the previous one. As will be shown, this aspect is clearly and easily captured by the Petri Net based model presented in this paper.

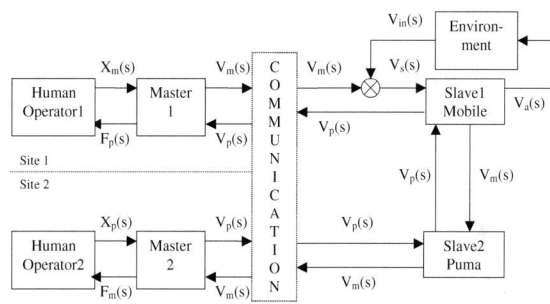

Figure 2: The block diagram of a multi-operator multi-site collaborative teleoperation system.

Block	Our Variables
Human Operator	$F_m, F_p \in \Re^3$: Applied force $X_m, X_p \in \Re^3$: Joystick pos.
Master	$V_m, V_p \in \Re^3$: Velocity desired
Slave	$V_a \in \Re^3$: Actual velocity
Environment	$V_{in} \in \Re^3$: Virtual contact $V_s \in \Re^3$: Velocity set

Table 1: Explanation of the various variables in Fig.2.

Master (Joystick):The dynamics of the joysticks are:

$$M_m \dot{V}_{mm}(t) = F_p(t) + F_{V_p}(s) \qquad (2)$$

$$M_p \dot{V}_{mp}(t) = F_m(t) + F_{V_m}(s) \qquad (3)$$

$$F_{V_m}(s) = C_m V_m(s) \qquad (4)$$

$$F_{V_p}(s) = C_p V_p(s) \qquad (5)$$

where M_m and M_p are the masses of the joysticks handles, V_{mm} and V_{mp} are velocity of joysticks movement, and F_m and F_p are as described earlier. F_{V_m} and F_{V_p} are the forces played by the joystick, which are simply the velocities V_m and V_p fed back from the robots scaled by the constants C_m and C_p respectively. The result of these dynamics is that the joysticks move to new positions $X_m(s+1)$ and $X_p(s+1)$. From these positions the desired velocities V_m and V_p are derived according to

$$V_m(s) = K_{m1} X_m(s) \qquad V_p(s) = K_{p1} X_p(s) \qquad (6)$$

where K_{m1} and K_{p1} are scaling constants, $X_m(s)$ and $X_p(s)$ are as before. $V_m(s)$ and $V_p(s)$ are the desired velocities of the mobile and the Puma respectively.

Communication Block (Internet): Resulting from event based control, the communication link is simply a delay element that plays no role in the modeling of the system. Since the advance of time does not affect the system and only the advance of the event s does, the system will remain stable when the connection is lost and will resume action only after the connection is re-established. This makes the system very robust since no initialization or synchronization is required.

Environment: Sensors on the mobile robot are used to detect objects. Based on the distance between the object and the robot, velocity is reduced. This is calculated according to a function of the distance from the object $f(d)$ that would give a velocity value $V_{in}(s)$. $V_{in}(s)$ is subtracted from the desired velocity $V_m(s)$ to give the velocity set for the robot $V_s(s)$,

$$V_s(s) = V_m(s) - V_{in}(s) \qquad (7)$$

As a result, the robot gets a velocity from the server that is less than the one desired by the operator.

Slave1 (Mobile robot): Once the robot receives $V_s(s)$, it will be commanded to move with that velocity, but it will actually move at velocity $V_a(s)$. The dynamics of the robot are described by the following equations:

$$M_s \dot{V}_a(t) = F_e(t) + \tau_s(t) \qquad (8)$$

Here M_s is mass of the robot, and F_e is the actual environment forces if any, and usually assumed very small. τ_s is the robot's internal servo control signal.

Slave2 (PUMA manipulator): The dynamic model for the robot arm can be written as

$$D(q)\ddot{q} + c(q,\dot{q}) + g(q) = u \qquad (9)$$

where q is the 6×1 vector of joint displacements, $\dot{q}$ is the 6×1 vector of joint velocities, u is the 6×1 vector of applied torques, $D(q)$ is the 6×6 positive definite manipulator inertia matrix, $c(q,\dot{q})$ is the 6×1 centripetal and coriolis torques, and $g(q)$ is the 6×1 vector of gravity term.

Let $Y \in \Re^6$ be a task space vector defined by $Y = (x,y,z,O,A,T)^T$. $(x,y,z)^T$ denotes the position of the end-effector in the Cartesian space, $(O,A,T)^T$ denotes an orientation representation(Orientation, Altitude and Tool angles). After applying non-linear feedback with $u = D(q)J^{-1}(-\dot{J}\dot{q}+v) + C(q,\dot{q}) + g(q)$ [17]. The Dynamic model of the arm can be simplified to: $\ddot{Y} = v$.

These dynamics and their interactions can be modeled as presented in Fig.3 using Petri Net, which is a directed graph that has an initial state (marking) [12].

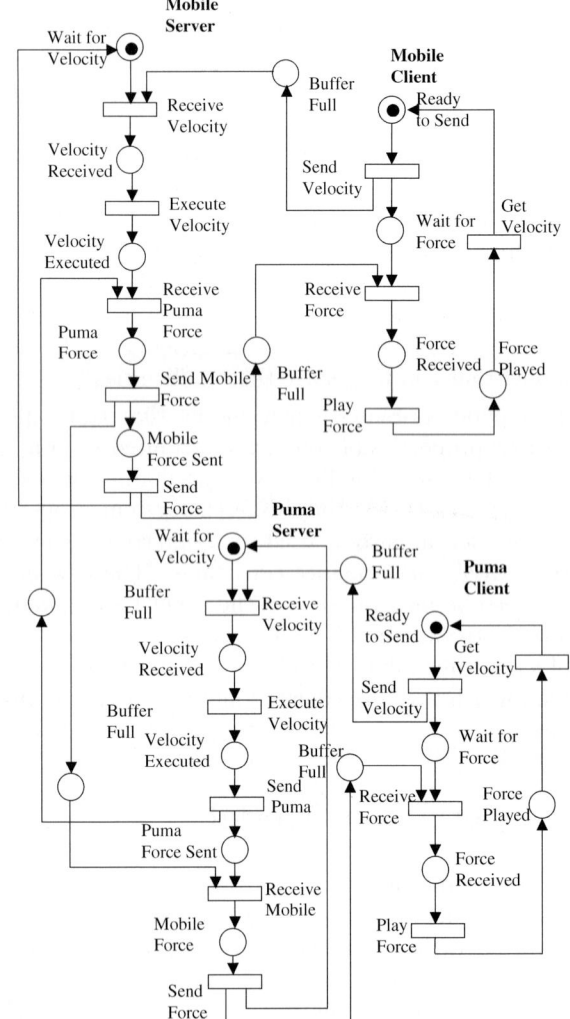

Figure 3: Petri Net model of the system shown in Fig.2.

Petri Net has several attractive features, which are based on its ability to describe and study systems that are characterized as being concurrent, asynchronous, distributed, parallel, non-deterministic and/or stochastic. Thus, it is a very adequate and efficient tool for studying Internet based telerobotic systems. Although Petri Nets have been used before for modeling and studying robotic systems but that was limited to manufacturing and scheduling related problems [13].

In addition, based on this model different properties of the system can be studied using Petri Nets analysis tools. Some of these properties can also be linked to physical properties.

3 Petri Net Analysis

Another strength of Petri Nets is their support for analysis of many important properties associated with the systems they model. These properties are divided into properties that are dependent on the initial marking (behavioral properties) and those properties which are independent of the initial marking (structural properties) [12] [14]. In this study interest is in the following behavioral properties:

-Boundedness: The model presented is k-bounded or simply bounded if the number of tokens in each place does not exceed a finite number k for any marking reachable from the initial marking. It is said to be safe if it is 1-bounded [12]. This implies that there is no accumulation of tokens in places at any time implying that the system is stable [16]. For this system stability means that the velocities of the Puma and the mobile base are tracking each other based on which operator is the master. So if the operator controlling the base is the master then the operator controlling the Puma would have to track the intentions of this operator, thus the Puma would follow the velocity of the mobile.

-Liveness: The model is said to be live if, no matter what marking has been reached from the initial state, it is possible to ultimately fire any transition of the net by progressing through some further firing sequence. This property guarantees deadlock-free operation [12]. This implies that, despite the non-determinism in the system, there is no case that would cause the system to stop operating normally.

To facilitate the analysis of the Internet based multi-operator at multi-site collaborative teleoperation system, while preserving properties of boundedness and liveness, the system model was reduced [12]. The coverability tree was built to study the boundedness and liveness of the system [12] [15].

The reduced system's coverability tree is shown in Fig.4. From this tree we can deduce that the system is bounded and live. Boundedness is deduced from the fact that all the nodes in the tree have ones and zeros in them. This implies that not only the system is bounded but also safe [12] [14] [15]. Which in other words means that the number of tokens in any place does not exceed one.

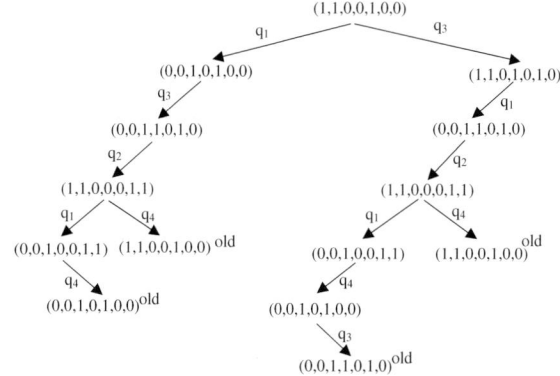

Figure 4: The coverability tree of the reduced model.

In addition, the coverability tree conveys that the system is dead-lock free since all the markings in the tree have enabled transitions [12]. So the system is both bounded (safe) and live. As for event-synchronization [9], since the system is safe it is clear that each place in the system can have only one token at most at any point in time. And since the passage of tokens in a place represents the advance of the event s for that place and since it is clear from the model that a place will not receive another token until all the other places have received a token then the system is event-synchronized. It is worth noting that in systems where delays are in the order of seconds synchronization becomes as important an issue as stability. Large random time delays would make any form of feedback, whether video or force, worthless if the different system parts are not synchronized. But the existence of random time delay makes time synchronization impossible to achieve. That is why event-synchronization was presented in [9]. Event-synchronization ensures that the feedback received corresponds to the most up to date status of the system and thus the right control signal would be sent.

4 Experimental Implementation and Results

The system consists of various operating systems and configurations. So the problem of interconnection had to be studied carefully. The joysticks used are programmable Microsoft SideWinder Force Feedback Pro. with 3 degrees of freedom. The mobile robot is a Nomadic XR4000 and the manipulator is a PUMA560. The software developed can be divided into five main parts: motion server, Puma server, Puma controller, motion client and Puma client.

Motion Server: Its service is moving the robot and receiving feedback. However the server does not execute requests blindly; it first checks the sensors and, based on input from them, makes a decision according to an obstacle avoidance algorithm. The obstacle avoidance algorithm first checks the different sensors on the robot, which are used to detect the distance to the closest object in the direction of motion. Once the motion server decides which velocity to set and sends it to the motors, it waits for the Puma server to send the feedback, which is the Puma desired velocity V_p. Then it would send the mobile desired velocity, V_m, to the Puma server and V_p to the motion client.

Puma server: This server is responsible for receiving the Puma velocity commands V_p from operator2. Then it forwards these commands to the Puma controller. After that the Puma server sends V_p to the motion server and receives V_m from it. Then V_m is forwarded to the Puma client.

Puma controller: This controller receives the desired velocity V_p and controls the Puma joint motors. The Puma controller is designed at task level. A singularity free hybrid motion controller is used to avoid the singularities of the robot arm [17].

Motion client: This client sends commands V_m to the motion server and relays force commands back to the joystick once feedback is received. Communication with the joystick is achieved with MS DirectX technology, and that with the server over the Internet.

Puma client: This program sends velocity commands V_p to the Puma server and relays force commands back to the joystick once feedback is received. Communication with the joystick is achieved with DirectX technology and with the server over the Internet.

Concerning the experiments the mobile manipulator (Robotics and Automation lab, Michigan State University), operator1 (Robot Control lab, Chinese University of Hong Kong) and operator2 (Nagoya University, Japan) were connected via the Internet. The delay experienced between these sites was random with no specific pattern or model. However, the system performance was stable and synchronized as will be shown in the experiments presented.

This experiment was master-slave operation, where one of the operators is requested to follow the force felt. This implies that the slave would eventually track the motion of the master. The results of one such experiment are seen in Fig.5, where the operator in Japan is controlling the mobile (slave) and the operator in Hong Kong is operating the Puma (master). The results show that the desired velocity of the mobile is tracking that of the Puma in real-time. The top row of Fig.5 shows a plot of time versus s, the event. The other plot in the top row shows the desired rotational velocity. The second row shows the desired velocities of the mobile in x and y directions, V_m. The Third row plots the desired velocities of the Puma in x and y directions, V_p. The plots of V_m and V_p also correspond to the forces fed back to operator2 and operator1 consecutively. The last row is the error between the mobile and the Puma desired velocities in both directions.

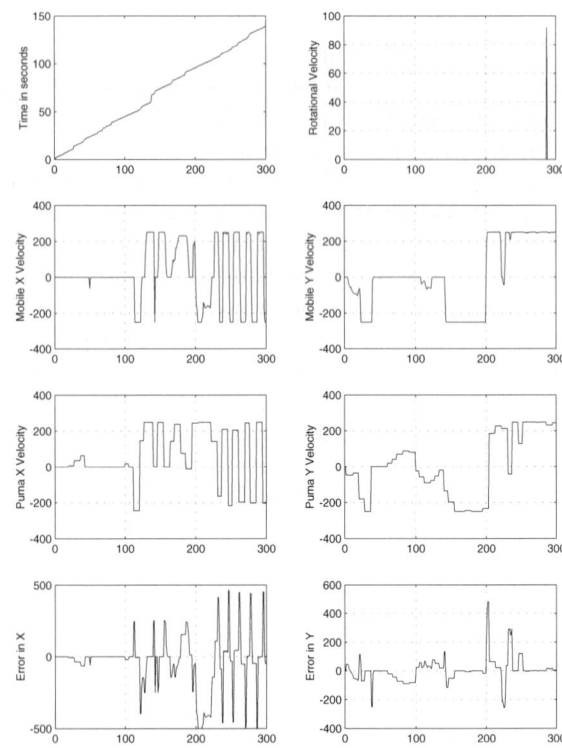

Figure 5: System Behavior while being controlled from Hong Kong and Japan.

The main points to note, are the synchronization and fast response. It is clear that both robots are event synchronized since the shift in direction occurs almost at the same s this confirms what the model depicted; no transition can fire out of order, that is why all the system parts are event-synchronized. Fast response is clear from the sharp decrease in the error between the velocities of the two robots. This implies that the desired velocity of the slave is tracking the desired velocity of the master, which was proven using Petri Net analysis and is a result of the system being safe.

5 Conclusion

Presented in this paper is a new modeling and analysis technique for Internet-based telerobotic systems. Specifically, multi-operator at multi-site bilateral teleoperation via the Internet was studied. The system was modeled using Petri Net and some of its properties were examined. Using Petri Net analysis techniques the event-based control system was found to be stable and event-synchronized .

In addition, implementation of the system was discussed and experimental results were presented. The experimental results confirmed the stability and synchronization of the system operation.

Petri Nets provided an efficient and simple tool for modeling complex telerobotic systems. It captured the concurrence, non-determinism and logical behavior of the system. In addition, several Petri Net analysis methods have been used to study the performance of the system. Therefore, Petri Nets promise to be a very efficient tool to model and analyze Internet-based telerobotic operations.

References

[1] K. Ohba, S. Kawabata, N. Y. Chong, K. Komoriya, T. Matsumaru, N. Matsuhira, K. Takase, K. Tanie, "Remote Collaboration Throught Time Delay in Multiple Teleoperation", IEEE/RSJ International Conference on Intelligent Robots and Systems, Korea, 1999.

[2] R. Anderson, M. Spong, "Asymptotic Stability for Force Reflecting Teleoperators with Time Delay", The Int. Journal of Robotics Research, 1992.

[3] W. Kim, B. Hannaford, and A. Bejczy, "Force-Reflection and Shared Compliant Control in Operating Telemanipulators with Time Delay", IEEE Trans. on Robotics and Auto., Vol. 8, April 1992.

[4] G. Leung, B. Francis, J. Apkarian, "Bilateral Controller for Teleoperators With Time Delay via μ-Synthesis", IEEE Trans. on Robotics and Automation, Vol 11, No. 1, February 1995.

[5] M. Otsuka, N. Matsumoto, T. Idogaki, K Kosuge, T. Itoh "Bilateral Tele-manipulator System With Communication Time Delay Based on Force-Sum-Driven Virtual Internal Models", IEEE Int. Conf. on Robotics and Automation, pp. 344-350, 1995.

[6] G. Niemeyer, J. Slotine, "Stable Adaptive Teleoperation", IEEE Journal of Oceanic Engineering, Vol 16, No. 1, January 1991.

[7] R. Riedi, M. Course, V. Ribeiro, R. Baraniuk, "A Multifractal Wavelet Model with Application to TCP Network Traffic", IEEE Trans. on Information Theory, pp. 992-1018, April 1999.

[8] W. Leland, M. Taqqu, W. Willinger, D. Wilson, "On the Self-Similar Nature of Ethernet Traffic", IEEE/ACM Trans. on Networking, Feb. 1994.

[9] I. Elhajj, J. Tan, N. Xi, W. K. Fung, Y. H. Liu, T. Kaga, T. Fukuda, "Multi-Site Internet-Based Cooperative Control of Robotic Operations", IEEE/RSJ Int. Conf. on Intelligent Robots and Systems, Japan, October 2000.

[10] Y. Zheng, "Human-Robot Coordination for Moving Large Objects", Workshop Note, 1997 ICRA.

[11] N. Hogan, "Multivariable Mechanics of The Neuromuscular System", IEEE Eight Annual Conf. of Engineering in Medicine and Biology Society, 1986.

[12] T. Murata, "Petri Nets: Properties, Analysis and Applications", Proceed. of the IEEE, April 1989.

[13] A. Moro, H. Yu, G. Kelleher, "Advanced Scheduling Methodologies for Flexible Manufacturing Systems using Petri Nets and Heuristic Search", IEEE Int. Conf. on Robotics and Auto., Arpil 2000.

[14] F. Baccelli, G. Cohen, G. Olsder, J. Quadrat, Synchronization and Linearity, John Wiley and Sons, England, 1992.

[15] J. Peterson, Petri Net Theory and the Modeling of Systems, Prentice-Hall, N.J., 1981.

[16] M. Song, "Integration of Task Scheduling, Sensing, Planning and Control in a Robotic Manufacturing Work-Cell", Doctoral Dissertation, Washington University, August 1997.

[17] J. Tan, N. Xi, "Hybrid System Design for Singularityless Task Level Robot Controllers", IEEE Int. Conf. on Robotics and Auto., 2000.

Supervisory Control for Systems of Vehicles in Path Networks

Elżbieta Roszkowska

Institute of Engineering Cybernetics
Wrocław University of Technology
ul. Janiszewskiego 11/17, 50–372 Wrocław, Poland
e-mail: ekr@ict.pwr.wroc.pl

Abstract

The paper is twofold. First we introduce a new concept - a partially directed graph, in which an edge can be undirected or have one of the two possible directions - and examine some of its structural properties. Then we assume that the edge attribute is variable and employ the concept to build a general, discrete event model of a system of vehicles moving in a path network. We examine some dynamic properties of the model and develop a flexible, suboptimal policy ensuring safe operation of the vehicle system.

1 Introduction

Early work dealing with conflict and congestion in AGV systems focused on eliminating or detecting vehicle conflicts during the vehicle route planning and scheduling phase. These methods typically require super polynomial complexity and the resulting plans are very sensitive to unexpected contingencies occurring during runtime. A more recently evolving approach [1, 2, 3, 4, 6] is to handle vehicle conflict in real-time by modeling the vehicle movement as a discrete event system in which the event occurrence is guarded by a policy guaranteeing that the system is safe, that is, that it does not deadlock.

In this paper we consider the problem of safe control for an AGV system with a bi-directional guide-path network and uni-directional vehicles, whose 'missions' are to visit specified sequences of workstations located in the network nodes. In [3] we discuss the same problem but for vehicles that circulate concurrently along some pre-determined paths. The model developed here follows the idea of [2] and allows real-time, dynamic path planning. Our work differs from the previous work both in the discrete-event representation of the vehicles' dynamics and the results obtained. For the modeling purposes we introduce a new concept - undirected graphs with two variable edge attributes: direction (including no-direction, to represent the absence of a vehicle in the zone) and mission. The values of the attributes change as a result of the occurrence of an event. Such an approach allows to model a vehicle system in a more natural way than e.g. Petri nets. We examine some structural properties of the partially directed graphs and employ the results to establish some properties of the dynamic model. As a consequence we develop a suboptimal policy ensuring safe operation of the vehicle system which, as we believe, is more flexible than those known so far.

2 Partially directed graphs

Definition 1 *A partially directed graph is a four-tuple $G = (V, E, in, \succ)$. The triple (V, E, in) is an undirected graph described by vertex set V, edge set E, and incidence function $in : E \to V \times V$, indicating the edges' vertices. $\succ : E \to \{-1, 0, 1\}$ is edge-direction function such that $\succ(e) = 0$ if edge e is undirected, $\succ(e) = 1$ if the direction is from vertex v to v', and $\succ(e) = -1$ if the direction is from v' to v, where $(v, v') = in(e)$.*

Definition 2 *For graph $G = (V, E, in, \succ)$ we define:*

1. *Edge e and vertex v are incident ($e \sim v$ for short) if $in(e) = (v', v'')$ and $v = v'$ or $v = v''$. If edge e is directed from vertex v to vertex v' then $v = tail(e)$ and $v' = head(e)$. Vertex v is called terminal if it is incident to only one edge, except possibly a self-loop edge. A start-vertex is a vertex with no predecessors, an end-vertex is a vertex with no successors.*

2. *A sequence $p = v_1, e_1, v_2, e_2, \ldots, e_{n-1}, v_n, e_n, v_{n+1}$ is a path in G iff for each $i = 1 \ldots n-1$, $v_i \sim e_i \sim v_{i+1}$, and either e_i is undirected or $tail(e_i) = v_i$ and $head(e_i) = v_{i+1}$. A path is directed (undirected) if it consists of only directed (undirected) edges. A*

path is *simple* if all edges and all vertices on the path, except possibly the first and last vertices, are distinct. A *cycle* is a simple path which begins and ends at the same vertex.

3. Vertices v and v' are *strongly connected* if $v = v'$ or there exist two paths, path p from v to v' and path p' from v' to v.

4. Edges e and e' are *strongly chain-connected* (or *strongly chained*) if $e = e'$ or there exists a sequence of cycles $c = c_1, c_2, \ldots, c_n$, such that edge e belongs to c_1, edge e' belongs to c_n, and for each $i \in 1, \ldots, n-1$, c_i and c_{i+1} have at least one common vertex.

5. Graph G is strongly connected if any given two vertices $v, v' \in V$ are strongly connected. Graph G is strongly chained if any given two edges $e, e' \in E$ are strongly chained. Graph $G = (U, \succ)$ is a tree if there are no cycles in U.

Note that both strong connectivity and strong chain-connectivity are reflexive, symmetric, and transitive. Thus, they partition their respective domains into disjoint sets, called equivalence classes.

Definition 3 For graph $G = (V, E, in, \succ)$ we define.

1. An equivalence class of strong connectivity is a subset of vertices $V_i \subseteq V$ such that any given two vertices $v, v' \in V_i$ are strongly connected and no vertex $v'' \in V - V_i$ is strongly connected to any $v \in V_i$. Subgraph $G_i = (V_i, E_i, in, \succ)$, where E_i is the set of edges which have their both vertices in V_i, is called a *strongly connected component* of G.

2. An equivalence class of strong chain-connectivity is a subset of edges $E_i \subseteq E$ such that any given two edges $e, e' \in E_i$ are strongly chained and no edge $e'' \in E - E_i$ is strongly chained to any $e \in E_i$. Subgraph $G_i = (V_i, E_i, in, \succ)$, where V_i is the set of vertices of edges in E_i, is called a *strongly chained component* of G. A component is *improper* if it consists of a single edge with two distinct vertices, and it is *proper* otherwise.

3. *v-condensation* $\mathcal{C} = \mathcal{C}(G)$ is the graph obtained from graph G through replacing each strongly connected component with a single vertex, called node.

4. *e-condensation* $\mathcal{B} = \mathcal{B}(G)$ is the graph obtained from graph G through replacing each proper strongly chained component with a single vertex, called node.

The following theorems provide some important information about the structure of partially directed graphs. Their proofs are presented in [5].

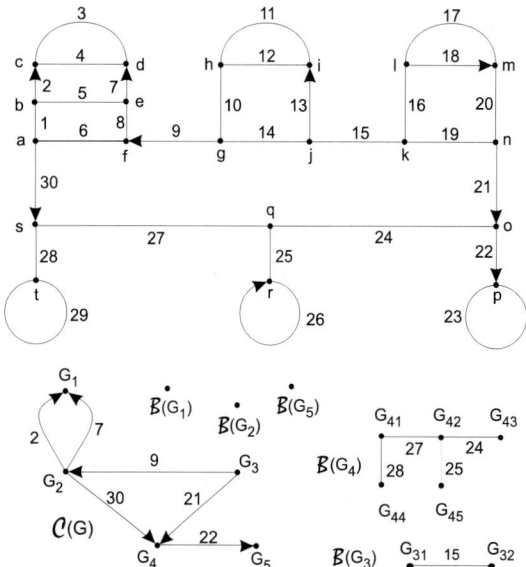

Figure 1: Illustration of Example 1.

Theorem 1 v-condensation $\mathcal{C}(G)$ of a partially directed graph G is a directed and acyclic graph.

Theorem 2 e-condensation $\mathcal{B}(G)$ of a partially directed, strongly connected graph G is an undirected tree.

As follows from the above theorems, a partially directed graph G can be partitioned into disjoint strongly connected components G_i joined by directed edges. Each subgraph G_i can be further partitioned into disjoint, proper strongly chained components G_{ij} joined by undirected edges.

Example 1 Fig. 1 shows an example of a partially directed graph $G = (V, E, in, \succ)$. The vertices are denoted by letters, the edges by numbers, and functions in and $\succ$ are given graphically. Graph G has five strongly connected components: G_1 with $V_1 = \{c, d\}$ and $E_1 = \{3, 4\}$, G_2 with $V_2 = \{a, b, e, f\}$ and $E_2 = \{1, 5, 6, 8\}$, G_3 with $V_3 = \{g, h, i, j, k, l, m, n\}$ and $E_3 = \{10, \ldots, 20\}$, G_4 with $V_4 = \{o, q, r, s, t\}$ and $E_4 = \{24, \ldots, 29\}$, G_5 with $V_5 = \{p\}$ and $E_5 = \{23\}$. G_1, G_2, and G_5 are strongly chained, so $G_{11} = G_1$, $G_{21} = G_2$ and $G_{51} = G_5$. G_3 has three strongly chained components: two proper components G_{31} with $V_{31} = \{g, h, i, j\}$ and $E_{31} = \{10, 11, 12, 13, 14\}$, and G_{32} with $V_{32} = \{k, l, m, n\}$ and $E_{32} = \{16, 17, 18, 19, 20\}$, and one improper component G_{33} – edge 15. Each edge of G_4 is a separate strongly chained component. Edges 26 and 29 are proper and 24, 25, 27, 28 are improper components.

Polynomially time-bounded algorithms to calculate condensations $\mathcal{C}(G)$ and $\mathcal{B}(G_i)$ can be found in [5].

3 Discrete event model of the vehicle system

To model the vehicle system we use an undirected graph PG with two edge attributes: network-state $\succ(e)$ and vehicle mission $m(e)$. Both functions $\succ(e)$ and $m(e)$ are variables which define the model's state. For each particular network-state $\succ$, the pair $(PG, \succ)$ is a partially directed graph. The undirected edges model empty zones and the direction of a directed edge e indicates the movement direction of the vehicle located in zone e. As a result of an event t, an edge e can turn from undirected to directed, which means that a vehicle has entered zone e, and from directed to undirected when the vehicle has left the zone. The events' occurrence is guarded by feasibility function $\varphi(s,t)$ and the resulting state-transition is given by next-state function $\delta(s,t)$.

Definition 4 Vehicle system is a five-tuple $VS = (PG, S, s_0, T, \varphi, \delta)$, where:

1. $PG = (V, E, in)$ is a connected undirected graph, called *path graph*; the edges correspond to the zones and the vertices to the nodes (intersections and zone ending points) of the vehicle path network. Each terminal vertex has a self-loop (so that a vehicle visiting a terminal node can change direction and go back along the same terminal zone it has come). There is one particular vertex $v^d \in V$, called docking station.

2. $S = D \times M$ is the state space of VS, with elements $s = (\succ, m)$ such that:
   - $\succ: E \to \{-1, 0, 1\}$ is network-state function
   - $m : E \to (V^n \times \{v^d\}) \cup \{null\}$, $n = 0, 1, \ldots$, is vehicle-mission function such that $m(e) = null$ iff $\succ(e) = 0$.

   $s(e) = (\succ(e), m(e))$ describes the state of the zone corresponding to edge e (or of zone e, for short). If $\succ(e) = 1$ then there is a vehicle in zone e traveling in the direction of $in(e)$. If $\succ(e) = -1$ then there is a vehicle in zone e traveling in the direction opposite to $in(e)$. The vehicle's mission $m(e) = [v_1, v_1, \ldots, v_n, v^d]$ is a vector of vertices (associated with the workstations located in the network nodes) that the vehicle located in zone e has to visit before it can exit the system through the docking station v^d. If $\succ(e) = 0$ then there is no vehicle in zone e (zone e is empty) and vehicle-mission function $m(e) = null$.

3. $s_0 = (\succ_0, m_0)$ is the initial state of VS given by $\succ_0: E \to \{0\}$ and $m_0 : E \to \{null\}$, which indicate that no vehicles are present in the system.

4. $T = T^* \cup T^e \cup T^m \cup T^{nm} \cup T^\dagger$ is the set of system events representing changes in the vehicles' location or mission, where:

   (a) $T^* = \{(e, m(e)) \mid e \sim v^d \ \& \ m \in M\}$ includes the events: a vehicle leaves docking station v^d for zone e with a new mission $m(e)$.

   (b) $T^e = \{(e, v, e') \mid e, e' \in E \ \& \ v \in V \ \& \ e \sim v \ \& \ v \sim e'\}$ includes the events: vehicle located in zone e leaves for zone e' via node v.

   (c) $T^m = \{e \mid e \in E \ \& \ tail(e) \neq v^d\}$ includes the events: vehicle located in zone e performs a mission step.

   (d) $T^{nm} = \{(e, m'(e)) \mid e \in E \ \& \ m' \in M\}$ includes the events: vehicle located in zone e receives a new mission $m'(e)$.

   (e) $T^\dagger = \{e \mid e \in E \ \& \ e \sim v^d\}$ includes the events: vehicle located in zone e leaves for docking station v^d.

5. $\varphi : S \times T \to \{enabled, disabled\}$ is *feasibility function* establishing for each pair (s, t) whether in state $s = (\succ, m)$ event t can occur. Depending on the event type, $\varphi(s, t) = enabled$ iff:

   (a) $t \in T^* \ \& \ t = (e_t, m(e_t)) \ \& \ \succ(e_t) = 0$

   (b) $t \in T^e \ \& \ t = (e_t, v_t, e'_t) \ \& \ head(e_t) = v \ \& \ \succ(e'_t) = 0$

   (c) $t \in T^m \ \& \ t = e_t \ \& \ tail(e_t) = m_1(e_t)$

   (d) $t \in T^{nm} \ \& \ t = (e_t, m'(e_t)) \ \& \ m(e_t) = [v^d]$

   (e) $t \in T^\dagger \ \& \ t = e_t \ \& \ m(e_t) = [v^d]$

   Less formally, a vehicle can be assigned a mission and enter the system if there is an empty zone incident to the docking station, change zones if there is an empty zone incident to its current location, accomplish next mission step if it has reached the first vertex of its mission vector, receive a new mission when there remains one last vertex in its mission - the docking station, and leave the system when it has reached a zone incident to the docking station and its mission vector includes only v^d.

6. $\delta : S \times T \to S$ is a partial function defined for each pair (s, t) such that event t is enabled in state s, called *next-state function*. For each edge $e \in E$, function $\delta(s, t)$, $s = (\succ, m)$, determines the new state $s'(e) = (\succ'(e), m'(e))$ which, depending on the event type, is given by:

   (a) $t \in T^* \ \& \ t = (e_t, m(e_t))$, $s'(e) = $
   $\begin{cases} (1, m(e)) & \text{if } e = e_t \ \& \ in(e_t) = (v^d, v) \\ (-1, m(e)) & \text{if } e = e_t \ \& \ in(e_t) = (v, v^d) \\ s(e) & \text{otherwise} \end{cases}$

(b) $t \in T^e$ & $t = (e_t, v_t, e'_t))$, $s'(e) =$
$$\begin{cases} (0, null) & \text{if } e = e_t \\ (1, m(e)) & \text{if } e = e'_t \ \& \ in(e'_t) = (v_t, v'_t) \\ (-1, m(e)) & \text{if } e = e'_t \ \& \ in(e'_t) = (v'_t, v_t) \\ s(e) & \text{otherwise} \end{cases}$$
where: $v_t \sim e'_t \sim v'_t$

(c) $t \in T^m$ & $t = e_t$, $s'(e) = (\succ(e), m'(e))$, where
$m'(e) =$
$$\begin{cases} [m_2(e), \ldots, m_n(e), v^d] & \text{if } tail(e_t) = m_1(e_t) \\ m(e) & \text{otherwise} \end{cases}$$

(d) $t \in T^{nm}$ & $t = (e_t, m'(e_t))$
$$s'(e) = \begin{cases} (\succ(e), m'(e)) & \text{if } e = e_t \\ s(e) & \text{otherwise} \end{cases}$$

(e) $t \in T^\dagger$ & $t = e_t$
$$s'(e) = \begin{cases} (0, null) & \text{if } e = e_t \\ s(e) & \text{otherwise} \end{cases}$$

Definition 5 For a system VS we define:

1. If an event $t \in T$ occurs in state s then the fact that a new state s' is reached is denoted by $s\,[\,t >\,s'$
2. The reachability set $R(s_0)$ is the set of states defined inductively as follows:
   (a) $s_0 \in R(s_0)$
   (b) $s \in R(s_0)$ & $s\,[\,t >\,s' \Rightarrow s' \in R(s_0)$
3. State $s \in S$ is safe iff $s_0 \in R(s)$. The system is safe iff each state $s \in R(s_0)$ is safe.

4 Testing state safety

Since as it can be easily noticed, the vehicles in VS can deadlock, the system is not safe. In order to ensure safe operation of VS it is necessary to guard the events' occurrence with an additional function $\psi : R(s_0) \times T \to \{admissible, inadmissible\}$ that provides that only such states can be reached, from which there exists a guarded event sequence that leads to s_0. For the further discussion consider a system VS with a path graph $PG = (V, E, in)$ and two states: $s \in R(s_0)$, $s = (\succ, m)$, and $s' \in R(s)$, $s' = (\succ', m')$. Let $G = (PG, \succ)$ and $G' = (PG, \succ')$ be the partially directed graphs corresponding to state s and s', respectively. We will be interested in the influence of the graphs' structure on the safety of system VS.

Definition 6 Let $chained(G)$ be the total number of edges in the strongly chained components of G, and let $room(G)$ be the total number of undirected edges in the strongly chained components of G. The room-condition for G, $\varrho(G)$, holds if $chained(G) = 1$ or $chained(G) = 2$ & $room(G) > 0$ or $chained(G) > 2$ & $room(G) > 1$.

Lemma 1 If graph G is strongly chained and the room-condition $\varrho(G)$ holds, then any vehicle present in the system can reach any vertex of PG in such a way that the new-reached graph G' is strongly chained and $room(G') \geq room(G)$.

Outline of proof. The case of $chained(G) \leq 2$ is evidently true. If G is strongly chained, then for any directed edge $e \in E$ and any vertex $v \in V$, there exists a path $p = head(e), \ldots, v, \ldots, tail(e)$ such that for each edge e' that occurs on p more than once, e' is traversed each time in the same direction. If $room(G') > 1$ then the vehicles can be moved in the train-like way along such paths, which allows any vehicle to visit any vertex on p and keeps the graph strongly chained.

Lemma 2 If graph G is strongly connected and the room-condition $\varrho(G)$ holds, then any vehicle present in the system can reach any vertex of PG in such a way that the new-reached graph G' is strongly connected and $room(G') \geq room(G)$.

Outline of proof. From Def. 4, if graph G has one edge then it is a self-loop and if G has two edges then they have common vertices and form a loop, hence the cases of $chained(G) \leq 2$ are evidently true. Since by Theorem 2, e-condensation $\mathcal{B}(G)$ is an undirected tree then, taking into account Lemma 1, it can be demonstrated that any vehicle can reach any vertex and leave the system or remain in a strongly chained component of G so that $\mathcal{B}(G') = \mathcal{B}(G)$ and $room(G') \geq room(G)$.

Lemma 3 State s is safe iff there exists state $s' \in R(s)$ such that G' is strongly connected and the room-condition $\varrho(G')$ holds.

Outline of proof. If s' can be reached, then by induction it follows from Lemma 2 that $s_0 \in R(s')$, hence state s is safe. If s' cannot be reached then, since s_0 satisfies the assumptions for s', s is not safe.

Lemma 3 allows us to substitute the problem of reachability of the initial state s_0 with an equivalent problem, i.e. the reachability of a particular vehicles' arrangement in network PG. Our approach to the safe-control for system VS is based on an algorithm that repeatedly looks for a deadlock in the v-condensation graph $\mathcal{C}(G)$ and if it does not find one, the algorithm attempts to change the vehicles' location so that $\mathcal{C}(G)$ eventually becomes one node, that is graph G becomes strongly connected. In order to do that we introduce conditions sufficient for VS to be in a deadlock or a pre-deadlock state and define operations on G and $\mathcal{C}(G)$ that correspond to particular vehicles' moves in the path network PG. Each operation either removes some vehicles from the system (when they are able to complete their missions) or results in such a change of vehicles' location, that the new-reached graph G'

is 'closer to strongly connected' than graph G, that is some nodes in v-condensation $\mathcal{C}(G)$ turn into one node in v-condensation $\mathcal{C}(G')$. The safe-control policy for system VS is based on the following concept.

Definition 7 $test : R(s_0) \to \{safe, not\_safe, ?\}$ is a function such that: (i) $test(s) = safe$ implies that there exists state $s' \in R(s)$ that satisfies the conditions of Lemma 3, (ii) $test(s) = not\_safe$ implies that $s_0 \notin R(s)$, and (iii) $test(s) = ?$ if neither (i) nor (ii) can be determined, i.e. the operations assumed allow no further folding of graph $\mathcal{C}(G)$ and no deadlock can be found in $\mathcal{C}(G)$.

$\psi : R(s_0) \times T \to \{admissible, inadmissible\}$ is admissibility function defined by: (i) $\psi(s,t) = admissible$ if $\varphi(s,t) = enabled$ and $test(\delta(s,t),t) = safe$, and (ii) $\psi(s,t) = inadmissible$ otherwise.

Theorem 3 System VS guarded by admissibility function ψ is safe.

Outline of proof. By Def. 7, function ψ allows only such state transitions for which the next-state $delta(s,t)$ satisfies the safety condition of Lemma 3.

The concept presented defines in fact a class of safe-control policies for system VS that includes both optimal and suboptimal policies. Admissibility function ψ is optimal iff for each $s \in R(s_0)$ such that $s_0 \in R(s)$ and $\varphi(s,t) = enabled$ for some $t \in T$, $test(\delta(s,t)) \in \{safe, not\_safe\}$. However, although it has not been proved that the problem of the initial-state reachability in system VS is computationally hard, no polynomially time-bounded solution is known either. The test proposed here is suboptimal and employs the ideas given in the following lemmas.

Lemma 4 Let G_n be an end node in v-condensation $\mathcal{C}(G)$ such that the docking station v^d is not in G_n. If $room(G_n) = 0$ then state s is not safe.

Outline of proof. By Theorem 1, all edges incident to G_n are directed, and since G_n is an end-node, they point to G_n. In order to proceed in its route, a vehicle located on an edge incident to node G_n must enter and eventually leave G_n. By Theorem 2, e-condensation $\mathcal{B}(G_n)$ is an undirected tree. Since $room(G_n) = 0$, then there are no undirected edges in the nodes of $\mathcal{B}(G_n)$. Thus, no vehicle that enters G_n can do the U-turn and go backwards, hence it will eventually deadlock with either the vehicles located on a node of $\mathcal{B}(G_n)$ or with another vehicle that has entered G_n, or with the vehicle located on an edge incident to G_n.

Note that a deadlock in a node G_i of $\mathcal{C}(G)$ can occur only then, when a vehicle enters G_i and $room(G_i) = 0$. Thus, $room(G_i)$ can be viewed as a current capacity of a single node. The following definition describes the current capacity of the nodes on a path, i.e. the maximal number of vehicles that can enter each node so that no deadlock occurs on the path, and inverse capacity, i.e. the maximal number of vehicles that can enter each node and return later to their initial location.

Definition 8 Let $p = G_0, e_1, \ldots, e_n, G_n$ be a path in v-condensation $\mathcal{C}(G)$. Node capacity on path p, $cap_p(G_i)$, is defined inductively by: $cap_p(G_n) = room(G_n)$ and $cap_p(G_i) = room(G_i) + cap_p(G_{i+1}) - 1$, $i = 0, \ldots, n-1$. Inverse node capacity on path p, $inv_p(G_i)$, is defined inductively by $inv_p(G_0) = room(G_0)$ and $inv_p(G_i) = room(G_i) + cap_p(G_{i-1}) - 1$, $i = 1, \ldots, n$.

Lemma 5 Let G_n be an end node in v-condensation $\mathcal{C}(G)$ and let $D \subset E$ be a cutset for the start-nodes of $\mathcal{C}(G)$ and G_n such that G_n is in part $\mathcal{C}_1$ and the docking station v^d is in part $\mathcal{C}_2$ of graph $\mathcal{C}(G)$. If for each edge $e \in D$ there is exactly one path p_e from $head(e)$ to G_n and capacity $cap_{p_e}(head(e)) = 0$, then state s is not safe.

Outline of proof. Since v^d is in part $\mathcal{C}_2$, then it is necessary for s to be safe that the vehicles eventually leave $\mathcal{C}_1$. This requires the accommodation of the vehicles located on at least one path p_e in the path's nodes. Such an accommodation is either impossible or requires entering a vehicle into some node G_i such that $room(G_i) = 0$. In both cases a deadlock occurs.

Lemma 6 Let P be the set of all paths from node G_0 to node G_n in v-condensation $\mathcal{C}(G)$. If (i): there is a path $p \in P$, $p = G_0, e_1 \ldots, G_n$, such that for each node G_i, except possibly G_0, capacity $cap_p(G_i) > 0$, then set P can be folded, i.e. there exists state $s' \in R(s)$ such that P becomes a strongly connected component G'_P of G'. If (i) holds and (ii): for each node G_i inverse capacity $inv_p(G_i) > 0$, then s' is safe iff s is safe.

Outline of proof. If (i) holds then the vehicles located on p can get accommodated in the path's nodes, hence p becomes undirected and thus all paths in P become strongly connected. If (ii) holds then state s can be reached back from s', thus s' is safe iff s is safe.

Lemma 6 defines an operation of path folding in $\mathcal{C}(G)$ as well as the conditions that state when the operation can be performed. Note that, since the safe-control policy assumed is not optimal, it is not necessary for its proper work to provide that only a safe state s' can be reached from a safe state s as a result of the folding. Thus, it is possible to implement in the testing algorithm only condition (i) or both conditions - (i) and (ii). Each option has its advantages and disadvantages. Condition (ii) may be over-preventive and forbid any further folding of the v-condensation graph

$\mathcal{C}(G)$, even if it is possible to fold it eventually into one node. On the other hand, neglecting condition (ii) may result in reaching an unsafe state s' from a safe state s, which unavoidably results in the *inadmissible* value of function ψ, whereas when guarded by (ii), the folding may eventually turn $\mathcal{C}(G)$ into one node and let ψ return the value *admissible*.

Example 2 Consider graph G of Example 1 and the set $P = \{p_1, p_2\}$ of paths from G_3 to G_5 in $\mathcal{C}(G)$: $p_1 = G_3, 21, G_4, 22, G_5$ and $p_2 = G_3, 9, G_2, 30, G_4, 22, G_5$. P can be folded into one node G_P as p_1 satisfies both conditions of Lemma 6: $room(G_5) = 1$, $room(G_4) = 1$, $room(G_3) = 8$, hence $cap_{p_1}(G_5) = 1$, $cap_{p_1}(G_4) = 1$, $cap_{p_1}(G_3) = 8$ and $inv(G_3) = 8$, $inv(G_4) = 8$, $inv(G_5) = 8$. Thus it is possible to accomodate the vehicles located on edge 21 and edge 22 in G_4 (edge 29) and G_5 (edge 23), resp., so that G_P is strongly connected. Graph $\mathcal{C}(G')$ has two nodes G_P and G_1, joined by two parallel paths $p_3 = G_P, 2, G_1$ and $p_4 = G_P, 7, G_1$. Since $cap_{p_3}(G_1) = room(G_1) = 2$ and $inv_{p_3}(G_P) = room(G_P) = 18$, $\mathcal{C}(G')$ can be folded into one node $G_{P'}$. Since $chained(G_{P'}) = 23$ and $room(G_{P'}) = 20$, then by Lemma 3, the arrangement of vehicles in G is safe, no matter what their missions are or where the docking station v_d is located.

Next to the folding we introduce operations that remove from the system the vehicles that can visit all the vertices of their mission vectors.

Definition 9 Operation $terminate(G, m)$ returns a new graph G' in which each edge e becomes undirected if there is an undirected path in G from $head(e)$ to v_d that includes in the proper order the vertices of $m(e)$. Operation $complete(G, m)$ returns a new graph G' in which each edge e becomes undirected if it belongs to a node G_i of $\mathcal{C}(G)$ such that all vertices of mission $m(e)$ are in G_i and the room-condition $\varrho(G_i)$ holds.

To detect deadlocks or paths to be folded we scan graph $\mathcal{C}$ with the BFS (Breadth-First-Search) algorithm. Below is a pseudo-code of the test.

Algorithm 1 Testing state safety.
Input: path graph $PG = (V, E, in)$, state $s = (\succ, m)$
Output: value of variable $test \in \{safe, not\_safe, ?\}$

PROCEDURE $clean(G, \mathcal{C}, m)$
  REPEAT
    $old\_G := G$
    G := $terminate(G)$; G := $complete(G)$;
    $\mathcal{C} := v\_condensation(G)$
  UNTIL $G = old\_G$
BEGIN *main*
  $test := ?$
  $\mathcal{C} := v\_condensation(G)$
  REPEAT
    $clean(G, \mathcal{C}, m)$
    $old\_G := G$
    $scan\_with\_BFS(\mathcal{C})$
      CASE(*conditions*) OF
        Lemma_4-5 : $test := not\_safe$; EXIT
        Lemma_6 : $G := fold(G)$
    $\mathcal{C} := v\_condensation(G)$
  IF $num\_of\_nodes(\mathcal{C}) = 1$ THEN
    $clean(G, \mathcal{C}, m)$
    IF Lemma_3(G) THEN $test :=$ safe
    ELSE $test := not\_safe$
  UNTIL $G = old\_G$ OR $test \neq ?$
END

Theorem 4 System VS guarded by admissibility function ψ (as in Def. 7) and Algorithm 1 is safe.
<u>Outline of proof.</u> By Lemmas 1-6 Algorithm 1 calculates correctly the values of $test(s)$, as required in Def. 7. Thus by Theorem 3, Theorem 4 is true.

5 Conclusions

The paper presents a new approach to modeling a vehicle system that can be viewed as a further development of predicate/transition Petri nets, in which the underlying graph is undirected and tokens have a direction attribute. In the case when only one token is allowed in a place (as assumed here) we substitute the place and its incident edges by one edge with a variable direction, including no-direction. The proposed method for graph folding is one of the solutions allowed by the general concept of state safety testing.

References

[1] C. Lee and J. Lin. Deadlock prediction and avoidance based on Petri nets for zone-controlled automated guided vehicle systems. *Int. J. Prod. Res.*, 33:3249–3265, 1995.

[2] S. Reveliotis. Conflict resolution in AGV systems. *IIE Transactions*, 32:647–659, 2000.

[3] E. Roszkowska Joint approach to design and control of process flows to avoid deadlocks in flexible production systems. *SPIE*, 3517:334–345, 1998.

[4] E. Roszkowska and M. Lawley. Supervisory control for flexible assembly systems with vehicle transport. In *Proc. MMAR'2000*, Vol.2, p. 853-858.

[5] E. Roszkowska. Partially directed graphs. To be submitted for publication, 2001.

[6] M. Yeh and W. Yeh. Deadlock prediction and avoidance for zone-controlled AGVS. *Int. J. Prod. Res*, 36:2879–2889, 1998.

Sensor Based Online Path Planning For Serpentine Robots

Mustafa GEVHER
mgevher@hc.aselsan.com.tr
ASELSAN Electronics Industry,
Telecommunication Division,
06370,Ankara, Turkey

Aydan M. ERKMEN
aydan@metu.edu.tr
Department of Electrical and
Electronics Engineering
Middle East Technical University,
06531,Ankara, Turkey

Ismet ERKMEN
erkmen@metu.edu.tr
Department of Electrical and
Electronics Engineering
Middle East Technical University,
06531,Ankara, Turkey

Abstract

This paper presents the multisensor-based online path planning of a serpentine robot in the unstructured changing environment of earthquake rubbles during search of living bodies. The robot has 6 links and 16 controllable degrees of freedom and is equipped with an ultrasound sensor and a camera on its head and infrared sensors along the links of the robot for obstacle avoidance.

The serpentine robot locomotion is achieved as a combination of basic snakelike and wormlike gaits which are that of lateral undulation, rectilinear motion, flapping motion (right, left), and of torsional rotation (which has the pivot point at the center of the robot). The path planning algorithm is adopted from Distance Transform Based exploratory path planning [1] and modified to our serpentine robotic configuration. The adopted method is used for calculating the cost of the possible next configuration states so that the snake/worm like robot selects its next gait accordingly. Environmental data is obtained from the multiple sensors located along the body of the robot. The sensors are located in the following configuration: one ultrasound sensor for detecting the obstacles in front of the robot and constructing the map for free space, and 12 infrared sensors (6 pairs) which are located on the left and right of the joints of the robot. Simulation results are presented in the paper as an illustration of the efficiency of the method in complex unstructured environments.

Keywords: Serpentine robots, Online Path planning, rescue robots, path planning in unstructured environments.

1.Introduction

Serpentine motion has been the focus of several recent research works where seminal papers were published [2][3][4]. Snakelike robots attracted the attention of researchers for applications in environments not suitable for wheeled and legged robots such as in ruins of collapse buildings, or in narrow paths of pipe lines. The major component of serpentine locomotion in such environments, besides control is path planning. Existing path planning algorithms have emerged due to the necessity of modifications brought to classical path finding algorithms since these classical approaches assumed single segment robot or nonredundant robotic configurations while robot snakes are hyper redundant structures. The most frequently used methods are potential fields[5], voronoi graphs[6], and Distance Transform based exploratory path planning[7].

Potential field methods are used successfully in many implementations, but they suffer from the problem of local minima. Recent modifications to the basic method helped to overcome this problem using harmonic functions as potentials[8].

Voronoi graphs are used to construct safe geometrical paths among obstacles using distance values obtained from sensors. Some implementations are used for snake-like multi-joint robots[6].

In this paper we adopt the Distance Transform based exploratory path planning and modify it to our snakelike robotic structure.

Our robot is composed of 6 identical segments joined together through a 2 way, 2 degrees of freedom joint enabling yaw and pitch rotation (Fig 1.1.).

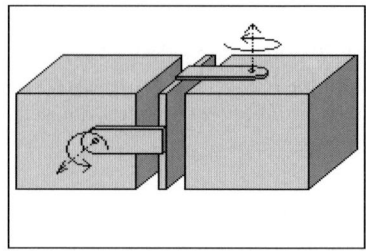

Fig 1.1.Configuration of each segment.

The whole robot is constructed using 6 identical segments. This results in 12 controllable degrees of freedom. An ultrasound sensor which is used for detecting the obstacles, and a thermal camera is located in the first segment in a dust free, antishock casting and operates intermittently when needed. This last segment functions as primary source of environmental data. 12 infrared sensors (6 pairs) are located on the left and right of the joints of the robot. (Figure 1.2.).

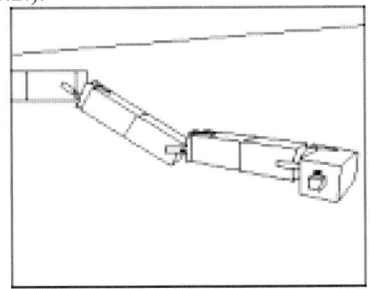

2. Mechanical layout of the front section of the snakelike robot.

asic contributions of this paper are:
1-instead of focusing to the global goal, our robot reduces its working area to a local one that evolves to meet the global goal. Many approaches have existed all along with geometrically predetermined intermediate goals. But here, our robot constructs a local map and using this local map, decides upon a suitable intermediate goal not in a predetermined manner.

2-Since the robot considers the local map for navigation, if the environment is changed during traversal either due to collision or due to loose pieces in rubbles, this change does not give rise to conflict for the path planning algorithm. The only required data in the global sense for the robot traversal is the direction of the ultimate goal. So changes of object interior of the global scale does not lead to confusion in the robots path planning algorithm.

3-Distance Transform based exploratory path planning is modified and applied to a multisegment robot configuration. The algorithm works fast and map building process does not need too many parameters. The computational and memory cost of this modified method is smaller compared to the original and to the other aforementioned methods.

2. Distance Transform Based Exploratory Path Planning

In this method the configuration space is first partitioned into cells. Cells which represents obstacles are assigned a high value, and the cell which represents the goal is assigned to zero. Such a parametric assignment generates an elevation map in the third dimension where obstacles create hills and the goal is nested within a basin. Other cells in the configuration space bear elevation values depending on distances to obstacles and/or the goal cell.
To find the value of a cell, the map is scanned twice in each pass; forward and reverse. In each pass the cell which has no value is given a value of one plus the minimum value of the four of its neighbors.
This process is done until there is no change in the values of the cells in the map (Fig 2.1).

8	9	10	9	10	11	12	13	14	15
7	8	9	8	9	10	11			14
6			7	8	9	10			13
5		5	6	7	8	9			12
4		4	5	6	7	8	9	10	11
3	2	3	4				10	9	10
2	1	2	3				9	10	9
1	0	1	2				8		
2	1	2	3	4	5	6	7		
3	2	3	4	5	6	7	8		

□ Goal   ■ Obstacle

Fig 2.1. Configuration space after Distance Transform is applied.

An 3D representation of the constructed map is shown in Fig 2.2.

Fig 2.2. Configuration space after Distance Transform is applied.

The original Distance Transform method is modified for our robot such that the cell representing goal is replaced with a line valley of cells with zero value within which the serpentine robot will match. This modification is done because we have an elongated snake robot in hand that has to fit a valley as a goal niche.

3. The Multisensor online serpentine path planning

3.1 Our Approach

In section 1 we had stated that the global goal is changed into an intermediate goal of a local map.
The ultrasound sensor scans for determining obstacles and free space and develop a local map. Thus, sensory data construct a local map within this sensor range. After the local map is obtained, possible next intermediate goals are found by considering points which are at the middle of the arcs representing free space. The intermediate goal is selected from the

candidate next states by considering the directions of the candidate states relative to the robot's head. In hardware applications, the direction which gives the highest energy of signal (thermal, chemical, sound) received from the goal (living being) is selected as intermediate goal. However, in the simulation here, we represent for illustrative purposes, the magnitude of the signals coming from the main goal as inversely proportional to the distance between sensor and goal. Thus this distance becomes minimum when the robot sensor faces the goal which is an emulation of the maximum signal energy coming with data from the goal, as it is the case in actual implementations. Among all directions generated in the scanning of the frontal sensor of the snake robot, the one closest to the direction of the main goal is selected as the intermediate goal. After the intermediate goal is found, The Modified Distance Transform Based Exploratory Path Planning (MDTBEPP) method is applied and the robot moves to this intermediate goal by using the serpentine gaits which are selected from the cost functions obtained through MDTBEPP. After reaching the intermediate goal, the robot makes a new scan and determines a next intermediate goal in this new local map. This process is repeated until the robot reaches the closest neighborhood of the main goal.

For example fig 3.1.1 gives a sample sequence of local maps formed by sensory data which contain the location of the closest detected obstacles through distance values of the obstacles relative to the frontal sensor of the robot. The distance values in each map are obtained in an array of angular span of 180 degrees. This data set must then be transformed into a local map which is made up of an array of 100x100 cells. The head, which is also the rotation center of the ultrasound sensor, is located in the center of the angular scan. Sensory data which are obtained through an angular scanning have naturally a polar coordinate representation in their obstacles and goal locations. These locations are then transformed into Cartesian coordinates before the MDTBEPP method can be applied.

In fig 3.1.1, for the first scan there are two candidates which are the middle points of the arcs representing free space. These states are shown as A_1 and B_1. Since state A_1 is closer to the main goal than B_1, A_1 is selected. After A_1 is reached by using MDTBEPP, the robot scans for next intermediate goal candidates. A_2, B_2, C_2 are the possible next states. C_2 is selected since it is closer to the direction of the main goal. Similarly after the third scan A_3 is selected. The last state which is also the main goal is A_4. The locomotion of the snake like robot to the intermediate sates is achieved by serpentine gaits which will be introduced in section 3.3.

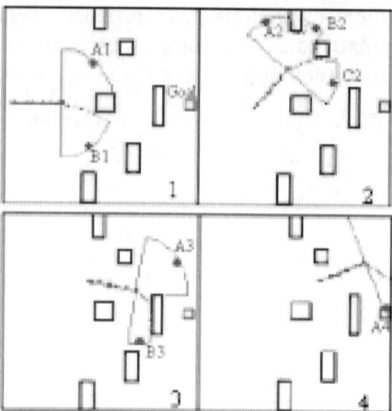

Fig 3.1.1. Robot determines intermediate goals/states to reach the ultimate goal(small box on the left center).

3.2 Map Building

Earthquake rubbles have an important property that underlines the complexity of this environment: During search, pieces of rubbles fall and thus change place so objects in the environment are considered to be dynamically changing.

The major aim of the serpentine search robot being that of finding, identifying and locking onto living beings under rubble, local map building is an essential component of our path planning approach.

Since the objects in the rubble environment are ready to change position and orientation, the local map is used to find the next position of the robot on its way to a goal which is the living being in an initially unknown but now, detected location. Using the local map provide the following benefits to our approach:

1. Building global map is computationally heavy when compared to building a local map. Thus a local map is computationally more accessible.
2. It is enough to know the direction of the location of the target and the location of the closest objects to construct the local map for path finding, thus a complete prior knowledge is not required. This allows our robot to operate in an environment initially unknown or minimally known during search of livingbeings by the snake-like robot.

Fig. 3.2.1 represents a sample of snake+environment interactions tracked by a simulation program while fig 3.2.2. shows the local map built by sensory data obtained for this snake+closest-environment interaction. In fig 3.2.1 the fishbone structure on the robot shows the line of sight of the infrared sensor pair located each side of the snake robot while the front radial line is the line of sight of the ultrasound sensor. The small squares in the middle of the arc are the candidates for the intermediate goal. The suitable goal is selected according to its direction relative to the main goal. As stated in section 3.1, the one which is closer to the main goal is selected as next intermediate goal.

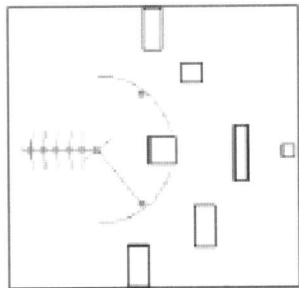

Fig 3.2.1. A sample environment used in simulation.

The cubic obstacle head-front from the snake robot in fig 3.2.1 is clearly seen in the local map of fig 3.2.2. The different gray levels in this figure represents the cost values obtained from MDTBEPP where darker regions represent minimum values, and brighter regions represent the higher cost values. Since the dimension of a local map is much smaller than that of a global map, the errors related to location and orientation of the robot is minimized when compared to finding location with a global map. When the intermediate goal is reached, the current local map is not needed anymore and a new local map is constructed and new intermediate goal is selected.

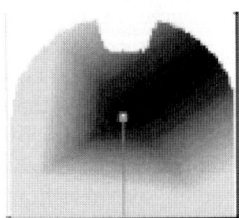

Fig 3.2.2. A local map obtained from the sensors.

3.3 Serpentine gaits of the search robot

Serpentine robots have advantage over wheeled and legged robots for their hyper-redundancy and flexibility. They can modifiy their bodies for more stability and they can provide more reliable locomotion compared to other mobile robots. Another benefit of a snakelike robit is that it has at least one stationary segment fixed in one location (a changing base) while the other segments are in motion, which results in minimization of the error of finding location and orientation. Also since some of the segments are fixed at one base location for certain motion, the robot is more robust to external disturbances.

The locomotion of the snake-like robot is achieved by adapting the natural snake motions to the multisegment robot configuration[4]. For the current implementation, our robot has 5 possible gaits which results in 5 possible next states:

1.Move forward:
In our algorithm robot moves forward with Rectilinear motion and Lateral Undulation..
Rectilinear motion: The segments displaces themselves as waves on the vertical axis. (Fig. 3.3.1)
Lateral Undulation :The segments follow the lines of horizontally propagating waves . (Fig. 3.3.2)

2.Move right/left:
Robot can move right/left with flapping motion (Flap right/left):
Flap righ/left: The front and the back segments of the robot move to the right/left side of the robot and then pull the center segments, which results a motion perpendiculular to the alignment of the robot. (Fig. 3.3.3)

3.Rotate right/left:
The robot undergoes a horizontal rotation to the right/left with respect to the pivot located near the middle of the robot. (Fig. 3.3.4)

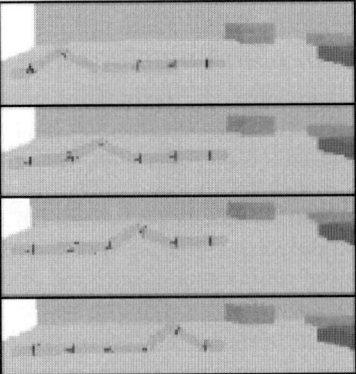

Fig 3.3.1.Rectilinear type gait.

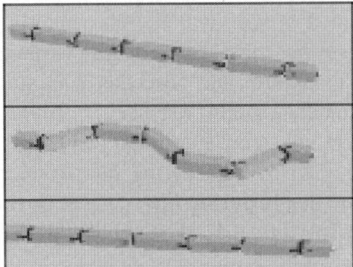

Fig 3.3.2.Lateral undulation gait .

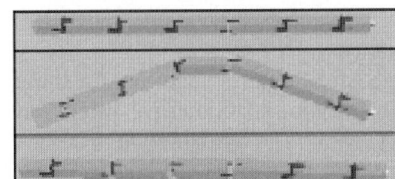

Fig 3.3.3. Flaping gait (right movement).

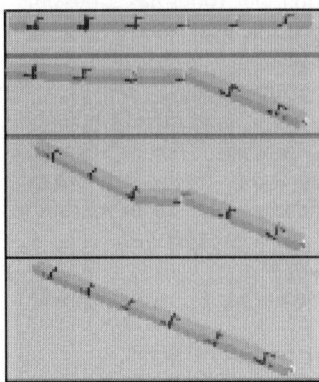

Fig 3.3.4. Rotational gait (right rotation)

In our simulation program the gaits are simulated in a 3D environment together with their physical interactions during traversal of the unstructured environment. These interactions include: collision simulation ,simulation of dynamic changes occuring to objects and to the robot, and simulation of the sensors.

4.Our path planning algorithm

Sensor data are limited by the range of the sensors. So the robot decides on an intermediate goal within the angular scan arc on the local map. This intermediate goal is determined according to the free space, and the direction of the ultimate goal position relative to the head of the robot. After the scanning of the robot frontal sensor, the locations which are in the middle of the each arc segments delimiting the free subspaces cut in between obstacles, are generated. The middle points of each arc are then the possible canditates for an intermediate goal. The one which is closest to the direction of the main goal is selected as intermediate goal as stated in section 3.1. Subsequently, the robot decides on the most suitable gait using Modified Distance Transform and proceeds with suitable gaits until intermediate goal is reached.

For a single link robot located at (x,y), the next move can be decided by considering the parametric values of the cells neighbouring the current position of the robot as in fig. 2.1. The cell with minimum parametric value is the most suitable next move for the robot. But for a multijoint hyperredundant robot, the remaining segments must also be considered. So all the segments must be included in the computation.

Since we have 6 control points on the snake robot for our path planning algorithm we would have to deal naturally with 12D configuration space. A practical simplification brought to the computations has been to projet the 12D configuration space into a 2D master configuration subspace where the local map is built and the intermedial goals are found and the first robot move is generated; and into 6 2D slave configuarton subspaces where the first robot move is propogated with modifications according to the snake gait selected..

For deciding which gait to choose among the gaits of our search robot that have been described in section 3.3, the algorithm takes into account the whole configuration of the robot on the possible next state. For the algorithm, 6 discrete control points are taken into consideration on the robot and are used for calculating a cost function for a gait. These control points are used to find the candidate cells where the robot segments would each possibly move after the next gait decision. So each of these cell values are multiplied with a weight value representing the possibility in candidacy of each cell and are added to the cost function. The cost function is formulated as:

$$F(s) = \sum_{i=1}^{6} w_i \cdot C(x_i, y_i)$$

Where $F(s)$ represents cost function of the possible next gait state s, w_i is the weight of the i'th control point, and $C(x_i,y_i)$ is the cost value obtained form Modified Distance Transform for i'th control point located at x_i and y_i. Weights of control points i depend on the ranking of the importance of contribution of each segment i to the snake displacement. This importance is a degree of constraint put on that segment during serpentine locomotion. A gait is selected such that it has minimum cost which is a way of demonstrating that this gait is the one that require the least of body energy in its realization in the corresponding local map.Thus we assign the weights for each control point such that front section has the maximum value and the end section has the minimum value. After each decided gait and before the next gait desicion making, the snake-robot body configuration resets. In the simulation the this reset is the lining up of the robot segments.

5.Simulation and Results

The modified algorithm is applied to the simulation program. Since the robot makes certain types of snakelike gaits, the most probable gait is chosen as the best move that minimizes the cost in the distance transform. A sample locomotion sequence is shown in Fig 5.1. Since the robot starts its next gait with initial line up (reset), and ends the gait with a final line up, the robot is shown as a line in these local maps. After the local map is built, the robot decides the next gait using MDTBEPP , then lines up and implements the selected gait. Intermediate goals are used to proceed towards the main goal. Sample simulation results are shown in Figure 5.2 which shows the displacement of snake-robot among obstacles.

This sample (fig 5.2) shows the path followed by the robot as composed of lines and arcs, which is the result of the serpentine gaits used by the snake-like robot. Straight lines in the direction of the robot is formed by rectilinear motion. Short lines deviating from the main path is formed by flaping motions. The arcs in the path are formed by rotational motions.

Figure 5.3. shows the number of gaits used for the second sample snake robot locomotion of fig 5.2 (the one to the right). Rectilinear gait is symbolized as R which is 8 times selected on the overall portion of the locomotion depicted here, Rotational gait is symbolized as Rt and is 4 times selected, and Flapping gait is symbolized as F and is 5 times selected.

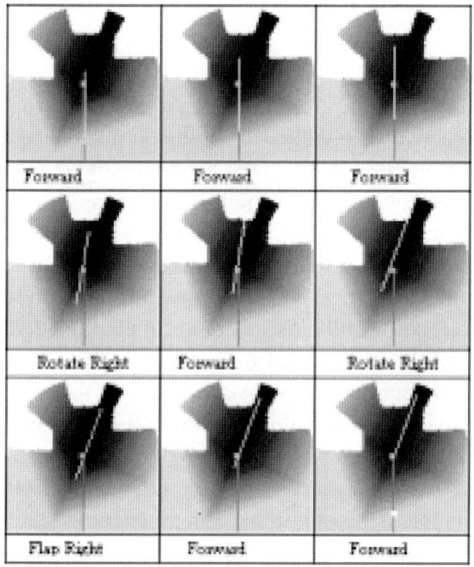

Fig 5.1. Robot reaches the intermediate state

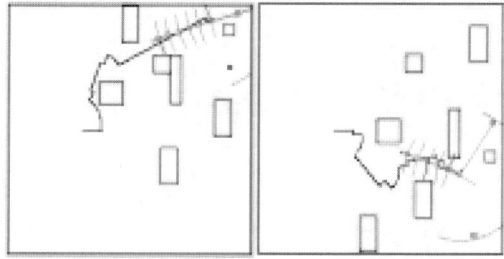

Fig 5.2. Two outputs of the simulation program

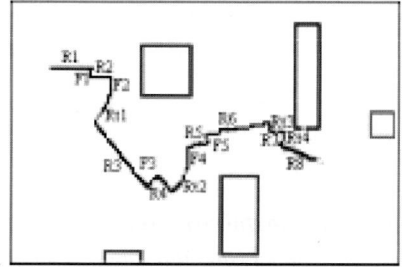

Fig5.3 Gaits selected by the algorithm.

6. Conclusion

In this paper, we introduce our sensor based online path planning for a serpentine robot which distinguishes itself in its optimal gait selection that would need smallest body energy possible in the corresponding local map. Body segments have been weighted according to their contribution ranking in the locomotion. The tail segments having small weights are found to frequently collide with objects in the environment. Therefore side sensors are used to avoid the collision of the body with the obstacles located along the body of the robot. By increasing the number of segments, we have found that our robot can curl around object by fixing some segments in a certain location, thus creating a changable base. Since the robot does not have base point, similar to shoulder for the arm, the robot must consider the end segments while locomotion. Most of the path planning algorithms for serpentine robots assumes the end segments of the robot are fixed, or at least concentrated in one location. Here we consider to take the serpentine gaits that move every segment of the robot. These are among the novelties of our snake robot and its path planner. In our algorithm we assume noisless data obtained from ultrasound and the infrared sensors is necesarry and is the focus of our present work.

7. References

[1] Jarvis, R.A. "collisison-Free Trajectory Planning Using Distance TransForm", Proceeding of National Conference and Exhibition on Robotics, Melbourne, August 1984, pp.20-24

[2] Hirose, S., Biologically Inspired Robots: Snake-like Locomotors and Manipulators, Oxford University Press, 1993, ISBN 0 19 856261 6.

[3] G.S. Chirikjian. "Theory and Applications of Hyper-Redundant Robotic Manipulators". PhD thesis, California Institute of Technology, Pasadena, CA, 1992.

[4] Dowling K. "Limbless Locomotion: Learning to Crawl with a Snake Robot". PhD thesis, Carnegie Mellon University, Pittsburgh, PA, 1997.

[5] Khatib, "Real-time Obstacle Avoidance For Manipulators and Mobile Robots", Int. J. of Robotics Research, Vol. 5(1), 90-98, 1986.

[6] H.Choset, J.W. Burdick. "Sensor Based Planning, Part I: Incremental Construction of the Generalized Voronoi Graph". In Proc. IEEE Int. Conf. On Robotics and Automotion, Nagoya, Japan 1995.

[7] K.S. Chong, L. Kleeman ,"Indoor Exploration Using a Sonar Sensor Array: A Dual Representation Strategy", IEEE/RSJ Int. Conf. on Intelligent Robots and Systems, 1997.

[8] C. I. Connolly, J. B. Bums, and R. Weiss, Path planning using Laplace's Equation, IEEE Int. Conf. on Robotics and Automation, 2101-2106, 1990.

Reactive Behaviours of Mobile Manipulators Based on the DVZ Approach

A.Cacitti (IEEE S.M.)   R.Zapata
Robotics Laboratory
LIRMM - UMR5506 CNRS
University of Montpellier, France

Abstract

This paper addresses the problem of reactive behaviours of mobile manipulators evolving in dynamic and unknown environments. The Deformable Virtual Zone (DVZ) method is proposed to resolve the problem through the reflex action theory. The formulation of the DVZ principle is given and the extension to mobile manipulators is detailed. This approach allows to implement fast control laws and can be seen as an efficient low level algorithm for controlling motions to avoid unscheduled obstacles collisions. Simulation results are discussed, showing the effectiveness of the proposed method.

1 Introduction

Artificial reflex actions for mobile robots can be defined as the ability to react when unscheduled events occur, for instance when they move in unknown and dynamic environments. The problem of designing reflex-oriented artificial systems has been partially solved by using a behavioural approach, consisting in directly relating inputs (stimuli) to outputs (actions) through state machines and to make these elementary machines communicate [1][2][3][4]. The ways these machines are programmed are various: fuzzy logic, neural networks, deterministic state machines and so on. Another design consists in considering a sensor-based approach which feedbacks sensory information to the robot control loop [5][6]. The most famous method of this family is the potential method developed by O. Khatib, twenty years ago [7].
In the first case, the stimuli are translated into virtual external forces which are simply added to the force due to the goal, when there is a goal, and all these forces produce an action, moving the system and locally modifying the world. It is quite clear that the second approach is not so different: the information provided by exteroceptive sensors generate the same virtual external forces which contribute to the system evolution, after a comparison with the forces due to the programmed goal.
For the last nine years, we have been interested in the problem of reactive behaviours for collision avoidance in the domain of mobile robotics. We have investigated the control of land autonomous vehicles, underseas robots, manipulators and walking machines. Many experiments have been carried out on real robots (wheeled mobile robots, legged robots and an AUV) and on simulated ones. We have developed a collision avoidance method called DVZ (Deformable Virtual Zone) that allows to control the reactive behaviours of a mobile robot [8]. In this paper we propose an extension of this method to the case of mobile manipulators.
This paper is organized into five sections. In section 2 the DVZ principle approach in its general frame is described. The extension to mobile robotic manipulators is presented in section 3, and in section 4 simulation results are discussed. Conclusions are given in sect. 5.

2 The DVZ Principle

This paragraph describes the DVZ principle in which a rigid body (the robot), evolving in an unknown environment, is supposed to be surrounded by a Deformable Virtual Zone (DVZ), the geometry of which depends on the body generalized coordinates (state) and whose deformations are due to the interaction with the environment.

Definition 1 *For a convex rigid body $R \subset \Re^3$, we define the **undeformed DVZ-set** $\overline{\Xi}_h$ as any convex surface surrounding R and verifying a one-one correspondence with the boundary ∂R of R.*

Definition 2 *The **undeformed DVZ** Ξ_h of a convex rigid body $R \subset \Re^3$, is the one-one map in the set of convex surfaces, relating ∂R to $\overline{\Xi}_h$ and formally defined by: $M \xrightarrow{\Xi_h} P_h$.*

Following the definition, $\overline{\Xi}_h = \Xi_h(\partial R)$.

Definition 3 *For a convex rigid body moving among obstacles, and for which is defined a DVZ-set $\overline{\Xi}_h$, we define the **deformed DVZ-set** $\overline{\Xi}$ as any convex surface surrounding R and included in $\overline{\Xi}_h$ and verifying a one-one correspondence with the boundary ∂R of R.*

Definition 4 *The **deformed DVZ** Ξ of a convex rigid body $R \subset \Re^3$, is the one-one map in the set of convex surfaces, relating ∂R to $\overline{\Xi}$ and formally defined by: $M \xrightarrow{\Xi} P$.*

Following the definition, $\overline{\Xi} = \Xi(\partial R)$.

Definition 5 *The **deformation** Δ is defined as the functional difference of Ξ and Ξ_h:*

$$\Delta = \Xi - \Xi_h \qquad (1)$$

The deformation Δ is a one-one map that associates the vector $P - P_h$ to the point $M \in \partial R$. It can therefore be considered as a vector field defined on ∂R.

The undeformed DVZ depends on the state vector π characterizing the motion capabilities of the body (its translational and rotational velocities for instance):

$$\Xi_h = \beta(\pi) \qquad (2)$$

Let M be a point of the boundary of R, $M \in \partial R$. The deformation vector $\Delta(M)$ depends on I, intrusion of the environment into the body space at M and on the undeformed DVZ at M:

$$\Delta(M) = \alpha(\Xi_h(M), I(M)) \qquad (3)$$

In the following we refer to (3) as:

$$\Delta = \alpha(\Xi_h, I) \qquad (4)$$

implicitly implying the vectors image of M through the correspondent map.
Differentiating equation (4) yields:

$$\dot{\Delta} = \nabla_{\Xi_h}[\alpha] \nabla_\pi[\beta] \, \phi + \nabla_I[\alpha] \, \psi \qquad (5)$$

where ∇_ξ is the derivation operator with respect to the variable ξ and where $\phi = \nabla_t \pi = \dot{\pi}$ and $\psi = \nabla_t I = \dot{I}$ are the two control vector of Δ.
The evolution of Δ is driven by a two-fold input vector $u = [\phi, \psi]^T$. The first control vector ϕ, due to the controller, tends to minimise the deformation of the DVZ. The second one, ψ, is unknown and is induced by the environment itself (and could, at most, try to maximize these deformations). The complete evolution of the deformation is modeled by a differential equation of the type:

$$\dot{\Delta} = A\phi + B\psi \qquad (6)$$

The DVZ control algorithm consists of choosing the desired evolution $\dot{\Delta}_{des}$ of the deformation and applying the following lemma:

Lemma 1 (DVZ Principle) *Given $\dot{\Delta}_{des}$, the best control vector $\check{\phi}$ in the sense of least-squares, that minimises function $\left\| \dot{\Delta}_{des} - \dot{\Delta} \right\|^2$ is obtained by inverting equation (6):*

$$\check{\phi} = A^+ (\dot{\Delta}_{des} - B\psi) \qquad (7)$$

where A^+ is the pseudo-inverse of A.

A simple and efficient control law consists of choosing the desired deformation as proportional to the real deformation and its derivative:

$$\dot{\Delta}_{des} = -K_p \Delta - K_d \dot{\Delta} \qquad (8)$$

where the two matrices K_p and K_d are respectively the proportional and derivative gain and are tuned in order to carry out the avoidance task.

2.1 Extension to Non-Rigid Bodies

We now extend the DVZ principle to the case of a non-rigid body R, i.e. a body presenting an internal configuration that can vary in time. Let η be the internal state of R. The internal configuration of R is completely described by η. We define the undeformed DVZ Ξ_h of R by applying the general definition 2.
The undeformed DVZ of a non-rigid body depends both on the state vector π characterizing the body motion capabilities and on the internal state η:

$$\Xi_h = \beta(\pi, \eta) \qquad (9)$$

Following the same conceptual development of the precedent paragraph, differentiating (4) yields:

$$\dot{\Delta} = \nabla_{\Xi_h}[\alpha] \nabla_\pi[\beta] \phi + \nabla_I[\alpha] \psi + \nabla_{\Xi_h}[\alpha] \nabla_\eta[\beta] \xi \quad (10)$$

where $\xi = \nabla_t \eta = \dot{\eta}$ is the evolution of the internal state.
The complete evolution of the deformation Δ for a non-rigid body is modeled by a differential equation of the type:

$$\dot{\Delta} = A\phi + B\psi + C \qquad (11)$$

This result represents the extension of (5) to non-rigid bodies. In this case the complete evolution of the deformation vector Δ depends as well on the internal configuration. We point out that $\nabla_\eta[\beta] = \nabla_\Theta[\beta]\nabla_\eta[\Theta]$ and $\nabla_\eta[\Theta] = \nabla_\eta[\Theta](\eta, \dot{\eta})$, being $\Theta = \Theta(\eta, \dot{\eta})$: the relation between $\dot{\Delta}$ and ψ is non linear. In general $\dot{\eta}$ is a subset of π.

The DVZ control algorithm does not change in substance and the DVZ principle is applied to (11), which takes in account the internal configuration additional term. We design the control law by choosing $\dot{\Delta}_{des}$ as in (8).

3 Mobile Manipulators

In this section we consider one link of the mobile manipulator and we aim to apply the DVZ principle. Firstly we define a meaningful shape for the link's DVZ and we express the deformation as a function of the mobile manipulator configuration and kinematic. Secondly we apply the extension of the DVZ control algorithm to generate the reflex command and let the reactive behaviour emerge.

A mobile manipulator can be characterized as a non-rigid body, presenting internal configurations defined by $\eta = (\theta, \gamma)$ where θ represents the joint angles of the manipulator and γ the absolute orientation of the mobile base. We expect to find the same differential equation (11) to describe the complete evolution of Δ.

3.1 Notation

We define the notation as follows: $\Sigma_L, \Sigma_B, \Sigma_S, \Sigma_0$ respectively the link, mobile base, sensors and absolute frames; ${}^2\mathbf{T}_1 = \begin{bmatrix} {}^2\mathbf{R}_1 & {}^2\mathbf{O}_1 \\ 0 & 1 \end{bmatrix}$, ${}^2\mathbf{R}_1$ and ${}^2\mathbf{O}_1$ respectively the homogeneous transformation matrice, rotation matrice and relative position from Σ_1 to Σ_2; ${}^0\mathbf{t}_{1,2} = \begin{bmatrix} {}^0\mathbf{v}_{1,2} & {}^0\boldsymbol{\omega}_{1,2} \end{bmatrix}^T$ the six-dimensional generalized velocity vector, referred to as twist, of Σ_1 relative to Σ_2, expressed in Σ_0; $\mathbf{P}$ a vector, $P = \|\mathbf{P}\|$ its norm, $\hat{\mathbf{P}} = \frac{\mathbf{P}}{P}$ its direction. It follows $\mathbf{P} = P \cdot \hat{\mathbf{P}}$.

In the following all variables are expressed with respect to Σ_L, link frame if not otherwise mentioned. Σ_L for link $i-th$ is defined as follows: axis $x \equiv$ symmetry axis of the link; origin $O_L \equiv$ center of joint $i+1$.

3.2 DVZ Definition

A natural choice to define the robotic arm's DVZ map Ξ_h of a mobile manipulator, is to associate the arm with a cylinder-like surface surrounding each link R. By doing so, a safe protective zone for each link (therefore for the whole arm) towards all directions is guaranteed. We define the undeformed DVZ set as:

$$\overline{\Xi}_h = \left\{ \mathbf{P}^h : \forall \mathbf{M} \in \partial R, \ P_x^h = M_x, \hat{\mathbf{P}}_{yz}^h = \hat{\mathbf{M}}_{yz}, \right.$$
$$\left. P_{yz}^h = \left(\sqrt{\hat{\mathbf{M}}_{yz}^T \cdot \mathbf{A} \cdot \hat{\mathbf{M}}_{yz}} \right)^{-1} + d_{min} \right\} \quad (12)$$

where

$$\mathbf{A} = \mathbf{V} \cdot \boldsymbol{\Lambda} \cdot \mathbf{V}^T, \boldsymbol{\Lambda} = \begin{bmatrix} \frac{1}{c_y^2} & 0 \\ 0 & \frac{1}{c_z^2} \end{bmatrix} \quad (13)$$

$$\mathbf{V} = \frac{1}{\|\mathbf{v_M}\rfloor_{yz}\|} \cdot \begin{bmatrix} \mathbf{v_M}\rfloor_y & -\mathbf{v_M}\rfloor_z \\ \mathbf{v_M}\rfloor_z & \mathbf{v_M}\rfloor_y \end{bmatrix} \quad (14)$$

$$c_y = (1-r) \cdot \|\mathbf{v_M}\rfloor_{yz}\|, \ c_z = (1-f) \cdot c_y \quad (15)$$

Following the definition, $\overline{\Xi}_h$ describes a generalized cylinder whose section is a variable ellipse either in size and orientation, refer to fig.1 and 2.

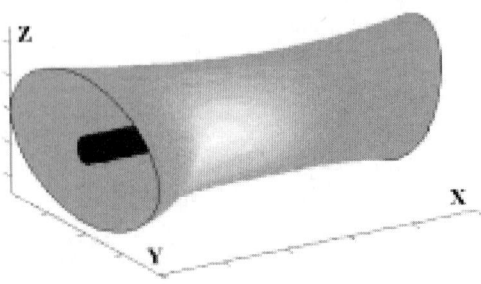

Figure 1: 3-D representation of link DVZ

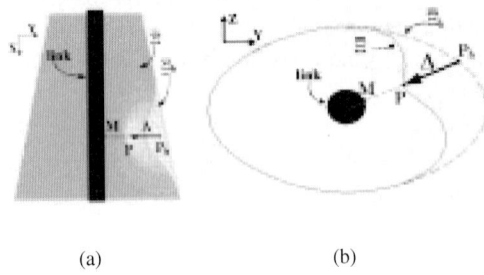

Figure 2: link DVZ cross sections

Vector $\mathbf{v_M}\rfloor_{yz}$ is the yz component of $\mathbf{v_M}$, velocity of point $\mathbf{M}$. Only the yz component of the velocity vector affects the DVZ since the x component represents a translation towards the link's axis and it will be duty of the DVZ of the successive or precedent link in the robotic arm chain (the end-effector or mobile base if a terminal link) to prevent collisions on that direction.

Parameter d_{min} defines a minimum safety zone around the link, inside which no intrusion is allowed. Normalized parameter f is the fineness of the DVZ, or width/length ratio. Normalized parameter r is the risk: a high value of r ($\cong 1$) means small DVZ, weakly dependent on state vector π, whereas a low value of r ($\cong 0$) means large DVZ, strongly dependent on state vector π. The three parameters (r, f, d_{min}) are referred to as the *intrinsic parameters* of the DVZ.

The velocity $\mathbf{v}_M$ is calculated by applying the direct kinematic to the mobile manipulator. We define the matrix operator $S(\mathbf{a})$ as:

$$S(\mathbf{a}) = \begin{bmatrix} 0 & a_z & -a_y \\ -a_z & 0 & a_x \\ a_y & -a_x & 0 \end{bmatrix}, \quad \mathbf{a} = \begin{bmatrix} a_x \\ a_y \\ a_z \end{bmatrix} \quad (16)$$

so that $\mathbf{b} \times \mathbf{a} = S(\mathbf{a}) \cdot \mathbf{b}$. Using the above expression, $\mathbf{v}_M$ is given as:

$$\mathbf{v}_M = \begin{bmatrix} {}^L\mathbf{R}_B & S(\mathbf{M}) \cdot {}^L\mathbf{R}_B \end{bmatrix} \cdot {}^B\mathbf{t}_{L,0} \quad (17)$$

${}^L\mathbf{R}_B$ represents the rotation matrice of Σ_B with respect to Σ_L and is function of the manipulator configuration θ. The six-dimensional vector ${}^B\mathbf{t}_{L,0}$ represents the twist of Σ_L with respect to Σ_0 expressed in Σ_B and is given as:

$${}^B\mathbf{t}_{L,0} = \mathbf{J} \cdot \dot{\theta} + \begin{bmatrix} \mathbf{I} & S({}^B\mathbf{O}_L) \\ \mathbf{0} & \mathbf{I} \end{bmatrix} \cdot {}^B\mathbf{t}_{B,0} \quad (18)$$

where $\mathbf{J}$ is the Jacobian matrix for the link.
The mobile base twist vector ${}^B\mathbf{t}_{B,0}$ is given as ${}^B\mathbf{t}_{B,0} = \chi(v_B, \dot{\gamma}_B, \gamma)$, where v_B and $\dot{\gamma}_B$ are the controlled velocities of the base. Function $\chi(\cdot)$ depends on the motion capabilities of mobile base, i.e. on its degrees of freedom, and can account for non-holonomic constraints. For instance, considering the case of an holonomic mobile base moving in the xy plane, v_B represents the absolute value of the linear velocity towards the front and the side of the base ($\lfloor v_B \rfloor_x$ and $\lfloor v_B \rfloor_y$) whilst $\dot{\gamma}_B$ the rotation velocity about the vertical axis z.
This result, combined with the above considerations, clearly underlines the dependence of the undeformed DVZ Ξ_h on the mobile manipulator motion capabilities, by means of the state vector defined as $\pi = \begin{bmatrix} \dot{\theta}, v_B, \dot{\gamma}_B \end{bmatrix}^T$, and on the mobile manipulator internal configuration η, that is $\Xi_h = \bar{\mathfrak{F}}_h(\pi, \eta)$.
The description of the environment surrounding the manipulator is given as I, proximity measurements by means of any sensorial equipment fitted on the mobile manipulator. Without loosing in generality, we consider I referred to the sensor frame Σ_S. The intrusion of the environment inside the link space is given by the deformed DVZ $\bar{\Xi}$, defined as:

$$\bar{\Xi} = \{\mathbf{P} : \mathbf{P}\ inside\ \bar{\Xi}_h, \mathbf{P} = {}^L\mathbf{R}_S(\theta) \cdot I + {}^L\mathbf{O}_S\} \quad (19)$$

Referring to fig.2, to each point $\mathbf{P} \in \bar{\Xi}$ corresponds, through Ξ, one point $\mathbf{M}$ on the boundary of the link such that:

$$\forall \mathbf{P} \in \bar{\Xi} \Rightarrow \mathbf{M} \in \partial R,\ M_x = P_x, \hat{\mathbf{M}}_{yz} = \hat{\mathbf{P}}_{yz} \quad (20)$$

Applying definition 5, the deformation vector field Δ is given as follows:

$$\Delta = \left\{ \mathbf{\Delta} : \forall \mathbf{M} \in \partial R,\ \Delta_x = 0, \mathbf{\Delta}_{yz} = \hat{\mathbf{P}}_{yz} \cdot (P_{yz} - P_{yz}^h) \right\} \quad (21)$$

It is easy to verify that $\Delta = \Delta(\Xi_h, I)$, as expected by the general DVZ theory for non-rigid bodies.

3.3 Generation of Reflex Command

We develop the fundamental equation to generate the reflex command by applying the DVZ control algorithm for non-rigid bodies. The goal of the control law is to minimise the norm of the deformation vector $\mathbf{\Delta}$, expressible from (21) as:

$$\Delta = P_{yz} - \left(\frac{1}{\sqrt{\hat{\mathbf{p}}_{yz}^T \cdot \mathbf{A} \cdot \hat{\mathbf{p}}_{yz}}} + d_{min} \right) \quad (22)$$

where we have substituted (12). Differentiating (22) yields:

$$\dot{\Delta} = \nabla_\pi[\Delta]\dot{\pi} + \nabla_I[\Delta]\dot{I} + \nabla_\eta[\Delta]\dot{\eta} \quad (23)$$
$$\dot{\Delta} = A \cdot \phi + B \cdot \psi + C \quad (24)$$

where $A = \nabla_\pi[\Delta]\dot{\pi}$, $B = \nabla_I[\Delta]$ and $C = \nabla_\eta[\Delta]\dot{\eta}$.
We assume that distance measurements (ρ) towards m directions ($\mathbf{u}$) of the three-dimensional space surrounding the mobile manipulator are available. We define $I = [\rho_1, \rho_2, \ldots \rho_m]^T$, where ρ_* is the measurement towards $\mathbf{u}_*$ and $\mathbf{u}_*$ is considered fixed in Σ_S. It follows that $\psi = \dot{I} = [\dot{\rho}_1, \dot{\rho}_2, \ldots \dot{\rho}_m]^T$. As already mentioned we have no control capabilities on ρ (i.e. I), being induced by the environment.

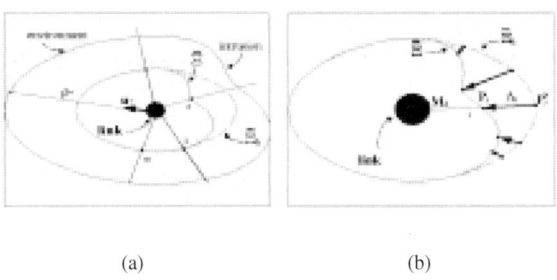

(a)   (b)

Figure 3: Sampled Deformation of DVZ

Therefore (24) represents a matrice equation corresponding to the system of equations below:

$$\begin{cases} \dot{\Delta}_1 = A_1 \cdot \phi + B_1 \cdot \psi_1 + C_1 \\ \dot{\Delta}_2 = A_2 \cdot \phi + B_2 \cdot \psi_2 + C_2 \\ \quad \vdots \\ \dot{\Delta}_m = A_m \cdot \phi + B_m \cdot \psi_m + C_m \end{cases} \quad (25)$$

Since we have knowledge of the environment towards m fixed directions, the $i-th$ equation of system (25) represents the DVZ principle applied to point M_i of the link, defined as in (20) for each of the m point $\mathbf{P}_i$ belonging to $\overline{\Xi}$ (refer to fig.3).

Considering each link constituting the manipulator, an equation on the form of (24) is calculated leading to the following system:

$$\begin{cases} \dot{\Delta}_{link1} = A_{link1} \cdot \phi + B_{link1} \cdot \psi + C_{link1} \\ \dot{\Delta}_{link2} = A_{link2} \cdot \phi + B_{link2} \cdot \psi + C_{link2} \\ \vdots \\ \dot{\Delta}_{linkn} = A_{linkn} \cdot \phi + B_{linkn} \cdot \psi + C_{linkn} \end{cases} \quad (26)$$

that we can compact in a matrice form as:

$$\dot{\Delta}_{arm} = A_{arm} \cdot \phi + B_{arm} \cdot \psi + C_{arm} \quad (27)$$

This result represents the complete evolution of the deformation for the whole manipulator. The DVZ control algorithm is next applied choosing the following control law:

$$\dot{\Delta}_{arm}^{des} = -K_p \cdot \Delta_{arm} - K_d \cdot \dot{\Delta}_{arm} \quad (28)$$

The collision avoidance control vector is given as:

$$\check{\phi} = A_{arm}^{+} \cdot \left(\dot{\Delta}_{arm}^{des} - B_{arm} \cdot \psi - C_{arm} \right) \quad (29)$$

and is then applied to the mobile manipulator.

4 Simulation Results

In this section we present some simulation results. A graphic simulator has been developed and implemented to test the effectiveness of the proposed approach for collision avoidance from either a computational and behavioural point of view. The simulator constitutes an interface that allows us to interactively change the position of the obstacles, the intrinsic parameters of the DVZ, the mobile manipulator configuration and initial state. It permits as well to visualize the three-dimensional motion of the mobile manipulator and the geometry of the DVZ. We discuss the behaviour of a simulated mobile manipulator looking at some typical situations.

In the following experiments, we consider an holonomic mobile base, moving in a two-dimensional space, together with a two-link planar manipulator. The mobile manipulator moves straight forward from initial point **A** to final point **B** at constant speed, keeping the arm still in its initial position. If an unscheduled obstacle is found on the path, the collision avoidance algorithm takes over, the reactive behaviour emerges and the configuration of the mobile manipulator changes in consequence. The mobile robot is equipped with 48 simulated ultrasonic sensors, 24 fitted on the mobile base and 24 on the terminal link, pointing to the surrounding two-dimensional space and measuring the distance towards a fixed direction. Manipulator interferences on sensor measurements are not taken in account. Ultrasonic sensors have already been used in experiments carried out on real mobile robots for reflexive collision avoidance using the DVZ algorithm[8]. The particular choice of sensorial equipment does not limit the application of the method. In fact, for mobile manipulators other systems of measurement could be used such as a stereoscopic-vision system, which measures the distance of the obstacles towards all directions within the scope of the cameras[9]. This is the actual ongoing research project of our laboratory.

Comparison between different situations are carried out. The figures show a mosaic of the mobile manipulator positions at every sample time. An intensification of darkness in the mobile base trail represents a decrease of speed.

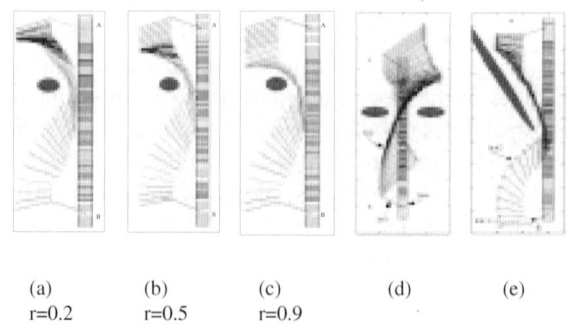

(a) r=0.2 (b) r=0.5 (c) r=0.9 (d) (e)

Figure 4: 2dof manipulator, 1dof base

In fig.4 the mobile base is allowed to move backward and forward only and the manipulator is two-link planar, where only the second link is shown for clearness. The control vector is therefore the three-fold vector $\pi = [\theta_1, \theta_2, v_x]^T$. The same simulation with different values of risk are compared. From 4(a) to 4(c) an increasing value of risk is chosen, consequently the size of the DVZ is smaller and the reflexive behavior emerges later (closer to the obstacle). In 4(d) two obstacles are considered, simulating a narrow passage as in entering through a door and in 4(e) a wall-like surface is introduced.

In fig.5 comparison between particular situations are presented. Fig.5(a) points out the need to attentively choose the sensing equipment, in order to avoid unreliable situation leading to undesired collisions. The big obstacle on the right shadows the mobile base ultrasonic sensors, preventing the manipulator's second link's DVZ to sense the intrusion. Equipping the robotic manipulator with (ultrasonic) sensors fitted on the end-effector, for example, avoids this

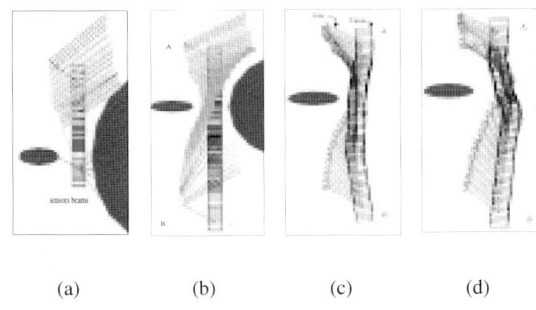

Figure 5: mobile manipulator passing through doors

situation leading to the correct behaviour 5(b). In fig.5(c) the base is allowed to rotate about the vertical axis z and the non-holonomic constraint is introduced. A one-link manipulator is considered. The results show that the reflexive behaviour due to intrusion in the arm's DVZ would cause a rotation of the base rather than the link, as reasonable, leading to this sort of attraction towards the obstacle that represents a dangerous situation. Thus, the mobile base DVZ is introduced and the correct behaviour emerges again, fig.5(d).

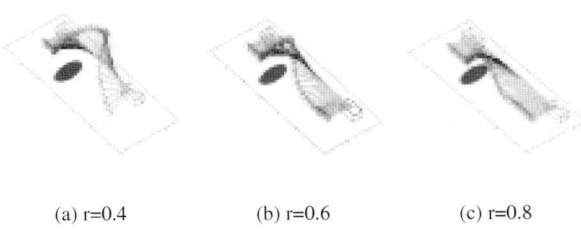

(a) r=0.4   (b) r=0.6   (c) r=0.8

Figure 6: Shoulder joint of antropomorph manipulator

Finally, fig.6 shows a one link manipulator with spherical joint(shoulder joint of an antropomorph robotic manipulator) avoiding an obstacle. The same simulation with different values of risk are compared confirming the good results and generality of the algorithm.

5 Conclusion

This paper has presented an algorithm for implementing collision avoidance reactive behaviours on mobile manipulators. The method is based on an extension of the DVZ principle and it has been developed by the authors. The approach consists of defining a Deformable Virtual Zone surrounding the mobile manipulator as a function of the motion capabilities of the robot. The DVZ is deformed by the interactions with the environment. The complete evolution equation of the deformation is calculated and the best control law is obtained. Simulation results show the effectiveness of the method. We plan to carry out experiments on real mobile manipulators using motion vision as a fundamental element in the perception of the environment in order to prove in practice the power of the proposed design. The proposed algorithm permits to pursue simultaneously both reactive behaviour, collision free interaction and short term target following in a sensor based environment. It must be underlined the DVZ principle cannot be implemented alone in the sense that it does not implement a planning procedure. These properties make the reflex algorithm amenable to control hierarchies and makes it possible to use it for preservation and as a building block in efficient path planning algorithms.

References

[1] T.L. Anderson, "Autonomous Robots and Emergent behaviours: A Set of Primitive behaviours for Mobile Robot Control", *IROS '90*, 1990, Tsukuba, Japon.

[2] R.A. Brooks, "Robot Beings", *International Workshop on Intelligent Robots and Systems '89*, September 1989, Tsukuba, Japon.

[3] D.T. Lawton, R.C. Arkin, J.M. Cameron, "Qualitative Spacial Understanding and Reactive Control for Autonomous Robot", *IROS '90*, 1990, Tsukuba, Japon.

[4] M. Soldo, "Reactive and Pre-Planned Control in a Mobile Robot", *Proc. IEEE Int. Conf. on Rob. and Aut.*, May 1990, Cincinnati, Ohio, USA.

[5] Holenstein, E. Badreddin, "Collision Avoidance in a Behaviour-Based Mobile Robot Design", *IEEE Int. Conf. on Rob. and Aut.*, April 1991 Sacramento, California, USA.

[6] T.S. Wilkman, M.S. Branicky, W.S. Newman, "Reflexive Collision Avoidance: A Generalized Approach", *IEEE Transactions on Robotics and Automation*, Vol. I, pp. 31-36, 1993.

[7] O. Kathib, "Real-Time Obstacle Avoidance for Manipulators and Mobile Robots", *Proc. IEEE Int. Conf. on Rob. and Aut.*, 1985, pp.500-505

[8] R. Zapata, P. Lepinay, "Reactive Behaviors of Fast Mobile Robots", *Journal of Robotic Systems,* January 1994.

[9] R.Zapata, P. Lepinay, L. Torres, J. Droulez, V.Creuze, "Prototyping a biologically-Plausible Vision System for Robotic Applications", *SBCCI 2000 13th Int. Symp on Integrated Circuits and Systems Design*, September 18-24, 2000, Manaus, Brazil.

The system development of unmanned vehicle for the tele-operated system interfaced with driving simulator

Duk-Sun Yun, Jae-Heung Shim, Min-Seok Kim, Young-Hoon Park, Jung-Ha Kim

Graduate School of Automotive Engineering, Kookmin University
861-1 Chungnung-dong, Sungbuk-gu,
Seoul, 136-702, KOREA

E-mail : yds@mecha.kookmin.ac.kr

URL : http://mecha.kookmin.ac.kr

Abstract

In this paper, the integration of driving simulator and unmanned vehicle by means of new concept for better performance is suggested. But, autonomous navigation system is one of the most difficult research topics from the point of view of several constrains on mobility, speed of vehicle and lack of environmental information. In these day, however, many innovations on the vehicle provide the appropriate automatic control in vehicle subsystem for reducing human error. This tendency is toward to the unmanned vehicle or the tele-operated vehicle ultimately.

The master system has host computer and simulator and slave system is electronic vehicle system. The slave vehicle system consists of three parts. First, laser sensor system for keeping the front sensory system and ultra sonic sensor system for keeping the side escaping collision. Second, acceleration system and brake control system for longitudinal motion control. Third, steering control system for lateral motion control. In this research, mechanical and electronic parts are implemented to operate unmanned vehicle as a whole-integrated system.

Driving simulator has a 6 degree-of-freedom motion-base and 3 channel fixed-based simulator. They are fully interactive, highly realistic and based on personal computer can be updated easily. Driving simulator is constituted by a vision system, a sound system, the system for control force loading and the 6-axis Stewart platform for motion base. This paper focuses on the integration of remote controlled unmanned vehicle and driving simulator. The vehicle mainly controlled lateral direction and longitudinal direction with actuators for controlling vehicle movement and sensors for closed-loop system we build intelligent system.

KEYWORD: Unmanned vehicle, Driving Simulator, Stewart platform, Tele-operated vehicle

1 Introduction

ITS(Intelligent Transportation System) is in fashion over the world rapidly. In addition to this fact, it has so many branches to be specialized in more detailed theme. To reincarnate AVCS(Advanced Vehicle Control System), the government of country related infrastructures, so called public facilities such as beacons, guided magnetic cores and radio stations for the wireless data communication are prerequisite. Autonomous vehicle can drive itself to the desired destination without aid of social infra and driver operation, so it is possible to move without help of other equipments all but sensor or actuator. This is an eventual objective of vehicle researches and developments of the implementation of autonomous vehicles. But the system stability is uncertain and the expenditure is so expensive that is cannot be go with itself yet. Although the performance of accident avoidance and maximization of road efficiency are attractive research theme, the problem of appropriate resolution of sensor, without disturbing computing power for the vehicle management and fast response of actuator for the control of vehicle motion must be solved. In this research, Laser radar or scanner detects long distance obstacles and makes out the front environment of a vehicle. Ultrasonic sensors can check the rear and side circumstances of the vehicle, which are relatively short distance. CCD camera can help autonomous vehicle keeping the lane, moreover; controlling the lateral motion of the vehicle, but it is used to gain the image data only in this research. Steering actuator also controls the lateral motion and D/A converter generates driving signal and DC motor mounted near the brake pedal are able to control the longitudinal motion of unmanned vehicle.

Driving Simulator(D.S.) developed in Kookmin University in Korea (KMU DS-1) has a 6 degree-of-freedom motion-base and 3 channel fixed-based simulator. They are fully interactive, highly realistic and based on Personal Computer can be updated easily.

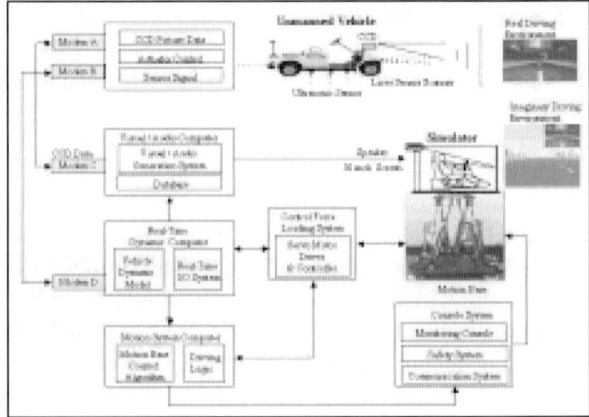

Fig. 1 Concept of tele-operated vehicle system for a driving simulator

Fig. 2 Motion platform (KMU DS-1)

It is developed for the vehicle system development, ITS, traffic safety evaluation and human factor (e.g., driving characteristic analysis of drunken drivers). It has a several research topic on real-time vehicle simulation, visual and audio system, motion control system, control force loading system and system integration, etc.

Fig. 3 System Setup of the unmanned vehicle

Fig. 3 shows the unmanned vehicle system ready for experiment, which is to drive vehicle itself and longitudinal control test, etc. Tele-operated vehicle system connected with driving simulator is a new research concept related to autonomous vehicle, but previous research has been just focused on monitoring of vehicle motion for the evaluation of suspension, steering, corner work and vehicle endurance test on the proving ground. In this paper, we proposed that tele-operated vehicle connected with driving simulator controlled by a driver on driving simulator. The motion cue generated by a driver is transmitted to the vehicle for the tele-operation via RF(Radio Frequency) module and it should be react immediately. Figure 1 shows the concept of tele-operated vehicle system, interfaced with a driving simulator. A data generated from the vehicle and real landscape image at the front of the vehicle should be propagated to driving simulator again, because of the real time motion control, enhancing realistic actuality. [1] [4] [7]

2 Hardware System

2.1 Longitudinal Control System

2.1.1 Velocity Control

There are two control methods for velocity control in this section. One is the linear position control of acceleration pedal, the other is the direct control of DC motor input signal. The former is easy to drive and monitor but hard to control, the later is easy to control, because the signal is generated by algorithm. Vehicle velocity calculated from the spinning wheel, which is produced by counting a rising signal edge at unit time. It is stored as a velocity data per unit time and displayed. On a basis of this data, the total driving distance is calculated and velocity profile of vehicle locus is generated for the path planning.

The basic notion of brake control is the master cylinder control linked with a brake pedal of passenger car. But, we choose DC motor as main actuator of brake control system on the test vehicle, indirectly. Because the master cylinder control is very difficult and dangerous, EV (Electronic Vehicle) is recommended.

In this paper, the vehicle for the experiment is electronic vehicle like a utility car, which can be seen in a golf ground or airport, etc.

Fig. 4 DC motor controller system

Fig. 4 shows DC Motor controller system for the drive actuator wired with DC Motor and this controller give an voltage signal to the motor, so it generates the torque, which make the wheel to rotate.[8][14] The displacement control of vehicle means wheel angle control, so the velocity can be calculated from the number of wheel spin at unit time. To meet with the requirement of safety, there are some prerequisite.

1) The vehicle should hold on desired velocity.

2) The vehicle should maintained a safety distance to the preceding vehicle and avoid the obstacle in front of the vehicle, showing abruptly.

3) The vehicle should have a riding comport, which is come from the limitation of acceleration.

2.1.2 Brake Control

To control vehicle velocity and acceleration and to stop the vehicle, the brake control is inevitable. The brake system is shown in figure 5, which shows that DC motor attached on the beneath side of vehicle floor and geared with brake pedal, wiring with brake drum in the rear wheel. The strategy of longitudinal control is mainly focused on the displacement, because of the time delay of signal processing. The displacement of locus at each unit time enable a velocity to be calculated with minimized time delay for the management algorithm. It is faster than the integration algorithm calculating vehicle displacement from velocity. In case of on/off point control of brake, the steel wire tied with brake pedal is simply pulled by the rotational force of DC motor, so that the unmanned vehicle stops. To derive the best performance and avoid discontinuity, on/off point control is executed as fast as possible. The evaluation of system performance is judged by comparing acceleration data generated by a driver and the brake control system. But, it requires so much computing power, yielding insufficient system resources.

Fig. 5 DC Motor Hardware System for Brake

In the braking control mode, PWM (Pulse Width Modulation) control is recommended, because the time of maintaining up level, which means ON signal of braking DC motor is defined by a parameter in algorithm, which is conformable to the velocity variation dependent on time. In addition to this fact, PWM enable the system to revise a current velocity. If the distance from the forward vehicle becomes less than 2/3 times of safety distance, unmanned vehicle will apply the brake control system and it should keep the safety distance from the forward vehicle for the safety. [5] [6]

2.2 Lateral Control System

2.2.1 Steering Control

Fig. 6 2Phase Step Motor System for Lateral Control

Figure 6 shows system setup for the steering control with step motor attached on the steering column, which has a reliable angular displacement and angular velocity, even acceleration. In order to keep the lane for the lateral control, unmanned vehicle needs standard signal as a reference for going straightly without leaning to one side. It might generate a topological error of navigation. A gyroscope can substitute the reference signal especially in case of a straight lane. Before activating the steering control system, unmanned vehicle must check the destination lane and verify the safety. Figs 7, 8 show that the changing values of heading angles are different from CW and CCW respectively, because the steering alignment is unstable and the calibrations of KVH Digital Gyro make the condition of steering test different. If any unbalance of the both graphs may occur, the whole increasing figures are similar with each graph. [1][10][12][13]

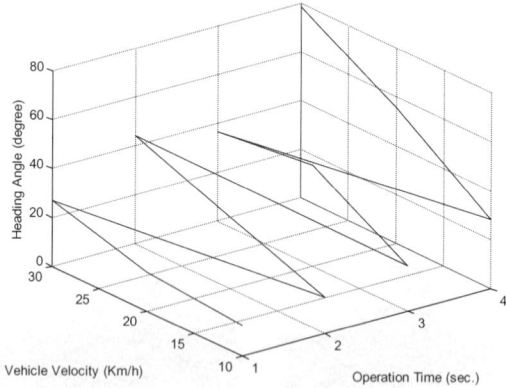

Fig. 7 Characteristic Graph of Steering System (CW)

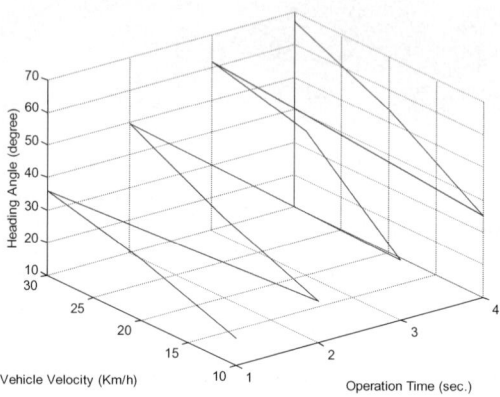

Fig. 8 Characteristic Graph of Steering System (CCW)

2.3 Sensory System & Algorithm

2.3.1 Laser Sensor Control

Laser sensor is used to sense the distance inter vehicles and detect an obstacle, appearing in front of the unmanned vehicle, suddenly. Figure 9 shows the laser sensor, which is manufactured by Silicon Heights Ltd., 300VIN model, attached on the front side of unmanned vehicle and. Laser sensor used for the longitudinal control, because it produces data stream for keeping the system safety distance from the collision with front vehicle and obstacle. But, it has so many error factors such as noise coming from electric magnetic interference, so called EMI, of D.C. motor. Figure 10 shows the characteristic of the laser sensor, which is to shoot the beam to the lower side from the sensor position, so that the sensor should be set on the vehicle, 1~1.5[m] from the ground. [9][10]

Fig. 9 System setup of Laser sensor

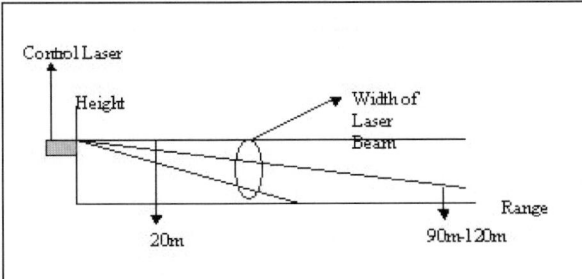

Fig. 10 Laser sensor system characteristic

2.3.2 Ultra Sonic Sensor Control

Fig. 11 Ultrasonic sensor mounting

Several sensors, which is a hall sensor, a laser range finder and ultrasonic sensors, help that the unmanned vehicle recognizes road situation. With a hall sensor, which is contacted on rear wheel dist drum, we can calculate vehicle velocity by counting the rising or the falling signal edges and dividing into the sampling time.
The laser range finder measures remote obstacle. (maximum 120[m]) A distance signal is processed with PCL-7122 I/O interfacing card. In this research, we use two ultrasonic sensors because the test field is straight corridor of inside building. So, the ultrasonic sensors detect the both side of wall around corridor, but it has so narrow space that the vehicle moves freely. Before we use ultrasonic sensor, we had to verify each sensor's characteristic. The distance signal of ultrasonic sensor is converted into digital data by A/D converter and interfaced to the main computer with PCL-7122.

2.3.3 System algorithm for unmanned vehicle

Figure 12 represents the system algorithm for the unmanned vehicle driving test. This algorithm can be extended to the inertial navigation (Dead Reckoning), because sensing the position is to be displayed on the digital map as information for driver. [11]

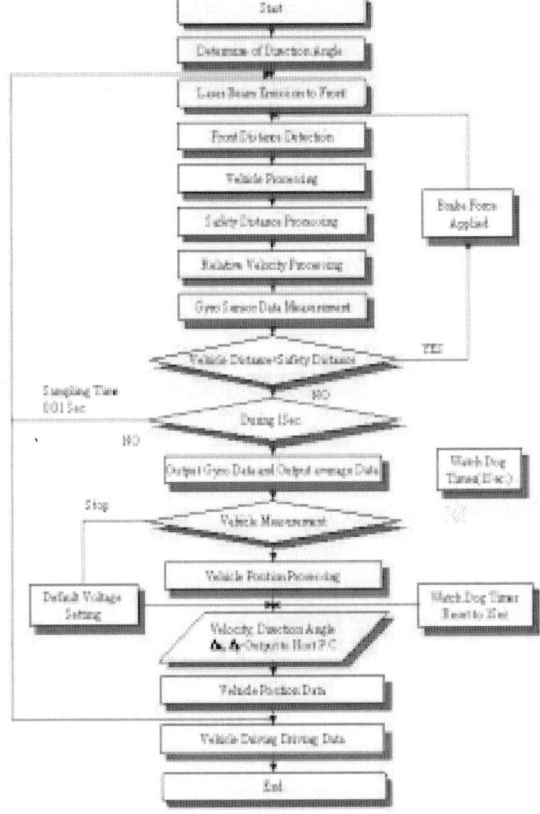

Fig. 12 Vehicle Collision Avoidance Algorithm Flowchart

3 Test Evaluation

The test is fully operated unmanned vehicle only and show the differences between unmanned vehicle driving mode with manual mode, an expert driver's driving mode is performed and scrutinized at the same test conditions. The experiment of vehicle motion, angular position, roll angle and pitching angle, is evaluated though a compass.

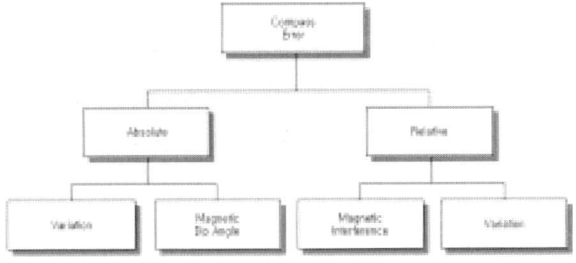

Fig. 13 The classification of compass error terms

The compass used in the experiment is the KVH-C100 Fluxgate Digital Compass made by KVH Industries. But, it has a few error term, which is classified greatly two parts, absolute and relative errors. [2] All of these errors should be eliminated by a filtering algorithm and hardware compensation, magnetic shielding, positioning the compass on the vehicle at its intended location, etc. [2]

Figure 14 shows the unmanned vehicle driving angular data in the test ground. The initial orientation of vehicle, so called initial heading angle is 220[deg] from real north side. In comparison with this result, Figure 14 and 16 show the vehicle angular data and Figure 16 and 17 show the angular velocity data trend of unmanned vehicle driving test and that of human driving test, individually. But, the difference of these test result show the response characteristic of a human driver and unmanned vehicle, managed by sensory system, algorithm and actuator's reaction. Unmanned vehicle is dependent on the performance of actuator response and sensitivity and sensory system, too. [1][3]

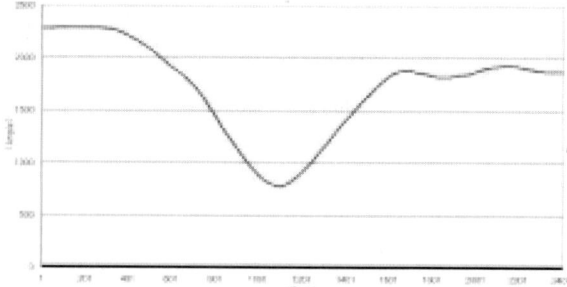

Fig.14 Angular locus of Unmanned Vehicle

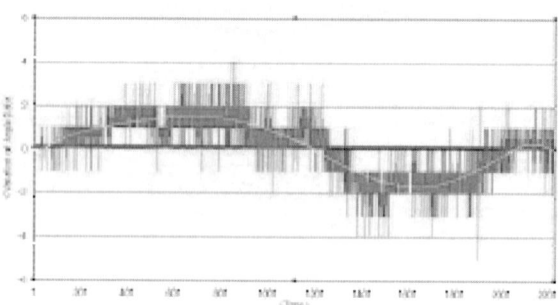

Fig. 15 Angular data trend and interpolation of data

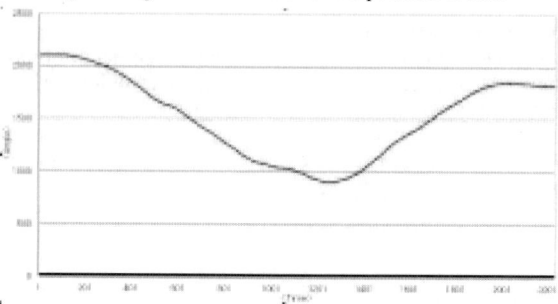

Fig. 16 Angular locus of manual driving test

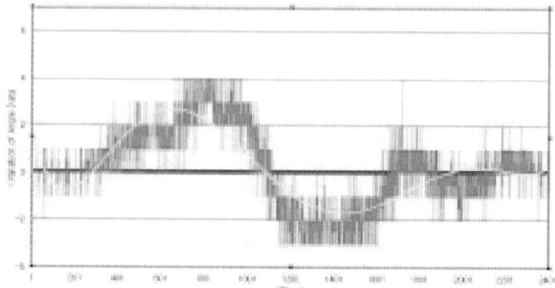

Fig. 17 Angular data trend and interpolation of data

In figure 18, the data show that the yaw motion of the vehicle is the most significant terms of each axis, however, the data of roll and pitching motion trend to be minute. Figure 20 show the acceleration data of the vehicle during the lane change motion, too. The data show the yawing motion (X axis) is significant while the vehicle changing the lane, but the pitching is so trivial that neglected. Pitching motion (Y axis) of the braking is so big that it can be recognized easily. The data means deceleration.

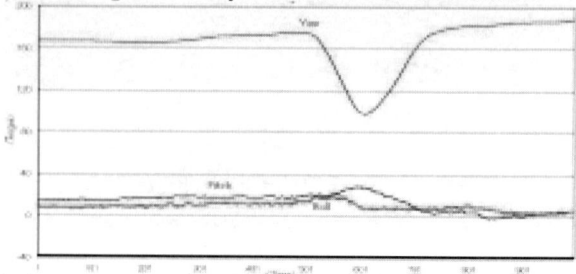

Fig. 18 Unmanned vehicle driving test - 3 axis
(Yaw, Pitch and Roll angle)

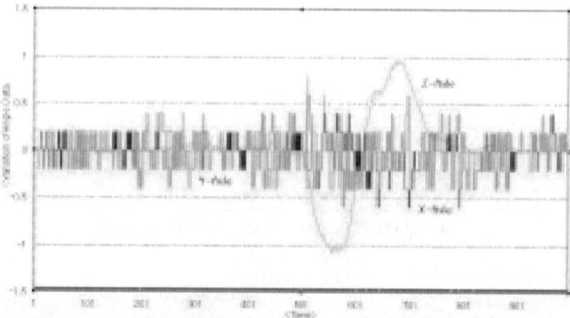

Fig. 19 Unmanned vehicle driving test showing the data trend of 3 axis

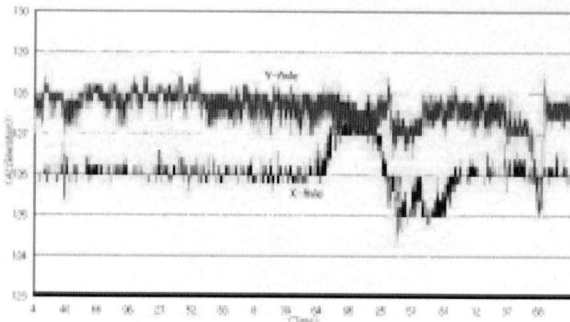

Fig. 20 Acceleration data of unmanned vehicle

Figs. 18, 19 and 20 show the experimental results of lane change mode test of unmanned vehicle. In fig. 18, the data show angular locus, which represents a absolute angle from real north, geographically. Pitching motion means nose down of the vehicle during the lane changing and roll motion means the characteristic of suspension, tire and linkage dynamics. These figures show accuracy of the sensors and how we adapted the sensors to circumstance for measuring. Fig. 21 shows the test result of laser range finder. We started to measure an obstacle in 25 meters ahead. The vehicle was moving to the obstacle at 8[km/h] speed. And we kept the measurement until the vehicle stopped in front of the obstacle. In that result, we could get the accuracy of the laser range finder. The error rate of laser range finer was ±4 percents.

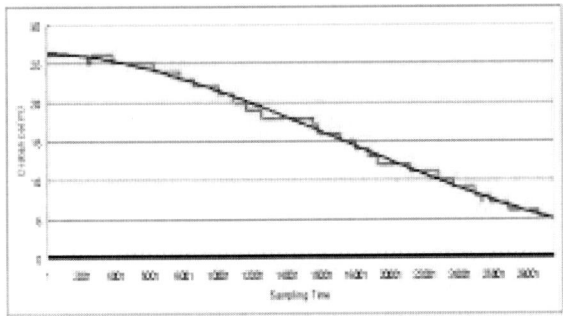

Fig. 21 Range change ratio in same speed

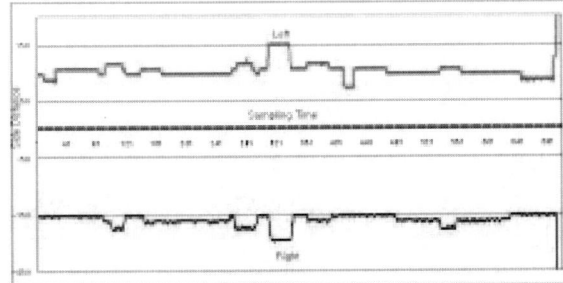

Fig. 22 Range change ratio in same speed

Fig 22 is the test road locus, which is made by ultrasonic sensors. We attached sensors to both sides of the vehicle. While the vehicle went through road, ultrasonic sensors got the distance data of a space between the vehicle and the wall. In case of ultrasonic sensor, Data are relatively precise. But the data interpolation should apply as each characteristic of ultrasonic sensor.

4 Conclusion

The work presented in this paper is mainly dealt with the basic concept, software and hardware implementation and development of unmanned vehicle. The analysis of experiment for each part is as follows:

Longitudinal Control: The wheel spinning data, streaming of pulse signal processing are combined for the implementation of vehicle velocity control. The electronic vehicle system has the advantage of easy manipulation and smart operation. Brake control should be applied carefully because of safety reason. On/Off control is applied and PWM control is experimenting. There is some lack of performance efficiency by On/Off point control. The mechanical structure of brake system is needed to upgrade for better performance. The bottom line is acceleration and deceleration range should be in the zone of $|\alpha| < 1.5$ for the safety.

Lateral Control: Steering control is the main system of lateral control and enables the vehicle to change lane and avoid collision. The normal lane change is easy but it should be careful whether the destination lane is safe or not, all the ways. So more sensory systems are needed to get the information of vehicle environment. For the collision avoidance, adaptive intelligent cruise control methods with laser range finder are applied. So unmanned vehicle can get the information of forward vehicle and maintain the reasonable safety distance.

By comparing with both results of test manual driving and unmanned vehicle driving, unmanned vehicle actuator response should be followed the manual driving mode, because the delay and monitoring time is slower than that of human. It can be solved by fuzzy logic. Several test make it possible to build the rule base and the vehicle to find the desired distance and velocity faster.

The sensor part: The laser range finder should higher resolution of laser range finder is demanded.
The ultrasonic sensor has to adjust gain value of each ultrasonic sensor. A Sensor for absolute vehicle headway direction is demanded for precise steering control without error.

Acknowledgements

This work was supported by the Brain Korea 21 Project.

References

1. Duk-Sun Yun, Jung-Ha Kim, "The system integration of unmanned vehicle & driving simulator for the tele-operated vehicle system", Intl. Conf. of IASTED on RA2000, Aug. 14, 2000, 197-201.

2. Lauro Ojeda, Johann Borenstein, "Experimental Results with the KVH-C100 Fluxgate Compass in Mobile Robots", Intl. Conf. of IASTED on RA2000, Aug. 14, 2000, 156-162.

3. Duk-Sun Yun, Jung-Ha Kim, "The Sensor Fusioning of Gyro Controller System for the Smart Vehicle", Intl. Conf. of IEEE on MFI, 1999, 189-194.

4. Woon-Sung Lee, "The Kookmin University Driving Simulators for Vehicle Control System Development and Human Factor Study", Driving Simulation Conference, 1999, 75-86.

5. Katsuhiko Ogata, *Modern Control Engineering*, Prentice Hall International, 1997.

6. Richard C. Dorf, *Modern Control Systems*, Addison-Wesley Publishing Company, 1995.

7. Gene F. Franklin, "Feedback Control of Dynamic Systems," Addison-Wesley Publishing Company, 1994.

8. Tak Kenjo, *Electric Motors and Their Controls*, Oxford University Press Inc., 1994.

9. Christopher O. Nwagboso, *Automotive Sensory systems*, Chapman & Hall, 1993, pp. 223-240.

10. Min-Cheul Kim, "Steering Control of a Vehicle Based on Kinematic Modeling," Pohang University of Science and Technology, 1993, pp. 16-20.

11. Mohammad Jamshidi, *Computer-Aided Analysis and Design of Linear Control Systems* (Prentice Hall, 1992)

12. R. w. Allen et. al., "Steady State and Transient Analysis of Ground Vehicle Handling", SAE Paper 87049, 1987.

13. Shibahata, et. al., "The Development of an Experimental Four-Wheel-Steering Vehicle, " SAE Paper 860623, 1986.

14. Benjamin C. Kuo and Jacob Tal, *DC Motors and Control Systems*, SRL Publishing Company, Champaign, Illinois, 1978.

Motion Planning for Humanoid Robots Under Obstacle and Dynamic Balance Constraints

James Kuffner, Koichi Nishiwaki, Satoshi Kagami, Masayuki Inaba, Hirochika Inoue

Dept. of Mechano-Informatics
The University of Tokyo
{kuffner,nishi,kagami,inaba,inoue}@jsk.t.u-tokyo.ac.jp

Abstract

We present an approach to path planning for humanoid robots that computes dynamically-stable, collision-free trajectories from full-body posture goals. Given a geometric model of the environment and a statically-stable desired posture, we search the configuration space of the robot for a collision-free path that simultaneously satisfies dynamic balance constraints. We adapt existing randomized path planning techniques by imposing balance constraints on incremental search motions in order to maintain the overall dynamic stability of the final path. A dynamics filtering function that constrains the ZMP (zero moment point) trajectory is used as a post-processing step to transform statically-stable, collision-free paths into dynamically-stable, collision-free trajectories for the entire body. Although we have focused our experiments on biped robots with a humanoid shape, the method generally applies to any robot subject to balance constraints (legged or not). The algorithm is presented along with computed examples using the humanoid robot "H6".

1 Introduction

Recently, significant progress has been made in the design and control of humanoid robots, particularly in the realization of dynamic walking in several full-body humanoids [11, 37, 27]. As the technology and algorithms for real-time 3D vision and tactile sensing improve, humanoid robots will be able to perform tasks that involve complex interactions with the environment (e.g. grasping and manipulating objects). The enabling software for such tasks includes motion planning for obstacle avoidance, and integrating planning with visual and tactile sensing data. To facilitate the deployment of such software, we are currently developing a graphical simulation environment for testing and debugging [19].

This paper presents an algorithm for automatically generating collision-free dynamically-stable motions from full-body posture goals. It expands upon the preliminary algorithm developed in [21], and has been tested and verified on the humanoid robot hardware platform "H6". Our approach is to adapt techniques from

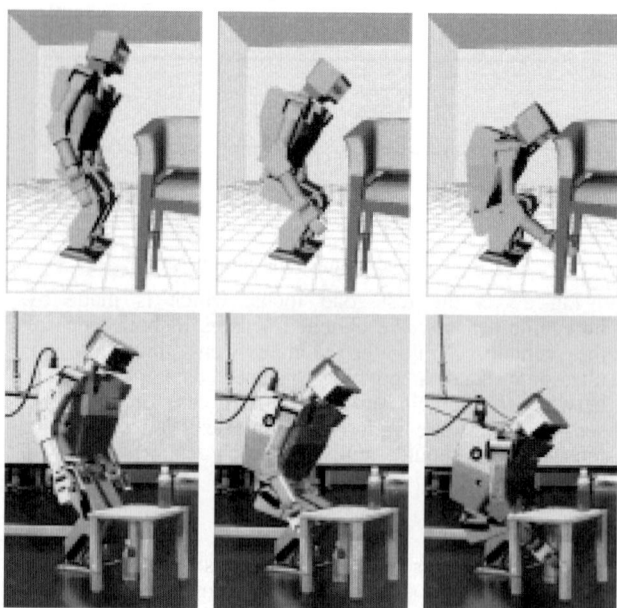

Figure 1: Dynamically-stable motion for retrieving an object (*top*: simulation, *bottom*: real robot hardware).

an existing, successful path planner [20] by imposing balance constraints upon incremental motions used during the search. Provided the initial and goal configurations correspond to collision-free, statically-stable body postures, the path returned by the planner can be transformed into a collision-free and dynamically-stable trajectory for the entire body. To the best of our knowledge, this paper represents the first general motion planning algorithm for humanoid robots that has also been experimentally confirmed on real humanoid robot hardware.

2 Background

Due to the complexity of motion planning in its general form [32], the use of complete algorithms [5, 12, 33] is limited to low-dimensional configuration spaces. This has motivated the use of heuristic algorithms, many of which employ randomization (e.g., [1, 2, 3, 4, 14, 15, 17,

20, 26]). Although these methods are incomplete, many have been shown to find paths in high-dimensional configuration spaces with high probability.

Motion planning for humanoid robots poses a particular challenge. Developing practical motion planning algorithms for humanoid robots is a daunting task given that humanoid robots typically have 30 or more degrees of freedom. The problem is further complicated by the fact that humanoid robots must be controlled very carefully in order to maintain overall static and dynamic stability. These constraints severely restrict the set of allowable configurations and prohibit the direct application of existing motion planning techniques. Although efficient methods have been developed for maintaining dynamic balance for biped robots [36, 31, 30, 16], none consider obstacle avoidance.

Motion planning algorithms that account for system dynamics typically approach the problem in one of two ways: 1) decoupling the problem by first computing a kinematic path, and subsequently transforming the path into a dynamic trajectory, or 2) searching the system state-space directly by reasoning about the possible controls that can be applied. The method presented in this paper adopts the first approach. Other methods using one of these two planning strategies have been developed for off-road vehicles[35, 6], free-flying 2D and 3D rigid bodies[23, 24], helicopters and satellites[8], and for a free-flying disc among moving obstacles[18]. None of these previous methods have yet been applied to complex articulated models such as humanoid robots. One notable exception is the VHRP simulation software under development[28], which contains a path planner that limits the active body degrees of freedom for humanoid robots for simultaneous obstacle avoidance and balance control. Since the space of possible computed motions is limited however, this planner is not fully general.

3 Dynamically-stable Motion Planning

Our approach is to adapt a variation of the randomized planner described in [20] to compute full-body motions for humanoid robots that are both dynamically-stable and collision-free. The first phase computes a statically-stable, collision-free path, and the second phase smooths and transforms this path into a dynamically-stable trajectory for the entire body. The planning method (RRT-Connect) and its variants utilize Rapidly-exploring Random Trees (RRTs) [22, 24] to connect two search trees, one from the initial configuration and the other from the goal. This method has been shown to be efficient in practice and converge towards a uniform exploration of the search space.

Robot Model and Assumptions: We have based our experiments on an approximate model of the H6 humanoid robot (see Figure 1), including the kinematics and dynamic properties of the links. Although we have focused our experiments on biped robots with a humanoid shape, the algorithm generally applies to any robot subject to balance constraints (legged or not). Aside from the existence of the dynamic model, we make the following assumptions:

1. **Environment model:** We assume that the robot has access to a 3D model of the surrounding environment to be used for collision checking.

2. **Initial posture:** The robot is currently balanced in a collision-free, statically-stable configuration supported by either one or both feet.

3. **Goal posture:** A full-body goal configuration that is both collision-free and statically-stable is specified. The goal posture may be given explicitly by a human operator, or computed via inverse kinematics or other means.

4. **Support base:** The location of the supporting foot (or *feet* in the case of dual-leg support) does not change during the planned motion.

3.1 Problem Formulation:

Our problem will be defined in a 3D world $\mathcal{W}$ in which the robot moves. $\mathcal{W}$ is modeled as the Euclidean space $\Re^3$ ($\Re$ is the set of real numbers).

Robot: Let the robot $\mathcal{A}$ be a finite collection of p rigid links $\mathcal{L}_i$ ($i = 1, \ldots, p$) organized in a kinematic hierarchy with Cartesian frames $\mathcal{F}_i$ attached to each link. We denote the position of the center of mass c_i of link $\mathcal{L}_i$ relative to $\mathcal{F}_i$. A *pose* of the robot is denoted by the set $\mathcal{P} = \{T_1, T_2, \ldots, T_p\}$, of p relative transformations for each of the links $\mathcal{L}_i$ as defined by the frame $\mathcal{F}_i$ relative to its parent link's frame. The *base* or *root* link transformation T_1 is defined relative to some world Cartesian frame $\mathcal{F}_{world}$. Let n denote the number of generalized coordinates or *degrees of freedom* (DOFs) of $\mathcal{A}$. A *configuration* is denoted by $q \in \mathcal{C}$, a vector of n real numbers specifying values for each of the generalized coordinates of $\mathcal{A}$. Let $\mathcal{C}$ be the *configuration space* or $\mathcal{C}$-space of $\mathcal{A}$. $\mathcal{C}$ is a space of dimension n.

Obstacles: The set of obstacles in the environment $\mathcal{W}$ is denoted by $\mathcal{B}$, where $\mathcal{B}_k$ ($k = 1, 2, \ldots$) represents an individual obstacle. We define the *$\mathcal{C}$-obstacle region* $\mathcal{CB} \subset \mathcal{C}$ as the set of all configurations $q \in \mathcal{C}$ where one or more of the links of $\mathcal{A}$ intersect (are in collision) with another link of $\mathcal{A}$, any of the obstacles $\mathcal{B}_k$. We also regard configurations $q \in \mathcal{C}$ where one or more *joint limits* are violated as part of the $\mathcal{C}$-obstacle region $\mathcal{CB}$. The open subset $\mathcal{C} \setminus \mathcal{CB}$ is denoted by $\mathcal{C}_{free}$ and its closure by $cl(\mathcal{C}_{free})$, and it represents the *space of collision-free configurations* in $\mathcal{C}$ of the robot $\mathcal{A}$.

Balance and Torque Constraints: Let $\mathcal{X}(q)$ be a vector relative to $\mathcal{F}_{world}$ representing the global position of the center of mass of $\mathcal{A}$ while in the configuration q. A configuration q is *statically-stable* if: 1) the projection of $\mathcal{X}(q)$ along the gravity vector g lies within the area of support $\mathcal{SP}$ (i.e. the convex hull of all points of contact between $\mathcal{A}$ and the support surface in $\mathcal{W}$), and 2) the joint torques Γ needed to counteract the gravity-induced torques $G(q)$ do not exceed the maximum torque bounds Γ_{max}. Let $\mathcal{C}_{stable} \subset \mathcal{C}$ be the subset of *statically-stable configurations* of $\mathcal{A}$. Let $\mathcal{C}_{valid} = \mathcal{C}_{stable} \cap \mathcal{C}_{free}$ denote the subset of configurations that are both *collision-free* and *statically-stable* postures of the robot $\mathcal{A}$. $\mathcal{C}_{valid}$ is called the set of *valid* configurations.

Solution Trajectory: Let $\tau : \mathcal{I} \mapsto \mathcal{C}$ where $\mathcal{I}$ is an interval $[t_0, t_1]$, denote a motion trajectory or *path* for $\mathcal{A}$ expressed as a function of time. $\tau(t)$ represents the configuration q of $\mathcal{A}$ at time t, where $t \in \mathcal{I}$. A trajectory τ is said to be *collision-free* if $\tau(t) \in \mathcal{C}_{free}$ for all $t \in \mathcal{I}$. A trajectory τ is said to be both *collision-free* and *statically-stable* if $\tau(t) \in \mathcal{C}_{valid}$ for all $t \in \mathcal{I}$. Given $q_{init} \in \mathcal{C}_{valid}$ and $q_{init} \in \mathcal{C}_{valid}$, we wish to compute a continuous motion trajectory τ such that $\forall t \in [t_0, t_1]$, $\tau(t) \in \mathcal{C}_{valid}$, and $\tau(t_0) = q_{init}$ and $\tau(t_1) = q_{goal}$. We refer to such a trajectory as a *statically-stable* trajectory.

Dynamic Stability: Theoretically, any statically-stable trajectory can be transformed into a dynamically-stable trajectory by arbitrarily slowing down the motion. For these experiments, we utilize the online balance compensation scheme described in [16] as a method of generating a final dynamically-stable trajectory after path smoothing (see Section 3.4).

Planning Query: Note that in general, if a dynamically-stable solution trajectory exists for a given path planning query, there will be many such solution trajectories. Let Φ denote the set of all dynamically-stable solution trajectories for a given problem. The query $Planner(\mathcal{A}, \mathcal{B}, q_{init}, q_{goal}) \longrightarrow \tau$, accepts as input the robot and obstacle models along with the initial and goal postures, and attempts to calculate a solution trajectory $\tau \in \Phi$. If no solution is found, τ will be empty (a null trajectory).

3.2 Path Search

Unfortunately, there are no currently known methods for explicitly representing $\mathcal{C}_{valid}$. The obstacles are modeled completely in $\mathcal{W}$, thus an explicit representation of $\mathcal{C}_{free}$ is also not available. However, using a collision detection algorithm, a given $q \in \mathcal{C}$ can be tested to determine whether $q \in \mathcal{C}_{free}$. Testing whether $q \in \mathcal{C}_{stable}$ can also be checked verifying that the projection of $\mathcal{X}(q)$ along g is contained within the boundary of $\mathcal{SP}$, and that the torques Γ needed to counteract gravitational torques $G(q)$ do not exceed Γ_{max}.

Distance Metric: As with the most planning algorithms in high-dimensions, a metric ρ is defined on $\mathcal{C}$. The function $\rho(q, r)$ returns some measure of the distance between the pair of configurations q and r. Some axes in $\mathcal{C}$ are weighted relative to each other, but the general idea is to measure the "closeness" of pairs of configurations with a positive scalar function.

For our humanoid robot models, we employ a metric that assigns higher relative weights to the generalized coordinates of links with greater mass and proximity to the trunk (torso): $\rho(q, r) = \sum_{i=1}^{n} w_i ||q_i - r_i||$. This choice of metric function attempts to heuristically encode a general relative measure of how much the variation of an individual joint parameter affects the overall body posture. Additional experimentation is needed in order to evaluate the efficacy of the many different metric functions possible.

Planning Algorithm: We employ a randomized search strategy based on Rapidly-exploring Random Trees (RRTs) [23, 20]. For implementation details and analysis of RRTs, the reader is referred to the original papers or a summary in [24]. In [21], we developed an RRT variant the generates search trees using a dynamics filter function to guarantee dynamically-stable trajectories along each incremental search motion.

The algorithm described in this paper is more general and efficient than the planner presented in [21], since it does not require the use of a dynamics filtering function during the path search phase. In addition, it can handle either single or dual-leg support postures, and the calculated trajectories have been verified using real robot hardware. The basic idea is the same as the RRT-Connect algorithm described in [20]. The key difference is that *instead of searching $\mathcal{C}$ for a solution path that lies within $\mathcal{C}_{free}$, the search is performed in $\mathcal{C}_{stable}$ for a solution path that lies within $\mathcal{C}_{valid}$*. In particular, we modify the planner variant that employs symmetric calls to the *EXTEND* function as follows:

1. The *NEW_CONFIG* function in the *EXTEND* operation checks balance constraints in addition to checking for collisions with obstacles (i.e. $q_{new} \in \mathcal{C}_{valid}$).

2. Rather than picking a purely random configuration $q_{rand} \in \mathcal{C}$ at every planning iteration, we pick a random configuration that also happens to correspond to a statically-stable posture of the robot (i.e. $q_{rand} \in \mathcal{C}_{stable}$).

Pseudocode for the complete algorithm is given in Figure 2. The main planning loop involves performing a simple iteration in which each step attempts to extend the RRT by adding a new vertex that is biased by a randomly-generated, statically-stable configuration (see Section 3.3). *EXTEND* selects the nearest vertex already in the RRT to the given configuration, q, with respect to the distance metric ρ. Three situations can occur: *Reached*, in which q is directly added to the RRT, *Advanced*, in which a new vertex $q_{new} \neq q$ is added to the RRT; *Trapped*, in which no new vertex is added due to the inability of *NEW_CONFIG* to generate a path segment towards q that lies within $\mathcal{C}_{valid}$.

NEW_CONFIG attempts to make an incremental motion toward q. Specifically, it checks for the existence of a short path segment $\delta = (q_{near}, q_{new})$ that lies entirely within $\mathcal{C}_{valid}$. If $\rho(q, q_{near}) < \epsilon$, where ϵ is some fixed incremental distance, then q itself is used as the new configuration q_{new} at the end of the candidate path segment δ (i.e. $q_{new} = q$). Otherwise, q_{new} is generated at

```
EXTEND(T, q)
1    q_near ← NEAREST_NEIGHBOR(q, T);
2    if NEW_CONFIG(q, q_near, q_new) then
3        T.add_vertex(q_new);
4        T.add_edge(q_near, q_new);
5        if q_new = q then Return Reached;
6        else Return Advanced;
7    Return Trapped;
```

```
RRT_CONNECT_STABLE(q_init, q_goal)
1    T_a.init(q_init); T_b.init(q_goal);
2    for k = 1 to K do
3        q_rand ← RANDOM_STABLE_CONFIG();
4        if not (EXTEND(T_a, q_rand) = Trapped) then
5            if (EXTEND(T_b, q_new) = Reached) then
6                Return PATH(T_a, T_b);
7            SWAP(T_a, T_b);
8    Return Failure
```

Figure 2: Pseudocode for the dynamically-stable motion planning algorithm.

a distance ϵ along the straight-line from q_{near} to q.[1] All configurations q' along the path segment δ are checked for collision, and tested whether balance constraints are satisfied. Specifically, if $\forall q' \in \delta(q_{near}, q_{new}), q' \in \mathcal{C}_{valid}$, then *NEW_CONFIG* succeeds, and q_{new} is added to the tree $\mathcal{T}$. In this way, the planner uses uniform samples of $\mathcal{C}_{stable}$ in order to grow trees that lie entirely within $\mathcal{C}_{valid}$.

Convergence and Completeness: Although not given here, arguments similar to those presented in [20] and [24] can be constructed to show uniform coverage and convergence over $\mathcal{C}_{valid}$.

Ideally, we would like to build a *complete* planning algorithm. That is, the planner always returns a solution trajectory if one exists, and indicates failure if no solution exists. As mentioned in Section 2, implementing a practical complete planner is a daunting task for even low-dimensional configuration spaces (see [13]). Thus, we typically trade off completeness for practical performance by adopting heuristics (e.g. randomization).

The planning algorithm implemented here is incomplete in that it returns failure after a preset time limit is exceeded. Thus, if the planner returns failure, we cannot conclude whether or not a solution exists for the given planning query, only that our planner was unable to find one in the allotted time. Uniform coverage and convergence proofs, though only theoretical, at least help to provide some measure of confidence that when an algorithm fails to find a solution, it is likely that no solution

exists. This is an area of ongoing research.

3.3 Random Statically-stable Postures

For our algorithm to work, we require a method of generating random statically-stable postures (i.e. random point samples of $\mathcal{C}_{stable}$). Although it is trivial to generate random configurations in $\mathcal{C}$, it is not so easy to generate them in $\mathcal{C}_{stable}$, since it encompasses a much smaller subset of the configuration space.

In our current implementation, a set $\mathcal{Q}_{stable} \subset \mathcal{C}_{stable}$ of N samples of $\mathcal{C}_{stable}$ is generated as a preprocessing step. This computation is specific to a particular robot and support-leg configuration, and need only be performed once. Different collections of stable postures are saved to files and can be loaded into memory when the planner is initialized. Although stable configurations could be generated "on-the-fly" at the same time the planner performs the search, pre-calculating $\mathcal{Q}_{stable}$ is preferred for efficiency. In addition, multiple stable-configuration set files for a particular support-leg configuration can be saved independently. If the planner fails to find a path after all N samples have been removed from the currently active $\mathcal{Q}_{stable}$ set, a new one can be loaded with different samples.

Single-leg Support Configurations: For configurations that involve balancing on only one leg, the set $\mathcal{Q}_{stable}$ can be populated as follows:

1. The configuration space of the robot $\mathcal{C}$ is sampled by generating a random body configuration $q_{rand} \in \mathcal{C}$.

2. Assuming the right leg is the supporting foot, q_{rand} is tested for membership in $\mathcal{C}_{valid}$ (i.e. static stability, no self-collision, and joint torques below limits).

3. Using the same sample q_{rand}, a similar test is performed assuming the left leg is the supporting foot.

4. Since most humanoid robots have left-right symmetry, if $q_{rand} \in \mathcal{C}_{valid}$ in either or both cases, we can "mirror" q_{rand} to generate stable postures for the opposite foot.

Dual-leg Support Configurations: It is slightly more complicated to generate statically-stable body configurations supported by both feet at a given fixed relative position. In this case, populating $\mathcal{Q}_{stable}$ is very similar to the problem of sampling the configuration space of a constrained closed-chain system (e.g. closed-chain manipulator robots or molecular conformations [25, 10]). The set $\mathcal{Q}_{stable}$ is populated with fixed-position dual-leg support postures as follows:

1. As in the single-leg case, the configuration space of the robot $\mathcal{C}$ is sampled by generating a random body configuration $q_{rand} \in \mathcal{C}$.

---
[1] A slight modification must be made for the case of dual-leg support. In this case, when interpolating two stable configurations, inverse kinematics for the leg is used to force the relative position between the feet to remain fixed. The same s technique is also used for generating random statically-stable dual-leg postures (see Section 3.3).

2. Holding the right leg fixed at its random configuration, inverse kinematics is used to attempt to position the left foot at the required relative position to generate the body configuration q_{right}. If it succeeds, then q_{right} is tested for membership in $\mathcal{C}_{valid}$.

3. An identical procedure is performed to generate q_{left} by holding the left leg fixed at its random configuration derived from q_{rand}, using inverse kinematics to position the right leg, and testing for membership in $\mathcal{C}_{valid}$.

4. If either $q_{right} \in \mathcal{C}_{valid}$ or $q_{left} \in \mathcal{C}_{valid}$, and the robot has left-right symmetry, additional stable postures can be derived by mirroring the generated stable configurations.

3.4 Trajectory Generation

If successful, the path search phase returns a continuous sequence of collision-free, statically-stable body configurations. All that remains is to calculate a final solution trajectory τ that is dynamically-stable and collision-free. Theoretically, any given statically-stable trajectory can be transformed into a dynamically-stable trajectory by arbitrarily slowing down the motion. However, we can almost invariably obtain a smoother and shorter trajectory by performing the following two steps:

Smoothing: We smooth the raw path by making several passes along its length, attempting to replace portions of the path between selected pairs of configurations by straight-line segments in $\mathcal{C}_{valid}$.[2] This step typically eliminates any potentially unnatural postures along the random path (e.g. unnecessarily large arm motions). The resulting smoothed path is transformed into an input trajectory using a minimum-jerk model [7].

Filtering: A dynamics filtering function is used in order to output a final, dynamically-stable trajectory. We use the online balance compensation scheme described in [16], which enforces constraints upon the zero moment point (ZMP) trajectory in order to maintain overall dynamic stability. The output configuration of the filter is guaranteed to lie in $\mathcal{C}_{stable}$. Collision-checking is used to verify that the final output trajectory lies in $\mathcal{C}_{valid}$, with the motion made slower in the case of collision.

Although this method has generated satisfactory results in our experiments, it is by no means the only option. Other ways of generating dynamically-stable trajectories from a given input motion are also potentially possible to apply here (e.g. [37, 29]). It is also possible to employ variational techniques, or apply algorithms for computing time-optimal trajectories [34]. Calculating the *globally-optimal trajectory* according to some cost functional based on the obstacles and the dynamic model is <u>an open problem, and</u> an area of ongoing research.

[2] When interpolating dual-leg configurations, inverse kinematics is used to keep the relative position of the feet fixed.

Task Description	Computation Time (seconds)			
	min	max	avg	stdev
Reach under chair	171	598	324	138
Lift leg over box	26	103	48	21
Reach over table	194	652	371	146

Table 1: Performance statistics ($N = 25$ trials).

4 Experiments

This section presents some preliminary experiments performed on a 270 MHz SGI O2 (R12000) workstation. We have implemented a prototype planner in C++ that runs within a graphical simulation environment [19]. An operator can position individual joints or use inverse kinematics to specify body postures for the virtual robot. The filter function can be run interactively to ensure that the goal configuration is statically-stable. After specifying the goal, the planner is invoked to attempt to compute a dynamically-stable trajectory connecting the goal configuration to the robot's initial configuration (assumed to be a collision-free, stable posture).

We have tested the output trajectories calculated by the planner on an actual humanoid robot hardware platform. The "H6" humanoid robot (33-DOF) is 137cm tall and weighs 51kg (including 4kg of batteries). Figure 1 shows a computed dynamically-stable motion for the H6 robot moving from a neutral standing position to a low crouching position in order to retrieve an object from beneath a chair. Figure 3 shows a different view of the real robot executing the same motion. Figure 4 shows a motion for positioning the right leg above the top of a box while balancing on the left leg. The motion in Figure 5 was not executed on the real robot, but is interesting in that it involves reaching for an object placed on top of a cabinet while avoiding both the cabinet and the shelves behind the robot. The robot is required to balance on one leg in order to extend the arm far enough to reach the obstacle on the table.

Each of the scenes contains over 9,000 triangle primitives. The 3D collision checking software used for these experiments was the RAPID library based on OBB-Trees developed by the University of North Carolina[9]. The total wall time elapsed in solving these queries ranges from under 30 seconds to approximately 11 minutes. A summary of the computation times for repeated runs of 25 trials each is shown in Table 1.

5 Discussion

This paper presents an algorithm for computing dynamically-stable collision-free trajectories given full-body posture goals. Although we have focused our experiments on biped robots with a humanoid shape, the algorithm is general and can be applied to any robot subject to balance constraints (legged or not). There are many potential uses for such software, with the primary one being a high-level control interface for automatically computing motions to solve complex tasks for humanoid robots that involve simultaneous obstacle-avoidance and balance constraints.

The limitations of the algorithm form the basis for our future work: 1) the current implementation of the planner can only handle a fixed position for either one or both feet. 2) The effectiveness of different configuration space distance metrics needs to be investigated. 3) We currently have no method for integrating visual or tactile feedback.

Acknowledgments: We thank Fumio Kanehiro and Yukiharu Tamiya for their efforts in developing the AutoBalancer software library. We are grateful to Steven LaValle and Hirohisa Hirukawa for helpful discussions. This research is supported in part by a Japan Society for the Promotion of Science (JSPS) Postdoctoral Fellowship for Foreign Scholars in Science and Engineering, and by JSPS Grant-in-Aid for Research for the Future (JSPS-RFTF96P00801).

References

[1] N. Amato and Y. Wu. A randomized roadmap method for path and manipuation planning. In *Proc. IEEE Int. Conf. Robot. & Autom. (ICRA)*, pages 113–120, 1996.

[2] J. Barraquand and J.-C. Latombe. Robot motion planning: A distributed representation approach. *Int. J. Robot. Res.*, 10(6):628–649, December 1990.

[3] R. Bohlin and L. Kavraki. Path planning using Lazy PRM. In *Proc. IEEE Int. Conf. Robot. & Autom. (ICRA)*, April 2000.

[4] V. Boor, M. Overmars, and A.F. van der Stappen. The gaussian sampling strategy for probabilistic roadmap planners. In *Proc. IEEE Int. Conf. Robot. & Autom. (ICRA)*, 1999.

[5] J.F. Canny. *The Complexity of Robot Motion Planning*. MIT Press, Cambridge, MA, 1988.

[6] M. Cherif and C. Laugier. Motion planning of autonomous off-road vehicles under physical interaction constraints. In *Proc. IEEE Int. Conf. Robot. & Autom. (ICRA)*, May 1995.

[7] T. Flash and N. Hogan. The coordination of arm movements: an experimentally confirmed mathematical model. *J. Neurosci.*, 5(7):1688–1703, 1985.

[8] E. Frazzoli, M.A. Dahleh, and E. Feron. Robust hybrid control for autonomous vehicles motion planning. Technical report, Laboratory for Information and Decision Systems, Massachusetts Institute of Technology, Cambridge, MA, 1999. Technical report LIDS-P-2468.

[9] S. Gottschalk, M. C. Lin, and D. Manocha. Obbtree: A hierarchical structure for rapid interference detection. In *SIGGRAPH '96 Proc.*, 1996.

[10] Li Han and Nancy M. Amato. A kinematics-based probabilistic roadmap method for closed chain systems. In *Proc. Int. Workshop Alg. Found. Robot.(WAFR)*, March 2000.

[11] Kazuo Hirai. Current and future perspective of honda humanoid robot. In *Proc. IEEE/RSJ Int. Conf. Intell. Robot. & Sys. (IROS)*, pages 500–508, 1997.

[12] H. Hirukawa, B. Mourrain, and Y. Papegay. A symbolic-numeric silhouette algorithm. In *Proc. IEEE/RSJ Int. Conf. Intell. Robot. & Sys. (IROS)*, November 2001.

[13] H. Hirukawa and Y. Papegay. Motion planning of objects in contact by the silhouette algorithm. In *Proc. IEEE Int. Conf. Robot. & Autom. (ICRA)*, pages 722–729, April 2000.

[14] T. Horsch, F. Schwarz, and H. Tolle. Motion planning for many degrees of freedom : Random reflections at c-space obstacles. In *Proc. IEEE Int. Conf. Robot. & Autom. (ICRA)*, pages 3318–3323, April 1994.

[15] D. Hsu, J.-C. Latombe, and R. Motwani. Path planning in expansive configuration spaces. *Int. J. Comput. Geom. & Appl.*, 9(4-5):495–512, 1997.

[16] S. Kagami, F. Kanehiro, Y. Tamiya, M. Inaba, and H. Inoue. AutoBalancer: An Online Dynamic Balance Compensation Scheme for Humanoid Robots. In *Proc. Int. Workshop Alg. Found. Robot.(WAFR)*, 2000.

[17] L. Kavraki, P. Švestka, J. C. Latombe, and M. H. Overmars. Probabilistic roadmaps for path planning in high-dimensional configuration space. *IEEE Trans. Robot. & Autom.*, 12(4):566–580, 1996.

[18] R. Kindel, D. Hsu, J.C. Latombe, and S. Rock. Kinodynamic motion planning amidst moving obstacles. In *Proc. IEEE Int. Conf. Robot. & Autom. (ICRA)*, April 2000.

[19] J.J. Kuffner, S. Kagami, M. Inaba, and H. Inoue. Graphical simulation and high-level control of humanoid robots. In *Proc. IEEE/RSJ Int. Conf. Intell. Robot. & Sys. (IROS)*, November 2000.

[20] J.J. Kuffner and S.M. LaValle. RRT-Connect: An efficient approach to single-query path planning. In *Proc. IEEE Int. Conf. Robot. & Autom. (ICRA)*, April 2000.

[21] J.J. Kuffner, S.Kagami, M. Inaba, and H. Inoue. Dynamically- stable motion planning for humanoid robots. In *IEEE-RAS Int. Conf. Human. Robot. (Humanoids)*, Boston, MA, September 2000.

[22] S. M. LaValle. Rapidly-exploring random trees: A new tool for path planning. TR 98-11, Computer Science Dept., Iowa State Univ. <http://janowiec.cs.iastate.edu/papers/rrt.ps>, Oct. 1998.

[23] S.M. LaValle and J.J Kuffner. Randomized kinodynamic planning. In *Proc. IEEE Int. Conf. Robot. & Autom. (ICRA)*, May 1999.

[24] S.M. LaValle and J.J Kuffner. Rapidly-exploring random trees: Progress and prospects. In *Proc. Int. Workshop Alg. Found. Robot.(WAFR)*, March 2000.

[25] S.M. LaValle, J.H. Yakey, and L.E. Kavraki. A probabilistic roadmap approach for systems with closed kinematic chains. In *Proc. IEEE Int. Conf. Robot. & Autom. (ICRA)*, 1999.

[26] E. Mazer, J. M. Ahuactzin, and P. Bessière. The Ariadne's clew algorithm. *J. Artificial Intell. Res.*, 9:295–316, November 1998.

[27] K. Nagasaka, M. Inaba, and H. Inoue. Walking pattern generation for a humanoid robot based on optimal gradient method. In *Proc. IEEE Int. Conf. Sys. Man. & Cyber.*, 1999.

[28] Y. Nakamura and et. al. V-HRP: Virtual humanoid robot platform. In *IEEE-RAS Int. Conf. Human. Robot. (Humanoids)*, September 2000.

[29] Y. Nakamura and K. Yamane. Interactive motion generation of humanoid robots via dynamics filter. In *Proc. of First IEEE-RAS Int. Conf. on Humanoid Robots*, September 2000.

[30] J. Pratt and G. Pratt. Exploiting natural dynamics in the control of a 3d bipedal walking simulation. In *In Proc. of Int. Conf. on Climbing and Walking Robots (CLAWAR99)*, September 1999.

[31] Marc Raibert. *Legged Robots that Balance*. MIT Press, Cambridge, MA, 1986.

[32] J. H. Reif. Complexity of the mover's problem and generalizations. In *Proc. 20th IEEE Symp. on Foundations of Computer Science (FOCS)*, pages 421–427, 1979.

[33] J. T. Schwartz and M. Sharir. On the 'piano movers' problem: Ii. general techniques for computing topological properties of real algebraic manifolds. *Advances in applied Mathematics*, 4:298–351, 1983.

[34] Z. Shiller and S. Dubowsky. On computing time-optimal motions of robotic manipulators in the presence of obstacles. *IEEE Trans. Robot. & Autom.*, 7(7), December 1991.

[35] Z. Shiller and R.Y. Gwo. Dynamic motion planning of autonomous vehicles. *IEEE Trans. Robot. & Autom.*, 7(2):241–249, April 1991.

[36] M. Vukobratovic, B. Borovac, D. Surla, and D. Stokie. *Biped Locomotion: Dynamics, Stability, Control, and Applications*. Springer-Verlag, Berlin, 1990.

[37] J. Yamaguchi, S. Inoue, D. Nishino, and A. Takanishi. Development of a bipedal humanoid robot having antagonistic driven joints and three dof trunk. In *Proc. IEEE/RSJ Int. Conf. Intell. Robot. & Sys. (IROS)*, pages 96–101, 1998.

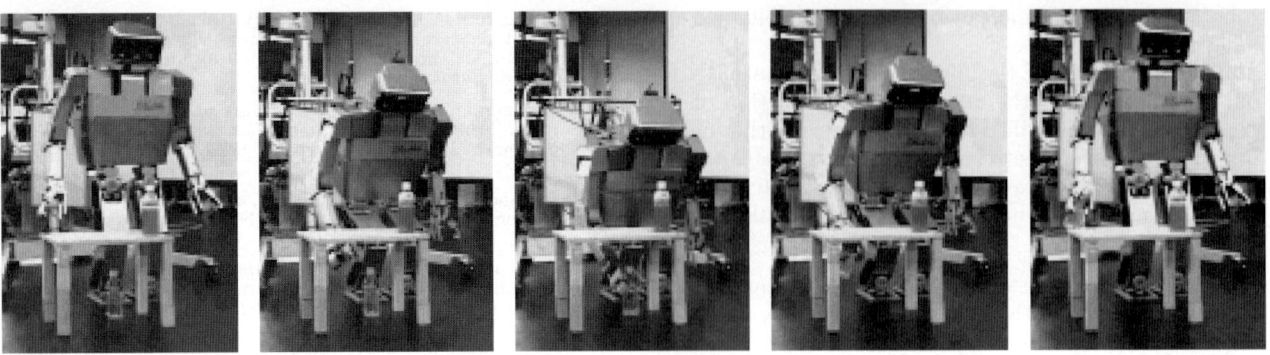

Figure 3: Dynamically-stable crouching trajectory for retrieving an object from beneath an obstacle

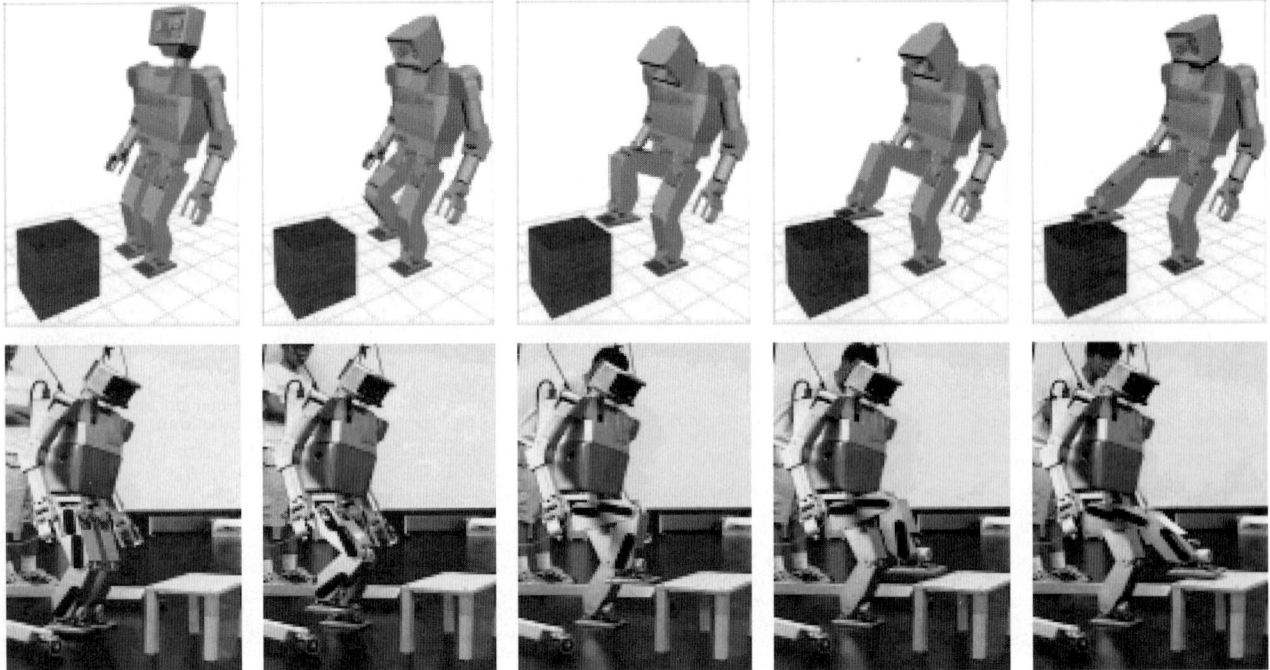

Figure 4: Positioning the right foot above an obstacle while balancing on the left leg. (*top*: simulation, *bottom*: actual hardware).

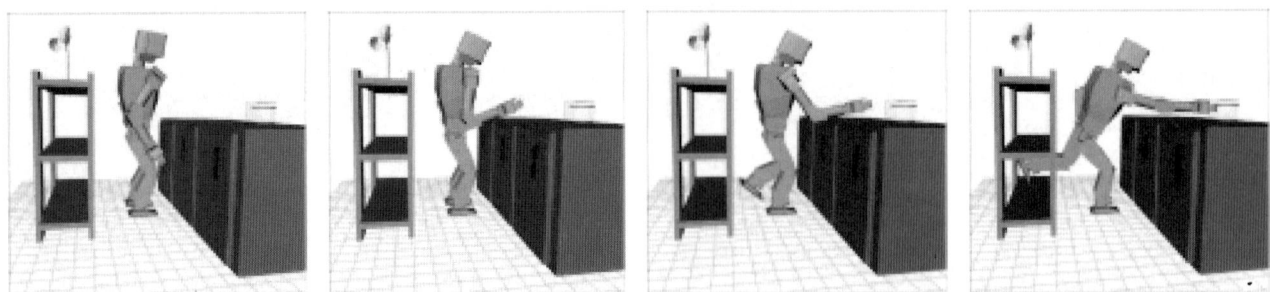

Figure 5: Reaching for an object atop a cabinet while avoiding obstacles and balancing on the right leg.

Exact Cellular Decomposition of Closed Orientable Surfaces Embedded in $\Re^3$

Prasad N. Atkar   Howie Choset   Alfred A. Rizzi   Ercan U. Acar

Carnegie Mellon University
Department of Mechanical Engineering and Robotics Institute
Pittsburgh, PA 15213

Abstract

*We address the task of covering a closed orientable surface embedded in $\Re^3$ without any prior information about the surface. For applications such as paint deposition, the effector (the paint atomizer) does not explicitly cover the target surface, but instead covers an **offset surface** — a surface that is a fixed distance away from the target surface. Just as Canny and others use critical points to look for changes in connectivity of the free space to ensure completeness of their roadmap algorithms, we use critical points to identify changes in the connectivity of the offset surface to ensure full surface coverage. The main contribution of this work is a method to construct unknown offset surfaces using a procedure, also developed in this paper, to detect critical points.*

1 Introduction

Conventional path planning determines a path between two points in the free configuration space $\mathcal{FS}$ of a system [14]. Recent path planning results describe *coverage path planning* algorithms that determine a path that enables an "effector" to pass over all points in the free configuration space [8], [12], [21]. Applications of this recent work include humanitarian de-mining, autonomous lawn mowing and floor cleaning, all of which are *planar* coverage tasks. This paper takes the first step towards lifting these planar coverage algorithms into three dimensions for applications such as the inspection of complicated surfaces (non-planar), material deposition, material removal, and CNC tool path planning. These applications require coverage of two-dimensional surfaces embedded in three dimensions. For applications such as visual inspection, the camera itself does not cover the target surface; to achieve visual coverage, the camera covers an *offset surface*, a surface that is a fixed distance away from the target surface. This paper describes how a robot equipped with a range sensor at its end effector, can incrementally construct (i.e., explore) an offset surface without any prior information about the target surface.

Our approach to coverage uses a *slice*, a co-dimension one surface (i.e., a plane), that is swept through the free space. We are interested in slices where topologi-

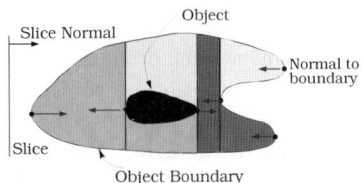

Fig. 1. Cellular decomposition for flat planar environments.

cal (e.g., connectivity) changes occur; these changes occur at points termed *critical points*. Just as Canny [3], [4], [5] uses critical points to ensure the connectivity of roadmaps, we use critical points to define cells in a cellular decomposition. In each cell, simple motions are sufficient to cover each cell and then *complete* coverage is achieved by visiting each cell in the decomposition. In this paper, we define a cellular decomposition for a large class of two-dimensional surfaces embedded in three dimensions in terms of critical points on the offset surface. To achieve sensor based coverage of the offset surface, we introduce a new method to detect critical points on the offset surface. This method assumes that the robot can measure distance from its effector to the target surface.

2 Prior Work

Our work rests on two tasks: coverage and generating offset surfaces. Most recent work in coverage is geared towards the plane and is either implicitly or explicitly, based on exact cellular decompositions. These algorithms decompose the space into different shapes such as rectangles [2], trapezoids [18], [19], "clumped trapezoids" [6], [8] or fine cells (grids) [10], [20], [21].

Choset et al [7] present a method to achieve an exact cellular decomposition of a planar environment that is formulated in terms of critical points. Between critical points, the number of connected portions of the slice in the free space remains the same. This means that each cell in the decomposition can be covered with simple back and forth motions and complete coverage is achieved by visiting each cell. Acar et al [1] sense critical points using distance information and show that the distance function gradient becomes perpendicular to the slice at the critical points (see Figure 1). We use an analogous approach to detect critical points on the offset surfaces.

Surveys [15], [17] of offset curves/surfaces literature indicate that there has been a vast amount of research in generating offset curves; however this field is still young. Farouki [9] gives procedures to determine a geometric representation for offset surfaces of simple solids, but these methods do not consider offset surfaces of concave objects or curved surfaces. Pham [16] describes methods to generate approximations of the offset surface for the NURB surface, but the method does not yield offset surfaces free from self intersections, discontinuities, and sharp ridges. There have been a few attempts to eliminate self intersections using non-analytical approaches. Kimmel and Bruckstein [13] use the wavefront approach in fluid dynamics to obtain the offset surface, while Gurbuz and Zeid [11] employ the approach of filling closed balls of a fixed radius at each point on the object. However, both methods are based on grid/cell decompositions; hence, their resolution and accuracy greatly depend upon the size of the grid. Since our work uses numerical tracing to construct the "sliced" offset surface of an arbitrarily shaped object, it is free from self intersections and discontinuities.

3 Offset Surface Coverage

A coverage offset surface has two dimensions and is embedded in $\Re^3$. We determine a path that covers this surface by repeatedly intersecting it with a two-dimensional slice. Each intersection generically is one or more loops; we term these loops as *coverage offset surface edges*, or COS_{edges}.

3.1 Coverage Offset Paths

We use a numerical technique that traces the roots of a function to generate the COS_{edge}. This function has two parts: an offset surface component and a slice component. Let x be a point in $\Re^3$. The distance between x and an object C_i is given by $d_i(x)$ ($d_i: \Re^3 \mapsto \Re$). The coverage offset surface, $\mathcal{COS}_i$ is

$$\mathcal{COS}_i = \{x \in \Re^3 : d_i(x) - \Omega = 0\}, \quad (1)$$

where $\Omega \in \Re$ is the desired fixed distance. By the pre-image theorem, $\mathcal{COS}_i$ is dimension two. Now, we intersect $\mathcal{COS}_i$ with a planar slice $\Sigma_{i,k} = \{x \in \Re^3 : \langle n_s, x \rangle - k = 0\}$, where $n_s \in \Re^3$ is the normal to the plane, and k defines the location of the plane. Varying k has the effect of sweeping the slice. The intersection of slice plane $\Sigma_{i,k}$ and $\mathcal{COS}_i$ defines the COS_{edge}, which can be represented by the pre-image of a function $G_{i,k,\Omega} \colon \Re^3 \mapsto \Re^2$,

$$G_{i,k,\Omega}(x) = \begin{pmatrix} d_i(x) - \Omega \\ \langle n_s, x \rangle - k \end{pmatrix}. \quad (2)$$

For concise notation, let $G(x)$ be $G_{i,k,\Omega}(x)$, and let $\mathcal{COS}$ denote $\mathcal{COS}_{\cup_i C_i}$. By the pre-image theorem, we know that for regular values, the pre-image of $G(x)$ is a one-dimensional manifold. We have shown, but omitted its

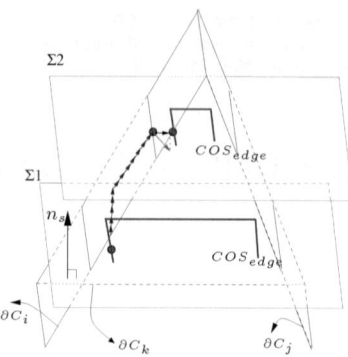

Fig. 2. Switching slice plane algorithm.

proof due to space limitations, that away from critical points and assuming a generic placement of objects, the offset edges form closed loops.

Once the planner traces the COS_{edge} loop, it must then shift to the next slice plane. Changing slice planes is not a trivial task. We have developed an algorithm that either determines a point on the COS_{edge} in the next slice plane or concludes that there is no COS_{edge} in the next slice corresponding to the COS_{edge} in the current plane. The algorithm has the following two steps:

1. Shifting the slice planes: The planner moves a point along the slice normal, until it reaches the next slice plane, or until it reaches the boundary of an object. In the latter case, the planner then starts moving along the boundary while increasing the distance to the current slice. In other words, it follows the slice gradient projected onto the surface boundary. See Figure 2. While following the boundary, if the planner comes across a point on the object surface such that it cannot move away from the previous slice, then the planner has reached a critical point on the object.

2. Moving onto a COS_{edge} or detecting that it does not exist: Once in the new slice plane, the planner moves a point away from the closest object while keeping the point in the plane, i.e., it follows $\Pi_\Sigma \nabla d_i(x)$, until it reaches a point on the COS_{edge} or a point on a two way equidistant sheet $SS_{ij} = \{x : d_i(x) = d_j(x), \nabla d_i(x) \neq \nabla d_j(x)\}$. In the latter case, the planner traces $SS_{ij} \cap \Sigma$ looking for a point on the COS_{edge}. After tracing the intersection completely, if the planner does not find any COS_{edge} point, then it concludes the absence of COS_{edge} in the new slice, corresponding to COS_{edge} in the old slice. This conclusion represents that the planner has passed a critical point on the offset surface lying between the new and the old slice. Such a critical point on $\mathcal{COS}$ appears when there is a corresponding critical point on the ob-

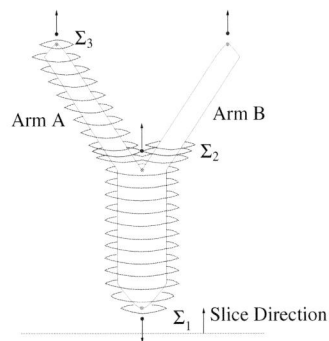

Fig. 3. Multiple loops in a single slice plane

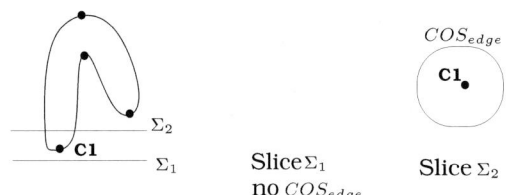

Fig. 4. Convex critical point: 0-to-1 connectivity change

ject, or when the object has a narrow "neck" which splits the offset surface into two disconnected parts.

3.2 Critical Points and the Adjacency Graph

For convex objects, there is at most one COS_{edge} in any slice plane; hence, repeatedly tracing a loop and then finding a seed point for the next loop in the next slice plane will cover the $\mathcal{COS}$ completely. However, for solids that are not convex, there may be more than one COS_{edge} loop in a single slice plane, therefore, the planner needs to know the number of COS_{edges} in a given slice plane to achieve complete coverage. In Figure 3, as we sweep the slice from bottom to top, at slice Σ_2, the number of COS_{edge} loops in a slice change from one to two. If the planner naively adopts the "trace the loop and jump to next slice" policy, it covers only one "arm" and fails to cover the other arm.

We use critical points on the offset surface to determine when the number of COS_{edge} loops changes. The change depends upon the "convexity" nature of a neighborhood of an offset critical point. Let CP be a critical point on the offset surface, and let $B_\epsilon(CP)$ be the neighborhood of CP, which is an open ball with center CP and of radius $\epsilon > 0$. Since the offset surface is a space-dividing surface, let the volume bounded by the offset surface that contains the object be denoted by $\mathcal{SO}$. Then, the critical points are divided into three classes : convex, concave, and semi-concave.

DEFINITION 3.1 *A critical point CP is*
- **convex** *if* $B_\epsilon(CP) \cap \mathcal{SO}$ *is convex,*
- **concave** *if* $B_\epsilon(CP) \cap (\mathcal{FS} \setminus int(\mathcal{SO}))$ *is convex,*
- **semiconcave** *if neither* $B_\epsilon(CP) \cap \mathcal{SO}$ *nor* $B_\epsilon(CP) \cap (\mathcal{FS} \setminus int(\mathcal{SO}))$ *is convex.*

The critical points correspond to nodes in our adjacency graph; edges correspond to cells whose boundaries are defined by two "adjacent" critical points. In this work, the planner incrementally constructs the adjacency graph by first covering a cell until it detects a critical point. The convexity nature of the critical point determines how many cells are associated with the critical point. In the next section, we show that a convex or concave critical point corresponds to a terminal node (a leaf) in the adjacency graph and a semi-concave critical point corresponds to a node that generically has three edges emanating from it. If the planner encounters a semi-concave critical point, the planner chooses one cell and covers it until the planner encounters another critical point. When the planner encounters a convex or concave critical point, then it returns to a semi-concave critical point with an "uncovered" cell associated with it. When all critical points have no uncovered cells, coverage is complete.

The challenge then becomes how to detect a critical point and what its type is.

4 Detecting the Critical Points

As we pass the critical point while covering $\mathcal{COS}$, *generically* only four kinds of local connectivity changes are possible: 0-to-1 (the number of loops changes from 0 to 1 as we pass the critical point), 1-to-0, 1-to-2 or 2-to-1.

Note that a slice is defined by the preimage of a scalar-valued function $\lambda: \Re^3 \mapsto \Re$. $\lambda^{-1}(c)$ is a slice, and $\lambda(x)$ denotes the value of the slice function evaluated at a point x on the $\mathcal{COS}$.

LEMMA 4.1 *At a convex critical point X, there is always a 0-to-1 or 1-to-0 change in the number of COS_{edge} loops (see Figure 4).*

Proof: By definition, $B_\epsilon(X) \cap \mathcal{SO}$ is convex. Since X is a critical point, it must be either a local minimum or maximum of the slice function evaluated on the offset surface. Since λ is a convex function and $B_\epsilon(X) \cap \mathcal{SO}$ is a convex set, $\forall x1, x2 \in (B_\epsilon(X) \cap \mathcal{SO})$, such that $x1, x2 \neq X$, either $\lambda(x1) < \lambda(X)$ and $\lambda(x2) < \lambda(X)$, or $\lambda(x1) > \lambda(X)$ and $\lambda(x2) > \lambda(X)$. Since $\mathcal{COS} = \partial\mathcal{SO}$, the same is true for all $x1, x2 \in B_\epsilon(X) \cap \mathcal{COS}$. Thus, there does not exist $x1, x2 \in (B_\epsilon(X) \cap \mathcal{COS})$ such that $\lambda(x1) > \lambda(X)$ and $\lambda(x2) < \lambda(X)$. Hence, all points in $B_\epsilon(X) \cap \mathcal{COS}$ must lie on one side of the slice that contains X, and there is no point in the $B_\epsilon(X) \cap \mathcal{COS}$ which lies on the other side of the critical slice. Therefore, there must be a 0-to-1 or 1-to-0 change at the convex critical point. ∎

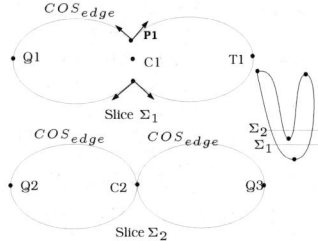

Fig. 5. Semiconcave critical point: 2-to-1 connectivity change

LEMMA 4.2 *At a concave critical point Y, there is always a 0-to-1 or 1-to-0 change in the number of COS_{edge} loops.*

Proof: Here, by definition, $B_\epsilon(Y) \cap (\mathcal{FS} \setminus int(\mathcal{SO}))$ is convex. Again, $\mathcal{COS}$ is the boundary of a convex set $B_\epsilon(Y) \cap (\mathcal{FS} \setminus int(\mathcal{SO}))$. The proof follows *mutatis mutandis*, as per Lemma 4.1. ∎

LEMMA 4.3 *There is a 1-to-m or m-to-1 change in the number of COS_{edge} loops at a semiconcave critical point Z, where $m > 1$ (see Figure 5).*

Proof: For a semiconcave critical point, both $B_\epsilon(Z) \cap \mathcal{SO}$ and $B_\epsilon(Z) \cap (\mathcal{FS} \setminus int(\mathcal{SO}))$ are not convex. Their intersection, $(B_\epsilon(Z) \cap \mathcal{SO}) \cap (B_\epsilon(Z) \cap (\mathcal{FS} \setminus int(\mathcal{SO})))$ is $B_\epsilon(Z) \cap \partial\mathcal{SO}$, which is the same as $B_\epsilon(Z) \cap \mathcal{COS}$. Hence, $\exists z1, z2 \in B_\epsilon(Z) \cap \mathcal{COS}$ such that $\lambda(z1) > \lambda(Z)$ and $\lambda(z2) < \lambda(Z)$. Thus, if $z1$ lies on one side of the slice plane that contains Z, then $z2$ must lie on the other. Without loss of generality, if there is only one COS_{edge} on the side where $z1$ lies, since there is a change in connectivity at the critical point, there must be $m(> 1)$ number of COS_{edge} loops on the $z2$ side. Hence, there must be a 1-to-m or m-to-1 change at the semiconcave critical point. A 1-to-m or m-to-1 connectivity change can be seen as $m - 1$ number of 1-to-2 or 2-to-1 connectivity changes at the same critical point. In this paper, we will always treat the 1-to-m change as $m - 1$ 1-to-2 changes and vice-versa. ∎

Now, we present methods to detect critical points on a closed, orientable and *connected* offset surface. When the planner is covering an unknown environment, only detecting a 1-to-0 connectivity change for the convex and concave critical points suffices. A 1-to-0 change is automatically detected by the switching slice plane algorithm. However, the 1-to-2 change at a semiconcave critical point is not very apparent to the planner. The planner uses cusps, or the non-smooth boundary points on the COS_{edge} where a discrete change occurs in the direction of the tangent to the COS_{edge}, to determine the semiconcave critical point. The planner looks for the cusps in the neighborhood of the critical point, and then traces the cusps to reach the critical point. Here,

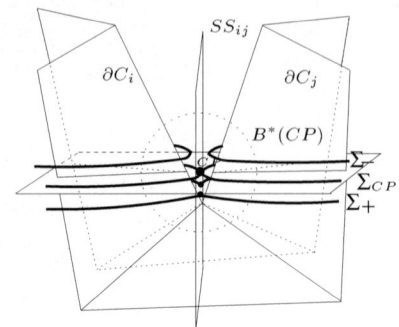

Fig. 6. Semiconcave critical point emanates cusps.

we consider a *deleted* neighborhood $B_\epsilon^*(Z) = B_\epsilon(Z) \setminus Z$ of the critical point Z.

LEMMA 4.4 *For every semiconcave critical point CP, there exists a cusp in its deleted neighborhood.*

Proof: From Lemma 4.3, we know that there is a 1-to-2 or 2-to-1 change in the number of COS_{edge} loops at a semiconcave critical point. Therefore, there exists a "critical slice" where two loops intersect non-transversely (i.e. the loops kiss each other). See Figure 6. Let Σ_{CP} denote the critical slice. Since we have assumed that a critical point is isolated, the loops intersect only at one point, the critical point. Clearly, both loops have at least one different convex object closest to them. Let C_i and C_j be the closest objects at the critical point. Then at the critical point, the distance of point CP from C_i and C_j is the same and is equal to Ω. Then, the equidistant sheet SS_{ij} passes through CP and locally splits the $\mathcal{COS}$. Note that the double equidistant sheet intersects the $\mathcal{COS}$ only on the "one loop side" of CP. Also, this intersection is one-dimensional and it locally separates the $\mathcal{COS}$. Let Σ_- be the slice plane with a slice value smaller than that of Σ_{CP}. Similarly, let Σ_+ have a larger slice value than that of Σ_{CP}. Without loss of generality, let the number of COS_{edge} loops change from 2 to 1 as the planner moves from Σ_- to Σ_+. Then, by continuity of the slice function, there exists a slice Σ_ϵ whose slice value is ϵ greater than that of Σ_{CP} such that $\Sigma_\epsilon \cap \mathcal{COS}$ is separated by $SS_{ij} \cap \mathcal{COS}$. Thus, SS_{ij} separates the COS_{edge} in slice Σ_ϵ into two parts (both parts exclude $COS_{edge} \cap SS_{ij}$) such that points belonging to one part are closer to a set of objects $\mathcal{C}_{A\epsilon}$, while points in the other part are closer to a different set of objects $\mathcal{C}_{B\epsilon}$, i.e., $\mathcal{C}_{A\epsilon} \neq \mathcal{C}_{B\epsilon}$. Hence at $SS_{ij} \cap \mathcal{COS} \cap \Sigma_\epsilon = \{K_i\}$, there is a *discrete* change in the gradient vector.

From Equation 2, we know that $Null\begin{pmatrix} \nabla d(x) \\ n_s \end{pmatrix}$ is the tangent to the COS_{edge}. Since there is a discrete change in $\nabla d(x)$ at K_i, the tangent at K_i has a discrete change in direction. Thus, COS_{edge} is non-smooth or,

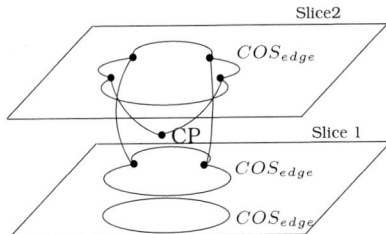

Fig. 7. Detecting the semiconcave critical point : tracing the cusps.

in other words, K_i is a cusp. This argument holds for sufficiently small values of ϵ, hence the proof. ∎

Thus, there exists a one-dimensional path, *EquiCOS*, which is the intersection of the offset surface and the equidistant sheet, such that the cusp points and the critical point (when it exists) lie on it. To guarantee that the semiconcave critical point is detected, the planner traces all the *EquiCOS* edges between the current and the previous slice planes. See Figure 7. Initially, it traces all *EquiCOS* edges emanating from the cusps in previous slice, and some of these cusps lead to cusps in the current slice plane. It then traces the *EquiCOS* edges from the remaining cusps in the current slice.

The *EquiCOS* is the pre-image of the function $EC(x)$:

$$EC(x) = \begin{pmatrix} d_i(x) - \Omega \\ d_i(x) - d_j(x) \end{pmatrix}, \quad (3)$$

where objects C_i and C_j are the first closest objects at the cusp. While tracing the *EquiCOS*, if at any point, the slice normal n_s lies in the convex hull of vectors $\nabla d_i(x)$ and $\nabla d_j(x)$ [8], then the planner has found the semiconcave critical point. If there is a 1-to-m change at this critical point, then m can be easily found by noting the starting point of each *EquiCOS* edge in the previous slice. If these *EquiCOS* starting points lie on n distinct COS_{edge} loops, then $m = n + 1$.

However, tracing all the *EquiCOS* edges lying between the previous and current slice planes may be computationally very expensive. A very useful heuristic uses the change in the number of cusps between the current and previous slices. Note that it is necessary that the number of cusps changes in the neighborhood of a critical point, but it is not sufficient. So, when such a change occurs, the positions of the cusps in the current slice are compared with the position of the cusps in the previous slice. Only the "new" cusps are traced. A 2-to-1 connectivity change increases the number of cusps in the current slice plane, while 1-to-2 connectivity change decreases the number of cusps. Thus, the planner can detect all three types of critical points and, therefore, can complete the offset surface coverage by constructing the adjacency graph.

5 Simulation

We simulate the critical point detection procedure using known polyhedral environments. For polyhedral environments, we use critical points on the target surface – the boundary of the polyhedral solid – to determine the critical points on the offset surface. We classify the critical points on the polyhedral object similarly to the offset critical points: convex, concave or semiconcave. For non-degenerate cases, the critical points of the target surface appear only at the vertices of the polyhedron. It is easy to determine which vertices are critical points by looking at the positive span of the surface normals that form the vertex. If the slice gradient lies in this positive span, then the vertex is a critical point [8]. Each critical point is then "lifted" to the offset surface, but this "lifted" critical point is not necessarily a critical point on the offset surface. We term these points as candidate critical points. If the distance between the candidate critical point and the closest point on the target surface is the offset distance, then the candidate critical point is indeed a critical point for the offset surface. It is worth noting that for convex, semi-concave, and concave critical points, there will be one, two, and three closest points respectively. Finally, not all critical points on the offset surface are derived from the target surface; these critical points, however, are detected while executing the switching slice plane algorithm. By looking at the target surface and by invoking the switching plane algorithm, we are guaranteed to encounter all critical points of the offset surface and hence ensure complete coverage.

The simulations are carried out by generating the offset surfaces for different offset distances and different slicing directions. We verify that the simulation yields the locations of critical points exactly as predicted. Figure 8 shows the coverage of the offset surface with semiconcave and convex critical points. The simulation shows the 2-to-1 connectivity change at the semiconcave corner. Figure 9 shows an object with a hole in it. For this object, again there are semiconcave and convex critical points on its offset surface. Thus, these simulations successfully demonstrate the critical point detection procedure for offset surfaces.

6 Conclusion

In this work, we introduce complete methods to cover an unknown closed, orientable and connected offset surface in $\Re^3$. The offset surface is the set of points that are a fixed distance away from a target surface (such as an automobile body). The offset surface is covered by incrementally tracing several paths on the offset surface using local numerical techniques. These paths are formed by repeatedly intersecting a slice with the cov-

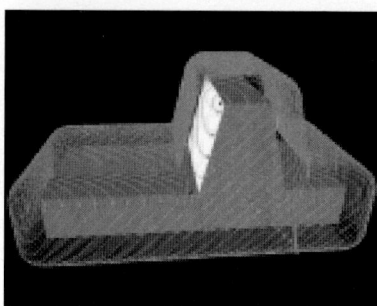

Fig. 8. Complete coverage of offset surface for a car-shaped object.

Fig. 9. Offset surface for an object with a hole.

erage offset surface and tracing the intersection. Note that these paths do not consider the kinematics of the robot, which will be considered in future work. We also assume that the offset surface is a separator (separating inside from outside); future work will consider scenarios where "obstacles" on the target surface will be not covered, yielding an offset surface that has "holes" in it. This is useful for applications such as paint stripping hulls of ships where the "obstacles" are port-holes.

The offset surface is decomposed into cells where the boundaries of the cells are defined by critical points of a slice function evaluated on the offset surface. Each cell thus generated is easy to cover using our offset path tracing procedure. The primary contribution of this paper is to provide methods which can detect critical points for offset surfaces of unknown environments. The result that semiconcave critical points "emanate" cusps can be particularly useful for surface coverage algorithms. Currently, we do not have a surface crawling robot to test our algorithms, so we have demonstrated the approach in this paper in known polyhedral environments.

The coverage algorithms presented in this work can be useful for a variety of applications such as robotic automobile-body spray painting, paint stripping, robotic inspection or CNC tool path generation. However, for applications like car painting, it is not only necessary that the target surface be covered completely, but it is also crucial that the target surface receive a uniform amount of paint. Our current coverage procedure guarantees complete coverage of an offset surface, but it does not take into account the effect of different deposition patterns. The cellular decomposition obtained by our method needs to be matched with decompositions based on geometrical aspects such as curvature. Our future work will include coverage procedures which consider these issues.

References

[1] E. Acar and H. Choset. Critical point sensing in unknown environments. In *Proc. of IEEE ICRA'00*, San Francisco, CA, 2000.

[2] Z. Butler, A. A. Rizzi, and R. L. Hollis. Contact-sensor based coverage of rectilinear environments. In *Proc. of IEEE Int'l Symposium on Intelligent Control*, Sept., 1998.

[3] J.F. Canny. *The Complexity of Robot Motion Planning*. MIT Press, Cambridge, MA, 1988.

[4] J.F. Canny. Constructing roadmaps of semi-algebraic sets i: Completeness. *Artificial Intelligence*, 37:203–222, 1988.

[5] J.F. Canny and M. Lin. An opportunistic global path planner. *Algorithmica*, 10:102–120, 1993.

[6] Z. L. Cao, Y. Huang, and E. Hall. Region filling operations with random obstacle avoidance for mobile robots. *Journal of Robotic systems*, pages 87–102, February 1988.

[7] H. Choset, E. Acar, A.A. Rizzi, and J. Luntz. Exact cellular decompositions in terms of critical points of morse functions. In *Proc. of IEEE ICRA'00*, San Francisco, CA, 2000.

[8] H. Choset and P. Pignon. Coverage path planning: The boustrophedon decomposition. In *Proceedings of the International Conference on Field and Service Robotics*, Canberra, Australia, December 1997.

[9] R.T. Farouki. Exact offset procedures for simple solids. *Computer Aided Geometric Design*, 2(4):257–80, 1985.

[10] Y. Gabriely and E. Rimon. Spanning-tree based coverage of continous areas by a mobile robot. *Annals of Mathematics and Artificial Intelligence*, Accepted, 2000.

[11] A.Z. Gurbuz and Zeid I. Offsetting operations via closed ball approximation. *Computer Aided Design*, 27(11):805–10, 1995.

[12] S. Hert, S. Tiwari, and V. Lumelsky. A Terrain-Covering Algorithm for an AUV. *Autonomous Robots*, 3:91–119, 1996.

[13] R. Kimmel and Bruckstein A.M. Shape offsets via level sets. *Computer Aided Design*, 25(3):154–62, 1993.

[14] J.C. Latombe. *Robot Motion Planning*. Kluwer Academic Publishers, Boston, MA, 1991.

[15] T. Maekawa. An overview of offset curves and surfaces. *Computer Aided Design*, 31(3):165–73, 1999.

[16] B. Pham. Offset approximation of uniform B-splines. *Computer Aided Design*, 20(8):471–474, 1988.

[17] B. Pham. Offset curves and surfaces: a brief survey. *Computer Aided Design*, 24(4):223–9, 1992.

[18] F.P. Preparata and M. I. Shamos. *Computational Geometry: An Introduction*. Springer-Verlag, 1985. p198-257.

[19] J. VanderHeide and N. S. V. Rao. Terrain coverage of an unknown room by an autonomous mobile robot. Technical Report ORNL/TM-13117, Oak Ridge National Laboratory, Oak Ridge, Tennessee, 1995.

[20] I.A. Wagner and Bruckstein A.M. Cooperative cleaners: A study in ant-robotics. Technical Report CIS-9512, Center for Intelligent Systems, Technion, Haifa, 1995.

[21] A. Zelinsky, R.A. Jarvis, J.C. Byrne, and S. Yuta. Planning Paths of Complete Coverage of an Unstructured Environment by a Mobile Robot. In *Proceedings of International Conference on Advanced Robotics*, pages pp533–538, Tokyo, Japan, November 1993.

Obstacle Detection Using Adaptive Color Segmentation and Color Stereo Homography

Parag H. Batavia and Sanjiv Singh
[parag/ssingh]@ri.cmu.edu

Carnegie Mellon University
Robotics Institute
Pittsburgh, PA 15213

Abstract

Obstacle detection is a key component of autonomous systems. In particular, when dealing with large robots in unstructured environments, robust obstacle detection is vital. In this paper, we describe an obstacle detection methodology which combines two complimentary methods: adaptive color segmentation, and stereo-based color homography. This algorithm is particularly suited for environments in which the terrain is relatively flat and of roughly the same color. We will show results in applying this method to an autonomous outdoor robot.

1. Introduction

This paper describes an obstacle detection algorithm for use in relatively flat areas where there is similarity in color. The method is robust to false positives and negatives through the use of two complimentary methods: color segmentation and color homography.

Color segmentation, as the name implies, uses color to classify image areas as "obstacle" or "freespace" The method we use is based on a training algorithm, in which examples of "freespace" are shown to the system, and it learns appropriate representations.

Stereo-based homography is often referred to as "poor man's stereo". Although computationally cheap, it does not provide depth information, as pure stereo does. Rather, it provides information on whether a particular image feature rises above the ground plane. In applications where a complete depth map is not needed, this can be a computationally cheap alternative. We extend the homography formulation to make use of color information, which improves robustness. This system is used to automatically train the color segmentation system.

In the rest of this paper, we describe how both methods are combined to form a robust obstacle detection system, followed by an example of its use on an outdoor mobile robot.

2. Obstacle Detection

Obstacle detection is a key component of an autonomous robot, particularly when dealing with large outdoor vehicles. The robot has to have very robust obstacle detection capabilities, since it is a heavy, potentially dangerous piece of equipment. The size of obstacles can vary, and the detection system has to operate reliably in various lighting conditions, along with light fog and rain, and at night as well. We have examined two methods for detecting obstacles, along with methods for integration of these two methods.

2.1. Color Segmentation

The basic idea behind color segmentation for obstacle detection is that pixels in an image are classified as "obstacle" or "freespace" based on color. When operating in domains in which traversable areas are of relatively constant color, such as grass, color segmentation works well.

Each pixel in a color image consists of a 3-tuple, representing the amount of energy contained in the red, green, and blue bands. Typically, each component of the tuple is a value between 0 and 255. Therefore, one simplistic method of color segmentation is a rule-based system. In such a system, various *rules* would be used to classify pixels, such as "*if red is between 100 and 175 and green is less than 25 and blue is more than 70,* **then** *classify as freespace*".

While conceptually simple and extremely fast, such methods are not general. They are specific to lighting conditions, and camera performance, and can easily be fooled by shadows and variations in grass condition. Therefore, we use a general, non-parametric representation similar to that used by Ollis [6] and Ulrich [12], but with extensions for automated training and adaptation. This approach uses a probabilistic formulation to classify pixels, based on a set of training images.

These images are not stored as standard (R,G,B) tuples. Rather, they are first converted to a different color space, known as Hue-Saturation-Value, or HSV. This is a cylindrical space, in which the H and S components contain the color information, in the form of a standard color wheel. The Hue is the actual color, or the angle of the point in the cylinder, the Saturation is the "purity" of the color, and is the radial distance of the point. The Value is the intensity, or brightness, and is the height of the point. This space has the advantage that if we ignore the Value component, we get additional robustness to shadows and illumination changes, along with a reduction in feature dimensionality.

Other researchers have addressed the problem of color segmentation. Hyams [3] uses a Spherical Coordinate Transform, which is a color space previously used in the medical domain, combined with a nearest-neighbor segmentation scheme to localize daughter vehicles with respect to a mothership. Shiji [9] uses a watershed algorithm along with extensions to avoid over-segmentation. McKenna [5] uses an adaptive mixture model to represent classes, which is a more compact representation than ours, but is computationally more expensive to train. The dichromatic reflection model, originally proposed by Shafer [8], is still used as well, as seen in [4].

2.1.1. Training

The training set is represented as a two dimensional histogram. The bins in the histogram are addressed based on the H and S values of a color pixel. The contents of the bin denote the number of occurrences of that particular H and S pair in the training set.

The color segmentation system has to "learn" what colors constitute traversable areas, such as grass. To train the classifier, we present it with several images of grass taken in various lighting conditions. For each pixel in the training image, the value of the corresponding histogram bin is incremented. Therefore, colors that occur often will have high values in the histogram. After training is done, the histogram is normalized by the total number of samples (i.e., the sum of the bin contents), so that bin contents now represent a probability. The top third of Figure 1 illustrates the training procedure.

The top-left image shows a set of training images of grass. The color values contained in these images are added to the histogram, which is shown in the top-right image. The histogram is represented as a color wheel. The area to the right of the center of the wheel, which corresponds to various shades of green, shows activity, which represents the colors in the training set.

The training time is linear in the number of image pixels, and in practice is extremely fast. The training can be done in a supervised manner by showing examples of freespace. Alternatively, the training set can be automatically acquired and adapted using homography, a complementary method

2.1.2. Run-Time

After training, the system is ready to classify pixels as obstacle or freespace. For each pixel, p, in a test image, we look up the bin value corresponding to the color of p. This provides us with a probabilistic measure, P, of p

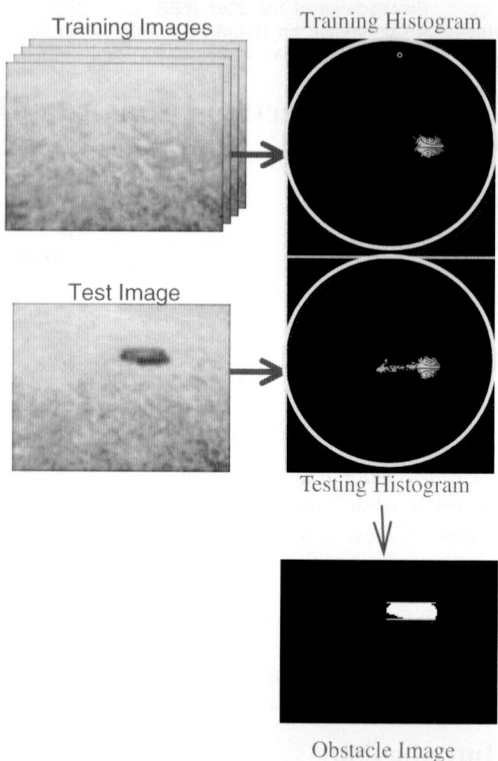

Figure 1: Color Segmentation training and histogram details.

being in the training set. If P is greater than a threshold, then we classify it as freespace. Else, it is an obstacle. I.e, anything which is not freespace is classified as an obstacle. This can lead to false positives, as we will discuss later.

The middle third of Figure 1 shows a test image, which contains grass and a bag. The figure on the middle-right shows the color distribution of the test image, again represented as a color wheel. Notice that although a large portion of the test image color distribution overlaps the training set color distribution, there is a significant portion which does not. This portion is due to the presence of the bag. The lower right figure shows an "obstacle image," in which white indicates obstacle and black indicates freespace. The bag is accurately detected, and there are no false positives.

2.1.3. Performance

Color segmentation relies on having a complete training set. As lighting changes, due to time of day or weather conditions, the appearance of grass and obstacle change as well, since the amount of incident sunlight changes. The color of grass is different under a cloud cover than under direct sunlight, and is different in the morning vs. mid-day. This can lead to false positives, if the system is only trained in one lighting condition, and then is used in another. Since training is so fast, and can be done on the fly, this is not a severe issue. If the environmental lighting changes, we can simply re-train and continue.

Similarly, color segmentation can classify flat objects, such as fall leaves, as obstacles, since their color is different from grass. In these cases, it is safe to drive over them. Therefore, there are four possible cases: 1) No obstacle, 2) true obstacle with significant height, 3) true flat obstacle, and 4) false obstacle due to lighting change. In cases 1 and 2, we do not need to modify the training set. In case three, we do not want to modify the training set, but want to recognize that it is safe to proceed. In case 4, we need to augment our training set to handle the new environmental conditions.

The next section describes stereo homography, which is a computationally cheap yet powerful method for detecting objects which rise above the ground plane. Homography can provide enough information to disambiguate between cases 2 and 3 or 4.

2.2. Color Homography

In conventional stereo, multiple cameras are used to find the range to image features. This range information comes at a steep computational cost. Computing depth using stereo is of the order $O(m*n*d)$, where m is the number of image pixels, n is the size of the correlation window, and d is the number of disparities searched, which is related to the range of depths which can be found. Another way to find obstacles is to use homography, which is linear in the number of image pixels. This is because homography does not compute range. Rather, it provides just enough information to determine whether a particular image feature is on the ground or above it.

The basic idea behind homography is this: If we know the extrinsic and intrinsic parameters of both cameras, and assume that all image features *lie on the ground plane*, we can solve the inverse perspective problem. I.e., any given image point in the [left/right] camera can now be back-projected into world coordinates. These world coordinates can then be forward-projected into the opposite camera. Using the left camera as an example, we can warp the left camera image to the right camera, and then compare the warped image against what the right camera actually sees. If all the image features actually *do* lie on the ground plane, then the warped image will match the actual right camera image. However, if certain image features lie above the ground plane, then our warped image will be incorrect in those areas, and this discrepancy can be detected.

Figure 2: Homography calibration images.

Previous work, as described in the next section, has made use of grey scale intensity images. Often, objects of different colors have the same intensity as the ground plane. In these cases, detecting ground plane violations through image subtraction fails. To avoid this, we use hue images, which capture the color properties of the background and potential obstacles.

2.2.1. Image Warping

We do a perspective warping of the left image to the right image, based on a 3x3 *homography matrix*. The equation to do the warping is:

$$x' = Hx \qquad (1)$$

Where x' is the (u,v,1) homogenous coordinate of the left image, x, and H is the homography matrix.

H is determined through a calibration procedure, described fully in [11], in which an image pair is taken, and a small set (usually four) of corresponding features in the left and right image are manually selected. These features must lie on the ground plane. A sample pair of calibration images is shown in Figure 2. Typical features to mark would be the corners of the white calibration markings.

Four corresponding features provides enough constraints to solve for the 8 free parameters in H (the 3,3 element of H is always 1). However, in practice, this yields a sub-optimal solution. Therefore, a levenberg-marquardt non-linear optimization step is applied to find the optimal value for H.

Once calibration is accomplished and H is found, obstacles can be detected. Figure 3 shows an example of homography being used for obstacle detection.

The top-left and top-right images are the left and right camera input images, respectively. The bottom-left image is the left image, warped as it would be seen from the right camera, given the assumption that *all* features lie on the ground. Note that there is a triangular black area in the warped image, on the right side. This is due to a lack of information, since the left camera field of view does not extend as far right as the right camera field of view. The bottom-right image is a thresholded

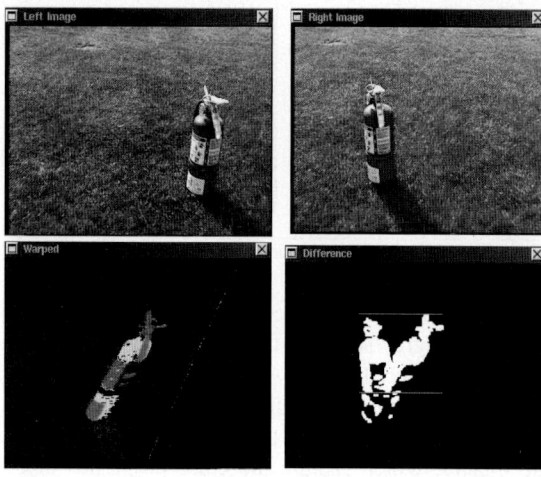

Figure 3: Homography example. The top images are input images. The bottom left image is a warped image. The bottom right image is the obstacle image.

difference image between the right camera image and the warped image. The two white areas in the difference image correspond to the portions of the obstacle which did not match the prediction, since it lies above the ground plane.

Previous work includes work by Storjohann [10], for indoor applications of homography and inverse perspective mapping. Batavia [1] has used a monocular version of homography, utilizing a single camera and known ego-motion, rather than two cameras, for highway obstacle detection. Santos-Victor [7] also used a monocular approach, but with a formulation that did not require knowledge of ego-motion. However, this approach requires the computation of normal-flow vectors. Bertozzi [2] used a stereo approach for detecting highway obstacles, with an on-line calibration-tuning ability.

2.2.2. Performance

This method is robust to the types of false positives which confound color segmentation. The sensitivity to true obstacles is determined by the image resolution, calibration accuracy, and field of view. This can be improved by increasing the image resolution and/or narrowing the field of view. In general, the same issues which affect stereo accuracy have an impact on homography accuracy.

Another issue is sensitivity to pitch variations. This variation can come from platform vibration, or it can come from a change in the terrain slope, which breaks the ground plane assumption. Large deviations in pitch from the calibrated conditions can lead to false positives, as the image warping process is dependant on a pre-determined pose.

2.3. Integration

Given two working, complimentary, obstacle detection methods, integration is an issue. Both methods have different strengths and weaknesses, and it is important to integrate them in such a way that the strengths of each are used to offset the respective weaknesses. For instance, color segmentation, although able to detect small obstacles and changes in color, is not sensitive to obstacle geometry, such as height. Homography is only able to find obstacles that are over a certain height. Both methods produce a list of candidate obstacles and centroid locations. It would be possible to just combine them in an 'OR' fashion. Alternatively, they could be 'AND'ed, so that both methods would have to detect a particular obstacle.

We use homography to act as a 'false positive' filter for color segmentation. When color segmentation detects an obstacle, homography is used to decide whether the obstacle is rising above the ground or not. If it is, then the object is classified as an obstacle. If it is not, then a decision has to be made whether to adaptively re-train the color segmentation system (in the case of global lighting change) or whether to simply ignore the object, assuming it is a temporary obstacle (such as a leaf). This decision is made based on the size of the obstacle. If it is extremely large, subtending most of the image, then it is likely that there is no object at all, and it is a global lighting change, and we re-train. Using homography as a filter allows us to adaptively re-train on the fly, without operator intervention.

3. Experimental Results

We have tested the fully integrated obstacle detection system offline, on image sequences, and online, using a mobile robot. We have also done extended duration testing using only color segmentation. The platform we use for online testing is described in the next section

3.1. Platform Description

The platform used for these experiments is a riding lawnmower and is pictured in Figure 4. The front bumper contains two CCD cameras and a SICK laser scanner.

Currently, an integrated PC, mounted in the rear, is used to communicate with the sensors and control the vehicle. The steering and throttle are hydraulically controlled, and are both actuated. A serial protocol is used to set steering and throttle positions. The planner and trajectory generation module takes pose information as input, and generates trajectory commands as output, and

Figure 4: Riding lawnmower with cameras and laser range finder.

passes them to the command and safety arbiter, which executes commands, based on input from the obstacle detection subsystem.

A dead-reckoning system is used for navigation, which integrates odometry information from two wheel-mounted encoders, along with heading information from a fiber optic rate gyro. Dead-reckoning will accumulate error over time. Over large distances, some form of globally-referenced localization will be required to bound the dead-reckoning error.

3.2. Results

The first test involves offline processing of two image sequences. The first sequence contains one true obstacle -- a small red fire extinguisher, and is the same sequence of Figure 3. The second sequence contains a false positive -- a large area of discolored grass and sand.

Figure 5 shows results from the two sequences. The top two images are the left and right camera images of the sand sequence. The middle left image is the color segmentation output on the left input image. The middle right image is a histogram of the *homography difference* image. Recall, the difference image is an absolute difference between the hue of the actual right input and the hue of the warped image. The bin centers are difference values, and the counts indicate the number of pixels in the difference image which fell into the corresponding bin. Peaks at high difference values indicate the presence of an obstacle. The middle right figure shows only a peak at very low difference values, indicating that homography does not detect an obstacle. Therefore, the output of the color segmentation is actually a false positive.

In contrast, the bottom two figures show an input image from the fire extinguisher sequence, and the corresponding difference histogram. Note the peak at higher difference values. This indicates the presence of an obstacle. The output of the color segmentation is true in this case.

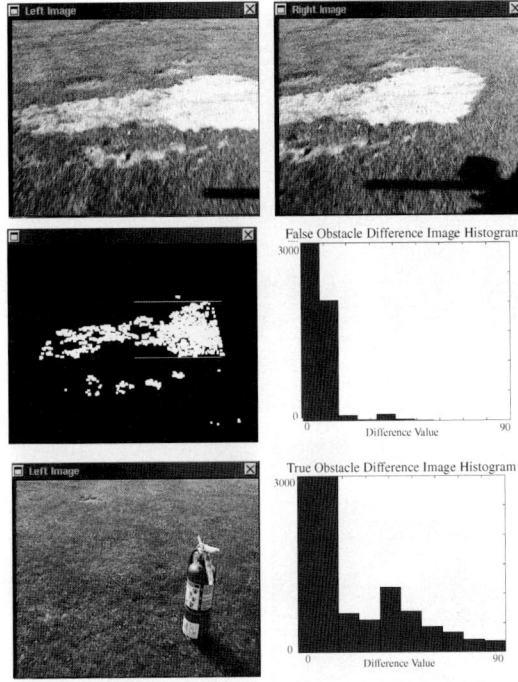

Figure 5: Offline homography results on true obstacle vs. false positive.

Our online testing made use of the lawnmower platform. The robot navigated autonomously through an oval pattern, travelling a total of 200m over 7 minutes. The color segmentation system starts with an initial training set of one image of grass. During this period, a "false obstacle," in the form of a green sheet, was set in front of the path. Figure 6 shows this. This sheet is meant to represent an area of grass which is of different color than the grass initially used for training. If operating alone, a human operator would have to decide whether or not to augment the training set with the new grass, since it appears as an obstacle.

Instead, color homography is used to validate the color segmentation output. The sheet is flat on the ground, so the homography prediction matches what is actually observed. Therefore, the object is declared a false positive. The decision to add this image to the color segmentation training set is made because the object is larger than a thresholded size. If the object had been smaller, it would have been ignored, and the mower would have continued.

3.3. Extended Duration Results

We conducted another test during a recent demonstration of the color segmentation system. In this test, only color segmentation was used. The mower runs in straight swaths of about 10 meters, then turns 180

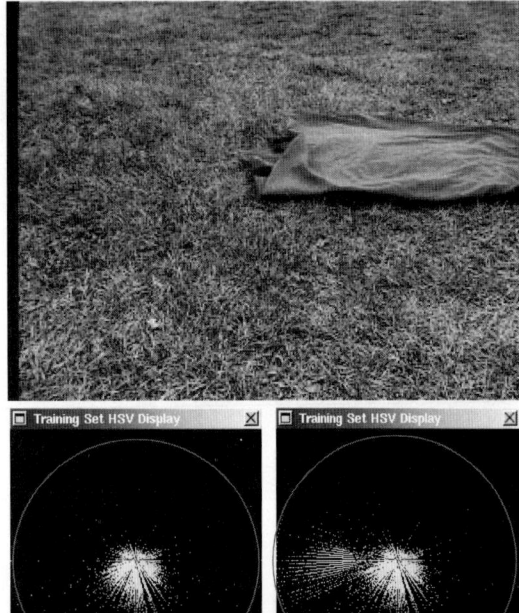

Figure 6: A green sheet simulating an area of new grass. The lower left image is a depiction of the training set before the new grass was added. The lower right is the training set after it was added.

degrees and repeats the pattern. The total distance travelled in this case is about 60 meters, and the total area mowed is about 80 square meters. Over a recent four day period, we repeated this pattern approximately 50 times, resulting in a total distanced travelled of about 3 km., and a total area covered of about 4 square km. Over these 50 trials, false positives were encountered on average of once every 250m of travel. True obstacles were also placed in its path, with 100% detection.

4. Conclusion and Future Work

We have demonstrated a novel integration of two vision-based obstacle detection methodologies: color segmentation, and color homography. Each method has strengths which compensate for the other's weaknesses, resulting in a robust method for obstacle detection. Furthermore, the homography is used to autonomously train the color segmentation system, allowing unsupervised training.

Future work includes further improving the robustness of the obstacle detection system and navigation system to allow for unattended operation over larger areas, on the order of 10 to 20 square kilometers. In particular, the homography approach is not limited to flat ground. Given terrain information, in the form of a digital map, along with accurate localization, the inverse perspective equations are solvable for arbitrary terrain. Also, the color segmentation can be improved through additional color-constancy work. In particular, accounting for the spectral contribution of varying sunlight should greatly reduce the number of false positives.

References

[1] P. Batavia and D. Pomerleau and C. Thorpe, "Overtaking Vehicle Detection Using Implicit Optical Flow," Proceedings of the IEEE Intelligent Transportation Systems Conference, Boston, MA, 1997.

[2] M. Bertozzi and A. Broggi and A. Fascioli, "Stereo Inverse Perspective Mapping: Theory and Applications," Image and Vision Computing, Vol 16, pp. 585-590, 1998.

[3] J. Hyams and M. Powell and R. Murphy, "Cooperative Navigation of Micro-rovers using Color Segmentation,' Journal of Autonomous Robots, Vol. 9, Num. 1, pp 7-16, August, 2000.

[4] V. Kravtchenko and J. Little, "Efficient Color Object Segmentation Using the Dichromatic Reflection Model," Proceedings of the IEEE Pacific Rim Conference on Communications, Computers, and Signal Processing, 1999.

[5] S. McKenna and Y. Raja and S. Gong, "Tracking Color Objects Using Adaptive Mixture Models," Image and Vision Computing 17, pp. 225-231, 1999.

[6] M. Ollis, "Perception Algorithms for a Harvesting Robot," Carnegie Mellon University Doctoral Dissertation, CMU-RI-TR-97-43, August, 1997.

[7] J. Santos-Victor and G. Sandini, "Uncalibrated Obstacle Detection Using Normal Flow," Machine Vision and Applications, Vol 9, pp 130-137, 1996.

[8] S. Shafer, "Using Color to Separate Reflection Components, "Color Research, Vol. 10, Num. 4, pp 210-128, 1985.

[9] A. Shiji and N. Hamada, "Color Image Segmentation Method Using Watershed Algorithm and Contour Information," Proceedings of the International Conference on Image Processing, Kobe, Japan, October, 1999.

[10] K. Storjohann and T. Zielke and H.A. Mallot and W. von Seelen, "Visual Obstacle Detection for Automatically Guided Vehicles," Proceedings of the International Conference on Robotics and Automation, pp 761-766, 1990.

[11] T. Williamson, "A High-Performance Stereo System for Obstacle Detection," Carnegie Mellon University Doctoral Dissertation, CMU-RI-TR-98-24, September, 1998.

[12] I. Ulrich and I. Nourbakhsh, "Appearance-Based Obstacle Detection," AAAI National Conference on Artificial Intelligence, Austin, TX, August 2000.

Acknowledgements

The authors would like to acknowledge the support of Iwan Ulrich, whose work provided the foundation for the color segmentation-based obstacle detection system described here, along with lower level camera driver support.

Visual Servoing Based on Multirate Sampling Control
– Application of Perfect Disturbance Rejection Control –

Hiroshi Fujimoto and Yoichi Hori

Department of Electrical Engineering, The University of Tokyo
E-mail: fuji@hori.t.u-tokyo.ac.jp, hori@hori.t.u-tokyo.ac.jp

Abstract

In this paper, novel multirate controllers are proposed for digital control systems, where it is restricted that the sampling period of plant output is comparatively longer than the control period of plant input. The proposed controllers can assure perfect disturbance rejection at M intersample points in the steady state. Moreover, the novel scheme of repetitive control is proposed based on the open-loop estimation and switching function, which enables to reject periodical disturbance without any sacrifice of the feedback characteristics. The proposed methods are applied to the visual servo system, and the advantages of these approaches are demonstrated by simulations.

1 Introduction

A digital control system usually has two samplers $\mathcal{S}$ for the reference signal $r(t)$ and the output $y(t)$, and one holder $\mathcal{H}$ of the input $u(t)$, as shown in Fig. 1. Therefore, there exist three time periods T_r, T_y, and T_u which represent the period of $r(t), y(t)$, and $u(t)$, respectively. The input period T_u is generally decided by the speed of the actuator, D/A converter, or the calculation on the CPU. On the other hand, the output period T_y is also determined by the speed of the sensor or the A/D converter. Practical control systems usually hold the restrictions on T_u and/or T_y. Thus, the conventional digital control systems make these three periods equal to the longer period between T_u and T_y.

In this paper, the digital control systems with longer sampling period $(T_u < T_y)$ are considered. This restriction may be general because D/A converters are usually faster than the A/D converters. The first example which has this restriction is the visual servo system of robot manipulator [1, 2]. Although the sampling period of the vision sensor such as a CCD camera is comparatively slow (over 33 [ms]), the control

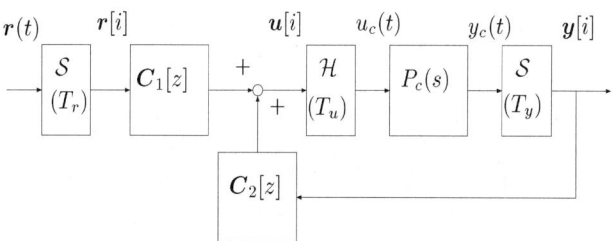

Figure 1: Two-degree-of-freedom control system.

period of joint servo is fast (under 1 [ms]). Therefore, the multirate controllers have been developed and implemented to the visual servo system [3, 4, 5].

The second example is the head positioning system of hard disk drives, in which the head position is detected by the discrete servo signal. Thus, the sampling frequency is restricted because it is determined by the rotation frequency and number of the servo signals. On the other hand, the control frequency of the actuator (voice coil motor) can be set faster than the sampling frequency of the head position. Therefore, the multirate estimation and control have been applied to hard disk drives in [6].

The third example is the velocity or position control of industrial motors with low precision encoder. In these systems, the sampling period cannot be set so fast, because the velocity information is not able to be detected precisely in the low speed region. Therefore, [7] has developed the instantaneous speed observer, which estimates the intersample velocity in use of the discrete-time observer.

For these systems with longer sampling periods, it is difficult to reject disturbance in high frequency region because the Nyquist frequency is relatively low. In this paper, multirate sampling control is introduced, in which the control input is changed N times during one sampling period. Using this scheme, novel multirate feedback controllers are proposed, which achieve disturbance rejection even in the semi-Nyquist frequency region. Moreover, the proposed methods are applied to the visual servo system of robot manipulator. In [6],

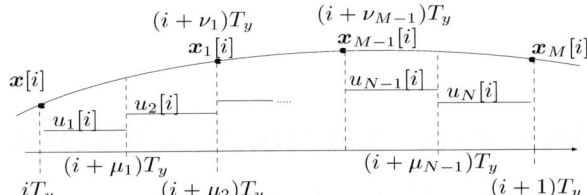

Figure 2: Multirate sampling control.

a novel multirate feedforward controller is proposed which assures the perfect tracking at M intersample points. Thus, this paper deals with only the feedback controller.

In the repetitive control system [8], conventional single-rate controllers do not have enough intersample performance to reject disturbance in the semi-Nyquist frequency region [9]. On the other hand, authors proposed a novel multirate feedback controller, which achieves the perfect disturbance rejection at M intersample points [6]. In this paper, the proposed approach is modified to be applicable to the visual servo system, in which the target object moves fast and periodically.

Repetitive feedback controllers based on the internal model principle have disadvantages that the closed-loop characteristics become worse and difficult to assure stability robustness [10]. Therefore, this paper proposes novel approach which never has these problems, based on the open-loop estimation with switching function and disturbance rejection by feedforward approach.

2 Design of the multirate feedback controller

In this section, the multirate feedback controller is proposed, which guarantees the perfect disturbance rejection at M intersample points at the steady state.

In the proposed multirate scheme, the plant input is changed N times during T_y and the plant state is evaluated M times in this interval, as shown in Fig. 2. The positive integers M and N are referred to as input and state multiplicities, respectively. N is determined by the hardware restriction. In this paper, the state multiplicity is defined as $M = N/n$, where n is the plant order.

In Fig. 2, $\mu_j (j = 0, 1, \cdots, N)$ and $\nu_k (k = 1, \cdots, M)$ are the parameters for the timing of the input changing and the state evaluation, which satisfy the conditions (1) and (2).

$$0 = \mu_0 < \mu_1 < \mu_2 < ... < \mu_N = 1 \quad (1)$$

$$0 < \nu_1 < \nu_2 < ... < \nu_M = 1 \quad (2)$$

If T_y is divided at same intervals, the parameters are set to $\mu_j = j/N, \nu_k = k/M$.

For simplification, the continuous-time plant is assumed to be SISO system in this paper. The proposed methods, however, can be extended to deal with the MIMO system by the same way as [11].

2.1 Plant discretization by multirate sampling

Consider the continuous-time plant described by

$$\dot{x}(t) = A_c x(t) + b_c u(t), \quad y(t) = c_c x(t). \quad (3)$$

The discrete-time plant discretized by the multirate sampling control of Fig. 2 becomes

$$x[i+1] = Ax[i] + Bu[i], \quad y[i] = Cx[i], \quad (4)$$

where $x[i] = x(iT_y)$, and matrices A, B, C and vectors u are given by

$$\left[\begin{array}{c|c} A & B \\ \hline C & O \end{array}\right] := \left[\begin{array}{c|ccc} e^{A_c T_y} & b_1 & \cdots & b_N \\ \hline c_c & 0 & \cdots & 0 \end{array}\right], \quad (5)$$

$$b_j := \int_{(1-\mu_j)T_y}^{(1-\mu_{(j-1)})T_y} e^{A_c \tau} b_c d\tau, \quad u := [u_1, \cdots, u_N]^T,$$

where $u_j[i]$ is the control input for $(i+\mu_{j-1})T_y \leq t < (i+\mu_j)T_y$ $(j = 1, \cdots, N)$. The intersample plant state at $t = (i+\nu_k)T_y$ is represented by

$$\tilde{x}[i] = \tilde{A}x[i] + \tilde{B}u[i], \quad (6)$$

where $\tilde{x}[i]$ is a vector composed of the intersample plant state $x_k[i] := x((i+\nu_k)T_y)$ of Fig. 2.

$$\begin{aligned}\tilde{x}[i] &:= [x_1^T[i], \cdots, x_M^T[i]]^T \\ &= [x_1^T((i+\nu_1)T_y), \cdots, x_M^T((i+1)T_y)]^T\end{aligned} \quad (7)$$

The coefficient matrices of (6) are given by

$$[\tilde{A} \mid \tilde{B}] := \left[\begin{array}{c|ccc} \tilde{A}_1 & \tilde{b}_{11} & \cdots & \tilde{b}_{1N} \\ \vdots & \vdots & & \vdots \\ \tilde{A}_M & \tilde{b}_{M1} & \cdots & \tilde{b}_{MN} \end{array}\right],$$

$$\tilde{A}_k := e^{A_c \nu_k T_y},$$

$$\tilde{b}_{kj} := \begin{cases} \mu_j < \nu_k : & \int_{(\nu_k-\mu_j)T_y}^{(\nu_k-\mu_{(j-1)})T_y} e^{A_c \tau} b_c d\tau \\ \mu_{(j-1)} < \nu_k \leq \mu_j : & \int_0^{(\nu_k-\mu_{(j-1)})T_y} e^{A_c \tau} b_c d\tau \\ \nu_k \leq \mu_{(j-1)} : & 0 \end{cases}$$

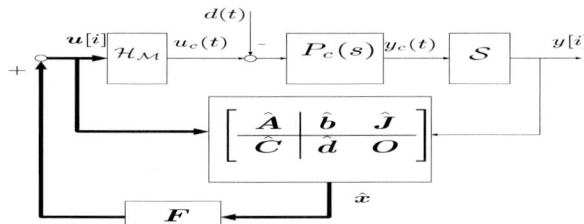

Figure 3: Multirate control with disturbance observer.

2.2 Design of the perfect disturbance rejection controller

In this section, the perfect disturbance rejection controller is proposed based on the state space design of the disturbance observer.

Consider the continuous-time plant model described by

$$\dot{x}_p(t) = A_{cp}x_p(t) + b_{cp}(u(t) - d(t)) \quad (8)$$
$$y(t) = c_{cp}x_p(t), \quad (9)$$

where $d(t)$ is the disturbance input. Let the disturbance model be

$$\dot{x}_d(t) = A_{cd}x_d(t), \quad d(t) = c_{cd}x_d(t). \quad (10)$$

For example, the step type disturbance is modeled by $A_{cd} = 0, c_{cd} = 1$, and the sinusoidal type disturbance with frequency ω_d is also modeled by

$$A_{cd} = \begin{bmatrix} 0 & 1 \\ -\omega_d^2 & 0 \end{bmatrix}, \quad c_{cd} = [1, 0]. \quad (11)$$

The continuous-time augmented system consisting of (8) and (10) is represented by

$$\dot{x}(t) = A_c x(t) + b_c u(t) \quad (12)$$
$$y(t) = c_c x(t) \quad (13)$$

$$A_c := \begin{bmatrix} A_{cp} & -b_{cp}c_{cd} \\ O & A_{cd} \end{bmatrix}, b_c := \begin{bmatrix} b_{cp} \\ 0 \end{bmatrix}, x := \begin{bmatrix} x_p \\ x_d \end{bmatrix}$$

$$c_c := [c_{cp}, 0].$$

Discretizing (12) with the multirate hold $\mathcal{H}_\mathcal{M}$, the intersample plant state at $t = (i + \nu_k)T_y$ can be calculated from the kth row of (6) by

$$x[i + \nu_k] = \tilde{A}_k x[i] + \tilde{B}_k u[i] \quad (14)$$

$$\tilde{A}_k = \begin{bmatrix} \tilde{A}_{pk} & \tilde{A}_{pdk} \\ O & \tilde{A}_{dk} \end{bmatrix}, \tilde{B}_k = \begin{bmatrix} \tilde{B}_{pk} \\ O \end{bmatrix}.$$

For the plant (12) discretized by (4), the discrete-time observer on the sampling points is obtained from the Gopinath's method by

$$\hat{v}[i+1] = \hat{A}\hat{v}[i] + \hat{b}y[i] + \hat{J}u[i] \quad (15)$$
$$\hat{x}[i] = \hat{C}\hat{v}[i] + \hat{d}y[i]. \quad (16)$$

As shown in Fig. 3, let the feedback control law be

$$u[i] = F_p \hat{x}_p[i] + F_d \hat{x}_d[i] = F\hat{x}[i], \quad (17)$$

where $F := [F_p, F_d]$. Letting e_v be the estimation errors of the observer ($e_v = \hat{v} - v$), the following equation is obtained.

$$\hat{x}[i] = x[i] + \hat{C}e_v[i]. \quad (18)$$

From (14) to (18), the closed-loop system is represented by

$$\begin{bmatrix} x_p[i + \nu_k] \\ x_d[i + \nu_k] \\ e_v[i+1] \end{bmatrix} = \begin{bmatrix} \tilde{A}_{pk} + \tilde{B}_{pk}F_p & \tilde{A}_{pdk} + \tilde{B}_{pk}F_d & \tilde{B}_{pk}F\hat{C} \\ O & \tilde{A}_{dk} & O \\ O & O & \hat{A} \end{bmatrix} \begin{bmatrix} x_p[i] \\ x_d[i] \\ e_v[i] \end{bmatrix}.$$

Because full row rank of the matrix $\tilde{B}_{pk}$ can be assured [12], F_d can be selected so as to the (1,2) element of the above equation becomes zero for all $k = 1, \cdots, M$.

$$\tilde{A}_{pdk} + \tilde{B}_{pk}F_d = O \quad (19)$$

The simultaneous equation of (19) for all k becomes

$$\tilde{A}_{pd} + \tilde{B}_p F_d = O, \quad (20)$$

$$[\tilde{A}_{pd} \mid \tilde{B}_p] := \begin{bmatrix} \tilde{A}_{pd1} & \tilde{B}_{p1} \\ \vdots & \vdots \\ \tilde{A}_{pdM} & \tilde{B}_{pM} \end{bmatrix}. \quad (21)$$

From (20), F_d is obtained by

$$F_d = -\tilde{B}_p^{-1}\tilde{A}_{pd}. \quad (22)$$

By (19), the influence from disturbance $x_d[i]$ to the intersample state $x_p[i + \nu_k]$ at $t = (i + \nu_k)T_y$ can become zero. Moreover, $x_p[i]$ and $e_v[i]$ on the sampling point converge to zero at the rate of the eigenvalues of $\tilde{A}_{pM} + \tilde{B}_{pM}F_p$ and $\hat{A}$ (the poles of the regulator and observer). Therefore, the perfect disturbance rejection is achieved ($x_p[i + \nu_k] = 0$) in the steady state. The poles of the regulator and observer should be tuned by the tradeoff between the performance and stability robustness.

Substituting (15) for (17), the feedback type controller is obtained by

$$\begin{bmatrix} \hat{v}[i+1] \\ u[i] \end{bmatrix} = \begin{bmatrix} \hat{A} + \hat{J}F\hat{C} & \hat{b} + \hat{J}F\hat{d} \\ F\hat{C} & F\hat{d} \end{bmatrix} \begin{bmatrix} \hat{v}[i] \\ y[i] \end{bmatrix}. \quad (23)$$

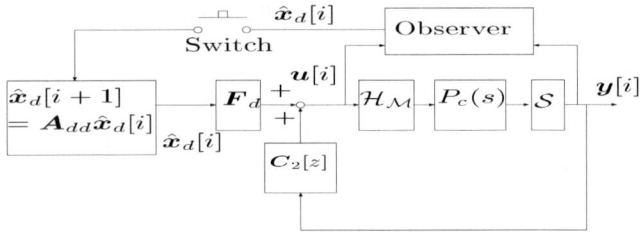

Figure 4: Feedforward repetitive control.

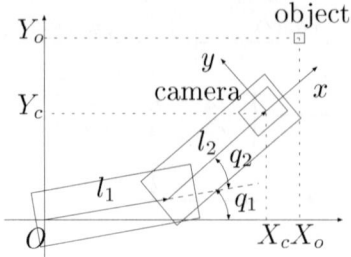

Figure 5: Two-link DD robot with camera.

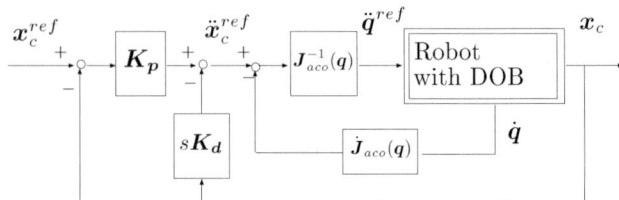

Figure 6: Workspace controller (inner-loop).

3 Repetitive control based on multirate control

In this section, two multirate repetitive controllers are proposed, which are 1) feedback approach based on internal model principle and 2) feedforward disturbance rejection approach based on the open-loop estimation and switching function.

3.1 Feedback repetitive control

The periodic disturbance of $T_0 := 2\pi/\omega_0$ is represented by

$$d(t) = a_0 + \sum_{k=1}^{\infty} a_k \cos k\omega_0 t + b_k \sin k\omega_0 t, \quad (24)$$

where ω_0 is known, a_k and b_k are unknown parameters. Letting the disturbance model (10) be (24), the repetitive feedback controller is obtained by (23), which has internal model $s^2 + (k\omega_0)^2$ in discrete-time domain. Moreover, the repetitive disturbance is perfectly rejected ($\boldsymbol{x}_p[i+\nu_k] = 0$) at M intersample points in the steady state.

3.2 Feedforward repetitive control

The repetitive feedback control based on the internal model principle has disadvantages that the closed-loop characteristics become worse and difficult to assure stability robustness [10]. Therefore, in this section, novel repetitive controller based on the open-loop estimation and feedforward disturbance rejection is proposed, as shown in Fig. 4.

The repetitive disturbance is estimated by the open-loop disturbance observer. When the estimation converges to the steady state, the switch turns on at $t = t_0$. After that, the switch turns off immediately. The repetitive disturbance is calculated by (25) from the initial value $\hat{\boldsymbol{x}}_d[t_0]$ which has the amplitude and phase information of the disturbance.

$$\hat{\boldsymbol{x}}_d[i+1] = \boldsymbol{A}_{dd}\hat{\boldsymbol{x}}_d[i], \boldsymbol{A}_{dd} = e^{\boldsymbol{A}_{cd}T_y} \quad (25)$$

Because the feedforward gain $\boldsymbol{F}_d$ is obtained by (22), the perfect disturbance rejection is achieved at M intersample points. The advantage of this approach is that the stability robustness can be guaranteed easily only by the conventional feedback controller $\boldsymbol{C}_2[z]$ which does not have the internal model.

4 Applications to visual servo system

In this section, the visual servo problem is considered, in which the camera mounted on the robot manipulator is controlled to track the moving object, as shown in Fig. 5. It is assumed that the motion of the object is periodic, and repetitive disturbance rejection control is applied, which was developed in section 3.

Because the sampling period of the camera is longer than the control period of the joint servo, the proposed approach is applicable. In order to focus on the dynamical problems of the multirate system, the kinematical problems of the visual servo system are assumed to be simple; the object movement is in two-dimensional plane, and the depth information between the camera and the object z is known.

4.1 Modeling of visual servo system

First, the workspace position controller is designed in order to control camera position, as shown in Fig. 6 [13]. Because this controller employs the robust disturbance observer (DOB) in the joint space, each joint axis is decoupled. Therefore, if the non-singularity of the Jacobian $\boldsymbol{J}_{aco}$ is assured, the transfer function from the work space acceleration command $\ddot{\boldsymbol{x}}_c^{ref}$ to the work space position $\boldsymbol{x}_c(=[X_c, Y_c]^T)$ can be regarded

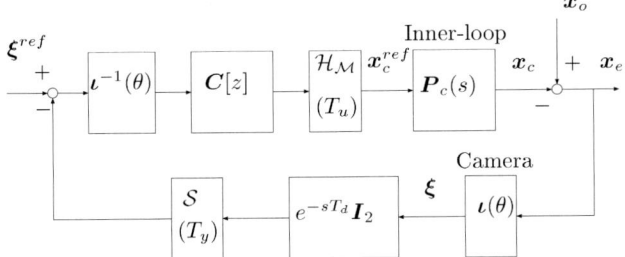

Figure 7: Visual servo system.

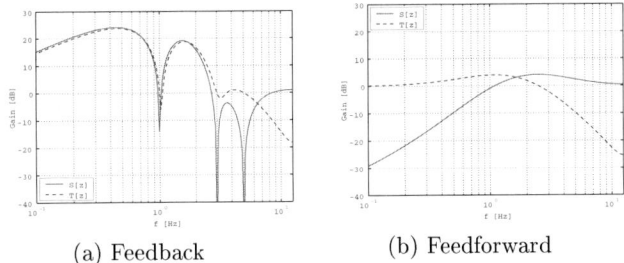

(a) Feedback   (b) Feedforward

Figure 8: Frequency responses $S[z]$ and $T[z]$.

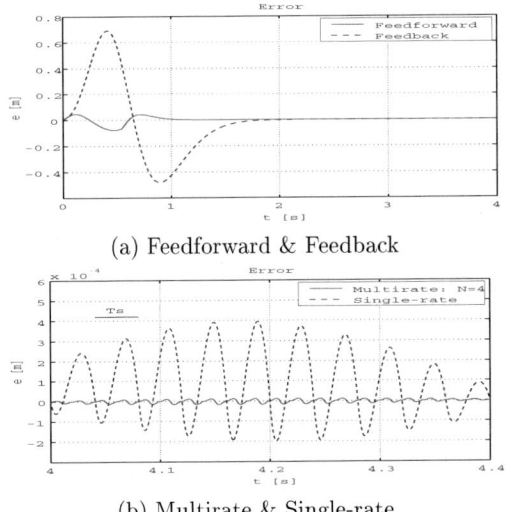

(a) Feedforward & Feedback

(b) Multirate & Single-rate

Figure 9: Position error $X_o - X_c$.

as a double integrator system in the frequency region under the cut-off frequency [13]. Letting x_c^{ref} be the control input u of the outer visual servo system, the plant is modeled by the analog system (26) because the sampling period of the inter-loop is very short (under 1 [ms]).

$$x_c(s) = P_c(s)u(s), \quad P_c(s) := \frac{K_p}{s^2 + K_d s + K_p} I_2 \quad (26)$$

In Fig. 6, the parameters of the position controller are set to $K_p = diag\{900, 900\}$ and $K_d = diag\{60, 60\}$.

Next, the perspective model of the camera is derived. In Fig. 6, the object position (x, y) on the camera coordinate system is determined only by the relative position between the camera position x_c and object position x_o. Therefore, the following model is obtained because the (x, y) is mapped to the feature point ξ on the image plane [3].

$$\xi = \frac{f}{z}\begin{bmatrix} x \\ y \end{bmatrix} = \frac{f}{z}\begin{bmatrix} \cos\theta & \sin\theta \\ -\sin\theta & \cos\theta \end{bmatrix}\begin{bmatrix} X_o - X_c \\ Y_o - Y_c \end{bmatrix} \quad (27)$$

Here f is the focus distance, z is the distance between the object and camera in $Z-$axis direction, and $\theta := q_1 + q_2$. (27) is defined by $\xi = \iota(\theta)(x_o - x_c) = \iota(\theta)x_e$.

Fig. 7 shows the proposed control system. In this paper, the desired feature ξ^{ref} is set to zero because the camera is controlled to be positioned just below the object. The movement of the object can be modeled as the output disturbance x_o. Therefore, the proposed method can achieve high performance tracking because the periodic motion can be rejected by the proposed method. Moreover, the control system of Fig. 7 is linearized and diagonalized by the inverse transformation $\iota^{-1}(\theta)$ of (27)[1]. Thus, the controllers can be designed independently in the x and y axes. The sampling period of the image and the control period of the position command x_c^{ref} are set to 40 [ms] and 10 [ms], respectively. Because the input multiplicity is

---
[1] In case of the setup of Fig. 7, $\iota^{-1}(\theta)$ is easily obtained from the inverse matrix of (27). In general case, it can be calculated from the inverse Jacobian [2].

$N = 4$ and the order of plant (26) is $n = 2$, perfect disturbance rejection is assured at $2(= N/n)$ intersample points. The T_d represents the time delay caused by the image processing. This delay generates the difficulty in control system. However, the proposed multirate sampling control is applicable to the system with time delay [14].

4.2 Simulation results

The repetitive disturbance is modeled on the $k = 1, 3, 5$ order[2]. The period of the object movement is $T_0 = 1[s]$.

Fig. 8 shows the sensitivity and complementary sensitivity functions $S[z]$ and $T[z]$ both of the feedback (Fig. 3) and the feedforward (Fig. 4) approaches. Fig. 8(a) indicates the disadvantages of the feedback repetitive controller, where the closed-loop characteristics become worse and difficult to assure stability robustness. On the other hand, in the proposed feedforward repetitive control (Fig. 4), the closed-loop characteris-

---
[2] These modes should be selected from the experimental analysis of the target motion.

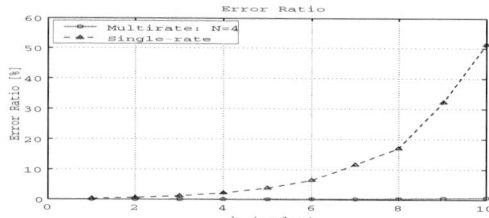

Figure 10: Error ratio $E_R(k)$.

tics depend only on $C_2[z]$ which does not need to have the internal model of repetitive disturbance. Therefore, the feedback characteristics are better than the feedback approach as shown in Fig. 8(b).

Fig. 9 shows the simulated results of position errors $X_o - X_c$ under the circular movement of the object. As shown in Fig. 9(a), the position error of the feedforward controller converges quickly after the switching action at $t_0 = 0.5$[s], while that of the feedback controller has large transient error. In the steady state, the errors of the plant position and velocity become zero at every $T_y/2$ by the proposed controller, as shown in Fig. 9(b). The intersample position error of the proposed multirate method is much smaller than that of the single-rate controller.

Fig. 10 shows analyzed results of the error ratio $E_R(k)$ for the disturbance order k. Considering the intersample response, the error ratio is calculated by

$$E_R^2(k) := \frac{\int_{t_s}^{t_s+kT_0}(X_o(t) - X_c(t))^2 dt}{\int_{t_s}^{t_s+kT_0} X_o^2(t) dt}, \quad (28)$$

where $X_o(t) = \sin k\omega_0 t, \omega_0 = 2\pi/T_0$, and t_s is selected as 2 [s] in order to evaluate the steady state. In the high frequency region close to the Nyquist frequency, the disturbance rejection performance is much improved by the proposed multirate control, compared with the single-rate controller. Therefore, it is found that the proposed method can demonstrate much effective performance for high-order disturbance.

5 Conclusion

In this paper, the repetitive disturbance rejection controllers were applied to the visual servo system of robot manipulator based on the multirate sampling control. Because the proposed control system assured the perfect disturbance rejection at M intersample points, the control system has achieved high tracking performance. Next, the novel scheme of repetitive control was proposed based on the open-loop estimation and switching function, which enabled to reject periodical disturbance without any sacrifice of the feedback characteristics. The experimental results of the proposed method will be presented in the next opportunity.

Finally, the authors would like to note that part of this research is carried out with a subsidy of the Scientific Research Fund of the Ministry of Education, Science, Sports and Culture of Japan.

References

[1] L. E. Weiss, A. C. Sanderson, and C. P. Newman, "Dynamic sensor based control of robots with visual feedback," *IEEE J. Robotics and Automation*, vol. 3, no. 5, pp. 404–417, 1987.

[2] S. Hutchinson, G. D. Hager, and P. I. Corke, "A tutorial on visual servo control," *IEEE Trans. Robotics and Automation*, vol. 12, no. 5, pp. 3745–3750, 1996.

[3] K. Hashimoto and H. Kimura, "Visual servoing with nonlinear observer," *IEEE Int. Conf. Robotics and Automation*, pp. 484–489, 1995.

[4] M. Nemani, T. C. Tsao, and S. Hutchinson, "Multi-rate analysis and design of visual feedback digital servo-control system," *ASME, J. Dynam. Syst., Measur., and Contr.*, vol. 116, pp. 44–55, March 1994.

[5] J. T. Feeddma and O. R. Mitchell, "Vision guided servoing with feature-based trajectory generation," *IEEE Trans. Robotics and Automation*, vol. 5, no. 5, pp. 691–700, 1989.

[6] H. Fujimoto, Y. Hori, T. Yamaguchi, and S. Nakagawa, "Proposal of perfect tracking and perfect disturbance rejection control by multirate sampling and applications to hard disk drive control," in *Conf. Decision Contr.*, pp. 5277–5282, 1999.

[7] Y. Hori, T. Umeno, T. Uchida, and Y. Konno, "An instantaneous speed observer for high performance control of dc servomotor using DSP and low precision shaft encoder," in *4th European Conf. on Power Electronics*, vol. 3, pp. 647–652, 1991.

[8] S. Hara, Y. Yamamoto, T. Omata, and M. Nakano, "Repetitive control system – a new-type servo system," *IEEE Trans. Automat. Contr.*, vol. 33, pp. 659–668, 1988.

[9] H. Fujimoto and Y. Hori, "Vibration suppression and optimal repetitive disturbance rejection control in semi-nyquist frequency region using multirate sampling control," in *Conf. Decision Contr.*, pp. 691–700, 2000.

[10] C. Kempf, W. Messner, M. Tomizuka, and R. Horowitz, "Comparison of four discrete-time repetitive algorithms," *IEEE Contr. Syst. Mag.*, vol. 13, no. 5, pp. 48–54, 1993.

[11] H. Fujimoto, A. Kawamura, and M. Tomizuka, "Generalized digital redesign method for linear feedback system based on N-delay control," *IEEE/ASME Trans. Mechatronics*, vol. 4, no. 2, pp. 101–109, 1999.

[12] M. Araki and T. Hagiwara, "Pole assignment by multirate-data output feedback," *Int. J. Control*, vol. 44, no. 6, pp. 1661–1673, 1986.

[13] T. Murakami and K. Ohnishi, "A study of stability and workspace decoupling control based on robust control in multi-degrees-of-freedom robot," *Trans. IEE of Japan*, vol. 113-D, no. 5, pp. 639–646, 1993. (in Japanese).

[14] H. Fujimoto, *General Framework of Multirate Sampling Control and Applications to Motion Control Systems*. PhD thesis, The University of Tokyo, December 2000.

Stacking Jacobians Properly in Stereo Visual Servoing

P. Martinet

LASMEA - GRAVIR
Blaise Pascal University of Clermont-Ferrand
63177 Aubière - Cedex, France
martinet@lasmea.univ-bpclermont.fr

E. Cervera

Robotic Intelligence Laboratory
Jaume-I University
12071 Castelló, Spain
ecervera@inf.uji.es

Abstract

Most visual servoing applications are concerned with geometrically modeled objects. In this paper, the problem of controlling a motion by visual servoing around an unknown object with a stereovision system is addressed. The main goal is to move the end-effector around the object in order to observe several viewpoints of the object for other tasks, e.g. inspection or grasping. The present work uses the well-known image-based visual servoing approach with a point, but the importance of the relationship between the end-effector and camera frames is clarified and emphasized. This relationship is needed for properly stacking the Jacobians or interaction matrices of each camera. A comparison with a visual servoing approach with a direct stacking of the Jacobians is presented. The centroid of a region, obtained by color segmentation, is used to move around the observed object. Experiments are developed on a PA-10 robot connected to a real time stereovision system, with two cameras mounted on the end-effector. Experimental results demonstrate the importance of a proper definition of the stacked Jacobians, to avoid undesired motions in the servoing task. Particularly, when turning around an unknown object, undesired motions on roll angle of the stereovision system can be avoided.

1 Introduction

Visual servoing applications have grown significantly since the last decade. Though in the first approaches, the scene observed by the camera was relatively simple, many works concerning unknown and complex objects have been developed recently.

Some methods require an initial learning step to obtain information characterizing the interaction between the sensor apparatus and the environment [3, 6]. In this case, it is necessary to get information from a predefined trajectory. To do so, the method proposed by Berry et al. [1] performs automatic motions around an unmodeled object in order to learn this interaction.

In this paper, the problem of moving around an object is addressed: no geometric model is needed and a stereovision system is used. Many works have been done in the field of visual servoing using a stereovision system [2, 5, 10]. Most of them use the stereovision system to recover the depth. Others use the epipolar constraint in order to execute the point to point matching process.

Recent developments in stereo-visual servoing have proved the theoretical soundness of the approach. Lamiroy et al. [7] present a solution to integrate the epipolar constraint directly in the control law. They rewrite the minimization problem under the optimization of the epipolar constraint, and show that, in the noiseless case and using rigid control points, both the classical and constrained approaches are identical.

Malis et al. [8] have formalized a *multi-cameras* visual servoing approach. They consider a system with N cameras which delivers a set $\mathbf{s} = (\mathbf{s}_1^T \mathbf{s}_2^T \cdots \mathbf{s}_N^T)^T$ of sensor signals ($dim(\mathbf{s}_i) = n_i$). Assuming that each sensor signal can control all the end-effector d.o.f m ($m \leq n_i$), they rewrite the global interaction relationship as:

$$\dot{\mathbf{s}} = \begin{pmatrix} \mathbf{L}_1 & 0 & \cdots & 0 \\ 0 & \mathbf{L}_2 & \cdots & 0 \\ \vdots & \vdots & \ddots & \vdots \\ 0 & 0 & \cdots & \mathbf{L}_N \end{pmatrix} \begin{pmatrix} {}^1\mathbf{M}_e \\ {}^2\mathbf{M}_e \\ \vdots \\ {}^N\mathbf{M}_e \end{pmatrix} {}^e\mathbf{v}$$

$$= \mathbf{L}\mathbf{M}_e {}^e\mathbf{v} \qquad (1)$$

where $\mathbf{L}_i$ represents the interaction matrix (or Jacobian matrix) of the ith camera, and ${}^i\mathbf{M}_e$ the transformation matrix between the velocity of the ith camera and the robot end-effector velocity. They define a global task function $\mathbf{e} = \mathbf{C}\dot{\mathbf{s}} = \sum_{i=1}^{N} k_i \mathbf{e}_i$ as a weighted mean of the task function relative to each camera, and demonstrate some properties in convergence and stability.

In our work, this scheme is applied to a stereo rig composed of two cameras. We show that less d.o.f.

can be controlled, by defining an appropriate hybrid task.

The paper outline is as follows: first, the modeling aspect is developed; secondly, the *task-function* approach is applied to obtain the control law. Next, results obtained at video rate with our robotic platform vision system are shown. Finally, some conclusions and possible extensions are presented.

2 Modeling

The main goal of this work is the positioning of the end-effector with respect to a fixed object, and to perform motions around it. A stereo rig is rigidly attached to the end-effector. No model of the object is known a-priori, thus limiting the choice of features for tracking [1]. Nevertheless, the position of the cameras in the end-effector frame and its intrinsic parameters are roughly known, without any special calibration.

The proposed approach uses 2D features extracted from regions in the image, segmented by color. Such features can be the centroid of the region, its size, the aspect ratio, and the angle of its first axis of inertia.

In this first work, only a point feature (the centroid of the blob) is used. Its observation by the stereo pair mounted on the end-effector makes possible the 3D positioning task. At the same time, the 3 remaining d.o.f are used for a secondary task, e.g. moving the end-effector around the object while keeping the fixed relative position.

Our stereovision system is composed of two parallel cameras. Figure 1 illllustrates the case when both cameras observe a 3D point P.

Let us define $\mathcal{F}_e$ as the Cartesian frame attached to the end-effector, $\mathcal{F}_l$ as the frame attached to the left camera, and $\mathcal{F}_r$ as the frame attached to the right camera.

The feature vector is defined as $\mathbf{s} = (u_l, v_l, u_r, v_r)^T$ where $(u_l, v_l)^T$ and $(u_r, v_r)^T$ are the image coordinates of the point, observed by the left and right cameras respectively.

2.1 First control law: *real stereo*

Let $^e\mathbf{v}$ be the kinematic screw applied to the robot end-effector. According to the multi-cameras visual servoing formulated in equation (1), the relationship between the time derivative of the feature vector and the end-effector screw is

$$\dot{\mathbf{s}} = \begin{pmatrix} \mathbf{L}_l\,^l\mathbf{M}_e \\ \mathbf{L}_r\,^r\mathbf{M}_e \end{pmatrix}\,^e\mathbf{v} = \mathbf{L}_{st}\,^e\mathbf{v} \qquad (2)$$

where

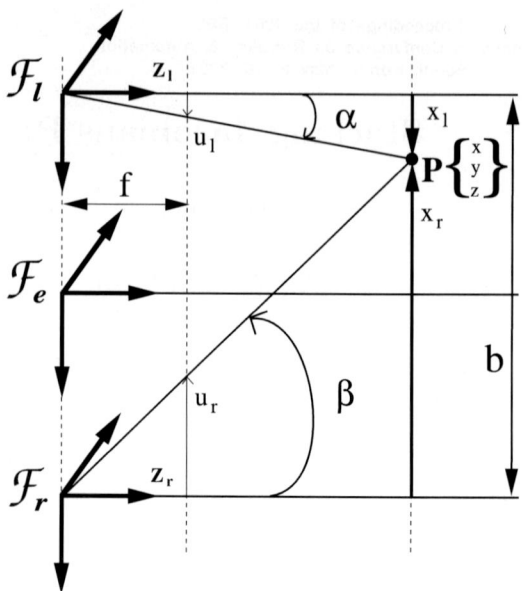

Figure 1: Stereovision: Case of a 3D Point.

- $\mathbf{L}_l$ and $\mathbf{L}_r$ are the interaction matrices relative to the left and right cameras respectively, defined by ($i = r$ or l) :

$$\begin{pmatrix} -\frac{F_u}{z} & 0 & \frac{u_i}{z} & \frac{u_i v_i}{F_v} & -F_u - \frac{u_i^2}{F_u} & \frac{v_i F_u}{F_v} \\ 0 & -\frac{F_v}{z} & \frac{v_i}{z} & F_v + \frac{v_i^2}{F_v} & -\frac{u_i v_i}{F_u} & -\frac{u_i F_v}{F_u} \end{pmatrix}$$

- $^l\mathbf{M}_e$ and $^r\mathbf{M}_e$ are the transformation matrices of the screw between the left and right camera frames and the end-effector frame. Given frames $\mathcal{F}_e$ and $\mathcal{F}_i$, the relationship between the screws is

$$^i\mathbf{v} = {^i\mathbf{M}_e}\,^e\mathbf{v} \qquad (3)$$

where the transformation matrix $^i\mathbf{M}_e$ is

$$^i\mathbf{M}_e = \begin{pmatrix} ^i\mathbf{R}_e & [^i\mathbf{t}_e]_\times\,^i\mathbf{R}_e \\ \mathbf{O}_3 & ^i\mathbf{R}_e \end{pmatrix} \qquad (4)$$

Though the resulting interaction matrix $\mathbf{L}_{st}$ is the same as that obtained by Maru *et al.* [10], our development is somewhat simpler and it is easier to generalize to other configurations of the cameras.

2.2 Second control law: *stacked-mono*

It is widely accepted in monocular visual servoing that the interaction matrix (the jacobian) of a set of points is constructed by stacking every interaction matrix of each single point.

One is tempted to apply this method directly to stereo vision, and thus, a simpler interaction matrix is obtained, if both matrices $^l\mathbf{M}_e$ and $^r\mathbf{M}_e$ are neglected.

In this case, the fusion of the sensor information is processed directly in the interaction matrix despite of the frame where they are defined. So, a more *classical* form is obtained:

$$\dot{\mathbf{s}} = \begin{pmatrix} \mathbf{L}_l \\ \mathbf{L}_r \end{pmatrix} {}^e\mathbf{v} = \mathbf{L}_{sm} {}^e\mathbf{v} \qquad (5)$$

The interaction matrix is similar to that obtained by stacking the matrices of several points, hence the name of the control law.

2.3 Theoretical comparison

It can be shown that the null space of the stereo interaction matrix $\mathbf{L}_{st}$ is always spanned by the three vectors

$$\begin{pmatrix} 0 \\ z \\ -y \\ 1 \\ 0 \\ 0 \end{pmatrix} \begin{pmatrix} -z \\ 0 \\ x \\ 0 \\ 1 \\ 0 \end{pmatrix} \begin{pmatrix} y \\ -x \\ 0 \\ 0 \\ 0 \\ 1 \end{pmatrix} \qquad (6)$$

which are in fact the same for the null space of the interaction matrix associated to a 3D point $\mathbf{p}_e = (x, y, z)^T$ in the robot end-effector frame

$$\begin{aligned}\mathbf{L}_{3D} &= \begin{pmatrix} -\mathbf{I}_3 & [\mathbf{p}]_\times \end{pmatrix} \\ &= \begin{pmatrix} -1 & 0 & 0 & 0 & -z & y \\ 0 & -1 & 0 & z & 0 & -x \\ 0 & 0 & -1 & -y & x & 0 \end{pmatrix} \end{aligned} \qquad (7)$$

The interaction matrix $\mathbf{L}_{st}$ can be rewritten as:

$$\mathbf{L}_{st} = \mathbf{L}_{3D}^{st} \mathbf{L}_{3D} \qquad (8)$$

where the matrix $\mathbf{L}_{3D}^{st}$ is defined as:

$$\mathbf{L}_{3D}^{st} = \begin{pmatrix} \frac{\partial \mathbf{s}_l}{\partial^l \mathbf{p}} {}^l\mathbf{R}_e \\ \frac{\partial \mathbf{s}_r}{\partial^r \mathbf{p}} {}^r\mathbf{R}_e \end{pmatrix} \qquad (9)$$

thus it is composed of the partial derivatives of the image points with respect to the velocity of the 3D point. The matrix is full rank, i.e., its null space is empty, since there is no motion of the 3D point which leaves both images unaffected. Indeed, only the motion along the projection ray leaves an image constant, but, since the cameras are not coincident, their projection rays are obviously different. Consequently, the null space of matrix $\mathbf{L}_{st}$ is the same as that of $\mathbf{L}_{3D}$.

In our experiments, only one point is used, thus 3 d.o.f. of the end-effector remain free for moving around the object.

In the second control law, excepting in some singular cases, the dimension of the null space of the interaction matrix $\mathbf{L}_{sm}$ is always 2. The reason is that the 3D relationship between both image points has been lost. As a consequence, in the equilibrium state (when the 3D point is centered in regard with the stereovision sensor apparatus), only 2 d.o.f. are available to perform motions around the object.

We show in the experiments that an undesired rotation around the z axis is present as a side-effect, due to the wrong dimension of the null space of the interaction matrix.

3 Control

The control law used in this study is based on the *Task function* formalism [11], firstly applied to visual servoing by Espiau *et al.* [4]. In this approach, the control is directly specified in terms of regulation in the image. It may be noted that this approach has the advantage of avoiding the intermediate step of the 3D estimation of the target with regard to the end effector [9, 12]. For a given robotics task, a *target image* is built, corresponding to the desired position of the end effector with regard to the environment. If the image jacobian is not full rank (number of d.o.f > number of independent visual features), it is possible to use an hybrid task. In an hybrid task, the primary task $\mathbf{e}_1$ maintains a visual constraint during the trajectory, while the secondary task $\mathbf{e}_2$ can be seen as representing a minimization of a secondary cost h_s.

A global *task function* $\mathbf{e}$ is then defined as:

$$\mathbf{e} = \mathbf{W}^+ \mathbf{e}_1 + \gamma(\mathbb{I}_n - \mathbf{W}^+\mathbf{W}) \frac{\partial h_s}{\partial r}^T \qquad (10)$$

where $\mathbf{W}^+$ and $(\mathbb{I}_n - \mathbf{W}^+\mathbf{W})$ are two projection operators which guarantee that the camera motions due to the secondary task are compatible with the regulation of $\mathbf{s}$ to $\mathbf{s}^*$. $\mathbf{W}$ is a full rank matrix *with the same null space* as that of the interaction matrix. The parameter γ is used to tune the preponderance between the primary and the secondary task.

Considering a motionless environment, the control law has the following expression:

$$\mathbf{v} = -\lambda \mathbf{e} - \gamma(\mathbb{I}_n - \mathbf{W}^+\mathbf{W}) \frac{\partial}{\partial t} \left(\frac{\partial h_s}{\partial r} \right)^T \qquad (11)$$

This control law is applied to both presented modelings, where matrix $\mathbf{W}$ is defined as follows for each control law:

	Real stereo	Stacked-mono
$\mathbf{W}$	$\mathbf{L}_{3D}^*$	$\mathbf{L}_{sm}^*$

The symbol * is used to precise that the corresponding expression is evaluated at the equilibrium situation.

4 Experimental results

The stereo system consists of two NTSC color cameras mounted on the end-effector of a Mitsubishi PA-10 manipulator. Cameras are coarsely positioned, being approximately mounted with the same orientation, and at equal distances from the end-effector's origin. No calibration procedure has been used.

A video-rate color segmentation system is used which extracts colored regions from an image and delivers the coordinates of its centroid, its aspect ratio, and the orientation of its major axis of inertia.

Each camera is connected to one of such image processing systems. Though the system is capable of sustaining a 60 Hz frame rate, only even or odd frames are used, due to alignment problems with interlaced frames, thus reducing the frame rate to 30 Hz.

An overview of the stereovision system is depicted in Fig. 2.

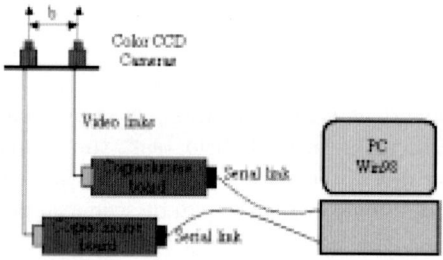

Figure 2: Overview of the stereovision system.

Though our setup is equivalent to the presented by Maru *et al.* [10], it must be noted that our task is under-constrained, and a secondary task has been defined for the motion around the object. Their object is modeled as a square and the feature vector is composed of four points (the corners).

4.1 Estimation of depth

In Fig. 1, f represents the focal length. We can write:

$$\begin{cases} sin(\alpha) = \frac{u_l}{F_u} = \frac{x_l}{z_l} \\ sin(\beta) = \frac{u_r}{F_u} = \frac{x_r}{z_r} \end{cases} \quad (12)$$

and finally:

$$x_l = x_r + b \quad (13)$$

where $F_u = \frac{f}{du}$ is the focal length along of the u axis.

With the relations 12 and 13, the depth of the observed point can be estimated as:

$$z = z_r = z_l = b.\frac{F_u}{u_l - u_r} \quad (14)$$

As a result, it is very simple to show one of the main advantages of a stereovision system in regard with a monocular vision system: the estimation of the depth. This estimation can be provided by:

$$\hat{z} = b.\frac{F_u}{u_l - u_r}$$

The Mitsubishi PA-10 manipulator has 7 d.o.f and is mounted on a mobile platform (XR4000 from Nomadic Inc.). In this implementation, the arm is only used and controlled as a Cartesian frame with 6 d.o.f.

The experimentation has been split in two steps. In the first step, a positioning task is executed during 300 iterations (one iteration corresponds to 33 ms). Then, the second step consists in a secondary task using a sinusoidal wave translation signal in x and y direction ($T_x = A_x.\omega_x.cos(\omega_x.t)$, $T_y = -A_y.\omega_y.sin(\omega_y.t)$ with $A_x = A_y = 0.6\ m$ and $\omega_x = \omega_y = 0.2\pi\ rd/s$). The aim of the secondary task is to describe a circle trajectory on a sphere while fixing the object centered in the image plane at a given distance.

The following table shows the different parameters (intrinsic and extrinsic) of both cameras, which have been roughly estimated:

F_u	F_v	b
300	450	118mm

The gains in the control laws are fixed to 1 for λ and 1/5 for γ.

4.2 Positioning task

In this paragraph, some results obtained in the real context when using both laws (*real stereo* and *stacked-mono*) are compared. The curves or graphs on the left side correspond to the *stacked-mono* control law, while those on the right side correspond to the *real stereo*. The reference feature to reach at the equilibrium is arbitrary fixed to $\mathbf{s}^* = (40, 0, -40, 0)^T$.

Figure 3 presents the servoing task during all the experimentation (both positioning and moving around the object).

Figure 4 gives the trajectories of the $2D$ points (left and right) in the same image plane.

Figures 5, 6, and 7, present the servoing task **only** during the positioning task. The sensor signals and the control vector have an exponential decay, but there is a persistent offset at the equilibrium. In fact, the sensor apparatus is not well calibrated, and the equilibrium sensor vector has been defined without taking into account this fact. To solve this problem, one way is to learn the desired sensor vector at the equilibrium with the uncalibrated

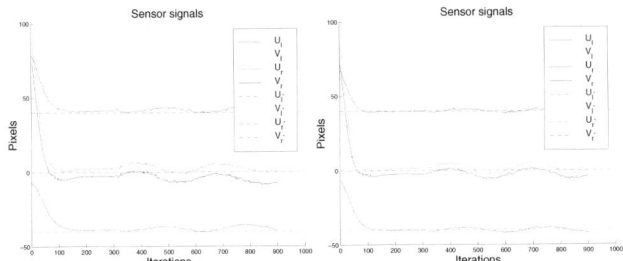

Figure 3: Sensor signals during the whole task

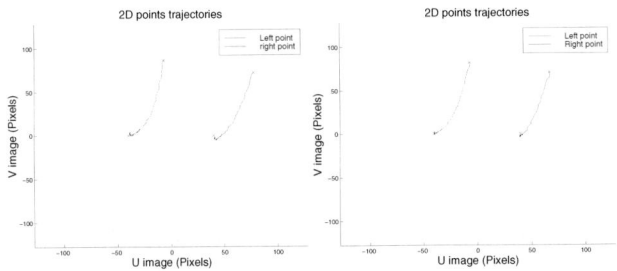

Figure 4: Image point trajectories

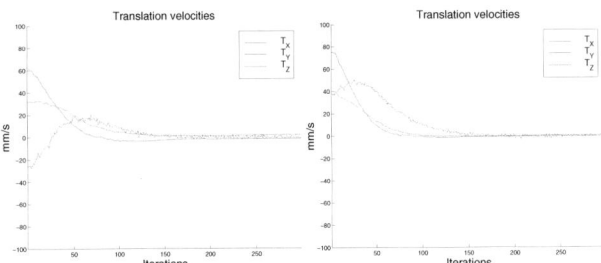

Figure 6: Translation velocities

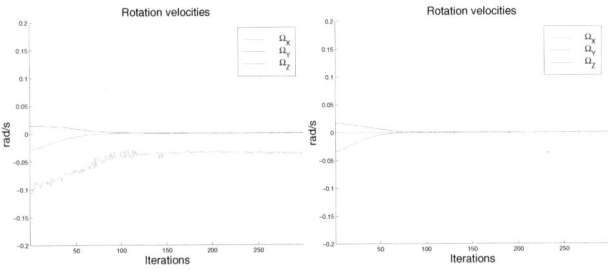

Figure 7: Rotation velocities

sensor. This is the reason why, in the *stacked-mono* approach, there exists a persistent rotation velocity in z direction (Ω_z). However, in the *real stereo*, this motion is cancelled due to the proper choice of matrix $\mathbf{W}$.

4.3 Secondary task

Figures 8, 9, and 10, present the servoing task during the secondary task.

The sensor signals do not remain in their equilibrium values: an offset due to the tracking error is present. In addition, in figures 9 and 10 the effect of the secondary task can be verified: the translation velocities T_x and T_y produce rotation velocities on Ω_y, Ω_x and a translation velocity on T_z (this corresponds to the vectors $\begin{cases} v_1 = (\quad 0, \quad z, \quad -y, \quad 1, \quad 0, \quad 0\quad)^T \\ v_2 = (\quad -z, \quad 0, \quad x, \quad 0, \quad 1, \quad 0\quad)^T \end{cases}$ of the kernel of the image jacobian).

Finally, in figure 10 (right side) the effect of the

choice of the matrix $\mathbf{W}$ which allows to suppress the rotation velocity Ω_z, can be verified. This fact demonstrates the main advantage of the *real stereo* control.

5 Summary and Conclusions

This paper has presented for the first time the application of stereo vision in an under-constrained visual servoing task. It has been shown that problems can appear if the relationships between frames are not properly taken into account. Particularly, it concerns some uncontrolled motions which can bring the robot in its joints limits.

On the contrary, when the modeling is correctly done, the use of a $3D$ point observed by a stereo vision system is sufficient to perform motions around an unknown and complex object.

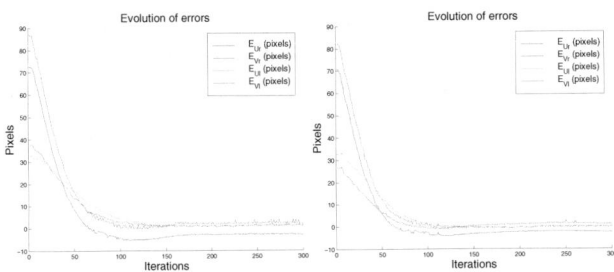

Figure 5: Evolution of the errors

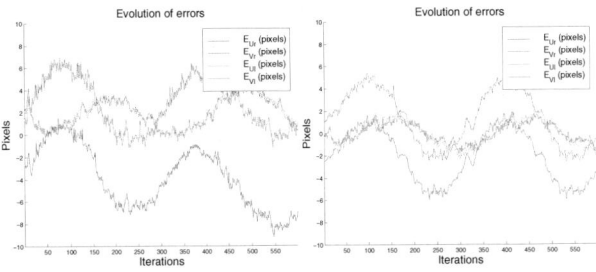

Figure 8: Evolution of the errors

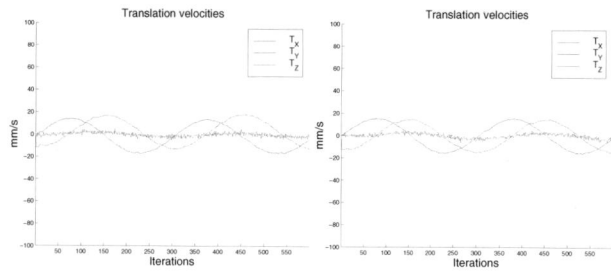

Figure 9: Translation velocities

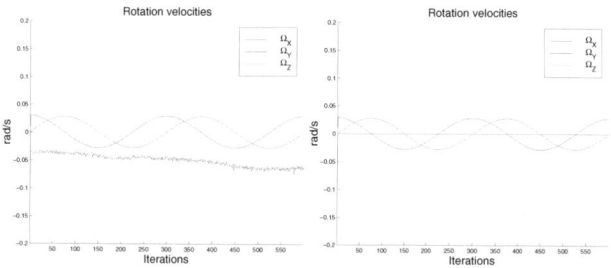

Figure 10: Rotation velocities

The choice of the centroid of the blob in both image planes is not the ideal invariant feature to perform this kind of task. As explained in [1], the center of a global bounding box is more relevant when using complex object. Future developments will concern the extension of the approach to a set of points, and other visual features (orientation and size of the blob).

First theoretical studies indicate that, when using two points in stereo, the null space of matrix $\mathbf{L}_{st}$ remains the same as that of $\mathbf{L}_{3D}$, which corresponds to a rotation around the line joining both points. With additional points, the null space is empty, thus we are interested in finding out which other properties are shared by both interaction matrices.

Acknowledgement

This work is partially funded by the Valencian Government under grant GV99-67-1-14, and a grant for a temporal stay of P. Martinet at Jaume-I University.

References

[1] F. Berry, P. Martinet, and J. Gallice. Real time visual servoing around a complex object. *IEICE Transactions on Information and Systems, Special Issue on Machine Vision Applications*, E83-D(7):1358–1368, July 2000.

[2] J. Crowley, M. Mesrabi, and F. Chaumette. Comparison of kinematic and visual servoing for fixation. In *Proc. IROS'95*, volume 1, pages 335–341, Pittsburgh, USA, 1995.

[3] K. Deguchi and Takhashi I. Image based simultaneous control of robot and target object motions by direct image interpretation method. In *Proceedings of the IEEE International Conference on Intelligent Robots and Systems*, volume 1, pages 375–380, Kyongju, Korea, 17-21 October 1999. IROS'99.

[4] B. Espiau, F. Chaumette, and P. Rives. A new approach to visual servoing in robotics. *IEEE Transactions on Robotics and Automation*, 8(3):313–326, 1992.

[5] G. Hager, W. C. Chang, and A. S. Morse. Robot hand-eye coordination based on a stereo vision. *IEEE Control Systems Magazine*, 15(1):30–39, 1995.

[6] M. Jägersand, O. Fuentes, and R. Nelson. Experimental evaluation od uncalibrated visual servoing for precision manipulation. In *Proceedings of the IEEE International Conference on Robotics and Automation*, volume 3, pages 2874–2880, Albuquerque, USA, 1997. ICRA'97.

[7] B. Lamiroy, B. Espiau, N. Andreff, and R. Horaud. Controlling robots with two cameras: How to do it properly. In *Proceedings of the IEEE International Conference on Robotics and Automation*, pages 2100–2105, San Francisco, California, USA, 24-28 April 2000. ICRA'2000.

[8] E. Malis, F. Chaumette, and S. Boudet. Multicameras visual servoing. In *Proceedings of the IEEE International Conference on Robotics and Automation*, volume 4, pages 2759–2764, San Francisco, California, USA, 24-28 April 2000. ICRA'2000.

[9] P. Martinet, J. Gallice, and D. Khadraoui. Vision based control law using 3d visual features. In *Proceedings of the Second World Automation Congress*, volume 3, pages 497–502, Montpellier, France, May 1996. ISRAM'96.

[10] N. Maru, H. Kase, S. Yamada, A. Nishikawa, and F. Miyazaki. Manipulator control by visual servoing with stereo vision. In *Proc. IROS'93*, pages 1866–1870, Yokohama, Japan, 1993.

[11] C. Samson, M. Le Borgne, and B. Espiau. *Robot Control. The task function approach*. ISBN 0-19-8538057. Clarendon Press, Oxford, 1991.

[12] W. J. Wilson, C. C. Williams Hulls, and G. S. Bell. Relative end-effector control using cartesian position based visual servoing. *IEEE Transactions on Robotics and Automation*, 12(5):684–696, October 1996.

Visual Servoing: Path Interpolation by Homography Decomposition

Justin A. Borgstadt and Nicola J. Ferrier*
Department of Mechanical Engineering
University of Wisconsin-Madison
borgstad@robios6.me.wisc.edu   ferrier@robios6.me.wisc.edu

Abstract

In order to successfully perform visual servoing of robot manipulators the control algorithm must account for the physical limitations of the manipulator. These constraints define the robot workspace boundary. Any visual servoing control algorithm must avoid this boundary if stability of motion is to be ensured. In this paper, a method is developed by which a desired object path can be interpolated between two arbitrary object poses based on image feature extraction. This path is defined to be a continuous change in object pose between the initial and final desired poses which avoids the workspace boundary. The development of this method involves the parameterization of the 2-D displacement transformation or homography. By decomposing this homography a set of object path poses can be interpolated in either the image plane or reconstructed in the 3-D workspace. Implementation of a visual servoing procedure confirms the validity of the interpolated path with respect to workspace boundary avoidance.

Keywords: *visual servo, path, homography*

1 Introduction

Visual servoing is the process by which machine vision is used in feedback control to manipulate the robot end-effector to a desired pose. Systems of a "look-then-move" architecture can be grouped into two general control type categories: position-based control and image-based control [9]. Position-based control combines image feature information, the task space manipulator jacobian, and a camera calibration to estimate the 3-D pose of the end-effector. Feedback control is computed to reduce errors between the current pose of the end-effector and the desired target pose. While clearly separating the vision and robot control issues position-based control is calibration sensitive

---

*Person to whom correspondence should be addressed. This research supported in part by NSF IRI-9703352.

and does not guarantee image features will stay in the field of view [2]. In image-based control the error function is computed in the image directly. Camera calibration errors are eliminated but the image jacobian used to define the control law is non-linear and requires depth estimation [9]. Additionally, the task space jacobian is still needed to accurately manipulate the end-effector.

Any type of motion control is ultimately determined by the available workspace of the robot. The robot model not only describes the physical 3-D link parameters but also describes the joint types, joint placement, and physical limitations of the actuators. The workspace boundary is broadly defined as the limit to which the end-effector can be manipulated with the robot fully stretched out [7]. For all joint positions corresponding to the workspace boundary the task space jacobian is singular. These configurations can be termed workspace boundary singularities [7]. Additionally, a loci of robot configurations lies within the work-space boundary where again the task space jacobian is singular and are defined as interior workspace singularities [7]. Any motion command attempting to move the end-effector must account for these singularity configurations to ensure stability and convergent motion.

1.1 Previous Work

Image-based visual servoing methods have been proposed which rely on estimating the image jacobian at each iteration of the control law from measured image features and their respective depth estimations [2, 5, 9]. This type of linear control commands the path of each image feature to be a straight line from the current image pose to the desired image pose. For large displacements particularly involving large rotations this approach can command undesirable motion of the end-effector beyond the workspace boundary.

Several methods have been proposed to avoid workspace boundary singularities. One image-based approach partitions the image jacobian into components containing purely image feature parameters and components containing depth parameters [6]. This method is designed to effectively control large motions of the end-effector about the optical axis. Other hybrid methods combine aspects of both position based and image based control [4, 10, 15]. These methods decompose the 2-D displacement transformation or homography computed from displaced co-planar image features into rotational and translational components. These components are then used to construct an error function of translations and rotations defined in 3-D or a combination of 2-D and 3-D. For these methods the image feature paths are not obvious and there is no guarantee the image features will stay in the camera field of view during the motion. In [12] a method is proposed where the image feature point paths are confined to the image space by implementing a potential function which "repels" image features away from the edges of the image. Another possible solution is presented in [3] where the end-effector trajectory follows a straight line in the task space. Methods have also been presented which avoid any motion of the robot in the vicinity of its joint limits [11] and kinematic singularity positions [13] through functions based on knowledge of the robot jacobian at each pose. In [14] an end-effector representation is used in conjunction with parameterized projective translation and rotation transformations. A projective jacobian is proposed to allow both the error function and the commanded joint space motion of the manipulator to be computed entirely in projective space.

1.2 Goals

In this paper, an End-point Closed Loop (ECL) architecture [9] is implemented where a single fixed camera is used to view a rigid object attached to the end-effector. The main goal is to develop a method by which a "desired viable object path" can be interpolated between two arbitrary image poses, where "desired viable object path" refers to a path along which the object can be manipulated where the change in object pose is continuous and workspace boundary singularities are avoided. A desired path can be uniquely defined by choosing a particular decomposition of the 2-D homography or collineation mapping the projective displacement of the object features between the initial and final image poses. Using a known object model the interpolation of this desired path can then be represented in the task space by a 3-D reconstruction *or* mapped directly to the image space. A visual-servoing procedure is used to validate the path interpolation method as a means to stabilize motion control with respect to workspace boundary singularities.

2 Image Displacements

Interpolating a viable object path for a given object displacement requires knowledge of the initial and final poses as well as how the object is to be displaced. A homography is a mapping from 2-D projective space to 2-D projective space, which is used here to define the 2-D displacement transformation between two object poses in the image. A specific form of the homography is derived and decomposed to interpolate a unique path. This specific form is obtained by directly mapping a 3-D displacement transformation to the image plane using a pinhole camera model.

2.1 Object Representation

Points are chosen here to represent a set of recognizable and distinguishable object features. An object pose can be represented by a matrix consisting of the object feature point vectors as follows:

$$P_k = \begin{bmatrix} M_{0,k} & M_{1,k} & M_{2,k} & ... & M_{j,k} \end{bmatrix} \quad (1)$$

where the point $M_{j,k}$ is the *jth* object feature point in the *kth* object pose described in the task space. The projection of this object pose into the 2-D image pixel frame $\{p\}$ can be found using a generalized projection transformation as follows

$$\tilde{p}_k = \tilde{H}\tilde{P} \, ^c T_h P_k \quad (2)$$

Here $\tilde{P}$ represents the 3x4 normalized perspective projection matrix and $\tilde{H}$ is the intrinsic camera parameter matrix [8]. The 4x4 matrix ^cT_h represents the transformation from a designated home frame $\{h\}$ to the camera frame $\{c\}$. This frame can be chosen to have the following form

$$^cT_h = \begin{bmatrix} 1 & 0 & 0 & 0 \\ 0 & 1 & 0 & 0 \\ 0 & 0 & 1 & ^cZ_h \\ 0 & 0 & 0 & 1 \end{bmatrix} \quad (3)$$

where ^cZ_h is the optical axis distance between $\{c\}$ and $\{h\}$. The pose P_0, referred to as the *virtual home position* of the object, is used to denote the pose at which the relationship between all object features is known *a priori* in $\{h\}$.

The true image pose of the object in the pixel frame p_k is given by

$$p_k = \begin{bmatrix} m_{0,k} & m_{1,k} & ... & m_{j,k} \end{bmatrix} \quad (4)$$

where $m_{j,k}$ is the normalized projection of object feature point $M_{j,k}$.

Eq. 2 can be used to compute the virtual home position pose p_0 providing a reference position from which to describe all image displacements. In this paper P_0 is chosen to be object centered in $\{h\}$.

2.2 Homographies: Image Mapping

The displacement of an object from P_0 to an arbitrary pose P_k can be represented by

$$P_k = \mathrm{T}_{k0} P_0 \quad (5)$$

where T_{k0} is a 4x4 homogeneous matrix of the form

$$\mathrm{T}_{k0} = \begin{bmatrix} \mathrm{R}_{k0} & \tau_{k0} \\ 0 & 1 \end{bmatrix} \quad (6)$$

The 3x3 matrix R_{k0} describes the change in orientation of the object about the 3 coordinate axes of $\{h\}$ where $\mathrm{R}_{k0} \in SO(3)$. This matrix is chosen to be parameterized by three Euler angles α_k, β_k, and γ_k defined to be positive counterclockwise about the positive ^hX, ^hY, and ^hZ axes of $\{h\}$ respectively and is given by $\mathrm{R}_{k0} =$

$$\begin{bmatrix} c\beta_k\, c\gamma_k & -c\beta_k\, s\gamma_k & s\gamma_k \\ c\alpha_k\, s\gamma_k + s\alpha_k\, s\beta_k\, c\gamma_k & c\alpha_k\, c\gamma_k - s\alpha_k\, s\beta_k\, s\gamma_k & -s\alpha_k\, c\beta_k \\ s\alpha_k\, s\gamma_k - c\alpha_k\, s\beta_k\, c\gamma_k & s\alpha_k\, c\gamma_k + c\alpha_k\, s\beta_k\, s\gamma_k & c\alpha_k\, c\beta_k \end{bmatrix} \quad (7)$$

Taken in the order given these angles uniquely describe an arbitrary change in orientation of the object with respect to P_0 [16]. The 3x1 column vector τ_{k0} represents the translational changes in displacement in the ^hX, ^hY, and ^hZ directions.

A specific transformation describing the displacement from p_0 to an arbitrary pose p_k in the image can be derived by using Eq. 2 to solve for P_k and P_0. Substitution of these poses back into Eq. 5 yields the following

$$\tilde{p}_k = \tilde{H} \mathrm{H}_{k0} \tilde{H}^{-1} \tilde{p}_0 \quad (8)$$

The 3x3 matrix H_{k0} is defined as the homography in the image centered frame $\{i\}$ and is given by

$$\mathrm{H}_{k0} = \tilde{P}\, ^c\mathrm{T}_h \mathrm{T}_{k0}\, ^c\mathrm{T}_h^{-1} \tilde{P}^{-1} \quad (9)$$

Since $\tilde{P}$ and $^c\mathrm{T}_h$ are constants the homography becomes dependent only on the 6 independent parameters of T_{k0}. Eq. 9 is merely symbolic since the perspective projection matrix $\tilde{P}$ is a 3x4 matrix and has no unique inverse. Therefore, a different approach must be used in order to parameterize H_{k0}.

2.3 Homography Parameterization

The transformation between two object poses in the image plane with *unit focal length* can be now be represented as

$$\tilde{p}_k = \mathrm{H}_{k0} \tilde{p}_0 \quad (10)$$

where $\tilde{p}_k$ and $\tilde{p}_0$ are the $\{i\}$ coordinates of the object. Using Eq. 5 the *j*th object feature point can be written for two object poses P_0 and P_k explicitly as

$$M_{j,0} = \begin{bmatrix} X_{j,0} \\ Y_{j,0} \\ Z_{j,0} \\ 1 \end{bmatrix} \quad M_{j,k} = \begin{bmatrix} X_{j,k} \\ Y_{j,k} \\ Z_{j,k} \\ 1 \end{bmatrix} \quad (11)$$

By combining Eqs. 5, and 10 the coordinates of $M_{j,0}$ and $M_{j,k}$ in $\{i\}$ can be found as follows

$$\tilde{m}_{j,0} = \tilde{P} M_{j,0} \quad \tilde{m}_{j,k} = \tilde{P}\, \mathrm{T}_{k0}\, M_{j,0} \quad (12)$$

By substituting Eq. 12 into Eq. 10 three equations can be written explicitly by matrix multiplication. Setting the coefficients of the variables $X_{j,0}$ and $Y_{j,0}$ equal to each other a *particular* solution for H_{k0} can be found as follows

$$(\mathrm{H}_{k0})_j = \begin{bmatrix} R_{k0}(1,1) & R_{k0}(1,2) & \frac{R_{k0}(1,3) Z_{j,0} + X_{k0}}{Z_{j,0} + {}^cZ_h} \\ R_{k0}(2,1) & R_{k0}(2,2) & \frac{R_{k0}(2,3) Z_{j,0} + Y_{k0}}{Z_{j,0} + {}^cZ_h} \\ R_{k0}(3,1) & R_{k0}(3,2) & \frac{R_{k0}(3,3) Z_{j,0} + Z_{k0} + {}^cZ_h}{Z_{j,0} + {}^cZ_h} \end{bmatrix} \quad (13)$$

The three elements of the last column of $(\mathrm{H}_{k0})_j$ include the term $Z_{j,0}$ which is the ^hZ-coordinate of the *j*th object feature point. As indicated by the subscript j, the general form of the homography is object feature point dependent. Therefore, if a 3-D object is described using object feature points which are chosen such that they are not coplanar in $\{h\}$, the homography H_{k0} between two object image poses becomes the set of image feature point homographies as follows

$$\mathrm{H}_{k0} = \{\ (\mathrm{H}_{k0})_0,\ (\mathrm{H}_{k0})_1,\ \ldots\ (\mathrm{H}_{k0})_n\ \} \quad (14)$$

Here n is the total number of object features which are measurable between two object image poses.

If the object can be represented by a set of coplanar object feature points and the plane described by these points is $\perp\ ^hZ$ such that $Z_{j,0} = 0$ for $j = 1, 2, .., n$ then the general form of H_{k0} is

$$\mathrm{H}_{k0} = \frac{1}{div_{k0}} \begin{bmatrix} div_{k0}\, R_{k0}(1,1) & div_{k0}\, R_{k0}(1,2) & div_{k0}\, x_{k0} \\ div_{k0}\, R_{k0}(2,1) & div_{k0}\, R_{k0}(2,2) & div_{k0}\, y_{k0} \\ div_{k0}\, R_{k0}(3,1) & div_{k0}\, R_{k0}(3,2) & 1 \end{bmatrix} \quad (15)$$

The inverse of element $H_{k0}(3,3)$ defines the divergence, div_{k0}, and . The divergence, a measure of how the object is scaled in the image, is given by

$$div_{k0} = \frac{{}^cZ_h}{{}^cZ_h + Z_{k0}} \quad (16)$$

and is nonzero for $^cZ_h > 0$. In practice ^cZ_h should always be chosen to be meet this condition since $^cZ_h \leq 0$ would physically mean p_0 was at the focal point with infinite divergence or inverted behind the focal plane. If $Z_{k0} \leq (-\ ^cZ_h)$ this would indicate the object had been displaced to a position at or behind the focal plane. Choosing ^cZ_h to be positive and assuming all displacements of the object are measurable in the image the divergence parameter is never zero. Additionally, the parameters x_{k0} and y_{k0} are image plane translations corresponding to the perspective projections of X_{k0} and Y_{k0}.

3 Path Interpolation

By measuring p_k in the pixel frame and using the pre-computed virtual home pose p_0 a numerical least squares method [1] is used to solve for the elements of the normalized $\mathbf{H}_{k0}$ given in Eq. 15. The symbolic form of the homography is then decomposed between any two arbitrary object image poses.

3.1 Homography Decomposition

The numerical solution for the elements of $\mathbf{H}_{k0}$ is represented here as

$$\mathbf{H}_{k0} = \begin{bmatrix} a & b & c \\ d & e & f \\ g & h & 1 \end{bmatrix} \qquad (17)$$

and is used to decompose the parameterized form into the 6 independent parameters: γ_k, α_k, β_k, x_{k0}, y_{k0}, and div_{k0}. Referring to Eqs. 7 and 15 the angle γ_k can be found from

$$\gamma_k = \arctan\left(-\frac{b}{a}\right) \qquad (18)$$

where the particular quadrant of γ_k can be determined by the corresponding signs of a and b.

The divergence div_{k0} and image translations x_{k0} and y_{k0} are given as follows:

$$div_{k0} = \left(\frac{a^2 + b^2 + d^2 + e^2 + g^2 + h^2}{2}\right)^{\frac{1}{2}} \qquad (19)$$

$$x_{k0} = \frac{c}{div_{k0}} \qquad y_{k0} = \frac{f}{div_{k0}} \qquad (20)$$

The angle α_k is given as

$$\alpha_k = \arccos\left(\left(\frac{div_{k0}^2 - (g^2 + h^2)}{a^2 + b^2}\right)^{\frac{1}{2}}\right) \qquad (21)$$

where $0 \leq \alpha_k < 90°$. A degenerate case exists if $a^2 + b^2 = 0$ or when $\alpha_k > 90°$ corresponding to self-occluded planar object feature points in the image.

The angle β_k is given by

$$\beta_k = \arccos\left(\left(\frac{a^2 + b^2}{div_{k0}^2}\right)^{\frac{1}{2}}\right) \qquad (22)$$

which gives $0 \leq \beta_k < 90°$ where a degenerate case exists if $div_{k0} = 0$. This can be avoided by choosing an appropriate value for ^cZ_h as previously mentioned.

The range of α_k and β_k can be increased from $-90°$ to $90°$ including negative rotations by a simple optimization using the numerical values in Eq. 17 and the symbolic formulas for the elements of Eq. 13.

3.2 Path Algorithm

By choosing to express the displacement of an object from p_0 to p_k by the particular parameterization of $\mathbf{H}_{k0}$ given in Eq. 15 a unique path m can be defined between any two arbitrary poses of the object. An object image pose lying on the path can be defined as $p_{k,m}$. This unique path m can be represented discretely in the following form

$$s_m = \{p_{1,m}, p_{2,m}, ..., p_{k,m}, ..., p_{N,m}\} \qquad (23)$$

where s_m is the set of N image poses of the object between and including the end poses $p_{1,m}$ and $p_{N,m}$ for k = 1,2,...,N. The pose $p_{k,m}$ can be found by from Eq. 10 where $\mathbf{H}_{k0}$ is computed by interpolating the rotation angles α_k, β_k, and γ_k, the translations x_{k0}, y_{k0}, and the div_{k0} between the values found at k=1 and k=N by decomposition.

The general procedure for finding the discrete path s_m in the image can now be stated as follows:

Assume: p_0 is known *a priori*

Given: p_1 and p_N for a particular displacement

Procedure: Decomposing the homographies $\mathbf{H}_{10}$ and $\mathbf{H}_{N0}$ into the respective rotation angles, translations, and divergence from p_0 to p_1 and from p_0 to p_N, the discrete path s_m can be found by choosing a satisfactory value for N and linearly interpolating these rotation angles and translations between p_1 and p_N using Eq. 10.

The particular path s_m found using this method depends entirely on the particular parameterization chosen for $\mathbf{H}_{k0}$. This method for path interpolation depends on having *a priori* knowledge of the object either in the form of a 3-D model or measured points in the image. The path can also be interpolated in the 3-D task space. This can be done by finding X_{k0}, Y_{k0}, and Z_{k0} from x_{k0}, y_{k0}, and div_{k0} respectively by using the perspective projection equations and Eq. 16. The decomposed parameters can be used to reconstruct the 3-D displacement transformation $\mathbf{T}_{k0}$ and a similar path interpolation procedure can be performed in $\{h\}$.

4 Experimental Results

Our path interpolation and visual servoing along the interpolated path methods were tested in a lab setting. Results are presented from a simulation to test the validity of the path interpolation method between any two arbitrarily specified object poses. The simulation is then constrained to include a model of the robot manipulator thereby restricting the motion of the object

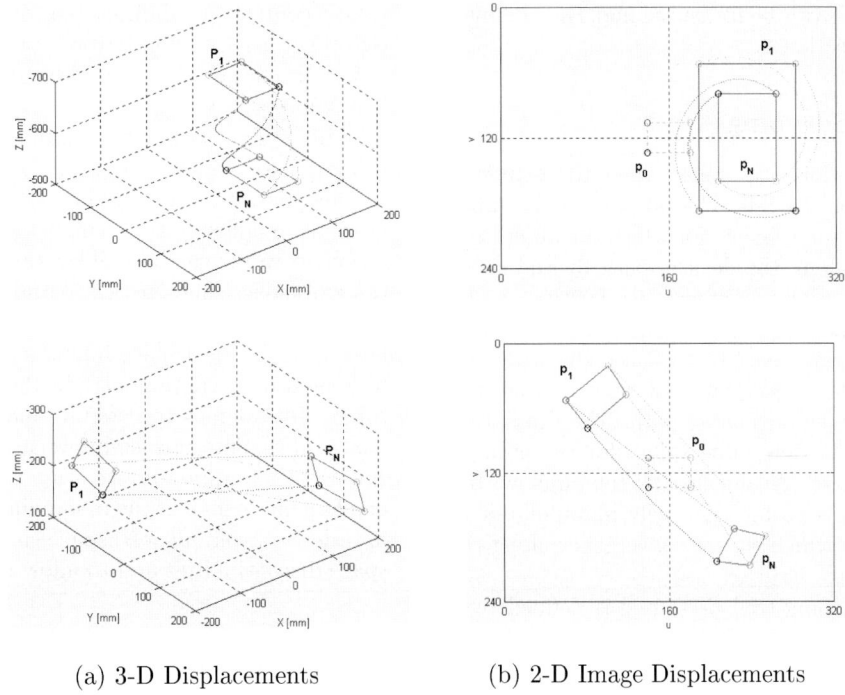

(a) 3-D Displacements (b) 2-D Image Displacements

Figure 1: Path Simulations

to the available workspace. The simulated and experimental results of one visual servoing trial are presented here. In this paper object feature points in the image were measured using a snake tracker. A rectangular box was chosen as the object where the features are all points on one face of the box (coplanar points to ensure a numerical solution for the homography). The box snake consists of a B-spline curve defined by 12 total control points with 4 knots of multiplicity 2 forming a closed loop. The vision system consists of a CCD camera mounted approximately 950 mm directly over the robot. The robot manipulator is a 6 degree of freedom Robix RCS-6 model where the rectangular box being manipulated is affixed to the end-effector. A primary PC handles all image I/O and system management. The Robix RCS-6 is run from a secondary PC via serial port communication.

4.1 Path Simulation

The path simulation takes as input a set of coplanar object feature points given at home position P_0 of the object in $\{h\}$. Our simulation takes as input 2 sets of screw displacement coordinates (3 displacement rotations and 3 displacement translations each) describing the displacement of the object from P_0 to P_1 and from P_0 to P_N in $\{h\}$. The object pose projections p_0, p_1 and p_N are computed in the image frame using Eq. 2

where $^cZ_h = 1000$ mm. The homographies H_{10} and H_{N0} are computed using the numerical solution and the path interpolation method described in Section 3.

Two sets of object displacements and the corresponding interpolated paths are shown in Figure 1. The left-hand column is the input 3-D displacement and 3-D interpolated path. The right hand column is the corresponding 2-D displacement of the object and the 2-D interpolated path. The corners of the box are the image features used to indicate each pose along the path. The first displacement is characterized by a 180° rotation about the ^hZ-axis and a translation in ^hX and ^hZ. The second displacement is characterized by 60° rotation about the ^hZ-axis and a translation in ^hX and ^hY. The *only* correspondence between the 3-D and 2-D interpolated paths is the projection of P_1 and P_N in $\{h\}$ to p_1 and p_N in the image pixel frame p where the 3-D path and 2-D paths were computed independently. However, the *projection* of the 3-D path into $\{p\}$ matches the 2-D path produced by our method within machine precision. This verifies that each object pose along the interpolated path in the image corresponds to an actual physical object pose in 3-D along the screw axis between P_1 and P_N. Additionally, for the second displacement the object was oriented such that p_k was not coplanar with p_0 for k = 1,2,...,N, demonstrating that the camera can be in any orientation relative to the object as long as the object features can be measured and P_0 is defined as described in Section 2.

4.2 Visual Servoing

The path simulation was constrained to include a model of the Robix RCS-6. The purpose of the simulation is to find displacements for which the object *can* be manipulated along the desired path by the robot while avoiding interior workspace singularities. An iterative servoing method is implemented here which involves reconstructing the 3-D displacement screw coordinates from the objects current pose to the next desired pose on the interpolated path. The simulation is also used here to determine values for the number of interpolated poses N and for the tolerance within which the object is to be iteratively manipulated to each desired pose which ensure convergence along the entire path.

The visual servoing trial presented is defined by a single rotation about joint 1 of the robot where $\Delta\theta_1 = 45^o$ and $N = 31$. The motion is represented in the left-hand column of Figure 2 by four simulated object image poses (p_1, p_{14}, p_{23}, and p_{31}). The right-hand column of Figure 2 shows four images (p_1, p_{14}, p_{23}, and p_{31}) of the actual visual servoing manipulation of the object from p_1 to p_{31}. Also indicated in both columns is the virtual home image pose p_0 and the desired image path between p_1 and p_{31}.

The joint positions corresponding to the robot configurations where the object pose p_k was manipulated to p_k^* within the desired tolerance for each value of k = 1,2,...,31 are shown in Figure 3.

Comparison of the joint positions for the entire motion between the simulation and the actual experimental results indicates a strong correlation. This verifies the path interpolation method combined with the visual servoing technique used here is a viable solution for this particular object displacement.

Some noise does exist which can be partially attributed to errors in calibration of the camera intrinsic parameters and the robot model. Most of this error seems to be attributable to the snake tracker which limits the minimum tolerance within which the object can be manipulated to each desired path pose. As can be seen in Figures 2 and 3 these sources of error did not play a significant role in the convergence of the object along the path from the initial pose to the desired final pose.

The simulation including the robot model was originally designed to aid in finding object paths which avoided interior workspace singularities as described in section 1. However, object poses have been observed to exist along certain interpolated paths which create control instabilities but do not represent interior workspace singularities. For these poses the object frame jacobian is well-conditioned and invertible but yet the next desired path pose is not attainable. The path interpolation method has been validated as a means to avoid workspace boundary singularities. However, observations suggest path dependent singularity-type conditions can exist in addition to workspace boundary singularities and interior workspace singularities. Although, the path is defined to be continuous and all interpolated path poses lie within the workspace boundary the path may not be represented continuously in the joint space of the robot. In order to successfully manipulate an object along interpolated paths between any two object image poses additional information possibly in the form of a kinematic map of the manipulator will be required in order to avoid *all* workspace singularities and joint space discontinuities and to allow true path planning.

5 Conclusions and Future Work

In this paper, a method for interpolating a viable object path between two arbitrary poses based on image feature extraction was presented. This was done by parameterizing and decomposing the 2-D displacement transformation matrix computed from measured object image features at an initial and desired final pose. Decomposing the parameterized form of the homography into rotational and translational displacement components an object path can be interpolated in either the image plane or in the camera frame relative to a known *a priori* home position. This path was validated through path simulations and experimental visual servoing trials where an object was manipulated along the interpolated path between the initial and final desired object poses.

The goal of developing a method by which a desired viable object path could be used in visual servoing to avoid workspace boundary singularities has been achieved . However, this method is not sufficient to guarantee visual servoing convergence for all types of object displacements. Other issues such as interior workspace singularities, discontinuities in the joint space, and special cases where the image feature paths may lie outside the camera's field of view are currently being addressed. Additionally, the subject of future papers will address the problem of handling 3-D objects and moving away from the 2-D planar case by implementing the general form of the homography derived in section 2.

The development of the path interpolation method in this paper is the first step in the development of a

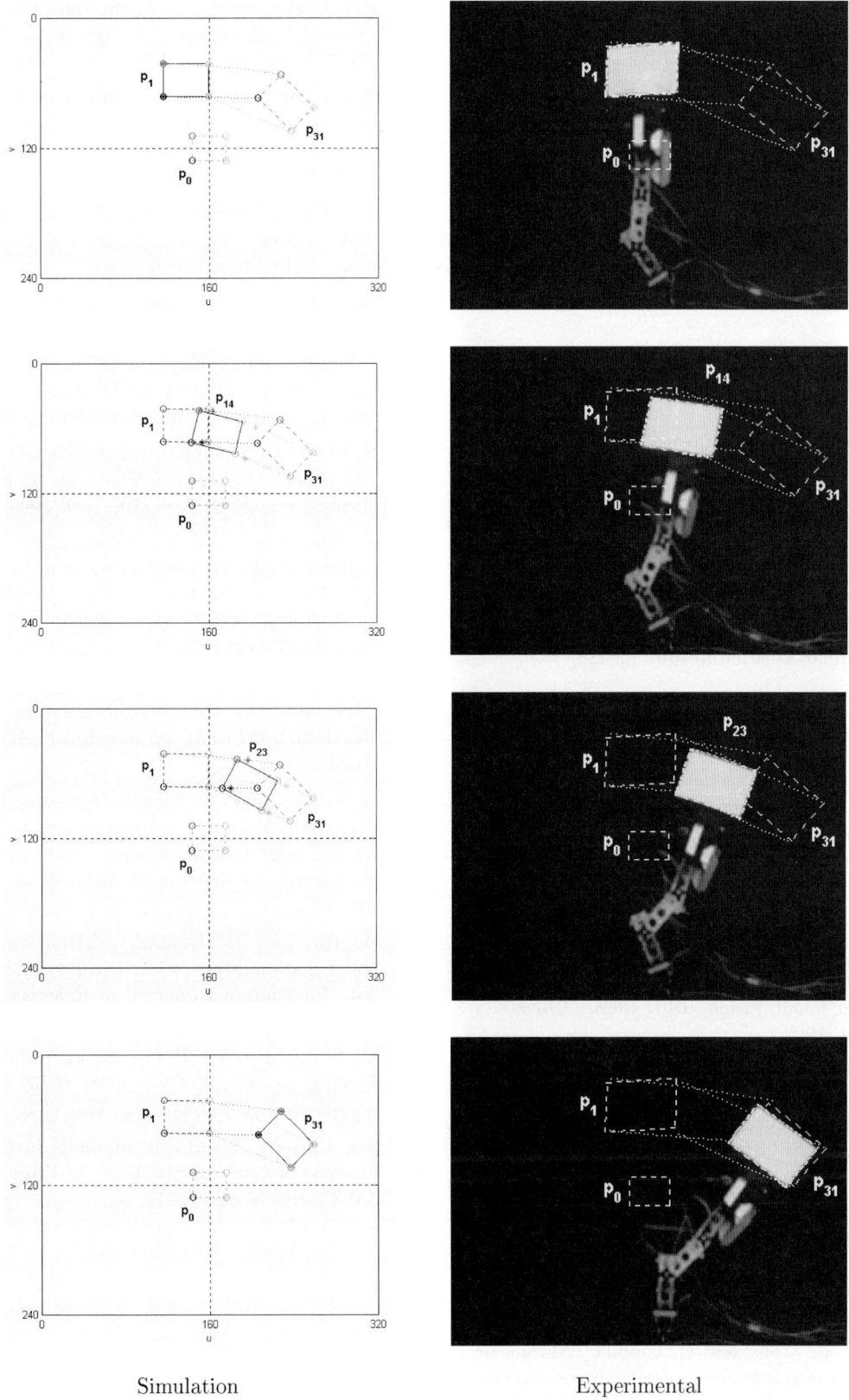

Simulation Experimental

Figure 2: Visual Servoing Results for $\Delta\theta_1 = 45^o$

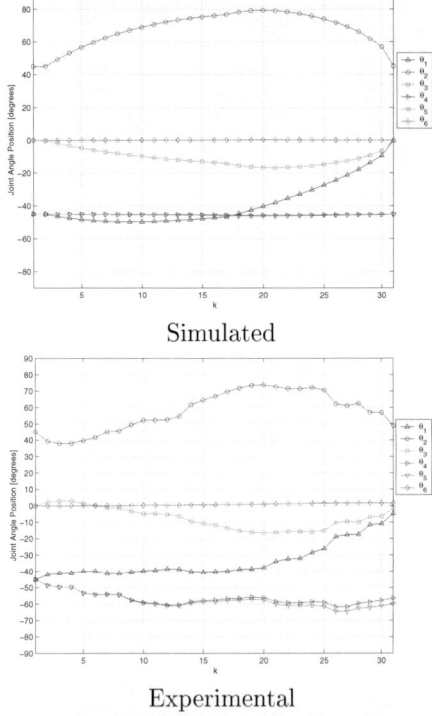

Figure 3: Joint Position Results for $\Delta\theta_1 = 45°$

true path planning method. A general procedure for visual servoing needs to be developed which incorporates the path interpolation method with knowledge of the reachable workspace and joint space of the robot manipulator in order to more intelligently plan a viable object path.

References

[1] Paul A. Beardsley. *Applications of Projective Geometry to Robot Vision*. PhD thesis, University of Oxford, 1992.

[2] F. Chaumette. Potential problems of stability and convergence in image-based and position-based visual servoing. In *The Confluence of Vision and Control*, LNCS (237), pp. 66–78. Springer Verlag, 1998.

[3] F. Chaumette and E. Malis. 2 1/2 d visual servoing: A possible solution to improve image-based and position-based visual servoings. *Proc. of the IEEE Int'l Conf. on Robotics & Automation*, 2000.

[4] G. Chesi, E. Malis, and R. Cipolla. Automatic segmentation and matching of planar contours for visual servoing. *Proc. of the IEEE Int'l Conf. on Robotics & Automation*, 2000.

[5] P. I. Corke and S. A. Hutchinson. Real-time vision, tracking and control. *Proc. of the IEEE Int'l Conf. on Robotics & Automation*, April 2000.

[6] P. I. Corke and S. A. Hutchinson. Recent results in visual servo control. In *Workshop on Integrating Sensors with Mobility and Manipulation*, April 2000.

[7] J. J. Craig. *Introduction to Robotics: Mechanics and Control*. Addison-Wesley Publishing Company, 2nd edition, 1989.

[8] O. Faugeras. *Three Dimensional Computer Vision*. MIT Press, Cambridge, MA, 1993.

[9] S. Hutchinson, G. Hager, and P. Corke. A tutorial on visual servo control. *IEEE Trans. on Robotics and Automation*, 12:651–670, Oct. 1996.

[10] E. Malis, F. Chaumette, and S. Boudet. 2-1/2d visual servoing. *IEEE Trans. on Robotics and Automation*, 15(2):238–250, April 1999.

[11] E. Marchand and F. Chaumette. A new redundancy-based iterative scheme for avoiding joint limits application to visual servoing. *Proc. of the IEEE Int'l Conf. on Robotics & Automation*, April 2000.

[12] Y. Mezouar and F. Chaumette. Path planning in image space for robust visual servoing. *Proc. of the IEEE Int'l Conf. on Robotics & Automation*, April 2000.

[13] B. Nelson and P. K. Khosla. Increasing the tracking region of an eye-in-hand system by singularity and joint limit avoidance. *Proc. of the IEEE Int'l Conf. on Robotics & Automation*, pp. 418–23, 1993.

[14] A. Ruf and R. Horaud. Visual servoing of robot manipulators part i: Projective kinematics. *The International Journal of Robotics Research*, 18(11):1101–1118, November 1999.

[15] C. J. Taylor and J. P. Ostrowski. Robust vision-based pose control. *Proc. of the IEEE Int'l Conf. on Robotics & Automation*, April 2000.

[16] J.J. Uicker., Jr. Matrix methods in the design analysis of mechanisms. Univ. of Wisconsin, ME 751 Course Notes, 1997.

Design and Tracking of Desirable Trajectories in the Image Space by Integrating Mechanical and Visibility Constraints

Youcef Mezouar
Youcef.Mezouar@irisa.fr

François Chaumette
Francois.Chaumette@irisa.fr

IRISA - INRIA Rennes
Campus de Beaulieu,
35042 Rennes Cedex, France

Abstract

Since image-based visual servoing is a local feedback control solution, it requires the definition of intermediate subgoals in the sensor space at the task planning level. In this paper, we describe a general technique for specifying and tracking trajectories of an unknown object in the camera image space. First, physically valid C^2 image trajectories which correspond to quasi-optimal 3D camera trajectory (approaching as much as possible a straight line) are performed. Both mechanical (joint limits) and visibility constraints are taken into account at the task planning level. The good behavior of Image-based control when desired and current camera positions are closed is then exploited to design an efficient control scheme. Real time experimental results using a camera mounted on the end effector of a six d-o-f robot confirm the validity of our approach.

1 Introduction

Classical approaches, using visual information in feedback control loops [6, 8], are point to point-based, i.e the robot must reach a desired goal configuration starting from a given initial configuration. In such approaches, a globally stabilizing feedback control solution is required. However if the initial error is large, such a control may produce erratic behavior especially in presence of modeling errors. For a very simple case, Cowan and Koditschek describe in [3] a globally stabilizing method using navigation function. By composing the error function of 3D Cartesian features and image features, Malis et al propose a globally stabilizing solution called *2 1/2 D* visual servoing for general setup [11]. Classical image-based visual servoing is a local control solution. It requires the definition of intermediate subgoals in the sensor space at the task planning level if the initial error is large. In this approach the robot effector is controlled so that the image features converge to the reference image features. The robot effector trajectory in the 3D Cartesian space is not controlled. Such a control can thus provide inadequate robot motion leading to no optimal or no physically valid robot trajectory [1]. However, it is well known that image-based control is locally stable and robust with respect to modeling errors and noise perturbations. The key idea of our work is to use the local stability and robustness of image-based servoing by specifying trajectories to follow in the image. Indeed, for a trajectory following a local control solution works properly since current and desired configurations remain close. Only few papers deal with path planning in image space. In [7] a trajectory generator using a stereo system is proposed and applied to obstacle avoidance. An alignment task using intermediate views of the object synthesized by image morphing is presented in [16]. A path planning for a straight-line robot translation observed by a weakly calibrated stereo system is performed in [14]. In previous work [12], we have proposed a potential field-based path planning generator that determines the trajectories in the image of a set of points lying on a planar target. In this paper, we propose to plan the trajectory of an **unknown and not necessarily planar** object. Both **mechanical (joint limits) and visibility** constraints are taken into account. Contrarily to others approaches [2, 13] exploiting the robot redundancy, the mechanical and visibility constraints can be ensured even if all the robot degrees of freedom are used to realize the task.

More precisely, we plan the trajectory of $\mathbf{s} = [\mathbf{p}_1^T \cdots \mathbf{p}_n^T]^T$, composed of the $2 \times n$ image coordinates of n points $\mathcal{M}_j$ lying on an unknown target, between the initial configuration $\mathbf{s}_i = [\mathbf{p}_{1i}^T \cdots \mathbf{p}_{ni}^T]^T$ and the desired one $\mathbf{s}^* = [\mathbf{p}_1^{*T} \cdots \mathbf{p}_n^{*T}]^T$. Our approach consists of three phases. In the first one, the discrete geometric camera path (that ensures the physical validity of the image trajectories) is performed as a sequence of N intermediate camera poses which approaches as much as possible a straight line in the Cartesian space. In this phase, the mechanical and visibility constraints are introduced. In the second one, the discrete geometric trajectory of the target in the image and the discrete geometric trajectory of the robot in the joint space are obtained from the camera path. Finally, continuous and derivable geometric paths in the image with an associated timing law $\mathbf{s}^*(t)$ is generated and tracked using an image-based control scheme.

The paper is organized as follows. In Section 2, we recall some fundamentals. The method of path planning is presented in Section 3. In Section 4, we show how to use an image-based control approach to track the trajectories. In section 5, a timing law is associated with the geometric path. The experimental results are given in Section 6.

2 Potential field method

Our path planning strategy is based on the potential field method [9, 10]. In this approach the robot motions are under the influence of an artificial potential field (V) defined as the sum of an attractive potential (V_a) pulling the robot toward the goal configuration (Υ_*) and a repulsive potential (V_r) pushing the robot away from the obstacles. Motion planning is performed in an iterative fashion. At each iteration an artificial force $\mathbf{F}(\Upsilon)$, where the 6×1 vector Υ represents a parameterization of robot workspace, is induced by the potential function. This force is defined as $\mathbf{F}(\Upsilon) = -\vec{\nabla} V$ where $\vec{\nabla} V$ denotes the gradient vector of V at Υ. Each segment of the path is oriented along the negated gradient of the potential function:

$$\Upsilon_{k+1} = \Upsilon_k + \varepsilon_k \frac{\mathbf{F}(\Upsilon_k)}{\|\mathbf{F}(\Upsilon_k)\|} \quad (1)$$

where k is the increment index and ε_k a positive scaling factor denoting the length of the kth increment. In our case, the control objective is formulated to transfer the system to a desired point in the sensor space and to provide robot motion satisfying the following constraints:

1. all the considered image features remain in the camera field of view

2. the robot joint positions remain between their limits

To deal with the first constraint, a repulsive potential $V_{rs}(\mathbf{s})$ is defined in the image. The second constraint is introduced through a repulsive potential $V_{rq}(\mathbf{q})$ defined in the joint space. The total force is given by:

$$\mathbf{F} = -\vec{\nabla} V_{ak} - \gamma_k \vec{\nabla} V_{rsk} - \chi_k \vec{\nabla} V_{rqk}$$

The scaling factors γ_k and χ_k allow us to adjust the relative influence of the different forces.

3 Trajectories planning

We consider that the target model is not available. In this case the camera pose can not be estimated. Only a scaled Euclidean reconstruction can be obtained by performing a partial pose estimation as described in the next subsection. This partial pose estimation and the relations linking two views of a static object are then exploited to design a path of the projection of the unknown object in the image.

3.1 Scaled Euclidean reconstruction

Let $\mathcal{F}^*$ and $\mathcal{F}$ be the frames attached to the camera in its desired and current positions. The rotation matrix and the translation vector between $\mathcal{F}$ and $\mathcal{F}^*$ are denoted $^*\mathbf{R}_c$ and $^*\mathbf{t}_c$, respectively. A target point $\mathcal{M}_j$ with homogeneous coordinates $\mathbf{M}_j = [X_j\, Y_j\, Z_j\, 1]$ (resp. $\mathbf{M}_j^*$) in $\mathcal{F}$ (resp. $\mathcal{F}^*$) is projected in the camera image onto a point with homogeneous normalized and pixel coordinates $\mathbf{m}_j = [x_j\, y_j\, 1]^T$ (resp. $\mathbf{m}_j^*$) and $\mathbf{p}_j = [u_j\, v_j\, 1]^T = \mathbf{A} m_j$ (resp. $\mathbf{p}_j^*$), where $\mathbf{A}$ denotes the intrinsic parameters matrix of the camera.

Consider a 3D reference plane Π given in $\mathcal{F}^*$ by $\pi^T = [\mathbf{n}^* - d^*]$, where $\mathbf{n}^*$ is its unitary normal in $\mathcal{F}^*$ and d^* is the distance from Π to the origin of $\mathcal{F}^*$ (see Fig. 1). It is well known that there is a projective homography matrix $\mathbf{G}_\pi$ such that:

$$\alpha_j \mathbf{p}_j = \mathbf{G}_\pi \mathbf{p}_j^* + \beta_j \mathbf{e} \quad \text{with} \quad \mathbf{e} = -\mathbf{A}\, ^*\mathbf{R}_c^T\, ^*\mathbf{t}_c \quad (2)$$

where α_j is a positive scaling factor and β_j is a scaling factor null if the target point is linked with Π. More precisely, if we define the signed distance $d(\mathcal{M}_j, \Pi) = \pi\, \mathbf{M}_j^*$, we have:

$$\beta_j = -\frac{d(\mathcal{M}_j, \Pi)}{Z_j^* d^*} \quad (3)$$

If at least four matched points belonging to Π are known, $\mathbf{G}_\pi$ can be estimated by solving a linear system. Else, at least height points (3 points to define Π and 5 outside of Π) are necessary to estimate the homography matrix by using for example the linearized algorithm proposed in [11]. Assuming that the camera calibration is known, the Euclidean homography $\mathbf{H}_\pi$ of plane Π is estimated as follow:

$$\mathbf{H}_\pi = \mathbf{A}^{-1} \mathbf{G}_\pi \mathbf{A} \quad (4)$$

and it can be decomposed into a rotation matrix and a rank 1 matrix [5]:

$$\mathbf{H}_\pi = \,^*\mathbf{R}_c^T - \,^*\mathbf{R}_c^T \mathbf{t}_{d^*} \mathbf{n}^{*T} \quad \text{where} \quad \mathbf{t}_{d^*} = \frac{^*\mathbf{t}_c}{d^*} \quad (5)$$

From $\mathbf{H}_\pi$ and the image features, it is thus possible to determine the camera motion parameters (that is the rotation $^*\mathbf{R}_c$ and the scaled translation $\mathbf{t}_{d^*}$) and the vector n^* [5]. The structure of the observed scene can also be determined. For example, the ratio between the Z-coordinates of a 3D point $\mathcal{M}_j$ expressed in $\mathcal{F}$ and the distance d^*, $\rho_j = Z_j/d^*$ can be obtained from $^*\mathbf{R}_c$, $\mathbf{t}_{d^*}$ and the image features [11].These parameters are important since they are used in the path planning generator and in the control scheme.

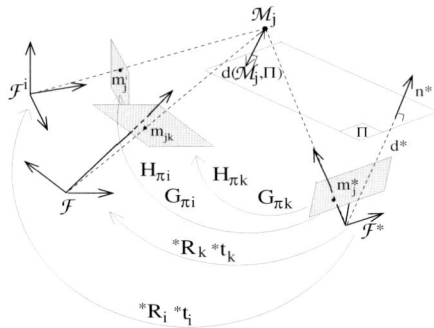

Figure 1: Scaled camera trajectory

3.2 Scaled 3-D Cartesian trajectory

The homography $\mathbf{G}_{\pi i}$ is computed from $\mathbf{s}^i$ and $\mathbf{s}^*$. According to (4), we obtain $\mathbf{H}_{\pi i}$. Then $\mathbf{n}^*$ is estimated, as well as the rotation $^*\mathbf{R}_i$ and the scaled translation $\mathbf{t}_{d^*i} = {}^*\mathbf{t}_i/d^*$ between $\mathcal{F}^i$ (frame linked to the camera in its initial position) and $\mathcal{F}^*$. If we choose as partial parameterization of the workspace $\mathbf{\Upsilon} = [\mathbf{t}_{d^*}^T \; (\mathbf{u}\theta)^T]^T$ where $\mathbf{u}$ and θ are the normalized rotation axis and the rotation angle extracted from $^*\mathbf{R}_c$, we obtain at the initial and desired robot configurations $\mathbf{\Upsilon}_i = [\mathbf{t}_{d^*i}^T \; (\mathbf{u}\theta)_i^T]^T$ and $\mathbf{\Upsilon}_* = \mathbf{0}_{6\times 1}$.
We thus have to determine a path starting at the initial configuration $\mathbf{\Upsilon}_{k=0} = \mathbf{\Upsilon}_i$ and ending at $\mathbf{\Upsilon}_* = \mathbf{0}_{6\times 1}$.

3.3 Trajectories in the joint space

To anticipate the possible encounter of a joint limit and to avoid it, we have to estimate the trajectory of the robot in the joint space. Indeed, the value of the joint coordinates at each iteration is used in the computation of the repulsive potential related to the joint limits avoidance. If the manipulator position in the joint space is represented by $\mathbf{q} = [q_1 \cdots q_m]^T$, we have:

$$\frac{\partial \mathbf{q}}{\partial \mathbf{\Upsilon}} = \frac{\partial \mathbf{q}}{\partial \mathbf{r}} \frac{\partial \mathbf{r}}{\partial \mathbf{\Upsilon}} = \mathbf{J}^+(\mathbf{q}) \mathbf{M}_{\mathbf{\Upsilon}}(d^*) \quad (6)$$

where $\mathbf{J}(\mathbf{q})$ and $\mathbf{M}_{\mathbf{\Upsilon}}(d^*)$ denote the robot Jacobian and the parameterization Jacobian respectively. The parameterization Jacobian can be computed directly from $\mathbf{\Upsilon}$ and d^* [12]:

$$\mathbf{M}_{\mathbf{\Upsilon}}(d^*) = \begin{bmatrix} d^* \, {}^*\mathbf{R}_c^T & \mathbf{0}_{3\times 3} \\ \mathbf{0}_{3\times 3} & \mathbf{L}_w^{-1} \end{bmatrix}$$

The computation of $\mathbf{L}_w^{-1}$ can be found in [11][1]:

$$\mathbf{L}_w^{-1} = \mathbf{Id}_{3\times 3} + \frac{\theta}{2}\mathrm{sinc}^2\left(\frac{\theta}{2}\right)[\mathbf{u}]_\wedge + (1 - \mathrm{sinc}(\theta))[\mathbf{u}]_\wedge^2$$

The trajectory of the robot coordinates in the joint space are then obtained from the trajectory of $\mathbf{\Upsilon}$ by a linearization of

---
[1] $[\mathbf{u}]_\wedge$ denotes the antisymmetric matrix associated to the vector $\mathbf{u}$

(6) around $\mathbf{q}_k$:

$$\mathbf{q}_{k+1} = \mathbf{q}_k + \mathbf{J}^+(\mathbf{q}_k) \mathbf{M}_{\mathbf{\Upsilon}_k}(d^*)(\mathbf{\Upsilon}_{k+1} - \mathbf{\Upsilon}_k)$$

3.4 Image trajectories

The homography matrix $\mathbf{G}_{\pi k}$ of plane Π relating the current and desired images can be computed from $\mathbf{\Upsilon}_k$ using (4) and (5):

$$\mathbf{G}_{\pi k} = \mathbf{A}({}^*\mathbf{R}_k^T - {}^*\mathbf{R}_k^T \mathbf{t}_{d^*k} \mathbf{n}^{*T})\mathbf{A}^{-1} \quad (7)$$

According to (2) the image coordinates of the points $\mathcal{M}_j$ at time k are given by:

$$\mu_{jk}\mathbf{p}_{jk} = \mathbf{G}_{\pi k}\mathbf{p}_j^* + \beta_j \mathbf{e}_k \quad (8)$$

where (refer to (2) and (3)):

$$\beta_j \mathbf{e}_k = \frac{d(\mathcal{M}_j, \Pi)}{Z_j^*} \mathbf{A}\,{}^*\mathbf{R}_k^T \mathbf{t}_{d^*k}$$

using the previous relation, (8) can be rewritten:

$$\mu_{jk}\mathbf{p}_{jk} = \mathbf{G}_{\pi k}\mathbf{p}_j^* + \frac{d(\mathcal{M}_j,\Pi)}{Z_j^*}\mathbf{A}\,{}^*\mathbf{R}_k^T \mathbf{t}_{d^*k} \quad (9)$$

Furthermore, if the relation (9) is applied between the desired and initial camera positions, we obtain [2]:

$$\frac{d(\mathcal{M}_j,\Pi)}{Z_j^*} = \mathrm{sign}\left(\frac{(\mu_j\mathbf{p}_{ji} - \mathbf{G}_{\pi i}\mathbf{p}_j^*)_1}{(\mathbf{A}\,{}^*\mathbf{R}_i^T\mathbf{t}_{d^*i})_1}\right) \frac{\|\mathbf{G}_{\pi i}\mathbf{p}_j^* \wedge \mathbf{p}_{ji}\|}{\|\mathbf{A}\,{}^*\mathbf{R}_i^T\mathbf{t}_{d^*i} \wedge \mathbf{p}_{ji}\|} \quad (10)$$

The equations (7), (9) and (10) allow to compute $\mu_{jk}\mathbf{p}_{jk}$ from $\mathbf{\Upsilon}_k$ and the initial and desired visual features. The image coordinates $\mathbf{p}_{jk}$ are then computed by dividing $\mu_j\mathbf{p}_{jk}$ by its last component. Furthermore, the ratio ρ_{jk}, which will be used in the repulsive force and in the control law, can easily be obtained from $\mathbf{\Upsilon}_k$ and $\mathbf{m}_{jk} = \mathbf{A}^{-1}\mathbf{p}_{jk}$.

3.5 Reaching the goal

The potential field V_a is defined as a parabolic function in order to minimize the distance between the current position and the desired one: $V_a(\mathbf{\Upsilon}) = \frac{1}{2}\|\mathbf{\Upsilon} - \mathbf{\Upsilon}_*\|^2 = \frac{1}{2}\|\mathbf{\Upsilon}\|^2$. The function V_a is positive or null and attains its minimum at $\mathbf{\Upsilon}_*$ where $V_a(\mathbf{\Upsilon}_*) = 0$. It generates a force $\mathbf{F}_a$ that converges linearly toward the goal configuration:

$$\mathbf{F}_a(\mathbf{\Upsilon}) = -\vec{\nabla} V_a = -\mathbf{\Upsilon} \quad (11)$$

When the repulsive potentials are not needed, the transition equation can be written (refer to (1) and (11)):

$$\mathbf{\Upsilon}_{k+1} = \left(1 - \frac{\varepsilon_k}{\|\mathbf{\Upsilon}_k\|}\right)\mathbf{\Upsilon}_k$$

---
[2] $(\mathbf{v})_j$ is the jth components of $\mathbf{v}$

Thus, $\mathbf{\Upsilon}_k$ is lying on the straight line passing by $\mathbf{\Upsilon}_i$ and $\mathbf{\Upsilon}_*$. As a consequence, the translation performed by the camera is a real straight line since $\mathbf{\Upsilon}_k$ is defined with respect to a motionless frame (that is $\mathcal{F}_*$). However, the object can get out of the camera field of view and the robot can attain its joint limits along this trajectory. In order that the object remains in the camera field of view and that the robot remains in its mechanical limits, two repulsive forces are introduced by deviating the camera trajectory when needed.

3.6 Mechanic and visibility constraints

A. Joint limits avoidance. The robot configuration $\mathbf{q}$ is called acceptable if each of its components is sufficiently far away from its corresponding joints limits. We note $\mathcal{L}$ the subset of the joint space of acceptable configurations. The repulsive potential $V_{rq}(\mathbf{q})$ is defined as:

$$\begin{cases} V_{rq}(\mathbf{q}) = \sum_{j=1}^{m} \dfrac{1}{(1-\frac{q_j}{q_{jmax}})(1-\frac{q_j}{q_{jmin}})} & \text{if } \mathbf{q} \notin \mathcal{L} \\ = 0 & \text{if } \mathbf{q} \in \mathcal{L} \end{cases}$$

q_{jmin} and q_{jmax} being the minimum and the maximum allowable joint values for the jth joint. The artificial repulsive force deriving from V_{rq} is:

$$\mathbf{F}_{rq}(\mathbf{\Upsilon}) = -\left(\frac{\partial V_{rq}(\mathbf{q})}{\partial \mathbf{\Upsilon}}\right)^T = -\left(\frac{\partial V_{rq}(\mathbf{q})}{\partial \mathbf{q}} \frac{\partial \mathbf{q}}{\partial \mathbf{r}} \frac{\partial \mathbf{r}}{\partial \mathbf{\Upsilon}}\right)^T$$

The previous equation can be rewritten:

$$\mathbf{F}_{rq}(\mathbf{\Upsilon}) = -\mathbf{M}_{\mathbf{\Upsilon}}^T(d^*)\mathbf{J}^{+T}(\mathbf{q})\left(\frac{\partial V_r(\mathbf{q})}{\partial \mathbf{q}}\right)^T$$

B. Visibility constraint. One way to create a potential barrier around the camera field of view is to define V_{rs} as an increasing function of the distance between the object projection and the image limits. A general description of such a function is given in [12].

4 Performing C^2 timing law

In the previous subsection we have obtained discrete trajectories. In order to design continuous and derivable curves and thus to improve the dynamic behavior of the system, we use cubic B-spline interpolation. The spline interpolation problem is usually stated as: given data points $\mathcal{S} = \{\mathbf{s}_k / k \in 1 \cdots N\}$ and a set of parameter values $\mathcal{T} = \{t_k / k \in 1 \cdots N\}$, we have to determine a cubic B-spline curve $\mathbf{s}(t)$ such that $\mathbf{s}(t_k) = \mathbf{s}_k, \forall t_k$. In practice, parameter values are rarely given. In our case, we can adjust them to the distribution of the vector of image features $\mathbf{s}_k$ or using the distribution of the camera positions $\mathbf{\Upsilon}_k$. In order to control efficiently the camera velocity, it is more reasonable to use the distribution of the camera positions. The time values are thus chosen spacing proportionally to the distances between camera positions (see Fig. 2):

$$\frac{\Delta t_k}{\Delta t_{k+1}} = \frac{t_{k+1} - t_k}{t_{k+2} - t_{k+1}} = \frac{\|\mathbf{\Upsilon}_{k+1} - \mathbf{\Upsilon}_k\|}{\|\mathbf{\Upsilon}_{k+2} - \mathbf{\Upsilon}_{k+1}\|}$$

Considering the transition equation (1), we obtain:

$$\Delta t_{k+1} = \frac{\varepsilon_k}{\varepsilon_{k+1}} \Delta t_k \quad \text{with} \quad \Delta t_0 = T$$

T being the time between two consecutive frames (chosen for example as the video rate). In practice, ε_k is chosen constant, we thus have $t_k = kT$. Given the data vectors $\mathbf{s}_k$ and the parameters values t_k, the image data can be interpolated by using a natural cubic B-spline interpolation and we obtain a C^2 function $\mathbf{s}(t)$ defined for $(k-1)\Delta T \leq t \leq k\Delta T$ by:

$$\mathbf{s}(t) = \mathbf{A}_k t^3 + \mathbf{B}_k t^2 + \mathbf{C}_k t + \mathbf{D}_k \qquad (12)$$

where the $n \times n$ diagonal matrices $\mathbf{A}_k, \mathbf{B}_k, \mathbf{C}_k, \mathbf{D}_k$ are obtained from $\mathcal{S}$ and $\mathcal{T}$.

Finally, the ratio ρ appears in the control law. By using the same process, $\mathbf{\Gamma}(t) = [\rho_1(t) \cdots \rho_n(t)]$ is computed from $\mathcal{R} = \{\mathbf{\Gamma}_k / k \in 1 \cdots N\}$ and $\mathcal{T}$.

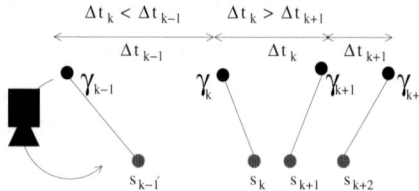

Figure 2: Controlling the time along the camera trajectory

5 Tracking the trajectories

To track the trajectories using an image-based control scheme, we use the task function approach introduced by Samson et al in [15]. A vision-based task function $\mathbf{e}$ to be regulated to $\mathbf{0}$ is defined by [4]:

$$\mathbf{e} = \widehat{\mathbf{L}}^+(\mathbf{s}(\mathbf{r}(t)) - \mathbf{s}^*(t))$$

The time varying vector $\mathbf{s}^*(t)$ is the desired trajectory of $\mathbf{s}$ computed in the previous sections. $\mathbf{L}$ denotes the well known interaction matrix (also called image Jacobian) and the matrix $\widehat{\mathbf{L}}^+$ represents the pseudo-inverse of a chosen model of $\mathbf{L}$. More precisely, when $\mathbf{s}$ is composed of the image coordinates of n points, two successive rows of the image Jacobian are given by:

$$\begin{bmatrix} -\frac{1}{d^* \rho_j} & 0 & \frac{x_j}{d^* \rho_j} & x_j y_j & -(1+x_j^2) & y_j \\ 0 & -\frac{1}{d^* \rho_j} & \frac{y_j}{d^* \rho_j} & (1+y_j^2) & -x_j y_j & -x_j \end{bmatrix}$$

The value of $\mathbf{L}$ at the current desired position is used for $\widehat{\mathbf{L}}$, that is $\widehat{\mathbf{L}} = \mathbf{L}(\mathbf{s}^*(t), \mathbf{\Gamma}^*(t), \widehat{d}^*)$, $\widehat{d}^*$ being an estimated value of d^* and $\mathbf{\Gamma}^*(t) = [\rho_1^*(t) \cdots \rho_n^*(t)]$ is the obtained trajectory for $\mathbf{\Gamma}$. An exponential decay of $\mathbf{e}$ toward $\mathbf{0}$ can be obtained by imposing $\dot{\mathbf{e}} = -\lambda \mathbf{e}$ (λ being a proportional gain), the corresponding control law is:

$$\mathbf{T}_c = -\lambda \mathbf{e} - \frac{\partial \mathbf{e}}{\partial t}$$

where $\mathbf{T}_c$ is the camera velocity sent to the robot controller. If the target is known to be motionless, we have $\frac{\partial \mathbf{e}}{\partial t} = -\widehat{\mathbf{L}}^+ \frac{\partial \mathbf{s}^*}{\partial t}$ and the camera velocity can be rewritten:

$$\mathbf{T}_c = -\lambda \mathbf{e} + \widehat{\mathbf{L}}^+ \frac{\partial \mathbf{s}^*}{\partial t}$$

where the term $\widehat{\mathbf{L}}^+ \frac{\partial \mathbf{s}^*}{\partial t}$ allows to compensate the tracking error. More precisely, we have from (12):

$$\frac{\partial \mathbf{s}^*}{\partial t} = 3\mathbf{A}_k t^2 + 2\mathbf{B}_k t + \mathbf{C}_k \quad \text{for} \quad (k-1)\Delta T \le t \le k\Delta T$$

This control law posses nice degrees of robustness with respect to modeling errors and noise perturbations since the error function used as input remains small and is directly computed from visual features.

6 Experimental Results

The proposed method has been tested on a six d-o-f eye-in-hand system. The specified visual task consists in a positioning task with respect to an unknown object. The target is a marked object with nine white marks lying on three differerent planes (see Fig. 3). The extracted visual features are the image coordinates of the center of gravity of each mark. The images corresponding to the desired and initial camera positions are given in Figs. 3(a) and 3(b) respectively. The corresponding camera displacement is very important ($t_x = 820mm$, $t_y = 800mm$, $t_z = 450mm$, $(u\theta)_x = 37dg$, $(u\theta)_y = 45dg$, $(u\theta)_z = 125dg$). In this case classical image-based and position-based visual servoing fail.

On all the following plots, joint positions are normalized between [-1;1], where -1 and 1 represents the joint limits.

In order to emphasize the importance of the introduced constraints in the trajectories, we first perform the path planning without repulsive potential. The results are given in Fig. 3. We can see that the visual features get out largely of the camera field of view (Fig.3(c)) and the axis q_5 attains its joint limit (Fig.3(d)). Then, only the repulsive potential associated to the visibility constraint has been activated (see Fig. 4). In that case, even if the visibility constraint is ensured (Fig 4(a)) the servoing can not be realized because the axis q_5 reaches its joint limit (Fig 4(b)). In Fig. 5, the two repulsive potentials are activated. The target remains in the camera field of view (see Fig. 5(a) and 5(c))

and all axes avoid their joint limit (see Fig. 5(b) and 5(d)). We can notice that the planned trajectories and the realized trajectories in the image are almost similar, that shows the efficiency of our control scheme. The error on the image coordinates of each target point between its current and desired location is given in Fig. 5(f). We can note the convergence of the coordinates to their desired value, which demonstrates the correct realization of the positioning task.

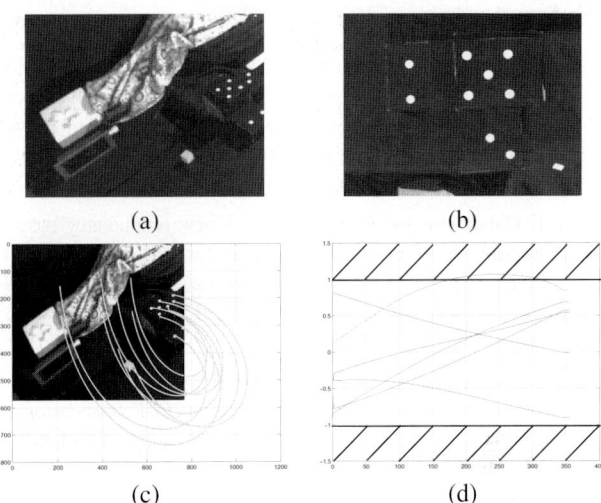

Figure 3: (a) Initial and (b) Desired images; Planned trajectories without any repulsive potential (c) in the image, (d) in the joint space

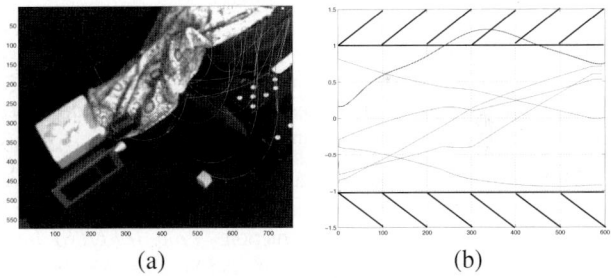

Figure 4: Planned trajectories without repulsive potential associated to the joint limits avoidance (a) in the image, (b) in the joint space

7 Conclusion

In this paper, we have presented a method ensuring the convergence for all initial camera position. By coupling an image-based trajectory generator and an image-based servoing, the proposed method extends the well known stability of image-based servoing when initial and desired camera location are close to the case where they are distant. The obtained trajectories provide some good expected properties. First, along these trajectories the target remains in the camera field of view and the robot remains in its mechanical limits. Second the corresponding robot motion is physically realizable and the camera trajectory is a straight line outside the area where the repulsive forces are needed. Experimental results show the validity of our approach. Future work will be devoted to generate the trajectories in image space of complex features in order to apply our method to complex objects.

References

[1] F. Chaumette. Potential problems of stability and convergence in image-based and position-based visual servoing. *The Confluence of Vision and Control D. Kriegman, G. Hager, A. Morse (eds), LNCIS Series, Springer Verlag*, 237:66–78, 1998.

[2] F. Chaumette and E. Marchand. A new redundancy-based iterative scheme for avoiding joint limits: Application to visual sevoing. In *Proc. IEEE Int. Conf. on Robotics and Automation*, volume 2, pages 1720–1725, San Francisco, California, April 2000.

[3] N.J. Cowan and D.E. Koditschek. Planar image based visual servoing as a navigation problem. *IEEE International Conference on Robotics and Automation*, pages 611–617, May 1999.

[4] B. Espiau, F. Chaumette, and P. Rives. A new approach to visual servoing in robotics. *IEEE Trans. on Robotics and Automation, 8(3) : 313-326*, 1992.

[5] O. Faugeras and F. Lustman. Motion and structure from motion in a piecewise planar environment. *Int. Journal of Pattern Recognition and Artificial Intelligence*, 2(3):485–508, 1988.

[6] K. Hashimoto. *Visual Servoing: Real Time Control of Robot Manipulators Based on Visual Sensory Feedback*. World Scientific Series in Robotics and Automated Systems, Vol 7,World Scientific Press, Singapor, 1993.

[7] K. Hosoda, K. Sakamoto, and M. Asada. Trajectory generation for obstacle avoidance of uncalibrated stereo visual servoing without 3d reconstruction. *Proc. IEEE/RSJ Int. Conference on Intelligent Robots and Systems*, 1(3):29–34, August 1995.

[8] S. Hutchinson, G.D. Hager, and P.I. Corke. A tutorial on visual servo control. *IEEE Trans. on Robotics and Automation*, 12(5):651–670, octobre 1996.

[9] O. Khatib. Real time obstacle avoidance for manipulators and mobile robots. *Int. Journal of Robotics Research*, 5(1):90–98, 1986.

[10] J. C. Latombe. *Robot Motion Planning*. Kluwer Academic Publishers, 1991.

[11] E. Malis and F. Chaumette. 2 1/2 d visual servoing with respect to unknown objects through a new estimation scheme of camera displacement. *International Journal of Computer Vision*, June 2000.

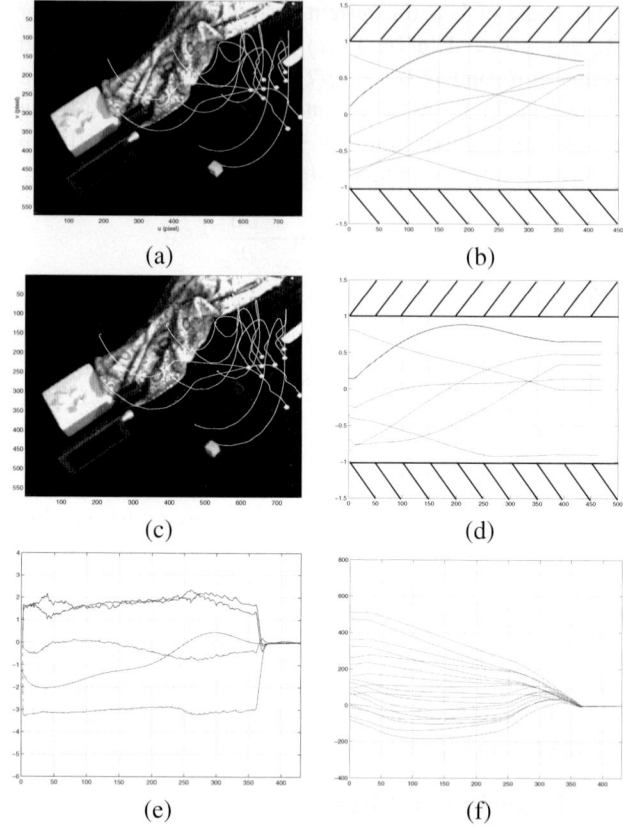

Figure 5: Planned trajectories with both repulsive potential (a) in the image, (b) in the joint space; realized trajectories (c) in the image, (d) in the joint space, (e) camera translational (cm/s) and rotational (dg/s) velocities versus iteration number, (f) errors in the image versus iteration number

[12] Y. Mezouar and F. Chaumette. Path planning in image space for robust visual servoing. *IEEE Int. Conference on Robotics and Automation*, 3:2759–2764, April 2000.

[13] B.J. Nelson and P.K. Khosla. Strategies for increasing the tracking region of an eye-in-hand system by singularity and joint limits avoidance. *Int. Journal of Robotics Research*, 14(3):255-269, June 1995.

[14] A. Ruf and R. Horaud. Visual trajectories from uncalibrated stereo. *IEEE International Conference on Intelligent Robots and Systems*, pages 83–91, 1997.

[15] C. Samson, B. Espiau, and M. Le Borgne. *Robot Control : The Task Function Approach*. Oxford University Press, 1991.

[16] R. Singh, R. M. Voyle, D. Littau, and N. P. Papanikolopoulos. Alignement of an eye-in-hand system to real objects using virtual images. *Workshop on Robust Vision for Vision-Based Control of Motion, IEEE Int. Conf. on Robotics and Automation*, May 1998.

A visual servoing algorithm based on epipolar geometry

G. Chesi, D. Prattichizzo, A. Vicino

Dipartimento di Ingegneria dell'Informazione - Università di Siena
Via Roma, 56 - 53100 Siena, Italy
email: {chesi,prattichizzo,vicino}@dii.unisi.it

Abstract

A visual servoing algorithm for mobile robots is proposed. The main feature of the algorithm is that it exploits object profiles rather than solving correspondence problems using object features or texture. This property is crucial for mobile robot navigation in unstructured environments where the 3D scene exhibits only surfaces whose main features are their apparent contours. The framework is based on the epipolar geometry, which is recovered from object profiles and epipolar tangencies. Special symmetry conditions of epipoles are used to generate the mobile robot control law. For the sake of simplicity, mobile robot kinematics is assumed to be holonomic and the camera intrinsic parameters are assumed partially known. Such assumption can be relaxed to extend the application field of the approach.

1 Introduction

This paper deals with the problem of controlling the pose of a mobile robot with respect to a target object by means of visual feedback.

Visual servoing has been applied recently to mobile robotics, see e.g. [6, 8, 4]. In visual servoing, the control goals and the feedback law are directly designed in the image domain. Designing the feedback at the sensor level increases system performance especially when uncertainties and disturbances can affect the robot model and the camera calibration [7].

In [7] the authors presented a classification of visual servoing systems. The approach used in this paper is known as image-based visual servoing, where the error between the robot pose and a target object or a set of target features is computed directly from image features.

Visual servoing algorithms make use of object cues whose image plane projections are controlled to desired positions through the visual servoing process. Usually, these cues are distinctive textures, like corners, of objects in the 3D scene.

However, it may happen that the 3D scene does not exhibit any appropriate textures but only smooth surfaces whose main features consist of their apparent contours, defined as the projection of the contour generators of objects' surfaces [3]. As pointed out in [10], if the object surface does not have any noticeable texture, the object profile is the only information available to estimate the structure of the surface and the motion of the camera.

The aim of this work is to exploit object profiles to synthesize a visual servoing algorithm. It is worthwhile to notice that, in general, tracking object profiles instead of textures can be performed in a more robust way since solutions of correspondence problems are not required. Exploiting profiles in visual servoing is crucial in outdoor navigation where objects in the scene are highly unstructured (hills, trees, etc.) and solving correspondences is a difficult task which usually gives rise to poor results.

Recent results on using apparent contours and profiles to reconstruct object surfaces and recover camera motion are due to Cipolla and his colleagues, see [3, 1] for example. In [1] visual servoing was based on the estimation of the homography between initial and final viewed profiles but the algorithm worked with planar closed contours and required a correspondence optimization procedure.

2 Visual modeling

Assume that a pinhole camera is fixed to a mobile robot moving on a plane. Let z_c be the optical axis of the camera-robot frame $<c>$. The configuration space of the mobile robot (or of the camera) is $R^2 \times SO(2)$, where $SO(2)$ is the special orthogonal group of 2×2 rotation matrices. Let $(Z_c\ 0\ X_c)^T$ be the camera center position in the base frame $<b>$, and α_c be the rotation angle of the camera-robot with respect to the z-axis of the base frame (see Fig. 1).

For the sake of simplicity a holonomic mobile robot is

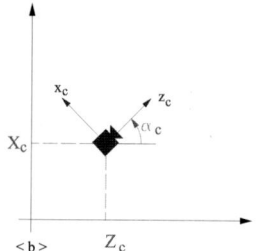

Figure 1: Mobile robot with a fixed camera.

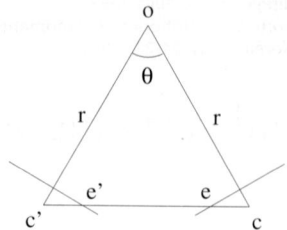

Figure 2: Symmetric camera displacement.

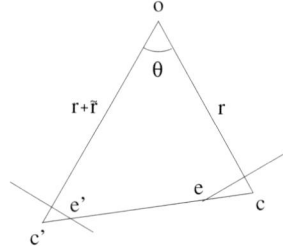

Figure 3: General cameras displacement.

considered. Thus system degrees of freedom are fully actuated by

$$\begin{cases} \dot{Z}_c &= u_z; \\ \dot{X}_c &= u_x; \\ \dot{\alpha}_c &= \omega \end{cases} \quad (1)$$

where $(u_z, u_x)^T$ is the linear velocity of the camera-robot on the plane and ω is the angular velocity about the y-axis.

2.1 Epipolar geometry

Before describing the proposed visual servoing algorithm, we introduce some concepts of epipolar geometry [5, 3]. Consider a pair of cameras with optical centers c', c, optical axes a', a and image planes q', q. The segment $c'c$ is called the *baseline* and its intersections with the image planes define the *epipoles*. The image line passing through the epipole and the image center is called the horizon line, while any plane containing the baseline is called an epipolar plane (see Fig. 2).

Given a pair of views of a scene and a set of corresponding points p'_i, p_i in homogeneous coordinates, there exists a matrix $F \in \mathcal{R}^{3\times 3}$, called the *fundamental matrix* [5], such that:

$$p'_i{}^T F p_i = 0 \ \forall i \quad (2)$$

For any point p_i (p'_i) in one view, the product Fp_i ($F^T p'_i$) defines a line, called the epipolar line, in the other view such that the corresponding point p'_i (p_i) belongs to this line. Moreover, the null right vector of F (F^T) represents the epipole e (e') on the image plane.

Consider now the situation depicted in Fig. 2. Two images are taken by the same camera, which undergoes a rotation θ about the axis o. The optical centers c', c are displaced at the same distance r from the intersection point o of the optical axes. Moreover, image planes and camera rotation axes are perpendicular to the epipolar plane containing o. Under such assumptions and for a camera intrinsic matrix $K = I$, the fundamental matrix F is given by (see [9, 10] for details):

$$F = \begin{pmatrix} 0 & \cos\theta - 1 & 0 \\ \cos\theta - 1 & 0 & \sin\theta \\ 0 & -\sin\theta & 0 \end{pmatrix} \quad (3)$$

where θ is the angle between the optical axes a' and a, and the epipoles e', e are given by

$$\begin{aligned} e' &= \left(-\tfrac{1}{\tan(\theta/2)},\ 0,\ 1\right)^T, \\ e &= \left(\tfrac{1}{\tan(\theta/2)},\ 0,\ 1\right)^T. \end{aligned} \quad (4)$$

Remark 1 *For the circular displacement in Fig. 2, a special symmetry condition holds: the x-coordinate of the two epipoles in (4) have the same magnitude and opposite sign. Such a symmetry will play a key role in designing the visual servoing algorithm.*

Symmetry is not preserved in the general configuration shown in Fig. 3 where the camera c' is shifted along its optical axis a distance $\tilde{r}$. In this case, the fundamental matrix F assumes the form

$$F = \begin{pmatrix} 0 & \beta\cos\theta - \gamma\sin\theta & 0 \\ -\beta & 0 & \gamma \\ 0 & -\beta\sin\theta - \gamma\cos\theta & 0 \end{pmatrix} \quad (5)$$

where

$$\beta = 1 - \cos\theta - \frac{\tilde{r}}{r+\tilde{r}}, \quad (6)$$

$$\gamma = \sin\theta \quad (7)$$

and the epipoles are $e' = (\alpha', \ 0, \ 1)^T$ and $e = (\alpha, \ 0, \ 1)^T$ with

$$\begin{array}{rcl} \alpha' & = & -\left[\tan(\theta/2)\left(1 + \frac{\tilde{r}}{r} \cdot \frac{1}{1-\cos\theta}\right)\right]^{-1}, \\ \alpha & = & \left[\tan(\theta/2)\left(1 - \frac{\tilde{r}}{r+\tilde{r}} \cdot \frac{1}{1-\cos\theta}\right)\right]^{-1}. \end{array} \qquad (8)$$

Remark 2 *The assumption of known camera intrinsic matrix ($K = I$) has been introduced for clarity of presentation. The symmetry property still holds for unknown focal lengths f_x and f_y, and henceforth we will assume that the intrinsic matrix is partially known:*

$$K = \mathrm{diag}(f_x, f_y, 1).$$

Under this assumption, the epipoles are given by

$$\begin{array}{rcl} e' & = & (f_x \alpha', \ 0, \ 1)^T \\ e & = & (f_x \alpha, \ 0, \ 1)^T \end{array} \qquad (9)$$

that is, they are obtained by scaling α and α' in (8) by the positive factor f_x.

Finally, note that the camera intrinsic matrix can be assumed completely unknown at the cost of a more involved discussion that, however, would not improve the problem insight.

3 Circular motion

In this section the visual servoing algorithm is described for circular motion. Extensions to general planar motion will be discussed in Section 4.

The core procedure of the proposed visual servoing algorithm consists in exploiting profiles to estimate the epipoles of current and desired images.

Consider the problem of moving a camera from an initial position to a desired position following a circular trajectory on the plane and exploiting only information derived from the desired image and the initial image.

As in Section 2, suppose that the image plane and the rotation axis of the camera are perpendicular to the motion plane, and that the optical axis intersects the trajectory center (see Fig. 2). Two different cases must be analysed.

3.1 Case I: known radius

Assume that the circular trajectory radius r is known. The image of the axis of rotation projects to the vertical line passing through the image center in each view of the scene. According to Section 2.1, the epipoles lie on the horizon line and their position is symmetric with respect to the rotation axis:

$$e' = (-w, \ 0, \ 1)^T; \quad e = (w, \ 0, \ 1)^T$$

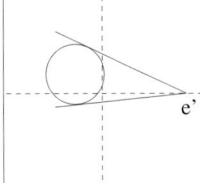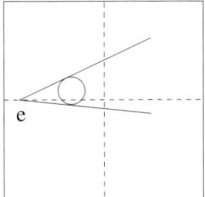

Figure 4: Epipolar tangencies for the symmetric cameras displacement case.

where

$$w = f_x / \tan(\theta/2). \qquad (10)$$

This means that the x-coordinates of the epipoles provide informations about the angle θ even when the cameras are uncalibrated.

In what follows, profiles will be exploited to estimate the epipoles of the initial and desired images. The basic idea is to use corresponding profiles in the images in order to estimate the epipole positions. Let obj be an object in the scene visible in both views and consider an epipolar plane p (distinct from the horizon plane) tangent to object obj. The tangent point between plane p and object obj is called *frontier point*: this point has the property of belonging to the apparent contour of object obj in each view [3]. Moreover, the epipolar line corresponding to this point is tangent to the apparent contour in each view (see Fig. 4) and, due to the chosen circular motion, the angle between the epipolar tangent and the horizon line is the same in each image.

This suggests that the epipole positions can be found by minimizing the sum of distances between the apparent contour and the corresponding epipolar tangency:

$$w = \mathop{\mathrm{argmin}}_{\tilde{w}} \left[\mathrm{Dist}(l'(\tilde{w}), C') + \mathrm{Dist}(l(\tilde{w}), C)\right] \qquad (11)$$

where $l'(\tilde{w}), l(\tilde{w})$ are the corresponding epipolar lines depending on the epipole positions $\tilde{w}$ and $-\tilde{w}$, C', C are the apparent contours and $\mathrm{Dist}(\cdot, \cdot)$ is the distance between the epipolar line and contour. In other words, we are looking for the epipole positions w and $-w$ such that the corresponding epipolar tangent in the current view is an epipolar tangent in the desired view.

Observe that the optimization problem in (11) has only one free parameter w. The visual servoing, leading the robot towards its final position algorithm, is designed on the basis of the estimated parameter w.

Let the circular trajectory of the camera be parameterized as follows (see Fig. 5):

$$\begin{array}{rcl} Z_c(t) & = & r\cos\varphi(t), \\ X_c(t) & = & r\sin\varphi(t), \\ \alpha_c(t) & = & \varphi(t) + \pi \end{array} \qquad (12)$$

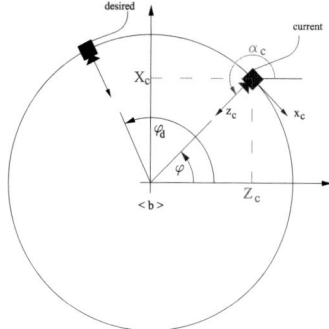

Figure 5: Parameterization of a circular motion centered on the origin of the base frame and with radius r.

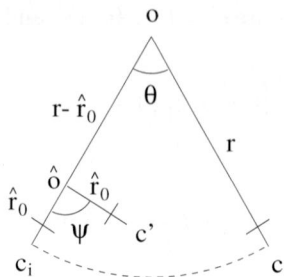

Figure 6: Robot motion under a (circular) control law with a wrong estimate of the radius.

where φ is the current camera position angle. The differential kinematics of the system is hence described by:

$$\begin{aligned} \dot{Z}_c(t) &= -X_c(t)\dot{\varphi}(t), \\ \dot{X}_c(t) &= Z_c(t)\dot{\varphi}(t), \\ \dot{\alpha}_c(t) &= \dot{\varphi}(t) \end{aligned} \quad (13)$$

where $\dot{\varphi}(t)$ is the control parameter steering the linear and angular velocity in (1).

To design the visual servoing algorithm, the control parameter $\dot{\varphi}(t)$ must be computable from the image measurement w. Observe that $\theta = \varphi_d - \varphi$, thus when φ approaches the desired value φ_d, $\frac{1}{w}$ decreases and goes to zero for $\varphi = \varphi_d$ as shown in (10), see also Fig. 5.

Therefore, a simple proportional control law of the visual measurement $\frac{1}{w}$ is given by:

$$\dot{\varphi}(t) = \frac{\lambda}{w(t)} \quad (14)$$

for some $\lambda > 0$.

3.2 Case II: unknown radius

Now suppose that the only a priori knowledge of the motion of the camera-robot is that a circular displacement occurs between the desired and the initial positions about an axis perpendicular to the motion plane and passing through an unknown point of the optical axis z_c. The trajectory radius r is unknown.

Let the initial configuration c_i and desired camera position c be as given as shown in Fig. 6. Starting from an initial guess $\hat{r}_0$ for the trajectory radius, apply controls ω, u_z and u_x (angular and linear velocities) as in (13) and (14). If $\hat{r}_0 \neq r$, the camera leaves the circular trajectory of radius r and reaches the new configuration c', after some amount of time, as shown in Fig. 6. The desired image and the current one (that taken by the camera in c') do not exhibit the property of symmetry discussed in Remark 1. In this new configuration (c', c), the epipoles are not symmetric with respect to the rotation axis and their positions are given by (9).

Two parameters, e_u and e_s, can be defined as:

$$e_u = \frac{1}{f_x \alpha'} + \frac{1}{f_x \alpha}, \quad (15)$$

$$e_s = \frac{1}{f_x \alpha'} - \frac{1}{f_x \alpha}. \quad (16)$$

The parameter e_u is defined as the sum of the inverse of the x-coordinates of the two epipoles. It accounts for the unsymmetric part of the displacements between the two views. On the other hand, parameter e_s accounts for the angle θ and as in Section 3.1 will steer the camera along the circular trajectory with known radius.

The following properties hold and are relevant for the design of the visual servoing procedure:

$$\begin{aligned} e_u \tilde{r} &\leq 0, \quad (17) \\ e_s(\varphi - \varphi_d) &\geq 0. \quad (18) \end{aligned}$$

In order to define the visual servoing procedure, we must compute the camera position c', obtained by rotating the camera of the angle ψ about an estimate of the rotation center $\hat{o}$ (Fig. 6).

The general camera position and orientation c' along the (unknown radius) trajectory with respect to the initial position c_i can be written as

$$\begin{aligned} Z_c(t) &= (r - \hat{r}(t))\cos\varphi_i + \hat{r}(t)\cos(\varphi_i + \psi(t)), \\ X_c(t) &= (r - \hat{r}(t))\sin\varphi_i + \hat{r}(t)\sin(\varphi_i + \psi(t)), \\ \alpha_c(t) &= \varphi_i + \psi(t) + \pi \end{aligned} \quad (19)$$

where $\hat{r}(0) = \hat{r}_0$, $\psi(0) = 0$ and φ_i identifies the initial camera position c_i on the plane. Note that the camera orientation at c' is such that the optical axis intersects $\hat{o}$. The corresponding differential kinematics are obtained

Figure 7: Trajectories followed for different initial estimates $\hat{r}_0$: $0.1r$, r, $2r$ and $3r$ (decreasing order).

by differentiating:

$$\begin{aligned}
\dot{Z}_c(t) &= \dot{\hat{r}}(t)\left[\cos\left(\varphi_i + \psi(t)\right) - \cos\varphi_i\right] \\
&\quad - \hat{r}(t)\dot{\psi}(t)\sin\left(\varphi_i + \psi(t)\right), \\
\dot{X}_c(t) &= \dot{\hat{r}}(t)\left[\sin\left(\varphi_i + \psi(t)\right) - \sin\varphi_i\right] \\
&\quad + \hat{r}(t)\dot{\psi}(t)\cos\left(\varphi_i + \psi(t)\right), \\
\dot{\alpha}_c(t) &= \dot{\psi}(t).
\end{aligned}$$

From (17) and (18) it can be easily shown that the simple proportional control law

$$\begin{aligned}
\dot{\hat{r}}(t) &= \lambda_r e_u(t), \\
\dot{\psi}(t) &= -\lambda_a e_s(t)
\end{aligned} \qquad (20)$$

is such that (for suitable positive λ_r and λ_a)

$$\lim_{t \to \infty} c'(t) = c.$$

Fig. 7 shows the trajectories followed by the camera for four different initial estimates $\hat{r}_0$ of the unknown circular radius $r = 1$. Control parameters were set to $\lambda_r = 1$, $\lambda_a = 0.1$. Further details on the controller design can be found in [2].

Observe also that in this case visual servoing (20) is entirely defined in terms of image measurements. The estimation of the epipoles is given by the solution of an optimization problem similar to (11) which has two free parameters (the two x-coordinates of the epipoles) instead of one and is constrained by the tangency condition in each view.

4 General planar motion

In this section the case of general cameras displacement is discussed. No a priori knowledge of the rotation and translation between the initial and the final positions is given. Assume only that the optical axes of the camera-robot in the initial and final configurations intersect at a point o. This case is that of general cameras displacement depicted in Fig. 3, where the property of symmetry discussed in Remark 1 does not apply.

Consider the trajectory consisting of a translation along the optical axis and a rotation about the axis through

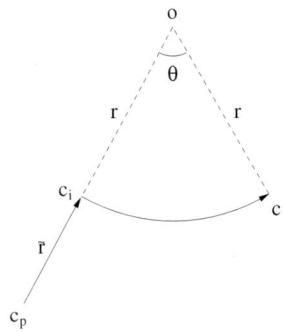

Figure 8: Trajectory followed by the robot in the case of general planar motion.

o, leading the camera from the initial configuration to the final one.

In order to reach the desired position, the visual servoing algorithm steers the robot along this trajectory in two steps:

1. the robot starts translating along the optical axis to reach a distance from o equal to r (i.e. to make $\tilde{r} = 0$ in Fig. 3);

2. the robot moves to the desired position with a circular motion.

Observe that the first step does not require knowledge of the current or desired radius. In fact it consists of a simple translation along the optical axis and the stopping condition occurs when parameter e_u (15), accounting for non-circular displacements, goes to zero. A simple proportional control law can be chosen as

$$\dot{\hat{r}}(t) = \lambda e_u(t) \qquad (21)$$

for some $\lambda > 0$ and, consequently, the robot differential kinematics become

$$\begin{aligned}
\dot{Z}_c(t) &= \lambda e_u(t) Z_c(t), \\
\dot{X}_c(t) &= \lambda e_u(t) X_c(t), \\
\dot{\alpha}_c(t) &= 0.
\end{aligned}$$

The second step brings the robot to the desired position following a circular trajectory as described in the previous section. Observe that neither knowledge nor estimation of the trajectory radius is required. Fig. 8 shows the complete trajectory followed by the camera in the case of general camera displacement from an initial configuration c_p to the final configuration c.

5 Simulations

Simulation results are reported to validate the proposed visual servoing algorithm for holonomic mobile robots.

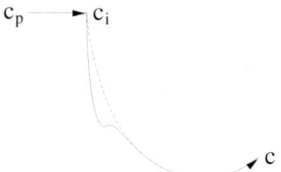

Figure 9: General planar motion: $\hat{r}_0 = 3r$ (solid) and $\hat{r}_0 = 0$ (dashed).

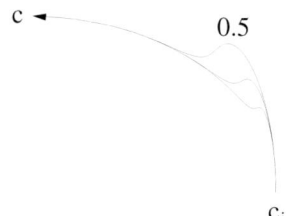

Figure 10: Trajectory for $\hat{r}_0 = 3r$ and different values of λ_r (0.5, 1, 2).

General planar motion was tested. The initial and final robot-camera configurations are

$$\begin{pmatrix} Z_{c_i} \\ X_{c_i} \\ \alpha_{c_i} \end{pmatrix} = \begin{pmatrix} -0.866 \\ -0.5 \\ \pi/3 \end{pmatrix} ; \quad \begin{pmatrix} Z_c \\ X_c \\ \alpha_c \end{pmatrix} = \begin{pmatrix} -0.866 \\ 0.5 \\ 2\pi/3 \end{pmatrix}.$$

Fig. 8 shows the ideal translational and rotational trajectories followed by the visual servoing. The pure translation moves the robot along the optical axis from c_p to the intersection c_i with the circle passing through c and centered on o. From this point the robot-camera starts to rotate as described in Section 3.2. This second part of the trajectory, which steers the mobile robot to c, strongly depends on the initial guess of the unknown radius r. Simulations are reported in Fig. 9: the solid line (dashed line) corresponds to an initial guess which is three (zero) times the true value. Control parameters are set to $\lambda = 1$, $\lambda_r = 1$ and $\lambda_a = 0.1$.

A second simulation was run to show system behavior for different control parameters. Circular motions with unknown radius were considered. Fig. 10 shows the trajectories followed for $\hat{r}_0 = 3r$, $\lambda_a = 1$ and different values of λ_r.

6 Conclusions

Epipolar geometry was exploited to design an image-based visual servoing algorithm for a mobile robot with a fixed camera. For the sake of simplicity mobile robot kinematics was assumed to be holonomic and the camera intrinsic parameters were assumed partially known. These assumptions can be relaxed to less restrictive conditions at the cost of a more involved discussion that, however, would not improve the problem insight.

The visual servoing algorithm is based on a measure of the symmetry of the epipolar geometry which is retrieved using image contours and tangency constraints but without solving any correspondence problem. Exploiting profiles in visual feedback is crucial in outdoor navigation where objects in the scene are highly unstructured and solving for correspondences is difficult.

Work is in progress to test the visual servoing algorithm on an experimental platform, the XR4000 by Nomadic Inc.

Acknowledgments

The authors wish to thank the student Massimo Pasqualetti of the University of Siena for his support in running simulations.

References

[1] G. Chesi, E. Malis, R. Cipolla: Automatic segmentation and matching of planar contours for visual servoing. In *Proc. Int. Conf. Rob. Autom.*, San Francisco, Cal., 2000.

[2] G. Chesi, D. Prattichizzo, A. Vicino. "Exploiting profiles in visual servoing", Int. Rep., Univ. of Siena, 2000.

[3] R. Cipolla and P.J. Giblin: *Visual Motion of Curves and Surfaces*. Cambridge University Press, 2000.

[4] F. Conticelli, D. Prattichizzo, A. Bicchi, F. Guidi, "Vision-Based Dynamic Estimation and Set-Point Stabilization of Nonholonomic Vehicles". In *Proc. Int. Conf. Rob. Autom.*, San Francisco, Cal., 2000.

[5] O. Faugeras: *Three-Dimensional Computer Vision: A Geometric Viewpoint*. MIT Press, Cambridge, 1993.

[6] K. Hashimoto and T. Noritsugu. "Visual servoing of nonholonomic cart". In *Proc. Int. Conf. Rob. Autom.*, Albuquerque, New Mexico, 1997.

[7] S. Hutchinson, G.D. Hager, and P.I Corke. "Tutorial on visual servo control". *IEEE Trans. Rob. Autom.*, 12(5):651–670, 1996.

[8] G.D. Hager, D.J. Kriegman, A.S. Georghiades, and O. Ben-Shahar. "Toward domain-independent navigation: dynamic vision and control". In *Proc. Conf. on Dec. and Control*, Tampa, Fl., 1998.

[9] P.R.S. Mendonca and R. Cipolla: Estimation of epipolar geometry from apparent contours: affine and circular motion cases. *Proc. IEEE Conf. Comp. Vis. Patt. Rec.*, Fort Collins, Col., I:9–14, 1999.

[10] P.R.S. Mendonca, K-Y.K. Wong and R. Cipolla: Circular motion recovery from image profiles. *Proc. of Workshop on Vision Algorithms: Theory and Practice*, Corfu, Greece, pages 119–126, 1999.

A Solution to the Adaptive Visual Servoing Problem

Alessandro Astolfi
Electrical Engineering Department
Imperial College
Exhibition Road
London SW7 2BT, UK
E-mail: a.astolfi@ic.ac.uk
FAX: +44 20 75946282

Liu Hsu
Electrical Engineering Department
COPPE/UFRJ
P. O. Box 68504
21945-970 Rio de Janeiro, Brazil
E-mail: liu@coep.ufrj.br
Fax +55 21 290 6626

Mariana Netto,* Romeo Ortega†
Lab. des Signaux et Systémes
Supelec
Plateau du Moulon
91192 Gif-sur-Yvette, France
E-mail: ortega@lss.supelec.fr
FAX: +33 1 69851765

Abstract

In this paper we address, and solve, a basic adaptive visual servoing problem whose solution was hampered by the nonlinear dependence of the system dynamics on the unknown parameters. Namely, we present a globally convergent vision–based position controller for a planar two-links manipulator in the so-called fixed–camera configuration, where the camera orientation and scale factor are considered unknown. The controller design technique of immersion and invariance, recently proposed in the literature, is used to derive a smooth adaptive scheme that ensures global asymptotic regulation without overparametrization, projections or persistency of excitation assumptions. In the case of tracking, we establish error bounds that depend on the speed of the reference trajectory, and can be reduced (eventually to zero) with improved prior knowledge on the camera scale factor. The efficacy of the approach is shown through simulations.

1 Introduction

Incorporating visual information in feedback control loops represents an attractive alternative for position and motion control of autonomous robot manipulators evolving in unstructured environments [6]. In this brief note we consider the problem of visual servoing of planar robot manipulators under a fixed–camera configuration with *unknown orientation and scale factor*. The control goals are to place the robot end-effector in some desired constant position—or to make it track a (slowly–moving) trajectory—by using a vision system equipped with a fixed camera.

It was shown in [2] that a (fixed parameter) PD–like controller ensures asymptotic *set–point regulation* of the full robot dynamics in spite of the uncertainty on the orientation parameter, which should however not be greater than $\pi/2$. See also [3, 4]. It is well-known that transient performance of PD–like schemes can be improved, particularly in tracking applications, adding an adaptation feature. The design of an adaptive controller is unfortunately complicated by the fact that the unknown parameters enter nonlinearly into the system dynamics. One way to bypass the nonlinearity obstacle is to overparametrize the system, see e.g. [5, 7, 8, 12]. However, this approach suffers from the well-known shortcoming of robustness degradation due to the slower convergence—intrinsic to a search in a bigger space—and the loss of identifiability in the absence of persistency of excitation. In [10] a globally convergent adaptive scheme that does not require overparametrization was proposed, but it uses a switching law that has shown to yield below–par performances.

The main contribution of our paper is the develop-

---

*The work of M. Netto is sponsored by the Brazilian foundation CAPES.

†Author to whom all correspondence should be addressed.

ment of a new (non–overparametrized) smooth adaptive scheme that, with the only prior knowledge of a lower bound on the scaling factor, ensures that the tracking error globally asymptotically converges to a residual set, whose size reduces to zero as the speed of the reference signal goes to zero. We also show that, if the scaling factor is known, then exact asymptotic global tracking of arbitrary (bounded) references is possible. Instrumental for our work is the utilization of the new controller design technique of immersion and invariance recently proposed in the literature [1].

2 Problem formulation

We consider a two-degrees of freedom robot manipulator that evolves in a plane. The vision system consists in a TV camera of CDD type that is fixed perpendicular to the plane where the robot evolves providing an image of the whole robot workspace, what includes the robot end-effector and the target. The image acquired by the camera supplies a two–dimensional array of brightness values from a three–dimensional scene. This image may undergo various types of computer processing to enhance image properties and extract image features [6]. As in [2], see also [3, 4, 5, 7, 8], we assume that the image features are the projection into the 2D image plane of 3D points in the scene space, hence we model the *action of the camera* as a static mapping from the joint robot positions $q \in \mathbb{R}^2$ to the position (in pixels) of the robot tip in the image output, denoted $y \in \mathbb{R}^2$. This mapping is described by

$$y = ae^{J\theta}[k(q) - \vartheta_1] + \vartheta_2 \qquad (1)$$

where $\theta \in \mathbb{R}$ is the orientation of the camera with respect to the robot frame, $a \in \mathbb{R}_+$ and $\vartheta_1, \vartheta_2 \in \mathbb{R}^2$ denote intrinsic camera parameters (scale factor, focal length and center offset, respectively). The function $k : \mathbb{R}^2 \to \mathbb{R}^2$ defines the robot direct kinematics, also, for compactness, we use the matrices

$$J = \begin{bmatrix} 0 & -1 \\ 1 & 0 \end{bmatrix}, \quad e^{J\theta} = \begin{bmatrix} \cos(\theta) & -\sin(\theta) \\ \sin(\theta) & \cos(\theta) \end{bmatrix}.$$

Invoking standard time–scale separation arguments, we assume an inner fast loop for the robot velocity control, and concentrate on the kinematic problem where we must generate the references for the robot velocities. The *robot* dynamics are then described by a simple integrator $\dot{q} = \tau$, where τ are the applied joint torques. The direct kinematics yields

$$\dot{k} = \mathcal{J}(q)\dot{q},$$

where $\mathcal{J}(q) := \frac{\partial k}{\partial q}(q) \in \mathbb{R}^{2\times 2}$ is the analytic robot Jacobian. As usual [2], we assume the robot operates far away from singular configurations, hence $\mathcal{J}(q)$ is nonsingular. Differentiating (1), and replacing the latter expression, we obtain the dynamic model of the overall system of interest

$$\dot{y} = ae^{J\theta}u, \qquad (2)$$

where we have introduced the input change of coordinates $u := \mathcal{J}(q)\tau$.

Control Problem Given the vision system (2), with measurable y, and a bounded target trajectory y_* with known bounded first and second order derivatives $\dot{y}_*, \ddot{y}_*$, respectively. Assume known a *prior estimate* on the scale factor a which satisfies the bound $a \geq a_m > 0$. Then, find a control signal u such that y asymptotically tracks, as close as possible, the reference trajectory, in spite of the lack of knowledge of a and θ. In particular, if $y_* = const$, or if $a = a_m$, we would like to ensure that y converges to y_* from all initial conditions $y(0)$.

In spite of the simplicity of the system dynamics, the task is, of course, complicated by the highly nonlinear dependence on the unknown parameters. We will show below that θ is a particularly critical parameter, hence an estimation law will be implemented for it. Unfortunately, we are unable at this point to estimate also a, but we will prove that the proposed controller is insensitive to this parameter in regulation, while for tracking applications the controller enjoys some nice robustness properties.

3 Main Result

In this paper we use the immersion and invariance (I&I) approach proposed in [1] to design an adaptive controller that solves the problem stated in the former section. Although the present work is self–contained, we refer the interested reader to that paper for further details and motivation of the I&I method.

To simplify the presentation of our main result we will find convenient to define the 2×2 constant rotation matrix[1]

$$R := -\frac{a}{a_m} e^{-J \arccos(\frac{a_m}{a})} \qquad (3)$$

which is parametrized by $\frac{a}{a_m}$ and satisfies the properties that R is Hurwitz for all $\frac{a}{a_m}$, and furthermore

$$\lim_{\frac{a}{a_m} \to 1} R = -I.$$

---
[1]Notice that, according to the problem formulation, $0 < \frac{a_m}{a} \leq 1$.

We also introduce the two-dimensional linear time–invariant dynamical system

$$\dot{w} = Rw - (R+I)\dot{y}_* \qquad (4)$$

For all initial conditions, the solutions $w(t)$ of this system satisfy the bound

$$|w(t)| \leq M(\frac{a}{a_m}, |\dot{y}_*|) + \epsilon_t$$

where ϵ_t converges exponentially to zero and the (smooth) function $M(\frac{a}{a_m}, |\dot{y}_*|)$ converges to zero if, either $\frac{a}{a_m} \to 1$, or $|\dot{y}_*| \to 0$.

In the proposition below we will present an adaptive I&I controller that ensures the tracking error asymptotically converges to w. In words, this means that we can ensure a small tracking error if *either* we have a good prior estimation of a, *or* the reference trajectory is slowly varying.

Proposition 1 *Consider the system (2) in closed–loop with the I&I adaptive controller*

$$u = -\frac{1}{a_m} e^{-J(\hat{\theta}+\frac{1}{2}|s|^2)} s \qquad (5)$$

$$\dot{\hat{\theta}} = s^\top (\tilde{y} + \ddot{y}_*) \qquad (6)$$

$$s = \tilde{y} - \dot{y}_* \qquad (7)$$

where $a \geq a_m > 0$ and $\tilde{y} := y - y_*$ is the tracking error. Then, for all initial conditions $y(0) \in \mathbb{R}^2, \hat{\theta}(0) \in \mathbb{R}$ and all bounded reference trajectories y_*, with bounded first and second order derivatives $\dot{y}_*, \ddot{y}_*$, we have that all trajectories of the system are bounded and the tracking error satisfies

$$\lim_{t\to\infty} |\tilde{y}(t) - w(t)| = 0, \qquad (8)$$

with w the solution of (4) with initial conditions $w(0) = \tilde{y}(0) \in \mathbb{R}^2$. In particular, if either $a_m = a$, or $|\dot{y}_*| \to 0$, then we have $\lim_{t\to\infty} |\tilde{y}(t)| = 0$.

Proof. First, we observe that θ is the critical parameter needed to achieve the convergence properties of the proposition. More precisely, if θ is known, we can design a stabilizing law for (2) using the lower bound estimate[2] of the uncertain parameter a. Indeed, the constant parameter feedback $u = -\frac{1}{a_m} e^{-J\theta} s$, where s is defined in (7), yields the target closed–loop dynamics

$$\dot{\tilde{y}} = -\frac{a}{a_m} s - \dot{y}_* \qquad (9)$$

---
[2] Actually, any fixed estimate of a will do the job if θ is known, as will become clear later, the lower bound is needed for the adaptive problem.

whose trajectories converge to zero if, either $\frac{a}{a_m} = 1$, or $|\dot{y}_*| \to 0$. Therefore, following the I&I adaptive controller procedure, we will take our control law of the form

$$u = -\frac{1}{a_m} e^{-J(\hat{\theta}+\beta_1(s))} s$$

where, $\beta_1(s)$ is a function to be determined—that, for convenience, we have defined explicitly dependent on s. Introducing the change of coordinates

$$z = \hat{\theta} - \theta + \beta_1(s) \qquad (10)$$

we obtain the first error equation

$$\dot{\tilde{y}} = -\frac{a}{a_m} e^{-Jz} s - \dot{y}_* \qquad (11)$$

The second error equation describes the dynamics of z which, as seen from (10), is defined by our choice of the function $\beta_1(s)$ and the parameter update law. In standard applications of the I&I approach, these two functions are selected to drive z to zero. Notice that, in this case, (11) reduces to the target dynamics (9). As will become clear below, due to the uncertainty in the parameter a, we will not obtain $z \to 0$, but will drive it anyway to a value that ensures the desired stability properties. Towards this end, we select

$$\beta_1(s) = \frac{1}{2}|s|^2 \qquad (12)$$

which yields (5), and $\dot{\hat{\theta}}$ as given by (6). Differentiating (10), using (6) and (12), and doing some simple calculations with (7) and (11) we get the second error equation

$$\dot{z} = -|s|^2 [\frac{a}{a_m} \cos(z) - 1] \qquad (13)$$

where we have used the fact that $s^\top e^{-Jz} s = |s|^2 \cos(z)$. The dynamics of the closed–loop system is fully characterized by the error system (11), (13).

Before completing our proof of the proposition, let us take a brief respite to unveil the stabilization mechanism that motivated our choice for the z dynamics. This can be better explained referring to Fig. 1, where we plot the function $f(z) = 1 - \frac{a}{a_m} \cos(z)$. From the figure, and the fact that $\dot{z} = |s|^2 f(z)$, we see that the trajectories of (13) remain *bounded*, and the dynamics—which is clearly 2π–periodic—exhibits alternating stable ($\bar{z}_s \mod(2\pi)$) and unstable ($\bar{z}_u \mod(2\pi)$) equilibria. From observation of (13) we also conclude that, in general, $z \to \bar{z}_s \mod(2\pi)$, and in this case $\tilde{y}$ will converge to the trajectories of (4), as desired. (Notice that $\bar{z}_s = \arccos(\frac{a_m}{a})$.) The qualifier "in general" is needed because, although highly improbable, we cannot rule out the occurrence of the case $s \to 0$, hence it needs to be considered. We will prove later that, if $s \to 0$, the tracking error actually goes to zero—which is more than expected! Let us

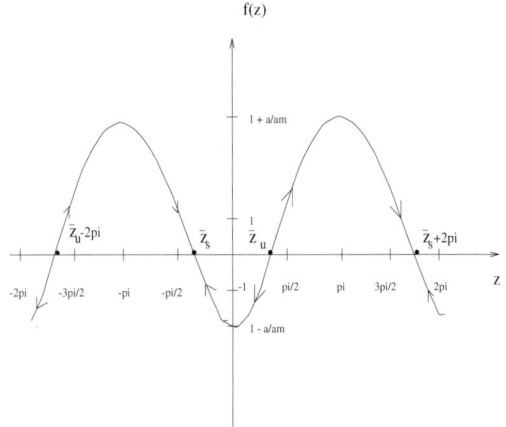

Figure 1: Plot of the function $f(z) = 1 - \frac{a}{a_m}\cos(z)$.

now continue with the stability analysis of (11), (13). Once we have shown that z is bounded, we prove now that $\tilde{y}$ is also *bounded*. To this end, we find convenient to write the $\tilde{y}$ dynamics in terms of s as

$$\dot{s} = -\frac{a}{a_m}e^{-Jz}s - \dot{y}_* - \ddot{y}_* \qquad (14)$$

and consider the function

$$W(s,z) = \frac{a_m}{2a}|s|^2 + \sin(z) + 1$$

Taking the derivative along the trajectories of (11), (14) yields vspace-0.2cm

$$\begin{aligned}\dot{W} &= -\frac{a}{a_m}|s|^2\cos^2(z) - \frac{a_m}{a}s^\top(\dot{y}_* + \ddot{y}_*) \\ &\leq -\frac{a}{a_m}|s|^2\left[\cos^2(z) - \frac{\alpha a_m^2}{2a^2}\right] + \frac{M_1}{\alpha}\end{aligned}$$

where $M_1 \geq \frac{a_m}{2a}|\dot{y}_* + \ddot{y}_*|^2$, and we have used the inequality

$$|s||\dot{y}_* + \ddot{y}_*| \leq \frac{\alpha}{2}|s|^2 + \frac{1}{2\alpha}|\dot{y}_* + \ddot{y}_*|^2$$

which holds for arbitrary $\alpha > 0$.

Now, from Fig 1 we see that the trajectories of (11) are not just bounded, but also monotone: nondecreasing in the interval $[\bar{z}_u - 2\pi, \bar{z}_s)$ and nonincreasing in the interval $[\bar{z}_s, \bar{z}_u)$.[3] Consequently, z *converges* to a finite

[3] As the dynamics is 2π–periodic, without loss of generality, we will restrict the analysis to the interval $\bar{z}_u - 2\pi, \bar{z}_u]$.

limit, that we denote z_∞. We will split our analysis into the two possible situations: $z_\infty = \frac{\pi}{2}$ or $z_\infty \neq \frac{\pi}{2}$.[4]

- Let us first assume that $z_\infty \neq \frac{\pi}{2}$. Then, there exists $\epsilon_1 > 0$, satisfying $|\cos(z_\infty)| \geq \epsilon_1$, and $T_1 \geq 0$ such that, for all $t \geq T_1$, we have

$$\dot{W} \leq -\beta(\alpha)|s|^2 + \frac{M_1}{\alpha} \qquad (15)$$

where we have defined the constant

$$\beta(\alpha) := \frac{a}{a_m}\left(\epsilon_1^2 - \frac{\alpha a_m^2}{2a^2}\right)$$

Remark that, for all ϵ_1, a and a_m, $\beta(\alpha)$ can be made positive choosing a sufficiently small α. Now, using the fact that $W \leq \frac{a_m}{2a}|s|^2 + 2$, we get the differential inequality vspace-0.2cm

$$\dot{W} \leq -2\frac{a}{a_m}\beta(\alpha)W + \bar{M}_1$$

where $\bar{M}_1 = \frac{4a}{a_m}\beta(\alpha) + \frac{M_1}{\alpha}$, from which we immediately conclude that W, and consequently s and $\tilde{y}$, are bounded.

- Let us now consider the case when $z_\infty = \frac{\pi}{2}$. We evaluate the integral

$$\lim_{t\to\infty}\int_0^t \dot{z}(\tau)d\tau =$$

$$= \lim_{t\to\infty}\int_0^t |s(\tau)|^2[1 - \frac{a}{a_m}\cos(z(\tau))]d\tau \leq M_2$$

where the bound M_2 follows from boundedness of z. Now, as we assumed $z_\infty = \frac{\pi}{2}$, there exists $T_2 \geq 0$, and $\epsilon_2 > 0$, such that

$$1 - \frac{a}{a_m}\cos(z(t)) > \epsilon_2, \ \forall \ t \geq T_2$$

Splitting the integral above into the intervals $[0, T_2]$ and $[T_2, \infty)$, and replacing this bound in the second integral we obtain vspace-0.1cm

$$M_2 > \int_0^{T_2} |s(\tau)|^2[1 - \frac{a}{a_m}\cos(z(\tau))]d\tau +$$

$$+ \epsilon_2 \lim_{t\to\infty}\int_{T_2}^t |s(\tau)|^2 d\tau$$

As the first right hand term is bounded, we conclude from this inequality that, if $z_\infty = \frac{\pi}{2}$, then s is *square integrable*.

[4] It is important to note that, as s may converge to zero faster than z, then we cannot conclude that $z_\infty = \bar{z}_s$. Of course, as pointed out already above, having $z_\infty \neq \bar{z}_s$ is highly improbable and it is treated only for the sake of completeness.

On the other hand, from (14), and the fact that e^{Jz} is unitary, we see that $\dot{s}$ satisfies the bound

$$|\dot{s}| \leq \frac{a}{a_m}|s| + M_3$$

where $M_3 \geq |\dot{y}_* + \ddot{y}_*|$, which means that s is a (so-called) *regular* signal [11]. Finally, it is possible to prove that regular signals which are square integrable, are bounded (see appendix). This completes our proof of boundedness of $\tilde{y}$.

We will now establish the *convergence* result (8). First, we see from (14) and boundedness of s and $\dot{y}_*, \ddot{y}_*$, that $\dot{s}$ is also bounded. Similarly, from (13), and boundedness of s, we have that $\dot{z}$ is bounded. Now, differentiating (13), and using boundedness of $s, \dot{s}, \dot{z}$ we prove that $\ddot{z}$ is also bounded, hence $\dot{z}$ is uniformly continuous. From convergence of z and Barbalat's lemma [11] we can then conclude that $\dot{z} \to 0$ as $t \to \infty$. Referring to (13), this establishes that

$$\lim_{t \to \infty} |s(t)|^2 [1 - \frac{a}{a_m} \cos(z(t))] = 0$$

which implies that, either s, or the signal in brackets converges to zero. We consider first the case when $\lim_{t \to \infty}[1 - \frac{a}{a_m}\cos(z(t))] = 0$, from which we immediately have that $z \to \bar{z}_s$. This means that the dynamics (11) can be written in the form

$$\dot{\tilde{y}} = [R + B(t)]\tilde{y} - [R + B(t) + I]\dot{y}_*$$

for some bounded matrix $B(t) \to 0$ as $t \to \infty$ and R defined in (3). The convergence result (8) then follows subtracting (4) from the equation above.

Finally, we will prove that, in the (highly improbable) case of $\lim_{t \to \infty} |s(t)| = 0$, we have $\tilde{y} \to 0$. From (7), we see that $s \to 0$ implies that $\tilde{y} \to \dot{y}_*$. On the other hand, from (11), it also implies that $\dot{\tilde{y}} \to -\dot{y}_*$. Combining these two limits we have that $\dot{\tilde{y}} + \tilde{y} \to 0$, which proves that $\tilde{y} \to 0$.

This completes the proof of the proposition. ◁

4 Simulations

The proposed adaptive controller given by (5)-(6) was tested through simulations. The two-link robot direct kinematics transformation is given by

$$y_1 = L_1 cos(q_1) + L_2 cos(q_1 + q_2) + O_1 \quad (16)$$
$$y_2 = L_1 cos(q_1) + L_2 cos(q_1 + q_2) + O_2 \quad (17)$$

where, L_1 and L_2 are the link lengths. The index 1 stands for the base of the manipulator, $y_1 = O_1$, $y_2 = O_2$ are the base coordinates in the workspace frame. For sake of comparativeness, the simulations were carried out in the same conditions of that in [7] where it is considered $L_1 = 0.8$m; $L_2 = 0.5$m; $O_1 = -0.666$m ; $O_2 = -0.333$m. A case of extreme misorientation was taken into consideration ($\theta = 1 rad$), where $\hat{\theta}(0) = 0$. The scaling factor a and its lower bound a_m were chosen as $a = 0.7$, $a_m = 0.5$. The initial conditions of the manipulator are $q_1(0) = 1.3 rad$ and $q_2(0) = -1.3 rad$. For the set-point control case, with $y_{1*} = y_{2*} = 0.1$, the convergence is observed in 5 seconds (within 4 percent of the final value). The tracking case was also tested where the reference trajectory was generated by a first-order filter $\dot{y}_* = -\lambda y_* + r$ with $r_1 = a sin(w_r t) + c + d sin(1.5 w_r t)$, $r_2 = b sin(w_r t + \psi) + c + d sin(1.5 w_r t + \psi)$, $a = b = d = 0.04$, $c = 0.1$, $\psi = 1 rad$, $w_r = 0.03 rad/sec$ and $\lambda = 1$. The simulation was carried out for 300 seconds. The phase-plane is shown in figure 2 where in spite of the extreme misorientation, we can observe that the tracking is achieved. The amplitudes in absolute value of the control torques (not shown) remain smaller than 0.6Nm fot all t. [5].

5 Conclusions

In this paper we have applied the I&I adaptive control design technique to fully solve the basic visual servoing problem of set–point regulation of a planar robot when the orientation and the scale factor of the camera are unknown. The controller also ensures bounded tracking with a residual error proportional to the reference speed and the *a priori* error on the scale factor.

Instrumental for the establishment of our results has been the utilization of the I&I approach. As clearly illustrated in this paper, I&I adaptive control constitutes a major departure from standard adaptive control techniques, where certainty equivalence controllers and separable Lyapunov functions are typically utilized. Abandoning certainty equivalence and employing Lyapunov functions that mix the plant and the estimator states—features which are intrinsic to the definition of adaptive I&I—seems to be necessary to tackle this class of nonlinearly parametrized problems.

In the problem at hand it has been possible to apply I&I to adapt the orientation parameter. However, we have been unable to estimate the scaling factor, a property that is required to achieve exact global tracking. In spite of this negative result we believe that, in contrast with other *ad-hoc* methods, the deep underlying geometric motivation of adaptive I&I makes it a promising candidate to survive scaling up to more

---

[5] the applied joint torques are computed from the inverse transformation: $\tau = \mathcal{J}^{-1}(q)u$

complicated examples.

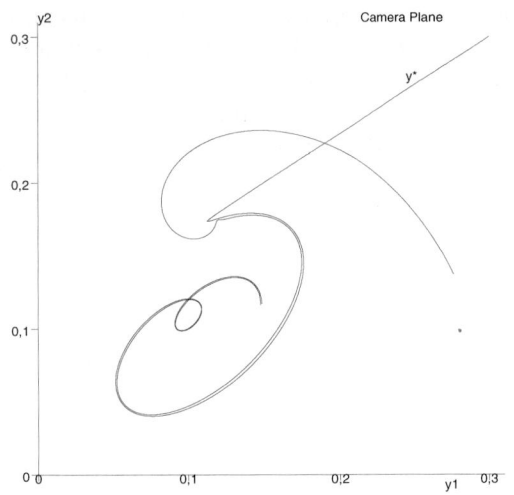

Figure 2: Convergence of the outputs to the desired trajectories.

References

[1] Astolfi, A. and R. Ortega, Immersion and invariance (I&I): A new tool in nonlinear control design, *European Control Conf ECC 2001*, Porto, P, September, 4-7, 2001, (submitted).

[2] Kelly, R., Robust asymptotically stable visual servoing of planar robots,*IEEE Trans. Robotics and Aut.*, Vol. 12, No.5, pp.759-766, October 1996.

[3] Coste-Manière, E., Couvignou, P. and Khosla, P.K. Visual servoing in the task–function framework: A contour following task. *Journal of Intelligent and Robotic Systems*, vol.12, pp. 1-21, 1995.

[4] Kelly, R., Shirkey, P. and Spong, M.W. Fixed camera visual servo control for planar robots. In *Proc. IEEE Int. Conf. on Robotics and Automation*,vol.3, Minneapolis, MN., pp.2643-2649, April 1996.

[5] Bishop, B.E., Spong, M.W., Adaptive calibration and control of 2D monocular visual servo systems, *Symp. on Robot Control*, Nantes, France., Sep. 3-5 1997.

[6] S. Hutchinson, G.D. Hager, and P.I. Corke. A tutorial on visual servo control. *IEEE Trans. on Robotics and Automation*, 12:651–670, October 1996.

[7] Hsu, L. and Aquino, P.,Adaptive visual tracking with uncertain manipulator dynamics and uncalibrated camera, *Proc. 38th Conference on Decision and Control*, pp.1248–1253, Phoenix, Dec 1999.

[8] Hsu, L., Costa, R.R. and Aquino, P.,Stable adaptive visual servoing for moving targets, *Proc. of the American Control Conference (ACC)*, Chicago, June 2000.

[9] Khalil, H.K. *Nonlinear Systems*, Prentice–Hall,Inc., New York, 1996.

[10] Lefebre, E., R. Kelly, R. Ortega, and H. Nijmeijer. On adaptive calibration for visual servoing. In *IFAC Symp Nonlinear Control Systems Design, Enschede, NL*, pages 1–3, 1998.

[11] S. Sastry and M. Bodson, *Adaptive Control: Stability, Convergence and Robustness*, Prentice–Hall, 1989.

[12] Dixon, W.E., Dawson, D.M., Zergeroglu, E. and Behal, A. , "Adaptive Tracking Control of a Wheeled Mobile Robot via an Uncalibrated Camera Sytem", IEEE Transactions on Systems, Man, and Cybernetics - Part B: Cybernetics, 2000.

Appendix A

Lemma 1 *A regular signal s that belongs to $\mathcal{L}_2$, also belongs to $\mathcal{L}_\infty$.*

Proof. We will prove Lemma 1 by absurd. We suppose that $s \notin \mathcal{L}_\infty$ and prove that this implies $s \notin \mathcal{L}_2$, leading to the absurd.

If $s \notin \mathcal{L}_\infty$, then there exists an unbounded sequence $t_1 < t_2 < \cdots$ such that $t_i - t_{i-1}$ is larger than any constant T and such that the sequence $|s(t_1)| < |s(t_2)| < \cdots$ is strictly increasing and unbounded. We now show that this is impossible.

If we prove that a lower bound $\rho(t)$ for $|s(t)|$, with $\rho(t) \geq 0$, $\forall t \geq 0$, is such that $\rho(t) \notin \mathcal{L}_2$ we can conclude that $s \notin \mathcal{L}_2$, leading to the absurd.

In the following, we will use the regularity property of the signal s to find the lower bound ρ for $|s|$. If s is regular then it satisfies $|\dot{s}| \leq L||s_t||_\infty + C_1$, where L, C_1 are nonnegative constants and $||s_t||_\infty = \sup_{\tau \leq t} u(\tau)$. It also implies that $|s|$ cannot decrease faster than an exponential signal. And this is the key point, that makes possible the construction of a lower bound ρ for $|s|$ in each interval $[t_{i-1}, t_i]$. In the sequel we detail this construction.

The lower bound ρ for $|s|$ in each interval $[t_{i-1}, t_i]$ can be given by a positive triangular pulse starting at t_{i-1} with value $|s(t_{i-1})|$ and decreasing with constant slope (independent of the interval). The slope is the initial one of an exponential signal of the form $[|s(t_{i-1})| - C_2]e^{-Lt} + C_2$ (the initial slope is equal to $-L[|s(t_{i-1})| + C_2]$), where C_2 is a constant that depends on L qnd C_1. More explicitly, the lower bound in the interval $[t_{i-1}, t_i]$ is

$$\begin{aligned} \rho(t) &= f(t) \quad, \text{if } \rho(t) > 0 \\ \rho(t) &= 0 \quad, \text{if } f(t) \leq 0 \end{aligned}$$

where $f(t) = |s(t_{i-1})| - L[|s(t_{i-1})| + C_2][t - t_{i-1}]$ is simply the line passing in the point $(t_{i-1}, |s(t_{i-1})|)$ with slope equal to $-L[|s(t_{i-1})| + C_2]$.

$\rho(t) = 0$ will happen inside the interval since T can be chosen arbitrarily large. Thus, the duration of the pulses can be easily calculated to be $\Delta = |s(t_{i-1})| / L[|s(t_{i-1})| + C_2]$. Now, it is easy to conclude that the areas under $\rho(t)$ in each interval adds up to infinity. Then, we have that $\int_0^\infty \rho(t)dt$ is unbounded. This directly implies that $\int_0^\infty \rho(t)^2 dt$ is also unbounded. Consequently, $\rho \notin \mathcal{L}_2$ leading to the absurd.

◁

An Industrial Application of Behavior-Oriented Robotics

Andreas Birk and Holger Kenn

Vrije Universiteit Brussel, Artificial Intelligence Laboratory,
Pleinlaan 2, 1050 Brussels, Belgium

Abstract

The so-called RoboGuard is a mobile security device which is tightly integrated into the existing surveillance framework developed and marketed by Quadrox, a Belgian SME. RoboGuards are semi-autonomous mobile robots providing video streams via wireless Intranets to existing watchguard systems, supplemented by various basic and optional behaviors. RoboGuards fill several market-niches. Especially, they are a serious alternative to the standard approach of using Closed Circuit Television (CCTV) for surveillance. Low production cost, user friendliness, and support for easy add-on functionalities were important design targets, which were achieved by following the AI paradigm of behavior-oriented robotics.

1 Introduction

The so-called RoboGuard is a joint development between Quadrox [17], a Belgian security SME, and two academic partners, the AI-lab of the Flemish Free University of Brussels (VUB) and the Interuniversity Micro-Electronics Center (IMEC). A RoboGuard allows remote monitoring through a mobile platform using onboard cameras and sensors. RoboGuards are supplements and often even alternatives to standard surveillance technology, namely Closed Circuit Television (CCTV) and sensor-triggered systems. RoboGuards are tightly integrated into the existing range of products of Quadrox. This is an important aspect for the acceptance of any novel technology in well-established markets as customers are usually not willing to completely replace any existing infrastructure.

For efficiency and security reasons, the RF-transmitted video-stream of the on-board cameras is compressed using a special wavelet-encoding [11]. The IMEC is the responsible partner for this feature of RoboGuard. The mobile base and its control, which are the main focus of this paper, are at the hands of the VUB AI-lab. The VUB AI-lab has a long tradition since the mid-1980s in basic research in the domain of behavior-oriented robotics. The expertise of the VUB AI-lab in this domain includes conceptual aspects as well as technological know-how, which are both incorporated in the RoboGuard system.

In accordance with recent interest in service robotics [12], there has also been previous work on security robots. This work is widely scattered, ranging from unmanned gunned vehicles for military reconnaissance operations [1] to theoretical research on reasoning within decision-theoretic models of security [14]. Our approach deals with a system operating in semi-structured environments under human control and which is a product, i.e., it must be competitive to existing alternative solutions for the task.

The rest of this paper is structured as follows. In section two, the relation of the field of behavior-oriented AI to the RoboGuard project is explained. Section three presents the hardware aspects of a RoboGuard. In section four, the software side of the RoboGuards is introduced. Section five concludes the paper.

2 Embodied Intelligence

"Classical" robots as we find them today in industry rely very much on precise mechanics. This is necessary as these robots rely on exact models, for example to describe and compute their kinematics. The need for extreme precision puts high demands on the parts, the assembling, and the maintenance of these robots. The field of behavior-oriented robotics or the "artificial life road to AI" [19] on the other hand has come up with robotic systems which are much less demanding in this respect. The model-based control schemes are substitute by reactive ones [9, 18], which establish close, dynamic couplings between sensors and motors. Instead of using complex models, sensor information is exploited as, as Rodney Brooks put it, "the world is its best own model" [10]. In addition to the much lower need for precision, there is a lower need for computation power as well, as reactive control schemes are usually computationally much cheaper than model-based control schemes like for example inverse kinematics.

Behavior-oriented robotics profits mainly from recent advances in sensor technology as there are more and more mass-produced, inexpensive sensors available for a multitude of different applications. Especially, advances in vision related hardware allow for some various sensing tasks at low cost. In combination with the tremendous savings at the mechanical part of the device, a behavior-oriented robot can be produced much cheaper than its classical counterpart [3]. This cost effectiveness is an important feature for RoboGuard.

In addition, RoboGuard needs several functionalities which are directly related to artificial life (Alife) [13] and research on animats (= animal + robot) [22]. For example, RoboGuard has to be self-sufficient, i.e., it has to be able to sustain itself over extended periods of time in respect to energy [16]. In this context, basic research in an artificial robotic ecosystem [20, 15, 2, 4] turned out to be very useful for this practical application.

A further advantage of employing behavior-oriented concepts in RoboGuard is the facilitation of semi-autonomy. In contrast to the naive intuition, including a human operator in the control loop of the base can make the task more complex. It is very difficult, if not impossible for a human teleoperator to efficiently steer a mobile base with video-streams from a on-board camera only. Operators do not take the current speed and momentum of the base into account, they neglect possible delays, they have difficulties to develop a feeling for the size of the base, and so on. In addition, the mobile base has to be protected from accidental or malicious misuse.

Shortly, the mobile base needs an advanced system for navigation and steering support including obstacle avoidance. The fusion of operator steering commands, autonomous drive and navigation functionality, as well as domain-specific plausibility and safety checks is a non-trivial task. For this purpose, the modular approach of using behaviors is especially suited. It also turned out that we could strongly benefit in this respect from insights gathered in the domain of robot soccer [6, 7, 8].

3 The Mobile Base

3.1 The RoboCube as Controller-Core

The basic Sensor-Motor-Control of RoboGuard centers at the side of electronics around the RoboCube, a

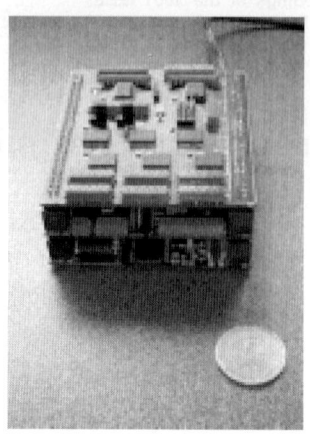

Figure 1: A picture of the RoboCube, an extremely compact embedded computer for robot control.

special robot control hardware developed at the VUB AI-lab. The VUB AI-lab has quite some tradition in developing flexible robot control hardware. Various experimental platforms have been build in the lab starting from the mid-eighties up until now. Over the years, experiences with approaches based on embedded PCs and different micro-controllers were gathered and lead to the *Sensor-Motor-Brick II (SMBII)*[21]. The SMBII is based on a commercial board manufactured by Vesta-technology providing the computational core with a Motorola MC68332, 256K RAM, and 128K EPROM. Stacked on top of the Vesta-core, a second board provides the hardware for sensor-, motor-, and communication-interfaces.

The so-called RoboCube (figure 1) is an enhanced successor of the SMBII. In RoboCube, the commercial computational core is replaced by an own design, also based on the MC68332, which saves significant costs and simplifies its architecture. In addition, the physical shape of RoboCube is quite different from the one of the SMBII. First, board-area is minimized by using SMD-components in RoboCube. Second, three boards are stacked on each other leading to a more cubic design compared to the flat but long shape of the SMBII, hence its name Robo*Cube*.

RoboCube has a open bus architecture which allows to add "infinitely" many sensor/motor-interfaces (at the price of bandwidth). But for most applications the standard set of interfaces should be more than enough. RoboCube's basic set of ports consists of 24 analog/digital (A/D) converter, 6 digital/analog (D/A) converter, 16 binary Input/Output (binI/O), 5 binary Inputs, 10 timer channels (TPC), and 3 DC-motor controller with pulse-accumulation (PAC). The RoboCube is described in more detail in [6, 5].

3.2 Components and Integration of the Mobile Base

When developing and integrating the different hardware components of the mobile base, it was necessary to engineer specific aspects through several iterated test and developments. For example, it is necessary to exactly adapt the drive-units (with motors, gears, encoders, wheels, etc.) to achieve a maximal performance at minimal cost. The same holds for the power-system and all other sub-units of the base. The following two bases serve as an example of this process of constant adaption and improvement. At the moment, the second base is produced in a small series to be used as RoboGuards.

The basic hardware aspects are the mobile platform including the RoboCube controller, the motor drivers, the support frame, the power system and the energy management. All these factors are strongly interdependent. In addition, they are strongly affect by the type of main-computer supplementing the RoboCube as this main-computer strongly affects the power consumption. The main-computer is used for the high-level computations, especially the image acquisition, the image compression and the communication on board of the robot. Due to an adaptation to the developments of the computer market, the type of main-computer on the robot was changed and therefore there were significant changes within the base-design between the first and the second version.

The most significant feature of the first version of the base (figure 2) is the usage of a network computer, namely the Corel Netwinder. At the beginning of the project, network computers seemed to be a promising technology especially in respect to this project. The Corel netwinder is very compact, offers many default interfaces, and it has a very low power-consumption.

But its computing power is not sufficient for the needs of this project. Furthermore, it is questionable if this trait of computers will survive the fast current developments in the market. To guarantee availability and increase in performance for the future, it was seen necessary to switch to a PC-based approach. This implied that the drive- and power-system of this first base were much too small. They had to be severely adapted for the next version. But the general development of motor-drivers and the control-electronics were already successfully completed on this base.

Figure 2: The first version of the RoboGuard base includes a network-computer, the Corel Netwinder.

Figure 3: The inside core of the second version of the RoboGuard base. It includes a mobile PC-board and four color-cameras allowing full 360 degrees surveillance. Optional components include two modules with 5 Sonar- and 6 active Infrared sensors, respectively.

The second version of the mobile platform (figure 3) and base was developed with several intermediate tests and changes. It is already a very matured version, i.e., there will be no or only minor changes to its low-level functionality for future versions. As mentioned above, a small series of these bases is produced at the moment to be used in RoboGuards.

4 The Control Software

4.1 Overview

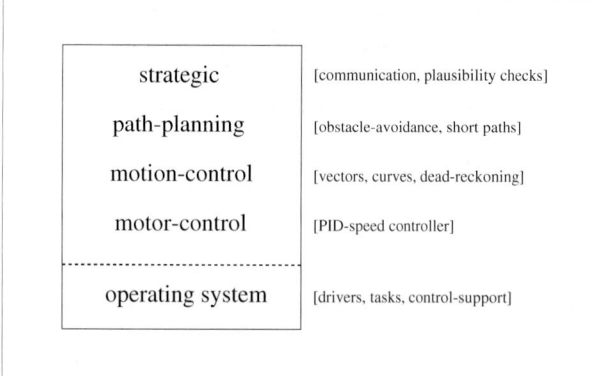

Figure 4: The software architecture of RoboGuard's mobile base. It is a layered architecture where several modules or behaviors run in simulated parallel.

The RoboGuard control software's task is the low-level control of the RoboGuard Base as well as several forms of support for the operator. Ideally, the operator has the impression that he or she is in full control while the system autonomously takes care of crucial tasks like obstacle avoidance, keeping on a trajectory, emergency stops, and so on. The software architecture is structured into several layers (figure 4), each allowing several modules or behaviors to run in (simulated) parallel.

4.2 RoboCube Software Drivers and Operating System Support

The RoboGuard control software relies on the RoboCube controller platform and on it's CubeOS operating system to implement the control application. The CubeOS nanokernel contains real-time multi-threading, abstract communication interfaces and thread control primitives. On top of the nanocore, a set of software drivers provides an application programming interface to the RoboCube's hardware.

The RoboGuard controller makes use of these drivers to control the MC68332 TPU on the RoboCube. The TPU is the main hardware interface for motor control and odometry. The TPU has 16 independent Channels which can be programmed to various functions. In the RoboGuard, four of these channels are used for odometry, two channels form a quadrature decoder. This quadrature decoder represents an up/down impulse counter that is controlled by the encoders on the motor axis. The CubeOS TPU driver configures two TP channels to form the decoder by linking them together in QDEC mode. Motor control is implemented by the TPU's pulse width modulation function. Again, the CubeOS TPU driver prepares one TPU channel per motor to generate a fixed-frequency square waveform with variable duty cycle that is controlled by the controller application.

The communication with the onboard PC makes use of the serial communication driver in CubeOS. It provides queued input and output to the application as well as platform-independent data encoding (XDR). Upon initialization, the controller application initializes the TPU driver which in turn initializes the TPU hardware and sets up the channel functions. The control information for the mobile base state and the odometry position is reset to orientation 0 degrees, position (0,0), speed 0.

Then, the control thread is configured to be restarted every 25 msec. This function is also executed by the TPU. Then the communication thread starts up and waits for incoming packets on the serial communication link that is connected to the onboard PC of the RoboGuard mobile base. Upon proper reception, the content of each packet is translated into control commands for the control task.

The control task is invoked every 25 msec. It first reads the QDEC decoder TP channels and computes a new base position and orientation. Then, the control command from the communication thread is compared with the current state of the base and correction values are computed which are then used to change the PWM settings of the motors. The control task then exits.

4.3 Motion- and Motor-Control

The motion layer interacts with the higher layers via a shared memory buffer that is written by a higher thread and is read by the motion thread (figure 4.3). The write operation is made atomic by delaying the

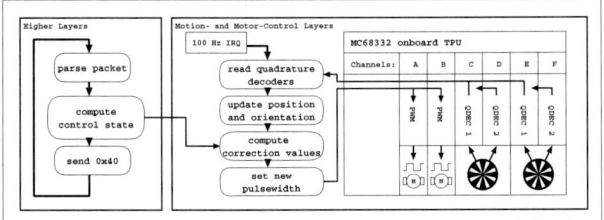

execution of the motion thread during write operations. So, target-values in the motion-controller can be asynchronously set by higher level behaviors. The motion-controller so-to-say transforms the target-values on basis of odometric data to appropriate target-values for the motor-control. The motion- and motor-control layers are based on generic software modules for differential drive robots, featuring

- PID-control of wheel speed
- odometric position- and orientation-tracking
- rotational and translational control

4.4 The Strategic and Path-Planning Layers

A core function on these layers is operator communication, i.e., the transmission of control states from the operator's console or so-called cockpit to the control hardware. To ensure a low-latency operation over the Internet link, a protocol based on UDP packets has been implemented. The packets are formed at the cockpit by synchronous evaluation of the control state and transmission to the onboard PC of the RoboGuard platform via Internet. Here, they are received and transmitted to the RoboCube via the serial port. The communication behavior parses the packets and makes its content available to other behaviors via shared memory.

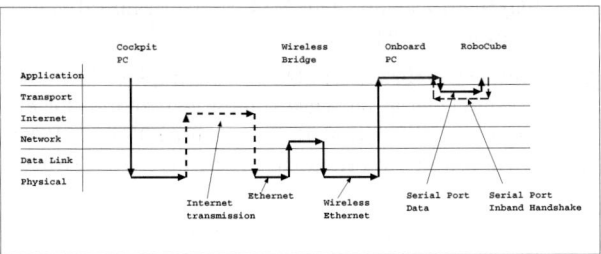

To ensure low-latency-operation, there is no retransmission on lost packets although UDP does not guarantee successful delivery of packets. However, since packets are transmitted synchronously and are only containing state information, there is no need to resend a lost packet since the following packet will contain updated state information. By exploiting this property of the protocol, low-latency operation can be assumed.

The communication between the RoboCube and the onboard PC uses inband handshaking to prevent buffer overruns in the RoboCube software. The communication layer software in the RoboCube confirms every packet with a 0x40 control code. Only if this control code has been received, the onboard PC communication layer software transmits the next packet. If the RoboCube communication layer software did not yet confirm a packet when a new packet arrives from the Internet transport layer, this packet is discarded so that the control layer software only receives recent packets, again ensuring low-latency operation.

Plausibility checks on the same layer can be used to discard packets or to modify the implications of the information they contain. This is done in a rule-based module. This functionality is optional and allows a convenient incorporation of background knowledge about particular application domains. The strategic layer also includes self-sufficiency behaviors like energy-management. Depending on the preferences of the customer, the arbitration of these behaviors can be handled in the rule-base. For example, a low-priority mission could be autonomously aborted if the base is likely to run out of energy during its execution.

The path-planning layer handles functionality to facilitate the operation of the base. It can incorporate world-knowledge on different scales, again depending on the preferences of the customer. Its simplest functionality consists path stabilization, i.e., jitters from the manual control can be smoothed away by temporal filtering. Behaviors for obstacle avoidance protect the system from accidental or malicious misuse, and help to move along narrow hallways and cluttered environments. Last but not least, it is possible to let the base navigate completely on its own when detailed world-knowledge in form of maps is provided.

5 Conclusion

The paper described the mobile base of RoboGuard, a commercial surveillance device. The base is developed with concepts and technology from behavior-oriented robotics, leading to low production costs and high user friendliness. Unlike "classical" robots, the base does not rely on extremely precise mechanics and detailed models. Instead, tight couplings between sensors and motors through software modules or behaviors running in (virtual) parallel are used. This allows a smooth fusion of steering commands from a human operator with autonomous functionalities like obstacle avoidance and motion stabilization. Furthermore, the generic modularity of a behavior-oriented approach facilitates customization to the specific needs of a particular customer.

References

[1] W.A. Aviles, T.W. Hughes, H.R. Everett, A.Y. Umeda, S.W. Martin, A.H. Koyamatsu, M.R. Solorzano, R.T. Laird, and S.P. McArthur. Issues in mobile robotics: The unmanned ground vehicle program teleoperated vehicle. In *SPIE Mobile Robots V*, pages 587–597, 1990.

[2] Andreas Birk. Autonomous recharging of mobile robots. In *Proceedings of the 30th International Symposium on Automative Technology and Automation*, 1997.

[3] Andreas Birk. Behavior-based robotics, its scope and its prospects. In *Proc. of The 24th Annual Conference of the IEEE Industrial Electronics*. IEEE Press, 1998.

[4] Andreas Birk and Tony Belpaeme. A multi-agent-system based on heterogeneous robots. In Alexis Drogoul, Milind Tambe, and Toshio Fukuda, editors, *Collective Robotics, CRW'98*, LNAI 1456. Springer, 1998.

[5] Andreas Birk, Holger Kenn, and Thomas Walle. Robocube: an "universal" "special-purpose" hardware for the robocup small robots league. In *4th International Symposium on Distributed Autonomous Robotic Systems*. Springer, 1998.

[6] Andreas Birk, Holger Kenn, and Thomas Walle. Onboard control in the robocup small robots league. *Advanced Robotics Journal*, 14(1):27 – 36, 2000.

[7] Andreas Birk, Thomas Walle, Tony Belpaeme, and Holger Kenn. The vub ai-lab robocup'99 small league team. In *Proc. of the Third RoboCup*. Springer, 1999.

[8] Andreas Birk, Thomas Walle, Tony Belpaeme, Johan Parent, Tom De Vlaminck, and Holger Kenn. The small league robocup team of the vub ai-lab. In *Proc. of The Second International Workshop on RoboCup*. Springer, 1998.

[9] Rodney Brooks. Achieving artificial intelligence through building robots. Technical Report AI memo 899, MIT AI-lab, 1986.

[10] Rodney Brooks. Intelligence without reason. In *Proc. of IJCAI-91*. Morgan Kaufmann, San Mateo, 1991.

[11] S. Dewitte and J. Cornelis. Lossless integer wavelet transform. *IEEE Signal Processing Letters 4*, pages 158–160, 1997.

[12] Joseph F. Engelberger. *Robotics in Service*. MIT Press, Cambridge, Massachusetts, 1989.

[13] Christopher C. Langton. Artificial life. In Christopher G. Langton, editor, *Proceedings of the Interdisciplinary Workshop on the Synthesis and Simulation of Living Systems (ALIFE '87)*, volume 6 of *Santa Fe Institute Studies in the Sciences of Complexity*, pages 1–48, Redwood City, CA, USA, September 1989. Addison-Wesley.

[14] N. Massios and Voorbraak F. Hierarchical decision-theoretic robotic surveillance. In *IJCAI'99 Workshop on Reasoning with Uncertainty in Robot Navigation*, pages 23–33, 1999.

[15] David McFarland. Towards robot cooperation. In Dave Cliff, Philip Husbands, Jean-Arcady Meyer, and Stewart W. Wilson, editors, *From Animals to Animats 3. Proc. of the Third International Conference on Simulation of Adaptive Behavior*. The MIT Press/Bradford Books, Cambridge, 1994.

[16] Rolf Pfeifer. Building "fungus eaters": Design principles of autonomous agents. In *From Animals to Animats. Proc. of the Fourth International Conference on Simulation of Adaptive Behavior*. The MIT Press/Bradford Books, Cambridge, 1996.

[17] The quadrox website. http://www.quadrox.be.

[18] Luc Steels. Towards a theory of emergent functionality. In Jean-Arcady Meyer and Steward W. Wilson, editors, *From Animals to Animats. Proc. of the First International Conference on Simulation of Adaptive Behavior*. The MIT Press/Bradford Books, Cambridge, 1991.

[19] Luc Steels. The artificial life roots of artificial intelligence. *Artificial Life Journal*, 1(1), 1994.

[20] Luc Steels. A case study in the behavior-oriented design of autonomous agents. In Dave Cliff, Philip Husbands, Jean-Arcady Meyer, and Stewart W. Wilson, editors, *From Animals to Animats 3. Proc. of the Third International Conference on Simulation of Adaptive Behavior*. The MIT Press/Bradford Books, Cambridge, 1994.

[21] Dany Vereertbrugghen. Design and implementation of a second generation sensor-motor control unit for mobile robots. Technical Report Thesis, Tweede Licentie Toegepaste Informatica, Vrije Universiteit Brussel, AI-lab, 1996.

[22] Stewart W. Wilson. The animat path to ai. In *From Animals to Animats. Proc. of the First International Conference on Simulation of Adaptive Behavior*. The MIT Press/Bradford Books, Cambridge, 1991.

Automated Container-Handling System for Container Production Nurseries

Hagen Schempf
hagen+@cmu.edu

Todd Graham
tgraham@rec.ri.cmu.edu

Robert Fuchs
rfuchs@rec.ri.cmu.edu

Chris Gasior
cgasior@rec.ri.cmu.edu

Carnegie Mellon University
Robotics Institute
5000 Forbes Ave.
Pittsburgh, PA 15213

I. ABSTRACT

Production of nursery crops in the US is accomplished in container- and field-growing conditions, with propagation and seedling-rearing carried out in greenhouses. Container-grown crops represent 60% of the US market and represent a highly labor-intensive and thus costly segment of ornamental crop production. The USDA, NASA and the ANLA have collaborated to develop an automated in-field container-handling system for reducing dependence on foreign labor while also increasing productivity. A first-generation system was developed at CMU, capable of automatically lifting and conveying plants from the ground (in a variety of regular patterns) onto trailers, and vice-versa. The system is capable of handling a vast array of container-designs from different manufacturers, and spans the size-range from #1 to #5 (approximate equivalence to gallons). The system is designed to handle 45,000 containers per 8-hour day with one to two operators. Testing currently underway indicates that the system approach is valid, with the next incarnation requiring reduction in size and weight, as well as reduction in manufacturing costs; a new design is currently underway, with as-built designs and experimental results for the first prototype's performance presented herein.

II. INDUSTRY OVERVIEW

US ornamental horticulture is a rapidly growing, $11 billion dollar a year industry (about 10% of the gross agricultural output of the US alone), tied to a dwindling migrant work force, working in outdoor conditions in very large acreage areas (see Figure 1).

Figure 1 : Typical container nursery view & labor task

Unskilled labor is becoming more costly and harder to find, while it is still needed to move potted plants - this represents a manual handling task of at least 450 million units per year, each handled 3 to 4 times a year. The nursery industry must address this problem if it is to survive and continue to flourish in the next millennium.

Nursery production automation is a growing field worldwide. At the highest level there are three main areas, namely greenhouse operations, container yards and field nurseries. Within these groupings, there are several areas that lend themselves to automation (see Table 1):

AREA	AUTOMATION-FRIENDLY
Greenhouse	*Seed/Propagate, Pick/Ship, Gather, Transplant/Set*
Container Yard	*Field Movement. Upshifting, Order-Picking, Shipping*
Field Nursery	*Dig, Plant, Stake, Harvest, Container Handling*

Table 1 : Automation Areas for Nursery Industry

In these areas it was judged [2] that automation has achieved different levels of automation-penetration worldwide - the levels are expressed in the histogram shown in Figure 2:

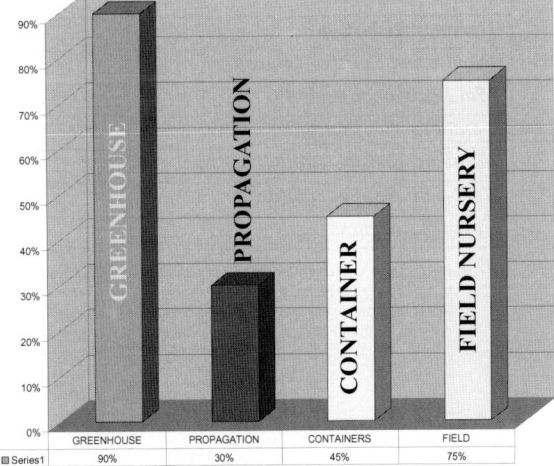

Figure 2 : Automation Levels in Nursery Industry

Current automation in the nursery production field worldwide are very wide-scattered, depending on the specialty-area. Holland [1], Germany, Italy and England are world leaders in the automation of greenhouse production, whether that be for woody ornamentals, flowers or even vegetables. Equipment from suppliers in these countries, and the associated infrastructure, are in use worldwide. Equipment ranges in variety from plug-planting machines, potting machines and flower imaging and quality segregation to automated soil-mixers, -fertilizers and washing machines (excerpt in Figure 3).

Figure 3 : Greenhouse automation equipment

Field-nursery production automation is primarily limited to the development of assistive tools, which assist in the excavation of field-grown trees and inventorying systems for such trees (see Figure 4).

Figure 4 : Field Nursery automation equipment

Container-handling devices for field-use, which is where most of the labor-costs are expended, have not seen any intensive development in the US nor abroad - for good reasons; systems that are available are shown in Figure 5.

Figure 5 : Container Nursery automation equipment

The variety in container-types and surface-conditions and nursery layouts is vast - they way greenhouse automation was made successful is through standardization of the infrastructure (containers, conveyors, irrigation, etc.), which is still sorely lacking in the US. Talks about standardized containers has been ongoing for many years, but due to the nature of the business, has not taken a foothold in the US. Differentiation amongst growers, climatic conditions and simple opinion-variability amongst growers results in methods and principles that make it hard to apply automation broadly in this vast market, without requiring dozens of various dedicated machines for different growers/growing-regions (thereby reducing the attractiveness to equipment developers).

III. PERFORMANCE REQUIREMENTS

The motivation to automate being obvious, it becomes important to realize that the US market has a high affinity to price and performance. The performance requirements that were derived for the proposed container-handling

DESCRIPTOR	TARGET	VALUE
Containers moved in the field per hour	Meet/Exceed 4-person daily rate	25,000/day[a]
System Design	Stand-alone System	N/A
Trailer Compatibility	Compatible with typical trailer	4' x 10'
Operator Reduction	Single-operator for system	1 Operator
Quality and Control Assurance	No extra plant/container damage	N/A
Multi-container usability	Adaptable[b] to #1, 2, 3 & #5s	Yes
Container Configurations	Can-to-Can, Can-tight, Spaced[c]	Yes[c]
Multi-surface operability	Gravel, Geotextile - NO Poly!	Yes
Cold-Frame Compatibility	Access into/sideways frames	Yes[d]
Cost-Effectiveness	Typical stand-alone system	$50K to $75K

a. Refers to #1 containers in an 8-hour workday with a single operator, or about 2,500 containers/hour!
b. manually adjustable over a range or usage of a different tool-head
c. in a follow on system adapted based on the baseline system
d. possibly with minor modifications in the door and/or hoop-structure and irrigation-location

Table 2: Performance Metrics

system, focussed around several key areas, namely (i) throughput (containers/day), (ii) applicability to existing infrastructure (containers, groundcover), (iii) compatibility with existing equipment (trailers, cold-frames), (iv) manpower reduction, (v) job-quality (compared to manual), and (vi) cost-effectiveness (ROI-based). The system has to be able to pick-up and drop-off in can-to-can and can-tight, as well as diamond-spaced configurations, and do so at a rate to pay back for the system in terms of labor-savings within as few seasons as possible. Performance variables and the expected value for each are shown in Table 2.

IV. SYSTEM DESCRIPTION

The design developed for the automated field-container handling system represents a self-mobile outdoor platform powered by an IC engine, perceiving containers through a set of ranging IR-sensors, controlled through on-board PLC-based ladder-logic computers, and actuated through a set of electro-hydraulic actuation systems. A CAD-rendering of the developed system is shown below in Figure 6:

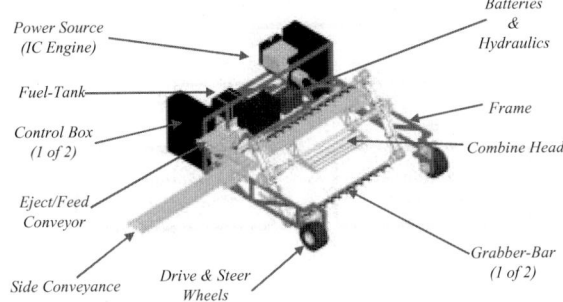

Figure 6 : CAD image of the container handling system

The above design relies on a self-powered skid-steered platform with rear floating rocker-arm caster-axle. The containers are now grabbed using a squeeze-pinch and moved in a circular arc fashion to a conveyor that speeds them off to the side (onto a waiting trailer); the operation is run in reverse for setting down and spacing out containers. Two grabbers ensure that a full 6-foot wide row of containers is moved every 8 seconds. Sensor-guided driving aligns the frame to pick up or drop off in any desirable configuration.

The overall system can thus be seen to consist of several major elements, including (i) frame, (ii) drive & steer, (iii) container grabber & handler, (iv), and power & control systems. The roles and interconnections of each of the above modules can be generically described as detailed below:

FRAME: The frame consist of an open U-shaped weldment, upon which rest the IC power-plant, hydraulic drive system, power and control electronics, locomotion and steering system, as well as the container grabbing and handling head and its associated conveyors. The system was oversized so as to allow for laboratory testing of all possibly useful features, which are then to be evaluated for inclusion in the commercial prototype (see Figure 7):

Figure 7 : Frame undergoing assembly

POWER & CONTROL: The main power source for the system consists of an internal combustion-engine mounted on the frame, providing both electrical power via a generator, and hydraulic power through a direct-coupled pump. The power is regulated through a dedicated cabinet, while the electronics and controls for the PLC and the relays and valves are housed in a separate compartment. Fuel-tanks and cooling radiators are mounted on the frame as well. A picture of the subsystems is shown in Figure 8:

Figure 8 : Power & Control Subsystems

HYDRAULIC DRIVE & STEERING: The hydraulic system is used to provide driving/steering power to the wheels, as well as articulating the grabber arms and pinching endeffectors. The drive and steering for the handling system is achieved by driving the two front wheels in a differential manner, while letting the system follow based on a simple rear-mounted rocker-arm caster-axle. The individual systems are shown in Figure 9 in different stages of assembly.

CONTAINER GRABBER: The method used to grab containers reliably, without requiring any dedicated container design, is based on a simple double half-moon friction-clamp design. By ganging these pinch-grabbers along an actuated rail, a whole row of containers can be grabbed at once and moved around. The bar-mounted pinch-grabber arms are mounted on a set of two rotating side-arms, allowing a combine-like full rotation of each row that has been grabbed; and internal gearing-pass ensures that the containers remain level during any part of the rotation. The sensory system used to control the machine-heading, combine-head rotational position, arm-extension and pincher open-close states, is based on the processing of infrared range-measurements from embedded sensors [3].

Figure 9 : Locomotion and Steering Subsystems

Preliminary testing on a single pincher and a full pinch-grabber bar basis have determined that this approach is reliable and robust to container types and outdoor conditions. Images depicting the pincher system during indoor testing, are shown in Figure 10.

Figure 10 : Container Grabber Subsystems

ELECTRONICS: The electronics and control system is based on commercial-off-the-shelf industrial automation hardware. The hardware architecture is shown in Figure 11. The control system hardware consists of an Allen-Bradley SLC-500 PLC, two Delta Computer Systems motion controllers and a variety of other components (e.g. sensors, relays, and contactors, etc.). As noted in Figure 11, the PLC communicates with the motion controllers via Ethernet. The PLC performs all supervisory and discrete device control. The PLC chassis houses the CPU and several I/O modules for: a) discrete and analog sensors inputs (e.g. proximity switches, IR sensors for container localization, etc.), and b) discrete outputs (e.g. solenoid valves, indicator lights, etc.). The motion controllers coordinate and control all 10 axes. The ten axes include: (i) two (2) drive wheels for locomotion, (ii) combine head rotation, (iii) four (4) hydraulic cylinders for the telescoping tubes, and (iv) the three conveyors. The system operator will interact and control the system via buttons, switches, and a joystick (see Figure 8).

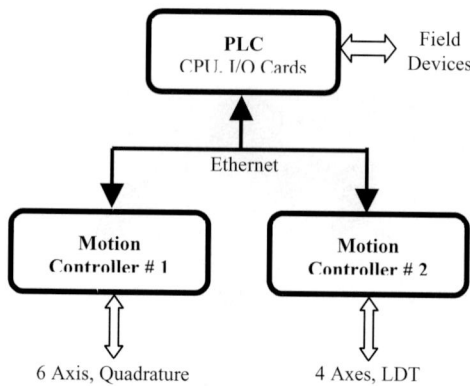

Figure 11 : High-Level Electronics Architecture

CONTAINER SENSING: In order to perform up-close positioning of the grabber-head so as to achieve 'proper' alignment with the containers for a full-row pick-up, despite the potential misalignment of the tool system itself, the misplacement of containers, etc., requires the use of an integrated sensing system. The most suitable candidate for simplicity, ruggedness and reliability turned out to be a non-contact infrared ranging system (see Figure 12). To build the range-imager, we integrated several of these relatively short-range (4 inches to 2 feet depending on IR diode-power) sensors onto the side of the frame along the pickup-line. This allows us to not only achieve a good 'average' sensory-alignment reading, but to also have a much better idea of the alignment of the container-row in the field, which will be useful if we are to properly space containers in the field. The test-setup we used (see Figure 12), includes a suite of several IR sensors, which are multiplexed through a computers' I/O port (parallel in the experimental setup's case) to obtain range-readings from each sensor at a rate of 10 per second. These readings are then processed based on the calibration-curve for each sensor, and then a range-map is built.

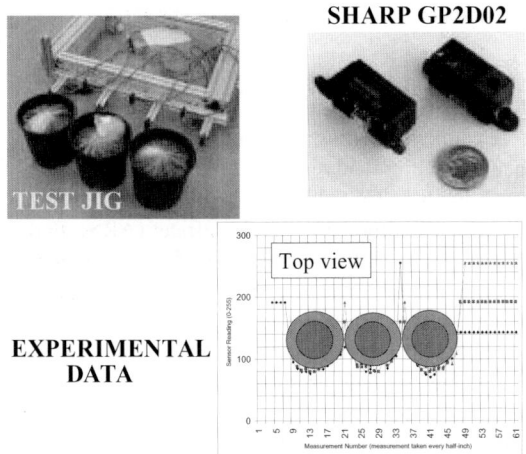

Figure 12 : IR Sensing System & Setup

If the sensor-array is moved laterally and in front of a row of pots, an image can be generated which a computer interprets so as to determine the inter-container spacing, which in turn can be used to determine the proper location of the gaps between the containers. This process makes the accurate placement of he grabber-head possible so as to provide final alignment through heading control and grabber-arm extensions. The block-diagram of the software that was developed in order to perform the ranging, computation and grabber-head alignment (including gross alignment by way of heading and displacement of the entire system), can be depicted as shown in Figure 13:

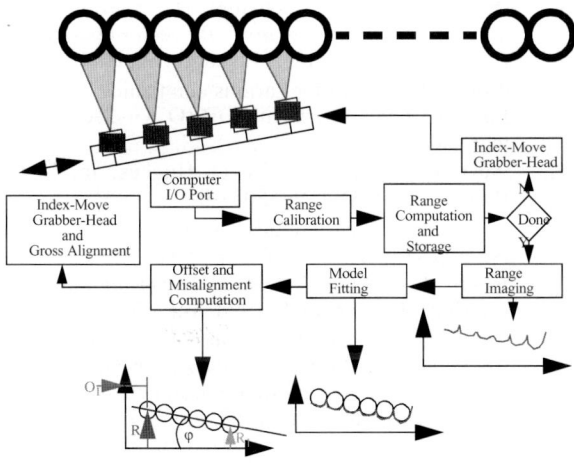

Figure 13 : Software Sensor-Control Diagram

SOFTWARE: The control system logic is implemented via Allen-Bradley's ladder-logic programming language. The RSlogix development environment was the primary tool for development of the ladder-logic control program. Unlike most industrial ladder logic programs, the software/control program was written using a modular, systematic approach. This systematic approach makes the code more reliable and easier to debug and maintain. The software architecture is shown in Figure 14. The program consists of a main program, device control, input references, output references and several processes. The main program provides overall control. The device control is the only place where devices are controlled. The input and output references map all internal software variables to the I/O hardware. The processes are where the majority of the control logic is implemented. These processes represent basic functionality of the system's various sub systems and are where the machine operation is sequenced. For example, one of the processes is for loading/queuing of the transfer conveyor so a grabber arm can pick up the containerized plants and set them on the ground. Another process, for example, is for unloading the conveyors after a grabber arm has placed containerized plants on the conveyor.

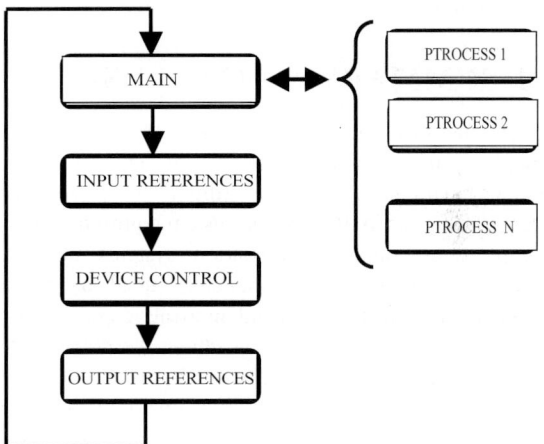

Figure 14 : Software Architecture Layout

SYSTEM: A fully assembled locomotion platform of the container handling system is shown in Figure 15 during locomotion trials on the experimental nursery at CMU's National Robotics Engineering Consortium experimental nursery:

Figure 15 : Fully integrated container handling system

V. FIELD TESTING

The handling system shown in Figure 15, was tested at REC's experimental nursery. The system performance was measured over a 6-foot wide and 50 foot long bed using a variety of #1 containers and different plant-types and weights. Initial testing indicates that the sensing scheme was able to position the system accurately enough (to within 0.15m), yet the closed-loop speed needs to be increased to achieve a productivity increase of about 30% (currently at ~ 20,000 containers per 8-hour day). The time spent between grabbing containers off the ground and hand-off onto the conveyor needs to be sped up to increase cycle-time as well. Large steering corrections did not result in expected behavior, due to the variability in traction we re seeing due to different groundcover and the fact that the weight distribution between the front driving wheels and the rear caster-axle is 30/70. The operator interface was found to be simple enough to use, even when manual reset and resumption of automated handling was required. Minor improvements in mounting and cooling for certain subsystems are being undertaken to complete the testing program before the onset of winter.

VI. SUMMARY & CONCLUSIONS

The container handling system presented herein represents a major step towards automation of labor-intensive container-handling tasks in medium to large-sized container nurseries in the US. The system represents a new class of smart outdoor automation systems utilizing existing hard-automation components, aided by smart sensors, intelligent software and innovative mechanism design. Testing of the system has shown its potential to achieve the desirable productivity of 25,000 to 45,000 #1 containers per day with one to two operators, without regard to the type of hauling-trailer. The system is capable of handling a large variety of containers available through US manufacturers. Groundcovers suitable for the machine and tested to date, include gravel and stone/asphalt/concrete. The current system needs to be reduced in weight and size (mostly width), as well as re-engineered for lower manufacturing costs. System maintenance requirements are expected to be reduced by switching to exclusively electrical power on board.

VII. FUTURE WORK

The system presented herein represents the first generation of field-container handling systems. We are currently simplifying and shrinking the system design, and expect to prototype an improved commercial prototype by the end of 2001. In addition, CMU is developing other more simple manual tools for assisting growers that have switched to growing trees in the ground in large containers (pot-in-pot) and those retailers involved in the landscape and garden-center sectors in this industry. Licensing arrangements are being sought to ensure the technology gains wide acceptance in the US and abroad.

VIII. ACKNOWLEDGEMENTS

The container handling system was jointly funded at Carnegie Mellon University (CMU), by NASA under research-grant #NCC5-223, the US Dept. of Agricultures' (USDA) Agricultural Research Office (ARS) under a SCA (#58-1230-8-101/58-3607-0-130), and a grant from the Horticultural Research Institute (#1999-128/2000-163), the research-arm of the American Nursery and Landscape Association (ANLA). We wish to further acknowledge the help from many growers and nursery/agricultural equipment/supply providers (MidWest Groundcover, John Deere, Toro, Lerio, Nursery Supplies, etc.) in the industry, that have gone out of their way to assist in the development through advice and equipment and supplies donations.

The above-described system and process has been described to the USPTO as part of a patent-filing - it has a patent-pending status.

IX. REFERENCES

[1] "*VISSER - Product Descriptions.*", Company Catalog and CD, November 1999

[2] Schempf, H., "*Automation and Mechanization: The Future of the Nursery Industry in the US*", NEGrows Conference, Boston, MA, January 21-27, 2000

[3] Dias, B., Stentz, A., Schempf, H., '*Sensory-based nursery container detection*', RI Tech Report Draft, Carnegie Mellon Univ., Pittsburgh, September 2000

[4] Product Literature for various companies: Bouldin-Lawson, Javo, Baertschi-FOBRO, Goetsch, Urdinati, etc.

[5] '*Systems for Success*', by D. Hughes, Jr., Cedar Rapids, IA, ISBN 0-9655037-0-4

[6] "*New Ideas*", Bi-monthly Newsletter, Wholesale Nursery Growers of America, 1991 - 1999.

[7] Jagers, F. et al, "*Hi-Tech take-over of pot-plant grading*", FlowerTECH 1998, Vol.1/No.1

[8] Adrain, J.L., et al, "*Cost Comparisons for Infield, Above Ground Container and Pot-in-Pot Production Systems*", Journal of Environmental Horticulture, Vol.16, No.2, June 1998, p.65

[9] '*American Standard for Nursery Stock*', American Association of Nurserymen, ANSI Z60.1-1996, Nov. 6, 1996

Actively Steerable Inpipe Inspection Robots for Underground Urban Gas Pipelines

S. G. Roh, S. M. Ryew, J. H. Yang, H. R. Choi
School of Mechanical Engineering,
Sungkyunkwan University
300, Chonchon–dong, Jangan–gu, Suwon, Kyonggi–do,
Korea, 440–746, hrchoi@mecha.skku.ac.kr

Seoul City Gas R&D Center

Abstract

In this paper we introduce robots called MRINSPECT (Multifunctional Robotic crawler for INpipe inSPECTion) III and IV which are under development for the inspection of urban gas pipelines. The proposed robots can freely move along the basic configuration of pipelines such as horizontal or vertical pipelines. Moreover it can travel along reducers, elbows, and steer in the branches by using steering mechanisms. Especially, their three dimensional steering capability provides outstanding mobility in navigation that is a prerequisite characteristic in urban gas pipelines. Their critical points in the design and construction are introduced with preliminary results of experiments.

1 Introduction

There are a wide variety of pipelines such as urban gas, sewage, chemical plant, nuclear power plant etc., which are indispensable in our life. Also, pipelines are the major tools for transportation of oils and gases and a number of countries employ pipelines as the main facilities for transportation. In our country, the urban gas pipelines currently go up to 13,000 Km long but since most of them have been constructed in 1980's, there happen a lot of troubles caused by aging, corrosion, cracks, and mechanical damages from third parties. Continuous activities for inspection, maintenance and repair should be performed from now on. However, those activities need enormous budgets that may not be easily handled by gas companies as they are mostly small and medium in size. Efficient equipments for inspection and integrated maintenance program are required in gas industries.

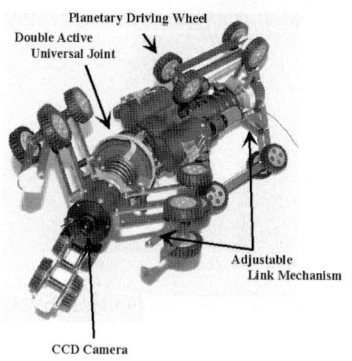

Figure 1: MRINSPECT III

Up to now various inpipe robots have been reported [1,2] but those robots just have the capability of traveling along straight pipelines with elbows. Few of them have been designed under the consideration of branched pipelines. In fact branched pipelines are one of the most popular configuration in underground urban gas pipelines and thus to conduct efficient traveling inside those pipelines steering capability is prerequisite. For last several years, we have developed several inpipe inspection robots with steering capability, and this paper introduces the robots named MRINSPECT III and IV among them which have been recently developed(refer to [3] as for MRINSPECT I and II). Both robots have totally different features in steering that may be regarded as the original ones, respectively. In this paper MRINSPECT III is briefly overviewed and a new robot MRINSPECT IV is described in detail. Several specific features are addressed concentrating on the steering and preliminary results of experiments are outlined.

This paper is organized as follows. In the next section we briefly introduce MRINSPECT III. Section 3 cover the issues related to the development of MRINSPECT IV and preliminary experiments. Finally we conclude with summary in section 4.

2 MRINSPECT III

As illustrated Fig. 1 MRINSPECT III consists of two vehicle segments and a steering mechanism called *Double Active Universal Joint(DAUJ)* between the segments. This robot is configured as an articulated type where two independent vehicles are connected via a double active universal joint providing omni–directional steering capability. DAUJ acts like a stiffness control-

Figure 2: Inpipe Inspection System

lable two–dof joint and it makes it possible to control the compliance of active joints in steering [4]. Each articulated body of the robot has three wheeled legs located circumferentially 120° apart. The legs employ a pantograph mechanism with a sliding base that ensures natural folding and unfolding of the body. With the proposed mechanism the legs just contract or expand along the radial direction when they are pressed. It is a quite advantageous feature because undesirable forces causing distortion does not exert on the body when the robot goes over obstacles such as steps, reducers, protrusions inside the pipelines. The driving motor of the vehicle is included in the rear articulated body which gives the major driving force to the system. The front body does not have any power and it just guides the motion. The wall pressing forces are obtained by the reflective forces of the spring that supports the moving base of the pantograph mechanism. Thus, the wall pressing forces can be easily preset depending on the payload by adjusting the spring constant and initial deflection. Fig. 2 shows the whole inspection system utilizing MRINSPECTION III which is composed of two MRINSPECT III's, control modules, and inspection tools.

3 MRINSPECT IV

3.1 Mechanism Overview

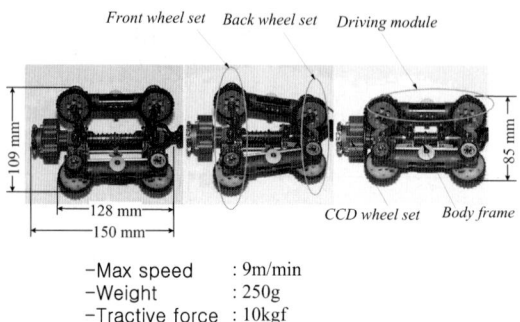

- Max speed : 9m/min
- Weight : 250g
- Tractive force : 10kgf

Figure 3: Construction and specification of MRINSPECT IV

MRINSPECT IV shares several aspects with MRINSPECT III but most of the mechanism has been renewed to miniaturize the robot as shown in Fig. 3. The robot is largely composed of 1) a body frame that mounts a CCD camera assembly and driving modules with foldable linkages, 2) a camera assembly which is for the navigation and the visual inspection of the pipelines, and 3) three modularized driving modules illustrated in Fig. 4 which are located circumferentialy with 120° apart.

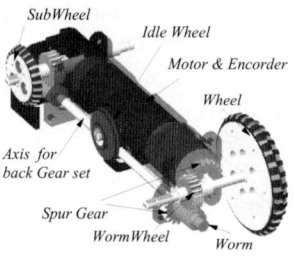

Figure 4: Driving module

A driving module consists of a DC motor with an encoder and a reducer, several wheels and casings. As it can be easily disassembled from the body frame, it ensures the convenience in maintenance. The driving units can be controlled independently and thus they amplify the tractive forces as well as provide steering capability. To provide the sufficient tracting forces and flexibility in navigation we designed a body frame illustrated in Fig. 5. At the end of the legs on the body frame three driving modules are fixed 120° apart circumferentially.

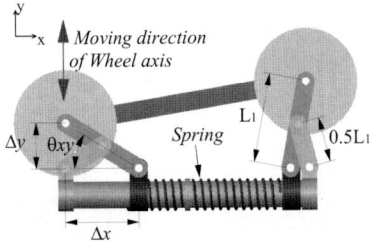

Figure 5: Link mechanism

Two wheels of the driving module move independently along the radial direction due to constraints of links and interaction between elastic force at the main spring of the body frame and the reaction forces of the wall. Distance between the main shaft of the robot and the wheel changes according to the link construction, and it makes the wheel have effective contact with the wall inside the pipelines whatever diameters changes. It assures stable traveling as long as providing sufficient traction forces. Front wheel of each driving module have been synchronized each other. As shown in Fig. 3, wheels in front and back of each driving module are named the front wheel set and the back wheel set, respectively. Constraint of the wheels inside each wheel set makes the robot capable of neglecting the effect of gravity, which makes the central axis of the robot always coincide with that of the pipelines.

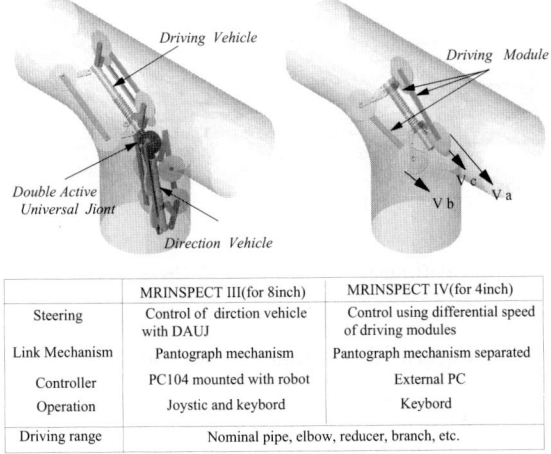

Figure 6: Comparision of MRINSPECT III and IV

MRINSPECT IV has upgraded MRINSPECT III in many aspects, especially in the steering mechanism. MRINSPECT III performs three-dimensional steering with a double active universal joint, while MRINSPECT IV does with speed differences of each driving module. Fig. 6 illustrates the distinction of the steering action between MRINSPECT III and IV. MRINSPECT III turns the front body by using DAUJ like a manipulator. In this case the robot should not rotate along the driving direction due to the contact constraint between the wheels of the robot and the wall, which is coped with DAUJ. On the contrary MRINSPECT IV steers its own body with the velocity differences among the driving modules. Thus, though it does not require complicated mechanisms, its control becomes quite difficult. The analysis related to the steering control is addressed in the later section.

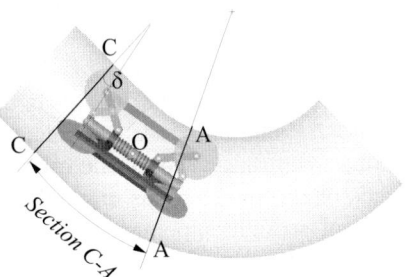

Figure 7: Straight to elbow

3.2 Navigation in Elbow

Basically the size of the robot is the critical points in the design in realizing the movement in the elbow. The basic relations and analysis can be found in the reference [3] and in this section we focus on the additional features in the navigation along the elbow.

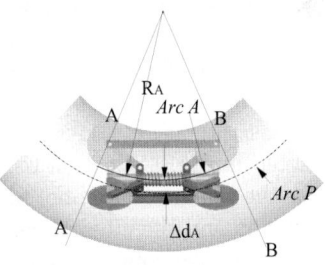

Figure 8: Robot in elbow

To travel smoothly along the elbow it should be considered the curvature of the elbow varies depending upon the inner area of the pipelines having contact with the robot.

When speeds of all of the wheels are same, the wheels in the outer side may be caused to slide, which

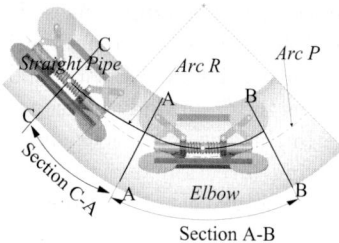

Figure 9: Moving trajectory in elbow

gives overload to the driving system. Thus each driving module of the robot can adjust its speed according to the curvature but it is a control work requiring careful observation. When the robot turns to the elbow from the straight pipelines the front wheel set of the robot is to be placed at inner area of the elbow, while back wheel set is to be located at the straight pipeline shown in Fig. 7. The cross section connecting contact points between rear wheel of the robot and walls with the axis of the pipeline has an angle δ with the cross section of straight pipeline. As the robot enters the elbow, the angle δ increases. When the robot

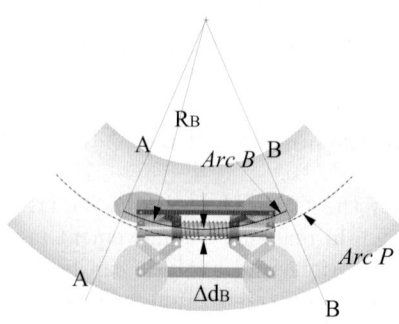

Figure 10: Dependence on posture

completely gets into the elbow, its center moves along Arc A having the curvature R_A from the center of the curvature, and has the offset Δd_A with the center of curvature depicted in Fig. 8. Section C-A in Fig. 9 is the range where the center of the robot moves between the straight pipeline and the elbows, while section A-B is the range where the robot is completely placed in the elbow. Arc R is the route where the center of the robot moves as depicted in Fig. 10. Because the robot is driven by the independent driving modules located 120° apart, the contact area varies depending on the posture the robot. Thus offset distance between the center of the robot and the pipeline axis varies depending on δ. Based on the those observations we should calculate the route of travel, which is the basic for modulating the speed differences.

3.3 Navigation in Branch

As shown in Fig. 11 branches can be considered to consist of two elbows and V-shaped area between the elbows. The V-shaped area has a flat surface rather

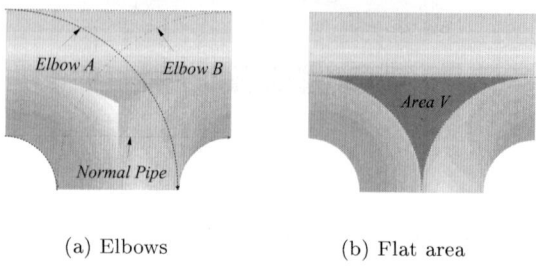

(a) Elbows   (b) Flat area

Figure 11: Formation of branch

than curved one and the flat surface can be found out only in the V-shaped area throughout the pipeline. Consequently, the V-shaped area may be an obstacle to the robot because it has been designed to be suitable for the curved surface. To travel in the branch the robot basically follows the method similar to the method in the elbow but there are several characteristic features that makes it difficult to control its motion. Those can be summarized as follows. First, as the robot moves along the branch, it meets a variety of cross sections depending on the posture of the wheel as shown in Fig. 12. In the second, the robot has

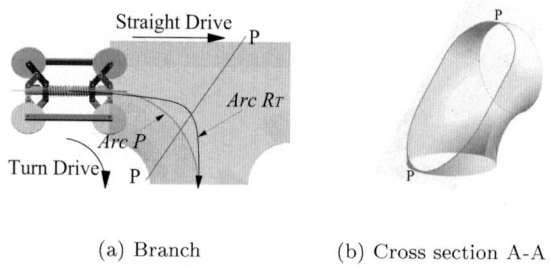

(a) Branch   (b) Cross section A-A

Figure 12: Change of cross section

been designed to move while making six wheels have contact with the inner side of the pipe radially. Nevertheless when the robot enters a branch, some of the wheels may not have contact with the wall. According to Fig. 12 the cross section does not have a circular shape any more in the branch and some of the wheels may carry idle rotations. In this case the robot loses its degree of motion and consequently we may not able to control its travel route exactly. When the robot turns in a branch, its center is assumed to move along the curve like $Arc\ R_T$ as illustrated in Fig. 12 that has different meaning from $Arc\ R$, $Arc\ A$ and $Arc\ B$ of the elbow because these Arcs of the elbow are a deterministic path produced by the robot while its wheels keep contact with the inner walls. When the robot enter the branch, initially the diameter of the pipeline does not change a lot until the front wheel set reaches section B-B after passing through section A-A illustrated in Fig. 13, but the robot cannot rotate in this region whatever speed differences are given.

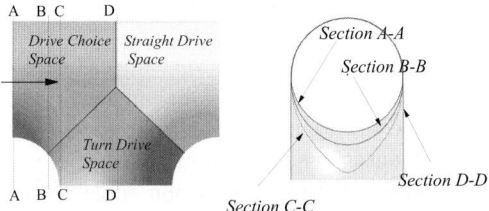

Figure 13: Constraint space in branch

When the front wheel set comes to the section C-C, the diameter of the pipeline changes greatly to let the robot travel toward *turn drive space* represented in Fig. 13. Even at this region, the robot can not rotate by itself with speed differences. In this region, the robot travels toward the *turn drive space* because the front wheel set is still placed close to inner side of the pipeline. Also, the front wheel set of the robot has still contact with the inner side of the pipeline and as the wheel set is confined absolutely up to this section, it cannot rotate by using speed differences. In fact, despite speed difference of the robot wheel only slip is produced between the wheels and inner sides of the pipelines, and the robot cannot rotate to the direction that it has to move. As a result, in this region the robot can still select either straight travel or rotation and thus this space is said to be *drive choice space*. When the front wheel set is close to section D-D, either one or two wheels, which are placed at the *turn drive space*, do not contact inner side of the pipe. Such a phenomenon indicates that there exists no inner side of the pipe contacting the wheel, which may prevent the robot from moving to the *turn drive space*. In this region, the robot can turn along the desired direction with speed differences. Thus as soon as the robot enters the branch, it decides the direction of rotation using CCD camera attached in the front of the robot, and then adjusts the rotational speed of each three driving modules. In Fig. 14 P_a, P_b and P_c indicate the points where each wheel contacts the inner side of the pipeline(this is based on the assumption that each wheel contacts the points).

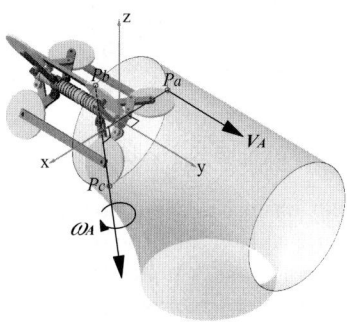

Figure 14: Three dimensional representation of related velocity vectors

It also shows the rotational direction of the robot when only the wheel at P_a rotates while two wheels remain without rotating. When speeds at P_a is $\boldsymbol{V}_A$ and speeds at P_b, P_c are 0 respectively, the robot turns with $\boldsymbol{\omega}_A$ of angular speed along the direction of the vector connecting P_b and P_c. As explained, the robot actually starts to rotate at section D-D, and the speed differences does not have a lot of influence on the rotation in front of the section D-D. Thus, we just need to start modulating the speeds of the wheels at the section D-D and to turn the robot to the desired direction, the rotation speed should be controlled accordingly as shown in Fig. 15. To turn the robot, first $\boldsymbol{V}_A$ at P_c assuming that it is the place closest to the rotational direction $-z$, is set to be 0. Using the CCD camera we already know the angles θ, $\theta + 120°$, and $\theta + 240°$ between z-axis and wheels, respectively. $\boldsymbol{\omega}_P$

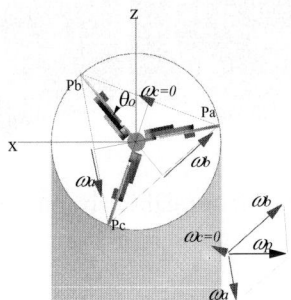

Figure 15: Turning at section D-D

of the angular speed can be freely selected to let the robot rotate, but $\boldsymbol{V}_A$, and $\boldsymbol{V}_B$ have to be decided to select the robot's travel speed at contact points between each wheel and inner wall of the pipeline. The following equation can be used to derive V_A, and V_B.

$$\omega_a = V_A \times \frac{3}{2} P_a \quad (1)$$
$$\omega_b = V_A \times \frac{3}{2} P_b \quad (2)$$
$$P_b \cdot \omega_b = 0 \quad (3)$$
$$P_b \cdot \omega_a = |P_b||\omega_a|\cos 150° \quad (4)$$

Those are the basic equations for navigation in the branch and in reality a lot of unexpected patterns of movements can be met. It need further research to completely analyze the motion in the branch.

4 Preliminary Experiments

In the experiments we demonstrated the movement in the elbow with 90° and the steering in the branch. Fig. 16 shows the scene of the robot running along the elbow and Figs. 17 and 18 depicts the steering in the branch which has 90° of the curvature angle.

Figure 16: Robot in elbow

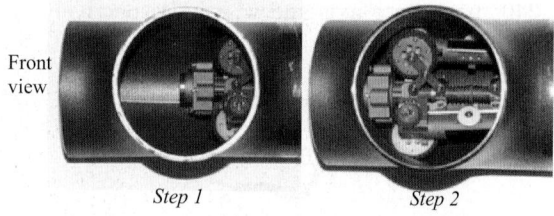

Figure 17: Straight drive in branch

5 Conclusion

In this paper we introduces two inpipe inspection robots, MRINSPECT III and IV. The system with MRINSPECT III has already been developed as shown in Fig. 2 but the system for MRINSPECT IV is under development. Both robots has the characteristic features in steering which are not discussed in the

Figure 18: Steering in branch

other robots. The steering is very important in underground pipelines, especially in urban gas pipelines and the proposed robots proves their effectiveness through real implementations. As the further work, we are going to implement NDT inspection tools on our robot and related works will be continued.

Acknowledgments

The authors are grateful for the support provided by a grant from Seoul City gas R&D center and the Safety and Structural Integrity Research Center at the Sung Kyun Kwan University.

References

[1] Y. Kawguchi, I. Yochida, H. Kurumatani, and T. Kikuta, "Development of an In–pipe Inspection Robot for Iron Pipes," *J. of the Robotics Society of Japan*, Vol. 14, No. 1, pp. 137–143, 1996.

[2] S. Hirose, H. Ohno, T. Mitsui, and K. Suyama, "Design of In–pipe Inspection Vehicles for $\phi 25, \phi 50, \phi 150$ pipes", *Proc. of IEEE Int. Conf. on Robotics and Automation*, pp.2309-2314, 1999.

[3] H. R. Choi, S. M. Ryew, S. W. Cho, "Development of Articulated Robot for Inspection of Underground Pipelines", *Trans. of the 15th Int. Conf. on Structural Mechanics in Reactor Technology(SMiRT-15)*, Vol.3, pp.407-414, 1999.

[4] S. M. Ryew, S. H. Baik, S. W. Ryu, K. M. Jung, S. G. Roh, H. R. Choi, "Inpipe Inspection Robot System with Active Steering Mechanism" *IEEE Int. Conf. on Intelligent Robot and Systems(IROS 2000)*, pp. 1652-1657, 2000.

Design of Continuous Alternate Wheels for Omnidirectional Mobile Robots

Kyung-Seok Byun, Sung-Jae Kim, Jae-Bok Song

Department of Mechanical Engineering, Korea University
5-ga Anam-dong Sungbuk-gu, Seoul, 136-701, Korea.
E-mail: jbsong@korea.ac.kr

Abstract

Many types of omnidirectional wheels with passive rollers have a gap between rollers. This gap causing the wheel to make discontinuous contact with the ground, leads to vertical and/or horizontal vibrations. In this paper a novel design of continuous alternate wheel was proposed to minimize a gap between rollers. In the continuous alternate wheel, inner and outer rollers are arranged continuously, thus resulting in no gap between the rollers. This paper details the design process including the systematic approaches to determine the optimum number of rollers, the radii of rollers and the inclination angle of the inside of an outer roller for given design specifications. Finally, an actual continuous alternate wheel is constructed to verify validity of the design guidelines.

1. Introduction

Applications of wheeled mobile robots have recently extended to service robots for the handicapped or the aged and industrial mobile robots working in various environments. The most popular wheeled mobile robot is equipped two independent driving wheels. This robot can rotate about any point, but does not allow sideways motion. To overcome this drawback, mobile robots with steerable wheels were suggested. They allow both rotation and sideways motions but not simultaneously. If such robots are used as service robots, for example, they may get in the way of a person they assists, require unnecessarily large space or move along a complicated path when changing their direction.

To cope with these problems, omnidirectional mobile robots have been proposed. They are capable of arbitrary motion in an arbitrary direction without changing the direction of wheels, because they can achieve 3 DOF motion on a two-dimensional plane. Various types of omnidirectional mobile robots have been proposed so far; off-centered wheels [1], ball wheels [2], and universal wheels [3] are more popular among them.

The initial universal wheel design shown in Fig. 1a has multiple passive rollers whose axes are positioned tangent to the wheel circumference. Since this type of wheel makes discontinuous contact with the ground due to a gap between successive rollers, however, the robot platform suffers form vertical vibrations. To minimize the gap between rollers, various variations of universal wheels have been devised. In the Mecanum wheel [4] shown in Fig. 1b, rollers are arranged in such a way that the contact between the wheel and the floor is continuous. In the double wheels [5] shown in Fig. 1c, wheels are arranged in an overlapping way. These types of wheels touch the ground continuously, but the points of contact with the ground are not continuous as seen in the figure. This discontinuous contact causes horizontal vibrations [6].

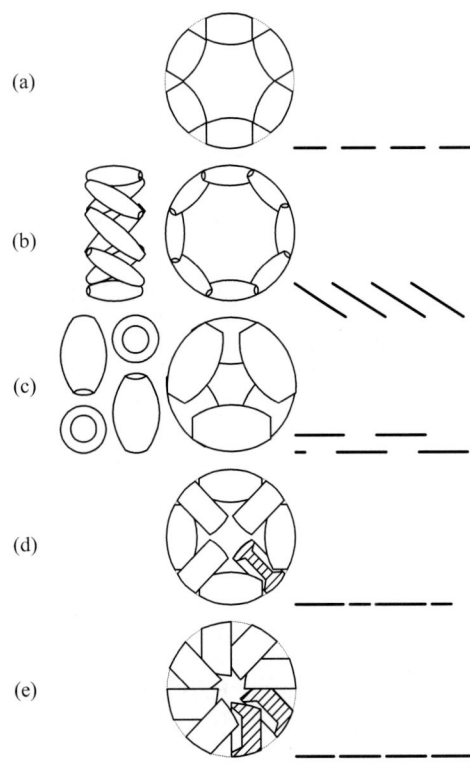

Fig. 1 Various wheel types using passive rollers and their traces; (a) classic, (b) Mecanum, (c) double, (d) alternate, and (e) half.

Other attempts have been made to minimize the gap between rollers for reduction in horizontal and vertical vibrations. In the alternate wheel mechanism [7] shown in Fig. 1d, the large and small rollers are alternated to reduce the gap size. In the half wheel mechanism [6] shown in Fig. 1e, half-divided rollers are arranged in an overlapping manner. However, the gap between rollers cannot be eliminated completely even in these wheels.

In this research a new type of universal wheel is proposed to cope with the problems caused by the gap between the rollers. This new wheel features virtually no gap between passive rollers. Because this wheel makes continuous contact with the ground and has alternating large and small rollers about the wheel, this wheel will be termed a continuous alternate wheel (CAW).

This paper presents details in the design procedure and construction of the continuous alternate wheel. Chapter 2 provides the detailed design in the number of rollers and the shape and size of rollers. Chapter 3 is concerned with the overall structure of a continuous alternate wheel. Finally, conclusions are drawn and future work is outlined in Chap.4.

2. Design of passive rollers

The rollers used in a universal wheel are barrel-shaped as shown in Fig. 2, because the surface contour of a roller should match the wheel circumference. Fig. 2 shows a universal wheel with smaller inner rollers and larger outer rollers arranged alternately around it. In each roller, the most convex radius in the middle is termed a maximum radius, and the radius at its ends is termed a minimum radius.

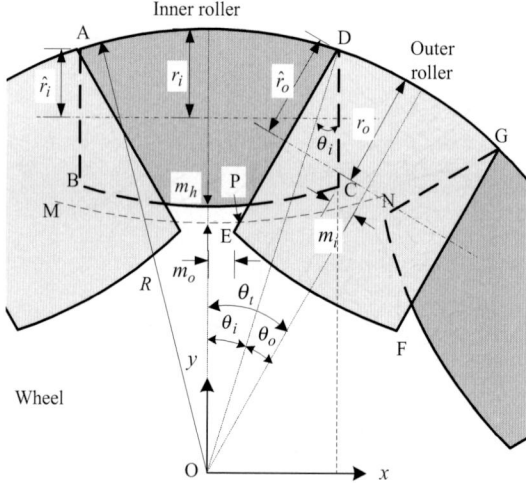

Fig. 2 Alternate inner and outer rollers

The height of a surmountable bump for a universal wheel with passive rollers depends on the size of the minimum radius (not on the wheel radius) and friction of a roller [6]. Therefore, it is preferable that the minimum radius of a roller be as large as possible and thus approach the maximum radius for a given wheel size. In this regard, the half wheel structure is not desirable because the minimum radius is small compared with the maximum one.

2.1 Number and radii of rollers

Fig. 2 shows the configuration of a continuous alternate wheel where the inner and outer rollers are alternated. As mentioned previously, since virtually no gap between rollers exists, the wheel contacts the ground continuously. First the conditions for no gap between rollers will be investigated below.

From the geometry in Fig. 2, the relationship between the wheel radius R and the half-angles θ_i and θ_o for the inner and outer rollers, respectively, becomes

$$2n(\theta_i + \theta_o) = 2\pi \quad (1)$$

where n represents the number of inner (or outer) rollers. If $n = 4$, for example, the wheel consists of 4 inner rollers and 4 outer rollers, and thus $\theta_t\ (=\theta_i+\theta_o)$ becomes $\pi/4$. The maximum radius r_i and the minimum radius $\hat{r}_i$ of an inner roller have the geometric relation

$$(R - r_i + \hat{r}_i)^2 + R^2 \sin^2 \theta_i = R^2 \quad (2)$$

and the maximum radius r_o and the minimum radius $\hat{r}_o$ of an outer roller have the similar relation

$$(R - r_o + \hat{r}_o)^2 + R^2 \sin^2 \theta_o = R^2. \quad (3)$$

Next, for no overlapping of inner rollers, the minimum radius of a inner roller should satisfy the following condition

$$(R \sin \theta_o - m_i) \geq 2\hat{r}_i \sin(\theta_i + \theta_o) \quad (4)$$

where m_i is the space margin between the inner rollers, which depends on the size of the supporting frame (refer to Section 3). The minimum radius of an inner roller is then obtained by

$$\hat{r}_i \leq \frac{R \sin \theta_o - m_i}{2 \sin(\theta_i + \theta_o)} \quad (5)$$

and thus the upper limit on $\hat{r}_i$ becomes

$$\hat{r}_{i\max} = \frac{R \sin \theta_o - m_i}{2 \sin(\theta_i + \theta_o)}. \quad (6)$$

Then the upper limit on r_i is easily computed by

$$r_{i\max} = \hat{r}_{i\max} + R(1 - \cos \theta_i). \quad (7)$$

Similarly, the condition for no overlapping of outer rollers on the minimum radius of the outer roller is obtained by

$$\hat{r}_o \leq \frac{R \sin \theta_i - m_o}{2 \sin(\theta_i + \theta_o)} \quad (8)$$

where m_o represents the margin between the outer rollers, which depends on the size of the supporting frame. The upper limits on the minimum and maximum radii of the outer rollers are given by

$$\hat{r}_{o\max} = \frac{R \sin \theta_i - m_o}{2 \sin(\theta_i + \theta_o)}, \quad (9)$$

$$r_{o\max} = \hat{r}_{o\max} + R(1 - \cos \theta_o). \quad (10)$$

The next step is to find the geometric conditions that the inner and outer rollers do not overlap. Since the rollers are solid as shown in Fig. 2, some portion of the inner roller interpenetrates inside of the outer roller. This problem will discussed in Section 2.2, and the condition to avoid overlap between surfaces of the inner and outer rollers is first considered here. Let the xy coordinates be defined at the center of the wheel as shown in Fig. 2. The equation for segment DE is given by

$$y = \frac{1}{\tan(\theta_i + \theta_o)}(x - R\sin\theta_i) + R\cos\theta_i \quad (11)$$

and the equation of the circle offset by the margin m_h from the surface of the inner roller (represented by the dashed arc MN) is described by

$$x^2 + \{y - (2R - 2r_i - m_h)\}^2 = R^2 \quad (12)$$

Solving Eq. (11) and (12) yields the intersecting point P whose coordinates are

$$x_P = \frac{-ae \pm \sqrt{a^2 - e^2 + 1}}{a^2 + 1}, \quad (13a)$$

$$y_P = a(x_P - b) + c \quad (13b)$$

where $a = 1/\tan(\theta_i + \theta_o)$, $b = R\sin\theta_i$, $c = R\cos\theta_i$, $d = 2R - 2r_i - m_h$, $e = -ab + c - d$.

To avoid overlap, the segment DP must be less than or equal to the minimum diameter DE of the outer roller. Hence

$$\hat{r}_{o\min} = \frac{\sqrt{(R\sin\theta_i - x_P)^2 + (R\cos\theta_i - y_P)^2}}{2} \quad (14a)$$

The lower limit on the maximum radius of the outer roller then becomes

$$r_{o\min} = \hat{r}_{o\min} + R(1 - \cos\theta_o) \quad (14b)$$

In summary, given a wheel radius R and the margins m_i, m_o, m_h, the upper and lower limits for radii of the inner and outer rollers can be determined as a function of θ_i (or equivalently θ_o) and n. For example, when m_i, m_o, m_h are set to 5.5mm, 4.5mm, 7.0mm, respectively, for a wheel radius of 10cm, the roller radii as a function of θ_i for n in the range of 3 to 8 are illustrated in Fig. 3. In Fig. 3, the maximum radius of an outer roller should be searched for in the region of $r_{o\max} \geq r_{o\min}$; otherwise, no solution exists. It is found that there is no solution for $n = 3$, since $r_{o\min}$ is greater than $r_{o\max}$ in the entire range of θ_i. Since the height of a surmountable bump is limited by the minimum radius of the inner roller, which is the smallest among four different radii, the roller shape should be decided so that the minimum radius of an inner roller has the largest value. It is observed in Fig. 3 that as θ_i increases, the radii of the outer roller increase, but those of the inner roller decrease. Therefore, the largest minimum radius of the inner roller corresponds to the point where $r_{o\max} = r_{o\min}$ in the plots. By equating Eq. (10) and (14b), the following equation is obtained.

$$\frac{\sqrt{(R\sin\theta_i - x_P)^2 + (R\cos\theta_i - y_P)^2}}{2} = \frac{R\sin\theta_i - m_o}{2\sin(\theta_i + \theta_o)} \quad (15)$$

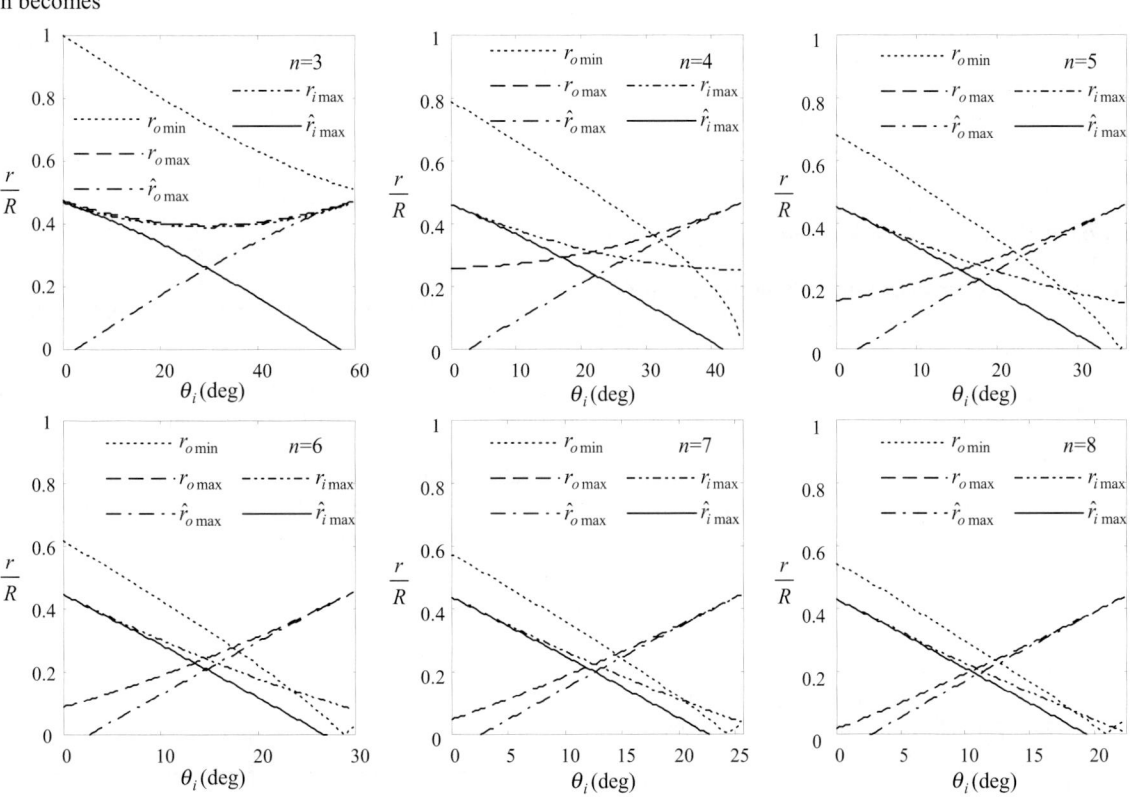

Fig. 3 Roller radii as a function of θ_i for various values of n

Since this equation cannot be solved explicitly, the solution by a numerical analysis technique should be obtained. Fig. 4 shows computational results of the roller radii as a function of n.

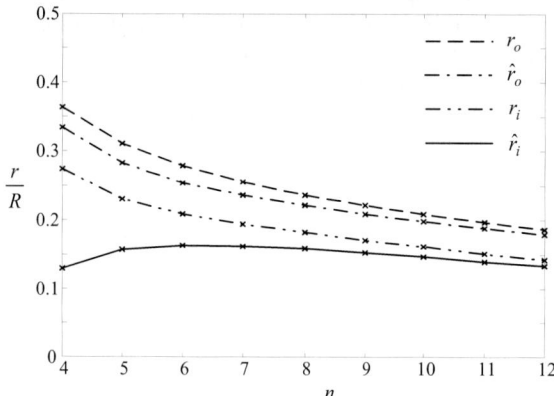

Fig. 4 Variation of radii of inner and outer rollers as a function of n

It is noted from Fig. 4 that for $n < 6$, the minimum radius of an inner roller becomes smaller while the other radii get larger. For $n > 6$, all the radii tend to decrease, which is not desirable. Considering all these facts, the number of rollers was chosen as 6 in the example design of a continuous alternate wheel.

2.2 Shape of inside of outer roller

Since the structure of the continuous alternate wheel involves slight interpenetration of the inner roller into the interior of the outer roller, the inside surface of the outer roller must have some inclination (denoted DH in Fig. 5) to avoid interference with the side surface (whose diameter is CD) of the inner roller. As the angle ϕ between the x axis and DH decreases, the inside surface of an outer roller can be thicker, thus leading to more solid structure.

The side circle of an inner roller whose diameter is CD can be found by the intersection of the sphere and the plane represented by

$$(x - \hat{r}_i \cos\theta_t)^2 + (y - \hat{r}_i \sin\theta_t)^2 + z^2 = \hat{r}_i^2, \quad (16)$$

$$y = \tan\theta_t x. \quad (17)$$

Combining Eq. (16) and (17) gives the following equation of circle

$$(\frac{x}{\cos\theta_t} - \hat{r}_i)^2 + z^2 = \hat{r}_i^2, \quad y = \tan\theta_t x. \quad (18)$$

The distance L between this circle and the axis (represented by the segment PH) that passes the point $(\hat{r}_o, 0, 0)$ and is parallel to the y-axis is expressed by

$$L^2 = (x - \hat{r}_o)^2 + z^2. \quad (19)$$

Substituting Eq. (17) and (18) into (19) yields

$$L(y)^2 = -y^2 - 2(\frac{\hat{r}_o \cos\theta_t - \hat{r}_i}{\sin\theta_t})y + \hat{r}_o^2 \quad (20a)$$

or, equivalently,

$$L(y) = \pm\sqrt{\hat{r}_o^2 - y^2 - 2(\frac{\hat{r}_o \cos\theta_t - \hat{r}_i}{\sin\theta_t})y}. \quad (20b)$$

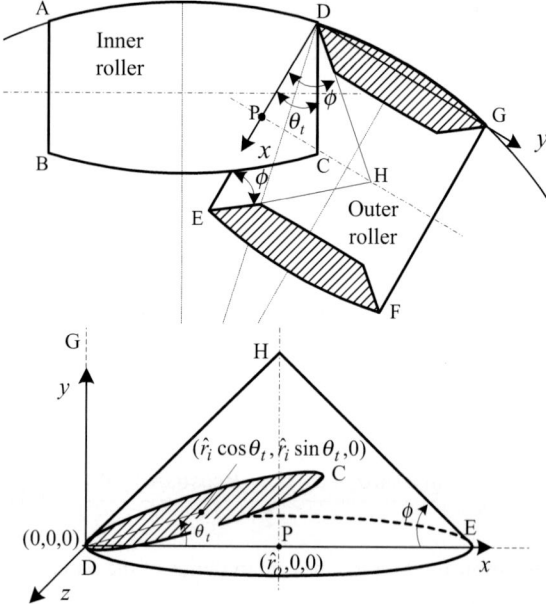

Fig. 5 Interpenetration of inner roller into outer roller

Partial differentiation of Eq. (20b) with respect to y gives

$$\frac{\partial L(y)}{\partial y} = -\frac{y + \frac{\hat{r}_o \cos\theta_t - \hat{r}_i}{\sin\theta_t}}{\sqrt{\hat{r}_o^2 - y^2 - 2(\frac{\hat{r}_o \cos\theta_t - \hat{r}_i}{\sin\theta_t})y}}. \quad (21)$$

From Eq. (21), the minimum angle ϕ to avoid interference can be found at $y = 0$

$$\frac{\partial L(0)}{\partial y} = -\frac{1}{\hat{r}_o}\frac{\hat{r}_o \cos\theta_t - \hat{r}_i}{\sin\theta_t} = \frac{\hat{r}_i/\hat{r}_o - \cos\theta_t}{\sin\theta_t}. \quad (22)$$

Table 1 shows the results of computation for a given example. Since the number of rollers was chosen as 6, the total roller angle becomes 30°, and θ_t is determined to be 17.4°.

Table 1 Design parameters for continuous alternate wheel

n	6	r_o	27.8mm
R	10cm	$\hat{r}_o$	25.4mm
θ_t	17.4°	r_i	20.9mm
θ_o	12.6°	$\hat{r}_i$	16.3mm
ϕ	24.3°		

3. Structure of continuous alternate wheel

An actual continuous alternate wheel was fabricated based on the design parameters given in Table 1. In the process of fabrication, more design factors other than the previously determined ones such as the number of rollers, roller radii, the inside inclination angle of an outer roller were considered. This section is concerned with the overall structure of a continuous alternate wheel.

3.1 Rollers

The shapes of inner and outer rollers designed in the previous section are capable of continuously contacting the ground without any interference between rollers.

Polyurethane was selected as the roller material since it provides a friction coefficient and mechanical properties suited to a universal wheel. However, a urethane roller is not solid enough to prevent deformation when it is subject to loads due to contact with the ground. Therefore, supporting frames in the form of a hollow cylinder are inserted inside the rollers to support urethane rollers.

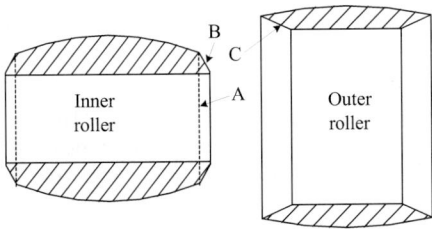

Fig. 6 Final shapes of rollers

Referring to Fig. 5, the fringe of the outer roller is relatively thin to avoid interference with the inner roller. However, this thin part cannot be supported by any supporting frame, since it interferes with the inner roller. As a result, this thin fringe of the outer roller is not solid enough to bear the external load due to contact with the ground. To overcome this problem, the final shapes of the rollers are designed as shown in Fig. 6. The inside surface of an outer roller forms a circular cone (denoted C in Fig. 6) with the inclination angle determined by the previous analysis. Note that another circular cone (denoted B) is added to the original sides of the inner roller (denoted A). These circular cones B and C do not contact each other when they are not subject to a load, since a small margin exists between the two surfaces. When the thin part of the outer roller makes contact with the ground, however, surface C is pressed against surface B. Then, surface B support surface C, and actually they rotate together about each roller axis. In this way, the problem of deformation of the fringe of the outer roller can be overcome without causing any interference with the inner roller.

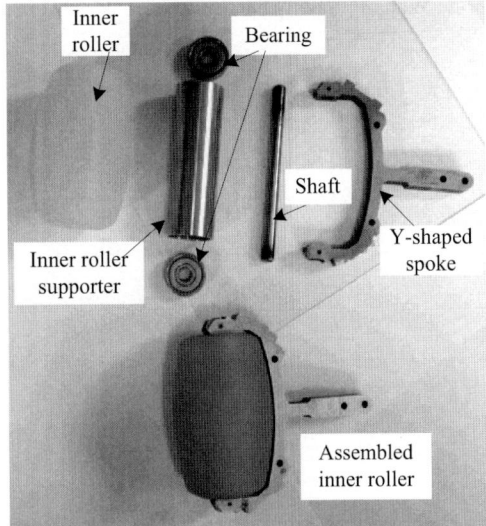

(a) Parts of inner roller

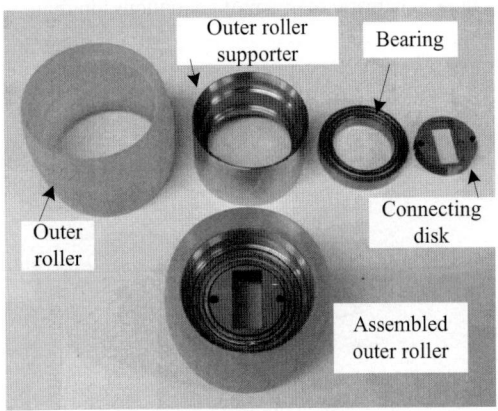

(b) Parts of outer roller

Fig. 7 Photo of roller parts and assembled rollers

3.2 Supporting structure

A supporting structure is required to place rollers around the wheel. The supporting structure for the inner roller is composed of a supporter, a spoke, a roller axis and bearings, while that for the outer roller consists of a supporter and bearings as shown in Fig. 7. One reason why other universal wheels have a gap between rollers is the space for the supporting structure. In the proposed wheel, however, no gap is required since the outer roller was designed to enclose the supporting structure and part of the inner rollers.

Fig. 8 shows schematic diagram and cutaway view of the final wheel and Fig. 9 is a photo of the continuous alternate wheel. As shown in Fig. 8, the wheel has a hub with radially disposed six Y-shaped spokes made of stainless steel. Each spoke supports one inner roller through the roller axis with bearings at both ends. Two consecutive spokes, on the other hand, support the outer roller together through bearings. Note that the margins m_i, m_o, and m_h were considered at the beginning of the design to provide appropriate space for spokes.

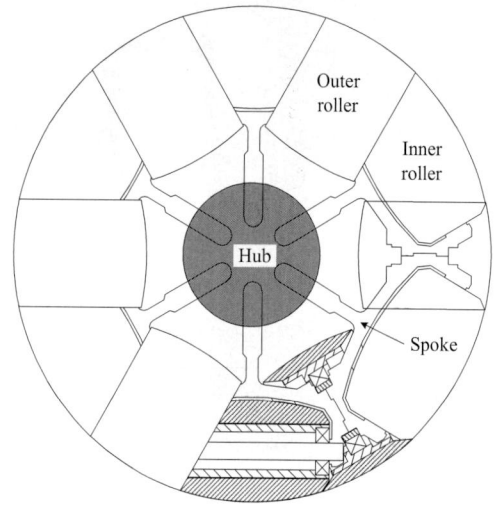

Fig. 8 Schematic diagram of continuous alternate wheel

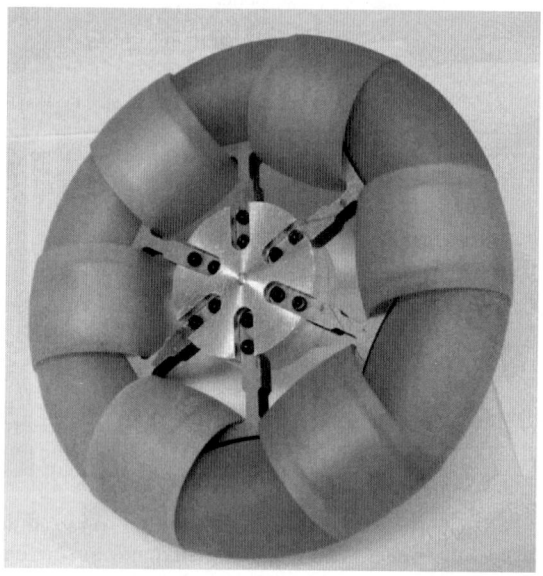

Fig. 9 Photo of continuous alternate wheel

4. Conclusion

In this research a novel design of continuous alternate wheel was proposed to minimize a gap between rollers, which causes vertical and/or horizontal vibrations in many types of omni-directional wheels with passive rollers. This wheel is an improved version of a conventional alternate wheel where inner and outer rollers are disposed alternately. In the continuous alternate wheel, however, the inner and outer rollers are arranged continuously, thus resulting in no gap between the rollers.

This paper details the design process of the continuous alternate wheel. The systematic approaches were presented to determine the optimum number of rollers, the radii of rollers, and the inclination angle of the inside of an outer roller for given design specifications

Using the polyurethane rollers, the actual continuous alternate wheel was constructed to verify validity of the design guidelines. The omnidirectional mobile robots equipped with these proposed wheels are now under construction.

References

[1] Wada, M, Mory, S, "Holonomic and omnidirectional vehicle with conventional tires," 1996 *Int. Conf. On Robotics and Automation*, pp.3671-3676, 1996.

[2] West, M., Asada, H., "Design of ball wheel mechanisms for omnidirectional vehicles with full mobility and invariant kinematics," *Journal of mechanical design*, pp.119-161, 1997.

[3] Blumrich, J. F., "Omnidirectional vehicle," *United States Patent* 3,789,947, 1974.

[4] Ilou, B. E., "Wheels for a course stable self-propelling vehicle movable in any desired direction on the ground or some other base," *United States Patent* 3,876,255, 1975.

[5] Asada, H, Sato, M, Bogoni, L, "Holonomic and omnidirectional vehicle with conventional tires," 1995 *Int. Conf. On Robotics and Automation*, pp.1925-1930, 1995.

[6] Ferriere, L., Raucent, B., and Campion, G., "Design of omnimobile robot wheels," 1996 *Int. Conf. On Robotics and Automation*, pp.3664-3670, 1996.

[7] Carlisle, B., "An omni-directional mobile robot," *Development in robotics*, Kempston, pp. 79-87, 1983.

Holonomic Omni-Directional Vehicle with New Omni-Wheel Mechanism

Riichiro DAMOTO, Wendy CHENG and Shigeo HIROSE

Tokyo Institute of Technology

Department of Mechano-Aerospace Engineering
2-12-1 O-okayama Meguro-ku, Tokyo 152-8552 Japan
Fax +81-3-5734-3982
E-mail damoto@mes.titech.ac.jp

Abstract : *In this paper, we focus on the "Vuton-II", a new omni-directional vehicle developed as a transport vehicle to operate within factories, hospitals, and warehouses. This vehicle is composed of three or more "Omni-Discs." The Omni-Disc mechanism ensures that individual wheels of the Omni-Disc assembly are always aligned in the same direction, and can always roll freely. The Vuton-II was based on the Vuton-I, an earlier omni-directional vehicle with similar targeted use and controls characteristics. The Vuton-II is designed to be low in cost, short in stature, and reasonably high in payload.*

KeyWords : *Omni-directional vehicle, Omni-Disc*

1. Introduction

Transport vehicles for factories, hospitals and warehouses need to possess high mobility to enable correct orientation to target locations while moving freely on narrow floor surfaces and to avoid repetitively switching drives and ensure smooth turning in any desired direction.

The authors have previously developed an omni-directional vehicle, the "Vuton-I", that fits the description above. Building off the Vuton-I, the authors have developed a new omni-directional vehicle, "The Vuton-II", which also fits the description above. This new vehicle is designed to be low in cost, short in stature, and reasonably high in payload. This paper will provide an introduction to the mechanisms of the vehicle and the principles of its actuator, as well as present the results of driving experiments.

2. Prior Omni-Directional Vehicles

Automobiles need to switch drives often when engaged in tasks such as parallel parking. The drive performance of ordinary automobiles, which includes neither rotating in place nor maneuvering in a sideways fashion, cannot be considered satisfactory for confined operational environments. For this reason, numerous efforts have been made to develop omni-directional vehicles with two independent translational degrees of freedom and one rotational degree of freedom, for a total of three degrees of motion freedom on a flat surface.

One potential solution is to install an independent steering mechanism in each wheel. The Nakano group has conducted leading research in this approach [1]. However, with this method, the turning motion cannot always be performed continuously. For example, when such a vehicle attempts a turn of 90 degrees, it must first stop and turn its wheels 90 degrees as a preparatory action.

To realize an instantaneous change in motion with three degrees of freedom on a flat surface requires holonomic motion characteristics. Holonomic motion characteristics are thus indispensable for omni-directional vehicles.

There have been attempts to produce this kind of omni-directional vehicle by using wheels on which many free rollers are arranged [2][3]. The basic form of such a wheel is shown in Figure 1. Omni-directional and holonomic motion can be achieved if three or more of these wheels are arranged in differing directions. However, this kind of mechanism raises the center of gravity of the chassis, and decreases the stability of the vehicle's motions.

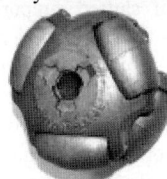

Fig.1 Example of free roller wheels for holonomic omni-directional vehicles

Other than these specialized wheels, omni-directional vehicles with crawler tracks, as shown in Figure 2, have also been proposed [4]. The crawler track is a system in which multiple spheres are supported along the crawler belts, and the spheres that touch the ground are simultaneously in contact with two rods located inside the crawler belts. Because the crawler has this kind of mechanical configuration, it generates ordinary forward motion in the direction of the belts by rotating the belts, and it generates sideways motion by rotating the pair of rods and driving the spheres sideways. In this mechanism, multiple spheres rotate at the same time, and translational omni-directional motion can be produced without slippage. However, attempts at rotational motions with sufficiently small rotational radii inevitably leads to slipping, decreased mobile efficiency, and potential floor surface damage.

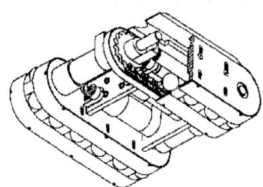

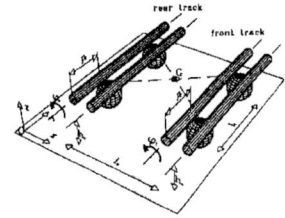

Fig.2 Omnitrack with balls along crawler

3. The Vuton Crawler

3.1 Mechanism of the Vuton Crawler

In order to resolve the problems associated with the conventional omni-directional mobile mechanisms described in Section 2, a novel type of crawler, called the "Vuton crawler", has been developed and is shown in Figure 3 [5]. The free rollers of the Vuton crawler are supported by square frame members with rotational motion independence, and the square frame members are connected to the chains at diagonal points separated by a horizontal distance t. The pair of chains is also offset with a distance t, and the chains are driven simultaneously. With this arrangement of shifted support, the Vuton crawler always maintains its free rollers in a horizontal posture.

Figure 4 shows a prototype Vuton crawler. One of the chains (2) is directly driven by a geared motor (1), and the other chain (3) is synchronously driven by transmitting the rotation of the motor (1) via a timing belt (4) to an axis offset from the geared motor by a distance t. The diagonal connecting sections (6) of the square frame (5) are linked into the chains via axle receiving units that substitute for chain units. The square frame (5) moves so as to contact the support rollers (7), which are arranged on the lower part of the crawler frame. This configuration transfers the support forces applied to the free rollers to the crawler frame. (8) is a wedge type mechanism which regulates the chain sprocket distance.

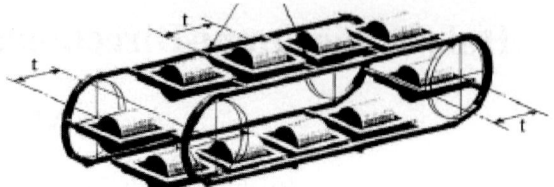

Fig.3 The mechanism of the Vuton crawler

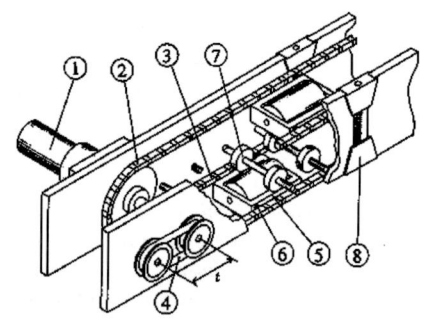

Fig.4 The driving mechanism of the Vuton crawler

3.2 Control of the Vuton Crawler

A vehicle equipped with three or more Vuton crawlers in mutually differing directions can realize omni-directional and holonomic mobility. The steering control is also very simple. When translational and rotational motion velocities are commanded to the vehicle, these instructions can be directly translated to the forward velocity of each crawler.

For example, consider the steering control of the omni-directional vehicle provided with four Vuton crawlers (shown in Figure 5). The velocity command values v_i (i=1-4) to the four crawlers may be calculated from the velocity command values (v_x, v_y) and the rotational angular velocity (ω) around the point of origin, all defined in the mobile body coordinate system (x,y):

$$\begin{bmatrix} v_1 \\ v_2 \\ v_3 \\ v_4 \end{bmatrix} = \begin{bmatrix} 0 & 1 & r_0 \\ -1 & 0 & r_0 \\ 0 & -1 & r_0 \\ 1 & 0 & r_0 \end{bmatrix} \cdot \begin{bmatrix} v_x \\ v_y \\ \omega \end{bmatrix} \quad (1)$$

Here, r_0 is the distance from the vehicle center point to each crawler, and the counter-clockwise rotation is assumed positive. The 4x3 matrix is the Jacobian matrix of the control of the omni-directional vehicle.

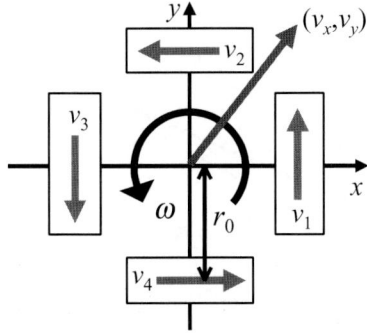

Fig.5 Control of the omni-directional vehicle

3.3 The Prototype Machine and Driving Tests

The Vuton-I was manufactured by arranging four Vuton crawlers, as shown in Figure 6. Its specifications are: it has a mass of 29.5 kg (without the battery); it has dimensions of 560 mm x 560 mm x 135 mm. It can be manually driven by a joystick, or automatically driven by a computer. The joystick produces velocity commands to the motors of each crawler by generating the relationships outlined in Equation (1) with an analog circuit.

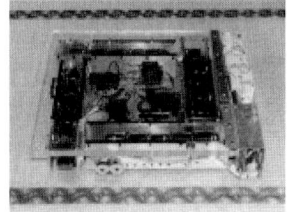

Fig.6 Photograph of the Vuton-I

4. The Omni-Disc

4.1 The principle and features of the Omni-Disc

In addition to the Vuton crawler mechanism and associated omni-directional vehicle Vuton-I, at this time the authors have also developed a new type of omni-directional wheel named the Omni-Disc and associated vehicle Vuton-II. This new vehicle is designed to be lower in cost and shorter in stature than the Vuton-I, but still maintain a reasonably high payload capacity.

The free roller wheel as shown in Figure 1 raises the center of gravity of the chassis, which may make its motion unstable. We first considered inclining the large wheel (9) on which many free rollers (10) are arranged (as shown in

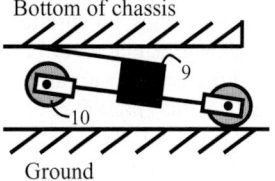

Fig.7 Inclined free roller wheel

Figure 7) to bring the chassis closer to the ground. But, in this mechanism, frictional forces prevent the motion of the chassis when two small free rollers not aligned to the chassis' movement direction touch the ground simultaneously, as shown in Figure 8.

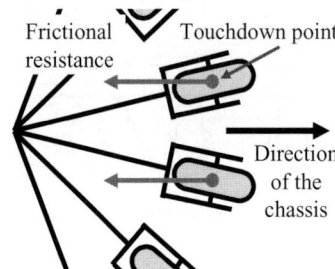

Fig.8 Two free rollers with touchdown points

In order to avoid this problem, we designed a new mechanism that keeps the direction of the free wheels always constant, as shown in Figure 9.

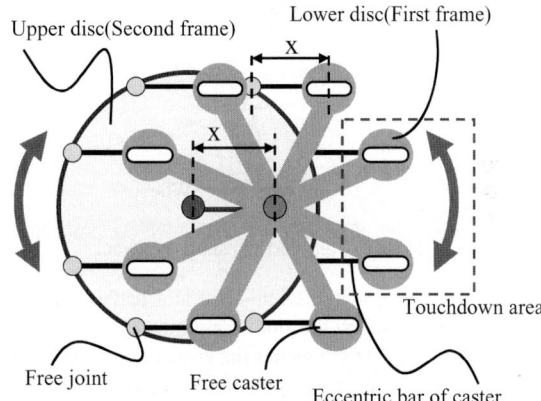

Fig.9 Principle of the Omni-Disc

As shown in Figure 9 and Figure 10, the two rotating discs in the Omni-Disc lie in separate planes, and are eccentrically fixed. The vertical shafts that support the multiple small free wheels are installed in the lower disc so that they can rotate freely. The tips of the vertical shafts of the free wheels, which have the same eccentricities as the two rotating shafts of the discs, are inserted in the holes in the upper disc as shown in Figure 11. This ensures that individual wheels of the assembly are always aligned in the same direction, and that the Omni-Disc casters can always roll freely in that direction; this resolves the problem shown in Figure 8. The two discs and multiple free wheels of the Omni-Disc constitute a kind of parallel mechanism. The whole Omni-Disc mechanism is inclined 4 degrees with respect to its supporting plate (11), so only one or two casters at the tip of the Omni-Disc touch the ground simultaneously.

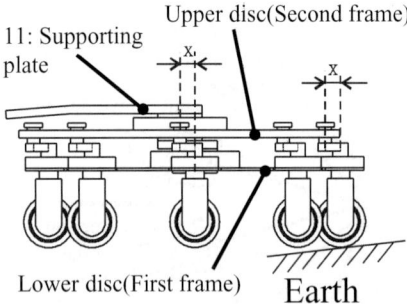

Fig.10 Side view of the Omni-Disc

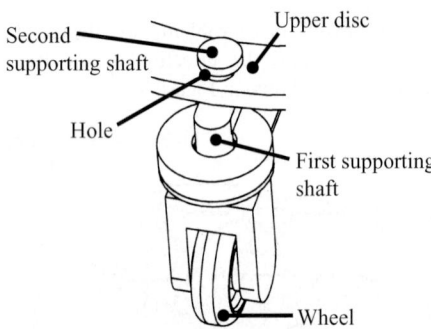

Fig.11 Hole to hold vertical shaft of wheel

The Lower disc of the Omni-Disc is a leaf spring that allows casters to touch the floor continuously. If only one wheel of the Omni-Disc touches the ground, the leaf spring suffers a displacement corresponding to 1/n of the weight of the vehicle (n is the number of Omni-Discs that support the load). When two wheels of the Omni-Disc touch the ground simultaneously, the distance between the wheels and center of the bottom of the chassis is shorter, and the displacement of the springs (one for each wheel) is also smaller, because each spring supports only 1/2n of the weight of the vehicle. This reduces the height variations associated with different touchdown combinations, and so the rotation of the disc is very smooth, as shown in Figure 12.

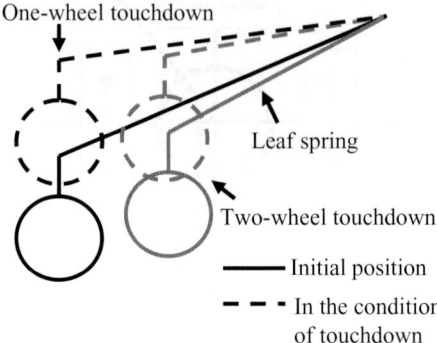

Fig.12 Two types of touchdowns

In Fig.13, if we define the tip of the Omni-Disc's touchdown area as the front and consider merely the rotational direction of the casters, it is acceptable for casters in the entire front half of the Omni-Disc mechanism to touch the ground. However, if the casters of the Omni-Disc touch the ground over a wide range, the casters at the edges of this range can't move at sufficiently high velocities toward the chassis, and they will produce a relative deceleration rather than an acceleration effect on the chassis.

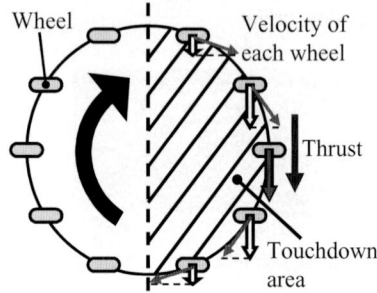

Fig.13 Touchdown area of Omni-Disc

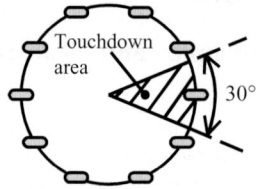

Fig.14 Desirable touchdown area

So, it is desirable that the casters only ground in a range of 30 degrees around the center of Omni-Disc's touchdown area, as shown in Figure 14. That's why the Omni-Disc mechanism is angled downwards by 4 degrees.
The Omni-Disc is much cheaper than the Vuton crawler

because it is composed of many commercial instead of customized components.

4.2 Passive Omni-Disc

The passive Omni-Disc, as a free caster, has an advantage over commercial free casters. An ordinary commercial caster can rotate freely around its vertical shaft, but generates frictional forces that might prevent straight motion of the chassis when the caster experiences large changes in its movement direction. In the case of the Omni-Disc, even though the wheel at the tip of the mechanism, which is touching the ground rotates around the center of the Omni-Disc, its alignment is kept constant by the parallel mechanism. Thus, when the direction of the chassis changes suddenly, the frictional forces generated by the Omni-Disc are much smaller, and do not impede the motion of the chassis. The overall view of the passive Omni-Disc as a free caster is shown in Figure 15.

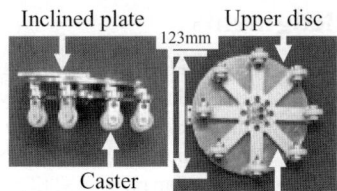

Fig.15 Passive Omni-Disc as free caster

Experiments measuring the motions of a differential drive vehicle (shown in Figure 16) outfitted with the Omni-Disc and a type of ordinary commercial free caster were carried out to compare the passive Omni-Disc with ordinary casters. In the experiment, commercial casters and the coordinate system as shown in Figure 16 were used. The diameter of

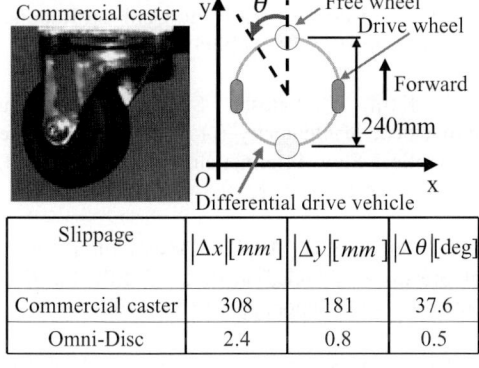

Slippage	$|\Delta x|[mm]$	$|\Delta y|[mm]$	$|\Delta \theta|[deg]$
Commercial caster	308	181	37.6
Omni-Disc	2.4	0.8	0.5

Fig.16 Comparison of Omni-Disc and Commercial Free Caster

the drive wheel is 50 mm and the power of the geared motor that actuates one drive wheel is 4.5 watts. During testing, the chassis was driven forward 0.8 m and backwards 0.8 m in the same posture and at a speed of 1.23×10^{-1} [m/s]. Comparing the average slippage distance from the initial position showed that the slippage of the chassis with the passive Omni-Disc was very small; the excellence the Omni-Disc's running performance was confirmed.

4.3 Active Omni-Disc

By driving the upper disc as a spur gear with a gear system, it is possible to obtain thrust in the direction perpendicular to the small casters' rotating direction, and a new type of omni-directional vehicle can be created. One actuator unit of the omni-directional vehicle "The Vuton-II" is shown in Figure 17, and the gear system that drives the wheel is shown in Figure 18.

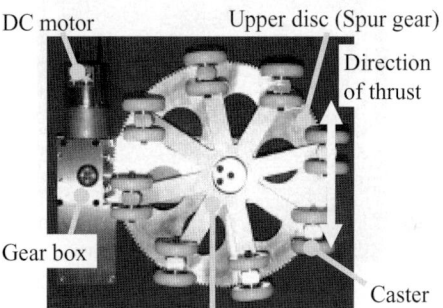

(a) Bottom view of the active Omni-Disc

(b) Side view of the active Omni-Disc

Fig.17 Active Omni-Disc

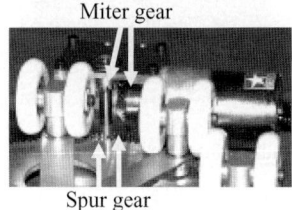

Fig.18 The gear system of the Omni-Disc

5. The Prototype of "The Vuton-II" with Active Omni-Discs

The overall view of the first prototype of the new Omni-directional vehicle "Vuton-II" is shown in Figure 19. The vehicle is equipped with four active Omni-Discs, so the positions of the actuators and the control method of this vehicle are identical to those described in Section 3.2 for the Vuton-I. Vuton-II's specifications are: it has a mass of 11.4 kg (without the battery); it has dimensions of 622 mm x 622 mm x 90 mm; and it has a carrying load capacity of 45 kg. The basic running experiment of this vehicle have been performed, as shown in Fig.20, and Vuton-II exhibited quite good performance, just like the Vuton-I.

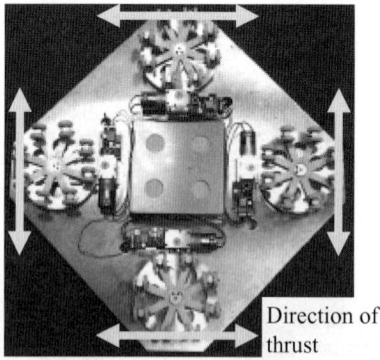

(a) Bottom view of the Vuton-II

(b) Side view of the Vuton-II

Fig.19 Overall view of the Vuton-II

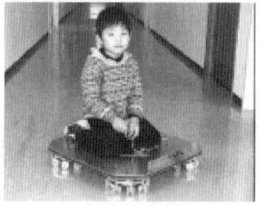

Fig.20 Manual control experiment of the Vuton-II

6. Conclusions

In this paper, after outlining the challenges facing conventional omni-directional vehicles, we proposed the Vuton crawler and the Omni-Disc as potential actuators for omni-directional transport vehicles. Then, we discussed the design, control, and testing of two omni-directional vehicles, one using four Vuton crawlers and one utilizing four Omni-Discs.

The new omni-directional vehicle described in this paper, the Vuton-II, uses Omni-Disc mechanisms that support multiple free casters horizontally. In principle, it can realize translational and rotational motion without slippage, and so it can be highly effective for practical applications that demand:
1) High mobility.
2) Smooth response to slight irregularities in the floor surface.

The development of the second prototype of the Vuton-II with improved Omni-Disc mechanism and much higher payload capacity will be carried out in near future.

Acknowledgments

This research is supported by The Grant-in Aid for COE Research Project of Super Mechano-Systems by The Ministry of Education, Culture, Sports, Science and Technology. Cheng was supported by the U.S. National Science Foundation and the Japanese JISTEC.

References

[1] E. Nakano, N. Koyachi, "An Advanced Mechanism of the Omni-Directional Vehicle (ODV) and Its Application of the Working Wheel chair for the Disabled," in Proc. of Int. Conf. of Advanced Robotics, pp.277-284 (1983).

[2] H. Asama, M. Sato, L. Bogoni, H. Kaetsu, A. Matsumoto, and I. Endo, "Development of an omni-directional mobile robot with 3 dof decoupling drive mechanism," in Proc. of IEEE Int. Conf. on Robotics and Automation, pp.1925-1930 (1995).

[3] N. Keiji, T. Satoshi, S. Makoto, T. Yutaka, "Improvement of Odometry for Omnidirectional Vehicle Using Optical Flow Information," in Proc. of IEEE/RSJ IROS, Takamatsu, Japan, pp.468-473 (2000)

[4] M. West and H. Asada, "Design of a Holonomic Omnidirectional Vehicle," in Proc. of IEEE Int. Conf. on Robotics and Automation, pp.97-103 (1992)

[5] S. Hirose and S. Amano, "The VUTON: High Payload, High Efficiency Holonomic Omni-Directional Vehicle," in Proc. of 6th Int. Symp. on Robotics Research, pp.253-260 (1993).

Dynamic Simulation of Actively-Coordinated Wheeled Vehicle Systems on Uneven Terrain

Min-Hsiung Hung

Dept. of Electrical Engineering
Chung Cheng Institute of Technology
National Defense University
Tao-Yuan, Taiwan, R.O.C.

David E. Orin

Dept. of Electrical Engineering
The Ohio State University
Columbus, Ohio, U.S.A.

Abstract

In this paper, a graphical dynamic simulator is developed that can simulate actively-coordinated wheeled vehicle systems on uneven faceted terrain. Based on the considerations of model fidelity and computational efficiency, a simple geometric model for wheel-terrain contact is proposed. In addition, a computationally-efficient algorithm for contact detection is developed. We also devise a contact force model based on soil mechanics. Simulation results of a case where the WAAV [1] is traversing a concave edge between facets are used to demonstrate the good performance of our contact model.

1 Introduction

Variably-configured vehicles with active coordination (also called actively-coordinated vehicles) generally possess larger numbers of independently-controlled actuators than those of motion degrees of freedom. The redundancy of actuation can be used to distribute and optimize the contact forces at the vehicle-terrain interface to improve contact conditions, such as reduction of slip and sinkage, and to enhance the system performance, such as minimization of power consumption. Therefore, their mobility and terrain-adaptive properties on unprepared terrain are generally superior to those of conventional vehicles. This type of vehicle has potential in many off-road applications including those in mining, agriculture, forestry, military locomotion, and exploration of planetary surfaces.

During the past decade, several prototypes of actively-coordinated wheeled vehicles have been developed: such as the Attached Scout Concept Rover [2] for Mars exploration by FMC, the Actively Articulated Six Wheeled Vehicle Concept [3] for the same purpose by Martin-Marietta, and the Wheeled Actively Articulated Vehicle (WAAV) [1] at The Ohio State University. However, in much of the work only preliminary research results have been obtained [1, 4, 5]. In order to bring actively-coordinated wheeled vehicles into applications, more advanced methods for coordination and control still need to be developed.

For the test and evaluation of coordination and control methods of vehicles, dynamic simulation is a very useful tool. By using numerical simulation with a reasonable execution speed, vehicle trials may be repeated at no risk. This will expedite the development of a system and save cost. In this work, a dynamic simulator (see Figure 1) with efficient algorithms is developed. It is able to simulate actively-coordinated, off-road wheeled vehicles with multiple modules operating on uneven terrain. In particular, this paper will focus on the development of the vehicle-terrain contact model.

Several issues are involved in the modeling of vehicle-terrain contact. Among them include (1) object modeling in a proper form based on the considerations of model fidelity and computational cost, (2) contact detection together with identification of contact locations, and (3) computation of the corresponding contact force. In the following sections, we propose efficient approaches to these problems.

The remainder of the paper is organized as follows. Section 2 describes the proposed geometric contact model. In Section 3, contact detection is addressed. In Section 4, the contact force model is formulated. Simulation results are given in Section 5. The paper ends with a summary and conclusions in Section 6.

2 Geometric Contact Model

With regard to object modeling, one approach is to model the wheels and terrain with polyhedra. There exists many efficient algorithms, such as the one in [6], to do collision detection for moving polyhedra. However, to model a smooth surface closely, such as for a wheel, a large numbers of faces are needed in the

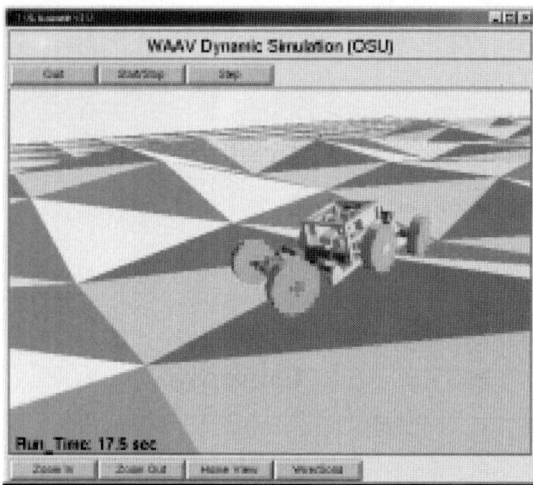

Figure 1: The dynamic simulator for WAAV, with triangular-facet terrain model.

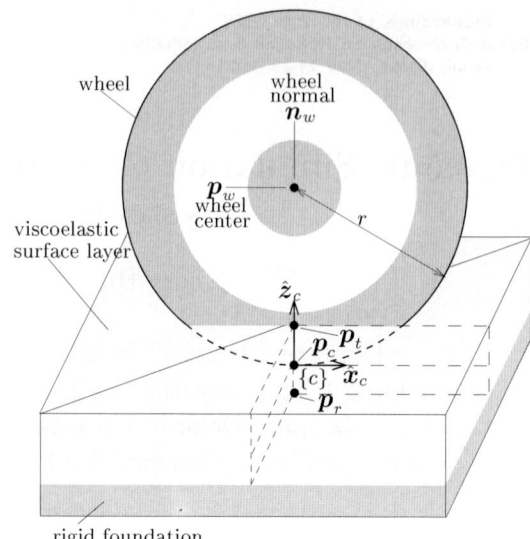

Figure 2: Geometric model for wheel-terrain contact.

polyhedron. This will significantly increase the computation time for contact detection.

In a dynamic simulation package, called *DynaMechs* [7], a polling strategy is used to detect contact. Evenly distributed points on the wheels of the vehicle system can be chosen as the polled points. In each time step, these points are computed relative to the environment to see whether contact occurs or not. Since this approach does not take full profiles of the wheel surface into consideration, a large number of points need to be defined to obtain a contact location close to the actual one. This also increases the computation time.

Based on the consideration of model fidelity, algebraic equations can also be used to represent the profiles of the wheels and terrain. For example, a wheel can be modeled as a cylinder or a torus [8]. Contact is then determined by solving a set of nonlinear equations. Generally, an analytical solution of this set of nonlinear equations is difficult to find, and an iterative algorithm must be applied to obtain the solution, which may not be computationally efficient.

To make a compromise between model fidelity and computational efficiency, we model a wheel as a thin disk that represents the middle cross section of the wheel. In addition, the terrain is modeled as a set of triangular facets; that is, the terrain is uneven and composed of piecewise planes with different slopes (see Figure 1). By increasing the number of facets, a smooth terrain surface can be approximated. We also model the wheel-terrain contact as a point contact. The deepest penetration point into the terrain is assumed to be the contact point on the wheel. For each wheel contacting a facet of the terrain, there is only one contact point although multiple contact points result when traversing an edge or vertex of the terrain.

In our wheel-terrain contact model, we assume that the wheel is nearly rigid, and the terrain is deformable under pressure. This is the case where rigid wheels or wheels with high tire pressure move on soil. Also, the terrain is modeled as a rigid foundation covered with a viscoelastic surface layer of uniform thickness [9]. The surface layer is used to account for the system's compliance (includes the compliance in the vehicle system, soil deformation, etc.). When a wheel and the terrain are in contact, it is assumed that the wheel will penetrate into the terrain. However, the penetration will never reach the inner rigid foundation. A geometric model for a wheel in contact with a terrain facet is illustrated in Figure 2.

Usually, a wheel contacts the terrain with a small contact area. However, we simulate the contact on the wheel with a net force acting at a single point, called the wheel contact point. For each wheel-terrain contact, a pair of spatial points are located: one on the wheel, p_c, and the other on the terrain, p_t. The relative movement (velocity and position) between p_c and p_t is used to model the deformation of the terrain and to develop the contact force model.

In this work, the contact point p_c on the wheel is assumed to be the deepest penetration point into the terrain. The corresponding closest point on the rigid foundation to the wheel is labeled as p_r. A contact coordinate system, $\{c\}$, is set at p_c. Its z-axis, $\hat{z}_c$, points up along the direction from p_r to p_c. Also, the x-axis, $\hat{x}_c$, of $\{c\}$ is directed normal to both the wheel axle and $\hat{z}_c$, while the y-axis, $\hat{y}_c$, is set to form a right Cartesian coordinate system. In addition, p_t

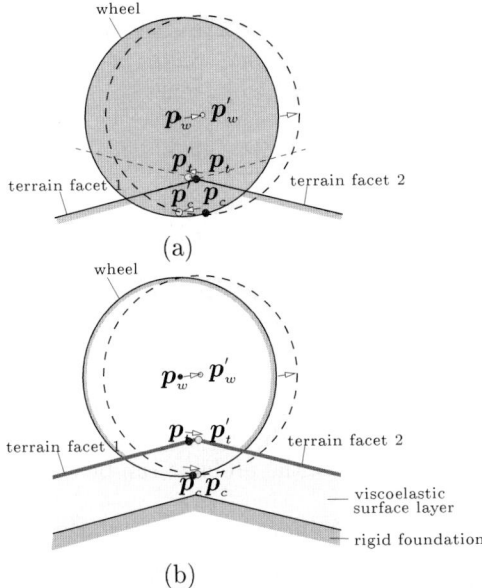

Figure 3: (a) Discontinuity of the contact point with least deepest-penetration approach. (b) Continuity of the contact point with elastic-surface-layer approach.

is initialized as the projection of p_c onto the surface of the viscoelastic layer along $\hat{z}_c$.

When traversing a convex edge, a wheel contacts both of the adjacent facets simultaneously. In [1], the contact point with less penetration into the plane is chosen as the correct one. When the wheel center is at p_w (see Figure 3(a)), the wheel has less penetration into facet 1 than into facet 2. Thus, the wheel is regarded as contacting facet 1, and the corresponding contact points are p_c and p_t. Subsequently, the wheel moves forward to the position with the center at p'_w. At this moment, the wheel has less penetration into facet 2 than into facet 1. The wheel is regarded as contacting facet 2, and the corresponding contact points are p'_c and p'_t. It is noted that the contact points are suddenly changed to backward locations while the wheel is moving forward during the edge crossing. This discontinuity does not simulate the edge traversal of the wheel in a realistic manner.

On the other hand, the introduction of the viscoelastic surface layer [9] for the terrain model combined with the assumption that a contact occurs at the deepest penetration point overcome this problem. When a wheel penetrates into the viscoelastic surface layer, the deepest penetration point on the wheel is the closest point to the rigid foundation. During the edge or vertex traversal, there is always only one closest point to the rigid foundation. Therefore, the contact point on the wheel can be uniquely determined. Also, the contact points p_c and p_t are continuous (see Figure 3(b)), which is more realistic.

Our strategy for modeling the wheels and terrain make the computation of contact detection relatively simple and fast, which is shown later. Specifically, in the case where a wheel contacts a facet of the terrain, the solution can be found analytically. Even in the case where a wheel traverses an edge or a vertex of the terrain, the computation is still relatively fast compared to other approaches mentioned previously.

3 Contact Detection

There are three possible cases for a wheel traversing uneven faceted terrain: traversing a plane, an edge, and a vertex. The computation of the corresponding contact locations for each case is presented in the following. The components of all vectors are referenced to an earth-fixed coordinate system $\{E\}$.

3.1 Traversing a Planar Facet

Figure 2 shows the case of a wheel traversing a planar faceted terrain. In the figure, p_c, p_t, p_r, p_w, $\hat{x}_c$, and $\hat{z}_c$ have been defined previously. In addition, n_w is the unit normal vector of the wheel plane and is parallel to the wheel axle; r is the wheel radius; and n_t is the unit normal vector to the rigid foundation facet and is parallel to $\hat{z}_c$ in this case.

Given n_w, p_w, r, and the equation of the rigid foundation plane, p_c can be analytically computed as follows. First, p_c is in the wheel plane, and the distance between p_c and p_w is r:

$$n_w \cdot (p_c - p_w) = 0, \quad (1)$$
$$\|p_c - p_w\|^2 = r^2, \quad (2)$$

where $\cdot$ is the inner product operation, and $\|*\|$ represents the 2-norm of a vector.

Next, the vector from p_w to p_c is perpendicular to the common normal vector of n_w and n_t:

$$\frac{n_w \times n_t}{\|n_w \times n_t\|} \cdot (p_c - p_w) = 0, \quad (3)$$

where $\times$ denotes the cross product.

Combining Eqs. (1) and (3) which are linear, we can express the x and y coordinates of p_c as a function of its z-coordinate and substitute them into Eq. (2). Equation (2) then becomes a second-order equation in the z-coordinate, which possesses two solutions. The correct z-coordinate is the one with the lower value (deepest penetration). In turn, the x and y coordinates of p_c can be obtained.

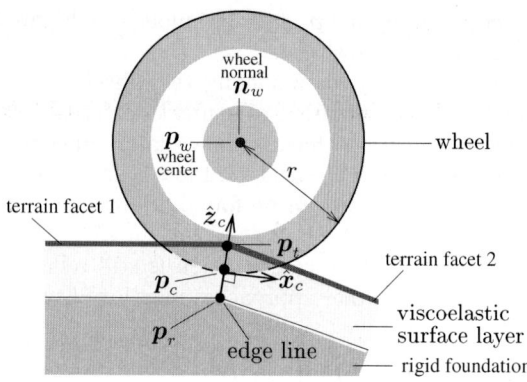

Figure 4: A wheel traverses an edge.

Since p_r is the perpendicular projection point of p_c on the rigid foundation, it can be expressed as

$$p_r = p_c + s\, n_t, \qquad (4)$$

where s is a scalar. Also, p_r must satisfy the plane equation of the rigid foundation:

$$n_t \cdot (p_r - p_0) = 0, \qquad (5)$$

where p_0 is any point other than p_r on the rigid foundation plane. By substituting Eq. (4) into Eq. (5) and using the fact that $\|n_t\| = 1$, we can compute s as follows:

$$s = n_t \cdot (p_0 - p_c). \qquad (6)$$

In turn, p_r is obtained. If the distance between p_r and p_c is larger than the the uniform thickness d_e of the viscoelastic surface layer, or p_r is outside the rigid foundation facet, then there is no contact. Otherwise, the wheel contacts the terrain, and the penetration depth d_p of the wheel into the terrain is

$$d_p = d_e - \|p_c - p_r\|. \qquad (7)$$

Also, p_t is the projection of p_c onto the surface of the viscoelastic layer along n_t and can be computed by

$$p_t = p_c + d_p\, n_t. \qquad (8)$$

Finally, $\hat{z}_c$ is set to n_t, and $\hat{x}_c$ is set to the common normal vector of n_w and n_t:

$$\hat{x}_c = \frac{n_w \times n_t}{\|n_w \times n_t\|}. \qquad (9)$$

3.2 Traversing an Edge Between Facets

When a wheel traverses an edge (see Fig. 4), p_r is located on the edge line. It can be expressed as follows:

$$p_r = p_1 + t\, u, \qquad (10)$$

where p_1 is a point on the edge line, t is a scalar, and u is the unit directional vector of the edge line. Also, the vector from p_r to p_c must be perpendicular to u:

$$u \cdot (p_c - p_r) = 0. \qquad (11)$$

By combining Eqs. (10) and (11) and using the fact that $\|u\| = 1$, the scalar t is:

$$t = u \cdot (p_c - p_1). \qquad (12)$$

To find the closest points between the wheel and the edge line, we must find the smallest distance between p_r and p_c under the constraint that p_c is located on the circumference of the wheel (expressed in Eqs. (1) and (2)). Thus, the problem can be described as

$$\begin{aligned}\text{minimize} \quad & \|p_r - p_c\|^2 \\ \text{subject to} \quad & n_w \cdot (p_c - p_w) = 0 \\ & \|p_c - p_w\|^2 = r^2 \end{aligned} \qquad (13)$$

By replacing p_r with Eqs. (10) and (12), Eq. (13) becomes a nonlinear optimization problem of three variables, the coordinates of p_c, subject to one linear and one nonlinear constraint. This nonlinear-constraint minimization problem can be solved by an algorithm that uses a successive quadratic programming method to solve the general nonlinear programming problem, such as the "NCONG" routine in the IMSL library. After p_c is computed, p_r can be obtained. If the distance between p_r and p_c is larger than the the uniform thickness d_e of the viscoelastic surface layer, or p_r is outside the edge-line segment of the adjacent facets, then there is no contact. Otherwise, the wheel contacts the terrain, and the penetration depth d_p of the wheel into the terrain is computed using Eq. (7).

For the edge contact, n_t is assigned to be the vector from p_r to p_c, and $\hat{z}_c$ is set along this normal vector:

$$\hat{z}_c = \frac{p_c - p_r}{\|p_c - p_r\|}. \qquad (14)$$

The unit vector $\hat{x}_c$ is set to be the common normal vector of n_w and n_t (Eq. (9)). Finally, p_t is initialized as the projection of p_c along $\hat{z}_c$ onto the surface of the viscoelastic layer, which can be computed with Eq. (8).

3.3 Traversing a Vertex Where Facets Intersect

When a wheel traverses a vertex where several facets intersect, p_r is located at the vertex, as shown in Figure 5. In this case, the underlying problems of finding the closest points and assigning the contact coordinate system are the same as those in the edge-contact case,

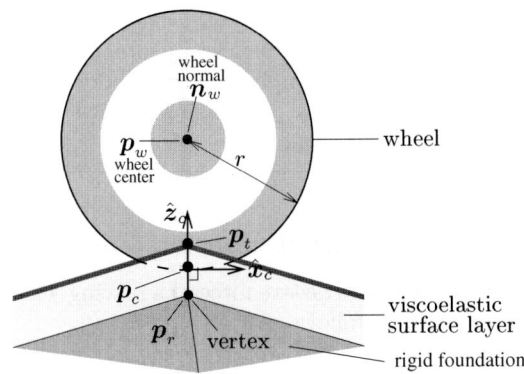

Figure 5: A wheel traverses a vertex.

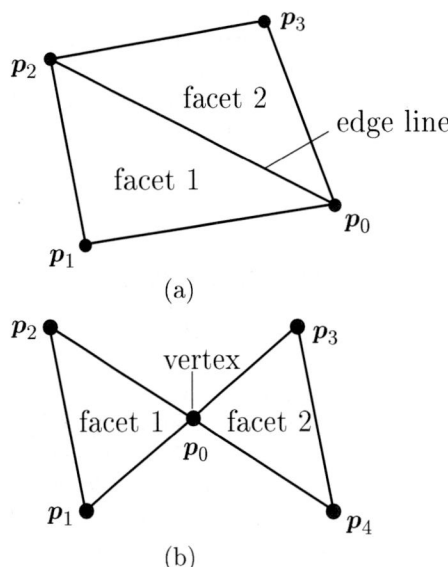

except that p_r in Eq. (13) is a known vertex. After p_c is computed, if the distance between p_c and p_r is larger than the the uniform thickness d_e of the viscoelastic surface layer, then there is no contact. Otherwise, the wheel contacts the terrain, and the penetration depth d_p of the wheel into the terrain is computed using Eq. (7). Finally, p_t is initialized as the projection of p_c along n_t ($\hat{z}_c$) onto the surface of the viscoelastic layer, which can be computed with Eq. (8).

3.4 Contact Detection Algorithm

The contact detection algorithm can be described in the following steps:

Step 1 *Create set L of possible contacting terrain facets*: From the forward dynamics of the vehicle system [7], the position of the wheel center p_w is given. We then create a circular area A on the x-y plane of $\{E\}$ with center at p_w^{xy} and rad where p_w^{xy} is the projection of p_w on the x-y of $\{E\}$. Set L consists of facets whose proje on the x-y plane of $\{E\}$ intersects A. For facets outside set L, it is impossible to co the wheel, and these facets can be ignored t computation time.

Step 2 *Do plane-contact computation as in tion 3.1 for each facet in L*: The contact status variable C_{status} for each facet is set as follows: $C_{status} = 1$ when plane contact occurs; $C_{status} = 2$ when the wheel penetrates a facet plane, but p_r is outside the edges of the associated rigid foundation facet; and $C_{status} = 3$ when the wheel is off the terrain ($d_p > d_e$). If plane contact occurs ($C_{status} = 1$), the corresponding contact data, p_c, $\{c\}$, p_r, and p_t, are recorded.

Step 3 *Do edge-contact computation as in Section 3.2 for each pair of adjacent facets in L if they are convex, and both of their values of*

Figure 6: (a) A pair of adjacent facets. (b) A pair of facets that intersect at a vertex.

C_{status} *are 1*: For a pair of adjacent facets, the edge can be flat, concave, or convex. In Figure 6(a), points (p_0,p_1,p_2) compose facet 1, while points (p_0,p_2,p_3) compose facet 2. If we assume that the normal vector to each facet is upward, then the normal vector n_1 to facet 1 is

$$n_1 = \frac{(p_2 - p_0) \times (p_1 - p_0)}{\|(p_2 - p_0) \times (p_1 - p_0)\|}, \quad (15)$$

and the equation for facet 1 is

$$n_1 \cdot (p - p_0) = 0. \quad (16)$$

ationship of the adjacent facets can be de :d as follows: They are flat if p_3 is on ne of facet 1, i.e. $n_1 \cdot (p_3 - p_0) = 0$; e concave if the direction from the plane 1 to p_3 is the same as that of n_1, i.e. $- p_0) > 0$; otherwise, they are convex, i.e. $n_1 \cdot (p_3 - p_0) < 0$. Edge contact can happen only on the edge between convex adjacent facets. When edge contact occurs, both of their values of C_{status} are updated to 4, and the corresponding contact data, p_c, $\{c\}$, p_r, and p_t, are recorded.

Step 4 *Do vertex-contact computation as in Section 3.3 for each pair of adjacent facets in L, which intersect at a vertex if they are convex, and both of their values of C_{status} are 1*: Vertex contact may occur only when two facets intersecting at a vertex are convex. Figure 6(b) shows a pair of facets that intersect at a vertex. The normal vector and the plane equation of facet 1 can be computed using Eqs. (15) and (16), respectively. If both of the directions from facet 1 to p_3 and p_4 are opposite to that of n_1, facets 1 and 2 are convex. When vertex contact occurs, both of the values of C_{status} are updated to 5, and the associated contact data, p_c, $\{c\}$, p_r, and p_t, are recorded.

From the variable C_{status}, the contact status is known.

4 Contact Force Model

Broadly, two approaches can be used to estimate the contact forces: the hard contact approach that estimates forces by assuming that the contacting bodies are infinitely rigid and do not penetrate each other, and the soft contact approach that computes forces by modeling the localized deformation in the vicinity of the contact. The hard contact approach usually has problems with the existence and uniqueness of the solutions when friction is present [10]. In addition, compliance normally exists for an off-road vehicle-terrain system. Thus, the soft contact approach is taken in this work. It is worth noting that with a soft contact model, kinematics loops are broken, and the dynamics computation is simplified.

In [4], the author models tractive, lateral, normal, and motion resistance forces for driving wheels on soil, based on vehicle-soil mechanics. In this work, we will extend the development in [4] to also simulate impact forces when wheels first contact the terrain, and retain static forces when the vehicle stands still on a slopped terrain.

4.1 Normal Contact Force

The normal contact force based on soil mechanics has been empirically estimated as follows [11]:

$$N = \frac{3-n}{3}(K_c + bK_\phi)\sqrt{D}\, z_l^{\frac{2n+1}{2}}, \quad (17)$$

where N is the normal force, (N); n the dimensionless exponent of soil deformation, z_l the depth of sinkage solely due to the normal load, (m); D the diameter of the wheel, (m); b the width of the wheel's ground contact area, (m); K_c the cohesive modulus of terrain deformation, (N/m^{n+1}); and K_ϕ the frictional modulus of terrain deformation, (N/m^{n+2}).

To model the impact force, a nonlinear damping compliant model [12] is involved, and the proposed normal contact force f_c^z is computed as follows:

$$f_c^z = \sigma(N - \lambda_n s_z v_c^z), \quad (18)$$

where N is computed by Eq. (17); σ is equal to 1 if the wheel contacts the terrain; otherwise it is equal to 0; λ_n is the damping constant along the contact normal direction; s_z is the penetrating displacement along the contact normal; and v_c^z is the penetrating velocity along the contact normal.

4.2 Tangential Contact Force

The tangential tractive force of a driving wheel can be computed as follows [13]:

$$H = (c\,b\,l + f_c^z \tan(\phi))\left[1 - \frac{K}{s\,l}\left(1 - \exp\frac{-s\,l}{K}\right)\right], \quad (19)$$

where H is the tangential tractive force, (N); f_c^z the normal force, (N); b the width of the wheel's ground contact area, (m); l the length of the wheel's ground contact area, (m); c the cohesion of the soil, (N/m^2); ϕ the soil friction angle, (deg.); K the soil deformation modulus, (m); and s the slip.

When a wheel stops on an inclined terrain, its slip is zero. In this case, the tractive force according to Eq. (19) is zero. This does not simulate the tractive force correctly. To add static-force effects, we involve spring and damper terms in the equation [9]:

$$f_c^t = H - k_x s^x - B_x v_c^x, \quad (20)$$

where k_x is the spring coefficient along the tangential direction, (N/m); s^x the contact displacement along the tangential direction, (m); B_x the damping coefficient along the tangential direction, (N-s/m); and v_c^x the contact velocity along the tangential direction, (m/s). Finally, the tangential force is the tractive force minus the motion resistance:

$$f_c^x = f_c^t - f_r \quad (21)$$

4.3 Lateral Contact Force

A similar formulation for the lateral contact force of a driving wheel, based on empirical results, has been developed as follows [13]:

$$L = (c\,b\,l + f_c^z \tan(\phi))(1 - \exp^{-B\alpha}), \quad (22)$$

where L is the lateral cornering force, (N); B the constant coefficient dependent on the wheel parameters and the soil conditions, (1/deg.); α the slip angle, (deg.); and other definitions of parameters were given previously.

To add the static-force effects, we involve spring and damper terms in the equation:

$$f_c^y = L - k_y s^y - B_y v_c^y \quad (23)$$

where k_y is the spring coefficient along the lateral direction, (N/m); s^y the contact displacement along the lateral direction, (m); B_y the damper coefficient along the lateral direction, (N-s/m); and v_c^y the contact velocity along the lateral direction, (m/s). A variety of soil parameters can be found in [14].

5 Simulation Results

A graphical dynamic simulator used to test our algorithm is shown in Figure 1. The simulator broadly consists of four parts: the forward dynamics of the vehicle, which simulates the vehicle system; the contact model that estimates the contact forces; the set of triangular facets that represents the terrain; and the graphic viewer that shows the animated motion of the vehicle system contacting the uneven faceted terrain.

The forward dynamics of the vehicle system is implemented using the simulation package, *DynaMechs* [7], a general-purpose dynamic simulation tool, which is able to simulate general tree-structured vehicles, such as the WAAV. The three dimensional (3D) graphical viewer that displays the animated motion of the simulation is implemented with a graphical package, called *GLAnimate*, based on *XAnimate* which was developed in [15].

To demonstrate the good performance of the proposed contact model, the results of the case where the WAAV crosses concave facets are shown as follows. The WAAV is crossing concave facets at a constant speed along the x direction, shown in Figure 7 (a). The trace of the resultant contact points $\boldsymbol{p}_r$ and $\boldsymbol{p}_c$ of the right front wheel is shown in Figure 7(b). In this case, the edge line is located at $x = 8.0$ m. As we can see in the figure, the wheel contacts only one facet before $time = 2.84$ sec. At $time = 2.84$ sec, contact with the second facet occurs. During the period of $time = 2.84$ sec to $time = 2.90$ sec, the wheel contacts both facets, simultaneously. At $time = 2.90$ sec, contact with the first facet is released, and the wheel contacts only the second facet afterward. Note in all cases that the movement of the contact point is continuous.

Next, we test the WAAV at different constant speeds along the x direction: 1 m/s, 0.7 m/s, 0.4 m/s, and 0.1 m/s, respectively. The vertical contact forces of the right front wheel are shown in Figure 7(c). As we can see, the vertical contact force is constant before the second contact occurs. During the period of dual contact, the wheel impacts the second facet such that the vertical force increases. When the wheel releases the first contact and moves steadily on the second facet, the vertical contact force decreases. This is because the front module goes up a hill, and the cen-

(a)

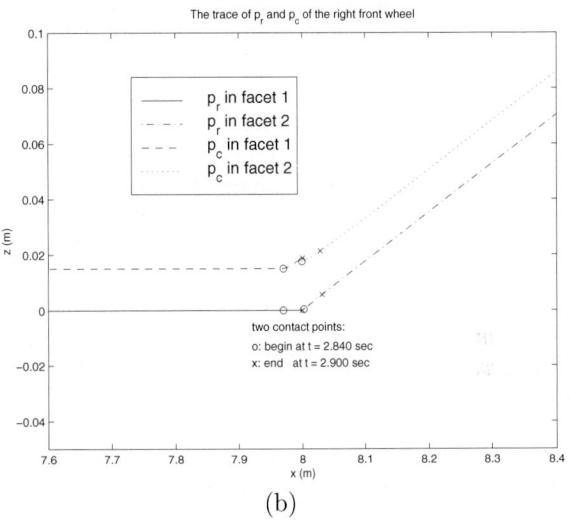

(b)

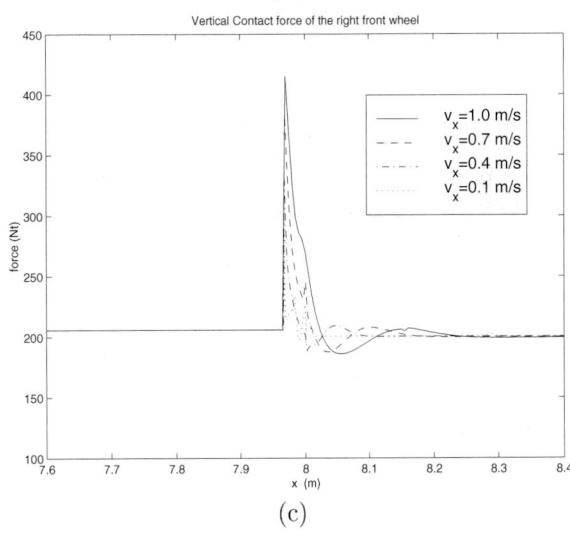

(c)

Figure 7: Results of WAAV crossing concave facets. (a) WAAV is crossing concave facets. (b) Trace of $\boldsymbol{p}_c$ and $\boldsymbol{p}_r$ of the right front wheel. (c) Vertical force of the right front wheel with different speeds.

ter of gravity of the whole vehicle system shifts from the original position to the end of the rear module. It is also noted that the higher the forward speed of the vehicle, the larger the amount of increase in vertical contact force. This is because the wheel has a greater impact with the terrain with a higher forward speed.

6 Summary and Conclusions

In this paper, a dynamic simulator is developed that can simulate actively-coordinated wheeled vehicle systems on uneven faceted terrain. Based on the considerations of model fidelity and computational efficiency, a simple geometric model for wheel-terrain contact is proposed. Also, a computationally-efficient algorithm for contact detection is developed. The introduction of the viscoelastic surface layer for the terrain model, combined with the assumption that a contact occurs at the deepest penetration point, enables the contact point to be uniquely determined when the wheel traverses a convex edge or crosses a vertex. Furthermore, the contact point is guaranteed to be continuous during the traversal of a convex edge.

A contact force model is also developed based on soil mechanics. Three forces at the contact interface are modeled: the normal force which supports the normal load, the tractive force which is generated from the forward friction force, and the lateral (cornering) force which exists when the vehicle makes a turn. In addition, impact forces are modeled.

Simulation results of a case where the WAAV is traversing a concave edge are used to demonstrate the good performance of the contact model. It is believed that the proposed contact model and dynamic simulator can contribute to the development of a variety of actively-coordinated wheeled vehicles in the future.

7 Acknowledgments

This work was supported by Grant No. IIS-9907121 from the National Science Foundation to The Ohio State University and by a fellowship from The Chung Cheng Institute of Technology, Taiwan, R.O.C.

8 References

[1] S. V. Sreenivasan, *Actively Coordinated Wheeled Vehicle Systems*. PhD thesis, The Ohio-State University, Columbus, Ohio, 1994.

[2] L. S. McTamaney *et al.*, "Final Report for Mars Rover/Sample Return (MRSR) Studies of Rover Mobility and Surface Rendezvous," tech. rep., FMC Corp., Corporate Technology Center, December 1989.

[3] A. J. Spiessbach and W. R. Woodis, "Final Report for Mars Rover/Sample Return (MRSR) Rover Mobility and Surface Rendezvous Studies," tech. rep., Martin Marietta Space Systems, Denver, March 1989.

[4] S. C. Venkataraman, *Active Coordination of Wheeled Vehicle Systems for Improved Dynamic Performance*. PhD thesis, Ohio State Univ., Columbus, OH, 1997.

[5] M. H. Hung, D. E. Orin, and K. J. Waldron, "Efficient Formulation of the Force Distribution Equations for General Tree-Structured Robotic Mechanisms with a Mobile Base," *IEEE Transactions on Systems, Man, and Cybernetics," Part B: Cybernetics*, vol. 30, pp. 529–538, August 2000.

[6] P. G. Xavier, "Fast Swept-Volume Distance for Robust Collision Detection," in *Proc. of the IEEE International Conf. on Robotics and Automation*, vol. 2, (Albuquerque, NM), pp. 1162–1169, April 1997.

[7] S. McMillan, D. E. Orin, and R. B. McGhee, "A Computational Framework for Simulation of Underwater Robotic Vehicle Systems," *Journal of Autonomous Robots*, Kluwer Academic Publishers, Boston, pp. 253–268, 1996.

[8] C. J. Hubert, "Coordination of Variably Configured Vehicles," Master's thesis, The Ohio-State University, Columbus, Ohio, 1998.

[9] C. A. Tenaglia, D. E. Orin, R. A. LaFarge, and C. Lewis, "Toward Development of a Generalized Contact Algorithm for Polyhedral Objects," in *Proc. of the IEEE International Conf. on Robotics and Automation*, (Detroit, Michigan), May 1999.

[10] P. E. Dupont, "The Effect of Friction on the Forward Dynamics Problem," *International Journal of Robotics Research*, vol. 12, pp. 164–179, April 1993.

[11] M. G. Bekker, *Introduction to Terrain-Vehicle Systems*. Univ. of Michigan Press, Ann Arbor, 1969.

[12] D. W. Marhefka and D. E. Orin, "A Compliant Contact Model with Nonlinear Damping for Simulation of Robotic Systems," *IEEE Transactions on Systems, Man, and Cybernetics, Part A: Systems and Humans*, vol. 29, pp. 566–572, November 1999.

[13] D. A. Crolla and A. S. A. Et-Razaz, "A Review of the Combined Lateral and Longitudinal Force Generation of Tyres on Deformable Surfaces," *Journal of Terramechanics*, vol. 24, no. 3, pp. 199–225, 1987.

[14] J. Y. Wong, *Theory of Ground Vehicles*. John Wiley & Sons, Inc., 1978.

[15] D. W. Marhefka and D. E. Orin, "XAnimate: An Educational Tool for Robot Graphical Simulation," *IEEE Robotics & Automation Magazine*, vol. 3, pp. 6–14, June 1996.

A Methodology for Software Development Cost Analysis in Information-Based Manufacturing

High-Way Sun, MengChu Zhou*, and Carl Wolf

Department of Industrial and Manufacturing Engineering
*Department of Electrical and Computer Engineering
New Jersey Institute of Technology
University Heights, Newark, NJ 07102-1982, USA
shw@puma.att.com, zhou@njit.edu, carl.wolf@njit.edu

Abstract-*The long lead-time response is always a problem for supply chain distribution systems, especially in the perishable food industry and rapidly advancing computer industry. The Internet common base environment can support suppliers and customers with on-line information and provide quick response time. This paper is based on a project between the New Jersey Institute of Technology (NJIT) and the Tropicana Juice Distribution Center in Jersey City, New Jersey. One goal of the project is to use information technology to automate the present system and thus shorten the response time from seven to three days. The existing structure uses Electronic Distribution Interface, fax, telephone or email to place orders. The project proposes an integrated Internet working environment for all the members of the supply chain. This paper focuses on the cost analysis of its related software that consists of in-house development and outsourcing. Activity Based Cost (ABC) and lifecycle engineering methods are used to catalog all the cost expenses to determine how much effort the software development takes.*

1. INTRODUCTION

Information-based manufacturing is critical for manufacturers to compete for today's global market (Huang et al., 1999; Seybold and Marshak, 1999; Viswanadham, 2000). The development cost analysis becomes an important issue and of industrial relevance for a software project involving both in-house development and outsourcing expenses. This work is motivated by the need to implement information-based manufacturing for Tropicana Jersey City Distribution Center. The project is based on the architecture of the Jersey City Distribution Center. The center is responsible for supplying Tropicana juices in all states of the Northeast USA, as well as in Canada. Fresh orange juice is the major portion of the business, and has a shelf life of approximately 65 days. At present, five juice trains are delivered weekly from Florida to the distribution center, and trucks from the owners' served region arrive at the facility to pick up orders. This facility consists of an Automated Storage and Retrieving System, which is fully integrated into an Automated Warehousing System. In order to shorten the customers' order process time from seven to three days, we propose to use the Internet to streamline the distribution process. According to Seybold et al. (1999),

e-business is defined as the use of the Web, e-mail and other devices to streamline the business process, improve customer service time and reduce cycle time, etc. To succeed in implementing an e-business strategy, the proposed team includes about 97 skilled software developers: 57 regular employees in-house and about 40 contractors from an outsourcing software consultant house. The focus of this paper is to report a cost analysis method and the results obtained for such a large software package development by using such a method.

Tropicana's current practice faces three problems. First, its Internet software is not connected to all of the supply chain. Consider orders from individual retailers. Currently, Tropicana managers control about 15%-20% of inventory orders from retailers through Tropicana's Continuous Replenishment Program (CRP) system. Meanwhile, individual stores randomly place the other 80%-85% of orders, which are not under Tropicana's control. The Tropicana customer service department administers orders from those individual customers. An Internet interface between customer service and customers, from the supply chain perspective, can benefit to both the customers and the warehouse. The advantage to the warehouse is that it is able to centralize demand information for individual stores as they decide to ship more juice to Jersey City from Florida. The retailers benefit from in-time delivery and less stock out cost. This helps reduce random variation and hence uncertainties in demand on the warehouse.

The second problem is the central ordering of juices that are shipped to the distribution center from Florida. Currently there are five juice trains scheduled to arrive weekly from Florida. The cargoes combine different classes of juice. The company never ships partially filled trains from Florida. The Jersey City distribution center sometimes builds up inventory of certain classes of juices that are close to their expiration date, and the company has to get rid of them either at very low price with sales promotion or donate them as charity. A carefully designed and sophisticated coordination in ordering policies will reduce the chances for these problems and result in savings. One approach would be to create an incentive for the customers to synchronize their order function with Tropicana. This is the so-called supplier-retailer coordination problem. A carefully designed coordinated system will benefit each and every player in the supply chain network. Shin, Collier and Wilson (2000) discusses the four factors to judge supplier quality: cost, quality, delivery and flexibility. This may require the design of contracts or cost sharing agreements with customers.

The third problem is how to combine marketing strategies with inventory levels and other factors. Marketing strategies such as sales incentives can influence demand. Foreseeing an inventory buildup problem, the company can use price as a tool to increase demand or reduce demand when insufficient inventory is on hand. Chen, F. et al. (2000) indicate a company can use promotional initiatives to catch the emerging orders and gain larger profit.

An Internet order system can help eliminate the above problems. The objective of this paper is to evaluate the Internet software development cost and uses the Tropicana Jersey City Distribution Center as a case study example. The proposed method is applicable to other industries. Section 2 will introduce the existing structure and explain the re-engineering model and cost analysis methodology. Section 3 will discuss the ABC cost analysis of the re-engineering process. Section 4 concludes the paper.

2. LIFE CYCLE AND ACTIVITY BASED COST (ABC) FOR PROJECT EVALUATION

2.1) Existing order processes and Internet-based structure

The existing Tropicana data flow is shown in Fig. 1, 80%-85% of the customer orders come from the directly orders through the customers; 15%-20% generated from the Continuous Replenishment Program (CRP) that provides the information of customer inventory and demand through the Electronic Distribution Interface. Then the Customer Service Department uses Bradentons' Order and Inventory Management System to maintain the orders and assign the carriers. The Traffic department picks up orders from the remote printer and manually assigns a pickup date, time, and carrier to each order ticket. Finally the Dispatch controls the logistics and operation of inbound and outbound trucks.

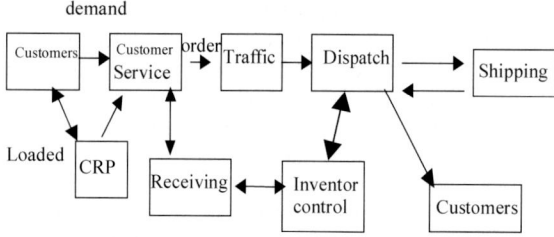

Fig 1 The existing Tropicana data

To shorten the processes time and provide a common work environment for all the parties in the supply chain, it is suggested to create an Internet E-Business environment as shown in Fig 2. Internet can link all the existing software and database. These parties including the customers, Customer Service Department, Warehouse, Traffic Department, Manufacture Plants and top mangers can view the ordering processes through a clear user interface. It thus saves the communication time and prevents mistakes. Hoque and Lohse (1999) considered the search cost of the interfaces for electronic commerce and indicated the importance to understand the human information process from electronic media. Hence, this is the key to develop an on-line retail stores.

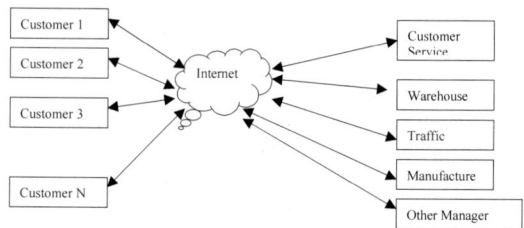

Fig 2 Picture of the Internet Architature

2.2) ABC methodology

Before a company's move to the information-based manufacturing environment, it is essential to carefully analyze the related cost, especially the software development cost. Recently, Sueyoshi (1999) and other researchers used the Data Envelopment Analysis method to measure the decision entities, and then built the linear programming model to make the optimal cost decision. Similarly, Dahlstrom and Nygaard (1999) used the mailing surveys to get the data of the bargaining costs, monitoring costs, and the maladaption costs which are caused by the interorganizational transaction. This paper proposes a cost analysis methodology based on the Cost Management method (Hansen and Mowen, 1997), which is basically Activity-Based Cost (ABC) methodology. The first stage of ABC is to identify activities and then the get the cost information from the associated individual activities. The last step is to separate the cost into homogeneous sets. To streamline the process, activities are grouped together in homogeneous sets based on similar characteristics:
(1) They are logically related, and
(2) They have the same consumption ratios for all products.

The collection of overhead costs associated with each set of activities is called a homogeneous cost pool. Once the pool is defined, the cost per unit of the activity driver's practical capacity is determined. This is called the pool rate. Computation of the pool rate completes the first stage. Thus the first stage produces four outcomes:
(1) activities are identified,

(2) costs are assigned to activities,
(3) related activities are summed to defined homogeneous cost pools, and
(4) pool (overhead) rates are computed.

Thus, the overhead assigned from each cost pool to each project is computed as follows:
Applied overhead (to a project) = Pool rate x Activity usage

2.3) The Consistent Software Life Cycle of the Project

The below life stages characterize a software development project:
- **Define** – The Business to Business Internet platform and the internal requirement from the customer service, continuous replenishment program, the logistical department and the dispatch center. The outside environment should consider the retailers' order interface, the shipping contractors and so on.
- **Design** – The software/hardware architecture and the data flow transmission design.
- **Construct** – The coding processes by the programmers.
- **Test** – Software/hardware system test from different angles and using different cases.
- **Install** – Install inside Tropicana and all members in its supply chain.
- **Maintain** – Debug the errors and keep the system running.

Each version consists of 6 steps in its life cycle as shown in Fig 3.

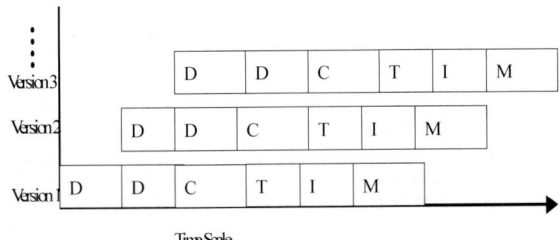

Fig 3 The Software Development Life Cycle

The combination of ABC and life-cycle engineering concept allows one to perform accurate cost analysis for a software project.

3. TROPICANA CASE STUDY

3.1) The Reengineering Cost Analysis from ABC Method and Life Cycle Point of View

In our method, we need to first separate different activities into different sub-categories. They include supervision, system engineering, software development and overhead staffs. Three cost-driving factors involved to determine activities are:
- Staffs
- Software
- Hardware

3.2) Staff cost analysis *
The Staff Factor is composed of:
- The developers' level of education.
- The developers' experience with application.
- The developers' knowledge of languages.
- The developers' familiarity with the infrastructure environment.

Table 1 The Estimated Annual Personnel Cost in Project

Personnel Resource	Number	Salary	Amount
TROPICANA employee			
Supervisor			
Senior	3	$120,000	$360,000
Junior	3	$100,000	$300,000
System Engineer			
Senior	5	$100,000	$500,000
Junior	5	$80,000	$400,000
System Analysis			
DB analysis	3	$75,000	$225,000
Tool analysis	3	$60,000	$180,000
Software developer			
15 years and higher	5	$90,000	$450,000
10 years and higher	10	$80,000	$800,000
5 years and higher	10	$65,000	$650,000
Network and System Support			
Database administrator	2	$120,000	$240,000
WEB administrator	2	$100,000	$200,000
Technical Document Writer			
20 years experience	3	$60,000	$180,000
Customer Support Experience	3	$100,000	$300,000
Subtotal	57		$4,785,000
Benefits (insurance, 401K and Stock option)			x 1.4
Tropicana TOTAL			**$6,699,000**
Contractors from software consultant companies			
Software Developers			
Senior	10	$120,000	$1,200,000
Junior	10	$80,000	$800,000
Testers			
Senior	10	$80,000	$800,000
Junior	10	$50,000	$500,000
Subtotal	40		$3,300,000
Total	97		**$9,999,000**

From Table 1, the personnel cost is:
Tropicana $6,699,000/9,999,000 = 67.0%
Software consultant $3,300,000/9,999,000 = 33.0%
The average personnel cost
 Tropicana $6,699,000/57 = $117,526
 Software consultant $3,300,000/40 = $82,500
According to their characteristics, the activities can be classified into six pools as shown in Table 2.

---
* The data come from the current job market survey. We modified some of the data to protect the source.

Table 2 The cost analysis via ABC method

Personnel Resource	Number	Salary	Amount
Pool 1 : Supervisor			
Senior	3	$120,000	$360,000
Junior	3	$100,000	$300,000
Subtotal	6		*$660,000*
Benefit Factors (x 1.4)			$924,000
Average			*$154,000*
Pool 2 : System Engineers			
System Engineer			
Senior	5	$100,000	$500,000
Junior	5	$ 80,000	$400,000
System Analysis			
DB analysis	3	$75,000	$225,000
Tool analysis	3	$60,000	$180,000
Subtotal	16		*$1,305,000*
Benefit Factors (x 1.4)			$1,827,000
Average			*$114,187*
Pool 3 : Software Developer			
Software Developer			
15 years and higher	5	$90,000	$450,000
10 years and higher	10	$80,000	$800,000
5 years and higher	10	$65,000	$650,000
Network and System Support			
Database administrator	2	$120,000	$240,000
WEB administrator	2	$100,000	$200,000
Subtotal	29		*$2,340,000*
Benefit Factors (x1.4)			$3,276,000
Average			*$112,965*
Pool 4 : Supporting Overhead			
Technical document writer			
20 years experience	3	$60,000	$180,000
Customer Support			
Experience	3	$100,000	$300,000
Subtotal	6		*$480,000*
Benefit Factors (x 1.4)			$672,000
Average			*$112,000*
Pool 5 : Contractors from software consultant companies			
Software Developers			
Senior	10	$120,000	$1,200,000
Junior	10	$80, 000	$800,000
Subtotal	20		*$2,000,000*
Average			*$100,000*
Pool 6 : Testers			
Senior	10	$80,000	$800,000
Junior	10	$50,000	$500,000
Subtotal	20		*$1,300,000*
Average			*$65,000*

The six pools of the activities are summarized in Tables 3.

Table 3 From the ABC method to analyze the cost

Pool 1	Supervisor	Average	$154,000
Pool 2	System Engineer	Average	$114,187
Pool 3	Software Developer	Average	$112,965
Pool 4	Supporting Overhead	Average	$112,000
Contractors from software consultant companies			
Pool 5	Software Developers	Average	$100,000
Pool 6	Testers	Average	$65,000

From Tables 3, if we consider the similar computer skill level staffs the supporting people from Tropicana and the testers from the consulting company can replace each other. The outsourcing staffs bear much lower cost. For example, the average supporting staff cost from Tropicana is $112,000. Compared with the testers from a consultant house $65,000, if 6 people are replaced, the cost saving benefit is 6 x ($112,000-$65,000) = $282,000. We can also conduct a benefit comparison for software programmers from the in-house and outsourcing developers. If 29 Tropicana's developers are replaced with the outsourcing ones, the project can save 29 x ($112,695-100,000) = $368,155. Therefore, managers needs to decide whether the outsourcing portion should be increased in order to reduce the total project cost.

3.3) The Software/Hardware Cost

It is composed of:
- Compatibility between development computers and target computers.
- Computing conditions during the development.
- Availability of the development computers.
- Size of the development database.
- Response time of the development computers.
- Equipment available in the networked place.

The cost is summarized in Table 4.

Table 4 The Annual Hardware, Software Cost and Office Space Analysis

Personnel Resource	Number	Cost	Amount
Depreciation equipment			
HP workstation	100	$6,500	$650,000
PC Pentium III 800MHz	100	$2,900	$290,000
Notebook Toshiba 600 MHz	50	$3,000	$150,000
MS Office	100	$950	$95,000
Subtotal			*$1,185,000*
Suppose 2 years depreciation x 0.5			$592,500
Annual expense fee			
Query Language Oracle	100	$100	$10,000
Office Space	100	$7,200	$720,000
Subtotal			*$730,000*
Total equipment Annual fee			**$1,322,500**

3.4) Total Cost Summary

Table 5 presents the total staffs and equipment costs.

Table 5 Total Cost Analysis

Personnel Resource	Number	Pool Avg	Amount
Supervisor	6	$154,000	$924,000
System Engineers	16	$114,187	$1,827,000
Software Developer	29	$112,965	$3,276,000
Supporting Overhead	6	$112,000	$672,000
Subtotal			$6,699,000
Contractors from software consultant companies			
Software Developers	20	$100,000	$2,000,000
Testers	20	$65,000	$1,300,000
Subtotal			$3,300,000
Equipment and Space			$1,322,500
Total Estimated Annual Cost for Tropicana Project			**$11,321,500**

According to Table 5, we have the in-house development cost (57 staffs):
$6,699,000 + (57/97) x $1,322,500 = $7,476,139
The outsourcing cost (40 staffs):
$3,300,000 + (40/97) x $1,322,500 = $3,845,361

Consequently, we derive the following conclusions:

1. The in-house expense takes 66% and the outsourcing 34% of the total project cost.
2. The average supporting staff from Tropicana costs $112,000 while the same computer skill level tester from a consultant house costs $65,000. Hence, outsourcing this test portion can reduce the cost significantly.
3. Replacing the 29 in-house software developers with consultants can also cut the cost significantly.
4. The personnel expense of $9,999,000 takes 88% of the total project cost; and
5. The equipment and office space costs $1,322,500 or 12% of the project.

4. CONCLUSIONS

This paper presents a systematic approach to analyzing software development cost encountered in conversion of traditional manufacturing into information-based manufacturing. It is based on the concept of the ABC method and lifecycle engineering to obtain whole pictures of the expenses. Internet software systems are expensive to develop. To protect the software staffs of the company, fully utilize their deep understanding of the company's operations, and reduce the development cost, this work suggests to join in-house and outsourcing development forces. Proper budget control is an important key in order to obtain the support from high level managers and stockowners. The project managers should tighten control of overhead expenses. Enlarging the outsourcing portion of a project can often lower the total cost. To invite the customers to share the expense can also help. Bock and Geoffrey (1999) also emphasized customer relations as the most important aspect of E-Business.

Obviously Internet technology updates very quickly. Today's super computer may be obsolete shortly. The life cycle of the hardware and software are shortened rapidly. To move the e-business we need continue to invest in it. Managerial staffs must know the cost and use the budget wisely. Therefore, the presented methodology can be used to help them make the best decisions by offering detailed and accurate costs analysis results for a complex software project.

5. REFERENCES

1. Hansen, D. R. and Mowen M. M. (1997), *Cost Management.* South-Western College Publishing, pp. 300-351.
2. Hoque, A. Y. and Lohse, G. L. (1999), "An Information Search Cost Perspective for Designing Interface For Electronic Commerce," *Journal of Marketing Research*, Vol. XXXVI, pp. 387-394.
3. Dahlstrom, R. and Nygaard (1999), "An Empirical Investigation of Ex Post Transaction Costs in Franchised Distribution Channel," *Journal of Marketing Research,* Vol. XXXVI, pp. 160-170.
4. Huang, E., F.-T. Cheng, and H.-C. Yang (1999), "Development of a Collaborative and Event-Driven Supply Chain Information System Using Mobile Object Technology," in *Proc. 1999 IEEE Int. Conf. on Robotics and Automation,* Detroit, Michigan, U.S.A., pp. 1776-1781.
5. Zornes, J. A., Wilderman, B., Gilmore, D. and Lehman C. (1999), "Application Delivery Strategies; Electronic Business Strategies," *Joint Trend Teleconference –Supply Chain Dynamics for the 21st Century Combining Operations, Planning and Externalization*, pp. 1-27.
6. Sueyoshi, T. (1999), "DEA Duality on Returns to Scale (RTS) in Production and Cost Analyses: An Occurrence of Multiple Solutions and Differences Between Production-Based and Cost-Based RTS Estimations," *Management Science*, Vol.45 No.11, pp. 1593-1608.
7. Shin, H., Collier D. A. and Wilson, D. (2000), "Supply management orientation and supplier/buyer performance," *Journal of Operation Management*, 18, pp. 317-333.
8. Tsay A.(1999), "The Quantity Flexibility Contract and Supplier-Customer Incentives," *Management Science*, Vol. 45, No.10, pp. 1339-1358.
9. Lee, H., So, K. Tang, C. (2000), "The Value of Information Sharing in a Two-Level Supply Chain," *Management Science*, Vol.46, No.5, pp. 626-643.
10. McCutcheon, D. and Staurt, F. (2000), "Issues in the choices of supplier alliance partners," *Journal of Operations Management*, 18, pp. 279-301.
11. Alfredesson, P. and Verrijdt, J. (1999), "Modeling Emergency Supply Flexibility in a Two-Echelon Inventory System," *Management Science,* Vol. 45, No.10, pp. 1416-1431.
12. Chen, F. et al. (2000), "Quantifying the Bullwhip Effect in a Simple Supply Chain: The Impact of Forecasting, Lead Times, and Information," *Management Science,* Vol. 46, No.3, pp. 436-443.
13. Cachon G. (1999), "Managing Supply Chain Demand Variability with Scheduled Ordering Polices," *Management Science*, Vol. 45, No. 6, pp. 843-856.
14. Bock, G. (1999), "Getting Started with an E-Business Strategy," *E-Business Strategies & Solution,* Jan. 1999, pp.10-12.
15. Seybold, P. B. and Marshak, D. S. (1999), "Particia Seybold Group's E-Commerce Service Evolves into E-Business Strategies & Solution," *E-Business Strategies & Solution,* Jan. 6, 1999, pp.2-4.
16. Viswanadham, N. (2000), "Supply chain engineering and automation," *Proc. 2000 IEEE Int. Conf. on Robotics and Automation*, San Francisco, CA, pp. 408-413.

OPTIMAL FIXTURE LAYOUT DESIGN IN A DISCRETE DOMAIN FOR 3D WORKPIECES

Yu Michael Wang and Diana Pelinescu
Department of Mechanical Engineering
University of Maryland
College Park, MD 20742 USA
E-mail: yuwang@eng.umd.edu

Abstract

This paper addresses two major issues in fixture layout design: (1) to determine the feasible fixture configurations that satisfy fundamental requirements such as kinematic localization and total fixturing (form-closure) and (2) to evaluate the acceptable fixture designs on several quality criteria and select the optimal fixture appropriate with practical demands. The performance objectives considered include the workpiece localization accuracy, and the norm and distribution of the locator contact forces. An efficient automated tool based on an interchange algorithm is developed for designing optimal fixture layout for arbitrary 3D parts. A thorough analysis is performed on the fixture characteristics during the single and multiple-criteria optimization process for different frequent cases, and on the inter-relationship between locators and clamps, leading to conclusions and strategies for performing fixture synthesis.

1. Introduction

Proper fixture design is crucial to product quality in terms of precision and accuracy in part fabrication and assembly. Fixturing systems, usually consisting of clamps and locators, must be capable to assure certain quality performances, besides of positioning and holding the workpiece throughout all the machining operations. Although there are a few design guidelines such as 3-2-1 rule, automated systems for designing fixtures based on CAD models have been slow to evolve.

This article describes a research approach to automated design of a class of fixtures for 3D workpieces. The parts considered to be fixtured present an arbitrary complex geometry, and the designed fixtures are limited to the minimum number of elements required, i.e. six locators and a clamp. Furthermore, the fixels are modeled as non-frictional point contacts and are restricted to be applied within a given collection of discrete candidate locations. In general, the set of fixture locations available is assumed to be a potentially very large collection; for example, the locations might be generated by discretizing the exterior surfaces of the workpiece. The goal of the fixture design is to determine first, from the proposed discrete domain, the feasible fixture configurations that satisfy the form-closure constraint. Secondly, the sets of acceptable fixture designs are evaluated on several criteria and optimal fixtures are selected. The performance measures considered in this work are the localization accuracy, and the norm and distribution of the locator contact forces. These objectives cover the most critical error sources encountered in a fixture design, the position errors and the unwanted stress in the part-fixture elements due to an overloaded or unbalanced force system.

The optimal fixture design approach is based on a concept of optimum experiment design. The automated developed algorithm enumerates efficiently the admissible designs exploiting the part geometry and performing a force analysis, and selects further the optimal fixture design that will assure a certain desired quality.

2. Related Work

Literature on general fixturing techniques is substantial, e.g., [1]. The essential requirement of fixturing is the century-old concept of *form closure* [2], which has been extensively studied in the field of robotics in recent years [3, 4]. There are several formal methods for analyzing performance of a given fixture based on the popular screw theory, dealing with issues such as kinematic closure [5], contact types and friction effects [6].

A different analysis approach based on the geometric perturbation technique was reported in [7]. An automatic modular fixture design procedure based on this method was developed in [8] to include geometric access constraints in addition to kinematic closure. The problem of designing modular fixtures gained more attention lately [9]. There has also been extensive research in fixture designs, focusing on workpiece and fixture structural rigidity [6], tool accessibility and path clearance [7].

The problem of fixture synthesis has been largely studied for the case of a *fixed* number of fixture elements (or fixels) [8, 10], particularly in the application to robotic manipulation and grasping for its obvious reasons [3, 4].

This article aims to be an extension of the results on the fixture design issues previously reported in [14].

3. Fixture Model

The fundamental performance of a fixture is characterized by the kinematic constraints imposed on the workpiece being held by the fixture. The kinematic conditions are well understood [3, 4, 5, 7, 12]. For a fixture of n locators ($i = 1, 2, \ldots, n$), the fixture can be represented by:

$$\delta y = G^T \delta q \qquad (1)$$

where $\delta y = \{\delta y_1 \delta y_2 ... \delta y_n\}^T$ and $\delta q = \{\delta r\ \delta\theta\}^T$ define small perturbations in the locator positions and the location of the workpiece respectively. The fixture design is defined by the *locator matrix* $G = [h_1 h_2 ... h_n]$ where $h_i^T = -\{n_i^T (r_i \times n_i)^T\}$ and n_i and r_i denote the surface normal and position at the ith contact point on the workpiece surface. The problem of fixture design requires the synthesis of a fixturing scheme to meet a given set of performance requirements.

4. Quality Performance Criterions for a Fixture

4.1. Accurate Localization

An essential aspect of fixture quality is to position with precision the workpiece into the fixturing system. In general the workpiece positional errors are due to the geometric variability of the part and the locators set-up errors. This paper will focus only on the workpiece positional errors due to the locator positioning errors.

As an extension of the fixture model equation (eq.1), the locator positioning errors δy can be related with the workpiece localization error δq as follows:

$$\|\delta y\|^2 = \delta q^T (GG^T) \delta q \quad (2)$$

Clearly, for given source errors the workpiece positional accuracy depends only on the locator locations being independent from the clamping system, the Fisher information matrix $M = GG^T$ characterizing completely the system errors. It has been shown [12] that a suitable criterion to achieve high localization accuracy is to maximize the determinant of the information matrix (D-optimality), i.e., $\max(\det M)$.

4.2. Minimal Locator Contact Forces

Another objective in planning a fixture layout might be to minimize all support forces at the locator contact regions throughout all the operations with complete kinematic restraint or force-closure. Locator contact forces in response to the clamping action are given as:

$$\alpha_c = -G^T (GG^T)^{-1} h_c \lambda_c = -G^T M^{-1} h_c \lambda_c \quad (3)$$

Normalizing these forces with respect to the clamping intensity we obtain:

$$p_c = \alpha_c / \lambda_c = -G^T M^{-1} h_c, \quad \text{where}$$
$$p_{ci} = -h_i^T M^{-1} h_c, (i = 1, 2, ..., n) \quad (4)$$

The force-closure condition requires these forces to be always positive for each locator i of a set of n locators:

$$p_{ci} > 0, \quad \text{for}\ \lambda_c > 0 \quad (5)$$

Computing the norm of the locator contact forces:

$$\|p_c\|^2 = \sum p_{ci}^2 = h_c^T M^{-1} h_c \quad (6)$$

leads to an appropriate design objective, i.e. $\min(\|p_c\|)$.

Note that this objective indicates both locator and clamp positions to be determinant in the optimization process.

4.3. Balanced Locator Contact Forces

Another significant issue in designing a fixture is that the total force acting on the workpiece have to be distributed as uniformly as possible among the locator contact regions. If $\bar{p}$ represents the mean reactive force in response to the clamp action, then we define the dispersion of the locator contact forces as:

$$d = \frac{1}{n}\sum_{i=1}^{n}(p_{ci} - \bar{p})^2 \quad \text{where}\ \bar{p} = \frac{1}{n}\sum_{i=1}^{n} p_{ci} \quad (7)$$

Therefore, minimizing the defined dispersion represents an objective for a balanced force-closed fixture: $\min(d)$.

5. Optimal Fixture Design with Interchange Algorithms

As mentioned earlier, by generating on the exterior surface of the workpiece to be fixtured a set of discrete locations defined as position and orientation, we create a potential collection for the fixture elements. For example, using the information contained in the part CAD model, a discrete vector collection (unitary, normal vectors) can be generated as uniformly as possible on those surfaces accessible to the fixture components (fig.1).

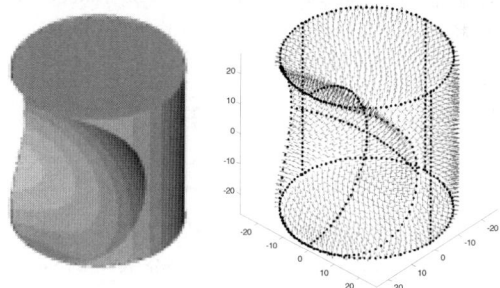

Figure 1: Part CAD model and global collection of candidate locations for the fixture elements.

The fixture design layout will select from this collection optimal candidates for locators and clamps with respect to the performance objectives and to the kinematic closure condition. Dealing with a large number of candidate locations the task of selecting an appropriate set of fixels is very complex.

As already introduced in [12, 14] an effective method for finding the desired fixture with regard to one of the previous quality objectives is the optimal pursuit method with an interchange algorithm. Due to its own limitations and to the fact that the objectives are functions with many extremes, the exchange procedure may not end up to a unique optimized fixture configuration, but to several improved designs depending on the initial layout. Therefore the solution offered by the *multiple interchange with random initialization algorithm* is overwhelming favorable, fact that recommends this procedure over the single interchange algorithms.

The algorithm can be described as a sequence of three phases:

Phase 1: Random generation of initial sets of locators.

The starting layout is generated by a random selection of distinct sets, each consisting from 6 locators out of the list of N candidate locations. If the clamp is pre-determined, a valid selection is obtained through a simultaneous check for all kinematic constraints. A big initial set of proposed locators is preferred, giving the opportunity of finding a convergent optimal solution. However from the efficiency point of view the designer has to balance the algorithm between the accuracy of the final solution and the computation time.

Phase 2: Improvement by interchange.

The interchange algorithm's goal is to pursue for an improvement of the initial sets of locators with respect to one of the objectives. Basically, this is done iteratively by exchanging one by one the proposed locators with candidate locations from the global collection. It is also essential to consider the form-closure restraint during the exchange procedure. The process will continue as long as an improvement of the objective function is registered.

Studying the effect of interchange on the proposed quality measures leads us to some efficient algebraic properties. For example, an interchange between a current locator j ($j = 1,2,...,6$) and a candidate location k ($k = 1,2, ... ,N-6$) yields changes in the optimized function such that:

$$\det M_{(j,k)} = p_{jk}^2 (\det M) \qquad (8)$$

$$\|p_c\|_{(j,k)}^2 = \|p_c\|^2 + \frac{p_{cj}}{p_{jk}^2}\left[p_{cj}(1+p_{kk})-2p_{ck}p_{jk}\right] = \|p_c\|^2 + \Delta p_c$$

where $p_{cj} = -h_j^T M^{-1} h_c$; $p_{ck} = -h_k^T M^{-1} h_c$;

$$p_{jk} = h_j^T M^{-1} h_k; \quad p_{kk} = h_k^T M^{-1} h_k; \qquad (9)$$

Thus, at each interchange the pair is selected such that the significant term that controls the function evolution is improving, e.g. $\max p_{jk}^2$ and $\min \Delta p_c$, easing the iterative process.

Phase 3: Selecting the optimal solution.

Applying the interchange algorithm for each initial set of locators we will end up with several distinct solutions on the configuration scheme of the fixture, the best fixture design corresponds evidently to the maximum improvement of the objective function. It should be emphasized that this algorithm can be used sequentially for different objective functions. Depending on the objective pursued the best solution can be evident (for a single objective) or might need the designer's final decision (for multiple objectives).

6. Fixture Locator Optimization

In many applications the clamp is already fixed given some practical considerations. Then with the clamp pre-defined, the *best fixture* with respect to a certain performance criterion is constructed by selecting a suitable set of locators such that a significant improvement of the objective-function is registered.

Using the *random interchange algorithm* we can analyze the impact of the optimization process on the fixture characteristics, as well as we can select the best optimized fixture solution for a specific criteria. In analyzing the effect of random interchange algorithm on several parts, there can be made the following statistical and empirical observations.

6.1. Single-objective Optimization

1. Accurate localization objective

Optimization for a precise localization $\max(\det M)$ objective would usually make significant improvement in the fixture quality. During the optimization process while the determinant is increasing, the general tendency of the norm or dispersion of the contact forces, with few exceptions, is a decreasing one (fig. 2). This indicates that benefic changes are recording for all objectives, seldom existing a conflict between them.

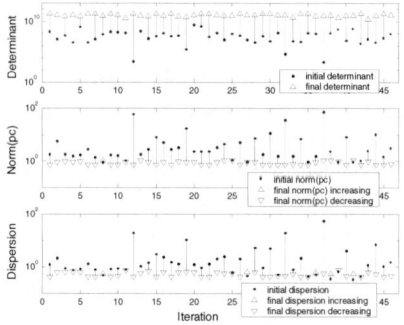

Figure 2: Random interchange for $\max(\det M)$ objective.

Furthermore, the best fixture solutions present locators spread away as far as possible from each other, gathering very close to the object boundaries (fig. 3). Starting with a large collection of initial sets of locators, the interchange design process converges to a few final solutions. The local optimal solutions usually cluster around the best solution (fig. 3).

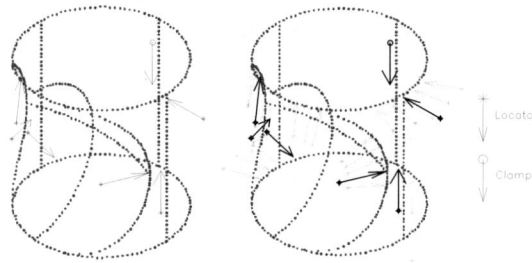

Figure 3: Best fixture solution; collection of optimized solutions cluster together around the best solution.

2. Minimal locator contact forces objective

Analyzing the impact of the optimization process for minimizing the locator contact norm forces $\min(\|p_c\|)$, some undesired changes are revealed. The determinant $\det(M)$ would often deteriorate significantly, affecting the localization quality of the part. Furthermore, we observed that one or more locators of the resulting fixture might have reactive forces close to zero, being almost inactive. To overcome this deficiency large acting forces must be imposed for the clamp, leading to excessive loads on workpiece and fixture. On the other hand, we noticed that the dispersion of the contact forces would generally decrease, being benefic for an equilibrated fixture design. This emphasis that there is no conflict between the locator norm and dispersion objectives (fig. 4).

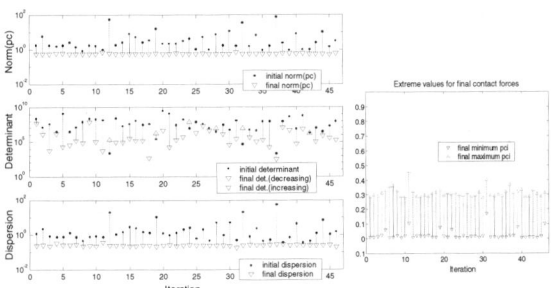

Figure 4: Random interchange for $\min(\|p_c\|)$ objective.

The best-optimized fixture configuration can efficiently be constructed using the random interchange algorithm. Applying this procedure on a large initial set of locators, we end up with a collection of optimized configurations of different aspects, presenting locators spread all over the part surface (fig. 5). However, there exist a strong preference in the fixels arrangement, one or more locators gathering on the opposite side of the part relative to the given clamp location.

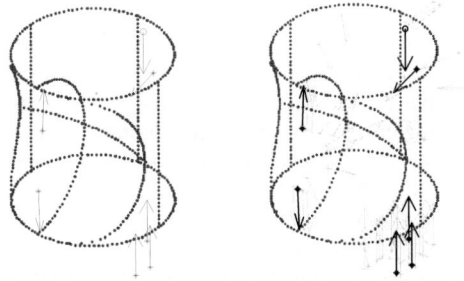

Figure 5: Best fixture solution; collection of optimized solutions obtained with random interchange.

3. Balanced locator contact forces objective

Trying to distribute more uniformly the reaction forces to the clamping action between the locators, i.e. $\min(d)$, we notice a better behavior of the fixture characteristics compared with the previous objective. Eventhough the evolution tendency of the determinant can not be predicted, the fluctuations are relatively small (fig. 6). Also, the minimum value of the forces is pulled up from zero, implying that all locators are active and as a consequence the clamp force may be relaxed a lot. Therefore, we recommend strongly as a second objective in fixture design the minimization of the force dispersion.

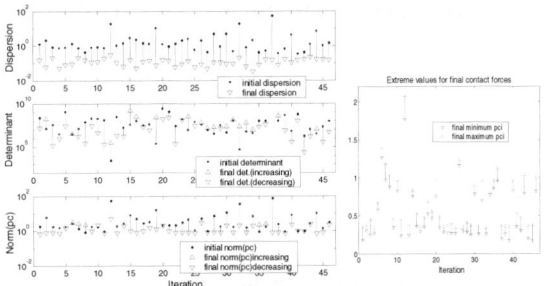

Figure 6: Random interchange for $\min(d)$ objective.

Fig. 7 illustrates the best-optimized solution for this specific objective and in addition presents also the collection of optimized solutions generated by the random interchange procedure. The locators are spread over the part surface tending to migrate towards surface interior.

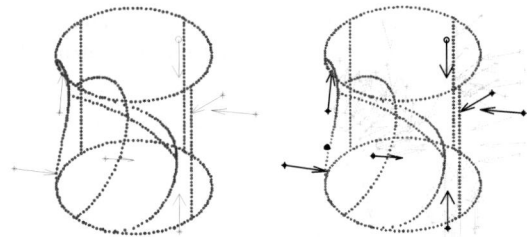

Figure 7: Best fixture solution; collection of optimized solutions obtained with random interchange.

6.2. Multiple-objective Optimization

1. Multi-objective trade-offs

In some applications both localization quality and a minimum force dispersion are important. In this case we may have to use a 2-step algorithm: first $\max(\det M)$ and secondly $\min(d)$. The proposed order is a consequence of the above observations. First, maximizing the determinant will automatically decrease the dispersion, fact benefic; proceeding next, a decreasing in dispersion leads in a decreasing in determinant value. Therefore, during the second phase of the algorithm *tradeoffs* between the two objectives occur, fact expressed also through the *Pareto-line* plot (fig. 8). Thus, the designer decision is determinant in finalizing the fixture scheme.

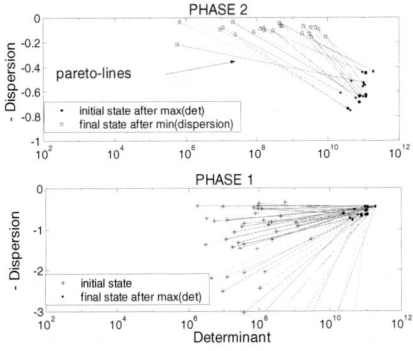

Figure 8: Behavior during a 2-step random interchange algorithm for a collection of locator sets.

2. Designer decision in finalizing the fixture

During the second phase of the algorithm a fairly significant decrease in the determinant value is registered, so few solutions will be acceptable for the multi-objective problem. In order to overcome these problems, *an active designer control* during $\min(d)$ interchange procedure is recommended. Essentially, the modifications consist in controlling the exchange procedure, such that the determinant of the improved locators must be permanently greater than a certain bound, simultaneously with the check for the form-closure condition. Even considering a tight bound for the determinant, more solutions are acceptable for the design than in the uncontrolled $\min(d)$ optimization case (fig. 9).

As an example, the behavior of a single set of locators is studied during the interchange process of a 2-step algorithm controlled for two different bounds of the determinant value, emphasizing the fact that in the trade-off zone the designer decision is decisive in finalizing the fixture configuration (fig. 10).

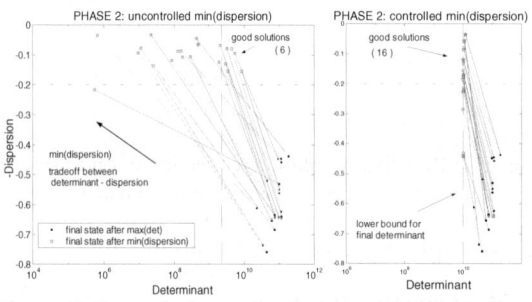

Figure 9: Second phase of a 2-step random interchange algorithm: uncontrolled $\min(d)$; controlled $\min(d)$.

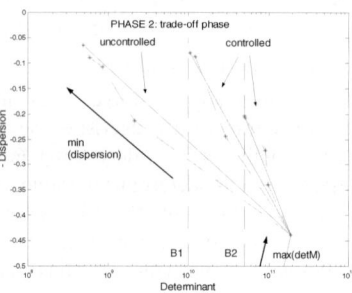

Figure 10: General behavior during a 2-step algorithm applied on a single set of locators.

7. Fixture Optimal Clamping

This section deals with a more complicated problem: to *search simultaneously for the optimal clamp and locators* in order to achieve a required fixture quality. Varying the clamp, it is obvious that the number of combinations for possible clamp-locators candidates is increasing very much. It will be shown that this problem is manageable for the precise localization objective. For the other objectives we will have to restrain the search of the optimal clamp inside of a small set of proposed locations, such that the optimization procedure could be handled.

7.1. Feasible Clamping Regions

Applying an automated procedure based on the derived conditions (eq.5) for given sets of locators we may identify the entire *feasible regions of clamp locations* that would satisfy the form-closure requirement. Fig. 11 illustrates the feasible clamping regions for different sets of locators. It can be noticed that the feasible clamps, if there exist, are gathered in compact regions over the part surface. Their number, distribution and aspect are completely determined of the locator scheme configuration. It is obvious that the locators that correspond to no clamp solution are excluded to form a feasible fixture.

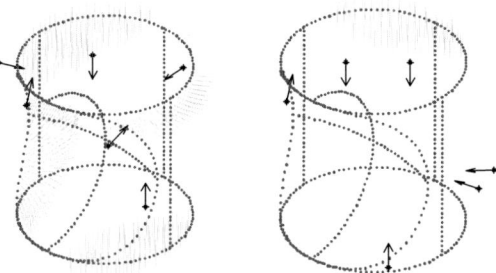

Figure 11: Feasible clamping regions.

7.2. Optimal Clamping for Precise Localization

For precise applications the main concern in fixture design is the precision of workpiece localization. Thus, the primary objective is $\max(\det M)$ while secondary objectives may include $\min(\|p_c\|)$ or $\min(d)$.

As shown previously, a highly accurate localization of a part is dependent only on locator positions. Without considering clamps (i.e., the form-closure constraint), the interchange algorithm for $\max(\det M)$ would give us the best solution for optimal localization. Once the locators are determined, we may identify the entire feasible clamping regions for a form-closed fixture. These will represent the *optimal clamping regions* that assure the highest accurate localization of the part. If such regions exist, we have the opportunity to choose a clamp for a desired second objective completing the design (fig.12). When no feasible clamp regions exist, this hierarchical design process would fail, and we have to compromise between the precise localization and feasible clamping.

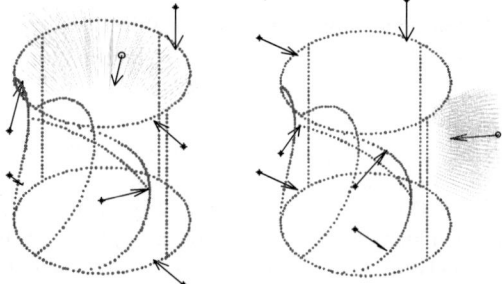

Figure 12: Symmetrical solutions for best locators and the corresponding optimal clamp regions for $\max(\det M)$.

7.3. Optimal Clamp from a Set of Clamps

In some applications the clamps have certain preferred locations, therefore the need to choose the best clamp from a proposed collection might be raised.

For example, let's consider that a collection of preferred clamps is given, and an optimal fixture design with respect to the highly precise localization objective is needed. It is obvious that applying a random interchange procedure successively for each clamp, we find optimal fixture configurations for each specified clamp. Comparing the determinant values offered by these fixture schemes (fig. 13), we end up by selecting an optimal clamp and its corresponding locators, constructing the *best- improved fixture design* (fig. 14).

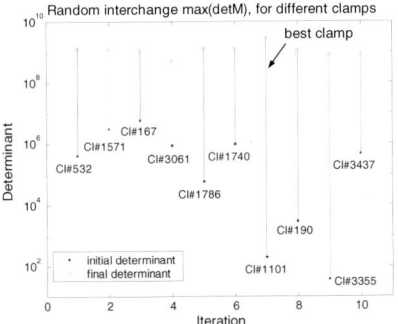

Figure 13: Clamp selection from a collection of clamps for single-objective design.

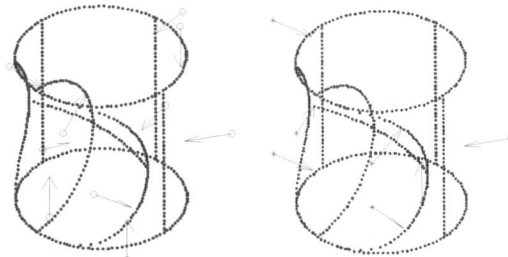

Figure 14: The initial collection of proposed clamps; the best clamp and the corresponding locators.

Furthermore, by extension, the selection of the optimal clamp from a set of proposed locations with regard to the multi-objective design problem can be considered. It will consist mainly in applying the random 2-step interchange algorithm *consecutively for each proposed clamp*.

By collecting the results after applying this procedure for all the clamps, we can compare their different behavior, and select the most appropriate one. It is obvious that an optimal clamp allows only small fluctuations of the determinant while the force dispersion is decreasing significantly (fig. 15). As an example, fig. 16 illustrates the final fixture design consisting of the best clamp selected from a proposed collection with respect to the multi-objective problem, and the corresponding optimal locators.

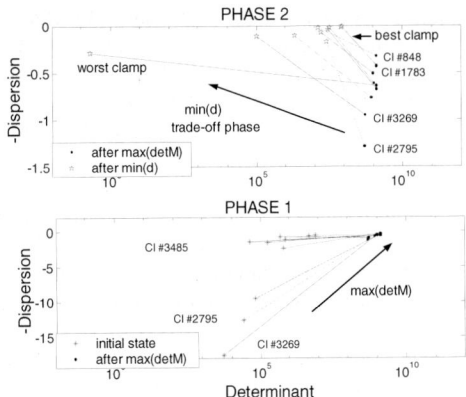

Figure 15: Clamp selection for multi-objective design.

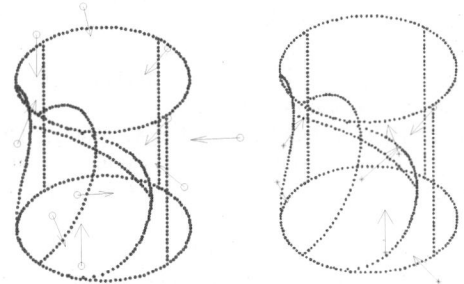

Figure 16: The initial collection of proposed clamps; the best clamp and the corresponding locators.

8. Conclusions

This article focuses on optimal design of fixture layout for 3D workpieces with an optimal random interchange algorithm. The quality objectives considered include accurate workpiece localization, minimal and balanced contact forces. A thorough analysis is performed on the optimization process with respect to the performance objectives. Several main directions for the 3D-fixture synthesis, such that *single-criteria optimal design* with clamp pre-defined and *multi-criteria optimal design* with hierarchical approach and combined-objective approach, have been solved. Examples are used to illustrate some empirical observations with respect to the design approach and its effectiveness.

The work described here is just a tip of the iceberg. The inter-relationship between the locators and the clamps has a determinant role on the fixture quality measures. A more coherent and complete approach to study the influence of the clamp and on an unlimited search of the optimal clamp position must be subject of further studies.

References

1. P.D.Campbell, *Basic Fixture Design*. New York: Industrial Press, 1994.
2. F. Reuleaux, *The Kinematics of Machinery*. Dover Publications, 1963.
3. B. Mishra, J. T. Schwartz, and M. Sharir, "On the existence and synthesis of multifinger positive grips",

Robotics Report 89, Courant Institute of Mathematical Sciences, New York University, 1986.

4. X. Markenscoff, L. Ni, and C. H. Papadimitriou, "The geometry of grasping", *International Journal of Robotics Research*, vol. 9, no. 1, pp. 61-74, 1990.
5. Y.-C. Chou, V. Chandru, and M. M. Barash, "A mathematical approach to automated configuration of machining fixtures: Analysis and synthesis", *Journal Engineering for Industry*, vol. 111, pp. 299-306, 1989.
6. E. C. DeMeter, "Restraint analysis of fixtures which rely on surface contact", *Journal of Engineering for Industry*, vol. 116, no. 2, pp. 207-215, 1994.
7. H. Asada and A. B. By, "Kinematics analysis of workpart fixturing for flexible assembly with automatically reconfigurable fixtures", *IEEE Journal Robotics and Automation*, vol. RA1, pp. 86-93, 1985.
8. R. C. Brost and K. Y. Goldberg, "A complete algorithm for designing modular fixtures for polygonal parts", Tech. Rep. SAND93-2028, Sandia National Laboratories, 1994.
9. Y. Zhuang, K. Goldberg, and Y.-C. Wong, "On the existence of solutions in modular fixturing", *International Journal of Robotics Research*, vol. 15, no. 5, pp. 5-9, 1996.
10. W. Cai, S. J. Hu, and J. Yuan, "A variational method of robust fixture configuration design for 3-d workpieces", *Journal of Manufacturing Science and Engineering*, vol. 119, pp. 593-602, November 1997.
11. D. Baraff, R. Mattikalli, and P. Khosla, "Minimal fixturing of frictionless assemblies", CMU-RI TR-94-08, The Robotics Institute, Carnegie Mellon University, Pittsburgh, PA, 1994.
12. Y. Wang, "Automated fixture layout design for 3d workpieces", in Proceedings of 1999 IEEE Int'l Conf. On Robotics and Automation (CD-ROM), Detroit, MI, May 1999.
13. A. Atkinson and A. Doney, *Optimum Experimental Designs*. New York: Oxford University Press, 1992.
14. Y. Wang and D. Pelinescu, "Precision localization and robust force closure in fixture layout design for 3D workpieces", IEEE Int'l Conf. on Robotics and Automation, San Francisco, April 2000.

Holonic Robot System: A Flexible Assembly System with High Reconfigurability

MASAO SUGI YUSUKE MAEDA YASUMICHI AIYAMA[†] and TAMIO ARAI

DEPARTMENT OF PRECISION ENGINEERING,
THE UNIVERSITY OF TOKYO
7-3-1 HONGO, BUNKYO-KU, TOKYO 113-8656, JAPAN

[†]INSTITUTE OF ENGINEERING MECHANICS AND SYSTEMS,
UNIVERSITY OF TSUKUBA
1-1-1 TENNODAI, TSUKUBA CITY, IBARAKI PREF. 305-8577, JAPAN

{sugi,maeda,arai}@prince.pe.u-tokyo.ac.jp
aiyama@esys.tsukuba.ac.jp

Abstract

This paper proposes a flexible assembly system that manages production by negotiating with autonomous manufacturing devices. With this mechanism, physical layout of the system is independent of the type of products to be assembled, and the system can asynchronously process multiple parallel production tasks. This system also supports Plug & Produce; a system function that enables easy reconfiguration. Thus the system can be quickly accommodated to breakdowns and changes in production quantities. This paper also deals with an index for general evaluation of the manufacturing systems. With this index, we can plan appropriate improvement of the system layouts. Then we reconfigure the system using Plug & Produce function, in order to accommodate it to the changes in the manufacturing environments.

1 Introduction

In recent years, the demands of the consumers have been diversifying, and the manufacturing environments have been getting no more static but dynamic. The manufacturers are expected to keep varieties of their product lineups, and quickly produce appropriate types of products in adequate quantities timely. For this reason, not only productivity, but flexibility against the changes of the manufacturing environments is also required for manufacturing systems. Applying distributed autonomous systems to manufacturing has been become very important.

Manufacturing system with this architecture has been widely studied. A common feature of those studies is the hierarchy of agents performing task assignments by competitive bidding [3, 7, 8]. Such hierarchical architectures, where agents achieve both autonomy and cooperation in balance, agree with the concept of Holon and Holarchy (i.e. holonic hierarchy) proposed by A. Koestler [4]. Most of the existing researches have, however, focused on the distributed scheduling or the negotiation between agents, and little attention has been paid to the implementation onto the actual manufacturing systems.

It is very important to estimate the expected performance at the designing stage, and to evaluate the current status in manufacturing systems [5, 6, 9]. Evaluation of conventional systems have been attempted so far, but that of distributed autonomous systems still remains to be developed.

This paper proposes Holonic Robot System that aims at assembling activities of real manufacturing devices. This system assembles products by negotiations between production management agents and device agents. Furthermore, this paper proposes **"Plug & Produce"**, a system function which enables easy addition/removal of devices. Then Plug & Produce function is introduced to Holonic Robot System. Evaluation of the proposed system is also dealt with in this paper, since it is essential for effective utilization of system's high reconfigurability. We propose the evaluation index suitable for distributed autonomous manufacturing systems. We also describe how to calculate the index.

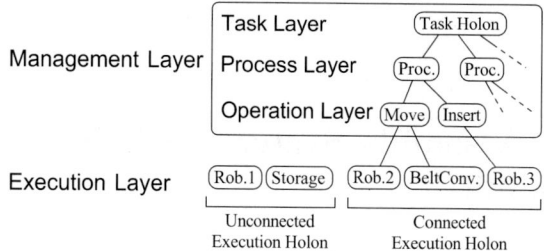

Figure 1: Holarchy of Management and Execution Layers

2 Holonic Robot System (HRS)

In this section, Holonic Robot System (HRS) is outlined. Holonic Robot System, based on the concept of Holon by A. Koestler [4], is an assembly system with high flexibility [1, 2].

2.1 Holarchy

Holonic Robot System consists of two different kinds of layers as shown in Figure 1. *Execution layer* is located at the lower level and represents the manufacturing devices. It controls devices. In this layer, holons exist statically since they correspond to actual devices. *Management layer* is the upper one to manage assembly processes. Inside of the management layer, there are three kinds of layers; task layer, process layer and operation layer.

Each component of the layers is a holon. An upper holon asks lower holons to execute an order and lower holons reply by bids. These three layers form a kind of hierarchy, called *holarchy*. We will show the details of the management holons and the execution holons.

2.2 Management holons

The management holons aim at:

- Managing assembly processes of a product: The management holons look after assembly processes according to a product database that describes how to assemble a product with parts.
- Assigning devices to assembly process: According to the sequence of assembly, the holons ask assembly devices and make contracts with them.

Here an assembly "device" means equipment such as manipulators, belt-conveyors, warehouses and so forth. These two characteristics are realized by introducing three layers. Holons in each layer have different functions as follows;

(I) Task holon: It manages the whole assembly process of a product. There exists only one task holon for the manufacturing of a product. It works as the top holon of the assemble holarchy.

(II) Process holon: It manages one assembly process that forms a part of the whole assembly task. A process holon is dynamically created by the task holon.

(III) Operation holon: One assembly process may include a series of motion of devices. According to the request from a process holon, an operation holon makes devices move.

2.3 Execution holons

Execution holons exist *statically* and *originally* in order to control existing assembling devices such as manipulators, belt-conveyors and warehouses. Note that one device is usually subject to its own execution holon. When a device is mounted onto a system, the existing "Execution holon" is also installed onto the holarchy.

It bids for a movement request from an operation holon and negotiates about the movement with its upper holons. When it receives an order, it controls motions of the device and reports results. It observes the status of devices, and it also detects and reports errors.

2.4 Experimental implementation

An assembly cell is installed according to the proposed architecture. We made an assembly cell that has three manipulators, one bi-directional belt-conveyor and two warehouses as shown in Figure 2. The task of a cell is to assemble parts into a product. The parts are initially set in warehouses and assembled in the cell, then the product is transferred back to one of the warehouses.

Each execution holon has a configuration file containing information on:

(a) relations between its local coordinates and the world coordinates,

(b) relay points to its neighboring devices that used for transferring parts from device to device, and

(c) data of objects stored in a buffer of each device.

In this system, every product is described as an assembly tree, i.e. precedence graph. Assembly trees are decomposed by the management holons into simpler operations ("move" or "insert"). An operation holon negotiates with execution holons, i.e. manufacturing devices, and assigns the operation to proper

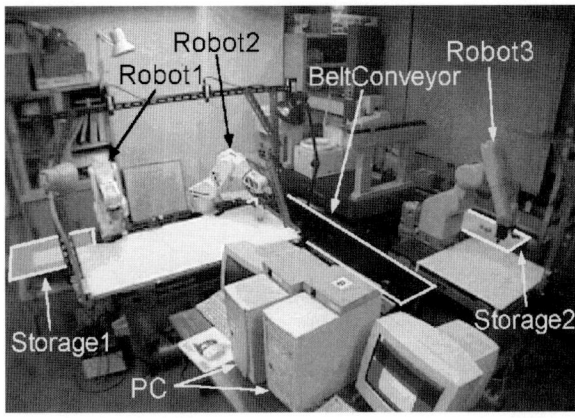

Figure 2: Overview of Holonic Assembly Cell

devices. Neither specified cell layouts nor control programs for each product type are necessary. If only a new assembly tree is prepared, the system can start to assemble new products. It means that the proposed system is flexible against changes in product lineup.

Using this assembly cell, we carried out experiments to assemble some kinds of products. Detailed description of the experiments is shown in our previous paper [1]. The results show that the system can cope with multiple parallel tasks asynchronously.

3 Plug & Produce

3.1 Outline of the concept

The physical reconfiguration of a usual manufacturing system is very expensive, because the inevitable information renewal by the physical change is very troublesome. Thus the reconfiguration is preferred to be avoided as much as possible. If the reconfiguration can be carried out easily, the system can deal with the change of production quantity and the occurrence of machine breakdowns rapidly, by adding or removing some machines in the system.

The authors consider the assurance of easy reconfiguration necessary for the manufacturing systems, and call this function as "**Plug & Produce**," on the analogy of the Plug & Play on personal computers.

Plug & Play enables easy addition/removal of peripherals by automating information renewals, i.e. installation of the device drivers or modification of the system configuration file, which has been manually managed. In Plug & Produce, information renewals associated with physical changes are automated in order to reduce costs at the reconfiguration. This method is equivalent to that in Plug & Play.

Applying Plug & Produce function along with distributed autonomous architecture, manufacturing systems will achieve much higher flexibility.

3.2 Requirements for Plug & Produce

In this subsection, let us consider what is necessary for the distributed autonomous manufacturing system in introducing Plug & Produce function.

Plug & Produce and Plug & Play have the same target and they use the same approach to the aim. However, system requirements are rather different. This is caused by the difference of system architectures and existence of cooperation.

In Plug & Play, there is no cooperation among the peripherals, and resource conflict occurs in the IRQ (interrupt request) number or the I/O port addresses, and is solved by the central management of the PC.

On the other hand, manufacturing devices cooperate with each other, and they conflict on their working spaces. Furthermore, this conflict should be solved by means of distributed coordination, since we presume the architecture of the system not to be centralized but to be distributed.

In this distributed solution of the problem, heterogeneity of the system should be considered. The abilities of the manufacturing devices are not uniform in the view of degrees of freedom (e.g. manipulators have three or more degrees of freedom, but belt-conveyors have only one).

Manufacturing systems can realize the Plug & Produce functionby satisfying these requirements.

3.3 Implementation of Plug & Produce

In the previous subsection, we considered the requirements for introducing Plug & Produce onto manufacturing systems with distributed architecture. Here Plug & Produce function is implemented onto HRS.

As mentioned in the subsection 3.2, mutual conflict of devices occurs on their working spaces. There are some available methods to solve this problem. For example, if each device knows the position of the neighbors at any moment, it can evade collision with others. Such a method is, however, too much rigorous for assembly system, where movement of the manufacturing devices are rigidly limited to several patterns of the motion.

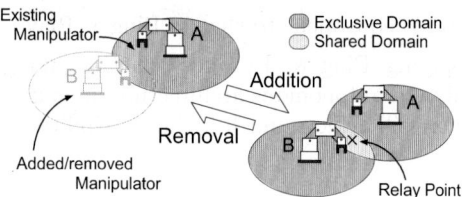

Figure 3: Changes in Information in Addition/Removal of a Device

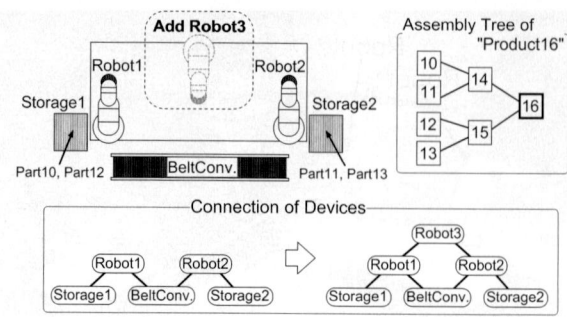

Figure 4: Schematic View of Cell in Simulations

The working space of each device is divided into two kinds of domains, and the rules of device motion are introduced as follows:

- **Shared Domain**: A domain where the neighboring devices can enter. This domain is used only for transferring parts or products. Relay points of two neighboring devices are defined in this domain.

- **Exclusive Domain**: A domain where other devices cannot enter. This domain is used for assembly and keeping parts or products.

Each device memorizes the allocation of exclusive domain and shared domain in its own working space. This allocation will be modified if the number of neighboring devices changes, as shown in Figure 3. If a new device comes, a new shared domain will be defined for a new comer and the exclusive domain will be decreased. On the other hand, if a neighbor is removed, the shared domain with the removed device will be incorporated into the exclusive domain.

Here, let us settle the procedure for adding new devices. As preconditions for each device, the reachable area and the relation between its local coordinates and the world ones are given. Then the procedure is settled as follows;

1. A human operator sets the new device in the cell.

2. The operator launches an execution holon that corresponds to the new device.

3. The new holon asks all of the existing holons about their location and the shape of their working space. Then the new holon understands its neighbors.

4. The new holon secures all of the neighbors and makes them halt.

5. The new device negotiates with each neighbor to decide shared spaces and relay points.

6. If all the neighboring holons have finished the renewal of their information, the new holon releases them. Then the procedure finishes.

In the above procedures 4–6, the information stored in multiple devices are renewed simultaneously and guaranteed no contradiction to each other.

In procedure 5, the relay point negotiations between the new device and each of its neighboring devices are held. As mentioned in the subsection 3.2, not the new device but a device with lower degree of freedom seizes the initiative in this negotiation. For example, in a negotiation between a manipulator and a belt-conveyor, the conveyor decides the relay points and the manipulator obeys it. If the manipulator determines the relay points, these points might be inconvenient for the belt-conveyor, because the manipulator has three (or more) degrees of freedom but the conveyor does only one.

In the case of removal of a device, similar procedure is used. The device to be removed secures all of the neighboring devices, asks them to renew information, then releases them and eliminates itself.

3.4 Simulation about addition or removal of devices

The authors have constructed a simulator of the HRS with the Plug & Produce function by the above-mentioned procedures. Using this simulator, addition/removal of a device is examined.

An assembly cell composed of two manipulators, one belt-conveyor and two warehouses is assumed, as shown in Figure 4. In the midst of the assembly of "Product16" that consists of four parts, a new manipulator "Robot3" is added into the cell.

The result is shown in Figure 5. At 28[sec] from the task started, a new execution holon "Robot3" is launched. Robot3 collects information of the existing devices and concludes that the neighbors are Robot1 and Robot2. Robot3 tries to secure both of them. Robot1 and Robot2 are secured at 31[sec], then information renewal starts. At 33[sec], the renewal finishes. Robot3 releases Robot1 and Robot2 and joins

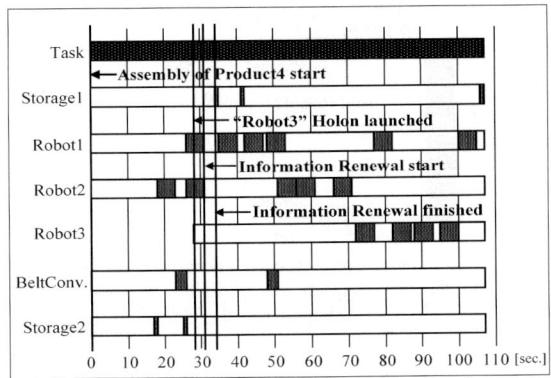

Figure 5: Time Table of Devices in Simulation of Addition of a New Manipulator

the bids. The time spent for the whole procedure was 6[sec]. From 72[sec] to the end, Robot3 contracts for three operations and executes them. This means that the added manipulator works regularly.

We have also examined the case of the removal of the belt-conveyor in the middle of the assembly of "Product16". The assembly continues to function and the reconfiguration is completed.

In this way, easy and quick addition/removal of devices is realized by automated information renewal of devices.

4 Evaluation of the System

In HRS, physical layout of the system is totally independent of product types. Thus we can arbitrarily compose manufacturing devices. If some of the devices may break down, the system can maintain manufacturing activities using remaining devices. We can also add new devices or remove the existing ones freely, even if manufacturing tasks are ongoing. Nevertheless, these functions are only the necessary condition for system to be flexible, and not the sufficient condition. This may be due the facts that we don't have any evaluation criterion of determining optimal layouts, or we cannot find out where to add a new device. An index is necessary for evaluating total performance of the system.

In addition, predictive evaluation is necessary; if the system is finally turned out to be inadequate after reconfiguration, such an "observe-only" evaluation is useless. Before the system reconfiguration, we must predict the performance of the modified system for every alternative layout plan, reject inadequate plans and choose the optimal configuration.

In a word, it is necessary to introduce an index to evaluate the system and a method to predict the future performance using that index. With both of the index and the prediction method, HRS will really obtain high flexibility.

4.1 System evaluation index

In this subsection, we consider what item should be included in the general index, and determine the index.

Productivity is regarded as the most important ability of manufacturing systems. This can be simply measured as production quantity per unit time (or as its reciprocal, necessary time per one product). Flexibility has been often neglected so far, but it is also an important ability. In this paper, flexibility is considered equal to fault-tolerance. Fault-tolerance can be calculated using theory of reliability.

Koren [5] proposed an index "Expected Productivity", which is expressed by productivity and reliability. Concretely, expected productivity $E[P]$ is defined as follows;

$$E[P] = \sum_{\lambda \in \Lambda} Pr(\lambda) V(\lambda), \qquad (1)$$

where Λ is a set of possible fault patterns. A system composed of n-devices has 2^n patterns. $Pr(\lambda)$ is probability for occurrence of a fault pattern λ. $V(\lambda)$ is the productivity under the pattern λ.

We think this index is appropriate, since it contains both flexibility and productivity. In this research, expected productivity is used for the evaluation.

4.2 Calculation method of the index

Expected productivity assumes the system to be conventional manufacuturing line. In order to use this index for evaluation of HRS, we must solve some problems.

Productivity of manufacturing lines are easily calculated using their tact times. On the other hand, HRS is asynchronous and dynamic system, and can execute multiple tasks in parallel, which is mentioned in section 2. Even if all the conditions of the two assembly cells are the same, their total processing times might be different, since their behavior is not always the same. Thus, it is difficult in HRS to predict exact productivity of the cell. In this paper, the following approximation is used for prediction.

1. Prepare an assembly cell in the simulator, where the condition of the imaginary cell reflects that of the real system.

2. Put in more than one assembly tasks in the simulator. Using the total processing time resulted, calculate processing time per one task. As the number of the tasks increases, processing time per task will be reduced through the system's parallel processing ability. However this reduction of time will finally reach a critical point, since the number of tasks executed in parallel might be limited.

3. Increasing the number of tasks gradually, find out the minimum value of time per task. Calculate the productivity (i.e. quantity per unit time) as the reciprocal of this minimum value.

4. Repeat the same simulation and get the average of the productivity.

Another problem is how to take into account the heterogeneity of the system. In ref. [5], manufacturing devices are treated as versatile machines that is not so concrete an entity. In this model, whether the system can continue the production or not is depends only on the preservation of the connection; if the system separated, the production is no more feasible. We deal with heterogenious system, where manipulators (for assembling parts or moving parts), belt-conveyors (only for moving parts), and warehouses (for feeding parts or keeping products). We must consider whether these abilities (assembling, feeding, etc.) is preserved in the system, in order to continue the production.

4.3 Sample case: real calculation of the index

In this subsection, we calculate the expected productivity on a sample problem.

Figure 6 shows an assembly cell with two manipulators, one belt-conveyor and two warehouses. The third manipulator "Robot3" is added in the cell. Here let us name the original cell with two manipulators "Cell-1" and the enhanced cell "Cell-2". We will evaluate these two cells.

Let us consider the expected productivity of "Cell-1". As shown in Figure 6, "Cell-1" has a linear connection. If one of the devices breaks down, "Cell-1" cannot continue assembly. In breakdowns of a manipulator or a belt-conveyor, the cell is separated into two. In the case of a warehouse, the parts necessary for the assembly cannot be supplied any more. Therefore, the expected productivity of the "Cell-1" $E[P]$(Cell-1) is expressed as the multiplication of productivity V(Cell-1) and reliability of each devices,

$$E[P](\text{Cell-1}) = V(\text{Cell-1}) \cdot R_{Storage1} \cdot R_{Robot1} \\ \cdot R_{BeltConv} \cdot R_{Robot2} \cdot R_{Storage2}. \quad (2)$$

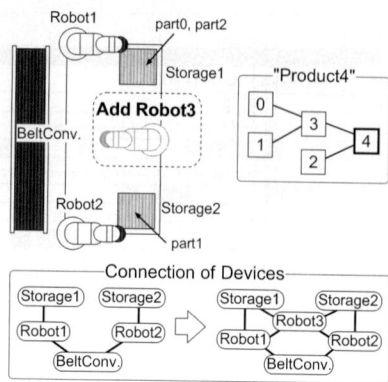

Figure 6: Benchmark Environment

Table 1: Usable Fault Patterns and Their Probabilities

Pattern	S1	S2	R1	R2	R3	B	$Pr(\lambda_i)$
λ_1							5.99×10^{-1}
λ_2						×	3.57×10^{-2}
λ_3					×		7.54×10^{-2}
λ_4				×			7.54×10^{-2}
λ_5				×	×		3.97×10^{-3}
λ_6			×				7.54×10^{-2}
λ_7			×		×		3.97×10^{-3}
λ_8			×	×			8.38×10^{-3}
λ_9			×	×	×		4.41×10^{-4}

In order to calculate $E[P]$(Cell-1), we have to know the reliability of each manufacturing devices. In this paper, each reliability is given by,

$$R_{Robot1} = R_{Robot2} = R_{Robot3} = 0.9, \\ R_{Storage1} = R_{Storage2} = 0.99, \quad (3) \\ R_{BeltConv} = 0.95.$$

Substituting these values in the equation (1), we get

$$E[P](\text{Cell-1}) = V(\text{Cell-1}) \cdot 0.99 \cdot 0.9 \cdot 0.95 \cdot 0.9 \cdot 0.99 \\ = V(\text{Cell-1}) \times 0.754, \quad (4)$$

where V(Cell-1) will be obtained through the simulation.

Next let us consider the expected productivity of "Cell-2". "Cell-2" has network connection, while "Cell-1" has linear one. Consequently, "Cell-2" can continue production even if some devices are out of order. Here we list all the fault patterns in which the cell can continue the assembly. Calculating probability of each usable pattern, we obtain the result shown in Table 1.

Table 2: Productivity and Probability of Each Fault Pattern

Pattern	$V(\lambda_i)$ (1/min.)	$Pr(\lambda_i)$
λ_1	1.418	5.99×10^{-1}
λ_2	1.418	3.57×10^{-2}
λ_3	1.066	7.54×10^{-2}
λ_4	1.463	7.54×10^{-2}
λ_5	1.463	3.97×10^{-3}
λ_6	1.281	7.54×10^{-2}
λ_7	1.281	3.97×10^{-3}
λ_8	1.289	8.38×10^{-3}
λ_9	1.289	4.41×10^{-4}

Here we must determine the productivity $V(\lambda_i)$ of each pattern λ_i ($i = 1, \ldots, 9$) through simulations. Using the method described in subsection 4.2, we obtain the result shown in the Table 2. Using Table 2, we obtain the expected productivity of "Cell-2",

$$E[P](\text{Cell-2}) = \sum_{i=1}^{9} Pr(\lambda_i) V(\lambda_i) = 1.09. \quad (5)$$

Here let us obtain the expected productivity of "Cell-1". Hence "Cell-1" is equal to the pattern λ_3 of "Cell-2",

$$\begin{aligned} E[P](\text{Cell-1}) &= V(\text{Cell-1}) \times 0.754 \\ &= V(\lambda_3) \times 0.754 \\ &= 1.066 \times 0.754 = 0.804. \end{aligned} \quad (6)$$

The expected productivity of "Cell-2" is 1.09 and that of "Cell-1" is 0.804. This means that "Cell-2"is superior to"Cell-1". Thus we have derived the difference between two cells quantitatively, supporting what is expected. Moreover, using this index and simulations, we can predict the increase of system performance at reconfiguration. Now the system can deal with changes of environment, by adequate reconfiguration.

5 Conclusion

An assembly system with high flexibility has been studied as an approach to the total view of assembly systems in the future.

- Holonic robot system as an assembling system with a holonic architecture is proposed.
- A holonic robot system is implemented as an assembly cell with three manipulators, one belt-conveyor, and two warehouses. The cell can deal with multiple parallel tasks asynchronously.
- A system function Plug & Produce is proposed that supports easy and quick reconfiguration.
- Plug & Produce function is realized on holonic robot system. System shows quick response to physical reconfiguration.
- Applying expected productivity as general index that evaluates the system, the method to predict the system performance is presented. By this index and prediction method, we can add a new device to proper location quickly.

Acknowledgements

This research is supported by IMS international research program HMS (Holonic Manufacturing Systems). The authors thank the members of HMS, especially those of HANDS (Holonic Handling Systems) for their valuable comments.

References

[1] Arai, T., Aiyama, Y., Sugi, M. and Ota, J., "Holonic Assembly System with Plug & Produce," *Proceedings of the 2nd Workshop on Intelligent Manufacturing Systems (IMS) '99*, pp. 119–126, 1999.

[2] Arai, T., Sugi, M., Aiyama, Y. and Ota, J., "Holonic Robot System for Assembly," *Proceedings of the 6th International Conference on Intelligent Autonomous System (IAS-6)*, pp. 371–378, 2000.

[3] Iwata, K., Onosato, M. and Koike, M., "Random Manufacturing System," *Annals of the CIRP*, Vol.43, no.1, pp. 379–383, 1994.

[4] Koestler, A., *The Ghost in the Machine*, Arkana, 1967.

[5] Koren, Y., Hu, S. and Weber, T., "Impact of Manufacturing System Configuration on Performance," *Annals of the CIRP*, Vol.47, no.1, pp.369–372, 1998.

[6] Makino, H., "Versality Index —An Indicator for Assembly System Selection," *Annals of the CIRP*, Vol.39, no.1, pp. 15–18, 1990.

[7] Márkus, A., Kis Váncza, T. and Monostori, L., "A Market Approach to Holonic Manufacturing," *Annals of the CIRP*, Vol.45, no.1, pp. 433–436, 1996.

[8] Ramos, C., "A Holonic Approach for Task Scheduling in Manufacturing Systems," *Proceedings of the 1996 IEEE International Conference on Robotics and Automation*, pp. 2511–2516, 1996.

[9] Valckenaers, P., Van Brussel, H., Bongaerts, L. and Wyns, J., " Holonic Manufacturing Systems," *Integrated Computer-Aided Engineering*, Vol.4, John Wiley & Sons. Inc., pp. 191–201, 1997.

New Architecture for Corporate Integration of Simulation and Production Control in Industrial Applications

E. Freund, A. Hypki, D. Pensky

Institute of Robotics Research (IRF), University of Dortmund
Otto-Hahn-Str. 8
D-44227 Dortmund, Germany
email: {freund, hypki, pensky}@irf.de

Abstract

These days systems for 3D simulation and flexible control of robot-based workcells are in common use. However especially small and midsize companies do not make use of the advantages these technologies imply. Reasons for this behavior are the costs of the soft- and hardware systems as well as the fact that you need experts to operate those systems. This article describes an innovative architecture that has been developed for a corporate integration of simulation and control systems into production processes. The coexistence and cooperation of PC-based simulation and control makes the cost-effective use of Virtual Reality (VR) methods and a wide range of novel industrial applications possible. By use of VR technologies complex systems can be operated by any user without long periods of training. Thus, small and midsize companies are enabled to use innovative technologies to increase their productivity and competitiveness.

Introduction

As a result of increasingly shorter development cycles for new products there is a growing demand for production systems to offer highest possible product flexibility. It is necessary to integrate new products without impairing the current production. Moreover simulation and control systems are needed, which permit a close-to-reality simulation in connection with the possibility to transfer the tested programs into real operation as fast as possible.

All through the development of simulation and control systems for industrial production the requirements of small and midsize companies have not been respected. These companies are not able to spend time and money for installing complex and difficult-to-use systems. Because of the lack of high-qualified employees software systems have to become more and more easy to use in general. PC- and Windows™-based systems are suited concerning their well-known user interfaces and low priced hardware.

The Institute of Robotics Research (IRF), Dortmund, Germany and its partners develop systems for the whole range of modern automation technology. The systems include planning and optimization of the layout, the programming of the automation components including the upper control hierarchies, the simulation of the entire workcell up to control and teleservice of the real cell.

The PC-based 3D simulation system COSIMIR® (Cell Oriented Simulation of Industrial Robots) and the flexible workcell controller LUCAS (Layered Universal Controller for Automation Systems) have been integrated into an innovative architecture to meet the needs especially of small and midsize companies.

The new type of architecture is described on the basis of two research projects.

Within the scope of the project DEMON a novel system structure has been developed for disassembly tasks executed by a robot-based workcell. Virtual Reality techniques make the operation of the control system easier.

The project FA-SiCo (Factory Automation – Simulation and Control) includes the development of methods for a so-called Parallel Simulation of production processes. Using this technology opens up new applications for corporately integrated simulation and control systems.

Architecture

In Figure 1 the structure of the integration of VR, simulation, and workcell control is presented. The left side of Figure 1 shows the different tasks realized by the usage of the workcell simulation system COSIMIR®. The 3D-simulation system COSIMIR®, which supports the simulation of entire workcells [1], was developed at the IRF [2]. It is PC-based and is running under Microsoft Windows™. Due to its ergonomic user interface supporting the complete range of Windows standards the training periods are minimized.

The whole process starts with the planning of the workcell layout, i. e. the generation of a model for further programming and simulation operations. This modeling is accomplished with the help of libraries, including for example industrial robots, PLCs, and mechanisms like conveyor belts and grippers. The 3D model of the workcell is used for the visualization within the Virtual Reality environment, being displayed for example in a head mounted display. The user selects the necessary tools and executes the desired tasks. These actions are interpreted by the VR system and are transferred into robot programs. These programs are tested in the simulation environment of COSIMIR®, especially taking care of aspects like the reachability of all positions and of movements without collisions.

The right area of Figure 1 shows the different tasks of the flexible workcell controller LUCAS [3]. During configuration time the setup, e. g. of the communication channel (TCP/IP, ProfiBus, RS232 etc.), for each single automation component integrated in the whole workcell is done. Afterwards the coordination of the different devices is programmed using the process plan language implemented in LUCAS.

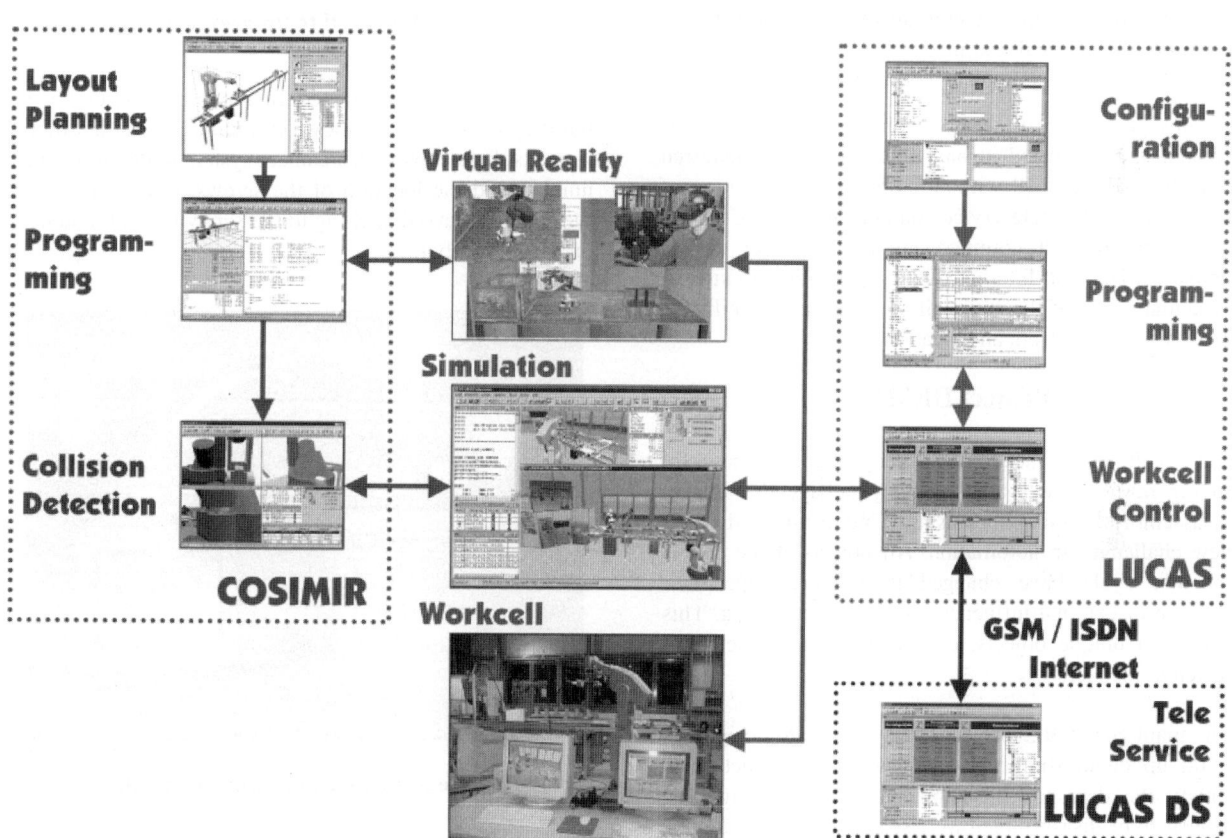

Figure 1: Connection Between Simulation and Control System

These steps create the basis for the first integration of simulation and workcell control, i.e. also the first interaction of COSIMIR® and LUCAS. LUCAS is responsible for loading the programs, developed and tested using COSIMIR® as described before, into the different simulated automation devices. Additionally LUCAS is also responsible for the control of the whole workcell environment. The behavior of the complete workcell is simulated. In addition to the simulation of single robot programs done, aspects like material flow, coordination of the devices, and the correctness of the sequences of the different processing steps are also simulated. By this integration of simulation and control, based on exactly those programs used for the real workcell later on, a very high transferability of the achieved results is guaranteed.

After successful tests within the simulation environment LUCAS switches between the simulated and the real world by downloading the programs for the robots, PLCs, etc. into the physical automation devices. Thus LUCAS takes control of the complete workcell.

During production time new robot programs may be created and simulated at any time and then transferred into the real workcell without stopping the operation of the workcell. The teleservice components of LUCAS are based on different communication layers like mobile phones, ISDN, and Internet. They support the monitoring of actual states of the workcell as well the possibility to modify programs from any location.

Project DEMON

Due to the rising necessity of recycling as a base for saving raw materials and avoiding garbage the area of disassembly becomes increasingly interesting and opens new challenges in automation. All objects at the end of the life cycle have changed their outer appearance because of external influences as well as of ageing. This results in unique objects, each one to be handled in a different manner.

The main goal of the project DEMON was to design and build up demonstrators for a new control structure to satisfy the demands of automated disassembling. The objects to be disassembled were selected from automotive industry; one task was to unscrew car wheels, the other example was the disassembly of fuel pumps.

Figure 2 shows the robot unscrewing a car wheel. The model for the simulation and the VR environment is automatically generated and permanently updated according to sensor information. The robot programs are created using this model, too e.g. to define the position of each single screw.

Figure 2: Workcell to unscrew car wheels

The workcell for automated fuel pump disassembly is build up of 3 robots as shown in Figure 3. The first robot inspects the pumps with a camera system to determine the number and the location of the screws, robot 2 polishes the screws to avoid sticking to the screwing tool mounted on the third robot.

Figure 3: Disassembling of fuel pumps

Robot 3 is equipped with a gripper exchange system to hold the screwing tool as well as a gripper to handle the parts of the pump and to sort them according to their material. Figure 3 also presents the user interfaces of COSIMIR® on the left and LUCAS on the right monitor.

Reality   Simulation

Figure 4: Close-to-Reality Simulation of the FA-SiCo Demonstrator with COSIMIR®

Project FA-SiCo

The goal of the project Factory Automation – Simulation and Control (FA-SiCo) was to demonstrate that by the use of innovative technologies especially small and midsize companies are able to meet new requirements because of decreasing development cycles of products. Therefore a fabrication system was built up to explain the efficient and cost-effective application of these technologies.

The manufacturing system shown in Figure 4 and consisting of a transfer system and two robots provides highest product flexibility as well as an intuitive user interface and optimized use of the integrated components. Several systems are integrated into the demonstrator's software including the modeling of the automation components and products, the simulation of all production processes, the workcell control, and teleservice.

Close-to-Reality Simulation

The 3D simulation system COSIMIR® is used for a close-to-reality simulation and the programming of the automation components within the scope of the project FA-SiCo (see Figure 4). For modeling of the demonstrator's simulation model existing model libraries were used. The product parts were created with a 3D-CAD system and they were imported into the simulation model after being converted to one of the graphic exchange formats supported by COSIMIR®. By the use of the simulation model the positions of the robots and the gripper exchange systems were determined relatively to the other workcell components.

The 3D simulation of robot-based workcells by COSIMIR® includes simulation of robot programs, PLCs, sensors, actuators, material flow, and simulation of handling and treatment processes.

All programs for the two Mitsubishi RV-E4NM robots are entirely developed within COSIMIR® using the programming language MELFA Basic. The download and upload of programs and position lists is performed via an RS-232 interface. Because of small differences between simulation model and reality, commonly different position lists for the simulated and the real robots are used.

Hierarchical Control

During the FA-SiCo project LUCAS was used for coordination of the automation components. Via communication interfaces to the robots (RS-232) and to the material flow system (CANOpen) the production controller starts process tasks in the components at the right time using a predefined process plan for each product. All production data is stored in a standard database and can be accessed by overlaid production planning and control systems. The integrated process visualization is used to provide an easy and intuitively to use interface to monitor and to operate all production processes.

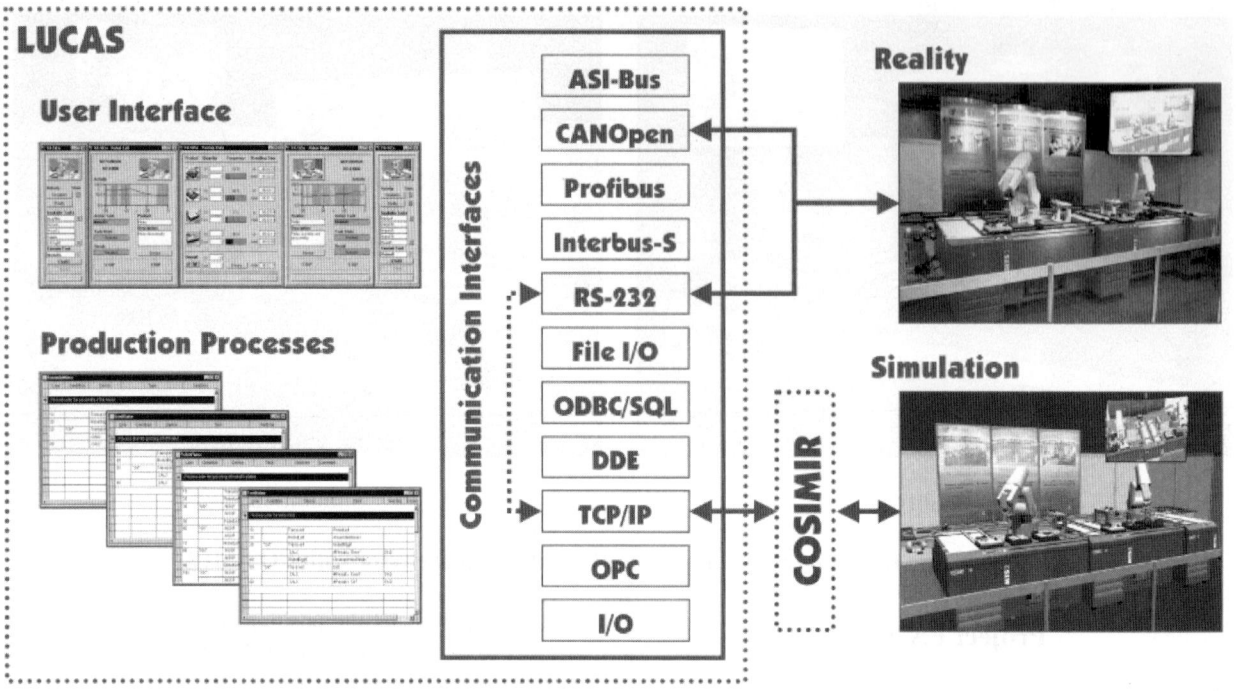

Figure 5: Connection Between Simulation and Control System for Parallel Simulation

The DS (Distributed System) functionality of LUCAS is used to integrate teleservice techniques into the controller system. An appropriate connection between the workcell controller and the teleservice station has to be established. Because the system uses standard network functionality, communication media like LAN or Internet via analog modem, GSM connection, or ISDN are suitable.

Control meets Simulation

For optimization of production processes a powerful connection between simulation and control system is mandatory. Thereby all overlaid programmable coordination procedures can be developed using the simulation. For that purpose the real components are not needed.

The systems COSIMIR® and LUCAS can be connected via a network based interface. Because it is based on TCP/IP the connection can be established over an ordinary Internet connection.

The different process tasks commonly being sent to the real components are executed by the simulated components. Using the close-to-reality simulation of COSIMIR® cycle times for production processes can be determined and optimized while executing the overlaid coordination processes by LUCAS. Moreover operators can be trained with the simulation using the same graphical user interface as for operating the real workcell. Figure 5 illustrates the described connection for the FA-SiCo demonstrator.

A wide range of industrial applications is supported by this powerful link between simulation and control methods:

- Offline-Programming of Control Structures, Batch Processes and Development of Graphical Operator Interfaces
- Operator Training and VR Operation
- Simulation in Advance
- 3D Process Visualization

While actual production is running simulation can be carried out in parallel. The parallel simulation is possible because of the controller's structure. Every process task can be sent to both simulation and reality. Thus, the connection of simulation and control system also provides a 3D visualization of production processes.

Summary

This paper presents a new approach to integrate workcell simulation and workcell control in combination with Virtual Reality components to facilitate workcell programming. New structures like the one described are necessary to meet the requirements of industrial automation processes. Especially in the field of automated disassembly the conventional robot programming by Teach-In does not lead to satisfying results.

The integrated system is based on the 3D simulation system COSIMIR® and the flexible workcell controller LUCAS. The Virtual Reality extensions for COSIMIR® are used as interface to the head mounted display and to the data glove on the one side. On the other side these extension modules fulfill the interpretation of user action and they are responsible for the automated generation of robot programs.

The VR and simulation system COSIMIR® is connected to the workcell controller LUCAS. By this integration of simulation and control, based on exactly those programs used for the real workcell later on, a very high transferability of the achieved results is guaranteed.

According to the developed structure three different workcells were built up to transfer the theoretical approach into practical environments. These realizations have proven the advantages of the concept and have shown a significant reduction in programming time. During operation these systems guarantee a reduction of faults combined with an increase in productivity.

Acknowledgement

Funding for this work was provided by the German Federal Ministry of Education and Research (BMBF) and the Ministry of Economics and Small Businesses, Technology and Transport (MWMTV) for North Rhine-Westphalia, Germany.

References

[1] Freund, E.; Hypki, A.; Uthoff, J. (1996). COSIMIR: Simulation of Complete Robotic Workcells. *Proceedings of the 12th International Conference on CAD/CAM Robotics and Factories of the Future*, London

[2] Freund, E.; Hypki, A.; Uthoff, J. (1997). New Strategies in Programming and Control of Flexible Manufacturing and Automated Disassembling Processes. *Proceedings of the 13th International Conference on CAD/CAM Robotics and Factories of the Future*, Pereira

[3] Freund, E.; Rothert, B.; Theis, K.; Uthoff, J. (1996). Comprehensive Approach to Control of Flexible Manufacturing Systems. *Proceedings of the 12th International Conference on CAD/CAM Robotics and Factories of the Future*, London

Vision-based Automatic Forming of Rheological Objects Using Deformation Transition Graphs

Shinichi Tokumoto, Takuya Saito, and Shinichi Hirai

Dept. of Robotics,
Ritsumeikan Univ., Kusatsu, Shiga 525-8577, Japan
E-mail: gr082982@se.ritsumei.ac.jp

Abstract

Manipulative operations of rheological objects can be found in many industrial fields such as food industry and medical product industry. Automatic operations of rheological objects are eagerly required in these fields. In this paper, we will realize a vision-based automatic forming of rheological objects using deformation transition graphs.

First, we will develop forming machine of rheological objects with multi degrees of freedom. Secondly, we will analyze the forming processes of rheological objects. Thirdly, we will introduce a deformation transition graph so that the forming processes can be described in a systematic manner. Finally, we will propose a forming control method of a rheological object based on the deformation transition graph.

1 Introduction

In food industry, we can find many manipulative operations that deal with deformable rheological materials such as dough, paste, jelly, and meat. Most these operations are done by humans. Especially, operations with large deformation of rheologic objects depend upon humans. For example, forming of pizza dough to a thin circular shape is performed by humans while extension of the dough can be done by mechanical stretchers. Thus, automatic operations by machines are strongly required for the purpose of cost reduction and cleanness of food.

Object deformation has been investigated in various research fields. Modeling of viscoelastic objects has been studied in computer graphics [1, 2] and virtual reality [3]. These researches focus on the deformation modeling of viscoelastic objects and forming operations of viscoelastic objects are out of consideration. Thus, we have no method to determine a forming strategy based on the object model. Material nature of rheologic objects is investigated extensively in rheology [4]. Unfortunately, deformation of rheologic objects in 3D space is not studied in rheology. Handling operations of deformable objects have been studied recently. Automatic handling of deformable parts in garment industry and shoe industry has been experimentally studied [5, 6]. Zheng and Chen have proposed a strategy to insert a deformable beam into a hole [7]. Wada et al. have proposed a control law for the positioning operation of extensible clothes [8]. These researches focus on handling of elastic objects and forming operations of rheologic objects are out of focus.

In this paper, we will focus on the automatic forming operation of rheologic objects. First, we will develop forming machine of rheological objects with multi degrees of freedom. Secondly, we will analyze the forming processes of rheological objects. Thirdly, we will introduce a deformation transition graph so that the forming processes can be described in a systematic manner. Finally, we will propose a forming control method of a rheological object based on the deformation transition graph.

2 Development of Forming Machine with Multi Degrees-of-Freedom

In this section, we will develop forming machine of rheological objects with multi degrees of freedom.

Figure 1 shows the configuration between a cylindrical roller and a planar table. The roller has six degrees of freedom relative to the table; three for translation and three for rotation. Traditional stretchers have two degrees of freedom; The five freedoms can be assigned to either a roller or a table. Note that feed motion T_1 is assigned to a table in stretchers. Translational freedom T_1 is thus assigned to a table. Rotation around a vertical axis is involved in planar motion in the table plane. Rotational freedom R_2 is thus assigned to the table. The other freedoms T_2 and R_1, R_3 are not planar motion in the table plane. Freedoms T_2 and R_1, R_3 are thus assigned to a roller. Consequently, a table should have three degrees of motion, T_1 and R_2, while a roller should have the other degrees of freedom, T_2 and R_1, R_3.

Based on the above investigation, we will develop a forming machine illustrated in Figure 2, where the five freedoms can be controlled to change the configuration between a roller and a table. A roller is supported by a link mechanism with rotational joints a_1, a_2, and a_3 and prismatic joints b_1 and b_2. All rotational joints are passive while two prismatic joints are driven by actuators. The distance between the roller and the table as well as the angle from the horizon of the roller can be changed by controlling the two actuators, which drive

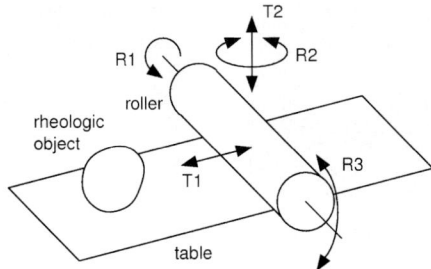

Figure 1: Motion freedoms of forming machine

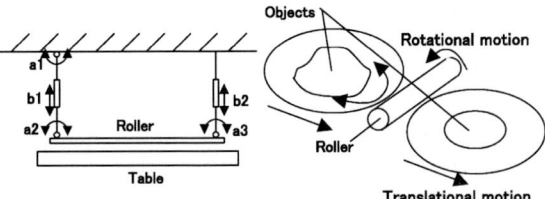

Figure 2: Mechanism of proposed forming machine

prismatic joints b_1 and b_2. As a result, translation T_2 and rotation R_2 are controllable in the proposed mechanism. A circular table can rotate around its axis and can translate reciprocally along the table plane. Namely, rotation R_2 and translation T_1 are controllable in the mechanism. These four effective freedoms are driven by DC servo motors, which are controlled from a PC through a D/A converter. Rotation R_1 and translation T_1 are synchronized by a driving belt. Thus, the tangential speed at the surface of the roller coincides with the translational speed of the table.

The developed forming system is shown in Figure 3. A CCD camera is installed in the system to capture the deformed shape of a rheological object. The captured image is sent to a PC and the motion of mechanism is determined based on the captured image.

3 Analysis of forming processes of rheological objects

In this section, we will investigate the forming process of rheological objects. Let us introduce an outline function to describe the deformed shape and develop a method to compute the outline function. Figure 4-(a) shows an example of grayscale images captured by the CCD camera. Note that the table surface is black and that the location of the table in a captured image is constant. Thus, we can easily extract the deformed shape of a rheological object, as shown in Figure 4-(b).

Let x_{max} be the width of the obtained image and y_{max} be its height. Let $p(i,j)$ be the grayscale at a lattice point (i,j). Then, the gravity center of the deformed shape (X_g, Y_g) is given by

Figure 3: Prototype of forming system

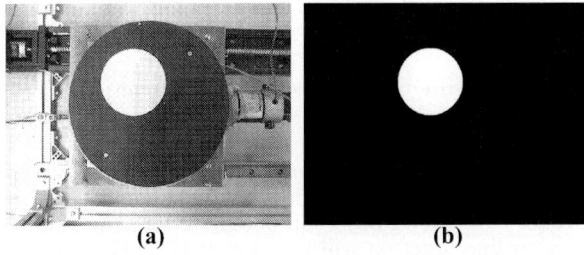

Figure 4: Images of rheological object :(a) Original image, (b) Extracted object image

$$\begin{bmatrix} X_g \\ Y_g \end{bmatrix} = \frac{1}{W} \sum_{i=0}^{x_{max}} \sum_{j=0}^{y_{max}} p(i,j) \begin{bmatrix} i \\ j \end{bmatrix}$$

where

$$W = \sum_{i=0}^{x_{max}} \sum_{j=0}^{y_{max}} p(i,j).$$

Let $f(\theta)$ be the distance from the gravity center to the contour of the deformed shape at angle θ from x-axis, as illustrated in Fig. 5. The deformed shape of a rheological object is then described by a function $f(\theta)(0° \leq \theta \leq 360°)$, which is referred to as *an outline function* of the deformed shape. Figure 6 shows an example of the outline function of a deformed shape. Figure 6-(a) is the deformed shape image and Figure 6-(b) is the outline function of that.

Let us investigate the forming processes of rheological objects. In the forming process, a cylindrical roller extend a rheological object, as illustrated in figure 7. Note that the location of the gravity center changes during a forming process, as illustrated in Figure 7. Thus, the distance between a point on the contour

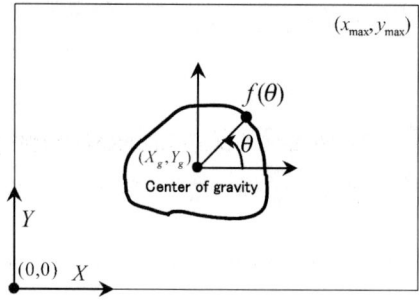

Figure 5: Outline function of deformed shape

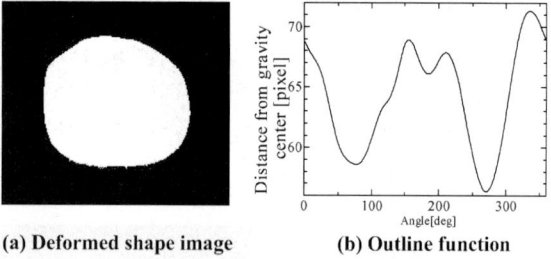

(a) Deformed shape image   (b) Outline function

Figure 6: Example of outline function

and the gravity center may change during a forming process even if the point does not move. Thus, when a roller extends an object in one direction, the value of an outline function will increase at the roller angle as well as the value at the opposite roller direction after the forming.

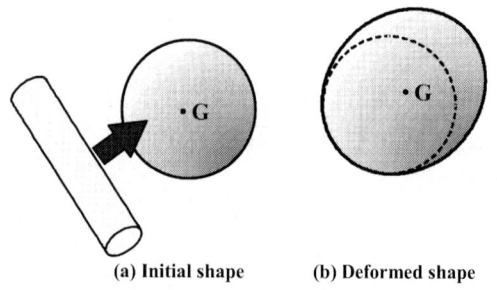

(a) Initial shape   (b) Deformed shape

Figure 7: Extentional forming of rheological objects

Figure 8 shows a forming process of a circular rheological object of radius 47 mm. Figure 8-(a) is an initial shape and Figure 8-(b) is a deformed shape. The roller direction is given by 90°. Outline functions corresponding to the initial shape and the deformed shape are plotted in Figure 8-(c). As shown in the figure, the distance increases at angle 90° as well as at angle 270°. Consequently, we find that the value of an outline function increases at the roller direction and at the opposite direction.

Next, we will investigate the forming process of sim-

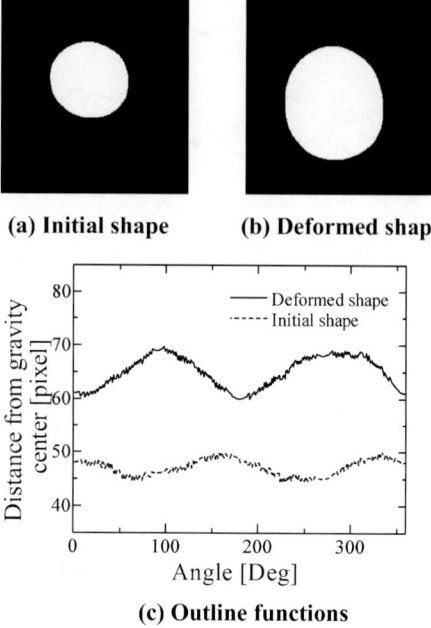

Figure 8: Forming of circular object of radius 47 mm

ilar initial shapes. Figure 9 shows a forming process of a circular rheological object of radius 57mm. Figure 9-(a) is an initial shape and Figure 9-(b) is a deformed shape. The roller direction is given by 90°. Outline functions corresponding to the initial shape and the deformed shape are plotted in Figure 9-(c). Initial shape given Figure 9-(a) is 1.2 times as large as the shape in Figure 8-(a). Let us normalize outline functions plotted in Figure 8-(c) by multiplying 1.2 so that they can be compared with the functions plotted in Figure 9-(c).

Normalized outline functions are plotted in Figure 10. Solid lines describe the forming process of a circular object with radius 47 mm. Broken lines describe the forming process of a circular object with radius 57 mm. From the figure, we find that the deformed shapes are similar each other when the initial shapes are similar. Consequently, it turns out that the forming processes preserve the similarity of deformed shapes.

4 Deformation Transition Graphs
4.1 Descriptiom of Forming Processes

In this section, we will introduce a deformation transition graph to describe the forming process of rheological objects. Since the proposed forming machine has multi degrees-of-freedom, it is required to determine a series of actions of the forming machine to perform a given forming operation of rheological objects successfully. Deformation transition graphs are useful to determine a series of actions for the forming machine.

One action of the forming machine yields a tran-

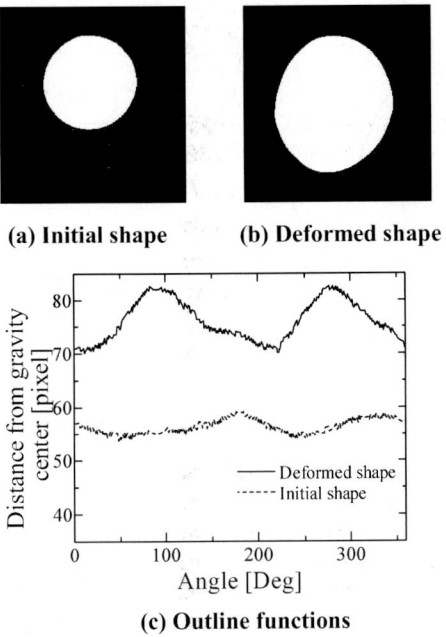

Figure 9: Forming of circular object of radius 57 mm

(a) Initial shape (b) Deformed shape

(c) Outline functions

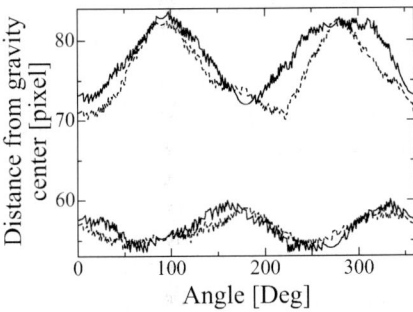

Figure 10: Normalized outline functions corresponding to forming processes of similar objects

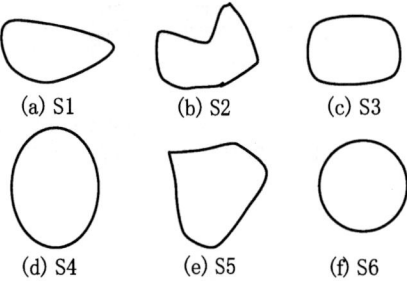

Figure 11: Examples of different shapes of rheological object

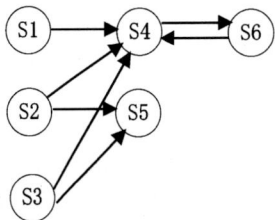

Figure 12: Deformation transition graph

sition from one shape to another of a rheological object. This implies that an action of the forming machine corresponds to a transition between one shape to another. Let us investigate whether an appropriate action of the forming machine yields a transition from one shape to another or not. For example, forming of a rheological object from the shape S_1 shown in Figure 11-(a) into the shape S_4 shown in Figure 11-(d) can be performed by moving the roller horizontally. On the other side, forming from the shape S_1 into the shape S_5 shown in Figure 11-(e) cannot be performed by any action of the forming machine. Namely, some transitions among shapes can be performed by actions of a forming machine and the other transitions are not possible.

Let us classify all actions of a forming machine into a finite number of categories. Each category is represented by one action involved in the category. Then, one shape of a rheological object can be formed into a finite number of shapes by representative actions. Let us describe the relationship between the representative forming actions and deformation of a rheological object using a graph, as illustrated in Figure 12. A node of the graph describes a shape of a rheological object and an arc between two nodes denotes a representative forming action, where the object deforms from one shape corresponding to the starting node into the other corresponding to the end node. This graph is referred to as a *deformation transition graph*. Determination of a series of actions from the initial shape to the goal shape results in path finding from one node of the graph corresponding to the initial shape into another node corresponding to the goal shape. Consequently, deformation transition graphs are useful to plan the forming operations of rheological objects.

4.2 Feature extraction of outline function

In this section, we will classify deformed shapes of a rheological objects to define the nodes of a deformation transition graph. In the extensional forming of a rheological object, a roller approaches to the object toward its concave regions or toward its convex regions, as illustrated in Figure 13. This implies that concave regions and convex regions of a deformed shape are essential in the extensional forming. Thus, we will classify deformed shapes by concavity and convexity of the shapes.

Let us explain the classification using a simple outline function $f(\theta)$ plotted in Figure 14-(a). Note that a concave region of a deformed shape corresponds to a local minimum of an outline function and its convex

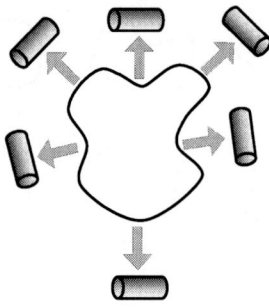

Figure 13: Possible directions of roller in extensional forming

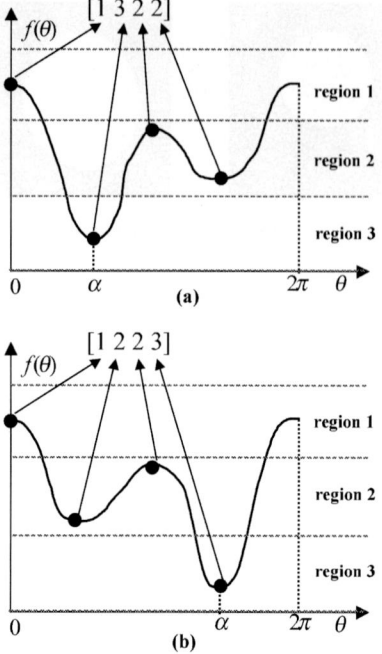

Figure 14: Symbolic representation of deformed shapes

region corresponds to a local maximum of the function. Here, we will focus on two local minimums, say, the smallest one and the second smallest one as well as on two local maximums, say, the largest one and the second largest one. Without losing the generality, we assume that an outline function takes its maximum value at $\theta = 0$ Assume that the function takes its minimum value at $\theta = \alpha$. Let us examine the value of the function at the two local minimums and at the two local maximums. In order to describe the function qualitatively, let us divide the range of the outline function as follows:

$$\begin{aligned} \text{region 1} &= [f(0) - d, f(0) + d], \\ \text{region 2} &= [f(\alpha) + d, f(0) - d], \\ \text{region 3} &= [f(\alpha) - d, f(\alpha) + d] \end{aligned}$$

where $d = (f(0) - f(\alpha))/4$. We will examine which region the value of the function is involved in at the two local minimums and at the two local maximums. For example, the value of an outline function given in Figure 14-(a) involved in region 2 at the second largest local maximum and at the second smallest local minimum. The numbers of regions corresponding to the four local extreme points are listed in order to describe the function qualitatively. The function plotted in Figure 14-(a) is then symbolized as [1322]. These symbols correspond to individual nodes of a deformation transition graph. Note that the maximum of an outline function is always involved in region 1. Consequently, we have $3^3 = 27$ nodes in a deformation transition graph. Figure 14-(b) shows an outline function, which is symbolized as [1223].

4.3 Determination of forming actions

In this section, we will determine forming actions, which represent arcs in a deformation transition graph. Here, we will focus on the angle of the table of a forming machine, ψ, and the height of the roller, D. As mentioned in the previous section, the direction of a roller corresponds to four local extremes of an outline function. Thus, we will determine angle ψ so that the direction of a roller satisfy this condition.

Next, let us investigate how to determine the height of a roller, D. As mentioned in Section 3, a forming action can preserve the similarity of deformed shapes. Forming actions that preserve this similarity are useful in the determination of forming actions since we do not have to deal with the absolute values of action variables, say, ψ and D. Thus, we will focus on forming actions that preserve the similarity of deformed shapes.

Let $D_{present}$ be the present height of a roller and D_{next} be the height in the following action. Note that D_{next} must be determined so that the similarity is preserved. Thus, we will investigate an appropriate value of D_{next} corresponding to $D_{present}$ in advance experimentally. Namely, we will obtain function $D_{next} = D_{next}(D_{present})$ through forming experiments. Once the function is obtained, we can determine the height D_{next} uniquely corresponding to the present height $D_{present}$.

5 Forming Control Using Deformation Transition Graphs

In this section, we will propose a control method for forming rheological object using a deformation transition graph.

Here, we will investigate a forming of pizza dough into a circular shape. Goal node of a deformation transition graph in this forming is denoted by [1111], as illustrated in Figure 15. One action should be determined at any node except the goal node so that the deformed shape can be guided to the goal shape. In this forming, an action where the roller approaches toward the smallest local minimum is selected at any node but the goal node.

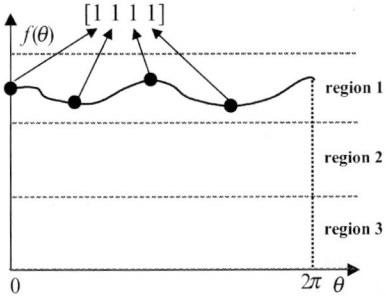

Figure 15: Outline function at goal node

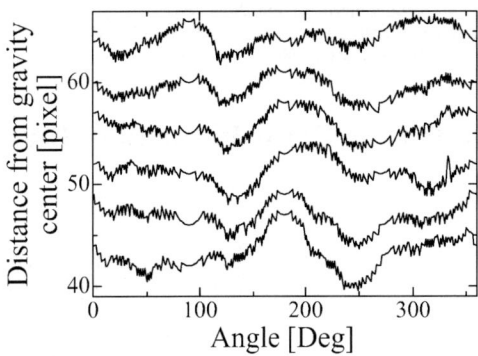

Figure 17: Outline functions at individual iterations

Figure 16 shows an example of the forming. Figure 16-(a) through (f) describe the deformed shapes of a rheological object at individual iterations during the forming. The object consists of wheat flour and water in weight ratio 1:3. Figure 17 provides the corresponding outline functions. From these figures, we find that the rheological object can be formed into a circular shape successfully.

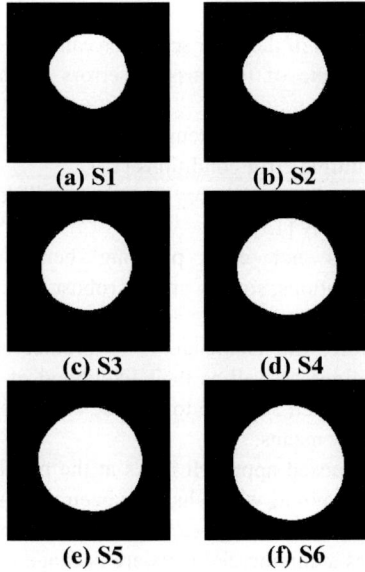

Figure 16: Deformed shapes of dough at individual iterations

6 Concluding Remarks

In this article, we have proposed a control method in the vision-based forming of rheological objects. First, we have developed forming machine of rheological objects with multi degrees-of-freedom. Secondly, we have analyzed the forming processes of rheological objects. Thirdly, we have introduced a deformation transition graph so that the forming processes can be described in a systematic manner. Finally, we have proposed a forming control method of a rheological object based on the deformation transition graph.

Future issues include (1) experimental evaluation of the robustness of the proposed method against the variation of object properties, (2) forming of rheological objects into a shape except circle, and (3) optimization of series of forming actions.

References

[1] Terzopoulos, D., and Fleisher, K., *Modeling Inelastic Deformation: Viscoelasticity, Plasticity, Fracture*, Computer Graphics, Vol.22, No.4, pp.269–278, Alberquerque, May, 1988

[2] Joukhader, A., Deguet, A., and Laugie, C., *A Collision Model for Rigid and Deformable Bodies*, Proc. IEEE Int. Conf. on Robotics and Automation, pp.982–988, Alberquerque, May, 1998

[3] Chai, Y., and Luecke, G. R., *Virtual Clay Modeling Using the ISU Exoskeleton*, Proc. IEEE Virtual Reality Annual Int. Symp., pp.76–80, 1998

[4] Barnes, H. A., Hutton, J. F., and Walters, K., *An Introduction to Rheology*, Elsevier Science Publishers, 1989

[5] Taylor, P. M., eds., *Sensory Robotics for the Handling of Limp Materials*, Springer–Verlag, 1990

[6] Henrich, D., and Wörn, H. eds., *Robot Manipulation of Deformable Objects*, Springer-Verlag, Advanced Manufacturing Series, 2000

[7] Zheng, Y. F., Pei, R., and Chen, C., *Strategies for Automatic Assembly of Deformable Objects*, Proc. IEEE Int. Conf. on Robotics and Automation, pp.2598–2603, 1991

[8] Hirai, S., Wada, T., *Indirect Simultaneous Positioning of Deformable Objects with Multi Pinching Fingers Based on Uncertain Model*, Robotica, Millennium Issue on Grasping and Manipulation, Vol.18, pp.3-11, 2000

[9] Tokumoto, S., Fujita, Y., and Hirai, S., *Deformation Modeling of Viscoelastic Objects for Their Shape Control*, Proc. IEEE Int. Conf. on Robotics and Automation, Vol.1, pp.767–772, Detroit, May, 1999

Off-Line Error Prediction, Diagnosis and Recovery using Virtual Assembly Systems

Cem M. Baydar and **Kazuhiro Saitou**
({cbaydar, kazu}@umich.edu)

Department of Mechanical Engineering
The University of Michigan, MI, 48109, USA.

ABSTRACT

Automated assembly systems often stop their operation due to the unexpected failures occurred during their assembly process. Since these large-scale systems are composed of many parameters, it is difficult to anticipate all possible types of errors with their likelihood of occurrence. Several systems were developed in the literature, focusing on on-line diagnosing and recovering the assembly process in an intelligent manner based on the predicted error scenarios. However, these systems do not cover all of the possible errors and they are deficient in dealing with the unexpected error situations. The proposed approach uses Monte Carlo simulation of the assembly process with the 3D model of the assembly line to predict the possible errors in an offline manner. After that, these predicted errors can be diagnosed and recovered using Bayesian reasoning and Genetic Programming. A case study composed of a peg-in-hole assembly was performed and the results are discussed. It is expected that with this new approach, errors can be diagnosed and recovered accurately and costly downtime of robotic assembly systems will be reduced.

1. INTRODUCTION

Automation is one of the unavoidable concepts nowadays. The developments in the robotics area enable to use robots in large-scale assembly operations for high productivity. However, robotic assembly systems are so sensitive to any perturbation during their operation and this makes the system open to unexpected failures. The cost for excessive maintenance of these systems to recover from these unexpected failures was identified as 200 billion dollars in the USA in 1990 [15].

The unexpectedness of the errors for these systems arises from the fact that most of the errors are *unforeseen* to human design experts before the operation of the line. This is natural since these systems are composed of many working parameters such as dimensional variations of the products, fixtures, sensor capabilities and robot repeatability and when these working parameters are coupled with the 3D workspace, it is difficult to anticipate the conditions, likelihood of occurrence and 3D state of all of the failures [2].

Previous approaches on the error diagnosis and recovery are focused on either "on-line" investigation of error followed by a manual recovery when an error is detected or providing automated intelligent means (i.e., expert systems) to diagnose and patch the process. However, these systems are deficient to deal with most of the errors because of the following reasons:

- Not all of the error scenarios can be predicted,
- 3D-state of the possible errors is not included [2].
- Most of the systems are deficient in dealing multiple error conditions [17],
- Mapping the sensory domain to failure domain is not easy [13],
- It is not easy planning heuristics to all conditions, so they are not robust [13].

Therefore, the challenge is to predict all possible error conditions as well as their likelihood of occurrence and the associated 3D-state to provide efficient and robust error recovery means.

The proposed approach looks at the problem from a different viewpoint, which has not been used so far. It is called *"off-line error prediction and recovery"*. The method uses a commercial software package to model the assembly environment virtually in 3D. After that, the possible error situations and their likelihood of occurrence are predicted by using Monte Carlo simulation of the assembly process. Having the sensory symptoms and their associated failure type and 3D state, these conditions are stored and used for the diagnosis using Bayesian reasoning. Next step is using an off-line error recovery system to generate *robust* recovery plans [4] that can deal with multiple error conditions of similar nature using Genetic Programming [3,12] and the 3D model of the assembly system. Finally, this offline recovery system can be downloaded to the controller of the robotic system to patch the process.

It is expected that this approach will provide efficient means of gathering information about the probable error situations during an assembly process and use this information correctly to develop robust recovery plans. The outcomes of this approach have impact on the industry to reduce costly downtime and maintenance expenses.

2. PREVIOUS WORK

Past research on error recovery in automated assembly lines has focused on using failure trees, expert systems or other intelligent reasoning methods. Among these people, Srinivas [18] is one of the earliest researchers who investigated error detection and recovery strategies. His approach was considering the tasks decomposable into a sequence of transformations from the initial state to a goal state. The next step is building a failure tree and generating an error recovery plan.

Expert systems are also one of the most popular tools used in the error diagnosis and recovery in flexible assembly systems. Several systems were developed in the literature [1] to provide diagnosis and recovery. These are knowledge-based systems, which try to provide recovery plans for possible error conditions for multi-station assembly systems. However, since they are at an abstract level and do not include all of the possible scenarios, they are deficient in handling the unexpected error situations.

Manipulating PLC codes is another approach. Zhou and DiCesare [21], proposed four argumentation methods of process control logic code with error recovery codes: input conditioning, alternate path, feedback error recovery and feedforward error recovery. Fuzzy reasoning was also used in conjunction with the Fuzzy Petri-Nets [6,10] or with expert systems [11,19] to provide probabilistic reasoning on the error diagnostics.

However, those approaches are deficient in handling geometric features of the assembly line, which is essential to make predictions of error scenarios. Since those 3D error states are missing, generated recovery plans' robustness are questionable since some unanticipated error states for the same error type (i.e., collision) may require different plan of recovery.

The need for a robust plan was first discussed in [9]. They developed an automated compliant motion planner based on geometric theory of error detection and recovery. However, the model is limited on modeling the configuration space with all of the properties such as kinematics or motion planning of the robots and 3D positional change of the products. Therefore, it is limited to one-station only. Consequently, a gap was formed between this type of concrete approach (i.e., prediction of the 3D error states) and the abstract approach that was followed by the expert systems.

The following illustration in Figure 1 shows the mapping of the two approaches and this gap between these two approaches. Therefore, a different approach is needed to fill this gap and combine the two different types of approaches discussed above in order to provide efficient means of error recovery. The proposed method discussed in the next section aims to satisfy this need by combining latest developments in the robotics simulation technology with the intelligent reasoning and recovery planning.

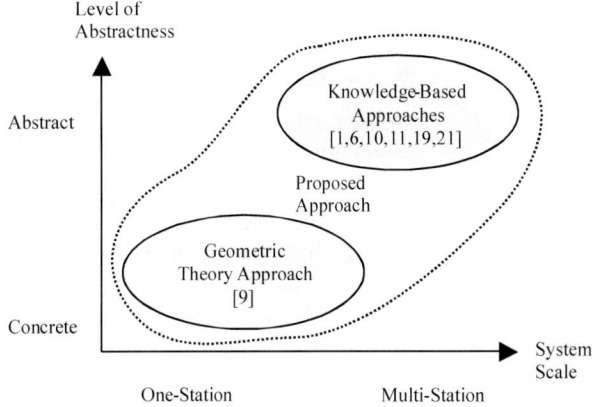

Figure.1: Station Level and System Level Approach

3. PROPOSED METHOD

Recent developments in the computer aided robotic simulation field revealed a new concept called "off-line programming". In *off-line programming*, any robotic system can be modeled virtually in 3D and the performance of the system can be evaluated accurately from the simulations.

The proposed method takes the advantage of off-line programming to predict the possible error scenarios with their 3D-geometric state. The first step is the three-dimensional geometry-based modeling of an entire assembly line using a commercial modeling software. At this step Workspace [22] from FLOW Software Inc. is used. After the system is modeled virtually, next step is the prediction of probable error scenarios by Monte Carlo simulation of assembly processes with the three-dimensional model, based on the statistical model of the dimensional and functional errors in sensors, actuators, products and fixtures.

After probable symptoms and their likelihood is obtained, the next step is the *off-line* logic synthesis for error diagnosis and recovery from the predicted error scenarios based on the three-dimensional model using Bayesian reasoning and Genetic Algorithms. The use of Genetic Algorithms to generate recovery logic is discussed in our previous works in detail [2,3,4].

The final step is building a library of recovery logic and implementing this library to the robot controller in the assembly system to patch the process against unexpected error situations. The following sections give information on the details of the each step.

The following figure summarizes the logic of the proposed approach.

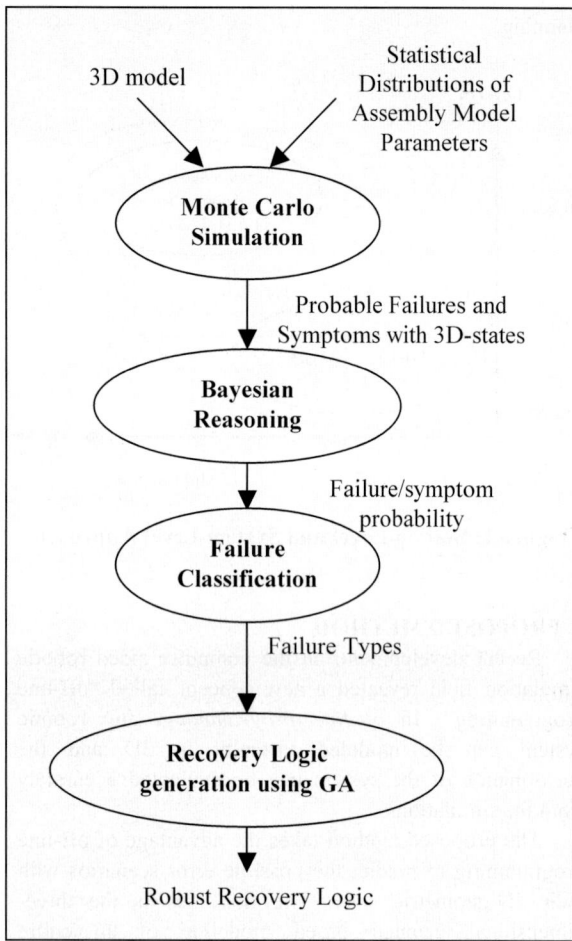

Figure.2: Working mechanism of the proposed approach

3.1 Prediction of Error Scenarios:

A widely used technique for simulating the errors is applying statistical methods to tolerance analysis of mechanical assemblies. At this step, Monte Carlo simulation is used to predict the possible errors. Process parameters are sampled from the appropriate distributions and simulations are performed. The main drawback of this method is that, to get accurate estimates, it is necessary to generate very large samples, which is computationally expensive. However, with the availability of high-speed computers this was overcome. For each simulated error with relatively high likelihood of occurrence, error diagnosis logic will be synthesized for effective error recovery from the error. Fundamental difference from the diagnosis in on-line cases is that a complete sequence of the events, which caused the detected error, is readily available in terms of the sampled parameters.

3.2 Error Diagnosis:

Since only providing error recovery logic is not adequate for the complete recovery process, a diagnosis system is necessary to identify the correct source(s) of error. Error diagnosis is implemented in the following way. First, from the simulation results conditional probabilities of each error situation is obtained for the sampled parameters. Since Monte Carlo simulation is being used, complete sequence of the events that caused an error and their likelihood are readily available. A reasoning engine is developed based on the symptoms (outputs from the sensory values) and the probable error conditions as it is suggested in [13]. This engine will process each possibility of failure and come up with most probable one (or multiple) of the five error classifications. The belief value of each type of failure is calculated by using Bayesian reasoning using the following formula:

$$Bel(F_k) = \frac{P(Y_o \setminus F_k) * P(F_k)}{\sum_{\forall l} P(Y_o \setminus F_l) * P(F_l)}$$

In the above formula, Y_o indicates the given symptoms from the sensor array. The sensor values can only get 0 or 1 depending on their activeness. F_k is the type of failure from the following failure array given in the table below. Depending on the number of sensors or the assembly process each failure type can take any numerical value.

Table.1: Failure Array

Failure Array = {d, e, f, g, h}
d= Grasping Error
e= Collision Error
f= Sensor Failure
g= Misplacement Error
h= Flawed Parts

3.3 Error Recovery:

The proposed approach provides the generation of the error recovery logic using a method called Genetic Programming (GP). The term "Genetic Programming" was first introduced by Koza [12] and it enables a computer to do useful things by automatic programming. It uses the working principles of Genetic algorithms (GAs). In Genetic Programming, each member in the population is a computer program for the solution of the problem. Using an error situation obtained with the sampled parameters, a fitness function based on the allowed recovery criteria can be defined. After the definition of this fitness function, genetic programming can be used to explore an efficient recovery algorithm.

The performance of the error recovery logic can be tested in a *generate and test* fashion [2] such that, several recovery logic algorithms are generated with the genetic programming engine and tested with the commercial software package, Workspace. The results of this evaluation will then be inputted to the GP engine and improved recovery logic is generated based on the obtained results [3,4].

4. EXPERIMENTAL RESULTS

A sample assembly process, which was mainly focused on inserting a peg in a hole, is experimented. The sampled parameters are statistical variations in the *dimensions of peg* and *hole*, *robot repeatability* (both translational ability and wrist repeatability effect), *grasping ability* and sensor reliability for the *grasping sensor* in the gripper and the *position sensor* located above the peg. Figure.3 shows the studied system.

The assembly process is as follows: First, a peg is grasped from the table. During this process, a camera is examining the position of peg, detecting whether it was grasped correctly or not. A sensor in the gripper is also detecting whether the peg is in the gripper or not. After peg is grasped, it is inserted into the hole. During the insertion, process torque/force sensors indicate whether a collision is occurred or not. In addition, during the releasing step of the peg, gripper sensor detects the whether the peg is released correctly or not.

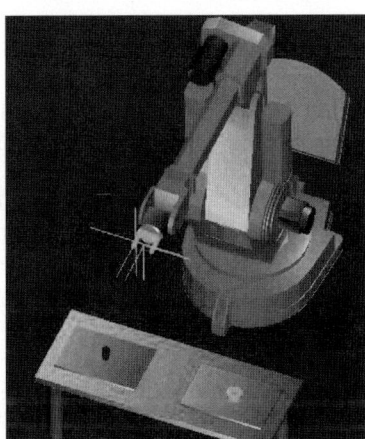

Figure.3: Monte Carlo Simulation of a Peg-in-Hole Problem

During the Monte Carlo simulation, normal distribution is assumed for the variations of peg, hole and robot repeatability, as it is suggested in [7]. Robot repeatability is a measure of positional deviation from the average position achieved by repetitive motion commands from a start position to target position. The parameters and tolerances are as follows:

Table.2: Tolerances of the Sampled Parameters

Sampled Parameter	Value
Hole diameter	50.3 ± 0.0508 mm
Peg diameter	49.92 ± 0.03175 mm
Robot Repeatability	0.2 mm
Wrist Angle Repeatability	0.006 deg

For calculating the mean and standard deviation of the normal distribution, it is assumed that the tolerances are taken in 3σ-range as suggested in [7]. Therefore, the distribution values of the sampled parameters are given in the table below:

Table.3: Distribution values of sampled parameters

Sampled Parameter	Distribution type
Peg	Normal (0, 0.0106mm)
Hole	Normal (0, 0169 mm)
Robot Repeatability	Normal (0, 067 mm)
Wrist Angle Repeatability	Normal (0, 0.002 deg)
Grasping ability	Uniform (0.9 probability)
Position sensor	Uniform (0.95 probability)
Gripper sensor	Uniform (0.95 probability)

The number of simulations is taken as 1000 and the process capability is investigated. Out of 1000 simulations 317 errors are occurred. The observed results for each error type are shown in the table below:

Table.4: Failure Types and percentage

Failure Type:	Percentage:
Collision Error	36.6 %
Grasping Error (Picking)	27.44 %
Grasping Error (Releasing)	27.1 %
Misplacement Error	0
Sensor Failure	0
Flawed Parts	0
Collision + Sensor Failure	5.397 %
Gripper + Sensor Failure	3.45 %

It is realized that none of the pure sensor failures is observable. This does not mean that no sensor failures occurred, the reason is sensor failures cannot be detected at this stage and may be propagated with the later steps of the assembly process to produce detectable errors. Also there is coupled collision and sensor failures that are detected as pure collision errors from the detection system, however Monte Carlo simulation reveals that the percentage of this coupled error type is 5.397 %.

A failure diagnosis system is developed based on a probabilistic Bayesian reasoning discussed. The system uses the symptoms obtained from the assembly system and calculates the likelihood of each possible failure based on these symptoms, coming up with most probable failure type(s). The sensor and failure arrays are given in Table.5.

The sensor values can only get 0 or 1 depending on the activeness. The values for the failure array are given in Table.6 below:

Table.5: Sensor and Failure Arrays

Sensor Array {a, b, c}	Failure Array {d, e, f, g, h}
a = Gripper Sensor	d = Grasping Error
b = Torque/Force Sensor	e = Collision Error
c = Camera	f = Sensor Failure
	g = Misplacement Error
	h = Flawed Parts

Table.6: Failure Types and Associated Values

Failure Type:	Values:
Grasping Error	0-none, 1-picking, 2-releasing
Collision Error	0-none, 1-collision
Sensor Failure	0-none 1-gripper sens., 2-camera, 3-both
Misplacement Error	0-none, 1-misplacement
Flawed Parts	0-none, 1-flawed part

For each failure type, a belief value is calculated using the equation discussed before. The following results are obtained with the given input symptoms as shown in Table.7:

Table.7: Results of Diagnosis Process

Symptom	Failure Type/Probability
(1,0,0)	(2,0,0,0,0) / 0.996
(0,1,0)	(0,1,0,0,0) / 0.989
(0,0,1)	(1,0,0,1,0) / 1
(1,0,1)	(1,0,0,0,0) / 1

It is realized that having a symptom from torque/force sensor does not mean that the error is due to pure collision. There is a possibility of having this error coupled with one/all sensor failures but the most probable action is using the error recovery strategy for collision.

A virtual recovery system is developed based on the obtained errors and their states in 3D Workspace. The system uses Genetic Algorithms to provide error recovery logic and based on a system developed before and discussed in [3,4]. A robust recovery code for collision error is generated in RAPID language. The following lines are belong to the generated code for the collision recovery only:

```
MODULE collision recovery
CONST   robtarget NewGP005   :=   [[1553.09,
530.71, 770.341], [0.707107, -8.65927E-017,
0.707107, 8.65927E-017], [0, 0,-1, 0],
[9E9, 9E9, 9E9, 9E9, 9E9, 9E9]];
CONST   robtarget NewGP006   :=   [[1553.09,
530.89, 770.341], [0.707107, -8.65927E-017,
0.707107, 8.65927E-017], [0, 0,-1, 0],
[9E9, 9E9, 9E9, 9E9, 9E9, 9E9]];
CONST   robtarget NewGP007   :=   [[1553.09,
530.89, 722.85], [0.707107, -8.65927E-017,
0.707107, 8.65927E-017], [0, 0,-1, 0],
[9E9, 9E9, 9E9, 9E9, 9E9, 9E9]];
PERS tooldata t_Gripper := [ TRUE, [[0,
2.27777E-014,      93],[1,       0,       0,
0]],[0.01,[0.01,   0.01,   0.01],[1, 0, 0,
0],0, 0, 0]];
PERS  tooldata  t_Nil  :=  [  TRUE,  [[0,  0,
0],[1,   0,   0,   0]],[0.01,[0.01,   0.01,
0.01],[1, 0, 0, 0],0, 0, 0]];
PERS   wobjdata   w_Nil   :=   [   FALSE, TRUE,
"",[[0, 0, 0],[1, 0, 0, 0]],[[0, 0, 0],[1,
0, 0, 0]]];

PROC Path1()
MoveJ NewGP005, v1000, z1, t_Gripper;
MoveJ NewGP006, v1000, z1, t_Gripper;
MoveL NewGP007, v1000, z1, t_Gripper;
!-    ThisDocument.RunBehavior    "Gripper",
"OpenGripper", ""

!-    ThisDocument.RunBehavior    "Gripper",
"UnGrasp", ""
ENDPROC

PROC main()
Path1;
ENDPROC
ENDMODULE
```

This case study demonstrated the validity of the proposed approach. The sensory information is mapped on the failure domain efficiently, to predict the probable failures and their 3D state. Another advantage is the generation of the recovery code using the virtual assembly system saves the time on finding a robust recovery logic algorithm for the system.

5. CONCLUSIONS

A new approach on the investigation of error diagnosis and recovery is discussed in this paper. Current systems use intelligent reasoning on the diagnosis and recovery for the automated assembly lines. However, they are deficient in anticipating all errors and they leave the 3D state of the possible errors out of consideration, which makes the generated recovery codes non-robust. Because of these facts, the need is identified as to predict all possible error conditions as well as their likelihood of occurrence and the associated 3D-state to provide efficient and robust error recovery means.

The proposed system uses a commercial software package for robotic simulation for the prediction, diagnosis and recovery of the possible failures during an assembly process in four steps:

- Modeling of the assembly system using a commercial off-line robotic software package.

- Monte Carlo simulation of assembly processes to predict the possible error conditions and their likelihood of occurrence.

- Logic synthesis for error diagnosis and recovery from the predicted error scenarios based on the three-dimensional model using Bayesian reasoning and Genetic Algorithms.

- Downloading the developed recovery codes to the robotic controller to patch the assembly process against the unexpected errors.

A case study was conducted by developing a virtual assembly system, which is responsible from a peg-in-hole assembly process. The obtained results showed that, the system is capable of identifying the possible failures and their likelihood as well as their 3D geometrical state. The system is also capable of generating robust recovery codes as in discussed in our previous works [2,3,4].

The proposed method aims to cover the gap between the configuration space approach and the abstract-level knowledge-based approach. Future studies will be performed on more complicated assembly process composed of multiple assembly stations. It is expected that with this new approach, errors can be diagnosed and recovered accurately and costly downtime of robotic assembly systems will be reduced.

References

[1] Abu-Hamdan M. G., El-Gizawy A. S., *"Computer Aided Monitoring System for Flexible Assembly Operations"*, Computers in Industry, Vol. 34, pp. 1-10, 1997.

[2] Baydar C., Saitou K., *"Off-line error recovery logic synthesis in automated assembly lines by using genetic programming"*, Proceedings of the 2000 Japan-USA Symposium on Flexible Automation, 2000.

[3] Baydar C., Saitou K., *"A Genetic Programming Framework for Error Recovery in Robotic Assembly Systems"*, Genetic Evolution and Computation Conference 2000, 2000.

[4] Baydar C., Saitou K., *"Generation of robust recovery logic in assembly systems using multi-level optimization and genetic programming"*, Proceedings of the ASME-DETC2000/CIE Conference, 2000.

[5] Brnjolfsson S., Arnstrom A., *"Error Detection and Recovery in Flexible Assembly Systems"*, International Journal of Advanced Manufacturing Technology, Vol.5, pp. 112-125, 1990.

[6] Cao T.C., Sanderson A.C., *"Sensor-based error recovery for robotic task sequences using fuzzy petri-nets"*, Proceedings of the 1992 IEEE International Conference on Robotics and Automation, Vol.2, pp.1063-1069, 1992.

[7] ElMaraghy H. A., ElMaraghy W. H., Knoll L., *"Design Specification of Parts Dimensional Tolerance for Robotic Assembly"*, Computers in Industry, v10, pp: 47-59, 1988.

[8] Evans E.Z., Lee S.G., *"Automatic Generation of Error Recovery Knowledge Through Learned Activity"*, Proceedings of the 1994 IEEE International Conference on Robotics and Automation, Vol. 4., pp. 2915-2920, 1994.

[9] Jennings J., Donald B., Campbell D., *"Towards Experimental Verification of an Automated Compliant Motion Planner Based on a Geometric Theory of Error Detection and Recovery"*, IEEE International Conference on Robotics and Automation, pp. 632-637, 1989.

[10] Jing Q., Xisen W., Zhihua P., Youngcheng X., *"A Research on Fault Diagnostic Expert System Based on Fuzzy Petri Nets for FMS Machining Cell"*, Proceeding of IEEE International Conference on Industrial Technology, pp. 122-125, 1996.

[11] Kang L., Wenhan Q., *"Fuzzy Expert System in Robotic Assembly Workcell"*, Proceedings of IEEE TENCON, pp. 738-741, 1993.

[12] Koza J. R., "Genetic Programming: On the Programming of Computers by Natural Selection", MIT Press, Cambridge, MA, 1992.

[13] Lopes L.S., Camarinho-Matos L.M., *"Towards Intelligent Execution Supervision for Flexible Assembly Systems"*, IEEE International Conference on Robotics and Automation, pp.1225-1230, 1996.

[14] Lunze J., Schiller F., *"An Example of Fault Diagnosis by means of Probabilistic Logic Reasoning"*, Control Engineering Practice, Vol.7, pp. 271-278,1999.

[15] Luxhoj J.T., Riis J. O., Thorsteinsson U., *"Trends and Perspectives in Industrial Maintenance Management"*, Journal of Manufacturing Systems, Vol.16, No.6, 1997.

[16] Lam R. K., Pollard N. S., Desai R. S., *"Studies in Knowledge-Based Diagnosis of Failures in Robotic Assembly"*, IEEE Conference on Robotics and Automation, pp. 60-65, 1990.

[17] Sampath M., Sengupta R., Lafortune S., Sinnamohideen K., Teneketzis C., *"Failure Diagnosis using Discrete-Event Models"*, IEEE Transactions on Control Systems Technology, v4, n2, pp. 105-124, 1996.

[18] Srinivas S., "Error Recovery in Robot Systems", Ph.D Thesis, California Institute of Technology, 1977.

[19] Tzafestas S. G., Stamou G. B., *"Concerning Automated Assembly: Knowledge-Based Issues and a Fuzzy System for Assembly under Uncertainty"*, Computer Integrated Manufacturing Systems, Vol. 10, No.3, pp. 183-192, 1997.

[20] Visinsky M.L., Cavallaro J.R., Walker I. D., *"Expert System Framework for Fault Detection and Fault Tolerance in Robotics"*, Computers in Electrical Engineering Vol. 20, No.5, pp 421-435, 1994.

[21] Zhou M. C., DiCesare F., *"Adaptive design of petri-net controllers for error recovery in automated manufacturing systems"*, IEEE Transactions on Systems, Man and Cybernetics, Vol.19, No.5, pp.963-973, 1989.

[22] Workspace 5 User-Manual, 2000.

Network Distributed Virtual Design using Coevolutionary Agents

Raj Subbu
subbu@eamri.rpi.edu

Arthur C. Sanderson
acs@eamri.rpi.edu

Electronics Agile Manufacturing Research Institute and
Electrical, Computer, and Systems Engineering Department
Rensselaer Polytechnic Institute, Troy, New York 12180-3590

Abstract

In an increasingly networked global marketplace, products and services are seldom created in isolation and are instead being realized through strategic and dynamic partnerships between suppliers, contract manufacturers, and customers. Superior design-supplier-manufacturing decisions are critical to the survival of enterprises that seek to compete in this environment. A novel evolutionary decision support framework (coevolutionary virtual design environment) particularly suited to distributed network-enabled organizations is introduced. In this framework an electronic interchange of design, supplier, and manufacturing information facilitates concurrent, network distributed decision-making based on evolutionary computation. The approach utilizes distributed evolutionary agents and mobile agents as principal entities that support a network-efficient exploration of planning alternatives in which successive populations systematically select planning alternatives that reduce cost and increase throughput.

I. INTRODUCTION

Advances in information technologies are driving fundamental changes in the processes and organizations of global enterprises. Innovations in software, networks, and database systems enable widely distributed organizations to integrate activities, share information, collaborate on decisions, and execute transactions. As a result it is becoming increasingly uncommon for the creation of products and services in isolation, and they are being realized based on the creation of strategic and dynamic partnerships between suppliers, contract manufacturers, and customers. This trend serves as much of the motivation for *business-to-business electronic commerce*. However, as the numbers of these distinct entities increase and they get more distributed, the complexity of forming efficient partnerships grows; it becomes more difficult to make ideal assignments (partnerships) with respect to multiple criteria including cost, time, and quality. Fundamental to this complexity is that each assignment has the potential to affect overall product cost, and product realization time, and therefore assignments cannot be considered independent of one another. Due to this complexity it is increasingly impossible to make these dynamic partnerships purely on the basis of prior experience, and it becomes necessary to develop efficient decision-making systems that can automate significant portions of the overall decision task. Such decision-making systems access information available at multiple, logically interrelated, distributed databases in order to evaluate the consequences of the assignments.

In this paper, we present a novel approach to design-supplier-manufacturing decision-making that is particularly suited to distributed network-enabled organizations. In this *Coevolutionary Virtual Design Environment*, an electronic interchange of design, supplier, and manufacturing information supports a cooperative distributed decision-making framework based on evolutionary computation. The principal entities in the coevolutionary virtual design environment are distributed evolutionary agents and mobile agents that support a *network-efficient*, concurrent, cooperative exploration of the space of design, supplier, and manufacturing decisions.

In Section II, we provide background information on the *Virtual Design Environment*, and distributed coevolutionary computation. In Section III, we discuss a theoretical foundation for distributed coevolutionary algorithms. Section IV presents a description of the design-supplier-manufacturing planning decision problem, while Section V describes the components of the *Coevolutionary Virtual Design Environment*. In Section VI, we present simulation results that highlight the network efficiency and search quality of distributed coevolutionary algorithms, and conclude in Section VII.

II. BACKGROUND

In this section, we present background information on the Virtual Design Environment, and distributed coevolutionary computation. This discussion assumes that the reader has a background in evolutionary algorithms.

A. Virtual Design Environment

The *Virtual Design Environment* (VDE) [6; 8; 9], a distributed, heterogeneous information architecture, is a novel approach to integrated design-supplier-manufacturing decision-making. The VDE utilizes evolutionary agents, modular computer programs that generate and execute queries among distributed computing and database resources and support a global optimization of integrated design-supplier-manufacturing planning decisions. The prototype VDE has been evaluated using design and supplier data based on a real commercial electronic circuit board product (Pitney Bowes) and data from three commercial manufacturing facilities. The VDE simultaneously selects parts for a design, selects suppliers for these parts, and makes manufacturing selections. Suppliers and manufacturing resources are distributed, and information about parts, suppliers and manufacturing resources is available through network databases. During the course of the evolutionary optimization, the VDE generates *virtual designs* (complete integrated planning decisions) that are evaluated against an evaluation function based on cost and time models. These computations require information collected dynamically over the network. As the evolution proceeds, successive generations of virtual designs are created, and the population systematically converges towards promising integrated planning decisions.

B. Distributed Coevolutionary Computation

Parallel and distributed evolutionary computational techniques include the coarse-grained approach of evolving independent populations on multiple nodes and occasionally migrating individuals between nodes, and the fine-grained approach of distributing individuals among multiple nodes and allowing localized interactions. In these methods (see for example [1; 3; 10]), each node can potentially directly manipulate variables in all n dimensions. These methods have been pursued primarily for speeding up computations in large scale problems and for simultaneously alleviating the effects of premature convergence. Recently, a class of distributed evolutionary computation has been proposed in which the problem is solved by multiple cooperating algorithm components that concurrently explore subspaces of the feasible space, and achieve coevolutionary collaborative problem solving by exchanging partial results [4; 5; 7].

Subbu and Sanderson [7] present a model of distributed coevolutionary algorithms applied to optimization problems in which the variables are *partitioned* among p nodes (see Figure 1). An evolutionary algorithm at each of the p nodes performs a local evolutionary search based on its own set of primary variables using *local and rapidly* accessible information (from a local database) while the secondary variable set at each node is clamped during this phase. An intercommunication between the nodes updates the secondary variables at each node. The local search and intercommunication phases alternate, resulting in a cooperative search by the p nodes.

The VDE functions in a network environment with distributed information sources that need to be accessed for decision-making. In this network environment, inter-node communication delays are a primary factor in sys-

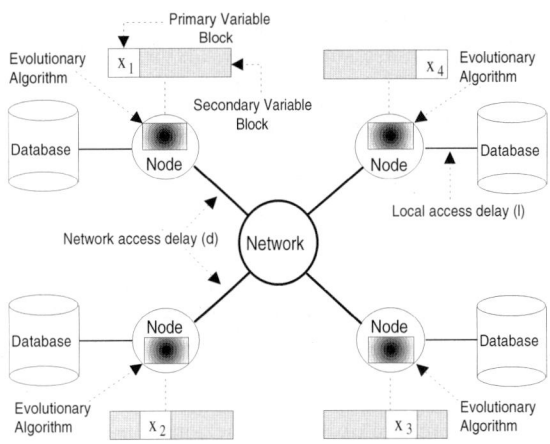

Fig. 1. Distributed coevolutionary computation.

tem performance. The prototype VDE [6; 8; 9] is conceptually based on a centralized optimization model, where computation is performed at one node while information resident at various network nodes is frequently accessed in order to perform fitness evaluations of the alternative planning decisions. Such a centralized planner that requires information from multiple databases is prone to large delays due to remote access latencies.

In this paper, a distributed coevolutionary algorithm is introduced to search for a global solution to the design-supplier-manufacturing planning problem. The approach uses network distributed evolutionary agents and mobile agents to support a concurrent and cooperative exploration of different subspaces of the entire space of integrated design, supplier, and manufacturing decisions, while simultaneously economizing inter-node communications.

III. THEORY OF COEVOLUTIONARY ALGORITHMS

In this section, we summarize the theory of distributed coevolutionary algorithms due to Subbu and Sanderson [7]. The theoretical results in [7] are developed in the space of reals to facilitate a tractable and compact mathematical analysis.

We consider the general nonlinear problem $\min_{x \in \mathcal{X}} \psi(x)$ where $\mathcal{X} \subset \mathbb{R}^n$ is a closed convex space of reals and $\psi : \mathcal{X} \longrightarrow \mathbb{R}_+$ is in general nonlinear and not separable. In general, we will be satisfied with efficiently approximating the value $\psi^* = \min_{x \in \mathcal{X}} \psi(x)$ and finding the corresponding global optimizer $x^* = \arg\min_{x \in \mathcal{X}} \psi(x)$, where $\psi^* = \psi(x^*)$. The variable vector x consists of p blocks $(x_1, x_2, \cdots, x_p)$, where $x_i \in \mathbb{R}^{n_i}, n = \sum_{i=1}^{p} n_i$, and we distribute the p blocks among p nodes. At any node i, x_i is its primary variable set, while $\bar{x}_i$ is its secondary variable set. Given a feasible space $\mathcal{X}$ and a variable distribution, each node i performs a local evolutionary search in its primary subspace $\mathcal{X}_i$, and so $\mathcal{X}$ is the product space $\mathcal{X} = \prod_{i=1}^{p} \mathcal{X}_i$. $(x_i^*|\overline{x}_i) = \arg\min_{x_i \in \mathcal{X}_i} \psi(x_i|\overline{x}_i)$ is the optimizer in the restricted space $(\cdot|\overline{x}_i)$.

The evolutionary search in the primary subspace of each node i utilizes proportional selection and stochastic variational operations, and is described by the following scheme.

1. Set $g = 0$. Choose a probability distribution $P_{(i,g)}$ on $\mathcal{X}_i$, and sample N times to obtain points $(x_{(i,g)}^{(1)}|\overline{x}_i), (x_{(i,g)}^{(2)}|\overline{x}_i), \cdots, (x_{(i,g)}^{(N)}|\overline{x}_i)$.

2. Let $\psi'(x_{(i,g)}^{(j)}|\overline{x}_i) = M - \psi(x_{(i,g)}^{(j)}|\overline{x}_i)$, where M is an upper bound of objective function values. Construct the distribution $R_{(i,g)}$ on $\mathcal{X}_i$ by selecting N points based on their fitness. The probability of selecting a point is given by

$$p_{(i,g)}^{(j)} = \frac{\psi'(x_{(i,g)}^{(j)}|\overline{x}_i)}{\sum_{k=1}^{N} \psi'(x_{(i,g)}^{(k)}|\overline{x}_i)} \quad (1)$$

3. Construct the distribution at the next generation $P_{(i,g+1)}$ on $\mathcal{X}_i$ according to

$$P_{(i,g+1)}(dx_i) = \sum_{j=1}^{N} p_{(i,g)}^{(j)} Q_{(i,g)}(x_{(i,g)}^{(j)}|\overline{x}_i, dx_i) \quad (2)$$

$Q_{(i,g)}(\cdot,\cdot)$ is a measurable, nonnegative function with respect to the first argument and a probability measure with respect to the second argument. A desired sample $(x_i|\overline{x}_i)$ in the distribution $P_{(i,g+1)}$ is obtained by first sampling $R_{(i,g)}$ and then sampling $Q_{(i,g)}$.

4. Set $g \longleftarrow g + 1$, and repeat from Step 2 until some stopping criterion is met.

The construction of the distribution $R_{(i,g)}$ takes into account the global aspect of the search strategy, whereby a point z from all $\mathcal{X}_i$ is chosen, while the distribution $Q_{(i,g)}$ comprises the local aspect of the search strategy, whereby a point in the neighborhood of z is chosen. Then, as $N \to \infty$ the distribution of samples in a local search would follow the sequence

$$P_{(i,g+1)}(dx_i) = \frac{\int_{\mathcal{X}_i} P_{(i,g)}(dz_i) \psi'(z_i|\overline{x}_i) Q_{(i,g)}(z_i, dx_i)}{\int_{\mathcal{X}_i} P_{(i,g)}(dz_i) \psi'(z_i|\overline{x}_i)} \quad (3)$$

and under mild conditions (see [7]) the above distribution sequence converges to the distribution concentrated at $(x_i^*|\overline{x}_i)$ as $g \to \infty$. Importantly, it is shown in [7] that for a unimodal Gaussian objective and a bimodal objective that is a linear combination of Gaussians, the local population distribution at each node i *converges geometrically* to $(x_i^*|\overline{x}_i)$. In the event that $(x_i^*|\overline{x}_i)$ is not unique, the theoretical population distribution would converge to multiple points. In practice, we may assume without loss of generality that the distribution converges to only one such point.

The local search described above is initialized with a randomly selected complete vector of variables x_g. This vector is broadcast to all nodes. The local search starting from this point may be represented as a mapping $T_i : \mathcal{X} \longrightarrow \mathcal{X}_i$ that generates the sequence

$$x_{(i,g+m+1)} = T_i(x_{(1,g)}, \cdots, x_{(i-1,g)}, x_{(i,g+m)},$$
$$x_{(i+1,g)}, \cdots, x_{(p,g)}) \quad m \geq 0$$

Then, as m increases

$$x_g^{(i)} = (x_{(1,g)}, \cdots, x_{(i-1,g)}, x_{(i,g+m)}, x_{(i+1,g)}, \cdots, x_{(p,g)})$$

and $x_g^{(i)}$ would converge geometrically to $(x_i^*|\overline{x}_i)$, where $x_g^{(i)}$ is the result of m generations of local search at node i, starting from point x_g. However, global convergence of the distributed search to the point x^* is the overall goal. For this purpose, let $\mathcal{Z}_g = \left\{ x_g, x_g^{(1)}, \cdots, x_g^{(p)} \right\}$, and let $S : \mathcal{X} \longrightarrow \mathcal{X}$ represent the computation that selects that vector from $\mathcal{Z}_g - x_g$ which has the highest fitness and makes it the new iterate x_{g+1} only if its fitness is greater than that of x_g (else $x_{g+1} = x_g$). The computation $x_{g+1} = S(x_g)$ represents a *global iteration* that encapsulates the combined m-step local search at each node and the intercommunication operation that facilitates selection and update of new iterates. It can be easily seen that the mapping S generates a non-decreasing sequence $\{\psi'(x_g)\}$.

It is shown in [7] that for a unimodal Gaussian objective the sequence $\{\psi'(x_g)\}$ generated by the mapping S *converges geometrically* to $\psi'(x^*)$. It is also shown that for a bimodal objective based on a linear combination of Gaussians, and whose optima are favorably aligned with respect to the search coordinates, the global convergence characteristics are similar to that for the unimodal Gaussian objective. In general instances, when a favorable transformation of coordinates (or additional problem specific information) is unavailable, randomization is introduced in the *coordination* (selection of new iterates) in order to achieve global convergence.

IV. PROBLEM DESCRIPTION

In this section, we present a high-level description of the design-supplier-manufacturing optimization problem applied to printed circuit assembly design and manufacturing. A detailed formulation of this problem appears in Subbu et al. [9]. Earlier versions of this formulation appear in [6] and [8].

A pictorial model of the problem of integrated design, supplier, and manufacturing planning for modular products where suppliers and manufacturing resources are

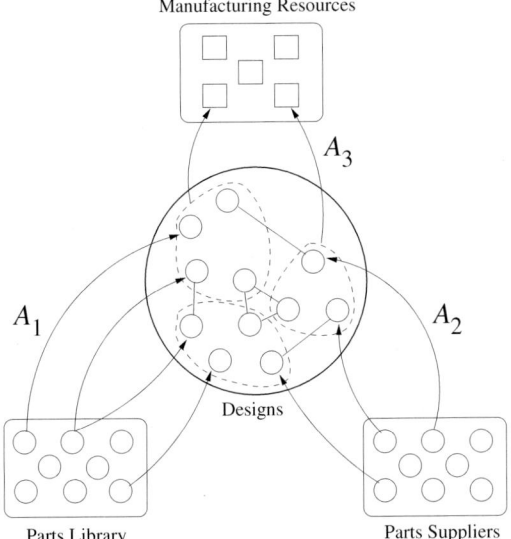

Fig. 2. Structure of the design, supplier, manufacturing planning decision problem. Lines with arrowheads indicate assignments. Dashed lines indicate aggregates. Identical parts in various designs have solid lines between them.

network distributed is shown in Figure 2. The mathematical structure of this planning task is given by $\min\{\psi(x) : \mathbf{A}x = b, x \in \mathbb{Z}_+^n\}$, where x represents a complete decision vector, $\psi(\cdot)$ is a nonlinear objective function, $\mathbf{A}$ is a constraint matrix, and b is a constraint vector. A decision problem in this formulation consists of three assignment problems (A_1, A_2, A_3). The assignment problem A_1 is the assignment of parts (from a parts library) to a design that satisfies a predetermined functional specification. Multiple designs that satisfy the functional specification are possible. The assignment problem A_2 is the assignment of suppliers (from a list of available suppliers) who will supply the parts in a design, and the assignment problem A_3 is the assignment of designs to available manufacturing resources. Each of these assignments contributes to overall product cost and product realization time, and has nonlinear (cannot be evaluated as weighted sums) effects on these measures. Therefore, the assignment problem triple (A_1, A_2, A_3) constitutes a set of highly coupled problems and each of the assignments cannot be considered independent of one another. Product cost is computed as an aggregate of the cost of parts in a given design and cost of manufacturing the design, while product realization time is computed as an aggregate of the parts supply lead time and time to manufacture the design. The overall objective function that is to be minimized is a heuristic weighting of the product cost and an exponential function of the product realization time, and is given by

$$\psi(x) = C(x)e^{(T(x)-\alpha)/\beta} \quad (4)$$

$C(x)$ and $T(x)$ respectively represent the product cost and product realization time for a complete design-supplier-manufacturing assignment x, and α, β are non-zero constants.

V. COEVOLUTIONARY VIRTUAL DESIGN ENVIRONMENT

The *Coevolutionary Virtual Design Environment* is a distributed information architecture that supports coevolutionary optimization of the design-supplier-manufacturing planning problem applied to printed circuit assembly design and manufacturing. In this section, we describe its architecture.

Figure 3 shows a high-level configuration of the networked environment that consists of several logical clusters of network nodes and a product design node. Nodes

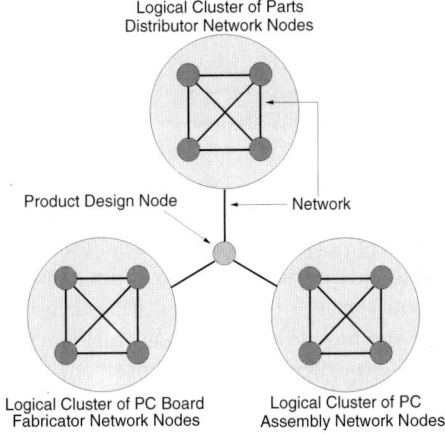

Fig. 3. High-level configuration of the networked environment.

in a logical cluster correspond to a class of functionally equivalent resources, and in general are physically distributed over a network. In this prototype environment we consider three logical clusters of network nodes: (a) *Parts Distributor Nodes:* Each node in this cluster corresponds to a parts distributor (parts warehouse) that stocks parts from several manufacturers, (b) *Printed Circuit Board Fabricator Nodes:* Each node in this cluster corresponds to a printed circuit board manufacturer that may have several alternative board manufacturing lines, (c) *Printed Circuit Assembly Nodes:* Each node in this cluster corresponds to a manufacturing facility with alternative manufacturing lines, each of which is capable of manufacturing printed circuit assemblies given a design (collection of parts) and an associated printed circuit board. The *product design* node generates functional specifications that serve as partial templates for virtual designs. While the search at a parts distributor node is over the space of functionally equivalent designs and is achieved by selecting alternative parts and suppliers for those parts, the search at a printed circuit fabricator node is over the space of available board manufacturing resources, and the search at a printed circuit assembly node is over the space of available assembly resources.

Figure 4 shows the computational components of a network node. Each such node is composed of a net-

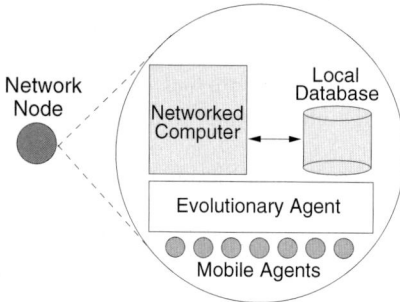

Fig. 4. Computational components of a network node.

worked computer, a local database, an evolutionary agent and several mobile agents that execute on the networked computer. An *evolutionary agent* implements a local evolutionary algorithm that searches over the subspace corresponding to locally available information, initializes with appropriate information that allows for decision-making, generates and executes queries among the local database and applications, coexists in a pool of evolutionary agents, and participates in coordinating the global computation through interactions with other evolutionary agents and mobile agents. An evolutionary agent at a node i is able to solve only the subproblem $\min_{z \in \mathcal{X}_i} \psi(z|\overline{x}_i)$, and an intercommunication with the other distributed evolutionary agents updates its secondary variable set $\overline{x}_i$, consistent with the description in Section III. However, for the design-supplier-manufacturing planning problem described in Section IV, there is a nonlinear coupling between all the variables and for a local variable assignment $z \in \mathcal{X}_i$ an evaluation $\psi(z|\overline{x}_i)$ cannot be completed without network-based information requests from other nodes. This interdependency between subproblems arises because it is not possible to neatly decompose the overall evaluation into a set of subproblem evaluations such that there is an exact fit between a subproblem evaluation and location of information. This problem is important and also appears in the context of multiagent systems research [2]. As a solution to this problem and to facilitate a network-efficient computation of the evaluation function at each node, we introduce *mobile agents* in the architecture. These mobile agents migrate to nodes during intercommunication operations carrying updated information from the nodes they migrate and provide the missing computational functionality at a node. For instance, an agent that migrates from a printed circuit assembly node to a parts distributor node would carry with it a model of resources and information of a specific assembly line (at the printed circuit assembly node), such that when the evolutionary agent at the parts distributor node selects an alternative set of parts, the associated cost and time of assembly at that manufacturing line can be computed.

VI. SIMULATION RESULTS

In this section, we experimentally evaluate the performance of distributed coevolutionary algorithms executing over a simulated network environment, under a variety of conditions, and statistically study their search quality and network efficiency relative to a centralized evolutionary algorithm. For these evaluations we have used a number of network environment configurations and printed circuit assembly design-supplier-manufacturing optimization problems. The problems and corresponding underlying data and models are generated using a combination of real design and manufacturing data and randomization.

When more than one node exists in a logical cluster of network nodes (see Figure 3), in the context of a coevolutionary search the subproblems solved at each node in a cluster are different in spite of being functionally similar. The subproblems are different because of differences in resources available at each node. The coevolutionary algorithm has no direct means (while the centralized algorithm does) to search the larger subproblem that is implicitly created by an augmentation of the subproblems at each node in a logical cluster. In order to provide this means to the coevolutionary algorithm, we introduce an information splicing operation whose principal function is to stochastically combine information from nodes.

We study the effects of two primary factors on coevolutionary algorithm performance—(a) effect of number of generations between nodal intercommunications (communication interval), and (b) the effect of the *delay ratio*, $\frac{l}{d}$, which is the ratio of the local access delay and network access delay for a given networked environment (see Figure 1). Global coordination (see Section III) of the distributed computation is achieved through a structured competition between local solutions from each of the nodes and randomized solutions created using the information splicing operation described above.

Algorithm performance is evaluated based on two metrics: (a) *Percentage Convergence Error:* which measures the percentage deviation of the average performance of a coevolutionary algorithm relative to the average performance of a centralized algorithm, and (b) *Percentage Computational Advantage:* which measures the percentage time advantage of the average performance of the distributed computation over the average performance of the centralized computation when network and local access delays are considered. For these evaluations, average performance of an algorithm given a parameter tuple is computed using 20 independent trials.

Figure 5 shows the average (and standard deviation) percentage convergence error with respect to the communication interval, while Figure 6 shows the average (and standard deviation) percentage computational advantage with respect to the communication interval and delay ratio $\frac{l}{d}$. The statistical moments corresponding to each communication interval are derived using a set of 10 increasingly complex printed circuit assembly problems for which the number of network nodes range from 3 through 30.

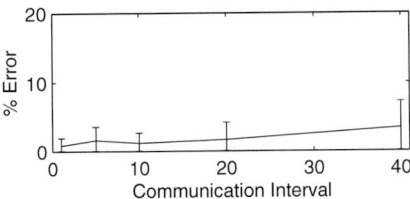

Fig. 5. Average (and standard deviation) of the percentage convergence error with respect to communication interval.

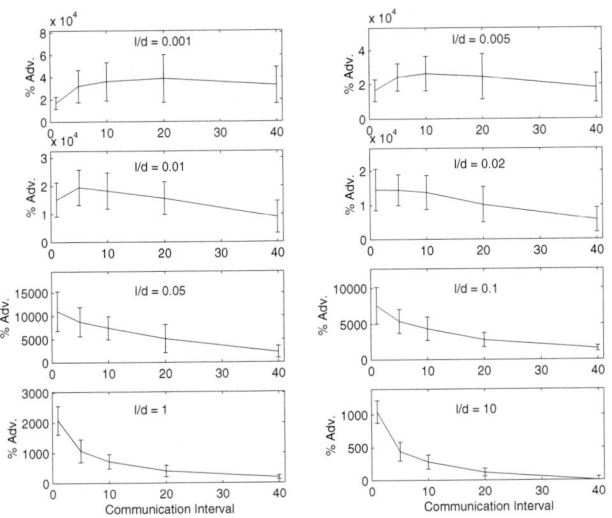

Fig. 6. Average (and standard deviation) of the percentage computational advantage with respect to communication interval and delay ratio $\frac{l}{d}$.

The percentage convergence error of the distributed computation is higher when the communication frequency is small (or communication interval is large). This can be explained by noting that when a local search at a node executes for a large number of generations before an intercommunication, the local population at that node would be less diverse and would contribute less diverse solutions during global coordination. The percentage convergence error is smallest when the communication frequency is highest, since higher the communication frequency, greater is the chance that the coevolutionary algorithm quickly finds good solutions. However, a higher intercommunication frequency is computationally advantageous only for higher delay ratios (> 0.02), corresponding to the cases when local access delays are a larger fraction of the network access delays. An important observation is that for network environments with small delay ratios (< 0.02) it is systematically more advantageous to communicate less frequently. Also, as the delay ratio gets very large, the advantage of distributed computation reduces.

VII. CONCLUSIONS

In this paper, we have described the Coevolutionary Virtual Design Environment that supports network-efficient design, supplier, and manufacturing planning optimization for printed circuit assemblies. The coevolutionary algorithms in this environment are presented using a mathematical foundation, and are experimentally evaluated using statistical techniques. Importantly, we have demonstrated the combination of distributed coevolutionary algorithms and a multiagent architecture as a network-efficient strategy for concurrent, cooperative exploration of a *highly coupled* space of design, supplier, and manufacturing decisions.

Acknowledgments: This work has been conducted in the Electronics Agile Manufacturing Research Institute (EAMRI) at Rensselaer Polytechnic Institute. The EAMRI is partially funded by grants DMI-9320955 and DMI-0075524 from the National Science Foundation.

References

[1] CANTÚ-PAZ, E. *Designing Efficient and Accurate Parallel Genetic Algorithms.* PhD thesis, University of Illinois at Urbana-Champaign, Urbana, Illinois, 1999.

[2] LESSER, V. R. Cooperative multiagent systems: A personal view of the state of the art. *IEEE Transactions on Knowledge and Data Engineering 11*, 1 (1999).

[3] MÜHLENBEIN, H. Evolution in time and space–the parallel genetic algorithm. In *Foundations of Genetic Algorithms*, G. J. E. Rawlins, Ed. Morgan Kaufmann, 1991.

[4] ORTEGA, J., BERNIER, J. L., DÍAZ, A. F., ROJAS, I., SALMERÓN, M., AND PRIETO, A. Parallel combinatorial optimization with evolutionary cooperation between processors. In *Proceedings of the Congress on Evolutionary Computation* (Washington D. C., July 1999).

[5] POTTER, M. A. *The Design and Analysis of a Computational Model of Cooperative Coevolution.* PhD thesis, George Mason University, Fairfax, Virginia, 1997.

[6] SUBBU, R., HOCAOĞLU, C., AND SANDERSON, A. C. A virtual design environment using evolutionary agents. In *Proceedings of the IEEE International Conference on Robotics and Automation* (Leuven, Belgium, May 1998).

[7] SUBBU, R., AND SANDERSON, A. C. Modeling and convergence analysis of distributed coevolutionary algorithms. In *Proceedings of the IEEE Congress on Evolutionary Computation* (San Diego, California, July 2000).

[8] SUBBU, R., SANDERSON, A. C., HOCAOĞLU, C., AND GRAVES, R. J. Distributed virtual design environment using intelligent agent architecture. In *Proceedings of the Industrial Engineering Research Conference* (Banff, Canada, May 1998).

[9] SUBBU, R., SANDERSON, A. C., HOCAOĞLU, C., AND GRAVES, R. J. Evolutionary decision support for distributed virtual design in modular product manufacturing. *Production Planning and Control 10*, 7 (1999). Special Issue on Agile, Intelligent, and Computer Integrated Manufacturing.

[10] TANESE, R. *Distributed genetic algorithms for function optimization.* PhD thesis, University of Michigan, Ann Arbor, Michigan, 1989.

MULTI-COMPONENT GENETIC ALGORITHM FOR GENERATING BEST BENDING SEQUENCE AND TOOL SELECTION IN SHEET METAL PARTS

CHITRA MALINI **THANAPANDI***, ARANYA **WALAIRACHT***, SHIGEYUKI **OHARA***

*Department of Electronics, School of Engineering, Tokai University
1117,Kitakaname, Hiratsuka-shi, Kanagawa-Ken, 259-1292,Japan
Phone:(+81)- 463-58-1211 Ext.4107, Fax: (+81)- 463-50-2031
E-mail: {0aeem013, 7jeed001, ohara}@keyaki.cc.u-tokai.ac.jp

Abstract

This paper investigates multi-component genetic algorithm for obtaining an optimal solution to the problem of generating bending sequence and tool selection in sheet metal parts. Finding bending sequence and assigning tools to it result to a combination explosion which emphasis probabilistic approach. It finds a feasible solution based on collision avoidance and user's desire. The multi-component genetic algorithm represents three important factors that determine the feasible solution the bend sequence, punch and die to be selected. In this paper, we analyze and define the problem and discuss our new approach towards solving it and the ideas are being implemented and evaluated using the real part data.

Keywords: *Automated Process Planning, Sheet Metal Bending, Combination Explosion, Collision Detection, User Desire Fitness Function, Multi-Component Genetic Algorithm.*

1. Introduction

In manufacturing industries, with the economic reasons the need to manufacture small sized batches and shorter delivery times are inevitable. Hence the process planning departments have to generate more process plans in a shorter period of time. With the complex parts, it becomes tedious as it involves combination (non-polynomial, exponential) problem, which cannot be solved within the required time and is a major bottleneck in manufacturing industries. Hence we believe that Automated Process Planning will significantly enhance the throughput and dramatically improve the performance.

In sheet metal industries, the required shape has to be bent from the flat layout. This bending process, the most complex process is due to the determination of bend sequence and assigning proper tools for each bend. Determining bend sequence is itself a combination problem and this when coupled with tool assignment results to a huge combination. And this clearly states a deterministic approach is impossible to solve the problem and Genetic Algorithm, a probabilistic approach is being proposed in this paper.

This paper is concerned with techniques for automated bending sequence and tool assignment for each bend. And furthermore, the system considers user's desire upon which the final optimal solution is decided.

In the following sections, the problem is defined, analyzed and are explained how they are mapped to Genetic Algorithms. Finally, the expected results are being discussed in the last section and the future ideas are being described.

2. Related Works

This bending problem has been researched and many AI techniques have been applied. L.J.de.Vin [1] has developed a process planning system for finding the optimal sequence. This system focuses on generating optimal bending sequence without part-tool collision and tolerance constraints. Heuristic rules have been used to reduce the search space. Shpitalni.M [2] uses A* algorithm and a set of heuristics to find the feasible solution with the lower cost without exceeding time limitation. J.R.Duflou [3] uses branch and bound algorithm which is characterized by the dynamic updateable penalty system. The knowledge is obtained through the analysis of partial sequences and heuristics are applied to refine the search space. As heuristics depend on part geometry, parts with new feature have more probability to remain unsolved.

In all these researches, the feasible solution are based on collision detection and importance for the operator's desires are not given much importance and are decided by the system itself. But since the human role is essential in bending process they should be given importance and need to be considered while evaluating the automatic bending process planning. And in this paper, we propose a new approach to obtain a feasible solution based on both collision detection and user's desire and is explained.

3. Problem Analysis & Definition

3.1. Problem Analysis

In the bending problem, the flat layout has to be bent along its bend lines to get the required shape. Bending sequence has to be determined and to bend them proper tools have to be assigned for each bend. The tools are a set of punch and die having their own characteristics. The punch can be defined and differentiated from other punches by its angle, its type it can bend and its radius. The type of bend means the shape it can bend (V, R, U...). The angle and the type of the bend and its depth, it can support, can define the die.

3.2. Reduction of Tool Search Space

Having tools of varying type, angle and radius any tools cannot be selected for any bend line. For a bend line, tools having the same angle or angle less than the bend line of the same bend type is only eligible. So considering all tools is meaning less and makes the search space so wider making the search time consuming. Hence we filter the tools by bend type and angles of the bend line. Tools to be selected for a bend line can be grouped by the angle of the bend line. That is all tools having angles the angles of bend line can be grouped and they are eligible to be selected. Hence for a product with N bend lines and n varying angles and t number of eligible tools for each angle then the combination order of the tools will be

$$\sum_{i=1}^{n} ni * ti!$$

And this when combined with bending sequence results to huge combination of

$$N!(\sum_{i=1}^{n} ni * ti!)$$

If there is a product with 7 bend lines with all 3 different angles and the eligible tools for each angle is 5,then the combination will be 4233600 combinations.

With the present researches focusing on generating feasible solution based on collision avoidance consideration of user desire is neglected. Bending operations as a manufacturing activity distinguish themselves from many other processes because of the physical effort that is required from the side of the machine operator. Even with the sophisticated CNC machine tools, robot systems still the need for the operator is conditio sine qua non. As bending is an important activity performed by means of conventional or general purpose CNC press brake machines, this human factor and his desire should be taken into consideration when evaluating a proposed process plan. The importance for the user's desire can be found in [4]. The following section explains the user's desire considered.

3.3. User's Desire

From the user (operator) point of view, user's desire can be classified to:
- *Safety:*

Part Orientation: This shows whether the part has to be oriented between the punch and die facing top or bottom.

Part Insertion: This shows whether the right part or the left part has to be inserted between the punch and die while bending.

The lesser the part orientation changes and smaller bend being inserted is safer for the operators.
- *Process Time Reduction:*

The number tools to be changed should be kept minimum so that the bending process can be done as fast as possible to meet the demands.
- *Risk:*

When accurate tools are being selected for a particular bend the risk for damaged products can be reduced. This means the angle of the tool and angle of the bend should be the same.
- *User's Knowledge:*

User's knowledge regarding the tool selection and bend sequence are given high performance and are selected with more probability.

3.4. Problem Definition

Closely viewing this problem we can see two problems are being embedded in it. The determination of bend sequence is a sequencing problem and assigning tools to each bend line is a scheduling problem. This can be expressed in a tabular form and a feasible solution can be obtained by optimizing the table satisfying the user's desire.

Thus the problem can be defined as: Given a set of bend lines for a desired shape, bend sequence has to be generated and tools have to be assigned for each bend in such a way that there is no collision between the part-part, part-tool, part-machine and also satisfies the user's desire.

And in this paper we focus on both reducing the search space as well as generating feasible solution based on user's desire. Fig.1 explains the problem being tabulated.

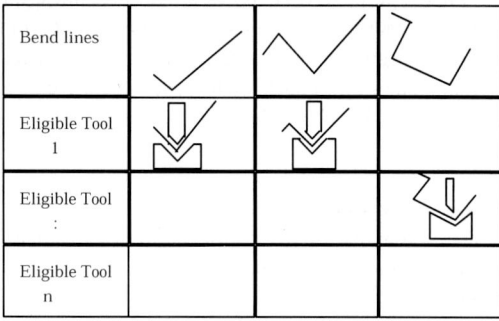

Fig.1.Tabulating the problem

4. Automatic Bending Process Planning Using Genetic Algorithms

4.1. Genetic Algorithms

Genetic Algorithms (GA) are evolutionary intelligent search algorithms [5] working with the coding of set of parameters. The set of parameters, simply said genes are to be coded into finite length of string. GA's work on the principle of survival of the fittest among the strings. It maintains a population of P candidate solutions over many generations, known as chromosomes. These chromosomes are selected according to their fitness value. The fitness value of a chromosome is the payoff value that is associated with the chromosome. For widening the search space, variation is introduced into the population using the two GA operators. The first operator crossover exchanges the information of genes in the chromosomes between the selected chromosomes. Mutation is the second operator that prevents premature loss of important information by randomly flipping the genes in the chromosome. The following shows the flow of Genetic Algorithm.

Function Genetic Algorithms (population, fitness function)
If (individual is fit) Exit;
Else
Repeat till the fit individual exist.
   {   Do selection;
      Do crossover;
      Do mutation;
   }

4.2. Multi-Component Genetic Algorithms

In this bending problem, multi-component Genetic Algorithm is being used and is explained in this section. The important parameters, bend sequence, punch to be used and die to be used for each bend line are represented as components of the chromosome. The punch and the die component depend on the geometry of the parts. The GA operators are applied to these components individually.

4.3. Structure of the chromosome

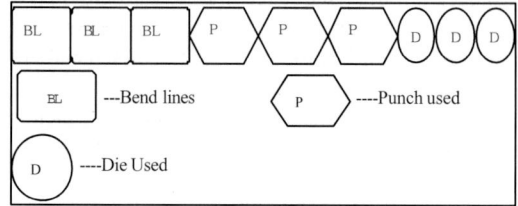

Fig.2. Multi-component chromosome structure

The chromosome is a combination of three components, bend sequence, punch and die to be assigned. The length of each component of the chromosome is dependent and limited to the number of bend lines of the part. Fig.2 shows the chromosome structure of a problem of a part with 3 bend lines.

Bend sequence chromosomes (BL) represent the bend lines to be sequenced. It is represented by its characters of bend line number, bend angle, bend type, part Orientation and part Insertion. The bend line number is randomly selected and Part Orientation is set to default as top. The Part Insertion is set to right. The bend angle and bend type information is read from the part data.

The punch (P) and the die (D) chromosome represent the punch and the die to be selected for the respective bend line in the bend sequence chromosome. The punches are classified into families based on their bend types. They are further classified into parent and child. The punches of the same family are grouped upon the angles it can bend and make a parent. Each punch in the parent group differs from other by their tool property width (radius) and length and are said to be child. Die are also classified as same as punches, but the child is differentiated by the depth and length. The punch and the die which fits the bend type and bend angle of the bend line in the bend sequence chromosome are selected randomly.

4.4. Selection

The chromosomes are evaluated by their fitness value and checked whether optimal solution exists. If there are no optimal solutions generated then, these chromosomes are selected for reproduction. Roulette selection method is used to select the chromosomes for reproduction. In selection the fitness values are treated as probabilities and chromosomes with higher fitness value have more probability for being selected.

4.5. Crossover

The chromosomes selected are given a chance to crossover to create new chromosomes with variation. In crossover chromosomes pairs are selected one being mother and another being father. A single crossover point is selected randomly and the genes of the chromosomes after the crossover point are exchanged between the mother and father chromosomes. As no bend lines can be repeated or omitted (bend lines numbers have to be maintained) in bending problem traditional crossover cannot be applied. Hence a single point position based crossover is being used. In single point position based crossover a point in which the genes have to be exchanged is selected randomly. If the gene of the mother chromosome is same as the gene of the father chromosome no exchange is done. If the gene of the mother chromosome is not as same as that of the father then the genes are exchanged and the genes in that crossover point is replaced in the position of the exchanged gene.

Position based crossover is shown below:
Chromosome A: 2 5 4 1 3 (Mother)
Chromosome B: 3 1 2 5 4 (Father)

If the crossover point selected by random is 2,then the bend line 1 and 5 are exchanged. And in Chromosome A the bend line 5 is replaced in the position of bend line 1 and in Chromosome B the bend line 1 is replaced in the position of bend line 5.
Chromosome A: 2 1 4 5 3
Chromosome B: 3 5 2 1 4
Double lines show the exchange of bend lines and single lines shows the replaced positions. Continuing in this same fashion the chromosome results in
Chromosome A: 4 1 2 5 3
Chromosome B: 3 5 4 1 2

Chromosome A: 4 5 2 1 3
Chromosome B: 3 1 4 5 2

Chromosome A: 4 5 3 1 2
Chromosome B: 2 1 4 5 3
And final reproduced chromosomes will be
Chromosome A: 4 5 3 1 2
Chromosome B: 2 1 4 5 3
Depending on the bend lines the punch and the die are also exchanged in the same way. This is shown in the following example with the 3 bend lines.

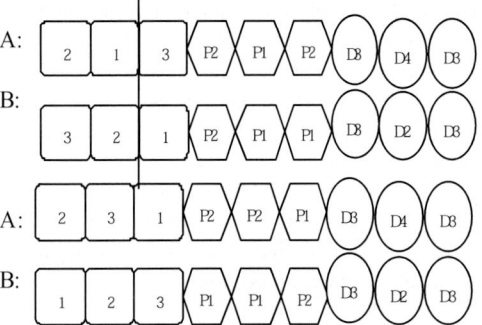

Fig.3. Chromosomes before and after crossover

If the random point selected is 2,then after crossover the chromosomes look like Fig.3.
The bend lines are exchanged on position basis and depending upon the bend lines the punch and die are also exchanged, but since the die chromosomes genes to be crossover are same no exchange is made.
The chromosomes are done crossover at a probability (p_m) of 0.8 per pair.

4.6. Mutation

The chromosomes and the genes to be mutated or flipped are selected randomly. And with the bend sequence component the gene that is selected for mutation, its characteristics are flipped. The bend sequence component's characteristic Part Orientation and Part Insertion are flipped. On mutation, Part Orientation is flipped from the default setting of top to bottom and the Part Insertion is also flipped from right to left.

With the tools (Punch and Die), any other tool is selected from the parent group. That is, another tool with the same angle but different property is selected. The chromosomes are done mutation at a probability (p_c) of 0.1 per bit.

4.7. Fitness Function

The chromosomes of the bending problem are evaluated based on the fitness function, which is a combination two-fitness function. The first fitness function is collision detection function and the second is User Desire function.

4.7.1. Collision Detection Function

It is a function that checks whether the candidate solution generated by the Genetic Algorithm has a collision. 2-D graphic approach is being used to check collision. The flange, which is bended, and the flange not bended are considered as separate objects and the punch and the die selected for the relative bends are also considered as objects. The areas of these objects are calculated and check is made if there is any intersection in these areas. If there is no intersection it is said there is no collision and if there results some intersection between any objects it is said that there is collision in the particular sequence (solution).

4.7.2. User Desire Function

It is a function that checks whether the solutions, which are collision free, satisfy the user's desire. The user's desire as explained in the section 3.3 are being evaluated by the respecting functions:

- *Safety:*

The part orientation changes are being calculated and lesser the changes the more the bonus is earned. With the part Insertion the distance of the flange right and left to the bend line is calculated and if the smallest flange is inserted bonus is added.

- *Process Time Reduction:*

The maximum number of tools that can be used for a product equals to the number of bend lines. With the number of tools to be used being reduced the bonus is increased.

Bonus=Min.ToolBonus+(Max.Tool – Tool used)*Extra Bonus where extra bonus is initialized with some fixed value eg.10.

- *Risk:*

When the angle of the tool and the angle of the bend line of the part match perfectly, we can say that the risk of imperfect shape can be reduced. Hence, Bonus is granted when:

Selected tool's angle==angle of the bend line

- *User's Knowledge:*

User preferred tools are given high preference and extra bonus is given when:

Selected tool= user preferred tool.

5. Implementation and Evaluation

The above-described ideas has been implemented in BBST system (Best Bending Sequence &Tool Selection) and evaluated with the real part data having 7 bend lines. This section explains our present system 's input and output and its results.

5.1. Input and Output of BBST System

The system's input and desired output is explained in the following sections:
The system is fed with the part geometry, tool library (database) and user's desire. The part geometry and the tool geometry are in dxf.files. The system uses genetic algorithm to generate the best bending sequence and tools to be assigned for each bend. The system generates best bending sequence and tools to be assigned for each bend with no collision and satisfying user desire.

5.2. Evaluated Results

The following part as shown in Fig.4 is implemented with the ideas and the results were evaluated. The part has 9 bend lines and we used the tool database, which has 15 tools to generate the best bending sequence and tool selection.

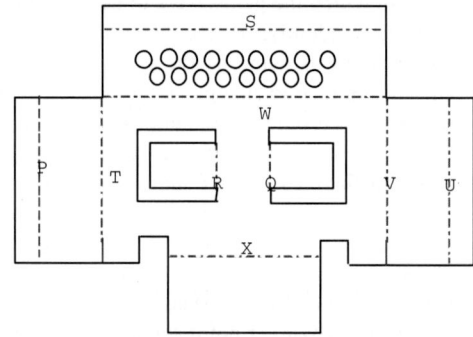

Fig. 5.Sample Part (Data_009) unfolded used for evaluation

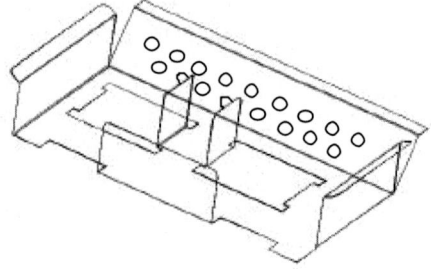

Fig.5.Sample Part (Data_009) folded to its final shape

If the above data were to be handled by the user, which has 1.73E+23 possible combinations, it would not be possible to get the results within the required time and at present only experts are able to handle these problems with their experience. But with a new complicated parts it too becomes difficult.

The sequence obtained by the BBST system is as follows and is shown in Fig.6. When the part was bend according to the sequence generated we were able to get the shape required as in Fig.5 without collision satisfying user desire.

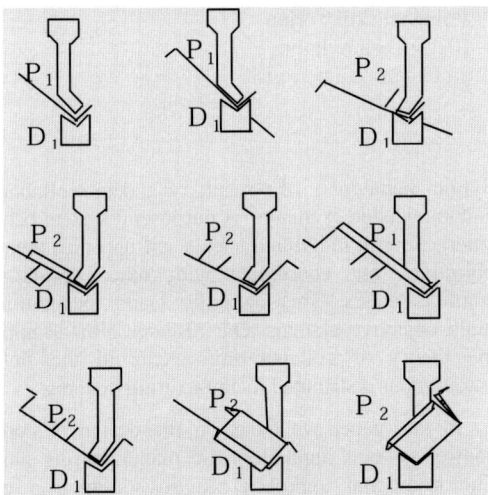

Fig.6. Bend sequences and tool assignment for the sample part (Data_009)

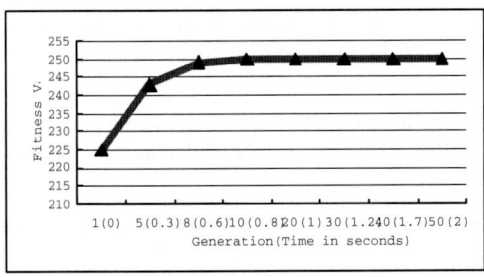

Fig.7. Convergence and performance of solution for the sample part (Data_009)

The number of tools used was kept minimum. The number of tool changes was 4 and 5 tools were selected for bending the part. The Part Insertion was done with the minimum distance towards the machine and this ensures the safety.

And our BBST system was able to get the sequence and tools selected within 50 generations. The result was obtained on a sample population of 100 chromosomes and was able to get results from its 10_{th} generation. After stable results continuing in its next 40 generations the search was stopped. The system was tested on Dell Computer Windows 2000 with 128 MB RAM and took 2 seconds to generate solution. Convergence and performance of the solution for the sample part Data_009 using Multi-Component Genetic Algorithms in BBST system is shown in Fig.7.

6. Conclusion

In this paper, we proposed a new approach for bending problem. We reduce the search space by filtering the tools and use Genetic Algorithm to generate the best bending sequence and assign proper tools to each bend. The feasible solution is based on collision detection and satisfying user's desire. The user's desire is defined and its fitness function is described. These ideas were implemented for real part data with 9 bend lines and the results are evaluated.

In our future research, we are planning it to be implemented and get our expected results with complicated parts. We are also considering of using expert rule database, which has the operator's knowledge on bend sequencing to fasten the search of sequencing.

References

[1] DeVin, J.deVries, A.H.Streppel, E.J.W.Klaassen and H.J.J.Kals, "The generation of bending sequences in a CAPP System for sheet metal components", Journal of materials processing technology, Volume 41 (1994), pp. 331-339.

[2] Radin, B. and Shpitalni, M., "Two-Stage Algorithm for determination of the bending sequence in sheet metal products", Proceedings of the ASME Design Automation Conference, Irvine, CA, (1996) USA, pp.1-12.

[3] J.R.Duflou, D.Van Oudheusden, J.P.Kruth and D.Cattrysee, "Methods for sequencing of sheet metal bending operations", International Journal of Production Research, Volume 37, (1999), pp.3185-3202.

[4] J.R.Duflou, "Ergonomics based criteria for manufacturability and process plan evaluation for bending processes", Proceedings of the 4th International Conference on Sheet Metal, Enschede, Volume 1(1996), pp.105-116.

[5] D.E.Goldberg,"Genetic Algorithms in search, optimization and Machine Learning," Addison Wesley, Massachusetts, 1989.

A Hybrid Software Agent Model For Decentralized Control

Fabio Balduzzi, Davide Brugali

Dpt. Automatica e Informatica - Politecnico di Torino
C.so Duca degli abruzzi, 24 – 10129 Torino, Italy
E-mail: {balduzzi, brugali}@polito.it

Abstract

In this paper we present a novel approach to the control of manufacturing systems via agent interactions. Based on a decentralized control strategy the agents can collaborate with each other conveniently. We provide a distributed architecture to model the elementary services (an unreliable machine coupled with a finite buffer) in terms of hybrid automata which are co-operating in a manufacturing system, and we show how the agents can efficiently make their control decisions using their local rules.

1 Introduction

In the past few decades, advances in Computer Science, Automatic Control, and Communications have enabled the development and control of large scale, highly complex manufacturing systems. For these systems the traditional centralized approach to automatic control is inadequate because of the deficiencies in robustness and flexibility. As a matter of fact, centralized architectures are highly sensitive to system failures, since the whole system depends on the availability of the single decision making component to work.

Furthermore, the design of centralized control systems tends to be hard and expensive as the physical systems are by their nature geographically distributed and coordinated in an autonomous and decentralized manner. Even when technologically feasible, a centralized decision point might require elevated cost of the hardware and software involved, especially for large systems. In addition, centralized control problems have to be solved with "ad hoc" techniques. Every new problem should be solved from scratch. These problems are frequently encountered in the domain of industrial systems, characterized by a large number of activities that have to be coordinated.

In this context, many researchers adopt the *multi-agent paradigm* [11] for the design and development of decentralized control systems is suitable for such a decentralized control. Multi-agent systems are heterogeneous collection of autonomous software agents that have different capabilities, tasks and objectives.

System behavior is the result of agents' collaboration and knowledge exchange. A purposeful system behavior emerges when individual agents self optimize their own objectives but coordinate with other agents when conflicts arises. This naturally leads to multi-agent, multi-objective systems. Our research aims at applying the theory of collaborating agents to the field of autonomous distributed manufacturing systems.

In this paper we focus on the design of complex control systems applied to the manufacturing domain. The traditional approach requires defining a global model of the physical system to be controlled. The model is expressed in terms of global state variables and dynamic, and abstracts from the physical system organization in sub-systems. This model is then used to design the control system that impose to the physical system a behavior that satisfies specific requirements. Usually, both the system model identification and the consequent control system design are complex tasks.

In this paper we propose an alternative approach that is based on the "divide and conquer" principle. We observe that:

- usually, a complex physical system is composed by a set of interconnected sub-systems; each sub-system is less complex than the whole system;
- the behavior of the whole system emerges from the interactions among the single sub-systems.

In this paper we argue that it is possible to impose a desired behavior to the whole system by imposing an adequate behavior to the single sub-systems and by defining an adequate protocol for the exchange of information between the agents. Although we are not able to prove it in the general case, we demonstrate that this approach is feasible in a variety of specific cases. This approach demands for a solution to the following problems:

- we must identify modeling rules that allow decomposing the original problem into a set of local problems for the single sub-systems;
- we must identify the information flow between the single sub-systems. The goal is to minimize the information exchange between sub-systems in order

to reduce dependencies among sub-systems with respect to changes in their behavior specification;

- we must identify developing mechanisms and techniques that allow a seamless transformation of the decentralized control problem specification into software programs.

Our research is supported by a current project, which aims to create a proof of concept agent-based prototype. The prototype will demonstrate that multiple interacting agents can solve complex tasks that have been traditional approached in a centralized manner.

2 The Manufacturing System Model

In this paper we consider manufacturing systems consisting of unreliable machines and buffers of finite capacity. The manufacturing process allows the simultaneous production of several parts with general service time distributions and routing policies. Each part has to perform its own orderly sequence of operations in order to be completed. The production cycle specifies for each part the sequence of machines it must visit and the operation performed by them. The same operation can be performed on alternative machines (routing).

The dynamics of a manufacturing system is described by splitting the discrete event processes into two hierarchical levels: at the lower level, the microscopic behavior of arrivals and departures of parts to(from) each machine is modeled using first-order fluid approximations [3,4]. Such model allows the mathematical formulation of the continuous dynamics of the system. At the higher level a discrete event model will represent the transitions through a sequence of macro-states upon the occurrence of a limited number of events. Both levels of the process can be represented by a hybrid automaton [4,5,13,15] which exhibits both continuous and discrete changes.

The proposed hybrid automaton formalism for describing elementary service can be also applied to the agents in charge of controlling the system. This yields significant advantages over all other formalisms, one for all the inherent capabilities analysis. As a matter of fact, several formal results can be applied in order to address problems of state reachability and safety verification of the system requirements.

Yet, hybrid automata have the expressive power to describe both agent dynamics as well as decision logic, usually modeled by discrete event systems. In addition, they are equipped with composition and abstraction operators in order to capture the distributed and hierarchical nature of manufacturing systems. Composition operators perform the proper interconnection and synchronization of subsystems whereas abstraction operators allow the ability of hiding unnecessary details at the higher level.

2.1 The Single-Server Model

The production process considered in this work consists of a set of n single-server stations, denoted M_i, for $i=1,\ldots N$, serving a single class of products. Parts move from machine M_i to M_j according to their production cycle and are queued in buffers, one for each machine, with the initial one (input buffer) acting as an unlimited supply of parts and the final buffer acting as a limited storage area for collecting finished products, thus representing the production target. The buffers have finite capacity C_i and the machines are unreliable. Machine service times are assumed independent random variables with identical distribution. The maximum average machine production rates are denoted V_i.

We assume that the system performs state transitions through a sequence of operational states, that we call *macro-states*, at the occurrence of a limited number of events that we call *macro-events*, i.e., machine starvation, blockage, breakdown and repair, material release times and due dates.

Let $\Delta_k=[t_k,t_{k+1}]$, for $k=0,1,2,\ldots$, be the interval of time between the occurrence of consecutive macro-events at time t_k and t_{k+1}, that we call *macro-period*, and let $v_{i,j}(k)$ be the constant average flow rates of parts from machines M_i to M_j. Let $pre: \mathcal{N} \rightarrow 2^{\mathcal{N}}$ and $post: \mathcal{N} \rightarrow 2^{\mathcal{N}}$ be set-valued functions that assign to each machine of index i its sets of upstream and downstream machine indices. The microscopic behavior of a production system during a macro-period can be approximated by the following three processes defined for each machine M_i.

- The buffer levels:

$$x_i(t) = x_i(t_k) + [v_{in,i}(k) - v_{out,i}(k)] \cdot (t - t_k). \qquad (1)$$

- The production volume processed by the machine since the last repair (used to evaluate the machine breaking time):

$$\chi_i(t) = \chi_i(t_k) + v_{out,i}(k) \cdot (t - t_k). \qquad (2)$$

- The time spent by the machine under repair since the last failure:

$$s_i(t) = s_i(t_k) + (t - t_k). \qquad (3)$$

For all these processes, it holds $t \in [t_k, t_{k+1})$, and

$$\begin{aligned} v_{in,i}(k) &= \sum_{h \in pre(i)} v_{h,i}(k) \\ v_{out,i}(k) &= \sum_{j \in post(i)} v_{i,j}(k) \end{aligned} \qquad (4)$$

are the inflow and outflow rates of parts of each machine.

We assume *operation-dependent* failures, thus the value of $\chi_i(t)$ in Equation (2) will be reset to 0 after each

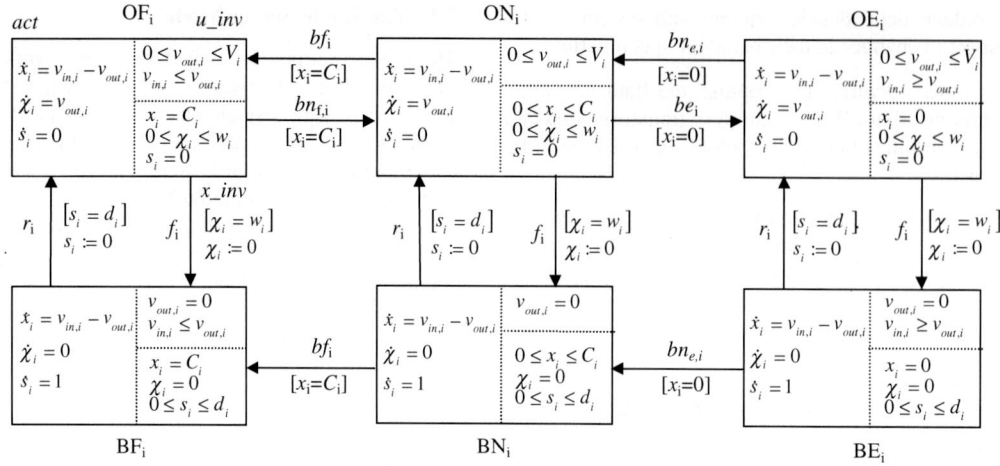

Figure 1: The hybrid automaton of an elementary service.

failure, i.e., failures occur after the production volume w_i, and its value will not increase until the machine is repaired. Similarly, the value of si(t) in Equation (3) will be reset to 0 after each repair, i.e., when a machine fails, it will be repaired

2.2 The Model of an Elementary Service

We now consider an elementary service composed of a buffer of finite capacity and of an unreliable machine. Such a service can be efficiently described by a hybrid automaton. In particular we consider a system with a continuous state vector $\mathbf{x} \in \mathcal{R}^n$ and a continuous input vector $\mathbf{u} \in \mathcal{R}^p$. The system may be in a finite number of discrete states called locations. In each location the state is constrained to belong to a subset of $\mathcal{R}^n$. As in the standard definition of hybrid automaton we consider a guard $g \in \mathcal{R}^n$ and a jump relation $j \in \mathcal{R}^n \times \mathcal{R}^n$. An edge is enabled when $\mathbf{x} \in g$ and the state jumps from $\mathbf{x}$ to $\mathbf{y}$ according to the jump condition, i.e., if $(\mathbf{x},\mathbf{y}) \in j$.

While in the standard definition of hybrid automata the activity function associates to each location a differential inclusion, in our definition we associate to each location a differential equation $\dot{\mathbf{x}} = \mathbf{f}(\mathbf{x},\mathbf{u})$. However, we allow the input vector $\mathbf{u}$ to take values in a convex set $S \subset \mathcal{R}^p$, defined by a set of linear inequalities expressing the machines and buffers capacity constraints. Thus we still have that the set of all possible derivatives is not a single vector but, for a given value of $\bar{\mathbf{x}}$ may take values in a subset of $\mathcal{R}^n$ defined by $\{\mathbf{f}(\bar{\mathbf{x}},\mathbf{u}) \mid \mathbf{u} \in S\}$.

The hybrid automaton of an elementary service is a structure $H_i = (L_i, \Sigma_i, act_i, x\_inv_i, u\_inv_i, E_i)$ where the continuous state vector is $\mathbf{x} = [x_i, \chi_i, s_i]^T$ while the input vector is $\mathbf{u} = [v_{in,i}, v_{out,i}]^T$.

The set of locations is denoted $L_i = \{OF_i, ON_i, OE_i, BF_i, BN_i, BE_i\}$ representing the joint machine status (O: *machine operational*, B: *machine broken*) and buffer status (F: *buffer full*, E: *buffer empty*, N: *buffer not full-not empty*). Each location represents the discrete state of the service. Let $stateOf(i) \in L_i$ denote the current macro-state of service i.

The alphabet $\Sigma_i = \{f_i, r_i, be_i, bf_i, bn_{e,i}, bn_{f,i}\}$ corresponds to the set of macro-events (f_i: *machine failure*, r_i: *machine repair*, bf_i: *buffer full*, be_i: *buffer empty*, $bn_{e,i}(bn_{f,i})$: *buffer not empty(not full)*). The dynamics at each location $act_i(l_i)$ can be obtained by differentiating Equations (1)-(3). The set x_inv_i of admissible state vectors changes from one location to another. As an example in location OE_i this set takes the form:

$$x\_inv_i(OE_i) = \begin{cases} x_i = 0 \\ 0 \leq \chi_i \leq w_i \\ s_i = 0 \end{cases} \quad (5)$$

since this location represents the discrete state in which the buffer is empty and the machine is operational. The set $u\_inv_i \subset S$ of admissible input vectors also changes from one location to another. It corresponds to a subset of the feasible region S for the average machine production rates. As an example in location OE_i this set takes the form:

$$u\_inv_i(OE_i) = \begin{cases} 0 \leq v_{out,i} \leq V_i \\ v_{out,i} \leq v_{in,i} \end{cases} \quad (6)$$

since this location represents the discrete state in which the buffer is empty and the machine is operational. Finally, $E_i \subset L_i \times \Sigma_i \times g_i \times j_i \times L_i$ represents the set of edges.

We can now prove the following proposition.

Proposition 2.2.1. *The hybrid automaton H_i of an elementary service is a linear hybrid automaton.*

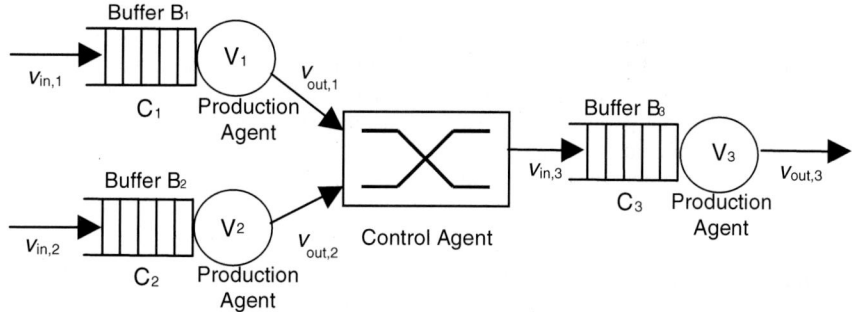

Figure 2: The interconnection of *Production* and *Control Agents*.

Proof. For each location $l_i \in L_i$ the invariants $x\_inv_i(l_i)$ and $u\_inv_i(l_i)$ are convex linear predicate on the variables, i.e., a system of linear inequalities over the variables x_i, χ_i, s_i, and the activity $act_i(l_i)$ is a linear combination of the vertices of the convex polyhedron S. For each edge $(l_i, \sigma_i, g_i, j_i, l'_i) \in E_i$ the guard g_i and the jump j_i are linear predicates on the variables. ∎

3 The System Control Model

In this paper we approach the problem of designing a control strategy that allows the production system to maximize the system throughput even in the presence of changing product class blend and production inconveniences (such as buffer blockages and machine breakdowns) while guaranteeing the machine load balance. This problem has been treated in a centralized way by several authors [2,4,9,12,16].

In particular in [4] the authors proposed a centralized solution where it is required the construction of a hybrid automata describing the whole network. A dynamic control policy was defined by assuming that in each location the choice of the input vector is obtained as the solution of a linear programming problem aimed at optimizing a given performance.

Therefore the centralized controller has to manage the following set of information:

i) the continuous state vector of the entire system;

ii) the set of linear constraints derived from the state and input invariants;

iii) the performance functional to be optimized.

The result of the control action is represented by the optimal machine production rates $v_{i,j}(k)$ to be expected for each macro-period. From a computational point of view this implies the evaluation of the $v_{i,j}$'s at each macro-event occurrences. In complex system the computational efforts may be very time consuming.

In this paper we address this control problem via a decentralized control architecture and policy. In particular we consider the goal of maximizing the system throughput and we show how a decentralized solution guarantees similar performances.

3.1 The Decentralized Control Architecture

The control problem previously described lends particularly well to a decentralized control solution that allows one to reduce the extent of the decision set and of the relevant information and data sets driving the decision procedure. The solution we propose is achieved via agent interactions [1,8], wherein each agent (e.g., an autonomous process that has the ability of communicating with its environment and can also control real equipment) manages a limited and consistent set of resources. We distinguish two types of agents.

- *Production Agents* manages stores and machines and enforce capacity and temporal precedence constraints on them. The current machine production rate and the current buffer level represent the state of a production agent.

- *Control Agents* manages the flows of parts among production agents in order to guarantee the desired production target and the balance of the machine loads.

The arrangement of the agents is locally hierarchical in nature. Each Control Agent receives information from and sends commands to the production agents in its neighborhood. However the hierarchy is not an orderly pyramid, and the agents at each level are involved in the process of setting the constraints within which they function.

Interactions among agents are established dynamically according to:

- the manufacturing production cycle;
- the current machines' availability.

In the traditional centralized systems the scope of the centralized controller includes all the active resources (production units, transports, assembly units, etc.).

In our system, each control agent supervises the flow of parts among adjacent production agents. Figure 2 shows an example of network of interacting agents, where the Control Agent is linked upstream to a set of production agents that provides independent flows of parts, and downstream to an output production agents that receives the combined flow of parts. Each agent manages an independent thread of control. Each control agent and has visibility only on the state of its input and output production agents.

3.2 The Model of a Decentralized Control Agent

In particular, at the occurrence of the macro-events, the evaluation of the desired machine production rates will be achieved via agent interactions and these reference values, that will become constant after a transient period, will be guaranteed by the distributed controller.

The arrangement of the agents is broadly hierarchical in nature. However the hierarchy is not an orderly pyramid, and the agents at each level are involved in the process of setting the constraints within which they function.

For the sake of simplicity we consider a network of elementary services wherein the inflow of each service of index i is provided by a finite set of upstream services, i.e., $pre(i) \subset \mathcal{N}$, card$\{pre(i)\} \leq N\}$, but the outflow is directed to only one downstream service, i.e., $post(i)=j$, card$\{post(i)=1\}$. We characterize this kind of service as a *multi-input single-output* service.

3.3 The Decentralized Control Strategy

A decentralized policy can be seen as a control law where the feedbacks are given in terms of agent interaction. It means that, in the general case, we will likely not able to find the optimality, that can be only achieved via a centralized control action. However we can achieve satisfactory performances, although suboptimal, but with less efforts and costs.
Let N be the index of final service in the system that supplies the output final buffer.

Proposition 3.3.1. *Let $l_i \in L_i$ denote the current location of service i. If every service $i=1,...,N$ applies the following local control law for each macro-period k:*

$$v_{out,i}^o(k) = \max\{v_{out,i}(k) \mid \mathbf{u}_i^*(k) \in u\_inv_i(l_i)\}, \quad (7)$$

where $\mathbf{u}_i^*(k)=[\ v_{out,i}(k),\ v_{in,i}^o(k)]^T$, then the objective function $J(k)=v_{out,N}(k)$ is maximized.

Proof. Let us first observe that $v_{in,i}(k)=v_{out,i-1}(k)$. Then, we can easily prove this results by showing that the maximization of $J(k)$ can be achieved in a centralized way by solving the following linear programming problem:

$$\max\left\{v_{out,N}(k) \mid (\mathbf{u}_1 \times \cdots \times \mathbf{u}_N) \in \bigcup_{i=1}^{N} u\_inv_i(l_i)\right\}. \quad \blacksquare$$

Let us now assume that the initial condition $\mathbf{x}(0)$ and $\mathbf{u}(0)$ are given. The above proposition implies the following control policy that is implemented in each agent as follows and computed at the occurrence of each macro-event at time t_k, i.e.,

- if $stateOf(i)$=ON $\Rightarrow \begin{cases} v_{out,i}(k)=V_i \\ v_{in,i}(k) \text{ any} \end{cases}$

- if $stateOf(i)$=OF $\Rightarrow \begin{cases} v_{out,i}(k)=V_i \\ v_{in,i}(k)=V_i \end{cases}$

- if $stateOf(i)$=OE $\Rightarrow \begin{cases} v_{out,i}(k)=v_{in,i}(k) \\ v_{in,i}(k) \text{ any} \end{cases}$

- if $stateOf(i)$=BN,BE $\Rightarrow \begin{cases} v_{out,i}(k)=0 \\ v_{in,i}(k) \text{ any} \end{cases}$

- if $stateOf(i)$=BF $\Rightarrow \begin{cases} v_{out,i}(k)=0 \\ v_{in,i}(k)=0 \end{cases}$

Note that in the case $stateOf(i)$=OF the outflow of the upstream services should be evaluated so that the following relation holds:

$$v_{in,i}(k) = \sum_{j \in pre(i)} v_{out,j}(k) \quad (8)$$

therefore we assume that the load balancing is satisfied if

$$v_{in,i}(k) = \alpha \sum_{\substack{j \in pre(i), \\ stateOf(j)=ON}} V_j + \sum_{\substack{j \in pre(i), \\ stateOf(j) \neq ON}} v_{out,j}(k-1) = V_i \quad (9)$$

from which we obtain $\alpha \in [0,1]$. Then we can evaluate $v_{out,j}(k)=\alpha V_j$.

3.4 Inter-Agents Information Exchange

Here we briefly outline the implementation of the information exchange process among agents.

We distinguish among *internal* and *external* macro-events. Looking at Figure 1 the transitions labeled bf_i, be_i, f_i, r_i correspond to macro-events occurring internally to a production agent, while transitions labeled $bn_{e,i}$, $bn_{f,i}$ correspond to external macro-events, i.e., macro-events occurring internally in some adjacent production agents. In particular the internal macro-events bf_i and be_i are caught respectively by the upstream control agents and the downstream control agent, which contextually update the states and outflow and inflow rates of their controlled production agents accordingly to the policy described in Proposition 1. The internal macro-event f_i and r_i are caught by both the upstream and downstream control agents.

4 Simulating a Production Agent Behavior

For the simulation of the agents' behavior we have developed an object-oriented tool where all the agents are represented as software objects and the actions are carried out by messages sent and received between objects.

A major challenge is to model the interactions between the agents in a coherent fashion. The fundamental issue in communicating is that some sort of visibility must exist among the agents. Visibility implies dependency with respect to changes, and hence it has consequences on the system's flexibility. In order to avoid the propagation of local changes to the entire system, it is better to avoid visibility loops between control agents and production. This is achieved by offering the agents two communication mechanisms where visibility and flow of information move in the same direction, and in the opposite direction, with respect to the communicating components [6,7].

The first mechanism, indicated here as Caller/Provider and deeply rooted in Object Oriented programming, is involved when an object invokes another object method. In our multi-agent system control agents have visibility of and ask for services from production agents by invoking their services through their interface. The second, the Broadcaster/Listeners mechanism, is achieved by giving the agents the capability of broadcasting and listening to events. In our system the production agents communicate with the control agents by broadcasting events.

The simulation tool uses a constraint satisfaction library [6,10], which supports interval arithmetic and allows the definition of linear relationships and differential equations. This library has been used to represent the agent's dynamic within a macro-period. Currently, the library implements the solving algorithm described in [14]. Each constrained variable is associated to an initial interval domain with values that may be not consistent with the defined constraints. The solving algorithm refines the initial interval domains as far as possible without loosing possible exact solutions and this is done so that the minimal consistent subinterval can be determined for each variable within its domain.

5 Summary and Conclusions

We have proposed a decentralized solution to the control problem of maximizing the system throughput of a production system consisting of unreliable machines and finite buffers. The solution has been implemented as a multi-agents system which interact with each other in an event-driven fashion. The behavior of each agent has been modeled and simulated as a hybrid automaton. Our future work will be to explore the feasibility of this approach in a more general setting.

References

[1] A. Aarsten, D. Brugali, G. Menga, "Designing Concurrent and Distributed Control System," *Communication of the ACM*, Oct., 1996.

[2] R. Akella, P.R. Kumar, P.R., "Optimal Control of Production Rate in a Failure Prone Manufacturing System," *IEEE Trans. Automatic Control*, Vol. 31, No. 2, pp. 116-126, 1986.

[3] F. Balduzzi, "Fluid Models and Hybrid Automata in Manufacturing," *Proc. IEEE Int. Conference on Robotics and Automation*, (San Francisco, California), April 2000, pp. 2647-2653.

[4] F. Balduzzi, A. Giua, G. Menga, "Optimal Control of Production Systems with Unreliable Machines and Finite Buffers," *Proc. IEEE Int. Conference on Robotics and Automation*, (Detroit, Michigan), May 1999, pp. 1462-1468.

[5] F. Balduzzi, A. Giua, C. Seatzu, "Modelling Automated Manufacturing Systems with Hybrid Automata," *Proc. Workshop on Formal Methods in Manufacturing*, (Zaragoza, Spain), September 1999, pp. 33-48.

[6] D. Brugali, "Hierarchical and Distributed Constraint Satisfaction Systems," *Implementing Application Frameworks: Object Oriented Frameworks at Work*, Fayad, Johnson, Schmidt (eds.) John Wiley&Co, 1999. ISBN#: 0-471-15012-0

[7] D. Brugali, G. Menga, A. Aarsten, "The Framework Life Span," *Communication of the ACM*, October 1997.

[8] D. Brugali, K. Sycara, "Towards Agent Oriented Application Frameworks," *ACM Computing Surveys*, to appear.

[9] D. Connors, G. Feigin, D. Yao, "Scheduling Semiconductor Lines Using a Fluid Network Model," *IEEE Trans. Robotics and Automation*, Vol. 10, No. 2, pp. 88-98, 1994.

[10] CCITT, "Specification and Description Language", Recommendation Z.100, 1986.

[11] E. Durfee, *Coordination of Distributed Problem Solvers*, Kluwer Academic Press, 1988.

[12] S.B. Gershwin, *Manufacturing Systems Engineering*, Prentice Hall, Inc., Englewood Cliffs, NJ, 1994.

[13] T.A. Henzinger, "The Theory of Hybrid Automata," *Proc. 11th Annual Symposium on Logic in Computer Science*, pp. 278-292, 1996.

[14] E. Hyvonen, "Constraint reasoning based on interval arithmetic: the tolerance propagation approach," *Artificial Intelligence*, 58, 71-112, 1992.

[15] A. Puri, P. Varaiya, "Decidable Hybrid Systems, " *Computer and Mathematical Modeling*, N. 23(11/12): pp. 191-202, 1996.

[16] S.P. Sethi, X.Y. Zhou, "Stochastic Dynamic Jobshops and Hierarchical Production Planning," *IEEE Trans. Automatic Control*, Vol. 39, No. 10, pp. 2061-2076, 1994.

A TIME WINDOW BASED APPROACH FOR JOB SHOP SCHEDULING

Haoxun Chen   Peter B. Luh   Lei Fang

Department of Electrical and Computer Engineering
University of Connecticut, CT 06269-2157, USA
Email: {hxchen, luh, fang}@engr.uconn.edu

1. INTRODUCTION

Scheduling is to allocate scare resources over time to accomplish a given set of tasks. For manufacturing scheduling, the goal is to deliver customer orders on time while minimizing the production and inventory costs. High quality schedules can improve delivery performance, reduce inventory costs, and are thus very important to manufacturers in today's time-based competition. To obtain a high quality schedule within an acceptable computation time, however, is extremely difficult because of the NP-hard nature and the large sizes of practical problems.

Recently, *constraint-based scheduling* has emerged as a new approach to solve scheduling problems [1, 2, 6]. The flexibility of the approach and its capability to solve notorious benchmark problems has attracted many researchers' attention in the scheduling community. Some commercial systems such as ILOG Scheduler [5] were developed based on the approach.

The constraint-based approach formulates a scheduling problem as one or a set of *Constraint Satisfaction Problems* (CSP) [2]. Each of them is to find a feasible solution subject to a set of constraints and solved by a search process like that of the branch and bound approach. *Constraint propagation* techniques, which exploit the constraints, are used to limit the search space so that the computation time to solve the CSPs can be greatly reduced [1, 6].

Most constraint-based scheduling algorithms were developed for problems to minimize the makespan, i.e., the completion time of all parts [1, 2, 6]. One reason for this is that these problems can be formulated as a set of CSPs to find a feasible schedule with the makespan less than or equal to a given value. For this kind of CSPs, powerful *edge-finding* techniques can be developed to carry out effective constraint propagation [1, 6].

Most practical problems, however, are due-date related where the scheduling objective is to deliver customer orders on time with low work-in-processing inventory. The objective is usually formulated as a weighted earliness and tardiness cost with the earliness term reflecting the work-in-process inventory cost and the tardiness term reflecting the penalty for late delivery of orders. Although these problems can also be formulated as a set of CSPs, the nonlinear nature of the objective function makes edge-find techniques difficult to apply.

The Lagrangian relaxation (LR) approach we recently developed can deal with the earliness and tardiness scheduling problem [4]. In the approach, machine capacity constraints are first relaxed by using Lagrange multipliers, and the relaxed problem is then decomposed into smaller subproblems, one for each part. These subproblems can be efficiently solved by using dynamic programming at the low-level [3, 8]. Multipliers are then iteratively adjusted at the high-level based on the degrees of constraint violation to maximize the dual objective derived from the relaxed problem. At the termination of such updating iterations, a list scheduling heuristic is applied to adjust subproblem solutions to obtain a feasible schedule satisfying all constraints.

The LR approach is quite effective for problems when the number of parts is not very large (e.g., less than 1000). When the number of parts is large, the performance of the approach may deteriorate because the heuristic propagates the beginning times of delayed operations in the scheduling process but the propagation may lose the useful information provided by the subproblem solutions.

Although the subproblem solutions are usually associated with an infeasible schedule even if the iterative process terminates, they should provide useful information for the time interval within which each part is processed to minimize the earliness and tardiness cost. Based on this observation, a time window based approach will be developed in this paper, which is effective for problems with a large number of parts. In this approach, LR is used to compute a time window for each part. Given the time windows for all parts, a constraint satisfaction algorithm or a heuristic will then be used to find a feasible schedule within the windows or a feasible schedule with less violation of the time window constraints. The time windows are refined as the Lagrange multipliers are updated in the process to maximize the dual objective.

The new approach is tested using randomly generated instances and compared with the LR approach. The results show that the new approach can generate better schedules and the improvement becomes very significant if the number of parts is big. The computation time of the new approach when the heuristic is used is similar to that of the LR approach.

2. PROBLEM FORMULATION

The job shop scheduling problem considered in this paper is to schedule N parts on H types of machines with time horizon K to minimize a weighted earliness and tardiness criterion [8]. Each machine type h ($h = 1, 2, ..., H$) has M_{hk} available machines at time k, $k = 1, 2, ..., K$, and the completion of each part i ($i = 1, 2, ..., N$) requires a series of N_i operations, denoted by $(i, 1)$, $(i, 2)$, ..., (i, N_i). Each operation (i, j) is non-preemptive and can be performed on a machine of a given machine type m_{ij}. For simplicity, all parts are assumed to be available at time 0. For the case where some parts are not available at time 0, the discrete-time, integer programming formulation and solution methodology developed in this paper are still applicable after a slight modification. The variables and symbols used in the formulation are listed below.

- B_i: Beginning time of part i;
- b_{ij}: Beginning time of operation (i, j);
- c_{ij}: Completion time of operation (i, j);
- C_i: Completion time of part i;
- D_i: Due date of part i;
- E_i: Earliness of part i, defined as max(0, $S_i - B_i$);
- M_{hk}: Number of available machines of type h at time k, $1 \leq h \leq H$, $1 \leq k \leq K$;
- p_{ij}: Processing time of operation (i, j);
- S_i: Desired raw material release time for part i;
- T_i: Tardiness of part i, defined as max(0, $C_i - D_i$);
- w_i: Tardiness weight for part i;
- β_i: Earliness weight for part i;
- δ_{ijhk}: 0-1 operation variable which is one if operation (i, j) is performed on machine type h at time k, and zero otherwise.

The constraints and objective function are briefly presented below.

Machine Capacity Constraints: The number of operations assigned to machine type h at time k should be less than or equal to M_{hk}, the number of machines available at that time, i.e.,

$$\Sigma_{ij}\, \delta_{ijhk} \leq M_{hk},\ k = 1, 2, ..., K,\ h = 1, 2, ..., H, \quad (1)$$

Operation Precedence Constraints: An operation cannot be started until its preceding operation has been finished, and it requires a specific amount of time for processing on the selected machine type, i.e.,

$$c_{i,j-1} + 1 \leq b_{i,j},\ i = 1, 2, ..., N,\ j = 1, 2, ..., N_i, \quad (2)$$

Processing Time Requirements: Each operation must be assigned the required amount of time for processing on its specified machine type, i.e.,

$$c_{i,j} = b_{i,j} + p_{ij} - 1,\ i = 1, 2, ..., N,\ j = 1, 2, ..., N_i, \quad (3)$$

Objective Function: The goals of on-time delivery and low work-in-process inventory are modeled as penalties (costs) on delivery tardiness and on releasing raw materials too early, and the objective function is

$$J = \Sigma_{i=1}^{N} w_i T_i + \Sigma_{i=1}^{N} \beta_i E_i \quad (4)$$

The problem is to determine operation beginning times b_{ij} for individual operations to minimize (4) subject to machine capacity constraints (1), operation precedence constraints (2), and processing time requirements (3).

3. SOLUTION METHODOLOGY

For the problem considered, it is not straightforward to directly apply edge-finding constraint propagation techniques since the objective is a nonlinear function. As an alternative, the approach to be developed will use LR to find the time windows within which each part is processed to minimize the earliness and tardiness cost. For the given time windows, a constraint satisfaction algorithm or a heuristic will then be used to find a feasible schedule within the windows or a feasible schedule with less violation of the time window constraints. The time windows are obtained by solving part level scheduling subproblems using dynamic programming after relaxing the machine capacity constraints, and are refined as the Lagrange multipliers are updated in the process to maximize the dual objective.

3.1. Lagrangian Relaxation

Within the LR framework [3, 8], the machine capacity constraints (1) are relaxed by using Lagrange multipliers λ_{hk}, and the relaxed problem is given by

$$\min_{\{b_{ij}\}} L,$$

s.t. (2) and (3),

with $L = \Sigma_{i=1}^{N} w_i T_i + \Sigma_{i=1}^{N} \beta_i E_i + \Sigma_{i,j} \Sigma_{k=b_{ij}}^{c_{ij}} \lambda_{m_{ij}k}$

$$- \Sigma_{h,k} \lambda_{hk} M_{hk}. \quad (5)$$

Since the formulation is separable, the relaxed problem can be decomposed into the following decoupled part subproblems for the given set of multipliers:

$$\min_{\{b_{ij}\}} L_i,\ \text{with}\ L_i = w_i T_i + \beta_i E_i + \Sigma_j \Sigma_{k=b_{ij}}^{c_{ij}} \lambda_{m_{ij}k}, \quad (6)$$

s.t. (2) and (3), $\quad i = 1, 2, ..., N.$

Each subproblem is to schedule the operations of a single part to minimize its tardiness and earliness penalties and the costs for using machines.

3.2. Determining the Time Windows for Parts

The time windows are determined by solving the part level subproblems using dynamic programming. In our previous work [3, 8], it has been shown that each part subproblem is a multistage optimization problem, which can be efficiently solved by using dynamic programming (DP) with pseudo-polynomial complexity. The dynamic programming algorithm can be performed in a forward or backward way. The forward DP will be used to determine the right side value (the maximum time) of the time window while the backward DP will be used to determine the left side value (the minimum time) of the window.

The forward DP (FDP) takes the completion times of operations as decision variables and starts with the first operation of a part. The algorithm then precedes to the second operation and finally to the last operation. At each stage, a cumulative cost is calculated for all possible decisions of this stage. For part i, the cumulative costs are obtained by recursively applying the following DP equations subject to operation precedence constraints (2) and processing time requirement constraints (3):

$$V_{i,1}(c_{i1}) = \beta_i E_i + \Sigma_{k=b_{i1}}^{c_{i1}} \lambda_{m_{i1}k}, \qquad (7)$$

$$V_{i,j}(c_{ij}) = \min_{c_{i,j-1}} \{ \Sigma_{k=b_{ij}}^{c_{ij}} \lambda_{m_{ij}k} + V_{i,j-1}(c_{i,j-1}) \},$$
$$j = 2, 3, \ldots, N_i - 1, \qquad (8)$$

$$V_{i,N_i}(c_{i,N_i}) = \min_{c_{i,N_i-1}} \{ w_i T_i + \Sigma_{k=b_{iN_i}}^{c_{iN_i}} \lambda_{m_{iN_i}k} + V_{i,N_i-1}(c_{i,N_i-1}) \}. \qquad (9)$$

The function $V_{i,j}$ is the cumulative cost for operation (i, j) and all operations preceding (i, j), and $\Sigma_{k=b_{iN_i}}^{c_{iN_i}} \lambda_{m_{iN_i}k}$ ($\beta_i E_i + \Sigma_{k=b_{i1}}^{c_{i1}} \lambda_{m_{i1}k}$ for $j = 1$ and $w_i T_i + \Sigma_{k=b_{iN_i}}^{c_{iN_i}} \lambda_{m_{iN_i}k}$ for $j = N_i$) is a "stage-wise" cost. The algorithm starts from the first stage and moves backward till the last stage is reached. The optimal subproblem cost is obtained as the minimal cumulative cost at the last stage. Finally, the optimal completion times of operations can be obtained by tracing forwards all the stages. The computational complexity of the above FDP algorithm is $O(N_i K)$.

The backward DP (BDP) is similar to FDP except that it takes the beginning times of operations as decision variables, starts with the last operation of a part, and moves backward till the first operation is reached.

Let $U_{i,1}(.)$ denote the cumulative cost function of part i at the first stage for BDP, $V_{i,N_i}(.)$ denote the cumulative cost function of the part at the last stage for FDP, and define

$$b_{i,1}^* = \arg \min_k \{ U_{i,1}(k) \mid 1 \leq k \leq K \}, \qquad (10)$$

$$c_{i,N_i}^* = \arg \min_k \{ V_{i,N_i}(k) \mid 1 \leq k \leq K \}. \qquad (11)$$

The beginning time $b_{i,1}^*$ and the completion time c_{i,N_i}^*, obtained by BDP and FDP, respectively, provide a time window $[b_{i,1}^*, c_{i,N_i}^*]$ for processing part i.

Sometimes, the above defined time windows are too tight so that no feasible schedule exists within the windows. In this case, a slight extension of the time windows becomes necessary.

The basic idea for this is to find a beginning time $\hat{b}_{i,1}$ less than $b_{i,1}^*$ and a completion time $\hat{c}_{i,N_i}$ larger than c_{i,N_i}^* so that the total cumulative cost $U_{i,1}(.)$ at $\hat{b}_{i,1}$ is within ε-neighborhood of the cost at $b_{i,1}^*$ and the total cumulative cost $V_{i,N_i}(.)$ at $\hat{c}_{i,N_i}$ is within ε-neighborhood of the cost at c_{i,N_i}^*, where $\varepsilon > 0$ is a parameter, i.e.,

$$\hat{b}_{i,1} = \min\{ k \mid k \leq b_{i,1}^* \text{ and } U_{i,1}(k') \leq (1+\varepsilon) U_{i,1}(b_{i,1}^*)$$
$$\text{ for any } k', k \leq k' \leq b_{i,1}^* \}, \qquad (12)$$

$$\hat{c}_{i,N_i} = \max\{ k \mid k \geq c_{i,N_i}^* \text{ and } V_{i,N_i}(k') \leq$$
$$(1+\varepsilon) V_{i,N_i}(c_{i,N_i}^*) \text{ for any } k', c_{i,N_i}^* \leq k' \leq k \}. \qquad (13)$$

The new beginning time $\hat{b}_{i,1}$ and completion time $\hat{c}_{i,N_i}$ provide an enlarged time window $[\hat{b}_{i,1}, \hat{c}_{i,N_i}]$ for part i.

3.3. Finding a Feasible Schedule within the Time Windows by Constraint Satisfaction

As soon as the time windows $[\hat{b}_{i,1}, \hat{c}_{i,N_i}]$ of all the parts are given, a constraint satisfaction algorithm will be used to find a feasible solution obeying the machine capacity constraints (1), operation precedence constraints (2), processing time requirement constraints (3), and the following time windows constraints:

$$\hat{b}_{i,1} \leq b_{i,1}, \qquad (14)$$

$$c_{i,N_i} \leq \hat{c}_{i,N_i}. \qquad (15)$$

The constraint satisfaction algorithm is developed based on ILOG Scheduler.

The scheduler is an extension of ILOG Solver (a solver for constraint programming) designed for applications requiring the management of resources over time. It comprises a library of C++ classes and functions that offer a natural object model for mathematical representation of finite capacity scheduling and resource allocation problems. The time window constraints (14) and (15) can be represented as temporal constraints defined in terms of activities in the library.

One powerful constraint propagation technique used in ILOG Scheduler is the *edge finding* technique. It consists of a set of rules that determines whether an operation must precede (or follow) a set of operations on the same resource. Most edge finding rules were developed for problems with the makespan criterion. However, these rules can also be applied to our problem to find a feasible schedule within given time windows.

Except for constraint propagation, the efficiency of a constraint satisfaction algorithm for the time window problem also depends on the search strategy of the algorithm. One key element of this strategy is the operation selection which determines whether an operation is selected to process first or last among all unscheduled operations, and what operation is actually selected. For simplicity of implementation, our algorithm always selects an operation to process first or always selects an operation to process last. The selection of the operation is based on a modified WSPT/CR rule.

The WSPT/CR rule [7] is a combination of the WSPT (Weighted Shortest Processing Time) rule and the CR (Critical Ratio) rule, which selects an operation with the maximal WSPT/CR ratio to dispatch next at each time. Suppose that the processing time of a dispatchable operation of a part is p, the due date, the priority weight, and the remaining processing time of the part (including the processing time of the operation) are D, w, and r, respectively, and the earliest possible starting time of the operation is t, the WSPT/CR ratio of the operation is defined as

$$\frac{w}{p\max(1,(D-t)/r)}, \quad (16)$$

where w/p and $(D-t)/r$ are the WSPT ratio and the critical ratio of the operation, respectively. By combining the advantage of the WSPT rule on minimizing the weighted complete time criterion (related to work-in-process inventory) and the advantage of the CR rule on minimizing due date related criterions, the WSPT/CR rule has been proved effective for just-in-time (earliness and tardiness) scheduling [7].

For our constraint satisfaction algorithm, if operations are always selected to process first, the WSPT/CR ratio of an operation (i, j) is defined as

$$WSPTCR_{ij} = \frac{w_i}{p_{ij}\max(1,(\hat{c}_{i,N_i} - t_{ij})/r_i)}, \quad (17)$$

where $\hat{c}_{i,N_i}$ is the right side value of the time window of part i, w_i is the tardiness weight of the part, r_i is the remaining time of part i (i.e., the total processing time of operation (i, j) and its succeeding operations), and p_{ij} and t_{ij} are the processing time and the earliest possible starting time of the operation, respectively. The modification of the critical ratio in the definition (17), which replaces the due date D_i of part i by $\hat{c}_{i,N_i}$, is to make partial schedules found in the search process of the algorithm to fit the time windows $[\hat{b}_{i,1}, \hat{c}_{i,N_i}]$ as much as possible. The operation with the smallest earliest start time is selected to schedule first and in case of ties the operation with the greatest WSPT/CR ratio is selected.

Symmetrically, if operations are always selected to process last, the WSPT/CR ratio of an operation (i, j) is defined as

$$WSPTCR_{ij} = \frac{\beta_i}{p_{ij}\max(1,(t_{ij} - \hat{b}_{i,1})/r_i)}, \quad (18)$$

where $\hat{b}_{i,1}$ is the left side value of the time window of part i, β_i is the earliness weight of the part, t_{ij} is the latest possible completion time of the operation, and p_{ij} and r_i are the same as those in (17). The operation with the largest latest completion time is selected to schedule last and in case of ties the operation with the greatest WSPT/CR ratio is selected.

For the backtracking when a search process reaches an infeasible partial schedule, a method similar to that in paper [6] is adopted. In the method, chronological backtracking is first used in order to solve the infeasibility at hand. If this does not lead to a feasible schedule after a given number of backtracks, the search process is restarted from a randomly generated point in the current path. A path that is the same as the current one before the point and different after that point will then be explored. The next operation after the point to be dispatched in the new path is taken as one that has the highest priority among all dispatchable operations that lead to an unexplored path. The priority is defined as that in the operation selection mentioned above. To avoid the repetition of the process, all previously explored paths are recorded. The process will be terminated after a given number of restarts.

3.4. Algorithm Framework for the Time Window Based Approach

With the above constraint satisfaction algorithm, our time window based approach is implemented as an iterative procedure. In the procedure, the Lagrange multipliers are iteratively updated (adjusted) to maximize the dual objective by using a subgradient method. At each iteration, the forward and backward DP algorithms are used to solve the relaxed subproblems of all parts to obtain the time windows, with the parameter ε adaptively determined according to the formula:

$$\varepsilon^{n+1} = \begin{cases} \varepsilon^n \rho, & \text{if the previous iteration finds a feasible schedule} \\ \varepsilon^n / \rho, & \text{if the previous iteration fails to find a feasible schedule} \end{cases}, \quad (19)$$

where n is the iteration number and $0<\rho<1$ is a parameter determined by numerical experiments.

Given the time windows, the constraint satisfaction algorithm will then be used to find a feasible schedule within the windows. The above iterative process repeats when a new set of multipliers is obtained.

3.5. Finding a Feasible Schedule by Heuristics

For large problems with thousands of parts and tens of machine types, the constraint satisfaction algorithm presented in last subsection is too time-consuming to be used in our time window based approach since the algorithm is invoked at each iteration of the approach. For these problems, a heuristics is developed to find a feasible schedule within the time windows or with less violation of the time window constraints.

Given the time windows [$\hat{b}_{i,1}$, $\hat{c}_{i,N_i}$], the heuristics performs a forward dispatch and a backward dispatch, and selects a schedule with less earliness and tardiness cost. The forward dispatch treats the left side time window constraints (14) as hard constraints and the right side time window constraints (15) as soft constraints. It starts with the first operations of parts. At each time, the operation with the earliest possible starting time is selected to dispatch next and in case of ties one with the greatest WSPT/CR ratio as defined in (17) is selected. The starting time of the operation is set to be its earliest possible starting time. Because the WSPT/CR ratio has taken account of the criticality of each part to meet its right side time window constraint (15), the forward dispatch is expected to derive a feasible schedule with less violation of the constraints.

By contrast, the backward dispatch treats the right side time window constraints as hard constraints and the left side time window constraints as soft constraints. It starts with the last operations of parts. At each time, the operation with the latest possible completion time is selected to dispatch next and in case of ties the operation with the greatest WSPT/CR ratio as defined in (18) is selected. The completion time of the operation is set to be its latest possible completion time.

The above described heuristics can be viewed as a special case of the constraint satisfaction algorithm where the search process is not backtracked and not terminated until a complete schedule is obtained no matter whether the time window constraints are violated.

4. NUMERICAL TESTING RESULTS

The approach developed in this paper has been implemented on a 450 MHZ PC using C++ language and tested using a set of instances randomly generated in the following way: The number of parts N and the number of machine types H are taken as two input parameters. The number of machines for each machine type is uniformly generated from integers 1 to 3 for 100×10 (100 parts and 10 machine types) problems, and from integers 1 to 10 for 1000×10 (1000 parts and 10 machine types) problems. The tardiness weights are uniformly generated from 1 to 10 and the earliness weights are uniformly generated from γ to 10γ, where γ is a parameter taken as 0.1 and 0.2. The parameter is taken much smaller than 1 because on-time delivery is usually more important than lower work-in-process inventory. Each part has H operations. For each of the operations, the machine type that the operation is performed on is uniformly generated from integers 1 to N, the indexes of all the machine types.

The processing time of the operation is uniformly generated from 1 to 10. The due dates are generated according to *flowtime*×U[1, N], where U[1, N] is a uniform distribution with the values from 1 to N, and the parameter *flowtime* is calculated as the maximal work load of all machine types divided by N. The work load of each machine type is defined as the total processing time of all operations performed on the machine type divided by the number of machines for the type. As soon as the due dates are determined, the desired raw material release time for each part will be generated as the due date of the part minus its total processing time. For each set of parameters, 10 problems are generated.

The costs of the schedules generated by the approach are compared with the costs of those generated by the LR approach in [8]. The comparison results for 100×10 problems are presented in Table 1, where the constraint satisfaction algorithm developed in Subsection 3.3 is used to generate a feasible schedule, ε is initially taken as 0.1, and ρ is taken as 0.95. In the table, the new approach is denoted by *TW*1 - Time Window Based Approach,

Version 1, and γ is the parameter for generating the earliness weights. The average computation time of the new approach for each problem in this table is 12.5 minutes.

From Table 1, we can see that compared with the LR approach, the new approach can generate better schedules for all problems and the improvement becomes more significant as the earliness weights increase.

Table 1: Test Results for 100×10 Problems

Prob. No.	$\gamma = 0.1$		$\gamma = 0.2$	
	LR	TW1	LR	TW1
1	10621.35	7682.53	20916.91	9138.23
2	18014.78	13818.84	20773.09	8429.93
3	9815.49	7183.74	20614.35	8416.53
4	17240.28	9825.43	21983.62	11549.52
5	12391.29	6263.03	32190.82	16213.10
6	18883.30	15319.79	24274.07	10673.83
7	21044.40	16333.74	30738.94	11864.25
8	22068.42	15283.51	14832.56	10433.89
9	18697.43	7380.49	27557.02	17857.37
10	24107.94	18134.56	23588.23	12043.57

Table 2 shows the comparison results for 1000×10 problems, where the heuristics developed in Subsection 3.5 is used to generate a feasible schedule, ε is initially taken as 1.0, and ρ is taken as 0.95. In the table, the new approach is denoted by TW2 - Time Window Based Approach, Version 2. The average computation time of the new approach for each problem in this table is 80 minutes, similar to that of the LR approach.

From Table 2, we can see that the performance improvement of the new approach over the LR becomes more significant as the problem size increases even if we use a heuristics to generate a feasible schedule. This demonstrates that our new approach can be used to solve practical scheduling problems with an improved performance.

Table 2: Test Results for 1000×10 Problems

Prob. No.	$\gamma = 0.1$		$\gamma = 0.2$	
	LR	TW2	LR	TW2
1	292762.34	102717.01	433738.36	161665.75
2	370953.13	131535.65	674216.03	146210.16
3	244733.34	156052.23	697044.60	145728.25
4	231005.37	107335.78	602330.85	239026.59
5	304829.26	84974.60	476988.23	130023.37
6	278151.29	120416.08	681479.95	130023.37
7	300733.03	210051.58	390219.08	254256.31
8	190323.35	98035.32	674660.31	156804.19
9	274785.41	79277.85	396350.67	142201.29
10	364924.22	143868.62	559224.35	100169.34

5. CONCLUSION

In this paper, a time window based approach is developed for job shop scheduling problems to minimize the weighted earliness and tardiness cost. With the time windows provided by Lagrangian relaxation within which parts are processed to minimize the cost and an effective algorithm to find a feasible schedule within or approximately within the windows, the approach can generate schedules better than those generated by the Lagrangian relaxation approach for large problems in a similar computation time. This demonstrates that our new approach can be used to solve practical scheduling problems with an improved performance.

REFERENCES

[1] Baptiste Philippe, Claude Le Pape, and Wim Nuijten, "Constraint-Based Optimization and Approximation for Job-Shop Scheduling," *Proceedings of the AAAI-SIGMAN Workshop on Intelligent Manufacturing Systems*, Montreal, Canada, 1995.

[2] Blazewicz J., W. Domschke, and E. Pesch, "The Job Shop Scheduling Problem: Conventional and New Solution Techniques," *European Journal of Operational Research*, Vol. 93, No. 1, 1996, pp.1-33.

[3] Chen, H., C. Chu and J. M. Proth, "An Improvement of the Lagrangian Relaxation Approach for Job Shop Scheduling: A Dynamic Programming Method," *IEEE Trans. on Robotics and Automation*, Vol.14, No.5, 1998, pp. 786-795.

[4] Hoitomt, D. J., P. B. Luh, and K. R. Pattipati, "A Practical Approach to Job Shop Scheduling Problems," *IEEE Transactions on Robotics and Automation*, Vol. 9, No. 1, February 1993, pp. 1-13.

[5] ILOG, Inc., "ILOG Scheduler 4.4: User's Manual,", 1999.

[6] Nuijten Wim and Claude Le Pape, "Constraint-Based Job Shop Scheduling with ILOG Scheduler," *Journal of Heuristics*, Vol. 3, No.1, 1998, pp. 271-286.

[7] Schafaei, R. and P. Brunn, "Workshop Scheduling Using Practical Data: Part 1: The Performance of Heuristic Scheduling Rules in a Dynamic Job Shop Environment Using a Rolling Time Horizon Approach," *Int. J. Prod. Res.*, Vol. 37, No. 17, 1999, pp. 3913-3925.

[8] Wang, J., P. B. Luh, X. Zhao, and J. Wang, "An Optimization-Based Algorithm for Job Shop Scheduling," *SADHANA*, a Journal of Indian Academy of Sciences, a Special Issue on Competitive Manufacturing Systems, Vol. 22, Part 2, April 1997, pp. 241-256.

FMS Scheduling Based on Timed Petri Net Model and RTA* Algorithm

*YoungWoo Kim, **Akio Inaba, *Tatsuya Suzuki, *Shigeru Okuma

*Dept. of Electrical Engineering, Graduate School of Nagoya University
Furo-cho, Chikusa-ku, Nagoya, 464-8603, Japan
Phone:+81-52-789-2778, FAX:+81-52-789-3140,
Email:kim,suzuki,okuma@okuma.nuee.nagoya-u.ac.jp

**Gifu Prefectural Research Institute of Manufacturing Information Technology,
4-179-1,Sue, Kagamihara, Gifu. Japan
Phone:+81-583-79-3300, FAX:+81-583-79-3001,
E-Mail:gifu-irtc@go.jp

Abstract

This paper presents a new scheduling method for manufacturing system based on a Timed Petri net model and a reactive fast search algorithm.

The following two typical problems are addressed in this paper. (1) Minimize the maximum completion time. (2) Minimize the total deadline over-time. As for problem (1), a search algorithm which combines RTA* and rule-based supervisor is proposed. Since both RTA* and rule-based supervisor can be executed in a reactive manner, machines and AGVs allocations can be scheduled reactively, and simultaneously. As for problem (2), original petri net model is converted to its reverse model and the algorithm developed in problem (1) is applied with regarding the due time as a starting time in the reverse model.

Keywords : Model Based Scheduling, Petri Net, RTA*, Supervisor, Due-Date problem

1 Introduction

Scheduling of manufacturing processes is the typical combinatorial optimization problem which decides the starting time and allocation of resources. Desirable characteristics for the scheduling can be stated as follows:

(1) It is easy to formulate a problem.

(2) Semi-optimal solution can be found within short time.

However, when we take into consider not only processes executed by machines but also transfer of works between machines, the formulation of the problem would be likely to be complicated and a search space would be enormous. As the result, previous researches on scheduling could not meet these characteristics. For example, the approach based on the mathematical programming[1] or network search meets characteristics (2), but it is usually difficult to formulate the problem. The approach based on the queuing theory or simulation-based method [2] meets characteristics (1), but it takes unreasonably long time to find semi-optimal solution. Moreover, in previous researches, machine allocation and AGV (used for transfer, Automatic Guided Vehicle) allocation problems were dealt with independently. These two allocation problems sometimes conflicted with each other. One of the promising ways to overcome these drawbacks is to make use of a formal model of FMS, such as Petri net model [6],[7]. The advantages of model based scheduling are: (1) It is easy to consider the various practical constraints which come from production environment, and (2) It is easy to monitor the current situation of the production system. The Petri net model based scheduling, however, has not always been satisfactory because it sometimes resulted in a combinatorial explosion even if the size of the problem is reasonable.

This paper presents a new scheduling method for manufacturing system based on a Timed Petri net model and a reactive fast search algorithm. The following two typical problems are addressed in this paper.

(1) Minimize the maximum completion time.

(2) Minimize the total deadline over-time.

The problem (1) can be regarded as to find an optimal path on the search space generated from the Petri net model. As for this problem, a search algorithm which combines RTA* (Real Time A* algorithm)and rule-based supervisor is proposed. Since both RTA* and rule-based supervisor can be executed in a reactive manner, machines and AGVs allocations can be scheduled reactively and simultaneously with avoiding combinatorial explosion. The problem (2) is not so straightforward to handle as the problem (1). In our model based approach, however, it can be easily dealt with as follows. First of all, the original Petri net model is converted to its reverse model in which the directions of all arcs are reversed. Secondly, the algorithm developed for the problem (1) is directly applied to the reverse model with regarding the due time as a starting time in the reverse model. This implies that the search proceeds backward. Finally, some minor refinements are done if there exist jobs of which end time (that is a commencement time in real system) is negative. The tractability of the problem (2) shows the great advantage of using model based approach. The usefulness of the proposed algorithm is also shown through some numerical experiments.

2 Modeling of FMS

2.1 Outline of Modeling

The FMS(Flexible Manufacturing System) environment considered in this paper is summarized as follows:

(1) The FMS consists of a load station, an unload station, machines, AGVs and an AGV pool.

(2) Each machine has an input buffer and an output buffer. Capacities of these buffers are finite. Loading and unloading of loads between an AGV and a machine are performed through buffers.

(3) An AGV can move along specified paths toward specified directions.

(4) An AGV moves toward an AGV pool when tasks are not allocated to it.

In this paper, the model of the FMS is described by a *system net* and a *process net*.

2.2 System net

The system net describes the physical behavior in the FMS based on colored timed Petri net. Places, transitions and tokens in the system net are defined as follows:

• Set of places P_S

Each place represents one of resources except AGVs. Subsets of P_S are defined as follows:

P_S^1 is a load station, P_S^2 is a unload station, P_S^3 is the set of input buffers, P_S^4 is the set of output buffers, P_S^5 is the set of machines, P_S^6 is the set of paths, and P_S^7 is the AGV pool.

Capacities of place for a machine and a path are supposed to be one, and those of others are more than one. Then, the set of places P_S is given by

$$P_S = \bigcup_{i=1}^{7} P_S^i. \quad (1)$$

• Set of transitions T_S

Each transition represents each action (transportation with an AGV and processing with a machine) in the FMS. There are nine kinds of action in the FMS as follows: (1), (2) loading an AGV with a job from the load station and the output buffer (T_S^1, T_S^2) respectively, (3), (4) unloading a job from an AGV to the unload station and the input buffer (T_S^3, T_S^4) respectively, (5) loading a machine with a job from the input buffer (T_S^5), (6) unloading a job from a machine to the output buffer (T_S^6) respectively, (7), (8) going toward and leaving the AGV pool (T_S^7, T_S^8) respectively, and (9) the movement of AGV along the path (T_S^9). Then, the set of transitions T_S is given by

$$T_S = \bigcup_{i=1}^{9} T_S^i. \quad (2)$$

• Color information for token c_S

In our model, each token has color information to represent the status of a corresponding AGV and/or

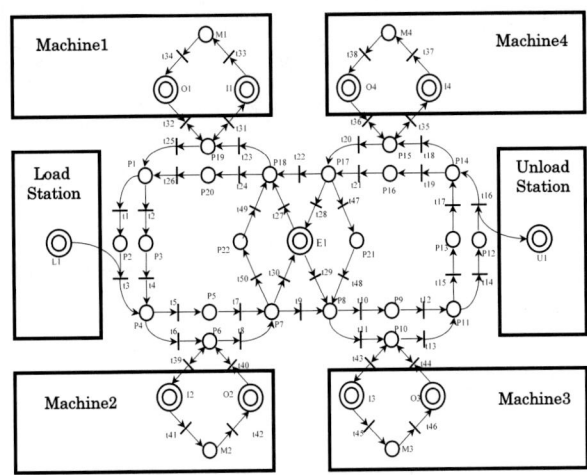

Figure 1: Example of System net

a job. Also, the location of each token represents the corresponding position of an AGV and/or a job. If the token represents the job, it indicates a type of a job and its stage in processing. If the token represents an AGV, it indicates a destination of the AGV and the information on the job which is carried. The color of token is defined as follows:

$$c_S = (c_d, c_j, c_o) \quad (3)$$

$c_d \in C_{Sd}, c_j \in C_{Sj}, c_o \in C_{So}.$

where
C_{Sd} is the set of destinations of an AGV
C_{Sj} is the set of indices for type of a job
C_{So} is the set of indices for a processing stage
(c_d, ϕ, ϕ) denotes an AGV which is not carrying any job, (ϕ, c_j, c_o) denotes a job and (c_d, c_j, c_o) denotes an AGV which is carrying a certain job.

As the result, the system net is defined as follows:

$$N_S = (P_S, T_S, C_S, I_S^-, I_S^+, \theta_S, M_{S0}) \quad (4)$$

where $C_S, I_S^-, I_S^+, \theta_S$ and M_{S0} are given in the following:

C_S is a color function.

I_S^- and I_S^+ are the backward and forward incidence relationship functions respectively.

$\theta_S(t)$ is a firing time.
M_{S0} is the initial marking.

Fig.1 shows an example of system net for the FMS including four machines. A double circle in Fig.1 denotes a place of which capacity is more than one. The system net enables us to analyze the status of the FMS visually, because the complete information of physical behavior in the FMS including competition among AGVs are included in it.

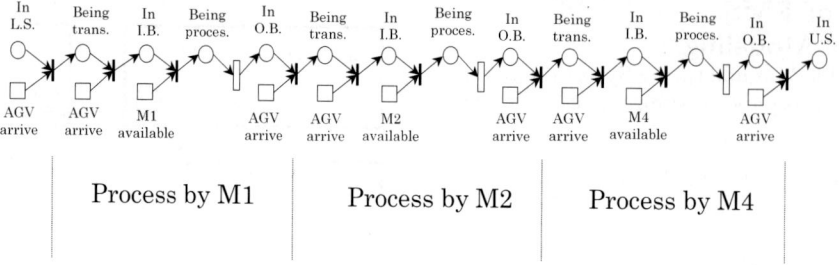

Figure 2: Example of process net

Table 1: Machine routings and processing time for job1

order	1	2	3
Job1	Machine1(5)	Machine2(10)	Machine4(2)

2.3 Process net

The process net describes processing information of jobs in the FMS. Fig.2 shows an example of process net for the job listed in Table 1. In Fig.2, circle, square, bar and rectangle indicate place for status of the job, place for status of the resource, immediate transition(i.e. firing time is equal to zero) and timed transition, respectively. Status of the job flows as follows : (1) 'being in the load station', (2) 'being transported', (3) 'being in the input buffer of the machine', (4) 'completion of process with the machine', (5) 'being in the output buffer'. After some iteration of this flow, it reaches at 'being in the unload station'. The processing time with each machine is represented by the firing time of corresponding transition (depicted by rectangle in Fig.2).

The process net is defined as follows:

$$N_P = (P_P, T_P, I_P^-, I_P^+, \theta_P, M_{P0}) \quad (5)$$

where P_P and T_P are set of places and transitions, I_P^- and I_P^+ are the backward and forward incidence relationship functions, M_{P0} is the initial marking, and $\theta_P(t)$ is firing time.

A relationship between the system net and the process net can be described as follows. Information on the arrival of AGVs and the availability of machines are sent to the process net based on the marking in the system net. On the contrary, information on the processing time with the machine, index number of job loaded on the AGV and destination of the AGV are sent to the system net based on the fired transition in the process net. Assignment of transportation task to the available AGVs in the system net is controlled based on a request from a load station and output buffers.

Our proposed two-stage modeling enables us to reduce a time required to both construct and modify the complete FMS model. Moreover, the competition between AGVs on the path and constraints imposed on the shop floor can be easily considered.

3 Proposed Scheduling Algorithm
3.1 Minimization of completion time

This subsection presents a model based scheduling algorithm for minimization of completion time. The proposed strategy consists of an RTA* algorithm and a supervisor. We call the algorithm introduced in this subsection 'forward scheduling algorithm' to distinguish from the algorithm introduced in the next subsetion. RTA* algorithm [4] was originally developed in the field of artificial intelligence and is one of the real-time graph search method. In the RTA*, solution is found by iterating an 'searching phase' and 'moving (executing) phase'. Different from original A*, the depth of search is set to be considerably small. This implies that the estimated cost $h(m_x)$ at each marking m_x plays an essential role. See [4] for more details about RTA* algorithm. It is, however, not straightforward to specify an appropriate $h(m_x)$, and in many cases, it depends on the applications. In this paper, $h(m_x)$ is calculated as the summation of the remaining cost, that is, the remaining transportation and processing time, as for jobs which are not in the unload stations at marking m_x.

In our proposed scheduling method, the RTA* algorithm is integrated with a supervisor [4],[5] in order to improve the quality of solution found by the RTA*. The role of the supervisor is to reduce the search space by eliminating obviously non-desirable markings. The operation of the supervisor can be realized by controlling the firing of transitions in the system net and the process net. In this paper, the following dispatching rules are adopted as the supervisory control logic:

(1) After the job i is processed in machine j, the transportation task of job i is assigned to an AGV which is nearest to machine j.

(2) If some tasks (processing with machines and transporting with AGVs) become enable, those tasks are executed simultaneously.

(3) A job loading at the load station or the output buffer is decided so as to avoid an overflow of input buffers.

The purpose of these rules are (1) to restrain an empty run of AGVs, (2) to make use of maximal number of available resources in the FMS, and (3) to avoid the deadlock. If the rule (3) is omitted, the search based on the RTA* often results in deadlock. The outline of the proposed reactive scheduler is depicted in Fig.3 and the detail of the algorithm is formulated in Appendix 1.

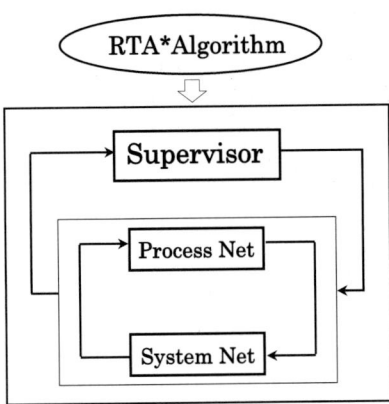

Figure 3: Outline of poposed algorithm

3.2 Minimization of total deadline over-time

In this subsection, a scheduling method for the minimization of total deadline over-time is introduced. Although the minimization of total deadline over-time is not so easy to handle as the minimization of completion time, the model based approach can provide a promising way as follows. First of all, the models developed in section 2 are converted to their reverse models and two-stage scheduling is applied.

Then, at first stage, the search agent tries to find the solution for minimization of maximum completion time by applying the forward scheduling algorithm to the reverse models with regarding the maximum due time as the starting time (We call this 'backward scheduling algorithm'). In this case, the search process starts from the final marking in the original model at maximum due time for all jobs, d_{max} and goes backward being driven by RTA* and supervisor. If the end time (commencement time in the original model) for all jobs, r_{end}, of this backward scheduling results in negative (obviously, this must be positive), some modification as for the priority which is assigned to each job will be made. This modification is taken care of in the second stage of the proposed scheduling.

The reverse system net N_S^I can be defined as follows:

$$N_S^I = (P_S, T_S^I, C_S, I_S^+, I_S^-, \theta_S, M_{S0}^I) \qquad (6)$$

where set of transitons T_S^I is given by

$$T_S^I = \{t_S^i \mid \bullet t_S^i = t_S \bullet, t_S^i \bullet = \bullet t_S, t_S \in T_S\}. \qquad (7)$$

The reverse model for the process net can be constructed by similar way with that of system net.

If the end time for all jobs r_{end} is negative, this implies that the corresponding solution is infeasible. Therefore, the scheduler tries to modify the priority of each job being loaded to make the solution feasible.

The objective of second stage is to obtain a positive $r_{end}(x)$ where x is a priority vector represented by a distribution of the probability defined as follows:

$$x = (x_1, x_2 \cdots x_n) \qquad (8)$$

where n denotes a number of job. The loading task at the unload station or the buffer of each machine is carried out by referring this priority vector, i.e. the job which has high priority is likely to be selected by AGV. Therefore, the problem for the second stage can be formulated as follows :

find x which makes $r_{end}(x)$ positive

The job i's priority at step $(k + 1)$ is decided by introducing intermediate variable $z_i(k)$ as follows:

$$x_i(k+1) = \frac{z_i(k+1)}{\sum_i z_i(k+1)} \qquad (9)$$

$$z_i(k+1) = x_i(k) - \alpha_i(k)\frac{\tau_i(k)}{\sum_i \tau_i(k)} \qquad (10)$$

where $\alpha_i(k)$ indicates the magnitude of improvement and is restricted in

$$0 < \alpha_i(k) < \delta, \qquad (11)$$

and $\tau_i(k)$ is given by

$$\tau_i(k) = \begin{cases} 0.01 & if \quad @r_i(k) < 0 \\ r_i(k) & otherwise \end{cases} \qquad (12)$$

where $r_i(k)$ represents the end time (i.e. the commencement time in the original model) of job i at stop k. Note that r_{end} is given by $\min_i r_i$.

Roughly speaking the job priorities $x_i(k)$ referring the end time, $r_i(k)$. By reducing the priority x_i of which corresponding job i has large due-time-margin r_i, the due time constraint is likely to be satisfied.

Note that conventional framework for the due time problem was very hard to apply to a complex (but realistic) system. Our approach, however, is easily applied to such system because the designer only have to regard the due time scheduling problem as the standard problem of minimization of completion time in the reverse model, and carry out some iterations with production priority.

The proposed two stage algorithm for minimization of total deadline over-time problem is formulated in Appendix 2.

4 Numerical Experiments

In order to confirm usefulness of the proposed algorithm, some numerical experiments are performed. The environment of FMS used for the numerical experiments is shown in Fig.1. PC STATION M366 (CPU:Pentium3 667MHz, memory:384MB, Sotec Ltd.) was used for the numerical experiments.

4.1 Case study for minimization of completion time

Fig.4 shows solutions (a Gantt chart) obtained by applying the algorithm developed in section3.1 The number of each job is supposed to be one and their processing times are listed in Table 2. In Fig.4, the

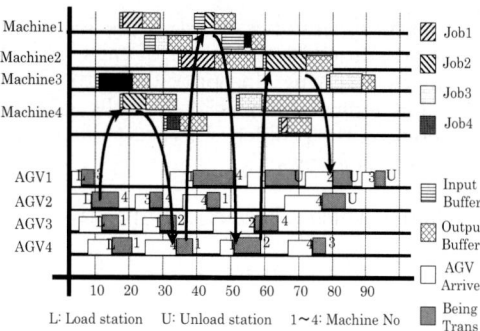

Figure 4: Solution by proposed algorithm

Table 2: Machine routings and processing times for Jobs

Process	1	2	3
Job1	M 1(5)	M 2(10)	M 4(2)
Job2	M 4(7)	M 1(3)	M 2(12)
Job3	M 1(3)	M 4(7)	M 3(10)
Job4	M 3(11)	M 4(4)	M 1(2)

horizontal axis represents time and the vertical axis represents resources such as machines and AGVs. For example, Job 2 is transported by the AGV 4 from the load station to the machine 4. After being processed with machine 4, Job 2 is transported to the machine 1. Finally, Job 2 is transported to the machine 2 by the AGV 1 and is processed with the machine 2. From this case study, feasibility of the solution has been confirmed.

Next, the solution and computational time obtained by applying the proposed algorithm are compared with those obtained by applying A* algorithm [3] with a supervisor. Conditions used for those simulations are exactly same. Table 3 shows solutions and computational times. They are also averages of ten trials and the number in parentheses denotes the capacity of the list $OPEN$. In Table 3, computational time required in the proposed algorithm is much smaller than one required in A* algorithm, although the solutions are comparable.

Table 3: Comparison with A* algorithm

Method	Completion time	Computational time(sec)
RTA*	104	1
A* (500)	114	16
A* (1000)	103	26
A* (1500)	95	49
A* (2000)	95	118

4.2 Case study for minimization of total deadline over-time

Table.4 shows the solution obtained by applying the two stage scheduling algorithm developed in section 3.2. The due time for each job is supposed to be 1200, 1000, 800 and 600, respectively. Also the number of each job is supposed to be 20. By means of 'X/Y' in Table 4, X represents the number of jobs which was commenced in the corresponding time slot, and Y represents the number of jobs which was terminated in the corresponding time slot. As listed in Table 4, all jobs satisfy their due-time requirements.

Table 4: Simulation of Due-time Problem

Time	Job 1	Job 2	Job 3	Job 4
0 `100	0/0	0/0	4/0	3/1
`200	0/0	0/0	3/4	2/3
`300	0/0	0/0	4/3	5/2
`400	0/0	0/0	4/4	4/4
`500	0/0	0/0	2/4	5/4
`600	0/0	4/0	0/1	1/6
`700	0/0	4/4	2/1	0/0
`800	0/0	8/4	1/3	0/0
`900	3/0	3/8	0/0	0/0
`1000	11/4	1/4	0/0	0/0
`1100	4/1	0/0	0/0	0/0
`1200	2/15	0/0	0/0	0/0

5 Conclusions

In this paper, a new scheduling method for the FMS based on the Timed Petri net model has been proposed. First of all, a new modeling method has been developed in which physical behavior of the FMS and the processing information on each job were described independently. Secondly, a new reactive scheduling method based on RTA* and the supervisor has been proposed and applied to the minimization of completion time. The proposed algorithm can obtain a semi-optimal solution with less computational time. Thirdly, two-stage scheduling method for minimization of deadline over-time has been proposed. In the first stage, the algorithm for the minimization of completion time was directly applied to the reverse model with regarding the due time as starting time, and in the second stage, priority vector was updated to make the solution feasible. Our two-stage algorithm can be easily applied to considerably complex problem. Finally, the usefulness of the proposed algorithm has been confirmed through some numerical experiments.

Appendix 1
Algorithm for minimization of completion time (MCT)

Step1 Put the initial marking m_0 in a list $EXECUTED$ D

Step2 Get a latest marking m_x from the list $EXECUTED$.

Step3 Assign the destination to available AGVs at marking m_x making a reference to the supervisor.

Step4 Find the firable transitions at the marking m_x. If there is no firable transition, terminate with failure.

Step5 Generate the next marking m_y for firable transitions referring the supervisor and set pointers from the next marking to m_x. Compute $c(m_x, m_y)$ for all m_y, and put m_y in a list $NEXT$.

Step6 For each marking m_y in the list $NEXT$, do the following.

  A: If m_y is already in the list $EXECUTED$, compute $\hat{f}(m_x, m_y) = h(m_y) + c(m_x, m_y)$.

  B: If m_y is not in the list $EXECUTED$, do the following :

    Step6B-1 Put the marking m_y in a list $OPEN$.

    Step6B-2 If the list $OPEN$ is empty, terminate **Step6B** with $\hat{f}(m_x, m_y) = \infty$, and initialize the list $OPEN$.

    Step6B-3 Remove the first marking m from the list $OPEN$.

    Step6B-4 If m is within depth $d-1$ from m_y, terminate **Step6B** with $\hat{f}(m_x, m_y) = \hat{f}(m_y, m) + c(m_x, m_y)$, and initialize the list $OPEN$.

    Step6B-5 Assign the destination to available AGVs at marking m referring to the supervisor.

    Step6B-6 Find the firable transition at marking m.

    Step6B-7 Generate the next marking m' for firable transitions referring the supervisor. Compute $h(m')$, $g'(m_y, m')$ and $\hat{f}(m_y, m') = h(m') + c'(m_y, m')$ for all m', and put m' in the list $OPEN$.

    Step6B-8 Reorder markings in the list $OPEN$ in the ascending order of $\hat{f}$.

    Step6B-9 Go to **Step6B-2**.

Step7 Reorder the marking in the list $NEXT$ in the ascending order of $\hat{f}(m_x, m_y)$.

Step8 Remove the first marking m_y from the list $NEXT$ (If there are some markings with same value of $\hat{f}(m_x, m_y)$, select a marking m_y randomly from them). If $\hat{f}(m_x, m_y) = \infty$, terminate this algorithm with failure.

Step9 Execute the action in real FMS which corresponds to the movement to m_y.

Step10 Put m_y in the list $EXECUTED$. If the marking m_y is the final marking m_e, terminate this algorithm.

Step11 If the list $NEXT$ is not empty, remove the first marking m_y from the list $NEXT$. Substitute $\hat{f}(m_x, m_y)$ for $h(m_x)$.

Step12 Initialize the list $NEXT$ and go to **Step2**.

Algorithm for minimization of total deadline over-time (MTDO)

Step1 Construct reverse model and set k_{last}.

Step2 Set $k = 1$.

Step3 If $k > k_{last}$, terminate it with failure.

Step4 Set all elements in priory distribution vector, x to be $1/n$, and substitute x for x_{best}.

Step5 Calculate the end time for job i, $r_i(x)$ for all jobs by applying MCT to the reverse model with regarding a due time as a starting time, and set $r_{end}(x) = \min r_i(k)$.

Step6 Do the following.

  A: If $r_{end}(x) > 0$, terminate with success.

  B: Do the following.

    Step 6B-1 Calculate
$$x_i(k+1) = \frac{z_i(k+1)}{\sum_i z_i(k+1)},$$
$$z_i(k+1) = x_i(k) - \alpha_i(k)\frac{\tau_i(k)}{\sum_i \tau_i(k)}$$
where,
$0 < \alpha_i(k) < \delta$,
and
$$\tau_i(k) = \begin{cases} 0.01 & if \ @r_i(k) < 0 \\ r_i(k) & otherwise. \end{cases}$$

    Step 6B-2 $k = k+1$, and go to **Step 3**.

References

[1] P. B. Luh et al.: 'Job shop schedulimg with group-dependent setups, finite buffers, and long time horizon', Preliminary version in Proc.34th IEEE Conf. on Decision and Control (1995)

[2] A.S.Kiran and M.L.Smith: 'Simulation Studies in Job Shop Scheduling- ', Computer Industrial Engineering, 8-2, 87-105, 1984

[3] R.E.Korf : 'Depth-First Iterative-Deepening: An Optimal Admissibal Tree Search', Artificial Intelligence, Vol.27, pp.97-109, 1995

[4] R.E.Korf : 'Real Time heuristic search', Artificial Intelligence, Vol.42, pp.189-211, 1990

[5] P.J.Ramadge and W.M.Wonham : 'Supervisory Control of A Class of Discrete-Event Processes', SIAM J. Contr. Opt., Vol.25, No.1, pp.206-230,1987

[6] P.Jean-Marie: 'A Class of Petri Nets for Manufacturing System Integration ', IEEE transactions on robotics and automation, Vol.13, No.3, pp.317-326, 1997

[7] W.Naiqi: 'Necessary and Sufficient Conditions for Deadlock-Free Operation in Flexible Manufacturing Systems Using a Colored Petri Net Model ', IEEE transactions on systems, man, and cybernetics -part C: applications and reviews, Vol.29, No.2, pp.192-3204, 1999

OPTIMAL CONFIGURATION AND PARTNER SELECTION IN DYNAMIC MANUFACTURING NETWORKS

N. Viswanadham, Roshan Gaonkar
The Logistics Institute – Asia Pacific

V. Subramaniam
Mechanical and Production Engineering

National University of Singapore
10 Kent Ridge Crescent, Singapore 119260

ABSTRACT

In this paper we develop a linear programming model for integrated partner selection and scheduling in a web-enabled global manufacturing network environment. We assume that all stakeholders in the supply chain share information on their capacities, schedules and cost structures. Based on this information, the model addresses the issue of partner selection for minimal cost of manufacturing and delivery. The model is solved using ILOG optimization tools.

1. INTRODUCTION

In this age of globalization, the ability of companies to meet rising customer expectations at lower operating costs has become a key competitive advantage. In order to enhance their competitiveness, companies no longer take ownership of all the assets and processes needed in delivering value to the customer. Instead they focus on their core competencies and partner with companies possessing complementary strengths. This together with a variety of other reasons has resulted in geographically distributed manufacturing, wherein, different components produced in different countries are assembled in another and the final product is customized at the customer's site in yet another country.

In this scenario, there is critical need for developing methods for partner selection and coordination among them and for integrated planning. Thankfully, the Internet provides the necessary platform to enable the seamless distribution of data, information, and knowledge across the entire value chain. The Internet thus has enabled effective monitoring of the activities executed by one's partners thereby further promoting and supporting the outsourcing of sub-processes and activities.

Our thinking in this paper is in line with emerging trends in collaborative e-commerce and the concept of value webs [1]. A value web is a dynamically changing network of independent companies, linked by the Internet to offer value to different customers [3]. The information linkages between companies are formed in real-time through the Internet, in response to market conditions and sometimes the formation of these linkages are facilitated by electronic marketplaces.

In the formation of effective value webs, the choice of partners for fulfillment of each and every order is important. This requires the development of optimization models and solutions, for partnership selection and value delivery, making full use of the information available on capacities, inventories, lead times, production-schedules and cost.

There is a significant amount of literature existing on partner selection in the operations research and management science literature. Weber and Current [4] discuss a multi-criteria analysis for vendor selection. They develop a model for minimizing total cost, late deliveries and supply rejection given the infrastructure constraints and constraints imposed by the company's policy. Arntzen et al [5] describe a global supply chain management model that was implemented at Digital Equipment Corporation. The model incorporates capacity constraints, import taxes, fixed charges, transportation constraints etc and recommends a production, distribution and supplier network. Erenguc et al [6] review and evaluates some of the relevant literature on production and distribution planning at each stage of the supply chain. The interested reader might find [7] useful for a comprehensive classification of publications on vendor selection criteria. Some other researchers have focused on the production scheduling aspects of the supply chain. Bretthauer and Cote [8] talk about a non-linear programming model for multi-period capacity planning.

Thus in the literature, most efforts have revolved around the selection of suppliers for a particular manufacturer. Our research here attempts to do much more. We select the supply chain configuration for every customer order and additionally provide the schedules for the manufacturing, assembly and inbound transportation within the supply chain. In fact, we embed the capacity availability and the fixed schedules of air/sea carriers into our algorithm and find the optimal component manufacture and assembly schedules. Our purpose in this paper is to develop a mathematical programming model for partner selection and supply chain design, and in the process build an integrated planning decision support system for channel-masters, supply chain process-owners and electronic market participants. We consider a four-tier supply chain with buyers, brand manufacturers, sub-assembly manufacturers, component suppliers and logistics service providers. The model determines the optimal order quantities to be allocated to each of the manufacturers, suppliers and logistics service providers and determines the production and delivery schedules for each of them.

In the remaining four sections of this paper we develop a linear programming model for integrated partner selection and scheduling in a web-enabled global manufacturing network environment. We begin by describing the problem we wish to address. We also formulate a linear programming model for integrated partner selection and scheduling. In the subsequent section we briefly describe the ILOG tools used in solving the model. We then proceed to present and discuss some of our results under the section on computational results. And finally we conclude by jotting down some of our observations in the field of electronic supply chains.

2. PROBLEM FORMULATION

2.1 Problem Description

We assume that there are a number of component suppliers, sub-assembly manufacturers, brand manufacturers and logistics service providers in different geographical locations. They all share information on their production schedules, capacity, cost, quality, etc. We also assume that there are a number of buyers with orders for a range of finished goods. These orders can be fulfilled by different sets of manufacturers and suppliers at different costs and in different lead times with the support of the logistics service providers. The logistics service providers have their own costs, capacity constraints and fixed shipping schedules. With access to operational information on all the participants in an electronic supply chain the challenge is, how best to meet the demands of the buyers, using a combination of sellers and logistics providers with minimal operational cost. In particular, a collaborative approach in supply chain management and coordination is required to form an effective and efficient value web. The Internet has enabled economically viable real-time supply chain coordination in value webs as shown in Figure 2.

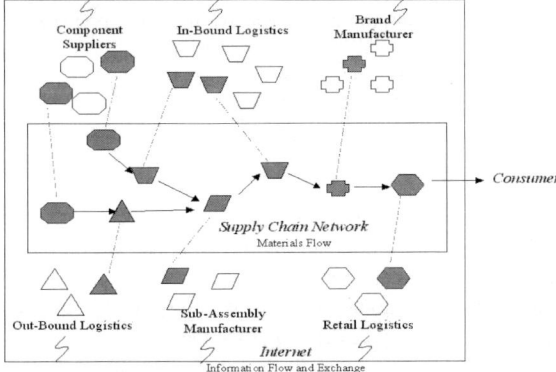

Figure 2: Supply Chain Configuration and Coordination between a set of partners using the Internet.

The challenge for a value web is the selection of the suppliers, assemblers and logistics service providers who can collectively meet the deadlines of the buyers and maximize the value delivered. Apart from incorporating the capacity constraints in the supply chain decisions, production activities need to be synchronized with the schedules of the logistics service providers, so that items can be ready for pickup in a just-in-time manner, instead of having to wait in inventory. There can be significant cost-savings in this exercise, through reduced inventory levels.

2.2 Notation

For development of a mathematical model for the above scenario, the following notations were used.

Identifiers

r : Component type identifier.
R : Number of component types.
v : Component supplier identifier.
V : Number of component suppliers.
i : Sub-assembly type identifier.
I : Number of sub-assembly types.
j : Sub-assembly supplier identifier.
J : Number of sub-assembly suppliers.
k : Manufacturer Identifier.
K : Number of Manufacturers.
m : Buyer Identifier.
M : Number of Buyers.
l : Brand Identifier.
L : Number of Brands.
t : Time Period identifier.
T : Total time horizon of the model.

Parameters

C_{abt} : Maximum production capacity for component/sub-assembly/brand of type a offered by Component Supplier/Sub-Assembly Supplier/Manufacturer b in time period t.

P_{ab} : Unit cost price for component/sub-assembly/brand of type a procured from Component Supplier/Sub-Assembly Supplier/Manufacturer b.

T_{abdt} : Maximum transportation capacity for shipment of component/sub-assembly/brand of type a from Component Supplier/Sub-Assembly Supplier/Manufacturer b to its customer d in time period t.

U_{abdt} : Unit transportation cost for shipment of component/sub-assembly/brand of type a from Component Supplier or Sub-Assembly Supplier or Manufacturer b to its customer d in time period t.

W_{ab} : Unit inventory cost incurred for component/sub-assembly/brand of type a in the possession of Component Supplier or Sub-Assembly Supplier or Manufacturer b.

Q_{ab} : Quantity for model type a required by Buyer b.

D_{ab} : Time period by which required quantity for model type a is to be delivered to Buyer b.

R_{ab} : Units of component type a required in the production of one unit of sub-assembly b.

M_{ab} : Units of sub-assembly type a required in the production of one unit of model b.

Variables

Q_{abt} : Quantity produced for component/sub-assembly/brand a by Component Supplier/Sub-Assembly Supplier/Manufacturer b in t.

I_{abt} : Inventory of component/sub-assembly/brand a with Component Supplier or Sub-Assembly Supplier or Manufacturer b in time period t.

S_{abdt} : Quantity shipped of component/sub-assembly/brand of type a from Component Supplier or Sub-Assembly Supplier or Manufacturer b to its customer d in time period t.

2.3 LP Model

We now develop a linear programming model for a dynamic manufacturing network. The objective of the model was to maximize the profit earned by the configurable manufacturing network subject to various capacity, production and logistics schedules and flow balancing constraints.

The profit was calculated, as given in Eqn. 1, as the sum of the revenue made from sales to the buyers, less the production costs and the costs incurred in the operation of the supply chain network, specifically transportation and inventory costs.

$$MaxPROFIT = \sum_{l=1}^{L} \sum_{m=1}^{M} P_{lm} Q_{lm}$$
$$- \begin{bmatrix} \sum_{r=1}^{R} \sum_{v=1}^{V} \sum_{j=1}^{J} \left(\sum_{t=1}^{T} S_{rvjt} U_{rvjt} + P_{rv} \sum_{t}^{T} S_{rvjt} \right) \\ + \sum_{i=1}^{I} \sum_{j=1}^{J} \sum_{k=1}^{K} \left(\sum_{t=1}^{T} S_{ijkt} U_{ijkt} + P_{ij} \sum_{t}^{T} S_{ijkt} \right) \\ + \sum_{l=1}^{L} \sum_{k=1}^{K} \sum_{m=1}^{M} \left(\sum_{t=1}^{T} S_{lkmt} U_{lkmt} + P_{lk} \sum_{t}^{T} S_{lkmt} \right) \end{bmatrix}$$
$$- \sum_{t=1}^{T} \begin{bmatrix} \sum_{r=1}^{R} \sum_{v=1}^{V} W_{rv} I_{rvt} + \sum_{r=1}^{R} \sum_{j=1}^{J} W_{rj} I_{rjt} \\ + \sum_{i=1}^{I} \sum_{j=1}^{J} W_{ij} I_{ijt} + \sum_{i=1}^{I} \sum_{k=1}^{K} W_{ik} I_{ikt} \\ + \sum_{l=1}^{L} \sum_{k=1}^{K} W_{lk} I_{lkt} + \sum_{l=1}^{L} \sum_{m=1}^{M} W_{lm} I_{lmt} \end{bmatrix}$$
... (1)

There are various capacity constraints in the virtual supply chain that make the solution non-trivial.

The component suppliers cannot produce more than their maximum production capacity. Hence,

$$Q_{rvt} \leq C_{rvt} \quad forall \quad r \in R, v \in V \ \& \ t \in T$$
... (2)

The components produced are held at the component suppliers end until they are shipped off to sub-assembly manufacturers. The production of new components adds to the inventory held by the supplier at the end of each time, while the products sold and shipped to the sub-assembly manufacturers in each time period reduces the suppliers inventory.

$$I_{rv(t-1)} + Q_{rvt} = \sum_{j=1}^{J} S_{rvjt} + I_{rvt} \quad forall \quad r \in R, v \in V, j \in J \ \& \ t \in T$$
... (3)

However, the quantity that can be transported in a single period is constrained by the maximum capacity of the transportation infrastructure. Considering our scenario with fixed shipping schedules, in time-periods when the service is available the transportation capacity is non-zero. However, for time-periods where particular flights or shipments are not scheduled the transportation capacity is zero. Hence the transportation of the component types from the component suppliers to the sub-assembly manufacturers site is bound by the below constraint.

$$S_{rvjt} \leq T_{rvjt} \quad forall \quad r \in R, v \in V, j \in J \ \& \ t \in T$$
... (4)

Once the components reach the sub-assembly manufacturer from the component supplier it adds to the manufacturer's inventory, which is then consumed by the production process. However before the manufacturing process can start and the component type can be consumed, the sub-assembly manufacturer will need to check adequate availability of all components that will be used in the sub-assembly production process. This imposes the following constraint on the component availability and the sub-assembly production.

$$I_{rj(t-1)} \geq \sum_{i=1}^{I} R_{ir} Q_{ijt} \quad forall \quad r \in R, i \in I, j \in J, t \in T$$
... (5)

However once the production process begins the inventory drops. The inventory status for component types with the manufacturer can be determined as given below in Eqn. 6.

$$I_{rj(t-1)} + \sum_{v=1}^{V} S_{rvjt} = \sum_{i=1}^{I} R_{ir} Q_{ijt} + I_{rjt}$$
$$forall \ r \in R, v \in V, j \in J, i \in I \ \& \ t \in T$$
... (6)

The capacity constraints and the inventory constraints that apply to the component suppliers apply to the sub-assembly manufacturers as well. The maximum production of sub-assemblies is constrained by the production capacity of the sub-assembler.

$$Q_{ijt} \leq C_{ijt} \quad forall \quad i \in I, j \in J \ \& \ t \in T$$
... (7)

The inventory of sub-assemblies at the sub-assembly manufacturers end increases at the end of each period by the quantity produced and decreases by the amount of sub-assemblies shipped out to customers, in that time period.

$$I_{ij(t-1)} + Q_{ijt} = \sum_{k=1}^{K} S_{ijkt} + I_{ijt} \quad forall \quad i \in I, j \in J, k \in K \& t \in T$$
... (8)

The quantity of sub-assemblies that can be shipped is constrained by the capacity of the transportation infrastructure.

$$S_{ijkt} \leq T_{ijkt} \quad forall \quad i \in I, j \in J, k \in K \& t \in T$$
... (9)

The shipped sub-assemblies will be stored at the manufacturer's end. Only in the case of sufficient availability of all the needed sub-assemblies, will production of the brands begin.

$$I_{ik(t-1)} \geq \sum_{l=1}^{L} M_{li} Q_{lkt} \quad forall \quad i \in I, l \in L, k \in K, t \in T$$
... (10)

As regards the inventory levels of sub-assemblies at the manufacturer's end incoming stocks will add to the inventory and sub-assemblies stocks will be used in the production of the brands. The inventory status for sub-assembly types with the manufacturer can be determined as given below in Eqn. 11.

$$I_{ik(t-1)} + \sum_{j=1}^{J} S_{ijkt} = \sum_{l=1}^{L} M_{li} Q_{lkt} + I_{ikt}$$
$$forall \quad i \in I, j \in J, k \in K, l \in L \& t \in T$$
... (11)

The manufacturer cannot manufacture the different brand types in a quantity more than their maximum capacity.

$$Q_{lkt} \leq C_{lkt} \quad forall \quad l \in L, k \in K \& t \in T$$
... (12)

The manufactured brands are stored at the assemblers warehouse so that they may then be delivered to the buyer on his requested date and time.

$$I_{lk(t-1)} + Q_{lkt} = \sum_{m=1}^{M} S_{lkmt} + I_{lkt}$$
$$forall \quad l \in L, k \in K, m \in M \& t \in T$$
... (13)

The transportation capacity constraint for the movement of the models from the assembly sites to the buyers will be governed by the below transportation capacity constraint.

$$S_{lkmt} \leq T_{lkmt} \quad forall \quad l \in L, k \in K, m \in M \& t \in T$$
... (14)

The models will be stored at the buyer's warehouse or will be kept aside by the assembler within his own premises for subsequent delivery.

$$I_{lm(t-1)} + \sum_{k=1}^{K} S_{lkmt} = I_{lmt} \quad forall \quad l \in L, k \in K, m \in M \& t \in T$$
... (15)

Finally, the entire process of procuring the product types, assembling and delivering the finished brands to the buyers needs to be completed by the date specified by the buyer, or equivalently the inventory level at the buyer's end on the due date for his order should equal the quantity ordered by the buyer.

$$I_{lm(t+D_{lm})} = Q_{lm} \quad forall \quad l \in L, m \in M \& t \in [1, T - D_{lm}]$$
... (16)

The solution of this model determines the selection of suitable suppliers and assemblers for each order and also provides a schedule for production and assembly activities within the supply chain.

With the above mathematical model any of the available optimization toolkits might be used in order to determine the firms involved in the dynamic supply chain for a given set of orders.

3. SOLUTION

3.1 ILOG OPL Studio

The above linear model was developed and optimized in the commercial optimization program, OPL Studio available from ILOG. ILOG provides a very comprehensive library of optimization algorithms implemented in C++. These algorithms can be used for the solution of a varied number of large-scale linear, integer or constraint programming models. ILOG also incorporates a set of modeling concepts, such as activities and resources, which are very useful in the solution of scheduling and allocation problems. ILOG studio utilizes the Optimization Programming Language (OPL) for modeling of problems. User-defined search strategies for each model can be specified in order to reduce the computational power required for the solution.

3.2 Computational Complexity

The above model LP model was developed in ILOG and solved for a scenario with 3 component suppliers, 5 sub-assembly manufacturers supplying 2 different product types to 3 manufacturers, who sell 2 different model types to 2 buyers. Not all manufacturers manufacture all models or all suppliers supply all product types. The time horizon for the model was taken as 12 periods. The number of variables that were encountered was 2535 and the constraints numbered 2899. The solution time was less than 10 seconds.

4. COMPUTATIONAL RESULTS

In order to verify the dynamic nature of the model that was developed in earlier sections, the model was solved for orders placed by each of the buyers and the supply chain configuration for each was observed and compared.

It was assumed that each buyer required 500 units of a model at the end of the 10th time-period and was willing to pay the price of $400, same as the rest of the buyers. The orders were to be fulfilled by the manufacturing network part of which is described in Appendix A. Each unit of the finished model requires 1 unit of sub-assembly 1 and 2 units of sub-assembly 2. Both the sub-assemblies are in turn made of 1 unit of component 1 and 1 unit of component 2. The manufacturing network can be designed to output more than one models. In such a situation is it very much possible to gain from economies of scale in the collective ordering and transportation of materials, which are used in the manufacture of multiple models. However, for ease of explanation the manufacture of only model has been considered in this experiment.

The optimal supply chain configuration for the fulfillment of 500 units of the finished goods required by buyer 1, consolidated over the entire time horizon, is obtained as given in Figure 3 below.

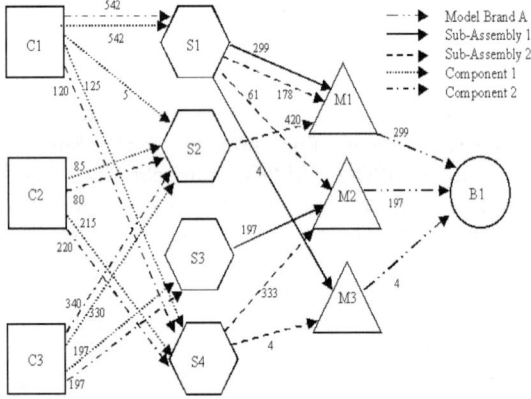

Figure 3. Configuration to meet Buyer 1 demand.

Similarly, the manufacturing network configuration for the fulfillment of any combination of orders from Buyer 1, Buyer 2 and Buyer 3 may also be obtained.

From the three scenarios it is noticed that, depending on where the buyer is, an appropriate manufacturing center is selected to fulfill the order. In case the demand is more than the quantity the manufacturer's supply chain is able to handle, the remainder of the demand will be fulfilled through other manufacturers. One of the bottlenecks in the supply chain that might arise is that the manufacturers are not able to manufacture at full capacity due to the lack of adequate sub-assembly and component supply from the suppliers, who cannot produce any more sub-assemblies or components. Consideration is also given to the schedules of the logistics service provider, so that item are produced just in time for pickup and delivery, instead of having to wait in the inventory.

The profit earned through the operation of the supply chain in each of the three cases is presented below.

Table 2: Profits in sales made to each of the three buyers

Revenues for sales	Profit from supply chain operations
Buyer 1 purchase 500 units @ $400	$85,724
Buyer 2 purchase 500 units @ $400	$87,935
Buyer 3 purchase 500 units @ $400	$79,593

Hence, the model suggests that it would be most profitable to accept orders from Buyer 2 as compared to orders from Buyer 1 or Buyer 3.

In the lack of any capacity constraints at the suppliers' and manufacturers' facility and the availability of transportation infrastructure the problem leads to the trivial solution where the cheapest complete link from the supplier to the buyer is chosen.

The solution of the LP model discussed earlier provides a breakdown of the optimum raw material production quantity, inventory holding and manufacturing capacity utilization for each time period. This information is key to scheduling supply chain activities to perform at optimal levels. Hence, the LP model provides an integrated strategic level partnership tool and a low level operational scheduling tool as well.

In order to simulate the large number of participants simultaneously trading in a marketplace, a solution was obtained for the manufacturing network configuration for multiple buyer requests. With multiple buyers trading on the marketplaces, the supply chain gets more complicated, with larger number of interconnections between the various participants in the value web.

5. CONCLUSIONS

In this paper, we have formulated and solved the partner selection problem in global manufacturing networks. This problem is very important in the current time of globalized manufacturing, proliferating electronic marketplaces and Internet enabled collaborative commerce. We specifically demonstrate how integrated supply chain planning can be conducted using standard optimization tools. We are developing a decision support tool for use in electronic marketplaces.

Our formulation here is linear and uses a LP model. We are planning to solve supply chain problems where the number of buyers and sellers are large and there are more tiers in the chain. We also want to impose more realistic constraints such as those forced by transportation lead times and solve the resulting mixed integer-programming problem.

6. REFERENCES

[1] David Bovet and Joseph Martha, "Value Nets", June 2000.
[2] N. Viswanadham, "Analysis of Manufacturing Enterprises – An Approach to Value Delivery Processes for Competitive Advantage", 1999.
[3] Dorian Selz, "Emerging Business Models: Value Webs", *Proceedings of the 19th ICIS Conference, Helsinki*, 12/98.
[4] Charles A. Weber and John R. Current, "A multiobjective approach to vendor selection", *European Journal of Operational Research*, 68 (1993) 173-184.
[5] Bruce C. Arntzen, Gerald G. Brown, Terry P. Harrison, Linda L. Trafton, "Global Supply Chain Management at Digital Equipment Corporation", *Interfaces*, 25:1 Jan-Feb 1995 69-93.
[6] S. Selcuk Erenguc, N.C. Simpson, Asoo J. Vakharia, "Integrated production/distribution planning in supply chains: An invited review", *European Journal of Operational Research*, 115 (1999) 219 – 236.
[7] Charles A. Weber, John R. Current and W.C. Benton, "Vendor selection criteria and methods", European Journal of Operational Research, 50 (1991), 2-18.
[8] Kurt M. Bretthauer and Murrray J. Cote, "Non-linear programming for multiperiod capacity planning in a manufacturing system", *European Journal of Operational Research*, 96 (1996) 167-179.

7. Appendix A

Presented below is representative data on the manufacturing capacity, transportation capacity, production costs, transportation costs and inventory holding costs for a particular brand delivered by the supply chain network. Similar information is also assumed to be available for all the supply chain participants for all the components and sub-assemblies and components used.

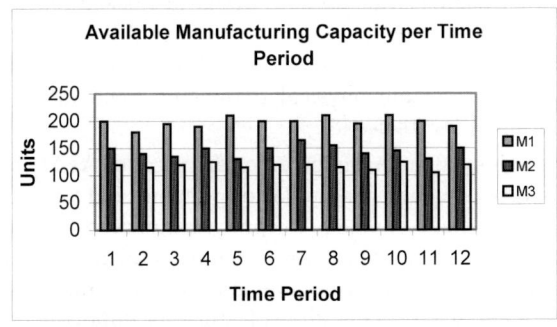

Fig. A-1: Manufacturer capacity information

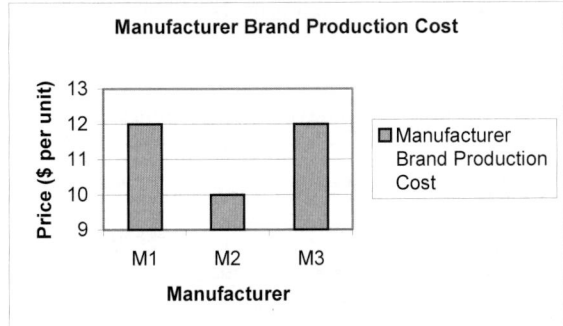

Fig A-2: Brand production costs

Fig A-3: Brand Inventory Holding Cost

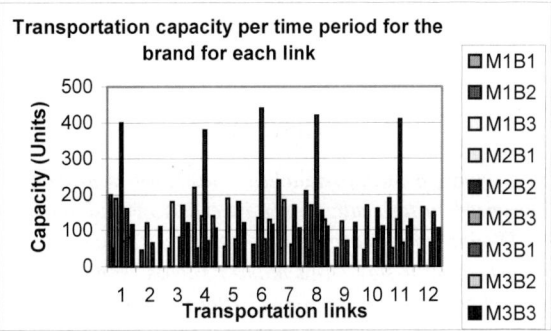

Fig A-4: Transportation capacity per time period for the brand for each link from the brand manufacturers to the buyers.

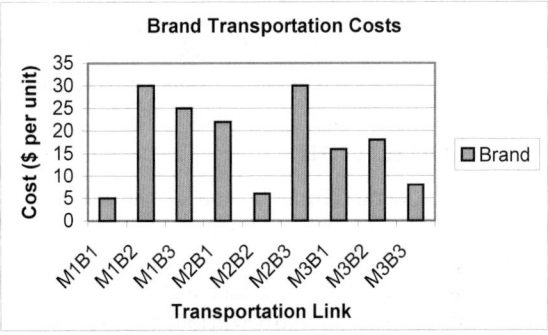

Fig A-5: Brand transportation cost for each link from the brand manufacturers to the buyers.

Development of a Scaled Teleoperation System for Nano Scale Interaction and Manipulation

Metin Sitti[1], Baris Aruk[2], Hiroaki Shintani[2], and Hideki Hashimoto[2]

[1] Dept. of EECS, University of California, Berkeley, CA 94720, USA
[2] Inst. of Industrial Science, University of Tokyo, Roppongi, 108-8558, Tokyo, Japan
sitti@eecs.berkeley.edu, hashimoto@iis.u-tokyo.ac.jp

Abstract

In this paper, a human-machine interface is proposed for teleoperated nano scale object interaction and manipulation. Design specifications for a bilateral scaled teleoperation system with slave and master robots, sensors, actuators, and control are discussed. Phantom and home-made haptic devices are utilized as the master manipulator, and piezoresistive MEMS fabricated probe is selected as the slave manipulator, and topology and force sensor. Force reflecting servo type teleoperation control is chosen, and initial experiments are realized for interacting with silicon surfaces and nano structures. It is shown that fine structures can be felt on the operator's finger successfully.

1 Introduction

One of the important future trend of the robotics field is the handling and interacting with smaller and smaller size objects where the human sensing, precision, and direct manipulation capabilities lack at those scales. Since the present trend in miniaturization goes down to the nanotechnology, handling and interacting with materials at the nano scale have been a new challenging issue for the robotics field [1]. In one side, some researchers are trying to understand more about the nano world physics and chemistry, and, on the other side, robotic researchers are attempting to construct new tools, new control and sensing technologies, and human-machine interfaces specific to the nano world.

Going to the nano scale world, manipulation and interaction systems can be grouped based on the utilized starting point, manipulator and object interaction process, utilized nano manipulator interaction, and operation techniques as given in Figure 1. For the operation techniques, teleoperation or automatic manipulation approaches are utilized. In the former approach, a human operator manipulates the nano objects by using a man-machine user interface. The operator controls the nano robot directly or sends task commands to the nano robot controller as shown in Figure 2. In the direct teleoperation system, the user interface can consist of visual, tactile or force feedback. Hollis et al. [2] used only tactile feedback from the nano world. In [3],[4], force feedback and 3D real-time Virtual Reality graphics interface are utilized during manipulation. Direct teleoperation approach can realize tasks requiring high-level intelligence and flexibility. However, it is slow, not precise, not exactly repeatable, and engaged in many complex and challenging scale difference problems. On the other hand, the task-oriented approach avoids these problems by executing only the given tasks in a closed-loop autonomous control. There is no report on task-oriented approach in the nano scale. In the other manipulation group, the automatic approach, nano robot has a closed-loop control using sensory information. It is utilized by Ramachandran et al. [5], Sitti et al. [6], Schaefer et al. [7]. However, the automatic control in the nano world seems to be very challenging and not reliable at present due to the complexity of the nano dynamics, difficulties in accurate nano positioning and real-time visual feedback, changing and uncertain physical parameters, and insufficient intelligent strategies [8].

In this paper, Atomic Force Microscope (AFM) probe based sensing and manipulation is connected with direct teleoperation technology for putting human operators inside of the nano world for touching, and physically interacting with surfaces, and manipulating materials at the nano scale. This kind of study is important especially for reliable tele-manipulation of nano deformable surfaces such as biological objects, carbon nanotubes, or polymers.

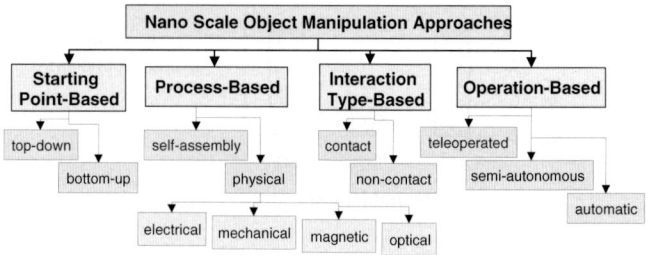

Figure 1: Grouping of nano manipulation systems depending on starting point, utilized process, manipulator and object interaction, and utilized operation technique.

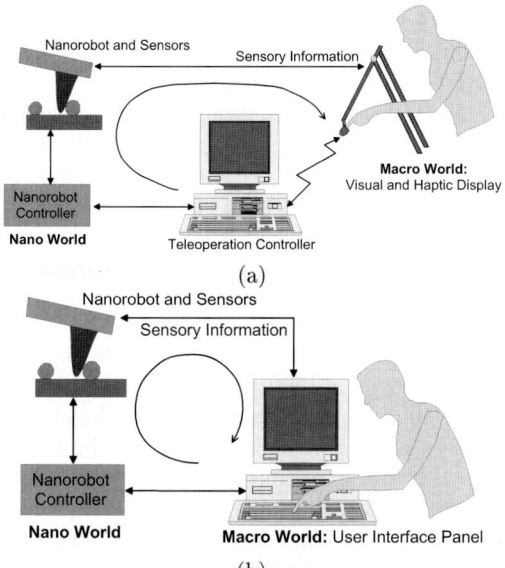

Figure 2: Teleoperation control approaches: (a) direct teleoperation; (b) task-oriented (semi-autonomous) teleoperation.

2 Design Specifications

2.1 Slave Site: Nano World

2.1.1 Nano Manipulator

For realizing mechanical, electrical or magnetic manipulation tasks, contact or non-contact type of manipulators should be designed. In this study, contact type mechanical manipulation is aimed which is an extension of macro scale robotics to the nano scale by utilizing nano physics and chemistry. Desired tasks for macro manipulation are illustrated in Figure 3. However, these tasks are for just 1D or 2D case. For 3D case such as pick and place task, new gripper or chemisty based manipulation strategies are needed to be developed. A possible mechanical 3D manipulation system is proposed in Figure 4. As nano manipulator, AFM probes are very popular [7],[9]. A piezoresistive MEMS fabricated silicon probe is selected in this study as shown in Figure 5. It has a very sharp pyramidal or spherical tip at the end, and is flexible in order to measure the nano forces on the probe tip. Probe bending is measured using a Wheatstone bridge [6]. For this type of manipulator, following parameters should be determined carefully:

- Mechanical properties of the probe are important for the forces can be applied on the objects during manipulation or interaction. Probe stiffness k_c is the main parameter to design for the applied forces, and, for a rectangular beam, it is given as:

$$k_c = \frac{t_c^3 L_x}{4 L_y^3} E \quad (1)$$

where E is the Young modulus, t_c is the thickness, L_x and L_y are the width and length of probe as illustrated in Figure 6. In the case of touching to the nano scale objects, k_c should be small for soft samples such as biological objects. However, for manipulation, it should be large for having sufficient loading force.

- Probe tip size and shape important such that sharp pyramidal or cylindrical tips are proper for indenting or cutting applications while very small size spherical tips are desirable for pushing nano objects on the surfaces (point contact is guaranteed). Furthermore, large tip size increases the adhesion between the tip and samples, and can cause sticking problems.

- Same probe is used as a 3D topology and force sensor. Topology sensing is mostly done in tapping mode [10] which needs a beam with high resonant frequency f_r and quality factor Q, which means large k_c, for increased sensitivity. Then there is a trade-off between topology and force sensing such that k_c should be large for imaging while the force measurement sensitivity is reduced. For a hard object manipulation, this problem does not exist since force values are large, but for soft object manipulation case careful design is necessary.

- The probe is actuated with piezoelectric x-y-z stack actuators (piezotube or precision stage) with a limited range (in the order of micrometers), very high resolution (down to pm), high bandwidth (0.1-100s of KHz), and closed loop sensors. Besides of this high resolution stage, a coarse long-range x-y-z stage is also needed for searching samples.

- During manipulation or interaction, sample or the probe can be moved by the piezo actuator depending on the design. Moving the probe is simpler for teleoperation control while its mechanical design is more challenging.

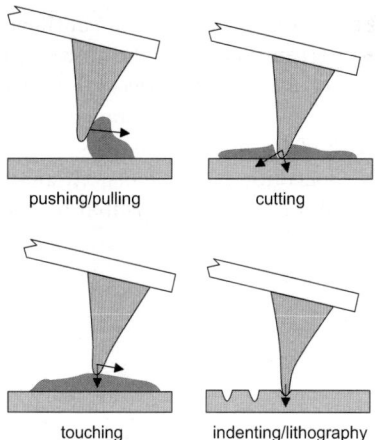

Figure 3: Possible mechanical manipulation tasks using an AFM probe as nano manipulator.

2.1.2 Force Sensing

AFM probe is used for **2(1/2)** d.o.f. force sensing. Taking the force vector as $\mathbf{F^c} = [F_x^c \ F_y^c \ F_z^c]^T$, and assuming the probe with the normal stiffness of k_c is tilted by an

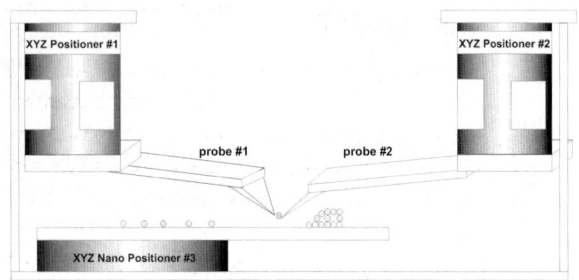

Figure 4: Conceptual drawing of the proposed 3D mechanical pick and place manipulation setup using two probes.

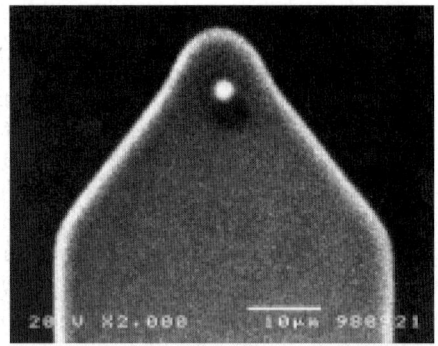

Figure 5: Bottom view of the piezoresistive cantilever (Park Scientific Instruments Inc.) by SEM where the silicon tip with apex radius of $20-30\ nm$ is located around $10\ \mu m$ inside.

angle of α along the x-axis for preventing the collision between the probe and substrate, the deflection is determined by the forces as follows [11]:

$$\begin{bmatrix} \zeta_x \\ \zeta_y \\ \zeta_z \end{bmatrix} = \frac{1}{k_c} \begin{bmatrix} c_1 & 0 & 0 \\ 0 & c_2 & c_3 \\ 0 & c_3 & 1 \end{bmatrix} \begin{bmatrix} F_x^c \\ F_y^c \\ F_y^c \end{bmatrix} \quad (2)$$

where $c_1 = 2L_z^2/L_y^2 + t_c^2/L_x^2$, $c_2 = 3L_z^2/L_y^2$, $c_3 = 3L_z/(2L_y)$, L_x, L_y and L_z are the cantilever lengths along the x-y-z axes, and t_c is the cantilever thickness. Thus, the important point is, the cantilever cannot be modeled as three decoupled springs for an accurate modeling where y and z axes are coupled. Therefore, the force sensing d.o.f. could be defined as 2(1/2). Using our piezoresistive deflection measurement system, only ζ_z can be measured (also ζ_x can be measured by the optical de-

tection method or a new piezolever design which is our future work). Thus, as F_y^c lateral force has also affect on ζ_z, it can be measured using ζ_z.

As important conclusion from here, if decoupled force feedback to the teleoperation system is desirable, the probe should be moved in x-z direction. Thus, F_y^c and F_z^c coupling can be eliminated. However, if 3 d.o.f. force feedback is desirable during x-y-z motion, these forces should be decoupled by modeling of the surface friction.

2.1.3 Vision Sensor

Possible microscopes for nano scale manipulation and interaction are given in Figure 7. AFM is the most promising one in the sense of broad range of object and environment types during imaging, and its 3D high resolution imaging. However, real-time imaging is the main problem of AFM. Moreover, the range of AFM images is limited by the scanner range and scanning time. Therefore, connecting AFM with optical microscope would be a good solution for a long range coarse to fine approach for the nano manipulation experiments.

Microscope Properties	AFM	STM	SEM	OM
Visible Object Sizes	> 0.1 nm	> 0.1 nm	> 5 nm	> 100 nm
Visible Object Type	all	conductors, 1/2conductors	conductors, 1/2conductors	all
Imaging Type	near-field	near-field	far-field	far-field
Interaction with Object	contact or non-contact	non-contact	non-contact	non-contact
Imaging Environment	all	vacuum or air	vacuum	air or liquid
Imaging Principle	interatomic forces	tunneling current	electron emission	light-matter interaction
Imaging Dimension	3-D	3-D	2-D	2-D

Figure 7: Main microscopes utilized for nano manipulation and interaction, and their properties.

2.1.4 Nanoforces

Forces at the nano scale can be given as van der Waals, surface tension, electrostatic, magnetic, electrodynamic, double layer, friction, contact repulsion, etc. [12]. These forces are highly nonlinear, attractive and/or repulsive, and can be long and short range. Thus, even without contacting to the surfaces, attractive or repulsive forces exist at the nano scale, and they should be felt with the teleoperation system. Depending on the environment (air, vacuum and liquid), the dominant forces change, and these forces also depend on the probe and nano object shape, material type, and separation distance or contact area.

2.2 Master Site: Macro World
2.2.1 Master Manipulator

Using a master manipulator, the motion of the nano manipulator is directed by the operator motions while the

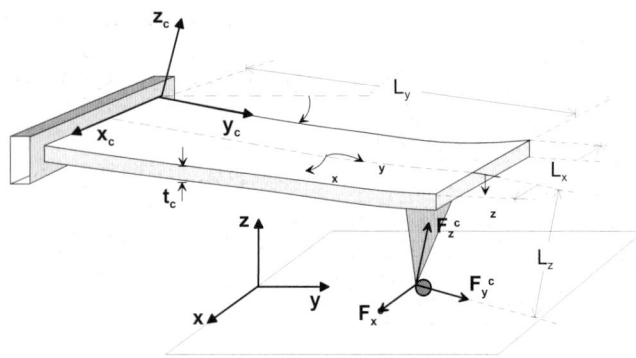

Figure 6: AFM cantilever bending along x-y-z axes during touching to nano objects.

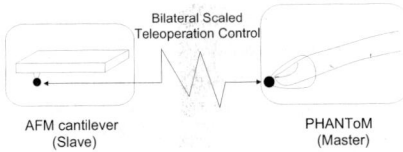

Figure 8: Design of the force feedback point on the probe, and haptic master device.

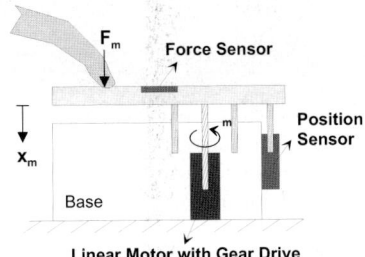

Figure 9: Home-made 1 d.o.f. force feedback master device for tele-nanorobotics applications [1].

scaled nano scale forces are felt by the operator. As the desired design specifications of the master haptic device: (1) high bandwidth ($0.1 - 2\ KHz$), (2) minimized friction and backlash for feeling the tiny nano force changes, (3) 2-3 d.o.f. with force feedback, (4) selecting the force feedback point as shown in Figure 8.

In our setup, a 1 d.o.f. home made haptic device [1] as shown in Figure 9, and 3 d.o.f. Phantom^{TM} as displayed in Figure 10 haptic devices are utilized. The former has the advantage of fully open control structure, and it is controlled by measuring the operator force (*admittance* type) at $33\ Hz$ bandwidth. The latter has few friction and backlash and a high bandwidth (around $1\ KHz$), and operator force is not measured (*impedance* type). Therefore, it could be ideal for tele-nanorobotic applications.

2.2.2 Visual Interface

Visual feedback is the most direct and useful feedback for teleoperation applications with on-line or off-line 2D or 3D displays. During teleoperated nano manipulation, on-line and simultaneous 2D and 3D images are the most useful ones since the nano objects can change by the en-

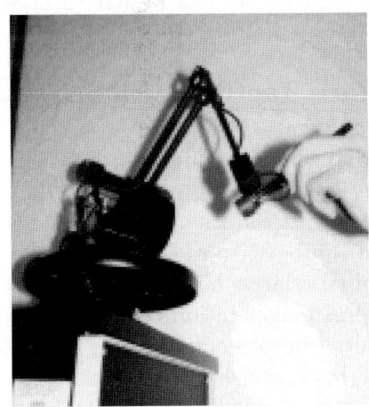

Figure 10: Phantom^{TM} as the 3 d.o.f. force feedback master device.

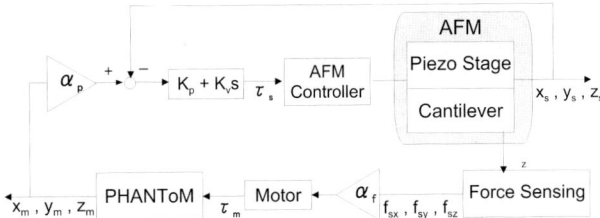

Figure 11: Teleoperation control structure for the Phantom haptic device.

vironmental conditions or during interactions unexpectedly. Furthermore, depending on the surface topology 2D or 3D images are preferable [13] (if the surface is flat and there are objects scattered on the surface, 2D image is easier to visualize while for rough and 3D complex surfaces 3D image would show better details). Using an AFM tapping mode imaging, 2D and 3D images can be held while they can be just off-line. For on-line imaging, Scanning Electron Microscope (SEM) can be utilized in vacuum environment applications [14].

2.3 Bilateral Teleoperation Control

Bilateral teleoperation controller generates the inputs for both slave and master manipulators in order to track scaled forces and positions. Since the nano forces can be nonlinear and abruptly changing by a fast motion, stability and robustness of the controller should be checked. As a rule of thumb, operator should move slowly and smoothly for undesired instabilities.

From the manipulation model, it can be seen that the operator should control the cantilever contact z-position while feeling the normal interaction force between the tip and particle on his/her hand using the haptic device. Also the operator will decide the direction of the horizontal x-y motion of the substrate using the Phantom device in 3 d.o.f. case, or mouse cursor for 1 d.o.f. haptic device. Teleoperation control structure for Phantom device is given in Figure 11. For home made haptic device, teleoperation control structure can be seen in Figure 12. In the latter, the aim of the teleoperation system is taken as to feel the scaled normal nano force $F_z(t)$ on the operator hand while the operator determines the z-position $x_s(t)$ of the cantilever. The teleoperation problem can be defined as to design a controller [15] such that as $t \to \infty$:

$$\begin{aligned} x_s &\to \alpha_p x_m\ , \\ f_m &\to \alpha_f f_s\ , \end{aligned} \qquad (3)$$

where $F_m(t) = F_{op}(t)$ and $F_s(t) = F_n(t)$ are the master and slave forces, $x_m(t)$ and $x_s(t)$ are the master and slave positions, and α_p and α_f are scaling constants for the position and force respectively.

As the controller, a force-reflecting servo type controller is selected [4] where:

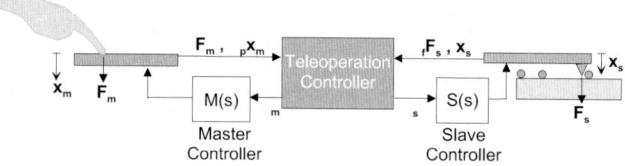

Figure 12: Bilateral force feedback teleoperation control system for home made 1 d.o.f. haptic device.

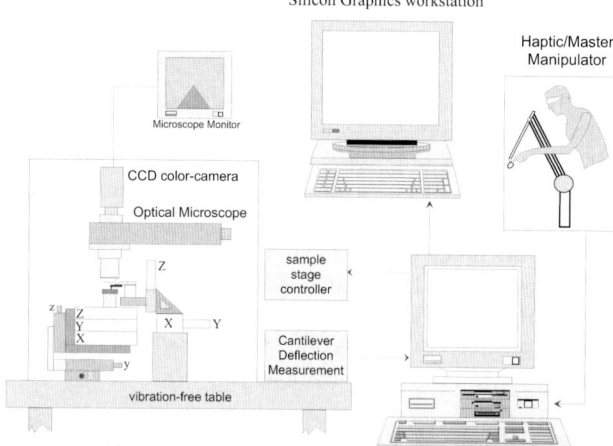

Figure 13: Overall experimental setup for the scaled teleoperation control.

$$\tau_m = -\alpha_f f_s - K_f(\alpha_f f_s - f_m)$$
$$\tau_s = K_v(\alpha_p \dot{x}_m - \dot{x}_s) + K_p(\alpha_p x_m - x_s) , \quad (4)$$

K_p and K_v are position control parameters, and K_f is the force error gain.

3 System Setup

Considering the described design specifications, a teleoperation control setup using the master devices and piezoresistive AFM probe is proposed as illustrated in Figure 13. 2D optical microscope and AFM images, and 3D AFM topology images are displayed on Silicon Graphics workstation monitor, and PC monitor while the Phantom or home made haptic device enables real-time force and tactile feedback.

4 Experiments

4.1 AFM Imaging

At first, the home made AFM system as shown in Figure 14 is tested for getting nano scale 3D topology images. As an example, a silicon substrate (TGT01 grating, Silicon-MDT Inc.) with 3 μm distance sharp conical tips is scanned in the tapping mode imaging as shown in Figure 15. From the image, it can be seen that topologies down to 10s of nanometer can be observed with our AFM setup.

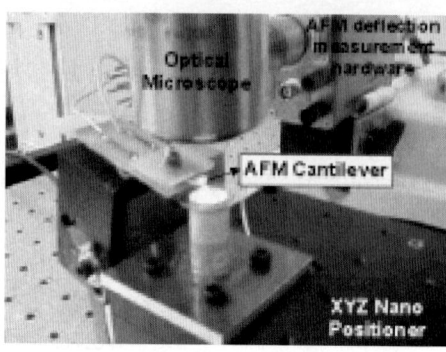

Figure 14: Photo of the home made AFM system setup with top-view high resolution Optical Microscope, piezoresistive probe, [16], nano-positioner, and vibration isolated table.

4.2 Point Contact Touching

For the teleoperated point touch to a surface using the AFM tip and the home made haptic device, the silicon AFM tip approached and retracted to a silicon flat substrate with the control parameters of $\alpha_p = 15 \times 10^{-9}$, $\alpha_f = 4 \times 10^6$ and $K_f = 100$. The result of the operator and scaled nano force, and master and scaled slave position data is given in Figure 16. Forces and positions are tracked successfully, and the effect of adhesion forces is observed as different from macro scale touching. Even in some experiments, very high capillary forces make the tip hardly detached from the substrate.

Using the Phantom haptic device, 2D/3D touching experiments are realized. A previously scanned silicon nanosurface is displayed in a Virtual Reality graphics environment, and the AFM tip is moved in $x-y-z$ coordinates to get the touch feedback. For the nanoforces, simulated models in [1] are utilized to generate the attractive (negative) and repulsive (positive) nanoforces. At first, the AFM tip is touched to the surface in z direction only, and the resulting forces on the operator finger are shown in Figure 17. It can be seen that the operator felt attactive forces in the order of 80 nN during retracting from the surface. Secondly, the tip is touched to the surface at the $x-y$ direction as shown in Figure 18 where the felt $x-y$ nanoforces can be seen. Thus, attractive and repulsive nanoforces can be felt successfully on the surfaces at 3D.

4.3 Surface Tactile Feedback

In this mode, the measured sample topology is felt on the operator's hand while scanning a sample such as in [2]. At first, the cantilever tip is approached to the surface, and put in contact with the substrate. Then, by controlling the x-y AFM cantilever motion, the operator can feel the scaled measured relative height of the surface and forces by the bilateral teleoperation controller. As the experimental result, silicon square gratings (TGZ03 grating, Silicon-MDT Inc.), which are seen in the optical microscope top view image as in Figure 19a, is scanned along the line shown in the figure. The resulting tactile-sensing

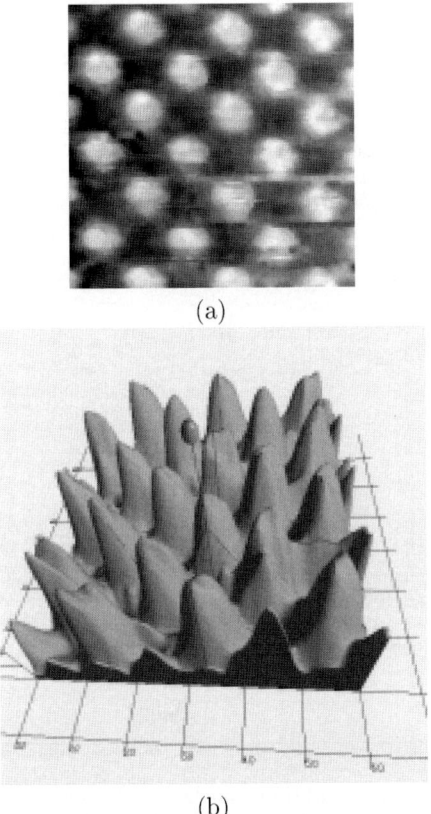

(a)

(b)

Figure 15: Experimental AFM topology image ($10 \times 10 \times 0.7\ \mu m^3$ size) of sharp silicon tips by the home made AFM: (a) 2D grey scale image, and (b) 3D Virtual Reality image.

with the scale parameters of $\alpha_p = 10^{-6}$ and $\alpha_f = 1 \times 10^6$, and the force error gain of $K_f = 10$ is given in Figure 19b. The upward jumps stand for the silicon grid structures of 480 nm height. In this mode, the nano objects are assumed to be fixed on the substrate since the touching operation is held in contact mode which can move not well-fixed samples on the surfaces.

4.4 Telemanipulation

As the telemanipulation application, a target material is carbon nanotubes (CNT) with around 50 nm diameter as given in Figure 20. They are very promising novel nanotechnology materials, and by pushing and cutting, their mechanical properties can be understood [14], [17].

5 Discussions

From the experimental results, it can be observed that the proposed bilateral scaled teleoperation control system is successful for providing tactile and force feedback at the nano scale. The important parameters for the reliable sensation are:

- Scanning speed: taking the bandwidth effect problem into consideration, the scanning speed should be small for feeling the small features easily.
- The stage positioning accuracy is around 15 nm during the experiments which added limitations to the

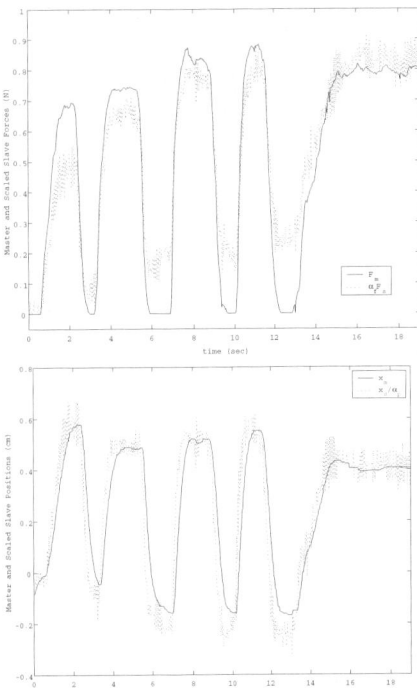

Figure 16: Experimental master and scaled slave forces (upper) and positions (lower) during touching to a silicon surface using the home made haptic device.

resolution of the felt nanoforces. This rarely caused instability problems in the case of very hard surfaces where vibration can occur if a large K_f is selected. Therefore, the accuracy of the stage should be increased to few nanometer for more stable and high resolution force feedback.

- During the tele-nano contact, depending on the sample and tip adhesion forces, the selection of the force scaling factor α_f differs such that: for high surface adhesion, i.e. large tip radius (R_t) and high adhesion energy, $\alpha_f = 1/\alpha_p$ gives better feedback while for small adhesion forces, $\alpha_f = 1/\alpha_p^2$ is a better selection which coincide with the theoretical calculations of Goldfarb [18]. Moreover, depending on the softness of the samples, these rules do not give good results such that the interaction forces become exponentially very small while the motion range increases.

- Electrostatic forces seem to be negligible for previously non-charged surfaces and grounded substrates. However, for a non-conducting sample, even the tip is conductive or semi-conductive, charges can be stored after the contact interaction, and these charges can increase the tip charge such that after a long time contact interaction with a sample, the electrostatic forces can increase as well as the tip can be deformed and the tip radius can increase. Thus, the adhesion forces could *increase with time* [19].

- During the approach process, the attractive forces becomes large only in the close vicinity to the sur-

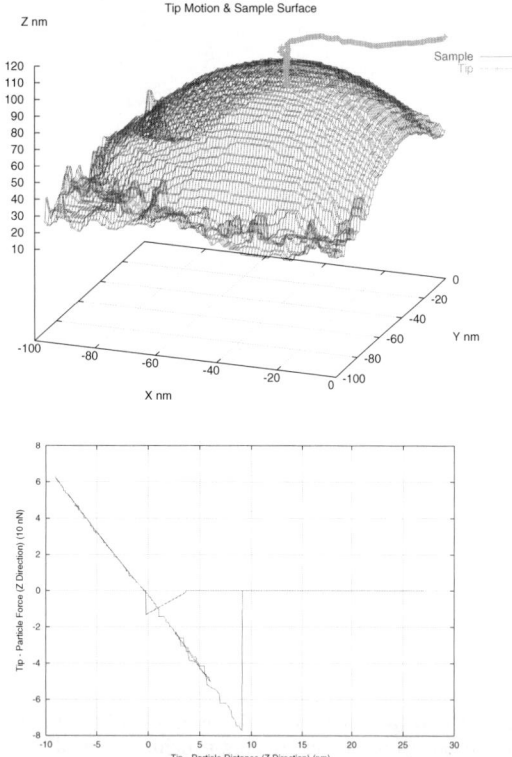
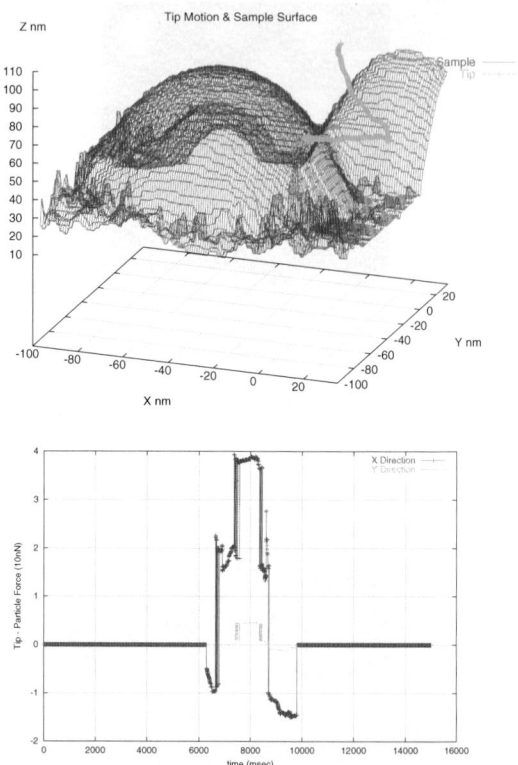

Figure 17: The AFM tip is touched to the surface at the z direction by the Phantom haptic device control (upper), and resulting nanoforces are shown (lower).

Figure 18: The AFM tip is touched to the surface at the $x-y$ direction (upper), and felt nanoforces are shown (lower).

face. Therefore, the approach speed should be slow and the positioning resolution of the stage should be high for a *repeatable* approach attractive force feedback.
- Water layer on the surfaces and tip can cause the instable jump-in-contact problem during approaching to surfaces with soft cantilevers, and increases the adhesion meniscus bridge length during the retraction. Thus, environmental conditions during the experiments are important for understanding which forces we are feeling during the experiments.
- Cantilever parameters such as the stiffness k_c, tip size and shape R_t, resonant frequency f_r, and quality factor Q_c should be measured for each probe to be used in the experiments for realistic theoretical predictions.

6 Conclusion

In this paper, design of a teleoperated nano scale interaction and manipulation system is proposed. Using 1 d.o.f. home made and 3 d.o.f. Phantom haptic devices with a force-reflecting servo type scaled teleoperation controller, preliminary experiments for interacting with the silicon samples are realized. Experimental and simulation results show that the proposed system can be used for nano scale interaction and manipulation applications. During the experiments it is observed that only visual feedback is not enough for interacting the nano world since the visual feedback includes many source of errors such as positioning drifts, nonlinearities, noises, and etc., and there is no real-time visual feedback possibility at present in ambient conditions for nano scale observation. Therefore, realtime force feedback is one of the solutions for compensating these errors and problems. As the future works, more Phantom device tele-nanomanipulation experiments will be conducted, and biological object manipulation experiments will be realized in liquid environments for sensing their elasticity and viscosity, and cutting, pushing, etc. type of mechanical manipulation tasks.

References

[1] M. Sitti and H. Hashimoto, "Tele-nanorobotics using atomic force microscope," in *Proc. of the IEEE/RSJ Int. Conf. on Intelligent Robots and Systems*, pp. 1739–1746, Victoria, Canada, Oct. 1998.

[2] R. L. Hollis, S. Salcudean, and D. W. Abraham, "Toward a tele-nanorobotic manipulation system with atomic scale force feedback and motion resolution," in *Proc. of the IEEE Int. Conf. on MEMS*, pp. 115–119, 1990.

[3] M. Guthold, M. Falvo, W. Matthews, S. Paulson, and et al., "Controlled manipulation of molecular sampels with the nanomanipulator," *IEEE/ASME Trans. on Mechatronics*, vol. 5, pp. 189–198, June 2000.

[4] M. Sitti, S. Horiguchi, and H. Hashimoto, "Tele-touch feedback of surfaces at the micro/nano scale: Modeling and experiments," in *Proc. of the IEEE/RSJ Int. Conf.*

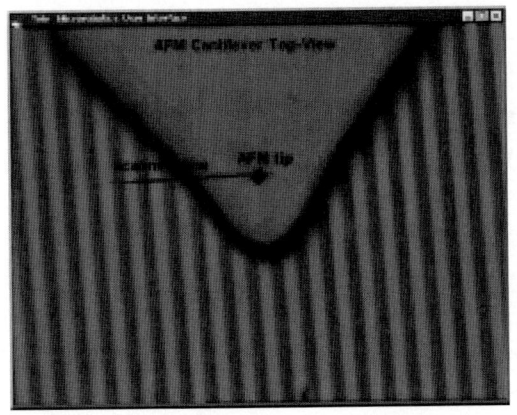

(a)

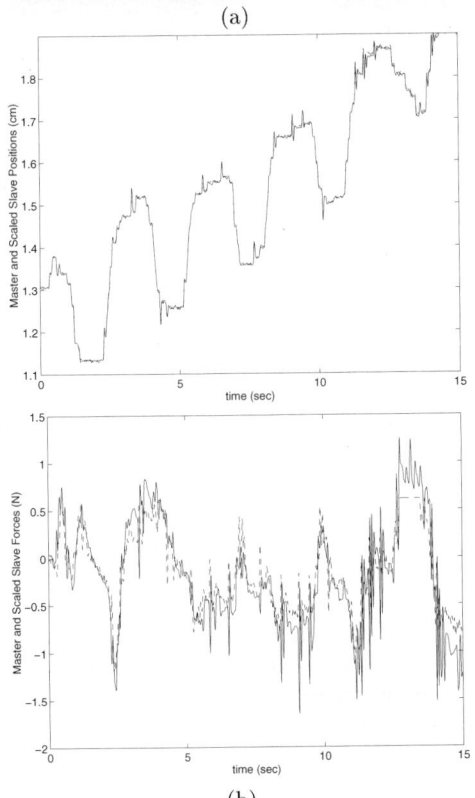

(b)

Figure 19: (a) Silicon grids with 480 nm height are felt along the black line, and (b) resulting topology and force feedback using the force reflecting servo type controller.

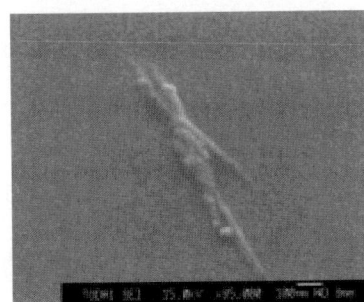

Figure 20: SEM image of the carbon nanotubes (courtesy of University of Tokyo) as a target manipulation material.

on *Intelligent Robots and Systems*, pp. 882–888, Korea, Oct. 1999.

[5] T. R. Ramachandran, C. Baur, A. Bugacov, and et al., "Direct and controlled manipulation of nanometer-sized particles using the non-contact atomic force microscope," *Nanotechnology*, vol. 9, pp. 237–245, 1998.

[6] M. Sitti and H. Hashimoto, "Controlled pushing of nanoparticles: Modeling and experiments," *IEEE/ASME Trans. on Mechatronics*, vol. 5, pp. 199–211, June 2000.

[7] D. Schafer, R. Reifenberger, A. Patil, and R. Andres, "Fabrication of two-dimensional arrays of nanometric-size clusters with the atomic force microscopy," *Appl. Physics Letters*, vol. 66, pp. 1012–1014, Feb. 1995.

[8] M. Sitti and H. Hashimoto, "Teleoperated nano scale object manipulation," in *Recent Advances on Mechatronics*, pp. 322–335, ed. by O. Kaynak, S. Tosunoglu and M.J. Ang, Springer Verlag Pub., Singapore, 1999.

[9] M. Sitti, "Teleoperated 2-d micro/nanomanipulation using atomic force microscope," *Ph.D. Thesis*, Dept. of Electrical Engineering, University of Tokyo, Tokyo, Sept. 1999.

[10] M. Sitti and H. Hashimoto, "Tele-nanorobotics using atomic force microscope as a manipulator and sensor," *Advanced Robotics Journal*, vol. 13, no. 4, pp. 417–436, 1999.

[11] J. Colchero, E. Meyer, and O. Marti, "Friction on atomic scale," *Handbook of Micro/Nano Tribology, Second Ed.*, CRC Press, pp. 273–333, 1999.

[12] J. Israelachvili, *Intermolecular and Surface Forces*. 2nd Ed., Academic Press, London, 1992.

[13] R. Taylor, II, J. Chen, S. Okimoto, and et al., "Pearls found on the way to the ideal interface for scanned probe microscopes," in *Proc. Visualization '97*, pp. 467–470, Phoenix, USA, 1997.

[14] M. Yu, M. Dyer, G. Skidmore, and et al., "Three-dimensional manipulation of carbon nanotubes inder a scanning electron microscope," *Nanotechnology*, vol. 10, pp. 244–252, Sept. 1999.

[15] Y. Yokokohji and T. Yoshikawa, "Bilateral control of master-slave manipulators for ideal kinesthetic coupling–formulation and experiment," *IEEE Trans. on Robotics and Automation*, vol. 10, pp. 605–619, Oct. 1994.

[16] M. Sitti and H. Hashimoto, "Two-dimensional fine particle positioning using a piezoresistive cantilever as a micro/nano-manipulator," in *Proc. of the IEEE Int. Conf. on Robotics and Automation*, pp. 2729–2735, Detroit, USA, May 1999.

[17] M. Falvo, R. T. II, and et al., "Nanometer-scale rolling and sliding of carbon nanotubes," *Nature*, vol. 397, pp. 236–238, 21 Jan. 1999.

[18] M. Goldfarb, "Dimensional analysis and selective distortion in scaled bilateral telemanipulation," in *Proc. of the IEEE Int. Conf. on Robotics and Automation*, pp. 1609–1614, Leuven, Belgium, 1998.

[19] J. Chen, C. Dimattia, and et al., "Sticking to the point: A friction and adhesion model for simulated surfaces," in *Proc. of teh Annual Symp. on Haptic Interfaces for Virutal Env. and Teleoperator Systems*, 1997.

Haptic Display Device with Fingertip Presser for Motion/Force Teaching to Human

Ryo Kikuuwe and Tsuneo Yoshikawa

Department of Mechanical Engineering, Kyoto University
Kyoto, 606-8501 Japan
{t60y0129@ip.media, yoshi@mech}.kyoto-u.ac.jp

Abstract

The purpose of this study is to establish a method of virtual-reality-mediated motion/force teaching from human teacher to human "teachee". For effective teaching, position and force information of the teacher's action have to be communicated to the teachee. Position information is easy to display, but force information is difficult to display by conventional haptic devices. In this paper, we propose a new type of haptic interface with fingertip presser which could make it possible to display both position and force information. The teacher's finger force of pushing a surface is displayed as a mechanical pressing force onto the teachee's fingertip, and position is displayed by active multijoint linkage. The results of basic experiments show that the proposed method is effective as a media of motion/force teaching, but reveal some limitations.

1 Introduction

Human motions are difficult to describe in words. In cases of training of motor skills or directing unskilled operators, verbal languages, gestures, or physical guidance by human instructors are traditionally used as media of communication.

Virtual reality (VR) technology has a potential to be a new kind of communication medium between teacher and "teachee", which could substitute a human instructors who physically guides the teachee's arms or fingers. VR-mediated motion teaching is advantageous over traditional methods for the following reasons: (i) Information is communicated correctly.

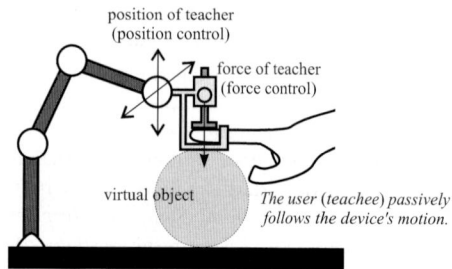

Figure 1: Proposed Mechanism — Haptic Device with Fingertip Presser

In cases of physical guidance such as tennis or golf coaching, the interference between the teacher's and the teachee's bodies obstructs the teacher's natural motion. This kind of problems can be solved by using haptic devices. (ii) Temporal or spatial distance do not affect. The teacher and the teachee do not have to be at the same place at the same time. Besides, instruction from one person to two or more persons is possible. (iii) Recorded data can be modified on demand. It is possible to emphasize specific aspects of the teacher's action.

For effective teaching, position and force information of teacher's action have to be communicated to teachee. Position information is easy to display, since it can be displayed by physical guidance using a haptic device (or generally, an active multijoint linkage). Force information, however, cannot be displayed by guidance. In this paper, we propose a new type of haptic interface, shown in Fig.1, composed of a multijoint arm and a fingertip presser attached on its end-effector. This mechanism makes it possible to display both position and force information. The results of basic experiments show that the proposed method is effective as a media of motion/force teaching, but reveal some limitations.

2 Previous Works on VR-Mediated Teaching

Various researches have been done for displaying virtual environment where an operator can experience a variety of physical behaviors[3]. Some of them are developed for skill training mainly in the area of medicine, and flight simulator is a typical example of VR-training that has been successful. However, those are simulators for mimic operations for learners and not meant to be media of teaching from expert(teacher) to novice(teachee).

Kuzuoka et al.'s GesureCam[4] and Mikawa et al.'s CTerm[5] could be mentioned as media of remote teaching. Each of them is a remote actuator on which a small camera and a laser pointer are mounted, and instructions are given by pointing objects by the laser pointer. Those kinds of systems could complement verbal direction by substituting finger pointing by human instructor in real-world collaborating task. However, since the signal is very simple, the teachee must have sufficient preliminary knowledge to accomplish a complicated task.

By using augmented reality (AR) technologies [6], more complicated graphic images can be drawn (or superimposed) on real environment surfaces, and a larger amount of information can be displayed to users. AR could make instructions easier to understand for teachee and could complement ambiguity of verbal direction. Suenaga et al.'s tele-instruction system for ultrasound tele-diagnosis[7] is a medium of teaching from a medical specialist to an ultrasound diagnostic device operator. Several graphic symbols are projected on the surface of patient's body at the request of the medical doctor, and the operator can know where to put and how to move the device.

As researches on teaching media involving haptics, Henmi and Yoshikawa's "virtual calligraphy system"[1] and Yokokohji et al.'s experiments using WYSIWIF display[2] could be mentioned. Henmi and Yoshikawa's idea was basically "record-and-replay" strategy, recording and replaying position and force information of teacher's action. Yokokohji et al. enhanced the previous Henmi's research, and investigated several control methods of haptic device. Since it is impossible to display both position and force information at one time through a haptic device, those systems depend heavily on visual cues.

3 Haptic Display Device with Fingertip Presser

The problem which surfaced in researches[1][2] is that it is impossible to display both position and force information at one time through a haptic device. A haptic device, generally an active multijoint linkage, could be used to communicate position information of teacher's action by physically guiding the teachee's arms or fingers. Force information, however, cannot be communicated in this method. It means that when the target task was to manipulate objects by hands, the teachee could understand how s/he should move his/her hands and fingers, but could not understand how strongly s/he should press the object.

To display force information, an additional channel is required. And, it has to generate a sensation which is equivalent to what the teacher feels when s/he is pushing a surface with his/her fingers. In this paper, we propose to use a fingertip pressing device as a medium of displaying force information. Mounting the fingertip presser on the end-effector of a multijoint linkage, as shown in Fig.1, both position and force information could be displayed to the teachee. Using this method, the teachee can understand how strongly the teacher is pushing the environment surface without preliminary knowledge about the impedance of the surface. The position and force information of the teacher's actions and information on environment are communicated through proprioception and haptic/tactile sensations, without help of vision. Besides, when the fingertip pressers are attached to more than one finger, internal force of grasping can be displayed.

The proposed method is expected to be useful for tasks of manipulating hard objects where delicate force control is required, such as planing or polishing, and tasks of manipulating soft objects, such as surgery or ceramic arts, and effective especially for visually impaired people since information is communicated without vision. Applicable usage of the proposed method is not limited to the field of virtual reality; training of delicate telerobotic operations, such as robotic surgery, will also be in the range of application.

4 Experiments

In order to evaluate the validity of the proposed method, we built an apparatus and conducted two experiments. Experiment I was conducted to test the availability of fingertip pressing function as a medium of teaching force. Experiment II was conducted to test the composite method of fingertip pressing and displaying position.

4.1 Apparatus

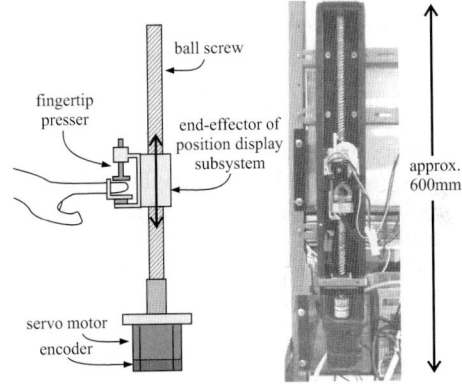

Figure 2: Apparatus

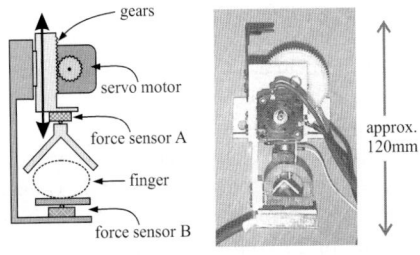

Figure 3: Apparatus's Fingertip Pressing Subsystem

Figure 4: Arrangement of Experiments

The apparatus is composed of a 1-DOF position display subsystem and a fingertip pressing subsystem, as shown in Fig.2 and Fig.3. These are connected to a control PC through D/A and A/D converters and digital I/O.

The fingertip pressing subsystem has a DC servo motor and 2 force sensors below and above the finger. The components which directly contact with the user's finger are aluminum roof-shaped (angle of 90°) part on the back of finger and a smooth acrylic board on the fingerpad. Force sensor A measures the user's voluntary generated pushing force f_U. Force sensor B measures mechanical pressing force F_D and allows feedback control of the force.

The position display subsystem is driven by AC servo motor and ball screw, and an encoder is attached to the motor. Position of its end-effector is controlled by processing velocity command to the dedicated amplifier of the AC servo motor.

4.2 Experiment I

The proposed method requires human user to make correspondence between the force F_D mechanically applied on his/her finger and the force f_U s/he voluntarily generates, though feeling of being pressed is quite different from that of pushing a surface. The purpose of Experiment I is to find out a relation between F_D and f_U.

4.2.1 Protocol

Each trial of this experiment was carried in the following procedure:
1. The subject placed his right-hand index finger on the apparatus as shown in Fig.4.
2. A randomly chosen force F_D was applied on the subject's finger for 5 seconds.
3. The mechanical force was removed for 2 seconds.
4. The subject pushed the apparatus "as strongly as" he was pressed before.
5. The subject's pushing force was recorded as f_U.

F_D was randomly chosen out of 5 ranks(2.5N, 3.5N, 5.1N, 7.4N and 11N), and 1 "set" of experiment consists of 25 trials (=[5 trials]× [5 ranks]). 6 male subjects, between 22 and 30 years of age, were recruited and are referred to as A, B, $\cdots$, F. For each subject, 4 sets were run at regular intervals of 3 to 6 days. Each set is named as A0, A1, $\cdots$, F2, F3.

4.2.2 Results

The results are shown in Fig.5, where geometric average and geometric standard deviation of voluntary pushing force f_U of every rank, every set are presented. Note that both axes of every figure are logarithmic. The estimated regression line of every set are drawn as a dotted line. We can see that F_D and f_U are not equal. Instead, they have a relation

$$f_U = CF_D{}^m. \quad (1)$$

In the following analysis, let us use an independent variable $X = \ln F_D - \ln \tilde{F}_D$ and a dependent variable $Y = \ln f_U$, where $\tilde{F}_D$ is the geometric mean of F_D used in this experiment ($\tilde{F}_D = 5.1[N]$). We obtained the following results: (i) Within subjects, there were no inter-set differences in regression slope ($F(18, 552) = 1.10$; $p = 0.349$). However, assuming intra-subject homogeneity of regression and removing the effect of X, significant inter-set differences were found in Y's averages ($F(18, 570) = 7.71$; $p < 1.0 \times 10^{-16}$). To put it shortly, a subject's regression line slides up and down without changing its inclination. (ii) Assuming intra-subject homogeneity of regression, there were significant inter-subject differences in regression ($F(5, 588) = 5.20$; $p = 0.0111$). (iii) Analyzing intercepts of regression lines (which correspond to f_U at the midpoint of the measurement range $F_D = 5.1[N]$), inter-subject variability was significantly greater than intra-subject one ($F(5, 18) = 17.0$; $p < 1.0 \times 10^{-5}$).

The statistic results above lead us to the conclusion that mechanical pressing force F_D and its correspondent voluntary pushing force f_U have a relation $f_U = CF_D{}^m$, where C and m values vary among individuals, and C value changes even in one person. Therefore, it would be desirable to calibrate C and m values every time prior to using the system.

4.3 Experiment II

Experiment II is conducted to evaluate the validity of the combined method of physical guidance and pressing fingertip, comparing a teaching effect of displaying both position and force to that of displaying position only. The target task chosen is very simple; that is to push an elastic object down to a specified position.

4.3.1 Protocol

The subject put his right-hand index finger on the apparatus as shown in Fig.4. Position was displayed by its movement and force by fingertip pressing force. The subject's hand were hidden by a board to prevent him from recognizing its movement visually. Therefore, the subject is supposed to perceive position by proprioception and force by sense of being pressed in his fingertip.

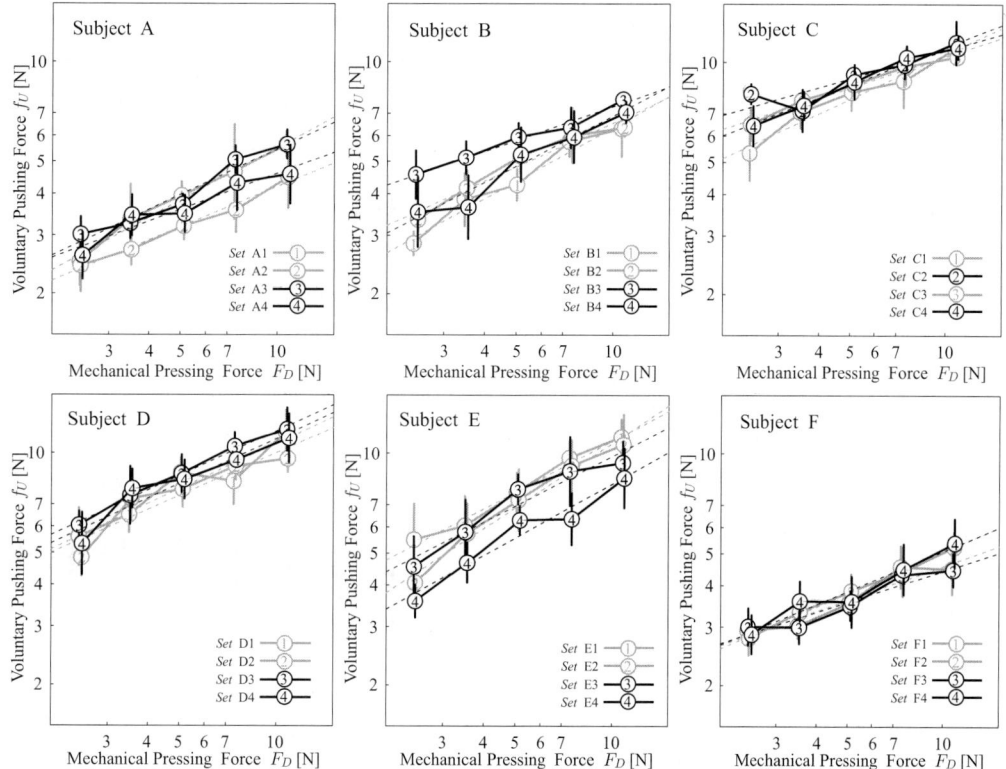

Figure 5: F_D versus f_U in Experiment I

The virtual object to push was a simple spring-mass-damper system shown in Fig.6(a), where $M_E = 0.04$[kg], $V_E = 0.01$[N·s/mm] and K_E varies by trials.

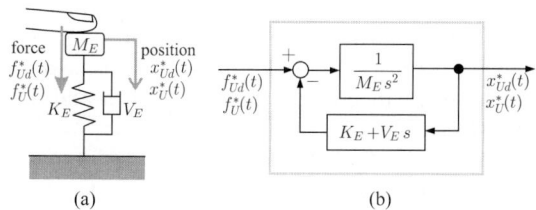

Figure 6: Virtual Environment

One trial in this experiment consists of a one-time teaching and a one-time execution. In teaching phase, an example trajectory of position, or trajectories of both position and force, was/were displayed to the subject by the apparatus. In the following executing phase, the subject pushed the apparatus as deeply and as strongly as he was instructed, and the subject's trajectories of position and force were recorded.

Time spans for each phase was 1.5 second, and an interval of 1 second was placed between the phases. Let us define several symbols as follows; $x^*_{Ud}(t)$ and $x^*_U(t)$ are the position trajectories in teaching and executing phases respectively, which are measured positive downward from the equilibrium of the virtual spring. $f^*_{Ud}(t)$ and $f^*_U(t)$ are the force trajec-

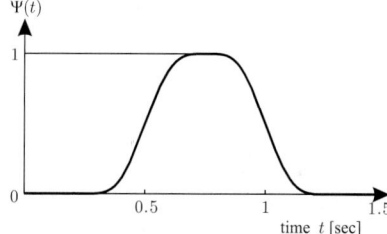

Figure 7: A Time Function $\Psi(t)$

tories in teaching and executing phases respectively. $f^*_{Ud}(t)$ is generated by the computer, while $f^*_U(t)$ is generated by human subjects. f_{Ud}, x_{Ud}, f_U and x_U are the peak values of $f^*_{Ud}(t)$, $x^*_{Ud}(t)$, $f^*_U(t)$ and $x^*_U(t)$ respectively. $\Psi(t)$ is a time function used to generate $f^*_{Ud}(t)$ which is shown in Fig.7 and described as;

$$\Psi(t) = \begin{cases} \dfrac{3}{2} \int_{0.25}^{t} \sin^3 2\pi(\tau - 0.25)d\tau \\ \qquad : \text{if} \quad 0.25 \leq t \leq 1.25 \\ 0 \qquad : \text{otherwise} \end{cases} \quad (2)$$

$F^*_D(t)$ is the trajectory of mechanical fingertip pressing force in teaching phase.

The simulator of the object is described as a block shown in Fig.6(b), which receives force input $f^*_{Ud}(t)$ or $f^*_U(t)$, and produces position output $x^*_{Ud}(t)$ or $x^*_U(t)$. Data flows in teaching and executing phase

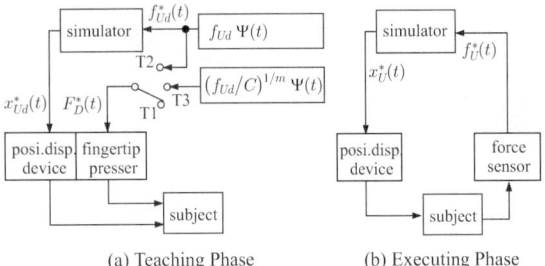

Figure 8: Data Flow

Table 1: V_n and E_n of each method.

method	imprecision	inaccuracy
T1a (T1 in Ua)	$V_{1a} = 2.78$	$E_{1a} = 0.0567$
T1b (T1 in Ub)	$V_{1b} = 2.43$	$E_{1b} = 0.127$
T2	$V_2 = 1.13$	$E_2 = 0.135$
T3	$V_3 = 0.745$	$E_3 = 0.107$

are shown in Fig.8(a) and Fig.8(b) respectively. The simulators used in both phases are identical to each other. The values of functions and variables are determined in the following procedure;

1. A pair of position and force (x_{Ud}, f_{Ud}) is randomly chosen. x_{Ud} is chosen out of 4 ranks; 1, 2, 4 and 8[mm], and f_{Ud} out of 4 ranks; 2.9, 4.4, 6.6 and 9.9[N]. Therefore the "task" (x_{Ud}, f_{Ud}) is chosen out of 16 candidates.
2. K_E is set to be $K_E = f_{Ud}/x_{Ud}$.
3. $f^*_{Ud}(t)$ is generated as $f^*_{Ud}(t) = f_{Ud}\Psi(t)$
4. In teaching phase, $f^*_{Ud}(t)$ is handed to the simulator and consequently $x^*_{Ud}(t)$ is produced. The position of the subject's finger is controlled to be $x^*_{Ud}(t)$. (The peak value of $x^*_{Ud}(t)$ necessarily becomes x_{Ud})
5. In executing phase, $f^*_U(t)$ is measured by force sensor B (see Fig.3). The simulator receives $f^*_U(t)$ and produces $x^*_U(t)$ in real-time. The position of the subject's finger is controlled to be $x^*_U(t)$.
6. After trial, x_U and f_U are set to be the maximum values of the recorded $x^*_U(t)$ and $f^*_U(t)$ (They necessarily satisfy $f_U = K_E x_U$).

Considering the results of Experiment I, three teaching methods were defined;

- method T1: displaying position only
- method T2: displaying both position and force *without* compensating for (1)'s relation; $F^*_D(t)$ is set to be $F^*_D(t) = f^*_{Ud}(t)$.
- method T3: displaying both position and force *with* compensating for (1)'s relation; $F^*_D(t)$ is set to be $F^*_D(t) = (f_{Ud}/C)^{1/m} \Psi(t)$.

When method T1 is used, the subject has no way to estimate the applied force from the displacement of the object, since he was not informed about K_E in advance. In contrary when method T2 or T3 is used, the subject can perceive f_{Ud} by mechanical pressing force in his fingertip.

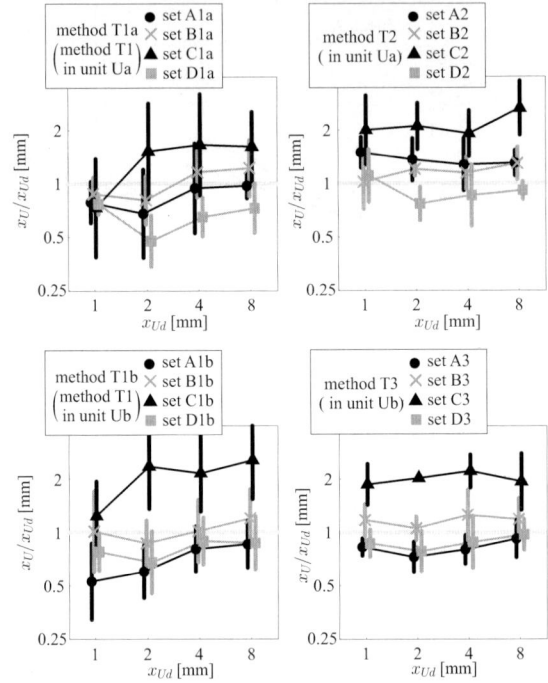

Figure 9: Result of Experiment II

C and m values of every subject was calibrated in advance by conducting 10 trials in the same procedure as Experiment I.

2 methods were tried in 1 "unit" of experiment, and 2 types of units were defined (unit Ua: T1 and T2, unit Ub: T1 and T3). 1 unit is composed of 32 trials(=[1 trial]×[2 methods]×[16 tasks]). Within each unit, methods were switched after every 4 trials. The reason why the both of Ua and Ub include method T1 is to detect unexpected inter-unit factors. For brevity, we refer to data sets obtained out of method T1 in unit Ua and Ub as T1a and T1b respectively.

4 male subjects, between 22 and 30 years of age, were recruited and are referred to as A, B, C and D. For each subject, 2 units Ua and Ub were run at a interval of more than 1 days.

The result obtained out of method Tn (n={1a, 1b, 2, 3}) of subject X (X={A, B, C, D}) is referred to as "set" Xn. Out of 1 subject, 4 sets(=[1 set]×[2 methods]×[2 units]) were obtained, and 1 set is composed of 16 trials(=[1 trial]×[16 tasks]).

4.3.2 Results

We analized the result by using the error in logarithm of position; $\varepsilon = \ln(x_U) - \ln(x_{Ud}) = \ln(x_U/x_{Ud})$. In Fig.9, geometric average and geometric standard deviation of x_U/x_{Ud} of every rank of x_{Ud} of every set are presented. In Table 1, V_n and E_n are defined as $V_n = V_{An} + V_{Bn} + V_{Cn} + V_{Dn}$ and $E_n = \bar{\varepsilon}_{An}^2 + \bar{\varepsilon}_{Bn}^2 + \bar{\varepsilon}_{Cn}^2 + \bar{\varepsilon}_{Dn}^2$ respectively, where $\bar{\varepsilon}_{Xn}$ is the intra-set average of ε of set Xn, and V_{Xn} is intra-set sum of $(\varepsilon - \bar{\varepsilon}_{Xn})^2$ of set Xn.

Firstly, we made a comparison between T1a and

T1b and found no significant differences between V_{1a} and V_{1b} ($p = 0.301$), or between E_{1a} and E_{1b} ($p = 0.223$). Therefore we supposed that units Ua and Ub were homogeneous, and that the only factor which distinguishes T3 from T2 was compensation for F_D-f_U relation. The statistic analysis, assuming the homogeneity between Ua and Ub, revealed the following facts: (i) Fingertip pressing, with or without compensation for F_D-f_U relation, made V_n significantly smaller ($p < 0.001$). (ii) Compensation for F_D-f_U relation made V_n more smaller with a borderline significance ($p = 0.055$). However, excluding trials of $X_{Ud} > 8$[mm], V_n became significantly smaller ($p = 0.0195$). (iii) There were no significant changes among E_n's.

The statistic results above lead us to the following conclusions:

1. Fingertip pressing improves the precision of reproducing a target position. This effect is enhanced by compensation for (1)'s relation especially when the motion is small.
2. However, fingertip pressing does not improve the accuracy.

To put it simply, fingertip presser helps the teachee to perceive the teacher's action *clearly* but not *correctly*.

Solutions for improving the accuracy will have to be sought in future researches. Several past studies on kinesthesia have been revealed that every human subjects had a particular inaccuracy of matching or reproducing targe positions, and that the inaccuracy could vary over time and between subjects [9]. Therefore, it is inferred that the inaccuracy observed in this experiment is rooted in the methodology of displaying position, not in that of fingertip pressing.

4.3.3 Discussion

To take (1)'s relation into consideration rigorously, $F_D^*(t)$ should have been

$$F_D^*(t) = \left(\frac{f_{Ud}^*(t)}{C}\right)^{\frac{1}{m}} = \left(\frac{f_{Ud}\Psi(t)}{C}\right)^{\frac{1}{m}}. \quad (3)$$

However when (3) was tried, we received an apparently unnatural feeling because the pressing force raises behind the moment when the movement starts. Therefore, in this paper, we used the equation

$$F_D^*(t) = \left(\frac{f_{Ud}}{C}\right)^{\frac{1}{m}} \Psi(t), \quad (4)$$

knowing it greatly lacks generality.

Cutaneous mechanoreceptors respond to mechanical stimuli in two distinct manners; all kinds of receptors respond during the intensity of stimulus is changing, while some of them respond to static maintenance of the stimulation as well [8]. Because (1) is only the 'static' relation, a 'dynamic' relation have to be sought in future research. A better substitution for (4) would be described as

$$F_D^*(t) = \left(\frac{f_{Ud}^*(t)}{C}\right)^{\frac{1}{m}} + \Phi\left(\frac{d}{dt}f_{Ud}^*(t)\right), \quad (5)$$

using an unknown function $\Phi(\cdot)$.

5 Conclusion

A new mechanism of a haptic device as a medium of motion/force teaching from human to human has been proposed. This mechanism has a fingertip presser, and the teacher's finger force of pushing a surface is displayed as a mechanical pressing force onto the teachee's fingertip. It can complement conventional physical guidance which displays position information.

Experimental results led us to some conclusions. First, mechanically applied pressing force F_D and its correspondent voluntary pushing force f_U have a relation $f_U = CF_D{}^m$, and that C and m vary among individuals. Therefore, to compensate for the relation, C and m have to be calibrated every time prior to using the system. Second, fingertip presser helps the teachee to reproduce the teacher's action "precisely", but not "accurately".

References

[1] K. Henmi and T. Yoshikawa: "Virtual Lesson and Its Application to Virtual Calligraphy System", *Proc. of the 1998 IEEE Int. Conf. on Robotics and Automation, pp.1275-1280*, 1998

[2] Y. Yokokohji et al.: "Toward Machine Mediated Training Of Motor Skill — Skill Transfer from Human to Human via Virtual Environment —", *Proc. of IEEE Int. Symp. on Robot and Human Communication (RO-MAN'96), pp.32-37*, 1996

[3] G. C. Burdea: "Force and Touch Feedback for Virtual Reality", *John Wiley & Sons, Inc.*, 1996

[4] H. Kuzuoka et al.: "Can the GesuterCam be a Surrogate?", *Proc. of ECSCW'95, pp.181-196*, 1995

[5] M. Mikawa et al.: "New Telecommunication using CTerm", *IEEE Robotics and Automation Conference Video Proceedings*, 1999

[6] R. Azuma: "A Survey of Augmented Reality", *Presence, 6(4)*, 1997

[7] T. Suenaga et al.: "A Tele-Instruction System for Ultrasound Tele-diagnosis", *9th Int. Conf. on Artificial Reality and Tele-Existence, pp.84-91*, 1999

[8] A. Iggo: "Sensory Receptors In The Skin of Mammals and Their Sensory Functions", *Revue Neurogique (Paris), 141, 10, pp.599-613*, 1985

[9] K. R. Boff et al.(eds.): "Handbook of Perception and Human performance", *A Wiley-Interscience publication*, 1986

A Framework for Haptic Rendering System Using 6-DOF Force-Feedback Device

Dae Hyun Kim[o], Nak Yong Ko[†], Geum Kun Oh[†], and Young Dong Kim[†]

[o] Graduate School of Control & Instrumentation Engineering,
[†] Dept. of Control & Instrumentation Engineering,
Chosun University, Kwangju, Korea
(E-Mail : conin@shinbiro.com)

Abstract

In this paper, we describe the design of the hardware and software for haptic rendering system. A haptic interface allows a human to explore and to interact with virtual environments using the sense of touch. Haptic rendering systems include three main components: detecting collision, computing the penetration depth, and restoring force and torques. We present novel algorithms for fast proximity queries for haptic interaction with a hybrid hierarchical representation, consisting of bounding axis-aligned bounding box trees(AABB-Trees) and tight fitting oriented bounding box trees(OBB-Trees). The resulting contact information is used for computing the restoring forces and torques. The rendering algorithms have been interfaced with 6-DOF force-reflecting device and applied to a number of polygonal models. This framework has been implemented in a system that is able to haptically render complex virtual environments.

1. Introduction

In recent times there has been an increasing interest on using computers to replicate simulations, which would be too costly or too dangerous to create otherwise. These so called virtual environments are expected to supply all of the sensory information that would be accessible to the user and allow the control the user would expect to have in such an environment. Virtual environments require natural interaction between interactive computer systems and users. Compared with visual and auditory information, haptic display is not as well developed. Haptic rendering as an augmentation to visual display can improve perception and understanding both of force fields and of world models in the synthetic environments[4]. Haptic rendering systems include three main components: detecting collision, computing the penetrating depth, and restoring force and torques.

Collision detection problems and their variants are of vital importance in many fields[3,6,9,15,23]. Numerous algorithms and techniques have been proposed. In order to meet the stringent requirement of haptic rendering algorithms, a specialized system implementation need to be developed to substain 1000 updates/second in order to maintain a stable system. Some methods to detect collision usually decompose the object into a bounding volume hierarchical structure, so that concave polyhedral model can be handled. Those methods include octree[2], BSP tree[1], C-Tree[7], sphere tree[11], strip trees and box trees[13], R-trees[5], axis-aligned bounding boxes (AABB)[14], Oriented Bounding Box Tree (OBB-Tree) [19] and discretely oriented polytopes(K-DOPs)[18,22]. Among them, AABB tree tend to have fast overlap test, but require many such tests to perform collision detection. The OBB-Tree is more efficient than other hierarchical trees because of its tightest bounding box but the efficiency of hierarchy is affected by the choice of a bounding volume.

Given two polyhedral models, algorithms for intersection computation and boundary evaluation have been extensively studied in solid modeling. Spatial partitioning approach [8] was proposed to compute the contact region. Hierarchical representations were used to detect contacts between polygonal models[16]. Synder[21] used temporal coherence to track penetration depth on spline models. All these algorithms rely on a surface-based representation of the model. Their performance varies as a function of the size and relative configuration of two models.

If the collision was detected, the collision response for force rendering is considerable work[3,12,17]. However, these algorithms do not guarantee real-time performance. *Constraint-Based* methods were first proposed for haptic applications[17]. A god-point was constrained by the objects in the environment. A topology of the rendered object was required to prevent the god-point from falling through small gaps commonly found in graphical models. This computationally expensive preprocessing step limited the interactivity of the system. Other algorithm was proposed for computing the restoring force based on *virtual proxy* extended to 6-DOF force-feedback device [20].

In this paper, we describe a novel framework of haptic

rendering system for haptic interaction of complex polygonal models using a 6-DOF force-feedback device. In the implementation of haptic rendering, we use the hybrid hierarchical technique with sphere-trees and OBB-trees. We will compute the restoring forces based on simple impedance control techniques. To prevent sudden jumps and sharp discontinuities between successive frames, we also use an interpolation method for smooth force shading effects with *virtual proxy* algorithm[20]. In addition, we used a simple linear smoothing scheme to minimize discontinuity in force and torque display between successive frames[24].

The rest of the paper is organized as follows. Section 2 briefly describes our system. Section 3 presents the algorithms for detecting collisions using hybrid hierarchical method. Section 4 describes computation of the restoring force and torques and the interpolation for surface normal. In section 5, we will describe the implementation of our rendering system to perform complex polygonal models.

2. System Overview

In this section, we describe a framework of our haptic rendering system. Due to the stringent update requirements for real-time haptic display, we run a stand-alone haptic server on a PC connected to the force-feedback device. The client application runs on another desktop PC that manages virtual environments. Through direct connection with UDP packet, the client application sends the haptic server the geometric features of the scene to be haptically displayed, and the server sends back device information such as the position and orientation of the force-feedback device' probe.

The hardware system used in this work a 6-DOF force-feedback device. This device has three active joints and three passive joints, each of which is equipped with an encoder for position sensor, and DC servomotors for active joints as shown in Figure 1. And the stable workspace has the dimensions of 20 cm by 20 cm by 20 cm.

Figure 1. 6-DOF force-feedback device

The rendering system consists of routines for hardware interface, collision detection and computation of contact regions and contact force. The overall architecture of the system is shown in Figure 2. As shown in Figure 2, the computation for haptic rendering at each time frame involves the following steps:

Collision Detection:
The algorithm detects intersections, which occurred between a probe held by user and the virtual environments.

Computation of contact region:
If an intersection occurred, then the algorithm computes the contact points from the contact region.

Computing the penetration depth:
Penetration depth along the contact normal direction is computed from the contact region.

Computing restoring forces and torques:
A restoring force is calculated with the penetration depth. Given the force and the contact region, restoring torques can be easily computed using Jacobian theorem.

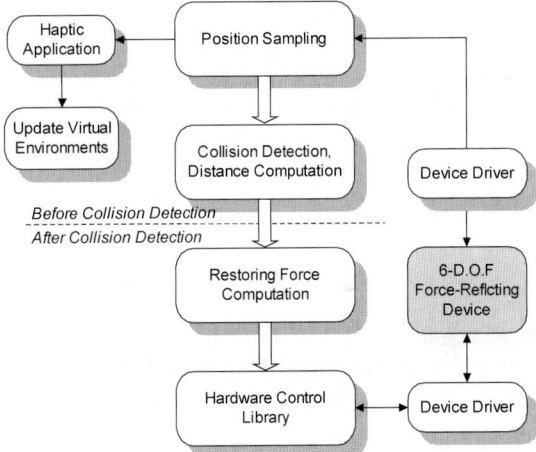

Figure 2. Overall architecture of the haptic display system

The contact region is the set of all points where the two objects come into contact with each other. An accurate computation of the contact region at each frame helps in smoothing the transition of force display from one frame to the next in a haptic simulation when there are multiple contacts between the probe and the simulated environment.

3. Collision detection

In order to create a sense of touch between the user's

hand and a virtual object, restoring forces are generated preventing penetration into the virtual object. If a collision or penetration has occurred, then determining the contact point on the object surface compute this.

3.1 First Level Overlap Test

An intersection test between two models is done by recursively testing pairs of nodes. For each visited pair of nodes, the minimal AABBs [14] are tested for overlap.

Given an AABB, B, defined by $\mathbf{b}^{min}$ and $\mathbf{b}^{max}$, four different diagonals can be constructed. These all pass through the center of B and have endpoints at the vertices B. For the intersection test, first find out which of the box diagonals is most aligned with the plane normal, $\mathbf{n}$. After the most aligned diagonal is found, the diagonal's two AABB vertices, called $\mathbf{v}^{min}$ and $\mathbf{v}^{max}$, are inserted into the plane equation π. This is illustrated in Figure 3. This overlap test can be improved upon by the notion that if $\mathbf{v}^{min}$ is in the positive half-space of the plane, then $\mathbf{v}^{max}$ will also be in the positive half-space [25]. In this case the $\mathbf{v}^{max}$ point need not be tested against the plane.

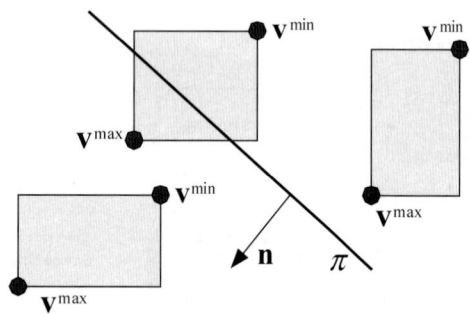

Figure 3. Example of AABB overlap test

As shown in Figure 3, if the pair of these vertices is on the same side of plane, then the AABB does not intersect the plane; otherwise, it does. Note that if $\mathbf{v}^{min}$ is tested against the plane first and is found to be on the same side as the plane, then we can immediately reject the intersection. Only pairs of objects whose bounding boxes overlap in all three dimensions are passed to the pairs test module of the system.

3.2 Second Level Overlap Test

The next level detects pair-wise collisions. Our implementation computes a tree of OBBs for every object, with a box containing the entire objects as the root and boxes only containing one or a very few primitives as the leaves. A fast collision detection routine[19, 26] will be derived for testing whether two OBBs. The algorithm uses the separating axis theorem, and is about an order of magnitude faster than previous method, which uses closest features[6] or linear programming. There are 15 potential separating axes: 3 for the independent faces of the first OBB, 3 for the independent faces of the second OBB, and 9 generated by an edge from the first OBB and an edge from the second OBB. If the objects are disjoint, then a separating axis exists, and one of the 15 axes mentioned above will be the separating axis. If the objects are overlapping, then clearly no separating axis exists. So, testing the 15 given axes is a sufficient test for determining overlap status.

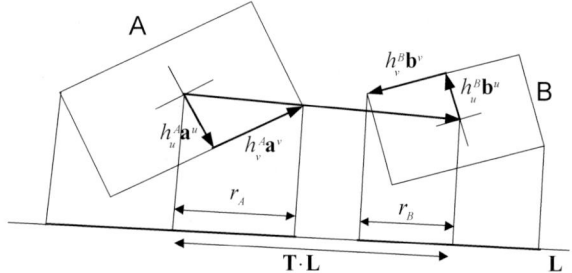

Figure 4. The separating axis theorem illustrated. The two OBBs, A and B, are disjoint, since the projections of their "radii" on the axis determined by L are not overlapping[19].

4. Computing Contact Forces and Torques

4.1 Restoring Force and Torques

We compute the restoring forces based on the penalty methods. Once the contact point is determined, simple impedance control techniques can be used to calculate a force to be displayed. Stiffness and damping elements will model the local material properties. Hook's law transfers directly into Equation (1) where k is the stiffness of the surface and D_p is the depth of penetration.

$$F_{stiffness} = k \cdot D_p \qquad (1)$$

The damping force should be based on the motion of the force-feedback device's probe relative to the motion of the virtual proxy. Since we don't want to impede motion tangential to the surface, only the motion along the surface normal $\hat{N}$ should be used. Equation (2), in which c is the damping coefficient, computes this force. The damping force should only be applied when it stiffens the surface.

$$F_{damping} = c(\dot{D}_p \cdot \hat{N})\hat{N} \qquad (2)$$

$$F_{normal} = F_{stiffness} + F_{damping} \qquad (3)$$

The computed restoring force F_{normal} is applied to the contact region along the contact normal direction to resolve the penetration, thereby generating a sense of touch. When the restoring force is applied at points cp_i ($i = 1, \cdots, n$), the torques can be summed to yield a restoring force.

$$\Gamma_r = \sum_i (cp_i - x_i) \times F_{normal, i} \quad (4)$$

where x_i is the center of mass.

4.2 Interpolated Surface Normal

The displayed force is computed as a function of penetration depth, contact region and contact normal direction. When the force is computed, an additional step must be taken to determine the direction of the force. It is possible that the magnitude of contact forces can vary; thereby creating sudden jumps and introduces sharp discontinuities between successive frames. We use an interpolation method for smooth force shading effects.

Each polygon has associated with it a normal vector $\vec{n}_i$. Consider an individual triangle in the polygon mesh, with three adjacent neighboring triangles, as shown in Figure 4. The normal vector for the triangle in the figure 5 is $\vec{n}_0$, and the adjacent polygons have normal vectors $\vec{n}_1$, $\vec{n}_2$, $\vec{n}_3$.

$$\vec{n}_{01} = \frac{\vec{n}_0 + \vec{n}_1}{2}, \quad \vec{n}_{02} = \frac{\vec{n}_0 + \vec{n}_2}{2}, \quad \vec{n}_{03} = \frac{\vec{n}_0 + \vec{n}_3}{2}. \quad (5)$$

With distances l_1, l_2, and l_3 from the edges of the polygon, the force direction can be computed as

$$\vec{n}(\vec{x}) = \frac{l_1(\vec{n}_{02} + \vec{n}_{03}) + l_2(\vec{n}_{01} + \vec{n}_{03}) + l_3(\vec{n}_{01} + \vec{n}_{02})}{2(l_1 + l_2 + l_3)} \quad (6)$$

In addition, we used a simple linear smoothing scheme to minimize discontinuity in force and torque display between successive frames [24].

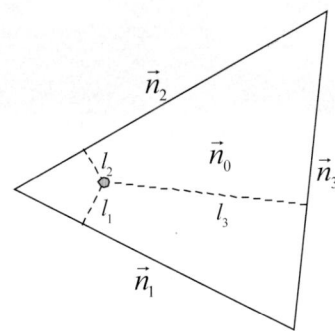

Figure 5. Surface normal interpolations

5. Implementation and Performance

We have implemented haptic rendering algorithms as part of a general-purpose framework for performing haptic interaction in virtual environments.

5.1 Hardware Configuration

We designed the PC-based controller with dedicated single-board. This would make the system more mobile and allow multiple devices to be run simultaneously [27]. The developed controller includes six channel encoder counters, a 12bit digital-to-analog converter, digital I/O, and a fast-speed analog-to-digital converter for current measurement. The joint position measured by encoder are acquired through 24 bit counter chips, and the servo values for current amplifiers are controlled through 12 bit D/A converters so that the joint torque is set to the desired value. Figure 6 shows the block diagram consist of the controller.

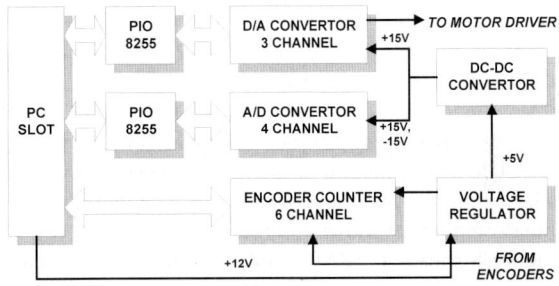

Figure 6. Block diagram for integrated controller

Small amplifiers constructed from TL072 power operational amplifiers power the motors. The two different signals drive the gates of the H-bridge driving circuit, which consists of four TIC41C power transistors. The diagram is shown in Figure 7. With power supply of ±15 volts and D/A output of ±10V, the amplifiers generated a maximum current of 1.5 amps, resulting in a maximum motor torque of 0.029 Nm for each motor.

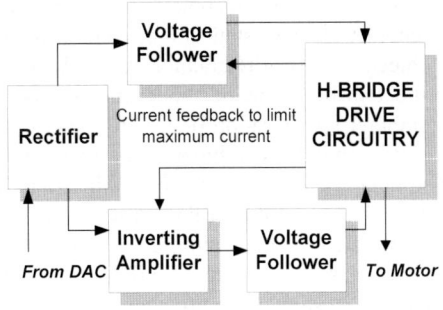

Figure 7. Circuit diagram for current amplifiers

5.2 Networking

Because of the real-time interaction of the force-feedback device with the polygonal model is critical, networking design and throughput are very important. The general guidelines followed are:

- UDP is faster than TCP, and the associated packet loss is acceptable for state update that occurs continuously.
- TCP is needed for guaranteed update of models or state changes that are sent only one.
- Polygonal model updates are not always necessary.

In our implementation, it was found necessary to utilize two-computer system. One is a Pentium II 350MHz for performing haptic rendering algorithms and the other is a Pentium III 750MHz dual for graphical display.

5.3 Performance

The framework has been implemented in C++ and runs under Windows NT operating system. The data structure of objects is VRML file format [29], which can be generated by various simulation tools such as 3D Studio MAX. The current rendering library supports a set of functions provided by OpenGL graphic library, GLUT, and GLUI for graphical user interface. The current system is adept in modeling a large number of geometric models. Some examples are shown in Figure 8, 9 and 10.

Figure 8 shows a VRML model of the engine-cylinder containing 15,230 polygons. The high level interface simplifies the implementation of applications like VRML browsers. It is composed of moving piston and we check for collisions and distances with the cylinder. Figure 9 shows a VRML model of the classic teapot, composed of 9,218 triangular surfaces generated using 3D Studio MAX. Figure 10 shows the dynamic simulation uses classic torus of the same number of triangular surfaces, 3,240, moving in closed environment. Each torus has their transnational and rotational velocity obtained by randomly sampling values.

Experiments have been carried out to investigate force and torque display to several applications: engine-cylinder pushing, interaction with static object and with multiple moving objects. Their performance varies based on the models, the configuration of the probe relative to the model, computer specification (e.g. CPU type, cache and memory size), and the combination of techniques. The engine-Cylinder model contains long thin triangles that are aligned with the world coordinate axes. AABBs bound these triangles tightly, their simple structure allows for fast overlap tests. Therefore, AABBs was better on models with axis-aligned components than on models with non-aligned components. In case of teapot model, OBB overlap test has good performance. The dynamic simulation uses 4 polygonal models of the same sharp in

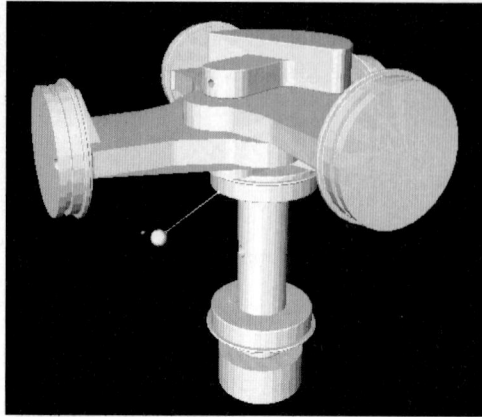

Figure 8. Engine-Cylinder (15,320 polygons)

Figure 9. Classic teapot (9,218 polygons)

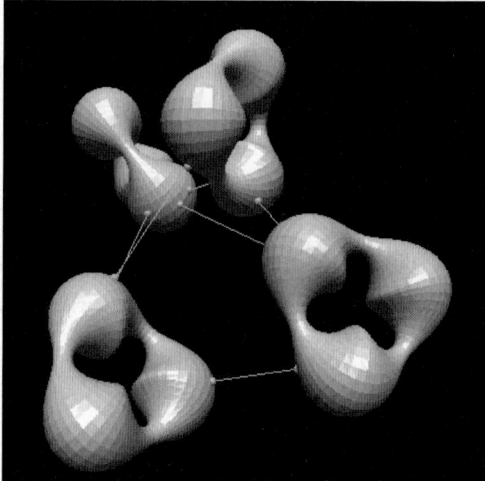

Figure 10. Dynamic simulation (3,240 polygons each)

environment. During the simulation, 1000 steps, we used the transnational and rotational velocity obtained by randomly.

6. Conclusions

In this paper, we have described the framework of haptic rendering system using 6-DOF force-feedback device. In terms of haptic display of complex polygonal models, a major interest is to accurately compute all the contacts and the penetration depth between general polygonal models in with more 1KHz update rate. Taking advantage of using hybrid-bounding volumes in a dynamic environment, collision was detected more efficiently. The AABB tree do not tend to fit their underlying geometry as tightly as do more complex shapes such as OBBs but tend to have quick overlap states. The advantage of OBB tree is that they can bound their enclosed geometry more tightly for computing collision pairs.

There are many open issues for future works. We would like to classify applications where different BVs work well. Based on this understanding, we need to explore hybrid hierarchies in more detail. We would also like to develop efficient overlap testing algorithms and analyze relationships for penetration depth and restoring force computation as well.

Reference

[1] W. Thibault and B. Naylor, "Set Operations on Polyhedra using Binary Space Partitioning Trees," *ACM Computer Graphics*, Vol. 4, pp. 153-162, 1987.

[2] M. Moore and J. Wilhelms, "Collision Detection and Response for Computer Animation," *ACM Computer Graphics*, Vol. 22, No. 4, pp. 289-298, 1988.

[3] D. Baraff, "Curved Surface and Coherence for non-penetrating rigid body simulation," *ACM Computer Graphics*, Vol. 24, No. 4, pp. 19-28, 1990.

[4] Frederick P. Brooks, Jr. Ming Ouh-Young, James J. Batter, and P. Jerome Kilpatrick, "Project GROPE : Haptic Displays for Scientific Visualization," *Computer Graphics(SIGGRAPH '90 Proceeding)*, Vol. 24, pp. 177-185, 1990.

[5] N. Beckmann, H.Kriegel, R.Schncider, and B. Seeger, "The R*-Tree : An Efficeint and Robust Access Method for Points and Rectangles," *Proceedings of SIGMOD Conference on Management of Data*, pp. 322-331, 1990.

[6] M.C.Lin and John. F. Canny, "Efficient algorithms for incremental distance computation," *In IEEE Conference on Robotics and Automation*, pp. 1008-1014, 1991.

[7] Ji-Hoon Youn and K. Wohn, "Realtime Collision Detection for Virtual Reality Applications," *IEEE First Annual Virtual Reality Symposium*, pp. 415-421, 1993.

[8] W. Bouma and G. Vanecek, "Modeling contacts in a physically based simulation," *Second Symposium on Solid Modeling and Applications*, pp. 409-419, 1993.

[9] Y. Yang and N. Thalmann, "An Improved Algorithm for Collision Detection in Cloth Animation with Human Body," *First Pacific Conference on Computer Graphics and Application*, pp. 237-251, 1993.

[10] David Baraff, "Fast Contact Force Computation for Non-penetrating Rigid Bodies," *In Proceedings of ACM SIGGRAPH*, pp. 23-24, 1994.

[11] S. Quinlam, "Efficient Distance Computation Between Non-Convex Objects," *Proceedings of International Conference on Robotics and Automation*, pp. 3324-3329, 1994.

[12] B. Mirtich and J. Canny, "Impulse-Based Simulation of Rigid Bodies," *In Proceeding of ACM Interactive 3D Graphics, Monterey, CA*. 1995.

[13] G. Zachman and W. Felger, "The Boxtree : Enabling real-time and Exact Collision Detection of Arbitrary Polyhedra," *Proceedings of SIVE '95*, 1995.

[14] H. Held, J.T. Klosowski, and J.S.B. Mitchell, "Evaluation of Collision Detection Methods for Virtual Reality Fly-Throughs," *In Canadian Conference on Computational Geometry*, 1995.

[15] Hahn JK, "Realistic Animation of Rigid Bodies," *Computer Graphics*, Vol. 22, No. 4, pp. 218-228, 1995.

[16] J. Synder, "An Interactive Tool for Placing Curved Surfaces without interpenetration," *In Proceedings of ACM SIGGRAPH*, pp. 209-218, 1995.

[17] Zilles C., Salisbury J. K, "Constraint based God Object Method for haptic Display," *Proc. of the IEEE/RSJ International Conference on Intelligent Robotics and System*, 1995.

[18] M. Held, J. Klosowski, and Joseph S.B. Mitchell, "Real-time Collision Detection for Motion Simulation within Complex Environments," *International Proceeding of ACM SIGGRAPH '96 Visual Proceedings*, pp. 151, 1996.

[19] S. Gottschalk, M. Lin and D. Manocha, "OBB-Tree : A Hierarchical Structure for Rapid Interfenece Detection," *International Proceeding of ACM SIGGRAPH '96*, pp. 171-180, 1996.

[20] D.C.Ruspini, K.Kolarov, and O.Khatib, "The haptic display of complex graphical environments," *Proc. Of ACM SIGGRAPH*, pp. 345-352, 1997.

[21] M. Ponamgi, D. Monacha, and M. Lin, "Increamental algorithms for Collision Detection between Polygonal models," *IEEE Transaction on Visualization and Computer Graphics*, Vol. 3, No. 1, pp. 51-67, 1997.

[22] J. Klosowski, M. Held, Joseph S.B. Mitchell, H. Sowizral, and K. Zikan, "Efficient Collision Detection using Bounding Volume Hierarchies of k-DOPs," *IEEE Trans. On Visualization and Computer Graphics*, Vol. 4, No. 1, pp. 21-37, 1998.

[23] Brain Mirtich, "V-Clip : Fast and Robust polyhedral Collision Detection," *ACM Transactions on Graphics*, Vol. 17, No. 3, pp. 177-208, 1998.

[24] A. Gregory, A. Mascarenhas, S. Ehmann, M. Lin and D. Manocha, "6-DOF Haptic Display of Polygonal Models," *IEEE Visualization Conference*, 2000.

[25] Hoff III and Kenneth E., "A Faster Overlap Test for a Plane and a Bounding Box," 1996.
http://www.cs.unc.edu/~hoff/research/vfculler/boxplane.html

[26] Gottschalk, Stefan, "Collision Queries using Oriented Bounding Boxes," Ph.D. Thesis, Department of Computer Science, University of North Carolina at Chapel Hill, 1999.

[27] D.H.Kim, J.S.Park, and Y.D.Kim, "A Personal Computer Based Force-Reflecting System: Design and Application", *Proc.of the 3rd Int. Workshop on Advanced Mechatronics*, pp.131-135, December, 1999.

[28] Web site, http://www.vrml.org/

A New Haptic Interface Device Capable of Continuous-Time Impedance Display within Sampling-Period: Application to Hard Surface Display

Masayuki Kawai and Tsuneo Yoshikawa

Department of Mechanical Engineering
Kyoto University
Kyoto, 606-8501 Japan
t60w0805@ip.media.kyoto-u.ac.jp and yoshi@mech.kyoto-u.ac.jp

Abstract

In this paper, we discuss displaying a virtual object with coupling impedance by a haptic interface device with long sampling period. For such system to be stable, coupling impedance has to be set low. But it means that the operator feels soft surface of the virtual object and it is not suitable for displaying a virtual object with hard surface. This is because the system is a discrete time system. So we propose a haptic interface device with analog springs and dampers that exert reaction force continuously. The method can decrease the influence of the sampling period and make haptic interface more stable. Finally we perform some experiments with two dimensional virtual environment to confirm the effect of the proposed method.

1 Introduction

Haptic interface is important for constructing virtual world that operators feel real. Keeping haptic interface stable is one of the most important problems, because haptic device has direct contacts with operators and instability causes damages to the operator and the system. But it is also a difficult problem to keep haptic interface stable, since the human dynamics is unknown.

A number of researchers have analytically studied stability issues in haptic interaction. Minsky et al. [1] discussed a tradeoff relation among simulation rate, virtual wall stiffness and viscosity of haptic display device for simple virtual environments. Colgate et al. [2] discussed haptic display of simple virtual wall and derived conditions under which the haptic display would exhibit passive behavior from an idea that the operator is replaced with an arbitrary passive impedance. Then, they led conditions of stability for tool-use haptic device for a dynamic virtual object with coupling impedance which they call the virtual coupling [3]. Such fundamental stability and performance issues were well discussed by Adams and Hannaford [4] with two-port model, which is a common method of analyzing stability and performance in bilateral tele-operation. In these studies, the roots of problems are that the system is a discrete time system. When a system measures and controls at a longer interval, it becomes more difficult to keep the system stable and virtual impedance is used between the real and virtual objects to secure stable display. We call it coupling impedance in this paper. When the system is a long-sampled system, it is possible to keep stable only with low coupling impedance and it means that the operator feels soft surface and it is not suitable for displaying hard surface.

In this paper, we propose a method to display hard surface for long-sampled system with impedance by electric analog circuit that exerts reaction force continuously. The haptic display device has joints with electric spring and damper which works as coupling impedance. Since the springs and dampers works continuously within sampling period, the method can decrease the influence of discrete sampling period. First, we discuss the basic concept of this method and the stability issues. Then, we perform some experiments with two dimensional virtual environment to confirm the effect of the proposed method.

2 Coupling Impedance in One Dimensional Environment

First, we explain coupling impedance conventionally used for 1-DOF(Degree Of Freedom) device and one dimensional virtual environment, then we propose a haptic display with a hardware that exerts impedance continuously.

2.1 Coupling Impedance

Fig.1 shows a 1-DOF haptic display system and the dynamic equation of the device is described as :

$$f_a = m\ddot{x} + b\dot{x} + f_h \qquad (1)$$

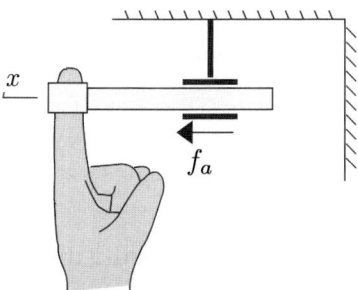

Figure 1: Figure of 1-DOF Haptic device

where x is the fingertip position, f_a is the exerted force by the actuator, m is the mass of the device, b is the viscosity and f_h is the force from the operator. When a virtual rigid wall is located at x_d, coupling impedance is described as:

$$f_a = -k_p(x^* - x_d) - k_d \dot{x}^* \quad (2)$$

where k_p, k_d are coefficients of virtual stiffness and damper and x^* express the discrete variable of x.

The impedance between the real and virtual object is discretely exerted by software and we call it "discrete-time impedance". The method is practically simple and useful. If the system becomes unstable and makes vibration, it is easy to make the system stable by decreasing the coupling impedance. But it means that when the system is a long-sampled system, the method can display objects with soft surface.

2.2 Haptic Display with Continuous-time Impedance

In order to display a virtual object with hard surface, we propose haptic interface with a device which exerts impedance continuously. In this section, we explain the basic concept of the proposed method. Fig.2 shows a 1-DOF haptic device with a spring and a damper attached to the system of Fig.1. The spring and damper are real hardware and exert the reaction force continuously. The dynamic equation is expressed

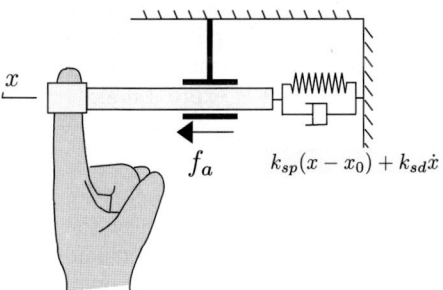

Figure 2: 1-DOF Haptic Device with Hardware Impedance

as:

$$f_a = m\ddot{x} + b\dot{x} + f_h + k_{sp}(x - x_0) + k_{sd}\dot{x} \quad (3)$$

where k_{sp}, k_{sd} are the coefficients of the real stiffness and the real damper of the device and x_0 is the position of the natural length of the spring. Haptic display utilizing the real impedance is described as:

$$f_a = k_{sp}(x_d - x_0) \quad (4)$$

It means that the equilibrium of the spring is moved to the contact point x_d on the surface of the virtual object. If the mass and the damping of the haptic device are slight, the impedance on the surface of the virtual object becomes:

$$f_h = k_{sp}(x - x_d) + k_{sd}\dot{x} \quad (5)$$

It is an equivalent of the real impedance with k_{sp}, k_{sd} located on the surface and the operator can feel the reaction force continuously. we call it "continuous-time impedance" in this paper.

In the case to display a static virtual object, x_d and x_0 are constant in the eqn.4, therefore the system has no influence of the sampling period. It is possible to display with high impedance during the contact despite of the sampling period. Moreover, it is effective to add the output of discrete-time impedance (eqn.2) and continuous-time impedance (eqn.4) as:

$$f_a = -\bar{k}_p(x^* - x_d) - \bar{k}_d \dot{x}^* + k_{sp}(x_d - x_0) \quad (6)$$

In this case, the impedance on the surface of the virtual object becomes:

$$f_h = (\bar{k}_p + k_{sp})(x - x_d) + (\bar{k}_d + k_{sd})\dot{x} \quad (7)$$

instead of eqn.5.

3 Stability Analysis

In this section, we discuss the stability issues for the methods described in the previous section.

3.1 System of Discrete-time Impedance

The system of discrete-time impedance, which is described in the section 2.1, is represented by Fig.3 under a condition that the position of the virtual object is $x_d = 0$. In the figure, T is the sampling period, $Z_0(s)$

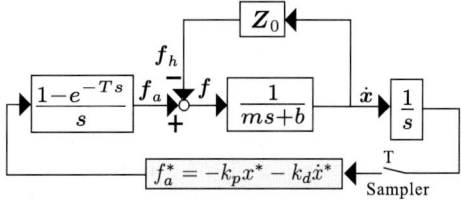

Figure 3: Model of 1-DOF Haptic Interface

is the transfer function that represents the dynamics

of the operator and $(1-e^{-Ts})/s$ means zero-order hold circuit. For analysis of systems containing human dynamics, passivity analysis is generally used and Colgate et al [2] derived the following condition from the concept that the dynamics of the operator is replaced with an arbitrary passive impedance:

$$b > \frac{k_p T}{2} + k_d \qquad (8)$$

3.2 System of Continuous-time and Discrete-time Impedance

The system of continuous-time impedance and discrete-time impedance, which is described in the eqn.6, is expressed by Fig.4. Since $k_{sp}x_0$ in the eqn.4 is

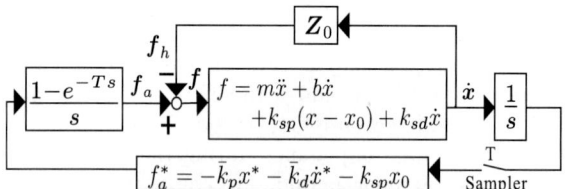

Figure 4: Model of 1-DOF Haptic Interface with Spring and Damper

constant and the spring which is installed to the joint can be moved to the operator Z_0, Fig.4 is equivalent to Fig.5. In the figure, Z_0' is a system containing the

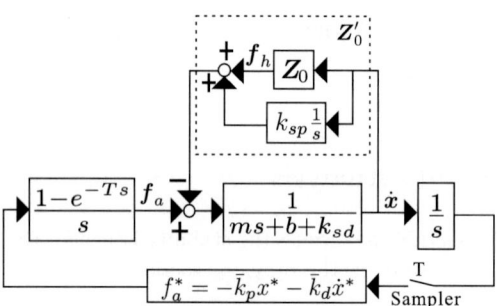

Figure 5: Modified Model of Fig.4

dynamics of the operator Z_0 and the real spring. If the dynamics of the operator Z_0 is passive in the Fig.5, the system Z_0' is obviously also passive. A condition of the passivity is derived as:

$$b + k_{sd} > \frac{\bar{k}_p T}{2} + \bar{k}_d \qquad (9)$$

It means that when the continuous-time impedance is used, the discrete-time impedance can be set higher according to the damping k_{sd} of the continuous-time impedance. Even if $k_{sd} = 0$ in the continuous-time impedance, the stiffness which is stably displayed to the operator increases to $k_{sp} + \bar{k}_p$, as shown in the eqn.7.

4 Continuous-time Impedance for Two or More Dimensional Environment

In this section, we extend the method of one dimensional environment described in the previous section to two or more dimensional environment with multi-linkage device.

4.1 Haptic Display in Multi-dimensional Environment

Multi-linkage device, as shown in Fig.6, is necessary for displaying in two or more dimensional environment. The peculiarity of such multi-linkage is that the finger-

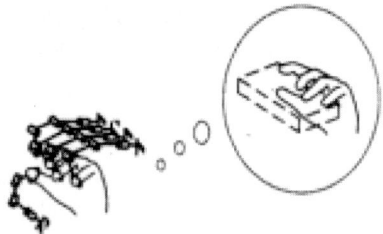

Figure 6: Example of Multi-dimensional Haptic Display

tip impedance depends not only on the impedance of each joint but also the angle of each joint. For keeping an arbitrary impedance at the fingertip, it is necessary that the continuous-time impedance of each joint is controllable.

4.2 Continuous-time Impedance For Multi-linkage Device

For changing the continuous-time impedance of each joint, we make use of a electrical circuit with controllable amplifier. Fig. 7 shows a part of the circuit for ith and jth joint. The potentiometers measure the angle of joints and the output enters a differentiation circuit and a differential amplifier circuit, which subtracts the output of the potentiometer from the output of the computer. the output from each circuit is feedback to motor driver of each joints through VCA(Voltage Control Amplifier), whose gain the computer can change. The kinematics and dynamics of the device with the circuit is expressed as:

$$\boldsymbol{r} = \boldsymbol{f}(\boldsymbol{q}), \quad \dot{\boldsymbol{r}} = \boldsymbol{J}\dot{\boldsymbol{q}} \qquad (10)$$

$$\boldsymbol{\tau} = \boldsymbol{M}\ddot{\boldsymbol{q}} + \boldsymbol{h} + \boldsymbol{K}_{sp}(\boldsymbol{q} - \boldsymbol{q}_0) + \boldsymbol{K}_{sd}\dot{\boldsymbol{q}} + \boldsymbol{J}^T \boldsymbol{F}_h \qquad (11)$$

where $\boldsymbol{r}$ is the fingertip position of the device, $\boldsymbol{q}$ is the angle of joint and $\boldsymbol{J}$ is the Jacobian matrix. $\boldsymbol{K}_{sp}$ and $\boldsymbol{K}_{sd}$ are controllable stiffness and damping by the circuit, which works as continuous-time impedance within a sampling period. $\boldsymbol{\tau}$ is the output force from the actuator and the relation between $\boldsymbol{\tau}$ and the output $\boldsymbol{u}$ from the computer is :

$$\boldsymbol{\tau} = \boldsymbol{K}_{sp}\boldsymbol{u} \qquad (12)$$

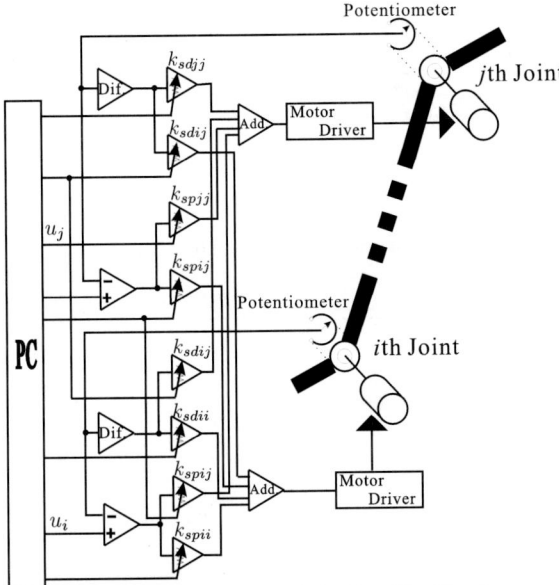

Figure 7: Outline of Analog Feedback

4.3 Haptic Display with Continuous-time Impedance Circuit

When an object coordinate frame Σ_C is defined on the contact point r_d located on the surface of the virtual object, as shown in Fig.8, and desirable fingertip

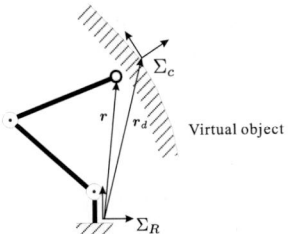

Figure 8: Object Coordinate

stiffness $\bar{K}_{sp}$ and damping $\bar{K}_{sd}$ along the object coordinates are given, the gain of the amplifier of each joint should be set as:

$$K_{sp} = J^T R_c \bar{K}_{sp} R_c^T J$$
$$K_{sd} = J^T R_c \bar{K}_{sd} R_c^T J \quad (13)$$

where R_c is the rotational matrix of Σ_c. The right terms are constructed as a symmetric matrix and so K_{sp}, K_{sd} can also be installed as a symmetric matrix, as shown in Fig.7. Haptic display with this system is described as:

[1] When the finger is not touching a virtual object, the stiffness and the damping are set as:

$$K_{sp} = 0, \quad K_{sd} = 0 \quad (14)$$

It means that the actuator makes no output and the operator feels only the dynamics of the device.

[2] When the finger is touching a virtual object and the contact point r_d is defined on the surface of the virtual object, the stiffness and the damping are set as eqns.13 and the output of the actuator is:

$$\tau = -J^T(\bar{K}_p(r - r_d) + \bar{K}_d \dot{r})$$
$$+ J^T R_c \bar{K}_{sp} R_c^T J(q_d - q_0) \quad (15)$$

where q_d is the angle where the fingertip position would be on the contact point r_d and K_{cp}, K_{cd} are software stiffness and damping. The first two terms are discrete-time impedance by software and the last term works for making continuous-time impedance by electric analog springs and dampings. The impedance on the surface of the virtual object is the sum of impedance of the discrete-time impedance and the continuous-time impedance.

$$(\bar{K}_p + \bar{K}_{sp})(r_d - r) + (\bar{K}_d + \bar{K}_{sd})\dot{r} \quad (16)$$

When the sampling period is long, it is impossible to set the part of the discrete-time impedance high. However, the continuous-time impedance can be set high regardless of the sampling period, as discussed in the previous section.

5 Experiment

In this section, we perform some experiments to check the effects of the proposed method with two dimensional virtual environment and 2-DOF haptic device.

5.1 Overview of Experimental Device

Fig.9 shows an overview of experimental setup. The haptic device is a 2-DOF robot with linkage whose length is shown in Fig.10. The device has potentiometer, DC servo motor and encoder on each joint. The potentiometers are used for the electrical circuit and the computer measures the angles of each joints by the encoders. A force sensor is installed at the fingertip of the robot and is used only for measuring the experimental result and not for control.

5.2 Experiment of Maximum Stiffness of Virtual Wall

First, we perform experiments to measure maximum stiffness for a virtual wall by each method: discrete-time impedance, continuous-time impedance and adding both impedance. Generally a target of haptic device is a human but it is difficult to keep constant conditions and to get quantitative results. In this experiment, we use a mass instead of a human. The

Figure 9: Overview of Device

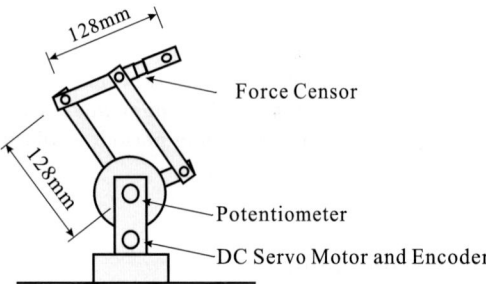

Figure 10: Length of Linkage

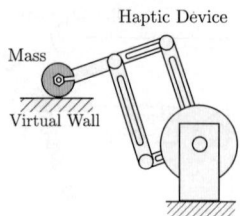

Figure 11: Experiment of Measuring Maximum Stiffness

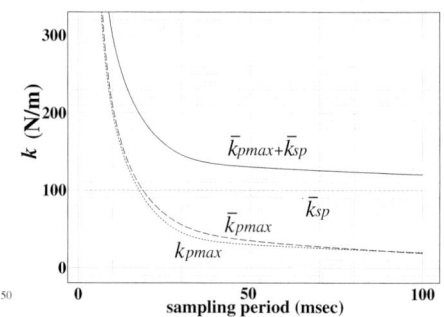

Figure 12: Maximum Stiffness

5.3 Experiment of Human Operation

Next, we perform some experiments that the operator traces on a virtual slope, whose angle is $\pi/4$, and we measure the fingertip force and position(Fig.13). The surface is supposed to be no-frictional and the impedance along the horizontal direction with the slope is set zero. Examples of results by the con-

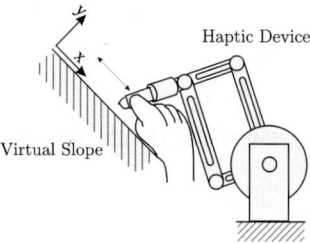

Figure 13: Experiment of Human Operation

ventional method only with discrete-time impedance are shown in Fig.14(a)(b). In this experiment, the discrete-time impedance is chosen the value which the operator feels stable. Fig.14(a) is the results along horizontal direction for the slope, Fig.14(b) is the results along vertical direction. In the figures, the solid line is the fingertip force and the dotted line is the fingertip position. Fig.15(a)(b) show results of the proposed method, adding continuous-time impedance to discrete-time impedance. The sink to the virtual wall, which is the dotted line in Fig.15(b), is shallower than Fig.14(b) and it is shown that continuous-time impedance can be stably set higher than the case of only discrete-time impedance.

Fig.16(a)(b) show results when stiffness of interference between joints is set zero and the continuous-time

mass is represented as a point in the virtual environment and is dropped on a virtual floor(Fig. 11). Since the virtual floor has impedance in the virtual environment, the mass bounds on the floor. If the impedance is high, the mass continues to bound. If it is low, the mass stops on the floor stably. We measure maximum value of stiffness at each sampling period from 5 to 100 ms. In order to make the comparison clear, the stiffness is only measured and the dampings are set zero ($k_d = \bar{k}_d = \bar{k}_{sd} = 0$). The result is shown in Fig. 12. In the figure, the abscissa is the sampling period and the ordinate is the stiffness.

The dotted line k_{pmax} is the maximum value of the discrete-time stiffness and the relation with the sampling period is trade off, which has ever derived by the previous works like [1][2]. The dashed line $\bar{k}_{pmax}$ is the maximum value of the discrete-time stiffness $\bar{k}_p$ when it is used at a time with the continuous-time stiffness $\bar{k}_{sp}$, which is the dashed line on 100 N/m in the figure. $\bar{k}_{pmax}$ becomes almost same as k_{pmax} and it means that continuous-time stiffness has no effect for the stability of discrete-time stiffness. The stiffness displayed to the real world is the solid line, which is the sum of the discrete-time stiffness $\bar{k}_{pmax}$ and the continuous-time stiffness $\bar{k}_{sp}$.

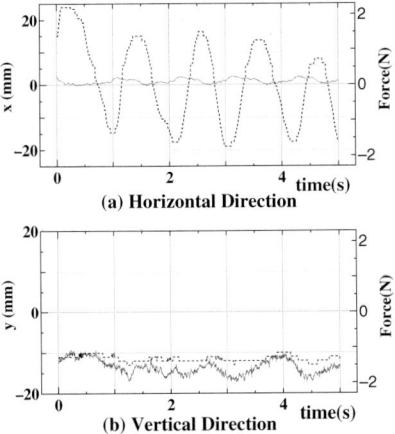

Figure 14: Result of Only Discrete-time Impedance

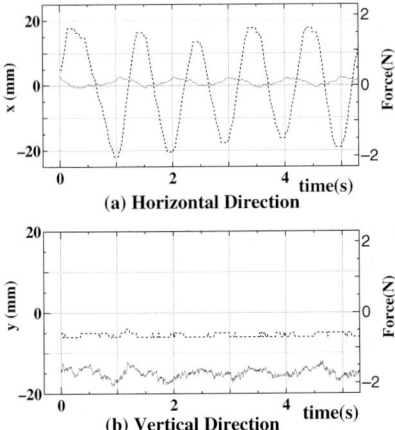

Figure 15: Result of Continuous-time Impedance with Discrete-time Impedance

stiffness K_{sp} is a diagonal matrix. It is equivalent to display using a PD control to the contact point r_d on each joint independently. In the figure, the fingertip force along the horizontal direction (solid line in Fig. 16(a)) is greater than the force in Fig.15(a). It means that haptic display with PD control independently for joint makes resistant force for the non-restricted direction. On the other hand, the proposed method can decrease such a resistant force.

6 Conclusion

In this paper, we proposed haptic interface with a device which exerts continuous-time impedance and a method to make use of the motion of the continuous-time impedance between sampling interval for decreasing instability caused by discrete sampling period. The method is possible to display harder surface during a contact with a virtual object and we analytically compared the stability of the proposed method with the conventional method of discrete-time impedance for a static virtual object. We proposed a structure of the

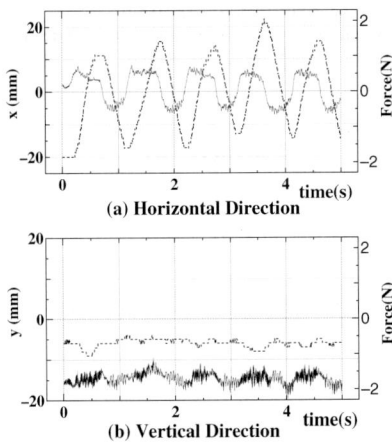

Figure 16: Result of Continuous-time Impedance Applied to Each Joint Independently

device by electric analog circuit with voltage control amplifier and performed some experiments to confirm the effects of the method.

However, even if the continuous-time impedance is used, there are some moment when the continuous-time impedance behaves as discrete-time system, like as a moment of collision or displaying a dynamic virtual object. Because contact point is suddenly defined or moving discretely in these case. Therefore, further research topics are to analyze influence of the discrete-time behavior and to improve the proposed method when a moment of collision or displaying a dynamic virtual object.

References

[1] M. Minsky, M. Ouh-young, O. Steele, F. P. Brooks, and M. Behensky, "Feeling and Seeing:Issues in Force Display" Comput. Graph, Vol.24,no.2,pp.235-243, 1990

[2] J. E. Colgate and G. Schenkel, "Passivity of A Class of Sampled-data Systems:Application to Haptic Interfaces" in Proc. of the American Control Conference, pp 3236-3240, 1994

[3] J. E. Colgate, M. C. Stanley and J. M. Brown, "Issues in the Haptic Display of Tool Use" IROS'95, 1995

[4] R. J. Adams and B. Hannaford, "Stable Haptic Interaction with Virtual Environment" IEEE Trans. on Robotics and Automation Vol.15, No.3, pp 465-474, 1999

[5] T. Yoshikawa and H. Ueda, "Haptic Virtual Reality: Display of Operating Feel of Dynamic Virtual Objects", Robotic Research, The Seventh Symposium, pp214-221, 1996

Design Of A New 6-DOF Parallel Haptic Device

J. H. Lee, K. S. Eom, B-J Yi, I. H. Suh

School of Electrical Engineering and Computer Science, Hanyang University, Korea

Email : bj@email.hanyang.ac.kr

Abstract *In this paper, a new 6-dof parallel haptic master device is proposed. Many existing haptic devices require large power due to having floating actuator and also have small workspaces. The proposed new mechanism is relatively light by employing non-floating actuators and has large workspace. Kinematic analysis and kinematic optimal design problem is performed for this mechanism. Dexterous workspace, global isotropic index, and global maximum force transmission ratio are considered as kinematic design indices. To deal with such multi-criteria optimization problem, composite design index is employed. Actuator sizing for this mechanism is also carried out.*

Keywords : Haptic Device, Parallel Mechanism, Optimal Design

1. Introduction

The application of virtual reality is expanding fast in various areas: medical, teleoperation, entertainment, etc. Virtual reality technology mainly consists of vision, haptic display, and audio. However, progress in the field of haptic interface has been much slower than vision and audio, relatively.

Since master-slave system proposed by Goertz in the 1950's[1], many researchers have developed various type of haptic display such as an exoskeleton type master arm by M. Bergamasco[5], PHANToM[3] by Massie and Salisbury, MagLev Wrist[4] by Hollis in Carnegie Mellon Univ., and the magnetic levitation haptic interface by Merkelman et al, etc. Although many haptic interfaces have been developed as above, they have generally three or less DOF which are not enough to display the real phenomenon. And they mostly have serial structures which can not achieve hard contact feeling. Other devices having parallel structure are heavy and have relatively small workspace, and required large power due to floating actuators[2,6,8].

General requirements of Haptic interface include large workspace for human operator, low apparent mass/inertia, low friction, high structural stiffness, backdriveability, low backlash, high force bandwidth, high force dynamic range, absence of mechanical singularities, compactness, an even 'feel' through the workspace, and so on[2]. These requirements can be fulfilled by employing a good mechanism in company with good control algorithm.

In this paper, we will propose a new 6-DOF parallel haptic interface which is actuated by non-floating actuators as shown in Fig. 1. This new mechanism has large workspace with no singularity. To maximize kinematic performances of the haptic interface, multi-criteria based optimization is carried out.

2. Kinematics of a new 6-DOF Haptic Device

2.1 Geometric Description

The proposed 6-DOF Haptic Device consists of a top plate, six actuators on the base, and three parallel chains connecting the top plate to the six actuators in Fig. 2. Let $\{B\}$ and $\{T\}$ be the base frame fixed to the ground with its origin at the center of the base and the local frame fixed to the top plate with its origin at the center of the top frame, respectively. Each of three ball-socket joints ($^m\underline{C}$, $m=1,2,3$) of the top plate are placed on the circle of radius (R_t) with 120° apart from each other. Three pairs of actuators are placed on the ground with 120° apart from each other. Each of three actuator pairs consists of the upper actuator (M1) that is placed on the circle of radius R_{B1} horizontal to the ground and the lower actuator (M2) that is placed on the circle of radius R_{B2} vertical to the ground. And H_B denotes the distance from upper actuator and lower actuator in the x-direction.

Each chain consists of upper closed-chain and lower closed-chain as Fig. 3. The upper chains connect the ball-socket joint ($^m\underline{C}$, $m=1,2,3$) of the top plate to the upper actuator (M1).

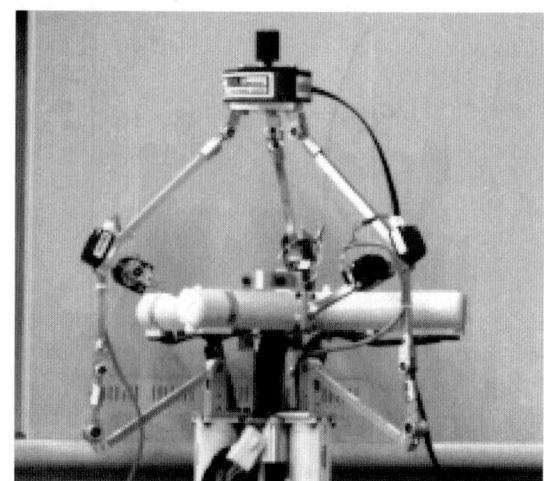

Fig. 1.   New 6-DOF Parallel Haptic Device

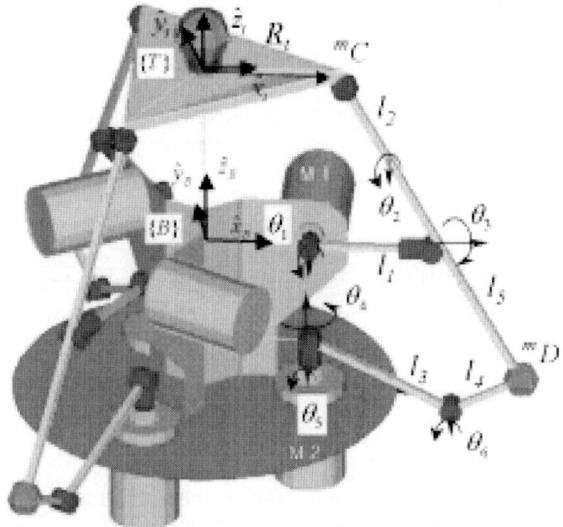

Fig. 2 Kinematic Structure of New 6 DOF Haptic Device

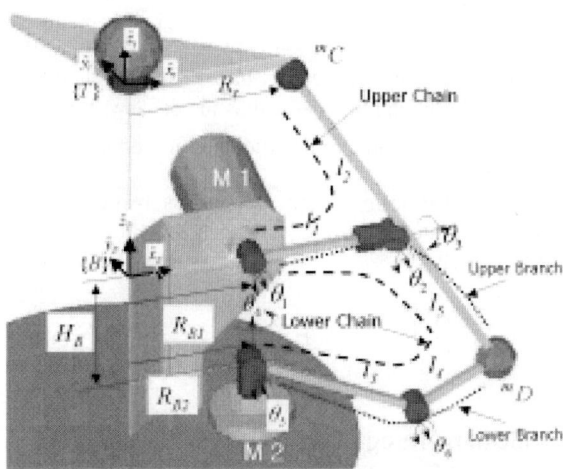

Fig. 3 Leg Structure of 6 DOF Parallel Haptic Device

And the lower chains connect the lower ball-socket joint ($^m\underline{D}$, $m=1,2,3$) to the lower actuator (M2). So upper actuators support gravity loads and generate the z-directional motion, and the lower actuators generate x-, and y-directional motions, respectively. Interaction of the three chains generates the Rotational motion.

We define the output displacement vector as

$$\underline{u} = (x_t, y_t, z_t, \theta_x, \theta_y, \theta_z)^T, \quad (1)$$

where (x_t, y_t, z_t) represent the positions of the origin of the top plate, and (θ_x, θ_y, θ_z) denotes $\hat{x}_t$-$\hat{y}_t$-$\hat{z}_t$ Euler angles equivalent to $[R_b^t]$ expressed by

$$[R_b^t] = [Rot(\hat{x}_t, \theta_x)][Rot(\hat{y}_t, \theta_y)][Rot(\hat{z}_t, \theta_z)]. \quad (2)$$

2.2 Forward Kinematics

General parallel mechanisms have many forward kinematic solutions, and this new 6-DOF parallel haptic device has 8-th order polynomial. So we place 3 additional encoders at passive joints(θ_2) of upper chains to obtain a unique forward solution.

2.3 Inverse Kinematics

We can solve the inverse kinematic solution of each upper chain from the position vectors of the upper ball-socket joint ($^m\underline{C}$), which are given from the position and orientation of the top plate. In the same manner, the inverse kinematic solution of each lower chain can be obtained from position vectors of the lower ball-socket joint ($^m\underline{D}$, $m=1,2,3$.), which are given from the forward kinematics of the upper chain.

2.4 First-order Kinematic Modeling

In the following, we describe the 1-*st* order KIC (Kinematic Influence Coefficient) as the relationship between the operational velocity ($\dot{\underline{u}}$) and the active joint velocity ($\dot{\underline{\theta}}_A$). Three upper contact points located at the upper ball-socket joints are denoted as

$$\underline{C} = (^1\underline{C}^T, {}^2\underline{C}^T, {}^3\underline{C}^T)^T, \quad (3)$$

where

$$^m\underline{C} = (^m_c x, {}^m_c y, {}^m_c z)^T. \quad (4)$$

Each upper contact point vector is expressed as

$$^m\underline{C} = \underline{u}_t + {}^m_c\underline{r}, \quad (5)$$

where

$$^m_c\underline{r} = [R_b^t] \, {}^m_c\underline{r}^{(t)}. \quad (6)$$

Differentiating $^m\underline{C}$ with respect to time results in

$$^m\underline{\dot{C}} = \underline{\dot{u}}_t + \underline{\omega} \times {}^m_c\underline{r}, \quad (7)$$

where

$$\underline{\dot{u}}_t = (\dot{x}_t, \dot{y}_t, \dot{z}_t)^T, \quad (8)$$

$$\underline{\omega} = (\omega_x, \omega_y, \omega_z)^T, \quad (9)$$

and

$$^m_c\underline{r} = (^m_c r_x, {}^m_c r_y, {}^m_c r_z)^T. \quad (10)$$

Eq. (7) can be written in a matrix form as

$$^m\underline{\dot{C}} = [^mG_u^c] \, \underline{\dot{u}}, \quad (11)$$

where

$$[^mG_u^c] = \begin{bmatrix} 1 & 0 & 0 & 0 & {}^m_c r_z & -{}^m_c r_y \\ 0 & 1 & 0 & -{}^m_c r_z & 0 & {}^m_c r_x \\ 0 & 0 & 1 & {}^m_c r_y & -{}^m_c r_x & 0 \end{bmatrix}, \quad (12)$$

$$\underline{\dot{u}} = (\dot{x}_t, \dot{y}_t, \dot{z}_t, \omega_x, \omega_y, \omega_z)^T. \quad (13)$$

Then the relationship between $\underline{\dot{u}}$ and $\underline{\dot{C}}$ can be described as

$$\underline{\dot{C}} = [G_u^c]\underline{\dot{u}}, \quad (14)$$

where
$$[G_u^c] = \left[[^1G_u^c]^T [^2G_u^c]^T [^3G_u^c]^T \right]^T. \quad (15)$$

The open-chain kinematics of each leg is described as

$$^m\underline{\dot{C}} = [^m_uG_\theta^c] \, ^m_u\underline{\dot{\theta}}, \quad (16)$$

where $^m_u\underline{\dot{\theta}}$ is the velocity vector of the joints in m-th upper chain.

Assuming no singularity in $[^m_uG_\theta^c]$, the first-order inverse kinematic formulation is obtained as

$$^m_u\underline{\dot{\theta}} = [^m_uG_c^\theta] \, ^m\underline{\dot{C}}. \quad (17)$$

Congregating from $m = 1$ to 3 yields

$$_u\underline{\dot{\theta}} = [_uG_c^\theta] \, \underline{\dot{C}}. \quad (18)$$

where

$$_u\underline{\dot{\theta}} = (\,^1_u\underline{\dot{\theta}}^T \;\, ^2_u\underline{\dot{\theta}}^T \;\, ^3_u\underline{\dot{\theta}}^T)^T,$$

and

$$[_uG_c^\theta] = \begin{bmatrix} [^1_uG_c^\theta] & 0 & 0 \\ 0 & [^2_uG_c^\theta] & 0 \\ 0 & 0 & [^3_uG_c^\theta] \end{bmatrix}.$$

2.5 Internal Kinematics

The lower closed chain consists of two branches which are constrained by the lower ball-socket joint ($^m\underline{D}$, $m = 1 \sim 3$), and the common velocity is described as

$$^m\underline{\dot{D}} = [^mG_{u\theta}^D] \, ^m_u\underline{\dot{\theta}} = [^mG_{l\theta}^D] \, ^m_l\underline{\dot{\theta}}, \quad (19)$$

where $[^mG_{u\theta}^D]$ and $[^mG_{l\theta}^D]$ represent the 1-st order KICs for the upper branch and the lower branch, respectively. From Eq. (18), we can obtain the first-order kinematic relationship between the upper branch and the lower branch as

$$^m_l\underline{\dot{\theta}} = [^mG_{l\theta}^D]^{-1}[^mG_{u\theta}^D] \, ^m_u\underline{\dot{\theta}} = [^mG_{u\theta}^\theta] \, ^m_u\underline{\dot{\theta}}, \; m = 1 \sim 3. \quad (20)$$

By using a row-column selection of Eq. (20), the first-order relationship between the active joint velocity ($^m\underline{\dot{\theta}}_A$: $^m\dot{\theta}_1, \, ^m\dot{\theta}_4, \, m = 1 \sim 3$) and the joint velocity ($_u\underline{\dot{\theta}}$) of the upper chains can be expressed as

$$\underline{\dot{\theta}}_A = [G_{u\theta}^A] \, _u\underline{\dot{\theta}}, \quad (21)$$

where

$$[G_{u\theta}^A] = \begin{bmatrix} 1 & 0 & 0 & 0 & 0 \\ [^1G_{u\theta}^\theta]_{1;} & 0 & 0 \\ 0 & 1 & 0 & 0 & 0 \\ 0 & [^2G_{u\theta}^\theta]_{1;} & 0 \\ 0 & 0 & 1 & 0 & 0 \\ 0 & 0 & [^3G_{u\theta}^\theta]_{1;} \end{bmatrix}_{(6\times 9)}, \quad (22)$$

$$_u\underline{\dot{\theta}} = [\,^1\theta_1 \, ^1\theta_2 \, ^1\theta_3 \, ^2\theta_1 \, ^2\theta_2 \, ^2\theta_3 \, ^3\theta_1 \, ^3\theta_2 \, ^3\theta_3]^T_{(9\times 1)}. \quad (23)$$

Substituting Eq. (17) into Eq. (21) yields the following relationship between the velocity of upper contact point the active joint velocity

$$\underline{\dot{\theta}}_A = [G_c^A] \, \underline{\dot{C}}, \quad (24)$$

where

$$[G_c^A] = [G_{u\theta}^A][_uG_c^\theta]. \quad (25)$$

And by substituting Eq. (14) into Eq. (24), the relationship between the operational velocity and active joint velocity is obtained as

$$\underline{\dot{\theta}}_A = [G_u^A] \, \underline{\dot{u}} \quad (26)$$

where

$$[G_u^A] = [G_c^A][G_u^c]. \quad (27)$$

Assuming no singularity in $[G_u^A]$, the forward Jacobian is obtained by matrix inversion as

$$\underline{\dot{u}} = [G_A^u] \, \underline{\dot{\theta}}_A. \quad (28)$$

2.6 Jacobian Scaling

General spatial motion involves both the translation and the rotation. Therefore, the Jacobian for this case has different units in the translational part and the rotational part. When the translational motion and the rotational motion are investigated separately, the result does not represent general 6-DOF motion characteristics. Many scaling techniques to treat the translational and rotational parts simultaneously have been proposed[9]. In this work, a Jacobian scaling technique based on the nominal link(l_{NL}) whose length is defined as the distance from the origin of the base frame to the midpoint of the top plate, is employed.

The scaled form of the Jacobian is related to the original jacobian as

$$[*G_A^u] = [S][G_A^u], \quad (29)$$

where

$$[S] = \begin{bmatrix} [I] & 0 \\ 0 & [l_{NL}]_{diag} \end{bmatrix}. \quad (30)$$

3. Kinematic Design Performance Index

Many aspects can be considered in the design of haptic device. Workspace, kinematic isotropy, and force transmission ratio are considered in this paper. Workspace means operating space for human operator, kinematic isotropy is a criteria which indicate how evenly the system moves in all directions. And force transmission ratio is also important to save input energy.

3.1 Workspace

One of the basic aspects of haptic device design is determining the workspace. The operating region or workspace of a manipulator is defined as a reachable and dexterous workspace. Also, a manipulator should be designed for the workspace property of well-connectedness, while allowing its end-effector to move from one regular point to another without passing through a critical value(i.e., singularity). A larger workspace enables the user to work in a wide range. This implies that this index should be maximized. The workspace volume is defined as

$$V = \int_V dV, \quad (31)$$

where the orientation angle of the top plate is defined as
$$-45° \leq \theta_x, \theta_y \leq 45°, \quad -90° \leq \theta_z \leq 90°. \quad (32)$$

3.2 Kinematic Isotropic Index

The Kinematic Isotropic Index is defined as
$$\sigma_I = \frac{\sigma_{min}(*[G_A^u])}{\sigma_{max}(*[G_A^u])}. \quad (33)$$

where σ_{min} and σ_{max} denote the minimum and the maximum singular values of $*[G_A^u]$, respectively. When σ_I approaches unity, the end-effector can generate uniform velocity in all directions. Also, global design index which represents the average of the manipulator's isotropic index over the whole workspace is defined as

$$\Sigma_I = \frac{\int_V \sigma_I dV}{\int_V dV}. \quad (34)$$

The greater Σ_I is, the better isotropy the mechanism has over the workspace. Therefore, Σ_I should be maximized.

3.3 Maximum Force Transmission Ratio

The maximum force transmission ratio implies the maximum magnitude of an actuator load required for the unit end-effector force, and it is defined as
$$\sigma_F = \sigma_{max}([*G_A^u]). \quad (35)$$

If σ_F becomes smaller, the actuator load will be reduced. This means that the manipulator can bear more weight with less actuators. The global design index for σ_F is defined as

$$\Sigma_F = \frac{\int_V \sigma_F dV}{\int_V dV}. \quad (36)$$

The smaller Σ_F is, the smaller capacity of the actuator is required. Thus, Σ_F should be minimized.

4. Optimum Design

In order to maximize the performances of the proposed haptic device, multi-criteria based design methodology based on genetic optimization algorithm are employed.

4.1 Composite Design Index(Cost function)

Several methodologies have been proposed to cope with multi-criteria based design. However various design indices are usually incommensurate concepts due to differences in unit and physical meanings, and therefore should not be combined unless they are transferred into a common domain. In consideration of this fact, a multi-criteria based design methodology employing a concept of composite design index is introduced[10]. This process consist of normalization which transfer various indices into a same domain and synthesis which combines several indices into one.

For V and Σ_I, the most favored preference is given the maximum value, and the least favored preference is given the minimum value of the criterion. Then, the preference design indices, $\tilde{V}$ and $\tilde{\Sigma}_I$ are expressed as

$$\tilde{V} = \frac{V - V_{min}}{V_{max} - V_{min}}, \quad (37)$$

$$\tilde{\Sigma}_{KSI} = \frac{\Sigma_{KSI} - (\Sigma_{KSI})_{min}}{(\Sigma_{KSI})_{max} - (\Sigma_{KSI})_{min}}, \quad (38)$$

where, "~" on each design index implies that it is transferred into the common preference design domain. Conversely, for $\tilde{\Sigma}_F$, the top preference is given the maximum value of the criterion. Then, the preference design indices, $\tilde{\Sigma}_F$ is expressed as

$$\tilde{\Sigma}_F = \frac{(\Sigma_F)_{max} - \Sigma_F}{(\Sigma_F)_{max} - (\Sigma_F)_{min}}. \quad (39)$$

To deal with this multi-criteria based design, a kinematic composite design index(KCDI) is employed which combines several individual preference design indices as a unique design index by using the max-min principle of fuzzy theory. KCDI is expressed as

$$KCDI = \min\{\tilde{V}, \tilde{\Sigma}_{KSI}, \tilde{\Sigma}_F\}. \quad (40)$$

KCDI is defined as the minimum value among the preference design indices calculated for a set of kinematic parameters. And a set of design parameters, which has the maximum value of the KCDI, is chosen as the optimal set of design parameters.

Table 1 Kinematic Constraint

Constant Parameters (mm)		
Motor Diameter	R_{B1}	$l_1 + l_2$
50	50	250
Variable Parameters (mm)							

l_2		l_3		l_4		l_5	
Min.	Max.	Min.	Max.	Min.	Max.	Min.	Max.
30	220	30	150	30	130	30	130

R_t		R_{B2}		H_B	
Min.	Max.	Min.	Max.	Min.	Max.
30	130	30	70	40	100

4.2 Design Parameters and Constraint

The Kinematic design parameters for the new 6-dof haptic device are the radius of the top plate, R_t, the link lengths of each leg, l_1, l_2, l_3, l_4, l_5, and the horizontal distance(R_{B2}) and vertical distance(H_B) from the base origin to the lower actuator($M2$).

These design parameters and space constraints are described in Table 1. Motor diameter, R_{B1}, and the sum of l_1 and l_2 are set to some constant values to provide the size limit of the whole system. And the other parameters are also bounded to guarantee the space of joint.

4.3 Optimization result

Genetic algorithm is employed to solve the nonlinear kinematic optimization for the new 6-DOF parallel haptic device[7]. We set the population size to 20. And in each generation we evaluate KCDI of each chromosome, select new population with respect to the probability distribution based on fitness values, and alter the chromosomes in the new population by mutation and crossover operators. After some number of generations, we obtain the best chromosome(parameter set) which represents an optimal solution. The optimization result is represented in Table 2. Figure 3 denotes the distribution of the kinematic indices resulting from optimization.

Table 2 Optimization Result

Design Parameters (mm)				
l_1	l_2	l_3	l_4	l_5
89.3	160.7	133.5	55.1	98.0
R_t	R_{B1}	R_{B2}	H_B	
37.7	50.0	50.0	50.0	
Performance				
Workspace (cm^3)	Average Kinematic Isotropy		Average Force Transmission Ratio	
1010.0	0.0158		0.00225	

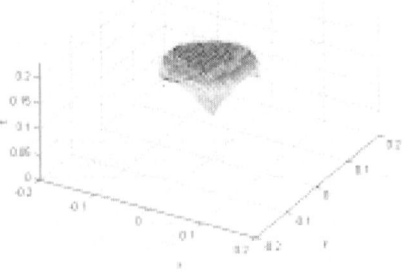

(a) Dexterous Workspace

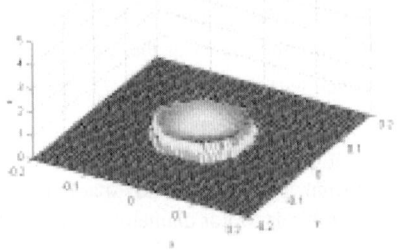

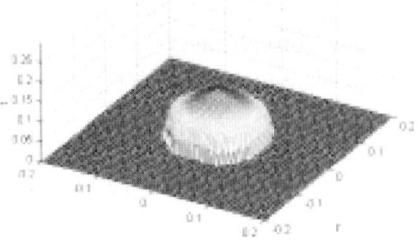

(b) Force Transmission Ratio

(c) Kinematic Isotropy

Fig. 4. Distribution of Performance Indices

4.4 Tolerance Analysis

General manipulators have inevitable joint tolerance due to manufacturing bottleneck. This tolerance causes undesirable motion or curtails performance, and thus we have to consider the design which minimizes its effect.

To analyze the effect of tolerance on the upper plate, most severe joints causing undesirable displacement have to be identified. For the designed device, the tolerance of the joint 4 of each chain is notable. Thus a Jacobian relating the output to the three pseudo-joints is expressed as

$$\delta \underline{u} = [G_{\theta_t}^u] \, \delta \underline{\theta}_t. \tag{41}$$

The displacement of upper plate due to tolerance at joint 4 is displayed in Fig. 6. In this example, the error range of the joint tolerance is $\pm 0.5°$, and the initial position of the upper plate is $u = [0 \ 0 \ 0.5]^T$. To minimize this tolerance effect, a design index representing the size of the tolerance effect should be incorporated into the optimization problem.

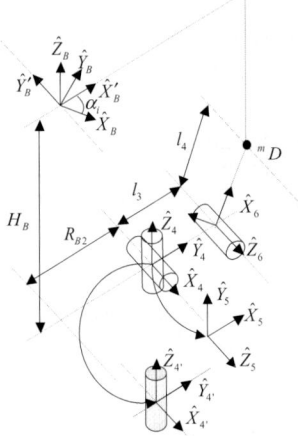

Fig. 5. Coordinate Including Joint with Tolerance

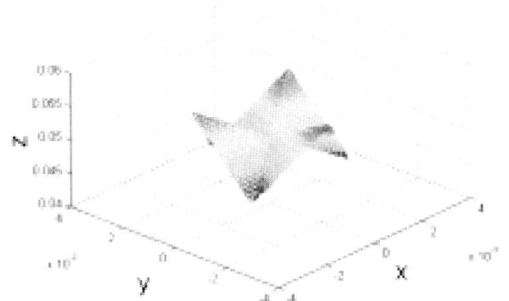

Fig. 6. Errors in Upper Coordinate by Joints with Tolerance

5. Actuator Sizing

Actuator size is defined as maximum joint torque enough to withstand the given external load and gravity, and can be computed as following[11].

The output load is bound as

$$T_u = \|\underline{T}_u\|_2 = (\underline{T}_u[W]\underline{T}_u)^{1/2} \leq (T_u)_{max}, \quad (42)$$

with $(\theta_i)_{min} \leq \theta_i \leq (\theta_i)_{max}$, $i = 1, 2, \cdots, N_p$ (43)

where $(T_u)_{max}$ is the maximum external load, θ_i is the angle of the i-th joint, and N_p is the number of all joints. The actuator torque due to the external load is expressed as

$$(T_{An})_{ext} = (T_u)_{max} ([G_A^u]_{,n}^T [W]^{-1} [G_A^u]_{,n})^{\frac{1}{2}}. \quad (44)$$

Consequently, the required torque at the n-th actuator is the sum of $(T_{An})_{ext}$ supporting the external load and T_{An}^G supporting the gravity load as

$$T_{An}^M = \max\left\{\left|T_{An}^G + (T_{An})_{ext}\right|, \left|T_{An}^G - (T_{An})_{ext}\right|\right\}. \quad (45)$$

The maximum T_{An}^M for the entire workspace is the appropriate actuator size. To generate the force of $20\,N$ and the torque of $1.0\,Nm$ at the end effector, the actuator sizes are calculated as in Table 3.

Table 3. Actuator Size

Payoad at the top plate	Actuator Torq.(Nm)	
Force(20.0 N)	T_{A1}	2.88
	T_{A2}	12.03
Torq.(1.0 Nm)	T_{A1}	3.79
	T_{A2}	6.79
Force(20.0 N) + Torq.(1.0 Nm)	T_{A1}	4.53
	T_{A2}	13.82

6. Conclusion

A new 6-dof parallel haptic device which is actuated by non-floating actuators is proposed. This device is adequate for haptic device due to light weight and high force-reflecting capability. To design the haptic mechanism, a multi-criteria based kinematic optimal design problem is formulated, which simultaneously considers various indices. And the optimal solution is found by employing Genetic Algorithm. Actuator sizing is also carried out to complete the design problem. More sophisticated design minimizing the loss of position accuracy due to the joint clearance has to be further studied. And application of the purposed haptic device to manipulation of virtual reality is ongoing subject.

Acknowledgement

This work was supported by grant No. (2000-2-30200-008-3) from the Basic Research Program of the Korea Science & Engineering Foundation.

Reference

[1] R.C. Goertz, "Fundamentals of General-Purpose Remote Manipulators," Journal of Nucleonics, Vol. 10, No. 11, pp.36-42, 1952.
[2] R.E. Ellis, O.M. Ismaeil, and M.G. Lipsett, "Design and Evaluation of High-Performance Haptic Interface", Robotica, Vol.4., 1996.
[3] T. Massie, K. Stalisbury, "PHANToM Haptic Interface: A Device for Probing Virtual Objects", ASME Journal of Dynamic System and Control, New York, NY, pp.295-299, 1994.
[4] P. J. Berkelman, R. L. Hollis, and S. E. Salcudean, "Interacting with Virtual Environments using a Magnetic Levitation Haptic Interface", Proceedings of IEEE International Conference on Intelligent Robots and Systems, pp. 117-122, piscataway, NJ, 1995.
[5] M. Bergamasco, et al., "An Arm Exoskeleton System for Teleoperation and Virtual Environments Applications," Proceedings of International Conference On Robotics and Automation, San Diego, California, pp. 1449-1454, 1994.
[6] M. Ishii and M. Sato, "A 3D Spatial Interface Device Using Tensed Strings," Presence-Teleoperators and Virtual Environments, Vol.3, No.1, MIT Press, Cambridge, MA, pp.81-86, 1994.
[7] Zbigniew Michalewicz, "Genetic Algorithms + Data Structures = Evolution Programs," Springer, 1996.
[8] J.M. Hollerbarch, "Some Current Issues in Haptic Research," Proceedings of International Conference On Robotics and Automation, pp. 757-762, 2000.
[9] Leo J. Stocco, S. E. Salcudean, and F. Sassani, "On the Use of Scaling Matrices for Task-Specific Robot Design", IEEE Transactions On Robotics And Automation, Vol. 15, No. 5, October 1999.
[10] J.H. Lee, B-J. Yi and S.R. Oh, "Optimal Design of a Five-Bar Finger with Redundant Actuation," Proceedings of International Conference On Robotics and Automation, pp. 2068-2074, 1998.
[11] M. Thomas, H.C. Yuan-Chou, and D. Tesar, "Optimal Actuator Sizing for Robotic Manipulators Based on Local Dynamic Criteria," ASME Journal of Mechanisms, Transmissions and Automation in Design, Vol. 107, pp.163-169, 1985.

Wearable Master Device
Using Optical Fiber Curvature Sensors for the Disabled

Kyoobin Lee*, Dong-Soo Kwon**

*: Dept. of Mech. Eng. KAIST (Tel: +82-42-869-3082; Fax: +82-42-869-3095; E-mail: leekb@robot.kaist.ac.kr)
**: Dept. of Mech. Eng. KAIST (Tel: +82-42-869-3042; Fax: +82-42-869-3210; E-mail: kwonds@me.kaist.ac.kr)

Abstract

This paper addresses a wearable master device for physically handicapped persons whose arms are disabled. Optical fiber curvature sensors are used to measure the human body motion. For the developed wearable master device, a calibration and mapping method of the sensors is proposed to extract 2-DOF human shoulder motions. An experiment shows that the wearable master device can be used for a 2-DOF input device effectively for handicapped persons. It has also shown that a subject can control a mobile robot with the wearable master device.

1. Introduction

In order to aid handicapped persons in the basic tasks of their daily lives, various assistant devices have been studied and developed. Such devices assist handicapped persons to move themselves to another place, to handle an object or to eat meal. Several rehabilitation robot manipulators have been developed[1][2] for handicapped persons. Although the most popular type of master device for rehabilitation robotic system is the joystick type, the joystick is not appropriate for physically handicapped persons who cannot move their arms. Therefore, it is necessary to develop a convenient and portable master device that can measure human body motion.

Optical fiber curvature sensors are used to measure human body motion. Although we can use image processing systems or absolute position sensor systems to measure the angle of body joints, these systems are expensive and bulky. Optical fibers seem to be well-suited for sensors that measure body motions of handicapped persons because of their flexibility, lightness, small size and low cost.

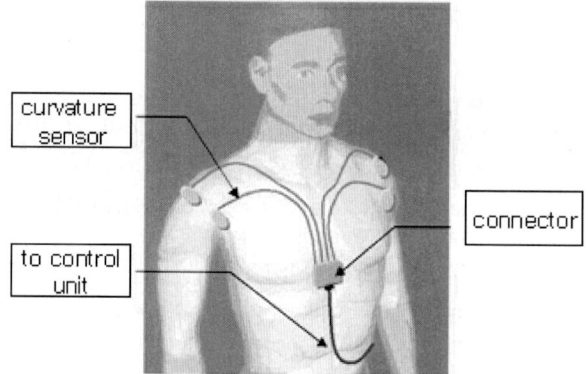

Figure 1. A conceptual sketch of wearable master device

A conceptual sketch is shown in Fig. 1. Optical fiber curvature sensors are attatched on an upper garment.

2. Optical fiber curvature sensor

Optical fiber curvature sensors are used for measuring human body motions. Transmitted light power in the optical fiber decreases exponentially when the fiber is curved [3][4]. This characteristic of the optical fiber is essential for curvature sensing. By using several optical fiber curvature sensors, it is possible to measure two or more degrees of freedom of motion.

The optical fiber curvature sensor designed for this research is composed of five parts: a light source, a light detector, an optical fiber, an electric circuit and a power source.

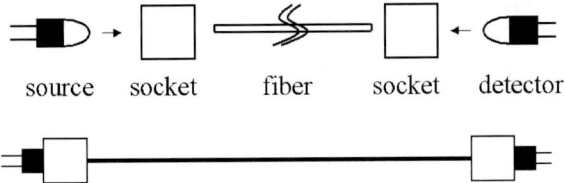

Figure 2. Assemblage of optic fiber curvature sensor

The light source and the detector are an IR LED and a phototransistor, respectively and the electric circuit consists of one OP-amp and several resistances and capacitors.

Power loss of the light in the optical fiber is related with the amount of bent angle, regardless of the bending direction. Methods to measure both the amount of bending and the bending direction are proposed.

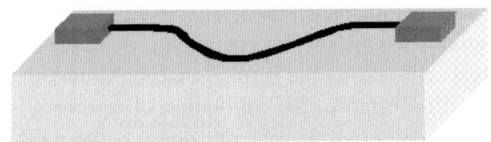

Figure 3. Configuration for measuring the bending direction

First, if the fiber is pre-bent as shown in figure 3, we can distinguish the bending direction.

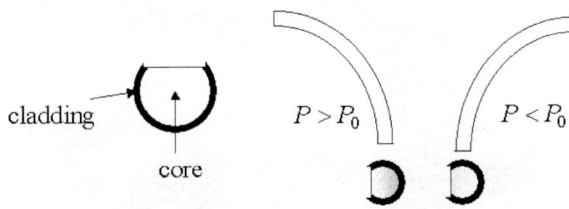

Figure 4. Surface treatment for measuring the bending direction

Second, if one side of the cladding is taken off as shown in Fig. 4, the amount of the transmitted light power varies according to the bending direction. When the fiber is bent to the harmed side, the transmitted light power is increased and when the fiber is bent to the opposite side, the transmitted light power is decreased[5].

3. A Wearable Master Device Using Optical Fiber Curvature Sensors

A tight upper garment is made to which sensors are attached and two sensors are attached as shown in Fig. 5. The positions of sensors were experimentally determined where maximum variation of the sensor outputs is met.

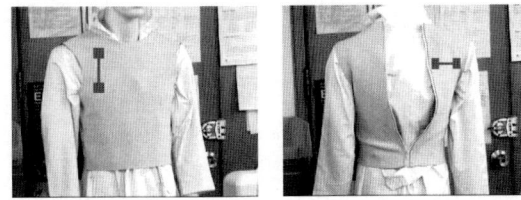

Figure 5. Locations of curvature sensors

The upper garment should be tight to transmit human body motion to the optical fiber curvature sensors. If the upper garment is loose, the position of the optical fiber curvature sensor relative to human body can be changed every moment. It makes it impossible to measure the human body motion from data of optical fiber curvature sensors.

The position of the optical fiber curvature sensor is selected by searching the position where the amount of change of sensor data is met.

4. An Artificial Neural Network for Mapping from Sensor Data to Available Position Commands

A simple artificial neural network is used for mapping function from sensor data measured from the human shoulder motion to available position commands. The neural network makes it possible that the sensor controller adjusts its mapping function according to different users. A system block diagram is shown in Fig. 6.

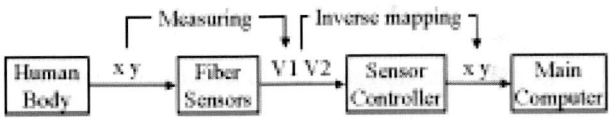

Figure 6. System block diagram

The ANN(artificial neural network) is used in the 'sensor controller' block. The sensor controller generates the position command from sensor data. Inputs of the ANN are two sensor data. Outputs of the ANN are coordinates in the workspace. The coordinates have to be continuous for proportional control. Although we can use only direction information for controlling 2-DOF motion, the amount of direction is useful for more efficient command inputs. For example, analogue joysticks are more helpful in computer tasks than digital joysticks (directions only). Therefore, the ANN must be the inverse model of the 'fiber sensors' block. Weights of the ANN are learned by the mechanism as shown in Fig. 7.

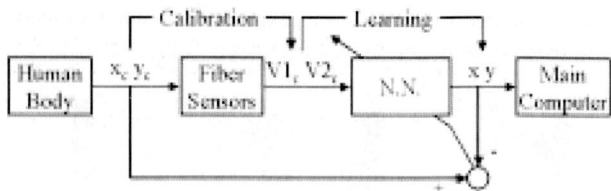

Figure 7. Learning of the artificial neural network

The ANN is a multi-layered perceptron. It has one hidden layer and 6 neurons in the hidden layer (Fig. 8).

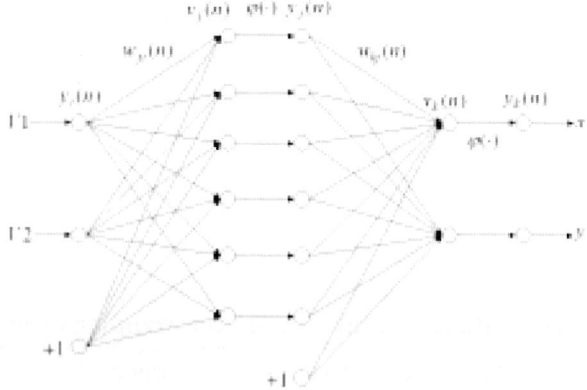

Figure 8. Signal-flow graph of the ANN

The activating function is as follows.

$$\varphi_j(v_j(n)) = \frac{1}{1+\exp(-v_j(n))} \quad (1)$$

where

$v_j(n)$ is the weighted sum of all synaptic inputs plus the bias.

Teachers of the ANN are position data and sensor data that are measured during the calibration process. The learning algorithm of the ANN is a steepest descent rule including a momentum term as follows[6].

$$\Delta w_{ji}(n) = -\eta \frac{\partial E(n)}{\partial w_{ji}(n)} + \alpha \Delta w_{ji}(n-1) \quad (2)$$

where

w_{ji} is the weight.

η is the learning-rate parameter.

α is the momentum constant

$$E(n) = \frac{1}{2}\sum_{j \in C} e_j^2(n)$$

$e_j(n) = d_j(n) - y_j(n)$, neuron j is an output node.

$d_j(n)$ is the j-th desired output.

$y_j(n)$ is the j-th output of the ANN.

5. Experiments

Two experiments have been carried out with the wearable master device (upper garment) shown in section 3. The first experiment is to move a cursor to five target points. The second experiment is to control a mobile robot.

In experiments, the first process is calibration. Every time the subject puts on the wearable master, the position of the wearable master w.r.t the human body is slightly changed. Since the amount of bending of a human shoulder is quite small, the change of the position of the wearable master causes the change of the coordinates of the position command dramatically. So calibration process is essential when the subject puts on the wearable master.

Short calibration time is important for human-machine interface. For instance, the calibration of joysticks is very short and easy. For the wearable master, the calibration points are chosen as shown in Fig. 9. In the calibration process, the subject moves his shoulder according to the instruction of the computer. The learning-rate parameter, η is set to 0.1. The momentum constant, α is set to 0.1. The number of epochs is 30000.

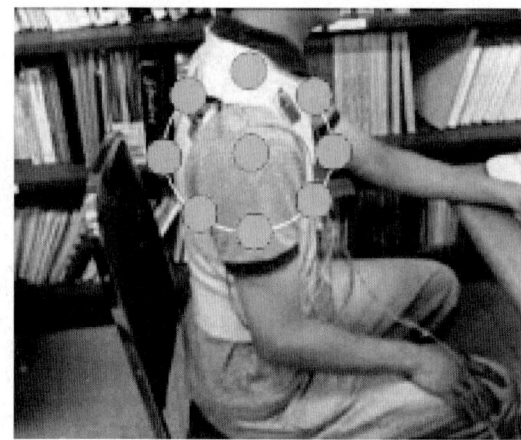

Figure 9. Calibration points

The first experiment is carried out as follows. To show the performance of the wearable master, a computer monitor displays five target points and one cursor point that corresponds to the movement of a right shoulder of a subject. The purpose of this experiment is that the subject should move his right shoulder to move the cursor to the target points. An experimental result is shown in Fig. 10.

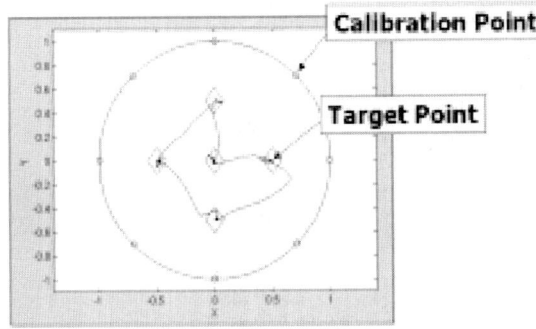

Figure 10. A result of an experiment on a human body

Fig. 10 shows that the trajectory of the cursor can be moved to the five target points very closely. The average position error is 0.32% of the workspace. The resolution is 0.14%. The elapsed time is 14 seconds. This result indicates with promise that the proposed wearable master device with optical fiber curvature sensors can be used as an input device for the disabled.

The second experiment is to control a mobile robot. The robot used in this experiment is a two-wheeled mobile robot developed in 'Intelligent Control' laboratory in KAIST as shown in Fig. 11. It has a radio frequency (RF) wireless communication module. The motors of the robot have encoders. The main computer sends velocity commands to the robot and a PID-controller in the robot drives the two motors in rate mode.

Figure 11. Two-wheeled mobile robot

Fig. 12 shows how to map from the human shoulder motion to the robot motion. When the subject moves his right shoulder up, down, forward and backward, the robot moves forward and backward and turns left and right, respectively. The number of direction is fixed by four. Of course, we can make the mapping function to control the direction finely, but it makes the subject to feel more difficult to control the mobile robot.

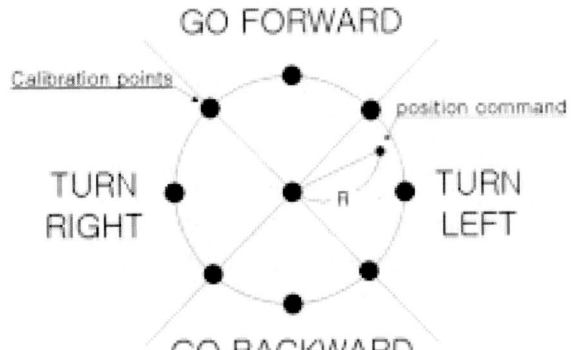

Figure 12. Mapping from position command to robot motion

The rate of the movement is proportional to exponent of the distance from neutral position, R as follows.

Linear velocity, $V = 0.5(e^{4R} - 1)\, cm/s$ (3)

Angular velocity, $\omega = 0.3(e^{4R} - 1)\, rad/s$ (4)

The exponential mapping makes it possible that the subject can control the robot finely.

The experimental result of Fig. 13 is made by overlaying frames captured every 1.2 second. The elapsed time is 24 seconds. The subject can control the speed and direction of the mobile robot comfortably.

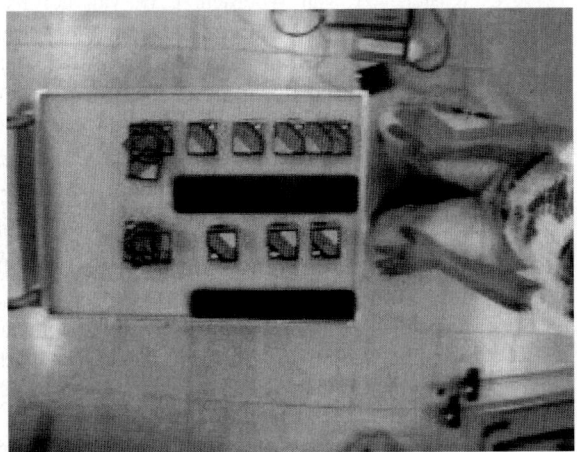

Figure 13. An experiment to control a mobile robot

6. Conclusions

The development of a wearable master device for handicapped persons depends on having sensors that are small, light, and attached simply to garments.

Optical fiber curvature sensors are used to measure

the human body motion. Despite the problems of ambiguousness and coupled output of optical fiber curvature sensors, useful 2 DOF motion is extracted with 2 sensors through a sensor controller using an artificial neural network. The resolution of the wearable master device is lower than 0.2% of the workspace. An experiment to control a two-wheeled mobile robot like a wheelchair is carried out successfully.

With these promising results, physically handicapped persons whose arms are disabled will be able to control motorized wheelchairs or rehabilitation robotic systems with lightweight and simple wearable master device that is equipped with optical fibers.

Acknowledgment

The authors are grateful for the support provided by MOST, National R&D Program (Critical Technology 21), Program ID: 99-J10-01-01-A (Development of Service Robot Technology).

REFERENCES

[1] Won-Kyung Song, He-Young Lee, Jong-Sung Kim, Yong-San Yoon, and Zeungnam Bien, "KARES: intelligent rehabilitation robotic system for the disabled and the elderly," Engineering in Medicine and Biology Society, 1998. Proceedings of the 20th Annual International Conference of the IEEE, Volume: 5 , 1998 , Page(s): 2682 -2685 vol.5

[2] J.C. Rosier, J.A. van Woerden, L.W. van der Kolk, B.J.F. Driessen, H.H. Kwee, J.J. Duimel, J.J. Smits, A.A. Tuinhof de Moed, G. Honderd, and P.M. Bruyn, "Rehabilitation robotics: the MANUS concept," Advanced Robotics, 1991. 'Robots in Unstructured Environments', 91 ICAR., Fifth International Conference on , 1991 , Page(s): 893 -898 vol.1

[3] P. Halley, "LES SYSTÈMES À FIBRES OPTIQUES", Editions Eyrolles.

[4] V. Louis, P. Le-Huy, J.M. Andre, M. Abignoli, and Y. Granjon, "Optical Fiber Based Sensor for Angular Measurement in Rehabilitation," Systems, Man and Cybernetics, 1993. 'Systems Engineering in the Service of Humans', Conference Proceedings., International Conference, Vol. 1 ,1993, pp.153-7

[5] A. Djordjevich and YuZhu He, "Thin structure deflection measurement," Instrumentation and Measurement, IEEE Transactions on, Volume: 48 3 , June 1999 , Page(s): 705 –710

[6] M. Hagiwara, "Theoretical derivation of momentum term in back-propagation," International Joint Conference on Neural Networks, vol. I, 1992, pp. 682-686, Baltimore

Maneuverability of A Flat-Streamlined Underwater Vehicle

Jenhwa. Guo, Forng-Chen Chiu
Department of Naval Architecture & Ocean Engineering
National Taiwan University
73 Chou-Shan Road. ,Taipei ,Taiwan, R.O.C.
jguo@ccms.ntu.edu.tw

Abstract - Maneuverability describes a vehicle's ability to change course or turn. Maneuverability of conventional underwater vehicles, such as torpedoes, can be determined by altering the position and length of control fins. To perform large-area surveying tasks, autonomous underwater vehicles (AUVs) generally require different maneuverability characteristics in their vertical and horizontal planes of motion. Furthermore, AUVs are significantly slower than torpedoes, and control fins are relatively ineffective at slow speeds. While relying solely on control fins to determine the maneuverability, this study investigates the maneuverability characteristics of a flat-streamlined underwater vehicle. A Planar Motion Mechanism (PMM) testing system is adopted to conduct a series of captive model tests in order to measure the stability derivatives of the vehicle, AUV-HM1. Stability and maneuvering indices are then derived from the measured data of stability derivatives. Finally, maneuvering criteria of a flat, baseline vehicle are evaluated using a prediction method.

Keywords: AUVs, Maneuverability, Stability

I. INTRODUCTION

Several "flat-fish" type autonomous underwater vehicles (AUVs) have been developed recently. The mission designation of PTEROA, the vehicle developed by the group at the University of Tokyo (Ura, 1989) is two fold: high gliding performance, and the ability to maintain a steady altitude while cruising over the sea bed. The vehicle AQUA EXPLORER developed at the KDD has been designed to track submarine telecommunication cables. (Asakawa et al., 1993). Furthermore, MARIUS was developed for environmental surveying and the acquisition of oceanographic data in coastal waters (Pascoal et al., 1997). Meanwhile, an AUV named VORAM is being developed at KRISO for observing and investigating the sea bed (Lee et al. , 1998). This work investigates the maneuverability of a flat, streamlined AUV, AUV-Hai-Min (Guo et al., 1995). The designated mission of the AUV-HM1 includes area searches and surveys in shallow water. A typical search/survey scenario requires good responsiveness to control in the horizontal plane of motion, and good ability to maintain altitude in the vertical plane. This study aims to further reveal the maneuverability and stability of vehicles with flat cross-sections, using experimental and theoretical methods.

Maneuverability measures the effectiveness of control inputs that are directly related to hull form and fin size. Maneuverability must be analyzed during the preliminary design stage. This work performed straight-line captive-model experiments in the vertical and horizontal planes of motion to find the stability and control characteristics of the autonomous underwater vehicle, AUV-HM1. The experiments were performed for the following configurations, excluding propellers: (1) the hull plus control surfaces in the vertical plane of motion, (2) the hull plus control surfaces in the horizontal plane. The experiments were performed in a model basin on a towing carriage (130m x 8m x 4m) using a Vertical Planar Motion Mechanism (VPMM). Captive model tests included angle of attack and elevator angle tests in the oblique-towing mode; and pure heave, pure pitch, and combined pitch/heave tests in the forced motion mode. The hydrodynamic force and moment measurements were non-dimensionalized using the total length of the vehicle body. Stability derivatives are referred to axes which originate at a point 1.0 meter aft of the nose on the hull centerline. The maneuvering behavior of the vehicle in its vertical and horizontal configurations is then determined using stability indices derived from the experimental data.

II. TESTING MODEL

The testing model has the same dimensions as the AUV-HM1, as listed in Table 1. Meanwhile, Figure 1 illustrates the horizontal and longitudinal center planes, with all cross sections being elliptical. Following Landweber et al. (1950), the longitudinal center plane is a sixth-degree polynomial form. Figure 2 presents configurations of the elevators and fixed vertical fins. The fin profiles selected are NACA0012, and the model is supported by a pair a vertical struts. The force gauge located beneath the struts was used to measure the longitudinal, normal, lateral forces, and rolling moment with respect to the body axes, and the pitching moment around the reference point was determined from the

Table 1 Principal particulars of the testing model

Length (body/overall)	2.0/2.0 m
Breadth (body/including elevators)	1.0/1.6 m
Height	0.6/0.6 m
Project area of body	1.791 m^2
Wetted surface area of body	4.38 m^2
Displaced volume of body	0.677 m^3
Centroid of body	0.909m from nose
Project area of horizontal fins	0.091 m^2
Project area of vertical fins	0.148 m^2

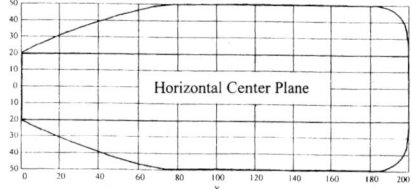

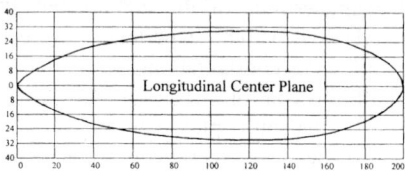

Fig. 1 Profile of the body

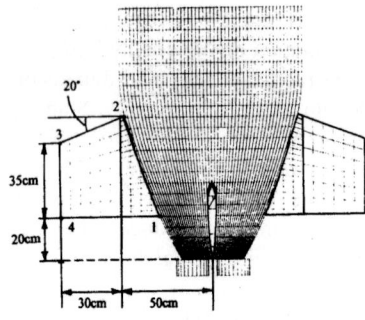

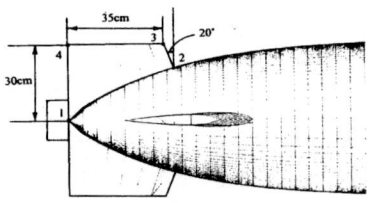

Fig. 2 Profile of fins and elevators

difference in the measured reaction forces at each strut multiplied by half of the distance between the struts. The distance between the struts is 1.0 m, the center of rotation is 0.778 m from the nose, and the model is submerged at a depth of 2.0 m.

III. EQUATIONS OF MOTION

1. Coordinate System

1.1 Configuration A

A set of body-fixed coordinate axes is used to describe the motion of the AUV-HM1. Figure 3 displays the positive direction of the axes and angles, while α denotes the angle of attack in the vertical plane. Furthermore, δ_e represents the angular deflection of the vertical fins. The z-axis is defined as being positive in the downwards direction, and the longitudinal and transverse horizontal axes of the AUV are represented by the x- and y-axes, respectively. For hydrodynamical analysis, the origin of the coordinate axes is located at the center of rotation of the test mechanism. For maneuvering analysis, stability derivatives are calculated with respect to the axes which originate from a point 1.0 m aft of the nose on the hull centerline using coordinate transformation.

1.2 Configuration B

To measure the horizontal plane of motion, the model is heeled starboard side down by 90 degrees. The coordinate axes defined in this configuration are illustrated in Fig. 4.

2. Linearized Maneuvering Equations of Motion

By reference to the body axes, the linearized equations of motion can be expressed as

$$(m+m'_z)\dot{w}' - Z'_w w' - (m'x'_G + m'_z x'_z)\dot{q}' \\ -\{Z'_q + (m'+m'_x)\}q' = Z'_\delta \delta \\ (I'_{yy} + J'_{yy})\dot{q}' - \{M'_q - (m'x'_G + m'_z x'_z)\}q' \\ -(m'x'_G + m'_z x'_z)\dot{w}' - M'_w w' = M'_\delta \delta \quad (1)$$

where the control mechanisms are
$Z_\delta = Z'_{\delta e}$, $M_\delta = M'_{\delta e}$, and $\delta = \delta_e$ for configuration A; and $Z_\delta = 0$, $M'_\delta = \dfrac{\ell'}{2}$, and $\delta = \Delta T$ for configuration B. The non-dimensional coefficients are defined as

$m' = m/0.5\rho L^3, m'_x = m_x/0.5\rho L^3, m'_z = m_z/0.5\rho L^3,$

$x'_G = x_G/L, x'_z = x_z/L, I'_{yy} = I_{yy}/0.5\rho L^5,$

$J'_{yy} = J_{yy}/0.5\rho L^5, Z'_w = Z_w/0.5\rho L^2 U,$

$M'_w = M_w/0.5\rho L^3 U, \dot{Z}'_q = Z_q/0.5\rho L^3 U,$

$M'_q = M_q/0.5\rho L^4 U, Z'_{\delta e} = Z_{\delta e}/0.5\rho L^2 U^2,$

$M'_{\delta e} = M_{\delta e}/0.5\rho L^3 U^2, \Delta T' = \Delta T/0.5\rho L^2 U^2, \ell' = \ell/L$

, and $\rho = 101.82 kg_f \sec^2/m^4$ for 20°C fresh water.

where m denotes the mass of the model with its enclosed water; x_G represents the x coordinate of the center of gravity; m_x is the added mass in x-axis; m_z is the added mass in the z-axis; x_z denotes the x coordinate of the center of gravity in the z-axis; I_{yy} represents the pitching moment of inertia including enclosed water; J_{yy} is the added moment of inertia in the pitch; Z_w, Z_q and M_w, M_q are derivatives of the normal force and moment components with respect to the linear velocity in z-direction w and angular velocity in y-direction q respectively; $Z_{\delta e}$ and $M_{\delta e}$ are the force and moment rate caused by control surface displacement δ_e; ΔT is the thrust force difference; ℓ is the distance between thrusters; and U is the forward speed.

IV. MODEL EXPERIMENTS

1. Test Apparatus and Test Conditions

The apparatus for measuring hydrodynamic force and moment is capable of performing ± 150mm in heaving amplitude, ± 15deg in pitching, 0.2 ~ 1.5Hz in oscillation frequency. The capacity of the force gauge is ± 300kgf. Test conditions are as follows:
a. forward speed : 1.414, 2.000, 2.828 (m/sec)
b. oscillation frequency:0.3,0.4,0.5 (Hz)
 pure heaving amplitude : 20, 40, 60(mm)
 pure pitching amplitude : 2.0, 4.0, 6.0 (deg.)
 combine pitching amplitude : 2.0, 4.0, 6.0 (deg.)
c. angle of attack for oblique towing tests : -6.0 ~ +14.0 (deg.)

2. Data Presentation

Non-dimensional data are presented in tabular form herein. Table 2 summarizes the results of oblique towing tests and PMM tests using the model in configuration A. The accuracy of experiments is discussed by Chiu et al. (1996, 1997). These derivatives are referred to axes which originate at the center of rotation of the test apparatus. Table 3 lists data obtained using the model in configuration B, while Table 4 presents the stability derivatives referred to the origin for maneuvering analysis.

Table 2 Results of oblique towing tests and PMM tests in configuration A
(a) Results of oblique towing tests

U(m/sec)	Z'_w	M'_w
1.414	-0.664	0.00521
2.000	-0.673	-0.00338
2.828	-0.655	-0.00197

(b) Added inertia obtained by PMM tests

m'_x	m'_z	x'_z	J'_{yy}
0.023	0.077	-0.107	0.0044

(c) Damping terms obtained by PMM tests

Z'_w	M'_w	Z'_q	M'_q
-0.0874	0.0194	-0.0424	-0.0260

(d) Results of elevator force tests

	$Z'_{\delta e}$			$M'_{\delta e}$		
α(deg) \ U(m/s)	0	3	6	0	3	6
1.414	-0.169	-0.167	-0.184	-0.0524	-0.0497	-0.0552
2.000	-0.176	-0.159	-0.178	-0.0531	-0.0478	-0.0575
2.828	-0.182	-0.165	-0.188	-0.0529	-0.0515	-0.0584

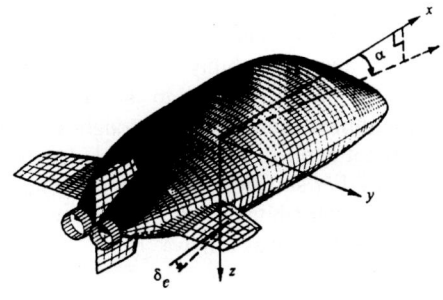

Fig. 3 Configuration A

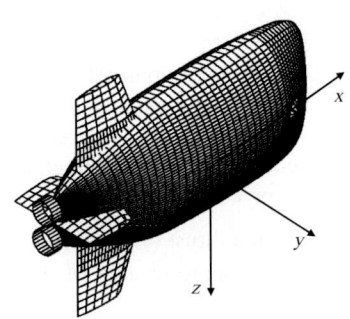

Fig. 4 Configuration B

Table 3 Results of oblique towing tests and PMM tests in configuration B
(a) Results of oblique towing tests

U(m/sec)	Z'_w	M'_w
1.414	-0.0998	0.0162
2.000	-0.0981	0.0242
2.828	-0.0902	0.0288
Average	-0.096	0.023

(b) Added inertia obtained by PMM tests

m'_x	m'_z	x'_z	J'_{yy}
0.023	0.077	-0.107	0.0044

(c) Damping terms obtained by PMM tests

Z'_w	M'_w	Z'_q	M'_q
-0.0874	0.0194	-0.0424	-0.0260

Table 4 Non-dimensionalized stability derivatives Obtained by coordinate transformation

.	Configuration A	Configuration B
m'_x	0.023	0.023
m'_z	0.239	0.077
x'_z	-0.022	0.004
J'_{yy}	0.0117	0.00352
Z'_w	-0.673(O.T.)	-0.0960(O.T.)
M'_w	0.0712(O.T.)	0.0337(O.T.)
Z'_q	-0.156	-0.0317
M'_q	-0.078	-0.0250
$Z_{\delta e}$	-0.176	----
$M_{\delta e}$	-0.0336	----

*Origin is located at 1.000m from nose
**O.T. represents results from oblique towing tests

Table 5 Non-dimensionlized stability and turning indices calculated from stability derivatives

Stability Indices	Configuration A	Configuration B
T_1	0.693	0.413
T_2	0.289	-2.987
T_3	-0.415	-2.552
T	1.397	-0.021
l_q	2.083	0.159
l_w	0.106	0.351
G	0.949	-1.21
K	-0.754	-3.27

*Origin is located at 1.000m from nose
$m = 0.168$; $x_G = 0$; $I_{yy} = 0.0113$

V. MANEUVERING CRITERIA

Equation (1) can be written as a decoupled second order equation, as follows:

$$T'_1 T'_2 \ddot{q} + (T'_1 + T'_2)\dot{q} + q = K'\delta' + K'T'_3 \dot{\delta}' \quad (2)$$

where

$$T'_1 T'_2 = \frac{(m' + m'_z)(I'_{yy} + J'_{yy}) - (m'x'_G + m'_z x'_z)^2}{[M'_q - (m'x'_G + m'_z x'_z)]Z'_w - \dot{M}'_w[Z'_q + (m' + m'_x)]}$$

$$T'_1 + T'_2 = \frac{-(m' + m'_z)[M'_q - (m'x'_G + m'_z x'_z)] - (I'_{yy} + J'_{yy})Z'_w}{[M'_q - (m'x'_G + m'_z x'_z)]Z'_w}$$
$$\frac{-(m'x'_G + m'_z x'_z)[Z'_q + (m' + m'_x)] - M'_w(m'x'_G + m'_z x'_z)}{- M'_w[Z'_q + (m' + m'_x)]}$$

$$K' = \frac{M'_w Z'_{\delta e} - M'_{\delta e} Z'_w}{[M'_q - (m'x'_G + m'_z x'_z)]Z'_w - M'_w[Z'_q + (m' + m'_x)]}$$

$$T'_3 = \frac{(m'x'_G + m'_z x'_z)Z'_{\delta e} - M'_{\delta e}(m' + m'_z)}{M'_w Z'_{\delta e} - M'_{\delta e} Z'_w}$$

T'_1, T'_2 are roots of the characteristic equation of Eq. (2), and are generally referred to as stability indices. Meanwhile, K' is the index of turning ability. For maneuvering at low frequencies, Eq.(2) can be approximated by a first order equation, with time constant $T' \equiv T'_1 + T'_2 - T'_3$,

$$T'\dot{q}' + q' = K'\delta' \quad (3)$$

Responsiveness to control inputs is an appropriate definition of maneuverability. Consequently, the maneuverability index is taken to be the extent of heading angle turned from an initial straight course, per unit control input applied, after the vehicle has traveled one body length. For situations in which the controls are fixed, the maneuverability index P (Nomoto and Norrbin, 1965) is defined as follows

$$P = \frac{\theta}{\delta} = K\left[1 - T + T e^{-1/T}\right]$$
$$\approx \frac{1}{2}\frac{K}{T} \quad (4)$$

where θ denotes the pitching response of the vehicle after moving one body length. The P value in configuration A is -0.27, implying approximately 2.7 degrees of heading change within one body length, when the elevator angle is over 10 degrees. For configuration B, the P value is 0.7, implying that given a thrust difference of 10kgf, the vehicle experiences a heading change of 0.25 degrees within a single body length.

The measure of dynamic stability can be defined as

$$G \equiv 1 - \frac{l'_w}{l'_q} \quad (5)$$

where

$$l'_q = \frac{-M'_q + (m'x'_G + m'_z x'_z)}{Z'_q + (m' + m'_x)}$$

$$l'_w = \frac{M'_w}{-Z'_w}$$

The system is unstable if G is negative, statically stable if G exceeds 1m and dynamically stable if G is between 0 and 1. Table 5 lists stability and turning indices of the AUV-HM1 in both configurations. Notably, configuration A is dynamically more stable.

VI. Prediction Of Stability And Maneuverability

Methods of predicting hydrodynamic force coefficients for slender bodies such as airplanes and torpedoes are well established, while similar methods for flat-streamlined underwater vehicles, such as the AUV-HM1, are neglected. Maeda et al. (1989) presented a method suitable for undersea vehicles and based on DATCOM (Hoak et al.,1978), a public-domain empirical method determining aircraft stability properties. Meanwhile, for estimating the stability derivatives of AUV-HM1, Chiu et al. (1996, 1997) have examined the validity of using empirical formula proposed for standard torpedoes by Botaccini. The efficacy of estimations that use a modified Botaccini method is confirmed by comparing the measured and estimated coefficients. Equivalent ellipsoids were applied herein to approximate the shape of the vehicle and its control fins. A theoretical method based on potential flow theory was used for estimating inertia terms, while the Botaccini method was used to predict the damping coefficients of the vehicle. The maneuverability and dynamic stability index is calculated herein using various body lengths, while maintaining the water displacement volume of the vehicle constant. Figure 5 displays the profile of the baseline vehicle used for the prediction, while Figs. 6 and 7 illustrate the variations of maneuverability and stability with increasing body length. Simulation results indicate that the flat body significantly influences maneuverability and dynamic stability. With an increasing body length, the cross-section becomes increasingly symmetrical, causing vehicle behavior to become increasingly similar in the lateral and horizontal modes.

VII. Conclusion

The AUV-HM1 is designed for area search and survey missions in shallow water. Search/survey scenarios generally require sufficient responsiveness to control in the horizontal plane of motion, with high stability for performing depth control in the vertical plane. The maneuverability of the AUV-HM1 is characterized using indexes G, P, representing stability, and responsiveness to control, respectively. Configuration A of the vehicle is found to be more stable. However, configuration B of the vehicle is found to have a superior turning performance. Maneuvering behavior in configuration A corresponds to the vertical plane of motion of the AUV-HM1, while that of configuration B corresponds to the horizontal plane. We conclude that this flat-streamlined design of the hull form is compatible with the operational requirements of the AUV-HM1. Performance of a flat, baseline vehicle was also predicted, and the G and P values of the baseline configuration are illustrated using different body lengths while maintaining the displacement volume of the vehicle constant. Flattening the body of the vehicle is found to reduce stability in the horizontal plane while increasing stability in the vertical plane. The flat body is more responsive in relation to axis-symmetric body in both the horizontal and vertical planes.

Acknowledgments

The authors would like to thank the National Science Council of R.O.C., for financially supporting this research under Contract number NSC84-2611-E-002-026 and NSC85-2611-E-002-012. Yi-Yuan Chang, Chieh-Chih Wang, Chien-Chen Huang, Chun-Chin Wei, Jen-Chao Chung, and Wei-Chen Tsai are appreciated for their assistance in the PMM experiments.

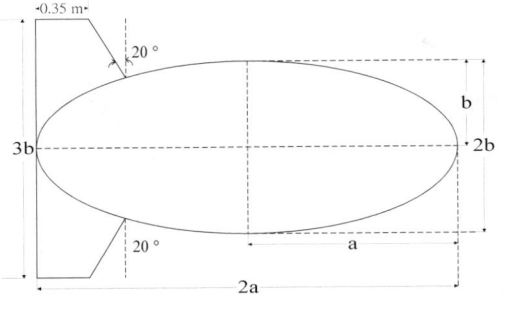

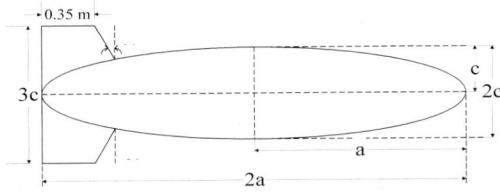

Fig. 5 Baseline arrangement of the body and fins used in the predictions herein. The upper plot corresponds to configuration A, and the lower plot corresponds to configuration B.

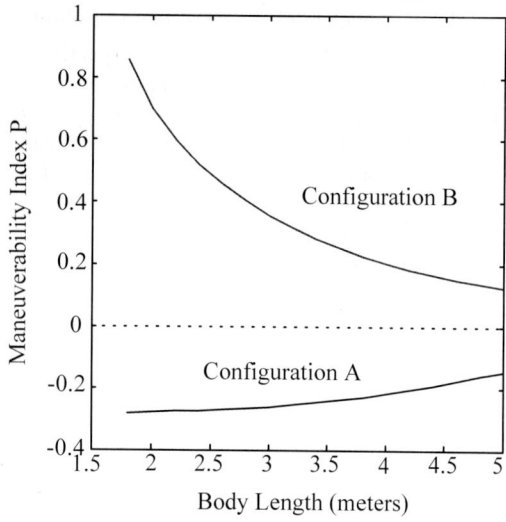

Fig. 6 Maneuverability with different body lengths

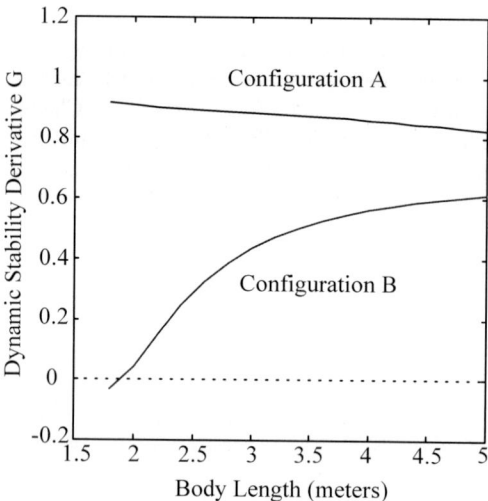

Fig. 7 Dynamic stability with different body lengths

REFERENCES

1. Asakawa, K., Kojima, J., Ito, Y., Shirasaki, Y., and Kato, N. (1993) "Development of Autonomous Underwater Vehicle for Inspection of Underwater Cables," *Proc. of Underwater Intervention '93*, pp. 208-216.
2. Chiu, F.C., Guo, J., Chang, Y.Y., Wang, C.C., Wang, J. P., (1996) "On the Linear Hydrodynamic Forces and the Maneuverability of Unmanned Untethered Submersible with Streamlined Body," *J. of Japanese Society of Naval Architecture*, No. 180. (In Japanese).
3. Chiu, F. C., Guo, J., Huang, C.C., Wang, J. P., (1997) "On the Linear Hydrodynamic Forces and the Maneuverability of Unmanned Untethered Submersible with Streamlined Body (2nd Report: Lateral Motions)," *J. of Japanese Society of Naval Architecture*, No. 182 (In Japanese).
4. Guo, J., J. F. Tsai and F. C. Chiu (1995) "Design, Simulation and Control of a Highly Maneuverable Autonomous Underwater Vehicle Testbed," *MARIENV '95*, Tokyo.
5. Hoak, D. E. and R. D. Finck (1978) *USAF Stability and Control DATCOM*, McDonnel-Douglas Corp.
6. Landweber, L. and M. Gertler (1950) *Mathematical Formulation of Bodies of Revolution*, The David W. Taylor Model Basin Report 719.
7. Lee, P.-M., Lee, C.-M., Jeon, B-H., Hong, S.-W., (1998) "System Design and Quasi-Sliding Mode Control of an AUV for Ocean Research and Monitoring," *Proc. Int'l Symp. on Underwater Technology*, Tokyo, pp.179-184.
8. Nomoto, K., and Norrbin, N. H., (1969) "A Review of Methods of Defining and Measuring the Maneuverability of Ships," *International Towing Tank Conference*, Appendix I, Report of Maneuverability Committee, Rome.
9. Maeda, H., Tatsuta, S., (1989) "Prediction Method of Hydrodynamic Stability Derivatives of an Autonomous Underwater Non-Tethered Submerged Vehicles," *Int'l Conf. On Offshore Mechanics and Arctic Engineering*, pp. 105-114.
10. Pascoal, A.Oliveira, P., Silvestre, C., Bjerrum, A., Ishoy, A. Pignon, J.-P., Ayela, G., and Petzelt, C., (1997) "MARIUS: An Autonomous Underwater Vehicle for Coastal Oceanography," *IEEE Robotics and Automation Magazine*, December, pp.46-59.
11. Ura, T., (1989) "Free Swimming Vehicle "PTEROA" for Deep Sea Survey," *Proc. ROV'89*, pp.263-268.

Closed Loop Time Invariant Control of 3D Underactuated Underwater Vehicles

M. Aicardi, G. Casalino , G. Indiveri

Abstract— A novel strategy to design time invariant motion controllers for underactuated mobile systems is applied to the position and attitude control of an underactuated 3D vehicle. The idea consists in defining a velocity vector field such that an ideal, fully actuated system would exponentially achieve the control objective by simply following such field. Then a steering law for the given underactuated system is designed such that it is exponentially stabilized parallel to the above mentioned velocity vector field. For the particular problem here addressed, due to the use of polar like coordinates, this method yields a discontinuous control law. Both the design process and the resulting solution have a most clear physical interpretation. Important practical requirements as approaching the target on a null curvature path, avoiding cusps in the whole path and moving in only one given forward direction are easily satisfied within this framework.

I. Introduction

It is a well known fact that because of Brocketts Theorem [1] the point stabilization problem of an underactuated 3D floating rigid body cannot be solved by time-invariant, smooth feedback. As for many other systems affected by Brocketts negative result, either time-varying or discontinuous solutions must be considered. Both approaches have advantages and disadvantages: while the convergence rate may be exponential in both frameworks, time-varying solutions are in general smooth and globally defined, but the resulting paths may exhibit cusps or very unnatural and perhaps practically unacceptable oscillations due to the trigonometric time dependence of the controller. On the other hand with discontinuous control solutions it may be much easier to satisfy "practical" constraints as driving in only one forward direction avoiding cusps and converging on the target on a null curvature path at the expense of isolated discontinuity points (including the target). Time-varying solutions for both the kinematic and dynamic model of an underactuated underwater vehicle have been proposed in [2] and [3].

The control design of discontinuous solutions is usually performed in polar coordinates so that the origin of the original cartesian coordinate system (target) is automatically a singularity for the controller as the polar coordinate angular variables are not defined in the origin. Indeed it is for this very reason that Brocketts theorem does not prevent the existence of a continuous law driving the polar-like configuration error to zero [4]. The discontinuous control method has been used to solve a wide number of motion control problems for underactuated systems, not only related to mobile robots (by example [5] for a recent reference). It has been extensively studied by A. Astolfi who suggested a general algorithm to design discontinuous controllers for nonholonomic systems in chained form [6]. Indeed a state transformation may be used to write the kinematic equations of a floating rigid body in chained form [2], but as such transformation is not defined for certain angular configurations (one of the direction cosines is assumed to be always strictly positive) the resulting controller would not be able to steer the vehicle from all the possible orientations. A discontinuous solution not suffering from such drawback has been presented in [7]. In spite of this results and of other time-invariant solutions as the one suggested in [8], there is yet no general method to design time-invariant position controllers for underactuated systems: the aim of this paper is to suggest, outlining a solution for the 3D underactuated underwater vehicle "parking" problem here addressed, a very natural and novel approach that has been successfully adopted to design a linear course tracking law for an underactuated vessel [9] and a point stabilizing controller for the unicycle model [10].

The control law design process is made of two major steps: first a velocity vector field is defined such that an ideal system free of any nonholonomic constraint subject to this velocity would exponentially converge to the desired configuration and then a steering law is computed such that the only actuated linear motion axis of the nonholonomic system is exponentially parallel to the previously defined velocity vector field. While the first step is trivial, the second may be technically more involved, but straightforward. Notice that the asymptotic characteristics of the vehicles motion are completely specified by the user defined velocity

---

M. Aicardi and G. Casalino are with the Department of Communications Computers and Systems Science, University of Genoa, Via Opera Pia 13, 16145 Genova, Italy, E-mail: {michele/pino}@dist.unige.it

G. Indiveri is with the GMD-AiS, German National Research Center on Information Technologies, Institute for Autonomous Intelligent Systems, Schloss Birlinghoven, 53754 Sankt Augustin, Germany

vector field, thus it is not difficult to satisfy important practical asymptotic constraints as, by example, reaching the target on a straight line. The application of this idea makes both the design process and the resulting solution extremely clear, it guarantees exponential convergence and it may allow to solve problems that cannot be solved and, to the knowledge of the authors, have not been solved otherwise [9][11]: this is the case of the linear course stabilization for an underactuated marine vessel.

II. Model definition

Consider an underwater vehicle having nonactuated sway, heave and roll which is by far the most common case in real applications. The controllability of such a kind of underactuated kinematic model is addressed in [8]. With reference to figure (1) consider a rigid body having a body fixed reference $<b>$ such that the linear and angular velocities obey

$$\underline{u} = u\,\underline{i}_b \quad (1)$$
$$\underline{\omega} \cdot \underline{i}_b = 0 \quad (2)$$

As for the planar unicycle case [4], [12], [13] and in spite of Brocketts Theorem it has been shown [7] that also in the 3D case a polar-like description of the system kinematic allows a time-invariant closed loop control law to be found that guarantees asymptotic convergence of the position and orientation error to zero as long as the system is not initially located in the target position. Moreover the planar version [4] of such controller has been most efficiently adopted to control the motion of the underwater vehicles Romeo and Roby2 of CNR-IAN (Genova, Italy) [14].

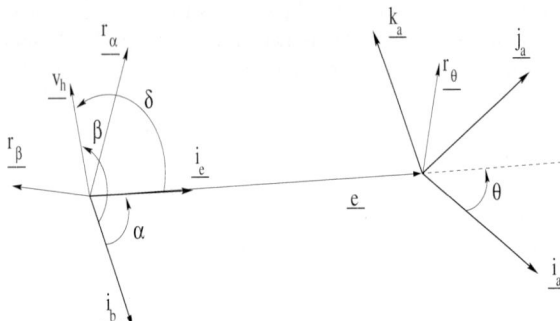

Fig. 1. The model

III. Problem definition and control law design

With reference to figure (1) consider the body and target fixed frames $<b> = \{\underline{i}_b, \underline{j}_b, \underline{k}_b\}$ and $<a> =$ $\{\underline{i}_a, \underline{j}_a, \underline{k}_a\}$ and the vector $\underline{e}$ pointing from $<b>$ to $<a>$. The angles between $\underline{i}_a$, $\underline{i}_b$ and $\underline{e}$ are θ and α such that when they are different from multiples of π the unit vectors $\underline{r}_\theta$ and $\underline{r}_\alpha$ parallel to $\underline{i}_a \wedge \underline{i}_e$ and $\underline{i}_b \wedge \underline{i}_e$ and normal to the planes $\{\underline{i}_a, \underline{e}\}$ and $\{\underline{i}_b, \underline{e}\}$ are well defined. Calling $\underline{v}_h$ a velocity vector to be yet defined, β is the angle between $\underline{i}_b$ and $\underline{v}_h$ and $\underline{r}_\beta \parallel \underline{i}_b \wedge \underline{v}_h$ the unit vector normal to the plane $\{\underline{i}_b, \underline{v}_h\}$ when $\beta \neq n\pi \ \forall \ n \in Z$.

Problem statement The control objective is to design a closed loop law for the linear and angular velocities $\underline{u}$ and $\underline{\omega}$ such that the origin of frame $<b>$ (the underactuated vehicle) is exponentially driven to the origin of the frame $<a>$ along the $\underline{i}_a$ axis while the constraints given by equations (1) and (2) are satisfied.

A. The velocity vector field $\underline{v}_h$

When the vehicle is not located in the origin of the target frame $<a>$, i.e. $e = \|\underline{e}\| \neq 0$ and $\underline{r}_\theta$ is well defined, i.e. $\theta \neq n\pi : n \in Z$, the reference $<e>$ may be defined as

$$<e> = \{\underline{i}_e = \underline{e}/\|\underline{e}\|, \ \underline{j}_e = \underline{r}_\theta \wedge \underline{i}_e, \ \underline{k}_e = \underline{r}_\theta\} \quad (3)$$

The vector field $\underline{v}_h$ is defined to be such that the projection of $\underline{v}_h$ on the reference $<e>$ is given by

$$\begin{aligned}{}^e\underline{v}_h &= {}^e[\,\gamma_e\,\underline{e} + \gamma_\theta\,\theta\,(\underline{r}_\theta \wedge \underline{e})\,] = \\ &= \gamma_e e \underline{i}_e + \underline{j}_e \gamma_\theta \theta e \quad : \quad \gamma_e, \gamma_\theta > 0. \end{aligned} \quad (4)$$

Given the nature of reference $<e>$, the field $\underline{v}_h$ is discontinuous when $\theta = \pi$. This is actually a drawback of the proposed strategy as when $\theta \underline{r}_\theta$ is not defined neither equation (4) is. As a consequence the field $\underline{v}_h$ is not defined when $\theta = \pi$, while for $\theta = 0$ the vector $\theta \underline{r}_\theta$ is well defined and equal to the null vector so that $\theta = 0 \Rightarrow \underline{v}_h = \gamma_e e \underline{i}_e$.

Remark 1: By construction the origin of frame $<b>$ subject to the velocity given by equation (4) would exponentially converge to the target provided that $e_{t=0} \neq 0$ and $\theta_{t=0} \in (-\pi, \pi)$. Moreover if $\gamma_\theta > \gamma_e$ the curvature of the path tends to zero.

Of course the problem is that in general the vector $\underline{v}_h$ will not satisfy the constraints given by equations (1) and (2). The basic idea is that of building a closed loop control law for the angular velocity $\underline{\omega}$ of the frame $<b>$ (nonholonomic vehicle) that drives $\underline{u} = u\underline{i}_b$ exponentially parallel to $\underline{v}_h$. Once that this will occur, the vehicle will exponentially converge to the target by the very definition of the field $\underline{v}_h$ (in particular u can be taken to be $u = \|\underline{v}_h\|$ as discussed in the following).

B. The steering law

Given the above definitions of $\underline{v}_h$ and β a closed loop control law for the angular velocity $\underline{\omega}$ that drives β to zero is to be found. Consider the more general situation in which two unit vectors $\underline{\nu} \equiv \underline{v}_h/\|\underline{v}_h\|$ and $\underline{i}_b$ are given such that their angular velocities are $\underline{\omega}_\nu$ and $\underline{\omega}$ and a law for $\underline{\omega}$ is searched such that the vector $\underline{\beta} \equiv \underline{r}_\beta \beta$ is asymptotically null. This control problem may be approached considering the Lyapunov function

$$V_\beta = \frac{1}{2} \underline{\beta} \cdot \underline{\beta} = \frac{1}{2} \beta^2 \qquad (5)$$

and its time derivative

$$\dot{V}_\beta = \underline{\beta} \cdot (\underline{r}_\beta \dot{\beta} + \underline{\dot{r}}_\beta \beta) = \underline{\beta} \cdot \underline{r}_\beta \dot{\beta} \qquad (6)$$

as $\underline{\dot{r}}_\beta = \underline{\omega} \wedge \underline{r}_\beta \perp \underline{\beta}$. By standard kinematic it follows that

$$\dot{\beta} = \underline{r}_\beta \cdot (\underline{\omega}_\nu - \underline{\omega}) \qquad (7)$$

so that in order to guarantee that equation (6) is negative definite it is sufficient, by example, that

$$\underline{r}_\beta \left(\underline{r}_\beta \cdot (\underline{\omega}_\nu - \underline{\omega}) \right) = -k \underline{\beta} \; : \; k > 0. \qquad (8)$$

By direct calculation it is found that equation (8) is satisfied choosing

$$\underline{\omega} = k \underline{\beta} + \underline{r}_\beta (\underline{r}_\beta \cdot \underline{\omega}_\nu) \qquad (9)$$

which guarantees exponential convergence of β to zero as can be seen replacing (9) in (7). Notice that the angular velocity $\underline{\omega}_\nu$ of a unit vector $\underline{\nu}$ is always given by $\underline{\omega}_\nu = \underline{\nu} \wedge \underline{\dot{\nu}}$ so that finally equation (9) may be expressed as

$$\underline{\omega} = k \underline{\beta} + \underline{r}_\beta \left(\underline{r}_\beta \cdot (\underline{\nu} \wedge \underline{\dot{\nu}}) \right). \qquad (10)$$

In order to implement equation (10) given that $\underline{\nu}$ is the unit vector parallel to $\underline{v}_h$, its time derivative $\underline{\dot{\nu}}$ must be explicitly computed. As shown in appendix A:

$$\underline{\dot{\nu}} = \frac{u}{e} \left(\left(1 + \frac{\gamma_e \gamma_\theta}{\gamma_e^2 + \gamma_\theta^2 \theta^2} \right) (\underline{r}_\theta \cdot \underline{r}_\alpha \sin \alpha) \underline{r}_\theta + \right.$$
$$\left. (\underline{i}_b \cdot \underline{r}_\theta)(\underline{r}_\theta \wedge \underline{i}_e) \right) \wedge \underline{\nu}. \qquad (11)$$

The control signal given by equation (10) satisfies the constraint (2), but may be not defined in correspondence of those states in which $\underline{r}_\beta$ or $\underline{\nu}$ are not defined. These cases will be more deeply discussed in a following section. The presence of the factor $1/e$ in (11) is a natural consequence of the polar-like description and may be "compensated" by a proper choice of u. Notice however that in spite of any choice for u equation (11) fails to be defined in $e = 0$ as $\underline{r}_\theta|_{e=0}$ is not defined.

C. Discontinuities in the steering law

For the sake of clarity and notational simplification the analysis is carried out assuming that all angles are in the range $[-\pi, \pi]$, the extension to wider ranges being obvious. By direct analysis of equations (10) and (11) it is apparent that the steering control law may not be defined if $\beta = \pi$, $\beta = 0$, $\theta = \pi$ or $\theta = 0$ while the cases $\alpha = 0, \pi$ should not worry as the term $\underline{r}_\alpha \sin \alpha$ in equation (11) is always well defined.

C.1 $\theta = \pi$

The most serious case is perhaps $\theta = \pi$ as in this situation the field $\underline{v}_h$ itself is not defined and so neither is the scalar β. As a consequence the steering control $\underline{\omega}$ will have to be discontinuous when $\theta = \pi$. A possible choice for $\underline{\omega}_{|\theta=\pi}$ is

$$\underline{\omega}_{|\theta=\pi} = \begin{cases} 0 & \text{if } \alpha \notin \{0, \pi\} \\ \varepsilon \underline{k}_b & \varepsilon > 0 \text{ otherwise.} \end{cases} \qquad (12)$$

Although without entering in the details of the analysis, it is worthwhile to examine at least qualitatively the consequences of this choice. Assuming that the vehicle moves in only one forward direction as will be apparent from the next section, the case $\theta = \pi \cup \alpha \in \{0, \pi\}$ corresponds to the situation in which the vehicle is moving along the positive part of the $\underline{i}_a$ axis of the target frame $<a>$ towards the target ($\alpha = 0$) or away from it ($\alpha = \pi$). In such situation any infinitesimal angular velocity driving the vehicle off the $\underline{i}_a$ axis as suggested by equation (12) will bring the system in a region where all relevant angles and unit vectors are properly defined for the implementation of equations (10) and (11). Moreover notice that due to the radial symmetry of the $\underline{v}_h$ field in any neighborhood of the $\underline{i}_a$ axis and due to the exponential convergence of β given by (10), the implementation of equation (12) when $\theta = \pi \cup \alpha \in \{0, \pi\}$ suggests that the $\underline{i}_a$ axis will not be crossed again once it is left. A similar reasoning justifies the choice $\underline{\omega}_{|\theta=\pi} = 0$ if $\alpha \notin \{0, \pi\}$ as in this situation the "singular" axis $\underline{i}_a$ is left behind by just moving in the given $\underline{i}_b$ direction.

C.2 $\theta = 0$

The case $\theta = 0$ is less problematic as actually in such case $\underline{v}_h$ as given by equation (4) is well defined and parallel to $\underline{i}_a$. Notice that when $\theta = 0$ the angles α and β are equivalent, i.e. $(\alpha_{|\theta=0}) \equiv (\beta_{|\theta=0})$ and $\underline{\nu}_{|\theta=0} \equiv \underline{i}_e$. A possible choice for $\underline{\omega}_{|\theta=0}$ is

$$\underline{\omega}_{|\theta=0} = \begin{cases} (k\beta + (u/e)\sin\alpha)\underline{r}_\beta & \text{if } \beta \notin \{0, \pi\} \\ \varepsilon \underline{k}_b : \varepsilon > 0 & \text{if } \beta = \pi \\ 0 & \text{if } \beta = 0. \end{cases} \qquad (13)$$

The first of the three cases given by equation (13) is obtained from equation (10) noticing that if $\underline{\nu} = \underline{i}_e$ as when $\theta = 0$, then $\underline{\dot{\nu}}_{|\theta=0} = (\underline{r}_\alpha \wedge \underline{i}_e)(u/e)\sin\alpha$. This specific choice of $\underline{\omega}_{|\theta=0}$ when $\beta \notin \{0, \pi\}$ guarantees that the Lyapunov function (5) and its time derivative (6) are continuous. The choice $\underline{\omega}_{|\theta=0} = \varepsilon\,\underline{k}_b : \varepsilon > 0$ when $\beta = \pi$ follows from the observation that in such situation the vehicle is moving along the negative part of the $\underline{i}_a$ axis in the opposite direction of the target. Any infinitesimal angular velocity will drive the system off this axis in a region where all relevant angles and unit vectors are suitably defined for the implementation of equations (10) and (11). Moreover the structure of the field $\underline{v}_h$ suggests that once this happens the vehicle will not cross the $\underline{i}_a$ axis with $\beta = \pi$ again. The last case $\theta = \beta = \alpha = 0$ corresponds to the trivial situation in which the vehicle is moving on the $\underline{i}_a$ axis towards the target and thus it may simply proceed on a straight line.

C.3 $\beta = 0$ or $\beta = \pi$

Finally it should be noticed that thanks to the exponential convergence of β to zero the unit vector $\underline{r}_\beta$ used to compute the angular velocity $\underline{\omega}$ will be actually well defined at all finite times as long as $\beta_{|t=0} \neq 0, \pi$. This suggests that in the case that the initial configuration of the vehicle should imply $\beta = 0$ or $\beta = \pi$ any angular velocity satisfying equation (2) may be applied at the time $t = 0$ in order to enter the region of the state space where equations (10) and (11) may be directly implemented.

D. The linear velocity law

Having computed a steering law that guarantees asymptotic convergence of β to zero, the most natural choice for the linear velocity is

$$u = \|\underline{v}_h\| = e\sqrt{\gamma_e^2 + \gamma_\theta^2 \theta^2} \quad . \tag{14}$$

With such law for u the vehicle will asymptotically follow the velocity field $\underline{v}_h$ with a well defined angular velocity as the ratio u/e required in equation (11) is defined in the whole state space at all finite times. This follows from the observation that the state equation for e is given by

$$\dot{e} = -u\cos\alpha \tag{15}$$

thus replacing equation (14) in (15) it is found that

$$e(t) = e_{|t=0}\exp\left[-\int_0^t \sqrt{\gamma_e^2 + \gamma_\theta^2 \theta(\tau)^2}\,\cos\alpha(\tau)d\tau\right] \tag{16}$$

showing that if $e_{|t=0} \neq 0$ then e is strictly positive at *all* finite times. Measurement noise in the vicinity of $e = 0$ will cause discontinuities in the control action and may result in high frequency chattering-like behaviour to be accordingly filtered. This is actually the price to be paid in order to reach a neighborhood of the target along a smooth "flat" (the asymptotic curvature is zero if $\gamma_\theta > \gamma_e$) path as opposed to the oscillating ones obtained by time-varying controllers. Moreover notice that as u given by equation (14) has constant sign it guarantees the absence of cusps in the vehicles path thus satisfying a major requirement for real world applications.

IV. CONCLUSIONS

The above presented approach has been tested by simulations one of which is here reported for reference. In figure (2) the path resulting from initial position

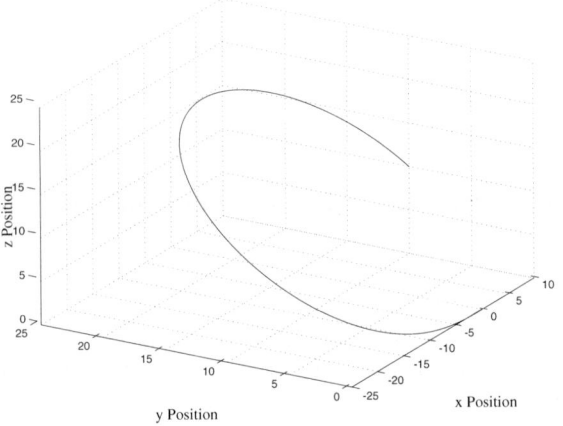

Fig. 2. Path resulting from the application of the proposed closed loop control strategy. Refer to the text for greater details regarding this example.

$(x_0, y_0, z_0) = (10, 10, 10)$ and orientation (roll, pitch, yaw) $(\phi_0, \theta_0, \psi_0) = (0, -\pi/4, \pi/2)$ with gains $k = 1$, $\gamma_e = 1/2$ and $\gamma_\theta = 3/4$ is reported. The time histories of the linear and angular velocities together with the one of β that converges exponentially as expected are displayed in figure (3). As apparent the starting configuration was such that no discontinuities are present. Strictly speaking the proposed solution does not solve the point stabilization problem for the 3D nonholonomic systems as the target configuration ($e = 0$) lies on the boundary of the polar-like state space domain, but is not part of it. Nevertheless this has virtually no importance from the practical point of view as the proposed strategy can be applied from *any* neighborhood of the null measure set $e = 0$. This is a clear and known consequence of the use of polar-like variables. It should be noticed that the controller discontinuities are due to the use of polar-like variables (necessary

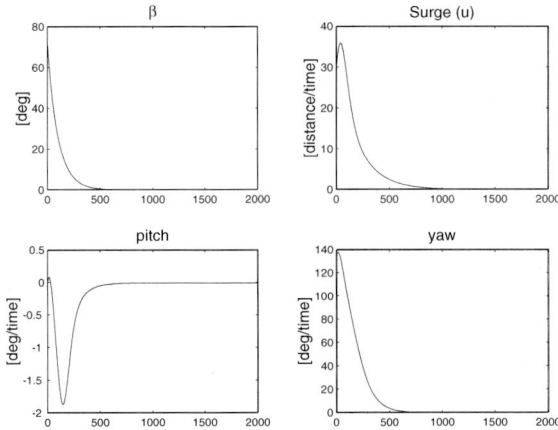

Fig. 3. Time history of β and of the control signals u (surge), ω_2 (pitch) and ω_3 (yaw) variables relative to the path displayed in figure (2).

in order to prevent Brocketts theorem obstruction if a time invariant solution is desired) and not to the strategy of converging on a filed $\underline{v}_h$. Indeed the method in itself of defining a smooth velocity filed $\underline{v}_h$ and steering the underactuated system parallel to it usually leads to smooth control solutions [9][11].

The proposed control design methodology exploits the prior definition of a velocity vector field that would exponentially drive the configuration error of a nonholonomic constraint-free systems to zero. It is actually sufficient to define a steering law that asymptotically orients the nonholonomic vehicle along the direction of this field to achieve the control objective. This simple and novel method makes both the design process and control solution very clear and straightforward, not to mention that it allows to design exponentially converging controllers that structurally satisfy major practical requirements as moving in only one forward direction, avoiding cusps in the paths and approaching the target on a null curvature path. Moreover the method has been shown [9][11] to be effective in solving motion control problems for underactuated system that have not been otherwise solved and *cannot* be solved by standard techniques as feedback linearization.

APPENDIX A: THE TIME DERIVATIVE OF $\underline{\nu}$

With reference to figure (1) and to the definition (4) of $\underline{v}_h$, $\underline{\dot{\nu}}$ may be generated at most by three different kinds of infinitesimal motions of the frame $<b>$, i.e. $\underline{\dot{\nu}} = \underline{\dot{\nu}}_1 + \underline{\dot{\nu}}_2 + \underline{\dot{\nu}}_3$:
- $\underline{\dot{\nu}}_1$ due to linear movements of $\underline{e}$ along $\underline{i}_e$
- $\underline{\dot{\nu}}_2$ due to rotations of $\underline{e}$ around $\underline{r}_\theta$
- $\underline{\dot{\nu}}_3$ due to rotations of $\underline{e}$ around $\underline{j}_e = \underline{r}_\theta \wedge \underline{i}_e$.

Given that both components of $\underline{v}_h$ are linear in e, $\underline{\dot{\nu}}_1 = 0$ as $\underline{\nu}$ is actually *not* affected by the first kind of infinitesimal motions, namely translations along $\underline{i}_e$ leave $\underline{\nu}$ unchanged. As far as the rotations around $\underline{r}_\theta$ are concerned it follows that

$$\underline{\dot{\nu}}_2 = (\dot{\delta} + \dot{\theta})(\underline{r}_\theta \wedge \underline{\nu})$$

being δ the angle between $\underline{i}_e$ and $\underline{v}_h$ as shown in figure (1). The time derivative of δ can be evaluated noticing that:

$$^e\underline{e} \cdot {}^e\underline{v}_h \equiv \|\underline{v}_h\|\|\underline{e}\| \cos \delta$$
$$= e^2 \sqrt{(\gamma_e^2 + \gamma_\theta^2 \theta^2)} \cos \delta = \gamma_e \, e^2 \quad (17)$$
$$^e\underline{e} \wedge {}^e\underline{v}_h \equiv {}^e\underline{r}_\delta \|\underline{v}_h\|\|\underline{e}\| \sin \delta =$$
$$= {}^e\underline{r}_\delta \, e^2 \sqrt{(\gamma_e^2 + \gamma_\theta^2 \theta^2)} \sin \delta = {}^e\underline{r}_\delta \, \gamma_\theta \, \theta e^2 (18)$$

from which it follows that $\cos \delta$ and $\sin \delta$ are given by

$$\cos \delta = \frac{\gamma_e}{\sqrt{(\gamma_e^2 + \gamma_\theta^2 \theta^2)}} \quad (19)$$
$$\sin \delta = \frac{\gamma_\theta \, \theta}{\sqrt{(\gamma_e^2 + \gamma_\theta^2 \theta^2)}}. \quad (20)$$

Finally differentiating equation (19) and using equation (20) the time derivative of δ is found to be

$$\dot{\delta} = \frac{\gamma_e \, \gamma_\theta}{\gamma_e^2 + \gamma_\theta^2 \theta^2} \dot{\theta}. \quad (21)$$

Assuming all the relevant unit vectors ($\underline{r}_\theta$, $\underline{r}_\alpha$, $\underline{\nu}$) to be well defined, the value of $\dot{\theta}$ can be computed noticing that θ changes if and only if there is an angular velocity parallel to $\underline{r}_\theta$ applied on $\underline{e}$. The only motion that can generate such angular velocity is the linear translation along $\underline{i}_b$ which, in fact, causes a rotation of $\underline{e}$ having an angular velocity $\underline{\omega}_e = \underline{r}_\alpha(u/e) \sin \alpha$. As a consequence of this angular velocity it follows that $\dot{\theta} = \underline{r}_\theta \cdot \underline{r}_\alpha (u/e) \sin \alpha$ and thus

$$\underline{\dot{\nu}}_2 = \frac{u \sin \alpha}{e} \left(1 + \frac{\gamma_e \, \gamma_\theta}{\gamma_e^2 + \gamma_\theta^2 \theta^2}\right)(\underline{r}_\theta \cdot \underline{r}_\alpha)(\underline{r}_\theta \wedge \underline{\nu}). \quad (22)$$

As far as the contribution $\underline{\dot{\nu}}_3$ to $\underline{\dot{\nu}}$ due to rotations around $\underline{j}_e = \underline{r}_\theta \wedge \underline{i}_e$ is concerned it must first be noticed that such motions do not affect the value of θ. This can be shown, by example, considering the effect of a constant rotation around $\underline{j}_e$, i.e. an angular velocity $\underline{\sigma} = \sigma \underline{j}_e$: the absolute time derivative of $\underline{i}_e$ due to such angular velocity would be given by

$$\frac{d_{<a>}}{dt} \underline{i}_e = \underline{\sigma} \wedge \underline{i}_e = \sigma \underline{j}_e \wedge \underline{i}_e = -\sigma \underline{r}_\theta$$

and as the absolute derivative of the fixed reference unit vectors is identically null ($\frac{d_{<a>}}{dt} \underline{i}_a = 0$) it follows that

$$\frac{d_{<a>}}{dt} \cos\theta = \frac{d_{<a>}}{dt}(\underline{i}_a \cdot \underline{i}_e) = -\sigma\, \underline{i}_a \cdot \underline{r}_\theta = 0$$

showing that the angle θ does *not* change in consequence of rotations of $\underline{e}$ around $\underline{j}_e$. This fact allows to compute $\underline{\dot{\nu}}_3$ as follows:

$$\underline{\dot{\nu}}_3 = \frac{\underline{u} \cdot \underline{r}_\theta}{e}\left(\underline{j}_e \wedge \underline{\nu}\right) \quad (23)$$

being $\underline{u} \cdot \underline{r}_\theta$ the only component of the systems velocity that may cause a rotation of $\underline{e}$ around $\underline{j}_e$. Recalling the above partial results, the time derivative of $\underline{\nu}$ is found to be:

$$\begin{aligned}\underline{\dot{\nu}} &= \frac{u}{e}\left(\left(1 + \frac{\gamma_e\,\gamma_\theta}{\gamma_e^2 + \gamma_\theta^2\,\theta^2}\right)(\underline{r}_\theta \cdot \underline{r}_\alpha \sin\alpha)\underline{r}_\theta + \right.\\ &\quad \left. (\underline{i}_b \cdot \underline{r}_\theta)\left(\underline{r}_\theta \wedge \underline{i}_e\right)\right) \wedge \underline{\nu}.\end{aligned} \quad (24)$$

Once again it is observed that this equation is defined only when the relevant unit vectors are well defined. It is a straightforward exercise to show that in the planar case, i.e. when $\underline{r}_\alpha = \underline{r}_\theta = \underline{r}_\beta = \underline{k}_a$, the resulting angular velocity $\underline{\omega}$ (10) is exactly the same one obtained for the $2D$ unicycle model [10] applying the same position control strategy.

References

[1] Brockett, Millmann, and Sussmann, eds., *Differential Geometric Control Theory*, ch. Asymptotic Stability and Feedback Stabilization, by Brockett, R. W., pp. 181–191. Birkhauser, Boston, USA, 1983.

[2] O. Egeland, E. Berglund, and O. J. Sørdalen, "Exponential stabilization of a nonholonomic underwater vehicle with constant desired configuration," in *IEEE Int. Conf. on Robotics and Automation, ICRA'94*, (S. Diego, CA, USA), pp. 20–25, May 1994.

[3] K. Y. Pettersen and O. Egeland, "Time-varying exponential stabilization of the position and attitude of an underactuated autonomous underwater vehicle," *IEEE Trans. on Automatic Control*, vol. 40, no. 1, pp. 112–115, 1999.

[4] M. Aicardi, G. Casalino, A. Bicchi, and A. Balestrino, "Closed loop steering of unicycle-like vehicles via lyapunov techniques," *IEEE Robotics and Automation Magazine*, pp. 27–35, March 1995.

[5] R. Mukherjee and M. Kamon, "Almost smooth time-invariant control of planar multibody systems," *IEEE Transact. on Robotics and Automation*, vol. 15, no. 2, pp. 268–280, 1999.

[6] A. Astolfi, "Discontinuous control of a nonholonomic system," *Systems & Control Letters*, vol. 27, pp. 37–45, 1996.

[7] M. Aicardi, G. Cannata, and G. Casalino, "Smoothness of a feedback control law for a non-holonomic 3D vehicle," in *IFAC 3rd Int. Workshop on Motion Control*, (Grenoble, France), pp. 239–244, September 21-23 1998.

[8] O. Egeland, M. Dalsmo, and O. J. Sørdalen, "Feedback control of a nonholonomic underwater vehicle with a constant desired configuration," *Int. Jou. of Robotics Research*, vol. 15, no. 1, pp. 24–35, 1996.

[9] G. Indiveri, "Linear course tracking for underactuated marine vehicles: a time-invariant nonlinear controller," Tech. Rep. GMD Report 83 (ISSN 1435-2702), GMD - German National Research Center on Information Technologies, Scloß Birlinghoven, D-53754 Sankt Augustin, Deutschland, December 1999.

[10] M. Aicardi, G. Cannata, G. Casalino, and G. Indiveri, "On the stabilization of the unicycle model projecting a holonomic solution," accepted at *8th Int. Symposium on Robotics with Applications, ISORA 2000*, (Maui, Hawaii, USA), June 11-16 2000.

[11] G. Indiveri, M. Aicardi, and G. Casalino, "Robust global stabilization of an underactuated marine vehicle on a linear course by smooth time-invariant feedback," submitted to 39th *IEEE Conference on Decision and Control CDC'2000*, (Sydney, Australia), December 2000.

[12] A. Astolfi, "Exponential stabilization of a wheeled mobile robot via discontinuous control," *ASME Jou. of Dyn. Sys. Mea. and Cont.*, March 1999.

[13] G. Indiveri, "Kinematic time-invariant control of a 2 D nonholonomic vehicle," in 38th *IEEE Conference on Decision and Control CDC'99*, (Phoenix, Arizona, USA), December 1999.

[14] M. Caccia, G. Casalino, R. Cristi, and G. Veruggio, "Acoustic motion estimation and control for an unmanned underwater vehicle in a structured environment," *Control Engineering Practice*, vol. 6, pp. 661–670, 1998.

Interoperability and Synchronisation of Distributed Hardware-in-the-Loop Simulation for Underwater Robot Development: Issues and Experiments

David M. Lane, G.J.Falconer, G.Randall
Ocean Systems Laboratory
Dept. Computing & Electrical Engineering
Heriot-Watt University
Edinburgh, Scotland, EH14 4AS, UK
dml@cee.hw.ac.uk   http://www.cee.hw.ac.uk/oceans

Ian Edwards
Halliburton Subsea Systems Ltd
Aberdeen, Scotland, UK

Jonathon Evans, Jeremy Smith
University of Liverpool
UK

Julia Hunter, John Standeven, Martin Colley
University of Essex UK

Abstract[1]

Development and integration of subsystems on advanced robots such as unmanned underwater vehicles, can benefit from the availability of a hardware-in-the-loop (HIL) simulation facility. Although complete interoperability of simulated and real subsystems appears desirable, substantial additional complexity of data flows and hardware can be introduced. Where non-real time simulations are involved, methods of synchronising subsystems running at different speeds must be employed, These should take account of the realities of starting and stopping real subsystem. The paper reviews some of these issues and presents the CORESIM distributed HIL system based around HLA. The system has been used to evaluate a timeslicing synchronisation approach, and to assist in development of docking, visual servoing and concurrent mapping & localisation systems for underwater vehicles.

1. Introduction

New developments in unmanned underwater vehicles (UUVs) and subsea robots are leading to increasingly sophisticated systems, encompassing mechanical hardware, electronics and software for propulsion, manipulation, navigation, control, communications, user interface and more (fig 1).

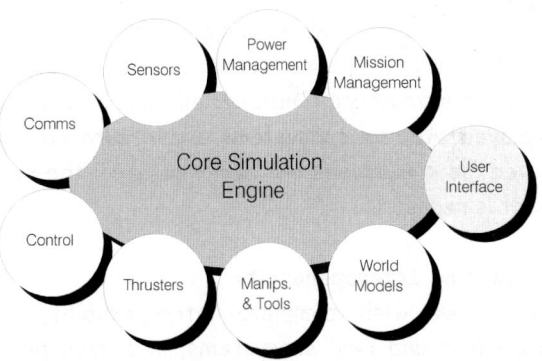

Fig 1: Unmanned Underwater Robot Subsystems

Multiple partners are often involved in developing different subsystems. Successful design, integration and test will benefit from availability of a hardware in the loop (HIL) simulation system to interoperably mix simulations and real subsystems. However, without substantial computing resources, some simulations may not run in real-time. Furthermore, both virtual and real worlds will be involved for subsystems such as manipulation. [1-10]

This paper examines issues of interoperability and synchronisation where real and logical-time subsystems are concurrently mixed with real and virtual worlds. CORESIM - an open framework for integrating vehicle subsystems and immersing them in a simulated environment - is also presented.

Our motivation is to interoperably combine simulations & real hardware to:
- Test & integrate remote subsystems
- Evaluate worktasks a priori
- Replay mission a posteriori
- Enable remote presence
- Facilitate downstream simulation based design and acquisition.

2. Hardware in the Loop Simulation

Fig 2 presents a logical view of the relationship between simulated and physical subsystems. To the left, simulated sensors, actuators & environment for initial testing and validation. In the centre prototyping physical sensors, actuators and environment for initial HIL subsystem testing. To the right, the real sensors/actuators, as deployed with the final system. We employ a common communications spine infrastructure. Initially, all simulations are employed. As real subsystems are integrated and tested, so simulations are removed, until the real robot is realised.

3. Seeking Interoperability

Ideally, we wish complete interoperability of simulations and real subsystems, in a true plug-and-play approach. However, not all combinations are possible or even useful [10]. We examine these in further detail.

Real-Sensor/Simulated Actuator

Consider the case where a selection of real sensors (and signal processing) have been integrated, but (some of) the robot actuators (and control systems) remain simulated. For proprioceptive sensors (e.g. compass, dvl, ins) to behave consistently with the simulated vehicle dynamics, they must be mounted on a motion platform and animated with the output of the vehicle model. This requires additional hardware (in our case a cartesian robot mounted over a test tank) with sufficient motion bandwidth to accommodate characteristic movements, and software integrated within the communication spine.

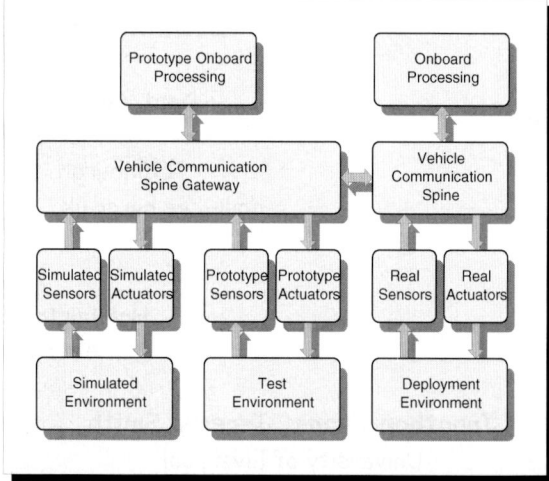

Fig 2: Hardware in the Loop Simulation Approach

The same applies to exterioceptive sensors (e.g. sonar, video). However, in addition, the sensed real world must remain consistent with the virtual world in which the dynamics simulation moves. This requires either complete a priori knowledge of the real world (i.e. not unstructured) or some additional sensing and processing to update the virtual world to remain consistent.

In both cases, the pure plug and play HIL architecture must be modified to include additional data flows between processes and hardware. Some of these only exist to support the HIL aspects, and will not form part of the final in-water robot. The complexity of these systems can be equal to, or surpass, that of the robot itself, questioning its utility.

Simulated Sensor/ Real Actuator

Consider now the inverse case, where proprioceptive and exterioceptive sensors are simulated, but some of the real robot actuators have been integrated (e.g. manipulator, real vehicle platform in water, thruster mounted on cartesian robot etc).

In the case of a real vehicle, with a simulated manipulator mounted on the front, reaction forces from the arm motion and contacts must be applied to the real vehicle, to obtain realistic behaviour. The obvious way is to use the thrusters, but this then limits the available thrust vector for control, and no longer models the final in-water system. Alternatives include an appropriate mechanical mechanism which does not limit the real vehicle's natural dynamic behaviour, or possibly a model based control strategy, where the model being followed is a coupled vehicle/arm system.

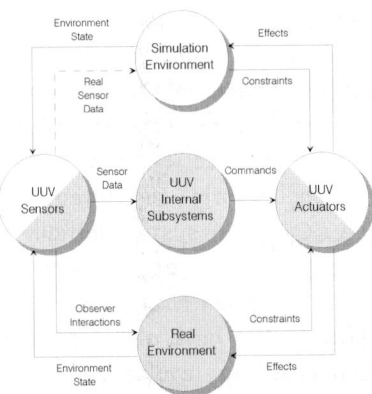

Fig 3: Hidden Data Flows for Interoperable HIL

For the case of a simulated vehicle and real manipulator, the base of the arm must be animated with the vehicle motion (as for the proprioceptive sensor above), using a stewart platform or similar. However, in addition, the reaction forces at the base of the arm must be measured and applied to the simulated vehicle dynamics. Fortunately, typical UUVs move slowly, and so hydrodynamic drag and other inertial effects on the arm which are caused by the vehicle motion will be small.

Once again, additional data flows, hardware and complexity is involved. Although do-able, the complexity of implementing a completely interoperable HIL involving all UUV subsystems with real and virtual world, may be more than practical considerations merit. Pragmatists may prefer to limit the levels of interoperability provided, according to need and resources.

4. Synchronisation

Ideally, all simulations would run in real time, and a real-time clock would be used throughout the system to keep all synchronised. In practice, however, some simulations require substantial computing resources to achieve this and remain faithful. Sonar devices (sidescan, forward look etc) are the prime example, involving physical models of sound propagation and reflection.

If we allow non-real time simulations within the interoperable HIL system, the issue of synchronising logical and real-time is raised. Two classical approaches are [1,3,4,5,7,8,9]

a) to run everything in *lock-step*. The simulation supervisor issues 'tic' to commence the next logical time interval, and awaits 'toc' from all subsystems before repeating

b) a more opportunistic strategy using *timeslicing*.

Lockstep is intuitively the easiest, but slows the progress of logical time to that of the slowest simulation. Real subsystems must also be interruptible and freezable over very short time intervals (milliseconds). This is not helpful where velocity control methods are used in a real subsystem e.g. in UUV and manipulator motion control.

Opportunistic strategies using time-slicing are more helpful, in that longer time intervals are used for interrupting and freezing a real subsystem while simulations catch up. In the classical approach, subsystems proceed at their own pace using their own local copy of logical time. When new input data is required but unavailable, fast local proxy simulations are used as an approximation, and the subsystems proceeds. If, when the correct data is available, there is a discrepancy with the proxy output, time is re-wound, anti-messages are sent out to negate interim outputs from the subsystem, and the system continues.

The presence of anti-messages makes the above prone to *thrashing*. Anti-messages continually re-wind time in adjacent subsystems, so that no progress is made, or logical time continually rewinds to the beginning. For our activities we have therefore employed a variation on time slicing, where fixed intervals are used to coarsely synchronise all.

Here, simulation subsystems are executed for a fixed period, while real subsystems remain frozen. Fast, local proxy simulations are used to predict the missing outputs. After the fixed period, the real subsystems are executed, and their outputs compared with those of the proxies. If there are discrepancies above a given threshold, logical time is rewound (rollback) in the simulated subsystems, and processing repeated using newly available data from the real subsystems.

The advantage of this approach is that real hardware can run for a reasonable length of time between frozen periods, and there should be no more that one logical time re-wind per subsystem in each fixed period (i.e. no thrashing). The disadvantage is that interaction between the real and simulated subsystems is not properly modelled during each fixed period. Determining the ideal period length is thus a balance of simulation accuracy (i.e. short periods) and real-subsystem run length (i.e., long periods)

In practice, the proxy simulations can be used to approximate either the real subsystems, or the HIL simulations. Pseudo code for using the proxy simulations to approximated the real subsystems is:

```
while() { pause_time = sim_time + T;
    while (sim_time < pause_time) {
        get sim_inputs;
        log sim_inputs;
        run sim;
        log sim_outputs;
        sim_time++;  }
    while (real_time < sim_time) {
        get logged inputs;
        run hardware;
        if (logged outputs invalid)
            ROLLBACK;
        else real_time++; }}  }
```

Pseudo code for using the proxy simulations to approximated simulated subsystems is:

Hardware process:
```
while() {get inputs from proxy;
    log inputs;
    run hardware;
    log output;}
```
Simulation process:
```
while() {get inputs;
    if (proxy inputs invalid)
        ROLLBACK;
    else get logged output;}
```

5. CORESIM Implementation

We have implemented concepts from the above across a communications software spine within a distributed computing environment. UNIX boxes, VMEcrate with VMEexec RTOS and PC/Windows have been concurrently supported, linked by TCP/IP. The CORESIM system initially supports limited interoperability of sensors, including animation of a cartesian robot over a laboratory test tank (fig 4). We have also implemented the basic mechanisms to evaluate our modified approach to timeslicing.

Fig 4: CORESIM: Cartesian Robot Carrying Sonar and Video Sensors during Docking Manoeuvre

The simulation infrastructure is provided by the DoD High Level Architecture (HLA) [14], with vehicle simulations expressed as 6 Federates, (pilot's interface, bottom station, propulsion, sensing, vehicle dynamics, simulation management). A drag and drop interface allows users assign processes to particular machines [11].

A subscription communication protocol has been implemented within the HLA run time infrastructure (RTI), to allow hot swapping of real and simulated HLA components at run time. The HIL simulation runs on a virtual machine, and processes publish data and write when available. Process also subscribe to data, and receive using call-back or polling mechanisms. The RTI is responsible for message routing, which is thus invisible to the programmer. Data can be annotated with meta-information defining priority, reliability and more.

For the virtual world modelling is carried out within DVise from Division (fig 5)

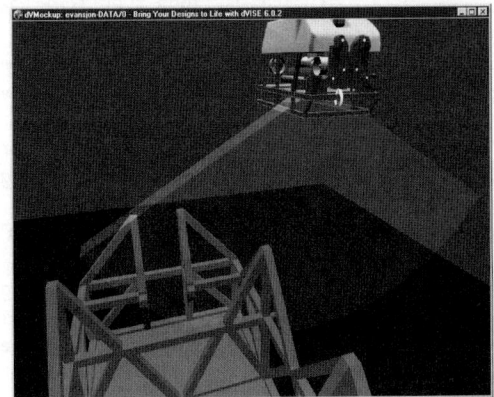

Fig 5: CORESIM: Graphical Display During Docking Manoeuvre

Fig 6: CORESIM: Sonar Data During Docking Manoeuvre

6. Application and Experimental Results

CORESIM has been used to evaluate the use of a 1.2 MHz profiling sonar as a docking sensor for an autonomous AUV shuttle within the EU Thermie SWIMMER project [13]. It has also been used to evaluate visual servoing for ROV stabilisation [16] and concurrent mapping & localisation algorithms with a forward look scanning sonar [17].

For evaluation of the basic timeslicing mechanism, we consider use of a flow metre to measure UUV velocity, mounted on the cartesian robot. The flow metre is interoperably swapped with a simulated velocity sensor. The timeslicing mechanism was employed to initiate rollbacks for varying error values between simulated and actual sensor values.

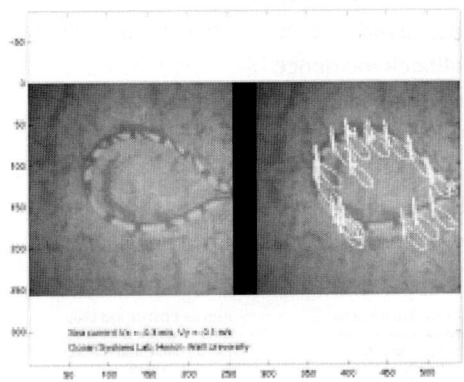

Fig 6: CORESIM: UUV Visual Stabilisation

Fig 7 shows the results of a typical run, plotting speed in X direction moving from the origin to position X=1m, Y=2m at a demanded speed of 0.1m/s. 4 rollbacks took place over a 13s period at 10.25, 10.5, 10.75 and 12.75 seconds.

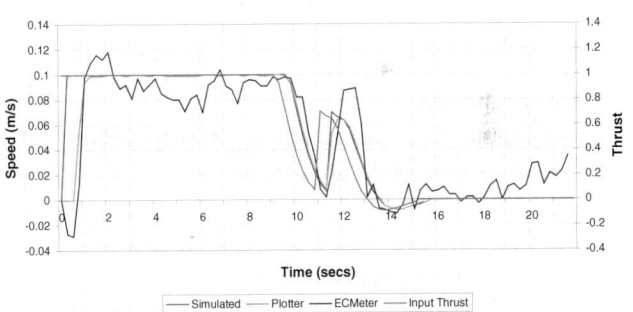

Fig 7: Speed in X direction during positioning – simulated, cartesian robot, current metre & thrust

Fig 8 shows plan view of the same run, performed twice at 0.5m/s with timeslice rollback error values of 5 and 20cm. Plots show position values from the the UUV simulation, the cartesian robot sensors and the current metre. Although runs are similar, there is some none determinisim caused by the noisy sensor data and the differing pattern of rollbacks.

From this, our practical experience of timeslicing can be summarised thus:

a) Rollbacks keep sub-systems synchronized, but inevitably destroy realism
b) Space scaling helps reduce incidence of rollback (eg cartesian robot moves x6 less than UUV simulation)
c) Increased error tolerance reduces rollback incidence
d) Rollback requires real subsystems must be resetable – ie they can be re-wound to an earlier (stored) state
e) Rollbacks and noisy sensors lead to some indeterminacy in simulation progression.

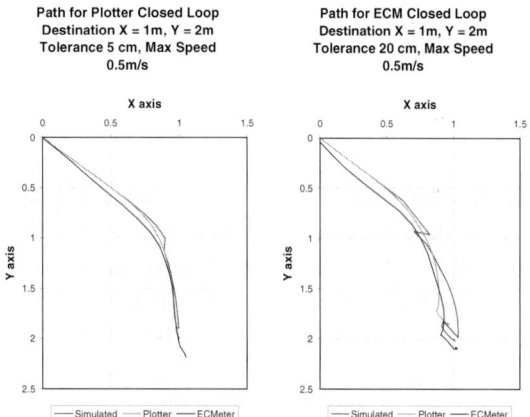

Fig 8: Varying UUV Trajectory with Varying Timeslicing Rollback Error Threshold

7. Conclusions

Although attractive, complete interoperability of simulated and real subsystems for subsea robot HIL is neither straightforward or necessary in practice. Where non-real time simulations are involved, timeslicing and rollbacks can be used to allow real subsystems to run for longer lengths of time. However, noisy sensors and rollbacks lead to indeterminacy, questioning the long term validity of the simulation. CORESIM has been a valuable tool in sensor processing development.

Acknowledgements

This work was funded by Shell UK Exploration & Production, Lockheed Martin, Halliburton Subsea Systems Ltd, Tritech International Ltd., Slingsby Engineering, Sea-Eye Marine and the Defence Evaluation and Research Agency (DERA) in conjunction with the UK Engineering & Physical Science Research Council (EPSRC) and the Centre for Marine & Petroleum Technology (CMPT) in the managed programme Technology for Unmanned Underwater Vehicles (TUUV), phase II - Core Simulation Engine 1997-99.

References

1. C. Frangos; Control System Analysis of a Hardware-In-The-Loop Simulation; IEEE Trans. Aerospace & Electronic Systems, Vol. 26, No. 4, July 1990.
2. Y. Kuroda, K. Aramaki, T. Ura; AUV Test Using Real/Virtual Synthetic World; Proc. 1996 IEEE Symp. Autonomous Underwater Vehicle Technology, pp365-372.
3. D. Maclay; Simulation gets into the loop; IEE Review, May 1997.
4. M. Raynal, M. Singhal; Logical Time: Capturing Causality in Distributed Systems; Computer, Feb 1996, pp49-56.
5. S. Alles, C. Swick, M. Hoffman et. al., A Real-time Hardware-in-the-Loop Vehicle Simulator for Traction Assist, Int. Journal of Vehicle Design, Vol.15, No. 6, 1994, pp597-625.
6. Donald P. Brutzman, Yutaka Kanayama and Michael J. Zyda, Integrated Simulation for Rapid Development of Autonomous Underwater Vehicles, 1992 IEEE Symp. On Autonomous Underwater Vehicle Technology.
7. David R. Jefferson "Virtual Time". ACM Transactions on Programming Languages and Systems Vol 7. No. 3 July 1985 pp404-425
8. J.I. Leivant, R.J. Watro "Mathematical Foundations for Time Warp Systems" ACM Transactions on Programming Languages and Systems vol 15 no 5 Nov 1993, pp771-794
9. R.M. Fujimot "Parallel Discrete Event Simulation" Communications ACM Oct 1990 vol 33, no 10 pp30-53
10. D.M. Lane, G.J. Falconer, G. Randall, et al "Mixing Simulations and Real Subsystems For Subsea Robot Development - Specification and Development of The Core Simulation Engine" IEEE Int. Conference OCEANS 98, Nice France, 29 Sept-2 October 1998.
11. J. Hunter "Core Simulation Engine API Version 2.1", University of Essex Dept Computer Science, May 1999
12. D.M.Lane, J. Smith, J Evans, J.Standeven, I Edwards "Achieving Hardware in the Loop Interoperability in Distributed Undersea Robot Simulation" Undersea Weapon Simulation Based Design Workshop, US Naval Undersea Warfare Centre, Rhode Island, USA. June 7-9 2000
13. J.C. Evans, J.S. Smith and K.M. Keller ASOP (Active Sonar Object Prediction): Short Range Positioning for the Autonomous SWIMMER Vehicle. Proceedings of the Fifth European Conference on Underwater Acoustics, ECUA 2000, ED. P. Chevret and M.E. Zakharia, Lyon, France, July 2000
14. High Level Architecture - Federation Development and Execution Process Model (FEDEP). US Dept of Defence, DMSO, version 1.2 May 1998. (http://www.dmso.mil).
15. D.M. Lane, A.G. McFadzean: Distributed Problem Solving and Real Time Mechanisms in Robot Architecture Engineering Applications of Artificial Intelligence Journal. Vol 7 No. 2, 1994.
16. J.F. Lots, D.M. Lane, E. Trucco "Application of 2.5D Visual Servoing to Underwater Vehicle Station Keeping". Proc IEEE Int Conf. OCEANS 2000. Rhode Island Convention Centre, USA. 11-14 Sept 00.
17. Y. Petillot, I. Tena Ruiz, D.M.Lane "Underwater Vehicle Obstacle Avoidance and Path Planning Using a Multi-Beam Forward Looking Sonar" Forthcoming IEEE Journal Oceanic Engineering, 2000.

Planar Motion Steering of Underwater Vehicles by Exploiting Drag Coefficient Modulation

M. Aicardi, G. Casalino, G. Indiveri

Abstract— An underwater planar vehicle, actuated by rear thrusters and equipped with longitudinal control surfaces which allow the drag coefficient modulation in the sway direction, is considered in a dynamic setting. The maneuvring controls for the vehicle, in order to reach the required final position and attitude are devised by exploiting both the rear thrusters actuation and the capability offered by the admitted presence of the longitudinal, modulable, control surfaces.

I. Introduction

The control of non-holonomic systems has been considerably put under attention in the last years even in connection with the development of the theory of control of non linear systems. Many results are available regarding either the theoretical aspects ([1], [2], [4]) and their application to terrestrial ([3], [5]), space or marine vehicles ([8] [9]). As to marine vehicles, specific results have been provided either regarding the task of following a predefined trajectory ([6] [9]) and for the stabilization of the position and orientation on a target point ([7] [8]).

In this paper, the second of the above problems (let us call it *parking problem*) is considered.

Following the same reasoning line that has been proven effective in [5] we shall consider a polar representation of the position and orientation of the vehicle. For the unicycle model such an approach has revealed that a smooth and time-invariant strategy exists that stabilizes the vehicle on the target point. Such a result does not contradict the Brockett's result since there exist a non-smooth transformation between polar and cartesian state variables. Nevertheless, the stabilization in the polar domain does not imply the stabilizazion in the cartesian domain (by the way a unicycle controlled by the strategy proposed in [5] is asymptotically stable in the polar domain and *unstable* in the cartesian domain in the sense of Lyapunov). On the other hand, polar coordinates seem to be more

M. Aicardi and G. Casalino are with the Department of Communications Computers and Systems Science, University of Genoa, Via Opera Pia 13, 16145 Genova, Italy, E-mail: {michele/pino}@dist.unige.it

G. Indiveri is with the GMD-AiS, German National Research Center on Information Technologies, Institute for Autonomous Intelligent Systems, Schloss Birlinghoven, 53754 Sankt Augustin, Germany

natural to represent the 'posture' of the vehicle (they mimic the 'human' choice).

Starting from this point of view, we shall consider a marine vehicle (therefore with the presence of a sway component of the velocity) and extend the result obtained in the *terrestrial* case to the new situation. To do this, one may be tempted to assume available lateral thrusters that can compensate the sway velocity.

In this paper we do not consider such a possibility but on the other hand we shall assume that the drag coefficient can be modified by means of a modulation of longitudinal control surfaces. More specifically, the role assigned to such control surfaces results in that of suitably controlling (withot any specific compensation) the sway velocity in order to allow the asymptotic convegence toward the desired final position and attitude for the vehicle. Anaytical proofs of stability and convergence are provided within the paper. Different simulation experiments are also provided and compared with the purely kinematic case showing the effectiveness of the proposed dyamic control technique. The so obtained satisfactory results actually encourage further studies devoted to their eventual extension to the more general 3D dynamic case.

II. Model and control problem

Consider the following figure:

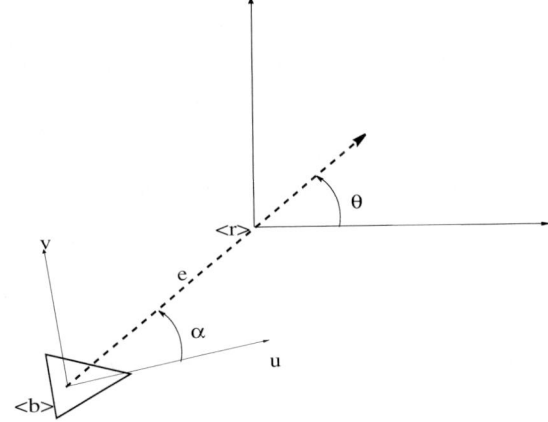

Fig. 1. The local vs. reference frame described by polar coordinates

The use of polar coordinates have already been successfully used in [5] where they have been shown to be suitable to encompass the limitations of the Brockett's theorem [1]. The state equation can be written (following [5] and [9]) as

$$\dot{e} = -u\cos\alpha - v\sin\alpha \quad (1)$$
$$\dot{\alpha} = -\omega + u\frac{\sin\alpha}{e} - v\frac{\cos\alpha}{e} \quad (2)$$
$$\dot{\theta} = +u\frac{\sin\alpha}{e} - v\frac{\cos\alpha}{e} \quad (3)$$
$$\dot{v} = -Dv - mu\omega \quad (4)$$

having denoted with u and v respectively the surge and sway velocity, with ω the angular velocity, and with D, m the drag and inertia parameters. In (4) we shall assume to be able to modify the term D modulating the quantity of underwater surface. Before entering into the details of the paper, it is worth observing that the model does not report the presence of the quadratic drag term $H|v|v$. However, as it is apparent the quadratic drag component can only help in reducing the magnitude of the sway velocity. Then, we omit such a term and will make some remark at the end of the paper.

As to the control problem, we want to determine suitable laws for $u, \omega,$ and v in order to park asymptotically the vehicle over the reference frame. In other words we want $e, \alpha, \theta,$ and v to converge to zero in an asymptotic way.

For a planar unycicle the appropriate control laws have already been determined in [5]. We recall them here (using the parameter h in [5] set to 1):

$$u = \gamma e \cos(\alpha) \quad (5)$$
$$\omega = k\alpha + (\alpha + \theta)\frac{\gamma\sin 2\alpha}{2\alpha} \quad (6)$$

A key role was played by the structure of u given in (5) that allowed to eliminate the dependence on e in the denominator of the state equations and of (6).

It is now somehow appealing to understand if and when the same laws can be used to control a marine vehicle. Obviousy, a-priori, we cannot say anything about the structure of v and then we could face a singularity in the state equations. However, as shown in the following section, a suitable structure for D exists such that also v will help to smooth the behaviour of the system. So, we shall divide the study in two steps:

- determine if there exist a structure for D suitable to make $v = e\,g()$
- show the performance of the system controlled by (5) and (6)

A. *Structure of v*

Assume
$$v = eg() \quad (7)$$
being $g()$ a suitable function not depending on e.

If (7) were true it would satisfy the differential relation given by (4), i.e.,
$$\dot{e}g + e\dot{g} = -Deg - muw \quad (8)$$

Now, considering the control laws given by (5) and (6) we have
$$uw = ef() \quad (9)$$
with $f()$ not depending on e. Then, by using (9) and (5) in (8) we have
$$(-\gamma e\cos^2\alpha - eg\sin\alpha)g + e\dot{g} = -Deg - mef \quad (10)$$

Equation (10) means that g must satisfy a differential equation *independent* of e given by
$$\dot{g} = (-D + \gamma\cos^2\alpha + g\sin\alpha)g - mf \quad (11)$$

The differential equation (11) is not so good. In fact, the quadratic dependence on g may generate a finite-escape-time of the solution. In fact, from a physical point of view, it may happen that e can be *zero* with the vehicle having a non null sway velocity v. Hence in this case, the two conditions $e = 0$ and $v \neq 0$ togheter with (7), would imply an infinite value for g. To overcome such a fact, it is simple to note that if D is chosen in the following way:
$$D = D' + \gamma\cos^2\alpha + g\sin\alpha \quad ; \quad D' > 0 \quad (12)$$

the differential equation for g turns out to be a linear one with the term mf acting as a forcing signal, i.e.,
$$\dot{g} = -D'g - mf \quad (13)$$

where the forcing term mf does not depend on e Using the control laws given by (5) and (6), and also the structure in (7) the closed-loop equations can be rewritten as:

$$\dot{e} = -\gamma e\cos^2\alpha - eg\sin\alpha \quad (14)$$
$$\dot{\alpha} = -k\alpha - \gamma\theta\frac{\sin 2\alpha}{2\alpha} - g\cos\alpha \quad (15)$$
$$\dot{\theta} = \gamma\frac{\sin 2\alpha}{2} - g\cos\alpha \quad (16)$$
$$\dot{g} = -D'g + \\ -m\gamma\cos\alpha\left(k\alpha + \gamma\frac{\sin 2\alpha}{2} + \gamma\theta\frac{\sin 2\alpha}{2\alpha}\right) \quad (17)$$

First note that the set of equations (15),(16), (17) form an autonomous subsystem whose solution is independent on e. This means that the time evolution of α, θ and g is invariant with respect to the absolute 'distance' of the vehicle from the target. The conjecture about the structure of v is then proven provided D is chosen as in (12).

Obviously, since D has to be always positive we shall take care about the choice of D'. This fact will be taken into account at the end of the study.

B. Use of the 'terrestrial' controls: performance analysis

The result given in the previous subsection guarantees that the state equations of the system do not have any singularity.

We can now use the stability theory to analize the behaviour of the system. In [10] it has been proven that the unycicle system controlled by (5) and (6) shows an exponentially asymptotically stable behaviour. Then, (see [11]) there exist a Lyapunov function $W_u(\alpha, \theta)$ and four strictly positive scalars z_1, z_2, z_3, z_4 such that:

$$\begin{aligned} z_1 \|\alpha, \theta\|^2 &\leq W_u(\alpha, \theta) \leq z_2 \|\alpha, \theta\|^2 \\ \dot{W}_u &\leq -z_3 \|\alpha, \theta\|^2 \\ \|\partial W_u/\partial \alpha \; \partial W_u/\partial \theta\| &\leq z_4 \|\alpha, \theta\| \end{aligned} \quad (18)$$

The important point of the above properties is, as will be apparent in the following, that

$$\lim_{\alpha, \theta \to 0} \frac{\|\partial W_u/\partial \alpha \; \partial W_u/\partial \theta\|^2}{\dot{W}_u} \quad (19)$$

is always well defined.

To the aim of the paper consider now the following 'candidate-Lyapunov-function':

$$W = W_u + \frac{1}{2} g^2 \quad (20)$$

Differentiating with respect to time,

$$\dot{W} = \frac{\partial W_u}{\partial \alpha} \dot{\alpha} + \frac{\partial W_u}{\partial \theta} \dot{\theta} + g \dot{g} \quad (21)$$

Substituting now (15),(16) in the above derivative we get, by virtue of the linear presence of g in such equations:

$$\dot{W} = \dot{W}_u - \left(\frac{\partial W_u}{\partial \alpha} + \frac{\partial W_u}{\partial \theta}\right) g \cos \alpha - D' g^2 - g m f \quad (22)$$

that we rewrite as:

$$\begin{aligned} \dot{W} =\; & -D' g^2 \\ & - [(\partial W_u/\partial \alpha + \partial W_u/\partial \theta) \cos \alpha + mf] g + \dot{W}_u \end{aligned} \quad (23)$$

note that, as can be seen by (17)

$$\lim_{\alpha, \theta \to 0} f = 0 \quad (24)$$

Now, consider (23) as a second order equation in g. Since the second order coefficient is surely negative, if such an equation would not admit real solutions than the overall function would be always non positive. With a straightforward application of the elementary properties of the second order equations we can compute the Δ as

$$\Delta^2 = [(\partial W_u/\partial \alpha + \partial W_u/\partial \theta) \cos \alpha + mf]^2 + 4 D' \dot{W}_u \quad (25)$$

The properties reported in (18) allows to say that Δ^2 is negative if

$$D' > \frac{[(\partial W_u/\partial \alpha + \partial W_u/\partial \theta) \cos \alpha + mf]^2}{4 z_3 \|\alpha, \theta\|^2} \quad (26)$$

To conclude the proof of the result we just have to say something about the structure of the r.h.s of (26). In particular, it is interesting to understand what happens as α and θ extremely small. It can be easily seen that, when the denominator of Δ^2 becomes small, also the numerator shows the same behaviour with a dependence not slower than the denominator itself. This is due to the third of (18) and from the fact that f is uniformly Lipschitz function in its arguments. Then, the upper bound given by the r.h.s. of (26) is always well defined. Finally, as to the positivity of D we can easily note that, whenever the system parameters has been fixed, the behaviour of g is surely upper-bounded so that we can always choose D' such that both (26) is satisfied and $D = D' + \gamma \cos^2 \alpha + g \sin \alpha > 0$.

At this point we have just shown the convergence of α, θ and g to zero.

As regards e we can say that, on the basis of (14)
- e cannot cross zero
- for small values of α the behaviour is exponentially decreasing

On the basis of such considerations we can say that there exist a time instant t^* (corresponding to the *practical* convergence of α, θ and g to zero, beyond which also e will monotonically and exponentially decrease to zero.

Note that, we cannot say that e always decreases (as would happen in the unicycle case) but just that it eventually exponentially converges to zero. This is however sufficient to our aim.

Remark. Coming back to the absence of the quadratic drag term, it should now be apparent that its presence would automatically increase the value of

D' and then help us for our objective. Nonetheless, if such a term would be present, we could not say any more that exactly $v = eg$. However, we could always say that $|v| \leq eg$, hence allowing alwasy to exclude singularities in the closed loop state equations.

III. SIMULATION RESULTS

In this section we report the simulation of the system under consideration compared with the behaviour of the terrestrial unycicle vehicle, in order to point out the effect of the sway velocity and the actions taken by the controller. The system's parameters are the following: $k = 8, \gamma = 2, m = 1, D' = 20$.

Figure 2 shows the behaviour if the trajectory, state variables and controls for a vehicle initially put in $\{0, -1, -\pi\}$ with null initial sway velocity v:

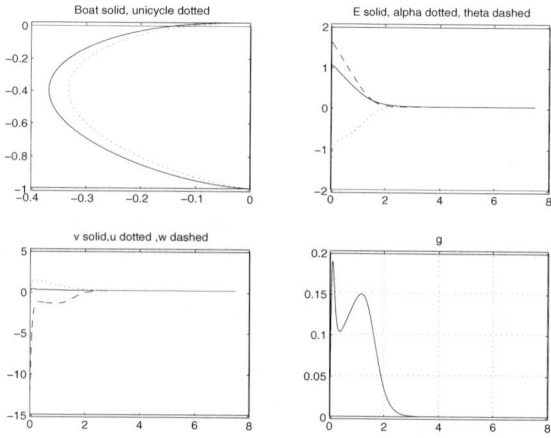

Fig. 2. Case of null initial v

Figure 3 shows the effect of a non zero initial v.

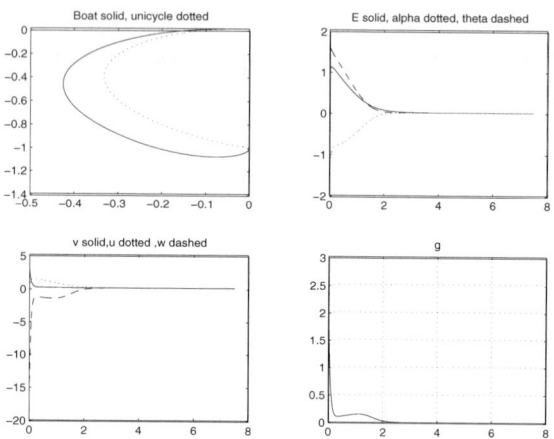

Fig. 3. Case of non-null initial v

To deal with 'far' manoeuvring, we report in figure 4 the same simulation as the one in Fig. 2 except for the fact that in the following case $y(0) = -100$. As it can be seen, the trajectories are the same, and more important function g has exactly the same pattern as the one in Fig. 2 as was expected from the fact that $\dot{g}$ does not depend on e as already noted.

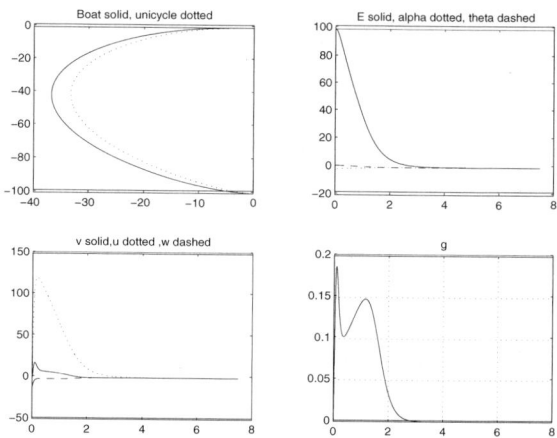

Fig. 4. Case of 'far' initial conditions

Figure 5 refers to a null initial absolute angle. As it can be seen the 'S' manoeuvring is accomplished quite well.

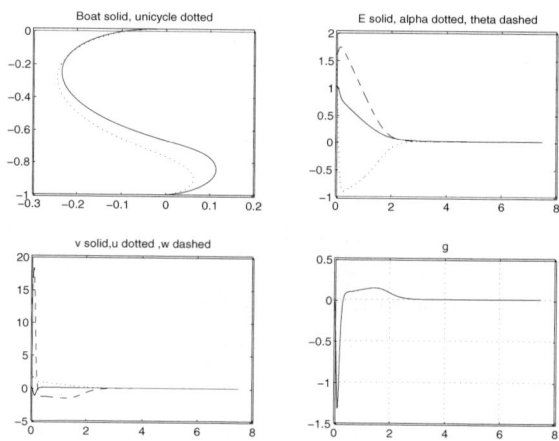

Fig. 5. 'S' manoeuvring with $D' = 20$

Let us finally choose $D' = 6$, and consider the same 'S' manoeuvring. Obviously, the trajectory in Figure 6 is very different than the previous one, but the general properties of the closed loop system are preserved. As it can be seen in Figure 7, in the very early phases of the navigation, the vehicle is not tangent to its trajectory showing the sliding due to the sway velocity.

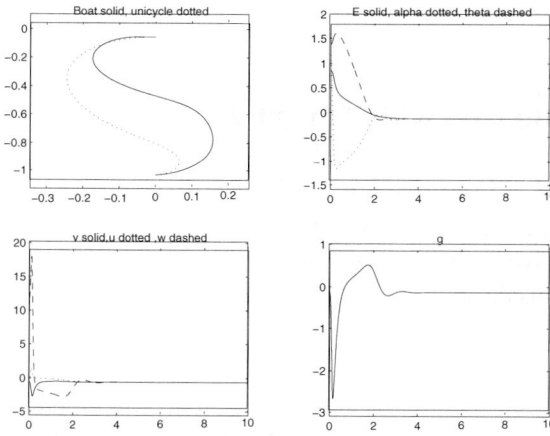

Fig. 6. 'S' manoeuvring with $D' = 6$

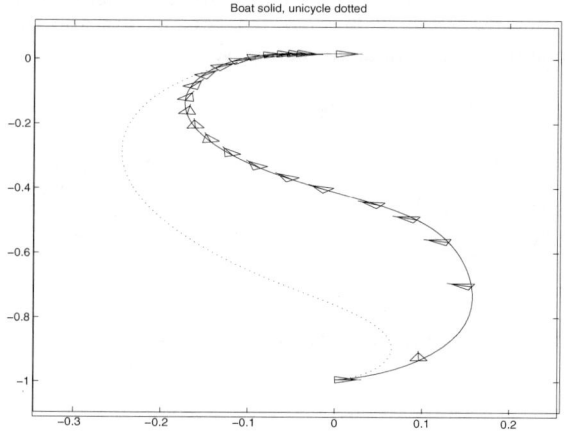

Fig. 7. Sliding with $D' = 6$

References

[1] R. W. Brockett, "Aymptotic Stability and Feedback Stabilization", in R. W. Brockett, R. S. Milman, and H. J. Sussmann, editors, *Differential Geometric Control Theory*, Birkhauser, Boston, pp. 181-193, 1983.

[2] I. Kolmanowsky and N. H. McClamroch, "Developments in Nonholonomic Control Problems", *IEEE Control Systems* pp. 20-36, December 1995.

[3] E. Badreddin and M. Mansour, "Fuzzy-tuned State Feedback Control of a Nonholonomic Mobile Robot", *Proc. 12th World Congress of the IFAC*, Sydney, Australia, 1993.

[4] A. Astolfi, "On the Stabilization of Nonholonomic Systems", *Proc. of the 33rd IEEE Conference on Decision and Control* pp. 3481-3486, 1994.

[5] M. Aicardi, G. Casalino, A. Bicchi, A. Balestrino, "Closed Loop Steering of Unicycle-Like Vehicles via Lyapunov Techniques", *IEEE Robotics and Automation Magazine*, pp. 27-35, March 1995.

[6] K. Y. Pettersen, O. Egeland, "Exponential Stabilization of an Underactuated Surface Vessel", *Proc. CDC'96*, Kobe, Japan, 1996.

[7] K. Y. Pettersen, H. Nijmeijer, "Global practical stabilization and tracking for an underactuated ship - a combined averaging and backstepping approach", *Proc. IFAC Conf. on Systems Structure and Control*, Nantes, France, 1998.

[8] M. Reyhanoglu "Exponential Stabilization of an Underactuated Autonomous Surface Vessel", *Automatica*, Vol. 33, No. 12, pp. 2249-2254, 1997.

[9] M. Aicardi, G. Casalino, G. Indiveri, "Nonlinear time-invariant feedback control of an underactuated marine vehicle along a straight course", *5th Ifac Conference on Manoeuvring and Control of Marine Crafts*, Aalborg, Denmark, August 23-25, 2000

[10] M. Aicardi, G. Casalino, A. Balestrino A. Bicchi, "Closed Loop Smooth Steering of Unicycle-like Vehicles", *IEEE CDC 1994*, Orlando, Florida, 1994.

[11] J. J. Slotine, W. Li, "'Applied Nonlinear Control'", Prentice-Hall, 1991

Underwater Cable Following by Twin-Burger 2

*Arjuna Balasuriya and **Tamaki Ura,
*School of EEE, Nanyang Technological University, Singapore,
**Institute of Industrial Science, University of Tokyo, Japan.
Email: earjuna@ntu.edu.sg

Abstract

In this paper, a sensor fusion technique is proposed for Autonomous Underwater Vehicles (AUVs) to track underwater cables. The work presented here is an extension of the vision based cable tracking system proposed in [1,2,3,4]. The focus of this paper is to solve the two practical problems encountered in vision based systems; namely 1) navigation of AUV when cable is invisible in the image, and 2) selection of the correct cable (interested feature) when there are many similar features appearing in the image. The proposed sensor fusion scheme uses deadreckoning position uncertainty with a 2D position model of the cable to predict the region of interest in the image. This reduces the processing data increasing processing speed and avoids tracking other similar features appearing in the image. The proposed method uses a 2D position model of the cable for AUV navigation when the cable features are invisible in the predicted region. An experiment is conducted to test the preformance of the proposed system using the AUV "Twin-Burger 2". The experimental results presented in this paper shows how the proposed method handles the above mentioned practical problems.

1. Introduction

Recently, vision based AUV navigation had got much attention However, due to undesirable optical behaviour underwater, there were many occasions where the cable is not visible enough for the vision processor to track the cable. In addition the environment itself makes the cable invisible with time, due to the growth of underwater plants etc..

This paper looks at this particular problem and proposes a system which could take care of the situation when the cable is invisible optically. The proposed system uses multi-sensors fused together to keep track of the cable even when it is invisible to the CCD camera mounted on the AUV. A rough layout model of the cable is used with uncertain deadreckoning data to predict a region where the cable is expected in the image. This confines the visual processing to a smaller area avoiding misinterpretations of other similar features and also decreasing the processing time due to the reduction in image data. The performance of the proposed algorithm is tested using the test-bed AUV, the "Twin-Burger 2" with a cable setting highlighting the above mentioned practical situations. It is shown that the proposed multi-sensor fusion method manages to over come the above difficulties which arise when using visual servoing.

Section II discusses the overview of the proposed system and highlights the fusion techniques used to overcome the above mentioned problems. Section III presents the experiment conducted to demonstrate the performance of the proposed algorithm and section IV shows the results obtained. The discussion and the conclusions are made in section V.

2. System Overview

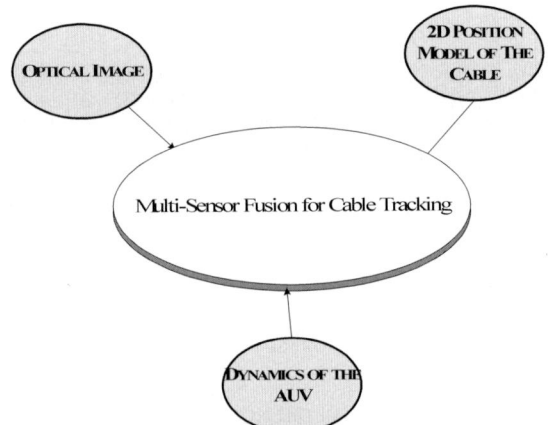

Fig. 1: Multi-Sensor Fusion

The system proposed is diagramatically shown in Fig. 1. In this technique, information other than the optical information is used in decision making for the navigation of the underwater vehicle. Position of different points on the cable is measured and this information is used as a rough position model of the cable. The rough position data model of the cable is used to predict the most likely region of the cable in the image, which reduces the amount of image processing data decreasing the processing time. Due to the narrowing of the region of interest in the image, the chances of misinterpretation of similar features appearing in the image can be avoided. In the proposed technique, dynamics of the AUV are also used to predict the position of the cable in the image [1,2,4]. This compensates the delays introduced by the image processing algorithm [4].

A. 2D Position Model of the Cable

A rough model of the layout of the cable is generated by taking the position (x_i, y_i) of a few points along the cable as shown in Fig. 2.

The line connecting x_i, y_i to x_{i+1}, y_{i+1} acts as the model of cable and is used to predict the most probable region in the image for image processing and is used for navigation command generation when the vision processor cannot recognize the cable in the environment.

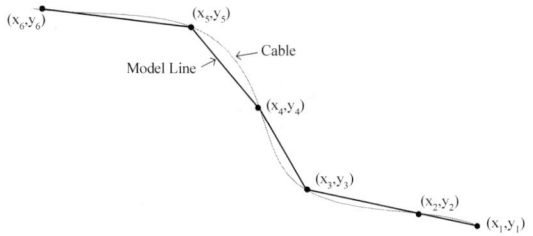

Fig. 2: Rough Layout Model of the Underwater Cable

The use of the model for cable pose prediction is explained in Fig. 3. The region of interest corresponds to the region of uncertainty of the rough model. Using the camera model, the corresponding position of the model line is determined and the interested region for image processing is selected according to this position. The interested region is selected according to the uncertainty of the cable model. Uncertainty is considered as the maximum possible deviation of the actual cable with respect to the model line. Narrowing of the region of interest in the image reduces the chances of misinterpretations of other similar features in the image and also increases the speed of processing due to the reduction in the input data for image processing. If the cable is not detected in this region then, the navigation is carried out by following the model line. The model line features (ρ, θ) are used in the Hough plane too as shown in Fig. 4.

The cable image introduces a high concentration of pixels in a particular direction forming a line in the image plane.

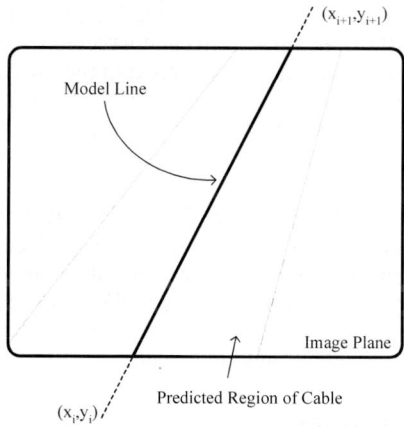

Fig. 3: Region Estimation for Image Processing

This line feature introduces a peak in the Hough space and its existing region can be predicted using the uncertainty of the model line as shown in Fig. 4. This technique avoids the extraction of other peaks in the Hough plane. In other words, it is possible to distinguish the cable of interest even when there are similar cables appearing in the image.

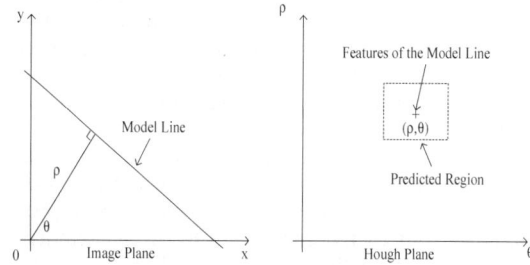

Fig. 4: Region Estimation in the Hough Plane

If a peak representing the cable cannot be found in the region predicted, the features of the model line is used for the navigation of the AUV. The integration of different components for cable tracking can be schematically represented as shown in Fig. 5. The low-level controller provides the dynamic state of the AUV and the optical image, captured by a CCD camera. The optical image is pre-filtered for optical noises using the SoLoG filter explained elsewhere [4]. The position of the AUV determined by deadreckoning is used to predict the region of interest in the image plane and that region is transformed into the Hough plane. Also using the position data of the AUV, the interested region in the Hough plane is predicted and the features of the cable image are extracted from the Hough plane. The size of the region varies from frame to frame and as a result, the image processing time will not be a constant.

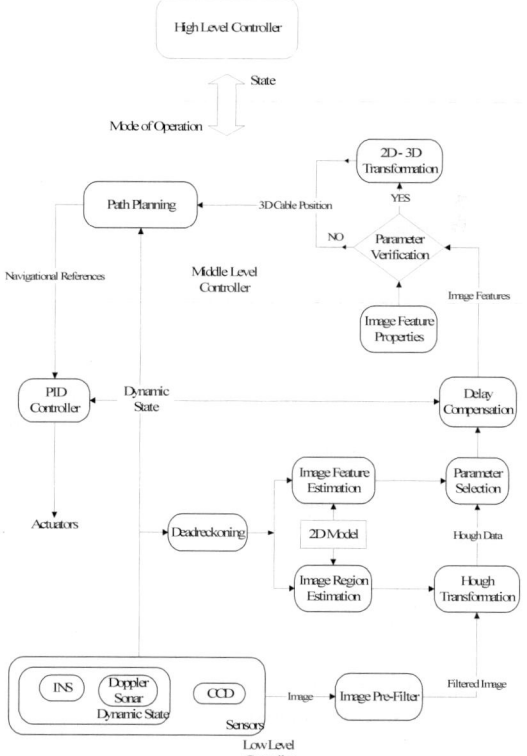

Fig. 5: The System Architecture

Therefore, in order to compensate the delays caused by image processing, a predictor based on the dynamics of the AUV which is explained in detail in [4] is used as the delay compensator. Once the instantaneous features of the cable is determined, these features are verified to check whether they represent the cable. The properties of the cable features are; 1) there are no abrupt changes, and 2) number of pixels on the line feature should be greater than a defined threshold. If the features extracted by image processing represents the properties of the cable then those 2D features are transformed into the vehicle coordinate system for determining the navigational references. Else the model line features are used for determining the navigational references. The generation of navigational references are discussed in [1,2,3,4].

The high-level controller explained in [4] is modified in this algorithm as shown in Fig. 6. It is important to use the same coordinate system for both 2D model of the cable and AUV positioning. This task is achieved by positioning the AUV initially with respect to a defined target (initial mark) using visual data. The initial mark is selected in such a way that it has two line features falling in the x and y axis of the coordinate system used for AUV positioning. The direction of the cable at the initial location is selected as the y axis and a perpendicular line feature (mark) to that direction is selected as the x axis as shown in Fig. 7. The intersection of these two lines is taken as the origin of the coordinate system.

Two modes of operation are carried out to initialize the position at the cross point of the mark shown in Fig. 7.

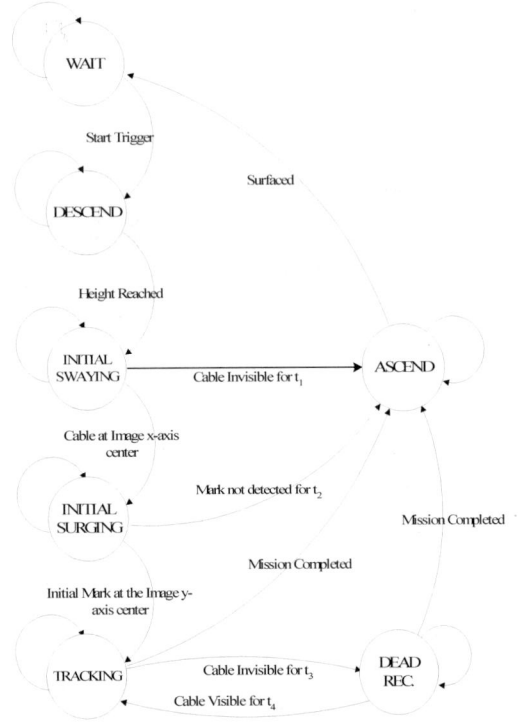

Fig. 6: High-Level Controller

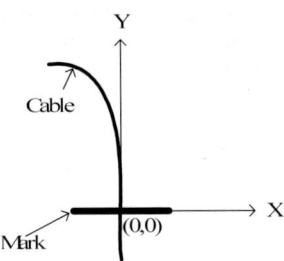

Fig. 7: Initial Positioning Mark

The initial dive of the AUV is done by keeping the heading direction, determined prior to the mission, along the Y axis. Once the defined height of the AUV with respect to the bottom is reached the AUV switch into the INITIAL SWAYING mode. Here, using visual feedback the AUV is navigated in the direction parallel to X axis only with sway actuator. At this instant heading and height references are kept constant. This mode is carried out until the image of the cable appears at the horizontal centre of the image. In this task the camera is kept in a forward-down looking position. The camera axis is parallel to the centre axis of the AUV. Once the cable is at the centre of the image the X-axis position of the AUV is made zero. Next, by keeping the same sway position, heading and height the AUV is moved forward by surging until the mark appears at the vertical centre of the image. It is called the INITIAL SURGING mode. Therefore, the mark-cable intersection point is taken as the (0,0) point of the coordinate system used for the 2D model of the cable, and for the positioning of the AUV. If the initial positioning is failed, the AUV ascends to the surface indicating that it could not recognize the initial mark. Else it starts tracking the cable and during tracking if the vision system cannot recognize the cable, the mode of operation is switched to the DEADRECK mode which uses the 2D model line for navigation until the cable is visible to the CCD camera mounted in the front of the AUV.

The proposed algorithm therefore, can handle the situations when the cable is invisible to the CCD camera and when there are many similar cables appearing in the image. The introduction of the 2D model of the cable improves the performance of the autonomous underwater cable tracking system.

3. Experiment

In order to demonstrate the performance of the proposed cable tracking algorithm an experiment is carried out using the test-bed AUV, "Twin-Burger 2" shown in Fig.8, available at the University of Tokyo, Japan. The dynamic state of this AUV is measured using the on-board sensors such as Rate Gyros, CCD camera, Doppler sonar, Compass, etc..

The experiment is carried out in a testing tank, whose depth is 3.5m and the setting of the cable is as shown in Fig. 9. The aim of the experiment is to test the performance of the proposed system to handle the two main practical points mentioned in section 2. The problem in vision processing to recognize the cable to be tracked when there are similar features appearing in the image is tested by laying a similar cable close to the cable of interest. The problem in vision based cable tracking to track a cable when it is invisible in the image is tested by discontinuing the cable. Also a similar cable, whose layout is different, is introduced to the image when the tracking cable is invisible in the image.

In this experiment a yellow colour hose pipe is used as the underwater cable to be tracked. The relative vertical distance between the cable and the AUV is controlled to be at 1m. An aluminium beam is used as the mark at the starting point.

Fig. 8: Test-Bed AUV the "Twin-Burger 2"

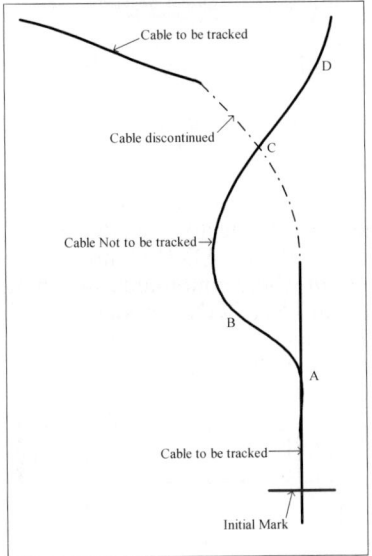

Fig. 9: The Experimental setting of the Underwater Cable

The CCD camera, mounted in the front of the AUV, is positioned with pan zero and with a tilt angle of 55° with respect to the horizontal. The surge, sway, yaw and heave motions are controlled for the autonomous underwater cable tracking mission. Path planning according to the visual features are carried out as explained in [4]. The cable is laid for a distance of about 27m and is invisible for a distance of about 8m.

Similar hose pipe is laid close to the interested hose in order to check the performance of the tracking algorithm. The surge speed control reference is kept at 0.1 m/s for vision based tracking and 0.2m/s for deadreckoning. Sway speed control reference is dependent on the change of position of the cable in the image in vision based tracking and is dependent on target position for deadreckoning. Yaw rate control reference is dependent on the position change of the cable in the image and is dependent on the position model line in the case of deadreckoning. Heave reference is generated to maintain a constant height with respect to the cable.

5. Results

The 2D position model of the cable is constructed by obtaining the position of eight points on the cable as shown in Table 1. The trajectory taken by the Twin-Burger in following the cable shown in Fig. 9 is presented in Fig. 10.

Table 1: The 2D Position Model of the Cable

X	Y
0.0	0.0
-0.1	5.0
-0.3	10.0
-0.8	15.0
-1.3	17.0
-2.5	19.0
-4.0	21.0
-5.0	22.0

The trajectory taken by the Twin-Burger is shown by the dark line in Fig. 10. Initially there is a deviation from the 2D model line due to the search of the initial mark. Position data is captured using the Doppler sonar on board Twin-Burger 2.

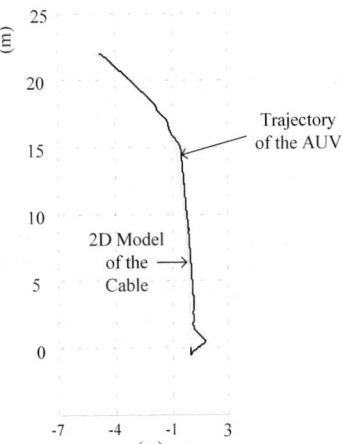

Fig. 10: Trajectory of Twin-Burger 2

Different modes commanded by the High Level Controller in following the cable is shown in Fig. 11. Heave controller uses the range to bottom to maintain a constant height with respect to the cable as shown in Fig. 13.

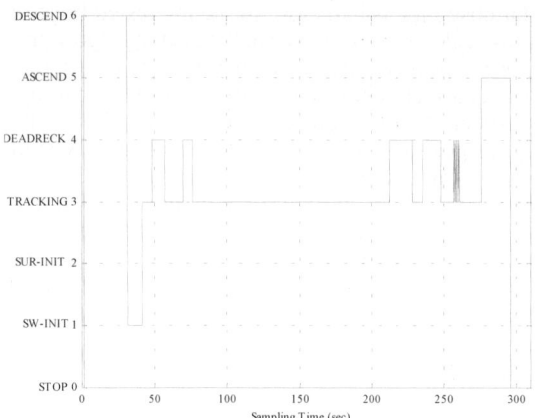

Fig. 11: Modes of Operations

DESCEND mode initializes the mission and will descend to the bottom until the desired height, in this case 1m, with respect to the cable is reached. As shown in Fig. 13, the height is reached around 30sec sampling time and as shown in Fig. 11 the mode changes to the SW_INIT where visual data is used to search for the cable. From Fig. 12 it can be seen that the cable is visible around the 30sec sampling time. The value 200 in Fig. 12 indicates that the cable is invisible. Once the cable in the image reaches the horizontal centre zone, in this case image parameter $\rho > 70$, $\rho < 120$ and $\theta < 30°$, $\theta > -30°$, the mission mode changes to the SUR_INIT as shown in Fig. 12.

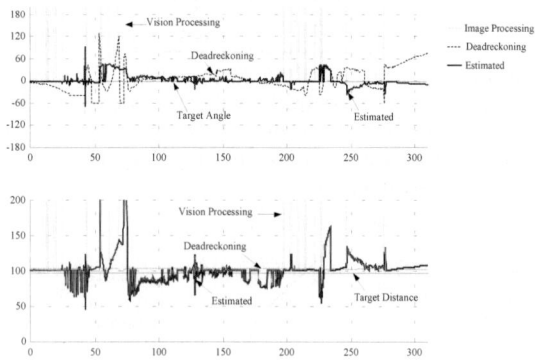

Fig. 12: Visual Parameters

The sway position is marked as the initial location or in other words as the origin of the x-coordinate used for position of the AUV. In the SUR_INIT mode the AUV is navigated with initial heading, and the sway control reference target is set to the sway position at which the mode of operation changed to SUR_INIT from SW_INIT and surge control reference is set to 0.05 m/s as shown in Figs. 14 - 16. It is clear from Fig. 12 that there is some difficulty for the vision processor to recognize the cable and as a result the mode of operation is changed to the DEADRECK mode. Due to the discontinuity of the cable, around sampling time 210 sec. the cable is invisible to the image processor as shown in Fig. 12 and as a result the mode of operation is changed to the DEADRECK mode and is back on TRACKING mode around the 250 sec. time.

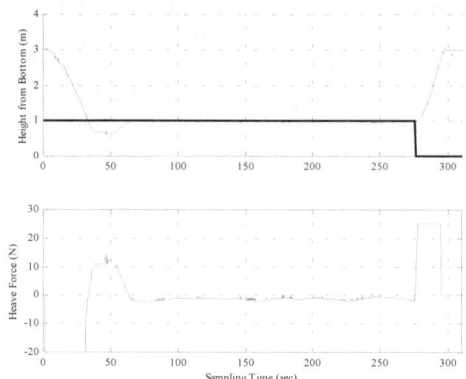

Fig. 13: Heave Controller

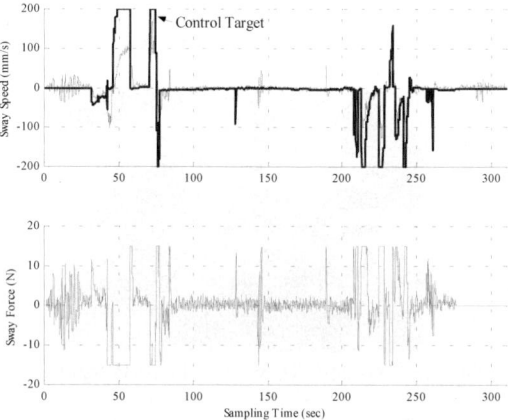

Fig. 14: Sway Controller

Considering the data points around sampling time 230sec. in Fig. 12, it can be seen that the vision processor recognizes a feature similar to the cable. However due to the difference between the deadreckoning based features and estimated features by dynamics, the mode switches back to the DEADREC mode. After 250sec. there is a region where continuous switching between DEADREC and TRACKING modes. This phenomena is due to the appearance of the cable just inside the region estimated in the image.

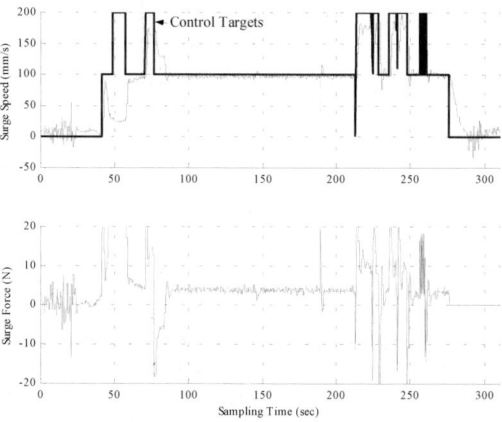

Fig. 15: Surge Controller

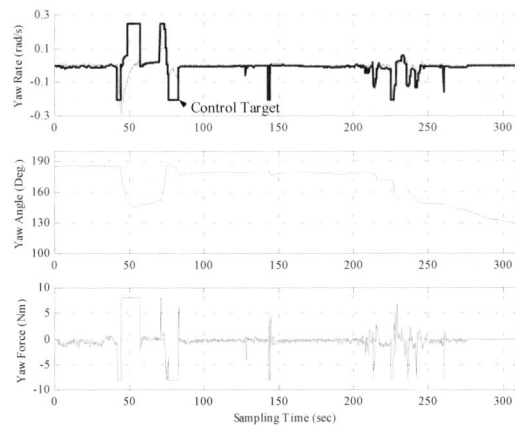

Fig. 16: Yaw Controller

6. Conclusions

In this paper, fusion of multi-sensors for the autonomous tracking of underwater cables is reported. This work is an extension of the previous work presented on vision based underwater cable tracking. Two important practical problems; 1) invisibility of the cable in the CCD image, and 2) selection of the cable to be tracked when there are many cables appearing in the CCD image, arising in vision based tracking are discussed. In this paper, image data, deadreckoning data and 2D position model data are fused together to overcome the above mentioned problems. The performance of the proposed algorithm is demonstrated by conducting an autonomous underwater cable tracking mission in a real-world environment using the AUV 'Twin-Burger 2'.

The Fig. 11 shows how the mode of operation is selected by the High Level Controller, dependent on different sensor data. The reliability of vision based tracking systems are enormously improved with the fusion of a position data model. The use of deadreckoning data for navigation only when the cable is invisible in the image will encounter less problems due to accumulated error compared to total navigation with deadreckoning. Uncertainty in position data by deadreckoning introduces a region in the image in which the cable is most likely be located. This helps the algorithm to reduce the amount of image processing data, decreasing the processing time. Prediction of a region avoid tracking of other similar features appearing in the image. These situation can be seen in the experimental results. The point A in Fig. 9 is a situation to demonstrate the case when there are similar features in the image and the section consisting C is a situation to demonstrate the case when the cable is invisible. Further test had been done by introducing a similar cable in the image when the tracking cable is invisible (point C). It is interesting to note that the image processor recognizes it as a cable as shown in Fig. 12 around sampling time 230. However, the algorithm detects it as a different cable and switches back to DEADREC mode as shown in Fig. 11.

Therefore, it can be concluded that the use of different sensors fused together, even with uncertainties, makes the performance of the entire system more reliable. The use of uncertain deadreckoning data for the prediction of visual parameters improves the visual processing and vision based tracking.

ACKNOWLEDGEMENT

The authors would like to thank the members of the Ura Laboratory, Institute of Industrial Science, University of Tokyo for the help given in conducting the experiments and the Ship Research Institute, Tokyo for allowing us to use their testing facilities for this experiment.

REFERENCES

[1] B.A.A.P Balasuriya and Tamaki Ura, "Vision Based Autonomous Underwater Vehicle Navigation: Underwater Cable Tracking", Proc. of OCEANS'97 IEEE/MTS, Nova Scotia, Canada, Octorber 1997.

[2] B.A.A.P Balasuriya and Tamaki Ura, "Autonomous Target Tracking by Underwater Robots Based on Vision", Proc. of IEEE UT'98, Tokyo, Japan, April 1998.

[3] B.A.A.P Balasuriya and Tamaki Ura, "Vision Based Object Following for Underwater Vehicles (part 2: collaborative control with pan and tilt camera mechanism)", Journal of the Society of Naval Architects of Japan, vol. 182, December 1998.

[4] B.A.A.P Balasuriya, "Computer Vision for Autonomous Underwater Vehicle Navigation", Doctoral Thesis, Univ. of Tokyo, April 1998.

[5] T. Fujii, T. Ura and Y. Kuroda, "Mission Execution Experiments with a Newly Developed AUV 'Twin- Burger'", Proc. of UUST'93, Durham, New Hampshire, September 1993.

[6] J. Kojima and Y. Shirasaki, "Acoustic Video-signal Transmission System for Autonomous Under water Vehicle", Proc. of AUV'96, Monterey, pp. 348-353, June 1996.

A Virtual Collaborative World Simulator for Underwater Robots using Multi-Dimensional, Synthetic Environment,

S.K. Choi and J. Yuh

Autonomous Systems Laboratory
Department of Mechanical Engineering
University of Hawaii at Manoa
2540 Dole Street #302
Honolulu, HI 96822 USA
schoi@hawaii.edu & yuh@eng.hawaii.edu

Abstract

***DVECS** (**D**istributed **V**irtual **E**nvironment **C**ollaborative **S**imulator) is used for testing of unmanned underwater vehicles (UUVs) where both real and simulated vehicles can interact and cooperate in a hybrid, synthetic, virtual environment. This virtual reality system can be used to determine: (1) the optimal performance and criteria for the cooperating vehicles and its relative application; (2) the determination of the advantages and disadvantages of collaborative application tasks between multiple UUVs; and (3) the optimal communication links between the cooperating vehicles and its remote control stations. It also utilizes a virtual reality projection and polarized eyewear systems, which immerses the user for optimal visualization.*

Introduction

Even with the increased interest in the development of underwater robotic technology, the design, fabrication and analysis of autonomous underwater vehicles (AUVs) are still very complex, expensive, and time-consuming. The unpredictable and hazardous underwater environment is extremely unforgiving and remote. With stark limitations in communication, an AUV must operate in a fully autonomous or near autonomous modes. This requirement immensely complicates the development, diagnosis, and evaluation of AUV's many subsystems. In order to ensure reliability in these systems, it is imperative to obtain and maintain accurate software and hardware data. For this purpose, it is necessary to test and re-test these systems under severe or extreme conditions in a controlled laboratory environment before operational or sea-trial deployment. In addition, since many military, scientific, and commercial tasks in open-oceans often require multi-national participation, it becomes a necessity to rehearse these operations before the actual operation, thus establishing operational strategy and ensuring the success of the operation without releasing proprietary or secured materials. Therefore, the Distributed Virtual Environment Collaborative Simulator (DVECS) becomes the ultimate tool for these needs.

Several universities have conducted research in the graphic simulator arena. To mention a few, they are: (a) the Naval Postgraduate School and their NPS AUV Integrated Simulator for their NPS AUV [1]; (b) the University of Tokyo and their Multi-Vehicle Simulator for their Twin-Burger AUV [2]; and (c) the Autonomous Undersea Systems Institute and their Cooperative AUV Development Concept [3]. Both the NPS and UT systems were developed on the IRIX environment of the Silicon Graphics workstation, while the AUSI system runs on the Win32 environment on an Intel based system. All systems, however, are running OpenGL graphic protocols, which is platform independent.

DVECS was developed with the sole objective of reducing the lead-time required for (1) the tedious aspects of pre-testing software and hardware before deployment and (2) the collaboration of various AUVs without having to consider the transportation of these vehicles. Much of DVECS is based and developed on the graphic test platform architecture for AUVs by Yuh, Adivi and Choi [4], and the SGI GL based 3-dimensional graphics by Choi, Yuh and Takashige [5].

DVECS utilizes a similar architecture of a combined hierarchical and heterarchical structure of the previous test platforms along with UT's MVS simulation system; but DVECS differs by utilizing a variety of different wireless communications methods – RF links, commercial cellular telephones, wireless Ethernet, wireless LAN, and asynchronous transfer mode – for data transfer. Finally, DVECS incorporates a projection VR system consisting of a RGB high-resolution, high-refresh-rate projector and polarized eyewear with an emitter, and creates an immersion effect.

ODIN

The Omni-Directional Intelligent Navigator (ODIN) is the University of Hawaii's experimental AUV as shown in Figure 1. The electro-mechanical components were designed for easy expandability and are in Table 1. [6]

Table 1: ODIN Specifications (condensed list)

Component	Specification(s)
Pressure Vessel	Near-sphere shape; Al 6061-T6.
Main Computer	Motorola MC68040/33; 4MB DRAM.
Com. Bus	3U-VME Bus; 12 slot chassis.
Ext. Memory	8 MB with 512 KB EPROM.
RTOS	VxWorks 5.2 OS software.
Power Supply	24 Lead Gel batteries.
Thrusters	8 Cylindrical shaped with hood; Al 6061-T6; Brushless motor; bi-directional propeller; torque-controlled.
Stepper Motors	X, Y & Z weight compensation control.
Manipulator	1-dof (90° motion); potentiometer.
Communication	RF Modem; open; RS232 connection.
Sonar Ranger	8 channel range data (0.1-14.4 m).
Inertia System	INS – 3-axis angle and heading rates, 3-axis linear acceleration; RS232 input.
Pressure Sensor	Calibrate between -1.5 to 8.2 V.

DVECS

The design, testing and operation of AUVs and their control systems can benefit immensely from interactive, 3-dimensional (3D) computer simulations. In particular, developing and testing complex systems that involve multiple autonomous underwater robots operating in an uncontrolled environment is considerably safer and cost-effective in a controlled synthetic environment than a real environment, since the research vehicles are not placed at risk of loss or damage. Mission planning, monitoring and analysis can also benefit from an interactive, 3D virtual environment since its performance can be tested prior to actual sea-trials. For these reasons, the DVECS is used in hybrid synthetic simulations for testing real and virtual vehicles in a common environment and, for mission collaboration, planning, monitoring and analysis of existing UUVs. (Figure 2)

Since DVECS and MVS are based on a similar concept and to achieve this collaboration effort, DVECS architecture is designed to operate in a networked environment such that each component of the simulation can be run on a separate system, processor or virtual machine within a single computer; thus, distributing the computation load.

This feature offers several advantages over a single system layout. First, it is possible to modify or create new components with minimal restrictions on internal architecture. As long as each component adheres to specifically prescribed requirements for communication, language and operating system, the design and implementation of each component is irrelevant to the rest of the system.

Second, users can design and test their AUV simulations across the Internet using common servers. This allows computationally intensive 3D simulation of interaction between multiple objects in the virtual world to be simulated on one or more centralized high performance computers, while the processing requirements for the user's AUV simulations are no more than what their physical AUVs would normally require. Thus, this optimizes the computation time and allows users to accurately evaluate their vehicle's computation performance and requirements.

Finally, multiple simulated or physical entities can interact over a networked environment without requiring them to share code or knowledge of each other's capabilities so proprietary or secured algorithms can be tested in a common environment without making them public. This allows for collaboration from many different sectors of the underwater community that wish to "evaluate" their AUV in conjunction with pre-tested AUVs, as shown in Figure 3.

Communication

One key feature of DVECS is that multiple simulated and real systems must be able to interact with one another. In order to accomplish this, communication between the various components is essential and critical. Within the DVECS environment, UDP socket DGRAMs or TCP socket streams convey messages from one component to another. These messages or data are used in the various sub-systems – vehicle graphic module, background graphic module, numerical module, navigation module, control module, & communication module – to exactly and accurately determine the position of the AUV within the virtual environment. In addition, a simple network reflector interface that allows a simulated entity within the DVECS to be controlled by telemetry data from an external component is used to garner data from ODIN, the UH AUV. This interface can be used to monitor a physical or simulated vehicle either directly over a network or with an additional software reflector if the communications system does not include network support.

Currently, messages are converted to the use of DGRAMs, since they offer a more optimal solution where the loss of data packets is not as critical as the delay – such as *"wait till lost packet is recovered"* – or the suspension of a stream causing noticeable latencies. The communication data, in DGRAMs or TCP packets, contains the following information:

< DESTINATION ADDRESS | TIME STAMP | ENTITY ID | MESSAGE TYPE | OPTIONAL DATA >

The *OPTIONAL DATA*, as indicated here, depends on message type and is usually the information about the vehicle itself. For example, the information to DVECS will be the 6 position and orientation data that includes X | Y | Z | Roll | Pitch | Yaw $(x, y, z, \phi, \theta, \psi)$ and an *ENTITY ID*, which is an integer ID for the entity within the DVECS universe – for example, ODIN typically uses "*21*". Messages can include things like resize, translate, rotate, position and orientation, range (for sonar), depth, etc.

Wireless communication between a vehicle in the field (ODIN) and DVECS, for use as a monitoring or testing system, can thus be accomplished either of two methods. First, it is possible by a serial data link that connects to an intermediate interface that relays data between our TCP packets and the serial link, or second, by a direct network connection over a wireless link with an interface either on board the AUV or at the test site. Either system is completely transparent to the DVECS and the communications delay created by adding additional reflectors is typically negligible.

Typically, for ODIN, the transfer of data between the test vehicle and the monitoring laptop is via a RF modem and limited to 9600 baud. The transfer of data from the monitoring laptop to the DVECS is via a cellular phone, microwave connection, or wireless LAN. The limitation of the wireless LAN is sight-to-sight.

Sensor Fusion

For interactive vehicle testing in a hybrid, simulated synthetic environment, combining the actual and the virtual sensor measurements generates the synthetic sensor data. This is easiest for range data, such as from sonar, in which the synthetic range is simply the minimum of the actual and virtual sensor data.

UUVs with even a modest number of sensors can generate an enormous amount of data that can be difficult for an operator or observer to easily interpret. For this reason, it is desirable to combine data from a variety of sensors and present it in a clear, easily understandable way. Within DVECS, it includes a sensor fusion system that can apply sensor data to surfaces in the virtual world that represents actual objects in the real environment of the AUV.

An obvious way to apply sensor data to the surfaces in the virtual world is to use tomographic methods, solving the inverse problem for the sensors. Unfortunately, even for sensors such as cameras under non-reflective and non-refractive conditions where the light paths are all straight lines between the sensor and the object being sensed, directly solving the inverse problem can be computationally intensive and require a significant amount of processing time. For other sensors, such as sonar in water, where multi-path effects become important and with refractive media, the paths are not straight lines and directly solving the inverse problem at reasonable resolution within a reasonable amount of time is impractical on current computers.

In order to overcome these problems, the use of texture mapping of images generated from the sensor data was selected. DVECS achieves this by generating a 'sensor space' coordinate system on the geometries, combining sensor data into texture images and applying the textures to the surfaces of the geometries.

There are several benefits of using texture mapping over directly solving the inverse problem. First, generating and manipulating the texture images can be done with commonly available image processing routines. This has the advantages that these routines can be highly optimized so processing time is minimal and that these routines are well tested, existing code considerably simplifying and shortening design and implementation.

Second, computing the sensor space parameterization of the geometries typically involves solving relatively simple equations at a small number of points on the geometry so does not require a significant amount of processing time. For a simple optical camera in non-refractive media imaging a non-reflective surface, the form of this parameterization is very simple. For simplicity, adopt a sensor space coordinate system where the origin is the focal point, forward distance from the focal point away from the vehicle is the positive z direction, up is positive y and when facing positive z and oriented with positive y up, horizontal left is positive x. Given a point, p, on geometry G, if the corresponding view plane then is parameterized by the coordinates $[u,v,L]$, for fixed L and point p has coordinates $[x,y,z]$ then the parameterization of p has the simple form,

$$[u,v]^T = \frac{L}{z}[x,y]^T.$$

Although the actual parameterization will take on a more complicated form for more complicated sensor problems, in many cases, this approximation can be used quite effectively. Unfortunately, obvious problems arise with discontinuities and multiple representation of a given area in multi-path problems or in media with variable refractive index. In principle, however, many of these problems may be dealt with by using image-processing techniques at the stage of generating the textures. The sensor fusion for ODIN within DVECS can be seen in Figure 4.

Current Simulation

The water current simulation in DVECS is very simple. It contains no turbulence and no curl, and is totally isotropic. Basically, it takes a fixed vector - an amplitude and a frequency - and generates an acceleration based on:

$$F_D = F_0 + Sin(\Omega t)v,$$

where F_D is the total disturbance, F_0 is the current direction, and v is the constant direction vector that specifies perturbation direction (+/-). This simplified assumption can be easily modified to include cross-sections. Random current simulation is being implemented; however, accurate vortices and turbulences may be difficult to generate.

Vehicle Dynamics

DVECS, when in a monitoring mode, does not consider vehicle dynamics since transmitted data directly reflects the motions of the vehicle in the real-world environment. However, when in the simulation mode, the interacting system must consider the basic underwater vehicle dynamics. The following vector equation:

$$M\dot{V} + A(V)V + h = F$$

where $V \in R^6$ is the linear and angular velocities in the vehicle coordinates; $M \in R^{6x6}$ is the inertia matrix; $A \in R^{6x6}$ includes all the nonlinear dynamic terms with velocity terms; and $h \in R^6$ is a vector representing other forces and torques except $F \in R^6$, which represents the forces and torques generated by the thruster forces. [7]

Virtual User Interface

As mentioned before, DVECS uses a Silicon Graphics workstation setup. This dual workstation setup comprises of an Onyx and an Indy. It interfaces with an Electrohome Virtual Reality Projection unit and Stereographics CrystalEyes eyewear and emitter system.

DVECS software is a multi-layered C++ program modularized by its subsystems and utilizes the inheritance properties. It uses OpenGL graphics libraries to generate the background, vehicle and obstacles, and uses Open Inventor 3D toolkit protocols to create the 3-dimensional, virtual images.

The DVECS consists of three windows. They are the Main View Window, Main Menu Bar Window, and Main Control Panel Window.

The Main View displays the virtual environment with the vehicle being tested. The background environment can be modified to represent an area being used, such as the UH dive well, or can used pre-mapped seafloor data to represent a specific deployment area. The window allows instantaneous change in viewpoints and magnification.

The Main Menu Bar allows access to the background, multiple vehicles or obstacles. This is a simple, pull-down menu layout allowing for access to a specific object's properties – dimension, location and attributes.

The Main Control Panel allows access to different environments or simulations, mission controls (start, stop, pause), placement of grid layout, modification of object labels, and control of the sensor and thruster data. It also allows monitoring of system messages, warnings, etc.

Finally, most commonly available high performance graphics hardware includes texture-mapping optimization, so scenes with complex geometries and high resolution, high detail textures are rendered in real time.

Conclusion

This paper presents a brief description of the **D**istributed **V**irtual **E**nvironment **C**ollaborative **S**imulator (**DVECS**) developed at the Autonomous Systems Laboratory of the University of Hawaii. Various tests have shown that the system can greatly help reduce the development time of underwater vehicle hardware and software testing and verification as well as issues on multi-agent problems. Further tests are scheduled to refine the program to allow higher degrees of robustness.

Acknowledgement

This project is partially supported by the National Science Foundation Grant Number INT-9603043 and partially supported by the Office of Naval Research Grant Numbers N00014-97-1-0961 and N00014-00-0629.

References

[1] D.P. Brutzman, Y. Kanayama & M.J. Zyda, "Integrated Simulation for Rapid Development of Autonomous Underwater Vehicles," Proc. of the IEEE Oceanic Engineering Society AUV92 Conference, Jun. 1992.

[2] Y. Kuroda, K. Aramaki, T. Fujii & T. Ura, "A Hybrid Environment for the Development of Underwater Mechatronic Systems," Proc. of the 1995 IEEE 21st International Conference on Industrial Electronics, Control and Instrumentation, Nov. 1995.

[3] S.G. Chappell, R.J. Komerska, L. Peng & Y. Lu, "Cooperative AUV Development

Concept (CADCON) – An Environment for High-Level Multiple AUV Simulation," Proc. of the 11th International Symposium on Unmanned Untethered Submersible Technology, Aug. 1999.

[4] J. Yuh, V. Adivi & S.K. Choi, "Development of a 3D Graphic Test Platform for Underwater Robotic Vehicles," Proc. of the 2nd International Offshore and Polar Engineering Conference, Jun. 1992.

[5] S.K. Choi, J. Yuh & G.Y. Takashige, "Omni-Directional Intelligent Navigator," Underwater Robotic Vehicle: Design and Control, TSI Press, NM, 1995.

[6] K. Kawaguchi, C. Ikehara, S.K. Choi, M. Fujita, W.C. Lee & J. Yuh, "Design of an Autonomous Underwater Robot: ODIN II," The Proc. of the First International Symposium on Intelligent Automation and Control, FRANCE, May 1996.

[7] S.K. Choi & J. Yuh, "Experimental Study of a Learning Control System with Bound Estimation for Underwater Robots," J. of Autonomous Robots, Mar. 1996.

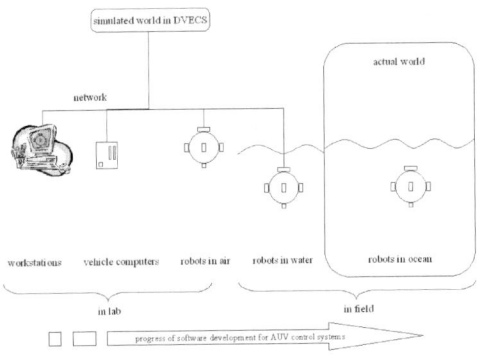

Figure 2: DVECS development environment

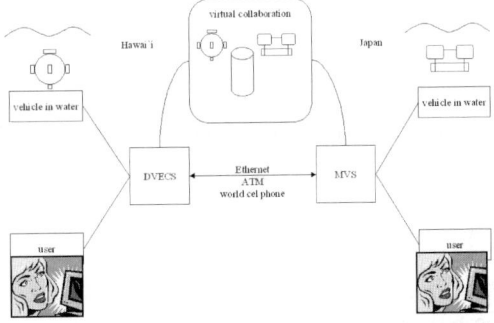

Figure 3: DVECS and MVS Collaboration

Figure 1: ODIN in UH Pool

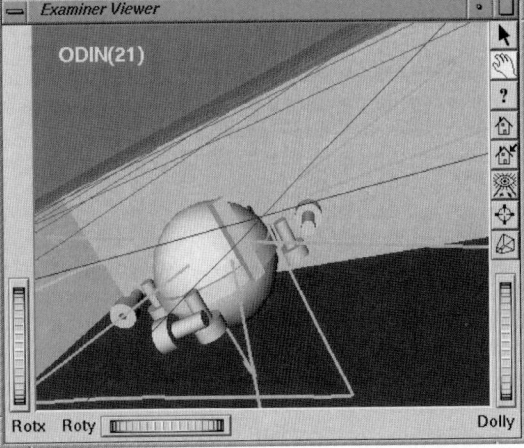

Figure 4: ODIN's sonar representation in DVECS.

Capturing Molecular Energy Landscapes with Probabilistic Conformational Roadmaps

Mehmet Serkan Apaydin[1], Amit P. Singh[2], Douglas L. Brutlag[2], and Jean-Claude Latombe[1]

Departments of Computer Science[1] and Biochemistry[2]
Stanford University, Stanford, CA 94305, USA

Abstract: Probabilistic roadmaps are an effective tool to compute the connectivity of the collision-free subset of high-dimensional robot configuration spaces. This paper extends them to capture the pertinent features of continuous functions over high-dimensional spaces. This extension has several possible applications, but the focus here is on computing energetically favorable motions of bio-molecules. Many bio-chemical processes essential to life require the interacting molecules to adopt different shapes over time. Computational tools predicting such motions can help better understand these processes and design useful molecules (e.g., new drugs). In this context, a molecule is modeled as an articulated structure moving in an energy field. The set of all its 3-D placements is the molecule's conformational space, over which the energy field is defined. A probabilistic conformational roadmap (PCR) tries to capture the connectivity of the low-energy subset of a conformational space, in the form of a network of weighted local pathways. The weight of a pathway measures the energetic difficulty for the molecule to move along it. The power of a PCR derives from its ability to compactly encode a large number of energetically favorable molecular pathways, each defined as a sequence of contiguous local pathways. This paper describes general techniques to compute and query PCRs, and presents implementations to study ligand-protein binding and protein folding.

1. Introduction and Motivation

An insight from research in biology is that the function of a bio-molecule follows from its form. For instance, to act as a potent inhibitor, a drug molecule must bind solidly against a protein's cavity (the binding site), which requires that the molecular surfaces in contact have close steric and energetic match [SK93]. In addition, molecules are neither static, nor rigid. In fact, chemical processes essential to life depend on the ability of certain molecules to adopt different shapes over time. For example, a drug molecule must both move to eventually reach a binding site and deform to achieve a conformation that fits well and lock into this site (docking process). These molecular movements occur under the influence of forces induced by energy fields [Hai92].

Computational models able to effectively simulate and predict molecular motions have important potential applications, notably in drug and protein design. For instance, being able to reliably simulate the ligand-protein docking process would make it possible to automatically extract promising drug candidates (leads) from large existing databases of ligands [LFKL00] and test the docking abilities of variants of these leads.

Molecules can be modeled as articulated structures made of spheres (representing atoms) connected by links (bonds between atoms). The main degrees of freedom (dofs) are torsional dofs about some links. Consider three consecutive links v_1, v_2, and v_3. The torsional dof around v_2 corresponds to varying the dihedral angle made by the plane containing v_1 and v_2 and the plane containing v_2 and v_3. The assignment of an angular value to each dof defines a *conformation* of the molecule (a concept similar to that of a configuration in robotics). The set of all conformations is the molecule's conformational space, which has as many dimensions as there are dofs.

While drug molecules typically consist of 10 to 50 atoms, with 5 to 15 torsional dofs, proteins contain thousands to hundreds of thousands of atoms, with hundreds to thousands of dofs. Energy fields are defined over the conformational spaces of the molecules (or the cross product of several such spaces, if we consider multiple molecules interacting and deforming simultaneously). Molecular movements are described as continuous pathways in a conformational space.

Many computational models assume that molecules are rigid structures. Such models may tell us that a ligand fits into a protein's active site, but give no indication of the conformational changes of the ligand and/or the protein that were required to achieve their final bound state. In reality, the docking may not even be possible because it would require a molecule to traverse high-energy conformations. The transition from static to dynamic models brings us into molecular dynamics [Hai92]. In theory, the energy fields causing molecular motions are well understood. But the precise simulation of these motions over the time periods during which the phenomena of interest take place is well beyond the capabilities of today's fastest computers [DK98]. This is primarily due to the fact that in order to ensure simulation accuracy, the time steps taken by molecular dynamics techniques are usually on the order of femtoseconds. Taking the solvent into account further adds to this complexity.

Researchers have developed approximate energy models that are less expensive to compute, e.g., by using principal component analysis to detect important dofs and "freeze" the others [TPK00], ignoring energy terms involving atoms that are some distance apart, and/or treating groups of atoms (e.g., rings, side-chains, α-helices) as single units. Simulation techniques connecting local minima of such models produce plausible pathways at a more reasonable cost. Some randomness may be introduced into the computation to account for model imperfection [KBK94].

Classical simulation techniques lead to generating individual pathways such as the one shown in Figure 1. However, the number of pathways that can be computed in practice is rather small. Instead, our goal is to capture the relevant characteristics of the "energy landscape" over the conformational space by a network of pathways. This network, called a *probabilistic conformational roadmap* (PCR), is a graph whose nodes and edges are respectively low-energy conformations and short weighted pathways. The weight of a pathway estimates the energetic difficulty for the molecule(s) to move between the two

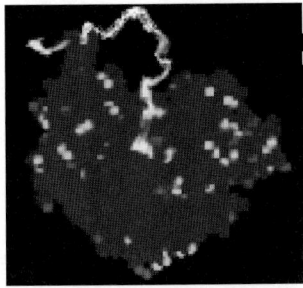

Figure 1. Ligand docking against a protein

conformations. By combining a large number of short paths, a PCR compactly encodes a large number of energetically plausible paths. It thus has the ability to represent the pertinent energy landscape over the conformational space in a form that is more directly exploitable than the original energy function. Once computed, a PCR may be queried in a variety of ways.

Probabilistic techniques (combined with optimization and clustering) have been used to sample conformational spaces of ligands and identify their low-energy conformations [FHK96]. But they do not attempt to connect the sampled conformations into a network. PCRs are a direct extension of the probabilistic roadmaps introduced in [KSLO96]. However, while robot configurations are admissible (e.g., collision-free), or not, a conformational space is the domain of a continuous energy function, where lower-energy conformations are more favorable. Therefore, while a classical probabilistic roadmap tries to capture the landscape of a binary function, a PCR has a similar goal, but with a more complex continuous energy function. The method in [KB99], which constructs a roadmap connecting local minima of a potential function, has some resemblance with ours. But it connects local minima using an up-hill search technique to climb out of local minima towards saddle points. This operation, which requires many evaluations of the potential function, would be too expensive in our case. The concept of a PCR was first introduced in [SLB99], along with its application to ligand-protein binding. The present paper extends this concept, provides new results for ligand-protein binding, and explores the application of PCRs to protein folding. Other ongoing research aimed at applying PCRs to ligand-protein binding and protein folding is reported in [BSA00, SA00].

The problem of capturing functional landscapes over complex spaces is one of general interest. For example, outdoor mobile robots must compute motion plans that take the navigability of the local terrain into account (e.g., muddy and steep areas are more difficult to traverse than flat hard terrain). The navigability of a terrain often depends on the heading of the robot and is best defined over the robot's configuration space. Algorithms have been proposed to compute such paths with acceptable or optimal characteristics [MM97]. Roadmaps similar to PCRs could better capture the pertinent properties of the navigability function over the configuration spaces. Another application is minimally-invasive surgical planning, where one must plan the paths of surgical instruments (e.g., scalpels, endoscopes) to minimize damage on healthy tissue, with some tissues (e.g., blood vessels) being more critical than others (e.g., fat).

Section 2 outlines the basic PCR framework. Sections 3 and 4 apply this framework to two problems, ligand-protein binding and protein folding.

2. Probabilistic Conformational Roadmap

2.1. Classical Probabilistic Roadmap

A classical probabilistic roadmap R is created over a robot's configuration space C [KSLO96]. R is a graph whose nodes are points of C (called *milestones*) and edges are short simple paths (*local paths*) between milestones. The local paths are usually straight-line segments. Points in C are either admissible (e.g., collision-free), or non-admissible. R should lie in the admissible subset C_A of C and capture the connectivity of C_A as well as possible. Ideally, there should be a one-to-one correspondence between the connected components of R and those of C_A, and every point in C_A should be connectable to a milestone by a simple path [HLMK99].

The roadmap R is computed as follows. The range of each dof parameter is normalized so that $C = [0,1]^n$, where n is the number of dofs. Points are picked at random in $[0,1]^n$ and the admissible ones are retained as milestones. Next, pairs of milestones that are sufficiently close to one another are considered and for each pair a local path connecting the two milestones is tested for admissibility. If this path lies in C_A, an edge of R is created between the two corresponding milestones. This basic scheme admits many variants. Points may be picked from $[0,1]^n$ uniformly, or using more sophisticated probabilistic distributions.

Theoretical analysis shows that under reasonable assumptions the probability that a probabilistic roadmap made of s milestones fails to correctly capture the connectivity of a given space C_A converges toward 0 as e^{-s}. In practice, probabilistic roadmaps have been used successfully to solve motion-planning problems in high-dimensional spaces and/or in the presence of complex admissibility constraints.

2.2. Probabilistic Conformational Roadmaps

Let C be the conformational space of a molecule or a group of interacting molecules. If we study protein folding, C may be the conformational space of the protein of interest. But for practical reasons, C may only encode a subset of the protein's dofs, e.g., by considering every amino-acid side-chain as a rigid unit. If we study ligand-protein binding, C may encode dofs of both the ligand and the protein, or it may only be the ligand's conformational space if we assume that the protein does not deform significantly during the docking process.

Let $E: C \rightarrow \mathbf{R}$ be a potential energy field over C. E may combine terms that express a molecule's own potential energy and terms that relate to the interaction between molecules. To illustrate, Figure 2a shows an imaginary function E over a two-dimensional space $C = [0,1]^2$. E varies between -47 and +52.5.

We need a metric over C, such as the maximal distance between any two corresponding atoms. Various other metrics could also be used.

A PCR is constructed by picking points from C uniformly at random. This is done by assigning random values to each coordinate of C, within its given range of possible values. For

each point q we compute $E(q)$ and we accept q as a milestone of the PCR at random with the following probability distribution:
- 0 if $E(q) > E_{max}$
- $(E_{max} - E(q))/(E_{max} - E_{min})$ if $E_{max} \geq E(q) \geq E_{min}$
- 1 if $E_{min} > E(q)$

The resulting milestone distribution is denser in low-energy regions of C.

Let s be the number of milestones selected as above. The next step is to connect every milestone by local paths to at most k other milestones, where k is selected roughly equal to the number of dimensions of C, so that the resulting PCR has size linear in s. The connection algorithm is the following:

For $i = 1, 2, ..., s-1$

1. Set Q to be the queue of the K milestones m_j ($j > i$) that are closest to m_i, sorted according to their distance to m_i.

2. While the number of edges at m_i is less than k and Q is not empty
   a. $m \leftarrow \text{extract}(Q)$
   b. If the straight-line segment (local path) between m and m_i lies in a low-energy region, then connect m and m_i by an edge

We implement Step 2.b by discretizing the segment into a series of points spaced by some small distance ε. An edge is generated if all these points have energy less than a given threshold. Hence, local paths that traverse a high-energy barrier are discarded. Step 2.b is potentially expensive, as it requires computing the energy function at multiple points. So, we bound the number of times it is executed by K for each milestone. K is set to 3 to 5 times k.

Figure 2b shows the projection of a PCR computed by this algorithm for the landscape of Figure 2a, with $k = 4$, $K = 12$, $E_{min} = -40$ and $E_{max} = 20$. The metric used here is the Euclidean distance in $\mathbf{R}^2$.

Finally, for every pair of connected milestones m and m', we estimate the likelihood of the molecule(s) to transit along the local path τ joining them. Let p be the number of discretized points generated at resolution ε along τ and $E_1, ..., E_p$ be the values of E already computed at these points. For any three successive points q_{i-1}, q_i, and q_{i+1} we use the following equation to estimate the probability of moving from q_i to q_{i+1}:

$$\Pr[q_i, q_{i+1}] = \frac{e^{-(E_{i+1} - E_i)/kT}}{e^{-(E_{i+1} - E_i)/kT} + e^{-(E_{i-1} - E_i)/kT}}.$$

We compute the weight of τ as: $w = -\Sigma_{i=1 \text{ to } p} \log(\Pr[q_i, q_{i+1}])$.

Local paths through higher-energy conformations have higher weights than those lying entirely in a low-energy area. Since the total weight is not the same in both directions, we compute and store both weights. During query processing, we select the weight corresponding to the direction in which the path is traversed.

The construction of a PCR requires choosing several parameters: s, k, K, E_{min}, E_{max}, and ε. The most difficult to select is the number s of milestones. We do not know how big should s be for the PCR to effectively capture the landscape of C. An exponential rate of convergence has been formally established for

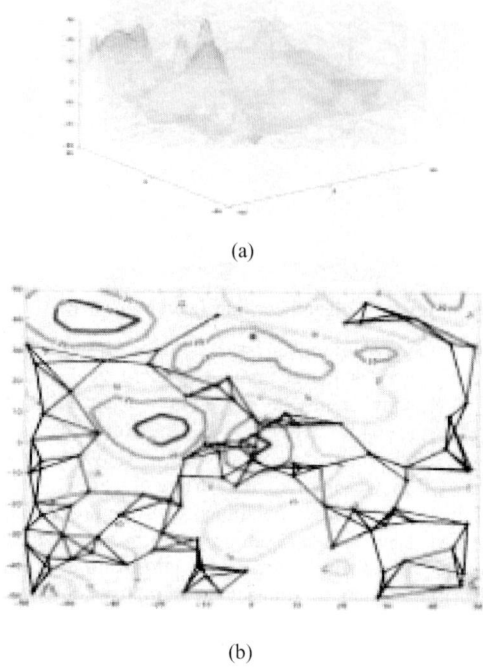

(a)

(b)

Figure 2: Fictitious energy function and computed PCR

classical probabilistic roadmaps [HLMK99], but no such result has been proven yet for PCRs.

2.3. Querying a PCR

It is important to note that the information contained in R cannot be better than the energy function E used to construct it. Furthermore, one could obtain individual pathways reflecting the energy function more precisely, by tracking the values of E at a fine resolution in the conformational space, rather than by sequencing local pathways contained in the PCR. The main advantage of a PCR is that it encodes a large number of paths scattered across the energetically favorable regions of C. Hence, though E is imperfect and milestones provide relatively low-resolution sampling of C, one may get significant and reliable information by collecting invariants and statistics from subsets of paths contained in the PCR. Query processing should take advantage of this strength, instead of relying on individual paths.

The most straightforward query is to determine if energetically favorable paths exist between two input conformations. Defining the weight of a pathway to be the sum of the weights of the local paths it contains, a search algorithm finds N best paths in the PCR between the two input conformations (for some given N). These paths can be visualized and statistics can be computed (e.g., number of milestones, average weight, energy profile). To avoid showing many similar results, the paths can be grouped into clusters using a similarity metric and only one path in each cluster may be output. If the same milestone m lies on several such paths, m may be considered as a likely intermediate, hence a potentially relevant biological structure.

Another query is to find N best paths that enter an input goal conformation, and display the milestones that are contained in those paths. Again, similar paths can be grouped into clusters, and milestones that lie on several distinct paths can be identified as likely intermediates. The average weight of these paths can be compared to the average path weight for other possible end conformations, in order to provide insight on why an end conformation is more likely to be attained than another.

One may use a PCR to perform stochastic simulations. Starting at some initial conformation, a simulation run proceeds step by step. At each step, it decides at random to either stay at the current milestone or transit to an adjacent milestone using a probability distribution based on the weights of the local paths. This type of simulation corresponds well to the modeled biological process since the molecules do not have the prior knowledge of their final conformations. Multiple simulation runs can be performed and statistics can be collected about the traversed milestones. Further analysis may also help discover funnels of attraction steering a molecule toward an end conformation.

2.4. Computational Enhancements

The cost of computing a PCR dominates that of performing many queries. It is therefore desirable to develop techniques that can produce good PCRs faster. One technique, which was successfully applied to classical roadmap [KSLO96], is to construct a roadmap in two stages. A first roadmap R_1 is created with $s_1 < s$ milestones. Then, $s_2 = s-s_1$ milestones (and the corresponding connections) are added to R_1 to form the final roadmap R. The new milestones are picked at random around milestones of R_1 that are the least connected to other milestones (e.g., the number of connections is less than k). Another technique is to evaluate the energy function at a few conformations around every low-energy milestone m and, if important energy variations are detected, to pick new milestones around m. This may help build denser PCRs around important states.

One may use multiple energy models of different complexity. Suppose that a function E' is available, which approximates the energy function E, but at a fraction of the computational cost of E. Let $H(E)$ and $H(E')$ designate the respective subsets of C over which E and E' take high values. We would like $H(E') \subseteq H(E)$ and the difference between the two subsets to be rather small. Since a molecular energy function such as E contains many terms, it is often possible to build E' by ignoring a large number of relatively small terms. We can use E' to build a roadmap R' of size s' much greater than s. Next, we re-consider the milestones and connections in R' and accept/reject them using E to produce the final roadmap R. If E' costs one or two orders of magnitude less time to evaluate than E, we can obtain a PCR of given size s in much less time than by generating it using the only function E.

Each time a point is picked at random in C, a gradient technique could track the steepest descent of E and generate a point of lower energy. This new point would then be the actual milestone candidate. Obviously, the cost of generating each milestone would be greater, and this cost would have to be weighted against that of generating more milestones (without local optimization). Local paths could also be improved by iteratively deforming them into curved ones in order to minimize their weights. Other improvements may use prior knowledge about the molecules and/or the molecular process. For instance, atomic symmetries that cause some torsional angles to have preferred values may be detected. Milestones can then be generated by selecting these angles using non-uniform distributions with peaks at the preferred values. If one knows in advance critical low-energy conformations, such as a ligand's binding conformation or a protein's folded state, these conformations can be input as milestones. A greater density of additional milestones may be generated around them since they often lie in convoluted low-energy passageways [LGH97], which may be difficult to capture by a standard sampling technique [HKL98].

3. Ligand-Protein Binding

3.1. Problem Statement

Biomolecular interactions, such as ligand-protein binding, are critical to the process of life. Ligands are usually small molecules (10-100 atoms) which bind to a specific site on a larger receptor protein. Ligands are used for signaling and regulation in virtually all cellular pathways. Most drug molecules are ligands that inhibit or enhance the activity taking place at the protein sites where they bind. For instance, it was discovered that a specific enzyme (protein) -- the HIV-1 protease -- cleaves the amino-acid chains produced by the HIV virus, hence playing an essential role in the life cycle of this virus. Drugs have been designed which bind to the active site of the HIV-1 protease and thus physically block the amino-acid chains produced by the HIV virus from entering this site.

Most techniques to predict ligand-protein binding attempt to compute the final conformation of the ligand by maximizing an energy score and do not explicitly study the dynamic or kinetic properties of the binding process. To study such properties, researchers have relied on Molecular dynamics, Brownian Dynamics and Monte Carlo simulation techniques. However, these techniques are computationally intensive, especially for ligands with many dofs, and usually provide a relatively small number of plausible ligand pathways.

3.2. Application of PCR

PCRs offer a novel approach to studying the dynamics and kinetics of the ligand-protein binding process by sampling from the space of *all* possible paths that a ligand may follow as it binds to the receptor protein. Hence, instead of simulating the binding process, we use a PCR to effectively guess several possible intermediate conformations of the ligand and obtain a distribution of energetically favorable paths to the binding site via these intermediate conformations.

In the following, we assume that the protein is rigid. This assumption allows us to generate PCRs in the ligand's conformation space. There are cases where deformations of the protein could not be ignored [TPK00]. In those cases, one must consider a conformation space encoding both the ligand's dofs and some protein's dofs.

We model the ligand as an articulated linkage made of spheres (atoms) connected by straight links (bonds). An arbitrarily chosen terminal atom is given 5 dofs, 3 specifying its center's coordinates and 2 specifying the orientation of its only bond. Bonds between non-terminal atoms are assigned a single torsional dof. Since bond angles and bond lengths usually undergo only small variations, we assume they are constant. Atomic rings are modeled as rigid units. Terminal hydrogen atoms are not explicitly modeled, but are accounted for by increasing the radius of the atoms they are bonded to. The 5 dofs of the root atom and the torsional dofs define the conformational space C.

Our energy model over C consists of two components: the energy of interaction of the ligand with the receptor and the internal energy of the ligand. For a given point in C, the energy of interaction is computed by first calculating the coordinates of the ligand atoms in a fixed coordinate system. The energy contributions of each ligand atom are computed based on the potential field created by the protein at the atom's coordinates. This field is calculated from the atom coordinates and charge distribution of the protein. It consists of the van der Waals potential and the electrostatic potential. The van der Waals potential represents the steric constraints on atomic interactions and is modeled using the following Lennard-Jones 12-6 function:

$$v(r) = \varepsilon\left\{\left(\frac{r0}{r}\right)^{12} - 2\left(\frac{r0}{r}\right)^{6}\right\},$$

where r is the distance between two atoms, r_0 is the distance at which the energy is minimum, and ε is the well depth, i.e., $-v(r_0)$, usually about 0.2 kcal/mol.

Since the standard Coulombic equation of electrostatic interaction is valid only for an infinite medium of uniform dielectric, it cannot be used here. The dielectric discontinuity between protein and solvent generates induced or reflected charges that can play a significant role in the binding process. Hence, we model electrostatics using the following Poisson-Boltzmann equation, which is a widely accepted model of electrostatic interactions in solution:

$$\nabla \cdot [\varepsilon(r) \nabla \cdot \varphi(r)] - \varepsilon(r)\kappa(r)^2 \sinh[\phi(r)] + 4\pi\rho^f(r)/kT = 0$$

where ϕ is the electrostatic potential in units of kT/q, k is the Boltzmann constant, T is the absolute temperature, q is the charge on a proton, ε is the dielectric constant, and ρ^f is the fixed charge density. We use the Delphi program [SH90] to solve the equation on a 3-D grid at a resolution of 0.5Å. The van der Waals potentials are computed at the same grid resolution by calculating for each grid point the potential contribution of all receptor atoms within a threshold distance of 10Å.

We compute the energy of interaction of every ligand atom with the protein by indexing the atom's center to the nearest grid point and retrieving the van der Waals and electrostatic potentials at this point. The total energy of interaction is computed by summing the contributions of each atom. The ligand's internal energy is computed by applying the standard van der Waals and Coulombic equations to each non-bonded pair of ligand atoms. (Since a ligand is small and flexible, we assume that its surface is not well defined and hence use the standard Coulombic equation, with a dielectric constant between 60-80.)

Milestones are generated as described in Section 2.2. In addition, extra milestones are generated by iteratively over-sampling regions of lowest energy in C. The final bound state of the ligand is also entered into the roadmap as a milestone.

3.3. Experimental Results
We have constructed PCRs for various ligand-protein complexes. Initial tests were performed on three complexes identified as 1ldm, 4ts1, and 1stp in the Protein Data Bank (PDB, http://www.rcsb.org). Further tests were carried out with proteins that appear to mediate ligand binding primarily by electrostatic effects (e.g., superoxide dismutase and acetylcholine esterase).

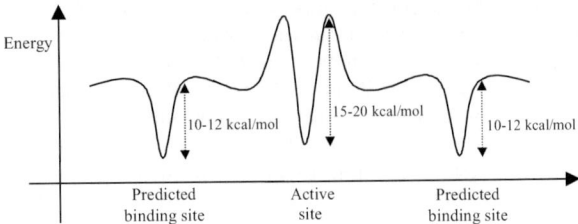

Figure 3: Illustration of energy barriers around an active site

The PCRs were constructed with between 5,000 to 100,000 milestones. On a 195-MHz MIPS R10000 processor, the average PCR construction time ranged from 3-8 minutes for smaller roadmaps to 0.5-3 hours for larger ones. In all runs, our software generated PCRs containing 2 to 5 connected components. As voids and narrow cavities do occur within a protein structure or on its surface, some milestones may be picked in these regions, thus yielding more than one connected component. In each run, over 98% of the milestones were in a single component also containing the bound state.

We studied if our software was able to distinguish the active site from other low-energy sites. The two attributes we used to distinguish between the active and other predicted binding conformations were the ligand's absolute energy and the average weight of the paths entering and leaving the conformation. The average weight was computed by generating minimum weight paths from several randomly selected initial conformations to the final conformation.

We observed that the absolute energy of the ligand was not a strong discriminating factor between the active site and other predicted sites. In two of our three test cases (1ldm and 1stp) the algorithm found ligand conformations outside the active site with energies equal to or even slightly lower than the ligand's energy in its active site conformation. Instead, using the average path weight, our software was able to distinguish between the active site and other predicted sites. The average weight of all paths entering and leaving the active site was on average 30% higher than the weights for all other low-energy sites. Therefore, while it is significantly more difficult for the ligand to *leave* the active site than the other low-energy binding sites, it is also more difficult for the ligand to *enter* the active site. We believe that this result indicates the presence of an energy barrier around the active site that traps the ligand within the site. Figure 3 shows a schematic of a possible energy contour that could yield a similar result. Our experiments also show that the average weight of paths entering the active site is of the same order as the weight of paths leaving the predicted sites (a result reflected in Figure 3). Hence, the difficulty of entering the active site is approximately equal to the difficulty of leaving the other binding sites.

Other tests have focused on analyzing the role of the electrostatic energy in binding. In one series of experiments, we have selectively eliminated one or more of the charges on the protein. When all charges are turned off, the results show that the energy barrier we previously detected is largely eliminated. Hence, not only do the energy minima in the binding site increase, but the energy barrier surrounding the binding site also seem to decrease, hence flattening the curve in Figure 3. These results indicate that the barriers to ligand docking are mainly of electrostatic nature, and not caused by van der Waals potentials. In addition, we have stochastically simulated the motion of the

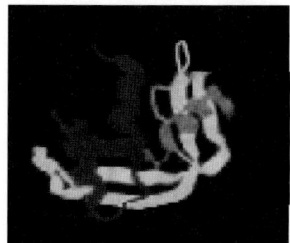

Figure 4. Secondary structure of the ribonuclease A

ligand in a PCR by selecting paths from each milestone based on the distribution of outbound local path weights. Initial results indicate that electrostatics steering is detectable by the PCR, but only at short distances from the molecular surface (5-7Å).

4. Protein Folding

4.1. Problem Statement

A protein is a sequence of amino acids that folds to generate a compact 3-D structure. This structure performs many functions, from building larger assemblies such as muscle fibers to providing specific binding sites for other molecules. The position of the atoms in a folded protein is referred to as the protein's tertiary structure. The primary structure is the amino-acid sequence, while the secondary structure refers to specific local arrangements of a few to a few dozen amino acids. There are two main types of secondary structure elements (SSE): α-helices and β-strands. These SSEs have regular structures, with repeating torsion angles and a constant pattern of hydrogen bonds. They are usually connected by *loops*, which have irregular shapes. An α-helix has a corkscrew shape, with the atoms on the backbone closely packed and the side-chains extended in a helical array. A β-strand is an almost fully extended series of 5 to 10 amino acids. Two or more β-strands often align side-by-side into a β-sheet held together by hydrogen bonds. Most folded proteins are a sequence of α-helices and β-strands connected by loops. Figure 4 shows the secondary structure of the ribonuclease A (a digestive enzyme). Note how intricately and compactly the SSEs are interwoven.

Recent advances in X-ray crystallography and NMR imaging have made it possible to elucidate the folded conformations of a rapidly increasing number of proteins. However, little is known today about the folding pathways that transform an extended string of amino acids into a compact and stable structure. So far it has only been possible to identify approximate intermediate conformations for few proteins. Some biological experiments track a particular property of the protein during folding, but they provide a limited way of following the folding pathway. The ability to predict pathways would help design proteins with desirable properties [KL99]. It could also help determine why relatively small alterations in amino acids may result in dramatic changes of a protein's folded state. Several diseases such as Cretzfeldt-Jakob's, Alzheimer's, and cystic fibrosis are believed to be the result of protein misfolding.

4.2. Application of PCR

The application of PCRs to protein folding is made complex by the large number of dofs. To simplify, we assume here that the SSEs of the protein have already formed and are given as inputs.

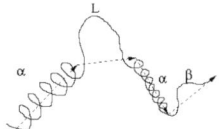

Figure 5: Representation of a protein

This assumption loosely corresponds to studying the folding process after the protein has acquired the so-called "molten globule" state, an observed intermediate for some proteins [PR97]. This state has nearly the same secondary structure as the final fold, but the tertiary structure is not as compact. The pathways provided by a PCR based on this assumption could help understand how α-helices and β-sheets interweave into a compact geometric arrangement.

In a similar way to [SB97], we represent the protein as a sequence of vectors, each representing an SSE (Figure 5). We consider the following dofs (Figure 6):

- A *revolute* dof is located at the extremity of each vector, except the last one. The corresponding parameter is the angle made by the vectors ending and starting at this point.
- A *dihedral* dof is associated to every three consecutive vectors. The parameter is the angle made by the plane containing the first two vectors and the plane containing the last two.
- A *twist* dof is associated to every α-helix and β-strand. A coordinate frame is attached to this SSE with its z axis aligned with the element vector. The dof parameter is the angle between the x axis of this frame and the orientation of the first amino acid on that vector. The twist of an α-helix or β-strand about its own axis does not affect the positions and orientations of other SSEs.
- A *prismatic* dof is associated with each loop. The parameter is the length of the loop vector, which is allowed to vary within a range that is a function of the number of amino acids in the loop.

Our potential function is taken from [STD95]. It has a hydrophobic-interaction and an excluded-volume part. Amino acids are categorized into two groups, hydrophobic (H) and hydrophilic (or polar, P). H-H contacts are favorable, whereas H-P or P-P contacts do not contribute to the energy. The exclusion term ensures that no two atoms are too close. There are also terms of a third type for β-sheets, which account for hydrogen bonding. These terms are a function of the distances between side-chain centroids. This model assumes that hydrophobic interactions drive the folding process and that the specific identity of the side-chains is only responsible for the fine-tuning of the fold. It is argued in [STD95] that the level of success of their function is comparable to that of functions with hundreds to thousands of parameters.

We generate each milestone of a PCR by sampling each dof of the protein's model at random. We explicitly input the folded state as a milestone. Lower energy conformations have a higher probability of being accepted. We only compare the exclusion energy component, rather than the total energy, in accepting a

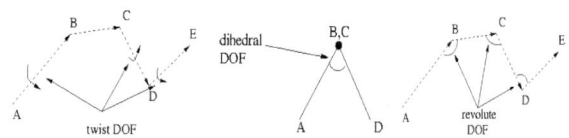

Figure 6: Degrees of freedom in our protein model

conformation. The reason for this is that H-H interactions may counterbalance the contribution of mutually close side-chains to the energy, thus resulting in self-colliding conformations. We set the exclusion energy threshold to prevent any two side-chains from coming closer than 3.8Å. We also take extra samples around the given folded state.

Each milestone is then connected to at most k milestones among the nearest ones, where k is the number of dof in our protein's model. The nearest neighbors of a milestone are found using ANN [AM93], with the Euclidean distance over the conformational space, after normalizing each dof parameter to lie between 0 and 1. We also tried the RMSD metric, but it was slower and did not give significantly different results. To discretize a local path (and eventually decide if it is part of the PCR, or not), we break the path into segments of equal length, such that the variation of every angular dof is less than $\pi/12$ and that of every length dof is smaller than 0.5Å. The path weights are assigned as described in subsection 2.2.

4.3. Experimental Results
After generating PCRs, we performed the following queries:

 1) Compute the minimum-weight path between an arbitrary start conformation and the folded conformation.

 2) Compute M near-optimal paths between the same start conformation and the folded conformation.

For 2), we computed 199 near-optimal paths between the start and goal conformations using the algorithm in [NB94]. For each milestone m in the best path, we computed the number of near-optimal paths that contain m or a milestone close to m.

We considered two proteins previously analyzed in [STD95]: 1hdd and 1le2. We obtained the description of the secondary structure from DSSP [KS83]. We took extra samples around the folded structure. The PCRs were constructed with 1500 to 5,000 milestones. On a 400-MHz Pentium II processor, the average PCR construction time ranged from 7 minutes for smaller roadmaps to 10 hours for larger ones. In all runs, the largest connected component contained more than 95% of all the nodes.

In the initial random sampling, each prismatic dof was uniformly assigned values between 0.5 and 6Å per amino acid on the loop. Each revolute dof was uniformly distributed in [0,π] and each dihedral and twist dof was uniformly distributed in [-π,π]. To generate extra samples around a given milestone, each angle was picked within ±π/6 of the corresponding angle in the milestone and each length was picked within ±0.5Å of the corresponding length. Figure 7 show the results for 1hdd. Energy vs. RMSD distribution is shown in (a); energy profile, rmsd profile, and the ratio of the number of times the milestones on the best path are also visited in the near-optimal paths.are shown in (b) and (c) For (a), red points correspond to samples taken around the native structure, whereas blue points are regularly sampled milestones.

For 1hdd, Figure 7 shows that there are milestones of lower energy than the folded state. This may be due to the various approximations made in the energy model. In (b), the energy profile shows a barrier just before reaching the folded state, similar to the profiles in [SA00]. But several runs led different plots, and (c) shows another profile, for a different PCR of the same size and for another random starting configuration. No barrier is observed before reaching the folded state and no node is visited extensively in the 200 best paths.

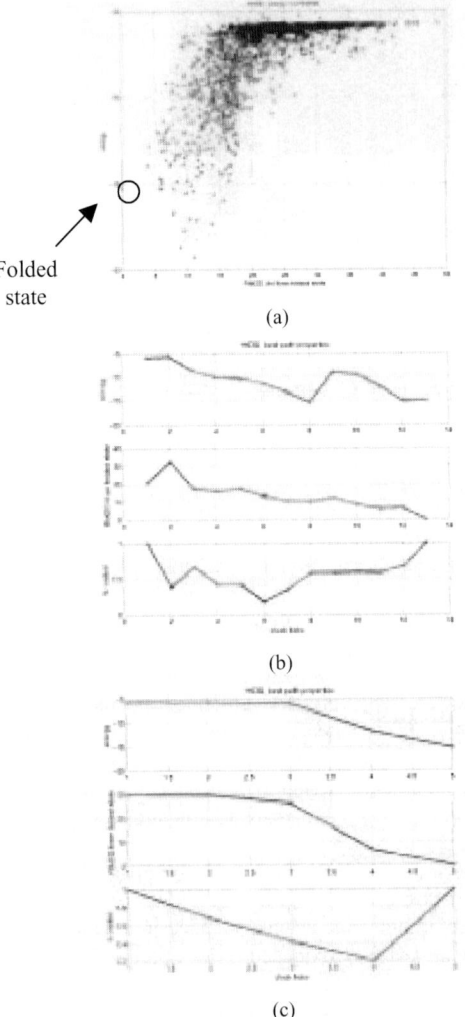

Folded state

(a)

(b)

(c)

Figure 7: Results for 1hdd

For 1le2, our PCR found a configuration which is visited in all 200 best paths. This configuration is displayed in Figure 8, along with the folded structure. Only the backbone atoms are shown. The solid vectors stand for α–helices, whereas the dashed vectors stand for loops. Both structures have the same topology, but the folded state is slightly more compact.

5. Current Research
We are pursuing our work on applying PCRs to ligand-protein binding and to protein folding. The kinematic and energetic models of the molecules need to be improved, especially for protein folding. Sampling techniques incorporating domain-specific heuristics must also be developed to take advantage of the most recent knowledge about biomolecular interactions. Concurrently, we are investigating new general techniques to produce pertinent roadmaps more quickly. Indeed, it is clear from our current work that large PCRs made of several 100,000 milestones, or more, will eventually have to be computed. Moreover, we believe that capturing function landscapes over high-dimensional spaces is a problem arising in several applications and that probabilistic roadmaps similar to PCR are a

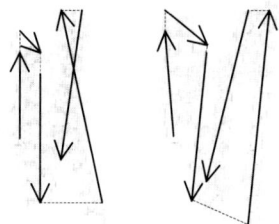

Figure 8. Folded state (l) and most visited state (r) in 200 best paths for 1LE2

promising tool to do it. Hence, any general improvement in computing PCRs can have a significant impact both in computational biology and beyond.

A remaining challenge in applying probabilistic roadmaps to robot motion planning is the so-called "narrow passage" issue [HKL98]. This issue also arises in protein folding, but with much greater acuity. The low-energy subset of the protein's conformation space tends to form a maze of very narrow passages [LGH97], with the (possibly unknown) folded conformation lying in one of them and the initial, extended conformation lying outside the maze. The technique proposed in [HKL98] for robot motion planning is to widen the narrow passages by allowing a small penetration of the robot into the obstacles In protein folding, using an energy function E' approximating the function E with a smaller domain of high values (see Subsection 2.4) has a similar effect. Investigating narrow passages for protein folding may eventually benefit robot motion planning.

Acknowledgements: This work has been partially funded by NSF-ITR grant CCR-0086013 and a Stanford's Bio-X program grant 127X406. A.P.S. and D.L.B. are supported by a grant from the National Human Genome Research Insitute HG02235. M.S.A was supported by the D.L. Cheriton Stanford Graduate Fellowship. This paper has greatly benefited from discussions with L. Guibas, M. Levitt, P. Koehl, and V. Pande (Stanford U.), N. Amato (A&M Texas), C. Camacho (Boston U.), A. Zell (U. of Tübingen) and L. Kavraki (Rice).

References

[AM93] S. Arya and D.M. Mount. Approximate Nearest Neighbor Searching. *Proc. 4th Ann. ACM-SIAM Symp. on Discrete Algorithms*, 271-280, 1993.

[BSA00] O.B. Bayazit, G. Song and N.M. Amato. *Ligand Binding with OBPRM and Haptic User Input: Enhancing Automatic Motion Planning with Virtual Touch*. TR00-025, Dept. of Comp. Sci., Texas A&M U., Oct. 2000.

[DK98] Y. Duan and P.A. Kollman. Pathways to a Protein-Folding Intermediate Observed in a 1-Microsecond Simulation in Aqueous-Solution. *Science*, 282:740-744, 1998.

[FHK96] P.W. Finn, D. Halperin, L.E. Kavraki, J.C. Latombe, R. Motwani, C. Shelton, and S. Venkatasubramanian. Geometric Manipulation of Flexible Ligands. In *Lecture Notes in Comp. Sc.*, 1148, M.C. Lin and D. Manocha (eds.), Springer, NY, 67-78, 1996.

[Hai92] J.M. Haile. *Molecular Dynamics Simulation: Elementary Methods*. Wiley, NY, 1992.

[HKL98] D. Hsu, L. Kavraki, J.C. Latombe, R. Motwani, and S. Sorkin. On Finding Narrow Passages with probabilistic Roadmap Planners. In *Robotics: The Algorithmic Perspective*, P.K. Agarwal, L.E. Kavraki, and M.T. Mason (eds.), A K Peters, Natick, MA,. 141-153, 1998.

[HLMK99] D. Hsu, J.C. Latombe, R. Motwani, and L.E. Kavraki. Capturing the Connectivity of High-Dimensional Geometric Spaces by Parallelizable Random Sampling Techniques. In *Advances in Randomized Parallel Computing*, P.M. Pardalos and S. Rajasekaran (eds.), Combinatorial Optimization Series, Kluwer, Boston, MA, 159-182, 1999.

[KS83] W. Kabsch and C. Sander. Dictionary of Protein Secondary Structure: Pattern Recognition of Hydrogen-Bonded and Geometrical Features. *Biopolymer*, 22(12):2577-637, 1983.

[KSLO96] L.E. Kavraki, P. Svetska, J.C. Latombe, and M. Overmars. Probabilistic Roadmaps for Path Planning in High-Dimensional Configuration Spaces. *IEEE Tr. Rob. and Autom.*, 12(4):566-580, 1996.

[KB99] S.W. Kim and D. Boley. *Building and Navigating a Network of Local Minima.*. TR 99-033, CSE Dept., U. of Minnesota, 1999.

[KBK94] R.M. Knegtel, R. Boelens, and R. Kaptein. Monte Carlo Docking of Protein-DNA Complexes: Incorporation of DNA Flexibility and Experimental Data. *Protein Eng.* 7(6), 761-7, 1994.

[KL99] P. Koehl and M. Levitt. De Novo Protein Design. I. In Search of Stability and Specificity. *J. Mol. Biol.*, 293:1161-1181, 1999.

[LFKL00] S.M. LaValle, P.W. Finn, L.E. Kavraki, and J.C. Latombe. A Randomized Kinematics-Based Approach to Pharmacophore-Constrained Conformational Search and Database Screening. *J. of Comp. Chem.*, 21(9):731-747, 2000.

[LGH97] M. Levitt, M. Gerstein, E. Huang, S. Subbiah, and J. Tsai. Protein Folding: The Endgame. *Ann. Rev. Biochem.*, 66:549-579, 1997.

[MM97] C. Mata and J. Mitchell. A New Algorithm for Computing Shortest Paths in Weighted Planar Subdivisions. *Proc. 13th Int. Annual Symp. On Comp. Geom.*, ACM Press, New York, 264-273, 1997.

[NB94] D. Naor and D.L. Brutlag.. On Near-Optimal Alignments of Biological Sequences. *J. Comp. Biol.*, 1(4):349-366, 1994.

[PR97] V.S. Pande and D.S. Rokhsar. Is the Molten Globule a Third Phase of Proteins? *Proc. of the Nat. Acad. of Science*, USA, 1997.

[SH90] K. Sharp and B. Honig, Electrostatic interactions in macromolecules: theory and applications. *Ann. Rev. Biophys. Chem.*, 19:301-32, 1990

[SLB99] A.P. Singh, J.C. Latombe, and D.L. Brutlag. A Motion Planning Approach to Flexible Ligand Binding. *Proc. 7th Int. Conf. on Intel. Sys. for Mol. Bio.*, AAAI Press, Menlo Park, CA, 252-261, 1999.

[SK93] B.K. Shoichet and I.D. Kuntz. Matching Chemistry and Shape in Molecular Docking. *Protein Eng.* 6(7) 723-32, 1993.

[SB97] A.P. Singh and D.L. Brutlag. Hierarchical Protein Structure Superposition Using Both Secondary Structure and Atomic Representations. *Proc. 5th Int. Conf. On Intell. Syst. for Mol. Bio.)*, AAAI Press, Menlo Park, CA, 284-293, 1997.

[SA00] G. Song and N.M. Amato. *Using Motion Planning to Study Protein Folding Pathways*. TR00-026, Dept. of Comp. Sci., Texas A&M U., Oct. 2000.

[STD95] S. Sun, P.D. Thomas, and K.A. Dill. A Simple Protein Folding Algorithm Using a Binary Code and Secondary Structure Constraints. *Protein Eng.* 8:769-778, 1995.

[TPK00] M. Teodoro, G. Phillips, and L.E. Kavraki. Singular Value Decomposition of Protein Conformational Motions: Application to HIV-1 Protease. *Currents in Comp. Mol. Bio.*, M. Satoru, R. Shamir, and T. Tagaki (eds.), Universal Academy Press., 198-199, 2000.

Physical Geometric Algorithms for Structural Molecular Biology*

Chris Bailey-Kellogg[†]   John J. Kelley, III[†‡]   Ryan Lilien[†]   Bruce Randall Donald[†‡§¶]

Abstract

A wealth of interesting computational problems arises in proposed methods for discovering new pharmaceuticals. This paper surveys our recent work in three key areas, using a Physical Geometric Algorithm (PGA) approach to data interpretation, experiment planning, and drug design:

(1) *Data-directed computational protocols for high-throughput protein structure determination.* A key component of structure determination through nuclear magnetic resonance (NMR) is that of assigning spectral peaks. We are developing a novel approach, called JIGSAW, to automated secondary structure determination and main-chain assignment. Jigsaw consists of two main components: graph-based secondary structure pattern identification in unassigned heteronuclear (^{15}N-labeled) NMR data, and assignment of spectral peaks by probabilistic alignment of identified secondary structure elements against the primary sequence.

(2) *Experiment planning and data interpretation algorithms for reducing mass degeneracy in mass spectrometry (MS).* MS offers many advantages for high-throughput assays (e.g. small sample size and large mass limits), but it faces the potential problem of mass degeneracy — indistinguishable masses for multiple biopolymer fragments (e.g. from a limited proteolytic digest). We are studying the use of selective isotopic labeling to substantially reduce potential mass degeneracy, especially in the context of structural determination of protein-protein and protein-DNA complexes.

(3) *Computer-aided drug design (CADD).* We are developing new CADD tools and applying them to the design of an inhibitor for the Core-Binding Factor-β oncoprotein (CBFβ-MYH11), a fusion protein involved in some forms of Acute Myelomonocytic Leukemia (AMML). Computational-structural studies of CBF help determine the molecular basis for its function and assist in the development of therapeutic strategies. A key issue in such studies is geometric modeling of protein flexibility; our approach attempts to account for flexibility by using an ensemble of structures representing low-energy conformations as determined by solution NMR.

Our long-range goal is the structural and functional understanding of biopolymer interactions in systems of significant biochemical as well as pharmacological interest. The research overviewed here represents a set of important steps towards that goal.

1 Introduction

The field of *Physical Geometric Algorithms (PGA)* studies computational processes that compute or reason about geometric or spatial relationships in the physical world, and their realization in application areas such as robotics and microelectromechanical systems. PGA research pursues the value proposition that, for such systems, predictions of behavior, arguments of correctness, and combinatorial precision devolve to a geometric analysis.

Some of the most challenging and influential opportunities for PGA arise in developing and applying information technology to understand the molecular machinery of the cell. Our recent work (and work by others) shows that many PGA techniques may be fruitfully applied to the challenges of computational molecular biology. PGA research may lead to high-throughput, automated systems that are useful in structural molecular biology.

Concomitantly, a wealth of interesting computational problems arises in proposed methods for discovering new pharmaceuticals. These problems include identifying the low-energy conformations of molecules, interpreting protein NMR (nuclear magnetic resonance) and X-ray data, inferring constraints on the shape of active drug molecules based on measurements of activity of related drug molecules, and docking candidate drug molecules to known protein targets. We survey our research on computer-aided drug design, new techniques for automated NMR data interpretation, and ex-

---

*This work is supported by the following grants to B.R.D.: National Science Foundation grants IIS-9906790, EIA-9901407, EIA-9802068, CDA-9726389, EIA-9818299, CISE/CDA-9805-548, IRI-9896020, and IRI-9530785, U.S. Department of Justice contract 2000-DT-CX-K001, and an equipment grant from Microsoft Research.

[†]Dartmouth Computer Science Department, Hanover, NH 03755, USA.

[‡]Dartmouth Chemistry Department, Hanover, NH 03755, USA.

[§]Dartmouth Center for Structural Biology and Computational Chemistry, Hanover, NH 03755, USA.

[¶]Corresponding author: 6211 Sudikoff Laboratory, Dartmouth Computer Science Department, Hanover, NH 03755, USA. Phone: 603-646-3173. Fax: 603-646-1672. Email: brd@cs.dartmouth.edu

periment planning and data interpretation for reducing mass degeneracy in mass spectrometry.

2 Computer-Aided Drug Design

PGA research in computational structural biology can assist in our long-range goal of understanding biopolymer interactions in systems of significant biochemical as well as pharmacological interest. One example is given by our work on Core-Binding Factor (CBF). CBF is a heterodimeric transcription factor involved in hematopoesis. Oncogenic translocations in CBF-α and -β are implicated in Acute Myelomonocytic Leukemia (AMML). The oncogenic form of CBF-β, fusion protein CBFβ-MYH11, oligomerizes with wild-type α-subunits to sequester them outside the nucleus, vitiating transcription. The ultimate goal of this research is to design a small-molecule inhibitor to disrupt the complex formed by the wild-type α with the oncoprotein CBFβ-MYH11. As a first step, we designed a ligand *in silico* to disrupt dimerization of the wild-type α and β subunits. Our inhibitor could be useful in itself, in that one could potentially disrupt the healthy transcription factor with a small ligand, allowing new *in vivo* studies of AMML. Our inhibitor could also serve as a lead compound to inhibit the oncogenic form of CBF-β.

A key issue was geometric modeling of protein flexibility. In our "Computational Screening Studies for Core-Binding Factor-β: Use of Multiple Conformations to Model Receptor Flexibility," [19] we present an approach in computer-aided drug design that attempts to account for a target protein's flexibility. Computational techniques were employed in docking a database of 70,000 ligands to an ensemble of structures representing low-energy conformations of CBF-β (as determined by solution NMR). Docking algorithms were used for each run and the top binding ligands were consolidated and screened in the wet-lab. Using our protocol, a small molecule inhibitor was designed to prevent dimerization of CBF. Our results — a ligand designed to disrupt the wild-type protein-protein interface — were validated in the wet-lab using electrophoretic mobility shift assays and SAR by NMR (^{15}N-HSQC chemical shift perturbation).

3 Algorithms for NMR Structural Biology

3.1 Introduction

High-throughput, data-directed computational protocols for Structural Genomics (or Proteomics) are required in order to evaluate the protein products of genes for structure and function at rates comparable to current gene-sequencing technology. We are pursuing a PGA approach known as the JIGSAW algorithm, a novel high-throughput, automated approach to protein structure characterization with nuclear magnetic resonance (NMR). For more details on this work, please see our recent papers in *The Journal of Computational Biology*, and *The International Conference on Computational Molecular Biology (RECOMB)* [3, 4, 16].

Jigsaw applies graph algorithms and probabilistic reasoning techniques, enforcing first-principles consistency rules in order to overcome a 5-10% signal-to-noise ratio. Jigsaw utilizes only four NMR experiments, none of which requires ^{13}C-labeled protein, thus dramatically reducing both the amount and expense of wet lab molecular biology and the total spectrometer time. Results for three test proteins demonstrate that Jigsaw correctly identifies 79-100% of α-helical and up to 65% of β-sheet NOE connectivities, and correctly aligns up to 90% of secondary structure elements. Jigsaw is very fast, running in minutes on a Pentium-class Linux workstation. This approach yields quick and reasonably accurate (as opposed to the traditional slow and extremely accurate) structure calculations. It could be useful for quick structural assays to speed data to the biologist early in an investigation, and could in principle be applied in an automation-like fashion to a large fraction of the proteome.

3.2 Algorithmic Approach

Jigsaw consists of two main components: (1) graph-based secondary structure pattern identification in unassigned heteronuclear (^{15}N-labeled) NMR data, and (2) assignment of spectral peaks by probabilistic alignment of identified secondary structure elements against the primary sequence. Deferring assignment eliminates the bottleneck faced by traditional approaches, which begin by correlating peaks among dozens of experiments.

The first key idea of Jigsaw (see Figure 1) is that regular protein secondary structure yields stereotypical through-space atom interactions, which are visible in a NOESY spectrum through the Nuclear Overhauser Effect (NOE). We can find such patterns in a spectrum *even if the positions in the primary sequence (assignments) are unknown*.

Jigsaw encodes NOESY data in a graph with nodes representing *unassigned* putative residues and edges representing possible interactions observed in the NOESY spectrum. This graph is very noisy (only about 10% signal) since many residues have approximately the same chemical shift for an interacting proton. However, buried within this graph is a set of edges that look like the canonical α-helix and β-sheet interactions above.

Jigsaw relies on the fact that the noise edges are evenly distributed, and thus that *it is unlikely that large groups of incorrect edges will conspire to form alpha/beta patterns*. Jigsaw imposes a set of constraints derived from the patterns in order to focus a graph search,

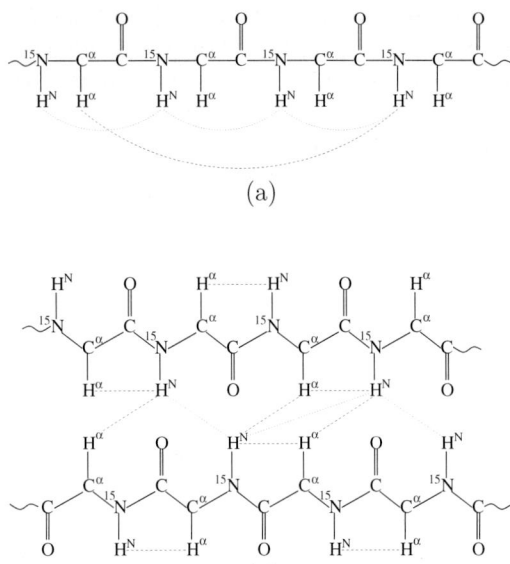

Figure 1: NOESY H^α-H^N (dashed) and H^N-H^N (dotted) interactions in (a) α-helices and (b) β-sheets.

working a "jigsaw puzzle" to find the correct secondary structure (see Figure 2). Of course, this jigsaw puzzle is somewhat different in that it has a very large number of extra pieces (and some missing ones, as well). However, we have shown that empirically the graph constraints serve to focus the search and avoid combinatorial explosion.

Jigsaw ranks secondary structures it discovers, based on criteria such as how well spectral peaks match, how many edges are missing, how many of the residues the graphs reach, and so forth. Figure 3 show β-sheets computed by JIGSAW for one example protein, CBF-β (discussed above).

The second step in Jigsaw is to align the α-helices and β-strands to substrings of the primary sequence. This relies on the use of a TOCSY spectrum, which has "fingerprints" of the protons on the side chains of the residues. These fingerprints in turn indicate probable amino acid types, which we locate in the primary sequence. This process is carried out probabilistically, assigning the probability that a strip is a certain AA type based on the results of a point-matching algorithm between observed and expected chemical shifts. Figure 4 shows both canonical proton shifts for the different amino acid types, culled from the BioMagResBank (BMRB), and observed proton shifts for some residues of Human Glutaredoxin. While most residues don't look that similar to the expected fingerprints, enough of them look enough like the expected fingerprints that a long helix or strand can be correctly aligned.

Given individual AA-type probabilities, the align-

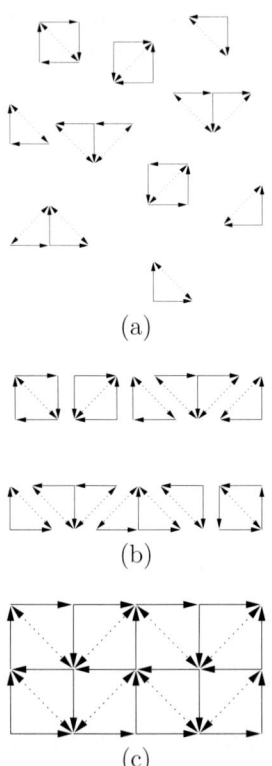

Figure 2: JIGSAW algorithm overview: (a) identify graph fragments, (b) merge them sequentially, and (c) collect them into complete secondary structure graphs. Only correct fragments are shown here. Graphs from experimental data also generate a large number of incorrect fragments, but mutual inconsistencies prevent them from forming either long sequences or large secondary structure graphs.

ment proceeds by computing the joint probability over a secondary structure string (helix or strand), starting at each location in the primary sequence. The best match is taken as the proper alignment.

This alignment process has proved effective in practice. Table 1 provides example results for the helices and strands of CBF-β. The first set of results uses fingerprints collected from a set of experiments; the second uses fingerprints observed by a single experiment, the 80 ms ^{15}N TOCSY. Results indicate both the rank of the correct alignment in the list of results, and its relative score — either the ratio of it to the second-best (if it's best) or the ratio of it to the best (if it's not). While the TOCSY alone yields good alignment results, the multi-exeriment results suggest that as TROSY-based pulse sequences improve, the results for a single experiment should get even better. In general, long sequences align better than short ones, although unusually noisy data can disrupt the alignment.

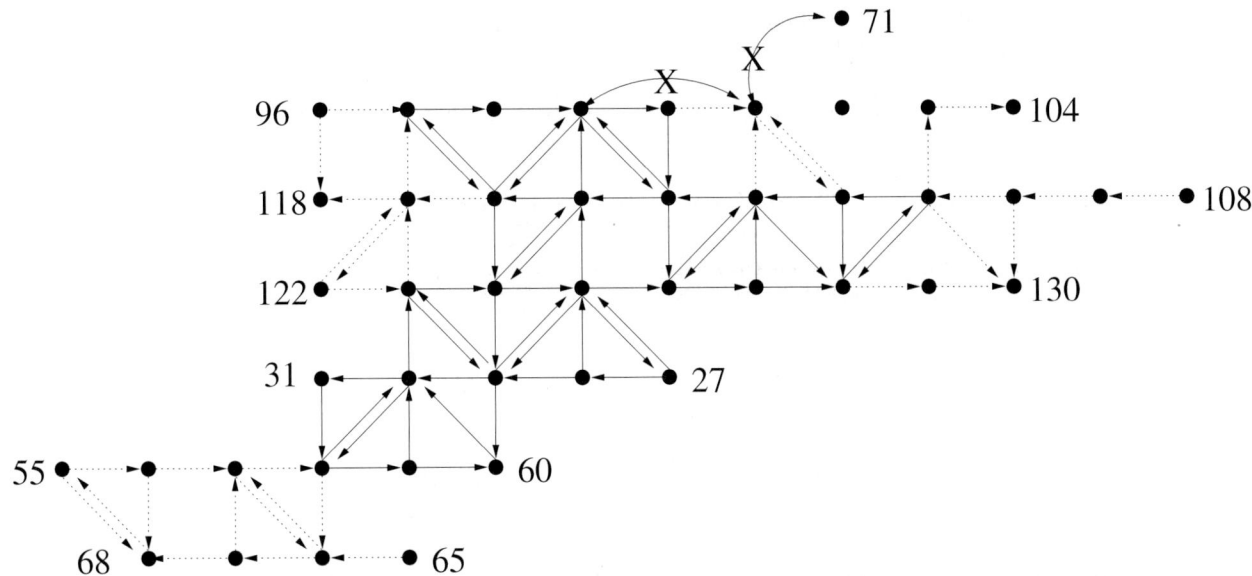

Figure 3: Some β-sheets of CBF-β computed by JIGSAW. Edges: solid=correct; dotted=false negative; X=false positive.

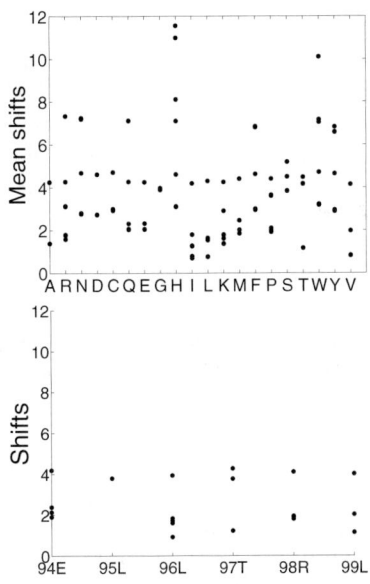

Sequence	Multi-experiment		^{15}N TOCSY	
	Rank	ρ	Rank	ρ
α_1:10–16	1	$9 \cdot 10^4$	1	$3 \cdot 10^2$
α_2:18–23	1	$2 \cdot 10^4$	17	$4 \cdot 10^{-6}$
α_3:34–36	1	$4 \cdot 10^1$	3	$7 \cdot 10^{-2}$
α_4:43–52	1	$1 \cdot 10^{13}$	1	$2 \cdot 10^4$
α_5:131–140	1	$7 \cdot 10^{14}$	1	$1 \cdot 10^{19}$
$\beta_{1,1}$:27–31	1	$4 \cdot 10^3$	5	$3 \cdot 10^{-2}$
$\beta_{1,2}$:55–60	1	$2 \cdot 10^6$	1	$2 \cdot 10^4$
$\beta_{1,3}$:65–68	1	$2 \cdot 10^1$	1	$1 \cdot 10^3$
$\beta_{2,1}$:96–104	1	$2 \cdot 10^1$	1	$7 \cdot 10^2$
$\beta_{2,2}$:108–117	1	$4 \cdot 10^{10}$	11	$3 \cdot 10^{-5}$
$\beta_{2,3}$:122–130	1	$3 \cdot 10^4$	5	$1 \cdot 10^{-1}$

Table 1: Fingerprint-based alignment results for α-helices and β-strands of CBF-β, with fingerprints obtained from a set of experiments or a single 80 ms ^{15}N TOCSY. ρ indicates the relative score of the alignment — relative to either the best alignment, if the correct one is not best, or else to the second-best alignment.

Figure 4: (top) BMRB ^{1}H mean chemical shifts over different amino acid types. These shifts define "fingerprints" for the different amino acid types. (bottom) Observed fingerprints for some residues of Human Glutaredoxin. Observed fingerprints don't exactly match expectations (e.g. 95 and 96 are both L), but yield enough information that joint probability across an entire α-helix or β-strand identifies the proper alignment in the primary sequence.

3.3 Summary

Jigsaw offers a novel approach to the automated assignment of NMR data and the determination of protein secondary structure. Since Jigsaw uses only four spectra and ^{15}N-labeled protein, it is applicable in a much higher throughput fashion than traditional techniques, and could be useful for applications such as quick structural assays and Structure-Activity Relation (SAR) by NMR. It demonstrates the large amount of information available in a few key spectra. Finally, Jigsaw formalizes NMR spectral interpretation in terms of graph algorithms and probabilistic reasoning techniques, laying the groundwork for theoretical analysis of spectral information. Jigsaw has been successfully applied to NMR data sets from (1) the glutaredoxin family of proteins, which play an important role in maintenance of the redox state of the cell as well as in DNA biosynthesis and (2) Core-Binding Factor (described above). These are just first steps in developing new PGA for NMR structural biology. Future work will extend the Jigsaw formalism to apply to larger proteins, develop faster and more accurate algorithms, and collaborate with NMR structural biologists to develop a useful suite of high-throughput tools. PGA will be important for spectral intrerpretation, conformational search, pattern recognition, kinematics, dynamics, and modelling. Computational approaches adapted from robotics and machine vision can be useful in solving key problems in NMR structural biology. New algorithms are required that can quickly extract significantly more structural information from sparse experimental data. For example, in [16], a novel approach to multidimensional NMR analysis is proposed in which the data are interpreted in the time-frequency domain, as opposed to the traditional frequency domain. Time-frequency analysis exposes behavior orthogonal to the magnetic coherence transfer pathways, thus affording new avenues of NMR discovery. In particular, we demonstrate the heretofore unknown presence of through-space inter-atomic distance information within ^{15}N-edited heteronuclear single-quantum coherence (^{15}N-HSQC) data. A biophysical model explains these results, and is supported by further experiments on simulated spectra.

4 Algorithms for Structure from Mass Spectrometry

4.1 Introduction

Mass spectrometry (MS) promises to be an invaluable tool for functional genomics, by supporting low-cost, high-throughput experiments. However, large-scale MS faces the potential problem of mass degeneracy — indistinguishable masses for multiple biopolymer fragments (e.g. from a limited proteolytic digest). In structural mass spec, mass peaks must be uniquely assignable in order to distinguish hypotheses. We are studying the PGA tasks of planning and interpreting MS experiments that use selective isotopic labeling, thereby substantially reducing potential mass degeneracy. Selective isotopic labeling allows, for example, all Leu and Ala residues in a protein to be labeled using either auxotrophic bacterial strains or cell-free synthesis. *Mass tags* — the mass differences between unlabeled and labeled proteins — can eliminate mass degeneracy by ensuring that potential fragments have distinguishable masses (see Figure 5). For more details on this work, please see our recent papers in *The Journal of Computational Biology* and *The International Conference on Intelligent Systems for Molecular Biology (ISMB)* [1, 2].

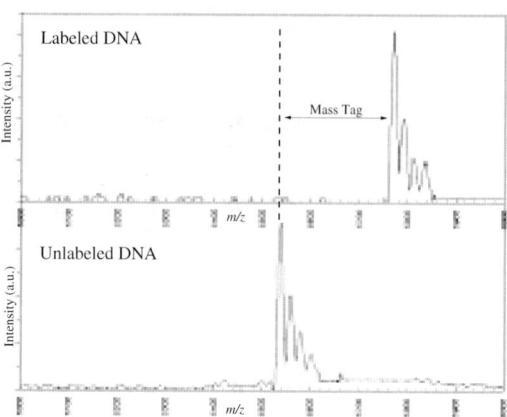

Figure 5: MALDI-TOF mass spectra of an 18 bp DNA oligonucleotide d(GACATTTGCGGTTAGGTC): (top) ^{13}C-, ^{15}N-labeled 18-mer; (bottom) ^{12}C,^{14}N-labeled 18-mer. The m/z difference between the two peaks is called the mass tag.

We have developed algorithms to support an experimental-computational protocol called *Structure-Activity Relation by Mass Spectrometry (SAR by MS)*, for elucidating the function of protein-DNA and protein-protein complexes (see Figure 6). In SAR by MS, a complex is first modeled computationally to obtain a set of binding-mode and binding-region hypotheses. Next, the complex is crosslinked and then cleaved at predictable sites (using proteases and/or endonucleases), obtaining a series of fragments suitable for MS. Depending on the binding mode, some cleavage sites will be shielded by the interface/crosslinking. Residues exposed in the isolated proteins that become buried upon complex formation are considered to be located either within the interaction regions or inaccessible due to conformational change upon binding. Thus, depending on the function, we will obtain a different mass spectrum. Analysis of the mass spectrum (and perhaps comparison to the spectra of the uncomplexed constituents) permits determination of binding mode and region, *provided that peaks are uniquely assignable*.

We have explored the PGA problem of eliminating mass degeneracy in SAR by MS, developing a computational experiment planning framework that seeks to maximize the expected information content of an SAR by MS experiment, and an efficient data analysis algorithm that interprets the resulting data.

4.2 Algorithmic Approach

For ease of exposition, we address proteins and protein-protein complexes here. A protein or protein-protein complex is digested by a protease, yielding a set of s possible *segments*. Any digestion site might be shielded,

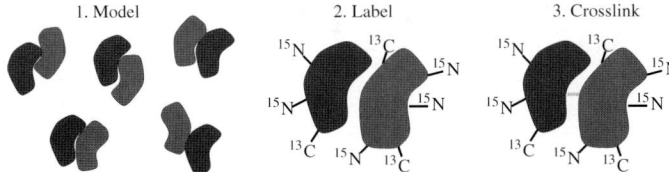

Figure 6: SAR by MS protocol overview. Mass peaks implicate residues in interaction site and nearby in space, assuming that mass peaks can be uniquely assigned. Isotopic labeling is employed to reduce mass degeneracy. Several different labelings may be used for 'multidimensional' MS (resulting in several different spectra, which must be correlated during data analysis).

^{13}C-labeled	^{15}N-labeled	χ	P(interp)
Unlabeled	Unlabeled	27	0.021
NDQEHILKSTWV	RCQHKMSTWYV	18	0.88
QGISWV	ACQEGIKPY	10	0.99
ANDCEGHILS	RCQGILMFPSWY	3	0.9998
ARNQEHKMSV	ACQGLMWY	1	0.99999
DCQEILSW	ANEGLKMFTWY	0	0.9999997

Table 2: Isotopically-labeled experiment planning results from the randomized algorithm for the protein UBC9. χ = number of remaining ambiguities. P(interp) is the probability that spectral differencing can eliminate all incorrect fragments.

yielding a set of $O(s^2)$ possible *1-fragments* in the sequential union of segments. Finally, any pair of these might be cross-linked, yielding a set of $O(s^4)$ possible *2-fragments* in the cross product of 1-fragments.

Our goal is to ensure that no pair of fragments has the same mass. The entails enforcing a system of linear inequalities of the form $f_{kl}(X) \neq 0$, where k and l are fragments and X is a labeling (a $\{0,1\}^2$ vector indicating whether or not each different amino acid type is ^{13}C and/or ^{15}N labeled). There are a quadratic number of such constraints, so a complex with s cleavage sites per protein has $O(s^8)$ such constraints to satisfy. However, it is clear that not all 1-fragment/1-fragment interactions are possible. Some may be excluded based on 1-fragment length. For example, it may be impossible to shield two cleavage sites that are t-apart with a single u-mer if $u \ll t$. Such reasoning requires careful modeling: for example, the longer strand may be heavily kinked. In general, the set of possible binding modes can be constrained by a variety of techniques, for example by docking studies, chemical shift mapping for protein-protein complexes, and docking algorithms, together with homology searching, DNA footprinting, and mutational analysis. When available, this information restricts the set of *a priori* fragment interpretations.

The goal of single-experiment planning is to find a labeling X that minimizes the amount of mass degeneracy. To do this, we attempt to minimize the number of constraint violations of the form $f_{kl}(X) = 0$. We have shown this problem to be NP-complete, even if restricted to ^{13}C labeling.

Even if we could solve single-experiment planning, the resulting labeling might have too much mass degeneracy. Therefore, we pursue a different approach, allowing experiment plans to use several different labelings. A necessary condition is that every pair of fragments be distinguishable in some labeling (else we could never determine which of the pair is present). However, this isn't sufficient if there are multiple experiments, since a fragment k could be mass degenerate with g_1 in experiment 1 and g_2 in experiment 2 (and thus distinguishable from g_2 in experiment 1 and from g_1 in experiment 2, satisfying the necessary condition), making it impossible to know whether k actually exists. A sufficient condition is that for every fragment, there is at least one labling in which it is distinguishable from every other fragment.

The sufficient condition is too strong in practice, since there are more potential than observed fragments, and (as discusssed below), we can leverage negative evidence. We have implemented a randomized algorithm to plan a set of labelings that satisfies the necessary condition. Table 2 demonstrates its effectiveness: the number of degenerate pairs goes to 0 after a small number of experiments, and the probability of interpretation (discussed below) converges to 1.

Given a set of mass spectra, we can leverage negative evidence to eliminate fragments not supported by a peak in each spectrum. An efficient (polynomial-time) algorithm for testing the existence of fragments builds a range tree for the fragments, with keys representing intervals around the predicted masses. This preprocessing step can be performed in parallel with the molecular biology. Then, given a set of spectra, simply look up each peak to find fragment explanations and intersect the sets.

A given experiment plan can be analyzed in a probabilistic framework that predicts how likely it is that the interpretation algorithm will be able to resolve all ambiguities. The key ideas are outlined below, with intuition in Figure 7.

1. Determine the *a priori* probability $\wp$ that a fragment hypothesis is incorrect (e.g. uniform $\wp = 1 - p^*/p$, or based on model). *Correctness* is a function of the biological ground truth.

2. A fragment f *appears* in an experiment i when something it is degenerate with (set $C(f,i)$ of size $c(f,i)$) is correct. *Appearance* is a function of our observations.

$$P(\text{appears}(f,i)) = 1 - \prod_{g \in C(f,i)} P(\text{incorrect}(g))$$
$$= 1 - \wp^{c(f,i)}$$

3. Spectral differencing can *eliminate* fragment f unless it appears in all experiments.

$$P(\text{elim}(f,L)) = 1 - \prod_{i \in L} P(\text{appears}(f,i))$$
$$= 1 - \prod_{i \in L}(1 - \wp^{c(f,i)}).$$

4. An experiment plan L is *interpretable* if all fragments are either correct or eliminatable.

$P(\text{interpretable}(L))$
$$= \prod_{f \in \mathcal{F}}(1 - P(\text{incorrect}(f)) \cdot (1 - P(\text{elim}(f,L))))$$
$$= \prod_{f \in \mathcal{F}}(1 - \wp \cdot \prod_{i \in L}(1 - \wp^{c(f,i)})).$$

P(interp) in Table 2 above shows that the probability of interpretability converges to 1 for an example protein. Fig 8 provides another example: how likely it is that randomly planned sets of labelings are correct. With 5 labelings, it is quite likely (practically guaranteed for UBL1) that the experiments will be interpretable.

4.3 Summary

We have tested our high-throughput techniques for obtaining structure from mass spec on the UBL1/UBC9 protein-protein complex. Yeast ubiquitin conjugating enzyme UBC9 has a functional human homolog UBEI2, which is critical for regulating the cell cycle. It complexes with UBL1 (a human ubiquitin-like protein) and associates with the RAD51/RAD52 proteins in their double-stranded DNA repair pathway. UBEI2/UBC9 is also involved in DNA recombination, and is essential for cell-cycle progression. These are just first steps in developing new PGA for structural MS. Future work will extend our algorithmic approach to SAR by MS, developing more efficient approximation algorithms, and generalizing our method to larger complexes by incorporating prior information into the probabilistic framework. In general, the set of possible binding modes can be constrained by a variety of techniques, for example by protein docking algorithms and NMR chemical shift mapping for protein-protein complexes, together with homology searching, DNA footprinting, and mutational

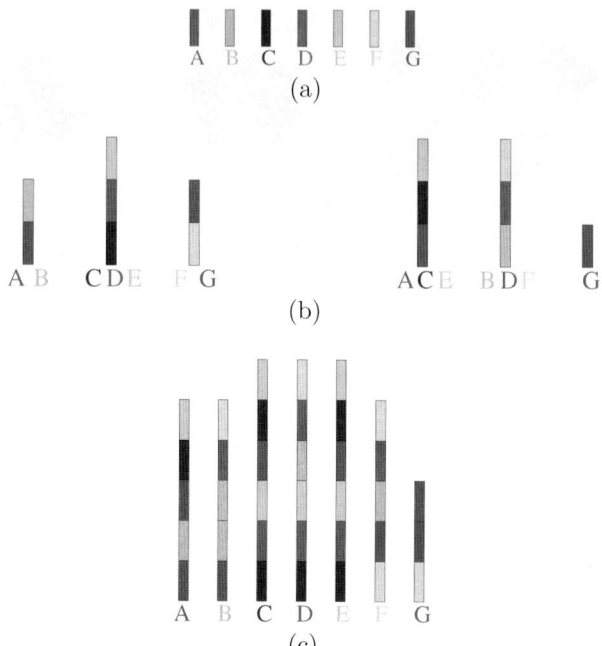

Figure 7: *This figure should be viewed in color. Postscript is available at* http://www.cs.dartmouth.edu/~cbk/papers/icra01.ps.gz. Log-space intuition for probablistic framework. (a) *A priori* fragment probabilities. (b) Probability of appearance depends on equivalence classes of mass-degenerate fragments in each experiments. (c) Probability of elimination depends on appearance in all experiments.

analysis. When available, this information restricts the set of *a priori* fragment interpretations. In turn, this should greatly help the combinatorics, since an experiment would only need to distinguish the fragments identified by hypothesis, and could allow degeneracy in unrelated fragments. In this model, predictions of docking and binding will be made on the computer, and labeling+MS would be performed as a way of screening these hypotheses to test which are correct.

References

[1] C. Bailey-Kellogg, J. J. Kelley III, C. Stein, and B. R. Donald. Reducing mass degeneracy in SAR by MS by stable isotopic labeling. In *The 8th Int'l Conf. on Intelligent Sys. for Mol. Bio. (ISMB-2000)*, pages 13–24, August 2000.

[2] C. Bailey-Kellogg, J. J. Kelley III, C. Stein, and B. R. Donald. Reducing mass degeneracy in SAR by MS by stable isotopic labeling. *J. Comp. Bio.*, 8(1):19–36, 2001. In press.

[3] C. Bailey-Kellogg, A. Widge, J. J. Kelley III, M. J. Berardi, J. H. Bushweller, and B. R. Donald. The NOESY Jigsaw: Automated protein secondary structure and main-chain assignment from sparse, unassigned NMR data. In *The 4th Int'l Conf. on Comp. Mol. Bio. (RECOMB-2000)*, pages 33–44, April 2000.

[4] C. Bailey-Kellogg, A. Widge, J. J. Kelley III, M. J. Berardi, J. H. Bushweller, and B. R. Donald. The NOESY Jigsaw:

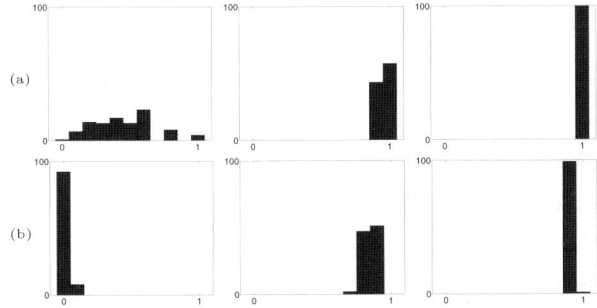

Figure 8: Interpretability of randomly planned sets of 1, 2, and 5 labelings (left to right), for (a) UBL1 and (b) UBC9. Each bar indicates how many sets, out of 100, have the given probability of interpretability.

Automated protein secondary structure and main-chain assignment from sparse, unassigned NMR data. *J. Comp. Bio.*, 7:537–558, 2000.

[5] H. J. Bohm and G. Klebe. What can we learn from molecular recognition in protein-ligand complexes for the design of new drugs? *Angew. Chem. Int. Ed. Engl.*, 35:2588–2614, 1996.

[6] J. Cavanagh, W.J. Fairbrother, A.G. Palmer III, and N.J. Skelton. *Protein NMR Spectroscopy: Principles and Practice.* Academic Press Inc., 1996.

[7] X. Chen, Z. Fei, L. M. Smith, E. M. Bradbury, and V. Majidi. Stable isotope assisted MALDI-TOF mass spectrometry allows accurate determination of nucleotide compositions of PCR products. *Anal. Chem.*, 71:3118–3125, 1999.

[8] X. Chen, S. V. Santhana Mariappan, J. J. Kelley III, J. H. Bushweller, E. M. Bradbury, and G. Gupta. A PCR-based method for large scale synthesis of uniformly ^{13}C/^{15}N-labeled DNA duplexes. *Federation of European Biochemical Societies (FEBS) Letters*, 436:372–376, 1999.

[9] X. Chen, L. M. Smith, and E. M. Bradbury. Site-specific mass tagging with stable isotopes in proteins for accurate and efficient peptide identification. *Anal. Chem.*, 2000. In press.

[10] H. A. Gabb, R. M. Jackson, and M. J. Sternberg. Modelling protein docking using shape complementarity, electrostatics and biochemical information. *J. Mol. Bio.*, 272:106–120, 1997.

[11] P. Güntert, C. Mumenthaler, and K. Wüthrich. Torsion angle dynamics for NMR structure calculation with the new program DYANA. *J. Mol. Bio.*, 273(1):283–298, October 1997.

[12] B.J. Hare and G. Wagner. Application of automated NOE assignment to three-dimensional structure refinement of a 28 kD single-chain T cell receptor. *J. Biomol. NMR*, 15:103–113, 1999.

[13] X. Huang, N.A. Speck, and J.H. Bushweller. Complete heteronuclear NMR resonance assignments and secondary structure of core binding factor β (1-141). *J. Biomol. NMR*, 12:459–460, 1998.

[14] J. J. Kelley III. *Glutaredoxins and CBF: The backbone dynamics, resonance assignments, secondary structure, and isotopic labeling of DNA and proteins.* PhD thesis, Dartmouth College, 1999.

[15] J.J. Kelley III and J.H. Bushweller. ^{1}H, ^{13}C, and ^{15}N NMR resonance assignments of vaccinia glutaredoxin-1 in the fully reduced form. *J. Biomol. NMR*, 12:353–355, 1998.

[16] C. Langmead and B. R. Donald. Extracting structural information using time-frequency analysis of protein NMR data. In *The 5th Int'l Conf. on Comp. Mol. Bio. (RECOMB-2001)*, April 2001. Accepted; in press.

[17] C. J. Langmead and B. R. Donald. Time-frequency analysis of protein NMR data. Poster, The 8th Int'l Conf. on Intelligent Sys. for Mol. Bio. (ISMB-2000), August 2000.

[18] R. H. Lathrop and T. F. Smith. Global optimum protein threading with gapped alignment and empirical pair score functions. *J. Mol. Bio.*, 255:651–665, 1996.

[19] R. Lilien, M. Sridharan, X. Huang, J. H. Bushweller, and B. R. Donald. Computational screening studies for core binding factor beta: Use of multiple conformations to model receptor flexibility. Poster, The 8th Int'l Conf. on Intelligent Sys. for Mol. Bio. (ISMB-2000), August 2000.

[20] J. A. Loo. Studying noncovelent protein complexes by electrospray ionization mass spectroscopy. *Mass Spectrometry Reviews*, 16:1–23, 1997.

[21] A. G. Marshall et al. Protein molecular mass to 1 da by ^{13}C, ^{15}N double-depletion and FT-ICR mass spectrometry. *J. American Chem. Soc.*, 119(2):443–434, 1997.

[22] C. Mumenthaler and W. Braun. Automated assignment of simulated and experimental NOESY spectra of proteins by feedback filtering and self-correcting distance geometry. *J. Mol. Bio.*, 254:465–480, 1995.

[23] R. Norel, D. Petrey, H. J. Wolfson, and R. Nussinov. Examination of shape complementarity in docking of unbound proteins. *Proteins: Structure, Function, and Genetics*, 36:307–317, 1999.

[24] A. Scaloni, N. Miraglia, S. Orrù, P. Amodeo, A. Motta, G. Maroni, and P. Pucci. Topology of the calmodulin-melittin complex. *J. Mol. Bio.*, 277:945–958, 1998.

[25] S.B. Shuker, P.J. Hajduk, R.P. Meadows, and S.W. Fesik. Discovering high-affinity ligands for proteins: SAR by NMR. *Science*, 274:1531–1534, 1996.

[26] T. Solouki et al. High-resolution multistage MS, MS2, and MS3 matrix-assisted laser desorption/ionization FT-ICR mass spectra of peptides from a single laser shot. *Anal. Chem.*, 68(21):3718–3725, 1996.

[27] C. Sun, A. Holmgren, and J. Bushweller. Complete ^{1}H, ^{13}C, and ^{15}N NMR resonance assignments and secondary structure of human glutaredoxin in the fully reduced form. *Protein Sci.*, 6:383–390, 1997.

[28] H. Takahashi, T. Nakanishi, K. Kami, Y. Arata, and I. Shimada. A novel NMR method for determining the interfaces of large protein-protein complexes. *Nature Structural Biology*, 7:220–223, 2000.

[29] R. A. Venters, W. J. Metzler, L. D. Spicer, L. Mueller, and B. T. Farmer. Use of H_N^1-H_N^1 NOEs to determine protein global folds in predeuterated proteins. *J. American Chemical Society*, 117(37):9592–9593, 1995.

[30] K. Wüthrich. *NMR of Proteins and Nucleic Acids*. John Wiley & Sons, 1986.

[31] M. M. Young, N. Tang, J. C. Hempel, C. M. Oshiro, E. W. Taylor, I. D. Kuntz, B. W. Gibson, and G. Dollinger. High throughput protein fold identification by using experimental constraints derived from intramolecular cross-links and mass spectrometry. *PNAS*, 97:5802–5806, 2000.

[32] F. Zappacosta, A. Pessi, E. Bianchi, S. Venturini, M. Sollazzo, A. Tramontano, G. Marino, and P. Pucci. Probing the tertiary structure of proteins by limited proteolysis and mass spectrometry: the case of minibody. *Protein Sci.*, 5:802–813, 1996.

[33] D.E. Zimmerman, C.A. Kulikowsi, Y. Huang, W. Feng, M. Tashiro, S. Shimotakahara, C. Chien, R. Powers, and G. Montelione. Automated analysis of protein NMR assignments using methods from artificial intelligence. *J. Mol. Bio.*, 269:592–610, 1997.

A Motion Planning Approach to Folding:
From Paper Craft to Protein Folding*

Guang Song   Nancy M. Amato

Department of Computer Science
Texas A&M University
College Station, TX 77843-3112
{gsong,amato}@cs.tamu.edu

Abstract

In this paper, we present a framework for studying folding problems from a motion planning perspective. Modeling foldable objects as tree-like multi-link objects allows us to apply motion planning techniques to folding problems. An important feature of this approach is that it not only allows us to study foldability questions, such as, can one object be folded (or unfolded) into another object, but also provides us with another tool for investigating the dynamic folding process itself. The framework proposed here has application to traditional motion planning areas such as automation and animation, and presents a novel approach for studying protein folding pathways. Preliminary experimental results with traditional paper crafts (e.g., box folding) and small proteins (approximately 60 residues) are quite encouraging.

1 Introduction

Folding is a very common process in our lives, ranging from the macroscopic level – paper folding or gift wrapping – to the microscopic level – protein folding. In most instances, while one desires a particular final state to be reached (e.g., the package is wrapped, or the protein's structure is obtained), the knowledge of the dynamic folding process used to reach a particular state is of interest as well. For this reason, we believe motion planning has great potential to help us understand folding. In particular, while motion planning does have the ability to answer questions about the reachability of certain goal states from other states, its primary objective is to in fact determine the motions required to reach the goal.

The problem of folding (and unfolding) is an interesting research topic and has been studied in several application domains. Lu and Akella [23] consider a carton folding problem and its applications in packaging and assembly. In computational geometry, there are various paper folding problems as well [25]. In computational biology, one of the most important outstanding problems is protein folding, i.e., folding a

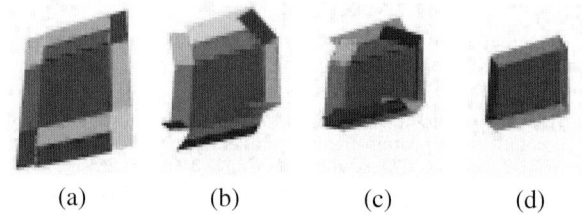

Figure 1: Snapshots of a carton folding.

one-dimensional amino acid chain into a three-dimensional protein structure.

There are large and ongoing research efforts whose goal is to determine the native folds of proteins (see, e.g., [15, 21]). In this paper, we assume we already know the native fold, and our focus is on the folding process, i.e., how the protein folds to that state from some initial state. Although there have been some recent experimental advances [10], computational techniques for simulating this process are important because it is difficult to capture the folding process experimentally.

Our approach is based on the successful *probabilistic roadmap* (PRM) motion planning method [17]. We have selected the PRM paradigm due to its proven success in exploring high-dimensional configuration spaces. A major strength of PRMs is that they are quite simple to apply, requiring only the ability to randomly generate points in C-space, and then test them for feasibility. The protein folding problem has a complication in that the way in which the protein folds depends on factors other than the purely geometrical constraints which govern the polygonal problems. Nevertheless, we show that these additional factors can be dealt with in a reasonable fashion within the PRM framework.

Our preliminary experimental results with traditional paper crafts and small proteins of approximately 60 residues, or 120 degrees of freedom, are quite promising. See Figures 1 and 2 for some path snapshots. Further results on protein folding can be found in [27].

---
[1] This research supported in part by NSF CAREER Award CCR-9624315, NSF Grants IIS-9619850, ACI-9872126, EIA-9975018, EIA-9805823, and EIA-9810937, and by the Texas Higher Education Coordinating Board under grant ARP-036327-017.

2 Related work

Paper Folding. Many problems related to the folding and unfolding of polyhedral objects have recently attracted the attention of the computational geometry community [25]. For example, [8] shows that every polyhedron can be 'wrapped' by folding a strip of paper around it, which addresses a question arising in three-dimensional origami, e.g., [1]. In most cases, origami problems cannot be modeled as trees since the incident faces surrounding a given face form a cycle in the linkage structure. Such cycles, often called closed chains, impose additional constraints on the motion planning problem (see, e.g., [13, 19]). In this paper we are interested in problems with tree-like linkage structures. There are still many interesting problems involving folding of tree-like linkages. For example, not every tree-like linkage in the plane can be 'straightened' (called 'locking'), that is, there are some pairs of configurations of the linkage which cannot be connected if the links are not allowed to cross [4]. In three dimensions, there exist open (and closed) chains that can lock [4, 5], while, in dimensions higher than three, neither open nor closed chains can lock [6].

Protein Folding. The protein folding problem is to predict a protein's three-dimensional conformation based solely on its amino acid sequence. Many different approaches for predicting protein structure have been explored. In folding simulations, several computational approaches have been applied to this exponential-time problem, including energy minimization, molecular dynamics simulation, Monte Carlo methods, and genetic algorithms (see [15] and references therein). Among these, molecular dynamics is most closely related to our approach. Much work had been carried out in this area [7, 9, 12, 20], which tries to simulate the true dynamics of the folding process using the classical Newton's equations of motion. The forces applied are usually approximations computed using the first derivative of an empirical potential function. Molecular dynamics simulations help us understand how proteins fold in nature, and provide a means to study the underlying folding mechanism, to investigate folding pathways, and can provide intermediate folding states.

Most of the proposed techniques have tremendous computational requirements because they attempt to simulate complex kinetics and thermodynamics. In this paper, we present an alternative approach that finds approximations to the folding pathways while avoiding detailed simulations. Our motion planning approach is based on the successful *probabilistic roadmap* (PRM) method [17] which has been used to study the related problem of ligand binding [3, 16, 26], which is of interest in drug design. The results were quite promising. Advantages of the PRM approach are that it efficiently covers a large portion of the planning space, in this case, the conformation space, and that it also provides an effective way to incorporate and study various initial conformations.

3 Preliminaries

C-spaces of folding objects. Both the paper polygon and the amino acid sequence are modeled as multi-link tree-like articulated 'robots', where fold positions (polygon edges or atomic bonds) correspond to joints and areas that cannot fold (polygon faces or atoms) correspond to links. The fold positions of the paper polygon are modeled as revolute joints. For the amino acid sequence of the protein, we consider all atomic bond lengths and bond angles to be constants, and consider only torsional angles (phi and psi angles), which we also model as two revolute joints (2 dof). Thus, in both cases, our models will consist of $n + 1$ links and n revolute joints.

The joint angle of a revolute joint takes on values in $[0, 2\pi)$, with the angle 2π equated to 0, which is naturally associated with a unit circle in the plane, denoted by S^1. Therefore, the configuration space of interest for our multi-link objects can be expressed as:

$$\mathcal{C} = \{q \mid q \in S^n\}.$$

Note that $\mathcal{C}$ simply denotes the set of all possible configurations, but says nothing about their feasibility. The validity of a point in $\mathcal{C}$ will be determined by collision detection for the polygon problems and by potential energy computations for the proteins.

Potential Function. The way in which the protein folds depends on the potential energy of the configurations. We start with:

$$\begin{aligned} U_{tot} &= \sum_{restraints} K_d\{[(d_i - d_0)^2 + d_c^2]^{1/2} - d_c\} \\ &+ \sum_{atom\ pairs} (A/r_{ij}^{12} - B/r_{ij}^6), \end{aligned}$$

which is similar to the potential used in [20]. The first term represents constraints which favor the known secondary structure through main-chain hydrogen bonds and disulphide bonds. The van der Waals interaction among atoms is considered in the second term. All parameters can be found in [20].

However, even for relatively small proteins (around 60 residues), there will be nearly one thousand atoms. Non-hydrogen atoms also number in the hundreds. Therefore, performing all pairwise van der Waals potential calculations (the second summation) can be computationally intensive. To reduce this cost, we use a step function approximation of the van der Waals potential component. This is computed by considering only the contribution from the side chains and modeling each side chain with a fixed-size rigid sphere (a further approximation). The side chain was chosen because it reflects the geometric configuration of a residue. By doing this, the computational cost is reduced by two orders of magnitude. Our results indicate that enough accuracy is retained to capture the main features of the interaction.

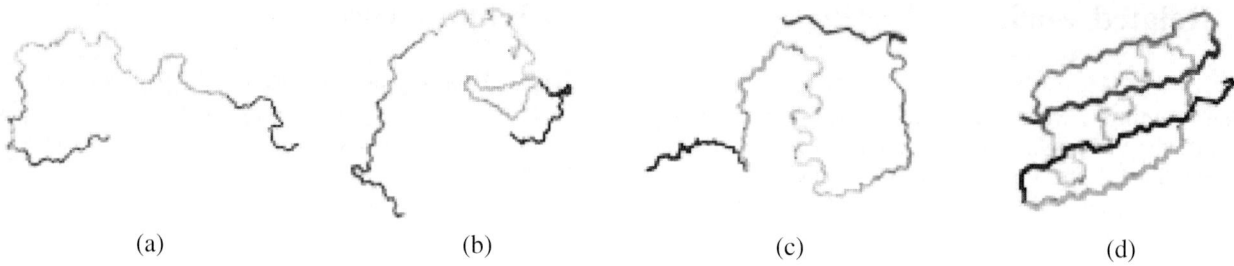

(a) (b) (c) (d)

Figure 2: Snapshots of protein GB1 folding.

4 PRMs for Folding Problems

As mentioned in Section 1, our approach to the folding problem is based on the PRM approach to motion planning [17].

The folding problems, especially protein folding, have a few notable differences from usual PRM applications. First, as our problems are not posed in an environment containing external obstacles, the only collision constraint we impose is that our configurations be self-collision free, and, for the protein folding problem, our preference for low energy conformations leads to an additional constraint on the feasible conformations. Second, in PRM applications, it is usually considered sufficient to find *any* feasible path connecting the start and goal. For our folding problems, however, we are interested not only in whether there exists a path, but we are also interested in the *quality* of the path. For example, for the paper folding problems, one is interested in a path which makes a minimal number of folds, and for the protein folding we are interested in low energy paths.

I. Node generation. As described in Section 3, since all joints are revolute, a configuration $q \in \mathcal{C}$ can be generated by assigning each joint angle a value in its allowable range. Once all the joint angles are set, the object's three-dimensional structure is fully determined.

For the paper folding, the configuration of each link is then calculated and self-collision among the links is checked. The node is discarded if any collision occurs.

For the protein molecular model, after the joint angles are known, the coordinates of each atom in the system are calculated, and these are then used to determine the potential energy of the conformation, as defined in Section 3. The node is accepted and added to the roadmap based on its potential energy E with the following probability:

$$P(E) = \begin{cases} 1 & \text{if } E < E_{\min} \\ \frac{E_{\max}-E}{E_{\max}-E_{\min}} & \text{if } E_{\min} \leq E \leq E_{\max} \\ 0 & \text{if } E > E_{\max} \end{cases}$$

This acceptance criterion was also used when building PRM roadmaps for ligand binding in [26]. This filtering helps us to generate more nodes in low energy regions, which is desirable since we are interested in finding the pathways that are most energetically favorable (low energy). In our case, we set $E_{\min} = 50000$ KJouls/mol and $E_{\max} = 89000$ KJouls/mol, which favors conformations with well separated side chain spheres.

II. Constructing the roadmap. The second phase of the algorithm is roadmap connection. For each node, we first find its k nearest neighbors in the roadmap (using Euclidean distance in C-space), for some small constant k, and then try to connect it to them using some simple local planner. For both the paper folding and protein folding models, each connection attempt performs feasibility checks for N intermediate configurations between the two corresponding nodes as determined by the chosen local planner. If there are still multiple connected components in the roadmap after this stage other techniques will be applied to try to connect different connected components (see [2] for details).

When two nodes are connected, the corresponding edge is added to the roadmap. We associate a weight with each edge. For the paper folding, the weight is simply N, the number of intermediate configurations on the edge. For the protein folding, the weight is:

$$Weight = \sum_{i=0}^{N-1} -log(P_i),$$

where the the probability P_i of moving from conformation i to $i+1$ is determined by:

$$P_i = \begin{cases} e^{\frac{-\Delta E}{kT}} & \text{if } \Delta E > 0 \\ 1 & \text{if } \Delta E \leq 0 \end{cases}$$

which keeps the detailed balance between two adjacent states. By assigning the weights in this manner, we can find the shortest or most energetically feasible path when performing subsequent queries. A similar weight function, with different probabilities, was used in [26].

III. 'Querying' the roadmap. The resulting roadmap can be used to find a feasible path between given start and goal configurations. For our folding problems, we first connect the start and the goal into the roadmap, just as was done for the other roadmap nodes in the connection phase. Dijkstra's algorithm is then used to find the smallest weight path between the start and goal configurations. For the protein folding, if the potential of some intermediate node is too large (as compared to some predetermined maximum), a failure is reported, otherwise the path is returned.

5 Validating Folding Pathways

For the protein folding pathways found by our PRM framework to be useful, we must find some way to validate them with known results. Even though the folding pathways provided by PRMs cannot be explicitly associated with actual timesteps, they do provide us with a temporal ordering. Therefore, we could study (i) the intermediate or transition states on the pathway, and the order in which they are obtained, or (ii) the formation order of secondary structures.

Folding intermediates have been an active research area over the last few years. It is thought that some, but not all, proteins go through intermediate states to reach the native conformation, see, e.g., [24]. Therefore, one possibility is to compare our folding pathways with experimental results known about folding intermediates.

The formation order of secondary structures is related to a fundamental question in protein folding: do secondary structures always form before the tertiary structure, or is tertiary structure formed in a one-stage transition? In this paper, we focus on validating our folding pathways by comparing the order in which the secondary structures form in our paths with results for some small proteins that have been determined by pulse labeling and native state out-exchange experiments [22].

6 Results and Discussion

We now describe results on paper folding and protein folding problems obtained using our PRM-based approach. For the paper folding problems we used the obstacle-based PRM called OBPRM [2], which generates nodes near constraint surfaces (C-obstacle surfaces). For the protein folding, the results presented follow the basic PRM approach [17] of uniform sampling in C-space. We used the RAPID [11] package for 3D collision detection. The experiments were performed on an SGI Octane R10000. In this paper we can only show path snapshots; movies can be found at http://www.cs.tamu.edu/faculty/amato/dsmft.

6.1 Models studied

We study two paper folding models: a *box* and a *periscope*. The periscope has 11 degrees of freedom (11 joints) and the box has 12. However, for the box, the number of dof can be reduced to five using symmetry arguments. Both foldings are non-trivial, and in fact, correspond to what are known as 'narrow passage' problems [14].

We present results for two small proteins. Protein GB1 has 56 residues (112 dof) and consists of one alpha helix and two beta-sheets. Its structure has been determined by both NMR and crystallography. Protein A has 60 residues (120 dof) and consists of three alpha helices. The pdb files used for the proteins were 1GB1.pdb and 1BDD.pdb, respectively, from the Protein Data Bank at http://www.rcsb.org/pdb/.

Paper Folding Roadmap Construction Statistics					
Model	dof	Gen	Con	#CC	#Nodes
Box	12(5)	38.7	201	1	1035
Periscope	11	13	177	1	883

Table 1: Roadmap construction statistics for the Box and Periscope models. The Box has 12 links, but its dof becomes 5 after symmetry is exploited. 'Gen' and 'Con' represent node generation and connection times in seconds, resp. #Nodes and #CC are the number of nodes and connected components, resp., in the resulting roadmap.

6.2 Paper folding results

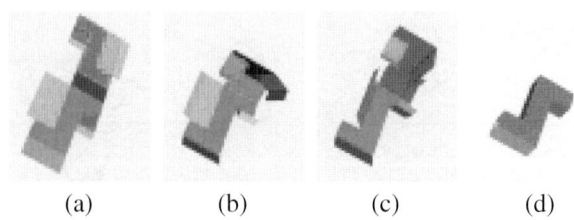

(a)   (b)   (c)   (d)

Figure 3: Snapshots of the periscope folding.

Some statistics regarding the roadmaps constructed for the paper folding problems are shown in Table 1. As can be seen, in both cases the problems were solved rather quickly with relatively small roadmaps. These results are really quite remarkable as the problems are actually considered to be quite challenging motion planning problems. Nevertheless, we see that just a few minutes were needed to construct roadmaps containing solution paths. We believe our success with these problems can be attributed to the tendency of the OBPRM roadmaps to contain nodes near the constraint surfaces (i.e., near self-collision configurations) which include configurations necessary for successful paths. For example, configurations in which the flaps of the box fold over other flaps. Snapshots of the folding paths found are shown in Figures 1 and 3 for the box and the periscope, respectively.

6.3 Protein folding results

The results for the protein folding examples are also very interesting. Some statistics regarding the roadmaps constructed for the protein folding problems are shown in Table 2. We provided the goal conformations beforehand, and then searched in the roadmap for the minimum weight path connecting the extended amino acid chain to the final three-dimensional structure. Snapshots of folding paths found by our planner for protein GB1 and protein A are shown in Figure 2 and Figure 4, respectively.

Validation of folding pathways. Protein GB1 has 56 residues (112 dofs), and consists of a central alpha helix and two beta-sheets, each composed of two beta strands. Pulse labeling experimental results [18, 22] indicate that the alpha helix and beta strand 4 form first and are protected during

Protein Folding Roadmap Construction Statistics								
Model	dof	Gen	Con	#N sam	#N ret	#N BigCC	#edges	#N path
Protein GB1	112	130	1500	5000	594	559	898	1
		500	5300	20000	2508	2381	3890	2
		2600	42300	100000	12392	11865	20433	3
Protein A	120	400	1300	5000	555	508	767	5
		1600	5800	20000	2308	2140	3352	4
		9500	50100	100000	11715	11057	17719	6

Table 2: Roadmap construction statistics for Protein GB1 and Protein A. 'Gen' and 'Con' represent the node generation and connection times in seconds, resp. '#N sam' is the number of sampled nodes and '#N ret' is the number of nodes retained after rejecting nodes with high potentials. '#N BigCC' is the number of the nodes in the biggest connected component of the roadmap, '#edges' is the total number of edges, and '#N path' is the number of roadmap nodes in the final folding path.

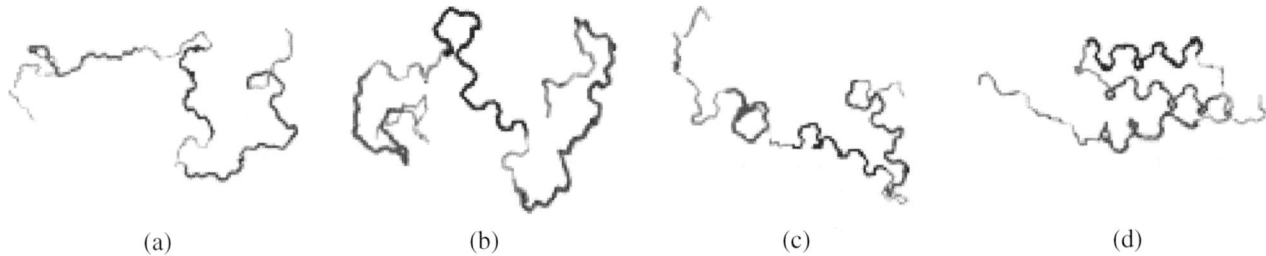

(a)   (b)   (c)   (d)

Figure 4: Snapshots of protein A folding.

hydrogen-deuterium exchanges. This was consistent with the path found by our method. For example, from the snapshots shown in Figure 2, one can see that the alpha helix in the middle forms first.

Protein A has 60 residues (120 dofs), and consists of three alpha helices. The pulse labeling results [22] show that the three alpha helices form at about the same time. As seen in the path snapshots in Figure 4, our paths seem to be consistent with these results.

In general, these results are very encouraging – in both cases, the formation order of the secondary structures seems to agree with the results of the pulse labeling experiments. Thus, while further investigation and tuning of the PRM technique for proteins is still needed, our preliminary findings show that this motion planning approach is a potentially valuable tool. For example, it could be used to study the secondary structure formation order for proteins where this has not yet been determined experimentally.

Analyzing folding pathways. By analyzing the paths found, we may be able to gain some insight into the natural folding process. Towards this end, we analyzed the profiles of the potential energies of the intermediate conformations on the folding paths. This is shown for proteins GB1 and A in Figure 5(a) and 5(b), respectively. We expect that as the number of nodes sampled increases (the sampling is denser), our roadmaps will contain better and better approximations of the natural folding path. Our results support this belief, and moreover, enable us to estimate how many nodes should be sampled. In particular, we can see in the plots that as the number of nodes, N, is increased, the paths seem to improve in quality, and have fewer and smaller peaks in their profiles.

Another interesting point is the similarity among the paths for all roadmap sizes. In particular, they all illustrate that there is a peak (or peaks) near the goal conformation. Some researchers believe such energy barriers around a folding state are crucial for a stable fold. Also, the profiles clearly show that the peak(s) right before the final fold are contributed by the van der Waals interaction, which is consistent with the tight packing of atoms in the native fold. The similarity among these paths also implies that they may share some common conformations, or subpaths, and this knowledge could be used to bias our sampling around these regions, hopefully further improving the quality of the paths.

7 Conclusion and Future Work

In this paper, we present a framework for studying folding problems from a motion planning perspective. Our approach, which is based on the PRM motion planning method, was seen to produce interesting results for representative problems in paper folding and protein folding. One of the most important benefits of this approach to folding problems is that it enables one to study the dynamic folding process itself. We believe that our results establish that this is a promising approach which deserves further investigation.

Acknowledgements

We would like to thank Jean-Claude Latombe for pointing out to us the connection between box folding and protein folding. We would also like to thank Marty Scholtz for

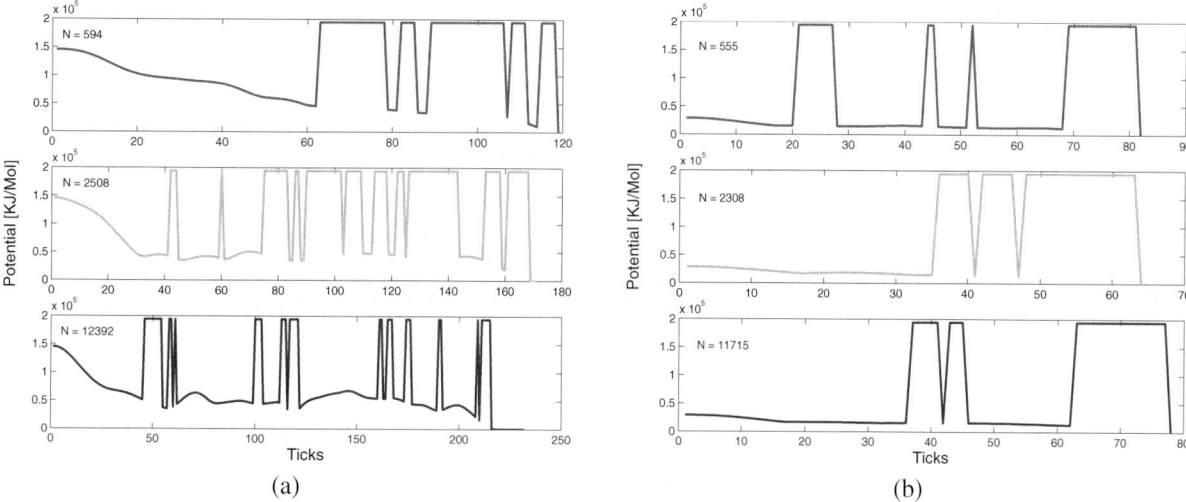

Figure 5: Potential along the folding path shown for each intermediate configuration on the path ('tick') for different sized roadmaps. (a) Protein GB1, roadmaps with $N = 594, 2508, 12392$ nodes (top to bottom). (b) Protein A, roadmaps with $N = 555, 2308, 11715$ nodes (top to bottom).

suggesting validation using the pulse labeling results, and Michael Levitt and Vijay Pande for useful suggestions.

References

[1] Jin Akiyama. Why Taro can do geometry. In *Proc. 9th Canad. Conf. Comput. Geom.*, page 112, 1997.

[2] N. M. Amato, O. B. Bayazit, L. K. Dale, C. V. Jones, and D. Vallejo. OBPRM: An obstacle-based PRM for 3D workspaces. In *Proc. Int. Workshop on Algorithmic Foundations of Robotics (WAFR)*, pages 155–168, 1998.

[3] O. B. Bayazit, G. Song, and N. M. Amato. Ligand binding with obprm and haptic user input: Enhancing automatic motion planning with virtual touch. In *Proc. IEEE Int. Conf. Robot. Autom. (ICRA)*, 2001. To appear. This work will be presented as a poster at RECOMB'01.

[4] T. Biedl, E. Demaine, M. Demaine, S. Lazard, A. Lubiw, J. O'Rourke, M. Overmars, S. Robbins, I. Streinu, G. Toussaint, and S. Whitesides. Locked and unlocked polygonal chains in 3D. In *Proc. 10th ACM-SIAM Sympos. Discrete Algorithms*, pages 866–867, January 1999.

[5] J. Cantarella and H. Johnston. Nontrivial embeddings of polygonal intervals and unknots in 3-space. *J. Knot Theory Ramifications*, 7:1027–1039, 1998.

[6] R. Cocan and J. O'Rourke. Polygonal chains cannot lock in 4D. In *Proc. 11th Canad. Conf. Comput. Geom.*, pages 5–8, 1999.

[7] V. Daggett and M. Levitt. Realistic simulation of naive-protein dynamics in solution and beyond. *Annu. Rev. Biophys. Biomol. Struct.*, 22:353–380, 1993.

[8] E. D. Demaine, M. L. Demaine, and J. S. B. Mitchell. Folding flat silhouettes and wrapping polyhedral packages: New results in computational origami. In *Proc. 15th Annu. ACM Sympos. Comput. Geom.*, pages 105–114, June 1999.

[9] Y. Duan and P.A. Kollman. Pathways to a protein folding intermediate observed in a 1-microsecond simulation in aqueous solution. *Science*, 282:740–744, 1998.

[10] W.A. Eaton, V. Muñoz, P.A.Thompson, C. Chan, and J. Hofrichter. Submillisecond kinetics of protein folding. *Curr. Op. Str. Bio.*, 7:10–14, 1997.

[11] S. Gottschalk, M.C. Lin, and D. Manocha. Obb-tree: A hierarchical structure for rapid interference detection. Technical Report TR96-013, University of N. Carolina, Chapel Hill, CA, 1996.

[12] J.M. Haile. *Molecular Dynamics Simulation: elementary methods*. Wiley, New York, 1992.

[13] L. Han and N. M. Amato. A kinematics-based probabilistic roadmap method for closed chain systems. In *Proc. Int. Workshop on Algorithmic Foundations of Robotics (WAFR)*, 2000.

[14] D. Hsu, L. Kavraki, J-C. Latombe, R. Motwani, and S. Sorkin. On finding narrow passages with probabilistic roadmap planners. In *Proc. Int. Workshop on Algorithmic Foundations of Robotics (WAFR)*, 1998.

[15] G. N. Reeke Jr. Protein folding: Computational approaches to an exponential-time problem. *Ann. Rev. Comput. Sci.*, 3:59–84, 1988.

[16] L. Kavraki. Geometry and the discovery of new ligands. In *Proc. Int. Workshop on Algorithmic Foundations of Robotics (WAFR)*, pages 435–448, 1996.

[17] L. Kavraki, P. Svestka, J. C. Latombe, and M. Overmars. Probabilistic roadmaps for path planning in high-dimensional configuration spaces. *IEEE Trans. Robot. Automat.*, 12(4):566–580, August 1996.

[18] J. Kuszewski, G.M. Clore, and A.M. Gronenborn. Fasting folding of a prototypic polypeptide: The immunoglobulin binding domain of streptococcal protein g. *Protein Science*, 3:1945–1952, 1994.

[19] S.M. LaValle, J.H. Yakey, and L.E. Kavraki. A probabilistic roadmap approach for systems with closed kinematic chains. In *Proc. IEEE Int. Conf. Robot. Autom. (ICRA)*, 1999.

[20] M. Levitt. Protein folding by restrained energy minimization and molecular dynamics. *J. Mol. Bio.*, 170:723–764, 1983.

[21] M. Levitt, M. Gerstein, E. Huang, S. Subbiah, and J. Tsai. Protein folding: the endgame. *Annu. Rev. Biochem.*, 66:549–579, 1997.

[22] R. Li and C. Woodward. The hydrogen exchange core and protein folding. *Protein Sci.*, 8:1571–1591, 1999.

[23] L. Lu and S. Akella. Folding cartons with fixtures: A motion planning approach. In *Proc. IEEE Int. Conf. Robot. Autom. (ICRA)*, pages 1570–1576, 1999.

[24] C.R. Matthews. Pathways of protein folding. *Annu. Rev. Biochem.*, 62:653–683, 1993.

[25] J. O'Rourke. Folding and unfolding in computational geometry. In *Proc. Japan Conf. Discrete Comput. Geom. '98*, pages 142–147, December 1998. Revised version submitted to LLNCS.

[26] A.P. Singh, J.C. Latombe, and D.L. Brutlag. A motion planning approach to flexible ligand binding. In *7th Int. Conf. on Intelligent Systems for Molecular Biology (ISMB)*, pages 252–261, 1999.

[27] G. Song and N. M. Amato. Using motion planning to study protein folding pathways. In *Proc. Int. Conf. Comput. Molecular Biology (RECOMB)*, pages 278–287, 2001.

Ligand Binding with OBPRM and User Input*

O. Burchan Bayazit   Guang Song   Nancy M. Amato
{burchanb,gsong,amato}@cs.tamu.edu
Department of Computer Science, Texas A&M University
College Station, TX 77843-3112

Abstract

In this paper, we present a framework for studying ligand binding which is based on techniques recently developed in the robotics motion planning community. We are interested in locating binding sites on the protein for a ligand molecule. Our work investigates the performance of a fully automated motion planner, as well as the effects of supplementary user input collected using a haptic device. Our results applying an obstacle-based probabilistic roadmap motion planning algorithm (OBPRM) to some protein-ligand complexes are encouraging. The framework successfully identified potential binding sites for all complexes studied. We find that user input helps the planner, and a haptic device helps the user to understand the protein structure by enabling them to feel the difficult-to-visualize forces.

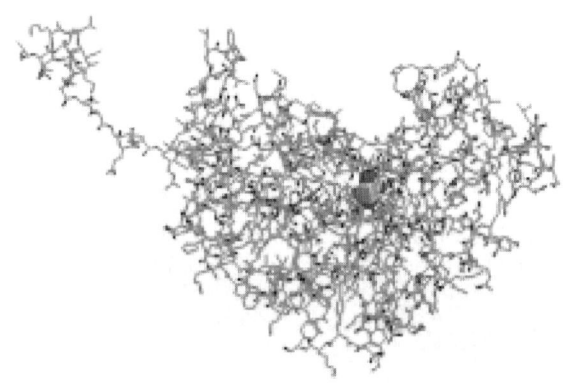

Figure 1: A protein in wireframe with a space-fill ligand. PDB file: 1LDM.pdb.

1 Introduction

Efficiency of a drug molecule is measured by its ability to find a specific position and orientation inside a protein. This process is called binding (or docking) and the drug molecule is often referred as a ligand. The binding configuration should satisfy some constraints based on geometric, electrostatic, and chemical reactions between the ligand and protein atoms. A good binding site should also be accessible to the ligand which must reach it from an outside location. This makes the path to the binding site important, and motivates the use of motion planning to study this problem.

Most researchers investigating automated docking methods simplify the problem by treating the ligand and protein molecules as rigid bodies. Our experiments show that this simplification negatively impacts our ability to identify the binding site. However, if the molecules are flexible (articulated bodies), the dimensionality of the problem becomes very high, making a deterministic search of the configuration space infeasible.

In this work, we apply *probabilistic roadmap* (PRM) motion planning methods to protein-ligand binding. PRMs have been extremely successful for problems with high-dimensional configuration spaces, and moreover, are very simple to apply, requiring only the ability to randomly generate points in C-space, and then test them for feasibility. Besides ligand binding [16], it has been applied also to study protein folding problems [17, 18] as well. The configuration of the ligand in the binding site has low potential energy, and so the usual PRM feasibility test (collision) is replaced by a test for low potential energy. In this study, we use an obstacle-based PRM called OBPRM [1] that generates configurations close to the surface of the protein. Since our goal, the binding site, is a local minima of the potential, we used a gradient-descent like algorithm to drive the roadmap nodes towards local minimum. To recognize potential binding sites, we construct 'local roadmaps' around (most of) the roadmap nodes and give each candidate node a score based on an analysis of its local roadmap. Our approach was able to generate configurations close to the true binding site in the three tested protein-ligand complexes. Two of these were studied in [16], where in one case their approach failed to generate a configuration in the binding site.

While our results are very encouraging, a fully automated approach suffers from the known problems of PRMs. The most significant of which is the *narrow passage* problem [3], so called because it is difficult for the planner to

---

*This research supported in part by NSF CAREER Award CCR-9624315 (with REU Supplement), NSF Grants IIS-9619850 (with REU Supplement), ACI-9872126, EIA-9975018, EIA-9805823, and EIA-9810937, and by the Texas Higher Education Coordinating Board under grant ARP-036327-017. Bayazit is supported in part by the Turkish Ministry of Education.

sample important configurations in a narrow C-space region. In molecular docking, the narrow passage becomes a passage through high potential areas. Another challenge with PRMs is that it is hard for the user to visualize and understand the progress made by the planner. In molecular docking, this problem becomes even more significant since the familiar three-dimensional workspace is replaced with potential energy landscape. To address these problems, we add a human operator who explores the energy landscape with a haptic device. The user selects some important configurations of the ligand with the haptic device which are passed to the planner for further processing. Since feeling the potential energy forces is not sufficient to understand the planner's progress, roadmap visualization methods similar to [2] are also employed.

In his paper we will use the terms configuration and conformation interchangeably.

2 Previous Work

Many automated docking algorithms have been proposed, such as, AutoDock [10], Dock [5], FlexX [15], and FTDock [4]. However useful they are, their success ratio is often relatively low (2-20%) [20]. Also, in most cases these algorithms make the simplifying assumption that the ligand is rigid.

Recently, a new approach which makes use of the PRM paradigm for protein-ligand binding was proposed in [16]. Their results were very promising, and are in large part the motivation for this work. They uniformly sample configurations, and resample more densely around a small subset of low potential configurations from the original sample. In essence, they randomly sampled configurations, identified those with low energy, and further explored those regions. Since the goal of the obstacle-based PRM used in our work is to generate, directly, configurations near the protein's surface, we believe that it will provide a better sample of this area than the uniform sampling approach of [16]. As will be seen in Section 7, some evidence in favor of our approach is our success in generating configurations in the binding site for a test case in which [16] failed.

Some researchers, such as SMD [6], STALK [7], and SCULPT [19], have added a human operator into the decision process. A revolutionary approach taken in [12] was to include a haptic interface and let the user feel the force applied on the ligand during the docking process. They also show in [12] that force feedback performed better than visual display in the docking process. In [13], an interactive molecular docking simulator was used for interactive molecular design. It is shown in [21] that SCULPT can be improved by adding a haptic device.

Unfortunately, there has not been much work on improving automated planners with human input. In [2], we used a haptic device together with a PRM planner on motion planning problems typically of maintainability studies in mechanical CAD designs. Our results showed that human insight improved the planner's ability to solve problems involving narrow passages in C-space.

3 Potential Energy

A key requirement in simulating the ligand binding process is to have a good model of the interaction between a ligand and its receptor, a protein. Proteins usually consist of thousands of atoms, whereas the ligand typically consists of tens. While both protein and ligand are in fact flexible, we will model the protein as a rigid body and the ligand as a flexible body as was done in [16]. In particular, the ligand is modeled as an articulated body with a free base, where the torsional movement of atomic bonds is modeled by revolute (1 dof) joints (bond angles and lengths are fixed). Thus, the base atom requires 6 dof and each additional torsional movement requires 1 dof.

During binding, the ligand can be viewed as moving in the potential field created by the protein's atoms, searching for a stable low potential configuration. Since a complete, accurate model of the potential is computationally infeasible, we approximate the potential by the van der Waals terms only, excluding other components such as Coulomb potential and solvent effects. While this is an extreme simplification, the results obtained are quite impressive. In particular, the potential we use is

$$U = \sum_{atom\ pair\ i,j} (A/r_{ij}^{12} - B/r_{ij}^6),$$

where r_{ij} is the distance between protein atom i and ligand atom j. Parameters are taken from [8].

Even with our simplified potential, computing the potential energy of a ligand configuration is expensive since it includes a contribution for each atom pair. To optimize this computation we use a grid-based energy calculation. In this approach, the protein is covered by a three-dimensional grid, and each grid point stores the potential applied by all protein atoms at that point [14]. During potential energy calculations, the value stored at the closest grid point to each ligand atom is used as an approximation of potential for that atom. The accuracy of this method directly depends on the size of the grid. Our experiments indicate that a 0.5 Å grid is sufficient for realistic force feedback for our haptic device.

4 Haptic Interaction for Molecular Docking

Our prototype system consists of a PHANToM haptic device (6 dof) output) [9], and a Xeon 550Mhz NT Workstation. The operator uses the PHANToM to manipulate a *rigid* ligand around a virtual protein.

A challenge with the molecular docking is to compute the potential energies of the configurations fast enough for haptic interaction. To achieve this, we used a grid-based

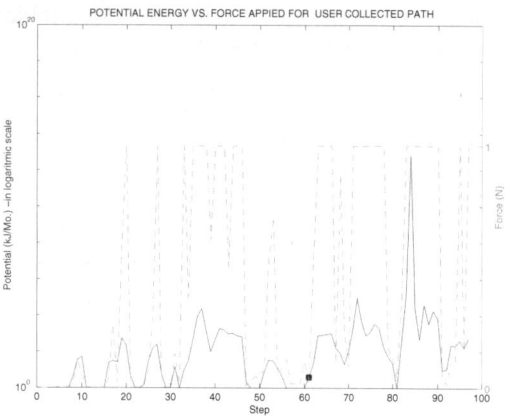

Figure 2: Correlation between potential energy (logaritmic left y-axis, in solid line) and user-felt force (right y-axis, in dashed line).

force calculation as described above, which was also used in [11]. The efficiency of this method can be observed in Figure 2, where both the real potential energy and the force the user felt (based on a grid) while collecting a path of ligand configurations are shown. In the figure, the potential is shown with the solid line and its (logarithmic) y-axis is represented on the left. Similarly the user felt force is shown with the dashed line and the its y-axis is represented on the right. To stay in the PHANToM's force feedback limits, any computed force greater than 1N was reduced to 1N without changing its direction, which resulted in a repulsive force in the correct direction. The binding configuration is represented by a small rectangle in both lines. The correlation between the figures shows the user feels low repulsive forces when the real potential energy is low. Note that at the binding configuration, the potential energy is very small and the user felt almost no force, hence with visual and haptic feedback the user can estimate the position of a binding site.

Our haptic system also lets the user investigate a path or roadmap generated by the automated planner. The user can visualize a path, select a configuration in the path, and investigate the potential field around that configuration. Similarly, the user can visualize a roadmap. Roadmap configurations are represented by points in the three-dimensional workspace and roadmap edges by lines connecting them [2]. The user can investigate a specific roadmap configuration by moving the haptic pointer close to a roadmap configuration causing the configuration to pop up visually.

5 Generating Binding Site Candidates

Good binding configurations are thought to be configurations at local minima in the potential landscape, and moreover, should be surrounded by high potentials so that the ligand will tend to stay in the site once it gets there [16].

Our goal is to automatically generate such configurations of the ligand. Our strategy is to first generate nodes near the potential energy surface, and then push the generated nodes to their local minima. Nodes can be generated automatically with our OBPRM planner, or manually by the user with the haptic device. Next, these nodes are connected into a roadmap that will be used to determine the accessibility of a site (this is very useful for eliminating 'decoy' nodes that are inaccessible, but otherwise appear to be good candidates). Finally, we will compute a weight for all low potential configurations contained in large connected components of the roadmap (see Section 6). Details of each step are discussed below.

Automatic Node Generation. OBPRM[1] is designed to generate contact and surface nodes. We model each protein atom as a unit cube (1 Å), and each ligand atom as a sphere with 1 Å radius. Considering only the base atoms of the articulated ligand model, we generate a collision-free (geometrically) node for the base. Then, we assign random values for the joint angles. Next, we evaluate the potential of the ligand's configuration and keep the node only if its potential is less than some fixed value $E_{\max}$. A fairly high value of 5,000 KJ/Mol was chosen for $E_{\max}$. However, after applying gradient descent to the nodes (described below), most remaining nodes have potentials below zero.

User Generated Nodes. We also experiment with operator collected nodes using a visual and haptic interface. Our previous work with CAD models has shown significant improvement when user input is given to the automatic planner [2]. A user would collect a node when he/she felt like it was local minimum in the potential and also saw it was in some sort of pocket on the protein. The ligand was restricted to a rigid body for the haptic interaction.

Approximate Gradient Descent. Since a binding configuration is commonly thought to be a local minima surrounded by higher potentials, it is natural to push our generated nodes to local minima. In particular, we perform an approximate gradient descent as follows. For each generated node, we sample uniformly twice the ligands dof nearby nodes, and select the lowest potential new node for iteration. This process is repeated until an iteration limit or a local minima is reached (when all sampled nodes have higher potential, usually in 2-4 iterations).

Node Connection. After gradient descent is performed, all local minima nodes are added to the roadmap (with the previously generated nodes). Then, we attempt to connect each node to its k nearest neighbors along the straight-line in C-space between them (we use $k = 10$). An edge between the node and its neighbor is added to the roadmap if it passes some validity check. The only difference here is that our validity check uses potential evaluation instead of collision detection.

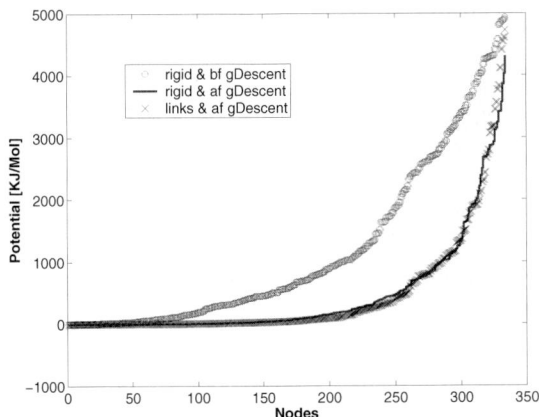

Figure 3: The potential distribution of all roadmap nodes before and after gradient descent for 1LDM protein-ligand complex.

6 Recognizing Binding Sites

In order to identify configurations in the binding site, as well as to determine promising configurations for further processing, we need to rank (score) configurations.

In the PRM based method presented in [16], they first identify the 10 roadmap nodes with the lowest potential, and then further evaluate them with a score function called average path weight. They successfully found the binding sites for two of the three examples they studied.

Our approach in this paper is similar to theirs, but has two significant differences.

First, after the gradient descent most of the roadmap nodes end up having very low energies, as shown in Figure 3, and the approximation used for the potential can easily change the relative order of the nodes. So potential energy alone cannot be used as a selection criteria. However, one factor that can be used to filter nodes is accessibility. We use the roadmap's connectivity to filter out inaccessible configurations, and in particular, only evaluate nodes contained in the largest connected component(s) of the roadmap.

Second, we use a different weight function. Since the binding site is a *local* feature, we observe there is no need to evaluate path weights globally as in [16]. Instead, we propose building a 'local roadmap' for each candidate node and evaluating path weights in it. In particular, we first uniformly sample n nodes within a fixed distance r from the candidate node, and then define the weight, or score, of the candidate as the average potential of all the sampled nodes. The potential is truncated to $5,000\ KJ/Mol$ if it has a larger value.

7 Experiments

We tested our approach on three different protein-ligand complexes obtained from the Protein Data Bank at http://www.rcsb.org/pdb/, Two of these complexes, 1LDM (M_4-Lactate Dehydrogenase-Oxamate) and 1STP (Streptavidin-Biotin), were studied in [16]. The third, 1A5Z (L-Lactate Dehydrogenase-Oxamate), was chosen based on the complexity of protein's potential field which created many isolated areas. Some details of these complexes are shown in Table 2.

In our experiments we investigate the following questions:

- Does the approximation of the ligand molecule as a rigid object affect our results?
- Can an OBPRM-based automated planner generate nodes in binding sites?
- Can a user provide helpful information to an automated planner?

To answer these questions we perform three different experiments. (i) In the first experiment, the ligand is treated as a rigid body. The planner generates configurations, and finds and ranks potential binding sites using these configurations. (ii) In the second experiment, the user collects configurations using the PHANToM by manipulating a rigid ligand. The automated planner then fine-tunes these configurations, now treating them as articulated bodies, and evaluates the possible binding sites. (iii) In the third experiment, we investigate the behavior of the automated planner when the ligand is flexible, i.e., articulated.

The running time and roadmap statistics are shown in Table 1. The number of configurations the user collected varies and can be found in Table 2.

7.1 Results

A summary of our results is shown in Table 3. For each experiment for a protein-ligand complex, we list the top five scoring configurations identified. The first column represents the rank of the configuration. The 0 ranked configuration is the binding configuration for the respective complex. The RMSD column shows the Root Mean Square Deviation with respect to the known binding site. The score of that configuration is computed as described in Section 6. The last column represents the potential energy of the complex at that configuration. Based on the rigid body experiments of 1A5Z and 1STP (Table 3(a) and 3(c)) we did not consider the rigid body representation for 1LDM.

As can be seen, in almost all cases, at least one the top five scoring configurations is very close (2-3 Å) to the true binding configuration. This suggests that our automated planner is effective at finding configurations close to the binding site. In addition, our success in identifying the binding site for 1STP shows an advantage of our OBPRM based approach over the results presented in [16], where the closest configuration identified was 13 Å from the true binding site.

Based on the automated results for the rigid body and articulated representations, for 1A5Z and 1STP in Table 3,

Roadmap Statistics						
Model	Gen. time	Conn. time	Total time	# node(init #, # local min added)	# edge	bigCC
1A5Z Rigid	55.	70	229.8	1429 (1000, 429)	3784	1182
1A5Z User	75	101	199	1684 (1666, 18)	3045	1304
1A5Z Artic.	75	203	657.9	2417 (1666, 751)	6253	2071
1LDM User	853	141	279	2011 (2000, 11)	4141	1643
1LDM Artic.	85	299	1089	2912 (2000, 912)	8472	2579
1STP Rigid	104	369	822	1771(1000, 771)	5589	1738
1STP User	162	163	343	1674 (1666, 8)	3115	1504
1STP Artic.	162	447	2095	3020 (1666, 1354)	7971	2909

Table 1: Roadmap statistics. Total time includes node generation time, connection time, and node score calculation time. All times are in seconds.

Pair ID	Protein # Atom	Ligand #Atom	dof	# Cfgs user collected
1LDM	2544	6	7	37
1A5Z	2416	6	7	129
1STP	1001	16	11	92

Table 2: Statistics for different protein-ligand complexes. The table also shows number of configurations that are collected by the user during the node generation phase.

it is clear that better results were obtained using the articulated representation, which did incur some additional computational costs as seen in Table 1. The articulated representation was slightly better for 1A5Z, where the difference in the degrees of freedom between the rigid and articulated versions is small. When these differences increase the performance gap increases, as with 1STP. In general, an articulated ligand can move easier then a rigid ligand in narrower areas by taking on the shape of the surrounding potential surface. Our comparison of the rigid and articulated representation is potentially relevant for other docking methods, since representing the ligand as a rigid body is a commonly used simplification.

The results for the fully automated articulated model and those starting from user generated nodes are comparable. In 1A5Z and 1LDM the user was able to find the closest configurations to the binding site. However, in 1STP the automated planner reached a closer configuration. This may be related to the fact that the user has to work with a rigid version of the ligand, while the automated planer takes advantage of the high-dimensional space. The large difference in the performance of the rigid body and the articulated body representations for the 1STP complex supports this idea.

An advantage of user input is that, as shown in Table 1, it was noticeably faster than the fully automated method since the user provided a much smaller sample of nodes. Fewer nodes were provided since the user only collected nodes in/near pockets. A disadvantage is that the user input suffers the users' bias. It is very tedious for the user to explore every corner of the protein to find a binding site, therefore he/she may miss the true binding site. In addition, a user may be misled by some 'decoy' sites based on only visual and/or force feedback as in 1A5Z where several nodes lie in a site far away.

Finally we remark that our energy filtering keeps many nodes that are in the flat region of the potential distribution plot. Therefore, we avoid throwing away nodes near binding sites even though their potential may be slightly higher, which could very easily be caused by inaccuracy of our potential function.

8 Conclusions and Future Work

Our results show that our approach to molecular docking is promising. In the examples we studied, we were able to find and recognize configurations in the true binding site. However, further processing, perhaps with a more accurate potential function, will be needed to find an exact binding configuration. For example, candidates identified by our method might be used as input for other docking programs that perform detailed simulations, such as molecular dynamics methods. We also saw that better results were obtained with an articulated representation for the ligand, as opposed to the commonly used rigid body simplification. User input was seen to improve efficiency, and moreover, haptic feedback was observed to help the user better understand the problem.

Our future work includes finding a better representation for potential energy formulations, concentrating on the binding sites and reaching the binding configurations, and improving the haptic interface by letting the user collect configurations for a flexible ligand.

References

[1] N. M. Amato, O. B. Bayazit, L. K. Dale, C. V. Jones, and D. Vallejo. OBPRM: An obstacle-based PRM for 3D workspaces. In *Proc. Int. Workshop on Algorithmic Foundations of Robotics (WAFR)*, pages 155–168, 1998.

[2] O. B. Bayazit, G. Song, and N. M. Amato. Enhancing randomized motion planners: Exploring with haptic hints. In *Proc. IEEE Int. Conf. Robot. Autom. (ICRA)*, pages 529–536, 2000.

[3] D. Hsu, L. Kavraki, J-C. Latombe, R. Motwani, and S. Sorkin. On finding narrow passages with probabilistic roadmap planners. In *Proc. Int. Workshop on Algorithmic Foundations of Robotics (WAFR)*, 1998.

[4] E. Katchalski-Katzir, I. Shariv, M. Eisenstein, A.A. Friesem, and C. Aflalo I.A. Vakser. Molecular surface recognition: Determination of geometric fit between proteins and their ligands by correlation

(a) 1A5Z

Rank	RIGID			USER INPUT (RIGID)			ARTICULATED (7dof)		
	RMSD(Å)	Score	Pot.	RMSD(Å)	Score	Potential	RMSD(Å)	Score	Pot.
0	0.00	3743	35.51	0.00	3743	35.51	0.00	3743	35.51
1	5.74	4694	0.09	2.08	4138	-8.783	5.82	4658	19.92
2	5.20	4462	24.81	1.94	3670	-5.19	7.31	4488	41.43
3	4.10	4422	0.81	15.90	3316	-5.209	2.31	4459	13.67
4	9.27	4376	30.36	16.89	2682	22.207	5.63	4393	-9.59
5	4.35	4290	-7.79	16.11	2413	-5.95	2.42	4342	19.88

(b) 1LDM

Rank	USER INPUT (RIGID)			ARTICULATED (7dof)		
	RMSD(Å)	Score	Pot.	RMSD(Å)	Score	Potential
0	0.00	4570	5.00	0.00	4570	5.00
1	1.52	4652	-6.06	5.03	4634	30.96
2	2.93	4564	-6.73	2.03	4593	23.35
3	3.79	4541	-7.04	3.07	4577	48.12
4	2.35	4499	-6.88	4.14	4475	20.95
5	2.11	4463	-8.70	1.66	4418	43.34

(c) 1STP

Rank	RIGID			USER INPUT (RIGID)			ARTICULATED (11dof)		
	RMSD(Å)	Score	Pot.	RMSD(Å)	Score	Pot.	RMSD(Å)	Score	Pot.
0	0.00	4552	-5.21	0.00	4552	-5.21	0.00	4552	-5.21
1	4.21	4127	48.79	5.50	3431	49.22	2.56	4609	0.37
2	11.40	3934	-12.73	3.77	3426	11.10	5.05	4284	36.44
3	10.63	3921	7.94	7.32	3425	-6.47	3.44	4151	-5.79
4	12.42	3894	-10.51	4.99	3287	48.80	4.11	4014	18.31
5	11.29	3876	38.19	6.90	3211	44.05	7.41	3952	22.68

Table 3: Performance of the scoring function on each protein. The tables show the results of three different kind of experiments; OBPRM with the ligand as a rigid body, OBPRM with the ligand as an articulated body and user collected configurations. For 1LDM, we only experimented on OBPRM with the articulated ligand and user collected configurations.

techniques. In *Natl. Acad. Sci. USA*, volume 89, pages 2195–2199, 1992.

[5] I.D. Kuntz. Structure-based strategies for drug design and discovery. *Science*, 257:1078–1082, 1992.

[6] Jonathan Leech, Jan F. Prins, and Jan Hermans. Smd: Virtual steering of molecular dynamics for protein design. *IEEE Computational Science and Engineering*, pages 38–45, 1996.

[7] D. Levine, M. Facello, P. Hallstorm, G. Reeder, B. Walenz, and F. Stevens. Stalk: An interactive system for virtual molecular docking. *IEEE Computational Science & Engineering*, April-June:55–66, 1997.

[8] M. Levitt. Protein folding by restrained energy minimization and molecular dynamics. *J. Mol. Bio.*, 170:723–764, 1983.

[9] T. H. Massie and J. K. Salisbury. The PHANToM haptic interface: A device for probing virtual objects. In *Int. Mechanical Engineering Exposition and Congress, DSC 55-1*, pages 295–302, Chicago, 1994. C. J. Radcliffe, ed., ASME.

[10] G.M. Morris, D.S. Goodsell, R.S. Halliday, R. Huey, W.E. Hart, R.K. Belew, and A.J. Olson. Automated docking using a lamarckian genetic algorithm and empirical binding free energy function. *J. Computational Chemistry*, 19:1639–1662, 1998.

[11] M. Ouh-young. *Force Display in Molecular Docking, TR90-004*. PhD thesis, Univ. of North Carolina at Chapel Hill, Chapel Hill, NC, USA, 1990.

[12] Ming Ouh-young, David V. Beard, and Frederick P. Brooks, Jr. Force display performs better than visual display in a simple 6-d docking task. In *Proc. IEEE Int. Conf. Robot. Autom. (ICRA)*, pages 1462–1466, 1989.

[13] Ruth Pachter and James A. Lupo. Virtual reality for materials design. In *The 14th Annual Dayton Section Symposium AESS/IEEE*, pages 51–56, 1997.

[14] N. Pattabiraman, M. Levitt, T. E. Ferrin, and R. Langridge. Computer graphics in real time docking with energy calculation and minimization. *J. of Computational Chemistry*, 6:432–436, 1985.

[15] M. Rarey, B. Kramer, T. Lengauer, and G. Klebe. A fast flexible docking method using an incremental construction algorithm. *J. Molecular Biology*, 261:470–489, 1996.

[16] A.P. Singh, J.C. Latombe, and D.L. Brutlag. A motion planning approach to flexible ligand binding. In *7th Int. Conf. on Intelligent Systems for Molecular Biology (ISMB)*, pages 252–261, 1999.

[17] G. Song and N. M. Amato. A motion planning approach to folding: From paper craft to protein folding. In *Proc. IEEE Int. Conf. Robot. Autom. (ICRA)*, 2001. To appear.

[18] G. Song and N. M. Amato. Using motion planning to study protein folding pathways. In *Proc. Int. Conf. Comput. Molecular Biology (RECOMB)*, pages 278–287, 2001.

[19] M.C. Surles, J.S. Richardson, D.C. Richardson, and F.P. Brooks, Jr. Sculpting proteins interactively: Continual energy minimization embedded in a graphical modeling system. *Protein Science*, 3:198–210, 1994.

[20] Maxim Totrov and Ruben Abagyan. Derivation of sensitive discrimination potential for virtual ligand screening. In *Third Annual International Conference on Computational Molecular Biology*, pages 312–320, 1999.

[21] Len Wanger. Haptically enhanced molecular modeling: A case study. In *Proceedings of the Third PHANTOM Users Group Workshop*, 1998.

Molecular Docking: A Problem With Thousands Of Degrees Of Freedom

Miguel L. Teodoro[1]
mteodoro@rice.edu

George N. Phillips Jr[2]
phillips@biochem.wisc.edu

Lydia E. Kavraki[3]
kavraki@rice.edu

1 Department of Biochemistry and Cell Biology and Department of Computer Science, Rice University
2 Department of Biochemistry and Department of Computer Science, University of Wisconsin-Madison
3 Department of Computer Science and Department of Bioengineering, Rice University

Abstract

This paper reports on the problem of docking a highly flexible small molecule to the pocket of a highly flexible receptor macromolecule. The prediction of the intermolecular complex is of vital importance for the development of new therapeutics as docking can alter the chemical behavior of the receptor macromolecule. We first present current methods for docking which however have several limitations. Some of these methods consider only the flexibility of the ligand solving a problem with a few tens of degrees of freedom. When the receptor flexibility is taken into account several hundreds or even thousands of degrees of freedom need to be considered. Most methods take into account only a small number of these degrees of freedom by using chemical knowledge specific to the problem. We show how to use a Singular Value Decomposition of Molecular Dynamics trajectories to automatically obtain information about the global flexibility of the receptor and produce interesting conformations that can be used for docking purposes.

1 Introduction

The application of computational methods to study the formation of intermolecular complexes has been the subject of intensive research during the last decade. It is widely accepted that drug activity is obtained through the molecular binding of one molecule (the ligand) to the pocket of another, usually larger, molecule (the receptor), which is commonly a protein. A complex of a protein with a therapeutical drug is shown in Figure 1. In their binding conformations, the molecules exhibit geometric and chemical complementarity, both of which are essential for successful drug activity. The computational process of searching for a ligand that is able to fit both geometrically and energetically the binding site of a protein is called molecular docking.

The rapid generation of quality lead compounds is a major hurdle in the design of therapeutics, so that accurate automated procedures would be of tremendous value to pharmaceutical and other biotechnology companies. However, designing a drug based on the knowledge of the target receptor structure as determined by current experimental techniques is a process prone to error. The two major reasons responsible for failures are

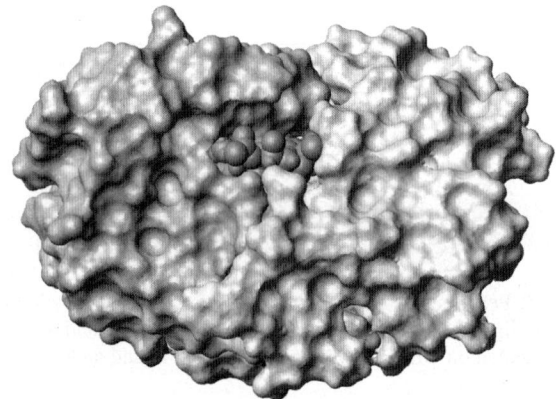

Figure 1 – Therapeutic drug molecule (small molecule towards the center of the figure) bound to protein receptor (HIV-1 protease). The drug molecule fits tightly in the binding site and blocks the normal protein function.

inaccuracies in the energy models used to score potential ligand/receptor complexes, and the inability of current methods to account for conformational changes that occur during the binding process not only for the ligand, but also for the receptor. Although this problem has been partially solved by incorporating ligand flexibility in search methods, predicting receptor structural rearrangements is a very complex problem which has not been solved. The docking problem is analogous to an assembly-planning problem where the parts are actuated by molecular forcefields and have thousands of degrees of freedom.

In this article we report on the current methods used to solve the docking problem and on some of the problems and possible solutions to incorporate protein flexibility in the docking process. Section 2 introduces some of the terminology and concepts relevant to this problem and Section 3 reports on some of the docking methods used in academia and industry. However, the models described in Section 3 follow the assumption of a rigid protein which limits their use. In Section 4 we describe some of the methods currently under development to model protein flexibility. We will also describe our own model which incorporates full protein flexibility in docking and can be easily automated.

2 Molecular Modeling

A molecule is characterized by a pair (A; B), in which A represents a collection of atoms, and B represents a collection of bonds between pairs of atoms. Information used for kinematic and energy computations is associated with each of the atoms and bonds. Each atom carries standard information, such as its van der Waals radius. Three pieces of information are associated with each bond: (i) bond length, is the distance between atom centers; (ii) bond angle, is the angle between two consecutive bonds; (iii) whether the bond is rotatable or not (for an illustration of rotatable bonds see Figure 2). Since bond lengths and angles do not change significantly, it is common practice to consider them fixed. Thus the degrees of freedom of the molecule arise from the rotatable bonds. The three dimensional embedding of a molecule defined when we assign values to its rotatable bonds is called the conformation of the molecule. Ligands typically have 3-15 rotatable bonds, while receptors have 1,000-2,000 rotatable bonds. The dimension of the combined search space makes the docking problem computationally intractable.

One key aspect of molecular modeling is calculating the energy of conformations and interactions. This energy can be calculated with a wide range of methods ranging from quantum mechanics to purely empirical energy functions. The accuracy of these functions is usually proportional to its computational expense and choosing the correct energy calculation method is highly dependent on the application. Computation times for different methods can range from a few milliseconds on a workstation to several days on a supercomputer.

In the context of docking, energy evaluations are usually carried out with the help of a scoring function and developing these is a major challenge facing structure based drug design[1]. No matter how efficient and accurate the geometric modeling of the binding process is, without good scoring functions it is impossible to obtain correct solutions. The two main characteristics of a good scoring function are selectivity and efficiency. Selectivity enables the function to distinguish between correctly and incorrectly docked structures and efficiency enables the docking program to run in a reasonable amount of time.

A large number of current scoring functions are based on forcefields that were initially designed to simulate the function of proteins[2,3]. A forcefield is an empirical fit to the potential energy surface in which the protein exists and is obtained by establishing a model with a combination of bonded terms (bond distances, bond angles, torsional angles, etc.) and non-bonded terms (van der Waals and electrostatic). The relative contributions of these terms are adjusted for the different types of atoms in the simulated molecule by adjusting a series of empirical parameters. Some scoring functions used in molecular docking have been adapted to include terms such as solvation and entropy[4]. A separate approach is to use statistical scoring functions that are

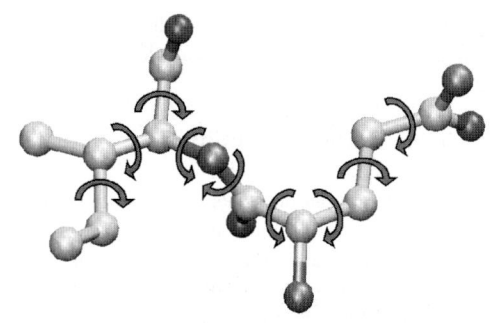

Figure 2 - A drug molecule. Spheres represent atoms and bonds connecting them are represented by sticks. Curved arrows represent the rotatable degrees of freedom around bonds.

derived using experimental data[5].

3 Rigid Protein Docking

Most of the docking methods used at the present moment in academic and industrial research assume a rigid protein. To illustrate the methodology used by these methods we will briefly discuss three of the most common programs used for docking: Autodock[4], Dock[6] and FlexX[7].

Autodock uses a kinematic model for the ligand similar to the one illustrated in Figure 2. The ligand begins the search process randomly outside the binding site and by exploring the values for translations, rotations and its internal degrees of freedom, it will eventually reach the bound conformation. Distinction between good and bad docked conformations is carried out by the scoring function. Autodock is able to use Monte Carlo methods or simulated annealing (SA) in the search process and in its last version introduced the ability to use genetic algorithms (GA). The routine implemented in the recent release is a Lamarkian genetic algorithm (LGA), in which a traditional GA is used for global search and is combined with a Solis and Wets local search procedure. Morris et al[4] show that the new LGA is able to handle ligands with a larger number of degrees of freedom than SA or traditional GA.

FlexX and Dock both use an incremental construction algorithm which attempts to reconstruct the bound ligand by first placing a rigid anchor in the binding site and later using a greedy algorithm to add fragments and complete the ligand structure. Although these programs are more efficient than Autodock in the sense that they require fewer energy evaluations there exist some tradeoffs. One of main problems is that it is not trivial to choose the anchor fragment and its choice will determine what solutions can be obtained. Also the greedy algorithm propagates errors resulting from initial bad choices that lead to missing final conformations of lower energy.

In order to solve the docking problem conformation methods using standard robotics techniques such as probabilistic roadmap planning have been recently described[8,9]. In addition to being successful in finding the correct docking conformation these methods are useful in identifying possible binding sites and in providing a computational efficient description of the dynamics of ligand binding.

4 Modeling Protein Flexibility

One of the greatest challenges facing current structure-based rational drug design is the integration of protein flexibility in docking methods used to screen databases of possible therapeutic compounds. In this section we present a survey of the approaches under research to model protein flexibility and at the end of this section we present our own solution to this problem.

Current docking methods follow the assumption that protein structures are rigid entities and that it is the ligand that during the binding process changes its three-dimensional structure to find the best spatial and energetic fit to the protein's binding site. This assumption follows the model of lock-and-key binding first proposed by Emil Fischer in 1890. However, a better description of the mechanism of interaction between a protein and its ligand was given by Koshland in 1958 with the induced-fit model. In this model both the protein and the ligand are flexible and when they interact to form a complex both structures change their conformation to form a minimum energy perfect-fit. Unfortunately doing an exact modeling of the flexibility available to the protein during the binding process is still far beyond our present computational capability. Whereas conventional ligand modeling techniques are able to handle up to approximately 30 degrees of freedom when searching for a docked conformation, modeling the full flexibility of the protein requires more than 1000 degrees of freedom, even for a small size protein.

Although the methods described in Section 3 have shown reasonable success in screening for candidate drugs, several studies have exposed their problems and limitations[10,11]. These problems are especially important when non-negligible changes in conformation are present during the binding process. This leads to final docking results that entirely fail to identify potential drug candidates or otherwise assign them very poor binding scores. To overcome limitations from the rigid protein assumption, several approximations have been used to model protein flexibility. These approximations can be divided in two groups: models which try to account for the flexibility of the protein in the binding region and models which simulate the flexibility as a whole.

4.1 Partial Protein Flexibility

The first approximation used in modeling partial protein flexibility was the soft-docking method first described by Jiang and Kim[12]. The principle underlying this method consists of decreasing the van der Waals repulsion energy term between the atoms in the binding site and those in the ligand. This method could result in final solutions that include physically impossible atom collisions. Nevertheless, due to the mobility available to the protein atoms in the binding site, it is possible that there is a low energy rearrangement of these that would eliminate collisions while maintaining the conformation of the ligand returned during its conformational search. This method has the advantage of being computationally efficient as it still describes the protein using fixed coordinates. The method is also easy to implement since it does not require changes to the energy evaluation function besides changing van der Waals parameters.

The most common approximation used to incorporate partial protein flexibility in modeling the binding process is to select a few degrees of freedom in the protein binding site and do a simultaneous search of the combined ligand/protein conformational space. Incorporating select degrees of freedom from the binding site in the conformational search process is based on the assumption that these degrees of freedom are the ones playing a major role in determining the conformational changes during the binding process. This choice requires deep chemical understanding of the system under study and is therefore difficult to automate. Furthermore, even for proteins which are considered relatively rigid, Murray *et al*[11] show that protein backbone changes often play a critical role and these are difficult to model using only a few degrees of freedom. The optimization techniques used for this approach are the same as for the rigid protein but are now required to handle a larger number of degrees of freedom resulting in overall less efficiency.

One of the earliest reports of using select degrees of freedom from the protein was described by Jones *et al*[13] and was implemented in the program GOLD (Genetic Optimization for Ligand Docking). This program improves on the rigid protein model by performing a conformational search on the binding site with the aim of improving the hydrogen bonding network between the protein and the ligand. Hydrogen bonds are local electrostatic interactions between pairs of atoms which play an important energetic role in ligand recognition and binding. GOLD selects the degrees of freedom in the binding site that correspond to reorientations of hydrogen bond donor and acceptor groups. These degrees of freedom represent only a very small fraction of the total conformational space that is available but should account for a significant difference in binding energy values. More recent studies have been reported[14-16] in which other degrees of freedom from aminoacid sidechains are also used in the conformational search. These are searched using stochastic methods with arbitrary step sizes or using rotamer libraries[17]. Rotamer libraries consist of discrete sidechain conformations of low energy which are usually determined from statistical analysis of structural data derived experimentally.

4.2 Full Protein Flexibility

Ideally ligand docking to a protein could be simulated using Molecular Dynamics (MD). This has the advantage that not only it takes into account all the degrees of freedom available to the protein but also enables an explicit modeling of the solvent. Furthermore, accurate energy calculations can also be carried out using the free energy perturbation method. Unfortunately, modeling proteins using MD is computationally expensive, and the computational power necessary to simulate the full process of diffusion and ligand binding without any approximations will be out of our reach for many years to come. Recently Mangoni et al[18] reported a modification to the standard MD protocol which reduces the computational time required for the docking simulation. The protocol consists of separating the center of mass motion of the ligand from its internal and rotational motions by coupling the different degrees of freedom to separate thermal baths. This optimization allows the ligand to sample the space surrounding the binding site faster while maintaining correct interactions with both protein and solvent.

An alternative approach to model full protein flexibility is to generate an ensemble of rigid protein conformations that together represent the conformational diversity available to the protein. These conformations can later be docked to a database of ligands using traditional rigid-protein/flexible-ligand methods. There are several possible methods to generate the ensembles, but unfortunately their accuracy is proportional to the difficulty in obtaining them. The most accurate ensemble is the one determined exclusively from experimental data. An example is the case where several structures of protein/ligand complexes are determined using X-ray crystallography bound to different candidate drugs. Under these circumstances it is usually possible to observe alternative binding modes directly[19]. Another less accurate option is to use the ensemble of structures that results from an experimental protein structure determination using the NMR (Nuclear Magnetic Resonance) technique. This docking methodology was first reported by Knegtel et al[20]. Finally, one can generate an ensemble using computational methods such as Monte Carlo (MC) or MD sampling. The accuracy of these alternatives is closely related to the accuracy of the force field used and is limited by the ability of these computational techniques to effectively sample the conformational space[21]. Docking to an ensemble of structures generated using MD was first reported by Pang and Kozikowski[22].

A different representation for full protein flexibility is to divide the protein in tightly coupled domains whose constituent atoms move collectively as one. Hinges connect the domains and the motion of the protein is simulated similarly to an articulated robot. Required conformational changes inside domains can be handled using minimization. An application of this model to the docking problem was reported by Sandak et al[23].

4.2.1 Modeling Protein Flexibility With Collective Modes Of Motion

The approach we are presently investigating to account for full protein flexibility while reducing the computational complexity of the problem is to use the concept of essential dynamics[24]. This formulation divides the conformational space accessible to the protein into two subspaces: (1) an essential subspace containing only a few degrees of freedom which correspond to major modes of anharmonic motion and describe most of the positional fluctuations; and (2) a nonessential subspace consisting of constrained harmonic motions. By using only the major modes of motion in the essential subspace of the protein it is possible to simulate an approximation to the interaction between a protein and its ligand in a conformational space of much lower dimensionality.

The mathematical formulation we use to determine the collective major modes of motion is the Singular Value Decomposition (SVD) of the displacement matrix derived from a molecular dynamics simulation[25,26]. As an alternative to MD data it is also possible to use ensembles of structures determined experimentally either by X-ray crystallography or by NMR.

The SVD of a matrix, A, is defined as:
$$A = U \Sigma V^T, \quad (1)$$
where U and V are orthonormal matrices and Σ is a nonnegative diagonal matrix whose diagonal elements, σ_i, are the singular values of A. The columns of matrices U and V are called the left and right singular vectors, respectively. Matrix A is constructed by the column-wise concatenation of atomic displacement vectors, for each time sample during a molecular dynamics run. The left singular vectors of the SVD of A will span the space sampled by the protein during the entire simulation. The left singular vectors corresponding to the largest singular values reflect the major modes of motion in the protein and span the essential subspace. The right singular vectors are projections of the dynamics trajectory along the left singular vectors. The advantage of this mathematical transformation is that it changes the basis of representation of our problem. Whereas initially all our degrees of freedom were identically important, using this method we are able to rank our collective degrees of freedom by the order of their eigenvalues. Moreover, this method does not require an intimate knowledge of the system in order to select a few degrees of freedom. The choice is determined by the eigenvalue rank.

The data used in the SVD computation was taken from a 500ps simulation of HIV-1 protease in a box of water molecules using periodical boundary conditions and full electrostatic computation. In Figure 3 we show the rank order eigenvalue spectrum of the SVD analysis of the coordinate data for all 3120 atoms in our protein system. In this plot only the first 30 out of 9360 eigenvalues are shown. The largest eigenvalue accounts for 18% of the cumulative eigenvalue sum and the first 20 account for 63%. It is clear from these values that on the new basis only a few degrees of freedom account for most of the conformational variation. Given this result we are able to approximate the most significant part of the motion in a space with significantly fewer dimensions. One problem with this approximation is that some of the motions in this new conformational space lead to high-

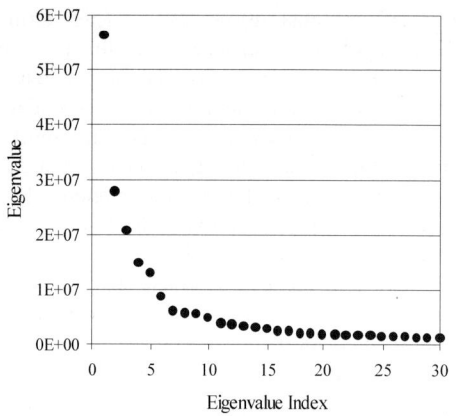

Figure 3 - Rank order eigenvalue spectrum of a SVD analysis of a 500ps MD trajectory.

energy conformations due to distortions in the internal structure of the ligand. We are currently dealing with this problem by performing standard energy minimization methods. We are also developing fast geometry correction methods for the internal structure of the protein which could help eliminate this problem.

In Figure 4 (center representation) we show the backbone representation for HIV-1 protease as determined by X-ray crystallography. The arrows show the mapping of the high dimensional first left singular vector motion into several Cartesian vectors on the backbone of the protein. The directions of the arrows indicate the direction of the motion at that position of the protein and the size of the arrows indicate the relative magnitudes of motion from one region to another. This mapping is in accordance to what would be expected for HIV-1 protease with most of the motion concentrated in flexible loops and on the two flexible flaps that cover the binding site. Using this experimental structure and the first left singular vector we can "actuate" the protein along this degree of freedom in one direction or the other (left and right representations). It is important to note that although the protein is moving as a whole, we are using only one degree of freedom of the new basis to describe that motion. This method can also be used to generate ensembles of structures for rigid docking as described in Section 4.2. In comparison with the ensemble extracted directly from the MD trajectory our method generates a more representative ensemble since the sampling is being done over the most significant degrees of freedom.

5 Discussion

Most of the docking programs presently being used simulate the binding of a flexible ligand to a rigid biological receptor. This model does not reflect the actual physical process of binding and limits or in some cases even prevents the correct identification of potential drug candidates. In this paper we reviewed some of the approaches under research to incorporate protein flexibility in the docking simulation. Some of these approaches have drawbacks such as high computational cost, limited sampling of the receptor conformational space, or require a deep understanding of the biological system making automation difficult. Here we described an alternative method to model protein flexibility based on the SVD of a molecular dynamics trajectory. This procedure is of general applicability, requires a practical amount of computational power and is easily automated.

Our discussion reveals the challenging representational and computational problems that need to be addressed to arrive to efficient molecular docking techniques. We believe that the work done in robotics on kinematics can help in the accurate simulation of protein flexibility and reduce the need of expensive energy minimizations. We also believe that the development of probabilistic path planners that can deal with many degrees of freedom robots will lead to the development of planners that lead to the docking of the flexible ligand in a flexible protein [9]

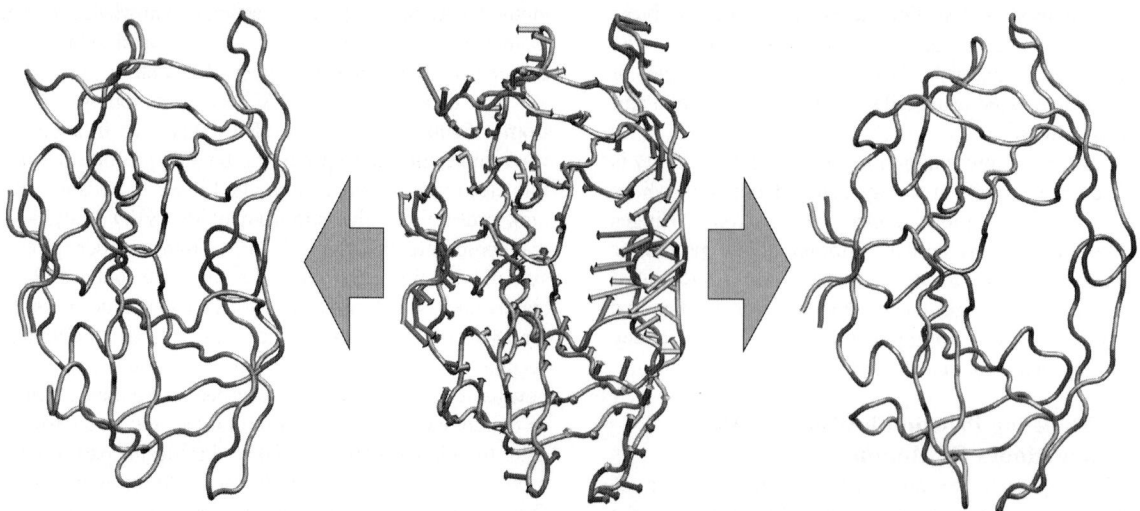

Figure 4 - First collective mode of motion for HIV-1 protease. The backbone structure of the protein as determined experimentally is shown in the center. Arrows indicate the mapping of the first left singular vector in the most significant collective motion. The structure can be "actuated" using this single degree of freedom (left and right structures).

Acknowledgements: Miguel Teodoro is supported by a PRAXIS XXI Pre-doctoral fellowship from the Portuguese Ministry of Science. George Phillips and Miguel Teodoro are partially supported by ATP 003604-0120-1999. Work on this paper by Lydia Kavraki is supported in part by NSF IRI-970228, NSF CISE SA1728-21122N, ATP 003604-0120-1999 and a Sloan Fellowship. All authors are affiliated with the W.M. Keck Center for Computational Biology.

References

1. Vieth, M., Hirst, J.D., Kolinski, A. & Brooks, C.L.I. Assessing energy functions for flexible docking. *J Comp Chem* **19**, 1612-1622 (1998).
2. Cornell, W.D. et al. A second generation force field for the simulation of proteins and nucleic acids. *J Am Chem Soc* **117**, 5179-5197 (1995).
3. MacKerell, A.D., Bashford, D., Bellot, M., Karplus, M. & al, e. All-atom empirical potential for molecular modeling and dynamics studies of proteins. *J Phys Chem B* **102**, 3586-3616 (1998).
4. Morris, G.M. et al. Automated docking using a Lamarkian genetic algorithm and an empirical binding free energy function. *J Comp Chem* **19**, 1639-1662 (1998).
5. Muegge, I. & Martin, Y.C. A general and fast scoring function for protein-ligand interactions: a simplified potential approach. *J Med Chem* **42**, 791-804 (1999).
6. Ewing, T.J.A. & Kuntz, I.D. Critical evaluation of search algorithms for automated molecular docking and database screening. *J Comp Chem* **18**, 1175-1189 (1997).
7. Kramer, B., Metz, G., Rarey, M. & Lengauer, T. Ligand docking and screening with FlexX. *Med Chem Res* **9**, 463-478 (1999).
8. Bayazit, O.B., Song, G. & Amato, N.M. Ligand Binding with OBPRM and Haptic User Input. in *2001 IEEE International Conference on Robotics and Automation (ICRA'01)* (Seoul, Korea, 2001).
9. Singh, A.P., Latombe, J.C. & Brutlag, D.L. A motion planning approach to flexible ligand binding. in *International Conference on Computational Biology, ISMB* (1999).
10. Brem, R. & Dill, K.A. The effect of multiple binding modes on empirical modeling of ligand docking to proteins. *Prot Sci* **8**, 1134-43. (1999).
11. Murray, C.W., Baxter, C.A. & Frenkel, A.D. The sensitivity of the results of molecular docking to induced fit effects: Application to thrombin, thermolysin and neuraminidase. *J Comput Aided Mol Des* **13**, 547-562 (1999).
12. Jiang, F. & Kim, S.H. "Soft docking": matching of molecular surface cubes. *J Mol Biol* **219**, 79-102. (1991).
13. Jones, G., Willett, P., Glen, R.C., Leach, A.R. & Taylor, R. Development and validation of a genetic algorithm for flexible docking. *J Mol Biol* **267**, 727-48 (1997).
14. Apostolakis, J., Pluckthun, A. & Caflisch, A. Docking small ligands in flexible binding sites. *J Comp Chem* **19**, 21-37 (1998).
15. Schnecke, V., Swanson, C.A., Getzoff, E.D., Tainer, J.A. & Kuhn, L.A. Screening a peptidyl database for potential ligands to proteins with side-chain flexibility. *Proteins* **33**, 74-87. (1998).
16. Totrov, M. & Abagyan, R. Flexible protein-ligand docking by global energy optimization in internal coordinates. *Proteins* **Suppl**, 215-20 (1997).
17. Leach, A.R. Ligand docking to proteins with discrete side-chain flexibility. *J Mol Biol* **235**, 345-56 (1994).
18. Mangoni, M., Roccatano, D. & Di Nola, A. Docking of flexible ligands to flexible receptors in solution by molecular dynamics simulation. *Proteins* **35**, 153-62 (1999).
19. Munshi, S. et al. An alternate binding site for the P1-P3 group of a class of potent HIV-1 protease inhibitors as a result of concerted structural change in the 80s loop of the protease. *Acta Crystallogr D Biol Crystallogr* **56**, 381-8. (2000).
20. Knegtel, R.M., Kuntz, I.D. & Oshiro, C.M. Molecular docking to ensembles of protein structures. *J Mol Biol* **266**, 424-40. (1997).
21. Clarage, J.B., Romo, T., Andrews, B.K., Pettitt, B.M. & Phillips, G.N., Jr. A sampling problem in molecular dynamics simulations of macromolecules. *Proc Natl Acad Sci U S A* **92**, 3288-92 (1995).
22. Pang, Y.P. & Kozikowski, A.P. Prediction of the binding sites of huperzine A in acetylcholinesterase by docking studies. *J Comput Aided Mol Des* **8**, 669-81. (1994).
23. Sandak, B., Wolfson, H.J. & Nussinov, R. Flexible docking allowing induced fit in proteins: insights from an open to closed conformational isomers. *Proteins* **32**, 159-74 (1998).
24. Amadei, A., Linssen, A.B. & Berendsen, H.J. Essential dynamics of proteins. *Proteins* **17**, 412-25 (1993).
25. Romo, T.D., Clarage, J.B., Sorensen, D.C. & Phillips, G.N., Jr. Automatic identification of discrete substates in proteins: singular value decomposition analysis of time-averaged crystallographic refinements. *Proteins* **22**, 311-21 (1995).
26. Teodoro, M.L., Phillips, G.N. & Kavraki, L.E. Singular Value Decomposition of Protein Conformational Motions. in *Currents in Computational Molecular Biology* (ed. Satoru, M., Shamir, R. & Tagaki, T.) 198-199 (Universal Academy Press, Inc., Tokyo, 2000).

Programmable Assembly at the Molecular Scale: Self-Assembly of DNA Lattices (Invited Paper*)

John H. Reif †Thomas H. LaBean and Nadrian C. Seeman ‡

Duke University   New York University

Abstract

DNA self-assembly is a methodology for the construction of molecular scale structures. In this method, artificially synthesized single stranded DNA self-assemble into DNA crossover molecules (tiles). These DNA tiles have sticky ends that preferentially match the sticky ends of certain other DNA tiles, facilitating the further assembly into tiling lattices. DNA self-assembly can, using only a small number of component tiles, provide arbitrarily complex assemblies. The self-assembly of large 2D lattices consisting of up to thousands of tiles have been recently demonstrated, and 3D DNA lattices may soon be feasible to construct. We describe various novel DNA tiles with properties that facilitate self-assembly and their visualization by imaging devices such as atomic force microscope. We discuss key theoretical and practical challenges of DNA self-assembly, as well as numerous potential applications. We briefly discuss the ongoing development of attachment chemistry from DNA lattices to various types of molecules, and consider application of DNA lattices (assuming the development of such appropriate attachment chemistry from DNA lattices to these objects) as a substrate for: (a) molecular robotics; for manipulation of molecules using molecular motor devices, (b) layout of molecular electronic circuit components, (c) surface chemistry, for example ultra compact annealing arrays, We also discuss bounds on the speed and error rates of the various types of self-assembly reactions, as well as methods that may minimize errors in self-assembly.

1 Introduction to Tiling Self-Assemblies

1.1 Self-Assembly: This is a process involving the spontaneous self-ordering of substructures into superstructures.

Biological Self-Assembly. We take inspiration from the cell, which performs a multiplicity of self-assembly tasks, including the self-assembly of cell walls (via lipids), of microtubules, etc. Many of these biological self-assembly processes

---

*A postscript version of this paper is at URL http://www.cs.duke.edu/~reif/paper/SELFASSEMBLE/nanoassemble.ps

†Contact address: Department of Computer Science, Duke University, Box 90129, Durham, NC 27708-0129. E-mail: reif@cs.duke.edu. Supported by Grants NSF/DARPA CCR-9725021, CCR-96-33567, NSF IRI- 9619647, NSF ITR EIA-0086015, ARO contract DAAH-04-96-1-0448, and ONR contract N00014-99-1-0406.

‡ Nadrian C. Seeman is at NYU and supported by NIH Grant GM-29554, ONR contract N00014-98-1-0093, DARPA/NSF grant NSF-CCR-97-2502, AFSOR contract F30602-98-C-0148 NSF grants CTS-9986512 and EIA-0086015.

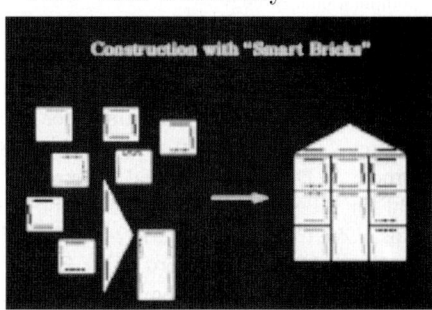

Figure 1: A tiling assembly using 'Smart Bricks' with affinity between colored pads.

utilize the specificity of ligand affinities to direct the self-assembly. We will focus instead on self-assemblies whose components are artificially constructed tiles.

1.2 Domino Tiling: These were defined by Wang [Wang61] (Also see the text [Grunbaum, et al, 87]). The input is a finite set of unit size square tiles, each of whose sides are labeled with symbols over a finite alphabet (the pads). Additional restrictions may include the initial placement of a subset of these tiles, and the dimensions of the region where tiles must be placed. Assuming arbitrarily large supply of each tile, the problem is to place the tiles, without rotation (a criterion that cannot apply to physical tiles), to completely fill the given region so that each pair of abutting tiles have identical symbols on their contacting sides. (See Figure 1.)

Domino tiling problems over an infinite domain with only a constant number of tiles was first proved by [Berger66] to be undecidable; proofs rely on constructions wherein tiling patterns simulate single-tape Turing Machines (see also [Berger66, Robinson71, Wang75]). Other results include reductions of NP-complete problems to finite-size tiling problems [LewisPapa81, Moore00]

1.3 Self-Assembly of Tiling Lattices. Domino tiling problems do not presume or require a specific process for tiling. However, we will presume the use of the self-assembly processes for construction of tiling lattices. In this self-assembly process, the preferential matching of tile sides facilitates the further assembly into tiling lattices. The sides of the tiles are assumed to have some methodology for selective affinity, which we call *pads*. Pads function as programmable binding domains, which hold together the tiles.

Since domino tiling problems are undecidable (see Section 3), tiling self-assemblies can theoretically provide arbitrarily complex assemblies even with a constant number of distinct tile types. As a very simple example, it possible to construct tiling assemblies with self-delimiting boundaries (e.g., rectangular boundaries of a give width w and length h), by use of a set of wh distinct tiles, with w distinct pads types on the bottom and top of a set of square tiles and a set of h pad types on the other sides of these tiles, and this can also be done with a constant number of distinct tiles. (See [Rothemund and Winfree, 2000b] for improved program-size complexity of self-assembled squares.)

Pad binding mechanisms for the preferential matching of

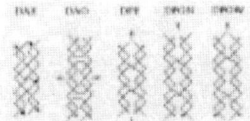

Figure 2: Double crossover isomers.

tile sides can be provided by various methods:
(i) *molecular affinity*, using for example hydrogen bonding of complementary DNA or RNA bases,
(ii) *magnetic attraction*, e.g., pads with magnetic orientations constructed by curing the polymer/ferrite composites in the presence of strong magnet fields, and also pads with patterned strips of magnetic orientations,
(iii) *capillary force*, using hydrophobic/hydrophilic (capillary) effects at surface boundaries that generate lateral forces,
(iv) *shape complementarity* (or conformational affinity), using the shape of the tile sides to hold them together.

There are a variety of distinct materials for tiles, at a variety of scales:

(a) Meso-Scale Tiling Assemblies have tiles of size a few millimeters up to a few centimeters. Whitesides at Harvard University has developed and tested multiple technologies [Zhao, et al, 98] [Xia et al, 98a,98b], [Bowden,et al 98], [Harder,et al 00] for meso-scale self-assembly, using capillary forces, shape complementarity, and magnetic forces (see http://www-chem.harvard.edu/GeorgeWhitesides.html). [Rothemund, 2000] also gave some meso-scale tiling assemblies using polymer tiles on fluid boundaries with pads that use hydrophobic/hydrophilic forces. A materials science group at the U. of Wisconsin also tested meso-scale self-assembly using magnetic tiles (http://mrsec.wisc.edu/edetc/selfassembly). These meso-scale tiling assemblies were demonstrated by a number of methods, including placement of tiles on a liquid surface interface (e.g., at the interface between two liquids of distinct density or on the surface of an air/liquid interface).

(b) Molecular-Scale Tiling Assemblies have tiles of size up to a few hundred Angstroms. Specifically, DNA tiles will be the focus of our discussions in the following sections.

1.4 Goals and Organization of this Paper. The goal of this paper is describe techniques for self-assembly of DNA tiling arrays and applications of this technology. We give in Section 2 a description of self-assembly techniques for DNA tilings and discuss the kinetics of self-assemblies and error control in Section 3. We discuss applications in Section 4. Section 5 concludes the paper.

2 DNA Self-Assembly of DNA Tilings

2.1 DNA as a Construction Material. Nanofabrication of structures in DNA was pioneered by the Seeman laboratory ([Seeman82, 94b, 96a]), who built a multitude of DNA nano-structures using DNA branched junctions [Seeman89, Wang91a, Du92]. These previous systems were flexible, so control over synthesis and proof of synthesis were both limited to the topological level, rather than the geometrical level (in contrast to the tiles described below). These DNA nano-structures included: DNA knots [Seeman93], Borromean rings [Mao97], a cube [Chen91], and a truncated octahedron [Zhang94] (reviewed in e.g. [Seeman98] [Seeman94b] and [Seeman96a]).

2.2 DNA Tiles Constructed from DX and TX Complexes. . The building blocks in the tiling constructions to be discussed are branched DNA complexes, which we call *DNA tiles*, consisting of several individual DNA oligonucleotides that associate with well-defined secondary and tertiary structure (see [Winfree, et al 98] and below description of DX and TX tiles). These associate with well-defined secondary and tertiary geometric structure (which is much more predictable and less flexible than DNA nano-structures using DNA branched junctions). These complexes come in a number of varieties that differ from one another in the geometry of strand exchange and the topology of the strand paths through the tile. The branched DNA complexes used for tiling assemblies include the double-crossover (DX) and triple-crossover (TX) complexes, The DX and TX complexes consists of two (three, respectively) double-helices fused by exchange (crossover) of oligonucleotide strands at a number of separate crossover points. Anti-parallel crossovers cause a reversal in direction of strand propagation through the tile following exchange of strand to a new helix. For example, DAO and DAE are double-crossover DX tiles with two anti-parallel crossovers[1]. (See Figure 2.)

DX complexes have been used successfully as substrates for enzymatic reactions including cleavage and ligation [Liu, Sha and Seeman,99]. The TX (see [LaBean,99]) TAO and TAE tiles, are similar except that they have three double-helices interlocked by exchange of oligonucleotide strands at four separate crossover points, two between the first pair of helices, two between the second (See the TAO in Figure 3 and the TAE in Figure 4. Both DX and TX motifs are useful for tiling assemblies; the DX (TX) complexes provide up to four (six, respectively) ssDNA pads [Liu, et al 99a] for encoding associations with neighboring tiles.

2.3 DNA Tiling Lattices. Tile assemblies, or simply tilings, can be defined as superstructures or lattices built up from smaller, possibly repetitive, component structures. Individual tiles interact by annealing with other specific tiles via their ssDNA pads to self-assemble into desired superstructures. These lattices can be either:
(a) *non-computational,* containing a fairly small number of distinct tile types in a repetitive, periodic pattern; or
(b) *computational,* containing a larger number of tile types with more complicated association rules which perform a computation during lattice assembly.

---

[1] The structure of the TAE resembles the TAO in that it is constructed of three double-helices linked by strand exchange, however, it contains an Even (rather than Odd) number of helical half-turns between crossover points. Even spacing of crossovers allows reporter strands (shown in black) to stretch straight through each helix from one side of the tile to the other. These three horizontal reporter segments are used for building up a long strand which records inputs and outputs for the entire assembly computations. A 3D confirmation of the TAE tile has been rendered by Brendon Murphy at Duke University; see http://www.duke.edu/~bkm2/taehtml/present.html .

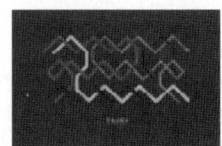 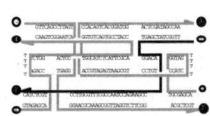

Figure 3: The TAO tile and a Strand and Sequence Trace through the TAO Tile.

DNA tiles are designed to contain several short sections of unpaired, single-strand DNA (ssDNA) extending from the ends of selected helices (often called 'sticky ends') that function as programmable binding domains, which are the *tile pads*.

Figure 4: The TAE Tile.

These DNA self-assembly procedures generally will be described as occurring in two distinct stages:
(i) annealing of ssDNA into tiles; and
(ii) assembly of tiles into superstructures.

However, direct assembly of DNA lattices from component ssDNA is also possible, and has in fact already been demonstrated for non-computational DNA lattices described below.

Programming Self-Assembly of DNA Tilings. Programming DNA self-assembly of tilings amounts to the design of the pads of the DNA tiles (recall these are sticky ends of ssDNA that function as programmable binding domains, and that individual tiles interact by annealing with other specific tiles via their ssDNA pads to self-assemble into desired superstructures). The use of pads with complementary base sequences allows the neighbor relations of tiles in the final assembly to be intimately controlled; thus the only large-scale superstructures formed during assembly are those that encode valid mappings of input to output. The papers [R97, WLW+98, WYS96, LYK+00, LWR99, RLS00, MLR+00] describe methods for executing computation using DNA Tilings (due to length limitations in this paper, we have omitted a detailed discussion of DNA computational tilings).

The Speed of Computing via DNA Tiling Assemblies (compared with silicon-based computing. The speed of DNA tiling assemblies is limited by the annealing time, which can be many minutes, and can be 10^{11} slower than a conventional computer. A DNA computation via self-assembly must take into account the fact that the time to execute an assembly can range from a few minutes up to hours. Therefore, a reasonable assessment of the power of DNA computation must take into account both the speed of operation as well as the degree of massive parallelism. Nevertheless, the massive parallelism (both within assemblies and also via the parallel construction of distinct assemblies) possibly ranging from 10^{15} to 10^{18} provides a potential that may be advantageous for classes of computational problems that can be

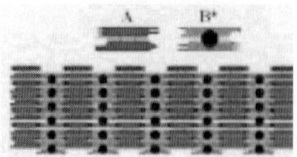

Figure 5: AB* Array. *Lattice formed from two varieties of DX, one containing an extra loop of DNA projecting out of the lattice plane, faciliting atomic force microscope imaging of the lattice.*

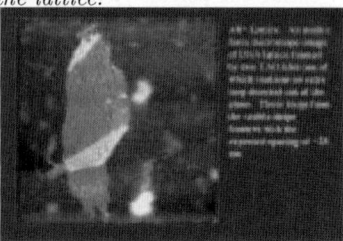

Figure 6: A non-computational DNA tiling formed by sets of two types of TAO tiles.
parallelized.

2.4 Two Dimensional DNA Tiling Assemblies. Recently Winfree and Seeman have demonstrated the use DX tiles to construct 2D periodic lattices consisting of upto a hundred thousand DX units [Winfree, et al 98] as observed by atomic force microscopy[2] The surface features are readily programmed [Winfree et al., 98; Liu et al., 99; Mao, et al 99]. (See Figure 5.) In addition, [LaBean et al, 99] constructed produced tiling arrays (see Figure 6) composed of DNA triple crossover molecules (TX); these appear to assemble at least as readily as DX tiles.

These tiling assemblies had no fixed limit on their size. [Reif97] introduced the concept of a *nano-frame*, which is a self-assembled nanostructure that constrains the subsequent timing assembly (e.g., to a fixed size rectangle). Alternatively, a tiling assembly might be designed to be *self-delimitating* (growing to only a fixed size) by the choice of tile pads that essentially 'count' to their intended boundaries in the dimensions to be delimitated.

2.5 Three Dimensional DNA Tiling Assemblies. There are a number of possible methods for constructing 3D periodic (non-computational) tilings. For example, stable tiling arrays with well-defined helices that come out of the plane (e.g., the TX tiling array constructed in [LaBean et al, 99]) may lead to ways to provide 3D tiling assemblies.

3 The Kinetics and Error Control in Self-Assembled Tiling Assemblies

3.1 Kinetics of Self-Assembled Tiling Assemblies. In spite of an extensive literature on the kinetics of the assembly of regular crystalline lattices, the fundamental thermodynamic and kinetic aspects of self-assembly of tiling assemblies For example, it is not yet known the affect of distinct tile concentrations and different rela-

---
[2]An atomic force microscope [AFM] is a mechanical scanning device that provides images of molecular structures laying on a flat 2D plate.

tive numbers of tiles, and the possible application of Le Chatelier's principle.

Winfree [W98] developed computer simulation of tiling self-assemblies. This software makes a discrete time simulation of the tiling assembly processes, using approximate probabilities for the insertion or removal individual tiles from the assembly. These simulations provide an approximation to the kinetics of self-assembly chemistry. Using this software as a basis, Guangwei Yuan at Duke developed improved simulation software with a Java interface (http://www.cs.duke.edu/~yuangw/project/test.HTML) for a number of example tilings, such as string tilings for integer addition and XOR computations. This simulation software was recently speed up by use of an improved method for computing on/of likelihood, as suggested by Winfree.

The meso-scale tiling experiments described in Section 1 have used mechanical agitation with shakers to provide a temperature setting for the assembly kinetics (that is, a temperature setting is made by fixing the rate and intensity of shaker agitation). These meso-scale tilings also have potential to illustrate fundamental thermodynamic and kinetic aspects of self-assembly chemistry.

3.2 Error Control in Self-Assembled Tiling Assemblies. As stated above, two dimensional self-assembled *non-computational* tilings have been demonstrated (and imaged via atom force microscopy) that involve up to a hundred thousand tiles. Certain of these appear to suffer from relatively low defect rates, perhaps in the order of less than a fraction of a percentage or less. The factors influencing these defect rates are not yet well understood and there are no known estimates on the error-rates for self-assembled *computation* tilings, since such tilings have been achieved only very recently and have only been done on a very small scale(error rates appear to be less than 5% [Mao et al 00]). There is reason (see the construction of a potential assembly blockage described in [Reif, 98]) to believe that in computational tilings, defect errors may be more prevalent; and moreover, they can have catastrophic effects on the computation. Experiments need to be done to determine the error rates of the various types of self-assembly reactions, both computational and non-computational.

There are a number of possible methods to decrease errors in DNA tilings. It is as yet uncertain which methods will turn out to be effective and it is likely that a combination of at least a few of the following methods will prove most effective.

(a) Error Control by Annealing Temperature Variation. This is a well known technique used in hybridization and also crystallization experiments. It is likely that this will provide some decrease in defect rates at the expense in increased overall annealing time duration. In the context of DNA tiling lattices, the parameters for the temperature variation that minimize defects have not yet been determined.

(b) Error Control by Improved Sequence Specificity of DNA Annealing. The most obvious methodology here is to improve the choice of the DNA words used for tile pads (that is, to decrease the likelihood of incorrect hybridizations from non-matching pads).DNA word design software for DNA tiles was developed by Winfree. This software has recently been improved by Horatiu Voicu at Duke University (see http://www.cs.duke.edu/~hvoicu/app.html) to include more realistic models of DNA hybridization, and was also speed up by use of a technique of Rajasakaran and Reif [RR95] known as nested annealing. Another possible approach is to use the observation and experimental evidence of [Herschlag 91] that stressed DNA molecules can have much higher hybridization fidelity (sequence specificity) than a relaxed molecule. This would entail redesigning DNA tiles, so their pads are strained single stranded loop segments with higher sequence specificity, or by the use of stressed DNA motifs known as "PX dumbbells" [Shen, Z. Ph.D. Thesis, NYU, 1999].

(c) Error Control by Redundancy. There are a number of ways to introduce redundancy into a computational tiling assembly. One simple method that can be developed for linear tiling assemblies, is to replace each tile with a stack of three tiles executing the same function, and then add additional tiles that essentially 'vote' on the pad associations associated with these redundant tiles. This results in a tiling of increased complexity but still linear size. This error resistant design can easily be applied to the integer addition linear tiling described above, and similar redundancy methods may be applied to higher dimension tilings.

(d) Error Control by Free Versus Step-wise Assembly. Self-assembly may be restricted so that certain assembly reactions can proceed only after others have been completed (*serial (or step-wise) self-assembly*). Alternatively, self-assembly reactions may be limited by no such restrictions *free self-assembly*). It is expected, but unproven, that free self-assembly is faster than serial self-assembly. [Reif, 98] suggested the use of serial self-assembly to decrease errors in self-assembly. There is not yet any experimental data on the error rates of self-assembly reactions and error control/repair of 'self-assembly' versus 'serial-assembly'. To decrease the human effort in serial assembly, assembly steps might be executed automatically with the use of a robotic machines (E.g., the Nanogen machine, which employs a chip that contains DNA sequences above electrodes. Tile components hybridized to these DNA sequences can be released in sequence by making the electrode sufficiently negative.).

(e) Use of DNA Lattices as a Reactive Substrate for Error Repair. DX complexes and lattices have been used successfully as substrates for enzymatic reactions including restriction cleavage and ligation of exposed hairpins attached to the tiles [Liu99a]. One approach is the use of DNA lattices to execute a broader class of reactions. For example, if restriction enzymes, topoisomerases or site-specific recombinases can be shown to operate on exposed portions of the DNA lattices, then it may be possible to modify the topology and geometry of the DNA lattice. This may aid in the DNA tiling computations described above, for example by providing mechanisms for error repair in DNA tiling computations. (Note: as mentioned above, this approach may also be of use for recycling of the component ssDNA for the next computation cycles.)

4 Applications of Non-Computational DNA Tiling Arrays

We now identify some further technological impacts of non-computational DNA assemblies; particularly their use as substrates for surface chemistry and molecular electronics, robotics. Many of these applications are dependant on the development of the appropriate attachment chemistry between DNA and the molecules attached to the arrays.

(4.1) Tiling Assemblies with Molecular Motors. Several types of molecules are known to couple chemical energy to the generation mechanical force, thereby functioning as molecular motors. Possible schemes for molecular motors include:

(a) Re-Engineering Biological Molecular Motors. Cells make use of a variety of such motor-like devices in processes such as mitosis. The best characterized of these fall into three categories.

(i) *ATP synthase and ADP* acts as a rotary motor, coupling proton flux through a membrane with the phosphorylation of ADP to ATP. The F1 component of ATP synthase can also be run in reverse, coupling hydrolysis of one ATP molecule to 120 B0 of rotation about the motor axis.

(ii) *Myosin* acts as a molecular running machine, skipping many steps along an actin filament with each molecule of ATP consumed. All of these motors are modular and can be re-engineered to accomplish linear or rotational motion of essentially any type of molecular component.

(iii) *Kinesin* acts as a molecular walking machine, translocating itself (and any attached components) in step-wise fashion along a microtubule. Each step along the microtubule consumes one ATP molecule.

Construction of these biological molecular motors and their linking chemistry to DNA arrays. These motors are composed of proteins with well known transcription sequences. There are also well known proteins (binding proteins) that provide linking chemistry to DNA. Hence it seems feasible that these molecular motors and attached linking elements may be synthesized from sequences obtained by concatenation of these transcription sequences. [Bachand, et al., 1999 and 2000], describes a biomolecular motor constructed of expressed ADP protein [Montemagno, et al, 1998 and 1999] with an attached [Soong, et al, 1999] silicon arm.

(b) DNA Motors. The Seeman laboratory made a DNA construction of a mechanical device capable of controlled movement [Mao, et al 99a]. This device consists of two DX molecules connected by a DNA double helix that contains a segment of DNA that can be converted to the left-handed Z-DNA structure. In B-promoting conditions, the two unconnected helices of the device are on the same side of the connecting helix, but they are on opposite sides in Z-promoting conditions. This results in an apparent rotary motion of about a half-revolution, leading to atomic displacements ranging from 2 to 6 nm, depending on the location of the atom relative to the axis of the stationary helix. This motion has been demonstrated by fluorescent resonance energy transfer (FRET). It is important to point out that the device based on the B-Z transition is only a prototype that was used to learn how to characterize motion in DNA systems. It lacks programmability, except to the limited extent that one can orient the two DX molecules at a variety of relative torsion angles in the B-state. Thus, all of the molecules must be in either the B-state or in the Z-state, assuming one has robust chemical control.

These molecular motors might be combined with the 2-D arrays, to achieve an array of devices (This has not been possible to do with the DX system, since it is most convenient there for the pivoting part of the system to point normal to the array. However, the TX system does not have this difficulty.). As an example, an array with attached kinesin may provide for the movement of objects across the surface of a two dimensional tiling array, similar to a conveyer belt, and this may be the basis of a transport system (*a molecular conveyer belt*) for molecular objects.

Programmable Sequence-Specific Control of DNA Mechanical Motion. However, such an array of molecular motors would be more useful if they can be selectively controlled. Such a system would lead to the ability to manipulate specific molecules and more generally, to do chemistry at chemically identical but spatially distinct sites. Because it couples a series of distinct structural states with programmability, such a system offers the potential of direct route to molecular robotics. The Seeman laboratory is developing a related system based on the paranemic crossover (PX) system, which leads to sequence-specific nanomechanical motion. It may be switched readily between two discrete states, PX or JX, in which the helices at one end of the molecule reverse positions in the transition between states. An array of these molecular devices would contain individually programmed PX/JX molecules, whose conformational state would be amenable to specific reversal (or not, depending on the program) from cycle to cycle. A DNA array with programmability of this sort may offer a mechanism

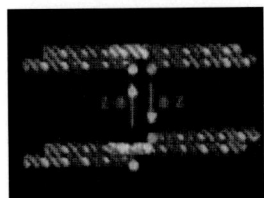

Figure 7: A prototype DNA nanomechanical device.

Figure 8: The Tour-Reed molecular electronic diode.

to do DNA computation of arrays whose elements (the tiles) hold state, as discussed in the next section.

4.2 Application to Layout of Molecular-Scale Circuit Components. Molecular-scale circuits have the potential of replacing the traditional microelectronics with densities up to millions of times current circuit densities. There have been a number of recent efforts to design molecular circuit components ([Petty et al 95] [Aviram,Ratner98]). Tour at Rice Univ. in collaboration with Reed at Yale have designed and demonstrated [Chen et al 99] organic molecules (see Figure 7) that act as conducting wires [Reed et al.97],[Zhou99] and also rectifying diodes (showing negative differential resistance (NDR), and as well as [CRR+,99], [RCR+,00], and have the potential to provide dynamic random access memory (DRAM) cells. These generally use $\sim 1,000$ molecules per device, but they have also addressed single molecules and recorded current through single molecules [BAC+96], [RZM+97]. These molecular electronic components make conformational changes when they do do electrical switching. One key open problem in molecular electronics is to develop molecular electronic components that exhibit restoration of a signal to binary values; one possible approach may be to make use of multi-component assemblies that exhibit cooperative thresholding.

The Molecular Circuit Assembly Problem: This key problem is to develop methods for assembling these molecular electronic components into a molecular scale circuit. Progress in the molecular circuit assembly problem could have revolutionary impact on the electronic industry, since it is one of the key problems delaying the development of molecular-scale circuits.

Top-down techniques versus bottom-up approaches. The usual approach of laying out circuits by top-down techniques (e.g., lithography) may not be practical at the molecular scale; instead bottom-up approaches (e.g., self-assembly) may need to be used. Hence this may be a key area of application of DNA tiling assemblies. There are a number of possible methods for the selective attachment of the molecular electronic components to particular tiles of the DNA tiling array, using annealing.

(i) linking chemistry between DNA and molecular electronics. Tour and Bunz recently prepared DNA-linked systems where the DNA could serve as a selective assem-

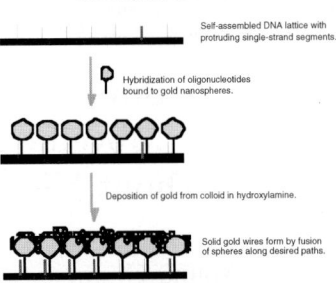

Figure 9: Diffusion of gold on beads to form molecular-scale gold wires.

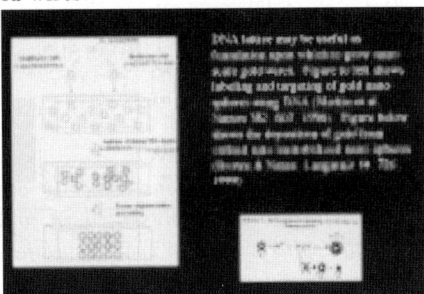

Figure 10: A scheme for molecular-scale gold wires.

bly glue for device configurations [WST+00].

(ii) The use of gold beads. In this approach, DNA strands attached to the gold beads can hybridize at selected locations of the arrays, and the molecular electronics components may self-assemble between the gold breads. Also, DNA lattices may be useful as a foundation upon which to grow nano-scale gold wires. This might be done by depositions of gold from colloid onto nanospheres immobilized on DNA tiling lattices. (See Figures 8 and 9.). Molecular probe devices may be used to test the electrical properties of the resulting molecular circuit attached to the DNA tiling array. *Computational* lattices (as opposed to regular, non-computational lattices), may also be employed to provide for the layout of highly complex circuits, e.g., the layout of the electronic components of an arithmetic unit. (Other related approaches for positioning of molecular electronic molecules without lithography include that of [Bumm, et al., 1999, 2000], which describes the use of directed self-assembly of molecular terrace structures in organic monolayers.)

5 Conclusion

We have discussed the potential advantages of self-assembly techniques for DNA computation; particularly the decreased number of laboratory steps required. We also discussed the potential broader technological impacts of DNA tiling lattices and identify some further possible applications. The chief difficulties are that of error control and predicable kinetics, as described in the previous section. Nevertheless, the self-assembly of DNA tilings seems a very promising emerging method for molecular scale constructions and computation.

References: See http://www.cs.duke.edu/~reif/paper/SELFASSEMBLE/references.ps

A Simple Algorithm for Determining of Movement Duration in Task Space without Violating Joint Angle Constraints

Young G. Jin

School of Electrical Engineering
Seoul National University
Seoul, Korea 151-742

Jin Y. Choi

School of Electrical Engineering
Seoul National University
Seoul, Korea 151-742

Abstract

In usual robot applications, velocity and acceleration constraints are represented in joint angle space, on the contrary that a goal is specified in task space such as the Cartesian coordinates. This fact means that a task trajectory planner for itself cannot resolve the constraint problem without working together with an inverse kinematics solving method. In this paper, we present a simple method to generate a joint angle trajectory without violating the joint velocity and acceleration constraints. The key idea is to determine movement duration to solve the constraint problem. The proposed method has a strong merit in that it can use the existing task trajectory planning method and inverse kinematics solving method without reformulating them for the constraints. The validity of the proposed method is shown by the simulation with a two-link serial manipulator executing a circle-drawing task.

1 Introduction

In usual, constraint is given or formulated in terms of joint velocity and acceleration limits, on the contrary that a lot of task specifications, such as end effector trajectories, are represented in task space(e.g., the Cartesian coordinates). This means that a task trajectory planner for itself cannot resolve the joint constraint problem. Thus, the task trajectory planning may be combined and formulated together with the inverse kinematics, and in some cases, with the control, considering the joint constraints(i.e., combining approach). Otherwise, another algorithm may give an interface between the task trajectory planning method and the inverse kinematics solving method so that the final joint angle trajectory should not violate the joint constraints(i.e., connecting approach).

So far, there have been many researches that adopt combining approach. Some works have used spline techniques [10, 17]. In their methods, spline joint angle trajectories are composed between given points or knots, and movement duration is optimized by determining time intervals between the points. Those methods are useful when the task is given as a path with some via points. If, however, the whole trajectory should be regarded as via points(e.g., a straight line in Cartesian task space), linear equations for finding the appropriate splines may get too complicated to solve.

The joint angle constraints may be incorporated into an inverse kinematics solving procedure [2, 13, 14]. In those methods, the inverse kinematics problem is formulated as a constrained optimization problem. Extra degrees of freedom are used to resolve the joint constraint problem, or inversely, the joint constraints are used to resolve the redundancy problem. The fact that redundant degrees are used, however, restricts the application of the methods just to redundant manipulators, not to general manipulators. Furthermore, excessive computation is required in the optimization procedure.

In order to include the nonredundant manipulator and to reduce the computational burden of optimization, a new algorithm for the inverse kinematics problem has been proposed on the basis of a time scaling technique [3]. In the method, whenever any joint reaches its limits of velocity or acceleration, virtual time is computed. This on-line time axis rescaling method, however, still requires an optimization procedure, even if the computational burden becomes light. In addition, the success of the algorithm depends too much on estimated initial movement duration. If the initial movement duration is chosen too small, the algorithm fails to find a new trajectory which satisfies the constraints. On the other hand, if the initial movement duration is chosen too large, the algorithm is not used and the manipulator cannot make a full use of its capability.

Some works have focused on the dynamic constraints of the manipulator and formulated the task trajectory planning method combined with control algorithms [4, 8, 15]. They have pointed out that torque/force constraints are more important than the velocity or acceleration constraints because the velocity/acceleration bounds vary with positions and payload conditions [4, 15].

Compared to such various techniques of combining approach, connecting approach which gives an interface algorithm is similar to divide-and-conquer approach and layered approach. Thus, it is likely to give a simpler solution to the constraint problem. In addition, other methods on the task trajectory generation and the inverse kinematics problem [1, 2, 6, 7, 12, 16], which are not formulated under the constraint, can be used without modification.

In this paper, following connecting approach, we present a simple interface or component algorithm, referred to as STICCON(Sampling TIme Calculation under the CONstraints). The algorithm determines movement duration under the joint constraints. Because the proposed method uses only basic numerical differentiation formulas, no complex optimization technique is needed. The validity of the method is demonstrated by the simulation with a two-link serial manipulator executing a circle-drawing task.

2 Basic Considerations and Their Influences on the Formulation of STICCON

There are two major facts that motivate the idea of STICCON. These facts can be often found in lots of works and applications related to trajectory generation. In this section, not only those facts and also the aspects of STICCON method on which each fact has an influence are discussed.

2.1 Need for Movement Duration: Goal of STICCON

The first fact to note is that movement duration plays an important part in some task trajectory generation schemes. The most typical example is a straight line trajectory. It is derived from the experimental results which show that the hand trajectory in point-to-point human reaching movement is approximately the straight line with a bell-shape velocity profile [5]. Suppose the movement starts and ends with zero velocity and acceleration in two dimensional plane. Then, the following hand trajectory is obtained:

$$\mathbf{X}(t) = \mathbf{X}_0 + (\mathbf{X}_f - \mathbf{X}_0) \cdot \left[6 \left(\frac{t}{t_f} \right)^5 - 15 \left(\frac{t}{t_f} \right)^4 + 10 \left(\frac{t}{t_f} \right)^3 \right] \quad (1)$$

where t_f is the movement duration, and $\mathbf{X}_0 = (x_0, y_0)$ and $\mathbf{X}_f = (x_f, y_f)$ are initial and final hand positions, respectively.

In equation (1), three variables must be determined to completely specify one trajectory. For human experiments, the initial and final positions are determined by the experimental setting and movement duration is measured in the experiment. For the robot application, however, it is required all the variables including the movement duration should be determined before the initiation of the movement to set the trajectory as a command to the robot manipulator.

The initial and final positions of the end effector are easily determined by the task. On the other hand, the movement duration is necessary to estimate so that it should not violate the joint constraints. Consequently, in a task trajectory generation method such as equation (1), it may be one of the most important problems to find the correct movement duration that satisfies the joint constraints.

This observation leads us to design a method for determining movement duration under the joint velocity and acceleration constraints. That is, it is the movement duration that is obtained by of STICCON method as the final goal or result. The computed movement duration is entered as a parameter into the task trajectory equation to obtain the final task trajectory without violating the joint constraints.

2.2 Off-line Trajectory Planning and On-line Tracking : Off-line Property of STICCON

Regarding trajectory planning and control, two types of approach may be considered. The first approach is to join trajectory planning together with control. The task is given as a path, not a trajectory, and the actual trajectory is realized by a controller. The control algorithm optimizes a criterion such as movement duration under the constraints like torque limits [4, 15]. In the approach, the concept of optimal control may be applied. Another approach is to separate trajectory planning and control. The constraint satisfaction is resolved in the level of trajectory planning, and a controller(e.g., PID controller) just plays a role of tracking [15]. Of the two types of approach, the latter is chosen, because the main issue of this paper is to find a trajectory used as the reference of a controller.

Then, the remaining part is whether the trajectory is planned off-line or on-line [11]. Off-line methods may get a globally optimized trajectory, but cannot use sensory feedback information to modify the preplanned trajectory. On the other hand, on-line methods may cope with dynamic situations such as unexpected disturbances. Local optimization is done based on a certain criterion. Those on-line methods, however, demand rather a big amount of computation.

If a trajectory is planned correctly, most situations requiring on-line feedback information are due to incomplete control and unexpected disturbances. They should be handled in the level of control where the problems happen, not in the level of trajectory planning. Therefore, if a trajectory without violating the constraints can be found before the movement is initiated, and if the goal(e.g., the target) is not changed, there is no reason to use on-line method in trajectory planning. In other words, it's not reasonable that a command trajectory is planned roughly so that it should be modified whenever any constraint is violated.

This fact has naturally resulted in "off-line" property of STICCON method. Here, "off-line" means that a command trajectory is generated before the initiation of the movement, and it is not changed during the movement except the situation such as a target change. The dynamic situations informed by the feedback sensors are handled by the tracking controller based on various control theories such as adaptive and robust control.

3 Algorithm of STICCON

When the proposed STICCON method is used, both the task trajectory planning method and the inverse kinematics solving method must be available. These techniques are used twice before and after STICCON method is applied. Thus, an initial joint angle trajectory is generated independently of the constraints and it is modified to satisfy the constraints after STICCON finds the correct movement duration.

As the first step of STICCON method, the command trajectory is represented in discrete form. For this purpose, the number of sampling points, denoted as N_s, should be determined. If the command trajectory is already in discrete form due to the use of a numerical inverse kinematics algorithm, N_s is set to the number of points in the joint trajectory.

If the command trajectory is represented in continuous form, e.g., in the form of mathematical equations, N_s should be selected under a certain criterion. If, fortunately, the frequency information of the movement might be available, well-known sampling theorem could be applied. Since it is not known in most cases, N_s is determined as great as possible within available computational power. Note that the amount of computation in STICCON is not increased exponentially with N_s since the sampling points are searched just once.

In computing joint velocity and acceleration, the basic formulas for numerical differentiation are used:

$$v_k^i = \frac{\theta_{k+1}^i - \theta_k^i}{\Delta t}, \quad (2)$$

$$a_k^i = \frac{\theta_{k+1}^i + \theta_{k-1}^i - 2\theta_k^i}{\Delta t^2}, \quad (3)$$

where θ_k^i, v_k^i, and a_k^i are the position, velocity, and acceleration of the ith joint at the kth time instant, respectively. Δt is the sampling time which STICCON should calculate such that specified joint angle constraints are not violated.

In order to compute velocity and acceleration at $k = 1$ and $k = N_s$, two temporary values are padded before and after the actual trajectory, i.e.,

$$\theta_0^i = \theta_1^i, \quad \theta_{N_s+1}^i = \theta_{N_s}^i. \quad (4)$$

As the velocity constraint, Δt should satisfy the following relationships for all $k = 1, 2, \cdots, N_s$:

$$\frac{\theta_{k+1}^i - \theta_k^i}{\Delta t} \leq V_{max}^i,$$
$$\frac{\theta_{k+1}^i - \theta_k^i}{\Delta t} \geq V_{min}^i, \quad (5)$$

where V_{max}^i and V_{min}^i are the upper and lower limits of velocity of the ith joint. And, as the acceleration constraint, Δt should satisfy

$$\frac{\theta_{k+1}^i - 2\theta_k^i + \theta_{k-1}^i}{\Delta t^2} \leq A_{max}^i,$$
$$\frac{\theta_{k+1}^i - 2\theta_k^i + \theta_{k-1}^i}{\Delta t^2} \geq A_{min}^i, \quad (6)$$

where A_{max}^i and A_{min}^i are the upper and lower limits of acceleration of the ith joint.

If Δt is found such that maximum and minimum values of velocity/acceleration for each joint do not violate the conditions (5) and (6), it is obvious that those conditions are also met at other points for that value of Δt. Thus, inequalities (5) and (6) can be transformed into

$$\frac{\max_k \left(\theta_{k+1}^i - \theta_k^i\right)}{\Delta t} \leq V_{max}^i,$$
$$\frac{\min_k \left(\theta_{k+1}^i - \theta_k^i\right)}{\Delta t} \geq V_{min}^i, \quad (7)$$

and

$$\frac{\max_k \left(\theta_{k+1}^i - 2\theta_k^i + \theta_{k-1}^i\right)}{\Delta t^2} \leq A_{max}^i,$$
$$\frac{\min_k \left(\theta_{k+1}^i - 2\theta_k^i + \theta_{k-1}^i\right)}{\Delta t^2} \geq A_{min}^i. \quad (8)$$

In inequalities (7) and (8), in purely mathematical sense, four combinations about sign must be considered(e.g., a left term has plus sign and a right term has minus sign, etc.). However, if the upper limit and the lower limit of velocity have the same sign, it means that the joint can move in only one direction, which shall not be the proper specifications of the robot manipulator. Thus, the upper and lower velocity limits shoud have plus sign and minus sign, respectively. Because of the similar reason, the upper and lower acceleration limits shall have plus sign and minus sign, respectively. Then, the relationships (7) and (8) can be rewritten as conditions for Δt:

$$\Delta t \geq \frac{\max_k \left(\theta_{k+1}^i - \theta_k^i\right)}{V_{max}^i},$$
$$\Delta t \geq \frac{\min_k \left(\theta_{k+1}^i - \theta_k^i\right)}{V_{min}^i}, \quad (9)$$

and

$$\Delta t^2 \geq \frac{\max_k \left(\theta^i_{k+1} - 2\theta^i_k + \theta^i_{k-1}\right)}{A^i_{max}},$$
$$\Delta t^2 \geq \frac{\min_k \left(\theta^i_{k+1} - 2\theta^i_k + \theta^i_{k-1}\right)}{A^i_{min}}. \qquad (10)$$

In conditions (9) and (10), it is easy to notice that the smallest sampling time which meets the equality is the best value, since smaller sampling time results in smaller movement duration which means that the manipulator is efficiently functioning under the constraints.

Thus, the tentative solutions for Δt obtained from the velocity and acceleration constraints are

$$\Delta t^i_{v,max} = \frac{\max_k \left(\theta^i_{k+1} - \theta^i_k\right)}{V^i_{max}},$$
$$\Delta t^i_{v,min} = \frac{\min_k \left(\theta^i_{k+1} - \theta^i_k\right)}{V^i_{min}}, \qquad (11)$$

and

$$\Delta t^i_{a,max} = \sqrt{\frac{\max_k \left(\theta^i_{k+1} - 2\theta^i_k + \theta^i_{k-1}\right)}{A^i_{max}}},$$
$$\Delta t^i_{a,min} = \sqrt{\frac{\min_k \left(\theta^i_{k+1} - 2\theta^i_k + \theta^i_{k-1}\right)}{A^i_{min}}}, \qquad (12)$$

where $\Delta t^i_{v,max}$, $\Delta t^i_{v,min}$, $\Delta t^i_{a,max}$, and $\Delta t^i_{a,min}$ are the tentative values for Δt obtained from the constraints of the ith joint. Note that,if the joint angle trajectory is not wrong, in equations (11) and (12), the numerators with '$\max_k$' must be positive, and the numerators with '$\min_k$' must be negative. Thus, it is guaranteed that $\Delta t^i_{v,max}$, $\Delta t^i_{v,min}$, $\Delta t^i_{a,max}$, and $\Delta t^i_{a,min}$ are all the positive quantities.

Then, let's denote the sampling time calculated by STICCON as Δt_{final}. As you can see in equations (11) and (12), four tentative values are obtained for each joint so that, in case of the manipulator of n degrees of freedom, the candidate set S is composed of $4n$ possible values for Δt_{final}, i.e.,

$$S = \{\Delta t^i_{v,max}, \Delta t^i_{v,min}, \Delta t^i_{a,max}, \Delta t^i_{a,min} | \\ i = 1, 2, \cdots, n\}. \qquad (13)$$

In the set S, the largest element is selected as the value for Δt_{final}:

$$\Delta t_{final} = \max\{\Delta t | \Delta t \in S\}. \qquad (14)$$

The last step of determining the movement duration t_f is just to multiply the final sampling time Δt_{final} in equation (14) and the number of intervals between the sampling points chosen in the first step:

$$t_f = \Delta t_{final} \times (N_s - 1). \qquad (15)$$

Generate a task trajectory with estimated t_f.
Transform it into a joint angle trajectory.

STICCON BEGIN

```
Determine the number of sampling points N_s.
Represent the trajectory in discrete form
  with selected N_s.
Initialize a candidate set(S = ∅).
for joint i=1 to n
  Compute tentative values for Δt_final,
    (Δt^i_{v,max}, Δt^i_{v,min}, Δt^i_{a,max}, Δt^i_{a,min})
  Add the computed candidates to S.
end for
Δt_final = max{Δt|Δt ∈ S}.
t_f = Δt_final × (N_s - 1).
```

STICCON END

Generate a task trajectory with calculated t_f.
Transform it into a final joint angle trajectory.

Figure 1: Summary of STICCON method for computing the movement duration.

Note that the number intervals is $N_s - 1$. With this, the calculation of the correct movement duration t_f is completed, which is the goal of STICCON method.

It should be reminded that STICCON method has assumed both task trajectory planning method and inverse kinematics problem solver are available. Thus, the last thing to do is, with the help of those techniques and the computed movement duration t_f, to generate a final joint angle trajectory without violating the constraints. The whole algorithm is summarized in Fig. 1.

4 Simulation

In this simulation section, the significance of the correct determination of the movement duration and the validity of the proposed STICCON method are presented. Simulations are done with a simple two-link serial manipulator executing a circle-drawing task. The dynamics are given by the equations

$$\tau_1 = [(m_1 + m_2)a_1^2 + m_2 a_2^2 + 2m_2 a_1 a_2 \cos\theta_2]\ddot{\theta}_1 \\ + [m_2 a_2^2 + m_2 a_1 a_2 \cos\theta_2]\ddot{\theta}_2 \\ - m_2 a_1 a_2 (2\dot{\theta}_1 \dot{\theta}_2 + \dot{\theta}_2^2)\sin\theta_2 \\ + (m_1 + m_2)g a_1 \cos\theta_1 + m_2 g a_2 \cos(\theta_1 + \theta_2),$$
$$\tau_2 = [m_2 a_2^2 + m_2 a_1 a_2 \cos\theta_2]\ddot{\theta}_1 + m_2 a_2^2 \ddot{\theta}_2 \\ + m_2 a_1 a_2 \dot{\theta}_1^2 \sin\theta_2 + m_2 g a_2 \cos(\theta_1 + \theta_2), \qquad (16)$$

where τ_i is the torque, m_i is the mass, a_i is the length, and $\theta_i, \dot{\theta}_i, \ddot{\theta}_i$ are the position, velocity, acceleration of the ith joint,respectively [9]. g is a gravitational constant. Note that, in STICCON, velocity and acceleration are denoted as v^i and a^i instead of $\dot{\theta}_i$ and $\ddot{\theta}_i$.

For a specific example, the following parameter values are used:

$$m_1 = m_2 = 5 \text{ kg},$$
$$a_1 = a_2 = 0.5 \text{ m}, \quad (17)$$
$$g = 9.8 \text{ m/s}^2.$$

The joints constraints are given the same as Chiaverini and Fusco [3],

$$\begin{array}{ll} V^1_{max} = 1.5, & V^2_{max} = 1.5, \\ V^1_{min} = -1.5, & V^2_{min} = -1.5, \\ A^1_{max} = 10, & A^2_{max} = 15, \\ A^1_{min} = -10, & A^2_{min} = -15, \end{array} \quad (18)$$

As the controller, PID controllers are used with

$$K^1_P = K^2_P = 6000,$$
$$K^1_I = K^2_I = 1000, \quad (19)$$
$$K^1_D = K^2_D = 1000,$$

where K^i_P, K^i_I, and K^i_D are proportional, integral, and derivative gains for the ith joint, respectively.

the parameters of a circle for the manipulator to follow are

$$r = 0.2 \text{ m}, \quad \mathbf{X}_c = \begin{pmatrix} x_c \\ y_c \end{pmatrix} = \begin{pmatrix} 0.5 \\ 0.0 \end{pmatrix} \text{ m}, \quad (20)$$

where r and $\mathbf{X}_c$ are a radius and a center, respectively. The initial and final positions of the circle are

$$\mathbf{X}_0 = \mathbf{X}_f = \begin{pmatrix} 0.3 \\ 0.0 \end{pmatrix} \text{ m}. \quad (21)$$

The circle is traversed from $-\pi$ to π so that the angle ϕ is given by the fifth-order polynomial

$$\phi(t) = \phi_0 + (\phi_f - \phi_0) \cdot \left[6\left(\frac{t}{t_f}\right)^5 - 15\left(\frac{t}{t_f}\right)^4 + 10\left(\frac{t}{t_f}\right)^3 \right] \quad (22)$$

where $\phi_0 = -\pi$ and $\phi_f = \pi$. Then, from equations (20) and (22), the task trajectory is given by the equations

$$\begin{array}{l} x(t) = x_c + r\cos\phi(t), \\ y(t) = y_c + r\sin\phi(t). \end{array} \quad (23)$$

The simulation setup described above has been shown in Fig. 2.

Because we don't have any information of the movement duration t_f, it should be estimated. Let us suppose that the estimated t_f is 2.8 seconds. Resulting task trajectory and profiles of joint velocity and acceleration are shown in Fig. 3. It can be observed that clippings have occurred at all joint

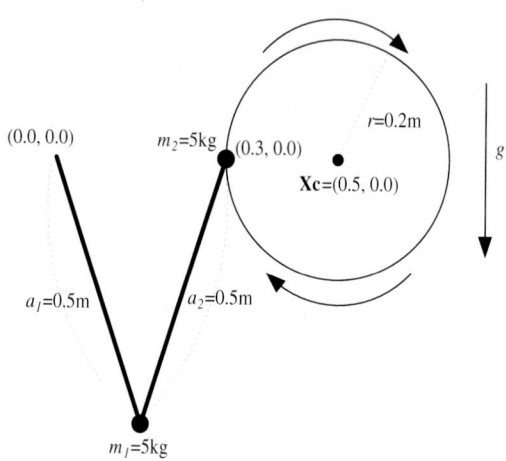

Figure 2: Description of simulation including parameters of both the manipulator and the task. The task is to draw a circle clockwise.

limits except the upper velocity limit of the first joint.

Actually, the commanded joint angle trajectory for PID controller to follow have the specification of

$$\begin{array}{ll} \max v^1_c = 1.1696, & \max v^2_c = 1.5742, \\ \min v^1_c = -1.6255, & \min v^2_c = -1.5742, \\ \max a^1_c = 5.7132, & \max a^2_c = 7.0826, \\ \min a^1_c = -4.0272, & \min a^2_c = -3.4734, \end{array} \quad (24)$$

where v^i_d and a^i_d are the velocity and the acceleration of the command trajectory of the ith joint. Note that, in comparison with equation (18), the violations happened only at the lower velocity limit of the first joint by 8.37%, and the upper/lower velocity limits of the second joint by 4.95%. This indicates the constraints are so coupled that the violation of some constraints resulted in successive violations of others. It is because of these successive violations that such an unsatisfactory trajectory in Fig. 3 is obtained. This fact suggests how it is important to determine the correct movement duration.

It is easy to see that a good result can be obtained if t_f is lengthened. Note that the capability of the manipulator will not be used to the full, if t_f becomes too longer. Then, how much should t_f become longer? STICCON gives an answer to this question. The movement duration which STICCON computes for this case is

$$t_f = 3.0343 \text{ sec}. \quad (25)$$

The results are shown in Fig. 4, The arrow indication in Fig. 4 means that the lower velocity limit of the first joint is a critical limitation in reducing movement duration. Note that it holds only to this circle-drawing task, and for other tasks, the critical constraint which determines the movement duration can be different. This 'critical constraint identification' is an additional feature of STICCON method.

5 Concluding Remarks

This paper has pointed out the importance of the correct determination of movement duration, and proposed a simple method, referred to as STICCON, to determine movement duration under the joint velocity and acceleration limits. It is simple in terms that no optimization technique is used. Only basic numerical differentiation formulas are used. The validity of the proposed method is proved by the simulation where a two-link serial manipulator with the joint constraints executes a circle-drawing task.

STICCON needs both the task trajectory generation method and the inverse kinematics solving method. Since STICCON is formulated independently of those methods, any method may adopt STICCON as a component algorithm, if the performance of the method is heavily affected by the accurate movement duration.

The followings are further researches:

- Finding a way for working together with methods based on torque constraint.

- Improvement of STICCON so that the same motion is executed in reduced time without violating the joint constraints.

Acknowledgments

This is supported by Brain Science and Engineering Program of Korea Ministry of Science and Technology and Brain Korea 21 Program of Korea Ministry of Education.

References

[1] R. L. Andersson, "Aggressive trajectory generator for a robot ping-pong player," *Int'l Conf. Robotics and Automation*, pp. 188–193, 1988.

[2] T. F. Chan and R. V. Dubey, "A weighted least-norm solution based scheme for avoiding joint limits for redundant manipulators," *IEEE Trans. Robotics and Automation*, vol 11, pp. 286–292, 1995.

[3] S. Chiaverini and G. Fusco, "A new inverse kinematics algorithm with path tracking capability under velocity and acceleration constraints," *Proc. 38th IEEE Conf. Decision and Control*, vol 3, pp. 2064–2069, 1999.

[4] O. Dahl and L. Nielsen, "Torque-limited path following by online trajectory time scaling," *IEEE Trans. Robotics and Automation*, vol. 6(5), ppl 554–561, 1990.

[5] T. Flash, and N. Hogan, "The coordination of arm movements: an experimentally confirmed mathematical model", *J. Neuroscience*, vol. 5, pp. 1688–1703, 1985.

[6] A. Ghosal, and B. Roth, "A new approach for kinematic resolution of redundancy," *Int'l J. Robotics Research*, vol. 7(2), pp. 22–35, 1988.

[7] G. Z. Grudic, and P. D. Lawrence, "Iterative inverse kinematics with manipulator configuration control," *IEEE Trans. Robotics and Automation*, vol. 9(4), pp. 476–483, 1993.

[8] J. M. Hollerbach, "Dynamic scaling of manipulator trajectories," *J. Dynamic Systems, Measurements, and Control*, vol. 108, pp. 163–171, 1986.

[9] F. L. Lewis, C. T. Abdallah, and D. M. Dawson, *Control of Robot Manipulators*, Macmillan Publishing Company, 1993.

[10] C. S. Lin, P. R. Chang, and J. Y. S. Luh, "Formulation and optimization of cubic polynomial joint trajectory for industrial robots," *IEEE Trans. Automatic Control*, vol. 28(12), pp. 1066–1074, 1983.

[11] J. Lloyd, and V. Hayward, "Trajectory generation for sensor-driven and time-varying tasks," *Int'l J. Robotics Research*, vol. 12(4), pp. 380–393, 1993.

[12] Y. Nakamura, and H. Hanafusa, "Inverse kinematic solutions with singularity robustness for robot manipulator control," *ASME J. Dynamic Systems, Measurements, and Control*, vol 108, pp. 163–171, 1986.

[13] J. Park, W.-K. Chung, Y. Youm and M. Kim, "Reconstruction of inverse kinematic solution subject to joint kinematic limits using kinematic redundancy," *Int'l Conf. Intelligent Robots and Systems*, vol 2, pp. 425–430, 1996.

[14] H. Seraji and R. Colbaugh, "Improved configuration control for redundant robots," *J. Robotic System*, vol 7, pp. 897–928, 1990.

[15] K. G. Shin, and N. D. McKay, "Minimum-time control robotic manipulators with geometric path constraints," *IEEE Trans. Automatic Control*, vol. 30(6), pp. 531–541, 1985.

[16] G. Tevatia, and S. Schaal, "Inverse kinematics for humanoid robots," *IEEE Int'l Conf. Robotics and Automation*, vol. 1, pp. 294–299, 2000.

[17] S. E. Thompson, and R. V. Patel, "Formulation of joint trajectories for industrial robots using B-splines," *IEEE Trans. Industrial Electronics*, vol. 34(2), pp. 192–199, 1987.

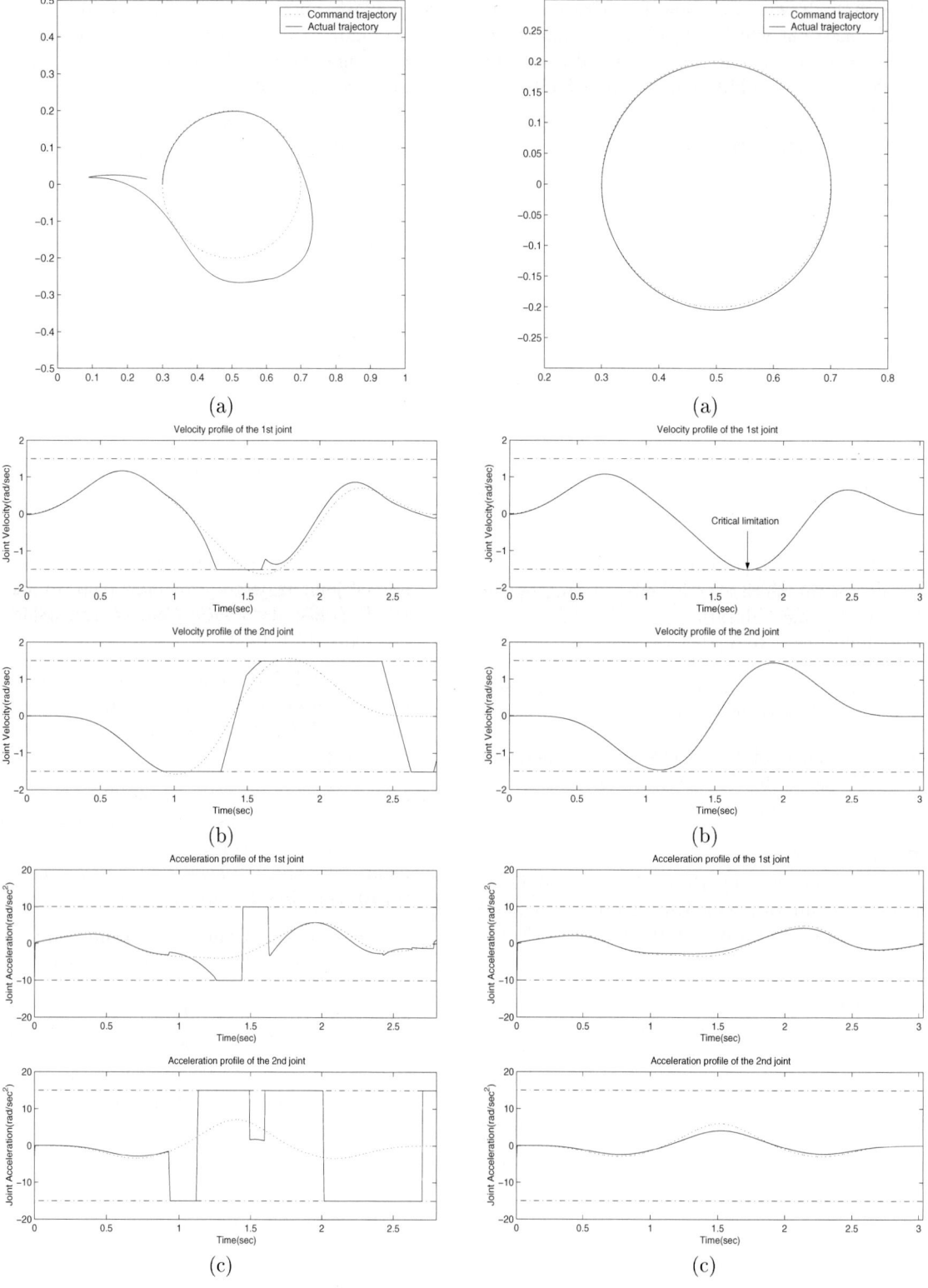

Figure 3: Results when the estimated movement duration($t_f = 2.8$ sec) is used. Commanded and actual trajectories are denoted as dotted and solid curves, respectively. Joint constraints are denotes as dash-dotted lines. (a) Trajectory in task space; (b) Joint velocity profile; (c) Joint acceleration profile.

Figure 4: Results when the movement duration calculated by STICCON($t_f = 3.0343$ sec) is used. Commanded and actual trajectories are denoted as dotted and solid curves, respectively. Joint constraints are denotes as dash-dotted lines. (a) Trajectory in task space; (b) Joint velocity profile; (c) Joint acceleration profile.

Design of Jerk Bounded Trajectories for On-Line Industrial Robot Applications

Macfarlane, S., Croft, E.A.
Industrial Automation Laboratory
Department of Mechanical Engineering
University of British Columbia
Vancouver, BC, Canada V6T 1Z4
e-mail: [sonja, ecroft]@mech.ubc.ca

Abstract

An on-line method for obtaining smooth, jerk-bounded trajectories has been developed and implemented. Jerk limitation is important in industrial robot applications, since it results in improved path tracking and reduced wear on the robot. The method described herein uses a concatenation of fifth-order polynomials to provide a smooth trajectory between two points. The trajectory is determined based on approximating a linear segment with parabolic blends (LSPB) trajectory. A sine wave approximation is used to ramp from zero acceleration to non-zero acceleration. This results in a controlled quintic trajectory which does not oscillate, and is near time-optimal given the jerk and acceleration limits specified. The method requires only the computation of the quintic control points, up to a maximum of seven points per trajectory way-point. This provides hard bounds for on-line motion algorithm computation time. Simulations and experimental results on an industrial robot are presented.

I. Introduction

It is well established that limiting jerk in manipulator trajectories is important for reducing manipulator wear, and improving tracking accuracy and speed [6, 9, 11, 12]. Limitation of jerk (and, proportionally, torque rate) results in smoothed actuator loads [11]. This effectively reduces the excitation of the resonant frequencies of the manipulator, and consequently reduces actuator wear [6]. Low jerk trajectories can be tracked faster and more accurately [5]. In addition, in certain applications, it is important that the manipulator motions be smooth. For example, a robot moving a tray of test tubes filled with fluid must move smoothly to avoid spillage. Fluid and aerosol deposition (e.g., gluing, spray painting) are further examples of applications in which smooth motion is of importance.

The method developed herein is intended for on-line use with industrial robots. Therefore, apart from trajectory smoothness, time-optimality is important, for economic reasons. As well, in order for the trajectory to be implemented in real-time, the method must have a low computational complexity.

Much work has been done in the area of smooth trajectory generation [2, 3, 4, 5, 8, 12]. Cubic splines are often used in smooth trajectory generation [3, 4, 12]. Higher order polynomials are generally not used due to their tendency to oscillate and, therefore, generate retrograde motion [4, 12]. Other methods are based on optimizing with respect to time, while limiting actuator torques or rates [2, 5, 14].

Cao et al. [3] optimized a piecewise cubic polynomial spline to obtain a smooth and time-optimal constrained motion. Chand et al. [4] used polynomial splines to interpolate between joint target points. These methods do not allow for the specification of a jerk limit. In [12], a global minimum-jerk trajectory was found using cubic splines. However, in this method the motion time must be known *a priori*. Jeon et al. [8] propose a method in which sine profiles are used to obtain the appropriate acceleration and deceleration characteristics. For predefined motions, coefficients are calculated and stored in advance. Using these coefficients, scaled velocity profiles are generated on-line. In [5], a method for calculating smooth and time-optimal path constrained trajectories was presented. The smooth and optimal motion was obtained by optimizing a base trajectory subject to actuator torque and torque rate limits. It was shown experimentally that trajectory smoothness improved tracking by the controller and the resulting motion was faster than both a quintic polynomial trajectory and a path constrained time optimal (PCTOM, [2]) trajectory. However, this method (as well as other optimization methods) has a large computational load and is not suitable for on-line use.

Implementation of pure time optimal motion with torque limits is problematic. Direct tracking of a time optimal trajectory leads to joint vibrations and overshoot of the nominal torque limits [5]. Dahl [7] used a secondary path velocity controller to modify a nominal minimum time velocity profile on-line, which resulted in good utilization of the available torque range and good path tracking. Kieffer et al. [10] developed a complex computed torque based controller to attempt to track the PCTOM while limiting actuator chatter. However, these controller-based approaches are not necessarily a practical choice for closed/proprietary industrial robots.

An alternative approach is to design smooth trajectories that approximate optimal motion yet do not inherently require the high torque rates associated with the switching points of torque bounded time optimal motion. This approach is effected by implementing jerk limitation.

In this paper, a methodology for achieving bounded jerk motion using a series of quintic control points is presented. This method computes, in real time, a point to point trajectory with up to seven quintic control points unique to the required path. Unlike approaches which use time scaled polynomials, this approach mimics the ideal (but unattainable with jerk limitations) behaviour of an optimal linear segment with parabolic blends (LSPB) trajectory for the path. It is shown that the maximum jerk, acceleration and velocity limits are respected while achieving near-optimal (LSPB) and on-line performance. The method takes advantage of the inherent smoothness of quintic polynomials and their fast computational properties, while avoiding their major drawback, namely, their tendency to oscillate. A number of simulations are presented to demonstrate the implementation of the method. Finally, we show comparative experimental results for two industrial robot trajectories.

II. Designing Trajectories

LSPB trajectories switch between maximum, minimum and zero accelerations to achieve time-optimal motion while respecting velocity limits. However, this type of motion results in infinite jerk, and as a consequence is not easily tracked, resulting in a longer motion time than expected. If this type of motion could be approximated using a smooth function that is at least C^2 continuous, the resulting motion would be fast and trackable.

Fifth-order polynomials (quintics) are the lowest order polynomials for which it is possible to specify the initial and final positions, velocities and accelerations (1),

$$p(t) = b_5 t^5 + b_4 t^4 + b_3 t^3 + b_2 t^2 + b_1 t + b_0. \quad (1)$$

This polynomial provides a quadratic jerk profile, as opposed to a cubic polynomial which provides a constant jerk profile. A cubic trajectory will result in jerk discontinuities at a minimum and potentially infinite jerk at trajectory way-points, unless computationally expensive spline fitting is done. Andersson [1] used quintics for stop point to stop point motion, and derived the quintic motion time such that velocity and acceleration limits are respected. It is also possible to ensure that jerk limits are respected. However, extending this work to cases where the initial and/or final velocities are non-zero would require repeatedly solving a third-order equation. Furthermore, the resulting trajectory may have a number of undesirable oscillations.

By assigning appropriate start and end conditions to the quintics, a smooth trajectory joining two specified points can be found, while satisfying acceleration continuity (limited jerk) at the trajectory end points. That is, a trajectory between two points is formed as a series of linked quintics, with each quintic having initial conditions equal to the end conditions of the previous quintic. Thus, such a trajectory has C^2 continuity.

Between the way-points, it is desirable to have the quintics follow a LSPB trajectory template without untoward oscillation effects. This requires careful selection of a series of control points between which the quintics will behave in a prescribed manner, respecting velocity, acceleration and jerk limits. The LSPB pattern is used to provide the "switching" points for the trajectory. That is, the trajectory is still formed as a series of acceleration ramps and velocity "cruise" segments, but the ramps are generated using a sine-shaped template.

The sine wave template provides a constant relationship between its amplitude and the maximum value of its derivative. Selecting the time for acceleration and deceleration ramps based on a sine wave with appropriate properties allows the jerk limits to be respected. The resulting acceleration/deceleration ramp is smooth and there is no positional oscillation.

IIa. General Approximation of a Square Acceleration Wave

Figure 1 shows the construction of a smooth trajectory using a concatenation of quintics and the sine wave approximation. In this example, the required change in position during the velocity ramp is smaller than the distance between the two specified points, allowing the robot to achieve the maximum cruise velocity, v_2.

In Figure 1, p_1 and p_2 are, respectively, the initial and final positions, and v_1 and v_2 are the initial and final velocities. D_1 is the distance covered during the ramp from zero acceleration to the maximum acceleration, a_{max}. V_a is the velocity reached at the end of this ramp, and dt_{max} is the time taken to ramp up to a_{max}. D_2 is the distance covered during the constant acceleration phase, and V_b is the velocity reached at the end of this phase. D_3 is the distance covered during the ramp down from a_{max} to zero acceleration. j_{max} is the maximum allowable jerk. Four quintics are necessary to describe the trajectory shown in Figure 1: the acceleration ramp-up quintic, the constant acceleration quintic, the acceleration ramp-down quintic, and the constant velocity quintic.

The sine wave template for acceleration has an amplitude equal to half the acceleration limit, $a_{max}/2$, and maximum slope equal to j_{max}, the maximum jerk. By following this ramp for one-half period, the quintic trajectory can ramp up smoothly without violating acceleration or jerk limits. The time allotted for the ramp is, therefore,

$$dt_{max} = \frac{\pi a_{max}}{2 j_{max}} \quad . \quad (2)$$

Using this approximation for all acceleration ramps limits the acceleration to the interval $[-a_{max}, +a_{max}]$ and the jerk to the interval $[-j_{max}, +j_{max}]$. The ramp up to a_{max} is followed by a constant acceleration phase and then by a ramp down to zero acceleration, which takes the same time as the ramp up. The ramp down begins when the remaining change in velocity is equal to dV_{rd}, where,

$$dV_{rd} = \frac{a_{max} dt_{max}}{2} \quad . \quad (3)$$

The acceleration ramp up and ramp down are symmetrical, which simplifies the integration process. The velocity ramp is also symmetrical about a straight line drawn from the start of the ramp to the end of the ramp.

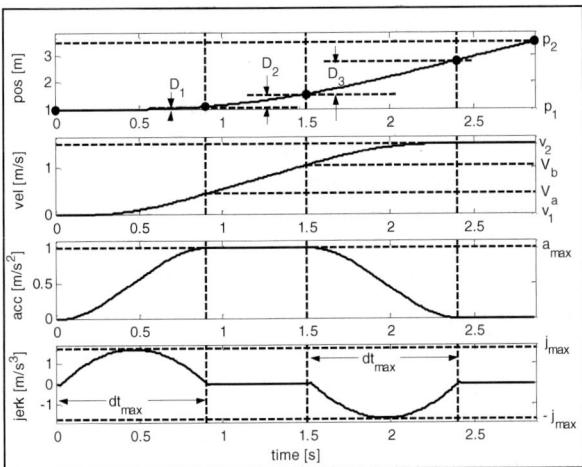

Figure 1. Smooth Trajectory Generated with a Sequence of Quintics (• indicate the quintic control points)

The corresponding LSPB trajectory time is smaller than the quintic trajectory time (by dt_{max}), but the quintic trajectory has bounded jerk, and is therefore trackable.

The quintic control points are determined as specified in the velocity ramp algorithm.

Velocity Ramp Algorithm (c.f. Figure 1)
1. Calculate the total point to point distance, $D = |p_2 - p_1|$
2. Calculate velocities V_a,

$$V_a = v_1 + \frac{a_{max} dt_{max}}{2} \quad , \quad (4)$$

and V_b,

$$V_b = v_2 - \frac{a_{max} dt_{max}}{2} \quad . \quad (5)$$

3. Calculate the distances covered during:
  • the ramp up to a_{max},

$$D_1 = a_{max} dt_{max}^2 \left(\frac{1}{4} - \frac{1}{\pi^2} \right) + v_1 dt_{max} \quad . \quad (6)$$

• the constant acceleration phase,

$$D_2 = \frac{V_b^2 - V_a^2}{2 a_{max}} \quad . \quad (7)$$

• the ramp down from a_{max} to zero acceleration

$$D_3 = a_{max} dt_{max}^2 \left(\frac{1}{4} - \frac{1}{\pi^2} \right) + V_b dt_{max} \quad . \quad (8)$$

4. If $(D_1 + D_2 + D_3 - D > 0)$
    Reduce v_2 based on D, and recalculate D_2, D_3 and V_b.
5. Assign the time, position, velocity and acceleration corresponding to the four quintic control points that form the non-zero acceleration portion of the motion.
6. If $(D_1 + D_2 + D_3 - D < 0)$
    Add a fifth quintic control point corresponding to the end of the constant velocity phase. Constant velocity quintics with zero acceleration end conditions collapse to linear segments.

IIb. Distance and Change in Velocity Limitations

If the distance to cover, D, is smaller than the distance that will be covered during a ramp up to a_{max} and then immediately down to zero acceleration, namely,

$$D_{min} = a_{max} dt_{max}^2 + 2 v_1 dt_{max}, \quad (9)$$

then the method used in Section IIa must be modified. This is also true if the required change in velocity, $(v_2 - v_1) < dV_{min}$, where,

$$dV_{min} = dt_{max} a_{max}. \quad (10)$$

dV_{min} is the change in velocity that will occur during the ramp up to and down from a_{max}.

In both cases, the maximum allowable acceleration cannot be reached without violating the velocity limit or overshooting D. The motion, in such cases, is based on a ramp up to a lower acceleration, a_{peak}, and then an immediate ramp down to zero acceleration. This is followed by a constant velocity phase, if necessary.

If $D < D_{min}$, the time necessary to ramp up to a_{peak} can be found by solving for dt in (11):

$$dt^3 + \frac{\pi v_1}{j_{max}} dt - \frac{\pi D}{2 j_{max}} = 0 \quad . \quad (11)$$

Eq. (11) has at only one positive real root in dt (for $v_1 > 0$), which can be found directly [13] or via a Taylor series approximation with an average of 1.5 iterations.

The peak acceleration, a_{peak}, is then

$$a_{peak} = \frac{2 j_{max} dt}{\pi} \quad . \quad (12)$$

If the change in velocity required is smaller than dV_{min}, dt and a_{peak} must be readjusted based on the required change in velocity, as shown in step 2 of the pulse algorithm below.

In the case of a motion limited by distance or velocity, three quintics are necessary to describe the motion between two points: the acceleration ramp up quintic, the acceleration ramp down quintic, and the constant velocity quintic.

Velocity Pulse Algorithm
1. Calculate dt from (11), and a_{peak} from (12).
2. If $(a_{peak}dt > v_2-v_1)$
   Recalculate dt:
$$dt = \sqrt{\frac{\pi(v_2 - v_1)}{2 j_{max}}}, \quad (13)$$
   and a_{peak} from (12).
3. Assign the time, position, velocity and acceleration corresponding to the three quintic control points that form the non-zero acceleration portion.
4. Assign the fourth quintic control point (corresponding to the end of the constant velocity phase) based on the remaining distance:
$$dD = D - (v_2 + v_1)dt. \quad (14)$$

III. Trajectory Considerations

The method described in Sections IIa and IIb can be used to determined a trajectory between two or more way-points. Each point provided has a specified position and velocity, while the acceleration at each of these points is set to zero, to ensure continuity and reduce the computational burden.

In order for the smoothness of a trajectory strung between a sequence of way-points to be respected, the acceleration at all the points must be continuous. The simplest manner of achieving this objective is to ensure that the acceleration be zero at each of the specified points. This minimizes the computational burden since it is not necessary to take the next trajectory segment into account when calculating a trajectory between two way-points. In addition, this requirement keeps the number of cases to a minimum.

IIIa. Algorithm for Trajectory Generation Between Two Points

A trajectory linking any two given points is calculated based on the distance between the points, the velocities that must be respected at each of the points, as well as the acceleration and jerk limits. Herein the velocity limit, v_{max}, is taken as the maximum of the velocities associated with the two way-points (unless both points are stop points, in which case a maximum velocity is specified).

Three cases (based on velocity) are used to generate trajectories between two points. In the following cases, let D be the distance between the two points, and let v_1, v_2 be the desired speeds at points 1 and 2 respectively.

Case 1: $v_2 \neq v_1$

First consider the case where $v_2 > v_1$. In order for the motion to be carried out as rapidly as possible, the motion begins with an acceleration phase. If $D > D_{min}$ (9) and $v_2-v_1 > dV_{min}$ (10), then the acceleration profile outlined in Section IIa is used. Otherwise, the acceleration profile outlined in Section IIb is used.

If $v_2 < v_1$, the speeds are interchanged such that the motion can be treated as an acceleration, as described above. The appropriate modifications are made at the end of the calculations so that the motion ends with a deceleration from v_1 to v_2.

Case 2: $v_2 = v_1 = 0$

If both v_1 and v_2 are equal to zero, then the resulting trajectory is composed, in the velocity space, of a ramp up to a peak velocity followed by a ramp down to zero velocity. A cruise at the peak velocity may also be needed depending on the magnitude of D. Two scenarios are possible based on the distance D:
- The maximum allowable velocity is not reached.
  In this case, the peak velocity is calculated based on D.
- The maximum allowable velocity is reached.
  In this case, the peak velocity is equal to v_{max}.

The trajectory is composed of an acceleration curve such that the peak velocity is reached (as described in Case 1). This is followed by a constant velocity quintic segment, if necessary. A deceleration curve, symmetrical to the acceleration curve, follows. Up to seven quintics are used to describe the trajectory linking the two points.

Case 3: $v_2 = v_1$

This case is trivial. It results in the coefficients of (1), b_5, b_4, b_3, b_2 being set equal to zero. That is, (1) becomes a first-order polynomial.

IIIb. Computational Considerations

Once the quintic control points (i.e. the start and end conditions of each quintic and the time corresponding to each quintic) have been calculated via the algorithms provided herein, it is necessary to determine the coefficients ($b_0, ..., b_5$ in (1)) corresponding to each quintic. By taking the first and second derivatives of (1) with respect to time, six equations can be derived. Since the position, velocity and accelerations are known at the start and the end of each quintic, the system of six equations and six unknowns (the quintic coefficients) can be solved. This system is shown in (15):

$$\begin{bmatrix} 1 & t_1 & t_1^2 & t_1^3 & t_1^4 & t_1^5 \\ 0 & 1 & 2t_1 & 3t_1^2 & 4t_1^3 & 5t_1^4 \\ 0 & 0 & 2 & 6t_1 & 12t_1^2 & 20t_1^3 \\ 1 & t_2 & t_2^2 & t_2^3 & t_2^4 & t_2^5 \\ 0 & 1 & 2t_2 & 3t_2^2 & 4t_2^3 & 5t_2^4 \\ 0 & 0 & 2 & 6t_2 & 12t_2^2 & 20t_2^3 \end{bmatrix} \begin{bmatrix} b_0 \\ b_1 \\ b_2 \\ b_3 \\ b_4 \\ b_5 \end{bmatrix} = \begin{bmatrix} x_1 \\ v_1 \\ a_1 \\ x_2 \\ v_2 \\ a_2 \end{bmatrix}, \quad (15)$$

where t_1, x_1, v_1, a_1 are the time, position, velocity and acceleration at the start of the quintic, similarly, t_2, x_2, v_2, a_2 are the time, position, velocity and acceleration at the end of the quintic, and $b_0,...,b_5$ are the unknown quintic coefficients. To simplify this system, t_1 is set to zero. The top of the matrix becomes diagonal and the system is reduced to a system of three equations and three unknowns, which is suitable for solving on a DSP-based trajectory planner.

There are no transcendental calculations involved in the calculation of the quintic control points needed to describe the trajectory between two way-points. The most complex operation necessary is a cubic root, used once per point only in the cases described in Section IIb. Thus, this method meets the low computational complexity requirement.

IV. Simulation Results

The trajectory generation algorithm presented herein has been implemented in MATLAB [15]. A simple straight line point to point motion in cartesian space is chosen to compare the simulated motion times of three different types of trajectories: LSPB, quintic concatenation and a single quintic.

The path is given, in parametric form, as
$$x(s) = 533.4$$
$$y(s) = 0$$
$$z(s) = 736.6 - 660.4s \quad (16)$$
$$s = 0, ..., 1.$$

The maximum allowable speed, v_{max}, is 444.5 mm/s, the maximum acceleration, a_{max}, is 2540 mm/s², and the jerk limit, j_{max}, is 79800 mm/s³.

Figure 2 shows the simulated trajectories corresponding to the z-coordinate of the robot end effector. The LSPB trajectory has the smallest motion time (Table 1), however the corresponding jerk is very large, and, therefore, the trajectory could not be tracked accurately by a robot. Using a single quintic that respects the acceleration and jerk limits results in a longer motion time than the quintic concatenation trajectory. The quintic concatenation trajectory is 0.05 s slower than the LSPB trajectory, but its jerk is limited, allowing the trajectory to be tracked, as will be shown in the next section.

Trajectory Type	Motion Time (s)
LSPB	1.66
Quintic Concatenation	1.71
One Quintic	2.79

Table 1. Motion Times for Trajectories of Figure 2.

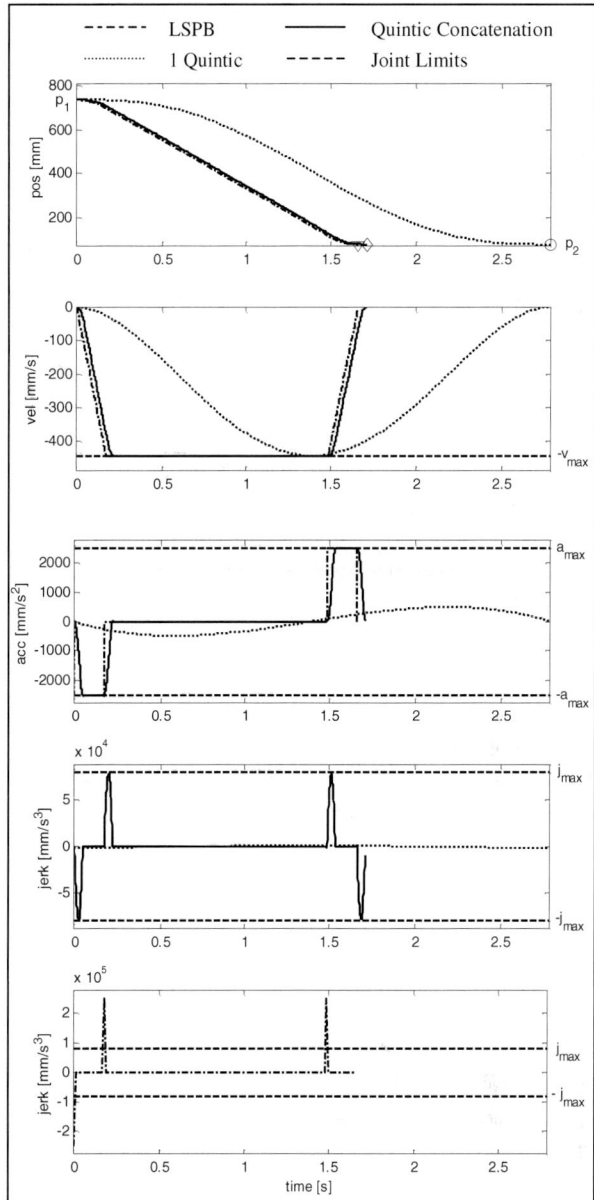

Figure 2. Comparison of Three Point to Point Trajectories

V. Experimental Results

The quintic concatenation trajectory for the path described in Section IV was implemented on a six degree of freedom CRS A465 robot. As can be seen in Figures 3 and 4, the path was tracked quite accurately. The maximum positional deviation is 0.33 mm. Comparison with a cubic splining method currently used in industry shows that although the positional deviation is comparable, the quintic concatenation trajectory provides a faster motion and a smoother velocity profile (a variance of 27.7 mm²/s² versus a variance of 1620 mm²/s² when at the maximum speed).

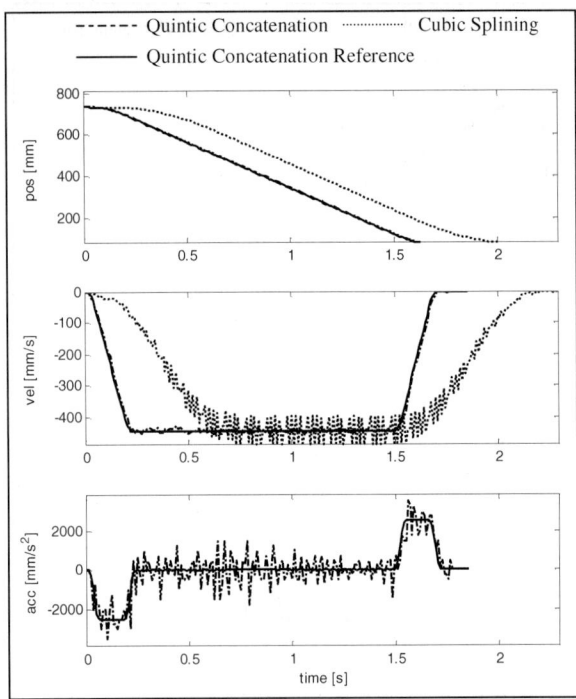

Figure 3. Reference and Experimental Position, Velocity and Acceleration of the End Effector Z-Coordinate.

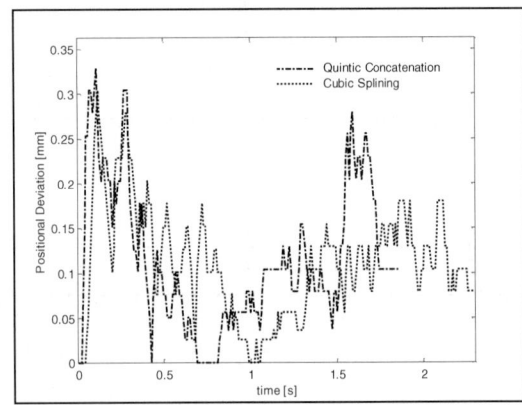

Figure 4. Path Deviation of End Effector

VI. Conclusions

The method presented herein provides a smooth, controlled trajectory for point to point motion with jerk limits. It is computationally efficient and has been successfully implemented on a real-time trajectory planner for a six-degree of freedom robotic manipulator. A concatenation of quintics is used to provide a controlled description of the point to point trajectory. A sine wave approximation for accelerations ensures that the quintics do not oscillate. The algorithms necessary for the implementation of this method have been provided, and simulation and experimental results have been presented.

Acknowledgements

The authors would like to thank the CRS Robotics R&D group for their helpful discussions and support for the implementation of this project. The financial support of NSERC is gratefully acknowledged.

References

[1] Andersson, R.L., *A Robot Ping-Pong Player: Experiment in Real-Time Intelligent Control*, MIT Press, 1988.

[2] Bobrow, J.E., Dubowsky, S. Gibson, J.S., "Time-Optimal Control of Robotic Manipulators Along Specified Paths", *Int. J. of Rob. Res.*, 1985, pp. 3-17.

[3] Cao, B., Dodds, G.I, Irwin, G.W., "Time-Optimal and Smooth Constrained Path Planning for Robot Manipulators", *Proc. IEEE Int. Conf. Rob. & Aut.*, 1994, pp. 1853-1858.

[4] Chand, S., Doty, K., "On-Line Polynomial Trajectories for Robot Manipulators", *J. Rob. Res.*, Vol. 4, No. 2, 1985, pp. 38-48.

[5] Constantinescu, D., Croft E.A., "Smooth and Time-Optimal Trajectory Planning for Industrial Manipulators along Specified Paths", *J. Rob. Sys.*, Vol. 17, No. 5, 2000, pp. 233-249.

[6] Craig, J.J., *Introduction to Robotics*, 2nd ed. New York, Addison-Wesley, 1989.

[7] Dahl, O. "Path-Constrained Robot Control with Limited Torques – Experimental Evaluation", *IEEE Trans. Rob. & Aut.*, Vol. 10, No. 5, 1994, pp. 658-669.

[8] Jeon, J.W., Ha, Y.Y., "A Generalized Approach for the Acceleration and Deceleration of Industrial Robots and CNC Machine Tools", *IEEE Trans. Ind. Elec.*, Vol. 47, No. 1, 2000, pp. 133-139.

[9] Jeon, J.W., "An efficient Acceleration for Fast Motion of Industrial Robots", *Proc. 1995 IEEE Int. Conf. Ind. Elec., Cont. & Instr.*, Part 2, 1995, pp. 1336-1341.

[10] Kieffer, J., Cahill, A.J., James, M.R., "Robust and Accurate Time-Optimal Path-Tracking Control for Robot Manipulators", *IEEE Trans. Rob. & Aut.*, Vol. 13, No. 6, 1997, pp. 880-890.

[11] Kyriakopoulos, K.J., Saridis, G.N., "Minimum Jerk Path Generation", *Proc. 1988 IEEE Int. Conf. Rob. & Aut.*, 1988, pp.364-369.

[12] Piazzi, A., Visioli, A., "Global Minimum-Jerk Trajectory Planning of Robot Manipulators", *IEEE Trans. Ind. Elec.*, Vol. 47, No 1, 2000, pp. 140-149.

[13] Press, W.H., Flannery B.P., Vetterling W.T., Teukolsky, S.A., *Numerical Recipes in C: The Art of Scientific Computing*, Cambridge University Press, 1992.

[14] Shin, K.G, McKay, N.D., "Minimum-Time Control of Robotic Manipulators with Geometric Path Constraints", *IEEE Trans. Aut. Cont.*, Vol. 30, No. 6, 1985, pp. 531-541.

[15] The MathWorks, Natwik, Massachusetts, *Matlab User's Guide*, 1995.

Control of Uncertain Flexible Joint Manipulator Using Adaptive Takagi-Sugeno Fuzzy Model Based Controller

Chang-Woo Park, Chang-Ho Hyun, Min-Sick Park, Chang-Hun Lee,
Jaehun Kim and Mignon Park

ICS Lab., Dept. of Electrical and Computer Eng., Yonsei Univ.
134, Shinchon-dong, Seodaemun-ku, Seoul, Korea
Tel : 82-2-361-2868   Fax : 82-2-312-3233
E-mail : cwpark@yeics.yonsei.ac.kr

Abstract

In this paper, in order to control a uncertain flexible joint manipulator, Adaptive Fuzzy Control (AFC) scheme via Parallel distributed Compensation(PDC) is developed for the multi-input/multi-output plant model represented by the Takagi-Sugeno(TS) model. The alternative AFC scheme is proposed to provide asymptotic tracking of a reference signal for the systems with uncertain or slowly time-varying parameters. The developed control law and adaptive law guarantee the boundedness of all signals in the closed-loop system. In addition, the plant state tracks the state of the reference model asymptotically with time for any bounded reference input signal. The suggested design technique is applied to tracking control of a flexible-link robot manipulator.

1. Introduction

Many of today's robots are driven by actuators with high gear ratios, such as harmonic drivers for high torque and low operation speed. Furthermore, mechanical damage to the robot and the environment can be minimized in an accidental collision involving the arm as the flexible joints and links can absorb a certain amount of impact force due to collision, and this will render the joint compliance sometimes be a desirable feature. Although the joint flexibility has demonstrated some potential merits, the difficulty with modelling and controlling such a flexible mechanical system with high performance made most robot designers prefer to manufacture mechanically rigid arms with stiff joints. Hence, in this paper, we will tackle the problem of controlling for flexible joint robots via fuzzy modelling and fuzzy model based controller and propose a complete solution to solving the problem of model uncertainty.

Fuzzy logic controllers are generally considered applicable to plants that are mathematically poorly understood and where the experienced human operators are available. However, the fuzzy control has not been regarded as a rigorous science due to the lack of the guarantee of the global stability and acceptable performance. To overcome this drawback, since Takagi-Sugeno(TS) fuzzy model[1] which can express a highly nonlinear functional relation in spite of a small number of fuzzy implication rules was proposed, there have been significant research on the stability analysis and systematic design of fuzzy controllers[2-4]. In their research, the nonlinear plant is represented by a TS fuzzy model and the control design is carried out based on the fuzzy model via the so-called Parallel Distributed Compensation(PDC) scheme.

In order to deal with the uncertainties of nonlinear systems, in the fuzzy control system literature, a considerable amount of adaptive schemes have been suggested[5-6]. An adaptive fuzzy system is a fuzzy logic system equipped with the an adaptive law. The major advantage of adaptive fuzzy controller over the conventional adaptive fuzzy controller is that the adaptive fuzzy controller is capable of incorporating linguistic fuzzy information from human operators. However most of them have consider only SISO plants and not dealt with adaptive control problem for TS fuzzy model has been given in a few cases[7]. In this paper, to control the flexible joint manipulator, we present alternative direct Adaptive Fuzzy Controller (AFC) for MIMO plants with poorly understood dynamics or plants subjected to parameter uncertainties. We utilize TS fuzzy model for uncertain flexible joint manipulator and PDC as the basis of our control scheme which is different structure from those used in the above studies based on TS fuzzy model. The adaptation law for adjusting the parameters in feedback and feedforward gain of PDC controller is designed so that the plant output tracks the reference model output. The developed control law and adaptive law guarantee the boundness of all signals in the closed loop system. In addition, the plant states track the states of the reference model asymptotically with time for any bounded reference input signal.

The proposed adaptive fuzzy control scheme is applied to tracking control of a uncertain flexible joint manipulator to verify the validity and effectiveness of the control schemes

2. TS Model Based Fuzzy Control

Consider the continuous-time nonlinear system described by the Takagi-Sugeno fuzzy model. The ith rule of continuous-time TS model is of the following form:

$$R^i : \text{If } x_1(t) \text{ is } M_1^i \text{ and } \cdots \text{ and } x_n(t) \text{ is } M_n^i \\ \text{then } \dot{x}(t) = A_i x(t) + B_i u(t) \quad (2.1)$$

where $x^T(t) = [x_1(t), x_2(t), \cdots, x_n(t)]$,
$u^T(t) = [u_1(t), u_2(t), \cdots, u_m(t)]$

Given a pair of input $(x(t), u(t))$, the final output of the fuzzy system is inferred as follows:

$$\dot{x}(t) = \frac{\sum_{i=1}^{l} w_i(t) \{A_i x(t) + B_i u(t)\}}{\sum_{i=1}^{l} w_i(t)} \quad (2.2)$$

where $w_i(t) = \prod_{j=1}^{n} M_j^i(x_j(t))$, and $M_j^i(x_j(t))$ is the grade of membership of $x_j(t)$ in M_j^i.

In order to design fuzzy controllers to stabilize fuzzy system (2.2), we utilize the concept of parallel distributed compensation (PDC). The PDC controller shares the same fuzzy sets with fuzzy model (2.2) to construct its premise part. That is, the PDC controller is of the following form:

$$R^i : \text{If } x_1(t) \text{ is } M_1^i \text{ and } \cdots \text{ and } x_n(t) \text{ is } M_n^i \\ \text{then } u(t) = -K_i x(t) \quad (2.3)$$

where $x^T(t) = [x_1(t), x_2(t), \cdots, x_n(t)]$ and $i = 1, \cdots, l$.

Given a state feedback $x(t)$, the final output of the fuzzy PDC controller (2.3) is inferred as follows:

$$u(t) = -\frac{\sum_{i=1}^{l} w_i(t) K_i x(t)}{\sum_{i=1}^{l} w_i(t)} \quad (2.4)$$

where $w_i(t) = \prod_{j=1}^{n} M_j^i(x_j(t))$

By substituting the controller (2.4) into the model (2.2), we can construct the closed-loop fuzzy control system as following:

$$\dot{x}(t) = \frac{\sum_{i=1}^{l} \sum_{j=1}^{l} w_i(t) w_j(t) \{A_i - B_i K_j\} x(t)}{\sum_{i=1}^{l} \sum_{j=1}^{l} w_i(t) w_j(t)} \quad (2.5)$$

A sufficient condition for ensuring the stability of the closed-loop fuzzy system (2.5) is given in Theorem 1, which was derived in [2].

Theorem 1

The equilibrium of a fuzzy control system (2.5) is asymptotically stable in the large if there exists a common positive definite matrix P such that

$$G_{ij}^T P + P G_{ij} = -Q_{ij} \quad (2.6)$$

for all $i, j = 1, 2, \cdots, l$
where $G_{ij} = A_i - B_i K_j$ and Q_{ij} is a positive definite matrix.

The design problem of model based fuzzy control is to select K_j ($j = 1, 2, \cdots, l$) which satisfy the stability conditions(2.6). In [4], the common P problem was solved efficiently via convex optimization techniques for LMI's (Linear Matrix Inequality). However, the fuzzy control (2.4) does not guarantee the stability of system in the presence of parameter uncertainty. Moreover, the design of the control parameters is not possible for the systems whose parameters are unknown. To overcome these drawbacks, in this research, an adaptive control scheme is developed for the plant models whose structures are known but parameters unknown.

3. MIMO Adaptive Fuzzy Control

In this section, a Adaptive Fuzzy Control scheme for MIMO TS fuzzy system is developed. Consider again the nonlinear plant represented by the TS model (2.1) or (2.2), where state $x \in R^n$ is available for measurement, $A_i \in R^{n \times n}$, $B_i \in R^{n \times q}$ ($i = 1, \cdots, l$) are unknown constant matrices and (A_i, B_i) are controllable. The control objective is to choose the input vector $u \in R^q$ such that all signals in the closed-loop plant are bounded and the plant state x follows the state $x_m \in R^n$ of a reference model specified by the system

$$\dot{x}_m = \frac{\sum_{i=1}^{l} \sum_{j=1}^{l} w_i(x) \mu_j(x) \{(A_m)_{ij} x_m + (B_m)_{ij} r\}}{\sum_{i=1}^{l} \sum_{j=1}^{l} w_i(x) \mu_j(x)} \quad (3.1)$$

where $(A_m)_{ij} \in R^{n \times n}$ ($i = 1, \cdots, l$) satisfy the stability condition of fuzzy system given in Theorem 1, $(B_m)_{ij} \in R^{n \times q}$, and $r \in R^q$ is a bounded reference input vector. The reference model and input r

are chosen so that $x_m(t)$ represents a desired trajectory that x has to follow.

3.1 Control Law

If the matrices A_i, B_i were known, we could apply the control law

$$u = \frac{\sum_{j=1}^{l} \mu_j(x)(-K_j^* x + L_j^* r)}{\sum_{j=1}^{l} \mu_j(x)} \quad (3.2)$$

where $\mu_j(x) = w_j(x)$, and obtain the closed-loop plant

$$\dot{x} = \frac{\sum_{i=1}^{l}\sum_{j=1}^{l} w_i(x)\mu_j(x)\{(A_i - B_i K_j^*)x + B_i L_j^* r\}}{\sum_{i=1}^{l}\sum_{j=1}^{l} w_i(x)\mu_j(x)} \quad (3.3)$$

Hence, if $K_j^* \in R^{q \times n}$, and $L_j^* \in R^{q \times q}$ are chosen to satisfy the algebraic equations

$$A_i - B_i K_j^* = (A_m)_{ij}, \quad B_i L_j^* = (B_m)_{ij} \quad (3.4)$$

then the transfer matrix of the closed-loop plant is the same as that of the reference model and $x(t) \to x_m(t)$ exponentially fast for any bounded reference input signal $r(t)$.

Let us assume that K_j^*, L_j^* in (3.4) exist, i.e., that there is sufficient structural flexibility to meet the control objective, and propose the control law

$$u = \frac{\sum_{j=1}^{l} \mu_j(x)(-K_j(t)x + L_j(t)r)}{\sum_{j=1}^{l} \mu_j(x)} \quad (3.5)$$

where, $K_j(t)$, $L_j(t)$ are the estimates of K_j^*, L_j^*, respectively, to be generated by an appropriate adaptive law.

3.2 Adaptive Law

By adding and subtracting the desired input term, namely,

$$\sum_{j=1}^{l} \mu_j(x)\{-B_i(K_j^* x - L_j^* r)\} / \sum_{j=1}^{l} \mu_j(x)$$

in the plant equation and using (3.4), we obtain

$$\dot{x} = \frac{\sum_{i=1}^{l}\sum_{j=1}^{l} w_i(x)\mu_j(x)(A_m)_{ij}}{\sum_{i=1}^{l}\sum_{j=1}^{l} w_i(x)\mu_j(x)} x$$
$$+ \frac{\sum_{i=1}^{l}\sum_{j=1}^{l} w_i(x)\mu_j(x)(B_m)_{ij}}{\sum_{i=1}^{l}\sum_{j=1}^{l} w_i(x)\mu_j(x)} r \quad (3.6)$$
$$+ \frac{\sum_{i=1}^{l}\sum_{j=1}^{l} w_i(x)\mu_j(x)B_i(K_j^* x - L_j^* r + u)}{\sum_{i=1}^{l}\sum_{j=1}^{l} w_i(x)\mu_j(x)}$$

Furthermore, by adding and subtracting the estimated input term multiplied by $\sum_i w_i B_i / \sum_i w_i$, that is,

$$\frac{\sum_i w_i B_i}{\sum_i w_i} \left\{ \frac{\sum_{j=1}^{l} \mu_j(x)\{(K_j(t)x - L_j(t)r)\}}{\sum_{j=1}^{N} \mu_j(x)} + u \right\}$$

in the reference model (3.1), we obtain

$$\dot{\hat{x}}_m = \frac{\sum_{i=1}^{l}\sum_{j=1}^{l} w_i(x)\mu_j(x)(A_m)_{ij}}{\sum_{i=1}^{l}\sum_{j=1}^{l} w_i(x)\mu_j(x)} \hat{x}_m$$
$$+ \frac{\sum_{i=1}^{l}\sum_{j=1}^{l} w_i(x)\mu_j(x)(B_m)_{ij}}{\sum_{i=1}^{l}\sum_{j=1}^{l} w_i(x)\mu_j(x)} r \quad (3.7)$$
$$+ \frac{\sum_{i=1}^{l}\sum_{j=1}^{l} w_i(x)\mu_j(x)B_i(K_j(t)x - L_j(t)r + u)}{\sum_{i=1}^{l}\sum_{j=1}^{l} w_i(x)\mu_j(x)}$$

By using the reference model (3.7), we can express (3.6) in terms of the tracking error defined as $e \triangleq x - x_m$, i.e.,

$$\dot{e} = \frac{\sum_{i=1}^{l}\sum_{j=1}^{l} w_i(x)\mu_j(x)(A_m)_{ij}}{\sum_{i=1}^{l}\sum_{j=1}^{l} w_i(x)\mu_j(x)} e$$
$$+ \frac{\sum_{i=1}^{l}\sum_{j=1}^{l} w_i(x)\mu_j(x)B_i(-\widetilde{K}_j x + \widetilde{L}_j r)}{\sum_{i=1}^{l}\sum_{j=1}^{l} w_i(x)\mu_j(x)} \quad (3.8)$$

where $\widetilde{K}_j = K_j(t) - K_j^*$ and $\widetilde{L}_j = L_j(t) - L_j^*$.
In the dynamic equation (3.8) of tracking error, B_i are unknown. We assume that L_j^* are either positive definite or negative definite and define $\Gamma_j^{-1} = L_j^* sgn(l_j)$, where $l_j = 1$ if L_j^* is positive definite and $l_j = -1$ if L_j^* is negative definite. Then $B_i = (B_m)_{ij} L_j^{*-1}$ and (3.8) becomes

$$\dot{e} = \frac{\sum_{i=1}^{l}\sum_{j=1}^{l} w_i(x)\mu_j(x)(A_m)_{ij}}{\sum_{i=1}^{l}\sum_{j=1}^{l} w_i(x)\mu_j(x)} e \quad (3.9)$$
$$+ \frac{\sum_{i=1}^{l}\sum_{j=1}^{l} w_i(x)\mu_j(x)(B_m)_{ij} L_j^{*-1}(-\widetilde{K}_j x + \widetilde{L}_j r)}{\sum_{i=1}^{l}\sum_{j=1}^{l} w_i(x)\mu_j(x)}$$

Now, by using the tracking error dynamics (3.9), we derive the adaptive law for updating the desired control parameters K_j^*, L_j^* so that the closed-loop plant model (3.6) follows the reference model (3.1). We assume that the adaptive law has the general structure

$$\dot{K}_j(t) = F_j(x, x_m, e, r), \quad \dot{L}_j = G_j(x, x_m, e, r) \quad (3.10)$$

where F_j and G_j ($i = 1, \cdots, l$) are functions of known signals that are to be chosen so that the equilibrium

$$K_{je} = K_j^*, L_{je} = L_j^*, \quad e_e = 0 \quad (3.11)$$

of (3.9), (3.10) has some desired stability properties.

We propose the following Lyapunov function candidate

$$V(e, \widetilde{K}_j, \widetilde{L}_j) = e^T P e + \sum_{j=1}^{l} tr(\widetilde{K}_j^T \Gamma_j \widetilde{K}_j + \widetilde{L}_j^T \Gamma_j \widetilde{L}_j) \quad (3.12)$$

where $P = P^T > 0$ is a common positive definite matrix of the Lyapunov equations $(A_m)_{ij}^T P + P(A_m)_{ij} < -Q_{ij}$ for all $Q_{ij} = Q_{ij}^T > 0$ $(i, j = 1, \cdots, l)$ whose existence is guaranteed by the stability assumption for A_m. Then, after some straightforward mathematical manipulations, we obtain the time derivative $\dot{V}$ of V along the trajectory of (3.9), (3.10) as

$$\dot{V} = -e^T \frac{\sum_{i=1}^{l}\sum_{j=1}^{l} w_i(x)\mu_j(x) Q_{ij}}{\sum_{i=1}^{l}\sum_{j=1}^{l} w_i(x)\mu_j(x)} e$$
$$+ 2tr\left\{ \frac{\sum_{i=1}^{l}\sum_{j=1}^{l} w_i(x)\mu_j(x) \widetilde{K}_j^T \Gamma_j (B_m)_{ij}^T sgn(l_j)}{\sum_{i=1}^{l}\sum_{j=1}^{l} w_i(x)\mu_j(x)} Pe\, x^T \right.$$
$$\left. + \sum_{j=1}^{l} \widetilde{K}_j^T \Gamma_j \dot{\hat{K}}_j \right\}$$
$$+ 2tr\left\{ \frac{\sum_{i=1}^{l}\sum_{j=1}^{l} w_i(x)\mu_j(x) \widetilde{L}_j^T \Gamma_j (B_m)_{ij}^T sgn(l_j)}{\sum_{i=1}^{l}\sum_{j=1}^{l} w_i(x)\mu_j(x)} Pe\, r^T \right.$$
$$\left. + \sum_{j=1}^{l} \widetilde{L}_j^T \Gamma_j \dot{\hat{L}}_j \right\}$$
$$(3.13)$$

In the last two terms of (3.13), if we let

$$\sum_{j=1}^{l} \widetilde{K}_j^T \Gamma_j \dot{\hat{K}}_j = -\frac{\sum_{i=1}^{l}\sum_{j=1}^{l} w_i(x)\mu_j(x) \widetilde{K}_j^T \Gamma_j (B_m)_{ij}^T sgn(l_j)}{\sum_{i=1}^{l}\sum_{j=1}^{l} w_i(x)\mu_j(x)} Pe\, x^T$$
$$(3.14a)$$

$$\sum_{j=1}^{l} \widetilde{L}_j^T \Gamma_j \dot{\hat{L}}_j = -\frac{\sum_{i=1}^{l}\sum_{j=1}^{l} w_i(x)\mu_j(x) \widetilde{L}_j^T \Gamma_j (B_m)_{ij}^T sgn(l_j)}{\sum_{i=1}^{l}\sum_{j=1}^{l} w_i(x)\mu_j(x)} Pe\, r^T$$
$$(3.14b)$$

we can make $\dot{V}$ to be negative, i.e.,

$$\dot{V} = -e^T \frac{\sum_{i=1}^{l}\sum_{j=1}^{l} w_i(x)\mu_j(x) Q_{ij}}{\sum_{i=1}^{l}\sum_{j=1}^{l} w_i(x)\mu_j(x)} e \le 0 \quad (3.15)$$

Hence, the obvious choice for adaptive law to make $\dot{V}$ negative is

$$\dot{\hat{K}}_j = \dot{K}_j(t) = \frac{\sum_{i=1}^{l} w_i(x)\mu_j(x) (B_m)_{ij}^T sgn(l_j)}{\sum_{i=1}^{l}\sum_{j=1}^{l} w_i(x)\mu_j(x)} Pe\, x^T$$
$$= \left\{ \frac{\sum_{i=1}^{l} w_i (B_m)_{ij}^T}{\sum_{i=1}^{l} w_i} \right\} \left\{ \frac{\mu_j}{\sum_{j=1}^{l} \mu_j} \right\} sgn(l_j) Pe\, x^T$$
$$(3.16a)$$

$$\dot{\hat{L}}_j = \dot{L}_j(t) = \frac{\sum_{i=1}^{l} w_i(x)\mu_j(x) (B_m)_{ij}^T sgn(l_j)}{\sum_{i=1}^{l}\sum_{j=1}^{l} w_i(x)\mu_j(x)} Pe\, r^T$$
$$= \left\{ \frac{\sum_{i=1}^{l} w_i (B_m)_{ij}^T}{\sum_{i=1}^{l} w_i} \right\} \left\{ \frac{\mu_j}{\sum_{j=1}^{l} \mu_j} \right\} sgn(l_j) Pe\, r^T$$
$$(3.16b)$$

Theorem 2 Stability of the AFC

Consider the plant model (2.2) and the reference model (3.1) with the control law (3.5) and adaptive law (3.16). Assume that the reference input r and the state x_m of the reference model are uniformly bounded. Then the control law (3.5) and the adaptive law (3.16) guarantee that

(ⅰ) $K(t)$, $L(t)$, $e(t)$ are bounded
(ⅱ) $e(t) \to 0$ as $t \to \infty$

Proof.

From (3.12) and (3.15), it directly follows that V is a Lyapunov function for the system (3.9), (3.10), which implies that the equilibrium given by (3.11) is uniformly stable, which, in turn, implies that the trajectory $\widetilde{K}(t)$, $\widetilde{L}(t)$, $e(t)$ is bounded for all $t > 0$. Because $e = x - x_m$ and $x_m \in \mathcal{L}_\infty$, we have that $x \in \mathcal{L}_\infty$. From (3.5) and $r \in \mathcal{L}_\infty$, we also have that $u \in \mathcal{L}_\infty$; therefore, all signals in the closed-loop are bounded.

Now, let us show that $e \in \mathcal{L}_2$. From (3.12) and (3.15), we conclude that because V is bounded from below and is nonincreasing with time, it has a limit, i.e.,

$$\lim_{t \to \infty} V(e(t), \widetilde{K}_j(t), \widetilde{L}_j(t)) = V_\infty < \infty \quad (3.17)$$

From (3.15) and (3.17), it follows that

$$\int_0^\infty e^T \left(\frac{\sum_{i,j=1}^{l} w_i \mu_j Q_{ij}}{\sum_{i,j=1}^{l} w_i \mu_j} \right) e \, d\tau = -\int_0^\infty \dot{V} d\tau = (V_0 - V_\infty) \quad (3.18)$$

where

$$V_0 = V(e(0), \widetilde{K}_j(0), \widetilde{L}_j(0))$$

On the other hand, from $0 \le w_i \le 1$, $0 \le \mu_j \le 1$, and

$$\lambda_{min}(Q_{ij}) \|e\|^2 \le e^T Q_{ij} e \le \lambda_{max}(Q_{ij}) \|e\|^2$$

we have

$$\{\lambda_{min}(Q_{ij})\}_{min} \|e\|^2 \le e^T \left(\frac{\sum_{i,j=1}^{l} w_i \mu_j Q_{ij}}{\sum_{i,j=1}^{l} w_i \mu_j} \right) e$$
$$\le \{\lambda_{max}(Q_{ij})\}_{max} \|e\|^2 \quad (3.19)$$

where

$$\{\lambda_{min}(Q_{ij})\}_{min} = \min\{\lambda_{min}(Q_{11}), \cdots, \lambda_{min}(Q_{ll})\}$$
$$\{\lambda_{max}(Q_{ij})\}_{max} = \max\{\lambda_{max}(Q_{11}), \cdots, \lambda_{max}(Q_{ll})\}.$$

After inserting (3.19) into (3.18), and straightforward manipulation, we have

$$\frac{(V_0 - V_\infty)}{\{\lambda_{min}(Q_{ij})\}_{min}} \le \int_0^\infty \|e\|^2 d\tau \le \frac{(V_0 - V_\infty)}{\{\lambda_{max}(Q_{ij})\}_{max}}$$

which implies that $e \in \mathcal{L}_2$. Because $e, \widetilde{K}_j, \widetilde{L}_j, r \in \mathcal{L}_\infty$, it follows from (3.9) that $\dot{e} \in \mathcal{L}_\infty$, which, together with $e \in \mathcal{L}_2$, implies that $e(t) \to 0$ as $t \to \infty$. □

4. Tracking Control of an Uncertain Flexible Joint Manipulator

In this section, the validity and effectiveness of the proposed AFC are examined through the simulation of tracking control for a flexible joint manipulator.

The control objective is to follow a given trajectory $q_d(t)$ and to produce a torque vector u such that the trajectory error approaches 0 as $t \to 0$. In the simulation, we examine the effects of parametric variation due to some internal uncertainties on behaviors of the closed-loop systems with the proposed TS model based AFC scheme, respectively.

4.1 Design of direct adaptive fuzzy control system

Consider the single link flexible joint manipulator whose dynamics can be written as

$$\dot{x}_1 = x_2$$
$$\dot{x}_2 = \frac{k}{J}(x_3 - x_1) + \frac{1}{J}u$$
$$\dot{x}_3 = x_4$$
$$\dot{x}_4 = -\frac{MgL}{I}\sin x_3 - \frac{k}{I}(x_3 - x_1) \quad (4.1)$$

where, I, J are, respectively, the link and the rotor inertia moments, M is the link mass, k is the joint elastic constant, L is the distance from the axis of the rotation to the link center of mass and g is the gravitational acceleration.

In order to apply the suggested AFC, we need a TS fuzzy model representation of the manipulator. By applying the Lyapunov linearization method at operating points $x_1 = 0, \pi$, we obtain the TS fuzzy model for the robot manipulator as followings.

Rule 1 : IF x_1 is about 0 THEN $\dot{x} = A_1 x + B_1 u$

Rule 2 : IF x_1 is about π THEN $\dot{x} = A_2 x + B_2 u$ (4.2)

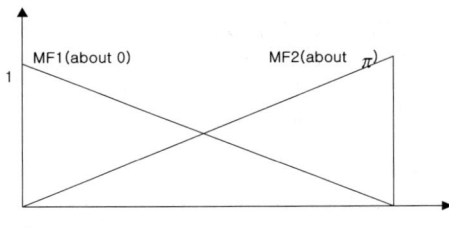

Fig. 1 Membership function for x1

The whole state space formed by state vector of the original nonlinear equations is partitioned into two different fuzzy subspaces whose center is located at the center of corresponding membership functions shown in Fig. 1.

To apply the proposed adaptive fuzzy control scheme, the reference model for the plant state x to follow should be specified. In this simulation, the closed-loop eigenvalues for each subsystem are chosen to be the same, which in turn make the reference model for each fuzzy subspace to be the same and linear one as following:

$$\dot{x}_m = \begin{bmatrix} 0 & 1 & 0 & 0 \\ 0 & 0 & 1 & 0 \\ 0 & 0 & 0 & 1 \\ -4 & -10 & -10 & -5 \end{bmatrix} x_m + \begin{bmatrix} 0 \\ 0 \\ 0 \\ 1 \end{bmatrix} r \quad (4.3)$$

The PDC controller shares the same fuzzy sets with fuzzy model to construct its premise part. That is, the PDC controller is of the following form:

R^i : If x_1 is MF_i,
then $u(t) = -K_i [x_1 x_2 x_3 x_4]^T + L_i r(t)$ (4.4)
$i = 1, 2$

The feedback control gains K_i and L_i of each fuzzy state feedback controller is updated by adaptive law so that the closed-loop plant follows the reference model (4.3).

Now by using (3.16), we derive the adaptive law for updating the elements of K_i and L_i so that the closed-loop plant follows the reference model.

$$\dot{K}_j(t) = \left\{ \frac{\mu_j}{\sum_{j=1}^{2} \mu_j} \right\} sgn(l_j) B_m^T P e \, x^T$$

$$\dot{L}_j(t) = \left\{ \frac{\mu_j}{\sum_{j=1}^{2} \mu_j} \right\} sgn(l_j) B_m^T P e \, r^T \quad (4.5)$$

where $B_m^T = [0\ 0\ 0\ 1]$, $P = \begin{bmatrix} 1.13 & 0 & 0.12 & 0 \\ 0 & 1.17 & 0 & 0.07 \\ 0.12 & 0 & 0.15 & 0 \\ 0 & 0.07 & 0 & 0.11 \end{bmatrix}$

4.2 Simulation results

The parameters of nominal plant model used in this simulation are as follows.

$M = 1Kg$, $I = 1Kg.m^2$, $L = 1m$, $k = 1N/m$, $J = 1Kg.m^2$, and $g = 9.8 m/s^2$

The initial values of K_i and L_i are designed from the nominal model parameters of the plant model (4.2) to be controlled.

To test the adaptation abilities of the proposed scheme, the mass of link is varied with time as

$m = 1 + 0.5 \sin 3\pi t$

The initial value for state x_3 is assumed as $x_3 = \frac{\pi}{3}$.

Figure 2,3,4 show the simulation results of tracking response of joint angle, control input and tracking error respectively. As shown in these figures, the tracking error approaches to almost zero in spite of some parameter perturbation.

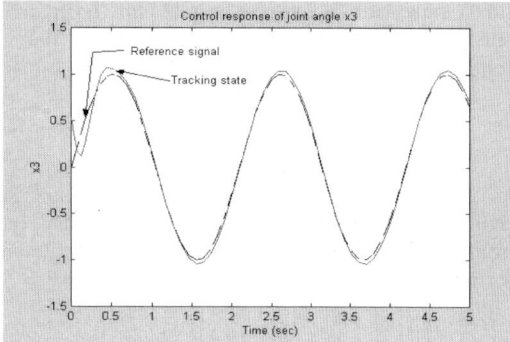

Fig. 2 Control response of joint angle x3

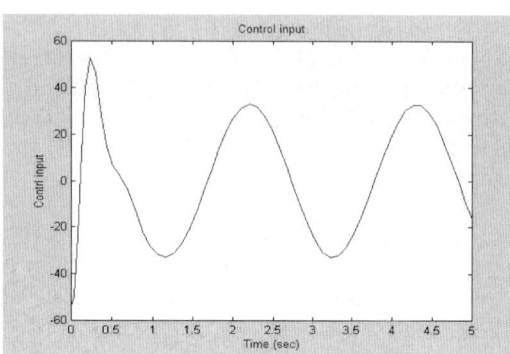

Fig. 3 Control input

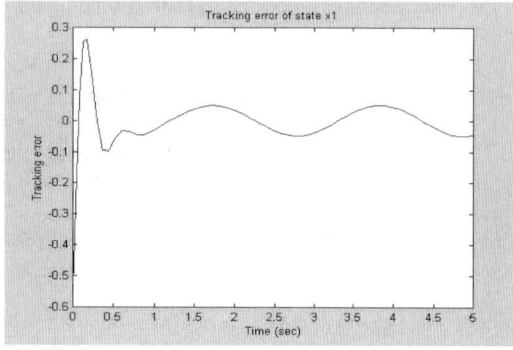

Fig. 4 Tracking error of state x3

5. Conclusions

In this paper, we have developed a alternative TS fuzzy model based adaptive control scheme for a flexible joint manipulators with parameter uncertainty in their fuzzy model. The adaptation law adjusts the controller parameters on-line so that the plant output tracks the reference model output. The developed adaptive law guarantees the boundedness of all signals in the closed-loop system and ensures that the plant state tracks the state of the reference model asymptotically with time for any bounded reference input signal. The proposed adaptive fuzzy control scheme was applied to tracking control of a single link flexible joint manipulator to verify the validity and effectiveness of the control scheme. From the simulation results, we conclude that the suggested scheme can effectively achieve the trajectory tracking in spite of parameter perturbation.

References

[1] T. Takagi and M. Sugeno, "Fuzzy identification of systems and its applications to modeling and control," *IEEE Trans. Systems, Man, and Cybernetics*, vol. 15, no. 1, pp. 116-132, 1985

[2] T. Tanaka and M. Sugeno, "Stability analysis and design of fuzzy control systems," *Fuzzy sets and system*, vol. 45, no.2, pp. 135-156, 1992

[3] K. Tanaka and M. Sano, "A robust stabilization problem of fuzzy control systems and its application to backing up control of a truck-trailer," *IEEE Trans. Fuzzy Systems*, vol. 2, no. 2, pp. 119-133, 1994

[4] H. O. Wang, K. Tanaka, M. F. Griffin, "An approach to fuzzy control of nonlinear systems: Stability and design issues," *IEEE Trans. Fuzzy Systems*, vol. 4, no. 1, pp. 14-23, 1996

[5] L.-X. Wang, *A Course in Fuzzy systems and Control*, Prentice-Hall International, Inc., 1997

[6] K. Fischle and D. Schroder, "An improved stable adaptive fuzzy control method," *IEEE Trans. Fuzzy Systems*, vol. 7, no. 1, pp. 27-40, 1999

[7] D. L. Tssay, H. Y. Chung and C. J. Lee, "The adaptive control of nonlinear systems using the sugeno-type of fuzzy logic," *IEEE Trans. Fuzzy Systems*, vol. 7, no. 2, pp. 225-229, 1999

[8] P.A. Ioannou, J. Sun, *Robust Adaptive Control*, Prentice Hall International Editions, 1996

Sensor Based Planning for Rod Shaped Robots in Three Dimensions: Piece-wise Retracts of $\mathbb{R}^3 \times S^2$

Ji Yeong Lee, Howie Choset and Alfred A. Rizzi, *Carnegie Mellon University*

Abstract

We describe a new roadmap, termed the rod-HGVG, for motion planning of a rod-shaped robot operating in a three-dimensional volume. This roadmap is defined in terms of work space distance enabling us to prescribe an incremental construction procedure. This allows the rod to explore its configuration space, $\mathbb{R}^3 \times S^2$, without ever explicitly constructing the configuration space. In fact, the rod robot need not know the work space ahead of time. We term the rod-HGVG a piecewise retract because it comprises many retracts. Homotopy theory asserts that there cannot be in general a one-dimensional retract of non-contractable five-dimensional space. Instead, we define an exact cellular decomposition on $\mathbb{R}^3 \times S^2$ and a retract in each cell. Next, we "connect" the retracts of each cell forming a piece-wise retract of the rod's configuration space.

1 Introduction

This work considers the sensor based planning of a rod robot in three dimensions. Although planning for rod-shaped robots is geometrically similar to planning for blimp robots, ultimately motion planning for highly articulated robots motivates this research and the motion planning of a rod robot is the first step toward this goal. In this paper, we will define a roadmap for a rod robot in three dimensions. This roadmap is the *rod hierarchical generalized Voronoi graph* (rod-HGVG). Recall from Canny's roadmap work [1] that for each connected component of free space, a roadmap has the following properties : (i) accessibility, (ii) departability, and (iii) connectivity. This means that if there exists a path between two configurations, we can determine a path by first finding a path to the roadmap from the start configuration, then following the roadmap, and then finding another path from the roadmap to the destination. If a planner can incrementally construct a roadmap, then it has in essence explored the configuration space.

Canny's original roadmap work required *a priori* information about the robot's configuration space, and hence was difficult to implement. The probabilistic community [5], [9] has successfully demonstrated the capabilities of probabilistic roadmaps for highly articulated robots; their approach does not construct the configuration space, but requires knowledge of the work space prior to the planning event. Rimon [8] first suggested a method for constructing roadmap using "critical point" sensors, but did not provide details on these sensors. Choset and Burdick developed a method to construct the Voronoi Graph in the plane and in three dimensions without requiring *a priori* workspace or configuration space information [2]. Later, Choset extended this result to a rod-shaped robot operating in the plane [3]. This method enabled a rod robot to map its configuration space without any prior information of its workspace (and configuration space). Loosely speaking, the rod-HGVG in $\mathbb{R}^3$ can be viewed as a combination of two prior roadmaps : (i) the point-HGVG, which is a roadmap for the point robot operating in a three-dimensional space, and (ii) the planar Rod-HGVG, which is a roadmap for rod robot operating in the plane. The rod-HGVG will be defined in terms of the distance information that sensors can easily provide, and it can be generated in an unknown environment.

1.1 The Generalized Voronoi Graph in $\mathbb{R}^2$ and $\mathbb{R}^3$

Ó'Dúnlaing and Yap [7] first applied the *generalized Voronoi Diagram* (GVD) to path planning of a disk-shaped robot. This result requires full knowledge of the environment to construct the roadmap and is restricted to the plane. Choset and Burdick extended this result beyond the plane by defining the *generalized Voronoi graph* (GVG) [2] which is a roadmap for a point robot operating either in the plane or in three dimensions. The GVG can be constructed using line of sight sensor data because it is defined in terms of the distance function,

$$d_i(r) = \min_{c \in C_i} \|r - c\|,$$

where $r \in FS$, $FS \subset R^2$ is the free space and C_i is an obstacle. A *point-two-equidistant face* is defined as

$$\begin{aligned} F_{ij} = \{ &r \in FS : d_i(r) = d_j(r), \\ &\nabla d_i(r) \neq \nabla d_j(r) \text{ and} \\ &d_i(r) \leq d_h(r) \; \forall h \neq i, j \} \end{aligned} \quad (1)$$

In planar case, F_{ij} is one-dimensional and the GVG is the union of F_{ij}'s, i.e., for an environment with n obstacles, the GVG is $\cup_{i=1}^{n-1} \cup_{j=i+1}^{n} F_{ij}$.

In $\mathbb{R}^3$, the *point-three-equidistant face* defines the GVG. That is,

$$\begin{aligned} F_{ijk} = \{ &r \in FS : d_i(r) = d_j(r) = d_k(r) \\ &\nabla d_i(r) \neq \nabla d_j(r), \nabla d_i(r) \neq \nabla d_k(r), \\ &\nabla d_j(r) \neq \nabla d_k(r), \text{and} \\ &d_i(r) \leq d_h(r) \forall h \neq i, j, k \} \end{aligned} \quad (2)$$

A three-equidistant face can also be defined as the intersection of two-equidistant faces, i.e., $F_{ijk} = F_{ij} \cap F_{jk} \cap F_{ik}$. However, we assume that the intersections are transversal [2] which is a different way of saying that the obstacles all lie in general position. The GVG is the union of the F_{ijk}'s, is one-dimensional, and not necessarily connected in $\mathbb{R}^3$. To connect the GVG, we will need to define additional structures, resulting in the *hierarchical generalized Voronoi graph* (HGVG) [2]. To distinguish the GVG from the rod-GVG, in later sections, we call this GVG defined in this section as point-GVG.

1.2 The Rod-HGVG in SE(2)

O'Dúnlaing, Sharir and Yap [6] extended their disk-result to rod-shaped robots, but this result also requires full knowledge of the environment and is restricted to the plane. Cox and Yap developed an "on-line" strategy for rod path-planning [4], but this result does not provide a roadmap of the rod robot's free space. In this section, we describe the *rod-hierarchical generalized Voronoi graph* (rod-HGVG) in the plane. This structure provides the groundwork for the rod-robot roadmap in three dimensions.

The rod in the plane has three degrees-of-freedom, and we represent a rod configuration as $q = (x, y, \theta)^T$. As with the point-GVG, we first define the distance function. Let q be a rod configuration, and $R(q)$ be the set of all points that the rod occupies at configuration q. Then, the distance between the obstacle C_i and the rod at q is defined as

$$D_i(q) = \min_{r \in R(q), c \in C_i} \|r - c\|.$$

The rod-GVG is the set of configurations equidistant to three obstacles. More formally, we first define the *rod-two-equidistant face* as follows

$$CF_{ij} = \{q \in SE(2) : D_i(q) = D_j(q) \leq D_h(q) \forall h \neq i, j \text{ and } \nabla D_i(q) \neq \nabla D_j(q)\}.$$

Then, the *rod-three-equidistance face* is defined as

$$CF_{ijk} = CF_{ij} \cap CF_{jk} \cap CF_{ik}.$$

In the planar case, we term the rod-three-equidistant faces as *rod-GVG edges*. The rod-GVG is simply the union of the rod-GVG edges, i.e., the set of rod configurations that are three-way equidistant. The rod-four-equidistance faces, which can be defined similarly, are termed *rod-meet points*, which correspond to configurations of the rod where rod-GVG edges intersect and terminate.

The rod-GVG edges are not necessarily connected, even in the planar case. To produce a connected structure we introduce another type of edge, called *R-edges*.

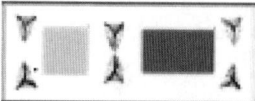 

Fig. 1. The Rod GVG-edge and R-edge in $\mathbb{R}^2$.

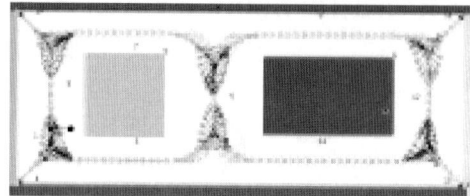

Fig. 2. The Rod-HGVG in $\mathbb{R}^2$.

An R-edge is the set of rod configurations defined as,

$$\begin{aligned}\mathcal{R}_{ij} = \{&q \in cl(RF_{ij})) : r \in F_{ij} \text{ and} \\ &(i)\ 0 \leq d_i(r) \leq d_i(r_1)\ \forall r_1 \in R(q) \text{ and} \\ &(ii)\ d_i(r) \leq D_h(q)\ \forall h \neq i, j\},\end{aligned}$$

where RF_{ij} is the collection of lines tangent to F_{ij} (which can be viewed as a "tangent bundle" of F_{ij}) [3]. Roughly speaking, R-edges connect the disconnected rod-GVG edges using point-GVG edges. The rod-HGVG then comprises Rod-GVG edges and R-edges. Figures 1 and 2 show an example of the rod-HGVG in $\mathbb{R}^2$.

We demonstrate connectivity of the rod [3] by defining a *piece-wise* retraction $H : \mathcal{FS} \times [0, 1] \to CF_{ijk}$. This H function also describes how the rod accesses the rod-GVG. The rod accesses the planar rod-GVG (and hence the rod-HGVG) via two gradient ascent operations: while maintaining a fixed orientation, the rod moves away from its closest obstacle and then while maintaining double equidistance and a fixed orientation, the rod moves away from the two closest obstacles until it reaches triple equidistance. Note that $\theta(q) = \theta(H(q, t))$ for all $t \in [0, 1]$, i.e., the rod arrives to the rod-GVG with a fixed orientation.

In order to "make" H continuous, we divide the configuration space $R^2 \times S^1$ into junction regions J_{ijk} where each junction region J_{ijk} is the pre-image of CF_{ijk} under H. This essentially guarantees that CF_{ijk} is a retract of J_{ijk}. Finally, note that if the rod is "small" enough, CF_{ijk} has one connected component and is diffeomorphic to S^1.

For each connected component of a junction region J_{ijk}, there is a connected component of CF_{ijk}. If motion planning should occur only in one junction region, then planning is trivial because CF_{ijk} is a retract of J_{ijk}. If the union of the CF_{ijk}'s formed a connected set, then planning is trivial again. In general, the CF_{ijk}'s will not form a connected set in $\mathbb{R}^2 \times S^1$, so we use the point-GVG to connect the CF_{ijk}'s of the junction regions.

The connections between junction regions are the rod configurations tangent to the point GVG, i.e., the R-edges.

Essentially, we are forming an exact cellular decomposition of $R^2 \times S^1$ where the junction regions are the cells. Planning in a cell is achieved with the rod-GVG edges, CF_{ijk}, and planning between the cells is achieved with the R-edges. We take an analogous approach to defining the rod-HGVG for the rod in three dimensions.

2 Rod-HGVG in $\mathbb{R}^3 \times S^2$, based on the point-GVG

In this section, we define the rod-HGVG. Since the rod's configuration space $\mathbb{R}^3 \times S^2$ has five dimensions, it is natural to first define a five-way equidistant structure which we term *rod-GVG edges*. Just like the planar rod-GVG, the three-dimensional rod-GVG (henceforth called the rod-GVG), is not necessarily connected. Just like the planar case, we decompose $\mathbb{R}^3 \times S^2$ into cells, also called *junction regions*, and "connect" them with *1-tangent edges*, structures that are analogous to R-edges. However, the rod-GVG edges by themselves are not retracts of the junction regions. Instead, the set of configurations equidistant to *four* obstacles forms a retract of a junction region. This four-way equidistant structure has two dimensions, and thus this structure, with the 1-tangent edges and rod-GVG edges do not form a roadmap. Therefore, we define an additional structures called *2-tangent edges*, which are four-way equidistant with an additional constraint. The rod-HGVG comprises rod-GVG edges, 1-tangent edges, and 2-tangent edges. This section formally defines these structures and the next two sections establish that the rod-HGVG is a roadmap.

2.1 Rod-GVG Edges

The rod-GVG is the set of configurations that are equidistant to five obstacles. The definition of the distance function in $\mathbb{R}^3 \times S^2$ is identical to that in $SE(2)$, i.e., $D_i(q) = \min_{r \in R(q), c \in C_i} \|r - c\|$, but the rod configuration q is parametrized by $q = (x, y, z, \theta, \varphi)^T$. Note that $\theta(q)$ and $\varphi(q)$ define the orientation of the rod. The rod gradient is derived in Appendix A. With distance and its gradient defined, we can define rod-equidistant faces, continuing from where we left off in the planar case. The *rod-four-equidistant face* is

$$CF_{ijkl} = CF_{ijk} \cap CF_{ikl} \cap CF_{jkl}.$$

Then the *rod-five-equidistant face*, which is a *rod-GVG edge* in the three-dimensional case, is defined as

$$CF_{ijklm} = CF_{ijkl} \cap CF_{iklm} \cap CF_{jklm}.$$

A *rod-meet point* is then CF_{ijklmn}, the zero-dimensional set of rod configurations that are six-way equidistant.

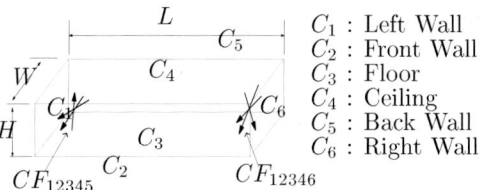

Fig. 3. Here, there are two set of rod configurations that are five-way equidistant. The configurations in CF_{12345} are equidistant to C_1, C_2, C_3, C_4 and C_5, and the configurations in CF_{12346} are equidistant to C_1, C_2, C_3, C_4 and C_6. Note that they are not connected to each other.

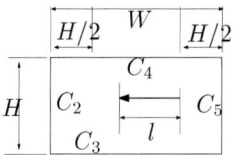

Fig. 4. Side view of Fig. 3 with $L > W, L > H$, but here, the length of the rod is smaller than $W - H$, thus, the rod cannot be four-way equidistant to C_2, C_3, C_4 and C_5. Therefore, five-way equidistant configurations do not exist either.

The rod-GVG edges can be generated by tracing the roots of the following function

$$G(q) = \begin{pmatrix} (D_i - D_j)(q) \\ (D_i - D_k)(q) \\ (D_i - D_l)(q) \\ (D_i - D_m)(q) \end{pmatrix}.$$

In the rod configuration space $\mathbb{R}^3 \times S^2$, the collection of rod-GVG edges is not necessarily connected, and in many environments, it may not even exist at all. Figure 3 shows a rectangular environment, where $L > W > H$ and CF_{12345} and CF_{12346} are the rod-GVG edges. If L is substantially larger than the length of the rod, there cannot be any six-way equidistant configurations, so these two component cannot be connected. Now, if W is also substantially larger than the length of the rod, the five-way equidistant configuration cannot exist (Figure 4).

2.2 1-tangent edges

The rod-GVG defined above will not necessarily be connected. Just like the rod-GVG in the planar case, we use the point-GVG to define additional structures that connect disconnected rod-GVG edge. These structures are similar to the R-edges where the rod is "tangent" to the point-GVG edge. More specifically, when C_i, C_j and C_k define a point-GVG edge F_{ijk}, we presume that

- The rod is "tangent" to F_{ijk} at r.
- At r, $d_n(r) < d_n(r_o)$ for all other points r_o on the rod, and for all three closest obstacles $C_n, n = i, j, k$, i.e, the point r is the closest to all three ob-

stacles.

The second condition asserts that $r_i = r_j = r_k$, where r_i, r_j and r_k are the closest points on the rod to each obstacles [3].

Instead of calling this an R-edge. we term this structure the *1-tangent edge* because it is tangent to a one-dimensional structure from the point-GVG. We can use the same technique as in the planar case to construct R-edge to construct this 1-tangent edge. The rod finishes the tracing a 1-tangent edge when it (i) reaches a four-way equidistant configuration, (ii) reaches an obstacle boundary, or (iii) detects the cycle. A cycle in the GVG is a disconnected edge diffeomorphic to S^1. Note that we consider only environments where the point-GVG is connected, and thus the condition (iii) will not occur.

2.3 2-tangent edges

Consider two 1-tangent edges that terminate at configurations equidistant to the same four obstacles. These two end point configurations cannot coincide with each other because we assume that the point-equidistant faces transversally intersect each other. In other words, two 1-tangent edges cannot intersect each other. The set of four-way equidistant configurations is two-dimensional and thus we need an additional constraint to define a one-dimensional structure that connects the two 1-tangent edges.

Consider the rod-four-equidistant face CF_{ijkl} with 1-tangent edge R_{ijk} and R_{ijl} each terminating at configurations in CF_{ijkl} (boundedness of the workspace assures this can happen). For R_{ijk}, the rod lies in the tangent space associated with F_{ijk}; restated, the rod simultaneously lies in the tangent spaces of F_{ij}, F_{ik}, and F_{jk}. Likewise, for R_{ijl}, the rod lies in tangent spaces F_{ij}, F_{il}, and F_{jl}. Note that in both cases, the rod lies in the tangent space associated with F_{ij}. Therefore, as our additional constraint, to travel from R_{ijk} to R_{ijl} along CF_{ijkl}, the rod must remain in the tangent space of F_{ij}. The edge formed on CF_{ijkl} with the additional constraint of staying in the tangent space of F_{ij} is termed a *2-tangent-edge* because the rod lies in a two-dimensional tangent space (as opposed to a one-dimensional tangent space with the 1-tangent edge). Note that for CF_{ijkl}, there are four 1-tangent edges that terminate on the set, and there can be six different 1-tangent edges on it.

As an example, the Figure 5 shows the case where F_{ij} is a plane. Here, the rod is "tangent" to $F_{ceiling, floor}$ when the rod is moving from q_2 to q_3, and "tangent" to $F_{front, floor}$ when moving from q_1 to q_2. Generally, when the rod is tracing four equidistant configuration from a configuration tangent to F_{ijk} to a configuration tangent to F_{jkl}, it must be in tangent space of F_{kl}.

Now we describe the tangent condition for 2-tangent edge $R_{ij/kl}$ more specifically, and demonstrate it is one-

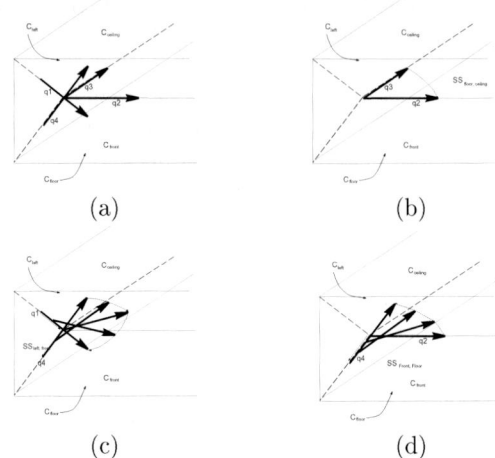

Fig. 5. In figure (a), q_1, q_2, q_3 and q_4 represent four-way equidistant configurations. Also they are terminal configurations. of different 1-tangent edges. Figures (b), (c) and (d) shows 2-tangent edges that connect some pairs of 1-tangent edges.

dimensional. First we need to define some additional structures. We define the set of rod configurations q in CF_{ij} that satisfy the conditions (i) q is tangent to F_{ij} (ii) $r_i = r_j$, where r_i and r_j are the points on the rod closest to each obstacles C_i and C_j as RC_{ij}, i.e.,

$$RC_{ij} = \{q \in cl(RF_{ij})) : r \in F_{ij} \text{ and }$$
$$(i)\ 0 \leq d_i(r) \leq d_i(r_1)\ \forall r_1 \in R(q) \text{ and}$$
$$(ii)\ d_i(r) \leq D_h(q)\ \forall h \neq i, j\},.$$

where, RF_{ij} is the set of planes tangent to F_{ij} (which is analogous to a tangent bundle of F_{ij}). We form the 2-tangent edge $R_{ij/kl}$ by intersecting RC_{ij} and CF_{ijkl}, i.e., $R_{ij/ik} = RC_{ij} \cap CF_{ijkl}$.

Note that $RC_{ij} \cap CF_{ijkl} = RC_{ij} \cap CF_{jkl}$ As shown in Appendix B, RC_{ij} is three dimensional. Since CF_{ikl} and RC_{ij} are each three dimensional, $R_{ij/kl}$ is generally one-dimensional by the pre-image theorem.

The $R_{ij/kl}$ can be constructed by tracing the roots of

$$G(q) = \begin{pmatrix} (D_i - D_j)(q) \\ (D_i - D_k)(q) \\ (D_i - D_l)(q) \\ (r_i - r_j)(q) \end{pmatrix}$$

Here, the first three elements ensures the four-way equidistance. The fourth element forces the two closest points on the rod to C_i and C_j to be the same and thus specifies the orientation of the rod. Since sensors can easily provide distance information and determine the closest points on the rod to the closest obstacles, the 2-tangent edge can be readily constructed without *a prior* information.

3 Accessibility

The rod accesses the rod-GVG (and hence the rod-HGVG) via four gradient ascent operations: the first three use a fixed orientation gradient $\tilde{\nabla} D_i(x)$ that directs the rod to increase distance to an object C_i while maintaining its orientation; the last gradient ascent operation uses the full gradient $\nabla D_i(x)$ which is derived in the Appendix. The first three gradient ascent operations implicitly define a function $H^3 : \mathcal{FS} \times [0,1] \to CF_{ijkl}$ that is analogous to the H mapping which enables a rod to access the planar rod-GVG. First, while maintaining a fixed orientation, the rod moves away from its closest obstacle (i.e., $\dot{c}(t) = \tilde{\nabla} D_i(c(t))$), then while maintaining double equidistance and a fixed orientation, the rod moves away from the two closest obstacles until it reaches triple equidistance (i.e., $\dot{c}(t) = \pi_{T_{c(t)}CF_{ij}} \tilde{\nabla} D_i(c(t))$, where $\pi_{T_{c(t)}CF_{ij}}$ is the projection operator), and then while maintaining triple equidistance and a fixed orientation, the rod moves away from the three closest obstacles until it reaches four-way equidistance (i.e., $\dot{c}(t) = \pi_{T_{c(t)}CF_{ijk}} \tilde{\nabla} D_i(c(t))$).

Finally, we define another mapping $HJ : CF_{ijkl} \to R_{ij/kl} \bigcup CF_{ijklm}$ that moves the rod away from the four closest obstacles, using the full gradient,

$$\dot{c}(t) = \pi_{T_{c(t)}CF_{ijkl}} \nabla D_i(c(t))$$

This step terminates at either of the following : (i) five-way equidistance, i.e., a rod-GVG edge, or (ii) a 2-tangent edge (proof appears in Appendix C). Once the rod has accessed the rod-HGVG, it can then begin incrementally constructing the rod-HGVG using the previously defined numerical techniques. If the rod-HGVG is connected, then numerically constructing it ensures complete exploration of a connected component of the rod's configuration space.

4 Connectivity

We will demonstrate that if the point-GVG is connected, then there exists a path between two rod configurations $q_1, q_2 \in \mathbb{R}^3 \times S^2$ if and only if there exists a path between two rod configurations q_1^*, q_2^* on the rod-HGVG in $\mathbb{R}^3 \times S^2$. Note that q_1^*, q_2^* are the "projected" configurations onto the rod-HGVG from q_1, q_2 via a mapping $HJ \circ H^3 : \mathbb{R}^3 \times S^2 \times [0,1] \to$ rod-HGVG.

The H^3 mapping is analogous to a retraction and is defined in a similar fashion as H in the planar case. In $\mathbb{R}^3 \times S^2$, we will take a similar approach as the planar case to "make" H^3 continuous; here, we define junction regions that are J_{ijkl} and the connections among them are the 1-tangent edges R_{ijk}, the set of rod configurations tangent to the point-GVG edge CF_{ijk}.

A junction region J_{ijkl} is the pre-image of CF_{ijkl} under $H^3 : \mathbb{R}^3 \times S^2 \times [0,1] \to CF_{ijkl}$. Again, H^3 is implicitly defined by a sequence of three fixed-orientation gradient ascent operations. Note that $\theta(q) = \theta(H^3(q,1))$ and $\phi(q) = \phi(H^3(q,1))$ and using a similar approach as in [3], H^3 can be shown to be continuous in each junction region J_{ijkl}. Therefore, CF_{ijkl} is a two-dimensional retract of J_{ijkl}. In fact, if the rod is "small" enough, CF_{ijkl} is diffeomorphic to a two-sphere, S^2. At this point, we can infer that if the point-GVG is connected in $\mathbb{R}^3$, then there exists a path between q_1 and q_2 in $\mathbb{R}^3 \times S^2$ if and only if there exists a path between $H^3(q_1,1)$ and $H^3(q_2,1)$ in the union of 1-tangent edges R_{ijk} and rod-four-equidistant faces CF_{ijkl}.

We have not defined a roadmap yet because CF_{ijkl} has two dimensions, not one. To define a one-dimensional structure on CF_{ijkl}, we define another mapping $HJ : CF_{ijkl} \to R_{ij/kl} \bigcup CF_{ijklm}$. This function is also implicitly defined via a gradient ascent operation. From Section 3, we showed that once the rod achieves four-way equidistance, continued gradient ascent (full gradient ascent) brings the rod either to a two-tangent edge $R_{ij/kl}$ or a rod-GVG edge CF_{ijklm}.

If the rod is "small" enough, then this last gradient ascent operation brings the rod only to a two-tangent edge. If the rod is "large" enough, the two-spheres CF_{ijkl} and CF_{ijkm} intersect and form an edge CF_{ijklm} and gradient ascent could bring the rod to a five-way equidistant configuration (but can also bring the rod to a 2-tangent edge). If the rod is even larger (or obstacles are appropriately shaped), a rod-meet-point occurs when three two-sphere intersect.

For the sake of discussion, lets assume the rod is "small" enough so no two-sphere intersect with each other. In other words, lets assume there are no rod-GVG edges and rod-meet-points on the rod-four-equidistant faces. If HJ were continuous on a rod-four-equidistant face, then the two-tangent edges would form a retract of the rod-four-equidistant faces and we are done.

Alas, this is not the case. Instead, we define another cellular decomposition on CF_{ijkl} where each cell is denoted $J_{ij/kl}$ and is defined by the pre-image of $R_{ij/kl}$ under the HJ mapping. For each cell $J_{ij/kl}$, HJ is continuous and therefore $R_{ij/kl}$ is a one-dimensional retract of $J_{ij/kl}$.

The remaining challenge is to establish that the two-tangent edges "link up" properly. This is easily shown because each end point of the 1-tangent edges coincides with the end point of three different 2-tangent edges. Consider the 1-tangent edge R_{ijk}; it has two types of end points: a boundary configuration and a four-way equidistant configuration (i.e., equidistant to C_l in addition to C_i, C_j and C_k.) Consider the end point that is four-way equidistant. By definition of the 1-tangent edge, its end point is tangent to the point-GVG edge F_{ijk}, which means that the end point is also tangent to

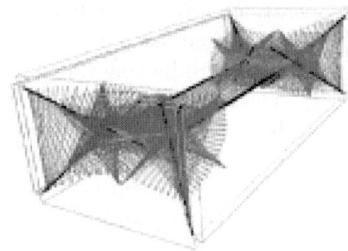

Fig. 6. Rod-HGVG in a rectangular box.

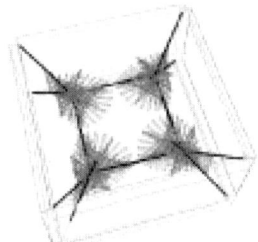

Fig. 7. Rod-HGVG in an square box. In this example the rod-HGVG does not exist, and junction regions are connected by 1-tangent edges

the point-two-equidistant faces F_{ij}, F_{jk} and F_{ik}. Therefore, this end point also belongs to the 2-tangent edges $R_{ij/kl}$, $R_{jk/il}$ and $R_{ik/jl}$. This shows that three of the six 2-tangent edges connect. By repeating this argument for the other combination of 1-tangent edges and 2-tangent edges, one can easily see how all of the 2-tangent edges link up.

5 Simulation Result

We performed a computer simulation using the algorithm described in this paper. Figures 6 and 7 shows the complete rod-HGVG in simple environments. For more simulation results, see http://voronoi.sbp.ri.cmu.edu.

6 Conclusion

This paper introduces a roadmap called *rod hierarchical generalized Voronoi graph* for a rod-shaped robot operating in a three-dimensional space. The rod-HGVG is defined in terms of workspace distance information, which sensors can easily provide. Using work space distance information, we can prescribe a sensor-based incremental method to achieve motion planning without constructing the configuration space. This is important for sensor based planning because we cannot construct configuration space without knowing the workspace first. Moreover, even when we have configuration space representation, it is still difficult to measure the distance in configuration space than in workspace. The rod-HGVG has three components : (i) rod-GVG edges, which are five-way equidistant, (ii) 2-tangent edges, which are four-way equidistant and (iii) 1-tangent edges, which are three-way equidistant.

The first contribution of this paper is the piece-wise retract: we defined an exact cellular decomposition of the rod's configuration space where in each cell, called a junction region, we defined a retract that makes motion planning in the junction region trivial. We then use the point-GVG to connect the retracts of each junction region, which is the second contribution of this paper. We were able to use connectivity of a structure defined in the robot's workspace to infer connectivity properties of the robot's configuration space. In other words, when connectivity of the point-GVG represents the connectivity of the works space, then connectivity of the rod-HGVG represents the connectivity of the rod's configuration space.

In general the point-GVG is not connected in $\mathbb{R}^3$, so future work will use the point-HGVG to connect the junction regions. We started with the point-GVG only to achieve a tractable subgoal. We also defined a piece-wise retract on CF_{ijkl} to guarantee connectivity of the rod-HGVG. This relied on HJ being continuous, which due to space limitations, was not detailed in this paper; a rigorous proof should be demonstrated.

Ideally, we would like to demonstrate this on a robot blimp. This work is a step towards the ultimate goal of sensor based planning for an articulated multi-body robot. The next step is to consider two rod robots, and then an n-rod robot.

References

[1] J.F. Canny. *The Complexity of Robot Motion Planning*. MIT Press, Cambridge, MA, 1988.

[2] H. Choset and J. Burdick. Sensor based motion planning: The hierarchical generalized voronoi graph. *International Journal of Robotics Research*, 19(2):96–125, February 2000.

[3] H. Choset and J.W.Burdick. Sensor Based Planning for a Planar Rod Robot. In *Proc. IEEE Int. Conf. on Robotics and Automation*, 1996.

[4] J. Cox and C.K. Yap. On-line Motion Planning: Case of a Planar Rod. In *Annals of Mathematics and Artificial Intelligence*, volume 3, pages 1–20, 1991.

[5] Jean-Claude Latombe, Lydia E. Kavraki, Peter Svestka, and Mark Overmars. Probablistic roadmaps for path planning in high dimensional configuration spaces. *IEEE Transactions on Robotics and Automation*, 12(4):566–580, 1996.

[6] C. Ó'Dúnlaing, M. Sharir, and C.K. Yap. Generalized Voronoi Diagrams for Moving a Ladder. I: Topological Analysis. *Communications on Pure and Applied Mathematics*, 39:423–483, 1986.

[7] C. Ó'Dúnlaing and C.K. Yap. A "Retraction" Method for Planning the Motion of a Disc. *Algorithmica*, 6:104–111, 1985.

[8] E. Rimon and J.F. Canny. Construction of C-space Roadmaps Using Local Sensory Data — What Should the Sensors Look For? In *Proc. IEEE Int. Conf. on Robotics and Automation*, pages 117–124, San Diego, CA, 1994.

[9] Steven A. Wilmarth, Nancy M. Amato, and Peter F. Stiller. Motion planning for a rigid body using random networks on

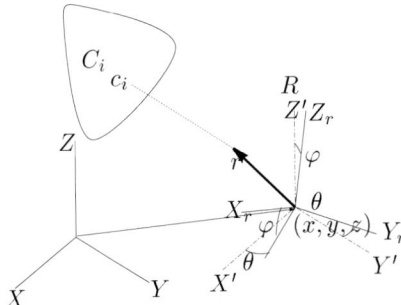

Fig. 8. World and Body coordinate system for the Rod

the medial axis of the free space. In *Proceedings of the 15th Annual ACM Symposium on Computational Geometry*, pages 173–180, June 1999.

Appendix
A Rod Gradient

Here, we derive the rod distance gradient. In Fig. 8, X, Y and Z denote the world coordinate frame, and X_r, Y_r and Z_r denote the body fixed coordinate frame on R (the rod). Let $(x, y, z)^T$ be the origin of the body fixed coordinate frame, and θ, φ be the orientation of the body fixed coordinate frame with respect to world coordinate frame. Let c be the closest point on the obstacle C_i to the robot and r be the closest point to C_i on the robot. Finally let $(a, b, c)^T$ be the coordinate of r in body fixed coordinate frame. Then the world coordinate of r is

$$r = \begin{bmatrix} x + a\cos\theta\cos\varphi - b\sin\theta + c\cos\theta\sin\varphi \\ y + a\sin\theta\cos\varphi + b\cos\theta + c\sin\theta\sin\varphi \\ z - a\sin\varphi + c\cos\varphi \end{bmatrix}$$

Then, following similar steps to the two dimensional case [3], the first three components of the distance gradient are

$$[\frac{\partial D}{\partial x}, \frac{\partial D}{\partial y}, \frac{\partial D}{\partial z}] = \frac{1}{D_i(q)}[(r_x - c_x), (r_y - c_y), (r_z - c_z)]$$

The remaining two rotational components are

$$\frac{\partial D}{\partial \theta} = \frac{1}{D_i(q)}\left\langle \begin{bmatrix} c_x - r_x \\ c_y - r_y \\ c_z - r_z \end{bmatrix}, \begin{bmatrix} aS\theta C\varphi + bC\theta + cS\theta S\varphi \\ -aC\theta C\varphi + bS\theta - cC\theta S\varphi \\ 0 \end{bmatrix} \right\rangle$$

$$\frac{\partial D}{\partial \varphi} = \frac{1}{D_i(q)}\left\langle \begin{bmatrix} c_x - r_x \\ c_y - r_y \\ c_z - r_z \end{bmatrix}, \begin{bmatrix} aC\theta S\varphi - cC\theta C\varphi \\ aS\theta S\varphi + cS\theta C\varphi \\ aC\varphi + cC\varphi \end{bmatrix} \right\rangle$$

B Dimension of RC_{ij}

In this section, we show that RC_{ij} is three-dimensional. Note that for any rod configuration $q \in CF_{ij}$, $R(q) \cap F_{ij}$ is not empty, and without loss of generality, we assume that $R(q) \cap F_{ij}$ is zero dimensional, i.e., a single point. For rod configurations in RC_{ij}, $r_i = r_j = R(q) \cap F_{ij}$. We call that point the "contact point" and simply denote it as r. We define the *constant distance curve* $cv_{ij}(\Omega)$ on F_{ij} as the set of all points on F_{ij} that have distance Ω to C_i, i.e.

$$cv_{ij}(\Omega) = \{x \in F_{ij} : d_i(x) = \Omega\}.$$

Now we consider the contact condition and dimensionality of RC_{ij}, in two separate cases, depending on the contact point r for each rod configurations: (I) $d_i(r)$ is at local mininum of the distance on F_{ij} to C_i on F_{ij}. (II) otherwise.

LEMMA B.1 *For rod configurations in CF_{ij} that are tangent to F_{ij} and $r_i = r_j$, such that $d_i(r_i)$ is not a local minimum of the distance (i.e., Case II above), (a) the contact point r must be either an end point P or Q, or (b) the rod must be in the tangent space of $cv_{ij}(d_i(r))$ and the contact point can be an interior point of the rod. Also, when the rod satisfies case (II) above, the rod has one degree of freedom in this set when the contact point is fixed.*

Proof: Note that, in a neighborhood about the origin of $T_r cv_{ij}$, $T_r cv_{ij}(d_i(r))$ separates the $T_r F_{ij}$ into two regions, such that all the points in one region have distance less than $d_i(r)$ to C_i, and the points in the other region have greater distance. for case (a), by hypothesis, the rod lies in a line that transversally intersects $T_r cv_{ij}(d_i(r))$. So, if $r \neq P, Q$, the rod itself intersects $T_r cv_{ij}(d_i(r))$ in its interior. So loosely speaking, P and Q are on opposite sides of $T_r cv_{ij}(d_i(r))$.

Let's define a set $L_i(\Omega)$, such that

$$L_i(\Omega) = \{x \in \mathbb{R}^3 : d_i(x) = \Omega\}.$$

This set is a two dimensional manifold in $\mathbb{R}^3$. Note that $cv_{ij}(d_i(r))$ can be seen as the intersection of $L_i(d_i(r))$ and F_{ij}. Because the rod intersects $T_r cv_{ij}(d_i(r))$ at r and r is on $L_i(d_i(r))$, the rod must "poke" into $L_i(d_i(r))$. This contradicts the condition that the point r is the closest point on the rod to each obstacles. So the rod cannot intersect the $cv_{ij}(d_i(r))$ in its interior, i.e., the intersection must occur at a rod's end point P or Q, when the intersection in transversal.

If the rod "touches" cv_{ij} at P or Q (case (a)), the rod cannot translate, but can rotate around the contact point as long as it "stays" on the tangent space $T_r F_{ij}$. When the rod is "tangent" to cv_{ij} (case (b)), the rod cannot change its rotation, but translate along its length. Therefore, when the contact point is fixed on F_{ij}, the rod has one degree of freedom in each cases. Because F_{ij} is two dimensional, the rod has three degrees of freedom in RC_{ij}. Figure 9 describe this case. ∎

LEMMA B.2 *For case (i) above, the contact point can be one of the interior points on the rod, and the rod has two degrees of freedom at each configuration in this set.*

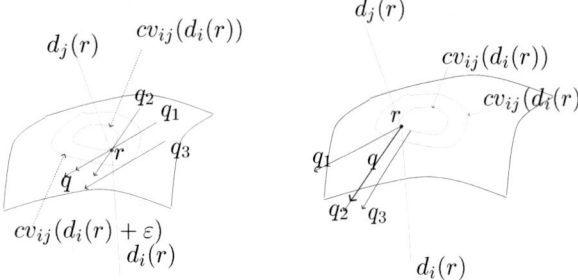

Fig. 9. Figure(a) : The rod configuration q touches F_{ij} at r, which is not a local minimum, and the rod is also tangent to $cv_{ij}(d_i(r))$. The three degrees of freedom that rod has are: (i) it can slide along its length (q_1) (ii) it can move "on" $cv_{ij}(d_i(r))$, maintaining the tangent condition to it (q_2). and (iii) it can slide normal to its length and become tangent to $cv_{ij}(d_i(r)+\varepsilon)$ (q_3). Figure(b) : The rod configuration q touches F_{ij} at r, which is not a local minimum, and the rod touches $cv_{ij}(d_i(r))$ at Q. The three degrees of freedom that the rod has are: (i) it can rotate around Q on the tangent plane (q_1). (ii) it can move "on" $cv_{ij}(d_i(r))$ (q_3) and (iii) it can move on F_{ij}, to $cv_{ij}(d_i(r))+\varepsilon(q_3)$.

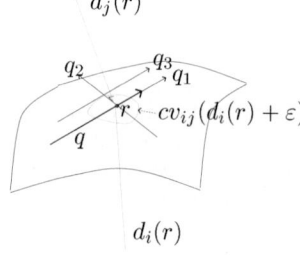

Fig. 10. When the rod touches F_{ij} at local minimum r, the rod has three degrees of freedom in the set R_{ij}. q is the original configuration. Then it can slide along its length (q_1), it can rotate around r (q_2) and it can slide without rotation so that it becomes tangent to $cv_{ij}(d_i(r)+\varepsilon)$ (q_3).

Proof: Let's call the set of rod configurations that satisfy case (i) as R_{ij}^1. This is a subset of RC_{ij}. In this case, $L_i(d_i(r)) \cap F_{ij}$ is a single point, which means that these two surfaces contact at a single point r, and share a common tangent plane at that point. Then, while the rod satisfies this condition, the rod is free to move on that plane, as long as $r \in R(q)$. That means that the rod has two degrees of freedom in this set R_{ij}^1, i.e. this set has two degrees of freedom. Because this set R_{ij}^1 is subset of R_{ij}, the rod configuration in the set R_{ij}^1 has at least two degrees of freedom in the set R_{ij}. Now, let's consider whether there are any additional degrees of freedom for the rod in the set R_{ij}. Because the dimension of the set R_{ij}^1 is two, if the rod has any additional degrees of freedom in R_{ij}, the movements in those directions will move the rod outside of the set R_{ij}^1, but not outside of the R_{ij}. And of course those movements must be independent from the movement we already considered. Recall that when the rod configuration belongs to $R_{ij} - R_{ij}^1$, the rod must be "tangent" to $cv_{ij}(d)$ for some value of d. Now if the rod moves outside the set R_{ij}^1, the distances to C_i and C_j increase. Let's denote that the new distance as $d_i(r)+\varepsilon$. Then, from the continuity of the distance function the $cv_{ij}(d_i(r)+\varepsilon)$ is a closed curve around the point r. The rod must be "tangent" to this curve, and there are only two point on that curve where the tangent is parallel to the rod. That means that the rod has (only) one more additional degree of freedom (which moves the rod outside of the R_{ij}^1, thus makes the rod lost contact at r, but still the rod stays in R_{ij}) at the given configuration in R_{ij}^1. Thus the rod has three degrees of freedom in this case. (Figure 10) ∎

C Accessibility Proof

In this section, we demonstrate that a rod starting at a four-way equidistant configuration will access either (i) a rod-GVG edge or (ii) a 2-tangent edge via full gradient ascent of distance to the four closest obstacles. Proving case (i) is trivial. Now, we prove case (ii), i.e., the gradient ascent will eventually direct the rod onto the $R_{ij/kl}$ when there are no five-way equidistant configurations. Figure 11(a) shows a rod configuration which intersects SS_{lk} nontraversally. Consider the plane formed by the rod vector PQ and the vector $r_k c_k$. Then let the global coordinate frame be located such that (i) the origin is located on P, (ii) the X and Y axes are on the plane formed by PQ and $r_k c_k$. Then the Z axis of the global coordinate frame is perpendicular to that frame. In this frame, the rod configuration is at $(0,0,0,0,0)$ and the body fixed coordinate of the r is $(a,0,0)$ with some positive value a. Note that in this coordinate system, the θ component of the gradient rotates the rod "in" the plane, ane the φ component of the gradient rotates the rod out of the plane. We interested in the θ component, which for the obstacle C_k is $\frac{\partial D_k}{\partial \theta} = \frac{1}{D_i(q)} \langle [c_k^x - r_k^x, c_k^y - r_k^y, 0]^T, [a_k \sin\theta, -a_k \cos\theta, 0]^T \rangle$. Then, the two vectors $(c_k^x - r_k^x, c_k^y - r_k^y, 0)^T$ and $(a_k \sin\theta, -a_k \cos\theta, 0)^T$ are parallel. So $\partial D_k / \partial \theta$ has positive value, and that means that the angle between the tangent plane on SS_{kl} and the rod increases as a result of gradient ascent. Now do the same analysis with $r_l c_l$; the vector $r_l c_l$ is in the "opposite" direction of $(a_k \sin\theta, -a_k \cos\theta, 0)^T$ and the coordinate value of a_l is larger because r_l is on the other side of SS_{kl}. This means that the gradient for C_l has a negative component of $\frac{\partial D_l}{\partial \theta}$ on plane defined by PQ and $r_l c_l$, with larger value than $\frac{\partial D_k}{\partial \theta}$, i.e., $\frac{\partial D_l}{\partial \theta} > \frac{\partial D_k}{\partial \theta}$. This decreases the angle between the tangent plane and the rod. Thus, the angle between the tangent plane and the rod will decrease during the gradient ascent, forcing the rod to become tangent.

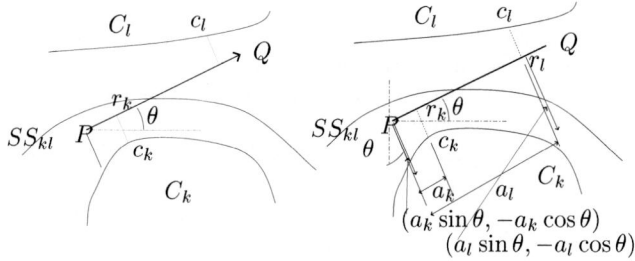

Fig. 11. The rod is two-way equidistant to obstacles C_k and C_l. The rod intersects $SS_k l$ transvesally. In second figure, the rod is shown on the plane defined by PQ and $r_k c_k$. Also in this figure $r_l c_l$ is on the same plane, but in general it will be on different plane.

Actor-Q Based Active Perception Learning System

Katsunari Shibata†, Tetsuo Nishino‡ & Yoichi Okabe‡

†Dept. of Electrical & Electronics Engineering, Oita Univ., 700 Dannoharu, Oita 870-1192, JAPAN
‡Res. Ctr. for Advanced Sci. & Tech., The Univ. of Tokyo, 5-3-1 Komaba, Meguro-ku, Tokyo 153-0041, Japan
shibata@cc.oita-u.ac.jp, okabe@okabe.rcast.u-tokyo.ac.jp

Abstract

An active perception learning system based on reinforcement learning is proposed. A novel reinforcement architecture, called Actor-Q, is employed in which Q-learning and Actor-Critic are combined. The system decides its actions according to Q-values. One of the actions is to move its sensor, and the others are to make an answer of its recognition result, each of which corresponds to each pattern. When the sensor motion is selected, the sensor moves according to the actor's output signals. The Q-value for the sensor motion is trained by Q-learning, and the Actor is trained by the Q-value for the sensor motion on behalf of the critic. When one of the other actions is selected, the system outputs the recognition result. When the recognition answer is correct, the Q-value is trained to be the upper limit of the Q-value, and when the answer is not correct, it is trained to be 0.0. The module to compute Q-value and the actor module are both consisted of a neural network, and are trained by Error Back Propagation. The training signals are generated based on the above reinforcement learning.

It was confirmed by some simulations using a visual sensor with non-uniform visual cells that the system moves its sensor to the place where it can recognize the presented pattern correctly. Even though the Q-value surface as a function of the sensor location has some local peaks, the sensor was not trapped and moved to the appropriate direction because the Q-value for the sensor motion becomes larger.

Key Words: Actor-Q Architecture, Reinforcement Learning, Neural Network, Active Perception, Visual Sensor

1 Introduction

Our living creatures obtain a variety of informations about the environment through our sensors, and utilize the information to generate our appropriate actions. However, the information of the environment is too huge, it is very inefficient to obtain all the detailed information. To solve this problem, we can move our sensors actively and obtain necessary information effectively. It is called "active perception".

When our visual sensor, which obtains the largest amount of information among our sensors, is observed, the distribution of the visual cells is non-uniform on the retina. We take a general view of the environment or target object by using whole the sensor, then move the dense part of the sensor to the appropriate place, and finally recognize the target correctly. The knowledge that tells us which part of the target should we focus on, cannot be thought inherited, but it is obtained by learning after our birth.

Actually, the system, in which a visual sensor moves, has been developed[1]. However, the system does not learn where its attention should be moved for appropriate recognition, but the main purpose is the pursuit of a moving object, and the captured image is processed by a given way.

On the other hand, reinforcement learning has been focused recently by its autonomous, adaptive, purposive learning. Mainly it is utilized to learn the action planning, but it is expected to be utilized to learn the whole process from sensors to motors including recognition, attention, and so on by employing neural networks[2]. The sensory signals are put into the neural network directly and the output signals are dealt as the motion commands.

It has been tried that the reinforcement learning is applied to active perception systems. In the system by Whitehead et al.[3], the block relocation task is employed, and "which block the attention frame should be moved to" is trained by Q-learning[4], Though the target block of the attention is selected using the Q-values, the motion of the sensor was not considered. Further, the recognition is not dealt with explicitly.

In the system by Shibata et al., visual sensory signals are put into a layered neural network directly, and the neural network generates the motion commands for the visual sensor and the recognition results[5]. However, there are three problems. At first, the reinforcement signal, which is a continuous scalar signal representing how the recognition outputs are close to the ideal ones, has to be given at every time step when the visual sensor makes a step motion. Secondly, the sensor is sometimes trapped at the place where the value function has a local peak, and the system makes a mistake. Third one is that the system does not make a recognition answer at the moment when it becomes to be able to recognize, but the timing when the system makes a recognition answer is fixed.

In this paper, an architecture is proposed not to

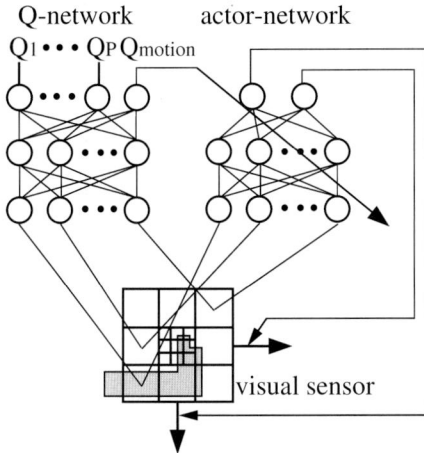

Figure 1: The active perception learning system based on Actor-Q architecture proposed in this paper.

be trapped at a local maximum on the value function surface, and to reach the global maximum. By this architecture, the timing to output the recognition result is also obtained through learning. The evaluation of the recognition result is not required at every time step, but the binary reinforcement signal, a correct (reward) or mistake (no reward), is given to the system only when the system outputs the result. That is similar to the experiments to examine the recognition ability of monkeys.

2 Actor-Q Architecture

Fig. 1 shows the active perception learning system based on Actor-Q architecture that is proposed in this paper. There are two layered neural networks, Q-network and actor-network, and the input of both networks are visual sensory signals. Because the input signals are the same, it is possible that only one neural network makes a role of the both. The outputs of the first network are used as Q-values. One of them is Q-value for the sensor motion and each of the others is Q-value for recognition of each pattern. This means that making a recognition output is considered as one action as well as moving the sensor, and one action is selected using these Q-values.

When it is selected to move the visual sensor, the sensor is moved according to the outputs of the actor network. The two outputs are used as the velocity of the sensor in the direction of x and y respectively. After the sensor motion, a new image caught by the visual sensor is put into the neural networks again, and an action is selected again according to the new Q-values. The Q-value for the sensor motion is trained by popular one-step Q-learning[4]. The training signal is computed as

$$Q_{training}(s(t), motion) = \gamma \max_a Q(s(t+1), a), \quad (1)$$

where γ: a discount factor. The neural network is trained by Error Back Propagation, but the other outputs of the Q-network are not trained. Note that the transformation between the output of the network and Q-value is necessary.

The velocity of the sensor along each of x and y axis is decided by the sum of the output $\mathbf{o}_m$ and the random number $\mathbf{rnd}$ as

$$\mathbf{m} = \beta(\mathbf{o}_m + \mathbf{rnd}), \quad (2)$$

where β: a constant. The actor-network is trained by the training signal as

$$O_{m,training} = O_m + \mathbf{rnd}(\gamma \max_a Q(s(t+1), a) - \max_a Q(s(t), a). \quad (3)$$

While in the actor-critic architecture[6], it is trained according to the change of the critic output.

When one of the other actions, that means one of the recognition outputs is selected, the recognition output is evaluated whether it is correct or not, and the trial finishes. If the output is correct, the corresponding output of the Q-network is trained to be $Q = 1.0$, and if not correct, it is trained to be 0.0. This corresponds that some juice is given to the monkey when it makes a correct answer, and some penalty is given when it makes a mistake. The flow chart of this learning is shown in Fig. 2.

3 Simulation

3.1 Task Setting

The visual sensor employed in this paper is as shown in Fig. 3. The sizes of the sensory cells are not uniform, such that it is small around the center of the sensor, and it is large at the fringe. The size of the small one is 0.5×0.5, while that of the large one is 1.5×1.5. The sensor has 9 small cells and 8 large cells, and totally 17 cells. The output of each visual sensory cell is the area ratio occupied by the projected pattern against its receptive field. When the signals are put into the neural network, they are linearly expanded from -1.0 to 1.0. The initial location of the sensor is chosen randomly under the condition that the sum of all the sensory signals is larger than 0.5.

The sets of presented patterns are shown in Fig. 4. In the first set, the difference among the 4 patterns exists around the upper-left corner not depending on the presented pattern. The smallest square in the pattern is just the same size as the small sensory cell. When the sensor catches the center of the presented patterns, it is difficult to identify the pattern. So the sensor is required to move its center at the upper-left corner of the pattern.

In the second set, in order to identify the pattern 1 or 2, the sensor should move its center to the upper-right corner of the pattern, while it should moved the

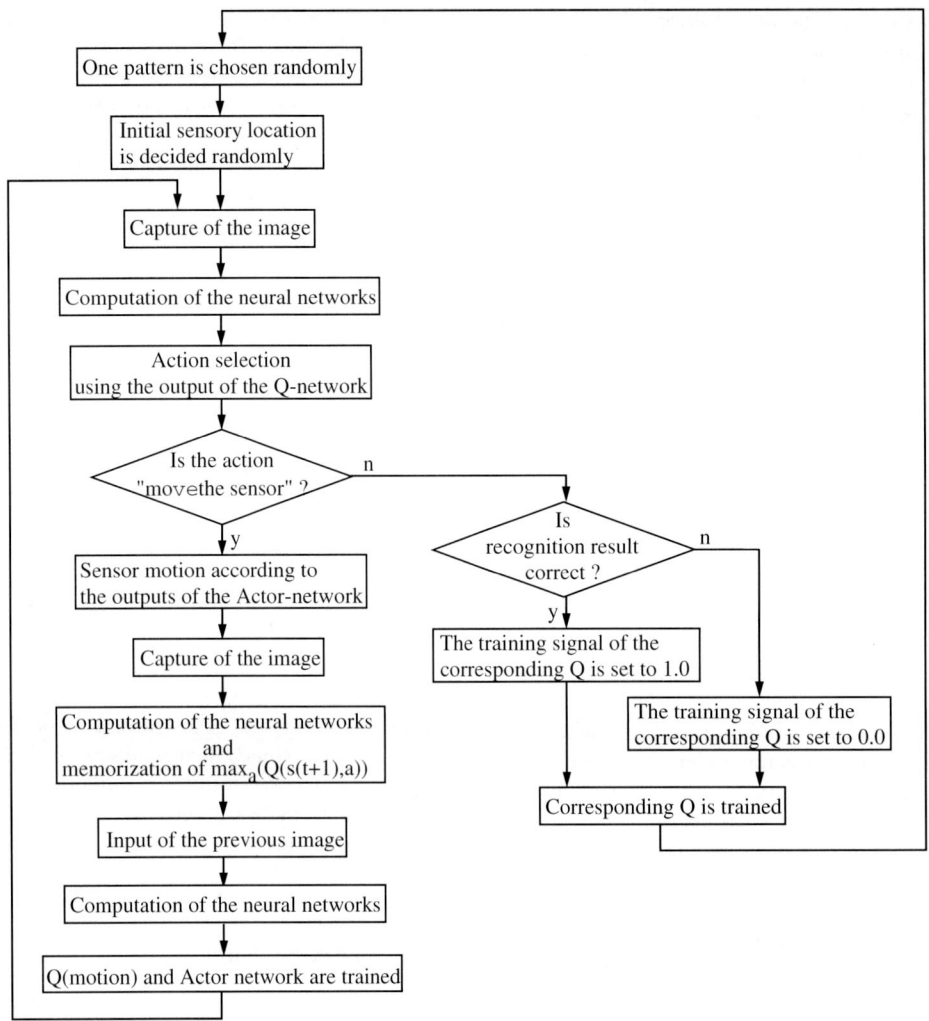

Figure 2: The flow chart of the proposed learning.

center to the upper-left corner to identify the pattern 3 or 4. So the visual sensor is required at first to know which group the presented pattern belongs to using whole the sensor, and then to move its center to the appropriate location.

Here the both neural networks, the Q-network and the actor-network have three layers. The number of hidden neurons is 30 in the Q-network, and 10 in the actor-network. The bias value is not introduced to the output layer in the both networks. That is because the bias sometimes leads to instability of learning or generates a constant flow of the sensor motion not depending on the sensor location.

Since the value range of the each neuron's output function is from -0.5 to 0.5, the training signal for the neural network is obtained from $Q_{training}$ in Eq. (1) and Eq. (3) by the transformation as

$$O_{training} = \alpha(Q_{training} - 0.5) \quad (4)$$

where α: a constant. While Q-value is transformed from the output of the network as

$$Q = O/\alpha + 0.5. \quad (5)$$

Here 0.8 is employed as α to avoid the saturation range of the output function. If the Q becomes less than 0.0, Q is set to be 0.0. As a discount factor γ in Eq. (1), 0.99 is employed.

In Eq. (2), 0.4 is employed as β. So the maximum motion step is 0.2 for each axis, while the minimum square of the pattern and the smallest sensor cell is 0.5×0.5. The random numbers **rnd** are the uniform random number powered by 3.0. The range is from -1.0 to 1.0, while the output range of the network is from -0.5 to 0.5.

One action is selected according to Boltzmann Distribution in the learning phase, and is selected according to the greedy method in the execution phase using the Q-values. The temperature is reduced gradually from 1.0 to 0.01 according to the progress of the learning as shown in Fig. 5. The number of the trial

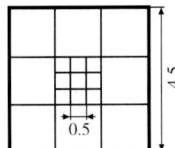

Figure 3: The visual sensor with non-uniform visual cells employed in this paper.

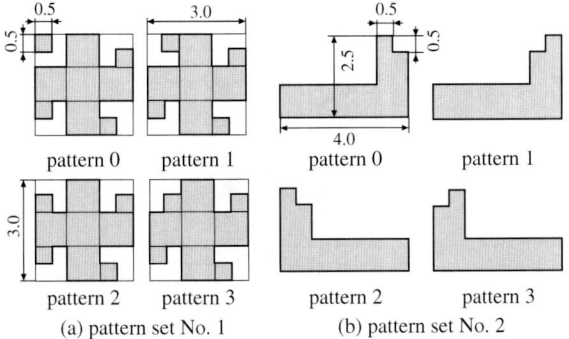

Figure 4: The presented pattern sets.

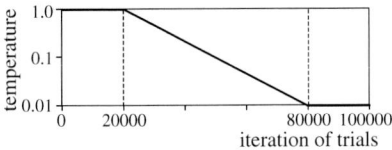

Figure 5: Temperature cooling schedule that is used in the action selection (log scale).

iterations is 100000.

3.2 Result

5 simulation runs are done for each pattern sets varying the initial weight values in the neural network, the initial sensor location, and the order of the presented patterns. In all the simulation runs, the system could identify the presented pattern after some motions. Fig. 6 shows an example of the trajectories of the visual sensor when the pattern set No. 1 is presented. It is seen that the system moves the center of its sensor to the upper-left corner of the presented pattern from given 132 initial sensor locations those are on the grid with 0.25 width. It is noticed that when the pattern 0, 1, or 2 is presented, the sensor moves and converges to almost one point, and the small square that is the difference from the other pattern is caught just by one of the small size sensory cells. This seems an effective and sure way of identification. While when the pattern 3 is presented, the convergence area is wider. That is the general property observed over the 5 simulation runs.

Fig. 7 shows an example of the trajectories of the visual sensor when the pattern set No. 2 is presented.

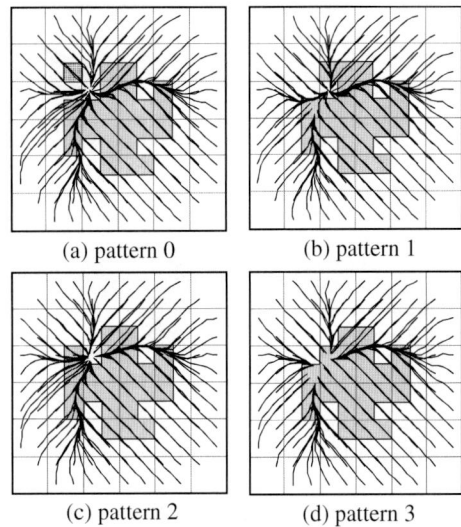

Figure 6: The trajectories of the visual sensor when the pattern set No. 1 is presented.

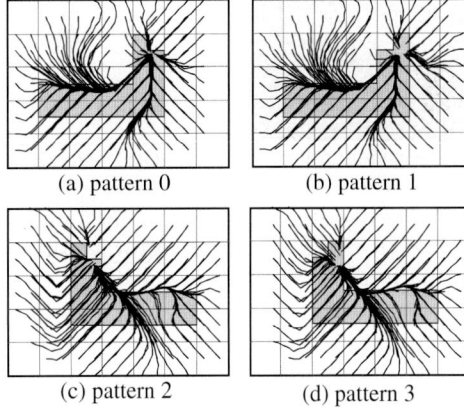

Figure 7: The trajectories of the visual sensor when the pattern set No. 2 is presented.

It can be seen that the system moves its sensor to the appropriate direction depending on the presented pattern. Concretely when the presented pattern is 0 or 1, the sensor moves to the upper-right corner, while when the pattern is 2 or 3, it moves to the upper-left corner. The sensor catches the small difference between the patterns on the center after a series of motions. Finally it could make a correct answer for each of the about 132 initial locations of the sensor.

Fig. 8 shows the distribution of the Q-values corresponding to each presented pattern. It is seen that the Q-value is large around the small difference area, and the area is the same as the location where the visual sensor converges as shown in Fig. 7. Fig. 9 shows the distribution of the Q-value for the sensor motion and the Q-value for the pattern 1 when the pattern 0 is presented. It is seen that the Q-value for the sensor

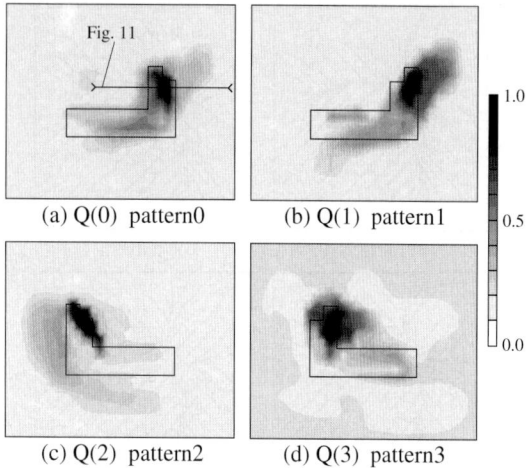

Figure 8: The distribution of the Q-values that is corresponding to the presented pattern.

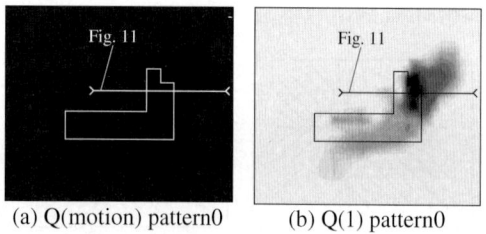

Figure 9: The distribution of the Q-value for the sensor motion and Q-value for the pattern 1 when the pattern 0 in the pattern set No. 2 is presented.

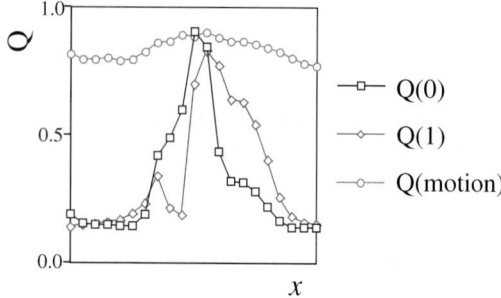

Figure 10: The one-dimension of distribution of the Q-values when the sections of the Q-value surfaces, Fig. 9(a), Fig. 10(a), and (b), are observed.

motion is always large not depending on the sensor location because the discount factor is close to 1.0. The Q-value for the pattern 1 is large around the upper-right corner.

Since it is difficult to see which value is larger at one sensor location, the section of the Q-value surface as shown in Fig. 8(a), Fig. 9(a),(b), is observed. Fig. 10 shows the distribution of the Q-values along the section. It can be seen that only at one point, the

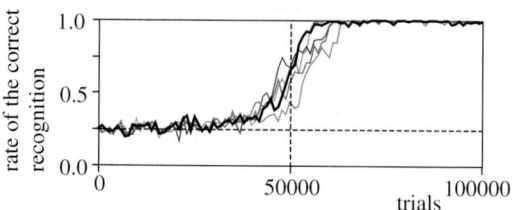

Figure 11: Learning curve when the pattern set No. 2 is presented. The y axis indicates the probability of the successful recognition.

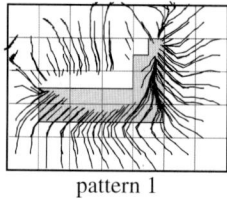

Figure 12: The trajectories of the visual sensor when the pattern 1 is presented after 50000 trials of learning.

Q-value for the pattern 0 is slightly larger than Q-value for the sensor motion, while the Q-value for the pattern 1 is always smaller than the Q-value for the motion. The Q-value for the sensor motion is reduced gradually from the maximum value, while the other Q value reduced suddenly. So even if the Q-value surface of one pattern has a local maximum, the Q-value for the sensor motion is larger, and the system selects to move the sensor. Accordingly the sensor is not trapped at the local maxima.

Fig. 11 shows the learning curve of 5 simulation runs when the pattern set No. 2 is presented. All the learning curve is similar, and the goal probability becomes large around 50000 trials. Fig. 12 shows the sensor trajectories after 50000 trials. The system makes an answer when the center of the sensor arrives on the pattern. In this case, the system makes the answer that the presented pattern is 0 even when the presented pattern is 1. When the sensor motion is selected, the Q-values for recognition are not trained. So the area of the sensor location where the Q-value for recognition is trained becomes smaller according to the progress of the learning. Then the Q-value surface for recognition becomes to have a strong peak. As above, the learning of Q-value and the learning of the motion make progress giving an effect with each other.

3.3 Simulation of Context Inputs

Next, it is examined that the system can generate the different series of sensor motions depending on the context inputs. The pattern set as shown in Fig. 13 is given. Each of the patterns cannot be identified even if the sensor goes to one of the corners of the pattern. For example, the difference between the pattern

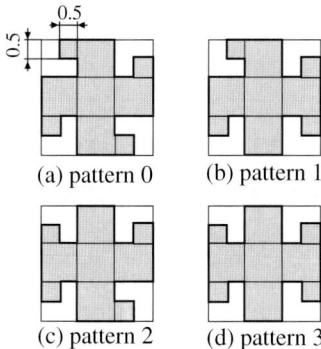

Figure 13: The pattern set in which the system requires the context inputs to identify each presented pattern.

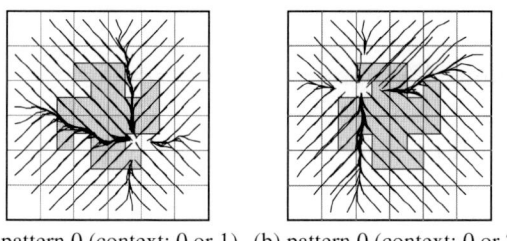

Figure 14: The difference of the sensor trajectories depending on the context inputs.

0 and 1 exists at the bottom-right corner of the pattern, while the difference between the pattern 0 and 2 exists at the upper-left corner. So when the sensor reaches the upper-left corner, it cannot identify whether the pattern is 0 or 1, while when the sensor reaches the bottom-right corner, it cannot identify whether 0 or 2. The context input that consists of 4 signals indicating the possibility that the presented pattern can be each of the 4 patterns, is also given to the neural network. In this case, only two of the signals are 2, which means a possible pattern, and the other two are -2. In this simulation, the number of the trial iterations is 2000000, and the number of the hidden neurons is 50 in the Q-network, and 20 in the actor-network. The discount factor γ is set to be 0.96. The temperature is reduced as Fig. 5, but the x axis is expanded linearly.

The difference of the sensor trajectories depending on the context inputs when the pattern 0 is presented is shown in Fig. 14. It can be seen that when the context inputs shows the possibility 0 or 1, the system moves its sensor to the bottom-right corner, while when the context shows the possibility 0 or 2, it moves its sensor to the upper-left corner. However, it is far more difficult to learn these motions than the previous simulation runs, and it needs 2000000 trials for learning. Fig. 15 shows the distribution of the Q-value for the pattern 0. It can be seen that the distribution is perfectly different from each other even if the visual sensory signals are completely the same.

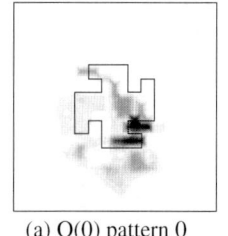 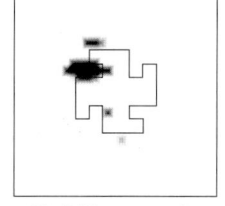

Figure 15: The difference in the Q-value distribution depending on the context inputs.

4 Conclusion

Q-actor architecture and the learning algorithm have been proposed for the active perception system based on reinforcement learning. Through the simulation using a visual sensor with non-uniform sensory cells, it was confirmed that the system becomes to move its sensor to the place where the difference between the patterns exists, and then to output the correct recognition result.

Acknowledgments   A part of this research was done under the Grant-in-Aid for Scientific Research on Priority Area No. 264 (System Theory of Function Emergence) supported by the Scientific Research Foundation of the Ministry of Education, Science, Sports and Culture of Japan.

REFERENCES

[1] S. Rougeaux and Y. Kuniyoshi, "Robust Real-Time Tracking on an Active Vision Head", *Proc. of IROS'97* (1997)

[2] Shibata, K., Okabe, Y. & Ito, K, "Direct-Vision-Based Reinforcement Learning in "Going to an Target" Task with an Obstacle and with a Variety of Target Sizes", *Proc. of NEURAP'98*, pp. 95–102 (1998)

[3] S.D.Whitehead & D.H.Ballard, "Learning to Perceive and Act by Trial and Error", *Machine Learning*, **7**, pp. 45–83 (1991)

[4] Watkins, C. J. C. H. and Dayan, P., "Q-Learning", *Machine Learning*, **8**, pp. 279–292 (1992)

[5] K. Shibata, T. Nishino & Y. Okabe, "Active Perception based on Reinforced Learning", *Proc. of WCNN '95*, **II**, pp. 170–173 (1995)

[6] Barto, A. G., Sutton, R. S. & Anderson C. W., "Neuronlike Adaptive Elements That Can Solve Difficult Learning Control Problems," *IEEE Trans. of SMC*, **13**, pp. 835–846 (1983)

Inverse Kinematics Learning by Modular Architecture Neural Networks with Performance Prediction Networks

Eimei OYAMA and Nak Young Chong
Intelligent Systems Laboratory
National Institute of Advanced Industrial
Science and Technology
Namiki 1-2, Tsukuba-shi, Ibaraki 305-8564 Japan
eimei@mel.go.jp

Arvin Agah
Department of Electrical Engineering
and Computer Science
The University of Kansas
Lawrence, KS 66045 USA

Taro MAEDA and Susumu TACHI
School of Engineering, The University of Tokyo
Hongo 7-3-1, Bunkyo-ku, Tokyo 113-8656 Japan

Abstract

Inverse kinematics computation using an artificial neural network that learns the inverse kinematics of a robot arm has been employed by many researchers. However, the inverse kinematics system of typical robot arms with joint limits is a multi-valued and discontinuous function. Since it is difficult for a well-known multi-layer neural network to approximate such a function, a correct inverse kinematics model cannot be obtained by using a single neural network. In order to overcome the discontinuity of the inverse kinematics function, we proposed a novel modular neural network system that consists of a number of expert neural networks. Each expert approximates the continuous part of the inverse kinematics function. The proposed system uses the forward kinematics model for selection of experts. When the number of the experts increases, the computation time for calculating the inverse kinematics solution also increases without using the parallel computing system. In order to reduce the computation time, we propose a novel expert selection by using the performance prediction networks which directly calculate the performances of the experts.

1 Introduction

The task of calculating all of the joint angles that would result in a specific position/orientation of an end-effector of a robot arm is called the inverse kinematics problem. An inverse kinematics solver using an artificial neural network that learns the inverse kinematics system of a robot arm has been used in many researches [1][2]; however, many researchers do not pay enough attention to the discontinuity of the inverse kinematics function of typical robot arms with joint limits. The inverse kinematics function of the robot arms, including a human arm with a wrist joint, is a multi-valued and discontinuous function. It is difficult for a well-known multi-layer neural network to approximate such a function. Therefore a novel modular neural network architecture for the inverse kinematics model learning is necessary.

A modular neural network architecture was proposed by Jacobs et al. and has been used by many researches [3][4][5]. However, the input-output relation of their networks is continuous and the learning method of them is not sufficient for the non-linearity of the kinematics system of the robot arm.

In order to learn a discontinuous inverse kinematics function, selecting one expert [6] has better performance than mixing all experts. The inverse kinematics function decomposes into a finite number of inverse kinematics solution branches [7][8][9][10]. DeMers et al. proposed the inverse kinematics learning method that a neural network learns each solution branch calculated by the global searches in the joint space [8][9][10]. However, the method is a purely off-line learning method and is not applicable for on-line learning, i.e. simultaneous or alternate execution of the robot control and the inverse model learning. Furthermore, the method is not goal-directed. There is no direct way to find an joint angle vector that corresponds to a desired hand position.

We proposed a novel modular neural network architecture for inverse kinematics learning based on DeMers' method [11][12]. The proposed modular neural network system consists of a number of experts, implemented by using artificial neural networks. Each expert approximates the continuous region of the inverse kinematics function. The proposed modular neural net system selects one appropriate expert whose output minimizes the expected position/orientation error of the end-effector of the arm calculated by using a forward kinematics model. The proposed system can learn a precise inverse kinematics model.

Since the proposed system uses the forward kinematics model of the robot arm for the calculation of the expected position/orientation error, the system requires the calculation of the outputs of all the experts and the calculation of the predicted position/orientation of the end-effector by using the forward kinematics model. When the number of the experts increases, the computation time for the calculation of the predicted errors of the experts also increases without using the parallel computing system.

In order to reduce the computation time, we propose a novel expert selection by using the performance prediction networks which directly calculate the performances of the experts. In order to evaluate the proposed architecture, numerical experiments of the inverse kinematics model learning were performed.

2 Modular Neural Net System with the Performance Prediction Networks

Let θ be the $m \times 1$ joint angle vector and x be the $n \times 1$ position/orientation vector of a robot arm. The relationship between θ and x is described by $x = f(\theta)$. f is a C^1 class function. Let $J(\theta)$ be the Jacobian of the robot arm, defined as $J(\theta) = \partial f(\theta)/\partial \theta$. When a desired hand position/orientation vector x_d is given, an inverse kinematics problem that calculates the joint angle vector θ_d satisfying the equation $x_d = f(\theta_d)$ is considered.

In this paper, a function $g(x)$ that satisfies $x = f(g(x))$ is called an inverse kinematics function of $f(\theta)$. The acquired model of the inverse kinematics system $g(x)$ in the robot controller is called an inverse kinematics model. Let $\Phi_{im}(x)$ be the output of the inverse kinematics model. Although $g(x)$ is usually a multi-valued and discontinuous function, the inverse kinematics function can be constructed by the appropriate synthesis of continuous functions [10][11][12].

2.1 Configuration of Proposed Inverse Kinematics Solver

Fig. 1 shows the configuration of the improved inverse kinematics solver with the modular architecture networks for inverse kinematics learning. Each expert network in Fig. 1 approximates the continuous region of the inverse kinematics function. The performance prediction network learns the performance of each expert. The expert selector selects one appropriate expert based on the outputs of the performance prediction networks as described in Section 2.2. The extended feedback controller calculates the inverse kinematics solution based on the output of the selected expert by using the output error feedback. When no precise solution is obtained, the controller performs a kind of global search, as shown in Section 2.3. The expert generator generates a new expert network based on the inverse kinematics solution.

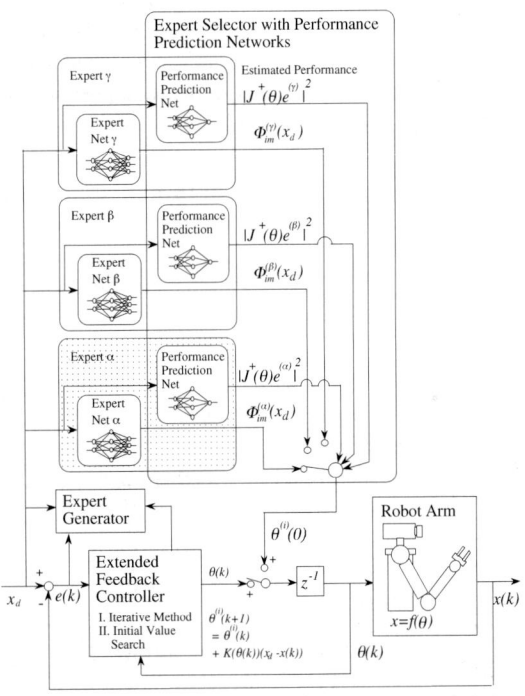

Fig. 1: Modular Neural Net System with Performance Prediction Networks

2.2 Configuration of the Expert and Selection by the Performance Prediction Networks

In order to cover the overall work space, each expert has its representative posture. The representative posture is the inverse kinematics solution obtained in the global searches by the extended feedback controller when the expert is generated. Let $\boldsymbol{\theta}_r^{(i)}$ be the representative posture of the i-th expert and $\boldsymbol{x}_r^{(i)}$ be the end-effector position/orientation that corresponds to $\boldsymbol{\theta}_r^{(i)}$. Let $\boldsymbol{\Phi}_{im}^{(i)}(\boldsymbol{x})$ be the output of i-th expert when the input of the expert is $\boldsymbol{x}$. Each expert is trained to satisfy the following equation:

$$\boldsymbol{x}_r^{(i)} = \boldsymbol{f}(\boldsymbol{\Phi}_{im}^{(i)}(\boldsymbol{x}_r^{(i)}))). \tag{1}$$

By changing the bias parameters of the output layer of the neural network, the above equation can easily be satisfied. Each expert approximates the continuous region of the inverse kinematics function in which the reaching motion can move the end-effector smoothly from its representative posture.

As stated in Section 1, the previously proposed modular net system with the forward kinematics model requires relatively large computation time. In order to reduce the computation time, we propose the use of the performance prediction networks that directly calculates the values which corresponds to the predicted end-effector position/orientation errors of the experts. The expert selector selects an expert with the best predicted performance among all the experts. If the performance prediction networks are accurate, the calculation of the output of only one selected expert instead of the outputs of all the experts is necessary for the inverse kinematics computation.

The idea of the performance prediction networks is based on a primitive reinforcement learning technique [13]. However, since the properties of each expert changes by learning, the careful construction of the learning algorithm of the proposed architecture is necessary. The learning of the performance prediction network will be described Section 2.4.

Let N_e be the number of the experts. Let $\boldsymbol{\Phi}_{im}^{(i)}(\boldsymbol{x}_d)(i=1,2,\ldots,N_e)$ be the output of the i-th expert and let $\boldsymbol{\Phi}_{pp}^{(i)}(\boldsymbol{x}_d)$ be the output of the performance prediction network which estimates the error of the i-th expert. When the desired end-effector position $\boldsymbol{x}_d$ is given, the performance prediction networks calculates the expected performances of all the experts $\Phi_{pp}^{(i)}(\boldsymbol{x}_d)(i=1,2,\ldots,N_e)$.

2.3 Extended Feedback Controller

The conventional on-line inverse model learning methods, such as Forward and Inverse Modeling proposed by Jordan [2] and Feedback Error Learning proposed by Kawato [14], are based on the local information of the forward system near the output of the inverse model. The desired output signal provided by these methods is not always in the direction that finally reaches the correct solution of the inverse problem [15]. An extended feedback controller avoids that drawback by employing a global search technique based on the multiple starts of the iterative procedure [16][15].

When a desired end-effector position $\boldsymbol{x}_d$ is given, the expert selector selects the expert with the minimum predicted error among all the experts. The extended feedback controller moves the arm to the posture that corresponds to the output of the expert and then improves the end-effector position/orientation by using the output error feedback, as described in Section 2.4. When no precise inverse kinematics solution is obtained, the other expert which predicted error is lower than an appropriate threshold r_{eim} is selected in increasing order of the predicted error and the iterative improvement procedure by the output error feedback is conducted. When no solution is obtained by the reaching motions from all the output of the selected experts, an expert is randomly selected and a reaching motion from the representative posture of the selected expert is conducted. an repeated until the reaching motion is successfully conducted or all the experts are tested. If a precise solution is obtained in the above procedural steps, the solution is used as the desired output signal for the expert, as shown in 2.4.

When no solution is obtained in the above procedural steps, the controller starts a type of global search. The controller repeats the initial joint angle vector generation by using a uniform random number generator and the reaching motion from the generated posture, as described in 2.4, until a precise solution is obtained. When a precise solution is obtained, a new expert is generated and the solution is used as the representative posture $\boldsymbol{\theta}_r$ of the expert.

2.4 Reaching Motion and Expert Learning

An illustration of the reaching motion, which is a kind of iterative improvement procedure, follows.

Let $\boldsymbol{\theta}(0)$ be the initial posture of the iterative procedure, which is the output of the selected expert $\boldsymbol{\Phi}^{(i)}(\boldsymbol{x}_d)$; the representative posture of the selected expert $\boldsymbol{\theta}_r^{(i)}$; or the randomly generated posture.

Let $\boldsymbol{x}_s$ be the initial end-effector position which is defined as $\boldsymbol{x}_s = \boldsymbol{f}(\boldsymbol{\theta}(0))$. The extended feedback controller conducts a reaching motion from $\boldsymbol{x}_s$ to $\boldsymbol{x}_d$ by using Resolved Motion Rate Control (RMRC) [17]. The reaching motion is conducted as the tracking control to the following desired trajectory of the end-effector $\boldsymbol{x}_d(k)(k = 0, 1, \ldots, T+1)$ described as follows.

Let T be an integer that satisfies $T - 1 \leq ||\boldsymbol{x}_d - \boldsymbol{x}_s||/r_{st} < T$. The desired trajectory $\boldsymbol{x}_d(k)$ is a straight line from $\boldsymbol{x}_s = \boldsymbol{f}(\boldsymbol{\theta}(0))$ to $\boldsymbol{x}_d$ which is calculated as follows:

$$\boldsymbol{x}_d(k) = \begin{cases} (1 - \frac{k}{T})\boldsymbol{x}_s + \frac{k}{T}\boldsymbol{x}_d & (0 \leq k < T) \\ \boldsymbol{x}_d & (k \geq T) \end{cases} \quad (2)$$

When the orientation is represented by the Direction Cosine Matrix or the Quaternion, the components of $\boldsymbol{x}_d(k)$ must be normalized.

We assume that a precise end-effector position feedback controller is already obtained by learning [18][19][20]. Otherwise, we assume that the controller can accurately estimate the coordinate transformation by the observation of the robot arm movement and the numerical differentiation technique [15].

Let $\boldsymbol{J}^+(\boldsymbol{\theta})$ be the pseudo-inverse matrix (Moore-Penrose generalized inverse matrix) of $\boldsymbol{J}(\boldsymbol{\theta})$ which is calculated as $\boldsymbol{J}^+(\boldsymbol{\theta}) = \boldsymbol{J}^T(\boldsymbol{\theta})(\boldsymbol{J}(\boldsymbol{\theta})\boldsymbol{J}^T(\boldsymbol{\theta}))^{-1}$. $\boldsymbol{J}^+(\boldsymbol{\theta})$ is used as the coordinate transformation gain of the output error feedback. Let $\boldsymbol{\theta}(k)$ be an approximate inverse kinematics solution at step k. When r_{st} is small enough, $\boldsymbol{\theta}(k)$ can be calculated as follows:

$$\boldsymbol{\theta}(k+1) = \boldsymbol{\theta}(k) + \boldsymbol{J}^+(\boldsymbol{\theta}(k))(\boldsymbol{x}_d(k+1) - \boldsymbol{f}(\boldsymbol{\theta}(k))). \quad (3)$$

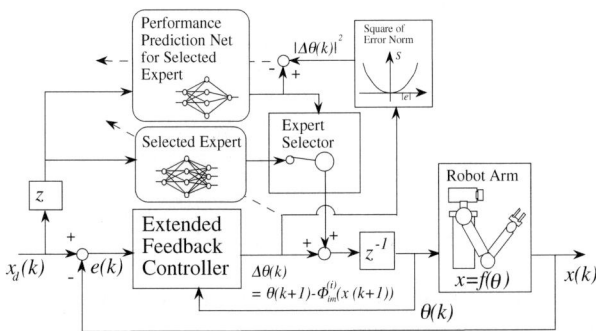

Fig. 2: Learning of Expert Network and Performance Prediction Network

Let $\boldsymbol{\Phi}_{im}^{\prime(i)}(\boldsymbol{x}_d)$ be the desired output signal for the i-th expert and $\boldsymbol{\Phi}_{pp}^{\prime(i)}(\boldsymbol{x}_d)$ be the desired output signal for the performance prediction network of the i-th expert. If a precise solution $\boldsymbol{\theta}(k)$, which end-effector position error norm $||\boldsymbol{x}_d(k) - \boldsymbol{f}(\boldsymbol{\theta}(k))||$ is lower than an appropriate threshold r_e, is obtained, the solution can be used for the selected expert learning as follows:

$$\boldsymbol{\Phi}_{im}^{\prime(i)}(\boldsymbol{x}_d(k)) = \boldsymbol{\theta}(k). \quad (4)$$

The learning of the performance prediction network is conducted as follows:

$$\boldsymbol{\Phi}_{pp}^{\prime}(\boldsymbol{x}_d(k)) = ||\boldsymbol{\theta}(k) - \boldsymbol{\Phi}_{im}^{(i)}(\boldsymbol{x}_d(k))||^2. \quad (5)$$

The above value is not the hand position error of the expert but directly corresponds to it. The learning of the selected expert network and the corresponding performance prediction network are illustrated in Fig. 2. When the controller cannot find a precise solution because of the singularity of Jacobian or the joint limits, the reaching motion is regarded as a failure.

3 Simulations

We performed simulations of the inverse kinematics model learning of a 7-DOF arm as shown in Figure 3. This arm is called TELESAR II (TELE-existence Slave Arm II), the original of which was developed for the experimental study on the remote robot control using the virtual reality [21]. The configuration of the arm is illustrated in Figure 3. The parameters $L_i(i = 1, 2, 3, 4)$ in Figure 3 is defined as $L_1 = 0.305m$, $L_2 = 0.260m$, $L_3 = 0.04m$ and $L_4 = 0.150m$.

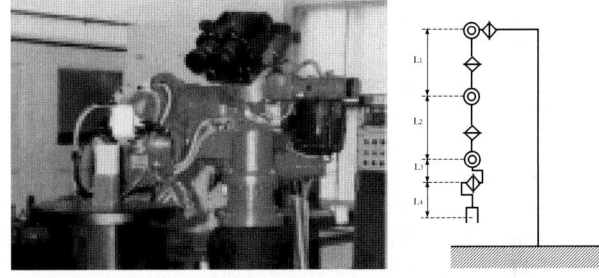

Fig. 3: TELE-existence Slave ARm II (TELESAR II)

For comparison, the modular neural network system which selects the expert with the minimum error among all the experts by using the complete forward model $\boldsymbol{f}(\boldsymbol{\theta})$ without error were tested. Furthermore, the system which uses the forward model consisting of a neural networks were also tested. Hereafter, MNPP indicates the Modular Neural network system with the performance prediction networks that consist of neural networks with no previous learning. MNCFM indicates the Modular Neural network system with a Complete Forward Model. MNFM indicates the Modular Neural network system with a learning Forward Model.

In the simulations, joint angle vectors were generated by using a uniform random number generator, and the end-effector positions that correspond to the generated vectors were used as the desired end-effector positions. In order to evaluate the performance of the solver, 16,384 desired end-effector positions were generated for the estimation of the root mean square (RMS) error of the end-effector position $e = x_d - f(\Phi_{im}(x_d))$. r_{eim} was $0.2m$, r_e was $0.0025m$, r_{st} was $0.02m$, and r_{jix} was 10^2.

and the 3rd layers of the performance prediction networks had 10 neurons each. The back-propagation method was utilized for the learning. The learning rate for the experts was set at 0.05. That for the performance prediction networks was set at 0.005. The momentum parameter was set at 0.5.

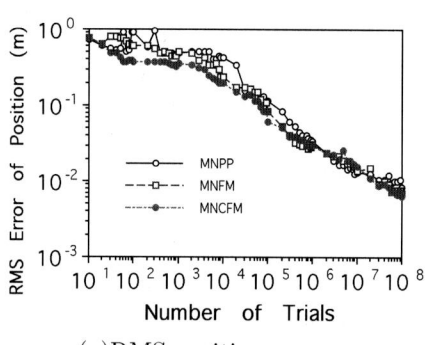

(a) RMS position error

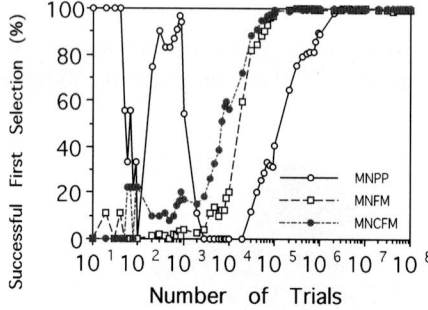

(b) Percentage of successful first selection

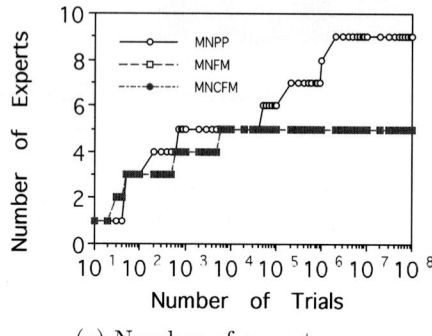

(c) Number of experts

Fig. 4: Performance Change of Proposed Inverse Kinematics Solver

A 4-layered neural network was used for the simulations. The 1st layer, i.e., the input layer, and the 4th layer, i.e., the output layer, consisted of linear neurons. The 2nd and the 3rd layers of the experts and the forward kinematics model had 25 neurons each. The 2nd

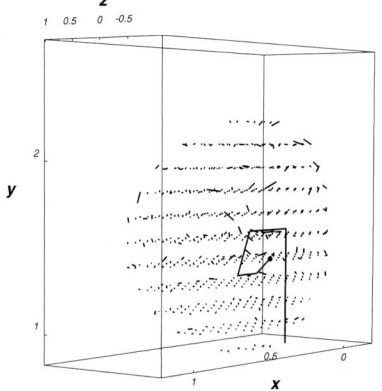

(a) Proposed Modular Neural Networks

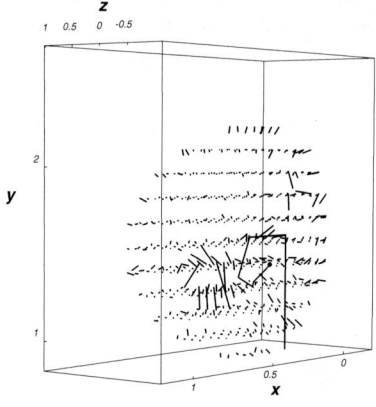

(b) Single Neural Network

Fig. 5: Position Error Vector of Inverse Kinematics Model of 7-DOF Arm

Fig. 4(a) shows the change of the RMS error of the end-effector position. It can be seen that the RMS error decreases and the precision of the inverse model becomes higher as the number of trials increases. The RMS end-effector position error of MNCFM, MNFM, and MNPP became lower than $1.5 \times 10^{-2} m$ after 10^7 learning trials. The precision of MNCFM is better than that of MNFM. The precision of MNFM is better than that of MNPP. However, there is not so much difference between them. We concluded that the proposed method succeeded in the inverse kinematics model learning of a 7-DOF arm.

Fig. 4(b) shows the percentage of the trials in which the posture generated by the first selected expert can successfully reach the desired position. The percentage of the successful first selection of MNCFM and MNFM was larger than 99% before 2×10^5 learning trials. On the other hand, that of MNPP was larger than 99% after 2×10^6 learning trials. The expert selection of MNCFM and MNFM was more precise than that of MNPP. Fig. 4(c) shows the number of the experts that constructs the inverse kinematics model. After 10^7 learning trials, MNFM and MNCFM generated 5 experts. MNPP generated 9 experts. The number of the experts of MNPP is larger than that of MNFM or MNCFM.

Simulations of the inverse kinematics model learning, consisting of a single neural network, using Forward and Inverse Modeling [2], were also performed. The learning was performed from 10 different initial states of the neural network. The minimum RMS error of the inverse kinematics model using a neural networks after 10^9 learning trials was $1.50 \times 10^{-2} m$. Fig. 5(a) shows the position error vectors of the proposed modular inverse kinematics model. Fig. 5(b) shows those of the inverse kinematics model which consists of single neural network learned by forward and inverse modeling. There are some regions where the inverse kinematics model is far from precise, which are caused by the discontinuity of the inverse kinematics function. The maximum error was larger than $0.4m$. An inverse kinematics model which consists of a single neural networks cannot obtain a precise inverse kinematics model. We also tested the mixture of experts used in [5]. However, we could not obtain a more precise inverse kinematics model by using the mixture of experts than by using a single neural network.

The selection of the expert is illustrated in Fig. 6(a). Each number in the figure indicates which expert is selected at each desired hand position. Fig. 6(b) shows the region where the predicted output error of the first expert is lower than $0.02m$. The graphics of the robot arm in Fig. 6(b) show the representative posture of the first expert. Fig. 6(c) shows the region where the predicted output error of the fifth expert is lower than $0.02m$.

The computation time of one kinematics solution of MNCFM with 5 experts was $0.43ms$ (Intel Pentium II 450 MHz) after 10^7 times learning trials, that of MNFM with 5 experts was $0.73ms$, and that of MNPP with 9 experts was $0.23ms$. The computation time improved using the performance prediction networks. However, the learning speed of the performance prediction networks was much slower than that of the forward kinematics model. The improvement of the learning method is necessary.

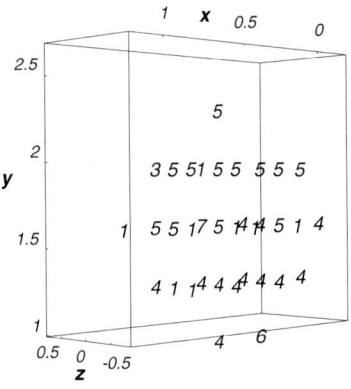

(a) Selected experts

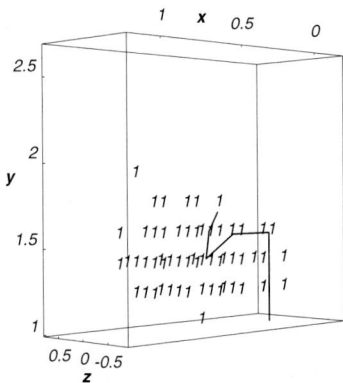

(b) Active region of first expert

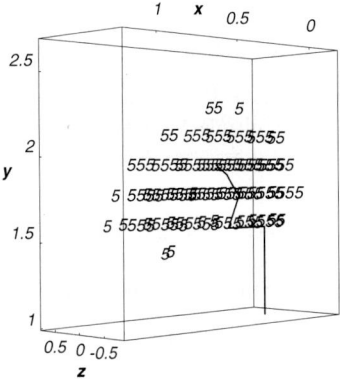

(c) Active region of fifth expert

Fig. 6: Configuration of Inverse Kinematics Model

4 Conclusions

In this paper, we proposed a novel modular neural network architecture with the performance prediction networks for the inverse kinematics model learning and confirmed the performance of the proposed system by numerical experiments. The computation time for calculating the inverse kinematics solution is reduced by the performance prediction networks. Although the proposed architecture has a number of limitations (for instance, the learning speed is very slow), we believe that the architecture can be used as a prototype of the inverse kinematics solver with learning function. The improvement for faster learning, the elimination of useless experts, and the utilization of the redundant degrees of freedom [22] will be reported in near future.

References

[1] M. Kuperstein,"Neural Model of Adaptive Hand-Eye Coordination for Single Postures," *Science*, Vol.239, pp.1308-1311, 1988.

[2] M. I. Jordan, "Supervised Learning and Systems with Excess Degrees of Freedom," *COINS Technical Report, 88-27*,pp.1-41,1988.

[3] R. A. Jacobs, M. I. Jordan, S. J. Nowlan and G. E. Hinton,"Adaptive Mixtures of Local Experts," *Neural Computation*,Vol.3, pp.79-87,1991.

[4] R. A. Jacobs and M. I. Jordan," Learning Piecewise Control Strategies in a Modular Neural Network Architecture," *IEEE Transactions on Systems, Man, and Cybernetics*, Vol.23, pp.337-345, 1993.

[5] H. Gomi and M. Kawato,"Recognition of Manipulated Objects by Motor Learning with Modular Architecture Networks," *Neural Networks*, Vol.6, pp.485-497, 1993.

[6] K. S. Narendra and J. Balakrishnan, "Adaptation and Learning Using Multiple Models, Switching and Tuning," *IEEE Control Systems Magazine*, Vol. 15, pp.37-51, 1994.

[7] Burdick, J., "Kinematics and Design of Redundant Robot Manipulators," Ph. D. Thesis, Dept. of Mechanical Engineering, Stanford University, 1988.

[8] D. DeMers and K. Kreutz-Delgado, "Learning Global Direct Inverse Kinematics," *Advances in Neural Information Processing Systems 4*, pp. 589-594, 1992.

[9] D. DeMers and K.Kreutz-Delgado, "Global Regularization of Inverse Kinematics for Redundant Manipulators," *Advances in Neural Information Processing Systems 5*, pp.255-262, 1993.

[10] D. DeMers and K. Kreutz-Delgado, "Solving Inverse Kinematics for Redundant Manipulators," in *Neural Systems for Robotics*, O. Omidvar and P. v. d. Smagt ed., Academic Press, 1997.

[11] E. Oyama and S. Tachi, "Inverse Kinematics Model Learning by Modular Architecture Neural Networks," *Proceedings of International Joint Conference on Neural Networks '99*, 1999.

[12] E. Oyama and S. Tachi, "Modular Neural Net System for Inverse Kinematics Learning," *Proceedings of International Conference on Robotics and Automation 2000*, 2000.

[13] A. G. Barto, "Reinforcement Learning," *Handbook of Brain Theory and Neural Networks*, M. A. Arbib ed., The MIT Press, pp.804-809, 1995.

[14] M. Kawato, K. Furukawa and R. Suzuki, "A Hierarchical Neural-network Model for Control and Learning of Voluntary Movement," *Biological Cybernetics*,Vol.57, pp.169-185, 1987

[15] E. Oyama and S. Tachi, "Inverse Model Learning by Using Extended Feedback System," *Proceedings of the fifth International Symposium on Measurement and Control in Robotics(ISMCR'95)*, pp.291-296, 1995.

[16] A. W. Moore,"Fast, Robust Adaptive Control by Learning only Forward Models," *Advances in Neural Information Processing Systems 4*, Morgan Kaufmann Publishers, pp.571-578, 1992.

[17] D. E. Whitney, "Resolved Motion Rate Control of Manipulators and Human Prostheses," *IEEE Trans. on Man-Machine System*, Vol.10, No.2, pp.47-53, 1969.

[18] F. H. Guenther and D. M. Barreca, "Neural Models for Flexible Control of Redundant Systems," in P. Morasso and V. Sanguineti (Eds.), *Self-organization, Computational Maps, and Motor Control*, Elsevier, pp. 383-421, 1997.

[19] E. Oyama and S. Tachi, "Coordinate Transformation Learning of Hand Position Feedback Controller by Using Change of Position Error Norm," *Advances in Neural Information Processing Systems 11*, pp. 1038-1044, 1999.

[20] E. Oyama, A. Agah, T. Maeda, S. Tachi and K. F. MacDorman, "Coordinate Transformation Learning of Human Visual Feedback Controller based on Disturbance Noise and Feedback Error Signal," *Proceedings of International Conference on Robotics and Automation 2001*, 2001.

[21] E. Oyama, N. Tsunemoto, Y. Inouse and S. Tachi, "Experimental Study on Remote Manipulation Using Virtual Reality," *PRESENCE*, Vol.2, No.2, pp.112-124, 1993.

[22] Y. Nakamura, H. Hanafusa and T. Yoshikawa, "Task-Priority Based Redundancy Control of Robot Manipulators," *The International Journal of Robotics Research*, Vol.6, No.2, pp.3-15, 1987.

Virtual Repulsive Force Field Guided Coordination for Multi-telerobot Collaboration

Nak Young Chong, Tetsuo Kotoku, Kohtaro Ohba, and Kazuo Tanie

Intelligent Systems Laboratory
National Institute of Advanced Industrial Science and Technology
1-2 Namiki, Tsukuba, Japan 305-8564
Email: nychong@ieee.org

Abstract

The Intelligent Systems Laboratory (ISL) has been developing coordinated control technologies for multi-telerobot collaboration in a common environment remotely controlled from multiple operators physically at a distance from each other. We have built a test bed and conducted a series of experiments, where we learned more about how the transmission delay over the network deteriorates the performance of telerobots. Previously, to overcome the problems arising from the throughput of the network such as the operator's delayed visual perception, we have suggested several coordination approaches in the local operator site. Likewise, this paper discusses the use of virtual repulsive force field in the on-line predictive simulator to assist the operator to cope with the collision between telerobots in remote environments. In the test bed, the operators control their master robot to get remote telerobots to work cooperatively with the other telerobots in a task. Specifically, the operator detects a priori the possibility of collision in the predictive simulator that runs in near real-time and the use of virtual force field prevents the telerobots from coming into collision. We have demonstrated various tasks by two telerobots and two operators via an Ethernet Local Area Network (LAN) subject to simulated communication delays and evaluated the validity of the virtual force field guided approach in Multi-Operator-Multi-Robot (MOMR) tele-collaboration.

1 Introduction

The current on-site work and maintenance that usually require a lot of travel is to be substituted for remote teleoperation over the network in a cost-effective way. Thus, we expect that the collaborative multi-telerobot system would be an alternative to support the coming society in which the working population decreases. In teleoperation with time delay, remote telerobot motions controlled from local operators would be visualized with round-trip time delays, thus the video camera image is often overlaid with graphics model predicted from local master control instructions. However, in the MOMR tele-collaboration as shown in Fig. 1, the telerobot under the control of the counterpart operator would not be straightforwardly predictable and accordingly the telerobots are most probably exposed to the possibilities of collision in remote environments. This seriously affects operators' decision-making and accordingly the performance of telerobots. Thus, to cope with the operator's delayed visual perception arising from the throughput of the network, we need to feed another supplementary information locally to the operator to safely steer remote telerobots through time delay [5].

To date, many works have been reported in the control of telerobot over the network and Sheridan extensively reviewed them in [13]. In addition, Goldberg et al. [7] [8] built systems that allow a robot to be teleoperated via the WWW. Brady et al. [1] proposed a new method for controlling telerobots over vast distances, where communication propagation delays exist. But most of the past works were applied only to a single telerobot controlled by a single operator. Recently, some efforts have been devoted to the teleoperation of multiple telerobots [9], [14]. But, they did not consider communication delays between local operators with large physical separation over the network.

In this work, we have built an experimental test bed to research the remote tele-collaboration through time delay between two distant operators. Specifically, to assist the operator suffering from the delayed visual

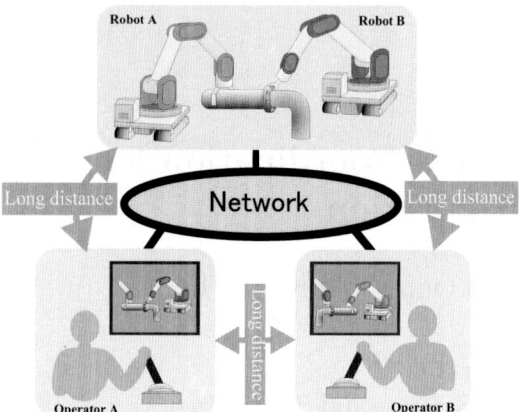

Figure 1: Network-based collaborative teleoperation.

perception on telerobots not under the operator's control, we developed an on-line graphics simulator that provides the operator with the work site view in near real-time. This simulator also feeds virtual force to the operator, which signals any possible collision between two telerobots. Thus, the operator can coordinate conflicting motions of multi-telerobots in remote sites even the counterpart operator is physically at a distance. The validity of the use of virtual force in the predictive simulator was verified through a series of experiments with two telerobots in a common work site controlled from two operators.

2 Use of Predictive Simulator

Over the past decades, the predictive display has been a well-tried approach for the time delay in teleoperation. It typically provides the operator with the immediate visualization of the master control commands where the real video image feedback from the remote site is delayed [10]. Likewise, the predictive simulator has played a major role in monitoring collision between telerobots and fine-tuning their end-effector position towards task goals when the network is subject to time delays [4]. But, to get the simulator-based approaches available, *a priori* models of remote environments are necessary. For this, recently, an interactive modeling tool based on the on-board sensor has been developed to build a reliable 3-D model as quickly as possible [6].

As is previously mentioned, telerobot motions not under the operator's control can not be predicted in local operator sites. Thus, the counterpart operator's robot motions are displayed with round-trip time de-

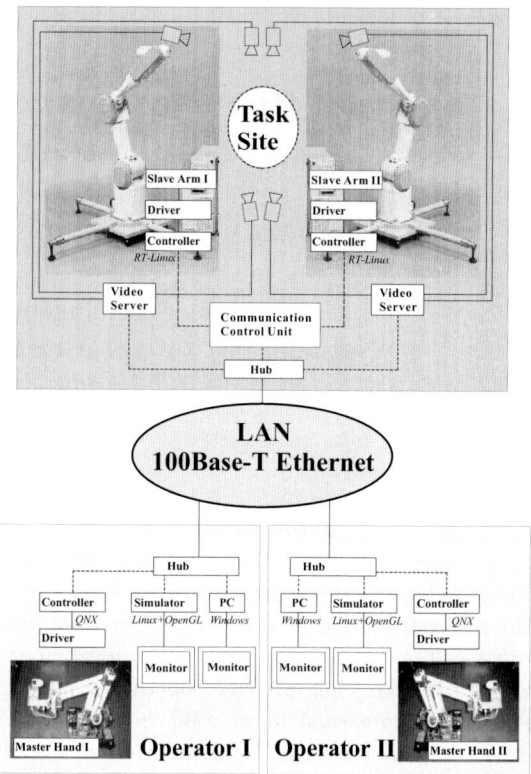

Figure 2: A collaborative teleoperation test bed.

lays and the operator's robot motions are predicted without time delay. Accordingly, there should be some mismatch between the graphics update of two cooperating telerobots in the simulator. To overcome this difficulty, we have already proposed several local coordination strategies and verified their validity through experiments using the graphics models of two planar robots [2] and real telerobots [3]. This work addresses another use of predictive simulator that feeds virtual force reflection to the operator to get the telerobots cooperating without any collision over the network with time delay.

3 Experimental Test Bed

This section goes into details about the test bed. (See Fig. 2)

3.1 Master control station

Fig. 3 illustrates a set of equipment in the operator's master control station. It consists of a prototype 6 DOF force-reflecting master system developed by the Toshiba R&D Center and an on-line graphics simulator running on a UNIX based operating system (Pentium II 450 MHz, Linux). Real video camera images from the task site is displayed in another PC (Pentium III 667 MHz, Windows) which has access to the video broadcasting server.

3.2 Transmission control

100Base-T Ethernet and dual-speed Ethernet hub are used to transmit information among two master stations and a task site. The communication control program in the Control Tower Station (Sun UltraSPARC 170, hereafter CTS) on the network gets all the communications between each site connected or disconnected. Likewise, the CTS also stores data in a buffer, which enables the communication over the LAN to simulate different time delays.

3.3 Video camera and broadcasting server

Local operators can have a better view of the task site if multi-camera views are incorporated. We mounted video cameras on the ceiling and the wall in the task site and also on the top of the gripper of each telerobot, respectively. To broadcast these camera images to local operators, we used a PC-less server (MegaFusion eWatch MD-100) that enables real-time streaming of video images over the Internet. Through incorporation of the system's plug-in software into a popular web browser, the system's video images can be viewed from a PC. The video image can be transmitted to the operator site at the maximum video rate 30 fps when one camera channel is input. The operator may have 4 different camera views from the task site, then the update rate becomes less than 2 fps which is impeded by source compression delay at the server and the image reconstruction delay at the client in addition to network transmission delays.

3.4 Telerobots and task environment

Two 7 DOF robots (PA-10, Mitsubishi Heavy Industries Ltd.) are positioned in opposite sides of a common working table as shown in Fig. 4. Different shaped and different colored acrylic plastic plates with a handle are on the table and are to be properly fitted into their pre-specified places by the cooperation of two robots. Some objects are out of reach of one robot. To proceed with the task, the objects should

Figure 3: Local master control station.

Figure 4: Telerobot systems and work environment.

be delivered within the reach of the robot by help of its counterpart robot.

4 Virtual repulsive force field guided approach

The throughput of the network restricts the operator's access to information about remote environments. To overcome the problems such as the operator's delayed visual perception, we have already proposed several coordination approaches in the local operator site [2]. Likewise, this paper discusses the use of virtual repulsive force field in the predictive simulator. We have the virtual force directly reflected to master operators to assist them to cope with collisions between telerobots in remote environments. It is well known that reflected force to the operator can improve the performance of task. Very recently, for the ground teleoperation of a space robot, reflected force was successfully used to improve the performance of the operator despite time delay [11]. Also, an algo-

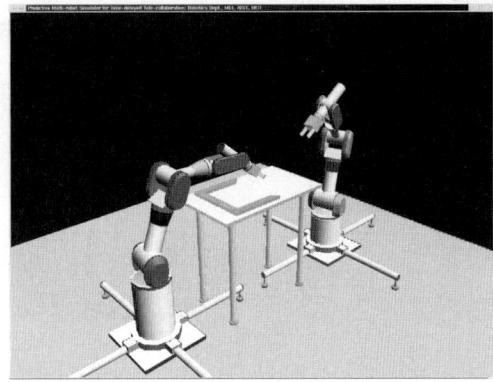

Figure 5: Predictive graphics simulator.

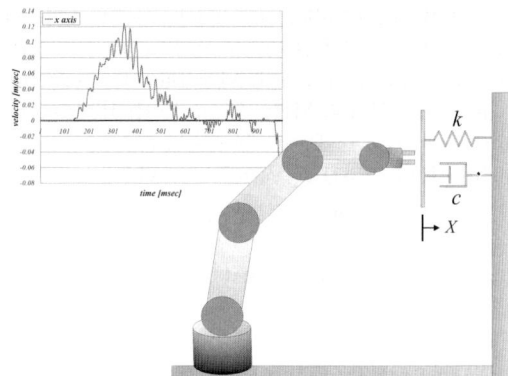

Figure 6: Teleoperator collides with a massless spring-damper buffer responding to a master's input velocity.

rithm has been implemented based on virtual springs in force-feedback teleoperation to keep a Stewart platform inside the useful workspace [12].

4.1 Predictive graphics simulator

We developed the predictive simulator using the OpenGL graphics system [15] that helped the operator visually verify their telerobot motions without time delay. (See Fig. 5) Specifically, the construction of 3-D graphics models of the telerobots and the task environment were performed. The robot graphics image is controlled by the master and its data are transmitted to the real robot. The CTS receives the robot data (e.g., joint configuration data) from the simulator and directly relays it to the counterpart operator's simulator. On the other hand, the CTS relays the same data to the real robot with time delay by storing the data in a ring buffer until the specified timer expires.

We installed a graphics accelerator (3dfx Voodoo3 3000 AGP) not to be caught up 3D graphics rendering burden. As an inter-process connector between the master controller that runs on the QNX and the graphics simulator on the Linux, a pair of cooperating sockets manage the communication via shared memory.

4.2 Force-reflecting master device

We first evaluate the fidelity of the force reflection at the master. We get the teleoperator approaching to a massless buffer with stiffness k and damping coefficient c in the predictive simulator as shown in Fig. 6. The input velocity of x-axis of the master is also shown in Fig. 6. Then, neglecting the inertial effect of

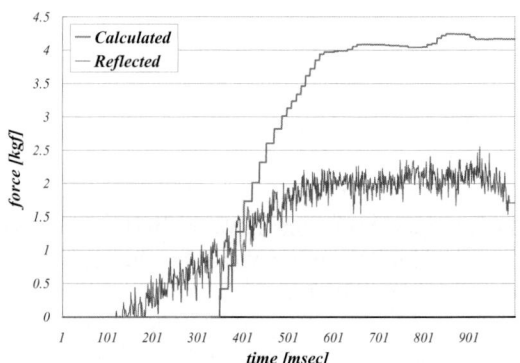

Figure 7: Reflected force measured at the master. ($k = 100 kg_f/m, c = 20 kg_f s/m$)

the teleoperator for simplicity, we can calculate the reaction force F_w at the end-effector of the teleoperator by

$$F_w = -kx - c\dot{x}. \quad (1)$$

Fig. 7 shows the calculated force F_w from Eq. 1 and reflected force measured at the master. It is noted that the reflected force conforms to the F_w and the rise of reflected force before contact comes up due to the friction and viscous damping of the master device.

4.3 Multi-telerobot coordination

In this work, the virtual reaction force F_w calculated in the simulator is directly fed back to the master controller. We employ this F_w to coordinate the motions of multiple telerobots by getting this force reflected to the master operator. The operator feels

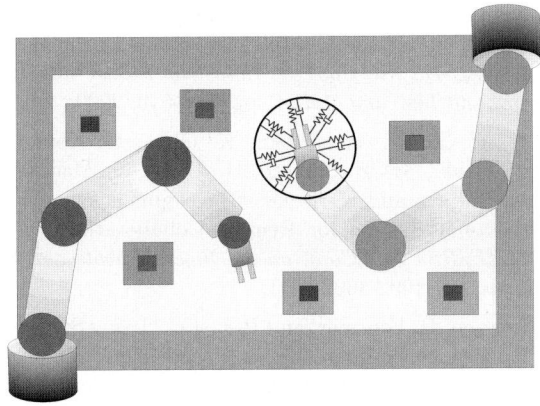

Figure 8: A virtual repulsive force field around the counterpart telerobot.

a barrier when making their telerobot approach its counterpart robot. Specifically, in the predictive simulator, a virtual repulsive force field is generated around the end-effector of the counterpart robot. (See Fig. 8) When one telerobot approaches its counterpart robot, the repulsive force F_w pushes the approaching robot back and keeps its distance from the robot. While the distance between two robots is secured enough, then the force field does not work any more.

Let $S_1(t)$ and $S_2(t)$ denote the sets in operational space occupied by each teleoperator. Also let $d(A,B)$ denote the minimum distance between two sets A and B. d_{col} implies the distance wherein the two telerobots are possible to collide. d_{col} can be calculated based on the maximum permissible velocity of telerobot over the communication delay from the CTS. Telerobots safely move out of the danger of collision if the distance is over d_{col}. Thus, according to the distance between two telerobots, the spring and damping coefficients $k_e = [k \ c]$ that generate the virtual force field can be given by

$$k_e = \begin{cases} 0 & : \ d(S_1(t), S_2(t)) \geq d_{col} \\ k_e & : \ d(S_1(t), S_2(t)) \leq d_{col}. \end{cases} \quad (2)$$

Here, for simplicity, we assume that the collision occurs only between the end-effectors of two robots.

5 Experimental Results and Discussion

We have several pieces of objects scattered on the table. The operators control their robot and collaborates with the counterpart on arranging the object

$TD[s]$	0		1	
Subj.	W/o simulator	W/ simulator	W/o simulator	W/ simulator
C & T	4m 05s	3m 35s	4m 30s	3m 51s
K & F	3m 20s	3m 00s	4m 49s	3m 31s

Table 1: Task completion time in 2 telerobot collaboration. (See [3].)

in order. We compared the task completion time with changing the network delays and also investigated how the predictive simulator works in collision detection without time delay. Several pairs of subjects evaluated the operation solely with the camera view. The same pairs also evaluated the trials with the predictive simulator as well as the camera view. Table 1 is cited from [3] where the audio-visual resources featured the simulator in collision detection.

Without the predictive simulator, operators usually control their robots at very low speed, otherwise they might fail to coordinate the conflicting motions of robots successfully. On the contrary, observing the predictive simulator, operators could make robots safely avoid the collision, since they were guided by an audio-visual information from the simulator. The simulator signals to the operator by getting the preserved audio file running and the original color changing when the robot comes in contact with its counterpart robot and/or the working table. In addition, from the simulator's overall view of robots and work site, the operator quickly make decision which direction they should move their robot and confidently move at a relatively high speed. Thus, the task completion time was reduced as shown in Table 1.

Our previous works showed that predictive simulators were effective in the collision detection thanks to its audio-visual characteristics. However, the operators should make strict observations about the predictive simulator to take precautions against collision between robots. The operator, moreover, might miss or neglect the advance notice of collision. Thus, in this work, we added telerobot contact force against a virtual barrier around the counterpart robot to the existing operator information. This force is directly delivered to the operator hand through the master controller. The virtual repulsive force field is not merely a precautionary information but also safeguard telerobots against approaching counterpart robots. This is more dependable level of operation whereby operators are almost completely released from the danger of collision. Thus, irrespective of whether novice operators make a mistake and control their telerobot toward its counterpart robot, the force field will not allow them to collide. In the simulator, the stiffness and damping coefficients of the force field are designed not to make the master device unstable at the instant of getting into/out of the force field.

6 Conclusion

Predictive simulators play an important role to improve the operators' performance in collaborative MOMR tele-operation systems. The use of virtual repulsive force field in the predictive simulator was described to help the operators overcome the delayed visual perception and securely ensure the collision-free cooperation between telerobots over the long-distance network. We have built an experimental test bed, where several example tasks were conducted with two operators and two telerobots over the LAN subject to simulated communication delays. Within the results of current experiment so far, the use of virtual force field satisfactorily steered a couple of telerobots to perform a task cooperatively without any collision and showed the feasibility to cope with the communication delay in MOMR tele-collaboration. Each operator could give larger master commands more confidently without having to consider collision of telerobots, because the virtual force field in the predictive simulator would constrain master commands where collision is likely to happen. Operators accordingly become released from the anxiety about the collision even the counterpart operator is physically at a distance from each other. This work hopefully will form the framework of MOMR tele-collaboration in the coming society as an alternative of costly on-site operations.

References

[1] K. Brady and T. J. Tarn, "Internet-Based Remote Teleoperation," *Proc. IEEE Int. Conf. on Robotics and Automation*, pp. 65-70, 1998.

[2] N. Y. Chong, T. Kotoku, K. Ohba, K. Komoriya, N. Matsuhira, and K. Tanie, "Remote Coordinated Controls in Multiple Telerobot Cooperation," *Proc. IEEE Int. Conf. on Robotics and Automation*, Vol. 4, pp. 3138-3143, 2000.

[3] N. Y. Chong, T. Kotoku, K. Ohba, H. Sasaki, K. Komoriya, and K. Tanie, "Use of Coordinated Online Graphics Simulator in Collaborative Multi-robot Teleoperation with Time Delay," *Proc. IEEE Int. Workshop on Robot and Human Interactive Communication*, pp. 167-172, 2000.

[4] N. Y. Chong, T. Kotoku, K. Ohba, H. Sasaki, K. Komoriya, and K. Tanie, "Audio-visual Guided Predictive Simulator in Multi-telerobot Coordination," *Proc. IEEE Int. Conf. on Industrial Electronics, Control and Instrumentation*, pp. 614-619, 2000.

[5] N. Y. Chong, T. Kotoku, K. Ohba, K. Komoriya, F. Ozaki, H. Hashimoto, J. Oaki, K. Maeda, N. Matsuhira, and K. Tanie, "Development of a Multi-telerobot System for Remote Collaboration," *Proc. IEEE/RSJ Int. Conf. on Intelligent Robots and Systems*, pp. 1002-1007, 2000.

[6] P. Even, R. Fournier, and R. Gelin, "Using Structural Knowledge for Interactive 3-D Modeling of Piping Environments," *Proc. IEEE Int. Conf. on Robotics and Automation*, pp. 2013-2018, 2000.

[7] K. Goldberg, M. Mascha, S. Gentner, N. Rothenberg, C. Sutter, and J. Wiegley, "Desktop Teleoperation via the World Wide Web," *Proc. IEEE Int. Conf. on Robotics and Automation*, pp. 654-659, 1995.

[8] K. Goldberg, B. Chen, R. Solomon, S. Bui, B. Farzin, J. Heitler, D. Poon, and G. Smith, "Collaborative Teleoperation via the Internet," *Proc. IEEE Int. Conf. on Robotics and Automation*, pp. 2019-2024, 2000.

[9] A. Kheddar, C. Tzafestas, P. Coiffet, T. Kotoku, S. Kawabata, K. Iwamoto, K. Tanie, I. Mazon, C. Laugier, and R. Chellali, "Parallel Multi-Robots Long Distance Teleoperation," *Proc. Int. Conf. on Advanced Robotics*, pp. 1007-1012, 1997.

[10] W. S. Kim and A. K. Bejczy, "Demonstration of a High-Fidelity Predictive/Preview Display Technique for Telerobotic Servicing in Space," *IEEE Trans. on Robotics and Automation*, Vol. 9, No. 5, pp. 698-702, 1993.

[11] L. F. Peñin, K. Matsumoto, and S. Wakabayashi, "Force Reflection for Time-delayed Teleoperation of Space Robots," *IEEE Int. Conf. on Robotics and Automation*, pp. 3120-3125, 2000.

[12] A. Rubio, A. Avello, and J. Florez, "On the Use of Virtual Springs to avoid Singularities and Workspace Boundaries in Force-Feedback Teleoperation," *IEEE Int. Conf. on Robotics and Automation*, pp. 2690-2695, 2000.

[13] T. B. Sheridan, *Telerobotics, Automation, and Human Supervisory Control*, Cambridge, MA: MIT Press, 1992.

[14] T. Suzuki, T. Fujii, K. Yokota, H. Asama, H. Kaetsu, and I. Endo, "Teleoperation of Multiple Robots through the Internet," *Proc. IEEE Int. Workshop on Robot and Human Communication*, pp. 84-89, 1996.

[15] M. Woo, J. Neider, and T. Davis, *OpenGL Programming Guide 2nd Edition*, Addison Wesley Developer's Press, 1997.

The Visual Acts Model for Automated Camera Placement During Teleoperation

B. G. Brooks, G. T. McKee
Department of Computer Science, The University of Reading, UK
P. S. Schenker
Jet Propulsion Laboratory, USA

Abstract

This paper describes the Visual Acts theory and architecture for automated camera placement during teleopration. The theory is motivated by the goal to provide operators with task relevant information in a timely manner. The Visual Acts architecture combines top-down deliberative task modelling with bottom-up reactive observation of an operator to select camera views which provide the operator with task-relevant information. The paper describes the reactive component of the architecture, its implementation in an agent-based architecture and experimental results that confirm the validity of the approach.

Keywords: Teleoperation, automated viewing, camera control.

1 Introduction

The term "teleoperation" denotes a remote control operation where the human operator and remote site are separated by some barrier. Typically teleoperation tasks can be performed in two ways. In the first case, a single operator has sole control. The complexity of managing and integrating remote viewing resources is sufficiently high, however, that a second operator is often required solely for this task [1]. In turn, this additional workload and personnel assignment is often unacceptable in resource–limited mission environments such as undersea and space telerobotic operations.

Reducing the complexity of the operator interface and improving the value of the sensor data returned from the remote environment are crucial to improving operator performance and, therefore, making teleoperation applicable to an ever more demanding set of tasks. In response to the first objective, researchers have been developing techniques to project the operator further and further into the remote environment [2] in a form that so closely matches his real–world experiences that he feels completely immersed [3; 4].

Within teleoperation, researchers have attempted to improve operator performance by providing tools to aid in the control of the remote manipulators [5], to aid the interpretation of the sensory feedback [6], or to improve the *value* of the sensor data for the operator [7]. Augmenting video feedback to the operator is a well known and effective technique for improving the presentation of sensory data from the environment [6]. Deliberately placing the sensors in the remote environment to maximize the value of the data returned is another effective technique for improving sensory data from the environment [8; 9]. Automating camera placement is one technique for doing this [10; 11; 12].

This paper describes a method for automated camera placement in a teleoperation environment. The system's ability to place cameras in the environment takes into account environmental constraints and the kinematics of the visual system, and can therefore select from a rich set of viewpoints. This in turn removes the burden of camera control from the operator and improves the quality of views offered. The core of this work is Visual Acts, a theory first presented in [11; 13; 14]. The following section outlines the Visual Acts theory. Section 3 describes the Visual Acts implementation, focusing on viewpoint evaluation. Section 4 describes the Visual Acts agent architecture employed in the implementation. Section 5 describes an experimental study employing a peg-in-hole task that was used to evaluate the model. Finally, section 6 provides our conclusions.

2 Visual Acts

In its simplest form Visual Acts can be described as a technique for computer assisted camera placement in a teleoperation environment. It is aimed at the dynamic placement of cameras that are mobile to some degree. Visuals Acts can roughly be divided into two topics; modelling and reasoning about the task being performed, and selecting camera viewpoints to deliver optimal views in the context of a particular operation.

2.1 Modelling the Task

Understanding the operator's intention and the task information they require is essential in order to offer effective assisted camera placement. A central tenet of Visual Acts is to position the cameras based on a reasoned model of the task. Visual Acts considers a task at four levels of abstraction, namely task, subtask, visual goal and viewpoint evaluation levels.

The *task* level specifies the goal state of the completed task in terms of geometric or topological constraints. Standard assembly planning techniques could be used to break the task down into a sequence of assembly subtasks. Visual Acts collects the set of all possible assembly sequences and then monitors the operator as the task proceeds to identify which subtask is being performed at any one time.

The *subtask* level specifies the goal state of each assembly operation, usually in terms of topological constraints. These tend to be classified as *operations* which define the general form of the subtask. Consider, for example, the subtask "put floppy disk in floppy drive." In this case the subtask would be classified as the operation *peg–in–hole*, in which the peg is the floppy disk, the hole is the floppy drive, and the "in" relationship defines the topological constraint.

The *visual goal* level identifies a set of simple manipulations that will need to be performed in order for an operation to be completed. Visual Acts models the operation in terms of the questions that the operator might be asking. The questions are deliberately selected for their visual relevance. For the *peg–in–hole* operation a valid question might be "how far is the peg from the hole?" Other *visual goals* are also deliberately selected for their pertinence to the current environmental conditions. For example, given a Cartesian manipulator, the operator might be more inclined to ask two separate questions; "how far is the peg from the hole in the x direction?" and "how far is the peg from the hole in the y direction?"

Finally, the *viewpoint evaluation* level is concerned with actual perceptual act that the operator uses to answer the questions posed by each visual goal. It defines an evaluation function for camera placement that takes into account the nature of the objects, the presentation of the data, and a knowledge of the operator's ability to process that data.

2.2 Selecting Camera Viewpoints

A significant portion of the Visual Acts model is the selection of camera viewpoints to deliver task-related data to the operator. The method it adopts for doing this is described in the following section. However, selecting camera viewpoints is more than just optimizing a viewpoint evaluation function. The core of Visual Acts includes:
- selection of which assembly subtask is active,
- selection of which visual goal is active, and
- evaluation of the viewpoint with respect to the visual goal.

In addition, automatic camera placement by definition removes camera control from the operator, which in turn may lead to problems. Therefore Visual Acts must also:
- reduce operator confusion/disorientation caused by cameras moving unexpectedly, and
- provide a method of converting master device controls according to the operator's current viewpoint.

There are also significant computational issues that need to be addressed. Visual Acts demands a method of searching for an optimal viewpoint taking into account the constraints of the environment (collision avoidance) and the kinematic constraints of the viewing system, and delivering this viewpoint in a *timely* fashion.

3 Visual Acts Implementation

Visual Acts starts out with three quantities that are given before the task can begin:
1. A geometric model of the remote workcell.
2. A kinematic model of the camera placement system, the remote teleoperated manipulator(s), and an optical model of the cameras.
3. A geometric description of the task.

Although the current implementation of Visual Acts demands a calibrated geometric model of the remote workcell, the intention is to reduce, and eventually remove, this requirement. The geometric model is manually calibrated to the real environment.

The Visual Acts implementation comprises a deliberative and a reactive component. The deliberative component decomposes the high-level task into a set of atomic operations, then decomposes these into visual goals, and represents the latter as viewpoint evaluation functions in the viewpoint selection algorithm. The decomposition does not rule out any possible variation: it neither dictates the sequence of operations nor the way in which operations are performed. The functions of the deliberative component were executed off-line. The reactive component concentrates on monitoring the operator to discover which task is being performed and which visual goal describes the operator's current actions, and evaluating proposed viewpoints.

3.1 Monitoring the Operator

The method of determining the operator's current intentions is to predict for each visual goal how the operator would behave were he to perform the task exactly as promoted by the goal, and then to measure how well this prediction matches his current activity. The latter is evaluated based on the motion of the object with respect to the target, which equates to the motion of the manipulator end effector.

3.2 Viewpoint Evaluation

The Visual Acts model evaluates task-related visual goals using a technique, visibility modelling, that assesses the degree of task completion that can be observed in the image. Visibility modelling aims to find viewpoints that ensure *sufficent* information for the operator to localize the objects, but also to optimize the viewpoint for evaluating the information pertaining to the task-related visual goals. Visual Acts takes the ability to identify the location of the axes of symmetry of the object in an image as a measure of the ability to localise the object. Two values need to be established:

1. How well the location of the axes of symmetry for each object can be determined from the image, which in turn involves
   (a) determining if the features needed to imply the location of the axes of symmetry are present, and
   (b) favouring images where these features are most prominent.
2. How well *task* information can be determined from the image.

The evaluation is based on knowledge of the objects' axes of symmetry and their surfaces (facets). The facets are listed in a database with two values according to the relative merits of each facet to the operator in determining the location and orientation of the object. The purpose for these values is to favour object features that provide significantly more information about the object than other features. The value of the viewpoint with respect to determining how well the location of the axes of symmetry for an object can be determined from the image is given by the equation:

$$L(o) = \sum_{f_i} S(f_i) K(f_i) \quad (1)$$

where the summation is over all the facets, f_i, of the object. $S(f_i)$ is the visible surface area of the facet in the image, and $K(f_i)$ is the value of the facet for the particular type of visual goal. For example, either end surface of a cylindrical object is more valuable for locating its longitudinal axis of symmetry than its cylindrical surface.

The operator needs to see *both* objects in the image in order to perform the task. The value of the viewpoint with respect to both objects is the minimum of the information from each:

$$E(o_1, o_2) = \min\left(L(o_1), L(o_2)\right) \quad (2)$$

where $L(o_i)$ is defined in equation 1. For example, a viewpoint showing end-surfaces of two cylindrical objects to be aligned is preferred to viewpoints in which either is viewed orthogonal to its longitudinal axis.

The current algorithm ignores this value except in ensuring that there is sufficient information in the image for the operator to locate the objects. The value of the viewpoint, therefore, is given by the equation:

$$V(o_1, o_2) = \begin{cases} T(o_1, o_2) & : \text{ if } E(o_1, o_2) > res \\ 0 & : \text{ otherwise} \end{cases} \quad (3)$$

where $T(o_1, o_2)$ gives a value for the task information content in the image, and *res* is a constant that enforces the *resolution* constraint (objects should occupy at least *res* area of the image).

The *task* information denoted by $T(o_1, o_2)$ in equation 3 depends on the type of visual goal: translational (such as "axes of symmetry aligned") or rotational (such as "axes of symmetry parallel"). For a translational visual goal, task information is proportional to the distance between the two objects' axes of symmetry in the image (l in figure 1). This is measured from the centre of the manipulated object (peg) along a line normal to the axis of symmetry of the target object (hole). For a rotational visual goal, task information is proportional to the angle between the axes of symmetry as projected in the image (θ in figure 1).

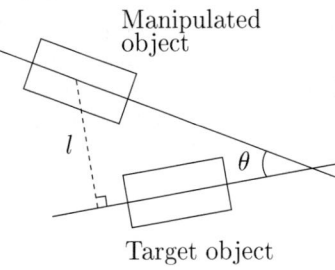

Figure 1: Task information for visibility modelling.

This viewpoint evaluation function is applied to each task related visual goal in turn.

3.3 Viewpoint Search Algorithm

The viewpoint search algorithm employed by Visual Acts was inspired by the MVP approach [15]. A number of seed viewpoints were pre-selected that were known to be within the constraints of the camera manipulator kinematics, visibility of the *workcell*, field–of–view and depth–of–field of the camera. These viewpoints were pre-selected for each experimental task and were deliberately chosen to cover the entire region as evenly as possible. The Visual Acts task–based algorithm was re-arranged into the form of an evaluation function suitable for a simple hill-climbing search method. The search executes each test roughly in the order of computational cost and selectivity. The tests for camera manipulator kinematics, visibility, field–of–view and depth–of–field simply return a boolean value indicating whether the viewpoint meets the minimum requirements of the constraint. If rejected by any test, the search was immediately discontinued in that area. A separate search was started for each pre-selected seed point and was restricted to a limited volume designed to reduce unnecessary overlap with neighbouring searches. This simple algorithm was considered acceptable since the Visual Acts function describes a fairly smooth search space.

4 The Visual Acts Architecture

A key choice in the development of Visual Acts was the adoption of an agent based approach in order to address two fundamental issues:

First, from the task description, and a knowledge of which operations have already been performed, Visual Acts can determine a set of feasible steps that might reasonably be performed next, and a set of feasible methods for performing those steps. Each feasible step implies an expected human behaviour. Visual Acts can employ an agent per probable feasible step; each agent monitoring the operator and evaluating his actions with respect to expected behaviour. This is similar to the concept of agents that produce an emergent behaviour, except in this case it is the operator producing the behaviour. Visual Acts is "reverse engineering" the behaviour to determine the active set of visual goals. The combination of these visual goals provides an acceptable description of the operator's immediate intentions.

Second, in many cases the operator may be attempting to perform more than one operation (or goal) in parallel, and therefore Visual Acts needs to be able to offer a view that may satisfy, to some degree, the requirements of a number of operations. However, even if the operator has a clearly identifiable operation in mind, the constraints on the camera (movement, occlusion, etc.) may make it difficult to provide an optimum view. In either case, there is no clear mapping between task and camera viewpoint. Several agent architectures exist where agents send votes to be combined rather than commands to be selected. Such architectures tend to encourange simultaneous fulfilment of multiple goals. This style of interaction can address the problem of matching resources to the operators' requirements.

The architecture developed for Visual Acts adopts features from the agent architectures of [16] and [17]. The aim is similar to the later in attempting to provide a viewpoint of value to all the visual goals, but includes the coordinator concept from the former such that the visual goal agents can alter the relative importance of their subordinate viewpoint evaluator agents according to their assessment of the operator's actions. In the Visual Acts architecture every visual goal, viewpoint evaluator, and camera resource is modelled as an agent:

1. The **visual goal agents** monitor the operator to see how closely the operator's actions match those suggested by their visual goal. The degree of match is used to set the subordinate viewpoint evaluator agents' fixed priority.

2. Each **viewpoint evaluator agent** searches for a viewpoint that provides the most information relevant to their visual goal. Then they send their viewpoint proposal to the camera resource agent.

3. Each viewpoint proposal received by the **camera resource agent**, plus the *current* viewpoint, are sent to *all* the viewpoint evaluator agents for re-evaluation.

4. The viewpoint evaluator agents re-evaluate each proposed viewpoint with respect to their own requirements, and returns the value (in the range 0 to 1) multiplied by their fixed priority (in the range 0 to 1) back to the camera resource agent.

5. Once all viewpoint proposals have been collected and evaluated, the camera resource agent sums the scores for each proposed viewpoint. It selects the one with the highest aggregate value, compares it to the value of the current viewpoint, and instructs the camera to move to the winning viewpoint.

6. In the case of multiple cameras, procedures 2 to 5 are repeated for every camera.

This algorithm was implemented and tested experimentally.

5 Experimental Results

An experiment was designed and performed to answer the simple question "Does Visual Acts work?" The sole criterion for evaluating Visual Acts in this respect is whether it can provide *sufficient* views to enable the operator to complete the task. The secondary aim was to measure its effect on operator performance.

The experimental task consisted of grasping a wooden peg and placing it in a hole. The operator was given one mobile camera and one robotic manipulator. The wooden peg was only slightly smaller than the hole or the gripper opening forcing the operator to conduct the task accurately. The robot manipulator was a 5 degrees of freedom CRS robot arm. The robot's waist and gripper were under joint control, but the wrist was under Cartesian control. This mixture of control was found to be the most natural and to feel most intuitive. A black and white camera on a pan-tilt head was mounted on the crane. The camera had automatic iris control and a fixed-focus lens, leaving the operator in control of the crane, pan and tilt. Figure 2 shows a diagram of the initial positions of the crane and the CRS robot arm. Figure 3 shows the operator console.

Figure 3: Operator console.

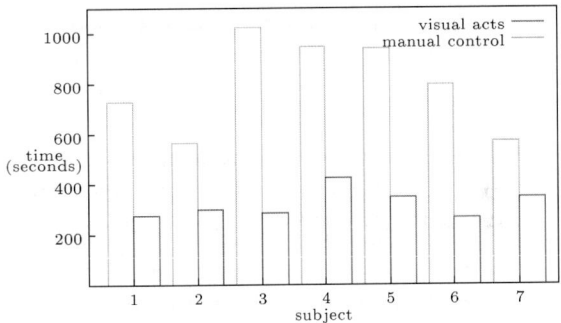

Figure 4: Timing measurements.

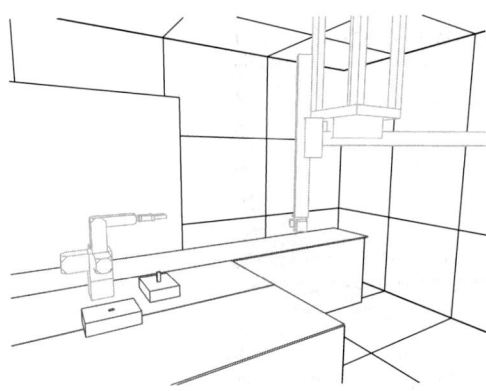

Figure 2: Initial layout of laboratory.

Operators were selected at random. Every other subject was asked to perform the task using manual control and then Visual Acts. Alternate subjects were asked to perform the task in the reverse order. The positions and velocities of all the robot joints in the laboratory, and the velocity instructions from the operator's control panel, were recorded in a log file. Figure 4 shows a bar graph of timings for each subject performing the task using Visual Acts and with the cameras under manual control. These results were fitted to a normal distribution, giving a mean and standard deviation of 797 and 184 respectively for manual control and 320 and 56 respectively for Visual Acts.

The task was designed such that it was impossible to complete without moving the cameras. Therefore, the first question posed, *"Does Visual Acts Work?"* would require Visual Acts to perform at least basic camera placement for the operator to complete the task. This question is trivially proven since all operators managed to complete the task using camera viewpoints selected exclusively by Visual Acts. The graph also shows that in all cases, operators performed the task more quickly using Visual Acts than under manual control. Statistical analysis of the data using a two tailed t-test confirmed the hypothesis that "Camera placement using Visual Acts improves operator performance."

Figure 5 shows a simulated view of the laboratory from a virtual camera positioned above the robot arm and looking down. The viewpoints are depicted in the simulation by lines. The end point of the lines closest to the task represents the actual view*point* selected, the orientation of the lines represents the viewing direction, and for manually selected viewpoints, the length of the line represents the length of time that the operator maintained that view.

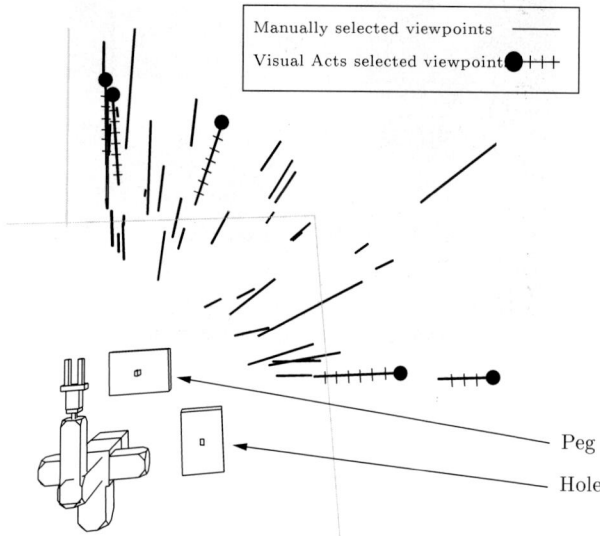

Figure 5: Viewpoints for grasping task.

It was expected that all operators would tend to select similar viewpoints for each component of the task, and that these viewpoints would appear to cluster together when visualized. However, the figure shows that the viewpoints selected by the human operators appear to be well spread out. This tends to confirm the observation that subjects were quickly getting frustrated at the effort of perfecting viewpoints and were tending to accept poor views in order that they could continue to make some (limited) progress at the task. Subjects tended to adapt the existing viewpoint with the aim of reducing the limitations of the current view rather than trying to find the best view for the next part of the task. In contrast, Visual Acts provided a smaller number of approximately orthogonal viewpoints.

6 Conclusions

The experimental results clearly show that Visual Acts works — it provides camera views of sufficient quality to allow the operator to perform the task. The more rigorous test of whether it improves teleoperator performance has also been shown. Visual Acts provides camera viewpoints that significantly reduce task completion time. Even removing the time taken to control the camera system, Visual Acts still significantly reduces the length of time spent manipulating the objects in the remote workcell.

References

[1] Paul S. Schenker and Gerard T. McKee. Human–machine interaction in telerobotic systems and issues in the architecture of intelligent and cooperative control. In *Proceedings of the IEEE International Symposium on Intelligent Control*, pages 385–390, Monterey, California, August 1995.

[2] F. Biocca and B. Delaney. Immersive virtual reality technology. In F. Biocca and M. R. Levy, editors, *Communication in the Age of Virtual Reality*, pages 57–124. Erlbaum, Mahwah, New Jersey, 1995. ISBN 0-805-81550-3.

[3] Thomas B. Sheridan. *Telerobotics, Automation, and Human Supervisory Control*. MIT Press, 1992. ISBN 0-262-19316-7.

[4] Larry Li, Brian Cox, Myron Diftler, Susan Shelton, and Barry Rogers. Development of a telepresence control ambidextrous robot for space apllications. In *Proceedings of the IEEE International Conference on Robotics and Automation*, pages 58–63, Minneapolis, Minnesota, April 1996.

[5] S. E. Everett and R. V. Dubey. Human–machine cooperative telerobotics using uncertain sensor or model data. In *Proceedings of the IEEE International Conference on Robotics and Automation*, pages 1615–1622, Leuven, Belgium, 16–20 May 1998.

[6] Ronald T. Azuma. A survey of augmented reality. *Presence: Teleoperators and Virtual Environments*, 6(4):355–385, August 1997.

[7] P. S. Schenker, S. F. Peters, E. D. Paljug, and W. S. Kim. Intelligent viewing control for robotic and automation systems. In Paul S. Schenker, editor, *Sensor Fusion VII*, volume 2355 of *Proceedings of the SPIE*, pages 301–313, Boston, Massachusetts, 31 October–1 November 1994.

[8] Konstantinos A. Tarabanis, Peter K. Allen, and Roger Y. Tsai. A survey of sensor planning in computer vision. *IEEE Transactions on Robotics and Automation*, 11(1):86–104, February 1995.

[9] Seth Hutchinson, Greg Hager, and Peter Corke. A tutorial on visual servo control. *IEEE Transactions on Robotics and Automation*, 12(5):651–670, May 1996.

[10] Yujin Wakita, Shigeoki Hirai, and Toshiyuki Kino. Automatic camera-work control for intelligent monitoring of telerobotic tasks. In *Proceedings of the IEEE/RSJ International Conference on Intelligent Robots and Systems*, pages 1130–1135, Raleigh, North Carolina, 7–10 July 1992.

[11] G. T. McKee and P. S. Schenker. Visual acts for goal–directed vision. In Paul S. Schenker, editor, *Sensor Fusion VII*, volume 2355 of *Proceedings of the SPIE*, pages 314–322, Boston, Massachusetts, 31 October–1 November 1994.

[12] Hari Das, Thomas B. Sheridan, and Jean-Jacques E. Slotine. Kinematic control and visual display of redundant teleoperators. In *Proceedings of the IEEE International Conference on Systems, Man, and Cybernetics*, pages 1072–1077, Cambridge, Massachusetts, 1989.

[13] Gerard T. McKee and Paul S. Schenker. Visual acts for remote viewing during teleoperation. In *Proceedings of the IEEE International Conference on Robotics and Automation*, pages 53–58, Nagoya, Aichi, Japan, 21–27 May 1995.

[14] Gerard T. McKee and Paul S. Schenker. Human–robot cooperation for automated viewing during teleoperation. In *Proceedings of the IEEE/RSJ International Conference on Intelligent Robots and Systems*, volume 1, pages 124–129, Pittsburgh, Pennsylvania, 5–9 August 1995.

[15] Konstantinos A. Tarabanis, Roger Y. Tsai, and Peter K. Allen. The MVP sensor planning system for robotic vision tasks. *IEEE Transactions on Robotics and Automation*, 11(1):72–85, February 1995.

[16] Luís Correia and A. Steiger-Garção. A model of hierarchical behaviour control for an automated vehicle. In P. Gaussier and J. Nicoud, editors, *From Perception to Action*, pages 396–399, Lausanne, Switzerland, 7–9 September 1994. IEEE Computer Society Press. ISBN 0-8186-6482-7.

[17] J. K. Rosenblatt. DAMN: A distributed architecture for mobile navigation. In H. Hexmoor and D. Kortenkamp, editors, *Proceedings of the AAAI Spring Symposium on Lessons Learned from Implemented Software Architectures for Physical Agents*, Stanford, California, 27–29 March 1995.

Sliding-Mode-Based Impedance Controller for Bilateral Teleoperation under Varying Time-Delay

Hyun Chul Cho and Jong Hyeon Park

School of Mechanical Engineering
Hanyang University
Seoul, 133-791, Korea
email: ando@hymail.hanyang.ac.kr

Kyunghwan Kim and Jong-Oh Park

Microsystem Research Center
Korea Institute of Science and Technology
Seoul, 130-650, Korea

Abstract

In the previous works, we have proposed new control schemes based on the sliding mode control and impedance control in order to cope with varying time delays. However, the previous controller needs local compliance to reduce contact forces between the slave and environment. In this paper, we modify the previous one to have a sliding surface for the slave include an impedance model. Since the nonlinear gain of the sliding controller is independent of the time delay as in the previous proposed controllers, it is not necessary to measure or estimate the time delay to implement the controller. The validity of the proposed control scheme is demonstrated by experiments with a 1-dof master/slave system connected through the Internet.

1 Introduction

Bilateral teleoperation systems transmit the information on the force at the slave to the master, which can considerably improve their task performance. However, these systems easily becomes unstable when communication time delays between the master and the slave exist. Anderson and Spong [1] pointed out that even a small constant time delay could make bilateral systems unstable and certainly degrade the operator's intuition and performance.

Recently, more and more computer networks such as the Internet are used as communication channels of teleoperation systems due to their availability and flexibility. With the computer networks, however, the communication time-delays between the master and the slave are not only significantly large but also *changing* depending on network congestions, which could make the overall system unstable.

Several control methods have been proposed to overcome the instability problem due to varying time-delays in communication channels: PD-control [5], scattering transformation [6, 8], wave variables [7, 13], H_∞ gain-scheduling [9, 11], environment predictive display [12], and master state prediction [14]. In order to implement these algorithms, however, some methods require time delay measurements using time stamps [5, 7, 9, 11, 14] and/or estimates to compensate data losses [5, 10]. And, as a conservative method, Kosuge et al. used "virtual time delay" to keep the time delays constant [6, 8, 12].

In our previous works [2, 3], we have proposed a new control scheme called the modified sliding-mode controller in order to cope with varying time delays. The proposed controller guarantees its nonlinear gain to be independent of the magnitude of the time delay. Thus, it is unnecessary to estimate or measure the varying time delay between the master and the slave. However, as any other position-based control scheme, it needs a kind of local compliance control to reduce impact at the collision between the slave and the environment. Control input changes according to a constrained condition may make the system be unstable [4], even though this is not the case in our previous experiments.

In this paper, we modify the sliding-mode controller for the slave to include an impedance model in its sliding surface. The proposed controller can be used for both constrained and unconstrained motion without control law changes in the case. Since the nonlinear gain for satisfying the sliding condition is not a function of the time delay and the scaling factors, we can freely design the desired impedance for the master and slave with scaling factors depending on applications.

This paper is organized as follows. Section 2 defines dynamic models for the master and slave, and delayed signals with scaling factors. Section 3 describes the proposed controller and its stability analysis. Experiments with a 1-dof teleoperation system and their results are shown in Section 4, followed by conclusions in Section 5.

2 Model Definitions

2.1 Dynamics for Master and Slave

The dynamics of the single dof master/slave system are modeled as a mass-damper system.

$$m_m \ddot{x}_m(t) + b_m \dot{x}_m(t) = u_m(t) + f_h(t) \quad (1)$$
$$m_s \ddot{x}_s(t) + b_s \dot{x}_s(t) = u_s(t) - f_e(t) \quad (2)$$

where x and u denote position and torque; m and b denote mass and viscous coefficient; subscript 'm' and 's' denote the master and the slave, respectively; f_h is the force applied at the master by the operator, and f_e is the force exerted on the slave by its environment.

2.2 Delayed Signals and Scaling Factors

A bilateral teleoperation system can be represented by three blocks as in Fig. 1, where x_m^d, $\dot{x}_m^d$, and f_h^d are the position, velocity commands and the operating force, respectively, transmitted from the master to the slave through the communication. Using the Internet as the communication line, the time delay between the master and slave varies differently with the signal direction as well as the network condition. Thus, the signals from and to the communication block are related as:

$$\begin{array}{ll} x_m^d(t) = x_m(t - T_1(t)), & \dot{x}_m^d(t) = \dot{x}_m(t - T_1(t)) \\ f_h^d(t) = f_h(t - T_1(t)), & f_e^d(t) = f_e(t - T_2(t)) \end{array}$$

where $T_1(t)$ is the time delay of the signal flowing from the master to the slave, and $T_2(t)$ is in the opposite direction. Note that these delays vary in time t and that the master velocity must be also transmitted to the slave as well as the position since the time derivative of x_m^d is needed in the slave controller but cannot be computed from x_m^d due to the varying time delay.

These delayed signal out of the communication block is then scaled down or up with some factors depending on the application. Using the scaling factors, the position/velocity command to the slave and the force command to the master are modified such that

$$x_s = k_p x_m^d, \quad f_h = k_f f_e^d$$

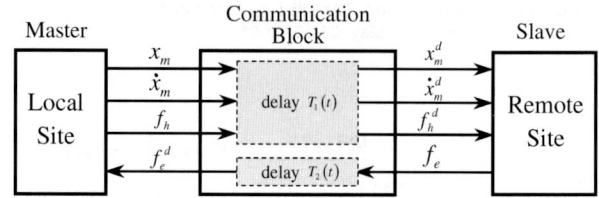

Figure 1: A block diagram of bilateral teleoperation.

where k_p and k_f are the position and force scaling factors, respectively.

2.3 Communication Protocol

The two main factors influencing Internet-based teleoperation are the transmission time delay between the master and slave and the loss of data packets owing to network congestion [5]. Before reaching its destination, data packets traverses several nodes, which route packets to different node or discard them according to the node congestion. This introduces an irregular communication time delay and possibly some loss of the data.

In the Internet-communication, TCP (Transmission Control Protocol) and UDP (User Datagram Protocol) are used as the standard communication protocol. Some researchers use the UDP for their teleoperation system [5, 11]. However, the orders of the sending data packets may be changed when transmitted with the UDP, and some kind of estimator is needed for compensating data losses. Thus, in this paper, the information exchanges are done by the TCP, which could prevent abrupt changes in the data using its *error recovery* and *reordering* capability.

3 Controller Design

In many telerobotic tasks, robot manipulators interact with their environments. Excessive contact force between the robot and the environment should be prevented to avoid system damages. Besides, good tracking in freespace in also important for task performance. It is well known that impedance control, which controls the relationship between the applied force and the position of the manipulator, is suitable for these continuously constrained and unconstrained tasks.

For this reason, the impedance control and the sliding-mode-based impedance control are used for the master and the slave, respectively. These controllers are designed based on so called position-force teleoperation. The force-controlled master device reflects to the human operator the contact force between the slave and its environment while the position-controlled slave follows the trajectory of the master.

3.1 Impedance Control for the Master

With the impedance control, the desired characteristics between the human force and the external force can be selected appropriately. Suppose that the target impedance for the master is specified by

$$\bar{m}_m \ddot{x}_m(t) + \bar{b}_m \dot{x}_m(t) + \bar{k}_m x_m(t) = f_h(t) - k_f f_e^d(t) \quad (3)$$

where $\bar{m}_m$, $\bar{b}_m$ and $\bar{k}_m$ are the desired inertia, damping coefficient, and stiffness, respectively.

Combining Eqs. (1) and (3) to remove acceleration $\ddot{x}_m$ results in the control input to the master:

$$u_m(t) = \left(b_m - \frac{m_m}{\bar{m}_m}\bar{b}_m\right)\dot{x}_m(t) + \left(\frac{m_m}{\bar{m}_m} - 1\right)f_h(t) \\ - \frac{m_m}{\bar{m}_m}\{k_f f_e^d(t) + \bar{k}_m x_m(t)\}. \quad (4)$$

3.2 Sliding-Mode-Based Impedance Control for the Slave

Sliding-mode-based impedance control is a kind of robust impedance control [15]. This type of controller uses the robust property of the sliding mode control against uncertainties such as parameter uncertainty and time delay. In this controller, the sliding surface includes an impedance model, and the slave exhibits its desired impedance behaviors when the controller satisfies the sliding condition.

A target impedance characteristic for the slave is specified such that:

$$\bar{m}_s \tilde{x}_s''(t) + \bar{b}_s \tilde{x}_s'(t) + \bar{k}_s \tilde{x}_s(t) = f_e(t) \quad (5)$$

where $\bar{m}_s$, $\bar{b}_s$ and $\bar{k}_s$ are the desired inertia, damping coefficient, and stiffness, respectively;

$$\tilde{x}_s''(t) := \ddot{x}_s(t) - k_p \ddot{x}_m^d(t) \\ \tilde{x}_s'(t) := \dot{x}_s(t) - k_p \dot{x}_m^d(t) \\ \tilde{x}_s(t) := x_s(t) - k_p x_m^d(t).$$

From this, the impedance error, $I_e(t)$, can be defined as

$$I_e(t) := \bar{m}_s \tilde{x}_s''(t) + \bar{b}_s \tilde{x}_s'(t) + \bar{k}_s \tilde{x}_s(t) - f_e(t). \quad (6)$$

Next, let us define sliding surface $s(t)$ as

$$s(t) := \frac{1}{\bar{m}_s} \int_0^t I_e(\tau) d\tau \\
= \dot{x}_s(t) + \frac{\bar{b}_s}{\bar{m}_s} x_s(t) + \frac{\bar{k}_s}{\bar{m}_s} \int_0^t x_s(\tau) d\tau \\
+ k_p \left(\frac{\bar{b}_m}{\bar{m}_m} - \frac{\bar{b}_s}{\bar{m}_s}\right) \int_0^t \dot{x}_m^d(\tau) d\tau \\
+ k_p \left(\frac{\bar{k}_m}{\bar{m}_m} - \frac{\bar{k}_s}{\bar{m}_s}\right) \int_0^t x_m^d(\tau) d\tau \\
- \int_0^t \left[\frac{k_p}{\bar{m}_m}\{f_h^d(\tau) - k_f f_e^{dd}(\tau)\} + \frac{1}{\bar{m}_s} f_e(\tau)\right] d\tau$$

where $f_e^{dd}(\tau) := f_e(\tau - T_1(\tau) - T_2(\tau - T_1(\tau)))$. Note that $\dot{s}(t)$ does not contain any jerk-term since the sliding surface is defined as the integration of $I_e(t)$.

The equivalent control [16] is obtained by using $\dot{s} = s = 0$. Substituting the delayed master impedance model of Eq. (3) by $T_1(t)$ and slave dynamics of Eq. (2) into $\dot{s} = 0$, and considering the uncertainty in the parameters give the control law for the slave.

$$u_s(t) = -k_p \frac{m_s}{\bar{m}_m}\{\bar{b}_m \dot{x}_m^d(t) + \bar{k}_m x_m^d(t) - f_h^d(t) \\ + k_f f_e^{dd}(t)\} - \frac{m_s}{\bar{m}_s}\left(\bar{b}_s \tilde{x}_s'(t) + \bar{k}_s \tilde{x}_s(t)\right) \\ + b_s \dot{x}_s(t) + \left(\frac{m_s}{\bar{m}_s} + 1\right) f_e(t) - K_g \cdot \text{sat}\left(\frac{s(t)}{\Phi}\right) \quad (7)$$

where K_g is the nonlinear gain; $\text{sat}(\cdot)$ is a saturation function; Φ is the thickness of the boundary layer used to reduce the chattering in the control input.

In the previous equations, the delayed external force signal, $f_e^{dd}(t)$ is needed for calculating $u_s(t)$ as well as $s(t)$. This delayed external force can be obtained easily by just sending it to the master and receiving it back from it. This procedure is depicted in Fig. 2. There is no need to measure the time delay in each direction or to store the external force data.

When the control input of Eq. (7) satisfies the sliding condition of $\dot{s}(t)s(t) \leq -\eta|s(t)|$, the states reach the sliding surface in a finite time. Applying the control input of Eq. (7) to the slave dynamics and expressing this in terms of $s(t)$ gives

$$\dot{s}(t) + K_g/m_s \cdot \text{sat}(s(t)/\Phi) = 0.$$

From this, the boundary of the nonlinear gain, K_g, for satisfying the sliding condition is obtained as

$$K_g \geq m_s \eta.$$

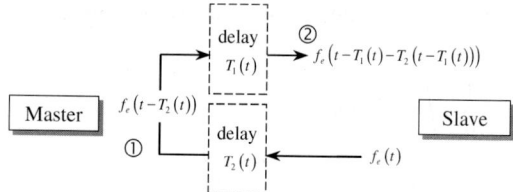

Figure 2: Delayed force signals.

Since the boundary consists of only a constant value, the gain can be selected independently of the time delay and the scaling factors.

With the proper gain, K_g, the state trajectory is kept in the region around the sliding surface. In this case, the slave shows the target impedance characteristic since $I_e = \bar{m}_s \dot{s} \approx 0$ [15]. The overall block diagram including the master, slave and the proposed controllers is shown in Fig. 3.

3.3 Stability Analysis

A passive network will always be absolutely stable; absolute stability do not, however, imply that the system is passive [18]. Therefore absolute stability can be a less conservative tool to treat stability problem. Absolute stability is defined as follows [18].

Definition: A linear two-port is said to be *absolutely stable* if there exists no set of passive terminating one-port impedances for which the system is unstable. If the network is not absolutely stable, it is potentially unstable.

From the properly controlled master and slave behaviors of Eqs. (3) and (5), its hybrid matrix [18] is defined as

$$\begin{bmatrix} F_h(s) \\ V_s(s) \end{bmatrix} = \begin{bmatrix} h_{11} & h_{12} \\ h_{21} & h_{22} \end{bmatrix} \begin{bmatrix} V_m(s) \\ -F_e(s) \end{bmatrix}$$

where $F_h(s)$, $V_m(s)$, $V_s(s)$ and $F_e(s)$ are the Laplace transforms of $f_h(t)$, $\dot{x}_m(t)$, $\dot{x}_s(t)$, and $f_e(t)$, respectively, and

$$h_{11} = \bar{m}_m s + \bar{b}_m + \frac{\bar{k}_m}{s}, \quad h_{12} = -k_f e^{-T_2 s},$$

$$h_{21} = k_p e^{-T_1 s}, \quad h_{22} = \frac{s}{\bar{m}_s s^2 + \bar{b}_s s + \bar{k}_s}.$$

Llewellyn's stability criteria [17], which provides necessary and sufficient conditions for absolute stability, is introduced for stability analysis. A two-port network is absolutely stable if and only if [18]

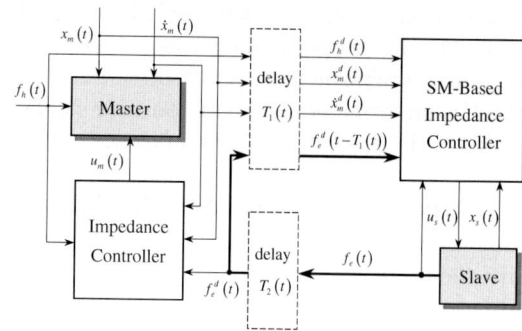

Figure 3: The overall system block diagram.

a) h_{11} and h_{22} have no poles in the right half plane;

b) Any poles of h_{11} and h_{22} on the imaginary axis are simple with real and positive residues;

c) For all real values of ω,

$$\text{Re } h_{11} \geq 0, \quad \text{Re } h_{22} \geq 0,$$
$$2\text{Re } h_{11}\text{Re } h_{22} - \text{Re}(h_{12}h_{21}) - |h_{12}h_{21}| \geq 0.$$

For the given two-port network, conditions (a) and (b), together with the first and second conditions in (c) are satisfied with positive impedance parameters. The last of condition (c) can be expressed by

$$k_p k_f \leq \frac{2\bar{b}_m \bar{b}_s \omega^2}{\left(\bar{k}_s - \bar{m}_s \omega^2\right)^2 + \left(\bar{b}_s \omega\right)^2} \cdot \frac{1}{1 - \cos(T_1 + T_2)\omega}$$

which is satisfied only if

$$k_p k_f \leq \frac{\bar{b}_m \bar{b}_s \omega^2}{\left(\bar{k}_s - \bar{m}_s \omega^2\right)^2 + \left(\bar{b}_s \omega\right)^2}, \quad \forall \omega \geq 0. \quad (8)$$

If the designed parameters satisfy Eq. (8), the teleoperation system will be stable for any set of passive human operators and environments.

4 Experiments

In this section, the performance of the proposed control scheme is investigated through the experiments with a 1-dof master/slave bilateral teleoperation system as shown in Fig. 4. At the master side, its human operator pulls and pushes a knob attached to its motor and the force exerted at the master is measured by a load-cell. At the other side, the slave can

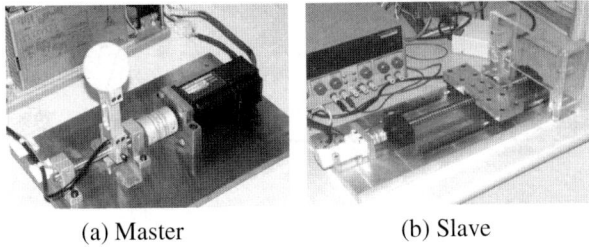

(a) Master (b) Slave

Figure 4: The master and slave system used in this experiment.

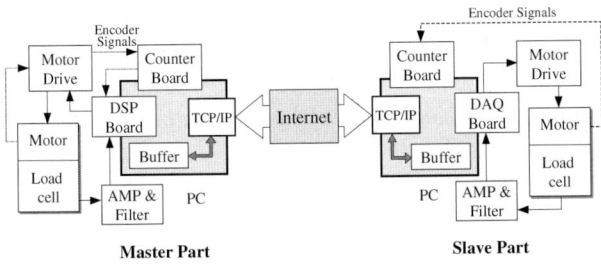

Figure 5: A detailed description of the master/slave system.

Table 1: The parameters for the experiments

k_p	0.05 [m/rad]	k_f	0.8×0.11 [m]
Φ	0.1 [m/s]	K_g	20.0 [N]
$\bar{m}_m$	0.02 [kgm^2]	$\bar{m}_s$	0.5 [kg]
$\bar{b}_m$	0.5 [Nms]	$\bar{b}_s$	70 [Ns/m]
$\bar{k}_m$	0.04 [Nm]	$\bar{k}_s$	600 [N/m]

move straight back and forth with a ball-screw mechanism and a load-cell is installed at its tip, which measures the contact force exerted by the environment. In the experiments, a wall with a some stiffness (about 31,000 [N/m]) is used as the environment of the slave.

In order to investigate the performance of the controller under a large time delay, the communication unit that introduces bilateral time delays is simulated with a memory buffer, in addition to the physical communication line. Data to be exchanged between the master and the slave is written to the buffer. It is sent to its intended destination only at the moment when the buffer is full. Thus, the time delay at the buffer depends on the size of the buffer. In the experiments, the buffer size is fixed at a constant value. Another source of the time delay is the physical communication line. Thus, the average of the time delay depends only on the size of the buffer, and its deviation is exactly that of the actual communication line. The detailed diagram of the entire system is given in Fig. 5.

Figures 6 and 7 show the experimental results when RTT (Round-Trip Time) is about 1.0 or 5.0 second. The parameters for the experiment are summarized in Table 1. In the table, the unit of k_f is selected to be [m] since the dimension of f_h and f_e is [Nm] and [N], respectively. In spite of a large time delay, the slave shows proper tracking performance, and stable contact behaviors.

5 Conclusions

In this paper, the sliding-mode-based impedance controller whose sliding surface includes an integration of an impedance model, is proposed. In the controller design procedure, we can design master and/or slave port impedance and select the nonlinear gain, which satisfies the sliding condition, independently of the irregular time delays in the communication channels. Experimental results show that the slave follows well commanded master trajectories, and maintains a stable contact even when there is a relatively large time delay.

References

[1] R. J. Anderson and M. W. Spong, "Bilateral Control of Teleoperators with Time Delay," IEEE Trans. on Automatic Control, Vol. 34, pp. 494–501, 1989.

[2] J. H. Park and H. C. Cho, "Sliding-Mode Controller for Bilateral Teleoperation with Varying Time Delay," IEEE/ASME Int. Conf. on Advanced Intelligent Mechatronics, pp. 311–316, 1999.

[3] J. H. Park and H. C. Cho, "Sliding Mode Control of Bilateral Teleoperation Systems with Force-Reflection on the Internet," IEEE/RSJ Int. Conf. on Intelligent Robots and Systems, pp. 1187–1192, 2000.

[4] R. Paul, "Problems and research issuses associated with the hybrid control of force and displacement," IEEE Int. Conf. on Robotics and Automation, pp. 1966–1971, 1987.

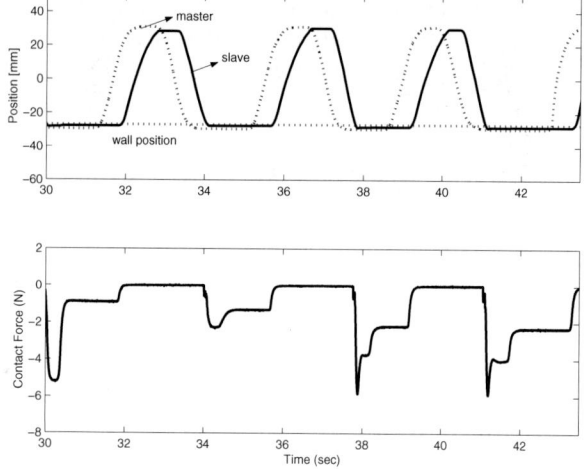

Figure 6: Scaled master/slave positions and external force when RTT≈1.0 s.

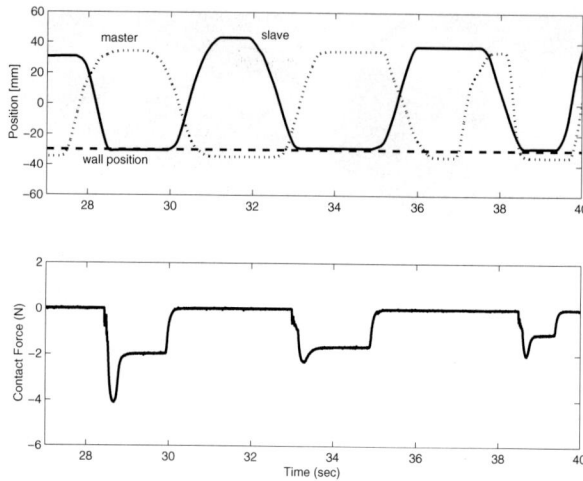

Figure 7: Scaled master/slave position and external force when RTT≈5.0 s.

[5] R. Oboe and P. Fiorini, "A Design and Control Environment for Internet-Based Telerobotics," Int. J. Robotics Research, Vol. 17, pp. 433–449, 1998.

[6] K. Kosuge and H. Murayama, "Teleoperation via Computer Network," Trans. of the Institute of Electrical Engineers of Japan-C, Vol. 117, pp. 521–527, 1997.

[7] Y. Yokokohji, T. Imaida and T. Yoshikawa, "Bilateral Control with Energy Balance Monitoring Under Time-Varying Communicaiton Delay," IEEE Int. Conf. on Robotics and Automation, pp. 2684–2689, 2000.

[8] K. Kosuge and H. Murayama, "Bilateral Feedback Control of Telemanipulator via Computer Network in Discrete Time Domain," IEEE Int. Conf. on Robotics and Automation, pp. 2219–2224, 1997.

[9] A. Sano, H. Fujimoto and M. Tanaka, "Gain-Scheduled Compensation for Time Delay of Bilateral Teleoperation System," IEEE Int. Conf. on Robotics and Automation, pp. 1916–1923, 1998.

[10] A. Lelevé et al., "Modeling and Simulation of Robotic Tasks Teleoperated through the Internet," IEEE/ASME Int. Conf. on Advanced Intelligent Mechatronics, pp. 299–304, 1999.

[11] A. Sano, H. Fujimoto and T. Takai, "Network-Based Force-Reflecting Teleoperation," IEEE Int. Conf. on Robotics and Automation, pp. 3126–3131, 2000.

[12] J. Kikuchi, K. Takeo and K. Kosuge, "Teleoperation System via Computer Network for Dynamic Environment," IEEE Int. Conf. on Robotics and Automation, pp. 3534–3539, 1998.

[13] G. Niemeyer and J.-J. E. Slotine, "Towards Force-Reflecting Teleoperation Over the Internet," IEEE Int. Conf. on Robotics and Automation, pp. 1909–1915, 1998.

[14] P. A. Prokopiou, S. G. Tzafestas and W. S. Harwin, "Towards Variable-Time-Delays-Robust Telemanipulation Through Master State Prediction," IEEE/ASME Int. Conf. on Advanced Intelligent Mechatronics, pp. 305–310, 1999.

[15] Z. Lu and A. A. Goldenberg, "Robust Impedance Control and Force Regulation: Theory and Experiments," Int. J. Robotics Research, Vol. 14, pp. 225–254, 1995.

[16] J.-J. E. Slotine and W. Li, *Applied Nonlinear Control*, Prentice-Hall, Inc., 1991.

[17] F. B. Llewellyn, "Some Fundamental Properties of Transmission Systems," Proc. IRE, Vol. 40, pp. 271–283, 1952.

[18] S. S. Haykin, *Active Network Theory*, Addison-Wesley, 1970.

Ground-Space Bilateral Teleoperation Experiment Using ETS-VII Robot Arm with Direct Kinesthetic Coupling

Takashi Imaida[1]*   Yasuyoshi Yokokohji[1]   Toshitsugu Doi[2]**
Mitsushige Oda[2]   Tsuneo Yoshikawa[1]

1: Department of Mechanical Engineering   2: National Space Development Agency of Japan
Graduate School of Engineering   Tsukuba Space Center
Kyoto University   2-1-1, Sengen, Tsukuba-shi
Kyoto 606-8501, Japan   Ibaraki 305-8505, Japan
{yokokoji|yoshi}@mech.kyoto-u.ac.jp
oda.mitsushige@nasda.go.jp

* currently with MITSUBISHI HEAVY INDUSTRIES, LTD.
** currently with TOSHIOBA CORPORATION

Abstract

A bilateral teleoperation experiment with ETS-VII was conducted on November 22, 1999. Round-trip time for communication between the NASDA ground station and ETS-VII was approximately six seconds. We constructed a bilateral teleoperator that is stable even under such a long time delay. Several experiments, such as slope tracing task and peg-in-hole task, were carried out. Experimental results showed that kinesthetic force feedback to the operator is helpful even under such long time delay and improves the performance of the task.

1 Introduction

Bilateral control provides important force information of the remote environment to the operator. It is well known, however, that even small communication delay may destabilize the system with conventional bilateral control methods, such as symmetric position servo and force reflecting servo[14]. Anderson and Spong[1] proposed a bilateral control law that maintains stability under the communication delay by using the scattering theory. Niemeyer and Slotine[11] studied further on this problem.

It has been assumed, however, that bilateral control would not be effective when the time delay becomes longer than about 1[sec]. For example, Kim *et al.*[5] described as "... However, this force-reflection technique can be utilized only up to an approximately 0.5- to 1-s communication time delay, since a long time delay in the force feedback loop causes the system to be unstable." Hirzinger *et al.*[2] mentioned that "In ROTEX the loop delays varied from 5-7sec. Predictive computer graphics seems to be the only way to overcome this problem." As Peñin *et al.*[13]

did, we also summarized previous works on teleoperation with force-feedback under the communication time delay in Table 1. All of them conducted real experiments. These previous works can be divided into two groups: (i) direct bilateral teleoperation without any models of the remote site and (ii) model-based teleoperation with pseudo force feedback from the local model of the remote environment. From the table, it seems that when the delay time is longer than about 1[sec], the model-based approach would be the only solution. However, we have been doubtful about this "1[sec] limitation" from the following reasons:

- Some of the observations were came from using a conventional bilateral controller with which stability is not guaranteed under the time delay condition. Probably, 1[sec] would be the limitation to stabilize such an unstable system by human operators.

- Bilateral control based on the scattering theory guarantees the system stability for any time delay. However, it tends to be sticky and heavy as the delay time becomes large. Again, 1-2[sec] would be the limitation for the operator to maneuver such a system comfortably[8]. However, the scattering theory is not the only solution to the time delay problem and some other types of bilateral controller can also guarantee the stability.

Instead of exactly drawing the limitation line at 1[sec], our claim is, in a sense, quite natural as follows: *"Time delay limitation depends on the difficulty of the task. Even if the delay time becomes longer than 1[sec], some tasks could be performed by direct bilateral teleoperation."*

Table 1: Amount of time delay in the previous works on teleoperation with force feedback

Author(s)	Model-Based?	Delay Time for Round Trip	Feature
[1] Anderson & Spong (1989)	No	80ms, 400ms, 4s	Scattering Theory
[11] Niemeyer & Slotine (1991)	No	1s	Wave Variables
[5] Kim et al. (1992)	No	1s	Shared Compliant Control
[8] Lawn & Hannaford (1993)	No	up to 1s	Comparison between Scattering Theory and Others
[6] Kosuge et al. (1996)	No	1.4s	Virtual Time Delay
[12] Obe & Fiorini (1998)	No	320ms	PD-type
[7] Kotoku (1992)	Yes	1s	Predictive Display with Force Feedback
[4] Funda et al. (1992)	Yes	3s	Teleprogramming
[15] Tsumaki et al. (1996)	Yes	5s	Velocity/Damping Control
[13] Peñin et al. (2000)	Yes	5-7s	Truss Structure Experiment on ETS-7

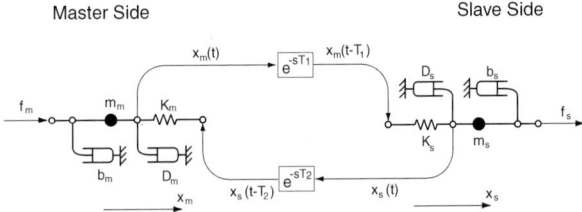

Figure 1: PD-type bilateral control

Actually, Ferrell[3] investigated the effect of time delay longer than 1[sec] in bilateral control. Although the conducted tasks were simple positioning with force feedback, he tested the delay time up to 3[sec].

In this paper, the result of ground-space teleoperation experiment using a robot arm mounted on ETS-VII (Engineering Test Satellite No.7) is shown. The experiment was conducted on November 22, 1999. Round-trip time for communication between the NASDA ground station and ETS-VII was approximately six to seven seconds. We constructed a bilateral teleoperator based on the PD-type controller that is stable even under such a long time delay. Several tasks, such as slope tracing task and peg-in-hole task, were carried out. All the tasks were completed by the direct bilateral control even without visual information. The experimental results demonstrate that kinesthetic force feedback to the operator is helpful even under such a long time delay.

2 Bilateral Controller with Time Delay

One of the well-know approach to time delay is to use scattering transformation, which was proposed by Anderson and Spong[1]. This approach was studied further by Niemeyer and Slotine[11], who introduced the notion of "wave variable". Besides the scattering-theory-based approach, there are several other approaches, which are less popular than the wave-variable approach. For example, Leung et al.[9] proposed a bilateral controller for time delay based on the H_∞-optimal control and μ-systhesis frameworks. Oboe and Fiorini[12] dealt with the time-varying delay problem over the Internet by using a simple PD-type controller.

We paid notice to this PD-type controller, which is shown in Figure 1. The dynamics of master and slave arms is formulated as follows:

$$\tau_m + f_m = m_m \ddot{x}_m + b_m \dot{x}_m, \quad (1)$$
$$\tau_s - f_s = m_s \ddot{x}_s + b_s \dot{x}_s, \quad (2)$$

where x_m and x_s denote positions of master and slave arms, and τ_m and τ_s are actuator driving forces, respectively. b_m and b_s represent viscous coefficients of the driving mechanism. f_m is the force that the operator applies to the master, and f_s denotes the force that the slave arm exerts to the environment.

The PD-type controller is given by the following equations:

$$\tau_m = -K_m \left(x_m(t) - x_s(t - T_2) \right) - D_m \dot{x}_m, \quad (3)$$
$$\tau_s = K_s \left(x_m(t - T_1) - x_s(t) \right) - D_s \dot{x}_s, \quad (4)$$

where K_m and K_s are position gains, and D_m and D_s are dumping gains. T_1 and T_2 denote delay times from master to slave and slave to master, respectively.

Oboe and Fiorini[12] analyzed the stability condition of this PD-type controller under time-varying delay conditions, but their analysis contains some errors and resultant condition is not true. We assumed constant time delays in both directions and derived the stability condition using Llewellyn's condition[10]. The derived condition is given by

$$(D_m + b_m)(D_s + b_s) \geq \frac{K_m K_s (T_1 + T_2)^2}{4}. \quad (5)$$

As the delay time becomes longer, the dumping gains should be increased, resulting in sticky feeling. Unlike the scattering-theory-based controller, however, the apparent

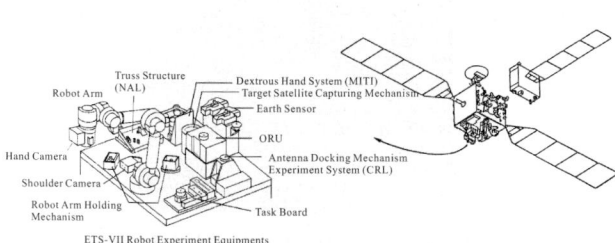

Figure 2: Robot system on ETS-VII

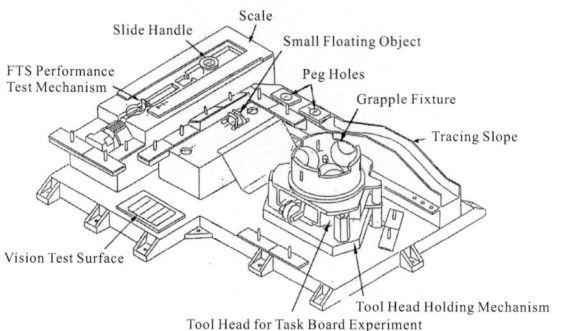

Figure 3: Task board

inertia keeps constant. Therefore, this PD-type controller is expected to be still useful even under a long time delay around the range of 5-7 seconds.

3 Overview of the Experiment

3.1 Purpose of the experiment

The purpose of the experiment can be summarized as follows. Detailed contents of the experiment will be described in section 4.

1. Basic performance check of the PD-type bilateral controller under 5-7[sec] time delay condition

2. Performance check in various tasks
   (a) accuracy of force command (pushing task)
   (b) accuracy of recognizing constraint surface slope (slope tracing task)
   (c) accuracy of recognizing contact state transitions (peg-in-hole task)
   (d) accuracy of recognizing unknown constraint direction (slide handle task)

3. Consideration of psychological aspect and operator's skill level

3.2 Experimental system

Figure 2 shows the robot experimental system on the ETS-VII. Figure 3 illustrates the task board used in the experiment. Figure 4 shows the configuration of the experimental system. Figure 5(a) shows the overview of the

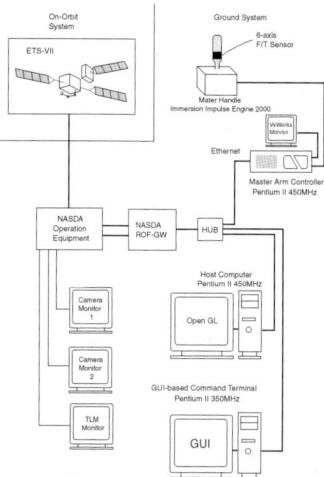

Figure 4: Configuration of experimental system

control station. A 2-DOF force feedback joystick (Impulse Engine 2000 by Immersion Co.), which is shown in Figure 5(b), was used for the master handle.

3.3 Modified bilateral controller

The PD-type bilateral controller discussed in section 2 assumes *grounded* dumper at both master and slave sides. Due to the limitation of the on-board arm controller specification of ETS-VII, however, we could not implement such a *grounded* dumper at the slave side. Instead, we reluctantly used a compliant controller where the dumping term is *relative* as shown in Figure 6.

We derived stability condition for this modified controller. The derived condition is to satisfy the following inequality for all $\omega \geq 0$:

$$(D_m + b_m)D_s > \frac{1}{2}\sqrt{\frac{K_m^2 K_s^2}{\omega^4} + \frac{K_m^2 D_s^2}{\omega^2}}$$
$$-\left(\frac{K_m K_s}{\omega^2}\cos\omega(T_1+T_2) + \frac{K_m D_s}{\omega}\sin\omega(T_1+T_2)\right). \tag{6}$$

Unfortunately the condition is not simple like eq.(5) and

(a) control station   (b) master handle

Figure 5: Experimental system.

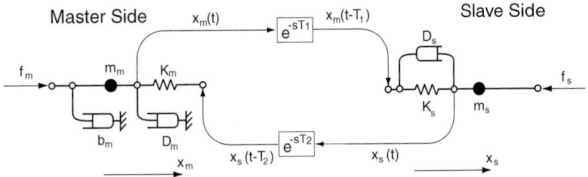

Figure 6: Modified bilateral controller

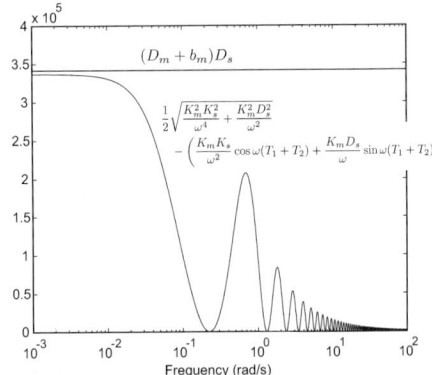

Figure 7: Plot of stability condition

we have to check the above inequality for all frequencies. Figure 7 illustrates left and right sides of this inequality with appropriate controller gains, and one can see that checking at $\omega = 0$ is sufficient. Although we found that the modified controller can be stabilized by applying enough amount of dumping gain, we could not increase the dumping gain at the master side large enough due to the hardware limitation of the master handle.

It should be noted, however, that the condition is, in a sense, conservative because it requires the passivity of the system to all passive class of environment and operator dynamics. With the maximum dumping gain available at the master side, we have checked overall system stability by classical Nyquist diagram. Assuming certain operator dynamics and environment dynamics (free motion and hard contact), we confirmed that the overall system is stable.

4 Detailed Contents of the Experiment

In this section, each experiment task will be described in detail. During the experiment, one can see two real im-

Figure 8: Shoulder camera image (slope tracing task)

Figure 9: Hand camera image (slope tracing task)

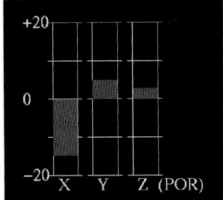

Figure 10: Bar graph of force telemetry

ages taken from the shoulder camera and the hand camera as shown in Figures 8 and 9, respectively. However, these two real images were shielded from the operator's position by putting masking boards as shown in Figure 5(a). In the following tasks, what the operator can see is only the computer screen showing telemetry force data or nothing, depending on the experimental conditions.

4.1 Pushing task

In the pushing task, the operator brings the tip of the robot arm into contact with the surface of the tracing slope shown in Figure 3. Then, he applies a rectangle force pattern 5[N]→15[N]→5[N] downwards without moving the arm. Since the force scaling factor between master and slave is five, the force pattern that the operator should actually apply is 1[N]→3[N]→1[N]. Settling time and errors were evaluated in the following three cases:

Case 1 (bilateral mode + force telemetry graph): The operator can get force feedback from the master handle. At the same time, he can monitor the telemetry force data displayed on the screen as shown in Figure 10.

Case 2 (bilateral mode): The operator must operate with force feedback alone and no visual information is provided.

Case 3 (unilateral mode + force telemetry graph): No force feedback is provided from the master handle. The telemetry force data on the screen is the only information fed back to the operator.

4.2 Slope tracing task

In the slope tracing task, the operator let the robot arm contour the sinusoidal slope, exerting a constant force (5[N]). As shown in Figure 8, a peg is attached to the tip of

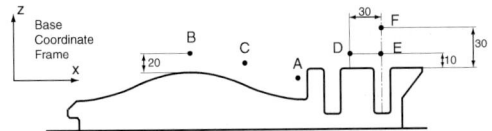

Figure 11: Starting points for slope tracing and peg-in-hole tasks

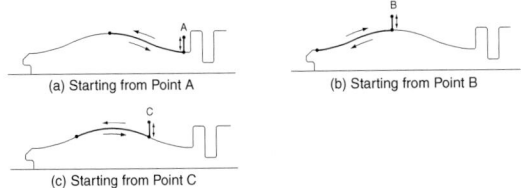

Figure 12: Three trajectory patterns in slope tracing task

the robot arm. The starting point, which is not informed to the operator, was chosen among points A, B and C shown in Figure 11. The operator was asked to move the arm down to the surface, then move 150[mm] left, and move back to the starting point. Depending on the starting point, the resultant trajectory will be one of the patterns shown in Figure 12.

In order to compare the task performance in equal condition, the operators were asked to complete the task preferably in three minutes and within a maximum of four minutes. The following two cases were tried:

Case 1 (bilateral mode + force telemetry graph)

Case 2 (unilateral mode + force telemetry graph)

Completion time and force errors were evaluated. In addition, the operator had to answer which starting point was selected when he finished each trial.

4.3 Peg-in-hole task

In the peg-in-hole task, the robot arm is initially placed at point D in Figure 11, 30[mm] left from the peg hole. The same peg used in slope tracing task is used again in this task. The diameter of the peg is 18[mm] and the hole has 0.4[mm] clearance. For smooth insertion, the peg tip was rounded and the hole was chamfered. The operator brings the peg into contact with the top surface, slides it horizontally until reaching the hole entrance (10[mm] below from point E), and inserts the peg into the hole. The operator is asked to avoid lateral force as much as possible when inserting the peg. He is also asked to identify the transition of the contact state, i.e., the instants when the peg starts to enter the hole and when it reaches the bottom of the hole, respectively. The following three cases were tried:

Case 1 (bilateral mode + force telemetry graph)

Case 2 (bilateral mode)

Case 3 (unilateral mode + force telemetry graph)

Completion time, the amount of lateral force during the insertion, and accuracy of recognizing the transition of the contact state were evaluated. In fact, we were not sure if this kind of task is possible under such a long time delay.

For the above three tasks (pushing, tracing, and peg-in-hole), 2D motion of the master handle was assigned to 2D translational motion of the robot arm in the vertical plane across the contouring slope and peg holes. Arm orientation and the remaining translational component were fixed by the on-board position controller.

4.4 Slide handle task

In the slide handle task, 2D motion of the master handle was assigned to 2D translational motion in the horizontal plane including the sliding direction. To make the sliding direction unknown to the operator, a certain amount of rotational coordinate transformation around the vertical axis of the plane was introduced.

At the initial stage, the peg attached to the tip of the robot arm is already inserted in the hole of the slide handle placed at the center of the slider guide. The operator should estimate the unknown sliding direction by probing the master handle. Then, he must move the robot to the end of the slider guide, then move to the other end, and finally move back to the center. The operator was asked to avoid lateral force (perpendicular to the sliding direction) as much as possible when moving the slide handle.

To complete the task, he must estimate the correct sliding direction and recognize when reaching the end of the slider guide. The operator should complete the task within three minutes. After the task, the operator should report his estimation of the sliding direction. The following three cases were tried:

Case 1 (bilateral mode + force telemetry graph)

Case 2 (bilateral mode)

Case 3 (unilateral mode + force telemetry graph)

Completion time, the amount of lateral force during the sliding motion, and accuracy of the estimation of the sliding direction were evaluated. Again, before we conducted this experiment, we were quite uncertain whether or not the operator can complete this task.

4.5 Skill level and other psychological factors

Due to the limited time allowed for us, most of the experiments were carried out by a single operator, who has been accustomed with the operation using the master handle. To see the effect of skill level, other two operators conducted some tasks. One is a NASDA operator, who has been accustomed with the operation of ETS-VII robot arm by NASDA's teleoperation facilities but is not familiar with the master handle used in this experiment. The other one

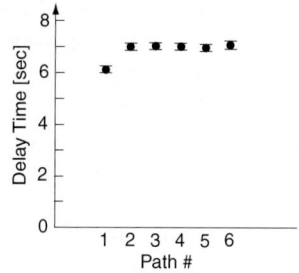

Figure 13: Measured delay time for round trip

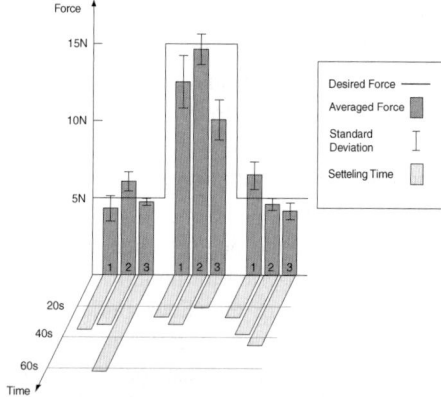

Figure 14: Result of pushing task (Task 1:B+T, Task 2:B, Task 3:U+T)

is a novice operator, who has not been trained with any device, but has a background of teleoperation.

To investigate psychological factors such as mental load and points of their attention during the task, operators were asked to fill out a questionnaire after the experiment.

5 Experimental Results

The experiment was conducted on November 22, 1999. Round-trip time for communication between the control station at the NASDA Tsukuba Space Center and ETS-VII, which is flying around an orbit 550[km] above, was approximately six to seven seconds. Figure 13 shows the measured delay time at each experimental unit called "path", corresponding the duration time (about 40[min]) when the ETS-VII is visible from the TDRS (Tracking Data Relay Satellite) on Geostationary orbit. One can see that delay time differs at each path, but does not fluctuate so much within each path. Figures 8 and 9 show the snapshots during the slope tracing task experiment.

5.1 Pushing task

Figure 14 shows the experimental result of the pushing task. One can see that the bilateral mode reaches the desired force more quickly and accurately than the unilateral mode.

Table 2: Actual starting points and declared points by the operators

Task Condition	Staring Point	Declared Point	Score
Task 1 (Bilateral + Telemetry)	A	C	×
Task 2 (Bilateral + Telemetry)	C	C	○
Task 3 (Unilateral + Telemetry)	C	C	○
Task 4 (NASDA Operator) (Bilateral + Telemetry)	C	C	○
Task 5 (longer operation time) (Bilateral + Telemetry)	B	B	○

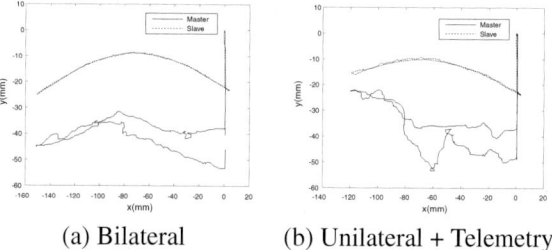

(a) Bilateral (b) Unilateral + Telemetry

Figure 15: Typical arm trajectories in slope tracing task

5.2 Slope tracing task

Table 2 shows the estimation result of the starting points. Except Task 1, the operators including the NASDA operator could estimate the starting points correctly. The reason of failure in Task 1 would be probably due to some psychological factors, such as a stress for the first trial. It should be noted, however, that in bilateral mode the shape estimation was confident and obtained with a little movement of the handle, whereas in unilateral mode the estimation was quite uncertain and obtained after the entire movement. Figure 15 explains the reason of this observation. In the bilateral mode, the trajectory of master handle reproduces the slope shape while in unilateral mode it is difficult to estimate the slope shape from the master handle trajectory.

Figure 16 shows the experimental results. The stroke

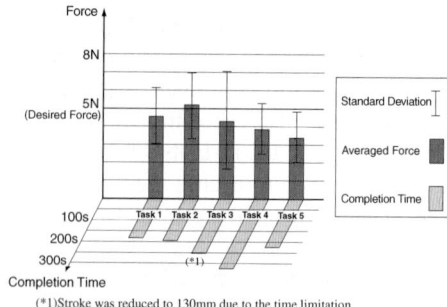

(*1)Stroke was reduced to 130mm due to the time limitation.

Figure 16: Result of slope tracing task (Task numbers correspond to those in Table 2.)

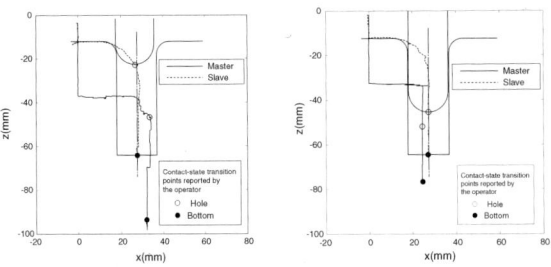

(a) Bilateral   (b) Unilateral + Telemetry
Figure 17: Arm trajectories during the peg-in-hole task

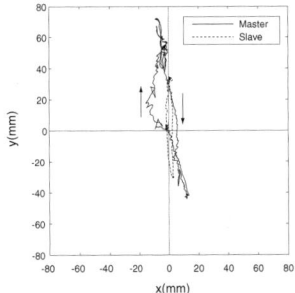

Figure 19: Trajectory of task 1 (Bilateral mode, rotated -75[deg])

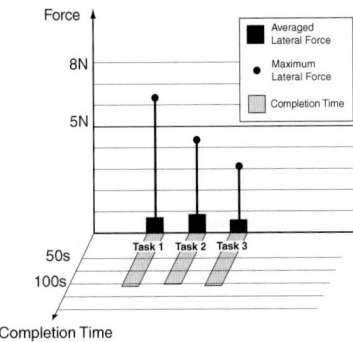

Figure 18: Result of peg-in-hole task (Task 1:B+T, Task 2:B, Task 3:U+T)

Table 3: Sliding directions and declared directions

Task	Rotation angle[deg]	Declared angle[deg]
Task 1 (Bilateral + Telemetry)	-75	-65
Task 2 (Bilateral)	0	0
Task 3 (Unilateral + Telemetry)	20	30
Task 4 (NASDA Operator) (Bilateral + Telemetry)	60	45
Task 5 (Novice Operator) (Bilateral + Telemetry)	-60	-45

in unilateral mode was reduced 20[mm] shorter than the initial plan to complete the task within the time limit. Like the pushing task, one can see that the bilateral mode keeps the desired force more accurately and complete the task faster than the unilateral mode.

It should be noted that the task performance of the NASDA operator, who used this system first time, was comparable to that of the skilled operator.

5.3 Peg-in-hole task

Figure 17 shows the arm trajectories in the peg-in-hole task. In the figure, the actual peg position when the operator judged as the starting point of the insertion is also drawn. One can see that in bilateral mode the operator could identify the transition of contact state accurately only from the force feedback information. On the other hand, the recognition of reaching to the bottom of the hole was better when using telemetry data than using force feedback from the master handle. Figure 18 shows the result of peg-in-hole task. Unexpectedly, unilateral mode gave smaller lateral force than bilateral mode. Probably the operator moved the arm more carefully in the unilateral mode. No significant difference was found in completion time.

5.4 Slide handle task

Table 3 shows the result of sliding direction estimation by the operators. In all cases, they could estimate the sliding direction with reasonable accuracies. It should be noted, however, that in bilateral mode the estimation was confident and obtained with a little movement of the handle, whereas in unilateral mode the estimation was quite uncertain and obtained after the entire movement, which is the same observation as in the slope tracing task. Figure 19 shows a typical example of the hand trajectory in the slide handle task. In the figure, the inserted rotational transformation was canceled so that master and slave trajectories coincide in the ideal situation. One can see that the operator moved the handle into a wrong direction but shifted to the correct direction, feeling the force feedback from the handle.

All operators including the NASDA operator and the novice operator could complete the task in bilateral mode. In unilateral mode, however, the operator could move the handle only in one way and could not complete the task within the assigned time. The NASDA operator performed only probing task to estimate the sliding direction, since we did not explain him the task procedure adequately.

Figure 20 shows the result of slide handle task. Lateral force could not be reduced in bilateral mode, because the operator exerted large lateral force when probing the sliding direction. After the direction was estimated, the lateral force was small.

5.5 Discussion

From the questionnaire survey after each task, the following observations are obtained:

- All three operators paid most attention to the force feedback from the master handle even when the

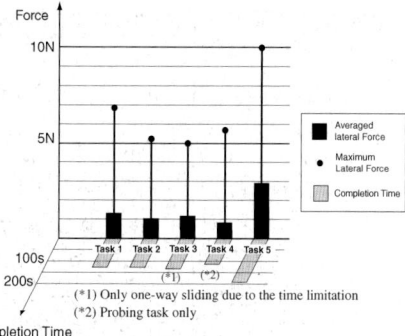

Figure 20: Result of slide handle task (Task numbers correspond to those in Table 3.)

telemetry force data was displayed on the screen.

- Using the kinesthetic force feedback information, the operators could recognize the shape of the contouring slope and the peg hole entrance with just a small movement. It is difficult, however, to recognize the instant of contact (e.g., when reaching the bottom of the hole), due to the low position gain.

- Telemetry force data was noisy and difficult to use for shape recognition. The data should have been pass through a low-pass filter.

- Even a novice operator could complete the task, showing that no specific skill is required.

It was surprising even for us that the kinesthetic force feedback information was still useful and could improve the task performance even under such a long time delay. Of course, the task should be performed slowly and the maneuverability would be poorer than the case without time delay. However, this experiment has proven that it is still possible to complete some tasks by direct bilateral control under 6-7[sec] time delay.

6 Conclusion

In this paper, the result of ground-space teleoperation experiment using a robot arm mounted on ETS-VII was shown. Stability condition of the PD-type bilateral controller was derived and the controller was modified due to the limitation of on-board robot controller of ETS-VII. Several tasks, such as wall following task and peg-in-hole task, were carried out under 6-7[sec] time delay. All the tasks were possible by the direct bilateral control even without visual information. The experimental results demonstrate that force feedback to the operator is helpful even under such a long time delay.

Time delay limitation depends on the difficulty of the task. Therefore, even if the delay time becomes longer than 1[sec], some tasks would be possible by direct bilateral teleoperation. This experiment is probably the first ground-space teleoperation by direct bilateral control.

Acknowledgement

The authors would like to thank Mr. Ryo Kikuue and Mr. Hiroyuki Iida, graduate students of Kyoto University, who joined the experiment as the SOP (Satellite Operation Procedure) commander and the telemetry observer, respectively.

References

[1] R.J.Anderson and M.W.Spong: "Bilateral Control of Teleoperators with Time Delay," *IEEE Trans. on Automatic Control*, vol.34, no.5, pp.494-501, 1989.

[2] G.Hirzinger et al.: "Sensor-Based Space Robotics –ROTEX and Its Telerobotic Features," *IEEE Trans. on Robotics and Automation*, vol.9, no.5, pp.649-663, 1993.

[3] W.R.Ferrell: "Delayed Force Feedback," *Human Factors*, pp.449–455, 1966.

[4] J.Funda et al.: "Teleprogramming: Toward Delay-Invariant Remote Manipulation," *PRESENCE, Teleoperators and Virtual Environments*, vol.1, no.1, pp.29–44, 1992.

[5] W.S.Kim and A.K.Bejczy: "Demonstration of a High-Fidelity Predictive/Preview Display Technique for Telerobotic Servicing in Space," *IEEE Trans. on Robotics and Automation*, vol.9, no.5, pp.698-702, 1993.

[6] K.Kosuge et al.: "Bilateral Feedback Control of Telemanipulators via Computer Network," *Proc. IEEE/RSJ International Conference on Intelligent Robots and Systems (IROS'96)*, pp.1380-1385, 1996.

[7] T.Kotoku: "A Predictive Display with Force Feedback and its Application to Remote Manipulation System with Transmission Time Delay," *Proc. IEEE/RSJ International Conference on Intelligent Robots and Systems (IROS'92)*, pp.239-246, 1992.

[8] C.A.Lawn and B.Hannaford: "Performance Testing of Passive Communication and Control in Teleoperation with Time Delay," *Proc. IEEE International Conference on Robotics and Automation (ICRA'93)*, vol.3, pp.776-83, 1993.

[9] G.M.H. Leung et al.: "Bilateral Controller for Teleoperators with Time Delay via μ-Synthesis," *IEEE Trans. on Robotics and Automation*, vol.11, no.1, pp.105-116, 1995.

[10] F.B.Llewellyn: "Some Fundamental Properties of Transmission Systems," *Proc. of the I.R.E.*, vol.40, no.5, pp.271-283, 1952.

[11] G.Niemeyer and J.J.E.Slotine: "Stable Adaptive Teleoperation," *IEEE J. of Oceanic Engineering*, vol.16, no.1, pp.152-162, 1991.

[12] R.Oboe and P.Fiorini: "A Design and Control Environment for Internet-Based Telerobotics," *Int. J. Robotics Res.*, vol.17, no.4, pp.433–449, 1998.

[13] Peñin et al.: "Force Reflection for Time-delayed Teleoperation of Space Robots," *Proc. IEEE International Conference on Robotics and Automation (ICRA 2000)*, pp.3120-3125, 2000.

[14] T.B.Sheridan: "Space Teleoperation Through Time Delay: Review and Prognosis," *IEEE Trans. on Robotics and Automation*, vol.9, no.5, pp.592 -606, 1993.

[15] Y.Tsumaki et al.: "Virtual Reality Based Teleoperation which Tolerates Geometrical Modeling Errors," *Proc. IEEE/RSJ International Conference on Intelligent Robots and Systems (IROS'96)*, pp.1023-1030, 1996.

A Numerical SC Approach for a Teleoperated 7-DOF Manipulator

Y. Tsumaki*, P. Fiorini**, G. Chalfant** and H. Seraji**

*Department of Aeronautics and Space Engineering
Tohoku University

**Jet Propulsion Laboratory

Abstract

To tackle the singularity problem, the SC (Singularity-Consistent) approach was introduced. It achieves very stable control at and around a singularity with feasible joint velocities and no directional error in the end-effector velocity. This approach is very suited for a direct manual teleoperation system because of its error-less character for the direction. Until now, it was applied for the teleoperation of a non-redundant manipulator. In this paper, the SC approach for a 7-DOF manipulator will be addressed. To derive the whole properties of the SC approach, analytical studies for both the adjoint and the determinant of the Jacobian are necessary. However, it is very difficult to realize this goal, since the kinematics of a 7-DOF manipulator is quit complicated. Therefore, here, we establish a method to apply the SC approach to a 7-DOF manipulator numerically without analyzing the kinematic properties. This method, though, cannot realize the whole properties of the SC approach but a stable control at and around the singularities is achieved which is the most important and demanded property of the SC approach. Moreover, this method can also be applied for any type of articulated manipulator if its Jacobian can be defined. The results of our approach have been confirmed by experiments with graphics model.

1. Introduction

During teleoperation, the operator is liable to move the slave arm to or around a singularity without perception. It is very dangerous since large joint velocities would occur or the system will become uncontrollable. Therefore the workspace of the manipulator is to be restricted not to allow it to reach the singularities practically. But, this method spoils a large part of the workspace. To tackle this problem, the DLS (Dumped Least Square) method was introduced [1], [2], [3]. The DLS method realizes stable motion around the singularities, but the end-effector deviates from the commanded direction. It means that the manipulator gets out of control for a while when it approaches a singularity. This phenomenon is not acceptable especially for the direct manual teleoperation. To tackle this problem, recently, the SC (Singularity Consistent) approach was introduced [4], [5], [6]. This approach can realize the reference direction exactly along with feasible joint velocities at and around the singularities. But, the end-effector cannot follow the exact commanded magnitude of the reference velocity. However, the directional error-less property is very suited for the manual teleoperation systems. Until now, this approach was applied to teleoperate the non-redundant 6-DOF manipulators [7].

In general, the redundancy is quite helpful to realize the complex tasks. Therefore, recently, the demand for the redundant manipulators is much increasing. But, unfortunately, their kinematics is complicated. Especially the kinematics of the redundant manipulators with displacement of joint axes (non-zero joint offsets) to fold themselves compactly is very complicated. The Robotics Research K-1207 arm which was developed for space applications is one of the representative examples [9]. It is well known that the redundant motions can be used to avoid the singular configurations. However, in teleoperation, even the redundant motion should be under the operator's control, since the arbitrary motions would induce serious problems like collisions against obstacles, etc. The concept of the arm plane [10], [11] is one of the solutions to handle the redundant motions. However, the arm plane induces some additional algorithmic singularities. Until now, the DLS method has been used to handle both the algorithmic and the kinematic singularities in spite of having the tracking errors.

On the other hand, the SC approach requires the analyses of both the Jacobian and its determinant. But, the analytical kinematic solution of a 7-DOF ma-

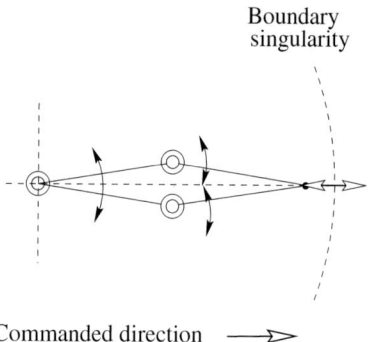

Figure 3. Chattering at the boundary singularity.

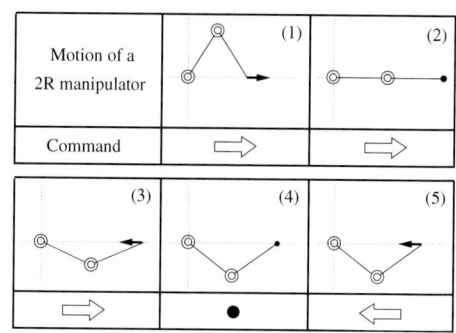

Figure 4. Motion through a singularity.

where $\boldsymbol{H}_p$ stands for the matrix $\boldsymbol{H}$ with pth column removed [5].

As a result, $(\operatorname{adj}\hat{\boldsymbol{J}}_A)\hat{\boldsymbol{u}}_r$ can be obtained from $\det \boldsymbol{H}_p$. It is easy to calculate the determinant of a matrix numerically by the well-known LU decomposition method.

3.3. Deciding the sign

One of the big advantages of the SC approach is its ability of reconfiguration using the Cartesian commands. To utilize this feature, the problem of chattering motions related to the decision of sign function σ should be solved. Fig. 3 shows the chattering motions that are typical problem in teleoperation systems. While calculating the inverse of Jacobian conventionally, the determinant of the Jacobian includes the sign information into itself. However, if σ is decided by the sign of this determinant, a chattering motion would occur at a boundary singularity, since the end-effector tries to keep up with the commanded direction. Of course this happens only in a discrete time system with digital servos. Therefore, maintaining the same sign while crossing through a singularity can be considered as a solution[5]. Fig. 4 illustrates this concept. σ succeeds the previous sign until the command approaches zero. As a result, the end-effector even moves in the opposite direction for a while, but, it can realize a smooth reconfiguration. Unfortunately, in the case of a 7-DOF manipulator, the situation becomes more complex since the manipulator is apt to meet singularities continually. As a result, it is difficult to find an effective rule to decide about the sign using only the determinant of Jacobian.

In order to avoid this situation, we introduce a simple but very strong method. The chattering motion is caused by drastic alternations in the joint velocities. Therefore, the proposed method finds out the maximum among all the joint velocity drifts for each sign in each sampling interval. Only then, among these t-

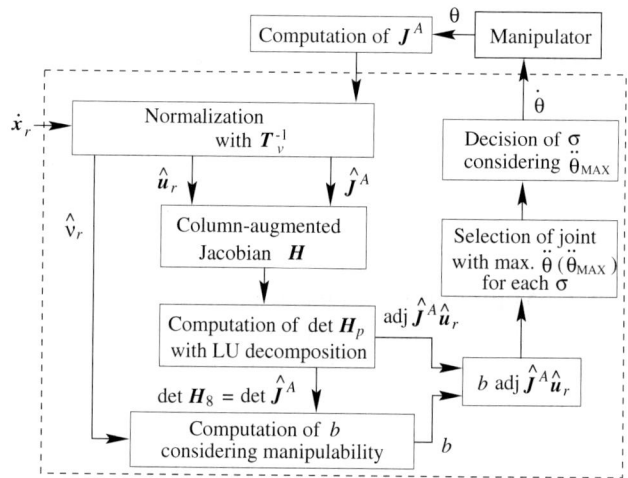

Figure 5. Algorithm for the numerical SC approach.

wo maximum known drifts, the sign of the one with lesser magnitude is selected. As a result, very smooth motions can be realized easily. We also adopt a strategy that adjusts the sign according to the one of the determinant after passing through a zero command. We would like to emphasize that this method can be applied for any type of manipulator without having knowledge of its kinematics.

3.4. The numerical SC approach algorithm

Fig. 5 summarizes the above mentioned procedures in the form of an algorithm. What needs to be emphasized is that the computations inside the dotted box are purely numerical. It means that this method can be applied to any type of articulated manipulator without any need to analyze its kinematics, but only if its Jacobian can be defined.

A Numerical SC Approach for a Teleoperated 7-DOF Manipulator

Y. Tsumaki*, P. Fiorini**, G. Chalfant** and H. Seraji**
*Department of Aeronautics and Space Engineering
Tohoku University
**Jet Propulsion Laboratory

Abstract

To tackle the singularity problem, the SC (Singularity-Consistent) approach was introduced. It achieves very stable control at and around a singularity with feasible joint velocities and no directional error in the end-effector velocity. This approach is very suited for a direct manual teleoperation system because of its errorless character for the direction. Until now, it was applied for the teleoperation of a non-redundant manipulator. In this paper, the SC approach for a 7-DOF manipulator will be addressed. To derive the whole properties of the SC approach, analytical studies for both the adjoint and the determinant of the Jacobian are necessary. However, it is very difficult to realize this goal, since the kinematics of a 7-DOF manipulator is quit complicated. Therefore, here, we establish a method to apply the SC approach to a 7-DOF manipulator numerically without analyzing the kinematic properties. This method, though, cannot realize the whole properties of the SC approach but a stable control at and around the singularities is achieved which is the most important and demanded property of the SC approach. Moreover, this method can also be applied for any type of articulated manipulator if its Jacobian can be defined. The results of our approach have been confirmed by experiments with graphics model.

1. Introduction

During teleoperation, the operator is liable to move the slave arm to or around a singularity without perception. It is very dangerous since large joint velocities would occur or the system will become uncontrollable. Therefore the workspace of the manipulator is to be restricted not to allow it to reach the singularities practically. But, this method spoils a large part of the workspace. To tackle this problem, the DLS (Dumped Least Square) method was introduced [1], [2], [3]. The DLS method realizes stable motion around the singularities, but the end-effector deviates from the commanded direction. It means that the manipulator gets out of control for a while when it approaches a singularity. This phenomenon is not acceptable especially for the direct manual teleoperation. To tackle this problem, recently, the SC (Singularity Consistent) approach was introduced [4], [5], [6]. This approach can realize the reference direction exactly along with feasible joint velocities at and around the singularities. But, the end-effector cannot follow the exact commanded magnitude of the reference velocity. However, the directional error-less property is very suited for the manual teleoperation systems. Until now, this approach was applied to teleoperate the non-redundant 6-DOF manipulators [7].

In general, the redundancy is quite helpful to realize the complex tasks. Therefore, recently, the demand for the redundant manipulators is much increasing. But, unfortunately, their kinematics is complicated. Especially the kinematics of the redundant manipulators with displacement of joint axes (non-zero joint offsets) to fold themselves compactly is very complicated. The Robotics Research K-1207 arm which was developed for space applications is one of the representative examples [9]. It is well known that the redundant motions can be used to avoid the singular configurations. However, in teleoperation, even the redundant motion should be under the operator's control, since the arbitrary motions would induce serious problems like collisions against obstacles, etc. The concept of the arm plane [10], [11] is one of the solutions to handle the redundant motions. However, the arm plane induces some additional algorithmic singularities. Until now, the DLS method has been used to handle both the algorithmic and the kinematic singularities in spite of having the tracking errors.

On the other hand, the SC approach requires the analyses of both the Jacobian and its determinant. But, the analytical kinematic solution of a 7-DOF manipulator is too complicated to analyze. Therefore in this article, we establish a numerical method to employ the SC approach to a teleoperated 7-DOF manipulator instead of analyzing its kinematics. Though, this method cannot realize all the salient features of the SC approach, yet a stable control at and around the singularities, which is the most demanded feature of the SC approach, is achieved. Furthermore, the proposed method can be applied to any type of articulated manipulator, if its Jacobian can be defined. The results of the proposed approach have been confirmed by experiments with a graphics model.

2. Fundamentals

2.1. SC approach

First of all, the fundamentals of the SC approach are addressed [7]. The inverse kinematics is written generally as:

$$\begin{aligned} \dot{\boldsymbol{\theta}} &= \boldsymbol{J}^{-1}\dot{\boldsymbol{x}} \\ &= \frac{1}{\det \boldsymbol{J}}(\text{adj}\boldsymbol{J})\dot{\boldsymbol{x}}, \end{aligned} \quad (1)$$

where $\boldsymbol{J}$ is the manipulator's Jacobian, $\det \boldsymbol{J}$ and $\text{adj}\boldsymbol{J}$ denote its determinant and adjoint, respectively. The cause of occurrence of a singularity is the determinant approaching to zero. Note, that the term $(\det \boldsymbol{J})^{-1}$ is a scaling factor that is related to the magnitude and the direction of the motion in the joint space. On the other hand, $(\text{adj}\boldsymbol{J})\dot{\boldsymbol{x}}$ determines the velocities of the individual joints.

In order to overcome the singularity problem, following modification in Eq. (1) are introduced [5].

$$\dot{\boldsymbol{\theta}} = \sigma b\,(\text{adj}\boldsymbol{J})\boldsymbol{u}_r \quad (2)$$

where $\boldsymbol{u}_r$ is the directional unit vector of the reference end-effector velocities, σ is a sign variable ($\sigma = \pm 1$), and $b \geq 0$ is a scalar variable. With a proper design of both σ and b, the manipulator can be controlled at singularities reducing the rank of Jacobian by one, and in its vicinity, without any error in the direction and with feasible joint velocities.

Here, we should note that there are two types of velocity relations at a kinematic singularity [6]

- *Type A:* $(\text{adj}\boldsymbol{J})\boldsymbol{u}_r \neq \boldsymbol{0}$
- *Type B:* $(\text{adj}\boldsymbol{J})\boldsymbol{u}_r = \boldsymbol{0}$.

In case of Type A relation, a special motion called *self-motion* is obtained. In this case, some of the joints are moving while the end-effector is motionless.

On the other hand, with Type B relation, all components of the vector $(\text{adj}\boldsymbol{J})\boldsymbol{u}_r$ vanish, and motions would stop entirely. Analysis shows, however, that in many cases, it is the common factor of $(\text{adj}\boldsymbol{J})\boldsymbol{u}_r$

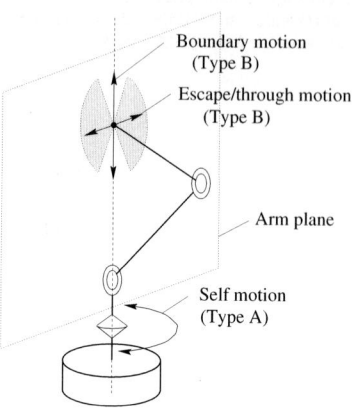

Figure 1. Type A and B relations at a shoulder singularity.

and $\det \boldsymbol{J}$ that yields this phenomenon. Hence, we can obtain a proper solution by eliminating this common factor [8]. This motion is related with a through singularity motion and an on singularity motion. But, it is necessary to analyze both the determinant and the adjoint matrix of the Jacobian to find out the common factor. In addition, type B relation requires very accurate direction which should lie in the arm plane. Such a precise command direction cannot be realized without computer assistance. It means that the computer system should have a knowledge of such precise command directions previously.

Note that the above two relations depend not only on the arm's configuration but also on the direction of the velocity command. An example of these relations are shown in Fig. 1.

2.2. Arm angle

In controlling a 7-DOF manipulator, the redundant motion needs to decided with some methods, e.g. minimum joint velocities [12], singularity avoidance [13], obstacle avoidance [14], etc. From the view point of teleoperation, the operator should be able to control all the motions always as he/she desires. In order to accomplish this target, the concept of arm angle was introduced [10], [11]. The arm angle is an angle between reference plane and arm plane which includes shoulder, elbow and wrist points. The definition of the arm angle ψ is illustrated in Fig. 2, where S, E, W are placed on shoulder, elbow and wrist, respectively [9]. $\boldsymbol{v}$ is an arbitrary fixed unit vector (e.g. the unit vector in the vertical direction of the base plane). ED is normal to the projection of SE onto SW, and $\boldsymbol{l}$ is an orthogonal unit vector to SW in the reference plane that contains both $\boldsymbol{v}$ and SW. Consequently, the arm

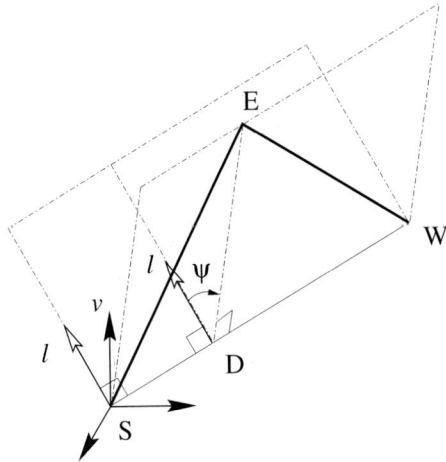

Figure 2. Illustration of the arm angle (ψ) concept.

angle ψ is the angle between ED and $\boldsymbol{l}$.

This arm angle, however, introduces some additional algorithmic singularities. It means that the operator is required to handle both the algorithmic and the kinematic singularities simultaneously.

2.3. Augmented Jacobian

The differential of the relationship between the end-effector, arm angle and the joint angle can be written as:

$$\begin{pmatrix} \dot{\boldsymbol{x}} \\ \dot{\psi} \end{pmatrix} = \begin{pmatrix} \boldsymbol{J}^{ee} \\ \boldsymbol{J}^{\psi} \end{pmatrix} \dot{\boldsymbol{\theta}} = \boldsymbol{J}^A \dot{\boldsymbol{\theta}} \quad (3)$$

where $\dot{\boldsymbol{x}} \in R^6$ is the end-effector velocity in the task space, $\dot{\psi} \in R^1$ is the arm angle's angular velocity, $\boldsymbol{J}^{ee} \in R^{6\times 7}$ is the end-effector Jacobian, $\boldsymbol{J}^{\psi} \in R^{1\times 7}$ is the arm angle's Jacobian, $\dot{\boldsymbol{\theta}} \in R^7$ is the joint velocity and $\boldsymbol{J}^A \in R^{7\times 7}$ is the augmented Jacobian [15]. As a result, the problem of both algorithmic and kinematic singularities becomes a common singularity problem of a non-redundant system with 7 parameters.

3. SC approach for a 7-DOF manipulator

Here, we apply the SC approach to a 7-DOF manipulator having displacements of joint axes (non-zero joint offsets) to fold itself compactly. The Robotics Research K-1207 arm, developed for the space applications, is taken as a representative example. The kinematics of such a manipulator is quite complicated. Therefore it is very difficult to handle the type B singularities, since it requires to analyze both the determinant and the adjoint of the Jacobian. This fact leads us to employ the numerical methods to apply the SC approach to this 7-DOF manipulator. Unfortunately, such a method cannot handle the type B singularities. However, a stable control at and around the singularities can be achieved without directional errors and it is the most important feature of the SC approach.

3.1. Basic concept of the SC approach for a 7-DOF manipulator

To apply the SC approach to a 7-DOF manipulator, it is necessary to consider a proper scaling procedure such that the orientation, position and arm angle variables can be treated in a uniform manner. For this purpose, we introduce the constants v_{max}, ω_{max} and ψ_{max} standing for the maximum translational, rotational and arm angle velocities, respectively. Using these constants, we can normalize the velocities, and obtain

$$\hat{\dot{\boldsymbol{p}}}_r = \hat{\boldsymbol{J}}^A \hat{\dot{\boldsymbol{\theta}}} \quad (4)$$

$$\hat{\dot{\boldsymbol{p}}}_r = \boldsymbol{T}_v^{-1} \dot{\boldsymbol{p}}_r \quad (5)$$

$$\hat{\boldsymbol{J}}^A = \boldsymbol{T}_v^{-1} \boldsymbol{J}^A \quad (6)$$

$$\boldsymbol{T}_v = \sqrt{3}\mathrm{diag}[v_{max}, v_{max}, v_{max}, \omega_{max}, \omega_{max}, \omega_{max}, \psi_{max}] \quad (7)$$

where $\dot{\boldsymbol{p}}_r$ is the reference velocity vector including $\dot{\boldsymbol{x}}_r$ and $\dot{\psi}_r$, $\hat{\dot{\boldsymbol{p}}}_r$ is the normalized one of $\dot{\boldsymbol{p}}_r$, and $\hat{\boldsymbol{J}}^A$ is the normalized augmented Jacobian. The maximum of the norm of $\hat{\dot{\boldsymbol{p}}}_r$ would be equal to 1 due to the introduction of the factor $\sqrt{3}$ in the above equation. According to the SC approach, the end-effector velocity is represented as

$$\hat{\dot{\boldsymbol{p}}}_r = \nu_r \boldsymbol{u}_r \quad (8)$$

where ν_r is a scalar, and $\boldsymbol{u}_r$ is a unit vector. Hence we can obtain

$$\dot{\boldsymbol{\theta}} = \sigma \hat{b}\,(\mathrm{adj}\hat{\boldsymbol{J}}^A)\boldsymbol{u}_r \quad (9)$$

where $\hat{b}$ is a function of the manipulability and the end-effector velocity ν [5].

3.2. A numerical SC approach

As we mentioned, analytical solution of Eq. (9) is quite complicated. Therefore, a numerical approach is utilized. Of course we suppose that the Jacobian can be defined. Unfortunately, it takes a lot of computing time to obtain the adjoint matrix numerically. Therefore, the following modification

$$\begin{array}{l}(-1)^n(\mathrm{adj}\hat{\boldsymbol{J}}_A)\hat{\boldsymbol{u}}_r = \boldsymbol{n}_H = [C_1, C_2, ...C_n]^T \\ C_p = (-1)^{p+1}\det\boldsymbol{H}_p \quad (p=1,2,...n)\end{array} \quad (10)$$

is made, where $\boldsymbol{n}_H$ is the null space vector of a column augmented Jacobian $\boldsymbol{H}$ which is defined as

$$\boldsymbol{H} = [\hat{\boldsymbol{J}}_A, -\hat{\boldsymbol{u}}_r] \quad (11)$$

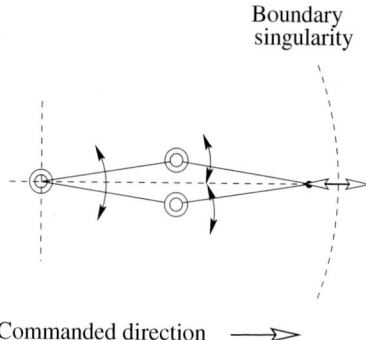

Figure 3. Chattering at the boundary singularity.

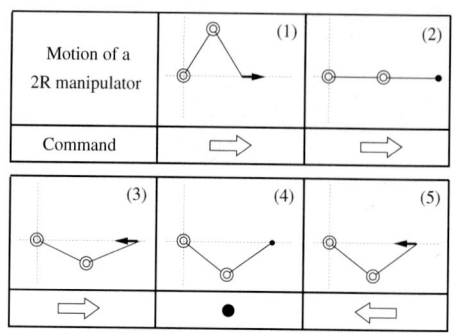

Figure 4. Motion through a singularity.

where H_p stands for the matrix H with pth column removed [5].

As a result, $(\operatorname{adj} \hat{J}_A)\hat{u}_r$ can be obtained from $\det H_p$. It is easy to calculate the determinant of a matrix numerically by the well-known LU decomposition method.

3.3. Deciding the sign

One of the big advantages of the SC approach is its ability of reconfiguration using the Cartesian commands. To utilize this feature, the problem of chattering motions related to the decision of sign function σ should be solved. Fig. 3 shows the chattering motions that are typical problem in teleoperation systems. While calculating the inverse of Jacobian conventionally, the determinant of the Jacobian includes the sign information into itself. However, if σ is decided by the sign of this determinant, a chattering motion would occur at a boundary singularity, since the end-effector tries to keep up with the commanded direction. Of course this happens only in a discrete time system with digital servos. Therefore, maintaining the same sign while crossing through a singularity can be considered as a solution[5]. Fig. 4 illustrates this concept. σ succeeds the previous sign until the command approaches zero. As a result, the end-effector even moves in the opposite direction for a while, but, it can realize a smooth reconfiguration. Unfortunately, in the case of a 7-DOF manipulator, the situation becomes more complex since the manipulator is apt to meet singularities continually. As a result, it is difficult to find an effective rule to decide about the sign using only the determinant of Jacobian.

In order to avoid this situation, we introduce a simple but very strong method. The chattering motion is caused by drastic alternations in the joint velocities. Therefore, the proposed method finds out the maximum among all the joint velocity drifts for each sign in each sampling interval. Only then, among these t-

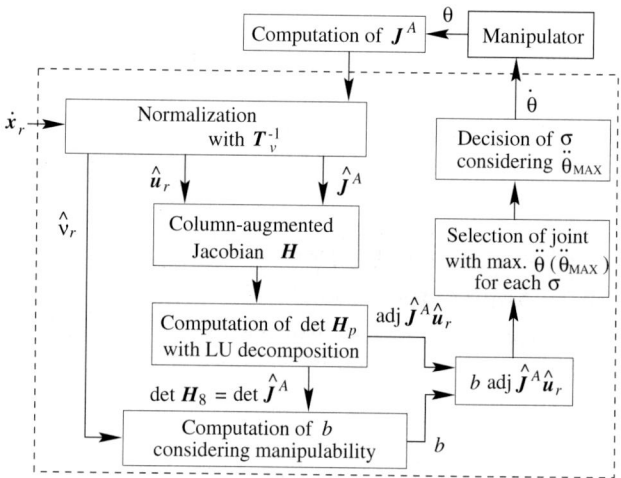

Figure 5. Algorithm for the numerical SC approach.

wo maximum known drifts, the sign of the one with lesser magnitude is selected. As a result, very smooth motions can be realized easily. We also adopt a strategy that adjusts the sign according to the one of the determinant after passing through a zero command. We would like to emphasize that this method can be applied for any type of manipulator without having knowledge of its kinematics.

3.4. The numerical SC approach algorithm

Fig. 5 summarizes the above mentioned procedures in the form of an algorithm. What needs to be emphasized is that the computations inside the dotted box are purely numerical. It means that this method can be applied to any type of articulated manipulator without any need to analyze its kinematics, but only if its Jacobian can be defined.

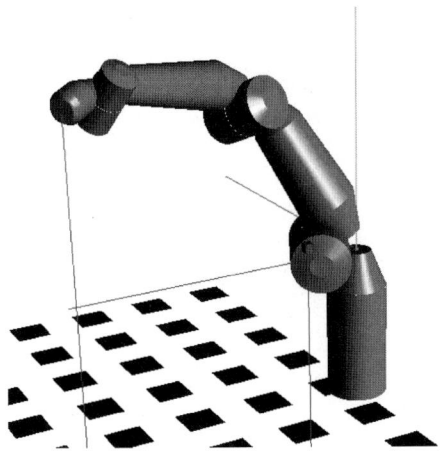

Figure 6. Computer graphics of the Robotics Research K-1207.

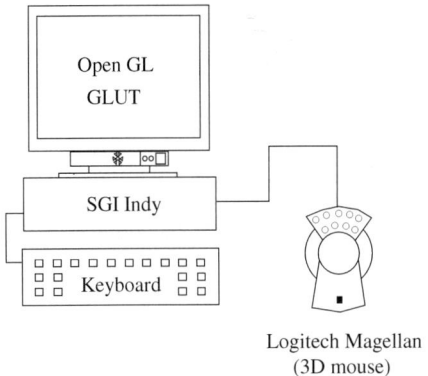

Figure 7. Experimental setup.

4. Experiments

We apply the proposed method to the graphical simulator of the Robotics Research K-1207. Fig. 6 shows an overview of the computer graphics of the manipulator. In this simulator, the operator can control the end-effector velocity of K-1207 using a 3D mouse that can generate the 6 axes commands simultaneously. However, the arm angle velocity is controlled by the keyboard. Fig. 7 shows an overview of the experimental setup. The DLS method is also applied to this manipulator to make a comparison with the proposed SC approach. One example experiment is shown in Fig. 8. A red (light) arrow and a blue (dark) arrow display the reference and the real end-effector velocities, respectively. It is easily notable that there is no directional error when the proposed SC approach is working. While in case of the DLS method, the manipulator even becomes uncontrollable for a while. In addition, we would like to emphasis that during self motions, there are no end-effector motions with the SC approach, but the end-effector does move while using the DLS method.

5. Conclusion

We established a numerical method to apply the SC approach to a teleoperated 7-DOF manipulator system. Though this method cannot really utilize all the properties of the original analytical SC approach, yet a stable control at and around the singularities, the most important character of the SC approach, is achieved without any directional errors. Furthermore, the proposed method can be applied to any type of articulated manipulator if its Jacobian can be defined. The validity of this approach has been confirmed by the experiments with a graphics model of a 7-DOF

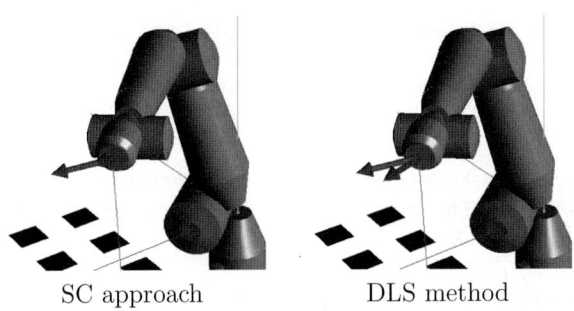

Figure 8. Difference between the SC approach and the DLS method.

manipulator.

Acknowledgment

The authors would like to thank Prof. Dragomir Nenchev for his very kind and useful advices.

References

[1] Y. Nakamura and H. Hanafusa, "Inverse Kinematic Solutions with Singularity Robustness for Robot Manipulator Control," *ASME J. Dynam. Sys. Measurement and Control*, Vol. 108, pp. 163–171, 1986.

[2] C. W. Wampler II, "Manipulator Inverse Kinematic Solutions Based on Vector Formulations and Damped Least-Squares Methods," *IEEE Trans. on Systems, Man and Cybernetics*, Vol. SMC–16, No. 1, pp. 93–101, 1986.

[3] S. Chiaverini, B. Siciliano and O. Egeland, "Review of the Damped Least-Squares Inverse Kinematics with Experiments on an Industrial Robot Manipulator," *IEEE Trans. on Control Systems Technology*, Vol. 2, No. 2, pp. 123–134, 1994.

[4] D. N. Nenchev, "Tracking Manipulator Trajectories with Ordinary Singularities: A Null Space Based Ap-

proach," *The Int. J. of Robotics Research*, Vol. 14, No. 4, pp. 399–404, 1995.

[5] Y. Tsumaki, D. N. Nenchev, S. Kotera and M. Uchiyama, "Teleoperation Based on the Adjoint Jacobian Approach," *IEEE Control Systems Magazine*, Vol. 17, No. 1, pp. 53–62, 1997.

[6] D. N. Nenchev, Y. Tsumaki and M. Uchiyama, "Singularity-Consistent Behavior of Telerobots: Theory and Experiments," *The Int. J. of Robotics Research*, Vol. 17, No. 2, pp. 138–152, 1998.

[7] Y. Tsumaki, S. Kotera, D. N. Nenchev and M. Uchiyama, "Advanced Experiments with a Teleoperation System Based on the SC Approach," *Proc. of the IEEE/RSJ Int. Conf. on Intelligent Robots and Systems*, pp. 1196–1201, 1998.

[8] D. N. Nenchev, Y. Tsumaki, M. Uchiyama, V. Senft and G. Hirzinger, "Two Approaches to Singularity-Consistent Motion of Non Redundant Robotic Mechanisms," *Proc. 1996 IEEE Int. Conf. on Robotics and Automation*, pp. 1883–1890, 1996.

[9] K. K. Delgado, M. Long and H. Seraji, "Kinematic Analysis of 7-DOF Manipulators," *The Int. J. of Robotics Research*, Vol. 11, No. 5, pp. 469–481.

[10] E. Nakano, "Mechanism and Control of Anthropomorphous Manipulator," *J. of the Society of Instrument and Control Eng.*, Vol. 15, No. 8, pp. 637–644, 1976 (in Japanese).

[11] J. M. Hollerbach, "Optimum Kinematic Design for a Seven Degree of Freedom Manipulator," *Robotics Research, The Second Int. Symp.*, pp. 215–222, 1984.

[12] A. Liegeois, "Automatic Supervisory Control of the Configuration and Behavior of Multibody Mechanisms," *IEEE Trans. on System, Man, and Cybernetics*, Vol. SMC-7, No. 12, pp. 868–871, 1977.

[13] T. Yoshikawa, "Analysis and Control of Robot Manipulators with Redundancy," *Robotics Research, The First Int. Symp.*, pp. 735–747, 1983.

[14] Y. Nakamura, H. Hanafusa, T. Yoshikawa, "Task-Priority Based Redundancy Control of Robot Manipulators," *The Int. J. of Robotics Research*, Vol. 6, No. 2, pp. 3–15, 1987.

[15] H. Seraji, "Configuration Control of Redundant Manipulators: Theory and Implementation," *IEEE Trans. on Robotics and Automation*, Vol. 5, No. 4, pp. 472–490, 1989.

Bilateral Controller Design for Telemanipulation in Soft Environments

Murat Cenk Çavuşoğlu[1], Alana Sherman[2], Frank Tendick[2,3]

[1] Department of Electrical Eng. and Comp. Sci., University of California, Berkeley, CA 94720
[2] Department of Bioengineering, University of California, Berkeley, CA 94720
[3] Department of Surgery, University of California, San Francisco, CA 94143
{*mcenk, alanas, tendick*} *@robotics.eecs.berkeley.edu*

Abstract

Previous research on teleoperation has focused on manipulation of hard objects. However, the design constraints are different in applications that involve manipulation of deformable objects, such as robotic telesurgery. In this paper a new measure for fidelity in teleoperation is introduced which quantifies the teleoperation system's ability to transmit changes in the compliance of the environment. This sensitivity function is highly appropriate for the application of telesurgery, where the ability to distinguish small changes in tissue compliance is essential for tasks such as tumor detection. The bilateral teleoperation controller design problem is then formulated as the optimization of this new metric with constraints on free space tracking requirements and robust stability of the system under environment and human operator uncertainties. The robust stability analysis can be applied to any teleoperator plant and guarantee stability given an uncertainty model. The analysis is also extended to evaluate effectiveness of using a force sensor in the teleoperation system.

Keywords — Bilateral Control Design, Haptics, Telemanipulation of Soft Objects, Teleoperation

1 Introduction

Previous research on teleoperation has focused on manipulation of hard objects. However, the design constraints are different in an application which involves manipulation of deformable objects. The stability-performance trade-off is the main determinant of the control design for teleoperation systems. Both performance and stability are inherently dependent on the task for which the system is designed. This paper addresses the issues in bilateral control design for telemanipulation of soft objects. The motivation behind this study is robotic telesurgery, where a surgical operation is performed by robotic instruments controlled by surgeons through teleoperation [2]. The goal of robotic telesurgery is to improve dexterity and sensation in minimally invasive surgery through the use of teleoperation technology.

As noted by Lawrence in [15], teleoperation controller architectures given in the literature can be classified in terms of the stability-performance trade-off. Control algorithms for ideal kinesthetic coupling [19] form one end of the spectrum, whereas passive communication based algorithms [1] form the other end. Conventional algorithms such as position error based force feedback and kinesthetic force feedback lie in the middle. There are also more recent controller designs using robust control theory. Kazerooni established an H_∞ based framework to design a teleoperation controller which transmits only force information and no position or velocity data [12]. Yan and Salcudean used H_∞ optimization to design controllers for motion scaling [18], and Hu et al. formulated the teleoperator control design as a convex H_∞ optimization problem [10]. Leung et al. used μ-synthesis to design controllers for teleoperation under time delay [16].

Operator performance is one of the important components of teleoperator design. Therefore experimental evaluation of control algorithms is crucial. Experimental studies at the NASA Jet Propulsion Laboratory [5, 13, 9] and by Lawn and Hannaford [14] compare various teleoperation algorithms within the context of operator performance. Human perceptual capabilities should also be considered. Jones and Hunter [11] performed experiments on determining human perceptual capabilities within the context of teleoperation. In a recent work, Daniel [4] takes into account considerations for improved stimulation of the tactile and kinesthetic receptors during teleoperator controller design by modifying the filter in the force feedback path. Colgate [3] introduced impedance shaping bilateral control as a means of "constructively altering the impedance of a task".

In the design of a teleoperation system controller, there are three considerations we believe to be important. First, it is important to have task-based performance goals rather than trying to achieve a marginally stable, physically unachievable ideal teleoperator response. Second, teleoperator controller design should be expressed explicitly as an optimization problem to accommodate task-based performance goals. Third, design of the teleoperation system must be oriented towards improving performance with respect to human perceptual capabilities. It is necessary to experimentally quantify human perceptual capabilities and to develop control design methodologies which will provide the means to include this in the control design.

In this paper a new measure for fidelity in teleoperation is introduced which quantifies the teleoperation system's ability to transmit changes in the compliance of the environment. This sensitivity function is highly appropriate for the application of telesurgery, where the ability to distinguish small changes in tissue compliance is essential for tasks such as tumor detection. The bilateral teleoperation controller design problem is then formulated as the optimization of this new metric with constraints on free space tracking requirements and robust stability of the system under environment and human operator uncertainties. The analysis is also extended to compare different sensor schemes within this context.

2 Formulation

The teleoperator can be modeled as a two-port network element relating force and position of the master manipulator, F_m and X_m, to the slave manipulator, F_s and X_s[1] (Fig. 1). We follow Hannaford [7] in using the hybrid parameters to characterize system behavior (Fig. 3)

$$\begin{bmatrix} F_m(s) \\ X_s(s) \end{bmatrix} = \begin{bmatrix} h_{11}(s) & h_{12}(s) \\ h_{21}(s) & h_{22}(s) \end{bmatrix} \begin{bmatrix} X_m(s) \\ F_s(s) \end{bmatrix} \quad (1)$$

Environment impedance transmitted through the teleoperator can be calculated as

$$Z_t = \frac{F_m}{X_m} = \frac{h_{11} + (h_{11}h_{22} - h_{12}h_{21})Z_e}{1 + h_{22}Z_e} \quad (2)$$

---
[1]In the literature, generally a force/velocity representation is used instead of a force/position representation. Although force/velocity representation has an advantage since the power is immediately given by the terminal variables of the two port, it introduces a pole/zero pair at the origin causing complications in stability analysis conditions, which is purely an artifact of the representation. Here, the force/position representation is used to avoid these complications.

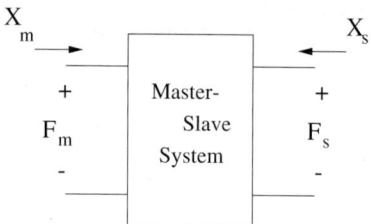

Figure 1: Two port input-output model of a teleoperation system.

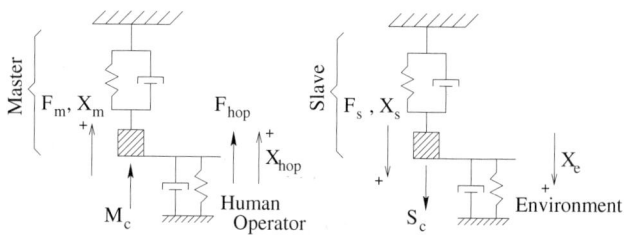

Figure 2: Physical model of the teleoperation system.

using the hybrid parameters. We will consider the model given in Fig. 2 as the underlying physical model throughout the analysis. This model is a linear model, and is only accurate locally. In Fig. 2, M_c and S_c are respectively the master and slave control inputs and F_{hop} is the human operator force.

3 Fidelity

Yokokohji defined an ideal response for teleoperation systems in [19]. With this, the goal of the control design was to match the position and forces at the master and slave manipulators exactly or through a virtual impedance. Lawrence defined transparency in [15] as the ratio between the transmitted and the environment impedances. Lawrence's design goal was to keep this ratio close to one over a maximal bandwidth. But, neither explicitly expressed their definitions of performance as measures that could be optimized.

We would like to explicitly distinguish the term *fidelity* in teleoperation. We define fidelity as a task dependent measure of performance which will be optimized during teleoperator controller design.

Transparency can be expressed as an optimization criterion, which will then be a specific choice of fidelity measure which quantifies how close the transmitted impedance is to the environmental impedance. However, if this particular choice of fidelity is best for a given task or not is another question.

In robotic telesurgery one would like to improve the ability to detect compliance changes in the environment

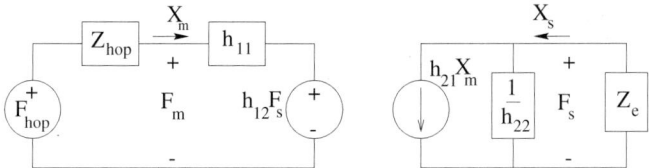

Figure 3: Hybrid parameters of a teleoperation system.

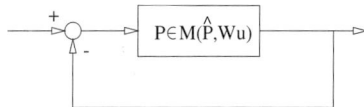

Figure 4: Closed loop system with multiplicative uncertainty.

in addition to the basic requirement of "good" tracking in free space and while in contact with tissue. This compliance detection is critical in a surgical application in two ways. First, the interaction of the needle with tissue during suturing, such as to feel when the needle punctures or leaves tissue, can be detected through the change in the compliance. Second, the structures hidden inside the tissue, such as blood vessels, major nerves, or tumors, can be located by noninvasively probing the tissue. In these cases, it is more desirable to have the ability to detect the changes in the environment impedance than simple position or force tracking between the master and slave manipulators. Therefore, it is necessary to introduce a fidelity measure which quantifies this ability.

The measure of fidelity proposed in this paper is the sensitivity of the transmitted impedance to changes in the environmental impedance. This can be defined as

$$\left\| W_s \frac{dZ_t}{dZ_e} \bigg|_{Z_e=\hat{Z}_e} \right\|_2 \quad (3)$$

where W_s is a frequency dependent weighting function, and $\hat{Z}_e$ is the nominal environment impedance.

In this study, a low pass filter with cutoff frequency of 40 Hz is used as the weighting function. This frequency was determined from our pilot experiments for determining human compliance discrimination thresholds. It is possible to improve this fidelity measure by including the frequency-dependent sensitivity of the human operator to compliance stimuli by incorporating it in the weighting function. A parallel research study is being conducted by our research group to determine this operator sensitivity function through psychophysics experiments [6].

4 Task-Based Optimization of the Teleoperation Controller

The controller to be used for the teleoperation system needs to satisfy some basic requirements such as stability under specified environment and operator variations. Once these are satisfied, the remaining freedom in the controller can be used to optimize a task dependent performance measure, in this case fidelity.

4.1 Stability

Any teleoperation system must maintain stability under operator and environment variations. Robust stability of the closed loop system under unstructured uncertainty can be used to check this by properly modeling the operator and environment variations as uncertainty in the system.

For stability analysis, we use a robust stability criterion for unstructured uncertainties as given in Zhou, Doyle, and Glover [20]. For SISO systems, the criterion is as follows.

Theorem 1 (Robust Stability Criterion)
Consider the closed loop system shown in Figure 4 with multiplicative unstructured uncertainty. The uncertainty is defined as

$$P \in M(\hat{P}, W_u) = \{\hat{P}(1 + W_u\Delta) : \Delta \in \mathcal{R}, \sup |\Delta(jw)| < 1,$$
$$\text{\# of rhp poles}(\hat{P}) = \text{\# of rhp poles }(\hat{P}(1 + W_u\Delta))\}, \quad (4)$$

where P is the loop gain, $\hat{P}$ is the nominal plant loop gain, W_u is the uncertainty weighting function, and $\mathcal{R}$ is the set of proper real rational functions. Then, the closed loop system shown is stable for all $P \in M(\hat{P}, W_u)$, if and only if it is stable for the nominal plant $\hat{P}$, and

$$\|W_u T\|_\infty \leq 1, \quad (5)$$

where $T = \frac{\hat{P}}{1+\hat{P}}$.

The uncertainty weighting function $|W_u(jw)|$ can be interpreted as the percentage uncertainty in $\hat{P}$ at the frequency w.

For the teleoperation system, the loop gain P is calculated in Hannaford [8] as

$$P = \frac{-h_{12}h_{21}Z_e}{(h_{11} + Z_{hop})(1 + h_{22}Z_e)} \quad (6)$$

where Z_e and Z_{hop} are respectively the environment and human operator impedances.

In this study we will consider the uncertainties in the human operator and the environment impedances.

First, consider the variation in the environment. Since Z_e appears as $\frac{Z_e}{1+h_{22}Z_e}$ in the loop gain expression, we proceed to put an upper bound to the variation in this term for the possible set of environments, $Z_e \in \mathcal{Z}_e$.

Start with some manipulation

$$P = \frac{-h_{12}h_{21}Z_e}{(h_{11}+Z_{hop})(h_{22}Z_e+1)} \quad (7)$$

$$= \underbrace{\frac{-h_{12}h_{21}}{(h_{11}+Z_{hop})}\frac{\hat{Z}_e}{h_{22}\hat{Z}_e+1}}_{\hat{P}} \underbrace{\frac{h_{22}\hat{Z}_e+1}{\hat{Z}_e}\frac{Z_e}{h_{22}Z_e+1}}_{1+W_{ue}\Delta} \quad (8)$$

Since we want to have the nominal environment $\hat{Z}_e$ for $\Delta = 0$, we pick

$$W_{ue}\Delta = \frac{1+h_{22}\hat{Z}_e}{\hat{Z}_e}\frac{Z_e}{1+h_{22}Z_e} - 1 \quad (9)$$

$$= \frac{1}{h_{22}\hat{Z}_e}\frac{Z_e - \hat{Z}_e}{\frac{1}{h_{22}}+Z_e} \quad (10)$$

then we pick an upper bound to

$$\left|\frac{Z_e - \hat{Z}_e}{\frac{1}{h_{22}}+Z_e}\right| < |\Phi(jw)| \quad (11)$$

for the possible environment values, which gives

$$W_{ue} = \frac{1}{h_{22}\hat{Z}_e}\Phi. \quad (12)$$

Φ can be a function of the controller values and other known variables present in h_{22}.

Similarly, for the operator impedance variation, we proceed to put an upper bound to the term $\frac{1}{h_{11}+Z_{hop}}$ for the possible set of operator impedances, $Z_{hop} \in \mathcal{Z}_{hop}$. We pick

$$W_{uh}\Delta = \frac{h_{11}+\hat{Z}_{hop}}{h_{11}+Z_{hop}} - 1 = \frac{\hat{Z}_{hop} - Z_{hop}}{h_{11}+Z_{hop}} \quad (13)$$

to have $\hat{Z}_{hop}$ for $\Delta = 0$. Then, we can pick an upper bound

$$\left|\frac{\hat{Z}_{hop} - Z_{hop}}{h_{11}+Z_{hop}}\right| < |W_{uh}(jw)| \quad (14)$$

which can be a function of the known variables present in h_{11}.

The two uncertainty terms can be combined to give a single multiplicative uncertainty weighting function as

$$W_u = W_{ue} + W_{uh} + W_{ue}W_{uh}. \quad (15)$$

4.2 Tracking Requirement

The tracking requirement is necessary to prevent the final controller parameter optimization from yielding trivial solutions. To illustrate this complication, consider the case of optimizing a controller for transparency at a given environment stiffness as operating point. The trivial solution to this optimization is to have a master controller which gives the master manipulator an apparent stiffness equal to the nominal environment stiffness, and have no feedback from slave to master or even not actuate the slave at all. The most natural constraint to prevent this kind of behavior is to require the teleoperation system to have sufficient tracking performance in free space. We will pose this tracking requirement as a condition on the disturbance sensitivity function of the forward position loop during motion in free space. In the hybrid parameter formulation of the teleoperator, this sensitivity function is given by

$$S = 1 - h_{21}. \quad (16)$$

Then the tracking requirement can be posed as

$$|S(jw)| < |b(jw)| \iff \|SW_p\|_\infty \leq 1, \; W_p = 1/b(jw) \quad (17)$$

which dictates a tracking error less than $|b(jw)|$ for a sinusoidal input with angular frequency w and magnitude 1. This effectively puts a condition on the slave position gain when the slave is controlled by the master position (position only loop in the forward direction).

4.3 Optimizing for Fidelity

The controller gains are chosen to optimize the fidelity among the set of controller values which satisfy stability and tracking requirements.

$$\arg\sup_{\substack{\|W_uT\|_\infty \leq 1 \\ \text{stable for } \hat{P} \\ \|W_pS\|_\infty \leq 1}} \inf_{\tilde{Z}_e \in \tilde{\mathcal{Z}}_e} \left\|W_s\frac{dZ_t}{dZ_e}\Big|_{\tilde{Z}_e}\right\|_2 \quad (18)$$

The fidelity term is slightly modified from (3) to be more general, optimizing the worst case fidelity for a given set of environment values, $\tilde{\mathcal{Z}}_e$. $\tilde{\mathcal{Z}}_e$ is the range of environments in which sensitivity of the transmitted impedance to environment impedance variations is desired. It is important to note that this is not a convex optimization since $\left\|W_s\frac{dZ_t}{dZ_e}\right\|_2$ is not convex in the controller parameters. Therefore, proper numerical techniques should be used during the computation.

5 Comparing Controller Architectures and Sensors

We would like to determine if the use of a force sensor is necessary on the slave manipulator of the telesurgical workstation for sufficient fidelity. For better performance, it is almost always desirable to put additional sensors on the manipulators, however, as this sensor will be located on the part of the instrument which will be inside the body, it is a source of complications in the manipulator design, sterilization requirements, and adds to the cost of the final product.

Within this context, we will compare three different control architectures: position error (PERR), kinesthetic force feedback (KFF), and position error plus kinesthetic force feedback (P+FF) (Fig. 5). In the PERR architecture, the force sent to the master is proportional to the position error between the master and slave manipulators. The KFF architecture uses a force sensor on the slave end to transmit forces back to the master. The P+FF architecture is a hybrid of KFF and PERR. In this architecture, the force fed back to the master is a linear combination of the position error and the interaction force between the slave and the environment. In all three controllers the master position is used to command the slave.

Essentially, the PERR and KFF architectures are the limit cases of the more general control architecture P+FF. Therefore it is possible to quantify the improvement due to using a force sensor for a given task by looking at how the fidelity of the P+FF architecture changes as the force gain is changed.

We define the *alpha*-curve is as the highest fidelity achievable with the P+FF controller as a function of the force gain α, subject to the stability and tracking constraints.

$$f(\alpha) = \sup_{\substack{\|W_u T\|_\infty \leq 1 \\ \text{stable for } \hat{P} \\ \|W_p S\|_\infty \leq 1 \\ G_m, G_s}} \inf_{\tilde{Z}_e \in \tilde{Z}_e} \left\| W_s \left. \frac{dZ_t}{dZ_e} \right|_{\tilde{Z}_e} \right\|_2 \quad (19)$$

The shape of this curve depends on the stability constraint as well as the fidelity measure being used. There are three different cases based on location of the maximum point of the curve (Fig. 6). If the PERR end is the maximum, use of a force sensor does not improve performance. If the KFF end is the maximum, then it is better to use purely the force sensor output as the source of force feedback. Finally, if the maximum is located at an intermediate point, it is possible to have better performance by using a combination of

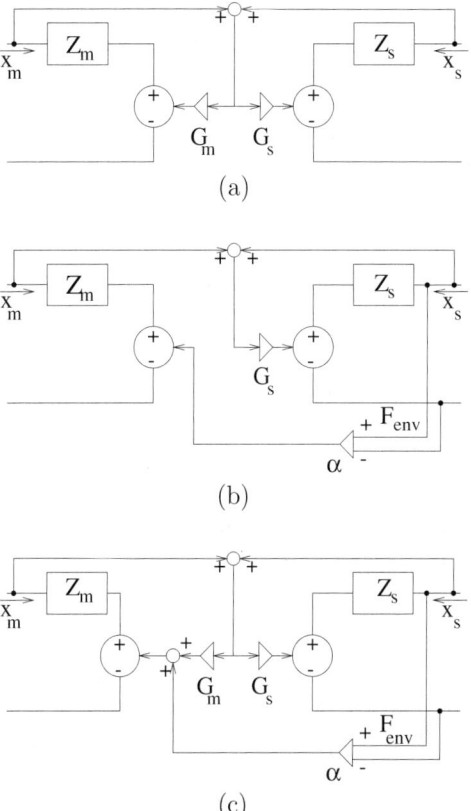

Figure 5: PERR, KFF and P+FF architectures

position error and the force measurements to generate force feedback. The relative value of the peak value of the curve to the PERR value can be used to judge if the amount of performance improvement justifies the use of the force sensor.

6 Case Study

The testbed used to evaluate the analysis described above is a teleoperation system with two identical three degree of freedom (DOF) robotic manipulators, Phantom v1.5 haptic interfaces (Sensable Technologies, Cambridge, MA) with custom motor drive electronics. The analysis here is carried out with a one DOF model, along the vertical direction, which is the axis orthogonal to the surface of the deformable body being manipulated. The local linear model of the manipulator in the vertical direction around the operating region is estimated as[2]

$$Z = \frac{1}{9.641e^{-5}s^2 + (0.002665 + D_x)s + 0.0322} \quad (20)$$

---
[2] All the units are in Newtons for force and mm for distance.

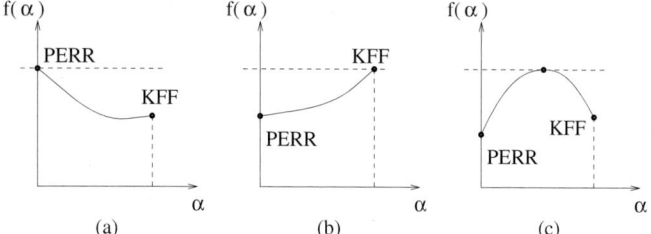

Figure 6: Possible cases for the shape of *alpha*-curve

where D_x is the active damping used. This model is constructed from black box system identification. In this study, active damping of $D_x = 5 \times 10^{-4}$ has been used on the slave side to improve the stability of the manipulator.

The following environment and operator impedance variations are considered

$$Z_e \in \{(B_e s + 1)K_e : B_e \geq 0.05, 0 \leq K_e < \infty\}, \quad (21)$$
$$Z_{hop} \in \{(0.0219s + 1)K_{hop} : 0.2 \leq K_{hop} \leq 2\} \quad (22)$$

with nominal impedances

$$\hat{Z}_e = 0.35(0.05s + 1) \quad (23)$$
$$\hat{Z}_{hop} = 1.51(0.0219s + 1). \quad (24)$$

The range of Z_e represents environments from 0 to infinite stiffness with a lower bound on damping. The nominal value of Z_e is the stiffness of the silicon gel used in experimental evaluation of teleoperation systems in [17]. The range and nominal value of Z_{hop} are experimentally determined from subjects using the haptic interface.

The following empirically determined upper bounds for the uncertainty terms of (11) and (14) are used in the stability analysis

$$\Phi(s) = 10^{3.8/20} \left(\frac{\frac{s}{19}+1}{\frac{s}{80}+1} \right)^2 \left(\frac{\frac{s}{125}+1}{\frac{s}{75}+1} \right)^3 \quad (25)$$

$$W_{uh}(s) = 10^{15.3/20} \left(\frac{\frac{s}{60}+1}{(s^2 + 20.760s + 60^2)/60^2} \right). \quad (26)$$

These upper bounds are determined by systematically varying the parameters G_s, B_e, K_e for (25) and G_m, K_{hop} for (26) within their specified limits (Fig.7). The upper bound used for tracking sensitivity function is

$$b(s) = \left(\frac{9.64 \times 10^{-5} + 3.66 \times 10^{-3}s + 0.032}{9.64 \times 10^{-5} + 3.66 \times 10^{-3}s + 0.232} \right) \\ \times \left(\frac{\frac{s}{70}+1}{\frac{s}{100}+1} \right)^8 \left(\frac{\frac{s}{138}+1}{\frac{s}{100}+1} \right)^8 . \quad (27)$$

This upper bound requires good position tracking at low frequencies where the voluntary hand movements

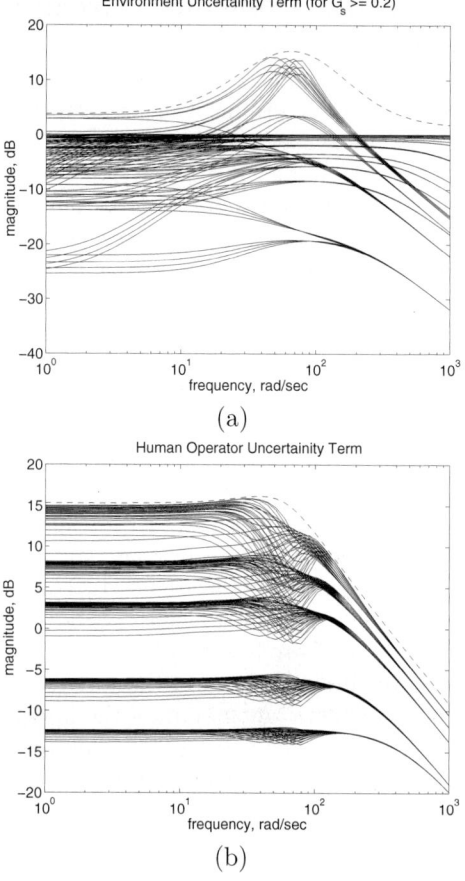

Figure 7: Uncertainty weighting functions: (a) Environment uncertainty term (b) Human operator uncertainty term. Dashed line is the upper bound for the uncertainty.

occur. The first term is chosen using the sensitivity function S at $G_s = 0.2$ and the remaining terms are chosen to accommodate underdamped behavior occurring for $G_s > 0.2$. The resulting $b(s)$ practically puts a lower bound on the slave position gain as $G_s \geq 0.2$.

It is important to note that the stability analysis performed with these upper bounds is conservative in the sense that it doesn't completely capture the dependence of the uncertainty weighting function on the known variables, such as controller gains. For example, the bound in (25) is chosen to be a constant transfer function, whereas it is actually possible to pick an upper bound which is a function of the controller gains. This dependence is a nontrivial function of controller gains, so a constant upper bound is used here.

It is also possible to find a single upper bound for the combined environment and operator uncertainties. However, the combined bound would have been completely independent of controller gains, whereas the

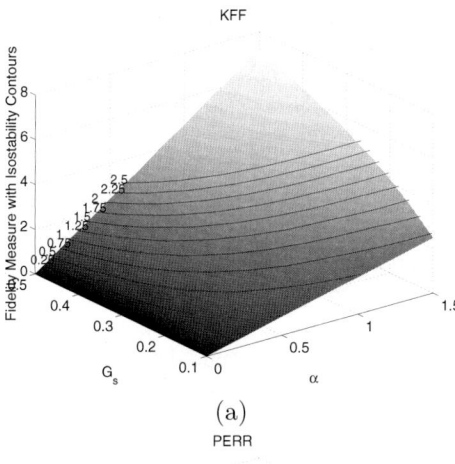

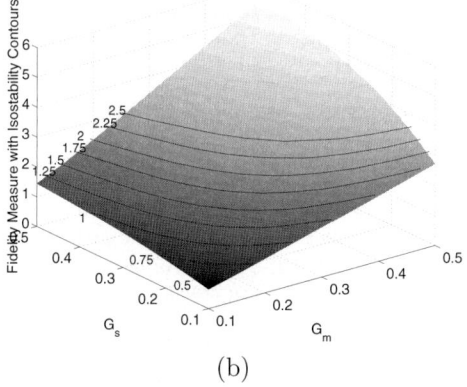

Figure 8: Fidelity of the PERR and KFF architectures as a function of controller parameters. Contours of constant stability are shown overlaid on the fidelity surface for comparison. Note that stability decreases as fidelity increases.

bound constructed from pieces have some (even though not complete) dependence from (12), since h_{22} is a function of controllers. This gives a less conservative upper bound than we would get with a single term.

The fidelity plots for the KFF and PERR controllers superimposed with isostability curves are shown in Fig. 8. The fidelity-stability trade-off can easily be observed on these plots, as the stability degrades as fidelity improves. The resulting *alpha*-curve is shown in Fig. 9. This curve predicts that using a force sensor will improve the performance and the KFF algorithm will perform best for the choice of the fidelity measure, tracking requirements, and the uncertainty bounds considered.

7 Discussion and Conclusion

As mentioned before, experimental evaluation of teleoperation systems is an important aspect of teleoperator controller design. Experiments comparing con-

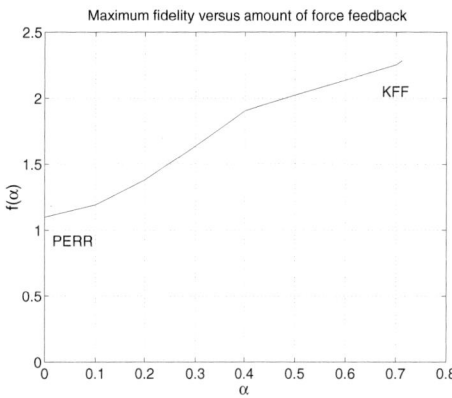

Figure 9: *Alpha*-curve for the teleoperation system studied

trollers designed following the methodology proposed in this paper are presented by the authors in [17], and it is one of the major thrusts of our future research.

It is important to note that the stability measure developed here is on the conservative side, mainly due to modeling errors in the weighting functions. It was possible to manually increase the gains of the physical setup and still maintain stability. It would be more appropriate to use a structured uncertainty model to best capture this kind of uncertainty. We are currently working on using linear fractional transformations to develop a better uncertainty model.

We are also working on a more detailed model of the system which includes the noise and the dynamic characteristics of the force sensor which were not modeled in the analysis. Including the non-idealities of the sensors is important to make a better comparison between the sensory schemes. For example, absence of noise in the force sensor model gives an unfair advantage to the KFF algorithm in the *alpha*-curve analysis. These modeling efforts will emphasize developing other quantitative means to compare sensory schemes.

As the final words, we would like to reemphasize three important points: 1) In design of teleoperation systems, it is important to have task-based performance goals rather than trying to achieve a marginally stable, physically unachievable ideal teleoperator response. 2) Teleoperator control design should be explicitly formulated as an optimization to accommodate task-based performance metrics. 3) Design of the teleoperation system must be based on human perceptual capabilities. For this, it is necessary to quantify human perceptual capabilities, and to have means to incorporate them into the control design (design methodology, tools, and proper formulation). This paper addresses these points by proposing a new fidelity measure for the compliance discrimination task, and developing a design methodology using robust control theory for task-based optimization of the teleoperation controller, focusing on telemanipulation of deformable objects. The analysis has also been used to evaluate the improvement from the use of a force sensor in the teleoperation system.

Acknowledgments

This work is supported in part by ONR under MURI grant N14-96-1-1200, ARO under MURI grant DaaH04-96-1-0341, and NSF under grant CISE CDA 9726362.

References

[1] R. J. Anderson and M. W. Spong. Asymptotic stability for force reflecting teleoperators with time delay. *International Journal of Robotics Research*, 11(2):135–148, April 1992.

[2] M. C. Çavuşoğlu, F. Tendick, M. Cohn, and S. S. Sastry. A laparoscopic telesurgical workstation. *IEEE Transactions on Robotics and Automation*, 15(4):728–739, August 1999.

[3] J. E. Colgate. Robust impedance shaping telemanipulation. *IEEE Transactions on Robotics and Automation*, 9(4):374–384, August 1993.

[4] R. W. Daniel and P. R. McAree. Fundamental limits of performance for force reflecting teleoperation. *International Journal of Robotics Research*, 17(8):811–830, August 1998.

[5] H. Das, H. Zak, W. S. Kim, A. K. Bejczy, and P. S. Schenker. Operator performace with alternative manual control modes in teleoperation. *Presence*, 1(2):201–218, Spring 1992.

[6] N. Dhruv and F. Tendick. Frequency dependence of compliance contrast detection. In *Proceedings of the Symposium on Haptic Interfaces for Virtual Environment and Teleoperator Systems, part of the ASME Int'l Mechanical Engineering Congress and Exposition (IMECE 2000)*, November 2000.

[7] B. Hannaford. A design framework for teleoperators with kinesthetic feedback. *IEEE Transactions on Robotics and Automation*, 5(4):426–434, August 1989.

[8] B. Hannaford. Stability and performance tradeoffs in bi-lateral telemanipulation. In *Proceedings of the IEEE International Conference on Robotics and Automation*, pages 1764–1767, 1989.

[9] B. Hannaford, L. Wood, D. A. McAffee, and H. Zak. Performance evaluation of a six-axis generalized force-reflecting teleoperator. *IEEE Transactions on System, Man, and Cybernetics*, 21(3):620–633, May/June 1991.

[10] Z. Hu, S. E. Salcudean, and P. D. Loewen. Robust controller design for teleoperation systems. In *Proceedings of the IEEE International Conference on Systems, Man and Cybernetics*, volume 3, pages 2127–2132, 1995.

[11] L. A. Jones and I. W. Hunter. Analysis of the human operator controlling a teleoperated microsurgical robot. In *Proceedings of 6th IFAC/IFIP/IFORS/IEA Symposium on Analysis, Design and Evaluation of Man-Machine Systems*, pages 593–597, 1995.

[12] H. Kazerooni, T.-I. Tsay, and K. Hollerbach. A controller design framework for telerobotic systems. *IEEE Transactions on Control Systems Technology*, 1(1):50–62, March 1993.

[13] W. S. Kim, B. Hannaford, and A. K. Bejczy. Force-reflection and shared compliant control in operating telemanipulators with time delay. *IEEE Transactions on Robotics and Automation*, 8(2):176–185, April 1992.

[14] C. A. Lawn and B. Hannaford. Performance testing of passive communication and control in teleoperation with time delay. In *Proceedings of the IEEE International Conference on Robotics and Automation*, pages 776–783, 1993.

[15] D. A. Lawrence. Stability and transparency in bilateral teleoperation. *IEEE Transactions on Robotics and Automation*, 9(5):624–637, October 1993.

[16] G. M. H. Leung, B. A. Francis, and J. Apkarian. Bilateral controller for teleoperators with time delay via mu-synthesis. *IEEE Transactions on Robotics and Automation*, 11(1):105–116, February 1995.

[17] A. Sherman, M. C. Çavuşoğlu, and F. Tendick. Comparison of teleoperator control architectures for palpation task. In *Proceedings of the ASME Dynamic Systems and Control Division, part of the ASME International Mechanical Engineering Congress and Exposition (IMECE 2000)*, volume 2, pages 1261–1268, November 2000.

[18] J. Yan and S. E. Salcudean. Teleoperation controller design using H-infinity optimization with application to motion-scaling. *IEEE Transactions on Control Systems Technology*, 4(3):244–258, May 1996.

[19] Y. Yokokohji and T. Yoshikawa. Bilateral control of master-slave manipulators for ideal kinesthetic coupling—formulation and experiment. *IEEE Transactions on Robotics and Automation*, 10(5):605–620, October 1994.

[20] K. Zhou, J. C. Doyle, and K. Glover. *Robust and Optimal Control*. Prentice-Hall, Inc., New Jersey, USA, 1996.

Authors Index

A

Abedelaziz, Benallegue	3465
Abiko, Satoko	441
Abolmaesumi, Purang	1549
Acar, Ercan U.	699
Adam, Amit	285, 291
Adan, Antonio	2541
Agah, Arvin	1006, 4186
Agarwal, Pankaj K.	2243
Aghili, Farhad	1130
Agrawal, Sunil K.	3338, 2480
Ahalt, Stanley C.	1671
Ahuactzin, Juan-Manuel	3722
Aicardi, Michele	915, 903
Aiyama, Yasumichi	799, 2474
Alami, Rachid	4256, 15, 2929, 33
Albro, Juanita V.	3630
Albu-Schaeffer, Alin O.	2852, 3356
Aldon, Marie-Jose	1585
Allen, Peter K.	2460
Altafini, Claudio	1265, 169
Amano, Takaharu	139
Amat, Josep	3250, 2779, 3600
Amato, Nancy M.	948, 3789, 21, 1500
	1940, 1475, 1376, 954
Anderson, Gary T.	2710
Andersson, Paul, H.	2002
Andou, Takayuki	3920
Ang, Marcelo H	2791
Angeles, Jorge	2873, 2407
Angeletti, Damiano	3220
Anticoli, Claud	626
Antonelli, Gianluca	447
Antoniali, Fabio M.	1591
Aoki, Takeshi	3772
Aoyama, Hisayuki	3429
Apaydin, Mehmet Serkan	932
Apostolopoulos, Dimitrios S	1742, 4174, 4158
Arai, Fumihito	139, 2486, 632, 577, 3195, 604
Arai, Hirohiko	1870, 2680
Arai, Tamio	3477, 799, 2474
Arai, Tatsuo	2998, 610
Aranda, Joan	3600
Arimoto, Suguru	2344, 2535, 97
Arkin, Ronald C.	1627, 477, 1975, 453
Armingol, Jose M	1124
Arras, Kai	1111
Arsicault, Marc	2642
Aruk, Baris	860
Asada, H. Harry	3831
Asada, Harry H	1857
Asada, Minoru	3987
Asahiro, Yuichi	3016
Asama, Hajime	2935, 3010, 3004
Asama, Kouichirou	2424
Asamura, Naoya	1851
Asano, Fumihiko	3139
Asoh, Hideki	1579
Astolfi, Alessandro	743
Atkar, Prasad N.	699
Atkeson, Christopher G.	1988
Aude, Eliana P.	2911
Avadhanula, Srinath	3901
Avizzano, Carlo A	1364
Axelsson, Mikael	4017
Ayache, Nicholas	1370
Azinheira, Jose R	4152

B

Bachem, A.	235
Bachiller, Margarita	2541
Bae, Byunghoon	145
Bagnell, J. Andrew	1615
Bahl, Vikas	175
Bai, Shaoping	2279
Bailey, Sean A.	3643
Bailey-Kellogg, Chris	940
Balaguer, Carlos	2668
Balaniuk, Remis	2337
Balasuriya, Arjuna P.	920
Balch, Tucker	2381, 1092
Balduzzi, Fabio E.	836, 40
Ban, Shigeki	348
Basch, Julien	1765
Batavia, Parag H.	705
Batlle, Joan	3250, 2779
Bayazit, O. Burchan	954
Baydar, Cem M.	818
Bayle, Bernard	1251
Begovic, Edvard	2255
Belousov, Igor R.	1878
Belta, Calin A.	1245

Bennewitz, Maren	271
Bentivegna, Darrin C.	1988
Bererton, Curt A.	2923
Bergamasco, Massimo	1364
Bergerman, Marcel	2114
Bergsten, Pontus	2980
Bernabeu, Enrique J.	3801
Berretty, Robert-Paul M.	1053
Bertozzi, Massimo	3698
Besnard, Sebastien	2859
Bessho, Yoshiharu	4023
Bessonnet, Guy	2285, 2499
Betaille, David	2045
Bhanu, Bir	491
Bi, Shusheng	127
Biagiotti, Luigi	1427
Bianco, Giovanni M.	2704
Bicchi, Antonio	2078, 1903, 2319
Bidaud, Philippe	3364
Bien, Zeungna	3662
Bing, Yang	3813
Birk, Andreas	749, 299
Bito, Janos F.	2002
Blanco, Dolores	4232, 2668
Boada, Beatriz L.	4232
Bobrow, James E	1433
Bobrow, Jim E.	3630
Bock, Hans Georg	4128
Bodner, Douglas	1813
Bombin, Carlos	3332
Bone, Gary M.	3061
Bonnifait, Philippe	1597
Bono, Riccardo	2369
Borenstein, Johann	3588, 2879
Borges, Geovany A.	1585
Borgstadt, Justin A.	723
Borovac, Branislav A.	2255
Borrelly, Jean-Jacques	2753
Bouron, Pascal	1597
Bouvet, Denis	2045, 3612
Bozzo, Tommaso	3220
Brady, Kevin	644
Branicky, Michael S.	1481
Braunstingl, Reinhard	1961
Bray, Marco	2319
Bredenfeld, Ansgar	205
Bretthauer, Georg	2437
Brock, Oliver	1469
Broggi, Alberto	3698
Brooks, Bernard G	1019
Brown, Keith	3232
Brown, Mark D.J.	4078
Brugali, Davide	836
Brutlag, Douglas L.	932
Bruyninckx, Herman	3344, 2523
Bruzzone, Gabriele	2369
Bu, Fanping	3459
Buehler, Martin	1130, 3153, 3650
Bueno, Samuel S	4152
Bullo, Francesco	3300
Bunschoten, Roland	499, 1174
Burdick, Joel W.	2716, 427, 2686
Burgard, Wolfram	271, 1665
Burschka, Darius	1707
Buskey, Gregg D	1635
Buss, Martin	103
Butterfass, Joerg	109
Byun, Kyung-Seok	767

C

Caccavale, Fabrizio	447, 435
Caccia, Massimo	2369
Cacitti, Alessio	680
Caffaz, Andrea	77
Campa, Ricardo	3523
Campolo, D.	3839
Campos, Mario	2992
Cannata, Giorgio	77
Cannon, David J.	2804, 3133
Canny, John F.	2235
Canou, Joseph	3999
Canuto, Enrico	2529
Carreras, Marc	3250
Carroll, James L.	517
Casal, Arancha	360
Casalino, Giuseppe	77, 3220, 915, 903
Casals, Alicia	3600
Caselli, Stefano	2147
Castano, Andres	3503
Cavusoglu, Murat Cenk	1045
Ceccarelli, Marco	4116, 1295
Cerrada, Carlos	2541
Cervera, Enric	717
Chaimowicz, Luiz	2992
Chalfant, Gene	1039
Cham, Jorge G.	3643
Chan, Vincent K	1313
Chang, Chih-Ching	1934
Chang, Eric C.-H.	3016
Chang, Pyung Hun	3955
Chang, Shi-Chung	535
Chang, Yi Cheng	1334
Chapelle, Frederic	3364
Chaumette, Francois	2773, 731, 247, 2767
Cheah, Chien C	2535

Cheboxarov, Victor V	2267
Chellali, Ryad	1878
Chen, Chun-Hung	535
Chen, Hao	842, 1439
Chen, Haoxun	842
Chen, I-Ming	2395, 2407
Chen, Jason R	4096, 1530
Chen, Jin-Liang	3795
Chen, Shih-Feng	3047
Chen, Wen-jia	3350
Chen, Wuwei	2968
Chen, Yifan	3127
Cheng, Fan-Tien	4060, 1832
Cheng, H.	3766
Cheng, Wendy	773
Cheng-Hsin, Tsai	3778
Chenut, Xavier	2846
Chesi, Graziano	737
Chesse, Stephane	2499
Chiaverini, Stefano	447
Chien, Yuh-Ren	2897
Chin-Po, Hwang	3778
Chirikjian, Gregory S.	1773
Chiu, F.C.	897
Cho, Changhyun	2628
Cho, Hye-Kyung	3618
Cho, Hyun C.	1025
Cho, Hyungsuck	1603
Cho, Sung-Kun	2956
Cho, Yong Jo	3577
Cho, Young Cheol	3571, 2419
Cho, Young-Jo	3618
Choi, Chong-Ho	2560
Choi, Hyoukryeol	761
Choi, J. W	2622
Choi, Jae-Joon	1452
Choi, Jin Y.	972
Choi, Joon Hyock	3412
Choi, Myoung H	4011
Choi, Sangeun	1352
Choi, Seung-gap	2547
Choi, Song K.	926
Choi, Sung Yug	1220
Choi, Youngjin	4041, 1142
Chong, Nak Young	1006, 1013, 4186
Choset, Howie	991, 699
Chou, Wusheng	2506
Chriette, Abdelhamid	1701
Chronis, George	485, 259
Chu, Jian	3582
Chu, Jiaxin	2968, 2723
Chung, Chi Youn	4066
Chung, Hakyoung	3588
Chung, Jaewoo	616
Chung, ShengLuen	541
Chung, Wan Kyun	1161, 4041, 1142, 3859, 121
Chung, Wen-Ya	1981
Clady, Xavier	1653
Clapworthy, Gordon J.	1878
Clark, Jonathan E.	3643
Clegg, Andrew C.	3226
Coletta, Paolo	2369
Collange, Franois	1653
Collewet, Christophe	247
Collins, Anne D.	2243
Company, Olivier	3256
Conticelli, Fabio	1463
Cook, Albert	1555
Coppelli, Alessandro	1903
Corke, Peter I.	2553, 3742
Cortes, Juan	1494
Costa, Ivan F.	2337
Coste-Maniere, Eve	386
Courty, Nicolas	223
Cowden, Chris	491
Cox, Daniel	3286
Cremers, Armin B.	1665
Croft, Elizabeth A.	979
Crubille, Paul	1597
Cufi, Xevi	2779
Cui, Youjing	1105
Cunningham, Christopher T.	3981
Cutkosky, Mark R.	589

D

Daachi, Boubaker	3465
Dai, Songtao	2165
Dale, Lucia K	1940
Damoto, Riichiro	773
Daney, David	3262
Dang, Anh X.	1257
Dario, Paolo	626
Das, Aveek K	157, 1714
Davidson, Morgan E	175
Davies, Bruce J.	2356
de la Escalera, Arturo	1124
de Lasa, Martin	3153
De Luca, Alessandro	2090
Decker, Michael W.	1257
Delingette, Herve	1370
Desai, Jaydev P.	3022, 4214
DeSouza, Guilherme N.	2171, 193
Despouys, Olivier	9
Di Febbraro, Angela	40
Dillmann, Rudiger	2596, 2578

DiMaio, S.P.	1549
Ding, Dan	2217
Do, Yongtae	2140
Doi, Toshitsugu	1891, 1031
Donald, Bruce R.	940
Donato, Gianluca	1757
Dong, Jianyu	1671
Dong, Lixin	632
Dornaika, Fadi	1194
Driankov, Dimiter	2980
Du, Yonghui	3061
Duan, Guanhong	2448
Dubey, Rajiv V.	374
Duckett, Tom	4017
Duindam, Vincent	427
Dunnigan, Matthew W.	3226

E

Ebert-Uphoff, Imme	1313, 1257
Ebihara, Daiki	1792
Edwards, Ian	909
Egerstedt, Magnus	3961
Ehrenmann, Markus	2596
Eisinberg, Anna	626
Eldershaw, Craig	3383
Elhajj, Imad H.	656, 662
Emery, Rosemary	2381
Emiris, Ioannis Z.	3262
Endo, Yoichiro	477
Engel, Dirk	2020
Eom, Kwang Sik	2331, 886
Erkmen, Aydan M.	674
Erkmen, Ismet	674
Esposito, Joel M	2818
Ezaki, Hideaki	4208

F

Fahlbusch, Stephan	3435
Falconer, Gavin J.	909
Fang, Chang-Jia	1201
Fang, Lei	842
Fang, Lijin	3535
Fascioli, Alessandra	3698
Fatikow, Sergej	3435
Faulkner, Gary	1555
Fearing, R.S.	3839
Feliu, Vicente B.	3853, 2541
Fernandez, Vicente	2668
Fernandez, Xavier	3600
Ferraro, Domenico	2147
Ferrier, Nicola J.	2140, 723
Fierro, Rafael O	157, 1714
Figliolini, Giorgio	4116
Fiorini, Paolo	2704, 1039
Fofi, David	3548
Forrester, Benjamin	259
Fourquet, Jean-Yves	1251
Fox, Dieter	1665
Fraichard, Thierry	3722
Freund, Eckhard	187, 1921, 806
Frintrop, Simone	2072
Frisoli, Antonio	1364
Froehlich, Edward M.	3643
Fu, Li-Chen	547, 3559, 2634, 1981, 229
Fuchiwaki, Ohmi	3429
Fuchs, Robert	755
Fuh, Jerry Y. H.	1340
Fujimoto, Hideo	380
Fujimoto, Hiroshi	711
Fujisawa, Kae	3772
Fujita, Masahiro	453
Fujiwara, Hisanaga	3529
Fukuda, Toshio	3668, 139, 2486, 632
	1518, 577, 3195, 604
Fukuyama, Junya	2037
Funahashi, Yasuyuki	597, 2466
Fung, Waikeung	662
Furuno, Seiji	1271
Furusho, Junji	3825
Furuta, Katsuhisa	3510
Fusco, Giuseppe	447

G

Gabriely, Yoav	1927
Gagov, Zavarin Vladislavov	2419
Gaines, James A.	133
Gao, Yan	2395
Gaonkar, Roshan S	854
Garcia de Quevedo, William	3423
Garcia, Andres H.	3853
Garcia, Eloisa	3523
Garcia, Gaetan	2045, 3612
Garcia, Rafael	2779
Gasior, Chris	755
Gautier, Maxime	2867
Gazeau, Jean-Pierre	2642
Ge, Q. Jeffrey	342
Ge, Shuzhi Sam	3871, 1105
Gevher, Mustafa	674
Gienger, Michael	4140
Gil, Thierry	3256, 1585
Ginhoux, Romuald	1226

Go, Seok Jo	2962
Goldberg, Ken	1065, 1053
Goldenberg, Andrew A.	3042, 3728
Goldenstein, Siome	3973
Goldfarb, Michael	1382
Goodwine, Bill	2811
Gorham, Barry J	2045
Gorman, Jason J.	2804
Goshozono, Toshihiko	407
Gosselin, Clement M.	1295
Goto, Satoru	2747, 3949
Graham, Todd	755
Grana, Manuel	2059
Gravagne, Ian A.	3877
Gravot, Fabien	2929
Grebenstein, Markus	109
Greiner, Russell	2134
Griffiths, John G	1561
Gruver, William	2517
Gu, Jason	1555
Guan, Yisheng	2197, 1909
Guere, Emmanuel	15
Guibas, Leonidas J	3973, 2903, 1765
Guo, Jenhwa	897
Gupta, Kamal	1948, 265
Gutmann, Jens-Steffen	1226
Gyorfi, Julius S.	1506

H

Haehnle, Matthias	3356
Hager, Gregory D	1707
Halliburton, William	3594
Hamel, Tarek	1701
Hamel, William R.	638, 393
Hamidzadeh, Babak	3326
Hamilton, Kelvin	3232
Hammond, Anthony	2140
Han, Chang-Soo	151
Han, Hyun-Yong	97
Han, Kuk-Hyun	217
Han, Sung Hyun	2622
Han, Z	2261, 2102, 1065, 1845, 3115
	3932, 2197, 1909, 3396, 2517
Hannaford, Blake	1863
Hara, Isao	1579
Hara, Takayuki	3477, 1239
Harada, Kensuke	3680, 2210, 2492, 3028
Harada, Tatsuya	3201
Harer, John L.	2243
Hariyama, Masanori	1168
Hasegawa, Naoya	3674
Hasegawa, Rika	453
Hasegawa, Tsutomu	4029
Hasegawa, Yasuhisa	1518, 3195
Hashimoto, Hideki	860
Hashimoto, Koichi	3529
Hashimoto, Minoru	3139
Hashizume, Kenichi	441
Hashizume, Takumi	1117
Hatakeyama, Takuro	3674
Hayakawa, Soichiro	3772
Hayashi, Kouki	2424
Hayashibe, Mitsuhiro	1543
He, Chao	1671
Hebbar, Ravi	325
Heidemann, John	2941
Hein, Andreas	2025
Helm, Chad	3423
Henry Phang, S. H.	4226
Herman, Przemys[3] aw	3819
Hermosillo, Jorge	3294
Hertzberg, Joachim	2072
Higuchi, Takuya	1567
Hirai, Shinichi	1233, 3807, 85, 812
Hirana, Kazuaki	2674
Hirata, Yasuhisa	3010, 3004
Hiroki, Takeuchi	3165
Hirose, Shigeo	3172, 2442, 773, 181
Hirsh, Robert L.	2171
Hirukawa, Hirohisa	4084
Hirzinger, Gerd	2852, 109, 3356
Hisseine, Dadi	3865
Ho, Antony W. T.	656
Ho, Minh Anh T.	1239
Ho, Teresa	1909
Hogg, Tad	360
Hol, Camile W.J.	1136
Hollerbach, John M.	1130
Hommel, Guenter	2698
Honegger, Marcel	2553
Hori, Yoichi	711
Horiuchi, Yohei	1728
Hornby, Gregory S.	4146
Horstmann, Sven	2590
Hoshino, Tasuku	3510
Hosoda, Koh	3987
Hosoe, Shigeyuki	3268
Howard, Ayanna M.	3067, 3084, 1413
Howe, David	1798
Howe, Robert D.	4214
Hsia, T. C.	3453
Hsieh, Bo-Wei	535
Hsu, David	1765
Hsu, Liu	743

Hu, Baosheng	3565
Hu, Jianbo	3582
Hu, Rong	2191
Hu, Xiaoming	4244, 3961
Hu, Yu Hen	2140
Hua, Wei	337, 3127
Huang, Han-Pang	1839, 1826
Huang, Jui-Hsian	2891
Huang, Shih-Shinh	3559, 1981
Huang, Tian	3280
Huang, Wesley H.	27
Huang, YiSheng	541
Huber, Eric	3078
Hung, Min-Hsiung	4060, 779, 1832
Hurtado, Jose F.	1086
Hwang, Jae-Chul	3274
Hyon, Sang-Ho	2741
Hypki, Alfred	806
Hyun, Chang-Ho	985

I

Iannitti, Stefano	2090
Iijima, Daisuke	523
Ikeuchi, Katsushi	465
Ikuta, Koji	2037, 3181
Imaida, Takashi	1031
Imin, Kao	3937, 3047, 3055
Inaba, Akio	848
Inaba, Masayuki	2120, 4110, 2307, 692, 1457, 4084
Inamura, Tetsunari	4208, 4220
Indiveri, Giovanni	205, 915, 903
Ingrand, Felix F.	9
Ingvast, Johan	1327
Inoue, Hirochika	2120, 4110, 2307, 1457, 4084
Inoue, Kenji	2998
Iocchi, Luca	4250
Iritani, Koji	2037
Ishida, Toru	4166
Ishiguro, Hiroshi	4166
Ishii, Hideki	3181
Ishii, Yohei	3189
Isukapalli, Ramana	2134
Itabashi, Kaiji	2674
Ito, Kazuyuki	2096
Ito, Ken	1207
Ito, Koji	3692, 505
Ito, Satoshi	1388
Itoigawa, Kouichi	577, 604
Ivanisevic, Igor	2649
Iwai, Ayako	597
Iwamura, Makoto	1271

Izumi, Kiyotaka	319, 2584

J

Jablokow, Kathryn W.	2804
Jakubiak, Janusz	2401, 2084
Jan, Y.J.	331
Jang, Gi-jeong	1677
Jang, Seong-Yong	3554
Jeng, MuDer	52, 541
Jensen, Bjoern	1609
Jeon, Jong Man	3577
Jeong, In-Soo	1603
Jewell, Geraint, W.	1798
Ji, H. zhang	1845
Ji, Qiang	2165, 2177, 2191
Ji, Xuerong	1512
Jia, Songmin	1915
Jin, Young G.	972
Jones, Andrew H.	193
Jongusuk, Jurachart	2885
Joni, Jeffry	1059
Joshi, Sameer A	1283
Ju, Ming-Yi	2897
Julier, Simon	4238
Jung, Kyu Dong	3412
Jung, Myung-Jin	3370
Jung, Seul	3453
Jung, Sunghwan	1452
Jurie, Frdric	1653

K

Kadmiry, Bourhane	2980
Kagami, Satoshi	2120, 401, 4110, 2307, 692, 2431
Kagami, Yoshiharu	1457
Kahmen, Andreas	199
Kajioka, Morimasa	1792
Kajita, Shuuji	2299, 3376
Kajitani, Makoto	3308
Kak, Avinash C.	2171, 193
Kakazu, Yukinori	523
Kamamichi, Norihiro	3139
Kameyama, Michitaka	1168
Kamiji, Norimasa	85
Kanai, Satoshi	211
Kanaoka, Katsuya	2840
Kanda, Takayuki	4166
Kanehiro, Fumio	4084
Kaneko, Makoto	3680, 4023, 2210, 2492, 3028
Kang, Bongsoo	2723
Kang, Hyosig	2031
Kang, Sung Jae	3208
Kang, Sungchul	1277

Kao, Imin	3937, 3047, 3055	Kim, Soo Hyun	413, 3734
Karavelas, Menelaos	3973	Kim, Sung-Jae	767
Kariya, Shingo	3668	Kim, Tae Won	2350
Kato, Tatsunori	3987	Kim, Whee Kuk	2413
Katsoulas, Dimitrios K	305	Kim, Won	2608
Katsuyama, Norikazu	1792	Kim, Yong-Jae	217
Kavraki, Lydia E.	1079, 1469, 960	Kim, Yoon Sang	2628
Kawabe, Hiroshi	407	Kim, Young Dong	874
Kawai, Masashi	610	Kim, Young Ho	3208
Kawai, Masayuki	880	Kim, YoungWoo	848
Kawaji, Akiko	604	Kimura, Hiroshi	465
Kawakami, Hiro	2014	Kimura, Hitoshi	2442
Kawamura, Sadao	3807, 2535, 85	Kinami, Masahiro	407
Kawasaki, Haruhisa	1388	Kinoshita, Genichiro	565
Kececi, Ferit	2159	Kircanski, Nenad	3728
Kee, Hongseok	145	Kishi, Jun	3189
Keller, James	485	Kishi, Kousuke	2014
Kelley, Jack	940	Kissner, Lea	2480
Kelly, Rafael	3523	Kitagawa, Ato	2454
Kenn, Holger	749, 299	Klandt, Jesko	2590
Khalil, Wisama	2859	Klavins, Eric	4200
Khatib, Oussama	386	Kluge, Boris	1683
Khosla, Pradeep K.	1994, 2923	Knoll, Alois	2217
Kiguchi, Kazuo	3668, 2584	Ko, Nak Yong	874
Kijimoto, Hirokazu	2474	Kobayashi, Hiroshi	1695
Kikuchi, Kohki	1695	Koch, Thorsten	187
Kikuuwe, Ryo	868	Koditschek, Daniel E.	4200, 3650
Kilchenman, Marcia	1382	Koehler, Christian	1683
Kim, Bong Keun	3859	Koenig, Sven	3594
Kim, Byoung-Ho	3034, 2614	Koishikura, Tarou	2466
Kim, Byung Hwa	253	Koji, Tatani	1968
Kim, Cheol-Taek	3314	Kojima, Kazuhiro	505
Kim, Dae Won	3577	Komoriya, Kiyoshi	1444, 1277
Kim, Dae-Hyun	874	Komsuoglu, Haldun	3650
Kim, Dae-Jin	3662	Kondak, Konstantin	2698
Kim, Do Hyung	2413	Kondo, Toshiyuki	3692
Kim, Hong Seok	3577	Koo, John	1757
Kim, Hyung Wook	2331	Kosaka, kio	2183
Kim, Jinsuck	3789	Kosmopoulos, Dimitrios I	305
Kim, Jin-Sung	3274	Kosuge, Kazuhiro	3010, 3004
Kim, Jong-Hwan	3847, 3370, 217	Kotoku, Tetsuo	1013
Kim, Jong-Sung	3662	Koutoku, Tetsuo	1277
Kim, Jongwon	3274	Koyachi, Noriho	610
Kim, Jung-Ha	686	Kozlowski, Krzysztof	3819
Kim, Jung-Hoon	3686	Kragic, Danica	2460
Kim, Ki-Tae	3554	Kress, Reid L.	393
Kim, Kyunghwan	1025	Kristensen, Steen	2590
Kim, M.S.	3109, 4066	Krose, Ben	499, 1174
Kim, Min-Seok	686	Ku, Peng-Jui	3417
Kim, Munsang	2628	Kubica, Eric G	2293
Kim, Nakhoon	145	Kubica, Jeremy	360
Kim, Seonho	145	Kubota, Takashi	3710, 1394
Kim, Shin	217	Kuffner, James J.	401, 692, 2431

Kumar, Saravana	3338
Kumar, Vijay	1245, 157, 2992, 2818, 3022, 1714, 2229
Kume, Shinji	3121
Kume, Youhei	3004
Kunii, Yasuharu	3710, 1394
Kunikatsu, Takase	1915
Kurazume, Ryo	3172
Kurimoto, Yujin	565
Kuroda, Yoji	3710, 1394
Kushida, Daisuke	2747
Kwak, Nojun	2560
Kwak, Yoon Keun	413, 3734
Kweon, In-so	1677
Kwon, Dong-Soo	1307, 1358, 3214, 892, 2313, 3238
Kwon, Guiryong	2486, 577
Kwon, Kihwan	1194
Kwon, Ohung	4134
Kwon, SangJoo	121
Kwon, Sung Min	1358
Kwon, Wook Hyun	1845, 3115, 3571, 2419
Kwon, Woong	3704
Kyriakopoulos, Kostas J.	3926
Kyung, Ki-uk	1358
Kyung, Min-Ho	1321
Kyura, Nobuhiro	2747, 3949

L

LaBean, Thomas H.	966
Lallemand, Jean-Paul	2642
Lam, H.K.	1736
Lamon, Pierre	1609
Lane, David M.	2767, 2785, 2375, 3232, 909, 3226, 2356
Large, Fred	3716
Latombe, Jean-Claude	932
Laumond, Jean-Paul	1494
Laurent, Guillaume	3914
LaValle, Steven M.	1481, 1954
LE, Minh-Duc	2108
Leang, Pat	491
Lee, Beomhee	3109, 4066
Lee, Byeung Leul	3412
Lee, Byoung-Ju	471
Lee, C. S. George	2986
Lee, Changho	1452
Lee, Chang-Hun	985
Lee, Changseung	616
Lee, Chong-Won	2628, 2313
Lee, Dong-Wook	3993
Lee, Doo Yong	1820
Lee, Ho-Hoon	2547, 2956
Lee, Hyun-Gu	471
Lee, J.H.	886
Lee, J.J.	3214
Lee, James B	1975
Lee, Jang Myung	1220, 253
Lee, Ji Yeong	991
Lee, Joo-Ho	471
Lee, Ju-Jang	2608, 3314
Lee, Kok-Meng	1059
Lee, Kyoobin	892
Lee, Man Hyung	253
Lee, Min Cheol	2962, 2622, 2325
Lee, Pan-Mook	3244, 3238, 2363
Lee, Sang Heon	413, 3734
Lee, Sang W.	3542
Lee, Sang Woo	3412
Lee, Seong-Whan	1659
Lee, Seung-Eun	217
Lee, Sooyong	2628, 3789
Lee, Sukhan	3704, 616
Lee, Sung-Uk	3955
Lee, Tae-Young	1301
Lee, Tat Hoi	1736
Lee, Tong Heng	3871
Lee, Wan-Chi	3795
Lee, Wongoo	2512
Lee, Woo W	4011
Lee, Yong-Beom	1659
Lefeber, Erjen	2084
Leung, F.H. Frank	1736
Lewis, D.	3408
Li, Jianfeng	2506, 2448
Li, Kai	2535
Li, Kejie	241
Li, Tsai-Yen	1934
Li, Wen J.	3656, 656
Li, Xudong	127
Li, Yanmei	3937, 3047, 3055
Li, Z.X.	2249, 3766
Likhachev, Maxim	1627
Likuan, Zhao	1845, 3115
Lilien, Ryan	940
Lilienthal, Achim J.	4005
Lim, Jong-Tae	650
Lim, Seungmo	616
Lin, Feng	3565
Lin, Jen-Yu	1832
Lin, Ming-Hung	547
Lin, Shir-Kuan	1201
Lin, Wei	2407
Lin, Yingqiang	491
Lipson, Hod	4146

Little, Jim	2051
Liu, Cheng-Yi	2634
Liu, Chun-Hung	3831
Liu, Da	2008, 229
Liu, David	229
Liu, Fuming	4029
Liu, G.F.	3748
Liu, Guangjun	1155, 2566
Liu, Hong	109
Liu, Jing-Sin	2897, 3795
Liu, Yun-Hui	3483, 2217, 241, 662
Lloyd, John E.	1884
Loeffler, Klaus	4140
Loh, Han Tong	1340
Lohmann, Boris	3865
Lohnert, Frieder	2590
Loisel, Philippe	247
Longman, Richard W	4128
Lopes, Ernesto P.	2911
Lorussi, Federico	2078
Lots, Jean-Francois	2767
Low, K.H.	3159
Low, Kin Huat	2279
Lowe, David	2051
Lu, Jilian	2261
Luan, Nan	3308
Luangjarmekorn, Poom	604
Luedemann-Ravit, Bernd	187
Lueth, Tim C.	2025
Luh, Peter B.	842
Lumelsky, Vladimir	2649
Lundin, Magnus	2147
Luo, Ren C.	1334, 4226
Lutticke, Tobias	2596
Lynch, Kevin M.	3300

M

MA, Shugen	4035, 3656
MacDorman, Karl F.	4186, 1968
Macfarlane, Sonja E.	979
Machado, Jose A. T.	3624, 4122
Maciel, Benedito C O	2114
Mae, Yasushi	2998
Maeda, Taro	1006
Maeda, Yusuke	3477, 799, 2474
Maekawa, Hitoshi	1444
Mahadevan, Sridhar	511
Mahony, Robert	1701
Maimone, Mark W.	1099
Makoto, Kaneko	3680, 4023, 2210, 2492, 3028
Malak, Richard J	1994

Mali, Amol	3016
Manocha, Karan A.	374
Marani, Giacomo	3220
Marchand, Eric	2773, 223
Marche, Pierre	3999
Marigo, Alessia	2078
Marquet, Frederic	3256
Martel, Sylvain	3423
Martin, Martin C.	1092
Martinet, Philippe	1653, 717
Masanao, Koeda	1968
Mascaro, Stephen A	1857
Mason, Richard J.	427
Mastrantuono, Domenico	4250
Masubuchi, Akihiro	1233
Masui, Tomohiro	3807
Masutani, Yasuhiro	421
Mata, Mario	1124
Matsakis, Pascal	485, 259
Matsumoto, Osamu	2299
Matsuno, Fumitoshi	2096, 1400
Matsushita, Motoshi	421
Matsuura, Hideo	2486
Matthies, Larry H.	1099
McBeath, Michael	1689
McKee, Gerard T	1019
McMordie, Dave	3650
Mei, Jiangping	3280
Meizel, Dominique	1597
Melchiorri, Claudio	1427
Melkote, Shreyes N.	1086
Meltzer, S	3408
Menciassi, Arianna	626
Meng, Max	3402, 1555
Meng, Yan	2797
Merlet, Jean-Pierre	1289
Metaxas, Dimitris	3973
Mezouar, Youcef	731
Micheli, Mario	1757
Miller, Andrew T.	2460
Miller, Shawna	1500
Mills, James K.	2968, 2723
Mimura, Nobuharu	597, 2466
Ming, Aiguo	3308
Ming-Fong, Chou	3778
Minguez, Javier	33
Mirolo, Claudio	3783
Misra, Arun K.	2873
Mita, Tsutomu	2735, 2741, 2885
Mitchell, Ken	1798
Mitsuishi, Mamoru	1567
Mittal, Gauri S.	163, 3402
Miura, Jun	1750, 1188

Miwa, Hiroyasu	2602, 459
Miyazaki, Fumio	421
Mizuno, Yuto	2466
Mochiyama, Hiromi	2223
Mohri, Akira	1271, 3489
Mok, Swee M.	313
Mombaur, Katja D	4128
Montano, Luis	33
Moon, Chanwoo	3109
Moon, Inhyuk	1188, 3208
Moore, Ned	3650
Moorehead, Stewart J	3098
Morales, Antonio	583
Morales, Eduardo	3092
Moreno, Luis	4232
Morgansen, Kristi A.	427
Mori, Taketoshi	3201
Morita, Hideyuki	2486
Morita, Satoshi	3692
Morizono, Tetsuya	1406, 1239
Morris, Daniel M.	325
Motai, Yuichi	2183
Motomura, Yoichi	1579
Mouaddib, El Mustapha	3548
Mouri, Tetsuya	1388, 597
Mueller, T.	235
Munasinghe, Sudath R	3949
Murphey, Todd D.	2716, 2686
Murray, Pamela	638
Mutsuo, Daito	1786
Myers, Donald R.	4078

N

Nacer, K. M'Sirdi	3465
Nagata, Fusaomi	319
Nagel, H.-H.	2159, 235
Nagy, Laszlo	2255
Najjaran, Homayoun	3728
Nakamura, Masatoshi	2747, 3949
Nakamura, Tatsuya	3920
Nakamura, Yoshihiko	348, 4208, 4220, 1647, 1, 2824, 2014, 1543
Nakaoka, Masaharu	1518
Nakaya, Koji	2442
Nakayama, Kanji	1388
Nam, Taek-Kun	2735, 2741
Nardi, Daniele	4250
Negahdaripour, Shahriar	2759
Nelson, Bradley J.	133, 620
Netto, Mariana S.	743
Neubert, Jeremiah J.	2140

Neumann, Mathias	3390
Newman, Wyatt S.	1352, 325, 571
Ng, Kuang-Chern	2791
Nguyen, An Thai	1765
Nguyen, Anh, Pham Thuc	2344
Nickels, Kevin M.	3078
Nijmeijer, Henk	2084
Nikolic, Milan	2255
Nishikawa, Kazufumi	2424
Nishino, Tetsuo	1000
Nishino, Yutaka	1180
Nishiwaki, Koichi	401, 4110, 692, 2431
Noborio, Hiroshi	1728, 1180
Nokata, Makoto	3181
Nourbakhsh, Illah	1609, 1111
Novales, Cyril	3999

O

O'Brien, Des J.	2375
O'Brien, John F.	354
Ochi, Kazuhiro	1695
Oda, Mitsushige	1891, 407, 1031
Ogawara, Koichi	465
Oh, Geum Kun	874
Oh, Jun-Ho	3686
Oh, Sang-Keon	3314
Oh, Sang-Rok	3034, 2614
Ohara, Shigeyuki	830
Ohashi, Kazushi	553
Ohba, Kohtaro	1013
Ohta, Atsuharu	2840
Ojeda, Lauro	3588
Okabe, Yoichi	1000
Okada, Kei	2120, 2307
Okada, Masafumi	348
Okamura, Allison M.	589
Okuma, Shigeru	2674, 3772, 848
Olson, Clark F.	1099
Olson, Kari	1481
Omata, Toru	103, 2203
Ong, Kenlip	313
Onogi, Yu	1695
Ootsuka, Takeo	3195
Orehek, Martin	1346
Orin, David E	779
Oriolo, Giuseppe	2090, 1591, 91
Ortega, Romeo	743
Ostrowski, James P.	157, 3022, 3396
Osumi, Hisashi	565
Otake, Mihoko	1457
Ottaviano, Erika	1295

Overmars, Mark H. 1053, 1488
Owens, Nancy E. 517
Oyama, Eimei 1006, 4186

P

Pagello, Enrico 3783
Pagilla, Prabhakar R. 3943
Pai, Dinesh K. 1884
Paiva, Ely C 4152
Paletta, Lucas 2072
Pallottino, Lucia 2319
Palopoli, Luigi 1463
Panin, Giorgio 77
Pao, Yoh-Han 571
Papadopoulos, Evangelos 3967
Papanikolopoulos, Nikolaos 2917
Park, Chang-Woo 985
Park, F.C. 3274
Park, Gwi-Tae 471
Park, Jaehyun 2512
Park, Jahng-Hyon 2974
Park, Jihwang 1452
Park, Jin-Woo 3554
Park, Jong Hyeon 1025, 4134
Park, Jong O. 1025
Park, Jonghoon 1161, 4041, 2210
Park, Jonghun 70, 1813
Park, Kyihwan 145
Park, Kyung Hoon 3516
Park, Min Kyu 2962, 2325
Park, Min-Sick 985
Park, Sangdeok 3859
Park, Seong-Jin 650
Park, Tong-Jin 151
Park, Yonmook 1621
Park, Young-Hoon 686
Paromtchik, Igor E. 2935
Parra-Vega, Vicente 3471
Parsa, Kourosh 2873
Pavlin, Gregor 1961
Pedersen, Liam 4174, 277, 4158
Pelinescu, Diana M. 792
Pensky, Dirk H. 806
Perdomi, Pierangelo 2319
Pernalete, Norali 374
Perrier, Michel 2773
Peterson, Todd S. 517
Petillot, Yvan R. 2785
Pfeiffer, Friedrich 4140
Pham, Minh Tu 2867
Phillips, George 960
Phillips, Roger 1561
Piat, Emmanuel 3914
Picinbono, Guillaume 1370
Piedboeuf, Jean-Claude 2832
Pierrot, Francois 3256
Plapper, Volker 3104
Pobil, Angel P. del 583
Poignet, Philippe 2867
Poisson, Gerard 3999
Pollack, Jordan B. 4146
Pollefeys, Marc 2153
Poo, Aun-Neow 2791
Poulakakis, Ioannis 3967
Prassler, Erwin 1683
Prattichizzo, Domenico 737
Pritchard, Michael J. 4078
Pylatiuk, Christian 2437
Pylkko, Heikki 1214

Q

Qi, Baohua 2710
Qi, Feng 3320
Qiang, Huang 4220
Qiao, Hong 1071
Qing, Wei 3932
Quarto, Francesco 1903

R

Ra, Jong Beom 1358
Racoceanu, Daniel I. 46
Raczkowsky, Joerg 2020
Raducanu, Bogdan 2059
Raghavan, N. R. Srinivasa 529
Rahn, Christopher D. 3877
Rajko, Stjepan 1954
Ramos, Josue J G 4152
Randall, Geoph 909
Rashid, Shahid 2948
Rauf, Abdul 2389
Recatala, Gabriel 583
Rehbinder, Henrik 4244
Reif, John H. 966
Reinkensmeyer, David J 1433
Remmler, Ian 21, 1475
Renaud, Marc 1251
Requicha, A. A. G. 3408
Resch, R. 3408
Reveliotis, Spyros A. 70, 1813
Reznik, Dan S. 2235
Ridao, Pere 3250

Ridderstr?, Christian	1327
Ridley, Peter R.	3742
Riekki, Jukka P	1214
Rimon, Elon	3636, 1927
Rives, Patrick	2753
Rivlin, Ehud	285, 291
Rizzi, Alfred A.	991, 3121, 699
Rizzo, Luigi	1903
Roberts, Jonathan M	1635
Roberts, Randy S.	3981
Robl, Christian	1346
Rodrigues, Carlos M. B.	3624
Rogalla, Oliver	2578
Roh, Dong Kyu	253
Roh, Se-gon	761
Rohanimanesh, Khashayar	511
Romero, Leonardo	3092
Roning, Juha J	1214
Ros, Lluis	2126, 3332
Rossmann, Juergen	1921
Rostami, Shadi	3326
Roszkowska, Elzbieta K.	668
Rouchon, Pierre	3294
Rovetta, Alberto F.	1575
Rudas, Imre J.	2002
Rutten, Eric P.	4104
Rybski, Paul E	2917
Ryew, SungMoo	761
Ryu, Jee-Hwan	1863, 3238
Ryu, Jeha	2389
Ryu, Se-Hee	2974

S

Sacks, Elisha P.	1524, 1321
Safaric, Riko	3417
Saffiotti, Alessandro	4017
Saigo, Muneharu	2299, 3376
Saito, Kei	1400
Saito, Takuya	812
Saitou, Kazuhiro	818
Sakaguchi, Masamichi	3825
Sakane, Shigeyuki	1207, 2661
Salcudean, S. E.	3760, 1549
Salemi, Behnam	4194
Salichs, Miguel A	2668, 1124
Salson, Cedric	2785
Salvi, Joaquim	3548
Samin, Jean-Claude	2846
Sanderson, Arthur C.	824
Sano, Akihito	380
Sanz, Pedro J.	583

Saranli, Uluc	3650
Sardain, Philippe	2285
Sastry, S Shankar	1720, 3885, 1641
Sato, Shuji	1406
Sato, Tomomasa	3201
Satoh, Hiroshi	139
Scalari, Giacomo	626
Schaefer, Ingo	3356
Scheinerman, Edward R.	1773
Schempf, Hagen	755
Schenato, Luca	3885, 1641
Schenker, Paul S	1019
Schlegl, Thomas	103
Schloeder, Johannes P	4128
Schmidt, Guenther	103
Schmoeckel, Ferdinand	3909
Schneider, Jeff G.	1615
Schoppers, Marcel	1099
Schulz, Dirk	1665
Schulz, Stefan	437
Schwarzer, Fabian	1537
Schweikard, Achim	1537
Se, Stephen	2051
Seeman, Nadrian C.	966
Segawa, Mitsuru	2454
Sekhavat, Sepanta	3716, 3294
Sekiguchi, Akinori	1647
Seng, Xinyan	3885, 1641
Sent, Danielle	1488
Seraji, Homayoun	3067, 3084, 1413, 1039
Serdeira, Henrique	2911
Sfakiotakis, Michael	2356
Shakernia, Omid	1720, 2948
Shamah, Benjamin	1742, 4174
Shang, X.	2517
Shapiro, Amir	3636
Sharp, Cory	2948
Sharp, Courtney S	1720
Sharp, Gregory C.	3542
Shen, Wei-Min	4194
Shen, Yantao	241
Sheng, Weihua	3127
Sherman, Alana	1045
Sherwood, Mark	3423
Shiang, Shen-Po	2897
Shibata, Katsunari	1000
Shibata, Takanori	2572
Shillcutt, Kimberly	1421, 1742, 4174
Shiller, Zvi	3716, 1
Shim, Jae-Heung	686
Shim, Jae-Kyung	1301
Shimamura, Koichiro	3920
Shimohira, Takahiro	1406

Shimshoni, Ilan	3605, 285, 291
Shin, Dong Hun	3516
Shin, Jae-Cheol	2313
Shin, Kang G.	4090, 553, 4072
Shinoda, Hiroyuki	1851
Shinohara, Tomoyuki	1851
Shintani, Hiroaki	860
Shirai, Tatsuya	3028
Shirai, Yoshiaki	1750, 1188
Shoham, Moshe	2273
Shoval, Shraga	2273, 3636, 2879
Siciliano, Bruno	435, 2729
Siegwart, Roland	1609, 1111
Sillitoe, Ian P.	2147
Silva, Filipe M.	4122
Silveira, Julio T.	2911
Sim, Kwee-Bo	3993
Simeon, Thierry	4256, 1494, 33
Simmons, Reid	3098
Simon, Carlo	58
Sin, Jeongsik	115
Singh, Amit P.	932
Singh, Sanjiv	705
Sinopoli, Bruno	1757
Sirouspour, Mohammad Reza	3760
Sitti, Metin	3893, 3901, 3839, 860
Skaff, Sarjoun	4180
Skubic, Marjorie A.	485, 259
Sokolov, Sergey M.	1780
Somolinos, Jose Andres S.	3853
Son, Kwon	2622, 2325
Son, Wookho	1376
Song, Guang	948, 1500, 954
Song, Jae-Bok	767
Song, Kai-Tai	2891
Song, Mumin	3127
Song, Peng	2229
Song, Se-Kyong	1307
Song, Won-Kyung	3662
Sonohara, Takayuki	1239
Southall, John B	1714
Speranzon, Alberto	169
Spindler, Fabien	2773
Staritz, Peter J.	4180
Steele, Jay W	3133
Stein, David	1773
Stephanou, Harry E.	115, 3417
Stern, Oliver	187
Storoshchuk, Orest	4072
Stramigioli, Stefano	3344
Stroupe, Ashley W.	1092
Su, Hongye	3582
Su, Kuo L.	4226
Su, Tong	2692
Subbu, Raj	824
Subramaniam, Velusamy	854
Subramanian, Devika	2065
Sucar, Enrique	3092
Sudsang, Attawith	1079
Sugar, Thomas G	2992, 3022, 1689
Sugi, Masao	799
Sugihara, Tomomichi	401, 4110, 2431
Suh, Il Hong	2331, 886, 3034, 2614
Suh, Jin-Ho	3495
Suh, John	3338
Suhara, Masaya	1394
Sukhatme, Gaurav S	2941
Suluh, Anthony	1689
Sun, Dong	2968, 3483
Sun, High-Way	787
Sun, Sang-Joon	3993
Sun, Yu	620
Sundaram, Sujay	1475
Sussner, Peter	2059
Suzuki, Ichiro	3016
Suzuki, Tatsuya	2674, 3772, 848
Svinin, Mikhail	3268
Swevers, Jan	2846

T

Tahara, Kenji	97
Tahboub, Karim A.	2655
Tahk, Min-Jea	1621
Takagaki, Tsuyoshi	3189
Takagi, Takeo	3010
Takagi, Tsuyoshi	453
Takahashi, Atsushi	3477
Takahashi, Katsumi	3674
Takahashi, Rie	2096
Takahashi, Yoshihiko	3674, 3189
Takai, Toshihito	380
Takanishi, Atsuo	2602, 459, 2424
Takanobu, Hideaki	2602, 459, 2424
Takesue, Naoyuki	3825
Takeuchi, Toshiki	1168
Takiguchi, Junichi	1117
Takubo, Tomohito	2680
Tam, K.S. Peter	1736
Tan , Da Long	1897
Tan, Jindong	3145
Tan, Min	3755
Tang, Lixin	3072
Tang, Wai Sum	4054
Tang, Xiaoqiang	2448

Tang, Ying	559	Uehiro, Kiyoshi	211
Tanie, Kazuo	2680, 2572, 1013, 3376	Ueyama, Tsuyoshi	1518
Tanikawa, Tamio	610	Uhlmann, Jeffrey	4238
Tanner, Herbert G.	3926	Umeda, Kazunori	565
Tanuki, Tomikazu	465	Umetani, Tomohiro	2998
Tar, Jszsef K.	2002	Umetsu, Tomohiko	2602, 459
Tarn, Tzyh-Jong	2102, 644	Ura, Tamaki	920
Taylor, Nick. K.	3232	Urmson, Chris	4180
Tchon, Krzysztof	2401, 2084		
Tena Ruiz, Ioseba	2785		
Tendick, Frank	1045		

V

Vallejo, Daniel	21
van der Stappen, A. Frank	1053
Van Geem, Carl	1494
Van Gool, Luc	2153
van Henten, Eldert J.	1136
van Straten, Gerrit	1136
van Willigenburg, Gerard	1136
Van Zwynsvoorde, Dominique	4256
Vassura, Gabriele	1427
Vaughan, Richard T	2941
Vendittelli, Marilena	91
Verdonck, Walter	2846
Vergauwen, Maarten	2153
Viant, Warren	1561
Vicino, Antonio	737
Victorino, Alessandro C.	2753
Vidal, Rene E	2948
Villani, Luigi	2729
Viswanadham, Nukala	854, 529
Vlassis, Nikos	499, 1579

Teodoro, Miguel	960
Terra, Marco H	2114
Tesar, Delbert	3286
Thanapandi, Chitra Malini	830
Theocharous, Georgios	511
Thomas, Federico	2126, 3332
Thomas, Susy	3847
Thrapp, Richard	2065
Thrun, Sebastian	271
Thukahara, Yasunori	577
Ting, Jen-Kuei	4060
Tischler, Neil A	3042
Tojo, Yoshiharu	1851
Tokumoto, Shinichi	812
Tomatis, Nicola	1111
Tomizuka, Masayoshi	3801
Tomura, Toyoaki	211
Tonani, Sergio	2529
Tornero, Josep	3801
Toru, Omata	103, 2203
Toshima, Iwaki	4208
Tovey, Craig	3594
Trifonov, Oleg V.	1780
Trinkle, Jeffrey C.	1376
Trucco, Emanuele	2767
Tsai, Lung-Wen	1283
Tso, Shiu Kit	2261
Tsuboi, Tatsuhiko	1233
Tsuda, Taishi	1567
Tsuji, Toshio	3680, 4023, 3028
Tsukagoshi, Hideyuki	2454, 181
Tsumaki, Yuichi	407, 1039
Tunstel, Edward	3067, 3084, 1413
Turro, Nicolas	386
Tzou, Jyh Hwa	1334

W

Wada, Takahiro	85
Wade, Eric	3831
Wagner, Michael D.	1742, 4174, 4158
Wakata, Koichi	1891
Walairacht, Aranya	830
Walker, Ian D.	3877
Wandel, Michael	4005
Wang, Che-Lung	1826
Wang, Chia-Yu E	1433
Wang, Danwei	3320
Wang, David	2293
Wang, Dongsheng	3755
Wang, Gwo-Chuan	3447
Wang, Honggua	3535
Wang, Jeen-Shing	2986
Wang, Jiabin	1798
Wang, Jinsong	2506, 2448
Wang, Qing Peng	1897

U

Uchibe, Eiji	3987
Uchiyama, Masaru	407, 3268
Ueda, Jun	1806

Wang, Qizhi	3755
Wang, Shige	4090, 4072
Wang, Tianmiao	2008
Wang, Y.T.	331
Wang, Youyi	1149
Wang, Yu M.	792
Wang, Yue Chao	1897, 3656
Wang, Yun Gan	1340
Wang, Zhuping	3871
Wang, Zigang	2008
Wang, Zongpei	3813
Warisawa, Shin'ichi	1567
Watanabe, Keigo	319, 3668, 2584
Watanabe, Tetsuyoh	1786
Weck, Manfred	3104, 199
Wehe, David K.	3542
Wei, Tao	1352
Wei, Yejun	2811
Weimar, Udo	4005
Wen, Hung	3559, 1981
Wen, John T.	2031, 354
Wenseng, Chang	3932
West, Michael E.	2363
Westbrook, Christian	2065
Whittaker, William L	1421, 3098, 4180
Whittaker, William R	1742, 4174, 4158
Wikander, Jan	1327
Will, Peter	3503, 4194
Winter, David	2293
Winther, Tobias	115, 3417
Winther, Tobias K.	3417
Woern, Heinz	2020, 3909
Wolf, Carl	787
Won, J.H.	3214
Wong, Ching-Chang	3778
Wood, Robert J	3901
Wu, Chi-haur	1506, 313
Wu, Naiqi	64
Wu, Wei Chung	3885
Wu, Weimin	3582
Wu, X.Z.	3748
Wu, Y.L.	3748
Wyeth, Gordon W	1635
Wysk, Richard A	3133

X

Xi, Ning	1897, 656, 662, 3145, 3127
Xia, Jun	342
Xiao, Jing	1512
Xie, Feng	2903
Xie, Xiaolan	52, 541
Xing, Keyi	3565
Xiong, Z.H.	2249, 3766
Xu, Dianguo	3813
Xu, Xun	2759

Y

Yamada, Takaaki	2584
Yamada, Takayoshi	597, 2466
Yamada, Yoji	1406, 1239
Yamaguchi, Mitsuharu	97
Yamakita, Masaki	3495, 3139
Yamamoto, Motoji	1271, 3489
Yamamoto, Shinya	1233
Yamamoto, Susumu	211
Yamane, Katsu	1, 2824
Yamashita, Masaya	2661
Yan, Joseph	3901, 3839
Yanai, Noritaka	3489
Yang, Ge	133
Yang, Gi-Hun	2313
Yang, Guilin	2407
Yang, Haw-Ching	1832
Yang, Jonghwa	761
Yang, Libo	1481
Yang, Simon X.	163, 3402
Yang, Yong	1340
Yanmei, Li	3937, 3047, 3055
Yao, Bin	3459
Yaroshevskiy, Victor S.	1780
Yashima, Masahito	2229
Yau, Wei Guan	229
Ye, Wei	2941, 2811
Yi, Byung-Ju	2331, 886, 3034, 2614
	2413, 413, 3734, 3286
Yim, Mark	3338, 2480, 3383
Yim, Sun Bin	3453
Yim, Woosoon	3441
Yin, Xuecheng	1059
Yiu, Y.K.	3748, 3766
Yoji, Miyazaki	1968
Yokoi, Hiroshi	523
Yokoi, Kazuhito	3376, 1277
Yokokohji, Yasuyoshi	1031
Yoneda, Kan	3172, 181
Yoo, Byung-Hoon	3554
Yoo, Ki Sung	2325
Yoon, Hyun Joong	1820
Yoon, Woo-Keun	407
Yoon, Yong-San	3214
Yoshida, Kazuya	441
Yoshida, Yasuo	368

Yoshikawa, Tsuneo	868, 1806, 1786, 880, 2840
Yoshimitsu, Tetsuo	3710
You, Bum-Jae	2413, 1659, 3618
Youm, Youngil	4041, 3859, 121
Yu, Biao	3943
Yu, Chih-Yuan	1839
Yu, Huiming	2692
Yu, Wen-Shyong	3447
Yu, Wenwei	523
Yu, Xiaoliu	3535
Yu, Yong	1948, 265
Yuan, Guangfeng	163, 3402
Yuh, Junku	3244, 2363, 926, 2350
Yun, Duk Sun	686
Yuta, Shin'ichi	3072

Z

Zapata, Rene	680
Zeghloul, S.	2642
Zelinsky, Alex	4096, 1530
Zell, Andreas	4005
Zemouri, Ryad A	46
Zerhouni, Noureddine	46
Zesheng, Tang	2008
Zhang, Dong	1439
Zhang, H.	1845, 2517
Zhang, Hong	2197, 1909, 3396, 1194
Zhang, Hui	3932
Zhang, Jianwei	2217, 241
Zhang, Jihui	3115
Zhang, Li	2903
Zhang, Mingjun	2102
Zhang, Tao	1065
Zhang, Yongde	2261
Zhang, Yongmian	2177
Zhang, Yue-Ming	4048
Zhang, Yuru	2506, 2517
Zhang, Zhong	3529
Zhao, Jing	4048
Zhao, Mingyang	3535
Zhao, Ming-yang	3350
Zhao, Side	3244
Zhao, Xingyu	3280
Zhao, Yonghong	571
Zhao, Zhanfang	2261
Zheng, Yuan F.	1671
Zhiqiang, Zheng	3932
Zhou, Chen	337, 1813
Zhou, Debao	3159
Zhou, Jianying	1149
Zhou, Lihua	3280
Zhou, MengChu	559, 64, 787
Zhou, Rujing	1149
Zhu, W.H.	1549
Zhu, Yonggen	1561
Zhuang, Hanqi	2797
Zollner, Raoul D.	2578
Zong, Guanghua	127